HANDBOOK OF HVAC DESIGN

Other Books of Interest

Baumeister and Marks • MARKS' STANDARD HANDBOOK FOR MECHANICAL ENGINEERS
Brady and Clauser • MATERIALS HANDBOOK
Bralla • HANDBOOK OF PRODUCT DESIGN FOR MANUFACTURING
Elliot • STANDARD HANDBOOK OF POWERPLANT ENGINEERING
Freeman • STANDARD HANDBOOK OF HAZARDOUS WASTE TREATMENT AND DISPOSAL
Gieck • ENGINEERING FORMULAS
Harris • HANDBOOK OF NOISE CONTROL
Harris and Crede • SHOCK AND VIBRATION HANDBOOK
Hicks • STANDARD HANDBOOK OF ENGINEERING CALCULATIONS
Juran and Gryna • JURAN'S QUALITY CONTROL HANDBOOK
Karassik et al. • PUMP HANDBOOK
Parmley • STANDARD HANDBOOK OF FASTENING AND JOINING
Rohsenow, Hartnett, and Ganic • HANDBOOK OF HEAT TRANSFER FUNDAMENTALS
Rohsenow, Hartnett, and Ganic • HANDBOOK OF HEAT TRANSFER APPLICATIONS
Rosaler and Rice • STANDARD HANDBOOK OF PLANT ENGINEERING
Rothbart • MECHANICAL DESIGN AND SYSTEMS HANDBOOK
Shigley and Mischke • STANDARD HANDBOOK OF MACHINE DESIGN
Tuma • ENGINEERING MATHEMATICS HANDBOOK
Tuma • HANDBOOK OF NUMERICAL CALCULATIONS IN ENGINEERING
Young • ROARK'S FORMULAS FOR STRESS AND STRAIN

HANDBOOK OF HVAC DESIGN

EDITORS

Nils R. Grimm, P.E.

Robert C. Rosaler, P.E.

McGRAW-HILL PUBLISHING COMPANY

New York St. Louis San Francisco Auckland Bogotá Caracas Hamburg Lisbon London Madrid Mexico Milan Montreal New Dehli Oklahoma City Paris San Juan São Paulo Singapore Sydney Tokyo Toronto

Library of Congress Cataloging-in-Publication Data

Handbook of HVAC design / editors, Nils R. Grimm, Robert C. Rosaler.
p. cm.
ISBN 0-07-024841-9: $96.00
1. Heating. 2. Ventilation. 3. Air conditioning. I. Grimm, Nils R. II. Rosaler, Robert C.
TH7011.H36 1990
697—dc20 89-78110
CIP

2 3 4 5 6 7 8 9 0 DOCDOC 9 8 7 6 5 4 3 2 1

ISBN 0-07-024841-9

The sponsoring editor for this book was Robert W. Hauserman, the developmental editor was Lester Strong, the designer was Naomi Auerbach, and the production supervisor was Suzanne W. Babeuf. It was set in Times Roman by the McGraw-Hill Publishing Company Professional & Reference Division composition unit.

Printed and bound by R. R. Donnelley & Sons Company.

We fondly dedicate this volume to our dear wives, Lillian Grimm and Shirley Rosaler, for whose patience and understanding we are very grateful. They shared in our problems and frustrations, and finally in our gratification from creating this work.

CONTENTS

Preface xi
General References xiii
List of Contributors xv

1. Conceptual and Preliminary Design *M. B. Herbert, P.E.* 1.1

2. Design Calculations *Sverdrup Corporation* 2.1

3. Pipe Sizing *Sverdrup Corporation* 3.1

4. Duct Sizing *Sverdrup Corporation* 4.1

5. Steam *Lehr Associates* 5.1

6. Hot-Water Systems *Lehr Associates* 6.1

7. Infrared Heating *Lehr Associates* 7.1

8. Electric Heating Systems *Lehr Associates* 8.1

9. Solar Space Heating *Lehr Associates* 9.1

10. Snow-Melting Systems *Lehr Associates* 10.1

11. Heat Tracing *Lehr Associates* 11.1

12. Heating Distribution Systems and Equipment for Unit Heaters *Modine Manufacturing Company* 12.1

13. Hydronic Cabinet Unit Heaters *Ted Reed Thermal, Inc.* 13.1

14. Heat Exchangers *ITT Standard, ITT Fluid Technology Corporation* 14.1

15. Radiation for Steam and Hot-Water Heating *Hydronics Institute* 15.1

16. Door Heating *The King Company* 16.1

17. Valance Heating and Cooling *Edwards Engineering Corporation* 17.1

18. Boilers *Cleaver-Brooks* 18.1

19. Fuels *Cleaver-Brooks* 19.1

20. Piping *Cleaver-Brooks* 20.1

21. Burners and Burner Systems *Cleaver-Brooks* 21.1

22. Safety and Operating Controls *Cleaver-Brooks* 22.1

23. Boiler Room Venting *Cleaver-Brooks* 23.1

24. Deaerators *Cleaver-Brooks* 24.1

25. Economizers *Cleaver-Brooks* 25.1

26. Air Preheaters *Cleaver-Brooks* 26.1

27. **Sootblowers** *Cleaver-Brooks* 27.1

28. **Good Practice in Inspection, Maintenance, and Operation of Boilers** *Cleaver-Brooks* 28.1

29. **Electric Boilers** *CAM Industries* 29.1

30. **Factory-Built Prefabricated Vents, Chimneys, and Stacks** *Van-Packer Company* 30.1

31. **Chilled Water and Brine** *Giffels Associates, Inc.* 31.1

32. **All-Air Systems** *Giffels Associates, Inc.* 32.1

33. **Direct Expansion Systems** *Giffels Associates, Inc.* 33.1

34. **Fans and Blowers** *Buffalo Forge Company* 34.1

35. **Coils** *Mario Coil Nuclear Cooling, Inc.* 35.1

36. **Air Filtration and Air Pollution Control Equipment** *Environmental Quality Sciences, Inc.* 36.1

37. **Air-Handling Units** *Tempmaster Corporation* 37.1

38. **Air Makeup (Replacement Air or Makeup Air)** *Aerovent, Inc.* 38.1

39. **Desiccant Dryers** *Gas Research Institute* 39.1

40. **Reciprocating Refrigeration Units** *Continental Products Company, Inc.* 40.1

41. Absorption Chillers *Gas Energy, Inc.* 41.1

42. Centrifugal Chillers *York International Corporation* 42.1

43. Screw Compressors *Sullair Refrigeration, Sundstrand Corporation* 43.1

44. Cooling Towers *The Marley Cooling Tower Company* 44.1

45. Applications of HVAC Systems *Giffels Associates, Inc.* 45.1

46. HVAC Applications for Cogenerating Systems *Brown and Root, Inc.* 46.1

47. Pumps for Heating and Cooling 47.1
Dresser Pump Division, Dresser Industries

48. Valves *Rockwell International* 48.1

49. Noise Control *Industrial Acoustics Company* 49.1

50. Vibration *Mason Industries, Inc.* 50.1

51. Energy Conservation Practice *Sverdrup Corporation* 51.1

52. Automatic Temperature, Pressure, Flow Control Systems 52.1
Honeywell, Inc.

53. Water Conditioning *Metropolitan Refining Company, Inc.* 53.1

Appendix A Engineering Guide for Altitude Corrections
Sverdrup Corporation A.1

Appendix B Metric Conversion Tables *Robert C. Rosaler, P.E.* B.1

PREFACE

Heating, ventilating, and air-conditioning (HVAC)—or creating a comfortable environment—is at once one of the oldest and one of the most modern technologies. It encompasses everything from the warming radiant heat of the caveman's flames to the comfortably cooled industrial complexes in the Sahara desert and the pressurized comfort of the Challenger space module. Today it is not unusual for an inhabitant of an advanced industrial country to live almost entirely within an artificially created environment. HVAC has turned many environmentally hostile regions into useful, productive areas.

The objective of the *Handbook of HVAC Design* is to provide a practical guide and a reliable reference for designing and operating HVAC systems. It details the necessary steps for planning, design, equipment selection, operation and maintenance. Included are the relevant associated disciplines and considerations necessary for a broad understanding of this subject, including economic factors, pollution controls, and the physiology of comfort.

Each topic is addressed by a leading organization or practitioner in the field.

Acknowledgments

The editors wish to acknowledge the valuable assistance and guidance of McGraw-Hill editors Robert Hauserman and Lester Strong.

Nils R. Grimm
Robert C. Rosaler

GENERAL REFERENCES

A project design program is essential to assure an economical, energy-efficient, maintainable, and flexible design that will not only be technically adequate but also meet the client's and/or user's needs within the allocated budget. Three good references for developing design criteria for the total project (all disciplines) are:

Architects Handbook of Professional Practice, 11th ed., Chapter 11, "Project Practices," American Institute of Architects, Washington, D.C., 1988.

Project Checklist, document D200, American Institute of Architects, Washington, D.C., 1982.

Guidelines for Development of Architect/Engineer (A/E) Quality Control Manual, National Society of Professional Engineers (NSPE), Washington, D.C., 1977.

LIST OF CONTRIBUTORS

Gary M. Bireta, P.E. *Project Engineer, Mechanical Engineering, Giffels Associates, Inc., Southfield, Michigan*
CHAPTER 31: Chilled Water and Brine

Richard T. Blake *Chief Chemist, Metropolitan Refining Company, Inc., Long Island City, New York*
CHAPTER 53: Water Conditioning

Edward A. Bogucz, P.E. *Edwards Engineering Corp., Pompton Plains, New Jersey*
CHAPTER 17: Valance Heating and Cooling

Nick J. Cassimatis *Gas Energy, Inc., Brooklyn, New York*
CHAPTER 41: Absorption Chillers

Cleaver-Brooks *Division of Aqua-Chem, Inc., Milwaukee, Wisconsin*
CHAPTER 18: Boilers; CHAPTER 19: Fuels; CHAPTER 20: Piping; CHAPTER 21: Burners and Burner Systems; CHAPTER 22: Safety and Operating Controls; CHAPTER 23: Boiler Room Venting; CHAPTER 24: Deaerators; CHAPTER 25: Economizers; CHAPTER 26: Air Preheaters; CHAPTER 27: Sootblowers; CHAPTER 28: Good Practice in Inspection, Maintenance, and Operation of Boilers

K. Coleman *Staff Engineer, Van-Packer Company, Manahawkin, New Jersey*
CHAPTER 30: Factory-Built Prefabricated Vents, Chimneys, and Stacks

Edward Di Donato *Flow Control Division, Rockwell International, Livingston, New Jersey*
CHAPTER 48: Valves

David F. Fijas *ITT Standard, ITT Fluid Technology Corporation, Buffalo, New York*
CHAPTER 14: Heat Exchangers

Warren Fraser *Dresser Pump Division, Dresser Industries, Mountainside, New Jersey*
CHAPTER 47: Pumps for Heating and Cooling

Ernest H. Graf, P.E. *Assistant Director, Mechanical Engineering, Giffels Associates, Inc., Southfield, Michigan*
CHAPTER 32: All-Air Systems; CHAPTER 45: Applications of HVAC Systems

Nils R. Grimm, P.E. *Section Manager—Mechanical, Sverdrup Corporation, New York, New York*
EDITOR CHAPTER 2: Design Calculations; CHAPTER 3: Pipe Sizing; CHAPTER 4: Duct Sizing; CHAPTER 51: Energy Conservation Practice; Appendix A: Engineering Guide for Altitude Corrections

Edward B. Gut *Honeywell, Inc., Arlington Heights, Illinois*
CHAPTER 52: Automatic Temperature, Pressure, Flow Control Systems

John C. Hensley *Marketing Services Manager, Marley Cooling Tower Company, Mission, Kansas*
CHAPTER 44: Cooling Towers

M. B. Herbert, P.E. *Consulting Engineer, Willow Brook, Pennsylvania*
CHAPTER 1: Conceptual and Preliminary Design

Martin Hirschorn *President, Industrial Acoustics Company, Bronx, New York*
CHAPTER 49: Noise Control

P. Hodson *Vice President and Manufacturing Manager, Van-Packer Company, Manahawkin, New Jersey*
CHAPTER 30: Factory-Built Prefabricated Vents, Chimneys, and Stacks

Robert Jorgensen *Buffalo Forge Company, Buffalo, New York*
CHAPTER 34: Fans and Blowers

Michael K. Kennon *The King Company, Owatonna, Minnesota*
CHAPTER 16: Door Heating

Ronald A. Kondrat *Product Manager, Heating Division, Modine Manufacturing Company, Racine, Wisconsin*
CHAPTER 12: Heating Distribution Systems and Equipment for Unit Heaters

Douglas Kosar *Project Manager, Cooling Systems, Gas Research Institute, Chicago, Illinois*
CHAPTER 39: Desiccant Dryers

Lehr Associates *Valentine A. Lehr, President, New York, New York*
CHAPTER 5: Steam; CHAPTER 6: Hot-Water Systems; CHAPTER 7: Infrared Heating; CHAPTER 8: Electric Heating Systems; CHAPTER 9: Solar Space Heating; CHAPTER 10: Snow-Melting Systems; CHAPTER 11: Heat Tracing

Robert L. Linstroth *Product Manager, Heating Division, Modine Manufacturing Company, Racine, Wisconsin*
CHAPTER 12: Heating Distribution Systems and Equipment for Unit Heaters

Chan Madan *Continental Products, Inc., Indianapolis, Indiana*
CHAPTER 40: Reciprocating Refrigeration Units

Ravi K. Malhotra, Ph.D., P.E. *President, RKM Associates, Inc., St. Louis, Missouri*
CHAPTER 35: Coils

Norman J. Mason *President, Mason Industries, Inc., Hauppague, New York*
CHAPTER 50: Vibration

Simo Milosevic, P.E. *Mechanical Engineering, Giffels Associates, Inc., Southfield, Michigan*
CHAPTER 33: Direct Expansion Systems

Kenneth Puetzer *Chief Engineer, Sullair Refrigeration, Subsidiary of Sundstrand Corporation, Michigan City, Indiana*
CHAPTER 43: Screw Compressors

James A. Reese *Tempmaster Corporation, North Kansas City, Missouri*
CHAPTER 37: Air-Handling Units

Robert G. Reid *Quality Assurance Manager, CAM Industries, Inc., Seattle, Washington*
CHAPTER 29: Electric Boilers

Richard D. Rivers *Environmental Quality Sciences, Inc., Louisville, Kentucky*
CHAPTER 36: Air Filtration and Air Pollution Control Equipment

Robert C. Rosaler, P.E. *Consultant, Santa Rosa, California*
EDITOR APPENDIX B: Metric Conversion Tables

Robert E. Ross *Hydronics Institute, Berkeley Heights, New Jersey*
CHAPTER 15: Radiation for Steam and Hot-Water Heating

Robert A. Russell, Jr. *Chief Engineer, Ted Reed Thermal, Inc., West Kingston, Rhode Island*
CHAPTER 13: Hydronic Cabinet Unit Heaters

John M. Schultz *Retired Chief Engineer, York International Corporation, York, Pennsylvania*
CHAPTER 42: Centrifugal Chillers

Joseph F. Schulz *Chairman, Van-Packer Company, Manahawkin, New Jersey*
CHAPTER 30: Factory-Built Prefabricated Vents, Chimneys, and Stacks

Walter B. Schumacher *Vice President, Engineering, Aerovent, Inc., Piqua, Ohio*
CHAPTER 38: Air Makeup (Replacement Air or Makeup Air)

Alan J. Smith *Brown & Root, Inc., Houston, Texas*
CHAPTER 46: HVAC Applications for Cogenerating Systems

Will Smith *Dresser Pump Division, Dresser Industries, Mountainside, New Jersey*
CHAPTER 47: Pumps for Heating and Cooling

ABOUT THE EDITORS

NILS R. GRIMM has more than thirty years of experience in the design of HVAC systems. Currently section manager for the Sverdrup Corporation, New York, Mr. Grimm is a Registered Professional Engineer in New York, New Jersey, and Connecticut and a member of the American Society of Heating, Refrigerating, and Air Conditioning Engineers (ASHRAE). He is author of the chapter titled "HVAC Systems" in McGraw-Hill's *Standard Handbook of Plant Engineering*.

ROBERT C. ROSALER is a consulting engineer with thirty years of experience as an engineering executive in design and manufacturing. Mr. Rosaler was editor-in-chief of the *Standard Handbook of Plant Engineering*, winner of the Association of American Publishers' Award as the outstanding engineering book of 1983. A member of ASHRAE, the American Institute of Plant Engineers, and the American Society of Mechanical Engineers, Mr. Rosaler is a Licensed Professional Engineer in New York and Massachusetts.

CHAPTER 1

CONCEPTUAL AND PRELIMINARY DESIGN

M. B. Herbert, P.E.
Consulting Engineer, Willow Grove, Pennsylvania

1.1 INTRODUCTION

Heating, ventilating, and air-conditioning (HVAC) systems are designed to provide control of space temperature, humidity, air contaminants, differential pressurization, and air motion. Usually an upper limit is placed on the noise level that is acceptable within the occupied spaces. To be successful, the systems must satisfactorily perform the tasks intended.

Most heating, ventilating, and air-conditioning systems are designed for human comfort. Human comfort is discussed at length in Ref. 1. This reference should be studied until it is understood because it is the objective of HVAC design.

Many industrial applications have objectives other than human comfort. If human comfort can be achieved while the demands of industry are satisfied, the design will be that much better.

Heating, ventilating, and air-conditioning systems require the solution of energy-mass balance equations to define the parameters for the selection of appropriate equipment. The solution of these equations requires the understanding of that branch of thermodynamics called "psychrometrics." Ref. 2 should be studied.

Automatic control of the HVAC system is required to maintain desired environmental conditions. The method of control is dictated by the requirements of the space. The selection and the arrangement of the system components are determined by the method of control. Controls are necessary because of varying weather conditions and internal loads. These variations must be understood before the system is designed. Control principles are discussed in Chap. 52 and in Ref. 3.

The recent proliferation of affordable computers with more than 500 kilobytes of random access memory has made it possible for most offices to automate their design efforts. Each office should evaluate its needs, choose from the available computer programs on the market, and then purchase a compatible computer and its peripherals.

No one office can afford the time to develop all its own programs. Time is also required to become proficient with any new program, including those developed "in-house."

Purchased programs are not always written to give the information required, thus they should be amenable to in-house modification. Documentation of purchased programs should describe operation in detail so that modification can be achieved with a minimum of effort. Refer to Chap. 2.

1.2 CONCEPT PHASE

The conceptual phase of the project is the feasibility stage; here the quality of the project and the amount of money to be spent are decided. This information should be gathered and summarized on a form similar to Fig. 1.1.

1.2.1 Site Location and Orientation of Structure

The considerations involved in the selection of the site for a facility are economic:

1. Nearby raw materials
2. Nearby finished-goods markets
3. Cheap transportation of materials and finished goods
4. Adequate utilities and low-cost energy sources for manufacturing
5. Available labor pool
6. Suitable land
7. Weather

These factors can be evaluated by following the analysis given in the *Handbook of Industrial Engineering and Management* (see Sec. 1.5). It is prudent to carefully evaluate several alternative sites for each project.

The orientation of the structure is dictated by considering existing transportation routes, obstructions to construction, flow of materials and products through the plant, personnel accessibility and security from intrusion, and weather.

1.2.2 Codes, Rules, and Regulations

Laws are made to establish minimum standards, to protect the public and the environment from accidents and disasters. Federal, state, and local governments are involved in these formulations. Insurance underwriters may also impose restraints on the design and operation of a facility. It is incumbent upon the design team to understand the applicable restraints before the design is begun. Among the applicable documents that should be studied are

1. Occupational Safety and Health Act (OSHA)
2. Environmental Protection Agency (EPA) requirements
3. National Fire Protection Association (NFPA), Fire Code (referenced in OSHA)

COMPANY ________ PROJECT ________ P.O. NO. ________ DATE ________

LOCATION ________ SPACE NAME ________ SIZE ________ SHEET NO. ________

ACTIVITY

	M	T	W	T	F	S	S
DAY OF WEEK							
NO. PEOPLE							
HOURS/DAY							

BUILDING CONSTRUCTION

FLOOR ________
WALLS ________
WINDOW GLASS ________ FRAME ________
SHADING ________
CEILING ________
ROOF ________
DOORS ________
PARTITIONS ________

CODES, BUILDING ________
MECHANICAL ________
PLUMBING ________
ELECTRICAL ________
FIRE ________

ENVIRONMENT

TEMPERATURE ____ ± ____ dB, ____ ± ____ WB, ____ ± ____ RH
VENTILATION ____ CFM, ____ AC/HR, ____ %OA
AIR FILTERS ____ TYPE, ____ %EFF. ____
AIR PRESSURE ____ IN.W.W., (+), (−)

LIGHTING TYPE ________, WATTS ________
ELECTRICAL CLASS ________
EMERGENCY-LIGHTING ________ POWER ________
TYPE CONTROL ________

TELEPHONE ________ INTERCOM ________
CCTV ________ COMPUTER ________
WORD PROCESSOR ________

HAZARDS & SAFETY

FIRE CLASS ________
HAZARDOUS MATERIALS & QUANTITIES ________

TYPE OF FIRE PROTECTION ________

REASONS ________

TYPE OF FIRE ALARM ________
SAFETY SHOWER & EYEWASH ________
FIRE BLANKET ________ STRETCHER ________

ITEM	EQUIPMENT LIST	QUATITY	SIZE			PLUMBING												PROCESS																	NOTES
						COLD WATER			HOT WATER			WASTE						GAS		AIR		VACUUM		STEAM		VENTILATION							ELECTRICAL		
																										EXH.		SUP.							
			LENGTH	WIDTH	HEIGHT	PRESSURE	PRESS. LOSS	GPM	TEMP./PRESS	PRESS. LOSS	GPM	SANITARY,GPM	PROCESS,GPM					PRESSURE	SCFM/SCFM	PRESSURE	SCFM	PRESSURE	SCFM	TEMP./PRESS	LB/HR	TEMP./ΔP	SCFM	TEMP./ΔP	SCFM				VOLTS/PHASE	KVA/KW	

FIGURE 1.1 Design information.

4. Local building codes
5. Local energy conservation laws, which usually follow the American Society of Heating, Refrigeration, and Air-Conditioning Engineers (ASHRAE) Standard 90-75

1.2.3 Concept Design Procedures

The conceptual phase requires the preparation of a definitive scope of work. Describe the project in words. Break it down to its components. Itemize all unique requirements, what is required, why, and when. Budgeting restraints on capital costs and labor hours should be included. A convenient form is shown in Fig. 1.2. This form is a starting tool for gathering data. It will suffice for many projects. For a major project, a more formal written document should be prepared and approved by the client. This approval should be obtained before proceeding with the design.

The method of design is influenced by the client's imposed schedule. Fast-tracking methods will identify long delivery items that might require early purchase. Multiple construction packages are not uncommon, since they appreciably reduce the length of construction time. Usually, more engineering effort is required to divide the work into separate bid packages. Points of termination of each contract must be shown on the drawings and reflected in the scope of work in the specifications. Great care in the preparation of these documents is required to prevent omission of some work from all contracts and inclusion of some work in more than one contract.

Some drawings and some sections of the specifications will be issued in more than one bid package. To prevent problems, the bid packages should be planned in the concept stage and carried through to completion of the project. All changes must be defined clearly for everyone involved in the project.

Every step of the design effort should be documented in written form. When changes are made that are beyond the scope of work, the written documents help recover costs necessitated by these changes. Also, any litigation that may be instituted will usually result in decisions favorable to those with the proper documentation.

After the scope of work has been accurately written down and approved, assemble the data necessary to accomplish the work:

1. Applicable building codes
2. Local laws and ordinances
3. Names, titles, addresses, and telephone numbers of local officials
4. Names, titles, addresses, and telephone numbers of client contacts
5. Client's standards

If the project is similar to previous designs, review what was done before and how well the previous design fulfilled its intended function.

Use check figures from this project to make an educated guess of the sizes and capacities of the present project. Use Figs. 1.3 and 1.4 to record past projects.

Every project has monetary constraints. It is incumbent upon the consultant to live within the monies committed to the facility. Use Figs. 1.5 and 1.6 to estimate the capacities and costs of the systems. Do not forget to increase the costs

COMPANY ______ PO NO. ______ DATE ______

LOCATION ______ SHEET NO. ______

SUBJECT ______

PROJECT BRIEF

COMPUTED BY ______ CHECKED BY ______

TYPE OF PROJECT ______

☐ HEATING ______
☐ VENTILATING, Comfort, Process, ______
☐ AIR CONDITIONING, Comfort, Process, ______
☐ PLUMBING, Sewage Treatment ______
☐ FIRE PROTECTION ______
☐ PROCESS PIPING ______
☐ ELECTRICAL, Power, Lighting, Control ______
☐ STRUCTURAL, Civil ______
☐ ARCHITECTURAL ______
☐ ______

DUE DATES:

Preliminaries ______, Cost Estimates ______, Final Documents ______,

SCOPE OF WORK

PROJECT ASSIGNMENTS: Proj. Mgr. Proj. Engr.

Discipline Engrs.

CONTACTS

Name & Title	Firm Name	Address	Telephone

FIGURE 1.2 Project brief.

from the year that the dollars were taken to the year that the construction is to take place.

Justification for the selection of types of heating, ventilating, and cooling systems is usually required. Some clients require a detailed economic analysis based on life cycle costs. Others may require only a reasonable payback time. If a sys-

JOB NAME SPACE NAME YEAR OF DESIGN TYPE OF SYSTEM	OUTSIDE DESIGN CONSIDERATIONS		INSIDE DESIGN CONSIDERATIONS		FLOOR AREA	CFM/SQ FT	% O A	BTU/HR-SQ FT (W/HR-SQ M)		LIGHT & POWER	SQ FT/PERSON	SQ FT/TON	IND APP DEW POINT
	DB °F/°C	WB °F/°C	DB °F/°C	WB °F/°C	SQ FT (SQ M)	(CMS/SQ M)		ROOM SENS	GRAND TOTAL	WATTS/SQ FT (WATTS/SQ M)	(SQ M/PERSON)	(SQ M/KW)	°F (°C)

FIGURE 1.3 Air-conditioning check figures.

JOB NAME SPACE NAME YEAR OF DESIGN TYPE OF SYSTEM	DESIGN CONSIDERATIONS		FLOOR AREA SQ. FT. (SQ. M)	VENTILATION		INFILTRATION		HEATING LOAD BTU/HR-SQ FT (W/HR-SQ M)	NOTES
	OUTSIDE °F (°C)	INSIDE °F (°C)		$\frac{CFM}{SQ\ FT}$ $\left(\frac{CMS}{SQ\ M}\right)$	AC/HR	$\frac{CFM}{SQ\ FT}$ $\left(\frac{CMS}{SQ\ M}\right)$	AC/HR		

FIGURE 1.4 Heating check figures.

COMPANY __________ PO NO. __________ DATE __________

LOCATION __________ SHEET NO. __________

SUBJECT __________

COMPUTED BY __________ CHECKED BY __________

ROOM NAME & SIZE TYPE OF SYSTEM	FLOOR AREA SQ FT (SQ M)	ROOM VOLUME CU FT (CU M)	HEAT REQUIRED			HEAT LOAD BTU/HR (KW)	ESTIMATED COST
			BTU/SQ FT (W/SQ M)	BTU/SQ FT (W/CU M)	BTU/SQ FT (W/CMS)		

FIGURE 1.5 Conceptual design estimate for air conditioning.

COMPANY ______________________ PO NO. __________ DATE __________

LOCATION ______________________ SHEET NO. __________

SUBJECT ______________________

____

COMPUTED BY ______________________ CHECKED BY ______________________

ROOM NAME & SIZE TYPE OF SYSTEM	FLOOR AREA SQ FT (SQ M)	ROOM VOLUME CU FT (CU M)	HEAT REQUIRED BTU/SQ FT (W/SQ M)	HEAT REQUIRED BTU/SQ FT (W/CU M)	HEAT REQUIRED BTU/SQ FT (W/CMS)	HEAT LOAD BTU/HR (KW)	ESTIMATED COST

FIGURE 1.6 Conceptual design estimate for heating.

tem cannot be justified on a reasonable payback basis, then it is unreasonable to expect the more detailed analysis of life cycle costs to reverse the negative results. A simple comparison between two payback alternatives can be made as follows:

$$\text{Payback years } N = \frac{\$ \text{ first cost}}{\$ \text{ savings, first year}} \tag{1.1}$$

This simple payback can be refined by considering the cost of money, interest rate i (decimal), and escalation rate e (decimal). The escalation rate is the expected rate of costs of fuel, power, or services. The actual number of years for payback n is given by

$$n = \frac{\log [1 + N(R - 1)/R]}{\log R} \tag{1.2}$$

where

$$R = \frac{1 + e}{1 + i}$$

and N is defined by Eq. (1.1). This formula is easily programmed on a hand-held computer. A nomographic solution is provided in Ref. 4.

There are many other economic models that a client or an engineering staff can use for economic analysis. Many books have been published on this subject from which the engineer may choose. Refer to Chap. 51.

1.3 PRELIMINARY DESIGN PHASE

The preliminary design phase is the verification phase of the project. Review the concept phase documents, especially if a time lapse has occurred between phases. Verify that the assumptions are correct and complete. If changes have been made, even minor ones, document these in writing to all individuals involved.

1.3.1 Calculation Book

The calculations are the heart of decision making and equipment selection. The calculation book should be organized so that the calculations for each area or system are together. Prepare a table of contents so anyone may find the appropriate calculations for a given system. Use divider sheets between sections to expedite retrieval. All calculations should be kept in one place. Whenever calculations are required elsewhere, make the necessary reproductions and promptly return the originals to their proper place in the calculation book.

1.3.2 Calculations

The calculations reflect on the design team. The calculations should be neat, orderly, and complete, to aid checking procedures. Most industrial clients require that the calculations be submitted for their review. Also when revisions are required, much less time will be spent making the necessary recalculations. All calculations made during this phase should be considered accurate, final calculations.

Many routine calculations can now be done more rapidly and more accurately with the aid of a computer. The computer permits rapid evaluation of alternatives and changes. If a computer program is not available for a routine calculation, the calculation should be done and documented on a suitable form. If a form does not exist, develop one.

All calculations should be dated and signed by the designer and checker. Each sheet should be assigned an appropriate number. When a calculation sheet is revised, a revision date should be added. When a calculation sheet is superseded, the sheet should be marked "void." Do not dispose of superseded calculations until the project is built satisfactorily and functioning properly.

List all design criteria on sheets such as Fig. 1.7, referencing sources where applicable. List all references used in the design at appropriate points in the calculations.

When you are doing calculations, especially where forms do not exist, always follow a number with its units, such as feet per second (meters per second), British thermal units (watts, foot-pounds, newton-meters), etc. This habit will help to prevent the most common blunders committed by engineers.

To avoid loose ends and errors of omission, always try to complete one part or section of the work before beginning the next. If this is impossible, keep a "things to do" list, and list these open ends.

1.3.3 Equipment Selection

From the calculations and the method of control, the capacity and operating conditions may be determined for each component of the system. Manufacturers' catalogs give extensive tables and sometimes performance curves for their equipment. All equipment that moves or is moved vibrates and generates noise. In most HVAC systems, noise is of utmost importance to the designer. The designer should know a lot about acoustics and vibrations. Read Chaps. 49 and 50 carefully. Beware of the manufacturer that is vague or ignorant about the noise and vibration of its equipment or is reluctant to produce certified test data.

Many equipment test codes have been written by ASHRAE, American Refrigeration Institute (ARI), Air Moving and Conditioning Association (AMCA), and other societies and manufacturer groups. A comprehensive list of these codes is contain in Chap. 13 and in ASHRAE handbooks. Manufacturer's catalogs usually contain references to codes by which their equipment has been rated. Designers are warned to remember that the manufacturer's representative is awarded for sales of equipment, and not for disseminating advice. Designers should make their own selections of equipment and should write their own specifications, based on past experience.

1.3.4 Equipment Location

Mechanical and electrical equipment must be serviced periodically and eventually replaced when its useful life has expired. To achieve this end, every piece of equipment must be accessible and have a planned means of replacement.

COMPANY ________________ P.O. NO. ________ DATE ________

LOCATION ________________ SHEET NO. ________

SUBJECT ________________

COMPUTED BY ________________ CHECKED BY ________________

OUTSIDE DESIGN DATA

Data for ________________ Elevation above mean sea level ________

Latitude ____ ° ____ ' ____ Latitude ____ ° ____ ' ____

Item	Winter	Summer
Temperature, DB/WB/DP†	/ /	/ /
Pressure, Total/Vapor	/	/
Humid. Ratio/%RH/Enthalpy	/ /	/ /
Specific Volume		
Mean Daily Temp Range		
Wind Velocity		
Hours Exceed Design, %		

Summer Design Day Temperatures

Hour	1	2	3	4	5	6	7	8	9	10	11	12	13	14	15	16	17	18	19	20	21	22	23	24
DB																								

Month	Cooling Out. Design		CLTD Corrections											
	DB	WB	To	N	NNE/NNW	NE/NW	ENE/WNW	E/W	ESE/WSW	SE/SW	SSE/SSW	S	Horiz.
JAN													
FEB													
MAR													
APR													
MAY													
JUN													
JUL													
AUG													
SEP													
OCT													
NOV													
DEC													

Heating Degree Days

Month	JAN	FEB	MAR	APR	MAY	JUN	JUL	AUG	SEP	OCT	NOV	DEC	YEAR
D.D.													

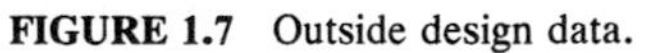

FIGURE 1.7 Outside design data.

The roof and ceiling spaces are not adequate equipment rooms. Placing equipment on the roof subjects the roof to heavy traffic, usually enough to void its guarantee. The roof location also subjects maintenance personnel to the vagaries of the weather. In severe weather, the roof may be too dangerous for maintenance personnel.

Ceiling spaces should not be used for locating equipment. Servicing equipment in the ceiling entails erecting a ladder at the proper point and removing a ceiling tile or opening an access door, to gain access to the equipment. Crawling over the ceiling is dangerous and probably violates OSHA regulations. No matter how careful the maintenance personnel are, eventually the ceiling will become dirty, the tiles will be broken, and if water is involved, the ceiling will be stained.

Also, the equipment will suffer from lack of proper maintenance, because no one on a ladder can work efficiently. This work in the occupied space is disruptive to the normal activities of that space.

Equipment should be located in spaces specifically designed to house them. Sufficient space should be provided so that workers can walk around pieces of equipment, swing a wrench, rig a hoist, or replace an electric motor, fan shaft, or fan belts. Do not forget to provide space for the necessary electrical conduits, piping, and air ducts associated with this equipment. Boilers and other heat exchangers require space for replacing tubes. Valves in piping should be located so that they may be operated without resorting to a ladder or crawling through a tight space. If equipment is easily reached, it will be maintained. Adequate space also provides for good housekeeping, which is a safety feature.

Provision of adequate space in the planning stage can be made only after the types and sizes of systems have been estimated. Select equipment based on the estimated loads. Lay out each piece to a suitable scale. Arrange the equipment room with cutout copies of the equipment. Allow for air ducts, piping, electrical equipment, access aisles, and maintenance workspace. Cutouts permit several arrangements to be prepared for study.

When you are locating the equipment rooms, be sure each piece of equipment can be brought into and removed from the premises at any time during the construction. A strike may delay the delivery of a piece of equipment beyond its scheduled delivery date. This delay should not force construction to be halted, as it would if the chiller or boiler had to be set in place before the roof or walls were constructed.

1.3.5 Distribution Systems

HVAC distribution systems are of two kinds: air ducts and piping. Air ducts are used to convey air to and from desired locations. Air ducts include supply air, return-relief air, exhaust air, and air-conveying systems. Piping is used to convey steam and condensate, heating hot water, chilled water, brine, cooling tower water, refrigerants, and other heat-transfer fluids. Energy is required to force the fluids through these systems. This energy should be considered when systems are evaluated or compared.

System Layouts. Locate the air diffusers and heat exchangers on the prints of the architectural drawings. Note the air-flow rates for diffusers and the required capacities for the heat exchangers. Draw tentative single-line air ducts from the air apparatus to the air diffusers. Mark on these lines the flow rates from the most remote device to the fan. With these air quantities, the air ducts may be sized. Use Chap. 4 or *ASHRAE Handbook, Fundamentals*, Chap. 32, or the *Industrial Ventilation Manual* to size these ducts. Record these sizes on a form similar to those shown there.

A similar method is used to size the piping systems; see Chap. 3. Remember, steam, condensate, and refrigerant piping must be pitched properly for the systems to function correctly. Water systems should also be pitched to facilitate draining and elimination of air.

Piping systems are briefly described in appropriate chapters of this book and in the *ASHRAE Handbook, Fundamentals*. A more substantial treatment is contained in Crocker and King's *Piping Handbook* (see Sec. 1.5).

1.4 REFERENCES

1. *1989 ASHRAE Handbook, Fundamentals*, ASHRAE, Atlanta, GA, 1989, chap. 8, "Physiological Principles for Comfort and Health."
2. *ASHRAE Handbook, Fundamentals*, chap. 6, "Psychrometrics."
3. John E. Hains, *Automatic Control of Heating and Air Conditioning*, McGraw-Hill, New York, 1953.
4. John Molnar, *Nomographs—What They Are and How to Use Them*, Ann Arbor Science Publishers, Ann Arbor, MI, 1981.

1.5 BIBLIOGRAPHY

ASHRAE: *Cooling and Heating Load Calculation Manual*, American Society of Heating, Refrigeration, and Air-Conditioning Engineers, Atlanta, 1979.

———: *Energy Conservation in New Building Design*, ASHRAE Standard 90-75, Atlanta, 1980.

———: Brochure on Psychrometry, ASHRAE, Atlanta, 1977.

———: *Simplified Energy Analysis Using the Modified Bin Method*, ASHRAE, Atlanta, 1984.

———: *1989 ASHRAE Handbook, Fundamentals*, ASHRAE, Atlanta, GA, 1989.

———: *1988 ASHRAE Handbook, Equipment*.

———: *1987 ASHRAE Handbook, HVAC Systems and Applications*.

———: *1986 ASHRAE Handbook, Refrigeration Systems and Applications*.

Baldwin, John L.: *Climates of the United States*, Government Printing Office, Washington, DC, 1974.

Crocker, S, and R. C. King: *Piping Handbook*, 5th ed., McGraw-Hill, New York, 1973.

Fan Engineering, 8th ed., Buffalo Forge Co., Buffalo, NY, 1980.

Hains, John E.: *Automatic Control of Heating and Air Conditioning*, McGraw-Hill, New York, 1953.

Handbook of Industrial Engineering and Management, 2d ed., Prentice-Hall, Englewood Cliffs, NJ, 1971.

Hydraulic Institute: *Pipe Friction Manual*, Hydraulic Institute, Cleveland, 1975.

Industrial Ventilation, A Manual of Recommended Practice, 17th ed., American Conference of Governmental Industrial Hygienists, Lansing, MI, 1982.

Kusuda, T.: *Algorithms for Psychrometric Calculations*, National Bureau of Standards, Government Printing Office, Washington, DC, 1970.

Molnar, John: *Facilities Management Handbook*, Van Nostrand Reinhold, New York, 1983.

———: *Nomographs—What They Are and How to Use Them*, Ann Arbor Science Publishers, Ann Arbor, MI, 1981.

NFPA: *National Fire Codes*, National Fire Protection Association, Batterymarch Park, Quincy, MA, 1985.

CHAPTER 2
DESIGN CALCULATIONS

Nils R. Grimm, P.E.
Section Manager, Mechanical, Sverdrup Corporation,
New York, New York

2.1 INTRODUCTION

One of the cardinal rules for a good, economical energy-efficient design is *not* to design the total system (be it heating, ventilating, air conditioning, exhaust, humidification, dehumidification, etc.) to meet the most critical requirements of just a small (or minor) portion of the total area served. That critical area should be isolated and treated separately.

The designer today has the option of using either a manual method or a computer program to calculate heating and cooling loads, select equipment, and size piping and ductwork. For large or complex projects, computer programs are generally the most cost effective and *should be used*. On projects where life cycle costs and/or annual energy budgets are required, *computer programs should be used*.

Where one or more of the following items will probably be modified during the design phase of a project, *computer programs should be used*:

- Building orientation
- Wall or roof construction (overall *U* value)
- Percentage of glazing
- Building or room sizes

However, for small projects a manual method should be seriously considered before one assumes automatically that computer design is the most cost-effective for all projects.

In the next section, heating and cooling loads are treated together since the criteria and the computer programs are similar.

2.2 HEATING AND COOLING LOADS

The first step in calculating the heating and cooling loads is to establish the project's heating design criteria:

- Ambient dry-bulb or wet-bulb temperature (or relative humidity), wind direction and speed
- Site elevation above sea level, latitude
- Space dry-bulb or wet-bulb temperature (or relative humidity), ventilation air
- Internal or process heating or cooling and exhaust air requirements
- Hours of operation of the areas or spaces to be heated or cooled (day, night, weekday, weekends, and holidays)

Even when the owner or user has established the project design criteria, the designer should determine that they are reasonable.

The winter outdoor design temperature should be based preferably on a minimum temperature that will not be exceeded for 99 percent of the total hours in the months of December, January, and February (a total of 2160 h) in the northern hemisphere and the months of June, July, and August in the southern hemisphere (a total of 2208 h). However, for energy conservation considerations, some government agencies and the American Society of Heating, Refrigeration, and Air-Conditioning Engineers (ASHRAE) Standard 90-75, *Energy Conservation in New Building Design*, require the outdoor winter design temperature to be based on a temperature that will not be exceeded 97.5 percent of the same total heating hours.

Similarly, the summer outdoor design dry-bulb temperature should be based on the lowest dry-bulb temperature that will not be exceeded 2½ percent of the total hours in June through September (a total of 2928 h) in the northern hemisphere and in December through March in the southern hemisphere (a total of 2904 h). For energy conservation reasons, some government agencies require the outdoor summer design temperature to be based on a dry-bulb temperature that will not be exceeded 5 percent of the same total cooling hours.

More detailed or current weather data (including elevation above sea level and latitude) are sometimes required for specific site locations in this country and around the world than are included in standard design handbooks such as Refs. 1 and 2 or computer programs such as Refs. 3 and 4 or from Ref. 5.

It is generally accepted that the effect of altitude on systems installed at 2000 ft (610 m) or less is negligible and can be safely omitted. However, systems designed for installations at or above 2500 ft (760 m) must be corrected for the effects of high altitude. Appropriate correction factors and the effects of altitudes at and above 2500 ft (760 m) are discussed in App. A of this book.

To avoid overdesigning the heating, ventilating, and air-conditioning system so as to conserve energy and to minimize construction costs, each space or area should be analyzed separately to determine the minimum and maximum temperatures that can be maintained and whether humidity control is required or desirable. For a discussion of humidity control see Chap. 39, "Desiccant Dryers," in this book.

The U.S. government has set 68°F (20°C) as the maximum design indoor temperature for personnel comfort during the heating season in areas where employees work. In manufacturing areas the *process requirements* govern the actual temperature. From an energy conservation point of view, if a process requires a space temperature greater than 5°F (2.8°C) above or below 68°F (20°C), the space should, if possible, be treated separately and operate independently from the general personnel comfort areas. The staff members working in such areas should be provided with supplementary spot (localized) heating, ventilating, and air-conditioning systems as the conditions require, in order to maintain personnel comfort.

The space's dry-bulb temperature, relative humidity, number of people, and ventilation air requirements can be established (once the activity to be performed in each space is known) from standard design handbook sources such as Refs. 2, 6 to 8, 10, and 22 for heating and Refs. 1, 6 to 22, 27, and 40 for cooling.

The normal internal loads generally produce a heat gain and therefore usually are not considered in the space heating load calculations but must be included in cooling load calculations. These internal loads, including process loads, are listed in standard design handbook sources such as Refs. 23 and 24.

The process engineering department or quality control group should determine the manufacturing process space temperature, humidity, and heating requirements. The manufacturer of the particular process equipment is an alternative source for the recommended space and process requirements.

The air temperature at the ceiling may exceed the comfort range and should be considered in calculating the overall heat transmission to or from the outdoors. A normal 0.75°F (0.42°C) increase in air temperature per 1 ft (0.3 m) of elevation above the breathing level [5 ft (1.5 m) above finish floor] is expected in normal applications, with approximately 75°F (24°C) temperature difference between indoors and outdoors.

There is limited information on process heating requirements in standard handbooks, such as Refs. 25 to 35, and on cooling requirements, such as Refs. 25, 27, and 29 to 35.

Usually the owner and/or user establishes the hours of operation. If the design engineer is not given the hours of operation for the basis of the design, she or he must jointly establish them with the owner and/or user.

The method of calculating the heating or cooling loads (manual or computer) should be determined next.

2.2.1 Manual Method

If the manual method is selected, the project heating loads should be calculated by following one of the accepted procedures found in standard design sources such as Refs. 21, 22, and 36 to 39. For cooling loads, see Refs. 21 to 24, 37 to 39, 41, and 42.

2.2.2 Computer Method

If the computer method has been chosen to calculate the project heating or cooling loads, one must then select a program to use among the several available. Two of the most widely used for heating and cooling are Trane's TRACE and other Customer Direct Service (CDS) Network diskettes and Carrier's E20-II programs.

Regardless of the program used, its specific input and operating instructions must be strictly followed. It is common to trace erroneous or misleading computer output data to mistakes in inputting the design data into the computer. *It cannot be overstressed that to get meaningful output results, the input data must be correctly entered and checked after entry before the program is run*. It is also a good policy, if not a mandatory one, to independently check the computer results the first time you run a new or modified computer program, to ensure the results are valid.

If the computer program used does not correct the computer output for the effects of altitude when the elevation of the project is equal to or greater than

2500 ft (760 m) above sea level, the computer output must be manually corrected by using the appropriate correction factors, listed in App. A of this book.

We outline the computer programs available with TRACE and other CDS diskettes and E-20-II in the remainder of this chapter. However, this is not to imply that these are the only available sources of programs for the HVAC fields. Space restraints and similarities to other programs are the main reasons for describing programs from only two sources.

2.2.3 Trane Programs

The following summary describes the programs available to the designer using Trane's TRACE and their other CDS computer programs to calculate heating and cooling loads.

Load Design—Ultra Edition (DSC-IBM-125). This is the workhorse of Trane's CDS Network package. It is designed to model the most complex load designs. It still has the flexibility, however, to do the quick runs with as few as 10 lines of input. It is capable of modeling

- Skylights
- Partitions
- Slab on grade
- Exposed floors, buildings, side fins, and overhangs
- Multiple exposures for walls and roofs
- Multiple system types
- Duplicate floors and zones
- Choice of units
- Unlimited walls, roofs, and floors per zone per run

The program runs completely on the microcomputer and is based entirely on ASHRAE algorithms and actual hour-by-hour weather tape data. All ASHRAE wall, floor, roof, and slab types are preloaded into the program. The user chooses either ASHRAE's total equivalent temperature difference (TETD) or cooling load temperature difference (CLTD) method. Output reports are chosen by the user.

Load Design, Weather Regions 1 through 10, Ultra Edition (DSC-IBM-151 through 160). These diskettes consist of weather tapes from the U.S. Weather Bureau (Natural Climatic Data Center) for use with design programs and/or energy analysis programs.

Energy Analysis Programs, TRACE Economic Program (DSC-IBM-116). This building energy analysis program is designed to calculate hourly loads throughout the year. It calculates the yearly energy consumption, operating costs, and equipment payback. This program runs an hour-by-hour simulation of data.

FANMOD or Fan System Economics (DSC-IBM-122). This program identifies and compares fan and air handler life cycle costs. It can help designers answer questions such as: "Which modulation method shows the lowest life cycle cost, forward-curved (FC) fans with inlet vanes or FC fans with frequency inverters?" or "What is the best mix of fan sizes and modulation methods for my project?"

Government Economics (DSC-IBM-303). This program is an automated method of performing life cycle cost analyses, which meet the Department of Defense (DOD) criteria for new, replacement, and renovation construction. The user merely specifies whether an Army, Navy, or Air Force study is desired, and the program automatically utilizes the correct economic criteria—building energy consumptions for the year, utility rate structures, total investment cost, and any other known operating and annual maintenance cost. The program generates a life cycle cost for each alternative, based on the total investment dollars and operating costs over the life of the study. It also compares each alternative to the baseline alternative by generating a saving-to-investment ratio (SIR) and a discounted payback period (DPP). TRACE is a comprehensive energy analysis program incorporating thermal storage, transfer functions, user-defined wall and roof types, temperature drift, peak-demand charge analysis, and proper treatment of solar loading on perimeter zones. Hundreds of different air-conditioning system and equipment types are available to analyze, including ice storage and chilled-water storage. Approved by all major government agencies, the program is inputted entirely on the microcomputer and then uploaded to the mainframe computer via a telephone modem. Output is then downloaded back to the user's microcomputer. Typical turnaround time is less than half an hour.

TRACE II Input/Edit (DSC-IBM-201). This is a simplified version of the TRACE program, featuring reduced input to as few as two pages, automatic zoning, predefined interior zones, and schedules based on building types. The TRACE II program runs exactly the same energy analysis program as TRACE. TRACE and TRACE II are the only programs in the CDS Network that need mainframe assistance to run. TRACE II also runs on the mainframe via the microcomputer and modem; costs must also be input. This program generates a life cycle cost for each alternative, based on the total investment dollars and operating costs over the life of the study. It also compares each alternative to the baseline alternative by generating a saving-investment ratio and a DPP.

Coil Selection Programs. There are a number of these:

Hot Water Coil Selection (DSC-IBM-101)
Chilled Water Coil Selection (DSC-IBM-102)
Refrigerant Coil Selection (DSC-IBM-103)
Steam Coil Selection (DSC-IBM-130)
Equipment Air Handling Unit Selection (DSC-IBM-104)
Fan Selection (DSC-IBM-106)
Large Commercial Roof Top Selection (DSC-IBM-120)
Fan Coil Selection (DSC-IBM-127)
Fan Coil Horizontal Data (DSC-IBM-128)
Fan Coil Low Vertical/Vertical Data (DSC-IBM-129)

The CDS Network programs provide the user with four coil selection programs; air handler, rooftop, and fan selection programs; VAV (variable air volume) box selection; and a fan coil selection program. All these programs are stand-alone, are easily updated via telephone modem, make multiple selections (from lowest cost to most efficient), generate sound power data, and are certified by the American Refrigeration Institute (ARI).

Spec Writers (DSC-IBM-119). This specification program is a word processor designed specifically for guide specification writing and editing of equipment manufactured by Trane. The designer can easily mold and change specifications provided by the Trane representative. In addition, it includes the ability to pick and choose paragraphs and adapt several different numbering schemes.

Chiller Economics. This program provides a quick but credible estimate of the annual operating cost of various chiller system configurations. These configurations may include

- Chiller type
- Chiller efficiency
- Free cooling
- Thermal storage
- Auxiliary condenser
- Pumping efficiency
- Tracer optimization strategies

The program performs an hour-by-hour equipment simulation by giving the program either a profile of ton-hours and peak ton-hours per month or tons and equivalent full-load hours. Using this information and weather file, the program can generate the hourly profile.

The program produces monthly energy consumption and utility cost tables. These tables include kilowatts, kilowatthours, therms, and water. Some of the economic comparisons given include installed cost, first-year analysis, investment, payback, returns, and net present value of returns. The resulting report is concise and provides the necessary information for an equipment and operating comparative analysis.

Acoustic. Among optional programs, this one automates the extensive arithmetic calculations necessary in acoustical studies of proposed and existing heating, ventilating, and air-conditioning (HVAC) systems. The program includes manufacturers' acoustic data on silencers, ceilings, walls, insulation, ductwork, and equipment. Alternative attenuation materials can be quickly analyzed. Inclusion of ASHRAE and manufacturers' data eliminates the need to reference ASHRAE manuals or search out manufacturer's sound data during acoustical evaluations.

Contractor Estimating. Also optional, this program features preloaded equipment files and complete piping and duct estimates. Labor, overhead, and miscellaneous material may be added to the menu to satisfy unique or changing procedures. Historical information can be stored on a data base. Output includes a complete bill of materials and system cost including overhead, material, profit, and labor in both hours and dollars.

CAD Interface with Ultra Edition Load Design and Duct Design. Also optional, this system integrates the entire computer-aided drafting and computer-aided engineering/design (CAD) processes. With the Trane CDS software and Sigma Design or Autocad CAD system, you can start with initial calculations and finish with final schedules. Begin with an architectural outline of the building, and the system measures lengths and areas of zones, generates reports, and provides input for the Ultra Edition Load Design program. This information is then fed into the Duct Design program, completing the duct design process, including schedules. The result is a new way to design HVAC systems. The designer is responsible for the concepts, judgments, and required changes. In other words, the system is a "silent partner" that handles the routine tasks.

2.2.4 Carrier Programs

The following summary describes the programs that are available to the designer using Carrier's E20-II programs to calculate heating and cooling loads.

Load Estimating, Commercial Version 1.3. This programs provides the following:

- Heating load and air-conditioning load calculation.
- Changes in input data are easily made (for example, *U* values, building configurations, building orientation, shading factors, window area, etc.) without rerunning the entire program.
- Program contains weather data for 324 U.S. and Canadian cities. Additional weather data can be stored in this program.
- Calculates block, zone, and room loads.
- Capability of 12-, 16-, and 24-h operation.
- Capability to rotate building and/or zones.
- Calculation of shading for reveals and/or overhangs.
- Able to store unlimited number of zones or blocks (up to 150 per disk) for later retrieval.
- Calculation of resultant relative humidity.
- Display of entering mixed-air conditions and coil selection parameters.
- Three different wall types per exposure.
- Ability to summarize maximum values of cubic feet per minute and/or maximum loads.
- Capable of combining any number of subzones into master zones and air-handling unit zones.
- Instant correction of input errors.
- Displays for every question the limits of input and the default value.
- Allows printing of all questions and input values.
- Capable of selecting room terminals.
- Calculates infiltration and slab heating loads.
- Allows mixing of different glass types and wall types on each exposure.

Load Estimating for Residential. This program is designed specifically for residential buildings, and its features are similar to those of the Commercial Version 1.3 program just described.

Life Cycle Costs, Version 2.1. This program provides the following:

- Obtains simple payback without the distraction of interest or inflation, or determines yearly cash flow and/or present-worth analysis for up to 50 years.

- Ability to include inflation and effect of interest rates.
- Ability to include fuel and its own escalation rate for one fuel.
- Ability to include maintenance at its own escalation rate and analysis using either cash flow or unified present worth.
- Ability to evaluate three different fuels at independent escalation rates.
- Ability to evaluate two different financial (debt) instruments.
- Capability of using variable escalation rates for all items.
- Determines equipment replacement cycles.
- Ability to store and compare different analyses.
- Program contains stored composite projected escalation rates of gas, oil, and electricity through the year 2030.
- Program contains stored composite projected inflation rates through the year 2030.

Operating Cost Analysis, Version 1.5. This program provides for the following:

- Capable of storing last building input data
- Allows input of one change and rerun of load profile or cost analysis
- Ability to produce graphs and profiles
- Ability to calculate energy savings of heat recovery systems
- Capable of using computer-generated weather data or input data
- Includes thermal flywheel effect
- Includes HVAC, lights, and domestic hot-water operating costs
- Capable of comparing calculated annual operating costs of alternative systems

Equipment Selection. This program selects air-handling units, coils, etc.

Specification Writing and Word Processing. This program enables the user to write, edit, and store specifications letters, proposals, etc.

Equipment Estimating. This program calculates the labor hours required for installation of equipment (everything from grilles, diffusers, fans, roof curbs, etc.).

2.3 REFERENCES

1. *1985 ASHRAE Handbook, Fundamentals*, ASHRAE, Atlanta, GA, 1985, chap. 24, "Weather Data and Design Conditions."
2. Carrier Corporation, *Handbook of Air Conditioning System Design*, McGraw-Hill, New York, 1965, part 1, chap. 2.
3. Loads Design Weather Region diskettes from the Trane Company, La Grosse, WI.
4. E20-II diskettes from Carrier Corp., Syracuse, NY.
5. National Climatic Data Center, Nashville, NC.
6. *1985 ASHRAE Handbook, Fundamentals*, chap. 8, "Physiological Principles for Comfort and Health," ASHRAE, Atlanta, GA, 1985.
7. Ibid., chap. 22, "Ventilation and Infiltration."

8. *Ventilation Standard*, ANSI/ASHRAE document 61-1981R, ASHRAE, Atlanta, GA, 1981.
9. *1982 ASHRAE Handbook, Systems and Applications*, ASHRAE, Atlanta, GA, 1982, chap. 18, "Retail Facilities."
10. Ibid., chap. 19, "Commercial and Public Buildings."
11. Ibid., chap. 20, "Places of Assembly."
12. Ibid., chap. 21, "Domiciliary Facilities."
13. Ibid., chap. 22, "Educational Facilities."
14. Ibid., chap. 23, "Health Facilities."
15. Ibid., chap. 25, "Aircraft."
16. Ibid., chap. 26, "Ships."
17. Ibid., chap. 27, "Environmental Control for Survival."
18. Ibid., chap. 30, "Laboratories."
19. Ibid., chap. 32, "Clean Spaces."
20. Ibid., chap. 33, "Data Processing System Areas."
21. Carrier Corp., *Handbook of Air Conditioning System Design*, part 1, chap. 1, McGraw-Hill, New York, 1965.
22. Ibid., chap. 6.
23. *1985 ASHRAE Handbook, Fundamentals*, chap. 26, "Air Conditioning Cooling Loads," ASHRAE, Atlanta, GA, 1985.
24. Carrier Corp., *Handbook of Air Conditioning System Design*, part 1, chap. 7, McGraw-Hill, New York, 1965.
25. *1985 ASHRAE Handbook, Fundamentals*, chap. 9, "Environmental Control of Animals and Plants."
26. Ibid., chap. 10, "Physiological Factors in Drying and Storing Farm Crops."
27. *1987 ASHRAE Handbook, Systems and Applications*, chap. 28, "Industrial Air Conditioning," ASHRAE, Atlanta, GA, 1987
28. Ibid., chap. 31, "Engine Test Facilities."
29. Ibid., chap. 34, "Printing Plants."
30. Ibid., chap. 35, "Textile Processing."
31. Ibid., chap. 36, "Photographic Materials."
32. Ibid., chap. 37, "Environment for Animals and Plants."
33. Ibid., chap. 39, "Air Conditioning of Wood and Paper Products Facilities."
34. Ibid., chap. 40, "Heating, Ventilating and Air Conditioning for Nuclear Facilities."
35. Ibid., chap. 42, "Underground Mine Air Conditioning and Ventilation."
36. *1985 ASHRAE Handbook, Fundamentals*, chap. 25, "Heating Load," ASHRAE, Atlanta, GA, 1985.
37. Ibid., chap. 23, "Design Heat Transfer."
38. Ibid., chap. 27, "Fenestration."
39. Carrier Corp., *Handbook of Air Conditioning System Design*, part 1, chap. 5, McGraw-Hill, New York, 1965.
40. *1985 ASHRAE Handbook, Fundamentals*, chap. 13, "Enclosed Vehicular Facilities," ASHRAE, Atlanta, GA, 1985.
41. Carrier Corp., *Handbook of Air Conditioning Systems Design*, part 1, chap. 3, McGraw-Hill, New York, 1965.
42. Ibid., chap. 4.

CHAPTER 3
PIPE SIZING

Nils R. Grimm, P.E.
Section Manager, Mechanical, Sverdrup Corporation, New York, New York

3.1 INTRODUCTION

Once the designer has calculated the required flows in gallons per minute (cubic meters per second or liters per second) for chilled-water, condenser water, process water, and hot-water systems or pounds per hour (kilograms per hour) for steam systems and tons or Btu per hour (watts per hour) for refrigeration, calculation of the size of each piping system can proceed.

3.2 HYDRONIC SYSTEMS

With respect to hydronic systems (chilled water, condenser water, process water, hot water, etc.), the designer has the option of using the manual method or one of the computer programs.

Whether the piping system is designed manually or by the computer, the effects of high altitude must be accounted for in the design if the system will be installed at elevations of 2500 ft (760 m) or higher. Appropriate correction factors and the effects of altitudes 2500 ft (760 m) and higher are discussed in App. A of this book.

The following is a guide for design water velocity ranges in piping systems that will not result in excessive pumping heads or noise:

Boiler feed	8 to 15 ft/s (2.44 to 4.57 m/s)
Chilled water, condenser water, hot water, process water, makeup water, etc.	4 to 10 ft/s (1.22 to 3.05 m/s)
Drain lines	4 to 7 ft/s (1.22 to 2.13 m/s)
Pump suction	4 to 6 ft/s (1.22 to 1.83 m/s)
Pump discharge	8 to 12 ft/s (2.44 to 3.66 m/s)

Where noise is a concern, such as in pipes located within a pipe shaft adjacent to a private office or other quiet areas, velocities within the pipe should not exceed 4 ft/s (1.22 m/s) unless acoustical treatment is provided. (Noise control and vibration are discussed in Chaps. 49 and 50 of this book.)

Erosion should also be considered in the design of hydronic piping systems, especially when soft material such as copper and plastic is used. Erosion can result from particles suspended in the water combined with high velocity. To assist the designer, Table 3.1 shows maximum water velocities that are suggested to minimize erosion, especially in soft piping materials.

TABLE 3.1

Annual operating hours	Maximum water velocity, ft/s	Maximum water velocity, m/s
1500	11	3.35
2000	10.5	3.20
3000	10	3.05
4000	9	2.74
6000	8	2.44
8000	7	2.13

If the manual method is selected to size the project's hydronic piping systems, it should be calculated by following one of the accepted procedures found in standard design handbook sources such as Refs. 1 to 5.

If the computer method is chosen to size the hydraulic piping systems, the designer must select a software program from the several that are available. Two of the most widely used are Trane's CDS Water Piping Design program and Carrier's E20-II Piping Data program. In addition to determining the pipe sizes, both programs print a complete bill of materials (quantity takeoff by pipe size, length, fittings, and insulation).

Whichever program is used, the specific program input and operating instructions must be strictly followed. It is common to trace erroneous or misleading computer output data to mistakes in inputting design data. *It cannot be overstressed that in order to get meaningful output data, input data must be correctly entered and checked after entry before the program is run*. It is also a good, if not mandatory, policy to independently check the computer results the first time you run a new or modified program, to ensure that the results are valid.

If the computer program used does not correct the computer output for the effects of altitude when the elevation of the project is equal to or greater than 2500 ft (760 m) above sea level, the computer output must be manually corrected by using the appropriate correction factors listed in App. A of this book.

The following describe the programs available to the designer using Trane's CDS Water Piping Design program for sizing hydronic systems.

Water Piping Design (DSC-IBM-123). This pipe-sizing program is for open and closed systems, new and existing systems, and any fluid by inputting the viscosity and specific gravity. The user inputs the piping layout in simple line-segment form with the gallons per minute of the coil and pressure drops or with the gallons per minute for every section of pipe. The program sizes the piping and identifies the critical path, and then it can be used to balance the piping so that the loops have equal pressure drops.

The output includes

- Complete bill of materials (including pipe sizes and linear length required, fittings, insulation, and tees)
- Piping system costs for material only or for material and labor
- Total gallons of fluid required

The following summary describes the program available to the designer using Carrier's E20-II Water Piping Design for sizing hydronic systems.

Water Piping Design (Version 1.0). This program provides the following:

- Enables the designer to look at the balancing required for each piping section, thereby permitting selective reduction of piping sizes or addition of balancing valves
- Calculates pressure drop and material takeoff for copper, steel, or plastic pipe
- Sizes all sections and displays balancing required for all circuits
- Sizes closed or open systems
- Corrects pressure drop for water temperature and/or ethylene glycol
- Calculates gallons per minute of total system
- Calculates total material required, including fittings
- Ability to store for record or later changes up to 200 piping sections
- Ability to change any item and immediately rerun
- Allows sizing of all normally used piping materials
- Allows balancing of system in a minimum amount of time
- Allows easy sizing of expansion tanks and determination of necessary gallons per minute of glycol for brine applications
- Estimates piping takeoff fitting by pipe size, quantities (linear feet, fittings, valves, etc.).

3.3 STEAM SYSTEMS

The author is not aware of any computer program for sizing steam piping for heating, ventilating, and air-conditioning (HVAC) systems. Unquestionably, one will become available within the next few years. Until then, the designer must use the manual method.

The effects of high altitude must be accounted for in the design when the system will be installed at elevations of 2500 ft (760 m) or higher. Appropriate correction factors and the effects of altitudes 2500 ft (760 m) and higher are discussed in App. A.

Table 3.2 is a guide for design of steam velocity ranges in piping systems that will not result in excessive pressure drops or noise. The steam piping systems

TABLE 3.2

Steam pressure		Velocity range	
lb/in^2	bar	ft/min	m/s
0–15	0–1.03	4000–6000	20.32–30.48
15–200	1.03–13.79	6000–10,000	30.48–50.08

should be sized by following one of the accepted procedures found in standard design handbook sources, such as Refs. 2, 3, and 5.

3.4 REFRIGERANT SYSTEMS

Here the designer has the option of using the manual method or at least one computer program.

Whether the piping system is designed manually or by computer, the effects of high altitude must be accounted for in the design when the system will be installed at elevations of 2500 ft (760 m) or higher. Appropriate correction factors and the effects of altitudes 2500 ft (760 m) and higher are discussed in App. A.

Liquid line sizing is considerably less critical than the sizing of suction or hot gas lines, since liquid refrigerant and oil mix readily. There is no oil movement (separation) problem in designing liquid lines. It is good practice to limit the pressure drop in liquid lines to an equivalent 2°F (1°C). It is also good practice to limit the liquid velocity to 360 ft/min (1.83 m/s).

The suction line is the most critical line to size. The gas velocity within this line must be sufficiently high to move oil to the compressor in horizontal runs and vertical risers with upward gas flow. At the same time, the pressure drop must be minimum to prevent penalizing the compressor capacity and increasing the required horsepower. It is good practice, where possible, to limit the pressure drop in the suction line to an equivalent temperature penalty of approximately 2°F (1°C). In addition to the temperature (pressure drop) constraints, the following minimum gas velocities are required to move the refrigerant oil:

Horizontal suction lines	500 ft/min (2.54 m/s) minimum
Vertical upflow suction lines	1000 ft/min (5.08 m/s) minimum

The velocity in upflow rises must be checked at minimum load; if it falls below 1000 ft/min (5.08 m/s), double risers are required. To avoid excess noise, the suction line velocity should be below 4000 ft/min (20.32 m/s).

The discharge (hot-gas) line has the same minimum and maximum velocity criteria as suction lines; however, the pressure drop is not as critical. It is good practice to limit the pressure drop in the discharge (hot-gas) line to an equivalent temperature penalty of approximately 2 to 4°F (1 to 2°C).

If the manual method is used to size the project, refrigerant piping systems should be calculated by following one of the accepted procedures found in standard design handbook sources such as Refs. 3, 6, and 7.

If the computer method is used to size the project hydraulic piping systems, the designer must choose a program among the several available. Two of the most widely used are Trane's CDS Water Piping Design program and Carrier's E20-II Piping Data program. In addition to determining the pipe sizes, both programs print a complete bill of materials (quantity takeoff by pipe size, length, fittings, and insulation). Whichever program is used, it is mandatory that the specific program's input and operating instructions be strictly followed. It is common to trace erroneous or misleading computer output data to mistakes in inputting design data into the computer. *In order to get meaningful output data, input data must be correctly entered and checked after entry before the program is run*. It is also a good, if not mandatory, policy to independently check the computer results the first time you run a new or modified program, to ensure that the results are valid.

If the computer program used does not correct the computer output for the effects of altitude when the elevation of the project is equal to or greater than 2500 ft (760 m) above sea level, the computer output must be manually corrected by using the appropriate correction factors, listed in App. A.

DX Piping Design (Version 1.0). Described in the following summary, this program is available to the designer using Carrier's E20-II DX Piping Design to size the refrigerant systems.

- This program will determine the minimum piping size to deliver the refrigerant between compressor, condenser, and evaporators while ensuring return at maximum unloading.
- This program is able to size piping systems using ammonia and Refrigerants 12, 22, 500, 503, 717.
- This program is capable of calculating low-temperature as well as comfort cooling applications.
- This program determines when double risers are needed, sizes the riser, and calculates the pressure drop.
- This program will include accessories in the liquid line and automatically calculates the subcooling required.
- This program permits entering, for all fittings and accessories, pressure drops in degrees Fahrenheit or pounds per square inch.
- This program will size copper or steel piping.
- This program can select pipe size based on the specific pressure drop.
- This program will calculate the actual pressure drop in degrees Fahrenheit and pounds per square inch for selected size.
- This program will estimate piping takeoff, listing by pipe size the quantities of linear feet, fittings, valves, etc.

3.5 REFERENCES

1. Cameron hydraulic data published by Ingersoll Road Company, Woodcliff Lake, NJ.
2. "Flow of Fluids through Valves, Fittings and Pipe," Technical Paper 410, Crane Company, New York.
3. *1985 ASHRAE Handbook, Fundamentals*, ASHRAE, Atlanta, GA, 1985, chap. 34, "Pipe Sizing."

4. Carrier Corp., *Handbook of Air Conditioning System Design*, McGraw-Hill, New York, 1965, part 3, chaps. 1, 2.
5. Ibid., part 3, chaps. 1 and 4.
6. Ibid., part 3, chaps. 1 and 3.
7. *Trane Reciprocating Refrigeration Manual*, Trane Company, La Crosse, WI, 1989.

CHAPTER 4
DUCT SIZING

Nils R. Grimm, P.E.
Section Manager, Mechanical, Sverdrup Corporation, New York, New York

4.1 INTRODUCTION

The function of a duct system is to provide a means to transmit air from the air-handling equipment (heating, ventilating, or air conditioning). In an exhaust system the duct system provides the means to transmit air from the space or areas to the exhaust fan to the atmosphere.

The primary task of the duct designer is to design duct systems that will fulfill this function in a practical, economical, and energy-conserving manner within the prescribed limits of available space, friction loss, velocity, sound levels, and heat and leakage losses and/or gains.

With the required air volumes in cubic feet per minute (cubic meters per second) determined for each system, the zone and space requirements known from the design load calculation, and the type of air distribution system [such as low-velocity single-zone, variable-air-volume (VAV) or multizone or high-velocity VAV or dual duct] decided upon, the designer can proceed to size the air ducts.

The designer must also choose one of three methods to size the duct systems: the equal-friction, equal-velocity, or static regain method. Of the three, the equal-friction and static regain methods are used most often. The equal-velocity method is used primarily for industrial exhaust systems where a minimum velocity must be maintained to transport particles suspended in the exhaust gases.

Static regain is the most accurate method, minimizes balancing problems, and results in the most economical duct sizes and lowest fan horsepower. It is also the only method that should be used for high-velocity comfort air-conditioning systems.

The equal-friction method is used primarily on small and/or simple projects. If manual calculations are made, this method is simpler and easier than static regain; however, if a computer is used, this advantage disappears.

Typical duct velocities for low-velocity duct systems are shown in Table 4.1. For high-velocity systems, typical duct velocities are shown in Table 4.2. The velocities suggested in Tables 4.1 and 4.2 may have to be adjusted downward to meet the required noise criteria. See Chap. 49 of this book for a discussion on noise and sound attenuation.

Whether the duct system is designed manually or by computer, the effects of

TABLE 4.1 Suggested Duct Velocities for Low-Velocity Duct System, ft/min (m/s)

	Main ducts		Branch ducts	
Application	Supply	Return	Supply	Return
Residences	1000 (5.1)	800 (4.1)	600 (3)	600 (3)
Apartments Hotel bedrooms Hospital bedrooms	1500 (7.6)	1300 (6.6)	1200 (6.1)	1000 (5.1)
Private offices Director's rooms Libraries	1800 (9.1)	1400 (7.1)	1400 (7.1)	1200 (6.1)
Theaters Auditoriums	1300 (6.6)	1100 (5.6)	1000 (5.1)	800 (4.1)
General offices Expensive restaurants Expensive stores Banks	2000 (10.2)	1500 (7.6)	1600 (8.1)	1200 (6.1)
Average stores Cafeterias	2000 (10.2)	1500 (7.6)	1600 (8.1)	1200 (6.1)
Industrial	2500 (12.7)	1800 (9.1)	2200 (11.2)	1600 (8.1)

TABLE 4.2 Suggested Duct Velocities for High-Velocity Duct System, ft/min (m/s)

	Main duct		Branch duct	
Application	Supply	Return	Supply	Return
Commercial institutions	2500–3800	1400–1800	2000–3000	1200–1600
Public buildings	(12.7–19.3)	(7.1–9.1)	(10.2–15.2)	(6.1–8.0)
Industrial	2500–4000	1800–2200	2200–3200	1500–1800
	(12.7–20.3)	(9.1–11.2)	(11.2–16.3)	(7.6–9.1)

high altitude must be accounted for in the design if the system will be installed at elevations of 2500 ft (760 m) or higher. Appropriate correction factors and the effects of altitudes of 2500 ft (760 m) and more are discussed in App. A.

4.2 MANUAL METHOD

If the manual method is used to size the project duct systems, they should be calculated by following one of the accepted procedures found in standard design handbooks such as Refs. 1 and 2. For industrial dilution, ventilation, and exhaust duct systems, they should be calculated and sized by the procedures set forth in Ref. 3.

When the equal-friction or equal-velocity method is used manually, the time to calculate duct sizes can be shortened by using Carrier's Ductronic Calculator or Trane's Ductulator. Both will size round or rectangular ducts in U.S. Customary System (USCS) or metric units.

4.3 COMPUTER METHOD

If the computer method is used to size the project's duct systems, one must select a program among the several available. Two of the most widely used are Trane's CDS Duct Design program and Carrier's E20-II Duct Layout program. In addition to determining the duct sizes, both programs print a complete bill of materials (quantity takeoff by pipe size, length, fittings, and insulation).

Whichever program is used, the specific program's input and operating instructions must be strictly followed. It is common to trace erroneous or misleading computer output data to mistakes in inputting design data. It cannot be overstressed that in order to get meaningful output data, the input data must be correctly entered and checked after entry before the program is run. It is also a good, if not mandatory, policy to independently check the computer results the first time you run a new or modified program to ensure that the results are valid.

If the computer program used does not correct the output for the effects of altitude when the elevation of the project is equal to or greater than 2500 ft (760 m) above sea level, then the output must be manually corrected by using the appropriate correction factors, listed in App. A.

4.3.1 Trane Programs

The following summary describes programs available to the designer using Trane's CDS Duct Design program to size the duct systems.

Varatrain (Static Regain) Duct Design (DSC-IBM-113). With this duct-sizing program, the user inputs the duct layout in simple line-segment form with the cubic feet per minute for the zone, the supply fan value of cubic feet per minute, and the desired noise criteria (NC) level.

The program sizes all the ductwork based on an iterative static regain procedure and selects all the VAV boxes when desired. It identifies the critical path and downsizes the entire ductwork system to match the critical-path pressure drop without permitting zone NC levels to exceed design limits.

The output of this program is an efficient, self-balancing duct design. It gives the designer a printout of the static pressure at every duct node, making troubleshooting on the jobsite a snap. The program will estimate the duct system and print a complete bill of materials, including schedule.

Equal-Friction Duct Design (DSC-IBM-108). This program outputs the total pressure as well as the pressure drop for each trunk section. The output also includes duct sizes, air velocity, and friction losses. The program can be used for fiberglass selection.

The program will calculate the metal gauges, sheet-metal requirements, and total poundage and provide a complete bill of materials.

4.3.2 Carrier Program

The following summary describes the program available to the designer using Carrier's E20-II Duct Design to size the duct system.

Duct Design. This program:

- Uses the static regain and equal-friction methods simultaneously
- Calculates round and rectangular ducts
- Allows for sound attenuation and internally insulated ducts
- Permits material changes in duct system for different sections
- Shows balancing requirements between circuits in same duct system
- Is capable of handling up to 200 sections of ductwork in one system
- Calculates sheet-metal poundage and material quantities and shows them in the summary

4.4 REFERENCES

1. *1985 ASHRAE Handbook, Fundamentals*, ASHRAE, Atlanta, GA, 1985, chap. 33, "Duct Design."
2. Carrier Corp., *Air Conditioning System Design*, McGraw-Hill, New York, 1965, part 2, chaps. 1–3.
3. Committee on Industrial Ventilation, *Industrial Ventilation—A Manual of Recommended Practice*, American Conference of Governmental Industrial Hygienists, Lansing, MI, 1989.

CHAPTER 5
STEAM

Lehr Associates
New York, New York

5.1 INTRODUCTION TO STEAM

Nearly any material, at a given temperature and pressure, has a set amount of energy within it. When materials change their physical state, i.e., go from a liquid to a gas, that energy content changes. Such a change occurs when water is heated to a gas—steam. When steam is used for heating, a cycle of different energy states occurs. First, water is heated in a boiler to its vaporization point, when it boils off as steam. The vapor is carried to the desired destination where it is allowed to cool, giving off heat. Usually, the water, now cooled back to a liquid, is returned to the boiler to be revaporized.

The heat content of water is usually measured in British thermal units (Btu's) or calories. Knowing the temperature is not sufficient to determine the energy content of steam—the pressure must also be known as well as the amount of actual vapor or condensate (moisture). "Steam" can exist as saturated (containing all the vapor it can), dry (at the saturation point or above), wet (below the saturation point), and superheated (capable of holding even more vapor). Wet steam—containing condensate—has less energy than dry steam.

These conditions are specified for water in a chart called a Mollier diagram (see Fig. 5.1). The Mollier diagram specifies the energy content for steam at various vaporization levels. On the two axes of the diagram are enthalpy (a measure of the heat content of a volume of steam) and entropy (a measure of the energy available for work). Rigorous analysis of the thermodynamics of a heating system involves measurements of the specific volume of steam available; its pressure, temperature, and moisture values; and the efficiencies of heat transfer of the elements of the heating system. Usually vendors of steam equipment provide details of their systems based on saturated-steam conditions, which simplifies their sizing and use. Saturated-steam tables (see Table 5.1) give the values that are necessary to determine the amount of energy the steam has available for heating.

To calculate the steam consumption of a heating device, the following equation should be employed:

$$Q = \frac{H}{\mathrm{SP}_{wv}(T_e - T_v) + h_{fg} + \mathrm{SP}_w(T_v - T_c)} \tag{5.1}$$

where H = heating load, Btu/h (W)
h_{fg} = latent heat of vaporization, Btu/lb (kJ/kg)
T_e = entering steam temperature, °F (°C)
T_v = steam temperature at vaporization, °F (°C)
SP_{wv} = specific heat of water vapor, Btu/(lb · °F) [cal/(g · °C)]
SP_w = specific heat of water, Btu/(lb · °F) [cal/(g · °C)]
T_c = leaving temperature of condensate, °F (°C)
Q = steam rate, lb/h (kg/h)

TABLE 5.1 Saturated-Steam Tables

Gauge pressure, in Hg vacuum	Absolute pressure, psia	Temperature, °F	Heat content: Sensible (h_f), Btu/lb	Heat content: Latent (h_{fg}), Btu/lb	Heat content: Total (h_g), Btu/lb	Specific volume of steam V_g, ft³/lb
27.9	1	101.7	69.5	1032.9	1102.4	330.0
25.9	2	126.1	93.9	1019.7	1113.6	173.5
23.9	3	141.5	109.3	1011.3	1120.6	118.6
21.8	4	153.0	120.8	1004.9	1125.7	90.5
19.8	5	162.3	130.1	999.7	1129.8	73.4
17.8	6	170.1	137.8	995.4	1133.2	61.9
15.7	7	176.9	144.6	991.5	1136.1	53.6
13.7	8	182.9	150.7	987.9	1138.6	47.3
11.6	9	188.3	156.2	984.7	1140.9	42.3
9.6	10	193.2	161.1	981.9	1143.0	38.4
7.5	11	197.8	165.7	979.2	1144.9	35.1
5.5	12	202.0	169.9	976.7	1146.6	32.4
3.5	13	205.9	173.9	974.3	1148.2	30.0
1.4	14	209.6	177.6	972.2	1149.8	28.0
0 psig	14.7	212.0	180.2	970.6	1150.8	26.8
1	15.7	215.4	183.6	968.4	1152.0	25.2
2	16.7	218.5	186.8	966.4	1153.2	23.80
5	19.7	227.4	195.5	960.8	1156.3	20.4
10	24.7	239.4	207.9	952.9	1160.8	16.5
15	29.7	249.8	218.4	946.0	1164.4	13.9
20	34.7	258.8	227.5	940.1	1167.6	12.0
25	39.7	266.8	235.8	934.6	1170.4	10.6
30	44.7	274.0	243.0	929.7	1172.7	9.5
40	54.7	286.7	256.1	920.4	1176.5	7.8
50	64.7	297.7	267.4	912.2	1179.6	6.7
60	74.7	307.4	277.1	905.3	1182.4	5.8
70	84.7	316.0	286.2	898.8	1185.0	5.2
80	94.7	323.9	294.5	892.7	1187.2	4.7
90	104.7	331.2	302.1	887.0	1189.1	4.3
100	114.7	337.9	309.0	881.6	1190.6	3.9
125	139.7	352.8	324.7	869.3	1194.0	3.2
150	164.7	365.9	338.6	858.0	1196.6	2.8
175	189.7	377.5	350.9	847.9	1198.8	2.4
200	214.7	387.7	362.0	838.4	1200.4	2.1

Note: Metric conversion factors are: 1 in Hg = 25.4 mm Hg; 1 lb/in² = 0.07 bar; °F = 1.8 × °C + 32; 1 Btu/lb = 554 cal/kg; 1 ft³/lb = 0.06 m³/kg.

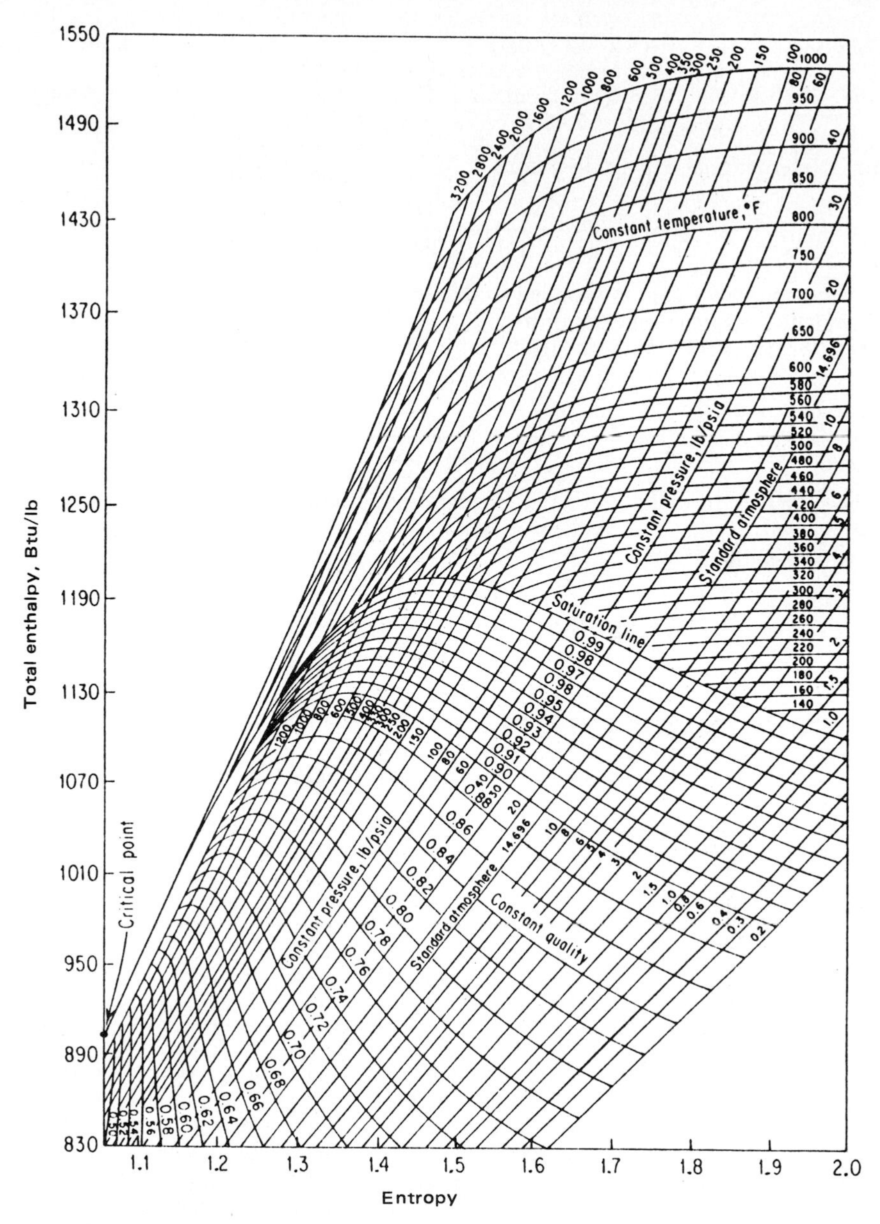

FIGURE 5.1 Mollier diagram.

So,

$$Q = \frac{H}{0.45(T_e - T_v) + h_{fg} + T_v - T_c} \tag{5.2}$$

or in International System (SI) units,

$$Q = \frac{H}{0.52(T_e - T_v) + 0.28h_{fg} + 1.16(T_v - T_c)} \tag{5.3}$$

When saturated steam is supplied to the heating unit, $T_e = T_v$, so $T_e - T_v = 0$. Normally T_c is maintained at or near T_v so that the factor $T_v - T_c$ can be omitted from the calculation without significantly affecting the outcome.

For a system supplying saturated steam we can simplify the calculation to

$$Q = \frac{H}{h_{fg}} \quad \text{or} \quad Q = \frac{H}{0.28h_{fg}} \quad \text{(SI units)} \tag{5.4}$$

The following formula converts the steam rate Q into gallons per minute (liters per second) so that the condensate will be in units normally associated with the flow of liquids:

$$\frac{Q}{500} = \text{gal/min} \quad \text{- or} \quad \frac{Q}{3600} = \text{L/s} \quad \text{(SI units)} \tag{5.5}$$

5.2 INTRODUCTION TO STEAM HEATING SYSTEMS

Steam systems are used to heat industrial, commercial, and residential buildings. These systems are categorized according to the piping layout and the operating steam pressure. This section discusses steam systems which operate at or below 200 psig (14 bar).

5.3 GENERAL SYSTEM DESIGN

The mass flow rate of steam through the piping system is a function of the initial steam pressure, pressure drop through the pipe, equivalent length of piping, and size of piping. The roughness of the inner pipe wall is a variable in determining the steam's pressure drop. All the charts and tables in this section that outline the performance of the steam transmitted through the piping assume that the roughness of the piping is equal to that of new, commercial-grade steel pipe.

5.4 PRESSURE CONDITIONS

Steam piping systems are usually categorized by the working pressure of the steam they supply. The five classes of steam systems are high-pressure, medium-pressure, low-pressure, vapor, and vacuum systems. A high-pressure system has an initial pressure in excess of 100 psig (6.9 bar). The medium-pressure system operates with pressures between 100 psig (6.9 bar) and 15 psig (1 bar). Systems that operate from 15 psig (1 bar) to 0 psig (0 bar) are classified as low-pressure. Vapor and vacuum systems operate from 15 psig (1 bar) to vacuum. Vapor systems attain subatmospheric pressures through the condensing process, while vacuum systems require a mechanically operated vacuum pump to attain subatmospheric pressures.

5.5 PIPING ARRANGEMENTS

The general piping scheme of a steam system can be distinguished by three different characteristics. First, the number of connections required at the heating device describes the system. A one-pipe system has only one piping connection which supplies steam and allows condensate to return to the boiler by flowing counter to the steam in the same pipe. The more common design is to have two piping connections, one for the supply steam and one for the condensate. This arrangement is known as a two-pipe system.

Second, the direction of the supply steam in the risers characterizes the piping design. An up-feed system has the steam flowing up the riser; conversely, a down-feed system supplies steam down the riser.

Third, the final characteristic of the piping design is the location of the condensate return to the boiler. A dry return has its condensate connection above the boiler's waterline, while a wet-return connection is below the waterline.

5.6 CONDENSATE RETURN

By analyzing how the condensate formed in the heating system is returned to the boiler, an understanding of how the system should operate is achieved. There are two commonly used return categories: mechanical and gravity.

If devices such as pumps are used to aid in the return of condensate, the system is known as a *mechanical return*. When no mechanical device is used to return the condensate, the system is classified as a *gravity return*. The only forces pushing the condensate back to the boiler or condensate receiver are gravity and the pressure of the steam itself. This type of system usually requires that all steam-consuming components be located at a higher elevation than the boiler or the condensate receiver.

With either mechanical or gravity return systems, the mains are normally pitched ¼ in (6.3 mm) for every 10 ft (3 m) of length, to ensure the proper flow of condensate. The supply mains are sloped away from the boiler, and the return mains are pitched toward the boiler.

5.7 PIPE-SIZING CRITERIA

Once the heating loads are known, the steam flow rates can be determined; then the required size of the steam piping can be specified for proper operation. The following factors must be analyzed in sizing the steam piping:

- Initial steam pressure
- Total allowable pressure drop
- Maximum steam velocity
- Direction of condensate flow
- Equivalent length of system

For different initial pressures, the allowable pressure drop in the piping varies. Table 5.2 gives typical values in selecting pressure-drop limits. To ensure that the parameters from the table are suitable for an application, check that the total system pressure drop does not exceed 50 percent of the initial pressure, that the condensate has enough steam pressure to return to the boiler, and that the steam velocity is within specified limits to ensure quiet and long-lasting operation.

When steam piping is sized, there is a trade-off between quiet, efficient operation and first-cost considerations. A good compromise point exists when the steam supply pipe is sized for velocities between 6000 and 12,000 ft/min (30.5 and 61 m/s). This allows quiet operation while offering a reasonable installed cost. If the piping is downsized so that the velocity exceeds 20,000 ft/min (101 m/s), the system may produce objectional hammering noise or restrict the flow of condensate when it is counter to the steam's direction. It is recommended that the piping be sized so that the velocity will never approach 20,000 ft/min (101 m/s) in any leg.

As condensate flows into the return line, a portion of it will flash into steam. The volume of the steam-condensate mixture is much greater than the volume of pure condensate. To avoid undersizing the return lines, the return piping should be sized at some reasonable proportion of dry steam. A maximum size would be to assume that the return is 100 percent dry steam. An acceptable velocity for the design of the return lines is 5000 ft/min (25.4 m/s).

5.8 DETERMINING EQUIVALENT LENGTH

The "equivalent length" of pipe is equal to the actual length of pipe plus the friction losses associated with fittings and valves. For simplicity's sake, the fitting and valve losses are stated as the equivalent length of straight pipe needed to produce the same friction loss. Values for common fittings and valves are stated in Table 5.3.

The equivalent length—*not the actual length*—is the value used in all the figures and charts for pipe sizing. Common practice is to assume that the equivalent length is 1.5 times the actual length when a design is first being sized. After the initial sizing and layout are completed, the exact equivalent length should be calculated and all the pipe sizes checked.

5.9 BASIC TABLES FOR STEAM PIPE SIZING

Figure 5.2 is is used to determine the flow and velocity of steam in Schedule 40 pipe at various values of pressure drop per 100 ft (30.5 m), based on 0 psig (1-bar) saturated steam. By using the multiplier tables, it may also be used at all satu-

TABLE 5.2 Pressure Drops for Steam Pipe Sizing

Initial steam pressure		Total pressure drop in supply piping		Pressure drop for mains and risers		Total pressure drop in return piping		Pressure drop for mains and risers	
psig	bar	lb/in²	bar	(lb/in²)/100 ft	bar/100 m	lb/in²	bar	(lb/in²)/100 ft	bar/100 m
Vacuum		1–2	0.069–0.138	⅛–¼	0.028–0.057	1	0.069	⅛–¼	0.028–0.057
0	0	⅙–¼	0.004–0.017	1/32	0.007	1/16	0.004	1/32	0.007
2	0.138	¼–¾	0.017–0.052	⅛	0.028	¼	0.017	⅛	0.028
5	0.345	1–2	0.069–1.38	¼	0.057	1	0.069	¼	0.057
15	1.03	4–6	0.276–0.414	1	0.228	4	0.276	½	0.114
30	2.07	5–10	0.345–0.069	2	0.455	5	0.345	1	0.228
50	3.45	10–15	0.069–1.03	2–5	0.455–1.14	10	0.69	2	0.455
100	6.90	15–25	1.03–1.72	2–5	0.455–1.14	15	1.03	2	0.455
150	10.3	25–30	1.72–2.07	2–10	0.455–2.28	20	1.37	2	0.455

TABLE 5.3 Length of Pipe to Be Added to Actual Length of Run—Owing to Fittings—to Obtain Equivalent Length

Size of pipe, in	Length to be added to run, ft*				
	Standard elbow	Side outlet tee†	Gate valve‡	Globe valve‡	Angle valve‡
½	1.3	3	0.3	14	7
¾	1.8	4	0.4	18	10
1	2.2	5	0.5	23	12
1¼	3.0	6	0.6	29	15
1½	3.5	7	0.8	34	18
2	4.3	8	1.0	46	22
2½	5.0	11	1.1	54	27
3	6.5	13	1.4	66	34
3½	8	15	1.6	80	40
4	9	18	1.9	92	45
5	11	22	2.2	112	56
6	13	27	2.8	136	67
8	17	35	3.7	180	92
10	21	45	4.6	230	112
12	27	53	5.5	270	132
14	30	63	6.4	310	152

*Metric conversion: 1 in = 2.54 cm and 1 ft = 0.31 m.
†Values given apply only to a tee used to divert the flow in the main to the last riser.
‡Valve in full-open position.
Example: Determine the length in feet of pipe to be added to actual length of run illustrated.

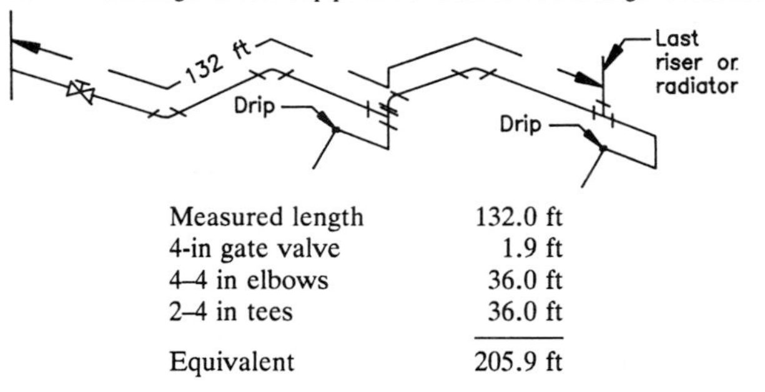

Measured length	132.0 ft
4-in gate valve	1.9 ft
4–4 in elbows	36.0 ft
2–4 in tees	36.0 ft
Equivalent	205.9 ft

Source: Reprinted by permission from *ASHRAE Handbook—1981 Fundamentals*.

rated pressures between 0 and 200 psig (1 and 14 bar). Figure 5.2 is valid only when steam and condensate flow in the same direction.

5.10 TABLES FOR LOW-PRESSURE STEAM PIPE SIZING

Table 5.4, derived from Fig. 5.2, gives the values needed to select pipe sizes at various pressure drops for systems operating at 3.5 and 12 psig (0.24 and 0.84

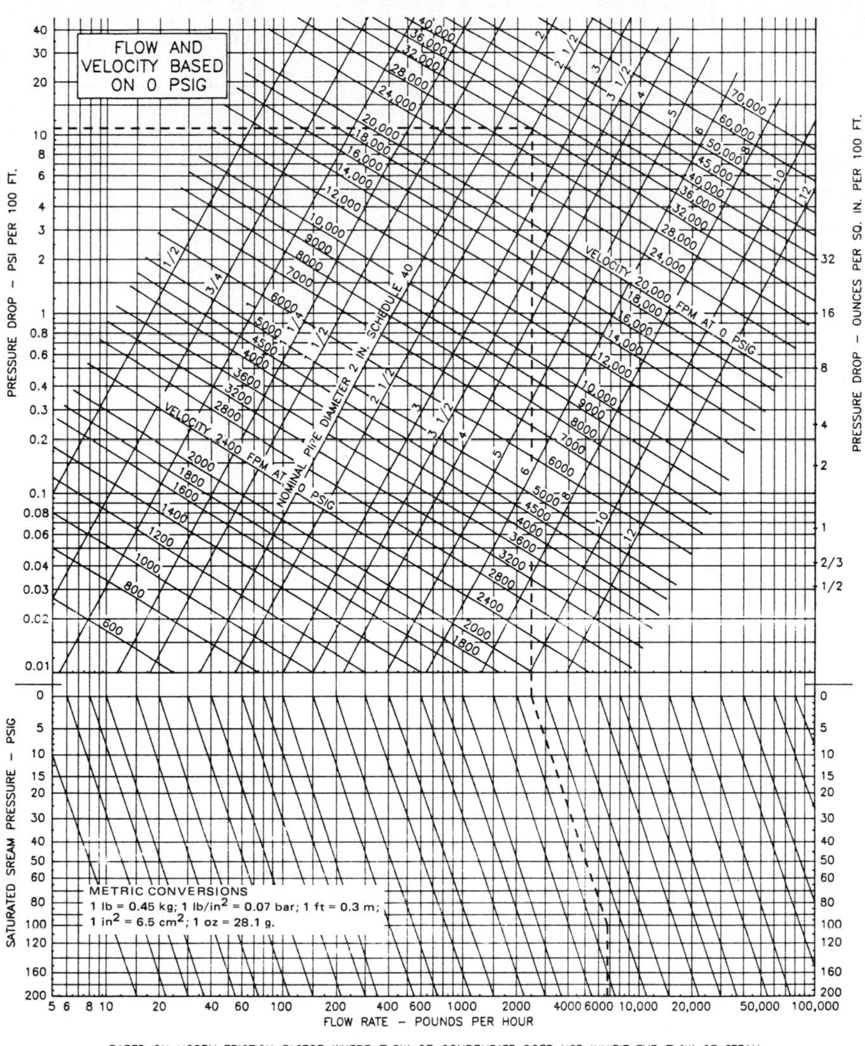

FIGURE 5.2 Basic chart for flow rate and velocity of steam in Schedule 40 pipe, based on saturation pressure of 0 psig (0 bar). *(Reprinted by permission from ASHRAE Handbook—1981 Fundamentals.)*

bar). The flow rates given for 3.5 psig (0.24 bar) can be used for saturated-steam pressures from 1 to 5 psig (0.07 to 0.34 bar), and those for 12 psig (0.84 bar) can be used for saturated pressures from 8 to 16 psig (0.55 to 1.1 bar) with an error not to exceed 8 percent.

Table 5.5 is used for systems where the condensate flows counter to the supply steam.

To size return piping, Table 5.6 is used. This table gives guidelines for return piping for wet, dry, and vacuum return systems.

TABLE 5.4 Flow Rate of Steam, lb/h, in Schedule 40 Pipe* at Initial Saturation Pressure of 3.5 and 12 psig†

Nom. pipe size, in	Pressure drop, lb/in² per 100-ft length‡													
	1/16 lb/in² (1 oz) Sat. press., psig		1/8 lb/in² (2 oz) Sat. press., psig		1/4 lb/in² (4 oz) Sat press., psig		1/2 lb/in² (8 oz) Sat. press., psig		3/4 lb/in² (12 oz) Sat. press., psig		1 lb/in² Sat. press., psig		2 lb/in² Sat. press., psig	
	3.5	12	3.5	12	3.5	12	3.5	12	3.5	12	3.5	12	3.5	12
1/4	9	11	14	16	20	24	29	35	36	43	42	50	60	73
1	17	21	26	31	37	46	54	66	68	82	81	95	114	137
1 1/4	36	45	53	66	78	96	111	138	140	170	162	200	232	280
1 1/2	56	70	84	100	120	147	174	210	218	260	246	304	360	430
2	108	134	162	194	234	285	336	410	420	510	480	590	710	850
2 1/2	174	215	258	310	378	460	540	660	680	820	780	950	1,150	1,370
3	318	380	465	550	660	810	960	1,160	1,190	1,430	1,380	1,670	1,950	2,400
3 1/2	462	550	670	800	990	1,218	1,410	1,700	1,740	2,100	2,000	2,420	2,950	3,450
4	640	800	950	1,160	1,410	1,690	1,980	2,400	2,450	3,000	2,880	3,460	4,200	4,900
5	1,200	1,430	1,680	2,100	2,440	3,000	3,570	4,250	4,380	5,250	5,100	6,100	7,500	8,600
6	1,920	2,300	2,820	3,350	3,960	4,850	5,700	5,700	7,000	8,600	8,400	10,000	11,900	14,200
8	3,900	4,800	5,570	7,000	8,100	10,000	11,400	14,300	14,500	17,700	16,500	20,500	24,000	29,500
10	7,200	8,800	10,200	12,600	15,000	18,200	21,000	26,000	26,200	32,000	30,000	37,000	42,700	52,000
12	11,400	13,700	16,500	19,500	23,400	28,400	33,000	40,000	41,000	49,500	48,000	57,500	67,800	81,000

*Based on Moody friction factor, where flow of condensate does not inhibit the flow of steam.

†The flow rates at 3.5 psig can be used to cover saturated pressure from 1 to 6 psig, and the rates at 12 psig can be used to cover saturated pressure from 8 to 16 psig with an error not exceeding 8 percent. The steam velocities corresponding to the flow rates given in this table can be found from the basic chart and velocity multiplier chart, Fig. 5.2.

‡Metric conversions: 1 in = 2.54 cm, 1 lb/in² = 0.07 bar, and 1 lb = 0.46 kg.

Source: Reprinted by permission from *ASHRAE Handbook—1981 Fundamentals*.

TABLE 5.5 Steam Pipe Capacities for Low-Pressure Systems, lb/h

For use on one-pipe systems or two-pipe systems in which condensate flows against the steam flow

Nominal pipe size, in	Two-pipe systems: Condensate flowing against steam		One-pipe systems		
	Vertical	Horizontal	Supply risers up-feed	Radiator valves and vertical connections	Radiator and riser runouts
A	B*	C†	D‡	E	F†
¾	8	7	6	. . .	7
1	14	14	11		7
1¼	31	27	20	16	16
1½	48	42	38	23	16
2	97	93	72	42	23
2½	159	132	116	. . .	42
3	282	200	200	. . .	65
3½	387	288	286	. . .	119
4	511	425	380	. . .	186
5	1,050	788	. . .	. . .	278
6	1,800	1,400	. . .	. . .	545
8	3,750	3,000			
10	7,000	5,700			
12	11,500	9,500			
16	22,000	19,000			

*Do not use column B for pressure drops of less than 1/16 lb/in^2 per 100 ft of equivalent run. Use Fig. 5.2 or Table 5.4 instead.

†Pitch of horizontal runouts to risers and radiators should be not less than ½ in/ft. Where this pitch cannot be obtained, runouts over 8 ft in length should be one pipe size larger than called for in this table.

‡Do not use column D for pressure drops of less than 1/24 lb/in^2 per 100 ft of equivalent run except on sizes 3 in and over. Use Fig. 5.2 or Table 5.4 instead.

Note: Steam at an average pressure of 1 psig is used as a basis of calculating capacities. Metric conversion factors of 1 in = 2.54 cm and 1 lb = 0.46 kg can be used.

Source: Reprinted from *ASHRAE Handbook—1981 Fundamentals*.

TABLE 5.6 Return Main and Riser Capacities for Low-Pressure Systems, lb/h

Pipe size, in	1/32 lb/in² or 1/3-oz drop per 100 ft			1/24 lb/in² or 2/3-oz drop per 100 ft			1/16 lb/in² or 1-oz drop per 100 ft			1/8 lb/in² or 2-oz drop per 100 ft			1/4 lb/in² or 4-oz drop per 100 ft			1/2 lb/in² or 8-oz drop per 100 ft		
	Wet	Dry	Vac.	Wet	Dry	Vac.	Wet	Dry	Vac.	Wet	Dry	Vac.	Wet	Dry	Vac.	Wet	Dry	Vac.
G	H	I	J	K	L	M	N	O	P	Q	R	S	T	U	V	W	X	Y
Return Main																		
3/4	...	...	...	...	...	42	...	...	100	...	...	142	...	...	200	...	...	283
1	125	62	...	145	71	143	175	80	175	250	103	249	350	115	350	...	...	494
1 1/4	213	130	...	248	149	244	300	168	300	425	217	426	600	241	600	...	...	848
1 1/2	338	206	...	393	236	388	475	265	475	675	340	674	950	378	950	...	...	1,340
2	700	470	...	810	535	815	1000	575	1,000	1400	740	1,420	2,000	825	2,000	...	...	2,830
2 1/2	1180	760	...	1580	868	1,360	1680	950	1,680	2350	1230	2,380	3,350	1360	3,350	...	...	4,730
3	1880	1460	...	2130	1560	2,180	2680	1750	2,680	3750	2250	3,800	5,350	2500	5,350	...	...	7,560
3 1/2	2750	1970	...	3300	2200	3,250	4000	2500	4,000	5500	3230	5,680	8,000	3580	8,000	...	...	11,300
4	3880	2930	...	4580	3350	4,500	5500	3750	5,500	7750	4830	7,810	11,000	5380	11,000	...	...	15,500
5	...	...	...	...	...	7,880	...	...	9,680	...	...	13,700	...	...	19,400	...	...	27,300
6	...	...	...	...	...	12,600	...	...	15,500	...	...	22,000	...	...	31,000	...	...	43,800
Riser																		
3/4	...	48	...	...	48	143	...	48	175	...	48	249	...	48	350	...	...	494
1	...	113	...	...	113	244	...	113	300	...	113	426	...	113	600	...	...	848
1 1/4	...	248	...	...	248	388	...	248	475	...	248	674	...	248	950	...	...	1,340
1 1/2	...	375	...	...	375	815	...	375	1,000	...	375	1,420	...	375	2,000	...	...	2,830
2	...	750	...	...	750	1,360	...	750	1,680	...	750	2,380	...	750	3,350	...	...	4,730
2 1/2	...	...	...	...	...	2,180	...	...	2,680	...	...	3,800	...	...	5,350	...	...	7,560
3	...	...	...	...	...	3,250	...	...	4,000	...	...	5,680	...	...	8,000	...	...	11,300
3 1/2	...	...	...	...	...	4,480	...	...	5,500	...	...	7,810	...	...	11,000	...	...	15,500
4	...	...	...	...	...	7,880	...	...	9,680	...	...	13,700	...	...	19,400	...	...	27,300
5	...	...	...	...	...	12,600	...	...	15,500	...	...	22,000	...	...	31,000	...	...	43,800

Note: This table is based on pipe size data developed through the research investigations of The American Society of Heating, Refrigerating and Air-Conditioning Engineers. Metric conversion factors of 1 in = 2.54 cm, 1 lb/in² = 0.07 bar, and 1 ft = 0.31 m can be used.

Source: Reprinted by permission from *ASHRAE Handbook—1981 Fundamentals*.

5.11 TABLES FOR SIZING MEDIUM- AND HIGH-PRESSURE PIPE SYSTEMS

Larger, industrial-type space-heating systems are designed to use either medium- or high-pressure steam at 15 to 200 psig (1.03 to 14 bar). These systems often involve unit heaters and/or air-handling units. Figures 5.3 to 5.6 provide tables for sizing steam piping for systems of 30, 50, 100, and 150 psig (2, 3.5, 6.9, and 10.5 bar).

5.12 AIR VENTS

The presence of air in the steam supply line impedes the heat-transfer ability of the system due to the high insulating value of air. Air also interferes with the flow of steam by forming pockets at the ends of runs that prevent the steam from reaching the system's extremities.

A valve that releases air from the system while restricting the flow of all other fluids is known as an "air vent." Air vents should be located at all system high

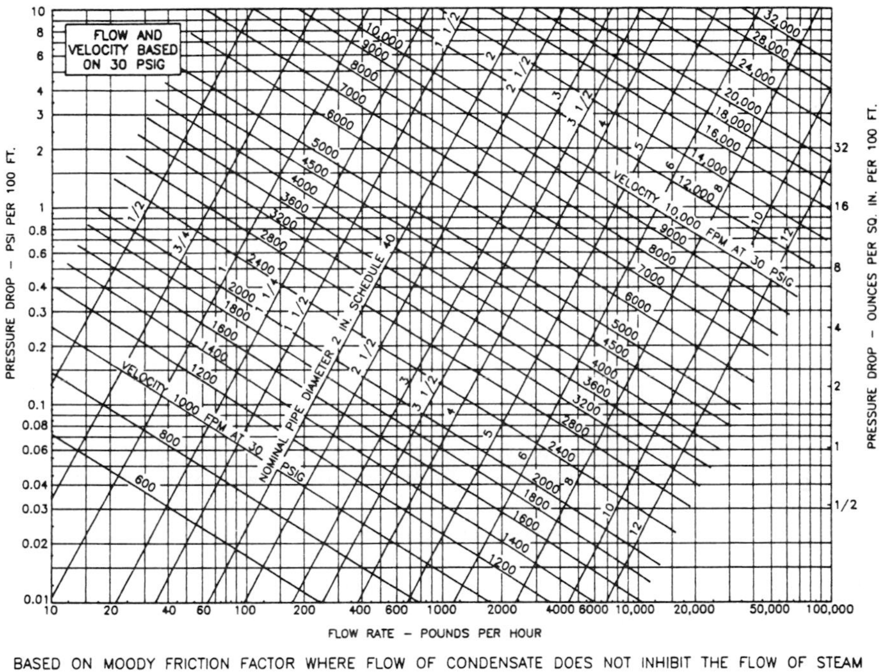

BASED ON MOODY FRICTION FACTOR WHERE FLOW OF CONDENSATE DOES NOT INHIBIT THE FLOW OF STEAM

(MAY BE USED FOR STEAM PRESSURE FROM 23 TO 37 PSIG WITH AN ERROR NOT EXCEEDING 9%)

METRIC CONVERSIONS:
1 lb = 0.45 kg; 1 lb/in^2 = 0.07 bar; 1 ft = 0.3 m;
1 in^2 = 6.5 cm^2; 1 oz = 28.1 g.

FIGURE 5.3 Chart for flow rate and velocity of steam in Schedule 40 pipe, based on saturation pressure of 30 psig (2.1 bar). *(Reprinted by permission from ASHRAE Handbook—1981 Fundamentals.)*

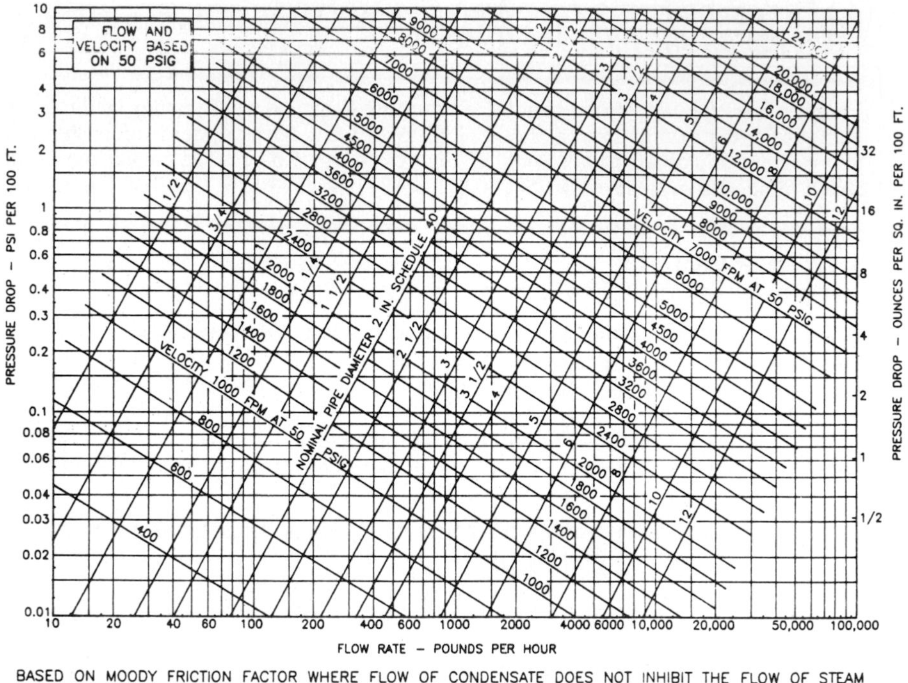

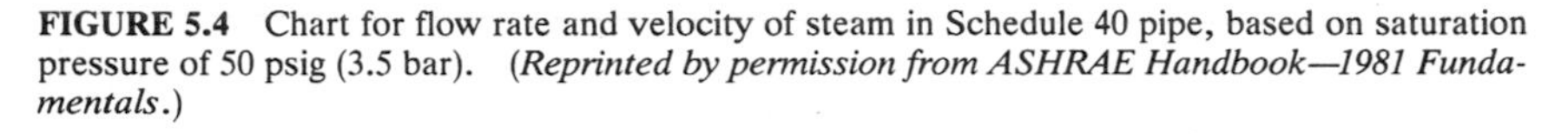

METRIC CONVERSIONS:
1 lb = 0.45 kg; 1 lb/in^2 = 0.07 bar; 1 ft = 0.3 m;
1 in^2 = 6.5 cm^2; 1 oz = 28.1 g.

FIGURE 5.4 Chart for flow rate and velocity of steam in Schedule 40 pipe, based on saturation pressure of 50 psig (3.5 bar). (*Reprinted by permission from ASHRAE Handbook—1981 Fundamentals.*)

points and where air pockets are likely to form. Venting should be done continually to prevent the buildup of air in the system.

Air enters the system by two means. First, when cold makeup feed water is supplied to the boiler, air is present in the water. As the water is heated, the air tends to separate from the water. Second, when the system is turned off, steam is trapped in the pipes. Eventually the steam cools and condenses. Since the volume of the condensate is negligible compared to the initial volume of the steam, a vacuum is formed in the piping. Air leaks into the system through openings in the joints until the internal pressure equalizes. Upon restarting the system, the air is swept along with the steam and becomes entrained in the system.

5.13 STEAM TRAPS

When steam is transmitted through the piping or the end-user equipment, it loses part of its heat energy. As heat is removed from saturated steam, a vapor-liquid mixture forms in the pipe. The presence of liquid condensate in the steam lines

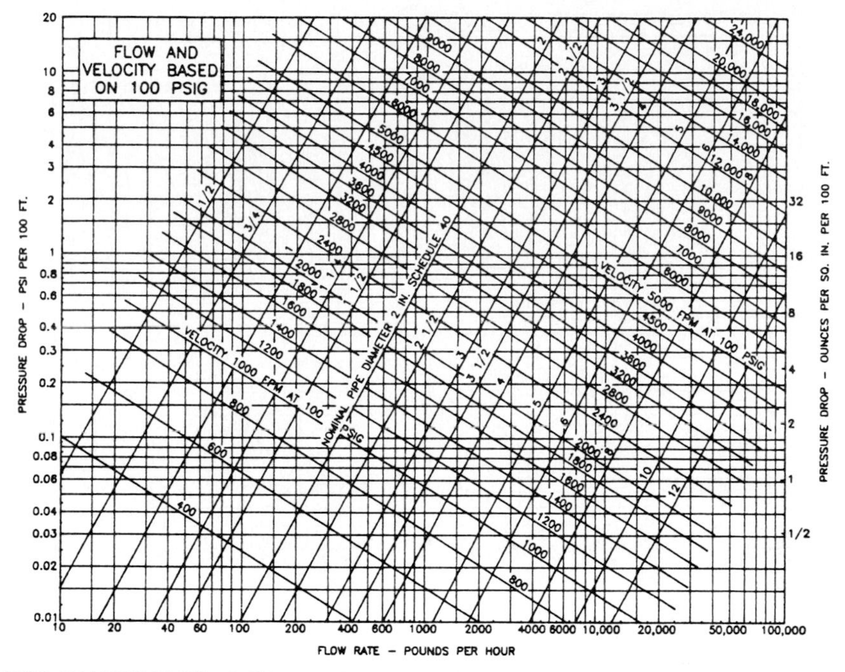

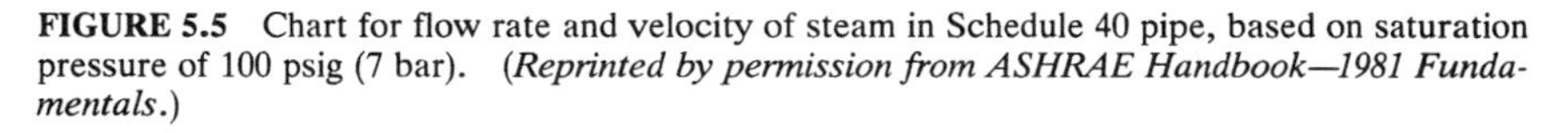

FIGURE 5.5 Chart for flow rate and velocity of steam in Schedule 40 pipe, based on saturation pressure of 100 psig (7 bar). (*Reprinted by permission from ASHRAE Handbook—1981 Fundamentals.*)

interferes with the proper operation of the system. Liquid condensate derates the system's heating capacity because water has a much smaller amount of available energy than steam does. Furthermore, the accumulation of water in the supply steam piping can obstruct the flow of the steam through the system.

A valve that permits condensate to flow from the supply line without allowing steam to escape is known as a "steam trap." All steam traps should be located such that condensate can flow via gravity through them. *Through mechanical means*, the steam trap recognizes when steam is present by sensing the density, kinetic energy, or temperature of the fluid at the trap. When conditions indicate that steam is absent, the trap opens and allows the condensate to drop to the return line. As soon as the trap senses the presence of steam, it slams shut.

5.14 STEAM TRAP TYPES

There are six types of steam traps normally employed in the heating, ventilating, and air-conditioning (HVAC) industry. Since traps differ in their operational characteristics, selection of the proper trap is critical to efficient operation of the

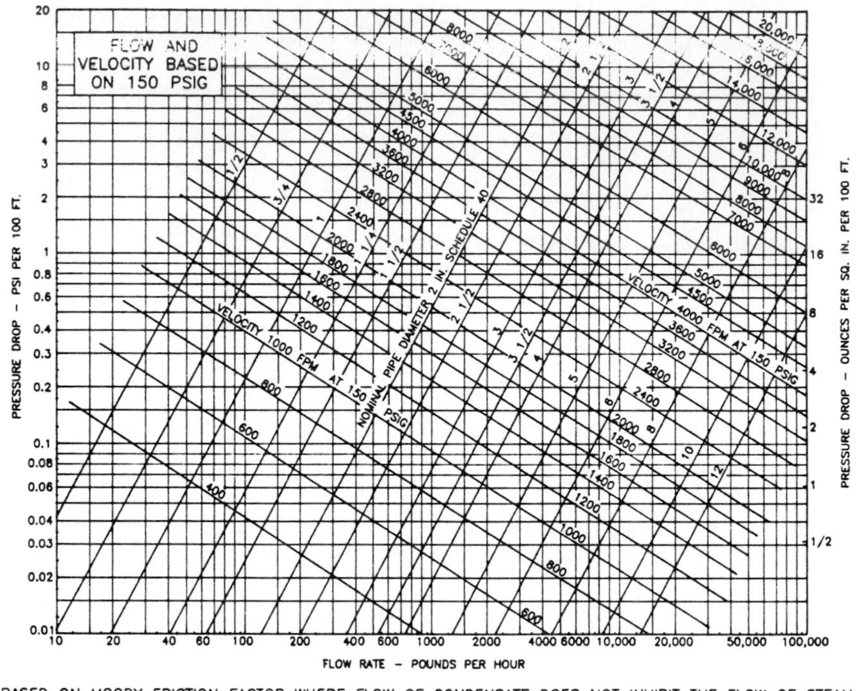

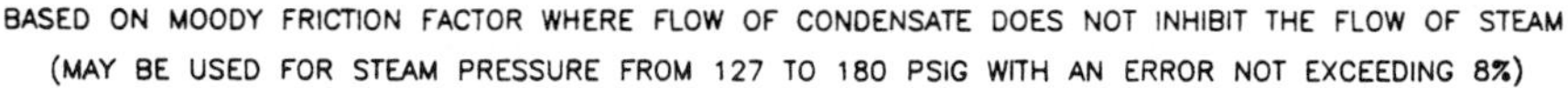

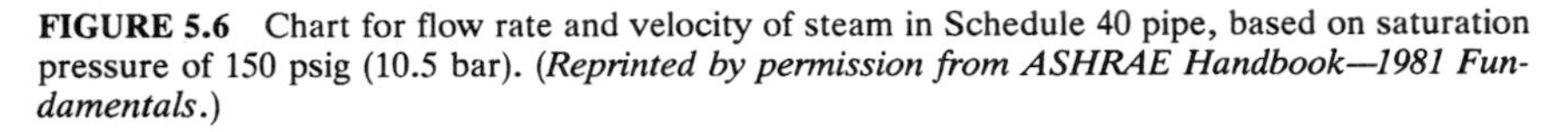

FIGURE 5.6 Chart for flow rate and velocity of steam in Schedule 40 pipe, based on saturation pressure of 150 psig (10.5 bar). (*Reprinted by permission from ASHRAE Handbook—1981 Fundamentals*.)

system. Different applications require specific types of traps, and no one type of trap will perform satisfactorily in all situations.

Three of the six basic types of traps operate thermostatically by sensing a temperature difference between subcooled condensate and steam: liquid-expansion, balanced-pressure thermostatic, and bimetallic thermostatic traps. Two other types—the bucket trap and the float-and-thermostatic trap—are activated by differences in density between steam and condensate. Finally, the thermodynamic steam trap operates on the differences in the velocity at which steam passes through the trap. This velocity difference can also be considered as a change in kinetic energy.

5.15 BALANCED-PRESSURE STEAM TRAPS

The balanced-pressure steam trap (Fig. 5.7) employs a bellows filled with a fluid mixture that boils below the steam temperature. When steam is present at the

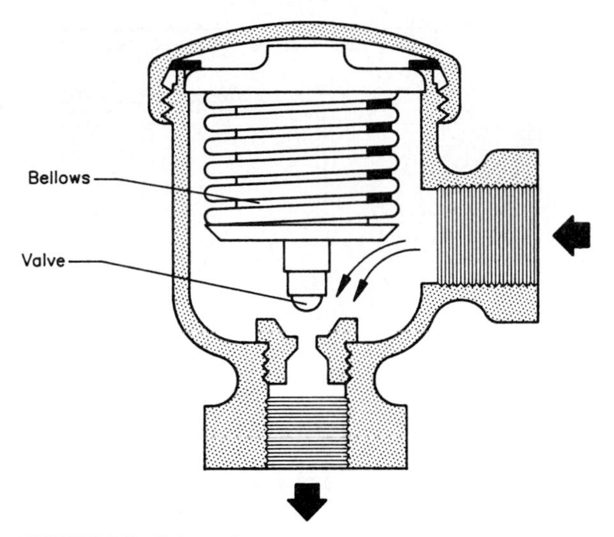

FIGURE 5.7 Balanced-pressure steam trap.

trap inlet, the liquid in the bellows is vaporized and expands to seal the trap. Condensate accumulates at the trap and starts to subcool. When the condensate cools enough to condense the fluid in the bellows, the trap opens and the condensate flows through the trap.

This type of trap has two possible drawbacks. First, it must allow condensate to subcool 5 to 30°F (2.8 to 16.7°C) below the steam temperature to operate. Second, it discharges condensate intermittently.

Advantages of the balanced-pressure trap are that it is freeze-proof, can handle a large condensate load, does a good job of air venting, and is self-adjusting throughout its operating range. These traps are typically used in conjunction with steam radiators and sterilizers.

5.16 BIMETALLIC THERMOSTATIC STEAM TRAPS

These traps operate on the same principle as the balanced-pressure steam trap. The bellows mechanism is replaced by a bimetallic strip formed from two dissimilar metals that have very different coefficients of expansion. As the bimetallic strip is heated, the difference in the expansion rate of the metals causes the strip to bend. The trap is fabricated so that when the strip is heated to the steam's temperature, there is enough movement to close off the valve. The bimetallic thermostatic trap (Fig. 5.8) has a slow response to load conditions, requiring as much as 100°F (55.5°C) of subcooling, and is not self-adjusting to changes in inlet pressure.

These traps are suited for superheated steam applications and situations where a great deal of condensate subcooling is required to prevent flashing in the return line. Normally these traps are applied to steam-tracing lines that can tolerate partial flooding.

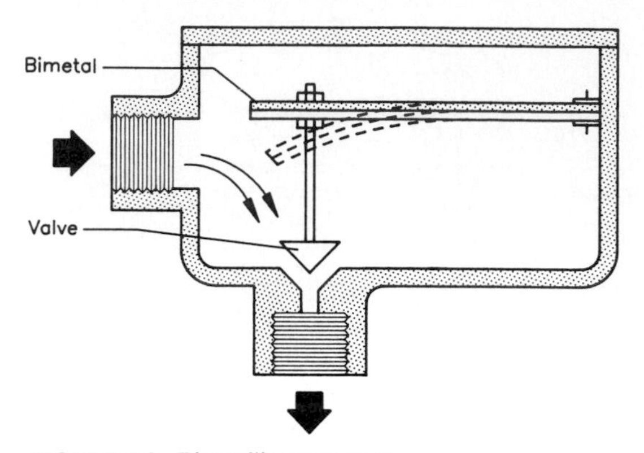

FIGURE 5.8 Bimetallic steam trap.

5.17 LIQUID-EXPANSION STEAM TRAPS

The liquid-expansion steam trap (Fig. 5.9) is designed with an oil-filled cylinder which drives a piston. When steam is present, the oil expands, thrusting the piston out. The end of the piston acts as the valve and seals the port to the return line. As condensate collects in the trap and cools, the oil starts to contract. The contraction of the oil causes the piston to move away from the port and permits the flow of condensate from the trap.

These traps are freeze-proof and are used for freeze protection of system low points and heating coils. Their limitations are that they are not self-adjusting to changes of inlet pressure and that they require condensate subcooling by 2 to 30°F (1.1 to 16.7°C).

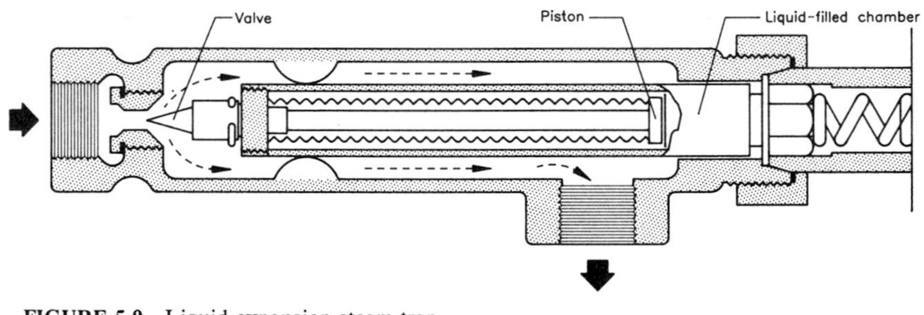

FIGURE 5.9 Liquid-expansion steam trap.

5.18 BUCKET STEAM TRAPS

Bucket traps operate by gravity, utilizing the density difference between liquid and vapor. When the body of the trap is filled with liquid and a vapor enters the bucket, the bucket will float. As the bucket fills with liquid, the bucket sinks. The bucket's movement activates a valve. If the bucket rises due to the vapor pressure, the valve closes; and when the bucket sinks, the valve opens, permitting condensate to flow from the trap. The most common type of bucket trap is the

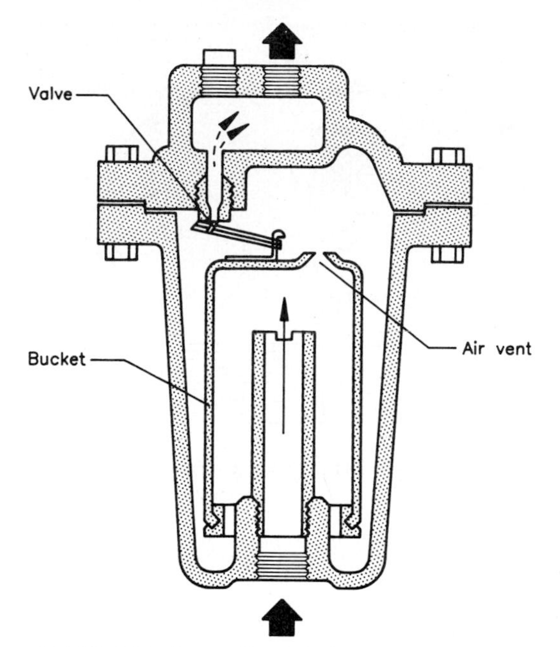

FIGURE 5.10 Inverted-bucket steam trap.

inverted bucket (Fig. 5.10), so named because the bucket has its open side facing down.

Bucket traps are capable of working at very high pressures, can discharge condensate at the saturated-steam temperature, and are resistant to water hammer. Unfortunately, if the water seal is lost, the bucket trap will continuously allow steam to pass through. Other disadvantages of these traps are their susceptibility to freeze-up, their lack of good air-venting capability, and their intermittent discharge.

Inverted-bucket traps are usually installed on high-pressure indoor steam main drips.

5.19 FLOAT-AND-THERMOSTATIC STEAM TRAPS

A float-and-thermostatic steam trap (Fig. 5.11) is actually two distinct traps in one unit. The balanced-pressure steam trap, outlined previously, is located at the top of the trap body and acts as an air vent. The rest of the unit consists of a float that rises and falls based on the level of condensate in the trap. The trap inlet is located above the outlet. The float position operates a valve that controls flow to the return line. As the condensate level rises above the outlet, the float causes the valve to open. If the condensate level drops enough, the float causes the valve to close. Since the float allows the valve to open only when the condensate level is above the outlet, a water seal is maintained to prevent steam from passing through the outlet when the valve is open.

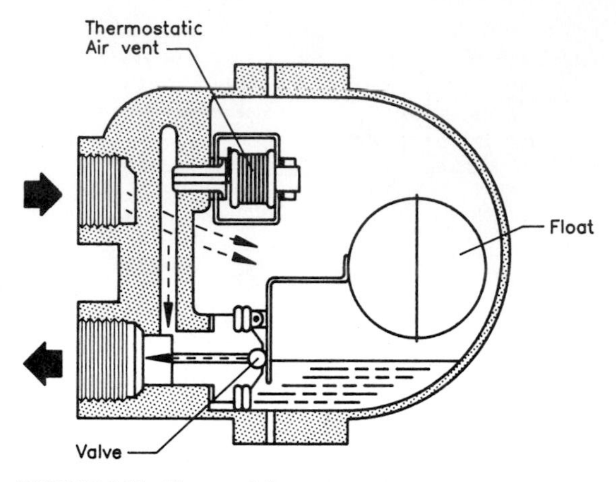

FIGURE 5.11 Float-and-thermostat steam trap.

The float-and-thermostatic steam traps cannot be used on a superheated-steam system unless they are modified and are usually not installed outdoors because they are subject to freeze-up. These types of traps will continuously vent air. They do not require subcooling of condensate and are unaffected by changes in system pressure. Typically float-and-thermostatic traps are used in conjunction with heating devices, such as unit heaters, water heaters, and converters.

5.20 THERMODYNAMIC STEAM TRAPS

The design of the thermodynamic steam trap (Fig. 5.12) is based on the theory that the total pressure of fluid passing through the trap will remain constant. Since the total pressure equals the sum of the static and dynamic pressures, any increase in dynamic pressure will cause a decrease in the static pressure, and vice versa. These traps have only one moving part, a disk that can seal off both the inlet and the outlet of the trap. Steam entering the trap accelerates radially over the disk, causing a reduction in static pressure under the disk. As the steam dead-

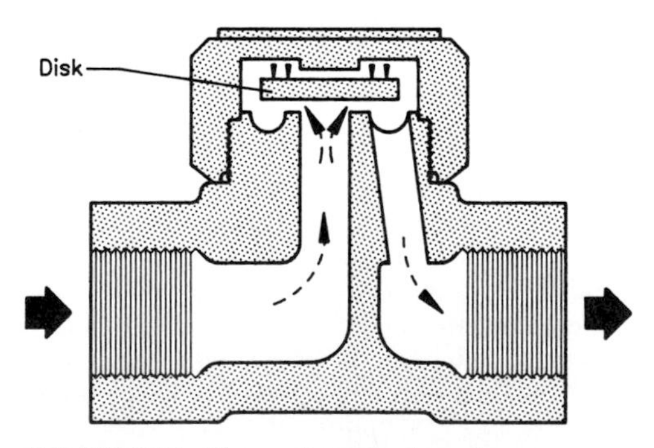

FIGURE 5.12 Thermodynamic steam trap.

ends above the disk, the static pressure above the disk increases. This difference in pressure induces the disk to seal off the trap's openings. The trap will remain closed until the steam in the trap condenses sufficiently to reduce the pressure above the disk to an amount less than the inlet steam pressure. At that point, the disk moves away from the inlet port.

Thermodynamic steam traps should not be used on systems operating below 5 psig (0.34 bar) or on those that have back pressures equal to or greater than 80 percent of their supply pressure.

These traps are compact and have a long life due to the simplicity of their design. They can operate under high pressures, responding quickly to load and pressure variations while discharging condensate without requiring subcooling. Thermodynamic traps are usually installed in main drips and steam tracer lines.

5.21 STEAM TRAP LOCATION

Steam traps are located either in the return line or in drip legs. A "drip leg" (shown in Fig. 5.13) is a piping assembly that hangs below the supply main; its purpose is to remove condensate and sediment from the main. Gravity allows

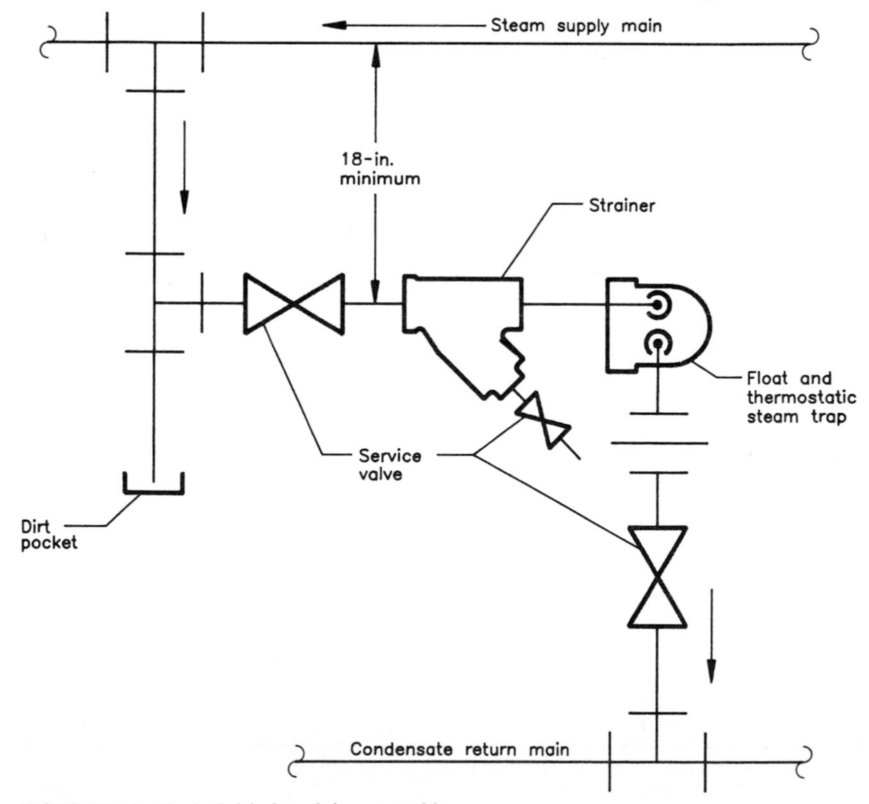

FIGURE 5.13 Typical drip-leg piping assembly.

condensate and sediment to leave the main and accumulate in the drip leg. When the condensate in the leg rises to the level of the trap intake, the trap fills and then discharges the condensate to the return line. The drip leg pipe should be of sufficient size to permit condensate to drain freely from the main. For mains of 4 in (102 mm) or less in diameter, the drip leg should be the same size as the main pipe. For mains larger than 4 in (102 mm), the pipe diameter of the drip leg should be half of the main's size, but not less than 4 in (102 mm). Where possible, all drip legs should be at least 18 in (45.7 cm) long. A trap should be installed in the return line after every steam-consuming device. Each device should have its own trap to prevent possible "short-circuiting" that could occur if multiple devices share a common trap. A drip leg should be located before risers, expansion joints, bends, valves, and regulators. System low points, end of mains, and untrapped supply runs of over 300 ft (100 m) are additional locations where drip legs should be installed.

5.22 STEAM TRAP SIZING

A steam trap must be properly sized to handle the full load of condensate. For heating devices, the method of determining the amount of condensate was discussed in Sec. 5.1. (See also Sec. 5.24.) Mains have their largest condensate loads during startup. Table 5.7 gives values for the condensate load of mains at startup.

The performance of a steam trap is affected by the inlet pressure and back pressure of the system. Therefore, when a trap is chosen, it is prudent to oversize the trap by a reasonable amount. Table 5.8 gives a guideline on how large to size traps. Grossly oversizing a trap will cause the system to operate improperly.

5.23 STEAM TRAP SELECTION

Once the size of the steam trap is known, the type of trap which will provide the best performance must be selected. When a trap is chosen, care must be taken to

TABLE 5.7 Startup Condensate Loads in Steam Mains, lb/h per 100-ft Length

Pipe size, in	Steam pressure, psig*†							
	0	5	15	30	50	100	150	200
2	6	7	8	9	10	13	15	16
2½	10	11	12	14	16	20	23	25
3	13	14	17	19	22	27	30	33
4	18	20	23	27	31	38	43	47
5	25	28	32	37	42	51	58	64
6	32	36	41	48	55	67	75	83
8	48	54	62	72	82	100	113	125
10	68	77	88	102	116	142	160	177
12	90	101	116	134	153	188	212	234

*Based on 70°F (21°C) ambient air, Schedule 40 pipe uninsulated.

†For metric equivalents, use the following conversion factors: 1 in = 2.54 cm = 25.4 mm; 1 lb/in^2 = 0.07 bar.

TABLE 5.8 Steam Trap Selection: Safety Factor

Trap type	Safety factor multiplier
Balanced-pressure thermostatic trap	3
Bimetallic trap	2.5
Liquid-expansion trap	3
Inverted-bucket trap	2.5
Float-and-thermostatic trap	2
Thermodynamic trap	1.5

select a type that will operate over the full range of pressures that the system will exert.

The best operating economy based on trap life and minimization of waste steam must be considered. If the trap will be subjected to low ambient temperatures, it should be of a freeze-proof design. For traps serving heating devices, continuous gas-venting capability is desirable. When the application is examined, the need for steam trap construction which is resistant to corrosion and water hammering should be considered.

5.24 DETERMINING CONDENSATE LOAD FOR A SYSTEM

The steam consumption of a system over time is equal to the amount of condensate formed during that period. Unfortunately, only when traps of the modulating type (such as float-and-thermostatic traps) are employed does the condensate return simultaneously equal the steam consumption.

If a blast type, say a bucket trap, is installed, the flow of condensate will be intermittent and equal to the trap's discharge rate, not the steam consumption rate. Since blast-type traps discharge intermittently, you can safely assume that not all the traps will discharge at once. For sizing purposes, the rule of thumb is that no more than two-thirds of the blast-type traps will discharge at any given time. This condensate load and the design steam consumption for the equipment utilizing modulating-type traps should be combined to determine the peak condensate load of the entire system. When the piping is sized, consider oversizing the condensate return main by one pipe size. This can be beneficial when future increases in the system's steam consumption are anticipated.

5.25 WATER DAMAGE

Water hammering is a phenomenon that occurs when condensate remaining in a pipe flashes into steam. The sudden expansion of the condensate causes a vibration in the pipe which can lead to premature failure of joints and can cause an objectional noise throughout the structure the pipe is serving. A more dangerous situation can develop if enough condensate accumulates in the pipe to block the passage of steam. The steam pressure behind the blockage will build up. Eventually the blockage may be transmitted through the pipe at a speed approaching the design velocity of the steam. When water travels at such a high velocity, it can damage the first obstruction it comes to, such as a valve or elbow. Both water hammering and damage from blockages can be prevented by proper trapping and pitching of the steam lines.

When certain gases, such as carbon dioxide (CO_2), are trapped in steam lines, the gases tend to mix with the condensate and form unwanted by-products, such as mild acids. These by-products will accelerate the rate of erosion in the system and cause premature failure in the system's components. Proper air venting will reduce the amount of gas in the system and increase its operating life.

5.26 WATER CONDITIONING

The formation of scale and sludge deposits on boiler heating surfaces creates a problem in generating steam. Water conditioning in a steam generating system should be under the supervision of a specialist. Refer to Chap. 52 of this handbook for a discussion of water treatment.

5.27 FREEZE PROTECTION

Whenever a steam system is servicing an area whose outdoor temperature will drop below 35°F (1.7°C), the designer must make provisions to prevent freezing. An alarm should be installed to alert the building operator of a loss of steam pressure or exceptionally low condensate temperatures. If air-handling units are used, the alarm should also terminate the supply fan's operation. The following recommendations will help to minimize freezing problems in steam systems:

1. Select traps of nonfreezing design if they are located in potentially cold areas.
2. Install a strainer before all heating units.
3. Do not oversize traps.
4. Make sure that condensate lines are properly pitched.
5. Keep condensate lines as short as possible.
6. Where possible, do not use overhead return.
7. If heating coils are used, allow only the interdistributing tube type.
8. Limit the maximum tube length of heating coils to 10 ft (3 m).
9. All coils and lines should be vented and drainable.

5.28 PIPING SUPPORTS

All steam piping is pitched to facilitate the flow of condensate. Table 5.9 contains the recommended support spacing for piping. The data are based on Schedule 40 pipe filled with water and an average amount of valves and fittings.

TABLE 5.9 Recommended Hanger Spacing

	Distance between supports, ft	
Pipe size, in	¼-in pitch for 10-ft length	½-in pitch for 10-ft length
1¾	4	3
1	6	4
1¼	10	5
1½	14	8
2	17	13
2½	19	15
3	22	18
3½	24	19
4	26	20
5	29	23
6	33	25
8	38	30
10	43	33
12	48	37
14	50	40

Note: Figures are based on Schedule 40 steel pipe filled with water including a normal amount of valving and fittings. These conversion factors can be used: 1 in = 2.54 cm and 1 ft = 0.3 m.

5.29 STRAINERS

Strainers (Fig. 5.14) should be located in the supply main before all steam-consuming devices and as part of the drip-leg assembly to collect particles and sediment carried in the system. Strainers located in areas not susceptible to freeze-up should extend down directly under the steam lines to allow sediment and particles to collect at the bottom of the strainer. In areas where freezing is possible, strainers should be installed at about a 20° angle below the horizontal

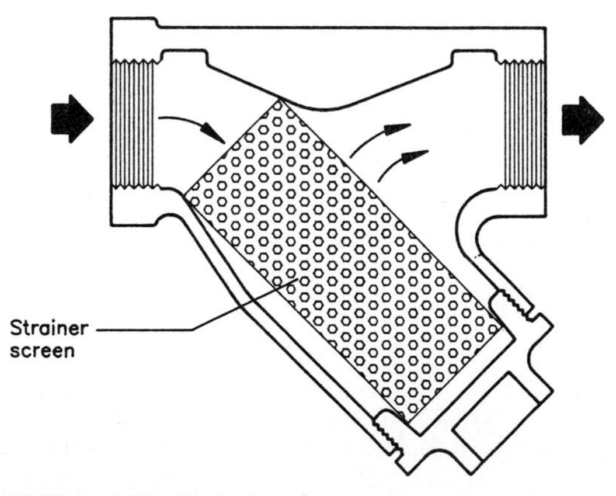

FIGURE 5.14 Typical strainer.

plane. This will form an air pocket which will allow for expansion if the water in the strainer freezes.

The strainers should be cleared regularly as part of a routine maintenance schedule.

5.30 PRESSURE-REDUCING VALVES

As steam pressure increases, the specific volume of the steam decreases as well as the heat of vaporization.

Many times the boiler is designed to operate at a higher steam pressure than the heating components. The higher boiler pressure allows the supply-main size to be reduced because of the smaller specific volume of the steam. At a convenient point in the main near the heating devices, a "pressure-reducing valve" is installed. This valve reduces the pressure and allows the steam to expand. As the steam expands, its heat of vaporization increases, allowing for greater system efficiency. The pipe size directly downstream of the pressure-reducing valve should be increased to accommodate the steam's expansion. This should be done even if the reducing-valve connections for the inlet and outlet are the same size.

5.31 FLASH TANKS

A reservoir where condensate accumulates at low pressure before it returns to the boiler is normally provided. Another name for this reservoir is the *flash tank*. As the hot condensate reaches a low-pressure area, some of the liquid will flash into steam.

At the top of the flash tank, a steam line routes the steam that has just formed back into the system to be utilized. The flash tank improves the efficiency of the system and guarantees that only liquid condensate is returned to the boiler.

5.32 STEAM SEPARATORS

The need for pure steam without the presence of water droplets is imperative to permit control devices to operate properly. A device that allows vapor to pass while knocking water droplets from the stream is known as a *steam separator*.

Steam separators should be installed before all control devices and anywhere else in the system where small water droplets cannot be tolerated. Obviously, steam separators are not required on superheated-steam installations.

CHAPTER 6
HOT-WATER SYSTEMS

Lehr Associates
New York, New York

6.1 INTRODUCTION

The predominant method of heating today's buildings, whether single-family dwellings or large structures, uses hot water to convey heat from a central generating source throughout the building. In nearly all new construction, the water is circulated through a piping distribution network by an electrically driven pump; this type of system is classified as a forced-circulation system. Heat from the circulating water is consumed by radiators, finned tubes, cabinets, or other types of terminal units (see Chaps. 13 to 18) distributed strategically throughout the structure.

Older systems used gravity to circulate the hot water, by utilizing the difference in weight between supply and return columns of the piping network. Since this type of system is rarely installed today, this chapter confines itself to forced-circulation systems. As a matter of fact, the latest American Society of Heating, Refrigeration, and Air-Conditioning Engineers (ASHRAE) guide refers readers to editions published before 1957 for details on designing gravity hot-water systems.

All hot-water heating systems rely on some form of central generating facility as the source of heat. This facility can be in the form of a boiler that consumes oil, gas, or electricity as the prime energy source or steam-to-water and water-to-water heat exchangers that derive heat from a utility or district-heating network.

This chapter gives details on the basic types of hot-water systems, as characterized by their temperature rating, general principles of system design, and special considerations of the equipment that comprises hot-water systems.

6.2 CLASSES OF HOT-WATER SYSTEMS

Hot-water systems are classified by operating temperature into three groups: low, medium, and high temperature. The *1987 ASHRAE Handbook* provides the following distinctions among these systems:

1. *Low-temperature water (LTW) system:* A low-temperature hot-water system operates within the pressure and temperature limits of the American Society of Mechanical Engineers' (ASME) *Boiler Construction Code* for low-pressure heating boilers. The maximum allowable working pressure for such boilers is 160 lb/in^2 (11 bar) with a maximum temperature of 250°F (121°C). The usual maximum working pressure for LTW systems is 30 lb/in^2 (2 bar), although boilers specifically designed, tested, and stamped for higher pressures frequently may be used with working pressures to 160 lb/in^2 (11 bar). Steam-to-water or water-to-water heat exchangers are often used, too.

2. *Medium-temperature water (MTW) system:* MTW hot-water systems operate at temperatures of 350°F (177°C) or less, with pressures not exceeding 150 psia (10.5 bar). The usual design supply temperature is approximately 250 to 325°F (121 to 163°C), with a usual pressure rating for boilers and equipment of 150 lb/in^2 (10.5 bar).

3. *High-temperature water (HTW) system:* When operating temperatures exceed 350°F (177°C) and the operating pressure is in the range of 300 lb/in^2 (20.7 bar), the system is an HTW type. The maximum design supply water temperature is 400 to 450°F (205 to 232°C). Boilers and related equipment are rated for 300 lb/in^2 service (21 bar). The pressure and temperature rating of each component must be checked against the system's design characteristics.

LTW systems are generally used for space heating in single homes, residential buildings, and most commercial- and institutional-type buildings such as office structures, hotels, hospitals, and the like. With a heat-transfer coil or similar device inside or near the boiler, LTW systems can supply hot water for domestic water supplies. Terminal units vary widely and include radiators, finned-tube fan-coil units, unit heaters, and others. Typically overall heat loads do not exceed 5000 to 10,000 MBtu/h (1.5 to 3 MW).

MTW systems show up in many industrial applications for space heating and process-water requirements. Overall loads range up to 20,000 MBtu/h (6 MW). Generally HTW systems are limited to campus-type district heating installations or to applications requiring process heat in the HTW range. System loads are generally greater than 20,000 MBtu/h (6 MW).

The designs of MTW and HTW systems resemble each other closely. The systems are completely closed, with no losses from flashing. Piping can run in practically any direction, since supply and return mains are kept at substantial pressures. Higher temperature drops occur in MTW and HTW systems, relative to LTW systems, while a lesser volume of water is circulated (depending on the heat load of the system). LTW systems lend themselves better to combined hot-water/chilled-water heating/cooling systems. Extra care and expense must be devoted to fittings, terminal equipment, and mechanical components, especially for HTW systems.

Finally, often a combined system is desirable: an MTW or HTW circuit for process heat and an LTW circuit for space heating. The hot water for the LTW system can be obtained via a heat exchanger with the main heating system.

6.3 DESIGN OF HOT-WATER SYSTEMS

Designing hot-water systems involves a complex interplay of heat loads and the type of generating system. A traditional starting point, primarily for residential LTW systems, was the assumption of a 20°F (11°C) temperature drop through the circuit, from which the overall flow rate could be determined. A more recent

TABLE 6.1 Typical Ratings of Wall Fin Elements

Element type	Rows	Hot-water capacity, Btu/(h · ft),* at 65°F (17.4°C), entering air with average water temperature of:					
		220°F 104.4°C	210°F 98.9°C	200°F 93.3°C	190°F 87.8°C	180°F 82.2°C	170°F 76.7°C
Steel, 1¼ in (32 mm)	1	1260	1140	1030	940	830	730
	2†	2050	1850	1680	1520	1350	1190
Copper-aluminum, 1 in (25.4 mm)	1	1000	900	820	740	660	580
	2†	1480	1340	1210	1100	970	860
Steel, grilled enclosure, 1 in (25.4 mm)	1	1310	1190	1080	980	860	760
	2†	2080	1880	1700	1540	1370	1210

*1 Btu/(h · ft) = 0.0768 kcal/(h · m).
†4-in (10.2-mm) center-to-center gap.

practice is to perform a rigorous analysis, because the 20°F (11°C) assumption can lead to oversized pipes and flow rates.

System design can be broken down into five elements:

1. Determining the heating load
2. Selecting terminal units or convectors based on the average water temperature and temperature drop and locating them on the architectural plan
3. Developing a piping layout, including the choice of return system
4. Locating mains, side branches, and other piping elements
5. Specifying mechanical components, the expansion tank, and the boiler

A good initial point is to run the flow main from the boiler to the terminal unit or units with the largest heat load and then to select branch runs to connect other terminal units. Common space-heating terminal elements are convectors or wall fins, both of which contain a length of finned tube over which air can be fanned if desired. The air entering temperature is usually assumed to be 65°F (18°C). Most manufacturers supply tables showing heat ratings of the convectors, based on the assumed temperature drop, and the average entering water temperature (AWT). See Table 6.1 for an example for finned-tube convectors.

An alternative approach is to assume a constant-temperature water flow (based on the leaving temperature of each class of terminal equipment) and to compute the required flow rate.

Both daily and annual variations in heat loads should be evaluated in order to arrive at a suitable design. This is especially true when LTW systems combining hot-water heating and cool-water cooling are envisioned. Figure 6.1 shows the seasonal effects of outside temperature on one type of piping design, the two-pipe system.

6.4 PIPING LAYOUT

Once a preliminary evaluation of heat load and terminal units has been performed, a piping layout can be undertaken. The usual starting-point options—running the flow main by the shortest and most accessible route to the largest heat loads—can be explored for the type of overall piping arrangement desired.

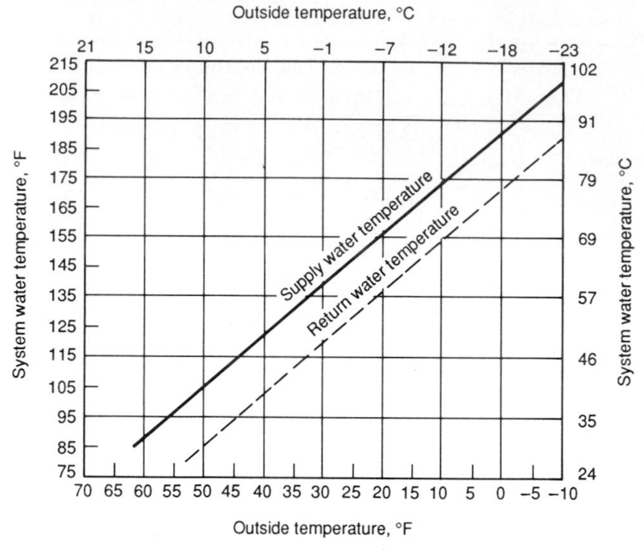

FIGURE 6.1 Seasonal operating characteristics of a two-pipe forced hot-water system. (*Courtesy of The Industrial Press.*)

Pipe circuits generally are organized into one- or two-pipe arrangements. One-pipe systems with radiators or similar terminal units often have a feed and return pipe that diverts water from the flow main to the radiator and back to the flow main; even though two pipes are present, the system is still considered a one-pipe arrangement (see Fig. 6.2). Finned-tube heating elements running along the outer walls of small residences—a common arrangement—are true one-pipe systems, as shown in Fig. 6.3. Each terminal unit in the circuit receives progressively

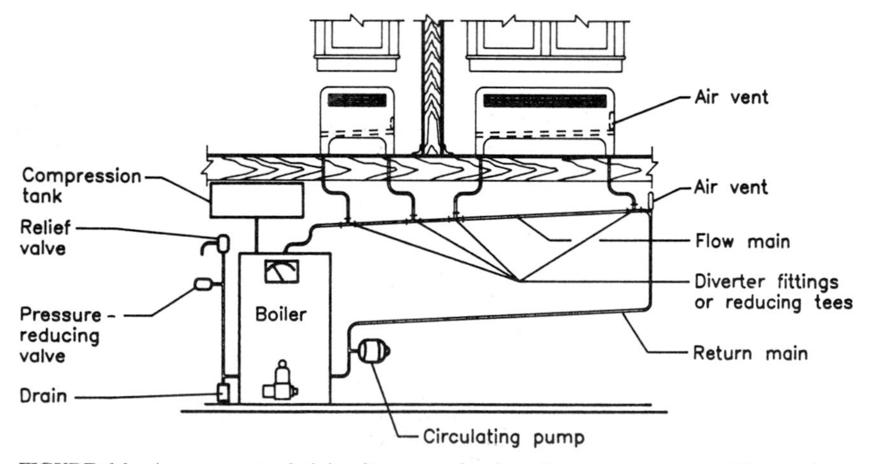

FIGURE 6.2 Arrangement of piping for a one-pipe forced hot-water system with closed expansion tank. (*Courtesy of The Industrial Press.*)

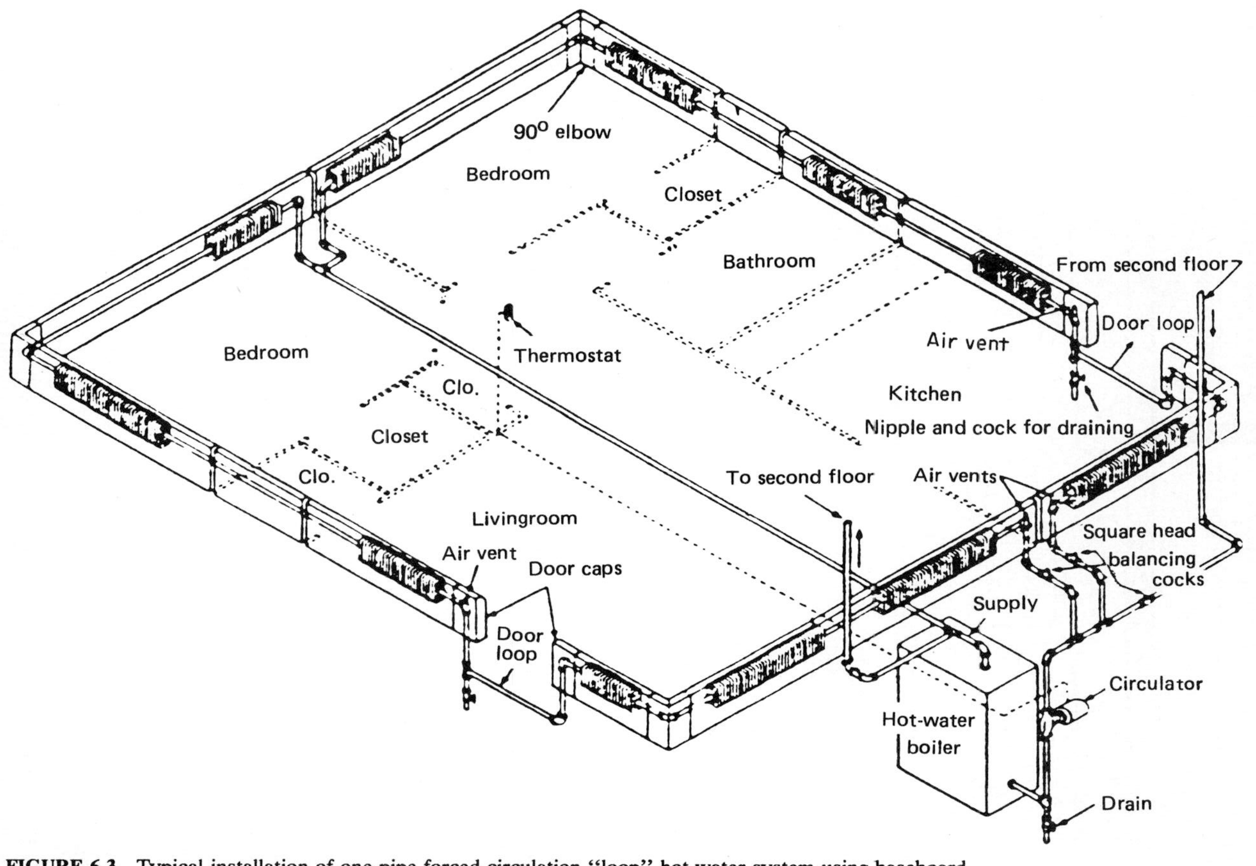

FIGURE 6.3 Typical installation of one-pipe forced-circulation "loop" hot-water system using baseboard radiators. (*Courtesy of The Industrial Press.*)

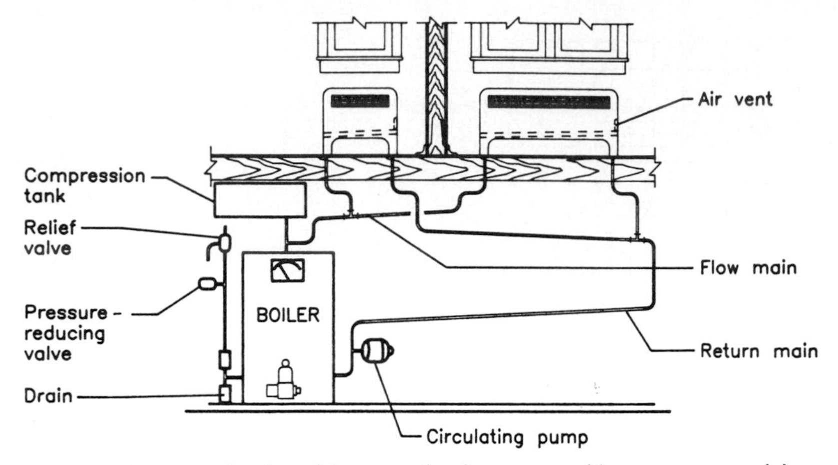

FIGURE 6.4 A two-pipe forced hot-water heating system with reverse-return piping. (*Courtesy of The Industrial Press.*)

lower water temperature; thus the units are sized larger as they are located farther from the heat source.

Two-pipe systems allow for parallel heating arrangements, whereby terminal units can receive hot water at roughly similar inlet temperatures. The cooled water returns via a second pipe. The flow of this pipe can be specified to run in direct or reverse fashion back to the heat generator. Choosing between these options allows for better balancing of heat supplies among various terminal units and for some variation in overall system capital cost. Reverse-return systems specify that the distance that the water travels to a particular unit is the same as the return distance from that unit (Fig. 6.4).

6.5 PRESSURE DROP AND PUMPING REQUIREMENTS

All hot-water systems require some type of pumping to overcome friction losses of the flowing water, because whatever head is developed by the height of the water system (static pressure) is offset by the return pressure. Some more complex systems are better served economically by two or more pumps strategically located, rather than one large pump.

Standard charts provide data on friction loss for runs of common types of piping (Fig. 6.5). To this should be added pressure losses from elbows, fittings, and other elements (Table 6.2). Similarly, manufacturers of radiators and other terminal units provide data on friction losses through their equipment.

Pump specifications are arrived at by first computing the overall pressure drop and the amount of desired water flow. "Pump curves"—charts which show the pressure developed by pumps as a function of the flow rate—can be used to arrive at the correct sizing. Many designers prefer to work with mass flow rate [pounds (kilograms) per hour] rather than gallons per minute (liters per second),

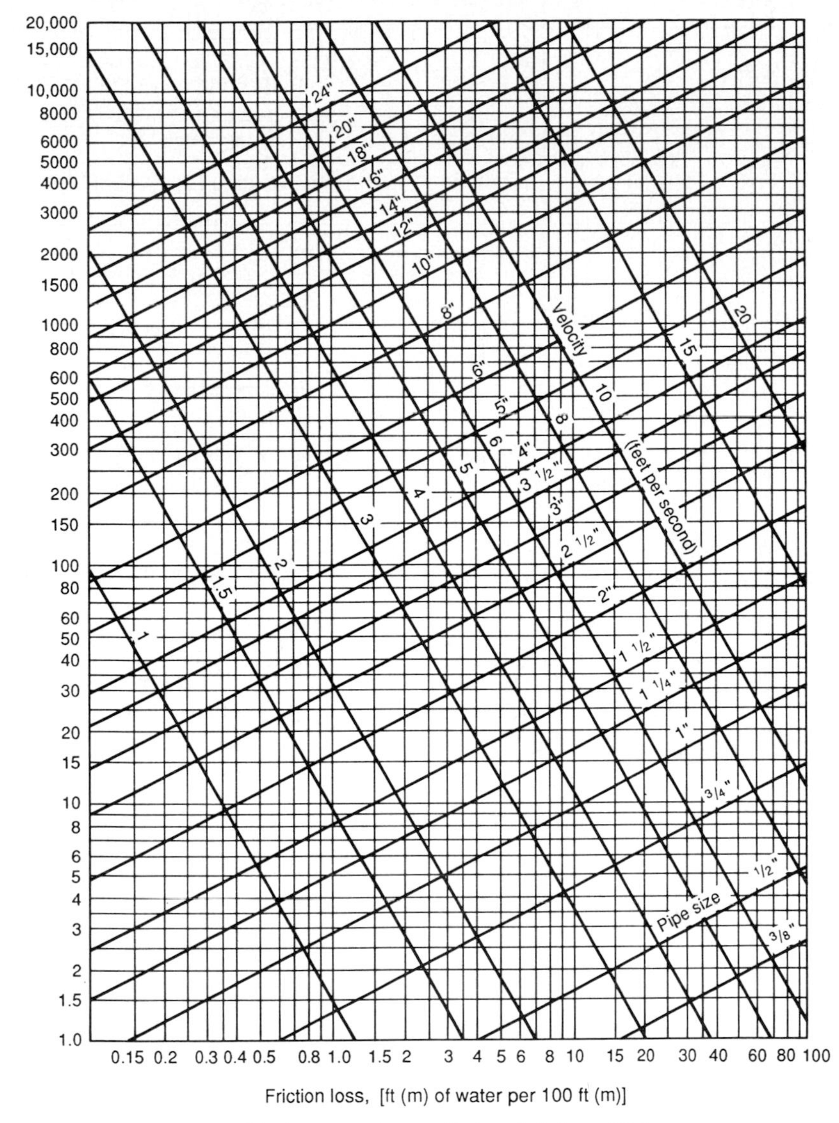

FIGURE 6.5 Friction loss for open-system piping. *(From Carrier Air Conditioning Company, Handbook of Air Conditioning System Design, McGraw-Hill, New York, © 1965. Used with permission.)*

TABLE 6.2 Fitting Losses in Equivalent Feet* of Pipe

Screwed, welded, flanged, flared, and brazed connections

Nominal pipe or tube size, in*	Smooth bend elbows						Smooth bend tees				Mitre elbows			
	90° Std.†	90° Long Rad.‡	90° Street†	45° Std.†	45° Street†	180° Std.†	Flow-through branch	Straight-through flow: No reduction	Straight-through flow: Reduced ¼	Straight-through flow: Reduced ½	90° Ell	60° Ell	45° Ell	30° Ell
⅜	1.4	0.9	2.3	0.7	1.1	2.3	2.7	0.9	1.2	1.4	2.7	1.1	0.6	0.3
½	1.6	1.0	2.5	0.8	1.3	2.5	3.0	1.0	1.4	1.6	3.0	1.3	0.7	0.4
¾	2.0	1.4	3.2	0.9	1.6	3.2	4.0	1.4	1.9	2.0	4.0	1.6	0.9	0.5
1	2.6	1.7	4.1	1.3	2.1	4.1	5.0	1.7	2.3	2.6	5.0	2.1	1.0	0.7
1¼	3.3	2.3	5.6	1.7	3.0	5.6	7.0	2.3	3.1	3.3	7.0	3.0	1.5	0.9
1½	4.0	2.6	6.3	2.1	3.4	6.3	8.0	2.6	3.7	4.0	8.0	3.4	1.8	1.1
2	5.0	3.3	8.2	2.6	4.5	8.2	10	3.3	4.7	5.0	10	4.5	2.3	1.3
2½	6.0	4.1	10	3.2	5.2	10	12	4.1	5.6	6.0	12	5.2	2.8	1.7
3	7.5	5.0	12	4.0	6.4	12	15	5.0	7.0	7.5	15	6.4	3.2	2.0
3½	9.0	5.9	15	4.7	7.3	15	18	5.9	8.0	9.0	18	7.3	4.0	2.4
4	10	6.7	17	5.2	8.5	17	21	6.7	9.0	10	21	8.5	4.5	2.7
5	13	8.2	21	6.5	11	21	25	8.2	12	13	25	11	6.0	3.2
6	16	10	25	7.9	13	25	30	10	14	16	30	13	7.0	4.0
8	20	13	. . .	10	. . .	33	40	13	18	20	40	17	9.0	5.1
10	25	16	. . .	13	. . .	42	50	16	23	25	50	21	12	7.2
12	30	19	. . .	16	. . .	50	60	19	26	30	60	25	13	8.0
14	34	23	. . .	18	. . .	55	68	23	30	34	68	29	15	9.0
16	38	26	. . .	20	. . .	62	78	26	35	38	78	31	17	10
18	42	29	. . .	23	. . .	70	85	29	40	42	85	37	19	11
20	50	33	. . .	26	. . .	81	100	33	44	50	100	41	22	13
24	60	40	. . .	30	. . .	94	115	40	50	60	115	49	25	16

*Conversion factors: 1 ft = 0.31 in; 1 in = 25.4 mm.
†R/D approximately equal to 1.
‡R/D approximately equal to 1.5.
Source: Carrier Air Conditioning Company, *Handbook of Air Conditioning System Design*, McGraw-Hill, New York, ©1965. Used with permission.

units common to pump curves. The conversion between the two is temperature-dependent; two quick conversions commonly used are

Water at 40°F (4.4°C): 1 lb/h = 0.002 gal/min (1 kg/h = 1.26 E − 4 L/s)

Water at 400°F (204.5°C): 1 lb/h = 0.0023 gal/min (1 kg/h = 1.45 E − 4 L/s)

The next step is to determine the system curve for the hot-water circuit. The following formula is employed:

$$\frac{H_1^{0.5}}{W_1} = \frac{H_2^{0.5}}{W_2} \tag{6.1}$$

where H_1 = known or calculated head, ft (m)
W_1 = design flow rate, gal/min (L/s)
H_2 = system curve head point, ft (m)
W_2 = system curve flow-rate point, gal/min (L/s)

With this equation, various system curve points can be plotted on the pump curve. The point where the system curve and the pump curve intersect is the operating point of the pump. Pump manufacturers specify optimum operating conditions (in terms of energy consumption, efficiency, and capacity of the pump) for their equipment.

6.6 PIPE SIZING

Hot-water system piping must be sized to carry the maximum desired amount of heating water throughout the system, while accounting for the static head of the elevation of the system and friction losses from pipe and fittings. Pipe sizes generally step down as water flows from the main(s) to branch circuits or individual heating units.

Once the overall heating demand and the operating temperature of the heating system are known, calculations can be made for pipe sizes. The relationship between Btu demand and water flow rate is

$$\text{Btu/h} = \text{gal/min}\ (500\ \Delta T\text{°F}) \tag{6.2}$$

A rough calculation of the overall friction head for the main can be done by measuring the longest main circuit and adding an equivalent length of 50 percent of the main to account for fittings. More accurate determinations are made by adding the equivalent pipe lengths of the fittings on the main to the length of the longest main. The manufacturer's literature usually includes charts similar to Table 6.2 showing equivalent lengths of common fittings.

Various methods have been worked out to determine the suitable pipe diameters to provide a sufficient flow rate. Usually the procedure must be iterated several times to select the best combination of flow rate, fluid velocity, and pressure drop. Table 6.3 shows these relationships for various pipe sizes if one assumes a maximum pressure drop of 4 ft per 100 ft (1.2 m per 30.5 m) and a maximum velocity of 10 ft/s (3 m/s). Once the pipe sizes have been determined, the system's pressure head should be compared to the head developed by the circu-

TABLE 6.3 Allowable Flow Rates for Closed System Piping, Standard-Weight Steel Pipe

Pipe size, in	Flow range, gal/min	Pressure drop range, ft per 100 ft
½	0–2	0–4
¾	3–4	2.5–4
1	5–7.5	2.0–4
1¼	8–16	1.25–4
1½	17–24	2–4
2	25–48	1.25–4
2½	49–77	2–4
3	78–140	1.5–4
4	141–280	1.25–4
5	281–500	1.5–4
6	501–800	1.75–4
8	801–1700	1.0–4
10	1701–2500	1.25–2.75
12	2501–3600	1.25–2.25
14	3601–4200	1.25–2.0
16	4201–5500	1.0–1.75
18	5501–7000	0.9–1.50
20	7001–9000	0.8–1.25
24	9001–13,000	0.6–1.00

Note: The above capacities are based on a maximum pressure drop of 4 ft per 100 ft and a maximum velocity of 10 ft/s. Conversions: 1 in = 25.4 mm, 1 gal = 3.8 L, and 1 ft = 0.31 m.

lation pump. The pump may have to be resized, necessitating another iteration of the pipe sizing.

6.7 VENTING AND EXPANSION TANKS

Hot-water systems require pressures greater than atmospheric at all times to prevent air infiltration. Flashing or boiling of water is also minimized by maintaining the system above the water vapor pressure—preventing this also minimizes water hammer.

Maintaining this pressure, as well as allowing for the expansion and contraction of water as it is heated and cools, is most frequently carried out by means of an expansion tank. The expansion of medium- or high-temperature water systems can be calculated by consulting steam tables. The specific volume of water at its initial conditions is subtracted from its volume at the highest temperature, to calculate the volume change. To a certain limited extent, the water's expansion and contraction are offset by the similar changes that system piping and heating units undergo. These changes can be calculated from coefficients of expansion of the materials of the piping.

The simplest type of expansion tank is open to atmosphere at an elevation that provides the pressurization (head) the system requires. Open tanks have the disadvantage of allowing air to enter the system via absorption in the water. Closed tanks are more common now, especially with larger systems. Three common types of expansion/pressurization tanks are in use today:

1. *Adjustable expansion tank.* This tank employs an automatic valve along with a closed tank that has water and air feeds. As the temperature in the system rises, the pressure rises. A control valve releases air in the tank to the atmosphere. When the pressure and the water level drop, high-pressure air is injected into the tank. High-temperature systems should use nitrogen rather than plain air to reduce corrosive effects.

2. *Pump-pressurized cushion tank.* This design involves a makeup tank which is fed by a pump and a back-pressure control valve. For small systems (depending on local codes and on the water pressure available) the pump is skipped and city water pressure is used to feed a makeup tank that pressurizes the heating circuit. In principle, either type of pressurized tank can be roughly sized by assuming the expansion and contraction rates of the water to be equal.

3. *Compression tank.* A compression tank employs a specified volume of gas within an enclosure. As the water temperature and volume increase, the pressure on the gas volume rises, causing that gas volume to decrease. In this manner, the tank accommodates changing water volumes while keeping the system within a specified range of upper and lower pressures.

In low-temperature systems, the compression tank is usually connected to the system through an air separator situated between the boiler exit and the suction inlet of the circulating pump. Air separated from the water will rise into the compression tank. When the compression tank is located at a system's high point, it can be smaller in volume since the pressure is at its lowest. Tank sizing is also dependent on the location of the circulation pump relative to the tank.

One commonly used formula for sizing the compression tank, when operating temperatures are below 160°F (71.1°C), is

$$V_t = \frac{EV_s}{P_0/P_1 - P_0/P_2} \tag{6.3}$$

where V_t = compression tank volume
V_s = volume of circulating system, exclusive of compression tank
E = coefficient of expansion from initial to operating temperature
P_0 = absolute pressure in compression tank prior to filling
P_1 = absolute static pressure after filling
P_2 = absolute pressure at system operating temperature

For operating temperatures between 160 and 280°F (71.1 to 137.8°C), this formula is used:

$$V_t = \frac{(0.00041t - 0.0466)V_s}{P_0/P_1 - P_0/P_2} \quad \text{(USCS units)} \tag{6.4}$$

$$V_t = \frac{(0.000738t - 0.03348)V_s}{P_0/P_1 - P_0/P_2} \quad \text{(metric units)} \tag{6.5}$$

where t = maximum operating temperature.

Compression tanks can be supplied with an impermeable membrane (diaphragm) to prevent air from being drawn into the circulating water when the system temperature drops. The diaphragm also allows thc compression tank to be smaller in volume.

Diaphragm compression tanks are equipped with sight glasses or similar devices to monitor the water level. Too low a water level prevents the air behind the diaphragm from affecting the system's pressure.

6.8 MECHANICAL AND CONTROL EQUIPMENT

Mechanical components for low-temperature systems are under less severe service than those for medium- and high-temperature units; correspondingly, the care with which components are specified should increase with the higher-temperature systems.

ASME and ASHRAE rules should be observed for dealing with pressure vessels. Specifically, the chemical condition of the circulating water in high-temperature systems should be checked periodically by an expert. Pressure gauges should be located at both ends of the circulation pump. Modulating combustion controls, rather than straight on/off controls, are necessary to minimize pressure swings that lead to flashing. Where compression-type expansion tanks are used, an interlock with the system's heat generator should be installed to prevent operation when the water level in the tank is too low or insufficient air is present to maintain the tank compression. Valves and fittings for high-temperature systems should be specified with materials that resist corrosion and erosion, such as stainless steel.

The primary control factor for a hot-water system is the operating temperature range, which in turn is based on outside air temperatures. Electronic thermosensors and thermostats function to keep the room air temperature within the desired range. The system should also be equipped with a manual on/off control.

The electronic control for moderating the room air temperature can be of several types. Most are based on a solenoid device, which sends a signal current on the basis of a temperature reading. The control can be a simple on/off device or can have various modulating schemes to minimize large temperature swings. Temperature controls can also be set for zone heating of certain rooms or areas within a large room, depending on the piping layout. In this case, the electronic control is connected with various flow control valves that will reduce or expand water flow to the heating units.

CHAPTER 7
INFRARED HEATING

Lehr Associates
New York, New York

7.1 INTRODUCTION

Infrared heating takes advantage of the fact that light can be used to convey heat, just as sunlight can warm a cold surface. Infrared rays heat the objects they strike rather than the air surrounding the objects. When infrared is used for total heating, the surrounding air conducts heat away from the objects and the space eventually reaches an equilibrium temperature. When it is desirable to heat only a small area within a larger one or in an open space (such as a construction site), infrared heating is unsurpassed because the energy required is only that used to warm objects or people, and not all the surrounding space.

7.2 TYPES OF HEATERS AND APPLICATIONS

A large variety of uses are suited to infrared heating, and many emphasize heating selected areas of large spaces. Examples include the following:

Bus shelters	Milking parlors
Canopies	Parking lots
Garages	Skating rinks
Gymnasiums	Snow melting
Indoor tennis courts	Stadiums
Loading docks	Work areas
Manufacturing areas	

Infrared heaters offer the user total flexibility in heating a building. Infrared systems can be used to maintain desired temperatures inside an entire building. They can also be used for supplementary areas or spot comfort heating within a building and are unmatched for outside area heating.

Infrared heaters are commonly designed to radiate energy by either burning a fuel or radiating heat from electrical-resistance wires. The details of common types are as follows:

7.2.1 Fueled Heaters

Gas-Fired. These units use either natural gas or propane, delivered under pressure. The gas flows along a tube or within a mesh-metal or ceramic "emitter." The emitter elements are mounted in a panel in parallel rows, each up to several feet in length. Tubular heaters are larger, with several tubes mounted in parallel and running up to 20 ft (6 m) in length. The heat output of a tubular heater can range upward of 100,000 Btu (29.3 kW), while panel-type heaters range from 20,000 to 200,000 Btu (5.9 to 58.6 kW). Both types are normally mounted high on walls or suspended by wires from ceilings. Small portable units, fueled from a pressurized liquefied petroleum gas (LPG) tank, can be floor-mounted.

7.2.2 Electric Heaters

Single- or Double-Element Tubular Units. These units come with reflectors that limit the radiation pattern to a 30°, 60°, or 90° angle. Double-element units can be symmetric or asymmetric. These heaters could be conduit- or chain-mounted, adjustable 45° in either direction. Metal-sheath, quartz tube, or quartz lamp elements may be used interchangeably.

Portable Units. These provide warm, sunlike heat for hard-to-heat spots or for construction or maintenance work outdoors. This type of heater is usually made of extruded aluminum, with gold-anodized reflectors, heavy-duty metal-sheath heating elements, and safety screens. The unit provides rapid heatup and is operated from power distribution centers, a disconnect switch, etc. Usually units up to 5 kW are provided with a stand; 6-kW units and up are mounted on two-wheeled carts.

Heavy-Duty Overhead Units Designed for Spot Heating Applications. These are operated manually from a disconnect switch or automatically controlled with a heavy-duty metal sheath heating element.

Glass-Panel Units. The heating element is fused onto specially tempered glass which will withstand a pressure of 27,000 lb/in^2 (1901.4 kg/cm^2). The fused-on heating element is covered with enamel for corrosion protection. The whole assembly is mounted in silicone rubber to completely eliminate hum and expansion noise.

7.3 PHYSIOLOGY OF INFRARED HEATING

Because infrared heating can be used so selectively—heating only a small space—its physiological effects are worth considering. A person under an infrared lamp is warmed by the absorption of the light; a portion of the light is reflected. Light reflected from surfaces around a person—the floor, furniture, and the like—is also absorbed by one's skin. Absorptance and reflectance are set by the color temperature (corresponding to the light's frequency) of the light and the properties of the surface on which the light falls. A person's skin has an absorptance of around 0.9 at 1200 K (color temperature), falling to around 0.8 at 2500 K.

A person feels comfortable or uncomfortable based on the surrounding temperature, humidity, air flow, and degree of activity. The "mean radiant temperature" (MRT) has been defined as the temperature at which an imaginary isothermal black enclosure would exchange the same amount of heat by radiation as in an actual environment. As room temperature falls below normal ranges, or as air flow increases heat flow away from a person, the MRT rises. The designer using infrared heating for comfort purposes can use a set of calculations by the American Society of Heating, Refrigeration, and Air-Conditioning Engineers (ASHRAE) to determine the MRT needed to provide desired comfort levels. ASHRAE has also established comfort charts that determine comfort levels on the basis of room temperature, relative humidity, and radiated heat.

Infrared heating can also be used, of course, for *total* space heating. Then comfort-level calculations become secondary to normal calculations of heat loss from a structure and the amount of heat provided by heating units.

7.4 SPACING AND ARRANGEMENT OF ELECTRIC HEATERS

7.4.1 Total Heating

To achieve total heating with infrared heat, calculate the normal heat loss of the building or room to be heated and supply an equivalent number of watts of infrared heat to replace what is being lost. Heaters should be arranged in a grid pattern to achieve the most even coverage of the floor area.

Infrared heaters heat the exposed floor and the objects in the room. The objects in turn radiate heat to the cooler surrounding air until an equilibrium is reached. Thermostats may be used to maintain the desired comfort level in the room.

Two parallel heaters aimed at an angle toward the section of a room are more efficient than a single heater positioned directly overhead. Two heaters can cover a greater area, with a more even coverage, than one heater.

7.4.2 Determining the Number and Size of Heaters

Determine the required watt density (WD) from Table 7.1.

EXAMPLE A tight, uninsulated building, such as a pole barn, steel shed, or concrete-block warehouse, located in an area with an ambient temperature of 45°F (7.22°C) would require 30 W/ft^2 (323 W/m^2) of occupied floor space to obtain a comfortable working temperature of 65°F (18.3°C) [assuming a 20°F (11.1°C) temperature rise].

From job conditions, determine the minimum mounting height available. The lower the mounting height, the greater the watt density attainable with the fewest heaters. However,

TABLE 7.1 Required Watt Density*

Desired† temperature rise, °F	Local spot and small-area heating				
	Tight, uninsulated building	Drafty‡ indoors or large glass area	Loading area, one end open	Outdoor, shielded; less than 5 mi/h wind	Outdoor,§ unshielded; less than 10 mi/h wind
20	30	40	50	55	60
25	37	50	62	70	75
30	45	60	75	85	90
35	52	70	87	100	105
40	60	80	100	115	120
45	67	90	112	130	135
50	75	100	125	145	150
60	90	120	150	175	N.R.
70	105	140	175	205	N.R.
80	120	160	200	235	N.R.
90	135	180	N.R.	N.R.	N.R.
100	150	200	N.R.	N.R.	N.R.

*Watts of input to heaters per square foot of occupied floor space.

†Temperature rise is the increase above the existing temperature in the room. For outdoor use, subtract temperature rise from 70° to determine lowest outside temperature at which comfort can be achieved with the watt density listed under the appropriate column.

‡Doors frequently open.

§Normal outdoor clothing; people stationary. N.R. = not recommended.

Note: Use the following conversion factors: °C = (°F − 32)/1.8; 1 mi/h = 1.609 km/h.

the lowest recommended mounting height is 10 ft (3.04 m), to prevent too high a watt density on the upper part of one's body. Higher mounting will provide coverage of a greater area at a lower watt density.

Using the selected mounting height, refer to *A* or *B* in Table 7.2 to obtain the watt density and area of coverage at three-fourths overlap or full overlap, respectively, at the indicated distance *S* between parallel heaters.

EXAMPLE At a mounting height of 14 ft (4.27 m), two heaters mounted with centerlines 8 ft (2.44 m) apart for full overlap would cover an area 14 ft (4.27 m) wide and 16 ft (4.88 m) long at a watt density of 22 W/ft^2 (237 W/m^2).

Table 7.2 is based on the use of 60° asymmetric heaters placed a distance *S* apart for three-fourths or full overlap.

7.4.3 Increasing the Watt Density

To increase the watt density, use additional heaters between the two parallel heaters. One additional heater equals 100 percent greater watt density; three additional heaters equal 150 percent greater watt density. Using Table 7.3, determine the watt density and coverage of the 60° and 60° asymmetric heater at the selected height. Add a sufficient number of heaters to attain the required watt density.

EXAMPLE At a 14-ft (4.27-m) mounting height, a watt density of 33 W/ft^2 (355 W/m^2) is required. With full-overlap spacing (*B* in Table 7.2), two asymmetric heaters provide only

TABLE 7.2 Spacing and Coverage

Mounting height, ft	A: WD	A: W × L, ft	A (3/4 overlap): W, ft	B: WD	B: W × L, ft	B (Full overlap): W, ft
10	33	13 × 12	8½	38	11 × 12	6
11	27	14 × 13	9	33	12 × 13	6½
12	24	15 × 14	10	29	12 × 14	7
13	21	16 × 15	10½	25	13 × 15	7½
14	18	17 × 16	11	22	14 × 16	8
15	16	18 × 17	12	20	15 × 17	8½
16	14	20 × 18	12½	18	16 × 18	9
17	13	21 × 19	13	16	17 × 19	9½
18	11	22 × 20	14	14	18 × 20	10
19	10	23 × 21	15	13	19 × 21	10½
20	9.6	24 × 22	15½	11	20 × 22	11
21	8.8	25 × 23	16	10	20 × 23	11½
22	8.0	26 × 24	17	9.6	22 × 24	12
23	7.2	27 × 25	17½	8.8	22 × 25	12½
24	6.8	28 × 26	18	8.4	23 × 26	13
25	6.4	30 × 27	19	8	24 × 27	13½

Note: 1 ft = 0.304 m. WD = watt density; *W* = width; *L* = length.

22 W/ft^2 (237 W/m^2) on 10-ft (3.04-m) centers, covering an area of 14 ft (4.27 m) by 16 ft (4.88 m). From Table 7.4, at a 14-ft (4.27-m) mounting height, a 60° heater will provide the necessary addition (11 W) over an area of 11 ft × 16 ft (3.35 × 4.88 m) when mounted between the two 60° asymmetric heaters. [*Note*: A 90° heater will add 10 W and cover an area of 16 × 16 ft (4.88 × 4.88 m). Select the heater that will best suit the needs of the application.]

7.4.4 Greater Length Coverage

To obtain coverage for lengths greater than those listed in Table 7.2, use additional parallel groups. Use a length centerline the same as the width centerline (*S*) listed, thereby increasing the length coverage by the same footage while retaining the same watt density.

EXAMPLE For the example above with a 14-ft (4.27-m) mounting height, a watt density of 33 W/ft^2 (355 W/m^2), but an area 14 ft × 24 ft (4.27 m × 7.32 m) to be covered, use an additional group of two parallel asymmetric heaters, plus the one "fill-in" heater, spaced on 8-ft (2.4-m) centers. This will provide an additional 8 ft (2.44 m) of coverage and maintain the watt density at 33 W/ft^2 (355 W/m^2).

7.4.5 Additional Width Coverage

Use side-by-side sets of parallel heaters. Very little overlap is required. At a 10-ft (3.04-m) mounting height, centerlines of back-to-back heaters should be 2 ft (0.61

TABLE 7.3 Single-Heater Watt Density and Coverage

	90°		60°		60° Asymmetric	
Mounting Height, ft	WD	*W* × *L*, ft	WD	*W* × *L*, ft	WD	*W* × *L*, ft
10	18	12 × 12	25	8½ × 12	19	11 × 12
11	15	13 × 13	22	9 × 13	16	12 × 13
12	13	14 × 14	18	10 × 14	14	12 × 14
13	11	15 × 15	16	10 × 15	13	13 × 15
14	10	16 × 16	14	11 × 16	11	14 × 16
15	9	17 × 17	13	12 × 17	9.6	15 × 17
16	8	18 × 18	11	12 × 18	8.8	16 × 18
17	7	19 × 19	10	13 × 19	8.0	17 × 19
18	6.4	20 × 20	8.8	14 × 20	7.2	18 × 20
19	5.6	21 × 21	8.0	15 × 21	6.4	19 × 21
20	5.2	22 × 22	7.6	15 × 22	5.8	20 × 22
21	4.8	23 × 23	7.0	16 × 23	5.4	20 × 23
22	4.4	24 × 24	6.2	17 × 24	4.8	22 × 24
23	4.0	25 × 25	5.8	17 × 25	4.5	22 × 25
24	3.7	26 × 26	5.4	18 × 26	4.1	23 × 26
25	3.5	27 × 27	5.0	19 × 27	3.9	24 × 27

Note: 1 ft = 0.304 m. WD = watt density; *W* = width; *L* = length.

m) apart and an additional 1 ft (0.3 m) apart for every 5 ft (1.52 m) of mounting height above 10 ft (3.04 m).

7.5 GAS INFRARED RADIANT HEATING

Gas-fired infrared units have many of the same advantages as electrically powered units, with the bonus of cost savings where the price of gas is less than the Btu equivalent amount of electricity. Gas-fired units can be used for both local (zone) heating of a workspace and total space heating.

Gas-fired units require that several precautions be taken to protect against fire hazards or damage from intense radiation. Generally these standards are detailed in local building codes. Among them are the following:

1. *Adequate combustion air:* Sufficient air must be available for complete combustion. In addition, the presence of flammable dusts or vapors must be kept low. Venting to prevent buildup of combustion gases must be allowed for; generally a positive air displacement of 4 ft^3/min per 1000 Btu/h (387 L/min per kW) of gas input is desired.
2. *Humidity control:* Among the combustion by-products is water vapor; the venting must be planned to allow for this.

3. *Clearance to combustibles:* Both the roof and articles or materials below and beside the heating unit must be kept a sufficient distance from the unit. This distance depends on the heat output of the unit and the flammability of the materials near it. Equipment manufacturers provide a clearance-to-combustibles chart that specifies minimum distances. For 50,000-Btu/h (14.65-kW) units, the top clearance is around 9 in (22.9 cm), the side clearance is around 10 in (25.4 cm), and the bottom clearance is 36 to 42 in (91.4 to 106.7 cm).
4. *Controls and shutoff:* Gas units are generally started by electric glow plugs or similar devices. Differential-pressure gauges are usually included to cut off gas flow in the event of a combustion air or flue blockage.

7.5.1 Design of Total-Heating Gas-Fired Systems

The selection of the proper heater size is determined by the available mounting height. There must be sufficient clearance from floor to roof to allow the heater to be installed at the recommended mounting height with enough clearance above the unit to prevent overheating of the roof.

From an economical standpoint, the largest heater is used for the available mounting height as long as the recommended maximum distance between heaters is not exceeded. Normally heaters are mounted 2 to 3 ft (0.6 to 0.9 m) below the crane rails, and the clearance to combustibles is maintained from the roof regardless of the combustibility of the roof.

7.5.2 Louvers

Inlet louvers may be necessary if the building is extremely tight and well insulated. If inlet louvers are used, they should be small and well distributed on at least two outside walls and on all outside walls in larger installations.

7.5.3 Annual Estimated Fuel Cost

Fuel consumption will be influenced by a number of variables over which the designer has little control. Consumption calculations should be treated as estimates.

A simplified formula to approximate the annual gas consumption in therms is:

$$\frac{HL \cdot DD \cdot 24}{TD \cdot 100{,}000}$$

where HL = heat loss, Btu/h (kW)
DD = degree days
TD = temperature differential, °F (°C)
100,000 = heating value of 1 therm of gas

7.5.4 Heater and Thermostat Placement

The placement and the spacing of the heaters are made easy by referring to Fig. 7.1 and Table 7.4. It is common practice to set in from the corner of a building 15 to 20 ft (4.57 to 6.10 m) with the first heater along an outside wall.

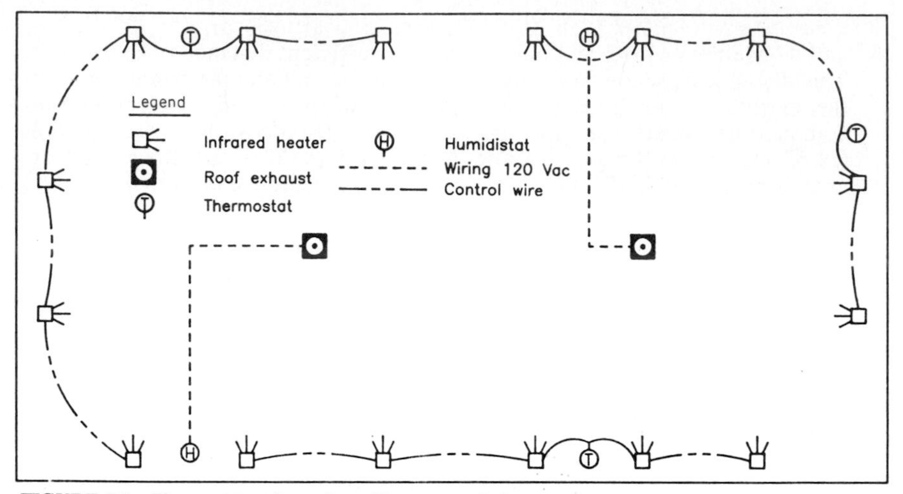

FIGURE 7.1 Placement and spacing of heaters and thermostats.

If clearances below the units necessitate mounting closer to the roof than suggested, a piece of noncombustible millboard is frequently utilized above the unit.

In most buildings a single perimeter loop system will satisfy the heat loss. If the maximum distance between heater rows is exceeded, an interior row or partial row may be needed.

In buildings where an interior row is indicated, the only heat losses are the roof and air losses. Center-row spacing is frequently double that used along the perimeter.

Control systems are available in three voltages: 120 V, 25 V, and millivolt. The selection of the voltage may be a matter of personal preference, or in some cases the code may dictate the system used. Millivolt control is used primarily for small systems since each heater has its own thermostat. The 25-V system is normally the most economical system to install. On most applications using the 25-V system, the 18/2 control wire is taped to the gas piping with a drop at each heater. If the code requires that 25-V control wiring be installed in conduit, then no savings advantage is experienced. Three thermostats for zone control are common practice with infrared systems and allow for flexibility of control. Thermostats should not be located directly in the rays of the infrared generators because they will not read the ambient temperature of the building correctly.

On 25-V systems, take special note of the volt-ampere ratings of the various control systems and the size of the transformers utilized. Some systems will allow the use of eight units on one 75-VA transformer while another system will set a limit of four units on the same transformer.

TABLE 7.4 Heater Placement Guide for Full Building Heat

	Input, Btu/h*							
	25,000	30,000	50,000	60,000	75,000	90,000	110,000	120,000
Mounting height, ft*	9–13	10–14	14–18	15–20	17–22	20–25	22–28	24–28
10°	7–10	7–12	11–15	12–16	14–19	16–22	18–25	18–26
30°								
Range between units, ft	7–18	8–20	12–25	15–30	18–34	20–42	25–46	28–50
Between rows ft* (max.)	60	60	80	80	100	100	110	110
Heater to wall, ft* (max.)	5	6	8	10	12	12	14	14
Required exhaust,† ft^3/min	100	120	200	240	300	360	440	480
	Clearance to combustibles, in							
Top	30	30	36	36	48	48	54	54
Below	72	72	88	88	104	104	120	120
Sides	30	30	36	36	42	42	48	48
Back	24	24	33	33	39	39	48	48

*Conversion factors: Btu/h × 0.293 = W; ft × 0.304 = m; ft^3 × 0.0283 = m^3.
†Air exhaust based on 1000 Btu/h of gas and 4 ft^3/min per 1000 Btu input.

CHAPTER 8
ELECTRIC HEATING SYSTEMS

Lehr Associates
New York, New York

8.1 INTRODUCTION

Electric energy is ideally suited for space heating. It is relatively simple to control and distribute. In many applications, the cleanliness and compactness of electric heaters are a very attractive design alternative. Such heaters require no fuel storage, produce no fumes or exhaust emissions, and, depending on the specific setup, present a safer alternative to combustion heaters.

Cost and energy conservation are the dominating design factors in electric heating. Usually, compared with other heating methods, electric heating has a lower installation cost, requires less maintenance, has lower insurance rates, and is easier to zone. Electric space heating is often used when minimal initial cost is a dominant factor. However, electricity is a relatively expensive form of power. The increase in energy costs during the 1970s has made electric energy cost-prohibitive in some cases. The operating costs of electric systems are normally higher than those of heating with gas or fossil fuels.

8.2 SYSTEM SELECTION

A careful analysis is necessary to select for a building a heating system that is both effective and cost-efficient. The effectiveness is determined by the ability of the system to meet the heating needs of the building. This is done by determining the power requirements to maintain an indoor design temperature over a range of outdoor temperatures. This power requirement is called the *load* on the system and is useful when presented in a load profile or load duration curve. A load profile curve shows the power required to maintain an indoor temperature plotted against the time of day. The load duration curve shows the same information in a different fashion. It plots the number of hours for which the system is at each load level. The total energy requirements can be determined by finding the areas under these curves.

Costs associated with the system must be estimated over the life cycle of the equipment. The initial expense of the equipment and its installation cost are obtainable, but the daily costs, long-term maintenance, and replacement value must be estimated. All these costs must be included in an efficiency analysis, if it is to be complete.

Operating costs are particularly difficult to determine. The estimated annual heating requirement is used along with estimates of electricity prices to determine yearly operating costs. The pricing structure of electricity must be accounted for, especially if demand changes or time-of-day rates apply. Local utility companies may provide estimates of future energy prices. The prices of other energy sources, such as oil and gas, should also be considered. The following cost comparison can be used to make an initial appraisal of the relative costs of electric versus gas or oil heating. (See Figs. 8.1 and 8.2.)

Electric heating systems are either centralized or decentralized. Centralized systems serve the heating needs of an entire structure. Hot-water, steam, and warm-air systems are examples of centralized systems. Decentralized systems come in various designs: natural-convection units, forced-air units, radiant units, or radiant panel-type systems. These designs are geared to the control of individual heating sections of a structure. The following systems are all presently in use:

I. Decentralized systems
- **A.** Natural-convection units
  - **1.** Floor drop-in units
  - **2.** Wall insert and surface-mounted units
  - **3.** Baseboard convectors
  - **4.** Hydronic baseboard convectors with immersion elements

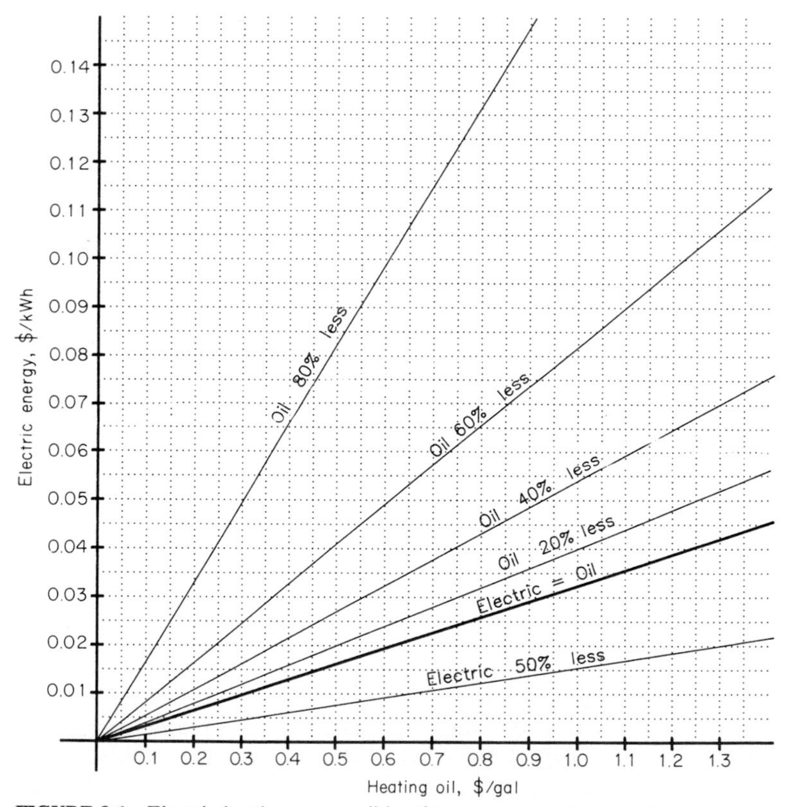

FIGURE 8.1 Electric heating versus oil heating—energy cost comparison.

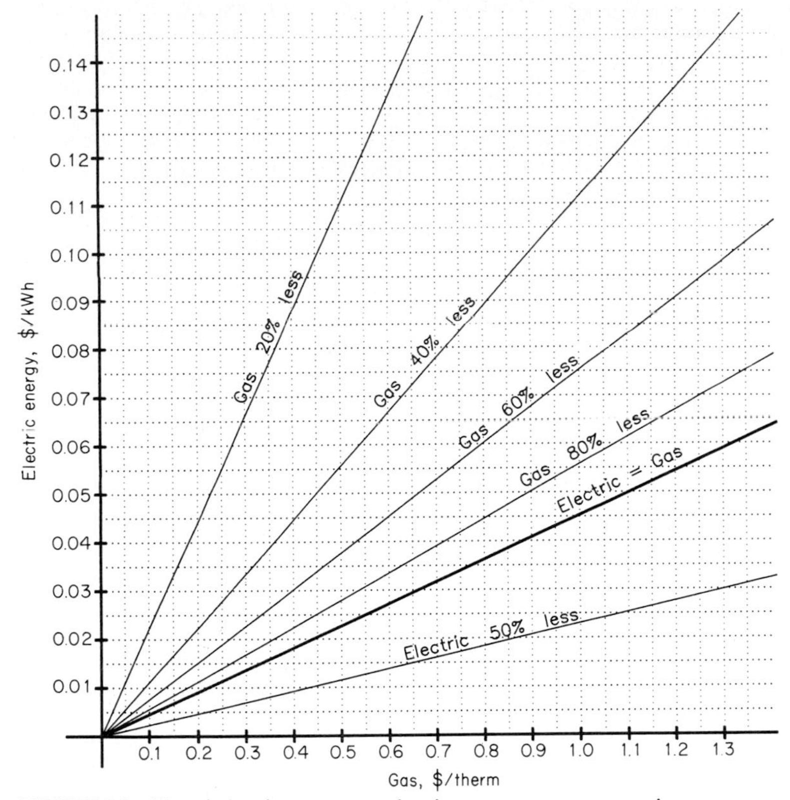

FIGURE 8.2 Electric heating versus gas heating—energy cost comparison.

B. Forced-air units
  1. Unit ventilators
  2. Unit heaters
  3. Wall insert heaters
  4. Baseboard heaters
  5. Floor drop-in heaters

C. Radiant units (high-intensity)
  1. Radiant wall, insert, and surface-mounted
  2. Metal-sheathed element with reflector
  3. Quartz-tube element with reflectors
  4. Quartz lamp with reflector
  5. Heat lamps
  6. Valance (cove) unit

D. Radiant panel-type systems (low-intensity)
  1. Radiant ceiling-mounted with embedded conductors
  2. Prefabricated panels
  3. Radiant floor-mounted with embedded conductors
  4. Radiant convector panels

II. Centralized systems
  A. Heated-water systems
    1. Electric boiler
    2. Electric boiler with off-peak storage
    3. Heat pumps
    4. Integrated heat-recovery systems
  B. Steam systems: electric boiler, immersion, or electrode
  C. Heated-air systems
    1. Duct heater
    2. Electric furnace
    3. Heat pump
    4. Integrated heat-recovery system
    5. Unit ventilator
    6. Self-contained heating and cooling unit
    7. Storage unit (ceramic, water)

8.3 CENTRAL HOT-WATER SYSTEMS

A central hot-water system comprises an electric boiler and a series of radiators, convectors, unit heaters, cabinet heaters, or heating and ventilating units. Water is heated in the boiler by elements or electrodes and is distributed throughout the structure. Multiple elements are combined to achieve the total capacity required. Multiple-element systems are arranged to energize or de-energize each unit in sequence to avoid large fluctuations in voltage. It is also important that the elements not be energized when the circulating pump is not operating.

Some boilers are designed for space efficiency and can be wall-mounted. A 20-kW boiler is approximately 1.5 ft^3 (0.04 m^3). Larger applications with capacities up to 5000 kW can be achieved with electrode boilers rated at 5 kV and higher. Electric boilers are space-efficient and clean. They have no need for flues, smokestacks, or fuel storage. See Chap. 29 for a detailed discussion of electric boilers.

Hot-water systems can be tailored to specific applications. Some boilers are particularly adaptable to multiple zoning. A control system energizes the number of elements needed to match the heat requirements of each zone. Hydronic water storage is also possible with hot-water systems. During normal system operation, water is stored at high temperature. This water can be used during off-peak operation. A large storage tank can hold water with temperatures of 200 to 275°F (93.3 to 135.0°C) at pressures up to 75 psig (5.2 bar). An automatic valve can supply water at a lower temperature by mixing it with cooler water from a main supply line. Steam can be obtained by withdrawing hot water into a low-pressure separating chamber.

8.4 WARM-AIR SYSTEMS

A central blower and system of ducts comprise a simple warm-air system. Electric heating units are installed in ducts near the blower to regulate the temperature of the entire structure. Individual room control may be achieved by heating air at the outlets. Warm-air systems provide a convenient means for fresh air intake, which ensures good ventilation. Other important functions of this type of system include circulation and filtering.

Heating elements for ductwork are fabricated in either slip-in or flanged designs. Electrical codes usually specify that at least 48 in (122 cm) of straight duct be installed between the element and fan outlets, duct elbows, baffles, or similar obstructions. This measure minimizes overheating of sections of the heating element that get uneven air flow.

The heating elements can be sized by looking at heat-transfer calculations. For easier design, nomograms have been developed that relate air flow, temperature rise, and kilowatt rating of the heating element (Fig. 8.3).

Electric furnaces consist of heating coils and a blower in an insulated housing. Air is drawn in through the bottom of the furnace and is filtered. The blower then forces

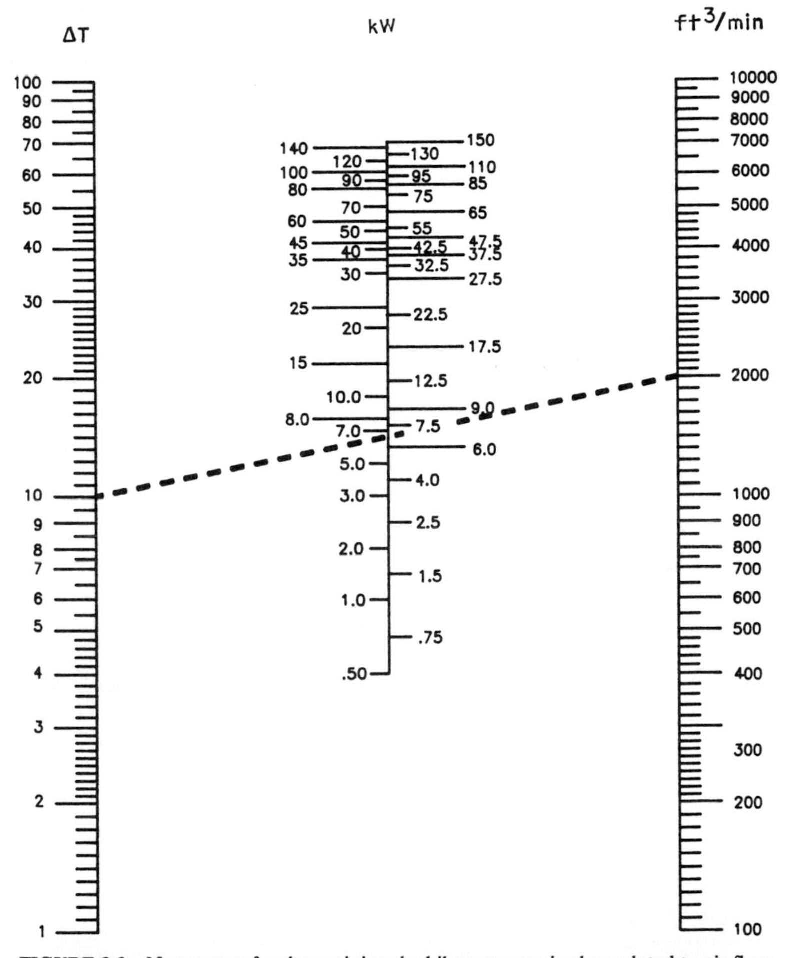

FIGURE 8.3 Nomogram for determining the kilowatts required as related to air flow and temperature rise. *(Courtesy of Industrial Engineering & Equipment Co., St. Louis, MO.)*

the air over the heating coils, transferring heat from the coils to the air. Electric furnaces are compact, requiring no flue or fuel storage.

8.5 CONVECTOR WITH METALLIC HEATING ELEMENT

Certain convective heating units use metallic heating elements. Heating is provided by bare-wire, low-temperature bare-wire, or sheathed elements. Air is heated by heat transfer from the elements. A reflector panel radiates heat away from the unit, minimizing the casing temperature rise and maximizing heat transfer to the room. The space between the elements and the reflector serves as an air passage that promotes convection. It is important that the location of electric convectors be such that air movement across the elements is not impeded.

Metallic heating element convectors can be wall-mounted, recessed, or surface-mounted. Wall- or surface-mounted types can be equipped with a circulating fan to force convection. Recessed versions do not use fans.

8.6 UNIT VENTILATORS

The function of a unit ventilator is to provide heat, ventilation, and cooling to an area. The ventilator comprises an inlet grille, air filter, fan, heating elements, damper, and diffuser. The total capacity of the unit is the sum effect of many small electric heating elements. The temperature is controlled by activating and deactivating individual elements. Unit ventilators use a certain percentage of outdoor air. This type of unit has many applications: classrooms, motels, offices, and nursing homes.

8.7 UNIT HEATERS

Unit heaters are available in three types: cabinet heaters, horizontal projection heaters, and vertical projection heaters. The cabinet type can be floor-, wall-, or ceiling-mounted. Recessed unit heaters are also available. Unit heaters are designed with a fan that circulates room air over the heating elements. Unit heaters require no venting or piping, which makes them a convenient source of supplemental heating. Unit heaters can be used to heat occupied rooms in unheated buildings or unattended equipment enclosures in which a certain temperature must be maintained.

8.8 BASEBOARD HEATERS

Electric baseboard heaters contain one or more horizontal heating elements in a metal casing. Elements are made of various materials and are available in many sizes. Some of these elements include finned, sheathed, cast grid, ceramic, extended-surface, and coated glass.

Baseboard heaters are placed on the floor along the base of a wall. Proper positioning is important to achieve uniform heating.

Electric hydronic baseboard heaters containing immersion heating elements and antifreeze are available.

8.9 INFRARED HEATERS

Infrared heaters use low-temperature elements to provide heating. Other designs use quartz tubes or lamps instead. Heaters are either suspended from the ceiling or mounted on the wall. Elements are placed in a trough-type casing surrounded by reflectors.

Infrared heaters have many applications. When convection heating is made impractical by high ventilation, infrared heaters can be used. They can be focused to heat specific sections of a room, and they provide comfort despite low ambient temperature. Outdoor spot heating is possible with infrared heaters. Industry uses modified designs for heating and industrial drying. See Chap. 7 for a discussion of infrared heating.

8.10 VALANCE, CORNICE, OR COVE HEATERS

These heaters are similar in shape to baseboard heaters. They are usually installed several inches below the ceiling on an upper outside wall. Heating elements come in various forms: metal-sheathed, coated-glass, or metal panels. Units are available up to 6 in (152 mm) high and project from the wall less than 3 in (76 mm). With these units the main source of heating is the room ceiling, which is heated by the convection flow of air over the heater. Also the heated panel provides some direct radiation into the room.

Hydronic valance units can provide either heating or cooling by circulation of either hot or cold water. Hydronic valance heating/cooling systems have a drain trough that collects and disposes of condensate during the cooling mode of operation. The fins are designed to allow the flow of condensate into a drain. See Chap. 17 for a discussion of valance heating and cooling.

8.11 RADIANT CONVECTOR WALL PANELS

Radiant convector wall panels have glass electric heating panels. The panels are supported on insulators within a metal frame, with a reflector arranged to provide space for air circulation. Protective guards are provided for safety.

By passing a current through a thin coat of conductive material on glass, heat is produced. The conductive material is usually sprayed-on aluminum or printed metallic oxide grids. Some radiant panels use tubular elements welded to extended aluminum panels. These units have emissivity characteristics similar to those of glass panels.

8.12 INTEGRATED HEAT RECOVERY

An integrated heating and cooling system takes account of all the heat sources and sinks within a structure. The system is modeled around these sources and sinks for maximum effectiveness and efficiency. It is becoming increasingly common for commercial and industrial settings to have a custom-designed system to meet their energy needs. A system considers all heat gains, including those from such sources as lights, people, machinery, and the sun. By transferring heat from hot areas to areas that require heating, the dual purposes of heating and cooling are served. Generally refrigeration equipment is incorporated into such systems to produce a cooling effect.

Integrated system design requires a careful accounting of the building's energy needs. Many sources of heat are available only during normal working hours, and supplemental heating is needed during off hours. Electric heaters are ideal for this purpose and can be used to provide heat for both personnel and machinery. In other applications heat is accumulated during periods of excess and stored for later use.

8.13 HEAT PUMPS

A heat pump is a device that operates in a cycle, requires work, and accomplishes the task of transferring heat from a low-temperature area to a high-temperature area. A heat pump uses the mechanical refrigeration cycle to cool and reverses the roles of the evaporator and the condenser to provide heating.

Heat pumps use electricity more efficiently than resistance heaters. For each kilowatt of electricity used by a heat pump, 1 kW of compression heat is produced plus a refrigeration effect. The refrigeration effect of the heat pump can vary from 10 to as high as 50 percent of the electric energy input. This effect is dependent on the temperatures involved. With resistance elements, 1 kW of heat is produced for each kilowatt of electricity.

8.14 COEFFICIENT OF PERFORMANCE

The "efficiency" of a heat pump is expressed by the coefficient of performance (COP). COP is used as an indicator of how a heat pump operates under specified temperature conditions. The most significant variable in the equation for the efficiency of the heat pump is the temperature of the heat sink. An equation for the COP is

$$\text{COP} = \frac{\text{heat of compression} + \text{refrigeration effect}}{\text{heat of compression}}$$

8.15 HEAT PUMPS AND BUILDING LOAD

To ensure proper equipment selection, some quantitative relationships must be determined and analyzed. Heating capacity versus outdoor air temperature

(OAT) is one such relationship. Figure 8.4*a* shows the output of a system operating at a series of outdoor air temperatures; the capacity drops off as the outdoor air temperature falls.

Another important relationship is heating load versus outside air temperature. Figure 8.4*b* indicates the load required to maintain an internal temperature at various outdoor air temperatures. With a reduction in outside air temperature, the heating load increases.

Applying this kind of analysis to a heat pump shows that under certain conditions a heat pump's capacity equals the heating load of a space. The temperature at this point is called the *balanced temperature*, determined graphically by the intersection of the capacity versus outside air temperature curve and the load versus outside air temperature curve (Fig. 8.4*c*).

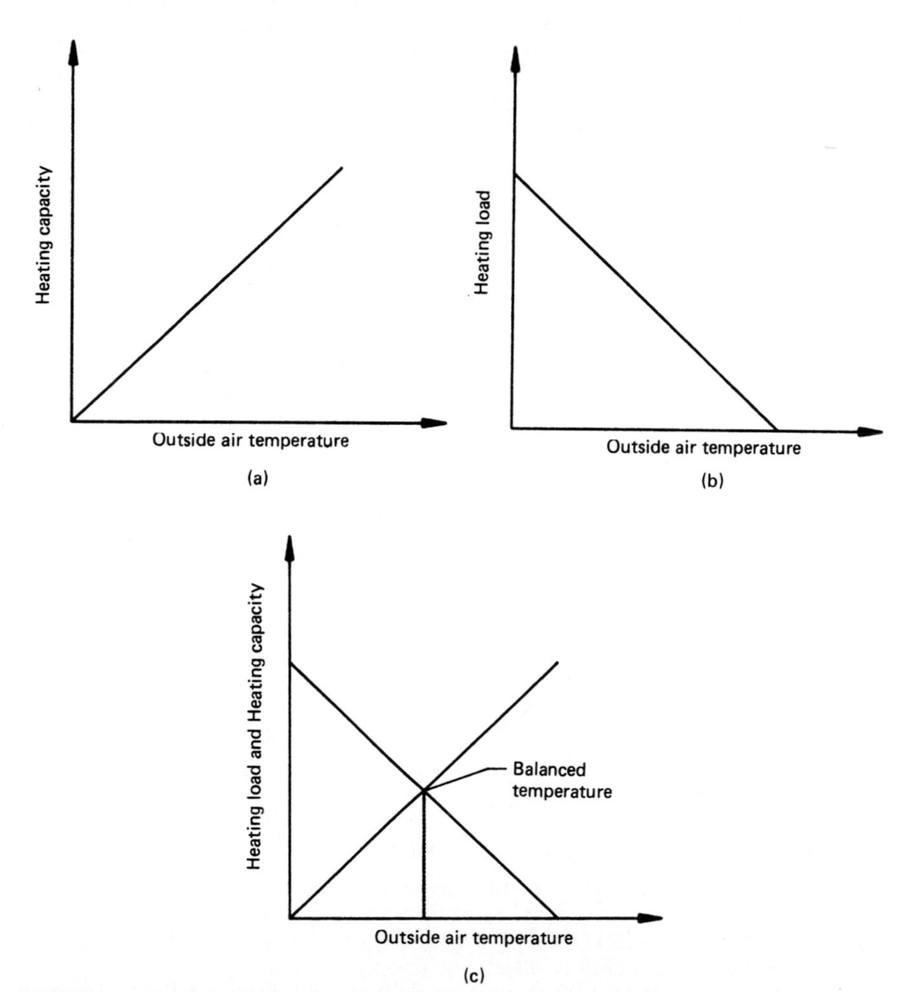

FIGURE 8.4 Relationships between heating capacity and heating load at various outside air temperatures.

When the outdoor air temperature is below the balanced temperature, the heating needs of a space will not be satisfied. Supplemental heating must be used, and its minimum capacity must make up the heating load of the space. The heat pump is most efficient when supplemental heat is not required. It is more important for a heat pump to satisfy an area's cooling requirement because supplemental heating can be easily provided.

8.16 HEAT PUMP TYPES

Heat pumps come in four types. They are designated by what they use as a heat source and sink. Air-to-air systems use outdoor air as a heat sink. The heat pump treats the air, and, when needed, a supplemental electric heater is activated to heat the supply air.

The air-to-water system uses a condenser water loop as the heat sink. Auxiliary devices such as a cooling tower and boiler are used to keep the temperature within limits, so that the heat pump will operate efficiently. When cooling is necessary, the tower is operated; and the boiler is used when heating is required. Heat is sometimes rejected to ground wells instead of cooling towers. In other instances, many heat pumps are attached to the same condenser water circuit; so when heating and cooling are required simultaneously in different areas, the heat added or rejected to the loop by the auxiliary devices is minimized.

A water-to-water heat pump is called a "heat reclaim chiller." It can provide both heating and cooling. It provides hot water at the condenser and cool water at the evaporator. The water-to-air heat pump is not very common. It is an air-cooled chiller that can reverse its cycle to produce warm water.

8.17 SPECIFYING ELECTRIC HEATING SYSTEMS

Generally, once the heat load of a building or room is known, the appropriate electric heating system or unit can be sized, just as for hydronic or other heating systems. The most efficient systems usually are decentralized ones, since the temperature conditions of each room can be adjusted individually to meet variations in air infiltration, heat generation by lights or appliances, or sunlight.

Heaters with direct electrical-resistance elements transform power to heat with 100 percent efficiency. Disregarding for the moment the subsequent inefficiencies of conveying that heat to air, water, or some other heating medium, the overall electricity consumption of an electric heating system can be estimated according to the following formula:

$$E = \frac{\text{HL(DD)(24)}}{\text{TD}(kV)} C_D$$

where E = energy required for auxiliaries, Btu (kWh)
HL = design heat loss including infiltration and ventilation, Btu/h (W)
DD = number of 65°F (18.3°C) degree-days for estimate period
TD = design temperature difference (indoor − outdoor), °F (°C)
k = heating value of fuel, units consistent with HL and E; for electric heat k can be approximated at 1

V = heating value of fuel, units consistent with HL and E; for electric heat, 3413 Btu/kWh (1000 W/kW)

C_D = empirical correction factor for heating effect versus 65°F (18.3°C) degree-days; determined by the graph in Fig. 8.5.

24 = constant used in unit conversion, that is, 24 h/day

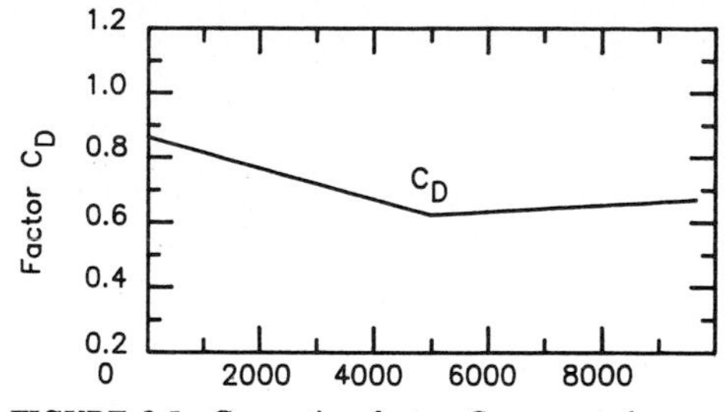

FIGURE 8.5 Correction factor C_D versus degree-days.

8.18 ELECTRIC CIRCUIT DESIGN

The National Electrical Code®, supplemented by local code restrictions, provides very strict and specific standards for connecting electric heating systems with a building's power supply.

Depending on the type of heating unit—particularly in the case of forced-air heaters—electrical codes usually require two types of overtemperature control. Power cutouts are the bimetallic type or liquid-filled bulbs. Both are designed to sense air blockage by means of a temperature rise, to interrupt the power circuit, and to reset automatically.

Many types of small-capacity heating units have integral power and thermostat controls. Thermostats carry line voltage to the unit and are designed to cycle the unit on a predetermined basis. Wall- or remote-mounted thermostats are supplied for larger heating units or for temperature control of large interior spaces. These thermostats are selected with a view toward how the room will perform during heatup and cooldown and whether radiant, convective, or forced-air heating is being employed.

CHAPTER 9
SOLAR SPACE HEATING

Lehr Associates
New York, New York

9.1 INTRODUCTION

Solar energy systems for heating, ventilating, and air-conditioning (HVAC) applications make use of incident energy from the sun to heat a transfer fluid or to energize photovoltaic materials. Photovoltaic materials are considered beyond the scope of this chapter; electricity generated by such solar panels can be used in many of the same ways as electricity from the local power grid, for electric heating, infrared heating, etc. Solar energy systems can be further separated into *passive designs*, which entail constructing and locating walls, roofs, windows, etc., in such a fashion to maximize the effects of incipient solar energy; or *active designs*, which involve transferring energy derived from sunlight to other parts of a structure or from one medium to another. Active systems are the focus of this chapter.

Solar energy is received by collection systems in several forms: direct radiation, reflected radiation, and diffuse radiation. The sum of these is termed "insolation." Direct radiation usually accounts for around 90 percent of the radiant energy. Tables of insolation totals for various latitudes, times of day and year, and collector tilts are available from standard handbooks. These tables allow estimates of how much energy is available to a solar collector, which in turn can be used to size the unit to meet the expected energy demand. Further corrections are made for weather patterns, collector efficiencies, and other factors.

Solar energy collected by active systems can be used for several purposes: fluctuating space heating, space heating plus hot-water heating, space heating plus hot-water storage, and other combinations. The more common type of system is the liquid-circulating flat-plate collector; this is the focus of the remainder of this chapter. Flat-plate collectors (henceforth called "collectors") employ a black absorber plate backed with insulation, tubes, or other passageways for heat-transfer fluid, transparent covers, and a frame. Plumbing connections at the back or side of the frame allow the unit to be hooked up to a circulation network running inside a house or other structure (see Fig. 9.1).

The collector operates by absorbing radiant energy from sunlight on the black absorber plate. Tubes carrying fluid pass along the surface of the plate, heating the fluid. Temperatures within the collector can reach a maximum of 400°F (204.4°C).

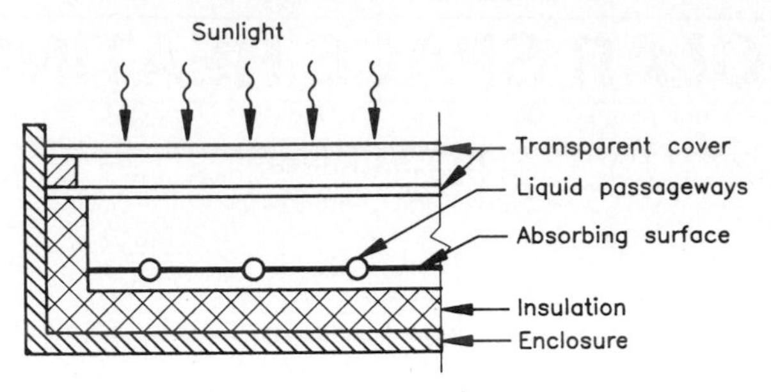

FIGURE 9.1 Flat-plate collector section.

9.2 TYPES OF DISTRIBUTION SYSTEM

Solar collectors can be arranged in a variety of configurations to suit various applications. Generally, the following types have undergone extensive development and commercialization:

1. *Thermosiphon systems:* This design is one of the simplest (Fig. 9.2) A storage tank is placed a certain height above the collector plate. As fluid in the collector warms, it becomes buoyant, causing a convective movement from the col-

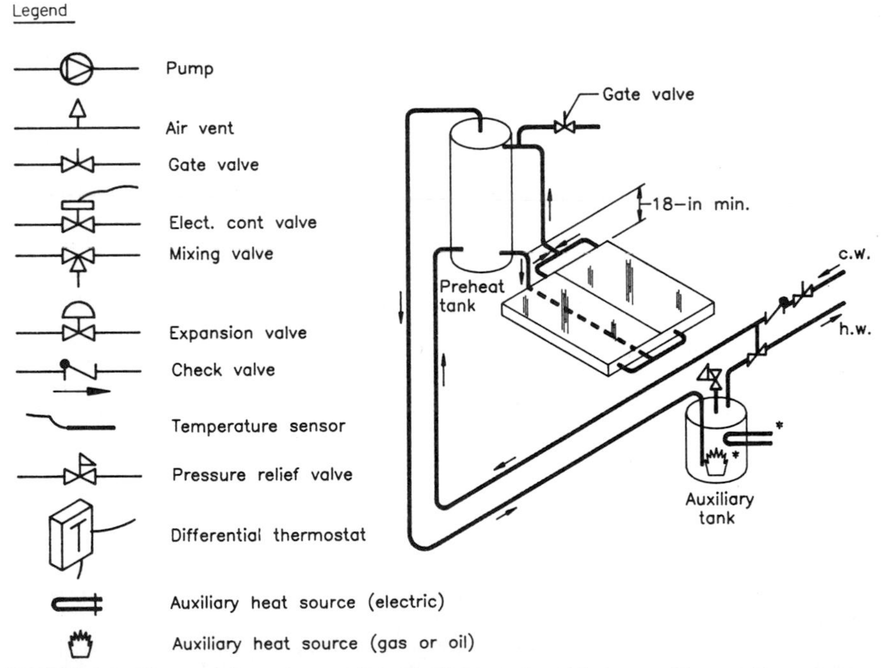

FIGURE 9.2 Thermosiphon system schematic diagram. Asterisk means either auxiliary energy source is acceptable. [*From Jan F. Kreider and Frank Kreith (eds.), Solar Energy Handbook, McGraw-Hill, New York, ©1981. Used with permission.*]

lector to the top of the storage tank (a pipe directs the fluid to that location). In time, the tank's reservoir becomes completely warmed, and the fluid can be released from the tank to space heating tubes below the tank or to an auxiliary storage tank in the basement of the building. Electrically driven pumps or subsequent heating with fuel causes the fluid to be distributed through the building's heating system.

2. *Direct forced-circulation systems:* Forced-circulation systems (Fig. 9.3) depend on a pump to move the fluid. A direct system uses solar energy in direct contact with the fluid of interest (usually water being heated). Such direct systems are often employed in climates where freezing is not a problem. However, various provisions for automatic drain-down or other measures can minimize freezing risks.

3. *Indirect forced-circulation systems:* These units are identical to direct systems except that a working fluid (i.e., an antifreeze solution) is used to collect solar energy (Fig. 9.4). A heat exchanger then transfers the energy to water or some other circulation fluid. Thus indirect systems are somewhat more complex and usually require a larger collector surface area to make up for inefficiencies in heat transfer.

4. *Combined collector-storage systems:* These units are indirect, forced-circulation heating systems with a heat pump or similar device (Fig. 9.5). The energy efficiency is improved by employing the heat pump to draw energy from the surroundings even when solar energy is limited. Another common feature is some means of heat storage, usually a supplementary hot-water tank or rock beds or similar media.

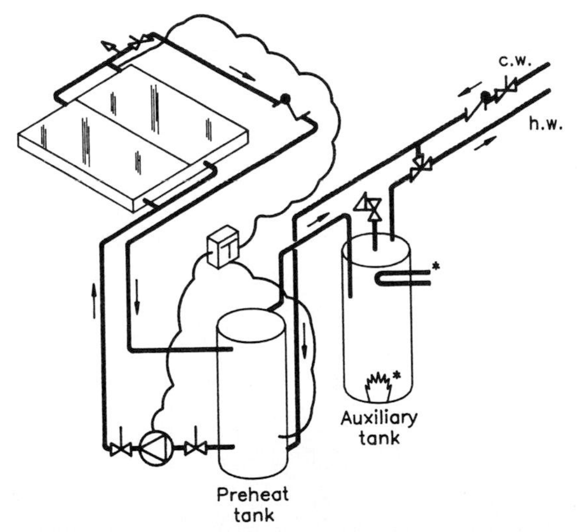

FIGURE 9.3 Direct forced-circulation water heating with preheat tank and separate auxiliary tank. Asterisk means either auxiliary energy source is acceptable. For explanation of other symbols, see Fig. 9.2. [*From Jan F. Kreider and Frank Kreith (eds.), Solar Energy Handbook, McGraw-Hill, New York, ©1981. Used with permission.*]

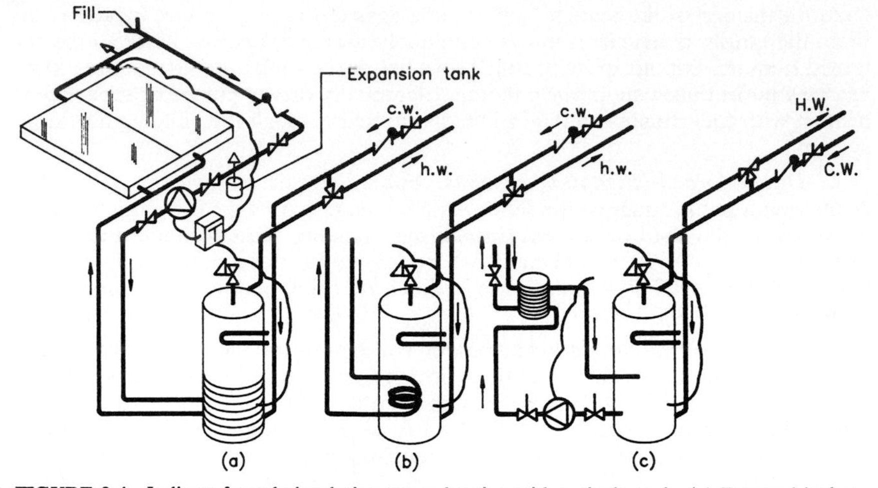

FIGURE 9.4 Indirect forced-circulation water heating with a single tank. (*a*) External jacket heat exchanger, (*b*) internal coil heat exchanger, and (*c*) external shell-and-tube heat exchanger with pump. For explanation of symbols, see Fig. 9.2. [*From Jan F. Kreider and Frank Kreith (eds.), Solar Energy Handbook, McGraw-Hill, New York, ©1981. Used with permission.*]

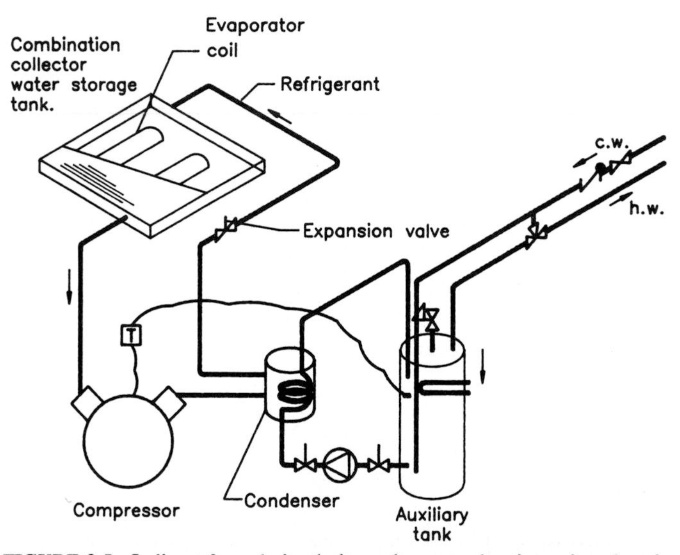

FIGURE 9.5 Indirect forced-circulation solar water heating using electric heat pump and combined collection/storage unit that is permitted to freeze. For explanation of symbols, see Fig. 9.2. [*From Jan F. Kreider and Frank Kreith (eds.), Solar Energy Handbook, McGraw-Hill, New York, ©1981. Used with permission.*]

The design of solar heating systems is composed of three interconnected design tasks: Solar panel selection, energy storage and heat transfer to the distribution system, and design of the primary heat distribution system. This section covers the distribution system design only. References to solar collectors and storage devices are made to clarify the design methodology and requirements.

9.3 GENERAL DESIGN

In general, the design of the solar distribution system is similar to that of the standard heating system. The solar heating system conforms with accepted plumbing, heating, ventilating, and air-conditioning engineering practices.

9.3.1 Pipe Sizing

The hot-water pipe should be big enough to provide heat to the building in the quantities specified by the load calculation. The maximum velocity in the system should not exceed 10 ft/s (3 m/s) at any time. Refer to Chap. 6 for a discussion of hot-water distribution pipe sizing.

9.3.2 Solar Hot-Water Operating Temperatures

A fundamental constraint of solar systems for space heating is that the maximum temperature of the solar unit is often below the normally used hydronic space-heating range. Obviously, if the solar system is to provide heating, the temperature it develops must be sufficient to deliver the necessary heat. A good rule of thumb is that in winter the maximum temperature of the solar unit should be 100°F (55.6°C) more than the existing outdoor ambient temperature. This rule assumes a 30 to 40 percent collector efficiency. Thus, when it is, say, 15°F (−9.4°C) outside, the solar collector can produce 115°F (46.1°C) water.

Such temperatures limit the operating range of space heating systems, most of which can operate well above 200°F (93.3°C). The system designer has several alternatives to consider:

1. The space heating system may be oversized to produce the same Btu/h (watt) output as a smaller, but hotter, conventional baseboard system.
2. The hot water can be used in conjunction with an air-heating fan-coil system. This coil can be sized appropriately for a required design supply water temperature to the order of 140 or 120°F (60 or 48.9°C). Alternatively, conventional floor panel systems (tubing embedded in the floor), which are usually designed to the order of 100°F (37.8°C) supply temperature, can use hot water from the solar collector at the required outdoor design temperature.
3. Slab-on-grade houses can utilize a modified hydronic floor panel perimeter loop in combination with the preheat air duct (supplementary standby) system. The perimeter loop (one or two copper tubes embedded in concrete close to the outside wall) prevents cold floors and provides for a relatively high percentage of required house heat at a low water temperature [on the order of 100°F (37.8°C)].

A hydronic system with a low water temperature has several advantages when viewed from the perspective of good solar-collector design:

1. A low required water temperature in the hydronic system will increase the range of outdoor temperatures at which the solar system will be operable.
2. A low temperature in the hydronic system will allow operation of the solar collector at a lower mean liquid temperature and will increase the collection efficiency.

9.4 HEAT-TRANSFER MEDIA

Liquid solar heating systems are generally more efficient than air-based ones, simply because of the greater heat capacity of liquids. Among liquid systems, there are a variety of constraints, such as the presence of freezing conditions or the need for certain temperature ranges, that dictate what type of heat-transfer medium to use.

Generally, water-based systems are most common, because of both the simplicity of the overall design and the design similarities to conventional hot-water heating systems. But water has the great disadvantage of freezing and harming the plumbing components of the solar system. Therefore, antifreeze-type solutions, such as ethylene glycol, are often utilized in solar space heating. A 50:50 water–ethylene glycol mixture is commonly specified. A heat-exchange coil is used to transfer heat from this solution to a hydronic space heating system.

In terms of the collector operating efficiency, the water system is best based on the following statements:

1. The glycol solution generally must run at a higher operating temperature to overcome the inefficiencies of the heat exchanger (while accomplishing the same ultimate heat output). Solar systems lose overall efficiency as their operating temperature rises.
2. All-water systems usually have a drainback feature to prevent freezing damage when the system is not running or it is night. The drainback allows all the energy collected by the system to be used. Systems without drainback have a certain amount of inherent reverse heat loss to atmosphere when they are not running.

Ethylene glycol is a toxic material. When glycol systems are used for domestic hot water, local building codes usually specify a double-walled containment between the glycol and water.

Other heat-transfer fluids that have been used in solar heating systems include propylene glycol, which is less toxic than ethylene glycol. Hydrocarbon oils, commonly used in industrial heat-transfer applications, have been tried but run the risk of igniting. Silicone fluids show low toxicity and reactivity but tend to be viscous and to have low specific heat.

9.5 WATER DRAINBACK SYSTEMS

All-water solar collector systems incorporate a drainback feature to prevent freezing damage. The overall system is usually pressurized. Collectors are pitched at angles that permit unimpeded downward flow when the system is shut off. Some means of air venting must be included to allow the system to refill upon being restarted. A compression/expansion tank (essentially, a vent tank) is used to store the drawn-down water. Overall, the system is equipped with differential thermostats that start a pump when a preset ΔT is reached between the exit and return points on the storage tank. When the return temperature reaches a low (i.e., when the onset of darkness prevents solar heating), the thermostat will automatically shut off the pump, allowing the drainback step to begin.

There are two types of drainback designs: open-drop and siphon return. Salient features of the two are as follows:

1. Open-drop designs use a pump to force water up to the solar collectors and then to an oversized vertical down-comer that returns the water to the storage tank. The down-comer is oversized by about two pipe sizes, and it has an air vent, to prevent any back-pressure filling of the down-comer. When the pump stops, the system automatically drains. Air-balancing valves are required at various points throughout the system to prevent bypasses or back pressures. The pump is sized and selected on the basis of overcoming the elevation difference between the collector and the storage tank.
2. Siphon returns are designed such that the down-comer fills rapidly. This setup balances the fluid head with the fluid in the down-comer, so that pumping requirements are essentially independent of system height. Balancing valves permit the eventual drainback when low temperatures are experienced. In addition, added insulation on plumbing parts near the solar collector minimizes ice formation.

9.5.1 Heat Exchangers and Multiple Heat-Transfer Fluids

Another alternative to freeze prevention is the use of a subsystem containing an antifreeze-based fluid (usually a glycol). Some designs call for the subsystem to be used only to keep the water flowing in the collectors warm during low-light, cold periods. Alternatively, the antifreeze subsystem can be used exclusively in the collector and related outdoor components, with the collected heat transferred to water indoors through a heat exchanger. In either case, the heat exchanger must be designed to work efficiently and must be properly sized.

In general, the heat exchanger should be sized generously to extract the maximum amount of heat from the fluid in the collector. An undersized exchanger will waste heat that has already been collected, decreasing the overall system efficiency. Glycol-to-water exchangers are commonly designed with straight-tube or U-tube configurations. The exchanger can also be placed in the storage tank at the base of the solar heating system, in which case glycol is flowing throughout the system's lines.

9.6 PUMPING CONSIDERATIONS

Glycol has an important effect on system flow rates.

A 50% glycol mixture has a slightly higher density—but lower specific heat—than water. The net relationship is that a greater fluid flow rate is required to sat-

TABLE 9.1 Increased Flow Requirement for Same Heat Conveyance for 50% Glycol Compared with Water

Fluid temperature, °F (°C)	Flow increase needed for 50% glycol compared with water
40 (4.4)	1.22
100 (37.8)	1.16
140 (60.0)	1.15
180 (82.2)	1.14
220 (104.4)	1.14

TABLE 9.2 Pressure-Drop Correction Factors for 50% Glycol Solution Compared with Water

Fluid temperature, °F (°C)	Pressure-drop correction with flow rates equal	Combined pressure-drop correction; 50% glycol flow increased per Table 9.1
40 (4.4)	1.45	2.14
100 (37.8)	1.10	1.49
140 (60.0)	1.00	1.32
180 (82.2)	0.94	1.23
220 (104.4)	0.90	1.18

isfy the system's design. The increased flow rate for a 50% glycol mixture as compared with water and for the same heat conveyance is shown in Table 9.1. As an example, suppose that a particular glycol collector subsystem is to be designed on the basis of 140°F (60°C) mean fluid temperature. The initial design base (stated to water) requires a flow of 10 gal/min (0.63 L/s). The new flow-rate requirement will then be 11.5 gal/min (0.73 L/s).

Above roughly 30°F (−1.1°C), the viscosity of a 50% glycol solution becomes higher than that of pure water, by a factor of around 8 to 16 cP (0.008 to 0.016 $N \cdot s/m^2$). This fact necessitates correction of the pressure drop at various operating temperatures. Table 9.2 shows the pressure drop iterated through two corrections; the first is the correction at equal water or glycol flow rates, and the second shows the correction at a glycol flow rate already corrected upward to account for the lower heat capacity of glycol solutions. In practice, the designer of a glycol system would then take a pure-water design and raise the flow rate by the factor indicated in Table 9.1 and raise the pressure drop by the "Combined factor" column in Table 9.2. The designer could then consult pump curves to specify the correct pump. Generally, the pump curve can be used as is (i.e., for pure water), as long as the flow-rate and pressure-drop corrections for glycol solutions have already been calculated.

9.7 ADDITIONAL FLUID SYSTEM CONSIDERATIONS

As with conventional hot-water heating systems, the performance of a solar heating system is improved by the addition of a compression tank. In particular, the compression tank helps minimize corrosion damage caused by dissolved oxygen, which is soon neutralized during operation. For an all-water system, this tank can be sized according to conventional procedures (see Chap. 6). A compression tank for an all-water system will experience greater vapor pressure changes than that for glycol systems, since the glycol system has a higher boiling point. Thus correct sizing is important. Conversely, a glycol mixture has a larger expansion rate than water alone. The general guideline is to size the tank 20 percent larger than if a water system were used.

9.7.1 Leak Protection

Leaks are a serious threat to the long-term operation of any solar heating system. To minimize their occurrence, welded or sweated joints should be used wherever possible, and threaded joints should be tightly wound with the appropriate sealing tape. Glycol systems should not contain a pressure-reducing valve, as this will lead to dilution of the mixture over time. Also pumps with mechanical seals, rather than gland seals, should be used.

9.7.2 Glycol Sludge

Sludge often forms where glycol is present. This is due to a number of factors, including reactions between the glycol and oils, fluxes, or other hydrocarbon contaminants; reactions with chromate-based water treatments (which should never be used with a glycol system) or with galvanized piping; and gradual degradation of the glycol solution. To minimize sludge generation, galvanized pipe should not be used and the system should be cleaned and flushed prior to startup.

9.7.3 Storage Subsystem

As has been mentioned, many solar heating systems employ storage tanks that contain heated water. The tank serves to moderate swings in system demands, to store heat for periods when few solar rays are available, and to provide a convenient means of heat exchanging with a glycol or other fluid subsystem. Storage subsystem sizing plays a significant role in the overall performance of a solar heating system. A common rule of thumb is to size the storage tank in the range 1.5 to 2.0 gal of water for each 1 ft^2 of solar collector area (61 to 81 L/m^2). Larger systems are rarely cost-effective.

Flows from the storage tank may be arranged to feed directly into space heating systems (a supplementary setup) or may be used as a standby system, operating through heat exchangers, air coils, or similar devices. The size of the storage system can be calculated more exactly by treating it as a means of storing surplus heat after space heating duties have been fulfilled. Thus, a large heat drawdown would allow for a larger storage unit.

9.8 MATERIALS AND EQUIPMENT

Equipment intended for heating the water or storing the hot water should be protected against excessive temperatures and pressures by approved safety devices and in accordance with one of the following methods:

1. A pressure relief valve and a temperature relief valve
2. A combination temperature/pressure relief valve
3. A pressure relief valve, a temperature relief valve, and an energy cutoff device
4. A combination temperature/pressure relief valve and an energy cutoff device

All safety devices specified by the plumbing designer should meet the criteria set forth by the American Society of Mechanical Engineers or Underwriters' Laboratories.

CHAPTER 10
SNOW-MELTING SYSTEMS*

Lehr Associates
New York, New York

10.1 INTRODUCTION

It has become common in recent years to use snow-melting systems in the vicinity of commercial installations, especially for such access areas as walkways, automobile ramps, and delivery areas. While obviating the need for snow shoveling, plowing, or blowing, snow-melting systems provide dry surfaces. These advantages also apply to residential applications. Usually no more than a standard hot-water boiler combined with the melting system is necessary. Wider use of residential snow-melting systems has been hampered by a lack of information and by a belief they are too expensive—but the costs need to be evaluated objectively before such a system is ruled out.

System costs are often inflated when the design is too conservative. Designers readily increase the capacity of the heating boiler or accessory pumping and piping equipment to cover poorly understood application criteria. Of course, this practice increases the cost, often unnecessarily. A better understanding of the application's parameters, combined with clearer knowledge of how snow-melting systems function, can be used to bring installation costs to a reasonable level.

Most snow-melting systems are one of three types: an embedded piping system carrying hot water or heating fluid (usually a glycol solution), embedded electrical-resistance heating elements, or radiative systems using infrared lamps. The last system can provide illumination as well as space heating, if desired.

Of the three types, the heated-fluid system provides the best explication of all the essential design elements for snow removal. We now examine this system in more detail. The basic design process comprises six steps:

1. Determine the snow melting load.
2. Decide on a piping layout.
3. Determine the gallon/minute (liter/second) flow rate, and specify the heat exchange unit.
4. Size and select the piping.

*Information contained in Tables 10.1, 10.2, and 10.3 and some of the text in this chapter used courtesy of ITT Fluid Handling Division.

5. Size and select a pump.
6. Determine the appropriate specialties.

10.2 DETERMINATION OF THE SNOW-MELTING LOAD

The performance of a snow-melting system depends on many factors, ranging from the type of installation to the predicted volume of snow. Other factors include the outside air temperature, snowfall rate and density, wind speed and exposure, humidity, and physical properties of the slab being heated. While it is possible to compute the effects of each of these factors, the calculations are laborious and time-consuming. A generalized method, presented here, offers adequate accuracy.

Basic physics tells us that the amount of heat needed to melt a load of snow is, first, the heat necessary to warm the slab to the snow-melting temperature. This value is called the "pickup load." The second amount is the heat needed to melt the snow itself. This value depends on the rate of snowfall, the snow's density, and, indirectly, the outside air temperature. When the temperature is very low, snow is usually less dense; when the temperature is higher, especially near the freezing point, the snow is heavier and denser. The proper heating capacity for the snow-melting system will be the larger of the pickup load or the melting load. Both must be calculated to make the appropriate determination.

To calculate the pickup load, two things must be known: the initial slab temperature and the amount of heat necessary to bring the slab up to 32°F (0°C), the snow-melting temperature. The initial slab temperature can often be considerably below the outside air temperature; a good initial temperature to use is 10°F (−12.2°C). The amount of heat necessary to bring the slab up to the freezing point will depend on the depth of bury of the piping that conveys the heating fluid.

A deeper bury for the piping embedment will require proportionately more heat because a greater mass of concrete is present. Thus it is desirable to bury the pipes only as deep as is necessary for structural reasons. Usually this can be interpreted as a minimum depth equal to the diameter of the pipes themselves. A customary depth of bury is 2 in (50.4 mm) below the surface of the slab. When the concrete that will become the slab is poured, take the precaution of securing the pipes so that they do not rise or move. Table 10.1 provides values for pickup loads on the basis of varying depths of bury and for a range of initial slab temperatures.

Good engineering judgment is the most direct way of calculating the snowfall rate. In the case of residential systems, a good nominal value is 1 to 1.5 in/h (25.4 to 38.1 mm/h) snowfall rate. Industrial systems can be sized with a design rate of 2 to 4 in/h (50.8 to 101.6 mm/h). This figure is usually looked on as the maximum rate; higher rates would be onerous to the system's cost. However, it may be advisable to check on local snowfall rate patterns.

Snow density depends on the air temperature during the snowfall; as previously mentioned, a higher temperature implies a denser snowfall. The average temperature during a snowfall is around 26°F (−3.3°C), which yields a density of about 6 lb/ft^3 (96.1 kg/m^3). Table 10.2 lists the Btu (watt) snow-melting heating quantities for a number of snowfall rates. This table implies an average snow density of 10 lb/ft^3 (160.2 kg/m^3).

A final step to sizing the snow-melting system is to obtain the area, in ft^2 (m^2), of the slab to be heated. Multiply this value by the appropriate value from Table

TABLE 10.1 Pickup Load, Btu/(h · ft^2)

Slab temperature at start, °F (°C)	Distance from centerline of tube to surface of slab, in (mm)											
	¾ (19)	1 (25.4)	1½ (38.1)	2 (50.8)	2½ (63.5)	3 (76.2)	3½ (88.9)	4 (101.6)	4½ (114.3)	5 (127)	5½ (139.7)	6 (152.4)
30 (−1.1)	7*	9	14	18	23	28	33	37	41	46	51	56
20 (−6.7)	41	55	83	110	138	166	194	221	249	276	304	332
10 (−12.2)	76	**101**	**152**	**203**	**303**	304	355	406	456	507	558	609
0 (−17.8)	110	**147**	**220**	**294**	**368**	442	575	588	662	735	808	881
−10 (−23.3)	145	**193**	**289**	**396**	**482**	578	675	771	867	964	1060	1156
−20 (−28.9)	179	239	359	478	598	718	838	958	1079	1195	1215	1335

*Values given in table are in Btu/(h · ft^2). The metric conversion factor is Btu/(h · ft^2) × 3.155 = W/m^2.
Note: Bold figures indicate normal design areas.

TABLE 10.2 Snow Melting

in/h	(mm/h)	Btu/(h · ft²)	(W/m²)	
½	(12.7)	60	(189.3)	
1	(25.4)	120	(378.6)	Normal used design range
1½	(38.1)	180	(567.9)	Normal used design range
2	(50.8)	240	(757.2)	Normal used design range
3	(76.2)	360	(1135.8)	Normal used design range
4	(101.6)	480	(1514.4)	Normal used design range
5	(127.0)	600	(1893.0)	
6	(152.4)	720	(2271.6)	

10.1 for the pickup load; perform the same multiplication of the area by the appropriate value from Table 10.2 to obtain the snow-melting load. The greater value (in Btu's or watts) is the appropriate one to use in sizing the overall system.

For our example, let us assume that the initial slab temperature is 10°F (−12.2°C) and that the tubes are buried such that the distance from the centerline of the tubes to the surface of the slab is 3 in (76.2 mm). From Table 10.1 the pickup load is 304 Btu/h · ft²) (959.1 W/m²).

From Table 10.2, based on an average 1.5 in/h (38.1 mm/h) snowfall rate, 180 Btu/h (567.9 W) is the required melting load. Since the pickup load is greater than the snow-melting load, the basis for our design will be the pickup load or 304 Btu/ft² (959.1 W/m²). The area to be melted is 1000 ft² (92.9 m²):

$$1000 \text{ ft}^2 \times 304 \text{ Btu/ft}^2 = 304{,}000 \text{ Btu}$$

$$(92.9 \text{ m}^2 \times 959.1 \text{ W/m}^2 = 89{,}100 \text{ W})$$

Our snow-melting requirement becomes 304,000 Btu/h (89,100 W).

10.3 PIPING LAYOUT

The same principles that are used in radiant panel systems apply to snow-melting systems. Grid or coil tubing patterns are suitable. A good rule of thumb is to use a tube spacing of 12 in (304.8 mm), center to center. When pipes are closer, installation costs go up; when they are spaced more widely, there may be incomplete melting.

10.4 DETERMINE THE GALLONS/MINUTE (LITERS/SECOND) REQUIREMENT AND SPECIFY A HEAT EXCHANGER

If a heat exchanger is used—as would be the case if a heating fluid other than pure water were considered—the lesser heat-carrying capacity must be factored in. For antifreeze glycol solutions, an increased amount of solution must be conveyed to deliver the same amount of heat. The relationship between the two is expressed by the following equation:

$$\frac{\text{Btu/h}}{8600} = \text{gal/min}$$

Note: 8600 is an English-unit constant based on a 50/50 glycol solution with a 20°F (−6.7°C) temperature difference. To use this equation in metric designs, first carry out the calculation in English units, and then convert the result to metric.

$$\frac{304{,}000 \text{ Btu/h}}{8600} = 35.3 \text{ gal/min}$$

Note: Expansion joints in concrete walks or driveways cause certain piping problems. Avoid concentrated stresses on the coils or grid panels and on piping mains. Piping mains should be covered with an insulation blanket below the level of the slab.

10.5 SELECT SPECIALTIES

Generally, there is no need to add the snow-melting load requirements to the boiler load, since U.S. Weather Bureau records indicate that 26°F (−3.3°C) is the average snowfall temperature. At this temperature most heating systems are operating at less than maximum capacity. The design condition is at a much lower temperature, in most cases. Table 10.3 shows the proportion of boiler capacity necessary for space heating at 25°F (−3.8°C) outside temperature.

TABLE 10.3 Percentage of Boiler Capacity Required for Space Heating When the Outside Temperature is 25°F (−4°C)

Design temperature of heating system				
20°F (−6.7°C)	10°F (−12.2°C)	0°F (−17.8°C)	−10°F (−23.3°C)	−20°F (−28.9°C)
90%	75%	65%	55%	50%

In our example, the snow-melting system requires 304,000 Btu/h (89,100 W) for snow melting alone. If a 500,000 Btu/h (146,541 W) boiler is required for the comfort heating system at a design condition of −10°F (−23.3°C), Table 10.3 shows that only 55 percent, or 275,000 Btu/h (80,598 W), is required by the comfort heating system at 25°F (−3.8°C). Therefore, a balance of 225,000 Btu/h (65,944 W) is available for snow melting. Subtracting this 225,000 Btu/h (65,944 W) from the total snow-melting load of 304,000 Btu/h (89,100 W) shows that only an additional 79,000 Btu/h (23,153.6 W) must be added to the boiler size.

Compression Tank. When a compression tank is sized, treat the snow-melting system should be treated in the same manner as a radiant panel system. The heat exchanger, when used, will take the place of the boiler and will be designed by following the guidelines for a flash-type boiler.

Refer to the manufacturer's literature for sizing criteria for the compression tank and fittings, boiler fittings, and relief valve.

10.6 ELECTRICAL SNOW MELTING

As in electrical space heating, resistance wires can be used to warm outdoor surfaces. The materials that deliver the heat are usually either mineral-insulated cable or resistance wires.

Heat load calculations for electric systems are the same as for hot-fluid designs. Electric heating inherently has a set heat output, so over- or underheating is minimized by the design. The amount of heat per unit area can be varied by the spacing of the heating elements. It is important that concrete slabs be properly designed when electric heating is used. When the elements are embedded in the slabs, expansion-contraction joints or other features create shear forces that could snap the elements. Slabs are sometimes poured in layers; in such cases, the uppermost layer of an element-containing slab should have a high degree of weatherability, including limited air content and aggregate size.

10.7 ELECTRIC HEAT OUTPUT

Designs for electric snow-melting systems differ slightly depending on whether mineral-insulated cable or embedded wires are used. Cable is laid out much as piping would be; embedded wires can be laid out in single strands or as prefabricated mats.

American Society of Heating, Refrigeration, and Air-Conditioning Engineers (ASHRAE) standards suggest cable layouts of 2 lin ft of cable per square foot of slab area (6.6 lin m/m^2) for concrete slabs and 3 lin ft/ft^2 (9.8 lin m/m^2) for asphalt surfaces. The difference is due to the heat conductance of the slab material.

The following procedure can be used to size the heating system with mineral-insulated cable:

1. Find the total power requirement W_t:

$$W_t = A \cdot w$$

2. Find the total resistance R:

$$R = \frac{E^2}{W_t}$$

3. Find the cable resistance per foot (meter) r_1:

$$r_1 = \frac{R}{L_1}$$

4. Calculate the required cable spacing S:

$$S = \frac{12A}{L}$$

5. Find the current I required for the system:

$$I = \frac{V}{R}$$

In these equations,

W_t = total power needed, watts
A = heated area of each heated slab, ft^2 (m^2)
w = desired watt density input, W/ft^2 (W/m^2)
V = voltage available, V
r_1 = calculated cable resistance, Ω/ft (Ω/m) of cable
L_1 = estimated cable length, ft (m)
R = total resistance of cable, Ω
L = actual cable length, ft (m)
r = actual cable resistance, Ω/ft (Ω/m) of cable
S = cable center-to-center spacing, in (mm)
I = total current per cable, amperes

The calculated cable resistance is compared to the resistances available from manufacturers. If there is a significant variation from what is calculated, the system is recalculated with the closest available cable (it is possible for large installations to custom-manufacture the cable). In addition, a cold lead—a length of cable that does not generate heat—must be allowed for, to reach an aboveground junction box. The junction box should be located so that it can be kept dry.

Embedded wires or mats follow similar design procedures. The manufacturer's specifications for the wire determine the necessary ohm/foot (ohm/meter) resistance and wire diameter. Prefabricated mats must be selected with an eye toward how the heated surface is being constructed; if it is made of hot-poured asphalt, beneath which the mat will be placed, a mat that can withstand that heat should be used.

Electric snow-melting cable or wire can also be used to prevent snow or ice buildup on gutters and downspouts. The cabling can be mounted in a zigzag fashion along the edge of a roof and along the length of a gutter, and it can be looped for some distance down the downspout. The *ASHRAE Handbook* recommends cabling rated at 6 to 16 W/lin ft (19.7 to 52.5 W/lin m).

10.8 INFRARED (RADIANT) SNOW MELTING

Infrared systems tend to be somewhat more expensive than embedded-cable or wire designs, in terms of both installed cost and operating cost. This additional expense is offset by the advantages of providing comfort heating, in areas such as entrances or walkways, as well as additional illumination. The lamps can be installed under entrance awnings, along building faces, or on poles.

The primary design considerations for these systems are heating capacity and fixture arrangement. If snow melting is the paramount criterion, fixtures with narrow beam patterns can be specified; if comfort heating is also important, wider patterns may be in order. Manufacturer's literature customarily provides ratings in watts per square foot (meter) for the fixtures. A common practice is to take the table value of average watt density for snow-melting requirements in various regions of the country, multiply it by the target area in square feet (meters), and multiply by a correction factor of 1.6. For more precise designs, the arrangement of infrared fixtures is first determined, and then a table is compiled by which the intensity of light (heat) incident on each section of the area being heated is determined. In this manner, stray energy losses at the edges of the area to be heated are kept to a minimum.

10.9 SYSTEM CONTROLS

Either manual or automatic controls can be installed with snow-melting systems. Manual systems require the operator to start the system some amount of time before precipitation begins to fall. This preheating depends on the type of installation that was specified originally (a low-duty residential system, where some snow buildup may be permissible, versus a high-duty commercial system, where no buildup is desirable), weather conditions, and the materials of construction of the slab or surface being heated. Automatic control systems are generally available that either sense the presence of snow (which would allow preheating) or that start when the surface temperature falls below a preset limit.

Since both electric and infrared snow-melting systems are powered by electricity, the control setups are similar. Usually it is a simple on/off setup. High-capacity snow-melting systems, which use substantial amounts of power, are often equipped with temperature sensors in the slab, permitting the system to cycle on and off during relatively warm, light-snow conditions. Finally, a number of fully automatic temperature and precipitation systems are becoming available.

CHAPTER 11
HEAT TRACING

Lehr Associates
New York, New York

11.1 INTRODUCTION

A common problem of many commercial and industrial fluid systems is keeping the fluid at a constant temperature while it is circulating through the system. Hot water from a boiler, for example, is often needed at elevated temperature a considerable distance from the boiler. In many industrial plants, piping is exposed to the elements, and there is danger of freeze damage or difficulties with increased viscosity or other properties. The solution to many of these problems is either to continually recirculate fluid within the system (while replenishing heat losses) or to provide heat tracing.

"Heat tracing" may be defined as *the supplemental heating of fluid system piping through extraneous means*. Generally, the heat is provided by a system completely separate from whatever device is providing overall heat for the fluid in the pipes. In the past, heat tracing was commonly implemented with supplementary steam lines in close proximity to the main process-fluid lines. Today, heat tracing is usually effected by electrical tracings—resistance wires wrapped along and around pipes and pipe fittings. A relatively recent innovation in the field is the use of variable-wattage (self-regulating) resistance wires. These wires are designed with the inherent capability of drawing current until a desired temperature at any point along the length of the tracing is reached. Wires can even cross over each other, a situation that would normally present a fire hazard.

In principle, electric heat tracing is very similar to electric heating with mineral-insulated (MI) cable. Overall calculations are made of the heat loss and the area to be warmed; then the desired power input is arrived at, and the general layout of the heating cable is designed. Electric heat tracing can offer substantial energy and capital-cost savings over continuously recirculating fluid systems, since the energy consumption need be only what is needed at specific times; additional heating units, balancing valves, and similar equipment are avoided. For some piping networks, electric heat tracing is the only practical method. The remainder of this chapter deals with electric heat tracing only.

11.2 BASIC DESIGN CONSIDERATIONS

A variety of applications are tailor-made for heat tracing:

1. Exterior pipes carrying fuel oil or water
2. Gutters, downspouts, and other rain-handling piping
3. Circulating water pipes on external cooling towers
4. Underground pipe carrying hot fluid for relatively long distances
5. Hot-water supplies for commercial dishwashing
6. On-demand hot water for bathing or other purposes
7. Prevention of condensation buildup in air lines
8. Freeze protection for instruments or other sensitive equipment

Heat-tracing duty can be continuous or intermittent. Commonly available tracings are rated for temperatures from ambient to around 300°F (148.9°C) for continuous use and 400°F (204.4°C) for intermittent use. General-purpose MI cable can also be used for some tracing applications; this material is rated for continuous use even above 1500°F (815.6°C). The power output of the tracing cable is usually in the range of 10 W/ft (32.8 W/m); MI cable is available at up to 200 W/ft (656.2 W/m). Tracing cable is also defined by the maximum length of one circuit. For 10-W/ft (32.8-W/m) cables, this length ranges from 100 to 1000 ft (30.5 to 305 m). (Longer system lengths can be handled by multiple circuits.)

Physically, the cable is a flat double run of copper wire encased in an insulating barrier material (Fig. 11.1*a* and *b*). The wires and insulation are then wrapped in metal braid and/or a plastic jacket for mechanical and chemical protection. MI cable (Fig. 11.1*c*) is usually circular in cross section, with a magnesium oxide insulation core and plastic or metal jacketing. Self-regulating or variable-wattage tracing cable (Fig. 11.1*a*) operates through the use of a special carbon-based conducting polymer. As the temperature drops, the polymer contracts, allowing for a higher flow of current between the two wires (in essence, the cable is an infinitely parallel circuit). As the temperature rises, the polymer expands, with reduced current flow and higher resistance. This design permits high current flow at one point along the length of the cable and low flow at another.

In low-duty applications, heat tracing is simply run along the process pipe or tube and attached with a cable tie or tape around the circumference of the pipe. Figure 11.2 shows some common tracing arrangements. At pipe elbows, it is generally desirable to run the tracing along the outer radius of the bend. At pipe tees or other unions, the tracing can be run optionally in several loops around the entire joint. A similar configuration is possible for valve bodies. (Self-regulating tracing permits overlaps around these fittings.)

For higher-duty applications, the tracing is wrapped in a helical fashion along the length of the pipe. Details specifying the pitch for such a helix are given in Sec. 11.3.

It is customary to apply the tracing next to the pipe or tube to be heated, beneath any insulation or wrapping on the pipe. The type of insulation used on the pipe then becomes a design factor in specifying the heat tracing. In underground applications, the tracing must also be sealed against water seepage. Usually this can be accomplished through the use of pipe insulation with waterproof jacketing.

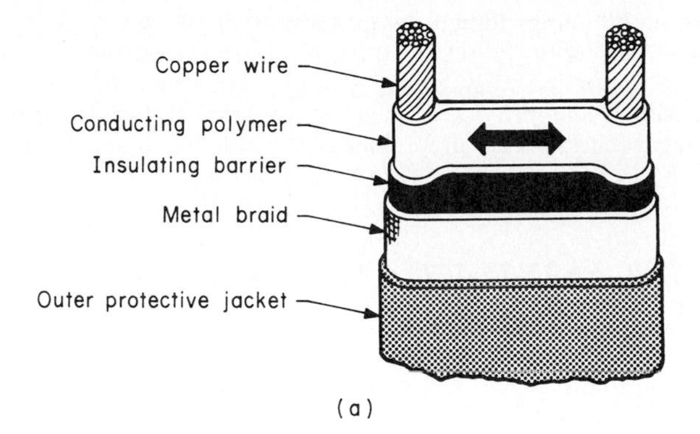

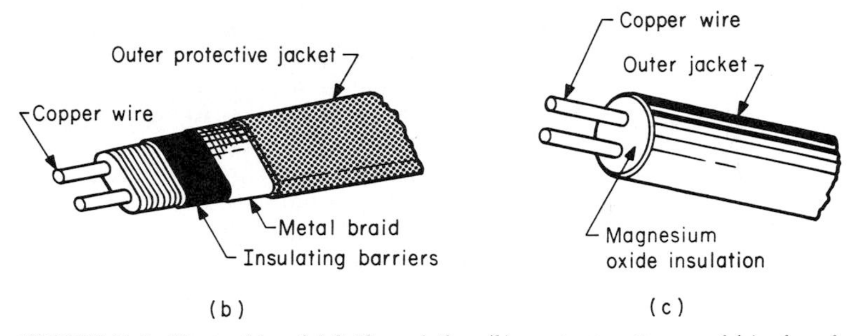

FIGURE 11.1 Heat cables. (*a*) Self-regulating, (*b*) constant-wattage, and (*c*) mineral-insulated. *(Courtesy of Chromalux, E. L. Wiegand Division, Emerson Electric Co. Used with permission.)*

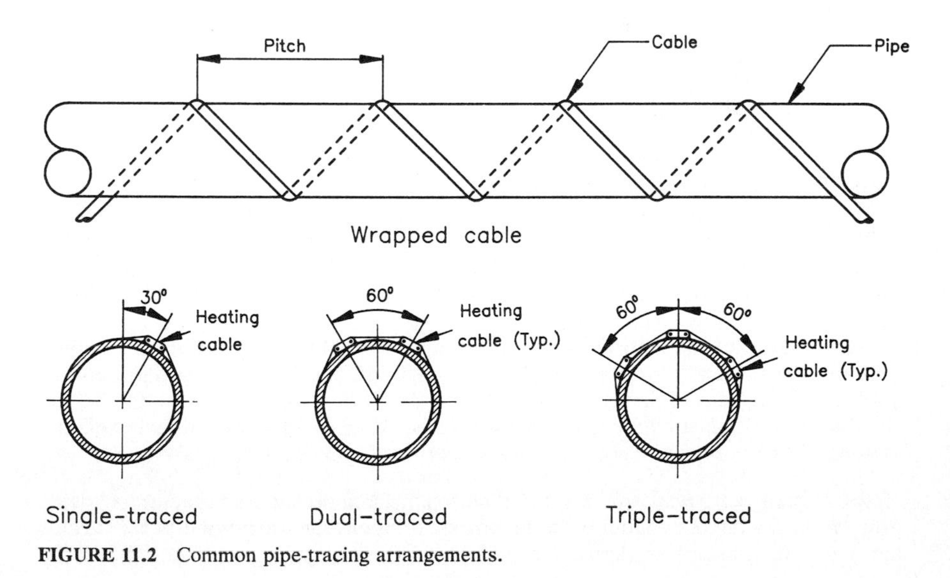

FIGURE 11.2 Common pipe-tracing arrangements.

Special heat-conducting aluminum tapes are available where heat is to be conducted from the tracing to the rest of the circumference of the pipe. Conducting tapes may also be applied to wrapping storage tanks or other large structures. Some manufacturers also provide special tee joints or end seals for field installation of the heat tracing; others pre-engineer the tracing to specified lengths and joints.

11.3 ELECTRIC HEAT-TRACING DESIGN

Specifying a heat-tracing installation follows the same general outline as for any type of heating system. The heat losses are calculated. The desired warming is analyzed, taking into account the type of piping and the fluid it is carrying. Finally, the physical dimensions of the system are detailed, and the ancillary electrical equipment is specified.

1. *Calculating heat demand:* A typical heat-tracing application requires that the pipe be kept at some temperature above ambient conditions. In outdoor environments, wind conditions should also be factored in.

As an example, a pipeline carrying fuel oil, outdoors, is to be traced. The desired maintenance temperature is 100°F (37.8°C). The minimum temperature prevailing at the pipe is 0°F (−17.8°C), the maximum wind velocity is 20 mi/h (32.2 km/h). For various pipe diameters and insulation thicknesses, check Table 11.1 for the heat loss in watts per foot per degree Fahrenheit of temperature differential. Table 11.1 is based on the following formula:

$$\frac{\text{W}}{\text{ft (of pipe)}} = \frac{2\pi k\ \Delta T}{Z \ln (D_o/D_i)} \tag{11.1}$$

where k = thermal conductivity of pipe insulation, Btu · in/(h · ft^2 · °F)
D_o = outside diameter of insulation, in
D_i = inside diameter of insulation, in
ΔT = temperature differential, °F
Z = 40.944 Btu · in/(W · h · ft)

(See Appendix B for metric conversions.)

The table assumes a k factor of 0.25 for fiberglass at 50°F (10°C) and adds in a 10 percent safety margin.

Multiply the value from Table 11.1 by the temperature differential [in the example, 100°F (37.8°C)] to determine the heat loss per foot (meter) of pipe. For a 1.5-in (3.8-cm) metal pipe with 2-in (5.1-cm) insulation, the Table 11.1 figure is 0.038 W/(ft · °F). Multiplied by the ΔT [100°F (37.8°C)], this leads to a heat loss of 3.8 W/ft (12.4 W/m) of pipe.

More rigorous calculation would determine the exact k factor of the insulation and add corrections based on higher or lower wind speeds. Underground piping would call for an additional safety factor of 25 percent, even with piping below the frost line. If plastic pipe is being used (rather than metal), higher heat capacity is needed to overcome the poorer heat-transfer characteristics of plastic pipe.

2. *Determining heat-tracing specifications:* Consult the manufacturers' literature to find a cable product with the proper temperature range of 100°F (37.8°C) service. An intermittent demand of 150°F (65.6°C) may be assumed. The cable

TABLE 11.1 Heat Losses from Insulated Metal Pipes, W/(ft · °F)*†
(For up to 20 mi/h wind speed)

Pipe size, in/s	Insulation ID, in	Insulation thickness, in							
		½	¾	1	1½	2	2½	3	4
½	0.840	0.054	0.041	0.035	0.028	0.024	0.022	0.020	0.018
¾	1.050	0.063	0.048	0.040	0.031	0.027	0.024	0.022	0.020
1	1.315	0.075	0.055	0.046	0.036	0.030	0.027	0.025	0.022
1¼	1.660	0.090	0.066	0.053	0.041	0.034	0.030	0.028	0.024
1½	1.990	0.104	0.075	0.061	0.046	0.038	0.034	0.030	0.026
2	2.375	0.120	0.086	0.069	0.052	0.043	0.037	0.033	0.029
2½	2.875	0.141	0.101	0.080	0.059	0.048	0.042	0.037	0.032
3	3.500	0.168	0.118	0.093	0.068	0.055	0.048	0.042	0.035
3½	4.000	0.189	0.133	0.104	0.075	0.061	0.052	0.046	0.038
4	4.500	0.210	0.147	0.115	0.083	0.066	0.056	0.050	0.041
	5.000	0.231	0.161	0.125	0.090	0.072	0.061	0.054	0.044
5	5.563	0.255	0.177	0.137	0.098	0.078	0.066	0.058	0.047
6	6.625	0.300	0.207	0.160	0.113	0.089	0.075	0.065	0.053
	7.625	0.342	0.235	0.181	0.127	0.100	0.084	0.073	0.059
8	8.625	0.385	0.263	0.202	0.141	0.111	0.092	0.080	0.064
	9.625	0.427	0.291	0.224	0.156	0.121	0.101	0.087	0.070
10	10.750	0.474	0.323	0.247	0.171	0.133	0.110	0.095	0.076
12	12.750	0.559	0.379	0.290	0.200	0.155	0.128	0.109	0.087
14	14.000	0.612	0.415	0.316	0.217	0.168	0.138	0.118	0.093
16	16.000	0.696	0.471	0.358	0.246	0.189	0.155	0.133	0.104
18	18.000	0.781	0.527	0.401	0.274	0.210	0.172	0.147	0.115
20	20.000	0.865	0.584	0.443	0.302	0.231	0.189	0.161	0.125
24	24.000	1.034	0.696	0.527	0.358	0.274	0.223	0.189	0.147

**Note*: Values in table are based on the formula below, plus a 10% safety margin. The k factor of 0.25 for fiberglass at 50°F is assumed.

$$\frac{\text{W}}{\text{ft of pipe}} = \frac{2\pi k\ \Delta T}{Z \ln (D_o/D_i)}$$

where k = thermal conductivity, Btu · in/(h · ft² · °F)
D_o = outside diameter of insulation, in
D_i = inside diameter of insulation, in
T = temperature differential, °F
Z = 40.944 Btu · N/(W · h · ft)

†For any desired metric equivalents, refer to metric conversion factors in App. B.
Source: Chromalox, E. L. Wiegand Division, Emerson Electric Co. Used with permission.

chosen must have a heat output equal to or in excess of the heat demand of the pipe system. In the example, since a 3.8 W/ft (12.5 W/m) heat demand is well within the range of most heat-tracing cables, a selection can be made from a variety of constant-wattage or self-regulating tracing.

If the heat demand were higher, one option would be to spiral-wrap the tracing around the pipe. Table 11.2 provides a determination of the pitch (spacing between spirals) that the tracing would require to equal a given heat demand. Read the table by dividing the system's heat demand by the heat output of the tracing. Locate that number or the next higher one in the row specifying the pipe diameter of the system. Then read the pitch in inches at the top of that column.

TABLE 11.2 Wrapping Factor,* ft cable/ft pipe†

Pipe	Pitch, in																	
size, in	2	3	4	5	6	7	8	9	10	11	12	14	16	18	24	30	36	42
½	1.90	1.47	1.29	1.19	1.14	1.10	1.08	1.06										
¾	2.19	1.64	1.40	1.27	1.19	1.14	1.11	1.09	1.07	1.06								
1	2.57	1.87	1.55	1.38	1.27	1.21	1.16	1.13	1.11	1.09	1.07							
1¼	3.07	2.18	1.76	1.53	1.39	1.30	1.24	1.19	1.16	1.13	1.11	1.08	1.06					
1½	3.43	2.41	1.92	1.65	1.48	1.37	1.29	1.24	1.20	1.16	1.14	1.10	1.08	1.06				
2	4.15	2.86	2.25	1.90	1.67	1.52	1.42	1.34	1.28	1.24	1.20	1.15	1.12	1.10	1.05			
2½	4.91	3.36	2.61	2.17	1.89	1.70	1.56	1.46	1.39	1.33	1.28	1.21	1.17	1.13	1.08	1.05		
3	5.88	3.99	3.06	2.52	2.17	1.93	1.76	1.63	1.53	1.45	1.39	1.30	1.23	1.19	1.11	1.07	1.05	
4	7.43	5.01	3.82	3.11	2.65	2.33	2.09	1.92	1.78	1.67	1.58	1.45	1.36	1.29	1.17	1.11	1.08	1.06
5	9.09	6.10	4.63	3.75	3.17	2.77	2.47	2.24	2.06	1.92	1.81	1.63	1.51	1.42	1.25	1.17	1.12	1.09
6	10.75	7.20	5.44	4.40	3.70	3.22	2.86	2.58	2.36	2.19	2.04	1.83	1.67	1.55	1.34	1.23	1.16	1.12
8	13.88	9.28	6.99	5.63	4.72	4.08	3.60	3.23	2.94	2.71	2.51	2.22	2.00	1.83	1.53	1.36	1.26	1.20
10	17.20	11.49	8.65	6.94	5.81	5.01	4.41	3.95	3.58	3.28	3.03	2.65	2.37	2.15	1.75	1.52	1.38	1.29
12	20.34	13.58	10.21	8.19	6.85	5.89	5.18	4.62	4.18	3.83	3.53	3.07	2.73	2.40	1.97	1.68	1.51	1.39
14	22.30	14.89	11.18	8.97	7.49	6.44	5.66	5.05	4.57	4.17	3.85	3.34	2.96	2.67	2.11	1.79	1.59	1.46
16	25.44	16.98	12.75	10.22	8.53	7.33	6.43	5.74	5.18	4.73	4.35	3.77	3.33	3.00	2.34	1.97	1.73	1.57
18	28.58	19.07	14.31	11.47	9.57	8.22	7.21	6.42	5.80	5.29	4.86	4.20	3.71	3.33	2.58	2.15	1.88	1.69
20	31.71	21.16	15.88	12.72	10.61	9.11	7.99	7.11	6.42	5.85	5.38	4.64	4.09	3.66	2.82	2.34	2.03	1.81
24	37.99	25.34	19.02	15.22	12.70	10.90	9.55	8.50	7.66	6.98	6.41	5.52	4.85	4.34	3.32	2.72	2.33	2.07

*To determine the wrapping factor, divide the calculated heat loss by the heat output of the cable. Locate the value that is equal to or the next highest in the row for the pipe size in your application. The value at the top of the column is the pitch or spacing from center to center of the cable along the pipe.

†For any desired metric equivalents, refer to metric conversion factors listed in App. B.

Source: Chromalox, E. L. Wiegand Division, Emerson Electric Co. Used with permission.

Where straight lengths of tracing are used, the overall length of the tracing will equal the pipe system's length, plus allowances for fittings. Roughly 1 ft (0.304 m) extra should be allowed for valves below 2-in (5.08-cm) diameter; above that, consult the manufacturer's literature. For flanges, allow for twice the flange diameter; for pipe hangers, allow 3 times the pipe diameter for each hanger.

11.4 ACCESSORY AND CONTROL EQUIPMENT

Like most electric HVAC systems, electric heat tracing offers relatively straightforward installation and control procedures. Care must be taken to prevent corrosion damage and to minimize fire or shock hazards. This section outlines the common industry practices.

The first control factor to consider with heat tracing is overcurrent protection. For constant-wattage units, the total current load can be calculated according to the following equation:

$$\text{Total current load (A)} = \frac{\text{cable length} \times \text{W/length at operating conditions}}{\text{operating voltage}} \tag{11.2}$$

The current load so determined should be multiplied by 125 percent to arrive at the correct rating for fuses, circuit breakers, or other overcurrent protection devices.

For self-regulating tracing, the calculation is different because the tracing experiences a large current inrush upon energization, followed by steadier current demand based on the system's operation. Inrush current is usually specified by the tracing supplier. The current load is then calculated according to the equation

$$\text{Total current load (A)} = \text{inrush current (A/length)} \times \text{cable length} \tag{11.3}$$

The temperature of the tracing line may be controlled by several means. Either the ambient temperature or the temperature of the tracing line can be monitored. Ambient sensing is normally recommended for freeze protection. The sensing device(s) should be placed where temperatures are expected to be lowest; additional sensors may be placed at points considered most representative of the system.

CHAPTER 12
HEATING DISTRIBUTION SYSTEMS AND EQUIPMENT FOR UNIT HEATERS

Ronald A. Kondrat
Robert L. Linstroth
Product Managers, Heating Division,
Modine Manufacturing Co.,
Racine, Wisconsin

12.1 INTRODUCTION

Although there may be various definitions for the term "unit heater," it is described here as a self-contained heating package suspended overhead or mounted from walls or columns, generally intended for installation in the space to be heated. Elements include a fan and motor, a heating element, and an enclosure. Depending on the type, additional items may include filters, power venters, directional air outlets, and duct collars.

Unit heaters have three principal characteristics: (1) relatively large heating capacities in compact casings, (2) the ability to project heated air in a controlled manner over a considerable distance, and (3) a relatively low installed cost per Btu (watt) of output. They are, therefore, usually employed in applications where the heating capacity requirements, the physical volume of the heated space, or both are too large to be handled adequately or economically by other means. By eliminating extensive duct installations, the space is freed for other use.[1]

Frequently, however, the unit heater's role is relegated to providing supplementary heating. Properly applied, the unit heater can be the *primary* source of heat in many types of buildings. The ideal applications for unit heaters exist in buildings having large open areas such as factories, warehouses, and other industrial and commercial buildings; showrooms, stores, and laboratories; and auxiliary spaces, where they may be used singly or in groups in corridors, vestibules, and lobbies or to blanket areas around frequently opened doorways; for spot heating; and for maintaining air temperatures year-round in greenhouse applications.

Unit heaters are also employed where filtration of the heated air is required. They may also be modified to provide ventilation by using outside air.

Unit heaters may be applied to a number of industrial processes, such as drying and curing, in which the use of heated air in rapid circulation with uniform distribution is of particular advantage. Unit heaters may be used for moisture absorption applications to reduce or eliminate condensation on ceilings or other cold surfaces of buildings in which process moisture is released. When such conditions are severe, exhaust fans and makeup air units may be required.[2] Today hydronic unit heaters are also being used with increasing frequency to utilize waste heat in order to conserve energy.

Consulting engineers, plant engineers, architects, contractors, and building owners often judge unit heaters on the basis of observed jobs. Sometimes these jobs are poorly engineered, and as a result the unit heater is unfairly assumed to be the problem contributing to an unsatisfactory heating system. Here are a few examples of common misapplications:

Unit heater motors are generally designed to operate at a maximum ambient temperature of 104°F (40°C). If the unit heater is so oriented that it does not discharge its heat into the space to be heated (away from the unit heater), heat builds up around the unit and its motor may fail prematurely.

If a unit heater's mounting height is specified for high-speed motor operation and the unit is operated at low speed, then the heated air may not reach floor level, thus sacrificing comfort and wasting heat.

Selection of the wrong unit heater type, over- or undersizing a unit, mounting the unit too high or too low, and other factors can contribute to poor comfort conditions and wasted heat.

The information contained in this chapter can be helpful to heating system design engineers, specifiers, and users in selecting and applying unit heaters for the overall upgrading of unit heater installations and in obtaining the most comfortable conditions.

12.2 UNIT HEATING SYSTEM DIFFERENCES

Steam, hot-water, gas-fired, gas-fired high-efficiency, oil-fired, and electric heating systems all have unique characteristics, and they can be compared only if the advantages and disadvantages of each are known.

If we compare initial unit heater costs by using comparable Btu (watt) capacities, oil-fired unit heaters are the most expensive and are followed in order by gas, electric, and steam/hot-water unit heaters. If we rank by installation costs, the order again begins with oil-fired units because of added piping, tanks, pumps, filters, valves, and venting. Similar installation costs accrue with steam/hot-water units requiring supply and return piping, traps, strainers, and valves. Gas-fired unit heaters, with one-way piping and venting, can be installed for about one-half the cost of hydronic units. Last, electric unit heaters with no piping or venting have lower installation costs.

No economic comparison would be complete without a review of operating costs projected to at least a year of future operation. Because operating or fuel costs vary locally, they must be interpreted individually. All local energy rates can be translated to cost per Btu (kWh). Natural or propane gas is normally sold by therms (100,000 Btu or 29.3 kWh) or by cubic feet (cubic meters). Your local utility or dealer will know your calculated rate of Btu/cubic foot of gas (usually

800 to 1150 for natural gas or 2500 for propane). Electric unit heaters deliver 3413 Btu/kW (3,600,000 J/kW). For oil heaters, as a reference, No. 2 fuel oil is about 140,000 Btu/gal (40×10^6 J/L). Overshadowing all these costs, however, is the fuel availability in each locality. Where fuel or power supplies are interruptible, the value of a dependable fuel supply could offset operating-cost advantages.

Negative pressures within a building will not affect the operation of hydronic or electric unit heaters because they are not connected to flue systems that lead to the outside of the building. But when vented gas or oil units are used in a building with negative pressure, there could be a problem. Since most gas-fired unit heaters are equipped with atmospheric burners, a negative pressure within the building will adversely affect the draft required for efficient operation of the unit. Negative building pressures can cause flue gases, which are normally vented to the outside, to spill into the building and endanger health and safety. Pilot flame outage is another result of negative pressure within a building. Where negative pressures exist, installation of mechanical draft systems may be required for applications using gas- or oil-fired unit heaters, or gas-fired "separated combustion" units can be used.

Hydronic unit heaters usually require some type of water treatment system. In existing steam or hot-water heating systems, water treatment has usually been provided. If the chemical condition of the condensate and/or supply water to be used in a heating system is unknown, it is recommended that the condensate and/or supply water be tested by a reputable testing firm and evaluated to determine if treatment is required.

The amount of maintenance a unit requires during its lifetime depends on the conditions under which the unit heater operates. Steam units, e.g., installed on high-pressure steam systems which supply steam for processing operations are often subject to corrosion from acids and oxygen present in makeup water introduced into the system. This is because processing operations require greater quantities of boiler makeup water. Proper installation and maintenance of deaerating equipment and adequate boiler-water treatment can do much to reduce corrosion in steam unit heater condensers as well as piping, valves, and traps. For similar reasons, adequate boiler-water treatment must be provided and maintained in hot-water boilers.

Certain airborne contaminants can cause metal parts to corrode. The corrosion attack can be rapid or slow, depending on the concentration of the contaminant. Gas- or oil-fired heat exchangers, flue or vent pipes, and metal chimneys are most vulnerable. When contaminating vapors exist, steps to correct or relieve the condition must be taken. For example, proper venting of degreasing or cleaning vats will reduce the distribution of the contaminants throughout the heated areas of the building.

Corrosive agents in the atmosphere can shorten the life of unit heaters (particularly gas- and oil-fired units). Minute particles of trichlorethylene and carbon tetrachloride, e.g., when heated and combined with water vapor in the products of combustion, produce an acid solution harmful to heat exchangers and other metal parts. Reduce corrosive atmospheric contaminates by installing exhaust hoods above tanks containing solvents and acids.

Oil, gas, and electric unit heaters are self-contained heating units; failure of one or more unit heaters will not affect the entire heating system. Failure of the entire system (except for an electric power failure or a cutoff of the main gas supply to the building) is highly improbable. Should a steam or hot-water boiler system fail, all the equipment supplied by that system is useless until the boiler is repaired and placed in operation. If there is only one boiler supplying the facility, then the entire facility will be without heat.

Properly installed, a single unit heater (oil, gas, hydronic, or electric) can be removed from the system for repair or replacement without shutting down the entire system.

Since gas and oil unit heaters draw combustion air from the building they heat, the air they consume must be replenished. This is not the case with hydronic or electric unit heaters.

In the event of a power failure, not only will electric unit heaters fail to operate, but so will gas, oil, and hydronic unit heaters since these units are powered by electric motors.

12.3 CLASSIFICATION OF UNIT HEATERS

Unit heaters can be classified as follows:

- By type of air mover (either propeller or centrifugal blower)
- By physical configuration (vertical or horizontal air delivery)
- By heating medium [steam, hot water, gas, gas (high-efficiency), oil, or electricity]

By Type of Air Mover. In making your selection, be sure to consider both the heating capacity and the air delivery capacity of the unit, to be sure that the air is delivered precisely where you want it to go.

Propeller Fan Unit Heaters. Propeller fan units are classified as zero-static-pressure type and cannot be used with air discharge nozzles or in connection with ductwork. This unit heater has a fixed air delivery with a temperature rise (difference between inlet and discharge air temperatures) usually in the range of 55 to 65°F (30.5 to 36.0°C). It is the most widely used of all unit heaters primarily because of its broad application, versatility, lower cost, and compact size.

It is used in free-air delivery applications where the heating capacity and distribution requirements can best be met by units of moderate output, used singly or in multiples, and where filtration of the heated air is not required. Horizontal blow units are usually associated with low to moderate ceiling heights. Downblow units are used in high ceiling spaces and where floor and wall space limitations dictate an out-of-the-way location for the heating equipment. Downblow units may be supplied with adjustable diffusers.[2]

Centrifugal Blower Fan Unit Heaters. Designed for use with ductwork or discharge air nozzles, the blower unit is classified as a high-static unit [up to 0.4-in (1-cm) water column external static pressure]. Since its operation is generally quieter than a propeller-type unit heater having the same air delivery, it is frequently installed in areas where lower sound levels are desirable. Also this unit has a variable-pitch motor sheave which permits the air volume to be adjusted to accommodate a range in air temperature rise. Equipped with air discharge nozzles, the unit is ideal for use in spot heating applications. In general, a blower unit heater is classified as a heavy-duty unit.

By Physical Configuration. Unit heaters are produced in two basic types: horizontal air delivery and vertical air delivery. Each has its own distinctly different heat-throw and heat-spread characteristics. For any comfort application, selection of the proper type is very important.

Horizontal Delivery Unit Heaters. Characterized by its horizontal air discharge, this type is widely used for general industrial and commercial heating applications. Horizontally positioned louvers attached to the air discharge opening are usually standard and can be adjusted to lengthen or shorten heat throw. Adjustable vertical louvers (when available, they are optional) used in combination with horizontal louvers permit complete directional control of heated air output.

Vertical Delivery Unit Heaters. Due to their directly downward air discharge, vertical units are particularly desirable for heating areas with high ceilings and where craneways and other obstructions dictate high mounting of heating equipment. Air distribution devices (optional) which attach to the air discharge openings provide distinctly different heat-throw patterns to meet specific heat-throw and heat-spread requirements.

By Heating Medium (Steam, Water, Gas, Oil, or Electric). Selection is usually dictated by the type of energy which is most prevalent, more reliable, or least costly. For example, in an industrial plant where steam is generated by in-plant processing operations, steam unit heaters may be selected if the existing boiler capacity is adequate to handle the heating load. In most areas electricity is more costly than gas. In certain areas gas may be in critical supply, and electricity, though more expensive, may constitute a more dependable energy source. Selection is also determined by economics which involve an examination of initial cost, operating cost, and conditions of use, including the size, location, and environment of the proposed installation.

It is standard practice among trade associations to rate all types of unit heaters on the basis of the amount of heat delivered in Btu/hour (watts) at an entering air temperature of 60°F (15.6°C).[2] If the units are operated under other than rated conditions, the manufacturers' literature should be reviewed or the manufacturer contacted directly. Most manufacturers' literature also provides data on other than standard rated performances.

FIGURE 12.1 Steam/hot-water horizontal delivery unit heater. A propeller fan moves room air through a condenser which is heated by steam or hot water. Adjustable louvers positioned horizontally in the air discharge opening permit heated air to be directed down, up, or straight out. Vertical louvers (optional) permit complete directional control of heated air. Steam and hot-water unit heaters are also available in vertical air delivery models. *(Courtesy of Modine Manufacturing Co.)*

Steam or Hot Water. There is generally no major difference in the construction of a steam or hot-water unit heater, as shown in Fig. 12.1. Both use a condenser which consists of fins attached to tubes through which steam or hot water produced by a remotely located boiler is circulated. Air moved across the condenser by a fan removes heat from the fins and discharges it to the space being heated. Although there is usually no difference between a steam unit and hot-water unit, the dissimilarity in piping arrangements of the two systems is discussed elsewhere in this chapter.

Rating of steam unit heaters has been standardized according to the following basis: dry saturated steam at 2-psig (13.78-kPa) pressure at the heater coil, air at 60°F (15.6°C) and 29.92 inHg (101.04 kPa) barometric pressure entering the heater, and the heater operating free of external resistance to air flow.[2]

The capacity of a heater increases as the steam pressure or hot-water temperature increases, assuming the entering air temperature is constant. The capacity decreases as the entering air temperature increases. The heating capacity for any condition of steam pressure and entering air temperature, other than standard, may also be determined by means of a standard procedure.[2]

A standard for the rating of hot-water unit heaters has been established according to the following basis: entering water at 200°F (93.3°C), water temperature drop of 20°F (11.1°C), entering air at 60°F (15.6°C) and 29.92 inHg (101.04-kPa) barometric pressure, and the heater operating free of external resistance to air flow. Variations in entering water temperature, entering air temperature, and water flow rate will affect capacity. A standard method is used to translate the heating capacity, as obtained under standard rating conditions, to other conditions of air and water temperature.[2]

Gas (Conventional Type). Heat exchanger tubes (aluminized or stainless steel) are vertically positioned over a natural or propane gas burner (Fig. 12.2). Flames produced by the burner burn within the tubes of the heat exchanger. Heat transferred from the flames and conducted to exterior surfaces of the tubes is removed by the air stream produced by a fan or blower, passing over the heat exchanger tubes.

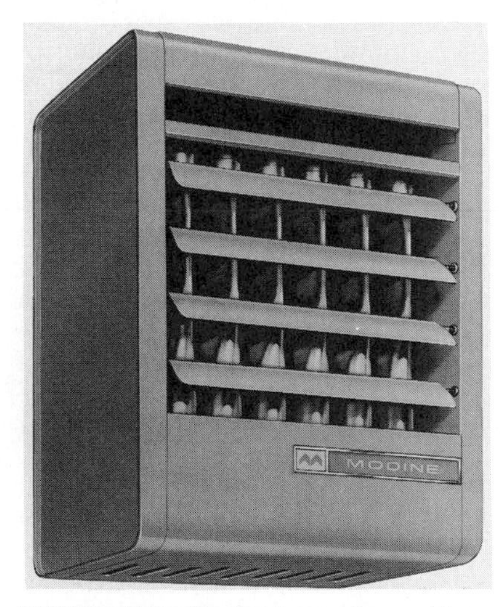

FIGURE 12.2 Gas-fired propeller or blower fan unit heater. A propeller or blower fan is the air mover pushing room air through the heat exchanger. Horizontally positioned adjustable air deflectors permit heated air to be directed up, down, or straight out. Vertically positioned deflector blades (optional) may be added for complete directional control of heated air. *(Courtesy of Modine Manufacturing Co.)*

Gas (High-Efficiency Type). Essentially the same as the conventional gas-fired unit heaters, these units can deliver almost 10 percent more average thermal efficiency compared with the conventional gas-fired units. The higher efficiency of at least one unit comes from reclaiming heat from combustion gases by drawing them through a secondary heat exchanger, rather than losing this heat through the vent pipe, as in most conventional gas-fired units. Advantages of this unit are that it cuts fuel consumption, trims electricity costs (because fewer operating hours are needed to satisfy the same heating requirements as conventional gas-fired units), and permits smaller and less expensive flue gas vents through the use of a power exhauster.

Gas (Separated Combustion Type). Gas-fired separated combustion units have the same basic design characteristics as conventional gas unit heaters except for the addition of (1) an enclosed burner compartment which is isolated from the heated space and (2) a power exhaust system. These units are designed to take all of their com-

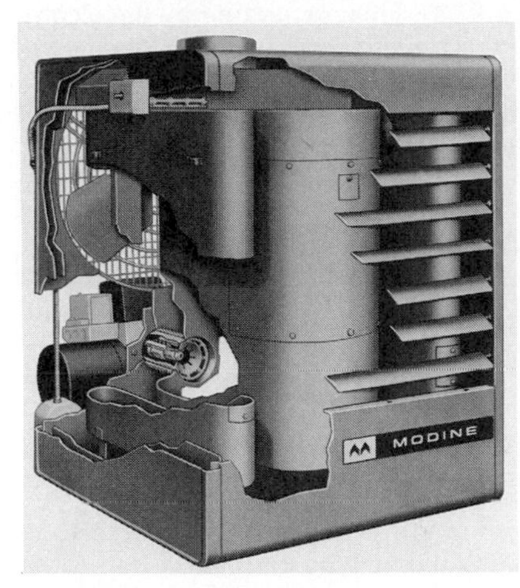

FIGURE 12.3 Oil-fired unit heater. A motor-driven propeller fan directs room air over the exterior surfaces of the heat exchanger. When the burner is ignited, a combination fan and limit control prevents fan operation until the heat exchanger has warmed up, and after the burner is shut down, it allows the fan to run until the heat exchanger has cooled. Horizontal and optional vertical louvers provide complete directional control of heated air. *(Courtesy of Modine Manufacturing Co.)*

FIGURE 12.4 Electric unit heater. Vertical and horizontal delivery units are available. Finned-tube heating elements located in the air stream heat the room air blown through the unit by the propeller fans. Horizontal delivery units are equipped with adjustable louver-type air deflectors. Vertical delivery models may be used with or without air deflector devices (optional). Three air deflector assemblies are offered, each with its own distinctive air distribution pattern. Heavy-duty, long-heat-throw, horizontal delivery models are also available. *(Courtesy of Modine Manufacturing Co.)*

bustion air directly from outside of the building. Because no indoor air is used for combustion, this type of unit is ideally suited for hard to heat applications which may include buildings with dusty or dirty atmospheres, high humidity, and negative pressure problems, public automobile repair facilities and parking garages, institutional buildings, and buildings where indoor air quality is of the utmost importance. Because separated combustion unit heaters are power vented, they can be vented through a side wall or vertically, usually using much smaller flue gas vent pipes than conventional unit heaters. High-efficiency models are available.

Gas-fired unit heaters are rated in terms of both input and output, in accordance with the approval requirements of the American Gas Association and the Canadian Gas Association.

Fuel Oil. In operation, No. 1 or No. 2 fuel oil is atomized under pressure by a gun-type burner and is mixed with high-velocity combustion air to burn in a solid-cone flame in an aluminized-steel heat exchanger (Fig. 12.3). Resultant radiated heat is transferred to the exterior surfaces of the heat exchanger and is removed by the fan-produced air stream passing over the surfaces of the exchanger.

Ratings of oil-fired unit heaters are based on heat delivered at the heater outlet in Btu/h (watts).[2]

Electricity. Heating elements consist of spirally wound fins on steel tubes which encase electrical-resistance wire (Fig. 12.4). Heat generated by the electric current is transferred through the tubes to the fin surfaces, where it is removed by the air movement produced by a fan.

Electric unit heaters are rated on the energy input to the heating element, expressed in terms of kilowatts, or Btu per hour.[2]

12.4 TYPICAL UNIT HEATER CONNECTIONS

These connections are illustrated in Figs. 12.5, 12.6, 12.7, 12.8, and 12.9.

12.5 CALCULATING HEAT LOSS FOR A BUILDING

Prior to selecting a unit heater, first determine the heat loss for the area to be heated. For a more exact method of calculating the heating load, see Chap. 2. An abbreviated method for calculating heat loss is listed on p. 12.10.

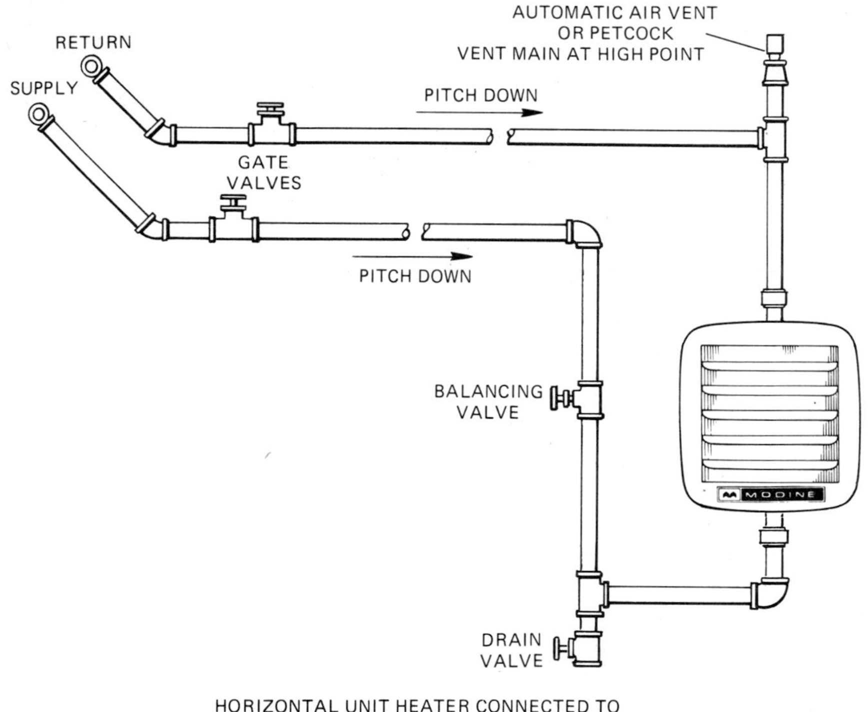

FIGURE 12.5 Hot-water heating. Piping should be sized to adequately accommodate the required water flow. Valves and unions on supply and return piping facilitate the takedown of the unit when necessary without shutting down the system. Return lines should be pitched for drainage. (*Courtesy of Modine Manufacturing Co.*)

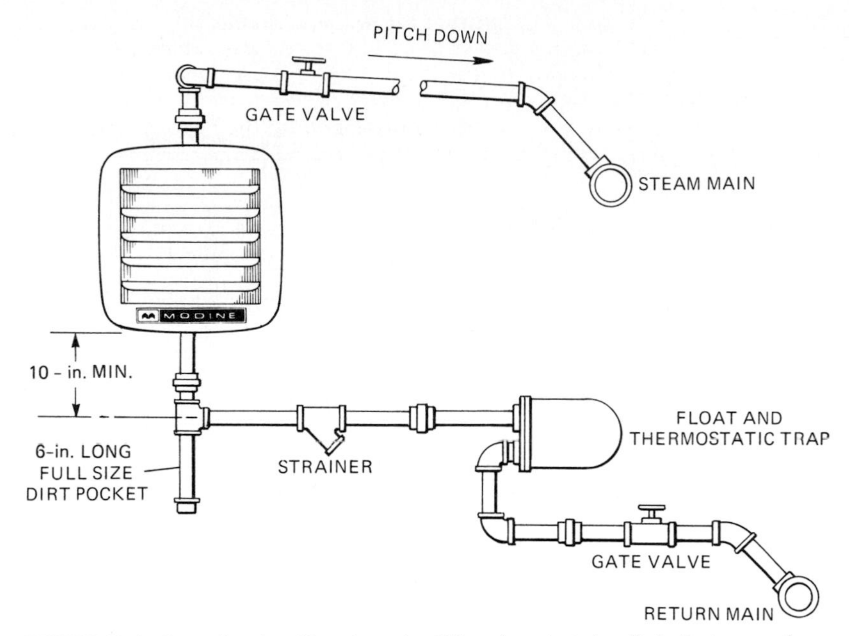

FIGURE 12.6 Steam heating. Pipe sizes should be adequate to handle both steam and condensate under maximum load conditions. Traps should be selected to handle the condensate when the unit heater is operating at maximum steam pressure and minimum entering air temperature. Provision should be made for elimination of entrapped air in system. (*Courtesy of Modine Manufacturing Co.*)

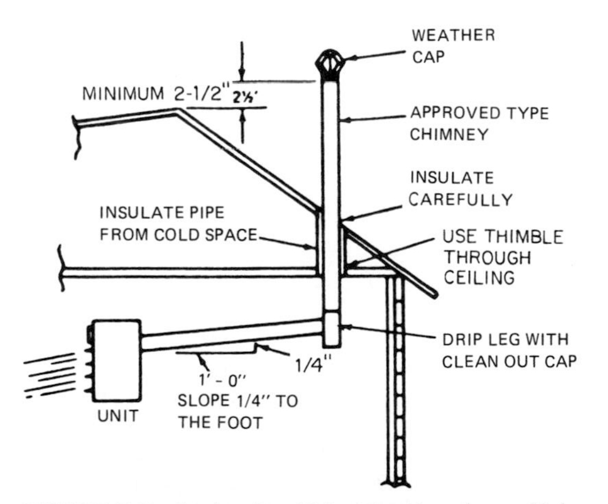

FIGURE 12.7 Gas heating. Piping should conform with local codes and requirements for type and volume of gas burned and pressure drop allowed in the line. Consult local codes for venting regulations. Vent pipe should be at least 6 in (15 cm) from combustible material and insulated at the junction with the roof or where the vent pipe passes through insulated spaces.

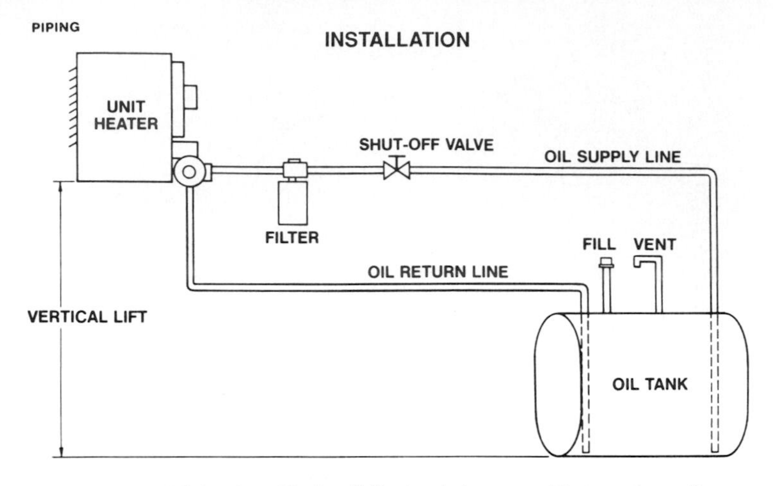

FIGURE 12.8 Oil heating. Single oil-fired unit heaters with two-stage oil pumps have a suction lift of 20 ft (6.1 m) on a two-pipe system and can be easily supplied from a separate tank. For multiple units, install a centralized oil distribution system with a separate booster pump connected to a common tank to feed fuel oil in either a looped or pressurized oil supply system. (*Note*: These circuits do not satisfy any particular local code requirements. Compliance with local codes is the responsibility of the installer.)

1. Determine both the inside temperature to be maintained and the design winter outside temperature for the locality. The difference between these two temperatures is called the "design temperature difference."
2. Calculate net areas in square feet (meters) of glass, floor, walls, and roof exposed to outside temperature or unheated spaces. Include all door areas as glass.
3. Select overall heat-transfer coefficients from Table 12.1 (or appropriate sections of Chap. 2), and compute the heat-transfer loss for each type of area in Btu/h (watts) by multiplying each area by its heat-transfer coefficient and by the temperature differential.
4. Add 10 percent to the heat-loss figures for areas exposed to prevailing winds.
5. Calculate the volume of the room or area in cubic feet (cubic meters), and multiply by the estimated number of air changes per hour due to infiltration (usually one or two). Determine the number of cubic feet per hour (cubic meters per second) of air exhausted by ventilating fans or industrial processes. Substitute the larger of these two figures into the formula to determine the heat required to raise the air from outside to room temperature.

Infiltration

$$\text{Btu/h} = [\text{volume (ft}^3 \times \text{number of air changes per hour} \times \Delta\ \text{TF}°]55$$
$$\text{W} = [\text{volume m}^3 \times \text{number of air changes per hour} \times \Delta\ \text{TC}°]/3$$

For exhaust air

$$\text{Btu/h} = \text{cfm} \times 1.08 \times \Delta\ \text{TF}°$$
$$\text{W} = \text{J/s} \times 1.2 \times \Delta\ \text{TC}°$$

FIGURE 12.9 Electric heating. Wiring must be in accordance with the National Electric Code® (ANSI CI) and applicable local electric codes. Power supply to unit heaters must be protected by a fused disconnect switch. Unit heater should not be located closer than 12 in (0.3 m) to combustible material. (*Courtesy of Modine Manufacturing Co.*)

6. The totals of Btu/h (watt) losses from steps 3, 4, and 5 will give the total Btu/h (watts) to be supplied by unit heaters. [*Note*: If processes performed in the room release considerable amounts of heat, this amount should be determined as accurately as possible and may be subtracted from the total Btu/h (watt) loss if they are operated 24 h/day during the peak heating season.]

12.6 SELECTING UNIT HEATERS

Once the building heat loss and the heating medium to be used have been determined, other factors must be considered during the unit heater selection process. First, establish the final air temperature and air delivery rate desired in cubic feet per minute (cubic meters per second).

The final air temperature, as stated earlier, should be between 55 and 65°F (30.5 and 36.0°C) higher than the desired room temperature that will be maintained. By locating final air temperatures of unit heaters operating at standard conditions in the manufacturers' literature, computation of these values is unnec-

TABLE 12.1 Common Heat-Transfer Coefficients*†

Building material	U factor‡	Building Material	U factor†
Walls		Wood, 1-in	
Poured concrete, 80 lb/ft^3§		(uninsulated) w/⅜-in built-up roofing	0.48
8 in	0.25	w/1-in blanket insulation	0.17
12 in	0.18	Wood, 2-in	
Concrete block, hollow cinder aggregate		(uninsulated) w/⅜-in built-up roofing	0.32
8 in	0.39	w/1-in blanket insulation	0.15
12 in	0.36	Concrete slab, 2-in	
Gravel aggregate		(uninsulated) w/⅜-in built-up roofing	0.30
8 in	0.52	w/1-in insulation board	0.16
12 in	0.47	Concrete slab, 3-in	
Concrete block, w/4-in face brick		(uninsulated) w/⅜-in built-up roofing	0.23
Gravel, 8 in	0.41	w/1-in insulation board	0.14
Cinder, 8 in	0.33	Gypsum slab, 2-in	
Metal (uninsulated)	1.17	(uninsulated) w/½-in gypsum board	0.36
w/1-in blanket insulation	0.22	w/1-in insulation board	0.20
w/3-in blanket insulation	0.08	Gypsum slab, 3-in	
Roofing		(uninsulated) w/½-in gypsum board	0.30
Corrugated metal (uninsulated)	1.50	w/1-in insulation board	0.18
w/1-in bolt or blanket	0.23	Windows	
w/1½-in bolt or blanket	0.16	Vertical, single-glass	1.13
w/3-in bolt or blanket	0.08	Vertical, double-glass, 3/16-in air space	0.69
Flat metal		Horizontal, single-glass (sky light)	1.40
w/⅜-in built-up roofing	0.90	Doors	
w/1-in blanket insulation under deck	0.21	Metal—single sheet	1.20
w/2-in blanket insulation under deck	0.12	Wood, 1-in	0.64
		2-in	0.43

*Heat loss through concrete floors on ground in large buildings may be ignored.
†Refer to *1985 ASHRAE Handbook, Fundamentals*, ASHRAE, Atlanta, GA, 1985, for expansion of this table.
‡Equivalent values for U factor: —Btu/(h)(ft^2)(°F) = 5.675 W/(m^2)(K) = U factor   1 K = 1°C   1 in = 2.54 cm = 0.254m
§For metric equivalents, see App. B.
Source: Modine Manufacturing Co. Used with permission.

essary, unlessthe condition varies from the standard. Should this be the case, consult the manufacturer or the manufacturer's representatives.

The ratings in cubic feet per minute (cubic meters per second), which are also based on standard conditions, are included in the manufacturers' literature. Again, if the conditions under consideration are other than standard, the manufacturer should be consulted because the volume of air that the unit heaters must provide should be a known factor.

An important point to consider with every heating medium is the change in entering air temperature. This change affects the total heating capacity in most unit heaters and the final air temperature in all units. The differences in temperature of the air entering the unit and of the air being maintained in the heated area are important, especially regarding downblow units.[2] As entering air temperatures increase, the outlet air temperature also increases, which makes the exiting air more buoyant and thus more difficult for it to penetrate the occupied zone near floor level.

Higher-velocity units and units with lower discharge air temperatures maintain lower temperature gradients than units with higher discharge temperatures. Valve-controlled or bypass-controlled units employing continuous fan operation maintain lower temperature gradients than units that employ intermittent fan operation. Lower temperature gradients lead to a general reduction in overall heat loss and improved comfort conditions. Directional control of the discharged air from a unit heater can also be an important factor in producing satisfactory distribution of heat and in reducing floor-to-ceiling temperature gradients.[2]

The use and character of the space to be heated influence the selection of the unit heater type(s) best suited to fulfill the heating requirement. Most industrial and commercial heating applications will benefit from the use of a combination of horizontal and vertical air delivery unit heaters. Each type differs principally in the method by which air and/or its direction (horizontal or vertical) is delivered.

The cost of equipment and its installation is less when fewer unit heaters are used, but comfort can be compromised. However, the ultimate system could utilize a greater number of units than might be economically practical. Generally, in large areas where maximum comfort is not necessarily the most important factor, it may be desirable to select the largest unit or units, such as in a warehouse-type building. In smaller areas or areas where the comfort of occupants is a prime consideration, specify more heaters of lower capacity.

A heated space occupied by a large number of people requires a greater number of air changes per hour than spaces occupied by relatively few people. As mentioned earlier, this should be considered when the heat loss is calculated for the space to be heated.

Mounting Height Is Critical. Improper mounting height is probably the cause of more unsatisfactory unit heater installations than any other single factor. When units are mounted higher than the maximum height recommended by the manufacturer, heated air may never reach the floor and thus drafts are created and fuel dollars are wasted. If units are mounted too low, hot blasts and excessive air velocities may cause discomfort to room occupants.

Horizontal delivery unit heaters usually are the best choice when ceilings are low and there are few obstructions in the path of air streams from the units. Standard louver-type air deflectors horizontally positioned in the air discharge openings of the unit heaters can be adjusted to direct the air up, down, or straight out (Fig. 12.2). If complete directional control of heated air is desired, add optional

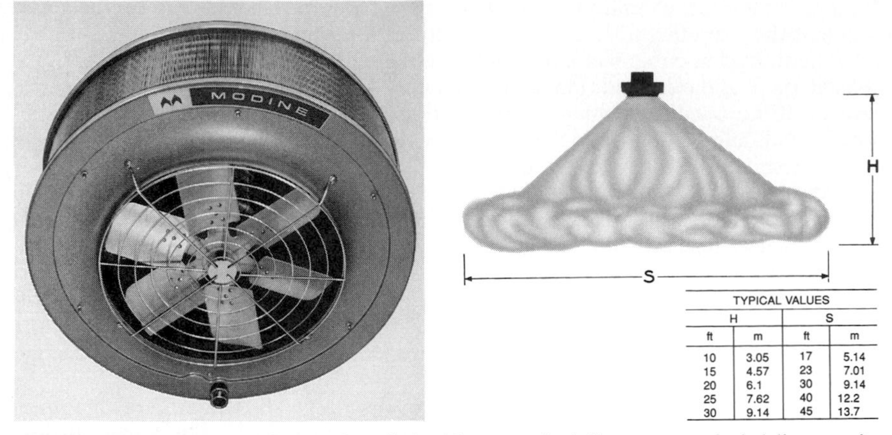

TYPICAL VALUES			
H		S	
ft	m	ft	m
10	3.05	17	5.14
15	4.57	23	7.01
20	6.1	30	9.14
25	7.62	40	12.2
30	9.14	45	13.7

FIGURE 12.10 Without deflector. Installed without an air deflector, a vertical delivery unit heater (*left*) has a heat spread similar to that shown at right. Where human comfort is not a major consideration and air control is not critical, vertical units are frequently installed without air deflectors. (*Courtesy of Modine Manufacturing Co.*)

louvers vertically positioned in the air discharge openings to achieve the desired effect.

Vertical delivery units are often used without optional air deflector assemblies (Fig. 12.10). However, several types of air deflectors are generally offered to accommodate the need for air distribution patterns not attainable when vertical units are installed without them. These air deflectors are illustrated together with the heat-throw and heat-spread patterns they produce in Figs. 12.11 to 12.14.

The use of air deflector assemblies influences heat throw and the recommended mounting heights of units. Therefore, when vertical delivery unit heaters are equipped with air deflectors, follow the manufacturer's mounting height recommendations for the unit heater with the specific deflector.

When it is necessary to install units at low mounting heights, select models with lower ft^3/min (m^3/min) ratings, because the greater volume of air handled by larger units can create excessive air movement.

Better air distribution and economy of heating system operation is realized when a greater number of smaller unit heaters are used (instead of fewer but larger units).

When standard steam unit heaters are installed on high-pressure systems, the temperature of the delivered air may be too high for comfort. Furthermore, high-temperature air is very buoyant, making proper heat distribution difficult. Some manufacturers offer a line of low-outlet-temperature unit heaters designed primarily for use on steam systems operating at steam pressures of about 30 lb/in^2 (2 bar) or more. These units supply heated air at more normal temperatures and are recommended for use on systems operated at high steam pressures, particularly in warehouses and storage areas, where adequate air volume and air velocity may be more important than maintaining temperatures high enough for maximum comfort. Low-pressure steam and conventional hot-water units are usually selected for smaller installations and for those concerned primarily with comfort heating.

For extremely hard-to-heat areas in a building, such as around frequently opened large doors and in other locations having extreme exposure, use a heavy-duty unit heater with a high-velocity horizontal air stream. This type of unit is

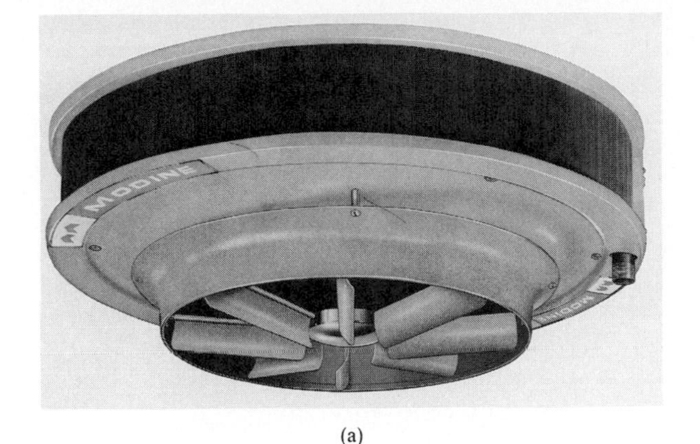

(a)

TYPICAL VALUES			
H		S	
ft	m	ft	m
10	3.05	12	3.66
15	4.57	11	3.35
20	6.1	15	4.57
25	7.62	19	5.69
30	9.14	22	6.70

(b) (c)

FIGURE 12.11 With cone-jet deflector. (*a*) Adjustable louver blades permit a variety of air distribution patterns, from (*b*) a high-velocity beam of heated air into a confined space on the floor to (*c*) a broad gentle cone of air to cover a much broader area. This permits heater to be mounted higher when louver blades are positioned vertically. (*Courtesy of Modine Manufacturing Co.*)

referred to as a "door heater." (For additional discussion on door heaters, refer to Chap. 16.) One such unit heater will often replace two or more smaller units. When a unit of this capability is installed, caution should be used in directing the high-velocity air stream away from room occupants. In addition, use blower-type gas unit heaters with nozzles equipped with adjustable air deflector blades to direct heat downward or obliquely into hard-to-heat areas.

12.7 WHEN QUIETNESS IS A FACTOR

Noise levels in workplaces may be a concern in some applications. Wherever fans and motors move air, sound is produced. Of course, such is the case with

(a)

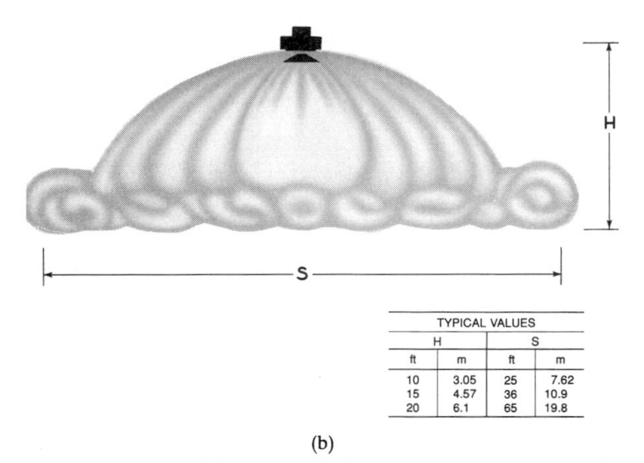

TYPICAL VALUES			
H		S	
ft	m	ft	m
10	3.05	25	7.62
15	4.57	36	10.9
20	6.1	65	19.8

(b)

FIGURE 12.12 With truncone deflector. (*a*) This cone-shaped air deflector may be raised or lowered to increase or decrease the heat spread of the unit heater. Generally it is used to provide (*b*) gentle air-flow patterns when unit heater is mounted low. (*Courtesy of Modine Manufacturing Co.*)

unit heaters. Sound emissions of certain large unit heater models may limit their use in applications where sound-level requirements may be critical. In such instances, smaller models should be selected which in total meet the heat load criteria of a larger single unit heater. The air velocity ratings are generally indicative of sound levels, i.e., the higher the ft^3/min (m^3/s) and ft/min (m/min) ratings of a unit heater, the greater the sound emission.

The absence of a standard sound testing and rating code in the unit heater in-

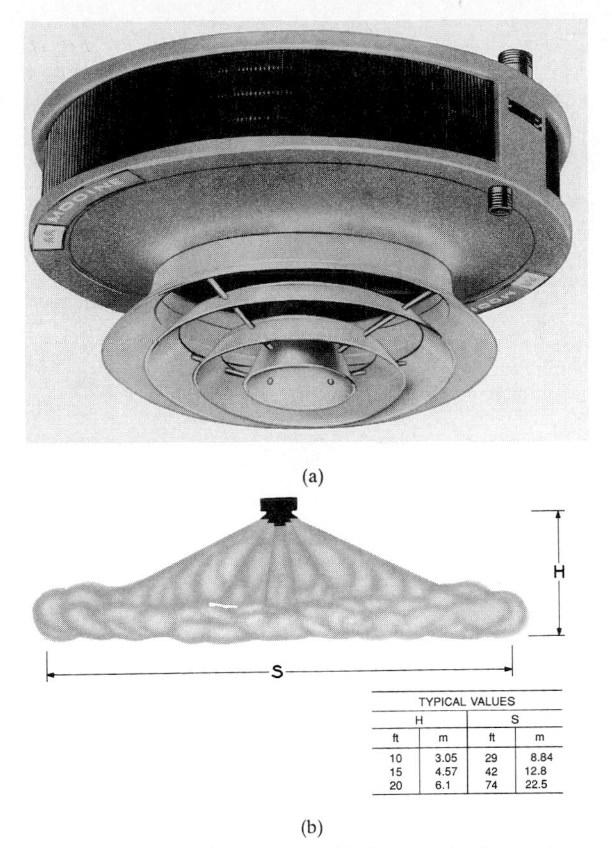

TYPICAL VALUES			
H		S	
ft	m	ft	m
10	3.05	29	8.84
15	4.57	42	12.8
20	6.1	74	22.5

FIGURE 12.13 With anemostat. (*a*) Comprised of several concentric cones in fixed position, this assembly produces (*b*) an exceptionally wide heat spread at very low air velocities. For example, a unit heater equipped with a four-cone anemostat may be mounted as low as 7 ft (2.1 m). (*Courtesy of Modine Manufacturing Co.*)

dustry makes it impossible to compare manufacturers' catalogued sound ratings.Unit heater manufacturers either develop their own procedures or adopt one of several sound testing standards developed by independent testing laboratories or other industry groups such as American National Standards Institute (ANSI S1.21-1972).

Publication 303, Application of Sound Power Level Ratings, by the Air Movement and Control Association (AMCA) states:

> Sound measurements cannot be made as precisely as those used to establish air movement or heat-transfer ratings. Within the present state of the art, differences in sound power levels ±2 dB or less are not considered significant. In comparing products of different manufacturers, it is good practice to disregard differences of less than 4 dB. This is particularly true in the first octave band where the difference of ±6 dB or less should be disregarded.[3]

(a)

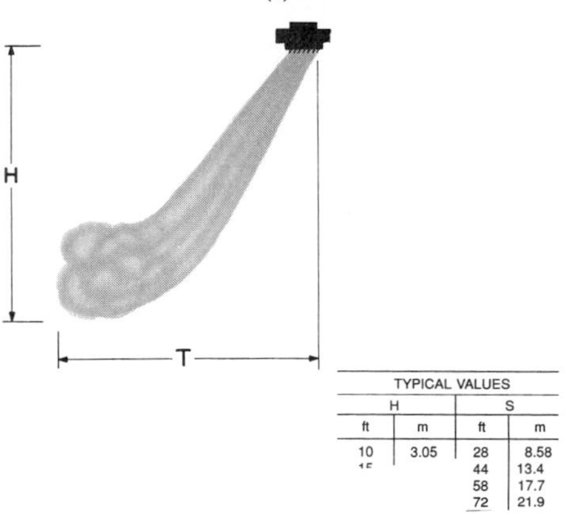

TYPICAL VALUES			
H		S	
ft	m	ft	m
10	3.05	28	8.58
[illegible]		44	13.4
		58	17.7
		72	21.9

(b)

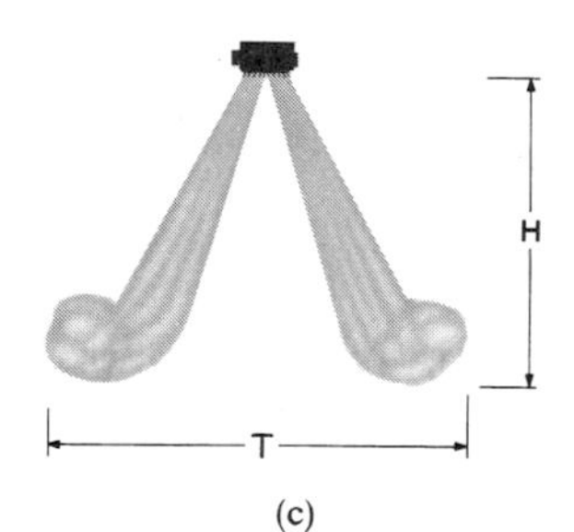

(c)

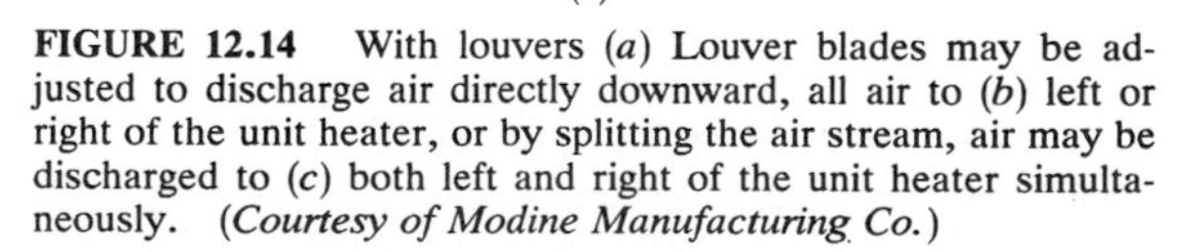

FIGURE 12.14 With louvers (*a*) Louver blades may be adjusted to discharge air directly downward, all air to (*b*) left or right of the unit heater, or by splitting the air stream, air may be discharged to (*c*) both left and right of the unit heater simultaneously. (*Courtesy of Modine Manufacturing Co.*)

Even though sound levels cannot be measured as accurately as air movement or as precisely as heat loss, the procedure for measuring sound is comparable to that for calculating the heat loss of a given space.

Wall, ceiling, floor surface, the kind of activity performed, and the type of equipment occupying the space all affect the heat loss. Similarly, these same factors affect the direct radiation of sound from its source to the human ear. All factors influence the sound absorption value of the space in which the unit heaters operate. Sound emissions from unit heaters are also affected by the type of mounting used to suspend them. Certain mounting arrangements will promote the transmission and amplification of operation noise and vibration through building structural members. Refer to Chaps. 49 and 50 for additional information on sound and vibration.

If sound criteria constitute a critical aspect in unit heater selection, complete details of the application requirement should be referred to the manufacturer of the unit heaters for evaluation and recommendation.

12.8 CONTROL OVER UNIT HEATER OPERATION

The control device common to all unit heaters is the thermostat. It cycles the flow of water or steam to unit heaters on some hydronic systems on which fans operate continuously. On other hydronic systems where steam or water flow is continuous, the thermostat activates unit heater fans intermittently when heating is required.

Oil, gas, and electric unit heating systems rely on a thermostat to activate the burners or the electric heating elements. On some units, once the heat exchangers are hot, a heat-sensing or time-delayed switch activates the fan. This may be the most desirable sequence since it prevents cold air from being initially discharged into the space to be heated.

When unit heater fans are operated only during the summer to provide air circulation, a manual summer/winter switch is required to prevent the heater from delivering unwanted heat.

Unit heaters lend themselves ideally to zone heating. Equipment and installation costs can be reduced by using a single thermostat to control the operation of several unit heaters.

The controls for a steam or hot-water unit heater can provide either *on/off operation* of the unit fan or *continuous operation* of the fan.

Continuous fan operation interrupts and lessens the vertical floor-to-ceiling air temperature stratification patterns which may occur.

One type of thermostat control used with unit heaters is designed to automatically return the warm air, which would normally rise to the ceiling, down to the occupied level. Two thermostats are required for this type of control arrangement.

The lower thermostat is placed in the zone of occupancy and is used to control the heating medium; it allows the unit heater to operate whenever the thermostat set point is not satisfied. The higher thermostat is placed near the unit heater at the ceiling or roof level, where the warm air tends to stratify. This thermostat allows the unit heater fan only to operate and deliver warm stratified air from the ceiling or roof level to the zone of occupancy, thus reducing heat stratification.

The fan will continue to run until the set-point temperature at the higher level is reached.

Numerous control options can be used to improve unit heater performance. Adding controls to a heating system will increase equipment and installation costs. However, the convenience, refinements in comfort, and energy savings often outweigh the extra costs. For example, when a pilot flame outage occurs, it is very inconvenient to manually relight the pilot flame on a gas unit heater mounted 20 ft (6 m) above the floor. In this instance an intermittent pilot ignition system is well worth the extra cost and can save the gas otherwise consumed by conventional standing pilot systems.

Another example: Low-voltage wiring need not normally be enclosed in conduit. This can be a real savings. But when line voltage is used, circuit wiring *must* be in conduit.

Because of the wide range of controls available, control selection can be confusing. Therefore, it is recommended that a unit heater or control manufacturer be consulted on this specialized subject. In addition, refer to Chap. 52.

12.9 LOCATING UNIT HEATERS

The following observations may be helpful:

- Use as few unit heaters as possible to give proper heat coverage of the area. The number of units selected will depend on the heat throw or heat spread of the individual heaters.
- More than any other single factor, improper mounting height is responsible for most unsatisfactory unit heater installations. When unit heaters are installed at heights higher than recommended, improper heat distribution results and comfort conditions are either difficult or impossible to maintain. But excessive air movement may cause discomfort when units are installed too low.
- Horizontal delivery unit heaters should be located so that the air streams of the individual units wipe the exposed walls of the building with either parallel or angular flow without blowing directly against the walls. Heaters should be spaced so that each supports the air stream from another heater. This sets up a circulatory air movement around the area to produce a blanket of warm air along the cold walls.
- It is advisable to locate unit heaters so that their air streams are subjected to a minimum of interference from columns, machinery, partitions, and other obstacles.
- Unit heaters installed in a building exposed to a prevailing wind should be located so as to direct a large portion of the heated air along the windward walls.
- Large expanses of glass, or large doors that are frequently opened, should be covered by long-throw unit heaters such as large horizontal delivery unit heaters or door heaters.
- In buildings having high ceilings, vertical delivery unit heaters equipped with the correct air distribution devices are recommended to produce comfort in central areas of the space to be heated. Horizontal delivery units are generally used to heat the peripheral areas of the same building.
- Horizontal delivery units should be arranged so they do not blow directly at

occupants. Their air streams should be directed down aisles, into open spaces on the floor, or along exterior walls of the building.

- When only vertical delivery units are used, they should be located so that exposed walls are blanketed by their warm air streams.
- To obtain the desired air distribution and heat diffusion, unit heaters are commonly equipped with directional outlets, adjustable louvers, or fixed or adjustable diffusers. For a given unit with a given discharge temperature and outlet velocity, the mounting height and heat coverage can vary widely with the type of directional outlet, adjustable louver, or diffuser.
- Several unit heaters may be operated by a single thermostat. In large, open spaces where similar activities are carried on, zonal heating will improve comfort and generally reduce fuel costs. Unit heaters may also be controlled individually, either manually or by a thermostat.
- For spot heating of individual spaces in larger unheated areas, single unit heaters may be used, but allowance must be made for unheated air from other spaces. The use of gas-fired infrared heaters is preferable for this situation. Other advantages of infrared heaters are that installation is easy, connection to the fuel supply is simple, installation costs are low, maintenance is inexpensive, and operation is noiseless. Infrared heating is fast and economical. Refer to Chap. 7 for additional information on infrared heaters.

Various spacing and arrangements of unit heaters for providing adequate heating coverage are demonstrated and described in Figs. 12.15 to 12.20. Installation costs will be simplified and reduced if unit heaters are properly placed in the space to be heated.

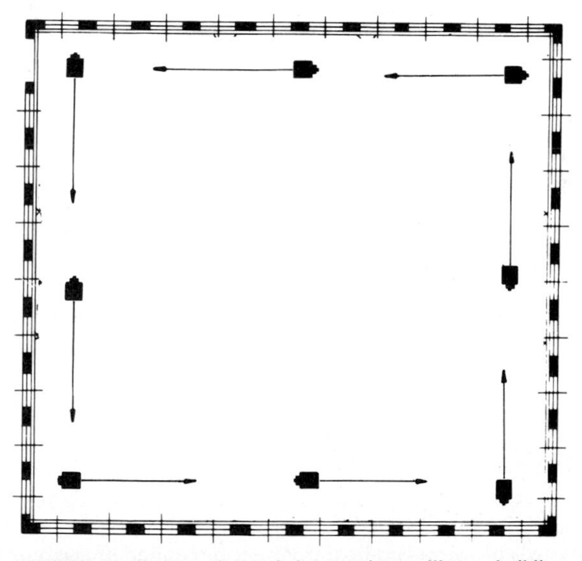

FIGURE 12.15 Locating unit heaters in a mill-type building. Here each horizontal delivery unit supports the air stream from another to produce circulatory air movement around the perimeter of the building where heat loss is greatest.

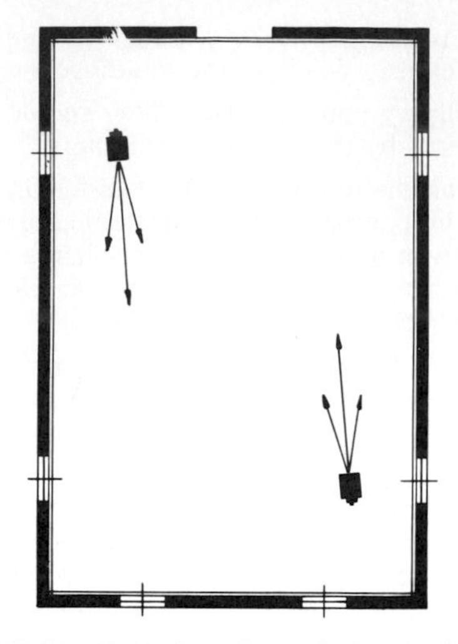

FIGURE 12.16 Locating unit heaters in a warehouse. Propeller unit heaters can provide maximum heat coverage of an area with a minimum number of units.

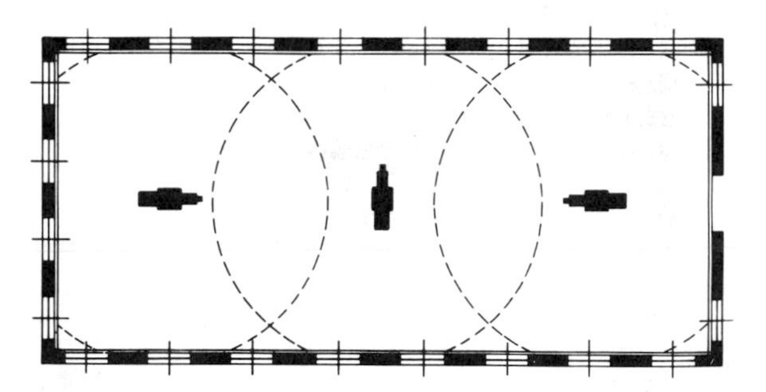

FIGURE 12.17 Locating unit heaters in a narrow building. An arrangement of vertical delivery unit heaters illustrates one of the advantages of a vertical (or downblow) air stream. Notice how heat spread from one unit overlaps the spread of another and how together they blanket cold outside walls with warm air to counter heat loss.

The following precautions should be taken:

- *Do not install gas- or oil-fired units in potentially explosive or flammable atmospheres laden with grain dust, sawdust, or similar airborne materials.* In such applications a blower-type heater is recommended in a separate room with ducting to the potentially explosive area.
- Consult piping, electrical, and venting instructions in the unit heater installation manual before installation.

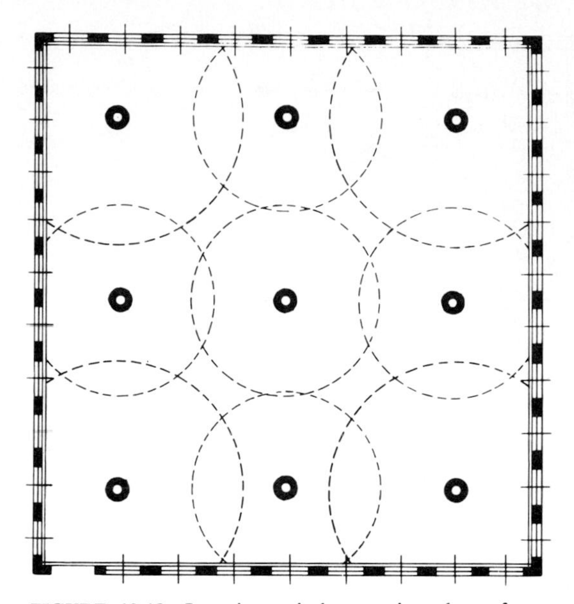

FIGURE 12.18 Locating unit heaters in a large factory area. Mounted high above machinery, assembly lines, and other obstructions on the floor, vertical delivery unit heaters beam heat directly downward to cover the entire floor area.

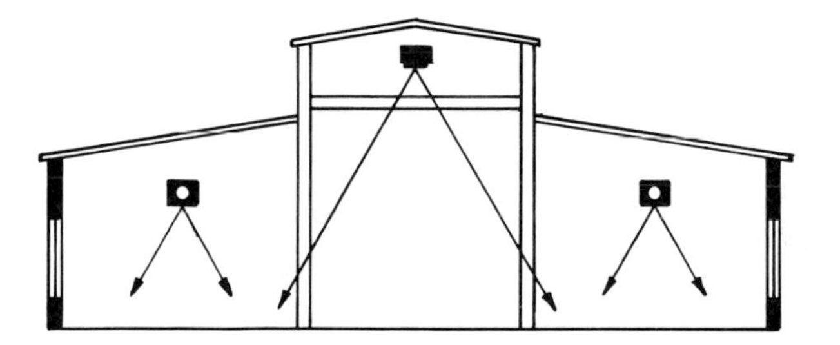

FIGURE 12.19 Locating unit heaters in a monitor-type building. Vertical delivery unit heaters installed in the high-ceiling central section of the building clear the craneway below them.

- *Do not locate any gas- or oil-fired unit heater in areas where chlorinated, halogenated, or acid vapors are present in the atmosphere.*
- Unit heaters in an occupied zone [less than 7 ft (2 m) above the floor level] *must* have fingerproof guards for all moving parts (fans, belts, sheaves, etc.). High-temperature surfaces, such as flue pipes, must be insulated or protected by safeguards to prevent body contact.
- Installation of gas-fired unit heaters in high-humidity or saltwater atmospheres may cause accelerated corrosion, reducing the normal lifespan of the units.

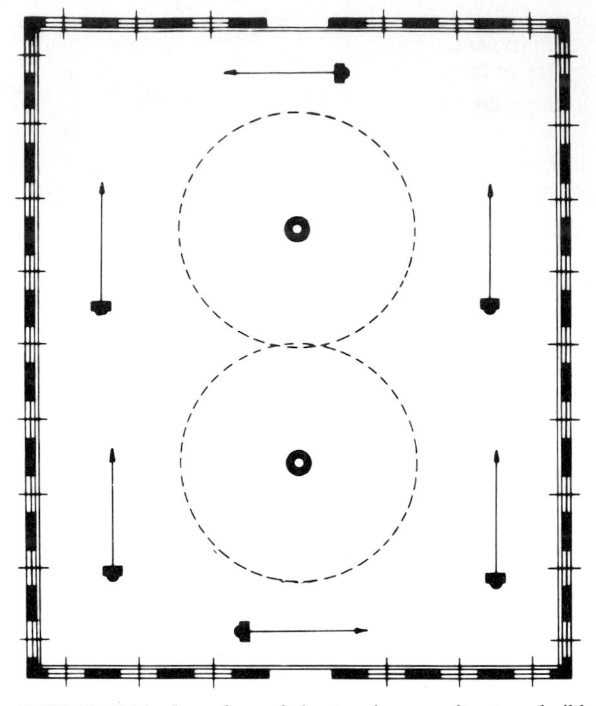

FIGURE 12.20 Locating unit heaters in a monitor-type building. In the same building, notice how horizontal delivery units in the two areas having low ceilings wipe outside walls with warm air.

12.10 SEVEN GOOD REASONS FOR REPLACING RATHER THAN REPAIRING UNIT HEATERS

1. Repairing a worn-out unit heater is costly. Labor and parts are not inexpensive. Repair the same unit twice, and chances are the price of a brand new unit heater will be exceeded, not including the added cost for takedown and reinstallation of the old unit heater each time it is repaired.
2. Repairs are unreliable. Repair a heat exchanger today, and perhaps next month, or on a cold day next winter, a motor or an overheat control on the same unit will need replacing.
3. It costs as much to reinstall a repaired unit heater as it does to install a new one. Labor costs can be a significant part of the bill. For a steam or hot-water unit, e.g., depending on the size and mounting height, it will take from 3 to 35 labor-hours to reinstall the unit. Add repair costs to installation costs, and the bill is substantial.
4. Replace an old, worn-out heater with a new unit heater, and many years of trouble-free service can be expected.
5. Replace an old "repairable" unit heater with a new unit, and get more usable

heat from fuel dollars. New units conserve fuel and cut heating bills because they are more efficient and throw heat farther—new motors are more efficient and use electricity frugally.

6. Replace old depreciated units with new, more efficient types to conserve energy.
7. Replace old units with new units of an alternate fuel type when production losses occur due to interruptible fuel supply.

12.11 REFERENCES

1. *1983 ASHRAE Handbook, Equipment*, American Society of Heating, Refrigerating and Air-Conditioning Engineers, Atlanta, GA, 1983, chap. 28, "Unit Ventilators, Unit Heaters and Makeup-Air Units," p. 28.5.
2. Ibid., p. 28.7.
3. AMCA Bulletin 303, Air Movement and Control Association, Arlington Heights, IL, June 1973, p. 1.

CHAPTER 13
HYDRONIC CABINET UNIT HEATERS

Robert A. Russell, Jr.
Chief Engineer, Ted Reed Thermal, Inc., West Kingston, Rhode Island

A cabinet heater is a specific type of unit heater which employs a centrifugal blower, a finned-coil heat exchanger, and an air filter. This package is housed in a "cabinet" which may be mounted on floors, walls, or ceilings. (See Fig. 13.1 which shows several types.) One of its chief benefits is that the cabinet heater can

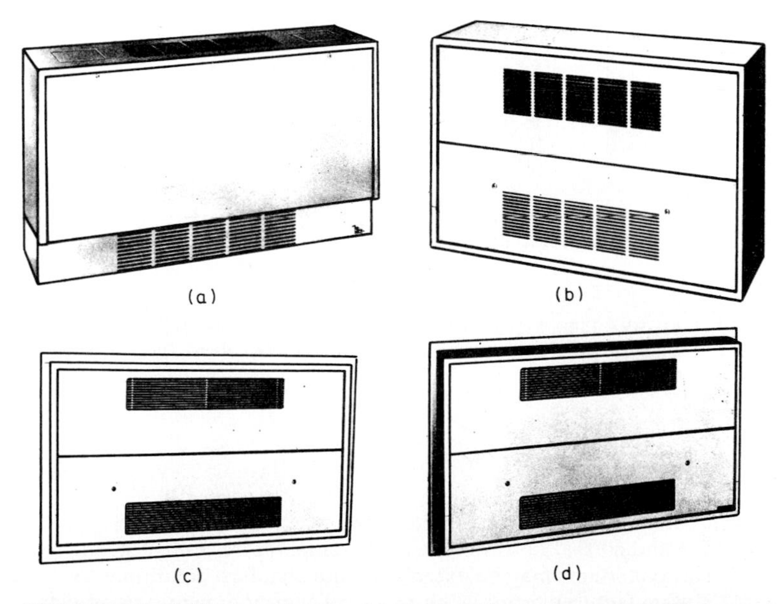

FIGURE 13.1 Cabinet unit heaters. (*a*) Exposed floor unit, model C; (*b*) exposed wall or ceiling unit, model CW; (*c*) fully recessed wall or ceiling unit, with aluminum bar grille, model CR; (*d*) partially recessed wall or ceiling unit, with aluminum bar grille, model CR. (*Courtesy of Ted Reed Thermal, Inc.*)

be used in building modernizations, where existing piping or wiring may be reused, to reduce construction costs.

The basic cabinet heater is primarily a heating-only appliance using 100 percent recirculated air. Modifications include cooling with chilled water (fan-coil units) and heating/cooling with 0 to 100 percent outside air (unit ventilators). Cabinet heaters may be exposed or recessed and can have a variety of air-flow arrangements through louvers, grilles, or ductwork.

13.1 CABINET UNIT HEATERS—HEATING ONLY

Figure 13.2 illustrates a floor-mounted cabinet heater with a pedestal mount. The steam/hot-water heating coil 1 consists of aluminum fins mechanically bonded to copper tubing. Air movement is via a centrifugal blower/fractional-horsepower motor package 2 mounted on a fan board 5 which serves as the barrier between the positive and negative pressure sections of the core. The air filter 3 can be disposable or a permanent, washable type. Air discharge is through the top louvers 4, and return air is through the pedestal 6. End compartments 7 provide space for piping connections and controls 8. Top-mounted doors 9 provide access to these compartments without removing the front panel.

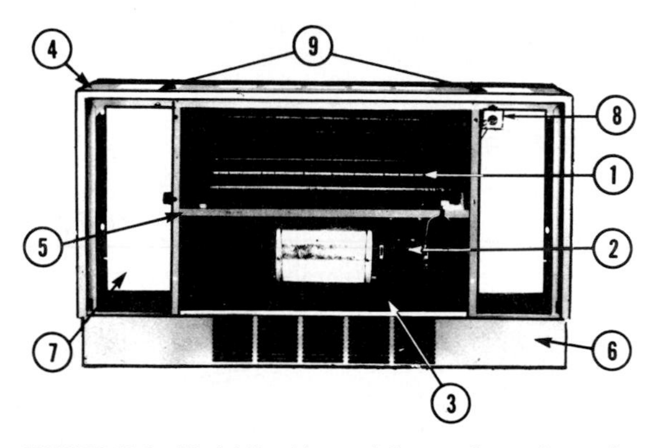

FIGURE 13.2 Model C cabinet unit heater. See main text for explanation of part numbers. *(Courtesy of Ted Reed Thermal, Inc.)*

13.1.1 Coil Types

Steam Coils. Steam coils are rated under standard conditions of 2 psig (13.8 kPa) steam pressure and 60°F (15.6°C) entering air temperature. Other steam pressure and entering air conditions may be extended from standard conditions as directed by the unit's manufacturer or by using classical thermodynamic principles. One method for testing air-heating coils is given in American Society of Heating, Refrigeration, and Air-Conditioning Engineers (ASHRAE) Standard 33-78.[1] Typical applications utilizing copper tubing are from 0 to 15 psig (0 to 103 kPa).

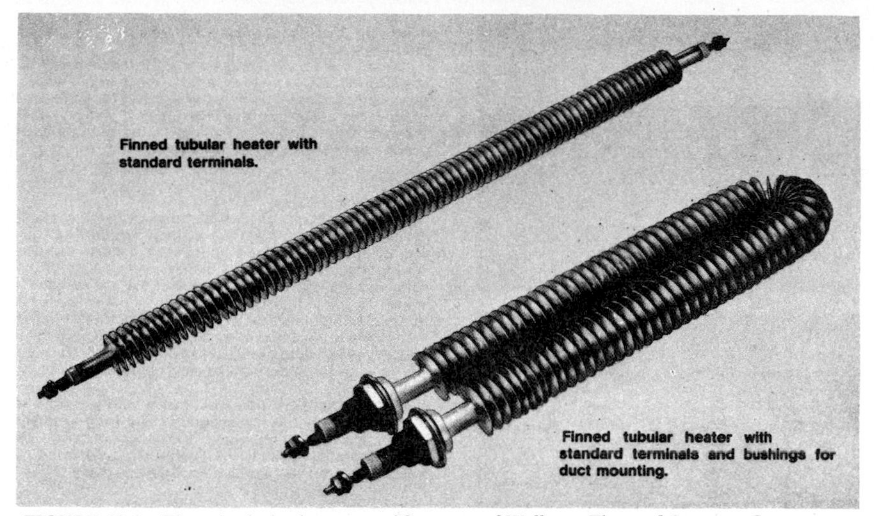

FIGURE 13.3 Finned tubular heaters. (*Courtesy of Wellman Thermal Systems Corporation*).

Hot-Water Coils. Standard ratings of hot-water cabinet heater coils are at the following conditions: 200°F (93.3°C) entering water temperature, 20°F (11.1°C) water temperature drop, and 60°F (15.6°C) entering air temperature. Capacities at various water flow rates and at entering air and water temperatures are extended from these conditions. Typical hot-water applications range from 160 to 240°F (71.1 to 115.6°C).

A common practice is to utilize a half-serpentine, single-row coil for both steam and hot-water use. This configuration has the benefit of minimal pressure drop across the coil in hot-water applications. Higher capacities require multiple-row, single-tube serpentine coils. This circuitry not only extracts the maximum amount of heat from the water, but also provides increased water velocity to ensure turbulent flow.

Electric Coils. Electric heating coil ratings are based on the energy input to the coil in kilowatts (1 W = 3.413 Btu/h). Capacities range from fractions of kilowatts to tens of kilowatts depending on the size and number of elements used. In the United States, single-phase voltages of 120, 208, 240, and 277 V are common for residential use, while commercial applications may require three-phase 208, 240, or 480 V.

As the primary heating mode, sheathed tubular elements may be inserted into finned copper tubes, much like a hydronic heating coil. This configuration provides greater heat-transfer surface. As an intermediate system, either straight or U-bent finned elements (Fig. 13.3) can be used in addition to hydronic coils.

13.2 FAN-COIL UNITS—HEATING AND COOLING

By varying the selection of coils, piping, and controls and with the addition of a condensate removal system, the cabinet heater may serve the dual purpose of heating and cooling. The control plant includes a water chiller, boiler or other heat source, circulating pumps, and temperature controls.

The cooling capacities of fan-coil units are rated at 80°F (26.7°C) dry-bulb (DB) temperature, 67°F (19.4°C) wet-bulb (WB) temperature, 45°F (7.2°C) entering water temperature (EWT), and a 10°F (5.6°C) water temperature rise. One method for testing and rating is given in American Refrigeration Institute's (ARI) Standard 440.[2]

Fan-coil units may use single or multiple coils and are adaptable to two- or four-pipe systems. Each combination has varying effects on initial costs, operational costs, and the amount of control provided.

13.2.1 Two-Pipe Single-Coil System

This system requires the selection of hot water or chilled water at the control plant. Advantages are lower installation costs and the ability to use lower hot-water temperatures. This is due to the oversized water coils that are required for the smaller temperature difference between chilled water and room air temperatures. Disadvantages include frequent changeovers during intermediate seasons, loss of control during mild weather when cooling is required during the day and heat in the morning and evening, and additional controls required to segregate chilled water from the boiler and hot water from the chiller.

13.2.2 Two-Pipe Double-Coil System

A common method of handling the intermediate seasons is to have an auxiliary electric heating coil which will provide for a small heating load while chilled water is available for cooling demand. Another variation is a dedicated chilled-water coil and a total electric heating coil.

13.2.3 Four-Pipe, Single- or Multiple-Coil System

This system eliminates the changeover difficulties of the two-pipe system by providing a continuous supply of both chilled and hot water. The coil configuration can be either two separate, dedicated coils or a split coil where two or three rows are used for cooling and a single row is used for heating.

Variations on this theme allow for cooling via chilled water or direct expansion and heating via hot-water, steam, or electric heating.

A typical control system for a chilled-water/hot-water four-pipe system would consist of a three-way motorized valve on the supply side of the coil and separate two-way motorized valves on the return side. A low-voltage heating/cooling thermostat with a fan control subbase would be used. On a call for heat, the three-way motorized valve opens to the hot-water supply, the hot-water return valve opens, the chilled-water return valve closes, and the blower is energized through a relay. As the room temperature increases to satisfy the thermostat, the hot-water return valve closes. With increasing room temperature, the three-way motorized supply valve moves from its normal position (open for hot water) to the cooling position (closed for hot water), the hot-water return valve is closed, and the chilled-water return valve opens. Blower operation can be either automatic or manual.

13.3 UNIT VENTILATORS—HEATING, COOLING, AND VENTILATING

The introduction of outdoor air through outside air dampers enables a fan-coil unit to heat, ventilate, or cool a room. Unit ventilators, sometimes called *classroom air conditioners*, are used in heavily occupied areas where a large amount of ventilation air is required. Damper operation can be either manual or motorized, with 0 to 25 or 0 to 100 percent of the unit's rated capacity introduced as outside air. Coils are sized to heat or cool the room loss plus ventilation air requirement.

The standard air rating of a unit ventilator is based on the delivery in cubic feet per minute (liters per second), converted to standard air at 70°F (21.1°C).

The three basic control cycles commonly used are cycle I, cycle II, and cycle III which are briefly defined as follows:

Cycle I. Except during warmup, 100 percent outdoor is admitted at all times.

Cycle II. A minimum amount of outdoor air (normally 25 to 50 percent) is admitted during the heating and ventilating stage. The percentage is gradually increased to 100 percent, if needed, during the ventilation cooling stage.

Cycle III. Except during the warmup stage, a variable amount of outdoor air is admitted as needed to maintain a fixed temperature of the air entering the heating element. This is controlled by the air-stream thermostat, which is set low enough, often 55°F (13°C), to provide ventilation cooling when needed.

13.4 SELECTION

Selection of a cabinet heater to match the calculated room design loads requires the following information:

Heating: Room heat load
Heating medium to be used and its availability
Entering air temperature
Required air circulation

Cooling: Indoor wet- and dry-bulb conditions

Ventilation: Outdoor wet- and dry-bulb conditions
Percentage of outside air needed for ventilation

In addition to the above input information, a manufacturer's ratings and specifications sheet is used (Table 13.1). The rating sheets provide unit capacities under the standard conditions listed previously in this chapter. Coil descriptions, blower configurations, motor information, and physical data are also provided. Note that sizes are based on nominal value in ft^3/min (L/s), say 200, 400, etc.

TABLE 13.1 Ratings and Specifications

UNIT SIZE ▶	2	3	4	6	8	10	12	14
① Heating Capacity — Hot Water								
Total MBH								
Standard Coil	12.4	22.1	28.1	44.3	48.8	55.7	72.3	79.2
High Capacity Coil	22.4	30.4	44.2	59.7	71.9	84.6	105.6	118.3
② Heating Capacity — Steam								
Total MBH — Standard Coil	19.8	28.3	42.5	52.1	67.7	76.0	95.4	102.6
Cond. Lb/Hr.	21.0	29.0	44.0	54.0	70.1	78.7	98.8	106.2
③ Heating Range — Hot Water								
MBH — Standard Coil	8-13	12-23	16-31	28-46	36-48	43-55	54-72	61-79
MBH — High Capacity Coil	16-24	21-31	28-44	36-60	54-72	65-84	81-105	91-118
④ Heating Range — Steam								
MBH — Standard Coil @ 2 PSI, 60° EAT	14-20	18-29	29-43	42-52	52-67	60-76	75-95	81-102
Coil								
Rows/FPI								
Standard Coil	1/11	1/11	1/11	1/11	1/11	1/11	1/11	1/11
High Capacity Coil	2/14	2/14	2/14	2/14	2/9	2/9	2/9	2/9
Face Area — Ft²								
Standard Coil	1.0	1.3	1.6	2.3	3.4	3.4	4.6	4.6
High Capacity Coil	1.1	1.5	1.8	2.7	3.6	3.6	4.8	4.8
Coil Conns.								
Standard Coil	¾" MPT	¾" MPT	¾" MPT	¾" MPT	1" MPT	1" MPT	1" MPT	1" MPT
High Capacity Coil	⅝" IDS	⅝" IDS	⅝" IDS	⅝" IDS				
Blowers								
No./Dia./Width In.	1/5¼/7	1/5½/7	2/5¼/7	2/5½/7	3/5½/7	3/5½/7	4/5½/7	4/5½/7
Drive	Direct	Direct	Direct	Direct	Direct	Direct	Direct	Direct
Speed RPM								
High	1090	1050	1090	1050	1090	1050	1090	1050
Medium	850	820	850	820	850	820	850	820
Low	650	625	650	625	650	625	650	625
CFM								
High	250	330	450	620	840	1050	1240	1430
Medium	195	265	350	485	655	820	970	1120
Low	150	195	270	370	545	685	805	930
⑤ Motor HP	1/30	1/30	1/15	1/15	1/10	1/10	1/8	1/8
Volts/Phase/Hertz	115/1/60	115/1/60	115/1/60	115/1/60	115/1/60	115/1/60	115/1/60	115/1/60
FL Amps, Standard Shaded Pole Motors	1.4	1.4	1.9	2.1	3.3	3.5	3.8	4.2
FL Amps, option #140	.5	.6	1.3	1.3	1.8	1.9	2.6	2.6
FL Amps, option #141	3.2	3.2	3.2	3.2	6.4	6.4	6.4	6.4
FL Amps, option #141A	4.4	4.4	4.4	4.4	8.8	8.8	8.8	8.8
Filters								
Type Supplied	Permanent	Permanent	Permanent	Permanent	Permanent	Permanent	Permanent	Permanent
Width/Length/Thickness	8½/21/½	8½/26/½	8½/31/½	8½/44/½	11/50/½	11/50/½	11/62/½	11/62/½
Length/Height/Depth Model C, CW, CR	39/25/9¾	44/25/9¾	49/25/9¾	62/25/9¾	72/28/12	72/28/12	84/28/12	84/28/12
Weights								
Shipping Wt. — Lbs.								
Model C	80	90	110	120	160	165	185	190
Model CW	90	100	120	130	170	175	195	200
Model CR	92	103	124	135	175	180	202	207

① Rating with 200° Entering Water Temperature, 60° EAT, and 20° TD at high fan speed.
② Rating with 2 lbs. steam and 60° EAT at high fan speed.
③ Range with 200° EWT at low to high fan speed.
④ Range at low to high fan speed.
⑤ Sizes 2-6 have 1 motor. Sizes 8-14 have 2 motors.
Horse power and amperage ratings represent both motors for sizes 8-14.

Notes: MBH = 100 Btu/h.
See App. B for conversion factors.

Source: Ted Reed Thermal, Inc. Used with permission.

13.4.1 Steam Selection

Steam capacities for cabinet unit heaters are based on standard conditions of 2 psig (13.8-kPa) steam and a 60°F (15.6°C) entering air temperature (EAT). Correction factors for other than standard conditions are listed in Table 13.2. Steam capacities are derived as follows:

$$\text{Unit capacity} = \text{capacity [at 2 psig (13.8 kPa) and 60°F (15.6°C) EAT]} \times \text{correction factor}$$

To simplify selection at nonstandard conditions, actual required capacities should be converted to the equivalent capacity at 2 psig (13.8 kPa) and 60°F (15.6°C) EAT. Selection is made from Table 13.3.

Determine the equivalent capacity at 2 psig (13.8 kPa) and 60°F (15.6°C) EAT as follows:

$$\text{Equivalent capacity} = \frac{\text{actual required capacity}}{\text{correction factor}}$$

The final air temperature should be greater than 100°F (37.8°C) to maintain a feeling of warmth. The final air temperature (FAT) at other than standard conditions is derived as follows:

$$\text{Final air temperature} = \frac{\text{actual capacity}}{1.085\ \text{(air flow)}} + \text{EAT}$$

where the air flow is in cubic feet per minute (liters per second).

EXAMPLE

Heating load, 29,000 Btu/h (8494 kW)
Entering air temperature, 60°F (15.6°C)
Entering steam pressure, 5 psig (34.5 kPa)

Solution From Table 13.2 the correction factor is 1.057.

$$\text{Equivalent capacity} = \frac{29{,}000}{1.057} = 27{,}500\ \text{Btu/h (8054.8 kW)}$$

Table 13.3 shows that selection of model 3 with 28,300 Btu/h is sufficient for application requirements.

$$\text{Actual capacity} = 28{,}300 \times 1.057 = 29{,}900\ \text{Btu/h (8757.7 kW)}$$

$$\text{Condensate rate} = \frac{\text{actual capacity}}{\text{latent heat of steam}}$$

$$= \frac{29{,}900\ \text{Btu/h}}{961\ \text{lb/h}} = 31\ \text{Btu/lb}$$

$$\text{Final air temperature} = \frac{29{,}900\ \text{Btu/h}}{1.085\ (330\ \text{ft}^3\text{/min})} + 60°\text{F} = 144°\text{F}$$

13.4.2 Hot-Water Selection

Hot-water heating capacities for cabinet unit heaters are based on standard conditions of 200°F (93.3°C) entering water temperature (EWT), 60°F (15.6°C) entering air temperature (EAT), and a 20°F (11.1°C) water temperature drop (WTD). Correction factors for other than standard conditions are listed in Table 13.4.

TABLE 13.2 Conversion Factors for Steam

Steam pressure, lb/in^2	Steam temperature, °F	Latent heat of steam, Btu/lb	Entering air temperature, °F										
			0	10	20	30	40	50	60	70	80	90	100
0	212.0	970	1.450	1.368	1.274	1.190	1.110	1.030	0.970	0.880	0.808	0.734	0.659
2	218.5	966	1.494	1.408	1.318	1.232	1.153	1.074	1.000	0.924	0.852	0.780	0.708
5	227.2	961	1.550	1.464	1.376	1.288	1.211	1.130	1.057	0.980	0.907	0.838	0.767
10	239.4	953	1.627	1.542	1.456	1.370	1.289	1.210	1.136	1.057	0.986	0.917	0.848
15	249.7	946	1.690	1.608	1.523	1.437	1.356	1.278	1.200	1.122	1.050	0.979	0.908

Note: See App. B for metric equivalents.
Source: Ted Reed Thermal, Inc. Used with permission.

TABLE 13.3 Steam Selection, Standard Coil, 2 psig, 60°F EAT

Unit size	ft^3/min	r/min	MBtu/h	EDR*	FAT, °F	Condensate, lb/h
2	250	1050	19.8	82	133	21
	150	625	14.6	61	156	15
3	330	1050	28.3	118	139	29
	195	625	18.0	75	145	19
4	450	1050	42.5	177	147	44
	270	625	29.3	122	160	30
6	620	1050	52.1	217	137	54
	370	625	42.7	178	166	44
8	840	1050	67.7	282	134	70
	545	625	52.9	220	149	55
10	1050	1050	76.0	317	126	79
	685	625	60.6	253	141	63
12	1240	1050	95.4	398	130	99
	805	625	75.1	313	146	78
14	1430	1050	102.6	428	126	106
	930	625	81.7	340	141	85

*Equivalent direct radiation.
Note: For each size, the top row indicates high fan speed and the bottom row indicates low fan speed. See App. B for metric equivalents.
Source: Ted Reed Thermal, Inc. Used with permission.

Hot-water capacities are derived as follows:

Unit capacity = capacity at 200°F (93.3°C) EWT and
60°F (15.6°C) EAT × correction factor

To simplify selection at nonstandard conditions, actual required capacities should be converted to the equivalent capacity at standard conditions. Then the model required can be quickly approximated from the capacity ranges in the rating and specification table (Table 13.1). Final selection is made from capacity tables, Tables 13.5 and 13.6. Capacities at 180°F (82.2°C) EWT and 60°F (15.6°C) EAT may be read directly from Tables 13.7 and 13.8.

EXAMPLE
Heating load, 27,000 Btu/h (7908.3 kW)
EAT, 60°F (15.6°C)
EWT, 210°F (98.9°C)
Water flow, 2 gal/min (0.126 L/s)

Solution From Table 13.4 the correction factor is 1.071. Thus

$$\text{Equivalent capacity} = \frac{27{,}000}{1.071} = 25{,}200 \text{ Btu/h (7381.1 kW)}$$

Table 13.5 shows selection of model 4 with 25,800 Btu/h (7556.8 kW) is sufficient for application requirements.

$$\text{Actual capacity} = 25{,}800 \times 1.071 = 27{,}600 \text{ Btu/h (8084 kW)}$$

TABLE 13.4 Hot-Water Correction Factors for Conditions Other than 200°F EWT and 60°F EAT

Entering water temperatures, °F	Entering air temperatures, °F										
	0	10	20	30	40	50	60	70	80	90	100
160	1.229	1.139	1.049	0.962	0.880	0.795	0.715	0.634	0.568	0.484	0.410
170	1.307	1.212	1.124	1.036	0.954	0.869	0.785	0.704	0.628	0.552	0.478
180	1.383	1.290	1.199	1.110	1.024	0.940	0.859	0.774	0.698	0.622	0.546
190	1.460	1.362	1.272	1.182	1.100	1.011	0.929	0.845	0.768	0.690	0.615
200	1.537	1.440	1.349	1.259	1.171	1.085	1.000	0.917	0.838	0.760	0.684
210	1.613	1.515	1.424	1.331	1.249	1.158	1.071	0.988	0.908	0.829	0.753
220	1.690	1.591	1.500	1.408	1.318	1.230	1.141	1.058	0.978	0.898	0.820
230	1.767	1.669	1.572	1.482	1.391	1.301	1.215	1.129	1.048	0.967	0.889
240	1.844	1.745	1.648	1.554	1.468	1.374	1.285	1.200	1.118	1.036	0.957

Note: See App. B for metric conversions.
Source: Ted Reed Thermal, Inc. Used with permission.

TABLE 13.5 Heating Capacities for Standard Coil, 200°F EWT, 60°F EAT

		Water pressure drop,	High fan speed				Low fan speed			
Unit size	gal/min	ft	ft³/min	MBtu/h	WTD, °F	FAT, °F	ft³/min	MBtu/h	WTD,* °F	FAT,† °F
	0.50	0.1		10.7	44.0	99		7.8	31.9	111
	1.00	0.1		12.0	24.7	104		9.1	18.6	119
2	1.28	0.1	250	12.4	20.0	105	150	9.5	15.3	122
	1.50	0.2		12.6	17.3	106		9.7	13.3	123
	2.00	0.2		12.9	13.3	107		10.2	10.5	127
	1.00	0.1		19.9	36.8	110		12.2	25.1	117
	1.50	0.2		20.1	27.6	116		13.6	18.7	124
3	2.23	0.3	330	22.1	20.0	121	195	14.9	13.8	130
	2.50	0.4		22.8	18.8	123		15.2	12.5	131
	3.00	0.5		23.0	15.7	124		15.6	10.7	133
	1.00	0.1		20.1	41.3	101		16.1	33.1	115
	2.00	0.3		25.8	26.6	113		19.2	19.7	125
4	2.87	0.5	450	28.1	20.0	117	270	20.5	14.7	130
	4.00	0.7		29.9	15.4	121		21.5	11.1	133
	5.00	0.9		31.0	12.8	123		22.1	9.1	135
	2.00	0.3		36.9	36.9	115		28.2	28.2	130
	3.00	0.6		41.0	27.3	121		30.4	20.3	136
6	4.43	0.9	620	44.3	20.0	126	370	31.8	15.9	139
	5.00	1.1		45.1	18.0	127		32.6	13.0	141
	6.00	1.4		46.3	15.4	129		33.2	11.0	143
	2.00	0.1		41.7	42.6	105		32.5	33.2	115
	4.00	0.2		47.5	24.2	112		35.5	18.1	120
8	5.00	0.4	840	48.8	20.0	113	545	36.2	14.8	121
	6.00	0.5		49.7	16.9	114		36.7	12.5	122
	8.00	0.9		51.3	13.1	116		37.2	9.5	122
	3.00	0.1		50.8	34.6	104		39.8	2.71	113
	5.00	0.4		54.9	22.5	108		42.7	17.5	117
10	5.10	0.5	1050	55.7	20.0	108	685	43.3	15.5	118
	7.00	0.7		57.1	16.7	110		44.5	13.0	119
	9.00	1.1		58.2	13.2	111		45.3	10.3	121
	4.00	0.2		68.1	34.8	110		51.4	26.3	118
	6.00	0.6		71.0	24.2	112		53.7	18.3	121
12	7.40	0.8	1240	72.3	20.0	113	805	54.7	15.1	122
	10.00	1.4		74.0	15.1	115		55.4	11.3	123
	12.00	1.8		74.9	12.8	115		55.9	9.5	124
	4.00	0.2		73.7	37.7	107		57.0	29.1	116
	6.00	0.6		77.2	26.3	109		59.4	20.2	118
14	8.10	1.0	1430	79.2	20.0	111	930	61.2	15.5	120
	10.00	1.4		80.5	16.5	111		62.0	12.7	121
	12.00	1.8		81.3	13.9	112		62.4	10.6	121

*Water temperature drop.
†Final air temperature
Note: See App. B for metric conversions.
Source: Ted Reed Thermal, Inc. Used with permission.

TABLE 13.6 Heating Capacities for High-Capacity Coil, 200°F EWT, 60°F EAT

Unit size	gal/min	Water pressure drop, ft	High fan speed				Low fan speed			
			ft³/min	MBtu/h	WTD, °F	FAT, °F	ft³/min	MBtu/h	WTD,* °F	FAT,† °F
2	1.00	0.7	250	20.1	41.3	134	140	13.5	27.8	148
	1.50	1.1		21.8	29.9	140		14.2	19.5	153
	2.00	1.7		22.7	23.4	144		14.5	15.0	155
	2.40	2.8		23.2	20.0	146		14.7	12.7	156
	3.00	4.3		23.8	16.4	148		14.9	10.3	158
3	2.00	1.4	325	31.4	32.1	149	195	21.2	21.7	160
	2.50	2.1		32.4	26.5	151		21.9	17.9	163
	3.00	2.8		33.0	22.5	154		22.3	15.2	165
	3.60	3.7		33.4	20.0	155		22.6	13.4	167
	4.00	4.6		33.8	17.3	156		22.9	11.7	168
4	2.00	1.5	440	41.9	42.9	148	270	28.7	29.4	158
	2.50	2.3		43.2	35.4	151		29.6	24.3	161
	3.00	3.0		44.1	30.1	152		30.2	20.6	163
	4.00	5.0		45.2	23.1	155		31.0	15.9	166
	4.70	8.8		45.8	20.0	156		31.4	12.7	168
6	2.00	1.4	615	49.8	49.8	135	370	35.9	35.9	149
	3.00	2.8		54.3	36.2	141		38.1	25.4	155
	4.00	4.8		56.8	28.4	145		39.3	19.6	158
	5.00	7.2		58.5	23.4	148		40.1	16.0	160
	6.00	10.1		59.7	20.0	150		40.6	13.5	161
8	5.00	1.0	835	67.0	27.4	134	540	50.9	20.8	147
	7.00	1.7		71.4	20.9	138		53.9	15.7	152
	7.40	1.9		71.9	20.0	139		54.4	15.0	153
	10.00	3.2		74.9	15.3	143		56.5	11.6	157
	12.00	4.3		76.5	13.0	144		57.6	9.8	158
10	6.00	1.4	1040	79.8	27.2	131	680	62.2	21.2	144
	8.00	2.2		83.6	21.4	134		65.0	16.6	148
	8.70	2.5		84.6	20.0	135		65.8	15.5	149
	12.00	4.3		88.0	15.0	138		68.1	11.6	152
	14.00	5.8		89.1	13.0	139		68.9	10.1	153
12	7.00	1.7	1220	99.8	29.2	135	790	77.2	22.6	150
	9.00	2.6		103.5	23.5	138		79.9	18.2	153
	10.80	3.6		105.6	20.0	140		81.5	15.4	155
	13.00	5.0		107.6	16.9	141		82.9	13.0	157
	15.00	6.5		109.1	14.9	142		83.9	11.4	158
14	9.00	2.6	1410	114.2	25.9	135	915	87.8	20.0	148
	11.00	3.7		117.1	21.8	137		89.9	16.7	151
	12.10	4.4		118.3	20.0	137		91.1	15.4	152
	15.00	6.5		120.2	16.4	139		92.4	12.6	153
	17.00	8.2		121.8	14.7	140		93.3	11.2	154

*Water temperature drop.
†Final air temperature
Note: See App. B for metric conversions.
Source: Ted Reed Thermal, Inc. Used with permission.

TABLE 13.7 Heating Capacities for Standard Coil, 180°F EWT, 60°F EAT

Unit size	gal/min	Water pressure drop, ft	High fan speed				Low fan speed			
			ft³/min	MBtu/h	WTD, °F	FAT, °F	ft³/min	MBtu/h	WTD,* °F	FAT,† °F
2	0.50	0.1	250	7.9	31.7	89	145	6.3	25.5	102
	1.00	0.1		10.4	20.8	98		7.8	15.7	112
	1.50	0.2		11.7	15.6	103		8.5	11.4	116
	2.00	0.2		12.5	12.5	106		9.0	9.5	119
3	1.00	0.1	330	14.4	28.8	100	195	12.3	24.6	118
	1.50	0.2		16.5	22.0	106		13.8	18.4	125
	2.00	0.3		17.9	17.9	110		14.7	14.7	129
	2.50	0.4		18.8	15.0	112		15.3	12.2	132
	3.00	0.5		19.5	13.0	114		15.7	10.5	134
4	1.00	0.1	450	17.9	35.9	97	270	14.5	28.9	109
	2.00	0.3		22.7	22.7	106		17.4	17.4	119
	3.00	0.6		25.1	16.7	111		18.7	12.4	124
	4.00	0.7		26.5	13.2	114		19.5	9.7	126
	5.00	0.9		27.5	11.0	116		20.0	8.0	128
6	2.00	0.3	620	31.2	31.2	106	370	23.9	23.9	119
	3.00	0.6		34.7	23.2	111		25.9	17.2	124
	4.00	0.8		36.9	18.4	115		27.0	13.5	127
	5.00	1.1		38.3	15.0	117		27.8	11.1	129
	6.00	1.4		39.4	13.1	118		28.3	9.4	131
8	2.00	0.1	840	35.0	42.6	98	545	27.3	27.9	106
	4.00	0.2		39.9	20.4	103		29.8	15.3	110
	5.00	0.4		41.0	16.8	105		30.4	12.4	111
	6.00	0.5		41.8	10.2	105		30.8	10.5	112
	8.00	0.9		43.1	11.0	107		31.3	8.0	112
10	3.00	0.1	1050	42.7	29.1	97	685	33.5	22.8	105
	5.00	0.4		46.1	18.9	100		35.9	14.7	108
	5.70	0.5		46.8	16.8	101		36.4	13.1	109
	7.00	0.7		48.0	14.0	102		37.4	10.9	110
	9.00	1.1		48.9	11.1	102		38.1	8.7	111
12	4.00	0.2	1240	57.2	29.3	102	805	43.2	22.1	109
	6.00	0.6		59.7	20.3	104		45.1	15.4	111
	7.40	0.8		60.8	16.9	105		46.0	12.7	112
	10.00	1.4		62.2	12.7	106		46.6	9.5	113
	12.00	1.8		62.9	10.7	106		47.0	8.0	113
14	4.00	0.2	1430	61.9	31.7	99	930	47.5	24.5	107
	6.00	0.6		64.9	22.1	101		49.5	17.0	109
	8.10	1.0		66.6	16.8	102		51.0	13.1	111
	10.00	1.4		67.7	13.8	103		51.6	10.7	111
	12.00	1.8		68.3	11.6	104		52.0	8.9	112

*Water temperature drop.
†Final air temperature
Note: See App. B for metric conversions.
Source: Ted Reed Thermal, Inc. Used with permission.

TABLE 13.8 Heating Capacities for High-Capacity Coil, 180°F EWT, 60°F EAT

Unit size	gal/min	Water pressure drop, ft	High fan speed				Low fan speed			
			ft³/min	MBtu/h	WTD, °F	FAT, °F	ft³/min	MBtu/h	WTD,* °F	FAT,† °F
	1.00	0.7		17.1	35.0	123		11.5	23.6	135
	1.50	1.1		18.6	25.4	128		12.1	16.6	139
2	2.00	1.7	250	19.4	20.0	132	140	12.4	12.8	141
	2.50	2.9		20.0	16.4	133		12.6	10.4	143
	3.00	4.3		20.3	13.9	135		12.8	8.8	144
	2.00	1.4		26.0	26.6	134		17.6	18.4	143
	2.50	2.1		26.8	22.0	134		18.1	15.1	145
3	3.00	2.8	325	27.4	18.7	137	195	18.5	13.1	147
	3.50	3.6		27.8	16.2	138		18.8	11.5	148
	4.00	4.6		28.1	14.4	139		19.0	10.0	150
	2.00	1.5		34.6	36.6	132		24.1	25.0	142
	2.50	2.3		35.7	29.4	134		24.6	20.7	145
4	3.00	3.0	440	36.5	24.9	136	270	25.4	17.5	147
	4.00	5.0		37.4	19.2	138		25.7	13.4	149
	4.70	8.8		37.9	16.6	139		26.0	9.9	151
	2.00	1.4		42.4	42.4	124		30.7	30.7	136
	3.00	2.8		46.4	30.9	129		32.6	21.7	141
6	4.00	4.8	615	48.6	24.3	133	370	33.6	16.8	144
	5.00	7.2		50.0	20.0	135		34.3	13.7	145
	6.00	10.1		51.0	17.0	136		34.7	11.5	147
	5.00	1.0		56.3	23.0	122		42.8	17.5	133
	7.00	1.7		60.0	17.5	126		45.3	13.2	137
8	7.40	1.9	835	60.9	16.7	127	540	45.7	12.6	138
	10.00	3.2		62.9	12.9	129		47.5	9.7	141
	12.00	4.3		64.3	11.0	131		48.4	8.2	142
	6.00	1.4		67.1	22.9	119		52.3	17.8	130
	8.00	2.2		70.3	18.0	122		54.6	14.0	134
10	8.70	2.5	1040	71.1	16.7	123	680	55.3	13.0	135
	12.00	4.3		74.0	12.6	125		57.2	9.8	138
	14.00	5.8		74.9	10.9	126		57.9	8.5	138
	7.00	1.7		83.9	24.5	123		60.7	17.7	131
	9.00	2.6		87.0	19.8	126		67.1	15.2	138
12	10.80	3.6	1220	88.7	16.8	127	790	68.5	13.0	140
	13.00	5.0		90.4	14.2	128		69.7	11.0	141
	15.00	6.5		91.7	12.5	129		70.5	9.6	142
	9.00	2.6		96.0	21.8	123		73.8	16.8	134
	11.00	3.7		98.4	18.3	124		75.6	14.0	136
14	12.10	4.4	1410	99.4	16.8	125	915	75.9	12.9	136
	15.00	6.5		101.0	13.8	126		76.9	10.6	137
	17.00	8.2		102.4	12.3	127		77.7	9.4	138

*Water temperature drop.
†Final air temperature
Note: See App. B for metric conversions.
Source: Ted Reed Thermal, Inc. Used with permission.

$$\text{FAT} = \frac{27{,}600}{450 \times 1.08} + 60°\text{F} = 117°\text{F}$$

13.4.3 Electric Heating Coils

Electric heating coils are chosen by matching the room heat loss with the equivalent Btu/hour (watt) rating of the coil. Table 13.9 provides capacities for auxiliary electric heating coils.

TABLE 13.9 Heating Capacities of Electrical-Resistance Coil

Unit size	Rating, kW	Equivalent rating, MBtu/h	Max. unit amperage @ 115/60/1
2	1.0	3.4	10
3	1.5	5.1	15
4	2.0	6.8	20
6	2.5	8.5	25

Note: See App. B for metric conversions.
Source: Ted Reed Thermal, Inc. Used with permission.

13.4.4 Cooling Coil Selection

EXAMPLE Here 100 percent recirculated air is used. The application requirements for cooling are:

Total cooling load, 6700 Btu/h (1962.4 kW)
Sensible cooling load, 5400 Btu/h (1581.7 kW)
Indoor design conditions, 80°F (26.7°C) dry bulb and 67°F (19.4°C) wet bulb
Entering water temperature, 45°F (7.2°C)
Ventilation provided by infiltration

13.4.5 Cooling Selection

With 100 percent recirculated air, coil entering-air conditions will be essentially 80°F (26.7°C) DB and 67°F (19.4°C) WB. Referring to the cooling capacity in Table 13.10, choose size 3 and by interpolation find that 1.05 gal/min will provide the required 5400 Btu/h (1581.7 kW) sensible capacity at 80°F (26.7°C) DB and 67°F (19.4°C) WB conditions. Total capacity is 5500 Btu/h (1610.9 kW), which is inadequate. To provide 6700 Btu/h (1962.4 kW) total capacity requires 1.25 gal/min (0.079 L/s) which provides 6100 Btu/h (1786.7 kW) sensible capacity.

13.4.6 Heating, Cooling, and Ventilation

EXAMPLE Here we use 25 percent outside air and 75 percent return air. The application requirements for cooling are:

Total cooling load, 23,750 Btu/h (6956.4 kW) (room only)
Sensible cooling load, 16,620 Btu/h (4868 kW) (room only)
Indoor design conditions, 80°F (26.7°C) DB, 67°F (19.4°C) WB

TABLE 13.10 Cooling Capacities for Two-Row Coil, 45°F EWT, High Fan Speed, Btu/h

Unit size	gal/min	Pressure drop, std. coil,* ft	Entering air temperature, °F											
			75 DB, 63 WB			78 DB, 65 WB			80 DB, 67 WB			84 DB, 69 WB		
			Sens. heat	Total heat	H_2O temp. rise	Sens. heat	Total heat	H_2O temp. rise	Sens. heat	Total heat	H_2O temp. rise	Sens. heat	Total heat	H_2O temp. rise
2	0.5	0.80	2.62	2.62	10.5	2.89	2.89	11.5	3.07	3.07	12.3	3.43	3.43	13.7
	1.0	1.55	3.17	3.17	6.3	3.49	3.49	7.0	3.71	3.71	7.4	4.15	4.15	8.3
	1.5	3.20	3.52	3.52	4.7	3.88	3.88	5.2	5.07	5.89	7.8	5.66	6.58	8.8
	2.0	5.50	4.40	5.01	5.0	4.84	5.69	5.7	5.12	6.39	6.4	5.71	7.15	7.1
	2.5	8.30	4.42	5.35	4.3	4.87	6.08	4.9	5.15	6.84	5.4	5.75	7.64	6.1
3	1.0	1.70	4.46	4.46	8.9	4.92	4.92	9.8	5.23	5.23	10.5	5.84	5.84	11.7
	1.5	3.60	4.98	4.98	6.6	6.53	7.27	9.6	6.91	8.17	10.9	7.72	9.14	12.2
	2.0	6.02	5.98	6.98	7.0	6.58	7.93	7.9	6.97	8.91	8.9	7.78	9.96	10.0
	2.5	9.30	6.02	7.47	6.0	6.62	8.49	6.8	7.01	9.55	7.6	7.83	10.67	8.5
	3.0	13.00	6.04	7.91	5.3	6.65	8.99	6.0	7.04	10.11	6.7	7.86	11.29	7.5
4	1.5	4.00	7.25	8.04	10.7	7.87	9.07	12.1	8.12	10.11	13.5	9.08	11.20	14.9
	2.0	6.80	7.64	9.02	9.0	8.31	10.19	10.2	8.61	11.39	11.4	9.61	12.63	12.6
	2.5	10.00	7.93	9.73	7.8	8.64	11.00	8.8	8.97	12.30	9.8	10.01	13.66	10.9
	3.0	14.40	8.16	10.26	6.8	8.89	11.61	7.7	9.25	13.00	8.7	10.32	14.44	9.6
	4.0	24.50	8.47	11.01	5.5	9.25	12.46	6.2	9.65	13.97	7.0	10.76	15.54	7.8
6	2.0	1.84	9.50	10.99	11.0	10.43	12.32	12.3	10.75	13.68	13.7	12.05	15.10	15.1
	3.0	3.82	10.38	12.98	8.7	11.31	14.58	9.7	11.72	16.24	10.8	13.11	17.96	12.0
	3.5	5.10	10.67	13.68	7.8	11.63	15.38	8.8	12.08	17.14	9.8	13.50	18.97	10.8
	4.0	6.52	10.91	14.26	7.1	11.90	16.04	8.0	12.37	17.89	8.9	13.82	19.81	9.9
	5.0	9.85	11.28	15.15	6.1	12.31	17.06	6.8	12.83	19.04	7.6	14.33	21.10	8.4
8	4.0	1.00	16.03	18.36	9.2	16.80	20.80	10.4	17.30	23.25	11.6	18.60	25.50	12.7
	5.0	1.50	16.69	19.98	8.0	17.60	23.00	9.2	18.10	25.60	10.2	19.50	28.00	11.2
	6.0	2.20	17.21	21.22	7.1	18.20	24.60	8.2	18.80	27.40	9.1	20.50	30.00	10.0
	8.0	4.00	17.96	23.02	5.8	19.10	26.60	6.7	19.50	29.50	7.4	21.50	32.80	8.2

Note: See App. B for metric conversions.
Source: Ted Reed Thermal, Inc. Used with permission.

Outdoor design conditions, 95°F (35°C) DB, 75°F (23.9°C) WB
Entering water temperature, 40°F (1.7°C)
Ventilation air required, 200 ft^3/min (94.4 L/s)
Outside air damper set at 25 percent outside air (oa) maximum, ft^3/min (L/s)

13.4.7 Cooling Selection

Since 25 percent outside air (oa) is specified, a size 8 fan coil [nominal 800 ft^3/min (377.5 L/s)] will be required to meet the 200 ft^3/min (94.4 L/s) ventilation requirement. Sensible heat (SH) gain from the outside air [abbreviated in what follows as SH (oa)] must be added to the room sensible heat load as follows:

$$\text{SH (oa)} = \text{ft}^3\text{/min (oa)} \times 1.08 \times (\text{DB°F outside} - \text{DB°F inside})$$

$$\text{SH (oa)} = 200\ \text{ft}^3\text{/min (oa)} \times 1.08 \times (95\text{°F DB outside} - 80\text{°F DB room})$$

$$\text{SH (oa)} = 3240\ \text{Btu/h outside air}$$

The total heat (TH) of the outside air is derived by

$$\text{TH (oa)} = \text{ft}^3\text{/min} \times 4.5\ (\text{Btu/lb outside} - \text{Btu/lb inside})\dagger$$

$$= 200\ \text{ft}^3\text{/min} \times 4.5 \times 7.0\ \text{Btu/lb oa (total heat of outside air} - \text{total heat of room air)}$$

$$= 6300\ \text{Btu/h}$$

$$\text{Unit sensible cooling load} = 16{,}620 + 3240 = 19{,}860\ \text{Btu/h}$$

$$\text{Unit total cooling load} = 23{,}750 + 6300 = 30{,}050\ \text{Btu/h}\ddagger$$

The conditions of the air mixture entering the coil are determined from a psychrometric chart to be 84°F (28.8°C) DB and 69°F (20.6°C) WB.

Referring to the cooling capacity table, Table 13.11, shows that the required sensible capacity at 84°F (28.8°C) DB and 69°F (20.6°C) WB will be obtained with 4.0 gal/min (0.25 L/s) [20,500 Btu/h (6004.4 kW)]. The total capacity is normally sufficient [30,000 Btu/h (8787.0 kW)].

13.5 APPLICATIONS

Due to the many configurations available, selecting a cabinet heater involves more than matching load requirements with capacities. Location is of prime importance in planning a project. Figure 13.4 shows some of the standard air-flow arrangements available. Some of these installation requirements also affect performance, e.g., mounting height and throw versus output.

Figure 13.5 indicates the performance of a single cabinet unit heater operating at high fan speed in a uniformly heated space at the single mounting height shown. Strong opposing drafts, large obstructions in the air stream of the unit,

†The Btu/lb value is determined from a psychrometric chart.
‡For metric equivalents, use conversion factors given in App. B.

TABLE 13.11 Cooling Capacities for Two-Row Coil, 40°F EWT, High Fan Speed, MBtu/h

Unit size	gal/min	Pressure drop, std. coil, ft	Entering air temperature, °F											
			75 DB, 63 WB			78 DB, 65 WB			80 DB, 67 WB			84 DB, 69 WB		
			Sens. heat	Total heat	H_2O temp. rise	Sens. heat	Total heat	H_2O temp. rise	Sens. heat	Total heat	H_2O temp. rise	Sens. heat	Total heat	H_2O temp. rise
2	0.5	0.80	3.02	3.02	12.1	3.28	3.28	13.1	3.46	3.46	13.8	3.82	3.82	15.3
	1.0	1.55	3.65	3.65	7.3	3.97	3.97	7.9	4.19	4.19	8.4	4.62	4.62	9.2
	1.5	3.20	4.05	4.05	5.4	5.43	6.16	8.2	5.70	6.81	9.0	6.29	7.49	9.9
	2.0	5.50	5.06	6.02	6.0	5.49	6.69	6.7	5.77	7.39	7.4	6.36	8.13	8.1
	2.5	8.30	5.10	6.43	5.1	5.53	7.14	5.7	5.81	7.89	6.3	6.41	8.68	6.9
3	1.0	1.70	5.14	5.14	10.3	5.60	5.60	11.2	5.90	5.90	11.8	6.51	6.51	13.0
	1.5	3.60	6.82	7.70	10.3	7.41	8.56	11.4	7.79	9.45	12.6	8.59	10.41	13.9
	2.0	6.02	6.89	8.38	8.4	7.49	9.32	9.3	7.87	10.29	10.3	8.68	11.33	11.3
	2.5	9.30	6.94	8.97	7.2	7.53	9.98	8.0	7.92	11.02	8.8	8.73	12.13	9.7
	3.0	13.00	6.97	9.50	6.3	7.57	10.56	7.0	7.96	11.66	7.8	8.78	12.83	8.5
4	1.5	4.00	8.08	10.09	13.5	8.69	11.13	14.8	8.93	12.19	16.3	9.87	13.30	17.7
	2.0	6.80	8.60	11.30	11.3	9.26	12.49	12.5	9.54	13.71	13.7	10.53	14.97	15.0
	2.5	10.00	8.97	12.17	9.7	9.67	13.46	10.8	9.99	14.79	11.8	11.02	16.17	12.9
	3.0	14.40	9.26	12.82	8.5	9.98	14.19	9.5	10.33	15.61	10.4	11.39	17.08	11.4
	4.0	24.50	9.67	13.73	6.9	10.44	15.21	7.6	10.83	16.74	8.4	11.93	18.34	9.2
6	2.0	1.84	10.67	13.68	13.7	11.49	15.03	15.0	11.79	16.41	16.4	13.07	17.86	17.9
	3.0	3.82	11.69	16.12	10.7	12.60	17.76	11.8	13.00	19.44	13.0	14.37	21.19	14.1
	3.5	5.10	12.06	16.98	9.7	13.01	18.72	10.7	13.44	20.51	11.7	14.84	22.37	12.8
	4.0	6.52	12.37	17.69	8.8	13.34	19.50	9.8	13.80	21.38	10.7	15.23	23.33	11.7
	5.0	9.85	12.85	18.77	7.5	13.87	20.71	8.3	14.37	22.73	9.1	15.86	24.82	9.9
8	4.0	1.00	17.94	22.98	11.5	18.30	25.40	12.7	19.00	27.70	13.8	20.50	30.00	15.0
	5.0	1.50	18.80	24.98	10.0	19.60	27.70	11.1	19.80	30.05	12.0	21.70	32.80	13.1
	6.0	2.20	19.47	26.52	8.8	20.40	29.30	9.8	21.00	32.10	10.6	22.70	35.05	11.7
	8.0	4.00	20.44	28.70	7.2	21.50	31.75	7.9	22.30	35.00	8.7	24.00	37.80	9.5

Note: See App. B for metric conversions.
Source: Ted Reed Thermal, Inc. Used with permission.

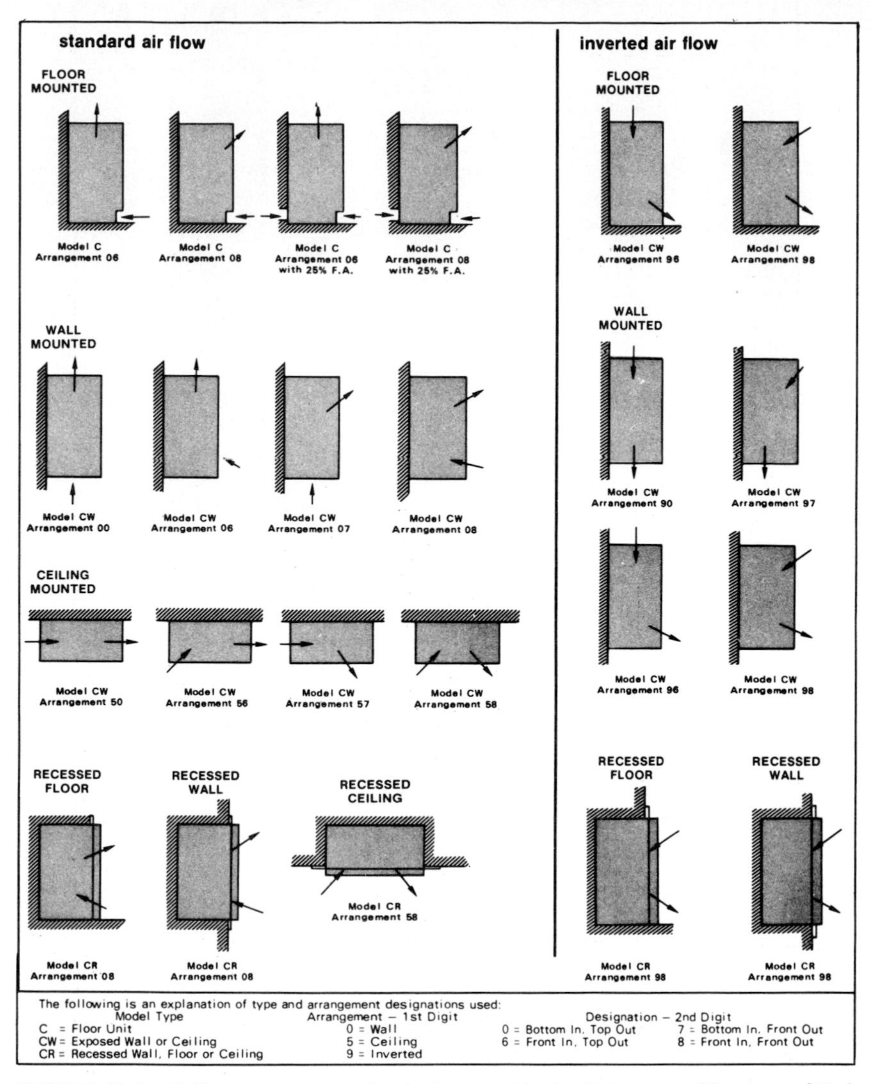

FIGURE 13.4 Air-flow arrangements for hydronic cabinet unit heaters. (*Courtesy of Ted Reed Thermal, Inc.*)

and higher than normal discharge air temperatures (resulting from high steam pressures) can prevent the heated air discharged by the cabinet unit from reaching the floor.

Under unfavorable conditions such as these, allowances must be made to ensure maintenance of desired comfort.

Correction Factor R *for Steam Units.* These correction factors are to be used as multipliers to correct the recommended mounting heights and throw of cabinet units installed under conditions other than 2 psi (0.14 bar) steam pressure and 60°F (15.6°C) entering air.

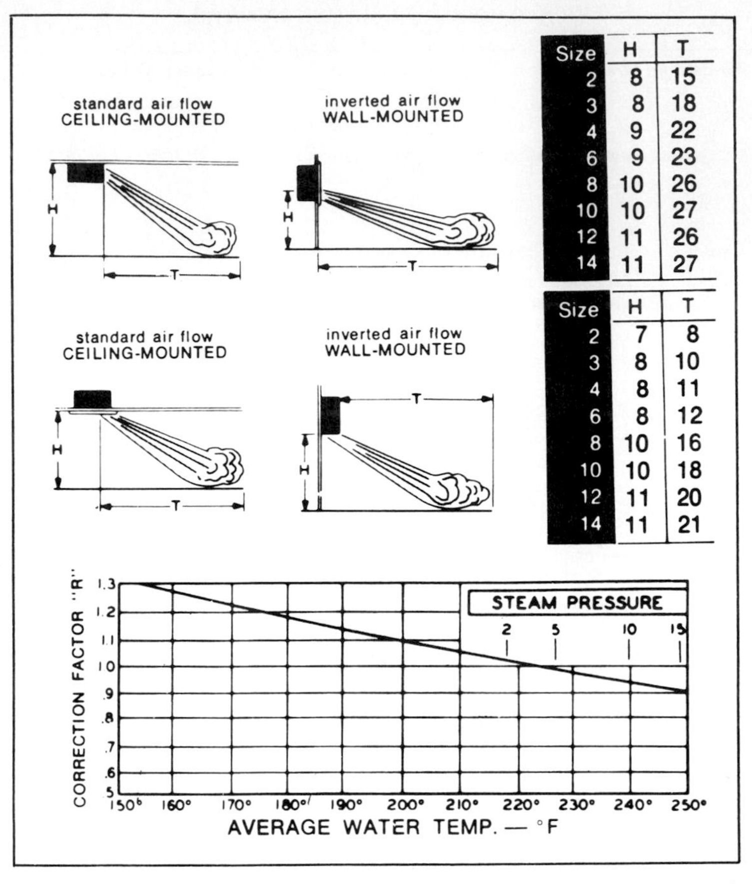

FIGURE 13.5 Mounting height and heat throw. (See App. B for metric conversions.) (*Courtesy of Ted Reed Thermal, Inc.*)

TABLE 13.12 External Static Pressure,* Standard Coil

	0.0 ESP*		0.05 ESP*		0.1 ESP*		0.125 ESP*	
Unit size	ft^3/min	Btu factor	ft^3/min	Btu factor	ft^3/min	Btu factor	ft^3/min	Btu factor
2	250	1.00	210	0.90	170	0.80	150	0.77
3	330	1.00	280	0.90	230	0.80	200	0.77
4	450	1.00	400	0.90	325	0.80	285	0.77
6	620	1.00	540	0.90	430	0.80	380	0.77
8	840	1.00	720	0.90	590	0.80	510	0.77
10	1050	1.00	905	0.90	735	0.80	640	0.77
12	1240	1.00	1065	0.90	870	0.80	755	0.77
14	1430	1.00	1230	0.90	1000	0.80	870	0.77

*For 0.1 ESP and for 0.125 ESP, specify high static motor.
Note: See App. B for metric conversions.
Source: Ted Reed Thermal, Inc. Used with permission.

Duct Applications. Recessed ceiling and wall units may be installed in a duct system. The external static pressure effect must be considered when the cabinet heater is sized. Table 13.12 provides conversion factors for hot-water applications. Information on high static motors is found on the ratings and specifications sheet, shown in Table 13.1 (option 141).

13.6 REFERENCES

1. *Method of Testing for Rated Forced Circulation Air Cooling and Heating Coils under Defrosting Conditions*, ASHRAE Standard 33-78, American Society of Heating, Refrigeration, and Air-Conditioning Engineers, Atlanta, GA, 1984.
2. *Room Fan Coil Air Conditioners*, ARI Standard 440, Air Conditioning and Refrigeration Institute, Arlington, VA.

CHAPTER 14
HEAT EXCHANGERS

David F. Fijas
ITT Standard, ITT Fluid Technology Corporation,
Buffalo, New York

14.1 INTRODUCTION

Today's heating and cooling systems contain a variety of heat exchangers for the heating and cooling of liquids, steam, and air. This chapter discusses the design aspects of various types of heat exchangers which might be found in the typical equipment room, including mechanical characteristics which might influence selection of various types and maintenance requirements which apply to the various designs.

The heat exchangers are broken down into three main categories of design:

- Shell-and-tube heat exchangers
- Plate-and-frame heat exchangers
- Air coils

These categories may contain several design features which are chosen for the particular application based on the expected service which the heat exchanger will be required to perform. Important parameters which influence the design of a heat exchanger include operating temperatures and pressures, fluid characteristics (clean, dirty, corrosive, etc.), thermal performance properties, duty cycle (continuous or repeated startups and shutdowns), and space limitations.

14.2 SHELL-AND-TUBE HEAT EXCHANGERS

Shell-and-tube heat exchangers (see Fig. 14.1) are by far the most common style found in equipment rooms. As the name implies, a bundle of smaller-diameter tubes is contained within an outer enclosure, the shell. The tubes are joined to tubesheets which prevent the shellside fluid from contacting the tubeside fluid. Heat transfer takes place through the tube wall.

The tubes are spaced apart and held in place by a series of plates (baffles). The baffles perform the additional function of directing the shellside fluid across the

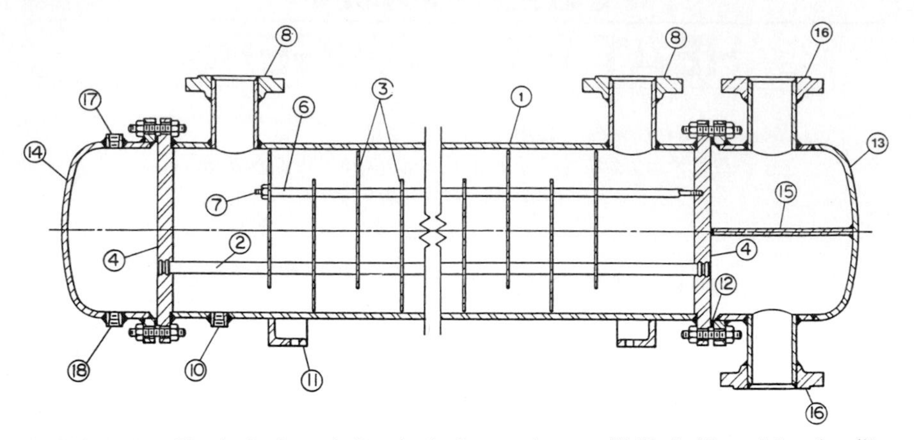

FIGURE 14.1 Fixed-tubesheet shell-and-tube heat exchanger. (1) Shell; (2) straight tube; (3) baffle; (4) fixed tubesheet; (5) floating tubesheet; (6) baffle spacers; (7) tie-rod; (8) shell nozzle; (9) shell vent; (10) shell drain; (11) support foot; (12) gasket; (13) in/out tubeside bonnet; (14) reversing tubeside bonnet; (15) pass rib; (16) tube nozzle; (17) tubeside vent; (18) tubeside drain; (19) tubeside channel; (20) channel cover; (21) U-tube; (22) packing; (23) lantern ring; (24) tapered channel; (25) floating head; (26) lifting ring; (27) removable shell end plate; (28) support plate; (29) clamp ring; (30) cover; (31) fins.

outside of the tube surfaces to achieve maximum heat transfer within allowable pressure losses.

Similarly, the tubeside fluid is directed through the inside of the tubes by appropriate design of headers and tube layouts. The use of pass partition plates in the headers can subdivide the tube layout. In this way the fluid will flow through only a portion of the tubes at one time, thereby making several passes through the length of the heat exchanger. Again, the number of passes is chosen to achieve a tube velocity appropriate to achieve the desired heat transfer within appropriate pressure-drop limits.

Construction of the shell-and-tube heat exchanger can be with either removable or nonremovable (fixed-tubesheet) tube bundles. Furthermore, removable bundles may be U-tube, packed floating head, packed floating tubesheet, or internal floating head designs.

14.3 NONREMOVABLE (FIXED-TUBESHEET) TUBE BUNDLES

Figure 14.1 illustrates a fixed-tubesheet heat exchanger. Note that each tubesheet on either end of the heat exchanger is joined to the heat exchanger shell by welding or brazing. This design tends to be the least expensive of the various designs due to its simplicity. Since more tubes may be obtained in a given shell diameter, a heat exchanger of smaller diameter than other types might be used. In most cases, the fixed tubesheet doubles as the shell flange to which the header is bolted. However, the flanges may be eliminated by welding the header to the tubesheet, as shown in Fig. 14.2.

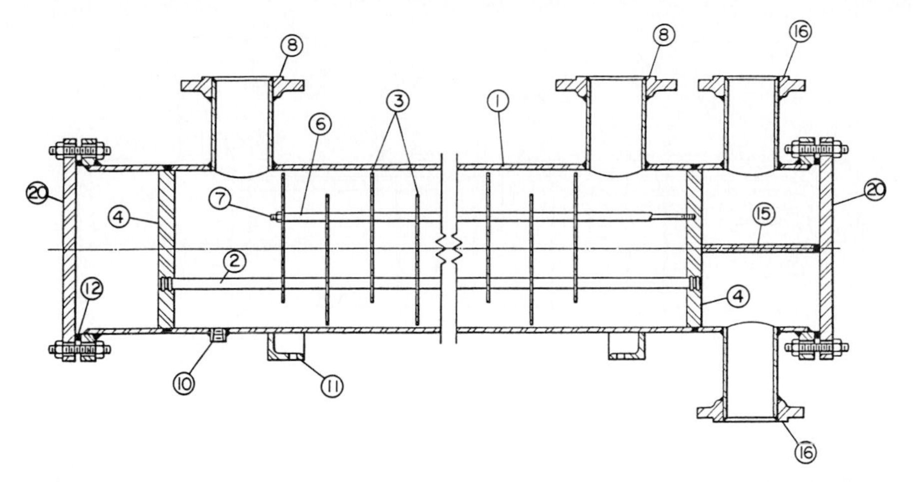

FIGURE 14.2 Fixed-tubesheet (nonremovable channel) shell-and-tube heat exchanger. See Fig. 14.1 for explanation of numbered parts.

Smaller sizes of fixed-tubesheet heat exchangers have been standardized by certain manufacturers. These units utilize hub construction in which the tubesheet, shell flange, and shell connection are included in one forged or cast piece. The hub is then brazed or welded to the shell to complete the shell assembly.

Since the tubes and shell are rigidly fixed relative to one another, this design does not provide for differential expansion between the shell and tubes. If the temperatures of these metals are significantly different during operation, the metals can expand at differing rates and thus highly stress the tube-to-tube sheet joint. These joints may, in fact, break loose, causing intermixing of the two fluids. Failure due to differential expansion may be avoided by using expansion joints in the shell. These function as an expansion bellows and flex as differential expansion occurs, thus preventing the tube joints from being overly stressed. Heat exchangers with expansion joints must be carefully handled since the joints are relatively thin and subject to failure if abused.

14.4 U-TUBE REMOVABLE TUBE BUNDLES

A U-tube bundle, illustrated in Fig. 14.3, utilizes tubes which are bent into a U shape. As such, the tube ends fit into opposite halves of a single tubesheet. This design reduces cost because only one tubesheet is used. The tubesheet is clamped between the shell and head flanges and is held stationary. The tubes are free to move independently of the shell, and thus the U-tube design permits differential thermal expansion to take place.

As with the fixed tubesheet, the header can be designed to provide a varying number of tubeside passes. However, the minimum is two passes, which means that a counterflow design is not possible.

Since the bundle can be removed from the shell, the shellside surface may be cleaned mechanically or, e.g., with a high-pressure water spray. U bends prevent

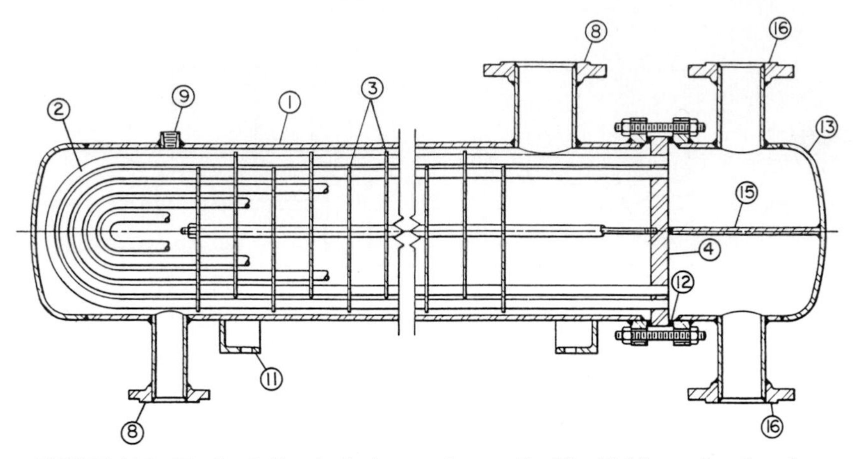

FIGURE 14.3 U-tube shell-and-tube heat exchanger. See Fig. 14.1 for explanation of numbered parts.

the tubes from being cleaned by rodding out, so chemical cleaning is usually used. Individual tubes cannot be removed (except in the outer row), so damaged tubes would not be replaced, but would be plugged instead.

Figure 14.3 shows a typical steam (in shell) to hot water (in tubes) heat exchanger. Note the steam inlet is greater in diameter than the condensate outlet. For water-to-water heat exchangers, the shell inlet and shell outlet have the same diameter.

14.5 PACKED FLOATING TUBESHEET REMOVABLE BUNDLES

Certain designs combine the straight-tube configuration of the fixed-tubesheet exchanger with the removable-bundle feature and differential expansion capability of the U-tube. These heat exchangers (see Fig. 14.4) contain one stationary tubesheet held between the shell and head flanges, while the second tubesheet is permitted to move (float). Sealing between the shell and tubeside fluids is accomplished by O rings or packing rings squeezed between the shell and head flanges and the tubesheet.

Some of these heat exchangers use double packings with a gland between. The gland is designed so that should one of the packings fail, the fluid no longer sealed will drip from between the flanges rather than mix with the other fluid. The bolting can be tightened to reseal the packing or the heat exchanger scheduled for repair without problems resulting from fluid contamination. The O ring or packing must be compatible with the fluids used in the heat exchanger. Often design pressures and temperatures are limited by this seal. In service, the high-temperature end of the heat exchanger should be at the stationary tubesheet end, so that the seal is not subjected to high temperatures with a resulting reduction in seal life.

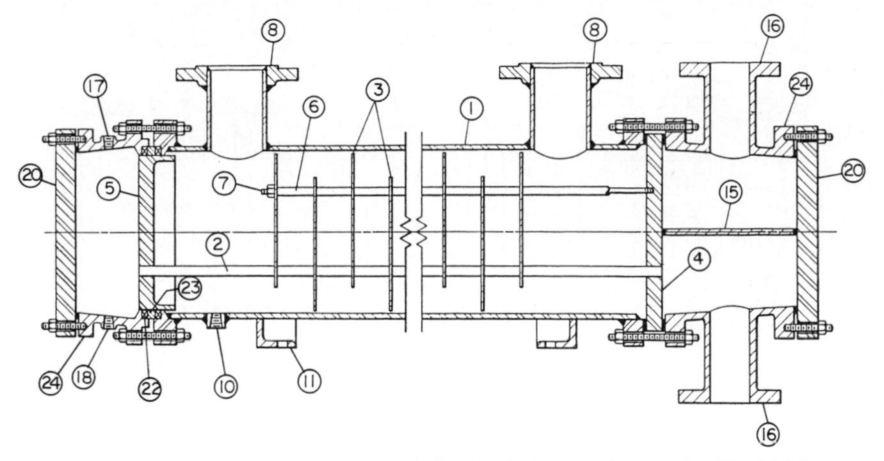

FIGURE 14.4 Packed floating tubesheet shell-and-tube heat exchanger. See Fig. 14.1 for explanation of numbered parts.

14.6 INTERNAL FLOATING HEAD REMOVABLE BUNDLES

Less often used, but important nonetheless, are heat exchangers in which the head and tubesheet float together inside the shell (Figs. 14.5 and 14.6). This allows the maximum thermal differential expansion between the tubes and shell, although the cost increases substantially. By using removable clamp rings (Fig. 14.6) the number of tubes in a given shell can be increased. However, the shell

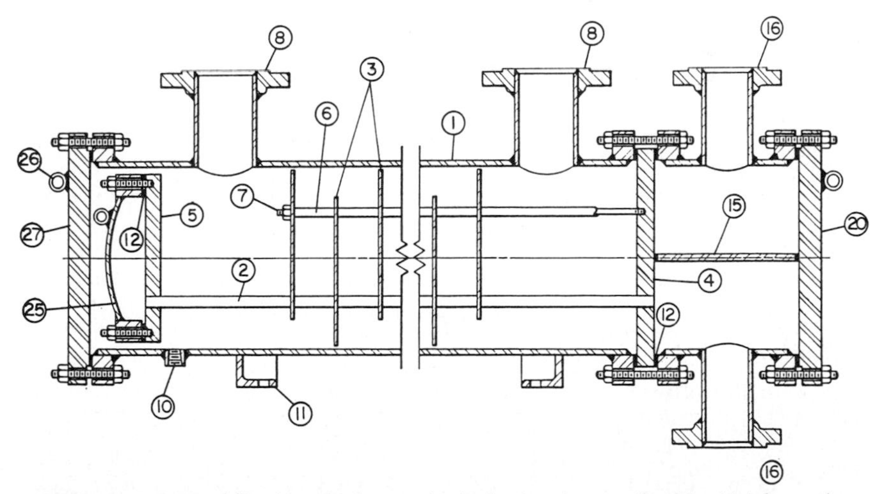

FIGURE 14.5 Integral floating tubesheet and tube heat exchanger. See Fig. 14.1 for explanation of numbered parts.

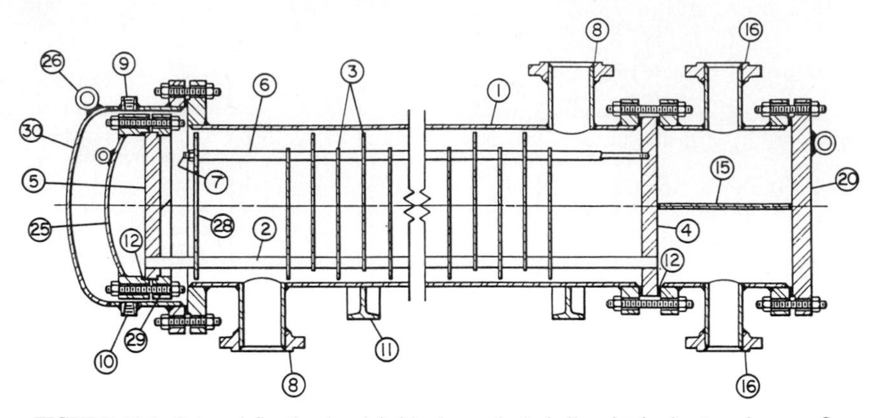

FIGURE 14.6 Integral floating head (with clamp ring) shell-and-tube heat exchanger. See Fig. 14.1 for explanation of numbered parts.

cover, clamp ring, and floating head must be removed before the bundle can be slid from the shell.

14.7 TUBES FOR SHELL-AND-TUBE DESIGN

Various types of tubes are found in shell-and-tube heat exchangers. Most are common plain tubes of ¼-in (6.35-mm) to ⅝-in (15.9-mm) outer diameter, although larger tubes are sometimes used. Wall thicknesses, or tube gauges, are usually a function of diameter and material. Typical values can be found in Table 14.1.

Certain fluids, such as oils or compressed air, do not have inherently good heat-transfer properties. Heat exchangers for these fluids often use tubes which are enhanced to improve heat transfer over plain tubes. When these fluids are on the shellside of a heat exchanger, finned tubes are often used. Figure 14.7 shows a typical integral fin tube which provides additional heat-transfer surface on the shellside, by machining fins in the outer wall of a heavy wall tube. The root diameter, or tube diameter, under the fin is approximately ⅛ in (3.18 mm) smaller than the ends of the tube joined to the tubesheet, so that if cleaning of the tubeside by mechanical means is desired, the rod must fit through the smaller in-

TABLE 14.1

Tube outer diameter		Tube gauge, BWG (wall thickness)					
		Copper or stainless steel			Steel		
in	mm	Gauge	in	mm	Gauge	in	mm
¼	6.35	27	0.016	0.41	24	0.022	0.56
⅜	9.53	23	0.025	0.64	22	0.028	0.71
⅝	15.90	18 or 20	0.049 or 0.035	1.24 or 0.89	16	0.065	1.65
¾	19.05	18 or 20	0.049 or 0.035	1.24 or 0.89	16	0.065	1.65

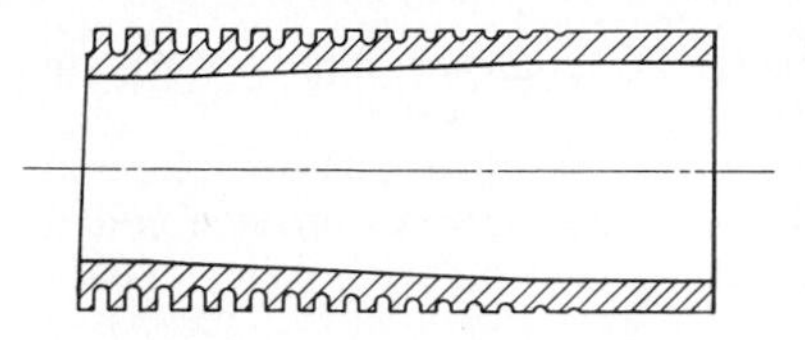

FIGURE 14.7 Integral low-fin tube.

ner diameter under the fin. A more recent configuration (Fig. 14.8) uses plate-type fins into which plain tubes are assembled. Due to the large amount of surface available, a very compact unit may be used.

If these poor heat-transfer fluids are used on the tubeside of an exchanger, e.g., compressed air in pipeline aftercoolers, the enhancement (Fig. 14.9) may be provided inside the tubes. Inserts may be mechanically or metallurgically bonded to the inside tube surface and can reduce the length of the heat exchanger significantly. However, mechanical cleaning is not possible due to the insert. Alternatively, designs are available with internal longitudinal fins to reduce tube length and still permit tube cleaning with a brush cleaner.

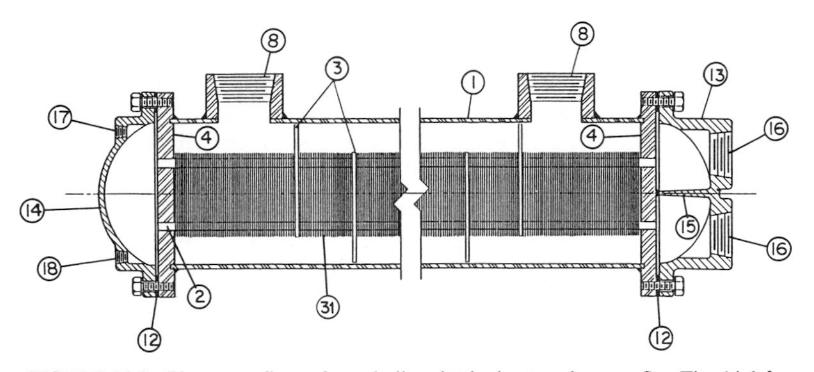

FIGURE 14.8 Plate-type fin surface shell-and-tube heat exchanger. See Fig. 14.1 for explanation of numbered parts.

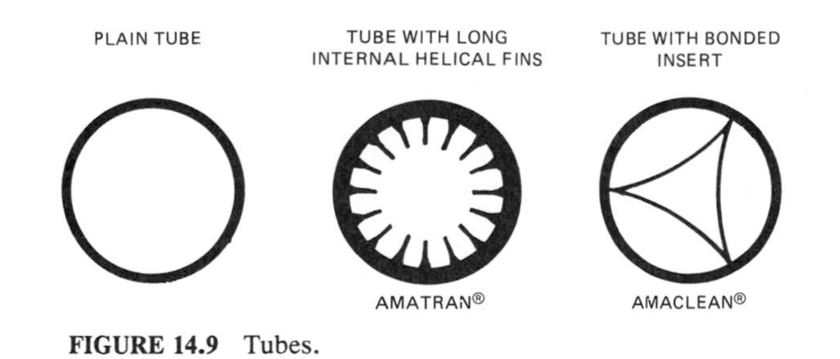

FIGURE 14.9 Tubes.

14.8 TUBE JOINTS

By far the most common method of joining tubes to tubesheets is by roller expansion. This method enlarges the diameter of the end of the tube so that residual stresses imposed on the tubesheet provide a pressure-tight mechanical joint. This type of joint is satisfactory in the great majority of cases. Should a tube joint of this type become loose for any reason, it is possible to repair the joint by rerolling the tube into the tubesheet.

For higher design temperatures and pressures, the tubesheet holes can be machined with one or more circumferential grooves. These grooves provide added strength when the tube wall is deformed to fit the grooves during roller expansion.

Where mixing of the fluids cannot be tolerated under any circumstances, seal welding of the rolled joints is possible. In certain small heat exchangers of brass or copper construction, the tube joints may be brazed. This permits tube layouts with more tightly spaced tubes to be used. However, should tube leaks occur at the brazed joints, repair is difficult.

14.9 HEADERS FOR SHELL-AND-TUBE DESIGN

Tubeside connections are made to headers of either bonnet or channel construction. Channels have a removable end plate, which permits access to the tube ends for inspection and repairs without disturbing the tubeside piping.

Bonnets, whether fabricated or cast, incorporate an end plate integral with the bonnet chamber and flanges. Tubeside piping must be disconnected so that the bonnet can be removed for inspection and repair of the tube ends.

When headers are bolted to the shell flanges, gaskets or O rings are used. These bolts should be tightened to the manufacturer's recommended torque, to ensure that the tubeside fluid does not leak. Flange thickness, bolting, and gasket combinations are designed to seal the tubeside fluid while keeping bolt and flange stresses within acceptable material limits. For O rings or full-face gaskets, high torques may be applied. However, care should be taken if the gasket is inside the bolt circle. Tightening puts a bending movement on the flanges. For some small-diameter thin flanges in cast iron, especially, cracking of the flanges can occur with overtightening.

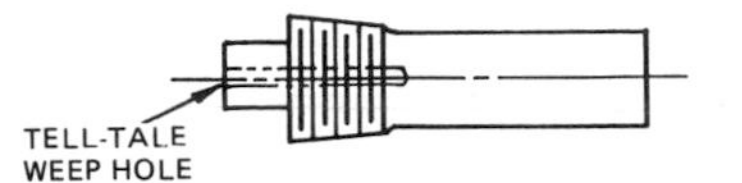

FIGURE 14.10 Sacrificial anodes.

Aggressive water, usually pumped on the tubeside of the heat exchangers, can cause corrosion problems. Heat-exchanger channels may include sacrificial anodes (Fig. 14.10) whose purpose is to react with the water so that the heat-exchanger metals do not. These usually have a telltale weep hole that results in a visible leak from the anode when the corrosion has reached a certain point. Then the anode should be removed and replaced.

14.10 PLATE-AND-FRAME HEAT EXCHANGERS

The plate-and-frame heat exchanger is used in liquid-to-liquid or steam-to-liquid applications. Its construction offers certain advantages and disadvantages compared to the shell-and-tube heat exchanger design. The plate-and-frame heat exchanger tends to be more compact in certain applications than a shell-and-tube design, can be disassembled and reassembled for cleaning and/or capacity changes, but is generally limited to lower design pressures and temperatures.

14.11 CONSTRUCTION

As the name implies, the plate-and-frame heat exchanger (Fig. 14.11) consists of heat-transfer plates carried and contained within a frame. The frame locates and supports the plates and includes two end plates between which the heat-transfer plates are compressed. The plates and connections are configured so that the hot and cold fluids flow on opposite sides of each heat-transfer plate in the slot formed by two adjacent plates.

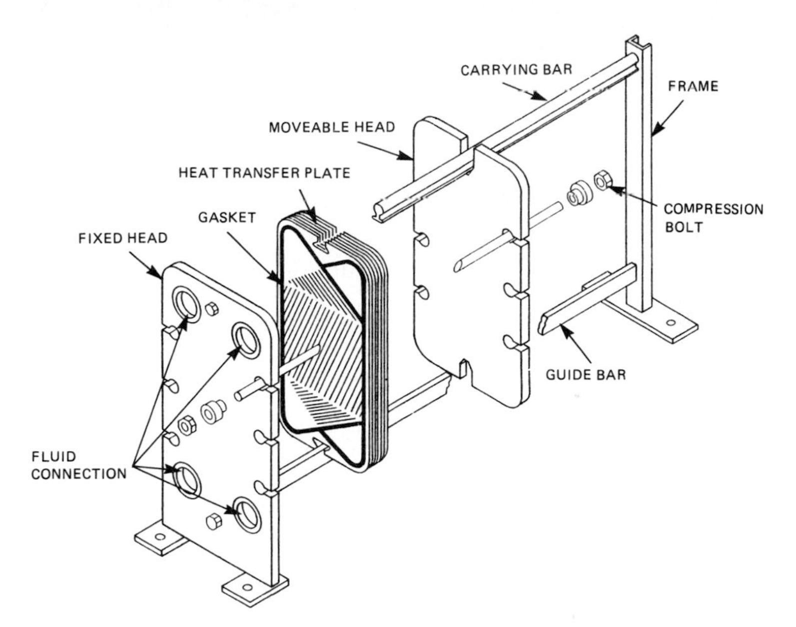

FIGURE 14.11 Plate-and-frame heat exchanger.

14.12 HEAT-TRANSFER PLATES

The heat-transfer plates (Fig. 14.12) are generally stamped to achieve a corrugated pattern of some kind. Typical designs use a chevron pattern or herringbone geometry for the corrugation, although different plate patterns may be found in specific models of various manufacturers. The corrugations serve two functions. First, the corrugations provide rigidity and strength to the plate so that relatively thin plate material may be used. When assembled, the corrugations provide a large number of contact points between plates. This further supports the fluid pressure even with thin plates. This is significant in minimizing resistance to heat transfer and reducing cost, a major plate-and-frame advantage when expensive alloy materials are used. Typical materials are

304 Stainless steel

316 Stainless steel

Titanium

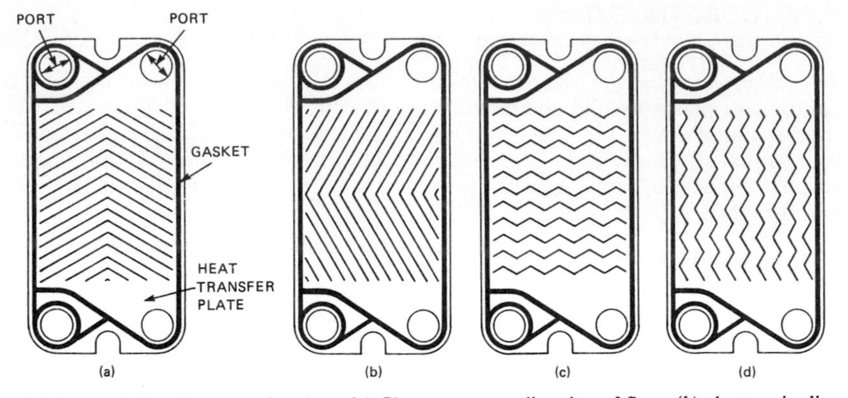

FIGURE 14.12 Heat-transfer plate. (*a*) Chevron across direction of flow; (*b*) chevron in direction of flow; (*c*) herringbone across direction of flow; (*d*) herringbone in direction of flow.

Hastelloy C
Inconel 825,600
Monel
Nickel 200
SMO 254

The second function of the corrugations is to induce turbulence in the fluid flowing across the plate. The turbulence generates relatively higher heat transfer for a given fluid velocity than is obtained in a shell-and-tube heat exchanger so that the number of plates (heat-transfer surface) can be minimized. Also the turbulence promotes a kind of self-cleaning action, so that the heat exchanger does not foul as easily.

The corrugations can be stamped so that the pattern goes across the direction of fluid flow or is aligned with the flow. Plate-and-frame heat exchanger designers choose the first type for high heat-transfer rates at higher pressure drops and the second for relatively lower heat-transfer rates and pressure drop. The two styles are sometimes mixed in the same heat exchanger to achieve the specified heat transfer and pressure drop.

The corners of the heat-transfer plates contain up to four holes, or ports, through which the fluids are distributed to the proper slots. By proper sequencing of plates, including some with fewer than four ports, the fluids can be forced to make one or more passes through the heat exchanger. This is done, as with the shell-and-tube heat exchanger, to achieve the proper fluid velocities to obtain the desired heat transfer and pressure loss.

It is important that the heat exchanger be assembled, e.g., after cleaning, with the plates in the same order as received. Often the plates are numbered sequentially to assist.

14.13 GASKETS

Each heat-transfer plate contains a gasket which, when compressed, enables the plate-and-frame heat exchanger to hold pressure, provides the proper distribution of the hot and cold fluids through the slots, and prevents mixing of the two fluids.

The gaskets are located in a groove which is stamped in the heat-transfer plate. The design of the groove and gasket combination is a critical factor in the functioning of the heat exchanger. Replacement gaskets must be obtained from the original manufacturer. Gaskets must be properly positioned within the grooves. The gaskets may be fixed by adhesive bonding or through the use of clips or studs.

In most cases, the gasket material limits the design temperatures and pressures of these heat exchangers. Common gasket materials are given in Table 14.2

TABLE 14.2

Gasket matcrials	Maximum temperature, °F (°C)
Nitrile butadiene rubber	230 (110)
Ethylene propylene rubber	302 (150)
Viton	365 (185)

and are normally chosen to match the specified design temperature and to ensure chemical compatibility with the two fluids flowing through the heat exchanger.

In some high-temperature applications of 350°F (177°C) to 600°F (315°C), compressed-fiber gaskets are used. However, these are very difficult to seal and require great care in assembly.

Most gaskets for plate-and-frame heat exchangers are designed so that leaks due to gasket failures can be seen by maintenance personnel before fluid intermixing occurs. Naturally, if a leak occurs at an outer edge, the fluid will leak to the outside and drip on the floor. A double gasket seal is used at the ports, and a weep hole is provided in the edge of the gasket between the double seal. Should one of the two seals fail, the fluid will fill the space between the two seals and leak from the weep hole. This fluid will also drip to the floor where it can be seen.

Since there is a large cumulative length of gasket seals in a typical plate-and-frame heat exchanger, safety shrouds are often included as part of the assembly. These provide a cover over the plate assembly so that if a gasket fails under high pressure, the leaking fluid will be retained under the shroud and therefore protect personnel in the area. The shrouds are usually screwed to the edge of the compression plates or are slid behind the compression bolts.

14.14 HEADS

One of the heads, called the "fixed head" or "frame plate," is an integral part of the frame, supporting the carrying bar and guide bar, and is bolted to the floor. The opposite movable head (or pressure plate) is hung, often by rollers, from the carrying bar. When the bolting is removed from the assembly, the movable head may be pushed back from the plate assembly, allowing access to the plates. The heads contain the connections for the two fluids.

Rather than fabricate the connection from an alloy material, carbon-steel flanged connections may be used with a rubber liner to prevent contact of the fluid with the carbon steel.

When possible, the heat exchanger is designed so that all four connections are on the fixed head. In this case, the movable head may be unbolted and moved for servicing without disturbing any of the piping. For multipass designs, however, two connections will be on the movable head. Then installation piping must be designed by the user to permit dismantling of the piping and movement of the head.

Compression of the large number of gaskets can require a substantial bolting force. Therefore, heads are generally relatively thick or are reinforced with ribs. Bolting of the heads should be done uniformly over all the available bolts. This eliminates overstressing of the bolts and/or heads and is required to ensure that gaskets remain properly seated and uniformly compressed.

14.15 CARRYING AND GUIDE BARS

Heat-transfer plates are typically designed to be hung from the overhead carrying bar. Then the lower guide bar is used to position the plate such that proper alignment of the plates and gaskets is achieved. Generally, plates can be removed by lifting the bottom edge above the guide bar and then twisting the plate to remove it from the carrying bar. Some manufacturers have separate hangers at the tops of the plates which must be removed before the plates can be taken out of the assembly.

14.16 CAPACITY CHANGES

Unlike shell-and-tube heat exchangers, the plate-and-frame design has the flexibility of changing the heat-exchange and flow capacities by adding or deleting heat-transfer plates. The original heat exchanger manufacturer can advise the specific changes which must be incorporated to achieve the new capacity. If a significantly large number of additional plates are required, the frame and bolting may require changes as well; so if a later expansion is anticipated, this should be specified at the time of initial purchase and a large enough frame provided.

14.17 ASSEMBLY, HYDROSTATIC PRESSURE TEST, AND STARTUP

The heads must be bolted tight enough to compress the gaskets so as to hold pressure. However, overtightening can damage the plates and actually reduce the pressure-holding capability. Most plate-and-frame heat exchanger nameplates give minimum and maximum dimensions between the heads. When the units are assembled, the compressed dimension should fall within these limits.

Like all elastomeric material, the gasket will tend to take a set after some time at operating temperatures and pressures. Should minor leaks occur, they may be stopped by retightening the bolts. A drip tray with a drain piped to a sewer may be useful to contain possible leaks and also to catch fluids trapped in the heat exchanger when it is dismantled for service.

After assembly, if the heat exchanger is hydrostatically tested for pressure, the pressure must not exceed the nameplate test pressures. Furthermore, certain plate-and-frame heat exchangers have a limit to the differential pressure which they can withstand. If so, this should be specified on the nameplate, and the pressures on both the hot and the cold sides of the heat exchanger should be increased together so that the difference between the two test pressures never exceeds the specified maximum differential.

During startup, to minimize internal stresses, differential expansion, and relative movement of parts, the temperature and pressures should be gradually increased to operating conditions. Temperatures and pressures should never exceed nameplate values.

14.18 COILS

Heat exchangers which utilize finned surface either to heat or to cool ambient air, or which use ambient air to cool a process fluid are termed "coils" (Fig. 14.13). The coil design is often a function of the required service, including pressure and temperature, maintenance, and corrosion resistance. Industrial coils can be fabricated with a variety of finned surfaces and header and casing designs.

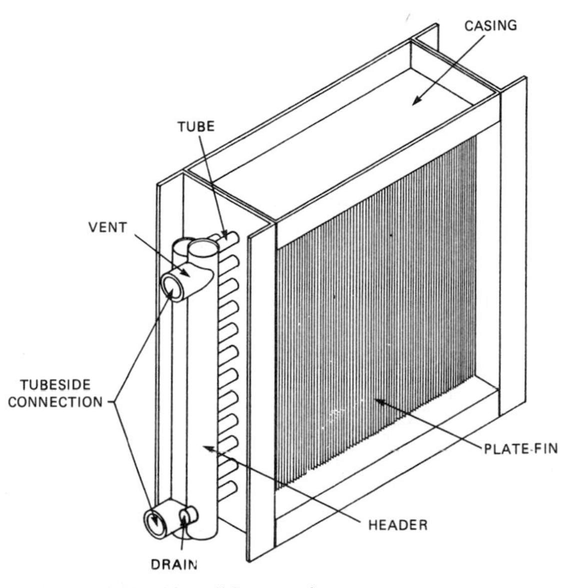

FIGURE 14.13 Air coil heat exchanger.

14.19 FINNED SURFACES

Coils may be designed so that fins are assembled on individual tubes, or fins may be one piece (plate fins) connecting several or all tubes in the coil. Coils are fabricated with various fin spacings, tube sizes, and layout geometries to meet the required performance of the particular application. Fins are typically made of aluminum or copper, but steel and stainless steel are also used. Tubes may be made of copper or copper alloys, steel, stainless steel, or aluminum.

Plate fins may be flat or corrugated. Flat fins offer the least resistance to flow at relatively lower heat transfer. Corrugated fins promote turbulence of the air stream, increasing the heat transfer and air pressure drop. Plate fins are normally attached to tubes by expanding the tube diameter outward along its entire length to make a pressure fit with the fin collars. The expansion may be accomplished hydraulically by high tubeside pressure, or the tubes may be expanded mechanically by pushing or pulling an oversized ball or bullet-shaped object through the tube. Heat transfer is a function of the quality of the pressure fit. Typically, coils are limited in temperature so as to maintain the bond between fins and tube. The manufacturer should be consulted for maximum temperatures.

Plate fins do not allow for movement of individual tubes in a coil. Some coils use fins that are helically wound around individual tubes before being assembled into the coil. Figure 14.14 shows three typical configurations which are used in various applications. Figure 14.14*a* is a wrapped-on L foot fin. Under tension, the fin is wrapped around the tube, and the ends of the fin are fastened to prevent loosening of the tension. The length of the foot sets the fin spacing. This mechanically bonded tube, similar in bond to the plate fin style, is typically used for service up to 350°F (177°C). A metallurgical bond can be achieved by helically winding a straight fin around a tube and then coating the entire surface with a braze or solder (Fig. 14.14*b*). Here the temperature is limited by the melting point of the braze [typically, 1200°F (649°C)] or solder [typically, 600°F (316°C)].

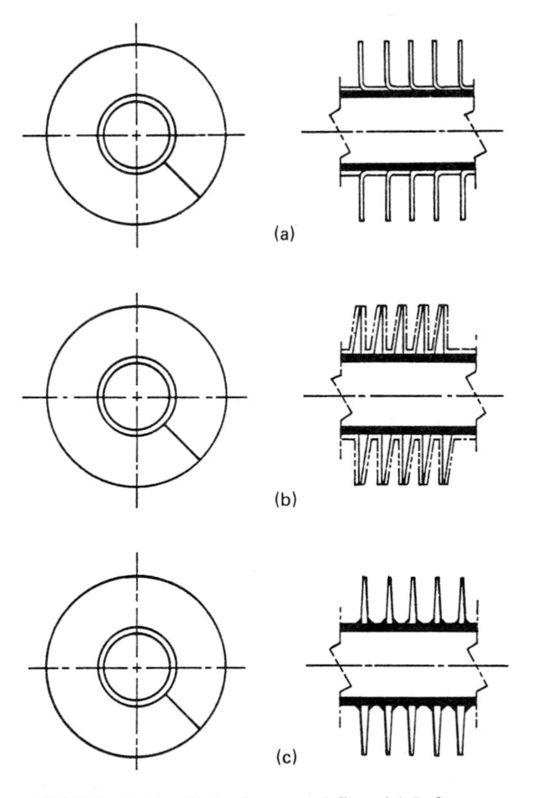

FIGURE 14.14 Helically wound fins. (*a*) L-foot mechanical bond; (*b*) solder or braze metallurgical bond; (*c*) embedded mechanical bond.

A better mechanical bond can be obtained by embedding the fin in a groove machined in the wall of the tube and deforming the tube wall against the base of the fin (Fig. 14.14*c*). Typically, these tubes are used in service up to about 750°F (400°C).

14.20 HEADERS FOR COIL DESIGN

Certain coils use tubesheet construction similar to that found in shell-and-tube heat exchangers where the tube joints are rolled or welded. Then cast or fabricated headers are bolted to the tubesheet. These removable headers permit access to the tube ends for maintenance and repair. Most coils use U bends to circuit the tubeside fluid through the coil, however, so that cleaning the tubes by rods is not usually possible.

More frequently, coils have pipe headers into which the tubes are welded or brazed. Tube ends cannot be serviced in these coils. When designed for high temperatures and pressures, these coils may have heavy wall carbon-steel pipe headers. Coils for lower-temperature and lower-pressure service may have watertube-type copper headers with soldered joints.

Depending on the circuiting of the coil, relatively few tubes may be connected to the header. Since the joints will be subjected to stress when the header connections are piped, care should be taken to minimize forces being applied to the header during piping installation.

14.21 TUBES FOR COIL DESIGN

Tubes have a smooth bore. Since plate-fin coils require expansion of tubes outward to increase the diameter, these tubes are often thinner than would be used in coils with helically wound fins. Embedded fin tubes must have heavier walls to accommodate the groove which is machined during the finning process.

In some coil applications, the heat-transfer capability is determined by the tubeside conditions. In these coils, where water velocities are low or where viscous fluids such as oils are in the tubes, turbulators are used. Several types are employed, including springs, spheres, and spiral tapes, to significantly reduce the size of the coils required for the application.

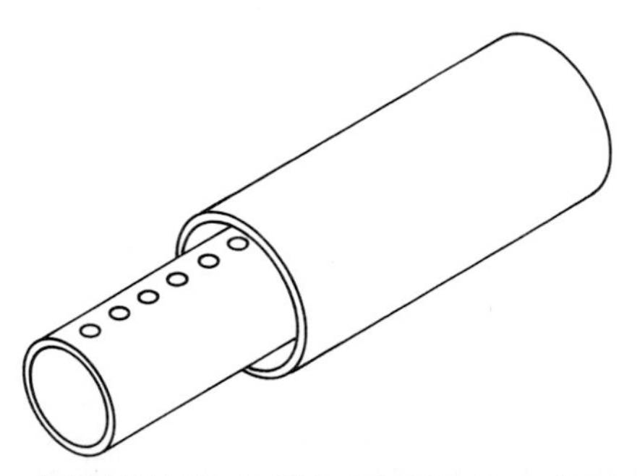

FIGURE 14.15 Orifice of double tube—distribution tube for heavy-duty coils.

Air heating coils using steam to preheat combustion air, e.g., may use distributing tube geometries (Fig. 14.15) to obtain uniform temperature profiles across the coil and to promote drainage of the condensate. These double-tube designs have an inner tube through which the steam enters the coil. Orifice holes in inner tubes distribute the steam evenly along the length of the coil into the surrounding outer tubes. Condensate flows through the annular space between the tubes and returns to the condensate header.

14.22 CIRCUITING

Coils are circuited to provide a flow path required to achieve an appropriate tube velocity. The tubeside fluid may make several passes back and forth through the coil. Coils with water are designed to avoid low pockets in the circuits which would prevent drainage. This is important in installations when the coil might be exposed to air temperatures below freezing. The exact method of circuiting depends on whether the air flow is horizontal or vertical. A coil which drains properly when oriented in a horizontal direction may not drain when installed in the vertical direction, and vice versa.

14.23 COATINGS

Some installations are such that the air flowing across the coils contains various kinds of contaminants. These might be in the general process atmosphere or in a process exhaust, as in heat recovery from a boiler flue. In systems where the contaminant might condense to form acids or otherwise corrode the coil materials, coatings may be used. Various kinds of coatings are available to provide resistance to particular chemicals contained in the flow across the coils. Care must be taken when the coils are cleaned so that the coatings are not chipped or scratched. Normally the coatings are limited to exposure temperatures, which should be checked against the particular installation requirements.

14.24 MAINTENANCE OF HEAT EXCHANGERS

Heat exchangers should be cleaned periodically to remove scale or sludge deposits from the surfaces. Such fouling decreases the heat-transfer capacity of the heat exchanger and increases the pumping power needed to pump fluids through the unit. The need for cleaning can be determined by monitoring pressure gauges and thermometers installed on the heat exchanger inlet and outlet piping. A significant increase in pressure drop or reduction in performance indicate cleaning is necessary.

Soft deposits may often be removed by circulating hot wash oils, water, or commercial cleaning compounds. Care should be taken that the metal is not cut or scratched in cleaning the metallic surfaces.

Removal of tube bundles from shell-and-tube heat exchangers and air coils from the ductwork casing should be done carefully. The weight of the bundle or coil should not be carried by individual tubes, since the tubes could be damaged. Instead tubesheets, baffle plates, or coil support plates should carry the weight. The heat exchanger manufacturer can provide assistance in recommending proper procedures for servicing and repairing heat exchangers.

14.25 BIBLIOGRAPHY

1. *Standards of Tubular Exchanger Manufacturers Association*, 7th ed., Tubular Exchanger Manufacturers Association, Inc., New York, 1988.
2. *Storage, Installation, Operation and Maintenance of Heat Exchangers*, Bulletin 104-17, ITT Standard, ITT Fluid Technology Corporation, Heat Transfer Division, Buffalo, NY, 1989.

CHAPTER 15
RADIATION FOR STEAM AND HOT-WATER HEATING

Robert E. Ross
Hydronics Institute, Berkeley Heights, New Jersey

15.1 INTRODUCTION

Comfort heating with steam or hot water from boilers is known as "hydronics" and usually includes heat-distributing radiation in each room to furnish warmth to the occupants. Free-standing cast-iron radiators, still being manufactured in the modern, small-tube style, emit both radiant and convective heat. The more popular types of radiation, fin tubes and convectors, provide heat primarily by natural convection of the air passing up over a finned heating element in an enclosure (Fig. 15.1).

Convectors in cabinets, either surface-mounted or recessed, provide concentrated heat in a short space of the wall, while fin-tube radiation occupies long stretches of the wall, spreading the heat output more evenly throughout the room. Both types are used with steam or hot water. They are usually furnished in a variety of standardized enclosures, but frequently are installed in architectural enclosures for decorative or protective purposes (Fig. 15.2).

15.2 PRODUCTION ENCLOSURES

For surface-mounted units, the slope-top enclosure (Fig. 15.3) is popular because it discourages sitting or blocking with packages. Other familiar production models are the flat-top front outlet (Fig. 15.4), flat-top top outlet (Fig. 15.5), and expanded metal cover for industrial applications (Fig. 15.6). Some models are available with dampers for temperature control by the room's occupants.

Fin-tube units can be provided as wall-hung installations with the bottom of the enclosures about 4 in (10 cm) above the floor or in the form of baseboard heaters at floor level, usually along the perimeter wall.

Steel enclosures, either in enamel or prime-coat finish, are available in 16 or 18 gauge (1.52 or 1.2 mm). They may be supported by the same bracket as the heating element or by separate back plates or hanger strips. Production covers

FIGURE 15.1 Fin-tube element in slope-top cover. *(Courtesy of Sterling Heating Equipment.)*

FIGURE 15.2 Cabinet convector. *(Courtesy of The Hydronics Institute.)*

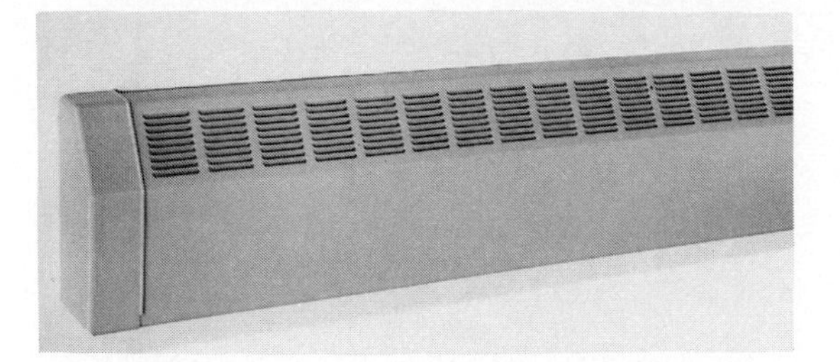

FIGURE 15.3 Slope-top enclosure. (*Courtesy of Sterling Heating Equipment.*)

FIGURE 15.4 Flat-top front outlet. (*Courtesy of Sterling Heating Equipment.*)

are available in even footages up to 8 ft 0 in (2.4 m), but can be ordered in intermediate lengths to fit site conditions.

Standard enclosures are generally open at the bottom and have an outlet at or near the top. In some cases a grille is available for the bottom inlet, and it is particularly useful where fin-tube radiation is installed above eye level. Convector enclosures are in the shape of self-contained boxes while fin-tube enclosures form a horizontal pattern along the wall, with sheet-metal accessories to provide access to valves and to adjust the length to fit specific spaces. Other accessories adjust the appearance to fit irregularities of the wall.

The covers are available in various heights, either to enhance the output of a heating element or to enclose two or three tiers of element for high output (Fig. 15.7). Covers are manufactured to fit most commercial and industrial dimension requirements, either as a self-contained unit with end caps (Fig. 15.8) or extending wall to wall, appearing visually as a continuous unit abutting the partitions (Fig. 15.9).

Baseboard heaters, for installation at floor level, are from 7 to 10 in (18 to 25 cm) high and about 3 in (8 cm) deep and are limited to one tier of element with maximum outputs below the maximums for wall-mounted fin tube (Fig. 15.10). They have a continuous slot along the bottom and along the upper front for convection, and usually they include a manual damper to reduce the output when desired by the occupants. Although most popular for residential heating, baseboard is also available in commercial-quality enclosures.

FIGURE 15.5 Flat-top top outlet. (*Courtesy of Sterling Heating Equipment.*)

15.3 ARCHITECTURAL ENCLOSURES

Where desired by the architect or engineer, fin-tube manufacturers can design and produce enclosures to match adjacent design details at the windows or to conform to other aesthetic or practical needs of the installation. Extruded shapes are often integrated into the enclosure design. Shorter, wider, or narrower configurations to meet almost any job requirement, instead of standard production enclosures, can be obtained given suitable lead time (Figs. 15.11 and 15.12).

Where low-profile high output is desired, two elements can be mounted side by side. Tamper-resistant covers, beyond simple locking devices, can be developed in consultation with the design engineer. Typical enclosures are made of 18 gauge (1.2 mm) steel, strongly braced to withstand physical damage, but they can also be obtained in 16 and 14 gauge (1.52 and 1.9 mm) where required for special protection.

FIGURE 15.6 Expanded-metal industrial cover. (*Courtesy of Sterling Heating Equipment.*)

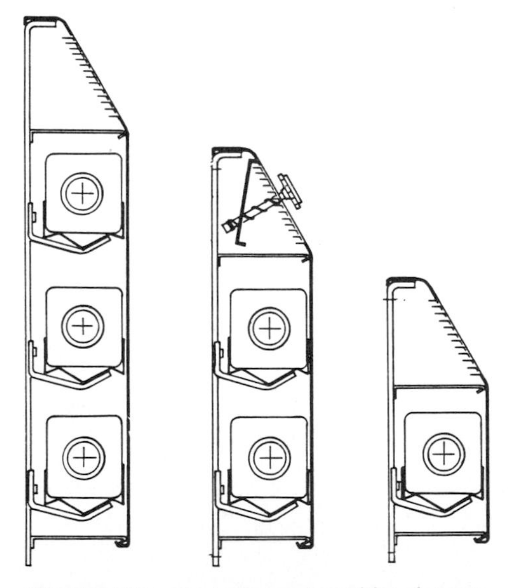

FIGURE 15.7 Single-tier and multitier elements. (*Courtesy of Slant/Fin Corp*).

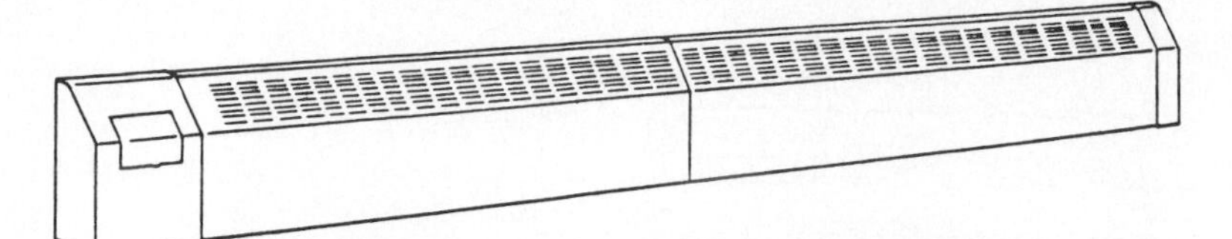

FIGURE 15.8 Unit with end caps. (*Courtesy of Slant/Fin Corp.*).

FIGURE 15.9 Wall-to-wall fin tube in front outlet enclosure. (*Courtesy of Sterling Heating Equipment.*)

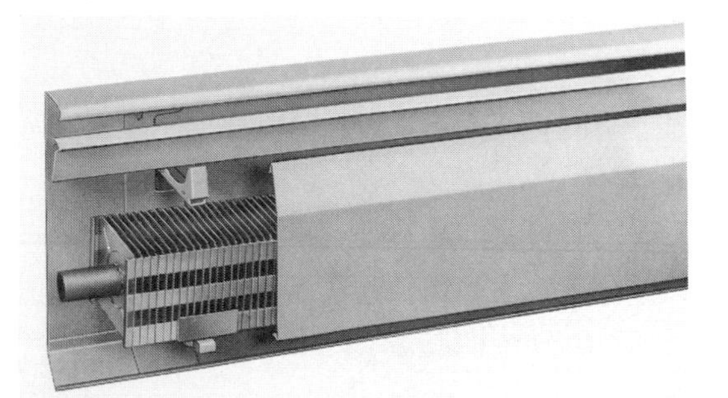

FIGURE 15.10 High-trim baseboard. (*Courtesy of Weil-McLain Corp.*)

FIGURE 15.11 Architectural fin-tube enclosure—high top. (*Courtesy of Sterling Heating Equipment.*)

FIGURE 15.12 Architectural fin-tube enclosure—low top. (*Courtesy of Sterling Heating Equipment.*)

15.4 HEATING ELEMENTS

Heating elements for cabinet convectors and for fin-tube radiation consist of pipe or tubing to carry the fluid and metal fins fixed to the tubing for extended heating surface. The "stack" effect created by the height of the enclosure draws cool room air into the bottom opening and out the top, through natural convection.

Elements for cabinet convectors (Fig. 15.13) consist of small-diameter copper or aluminum tubing [usually ⅜ in (1 cm)] arranged in parallel horizontal rows, with the number of rows determining the depth of the element and the cabinet. The tubing is fastened to headers at each end by the manufacturer. Tightly spaced aluminum fins are pierced by all rows of the tubing. Heating output is de-

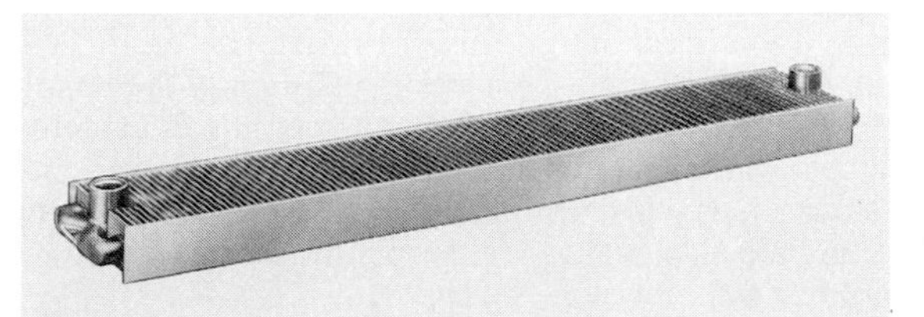

FIGURE 15.13 Convector elements. (*Courtesy of The Hydronics Institute.*)

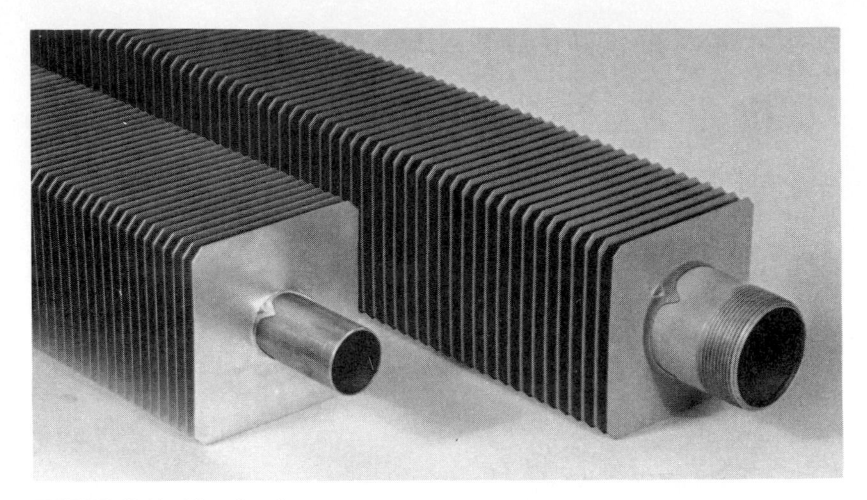

FIGURE 15.14 Fin-tube elements.

termined by the length and depth of the element and is affected by the height of the enclosure.

Access to the fittings and valves is either through an opening at the bottom or through an access door. Elements for hot water are installed level, while those for steam are pitched down in the direction of condensate flow.

Fin-tube elements (Fig. 15.14) vary extensively in physical dimensions and types of material. Tubing is usually ¾, 1, or 1¼ in (2, 2.5, or 3 cm) for copper tubing and 1¼ or 2 in (3 or 5 cm) for iron pipe, with standard lengths available up to 8 or 12 ft (2.4 or 3.7 m).

Although the configuration of the enclosure has an effect on the heat output, the basic determinants are the length of the element, the fin spacing and thickness, the height of the enclosure, and the choice between steel pipe and copper tubing (the latter creates greater heat transfer). A major difference in Btu (watt) output is developed by stacking the finned tubing above each other two or three tiers high.

The extended heating surface attached to the tube is either steel fins on iron pipe or aluminum fins on copper tubing. Fin dimensions vary in height and depth; the larger ones are used for greater heat output, and the narrower are useful where a slim enclosure is desired.

Fin spacing on the tube varies from 24 to about 60 fins per foot (79 to 197 fins per meter), with the larger quantity usually providing more heat transfer at greater material cost. In some cases, such as a low enclosure, the larger number of fins may produce a lower output because of resistance to air flow.

Piping connections to single-tier elements usually have the supply at one end and return at the other (Fig. 15.15), but the return piping is sometimes run back inside the enclosure to the supply end, to use the same vertical chase.

Multitier elements usually are arranged for parallel flow with headers fabricated at the job. They are also available in serpentine arrangement, which may be preferred for hot-water systems (Fig. 15.16).

Heating elements are supported by heavy steel brackets fastened directly to the wall or through a mounting plate which also supports the front cover. At-

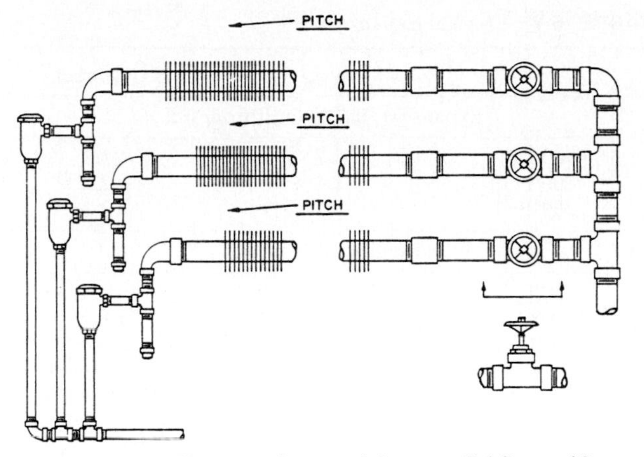

FIGURE 15.15 Elements of steam piping—parallel flow. (*Courtesy of Sterling Heating Equipment.*)

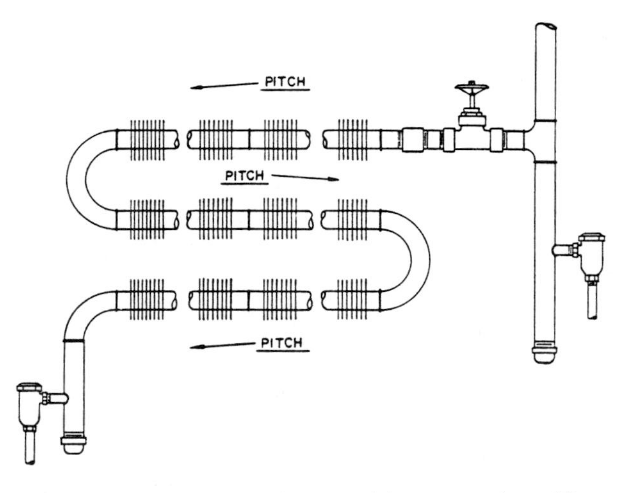

FIGURE 15.16 Elements of steam piping—serpentine. (*Courtesy of Sterling Heating Equipment.*)

tached to the wall brackets are supports spaced at appropriate intervals for physical support of the element, with a cradle between or around the fins. The cradle has provision for lateral slide to permit movement of the element due to expansion and contraction of the heating cycle. For steam systems the supports can be adjusted to provide proper pitch. See Table 15.1.

15.5 RATINGS

The output of radiation in Btu per hour (watts) varies with the type of element and the type of enclosure, and it should be obtained from manufacturers' cata-

TABLE 15.1 Thermal Expansion

	Steel	Copper
	Expansion, in/100 lin ft from 0°F	
100°F	0.7	1.1
200°F	1.4	2.2
300°F	2.1	3.3
	Expansion, cm/10 m from 0°C	
50°C	0.39	0.61
100°C	0.78	1.23
200°C	1.56	2.46

Note: Expansion is linear and may be extrapolated for other dimensions.

logs. The *I*=*B*=*R* ratings of baseboard and fin-tube radiation are based on actual tests of the equipment in the laboratory of The Hydronics Institute, and all such products currently certified by that organization are listed in their published ratings book.[1] In addition to the initial tests to establish the Btu/hour (watt) output, certified products are retested at specified intervals to ensure that the product continues to merit the published rating and is manufactured in accordance with catalog specifications.

Any manufacturer can submit a product for testing regardless of whether the company is a member of The Hydronics Institute. Certification tests are also performed at the laboratory on fin-tube units of special design when required by the engineer or architect for specific installations, to ensure the product will produce the output required.

Other determinants of Btu/hour (watt) output, in addition to the design and dimensions of the product, include circulating water or steam temperature, rate of flow, and incoming air temperature. The standards for these variables are specified in the "I=B=R Testing and Rating Code for Finned Tube Radiation." Standard tests are based on 1-lb (0.45-kg) steam at 215°F (102°C) with 65°F (18°C) entering air. Correction factors are listed for other steam and water temperatures in Table 15.2.

15.6 SELECTION

Two of the selection criteria are the heat output of the unit and the physical space requirements. For example, a single-tier copper-tube element in a low enclosure can have the same output per linear foot (meter) as a two-tier iron-pipe element in a higher cover. Considerations of the length of available wall and wall height beneath the windows, in relation to the required Btu/h (watt) rating, are some of the factors taken into account. The pipe diameter chosen will be mainly determined by permissible pressure drops.

Another major consideration in determining the Btu output per foot (meter) is the temperature drop of the circulating water. It has been the pattern to design the system on the basis of a 20°F (11°C) drop from the boiler supply to the boiler return, but greater drops of 30, 40, or 50°F (16, 22, or 28°C) may be used, with

TABLE 15.2 Correction Factors for Steam Pressures, Water Temperatures, and Air Temperatures Other than Standard

Steam pressures and air temperatures																
Steam pressure			Entering air temperature, °F (°C)													
Gauge	Abs., lb/in^2 (Pa)	Temp., °F (°C)	45 (7.2)	55 (12.7)	Std. 65 (18)	70 (20.9)	75 (23.7)	80 (26.4)	85 (29.1)	90 (32.2)	100 (37.7)	110 (43.3)	120 (48.4)	130 (53.9)	140 (59.4)	150 (64.9)
(Vac) 15 inHg	7.32 (0.27)	178.9	0.90	0.80	0.70	0.65	0.60	0.56	0.51	0.45	0.39	0.32	0.25	0.18	0.13	0.08
(Vac) 10 inHg	9.78 (0.37)	192.2	1.02	0.91	0.81	0.76	0.71	0.66	0.62	0.55	0.48	0.40	0.33	0.26	0.20	0.14
(Vac) 5 inHg	12.25 (0.45)	202.9	1.11	1.00	0.90	0.85	0.79	0.75	0.70	0.63	0.56	0.48	0.40	0.33	0.27	0.20
0 lb/in^2 (Pa)	14.696 (0.55)	212.0	1.19	1.09	0.97	0.92	0.87	0.82	0.77	0.70	0.63	0.54	0.46	0.38	0.31	0.25
0.899 (0.03)	15.595 (0.59)	215.0	1.22	1.11	1.00	0.95	0.90	0.84	0.80	0.75	0.65	0.57	0.48	0.40	0.33	0.26
5 (0.19)	19.70 (0.74)	227.1	1.34	1.22	1.11	1.05	1.00	0.95	0.90	0.81	0.75	0.66	0.57	0.49	0.41	0.34
10 (0.38)	24.70 (0.92)	239.4	1.45	1.33	1.22	1.17	1.11	1.05	1.00	0.91	0.85	0.75	0.66	0.58	0.50	0.42
15 (0.56)	29.70 (1.1)	249.8	1.55	1.43	1.31	1.26	1.20	1.14	1.09	1.00	0.94	0.84	0.75	0.66	0.57	0.49
20 (0.75)	34.70 (1.3)	258.8	1.63	1.52	1.40	1.33	1.28	1.23	1.17	1.07	1.02	0.92	0.82	0.73	0.64	0.55
25 (0.94)	39.70 (1.5)	266.8	1.71	1.59	1.47	1.41	1.36	1.30	1.25	1.15	1.09	0.98	0.89	0.80	0.71	0.62
30 (1.1)	44.70 (1.7)	274.0	1.78	1.66	1.54	1.48	1.42	1.37	1.31	1.21	1.15	1.05	0.95	0.85	0.76	0.68
40 (1.5)	54.70 (2.1)	286.7	1.91	1.79	1.66	1.61	1.54	1.49	1.43	1.32	1.27	1.16	1.06	0.97	0.87	0.78
50 (1.9)	64.70 (2.4)	297.7	2.02	1.90	1.77	1.71	1.65	1.60	1.54	1.42	1.37	1.26	1.16	1.06	0.96	0.87
60 (2.3)	74.70 (2.8)	307.3	2.10	2.00	1.87	1.81	1.75	1.69	1.63	1.51	1.47	1.35	1.25	1.15	1.05	0.95
70 (2.6)	84.70 (3.2)	316.0	2.20	2.09	1.95	1.89	1.83	1.77	1.71	1.59	1.55	1.44	1.33	1.23	1.12	1.03
80 (3.0)	94.70 (3.6)	323.9	2.27	2.17	2.03	1.97	1.91	1.85	1.80	1.69	1.63	1.52	1.41	1.31	1.20	1.10
90 (3.4)	104.70 (3.9)	331.2	2.36	2.24	2.11	2.05	1.98	1.93	1.87	1.74	1.70	1.59	1.48	1.38	1.28	1.17
100 (3.8)	114.70 (4.3)	337.9	2.43	2.31	2.18	2.11	2.05	2.00	1.94	1.81	1.77	1.65	1.54	1.44	1.33	1.23
125 (4.7)	139.70 (5.3)	352.9	2.59	2.47	2.33	2.27	2.21	2.16	2.10	1.96	1.92	1.80	1.69	1.59	1.48	1.38
150 (5.6)	164.70 (6.2)	365.9	2.73	2.62	2.47	2.43	2.35	2.29	2.23	2.08	2.05	1.94	1.82	1.72	1.61	1.51
175 (6.6)	189.70 (7.1)	377.4	2.86	2.74	2.60	2.54	2.47	2.41	2.35	2.21	2.17	2.05	1.95	1.85	1.73	1.63
200 (7.5)	214.70 (8.1)	387.8	2.95	2.85	2.71	2.63	2.58	2.52	2.47	2.31	2.29	2.17	2.06	1.96	1.84	1.75

TABLE 15.2 Correction Factors for Steam Pressures, Water Temperatures, and Air Temperatures Other than Standard (*Cont*)

Average water temperature, °F (°C)	Water temperatures and air temperatures														
	Entering air temperature, °F (°C)														
	45 (7.2)	55 (12.7)	Std. 65 (18)	70 (20.9)	75 (23.7)	80 (26.4)	85 (29.1)	90 (32.2)	95 (35.0)	100 (37.7)	110 (43.3)	120 (48.4)	130 (53.9)	140 (59.4)	150 (64.9)
90 (32.2)	0.19	0.13	0.08	0.06											
100 (39.7)	0.25	0.19	0.13	0.11	0.08	0.06									
110 (43.3)	0.31	0.25	0.19	0.16	0.13	0.11	0.08	0.06							
120 (48.4)	0.38	0.31	0.25	0.22	0.19	0.16	0.13	0.11	0.08	0.06					
130 (53.9)	0.45	0.38	0.31	0.28	0.25	0.22	0.19	0.16	0.13	0.11	0.06				
140 (59.4)	0.53	0.45	0.38	0.34	0.31	0.28	0.25	0.22	0.19	0.16	0.11	0.06			
150 (64.9)	0.61	0.53	0.45	0.42	0.38	0.34	0.31	0.28	0.25	0.22	0.16	0.11	0.06		
160 (70.4)	0.69	0.61	0.53	0.49	0.45	0.42	0.38	0.34	0.31	0.28	0.22	0.16	0.11	0.06	
170 (75.9)	0.78	0.69	0.61	0.57	0.53	0.49	0.45	0.42	0.38	0.34	0.28	0.22	0.16	0.11	0.06
180 (81.4)	0.86	0.78	0.69	0.65	0.61	0.57	0.53	0.49	0.45	0.42	0.34	0.28	0.22	0.16	0.11
190 (86.9)	0.95	0.86	0.78	0.73	0.69	0.65	0.61	0.57	0.53	0.49	0.42	0.34	0.28	0.22	0.16
200 (92.4)	1.05	0.95	0.86	0.82	0.78	0.73	0.69	0.65	0.61	0.57	0.49	0.42	0.34	0.28	0.22
210 (97.9)	1.14	1.05	0.95	0.91	0.86	0.82	0.78	0.73	0.69	0.65	0.57	0.49	0.42	0.34	0.28
215 (Std.) (100.7)	1.20	1.09	1.00	0.95	0.91	0.86	0.82	0.78	0.73	0.69	0.61	0.53	0.45	0.38	0.31
220 (103.4)	1.25	1.14	1.05	1.00	0.95	0.91	0.86	0.82	0.78	0.73	0.65	0.57	0.49	0.42	0.34
230 (108.9)	1.39	1.25	1.14	1.09	1.05	1.00	0.95	0.91	0.86	0.82	0.73	0.65	0.57	0.49	0.42
240 (114.4)	1.44	1.39	1.25	1.20	1.14	1.09	1.05	1.00	0.95	0.91	0.82	0.73	0.65	0.57	0.49
250 (119.9)	1.54	1.44	1.39	1.32	1.25	1.20	1.14	1.09	1.05	1.00	0.91	0.82	0.73	0.65	0.57
260 (125.4)	1.64	1.54	1.44	1.41	1.39	1.32	1.25	1.20	1.14	1.09	1.00	0.91	0.82	0.73	0.65
300 (130.9)	2.09	2.00	1.87	1.81	1.76	1.70	1.64	1.59	1.54	1.50	1.41	1.32	1.20	1.09	1.00

Note: Gauge pressure should be corrected for altitude.

possible reductions in pipe diameters for the mains. A design guide containing tables and worksheets is available from The Hydronics Institute.[2] The savings in piping costs must be balanced against the possible need for a higher-head circulating pump.

For example, with 180°F (82°C) boiler water supply and a 40°F (22°C) design drop, the average water temperature in the first third of the piping would be approximately 170°F (77°C) with 160 and 150°F (71 and 65°C) in the latter two-thirds, respectively. The active lengths of elements would be selected on the basis of those water temperatures. This applies to larger installations and to temperature drops of 30°F (17°C) or more on series-loop arrangements. The temperature drop has no effect on the selection of boiler size.

A 1°F (0.55°C) temperature drop per 1 gal/min (3.7 L/min) of water flowing emits 8.3 Btu (8755 J). In 1 h the output is approximately 500 Btu (527,000 J), and for a 20°F (11°C) drop each 1 gal/min (3.7 L/min) will provide 10,000 Btu/h (10.5 MJ/h or 28.3 W). If the system is designed for a 40°F (22°C) drop, the heat output per 1 gal/min (3.7 L/min) is twice as much, so for a given design output, only half the volume of water will have to be circulated. This permits a reduction in the pipe diameter of the main.

To ensure proper heat transfer through the elements, an adequate minimum flow rate must be maintained: for ¾-in (1.9-cm) diameter, 0.5 gal/min (1.9 L/min); 1-in (2.5-cm) diameter, 0.9 gal/min (3.4 L/min); and 1¼-in (3.2-cm) diameter, 1.6 gal/min (6.1 L/min).

15.7 APPLICATION

The choice of element and cover is related to the type of building and the purpose of the hydronic equipment. Fin-tube radiation is widely used for total comfort heating, particularly in buildings of moderate area for each floor, where the perimeter heat can properly reach interior areas. This also applies where floor space is divided into small areas.

For heating of industrial buildings and warehouses, it is good practice to use fin-tube elements in expanded-metal covers around the perimeter, supplemented at major loss areas (such as loading docks) with unit heaters. In cold climates where radiant floor panels are used in one-story buildings, fin-tube elements can provide quick pickup or supplement the radiant heat to supply the adequate amount of Btu's not provided by the floor panel.

Where a central ventilation system is required, tempered air at moderate temperature is introduced, often at the indoor design temperature of 70°F (21°C), and the perimeter fin-tube element supplies the necessary Btu's to satisfy the transmission heat loss.

Often in modern office buildings where air cooling is required in summer, the air system is used for core heating in winter, in conjunction with fin-tube radiation to satisfy the envelope losses. This is particularly useful for after-hours occupancy when the air system is shut down, because control of individual perimeter offices can be maintained via room thermostats and zone valves. This is especially advantageous because it eliminates the noise of blowers and ducts when all else in the building is quiet. In a hot-water system which has been purged of air, and with the usual provision for expansion and contraction, there is little expectation of sound coming from a proper fin-tube installation.

An economical operation for combination cooling and heating is the use of cabinet fan-coil units using chilled water in summer and the use of perimeter fin-tube elements, in conjunction with the cabinet units, using warm water in winter. The cabinet units can provide quick heat during morning pickup and individual office control for limited areas, while the fin-tube radiation is used to overcome downdrafts from the cold windows and walls all along the exterior.

Another combination widely used where quiet operation is required, such as in libraries and churches, is valance cooling in summer, using chilled water in finned elements installed behind enclosures along the upper wall near the ceiling, and the fin tube in its normal position near the floor for winter heating. The valance unit can also be used for heating, with low-output commercial baseboard installed in high-heat-loss areas at the floor line. This provides natural convection of air in both summer and winter, thereby eliminating the noises of forced air and the discomforts of unbalanced systems. See Chap. 17 for a detailed discussion of valance heating and cooling.

15.8 PIPING ARRANGEMENTS

Various convenient and economical piping arrangements depend on the configuration of the building, the degree of local temperature control desired, and many other considerations. Fin tube is eminently suitable for series-loop systems where each segment is piped directly to the next segment, room to room, or covering a wide space of wall in an industrial building.

In small buildings a single loop may suffice, running from the boiler and back to it directly. A larger building can have two or more individual loops connected to supply and return headers at the boiler or joining the return main at a convenient tee. Such loops can be balanced with a manual valve to adjust the water flow for proper heat emission, or they can be automatically controlled by zone valves or individual circulators to respond to temperature needs at various sides of the building. See Fig. 15.17. A typical piping layout for a small manufacturing plant is shown in Fig. 15.18.

Larger still are the buildings which use two-pipe reverse-return mains, or primary and secondary hookups. With two-pipe mains, each heating unit can be individually valved to throttle or turn off the flow. The reverse-return arrangment of the mains permits a balanced pressure drop through various parts of the system. Primary-secondary pumping includes a master pump to keep water circulating through the main, while the secondary circulators supply specific zones of the building, or individual risers and returns, in response to their respective thermostats. (See Fig. 15.19.)

15.9 AUTOMATIC CONTROL

Many control arrangements are possible with hydronic fin-tube installations, to provide comfort to the occupants and economy to the owner. These vary from simple on/off operation, to more efficient two-stage thermostat control, to modulating water temperature in response to outdoor conditions. The more sophisticated type senses indoor versus outdoor temperatures, maintains a steady condition of comfort at all occupied hours, and schedules appropriate setback periods for energy conservation.

The accurate sensing of modern thermostats is best served by the accurate re-

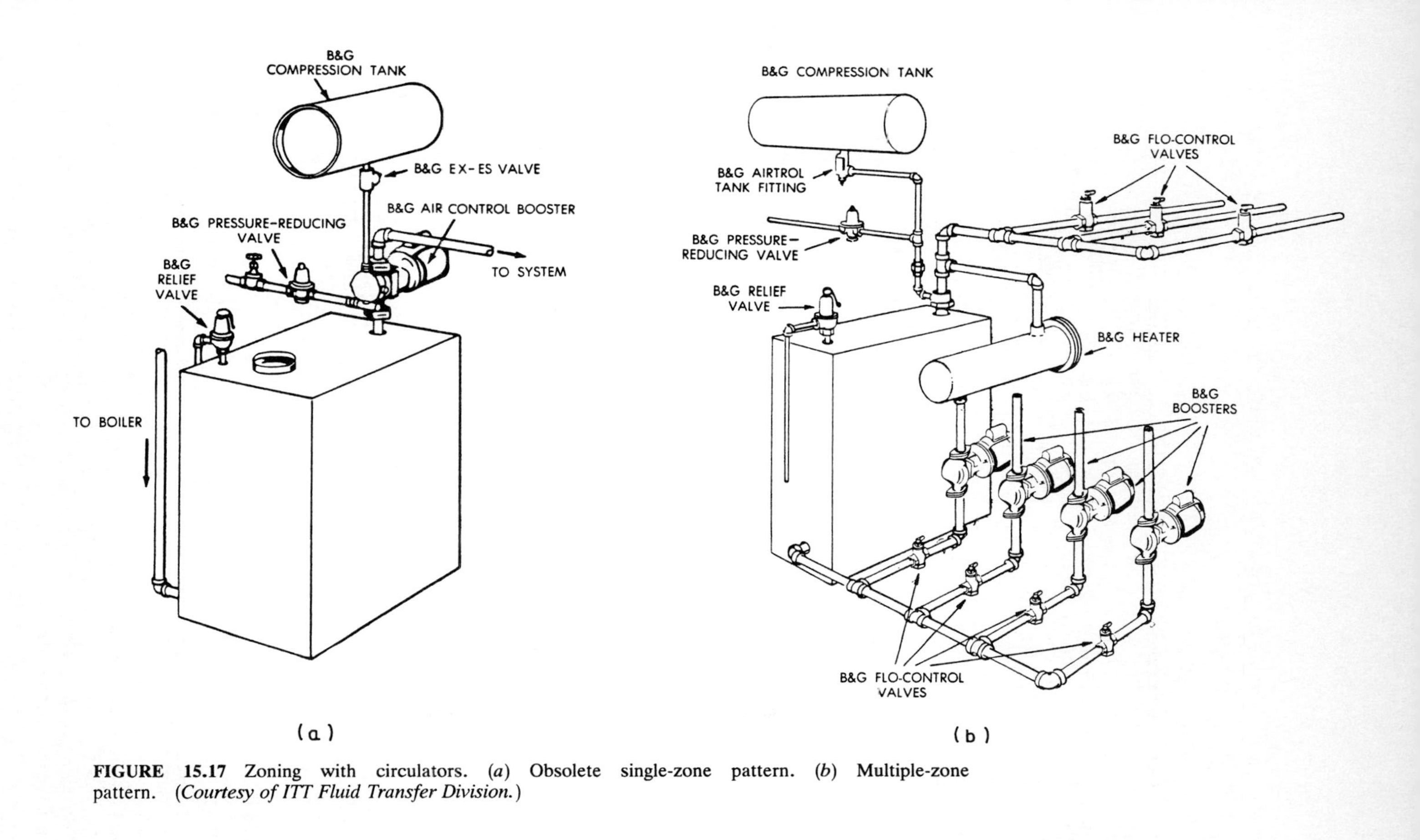

FIGURE 15.17 Zoning with circulators. (*a*) Obsolete single-zone pattern. (*b*) Multiple-zone pattern. (*Courtesy of ITT Fluid Transfer Division.*)

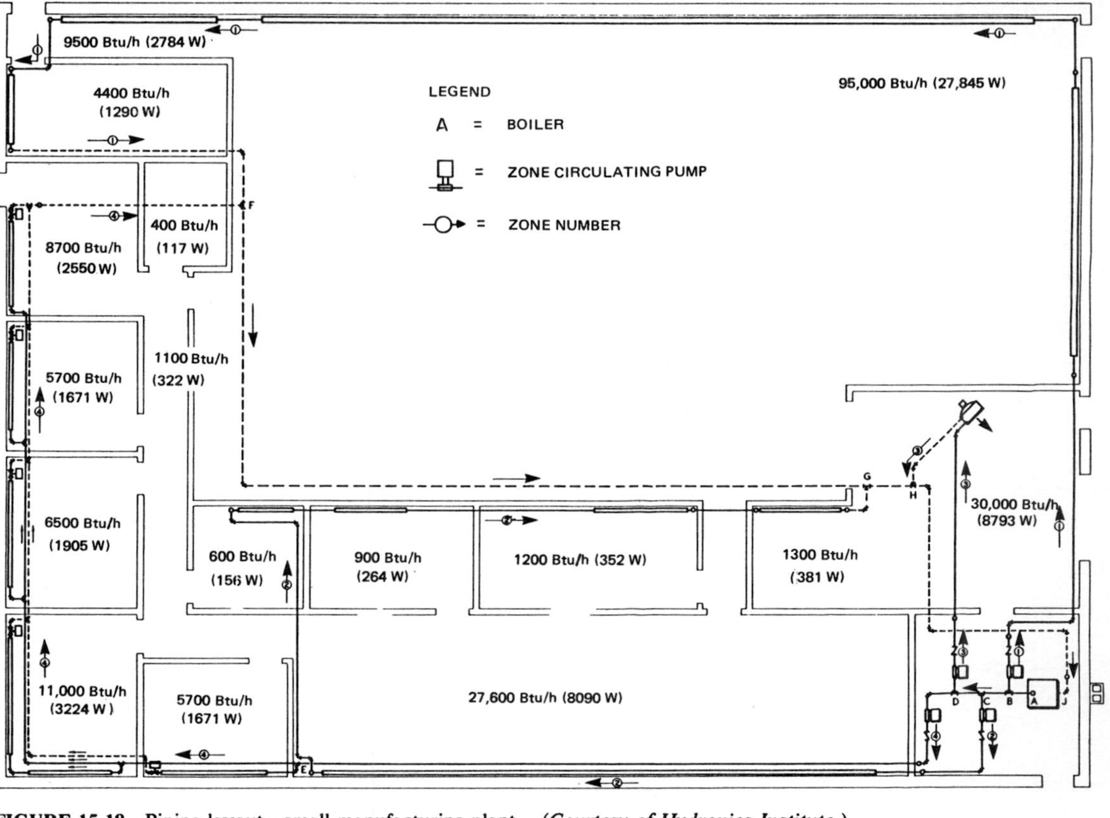

FIGURE 15.18 Piping layout—small manufacturing plant. (*Courtesy of Hydronics Institute.*)

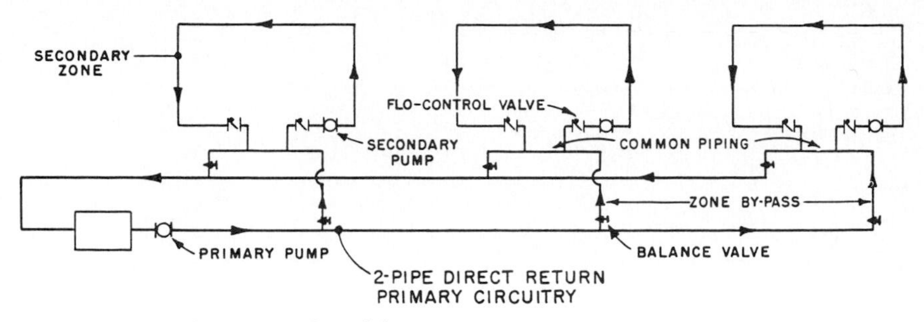

FIGURE 15.19 Primary-secondary piping.

sponse of circulating water. This is particularly effective where continuous circulating mains, with modulating water temperatures, automatically maintain a proper comfort balance between the heat output of the system, the heat loss of the space, and the fuel input to the boiler. (See Chap. 52, Automatic Control Systems.)

15.10 REFERENCES

1. *I=B=R Ratings for Baseboard Radiation*, The Hydronics Institute, Berkeley Heights, NJ, 1988.
2. *Advanced Installation Guide for Hydronic Heating Systems*, The Hydronics Institute, Berkeley Heights, NJ, 1973.

CHAPTER 16
DOOR HEATING

Michael K. Kennon
The King Company, Owatonna, Minnesota

16.1 INTRODUCTION

Door heating differs from the usual heating application because it involves an infiltration load. Compared to transmission loads, infiltration loads are erratic and difficult to predict. The best an engineer can do is to use appropriate techniques tempered with good judgment to arrive at solutions with a reasonable chance of success at a reasonable cost. It is impossible to predict exact performance of every system designed.

16.2 CHARACTERISTICS OF DOOR HEATING LOADS

Effective solutions to door heating problems recognize the differences between infiltration and transmission loads and respond accordingly. The differences include the following:

1. *Door heating loads undergo a sudden change in temperature.* With a transmission load, temperatures change slowly. Rarely would temperatures drop more than a couple of degrees per minute. With door heating loads, temperatures can drop 30 to 40°F (16° to 22°C) immediately.
2. *Door heating loads are compact.* Large differences in temperature exist over short distances. They can vary by 20°F (11°C) between two points 10 ft (2.5 m) apart.
3. *The location of a door heating load cannot be predicted.* The load will vary depending on the outside wind direction and how the building is being operated. Door heaters must be located in the immediate vicinity of the door.
4. *A door heating load is a moving air mass.* Transmission loads affect a constant volume of air. A door heating load consists of a mass of cold air pushing warm air out of its way. In order for the door heater to be effective, sufficient force must be imparted to the discharge air to penetrate the moving air mass. Because the cold air is moving, the control systems of a door heater must an-

ticipate the heating requirement. If a slow responsive control system like a thermostat is used, the load will move out of the range of the heater before the thermostat is activated.

5. *Door heating loads are of short duration.* A door heating load exists only as long as the door is open. Frequently the doors are open less than 5 min. Yet the heat loss during that time often exceeds the losses of several hours with the door closed. It should not be surprising that the capacities per hour of door heaters are high compared to those of space heaters.

In general, transmission loads involve slowly changing temperatures on a predictable, constant volume of air. Infiltration loads, however, are intensive loads that exist for short periods. Moreover, infiltration loads are concentrated in a relatively small area. A "door heater" can be defined as any device designed to respond to the special requirements of an infiltration load.

In order for infiltration to occur, exfiltration must also occur. Consistent success in solving door heating problems can be achieved by taking a "total building" approach, which uses an understanding of the infiltration-exfiltration dynamics of the building as a whole to develop a strategy for providing comfort in the door area.

16.3 TYPES OF DOOR HEATING EQUIPMENT AVAILABLE

16.3.1 Delivery Systems

Door heaters discharge air at relatively high temperatures (between 90 and 150°F (32 and 65°C) and with sufficient force to penetrate the infiltrating air mass. The door heater discharge slot is designed to keep the heated air in the immediate vicinity of the doorway. While thermostats are frequently used to control the capacity of the heating section, the blower sections are normally activated and deactivated by a door switch. Most types of door heaters are available for most heating media, as discussed below.

Space heaters are designed to distribute heat over a large area. Converting a space heater to a door heater requires a special discharge nozzle to confine the heat to the door area. Usually a 90° elbow is sufficient (Fig. 16.1). When the use of space heaters for door heating is discussed, converted unit heaters are usually referred to. When larger heaters are converted for door heating, there is virtually no difference between those and heaters designed for door heating.

Conventional door heaters usually consist of a fan, a heat exchanger, and a discharge nozzle (Fig. 16.2). Some provision is usually made for adjusting the discharge angle relative to the plane of the door. Frequently vanes are provided to adjust the lateral spread of the discharge air stream.

Combination door and space heaters are available. These units have two nozzles, one for distributing the air as a door heater and the other for distributing air as a space heater.

Air curtains are distinguished from door heaters in that their nozzles extend the full width of the door (Fig. 16.3). They provide comfort in the door area, first by reducing infiltration and then by adding sufficient heat to bring the air that does infiltrate up to an acceptable temperature.

Pant-leg air curtains have the discharge nozzle extending up both sides of the

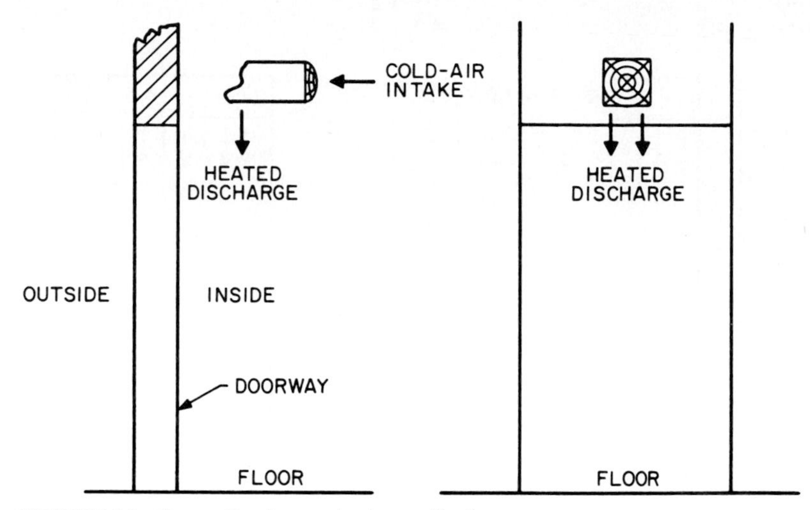

FIGURE 16.1 Conventional space heater application.

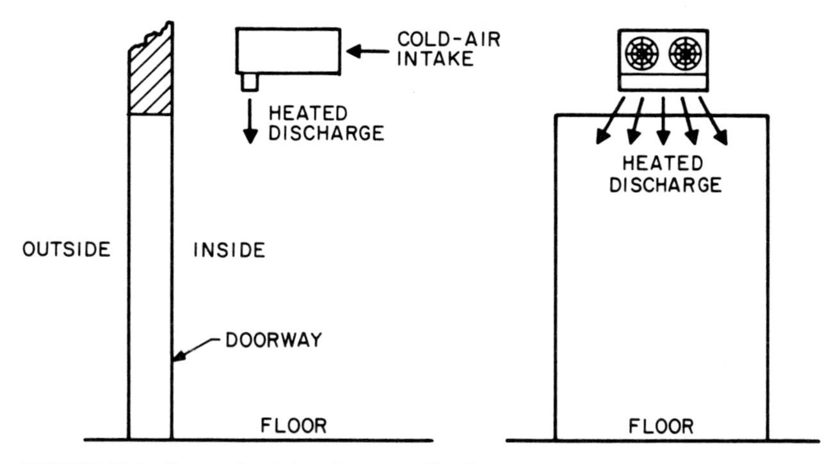

FIGURE 16.2 Conventional door heater application.

door, blowing toward the center (Fig. 16.4). This arrangement can be significantly more effective in maintaining comfortable conditions in the immediate vicinity of doors over 20 ft (5 m) high.

A *recirculating air curtain* has a discharge at the top of the door and a return in the floor (Fig. 16.5). In a few cases, recirculating air curtains discharge from one side of the door and return at the opposite side. Recirculating air curtains are frequently used in commercial applications because they substantially reduce air movement in the vicinity of the door. Recirculating air curtains in commercial applications have discharge nozzles up to 36 in (0.9 m) wide. The velocities usually do not exceed 1000 ft/min (31 m/min) compared to a minimum of 3000 ft/min (93 m/min) for industrial air curtains.

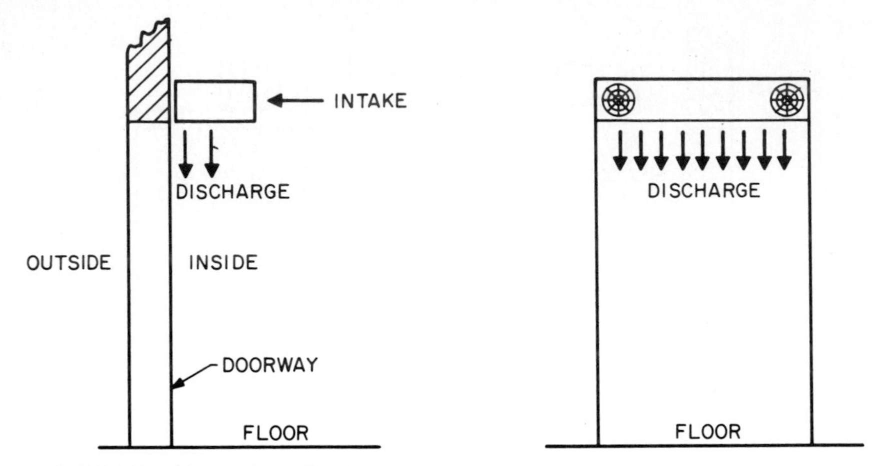

FIGURE 16.3 Air curtain application.

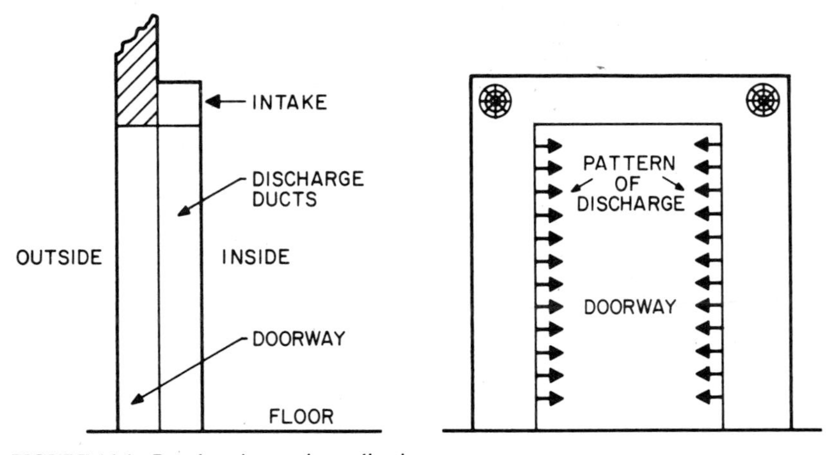

FIGURE 16.4 Pant-leg air curtain application.

Any of the above delivery systems can use *outside air* (Fig. 16.6). If the volume of outside air used is significant relative to the total infiltration, it will make a significant contribution to improving comfort in the door area. The amount of air that will come through the door will remain the same, whether it infiltrates or is brought in by a mechanical system. If air is brought in via a mechanical system, the designer can add more heat than is possible with most systems using ambient air. It is also the best design if a high degree of control is desired.

16.3.2 Heating Sections

Steam and Hot-Water Coils. Coils are the least expensive method for providing heat in a door heater. However, it is a stressful environment for coils, especially

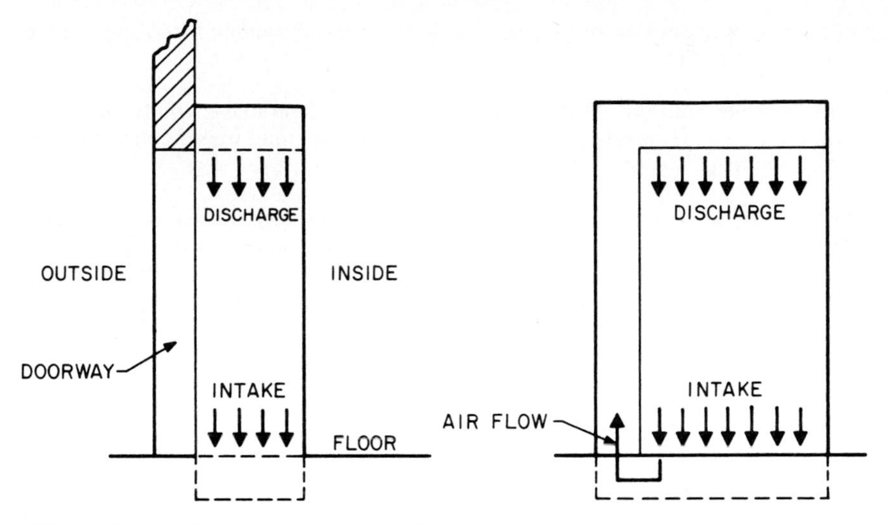

FIGURE 16.5 Recirculating air curtain application.

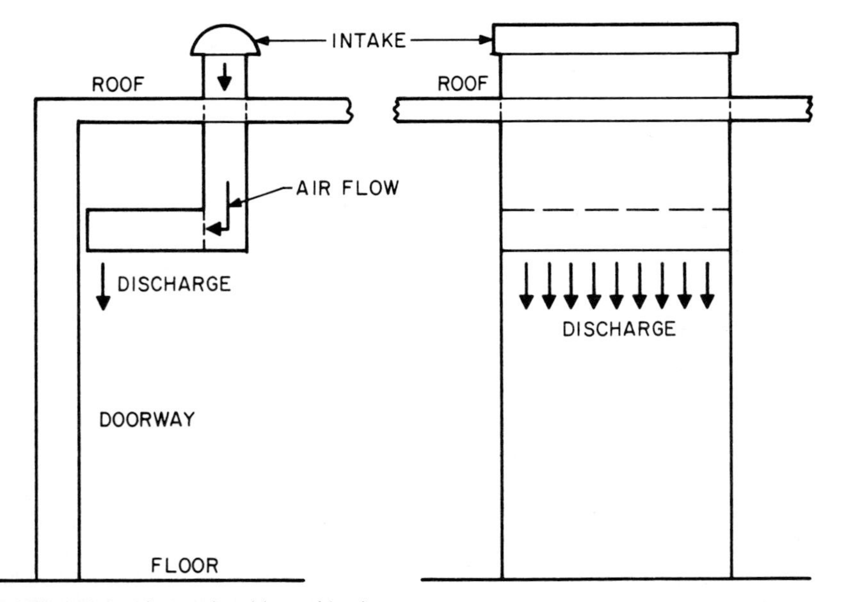

FIGURE 16.6 Air curtain with outside air.

on air curtains, where the tubes can be quite long. Therefore only industrial-grade coils [i.e., those with a minimum tube wall of 0.012 in (0.3 mm) should be used]. Although some method of controlling the amount of heat delivered should be used, the sensing devices will usually not be accurate enough to warrant an elaborate control system. In warmer climates, the capability to turn the heating section on and off should be sufficient. In cold climates, a two- or three-stage valve would be appropriate.

Electric Heating Sections. Electric heat is the least desirable method for industrial applications. However, it is a popular method for small door heaters in commercial buildings.

Indirect-Fired Gas. Indirect-fired gas door heaters usually consist of the heater blower section attached to one or more duct furnaces. Occasionally a door heater without a heating section is mounted with one or more indirect-fired gas unit heaters hung separately with their discharge directed into the air curtain inlet.

The amount of air moved by door heaters frequently interferes with the gravity exhaust of the unit heaters. This is usually the cause of a yellow flame. The problem can be corrected by using a fan in the vent.

Direct-Fired Gas. Most industrial heating applications require that direct-fired gas units be ducted to the outside. Under many circumstances it has been acceptable to use direct-fired gas door heaters without ducting to the outside, as long as they were controlled by a door switch. The assumption is that sufficient outside air would be induced into the discharge air to dilute the products of combustion. It is necessary to check local practice and regulations before using unducted direct-fired gas door heaters.

Air curtains need long burners to distribute the heat across the width of the door. As a result, they do not have the range of heating capacity most direct-fired gas installations do. Because of the high discharge temperatures, direct-fired gas air curtains should not be used on doors less than 12 ft (3.7 m) high.

16.4 CONTROLS AND CONTROL SYSTEMS

Door Switches. Door heaters are usually activated by limit switches on the door. A heavy-duty, roller-arm-type limit switch with a protective enclosure is a worthwhile investment. Usually a bracket that extends beyond the track is mounted to the door (Fig. 16.7).

Thermostats. Thermostats do not work as well as door switches as the primary control on door heaters for two reasons. (1) Infiltration loads are localized. Temperatures can vary 15°F (8°C) between two points 5 ft (1.5 m) apart. It is generally difficult to place a thermostat in a position where it will reliably read a heating requirement. (2) A door switch responds much faster, because it anticipates the load.

The door switch solution is not without problems, for it is inadequate for controlling the heating capacity. One method is to place a thermostat on the outside wall with the capacity of the heater based on the outside temperature. This works tolerably well as long as the volume of air infiltrating is relatively constant from day to day, where infiltration is primarily influenced by stack or mechanical pressures. Another method is to mount the thermostat on a column in the infiltrating air stream. This works well for direct-fired gas and electric units, but indirect-fired gas and coils respond too slowly. In some cases a two- or three-stage manual control works best.

A thermostat can also be used in parallel with the door switch. The door

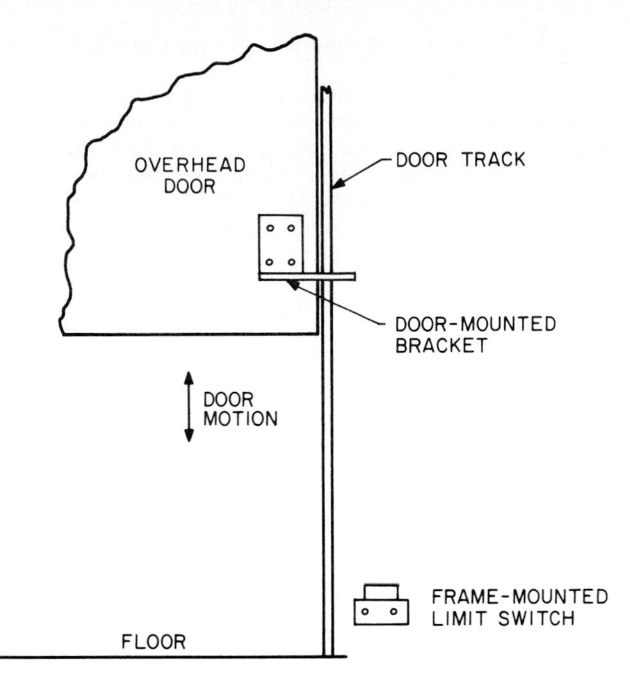

FIGURE 16.7 Application of a door limit switch.

switch will activate the unit, and the thermostat will keep it running after the door has been shut until the space reaches a comfortable temperature.

Automated Directional and/or Volume Controls. To be effective against winds, door heater discharges are normally angled against the wind. However, without the wind to push the heated air back into the building, the door heater discharges its heat outdoors. An automated directional control adjusts the nozzle direction in response to the pressure differential across the building wall. The unit would require a two-position motorized damper for directional control and a differential pressure switch, scaled in increments of 0.005 in (0.13 mm) water gauge (WG) with a deadband control to regulate the cycling of the motor. An automated volume control uses the same method to control the heater's air volume rather than the discharge direction. These systems, however, require a considerable amount of maintenance to keep functioning properly.

16.5 SELECTION OF DOOR HEATERS

16.5.1 Infiltration Load

The total-building approach recognizes that infiltration is caused by the pressure differential across the opening. Wind pressure is just one of the components. Stack pressure, mechanical pressure, and equilibrium pressure are just as important. In many cases they contribute more to the pressure differential than wind pressure.

Wind pressure is caused by the wind's impacting on the building. Stack pressure is caused by the difference in densities of the cold outside air and the warm inside air. Mechanical pressure is caused by a difference in the volumes of the exhaust and makeup air equipment. The "equilibrium pressure" is defined as the pressure assumed by the building to balance the rates of infiltration and exfiltration. It ensures that the infiltration air volume equals the exfiltrating air volume.

The exact solution of the volume of air moving through a door on a building without any internal partitions requires the solution of an equation of the form

$$\Sigma A_i C_i (W_i S_i - M - E)^{N_i} = \Sigma A_e C_e (W_e + S_e - M + E)^{N_e} \qquad (16.1)$$

where A = area of leakage
C = infiltration coefficient for type of opening
W = wind pressure, in WG (mmWG); positive on windward side and negative on leeward side
S = stack pressure, in WG (mmWG); negative if causing flow from outside to inside of building
M = mechanical pressure, in WG (mmWG); negative if volume of exhaust units exceeds that of makeup air units
E = equilibrium pressure, in WG (mmWG); positive when leakage area on windward side is greater than that on leeward side
i = windward openings
e = leeward openings
n = number of openings on leeward or windward sides

The equilibrium pressure is the unknown. Unfortunately the equilibrium pressure cannot be algebraically moved to one side of the equation. The solution requires an iterative method, a method that systematically tries different values for equilibrium pressure until one is found where the difference between the infiltration and exfiltration estimates is within acceptable limits. Normally a computer is required.

The general subject of infiltration is discussed in greater detail in Chap. 38.

16.5.2 Determining Design Conditions

The cost of door heating systems is sensitive to the design conditions selected. The engineer will need to select conditions for ambient temperatures with the door open and outdoor temperatures, correlated with speed and recovery time. Recovery time is the time required to return to normal ambient temperatures after the door has been closed.

There is no generally accepted method for specifying the design conditions for door heating. Our recommendations are presented here.

Ambient Temperature with the Door Open. It is unusual for the building's normal ambient temperatures to be maintained while the door is open. Door heating systems can be designed to maintain normal ambient temperatures, but this is rarely done because of the cost.

Normally it would be acceptable to allow temperatures in the immediate vicinity of the door to fall to 60°F (16°C) while the door is opened. For doors open relatively infrequently for short periods, the expense of equipment to maintain a 60°F (16°C) ambient temperature may not be reasonable. Under these conditions the client might consider a 50°F (10°C) design objective provided normal ambient temperatures can be recovered quickly once the door is closed. However, if the door is open for long periods (½ h) a number of times during the day, it may be necessary to design to a 65°F (20°C) ambient temperature.

Outdoor Temperature and Conditions. Selecting outdoor design conditions is equally difficult. If an engineer selects the 95 percent wind condition and the 95 percent winter temperature condition from a guide such as *ASHRAE Handbook, Fundamentals*,* the probability of the design conditions occurring is less than 0.09 percent. For a 180-day heating season, that is once every 6 years.

A reasonable design condition would be to use a temperature 15 percent higher than the winter design temperature and a wind speed between 12.5 and 17.5 mi/h (20 and 28 km/h).

Recovery Time. A third consideration is recovery time, or the amount of time needed to recover normal ambient temperatures after the door has been closed.

Recovery time R is given by

$$R = \frac{H/[1.08(N - C)A]}{60} \qquad (16.2)$$

where R = recovery time, min
H = capacity of heating units in excess of that needed for transmission load, Btu/h (W)
N = normal ambient temperature, °F (°C)
C = ambient temperature with door open, °F (°C)
A = area, ft^3 (m^3)

16.5.3 Calculating the Door Heating Load

Where infiltration simulations are not available, the following procedure will give adequate results for most applications. It tends to err by overstating infiltration and air curtain savings. Basically, it works by giving all the leakage area the characteristics (i.e., infiltration coefficient and infiltration exponent) of a large open area. This assumption enables reduction of the equilibrium pressure equation to

$$P_w - P_i = \frac{P_w - P_l}{1 + (A_w/A_l)^{1/N}} \qquad (16.3)$$

where P = wind pressure
A = leakage area
N = flow exponent
w = windward
i = inside building
l = leeward

*Published by Amer. Society of Heating, Refrigeration and Air Conditioning Engineers, Atlanta, GA.

This can be found in a number of older editions of the *ASHRAE Handbook, Fundamentals*.* A worksheet using this method is given at the end of this chapter.

The results of the procedure should be checked by comparing the measured velocity through the door with the velocity predicted under a given wind condition. A vane anemometer is the appropriate instrument because it averages velocities over the time of the test.

16.5.4 Comparing Door Heating Alternatives

The "effectiveness factor" is the ratio of the velocity of air moving through the door compared to the outside wind velocity. It can be measured for existing buildings or can be calculated by using an infiltration-estimating technique such as that described above.

Converted space heaters are adequate for mild infiltration conditions, i.e., infiltration conditions with effectiveness factors less than 0.35 for 8-ft (2.5-m) doors, 0.15 for 12-ft (3.7-m) doors, or 0.05 for 16-ft (5-m) doors.

When infiltration is more severe, equipment specifically designed for door heating is needed. The choice between conventional door heaters and air curtains is primarily an economic decision. An air curtain may have higher initial costs, but in many cases will reduce operating costs because it reduces building infiltration.

To make the comparison:

1. Calculate building infiltration with and without the air curtain.
2. Calculate the required heating capacity of the unit. If the air curtain reduces infiltration, you will not need as large a heater capacity as the conventional door heater.
3. Determine the comparative installed costs of the air curtain with its required heating capacity and the conventional door heater at its required heating capacity.
4. Calculate operating costs of the air curtain and of the conventional door heater. The air curtain will probably require a larger motor so its electrical consumption will be higher. The air curtain is likely to have more moving parts, so the maintenance costs would be expected to be higher.
5. Apply your organization's life cycle costing procedures. For the conventional door heater, you will need to consider the annualized initial costs plus the operating costs. For the air curtain, you will need to consider the annualized initial costs plus the operating costs less the energy savings resulting from the reduction in filtration.

Generally speaking, the more severe the infiltration problem, the greater the advantage of air curtains compared to door heaters. Large buildings, buildings with high-traffic doors on opposite or adjacent sides, old buildings, and buildings with doors open more then 6 h/day are good candidates for air curtains. Smaller buildings, buildings with high-traffic doors all on the same side, and buildings

*See, for example, *1977 ASHRAE Handbook, Fundamentals*, American Society of Heating, Refrigeration, and Air-Conditioning Engineers (ASHRAE), Atlanta, GA, 1977, eq. 5, p. 21.4.

with tight construction all possess characteristics that would tend to favor the conventional door heater.

16.6 ALTERNATIVES TO DOOR HEATING

One of the advantages of a total-building approach to door heating is that it suggests many options in solving a door heating problem. Obviously the first choice in solving the door heating problem would be to significantly reduce the infiltration.

As the infiltration load is evaluated, the engineer should be alert to significant, unnecessary paths for exfiltration. These include gravity vents, exhaust and makeup air equipment that is not being used, broken or open windows, and doors that are left open unnecessarily. Sometimes large sums of money have been unsuccessfully invested in door heating equipment, only to find that the solution to the problem was a very modest investment in exfiltration control.

There are also methods for controlling infiltration and improving the comfort in the area without adding heat to the area. Dock seals or dock shelters, strip doors, and vestibules can be used to reduced infiltration at the door.

Dock seals or dock shelters are intended to provide a seal between the truck body and the building wall. They can be used on loading dock doors where most of the traffic has a standard enclosure size. Dock seals or dock shelters are effective and relatively inexpensive. They should be used whenever possible.

Strip doors consist of plastic strips, 8 to 16 in (20 to 40 cm) wide, suspended from the top of the door. They are also an effective and inexpensive means of controlling mild infiltration conditions. In some applications, a potentially dangerous situation can develop as the strips accumulate dirt and visibility becomes impaired.

Vestibules are an enclosed pair of doors situated so that the first door is closed before the second is opened. They are effective as long as traffic is not so heavy as to keep both doors open simultaneously. An enclosed truck dock is a form of vestibule.

16.7 DOOR HEATER INSTALLATION

Conventional door heaters are mounted between one-half the door height and the total door height back from the door. Their discharge should be as close to the door height as possible without running the risk of their being hit by traffic in the vicinity.

Ideally air curtains are mounted with the discharge slots flush with the bottom of the door sill and adjacent to it. Frequently the site does not permit this arrangement. If the air curtain cannot be mounted so that the discharge is flush with the bottom of the door sill, care must be taken to ensure that the air curtain is mounted far enough away from the door that air discharged at a 15° angle will clear the sill. If the air curtain is mounted more than 3 ft (0.9 m) back from the door, side baffles should be provided.

With either conventional door heaters or air curtains, there must be sufficient clearance at the inlet. Objects cannot be closer than 3 times the short dimension of the inlet if air can pass around just one side or 1.5 times the short dimension if air can pass around both sides of the object.

16.8 DOOR HEATING WORKSHEET—EXPLANATION

Refer to par. 16.5.3, p. 16.9.

Windward Side of Building

C_0 = air curtain efficiency
C_1 = area of door to be heated
C_2 = area of other open doors
C_3 = area of windward wall
C_4 = leakage area of windward wall
C_5 = total windward leakage area without air curtains
C_6 = windward leakage area with air curtains

Leakage Area on Windward Side. The windward side of the building is the side facing the wind. There can be only one windward side; this procedure is not appropriate for oblique winds. "Other open doors" refers to the doors that would be open at the same time as the heated doors would be. A value of 0.001 in the equation for C_4 represents an average leakage area per square foot (meter) of wall surface in a typical industrial plant. Leakage areas can also be calculated by using the crack method, if greater precision is required.

The design air curtain efficiency is based on the ratio of infiltration with the air curtain versus infiltration without. An efficiency of 0.80 means that there is 80 percent less infiltration through the door with the air curtain than without. The procedure calculates the pressure difference, assuming the air curtain operates at the design efficiency. Once the pressure difference has been estimated, the specifications for the air curtain to deliver that efficiency can be calculated.

Leeward Side of Building

C_7 = area of leeward doors
C_8 = area of leeward walls
C_9 = leakage area of other walls
C_{10} = area of roof
C_{11} = leakage area of roof
C_{12} = total leeward leakage

Leakage Area on Leeward Side. The leeward sides are those surfaces of the building with a negative wind pressure, including the side opposite the windward side, the sides parallel to the wind, and the roof. A value of 0.0002 represents an average leakage area per square foot (meter) of roof area. The leakage is primarily through vents in the roof.

Total Leakage Area

C_{13} = total leakage area without air curtains
C_{14} = total leakage area with air curtains

Total Leakage Area. The sum of the windward and leeward leakage areas is the total leakage area.

Mechanical Pressure

C_{15} = capacity of exhaust systems
C_{16} = capacity of makeup air systems
C_{17} = difference between capacities
C_{18} = mechanical pressure without air curtains
C_{19} = mechanical pressure with air curtains

In this section the difference in makeup air and exhaust air volumes is divided by the leakage area to get the velocity. Then the velocity is converted to a pressure.

Stack Pressure

C_{20} = height of door
C_{21} = width of door
C_{22} = temperature of inside air
C_{23} = inside temperature factor
C_{24} = temperature of outside air
C_{25} = outside temperature factor
C_{26} = stack pressure constant
C_{27} = neutral level
C_{28} = stack pressure
C_{29} = infiltration due to stack pressure

This section calculates the pressure due to the difference in density of the cold outside air and the warm inside air. The procedure assumes that the neutral level of the building is at two-thirds the door height. This is reasonable for buildings with a few open doors. If needed, a more accurate estimate can be made of the neutral level based on the leakage areas for the wall and roof surfaces and the door areas and heights.

Wind Pressure

C_{32} = wind velocity
C_{33} = wind velocity, ft/min (m/min)
C_{34} = windward pressure
C_{35} = leeward pressure

This section converts a wind velocity in miles (kilometers) per hour to windward and leeward wind pressures. The effect of terrain and the aerodynamics of the building shell are accounted for.

Equilibrium Pressure

C_{30} = leakage area ratio without air curtains
C_{31} = leakage area ratio with air curtains
C_{36} = equilibrium pressure without air curtains
C_{37} = equilibrium pressure with air curtains

This procedure uses the difference in external pressures and the ratio of windward and leeward leakage areas to calculate equilibrium pressure. This is Eq. (16.1).

Pressure Difference across the Opening

C_{38} = without air curtains
C_{39} = with air curtains

This is the wind pressure less the equilibrium pressure plus the mechanical pressure and the stack pressure. The stack pressure is always positive. The mechanical pressure will be negative if the capacity of the exhaust systems exceeds that of the makeup air systems.

Infiltration through the Door

C_{40} = doorway velocities without air curtains
C_{41} = doorway velocities with air curtains
C_{42} = infiltration without air curtains
C_{43} = infiltration with air curtains

This section converts the pressure difference across the opening to a velocity and then to a volume. A value $1 - C_0$ accounts for the reduction in effective leakage area due to the air curtain.

Heating Capacity Required

C_{44} = acceptable temperature with door open
C_{45} = outdoor temperature
C_{46} = Btu/ft^3 (kJ/m^3) min required
C_{47} = heating capacity without air curtains
C_{48} = heating capacity with air curtains

This section calculates the heating capacity required to bring the infiltrating air up to the required temperature.

Equipment Sizing

Without Air Curtains

C_{49} = door heater nozzle width
C_{50} = base value
C_{51} = windspeed factor
C_{52} = efficiency factor
C_{53} = intermediate calculation
C_{54} = required velocity (use to specify)
C_{55} = required ft^3/min (m^3/min) (use to specify)
C_{56} = required Btu/h (W)

With Air Curtains

C_{57} = door heater nozzle width
C_{58} = base value
C_{59} = windspeed factor
C_{60} = efficiency factor
C_{61} = intermediate calculation
C_{62} = required velocity (use to specify)
C_{63} = required ft^3/min (m^3/min) (use to specify)
C_{64} = required Btu/h (W)

This section develops the required velocities, volumes, and heating capacities for the door heaters. For the conventional door heater, the velocity and volume are determined to ensure that the discharge air has sufficient force to reach the floor. In the case of air curtains, the velocity and volume express the force required to maintain the design efficiency. The required heating capacity can be provided by the door heater or by space heaters in the immediate vicinity.

16.9 DOOR HEATING WORKSHEET—SAMPLE FORM FOR USE

Windward Side of Building

C_0 = ______
C_1 = ______ ft^2 (m^2)
C_2 = ______ ft^2 (m^2)
C_3 = ______ ft^2 (m^2)
$C_4 = C_3\,(0.001)\ ft^2/ft^2$ = ______ ft^2 (m^2)
$C_5 = C_1 + C_2 + C_4$ = ______ ft^2 (m^2)
$C_6 = C_5 - C_1C_0$ = ______ ft^2 (m^2)

Leeward Side of Building

C_7 = ______ ft^2 (m^2)
C_8 = ______ ft^2 (m^2)
$C_9 = C_8(0.001)$ = ______ ft^2 (m^2)
C_{10} = ______ ft^2 (m^2)
$C_{11} = C_{10}(0.0002)$ = ______ ft^2 (m^2)
$C_{12} = C_7 + C_9 + C_{11}$ = ______ ft^2 (m^2)

Total Leakage Area

$C_{13} = C_5 + C_{12}$ = ______ ft^2 (m^2)
$C_{14} = C_6 + C_{12}$ = ______ ft^2 (m^2)

Imbalance of Mechanical Systems

C_{15} = ______ ft³/min (m³/min)

C_{16} = ______ ft³/min (m³/min)

$C_{17} = C_{15} - C_{16}$ = ______ ft³/min (m³/min)

Pressure Implied by Imbalance of Mechanical Systems

$$C_{18} = \left(\frac{\text{ABS}(C_{17}/C_{13})}{4005}\right)^{1.6667} \frac{C_{17}}{\text{ABS}(C_{17})} = \text{\_\_\_\_\_\_ inWG (mmWg)}$$

$$C_{19} = \left(\frac{\text{ABS}(C_{17}/C_{14})}{4005}\right)^{1.6667} \frac{C_{17}}{\text{ABS}(C_{17})} = \text{\_\_\_\_\_\_ inWG (mmWg)}$$

Stack Pressure

C_{20} = ______ ft (m)

C_{21} = ______ ft (m)

C_{22} = ______ °F (°C)

$$C_{23} = \frac{1}{C_{22} + 460} = \text{\_\_\_\_\_\_ °F(°C)}$$

C_{24} = ______ °F (°C)

$$C_{25} = \frac{1}{C_{24} + 460} = \text{\_\_\_\_\_\_°F(°C)}$$

$C_{26} = 7.642(C_{25} - C_{23})$ = ______°F(°C)

$C_{27} = C_{20}(0.6667)$ = ______ ft (m)

$C_{28} = C_{27} - C_{20}/2)C_{26}$ = ______ inWG (mWg)

$C_{29} = C_{28}(0.5)(4005C_{20})(0.6667C_{21})$ = ______ ft³/min (m³/min)

Wind Pressure

C_{32} = ______ mi/h (km/h)

$C_{33} = C_{32}(88)$ = ______ ft/min (m/min)

$$C_{34} = \left(\frac{0.54C_{33}}{4005}\right)^2 = \text{\_\_\_\_\_\_ inWG (mmWg)}$$

$$C_{35} = \left(\frac{0.31C_{33}}{4005}\right)^2(-1) = \text{\_\_\_\_\_\_ inWG (mmHg)}$$

Equilibrium Pressure

$$C_{30} = \left(\frac{C_5}{C_{12}}\right)^{1.6667} = \text{\_\_\_\_\_\_ ft}^2\text{(m}^2\text{)}$$

$$C_{31} = \left(\frac{C_6}{C_{12}}\right)^{1.6667} = \text{\_\_\_\_\_\_ ft}^2\text{(m}^2\text{)}$$

$$C_{36} = C_{34} - \frac{C_{34} - C_{35}}{1 + C_{30}} = \text{\_\_\_\_\_\_ inWG(mmWg)}$$

$$C_{37} = C_{34} - \frac{C_{34} - C_{35}}{1 + C_{31}} = \text{\_\_\_\_\_\_ inWG (mmHg)}$$

Pressure Difference across Opening

$$C_{38} = C_{34} - C_{36} + C_{28} + C_{18} = \_\_\_\_\_ \text{ inWG (mmWg)}$$

Infiltration through the Door

$$C_{40} = 4005 \frac{C_{38}}{\text{ABS}(C_{38})} \text{ABS}(C_{38})^{0.5} = \_\_\_\_\_ \text{ ft/min (m/min)}$$

$$C_{41} = 4005 \frac{C_{39}}{\text{ABS}(C_{39})} \text{ABS}(C_{39})^{0.5} = \_\_\_\_\_ \text{ ft/min (m/min)}$$

$$C_{42} = C_1 C_{40} = \_\_\_\_\_ \text{ ft}^3\text{/min (m}^3\text{/min)}$$

$$C_{43} = C_1 C_{41}(1 - C_0) = \_\_\_\_\_ \text{ ft}^3\text{/min (m}^3\text{/min)}$$

Heating Capacity Required

$$C_{44} = \_\_\_\_\_ \text{ °F (°C)}$$

$$C_{45} = C_{24} = \_\_\_\_\_ \text{ °F (°C)}$$

$$C_{46} = (C_{44} - C_{45})(1.08) = \_\_\_\_\_ \text{ Btu/(ft}^3 \cdot \text{min) [kJ/(m3} \cdot \text{min)]}$$

$$C_{47} = \frac{C_{46}}{1000} C_{42} = \_\_\_\_\_ \text{ kBtu (kJ)}$$

$$C_{48} = \frac{C_{46}}{1000} C_{43} = \_\_\_\_\_ \text{ kBtu (kJ)}$$

CHAPTER 17
VALANCE HEATING AND COOLING

Edward A. Bogucz, P.E.
Edwards Engineering Corp., Pompton Plains, New Jersey

17.1 DESCRIPTION

The "valance terminal unit" is a natural-convection heating and/or cooling unit which does not use a fan or blower to produce heating or cooling. The unit consists of a finned heat-transfer element housed in an insulated enclosure positioned in a room near the juncture of the ceiling and an adjacent wall. The heat-transfer element of the valance is heated or chilled by fluid or refrigerant circulated through tubes in the element.

17.2 FEATURES

Since the valance terminal unit operates as a convection device, these operating features are realized: electric energy is not consumed, operation is extremely quiet, no forceful blower drafts are created, maintenance is minimal and, in the conditioned space, the temperature in the comfort zone is exceptionally uniform. Since direct humidification and introduction of fresh air for ventilation are not possible with the basic valance terminal unit, supplementary systems must be provided when humidification and/or ventilation is required. In the cooling mode, the convected air is dehumidified during the operation. Additionally, no ductwork is required, only thermostat wiring is needed to the control valves, and maximum usable floor space is available. Rapid installation is possible, and in most cases both furring and plastering are eliminated.

17.3 CONSTRUCTION

Basic Components. The valance unit is a natural-convection heat-transfer device similar to a baseboard convector, but unlike the convector, the valance unit is

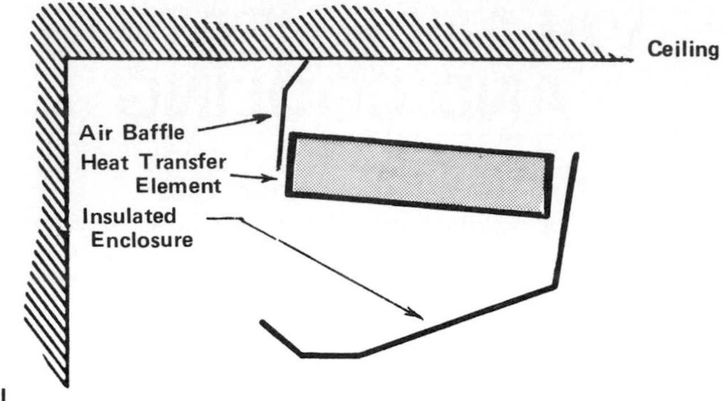

FIGURE 17.1 Basic components of the valance unit (cross section). *Note*: When used for cooling, the insulated enclosure is modified to include drain fitting for condensate drainage. *(Courtesy of Edwards Engineering Corp.)*

placed at the ceiling, not the floor. The three major components of the valance are (Fig. 17.1)

1. Heat-transfer element, consisting of tubes with external plate fins
2. Air baffle, to channel air flow over the heat-transfer element
3. Insulated enclosure, to house the element and to drain the condensate dropped from the element during the cooling operation

Heat-Transfer Element. The heat-transfer element is a plate-fin coil with aluminum fins on copper tubes. On two-pipe systems, a single set of tubes is provided through which both the hot and cold fluids may circulate as required. On four-pipe systems, a dual set of tubes enables the flow of cold fluid through one set and the flow of hot fluid through the second set. See Fig. 17.2 for typical heating/cooling four-pipe valance unit.

Air Baffle. The air baffle presses against the ceiling and is fastened to the upper rear edge of the casing which holds the element. The baffle prevents the air from bypassing the heat-transfer element as the air flows through the valance unit.

Enclosure. The enclosure is constructed of aluminum sheet, bent, formed, and lined with a layer of closed-cell, watertight plastic insulation. The enclosure, upon installation, is pitched in the direction of the condensate drain by adjusting the enclosure support hangers.

Accessories. Physical accessories supplied with valance units consist of wall supports, ceiling hangers, support hooks, wall trim pieces, and splicers.

17.4 OPERATION

17.4.1 Cooling

When the unit is operating in the cooling mode, chilled fluid or refrigerant gas is directed to flow through the valance element. Warm air in contact with the ele-

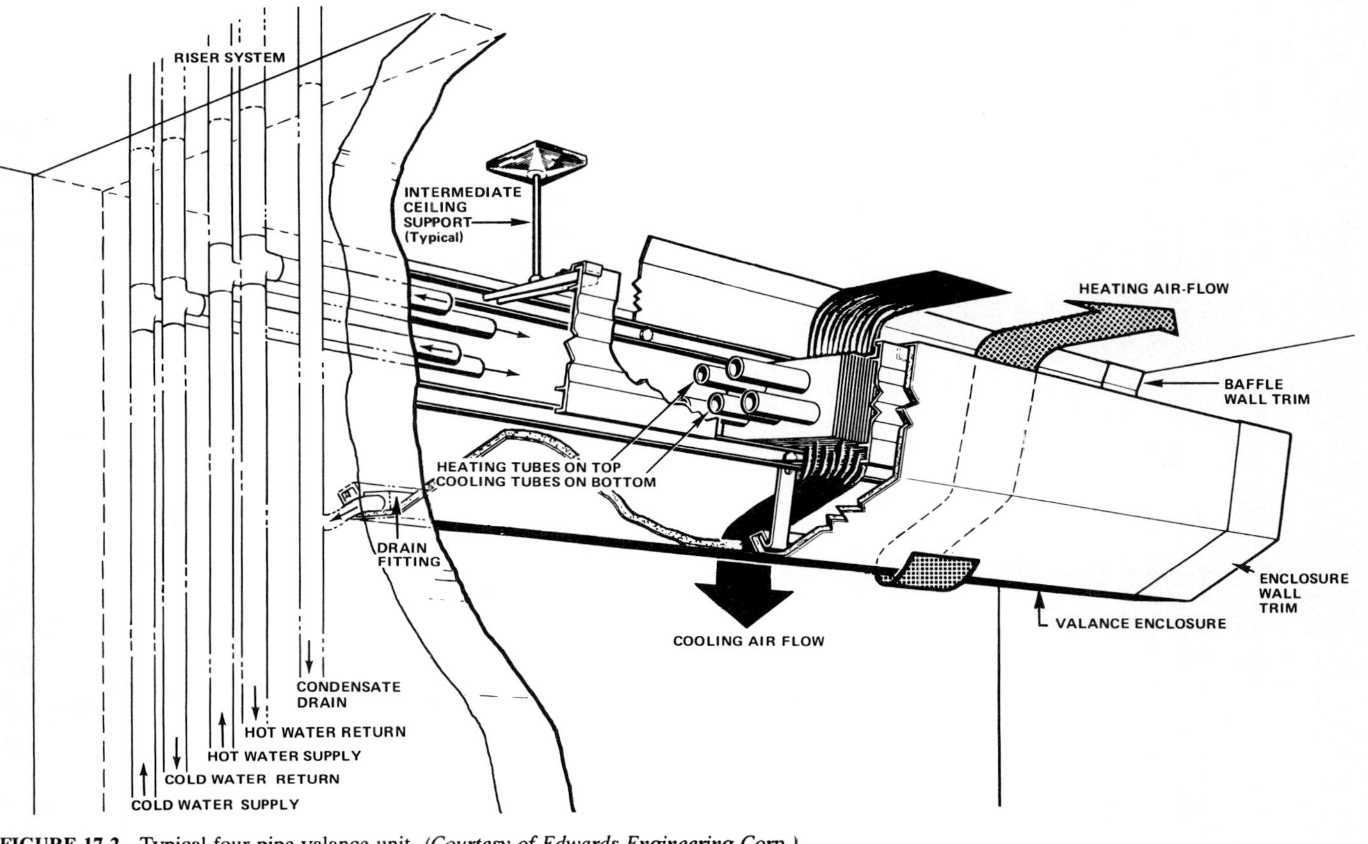

FIGURE 17.2 Typical four-pipe valance unit. *(Courtesy of Edwards Engineering Corp.)*

ment is cooled. This denser cool air falls gently to the floor from the rear outlet of the valance. The cool air spreads evenly across the floor and upward throughout the room. The cooling process is continuous (as long as it is directed by thermostatic control): The valance initiates a continuous movement of cool air down the wall adjacent to the unit, across the floor, up and back across the ceiling to the valance unit. See Fig. 17.3.

Drafts are minimal as a result of the air's passing through a large area across the valance unit, which then results in a low outlet velocity on the order of 35 ft/min (10 m/min). The air outlet on the valance is the length of the room wall and a minimum of 6 in (15 cm) wide. The valance therefore functions as a wall-to-wall linear grille with an evenly distributed low air velocity.

As the air is cooled by the heat-transfer element, the air temperature drops below the dew point and the moisture in the air condenses on the fins of the element. The moisture drops off the element onto the waterproof liner within the valance enclosure. The condensate flows along the insulated liner to a drain connection where the condensate flows out of the unit.

During the cooling cycle, dehumidification within the valance is exceptionally good. An average of 16 ft^2 (1.5 m^2) of coil element face area per nominal ton (3517 W) (refrigeration) exists in the valance unit versus about 1 ft^2/ton (26.4 mm^2/W) in a forced-air terminal unit. This ratio means that the coil air velocity in the valance is about one-sixteenth as great as that in forced-air units. Because of this low velocity, the air passing through the valance is in intimate contact with the cooling surface for a longer time than is air which passes through a forced-air unit. Consequently, more moisture is removed with the valance unit, thus resulting in exceptionally good dehumidification.

Additionally, during the cooling cycle the valance provides air filtration. The aluminum fins are covered with moisture due to the dehumidification effect. Dust and lint in the room air pass slowly through the valance element, eventually contact the wet fin surfaces, and adhere to them. This filter action has the characteristic of being self-cleaning. The air flow, condensate flow, and force of gravity are all in a downward direction. The dust and lint are washed down with the condensate into the drain liner and out through the drain connection.

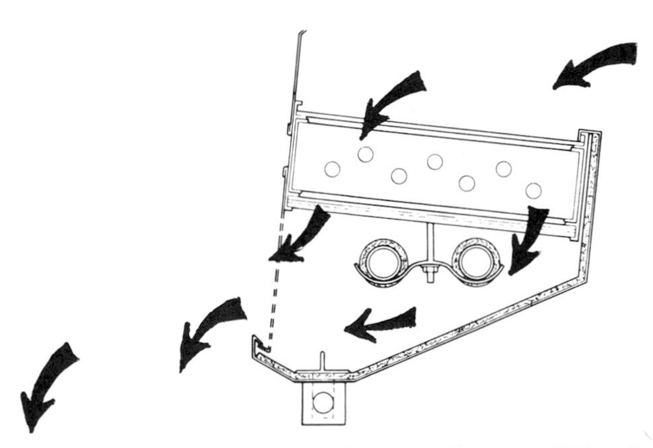

FIGURE 17.3 Air flow during cooling cycle. *(Courtesy of Edwards Engineering Corp.)*

17.4.2 Heating

When the unit is operating in the heating mode, hot heat-transfer fluid is controlled to flow through the valance element. Air in contact with the coil is heated. This less dense warm air rises and flows upward and outward through the front outlet of the valance unit. The warm air spreads across the ceiling of the room. This layer of warm air leaving the valance unit warms the ceiling of the room. The ceiling in turn radiates heat downward to the objects and floor of the room. When the layer of warm air from the valance strikes the wall opposite the valance, the air drops 4 to 5 in (10 to 12 cm), reverses its direction, and flows back to the valance for reheating. Smoke patterns show the air movement in Fig. 17.4.

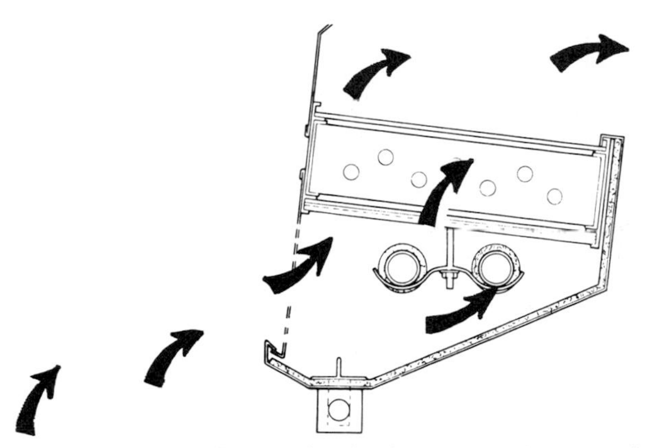

FIGURE 17.4 Air flow during heating cycle. *(Courtesy of Edwards Engineering Corp.)*

Since the floor is at right angles to the vertical rays radiating from the ceiling, the floor temperature experienced with valance heating is slightly warmer than the temperature for some distance above the floor (Fig. 17.5). This phenomenon contrasts with the cool floor temperatures noted with some other heating systems, such as forced hot air or baseboard convection.

The warm floor temperature experienced with valance heating retards the formation of drafts, even when all-glass exterior walls are encountered. Cold air may flow down the glass wall and to the floor, which has been warmed by energy radiated from the ceiling. The warm floor heated by radiation from the warm ceiling neutralizes and diffuses any cold draft flowing off the windows.

17.5 DESIGN OF THE VALANCE

17.5.1 Element Sizes

To accommodate various load requirements, element sizes can be fabricated with various numbers of tubes and wider fins. The fin height and fin spacing remain constant. The length of the element usually is selected to fit the wall-to-wall di-

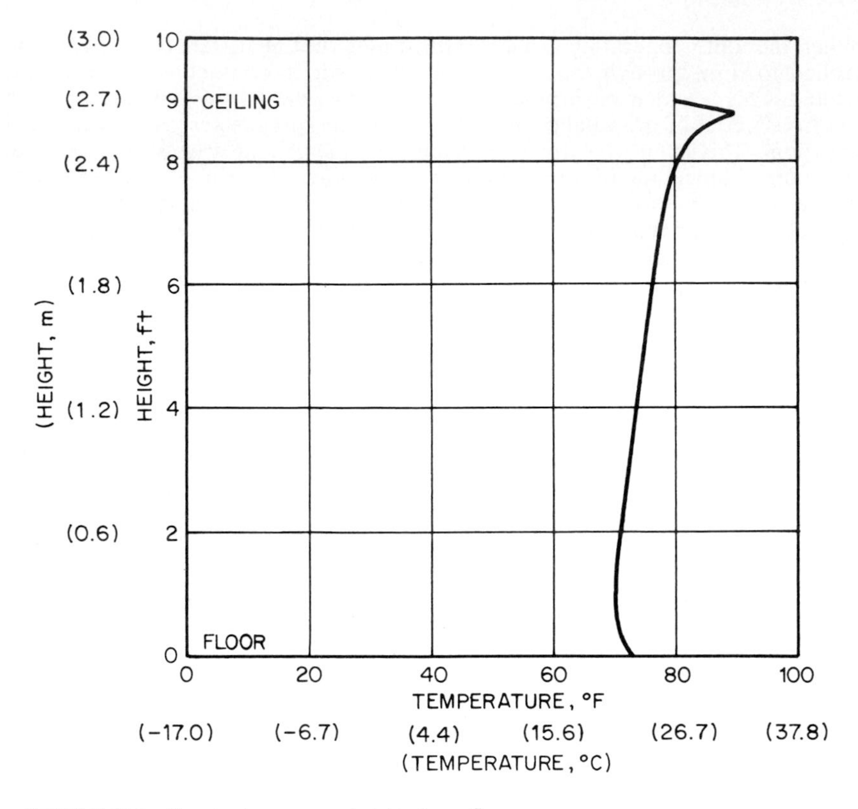

FIGURE 17.5 Temperature versus height above floor.

mensions of the room. A typical valance coil will involve a coil from two to eight tubes wide along with an appropriate tube circuiting or manifolding to secure a suitably designed pressure drop through the entire coil. The element tube circuiting may be designed for series flow, parallel flow, or suitable combinations to obtain desired fluid pressure drop.

17.5.2 Minimum Clearance

Adequate clearance must be provided for the air to flow into and exit from the valance enclosure. Refer to Fig. 17.6 for minimum dimensions.

17.5.3 Selection of Cooling Element

1. *Determination of the cooling load:* Determine the heat gain for the room, in Btu/h(w), using a standard procedure such as that recommended in Chap. 2, by

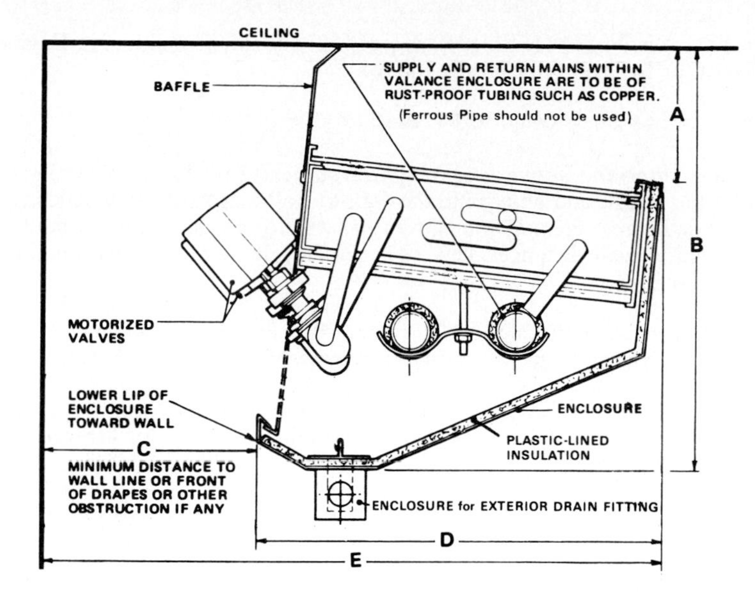

Rows Valance Units	Dimensions, in				
	A	B	C	D	E
2	3 1/2	10 3/4	3 13/16	7 3/16	11
3	3 3/4	11 13/16	5 5/8	9 1/4	15 1/4
4	4	12 5/8	6 1/2	11 13/16	18 5/16
5	4 1/4	13 1/2	7 1/2	14 3/8	21 7/8
6	4 1/2	14 7/8	9 3/4	16 13/16	26 9/16
8	6 1/8	20 3/8	12 3/4	22 9/16	35 5/16

Rows Valance Units	Dimensions, cm				
	A	B	C	D	E
2	8.9	27.3	9.7	18.3	27.9
3	9.5	30.0	14.3	23.5	38.7
4	10.2	32.1	16.5	30.0	46.5
5	10.8	34.3	19.1	36.5	55.6
6	11.4	37.8	24.8	42.7	67.5
8	15.6	51.8	32.4	57.3	89.7

FIGURE 17.6 Minimum installation allowances. *(Courtesy of Edwards Engineering Corp.)*

the American Society of Heating, Refrigeration, and Air-Conditioning Engineers (ASHRAE), or by the Hydronics Institute.

EXAMPLE Assume a heat gain of 12,000 Btu/h (3517 W).

2. *Establishing the active length of the valance:* Locate the position of the valance in the room, and determine the wall-to-wall length of the valance enclosure. Next determine the active length of the valance by deducting 12 in (30 cm) from the wall-to-wall distance. This 12-in (30-cm) deduction allows for the coil end connections.

EXAMPLE The wall-to-wall length is 15 ft (4.6 m). Deduct 12 in (30 cm) for coil end connections. The active coil length of the valance element then becomes 14 ft (4.3 m).

3. *Determination of the valance element cooling capacity required:* From the first two steps above, establish the required cooling capacity for the valance element in Btu/(h · ft) (W/m).

EXAMPLE 12,000/14 = 857 Btu/(h · ft) (818 W/m) of length.

4. *Selection of valance element size:* Refer to Table 17.1; for the average temperature of chilled fluid through the element, obtain the size of the valance cooling element.

TABLE 17.1 Cooling Ratings for Four-Pipe Dual-a-Matic Units

Ratings for average chilled water temperature t_c into valance at 70% sensible heat and 30% latent heat for 80°F dry-bulb room air temperature entering valance, Btu/(h · lin ft)

Valance tubes						
Cooling	Heating	t_c = 35°F	t_c = 40°F	t_c = 45°F	t_c = 50°F	t_c = 55°F
2	1	310	280	250	210	170
3	2	480	430	380	330	270
4	3	630	570	510	440	350
5	4	790	710	630	550	450
6	5	950	860	760	660	540
8	6	1270	1150	1020	880	710

Ratings for average chilled water temperature t_c into valance at 70% sensible heat and 30% latent heat for 26.7°C dry-bulb room air temperature entering valance, W/lin m

Valance tubes						
Cooling	Heating	t_c = 1.7°C	t_c = 4.4°C	t_c = 7.2°C	t_c = 10°C	t_c = 12.8°C
2	1	295.9	267.3	238.6	200.4	162.3
3	2	458.2	410.4	362.7	315.0	257.7
4	3	601.3	544.1	486.8	420.0	334.1
5	4	754.1	677.7	601.3	525.0	429.5
6	5	906.8	820.9	725.4	630.0	515.4
8	6	1212.2	1097.7	973.6	840.0	677.7

EXAMPLE At 40°F (4.4°C) average chilled fluid temperature, for 857 Btu/(h · ft) (818 W/m) a six-row-wide coil would be selected, since the tabular rating shows a capacity of 860 Btu/(h · ft) (820 W/m).

17.5.4 Selection of Heating Element

1. *Determination of the heating load:* Determine the heat loss for the room in Btu/h(w), using a standard procedure such as that recommended in Chap. 2, by ASHRAE, or by the Hydronics Institute.

EXAMPLE Assume a heat loss of 14,000 Btu/h (4103 W).

2. *Establish the active length of the valance:* Locate the position of the valance in the room, and determine the wall-to-wall length of the valance enclosure. Next determine the active length of the valance by deducting 12 in (30 cm) from the wall-to-wall distance. This 12-in (30-cm) deduction allows for the coil end connections.

EXAMPLE The wall-to-wall length is 15 ft (4.6 m). Deduct 12 in (30 cm) for the coil end connections. The active coil length of the valance element then becomes 14 ft (4.3 m).

3. *Determination of the required valance element heating capacity:* From the previous steps, establish the required heating capacity for the valance element in Btu/(h · ft) or (W/m).

TABLE 17.2 Heat Output for Four-Pipe Dual-a-Matic Units

Output for average water temperature T_{av} through valance, Btu/(h · lin ft)						
Valance tubes						
Cooling	Heating	130°F* 60°F	150°F 80°F	170°F 100°F	190°F 120°F	210°F 140°F
2	1	140	240	330	430	520
3	2	240	400	550	710	870
4	3	330	550	770	990	1210
5	4	420	710	990	1270	1550
6	5	500	860	1210	1550	1900
8	6	690	1170	1650	2120	2590
Output for average water temperature T_{av} through valance, W/m						
Valance tubes						
Cooling	Heating	54.4°C 15.6°C	65.6°C 26.7°C	76.7°C 37.8°C	87.8°C 48.9°C	98.9°C 60.0°C
2	1	133.6	229.1	315.0	410.4	496.3
3	2	229.1	381.8	525.0	677.7	830.4
4	3	315.0	525.0	735.0	945.0	1155.0
5	4	400.9	677.7	945.0	1212.2	1479.5
6	5	477.3	820.9	1155.0	1479.5	1813.6
8	6	658.6	1116.8	1575.0	2023.5	2472.2

**Note*: Top temperature is T_{av} and bottom is ΔT. ΔT is the difference between the average water temperature passing through the coil and a room temperature of 70°F (21.1°C) at 5-ft (1.5-m) occupancy level.

EXAMPLE 14,000/14 = 1000 Btu/(h · ft) (955 W/m) of length.

4. *Selection of valance element size:* Refer to Table 17.2 on the previous page; for the average temperature of the hot fluid through the element, obtain the size of the valance heating element.

EXAMPLE At 170°F (76.7°C) average water temperature for 1000 Btu/(h · ft) (955 W/m), a five-row-wide heating coil would be selected, since the tabular rating shows a capacity of 1210 Btu/(h · ft) (1155 W/m).

CHAPTER 18
BOILERS*

Cleaver-Brooks, Division of Aqua-Chem, Inc.,
Milwaukee, Wisconsin

18.1 INTRODUCTION

The term "boiler" applies to a device for generating (1) steam for power, processing, or heating purposes or (2) hot water for processing, heating purposes, or hot-water supply. Generally, a boiler is considered a steam producer. However, many boilers designed for steam are convertible to hot water.

Boilers are designed to transmit heat from a combustion source, (usually fuel combustion) to a fluid contained within the boiler vessel itself. If the fluid heated is other than water, e.g., Dowtherm®, the unit is classified as a vaporizer or a thermal liquid heater.

To ensure safe control over construction features, all insurable stationary boilers in the United States must be constructed in accordance with the ASME (American Society of Mechanical Engineers') *Boiler and Pressure Vessel Code*, known as the ASME *Boiler Code*. ASME publications related directly to the boiler industry include *Boiler and Pressure Vessel Code, Power Test Codes*, and *American Standards*.

18.2 BOILER TYPES

While many types of boilers are manufactured today that differ in size, configuration, materials, and construction methods, the popular types are firetube, watertube, and cast-iron sectional boilers.

18.2.1 Packaged Boilers

The one-piece package boiler is completely shop-assembled with fuel burner (usually oil, gas, or oil/gas), mechanical draft system (forced or induced), external insulation (usually with casing or jacket), refractories, trim (gauges, safety or relief valve, water column), panel-mounted controls, oil preheater (when residual

*Section 18.7, Cast-Iron Boilers, was written by Nils R. Grimm, co-editor of this handbook.

fuels are burned), and interconnecting piping and complete wiring. It is mounted on a structural-steel base, ready to be skidded or lifted into place on a single foundation.

The only connections required for operation are those leading to (1) sources of water, fuel, and electricity; (2) steam and condensate return piping; and (3) a stub stack for vent gases. The installation is sufficiently simple to be handled without extensive subcontracting.

The packaged boiler (also known as the *packaged steam generator*) has found wide application since World War II in every field requiring boilers except large central station units.

Advantages. A number of advantages are claimed for the packaged boiler. The most important is that responsibility for the entire unit is assigned to a single manufacturer. Standardized models provide the best possible manufacturing economy, and integration of custom-designed burner components results in uniformly high combustion efficiency.

The manufacturer of completely assembled packaged steam generators test-fires the equipment prior to shipping. The efficiency of the packaged boilers is relatively constant, between 25 and 100 percent of capacity. Because of standardization, complete piping and wiring information for the entire installation is available prior to purchase from the manufacturer.

Packaged units are space savers and do not require elaborate foundations. Adaptable in location, packaged units may be installed anywhere from penthouses to remote plants.

18.2.2 Firetube Boilers

In the firetube boiler concept, the combustion chamber and the flue gas passageways are tubes which are immersed in the boiler water. The tubes and water are contained in an enclosure referred to as a "shell." The heat released in the combustion process is transferred through the tube walls to the surrounding water. Usually the combustion chamber or furnace is a relatively large-diameter tube compared to those tubes that conduct the gaseous products of combustion to a point outside the shell or pressure vessel.

Many arrangements have been developed. Tubes have been placed in horizontal, inclined, and vertical positions, with one or more flue gas passes. The modern version of the firetube boiler is the modified Scotch marine type. This type of burner fires into a cylindrical furnace with the hot gases then going through one, two, or three successive banks of tubes to transfer as much heat as possible to the boiler water. (See Fig. 18.1.)

The modified Scotch boiler is probably the most popular power boiler manufactured today for capacities up to about 28,000 lb/h (12,600 kg/h) steam at pressures to 250 lb/in^2 (1724 kPa) or more. Its compactness is a design feature originally developed primarily for marine service.

18.2.3 Selecting a Packaged Boiler

Generally the decision to buy a packaged boiler is based on considerations of capacity, pressure, headroom, and space available for installation. Statistics show

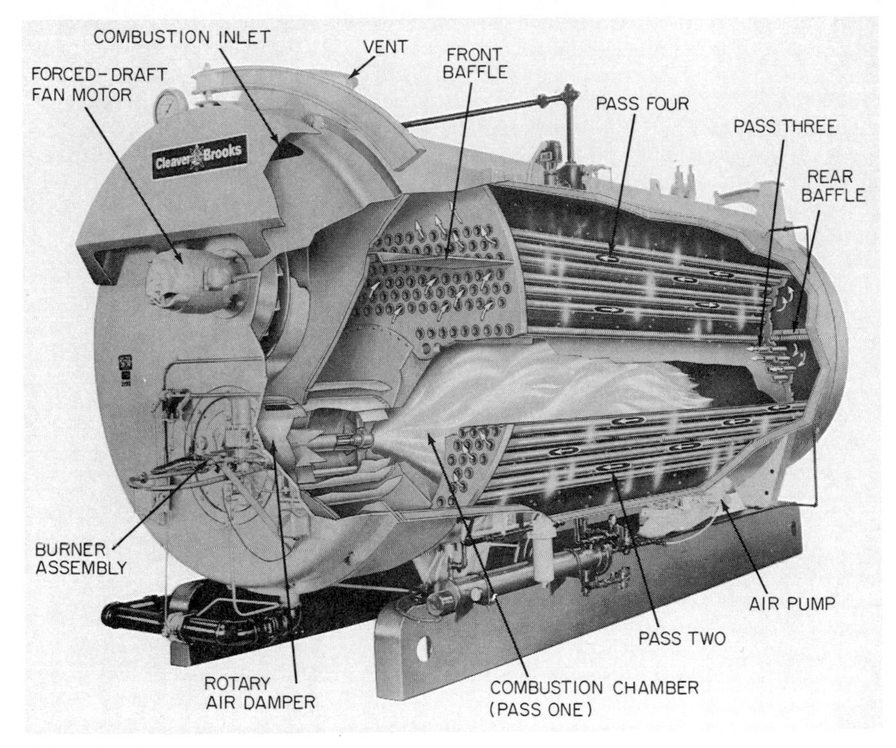

FIGURE 18.1 Firetube boiler. *(Courtesy of Cleaver-Brooks.)*

that firetube units are preferred for smaller loads up to about 25,000 lb/h (11,300 kg/h) and watertube units for loads over 25,000 lb/h (11,300 kg/h).

Multiunit installations of packaged boilers are being made in order to take advantage of the substantial savings gained over field-erected units. There are also operating and maintenance advantages that favor multiunit installations over larger field-erected boilers, such as operational flexibility to suit the load, permitting peak efficiency, and minimal production disruption due to outages.

18.2.4 Design Criteria

In the packaged firetube boiler, four basic internal construction features have been recognized as providing peak performance.

1. The *four-pass design* maintains high flue gas velocity for high heat transfer and greater operating economy. The advantage of the four-pass principle is expressed in a simple basic formula:

$$\text{Gas velocity} = \frac{\text{gas volume}}{\text{flow area}} \tag{18.1}$$

Maintaining a continuously high gas velocity is essential to good heat transfer. As hot gases travel through the tube passes, the gases transfer heat to the boiler water, and as the gases cool, they occupy less volume. The gas flow area (number of tubes) is reduced proportionately to maintain high gas velocity and produce constant and optimum heat transfer.

Gases moving rapidly through the four passes scrub the tube surfaces clean and increase the heat transfer by improving the gas film coefficient. Every unit area of tube surface becomes more effective in transferring heat to the water. The ultimate result is low-temperature exhaust gases and high overall efficiency. A four-pass arrangement is depicted in Fig. 18.2.

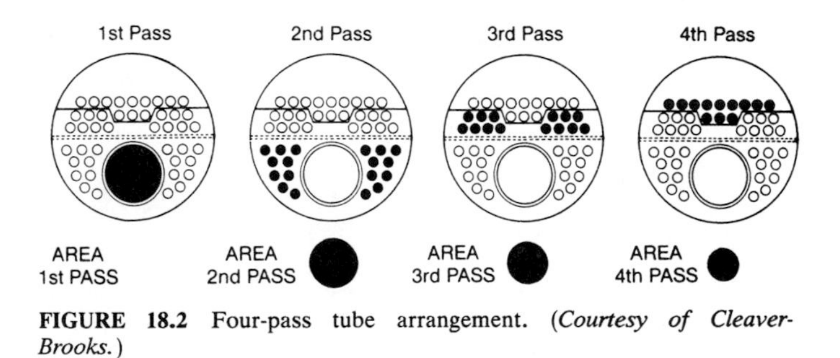

FIGURE 18.2 Four-pass tube arrangement. (*Courtesy of Cleaver-Brooks.*)

2. The *forced-draft design* provides air in required quantity and at the proper pressure. The forced-draft system supplies controlled quantities of clean, boiler room air—air of relatively constant density and volume. It ensures proper mixture of air and fuel for complete combustion with little excess air. The forced-draft fan supplies combustion air at the required pressure and eliminates the need for large-diameter breeching, high chimneys, induced-draft fans, barometric dampers, and sequencing draft controls.

The forced-draft fan moves fresh boiler room air instead of hot, corrosive exhaust gases, as is the case for induced-draft designs. Then air is directed through a diffuser plate which provides a controlled pattern, and air mixes thoroughly with the fuel. Metering of fuel and air is completely automatic in response to load demand. This feature is shown in Fig. 18.3.

3. Using 5 ft^2/bhp (0.0474 m^2/kW) of heating surface provides longer life and less maintenance (1 boiler horsepower = 1 bhp).

Heat transfer is the key to efficient and long-lasting equipment. Firetube systems with 5 ft^2 (0.46 m^2) of heating surface are noted for their ruggedness and ability to handle varying load conditions. With a realistic approach to heat-transfer rates, these systems offer long years of continuously uniform output at high efficiencies.

Adequate heating surface is necessary to transmit heat from combustion gases to the boiler water. The heating surface is made up of tubes, through which gases travel, and tube sheets. (See Fig. 18.1.) The greatest amount of heating surface offered in any packaged unit is 5 ft^2/bhp (0.0474 m^2/kW). This results in long life and low maintenance. See Fig. 18.1.

4. The *updraft construction* offers safer operation with a low furnace location. Low furnace location means safe operation and better combustion. The furnace, containing the hottest combustion gases, lies deep within the water level

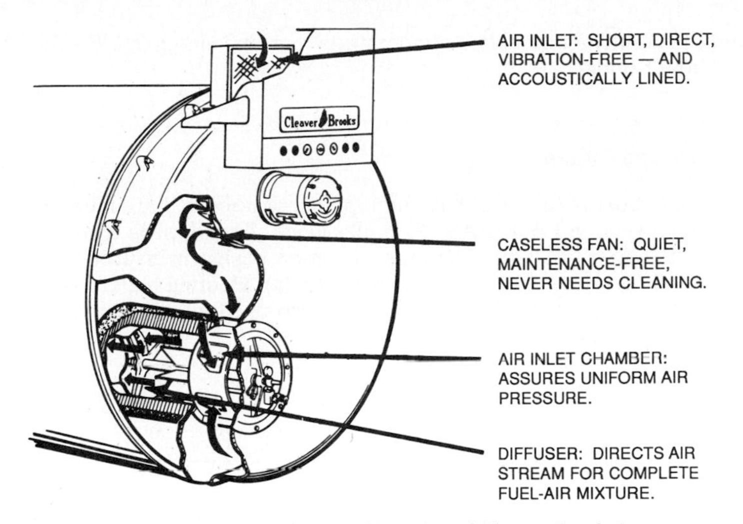

FIGURE 18.3 Forced-draft design. *(Courtesy of Cleaver-Brooks.)*

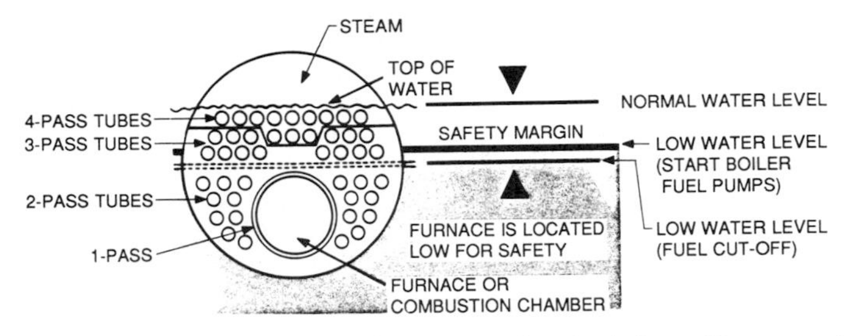

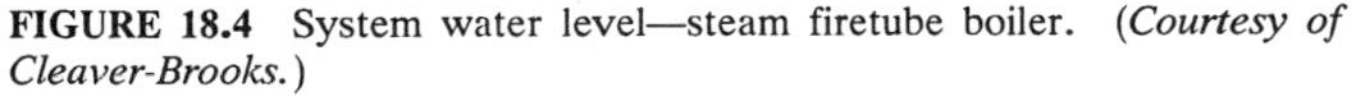

FIGURE 18.4 System water level—steam firetube boiler. *(Courtesy of Cleaver-Brooks.)*

and is designed with a generous clearance from the bottom of the shell to permit full circulation of water throughout the system.

See Fig. 18.4 to note the generous safety margin between the furnace and system water level. There is always the possibility with high furnace design that a low-water condition could expose the furnace. When the furnace is located well below the water level, an improved safety margin is provided, the foaming action of the steam decreases, and drier steam is produced.

18.3 FIRETUBE HEATING BOILERS

18.3.1 Operating Pressure

Low-pressure heating boilers in the United States are fabricated in accordance with section IV of the *ASME Code*, which limits the maximum operating pressure of low-pressure steam boilers to 15 psig (1 bar) maximum and hot-water boilers to

30 psig (2 bar) maximum and/or temperatures not exceeding 250°F (121°C) at or near the boiler outlet.

18.3.2 Process Steam

Process-steam boilers (often called "high-pressure boilers") are fabricated in accordance with section I of the ASME *Boiler Code*. This applies to all steam boilers above 15 lb/in^2 (103 kPa). The most common design pressures are 150, 200, 250, and 300 lb/in^2 (1035, 1379, 1724, and 2070 kPa). Heating boilers, as discussed in section IV of the ASME *Boiler Code*, are also used occasionally for process-steam requirements.

18.4 INDUSTRIAL WATERTUBE BOILERS

18.4.1 Steam Boilers

Watertube boilers are typically used for steam demands of 12,000 lb/h (5400 kg/h) and greater. Typical packaged units are available in sizes having a maximum output range from as low as 12,000 lb/h (5400 kg/h) to as high as 200,000 lb/h (91,000 kg/h). Burners, fuel-handling equipment, controls, and safety equipment are discussed in separate sections.

Steam is generated within the tubes in these units whereas in firetube boilers steam and water surround the tubes. See Table 18.1 for factors which affect evaporation. Watertube boilers are very suitable for high pressures and the generation of superheated steam. These units consist of tubes connecting an upper drum (steam drum) and a lower drum (water drum). Because of the tube and drum construction, high pressures, typically to 900 psig (62 bar) and above, are easily attainable.

Watertube boilers are also noted for their fast steaming capability because of the reduced water content compared to firetube boilers. A number of watertube boiler concepts are available from a variety of manufacturers.

18.4.2 Types of Watertube Boilers

1. *D-style:* This unit consists of a steam drum and a lower drum connected by tubes. The drums are located one above the other with the furnace offset, see Fig. 18.5.

The furnace can be offset either right or left depending on preference. On this type of unit, a single convection pass is used. Large D-shaped tubes form the top, bottom, and one side of the furnace, thus the name *D-style boiler*. The pitch on the D tubes is an important consideration and is discussed later in this section. D-style boilers are versatile in that they enable large combustion chamber volume and the addition of superheaters with relative ease.

2. *A-style:* Figure 18.6 shows a typical A-style boiler, consisting of a steam drum and two lower drums. This unit is symmetric in design, consisting of a centralized furnace with a convection pass on each side of the furnace. This type of boiler is typically used for steam outputs of 120,000 lb/h (55,000 kg/h) and above.

TABLE 18.1 Factor of Evaporation, lb/bhp Dry Saturated Steam

Feed-water temp., °F	Gauge pressure, psig																	
	0	2	10	15	20	40	50	60	80	100	120	140	150	160	180	200	220	240
30	29.0	29.0	28.8	28.7	28.6	28.4	28.3	28.2	28.2	28.1	28.0	28.0	27.9	27.9	27.9	27.9	27.9	27.8
40	29.3	29.2	29.1	29.0	28.9	28.7	28.6	28.5	28.4	28.3	28.2	28.2	28.2	28.2	28.2	28.1	28.1	28.1
50	29.6	29.5	29.3	29.2	29.1	28.9	28.8	28.8	28.7	28.6	28.5	28.5	28.4	28.4	28.4	28.3	28.3	28.3
60	29.8	29.8	29.6	29.5	29.4	29.2	29.1	29.0	28.9	28.8	28.8	28.7	28.7	28.6	28.6	28.6	28.6	28.5
70	30.1	30.0	29.9	29.8	29.7	29.5	29.4	29.3	29.2	29.1	29.0	29.0	28.9	28.9	28.9	28.3	28.8	28.8
80	30.4	30.3	30.1	30.0	30.0	29.8	29.6	29.6	29.5	29.3	29.2	29.2	29.2	29.2	29.1	29.1	29.1	29.0
90	30.6	30.6	30.4	30.3	30.2	30.0	29.9	29.8	29.7	29.6	29.5	29.5	29.4	29.4	29.4	29.3	29.3	29.3
100	30.9	30.8	30.6	30.6	30.5	30.3	30.2	30.1	30.0	29.8	29.8	29.8	29.7	29.7	29.7	29.6	29.6	29.6
110	31.2	31.2	30.9	30.8	30.8	30.6	30.4	30.3	30.2	30.0	30.0	30.0	30.0	30.0	29.9	29.9	29.8	29.8
120	31.5	31.4	31.2	31.2	31.1	30.8	30.7	30.6	30.5	30.4	30.3	30.3	30.2	30.2	30.2	30.1	30.1	30.1
130	31.8	31.7	31.5	31.4	31.4	31.1	31.0	30.9	30.8	30.7	30.6	30.6	30.5	30.5	30.4	30.4	30.4	30.4
140	32.1	32.0	31.8	31.7	31.6	31.4	31.3	31.2	31.1	31.0	30.9	30.8	30.8	30.8	30.8	30.7	30.7	30.6
150	32.4	32.4	32.1	32.0	31.9	31.7	31.6	31.5	31.4	31.2	31.2	31.2	31.1	31.1	31.0	31.0	30.9	30.9
160	32.7	32.7	32.4	32.4	32.3	32.0	31.9	31.8	31.7	31.5	31.4	31.4	31.4	31.4	31.3	31.3	31.2	31.2
170	33.0	33.0	32.7	32.6	32.6	32.3	32.2	32.1	32.0	31.8	31.7	31.7	31.7	31.6	31.6	31.6	31.5	31.5
180	33.4	33.3	33.0	33.0	32.9	32.6	32.5	32.4	32.3	32.2	32.1	32.0	32.0	32.0	31.9	31.9	31.8	31.8
190	33.8	33.7	33.4	33.3	33.2	32.9	32.8	32.7	32.6	32.5	32.4	32.4	32.3	32.3	32.2	32.2	32.1	32.1
200	34.1	34.0	33.7	33.6	33.5	33.2	33.1	33.0	32.9	32.8	32.7	32.6	32.6	32.6	32.6	32.5	32.4	32.4
212	34.5	34.4	34.2	34.1	33.9	33.6	33.5	33.4	33.3	33.2	33.1	33.0	33.0	33.0	32.9	32.9	32.8	32.8

Note: These metric conversion factors can be used: 1 psig = 6.9 bar, 1 lb = 0.45 kg, and °C = 5⁄9 (°F − 32).

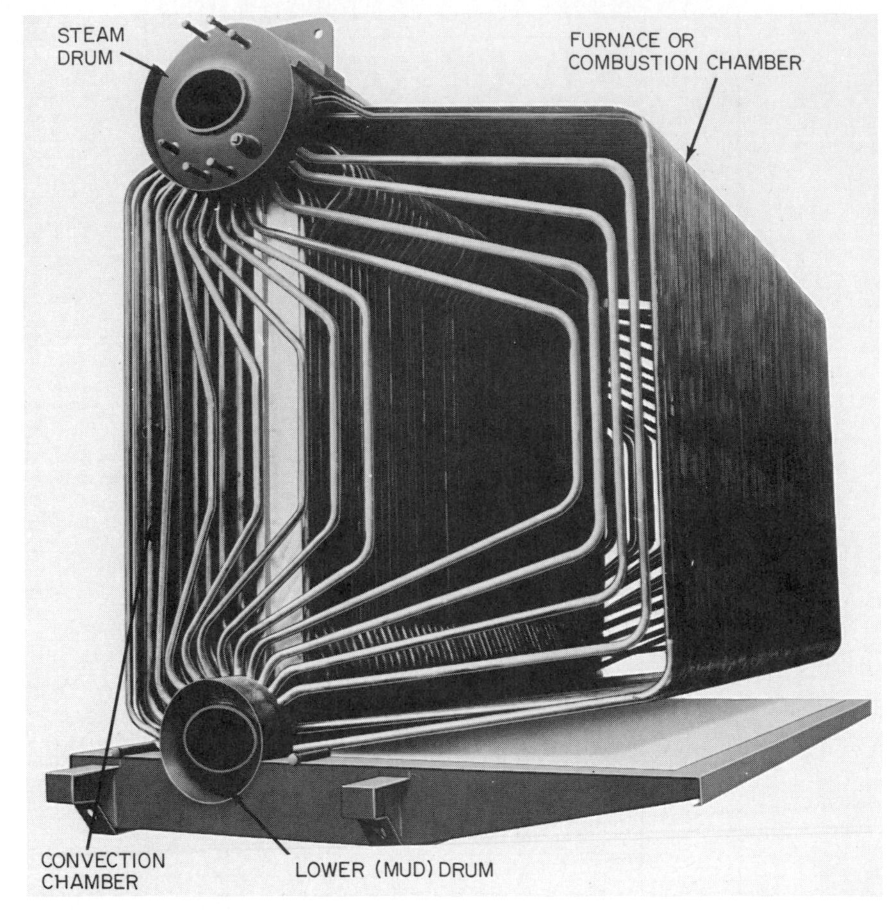

FIGURE 18.5 D-type boiler. *(Courtesy of Cleaver-Brooks.)*

In smaller sizes it does not offer the economy of materials compared to the D-style because it requires two lower drums. It also is not as versatile for incorporating superheater designs.

3. *O-style:* Boilers utilizing the O design are similar to A designs in that they are symmetric but have only one lower drum (see Fig. 18.7). Steam capacities are normally equal to those of D-style units. The configuration of the convection pass of this unit does not lend itself to good flue gas distribution between the two gas convection passes.

18.4.3 Principles of Operation

Heat Input Areas

Furnace. The furnace of the boiler is the combustion chamber, which allows for complete combustion of the fuel being burned. Being surrounded completely

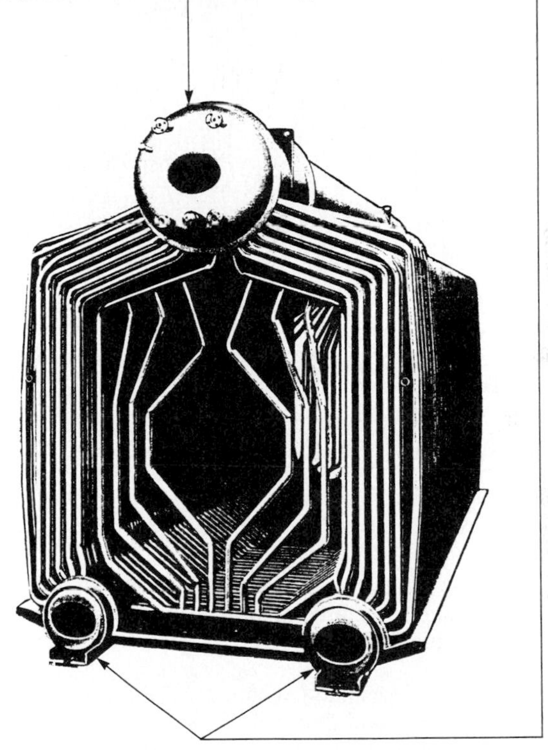

FIGURE 18.6 A-style boiler. (1 in = 2.54 cm) (*Courtesy of Cleaver-Brooks.*)

by tubes, it is the hardest-working section of the boiler because these tubes see radiant heat directly from the flame. Typically, 60 percent of the heat transfer will occur in the furnace while less than 20 percent of the total heating surface in the boiler is used.

Convection Zone. Products of combustion, often called "flue gas," pass through the convection zone. The purpose of the convection zone is twofold. First, it provides a zone of high gas turbulence where the flue gases are cooled (and steam is generated) so that acceptable efficiencies can be obtained. Second,

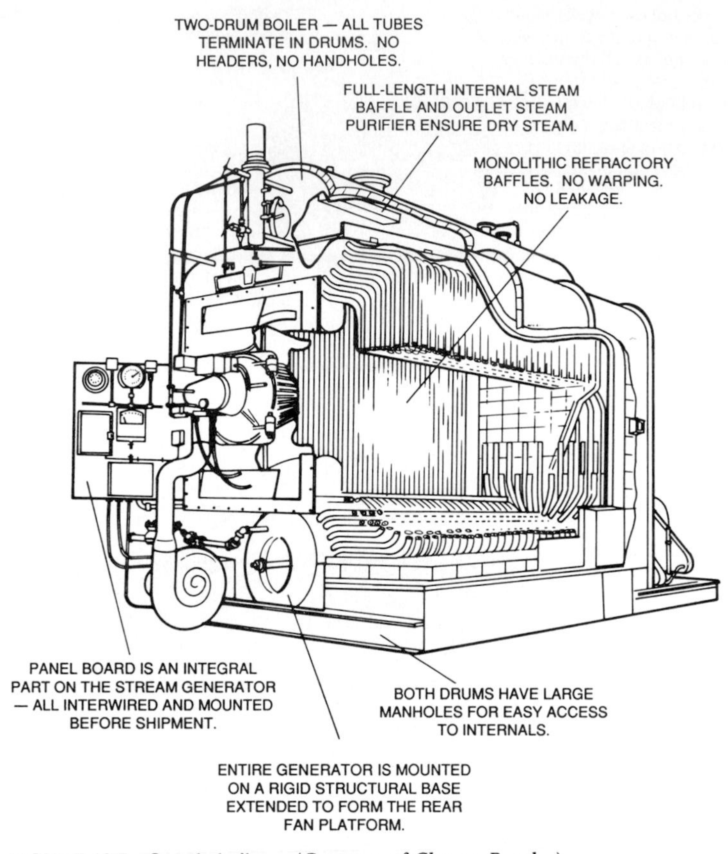

FIGURE 18.7 O-style boiler. (*Courtesy of Cleaver-Brooks.*)

the convection zone in the area at the flue gas outlet has a low heat input to allow for circulation in the boiler.

Steam Production. Circulation within a boiler is very important so that boiler tubes do not overheat and thus fail. This requires steam-generating tubes to always have a steam/water mix. This is referred to as the "circulation ratio" and is based on a maximum percentage of steam by volume exiting from the steam-generating tubes. This ratio is a function of the operating pressure. Typically packaged units are designed for a minimum operating pressure of 75 psig (517 kPa). As the operating pressure increases, the percentage of steam by volume will decrease.

To obtain the required circulation in a watertube boiler, the unit has a definite flow path for the steam/water mix. These are defined as riser tubes and downcomer tubes.

Riser Tubes. Tubes generating steam are called "risers." In these tubes the water moves from the lower drum to the steam drum, and along this path the heat input causes the generation of steam. As steam bubbles form in these tubes, the steam/water mix, being lighter, is replaced with relatively cool water. Thus upward circulation results.

Downcomer Tubes. These tubes allow the flow of water from the steam drum to the lower drum. They are located in the coolest section of the boiler so as not to generate steam, which would impair the downward flow of water. These tubes must accommodate enough water to satisfy the demands of the riser tubes and thus allow for continuous circulation. The downcomer area is normally the governing area for circulation. If the tubes do not permit sufficient water to reenter the lower drum and thus the riser tubes, the percentage of steam by volume exiting the riser tubes will be excessive, causing possible overheating of the riser tubes.

18.4.4 Boiler Design

Pressure Vessel

Drums. Watertube boilers use drums rolled from plate or pipe. They have spherical heads with manways for access at each end. The material thickness and the type of material used are dependent on the design pressure required.

All drums are of welded construction in conformance with the ASME *Power Boiler Code*. All drums are stress-relieved, and welded seams are radiographed and hydrostatically tested in the shop at 1.5 times the design pressure. The upper drum is wholly supported by the tubes themselves, and generally no extra supporting steel is required. All tubes are expanded into the drums. Tube holes must be bored to close tolerances. For maximum contact, surface tube holes are not normally serrated or grooved. This manufacturing technique is sometimes used for boilers designed for operation at pressures in excess of 500 psig (34.5 bar).

Tubes and Tube Attachments. The most common tubes used in watertube boilers are SA-178 with a wall thickness of 0.105 in (2.67 mm). These tubes have a working pressure in excess of 900 psig (62 bar). The tubes are formed by bending, as required for connection between the steam drum and the lower drum. Good design and manufacturing techniques provide tubes that remain relatively round in the bends, and the tube ends are not swagged down at connections to the drums.

Furnace Design (Six-Wall Cooling). Furnace construction is an important consideration in the design of the walls used to form the furnace.

Tangent Tube Walls. Many watertube boilers use tubes tangent to each other. This type of heat-transfer surface gives maximum water-backed surface in the furnace area. Due to the high heat absorption rates in the furnace, tubes placed in a tangential arrangement provide as much water-cooled surface as possible. Fins, studs, or other forms of extended metallic surface may result in shortened boiler life.

Fin-Tube Walls. These consist of fins welded to the tubes and used as an extended heating surface. The tube wall temperature is higher with this type of wall because less cooling water is available per unit of heat-absorbing surface.

Membrane Tube Walls. Solid fins are welded between tubes with this type of construction. Once again, the tube wall temperature is higher than the tangent tube design, as with fin-tube walls.

Furnace Floor. While the floor of a furnace will be one of the wall types described above, an added requirement of floor tile is needed. A media to reduce heat transfer on the floor is required because of a phenomenon called *top dry-*

ness. This occurs in tubes that are nearly horizontal in pitch because steam bubbles tend to rise to the top. This reduces the wetting effect from water at that surface, resulting in nonuniform tube wall temperature which makes the tubes more susceptible to failure.

Front and Rear Walls. The furnace construction discussed so far refers to the side, roof, and floor. Front and rear walls must be constructed in another manner. These walls are refractory walls with tubes for cooling to increase refractory life. See Fig. 18.8.

Tube Pitch. A very important consideration in furnace design is the slope of the tubes. Tubes that run perfectly horizontal will have poor circulation. All furnace floor tubes must have a minimum slope of 6.5° to the horizon to achieve good circulation and drainage. All furnace roof tubes must have a minimum slope of 7.5° to the horizon to permit good circulation and maximum steam-relieving capacity.

Convection Heating Surface. This part of the boiler consists of many tubes with relatively close spacing. Figures 18.8 and 18.9 illustrate a typical convection zone in a boiler. Considerations in design are the effective use of heat-transfer surface, tube removal, flue gas pressure drop, and cleaning of the tubes. The tube arrangement must be such that usually a tube can be removed without requiring the removal of adjacent tubes.

Sootblowers are normally installed or provided for, in the convection area of the boiler, when solid fuels or residual fuel oils are burned. Openings for inspection of the blower system are a design necessity.

Boiler Casing and Insulation. Modern watertube boilers use a pressurized furnace because a forced-draft fan is provided with the burner. Two types of construction are normally used to provide casings that do not leak: double-wall and membrane.

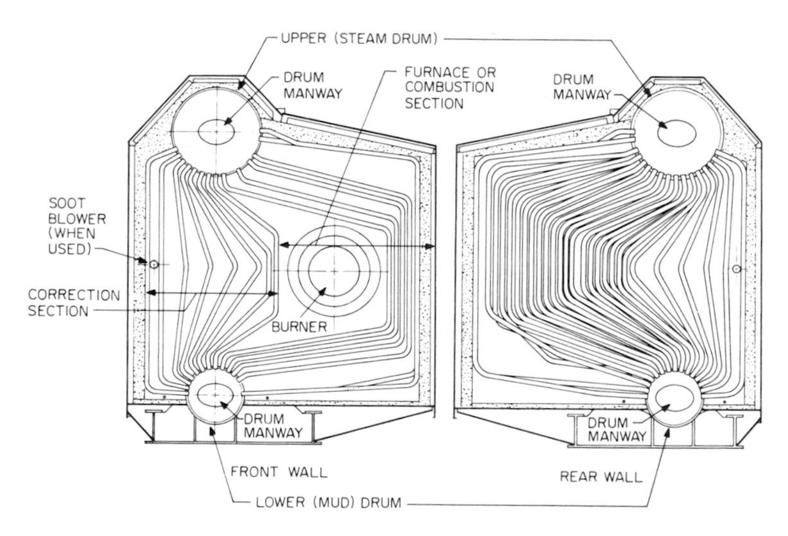

FIGURE 18.8 Front and rear walls of a D-type boiler. (*Courtesy of Cleaver-Brooks.*)

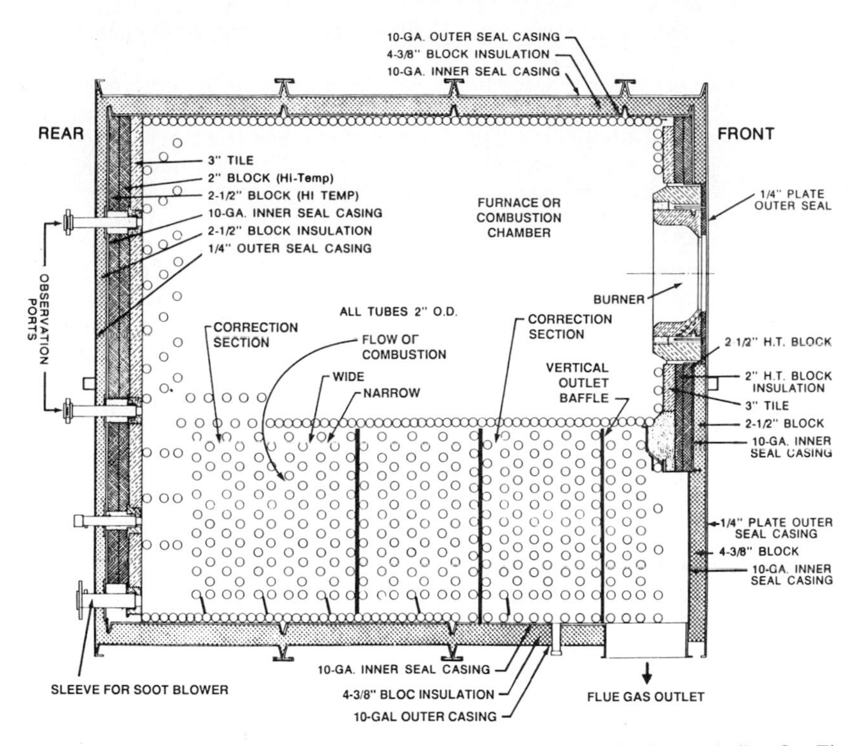

FIGURE 18.9 Double-wall construction. *Note*: This is the plan of a D-type boiler. See Fig. 18.5 for sections. (*Courtesy of Cleaver-Brooks.*)

Double-Wall Construction. Double-wall construction consists of an inner and outer casing with insulation sandwiched in between to reduce heat losses. See Fig. 18.9.

The inner casing is seal-welded to form a leakproof barrier to contain the flue gas. On the side walls and roof, the inner casing is laid tight up against the tubes but not attached to the tubes. The fact that the tubes are at saturated steam temperature enables standard-grade carbon steels to be used in this application. The tubes are not attached to the casing so that expansion of tube or casing can take place without stressing either part. The floor casing, while not tight against the tubes, experiences the same temperatures as the sides and roof.

The front and rear walls do not have the same amount of cooling and require different construction. Because of the radiant heat to which they are exposed, a refractory is used to form the innermost surface of the walls. Insulation covers the outer refractory surface to reduce the temperature. At this point the inner casing is applied. Because of the insulation a standard-grade carbon steel is used.

The inner casing is normally pressure-checked at 8 to 10 in water column (WC) (2 to 2.5 kPa) to ensure that no leaks occur before the final insulation is applied over it. Insulation is laid over the inner casing to reduce heat losses. The outer casing provides additional strength, a covering for the insulation, and an aesthetic appearance.

Both the inner and outer casings are commonly constructed from 10 gauge

(3.42 mm) steel. Expansion joints are provided over the length of the unit for expansion.

Membrane Construction. Membranes between tubes in the outermost tube rows serve the same purpose as the inner casing used in double-wall construction, i.e., to contain the hot gases. Construction of rear and front walls with membranes requires very special attention in the design. Complete membrane wall construction is not common because of the excessive amount of labor required for welding.

Steam Separation and Drum Internals (Steam Quality). Carryover of excessive moisture in steam is not acceptable. In most applications carryover can cause steam hammer in steam lines, reduce heat transfer in heating systems, and damage steam turbines. Boiler steam drum internals are required to eliminate these problems.

The standard requirement for steam quality is 99.5 percent, also referred to as ½ percent moisture. To accomplish this, special baffling and separating devices are usually needed in the steam drum. See Fig. 18.10. The purpose of the baffling is to produce a calm water surface so that the moisture can drop from the steam before it enters the separating device.

Some special applications require as little as 1 ppm (1 mg/L) moisture carryover. One common use is in steam-turbine applications. Special devices called "purifier cartons" are required for this. See Fig. 18.11.

Properly designed drum internals and separation equipment are one critical factor in obtaining good steam quality. Proper water treatment is also very important. See Chap. 53.

Blowdown. Removal of impurities that result from steam generation is important. Two types of blowdown are recommended: surface blowdown and bottom blowdown. The surface blowdown removes impurities that rise and float on the top of the water in the upper or steam drum. See Chap. 53 for further discussion of blowdown.

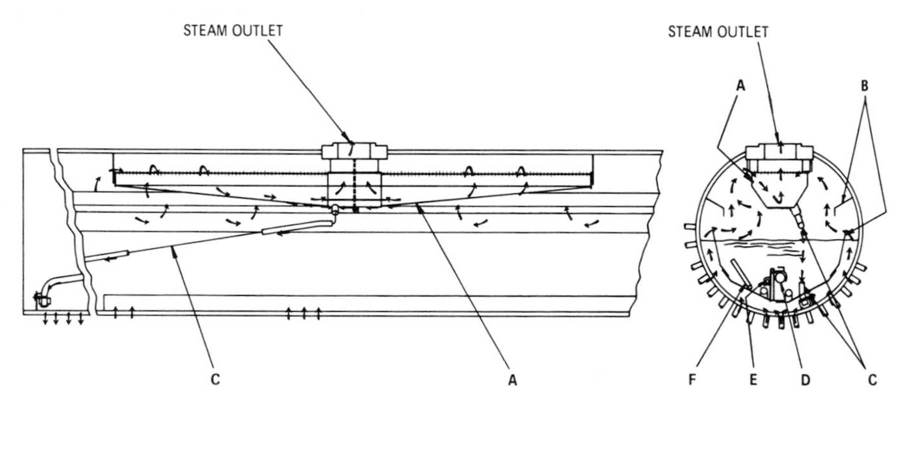

A. Dry Pan Carton
B. Flow Control Baffles
C. Dry Pan Drain
D. Feedwater Pipe
E. Chemical Feed Pipe
F. Continuous Blowdown Pipe

FIGURE 18.10 Drum internals and separator for 99.5 percent dry steam. (*Courtesy of Cleaver-Brooks.*)

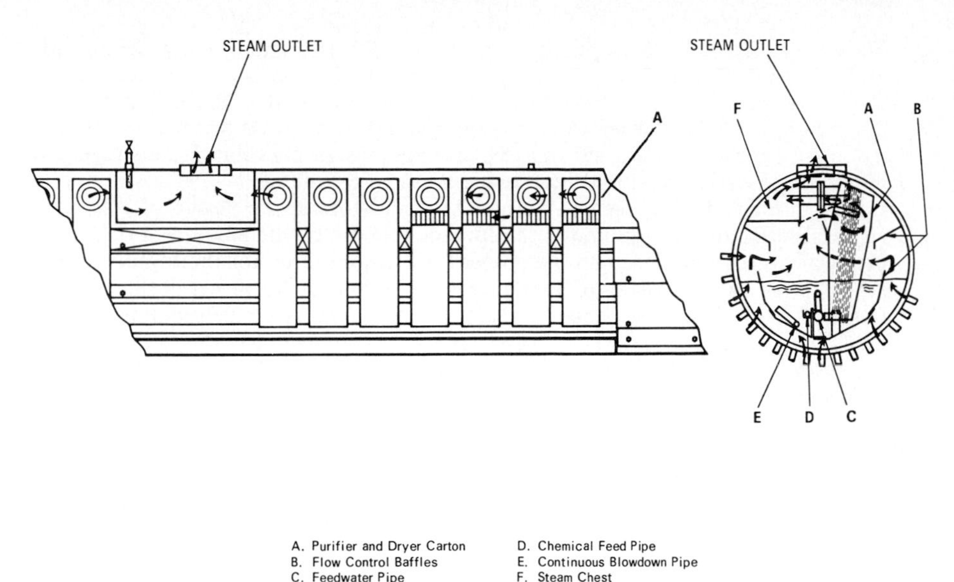

FIGURE 18.11 Drum internals and steam purifiers for 1-, 3-, and 7-ppm solids. (*Courtesy of Cleaver-Brooks.*)

Refractory. Refractory is commonly used on the furnace front and rear walls and in the area called the "throat tile," where the burner is attached. Boilers are designed with high-temperature refractories to withstand the temperatures to which they will be exposed. Inspections and maintenance should be performed according to the boiler manufacturer's recommendations. The throat tile is designed for the burner provided.

18.4.5 Boiler Heating Surface

Since there continue to be discussions on what actually constitutes a heating surface, the engineer must define this term. In the United States this is generally defined per the ASME *Power Test Code*, Stationary Steam Generating Units, PTC4-1946, p. 23, sec. 5, par. 116, or per the American Boiler Manufacturers' Association (ABMA) *Industry Standards and Engineering Manual*, 5th ed., 1958, sect. G, p. 4-2.11.

The term "effective projected radiant surface" (EPRS) also has different meanings to many people and should be defined. It is recommended that EPRS be defined as the projected waterwall surface minus the projected heating surface covered by refractory (tile) plus the area of furnace exit, to avoid any misunderstanding.

18.4.6 Furnace Heat Release

The furnace heat release per cubic foot (meter) of furnace volume has been a governing factor for many years in the selection of boilers. Some years ago, 25,000 Btu/(h · ft^3) (258.7 kW/m^3) was considered a sacred number beyond which nobody could trespass. Now 50,000, 60,000, and 809,000 Btu/(h · ft^3) (517.5, 621, and 8373 kW/m^3) have all been advanced by various sources as limiting factors. Marine boilers are known to have heat releases of 200,000 to 300,000 Btu/(h · ft^3) (2070 to 3105 kW/m^3). Therefore, it is the general feeling that the number itself is not a governing criterion and that the furnace design plus the flame shape and length characteristics within the furnace must be examined.

It is the general feeling that the governing factors in the design of a furnace (and subsequently the volume of the furnace) should be the amount of true cooling surface installed based on the amount of heat required for the furnace to absorb and the characteristics of the flame in the furnace. The burner design and the flame characteristic generated are also governing factors for determining the acceptability of a furnace design. The shape of the flame envelope is dependent on the proper burner selection, appropriate atomizer, and correct diffuser arrangements for fuel-air mixing and burning.

18.4.7 High-Temperature Water Systems

High-temperature water (HTW), medium-temperature water (MTW), and low-temperature water (LTW) systems use generators of various designs. HTW systems are generally operated from 350 to 430°F (177 to 221°C) with corresponding generator operating pressures between 200 and 525 psig (13.8 and 36.2 bar). MTW systems are generally operated from 250 to 325°F (121 to 163°C) with corresponding generator operating pressures between 50 and 160 psig (3.4 and 11 bar). LTW systems operate from 180 to 250°F (82 to 121°C) with corresponding generator operating pressures between 10 and 45 psig (0.7 and 3.1 bar).

Boilers operated as hot-water generators must be pressurized to prevent the formation of steam. The pressure level required for various operating temperatures can be found in Fig. 18.12. A safety factor is included in these data to offset slight deviations in system operating conditions.

The generator produces the heat to raise the temperature of the "return" water to the required temperature of the system "supply."

System and generator pumps force the water through the generator, the piping, and the heat-using apparatus. The generator, being in the pumped-flow circuit, is therefore part of the forced-flow circuit. Proper water and gas flow direction provisions inside the generator vary with the design.

System pressurization is accomplished through the use of an appropriately designed expansion tank. The design of this tank is crucial to proper system function and depends on system size and operating temperatures and pressures.

18.4.8 Watertube Generators

Industrial watertube HTW generators are available in several design concepts:

- Pumped jet-induced circulation (see Fig. 18.13)
- Forced circulation with or without separation drum
- Once-through, orifice-controlled, drumless, forced circulation (Fig. 18.14)

18.4.9 Jet-Induced Circulation

The D-type two-drum watertube HTW generators have proved their dependability by operating troublefree for over thirty years. They are economical in first

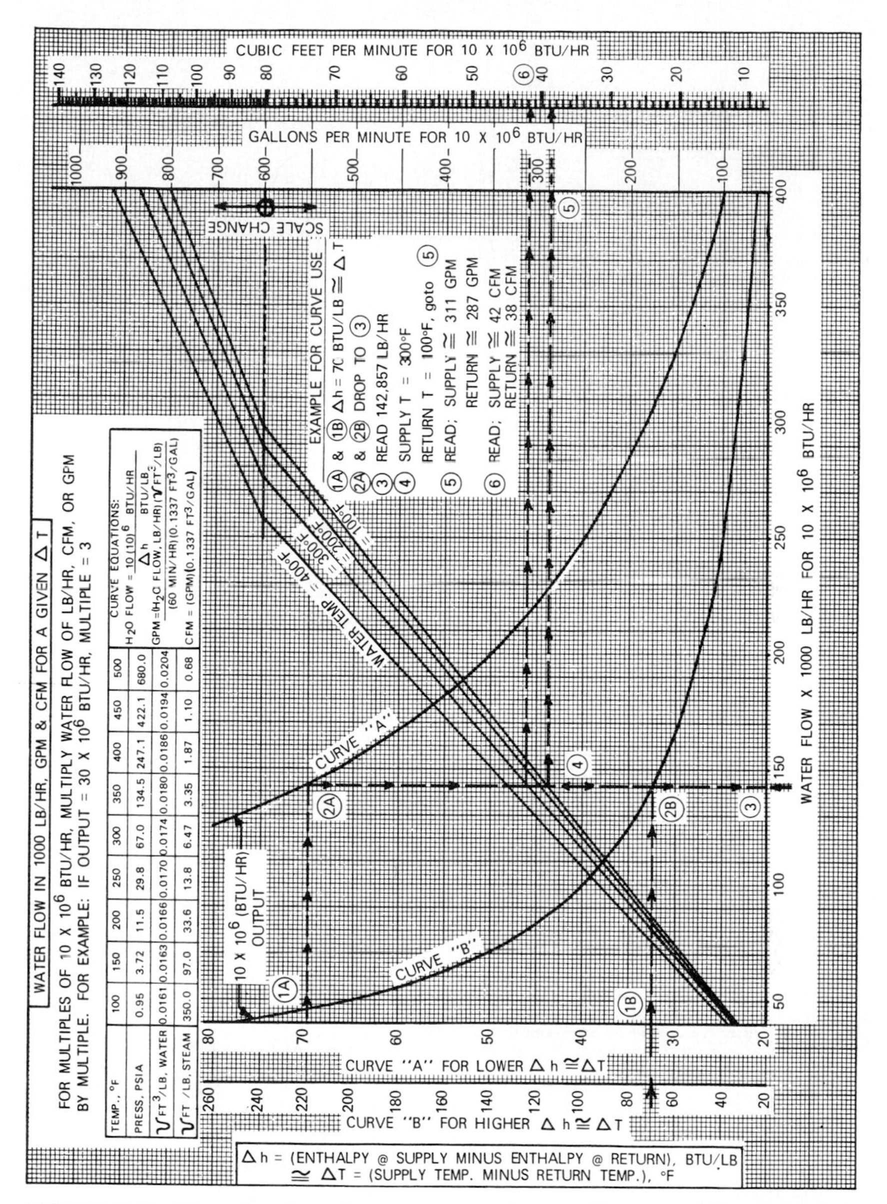

FIGURE 18.12 Water flow for a given temperature change. (*Courtesy of Cleaver-Brooks.*)

cost, low operating cost, low maintenance, and dependability. The output capacities range up to 65 MBtu/h (19 MW) for pressures to suit water temperatures up to 430°F (221°C), with supply-to-return temperature differences up to 200°F (93°C).

Figure 18.14 shows the water circulation pattern through the HTW unit. The return water flows from the circulation pump through the internal distribution

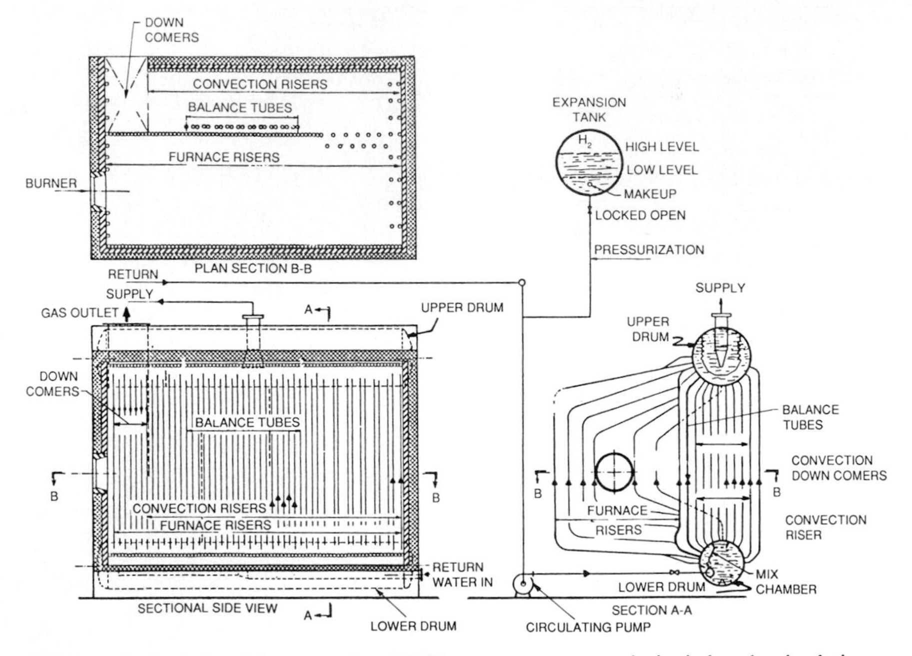

FIGURE 18.13 Industrial watertube HTW generator—pumped jet-induced circulation. (*Courtesy of Cleaver-Brooks.*)

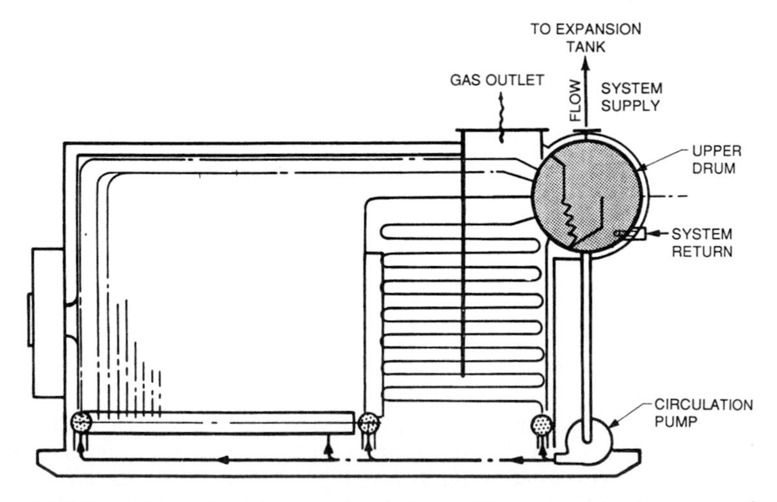

FIGURE 18.14 Industrial watertube boiler—orifice-controlled. (*Courtesy of Cleaver-Brooks.*)

pipe into the plenum chamber and then rises through all furnace tubes to the top drum. Here water mixes with the drum water. From there a part is taken out as the supply flow requires. The rest circulates down to the lower drum through the downcomer tubes in the gas outlet pass. The flow through the upper drum provides for free longitudinal drum flow.

Large, Jet-Induced Circulation, Multidrum Units. Shop-assembled, large-capacity HTW generator units are an advanced multidrum type of design. Figures 18.15 and 18.16 show units for capacities of 70 to 150 TBtu/h (20.5 to 44 MW) and 430°F (221°C) supply water temperature.

These large units have two lower drums and one upper drum, whereas the medium-size units have two drums. The return water enters the lower drums through a jet pipe that discharges into the mixing chamber of both lower drums. The "return" water is forced by the system or the boiler circulation pumps directly into the bottom drums (Fig. 18.15). An internal distributing pipe with multiple jet outlets discharges into a plenum chamber around the furnace tube inlets. The multiple-jet action of the return water induces the boiler water already in the drum to flow into the plenum chamber. There both waters mix and enter the furnace tubes, flow up through the tubes, and discharge into the upper drum, completing the first stage of circulation.

The second stage of circulation results from controlling the gas flow by gas pass baffling. The baffles define a hot-gas pass in which the watertubes are risers and a cold-gas pass in which the watertubes are downcomers. The combined action of return jets, plenum chamber, water movement in the lower drum, and hydraulic head self-balancing effect increases the downflow of water in the downcomers located in the cold-gas pass. Unheated downcomers at the front and rear of the boiler for large units use additional circulation safeguards.

Breakup screens and separation baffles mix the discharged water from the furnace tubes with the water present in the upper drum. The water leaving the upper

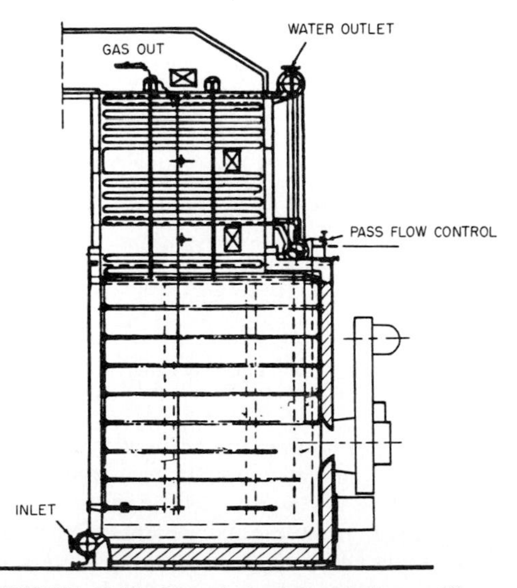

FIGURE 18.15 Forced-circulation boiler—orifice-controlled. (*Courtesy of Cleaver-Brooks.*)

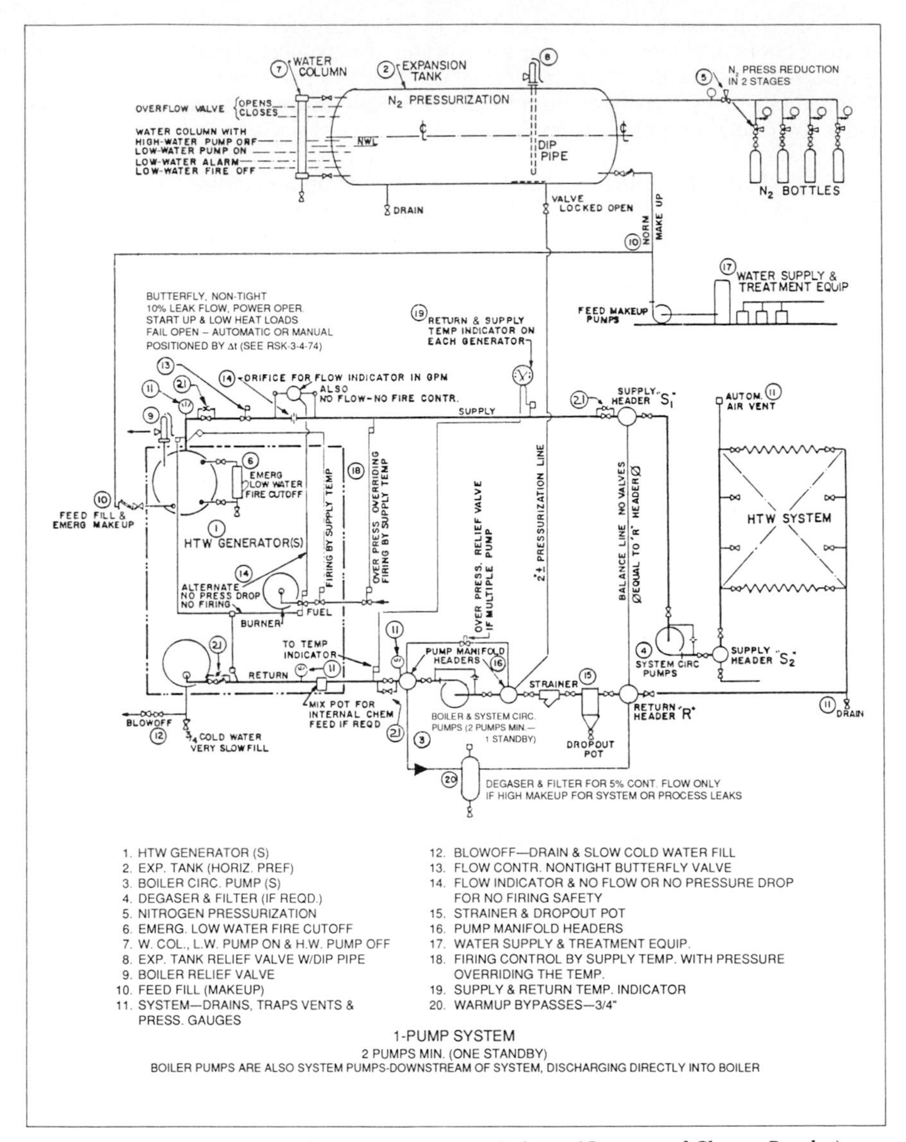

FIGURE 18.16 HTW system elements—one-pump design. (*Courtesy of Cleaver-Brooks.*)

drum of the generator does not contain steam because the operating drum pressure is higher than the operating water temperature saturation pressure. The difference is usually provided by nitrogen pressurization of the expansion tank. If the boiler or system pumps fail, the plenum chamber baffle and nozzle arrangement in the lower drums will allow natural water circulation by difference of the density of the water in riser tubes and downcomer tubes. Furnace tubes, therefore, are never without a flow of water. Separation baffles in the upper drum provide a channel for the longitudinal return flow toward the downflow end of the upper drum.

18.4.10 Forced-Circulation Boilers with Release Drum

Figure 18.14 illustrates a unit where the water is continuously circulated through the HTW generator by pumps which take the water from the upper drum and force it through single-pass or multipass, continuous-tube heat-absorption surfaces. These surfaces can form the furnace and the convection section, either in parallel or in series. Pump capacity must be sufficient to provide the required water velocity and mass flow in the tubes.

Multitube passes generally require orifice or flow control valves to produce, by pressure difference, equal flow through all tubes. Some designs do not incorporate orifices, but require relatively large boiler circulation pumps to maintain nearly equal water velocities through all tubes of each water pass. Minimum water velocities usually are kept at or above 10 ft/s (3 m/s) during normal loads. Pumps generally are single-speed drives and require constant pump horsepower at all loads, resulting in high power costs. Flow variations through these multitube boilers must be minimized and should not vary by more than 10 percent of design values.

18.4.11 Once-Through Orifice-Controlled Forced-Circulation Boilers—Drumless

Orifice-controlled once-through forced-circulation boilers are available in many designs. Typical designs are La-Mont, corner-tube, multipass, and combinations thereof.

The boiler convection section can be in back of or above the furnace (Fig. 18.15).

Some units employ multipass tube groups. Each group is equipped for automatic flow control by use of aquastats located in manifold blocks, which actuate the control valves.

Water is forced by pump pressure through the tubes forming the heat-absorbing surfaces in the furnace and in the convection section. These pumps can be either the boiler circulation pumps or the system pumps. The water flow quantity through these type generators cannot be varied by more than 10 percent from design value.

18.4.12 Tube Sizes and Orifices

The jet-induced first-stage forced-circulation watertube HTW generators use tubes 2 in (51 mm) in diameter and larger. The tubes have unrestricted internal flow area. Orifices in tube inlets or tube outlets are not used as in some generators.

Once-through orifice-controlled forced-circulation HTW generators may use tubes 2 in (51 mm) in diameter, but generally they are smaller. To ensure equal flow through each tube of a multitube group, inlet (or outlet) orifices are used. Such orifices may be in each tube or in the inlet or outlet manifold of a group of tubes. An orifice anywhere in fluid-carrying elements—whether individual tubes, groups of tubes, or manifold headers—can be obstructed by foreign material. Dirt or scale in a system can build up at an orifice and restrict or stop flow, causing overheating of tube metals and eventual tube failure.

18.4.13 Problems and Solutions

Casing Corrosion. Casing corrosion which occurs when high-ash and sulfur fuels are fired can be eliminated by seal-welded inner casings.

Orificed Generator. In an orificed generator, the tube distribution headers provide a dropout point for solids which, after some accumulation, flow into the orifices, where they can cause plugging. Unless firing is stopped at once (within seconds), plugged tubes will be damaged or will even rupture. A water-flow interruption in an orifice-controlled once-through flow generator would, of course, lead to furnace tube rupture if the fire is not immediately put out.

Natural-circulation provisions should be provided in the generator so as to give the flow control ample time to shut off the burner(s) if a no-flow incident by pump or generator should occur.

An HTW generator designed to accommodate natural circulation is less likely to be damaged by plugging, can be cleaned independent of the external system, can be easily inspected internally, can be easily isolated for short- or long-term storage, and is much less susceptible to damage caused by failure of external system components. One further advantage of such generator design is that the system circulation flow rate can be adjusted to meet varying system requirements without causing any operational difficulty in the boiler.

Makeup Water. Equipment to provide properly treated and deaerated makeup water should include pumps, level regulator, storage, deaerator, chemical treatment, etc. This equipment should be located close to the HTW generator. (For a detailed discussion of water treatment see Chap. 53.) In a plant where steam of 5 to 10 psig (0.34 to 0.68 bar) pressure is available, a small deaerator for makeup water is advisable.

Although an HTW system is a closed system and theoretically would require no makeup water, in practice provisions must be made for the supply and treatment of 1 to 3 percent of the system's water content as the hourly makeup water capacity.

When no plant steam is available for deaeration, chemical means must be used to rid the makeup water of oxygen. For a large system, a small, low-pressure steam boiler for steam to a small deaerator may be necessary.

Generator Arrangements within the System. The system can be a one-pump or two-pump type. Figure 18.16 shows the one-pump design where the water is pumped directly into the bottom drum of the generator. The pressure of the pump(s) moves the water through the generator, into the system's supply line, through the system load, and back through the return line to the inlet of the pump(s). One pump (group of pumps) provides the system flow. We thus have a one-pump system design.

Figure 18.17 shows a two-pump system. The generator circulating pumps move the water from the return line through the generator and into the supply line. A second pump or group of pumps takes the water from the supply line leaving the generator and pushes the water through the system into the return line.

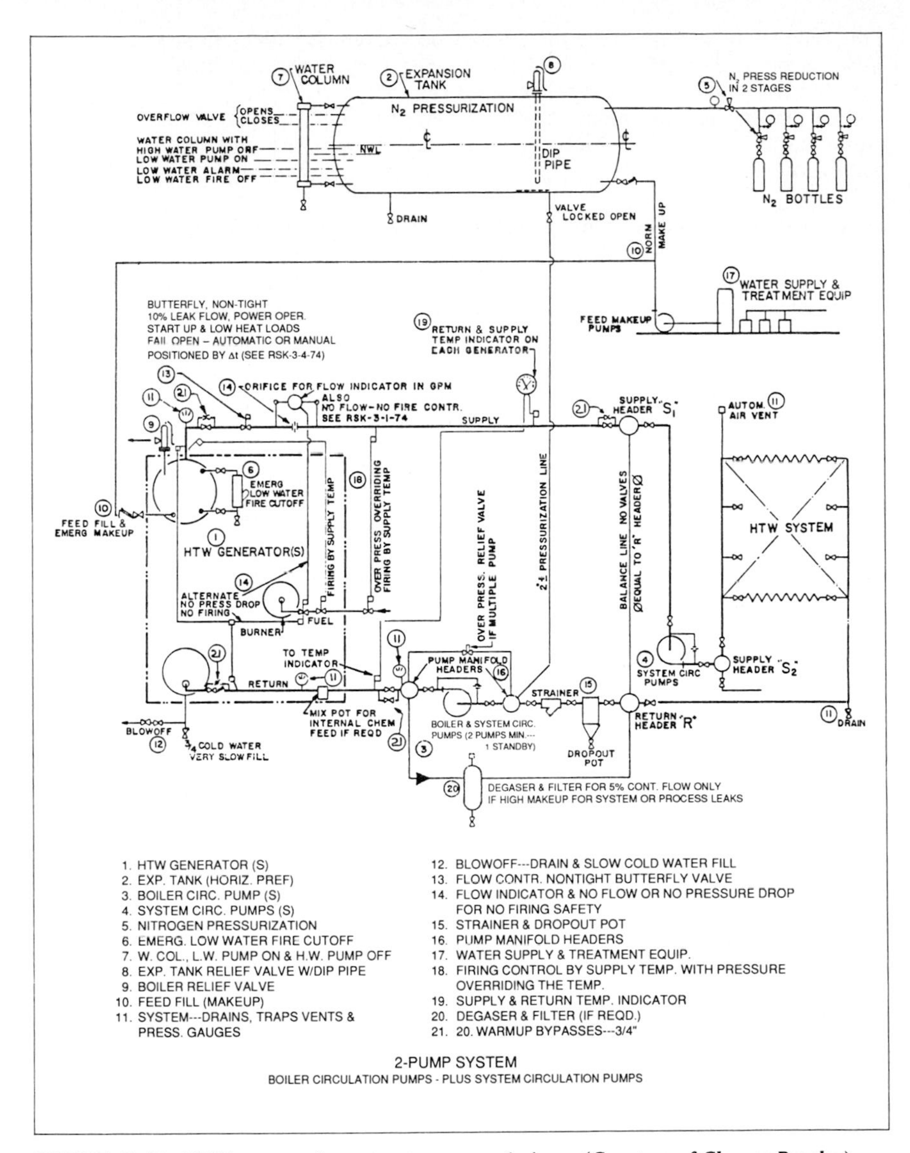

FIGURE 18.17 HTW system elements—two-pump design. (*Courtesy of Cleaver-Brooks.*)

This type of system divorces the system flow from the boiler flow requirement. Either system is satisfactory for a jet-induced flow generator, but is likely to be required for orifice flow control generators.

For either the one-pump or the two-pump system, it is advisable to use pump manifold headers which permit operation with any one pump to any one generator.

System Expansion. Expansion space may be provided as follows:

- Expansion space in the generator drum (this is not recommended and is not used except in very small systems), pressurized by steam or by inert gas, usually nitrogen (air, conducive to corrosion, is not recommended)
- System expansion tank, with or without overflow tank, pressurized by inert gas or by steam (Fig. 18.17), N_2 always preferred
- Overflow tank, atmospheric or pressurized
- Any combination of the above

The total water content of the system including generators, heat users, and all connecting piping will expand in volume as the water is heated from the system fill temperature to system operating conditions. This expansion must be accommodated. Large systems need large expansion tanks external to the generators. A small system, where the system water content is less than 2 times the water content of the boilers, may not need a separate expansion tank provided the generator is a drum-type unit and that the upper drum has sufficient space for system and generator water expansion at operating-temperature variations. The tabulation in Fig. 18.18 shows the available expansion space in the top drums of standard HTW jet-induced circulation, D-style and multidrum units.

Water expansion beyond the volume that the specific generator drum can accommodate will require a separate expansion tank. Operational difficulties of a few early systems generally were caused by insufficient expansion space, lack of expansion tanks, or a restricted permissible water-level variation within the generator drum or expansion tank.

Expansion Vessel. Expansion space for larger system volume expansions should be provided in a separate vessel and should be not less than 2 times the maximum volume of expansion of the water in the system, including the generator. Consideration must be given in this calculation to the probability that the generator may be operating with a greater temperature rise from inlet to outlet than the temperature change which occurs in the external system. The total water expansion volume must be taken by the expansion space provided in the expansion drum above normal drum water level.

The system water expansion under consideration would be only for normal operating-temperature variations and need not take into account the expansion from a cold-start temperature. For cold-start expansion, an overflow storage tank is recommended to save the treated water.

An inert-gas- or steam-pressurized expansion drum should pressurize the system on the generator circulation pump inlet side, and the bottom of such an expansion tank should be above the generator top-drum elevation. This is recommended so that a minimum static head continues to be available if a failure of the pressuring action occurs. The expansion tank should be arranged horizontally for minimum variation of water level and maximum hydraulic pump suction head.

EXPANSION SPACE IN			"DELTA" and "DL" JET-INDUCED CIRCULATION HTW GENERATORS																				
SIZE DL TYPE	SIZE DELTA TYPE	UPPER DRUM DIA. in ID	WATER CONTENT FILLED TO LWL			DRUM LENGTH (ft)	EXPANSION SPACE OF UNIT		VOLUME REQUIRED FOR WATER EXPANSION FROM 80° F to ...... (ft³.)				RATIO OF EXPANSION SPACE TO VOLUME REQUIRED FOR EXPANSION OF BOILER WATER WHEN WATER IS HEATED FROM 80° F to				PERMISSIBLE MAXIMUM WATER CONTENT OF SYSTEM WHEN WATER IS RAISED FROM 80° F to				RATIO OF DRUM EXPANSION SPACE TO WATER CONTENT FILLED TO LWL @80° F
			lb. @ 80° F	ft³.	gal		ft³.	gal	°F 300	350	400	450	°F 300	350	400	450	°F 300	350	400	450	
	26	36	6,480	104	780	8.2	36.2	270	8.85	12.4	16.55	21.4	4.10	2.9	2.20	1.70	2,380	1,500	920	530	0.35
	34	36	8,380	135	1,000	10.6	45.6	340	11.5	16.0	21.5	27.8	4.00	2.85	2.15	1.65	3,000	1,850	1,130	650	0.34
	42	36	10,270	165	1,230	13.0	55.0	412	14.0	19.7	26.2	34.0	3.90	2.80	2.10	1.60	3,610	2,240	1,365	770	0.33
	52	36	12,630	203	1,517	16.0	68.0	510	17.5	24.2	32.3	42.0	3.88	2.82	2.10	1.62	4,440	2,770	1,680	955	0.33
60	60	36	14,540	233	1,746	18.5	78.0	585	20.0	27.8	37.0	48.0	3.90	2.81	2.10	1.62	5,075	3,180	1,930	1,100	0.33
68	68	36	16,435	264	1,974	21.0	86.5	648	22.5	31.5	42.0	54.5	3.85	2.75	2.06	1.58	5,660	3,470	2,100	1,175	0.33
76	76	36	18,360	295	2,200	23.3	95.5	715	25.0	35.2	47.0	61.0	3.82	2.72	2.04	1.57	6,175	3,800	2,300	1,270	0.32
86	86	36	20,715	333	2,490	26.3	107	800	28.5	40.0	53.0	68.5	3.76	2.70	2.02	1.56	6,950	4,240	2,540	1,400	0.32
94	94	36	22,610	364	2,715	28.7	116	870	31.0	43.5	58.0	75.0	3.74	2.66	2.00	1.55	7,475	4,600	2,760	1,510	0.32

- FL Water Content = D Water Content × 1.17
- Table shows required and available expansion volumes for water at 80° F temperature heated to operating temperature. This is seldom so provided and to take care of this large temperature difference, overflow tanks should be provided.
- Expansion space generally is provided only for water temperature variations under operating conditions applied to boiler water, water in expansion drum, water in supply line, water in heat-absorbing apparatus, water in return line, and marginal water in the expansion drum.
- A supply water temperature of 400° F might vary between 350° F and 400° F, or about by 50°. The return line water temperature might vary for a return of 200° F from 150° F to 250° F, or by about 100°F. This ratio holds good for most of the HTW systems in that the water in the supply line and expansion drum may vary by some 50° F and the water in the return line may vary by some 100° F to 200° F.
- Note: When using metric, convert to USCS units before applying this chart. Use the following values: in = mm/25.4; °F = 1.8 °C + 32; ft³ = m³ × 35.311; ft = m × 3.28.

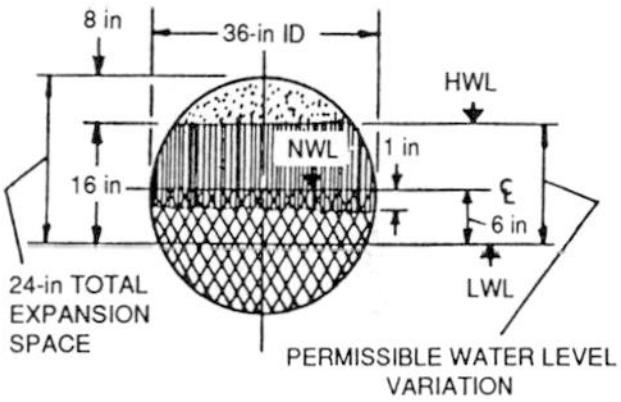

FIGURE 18.18 Expansion space in jet-induced HTW generator. *(Courtesy of Cleaver-Brooks.)*

The pressure in the drum maintains a positive suction head at the inlet to the circulation pumps and should be sufficient to provide the required pump suction head for the pumps selected; if it is less, cavitation will occur.

The system is pressurized by loading the expansion space of the expansion drum with steam or an inert gas such as N_2. The pressure is then transmitted by a small pipeline to the inlet side of the generator circulation pump. See Figs. 18.16 and 18.17. This pressurization line to the pump inlet connects to the bottom of the expansion drum with a locked-open shutoff valve near the drum. An inert gas such as nitrogen is used to eliminate the possibility of saturating the cool water contained in the drum with oxygen, thus minimizing corrosion of the generator and system components.

For good operational system and generator control, water-flow indicators should be used with a flow orifice located in the supply line from each unit where the water temperature is constant. A non-tight-closing motor or a manually actuated butterfly valve, in the supply line leaving the generator, is a desirable flow-changing means. For generator startup a ½- or ¾-in (12.7- or 19.1-mm) valve bypass line should be installed around generator shutoff valves to permit slow warmup of a cold generator.

Water-Flow Quantity. Figure 18.12 shows a curve from which we can read water weights and volumes for different water-flow rates through the system or the

HTW generator as they result from varying temperatures and enthalpies. The curve is based on 10 TBtu/h (2.9-MW) heat flow. Thus for 30 TBtu/h (8.8 MW), the flow values for water weight and volume must be multiplied by 3.

18.5 COMMERCIAL WATERTUBE BOILERS

18.5.1 General Information

Commercial boilers are defined, in this section, as boilers which produce steam or hot water for heating in commercial applications and for modest-size process applications. Commercial boilers come in a wide range of types, sizes, capacities, and design pressures and temperatures. Also commercial boilers can be, and are, designed to operate on a variety of fuels from natural gas to solid fuels.

Commercial boilers are normally furnished as packaged units, and should bear an acceptance label from an independent testing laboratory such as Underwriters' Laboratory (UL), the American Gas Association (AGA), etc., and the label should cover the entire package, not just the burner.

Operational safety is ensured by pressure relief valves, pressure and/or temperature controls, low water cutoff, and some type of flame-monitoring device. State-of-the-art commercial boilers are equipped with controls and safety features comparable to those found on industrial-size boilers. However, since unit capacities are lower for commercial boilers, the annual fuel costs are significantly less. Therefore, the fuel-burning equipment design may be less sophisticated than that found on industrial boilers and is usually capable of on/off or high/low/off operation.

18.5.2 Burners

Burner design for commercial boilers is classified according to two general categories. The first type of burner is referred to as "atmospheric" or "natural-draft," and the second as "power" or "forced-convection."

Atmospheric Burner. The atmospheric burner is the oldest and simplest type of burner which requires only an adequate draft and fuel pressure to function. This type of burner is limited in that it is capable of burning only gaseous fuels. Such gases normally are natural gas, propane, or manufactured gas. Ignition of the gas in an atmospheric boiler is accomplished by one of two methods.

1. Standing pilot (continuous pilot), initially lit by hand. These pilots are normally quite small, consuming less than several cubic feet per hour of gas. The pilot flame is monitored by a sensing device, which shuts off the supply of fuel if the flame is extinguished. Some burner systems may include more than one such pilot.
2. Spark ignition. This eliminates a standing pilot and normally will use an interrupted pilot, operating only during the required period for main burner operation.

Boilers equipped with atmospheric burners will be supplied with a draft correction device referred to as a "diverter." The diverter may be designed into the boiler or may be a separate assembly which is attached to the boiler casing at installation. The breeching which conducts the flue gas to the stack or chimney is attached to the diverter.

The diverter modifies the draft supplied by the chimney. In situations where high draft is created, the diverter allows boiler room air to flow into the chimney to satisfy the negative pressure. When conditions create downdrafts in the chimney, the flue gases from the boiler are relieved from the diverter openings into the boiler room. In any case, the combustion chamber is shielded from the direct effects of variable draft conditions that can occur in a chimney.

Forced-Convection Burner. A forced-convection or forced-draft burner utilizes a fan or blower to force combustion air into the boiler. This method provides better overall combustion and higher efficiencies than can be obtained with atmospheric burners. A forced-draft burner also permits oil to be used as a fuel by atomizing the oil prior to introducing it into the combustion chamber.

Some commercial boilers and burner systems of the forced-convection type may be designed to have a low draft loss. For those units, a draft-modifying device such as a barometric damper may be required to ensure acceptable combustion chamber conditions. This device is normally installed into the breeching.

Boilers of this size [up to 10 MBtu/h (2.9 MW)] are commonly used in installations where routine care and maintenance are not considered a high priority. Boilers designed for this market are generally equipped with adequate but simple control systems which require a minimum of routine adjustments and which eliminate the need for skilled operators.

If the proper selection of boilers is made, maintenance procedures can be simple. Fireside and waterside cleaning can be accomplished by relatively unskilled personnel.

Boilers of this type will provide efficient, dependable, and long-term operation if given proper care and maintenance. A typical installation of a pair of atmospheric gas-fired commercial watertube boilers is shown in Fig. 18.19.

18.6 FIREBOX BOILERS

Firebox-type boiler designs originated many years ago and are the basis for many of the modern boiler pressure vessel concepts. The need for conservation of space and improved energy conversion efficiencies have resulted in modifications to the early designs, but the basic functional principle remains unchanged.

This type of boiler consists of a large cylindrical pressure vessel fitted with tubes for the flow of flue gases from the combustion process. The pressure vessel is designed to be positioned over the top opening of a usually rectangular chamber. This chamber may be of double-wall construction with boiler water circulating between the walls or, as in many cases, may be constructed of refractory walls. One or more openings are provided in one wall of the chamber to permit the installation of a burner system and to provide access to the interior. The burner system initiates and controls the combustion of fuel in the chamber where the combustion process is completed. The products of combustion leave the chamber, flowing through the tubes in the upper pressure vessel. The bottom surface of the cylindrical pressure vessel is in direct contact with the flame and is subjected to the radiant heat from the combustion process.

FIGURE 18.19 Commercial watertube boiler. *(Courtesy of Cleaver-Brooks.)*

Tubes, tube sheets, and baffles may be arranged such that the gases flow through one or two banks of tubes to reach the boiler gas outlet. See Fig. 18.20.

Firebox boilers that include water-cooled sidewalls of flat plates are more difficult to construct for elevated working pressures. The thickness of the wall plates and the number of sidewall stays required to contain the internal pressures limit the availability of such vessels for operation at high pressures. Boilers of this type are more commonly used as heating boilers, generating low-pressure steam [up to 15 psig (1 bar)] or for environmental heating, operating at water outlet temperatures of up to 250°F (121°C).

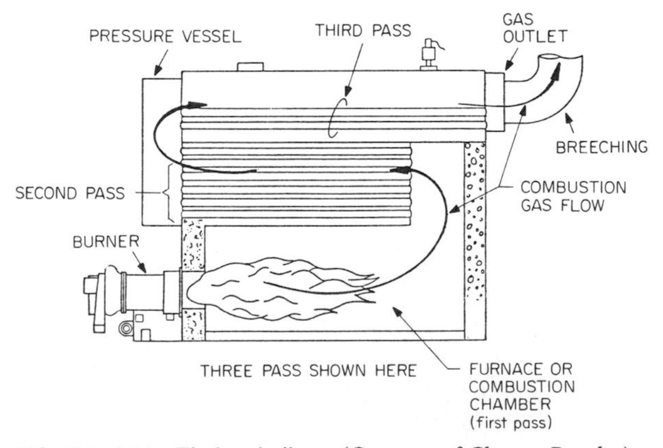

FIGURE 18.20 Firebox boiler. *(Courtesy of Cleaver-Brooks.)*

Firebox boilers with a flue outlet at the front and one bank of horizontal tubes returning the combustion gases from the rear of the boiler to the front are normally referred to as "horizontal return tubular (HRT) boilers." Figure 18.21 illustrates a boiler of this type. This firebox boiler design requires a greater field installation effort than other designs, which may affect the total cost of the boiler installation. An important design consideration is that the pressure vessel be limited to a cylindrical configuration, which can be efficiently constructed to generate steam at elevated pressures.

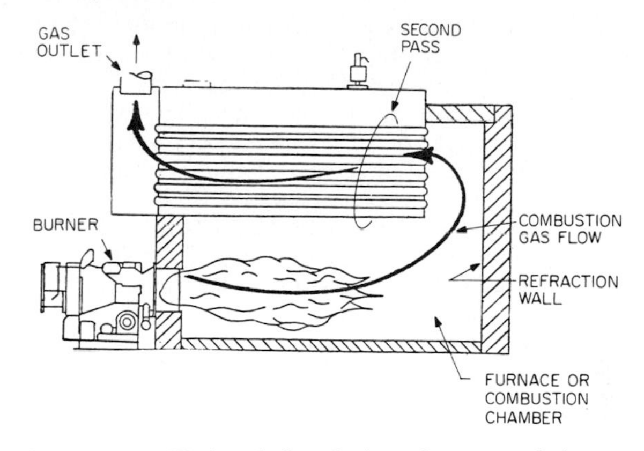

FIGURE 18.21 Firebox boiler—horizontal return, tubular. (*Courtesy of Cleaver-Brooks.*)

One benefit of firebox boilers is that a large combustion chamber volume is possible. This makes possible the burning of a wide variety of fuels which require significant time to complete the oxidation of all components before the gases enter the boiler tube passes. The residence time for the burning gases from the fuel may be 4 or more times longer than what might be possible in, say, a firetube boiler. Fuels such as coal, wood (both dry and wet), agricultural wastes, and others are routinely used in firebox boilers.

Firebox boilers are usually less efficient than firetube boilers since large-diameter tubes and fewer flue tube passes are used, causing low gas velocity (and the products of combustion leave the boiler at a higher temperature).

Firebox boilers are commonly supplied by the manufacturer less burner and controls. The boiler specifier or the boiler contractor will select a burner and control system appropriate for the available fuel or the needs of the owner. Under these conditions the installing contractor must have the skills and experience required to provide a good marriage of all parts from various suppliers for the final assembly to achieve the desired performance.

18.7 CAST-IRON BOILERS

18.7.1 Types

Cast-iron boilers can be classified as sectional with internal push nipples or sectional with external headers (drums).

The round cast-iron boiler, primarily used for residential or other small buildings, is no longer manufactured in the United States. Although the wood- or coal-fired, round cast-iron potbelly stove is still manufactured, it is not discussed further since its capacity, adequate for only one room (two rooms at best), is not applicable for central heating.

18.7.2 Use

Cast-iron boilers are used for low-pressure steam [15 lb/in^2 (1 bar) maximum] and hot water [30 lb/in^2 (2 bar), 250°F (121°C) maximum] heating systems for residential as well as small to medium institutional, commercial, and industrial buildings.

The installed cost of cast-iron boilers is generally greater than that of steel firetube boilers of equal capacity. Although the operating reliability and the maintenance costs are at least equal to those of firetube boilers, the average operating life of a cast-iron boiler is noticeably longer.

In the replacement boiler market, where access to existing boiler rooms is often restricted, cast-iron sectional boilers have a distinct advantage over other factory-assembled units and/or generators, since cast-iron boilers can be field-assembled easily and cost-effectively.

18.7.3 Design

General. Cast-iron boilers with automatic firing of gas and/or oil can be field-assembled if site restrictions require or factory-assembled—complete with burner, valving, trim, and prewired panel-mounted controls.

Coal-fired boilers can be either field-assembled or factory-assembled with trim and valving. However, when automatic stoker firing of anthracite or bituminous coal is required, the stoker unit is usually shipped separately and connected to the boiler at the site after the boiler is set.

Boiler Construction. Sectionalized cast-iron boilers inherently have low draft losses because of the relatively large combustion chamber volume and liberally sized flue passages. This low-draft-loss characteristic is primarily beneficial to boilers with atmospheric gas burners, since they will require the lowest chimney height of all types of boilers for the same rated heating capacity.

Cast-iron boilers, due to their relatively heavy (thick) sections, are intrinsically slow in heating up. However, their ability to retain heat (due to their mass) enables them to dampen the typical swings in heating demands. In addition, this same mass contributes significantly to their ability to resist corrosion and to thus have a long service life.

Sectional boilers are made up of vertical sections and look much like a loaf of sliced bread. The number of sections determines the boiler capacity, since the sections of each specific boiler model are identical. This design factor permits replacing damaged sections with new sections right in the boiler room.

The combustion gas flow in typical boilers (when any fuel is burned within the combustion chamber) is comprised of gases rising and entering the horizontal flue passages (cast within each section). Depending on the particular design, the combustion gases may pass horizontally once or twice before entering the breeching (boiler stack outlet). In general, external header boilers have more horizontal passes than internal push-nipple boilers.

Internal Push-Nipple Boilers. The cast sections of internal push-nipple boilers are joined with tapered nipples and held tightly together with tie-rods or bolts.

The nipples are sized to provide the required free circulation of water from section to section. In addition, in steam boilers the top push nipples must be large enough to provide unrestricted flow of steam.

To minimize the size of the combustion chamber and to maximize the heat-absorbing surface of gas-fired boilers, some manufacturers cast studs into each section.

Some smaller oil-fired boilers are built with water-filled spaces that completely surround the sides as well as the bottom of the combustion chamber. This type of boiler is called a *wet-base boiler*. The bottom of each section has legs filled with water that extend across the bottom of the boiler floor. This design permits unrestricted circulation of water within the base (water legs) of each section. The primary merits of a wet-base boiler are twofold: First, this type of boiler can be set directly on a wood or any combustible floor construction without danger of fire. Second, the available heat-absorption surface in the combustion chamber is increased.

External Header. Each cast section of the external header boiler is connected to the headers (drums) with screwed nipples. There are three headers: one is centered at the top, and the other two are at the bottom (one at each side). The top header serves as the supply manifold and is sized large enough to provide unrestricted flow of steam and/or hot water.

With this type of boiler construction, in the rare event that a section develops a leak or cracks, the following temporary expedient repair can be made within hours: When the boiler is taken out of service and drained, the damaged section's connections to the upper and lower headers are capped. With the damaged section isolated from the headers, the boiler can be refilled and brought back into service.

With this temporary repair completed, the boiler can remain in service until the next scheduled shutdown, at which time the damaged section can be replaced with a new section to be reconnected to the headers.

18.7.4 Burners and Fuels

Depending on the heating capacity (see Table 18.2), cast-iron boilers are designed for burning the following fuels: Automatic-firing natural gas or propane, light oil (No. 1 or No. 2), dual-fuel gas/oil (natural gas or propane and No. 1 or No. 2 oil), anthracite and bituminous coals.

Hand firing of anthracite coal, wood coal, and bituminous coal is generally not recommended for central heating boilers due to excessive smoking and sooting problems.

18.7.5 Efficiency

In general, cast-iron boilers in good working order and burner units can maintain the following efficiencies: 80 percent burning gas or light oil and 55 to 65 percent (stoker-fired) burning anthracite or bituminous coal. When hand firing is used, anthracite coal efficiencies less than 50 percent are normal.

The efficiencies noted above are the overall efficiency of the boiler-burner

TABLE 18.2 Typical Cast-Iron Sectional Boiler Capacity Range

	Nominal capacity range in $I = B = R$ net rating			
	Push-nipple boiler		Header (drum) boiler	
Fuel	Steam, lb/h (kg/h)	Hot water, MBtu/h (kW)	Steam, lb/h (kg/h)	Hot water, MBtu/h (kW)
Gas	25.2–4,035 (11.4–1,836.3)	26–4,160 (7.6–1,220)	1,008.8–11,349 (457.6–5,147.9)	1,040–11,700 (304.8–3,429.3)
Oil	76.5–407 (34.7–184.6)	79–420 (23–123)	1,008.8–11,349 (457.6–5,147.9)	1,040–11,700 (304.8–3,429.3)
Dual fuel, gas or oil	412–3,900 (186.9–1,769)	425–4,020 (125–1,180)	1,008.8–11,349 (457.6–5,147.9)	1,040–11,700 (304.8–3,429.3)
Anthracite coal	97–254 (44–115.2)	100–262 (29.3–76.8)	— —	— —
Bituminous coal	142.6–292 (64.7–132.5)	147–301 (43–88.2)	— —	— —

Note: Light oil, No. 1 or 2; gas: natural or propane; 1 lb/h = 0.970 MBtu/h. Coal heating value: anthracite, 12,500 Btu/lb (29 MJ/kg); bituminous, 12,000 Btu/lb (28 MJ/kg). $I = B = R$ net rating is equal to the gross output less an allowance for piping and pickup losses in accordance with $I = B = R$ testing and rating code for low-pressure cast-iron boilers.

unit. "Efficiency" is defined as the gross boiler heat output divided by the heat of the fuel burned during the same unit of time, under steady-state conditions.

18.7.6 Capacities

The nominal capacities of steam and hot-water sectional push-nipple and header-type boilers (manufactured in the United States) with respect to the fuel burned are summarized in Table 18.2.

18.8 SYSTEMS AND SELECTIONS

18.8.1 Load Analysis

Once the heating and/or process loads have been calculated, a decision must be made to use a steam or water system. A comparative study of original, operational, and maintenance costs of the overall system should always be made. The electric power for pumps and sizing of radiation should be considered.

The frequently cited advantages attributed to hot-water systems are:

1. Smaller pipe size for a given load.
2. Need for steam traps eliminated.
3. No condensate return systems required.
4. Lower operational temperatures.
   a. Reduced system heat losses.
   b. Improved boiler efficiency and reduced cost of fuel.
5. More uniform space heating environment.

6. Loss of system water reduced or eliminated and need for water treatment reduced.
7. Quieter system with lower piping expansion and contraction consideration.
8. Flexibility of system expansion. Extension of the system can be easily accommodated by use of booster pumps which will carry the heat through extension piping runs.

Advantages of steam systems are:

1. Higher temperature available at user
2. Constant temperature in user at condensing condition
3. Improved transfer of heat (coefficients) requiring smaller heat user equipment
4. Fast response to load change
5. Reduced mechanical stresses in users and operators
6. Greater familiarity due to past use
7. Minimized corrosion problem in generator caused by operation at reduced temperatures

Number of Boilers Required. Some jobs will be installed with a single boiler. This is a perfectly valid solution, provided that the loss of the boiler for a time will not seriously curtail the operation of the facility where the boiler is installed or that such a curtailment will be acceptable to the operators of the facility. In most cases, complete loss of heat and/or process steam cannot be tolerated. Then multiple boilers are recommended.

One of the most satisfactory methods of sizing two boilers is for each to have a capacity equal to two-thirds of the plant load. The two units will operate together during the full-load period. As the load demand drops to two-thirds of the design load, one boiler will carry the load with the capability of swinging to its low-load point before shutoff. Thus for most of the plant operating-load range, one boiler will carry the load.

If one boiler is down for service, the other boiler can carry nearly full plant load.

If loads are such that three boilers are desired, because of very low load situations when only one unit at relatively low rate is needed, use three units, each rated at one-half the design load. This will also provide service without interruption if one boiler becomes inoperative for any reason, since two boilers will deliver all the load. A variation of this three-boiler setup would be two units, each good for two-thirds the total load, plus a "pup" or small boiler selected to meet the lowest plant or "summer" requirement.

Heating Boilers

1. Total net load (connected load), Btu/h or W; building heat loss, Btu/h or W; process requirements, if any, Btu/h or W; domestic water heating, Btu/h or W; future expansion, if any, Btu/h or W
2. Pickup allowance, Btu/h or W
3. Piping losses Btu/h or W

The total of items 1, 2, and 3 is termed the "gross load."

Other factors affecting boiler load include:

1. *Air conditioning:* Some types of air conditioning systems use heat from the boiler, and in some parts of the United States this hot-weather load is greater than the winter heat requirement and must be used in determining the size of the boiler.
2. *Pickup allowance:* The pickup allowance covers sudden or quick pickup of increased-load or starting-load conditions and a period when outdoor temperature and other load factors are close to the design limits selected originally. If there are no unusual conditions, a maximum of 10 percent of the known loads may be added.
3. *Piping losses:* Allowance is made to cover nonuseful heat loss from piping, etc. If it is not calculated, a maximum of 15 percent of known loads may be added when there are no unusual circumstances.
4. *Maximum instantaneous demand:* This is another way of expressing sudden (unusually short-duration) peak-load requirements, excluding pickup allowance. All applications of any type of boilers (especially steam) should be investigated for any unusual maximum instantaneous demands which produce unusual peak conditions. The solution can take one of two forms:
   a. The boiler can be controlled, valved, or otherwise limited to keep the steam flow within the boiler's rated capacity.
   b. Additional boiler capacity should be provided to ensure safe pickup of this type of load.

General descriptions of the major factors determining the domestic hot-water requirements for typical buildings in each category are presented in Table 18.3 as a guide to designers: When unique hot-water requirements exist in a particular building, the designer should increase the recovery and/or storage capacity to account for this use.

General. Peak hourly and daily demands for various categories of commercial and institutional buildings are shown in Table 18.3. These demands for central-storage-type hot-water systems represent the maximum flows metered in over 160 buildings throughout the United States. Caution must be exercised in applying these figures to very small buildings.

Also shown in Table 18.3 are average hot-water consumption figures for these types of buildings. Averages for schools and food service establishments are based on actual days of operation, while averages for all others are based on total days.

Operating Pressure. Boilers cannot be operated at their design pressure because the safety valves must be set at the boiler design pressure or less by code and law. Boilers must be selected for operating at pressures greater than the maximum system operating pressure, to prevent safety valve leakage and to provide proper valve seating.

For 15 lb/in^2 (1-bar) boilers, the actual operating control pressure must be set lower so that when the burner control is satisfied, the maximum steam pressure will not exceed 15 psig (1 bar). Usually operating controls are set to maintain 12 psig (0.82 bar), or less; however, the boiler manufacturer and control manufacturer should be consulted if higher settings are desired.

For hot-water boilers the total operating pressure of static and velocity pressure must not exceed the 30 psig (2-bar) maximum pressure. Allowance for ve-

TABLE 18.3 Hot-Water Demands and Use for Various Types of Buildings

Type of building	Maximum hour	Maximum day	Average day
Men's dormitories	3.8 gal/student	22.0 gal/student	13.1 gal/student
Women's dormitories	5.0 gal/student	26.5 gal/student	12.3 gal/student
Motels			
20 units or less	6.0 gal/unit	35.0 gal/unit	20.0 gal/unit
60 units	5.0 gal/unit	25.0 gal/unit	14.0 gal/unit
100 units or more	4.0 gal/unit	15.0 gal/unit	10.0 gal/unit
Nursing homes	4.5 gal/bed	30.0 gal/bed	18.4 gal/bed
Office buildings	0.4 gal/person	2.0 gal/person	1.0 gal/person
Food service establishments			
Type A: Full-meal restaurants and cafeterias	1.5 gal/max. no. of meals/h	11.0 gal/max. no. of meals/h	2.4 gal/max. no. of meals/day*
Type B: Drive-ins, grills, luncheonettes, sandwich and snack shops	0.7 gal/max. no. of meals/h	6.0 gal/max. no. of meals/h	0.7 gal/max. no. of meals/day*
Apartment houses			
20 apts. or less	12.0 gal/apt.	80.0 gal/apt.	42.0 gal/apt.
50 apts.	10.0 gal/apt.	73.0 gal/apt.	40.0 gal/apt.
75 apts.	8.5 gal/apt.	66.0 gal/apt.	38.0 gal/apt.
100 apts.	7.0 gal/apt.	60.0 gal/apt.	37.0 gal/apt.
130+ apts.	5.0 gal/apt.	50.0 gal/apt.	35.0 gal/apt.
Elementary schools	0.6 gal/student	1.5 gal/student	0.6 gal/student*
Junior and senior high schools	1.0 gal/student	3.6 gal/student	1.8 gal/student*

Note: 1 gal = 3.7 L.
*Per day of operation.

locity head is usually not made if circulator pumps are found between boiler outlet and system resistance. For hot water above 30 lb/in^2 (2 bar), the boiler design pressure should be 10 or 20 percent above that for the system static plus velocity head.

In selecting the operating pressure of the boiler, the piping pressure drop must be added to the pressure required at the appliance. The boiler safety valve setting should be at least 10 percent higher than this operating total pressure, and the boiler design pressure must be equal to or greater than this. In some cases a higher design pressure is chosen to allow for possible future down-rating with age.

If the loads are calculated in Btu/h (W), the boiler size may be selected directly from the manufacturer's gross output at the boiler nozzle. If the load is given as pounds (kilograms) of steam per hour at a stated pressure in lb/in^2 (or bar) and a given feedwater temperature, it must be adjusted to "from and at 212°F (100°C)" conditions for comparison to the manufacturer's rating.

The appropriate factor of evaporation from Table 18.1 is used to adjust the rated boiler output [from and at 212°F (100°C)] to actual job or operating conditions. The available feedwater temperature and expected boiler operating pressure affect the manufacturer's standard boiler output ratings, which are based on "from and at 212°F (100°C)."

The following equation is used to correct the actual operating conditions from and at 212°F (100°C):

$$\text{Required boiler capacity from and at 212°F, bhp} = \frac{\text{design load at design temp., lb/h}}{\text{factor of evaporation from Table 18.1, lb/(h} \cdot \text{bhp)}}$$

Note: Metric units must be converted to English units before this equation is applied.

18.8.2 System Pressures

Steam boilers generate the pressure required to deliver the heat energy to the user. The boiler selected must be appropriate to operate at the pressure required by the system heat user. The pump which returns the condensate to the boiler must also be capable of delivering such condensate into the pressure vessel.

If the system design is for hot water, the system pressure is dictated by the building configuration and operating temperature. Sufficient pressure must be imposed on the generator to prevent the water from flashing into steam. Some safety margin in system pressure is required. The recommended procedure to establish the system pressure is to select the saturation pressure for 40°F (22°C) higher than the maximum expected system water temperature.

System pressure is created by the expansion tank which is either elevated to provide the needed pressure or is pressurized with a gas, preferably an inert gas such as nitrogen. Compressed air can be used with success, but the danger exists of oxygen absorption in the water in the expansion tank and corrosion of boiler internal piping systems. Most boiler manufacturers will provide a minimum boiler return water temperature level to prevent gas-side corrosion.

Most boilers are designed to provide generous internal circulation, taking advantage of the greater heat transfer that results. In general, relatively large system circulation rates can be used without significantly increasing system pump head. There are exceptions, however. Some boiler designs require forced circulation through single- or parallel-tube banks. Such designs may require higher pump discharge head or even a separate pump. Such factors should be thoroughly analyzed before final boiler selections are made.

Table 18.4 provides system header (and boiler) circulation rates for typical system temperature drops for installations requiring up to 1000 bhp (746 kW).

18.9 HEAT-RECOVERY BOILERS

18.9.1 Introduction

Heat-recovery boilers are used to reclaim heat from high-temperature exhaust gases resulting from thermal and/or chemical processes. Such regained heat is used to generate steam, heat water, or organic fluids. Table 18.5 lists a few processes and typical exhaust temperatures. Heat-recovery boilers are nor-

TABLE 18.4 Water Circulation Rates through Hot-Water Generators

bhp	Gross boiler output at nozzle, MBtu/h	System temperature drop, °F (°C)					
		10 (5.5)	20 (11.1)	30 (16.7)	40 (22.2)	50 (27.7)	60 (33.3)
		Circulation rate gal/min					
20	670	134	67	45	33	27	22
30	1,005	200	100	67	50	40	33
40	1,340	268	134	89	67	54	45
50	1,675	335	168	112	84	67	56
60	2,010	400	200	134	100	80	67
70	2,345	470	235	157	118	94	78
80	2,680	536	268	179	134	107	90
100	3,350	670	335	223	168	134	112
125	4,185	836	418	279	209	168	140
150	5,025	1005	503	335	251	201	168
200	6,695	1340	670	447	335	268	224
250	8,370	1675	838	558	419	335	280
300	10,045	2010	1005	670	503	402	335
350	11,720	2350	1175	784	587	470	392
400	13,400	2680	1340	895	670	535	447
500	16,740	3350	1675	1120	838	670	558
600	20,080	4020	2010	1340	1005	805	670
750	25,100	5025	2512	1675	1255	1006	838
800	26,800	5360	2680	1790	1340	1070	900
1000	33,500	6700	3350	2230	1680	1340	1120

Note: Conversion factors: 1 hp = 746 W and 1 gal = 3.8 L.

TABLE 18.5 Typical Process Exhaust Temperatures

Chemical and thermal process	Temperature, °F (°C)	
Chemical oxidation	1350–1475	(730–800)
Annealing furnace	1100–1200	(590–650)
Diesel-engine exhaust	1000–1200	(540–650)
Forge and billet heating furnace	1700–2200	(920–1200)
Refuse incinerator	1550–2000	(840–1100)
Open-hearth steel furnace, air-blown	1000–1300	(540–700)
Open-hearth steel furnace, oxygen-blown	1300–2100	(700–1150)
Basic oxygen furnace	3000–3500	(1650–1900)
Glass-melting furnace	1200–1600	(650–870)
Gas-turbine exhaust	700–1100	(370–590)
Fluidized-bed combustor	1600–1800	(870–980)

mally considered economical for exhaust gas temperatures above 600°F (315°C).

Heat-recovery boilers may be either firetube or watertube design. Selection depends on the type of gas, gas temperature, operating and design pressures, plant requirements, etc. Firetube boilers are preferred for lower steam or hot-water output and design pressure. Firetube design can be furnished to operate with higher waste gas pressures and will normally cause a greater pressure loss in the gas-flow stream. Watertube boilers are suggested for higher steam or hot-water output and design pressures and will cause lower gas pressure losses.

18.9.2 Design Considerations

Before any heat-recovery boiler is designed, an analysis of the economics involved should be conducted. Physical and chemical behavior of the hot gas stream and gas velocity should be studied to avoid erosion and corrosion of tubes. The pressure and temperatures of the waste gas stream are important, so as to avoid the need for filters, inducers, blending systems, etc. Possible effects on the basic process must be considered, and a value must be placed on the energy absorbed.

Tables 18.6 and 18.7 provide information related to types of heat-recovery boilers which are normally recommended for a variety of waste gas streams as well as a definition of waste materials and some problem constituents which result from their incineration.

Physical and Chemical Behavior of Gases. Incinerator flue gas may contain a significant amount of abrasive solid particulates which in time may damage the tube's surface and weaken the mechanical strength. Chemical behavior concerns involve the presence of corrosive or fouling gases. Incinerator gas, e.g., usually from the combustion of a variety of materials burned which may generate corrosive gases.

Table 18.8 indicates the flue gas composition for a municipal waste incinerator. The presence of SO_2, SO_3, or chlorine compounds is a serious concern unless metal surfaces are maintained above the dew point temperatures. The best operating range of boiler metal where such gases are present is from 350 to 450°F (175 to 230°C). Corrosion can occur above and below these temperatures.

Flue Gas Fluid Dynamic Behavior. Velocity distribution and the magnitude of the gas stream play a vital role in achieving high thermal performance and prolonged boiler life. Uniform velocity distribution through or around the tubes results in a uniform temperature gradient, proper internal circulation, and minimal stresses. Velocity control is important to avoid tube erosion and to maintain limited pressure drop. Table 18.9 lists recommended velocities for both firetubs and watertube boilers. The investigators suggest that a gas velocity less than 35 ft/s (11 m/s) allows particulates to deposit inside the tubes of firetube boilers, which compromises the effectiveness of the heat-transfer area. Similarly, insufficient gas velocities in a watertube design allow particulate matter to collect in the gas passageways and may affect the soot blowers' effectiveness.

Firetube heat-recovery boilers are designed in various configurations:

- Single-, double-, three-, and four-pass boilers for straight heat recovery from hot flue gases
- Double- or three-pass boilers with supplementary firing

TABLE 18.6 Heat Sources and Application of Heat-Recovery Boilers

Process	Type of boilers	Comments
Incinerator	Firetube and watertube	Care must be taken to prevent corrosion. Hoppers, soot blowers, and superheaters may be used. Fins are never used. Induced-draft fan is needed.
Gas turbine	Finned watertubes	Superheater, economizer, pressure drop critical. Must be within 10 to 12 inWG (2.5 to 3 kPa) exhaust gas pressure range.
Petroleum refinery (catalytic regeneration process)	Firetube or watertube	High-pressure gas requires special shell design. Produced steam used for the process itself. Installed space limited.
Heat-treat/metal refining processes (aluminum ore and copper ore refining)	Firetube and watertube	For small flow rates, firetube is preferred. Space limitation. Watertube boilers also used with soot blowers, greater tube thickness. Induced-draft fan needed.
Chemical processes (sulfur combustion)	Firetube and watertube	Corrosion prevention, metal temperatures above dew point. Induced-draft fan needed. Fan blades must be refractorized or of high metal alloy.
Rice hull combustion*	Firetube boilers	With furnace to cool down high-temperature combustion gas. Automatic to soot blowers required to clean tube. Gas velocity 45 to 50 ft/s (13 to 15 m/s) to avoid erosion.
Diesel engine*	Firetube or watertube	For small engines firetube boilers; for bigger engines, watertube with finned tubes or bare tubes. Superheater can be used in watertube design. Pressure drop critical.
Fluidized-bed combustor	Firetube and watertube	Hoppers, soot blowers, superheater, and economizer. Average gas velocity is 50 ft/s (15 m/s) for wood or coal fuels.
Glass melting*	Watertube and firetube	Slag buildup and fouling can be severe because of fluxing material (alkaline or acidic).
Miscellaneous (gasification, papermill)	Watertube and firetube	Gasified products cooled to lower temperature before being fired in any thermal process.

*See footnote at bottom of Table 18.2.

TABLE 18.7 Classification of Wastes to Be Incinerated

Waste type, description*	Principal components	Approximate composition, % by weight	Moisture content, %	Incombustible solids, %	Value of refuse as fired, Btu/lb (kJ/kg)
0, trash	Highly combustible waste, paper, wood, cardboard cartons. Including up to 10% treated papers, plastic or rubber scraps; commercial and industrial sources	Trash, 100	10	5	8500 (19.8)
1, rubbish	Combustible waste, paper, cartons, rags, wood scraps, combustible floor sweepings; domestic, commercial, and industrial sources	Rubbish, 80 Garbage, 20	25	10	6500 (15.1)
2, refuse	Rubbish and garbage; residential sources	Rubbish, 50 Garbage, 50	50	7	4300 (10)
3, garbage	Animal and vegetable wastes, restaurants, hotels, markets; institutional, commercial, and club sources	Garbage, 65	70	5	2500 (5.8)
4, animal solids and organic	Carcasses, organs, solid organic wastes, hospital laboratory, abattoirs, animal pounds, and similar sources	Animal and human tissue, 100	85	5	1000 (2.3)
5, gaseous	Industrial process wastes	Variable	Depends on predominant components	Variable according to wastes survey	Variable according to wastes survey
6, semisolid and solid wastes	Combustibles requiring hearth, retort, or grate	Variable	Depends on predominant components	Variable according to wastes survey	Variable according to wastes survey

*The figures on moisture content, ash, and Btu values as fired have been determined by analysis of many samples. They are recommended for use in computing heat release, burning rate, velocity, and other details of incinerator designs. Any design based on these calculations can accommodate minor variations.

TABLE 18.8 Composition of Flue Gas from Combustion of Solid Waste and Hazardous Materials

Chemical composition	Municipal solid	Hazardous
N_2	78–88%	78–88%
O_2	6–10%	6–12%
CO_2	6–12%	6–12%
SO_2	2–200 ppm	1200–1400 ppm
HCl	5–100 ppm	(Based on waste)
HF	0–1 ppm	(Based on waste)
CO	5–50 ppm	0–50 ppm
NO_x	50–100 ppm	100–200 ppm

Note: ppm = parts per million.

TABLE 18.9 Heat-Recovery Design Guidelines

	Maximum gas velocity, ft/s (m/s)						
	Firetube	In-line	Water-tube stag-gered	Fouling factor	Maxi-mum fins per in (mm)	Fin thick-ness, in (mm)	Soot blowers
Clean: Combustion products of natural gas	200 (61)	120 (36.6)	100 (30.5)	0.001	6 (0.24)	0.05 (1.27)	No
Average: Combustion products of No. 2 oil and wood, fume incinerators	170 (51.8)	100 (30.5)	80 (24.4)	0.002	5 (0.2)	0.05 (1.27)	Provision
Dirty: Combustion products of Nos. 5 and 6 oil, liquid incinerators, municipal sludge, fluid cat. cracker	130 (39.6)	80 (24.4)	65 (19.8)	0.004	4 (0.16)	0.06 (1.52)	Yes
Dirty and abrasive: Combustion products of coal	100 (30.5)	60 (18.3)	40 (12.2)	0.004	3 (0.12)	0.075 (1.91)	Yes

Watertube heat-recovery boilers are tailor-made to effectively absorb the heat available from the waste gas stream while maintaining appropriate gas velocities and minimizing erosion. Supplementary firing capabilities can be incorporated in this design, too.

Supplementary firing in waste heat-recovery boilers may be an economical alternative if the steam demand at the facility is greater continuously or periodically than that which can be generated from the waste heat recovered.

Supplementary firing control systems must be carefully integrated into the waste heat-recovery control system to provide the desired result. Damper controls frequently employed to modulate the flow of the waste gas stream through the boiler create changes in the gas dynamic within the exchanger. Supplementary fuel firing is accomplished by means of a forced-air burner system, which is effective and safe if the combustion environment is correct.

When two gas streams, each independent of the other, flow through the exchanger, the gas temperatures and pressures will vary as each flow changes. To achieve stable combustion from the supplementary system, such environmental changes need to be accurately sensed and appropriate changes made to the fuel and air flow at the burner.

Design calculations for heat transfer from waste gases follow the routine procedures well documented in other publications. For gases burdened with high levels of particulate matter, the gas-side coefficients may be reduced by as much as 10 percent in consideration of the surface fouling that is likely to occur. Overall coefficients between 5 and 16 Btu/(h · ft^2 · °F) [28.4 and 90.9 W/(m^2 · K)] are realistic. For gases containing erosive matter, the gas velocity must necessarily be low, therefore dictating a low coefficient. For very clean waste gas, designs can be based on heat-transfer coefficients which may approach 20 Btu/(h · ft^2 · °F) [113.6 W/(m^2 · K)].

For those instances where the gas particulate loading is very heavy, consideration should be given to the provision of a dust drop area. This may be installed in the ductwork carrying the gas to the boiler or within the boiler itself. The boiler or system designer should be consulted for these details before the total system design is begun.

Firetube boilers are employed more frequently in waste heat applications because flue gas velocities can be maintained at uniformly higher levels, which will result in higher heat-transfer rates per unit of surface area. Therefore the exchanger is smaller and less expensive. However, if the gas to be cooled contains little or no particulates, a watertube design which incorporates a finned-tube heat-transfer surface is possible and significant reductions in boiler size are achievable.

For installations where large quantities of gas are to be cooled [25,000 lb/h (11,300 kg/h) of gas or greater], a watertube design is likely to be specified unless the waste gas is available at the source of pressure of 1 psig (0.15 kPa) or more. Figures 18.22, 18.23, and 18.24 all depict watertube-type equipment. Note that superheaters (available only with this type of equipment) provide steam temperatures higher than saturation temperatures.

Economizers, although more regularly furnished with watertube-style boilers, may be used in conjunction with firetube design, to enhance the total recovery of heat.

18.10 SOLID-FUEL BOILERS

In recent times of increasing costs and uncertain supplies of oil and gas, coal- and other solid-fuel-fired boilers can offer important benefits. They rely on fuel sources that are stable, with ample reserves. Solid fuel such as coal, wood, etc., when readily available, can substantially reduce the cost of producing steam and hot water, now and in the future.

Figure 18.25 compares the cost of generating 1000 lb/h (453 kg/h) of steam by using various fuels.

Boilers designed to be used with solid fuels are usually significantly different

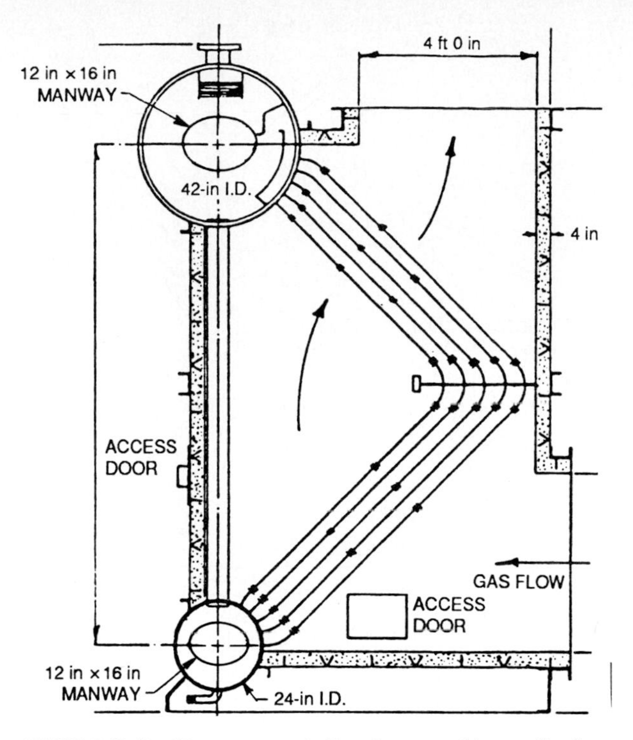

FIGURE 18.22 Heat-recovery boilers for gas-turbine application. (*Courtesy of Cleaver-Brooks.*)

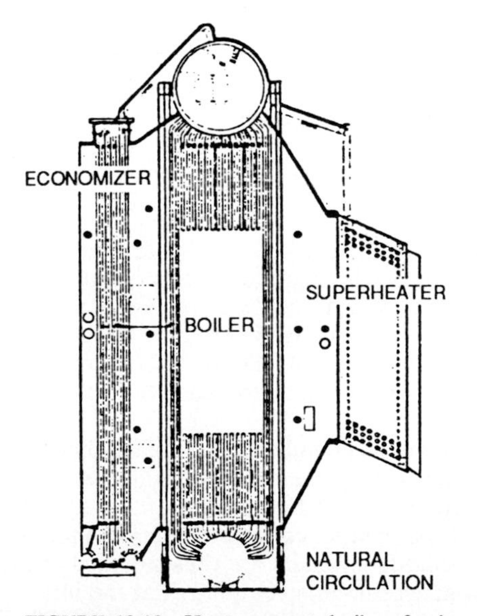

FIGURE 18.23 Heat-recovery boilers for incinerator application. (*Courtesy of Cleaver-Brooks.*)

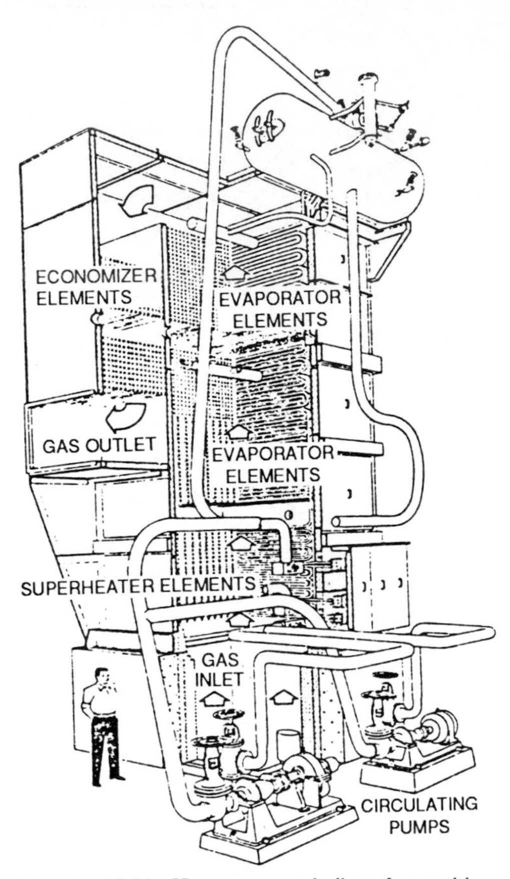

FIGURE 18.24 Heat-recovery boilers for positive waste heat application. (*Courtesy of Cleaver-Brooks.*)

from automatic fuel boilers. Because solid fuels burn much more slowly than oil or gas, a large combustion zone is required for equal heat release. To ensure complete combustion of the fuel, larger quantities of excess air are supplied, creating the need for larger flue gas flow areas. Therefore, the boilers are nearly always substantially larger for equal output. Solid fuels normally require a higher-temperature environment in the combustion zone to facilitate complete fuel combustion.

To accomplish this, often the chamber is completely lined with refractory, surrounded by insulation and seal casing. The convection zone or heat exchanger is normally mounted above this chamber. Because of this arrangement and because of the increased size of the components, the boiler cannot be easily handled or shipped as a package. Solid-fuel boilers are thus supplied to the site in a number of components, and final assembly is completed, at the installation.

A boiler with an external furnace is shown in Fig. 18.26. This boiler is also known as a horizontal return tubular (HRT) boiler. Although the gases are going

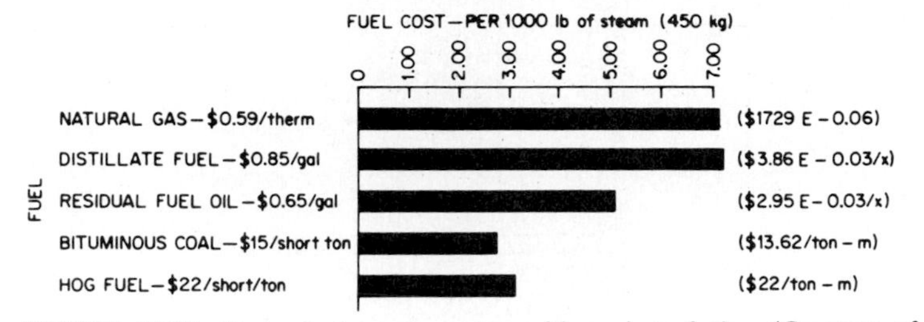

FIGURE 18.25 Cost of steam generator with various fuels. *(Courtesy of Cleaver-Brooks.)*

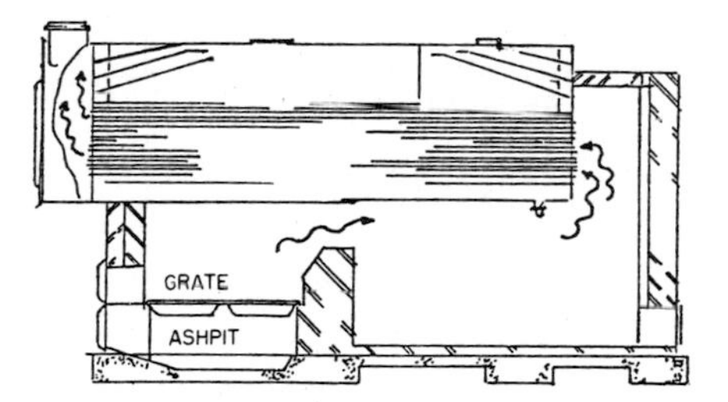

FIGURE 18.26 Horizontal return boiler—sectional sketch. *(Courtesy of Cleaver-Brooks.)*

through the horizontal tubes only once or in one pass, it could be done in two or three passes. This will cool the flue gases, decreasing the gas exit temperature and thus increasing boiler efficiency.

This design is the simplest of solid-fuel-fired boilers and is a throwback to very early boiler designs. The fuel-burning mechanism is located within the refractory combustion chamber. Access doors provide a means for ash removal, boiler cleaning, and inspection. Fuel-burning equipment can be automated as well as ash removal.

Because the fuel cannot be pumped from a tank or taken from a pipeline, but must be physically handled from a storage site for delivery to the burner equipment, and because ash results from the combustion of these fuels which must be frequently removed from the combustion zone, operators must be in frequent attendance.

For these reasons, solid-fuel-burning installations are normally found only at intermediate to large facilities. A boiler appropriate for an intermediate-size solid-fuel installation might be as shown in Fig. 18.27. This is a two-drum watertube unit, wherein the factory-completed heat exchanger section could be assembled to the combustion chamber at the site. The combustion zone extends upward into a water-cooled wall area of the exchanger, and ash hoppers are included at the boiler convection area as needed. Screen wall tubes, convection bank baffles, slag tubes, etc., are a part of the factory assembly. Superheater sections are frequently supplied as an option with this style boiler.

Boilers designed for solid fuels are usually tailored for the fuel to be used. Solid fuels vary significantly in how they can be handled, how they burn, and the quantity and nature of the residue. Fuel characteristics which strongly impact the

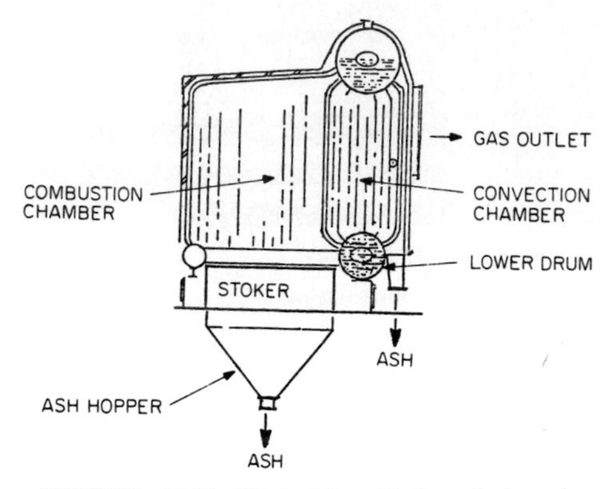

FIGURE 18.27 Watertube D-shaped two-drum boiler. *(Courtesy of Cleaver-Brooks.)*

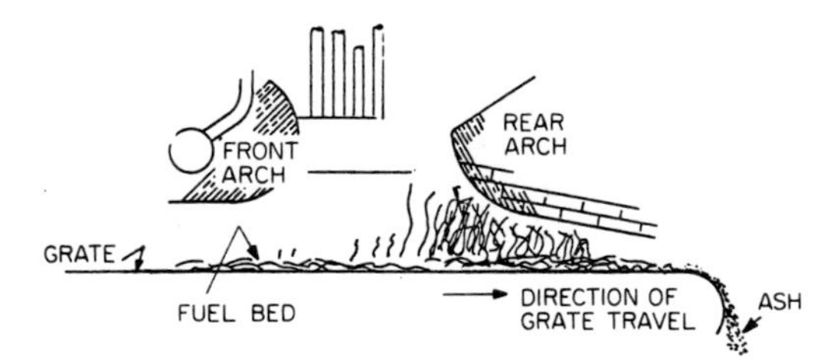

FIGURE 18.28 Arches for low-volatility fuels. *(Courtesy of Cleaver-Brooks.)*

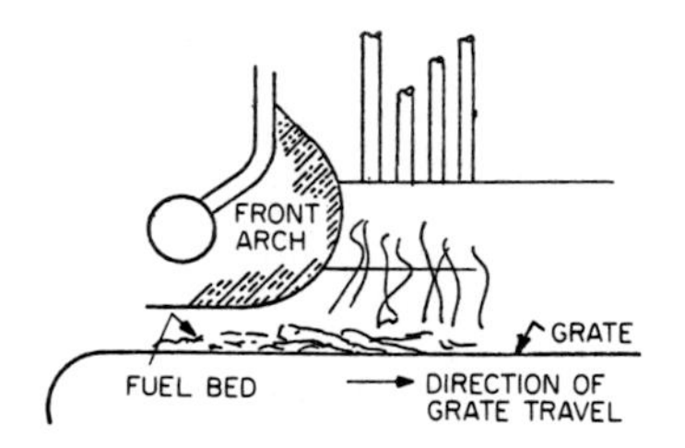

FIGURE 18.29 Arches for high-volatility fuels. *(Courtesy of Cleaver-Brooks.)*

design of the boiler, combustion chamber, burner equipment, and fuel and residue handling equipment are volatility, particle size, moisture content, ash content, and ash fusion temperature.

Each of the above characteristics affects the design required for a successful installation. The design of the combustion chamber is selected to provide the environment needed for the specific fuel. The boiler is then modified to complement the process and fit the combustor. For example, for anthracite fuels which are very low in volatility, extensive refractory arches are required to ensure continued ignition of the fuel. Relatively low levels of excess air are required, and the boiler assembly might be quite compact. See Fig. 18.28.

However, high-quality bituminous coals can be pulverized and injected into a furnace through properly designed burner hardware (Fig. 18.29). The boiler design can be quite similar to that intended for gas or oil except that modifications are needed for the removal of ash.

A detailed discussion of boilers and burner equipment for use with solid fuels is beyond the scope of this chapter. The above information, however, should encourage the reader to further investigate the possibilities of solid fuels for energy production, particularly for those installations in close proximity to a plentiful supply of coal, wood, agricultural wastes, or other combustible refuse.

CHAPTER 19
FUELS

**Cleaver-Brooks, Division of Aqua-Chem, Inc.,
Milwaukee, Wisconsin**

With the ever-changing cost and supply of standard boiler fuels, one needs to know what fuels are available and some of their basic properties. New boilers are being designed and existing ones retrofitted in order to burn fuels that in the past were not normally burned in a boiler. Some of the more popular new fuels used are solvents, alcohols, kerosene, used oils, digester gas, landfill gas, and wood. In the following pages various liquid fuels, gaseous fuels, and solid fuels are compared and discussed.

As shown in Table 19.1, several grades—Nos. 1, 2, 4, 5, and 6—of fuel oil are available commercially, and each has its own specific properties. These oils must pass particular tests specified by Commercial Standard CS12-48 or ASTM D-396. The remaining fuels shown in Table 19.1 are not normally considered boiler fuels. Because of this, standards like those mentioned above are not available.

The heating value or heat of combustion is very important to know in determining which fuel is most economical to burn. Fuels should be compared on a cost per Btu (kcal) basis.

The amount of heat given off during complete combustion of a known amount of fuel is the *heat of combustion*. For liquids and solids, the heat of combustion is usually given in units of Btu/lb (mJ/kg), while gaseous fuels are given in units of Btu/(s $\cdot$ ft^3) [kcal/(s $\cdot$ m^3)]. "Heating value," "calorific value," and "fuel Btu (kcal) value" are other terms frequently used.

The higher heating value (gross heating value) of a fuel is the heat obtained from combustion where the combustion products are cooled to a temperature at which all combustion-generated water is condensed. The lower heating value (net heating value) of a fuel is the heat obtained from combustion where the combustion products *are not cooled* to the point of condensation. In the United States, the higher heating value is the one normally reported and used in calculations.

The "flash point" of a liquid fuel may be defined as the temperature to which it must be heated to give off sufficient vapor to form a flammable mixture with air. This temperature varies with the apparatus and procedure employed, and consequently both must be specified when the flash point of an oil is stated. To determine the flash point, either the closed-cup (Pensky-Martens) test or the open-cup (Cleveland) test is used.

The sulfur present in any boiler fuel is a concern because of its corrosive and polluting effects. In boilers the sulfur dioxide and water vapor in the combustion

TABLE 19.1 Grades of Fuel Oil

Fuel	Heating value, Btu/lb (MJ/kg)	Specific gravity	Flash point, °F (°C)	Sulfur, wt %	Viscosity, SSU at 100°F (38°C)	Kinematic viscosity at 38°C, cSt	Ash, wt %
No. 1 fuel oil	19,670–19,860 (45.8–46.2)	0.805–0.845	100 (37.8)	0.5		1.4–2.2	—
No. 2 Fuel oil	19,170–19,750 (44.6–45.9)	0.855–0.876	100 (37.8)	0.5	32–38	1.8–3.6	—
No. 4 Fuel oil	18,280–19,400 (42.5–45.1)	0.887–0.910	130 (54.4)	0.3	60–300	10.3–64.6	0.1
No. 5 Fuel oil	18,100–19,020 (42.1–44.3)	0.922–0.946	130 (54.4)	0.3	20–40 @ 122°F (50°C)	39.5–81.3 @ 50°C	0.1
No. 6 Fuel oil	17,410–18,990 (40.5–44.2)	0.959–0.986	150 (65.6)	0.3	45–300 @ 122°F(50°C)	92–647 @ 50°C	0–0.3
No. 1 Diesel*	18,540 (43.1)	0.81–0.85					
No. 2 Diesel*	19,440 (45.2)	0.841					
JP-4*	18,540 (43.1)	0.87–0.90		0.4			
JP-5*	18,540 (43.1)	0.82	140 (60.0)	0.4			
Kerosene*	18,540 (43.1)	0.80					
Alcohols*	8,419–18,810 (196–438)	0.79–0.85					
Ketones*	10,400–16,100 (24.2–37.4)	0.80–0.81					
Solvents*	12,250–17,558 (28.5–40.8)	0.79–0.88					
Motor oil (used)*	18,460–19,270 (42.9–44.8)	0.84–0.96					

*Nonstandard liquid boiler fuel.

products may unite to form acids that can be highly corrosive to the breeching. The presence of some gaseous sulfur compounds may lower the dew point of water vapor in the flue gases, further aggravating corrosion problems.

The viscosity of an oil is the measure of its resistance to flow. Commercial oils have maximum limits placed on this property because of its effect on both the rate at which oil will flow through pipelines and the degree of atomization that may be secured in any given equipment.

Viscosity decreases rapidly as temperature increases, and preheating makes possible the use of oils of relatively high viscosity. The kinematic and Saybolt second universal viscosimeters are used for fuel oils of fairly low viscosity, and the Saybolt second Furol viscosimeter is used for more viscous oils. Viscosity versus temperature curves for commercial fuel oils are shown in Fig. 20.1.

The ash test is used to determine the amount of noncombustible impurities in the fuel. These impurities come principally from the natural salts present in the crude oil or from chemicals that may be used in refinery operations. Some ash-producing impurities in fuels cause rapid deterioration of refractory materials in the combustion chamber, particularly at high temperatures; some impurities are abrasive and destructive to pumps, valves, control equipment, and other burner parts.

Typical heating values and specific gravities for various gaseous fuels are given in Table 19.2, while Table 19.3 gives their chemical composition.

Coal is still the most widely used solid fuel for boilers. Properties of three classes of coal are given in Table 19.4. The composition of coal is reported by either the proximate analysis (Table 19.4) or the ultimate analysis, both expressed in weight percent.

The ash fusion temperature is a measurement of when the fuel's ash becomes molten. If the ash in the fuel has a low melting or fusion temperature, the ash will coalesce into masses and deposit on the firing surface or on boiler surfaces. Such deposits are difficult to remove and will interfere with boiler and burner operation.

New technology in the burning of wood in boilers is making wood a very economical boiler fuel in some instances. Normally the wood used for fuel is a waste or by-product of some process of manufactured products. Table 19.5 gives pertinent information for typical hardwood and softwood.

TABLE 19.2 Gaseous Fuels—Heating Values and Specific Gravity

Fuel	Heating value		Specific gravity
	Btu/(s · ft^3)	kJ/(s · m^3)	
Carbon monoxide	321	11,960	0.967
Coal gas	149	5,552	0.84
Coke oven gas	569	21,200	0.4
Digester gas	655	24,405	0.86
Hydrogen (H_2)	325	12,109	0.0696
Landfill gas	476	17,735	1.04
Natural gas	974–1129	36,290–42,065	0.60–0.635
Propane	2504–2558	93,296–95,308	1.55–1.77

TABLE 19.3 Chemical Composition of Gaseous Fuels

	Methane (CH_4)	Ethane (C_2H_6)	Propane (C_3H_8)	Butane (C_4H_{10})	Carbon monoxide (CO)	Hydrogen (H_2)	Hydrogen sulfide (H_2S)	Oxygen (O_2)	Carbon dioxide (CO_2)	Nitrogen (N_2)
Carbon monoxide	...	...	...	...	100					
Coal gas*	0.2	...	...	...	28.4	17.0	...		3.8	50.6
Coke oven gas	32.3	...	...	...	5.5	51.9	...	0.3	2.0	4.8
Digester gas†	64.0	...	...	...	—	0.7	0.8	...	30.2	2.0
Hydrogen H_2	...	...	...	...	...	100				
Landfill gas	47.5	...	...	...	0.1	0.1	0.01	0.8	47.0	3.7
Natural gas	82–93	0–15.8	...	...	0–0.45	0–1.82	0–0.18	0–0.35	0–0.8	0.5–8.4
Propane gas‡	...	2.0–2.2	73–97	0.5–0.8						

*Contains 0.02% sulfur.
†Contains 2.0% water.
‡Contains up to 24.3% C_3H_6.

TABLE 19.4 Composition of Coal (Proximate Analysis)

	Heating value, Btu/lb (MJ/kg)	Bulk density, lb/ft³ (kg/m³)	Fixed carbon, wt %	Moisture, wt %	Volatile matter, wt %	Sulfur, wt %	Ash, wt %	Ash fusion temp., °F (°C)
Anthracite	——	50–58 (800–929)	80.5–85.7	2.8–16.3	3.2–11.5	0.6–0.77	9.7–20.2	
Bituminous	——	42–57 (673–913)	44.9–78.2	2.2–15.9	18.7–40.5	0.7–4.0	3.3–11.7	2030/2900+ (1110/1543+)
Lignite	7440 (17.3)	40–54 (641–865)	31.4	39.1	29.5	0.4	4.2	1975/2070 (1079/1132)

TABLE 19.5 Composition of Wood

	Heating value, Btu/lb (dry) (kJ/kg)	Weight percent on a dry basis					Water, wt %	Fusion temp., °F (°C)
		Carbon	Hydrogen	Oxygen	Nitrogen	Ash		
Hardwoods	8732 (20,300)	50.9	6.3	41.9		0.84	32.7	2200–2750 (1204–1510)
Softwoods	8903 (20,700)	52.1	6.0	40.8	0.1	1.07	54.4	2200–2750 (1204–1510)

CHAPTER 20
PIPING

Cleaver-Brooks, Division of Aqua-Chem, Inc.,
Milwaukee, Wisconsin

The fuel oil piping system consists of two lines. The suction line is from the storage tank to the fuel oil pump inlet. On small burners the fuel oil pump is an integral part of the burner. The discharge line is from the fuel oil pump outlet to the burner. On systems that have a return line from the burner to the storage tank, this return line is considered part of the discharge piping when the piping losses are calculated.

20.1 OIL PIPING

20.1.1 Suction

Suction requirements are a function of

1. Vertical lift from tank to pump
2. Pressure drop through valves, fittings, and strainers
3. Friction loss due to oil flow through the suction pipe. This loss varies with:
   a. Pumping temperature of the oil, which determines viscosity
   b. Total quantity of oil being pumped
   c. Total length of suction line
   d. Diameter of suction line

To determine the actual suction requirements, two assumptions must be made, based on the oil being pumped. First, the maximum suction pressure on the system should be as follows:

No. 2 oil	12 inHg (305 mmHg)
No. 4 oil	12 inHg (305 mmHg)
Nos. 5 and 6 oil	17 inHg (432 mmHg)

Second, the lowest temperature likely to be encountered with a buried tank is 40°F (5°C). At this temperature the viscosity of the oil would be:

No. 2 oil 68 SSU* (12.5 cSt)

No. 4 oil 1000 SSU (21.6 cSt)

In the case of Nos. 5 and 6 oil, the supply temperature of the oil should correspond to a maximum allowable viscosity of 4000 SSU (863 cSt). This viscosity corresponds to a supply temperature of 110 to 225°F (43 to 105°C) for commercial grades of Nos. 5 and 6 oils. Then, using Fig. 20.1 and entering at 4000 SSU and going horizontally to the No. 5 fuel range, the maximum corresponding temperature is about 70°F (21°C). Likewise, the maximum corresponding temperature for No. 6 fuel is about 115°F (46°C).

The suction pressure limits noted above also allow for the following:

1. The possibility of encountering lower supply temperatures than indicated above, which would result in higher viscosities
2. Some fouling of suction strainers
3. In the case of heavy oil (Nos. 5 and 6), pump wear, which must be considered with heavy oils (See Figs. 20.3 to 20.6 for suction pressure curves.)

Strainers. It is a good practice to install suction-side strainers on all oil systems to remove foreign material that could damage the pump. The pressure drop associated with the strainer must be included in the overall suction pressure requirements.

Strainers are available as simplex or duplex units. Duplex strainers allow the ability to inspect and clean one side of the strainer without shutting down the flow of oil.

20.1.2 Discharge

Pumps. Pumps for fuel oil must be chosen based on several design criteria: viscosity of fuel oil, flow requirements, discharge pressure required, and fluid pumping temperature.

Viscosity. Charts for commercial grades of fuel oil are shown in Fig. 20.1. The pump must be designed for the viscosity associated with the lowest expected pumping temperatures.

Flow. Fuel oil pumps should be selected for approximately twice the required flow at the burner. The additional flow will allow for pressure regulation, so that constant pressure can be supplied at the burner.

Pressure. The supply pressure of the pump is based on the required regulated pressure at the burner.

A system utilizing a variable orifice for flow control typically requires from 30 to 60 psig (207 to 414 kN/m^2). The metering orifice type of system can be used on all grades of fuel oil. Burners utilizing an oil metering pump usually limit the supply pressure to prevent seal failure. As with metering orifices, there is no limitation on the grade of fuel oil used.

Temperature. The temperature of the oil must be considered, to ensure that the seals and gaskets supplied can withstand the fluid temperature.

---

*SSU is the abbreviation for standard Saybolt unit.

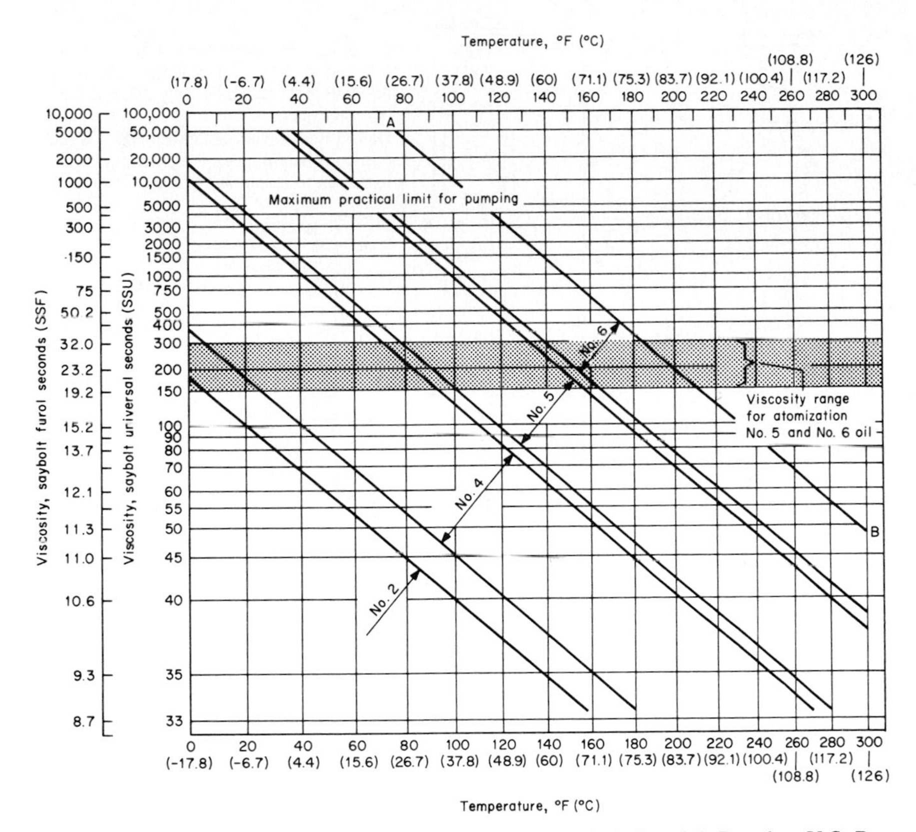

FIGURE 20.1 Viscosity-temperature curves for fuel oil Nos. 2, 4, 5, and 6. Based on U.S. Department of Commerce's Commercial Standard CS12-48. (*Courtesy of Cleaver-Brooks.*)

Pumping. The major difference between calculating hydronic and fuel oil piping systems is that the actual specific gravity of the oil being pumped must be accounted for.

The design pump head is equal to the suction lift, dynamic piping loss (including fittings, valving, etc.), and required supply pressure at the burner (if applicable).

Figure 20.2 should be used to determine the equivalent length of straight pipe that results in the same pressure drop as the corresponding pipe fitting or valve.

Figures 20.3 to 20.6 should be used to determine the appropriate dynamic piping losses with respect to type of oil being pumped, flow rate, and pipe size. The total equivalent length of straight pipe for fittings and valving, from Fig. 20.2, must be added to the total length of horizontal and vertical piping before multiplying by the appropriate piping loss factor.

The pressure loss for each strainer generally must be calculated separately and added to the total.

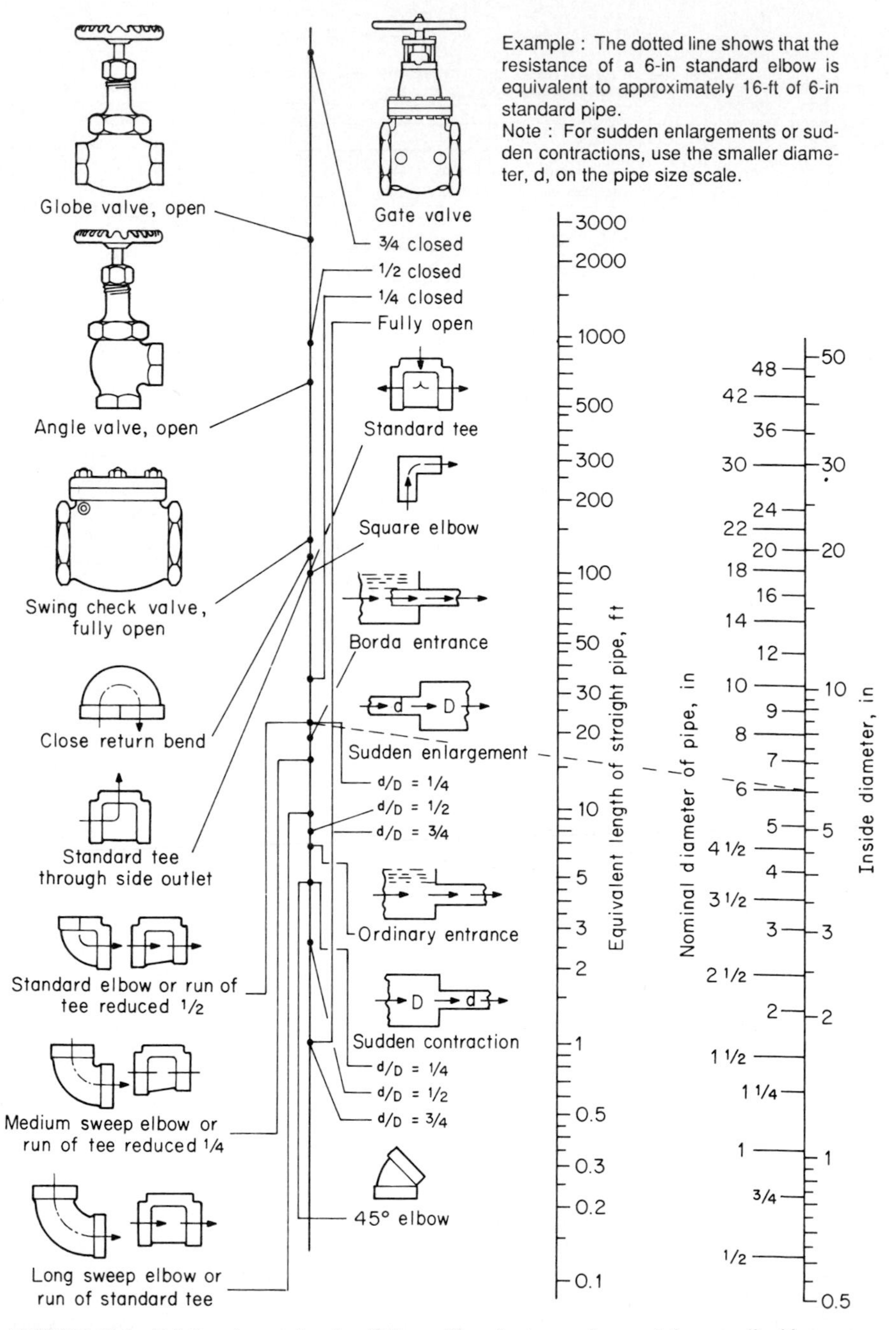

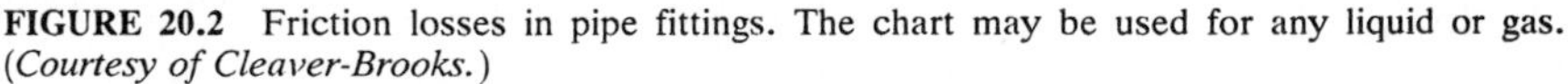

FIGURE 20.2 Friction losses in pipe fittings. The chart may be used for any liquid or gas. (*Courtesy of Cleaver-Brooks.*)

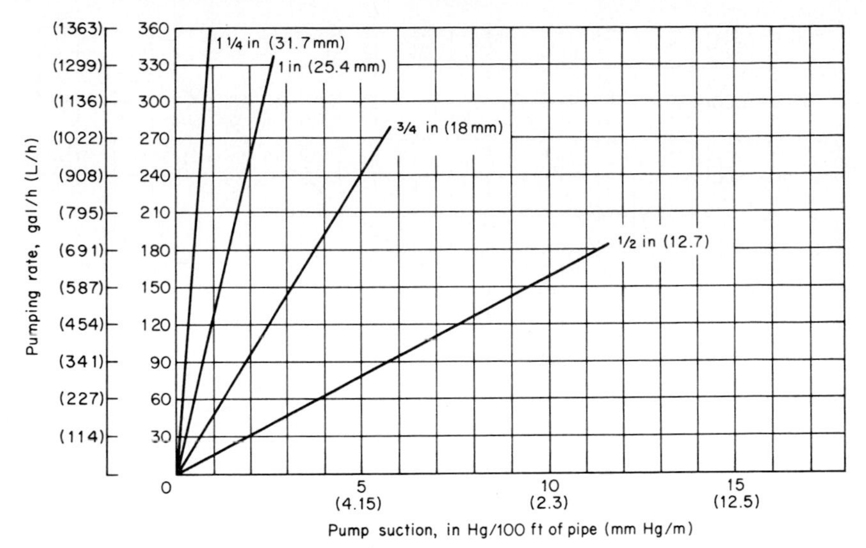

FIGURE 20.3 Pump suction curves for No. 2 fuel oil. Curves are based on a pumping temperature of 40°F (4.4°C), or 68 SSU. (*Courtesy of Cleaver-Brooks.*)

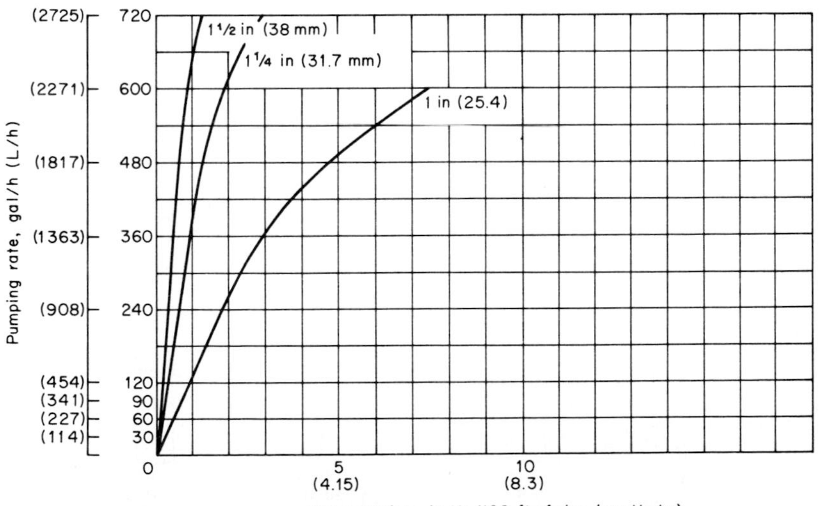

FIGURE 20.4 Pump suction curves for No. 2 fuel oil. Curves are based on a pumping temperature of 40°F (4.4°C), or 68 SSU. (*Courtesy of Cleaver-Brooks.*)

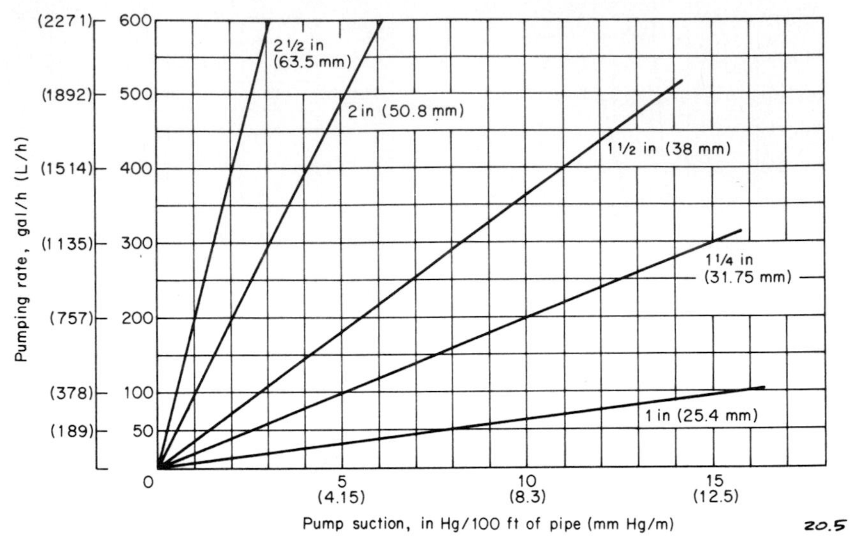

FIGURE 20.5 Pump suction curves for No. 4 fuel oil. Curves are based on a pumping temperature of 40°F (4.4°C), or 1000 SSU. (*Courtesy of Cleaver-Brooks.*)

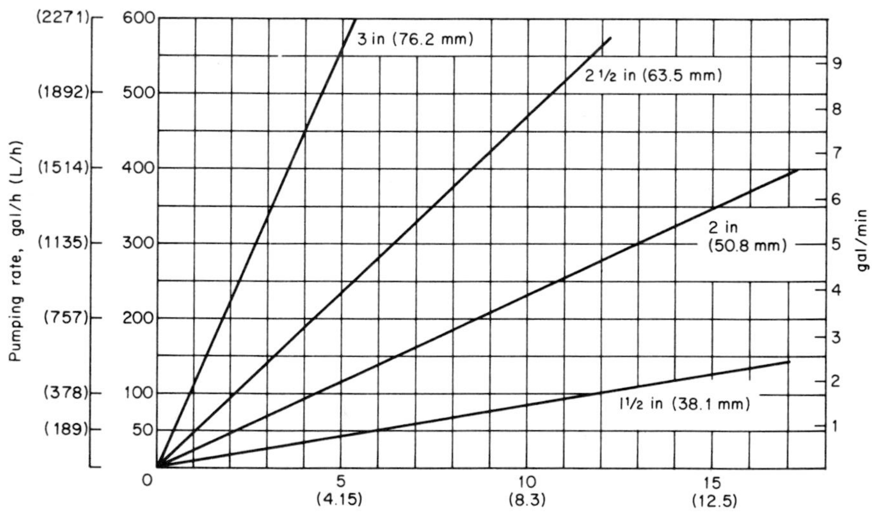

FIGURE 20.6 Pump suction curves for Nos. 5 and 6 fuel oils. Curves are based on a pumping limit of 4000 SSU. (*Courtesy of Cleaver-Brooks.*)

To obtain the suction lift in inches (millimeters) of mercury (Hg) from the bottom of the suction pipe (in the tank) to the boiler connection or pump suction centerline, multiply this vertical distance in feet (meters) by 0.88155 inHg/ft of water (73.428 mmHg/m of water) by the specific gravity of the oil being pumped.

For No. 2 oil with a specific gravity of 0.85 at maximum 40 SSU and 100°F (37.8°C):

$$\text{Suction lift} = Cd \tag{20.1}$$

Where the suction lift is inHg (J), C is in inHg/ft (mmHg/m), and d is in ft (m).

Heaters. Heaters are used to increase fuel oil temperatures, to provide the viscosity to atomize properly. Oil temperatures corresponding to a viscosity of 100 SSU [2 × 1.6 centistokes (cSt)] or less are recommended.

Heating can be accomplished by using hot water, steam, electricity, or a combination of these. Most packaged boilers have heaters that utilize electric elements for initial warmup and then transfer to either hot water or steam when the boiler has reached sufficient temperature and pressure. The heater sizing should be based on the supply pump design flow rate and temperature.

Electric heaters are commonly used to preheat heavy fuel oils on low-temperature hot-water boilers or on startup of a high-temperature hot-water or steam boiler.

The watt density of an electric heater should not exceed 5 W/in^2 (0.007 W/mm^2) because of dangers with vapor lock and coking on the heater surface. When steam is used as the heating medium for heavy oils, the steam pressure used should have a saturation temperature at least equal to the desired oil outlet temperature.

The flow of steam is controlled by using a solenoid valve that responds to a signal from the oil heater thermostat.

Some steam heaters include electric heating elements to allow firing of oil on a cold startup. When sufficient steam pressure is available, the electric heater is automatically de-energized.

Steam from the boiler is regulated to the desired pressure for sufficient heating. If the boiler pressure exceeds the steam heater pressure by 15 lb/in^2 (1 bar) or more, superheated steam will be produced by the throttling process. Steam heater lines should be left uninsulated to allow the steam to desuperheat prior to entering the heater. It is common practice to discharge the steam condensate leaving the oil heater to the sewer, to eliminate the possibility of contaminating the steam system in the event of an oil leak. The heat from the condensate is usually reclaimed prior to dumping it.

Excessive steam temperatures can also cause coking in the heater.

Hot-water oil heaters are essentially water-to-oil heat exchangers used to preheat oil. However, since the source of heat energy is boiler water circulated by the pump through the heater, any system leak could cause boiler water contamination. Therefore, safety-type heater systems are recommended for this service. Such an exchanger is frequently a double-exchange device using an intermediate fluid.

In cases where the oil must be heated to a temperature in excess of the hot-water supply temperature, supplemental heat must be provided by an electric heater. Tank heaters are commonly an insulated bundle of four pipes submerged in the oil tank. See Fig. 20.7. Tank preheating is required anytime the viscosity of the oil to be pumped equals 4000 SSU or greater.

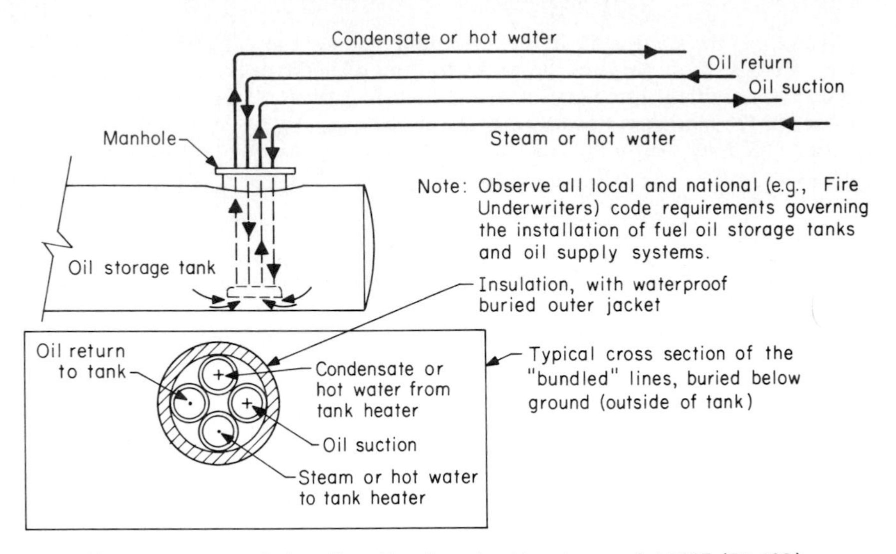

FIGURE 20.7 Tank heaters. (*Courtesy of Cleaver-Brooks.*)

Valves

Pressure Relief Valves. These are installed in the discharge line from the supply pump, to protect the pump and system from over pressure. Pressure relief valves are also commonly installed on oil heaters to relieve pressure so that oil may circulate even though the burner does not call for oil.

Pressure Regulators. These reduce system pressure and maintain a desired pressure at the burner.

Oil Shutoff. There are two commonly used styles of oil shutoff valves for burner service: electric coil and motorized. Electric coil solenoid valves are used on most small industrial and commercial burners. These valves are normally closed valves, and they control the flow of oil fuel to the burner. Two such valves for fuel shutoff are used on commercial and industrial boilers.

The second type of oil shutoff valve is a motorized valve that has a spring return to close. Motorized valves can be equipped with a proof-of-closure switch which ensures that the valve is in the closed position or prevents the burner from igniting if it is not. This type of switch is necessary to meet certain insurance requirements.

Manual Gas Shutoff Valves. Manual gas shutoff valves are typically a lubricated plug type of valve with a 90° rotation to open or close. The valve and handle should be situated such that when the valve is open, the handle points in the direction of flow.

The number of valves and their locations are based on insurance requirements. Typically, manual valves are installed upstream of the gas pressure regulator, directly downstream of the gas pressure regulator, and downstream of the last automatic shutoff valve.

Automatic Gas Shutoff Valves. Three types of automatic gas shutoff valves are used on burners: solenoid valves, diaphragm valves, and motorized valves.

Of the three automatic valves, the solenoid is the simplest and generally the least expensive. A controller opens the valve by running an electric current through a magnetic coil. The coil, acting as a magnet, pulls up the valve disk and allows the gas or oil to flow. Solenoid action provides fast opening and closing times, usually less than 1 s.

Diaphragm valves are frequently used on small to medium boilers. These valves have a slow opening and fast closing time. They are simple, dependable, and inexpensive. They are full-port valves and operate with little pressure loss.

Motorized shutoff valves are used for large gas burners that require large quantities of gas and relatively high gas pressures. There are two parts to a motorized valve: the valve and a fluid power actuator. A limit switch stops the pump motor when the valve is fully open. The valve is closed by spring pressure. The valve position (open or closed) is visible through windows on the front and side of the actuator. Motorized valves often contain an override switch which is actuated when the valve reaches the fully closed position. This proof-of-closure switch is needed to meet several different insurance company requirements.

Vent Valves. Vent valves are normally open solenoid valves that are wired in series and are located between two automatic shutoff valves in the main gas line or, in some cases, the pilot line. The vent valve vents to the atmosphere all gas contained in the line between the two valves.

Flow Control Valves

1. Butterfly valves are the most commonly used device for controlling the quantity of fuel gas flow to the burner. The pressure drop associated with a fully open butterfly valve is very low. Butterfly valves can be used for control of air flow and with special shaft seals can be used for all grades of fuel gas. Linkage arms are connected to the shaft of the valve and driven directly from the burner-modulating motor.

2. Modulating gas shutoff valves can be supplied with positioning motors that can operate on the on/off principle or high/low/off. In the case of the high/low/off shutoff valves, the air damper is controlled by the valve-modulating motor. This allows the valve position to dictate the amount of combustion air necessary for the gas input rate.

3. Pneumatic control valves are often butterfly valves that are driven by a pneumatic actuator. The signal to the pneumatic actuator is proportional to the combustion air flow and positions the valve to deliver the appropriate amount of gas. Often additional signals such as steam flow and combustion air flow are used to determine the signal to the valve and its corresponding position.

Gas Strainer. It may be advisable to use a strainer to protect the regulators and other control equipment against any dirt or chips that might come through with the gas.

Gas Compressors or Boosters. If the local gas utility cannot provide sufficient gas pressure to meet the requirements of the boiler, a gas compressor or booster should be used. *Caution*: The use of a gas compressor or booster must be cleared with the local gas utility prior to installation.

20.2 GAS PIPING

Figure 20.8 illustrates the basic arrangement for piping gas to boilers from street gas mains for a typical installation.

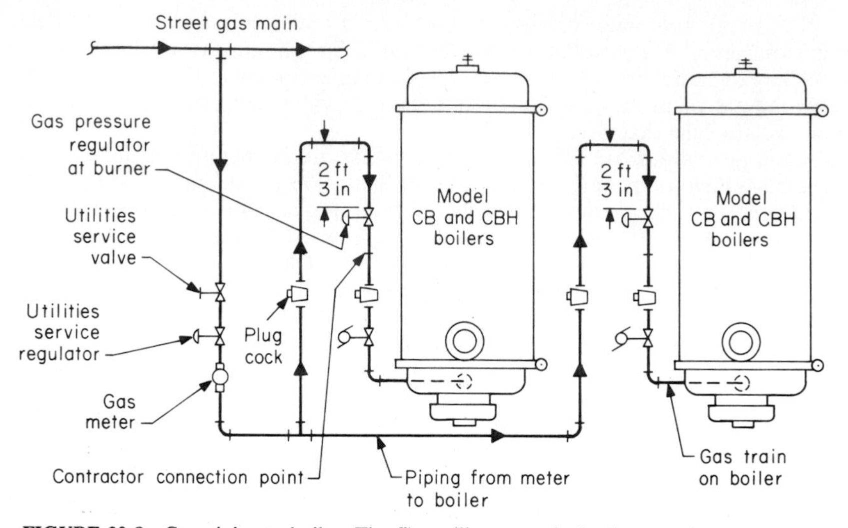

FIGURE 20.8 Gas piping to boiler. The figure illustrates the basic gas valve arrangement on boilers and shows the contractor's connection point for a typical installation. Actual requirements may vary depending on local codes and local gas company requirements, which should be investigated prior to both the preparation of specifications and construction. (*Courtesy of Cleaver-Brooks.*)

20.2.1 Line-Sizing Criteria

The first step in designing a gas piping system is to properly size components and piping to ensure that sufficient pressure is available to meet the demand at the burner. The boiler manufacturer should be consulted to determine the pressure required.

The gas service piping installed in the building must be designed, and components selected, to provide the required fuel gas flow to the boiler at the manufacturer's recommended pressure. The utility supplying gas to the facility will provide the designer with information on the maximum available gas pressure for the site. The gas piping design must be appropriate for the specific site conditions.

The gas train pressure requirements can be expressed as

$$P_s = P_R + P_C + P_P + P_F + P_B + P_{fp} \tag{20.2}$$

where P_s = supply pressure available
P_R = pressure drop across gas pressure regulator
P_C = pressure drop across gas train components
P_p = pressure drop associated with straight runs of pipe
P_F = pressure drop associated with elbows, tees, or other fittings
P_B = pressure drop across burner orifice or annulus
P_{fp} = boiler furnace pressure

Pressure drop calculations for regulators and valves are normally based on the C_v factor or coefficient of value capacity of air or in equivalent feet or diameters of pipe length.

The resistance coefficient k can be used to express the pressure drop as a number of lost velocity heads

$$k = \frac{PV^2}{2g}$$

Depending on the information available, the following equations can be used to determine the pressure drop through valves or across regulators:

$$k = f \cdot \frac{L}{D} \tag{20.3}$$

$$P = f \cdot \frac{L}{D} \cdot \frac{PV^2}{2g} \tag{20.4}$$

$$k = \frac{891d^4}{C_v^2} \tag{20.5}$$

$$H_v = 0.000228V^2 \text{ inWG} \qquad \text{for air} \tag{20.6}$$

$$P = \frac{k}{144} H_v \tag{20.7}$$

$$C_v = 0.0223(\text{ft}^3/\text{h}) \text{ @ 1-inWG drop})G \qquad \text{for 0- to 2-psig gases} \tag{20.8}$$

where k = resistance coefficient
f = Darcy friction factor
L = length of pipe or equivalent length of pipe for fitting, ft
D = diameter of pipe, ft
P = pressure drop or differential, lb/in^2
V = velocity, ft/s
C_v = valve conductance based on H_2O @ 1 lb/in^2 drop
g = acceleration of gravity
H_v = velocity head
G = gas gravity relative to air = $P/0.0765$
p = density of flowing fluid, lb/ft^3

Note: Metric units must be converted to English units before Eqs. (20.2) to (20.8) can be applied.

To determine the losses associated with straight runs of pipe (P_p) and pipe fittings (P_f), Eq. (20.2) can be used. Values for equivalent length of pipe or equivalent pipe diameter are listed in Fig. 20.2. The pressure drop for the burner orifice or annulus (P_B) can be calculated by using Eq. (20.4) and making the appropriate gas density corrections. The furnace pressure P_{fp} is a function of the furnace geometry, size, and firing rate. This pressure is often zero or slightly negative, but for some types of boilers and furnaces it can run as high as 15 in water column (inWC) (381 mm) positive.

20.2.2 Gas Train Components

Pressure Regulators. Pressure regulators or pressure-reducing regulators are used to reduce the supply pressure to the level required for proper burner operation. The regulated, or downstream, pressure should be sufficient to overcome line losses and deliver the proper pressure at the burner. Pressure regulators commonly used on burners come in two types: self-operated and pilot-operated.

In a self-operated regulator, the downstream, or regulated, pressure acts on one side of a diaphragm, while a preset spring is balanced against the backside of the diaphragm. The valve will remain open until the downstream pressure is sufficient to act against the spring.

Regulators for larger pipe sizes are normally the pilot-operated type. This class of equipment provides accurate pressure control over a wide range of flows and is sometimes selected even in smaller sizes where improved flow control is desired.

A gas pressure regulator must be installed in the gas piping to each boiler. The following items should be considered when a regulator is chosen:

1. *Pressure rating:* The regulator must have a pressure rating at least equivalent to that in the distribution system.
2. *Capacity:* The capacity required can be determined by multiplying the maximum burning rate by 1.15. This 15 percent over-capacity rating of the regulator provides for proper regulation.
3. *Spring adjustment:* The spring should be suitable for a range of adjustment from 50 percent under the desired regulated pressure to 50 percent over.
4. *Sharp lockup:* The regulator should include this feature because it keeps the downstream pressure (between the regulator and the boiler) from climbing when there is no gas flow.
5. *Regulators in parallel:* This type of installation would be used if the required gas volume were very large and if the pressure drop had to be kept to a minimum.
6. *Regulators in series:* This type of installation would be used if the available gas pressure were over 5, 10, or 20 psig (34.5, 68.0, or 137.9 kPa), depending on the regulator characteristics. One regulator would reduce the pressure to 2 to 3 psig (17.8 to 20.7 kPa), and a second regulator would reduce the pressure to the burner requirements.
7. *Regulator location:* A straight run of gasline piping should be used on both sides of the regulator to ensure proper regulator operation. This is particularly important when pilot-operated regulators are used. The regulator can be located close to the gas train connection, but 2 to 3 ft (0.6 to 0.9 m) of straight-run piping should be used on the upstream side of the regulator. *Note*: Consult your local gas pressure regulator representative. She or he will study your application and recommend the proper equipment for your job.

CHAPTER 21
BURNERS AND BURNER SYSTEMS

Cleaver-Brooks, Division of Aqua-Chem, Inc., Milwaukee, Wisconsin

21.1 INTRODUCTION

The primary purpose of a boiler is to transfer energy of the fuel to water, which can be for either increasing the water temperature or generating steam. The burner is used to convert the energy of the fuel to heat, through burning.

21.2 GAS BURNERS

The optimum air-fuel ratio for natural gas, at standard conditions, is approximately 9.5:1 (9.5 m^3 air/1 m^3 fuel) for perfect combustion. At 10 percent excess air, the air-fuel ratio is about 10.5:1. Furthermore, the fuel must be put into the furnace in such a way that all the gas comes in contact with enough air to burn it completely.

Also the air and gas must come together at the right time so that the mixture is ready for spontaneous ignition and a stable flame pattern is established.

The gas pressure requirements should be at a minimum, and the power requirements for the fan should be reasonable. The pressure developed by the fan should be high enough so that the amount of air delivered for combustion is not appreciably affected by changes in stack conditions. Thus, a change in pressure at the boiler stack outlet of as much as ± 0.5 in water column (in WG) (24.5 Pa) should not appreciably affect the air-fuel ratio. Significant changes in combustion settings can result from draft variations in boilers designed for low draft loss.

There are, of course, many controls, both operating and safety, in connection with the burner, and these are certainly important to good burner design. This section, however, deals primarily with the mechanical functioning and design of the burner itself.

The color of the flame in good gas burner design should be mentioned even though it is a controversial issue. Colors range from blue transparent to yellow luminous. Whatever the flame color, one must rely on instruments which analyze the stack gases. If the instruments show no unburned fuel in the stack and there

TABLE 21.1 Contents of Gases

Gas	Btu/ft^3 (kJ/m^3) gross	ft^3 air/ft^3 gas (m^3 air/m^3 gas) (no excess)	Ultimate CO_2, %	Specific gravity
Natural	1020 (38,000)	10.0 (2.83)	10.0	0.65
Blast furnace	92 (3,428)	0.68 (0.02)	25.5	1.02
Producer bituminous	163 (6,073)	1.24 (0.04)	18.6	0.86
Coke oven	569 (21,200)	5.45 (0.15)	10.8	0.40
Propane (refinery)	2504 (93,296)	23.20 (0.66)	14.0	1.77
Butane (refinery)	3184 (118,632)	30.00 (0.85)	14.3	2.00

TABLE 21.2 Flow Characteristics of Gases for 1 MBtu/h (14.3 kW)

Gas	Flow, ft^3/h (m^3/h)	Orifice size, in (mm)	Velocity, ft/s (MPa)	Pressure, in H_2O (kPa)
Natural	980 (27.7)	0.5 (12.7)	200 (1.02)	11.9 (2.96)
Propane	400 (11.3)	0.5 (12.7)	81 (0.41)	5.3 (1.32)
Butane	314 (8.9)	0.5 (12.7)	64 (0.33)	3.7 (0.92)

is no deposition on the boiler tube walls from minimal amounts of excess air, the flame is good—regardless of color.

Available Gases and Characteristics. The most common fuel gas is natural gas. Some other less common gases are listed in Table 21.1.

Flame Speeds and Temperatures. The most common gases are natural, propane, and butane. Fortunately the flame speed (ignition velocity) is about the same [2.3 ft/s (0.7 m/s)] for each of these gases for a mixture of the theoretical amount of air for complete combustion. The flame temperatures for these gases at around 3700°F (2030°C) are also similar. But these gases differ greatly in density and heating value. Thus the piping and porting must be different for each gas, to keep the same combustion characteristics. For example, the values in Table 21.2 show the variation in gas flow velocity and pressure, assuming that 1 MBtu/h (143 kW) is to be released through a ½-in (12.7-mm) orifice.

Although the values in Table 21.1 are approximate, they clearly show that the designer must use care in selecting burners which are to burn more than one gas.

21.3 OIL BURNERS

A well-designed oil burner will have the following functions and/or features:

- The air-fuel ratio must be held constant.
- All the oil must come in contact with air at the right time.
- There must be a proper flame pattern.
- Oil pressure and power requirements should be a minimum.

- The fan pressure should be adequate.
- The oil burner must atomize the oil sufficiently well.

21.3.1 Methods of Atomization

In oil burner design, probably no other subject is as controversial as the method of atomizing the fuel. There are many methods of breaking up the oil into fine particles, and every method has its staunch supporters. There is no one best method. As long as the final goal is reached, the method used is immaterial, even if the oil is swept into the combustion chamber with a broom.

21.3.2 Types and Characteristics of Oils

Fuel oils range from No. 6, a thick black oil, to No. 1, a very thin, colorless oil similar to kerosene. The characteristics of oils can be found in Table 21.3.

The heaviest and thickest oils must be heated to be pumped, and heated still more to reduce the viscosity for adequate atomization. But the lighter oils have their own handling and atomizing problems. For example, atomizing a light oil with steam might not work too well if the oil is stored outside in cold weather. Steam may condense in the gun. Our solution is to heat the oil, or to atomize with air, or to atomize mechanically (rotary cup, or pressure).

21.3.3 Basic Combustion Principles

For any fuel to burn, it must combine with a specific amount of air (actually oxygen) and the temperature of the mix must be sufficiently high to start the reaction, which then becomes self-sustaining. If there is insufficient air, the fuel-air mix will not burn, regardless of temperature. Similarly, if there is too much air, the fuel will not burn, regardless of temperature. However, between these limits the fuel will burn completely or partially, depending on how well the fuel and air are mixed and the amount of air in the mix.

The amount of oxygen left over after combustion is the "excess O_2" or "excess air." One objective in burner design is to have a minimum of excess air in the combustion products.

To find the theoretically correct quantity of air required to burn a fuel, the following equation may be used:

$$W = 11.52C + 34.56\left(H - \frac{O}{8}\right) + 4.32S \tag{21.1}$$

where W = lb (kg) air/lb (kg) of fuel
C = decimal percentage of carbon in fuel
H = decimal percentage of hydrogen in fuel
O = decimal percentage of oxygen in fuel
S = decimal percentage of sulfur in fuel

The air requirements are often expressed in (ft^3/min air)/(MBtu/h) or (m^3/s)/kW.

The calculated value for air required is in pounds of air per pound (kilogram) of fuel. On average, because of varying environmental conditions and fuel quality

TABLE 21.3 Commercial Standard CS12-48: Detailed Requirements for Fuel Oils[(A)]

No. Grade of Fuel Oil (B)	VISCOSITY Sayb. Univ. at 100 Degrees F		Sayb. Furol at 122 Degrees F		Kin. C.S. at 100 Degrees F		Kin. C.S. at 122 Degrees F		Gravity Degrees API Min.
	Max.	Min.	Max.	Min.	Max.	Min.	Max.	Min.	
1. A distillate oil intended for vaporizing pot-type burners and other burners requiring this grade. (D)					2.2	1.4			35
2. A distillate oil for general purpose domestic heating for use in burners not requiring No. 1.	40				(4.3)				26
4. An oil for burner installations not equipped with preheating facilities	125	45			(26.4)	(5.8)			
5. A residual type oil for burner installations equipped with preheating facilities		150	40			(32.1)	(81)		
6. An oil for use in burners equipped with preheaters, permitting a high-viscosity fuel			300	45			(638)	(92)	

(A) Recognizing the necessity for low sulphur fuel oils used in connection with heat-treatment, non-ferrous metal, glass and ceramic furnaces and other special uses, a sulphur requirement may be specified in accordance with the following table.

Grade of Fuel Oil	Sulphur, Max., Per Cent
No. 1	0.5
No. 2	1.0
Nos. 4, 5, and 6	no limit

Other sulphur limits may be specified only by mutual agreement between buyer and seller.

(B) It is the intent of these classifications that failure to meet any requirement of a given grade does not automatically place an oil in the next lower grade unless in fact it meets all requirements of the lower grade.

(C) Lower or higher pour points may be specified whenever required by conditions of storage or use. However, these specifications shall not require a pour point lower than 0 degrees F under any conditions.

(D) No. 1 oil shall be tested for corrosion in accordance with par. 15 for three hours at 122 degrees F. The exposed copper strip shall show no gray or black deposit.

(E) The 10 per cent point may be specified at 440 degrees F maximum for use in other than atomizing burners.

(F) The amount of water by distillation plus the sediment by extraction shall not exceed 2.00 per cent. The amount of sediment by extraction shall not exceed 0.50 per cent. A deduction in quantity shall be made for all water and sediment in excess of 1.0 per cent.

No. Grade of Fuel Oil (B)	Flash Point Degrees F Min.	Pour Point Degrees F Max.	Water and Sediment Per Cent Max.	Carbon Residue On 10 Per Cent Residuum, Per Cent Max.	Ash Per Cent Max.	Distillation Temperatures, Degrees F 10 Per Cent Point Max.	90 Per Cent Point Max.	End Point Max.
1. Distillate oil intended for vaporizing pot-type burners and other burners requiring this grade. (D)	100 or legal	0	trace	0.15		420		625
2. A distillate oil for general purpose domestic heating for use in burners not requiring No. 1	100 or legal	20	0.10	0.35		(E)	675	
4. An oil for burner installations not equipped with preheating facilities	130 or legal	20	0.50		0.10			
5. A residual type oil for burner installations equipped with preheating facilities	130 or legal		1.00		0.10			
6. An oil for use in burners equipped with preheaters permitting a high-viscosity fuel	150 or legal		2.00(F)					

(A) Recognizing the necessity for low sulphur fuel oils used in connection with heat-treatment, non-ferrous metal, glass and ceramic furnaces and other special uses, a sulphur requirement may be specified in accordance with the following table.

Grade of Fuel Oil	Sulphur, Max., Per Cent
No. 1	0.5
No. 2	1.0
Nos. 4, 5, and 6	no limit

Other sulphur limits may be specified only by mutual agreement between buyer and seller.

(B) It is the intent of these classifications that failure to meet any requirement of a given grade does not automatically place an oil in the next lower grade unless in fact it meets all requirements of the lower grade.

(C) Lower or higher pour points may be specified whenever required by conditions of storage or use. However, these specifications shall not require a pour point lower than 0 degrees F under any conditions.

(D) No. 1 oil shall be tested for corrosion in accordance with par. 15 for three hours at 122 degrees F. The exposed copper strip shall show no gray or black deposit.

(E) The 10 per cent point may be specified at 440 degrees F maximum for use in other than atomizing burners.

(F) The amount of water by distillation plus the sediment by extraction shall not exceed 2.00 per cent. The amount of sediment by extraction shall not exceed 0.50 per cent. A deduction in quantity shall be made for all water and sediment in excess of 1.0 per cent.

Source: Cleaver-Brooks, Division of Aqua-Chem, Inc.

differences, burners are nominally adjusted for fuel oil to operate in the 20 percent excess-air range. For air flow required for burning most oil fuels, the rule of thumb is 200 (ft^3/min)/(MBtu/h) [3.22E − 0.04 (m^3/s)/kW].

For most typical automatic fuels, Eq. (21.1) yields a value for W of 12 to 15. This value is also referred to as the "fuel-air ratio." For rough approximations, a fuel-air ratio of 15 is adequate.

The air required per gallon (liter) of fuel burned in a boiler can be readily worked out on an approximate basis. The turndown ratio is related to the air-fuel ratio. Suppose that a boiler is rated for 50 TBtu/h (14.655-kW) input. The air flow is about 50 × 200 (ft^3/min)/(TBtu/h) (at 20 percent excess air) or 10,000 ft^3/min (4.72 m^3/s). If now, with all air dampers closed, the fan delivers 2000 ft^3/min (0.94 m^3/s), the air turndown ratio is 10,000/2000 (4.72/0.94), or 5:1. If the air-fuel ratio stays constant, the low-fire boiler input is about 10 TBtu/h (2931 kW) for a fuel turndown of 5:1 also. Now if it is desired to have a lower input than 10 TBtu/h (2931 kW), this can be accomplished by reducing the low-fire fuel input. However, since the air flow cannot be reduced (the dampers are already closed), the air-fuel ratio will rise as low fire is approached. Therefore, air turndown for this boiler-burner package is always 5:1 maximum. However, fuel turndown may be much greater with only flame stability or perhaps flame chilling and subsequent unburned fuel generation as the lower limit.

21.3.4 Flame Stability

Every burner generates a flame front in the boiler furnace, and it is important for this flame front to remain stable for a wide range of firing rates and air-fuel ratios. A good burner should maintain a stable flame front with maximum air and minimum fuel without blowoff. A practical limit, however, may be the electronic flame scanner. The scanner may lose sight of a flame which is shrunk by excessive air.

21.3.5 Products of Combustion and Control of Pollutants

The basic products of combustion are carbon dioxide (CO_2) and water (H_2O). However, there are generally small amounts of carbon monoxide (CO), nitric and nitrous oxides (NO_x), unburned hydrocarbons, carbon particulates, sulfur, and ash. It is always good to keep these products to an absolute minimum, and this is a big part of the burner designer's task. These limits are generally spelled out by local, state, or federal authorities.

21.3.6 Control of Fuels

The basic devices used for fuel control are

- Pumps
- Relief valve
- Pressure-reducing valve

- Metering valve
- Fuel temperature controls

The final design of a fuel-burning system includes a blower with sufficient capacity to provide the air required for the design capacity at the pressure needed to overcome the pressure losses in the burner, boiler, and auxiliaries.

The designer must select or design fuel-flow controls which very closely meter fuel quantity flowing to the burner. Since the fuel input flow is only about one-fifteenth of the air flow, small variations in fuel flow will have a substantial effect on the fuel-air ratio. Therefore, these components which are selected to control such flow must singly or in combination provide a fairly precise and dependable function.

The pump selection is important in that the pump must deliver dependably a flow of fuel oil to the burner system in sufficient quantity to meet the burner maximum capacity, plus a small excess amount which returns to storage.

The fuel oil from the pump must be routed through the burner piping at a consistent pressure. This task is normally handled by carefully selected relief valves, back-pressure valves, or pressure-reducing valves. Quite often this is accomplished through the use of two or more such devices. The fuel, at constant pressure, is then allowed to flow through a metering valve to the burner nozzles. The metering valve is usually designed so that it is little affected by minor changes in oil viscosity. The valve incorporates a variable-size port which is actuated to provide varying fuel throughput to the burner in accordance with heat demand imposed on the boiler.

Figure 21.1 indicates the piping schematic for a metering orifice controller firing No. 6 oil with air atomization. The schematic also shows the nozzle purge system.

21.4 SOLID-FUEL BURNERS

During the era of bountiful supply of natural gas and oil fuels, it was not economically prudent to use solid fuels. Although solid fuels have always been available at a lower cost per unit of energy, the equipment required to deliver the fuel to the burner, as well as to control the combustion process, is more costly. Solid fuels normally contain significantly more noncombustible material (ash) which must be extracted from the combustion zone and disposed of. Larger boiler equipment is required to provide significantly greater combustion space for the slower burning process when solid fuels are used.

However, in recent years, a substantial effort has been made to improve solid-fuel burner systems, in an effort to economize on the cost of energy. There are vast improvements not only in burner systems but also in fuel-handling hardware. Also considerable research is being done on the concept of converting solid fuels to gaseous and liquid forms. In this section we discuss only presently available equipment of proven design.

Solid fuels, both coal and wood, are being burned in boilers in a number of system designs. The most popular and best known is the *stoker*. This combustion process is frequently referred to as "mass burning." The fuel is fed onto a grate from a fuel hopper from above or pumped upward onto a grate from below. The air required for the combustion of the fuel is normally delivered from below, be-

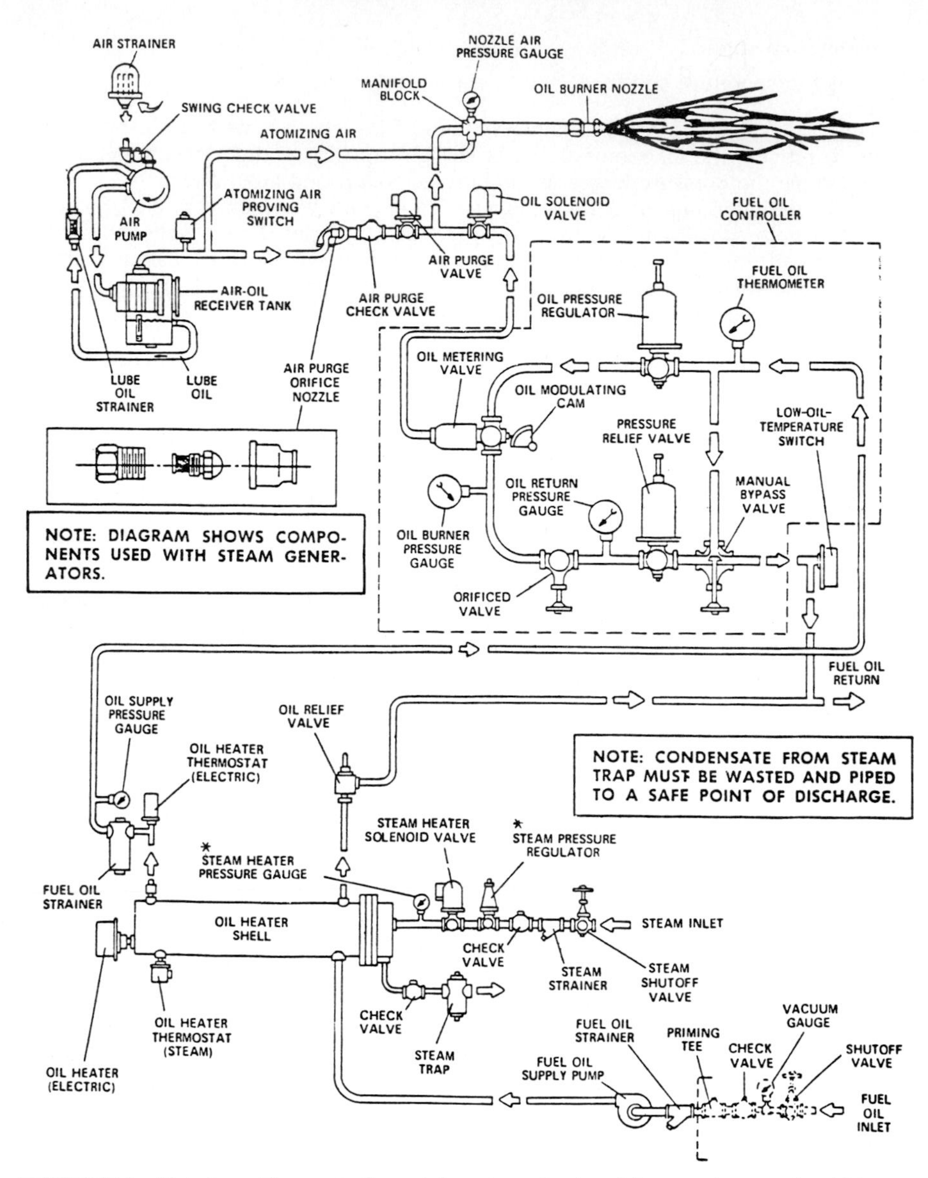

FIGURE 21.1 Metering orifice controller—piping schematic. Asterisk means item used on high-pressure generators only. (*Courtesy of Cleaver-Brooks.*)

ing forced through the bed of fuel on the grate surface. The fuel bed thickness varies and depends on the specific type of stoker system employed, type of fuel being used, and load on the boiler.

The most commonly used stoker system in small to intermediate industrial boilers is a chain grate or traveling grate. See Fig. 21.2. These stokers are formed

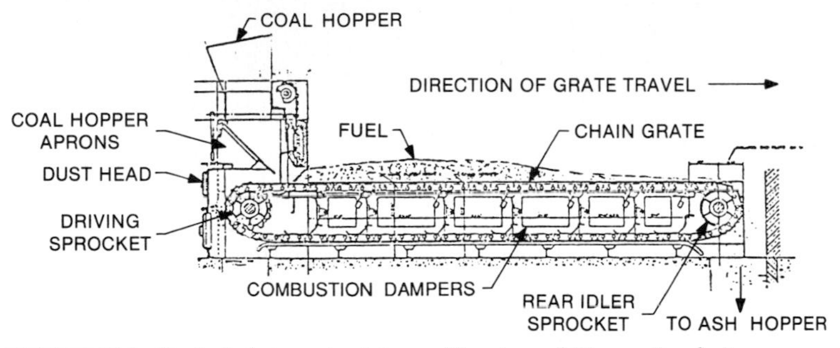

FIGURE 21.2 Typical chain grate stoker. (*Courtesy of Cleaver-Brooks*.)

of a series of endless chains of special configuration that provide a flat surface onto which the fuel is fed. The grate moves from the front boiler face toward the rear of the combustion zone. The fuel is fed onto the top surface of the grate in a layer of uniform thickness. The bed of fuel traverses the length of the furnace during which time it burns, releasing its energy to the heat-absorbing surfaces.

The speed at which the fuel bed travels is automatically adjusted under the control of the boiler pressure. The combustion air fed to a plenum chamber beneath the stoker flows upward through the fuel bed and is simultaneously controlled by fan dampers in such a manner as to provide a relatively constant air-to-fuel ratio. The depth of fuel carried on the stoker is normally controlled manually by adjusting the height of a gate at the fuel feed. The design of the system is such that the combustion process is completed before the fuel on the bed reaches the end of the stoker. The stoker links traverse a drum where the ash is dropped off into a hopper. The grates then move to the front or fuel feed end where a new bed of fuel is laid down. The grate heat-release rates for stokers of this type are moderately high.

There are many variations on this principle that employ mechanisms other than chain grates, where the fuel bed is moved through the combustion zone by reciprocating grates, vibrating grates, and sliding systems. Grate systems may be horizontally oriented within the combustion zone or inclined to allow the fuel to move downward toward the ash discharge zone. (See Fig. 21.3.)

Chain grates or traveling grates may have the fuel fed onto them by a spreader mechanism that throws the fuel onto the grate system. Where spreader systems are used, the fuel is injected from a point some distance above the grate bed. The traveling grate in this design usually travels from the rear of the combustion chamber toward the front, with the ash discharging into a hopper at this point. The solid fuels used for these systems may contain a higher level of fine particles, which will burn in suspension. The larger fuel pieces, being more dense, will fall to the grate, where the combustion process is completed. (See Fig. 21.4.)

A smaller grate surface is required for this system design, permitting the use of a combustion chamber of smaller floor dimensions. However, the height dimen-

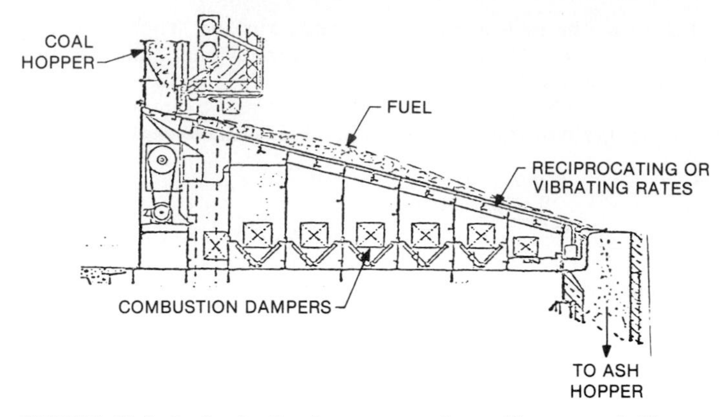

FIGURE 21.3 Inclined vibrating grate stoker. (*Courtesy of Cleaver-Brooks.*)

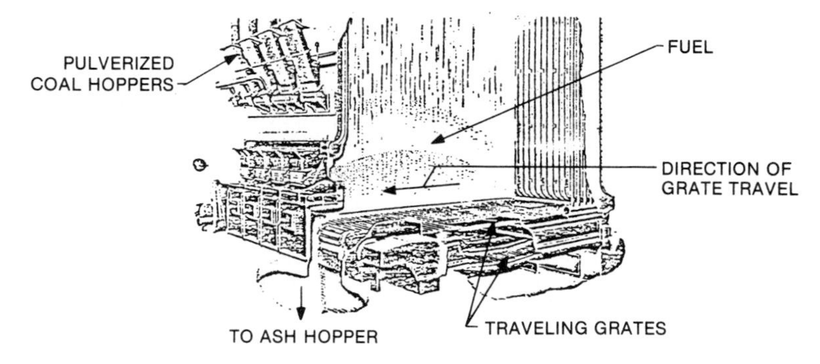

FIGURE 21.4 Full spreader or traveling grate. (*Courtesy of Cleaver-Brooks.*)

sion is usually greater, to provide sufficient volume in the combustion zone so that ample time is available to complete the combustion of all fuel delivered to the furnace. Spreader stoker systems are frequently used in boilers designed to generate steam in quantities of 75,000 lb/h (34,000 kg/h) and larger. Coal-burning systems of this design can be operated at relatively high heat-release rates per square foot (meter) of grate area.

Smaller boilers, having capacities of 10,000 lb/h (4536 kg/h) of steam or less, are frequently equipped with underfeed stokers, where the fuel is forced into the combustion zone by a ram or screw. (See Fig. 21.5.) When the fuel reaches the end of the feed conduit, the fuel moves upward over casting surfaces containing holes or slots through which the combustion air is delivered.

The combustion process is completed as the fuel moves in a sideward direction over the stoker surfaces, called "tuyeres." The ash from the spent fuel falls into channels or hoppers at each side of the stoker, from which it is easily removed.

The procedure for extraction of the ash varies and depends greatly on the quantity of ash generated, facilities available at the jobsite, quality of fuel to be burned, and means for disposal. Small boilers obviously will generate small quan-

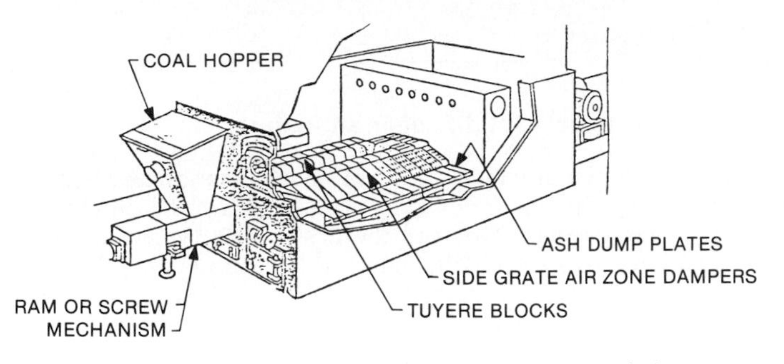

FIGURE 21.5 Underfeed stoker. (*Courtesy of Cleaver-Brooks.*)

tities of ash. It is not unusual to expect the boiler operator to manually remove the accumulated ash as required, from the points of collection, using simple tools. However, on large boilers, significant quantities of ash will collect in hoppers on the boiler. The number of hoppers and their location will vary according to boiler design. Ash can be extracted from these hoppers manually through seal valves, or automatic extraction systems can be provided. A variety of systems are available including pneumatic, water, and mechanical converters. These systems all include closed conduits which carry the material to a safe point of discharge. Such systems are normally installed in the field after the boiler has been installed on its foundation.

Many other solid-fuel burner systems are in use. For very large boilers typical of electric utilities, pulverized-fuel burners are common. For these systems, the fuel is prepared by grinding to powder form. The fuel is then blown into the combustion zone, along with the combustion air through air registers, not unlike large oil and gas burner systems. Efficient combustion with low levels of excess air is routinely achievable. These systems are normally restricted to very large boilers, since the cost of fuel preparation and equipment maintenance is high and can be justified only when large amounts of fuel are used. Systems of this type are complex, and they include fuel distribution and air-handling systems and elaborate control systems which require skilled boiler operating technicians. But in large plants the benefits are significant in terms of energy recovered per dollar of fuel, and the improved efficiency justifies the added equipment, maintenance, and labor costs.

Later technology and research have produced fluidized-bed systems and fuel gasification and liquefaction. The benefit of these systems is that they provide the potential for the burning of marginal-type fuels without polluting the atmosphere. Coals with high ash and/or sulfur content can be burned directly in a fluidized bed. Use of an appropriate bed material provides for the capture of sulfur compounds within the bed, minimizing the escape of acid-forming gases to the environment. The collection of particulate matter from the flue gases is a proven technology, making the fluidized-bed combustion systems a viable concept. However, control systems and fuel-handling and combustion air supply equipment are more complex than other available systems, making fluidized-bed equipment more expensive to install and operate.

Gasification and liquefaction designs are emerging from the laboratories holding promise for the future, but they will be broadly adapted only after more extensive commercialization.

21.5 CONTROLS

21.5.1 Control Systems—Solid-Fuel-Fired Boilers

The use of solid fuels can provide significant savings in purchased-fuel costs, but there are some other costs that must be considered when these fuels are burned. Fuel handling is an important factor in the design of a system if a dependable, efficient installation is to result. Even under the best conditions, some monitoring is required to ensure that fuel is being delivered dependably and at the appropriate rate to the boiler. Therefore, although control systems are provided to control the boiler automatically with optimum fuel-to-air ratios on smaller industrial installations, some degree of control sophistication may be omitted because an operator in periodic attendance is common.

The boiler water level control systems used are of the same type and quality as those on boilers intended for use with other fuels. Burner operating controls receive a signal from the boiler pressure for steam boilers or from the temperature control for hot-water boilers. The fuel feed rate is controlled to satisfy the boiler pressure sensor for steam or the temperature sensor for hot water. Depending on the firing system, the operating control(s) will increase the speed of the grate and the spreader feeders, if used, or will change the frequency of operating cycles of the fuel feed mechanism if an underfeed system is used. Simultaneously the dampers controlling the flow of combustion air on both overfire and underfire are adjusted to provide the appropriate air flow in a fixed relationship to the rate of fuel delivered. Solid-fuel-fired boilers are normally operated with a furnace pressure of zero to slightly negative. This condition is created by draft inducers and damper systems under the control of a furnace pressure sensor.

On grate burning systems, air delivered beneath the grate is ducted from the fan through an automated damper into a plenum chamber, which can be divided into a number of zones. Each zone is equipped with a manually adjusted damper to provide the optimum air to the underside of the grate, in accordance with the needs of the fuel bed. The need for frequent manual adjustment of these individual dampers is rare and will depend more on the specific fuel quality than on the steam load. (See Figs. 21.2 and 21.3.)

Stoker-fired boiler systems are well adapted to relatively stable, slowly changing steam loads. Greater time is required to bring a bed of solid fuel up to the temperature needed to increase the incidence of volatiles. The larger furnace envelope and boiler water content also have an impact on this process. Rapid startup and rapid shutoff are not easily achievable.

For more responsive steam supply using solid fuels, pulverized fuel may be employed. Burner systems designed for these fuels can provide a rapid increase or decrease of heat release in the boiler furnace, since this prepared fuel will volatilize very quickly in the combustion chamber environment. There is some lag in the fuel supply between pulverizer and burner, but this will cause few problems since only large boilers are used with this quality fuel and since the stored energy available in the boiler is usually sufficient to cope with minor load swings.

Pulverized-coal-burning systems are usually controlled by fairly sophisticated control networks for several reasons. The fuel is very combustible, and fast response flame sensors are needed to continuously monitor the flame activity, just as is required for oil or gas flames. Since the furnaces for coal-burning systems are usually large, multiple monitors are common. Also since the boilers using this fuel are large and the continuous energy release in the boiler is great, any sudden changes in the flows to the boiler, either fuel or air, are accompanied by a change in chamber pressure. The sidewalls of the combustion zone are usually formed of tubes carrying boiler water. Large volumes containing this combustion environment cannot be subjected to rapidly fluctuating internal pressures without running the danger of causing some failure. Therefore, the internal environment needs to be monitored with sensitive instrumentation to provide a signal to the boiler operator when upsets in the combustion process begin.

Clearly burners and control systems can vary from a very simple ram pushing fuel into a small chamber on cycles dictated by a steam-pressure-monitoring device, to a very complex computer control which monitors everything from the speed of conveyers delivering fuel to pulverizers to the temperature and pressure of the gas exiting the flue gas cleanup device.

Recommendations and requirements for control systems, particularly for larger boilers, are provided in a number of codes which are recognized by the American National Standards Institute (ANSI), insurance carriers, and equipment suppliers.

21.5.2 Combustion Fuel-Air-Ratio Controls

To achieve good combustion, the burner must provide a sufficient amount of air to burn the fuel. To improve combustion efficiency, the amount of air should be regulated, so that only enough air to completely burn the fuel is supplied. For safety reasons, the system should be designed and set up to provide more air than is actually required. If there is insufficient air, then the unburned fuel could be combined with additional air later and could be an explosive mixture.

To provide the proper amount of air for combustion, various fuel-air-ratio controls are used. The simplest control provides a fixed flow of air and fuel, and the burner simply turns on and off. Other types of burners are capable of varying the rate of fuel input and have a means of providing variable air and fuel flow rates based on heat requirements.

In addition to the basic types of controls, different systems are available which provide greater or lower accuracy in flow control. Feedback systems are sometimes added to more precisely control excess air and to prevent fuel-rich or fuel-smoking conditions. The more accurate controls and feedback systems allow the burner to operate closer to the optimum excess air level, which increases the efficiency of the unit.

21.5.3 Flow Control Equipment

Fuel-air-ratio control equipment can be broken down into three general types: atmospheric, fixed-flow, and variable-flow equipment.

Atmospheric Equipment. An atmospheric burner relies on the natural draft through a chimney to provide the required air flow. The proper design of the chimney is critical in obtaining the required air flow and preventing downdrafts or other conditions that may affect air flow. (See Chap. 30 for chimney design.) With this equipment, the air flow occurs at all times, whether the burner is on or off. The actual amount of air flow varies according to temperatures which determine the draft.

The fuel flow, except for solid fuels, is controlled by a valve, which is generally operated electrically to provide automatic operation. The fuel rate is regulated by a combination of a pressure control valve and an orifice.

The atmospheric burner is used in most homes for heating and hot water. Because few components are needed for this type of burner, it is both inexpensive and reliable. The accuracy of air-flow control is very poor, however, and because of this it is relatively inefficient.

One device that is sometimes added to increase efficiency is a draft control damper. This prevents air flow up the chimney when the burner is not operating and prevents the warm air in the room from escaping.

Fixed-Flow Equipment. Fixed-flow controls provide a specific flow of both air and fuel. A fan is used to supply combustion air, and the quantity can be regulated by adjusting a register or damper. The fuel flow is obtained in the same manner as for the atmospheric burner.

With this type of control, the quantity of air flow can be regulated with a fair degree of accuracy. The draft effect and changes in the draft are not important factors in determining air flow. (It is important, though, to provide proper designs to prevent large changes in stack pressure, which can prevent proper air flow.) The combustion air fan is usually cycled with the fuel valve, so that air flow only occurs when the burner is operating.

The fixed-flow control can be staged so that it can operate at more than one flow rate. For an oil burner, this can be done by using multiple valves and nozzles. Opening more valves increases the total fuel input. This can be done with single burner (one air fan) or multiple burners. Another method is to use a single nozzle and to vary the pressure, where the fuel flow increases with pressure. Again, this can be done with valves where the different valves are connected to different fuel pressure regulators. There are limits on the turndown available with pressure, since the square law requires large pressure variations to obtain small flow change. The multiple-nozzle approach can easily provide large turndowns.

The air-flow control for staged fixed input is generally achieved by positioning a damper. Typically an electric actuator is used to open or close the damper, to provide more or less air. In some cases, the air flow is fixed at the maximum input requirements, and only the fuel flow is adjusted. This approach is also used on some atmospheric burners.

Variable-Flow Equipment. Variable-flow or modulating controls provide the ability to infinitely vary both fuel and air flow between some minimum and maximum values. In most cases, variable-flow valves are used to regulate the flow of fuel and air. This valve is generally called an "air damper" when it is used to control air flow. The air is supplied by a fan which operates at a constant speed and is considered a constant-volume device.

The fuel is supplied to the valve at constant pressure, so that the area change in the valve causes a specific known change in fuel flow. The fuel system (both gas and oil) is designed so that there is a relatively large pressure drop across the fuel valve and a small drop in the nozzle. This allows the fuel valve to be the primary regulator of flow, with the nozzle sized to accommodate large changes in flow.

With natural gas, a butterfly valve is generally used to control flow. The valve is opened or closed to change flow rates. On oil, a number of different variable-orifice valves are used. Since the pressure drop at the nozzle must be low, air or steam is generally used to atomize the fuel.

21.5.4 Fuel-Air-Ratio Control Systems

The control system used to operate this equipment also varies by type of burner and accuracy requirements. Simple on/off burners (atmospheric or fixed-position) use switches that will cycle the burner on and off. These switches can be driven by any number of different parameters that reflect the need for heat. In a home furnace, the room temperature is used. For a boiler, the outlet water temperature or steam pressure is generally used. Because these burners are designed for simplicity and low cost, the operating controls tend to be very limited.

For staged input, a series of switches are used. The settings (temperature or pressure) are staged so that the input is increased as the need for heat increases.

Modulating systems come in three basic forms: single-point, parallel, and fully metered. Each system has advantages and disadvantages. All three systems, however, are limited in that they are used only for controlling between the minimum and maximum burner operating valves. Below this point, the burner operates as an on/off unit (described above).

The most common control is the single-point positioning system. A measurement of the error (actual minus desired) in temperature or pressure is used to operate a positioning actuator. The greater the error, the more the actuator moves. This actuator drives the fuel and air control valve via mechanical linkage. The linkage is adjusted so that the appropriate amounts of air and fuel are delivered to the burner at all firing rates. A jackshaft control system employs a shaft as the primary mechanical link between the fuel and air valves.

A parallel control system uses two actuators, one for air and one for fuel. These actuators operate from a single error measurement, and they independently drive the air and fuel valves. Operation of the system is adjusted so that the proper amounts of air and fuel are delivered to the burner at all firing rates. The primary difference between this system and the single-point positioning system is that the mechanical linkage between the air and fuel valve is replaced by an electric or pneumatic signal used to operate the individual actuators. Parallel positioning systems offer the advantage of allowing large differences between the air and fuel valve positions when mechanical linkage would be impractical.

The metering system is similar to the parallel positioning system except that it actually measures air and fuel flow rates. With both parallel and single-point positioning, the flow rate is assumed, based on the position of the valve. The metering system positions the valve to obtain the desired flow rate. The flow measurement accuracy, temperature, pressure, viscosity, and other values are added to increase the accuracy.

21.5.5 Feedback Systems

In the past few years, a number of improved stack measurement systems have been used either to provide fine tuning of the fuel-air ratio or to prevent the burner from operating in an improper manner. By directly measuring the products of combustion, all the possible variables that can affect the fuel-air ratio are taken into account.

The typical measurements, what the value represents, and how these values are used, are listed below. These systems are offered on new equipment and often are packaged for use on existing control systems. The two most commonly measured combustion products are oxygen and carbon monoxide (CO).

Oxygen. The oxygen measurement is the most common and is representative of the excess air in combustion. Products are offered for test instruments which are used to periodically monitor and set up the fuel-air ratio and for continuous control. As a controller, the oxygen trial system will provide a small change in either air flow or fuel flow to maintain the desired excess-air (or oxygen) level. *The primary flow rate control is obtained by one of the standard systems.*

Carbon Monoxide. The presence of this gas indicates incomplete combustion. A small amount of CO, measured in ppm (mg/L), is common. If this value increases, it represents a decrease in efficiency, since not all the fuel is converted to heat. Although CO has been used to control excess air in a similar way to oxygen, it is more frequently used as a monitor to detect problems. New low-excess-air burners do not experience a gradual change in CO with changes in excess air, which makes it difficult to use them for controlling. This excess-air value is also limited to gas fuel, with opacity being the measurement of incomplete combustion for liquid and solid fuels.

A modulating control generates the input necessary to match the load and is quick to respond to load changes. Its cost is obviously higher, but the major cost factor is the type of modulating control. The efficiency varies with the accuracy of the modulating control system.

A single-point positioning system is the lowest-cost modulating control. Generally the accuracy is the same as for good parallel positioning control and lower than that of a typical metering system. Parallel positioning costs more than the single-point method, but there is a dramatic increase in cost with the metering system.

A metering system that uses only a pressure differential does not offer any major increase in accuracy over the single-point control. Without correction for changes in air density and fuel density or makeup water, the same variable will limit the accuracy of all systems.

Oxygen trim systems have generally been recognized as the best method of increasing the accuracy of all systems, and they are used with all the different control systems. With oxygen trim, all variables that could change air or fuel flow are taken into account, and so the accuracy is limited only by the accuracy of the oxygen reading.

There has been a general shift away from metering systems to single-point positioning with oxygen trim. The cost of this package is well below that of a metering system, and the accuracy is the same. One advantage offered by some metering systems is the ability to correct for lead-lag that occurs when the boiler modulates. During modulation, the dynamics of the rate of change in air and fuel flow plus some mechanical factors result in a slightly different excess-air level. By having the air change lag the fuel change, the potential of going too low in excess air is eliminated.

In the final analysis, a number of factors determine performance, cost, and efficiency, so that this chapter is only as a guide in understanding the various controls and systems. Often the design of the process or heating system can overcome shortfalls, or they place tighter restrictions on the boiler. As with each component, it must be viewed with the total system and not by itself.

CHAPTER 22
SAFETY AND OPERATING CONTROLS

**Cleaver-Brooks, Division of Aqua-Chem, Inc.,
Milwaukee, Wisconsin**

22.1 INTRODUCTION

The purpose of safety controls is to minimize the possibility of an overpressure condition of the heating device and of the unintended ignition of a combustible mixture. The type and the number of devices used to protect and control heating equipment are, to a large extent, mandated in the United States by national standards such as the American Society of Mechanical Engineers' (ASME) *Boiler and Pressure Vessel Code*, American Gas Association (AGA), Underwriters' Laboratories (UL), National Fire Protection Association (NFPA), insurance carriers, and state and local authorities.

Operating controls determine the cycling and/or firing rate of the equipment based on the imposed load. In addition to the required devices, controls to meet a specific need may be necessary. Items such as efficiency, automation, response time, downtime, and economics affect the design of a control system.

22.2 SAFETY DEVICES

22.2.1 Flame Safeguard

Probably the most important safety device on a boiler is the flame safeguard control, whose primary function is to monitor the flame, to allow the fuel valves to remain open or closed, depending on the presence or absence of flame.

22.2.2 Fuel Interlocks

The temperature and pressure of the fuel(s) must be maintained within a narrow range to ensure proper combustion.

When gas is burned, two pressure switches are typically provided, one to detect a drop and the other to sense a rise in pressure. If either switch detects an improper pressure, the boiler is shut down. If the pressure were allowed to go above the normal operating range, the fuel-air ratio would go fuel-rich and/or overfire the boiler. If the pressure is too low, efficiency will be lowered and flame properties will be adversely affected, possibly causing flame outage.

The pressure of oil fuels must be maintained for the same reasons as for gas. But, in addition, low oil pressure may cause improper atomization of the oil, which in turn causes poor combustion.

22.2.3 Air Interlocks

The combustion air system should contain a device to shut down the burner when combustion air is not present. This is done with an air pressure proving switch. If combustion air is not present, the main fuel valves are prevented from opening. If combustion air is lost during operation (e.g., due to belt breakage), the main fuel valves are closed before a dangerous air-fuel ratio can develop. Dampers may be added to control the air-flow rate.

22.2.4 Low-Water Cutoffs

The water-level (water-column) control (optional on hot-water boilers) is a device which shuts down the burner when the water level approaches a dangerously low condition. On low-pressure firetube boilers, the low-water cutoff prevents firing when the waterline is at or below the lowest visible point of the water glass (*ASME Code* states that the lowest visible part of the water glass shall not be below the lowest safe waterline).

22.2.5 High-Limit Controls

On steam boilers, a pressure-sensitive switch monitors boiler pressure and will immediately shut the fuel valves(s) if an abnormally high pressure is detected. The switch is set at some value below that of the steam safety valve so that the burner shuts down before the safety valve "pops." However, it is set higher than any other device which controls steam pressure. Once the high-limit control locks out, it must be manually reset before the burner will fire. A similar function control is provided on low-pressure hot-water boilers. The control is responsive to boiler water temperature.

22.2.6 Safety and Relief Valves

The single device used to protect the pressure vessel is a *safety valve* for steam boilers or a *relief valve* for hot-water boilers. Details of the application of steam safety valves and water relief valves are given in the latest editions of the *ASME Boiler and Pressure Vessel Code*.

22.3 OPERATING CONTROLS

22.3.1 Fuel System Controls

Gas-fired boilers must have safety shutoff valves that respond to the action of the various limit controls and combustion safeguards. Safety shutoff valves are tested by recognized testing agencies for reliability and safety. Strainers may be added to ensure clean fuel gas. Hand-operated valves (such as lubricated plug cocks) are provided so that the gas supply to the burner can be shut off manually. Pressure regulators are added to provide regulated gas pressure at the burner. Metering valves and/or orifices may be added to control the fuel rate.

Snap-acting solenoid valves are used in gas pilot lines or as shutoff valves in the main line to the boiler; however, these are generally used on small boilers only. On larger boilers, motorized valves are used. These are slow-opening valves which produce an easy lightoff of the main burner and minimize pressure surges in the gas supply line. Some testing agencies (e.g., Underwriters Laboratories', Inc., Canadian Standards Association) or insurance carriers may require multiple safety shutoff valves and/or other valving arrangements.

Each gas-fired appliance should have a pressure-reducing valve (PRV) or regulator to maintain a constant gas pressure to the burner. Oil-fired boilers must have fuel oil safety shutoff valves that respond to the action of the various limit controls and combustion safeguard. Strainers to ensure clean fuel oil at the burner are recommended.

One or more solenoid valves control the flow of oil to the burner. However, motorized valves may be used to achieve certain lightoff characteristics or if valve-closure-proving switches are required. In either case the valve(s) chosen must have temperature and pressure ratings suitable for the type of oil being burned.

22.3.2 Limit Devices

A limit device is a temperature- or pressure-actuated switch which controls the cycling of the boiler. If the sensed boiler temperature or pressure rises above the setpoint, the switch opens and the boiler is kept off until the measured variable again falls below the setpoint. Next, the switch is closed, allowing the boiler to begin firing. The switch mechanism is automatically reset, unlike that described earlier under high-limit controls.

Each boiler must have a limit control in addition to the high-limit device. The high-limit setpoint is always higher than the limit control setpoint. If the boiler is designed for on/off operation, the limit control cycles the boiler on and off. If the burner is capable of off/low/high operation, a second limit control is used to control the firing rate between low fire and high fire. The setpoint of this control will be lower than that of the primary limit control.

If the boiler is equipped with a modulating burner, a device is required to detect changes in boiler temperature or pressure over a continuous range. This device transforms a changing process variable to a mechanical or electric output. Depending on the type of device, a potentiometer or slidewire output may be produced, or the output may be a voltage or current. These output signals are used to drive an actuator which positions air dampers and fuel-flow control valves.

22.3.3 Firing Rate Control

There are three major firing rate control methods: on/off, off/low/high, and modulating. In the on/off method, the burner is either off or firing at its rated capacity. This type is used primarily on small boilers when control of output pressure or temperature is not critical. The on/off/high mode allows the boiler to more closely match the load. Given a demand for heat, the boiler begins firing at approximately 30 to 50 percent of capacity (low fire). Upon a further increase in demand, the burner fires at maximum capacity (high fire). When the demand is satisfied, the firing rate is reduced and held at low fire. If the demand increases, the burner again goes to high fire; on a demand decrease, the burner will turn off.

The modulating firing rate control method provides the greatest flexibility in matching the boiler output to the load. On a call for heat, the boiler is brought on at low fire, which typically is 15 to 30 percent of capacity.

As a greater load is imposed on the boiler, an increasing amount of fuel and air is introduced to the burner. The output of the boiler varies continuously between low and high fire. In this way the boiler output matches any load between 25 and 100 percent of boiler capacity.

Gas-fired burners use a butterfly valve to vary the amount of gas to the burner. When oil is the fuel, a variable-orifice or characterizeable valve is used to meter the flow of oil. Typically, the butterfly valve or variable-orifice valve is mechanically linked to the combustion air damper to provide the proper fuel-air ratio at all firing rates.

22.3.4 Optional Features

Many controls can be added to the standard system to accomplish a specific task. Boiler temperature or pressure may be reduced at night or on weekends by adding a second limit control at a reduced setpoint which is selectable with a switch.

Each installation must be considered individually. When a control system is being designed, it is important that the load requirements be met, but at the same time the integrity of the boiler must be maintained by avoiding undue stresses or thermal shock.

22.3.5 Feedwater Control

Feedwater controls are used to maintain a safe and acceptable water level in the boiler. Too high a water level in the boiler can lead to water carryover with the steam. Too low a water level in the boiler can lead to a safety shutdown of the burner or a ruptured pressure vessel, should the low-water safety control fail to operate.

In the following paragraphs we discuss a few of the water level sensing controls and operating schemes used for controlling water level. Selection of the correct ones depends on load and operating conditions, type of boiler, and economics.

There are several different methods of sensing the water level in steam boilers. The most common uses a float located in an external chamber whose position changes in response to water-level variations. This float is either mechanically or magnetically connected to an electric or pneumatic device which actuates a feedwater valve and/or motor.

A second method of sensing the water level uses electric probes. When the probes make contact with the water, an electrical path is established which actuates a relay to power a feedwater valve or pump. This method is used only for on/off or open/close operation of a feedwater motor or valve.

A third method of sensing the water level involves inserting a special probe into the boiler shell or external chamber to measure the variance in electric capacitance as the water level changes. The variance in capacitance is converted to an electric or pneumatic output signal for control.

A fourth method of sensing the water level uses a thermohydraulic system that has a "generator" mounted at normal water level which sends a pressure signal to a feedwater valve in response to the amount of steam the generator is exposed to. This system is not generally used on low-steam-pressure applications.

In conjunction with the water-level sensing, several different control schemes are used to maintain the proper water level in the boiler. One way is to turn a feedwater pump on and off in response to a water-level signal. This system can be modified to include opening and closing a feedwater valve.

A second control scheme is to have the water-sensing element send a proportionate signal to a feedwater valve to throttle the valve in response to the water level. This system may or may not turn a feedwater pump on and off as the feedwater valve opens and closes.

On boilers where water levels are hard to maintain due to rapid and large load swings, firing rates, and low water volume, a more elaborate proportioning control scheme may be in order. Such a system would include additional sensors besides water-level sensors, such as steam flow or steam flow and feedwater flow sensors. The proportioning feedwater valve would be positioned in response to the two or three inputs it receives.

22.4 SUMMARY

A wide variety of control devices ensure the safe and reliable operation of heating equipment. The use of these devices, however, does not preclude the proper use and maintenance of the equipment. The wide selection also allows control systems to be designed which can achieve many operational objectives.

CHAPTER 23
BOILER ROOM VENTING

Cleaver-Brooks, Division of Aqua-Chem, Inc.,
Milwaukee, Wisconsin

23.1 BOILER ROOM AIR SUPPLY AND VENTILATION

The combustion process requires a supply of air at all times, and so it is essential that provisions be made to supply adequate air to the boiler room.

Installation codes normally require that boilers be installed in a separate fire-rated room, and it is not unusual for all fired appliances needed for the facility to be installed within the boiler room confines. If the boiler is installed in a room containing other air-handling equipment such as compressors, exhaust fans, incinerators, etc., then adequate fresh air must be provided for all these appliances, including the boiler.

In addition to fresh air for combustion, it is important to provide sufficient air to maintain the boiler at a reasonable temperature. Two or more fresh-air openings are usually provided to obtain air circulation within the boiler room. Generally boiler room temperatures at the boiler inlet above 100°F (38°C) and below 50°F (10°C) should be avoided. Proper insulation of steam lines, hot-water lines, oil lines, and stacks will assist in maintaining reasonable boiler room temperatures.

A boiler with nominal insulation will lose as much as 600 Btu/(h · bhp) (17.8 W/kW) to the room environment. Boiler rooms become intolerably hot without adequate insulation and ventilation. The control systems used normally have a maximum ambient temperature of 104°F (40°C).

The blower provided as a part of the boiler assembly is normally sized to provide sufficient air at 70°F (21°C). If the air in the room rises to 120°F (49°C), the oxygen delivered to the burner decreases by about 10 percent since the blower fan is a constant-volume device and the density decreases with increasing temperature. This reduction could upset the fuel and air adjustments made on the burner.

Numerous codes spell out the requirements for boiler installations, and they should be followed by the mechanical engineer for the building design. Local jurisdictions frequently have additional requirements which can go beyond the nationally recognized codes.

In the absence of code requirements and the boiler manufacturer's recommendations, the following rules of thumb apply:

Combustion air	8 to 10 $ft^3/(min \cdot bhp)$ [3.05E − 0.04 to 4.81E − 0.04 $m^3/(s \cdot kW)$]
Boiler room ventilation	2 $ft^3/(min \cdot bhp)$ [9.63E − 0.05 $m^3/(s \cdot kW)$]
Velocity access louvers	250 ft/min (1.27 m/s)

Combustion air and/or ventilation for all other equipment located within the boiler room must be added to the above.

The ventilation air should be designed to cross-ventilate the boiler room with the operating aisle and the electric equipment receiving the cool air first.

23.1.1 Special Precautions

1. Frequently the air that must be admitted to the boiler room for combustion is insufficient for proper ventilation in warm weather.

2. The ventilation requirements should consider all sources of heat in the boiler room. In addition to the heat given off by the boiler(s), heat is given off by breeching, condensate or deaerator systems, steam lines, and miscellaneous electric equipment.

3. If blowers are used to supply air to the boiler room, vent openings may not be required if such devices are approved by the authority having jurisdiction. If the air supply depends on a blower, an interlock switch should be provided to prevent the boiler from firing if the blower fails. Figure 23.1 shows the combustion air requirements for Cleaver-Brooks boilers and other similar boilers.

23.2 STACK (OR CHIMNEY)

It is good practice to connect the boiler gas outlet to a stack or chimney by using breeching ductwork. This ductwork conducts the flue gas to the chimney, using the draft that the chimney supplies efficiently and without building in excessive friction losses. See Chap. 30 for a discussion of stacks and breeching.

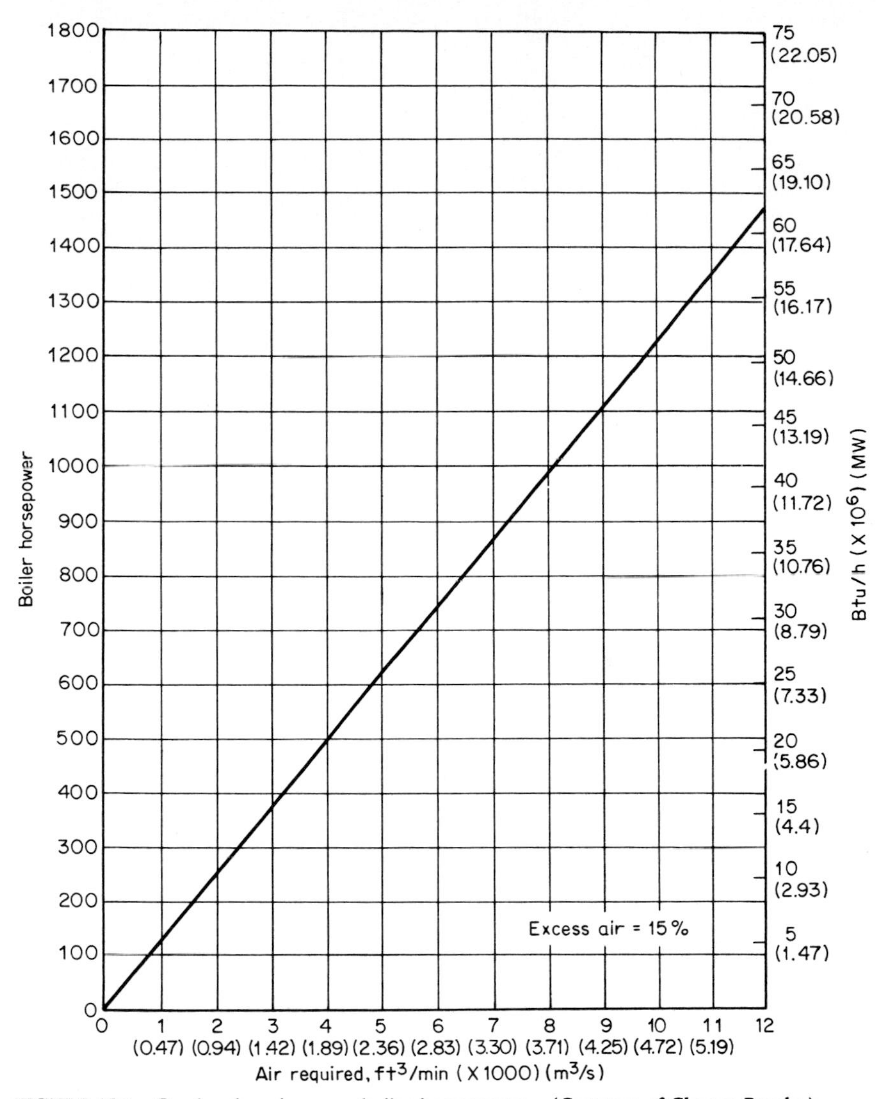

FIGURE 23.1 Combustion air versus boiler horsepower. (*Courtesy of Cleaver-Brooks.*)

CHAPTER 24
DEAERATORS

Cleaver-Brooks, Division of Aqua-Chem, Inc., Milwaukee, Wisconsin

Dissolved gases, particularly oxygen, are the most common causes of corrosion and pitting in the ordinary medium-pressure steam plant, and they are particularly aggressive in their action on feed lines, condensers, and heat exchanger surfaces. These dissolved gases enter with the feedwater and are liberated in the boiler. Carbon dioxide (CO_2) enters the boiler as a dissolved gas and is liberated on the application of heat. On the liberation in the boiler, the CO_2 passes out of the boiler with the steam.

Waterside corrosion and scaling seriously affect performance and boiler life. Many things contribute to corrosion, but the most important are dissolved oxygen and CO_2 in the feedwater for steam boilers. Removal of these two gases practically eliminates, or at least greatly reduces, corrosion attack. Pretreatment of feedwater makeup and returning condensate by deaeration is the most acceptable method.

Devices designed to remove the oxygen and carbon dioxide from boiler feedwater are commonly called "deaerators." Deaerators heat the boiler feedwater to the saturation temperatures at the deaerator pressure. A properly designed deaerator reduces oxygen to 0.005 cm^3/L or less (see American Society of Mechanical Engineers' [ASME] publication PTC 12.1.3-1958) since any oxygen in the feedwater is corrosive. Therefore it is very important to reduce the oxygen content to practically zero.

If the oxygen content is accurately controlled, little or no corrosion will result. Oxygen solubility in water varies with the temperature and pressure. As the temperature rises, the solubility decreases. So it is desirable to keep the feedwater at a high temperature and to vent the deaerator thoroughly, to reduce the partial pressure of air or oxygen in the vapor space over the water.

These are some of the benefits of a deaerator:

1. It reduces the cost of boiler water treatment by reducing the amount of chemicals required to neutralize oxygen and CO_2.
2. It reduces the amount of boiler blowdown by minimizing makeup water and treatment chemicals.
3. It reduces corrosion on return condensate lines.
4. It reduces maintenance repair and cleaning of steam traps.

5. It may provide an effective means for recovering the heat from exhaust steam.
6. It minimizes gases in the boiler drum which may interfere with maintaining pressure or temperature sensing.
7. It provides a reservoir to temporarily store and dampen variations in returning condensate.

Although there are no steam boiler plants in which feedwater deaeration would be unnecessary, the deaerator is particularly essential for any of the following situations:

- All boiler plants operating at 75 psig (5 bar) or more
- All boiler plants with little or limited standby capacity and where production depends on continuous boiler operation
- All boiler plants with 25 percent or more cold makeup water

One of the more popular types of pressurized deaerators capable of reducing dissolved oxygen to the desired levels is a spray type. The pressurized deaerator, shown in Fig. 24.1, is simple, troublefree, and quiet. Supply water enters the deaerator through a spring-loaded, self-cleaning nozzle, which sprays water into a steam-filled primary heating and vent concentration section. Here the water temperature and most of the gases are released. Then the water is collected in a conical water collector. From there it goes to an atomizing valve where high-velocity steam strikes it, breaks it down into a fine mist, and heats it to full steam saturation temperature. The mixture strikes a deflecting barrier which separates water and steam. Hot, gas-free water then drops to the storage compartment.

The steam and noncondensable material flow upward into the primary heating and vent section, where they contact the incoming water spray. The steam is condensed, and the gases released from the water flow upward to the vent outlet. Some steam also discharges through the vent to ensure thorough deaeration. The steam quantity used in this process normally amounts to less than 0.5 percent of boiler capacity. Usually the vent orifice is sized to vent the total amount of gases from 100 percent makeup water at 50°F (10°C). Water at higher temperatures releases less gas, but keeping the vent size maximum ensures thorough venting at all conditions, and this slight additional loss at higher temperatures is negligible.

The entire deaeration section in contact with undeaerated water should be made of stainless steel.

Other types of pressurized deaerators are the packed column type and the tray type. The functions of these two designs are essentially similar. In one case, the water to be deaerated flows in thin films over packing material, while in the other it flows over the edges of shelves or trays. In Fig. 24.2 the sequence of operation is as follows:

The supply or makeup water enters the deaerator through a spring-loaded self-cleaning spray nozzle, which sprays the water into the vent concentrating section. This is the first stage in raising the temperature of the water. The partially heated and deaerated water strikes the distributing baffle and falls onto the exchange packing. It then flows down through the packing in a very thin, turbulent film.

Steam enters the system at the top of the storage tank and sweeps the tank free of noncondensible gases. Steam then enters the bottom of the packed column

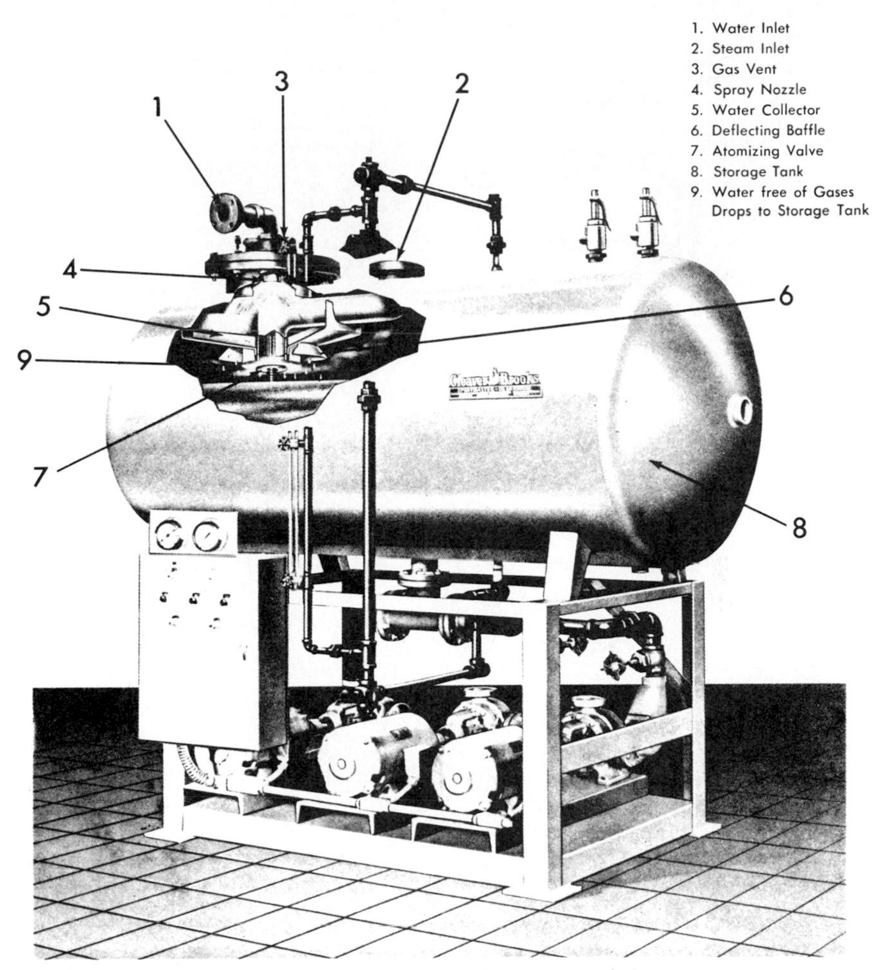

FIGURE 24.1 Pressurized deaerator. (*Courtesy of Cleaver-Brooks.*)

and flows upward through the voids in the exchange packing. The cleanest, hottest steam meets the hottest water flowing through the lowest part of the packing, completing the deaeration and heating of the water before it falls to the storage section.

Counterflow of steam and water in the deaerating section provides the best possible scrubbing of the dissolved gases from the water. The steam and noncondensible gases flow upward through the packing into the vent concentrating section where they meet the cold effluent water. Here steam is condensed, and the released gases are discharged into the atmosphere.

The packed column deaerator offers the highest-purity effluent, less than 0.005 cm^3/L oxygen and carbon dioxide by accepted test. It is also very effective at all loads. It has a long operating life because all parts coming in contact with

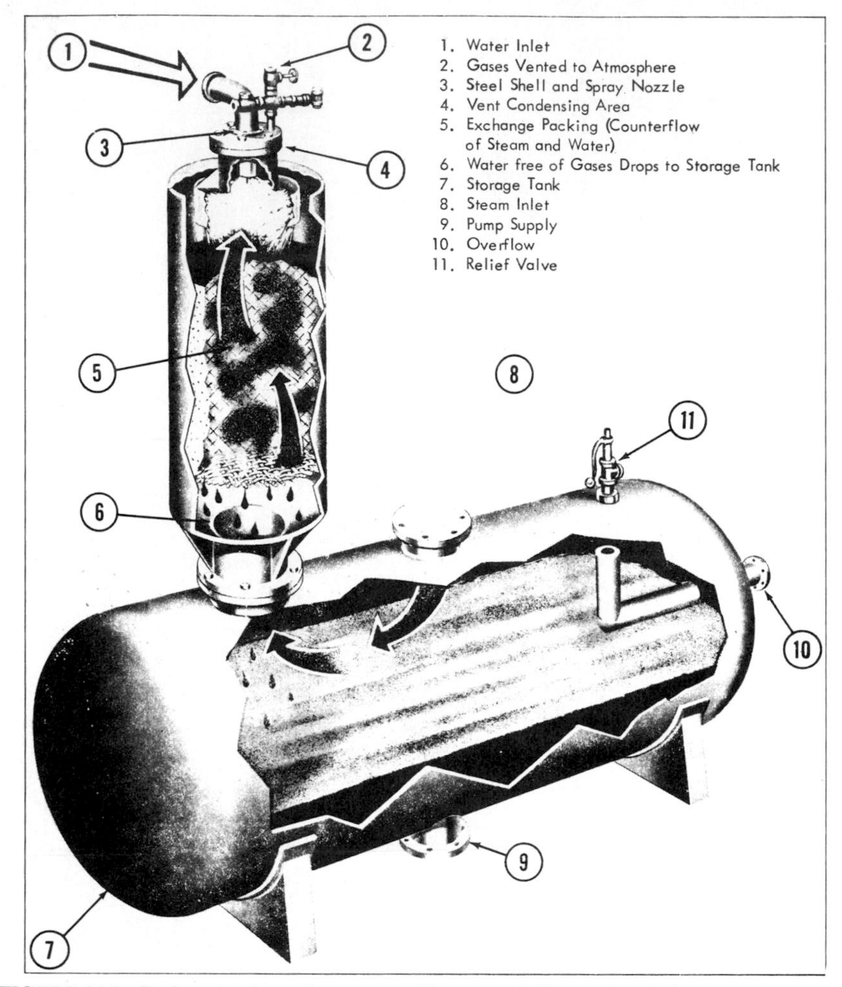

FIGURE 24.2 Packaged column deaerator. (*Courtesy of Cleaver-Brooks.*)

undeaerated water are made of corrosion-resistant stainless steel. Exchange packing is service-free, impervious to the effects of steam, water, or gases. The exchange packing is stainless-steel cylinders.

Figure 24.3 shows a typical deaerator setup. Under normal conditions, makeup water is added only if the condensate returning from the system is inadequate to generate the steam used.

In the case of the pressurized deaerator, if the high-temperature returns are less than 25 to 30 percent of the deaerator rating, they can be returned directly to the storage tank. "High-temperature returns" are defined as temperatures higher than the normal operating temperature of the deaerator.

For those applications in which the deaerator feed consists of 100 percent makeup water, the overflow drainer may be omitted.

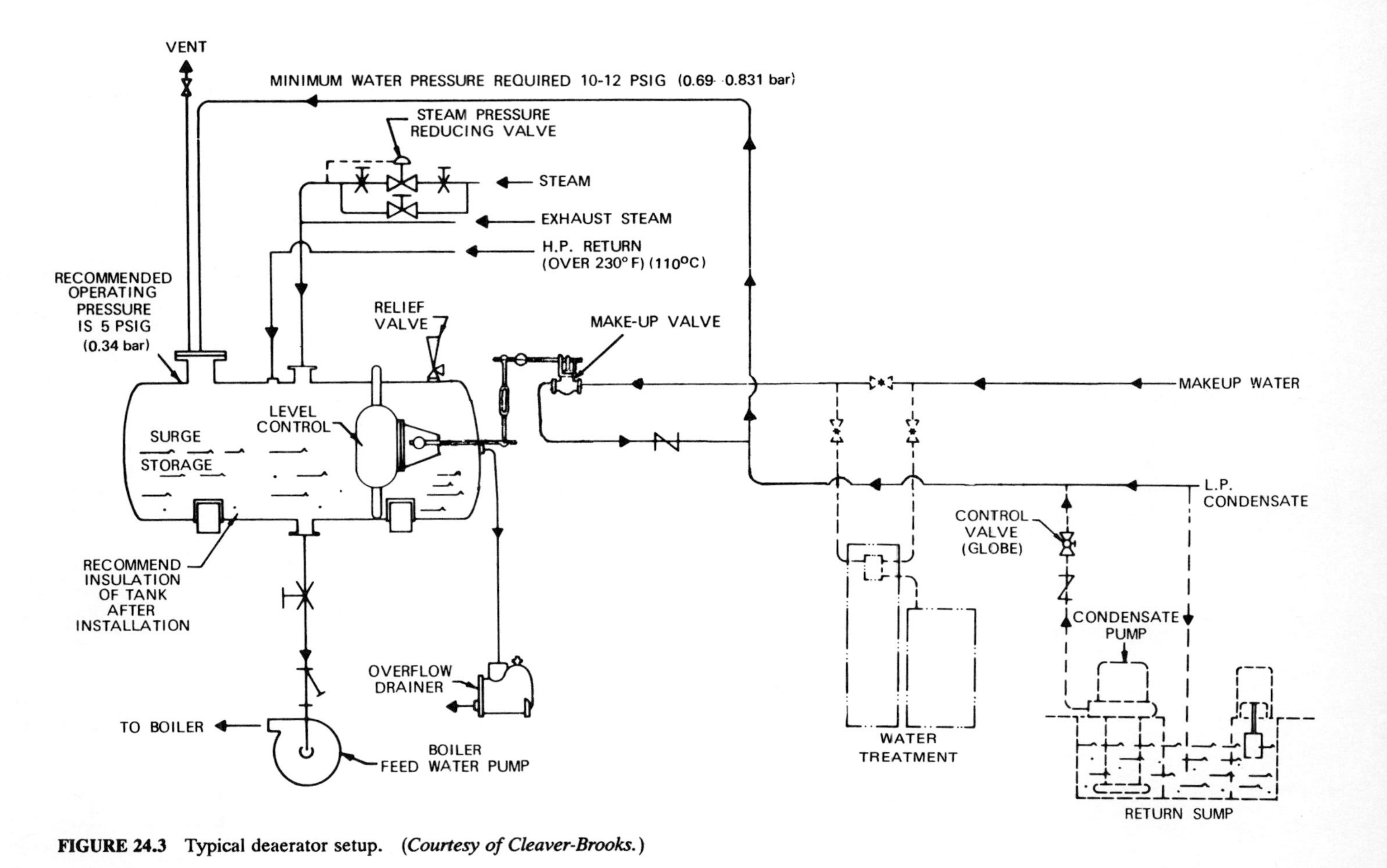

FIGURE 24.3 Typical deaerator setup. (*Courtesy of Cleaver-Brooks.*)

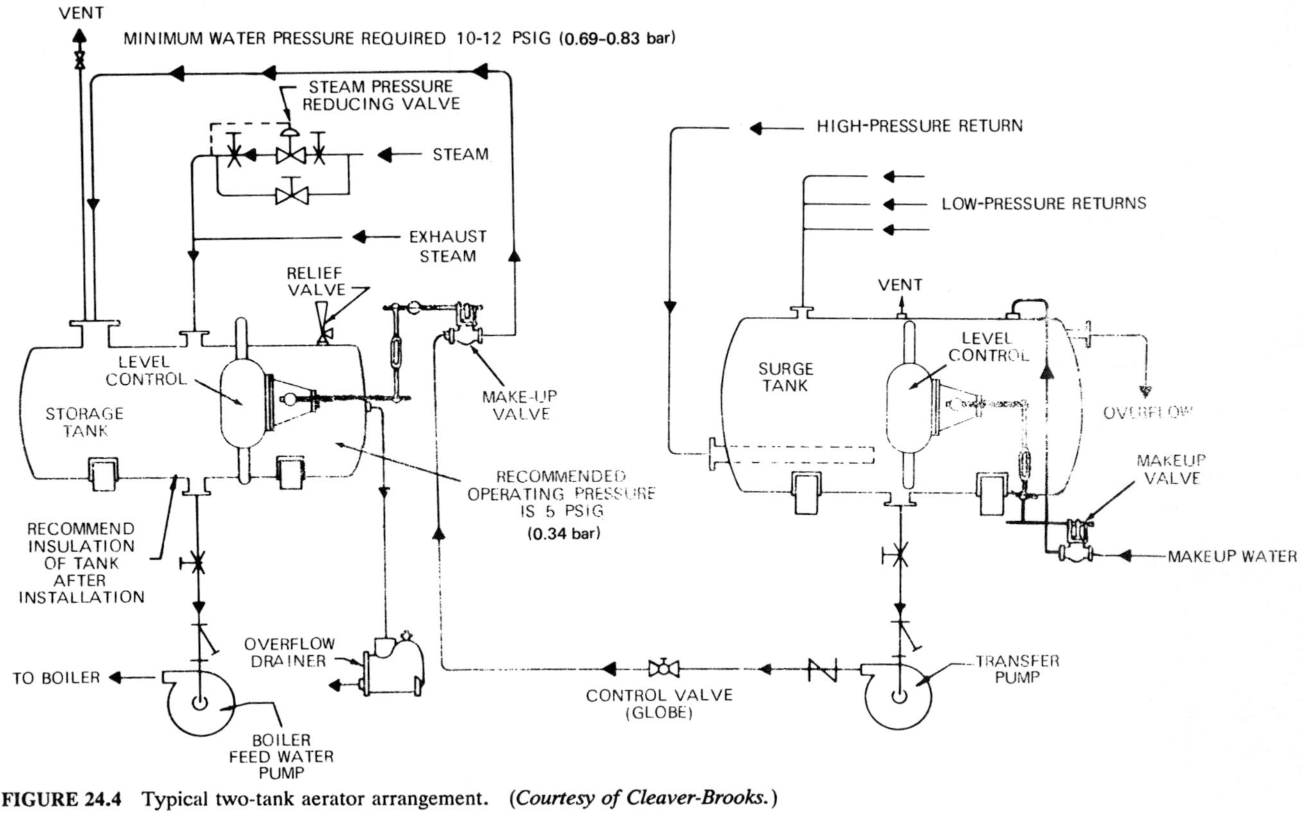

FIGURE 24.4 Typical two-tank aerator arrangement. (*Courtesy of Cleaver-Brooks.*)

Figure 24.4 shows a typical two-tank arrangement. In this layout, the high-pressure returns, low-pressure returns, and cold-water makeup are collected in a surge tank. A transfer pump is provided to move the water from the surge tank to the deaerator through a makeup valve. The transfer pump discharge must be limited to the maximum capacity of the deaerator. This is accomplished in the field by manual adjustment of a control valve.

This system is used primarily where there are many low-pressure returns from several sources in the plant that can only be collected conveniently in a single tank. This system can also be used where returns are at such low pressure that they cannot go into the pressurized deaerator directly.

CHAPTER 25
ECONOMIZERS

Cleaver-Brooks, Division of Aqua-Chem, Inc., Milwaukee, Wisconsin

Fuel savings and heat recovery have become a prime interest in today's high-pressure and high-temperature boilers. The use of economizers has shown the greatest increase in boiler efficiency. Boilers operating at steam pressures of 100 psig (6.9 bar) and above or hot-water boilers above 300°F (150°C) are viable prospects. A properly designed and installed economizer can increase overall boiler efficiency from 85 to 90 percent.

An economizer is basically a heat exchanger consisting of a bundle of tubes enclosed in a casing and installed on the flue gas exit side of the boiler.

In the economizer, the boiler feedwater is passed through tubes which are normally finned, while the stack gases pass around the outside of these tubes. The economizer reduces the temperature of the stack gases by transferring this heat to the boiler feedwater. The heat removed from the high-temperature stack gases is added to the feedwater, increasing its temperature; therefore less heat and fuel are needed in the boiler to heat the feedwater temperature to the saturated steam temperature.

An economizer at many installations can reduce fuel consumption by 5 to 10 percent. Typically, an economizer installation will pay for itself in less than 2 years—frequently in less than 1 year.

CHAPTER 26
AIR PREHEATERS

**Cleaver-Brooks, Division of Aqua-Chem, Inc.,
Milwaukee, Wisconsin**

26.1 INTRODUCTION

Air preheaters are used to heat combustion air by recovering heat from boiler flue gas. This results in increased fuel-to-steam or fuel-to-high-temperature efficiency.

26.2 EFFECT ON EFFICIENCY

Preheating the combustion air increases the efficiency by reducing the heat required to elevate the air to required combustion temperatures. For every 50°F (28°C) rise in combustion air entering the burner, the efficiency increases by 1 percent.

26.3 APPLICATION CONSIDERATIONS

Coal. Air preheaters are typically used for pulverized coal-fired units that require hot air for drying the fuel and for burning bituminous coal in large stoker-fired units.

Natural Gas and Fuel Oils. These burners are typically designed for ambient air temperatures of 60 to 100°F (15.5 to 38°C). When the air temperature to the burner is elevated, the burner and fan design must be considered. Burner wind boxes and equipment normally mounted on the burner must be designed for the temperatures encountered. Typically, wind boxes and air ducts must be insulated for safety precautions and to reduce heat losses to obtain maximum benefit.

A disadvantage to air preheat is the fan size required. The fan must be sized to overcome the preheater pressure drop and increased burner pressure drop. The increasing losses and lower air density will result in larger fan diameters and increased fan horsepower requirements.

26.4 TYPES OF PREHEATERS

There are two general classes of air preheater: the "regenerative" style, which incorporates a rotating disk filled with a medium alternately heated by flue gas and cooled by the passage of combustion air, and the "recuperative" design, which is a stationary heat exchanger that may use tubes or plates of various shapes to absorb the heat from the flue gas and transfer it to the air flowing on the opposite side of the surface.

Air heaters, like economizers, are subject to corrosion since they are employed to cool flue gases. The cooling medium is supplied by air to the combustion process. The air flowing over the surfaces may be at the boiler room temperature of 60 to 100°F (15 to 38°C), but could also be very cold in some environments. Although heat-transfer coefficients may be relatively low, it is possible to reduce the temperature of the flue gas below the dew point condensing temperature of some of the five exit gases.

Because the gas-to-gas coefficients are low, air preheaters are likely to have a large heat-transfer surface area, if they are to be effective. Since the preheater is not installed in the steam or water circuit, the preheater is not subjected to elevated pressures and is therefore not required to be constructed to any pressure vessel code. However, the exchanger designs offered may need to be complex to achieve the heat-transfer goals; therefore, the assembly cost may be high. Installation costs are also likely to be high because ductwork is required for both combustion air and flue gas. This ductwork must include expansion joints to allow movement of both duct networks caused by variable operating temperatures. In general, air preheaters are cost-effective only on very large installations that consume large quantities of fuel.

CHAPTER 27
SOOTBLOWERS

Cleaver-Brooks, Division of Aqua-Chem, Inc.,
Milwaukee, Wisconsin

27.1 INTRODUCTION

The purpose of sootblowers is to remove ash and soot buildup on convection pass tubes. Typically, soot builds up when residual-grade oils and coal are fired. These cause the greatest amount of soot deposits over a given time of firing. Combustion quality is a factor in soot buildup, but even the best combustion will result in some soot deposits.

Natural gas and No. 2 fuel oil are clean-burning fuels which do not cause deposits on tubes and therefore do not usually require the use of sootblowers.

27.2 BOILER DESIGN CONSIDERATIONS

Boilers must be designed with the removal of soot as a consideration. The design must allow for the installation of soot-blowing equipment and its proper operation. Typical packaged watertube boilers have lanes which allow the soot-blowing media air or steam to pass along so that the tubes can be scrubbed of soot. The distance that the cleaning media can travel effectively must also be considered. Industrial watertube boilers with narrow but high convection zones may require multiple sootblowers.

Firetube boilers, by design, are not conducive to soot blowing and typically require manual removal of soot by periodic shutdown and opening of boiler doors so that soot can be brushed away.

27.3 SOOTBLOWER OPERATION

Proper operation is required for sootblowers to be effective. Typically a fixed schedule is established based on the type of fuel being fired and operating experience. On installations where no previous experience is available, it is best to start with frequent operation and to reduce the frequency after some experience has been gained. Once a boiler has so much soot that it causes excessive stack temperatures, mechanical cleaning may be required. Mechanical cleaning of watertube boilers can require extensive work, usually involving removal of the boiler casing for access to tubes.

Generally sootblowers are operated twice a day, and the duration of soot blowing during each of these operations is the variable based on experience. It is very important that the boiler have a firing rate of at least 75 percent during sootblower operation so that the gas velocities are maintained at a high enough level to carry the dislodged soot from the boiler into the stack.

27.4 TYPES OF SOOTBLOWERS

These types of sootblowers are available:

1. Rotary steam: stationary and retractable
2. Rotary air puff: stationary and retractable
3. Ultrasonic

Industrial packaged watertube boilers normally use a rotary sootblower that is stationary. The rotary sootblower moves through a predetermined angle of rotation as required by the boiler design. By design these units incorporate a sootblower that can withstand the temperatures to which it is exposed. The sootblower element which is constantly in the unit is constructed of various materials over its length, so it can withstand the temperatures yet provide maximum economy in material use.

The retractable type of sootblower element is designed for use where the materials comprising it cannot continually withstand the temperatures to which it is exposed. In these cases the element is mechanically driven into proper soot-blowing position only during operation. The retracted sootblower extends outside the boiler casing and occupies additional boiler room space.

Rotary units use either steam or air as the blowing medium. Steam is most commonly used; it is normally supplied from the steam drum of the boiler. Air puff units are used when the boiler operating pressure is not sufficient to provide adequate velocities from the soot-blowing nozzles. Air is normally used on hot-water generators.

Ultrasonic sootblowers produce a sound wave which causes soot to vibrate loose from the areas where it is deposited. These units are not commonly used by boiler manufacturers. Little historic data is available on these devices. Questions unanswered at this time relate to the effectiveness and to possible deterioration of refractories and insulation within the boiler.

The results of proper sootblower operation should not be taken lightly given today's high-energy costs. Modern boilers are designed for optimum efficiency, and this can be achieved only with a clean boiler.

CHAPTER 28

GOOD PRACTICE IN INSPECTION, MAINTENANCE, AND OPERATION OF BOILERS

Cleaver-Brooks, Division of Aqua-Chem, Inc., Milwaukee, Wisconsin

28.1 INSPECTION

28.1.1 New Installations

Pressure Vessel. Before a new boiler is initially fired, it should be given a complete and thorough inspection on the fireside and the waterside. Examination of all the pressure parts, boiler, drums, tubes, water column, blowoff valves, safety and safety relief valves, baffles, nozzles, refractory setting, seals, insulation, casing, etc., is necessary. Any dirt, welding rods, tools, or other foreign material should be removed from the boiler. The waterside of the boiler should be thoroughly rinsed with water of not less than 70°F (21°C) and with sufficient pressure to effect a good cleaning.

Inspectors. During this period of initial inspection, it is a good time for the customer to call in an insurance inspector to check out the boiler. As a general rule, these inspectors like to inspect the boiler prior to the hydrostatic test. Any state or local inspectors (when required) should also be contacted.

Before any inspecting personnel are allowed to enter a boiler that has just recently been opened up, either the waterside or fireside, the boiler should be ventilated thoroughly. This eliminates any explosive gases or toxic fumes and ensures an adequate supply of fresh air.

Only approved types of waterproof low-voltage portable lamps with explosionproof guards, or battery-operated lamps, should be used within a boiler. Portable lights should be plugged only into outlets located outside the boiler.

Testing. When a hydrostatic test is required to comply with state, local, or insurance company requirements, all of the preparatory steps required and/or recommended by the manufacturer and authorities must be adhered to.

28.1.2 Existing Installations

An external examination of an existing installation can be made while the boiler is in operation. Close inspection of the external parts of the boiler, its appurtenances, and its connections should be made before the boiler is taken off the line and secured. Notes should be made of all defects requiring attention so that proper repairs can be made once the boiler has been shut down. When conditions permit, the shutdown of an existing installation can be timed to coincide with the annual or semiannual inspection, when the insurance and/or state inspectors can be present. Before an internal inspection can be made, the boiler must be secured in strict accordance with all the recommendations and requirements of the manufacturer and the authorities having jurisdiction.

28.1.3 Burner Controls

Before making any attempt to start a burner-boiler unit, read the manufacturer's instruction manual thoroughly, to get a good understanding of how the burner operates. The operator should familiarize herself or himself with the various parts of the burner and understand their functions and operation.

Parts. Describing all the different types of fuel-burning equipment is beyond the scope or intent of this chapter. However, the points covered here can, with slight variations, be applied to all types of equipment.

The burner should be checked over thoroughly to ascertain that all parts are in proper operating condition. These are some of the points to check:

1. All the fuel connections, valves, etc., should be gone over thoroughly. Make sure that all the joints are tight, pressure gauges and thermometers in place and tight, and valves operative. Fuel lines should be checked for leakage.
2. If the burner has a jackshaft, check that all the linkage is tight. If possible, check the linkage for travel and proper movement.
3. The electric system should be checked very carefully. Make sure that all connections are tight. Check all the terminal strip connections. Sometimes if these are not tight, the vibration from shipping can loosen them. Test plug-in controls and relays. Check the rotation of the fan, pump, and air compressor motors.

Safety Limits and Interlocks. All the various interlocks and limits should be checked to see whether they are operational. An actual operating test of these controls cannot be made until the burner is running. All or most of the following limits are found on the average packaged burner, depending on the insurance rules being complied with:

1. Operating and high-limit pressure control
2. Operating and high-limit temperature control (hot water)
3. Low-water cutoff
4. Flame failure (part of flame safeguard function)
5. Combustion air proving switch
6. Atomizing media (steam and/or air, proving switch for oil burner only)
7. Low-oil-temperature switch
8. Low-oil-pressure switch
9. High-oil-pressure switch
10. Low-gas-pressure switch

11. High-gas-pressure switch
12. Low-fire switch

Other operating limit interlocks can be added to the control system depending on the degree of sophistication; however, this list covers most of them.

28.1.4 Auxiliary Equipment

Deaerator. All the boiler room auxiliary equipment should be checked thoroughly to make sure everything is in good working order and ready to go.

Boiler Feed Pumps. Condensate pumps, transfer pumps, all the pumps in the system should be checked out. The coupling alignment should be checked to ensure that no misalignment exists. If the coupling alignment is not up to the pump or coupling manufacturer's recommendations, it must be corrected.

Piping connections to the pump suction and discharge ports should be checked. Adequate provisions should be made for proper expansion so that no undue strains are imposed on the pump casing. All drain lines, gland seal lines, recirculation lines, etc., should be checked to verify that they are in compliance with manufacturer's recommendations. After all these precautions have been taken, check the pump for proper rotation.

Heat-Recovery Equipment. Generally the heat-recovery equipment for a boiler consists of an air heater and/or an economizer. Simultaneous use of an air heater and an economizer on a boiler is rare; however, on some occasions use of both pieces of equipment can be economically justified.

Manufacturer's drawings and instructions should be followed when the inspection is made of the economizer and/or air heater. The inspection procedure is very similar for a tubular-type air heater and an economizer. Check for proper installation as well as breeching ductwork, arrangement, alignment, and expansion provisions. Make sure all internal baffling, tube arrangements, headers, tubesheets, etc., are installed in accordance with the manufacturer's recommendations. There should be no plugging or obstruction in the tubes and headers. In an economizer, check the feed piping to the inlet header and the piping between the economizer and the boiler. Shut off and check the valves, isolating valves, and relief valves. Test connections, thermometers, and pressure gauges should all be checked and inspected.

If the air heater is a regenerative type, only an operator who is completely familiar with such equipment can inspect and evaluate the appropriateness and condition of the assembly.

The air heater seals should be checked for proper clearance, expansion provisions, and clearances. The drive motor and gear-reducing unit must be lubricated and the rotation checked. All ducts, breeching, dampers, and expansion joints should be checked against installation drawings. Any debris, welding rods, rags, etc., should be removed.

28.2 STARTUP

After the initial inspection has been made, the new boiler is ready for startup. Startup time is a serious period. It can be the most critical period in the life of the boiler. However, it need not be, if proper planning and training have preceded this period. The manufacturer's instruction manual should be consulted and the contents thoroughly understood by all parties involved in the initial startup and eventual operation of the boiler(s). Owners and operating and maintenance personnel should be present during the initial startup period.

28.2.1 Checks

No attempt should be made to initially fire the boiler until the following points have been checked:

1. Feedwater supply must be ensured. The boiler should be filled to operating level with 70°F (21°C) or warmer water.
2. Steam lines must be hooked up and ready for operation. If the boiler is equipped with a superheater, then a proper vent should be provided to vent the superheater during boilout.
3. Blowoff and blowdown lines must be checked and ready for operation. Check all piping for adequate support and expansion provisions.
4. All fuel supply lines should be checked out. Make sure that the piping is correct. Check lines for tightness and leaks. Strainers are not normally supplied as part of the package (Cleaver-Brooks supplies strainers as standard on the oil supply line), but these are most important to the safe operation of gas- and oil-fired units. If the boiler is fired with coal, then the coal handling and burning equipment must be checked to make sure an adequate supply of fuel will be available.
5. The electric power supply should be reviewed. The power lines to the boiler-burner unit should be connected and ready for operation. Checking of the burner and controls is covered in Sec. 28.1.
6. Any walkways, platforms, stairs, and/or ladders needed to permit proper access to the boiler should be installed and ready for use.

28.2.2 Boilout

Once all the foregoing points have been checked, the next step is to prepare for the boilout. The newly installed boiler must be boiled out because its internal surfaces are invariably fouled with oil, grease, and/or other protective coatings. Boilout will also remove remaining mill scale and rust, welding flux, and other foreign matter normally incident to fabrication and erection.

Existing boilers which have had any tube replacement, rerolling, or other extensive repairs to the pressure parts should be boiled out. The lubricant used for rolling tubes plus the protective coating on new tubes must be removed by boiling out before the repaired boiler can be put back on line.

The boilout chemicals added to the water create a highly caustic solution which, upon heating, dissolves the oils and greases and takes them into solution. Through a series of blowoffs, the concentration of the solution is reduced. After a period of boiling and blowing down, the concentration is diluted enough that practically all the oils and greases and other matter have been eliminated. The boiler manufacturer's recommendations and requirements for boilout should be strictly adhered to.

28.3 ROUTINE OPERATION

Routine operation of a boiler room and all its associated equipment can become boring. Day by day, week in and week out, the same monotonous chores are performed over and over, but diligence must be maintained. It is dangerous to become lackadaisical and let these duties slide.

Safe and reliable operation depends to a large extent on the skill and attentiveness of the operators. Operating skill implies the following:

- Knowledge of fundamentals
- Familiarity with equipment
- Suitable background or training
- Operator diligence in performing these routine functions

28.3.1 Boiler Room Log

During operation certain procedures can and should be performed by the operating personnel. Actually the best way to keep track of these procedures and to help remind the operators to perform these functions is to keep a boiler room log.

In the interest of accident prevention and safe operating and maintenance practices, the Hartford Steam Boiler Inspection and Insurance Company and Cleaver-Brooks offer, without cost or obligation, copies of the boiler room log forms. Consult a Cleaver-Brooks agent or write to Hartford Steam Boiler Inspection and Insurance Company, 56 Prospect Street, Hartford, Connecticut 06102.

The Hartford Steam Boiler Inspection and Insurance Company, in their Engineering Bulletin No. 70, *Boiler Log Program*, explains why a log is desirable and cites the various tests to be performed on the equipment or apparatus. This bulletin recommends both the various tests to be performed and the frequency of log entries as follows:

Low-pressure steam and hot-water heating boiler logs	Weekly readings
High-pressure power boiler log for relatively small plants that do not have operators in constant attendance	Twice daily readings with more frequent checking during operation
High-pressure power boilers, larger units in large plants	Hourly readings, the preferred frequency for all high-pressure boilers

28.3.2 Safety and Relief Valves

Another extremely important safety check to be performed is the testing of all safety and relief valves. How frequently this should be done is an extremely controversial subject. Everyone seems to agree that safety valves and relief valves should be checked or tested periodically, but very few agree on how often.

The American Society of Mechanical Engineers' (ASME) book *Recommended Rules for Care and Operation of Heating Boilers*, in section VI, states that the safety or relief valves on steam or hot-water heating boilers should be tested every 30 days. A try-lever test should be performed every 30 days that the boiler is in operation or after any period of inactivity. With the boiler under a minimum of 5 lb/in^2 (34.5-kPa) pressure, lift the try lever on the safety valve to the wide-open position and allow steam to be discharged for 5 to 10 s (on hot-water boilers hold open for at least 5 s or until clear water is discharged). Release the try lever, and allow the spring to snap the disk to the closed position.

28.3.3 Burner and Controls

During normal operation, maintenance of the burner and controls is extremely important. If the boilers are firing continuously, most of the maintenance and repairs that can be done under these conditions is of a minor nature.

Spare oil guns should be kept cleaned and ready to be changed as necessary. Check all the valves in the fuel lines, feedwater, and steam lines for leaks and packing gland conditions. Repair, repack, and tighten where needed.

Check out the burner linkage, jackshafts, drive units, cams, etc. Make sure that all linkage, linkage arms, and connections are tight. Lubricate or grease where required. Note any wear or sloppiness in any of this equipment. If something cannot be repaired while the boiler is in operation, then this fact should be noted in the log so that proper repairs can be made when the boiler is taken out of service.

Operation of the burner should be observed. Is the burner firing properly? Check the flame shape and combustion. If the burner is not operating properly, have a qualified serviceperson work on the burner and put everything back into proper condition.

28.3.4 Feedwater Treatment

Improper water treatment is probably the largest direct cause of boiler failure. When the boiler feedwater and the boiler water do not receive proper preparation and treatment, scale and sludge deposits, corrosion, and pitting of the boiler surface result.

The boiler owner and operator must know that proper boiler feedwater treatment is an absolute necessity. If the boiler does not receive water of proper quality, the boiler's life will be needlessly shortened. (For detailed discussions of water treatment for boilers, see Chap. 53 on water conditioning.)

28.3.5 Boiler Blowdown

When a boiler is generating steam, the feedwater continuously carries dissolved minerals into the unit. This material remains in the drum, causing an increase in the total solids until some limit is reached beyond which operation is unsatisfactory. Many difficulties in boiler operation will occur because of excessive concentrations of sludge, silica, alkalinity, chlorides, or total dissolved solids.

When these constituents have reached a maximum limit (see Table 28.1), some of the water must be removed from the boiler and replaced with feedwater with a lower solids content.

The amount of blowdown or number of times the boiler is blown each day depends on the concentration of solids in the boiler water. Recommendations of the feedwater consultant should be followed regarding blowdown procedures.

TABLE 28.1 Specifications for Water Conditions

Pressure at outlet of steam-generating unit, lb/in² (bar)		Total solids, ppm (or mg/L)	Total alkalinity, ppm (or mg/L)	Suspended solids, ppm (or mg/L)
0–50	(0–3.4)	2500	500	150
51–300	(3.5–20.7)	3500	700	300
301–450	(20.8–31)	3000	600	250
451–600	(30.1–41.4)	2500	500	150
601–750	(41.5–51.7)	2000	400	100
751–900	(51.8–62)	1500	300	60
901–1000	(62.1–68.9)	1200	250	40
1001–1500	(69–103.4)	1250	250	20
1501–2000	(103.5–137.9)	750	150	10
2001 +	138 +	500	100	5

28.4 SHUTTING DOWN

28.4.1 Short Shutdowns

When a routine shutdown is scheduled, it should be planned so that there is time to perform certain operations in the shutdown procedure.

1. If the boiler is equipped with sootblowers and if any fuel other than natural gas has been fired, all the sootblowers should be operated before the boiler is taken off line.

All the recommended rules for operating the sootblowers should be followed. One of the most important is that the steam load on the boiler should be 50 percent of boiler rating or greater.

2. After the soot-blowing operation has been completed, the steam flow should be gradually reduced and the burner set to the low-fire position.

3. With the burner in the low-fire position, blow down the boiler along with the water column, gauge glass, and feedwater regulator. Turn off the burner in accordance with the burner manufacturer's instructions. If the boiler is equipped with a flue gas outlet damper, it should be fully closed, to allow the unit to cool slowly.

4. Remove and clean the burner oil gun. Place fuel supply equipment in standby condition (for gas, shut the main supply cock). Open the main electric switch, and take the feedwater regulator out of service. Hand-operate feedwater valves to keep the water level above one-half gauge glass.

5. When the boiler pressure falls below the line pressure, the boiler stop valve should be closed if the setting has cooled enough to prevent any pressure buildup. If the boiler is equipped with a nonreturn valve, this valve should close automatically when the boiler pressure drops below the line pressure. This, of course, isolates the boiler from other units remaining in service. As the drum pressure falls below 15 lb/in² (1 bar), the manual closing device (handwheel) of the nonreturn valve (if equipped) should be closed and the top drum vent opened. This prevents a vacuum on the boiler water side which will loosen well-set gaskets and cause future problems. While there is still a small amount of steam pressure available, the boiler should be blown down and filled back to a safe level with freshly treated hot water in preparation for the next startup.

If the boiler is being shut down just overnight or for the weekend, the previous procedure is generally all that is required. The primary concern is to ensure sufficient water in the boiler. If the boiler is going to be shut down only overnight, probably the boiler can be secured and will have pressure still showing the next morning.

28.4.2 Prolonged Shutdowns

Up to this point we have been discussing shutting down the boiler for just a few hours or a few days. If the boiler is going to be taken out of service and is expected to be out of service for several weeks or several months, then a different procedure must be followed.

There are two basic methods of laying up a boiler for protracted outages: wet storage and dry storage.

Wet Storage Method. If the unit is to be stored for not longer than a month and emergency service is required, wet storage is satisfactory. This method is not generally employed for reheaters or for boilers which may be subjected to freezing temperatures. Several alternative methods have been used.

1. The clean, empty boiler should be closed and filled to the top with water that has been conditioned chemically to minimize corrosion during standby. Water pressure greater than atmospheric should be maintained within the boiler during the storage period. A head tank may be connected to the highest vent of the boiler to keep the pressure above atmospheric.
2. Alternatively, the boiler may be stored with water at the normal operating level in the drum and nitrogen maintained at greater than atmospheric pressure in all vapor spaces. To prevent in-leakage of air, it is necessary to supply nitrogen at the vents before the boiler pressure falls to zero, as the boiler is coming off the line. If the boiler pressure falls to zero, the boiler should be fired to reestablish pressure and drums and superheaters should be thoroughly vented to remove the air before nitrogen is admitted. All partly filled steam drums and superheater and reheater headers should be connected in parallel to the nitrogen supply. If nitrogen is supplied to only the steam drum, the nitrogen pressure should be greater than the hydrostatic head of the longest vertical column of condensate that could be produced in the superheater.
3. Rather than maintain the water in the boiler at the normal operating level with a nitrogen cap, it is sometimes preferred to drain the boiler completely, applying nitrogen continuously during the draining operation and maintaining a nitrogen pressure greater than atmospheric throughout the draining and subsequent storage.

Dry Storage Method. This procedure is preferable for boilers out of service for extended periods or in locations where freezing temperatures may be expected during standby. It is generally preferable for reheaters.

1. The cleaned boiler should be thoroughly dried, since any moisture left on the metal surface will cause corrosion after long standing. After drying, precautions should be taken to preclude entry of moisture in any form from steam lines, feed lines, or air.

2. For this purpose, moisture-absorbing material may be placed on trays inside the drums or inside the shell. The manholes should then be closed, and all connections on the boiler should be tightly blanked. The effectiveness of the materials for such purposes and the need for their renewal may be determined through regular internal boiler inspections.

When boilers have been kept in dry storage with any of the aforementioned materials in the fireside and waterside sections, serious damage can be done to the boiler if these materials are not removed before the boiler is filled with water and fired.

We strongly recommend that large signs be placed in conspicuous places around the boiler to indicate the presence of moisture-absorbing materials. These signs could read similarly to the following:

IMPORTANT—Moisture-absorption material has been placed in the waterside and furnace areas of this boiler. This material must be removed before any water is placed in unit and before boiler is fired.

For long periods of storage, inspect every 2 or 3 months and replace with fresh and/or regenerated materials.

3. Alternatively, air dried external to the boiler may be circulated through it. The distribution should be carefully checked to be sure that the air flows over all areas.

4. If the boilers are to be stored in any place other than a dry, warm, protected atmosphere, steps should be taken to protect the exterior components also. Burner components that are subject to rust—jackshaft, linkage, valve stems, moving parts, etc.—should be coated with a rust inhibitor and covered to protect them from moisture and condensation. Electric equipment, electronic controls, relays, switches, etc., should be protected.

5. Pneumatic controls, regulators, and diaphragm- or piston-operated equipment should be drained or unloaded and protected so that moisture, condensation, rust, etc., will not damage it during a long period of storage. Feedwater lines, blowdown, the soot blower, drain lines, etc., should all be drained and dried out. Valve stems, solenoid valves, and the diaphragm should be protected by lubricant, rust inhibitors, plastic coverings, or sealants.

28.5 AUXILIARY EQUIPMENT

The term "auxiliaries" can be defined as all the attached or connected equipment, used in conjunction with a boiler, which should be operated and maintained in accordance with the regulations established by the equipment manufacturers. Adherence to these rules will help protect the operators from injury and prevent damage to the boiler being serviced and to the auxiliary equipment.

28.5.1 Deaerators

When a tray-type deaerator is involved, it can be cleaned with a light muriatic acid solution. The trays can be removed and brushed with the muriatic acid solution until all the scale deposit has dissolved.

Note: The operator should wear protective clothing, goggles, rubber gloves, and a rubber apron when using the acid solution.

After the trays have been cleaned, they should be thoroughly rinsed with water several times to remove all traces of the acid.

The packed column type of deaerator may or may not be cleaned with an acid solution depending on the type of packing in the column.

The spray-type deaerator will generally have a spray valve or spray pipe. With the spray valve it may be necessary to check the valve seat for wear or pitting. The spring compression dimension should be checked and adjusted to the manufacturer's recommendations. If the deaerator has a spray pipe, then the pipe should be inspected to see that all the spray holes are open. If any holes are plugged, they should be opened.

28.5.2 Strainers

Strainers in all lines should be cleaned at regular intervals determined by conditions and use; water gauge glasses should be kept clean and replaced if necessary. Tighten all packing glands and fittings to eliminate steam and water leaks. The overflow valve and makeup valve, with their associated level control mechanisms, should be checked. Refer to the appropriate manufacturer's literature for service recommendations of particular components.

28.5.3 Air Heaters

Tubular air heaters have no moving parts, and therefore one might assume that this type of air heater would need very little maintenance. This assumption would be correct if natural gas were the only fuel being fired. With natural gas or No. 2 fuel firing and a proper air-fuel ratio, the products of combustion are fairly clean. The heat-transfer surfaces will remain fairly clean and will normally not require any cleaning.

The air side of the air heater should be inspected to see that the baffles are all secure and in place. Sometimes air flow through these areas may cause a baffle to vibrate and break loose. Proper repairs should be made and the baffles welded or clamped in place.

Check for signs of corrosion in the low-temperature end. If the temperatures in the low-temperature end are below the dew point, the flue gases will condense and corrode the metal. Installations that are firing coal or heavy oil with high sulfur content are particularly subject to corrosion in this area. The higher the sulfur content of the fuel, the higher the dew point of the flue gases.

Where coal or residual oil is the main fuel, the air heater tubes become plugged with fly ash and soot. When this happens, brush or steam-clean the tubes.

Maintenance on a regenerative air preheater involves checking the radial and circumferential seals. The seals should be inspected for wear and proper clearance. If any of the seals are damaged or so worn that they cannot be adjusted for proper clearance, they should be replaced. Manufacturer's recommendations should be consulted regarding making the proper clearance adjustments.

28.5.4 Economizer

In general, the economizer is subject to the same types of troubles as the air heater. Corrosion in the low-temperature ends due to condensation of gases and erosion due to action of the flash or soot deposits are the more common problems.

28.5.5 Fuel Systems

For any boiler to be able to operate at peak efficiency and as free of trouble as possible, the fuel-handling system must be designed properly and must be kept in top condition at all times.

Full and effective use should be made of manufacturer's instruction books on operation and maintenance. Of special importance are written procedures prepared expressly for each installation. Fuel-burning equipment should never be operated below the safe minimum level at which a stable furnace condition can be maintained. In like manner, operation should not be tolerated at rates of fuel input which are excessive in relation to the available air supply.

This chapter cannot cover the size of industrial and utility boilers that normally use pulverized fuel. Those interested in obtaining further information on this type of equipment should consult the *ASME Boiler and Pressure Vessel Code*, section VII, "Recommended Rules for Care of Power Boilers." Two other books that cover the subject are *Steam—Its Generation and Use*, published by Babcock and Wilcox Co., and *Combustion Engineering*, published by Combustion Engineering, Inc. (see publication information addresses of companies listed in Secs. 28.7 and 28.8).

28.6 ABBREVIATIONS OF ORGANIZATIONS

ABMA	American Boiler Manufacturers Association
AGA	American Gas Association
AMCA	Air Moving and Conditioning Association (fans)
ANSI	American National Standards Institute
API	American Petroleum Institute
ASHRAE	American Society of Heating, Refrigeration, and Air-Conditioning Engineers
ASME	American Society of Mechanical Engineers
ASTM	American Society for Testing and Materials
AWS	American Welding Society
CSA	Canadian Standards Association
HI	Hydronics Institute
HSB	Hartford Steam Boiler Inspection and Insurance Company
IEEE	Institute of Electrical and Electronic Engineers
NBS	National Bureau of Standards

NEMA	National Electrical Manufacturers Association
NFPA	National Fire Protection Association
SBI	Steel Boiler Industry (an affiliate of HI)
SMA	Stoker Manufacturers Association
UL	Underwriters' Laboratories, Inc.

28.7 ACKNOWLEDGMENTS

The material contained in this chapter represents countless hours of preparation, including reading several books and sources of reference material. We wish to acknowledge those who contributed some of the information contained in this chapter. In addition, we wish to acknowledge the many Cleaver-Brooks employees for their knowledge, expertise, and dedication to this project.

American Boilers Manufacturer's Association (ABMA)
950 N. Glebe Road, Suite 160
Arlington, VA 22203

American Society of Mechanical Engineers (ASME)
United Engineering Center
345 E. 47th Street
New York, NY 10017

Babcock and Wilcox Company
161 E. 42d Street
New York, NY 10017

Combustion Engineering, Inc.
Windsor, CT 06095

Factory Insurance Association
85 Woodland Street
Hartford, CT 05102

Factory Mutual Insurance Association
1151 Boston-Providence Turnpike
P.O. Box 688
Norwood, MA 02002

Hydronics Institute
35 Russo Place
Berkeley Heights, NJ 07922

28.8 BIBLIOGRAPHY

ABMA: *ABMA Lexicon*, 4th ed., ABMA, Arlington, VA, 1976.

ASME: *ASME Boiler and Pressure Vessel Code* (Section I, "Power Boilers"; Section IV, "Heating Boilers"; Section VIII, "Pressure Vessels"), ASME, New York, 1972.

Babcock and Wilcox: *Steam: Its Generation and Use*, The Babcock and Wilcox Company, New York, 1972.

Cleaver-Brooks: *Cleaver-Brooks Packaged Firetube Boiler Engineering Manual (Q38)*, Cleaver-Brooks, Milwaukee, WI, 1986.

Fryling, Glenn R. (ed.): *Combustion Engineering: A Reference Book on Fuel Burning and Steam Generation*, Combustion Engineering, Inc., New York, 1967.

CHAPTER 29
ELECTRIC BOILERS

Robert G. Reid
Quality Assurance Manager, CAM Industries, Inc.,
Kent, Washington

29.1 ELECTRIC BOILER APPLICATIONS

Electric boilers are compact, clean, and very efficient sources of hot water or steam. Available in ratings from 5 to over 70,000 kW, they can provide sufficient heat for any heating, ventilating, and air-conditioning (HVAC) requirement for any application from humidification to a primary heat source.

No chimneys or vents, no combustion air, and no fuel storage areas are required. Electric boilers can be located in the building anywhere that heat or steam is needed or space is available.

Electric boilers should be considered in any project in which air-conditioning equipment is a major electric load and in which the use of an electric boiler as a heat source occurs only when the air-conditioning load is greatly reduced but has already created an electric demand charge. The boiler power consumption can be controlled so as not to create any new demand.

29.2 ELECTRIC BOILER CLASSIFICATIONS

Electric boilers are usually classified as either resistance or electrode type with further distinctions as high or low voltage and steam or hot water.

29.2.1 Resistance Boilers

Resistance-type boilers use metal-sheathed resistance elements which are submerged in the boiler water. The water is heated by convection from the surface of the sheaths of the heating elements. The electric current flows through a wire centered inside, but insulated from, the element sheath. The element sheath has no electric potential.

29.2.2 Electrode Boilers

Electrode boilers use the water as a resistor, and the electrodes only provide an electric contact with the water.

29.2.3 Voltage Classifications

Voltages on which boilers are designed to operate are classified as low, 600 V or less; or high, greater than 600 V. Resistance-type boilers are always low-voltage. Electrode boilers are available in both low- and high-voltage designs. Both resistance and electrode boilers are available in designs to generate either steam or hot water.

29.3 RESISTANCE BOILERS

Resistance boilers consist of a pressure vessel and resistance-type heating elements of sufficient number and rating to obtain the desired total boiler kilowatt rating. The boilers have controls to regulate the power needed by the steam or hot-water demand, a low-water cutoff, and other items as appropriate to the boiler's design (steam or hot water).

29.3.1 Resistance Steam Boilers

Resistance steam boilers (see Fig. 29.1) have vessels sized to provide a steam-to-water ratio of about 40:60 by volume. A feedwater regulator is provided to control a feed pump, solenoid valve, or float-operated valve to maintain the water at the desired elevation. A water-level gauge (sight glass) permits a visual check of the water level. A low-water cutoff is positioned to stop the power flow to the heating elements before the water level drops below the highest part of the heating elements.

Controls. The boiler steam output is controlled by a pressure-actuated switch or controller acting to energize or de-energize the contactors which control the electric power to the heating elements.

On/off controls are typical on steam boilers of 307,170 Btu/h (90 kW) or less. Larger steam boilers are typically provided with proportional control. All steam boilers should have a high-pressure-limit cutoff. Steam pressure gauges are required and must have a dial range 1.5 to 2 times the boiler safety valve setting.

Safety Valves. Safety valves must have a steam-relieving capacity of at least 3.5 lb/h (1.59 kg/h) for each kilowatt of capacity for which the boiler vessel is rated. The pressure setting of the valve cannot exceed the allowable working pressure marked on the boiler vessel. Two or more valves may be required to provide the required capacity.

Vessel Design Codes. Steam boiler vessels must be designed and fabricated in accordance with requirements of the boiler codes applicable where the boiler is to be installed. In the United States and Canada, the ASME (American Society of

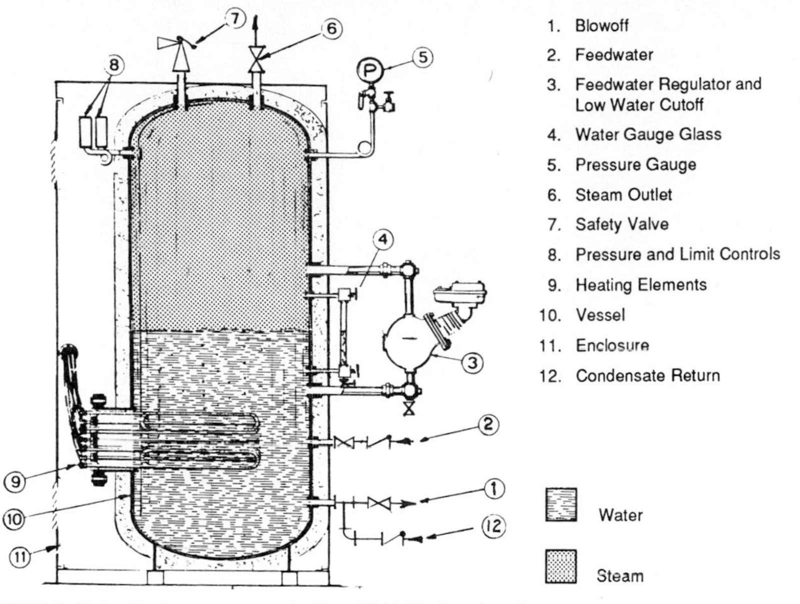

FIGURE 29.1 Resistance steam boiler. *(CAM Industries, Inc.)*

Mechanical Engineers') *Boiler and Pressure Vessel Code* [section I for greater than 15 lb/in^2 (103 kPa), or section IV for less than 15 lb/in^2 (103 kPa)] would apply.

Ratings Available. Resistance-type steam boilers are available in ratings from 5 to over 2500 kW at voltages from 208 to 600 V. Pressure ratings typically offered by manufacturers are 15 lb/in^2 (103 kPa) for low-pressure heating applications, 100 lb/in^2 (690 kPa) for "miniature" boilers (usually less than 100 kW), and 125 lb/in^2 (862 kPa) for boilers over 100 kW. Pressure ratings up to and over 2000 lb/in^2 (13,793 kPa) are available.

Selection. In selecting a resistance-type steam boiler, one must consider the steam output required, feedwater temperature, steam pressure needed, safety valve setting, feedwater supply pressure, rating of required electric supply circuit(s), and type of pressure control required or desired. Additionally, the feedwater quality and pretreatment required and the amount and disposal of boiler blowdown must be considered. Whether condensate can be recovered economically is also a factor.

29.3.2 Resistance Hot-Water Boilers

Boilers for hot water are similar to but smaller than steam boilers (for the same kilowatts). Controls consist of at least a low-water cutoff and temperature control. Figure 29.2 shows a typical resistance-type hot-water boiler.

Residential Boilers. Hot-water boilers used for residential heating are usually for single-phase 120/240-V supply and provide for energizing the heating elements in

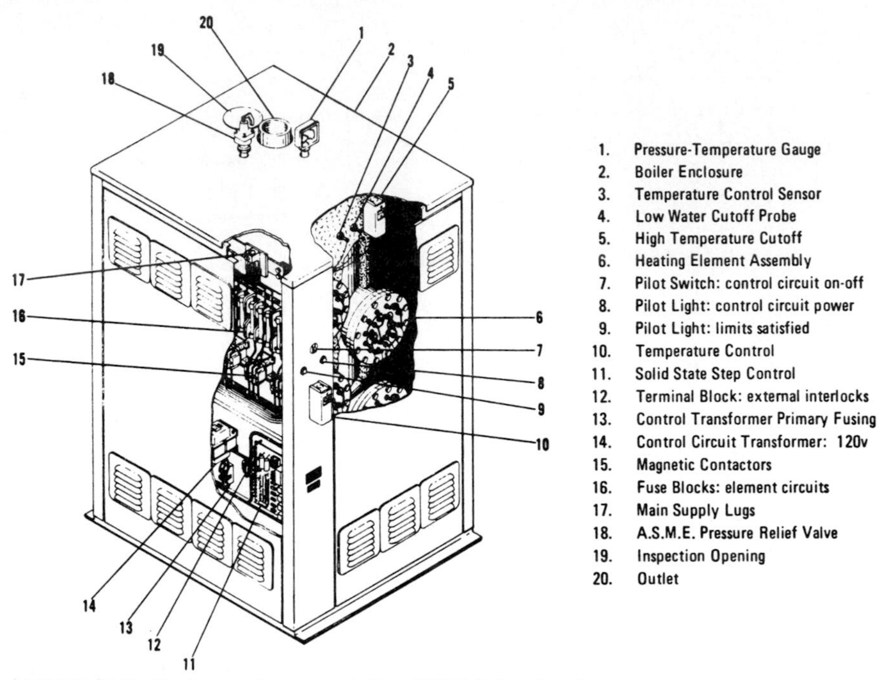

FIGURE 29.2 Resistance hot-water boiler. *(CAM Industries, Inc.)*

increments of 3 to 6 kW with a short (5- or 10-s) delay between increments. Such boilers are usually not over 30 kW.

Commercial and Industrial Boilers. Boilers designed for commercial and industrial installations usually subdivide the elements into groups of 15 to 36 kW and control the boiler in increments only when the total power exceeds 50 to 60 kW.

Controls. Temperature control systems available either are on/off or are proportional, using step controls. Controls can usually be designed to interface with building control systems to provide for outdoor reset or energy management system control.

Relief Valves. Pressure relief valves for hot-water boilers must have a relieving capacity equal to or greater than the boiler's rating in Btu/hour and a steam-relieving capacity of at least 3.5 lb/(h · kW) [1.59 kg/(h · kW).

Vessel Codes. Pressure vessels for these boilers usually are designed and fabricated to *ASME Boiler and Pressure Vessel Code*, section IV. Boilers for design temperatures above 250°F (121°C) and/or pressures above 160 lb/in^2 (1103 kPa) are designed to *ASME Boiler and Pressure Vessel Code*, section I. Maximum operating temperatures and pressures would be about 90 percent of design values.

Selection Considerations. Selection of an electric hot-water boiler should reflect the required heat output, flow rate, and temperature rise; pressure drop through the boiler (see Fig. 29.10); system pressure; supply voltage; ratings of required electric supply circuit(s); and type of control required or desired.

29.4 ELECTRODE BOILERS

Electrode boilers use water as a resistor. Protection against low-water burnout is inherent. They are generally much more compact than resistance-type boilers of equivalent kilowatt and voltage ratings. Only electrode boilers can be designed to operate on high voltages (above 600 V). This makes it possible and feasible to attain boiler ratings of 50,000 kW or more and to use voltages up to 25,000 V. Electrode boilers require properly conditioned water for successful operation. Water-quality requirements differ for each boiler design, voltage, and kilowatt rating. Water-quality requirements are critical. The treatment which may be needed to make the available water suitable for the boiler should be determined as part of the boiler selection process. Table 29.1 lists the water-quality limits typical for electrode boilers.

Controls for Electrode Boilers. Steam pressure and current flow (or kilowatts) are the variables used to control these boilers. Pressure is the primary variable with current flow being used as a feedback signal and as an override control variable, to limit the boiler power consumption to its rating or to some lower value.

Back-Pressure Valves (Steam Boilers). Because these boilers require steam pressure to force water out of the electrode chamber, a back-pressure valve is usually needed on the boiler steam outlet to ensure that the steam pressure will not drop below that required to operate the boiler. On high-voltage boilers, the back-pressure valve is sometimes used to control the boiler pressure and to isolate the boiler from steam line pressure variations caused by rapid fluctuations in the rate of steam consumption.

Electric Supply Systems. Electrode boilers are usually designed to operate on three-phase, four-wire electric systems having a grounded neutral. The boiler manufacturer should be consulted if an electric supply of a different type is to be used.

Vessel Codes. Boiler vessels for electrode boilers are designed and fabricated to requirements of the *ASME Boiler and Pressure Vessel Code*, section I or IV, as applicable for the operating temperatures and pressures required.

29.4.1 Electrode Boiler Types

There are three basic designs according to the control method used:

1. Totally immersed electrode with insulating-sleeve control
2. Variable-immersion electrode, water-level control
3. Jet or spray type with variable flow to electrodes

Totally Immersed Electrode Boilers. Totally immersed electrode boilers are available for both hot-water and steam applications and for both low and high volt-

TABLE 29.1 Typical Recommended Water-Quality Limits for Electric Boilers

Boiler type and water classification	Hardness, ppm maximum	pH range	p alkalinity, ppm	Iron, ppm maximum	Conductivity, MMho	TDS,† ppm maximum	Oxygen, ppm maximum	Notes
I. Resistance								
A. Steam								
1. Feedwater	5	≥7.5	*	*	*	*	0.05	
2. Boiler water	0	7.5–10.5	500	3	7000	3500	0.005	
B. Hot water								
1. Makeup water	5	≥7.5	*	*	*	*	0.05	
2. Boiler water	5	8.5–10.5	500	5	7000	3500	0.005	
II. Electrode								
A. Low-voltage								
1. Steam								
a. Boiler water	0	8.5–10.5	400	3	1500	750	0.005	‡
b. Feedwater	0.5	≥7.5	*	0.1	100	50	0.05	
2. Hot water								
a. Boiler water	5	8.5–11	400	2.5	1000	500	0.005	‡
b. Makeup Wtr	5	≥7.5	*	0.5	*	*	0.05	

B. High-voltage								
1. Totally immersed electrode								
a. Steam								
(1) Feedwater	1.0	≥7.5	*	*	*	*	0.05	
(2) Bolier water	0.1	8.5–11	125	0.5	<500	<250	0.005	‡
b. Hot water								
(1) Makeup	1.0	≥7.5	*	*	*	*	0.05	
(2) Boiler water	1.0	8.5–10	125	0.5	<500	<250	0.005	‡
2. Variable-immersion electrode								
a. Steam								
(1) Feedwater	0.1	≥7.5	*	*	*	*	0.05	
(2) Boiler water	0	8.5–10	125	0.5	<200	<100	0.005	‡
3. Jet type								
a. Steam								
(1) Feedwater	1	≥7.5	*	*	*	*	0.05	
(2) Boiler water	0.1	8.5–12	400	2	<3600	<2000	0.005	§

*Limit of this variable in feedwater is determined by boiler water limit and acceptable blowdown rate.
†TDS is total dissolved solids.
‡Conductivity limits vary with voltage and water temperature; consult the boiler manufacturer.
§Water must be nonfoaming.
Source: CAM Industries, Inc.

ages. Kilowatt ratings are determined by the voltage, number, and size of the electrodes; the conductivity of the boiler water at the design operating temperature; and the water circulation rate past the electrodes. Figure 29.3 shows a low-voltage hot-water boiler, Fig. 29.4 shows a high-voltage hot-water boiler, and Fig. 29.5 shows a high-voltage steam boiler.

Controls. Control systems for totally immersed electrode boilers employ a movable, insulating sleeve or shield which is interposed between the energized, cylindrical electrode and a concentric, metallic neutral assembly to which the electric current flows. The position of the insulating sleeve regulates the amount of current which will flow through the water and therefore the power consumed by the boiler. The position of the insulating sleeve is regulated by a temperature or pressure control system which matches the boiler output to the system load requirements.

Turndown Ratio. These boilers can be turned down to 5 to 10 percent of their rated maximum output. Further output reductions require that power to the boiler electrodes be interrupted by a contactor or circuit breaker rated for the anticipated frequency of operation.

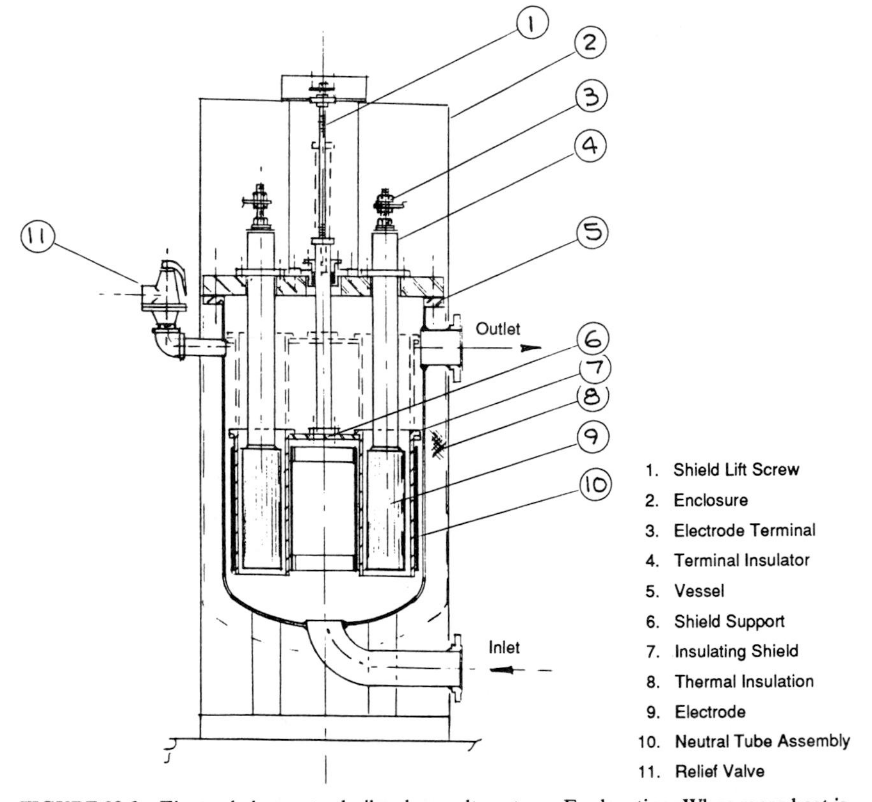

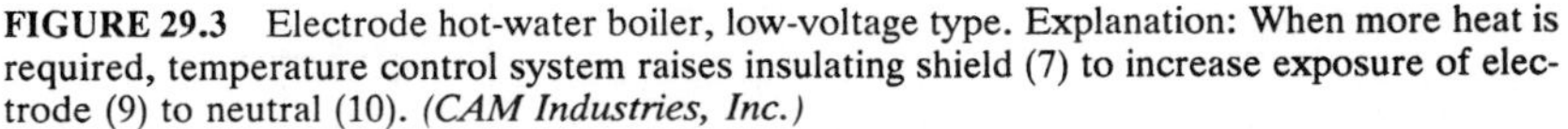
FIGURE 29.3 Electrode hot-water boiler, low-voltage type. Explanation: When more heat is required, temperature control system raises insulating shield (7) to increase exposure of electrode (9) to neutral (10). *(CAM Industries, Inc.)*

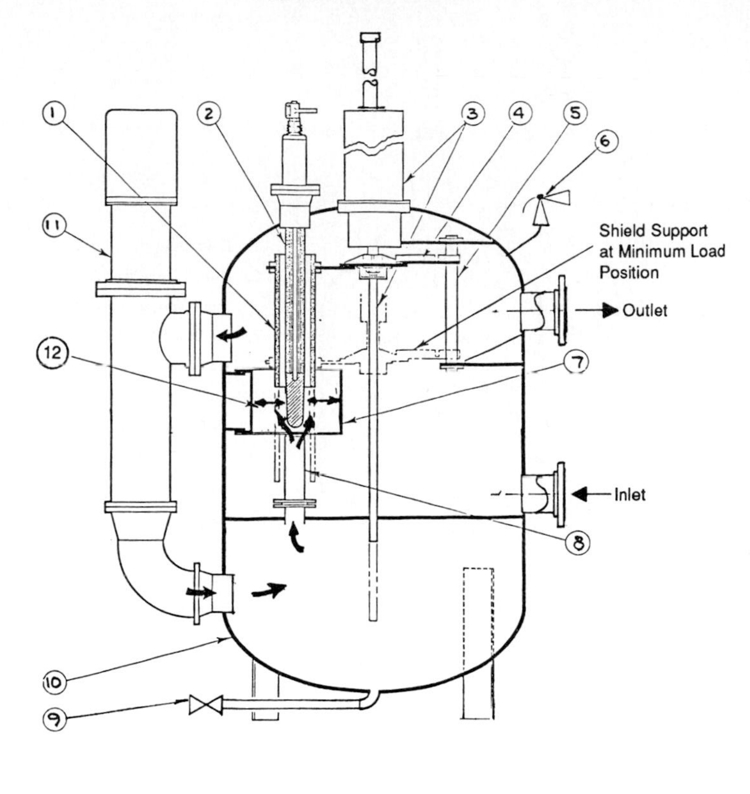

FIGURE 29.4 Electrode hot-water boiler, high-voltage type. Explanation: When heat is required, temperature control system causes insulating shields (1) to be lifted to expose electrodes (2) to neutral (7). The boiler circulating pump (11) operates continuously. System water is circulated by system pumps. *(CAM Industries, Inc.)*

Pressures and Ratings. High-voltage boilers of this design are well suited for use on voltages at the lower distribution ratings, that is, 2300 to 7200 V. Above 7200 V the design allows the use of higher-conductivity water than does the variable-immersion electrode boiler of the same voltage rating. Operating pressures are usually 75 lb/in^2 (517 kPa) or higher. Output ratings from 1 to 45 MW are available.

Variable-Immersion Electrode Boilers. Electrode boilers using a variable water level to change the depth of electrode immersion are used to generate steam. These boilers are available in both low- and high-voltage designs. The control

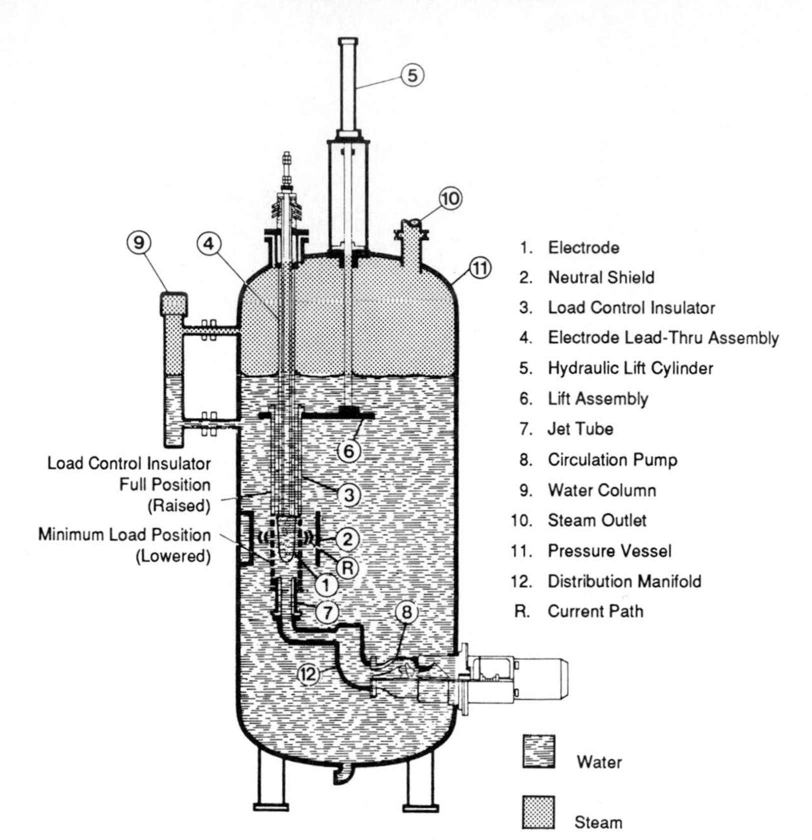

FIGURE 29.5 Electrode steam boiler with totally immersed electrode. Explanation: The rate of steam generation is controlled by raising or lowering the load control insulator (3) to change the exposure of the electrode (1) to the neutral shield (2). The boiler circulation pump (8) forces water past electrodes to carry steam bubbles away from electrodes. *(CAM Industries, Inc.)*

principle involved is that the flow of current through the water will be proportional to the wetted area of the electrodes, which is in turn controlled by the depth of electrode immersion in the water.

Control Systems. In all designs the electrode is fixed, and the water level is made to rise or fall as needed to develop the required steaming rate. Regulation of the water level is accomplished by transferring water between the chamber containing the electrodes and another holding chamber or vessel. The transfer is accomplished by causing a pressure difference to exist between the chambers. The two chambers can be in the same vessel, usually concentrically arranged, or in two or more vessels connected by piping.

Current flow in the low-voltage boilers of this type is from electrode to electrode. Figures 29.6 and 29.7 show the basic elements of the two systems as applied to low-voltage electrode steam boilers.

High-voltage electrode boilers using the variable-immersion principle are usually of single-vessel, two-chamber construction (see Fig. 29.8). Current flow is from electrode to neutral. Most high-voltage boilers of this type incorporate a cir-

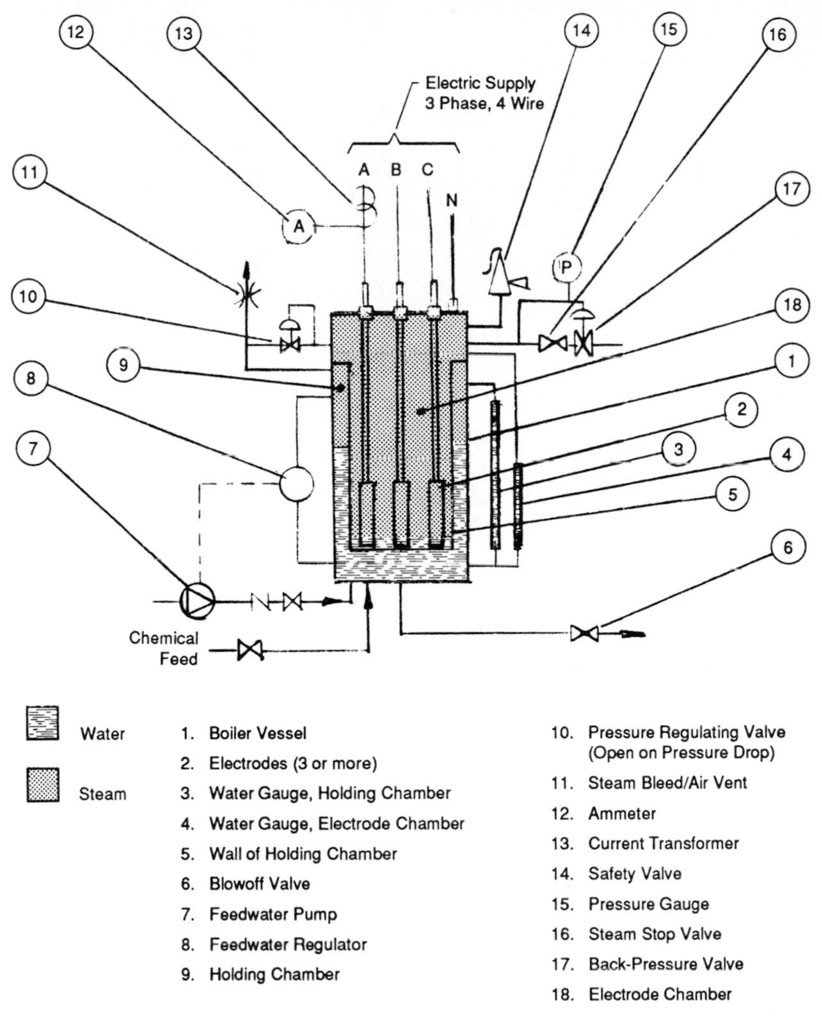

FIGURE 29.6 Single-vessel, low-voltage, electrode steam boiler. Boiler shown at high-pressure condition. Explanation: Water-level regulator (8) controls feed pump (7) to maintain a minimum water level in holding chamber (9). Pressure in top of holding chamber is controlled by pressure regulating valve (10) to be either equal to or less than the pressure in the electrode chamber (18). When electrode chamber pressure exceeds setting of valve (10), the valve throttles and pressure in holding chamber is reduced by bleeding through vent (11). Water from electrode chamber is displaced into holding chamber, reducing depth of electrode immersion and steam generated until pressure returns to set point. On pressure drop in electrode chamber, valve (10) opens to equalize pressure in both chambers and to increase depth of water on electrodes and steam output. *(CAM Industries, Inc.)*

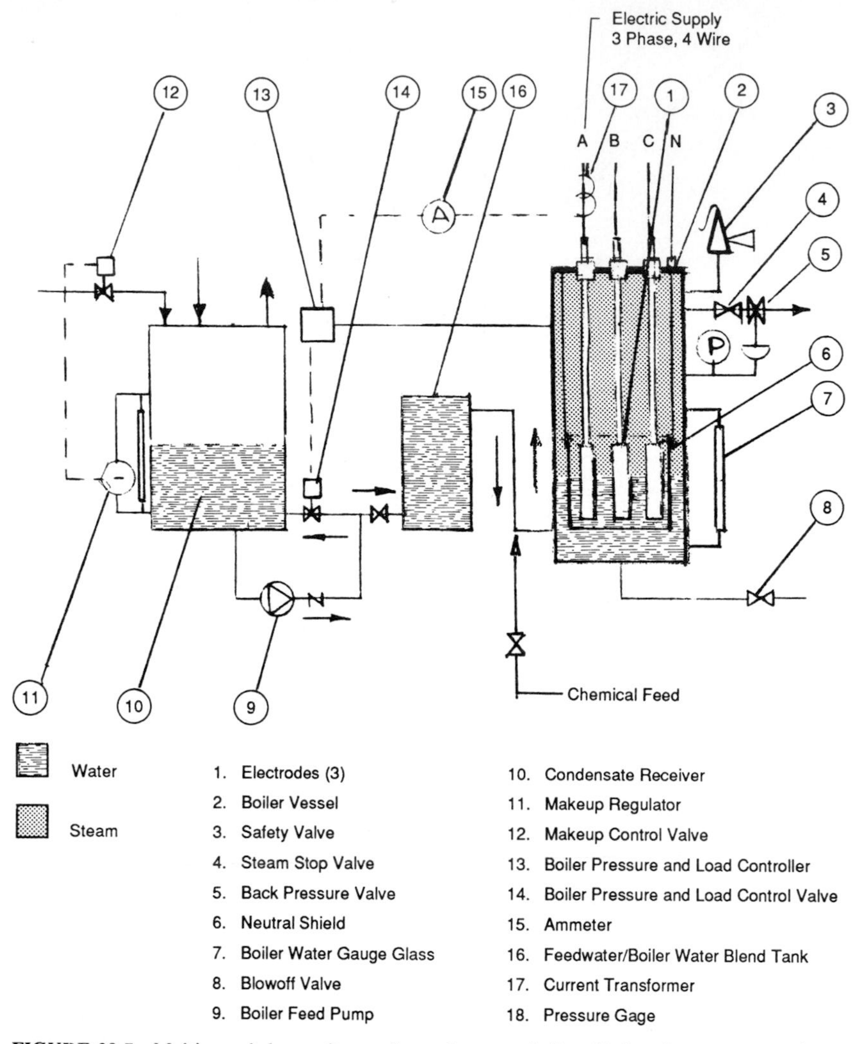

FIGURE 29.7 Multivessel, low-voltage, electrode steam boiler. Boiler shown at normal operating condition. Explanation: Pressure and load controller (13) controls boiler feed pump (9) to run continuously and causes valve (14) to bypass all feedwater in excess of steam generation rate back to condensate receiver (10). Water level in boiler is regulated to generate steam needed to maintain steam pressure or maximum kilowatts desired, as applicable. If steam demand decreases, water from boiler is displaced back to blend tank (16) and cold feedwater from blend tank is returned to condensate receiver (10). If no steam is needed, feed pump (9) stops. *(CAM Industries, Inc.)*

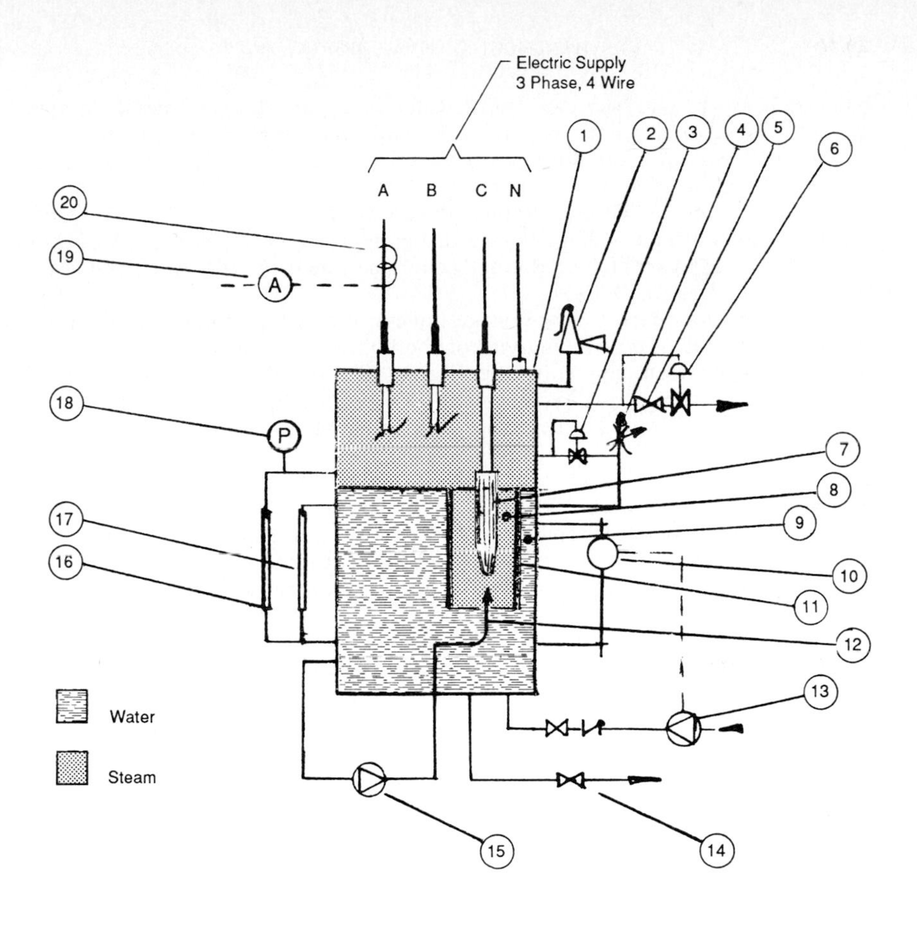

1. Boiler Vessel
2. Safety Valve(s)
3. Pressure Regulating Valve
4. Steam Bleed/Air Vent
5. Steam Stop Valve
6. Back-Pressure Valve
7. Electrodes (3)
8. Electrode Chamber (3)
9. Holding Chamber
10. Feedwater Regulator
11. Electrode Neutral (Concentric with Electrode) (3)
12. Water Circulation Outlet Pipe
13. Boiler Feed Pump
14. Blowoff Valve
15. Circulating Pump
16. Water Gauge, Electrode Chamber
17. Water Gauge, Holding Chamber
18. Pressure Gage
19. Ammeter
20. Current Transformer

FIGURE 29.8 Variable-immersion, single-vessel, high-voltage electrode steam boiler. Boiler shown in high-pressure operating condition. Explanation: Water-level regulator (10) controls feed pump (13) to maintain a minimum water level in holding chamber (9). Pressure in top of holding chamber is controlled by pressure-regulating valve (3) to be either equal to or less than pressure in electrode chamber (8). When electrode chamber pressure exceeds setting of valve (3), valve throttles and pressure in holding chamber is reduced by bleeding through vent (4). Water from electrode chamber is displaced into holding chamber (9), reducing depth of electrode immersion and rate of steam generation until pressure returns to set point. On pressure drop in electrode chamber, valve (3) opens to equalize pressure in both chambers and to increase depth of water on electrodes and steam output. Circulating pump (15) operates full-time to wash tips of electrodes via water circulation outlet pipe (12). *(CAM Industries, Inc.)*

culation pump to increase the rate of water flow past the electrodes. In some cases the pump is also used as part of the water-level regulation system.

Pressure and Kilowatt Ratings. Low-voltage boilers of this type are available for operation at a range of pressures from as low as 5 lb/in^2 (34 kPa) to as high as 1200 lb/in^2 (8275 kPa). Ratings cover the range of 15 to 2800 kW. Operating pressures of high-voltage boilers of this type are typically limited to about 75 lb/in^2 (517 kPa) or higher. Ratings cover the range of 2000 to 50,000 kW at voltages from 4.16 to 13.8 kV.

Water Requirements. Low-voltage boilers of this type require water which has been softened. The conductivity of the boiler water must be limited to the manufacturer's recommendation. See Table 29.1. High-voltage boilers of this type require very high-quality, low-conductivity feedwater. Demineralization is usually required. Feedwater flow should be modulated to suit steam flow rather than on/off.

Jet or Spray Electrode Steam Boilers. Jet or spray electrode boilers have electrodes located in the steam space above the boiler waterline. The boiler is fitted with one or more circulating pumps which draw water from the bottom of the boiler, forcing it up a central column and through nozzles directed at the electrodes. Each of the streams of water becomes a current path between the energized electrode and the neutral-potential nozzle header. The rate of circulation is many times the steaming rate. The excess water passes through the bottom of the electrodes and falls onto a perforated plate located a fixed distance below the electrode. This plate is also at neutral potential. A second current path is thus established.

Jet-type boilers are offered for high-voltage use only. The advantage of this boiler lies in the much greater electric resistance of the streams compared to that between immersed electrodes. As a result, the conductivity of the boiler water can be many times higher for jet-type boilers than for immersed-electrode designs.

Control Systems. Control of the boiler output is accomplished by varying the number of streams of water allowed to strike the electrode. The control occurs in one of two ways:

1. *Constant flow:* A regulating sleeve which is concentric with the nozzle header is raised or lowered. The sleeve shunts unwanted water streams back to the bottom of the boiler.
2. *Variable flow:* A valve is installed in the circulating pump discharge to regulate the water flow to the nozzle header.

The position of the regulating sleeve or flow control valve is controlled either to maintain the desired steam pressure or to generate steam at a preselected rate at constant kilowatts.

Jet-type boilers operate with a constant water level, and feedwater flow should be modulated at the same rate as the steam flow. High-water and low-water cutoffs and alarms should be provided as well as high-pressure cutoffs and alarms.

Conductivity controllers are required on all high-voltage electrode steam boilers and should be arranged to cause surface blowdown to occur when the conductivity is too high and to activate a chemical feed system when the conductivity is too low.

Water Requirements. The jet-type boiler was developed to allow the use of higher-conductivity water than other high-voltage designs. This reduces the

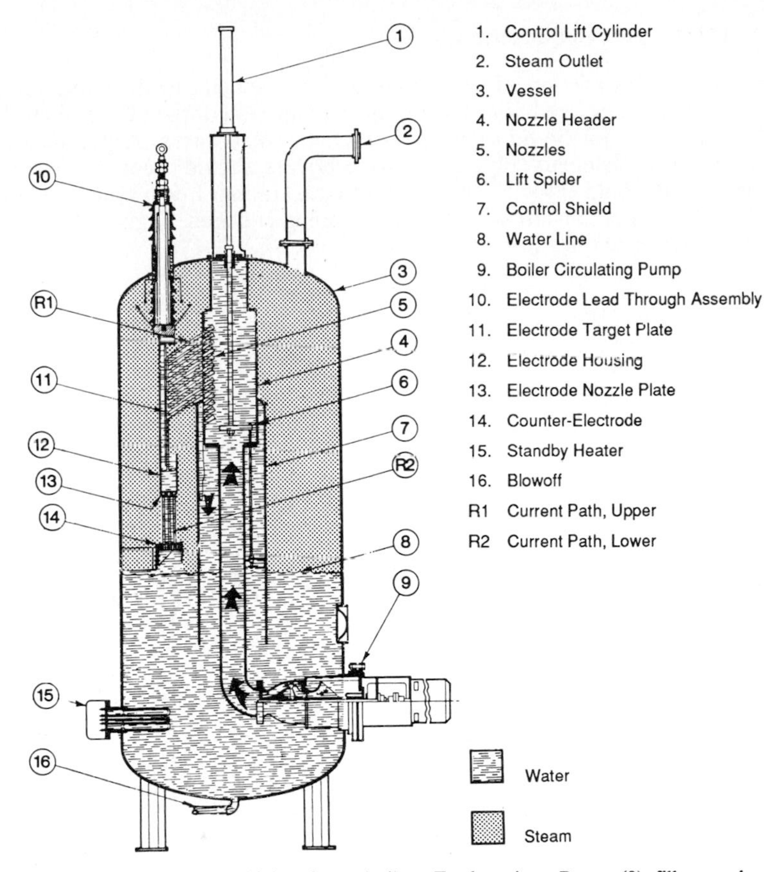

FIGURE 29.9 Jet-type high-voltage boiler. Explanation: Pump (9) fills nozzle header (4), water flows through nozzles (5) to strike electrode target plate (11), water falls to bottom of electrode housing (12) and through plate (13) to counterelectrode (14). R_1 and R_2 are current paths from electrode to nozzles (5) and counterelectrode (14), both of which are neutral. To control steam output, control shield (7) is lifted to divert nozzle discharge to inside of control shield. Steam output is proportional to water flow to electrode. *(CAM Industries, Inc.)*

blowdown requirements by a factor of 10 or more and permits the use of water which has been softened rather than demineralized. See Table 29.1.

Pressures and Ratings. Jet-type boilers operate best at pressures of 75 lb/in^2 (517 kPa) or higher. Operation at lower pressures requires special designs and significant derating. Jet-type boilers are offered for high voltage only. Ratings up to 78 MW at 25 kV are available. Figure 29.9 shows a jet-type boiler.

29.5 INSTALLATION

Planning the installation of electric boilers requires attention to both mechanical and electrical details, as well as to the water to be used in the boiler.

29.5.1 Mechanical Requirements

The mechanical installation of an electric boiler is subject to the same requirements as a fossil-fuel-fired boiler except for the omission of the vent and chimney and the provisions for liquid or gaseous fuel. In some cases, requirements for firewalls are less stringent. Piping for these boilers should meet the design requirements that would apply to any boiler of equal rating. Local boiler and building codes should be consulted before the planning is begun. Figure 29.10 shows pressure drops typical for hot-water boilers.

29.5.2 Clearances

Boilers must be installed with adequate clearances at the sides and top to permit access to all valves and piping and to meet the requirements of the local boiler or

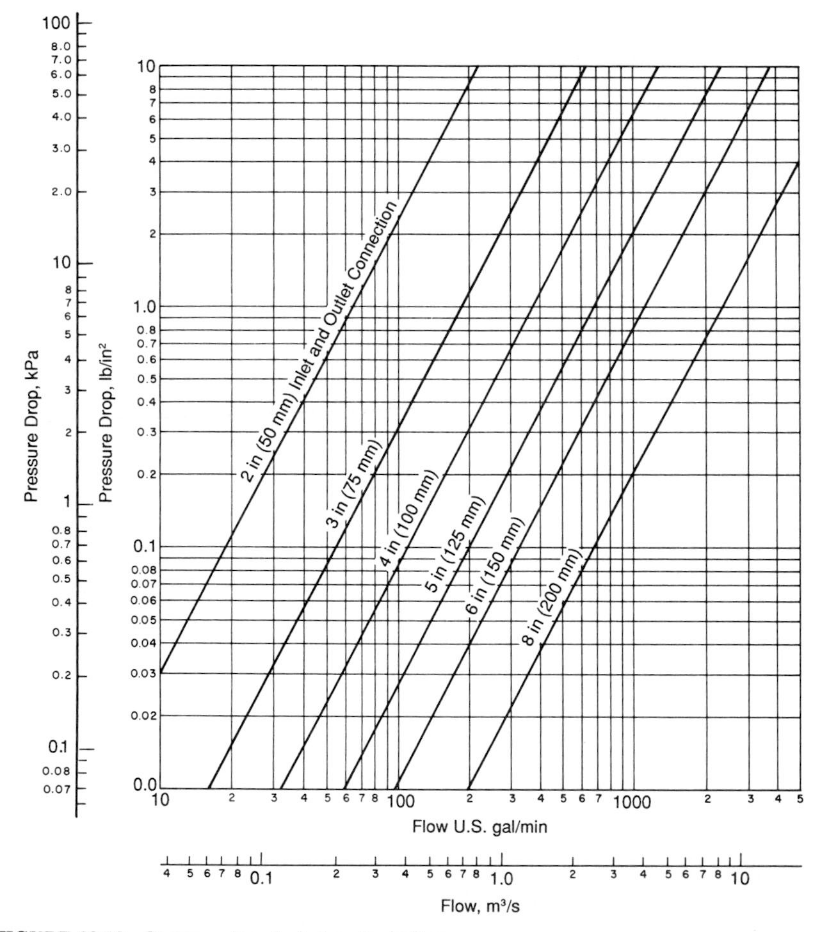

FIGURE 29.10 Pressure drop in hot-water boilers.

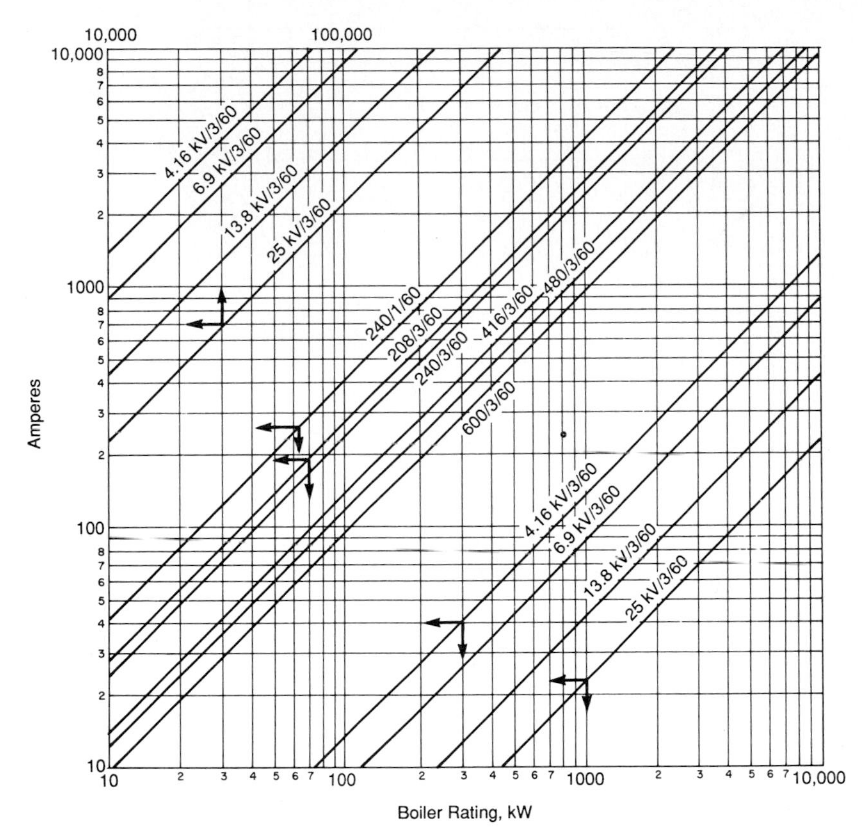

FIGURE 29.11 Boiler amperage draw.

building code. Clearances in front of electric cabinets are also prescribed by electrical and/or building codes.

29.5.3 Electric Wiring

Supply circuits for electric boilers *must meet the requirements of the local electrical code*. In the absence of a local code, the National Electrical Code® (NEC) (ANSI/NFPA 70) should be used as a guide; it is available from the National Fire Protection Association (NFPA). Conductors of the circuit supplying the boiler should be sized for 125 percent of the load served, as should the circuit protective devices.

Figure 29.11 shows the relation of amperes to kilowatts for the voltages commonly encountered in the United States. More than one supply circuit may be used to feed a boiler if the boiler load is (or can be) split and separate terminals provided for each supply circuit. Electrode boilers and high-voltage boilers have requirements for grounding, etc., which are, or may be, different from the requirements for resistance-type boilers. In all cases, the boiler manufacturer's recommendations should be followed.

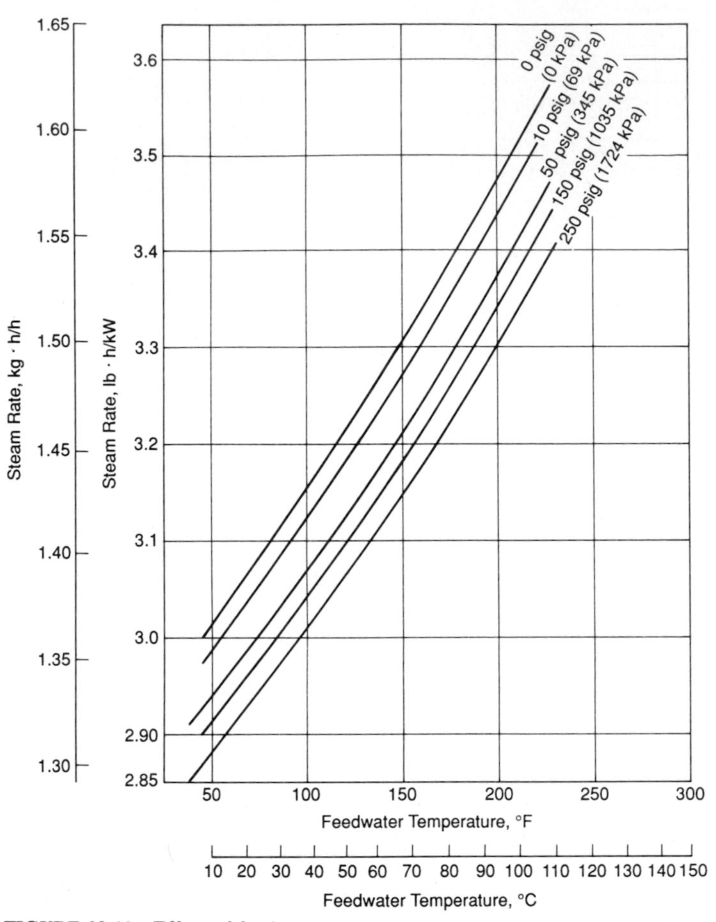

FIGURE 29.12 Effect of feedwater temperature on steam generated per kilowatt.

Water Quality. Table 29.1 shows the recommended water-quality limits for boiler feedwater and makeup water and for water in the boiler or heating system. Water-quality requirements are more critical for electrode boilers than for the resistance type. In addition, the minimum water-quality requirements for electrode boilers vary for the different designs and are usually more stringent as the voltage increases.

Recovery of condensate for return to the boiler is recommended both as an energy conservation measure (Fig. 29.12 shows the increase in pounds of steam per kilowatt with boiler feedwater temperature) and as a source of high-quality feedwater. Leakage from hot-water heating systems not only wastes heat but also increases corrosion in the system and can lead to scale deposits in the boiler.

CHAPTER 30
FACTORY-BUILT PREFABRICATED VENTS, CHIMNEYS, AND STACKS

J. F. Schulz
Chairman, Van-Packer Co., Manahawkin, New Jersey

P. Hodson
Vice President and Manufacturing Manager, Van-Packer Co., Buda, Illinois

K. Coleman
Staff Engineer, Van-Packer Co., Manahawkin, New Jersey

30.1 INTRODUCTION

This chapter covers all vents and chimneys that are fabricated off-site and assembled on-site to produce a finished product. All aspects are considered except structural engineering, which is referenced sufficiently to point designers in the proper direction for this information.

Underwriters' Laboratories (UL) listed gas vents and chimney test procedures are reviewed to provide an understanding of the limits imposed on listed products. The practical uses of these products are also discussed.

There is currently no UL listing for breechings; however, UL listed vents and chimneys may be used as breechings. A review of the criteria for breechings is presented to demonstrate what materials will provide the best service for specific conditions. Consideration is given to the thermal and corrosion analysis of chimney systems and the effect on longevity.

Most of the UL listed products, with the exception of refractory-lined medium-heat appliances, are suitable only for gas- and oil-fired appliances. Some residential metal chimneys are said to be suitable for wood-burning appliances. This area is discussed, and the controversy is reviewed.

Incineration coupled with waste heat recovery has stressed chimney linings considerably compared to previous incinerator chimney designs. The various wastes—hazardous, hospital, and municipal—are discussed, and the effectiveness of various linings is reviewed.

Factory-built steel stacks and precast chimneys demonstrate the many types in use, and their general characteristics are reviewed.

In Sec. 30.6, the subsections on chimney sizing, thermal and corrosion analysis of chimney systems, chemical loading, factors involved in the design of high-rise chimney and chute systems, and the self-pollution of buildings discuss the actual sizing procedures, with the other sections highlighting trouble areas.

30.2 LISTED FACTORY-BUILT CHIMNEYS AND VENTS

30.2.1 Overview and Important References

Listed Chimneys and Vents. Two important organizations that foster fire safety are the National Fire Protection Association (NFPA) and the Underwriters' Laboratories. The NFPA develops fire safety standards, and UL creates testing standards and lists products to accommodate the NFPA standards.

The first products listed were chimneys which utilized either stainless steel or refractory as flue liners to accommodate the temperatures and corrosion found in chimneys. The applicable NFPA standard is NFPA 211. Chimneys meeting this standard are tested in accordance with UL 103 and are capable of withstanding continuous firing at temperatures up to 930°F (517°C) above room temperature and infrequent brief periods of forced firing to no higher than 1330°F (739°C) above room temperature. These chimneys are suitable for low-heat appliances.

Gas vents conform to NFPA 211 and NFPA 54, are tested in accordance with UL 441, and are capable of withstanding continuous firing at flue temperature not exceeding 400°F (204°C) above room temperature. The combustion gases of natural gas are relatively noncorrosive, and thus aluminum flue linings are used with only a single metal jacket spaced a minimum of ¼ in (6.35 mm) from the flue; no insulation is required.

Medium-heat appliance chimneys conform to NFPA 211 and NFPA 82 and are tested in accordance with UL 959. They are capable of handling continuous flue temperatures up to 1730°F (980°C) above room temperature and intermittent excursions up to 1930°F (1070°C) above room temperature. Chimneys of this type may be used on low-heat appliances such as boilers on building heating appliances and on incinerators, furnaces, etc., within the temperature limitations. This was the first chimney used in the commercial/industrial market.

The latest chimney to be listed by UL is the 1400°F (760°C) equilibrium building heating appliance chimney. It conforms to NFPA 211 and is tested in accordance with a modified UL 103 standard with increases in the equilibrium and shock test temperatures. (Currently the Underwriters' Laboratories do not have a published standard to accommodate this product.) It can accommodate continuous firing up to 1330°F (739°C) above room temperature and 1730°F (980°C) for brief periods and can be used on all boiler types. It is not acceptable for incinerators of any type, due to not only temperature limitations but also the corrosive nature of incinerator flue gases.

Wood-Burning Appliance Chimneys. In the past, chimneys tested to UL 103 were recognized for use on conventional fireplaces and other wood-burning appliances. In 1984, NFPA 211 was upgraded to fully accommodate wood-burning appliances and to recognize the growing fire losses associated with creosote fires. Creosote fires operate at temperatures from 1700°F (927°C) to 2150°F (1177°C).

Underwriters' Laboratories Inc. (ULI, interchangeable with the abbreviation UL) of the United States and Underwriters' Laboratories of Canada (ULC) held diverse opinions regarding the intensity of creosote fires. The ULI currently tests chimneys in accordance with UL 103 but adds an optional temperature test consisting of three 10-min cycles at 2030°F (1128°C) above room temperature and allows the manufacturer to designate the product as Type HT. Many companies designate their products as 2100°F (1149°C) chimneys. This identification is not authorized by Underwriters' Laboratories and can be very misleading. The 2100°F (1149°C) value refers to a 10-min test. Use of these chimneys at temperatures of 2100°F (1149°C) for periods longer than 10 min can be dangerous and should be avoided.

The ULC uses a nominal 1200°F (649°C) continuous temperature to equilibrium, 1700°F (927°C) for 1 hour and 2100°F (1149°C) for 3- to 30-min cycles. These last values are said to approximate the temperatures of creosote fires. Unavoidably the use of high-grade metal and fabricating procedures to meet those requirements increases costs. The ULC standard for 650°C (1202°F) factory-built chimneys is identified as ULC S629-M1981. The Minnesota Reinsurance Association accepts the Canadian products for use in their area of influence.

Most chimneys are tested by Underwriters' Laboratories for temperature only. The only products tested for corrosion by acids that condense within the flue area of chimneys are the refractory-lined products. Note that this may be a satisfactory position to take at the temperatures tested; for the usual residential installation, however, corrosion does occur in stainless-steel chimneys at a rather rapid rate when flue surface temperatures fall below acid dew points, particularly on commercial/industrial chimneys. Problems may be minimal on short chimneys where the temperature loss is low; however, on poorly insulated or air-insulated chimneys, the loss in temperature from midflue to flue surface at any given cross section can be as high as 60 percent. Corrosion and thermal analyses should be considered on all commercial/industrial chimney installations during the design stage if one is to avoid premature chimney failures and replacement costs. Corrosion and thermal analysis are discussed further in Sec. 30.6.

Commercial/Industrial Chimneys. Commercial/industrial chimneys, unlike residential chimneys, are subject to a wide range of fuels from high-sulfur liquid fuels to coal and refuse. This brings not only sulfuric acid as a corrosive but also hydrochloric and hydrofluoric acid, which are extremely corrosive to metals even in small quantities as condensate at low temperatures or vapors at temperatures above 800°F (427°C).

The breeching runs and chimney heights are greater than for residential applications and thus lose more heat and could easily fall below the critical dew points of 320°F (160°C) for No. 6 fuel oil with 4 percent sulfur, 275°F (135°C) for No. 2 fuel oil with 0.6 percent sulfur, and 140°F (60°C) for hydrochloric acid. All temperatures are measured on the inside flue surfaces.

A thorough thermal analysis of commercial/industrial chimney systems is essential to a good design. This subject is discussed further in Sec. 30.6.

Publications of Interest. All the UL listed chimneys and vents may be found in the Underwriters' Laboratories *Gas and Oil Equipment Directory*, published annually.

The publications cited in this section may be obtained from:

Underwriters' Laboratories, Inc.
Publications Stock
333 Pfingsten Road
Northbrook, IL 60062

Underwriters' Laboratories of Canada
7 Crouse Road
Scarborough, Ontario M1R 3A9, Canada

National Fire Protection Association
Batterymarch Park
Quincy, MA 02269

Publications of interest include:

Underwriters' Laboratories standards for safety:

- *Chimneys, Factory Built, Residential Type and Building Heating Appliances*, UL 103
- *Gas Vents*, UL 441
- *Medium-Heat Appliance Chimney*, UL 959
- 1400°F (760°C) chimney (no published standard at this time but UL 103 is used as a guide)
- *Low-Temperature Venting Systems, Type L*, UL 641
- ULI's *Gas and Oil Equipment Directory*
- ULC's *List of Equipment and Materials*, vol. 1, *General*; vol. 2, *Building Construction*
- *Standard for 650°F (1202°F) Factory-Built Chimney*, CAN4-S629-M84
- *Standard for Factory-Built Chimneys*, CAN4-S602-M82
- *Standard for Type L Vents*, ULC-S609
- *Standard for Gas Vents*, CAN4-S605-77

National Fire Protection Association standards:

- *Chimneys, Fireplaces, Vents and Solid Fuel-Burning Appliances*, NFPA 211
- *Incineration and Waste Handling*, NFPA 82
- *National Fuel Gas Code*, NFPA 54
- *Installation of Oil-Burning Equipment*, NFPA 31

30.2.2 Venting Systems

Gas Vents (UL 441). Gas vents are installed as Type B (Fig. 30.1) or Type BW (Fig. 30.2) gas vents. Gas vents are to be installed in accordance with NFPA 54, *National Fuel Gas Code*, and NFPA 211, *Chimneys, Fireplaces, Vents and Solid Fuel-Burning Appliances*. They are tested in accordance with UL 441, *Gas Vents*.

Portions of the vent which extend through accessible spaces are to be enclosed with clearances as specified on their UL listed cards, to avoid personal contact with or damage to the vent.

The cross-sectional shape of gas vents may be round, oval, square, or rectangular. Most vents on the market are round or oval (Figs. 30.1 and 30.2). The appropriate designation of Type B or Type BW is marked on each listed vent pipe section and vent pipe fitting.

1. *Type B gas vents:* Type B gas vents (Fig. 30.1) are for venting gas appliance equipment with draft hoods listed for use with Type B vents. Type B gas vents shall not be used for venting recessed wall heaters listed for use with Type

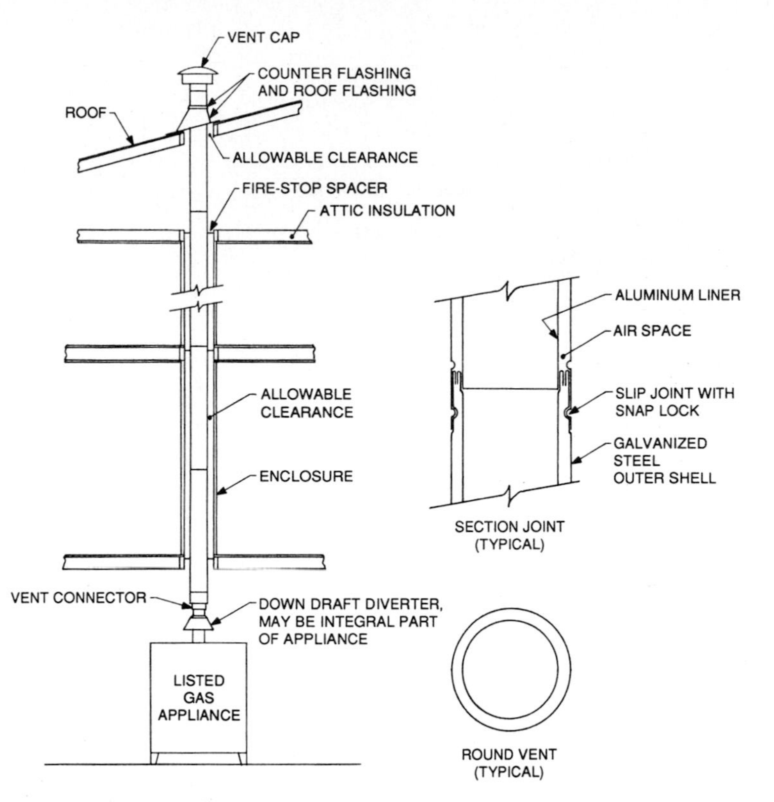

FIGURE 30.1 Type B gas vent.

BW gas vents only, for incinerators and appliances listed for use with chimneys only, or for combination gas/oil appliances and appliances which may be converted readily to the use of solid or liquid fuels.

2. *Type BW gas vents:* Type BW gas vents (Fig. 30.2) are for venting only; they are for approved vented recessed heaters having inputs not greater than specified in the individual listing. Type BW gas vents are designed to be installed within wall stud areas. See the UL standard for safety (UL 441) for the testing procedure and material temperatures for Type B and Type BW gas vents.

Type L Venting Systems (UL 641). Type L venting systems (Fig. 30.3) are for use with gas and oil appliances listed as suitable for venting with Type L vent systems. They may also be used where Type B gas vents are permitted. Type L venting systems may be used for chimney and vent connectors. Installation for oil-burning appliances is to be in accordance with the standards of the NFPA for the installation of oil-burning equipment (NFPA 31). NFPA 211, the standard for chimneys, fireplaces, vents, and solid-fuel appliances, identifies other permitted uses of Type L vents.

Portions of the venting system which extend through accessible spaces are to be enclosed, to avoid personal contact with or damage to the venting system.

The designation "Type L" venting system is marked on each listed vent pipe

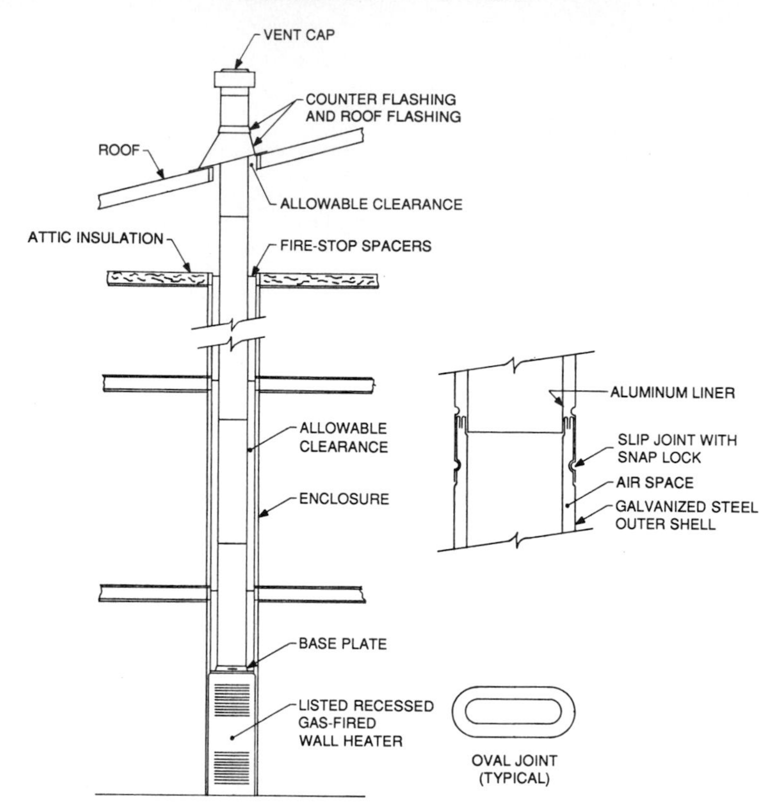

FIGURE 30.2 Type BW gas vent.

section and vent pipe fitting. The reader is referred to the UL 641 standard for safety for the testing procedure and material temperatures for Type L vents. For reference the maximum temperature rises above room temperatures for some materials as established by UL are shown in Table 30.4.

Connectors for Listed Gas Appliances. NFPA 211 specifies the following: Connectors for listed residential-type gas appliances installed in an attic shall be of Type B or Type L vent material. For listed gas appliances not installed in attics, connectors shall be of Type B or Type L vent material and metal pipe having a resistance to corrosion and heat not less than 0.016-in (0.406-mm) (28 gauge) galvanized steel.

Clearance to combustibles for listed Type B or Type L vents used as connectors shall be in accordance with their UL listing. Clearance to combustibles for single-wall galvanized connectors shall be 6 in (152 mm).

Connectors for Listed Type L Appliances. NFPA 211 specifies the following: Connectors for residential-type oil-fired appliances listed for use with Type L vents and installed in attics shall be of Type L vent material. For oil-fired appliances listed for use with Type L vents not installed in attics, connectors shall be of

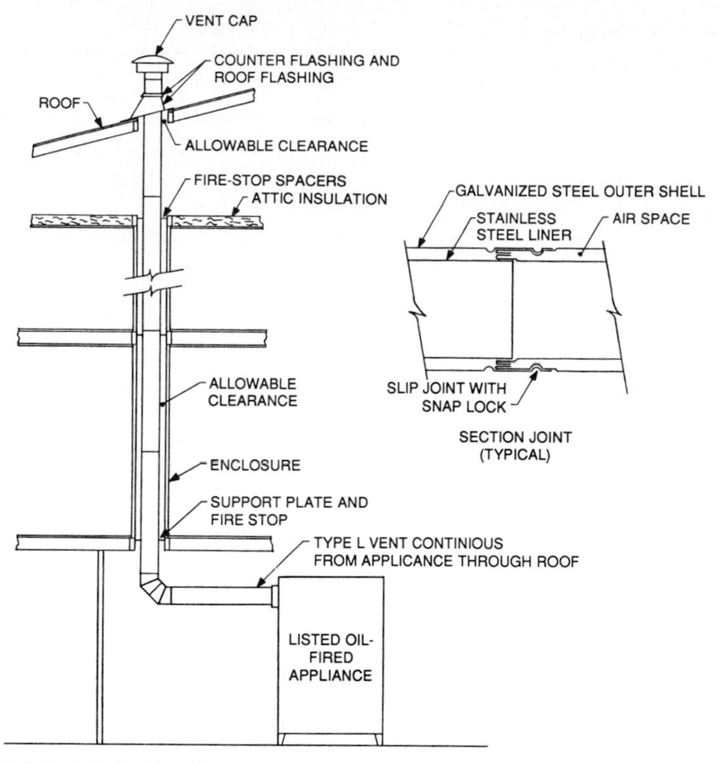

FIGURE 30.3 Type L vent.

Type L vent material or steel pipe not less than 0.016-in (0.4-mm) (28 gauge) galvanized steel having resistance to corrosion and heat.

Clearance to combustibles for listed Type L vents used as connectors shall be in accordance with their UL listing. Clearance for single-wall galvanized connectors shall be 9 in (229 mm).

30.2.3 Residential Low-Heat Appliance Chimney (UL 103)

Applications. Chimneys of this type are identified in the UL *Gas and Oil Equipment Directory* as "chimneys, residential type, and building heating appliances." Residential type and building heating appliance chimneys are intended for venting flue gases at a temperature not exceeding 1000°F (538°C), under continuous operating conditions from gas-, liquid-, and solid-fuel-fired residential appliances and building heating appliances specified in the chimney selection chart of NFPA Standard 211 (Table 30.1). *Note*: Such chimneys are listed by Underwriters' Laboratories of Canada as being suitable only for gas- or liquid-fuel-fired residential-type appliances and building heating equipment and designated as Type A chimneys. In the United States, Type 103 chimneys are still allowed on some solid-fuel-fired appliances; however, at this writing the controversy concerning wood-

TABLE 30.1 Chimney Selection Chart

Column I	Column II	Column III
	Chimney type	
1. Factory-built—residential type and building heating appliance 2. Masonry, residential type	1. Factory-built—residential type and building heating appliance 2. Masonry, low-heat type 3. Metal, low-heat type*	1. Factory-built—1400°F (760°C) 2. Masonry, low-heat type 3. Metal, low-heat type*
	Maximum continuous appliance outlet flue gas temperature	
1000°F (538°C)	1000°F (538°C)	1400°F (760°C)
	Types of appliances to be used with each type chimney†	
Residential type gas-, liquid-, and solid-fuel-burning appliances such as: 1. Dual-fuel furnaces 2. Fireplace inserts 3. Fireplace stoves 4. Fireplace stoveroom heater 5. Floor furnaces 6. Free-standing fireplaces 7. Hot-water heating boilers 8. Low-pressure steam heating boilers 9. Masonry fireplaces 10. Ranges 11. Residential incinerators 12. Room heaters 13. Stoves 14. Wall furnaces 15. Warm air furnaces 16. Water heaters	A. All appliances shown in column I B. Nonresidential-type building heating appliances for heating a total volume of space exceeding 25,000 ft^3 (708 m) C. Steam boilers operating at not over 1000°F (538°C) flue gas temperature; pressing machine boilers	All appliances shown in columns I and II and appliances such as: 1. Class A ovens or furnaces operating at temperatures below 1400°F (760°C) as defined in NFPA 86A 2. Annealing baths for hard glass (fats, paraffin, salts, or metals) 3. Bake ovens (in bakeries) 4. Candy furnaces 5. Core ovens 6. Feed drying ovens 7. Forge furnaces 8. Gypsum kilns 9. Hardening furnaces (below dark red) 10. Lead-melting furnaces 11. Nickel-plate (drying) furnaces 12. Paraffin furnaces 13. Restaurant-type cooking appliances using solid or liquid fuel 14. Sulfur furnaces 15. Tripoli kilns (clay, coke, and gypsum) 16. Wood-drying furnaces 17. Zinc-amalgamating furnaces

TABLE 30.1 Chimney Selection Chart (*Continued*)

Column IV	Column V
Chimney type	
1. Factory-built—medium-heat appliance 2. Masonry, medium-heat type 3. Metal, medium-heat type*	1. (No listing) 2. Masonry, high-heat type 3. Metal, high-heat type*
Maximum continuous appliance outlet flue gas temperature	
1800°F (982°C)	Over 1800°F (982°C)
Types of appliances to be used with each type chimney†	
All appliances shown in columns I, II, and III and appliances such as: 1. Alabaster gypsum 2. Annealing furnaces (glass or metal) 3. Charcoal furnaces 4. Cold stirring furnaces 5. Feed driers (direct fire-heated) 6. Fertilizer driers (direct fire-heated) 7. Galvanizing furnaces 8. Gas producers 9. Hardening furnaces (cherry to pale red) 10. Incinerators—commercial and industrial 11. Lehrs and glory 12. Lime kilns 13. Linseed oil boiling 14. Porcelain biscuit kilns 15. Pulp driers (direct fire-heated) 16. Steam boilers operating at over 1000°F (538°C) flue gas temperature 17. Water-glass kilns 18. Wood-distilling furnaces 19. Wood-gas retorts	All appliances shown in columns I, II, III, and IV and appliances such as: 1. Bessemer retorts 2. Billet and bloom furnaces 3. Blast furnaces 4. Bone calcining furnaces 5. Brass furnaces 6. Carbon point furnaces 7. Cement brick and tile kilns 8. Ceramic kilns 9. Coal and water gas retorts 10. Cupolas 11. Earthenware kilns 12. Glass blow furnaces 13. Glass furnaces (smelting) 14. Glass kilns 15. Open hearth furnaces 16. Ore roasting furnaces 17. Porcelain baking and glazing kilns 18. Pot-arches 19. Puddling furnaces 20. Regenerative furnaces 21. Reverberator furnaces 22. Vitreous enameling ovens (ferrous metals)

*Single-wall metal chimneys or unlisted metal chimneys shall not be used inside one- and two-family dwellings.

†For appliance types not listed in columns I through V, the appropriate chimney shall be selected on the basis of the appliance gas temperature when appliance is fired at its normal maximum input, and type of surroundings.

Source: Reprinted with permission from NFPA 211-1984, *Chimneys, Fireplaces, Vents and Solid Fuel-Burning Appliances*, Copyright© 1984, National Fire Protection Association, Quincy, MA 02269. This reprinted material is not the complete and official position of the NFPA on the referenced subject, which is represented only by the standard in its entirety.

burning appliances has not been resolved. It is prudent to check the latest code in your area. NFPA Standard 211 was revised in 1989. The reader is referred to UL 103, *Chimneys, Factory-Built, Residential Type and Building Heating Appliances*, for testing procedures and material temperatures for this type of chimney.

Low-Heat Appliance Chimneys in Current Use

1. *Solid Pac insulated type:* See Fig. 30.4. This very simple and highly efficient chimney utilizes an insulated material designed to give maximum insulation at high temperatures. Sections are installed by twisting the female end onto the male projection of the section below it.

Most products utilize both a stainless-steel flue liner and a stainless-steel outer jacket. They are produced in both the United States and Canada.

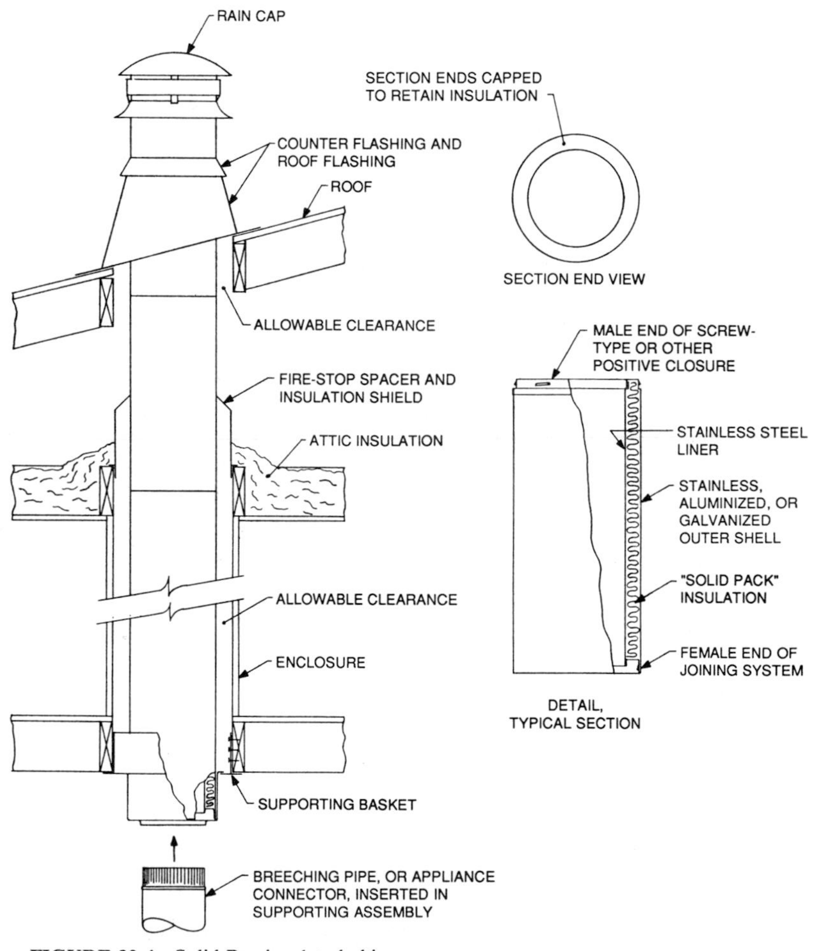

FIGURE 30.4 Solid Pac insulated chimney.

2. *Thermosiphon type:* See Fig. 30.5. Thermosiphon systems consist of three concentric cylindrical shells. The inner shell, or the flue, is fabricated from stainless steel. The two remaining shells are usually made of galvanized or aluminized steel.

The hot flue gases lose heat rapidly through the uninsulated flue wall. This heats the air in the inner annular area, causing it to rise. The two annular areas are connected at the bottom of the chimney by a plenum. The negative pressure produced in the inner annular area pulls cold air into the outer annular area, allowing a continuous flow of cool air over the exterior surface of the flue. The radiation effect of the two external shells also assists in lowering outer shell temperatures.

Sections are joined with friction-tight female and male joints on all three shells, to allow penetration of the flue and inner shell of the upper sections by those in the lower section, thus maintaining any condensate to the chimney interior. The outer shell of the upper section is penetrated by the lower section to keep atmospheric moisture out of the chimney.

3. *Modified thermosiphon type:* See Fig. 30.6. This type of chimney is a rel-

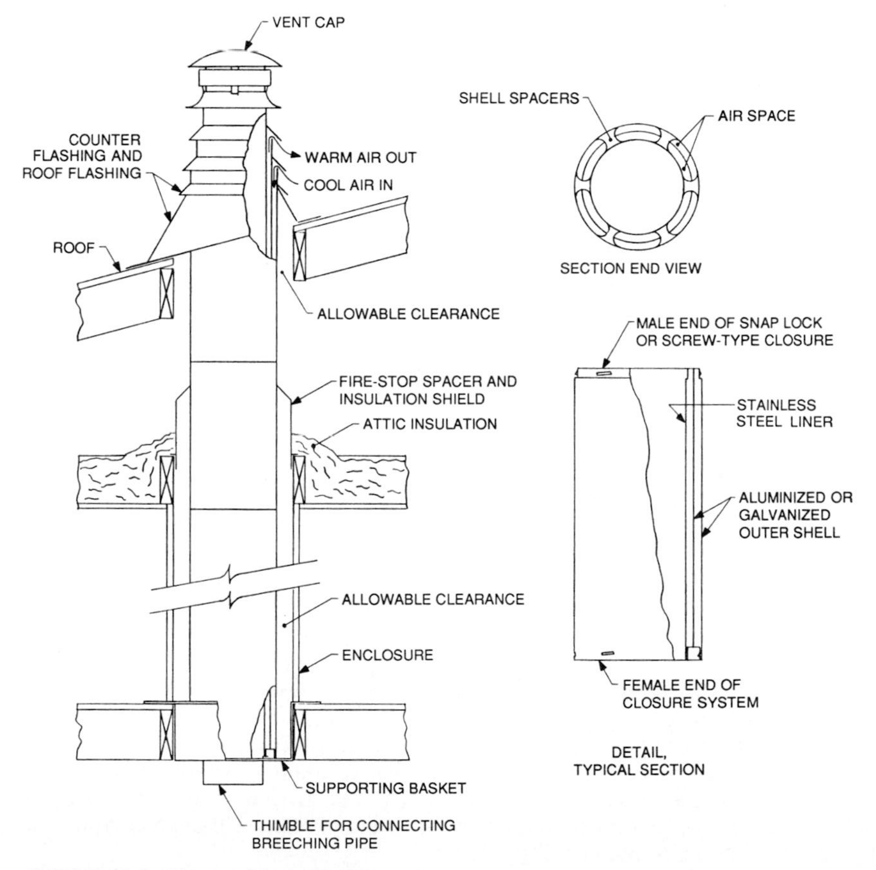

FIGURE 30.5 Thermosiphon chimney.

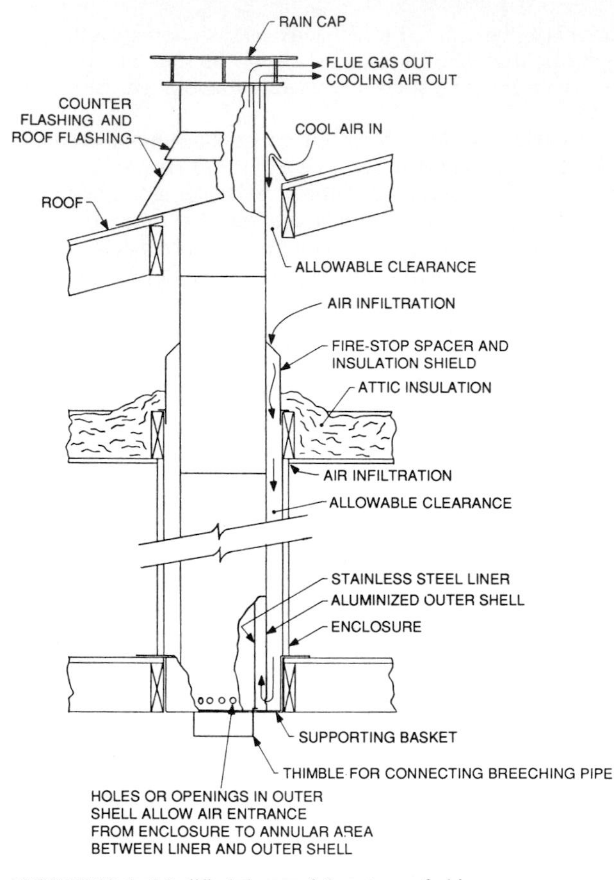

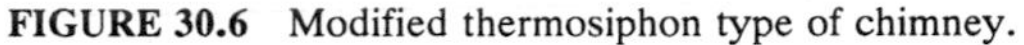

FIGURE 30.6 Modified thermosiphon type of chimney.

atively new chimney with a relatively new UL listing. It must be placed within a fire-rated enclosure, and Underwriters' Laboratories have allowed that enclosure to be, in effect, the outer chimney wall.

Air is drawn into this area through a ventilated thimble and travels downward in the enclosure to the bottom of the chimney, through holes in this outer jacket to the annular area between the jacket and the flue.

Warnings are required by UL indicating that the chimney is unsafe if the holes at the bottom of the jacket or opening in the ventilated thimble become clogged.

4. *Air-insulated chimney:* See Fig. 30.7. This chimney type utilizes the air movement within the annular areas, as does the thermosiphon; however, the movement is less defined. The flue, of course, has tight joints. The middle cylinder, however, is designed to allow air movement into both annular areas. The outer jacket combines a connecting device such as a screw-type joint with openings that enable air to enter the annular areas. The air flow may not be as well defined as in a true thermosiphon, but the overall movement, though complicated by various flow patterns, is still the same.

The radiation factors and surface coefficients also provide some reduction in

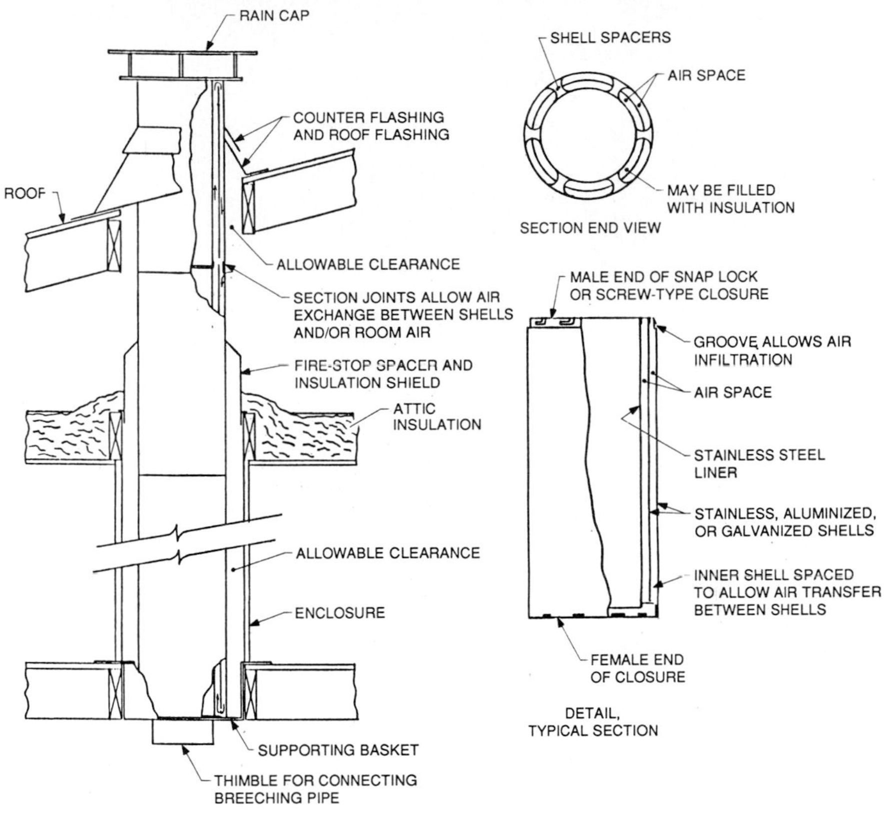

FIGURE 30.7 Air-insulated chimney.

outer surface temperatures. Some manufacturers have upgraded this chimney type to meet the more rigid requirements of UL 103 HT, by placing insulation in the annulus adjacent to the flue liner.

5. *Insulated refractory-lined type:* See Fig. 30.8. This unit is the original model 1 refractory unit produced in the United States for all low-heat appliances including those fueled with solid fuels.

6. *Rectangular insulated refractory-lined type:* See Fig. 30.9. This unit, too, is very similar to the original model 1 refractory unit produced in the United States but is rectangular and has also passed not only the UL 103 Type HT tests but also the ULC-S629-M1981 tests. It is manufactured in Canada but is also available in the United States.

This unit is unique as a chimney system since, unlike the refractory-lined type, only the flue and outer shell are factory-built. The insulating refractory is field-poured into the annular area, forming a continuous column of insulation.

7. *Combined thermosiphon/refractory-lined type:* See Fig. 30.10. This system combines features of an insulated system and a thermosiphon system. The refractory lining has been tested by Underwriters' Laboratories for 3- to 30-min firings at 2100°F (1155°C) and found to pass successfully. It has also been tested

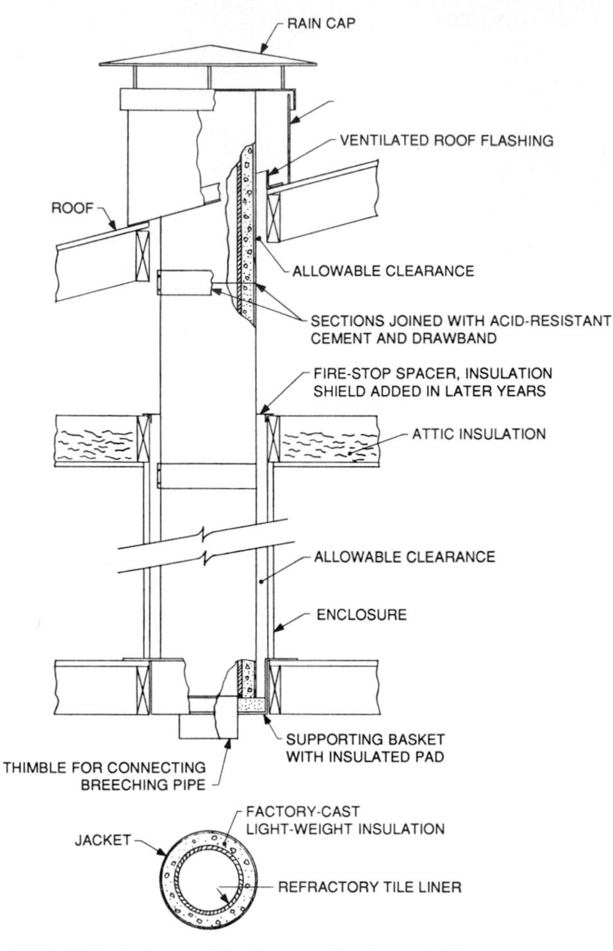

FIGURE 30.8 Insulated refractory-lined round chimney.

under actual creosote burnout testings and does contain flue fires without damage.

UL 103 Chimneys for Wood-Burning Appliances. For many years UL 103 chimneys were used primarily on oil-fired appliances. When wood-burning equipment became popular in the 1980s, problems arose that were previously unheard of. Soot and creosote collected within the chimney and ignited to cause severe chimney fires and often structural damage.

Creosote fires can develop temperatures as high as 2150°F (1177°C) within the chimney itself for short periods and 1400°F (760°C) to 1700°F (927°C) for substantially longer periods. These conditions are not addressed in UL 103 and have led to serious failures where the flue liner distorts to such a degree that the integrity of the chimney is lost and fires result.

Chimneys servicing wood-burning appliances are less forgiving of misin-

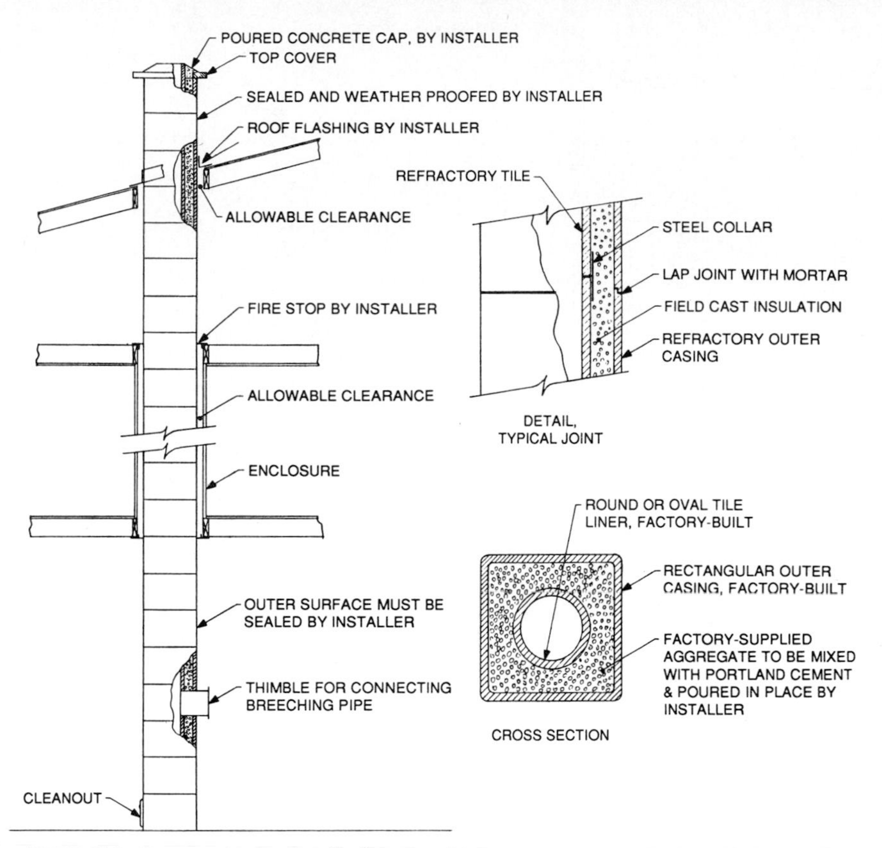

FIGURE 30.9 Rectangular insulated refractory chimney.

stallations than chimneys servicing oil-burning appliances because of the higher temperatures developed within the chimney by creosote fires. Other problems that become critical with wood-burning appliance chimneys have little effect on liquid- or gas-fired appliance chimneys:

- Reduced clearances to combustibles.
- Applying insulation up to chimneys. This most frequently occurs in attic areas, but the clearance spaces throughout should be free of insulation. Added insulation can raise temperatures on flue liners or other critical parts above allowable values.
- Construction of a chimney from products of different manufacturers. This procedure not only invalidates the UL listing but also could cause a problem with insurance claims.
- Field modification of factory-built metal chimneys. This invalidates the UL listing and may alter the design criteria of the product.

The industry's position has been to recommend frequent cleaning of the chimneys; however, as wood-burning chimneys achieve longer life and as more chimneys are installed, the data favor more rigid testing requirements.

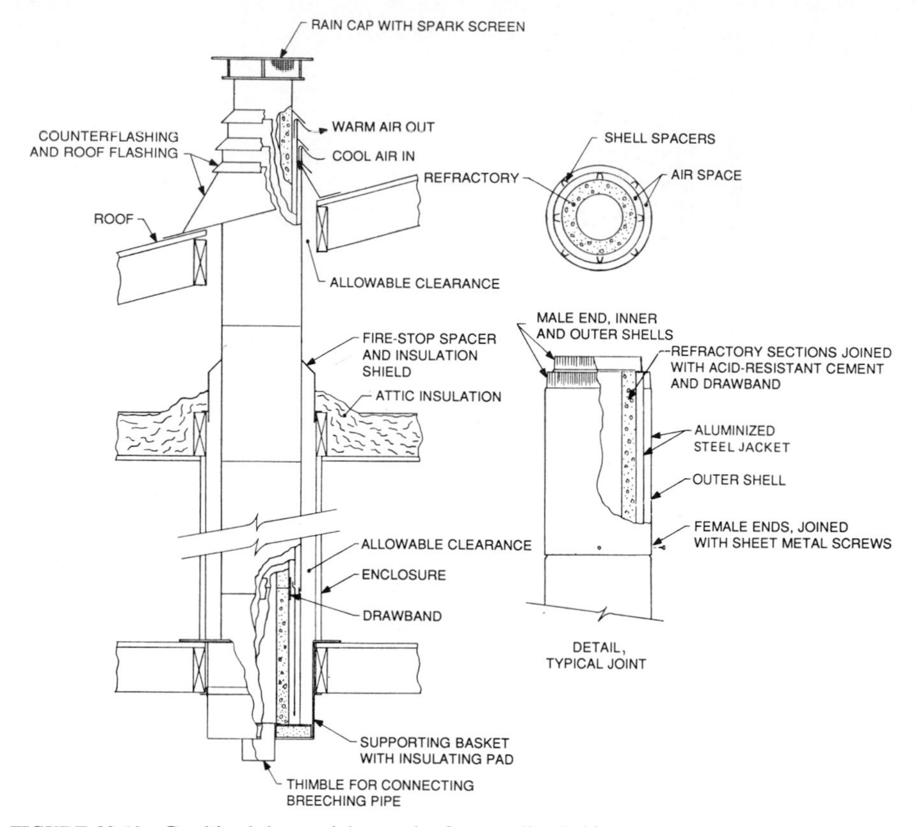

FIGURE 30.10 Combined thermosiphon and refractory-lined chimney.

Tests for Wood-Burning Appliance Chimneys. There is considerable controversy regarding chimneys for wood-burning appliances, and the approaches of ULI, the United States Underwriters, and ULC have differed to some extent.

ULI slightly upgraded the UL 103 test, calling it UL 103 Type HT. ULC developed a considerably more rigid test based on data collected from a field survey of over 1000 actual chimney fires.

1. *Underwriters' Laboratories Inc. Type HT:* Underwriters' Laboratories, recognizing the difference between wood- and gas- or oil-burning appliances, have added an optional 2100°F (1149°C) test for chimneys to be used on wood-burning appliances. They also required that printed warnings be shown in the installation instructions and placed on certain chimney parts.

The thermal shock and 1000°F (538°C) equilibrium tests remain the same, as does the 1400°F (760°C) test for 1 h. The 2100°F (1149°C) test is an option to the 1700°F (927°C) test and is run 3 times for 10 min for each cycle. The allowable temperature rise above ambient temperature on combustibles has been set at 175°F (79°C).

2. *ULC-S629 test for gas, liquid, and solid-fuel-fired equipment:* Underwriters' Laboratories of Canada recognized the problem and increased the maximum firing to 2100°F (1148°C) for three 30-min periods and increased the 1700°F (927°C) test runs for 1 h compared to 10 min for ULI. The ULC equilibrium test is run at 1200°F (649°C) with the inlet temperatures measured 3 diameters from the chimney inlet; ULI still uses 1000°F (538°C) for the equilibrium test with the inlet temperatures measured 5 chimney diameters from the chimney.

The ULC tests are substantially more severe, and some areas will accept chimneys tested to the Canadian standards. Underwriters' Laboratories of Canada identify their requirements in *Standard for 650°C (1202°F) factory-built chimneys* (CAN4-S629-M84). A list of companies that meet this standard can be found in the ULC's *List of Equipment and Materials*, volume 1: *General*.

3. *Controversial test standards for wood-burning appliance chimneys:* The Consumers Product Safety Commission (CPSC) ran a study on wood-burning appliance chimneys (*CPSC Final Report on FY 1984 Metal Chimney Project*, Washington, DC, Dec. 31, 1984) and found that a considerable potential for problems exists. It is stated that the UL 103 Standard "does not approach by several hundred degrees" the flue liner temperatures of actual chimney fires.

Unfortunately, nothing has been settled. As of January 1987, many areas of the country still allow UL 103 chimneys and simply recommend more frequent cleaning. Some areas accept products meeting ULC-S629 whereas others accept UL 103 Type HT. Nothing has been stabilized, nor will it be so for several years, until experience shows the proper direction.

An engineer called on to specify a chimney for a wood-burning appliance should consider UL 103 Type HT as minimum requirements. Standards for wood-burning appliance chimneys are predicted to become considerably more severe in the future if failures continue at their current rate.

Connectors for Low-Heat Appliance Chimneys. The following data have been taken from NFPA 211: Connectors for oil appliances and solid-fuel-burning appliances shall be of factory-built Type L vent material or steel pipe having resistance to corrosion and heat not less than that of galvanized pipe, as specified in Table 30.2. Clearance from connectors to unprotected combustible materials is 18 in (475 mm) for all low-heat oil-burning appliances.

TABLE 30.2 Metal Thickness for Galvanized-Steel Pipe Connectors

Diameter of connector, in (mm)	Galvanized sheet gauge no.	Minimum thickness in (mm)
<6 (<152)	26	0.019 (0.48)
6–10 (152–254)	24	0.023 (0.58)
11–16 (254–406)	22	0.029 (0.74)
>16 (>406)	16	0.056 (1.42)

Source: Reprinted with permission from NFPA 211-1984, *Chimneys, Fireplaces, Vents and Solid Fuel-Burning Appliances*, Copyright© 1984, National Fire Protection Association, Quincy, MA 02269. This reprinted material is not the complete and official position of the NFPA on the referenced subject, which is represented only by the standard in its entirety.

Effects of Corrosion. Most of these (low-heat appliance) chimneys are utilized in one- and two-story dwellings, and thus the chimney runs are relatively short.

TABLE 30.3 Sulfur in Fuel Oils

	No. 2	No. 3	Cold No. 5	No. 5	Bunker C or No. 6
Sulfur, % by weight	0.1–0.6	0.2–1.0	0.5–2.0	0.5–2.0	1–4
Dew point at maximum sulfur	275°F (135°C)	280°F (137.8°C)	295°F (146°C)	295°F (146°C)	320°F (160°C)

Over the years no real problems arose when No. 2 fuel oil was used; however, if fuels with higher sulfur contents are used, corrosion can occur.

1. *Solid and gaseous fuels:* Most metal chimneys are suitable for use with gas or No. 2 fuel oil where the flue surface temperatures will remain above 343°F (173°C) the acid dew point of No. 2 fuel oil with 0.6 percent sulfur, plus a 68°F (20°C) safety factor to accommodate variations of temperatures within chimney systems].

If fuel oil heavier than No. 2 is used, the increase in dew point will be in direct relationship to the sulfur content. Some manufacturers do not recommend the use of No. 6 fuel with stainless-steel-lined chimneys. If No. 6 fuel oil is to be used, it is very prudent to have a complete thermal analysis of the system. (See Table 30.3.) When breechings are longer than normal with any fuel, the heat loss could allow condensation within the breeching; thus thermal analysis is essential. Many listed products contain depressions at joints that will enhance corrosion by collecting condensed products at these points. Even though corrosion tests are not required by ULI for metal chimneys, it is good practice to require them if a corrosion analysis of the installation indicates a potential abnormal corrosion rate. Insulating refractory chimneys are tested by UL to resist flue acids (sulfuric) with a pH of 2 and have the capability of withstanding the whole range of fuel oils from No. 2 to No. 6.

2. *Corrosion in chimneys servicing solid-fuel appliances:* Corrosion here relates to the type of fuel:

- *Coal:* Coal presents a serious problem in metal chimneys of this type since hydrochloric acid (HCl) and sulfuric acid are condensed on the flue surfaces when these surfaces fall below 140°F (60°C). The refractory units will also show signs of corrosion but have a significantly longer life than metal ones.
- *Wood:* Wood presents a corrosion problem only if condensed creosote ignites within the flue and temperatures go over 800 to 1650°F (427 to 990°C); within this range intergranular corrosion degrades stainless steel and allows subsequent rusting. In double- or triple-wall systems, creosote fires can also oxidize the galvanized coatings.

A more detailed discussion of the corrosive nature of flue acids may be found in Sec. 30.6.

Summary of Various Chimney Tests

1. *UL 103 thermal shock test for 1700°F (906°C) flue gases:* This test is to be conducted prior to other thermal tests. It simply determines if the chimney can mechanically survive a higher-temperature excursion for a short time. No material temperatures are required.

2. *UL 103 equilibrium temperature test for 1000°F (538°C) flue gases:* This test starts at room temperature and is continued until equilibrium temperatures are attained on surfaces of the test chimney parts and the test enclosure.

The maximum temperature on surfaces of the test structure (such as ceilings, enclosures, floors, and joists) and on surfaces of chimney parts at points of zero clearance to the test structure shall not be more than 90°F (50°C) above room temperature during the period ending 4½ h from the start of the test and not more than 117°F (65°C) above room temperature for any subsequent period.

The temperature of any part of the chimney shall not exceed the maximum temperature specified for the material used. (See Table 30.4.)

TABLE 30.4 Maximum Temperature Rises

(The inclusion of a temperature limit for a material in this table is not indicative of the acceptability of the material if it does not otherwise conform to these requirements.)

	Maximum rise above room temperature			
	Column 1¶		Column 2¶	
Material	°F	°C	°F	°C
1. Aluminum alloys	330	183	430	239
1100 (2S)	430	239	530	294
3003 (3S)	530	294	630	350
2014, 2017, 2024, 5052*				
2. Aluminum-coated steel heat-resistant type†	1030	572	1275	708
3. Carbon steel-coated with Type A19 ceramic	1030	572	1130	628
4. Galvanized steel‡	480	267	630	350
5. Low-carbon steel, cast iron§	830	461	930	517
6. Stainless steel		686	1380	767
Types 302, 303, 304, 321, 347	1235			
Type 316	1200	667	1345	748
Type 309S	1560	867	1705	950
Types 310, 310B	1610	894	1755	975
Type 430	1310	728	1455	808
Type 446	1730	961	1875	1042

*These and other alloys containing more than 1.0 percent magnesium shall not be used when the reflectivity of the material is utilized to reduce the risk of fire.

†If the reflectivity of aluminum-coated steel is utilized to reduce the risk of fire, the maximum allowable temperature rise shall be 830°F (461°C).

‡The specified maximum temperature rise shall apply if the galvanizing is required as a protective coating or if the reflectivity of the surface is utilized to reduce the risk of fire.

§The specified maximum temperature rises apply to parts whose malfunction may cause the product to be unacceptable for use.

¶See par. 4, top of page 30.20.

Source: *Chimneys, Factory Built, Residential Type and Building Heating Appliances*, UL 103, Underwriters' Laboratories, Northbrook, IL, 19XX.

3. *UL 103 1400°F (760°C) flue gas test:* The maximum temperature attained on all surfaces and on points of zero clearance to the test structure shall not be more than 140°F (78°C) when the flue gas temperature is maintained for 1 h.

4. *UL 103 1700°F (927°C) flue gas test:* This test is conducted for 10 min after equilibrium temperatures are achieved on surfaces of chimney parts and test structures. The maximum temperature attained on all surfaces and on points of zero clearance to the test structure shall not be more than 175°F (97°C) when flue gas temperatures are maintained for 10 min.

For reference the maximum temperature rises above room temperature on any part of the chimney during the 1000°F (538°C) equilibrium test shall not exceed the maximum temperature specified for the material used. See column 1 of Table 30.4.

The maximum temperature rises above room temperature on any part of the chimney during the 1400°F (760°C) tests shall not exceed the maximum temperature specified for the material used. See column 2 of Table 30.4.

5. *Other UL 103 tests:* Various other mechanical tests are run to determine structural integrity:

- Draft test
- Vertical support test
- Strength test
- Wind load test
- Rain test
- Crushing test of nonmetallic flue gas conduit
- Resistance to action of acids of nonmetallic flue gas conduit
- Freezing and thawing test of water absorption of nonmetallic materials
- Cemented-joint test on flue gas conduit
- Sulfuric acid extraction test for porcelain-coated steel used for flue gas conduit

30.2.4 Chimneys for Building Heating Appliances "Only" (UL 103, Modified and Unpublished)

Applications. This type of chimney is not tested with combustible enclosures on all four sides, as the standard UL 103 chimneys are. This type is tested with closures on two sides only. Clearances determined by UL 103 thermal tests which appear on the individual listing cards are applicable only to clearances to combustibles in the appliance room. Above this room the chimney shall be enclosed within a fire-rated noncombustible enclosure. This type is intended for commercial/industrial building heating appliances only.

Chimneys which are designated in the individual listing and on the individual chimney parts as only a building heating appliance chimney are not for use in or on one- or two-family dwellings. These chimneys are to be installed as required for unlined uninsulated steel chimneys or smokestacks. They are not to be enclosed within combustible construction, but an interior chimney is to be enclosed in a noncombustible fire-resistive shaft with a 1-h fire rating for buildings of less than four stories and a 2-h fire rating for buildings of four stories or more where the chimney extends through any story of a building above that in which the connected appliance is located.

An unenclosed chimney in the appliance area may be placed adjacent to walls of combustible construction at the clearances established by Underwriters' Laboratories test procedures. See individual listings or cards for clearance.

Chimneys Tested to 1000°F (538°C) Equilibrium. Chimneys tested to this equilibrium temperature may be used only on those products as found in column II of Table 30.1.

Chimneys Tested to 1400°F (760°F) Equilibrium. Chimneys tested to this equilibrium temperature may be used on those products as found in columns II and III of Table 30.1.

Test Procedures. All test procedures except the 1400°F (760°C) equilibrium test are similar to the UL 103 tests required for low-heat appliance chimneys as previously outlined, but the test enclosure is on two sides only and greater clearances to combustibles are allowed.

The 1400°F (760°C) equilibrium test has been upgraded from the UL 103 1000°F (538°C) equilibrium test to allow this class of chimney to be used on a wider range of products. See Table 30.1, column III.

Building Heating Appliance Chimneys in Current Use. See Fig. 30.11. They come in three types: double-wall air-insulated, triple-wall air-insulated, and double-wall insulated.

30.2.5 Medium-Heat Appliance Chimneys for Commercial/Industrial Use (UL 959)

Applications. Medium-heat appliance chimneys are intended for venting all flue gas temperatures up to 1800°F (982°C) continuously and 2000°F (1093°C) intermittently and any appliances with temperatures below 1800°F, including all low-heat appliances and all appliances suitable for 1400°F (760°C) chimneys. These chimneys are used primarily in commercial/industrial construction on boilers, incinerators, fireplaces, grease ducts, process equipment—any process from ambient to 1800°F (982°C) continuous operation.

Chimneys of this type are tested under UL Standard 959 medium-heat appliance factory-built chimneys and are produced with up to 60-in (1524-mm) inside diameter. Applicable NFPA standards are NFPA 211, *Chimneys, Fireplaces, Vents and Solid Fuel-Burning Appliances*, and NFPA 82, *Incineration and Waste Handling*.

Medium-heat appliance chimneys have been tested at temperatures well above those established for creosote fires and are acceptable for use as fireplace chimneys in high-rise construction. In these situations the runs are usually very long and could lead to condensation of creosote on flue surfaces. The refractory-lined system sustains the fire and contains the creosote during fires with banded sections using silicone under the bands. Some medium-heat chimneys have been tested at 2300°F (1260°C) midflue temperature for 30 min. This test is repeated 3 times without failure and fully contains the very fluid creosote within the system. When medium-heat chimneys are used for wood-burning appliances, it is wise to determine if the product has been tested at the 3-cycle 2300°F (1260°C) 30-min test.

Some medium-heat appliance chimneys have also been listed for use with factory-built listed zero-clearance fireplaces.

Recognizing both the problems of fire fighting in high-rise construction above maximum fire ladder heights and the high cost of replacement of failed metal chimneys, many large cities are modifying their codes to require only refractory-lined chimneys suitable for medium-heat appliances for wood-burning appliances.

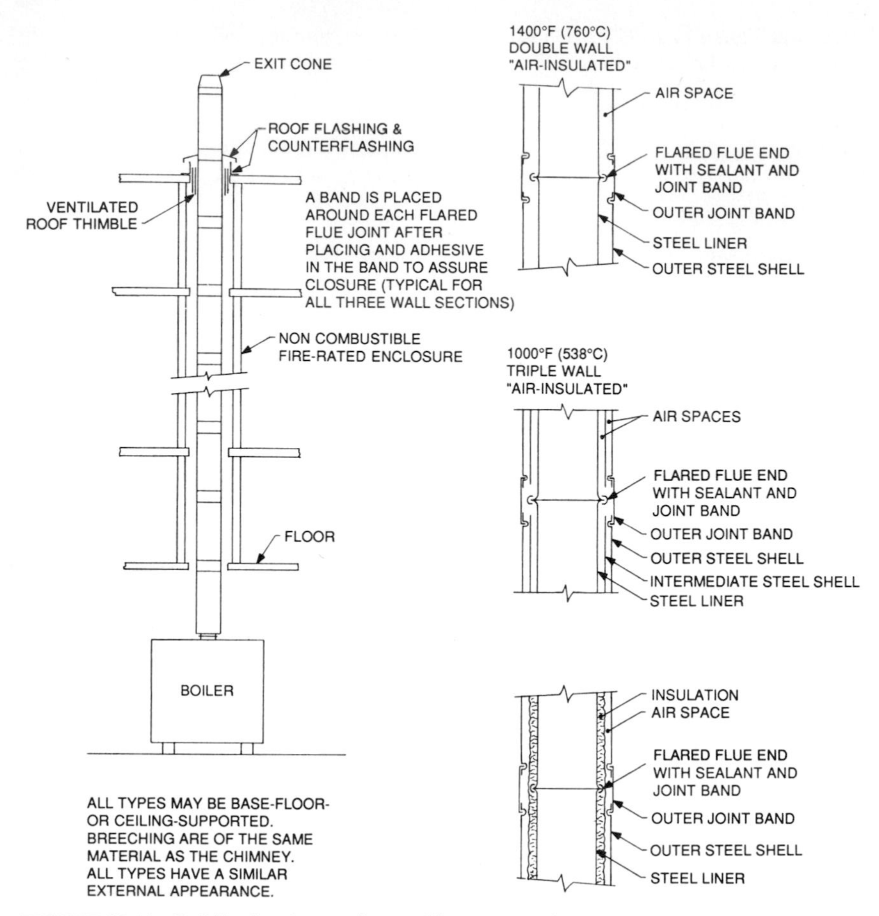

FIGURE 30.11 Building heating appliance chimney types in current use.

These chimneys are not to be enclosed within combustible construction, but an interior chimney is to be enclosed in a noncombustible fire-resistive shaft where the chimney extends through any story of a building above that in which the connected appliance is located. NFPA 211 requires that:

> The enclosure walls shall have a fire resistance rating of not less than 1 hour if the building is less than 4 stories in height. The enclosure walls shall have a fire resistive rating of not less than 2 hours if the building is 4 stories in height.*

Most model codes adhere to NFPA 211; however, it is best to consult the local authorities having jurisdiction. Some areas have more rigid requirements.

---

*Reprinted with permission from NFPA 211-1984, *Chimneys, Fireplaces, Vents and Solid Fuel-Burning Appliances*, Copyright © 1984, National Fire Protection Association, Quincy, MA 02269. This reprinted material is not the complete and official position of the NFPA on the referenced subject, which is represented only by the standard in its entirety.

UL 959 Thermal Tests

- 2000°F (1093°C) thermal shock test
- 1800°F (982°C) water shock test
- 1800°F (982°C) equilibrium test

Mechanical Strength Tests. After the thermal tests have been completed, the chimney assemblies are dismantled and the units are subjected to destructive testing. At the point of failure, the load is divided by 4, to achieve a 4X safety factor, and is reduced by an additional 15 percent for masonry variables. The remaining load is divided by the weight per foot (meter) of the chimney sections, to achieve the allowable height.

All load-bearing components are subject to this type of destructive testing to establish allowable loads.

Corrosion Tests. Refractory chimneys are the only chimney products subject to corrosion tests by flue acids. Underwriters' Laboratories have selected a sulfuric acid solution of $\frac{1}{50}$ *N* that gives a pH of 2 as being representative of normal chimney acids. A sample is immersed for 24 to 48 h between 21 and 32°C (70 and 90°F) and cannot lose more than 3 percent by weight.

Freeze and Thaw Test. Refractory chimneys are the only chimney products subject to this particular test.

Other Tests. Before a UL listing is allowed, guy spacing and maximum height above the top guy tests must be run. Both tests are run by subjecting the assemblies to a horizontal load of 30 lb/ft^2 (147 kg/m^2) to simulate the effect of a 100 mi/h (161 km/h) wind.

Medium-Heat Appliance Chimneys in Current Use. Figure 30.12 shows the only type of chimney listed as a medium-heat chimney at present. It consists of a refractory liner that ranges in thickness from 2 in (51 mm) on a 4-in (102-mm) inside-diameter (ID) unit to 6 in (152 mm) on a 60-in (1524-mm) ID unit. See Fig. 30.12.

The refractory of this type of chimney carries the entire structural load. The maximum height, refractory thickness, and chimney diameters may be obtained from UL listing cards or the manufacturer's catalogs.

Accessories, Components, and Support Systems Provided. See Figs. 30.13 and 30.14. Sections may be installed by placing one above the other and joining with a banded joint and joint cement. Sections may also be welded. An 11-gauge (minimum) jacket should be used for welded sections. Jacket materials may be 28-gauge aluminized steel, 11-gauge corten black iron or stainless steel, and heavier materials as required.

Free-standing Chimneys. Free-standing chimneys are not listed per se by UL, and they must be engineered to suit specific requirements by providing a jacket thickness to accommodate the stresses. Structural codes for free-standing chimneys do not require lugs, if the units have been tested and listed by UL. The design engineer must consider the allowable heights in the structural design. The American Society of Mechanical Engineers (ASME) recently issued a steel stack standard.

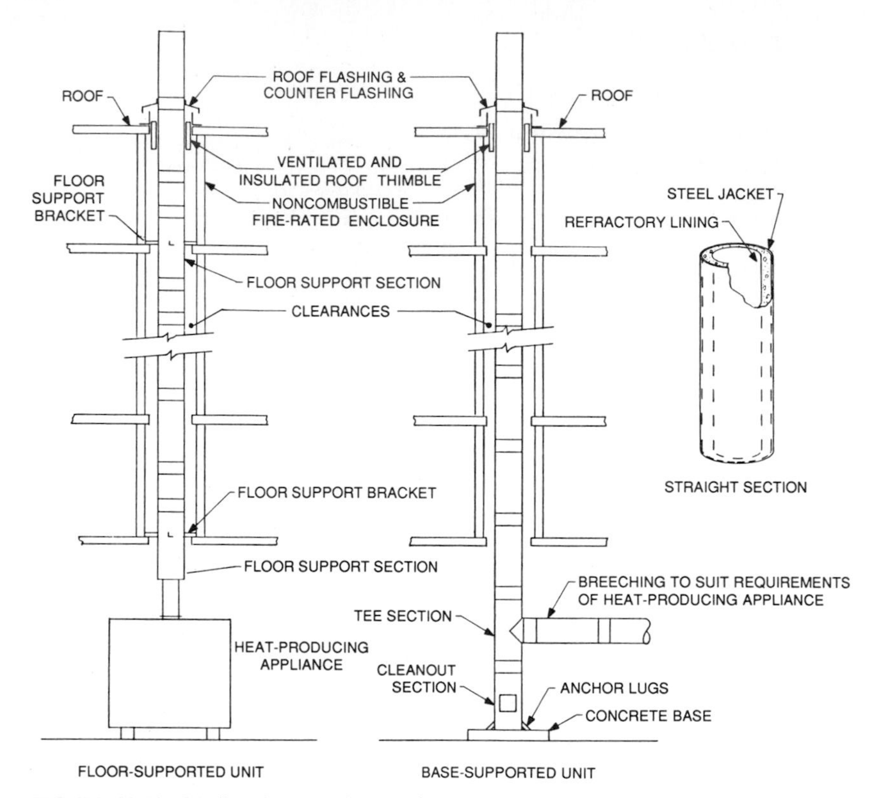

FIGURE 30.12 Medium-heat appliance chimney.

30.2.6 Chimney Connectors (Breechings) for Factory-Built, Listed Chimneys

Definition. The word "connectors," as used in NFPA 211, means the connection between the appliance and the chimney. This is satisfactory for the small sizes associated with the residential field; however, in the general chimney field, the accepted term is "breeching."

Environment. There are no Underwriters' Laboratories standards for breechings; none have been tested so far. Listed chimneys, however, have been used as breechings.

The horizontal nature of most breechings creates problems that are not considered for chimneys. If no condensation occurs in the breeching, there is little difference in operation between the two insofar as corrosion occurs; but if condensation does occur, even in small amounts, gravity carries these corrosive liquids to any depressions that exist in the chimney construction. Metal chimneys frequently have depressions at joints or in areas where a deep bead in the flue acts as a spacer for the outside shell. When small amounts of condensation occur in a breeching, they collect in all the depressions. Under these conditions the area where the condensate collects loses its passivity, and serious crevice corrosion may take place.

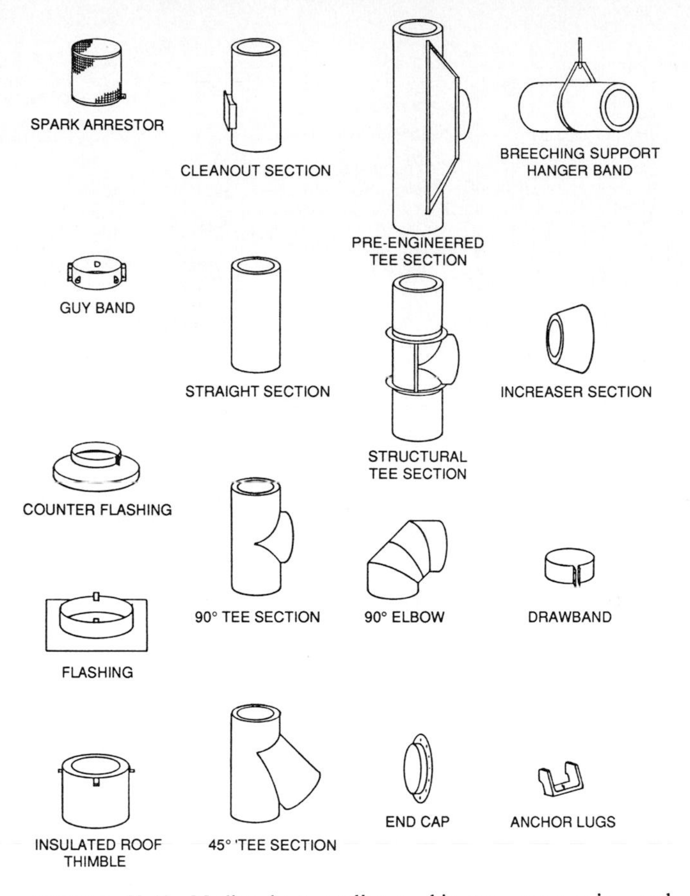

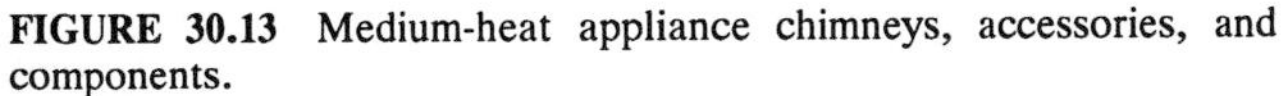

FIGURE 30.13 Medium-heat appliance chimneys, accessories, and components.

No metal chimney is designed for operation below the acid dew point since stainless steel and carbon steel are not suited for this condition. Both steels have about the same corrosion rate in chimney flue environments.

It is prudent to design all breeching systems to operate above acid dew points during all appliance cycles and under the minimum ambient conditions to which breechings will be subject.

Chimney Connectors (Breechings). NFPA requires that connectors be used to connect appliances to the vertical chimney or vent unless the chimney or vent is attached directly to the appliance.

Connectors (Breechings for Listed Gas Appliances Having Draft Hoods and Listed Type L Appliances). See Table 30.5.

Connectors (Breechings) for Appliances Using Listed Metal Chimneys. Connectors for gas-fired appliances without draft hoods, solid-fuel-fired appliances, and all

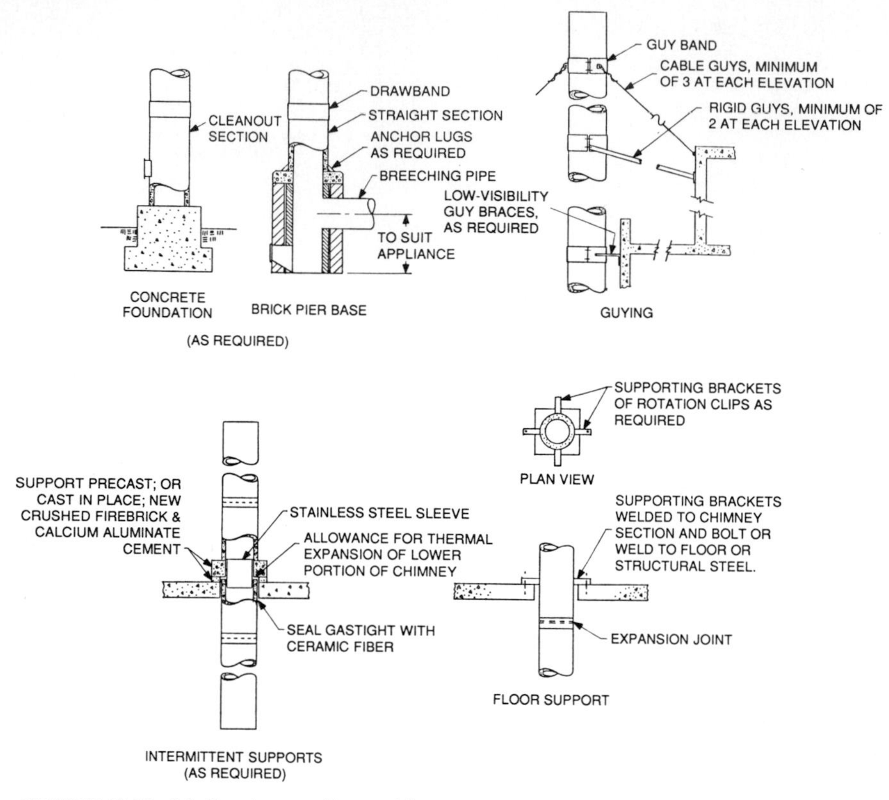

FIGURE 30.14 Medium-heat appliance chimney support systems.

TABLE 30.5 Materials suitable for connectors (breechings)

Appliance type	Attic installation	Appliances not installed in attics
Type B gas	Type B or Type L vent	0.016 in (0.406 mm) (28-gauge) single-wall galvanized steel
Type L	Type L vent	0.016 in (0.406 mm) (28-gauge) single-wall galvanized steel

oil-fuel-fired appliances except those listed for Type L when used in an attic installation may use the following breeching material:

- Factory-built chimney material
- Type L vent material
- Single-wall galvanized-steel pipe (see Table 30.6)

Unlined metal chimneys or chimney connectors are not suitable for any class of incineration.

TABLE 30.6 Metal Thickness for Galvanized-Steel Pipe Connectors

Diameter of connector, in (mm)	Galvanized sheet gauge no.	Minimum thickness in (mm)
<6 (<152)	26	0.019 (0.48)
6–10 (152–254)	24	0.023 (0.58)
11–16 (254–406)	22	0.029 (0.74)
>16 (>406)	16	0.056 (1.42)

Source: Reprinted with permission from NFPA 211-1984, *Chimneys, Fireplaces, Vents and Solid Fuel-Burning Appliances*, Copyright© 1984, National Fire Protection Association, Quincy, MA 02269. This reprinted material is not the complete and official position of the NFPA on the referenced subject, which is represented only by the standard in its entirety.

Breechings (Connectors) for Medium-Heat Appliance Chimneys. See Fig. 30.15. (From this point forward, the accepted terminology is "breeching.") This type of chimney is listed for the entire range of appliances and appliance temperatures, heating appliances, boilers, incinerators, furnaces, and all fuel types (gas, oil, and solid). Thus the choice of connectors from the appliance to the chimney will depend on the temperature at the appliance outlet.

1. *Breechings for appliance outlet temperatures of 600°F (316°C) to 1800°F (982°C):* Under these conditions the connector must be of the same refractory material as the chimney. Metal should not be used in contact with the flue gases since both stainless steel and carbon steel are subject to corrosion by HCl vapor above 600°F (316°C).

2. *Breechings for appliance outlet temperatures of 400°F (204°C) to 600°F (316°C):* Breechings maintained above 400°F (204°C) will be above the dew point for sulfuric acid, so corrosion is not a problem. It is only necessary to reduce heat loss to avoid a flue surface temperature below acid dew points.

3. *Breechings of double-wall steel chimney sections:* This product consists of a stainless-steel flue liner and an outer aluminized-steel shell spaced 1 in (2.5 cm) from the flue. This type utilizes only surface coefficients and the radiation effect and has no true insulation. The insulating value is less than that for a truly insulated breeching; however, if a thermal analysis shows temperatures above 400°F (204°C) and below 600°F (316°C) in the breeching, it is a very satisfactory breeching.

Insulated Carbon-Steel Factory-Built Breeching. See Fig. 30.16. Insulated carbon steels and stainless steels in the flue environment have the same corrosion resistance. Neither is satisfactory for conditions below the acid dew point; however, a 10-gauge carbon-steel breeching is 138 mils (35 mm) thick, or about 4 times thicker than the 35-mil (0.9-mm) breeching of stainless steel used in listed chimneys. Thus carbon steel will last 4 times longer at equivalent corrosion rates and be substantially cheaper. Also 1 in (2.5 cm) of insulation material provides a higher flue surface temperature than the double-wall steel breechings without insulation material do. In air-insulated chimneys, surface coefficients and radiation effects are the only elements providing insulation. See the material on thermal and corrosion analysis of chimney systems in Sec. 30.6.

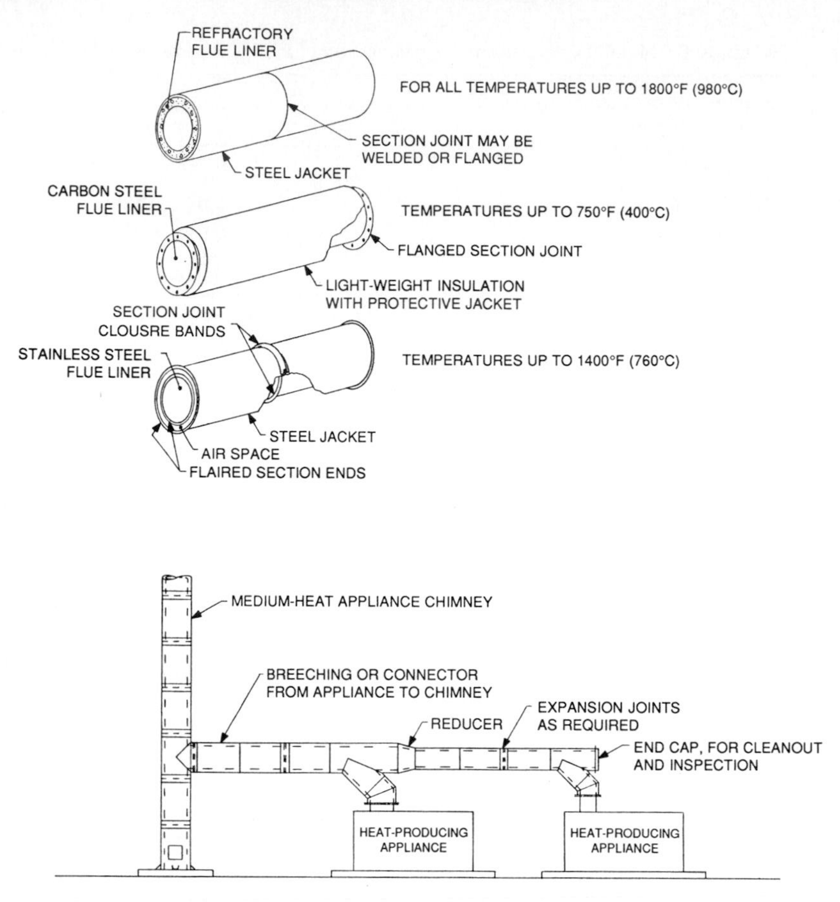

FIGURE 30.15 Breechings for medium-heat appliance chimneys.

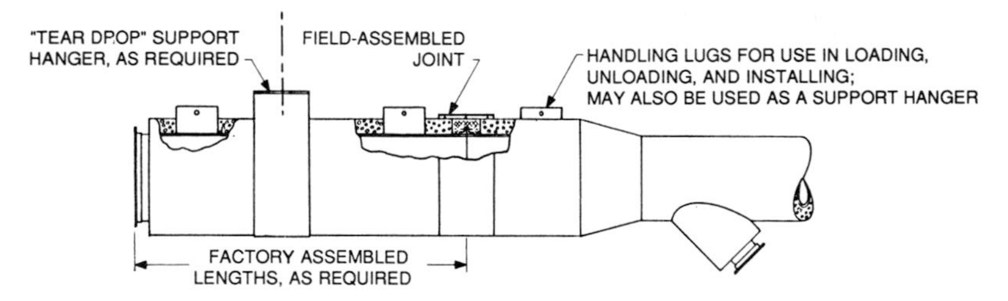

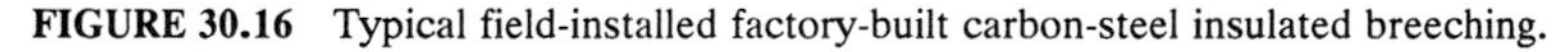

FIGURE 30.16 Typical field-installed factory-built carbon-steel insulated breeching.

Field-Installed Factory-Built Insulated Carbon-Steel Breechings

1. *Typical installation:* (See Fig. 30.16.) Material is shipped in lengths compatible with the installation equipment to be used on the job. Twelve-foot lengths have been selected as the most commonly used standard length.

2. *Field-assembled joints:* See Fig. 30.17. Sections can be provided with factory-installed flanged joints where welding is undesirable. The face of one joining flange is coated with silicone. The bolts provided are installed in the flange and tightened to provide a tight joint. Joints may be field-welded if required. In this case flanges will not be provided, but the steel flue will be beveled.

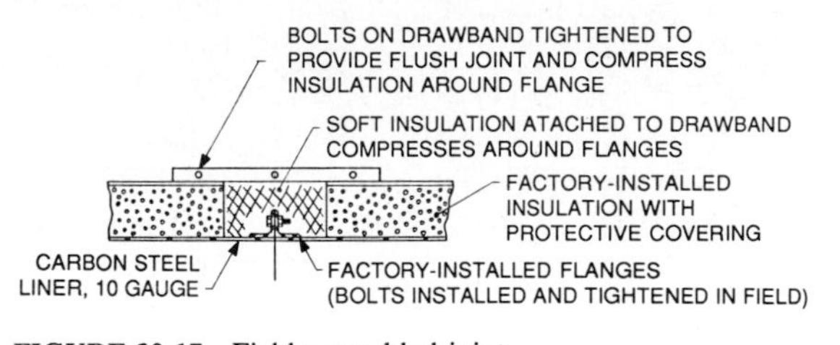

FIGURE 30.17 Field-assembled joint.

3. *Expansion joints:* See Fig. 30.18. This unit can tolerate a maximum expansion of 2 in (50.8 mm). When expansion joints are used, one portion must be secured to a fixed point. The thermal expansion is the same as that of carbon steel.

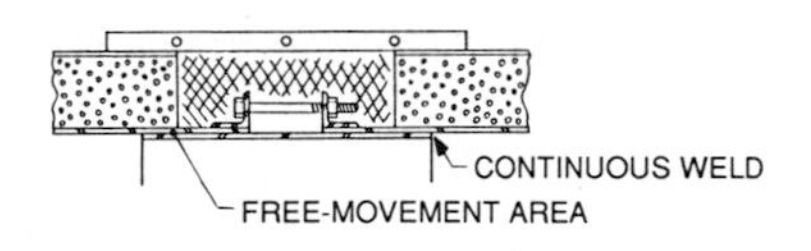

THIS UNIT CAN TOLERATE A MAXIMUM EXPANSION OF 2 IN (51 MM); WHEN EXPANSION JOINTS ARE USED, ONE PORTION MUST BE SECURED AS A FIXED POINT; THERMAL EXPANSION IS THAT OF CARBON STEEL

FIGURE 30.18 Expansion joints.

4. *Support systems:* "Teardrop" support hangers are provided and are spaced as required. Handling lugs used in loading and unloading and during installation may also be used as a support hanger. See Fig. 30.19.

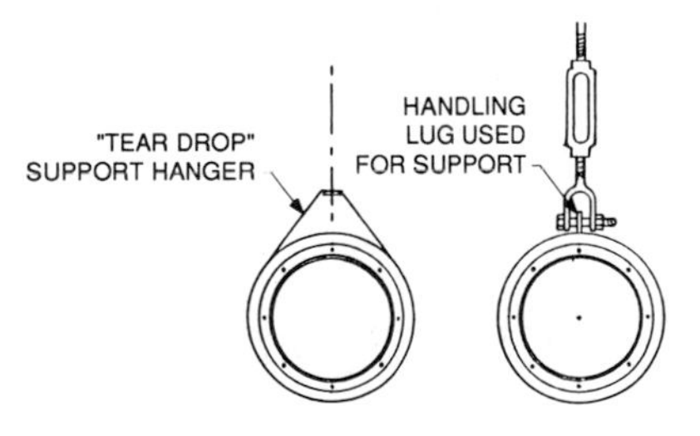

FIGURE 30.19 Support systems.

Connection of Factory-Built Carbon-Steel or Double-Wall Steel Breechings to Medium-Heat Appliance Chimneys. See Fig. 30.20. Medium-heat chimneys coupled with either carbon-steel or double-wall breechings provide systems of highest quality and lowest cost. A typical installation is shown in Fig. 30.21.

Underground Breechings. This is a completely separate class of breeching because of the electrolytic effect that ground currents can have on steel breechings. The breechings require completely water-resistant jackets and nonporous insulation (see Fig. 30.22). A typical direct-burial installation is shown in Fig. 30.23.

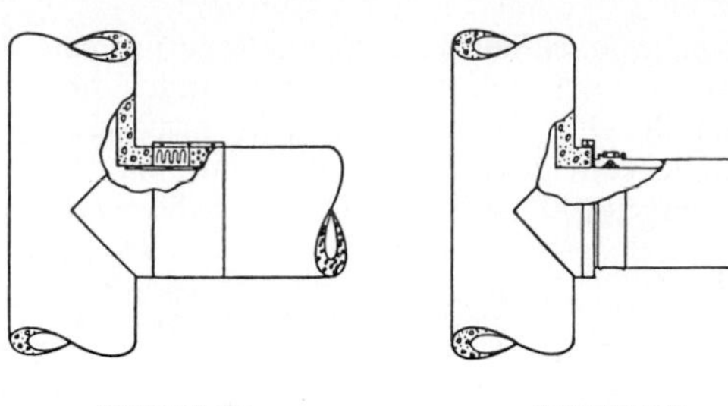

FIGURE 30.20 Connection of factory-built carbon-steel or double-wall steel breeching to medium-heat appliance chimney.

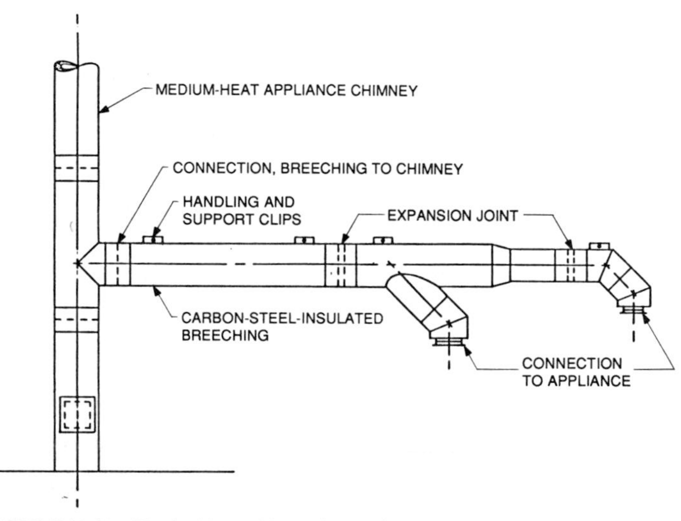

FIGURE 30.21 Typical breeching/chimney installation.

30.3 STEEL STACKS

30.3.1 Definition

It is general practice to call steel chimneys "stacks," no matter the size. Concrete chimneys are always called "chimneys."

30.3.2 Discussion and References

Most steel stacks with outside diameters suitable for over-the-road shipment are factory-fabricated and must be designed in accordance with good engineering practice, which is not covered in this section. However, the references for struc-

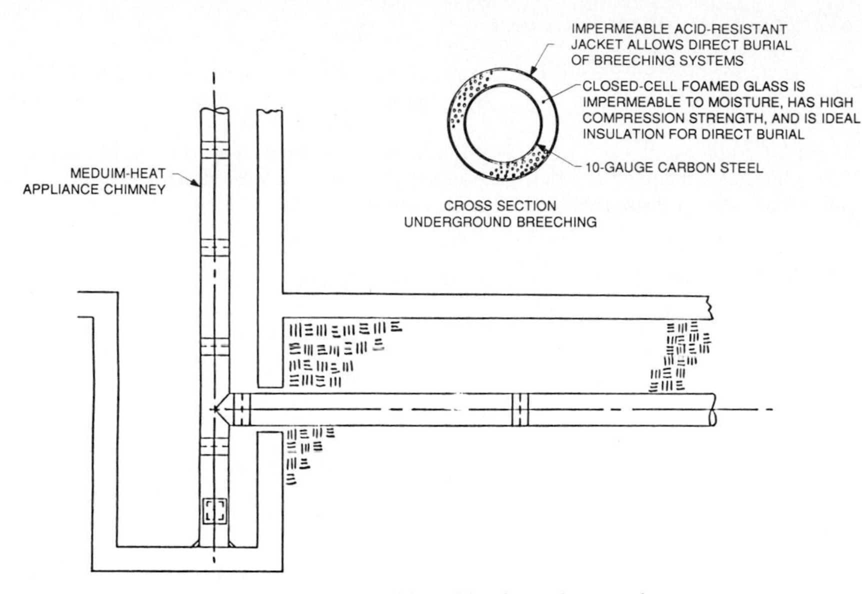

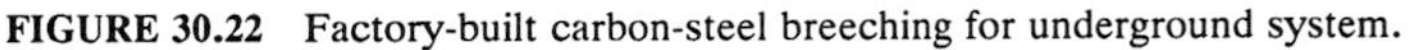

FIGURE 30.22 Factory-built carbon-steel breeching for underground system.

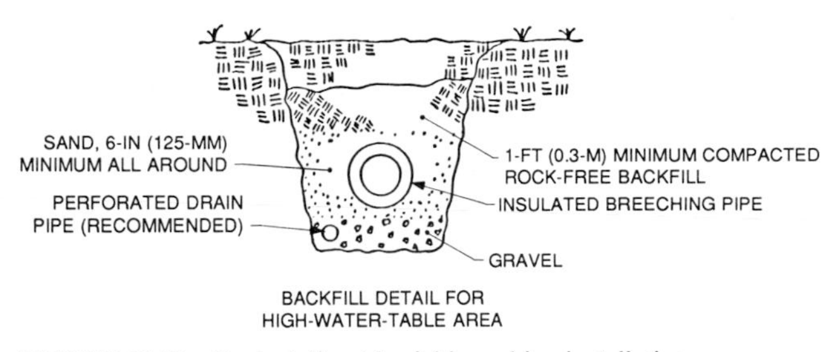

FIGURE 30.23 Typical direct-burial breeching installation.

tural design can be found in the following sources. The cost-effective height of stacks is usually the height of available cranes, or about 300 ft (100 m).

Free-Standing Individual Steel Stacks including Windscreen for Multiple Stacks

- American Society of Mechanical Engineers, *Steel Stack*, ASME, New York, NY 1986
- International Committee on Industrial Chimneys, *International Steel Stack Standard*, Comité International des Cheminées Industrielles, Brighton, UK, 1986 (available from CICIND, 136 North St., Brighton, UK BN1 1RG)
- *Guide for Steel Stack Design and Construction*, Sheet Metal and Air Conditioning Contractors National Association, Inc., Vienna, Va.
- American Society of Civil Engineering journals

Steel Flue Liners within a Windscreen

- American Society of Civil Engineers, *Design and Construction of Steel Chimney Liners*, ASCE, New York, NY, 1975.

Design. Steel stacks are considered tall, slender buildings. But rapid temperature changes through the shell and along its length, coupled with the corrosive nature of effluents, increase the design complexity.

30.3.3 Single-Wall Insulated But Unlined Stacks

Insulated steel stacks are quite suitable for continuous-firing situations such as gas- or oil-fired units that operate above 500°F (260°C) and when chemical loading is low.

When solid fuels are burned such as coal or refuse, the combustion process cannot be turned off until the charge is burned out. If during this burning process the flue gases fall below the acid dew points, then corrosion can occur.

Carbon steel is generally not suitable above 750°F (399°C) due to stress losses, but most chimney designers limit steel temperatures to 500°F (260°C) for free-standing chimneys with large stresses.

External insulation on steel chimneys increases the steel temperatures close to the midflue temperatures.

Note that field experience shows that waste heat boilers range from 600°F (316°C) to as low as 300°F (149°C) and allow flue surfaces to fall below acid dew points, particularly during burndown after the last charge. This cycling below the dew point accelerates stack corrosion.

30.3.4 Double-Wall Steel Stack with Single Flue

Uninsulated. Manufacturers of this type of chimney often call the units "air-insulated," but this is a definite misnomer. The air within the annular area is in motion; the air is in contact with the flue going up and with the outer shell going down. Some units are open at the top, bottom, and joints. This makes for a more positive upward flow, but countercurrents still exist. If units are hermetically sealed, the countercurrents operate within the units.

In all cases the only insulation value is that from the radiation effect of the outer shell and surface coefficients. There is some insulating effect, but it does not approach the insulating value of a lightweight refractory lining and is substantially below material with a proper thermal conductivity.

Good insulation can reduce heat loss and maintain a very low loss between midflue and flue surface temperatures, but it cannot raise temperatures. If flue surface temperatures fall below acid dew points, condensation will occur. High-efficiency boilers have been observed to have midflue temperatures as low as 190°F (31°C) at the appliance outlet. This is well below acid dew points, and condensation will occur no matter how good the insulation. This seems obvious, but field experience warns against overlooking this simple point.

Insulated. The various possible locations of insulated double-wall steel stacks have been cited as a benefit in their sales promotion, but testing and thermal analysis show them all to be about the same, with the thermal conductivity of the insulation being the controlling factor.

The inner shell is the flue which conducts the flue gases. The outer shell is usually the structural element, and it is designed to be isolated from flue gases. This condition exists only if the flue liner is not corroded. When this occurs, damage is hidden and may not surface until failure of the structural shell occurs.

Unlined double-wall steel chimneys of all types must have frequent inspections of both the flue liner and the outer shell, to avoid serious problems.

Pressurized or Ventilated Annular Areas. To avoid penetration of flue gases into annular areas after failure has occurred in the flue liner, some manufacturers apply a positive pressure to the annular area and require flexible closures between the flue and the outer shell. The positive pressure depends on the workmanship and porosity of the closure. Other manufacturers use a ventilated annular area with an opening at the bottom and top of the chimney. Advocates of both systems claim their product is better, but field experience shows that failures in all types of double-wall steel chimneys are numerous.

The outer shell, in effect, provides weather protection to the insulation, but it also creates a potential for hidden damage. If temperatures can truly be maintained above acid dew points, such as in gas- or oil-fired boilers, then double-wall steel chimneys are quite acceptable. Use of double-wall steel chimneys is questionable for coal-burning or incineration equipment without protection to the steel flue.

30.3.5 Typical Installations

Single-Wall Refractory-Lined Free-Standing Chimneys. See Fig. 30.24.

Single-Wall Chimney Lined with Acid-Resistant Borosilicate Block. See Fig. 30.25.

Free-Standing, Roof-Mounted, Refractory-Lined Chimneys with Helical Strakes to Counter Vibrational Forces. See Fig. 30.26.

Double-Wall Single-Flue Stacks. See Fig. 30.27.

Multiple Flues in a Structural-Steel Windshield. These systems usually have ventilated areas between the flue liners and the windscreen, but they could be under positive pressure.

Roof-Mounted Tower-Supported Chimney. Standard UL listed medium-heat appliance chimneys can be supported within a tower structure. This procedure is generally used when loads must be supported over a wide area of roof.

Cluster-Supported Chimneys. See Fig. 30.28. Architecturally pleasing clusters of chimneys can be designed when multiple flues are involved. When these structures are designed, the close relationship of the chimneys necessitates careful consideration of vibration effects.

30.4 PRECAST REINFORCED-CONCRETE CHIMNEYS

30.4.1 Introduction and References

Conventional reinforced-concrete chimneys of the large free-standing type, as used on power plants, etc., are built in accordance with the American Concrete

FIGURE 30.24 Single-wall refractory-lined free-standing chimney.

FIGURE 30.25 Single-wall free-standing chimney lined with acid-resistant borosilicate block.

FIGURE 30.26 Free-standing, roof-mounted, refractory-lined chimney with helical strakes to counter.

FIGURE 30.27 Vibrational forces in double-wall single-flue steel stacks.

Institute's (ACI) Standard 307 in the United States and Canada. This standard is concerned with only the structural design of tall, slender structures. ACI Standard 367 for precast concrete chimneys is being written.

When steel flue liners are used in these structures, the liners are usually designed in accordance with the American Society of Civil Engineers' (ASCE) *Design and Construction of Steel Chimney Liners*.

Both of these standards are concerned purely with the structural aspects. The effects of corrosion factors and the selection of lining materials are left to others. This area is the basis of this chapter.

The International Committee on Industrial Chimneys (CICIND) has developed a model code for concrete chimneys that has had great acceptance throughout the world and should be reviewed by anyone seriously interested in concrete chimneys. (For information on chemical loading, see CICIND Model Code for Steel Chimneys.)

Conventional reinforced-concrete chimneys are poured as monolithic shells, usually cylindrical, by either a slip-form technique or a jump-form procedure. With a slip form, the form is moved upward slowly as concrete is poured, allowing sufficient time for the concrete below to support itself. In the jump-form procedure, the form is placed and the concrete poured and allowed to set before the form is "jumped" to the next casting position. Reinforcing elements are placed both circumferentially and vertically in the outer area with sufficient cover over the rebar as required by the standards.

It is evident that this entire operation must take place at the job site for a considerable time until construction is complete.

Precast reinforced-concrete units are built to the same standards but are built off-site and are not carried to the job site until everything is ready. The foundation is cast (usually by others) with the rebar embedded securely in a configuration to suit the particular precast units used.

When construction begins, the units are placed one on the other until the structure is complete. This reduces the on-site time from months to weeks. Typically an 80-ft (25-m) high chimney can be constructed on-site in approximately 5 working days whereas a conventional cast-on-site chimney may take 5 weeks.

FIGURE 30.28 Cluster of three refractory-lined steel chimneys.

30.4.2 Precast Chimneys

Precast chimneys of all types are limited in height by the height of the tallest crane available in the area, or about 300 ft (90 m). Slip-form and jump-form techniques do not always require cranes, and so the chimneys can be built considerably higher. Chimneys over 300 ft (90 m) could be built with precast units; however, they compare less favorably to the conventional field-cast chimneys as the height increases.

Size Limitations. Precast units are factory-cast at facilities as close to the job site as possible, and this proximity determines the maximum sizes that can be handled as individual units. This height is usually about 12 ft (40 m) and depends on local regulations concerning oversized loads.

Sizes greater than those allowed by maximum overloads can be reduced by segmenting the units into two or three pieces to accommodate the shipping limitations.

Advantages. First, precast designs are limited only by the imagination. (See Fig. 30.29.) Precasting allows any variation of texture, color, or shape consistent with precast concrete and enables a chimney to become an architectural feature rather than an eyesore.

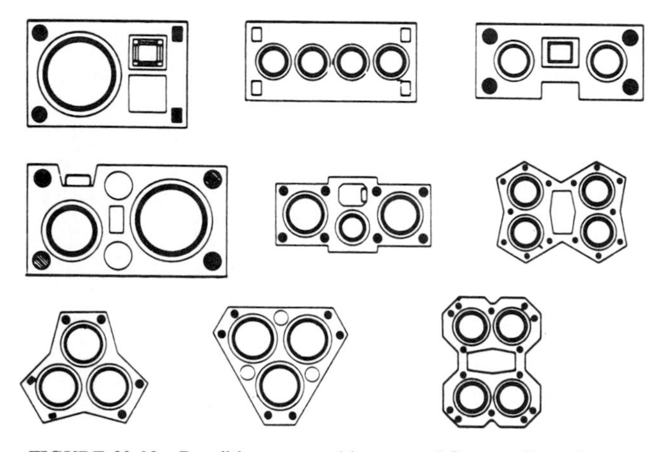

FIGURE 30.29 Possible precast chimney and flue configurations.

Second, precasting enables holes to be made to accommodate electric and other services to the chimney over its entire length, and precasting leaves areas for personnel lifts or ladders to carry service personnel to the various items of equipment on the chimney.

30.4.3 Types of Precast-Concrete Chimneys

Three types of construction have been in use for a number of years. The earliest type, which we call the "classical type," began to see use in the early 1950s. There are two patented systems, one invented by Kenneth Roy Jackson of the United Kingdom and the other invented by George Richter of West Germany, called the Jackson and Richter systems, respectively.

Classical Type

Discussion. This type of unit has been used in Great Britain since about 1950 and is produced by individual castings ranging from 30 in (75 cm) to 5 ft (1.6 m) high as a convenient size for handling. These basic units contain the required rebar in accordance with controlling standard and with adequate concrete cover over the rebar to accommodate texture or surface configurations.

FIGURE 30.30 Classical precast sections.

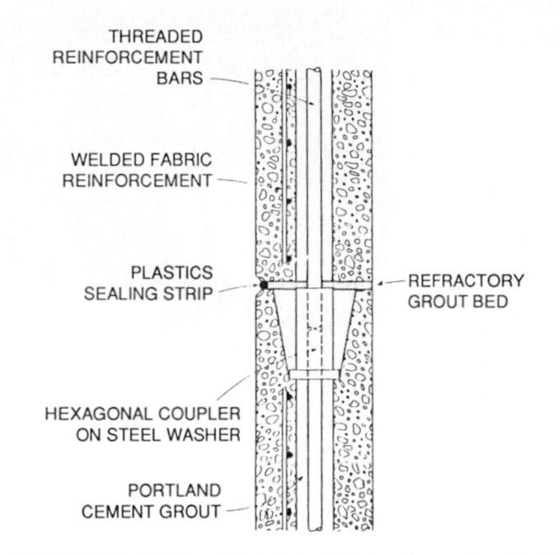

FIGURE 30.31 Tendon joints for classical system.

Installation. These shells of high-strength concrete utilize core holes (Fig. 30.30) in the outer shell to allow installation of steel tendons that connect the casting to each other. See Fig. 30.31. These are substantial steel rods with threaded ends. The first set of tendons is cast into the footing using a template designed for the specific unit to be used. These tendons are usually designed to be jointed by hexagonal couplers on steel washers at each joint.

Core holes to accommodate tendons are designed to allow superflow low-shrink grout to be placed in the annular space between the tendon and the casting.

Flue Support Floors. Flue liners are supported by a concrete floor that carries the weight of the flues. This type of chimney is designed most economically with all units of an identical shape and size.

Basic Concept. This type of structure, joined securely together with substantial tendons, is based on the concept that the joining system actually provides positive joining under controlled conditions. In jump-formed reinforced-concrete chimneys, the concrete has set before the next pour, and the joint is dependent upon surface conditions and the rebar penetrating the column to provide strength.

Liners. Liners for this type of structure can be free-standing acid brick structures or steel liners with whatever insulation and lining material are required for the chimney environment. When brick is used, it must be recognized that both the acid brick and the potassium silicate mortar used to join the bricks are porous. If positive pressures occur within the chimney, acid vapors could be forced through the flue wall into the area between the flue and windscreen. This may require protection of the portland cement surfaces. It is prudent to design chimneys with brick liners for negative pressure use only. Tongue-and-groove bricks provide tighter joints and minimize condensate seepage through joints.

Richter Type. U.S. patent 4,381,717, May 3, 1983, issued to Alfons Krautz.

Discussion. This patent covers the use of precast-concrete sections axially stacked one upon the other with flue apertures spaced within the casting to allow installation of free-standing flues. This chimney type was invented by George Richter of West Germany and was first issued as a German patent.

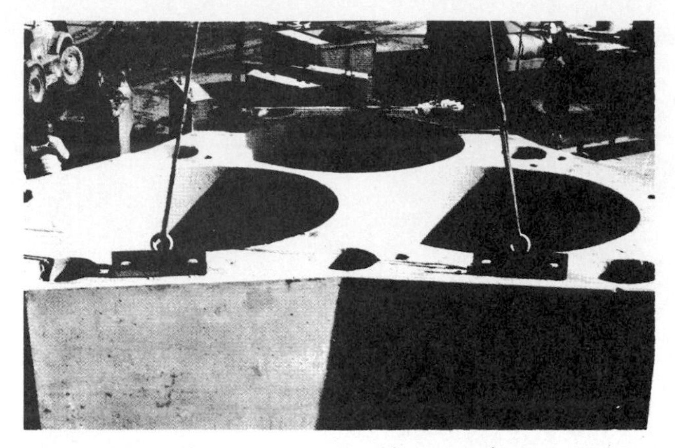

FIGURE 30.32 Richter-type precast chimney section.

FIGURE 30.33 Rebar guided through core holes (Kraut patent).

The difference between this type of structure and the classical type lies in the nature of the vertical reinforcing system. In this system, the core holes are increased in diameter to accommodate bundles of rebar. (See Fig. 30.32.) All other features are about the same.

Installation. The rebar is secured in the footing, bound together and guided through the core holes. (See Fig. 30.33.) With the rebar in place, plasticized low-shrink grout is placed in the holes and vibrated to fill all cavities in the rebar bundle and to engage the depressions in the core holes. As each section is placed, grout is installed on the horizontal surfaces of the castings. These procedures ensure a truly monolithic structure.

Additional castings are threaded with rebar until the structure is complete. Acid brick linings are installed as the castings are installed. Steel flues with various linings are installed when the structure is complete.

Flue Support Floors. Sections with vertical joints and flue liners are handled as in the classical units.

Basic Concept. The use of rebar bundles supports Richter's concept that these units in themselves can be compared to a tower designed of reinforced-concrete legs where the remaining concrete is used to handle lateral and other forces.

Liners. Liners are the same as in the classical type.

Jackson Type. U.S. patent 4,104,868, August 8, 1978, issued to Kenneth Ray Jackson, England.

Discussion. This concept is different from the preceding one in that no holes are cast in the shell to provide for in-situ plasticized concrete and rebar. Two shells are cast, an outer shell of reinforced portland cement, which has all the versatility of the classical and Richter systems, and an inner unreinforced acid-resistant refractory lining. The sizing of these shells is such that an annular area remains between the two cores. See Fig. 30.34.

FIGURE 30.34 Jackson precast chimney system.

This annular area is to be filled with rebar and in-situ structural concrete to form a reinforced-concrete column that meets ACI Standard 307. This column is the structural element and could stand by itself if the inner and outer shells were removed. In effect, the precast units are simply forms.

Jackson stresses that this system of casting utilizes the shrinking effect of the in situ concrete to apply pressure to the refractory flue liner, not only to transfer part of the weight of the flue (through friction) to the structural element but also to minimize thermal cracking that normally occurs in refractory concretes.

Shells with Vertical Joints. When vertical joints are required, horizontal reinforcement is needed as well as vertical reinforcement. This can be accomplished by securing connectors to the reinforcing of the cast segments. The horizontal reinforcing may be secured to these connectors.

FIGURE 30.35 Jackson footing.

Installation. Rebar is secured to the footing as with the other types and located in accordance with the particular design. See Fig. 30.35.

The base sections are grouted in place, and drains are properly installed. Two additional sets of castings 4 ft (1.2 m) high with female/male joints are installed, and in situ concrete is poured in the annular area. The chimney continues upward until completion.

30.4.4 Review of Precast Systems

The classical and Richter types are similar. In both cases the area between the shell and flue is open and can be relatively accessible. This area can be placed under positive pressure to ensure that flue vapors cannot penetrate porous flue structures. This area may also be ventilated to allow air flow into the bottom and out the top, to remove vapors. In the case of the ventilated airspace, the portland cement surfaces must be protected against acids since they can condense on this colder surface. It is most desirable to consider operating chimneys with porous flue liners under negative pressure. If positive pressures are required in the flue, the annular area can be pressurized at a greater pressure than the flue or a nonporous steel lining can be used.

The Jackson-type system is most effective as a single-flue, essentially circular structure. Multiple flues can be installed but will usually require more in situ concrete than is necessary.

If positive pressures are to be used in the flue, this will create thermal and pressure gradients through the entire structure. Corrosive flue acids can be forced through the porous refractory to the portland cement–refractory interface. This can be very damaging to the portland cement structural element, and damage could go unobserved until failure occurs. To avoid this problem with the Jackson system, the acid-resistant refractory flue liners can be coated with acid-resistant nonpermeable fiberglass-reinforced plastics. In the classical and Richter systems, the portland-cement interior surfaces of the castings can be protected with acid-resistant membranes if the system is to be placed under sustained positive pressures.

Designers should also investigate the compression strength of the lining material to determine if it can truly sustain its own weight. The frictional forces are factors to consider, but one must not forget the shear forces involved.

The most negative aspect of the Jackson system is the difficulty involved with repairing the flue linings if failure does occur.

Some rather spectacular structures have been built from precast units. See Figs. 30.36 and 30.37.

30.4.5 Bibliography

Pinfold, Geoffrey M.: *Reinforced Concrete Chimneys and Towers*, Cement and Concrete Association, London, 1975.

Jackson, Kenneth R.: *A Guide to Chimney Design*, I.P.C. Science and Technology Press Limited, Guildord, England, 1973.

30.5 CHIMNEYS FOR INCINERATORS

30.5.1 Discussion

In the past the use of refractory-lined chimneys for incinerators was unquestioned, and they worked exceptionally well. When heat-recovery units and air pollution control devices are added to the system, designers often have second thoughts about the chimney. Design factors showing temperatures above acid dew points and below critical high temperatures sometimes suggest that unprotected steel can be used.

It is unfortunate that design requirements are not always met; in fact, incinerator manufacturers guarantee incinerators to operate without problems only 85 percent of the time. Field data show that scrubbing systems operate only part of the time that the incinerator is burning and the boiler is functioning, due to equipment breakdown. All this may fall into an acceptable area insofar as air pollution control is concerned. However, little thought is given to what happens to the chimney environment during downtime of acid gas removal systems, or what happens when the 5 percent of acid gas passing through a 95 percent efficient scrubber condenses in the chimney. We discuss the chimney environment under normal operating conditions as well as optimum design conditions to determine how corrosive that environment is.

Many of the reference papers cover conditions of refractory or water walls in furnace and boiler tubes in heat-recovery units. Problems found in these areas

FIGURE 30.36 Multiflue Richter-type chimney.

FIGURE 30.37 Highly fluted Richter type.

may be reduced by air pollution control devices, but problems still arise in chimneys.

30.5.2 Overview

Corrosives. Incineration of waste products produces a wide range of material which can cause corrosion of flue liners, softening and flow of refractories below their design temperature, and other problems such as spalling of refractories if conditions within the flue reach critical temperatures and conditions.

An analysis of exhaust gases from municipal incineration[1] shows the presence of nitric oxide, chlorides, bromides, fluorides, phosphates, organic acids, alkalies, aldehydes and ketones, sulfur dioxide, zinc, lead, and other materials.

The prime corrosive materials are sulfuric acid, hydrochloric acid, and hydrofluoric acid.[12] Sulfuric acid is the first to condense, hydrochloric acid is the most common acid produced by halogens, and hydrofluoric acid is the most corrosive. Halogens are corrosive in both vapor and liquid phases.

These corrosives in flue gases can produce corrosion at temperatures over 600°F (316°C) in the case of halogen acids and below 400°F (204°C) for sulfuric and other condensed flue gases.

Zinc Chloride. Zinc chloride is volatile but readily condenses on cool furnace walls at 500°F (260°C) or less and can be carried to chimneys if scrubbing systems are bypassed or are not used in the system. Corrosion rates on carbon steel are high at elevated temperatures and continue at low temperatures.

Alkali Materials. The alkalies—soda (Na_2O), potash (K_2O), lithia (Li_2O), etc.—are introduced as salts and halides as plastics and other materials. Distress due to alkali attack may be seen as glazing or dripping of the refractory if the temperature is high enough or as shelling or decrepitation of the surface at lower temperatures.[5]

Case histories show that at 1800°F (982°C) to 2200°F (1204°C) a 3 to 10 percent alkali concentration in the slag can disrupt most refractories. Therefore the chemical composition of the fuel load is of extreme importance.

Incineration System Temperatures. Incineration temperatures range from 1400°F (760°C) to 2000°F (1093°C) at the incinerator outlet.[2] Chimney temperatures are reduced from these values by barometric dampers or other like devices. This range of temperatures has been the norm for many years. Medium-heat appliance chimneys with calcium aluminate bonded refractories are the standard for this range. The use of waste heat boilers to recover the energy of incineration reduces chimney temperatures. This temperature range provides considerable opportunity for the condensation of flue acids:

600°F (316°C) maximum

400°F (204°C) normal

300°F (149°C) minimum

Gas scrubbing devices can bring flue gases as low as 86°F (30°C), but typical temperatures range from 140°F (60°C) to 180°F (82°C) at saturation levels; all guarantee flue gases below acid dew points.[18]

As the presence of plastics increases in the mix to be incinerated, the incineration temperatures are usually increased from 1850°F (1010°C) to 2200°F (1316°C) with a 2-s dwell time. It is in this range that alkali problems become severe. Problems with refractories can occur at lower temperatures but are less likely to.

These changes from the norm of past years make it very important to understand the true effect of the incineration process on the exhaust system being designed.

30.5.3 Chemical Changes in Flue Gases from Incineration of General Waste

Tests on municipal solid waste run in 1975 and again in 1985 show some interesting changes.[1] See Table 30.7. Sulfur trioxide was not reported in either case, so

TABLE 30.7 Corrosives in Flue Gases from General Refuse Incineration

Corrosive	1975	1985
Water	10.3%	11.5%
Sulfur dioxide (SO_2)	104 ppm	400 ppm
Hydrochloric acid (HCl)	325 ppm	800 ppm
Hydrogen fluoride (HF)	Not reported	80 ppm

we assume that 2 percent of the sulfur dioxide was converted to SO_3 by using Fig. 10 in Ref. 12. The dew points in Table 30.8 were determined.

TABLE 30.8 Dew Point of Sulfuric Acid

	1975	1985
Sulfur trioxide	2.1 ppm	8 ppm
Dew point of H_2SO_4	267°F (131°C)	289°F (142°C)

Sulfuric Acid. The significant increases in corrosives in general waste from 1975 to 1985 clearly demonstrate the changing nature of general waste. The increase in H_2SO_4 is due to the increased amount of sulfur compounds in waste. We calculated the amount of SO_3 based on 2 percent of the SO_2 content and a 15 percent moisture content. An increase in water content and the presence of catalysts such as Fe_2O_3 and other similar materials could show increases even greater than the 2 percent selected. It is encouraging that the recent surge in resource recovery systems has brought to light the many problems that exist, and now many companies are reporting sulfur trioxide and halogens. Designers can establish the acid dew point and the concentrations of sulfuric acid by knowing the SO_3 content of flue gases. These can be determined by reviewing Sec. 30.6.2.

Hydrochloric Acid. Hydrochloric acid in flue gases shows a 246 percent increase from 1975 to 1985. This is due primarily to increases in plastics. However, in 1975 about 50 percent of the HCl formed was due to common table salt (NaCl, sodium chloride) and 50 percent due to plastics. The amount of table salt may have increased somewhat, but the amount of plastics in general refuse has increased significantly. Normal refuse in one series of tests showed about 550 ppm of hydrochloric acid. However, when 2 and 4 percent more plastics were added, the amount of hydrochloric acid in flue gases increased up to 3030 ppm at the 4 percent increase level.[1] This points to the importance of knowing not only the amount of plastics in an average load but also the amount in any given charge. Hydrochloric acid is a very powerful corrosive that reacts rapidly when in contact with most materials.

Federal and state regulations often require temperatures from 1850°F (1010°C) to 2200°F (1204°C) for burning quantities of plastics and other hazardous wastes.

It is prudent for incinerator designers to consider charging means that will provide a relatively uniform mix of materials in each charge.

Zinc Chloride. Zinc chloride is present in textiles, leather, and rubber.[20] Zinc oxide is the most common of the zinc compounds and is relatively nonvolatile and innocuous. However, in the flame, zinc oxide is readily reduced to metallic zinc which reacts with HCl to form $ZnCl_2$. Zinc chloride is volatile but readily condenses at temperatures of 500°F (260°C) or below. Zinc (and lead) chloride may be present as a liquid. Zinc chloride and sodium chloride form a eutectic which melts at 504°F (262°C). Zinc chloride destroys the protective oxide film on steel and facilitates the rapid corrosion by HCl.

Dew Points of Corrosives in Flue Gases. The actual acid dew points (Table 30.9) are of value, but since temperature measurements in chimneys are particularly difficult, it is common practice to add 68°F (20°C) as a safety factor in establishing a critical or signal temperature of flue surfaces below which condensation can take place.[14]

TABLE 30.9 Actual Dew Point of Flue Gas Corrosives

Corrosive	Dew point range (Ref. 14)
Sulfuric acid (H_2SO_4)	160°F (71°C)–330°F (166°C)
Hydrochloric acid (HCl)	80°F (27°C)–140°F (60°C)
Hydrobromic acid (HBr)	85°F (29°C)–155°F (69°C)
Hydrofluoric acid (HF)	Water dew point

30.5.4 Signal Temperatures for Corrosion Problems

The most important acids for "signal" purposes are, first, sulfuric acid (H_2SO_4), which is the first to condense as flue temperatures are lowered, and, second, hydrochloric acid, which is the most common halogen present. In both cases we add the 68°F (20°C) safety factor to the actual dew points. (See Table 30.10.) (We have elevated the hydrochloric acid dew point slightly to accommodate hydrobromic acid to provide signal temperatures—temperatures below which we can expect corrosion from condensed flue acids.)

TABLE 30.10 Signal Temperatures below Which Flue Acids May Condense

Flue acid	Signal temperature's actual dew point plus 68°F (20°C) safety factor
Sulfuric acid	398°F (186°C)
Halogen acids	223°F (89°C)

If a thermal analysis shows any portion of the flue surface below the signal temperatures, a more detailed thermal and corrosion analysis plus an analysis of the materials to be burned should be considered.

30.5.5 Combustion Problems

Heating Loads of Plastics. The value of municipal waste constituents ranges from about 1000 to over 20,000 Btu/lb (2300 to 46,000 kJ/kg) with an average of about 4577 Btu/lb (10,527 kJ/kg).[2] The highest Btu (kJ) values are found for plastics in Table 30.11 and for other materials such as rubber.[1] These can cause serious combustion problems with nonuniform firing rates.

TABLE 30.11 Heating Load of Four Common Plastics

Plastic	Btu	kJ/kg
Polyethylene	19,687	46,067
Polystyrene	16,816	39,468
Polyurethane	11,203	26,215
Polyvinylchloride	9,754	22,824

Care should be taken to establish a uniform mix of waste materials, particularly with hospital waste where chimneys are usually enclosed within walls surrounded by occupied rooms, which will make repairs difficult, or where loss of refractory could increase wall temperatures above acceptable limits.

Incomplete Combustion. Plastics can create a serious overfiring condition if they are not mixed uniformly with the refuse charge or if the unit is overfired. If, e.g., only plastics were charged, the Btu (kJ) value would increase substantially with the volume of combustion gases perhaps increasing to a volume greater than the incinerator design allows. This can cause two important problems that are often overlooked.

First, insufficient oxygen can create large quantities of carbon monoxide for sustained periods while pure plastics or high plastic charges are made. Second, the incompletely combusted materials will enter the flue system unburned until such a time as they obtain a source of oxygen. This can be at a point following a barometric damper or cleanout door. When combustion occurs within a chimney system, the chimney is subject to the more serious problems associated with combustion chambers. See Fig. 30.38.

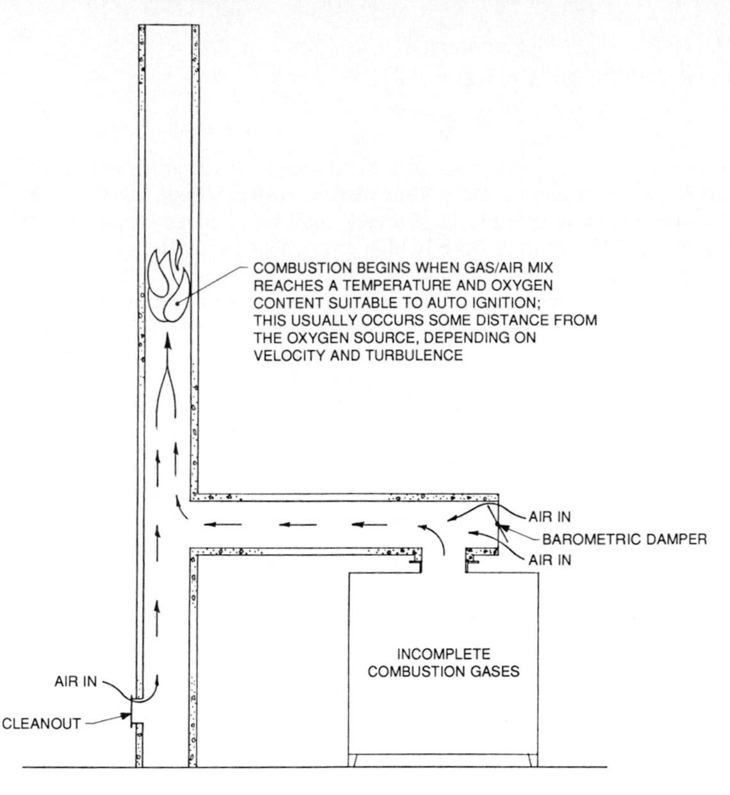

FIGURE 30.38 Combustion in chimney due to incomplete furnace combustion.

30.5.6 Chimney Problems

Alkalies in Refuse. Alkalies are introduced as salts and halides during the incineration process.[3] Many plastics, particularly polyethylene and polyvinylchlorides, contain from 0.0001 to 0.001 percent vanadium.[1] When burned, the combustion gases transport particles of these contaminants in a vapor, fused, or molten condition to various parts of the combustion chamber and adjoining flues.

It has been found[4] that vanadium pentoxide in itself does not react with the constituents of fireclay or high alumina when heated as high as 3100°F (1704°C). Rigby and Hutton[5] found that vanadium pentoxide in the presence of soda acts both as a flux and as a mineralizer to promote chemical change. Sodium aluminum silicates in the presence of soda and vanadium were formed at as low as 1472°F (800°C), whereas these silicates did not appear below 1652°F (900°C) when soda alone was present.

Studies of incinerator slags[6] have indicated that alkali materials such as Na_2O, K_2O, and Li_2O are found in the following quantities:

Soda (Na_2O)	0.6–11.6 percent
Potash (K_2O)	0.3–8.1 percent
Lithia (Li_2O)	0.03–0.13 percent

These materials are about equally destructive to refractory materials.[7] Alkalies form low-temperature melts that distress refractories, particularly those of low density. Distress due to alkali attack may be seen as glazing or dripping of the refractory if the temperature is high enough or as shelling or decrepitation of the surface at lower temperatures.

When free chlorides[3] are present with alkali, these compounds may condense within the lining, forming expansive alkali chloride phases. The presence of these phases can result in cracking of the refractory.

Case histories show that at temperatures of 1800°F (982°C) to 2200°F (1204°C) a 3 to 10 percent alkali concentration in the slag can disrupt most refractories.

Precautions for Incinerator Operators. Incinerator operators must carefully monitor the mixing of plastics with general waste, to avoid conditions that allow incomplete combustion in the incinerator. This could lead to afterburning in the chimney and breeching system and can seriously damage refractory materials. Note that if a unit is operated for substantial periods without proper combustion air, combustion can take place within the exhaust system. Slagging can occur at temperatures below 1500°F (816°C) with no evidence of slagging in the combustion chamber, particularly if vanadium compounds are present.

The presence of alkalies and vanadium compounds in waste materials makes it essential to observe the slagging tendencies within the incinerator itself. If slagging becomes evident to any degree, an analysis of the slag can determine the seriousness of the problem. Any amounts over 3 percent are suspect.

Refractories that are porous or lightweight may allow slag to enter the pores and create further damage. This problem may be abated somewhat by using a dense refractory.

Increased Combustion Temperatures. Increased temperatures up to 2200°F (1204°C) and dwell times up to 2 s, as often required by federal, state, or local regulations, can put severe stresses on both incinerator and chimney refractories.

Disruptive Effects from Carbon Monoxide. Carbon monoxide is not a contaminant, unlike the alkalies and halides, but it can be equally destructive.[8] Carbon deposition resulting from carbon monoxide contact can cause refractories to burst and fall apart. As early as 1910, Rhead and Wheeler[9] recognized that carbon monoxide may break up into carbon and carbon dioxide. They found that the reaction was relatively sluggish at temperatures around 800°F (427°C). Numerous investigations[10] have shown that this reaction is accelerated in the presence of certain catalysts, such as iron oxide, and so minor amounts of carbon monoxide over an extended period can result in considerable deposition and refractory disruption.

This situation becomes aggravated when relatively porous castable refractories are used, particularly when standard calcium aluminate cement is the binder. Iron contamination can occur in both the binder and the aggregate. If carbon deposition is suspected, it is prudent to use a denser refractory with an iron-free binder. The aggregate should also be selected for its minimal iron content.

Kaiser and Carotti[1] found iron in fly ash from 1 to 10 percent, and Criss and

Olsen[6] found slag to contain from 12.3 to 46 percent total iron (as Fe_2O_3). This clearly shows that the reduction of iron in refractories can be helpful but can be overcome by iron in fly ash or other portions of airborne combustion products.

30.5.7 Effect of Halogens and Alkalies on Refractories

Most refractories are not affected by halogen vapors and alkalies in the normal range of use of 1500 to 1800°F (816 to 982°C) when the concentrations of both are low, but it is prudent to use the most dense, least porous materials. Firebrick bonded with high-purity calcium aluminate cement has been found suitable for use with small amounts of alkalies and halogens from 1800 to 2200°F (982 to 1204°C) where glasses are formed on the surface to help seal the porosity. At higher temperatures the slag becomes free-flowing with a degregation of the refractory. Most castable materials are not recommended for incinerator systems burning hazardous wastes only; however, high-alumina, low-calcium-oxide (CaO) high-density castables have been successful.

30.5.8 Design Features to Handle Downtime of Waste Heat Boilers

Bypass Chimneys. In the event of a waste heat boiler breakdown, a power failure, or a water failure, the full 1600°F (871°C) temperature from the incinerator will exhaust into the chimney. In this case a bypass chimney suitable for temperatures up to 1800°F (982°C), a medium-heat appliance chimney, is used to handle these high temperatures. This is prudent since acid brick and other acid-resistant chimney liners may not tolerate sudden temperature excursions.

Bypass Boilers. In Europe it is common to use acid brick as flue liners for resource recovery systems. This brick can be susceptible to spalling with rapid temperature changes; thus the exhaust gases are bypassed into a second and sometimes a third standby boiler, to avoid introducing high temperatures into the chimney.

30.5.9 Reheat of Chimney Systems to Maintain Flue Surface Temperature above Acid Dew Points

To overcome the problem of corrosion by sulfuric, hydrochloric, and hydrofluoric flue acids as well as hydrobromic, nitric, and sulfurous acids, some companies maintain gas burners that are expected to maintain temperatures above 500°F (260°C) at the end of an incinerator burn, to protect chimneys and other equipment upstream of heat-recovery units. This can be deceptive, and care should be taken to carefully analyze this burn cycle to be sure all flue temperatures remain above the acid dew points. It is not smart to use unlined uninsulated steel stacks in an effort to offset the high cost of purging. This may work with precipitators but does not work in stacks where the heat loss is significant. An unlined uninsulated steel stack should never be used.

Insulated steel stacks may retain heat, but insulation cannot elevate temperatures. If flue surfaces fall below acid dew points, corrosion will occur.

If sulfuric acid is deposited on flue surfaces or retained in fly ash adhered to flue surfaces, a purging temperature of 500°F (260°C) will concentrate the acid but may not completely vaporize it. Sulfuric acid boils at as low as 350°F (167°C) and as high as 644°F (340°C), depending on the number of H_2O molecules involved. Sulfuric acid over about 225°F (107°C) has a corrosion rate of over 200 mils (5 mm) per year on both stainless and carbon steels.

30.5.10 Multiple Flues

When more than one furnace is used on a project, it is good to have a separate chimney for each furnace and to enclose these flues within a steel or concrete windshield. Single chimneys servicing multiple units are more susceptible to corrosion since flue velocities are reduced as units are removed from the system. This causes a much greater difference between midflue temperatures and flue surface temperatures.

30.5.11 Hospital Waste Incineration

Hospital waste contains large quantities of plastics which contain alkalies as pigments, fillers, etc., in the compounding with resins to provide the desired characteristics. In effect, we have a minihazardous waste situation but with reduced intensity, since the incineration temperature will range from 1400°F (760°C) to 1800°F (982°C) with a 0.25- to 2-s dwell or less in the combustion unit. If no scrubbers or waste heat-recovery boilers are used, flue temperatures may still be reduced owing to the mixing of ambient air with flue gases through barometric dampers.

Alkalies are present in significant quantities due to the high plastic content. So standard UL listed medium-heat appliance chimneys which have lightweight and porous calcium aluminate refractory plus an Fe_2O_3 content over 5 percent will have a shorter life than chimneys with a more dense refractory with reduced CaO and Fe_2O_3 contents and a higher Al_2O_3 content, they will provide more strength, and with a 2400°F (1316°C) service temperature they are more suitable for such incineration.

High temperatures may occur in both the combustion chamber and the chimney. If insufficient oxygen is present for complete combustion in the combustion chamber, air entering through barometric dampers or cleanout doors can provide the necessary air for autoignition of the superheated gases in the chimney. When this secondary combustion occurs in chimneys, temperatures up to 2200°F (1204°C) are not unusual. This is a range where alkali degradation becomes severe.

Hospital Waste Systems without Waste Heat Recovery or Scrubbers. If no scrubbers or heat-recovery systems are to be used, the chimney should be designed to handle up to 2200°F (1204°C) flue gases and be lined with a dense refractory with reduced Fe_2O_3 and CaO contents. Some refractories have been cobonded with potassium silicate and reduced amounts of calcium aluminate cement that have been acceptable for this use.

Systems with Heat-Recovery Units. When heat-recovery units are used, the chimney temperature is reduced to about 600°F (316°C) at the high end and 300°F (149°C) at the low end. Often a burner is installed with a timer and thermostatic controls to maintain a chimney temperature in the 500°F (260°C) range, particularly during the time when material in the refuse chamber is still smouldering. If this type of burner is not available or is turned off too soon, the combustion process will continue to send flue gases into the breeching and chimney at temperatures below acid dew points, causing potentially severe corrosion problems.

It is important to recognize that a chimney system is sized to accommodate both the products of combustion of refuse and the products of the burner assisting that combustion. When the intensity of combustion of the refuse is reduced, the flue velocity is also reduced and the heat losses increase; this increases the difference between the midflue temperature and the flue surface temperature. This difference becomes greater as the burning subsides. If the supplementary burner is also turned off, the difference can be major, since in effect the chimney is larger than needed at that time and so has more flue surface (for cooling), greater heat loss, and greater difference between midflue and flue surface temperatures.

Many operators are not aware that their units fall below dew points until the chimney system fails.

Heat-recovery units do not remove corrosives from flue gases; they remove only heat. Therefore when these units cause temperatures to fall below acid dew points, the units show rapid decomposition of chimney linings. The problem intensifies if there is no bypass chimney to accommodate the high temperatures when the heat-recovery boiler is down for repair or down because it is not required during the summer. The chimney must now also be able to handle the high temperatures with alkali decomposition as well as low temperatures with acid corrosion. This can be a major problem to designers if they mistakenly believe the units will not fall below acid dew points.

The designer who insists on a single chimney for a hospital incinerator must be aware that the construction will be a compromise, that decomposition will take place at both high temperatures and low temperatures depending on the duration at the two extremes. The most serious and rapid problem will be caused at the low-temperature end. The time below acid dew points, including shutdown and startup, can lead to serious corrosion to conventional calcium aluminate chimneys when that time exceeds as little as 25 h/year.

Systems with Bypass and Heat-Recovery Systems. Several acid-resistant refractories and coatings have been used successfully under wet acid conditions[16] with the urethane asphalt membrane, but none can tolerate high temperatures. When such refractories and coatings are required to perform both under wet acid conditions and at high temperatures, the acid resistance of the membrane can be destroyed.

If any of the following are used and if the maximum temperatures are not exceeded, service life under acid conditions of hospital waste should be good:

Potassium silicate concrete (without membrane)	1600°F (760°C)
Borosilicate block	950°F (510°C)
Urethane asphalt	160°F (71°C)

These products were introduced about ten years ago and still show no serious problems. Bypass chimneys will be required if the number of hours on bypass is less than 500 h/year a standard UL listed 1800°F (982°C) medium-heat appliance chimney is acceptable. If the heat-recovery unit is to be used only during the winter or if any other factor requires sustained use at high temperatures, a high-density calcium aluminate bonded 2400°F (1316°C) refractory should be considered.

Systems with Heat Recovery But No Bypass. When a single chimney is used to handle both high- and low-temperature problems, a compromise must be made in the selection of a lining system since the acid-resisting materials cannot handle the high temperatures. Since the combustion range in hospital incineration is 1400°F (760°C) to 1800°F (982°C), a potassium silicate product is tempting. This can be an acceptable system if the designer can ensure that a reducing atmosphere will not exist and that temperatures will not exceed 1600°F (871°C).

If temperatures go above 1600°F (871°C), serious cracking can occur and alkali corrosion may be severe. Any temperature above about 250°F (121°C) on the urethane asphalt membrane can damage it and allow corrosion of the steel jacket.

Another compromise is to use a 35 to 45 percent Al_2O_3 refractory with a low iron content that is cobonded with calcium aluminate and potassium silicate.

When dealing with hospital waste, the designer must recognize that he or she is dealing with a mini hazardous waste system where the type of loading is not likely to be controlled and that chimney selection will be a compromise unless a bypass chimney is used. Also beware of the potential processing of nuclear waste in hospital incinerators. This could require substantially increased temperatures in the furnace and thus the chimney.

Systems with Scrubbers and a Bypass System. In this situation we are assured that the system will be below acid dew points at all times since the usual temperature out of a scrubber is from about 140°F (60°C) to 180°F (82°C) and can go below 100°F (38°C). If no compromise is to be made, a bypass chimney must be installed. The main chimney can be lined with potassium silicate gunnite or cast potassium silicate concrete with a urethane asphalt membrane between the steel substrate and the potassium silicate. A second option is borosilicate block bonded to the steel substrate with a urethane asphalt membrane.

The bypass chimney can be a calcium aluminate bonded lightweight or dense refractory depending upon the temperature in the bypass system.

Scrubbers do present a serious low-temperature problem to chimney systems, but treated flue gases tend to have lower acid concentrations. This fact plus a high efficiency of scrubbing could lead engineers to feel that the acid problem has been eliminated. It may have been reduced, yes, but not eliminated. Flue acids which pass the scrubber can be below the acid dew point continuously, so the lining must resist acids continuously.

Another misjudgment that can be made with scrubbers is their high efficiencies of 90 percent or better in some cases. Unfortunately field tests on the downstream side show an approximate average of 80 percent efficiency. This should not be considered as collecting 80 percent of the acids but as allowing 20 percent to escape into chimney systems.

Systems with Scrubbers But No Bypass Chimney. If a bypass is not used, one can follow the same criteria as with waste heat-recovery systems, except that it may be prudent to install separate breeching systems for the scrubber mode and bypass mode. This will allow the use of acid-resistant linings in the horizontal breechings where gravity will force acids through a porous system. If potassium silicate concrete or gunnite is used, it should be applied over a urethane asphalt membrane.

The use of borosilicate block bonded to the steel shell with a urethane asphalt membrane provides the least porous system.

If potassium silicate is to be used in chimney systems without a bypass chimney, care should be taken to avoid high-temperature excursions above 1600°F

(871°C). Any temperature over 250°F (121°C) on the urethane asphalt membrane may reduce or eliminate the acid resistance of the urethane asphalt lining.

The high-density 40 to 60 percent alumina, low-Fe_2O_3, low-CaO, 2400°F (1316°C) castable cobonded with calcium aluminate cement and potassium silicate can give better high-temperature resistance than potassium silicate concrete and gunnite which usually use silica as the refractory and can be satisfactory with low acid concentrations, as found with scrubbers.

30.5.12 Characteristics of Lining and Chimney Systems

Introduction. Prior to the arrival of heat-recovery units and air pollution control devices, chimney selection was simple. One had only to establish the temperature exiting the combustion process. This was always above acid dew points. With the introduction of heat-recovery units and air pollution control devices, conditions changed. Chimney temperatures increased to approach previous furnace temperatures and came with associated problems such as alkali or halogen attack. The percentage of alkali increased, and chimneys began to have a limited life at the high temperatures.

Low chimney temperatures were introduced with air pollution control devices, thus increasing the corrosion problem in chimneys operating at design temperatures. These problems became severe as equipment such as reheaters, fans, water supply, heat-recovery units, etc., failed. Manufacturers only guaranteed an 85 percent operational rate, thus equipment failure was accepted.

Scrubber utilization factors showed the scrubber to be operating as little as 42 percent of the time. The incinerator burned with a norm of about 85 percent, still leaving a 15 percent equipment failure rate. Problems were occurring but were often overlooked when chimney linings were chosen.

Chimneys frequently fell below acid dew points during these periods of equipment failure or on shutdown and startup. Temperatures also surged to the full temperature of the furnace, allowing alkali and unremoved acid gases to have detrimental effects on chimney lining systems.

Some designers avoid problems by using a bypass chimney, thus allowing chimneys to be designed to suit a specific need. When a chimney must serve both wet acid conditions at the low temperatures exiting waste heat boilers and alkali conditions at the high temperatures of bypass conditions, the chimney can become unduly stressed.

The characteristics that make a chimney acid-resistant at low temperatures are not the same as those that enable alkali and high-temperature resistance. When one chimney is used as both a main chimney and a bypass chimney, the selection becomes a compromise.

The following discussion presents the characteristics of the various systems and hopefully allows designers to make selections to suit their needs. A price index shows the relative cost of each system. The price index, which shows the relative difference in price between the various systems, will vary depending on local labor, material thickness, system differences, etc. They are designed simply to indicate the order of magnitude of relative pricing.

Unlined Uninsulated Carbon- or Stainless-Steel Flue Liners or Chimneys. Carbon steel is limited to about 750°F (399°C) and stainless steel to 1200°F (649°C) to 1700°F (927°C) insofar as operating temperatures are concerned. Neither can tolerate condensed flue acids, and both are extremely susceptible to corrosion. The lack of insulation and high thermal conductivity can enable a midflue temperature at the appliance outlet to be 59 percent higher than the flue surface temperature given a 30°F (−1°C) ambient and 15 mi/h (24 km/h) wind velocity. Appliances with a midflue temperature of 500°F (260°C) would show a flue surface temperature of 205°F (96°C) which is well below the dew point of sulfuric acid from No. 2 fuel oil. This situation worsens as flue gases move up the chimney, continuously losing heat. Most chimneys of this type have short service lives. We assign a price index of 1 to this chimney.

Unlined Insulated Carbon- or Stainless-Steel Flue Liners or Chimneys. Insulation lowers heat loss considerably with the midflue and flue surface temperatures differing by about 2 percent. Operation with this type of chimney can be very effective if the temperature in any part of the system can be guaranteed to remain above acid dew points. A safe critical temperature is considered the dew point of No. 6 fuel oil, or 320°F (160°C), plus a temperature measurement safety factor of 68°F (20°C) or 388°F (195°C). A temperature below this value in any part of the system can lead to serious corrosion. Price index = 2.

Unlined Uninsulated Double-Wall Stainless-Steel Chimneys or Flue Liners. This product is a UL listed factory-built chimney with a stainless-steel flue, a 1-in (2.5-cm) annular airspace, and an outer aluminized-steel shell. The flue liner is flanged and joined by a V band filled with sealant. The product is listed for continuous service at temperatures above acid dew points and no more than 1400°F (760°C) on a continuous basis. Other nonpressure products are listed for service above acid dew points and no more than 1000°F (538°C) on a continuous basis. All are called "building appliance chimneys" and are very satisfactory products when used above acid dew points and within the testing limitations. Flue surfaces must not fall below acid dew points more than 25 h/year including startup and shutdown.

None of these products are listed for use with commercial/industrial incinerators, and they cannot tolerate even short periods below acid dew points, including startup and shutdown times. The only chimney listed for use on commercial/industrial incinerators is the medium-heat appliance chimney.

When refractory-lined products are used on incinerators, designers should consider all factors of the operation, as presented in the following data. Price index = 2.

Conventional Calcium Aluminate Bonded Lightweight Refractories. This group of chimneys includes the UL listed medium-heat appliance chimneys. The product usually uses a lightweight refractory. It contains about 9 percent CaO and about 6 percent Fe_2O_3. It provides good insulation with a 4 to 8 percent loss between midflue and flue surface temperatures.

It has been used successfully for many years without a problem on both gas- and oil-fired appliances using both No. 2 and No. 6 fuel oil and on incinerators.

Temperatures up to 1800°F (982°C) continuously to 2000°F (1093°C) intermittently are acceptable service conditions. This refractory tolerates intermittent condensation of flue acids in the 2 pH range with an acceptable weight loss, does not handle carbon deposits from carbon monoxide due to the high Fe_2O_3, and is susceptible to alkali degradation at the higher temperatures.

At higher concentration of H_2SO_4 and with HCl introduced in the system in concentrations below a pH of 3, decomposition can result. This product shows serious decomposition when subject to hydrochloric acid and sulfuric acid as condensed from incinerated refuse. Temperatures above 500°F (260°C) are acceptable and are often the design range of waste heat boilers or scrubbing systems. But designers must investigate the downtime and must run a thermal analysis on the system at low ambient temperatures and high wind velocities as well as at standard conditions, to ensure that chimneys remain above acid dew points at all points under all conditions including equipment failure. Serious corrosion can occur with as little as 50 h/year below acid dew points including startup and shutdown time while combustion occurs. Price index = 2.

High-Density Fireclay Castable Refractory Calcium Aluminate Bonded. This type of material is moderate in Al_2O_3 (35 to 45 percent), low in Fe_2O_3 (0.5 percent), with about 6 percent CaO, and it is a dense material suitable for service from 2000°F (1093°C) to 3000°F (1649°C). It resists alkali degregration and is less susceptible to carbon monoxide precipitation of carbon in reducing atmospheres.

The CaO content is too high and the Al_2O_3 content too low for hazardous waste and perhaps hospital waste, depending on the amount of chlorides and alkalics, but it is substantially more resistant to alkali than conventional lightweight castables. Corrosion can occur under wet conditions and low temperatures below the critical dew point temperature of about 388°F (198°C). The lower CaO content gives it slightly better acid resistance; however, under conditions found with hospital waste, serious corrosion can occur with as little as 100 h/year below acid dew points including startup and shutdown times. This material is porous and can transmit flue acids and vapors to the steel shell under positive pressures. Price index = 3 without membrane; price index = 3.5 with membrane.

High-Density, High-Al_2O_3, Low-Fe_2O_3, Low-CaO (2 percent or less), Castable Fireclay Refractory. This material is said to have been used with success on hazardous waste with good resistance to alkali. Corrosive chloride vapors may create problems in relatively large quantities. This material is usually used at high temperatures above 2000°F (1093°C) with the Al_2O_3 content increasing as temperatures increase.

When this material is recommended, the manufacturer usually suggests a temperature of 350°F (177°C) or above on the refractory/steel shell interface to avoid serious corrosion to the shell. Price index = 7.

High-Density Fireclay Castable Refractory Cobonded with Potassium Silicate. This refractory is designed as a compromise to service low-temperature acid problems during heat recovery and/or scrubbing operation, but it must also handle the high temperatures of the bypass cycle. It is designed to handle both with reasonable effectiveness. It is a fireclay-type refractory with low Fe_2O_3 (0.5 percent or less), about 3 percent CaO, and 35 to 45 percent Al_2O_3 cobonded with potassium silicate.

It has resisted alkali on hospital waste incinerator chimneys, and according to short-term laboratory tests it can tolerate 15 percent H_2SO_4 and 5 percent HCl for 2000 h or more. Its resistance increases as acid concentrations decrease. The low Fe_2O_3 content suggests fewer problems with carbon precipitation. Price index = 3.5 without membrane; price index = 4 with membrane.

Potassium Silicate Concrete. This group of materials utilizes silica aggregates and potassium silicate as the binder. The material is acid-resistant and is designed for long-term immersion in acids. The material is porous, and if positive pressures are to be used or if the material is installed in horizontal runs such as breechings, an acid-resistant membrane can protect the steel substrate.

This is one of the materials studied by Mitas et al.[16] We present here only those products that were fully successful. Designers wishing more detailed information on the remaining 82 materials can find it in Mitas et al.[16] Their report showed that "shrinkage cracks formed while the potassium silicate lining is curing but no further degradation was noted after installation. Linings are installed as protective topcoat over other membrane linings."

Services at low temperatures under acid conditions appear good. Manufacturers suggest that maximum temperatures in systems without a membrane can go as high as 1600°F (760°C). Temperatures above this value could cause serious cracking. The high silica content indicates that alkali or caustic resistance is low.

The high porosity suggests a membrane to protect the steel substrate, and the best one according to the "survey" is urethane asphalt. This requires a midflue temperature of less than 275°F (135°C). Price index = 5 without membrane; price index = 5.5 with membrane.

Urethane Asphalt Membrane. This membrane, when properly applied, is very resistant to condensed flue gases, but it should not be used with temperatures over 160°F (71°C). It is usually protected by other linings such as borosilicate block or potassium silicate. This is one of the materials found to be fully successful in the survey.

Borosilicate Block. This material is a foamed borosilicate glass with a density of 12 lb/ft^3 (19 kg/m^3). It is a closed-cell foam, thus impervious to both liquids and vapors. It resists condensed flue acids very well when used in conjunction with a well-applied urethane asphalt membrane that fully covers both the steel substrate and the joints between blocks, to fully secure the system to the steel.

The excellent insulating value of borosilicate block reduces heat loss significantly and protects the urethane asphalt substrate. Laboratory tests show that hardening of the urethane asphalt in hot areas does not affect the acid-resistant nature of the system to protect the steel substrate from corrosion. Price index = 5.5.

Acid Brick. This material is usually installed as a free-standing structure with or without corbeling or floor support sections within a reinforced-concrete windshield, but it could be installed within a steel windscreen. To achieve acid resistance, bricks are made as dense as possible, to resist acid penetration. They have been used successfully in wet acid environments throughout the world. Some bricks are manufactured with a tongue and groove on all sides, to allow ease of installation and to avoid holidays in the mortar. Other bricks grooved on only two sides serve a similar purpose, but most are flat on all sides.

Acid brick is porous, thus systems under positive pressure will create a differential pressure that forces vapors through the brick to the annular area between the windscreen and the brick flue. These corrective measures have been taken:

- Pressurize the annular area above flue pressures.
- Ventilate the annular area.
- Protect the interior surface of the windscreen from acid condensation.

An installed brick lining by itself can be less expensive than either potassium silicate or borosilicate block, if we assume that all are to be enclosed within a windscreen, thus requiring a steel shell plus the windscreen, which increases the cost. A brick lining must be enclosed within a windscreen whereas potassium silicate and borosilicate can be installed in a steel shell that becomes a free-standing structural element. When we compare this situation with a brick flue within a windscreen, the prices of the two approach each other. When we include protection to the interior of the windscreen against flue acids that penetrate acid brick, the difference is removed. Price index = 5.5.

30.6 DESIGN

30.6.1 Chimney Sizing

The following procedure for chimney sizing is a condensation of "Chimney, Gas Vent and Fireplace Systems," as found in the American Society of Heating, Refrigeration, and Air-Conditioning Engineers' *ASHRAE Equipment Handbook*, chapter 26. This chapter provides formula derivations and many more details. This condensation, however, is sufficient to work out most chimney-sizing problems.

Theoretical Draft. Chimney design involves balancing forces which tend to produce flow (draft) against those which tend to retard flow (friction). The force producing the flow in gravity or natural-draft chimneys is termed the "theoretical draft," defined as the static pressure resulting from the difference in density between a stagnant column of hot flue gases and an equal column of ambient air. The theoretical draft can be calculated by using the following equation:

$$D_t = C_1 BH\left(\frac{1}{T_o} - \frac{1}{T_m}\right) \tag{30.1}$$

where D_t = theoretical draft, in H_2O (Pa)
C_1 = 0.2554 (0.03413)
B = barometric pressure, inHg (mmHg)
H = effective height of chimney, ft (m)
T_o = outside temperature, °F + 460 (°C abs)
T_m = mean chimney temperature, °F + 460 (°C abs)

Refer to Table 30.12 for barometric pressure versus altitude. Refer to Table 30.13 for appliance outlet temperatures.

For a more simplified calculation, Table 30.14 can be used to determine the theoretical draft, assuming the density of the chimney gas is the same as that of air, the barometric pressure is at sea level, and the ambient temperature is 60°F (15.6°C). When the theoretical draft is found from Eq. (30.1), T_o must be calculated on the warmest ambient air temperature that the boiler and/or heat appliance will operate at.

Barometric Pressure versus Altitude. See Table 30.12.

Appliance Outlet Temperatures. See Table 30.13.

Theoretical Draft per Unit of Chimney Height. See Table 30.14.

TABLE 30.12 Barometric Pressure and Altitude

Altitude above sea level, ft (m)	Pressure	
	Hg	Pa
	29.92	101,293
2000 (610)	27.8	94,116
4000 (1219)	25.8	87,345
6000 (1829)	24.0	81,251
8000 (2438)	22.3	75,496
10,000 (3048)	20.6	69,740

TABLE 30.13 Appliance Outlet Temperatures

Appliance type	Outlet temperature	
	°F	°C
Natural-gas-fired heating appliance with draft hood	360	182
Liquefied-petroleum gas-fired heating appliance with draft hood	360	182
Gas-fired heating appliance, no draft hood	460	238
Oil-fired heating appliance, residential	560	293
Oil-fired heating appliance, forced draft over 400,000 Btu/h	360	182
Conventional incinerator	1400	760
Controlled air incinerator	1800–2400	982–1316
Pathological incinerator	1800–2800	982–1538
Turbine exhaust	900–1400	482–760
Diesel exhaust	900–1400	482–760
Ceramic kilns	1800–2400	982–1316

TABLE 30.14 Theoretical Draft

Mean chimney temp. T_m, °F	Theoretical draft D_t, inH_2O	Mean chimney temp. T_m, °F	Theoretical draft D_t, inH_2O
100	0.00105	850	0.00886
150	0.00215	900	0.00907
200	0.00312	950	0.00927
250	0.00393	1000	0.00946
300	0.00464	1050	0.00963
350	0.00526	1100	0.00979
400	0.00581	1200	0.01009
450	0.00629	1300	0.01035
500	0.00673	1400	0.01058
550	0.00713	1500	0.01079
600	0.00748	1600	0.01098
650	0.00780	1700	0.01115
700	0.00810	1800	0.01135
750	0.00837	1900	0.01145
800	0.00862	2000	0.01158

Note: This table is based on an ambient temperature of 60°F (15.6°C). For metric conversion: 1 Pa = 249 inH_2O and °C = 5/9 (°F − 32).

TABLE 30.15 Mass Flow Input Ratios

Fuel	Appliance	Mass flow input ratio	
		Btu	kJ
Natural gas	Draft hood	1.60	0.688
Natural gas	No draft hood	0.90	0.387
LP gas	Draft hood	1.64	0.705
Oil	All	1.24	0.533
Oil No. 2	Over 400,000 Btu/h (117,240 W)	0.85	0.366
Oil No. 6	Over 400,000 Btu/h (117,240 W)	0.86	0.370
Coal (bituminous)	All	1.54	0.662

Mass Flow of Combustion Products. Mass flow in a chimney or venting system may differ from that in the appliance depending on the type of draft control or number of appliances operating in a multiple system. The use of mass flow is preferable (rather than cubic feet or cubic meters) because it remains constant in any continuous portion of the system regardless of changes in temperature or pressure. For the chimney gases resulting from any combustion process, the mass flow W, in lb/h or kg/h, can be expressed as

$$W = IM \tag{30.2}$$

where I = appliance heat input, Btu/h (W)
M = mass flow input ratio, lb of combustion products per 1000 Btu of fuel burned (kg of combustion products per 1000 W of fuel burned)

Table 30.15 lists the mass flow input ratios for various fuels and appliances. If a Btu input rating is not given for the appliance, Table 30.16 lists conversion factors for other appliance ratings.

Mass flow within incinerator chimneys must account for the probable heating value of the waste, plus its moisture content, plus the use of additional fuel to initiate or sustain combustion. For incinerators, where constant burner operation accompanies the combustion of waste, the additional quantity of products due to

TABLE 30.16 Conversion Factors

Btu/h input	bhp @ 100% efficiency × 33,475
	bhp @ 80% efficiency × 42,000
	bhp @ 75% efficiency × 44,500
	bhp @ 70% efficiency × 47,800
	gal/h oil × 140,000
	gal/h oil × 150,000
	ft^3/h natural gas × 1000
	lb/h coal × 13,000
	W × 3.412
	lb steam × 0.9740 × 1000

Metric conversion: kW input = kW output/eff.; kW = 9.81bhp.

the additional fuel should be considered in the design process. For incinerator chimneys, the mass flow can be calculated from a slightly different formula:

$$W = \frac{\text{lb (kg) waste burned}}{\text{h}} \times \frac{\text{lb (kg) combustion products}}{\text{lb (kg) waste}} \tag{30.3}$$

Table 30.17 gives the values for pounds or kilograms of combustion products based on the type of waste being burned.

TABLE 30.17 Mass Flow for Incinerator Chimneys

		Combustion products	
Type of waste	Btu/lb (kJ/kg) waste	ft³/(min · lb waste) [L/(s · kg waste)] at 1400°F (760°C)	lb/(h · lb waste) [kg/(h · kg waste)]
Type 0	8,500 (19,805)	10.74 (11.17)	13.76 (13.76)
Type 1	6,500 (15,145)	8.40 (8.74)	10.80 (10.80)
Type 2	4,300 (10,019)	5.94 (6.18)	7.68 (7.68)
Type 3	2,500 (5,825)	4.92 (5.12)	6.25 (6.25)
Type 4	1,000 (2,330)	4.14 (4.31)	5.33 (5.33)

To determine the losses in the chimney, the velocity must be calculated first. The velocity can be computed from

$$V = \frac{W}{\rho_m C_2 d^2} \tag{30.4}$$

where V = velocity, ft/s (m/s)
W = mass flow, lb/h (kg/h)
ρ_m = flue gas density, lb/ft³ (kg/m³)
d = diameter, in (mm)
C_2 = 19.635 (0.785)

If the diameter is unknown, a reasonable estimate can be made by using a velocity of 17 ft/s (5.2 m/s).

If the volume of flow in the chimney is required, possibly for fan selection, the following formula can be used:

$$Q = AV \tag{30.5}$$

where Q = flow rate, ft³/min (m³/min)
A = area, ft² (m²)
V = velocity, ft/min (m/min)

Mass Flow Input Ratio. See Table 30.15.

Conversion Factors. See Table 30.16.

Mass Flow for Incinerator Chimneys. See Table 30.17.

System Losses. Flow losses due to friction may be estimated by several methods using formulas for flow in pipes or ducts. These include the equivalent-length method and the loss coefficient or velocity head method. Primary emphasis is placed here on the loss coefficient method, because in chimney systems, fittings usually cause the greater portion of the system pressure drop and conservative loss coefficients (which are practically independent of piping size) provide an adequate basis for system designs.

By using the velocity head method for resistance losses, a fixed numerical coefficient (independent of velocity), the k factor, is assigned to every turn in the flow circuit and to piping as well. Table 30.18 offers design values for the resistance loss coefficient for various fittings.

TABLE 30.18 Resistance Loss Coefficients

	Suggested design values, dimensionless	Estimated span and notes
Inlet acceleration		
Gas vent with draft hood	1.5	1.0–3.0
Barometric regulator	0.5	0.0–0.5
Direct connection	0.0	Also dependent on blocking damper position
Round elbow, 90°	0.75	0.5–1.5
Round elbow, 45°	0.3	
Tee or 90° breeching	1.25	1.0–4.0
Y breeching	0.75	0.5–1.5
Cap, top		
Open straight	0.0	
Low-resistance (UL)	0.5	0.0–1.5
Other	—	1.5–4.5
Spark screen	0.5	
Converging exit cone	$\left(\frac{d_1}{d_2}\right)^4 - 1$	System designed using d_1
Tapered reducer (d_1 to d_2)	$1 - \left(\frac{d_2}{d_1}\right)^4$	System designed using d_2
	$0.4 \frac{L \text{ ft (m)}}{d \text{ in (mm)}}$	Numerical coefficient from 0.2 to 0.5; see Fig. 30.39 for friction factors
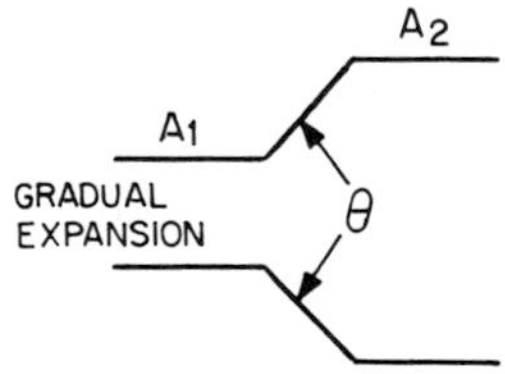 GRADUAL EXPANSION (A_1, A_2, θ)	θ	
	5° 0.17	
	7° 0.22	
	10° 0.28	
	20° 0.45	
	30° 0.59	
	40° 0.73	
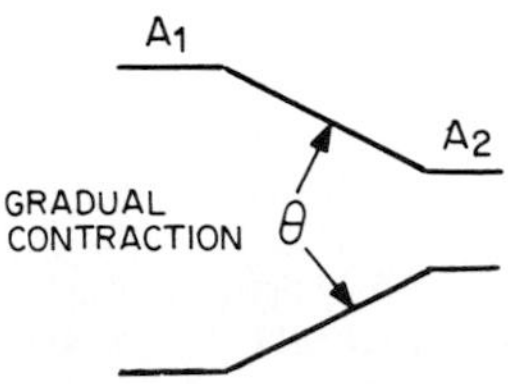 GRADUAL CONTRACTION (A_1, A_2, θ)	θ	
	30° 0.02	
	θ	
	45° 0.04	
	60° 0.07	

Once the loss coefficients for the system have been evaluated, the system losses can be calculated from

$$\Delta p = \frac{k \rho_m V^2}{5.2(2g)} \tag{30.6}$$

where Δp = system loss, inH_2O (Pa)
k = dimensionless loss coefficient
ρ_m = density of flue gas, lb/ft^3 (kg/m^3)
V = chimney velocity, ft/s (m/s)
g = gravitational constant, 32 ft/s^2 (9.7 m/s^2)

Refer to Table 30.19 for density versus temperature.

To determine the losses in a rectangular system, an equivalent circular diameter for equal friction and capacity can be used. A sample of these equivalent diameters is given in Table 30.20.

Resistance Loss Coefficients. See Table 30.18.

Friction Factor for Piping (Commercial Iron and Steel Pipe). See Fig. 30.39.

Density versus Temperature. See Table 30.19.

Masonry Chimney Liner Dimensions with Circular Equivalent. See Table 30.20.

Balancing the System. Equipment or appliances can be placed in three broad categories: negative pressure appliances, atmospheric appliances, and forced-draft appliances. *Negative pressure appliances* require a negative pressure at the outlet to induce combustion air flow into the combustion zone. *Atmospheric appliances* require a neutral pressure at the outlet without need for chimney draft, e.g., draft hood types of gas appliances in which the combustion process is iso-

TABLE 30.19 Density versus Temperature

Temperature		Density		Temperature		Density	
°F	°C	lb/ft^3	kg/m^3	°F	°C	lb/ft^3	kg/m^3
60	15.6	0.07656	1.22	300	149	0.05237	0.84
70	21.1	0.07512	1.20	350	177	0.04914	0.79
80	26.7	0.07373	1.18	400	204	0.04628	0.74
90	32.2	0.07238	1.16	450	232	0.04374	0.70
100	37.8	0.07109	1.14	500	260	0.04146	0.66
110	43.0	0.06984	1.12	550	288	0.03940	0.63
120	44.0	0.06864	1.10	600	316	0.03754	0.60
130	54.0	0.06747	1.08	650	343	0.03585	0.57
140	60.0	0.06635	1.06	700	371	0.03431	0.55
150	66.0	0.06526	1.04	800	427	0.03158	0.51
175	79.5	0.06269	1.00	900	482	0.02926	0.47
200	93.0	0.06031	0.96	1000	538	0.02725	0.44
225	107.0	0.05811	0.93	1500	816	0.02030	0.32
250	121.0	0.05606	0.90	2000	1093	0.01617	0.26
275	135.0	0.05415	0.87				

TABLE 30.20 Masonry Chimney Liners

Nominal liner size, in	Inside dimensions of liner, in	Inside diameter or equivalent diameter, in	Equivalent area, in^2	Typical outside dimensions of casing, in
4 × 8	2½ × 6½	4	12.2	
		5	19.6	
		6	28.3	
		7	38.5	
8 × 8	6¾ × 6¾	7.4	42.7	16 × 16
		8	50.3	
8 × 12	6½ × 10½	9	63.6	16 × 21
		10	78.5	
12 × 12	9¾ × 9¾	10.4	83.3	21 × 21
		11	95.0	
12 × 16	9½ × 13½	11.8	107.5	21 × 25
		12	113.0	
		14	153.9	
16 × 16	13¼ × 13¼	14.5	162.9	25 × 25
		15	176.7	
16 × 20	13 × 7	16.2	206.1	15 × 29
		18	254.4	
20 × 20	16¾ × 16¾	18.2	260.2	29 × 29
		20	314.1	
20 × 24	16½ × 20½	20.1	314.2	29 × 34
		22	380.1	
24 × 24	20¼ × 20¼	22.1	380.1	34 × 34
		24	452.3	
24 × 28	20¼ × 20¼	24.1	456.2	34 × 38
28 × 28	24¼ × 24¼	26.4	543.3	38 × 38
		27	572.5	
30 × 30	25½ × 25½	27.9	607.0	48 × 48
		30	706.8	
30 × 36	25½ × 31½	30.9	649.9	48 × 54
		33	855.3	
36 × 36	31½ × 31½	34.4	929.4	54 × 54
		36	1017.9	

Metric conversions: 1 in = 2.54 cm and 1 in^2 = 6.45 cm^2.

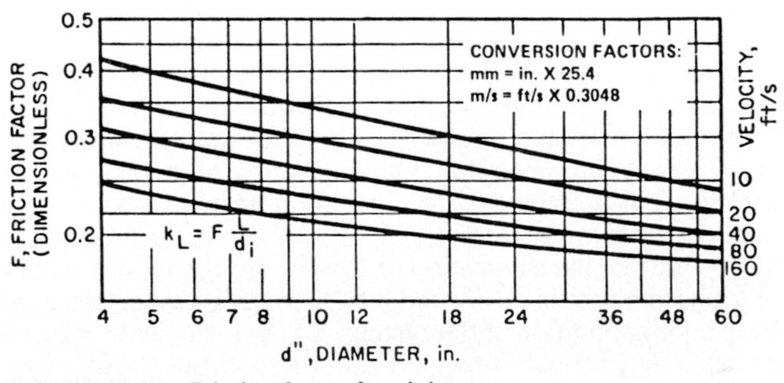

FIGURE 30.39 Friction factor for piping.

lated from chimney flow variations. *Forced-draft appliances* operate at above atmospheric pressure and have sufficient flue gas outlet static pressure that no chimney draft is needed. This forced-draft category, however, needs explanation.

Forced-draft appliances utilize a fan to force the products of combustion through the appliance, which at times results in a positive pressure at the breeching inlet. If a positive pressure exists, the overall draft is increased, therefore reducing the required stack diameter. Some boiler manufacturers state that there can be as much as 0.5 in water column (WC) (1.2 cm) of positive pressure at the boiler outlet while other manufacturers prefer the stack diameter to be calculated based on an outlet pressure of zero. To ensure proper stack sizing, the boiler manufacturer should be contacted regarding the issue of boiler outlet pressure.

Depending on the type of appliance, a different formula is needed to balance the system (see Table 30.21). The values previously calculated from equations for the system should be evaluated against these equations. In all three cases, if the system losses exceed the draft requirements, the system will not properly exhaust the combustion products. At this point, by using an iterative (trial-and-error) process, the stack and breeching diameters should be increased (or decreased) until the system balances. Another solution, for an undersized system, is to increase the chimney height or to include a draft inducer in the chimney system. For a system with a draft inducer, the static pressure supplied by the inducer should be added to the draft requirements in the pressure equation in Table 30.21.

TABLE 30.21 Pressure Equations for Δp

Appliance type	Pressure equation (loss = draft requirements)	Notes
Negative pressure	$\Delta p = D_t - D_o$	D_o is amount of negative pressure needed
Atmospheric	$\Delta p = D_t$	Neutral pressure at outlet
Forced draft	$\Delta p = D_t + D_o$	D_o is amount of positive pressure at outlet due to forced-draft system

Here Δp = system losses, D_t = theoretical draft, and D_o = appliance outlet pressure, all in inH_2O (Pa).

In reverse, a draft inducer can be selected based on the additional draft required to balance the pressure equation. Given the volume of flow in the system and the static pressure required, the fan selection for the inducer is simplified.

Multiple Systems. The most common configuration is the individual vent, stack, or chimney where one continuous system carries the products from appliance to terminus. Other configurations include the combined vent serving a pair of appliances, the manifold serving several appliances, and a branched system with two or more lateral manifolds connected to a common vertical system.

For the configurations where one system is used to vent several appliances, the sizing procedure involves the summation of losses through the various branches of the system. The system should be divided into sections based on the number of appliances venting into that portion of the system. The velocity and the loss coefficient are calculated separately for each branch. From these values the losses in each sec-

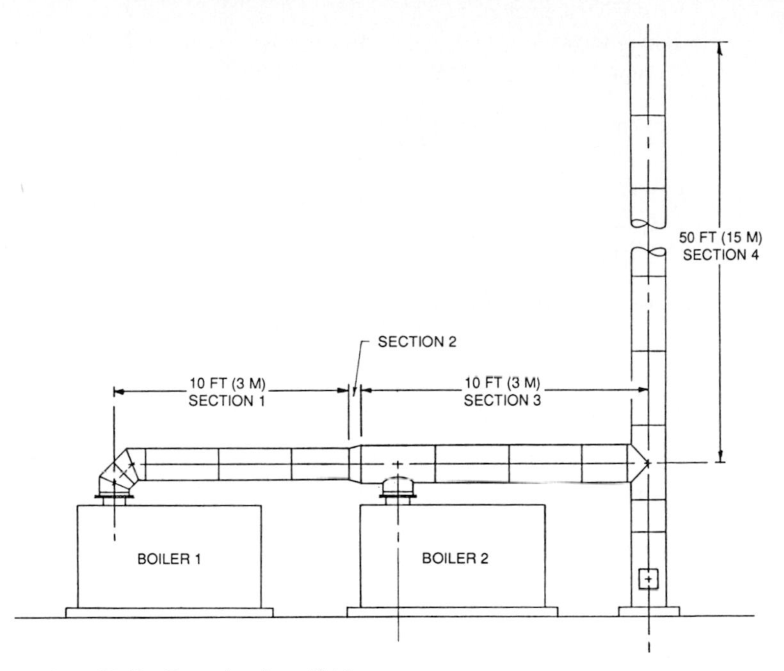

FIGURE 30.40 Example of manifold system.

tion can be computed. In the summation of the system loss, only the greatest branch loss of all the parallel branch losses should be considered.

The draft requirements of each appliance need individual consideration. A pressure equation for each appliance should be formulated to ensure that the system balances at each connection. If the system does not balance at any point, either a branch size or the vertical stack size must be changed. Once a size has been changed, the pressure equations should be recalculated to determine if the change leaves the system balanced.

EXAMPLE Determine if the system shown in Fig. 30.40 is balanced.

GIVEN:
- Two natural-gas-fired appliances with input ratings of 6,300,000 Btu/h each
- Exhaust temperature of 360°F (180°C)
- Pressure at boiler outlet of 0.0 in WG

Solution MASS FLOW RATE PER BOILER

$$W = IM = \frac{6{,}300{,}000(0.9)}{1000} = 5670 \text{ lb/h (2613 kg/h) flue gas} \tag{30.2}$$

SYSTEM LOSSES

SECTION 1:

$$V = \frac{W}{\rho_\eta(19.635)d^2} \tag{30.4}$$

$$= \frac{5670}{19.635(0.04857)(18^2)} = 18.35 \text{ ft/s (5.59 m/s)}$$

RESISTANCE COEFFICIENT

90° elbow $k = 0.75$

10-ft (3-m) piping $k = 0.4\left(\frac{10}{18}\right) = 0.22$

$k = 0.97$

$$p = \frac{k\rho_m V^2}{5.2(2g)} \tag{30.6}$$

$$= \frac{0.97(0.04857)(18.35)^2}{332.8} = 0.05 \text{ in WC } (12.45 \text{ Pa})$$

SECTION 2:

V upstream = 18.35 ft/s (5.59 m/s)

RESISTANCE COEFFICIENT

Increaser $k = 0.5$ (*see Table* 30.18).

$$p = \frac{k\rho_m V^2}{5.2(2g)} \tag{30.6}$$

$$= \frac{0.5(0.04857)(18.35)^2}{332.8} = 0.02 \text{ in WC } (4.98 \text{ Pa})$$

SECTION 3:

$$V = \frac{W}{C_2\rho_m d^2} \tag{30.6}$$

$$= \frac{5670(2)}{19.635(0.04857)(24^2)} = 20.64 \text{ ft/s } (6.29 \text{ m/s})$$

RESISTANCE COEFFICIENT

10-ft piping: $k = 0.4\left(\frac{10}{24}\right) = 0.17$

$$p = \frac{k\rho_m V^2}{5.2(2g)} \tag{30.6}$$

$$= \frac{0.17(0.04857)(20.64)^2}{332.8} = 0.01 \text{ in WC } (2.49 \text{ Pa})$$

SECTION 4:

V = 20.64 ft/s (6.29 m/s)

RESISTANCE COEFFICIENT

90° tee $k = 1.25$

50-ft piping: $k = 0.4\left(\frac{50}{24}\right) = 0.83$

$k = 2.08$

$$p = \frac{k\rho_m V^2}{5.2(2g)} \qquad (30.6)$$

$$= \frac{2.08(0.04857)(20.64)^2}{332.8} = 0.13 \text{ in WC } (32.37 \text{ Pa})$$

THEORETICAL DRAFT

$$D_t = 0.2554\, BH \left(\frac{1}{T_o} - \frac{1}{T_m}\right) \qquad (30.1)$$

$$= 0.2554(29.92)(50)(1/520 - 1/820) = 0.24 \text{ in WC } (59.76 \text{ Pa})$$

SYSTEM BALANCING

GIVEN: Neutral pressure at outlet.
Pressure equation:

$$\Delta p = D_t \qquad (30.6)$$

Total system losses = Δp = sec. 1 + sec. 2 + sec. 3 + sec. 4

= 0.05(12.45) + 0.02(4.98) + 0.01(2.49) + 0.13(32.37)

= 0.21 in WC (52.29 Pa)

Theoretical draft D_t = 0.24 in WC (59.76 Pa) exceeds the system losses; therefore the breeching and stack are properly sized.

Fireplace Chimney. Fireplaces with natural-draft chimneys obey the same gravity fluid-flow law as gas vents and thermal-flow ventilation systems. Mass flow of hot flue gases up to some limiting value is induced in a vertical pipe as a function of the rate of heat release, and this flow is regulated by the chimney area, chimney height, and system pressure loss coefficient. A fireplace may be treated analytically as a gravity duct inlet fitting having a characteristic entrance loss coefficient and an internal heat source. Proper functioning of a fireplace (prevention of smoking) is achieved by producing adequate intake or face velocity across those critical portions of the frontal opening, to nullify effects of external drafts or internal convection effects.

A minimum mean frontal inlet velocity of 0.8 ft/s (0.244 m/s) should control smoking adequately in conjunction with a chimney gas temperature at least 300 to 500°F (149 to 260°C) above ambient. For a reasonably conservative design, a frontal inlet velocity of 1.0 ft/s (0.305 m/s) and a temperature of 350°F (177°C) can be used.

To determine the volume of air entering the fireplace at 70°F (21°C), the following equation can be used:

$$Q_f = V_c A_f \qquad (30.7)$$

where Q_f = volume of air entering fireplace at 70°F (21°C), ft³/min (m³/min)
V_c = capture velocity, ft/min (m/min)
A_f = fireplace frontal area, ft² (m²)

The volume of flue gas in the chimney can be calculated from

$$Q_c = \frac{Q_f}{DCF} \tag{30.8}$$

where Q_c = volume of gas entering chimney at average chimney temperature, ft^3/min (m^3/min).
Q_f = volume of air entering fireplace at 70°F, ft^3/min (m^3/min)
DFC = density correction factor (see Table 30.22)

The velocity in the chimney is determined by:

$$V_c = \frac{Q_c}{A_c} \tag{30.9}$$

where V_c = velocity in chimney, ft/min (m/min)
Q_c = volume of gas entering chimney at average chimney temperature, ft^3/min (m^3/min)
A_c = area of chimney, $ft^3/(m^3)$

In most applications, the area of the chimney will fall between one-tenth and one-twelfth of the fireplace frontal area.

Once the velocity has been computed, the system losses can be calculated by following the procedure previously described. For a fireplace application, addi-

TABLE 30.22 Density Correction Factors

	Altitude above sea level, ft						
Temp., °F	0	1000	2000	3000	4000	5000	6000
0	1.15	1.11	1.07	1.03	0.99	0.95	0.91
70	1.00	0.96	0.93	0.89	0.86	0.83	0.80
100	0.95	0.88	0.88	0.85	0.81	0.78	0.75
150	0.87	0.81	0.81	0.78	0.75	0.72	0.69
200	0.80	0.74	0.74	0.71	0.69	0.66	0.64
250	0.75	0.70	0.70	0.67	0.64	0.62	0.60
300	0.70	0.65	0.65	0.62	0.60	0.58	0.54
350	0.65	0.60	0.60	0.58	0.56	0.54	0.52
370	0.64	0.59	0.59	0.57	0.55	0.53	0.51
400	0.62	0.57	0.57	0.55	0.53	0.51	0.49
450	0.58	0.54	0.54	0.52	0.50	0.48	0.46
500	0.55	0.51	0.51	0.49	0.47	0.45	0.44
550	0.53	0.49	0.49	0.57	0.45	0.44	0.42
600	0.50	0.46	0.56	0.45	0.43	0.41	0.40
650	0.48	0.44	0.44	0.43	0.41	0.40	0.38
700	0.46	0.43	0.43	0.41	0.39	0.38	0.37
750	0.44	0.41	0.41	0.39	0.37	0.36	0.35
800	0.42	0.39	0.39	0.37	0.36	0.35	0.33
850	0.40	0.37	0.37	0.36	0.34	0.33	0.32
900	0.39	0.36	0.36	0.35	0.33	0.32	0.31
950	0.38	0.35	0.35	0.34	0.33	0.31	0.30
1000	0.36	0.33	0.33	0.32	0.31	0.30	0.29

Metric conversion: °F = 9/5(°C)+32 and 1 ft = 0.3048 m.

TABLE 30.23 System Loss Coefficients

Loss	*K* factor
Loss to initiate flow	1.0
Inlet loss	
Cone-type fireplace	0.5
Masonry damper throat = 2 × flue area	1.0
Masonry damper throat = flue area	2.5

tional loss coefficients need to be considered. These values are found in Table 30.23.

The procedures for calculating theoretical draft and balancing the system follow those in previous sections. Draft inducers can also be used on fireplace chimneys to supplement the theoretical draft, if needed, to help prevent smoking.

Density Correction Factor. See Table 30.22.

System Loss Coefficients. See Table 30.23.

30.6.2 Chemical Loading for Unlined Steel Stacks*

Attack due to Sulfur Oxides. The most common form of internal chemical attack is due to acids formed by the condensation of sulfur oxides in the flue gas. Sulfur is found in all solid and liquid fuels to varying degrees and in gaseous fuels as well. During the combustion process, nearly all the sulfur in the fuel is oxidized to sulfur dioxide (SO_2) which is absorbed by condensing water vapor to form sulfurous acid.

A small quantity of sulfur dioxide is further converted to sulfur trioxide. Upon condensation, the SO_3 ions combine with water vapor to form sulfuric acid whose concentration can be as high as 85 percent.

Condensation of these acids occurs when the temperature of the flue gas falls below their acid dew point or when the flue gas comes into contact with a surface at or below the relevant acid dew point temperature.

The acid dew point temperature of sulfuric acid depends on the concentration of SO_3 in the flue gas (see Fig. 30.41). Provided the temperature of the flue gas and the surfaces with which it can come into contact are maintained at a minimum of 50°F (10°C) above the acid dew point estimated from Fig. 30.41, there is no danger of acid corrosion.

The acid dew point of sulfurous acid is about 149°F (65°C), a little above the water dew point. If the fuel contains other acids, such as hydrochloric acid and nitric acid, the fuel can be expected to condense in the same temperature range. Thus, even if fuel and combustion processes are chosen to minimize production of SO_3, severe corrosion can be expected if the temperatures of the flue gas or the surfaces with which it can come into contact fall below 140°F (60°C) or the acid dew point temperature relevant to hydrochloric acid. Again, a safety margin is recommended of 68°F (20°C) above the acid dew point temperature estimated from Fig. 30.41.

---

*This section is taken from J. F. Schulz and K. A. Schultz, "Thermal and Corrosion Analysis of Chimney Systems," *ASHRAE Transactions*, vol. 91, part 1A, no. 2883, 1985. Used by permission of the American Society of Heating, Refrigerating and Air-Conditioning Engineers, Inc., Atlanta, GA. For a detailed discussion of the topic, please refer to that publication.

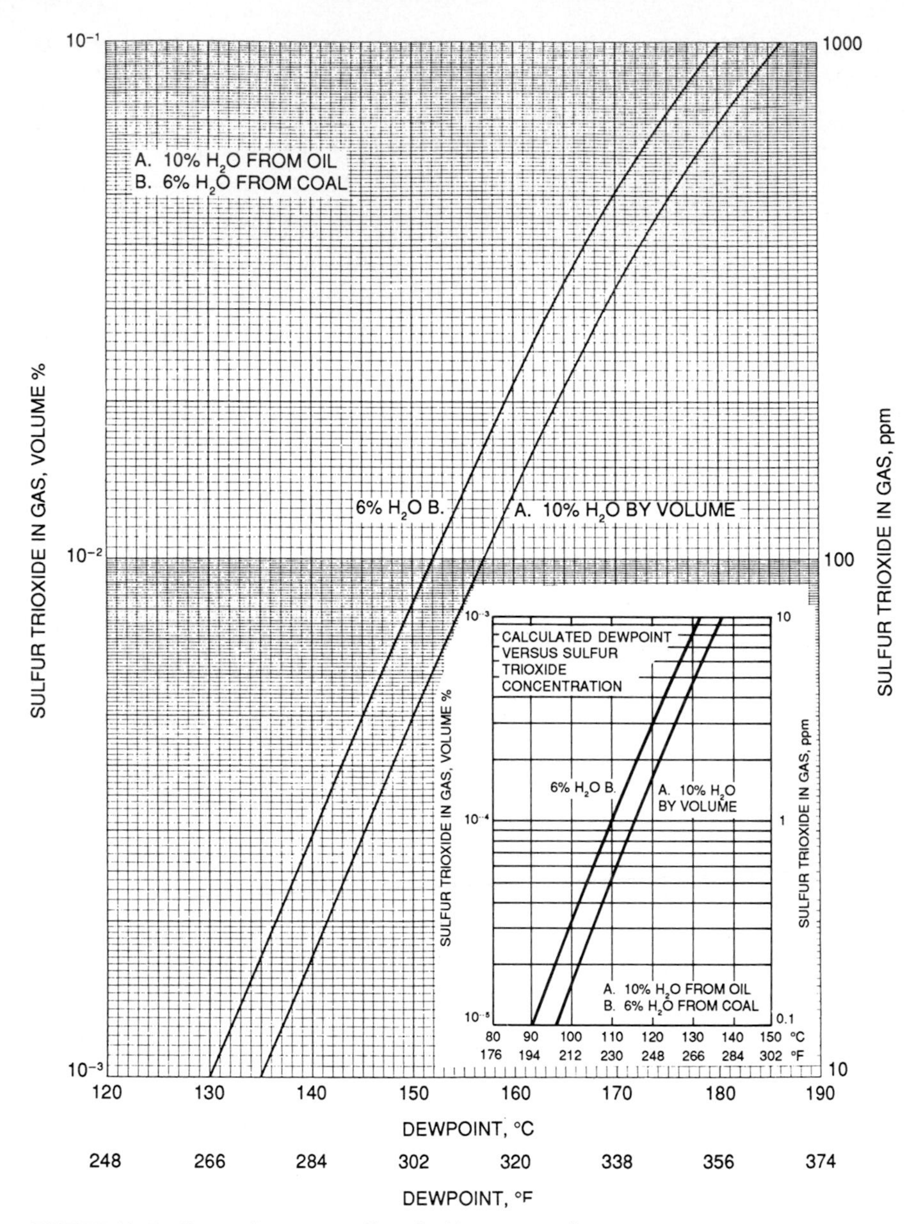

FIGURE 30.41 Dew point versus sulfur trioxide concentration.

Attack due to Chlorine, Chlorides, and Fluorides. Chlorides are found in all solid fuels, including refuse, and in many liquid fuels. Upon combustion chlorides are transformed to free chloride ions which on contact with water vapor are transformed to hydrochloric acid. The highest condensation temperature at which hydrochloric acid has been found is 140°F (60°C). Thus, when any flue surface temperature falls below the acid dew point, very serious corrosion will occur. This dew point is close to the water and sulfurous acid dew point temperatures. Even very small amounts of chlorides in combination with other condensed acids can cause serious corrosion problems.

Hydrogen chloride, hydrogen fluoride, and free chlorine in flue gases also become corrosive in their vapor stage. Stainless steels are attacked at temperatures above 600°F (316°C). Fluoride vapors are corrosive to stainless steels at temperatures above 480°F (250°C).

Chemical Effects. Limited exposure to acid corrosion conditions can be permitted in chimneys which are safe from chemical attack for most of the time.[11] Providing the flue gas does not contain halogens (chlorine, chlorides, fluorides, etc.), the degree of chemical load is defined in Table 30.24.

TABLE 30.24 Degree of Chemical Load for Gases Containing Sulfur Oxides

Degree of chemical load	Operating time when temperature of surface in contact with flue gases is below estimated acid dew point + 20°C, h/year
Low	< 25
Medium	25–100
High	> 100

Please note:

1. The operating hours in Table 30.24 are valid for an SO_3 content of 15 ppm. For different values of SO_3 content, the hours given vary inversely with SO_3 content. When the SO_3 content is not known, the chimney design should be based on a minimum SO_3 content amounting to 2 percent of the SO_2 content in the flue gas. It may also be estimated from Table 30.25.
2. In assessing the number of hours during which a chimney is subject to chemical load, account should be taken of startup and shutdown periods when the flue gas temperature is below its acid dew point.
3. While a steel chimney may generally be at a temperature above the acid dew point, care should be taken to prevent small areas from being subjected to local cooling and therefore running the risk of localized acid corrosion. Local cooling may be due to:
   - Air leaks
   - Fin cooling of flanges, spoilers, or other attachments
   - Cooling through support points
   - Downdraft effects at top of chimney

TABLE 30.25 Estimate of Sulfur Trioxide in Combustion Gas, ppm

Excess air, %	Oxygen in gas, %	Sulfur in fuel, %					
		0.5	1.0	2.0	3.0	4.0	5.0
Oil-fired units							
5	1	2	3	3	4	5	6
11	2	6	7	8	10	12	14
17	3	10	13	15	19	22	25
25	4	12	15	18	22	26	30
Coal-fired units							
25	4.0	3–7	7–14	14–28	20–40	27–54	33–66

Source: Adapted from an article by R. R. Pierce, *Chemical Engineering*, April 11, 1977.

4. The presence of chlorides or fluorides in the flue gas condensate can radically increase corrosion rates. In such cases the degree of chemical load should be regarded as high if the operating time below the acid dew point exceeds 25 h/year.
5. Regardless of temperature, the chemical load should be considered high if halogen concentrations exceed the following limits:
   - Hydrogen fluoride: 0.025 percent by weight (300 mg/m^3 at 20°C and 1-bar pressure)
   - Elementary chlorine: 0.1 percent by weight (1300 mg/m^3 at 20°C and 1-bar pressure)
   - Hydrogen chloride: 0.1 percent by weight (1300 mg/m^3 at 20°C and 1-bar pressure)

Corrosion in steel chimneys subject to 100 h below the acid dew point and high corrosion can be reduced by installing a refractory liner.

Refractory linings used on oil-fired appliances are usually bonded with calcium aluminate cement. Although these linings increase the service life of steel chimneys subject to sulfuric acid only, the situation changes when hydrochloric acid and/or higher sulfuric acid concentrations are introduced. For instance, if the fuel were refuse with temperatures below acid dew points, the HCl content could be as high as 7 percent and the H_2SO_4 content as high as 28 percent. Under these conditions serious corrosion occurs in as little as 2 h and is about equivalent to the example of 100 ppm SO_3, which is the actual amount in municipal waste.

30.6.3 Wind Effect on High-Rise Building Pressures

The wind effect is always considered when the heating load is calculated. Unfortunately, it is usually not considered for high-rise nonexposed chimney design. The same factors that affect the heating load affect the chimney. If these factors are not considered, the problem of smoke infiltration into the living areas will become serious.

We do not concern ourselves with maximum velocities because of their infrequent occurrence, but we must concern ourselves with velocities in the range of 20 to 30 mi/h (32 to 48 km/h).[29] This range occurs with fair frequency in most parts of the United States. Wind is affected by friction, thus surface winds maybe as much as 30 percent lower than those at higher levels. In high-rise construction, we must consider the wind pressures 100 ft (30.5 m) or more above the ground level or friction surface. We have elected to use 30 mi/h (48 km/h) as a practical design value. This should be increased if the area is subject to higher wind velocities.

Wind or moving air will tend to continue in the direction in which it is flowing until it encounters some external body or force. A mountain, a forest, a building, or some other object will direct the wind and change its direction. If we assume an open plain and a high-rise building, we have the least complicated situation. The air pattern is shown in Figs. 30.42 and 30.43.

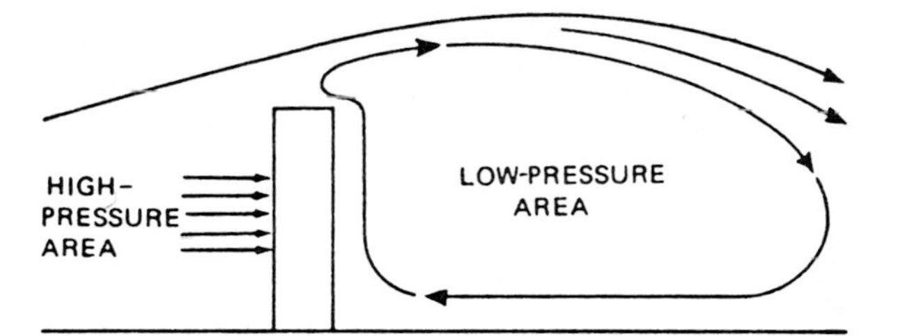

FIGURE 30.42 Wind pressure around buildings.

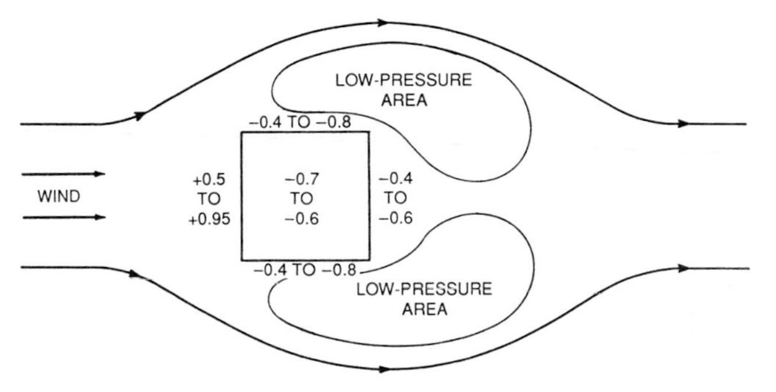

FIGURE 30.43 Pressure ratios compared to mean upstream wind velocities.

It is obvious that for a chimney terminated in the eddy area, the products of combustion or odors will be carried to the building on the leeward side in concentrations above acceptable limits,[24,32] depending on the relationship between the building pressures and the pressure in the eddy area. We must examine these pressures on all sides, since eddies involve the top, back and both sides of a structure.

A typical pressure differentiation on the various portions of a building, as determined by wind tunnel tests on models, shows positive pressure up to 0.95 of the wind pressure on the windward side and negative pressures up to 0.8 of the wind pressure on the leeward side.[27,29]

We have chosen 30 mi/h (48 km/h) as the design velocity. This provides an impact pressure of 0.434 inH_2O (11 cmH_2O). Thus,

$$\text{Positive pressure outside} = 0.95(0.434) = 0.41 \text{ inH}_2\text{O}$$
$$= 0.95(11) = 10.5 \text{ cmH}_2\text{O}$$
$$\text{Negative pressure outside} = 0.8(0.434) = 0.35 \text{ inH}_2\text{O}$$
$$= 0.8(11) = 8.8 \text{ cmH}_2\text{O}$$

Additional tunnel testing on models with various openings indicates that about 50 percent of the pressure on the outside will be transmitted to the interior of the building. So at 30 mi/h (48 km/h) the pressures due to wind are

$$\text{Inside positive pressure} = +0.21 \text{ inH}_2\text{O} \ (5.3 \text{ mmH}_2\text{O})$$
$$\text{Inside negative pressure} = -0.18 \text{ inH}_2\text{O} \ (-\ 4.4 \text{ mmH}_2\text{O})$$

These pressures must be considered when the exhaust system is designed for chutes, enclosures, or chimneys. For a detailed discussion of this topic, see Ref. 33.

30.6.4 Air Movement around Buildings and Eddy Zones

Air patterns around buildings have been subject to considerable investigation. Extensive wind tunnel tests and field confirmation have provided some general rules. For a single basic cube, i.e., a building where the length equals the width and the width equals the height, the height of the eddy above the ground will be 1.5 times the building width.[23] As we increase the number of cubes in height, we find that this cavity height above the roof remains unchanged and is approximately half the building width (Fig. 30.44).

Having established the eddy height, we may now review its effect. For a chimney terminated above the eddy area, there are no self-polluting problems, since the exhausts are carried away from the building (Fig. 30.45). With a chimney ter-

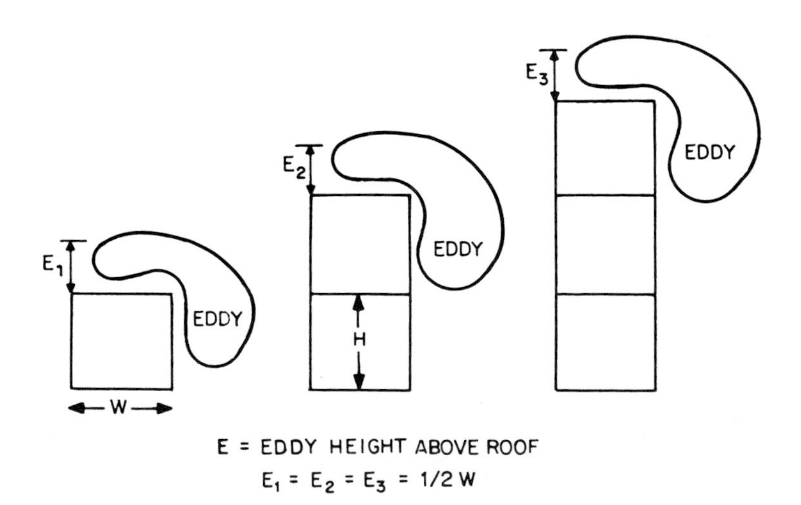

FIGURE 30.44 Eddy height remains constant with building.

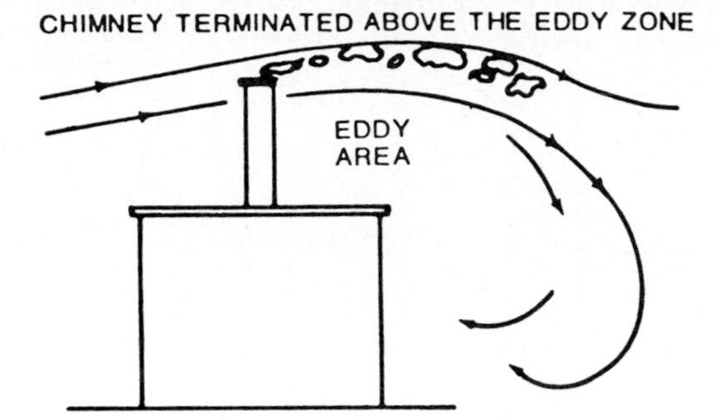

FIGURE 30.45 Chimneys terminated above the eddy area allow combustion products to be blown clear of buildings.

minated below the eddy zone, the exhaust products are carried back to the building to reenter through air intakes or by infiltration. During the heating season, pollutants enter the stories below the zero point; during the cooling season, they enter above the zero point (Fig. 30.46).

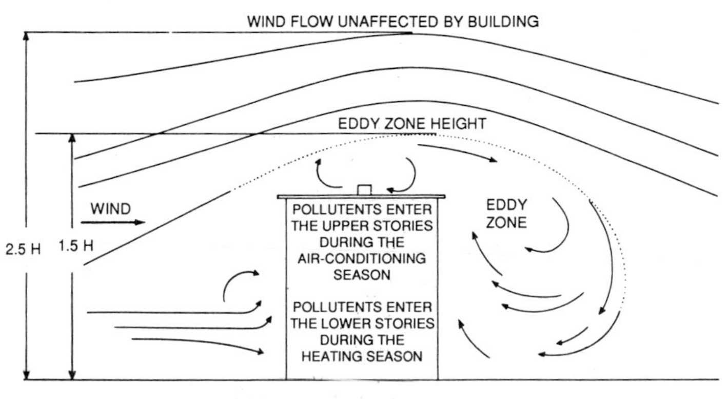

FIGURE 30.46 Chimneys terminated in eddy areas allow pollution to enter buildings.

The box-type building with flat roof and hidden roof appliances is popular. But to terminate a chimney or exhaust system above this area requires a 60-ft (18.3-m) chimney height for a 120-ft (36.6-m) wide building.[24] This is always objectionable to an architect and creates other problems. The chimney becomes a condenser, creating internal moisture problems. Both conditions may be solved by placing the chimney or exhaust duct within an enclosure as part of the mechanical room, to protect the chimney from cooling and to allow projection above the eddy (Fig. 30.47). We must remember that all building exhausts, boilers, incinerators, heating chimney, air conditioning, etc., reenter the building by infiltration, or intake air ducts, unless the pressures involved are overcome or the exhaust is terminated above the eddy area. The concentration of these effluents will determine the degree of recontamination.[25,26] Unfortunately, this concentration can be above acceptable levels and can be undetectable in the case of carbon monoxide.

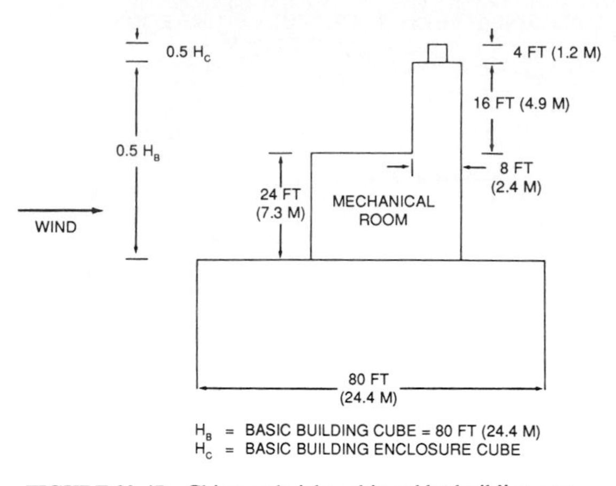

FIGURE 30.47 Chimney height achieved by building configuration.

The fenced roof and the parapet wall are aesthetically adequate (Fig. 30.48); however, the pressures and conditions allow recycling of effluents into air intake systems.[27] Acid effluents,[6] as from solid- or liquid-fuel heating systems, will be deposited on roof duct motors and roof equipment, to increase maintenance costs and the risk of system failure.

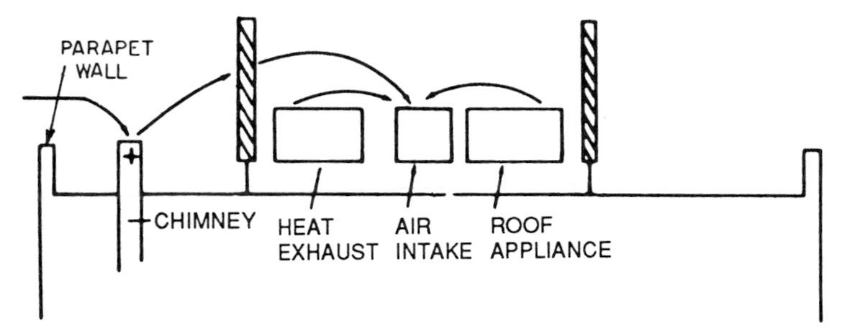

FIGURE 30.48 Parapet walls allow circulation of pollutants.

Exhaust systems terminated under a parapet wall will be under positive pressure with certain wind conditions.[29] For an incinerator or boiler chimney, this can have an adverse effect on the combustion process with a high carbon monoxide and odor output. This gas could be forced from the appliance to contaminate the appliance room and circulate throughout the building. Note that all appliances should have an adequate air supply if proper combustion is to occur. Open louvers are no answer, since they tend to be closed by personnel who object to cold air during the winter. A ducted air supply to the appliance or heated makeup air is a possible solution.

Figure 30.49 shows a chimney that may be terminated according to fire codes but is a frequent source of trouble for all types of appliances and exhaust systems. It is particularly a problem for fireplaces where downdrafts carry smoke and ashes into living quarters.

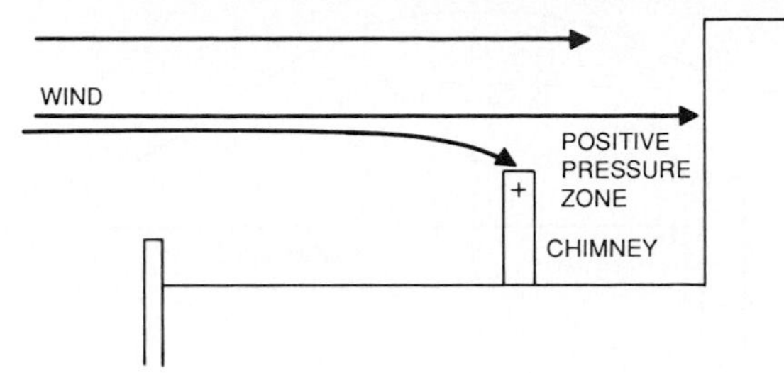

FIGURE 30.49 Obstruction creates positive chimney pressure.

A chimney terminated below the eddy area, below a parapet wall, or below a higher building will be subject to positive pressures. These pressures will force the products of combustion back into the building regardless of downdraft caps. These caps are only good to prevent downdrafts associated with wind direction.[30]

The effect of wind on an eddy area of low buildings is often overlooked in the determination of chimney heights. We tend to adhere to codes designed for fire safety only. Wind tunnel tests have shown the same air-flow effect as the high-rise building; however, in the case of low buildings, the width of the eddy area becomes important.

Wind builds up into an eddy area with a height equal to one-half of the building height, but moves closer to the roof as the width of the building exceeds 2 times its height H (Fig. 30.50). This places the building edge as the most critical portion

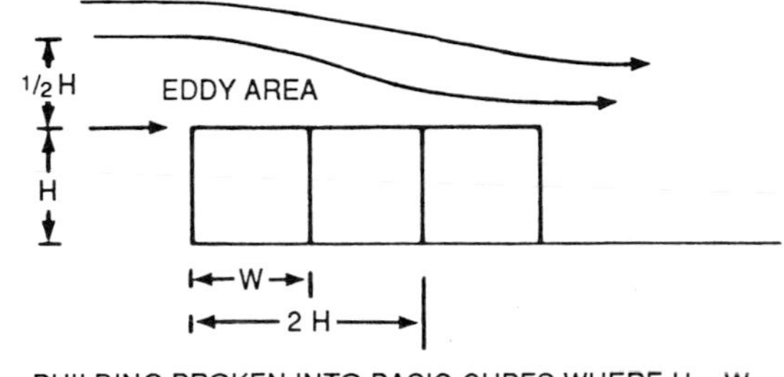

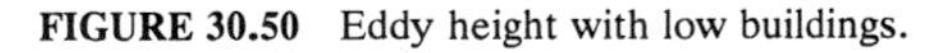

FIGURE 30.50 Eddy height with low buildings.

of low buildings. If a chimney is located on or within $2H$ of the edge of a 20-ft (6-m) high building, the chimney height must be 10 ft (3 m) above the parapet wall to be above the eddy area.[23] The level of the roof does not determine the height of the chimney. The height is determined by the height of the highest point in contact with the wind. Thus in Fig. 30.51 the building height, plus the parapet wall, becomes the controlling height. With this height as 20 ft (6 m), the chimney height must be 10 ft (3 m) above the parapet wall or 14 ft (4.3 m) above the roof.

It is in line with good design to locate a chimney above the eddy area (Fig. 30.52). This area becomes lower as we move from the edge; thus a lower chimney still above the eddy area is possible. Removal from the building edge has the further effect of raising the line of sight above the normal viewing height.

Termination of a chimney or exhaust system below the eddy area and close to the roof edge allows recycling of exhaust products into the air intake to contaminate the entire building (Fig. 30.53).

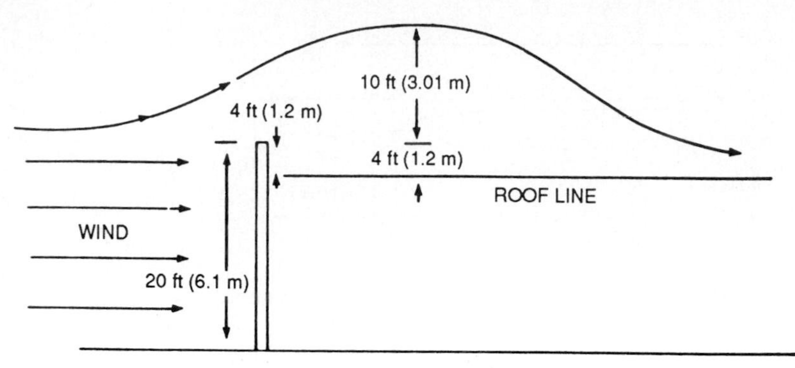

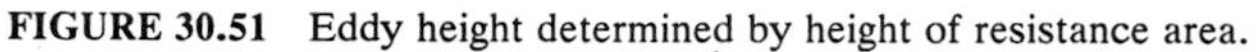
FIGURE 30.51 Eddy height determined by height of resistance area.

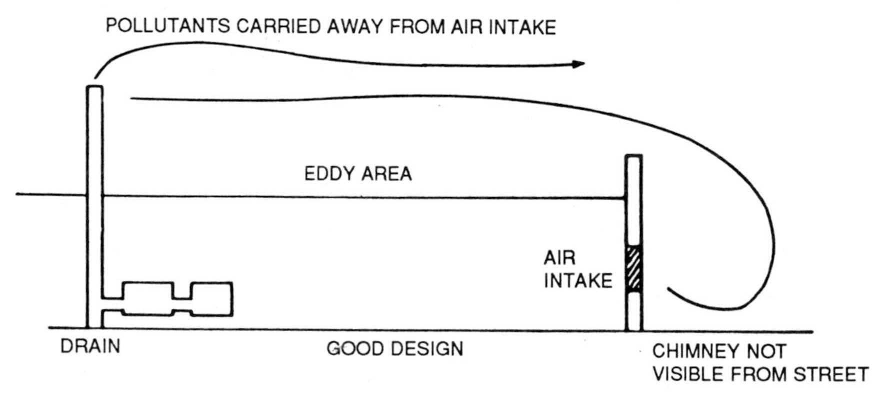

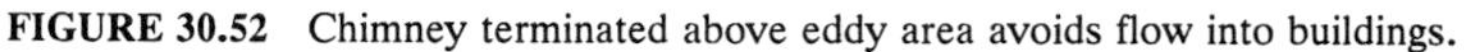
FIGURE 30.52 Chimney terminated above eddy area avoids flow into buildings.

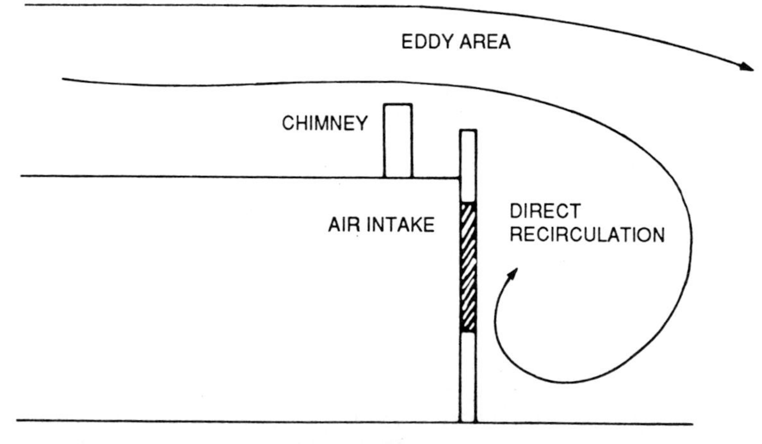

FIGURE 30.53 Chimneys terminated below eddy area allow recirculation of pollutants into building.

A coolie cap and most chimney caps defeat the proper functioning of a chimney even though designed to be above the eddy area (Fig. 30.54). The cap forces the pollutants below the eddy area and close to the roof, where they can be recirculated into the building by air intake ducts or infiltration.

It is good procedure to eliminate the rain cap and to use a tee-and-pier-supported chimney with a drain. This will remove both rain and condensation (Fig. 30.55). Chimneys may be supported from the roof where space is important.

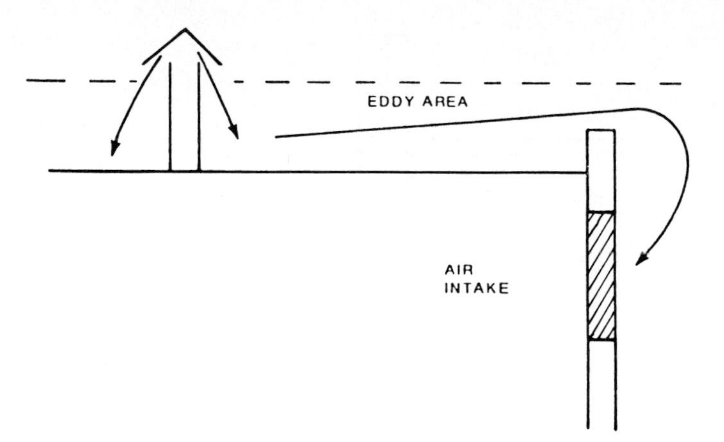

FIGURE 30.54 Rain cap forces pollutants toward building.

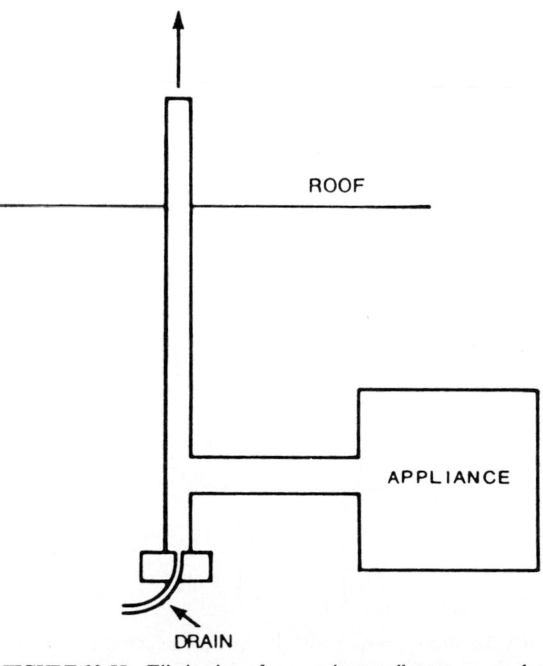

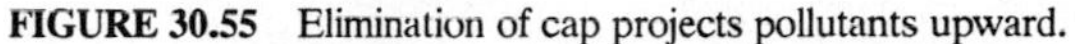

FIGURE 30.55 Elimination of cap projects pollutants upward.

The supporting member must be provided with a drain (Fig. 30.56). When a rain cap is absolutely necessary, a zero-loss cap may be used (Fig. 30.57). This provides an annular area around the inside of the chimney, for water removal. Figure 30.56 shows one design; others may be found in Clarke's paper[27] or in Ref. 31, which deals with industrial ventilation.

For a detailed discussion of this topic, see Ref. 34.

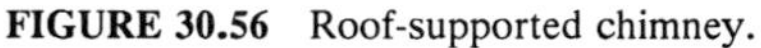

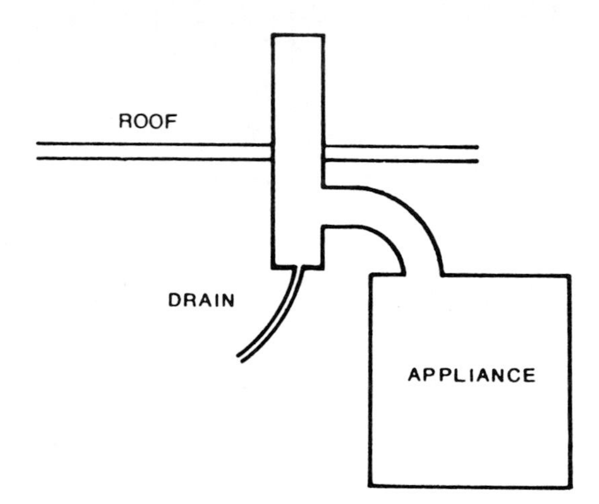

FIGURE 30.56 Roof-supported chimney.

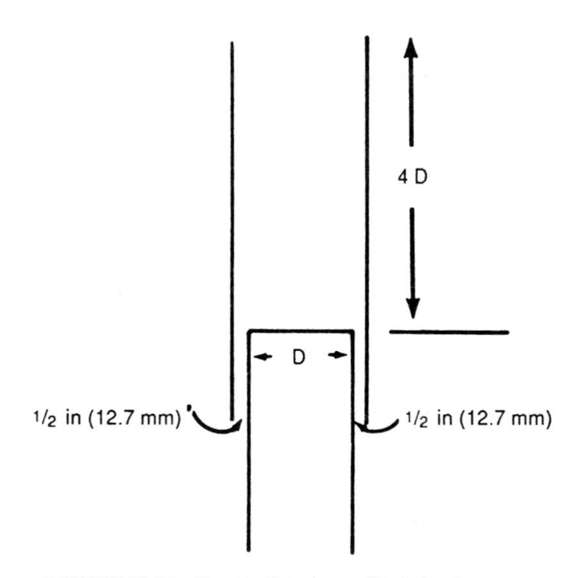

FIGURE 30.57 Zero-resistance roof termination.

30.7 REFERENCES

1. E. R. Kaiser and A. A. Carotti, "Municipal Incineration of Refuse with 2% and 4% Additions of Four Plastics," Report to the Society of the Plastics Industry, June 30, 1971, pp. 22, 24.
2. *Incinerator Standards*, Incinerator Institute of America, New York, 1968.
3. J. A. Caprio and E. Wolfe, "Refractories for Hazardous Waste Incineration—An Overview," *1982 National Waste Processing Conference*, ASME, New York, 1982.
4. J. R. McLaren and H. M. Richardson, "Action of Vanadium Pentoxide on Alumina-Silicate Refractories," *Transactions of British Ceramic Society*, vol. 58, no. 3, 1959, pp. 188–197.
5. G. R. Rigby and R. Hutton, "Action of Alkali and Alkali-Vanadium Oxide Slags on Alumina-Silica Refractories," *Journal of the American Ceramic Society*, vol. 5, no. 2, 1962, pp. 68–73.
6. G. H. Criss and A. R. Olsen, "The Chemistry of Incinerator Slags and Their Compatibility with Fireclay and Alumina Refractories," *1966 National Incinerator Conference*, ASME, New York, 1966.
7. G. H. Criss and A. R. Olsen, "Further Investigations of Refractory Compatibles with Selected Incinerator Slags," *1968 National Incinerator Conference*, ASME, New York, 1968.
8. G. H. Criss and R. R. Schneider, "Effect of Flue Gas Contaminants on Refractory Structures at Elevated Temperatures," Winter Meeting, ASME, New York, 1964.
9. T. F. E. Rhead and R. V. Wheeler, "Effect of Temperature on Equilibrium $2CO = CO_2 + C$," *Journal of the Chemical Society (London)*, vol. 97, no. 2, 1910, p. 278.
10. R. F. Patrich and R. B. Sosman, "Some Effects of Zinc and Carbon Monoxide on Fireclay Refractories," *Journal of American Ceramic Society*, vol. 32, no. 4, 1949, pp. 133–140.
11. *Steel Chimney Code*, International Committee on Industrial Chimneys, Brighton, UK, 1985. Available from CICIND, 136 North St., Brighton, UK BN1 1RG.
12. J. F. Schulz and K. A. Schultz, "Thermal and Corrosion Analysis of Chimney Systems." For full publication data, see footnote to Sec. 30.6.2.
13. Karen A. Harwood, "Thermal Shock Resistance of Fireclay Refractories," B.S. thesis, The Pennsylvania State University, University Park, 1985.
14. "Dew Point Corrosion by Inorganic Acids," paper 14, *Proceedings 1984 Air Pollution Seminar*, Air Pollution Control Association, Dec. 5–7, 1985.
15. James A. Caprio, "Refractory Practice Survey in Hazardous Waste Incinerators," paper presented at 76th Annual Meeting of the Air Pollution Control Association, June 19–24, 1983.
16. D. W. Mitas, D. O. Swenson, R. G. Gentner, and A. Zourides, "A Survey of Lining Materials Performance in Flue Gas Desulfurization Systems," paper presented at American Power Conference, Chicago, Apr. 22–24, 1985.
17. Jeffrey L. Hahn, "Innovative Technology for the Control of Air Pollution at Waste-to-Energy Plants," *1986 National Waste Processing Conference*, ASME, New York, 1986.
18. Klaus S. Feindler, "Long Term Results of Operating Ta Luft Acid Gas Scrubbing Systems," ibid.
19. Gary W. Schanche and Kenneth E. Griggs, "Features and Operating Experiences of Heat Recovery Incinerators," ibid.
20. P. L. Daniel, J. L. Barna, and J. D. Blue, "Furnace-Wall Corrosion in Refuse-Fired Boilers," ibid.

21. E. S. Domalski, A. E. Ledford, Jr., S. S. Bruce, and K. L. Churney, "The Chlorine Content of Municipal Solid Waste from Baltimore County, Maryland and Brooklyn, New York," ibid.

22. D. W. Mitas, D. O. Swenson, R. G. Gentner, and A. Zourides, "A Survey of Lining Materials Performance in Flue Gas Desulfurization Systems," paper presented at American Power Conference, Chicago, 1985.

23. Benjamin H. Evans, "Natural Air Flow around Building," Research Report 59, Texas A&M College, March 1957.

24. J. Halitsky, "Gas Diffusion near Buildings," paper presented at the ASHRAE 70th Annual Meeting, Milwaukee, WI, June 24–26, 1963.

25. J. Halitsky, "Estimation of Stack Height Required to Limit Contamination of Building Air Intakes," *American Industrial Hygiene Association Journal*, vol. 26, 1965.

26. P. S. Rummerfield, J. Cholak, and J. Kereiakes, "Estimation of Local Diffusion of Pollutants from a Chimney," *American Industrial Hygiene Association Journal*, vol. 28, no. 4, July-August 1967, p. 366.

27. J. H. Clarke, "Air Flow around Buildings," *Heating Piping and Air Conditioning*, May 1967, p. 145.

28. "Control of Corrosion and Deposits in Stationary Boilers, Burning Residual Fuel Oil," Government Research Report AD-422073, U.S. Department of Commerce, Office of Technical Services, Washington, DC.

29. F. S. Holdredge and R. H. Reed, "Pressure Distribution on Buildings," Summary Report 1 to the Department of the Army, Camp Detrick, MD, August 1956, Texas Engineering Experiment Station.

30. E. H. Perry and J. J. Vandeber, Jr., "A Study of Vent Cowl Performance and Location," *Research Report 1362, Heating and Air Conditioning Research*, American Gas Association.

31. Committee on Industrial Ventilation, *Industrial Ventilation: A Manual of Recommended Practice*, 10th ed., American Conference of Governmental Industrial Hygienists, p. 8–5.

32. P. S. Rummerfield, J. Cholak and J. Kereiakes, "Estimation of Local Diffusion of Pollutants from a Chimney," *American Industrial Hygiene Association Journal*, vol. 28, no. 4, July-August 1967, p. 366.

33. J. F. Schulz, "Factors Involved in the Design of High Rise Chimney and Chute Systems," *Proceedings of the 1968 National Incinerator Conference*, ASME, New York, 1968.

34. J. F. Schulz, "Self-Pollution of Buildings," *Proceedings of the 1972 National Incinerator Conference*, ASME, New York, 1972.

CHAPTER 31
CHILLED WATER AND BRINE

Gary M. Bireta, P.E.
Project Engineer, Mechanical Engineering, Giffels Associates, Inc., Southfield, Michigan

31.1 INTRODUCTION

Water systems are used in air-conditioning applications for heat removal and dehumidification. The two most common systems use chilled water and brine. Chilled water is plain water at a temperature from 40 to 55°F (4 to 13°C). Brine is a water/antifreeze solution at a temperature below 40°F (4°C). Here we describe the basic principles and considerations for chilled water. Additional considerations for brine follow.

31.2 SYSTEM DESCRIPTION

A chilled-water system works in conjunction with air-handling units or process equipment to remove the heat generated within a conditioned space or process. The terminal unit cooling coil(s) collect(s) the heat and then transfers it by conduction and convection to the water, which is conveyed through connecting piping to the evaporator side of a chiller. The chiller, which is a packaged refrigeration machine, internally transfers the heat from the evaporator to the condenser, where heat is discharged to the atmosphere by the condenser system. The chilled water leaving the evaporator is circulated back to the coils, where the heat-removal process repeats again. Figure 31.1 is a schematic diagram of the chilled-water system. In this chapter we discuss the basic principles and details regarding the chilled-water or evaporator side of the system. The designer should refer to Chaps. 40 to 44 for types and selection of the chiller and condenser system.

31.3 WHERE USED

Chilled-water systems are applicable when:

- The design considerations of a proposed air-conditioned facility require numerous separated cooling coils plus the restriction that the refrigeration system(s) of the facility be located in a single area.
- There is a need for close control of coil leaving air temperature or humidity

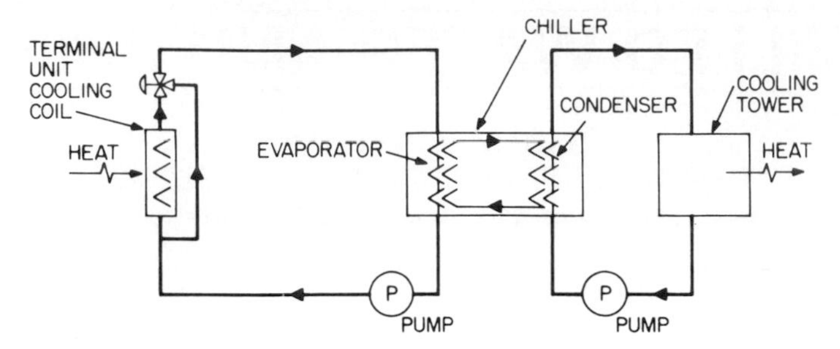

FIGURE 31.1 Schematic diagram of chilled-water system.

control. Control can be very smooth and exact because of the infinite modulating capability of the chilled-water valve.

- Future expansion will require additional cooling capacity. The additional capacity might be merely a matter of new terminal units and branch piping from the chilled-water mains to the coils, although this is limited by the unused capacity of the chillers and water distribution system.
- The coil leaving air temperature desired is 45°F (7°C) or higher. The leaving air will be at least 5°F (3°C) warmer than the coil entering water temperature.

31.4 SYSTEM ARRANGEMENT

The system designer must consider the cooling loads involved and the type and arrangement of the facility during the conceptual phase of a chilled-water system design. During initial design development, the designer should consider the impact of future system loads. System expansion costs can be reduced if space for additional equipment and the flow rates are planned for during the initial design. The module design concept adapts well for planning for future expansion.

Large facilities commonly consist of terminal units located near the areas they serve. The total combined loads of the facility result in a large peak demand with a wide operating range that is beyond the capability of a single chiller. A chilled-water arrangement for a large installation would commonly consist of multiple chillers centrally located with multiple cooling towers of the condenser system situated nearby outside. Figure 31.2 shows the evaporator side of a multiple-chiller arrangement. Installations of this type are typically arranged in modules with a chiller, cooling tower, and associated pumps dedicated to part of the peak load. A single distribution system transports the chilled water to the various areas and terminal units.

Two-way modulating control valves are used to vary the flow to the terminal units based on a signal from the conditioned space room thermostats. Two-way valves are preferred to three-way valves because the total system pumping cost is reduced during part-load conditions. Chiller manufacturers, however, demand a constant flow through the chillers for stable refrigeration control. In this situation a pump arrangement is needed that allows variable flow to the terminals while maintaining a constant flow through the chillers. This problem is solved by installing two sets of pumps in a primary/secondary arrangement. The secondary pumps can be controlled to match the demand of the terminals while the primary

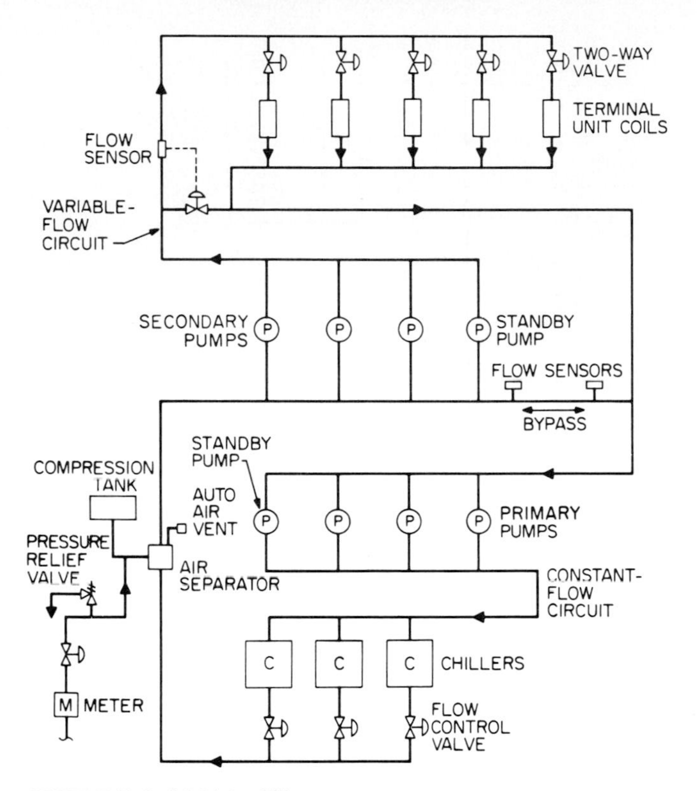

FIGURE 31.2 Multiple-chiller arrangement.

pumps maintain a constant flow through the chillers. The system bypass decouples the two pump sets which allows them to operate pressure-independent of each other. Reference 1 provides further explanation of primary/secondary pumping.

A small installation for an individual building or process may consist of a single chiller, cooling tower, pump, and small distribution system connected to nearby terminal units. The condenser system cooling tower would typically be located on the building roof or nearby, outside the building.

A three-way mixing valve modulates the terminal unit flow based on the cooling load demand while maintaining a constant flow through the chiller and pump. For small installations, the increased pumping cost is offset by the savings realized from fewer pumps and less complicated controls. Figure 31.3 shows the evaporator side of a single-chiller arrangement.

31.5 DISTRIBUTION SYSTEMS

There are two basic distribution systems for chilled water: the two-pipe reverse-return and the two-pipe direct-return arrangements. Figure 31.4 illustrates the direct-return and reverse-return configurations.

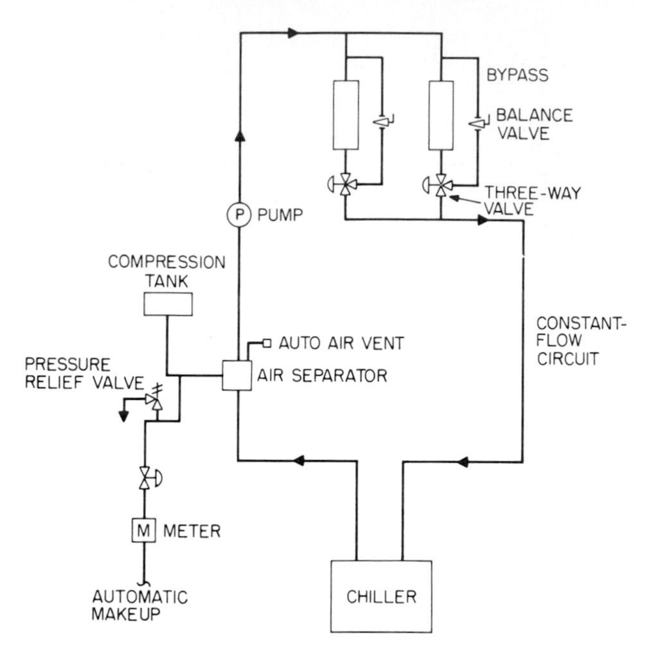

FIGURE 31.3 Single-chiller arrangement.

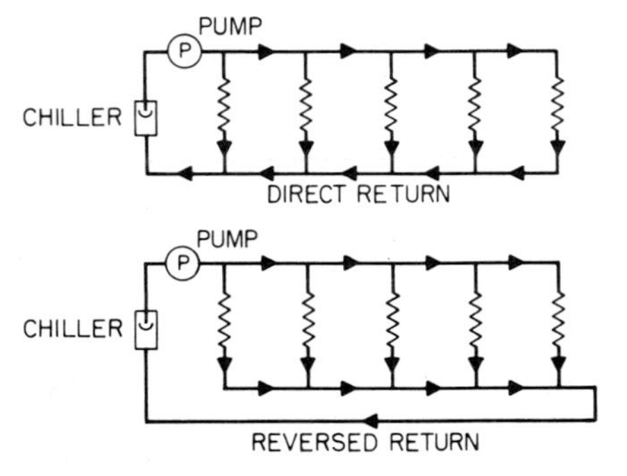

FIGURE 31.4 Piping distribution systems.

The reverse-return system is preferable from a control and balancing point of view, since it provides very close equivalent lengths to all terminals, resulting in closely balanced flow rates. In large installations, however, the additional piping for a reverse-return system is usually not economical.

The direct-return system is more commonly used. The system must be carefully analyzed to avoid flow-balancing problems. Balancing valves and flow meters should be provided at each branch takeoff and terminal unit. Control valves with high head loss are recommended and must be analyzed for varying "shutoff" heads through the system. In large systems, it is sometimes desirable to use a combination of a direct-return system for the mains with a reverse-return system for branch piping to sets of terminal units. This combination provides an economical main distribution segment with easier balancing within the branches. The overall system should be analyzed to determine the most economical distribution system for the application.

31.6 DESIGN CONSIDERATIONS

31.6.1 Design Temperatures

The chilled-water design temperatures must be established before the terminal unit flow rates can be ascertained. Chilled-water supply temperatures range from 40 to 55°F (4 to 13°C), but temperatures from 42 to 46°F (6 to 8°C) are most common. Temperature differentials between supply and return are in the 7 to 11°F (4 to 6°C) range for small buildings and the 12 to 16°F (7 to 9°C) range for conventional systems. Higher temperature differentials are preferred since they reduce the system flow rates, resulting in smaller piping and pumps, less pumping energy, and increased chiller efficiency. Before the design temperatures are finalized, the designer should ensure that the design temperature selected will result in terminal devices properly sized for their applications. The designer should refer to Chaps. 40 to 44 for design considerations and selection of chilled-water coils.

For large distribution systems, a terminal supply temperature approximately 1°F (0.5°C) higher than the leaving chiller temperature is sometimes assumed. The 1°F (0.5°C) increase accounts for pump and pipe heat gains between the chiller and terminal units. The additional load from these sources must be included during the sizing of the chiller(s).

If the system is subject to freezing, a water/glycol solution (brine) may be required. Refer to the discussion of brine for additional design considerations.

After the terminal units have been located and the flow rates established, a system flow diagram should be created. Several chiller manufacturers should be consulted for sizes, types, and operating ranges available. The designer must analyze the facility for an appropriate distribution system, pump arrangement, and control sequence for all components.

If continuous system operation is required, standby pumps are recommended to ensure system operation in the event of a pump failure. Standby chillers are not usually included because of their high initial cost and rare failure.

31.6.2 Piping

Figure 31.5 is provided as a general guide for selecting pipe sizes once the system flow diagram has been established. The shaded area provides economical pipe sizes as a function of flow, velocity, and friction loss. In situations where two pipe sizes are capable of handling the design flow, the larger of the two should be selected, in case of an unexpected increase in the flow rate. The system designer should carefully size the inlet and outlet connections to terminal units. If they were left unsized, these branches could be installed, the same size as the terminal unit connection, which might result in abnormally high pressure losses.

Corrosion inhibitors are commonly added to chilled-water systems to reduce corrosion and scale. Refer to Chap. 17 for recommended water conditioning. With proper water treatment and a closed system, the pipe interior should remain relatively free of scale and corrosion. The calculated pump head can be based on relatively clean pipe, although it is prudent to assume a minimum fouling factor. A 25 percent fouling factor is equivalent to using $C = 130$ in the Williams and Hazen formula for steel pipe.

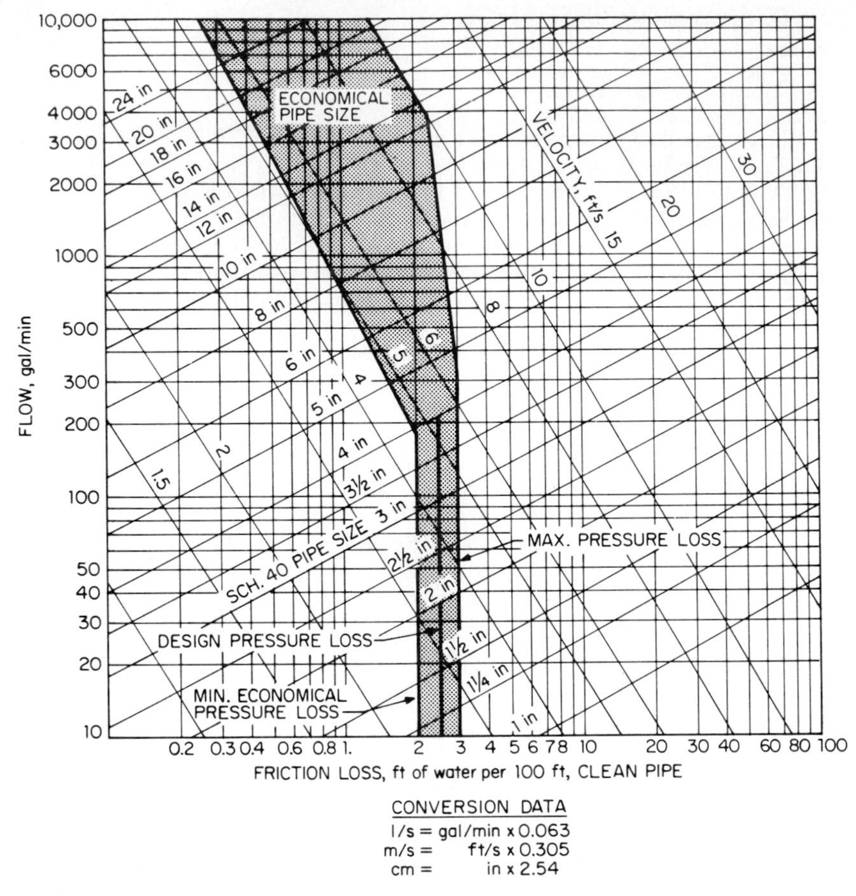

FIGURE 31.5 Pipe-sizing graph.

31.7 INSTALLATION CONSIDERATIONS

31.7.1 System Volume

Automatic makeup water with a positive-displacement meter is recommended for filling, monitoring volume use, and maintaining system volume. The systems should also be attached to an adequately sized compression or expansion tank and to a pressure relief valve. If the tank is not provided and the system has automatic makeup water, the relief valve will discharge with each rise in average water temperature, wasting water and chemicals.

31.7.2 Air Control

Manual or automatic air vents must be installed at system high points to vent air when the system is filled. Automatic vents should be provided with an isolation valve to enable replacement. Vent blowoff lines should be piped down to the closest waste drain. Air, in horizontal mains, can generally be kept out of the branch piping when the branch connections are in the bottom 90° arc of the main. A branch pipe with vertical downflow that connects to the bottom of a main can accumulate pipe scale and similar debris. Dirt legs, such as for steam drips, or strainers may be useful in these branches, especially for 2-in (51-mm) and smaller piping. Air in vertical piping will flow *down* with the water at 2 ft/s (0.6 m/s) or greater water velocity.

31.7.3 System Isolation

All equipment requiring maintenance and branch piping should be provided with manual isolation valves. Chain-wheel operators are recommended for frequently used valves located out of the operator's reach. Drain connections should be provided at low points to allow partial system drainage of isolated sections.

31.7.4 Coil Control

The coil capacity or degree to which an airstream is cooled or dehumidified as it passes through a chilled-water coil is generally controlled by varying the quantity of water flowing through the coil. (Capacity can also be controlled by varying the water temperature or by varying a portion of air which bypasses the coil and remixing it with the portion passed through the coil.) Ideally the control valve will vary the water flow continuously and uniformly as the valve strokes from full open to full closed; that is, 10 percent of stroke should cause a 10 percent change in flow. However, if a line-size butterfly valve is installed, it might pass 80 percent of full flow when only 20 percent open, which means the valve is trying to control 0 to 100 percent flow while 0 to 25 percent "open." The flow tends to be unstable in these situations, especially at low flow where the valve is apt to wiredraw, chatter, and hunt. Valve size is very important. In the interest of satisfactory control stability, the valves are frequently less than line size because the pressure loss through the fully open valve should be at least 33 percent of the total pressure loss of the circuit being controlled. For example, if the total coil circuit loss is 7 lb/in^2 (48 kPa), a valve sized for a drop of 3 to 5 lb/in^2 (21 to 34 kPa) would be satisfactory. Strainers are recommended upstream of control valves and any other piping elements which require protection against pipe scale and debris within the system.

31.8 SYSTEM MONITORING

Pressure gauges permanently installed in a system deteriorate over time from constant vibration. Gauges should be installed only at points requiring periodic monitoring. At points where infrequent indication is required, gauge cocks should be installed with a set of spare gauges provided to the operator. Thermometers are recommended at all terminal units and chillers.

For additional explanation and considerations for the design of a chilled-water system, see Ref. 2.

31.9 BRINE

The term "water" is used throughout this chapter for convenience, whereas it could be plain water or a brine. The term "brine" includes a water/glycol solution, a proprietary heat-transfer liquid, water and calcium or sodium chloride solution, or a refrigerant. The best choice of brine will depend on the parameters of the system, but plain water or a water/glycol solution is the overwhelming choice for comfort air-conditioning chilled-water systems. Propylene glycol is the least toxic of the glycols and should be used if there is any possibility (e.g., piping leaks) of contact with a food or beverages.

31.9.1 Where Used

An antifreeze must be added to the water, or a nonfreezing liquid must be used, whenever any portion of the water is subject to less than 33°F (0.6°C). Be sure to check the temperature of the refrigerant, or cooling medium, in the chiller. If it operates below 33°F (0.6°C), freeze protection is needed. In systems where the refrigerant is only a few degrees below 33°F (0.6°C), chiller freezeup can be precluded by maintaining chilled-water flow through the chiller for some period after the chiller has been shut off. This water flow will "boil" the refrigerant remaining in the evaporator.

31.9.2 Design Considerations

Adding an antifreeze to water will generally reduce the specific heat and conductivity and increase the viscosity of the solution. These, in turn, generally necessitate increased heat-transfer surface in the chiller and cooling coils, increased chilled-water flow, and increased pump head. See Fig. 31.6. For example, suppose a plain water system involves 1000 gal/min (63.1 L/s), 50 lb/in^2 (345-kPa) pressure drop for pipe friction, and 29 hp (21.6 kW) to overcome pipe friction. Then

- If 10% glycol is added, the parameters become 1008 gal/min (63.6 L/s), 53 lb/in^2 (365 kPa), and 30 hp (22.4 kW).
- If 40% glycol is added, the parameters become 1150 gal/min (72.6 L/s), 79 lb/in^2 (544 kPa), and 52 hp (38.8 kW).

The piping size and solution temperature rise are assumed the same as in the plain water and glycol systems.

The freezing point of aqueous glycol solutions can be found from charts similar to Fig. 31.7. Note that 40% glycol is needed to lower the freezing point to −10°F (−23°C). By definition, the freezing point is that at which the first ice crystal forms. Chilled-water piping has been protected from freeze damage with as little as 10% or even 5% glycol in 0°F (−18°C) weather. This is referred to as "burst protection." Ice crystals may form at 25 to 29°F (−4 to −2°C) in the 10% or 5% water/glycol solution, but the solution merely forms a slush and does not

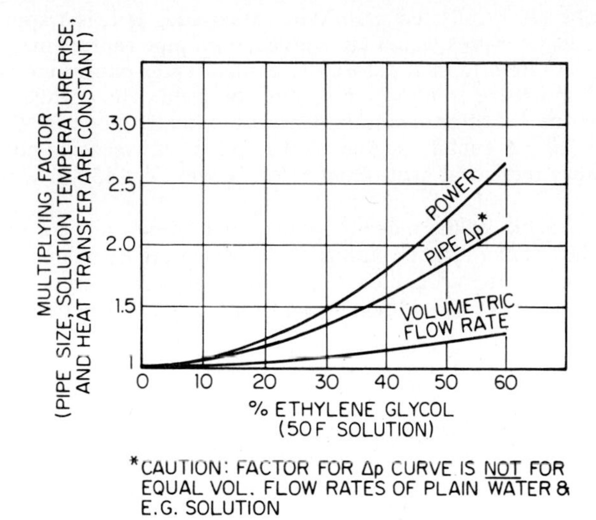

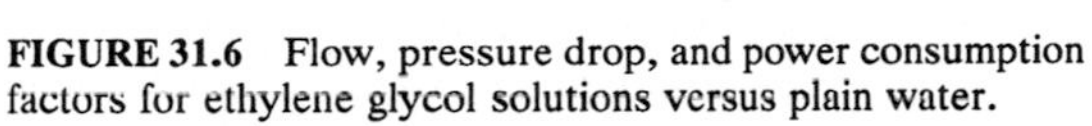

FIGURE 31.6 Flow, pressure drop, and power consumption factors for ethylene glycol solutions versus plain water.

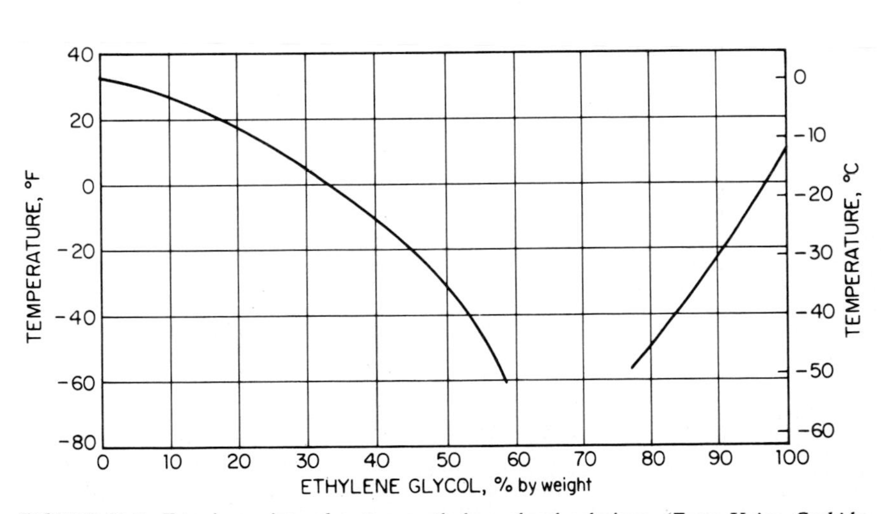

FIGURE 31.7 Freezing points of aqueous ethylene glycol solutions. *(From Union Carbide Corp., Ethylene Glycol Product Information Bulletin, Document F-49193 10/83-2M.)*

freeze solid. The slush must be permitted to expand. If it is trapped between shutoff valves, check valves, automatic valves, etc., pipe rupture may occur. The slush must be permitted to melt before the chilled-water pumps are started.

Automotive antifreeze solutions should not be used. The corrosion inhibitors added to automotive antifreeze solutions are specifically made for the materials encountered in an automobile engine. Automotive antifreeze is not meant for long life, whereas industrial heat-transfer fluids may last 15 years with proper care.

A chemical analysis of the makeup water must be checked for compatibility with the proposed chiller, pump, piping, and coil materials, chemical treatment, antifreeze, corrosion inhibitors, etc., to preclude the formation of scale, sludge, and corrosion. The water should be checked regularly for depletion of any components.

31.10 STRATIFIED CHILLED-WATER STORAGE SYSTEM

The chilled-water storage system is a conventional chilled-water system with the addition of a thermally insulated storage tank. Figure 31.8 shows a typical chilled-water storage arrangement. During the daily cooling cycle, the chillers operate to maintain cooling until the load exceeds the capacity of the system. At that point, the chillers and tank work in conjunction to handle the peak demand. As the load falls below the chiller's capacity, the chillers continue to operate to recharge the tank for the next day's demand.

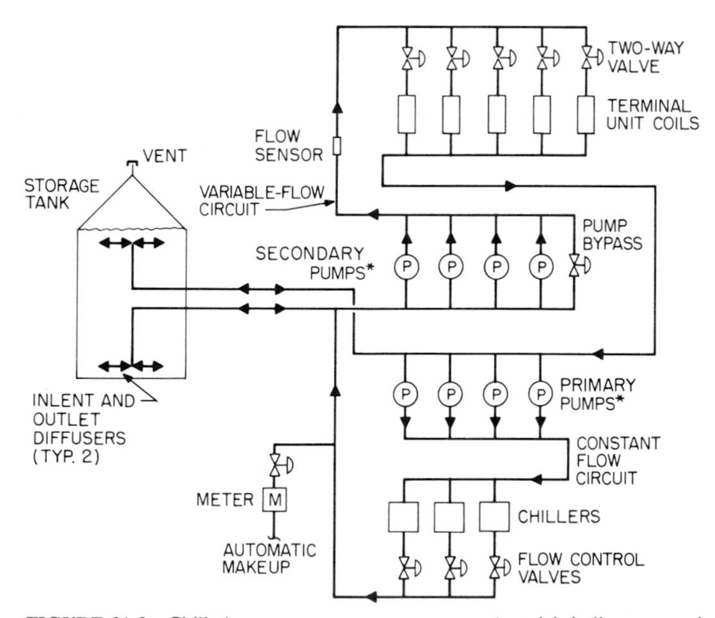

FIGURE 31.8 Chilled-water storage arrangement. Asterisk indicates standby pumps.

The advantage of this arrangement is that a portion of the equipment required for a conventional system to handle peak loads can be replaced by a less expensive storage tank. In addition, the owner's electric power rates are reduced since the tank has shaved the monthly peak power demand.

The system is classified as a stratified storage system because warm water and cold water within the storage tank remain separated by stratification. During operation, as a portion of chilled water is removed from the tank bottom for cooling, the identical portion of warm return water is discharged back into the tank at the top. A thermal boundary forms with the warmer, less dense water stratifying at the top and the denser, colder water remaining below. During periods of reduced load, the tank is recharged by removing the warm stratified water from the tank top, chilling it, and returning it to the tank bottom. During a daily cycle, the thermal boundary moves up and down within the tank, but the total water quantity remains unchanged.

31.10.1 Load Profile

Figure 31.9 shows a typical building cooling load profile utilized for storage applications. Curve *ABCDE* represents the cooling load profile during the day. Point *C* represents the maximum instantaneous peak. Line *FG* represents the installed chiller capacity. The area within *ABDE* represents the portion of cooling provided by the chillers, and area *BCD* represents the portion of cooling provided by the tank. The remaining areas, *FBA* and *DGE*, represent the chiller capacity available to recharge the tank. For storage applications, units of ton-hours (kWh) are used to determine the cooling load and storage requirements of the system.

31.10.2 Where Used

Chilled-water storage systems generally become economical in systems in excess of 400 tons (1407 kW) of refrigeration. In all cases, the economics of the installation should be the deciding factor for choosing a storage system.

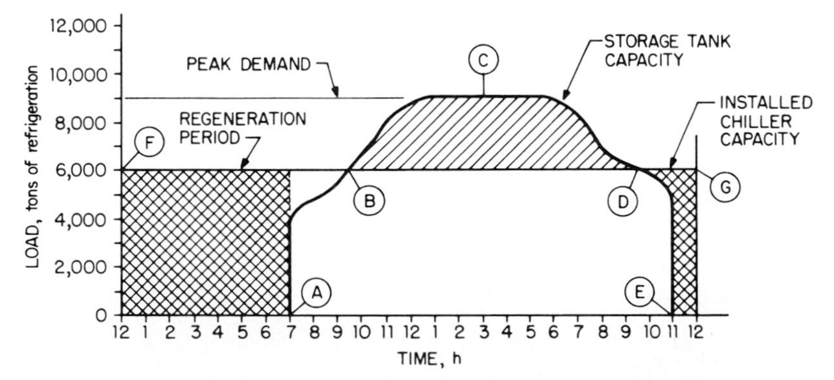

FIGURE 31.9 Cooling-load profile. See text for explanation of letter symbols.

31.10.3 Design Considerations

The design engineer must analyze the operating parameters of the facility in order to accurately predict the load-cycle hours for a given day. The number of hours required for cooling and available for tank regeneration must be determined. Once established, the daily cooling load can be calculated. If available, a computer cooling-load program capable of providing an hour-by-hour analysis is recommended for predicting the cooling load profile. Example 31.1 demonstrates the method for creating a load profile diagram and determining the refrigeration and tank capacity for a storage installation.

EXAMPLE 31.1

FIND: (1) Total refrigeration load
(2) Tank capacity

GIVEN: Cooling period, 8 a.m. to 8 p.m. (12 h)
Regeneration period, 8 p.m. to 8 a.m. (12 h)
Note: Hourly loads provided are not from Fig. 31.9 load profile

Daily cooling load, hour-by-hour analysis

Time	Total load	
8 a.m.	30,084 MBtu	(7.58×10^6 kcal)
9	36,972	(9.31×10^6)
10	45,144	(1.14×10^7)
11	59,268	(1.49×10^7)
12	72,168	(1.82×10^7)
1 p.m.	80,952	(2.04×10^7)
2	82,212	(2.07×10^7)
3	83,100	(2.09×10^7)
4	85,392	(2.15×10^7)
5	85,320	(2.15×10^7)
6	83,436	(2.10×10^7)
7	81,900	(2.06×10^7)
Total	825,948 MBtu	(2.08×10^8 kcal)

Total daily load (ton/h)

= 825,948 MBtu/[12 MBtu/(ton·h)]
= 68,829 ton/h

= (2.08×10^8 kcal/3022 kcal/ton·h)
= (68,829 ton/h)

Solution (1) Total Chiller Capacity

68,829 ton/h/24 hr = 2868 tons
(Select two 1500-ton chillers)

(2) Storage Tank Capacity

68,829 ton/h − (3000 ton × 12 h)
= 32,829 ton/h

Based on 20°F temp. diff (11.1K)

$$\frac{(7.48\ \text{gal/ft}^3\ (32{,}829\ \text{ton/h})\ [12{,}000\ \text{Btu/(h}\cdot\text{ton)}]}{(62.4\ \text{lb/ft}^3)\ (20°\text{F})\ [1.0\ \text{Btu/(lb}\cdot°\text{F)}]}$$

$$\left(\frac{(1000\ \text{L/m}^3)\ (32{,}829\ \text{ton/h})\ [3022\ \text{kcal/ton}\cdot\text{h})]}{(1000\ \text{kg/m}^3)\ (11.1\text{K})\ [0.998\ \text{kcal/(kg}\cdot\text{K)}]}\right)$$

capacity = 2,361,162 gal (8,955,680 L)

Increasing tank volume 20% for mixed zone and internal piping:

Total Capacity = 2,833,395 gal (10,746,817 L

31.10.4 Design Temperatures

As previously described for the conventional chilled-water system, the design temperatures of the system must be established. Temperature differentials for storage applications typically range from 16 to 22°F (−10 to −6°C). A higher temperature differential is preferred since it will reduce the size of the tank and system flows. Before the design temperatures are finalized, the terminal unit coils must be checked to ensure that they can be sized for proper operation. High differential temperature can result in low water velocities within the coil tubes, which may lead to poor part-load performance. Tube water velocities between 2 and 5 ft/s (0.6 and 1.5 m/s) are preferred for efficient heat transfer.

31.10.5 Tank Sizing

After the design temperatures have been established, the storage tank capacity can be determined. The capacity previously calculated in ton-hours can be converted to gallons, as illustrated in Example 31.1. The tank sizing must also allow both for unused space due to piping and related apparatus and for the mixed thermal boundary between warm and cold fluids.

31.10.6 Installation Considerations

Tank. Tanks for storage applications are field-fabricated of steel or concrete. Steel is generally preferred over concrete for stratified storage because steel readily absorbs and rejects the changes in water temperature without disturbing the thermal boundary. For example, when a concrete tank is recharged, the rising chilled water is warmed by the heat stored in the concrete.

The tank should have a roof, to keep out unwanted debris. The exterior of the tank must be insulated; spray foam or rigid board insulation is used for this type of installation. The tank should be equipped with provisions for access, filling and draining, venting, and overflow, with associated controls for temperature and level monitoring.

Diffuser. The size, number, and orientation of the diffusers within the tank depend on the design parameters of the system. The intent of the diffusers is to allow removal and replacement of the tank water without disrupting the stratified thermal boundary. The tank should be installed in parallel with the chillers. This arrangement will result in two diffusers, one at the top and one at the bottom of the tank.

Radial-type diffusers have been used with success for cylindrically shaped tanks. The diffuser consists of two steel plates with the inlet pipe located at the center. As the water enters the diffuser, the flow is distributed in all directions toward the outside wall. The diffuser should be designed for low outlet velocity. This is determined by designing for a Froude number of 1 or below. Example 31.2 demonstrates the method for sizing a radial diffuser.

EXAMPLE 31.2

FIND: Radial diffuser height between plates

GIVEN: Maximum flow = 3500 gal/min (220 L/s)
Diffuser diameter = 10 ft (3.048 m)
Water temperatures = 44°F (6.6°C), 64°F (17.7°C)

Solution The Froude number is dimensionless and defined by the following equation:

$$F = \frac{Q}{\sqrt{g(\Delta p/p)h^3}}$$

where F = Froude number = 1
Q = outlet/inlet flow ratio, defined as ft^3/s (m^3/s) flow divided by outlet perimeter in ft (m)
g = gravitational effect, ft/s^2 (m/s^2)
p = water density, lb/ft^3 (kg/m^3)
Δp = difference in water density between inlet and outlet, lb/ft^3 (kg/m^3)
h = distance between plates, ft (m)

$$1 = \frac{\dfrac{3500 \text{ gal/min}}{7.48 \text{ (gal} \cdot \text{min)/ft}^3 \times (60 \text{ s/min}) \times 10 \text{ ft} \times 3.14}}{\sqrt{32.2[(62.40 - 62.31)/62.40]h^3}}$$

or

$$1 = \frac{\dfrac{220 \text{ L/s}}{1000 \text{ L/m}^3 \times 3.048 \text{ m} \times 3.14}}{\sqrt{9.8 \text{ m/s}^2[(999.7 - 998.3)/999.7]h^3}}$$

Solving for h gives 1.13 ft (0.34 m).

Temperature Monitoring. The tank should be equipped with temperature controls to monitor the temperature gradient within the tank. Thermocouples are typically used and installed vertically along the tank wall. The spacing should be adequate to identify the location of the thermal layer at all times. Generally, spacing of 1 to 2 ft (0.3 to 0.6 m) should be adequate. However, judgment is required depending on the tank's dimension.

31.11 REFERENCES

1. *Primary Secondary Pumping Application Manual*, International Telephone and Telegraph Corp., Bulletin TEH-775, 1968.
2. The American Society of Heating, Refrigeration, and Air-Conditioning Engineers, Inc., *1987 ASHRAE Handbook, Systems*, ASHRAE, Atlanta, GA, chaps. 13 and 14.

CHAPTER 32
ALL-AIR SYSTEMS

Ernest H. Graf, P.E.
Assistant Director, Mechanical Engineering, Giffels Associates, Inc., Southfield, Michigan

Melvin S. Lee
Senior Project Designer, Giffels Associates, Inc., Southfield, Michigan

32.1 SINGLE-ZONE CONSTANT-VOLUME SYSTEM

The single-zone constant-volume system is the basic all-air system used in a temperature-controlled area (Fig. 32.1). This system will maintain comfort conditions in an area where the heating or cooling load is fairly uniform throughout the space.

The method of controlling the temperature in the area can vary based on the functions performed in the space. The basic general-use area will require only modulating the cooling or heating medium at the air-handling equipment to maintain desired space conditions. Areas requiring a closer temperature-humidity control will need a unit that can cool and reheat at the same time, to maintain a close temperature range.

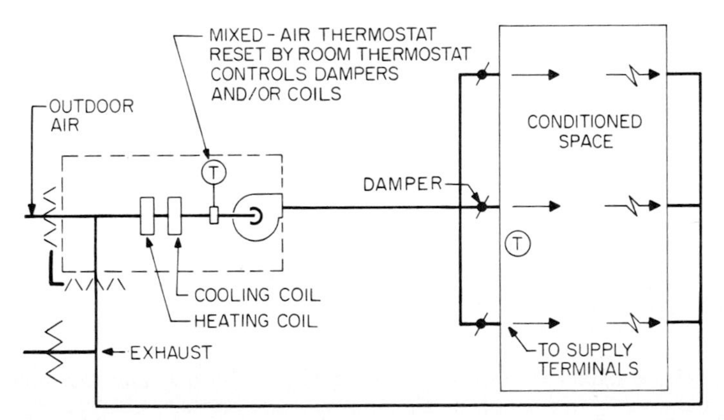

FIGURE 32.1 Single-zone system.

32.1.1 Central Equipment

A single-zone unit, shown in Fig. 32.1, consists of a supply fan, cooling coil, heating coil, filter section, and return-air/outdoor-air mixing plenum.

The combination of these components in the unit provides a system that can maintain a basic temperature-controlled environment with a change in either heating or cooling loads.

Variations and additions to these components can provide a system that can maintain a closely controlled temperature and humidity environment throughout the year. With the addition of a return-air fan to systems having long return air ducts, outdoor air can provide the required cooling medium during certain periods of the year. Adding a humidifier to the unit provides means to control and maintain a precise level of humidity to match the area function.

32.1.2 Ductwork System

The layout of the single supply duct to the conditioned space should be routed with a minimum of abrupt directional and size configuration changes.

The supply ductwork should be sized by using the static regain method to assist in a balanced air distribution in the duct system. The branch or zone mains should be provided with a balancing damper at the point of connection to the supply main. This will enable fine adjustments to be made to the air distribution system within the zone.

The supply air terminals in the space should be selected, sized, and located to provide even distribution throughout the space without creating drafts or excessive noise. Each terminal should have a volume damper to permit individual air balancing.

The return-air ductwork should be sized by using the equal-friction method from the space return registers back to the central equipment. The same ductwork configuration considerations and accessories should be used in laying out the return ductwork as listed for the supply duct system.

The location of the central equipment relative to the conditioned space should be considered when one is evaluating the need for acoustically lined ductwork or sound traps at the central equipment, to prevent transmission of noise through the duct system to the space.

32.1.3 Applications

The single-zone systems are generally used for small offices, classrooms, and stores. The single-package type of individual air-handling unit, complete with refrigeration and heating capabilities, can be roof-mounted or located in a mechanical space adjacent to or remote from the conditioned space.

32.2 SINGLE-ZONE CONSTANT-VOLUME SYSTEM WITH REHEAT

A single-zone constant-volume system with reheat has the same equipment and operating characteristics as the single-zone system, but has the advantage of being able to control temperatures in a number of zones with varying load conditions. See Fig. 32.2.

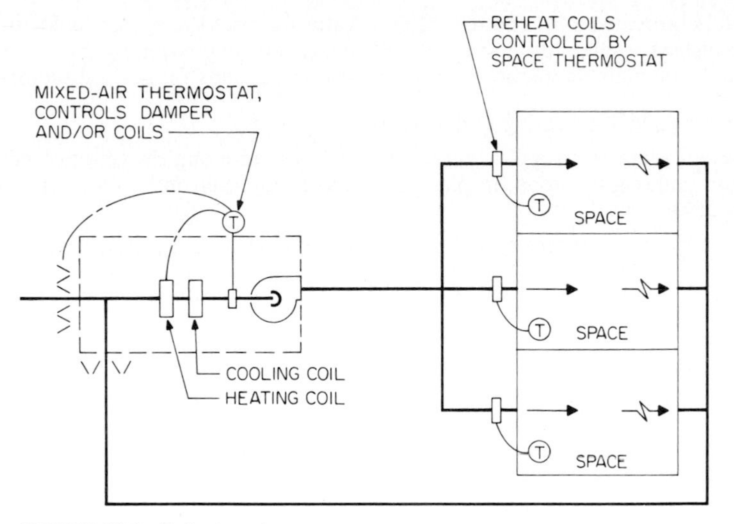

FIGURE 32.2 Reheat system.

Areas made up of zones with varying loads can be supplied by a single supply air system of constant volume and temperature. The air quantity and air temperature are based on the maximum load and comfort conditions established for the area. The individual rooms or zones within the area can be temperature-controlled with the addition of a reheat coil to the branch supply duct.

32.2.1 Central Equipment

The same equipment as described for the single-zone constant-volume system is used, except for the addition of reheat coils in the branch ducts serving zones with changing heating and cooling loads.

The heating coil can be electric, hot water, or steam. A space thermostat modulates the reheat coil to maintain the desired space temperature.

32.2.2 Ductwork System

The ductwork for the constant-volume system with reheat requires the same considerations as listed for the single-zone constant-volume system.

The addition of a reheat coil will require ductwork enlargement transition before the coil and a reducing fitting after the coil, to ensure proper air flow over the entire face area of the coil. Access doors should be provided in the ductwork on the entering and leaving sides of the coil for cleaning and inspection.

32.2.3 Applications

The single-zone system with reheat coils is used for small commercial facilities which may be divided into a number of areas and/or offices with varying internal and perimeter loads. These systems, which use reheat to maintain comfort, should be provided with controls to automatically reset the system cold-air supply to the highest temperature level that will satisfy the zone requiring the coolest air.

The leaving air temperature of a reheat coil depends on several factors:

- The design space heating temperature
- Whether there is a supplementary heating system along the exterior perimeter of the building (such as fin pipe convectors, fan coil units, etc.) for the zone served by the reheat coil
- Whether there is a space equipment cooling load during the heating season

For instance, one of the following conditions can determine the leaving air temperature of a reheat coil:

- *Condition 1:* When the space or zone does not have an exterior exposure or has a supplementary perimeter heating system and there is no equipment cooling load during the heating season, the reheat coil leaving air temperature should nearly equal the space design temperature.
- *Condition 2:* When the space or zone has an exterior exposure without a supplementary perimeter heating system and there is no equipment cooling load during the heating season, the reheat coil leaving air temperature should equal the space design temperature plus the temperature difference calculated to offset the space or zone heating loss from the exterior exposure.
- *Condition 3:* The space or zone is the same as for condition 1 except there is an equipment cooling load requirement during the heating season. Then the reheat coil leaving air temperature should be equal to the space design temperature minus the temperature difference calculated to offset the space equipment cooling load.
- *Condition 4:* The space or zone is the same as for condition 2 except there is an equipment cooling load requirement during the heating season. Then the reheat coil leaving air temperature should be equal to the space design temperature plus the temperature difference calculated to offset the space or zone heat loss from the exterior exposure minus the temperature difference calculated to offset the space equipment cooling load.

32.3 MULTIZONE SYSTEM

This type of system (Fig. 32.3) is used when the area being served is made up of rooms or zones with varying loads. Each room or zone is supplied by means of a single duct from a common central air-handling unit.

The central air-handling unit consists of a hot-air plenum and cold-air plenum with individual modulating zone dampers mixing hot and cold air streams and supplying the mixture through a dedicated duct to the space. A thermostat located in the occupied space modulates the zone dampers at the unit to achieve the desired temperature conditions.

32.3.1 Central Equipment

The multizone unit shown in Fig. 32.3 may be a factory-assembled packaged unit consisting of a mixing plenum, filter section, supply fan, heating coil, cooling coil

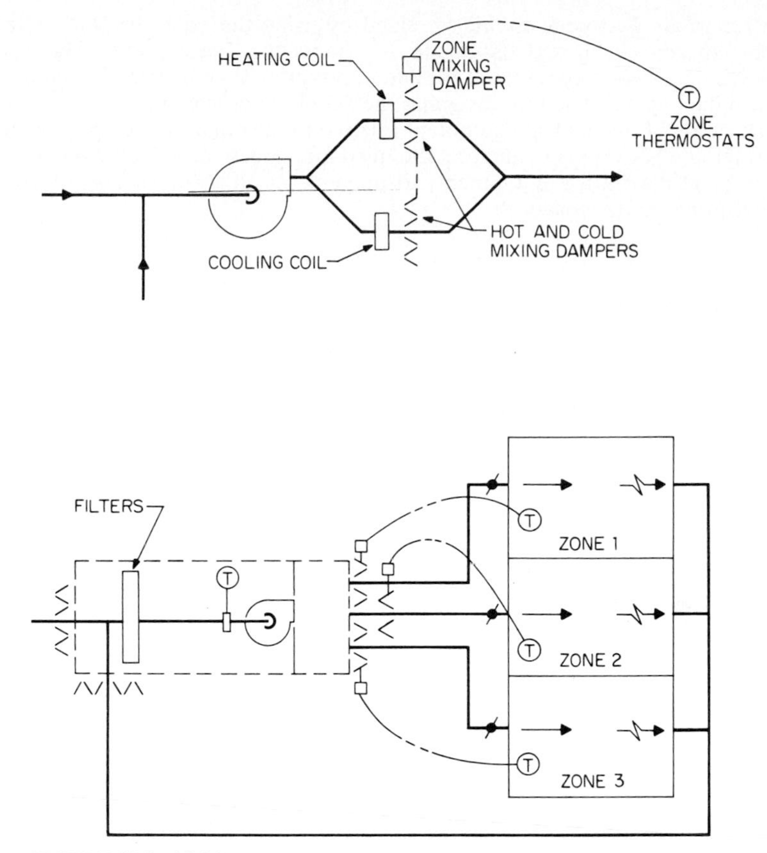

FIGURE 32.3 Multizone system.

and damper assemblies on the discharge side of coils. A humidifier can be added to the unit to maintain a winter humidity level.

32.3.2 Ductwork System

The supply ductwork for the multizone system originates at the central unit dampered discharge outlet from the hot and cold deck. Each zone will be supplied by a single duct with a number of supply air terminals. The supply ductwork should be sized by using the static regain method, to assist in a balanced air distribution in the duct system. Branch duct takeoffs from the duct mains should be provided with balancing dampers to permit fine adjustments to the air distribution system within the individual zones.

The supply air terminals in the space should be selected, sized, and located to provide even distribution throughout the zones without creating drafts or excessive noise. The supply air terminal should be provided with a volume damper to balance air quantities at the individual outlets.

The return-air ductwork should be sized by using the equal-friction method from the space return registers back to the central equipment. The same ductwork configuration considerations and accessories shall be used in laying out the return duct as was used in designing the supply duct system.

The location of the central equipment relative to the conditioned space should be considered when one is evaluating the need for acoustically lined ductwork or sound traps at the central equipment, to prevent transmission of noise through the duct system to the space.

32.3.3 Application

The multizone type of system is considered for office buildings, schools, or buildings with a number of floors and interior zones with varying loads.

The multizone system and dual-duct system, to an extent, will give similar performances inasmuch as the dual-duct system is sometimes described merely as a multizone system with extended hot and cold decks. However, the following real differences do exist:

- Packaged multizone air handlers are available with up to 14 zones whereas dual-duct systems have virtually no limit as to zones.
- Building configuration may be better suited to the numerous small ducts from a multizone system than to the two large ducts off a dual-duct air handler.
- The small zone off a multizone which also has large zones will have erratic air flow when the large zone dampers are modulating. The pressure-independent mixing boxes of a dual-duct system preclude this.
- The damper leakage at "economy" multizone units can be excessive, especially when maintenance is poor.
- It is undoubtedly more costly and cumbersome to add a zone to an existing multizone system than to use a dual-duct system.
- Packaged multizone systems are suitable for small systems and as such may include direct expansion cooling and gas-fired heating equipment. The step capacity control included with this equipment can result in noticeable cycling of space temperatures. The larger cooling and heating equipment generally accompanying dual-duct systems includes modulating capacity controls, and this precludes the space temperature cycling.
- The air in the short hot and cold plenums of multizone units can experience the same temperature gradient as that of a heating coil which has a "hot end," especially during low loads (and similarly for cooling coils). This temperature gradient can result in improper hot (or cold) air entering the zone duct. The long hot and cold ducts of the dual-duct system permit thorough mixing of air off the coils and eliminate the gradient.

32.4 INDUCTION UNIT SYSTEM

The induction unit system is used for the perimeter rooms in multistory buildings such as office buildings, hotels, hospital patient rooms, and apartments. See Fig. 32.4.

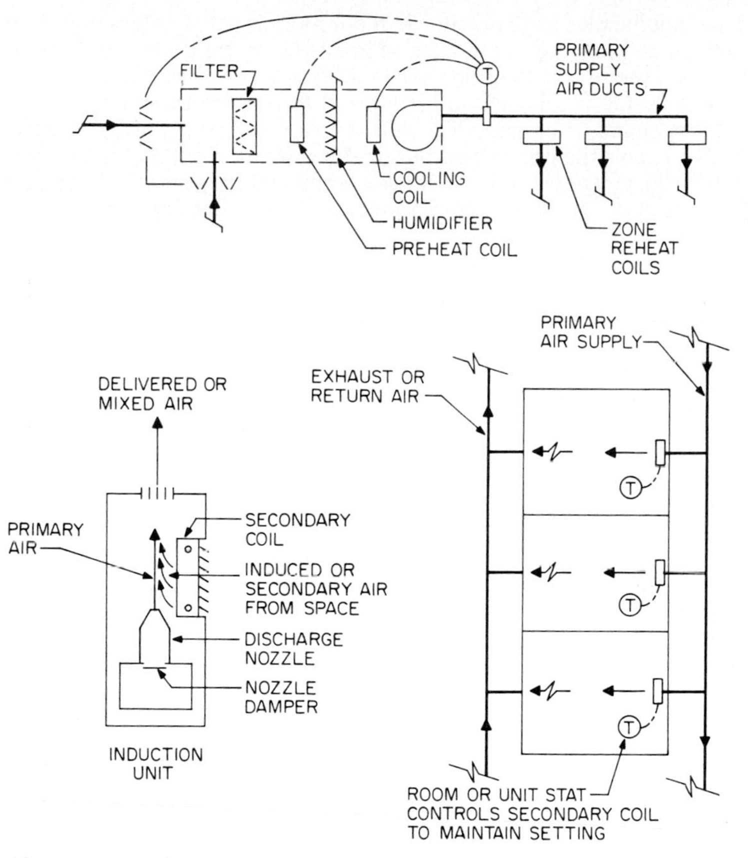

FIGURE 32.4 Induction unit system.

The system consists of a central air-handling unit which supplies primary air, heated or cooled to offset the building transmission loss or gain; a high-velocity duct system for conveying the primary air to the induction units; an induction unit with a coil for each room or office; and a secondary water system, which is supplied from central equipment. The secondary water system is heated or cooled depending on the time of year and the requirements of the space being served.

A constant volume of primary air is supplied from the central air-handling unit through a high-pressure duct system to induction units located in the rooms. The air is introduced to the room through the high-pressure nozzles located within the unit that cause the room air to be drawn over the unit coil. The induced air is heated or cooled depending on the secondary water temperature and is discharged into the room.

32.4.1 Central Equipment

The primary air supply unit for the induction system generally includes a filter, humidifier, cooling coil, heating coil, and fan. A preheat coil is also included when the unit handles large quantities of outdoor air which is less than 32°F (0°C). The heating coil may be in the form of zone reheat coils when the unit supplies induction units on more than one exposure (north, east, south, or west).

The supply fan is a high-static unit sized to provide the primary air requirements for each induction unit. The chilled water or refrigerant cooling coil dehumidifies and cools the primary air during the summer months. Primary air is supplied at a constant rate to the induction units and is generally 40 to 50°F (4 to 10°C) year-round. The final room temperature is maintained by the secondary coil.

32.4.2 Ductwork System

The air supply to the induction units originates from a central air-handling unit. The supply header ductwork should be routed around the perimeter of the building, with individual risers routed up through the floors supplying primary air to the induction units.

Limited available duct space frequently dictates that velocities in the risers be maintained at 4000 to 5000 ft/min (20 to 26 m/s). Rigid spiral ductwork is used, with elbows and takeoffs being of welded construction. Close attention must be paid to prevent noisy air leakage in the duct system.

A sound-absorbing section of ductwork should be provided at the discharge of the central air handler to absorb noise generated by the high-pressure fan.

The supply header and risers should be thermally insulated to prevent heat gain and sweating during summer operation and heat loss during winter operation.

The supply ductwork system should be sized by using the static regain method.

32.4.3 Application

The induction unit system is well suited to the multistory, multiroom buildings with perimeter rooms that require individual temperature selection.

The benefits in using the induction system in these types of buildings is in the reduced amount of space required for air distribution and equipment. The secondary coil in the induction unit is frequently connected to a two-pipe dual-temperature system which provides the coil with hot water during the winter and chilled water during the remaining seasons. The thermostat modulates water flow and therefore varies the temperature of the delivered or mixed air to compensate for the room heat loss or heat gain.*

32.5 VARIABLE-AIR-VOLUME SYSTEM

This system is used primarily where a cooling load exists throughout the year, such as the interior zone of office buildings. This air supply system uses varying amounts of constant-temperature air introduced to a space to offset cooling loads and to maintain comfort conditions. The system operates equally well at exterior

*For more details see *1987 ASHRAE Handbook, HVAC Systems and Applications*, American Society of Heating, Refrigeration, and Air-Conditioning Engineers, Atlanta, GA, 1984.

zones during the cooling season, but during the heating season supplementary equipment such as reheat coils, finned radiation, etc., must be provided at all spaces with an exterior exposure. When a reheat coil is added to a variable-air-volume (VAV) box, the temperature controls should reduce air flow to the minimum acceptable level for room air motion and makeup air and then activate the reheat coil.

The system typically consists of a central air-handling unit with heating and cooling coil, single-duct supply system, VAV box, supply duct with air diffuser, return air duct, and return air fan.

Constant-temperature air is provided from the central air-handling unit through a single-supply air duct to the individual VAV box which regulates supply air to zone to offset cooling load requirements.

32.5.1 Central Equipment

The air supply unit consists of a supply fan with variable inlet vanes, variable speed control or discharge dampers, cooling coils using refrigerant or chilled water, filters, heating coil using steam or hot water for morning warmup, return air fan which is modulated through controls to match supply fan demands, and mixed air plenum to provide outdoor air requirements to the system.

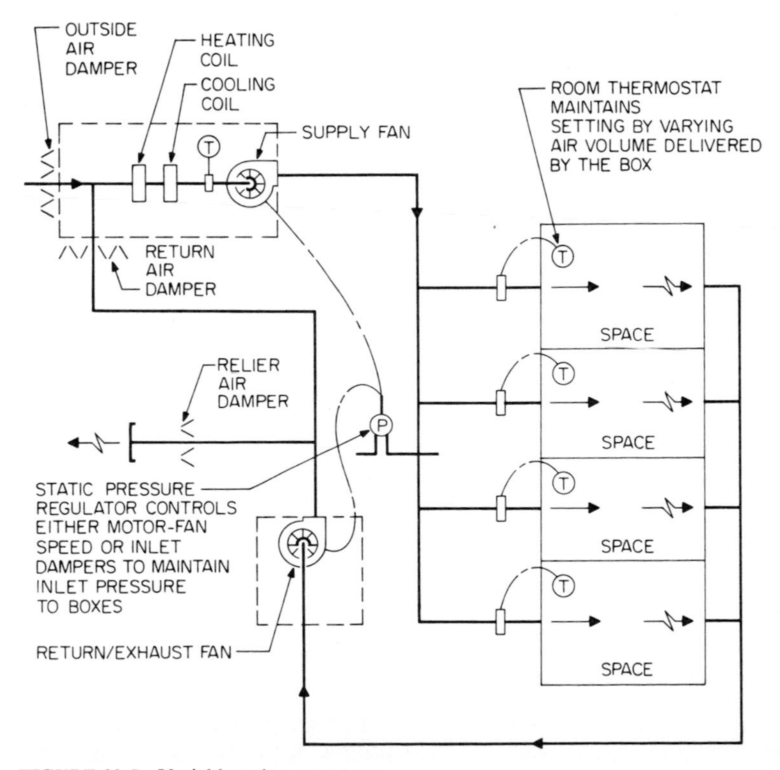

FIGURE 32.5 Variable-volume system.

The supply fan should be selected for the calculated load and system static pressure. During system operation, the supply air demand varies with space load requirements. To meet this demand, the supply and return fans' discharge air quantities must be modulated in unison with variable inlet vanes, variable speed control, or discharge dampers.

The cooling coil, being either a direct-expansion or chilled-water type, automatically controls the discharge air temperature for the unit. During the winter cycle, the mixed air damper and return air damper are modulated to maintain the discharge air temperature.

The following items may need special consideration when VAV systems are designed:

Minimum Outdoor Air. The outdoor air drawn into a building will tend to reduce as the supply fan volume reduces. This can become detrimental when the supply fan has a large turndown from its maximum flow and/or when the VAV system provides makeup air for constant-flow exhaust systems. The minimum outdoor air can be maintained by providing a short duct with a flow sensor downstream of the minimum outdoor air damper. The flow sensor modulates the outdoor and return air dampers to maintain the required minimum.

Building Static Pressure. It is important that return fan volume be properly reduced when the supply fan volume reduces. The return should reduce at a greater rate so as to leave a fixed flow rate for the constant-flow exhaust systems and building pressurization. Flow sensors at the supply and return fans can monitor and maintain a constant *difference* between supply and return air by modulating the return air and exhaust air flow.

Room Air Motion. Select air diffusion devices for proper performance at minimum as well as maximum flow to preclude "dumping" of air. See Sec. 45.1.11.

Building Heating. Calculations frequently show that the internal heat gain (lights, equipment, people) during occupied hours of basement, interior, and sometimes perimeter spaces is more than sufficient to keep these spaces warm. So it may appear that a "mechanical" heat source is not required. But these heat gains might not exist during unoccupied nights, weekends, and shutdown periods, and the spaces will cool down even when the only exposure is a well-insulated wall or roof. The central equipment of VAV systems is sometimes designed without heating coils and in itself cannot heat the building (it "heats" by providing less cooling). Unit heaters, radiation elements, convectors, or heating coils including controls coordinated with the VAV system at zero outdoor air are required for a timely morning warmup and heat when the space is unoccupied.

Calculations for winter usually show a need for heat at perimeter spaces. If the VAV boxes have "stops" for minimum air supply, there must be sufficient heat to warm this minimum air [usually 55 to 60°F (13 to 15°C)] in addition to that required for transmission losses.

32.5.2 Ductwork System

The supply air mains and branch ducts can be sized for either low or high velocity depending on the space available in the ceiling. A low-velocity design will result in lower operating costs.

Supply ducts should have pressure relief doors, or be constructed to withstand full fan pressure in the event of a static pressure-regulator failure concurrent with closed VAV boxes.

32.5.3 Application

The variable-air-volume system is considered for a building with a large interior zone requiring cooling all year. The varying amounts of cool air match the varying internal loads as people and lighting loads change throughout the day and night. When used in conjunction with perimeter heating systems, the VAV system can be used for the perimeter zone of a building also. These conditions describe the typical operations of office buildings, schools, and department stores which are the prime users of this type of system.

32.6 DUAL-DUCT SYSTEM

This type of system is used when the area served is made up of rooms or zones with varying loads with the entire area being supplied from a central air-handling unit. See Fig. 32.6. The central unit supplies both cold and hot air through separate duct mains to a mixing box at each zone. The zone box controlled by the space thermostat mixes the two air streams to control the temperature conditions within the zone.

32.6.1 Central Equipment

The dual-duct unit may be a factory-assembled packaged unit consisting of a mixing plenum, filter section, supply fan, heating coil, cooling coil, and discharge plenums. A humidifier can be added to maintain winter humidity level.

32.6.2 Ductwork System

The hot and cold supply ductwork headers originate at the unit discharges and run parallel throughout the building, connecting to the individual mixing boxes which sypply the zone ductwork.

The cold duct header is sized to carry the peak air volumes of all zones. The hot duct header is sized to carry a certain percentage of the cold air, usually 70 percent.

The zone mixing box responds to a space thermostat and modulates and mixes quantities of cold and hot air and delivers a constant volume of air to the space to maintain desired temperature levels. The size of the mixing box is based on the peak air volume of the zone or room.

The mixed air from the mixing box is discharged through a single duct terminating with a number of supply air diffusers, chosen and sized to provide even distribution throughout the zone.

Supply ductwork may be sized by the static regain or equal-friction method with aerodynamically smooth fittings and velocities not exceeding 3000 ft/min (15 m/s).

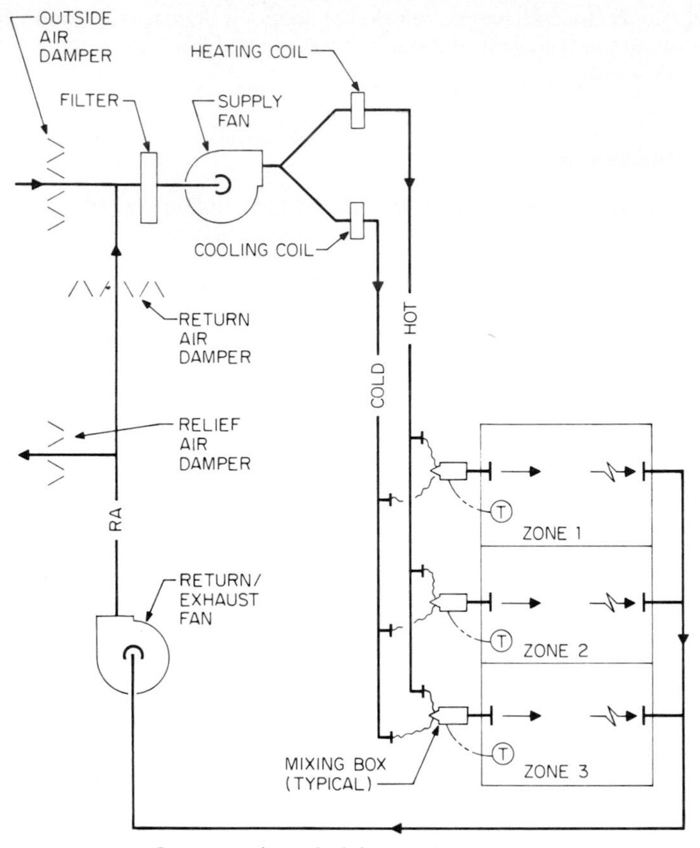

FIGURE 32.6 Constant-volume dual-duct system.

The return-air ductwork is often sized by the equal-friction method, but it does not exceed 1500 ft/min (7.6 m/s). The routing and configuration of the supply headers must satisfy the space limitations in ceiling and shaft areas. Access and space requirements for the mixing box should also be considered when the routing of the duct system is laid out.

The location of the central equipment relative to the conditioned space should be considered when deciding the need for acoustically lined ductwork or sound traps at the central equipment, to prevent transmission of noise through the duct system to the space.

32.6.3 Application

The dual-duct type system is considered for office buildings, schools, or buildings with a number of floors and zones with varying loads. Generally, however, this system has been "replaced" by the VAV system because of higher operating and first costs and increased duct space requirements.

CHAPTER 33

DIRECT EXPANSION SYSTEMS

Simo Milosevic, P.E.
Project Engineer, Mechanical Engineering, Giffels Associates, Inc., Southfield, Michigan

33.1 SYSTEM DESCRIPTION

To air-condition the interior space of a building, recirculating and makeup air is commonly cooled and simultaneously dehumidified while passing across a cooling coil. This coil could be a direct expansion (DX) type.

DX refrigeration utilizes refrigerant (working fluid of the cycle) fluid temperature, pressure, and latent heat of vaporization to cool the air. To evaporate liquid refrigerant to a vapor, the latent heat of vaporization has to be applied to the liquid. The quantity of heat necessary to evaporate 1 lb (0.45 kg) of liquid refrigerant varies with the thermal characteristics of different refrigerants. The boiling point of an ideal refrigerant has to be below the supply air temperature and above 32°F (0°C), so as to not freeze the moisture condensed from the building supply air. Most refrigerants in use today have a relatively low boiling point and nonirritant, nontoxic, nonexplosive, nonflammable, and noncorrosive characteristics for use in commercially available piping materials.

Early air-conditioning and refrigeration systems used toxic and hazardous substances such as ether, chloroform, ammonia, carbon dioxide, sulfur dioxide, butane, and propane as refrigerants. The most commonly used refrigerants today are numbers 11, 12, 22, and 502 (for low-temperature applications such as food freezing).* Currently there is concern that these refrigerants may be damaging the earth's ozone layer.

A simple refrigeration system is illustrated in Fig. 33.1. Basic elements of the system are the expansion valve (pressure-reducing valve), evaporator (cooling coil or DX coil), compressor, condenser, and interconnecting piping. The compressor and expansion valve are points of the system at which the refrigerant pressure changes. The compressor maintains a difference in pressure between the suction and discharge sides of the system (between the DX coil and condenser), and the expansion valve separates high- and low-pressure sides of the system. The function of the expansion valve is to meter the refrigerant from the high-pressure side (where it acts as a pressure-reducing valve) to the low-pressure side

*These refrigerants can be found on the market under the various trade names of Freon (registered trademark of E. I. du Pont de Nemours Co.), Genetron (registered trademark of Allied Chemical Corp.), and Isotron (registered trademark of the Pennsalt Chemicals Corp.).

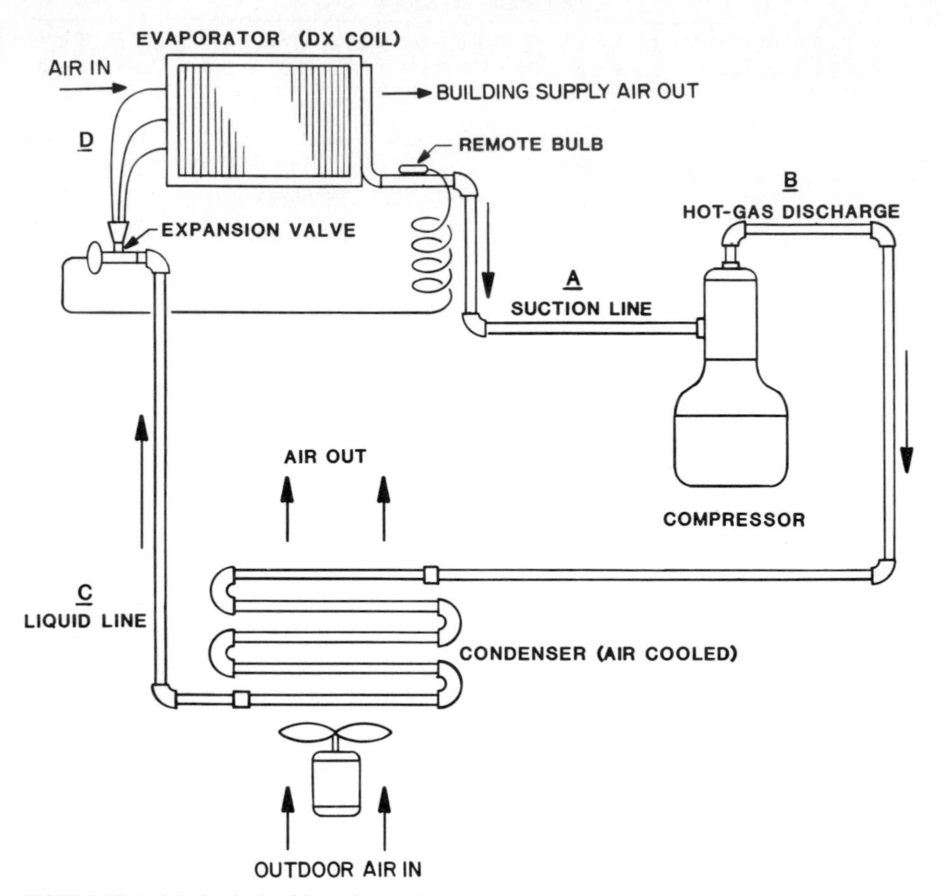

FIGURE 33.1 Mechanical refrigeration system.

(where it undergoes a phase change from a liquid to a vapor during the process of heat absorption).

The compressor draws vaporized refrigerant from the evaporator through a suction line *A*. In the compressor, the refrigerant pressure is raised from evaporation temperature and pressure to a much higher discharge pressure and temperature. In the discharge line *B*, refrigerant is still in the vapor state at high temperature, usually between 105 and 115°F (40 and 46°C). A relatively warm cooling medium (water or air) can be used to condense and subcool the hot vapor. In the condenser, heat of vaporization and of compression is transferred from the hot refrigerant gas to the cooling medium through the walls of the condenser heat-exchange surfaces while the gas becomes liquefied at or below the corresponding compressor discharge pressure *C*.

The expansion valve separates the high-pressure or condenser side of the system from the low-pressure or evaporator side. The purpose of the expansion valve is to control the amount of liquid entering the evaporator such that there is a sufficient amount to evaporate but not flood the evaporator *D*.

In the evaporator, liquid refrigerant is entirely vaporized by the heat of the building supply air. Heat equivalent to the latent heat of vaporization has been transferred from building air through the walls of the evaporator to the low-temperature refrigerant. Thus the building supply air is cooled and dehumidified. The boiling point (temperature of evaporation) at the evaporator pressure is usually between 34 and 45°F (1 and 7°C) for refrigerants 11, 12, and 22 and lower for refrigerant 502.

From the evaporator, vaporized refrigerant is drawn through suction piping to the compressor, and the cycle is repeated.

All refrigerants have different physical and thermal characteristics. Depending on the available condensing temperature, required evaporation temperature, and cooling capacity, different refrigerants are used for different applications.

Figure 33.2 illustrates the theoretical refrigeration cycle (without pressure

TEMPERATURE, °F (°C)
SP. VOLUME, FT^3/lb (m^3/kg)
ENTHALPY, BTU/lb (kJ/kg)
ENTROPY, BTU/lb • R (kJ/kg • K)
PRESSURE, lb/in^2 (kPa)
QUALITY, %

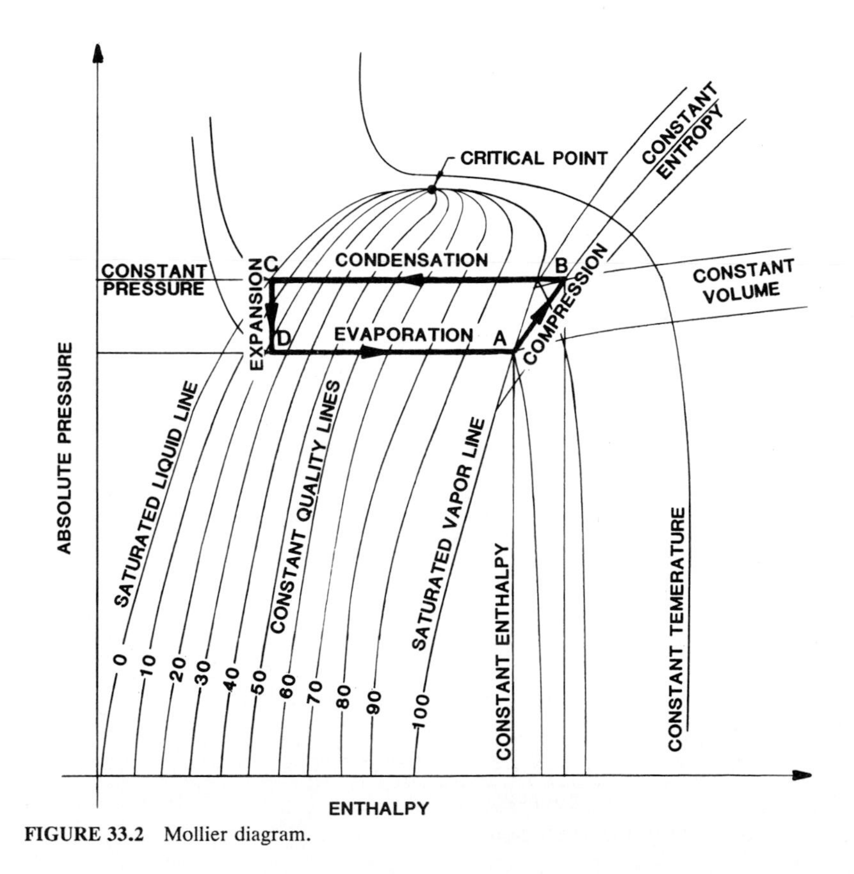

FIGURE 33.2 Mollier diagram.

losses in the system and without subcooling of the liquid or superheating of the vapor) shown on a Mollier or pressure-enthalpy diagram.

33.2 EQUIPMENT

The purpose of this section is to describe the components of a DX refrigeration system and their functions. In addition to the basic elements already mentioned (compressor, condenser, expansion valve, evaporator, and refrigerant piping), the typical refrigeration system incorporates other components or accessories for various purposes. Figure 33.3 illustrates a simple DX refrigeration system with an air-cooled condenser designed for operation in cold weather.

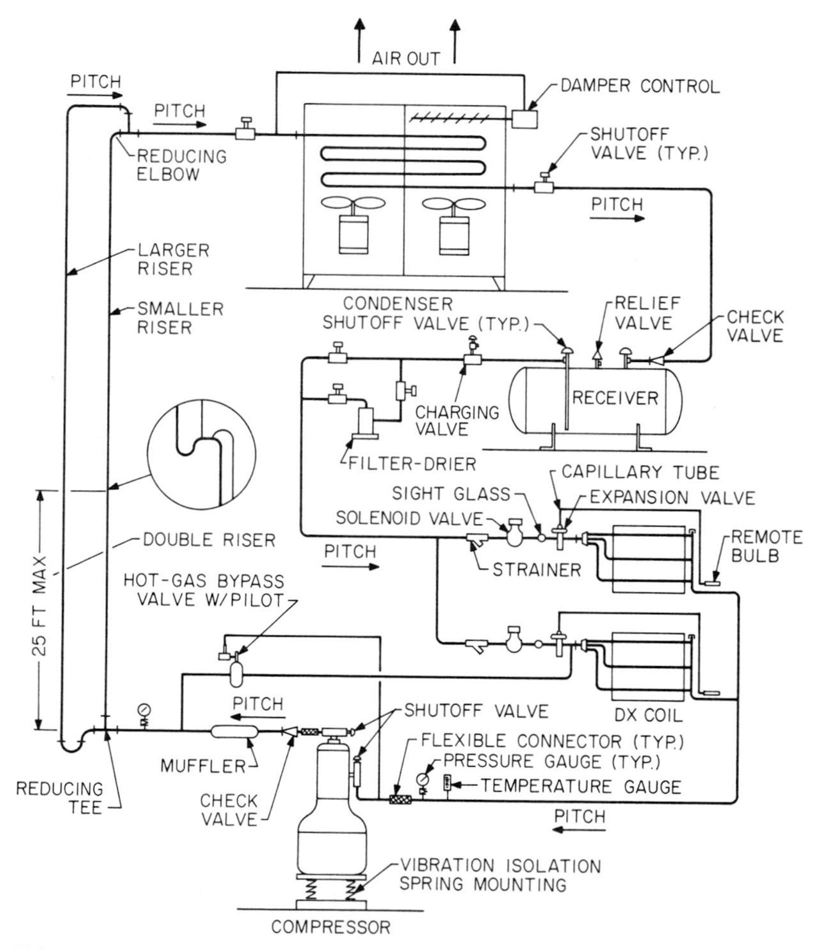

FIGURE 33.3 Reciprocating DX system.

33.2.1 Compressor

This is a vapor-phase fluid pump which maintains a difference in refrigerant gas pressure between the DX coil (low-pressure or suction side) and the condenser (high-pressure or discharge side) of the system. Compressors can be categorized as to construction, i.e., hermetic, semihermetic, and open (direct- or belt-driven). They also can be categorized by the type of machine, i.e., reciprocating, centrifugal, and screw. More about compressors can be found in Chaps. 40 to 44.

33.2.2 Condenser

This is a heat-exchange device where heat of vaporization and compression is transferred from hot refrigerant gas to the cooling medium in order to change the refrigerant from a superheated vapor to a liquid state and sometimes to subcool the refrigerant. Condensers can be air-cooled, where outdoor air is used to condense and subcool the refrigerant, or water-cooled, where city water or cooling tower water is used as the cooling medium. Evaporative condensers use both water and air to condense and subcool refrigerant: recirculating water is sprayed over tubes containing hot refrigerant and is evaporated by moving outdoor air, thus removing heat from the refrigerant.

Air-cooled condensers can be single- or multifan types. Axial fans are most commonly used because axial fans economically handle large air volumes at low static pressure. Centrifugal fans, which are capable of generating higher static pressures, are used in certain applications.

Condensers can also be categorized as single- and multicircuit, according to whether they are connected to one or more compressors.

33.2.3 Expansion Valve

This is a throttling or metering device with a diaphragm operator. The space above the diaphragm is connected to a remote sensor bulb with capillary tubing and filled with the same refrigerant as is used in the system. The valve controls flow of fluid refrigerant to maintain a set-point pressure in the evaporator.

The remote temperature-sensing bulb is normally strapped or soldered to the suction line (leaving evaporator) for maximum surface contact. An increase in heat load on the evaporator is sensed by the bulb, causing a corresponding increase in vapor pressure within the bulb, capillary tube, and space above the diaphragm. This pressure, transmitted by the diaphragm, moves the valve off its seat, to admit more liquid refrigerant into the evaporator, for evaporation by the increased heat load.

When the cooling requirements are satisfied, the process reverses.

33.2.4 Evaporator (DX Coil)

This is an extended-surface (finned tube) device where heat exchange occurs between building supply air and the liquid refrigerant in the coil tubes, causing the refrigerant to vaporize.

DX coils are either dry or flooded (with refrigerant liquid). The coils can have 20 or more parallel circuits and are of one- or multirow construction.

33.2.5 Refrigerant Piping

Typically, Type L copper tubing is used for handling chlorinated fluorocarbon refrigerants, discussed earlier.

33.2.6 Hot-Gas Bypass Control

This control is a way to maintain a reasonably stable evaporator suction pressure when a refrigeration system is operating at minimum load. Two hot-gas bypass methods are in use today on direct expansion systems: hot-gas bypass to the evaporator inlet and to the suction line.

Hot-gas bypass to the evaporator inlet introduces compressor discharge vapor to the DX coil after the expansion valve (see Fig. 33.3). This acts as an artificial heat load on the DX coil and raises the temperature at the coil outlet. The remote bulb of the expansion valve senses this temperature rise and opens the valve to increase the flow of refrigerant through the coil, resulting in a rise of the suction pressure and stabilization thereof.

The effectiveness of this method is a function of the distance between the (compressor) and the DX coil. This method should not be used when the distance is greater than 50 ft (15 m) for hermetic compressors (hot gas might start condensing, resulting in oil "holdup" and compressor lubrication problems).

Hot-gas bypass to the suction line introduces hot vapor from the compressor discharge to the inlet (suction) side of the compressor (see Fig. 33.4). This method requires an additional liquid line solenoid valve and expansion valve. When low suction pressure is sensed because of reduced heat load, hot gas is introduced to the compressor inlet through the hot-gas bypass valve. This causes the supplementary expansion valve to open and to introduce liquid refrigerant

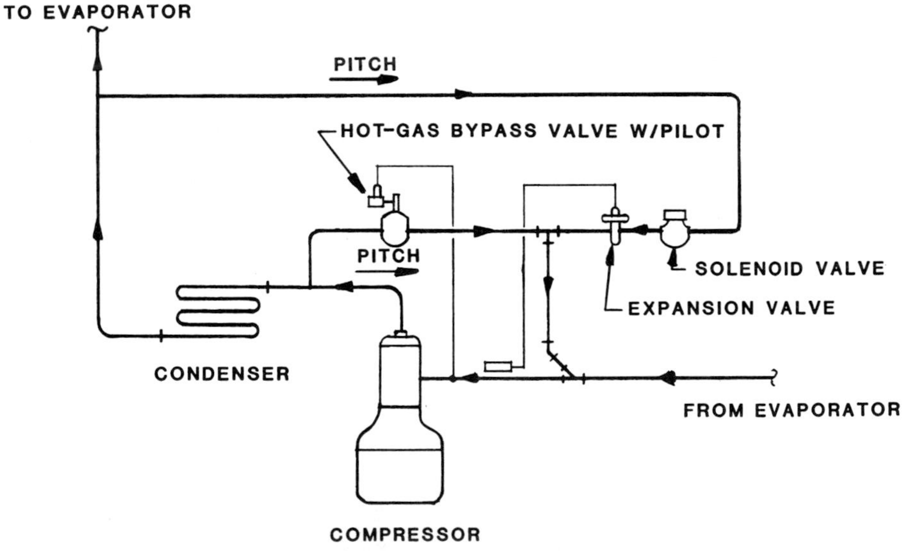

FIGURE 33.4 Hot-gas bypass to suction line.

into the hot gas. The liquid refrigerant evaporates and increases compressor suction pressure to stabilize operation.

Disadvantages of this method include the additional cost of expansion and solenoid valves, oil trapping in the DX coil, and the possibility of liquid slugs entering the compressor.

33.2.7 Suction and Hot-Gas Double Riser

This is a pipe assembly which promotes oil movement to the compressor on the suction side and from the compressor on the hot-gas side. A double-riser arrangement is used in a vertical piping layout when the compressor is below the condenser and/or when the compressor is above the DX coil. Figure 33.3 illustrates double riser on the hot-gas side. A similar setup would be provided on the suction side if the compressor were located above the DX coil.

Operation at minimum compressor capacity (with the compressor unloaded or one compressor running in multiple-compressor installations) reduces oil conveying velocities in the system which causes compressor oil to fill the trap, thus directing gas flow to the smaller riser. This riser is sized to produce a velocity of 1000 ft/min (5 m/s) which is sufficient to convey oil upward for return to the compressor. When full-load capacity is restored, pressure clears the trap and flow is established through both risers. The larger riser is sized for velocities between 1000 and 4000 ft/min (5 and 20 m/s) at full compressor load.

The maximum vertical rise of a double riser should not exceed 25 ft (8 m). If greater than 1 25-ft (8 m) rise is required, an intermediate trap should be incorporated every 25 ft (8 m) of rise (see Fig. 33.3).

33.2.8 Filter-Drier

This is usually installed in the liquid line to protect the expansion valve from dirt or moisture that may freeze in the expansion valve and to protect motor windings from moisture. The filter-drier core has an affinity for and retains water while simultaneously removing foreign particles from the liquid refrigerant (see Fig. 33.3).

33.2.9 Condenser Pressure Control

This control is necessary with lower outdoor air temperatures when the air-cooled condenser capacity increases and system load decreases, causing low condenser pressure. This is controlled by modulating air flow through the condenser with an outlet damper whose operator is driven by condensing pressure. In multifan condensers, cycling of the fans by outdoor air temperature thermostat provides step control of the air flow through the condenser. The last operating fan might have an outlet damper operated by condensing pressure.

In water-cooled condensers, water flow is usually controlled by a flow-modulating valve controlled by condensing pressure.

33.2.10 Hot-Gas Muffler

This muffler is usually installed on the discharge side of reciprocating compressors with long piping systems. This reduces gas pulsation and noise produced by reciprocating equipment.

33.2.11 Solenoid Valves

These valves are electrically operated two-position valves. They permit isolation of coil circuits to reduce the cooling produced and pumpdown of the low pressure side for eventual compressor shutoff when heat load is zero.

33.2.12 Sight Glasses

These should be installed in every system in front of the expansion valve. The operator can verify the flow of liquid and absence of gases or vapors upstream of cooling coils.

33.2.13 Shutoff Valves

These valves are usually the capped, packed, angle type mounted directly on the compressor or liquid receiver. The purpose of shutoff valves is to isolate portions of the refrigeration circuit to enable maintenance or repair.

33.2.14 Charging Valve

This is the point at which refrigerant is introduced (charged) into the system. Normally the charging valve is installed in the liquid line after the condenser or after the liquid receiver, if one is used.

33.2.15 Relief Valves and Fusible Plugs

These devices protect the refrigeration system from excessive pressure buildup. In the case of fusible plug activation, all refrigerant charge is released when the plug melts because of excessive temperature.

33.2.16 Check Valves

These are usually used in front of the liquid receiver and after the compressor, to prevent vapor migration from the receiver to the condenser or liquid migration from the condenser to the discharge of the compressor, after the system shutdown. This is especially important in systems where the receiver is located in a hot space or the compressor is located in a space cooler than the condenser.

33.2.17 Strainers

These are installed in liquid lines to protect solenoid and expansion valves from dirt.

33.2.18 Liquid Receiver

When condensers (evaporative, air-cooled) which inherently have a small refrigerant storage volume are used and the system is sufficiently large, a liquid receiver is installed after the condenser to collect and hold the system liquid refrigerant until it is required by the peak load.

33.2.19 Pressure and Temperature Gauges

These are used to indicate suction and discharge compressor pressures and temperatures, condenser water temperature, and compressor lubricating oil pressure.

33.3 WHERE USED

Direct expansion air-conditioning systems are available as window-type units with capacities from less than ½ ton (1.8 kW) to large packaged or built-up units of over 100 tons (352 kW) of refrigeration. They can be of self-contained construction (rooftop) where all elements, including the controls, are built in one cabinet, or they can be a split system where the condenser and compressor are located in one cabinet outside the building (condensing unit) and the DX coil and expansion valve are located inside the building in the air-handling unit. In the latter case, interconnecting refrigerant piping (liquid and suction lines) has to be field-installed and -insulated.

Capacities of split and self-contained air-conditioning systems (excluding window air conditioners) normally start at 2 tons (7 kW) and for self-contained systems range to 100 tons (352 kW), while split systems range to 120 tons (422 kW) of refrigeration.

Split and self-contained DX systems are normally used in situations where individual temperature and humidity control is required for numerous small spaces within a large building. Typical applications are apartment buildings, condominiums, small shopping malls (strip stores), office buildings with multiple tenants, medical buildings (doctors' and dentists' offices), and various departments within a manufacturing building. These systems are also used in nonair-conditioned plants where it is necessary to air-condition in-plant offices and spaces. In this case, packaged systems are frequently located on the roof of the in-plant space with heat rejection to plant space. These air-conditioning systems can be individually shut off when not needed, thereby saving energy. Also the systems can be individually metered to facilitate allocating the operating cost directly to the tenants or departments. In case of failure in one system, only the space being served will be affected whereas failure of a large, central built-up system would affect the entire building or buildings.

Typical use of DX system air conditioning is found in churches and restaurants where zoning of different spaces is important (different temperatures or times of use of different spaces, such as bar, kitchen, dining area, recreation hall, and church area).

The most common use of small DX systems is in residential spaces. Capacities of these systems start at 2 tons (7 kW) of refrigeration, which is enough for a small home. When two or more small DX systems are installed in larger homes, separate zones with independent temperature control and operating periods are established, e.g., sleeping areas which are cooled only at night, living quarters cooled only during the day, rooms with west exposure cooled only in late afternoon or evening, etc.

Special fields of use of DX cooling systems include computer rooms and vehicles. Self-contained, water-cooled condenser units are often used for computer rooms. They are normally designed for recirculation air only with bottom (under floor) discharge and sized to handle large sensible loads. However, computer rooms which have uniform heat release throughout the room can be conditioned by ceiling air distribution systems. For transportation vehicles (subway cars, public buses, and cars), modular systems with air-cooled condensers are used. All three major parts of the system (condenser, evaporator, and compressor) are in different locations in the vehicle and are connected with insulated piping.

Split systems are applicable as a retrofit or an option to standard air-handling units. A typical case is a residential furnace where space for a future DX coil is provided.

33.4 DESIGN CONSIDERATIONS

During initial design development, the designer must consider the type and function of facility to be air-conditioned, cooling loads involved, building layout, provisions for future expansions, and degree of required temperature and humidity control. If the designer decides that a DX system is suitable for the project, the next steps include evaluation of available condenser cooling media, type of system to be used, and location of the condenser and air handler if the system is comprised of multipackage units.

The simplest approach is to provide an air-cooled, single-package, rooftop-mounted air-conditioning unit. This system is completely self-contained including controls, so that the designer has only to connect ductwork to the unit and to bring in electric power and thermostat wiring. With restrictions on water use and the high cost of water in many areas of the country, air-cooled condensers have long been popular.

In general, air-cooled condensers have lower initial cost, they are lighter, maintenance is easier, and there is no liquid disposal problem. However, there are certain disadvantages and design considerations that the designer has to recognize before choosing a type of condenser. Air-cooled condensers require large amounts of relatively cool air, which could be a problem, especially with an indoor location of the condenser. Axial-flow fan condensers can be noisy. They require relatively clean air (condenser plugging problem). Startup difficulties at low outdoor air temperatures, capacity reduction on high outdoor temperatures, and operating problems at part load are common problems. Air-cooled condensers require locations free of any obstructions on both inlet and outlet sides. Usually clearance of 1.5 times the condenser height is required around the condenser. If a possibility of air short-circuiting (recirculation of hot air) occurs, the designer should consider condenser fan discharge stacks. Since the north side of the building is cooler and is in shade for most of the day, the condenser should be located in this area, if possible.

When a system operates for a longer period on minimum load, the suction pressure drops, as does the corresponding temperature. This can result in frost or ice on the cooling coil, restricting air flow through it. Also reduced refrigerant flow through the system may cause compressor lubrication problems and motor cooling problems in hermetic compressors.

In general, capacity control in a reciprocating compressor DX system is a problem. Control is achieved in steps, either with multiple-compressor arrangements or by compressor valve control (unloading compressor cylinders). In any case, this is step (not modulating) control, therefore, precise temperature control cannot be expected from DX systems. For more precise capacity control, multispeed and variable-speed motors are usually considered.

Temperature and humidity control can be achieved with parallel- and series-arrangement DX coils. A parallel coil arrangement is less expensive and provides better humidity control, but maintenance of constant leaving air temperature is difficult. Therefore, parallel coil arrangements are not recommended for reheat air distribution systems where a constant air temperature in front of reheat coils is important. Coils arranged in series are usually split to carry half the capacity each and are connected to separate compressors of the same capacity (two circuits). This division is done so that the first coil has one-third of the total number of rows and the second coil has the remaining two-thirds, because the first coil has greater air temperature differences and still will carry one-half of the total cooling load. The disadvantage in this arrangement is that one compressor (the one connected to the upstream coil) is always leading on load demand and is therefore wearing faster.

Air velocity through the cooling coil is limited to 550 ft/min (2.8 m/s) maximum because of condensate moisture carryover from coil fins.

Part-load system operation can increase lubricating oil migration problems. On long vertical piping runs, this is solved with double-riser piping arrangements, as discussed earlier.

If a split system is selected, the designer must consider the distance between the condensing unit and DX coil. This distance is limited to 50 ft (15 m) total length of piping for hermetic compressors of 20 tons (70 kW) of refrigeration capacity and under and to 150 ft (46 m) for semihermetic compressors with capacity of over 20 tons (70 kW) of refrigeration.

When modular systems are used, compressor vibration and noise factors must be recognized. These disadvantages can be mitigated by installing vibration isolators under the compressor and by providing muffler and flexible connectors at the compressor. Piping flexibility can be improved by using two or three 90° elbows in the piping near the compressor.

If the air-conditioning unit is not easily accessible, remote panel indication of air filter status, different pressures, and temperatures should be considered.

All equipment requiring maintenance should be provided with manual shutoff valves.

Some municipalities require licensed operators for compressor motors above certain sizes. This can be avoided by use of multiple compressors of smaller size.

CHAPTER 34
FANS AND BLOWERS*

Robert Jorgensen
Buffalo Forge Company, Buffalo, New York

34.1 FAN REQUIREMENTS

Fans provide the energy to move the air through the ducts and other apparatus that form the air side of a heating, ventilating, and air-conditioning (HVAC) system. Each system may be served by one or more fans. Sometimes fans are an integral part of a unit containing coils, filters, and other devices. At other times, free-standing fans are used.

34.1.1 Pressure

The energy transmitted to the air by the fan must equal exactly the energy lost by the air in moving through the system. Fan requirements and system losses are usually expressed in terms of energy per unit volume of gas flowing, which is known as pressure.

In fan engineering it is necessary to distinguish various pressures according to the method by which they can be measured or by the kind of energy with which they can be identified as follows:

1. The total pressure p_T at a point in a gas stream is the force per unit area which can be measured by a manometer connected to an impact tube that points directly upstream. It is equivalent to the sum of the pressure energy[†] and kinetic energy of a unit volume of gas and exists by virtue of the gas density, velocity, and degree of compression.
2. The static pressure p_S at a point in a gas stream is the force per unit area which can be measured by a manometer connected to a small hole in the duct wall or other boundary, the surface of which must be parallel to the path of the stream. It can be considered equivalent to the pressure energy of a unit volume of gas and exists by virtue of the gas density and degree of compression.

*All the illustrations and tables in this chapter appear with the permission of the Buffalo Forge Company, Buffalo, NY. The text is condensed from Robert Jorgensen (ed.), *Fan Engineering*, 8th ed., Buffalo Forge Co., Buffalo, 1983, and is also used with permission.

[†]The concept of pressure energy is a convenience in fan engineering. It derives from the flow work term in the general energy equation.

3. The velocity pressure p_V at a point in a gas stream is the force per unit area which can be measured by a manometer, one leg of which is connected to an impact tube pointing directly upstream and the other leg connected to a small hole in the duct wall (or its equivalent). It is equal to the kinetic energy of a unit volume of gas and exists by virtue of the gas density and velocity.

Pressures in fan systems are usually measured with some form of water column gauge (WG) such as the vertical U-tube or the inclined manometer. The unit of measurement is most commonly the inch water gauge (kilopascal), abbreviated as in WG (kPa).

The Air Movement and Control Association (AMCA) and the American Society of Heating, Refrigeration, and Air-Conditioning Engineers (ASHRAE) jointly published a test code, AMCA 210-74/ASHRAE 51-75, which is the standard of the industry. In it fan pressures are defined as follows:

1. The fan total pressure p_{FT} is the difference between the total pressure at the outlet of the fan p_{T2} and the total pressure at the inlet of the fan p_{T1}. If there is no inlet connection, p_{T1} must be assumed equal to zero.

$$p_{FT} = p_{T2} - p_{T1} \tag{34.1}$$

2. The fan velocity pressure p_{FV} is the velocity pressure corresponding to the average velocity through the fan outlet p_{V2}.

$$p_{FV} = p_{V2} \tag{34.2}$$

3. The fan static pressure p_{FS} is the difference between the fan total pressure and the fan velocity pressure:

$$p_{FS} = p_{FT} - p_{FV} \tag{34.3}$$

The symbols used above are not always those encountered in fan engineering. Frequently F_{TP}, F_{VP}, and F_{SP} are substituted for their obvious counterparts. The test standard noted above does not use either nomenclature, choosing instead to use a capital P for pressure and omitting the subscript F. The reader should have no trouble with any of these systems provided the distinction between fan pressures and pressures at a point or across a plane are carefully recognized.

34.1.2 Capacity

In the standard test code, "fan capacity" is defined as the volume rate of flow measured at the fan inlet conditions. Capacities in fan systems are determined from pressure measurements. These may be velocity pressure traverses of the duct itself or total pressure or static pressure measurements associated with an orifice or nozzle. Such measurements may be converted to a volume rate of flow by proper consideration of air density and duct or nozzle geometry. The unit of measurement for volume rate of flow is most commonly cubic foot per minute (cubic meter per second), abbreviated ft^3/min (m^3/s). Flow rate is conventionally designated by the symbol Q, and fan flow rate can be designated as either Q_F or Q_1. The latter reflects the standard definition. It is not unusual to refer to the fan flow rate as the cubic feet per minute (CFM) of the fan.

34.1.3 Density

Air density may be determined by measuring dry-bulb temperature, wet-bulb temperature, and barometric pressure and referring to a psychrometric density chart, such as that illustrated in Fig. 34.1.

Most performance data are published for the conditions that would be obtained if the fan were handling air at the standard density of 0.075 lbm/ft^3 (1.2 kg/m^3). This is substantially the density of dry air at 70°F (20°C) and 14.7 psia (101.3 kPa). If other conditions prevail, corrections must be made according to the fan laws.

34.1.4 Power Formulas

The formula for power H in terms of the total efficiency η_T, total pressures at inlet and outlet p_{T1} and p_{T2}, respectively, in in WG (kiloposcals), inlet capacity Q_1 in cubic feet per minute is

$$H = \frac{Q_1(p_{T2} - p_{T1})K_p}{6362\eta_T} \tag{34.4}$$

The power in terms of the static efficiency η_S, static pressure at the outlet p_{S2}, and total pressure at the inlet p_{T1} is

$$H = \frac{Q_1(p_{S2} - p_{T1})K_p}{6362\eta_S} \tag{34.5}$$

The compressibility coefficient K_p generally can be taken as unity for fans, particularly those that generate less than 10 in (2.5 kPa) WG.

The above equations for power are based on U.S. Customary System (USCS) units and yield the power in horsepower. (The standard test code uses the symbol H to designate this quantity and the compound symbol HP is frequently found.) If Q_1 is measured in cubic meters per second and pressures are measured in kilopascals, then power will be in kilowatts when the factor 6362 is changed to 1.0.

34.2 FAN TYPES

The various aerodynamic types of fans can be distinguished by the direction in which the air flows through the impeller when the energy is being transmitted by the blades or working surfaces. The principal types are

1. *Axial-flow fans,* including propeller fans, through which the air flows substantially parallel to the shaft
2. *Centrifugal fans,* which might be called radial-flow fans because the air flows in a radial direction relative to the shaft
3. *Mixed-flow fans,* through which the air flows in a combined axial and radial directions

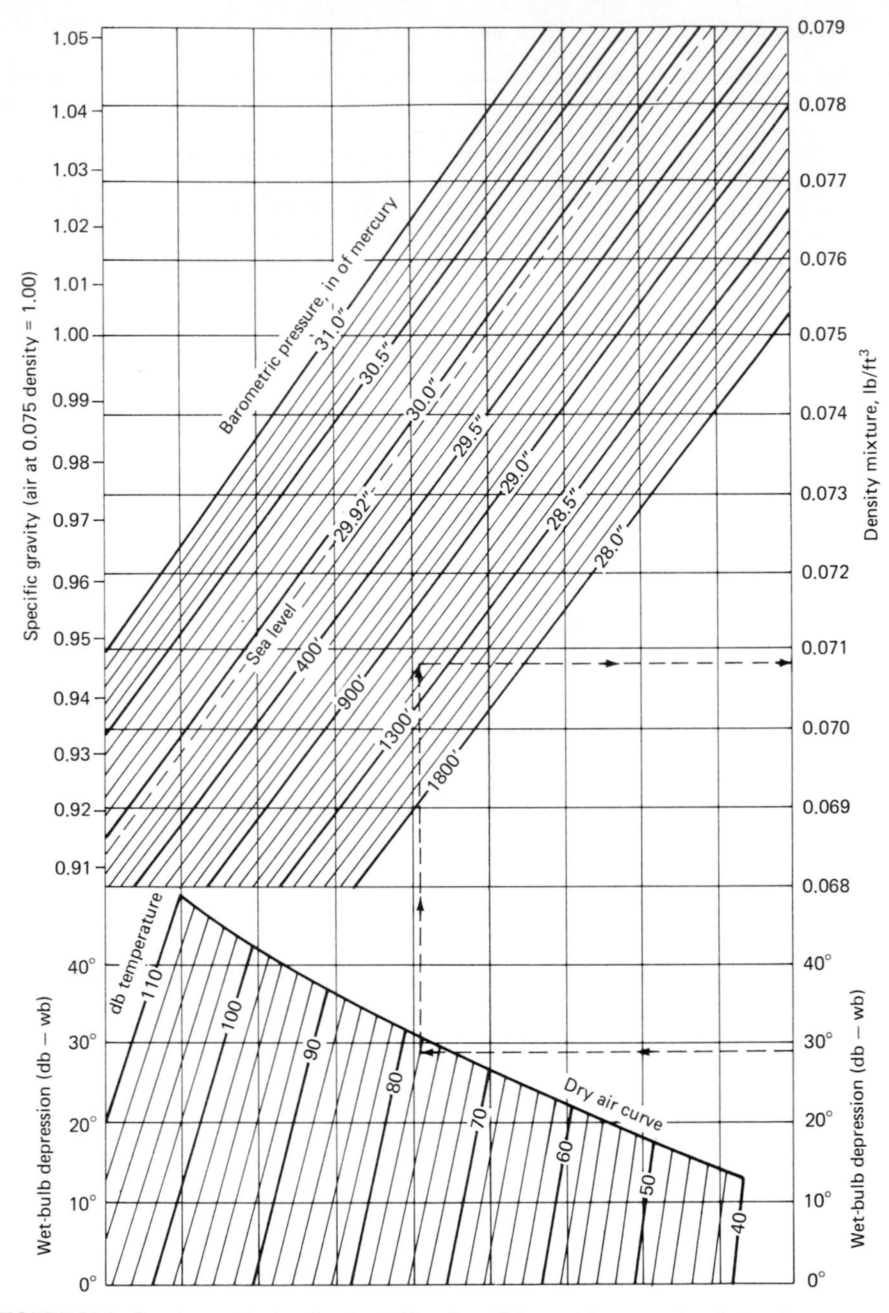

FIGURE 34.1 Psychrometric density chart. (See App. B for metric conversion factors.)

4. *Cross-flow fans,* through which the air flows in an inward radial direction and then in an outward radial direction.

34.2.1 Axial-Flow Fans

There are numerous aerodynamic types of fans within the axial-flow category. One of the most important distinguishing features is the *hub ratio*, i.e., the ratio between the diameter of the hub (or root of the blade) and the tip of the blade. Generally, the higher the hub ratio, the higher the inherent pressure capability of the impeller. Similarly, the higher the pressure requirements of the fan, the more complex the other parts of the fan must be, to achieve acceptable efficiencies.

Propeller fans are perhaps the simplest and best known of all fan types. For all practical purposes, they have a zero hub ratio, having only enough hub to satisfy the mechanical requirements for driving the fan. Even those fans with the most elaborately contoured blades are only capable of efficiently developing total pressures up to approximately 1 in WG (2.49 kPa). Most propeller fans are used for static pressures close to zero.

Propeller fans may be either direct- or belt-driven. They may be free-standing, wall- or ceiling-mounted, or incorporated in a roof ventilator (see Figs. 34.2 through 34.5).

Accessories such as bird screens and louvers may be employed. Motors and drives are usually sized for the power requirements at the operating point which will be at or near the best efficiency point. If this is the case, operation at shutoff or low flows should be prevented. Shutters should not be allowed to freeze closed; filters should not be allowed to get too dirty. Figure 34.6 shows the performance of a typical propeller fan.

FIGURE 34.2 Free-standing, direct-drive propeller fan.

FIGURE 34.3 Wall-mounted, direct-drive, heavy-duty propeller fan.

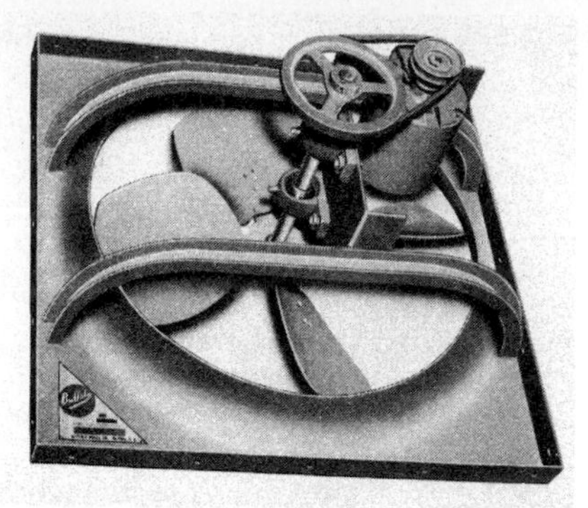

FIGURE 34.4 Ceiling-mounted, belt-drive propeller fan.

FIGURE 34.5 Direct-drive propeller fan in a roof ventilator.

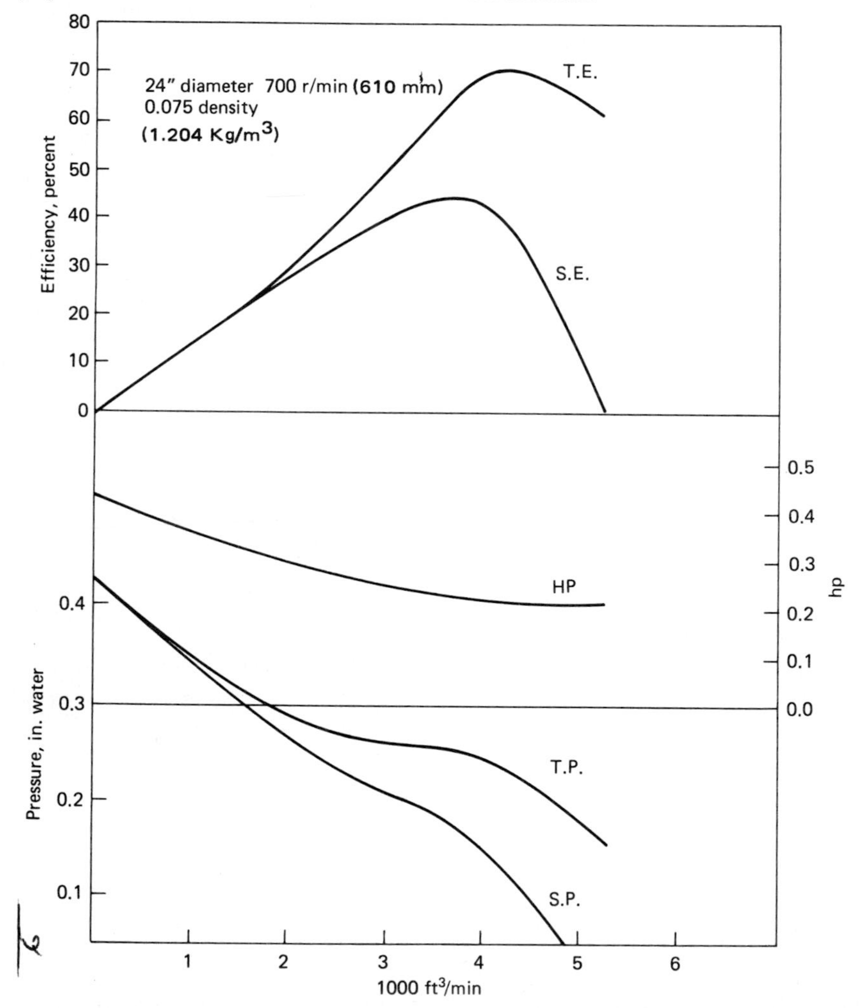

FIGURE 34.6 Performance for typical propeller fan. TE = total efficiency; SE = static efficiency; HP = horsepower; TP = total pressure, inH_2O; SP = static pressure, inH_2O. (See App. B for metric conversion factors.)

FIGURE 34.7 Adjustable-blade, direct-drive vane axial fan.

FIGURE 34.8 Fixed-blade, belt-drive vane axial fan.

Axial-flow fans are generally built with hub ratios ranging from 0.25 up. It is usually desirable to house the impeller in a cylinder together with a set of guide vanes, located on either the upstream or downstream side of the impeller. When this is done, the fan is generally described as a *vane axial fan*. When lower hub ratios are used, the vanes can be omitted with only a limited sacrifice in efficiency. The resulting fan is usually called a *tube axial fan*.

Axial-flow fans may be constructed for either direct or belt drive, as seen in Figs. 34.7. and 34.8. Variable-inlet vanes, suction boxes, inlet bells, discharge cones, sound attenuators, etc., may be used. Figures 34.9 and 34.10 illustrate some special features available for vane axial fans. The impeller may have fixed

FIGURE 34.9 Vane axial fan with clamshell-type access doors.

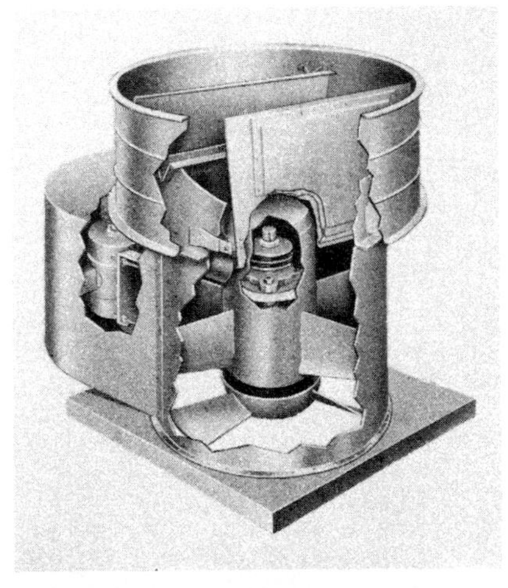

FIGURE 34.10 Belt-drive vane axial fan in a roof ventilator.

or adjustable blades. Blade adjustment may be possible when the fan is shut down or while the fan is in operation. The latter is called a variable-pitch or controllable-pitch fan. Depending on the hub ratio, speed, and other design factors, total pressures up to 10 to 20 in (2.55 to 5.0 kPa) WG can be developed in one stage. Multistage fans and fans in series can be used for higher pressures.

Most axial-flow fans require more power at low flows than at high flows at the same rotative speed. Motors need not be sized for the maximum power, if pro-

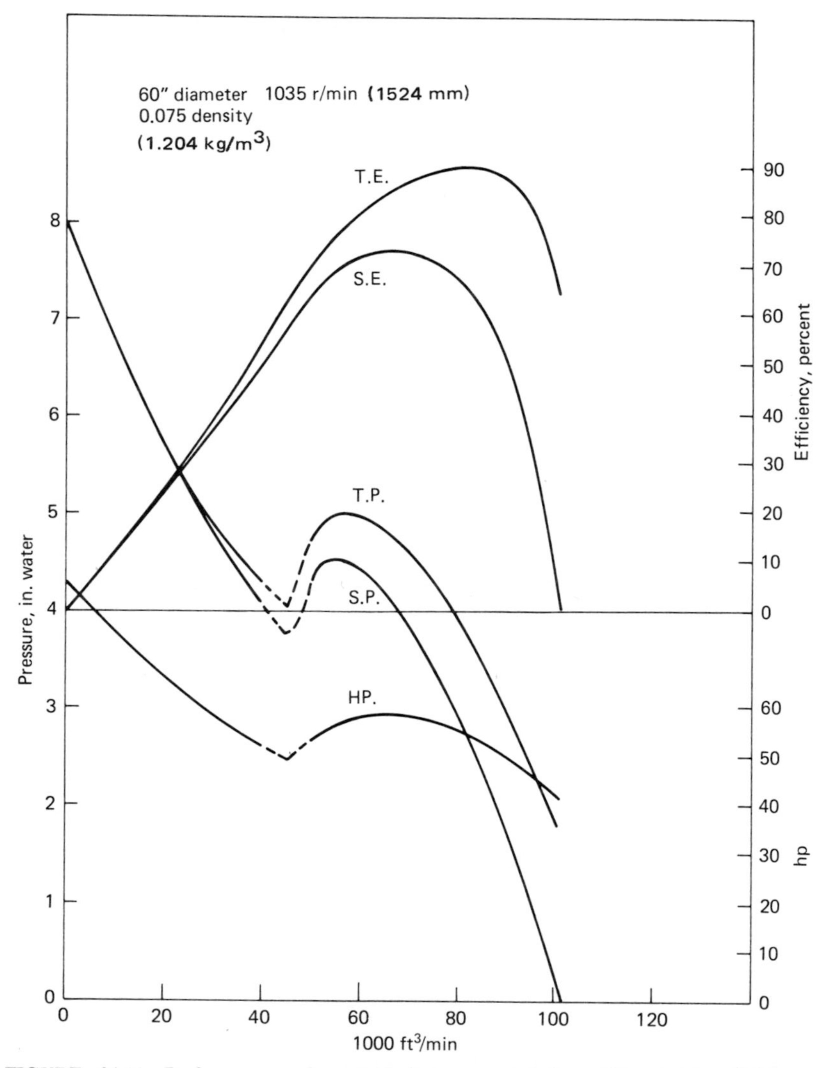

FIGURE 34.11 Performance of a typical vane axial fan. TE = total efficiency; SE = static efficiency; HP = horsepower; TP = total pressure, inH_2O; SP = static pressure, inH_2O. (See App. B for metric conversion factors.)

visions are made to avoid operation at or near the shutoff condition. Figure 34.11 gives the performance of a typical vane axial fan.

34.2.2 Centrifugal Fans

Within the centrifugal or radial-flow category, there are numerous aerodynamic types. Among the various distinguishing features are blade shape and blade depth (see Fig. 34.12). Generally, all centrifugal-type impellers should have blades curved at the heel, but this may be omitted on the deepest blade designs with very little sacrifice in efficiency. When curved, the heel should be curved forward in the direction of rotation. The blade tips may be curved in either a backward or

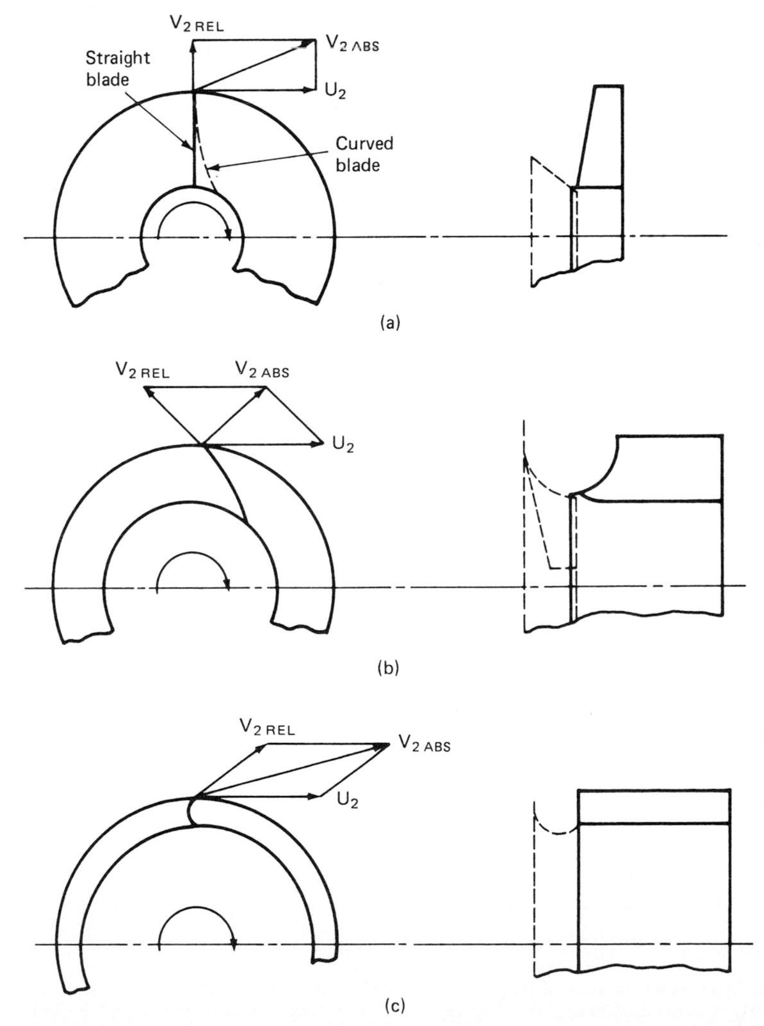

FIGURE 34.12 Blade designs: (*a*) radial discharge; (*b*) backward-curved discharge; (*c*) forward-curved discharge.

FIGURE 34.13 Centrifugal fan with inlet vanes and backward-curved impeller.

forward direction, or they may be radial. In general, the backward-curved types are the most efficient and most stable.

A typical backward-curved fan and its performance are shown in Figs. 34.13 and 34.14. Greater efficiencies can be obtained by employing airfoil-shaped, backward-curved blades. Performance for a typical airfoil-blade fan is shown in Fig. 34.15. Fans with forward-curved blades are generally smaller for a given duty than the other types. A small vent set with forward-curved blades is shown in Fig. 34.16. The performance of a larger forward-curved blade fan is shown in

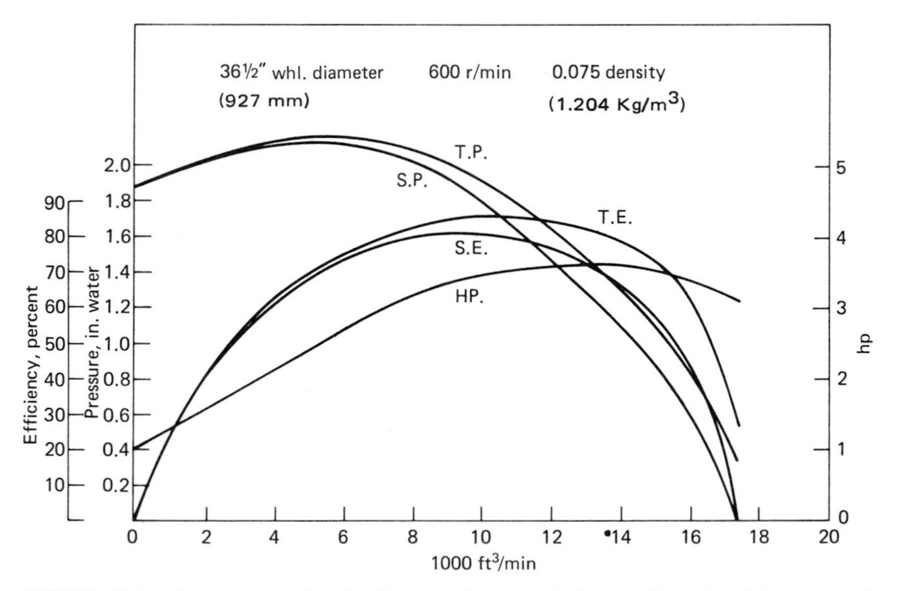

FIGURE 34.14 Performance of typical backward-curved-blade centrifugal fan. TE = total efficiency; SE = static efficiency; HP = horsepower; TP = total pressure, inH_2O; SP = static pressure, inH_2O. (See App. B for metric conversion factors.)

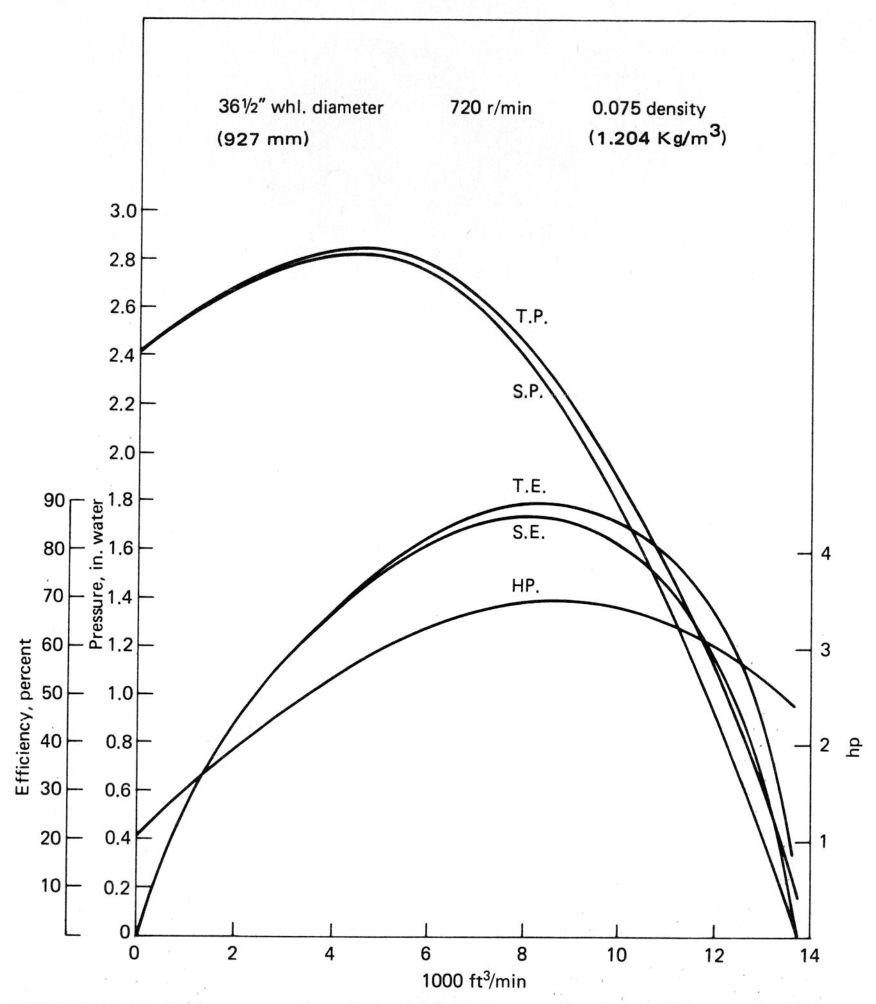

FIGURE 34.15 Performance of typical airfoil-blade centrifugal fan. TE = total efficiency; SE = static efficiency; HP = horsepower; TP = total pressure, inH_2O; SP = static pressure, in H_2O. (See App. B for metric conversion factors.)

Fig. 34.17. Radial blades are usually used in industrial applications where erosion or corrosion is a consideration. The deeper the blade, the more pressure the impeller will be capable of developing. Figure 34.18 shows a typical exhauster which can be equipped with various impeller types. Figure 34.19 shows a typical pressure blower, and Fig. 34.20 illustrates the performance of a larger radial-blade fan.

Centrifugal fans at constant speed and density have their lowest power requirements at the shutoff condition and so are often started with dampers closed. Both radial and forward-curved blades require considerably more power at high flows for a given speed than at the best efficiency point. Accordingly, some pro-

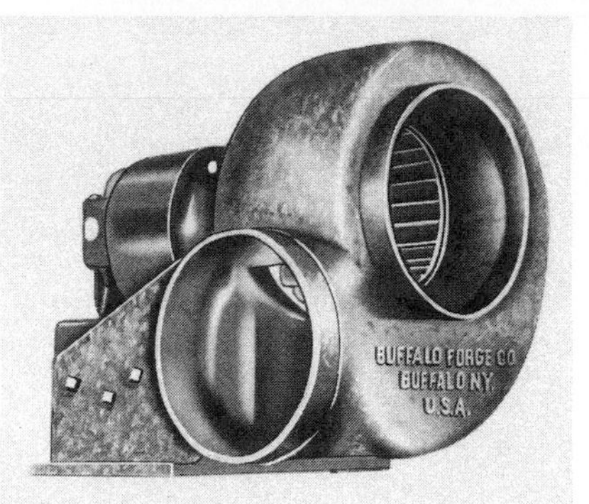

FIGURE 34.16 Cast-iron-housed vent set with forward-curved impeller.

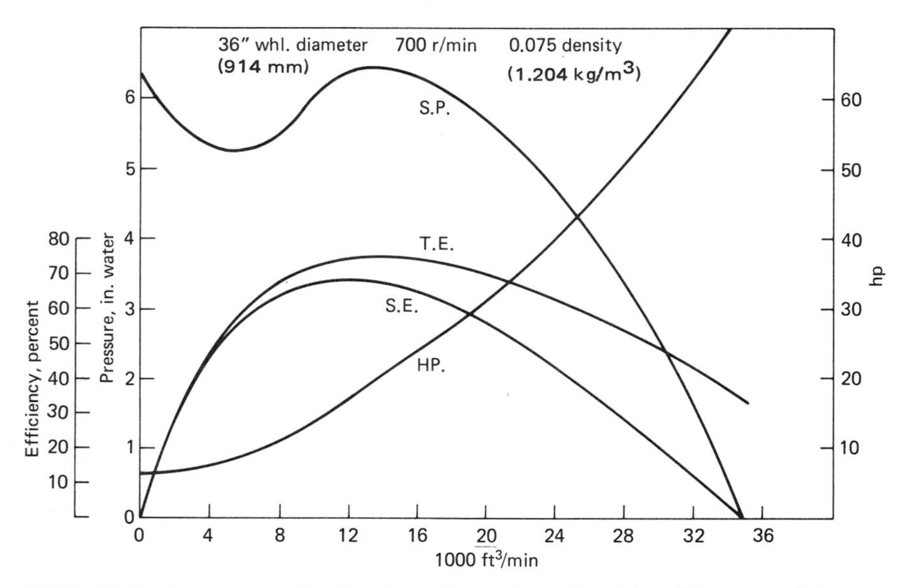

FIGURE 34.17 Performance of typical forward-curved centrifugal fan. TE = total efficiency; SE = static efficiency; HP = horsepower; TP = total pressure, inH_2O; SP = static pressure, inH_2O. (See App. B for metric conversion factors.)

tection should be provided to prevent the system resistance from falling below the design value, unless motors are oversized. Overloads can occur if the system resistance is overestimated. By distinction, the backward-curved types exhibit a horsepower characteristic such that if a fan is rated at or near the best efficiency point, no other point of operation at the same speed and density will require very much more power. This characteristic is variously described as nonoverloading, power-limiting, or Limit-Load.*

*Registered trademark of Buffalo Forge Co.

FIGURE 34.18 Arrangement 9 industrial exhauster with air, material, and open impellers.

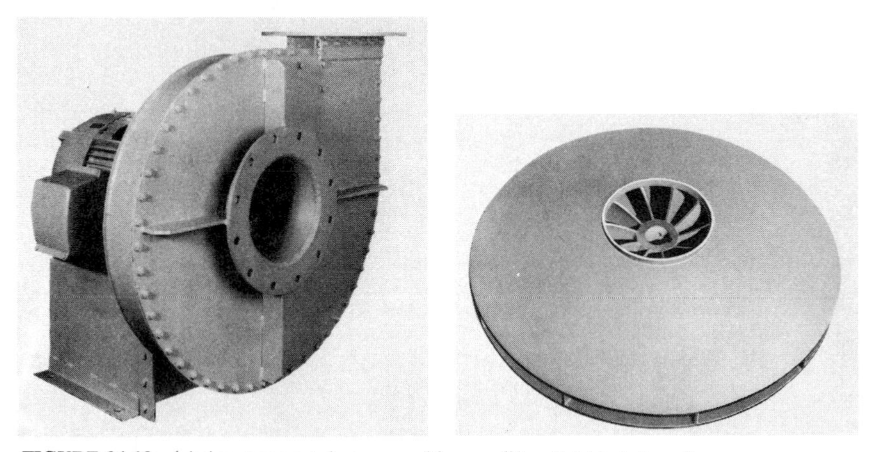

FIGURE 34.19 (*a*) Arrangement 4 pressure blower; (*b*) radial-blade impeller.

Centrifugal fans may be built for either belt or direct drive. Variable-inlet vanes, inlet boxes, inlet box dampers, outlet dampers, sound attenuators, etc., may be used.

34.3 FAN SYSTEMS

Fan systems may consist of any combination of fans, duct elements, heat exchangers, air cleaners, or other equipment through which all or part of the total flow must pass.

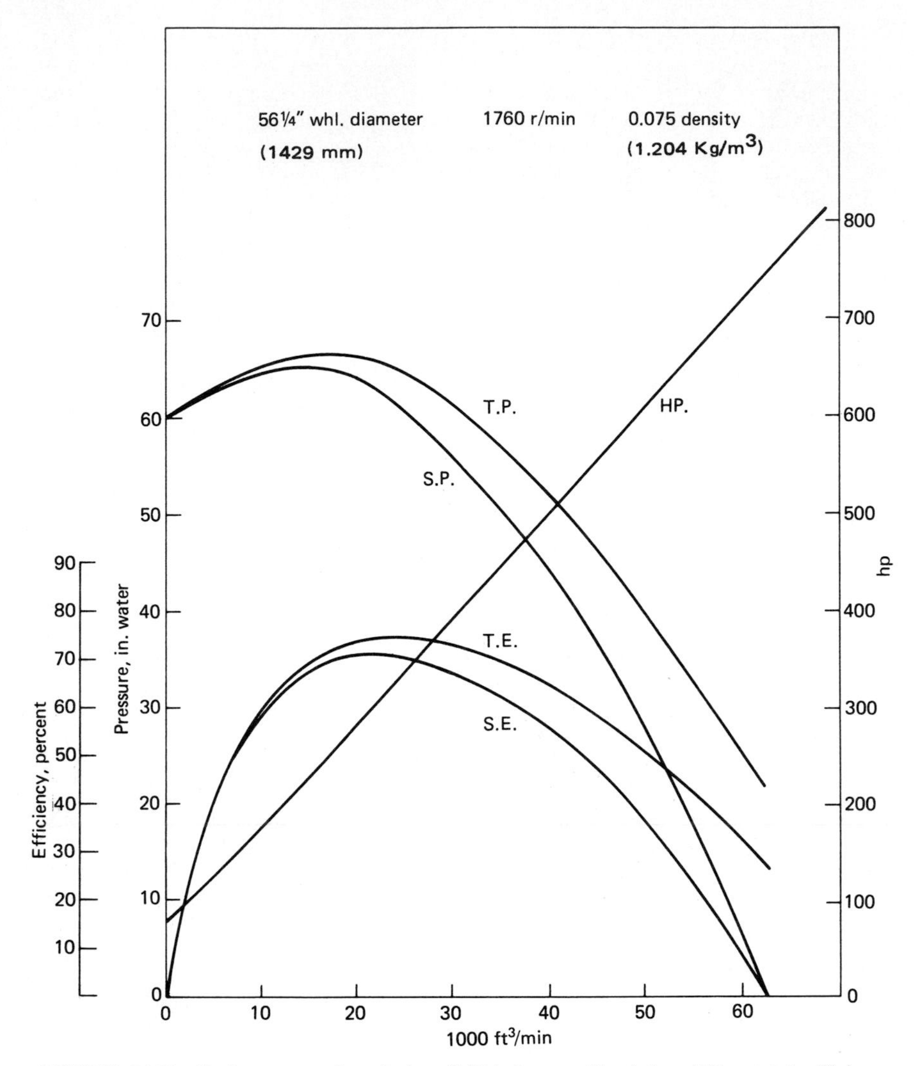

FIGURE 34.20 Performance of typical radial-blade centrifugal fan. TE = total efficiency; SE = static efficiency; HP = horsepower; TP = total pressure, inH_2O; SP = static pressure, inH_2O. (See App. B for metric conversion factors.)

34.3.1 Fan Location

The fan location in any system is dictated by the direction of flow and desired pressure relations. That is, a supply fan may be used to pump air into a space, or an exhaust fan may be used to draw air out of a space. The same through-flow conditions are obtained, but the pressure relations are different. In the first case there is a buildup of pressure in the space, and in the second case a reduction occurs in the space pressure. Both supply and exhaust fans may be used. Then the space pressure will depend on the relative amounts of air handled by each fan; i.e., space pressure will be positive if there is an excess of supply over exhaust and negative if there is an excess of exhaust over supply. Assuming the same capacities and end pressures for all cases, the total energy delivered by the fan or fans to the air passing through a given system is fixed, whether a supply fan, an exhaust fan, or both are used.

Obviously, if one fan is to supply air to several spaces, it must be located upstream of each space. A downstream location relative to each space is required if a fan is to exhaust air from several spaces.

When there must be several branches, the fan should be located as centrally as possible so that each particle of air will require approximately the same amount of energy for transport. Only one pressure can exist at a single point in space or in a duct system. This applies at the junction of any two branches regardless of differences in branch size, length, or configuration. Accordingly, the pressure drop along one branch must equal that along the other. If the branches are not designed to provide equal pressure drops at the required flow rates, the flow rates will differ from the design values. When the available pressure exceeds that required, the flow through a branch can be reduced to the design value by dampering. This is a waste of energy which sometimes cannot be avoided. In many cases, the size of the duct can be reduced to balance the pressure drops.

The fan must be chosen for a pressure sufficient to overcome the total losses based on the flow through the longest run. In other words, the same pressure must be dissipated by each portion of the air flowing, regardless of the lengths of the runs. Balanced design may be accomplished by using balancing dampers or by appropriate duct sizes in all branches.

34.3.2 Fan and System Matching

Fan performance must match system requirements. The only possible operating points are those where the system characteristic intersects the fan characteristic. At such points the pressure developed by the fan exactly matches the system resistance, and the flow through the system equals the fan capacity. If the flow rate is not equal to specifications, either the fan characteristic must be altered (by a change in speed, size, or whirl) or the system characteristic must be changed (by altering components or damper settings).

If more than one fan serves a system, their combined characteristics must be used to determine the capacity and performance characteristics of the system.

In any case, fan characteristics should be based on the gas density at the proposed fan location. System characteristics must be based on the individual densities in each of the system components. The overall system resistance is equal to the sum of the pressure losses for the individual components along any one flow path.

34.3.3 Two-Fan Systems

More than one fan may be required on an HVAC system. Supply and exhaust (return) fans are required to avoid excessive pressure buildup in the spaces being served. This combination may be considered a series arrangement in that both fans handle the same air. In a true multistage arrangement, both stages handle the same mass flow rate of air. However, as noted above, supply and exhaust rates may be adjusted to control the pressure in the space being served. The pressure requirements for each of the fans depend on the pressure in the space as well as the ductwork between the space and the fan.

Fans may also be installed in a parallel arrangement. Double-width, double-inlet fans are essentially two fans in parallel in a common housing. Sometimes two separate fans may be used because they fit the available space better than one larger fan. When such fans are used without individual ductwork and discharge into a common plenum, their individual velocity pressures are lost since they will merge into a common plenum pressure. Therefore the fan should be selected to produce the same fan static pressures.

Multiple-fan operation can be troublesome on some complicated systems. However, most HVAC systems are relatively uncomplicated, and most fans will come on-line and operate satisfactorily without special precautions. Fans which are grossly misapplied may stall or cause other problems.

34.3.4 Systems with Mass or Heat Exchange

If the air passing through a system is heated or cooled, its density will be decreased or increased, respectively, and assuming no change of mass, its volume rate will be increased or decreased accordingly. The resistance of any component should be calculated for the actual gas density, volume, and velocity through it. The overall system resistance is the sum of these individual resistances along any one flow path.

The mass rate of flow may be different at various locations. Multiple intakes or outlets have been discussed earlier. The mass rate may also vary due to a change of state, such as when water is evaporated into a gas stream. Conversely, water vapor may be condensed out of an air/vapor mixture, and in any combustion process additional gas may be generated. In any case, it is necessary to determine the rate of mass gain or loss from the appropriate relation for the physical or chemical process involved. The actual densities, volumes, and velocities should be used to determine individual resistances; then overall resistances can be determined by simple addition.

The fan for any system, including one with mass or heat exchange, must develop enough pressure to overcome the system resistance. This pressure requirement is the same regardless of the fan location. The best fan location based on power requirements is at the point where the volume entering the fan is minimized. Naturally, the fan will have to be downstream of all intakes on exhaust systems and upstream of all discharge openings on supply systems. Volume flow rate varies inversely with density when the mass flow rate is constant, and volume flow rate is proportional to weight flow when the density is constant. Both density and weight flow may vary, in which case volume flow will vary accordingly.

The difference in power requirements for various fan locations can be calculated from Eq. (34.4). For constant system resistance, the pressure required of the fan ($p_{T2} - p_{T1}$) will be constant regardless of its location. The volume flow Q will vary with position, as noted above. Assuming constant efficiency, the fan location requiring the least power is that place where the density is greatest. Density changes are not usually very great in HVAC systems. It would be appropriate to take advantage of the high-density location, but many other factors influence the choice of where the fan(s) will be located in the system.

34.3.5 Mutual Influence of Fan and System

Fan performance data are generally based on tests in which the air approaches the inlet with a uniform velocity, free of whirl. Duct element losses, except where noted otherwise, are based on similar flow conditions.

Elbows, unless provided with adequate turning vanes or splitters, produce uneven velocity patterns that may persist for considerable distances in subsequent straight ducts. Nonuniform inlet velocities may alter fan performance since different portions of the impeller will be loaded differently. Nonuniform velocities may also produce whirls in the inlet flow which affect fan performance. Every reasonable precaution should be taken to ensure uniform flow from all elbows.

The velocity pattern at fan discharge will vary with the design. Performance data are usually based on tests with a straight discharge duct. If an elbow is located close to the discharge, there may be some loss in fan performance due to a reduction in static-pressure regain. If the velocity along the inside radius of the elbow is higher than that along the outside radius, the loss will be higher than normal; but it will be less than normal if the velocity is higher along the outside radius than along the inside radius.

The effects of air-flow conditions different from those that would exist during a standard laboratory test are known as *system effects*. System-effect factors have been developed by various fan manufacturers and by the Air Movement and Control Association.[1] The former should be used in preference to the latter wherever possible because the system effects will vary with individual fan designs. Nevertheless, the AMCA data can be very useful, particularly if the system designer avoids the conditions where high system-effect factors tend to prevail. The most common reason for systems' failing to perform as projected is poor duct conditions near the fan.

34.3.6 Capacity Control

If a fan system must operate over a range of capacities, some means of adjusting either the fan characteristics or the system characteristics must be provided.

System characteristics may be altered by inserting additional resistance or by providing an additional flow path. Gradual adjustment can be provided with movable dampers either in the main passage or in the bypass.

Fan characteristics may be modified by altering blade position, rotative speed, or inlet whirl. Gradual adjustment is usually possible within the limits of the device used, whether it be a variable-pitch rotor, a variable-speed motor or transmission, or variable-inlet vane. Step-by-step adjustment is possible with multispeed motors and multiple-fan arrangements.

The choice of a specific control method should be based on an economic evaluation. Before such an evaluation can be made, it is necessary to estimate the operating times at various capacities. Each of the methods entails a loss in efficiency over most of its operating range. The most inefficient methods are usually the least expensive to install originally. Only a complete evaluation on the basis of both first cost and operating cost will reveal the best method.

The effects of various methods of capacity control on fan operation are illustrated in Fig. 34.21. The first three sets of curves are drawn for the same fan (a backward-curved centrifugal) operating at the same full-load speed. The system resistance varies as the square of the capacity.

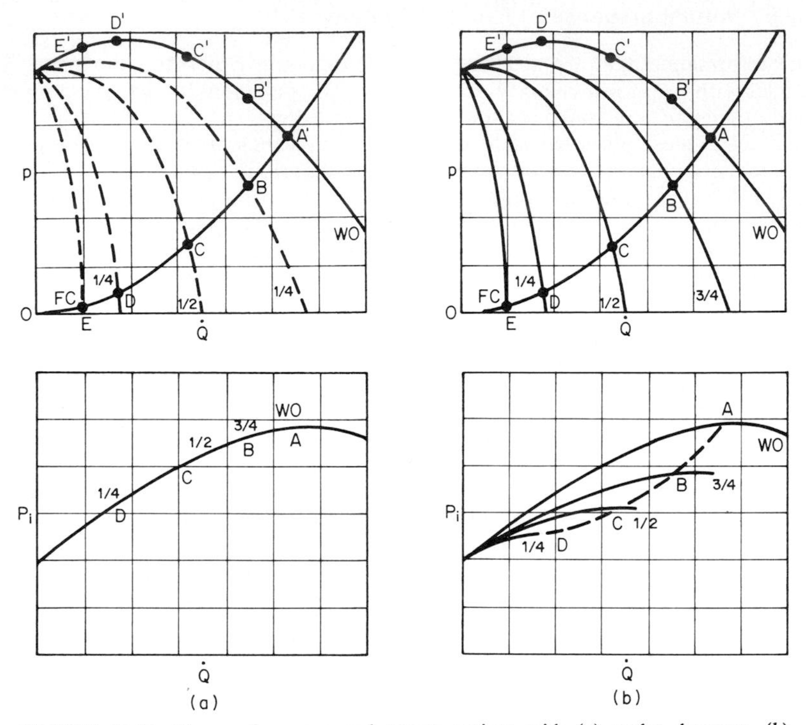

FIGURE 34.21 Fan performance and power savings with (*a*) outlet dampers, (*b*) variable-inlet vanes, (*c*) variable pitch, and (*d*) variable speed; (*e*) comparison of power savings. Details for inlet box dampers are not shown since they are quite similar to those for variable-inlet vanes. OD = outlet damper; IBD = inlet box damper; VIV = variable-inlet vanes; VS_i = input to variable-speed device; VS_o = output from variable-speed device; VP = variable pitch. (See App. B for metric conversion factors.)

Figure 34.21*a* is drawn for damper control. The pressure required of the fan is higher at reduced ratings than at design values because the damper increases system resistance. The dotted system curves include this increase, which also varies as the square of the capacity. The powers at one-half and three-fourths load are less than at full load because of the nature of the fan characteristics. This would not be true if the characteristics of an axial fan were used instead of those for a centrifugal fan.

Figure 34.21*b* is drawn for variable-inlet-vane (VIV) control. Three fan curves are drawn for three different positions or settings of the variable-inlet vanes. The powers required by the fan at reduced capacities are lower than those for damper control.

Figure 34.21*c* is drawn for speed control. Instead of three system curves, three fan curves are drawn, one each at full, three-fourths, and one-half speed. There is also a power curve for each speed. The powers at reduced capacities are less than the corresponding powers for either damper or VIV control, and no elements add resistance to the system.

Figure 34.21*d* is drawn for variable-pitch control. Three fan curves are drawn

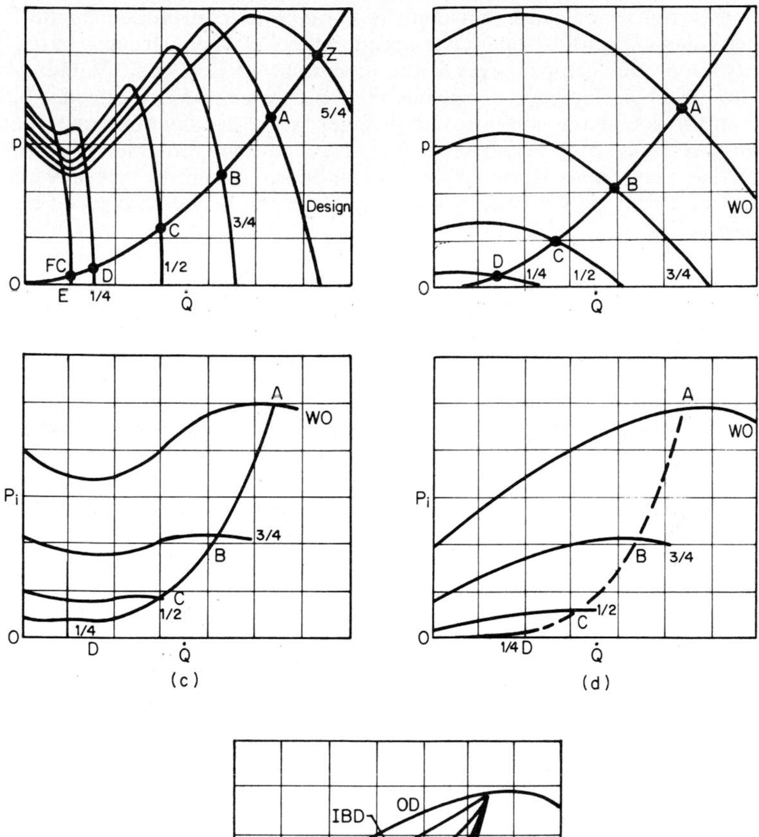

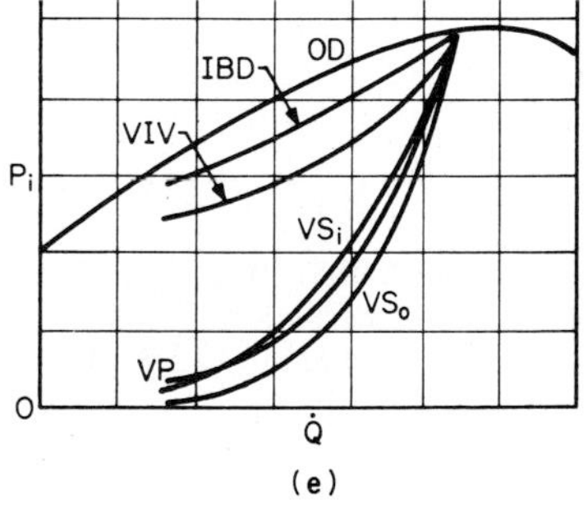

FIGURE 34.21 (*Continued*).

for three different blade positions or pitch settings. The power requirement is very close to that for variable speed.

Figure 34.21*e* shows the power relations for all four methods. The curves for damper and vane control are exactly as in Fig. 34.21*a* and *b*; however, two curves are shown for speed control. The extra curve 2 includes the efficiency of a hydraulic coupling used as a variable-speed transmission. Curve 2 therefore represents the input to the transmission rather than to the fan as for all other curves. This figure indicates that there is a range of capacities near full load where inlet vane control is superior to speed control on a power input basis. The simplicity of a duct damper gives it the advantage in first cost. Variable-inlet vanes are considerably less expensive than most methods of speed control. Speed control does have additional advantages when considerable operation at less than maximum speed is expected. The accompanying reduction in noise and increased life may also be appreciable. As can be seen from the typical variable-pitch-control curve, power savings approximate those of variable-speed control very closely.

34.4 FAN LAWS

The fan laws relate the performance variables for any homologous series of fans. The variables involved are fan size D, rotative speed N, gas density σ, capacity Q, pressure p, power P, sound power level L_W, and efficiency η.

The fan laws are the mathematical expression of the fact that when two fans are both members of a homologous series, their performance curves are homologous. At the same point of rating, i.e., at similarly situated points of operation, efficiencies are equal. The ratios of all the other variables are interrelated.

The fan laws can be expressed in terms of various combinations of dependent and independent variables. The most frequently used combination is fan law 1 where size, speed, and density are considered the independent variables and capacity, pressure, power, and sound power level are considered the dependent variables. Accordingly, the effects of change in speed or density can be calculated for a fan that has already been installed and such changes need to be evaluated. Table 34.1 gives fan law 1. The subscript a denotes that the variable is for the fan under consideration. The subscript b indicates that the variable is for the tested fan.

34.4.1 Fan Law Restrictions

It cannot be emphasized too strongly that the fan laws should be used only under very special circumstances. The most frequent error made in using the fan laws is trying to predict performance after changing some physical aspect of the system to which the fan is attached or even a damper setting.

In fan engineering terminology, the fan laws are only applicable to exactly similar points of rating. When a system is changed by adding a coil or some other means, the point of rating has to change. Of course, there are ways to predict the effects of these kinds of changes, but they involve the calculation of more than one point on the performance curve.

34.4.2 Equivalent Static Pressure

The use of the fan laws can sometimes be simplified by employing a device called *equivalent static pressure (ESP)*. The "equivalent static pressure" can be defined as the pressure that would be developed by a fan operating at standard air density instead of the actual pressure developed when operating at actual density. This is

TABLE 34.1 Fan Laws

For all fan laws: $\eta_a = \eta_b$ *and (point of rating)$_a$ = (point of rating)$_b$*

No.	Dependent variables			Independent variables				
1*a*	CFM_a	$= CFM_b$	×	$\left(\frac{SIZE_a}{SIZE_b}\right)^3$	×	$\left(\frac{RPM_a}{RPM_b}\right)^1$	×	(1)
1*b*	$PRESS_a$	$= PRESS_b$	×	$\left(\frac{SIZE_a}{SIZE_b}\right)^2$	×	$\left(\frac{RPM_a}{RPM_b}\right)^2$	×	$\left(\frac{\delta_a}{\delta_b}\right)$
1*c*	HP_a	$= HP_b$	×	$\left(\frac{SIZE_a}{SIZE_b}\right)^5$	×	$\left(\frac{RPM_a}{RPM_b}\right)^3$	×	$\left(\frac{\delta_a}{\delta_b}\right)$
1*d*	PWL_a	$= PWL_b$	+ 70 log	$\left(\frac{SIZE_a}{SIZE_b}\right)$	+50 log	$\left(\frac{RPM_a}{RPM_b}\right)$	+20 log	$\left(\frac{\delta_a}{\delta_b}\right)$

a useful tool when one is selecting a fan from published data which are based on operation at standard density. Example 34.1 (p. 34.37) will illustrate this.

$$\text{ESP} = p_{FS}\frac{\rho_{\text{std}}}{\rho_{\text{act}}} \tag{34.6}$$

In fact, the concept of equivalent static pressure can be extended to any pair of densities. Since most published data for a fan are based on standard air density, we do not explore the general case any further.

34.5 FAN NOISE

Fan noise consists of a series of discrete tones superimposed on a broadband component. The former, which may be called the *rotational component*, may be traced to the process of energy transfer that also leads to the development of head. The latter, which may be called the *vortex component*, may be traced to the formation of turbulent eddies of one kind or another that usually lead to losses of head.

34.5.1 Centrifugal Fan Noise

The predominant tone of the rotational component of fan noise in centrifugal apparatus is usually that at the blade-passing frequency. In very narrow blade designs, the higher harmonics may be of equal intensity. Widening the blades progressively weakens the higher harmonics.

Vortices can be created at the leading or trailing edges of the blades, along the sides of the blades, or at locations remote from the blades. In general, the size, rate of growth and decay, and point of origin and movement of these vortices will be random and the resulting noise will have a broadband spectrum.

Streamlining the leading edges of the blades minimizes vortex formation at that location. At the design capacity, both thin blades with rounded edges and thick blades with airfoil sections are quite effective in reducing vortex formation. The airfoil-shaped blade may enjoy some advantage, particularly when the leading-edge angle does not match the entering flow angle across the entire width of the blade.

Large eddies may be formed in the blade passages due to flow separation from a boundary. The greatest benefit to be derived from the use of airfoil-shaped blades is that of reducing separation. This is somewhat offset from a noise standpoint by the decrease in the optimum number of blades compared to that for thin blades. The thickness of the blade apparently has very little effect in centrifugal fans.

The speed of sound so greatly exceeds the airspeed in most fans that the noise is propagated upstream and downstream with equal facility. The acoustical impedances of the inlet and outlet openings are so nearly equal that in most cases the sound power radiated through the outlet can safely be assumed to be equal to that radiated through the inlet. The transmission through the casing walls is so small by comparison that when the total sound power output of a fan is mea-

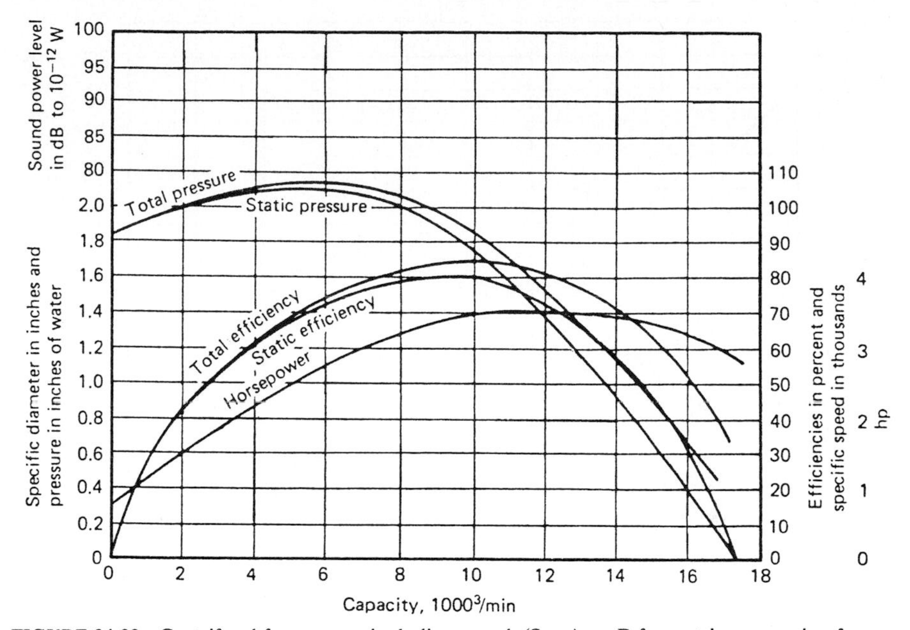

FIGURE 34.22 Centrifugal fan curves, including sound. (See App. B for metric conversion factors.)

sured, the portions radiated through the outlet and inlet are each reported as one-half of that total. The corresponding sound power levels are therefore each 3 dB less than the total sound power level.

The sound power level curve for a backward-curved thin-bladed centrifugal fan is shown in Fig. 34.22 together with other performance characteristics. This curve is typical of that for all centrifugal fan types. The overall shape of the sound power level curve indicates that the sound power output of a fan is a function of both capacity and pressure. Tests indicate that sound power outputs are proportional to the capacity ratio multiplied by the square of the pressure ratio, all other conditions being equal. The spectrum for a centrifugal fan may be approximated in most cases by subtracting 8, 4, 8, 9, 11, 16, 20, and 25 dB from the overall level to obtain the levels in the 63-, 125-, 250-, 500-, 1000-, 2000-, 4000-, and 8000-Hz octave bands, respectively, provided the blade-passing frequency falls in any other band.

34.5.2 Axial-Flow Fan Noise

The noise characteristics of axial-flow fans are very similar to those of centrifugal fans. The division of fan noise into rotational tones and broadband vortex components applies to both types.

In axial-flow fan apparatus, the predominant tone of the rotational component may be one of the higher harmonics rather than the fundamental blade-passing frequency, if the fan is used to develop appreciable pressure.

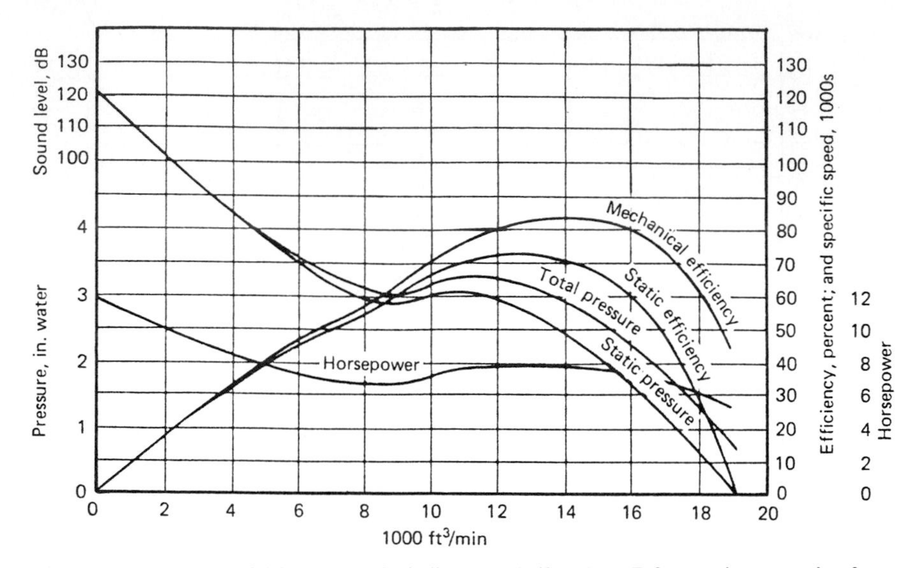

FIGURE 34.23 Vane axial fan curves, including sound. (See App. B for metric conversion factors.)

Increases in the number of blades generally have a beneficial effect on axial fan noise. The number of blades should differ from the number of guide vanes, to prevent strengthening of the fundamental tone.

The effects of streamlining on leading-edge vortices and side-separation eddies are the same for both axial and centrifugal fans. The effect of trailing-edge thickness is more pronounced in axial fans. There may be a noticeable increase in noise if the wake from one blade is cut by succeeding blades.

The sound power level curve for a vane axial fan is shown in Fig. 34.23 together with other performance characteristics. This curve is typical for all axial-flow fan types. The overall shape of the curve is slightly different from that for a centrifugal fan, consistent with the difference in pressure-capacity curves. The spectrum for a high-pressure vane axial fan may be approximated in most cases by subtracting 11, 13, 7, 5, 7, 11, 16, and 19 dB from the overall level to obtain the levels in the 63-, 125-, 250-, 500-, 1000-, 2000-, 4000-, and 8000-Hz octave bands, respectively. The corresponding values for a low-pressure propeller fan are 5, 5, 7, 9, 15, 18, 25, and 30 dB.

34.5.3 Sound Power Level and the Fan Laws

The ratio of the sound power levels for two similar fans can be predicted from the fan laws. This characteristic, like the other dependent variables, is predictable only if the fans have the same point of rating and both have good balance, good bearings, etc. Complete test data on one of the fans must be available.

The noise spectra for any two homologous fans at the same point of rating may be considered similar in most cases. The ratio of the sound power levels at any pair of corresponding frequencies will equal the ratio of the overall sound power levels. Corresponding frequencies may be the two blade frequencies or any two harmonics thereof. If the rotative speeds and thus the blade frequencies are equal, the fan laws may be used to predict the sound power levels in each of the standard octave bands, provided the corresponding test data are available. If rotative speeds are not equal, only the overall sound power level can be predicted.

Test data on sound power level will usually be based on measurements of the total noise radiated from the inlet and outlet and the casing. If the inlet and outlet have the same size and shape, there is a good chance that the noises radiated in the two directions will be equal. This is approximately true even in centrifugal fans having round inlets and rectangular outlets.

Fan law 1*d* (see Table 34.1) indicates that the ratio of sound power outputs varies as the seventh power of the size ratio, the fifth power of the speed ratio, and the square of the density ratio. These relationships have been verified by Madison and Graham[2] in the Buffalo Forge Company laboratory.

34.6 FAN CONSTRUCTION

34.6.1 Standard Designations and Arrangements

The fan industry through the Air Movement and Control Association has devised certain standard designations. Those for rotation, discharge, inlet box position, drive arrangement, and motor position are excerpted here. For a complete set refer to AMCA Publication 99-83.

The method of specifying rotation is to view the fan from the drive side and to indicate whether the rotation is clockwise or counterclockwise. The drive side of a single-inlet centrifugal fan is considered to be the side opposite the inlet, even in those rare cases where the actual drive location may be on the inlet side. It is necessary to specify which of the drives is used for reference on dual-drive arrangements. The rotation of a propeller or axial-flow fan is usually immaterial and a matter of individual design. There is no official designation of drive sides for axial fans, so if it is necessary to specify rotation, the direction from which the fan is viewed should also be specified.

The method of specifying discharge position is indicated in Fig. 34.24. If the fan is to be suspended from the ceiling or a side wall, discharge should be specified as if the fan were floor-mounted. The intended mounting arrangement should also be given. An angular measure is required for angular positions.

The various drive arrangements have been assigned a number, as indicated in Fig. 34.25. Designations for axial fans are consistent with standards for centrifugal fans. The official arrangement numbers may be used for fans with a bearing on the housing or subbase as shown or for pedestal-mounted bearings. Arrangements involving a bearing in the inlet should be avoided for small fans.

The method of specifying the inlet box position is to view the fan from the drive side (same as for rotation) and indicate the position of the intake opening. Angularity is referred to a horizontal centerline, as shown in Fig. 34.26. The various motor positions have been assigned letter designations, as indicated in Fig. 34.27.

34.6.2 Heat-Resistant Materials

Protecting fans in high-temperature gas streams involves both corrosive and structural considerations. Ultimate strengths of steels may improve slightly at

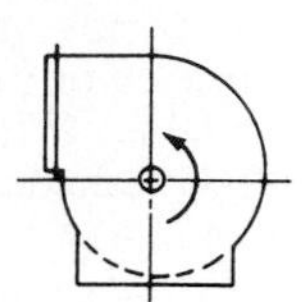

Counterclockwise
Top horizontal

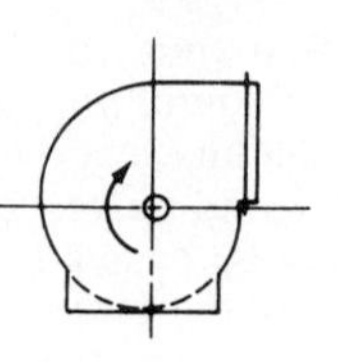

Clockwise
Top horizontal

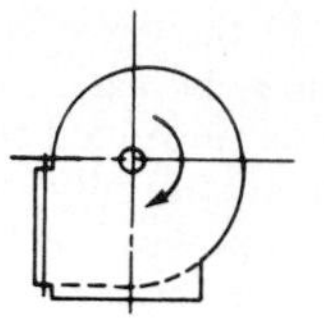

Clockwise
Bottom horizontal

Counterclockwise
Bottom horizontal

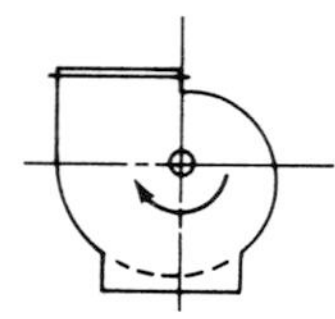

Clockwise
upblast

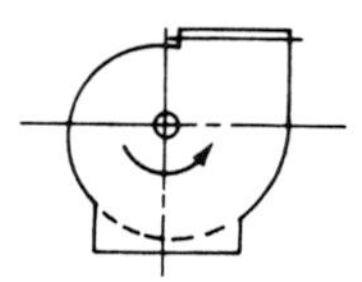

Counterclockwise
upblast

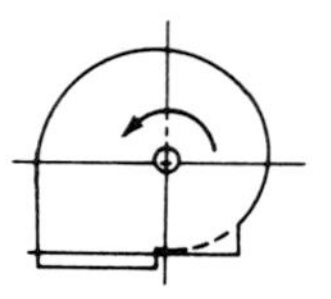

Counterclockwise
downblast

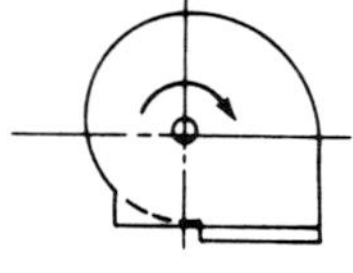

Clockwise
downblast

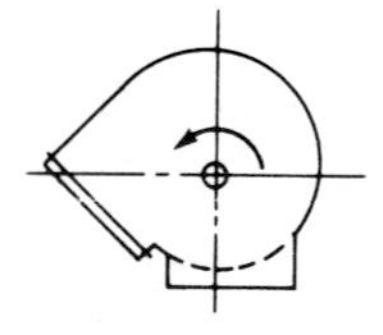

Counterclockwise
Top angular down

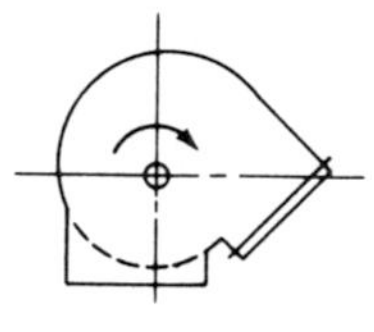

Clockwise
Top angular down

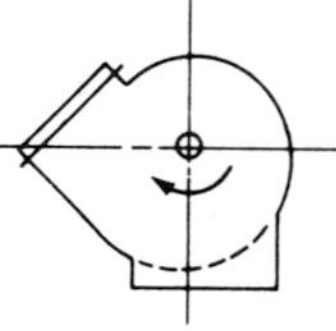

Clockwise
Bottom angular up

Counterclockwise
Bottom angular up

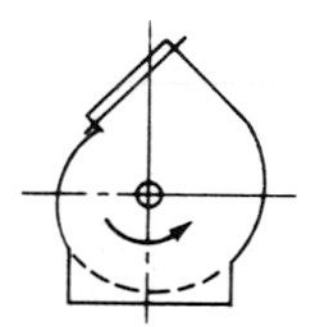

Counterclockwise
Top angular up

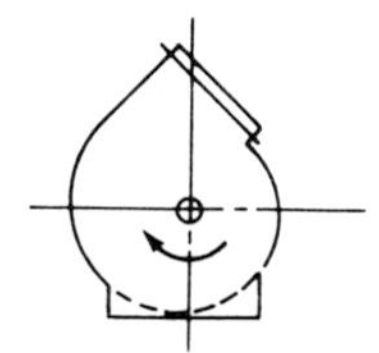

Clockwise
Top angular up

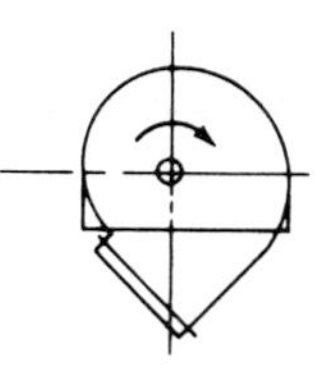

Clockwise
Bottom angular
down

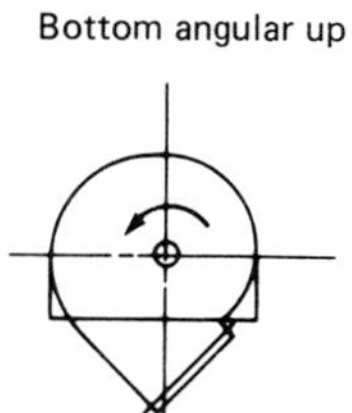

Counterclockwise
Bottom angular down

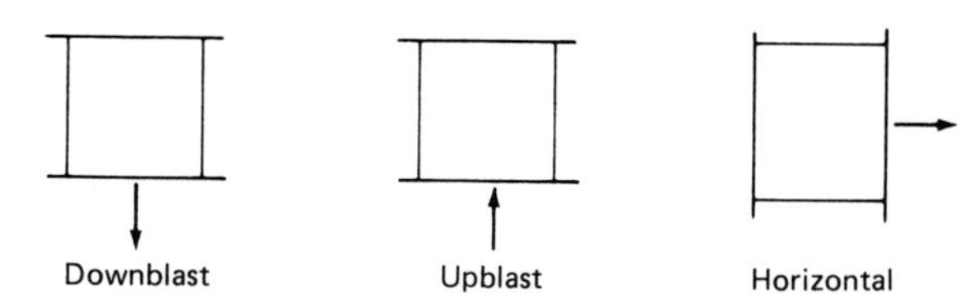

FIGURE 34.24 Standard rotation and discharge designations.

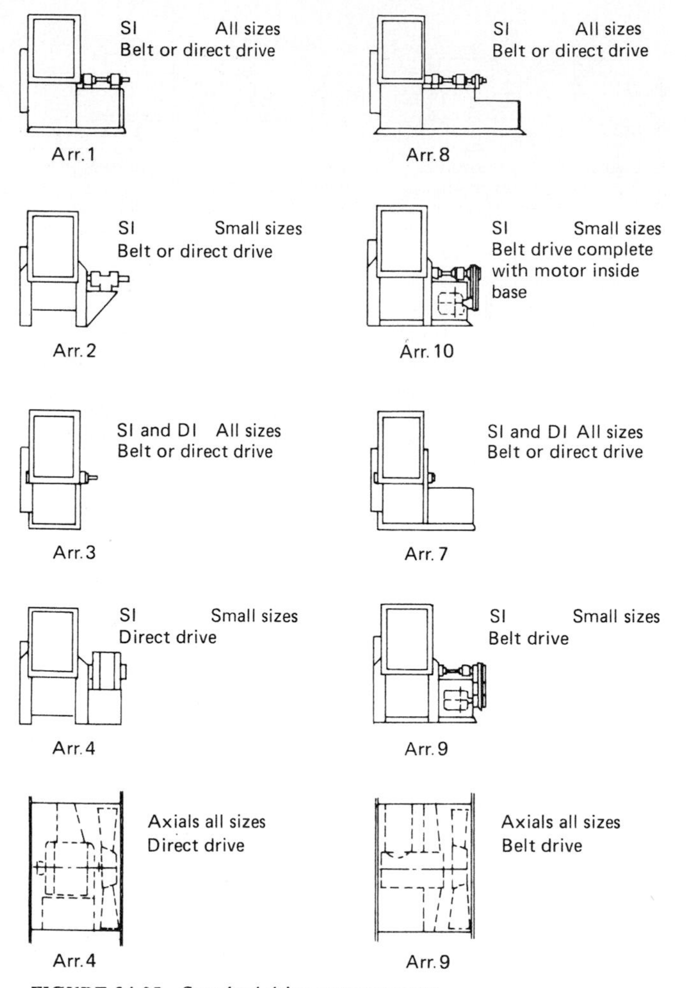

FIGURE 34.25 Standard drive arrangements.

moderate temperatures but eventually decrease rapidly. Yield strengths decrease with temperature even when the ultimate strengths are apparently unaffected. The design criteria may be the ultimate strength, yield strength, creep strength, or rupture strength, depending on the temperature, nature of the stressed part, and service requirements.

Mild steel scales rapidly at temperatures above 900°F (470°C) in normal atmospheres. A process involving a metal spray and heat treatment to form a steel-aluminum alloy at the surface may extend this operating range. Heat-resistant paints may provide some protection, but temperatures are limited by the binder used. Various low-alloy steels, stainless steels, and high-nickel alloys provide suitable strength and corrosion resistance in air.

Problems due to expansion and problems of cooling bearings also face the designer of high-temperature fans. In general, bearings should be kept out of the air

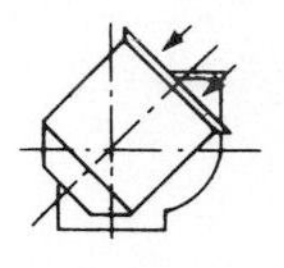

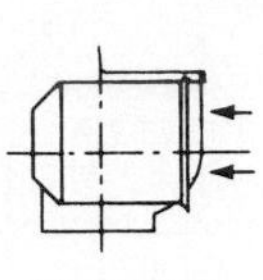

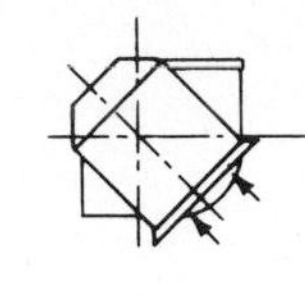

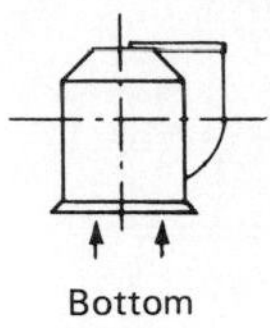

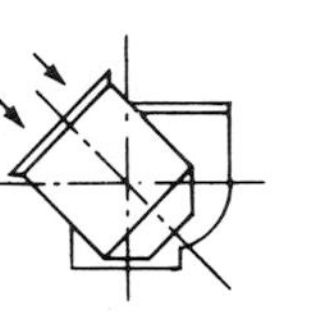

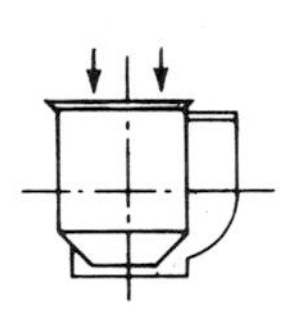

FIGURE 34.26 Standard inlet box positions.

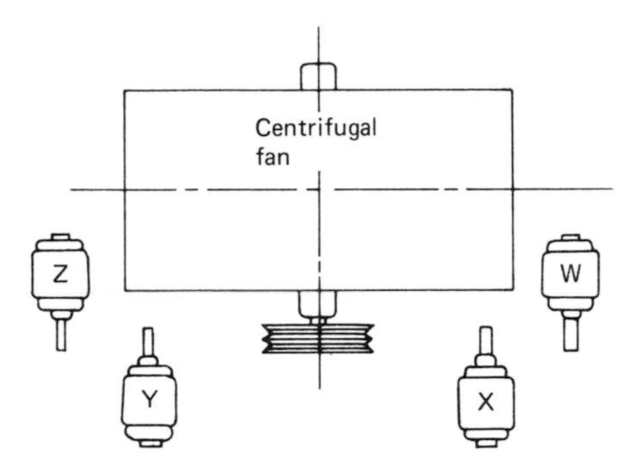

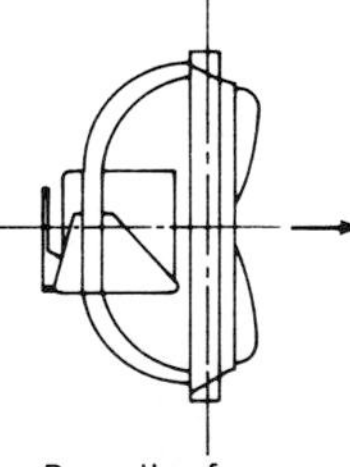

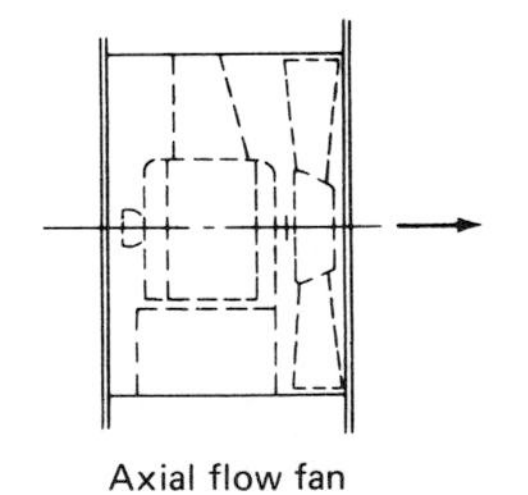

FIGURE 34.27 Standard motor positions.

stream on all but room-temperature applications. Grease lubrication is satisfactory up to 200°F (93°C) on many types of antifriction bearings. Oil lubrication is required for higher temperatures. Above 350°F (175°C) some sort of heat slinger or cooling disk is usually required on the shaft between the fan and bearing. Above 700°F (425°C) the bearing subbase should be physically separated from the fan housing, to prevent the direct conduction of heat. Fans and ductwork should be insulated.

34.6.3 Corrosion-Resistant Materials

The choice of a particular material or coating to protect a fan from attack by a corrosive gas is an economic matter. Ordinarily a fan is constructed of the materials which provide the necessary strengths and contours most economically. If the life of such a fan handling a corrosive gas is limited, the extra first cost of a special material or coating may be justified by the increase in life obtained.

Such an evaluation is easily made for any two materials if actual operating experience or test data are available. Uhlig[3] has compiled a list of corrosive agents and materials with various degrees of resistance. The accuracy of any such prediction is limited. The effects of temperature, local concentration, velocity, impurities, and fabrication must be considered when one is translating test results to terms of actual performance.

When corrosion takes place, two types of attack may be involved. Direct chemical attack is generally limited to high temperatures or highly corrosive environments or both. The rapid scaling of steel at temperatures above 900°F (525°C) and the effect of concentrated acids or alkalies are examples. Such reactions may be prevented by controlling the temperature or the concentration of the corrosive substance or by application of inert protective coatings.

Electrochemical attack is much more common. The requirements for such a reaction are that there be discrete anodic and cathodic regions connected by solid material submerged in an electrolyte. Such cells not only are produced by coupling two materials but also arise from minor variations in stress, surface composition, deposits of dissimilar metals, and condensation of an electrolyte.

Frequent cleaning will remove deposits and thus promote better life for any material. The composition of weld deposits and the heat-affected zones in welding as well as stress relief and surface treatment should be evaluated for their effect on corrosion resistance.

The formation of corrosion products may serve to retard or accelerate the reaction. Rust on mild steel tends to promote corrosion, while rust that forms on low-alloy steels may protect the metal from further corrosion. The film that forms on the surface of stainless steel is protective in many corrosive environments, as is that which forms on aluminum.

The protection afforded by many metallic coatings is due to their sacrificial action. Zinc on galvanized products and the anodic covering on aluminum-clad and cadmium-plated materials protect scratches and even sheared edges by this means. However, to avoid undesirable sacrificial action, the formation of galvanic couples by joining dissimilar metals should be avoided.

The protection provided by other coatings such as lead, rubber, or plastics is due primarily to their inertness to corrosive media. The optimum thickness varies with the corrosive medium, but the values in Table 34.2 are generally suitable. Temperature and tip-speed limits are shown.

Lead linings may be "burned" on or mechanically attached to the surfaces of

TABLE 34.2 Protective Coverings

	Thicknesses	Limitations
Lead linings	⅛–¼ in	12,000 ft/min at 200°F (94°C) 9,000 ft/min at 400°F (204°C)
Hard rubber coverings	⅛–3/16 in	6×10^6 RN^2,* 150°F (66°C) max
Soft rubber coverings	⅛–3/16 in	3×10^6 RN^2,* 150°F (66°C) max
Epoxy coatings	5–10 mil	200°F (94°C) max
Phenolic coatings	5–10 mil	350°F (177°C) max
PVC lacquers	5–10 mil	180°F (82°C) max
PVC plastisols	30–100 mil	160°F (71°C) max

*RN^2 = radius in ft × (r/min)^2.

Note: Conversion factors to the metric system are: in × 2.54×10^{-5} = mm; mil × 2.54×10^{-7} = mm; ft/min × 5.08×10^{-3} = m/s; $6 \times 10^6 RN^2$ [ft · (r/min)^2] = $1.829 \times 10^6 RN^2$ [m · (r/min)^2]; $3 \times 10^6 RN^2$ [ft · (r/min)^2] = $0.915 \times 10^6 RN^2$ [m · (r/min)^2].

fan wheels or housings made of steel. Costs limit use to applications where other means of protection are not particularly suitable.

Rubber coverings, in the thicknesses shown in Table 34.2, are sheets adhered to exposed steel surfaces and vulcanized in place. Rubber-coated parts may also be sprayed or dipped. Various natural and synthetic elastomers may be used including neoprene, Saran, and silicone rubber for increased temperature.

Plastic coatings are generally sprayed on. To obtain the optimum thickness, multiple layers are needed. Phenolics and plastisols require baking at moderate temperatures [375 to 450°F (191 to 232°C)], but vinyl lacquers and epoxies may be air-dried. Only the more flexible types of plastics may be used on fan wheels. Even so, standard fan construction must often be modified to eliminate all gaps and voids. Continuous welds and rounded edges as well as clean surfaces are required. Similar provisions must be made for ceramic and metal sprays.

FIGURE 34.28 Fiberglass-reinforced plastic vane axial fan.

Fiberglass-reinforced plastic (FRP), both polyesters and epoxies, is used for fan housings and impellers. Figure 34.28 shows a vane axial fan made with FRP. The portion of the shaft which will be exposed to the gas is encapsulated. Both manual lay-up and pressure-molding techniques are employed. The fiberglass should be covered with a sufficient thickness of a chemical-resistant resin to protect it where the gas might attack glass, as in the case of hydrogen fluoride. Rigid polyvinyl chloride (PVC) is also used to fabricate impellers and housings. It is generally limited to a fraction of the speeds for FRP equipment.

Various metals and alloys have been used to fabricate corrosion-resistant fans. Aluminum-base alloys are not affected by most gases in the absence of water at or near room temperature. Acid gases and SO_2 in the presence of water have some corrosive action on aluminum. All alkalies will attack aluminum. Many industrial fumes and vapors attack aluminum surfaces.

Copper at or near room temperature is not affected by dry halogen gases, but absolutely no moisture may be present, even with halogenated refrigerants. Normal and industrial atmospheres do not usually cause corrosion beyond the formation of a protective oxide film.

Copper-zinc, copper-nickel, copper-tin, and copper-silicon alloys are unaffected by most dry gases at ordinary temperatures. The presence of moisture increases the corrosiveness of the halogens, SO_2, and CO_2 considerably. Generally H_2S reacts even at low humidity.

34.6.4 Spark-Resistant Construction

The Air Movement and Control Association, in its Standard 99-0401-66[4], outlines three types of spark-resistant construction:

- *Type A* construction requires that all parts of the fan in contact with the air or gas being handled be made of nonferrous metal.
- *Type B* construction requires that the fan have an entirely nonferrous wheel and a nonferrous ring about the opening through which the shaft passes.
- *Type C* construction specifies that the fan must be so constructed that a shift of the wheel or shaft will not permit two ferrous parts of the fan to rub or strike.

In all three types, bearings should not be placed in the air or gas stream, and the user must electrically ground all fan parts.

Bronze and aluminum are commonly used where nonferrous parts are specified. Stainless steels have been allowed in certain instances, even though they are ferrous.

34.7 FAN SELECTION

In the majority of fan applications, it is neither necessary nor desirable to design a completely new fan for the specified job requirements. Standard designs are available in each of the various aerodynamic types. Fan selection is usually a matter of choosing the best size and type from those available.

Selecting a fan begins with the specification of requirements and ends with the evaluation of alternative possibilities. Of the many fans capable of satisfying particular capacity and pressure requirements, the best choice is the one that does the job most economically. First costs, operating costs, and maintenance costs must all be considered.

34.7.1 Specification of Requirements

A fan specification should give the fan supplier all the pertinent information regarding performance, service, evaluation, arrangement, etc., so that the best se-

TABLE 34.3 General Fan Specifications

General:______
Number of fans______
Aerodynamic type______
Size of connecting ductwork______
Service______
Capacity of each fan—Specify maximum and reduced ratings______
ft^3/min (m^3/s) at inlet or lb/h (kg/h)______
Fan pressure at each capacity:______
Inches water gauge (kilopascals)______
State whether static or total______
Indicate distribution between inlet and outlet______
Gas composition and conditions at each capacity:______
Name or gas analysis______
Molecular weight or specific gravity referred to air______
Ambient barometer in inHg (kPa) or elevation in ft (m)______
Temperature at inlet in °F (°C)______
Relative humidity in %______
Dust loading through fan______
Power evaluation factors:______
Expected life in years______
Expected operation at each rating in h/year______
Power rate in $/kWh______
Demand charge in $/kW or $/hp______
Physical data:______
Number of inlets______
Type of drive______
Arrangement number______
Direction of entry for inlet boxes______
Rotation, discharge, and motor position______
Construction details:______
Appurtenances______
Special materials______
Type and mounting of bearings______
Motor data:______
Electrical characteristics______
Type and enclosure______

lection can be made. Most of the important items are listed in Table 34.3. Explanatory notes are given for many of these items in the following paragraphs.

The number of fans and their aerodynamic type are items which should be specified only after the various possibilities have been compared. The type of service for which the fan is intended should be specified, to warn the supplier of any unusual conditions. In some instances, a duct layout should be included with the specifications. In any case, the sizes of the connecting ductwork should be indicated.

The capacity of the fan must be specified by the system designer. Due to the nature of the application, the designer may find it convenient to calculate the capacity as a weight rate of flow. Fan capacity should be expressed as a volume rate of flow and specified in cubic feet per minute (cubic meters per second) at inlet conditions. So to determine the appropriate fan capacity in ft^3/min (m^3/s), divide the weight rate of flow in pounds per minute by the inlet density in pounds per cubic foot:

$$\frac{\text{ft}^3}{\text{min}} = \frac{\text{lb/min flowing}}{\text{lb/ft}^3 \text{ at inlet}} \qquad \left(\frac{\text{m}^3}{\text{s}} = \frac{\text{kg/s}}{\text{kg/m}^3}\right) \tag{34.7}$$

If an adjustable flow rate is contemplated, the minimum capacity and any intermediate capacities at which power requirements are to be evaluated should be specified.

The fan must provide the air or gas with sufficient energy to overcome the losses encountered in passing through the system. The usual method of specifying this energy requirement is to stipulate the static or total pressure in inches water gauge (kilopascals) which the fan must develop at each capacity. In any case, the distribution of pressure between suction and discharge should be specified.

When a fan is located at the downstream end of the system, i.e., exhausting, the static pressure required is equal to the sum of the total-pressure losses in all the system components plus any difference in barometric pressure between the two ends of the system. For the duct system calculation, see Chap. 4. When the fan is located at any position other than the downstream end, i.e., blowing or boosting, an amount equal to the fan outlet velocity pressure should be subtracted from the requirement as stated above. The fan static pressure p_{FS} may thus be expressed in terms of the sum of the total-pressure losses $\Sigma p_{T_{loss}}$, *the difference between the barometric pressure at the system exit* p_{Bx}, and the barometric pressure at the system entrance p_{Be}, and the fan velocity pressure p_{FV} as

$$p_{FS} = \Sigma p_{T_{\text{loss}}} + p_{Bx} - p_{Be} - p_{FV} \tag{34.8}$$

with the proviso that the p_{FV} term is zero when the fan is located at the downstream end of the system.

Fan velocity pressure depends on the fan size, and since this is not usually known in advance, many designers ignore the p_{FV} term regardless of the fan location. This is usually justified as constituting a factor of safety. Better accuracy, when warranted, can be obtained by using p_{FV} for a trial size and correcting to the actual value in the final selection.

The sum of the total-pressure losses through the system should include an allowance for any elements required to connect the fan to the system. This will be a small amount unless the size of the fan opening differs greatly from the size of the connected ductwork.

In the usual case there is no difference in barometric pressure between the entrance and exit of the system. The exceptions occur whenever there is more than one device supplying energy to the air.

The total pressure required of the fan is equal to the sum of the total-pressure losses plus any difference in barometric pressure (as noted above):

$$p_{FT} = \Sigma p_{T_{\text{loss}}} + p_{Bx} - p_{Be} \tag{34.9}$$

The performance of a fan is a function of the density of the air or gas at the fan inlet. The inlet density determines not only the volumetric capacity for a specified weight rate of flow but also the pressure which the fan is able to develop. The factors which affect density (and should therefore be specified) are the barometric pressure, temperature, and relative humidity at the inlet, as well as the name or composition of the gas. The ambient barometer and the gauge pressure at the fan inlet may be specified in lieu of the inlet barometer.

The composition of the gas and information about any entrained material (dust loading, etc.) should be specified so that the fan supplier can offer the best selection based on any previous experience. See Sec. 34.7.2.

Whenever the gas composition and conditions are not specified, the fan supplier usually assumes air at standard conditions. Standard conditions for the fan industry are dry air at 70°F and 29.92 inHg barometer (20°C and 101.3 kPa). The density corresponding to these conditions is 0.075 lbm/ft^3 (1.2 kg/m^3). If the fan requirements are specified in terms of other standards, both the actual and the standard conditions should be given in detail. If the weight rate of flow corresponding to a certain number of standard cubic feet per minute (cubic meters per second) is required, sufficient information to determine both the actual and standard density should be listed.

Power requirements should be evaluated to obtain the best selection. The fan supplier should be advised of the method of evaluation. The usual method is to reduce the expected cost of power during the useful life of the equipment to the present value of an annuity sufficient to yield the annual expenditure. The annual expenditure will depend on the power rate and the expected operating schedule during the life of the equipment as well as the power requirements at the various operating conditions. The size of the hypothetical investment will depend on the expected life and rate of interest which could be obtained. Accordingly, the cost of the expected operation reduced to present value ($\$_{RPV}$) may be determined from the power rate (\$/kWh) and the expected annual power consumption (kWh/year):

$$\$_{RVP} = \frac{\$}{\text{kWh}} \frac{\text{kWh}}{\text{year}} \left[\frac{1 - (1 +)i^{-n}}{i} \right] \tag{34.10}$$

The bracketed term, which is the present value of an annuity to yield \$1 annually for n years if invested at rate of interest i, may be determined from any standard interest table.

Inefficiency may be penalized by charging the fan a flat amount for each horsepower or kilowatt required. In the power generation industry such a demand charge, which reflects the loss of power available for sale at peak load, may be applied in addition to any operating charge based on the cost of fuel.

Operating costs, as determined by Eq. (34.10) or any other method, should be added to the first costs and expected maintenance costs for each possible selection. The best selection is the fan having the lowest total cost. Maintenance costs are not determined as readily as first and operating costs. In many cases, maintenance costs can be assumed equal for alternative selections. All pertinent engineering factors should be examined to justify any such assumptions.

The items listed in Table 34.3 under "Physical data" and "Construction details" should be specified because the user is usually in a better position than the supplier to decide such issues. Certain of these items, such as rotation and discharge, have no influence on cost. Other items, such as arrangement number and appurtenances, may have great influence on first cost but no effect on the size and type of fan which should be selected. Still other items, such as the number of inlets and type of drive, largely determine the size and type of fan which should be selected.

If there is no connecting ductwork on the inlet side of the fan, either a single- or double-inlet fan can be used in many cases. The first cost is generally lowest for the double-inlet fan, particularly when relatively large capacities and low pressures are involved. Double-inlet fans require less head room but more floor space than a single-inlet fan for the same rating. Single-inlet fans are generally favored for high-pressure, low-capacity ratings. The advantages of providing only one inlet connection rather than two are obvious.

Direct-drive specifications limit the fan speeds to available motor speeds, and this in turn limits the number of possible fan selections except where adjustable-blade fans are considered. It is quite unlikely that a standard-size fan will be able to satisfy performance requirements exactly at a direct-connected speed. Accordingly, either requirements must be relaxed or a nonstandard fan must be used. The performance of a standard fan can sometimes be changed sufficiently by modifying the wheel diameter or width. At other times an odd-sized fan may be furnished. When warranted, an entirely new design may be furnished. Direct drives generally require less maintenance and involve less power transmission loss than belt drives.

Adjustable-blade and belt-drive specifications make a large number of fan selections possible, and standard-size fans may be used. Should requirements be altered slightly after installation, usually it is a comparatively easy and inexpensive matter to change the blade setting or belt drive.

The standard designations for arrangement number, direction of entry for inlet boxes, rotation, discharge, and motor position were given earlier. Arrangements with the impeller mounted between bearings are generally less expensive than those with overhung impellers. In the smaller-size fans, arrangements with bearings in the inlet are generally avoided because a closely situated bearing may block an appreciable portion of the inlet. Overhung impeller arrangements are frequently used to protect the bearings whenever the fan must handle hot, dirty, or corrosive gas. The alternative is to use inlet boxes and increase the center distance between bearings accordingly. Overhung pulleys or sheaves are preferred for easy maintenance, but jack shafts may be required on some larger drives.

Various appurtenances may be required, including vibration isolation bases, belt guards, drains, access or inspection doors, flanged connections, inlet screens, stack bracing, evases, stuffing boxes, shaft seals, heat slingers, outlet dampers, and variable-inlet vanes.

Standard fans are usually steel-plate products. However, certain standard lines are housed in cast iron. Under certain conditions special materials or special methods of construction may be justified. When corrosion resistance is required, the materials of construction should be specified by the user, if possible. Several classes of spark-resistant construction are generally available. Abrasion resistance is very difficult to achieve in a fan. Additional thicknesses of material or special materials or both may be specified. Centerline support may be required to maintain alignment when high-temperature gases are handled. Special materials may be required to prevent rapid oxidation at elevated temperatures.

Certain fan lines are normally furnished with antifriction bearings, others with sleeve bearings. Any preference should be specified together with details on lubrication system, cooling media, and type of mounting.

The type of motor, its enclosure, etc., should be specified, particularly if furnished by the user. In any case, the electrical characteristics should be listed.

34.7.2 Selecting the Proper Size and Type of Fan

Theoretically, almost any size fan of any type could be used to satisfy the maximum requirements of a particular job. Practical engineering and economic considerations reduce the possibilities to a relatively narrow range of sizes and to a few types.

Certain types or designs of fans are designated according to their usual field of application. There are ventilating fans, mechanical draft fans, industrial exhausters, and pressure blowers, with subclassifications in each case.

Ventilating fans are designed for clean-air service at normal temperatures. Some heavy-duty ventilating fans may be used for more severe conditions. Both centrifugal and axial designs are available. Centrifugal types may have either backward- or forward-curved blades. Maximum efficiencies are obtained with the former, particularly when airfoil-shaped blades are used. Forward-curved blade types are used when space and price are more important than efficiency. Belt drives are normally used so that any rating can be obtained with a standard-size fan. Propeller fans are usually designed for free delivery operation, but may be used up to 1 in (0.25 kPa) WG static pressure in some cases.

After the type or types of fans suitable for an application are determined, it is necessary to choose the best size fan in each type. There is only one size fan in each type that will operate at the point of maximum efficiency for any given rating. This optimum-size fan must be operated at a certain speed to produce the required rating. A smaller-size fan could be selected that would have to operate at higher speed, or a larger-size fan could be selected that would have to operate at a lower speed. In either case efficiency would be lower than that for the optimum size.

Fans which rate to the right of peak efficiency may be called *undersized fans*, and those which rate to the left of peak efficiency may be called *oversized fans*. To the left means lower capacity, and to the right means higher capacity on the base curve. Oversized fans are hard to justify unless future increases in capacity are envisioned. Occasionally the required operating speed of an oversized fan will match a motor speed. Slightly undersized fans are usually chosen because optimum sizes are rarely standard sizes. Ratings slightly to the right of peak efficiency usually have more stable operating characteristics, i.e., have steeper slopes than those at or to the left of peak efficiency. Sometimes a fan which is considerably undersized is the best choice. In such cases, the additional operating costs occasioned by the lower efficiency must be offset by the savings in first cost or some other engineering or economic factor.

34.7.3 Rating Fans from Published Data

Rating data are generally published in the form of tables or charts for each size fan of a given type. The average user finds such presentations convenient. Two typical methods of presenting rating data and examples of their use are illustrated below.

Multirating tables are probably the most common type of published data. Portions of three pages from a typical multirating table are illustrated in Fig. 34.29. Such tabulations are almost always based on standard air. To use such a table, enter with the required CFM and ESP and read the RPM and BHP on the appropriate line in the appropriate column. If the requirements do not match the listed values of CFM or ESP exactly, linear interpolations will give accurate results. The table value of RPM is the required operating speed. The table value of BHP must be multiplied by the ratio of actual density to standard density, to obtain the required operating horsepower. Example 34.1 illustrates the use of such tables.

SIZE 805 SINGLE INLET **Buffalo Limit-Load Type "BL" Fans** SINGLE INLET SIZE 805

Wheel Diameter 40¼". Limit Load H.P. $= 26.8 \times \left(\frac{\text{R.P.M.}}{1000}\right)^3$ Outlet Area 9.32 sq. ft. inside.

29,824 / 30,756 — 3200 / 3300

1076 / 1095 — 33.3 / 34.9 — 1102 / 1120 — 35.8 / 37.3

CFM	Outlet Velocity	2" S.P.		2¼" S.P.		2½" S.P.		3" S.P.		3½" S.P.		4" S.P.		4½" S.P.		5" S.P.		5½" S.P.		6" S.P.	
		RPM	BHP	RPM	BHP	RPM	BHP	RPM	BHP	RPM	BHP	RPM	BHP	RPM	BHP	RPM	BHP	RPM	BHP	RPM	BHP
13,048	1400	583	5.16	608	5.77	632	6.40	675	7.71	720	9.11	765	10.6	808	12.0	846	13.5	885	15.1	920	16.7
13,980	1500	596	5.61	622	6.23	643	6.89	687	8.22	728	9.65	772	11.2	812	12.7	851	14.2	890	15.9	926	17.5
14,912	1600	611	6.08	636	6.74	657	7.41	700	8.78	738	10.2	779	11.8	818	13.4	857	15.0	896	16.7	932	18.4
15,844	1700	627	6.59	649	7.26	671	7.97	713	9.38	750	10.9	787	12.4	828	14.1	864	15.7	901	17.5	938	19.2
16,776	1800	643	7.12	664	7.85	686	8.59	726	10.0	764	11.6	800	13.1	839	14.8	872	16.5	907	18.4	943	20.2
17,708	1900	659	7.66	680	8.46	701	9.22	740	10.7	777	12.3	[illegible]	[illegible]	[illegible]	[illegible]	[illegible]	[illegible]	[illegible]	[illegible]	[illegible]	21.0
18,640	2000	676	8.25	696	9.06	716	9.89	754	11.5	791	13.1	[illegible]	[illegible]	[illegible]	[illegible]	[illegible]	[illegible]	[illegible]	[illegible]	[illegible]	22.0
19,572	2100	695	8.91	713	9.73	733	10.6	768	12.3	805	14.0	[illegible]	[illegible]	[illegible]	[illegible]	[illegible]	[illegible]	[illegible]	[illegible]	[illegible]	23.0
20,504	2200	713	9.56	731	10.4	750	11.3	785	13.1	821	14.8	[illegible]	[illegible]	[illegible]	[illegible]	[illegible]	[illegible]	[illegible]	[illegible]	[illegible]	24.1
21,436	2300	[illegible]	[illegible]	[illegible]	[illegible]	[illegible]	12.1	803	13.9	837	15.7	[illegible]	[illegible]	[illegible]	[illegible]	[illegible]	[illegible]	[illegible]	23.2	993	25.3
22,368	[illegible]	[illegible]	[illegible]	[illegible]	[illegible]	[illegible]	12.9	820	14.7	854	16.7	886	18.6	916	20.6	[illegible]	22.5	977	24.5	1006	26.5
23,300	[illegible]	[illegible]	[illegible]	[illegible]	[illegible]	[illegible]	13.7	838	15.7	871	17.6	902	19.7	930	21.8	[illegible]	23.7	990	25.8	1019	27.9
24,232	[illegible]	[illegible]	[illegible]	[illegible]	[illegible]	[illegible]	14.7	855	16.7	887	18.6	919	20.8	947	22.9	[illegible]	24.9	1004	27.0	1033	29.2
25,164	[illegible]	[illegible]	13.6	823	14.6	841	15.7	873	17.6	904	19.7	936	21.9	963	24.1	[illegible]	26.3	1019	28.3	1046	30.5
26,096	2800	[illegible]	14.6	842	15.6	859	16.7	892	18.7	922	20.9	953	23.1	980	25.3	1008	27.6	1036	29.8	1061	32.0
27,028	[illegible]	844	15.6	861	16.7	877	17.8	910	20.0	940	22.0	971	24.2	997	26.5	1025	[illegible]	1052	31.3	1077	33.6
27,960	[illegible]	864	16.7	880	17.8	896	18.9	929	21.2	959	23.2	988	25.5	1014	28.8	1042	[illegible]	1068	32.7	1094	35.2
28,892	3100	884	17.6	900	18.9	916	20.1	948	22.4	978	24.7	1007	27.0	1032	30.1	1059	[illegible]	1085	34.3	1110	36.7
[illegible]	[illegible]	903	18.9	919	20.0	936	21.3	966	23.6	997	26.0	1025	28.3	1050	31.3	1076	33.3	1102	35.8	1127	38.3
30,756	3300	924	20.1	939	21.3	955	22.5	985	24.9	1015	27.5	1042	29.9	1068	32.6	1095	34.9	1120	37.3	1143	40.0
31,688	3400	946	21.3	959	22.6	975	23.8	1004	26.4	1033	29.0	1061	31.4	[illegible]	[illegible]	[illegible]	[illegible]	[illegible]	[illegible]	[illegible]	[illegible]
32,620	3500	967	22.5	980	24.0	[illegible]	[illegible]	[illegible]	[illegible]	[illegible]	[illegible]	[illegible]	[illegible]	[illegible]	[illegible]	[illegible]	[illegible]	[illegible]	[illegible]	[illegible]	[illegible]

SIZE 890 SINGLE INLET **Buffalo Limit-Load Type "BL" Fans** SINGLE INLET SIZE 890

Wheel Diameter 44½". Limit Load H.P. $= 44.2 \times \left(\frac{\text{R.P.M.}}{1000}\right)^3$ Outlet Area 11.38 sq. ft. inside.

29,588 / 30,726 — 2600 / 2700

883 / 897 — 30.5 / 32.1 — 908 / 922 — 33.0 / 34.6

CFM	Outlet Velocity	2" S.P.		2¼" S.P.		2½" S.P.		3" S.P.		3½" S.P.		4" S.P.		4½" S.P.		5" S.P.		5½" S.P.		6" S.P.	
		RPM	BHP	RPM	BHP	RPM	BHP	RPM	BHP	RPM	BHP	RPM	BHP	RPM	BHP	RPM	BHP	RPM	BHP	RPM	BHP
15,932	1400	527	6.30	550	7.04	572	7.82	610	9.42	651	1.11	692	12.9	731	14.7	765	16.5	801	18.4	833	20.4
17,070	1500	539	6.85	563	7.61	582	8.41	622	10.0	659	[illegible]	[illegible]	[illegible]	[illegible]	[illegible]	[illegible]	[illegible]	[illegible]	[illegible]	837	21.4
18,206	1600	553	7.43	575	8.23	594	9.05	633	10.7	668	[illegible]	[illegible]	[illegible]	[illegible]	[illegible]	[illegible]	[illegible]	[illegible]	[illegible]	843	22.4
19,346	[illegible]	[illegible]	[illegible]	[illegible]	[illegible]	[illegible]	[illegible]	645	11.5	678	[illegible]	[illegible]	[illegible]	[illegible]	[illegible]	[illegible]	[illegible]	[illegible]	[illegible]	848	23.5
20,484	[illegible]	[illegible]	[illegible]	[illegible]	[illegible]	[illegible]	[illegible]	657	12.3	691	[illegible]	[illegible]	[illegible]	[illegible]	[illegible]	[illegible]	[illegible]	[illegible]	[illegible]	853	24.7
21,622	[illegible]	[illegible]	[illegible]	[illegible]	[illegible]	[illegible]	[illegible]	669	13.1	703	15.0	736	17.1	769	19.2	797	21.2	828	23.5	859	25.7
22,760	[illegible]	[illegible]	[illegible]	[illegible]	[illegible]	[illegible]	[illegible]	682	14.0	715	16.0	749	18.0	778	20.2	[illegible]	22.3	835	24.5	867	26.9
23,898	2100	[illegible]	[illegible]	[illegible]	[illegible]	[illegible]	[illegible]	695	15.0	728	17.1	762	19.2	790	21.2	[illegible]	23.5	847	25.7	877	28.1
25,036	2200	645	11.7	661	12.7	678	13.8	710	16.0	742	18.1	774	20.2	802	22.4	[illegible]	24.8	860	27.0	887	29.4
26,174	2300	661	12.6	[illegible]	13.7	694	14.7	726	16.9	757	19.2	787	21.4	815	23.8	844	26.2	872	28.4	898	30.9
27,312	2400	677	[illegible]	696	14.6	711	15.8	741	18.0	773	20.4	801	22.7	828	25.1	857	[illegible]	883	29.9	910	32.4
28,450	2500	694	14.5	711	15.6	728	16.8	758	19.2	787	21.5	816	24.1	842	26.6	870	[illegible]	896	31.5	923	34.0
[illegible]	[illegible]	[illegible]	15.6	728	16.8	744	18.0	773	20.4	802	22.7	831	25.4	856	27.9	883	30.5	908	33.0	934	35.7
30,726	2700	728	16.6	745	17.8	760	19.2	790	21.5	818	24.1	846	26.7	871	29.4	897	32.1	922	34.6	947	37.3
[illegible]	[illegible]	746	17.8	761	19.0	777	20.4	807	22.9	834	25.6	860	28.2	887	30.9	912	33.7	937	36.4	960	39.1

SIZE 980 SINGLE INLET **Buffalo Limit-Load Type "BL" Fans** SINGLE INLET SIZE 980

Wheel Diameter 49". Limit Load H.P. $= 71.1 \times \left(\frac{\text{R.P.M.}}{1000}\right)^3$ Outlet Area 13.81 sq. ft. inside.

29,001 / 30,382 — 2100 / 2200

745 / 755 — 28.5 / 30.1 — 769 / 781 — 31.2 / 32.8

CFM	Outlet Velocity	2" S.P.		2¼" S.P.		2½" S.P.		3" S.P.		3½" S.P.		4" S.P.		4½" S.P.		5" S.P.		5½" S.P.		6" S.P.	
		RPM	BHP	RPM	BHP	RPM	BHP	RPM	BHP	RPM	BHP	RPM	BHP	RPM	BHP	RPM	BHP	RPM	BHP	RPM	BHP
19,334	1400	[illegible]	[illegible]	[illegible]	[illegible]	[illegible]	[illegible]	[illegible]	11.4	591	13.5	629	15.6	664	17.8	696	20.0	727	22.3	756	24.7
20,715	1500	[illegible]	[illegible]	[illegible]	[illegible]	[illegible]	[illegible]	[illegible]	12.2	598	14.3	634	16.5	667	18.7	699	21.1	731	23.6	761	26.0
22,096	1600	[illegible]	[illegible]	[illegible]	[illegible]	[illegible]	[illegible]	[illegible]	13.0	606	15.2	640	17.4	672	19.8	704	22.2	736	24.7	766	27.2
23,477	1700	[illegible]	[illegible]	[illegible]	[illegible]	[illegible]	[illegible]	[illegible]	13.9	616	16.1	647	18.4	681	20.9	710	23.2	740	26.0	770	28.5
24,858	1800	[illegible]	[illegible]	[illegible]	[illegible]	[illegible]	[illegible]	[illegible]	14.9	627	17.1	657	19.5	689	22.0	717	24.5	745	27.2	775	29.9
26,239	1900	542	11.4	559	[illegible]	576	13.7	608	15.9	638	18.2	668	20.7	698	23.2	724	25.8	752	28.5	780	31.2
27,620	2000	556	12.2	[illegible]	13.4	588	14.7	619	17.0	650	19.5	680	21.8	707	24.5	734	27.0	758	29.7	787	32.6
29,001	2100	571	[illegible]	585	14.4	602	15.6	631	18.2	661	20.7	692	23.2	717	25.8	745	28.5	769	31.2	796	34.1
30,382	2200	[illegible]	14.2	600	15.5	616	16.7	645	19.5	674	22.0	703	24.5	729	[illegible]	755	30.1	781	32.8	806	35.7
31,763	2300	600	15.2	617	16.6	630	17.9	659	20.5	688	23.2	714	26.0	740	[illegible]	767	31.7	792	34.4	816	37.5
33,144	2400	615	16.4	632	17.7	646	19.1	673	21.8	702	24.7	728	[illegible]	[illegible]	30.5	778	33.3	802	36.2	826	39.3
34,525	2500	630	17.6	[illegible]	[illegible]	[illegible]	[illegible]	[illegible]	[illegible]	[illegible]	[illegible]	[illegible]	[illegible]	764	32.3	790	35.1	813	38.2	837	41.3
35,906	2600	646	18.9	661	[illegible]	[illegible]	[illegible]	[illegible]	[illegible]	[illegible]	[illegible]	[illegible]	30.8	778	33.9	802	36.9	825	40.0	848	43.3
37,287	2700	661	20.2	[illegible]	[illegible]	[illegible]	[illegible]	[illegible]	[illegible]	[illegible]	[illegible]	[illegible]	32.4	791	35.7	815	38.9	837	42.0	860	45.2
38,668	2800	677	21.6	691	[illegible]	[illegible]	[illegible]	[illegible]	[illegible]	[illegible]	[illegible]	[illegible]	34.2	806	37.5	828	40.9	851	44.2	872	47.4
40,049	2900	694	23.1	707	24.7	720	26.3	748	29.6	772	32.6	[illegible]	35.9	819	39.3	842	42.9	864	46.3	885	49.7
41,430	3000	710	24.7	723	26.3	736	27.9	764	31.4	788	34.4	812	37.8	833	42.7	856	44.9	877	48.5	898	52.1
42,811	3100	726	26.1	739	27.9	752	29.7	778	33.2	803	36.6	827	40.0	848	44.5	870	47.0	891	50.8	912	54.4
44,192	3200	742	27.9	755	29.6	769	31.5	793	35.0	819	38.6	842	42.0	863	46.3	884	49.4	906	53.0	926	56.8
45,573	3300	759	29.7	772	31.5	784	33.3	809	36.9	834	40.7	857	44.3	877	48.3	899	51.7	920	55.3	939	59.3

Ratings are at 70° F. and 29.92" barometer.

FIGURE 34.29 Typical rating table for a centrifugal ventilating fan.

EXAMPLE 34.1: MULTIRATING TABLE SELECTION

Given: Multirating table for a specific design according to Fig. 34.29.

Required: Pick a fan to deliver 30,000 ft^3/min of air against 5 inWG static pressure. Conditions at the fan inlet are 100°F dry bulb, 75°F wet bulb, and 29.5 inHg barometer.

Solution Calculate the inlet density and equivalent static pressure (note that the tables are drawn up for standard air):

$$\rho_a = 0.0693 \text{ lbm/ft}^3 \qquad \text{from Fig. 34.1}$$

$$\text{ESP} = 5\left(\frac{0.075}{0.0693}\right) = 5.41 \text{ inWG} \qquad \text{from Eq. (34.6)}$$

To select a trial size and determine rating, examine the table for size 805. Note that interpolation is required between 5 and 5½ inWG and between 29,824 and 30,756 ft^3/min.

$$1076 + \frac{5.41 - 5.00}{5.50 - 5.00}(1102 - 1076) = 1097 \text{ r/min}$$

$$1095 + \frac{5.41 - 5.00}{5.50 - 5.00}(1120 - 1095) = 1115 \text{ r/min}$$

$$1097 + \frac{30{,}000 - 29{,}824}{30{,}756 - 29{,}824}(1115 - 1097) = 1100 \text{ r/min}$$

$$33.3 + \frac{5.41 - 5.00}{5.50 - 5.00}(35.8 - 33.3) = 35.4 \text{ hp}$$

$$34.9 + \frac{5.41 - 5.00}{5.50 - 5.00}(37.3 - 34.9) = 36.9 \text{ hp}$$

$$35.4 + \frac{30{,}000 - 29{,}824}{30{,}756 - 29{,}824}(36.9 - 35.4) = 35.7 \text{ hp}$$

The required operating power is less than the table value since the operating density is less than the standard density for which the table was prepared.

$$35.7\left(\frac{0.0693}{0.075}\right) = 33.0 \text{ hp}$$

Performing similar interpolations for the other two sizes shown in Fig. 34.29 and tabulating will yield the following:

Size 805	Size 890	Size 980
1100 r/min	909 r/min	773 r/min
35.7 hp at 0.075	33.2 hp at 0.075	31.9 hp at 0.075
33.0 hp at 0.0693	30.7 hp at 0.0693	29.5 hp at 0.0693

Note that the larger fan must run slower and the smaller fan faster than the original trial size. The power requirements are all fairly close to 30 hp (22 kW). Even allowing for a 3 percent power loss in the belt drive, the size 805 fan could be driven by a 30-hp (20 kW) motor without exceeding the normal service factor (1.15) of an open motor (1.03 × 33.0 = 34.0 and 1.15 × 30.0 = 34.5). The cost of the extra power and the reduction in motor life should be weighed against the savings in first cost for the smaller fan.

Assuming that the fan is expected to operate 7500 h/year for 20 years and that power costs will average 0.04 $/kWh while interest rates average 15 percent, the present value of the difference in cost for power to drive the 33.0- and 30.7-hp fans can be determined. Assuming equal motor and drive and efficiencies, the power difference is 0.746(33.0 − 30.7), or 1.72 kW. The cost of the extra power for 1 year is 0.04(1.72)(7500), or $516. The present value of an annuity which would yield this amount for 20 years if invested at 15 percent may be determined from interest tables or Eq. (34.10) as $516(6.26), or $3229.82. This will more than pay for the difference in first cost of the fans without even considering the reduction in motor life.

There are numerous other methods of presenting rating data in use. Figure 34.30 illustrates a method of showing the performance of an adjustable-blade fan. As usual, the sample zone curves are drawn for standard air. To use such a chart, enter with the required CFM and ETP and read blade position, efficiency, and

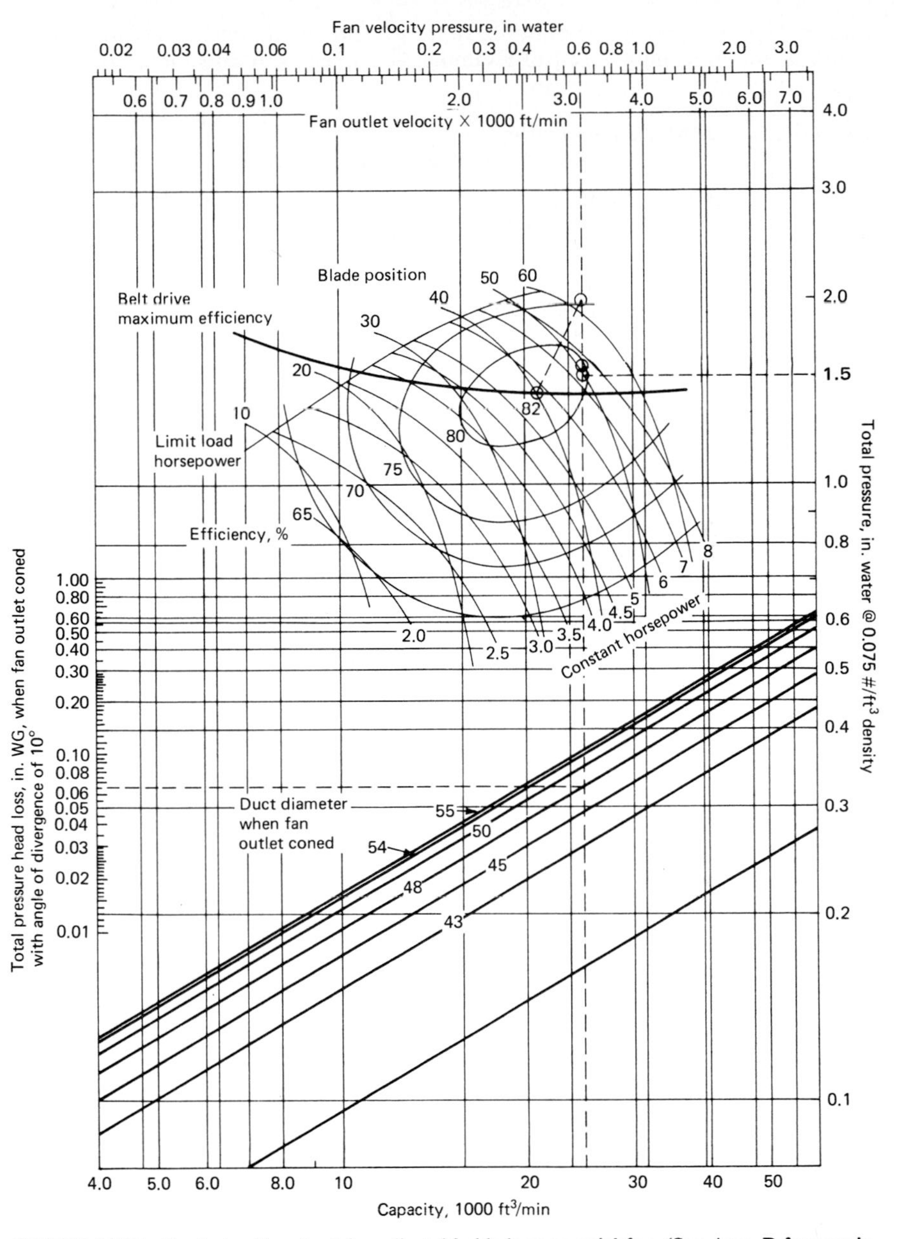

FIGURE 34.30 Typical rating chart for adjustable-blade vane axial fan. (See App. B for metric conversion factors.)

horsepower by interpolating between appropriate lines. If nonstandard air is to be handled, the procedures used in Example 34.1 must be used to determine ETP and BHP.

EXAMPLE 34.2: ZONE CURVE RATING

GIVEN: Zone curves for a specific design according to Fig. 34-30.

REQUIRED: Determine operating characteristics of a 38-A fan to deliver 25,000 ft^3/min of air against 1.5-inWG total pressure for standard air.

Solution Enter the chart at bottom of 25,000 ft^3/min. Note outlet velocity pressure of 3200 ft/min at top. Velocity pressure is 0.62 inWG at top. Enter chart at right at 1.5 inWG total pressure. Note intersection with capacity line. Blade position is 50. Note horsepower by interpolation is 7.3 hp. Note total efficiency is 80.5 percent.

34.8 REFERENCES

1. *Fans and Systems*, AMCA publication 201, Air Movement and Control Association, Arlington Heights, IL, 1982.
2. R. D. Madison and J. B. Graham, "Fan Noise Variation with Changing Fan Operation," *Transactions of the ASHRAE*, vol. 64, 1958, pp. 319–340.
3. H. H. Uhlig, *Corrosion Handbook*, John Wiley & Sons, Inc., New York, 1948, pp. 747–799.
4. Standard 99-0401-66, Air Movement and Control Association, Arlington Heights, IL, 1976.

CHAPTER 35
COILS

Ravi K. Malhotra, Ph.D., P.E.
President, RKM Associates, Inc.,
St. Louis, Missouri

35.1 COILS

Coils are used for cooling and heating an air stream under natural or forced convection. This chapter will cover design and application under forced convection. Coils are used as components in room air conditioners, central station air-handling units, and various factory-assembled air heating, cooling, and refrigeration units. The applications of each type of coil are limited to the design, code regulation, and materials of construction.

35.2 COIL CONSTRUCTION AND ARRANGEMENT

Basically there are two types of coils, one consisting of bare tubes and the other having extended fin surfaces. Bare-tube coils generally are made with copper, steel, stainless-steel, and aluminum tube material. Material selection depends on the type of application. Bare-tube coils are used mainly for evaporative cooling and sprayed coil dehumidifiers. The design and tube arrangement of bare-tube coils vary with the application and manufacturing capabilities. This chapter deals with extended fin surfaces only.

Several designs and arrangements of extended fin surface coils are available. The construction of extended fin surface coils involves the following consideration:

1. *Tube diameter:* Generally ¼- to 1-inch (6- to 25-mm) diameter tube coils are used in most air cooling and heating applications.
2. *Tube arrangement:* Coils are made with staggered or in-line tube arrangement. Staggered-tube coils have a higher airside heat-transfer coefficient compared to in-line tube coils. However, staggered-tube coils also have higher air pressure drop.

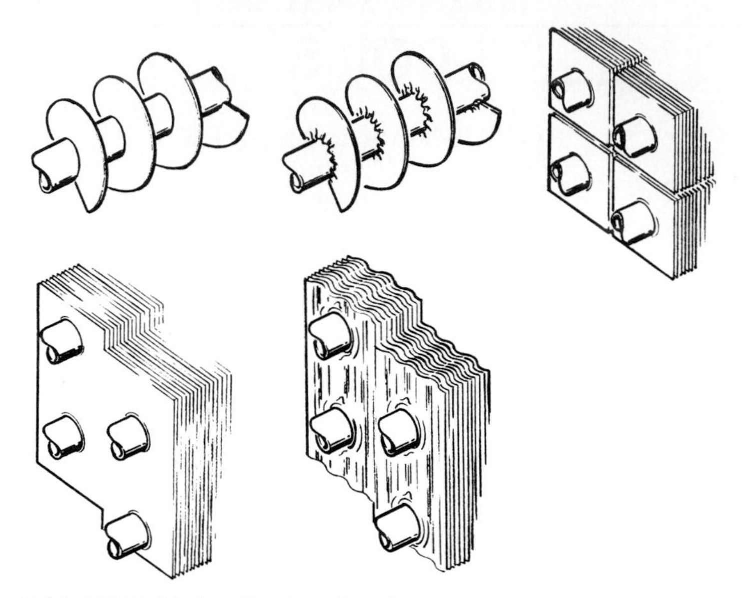

FIGURE 35.1 Various fin-tube coil arrangements.

3. *Fin type:* Fins are generally spiral or continuous plate made out of aluminum, copper, steel, copper-nickel, or stainless steel. Continuous-plate fins are more commonly used in air-conditioning applications. Fins are flat, corrugated, or louvered. Corrugated or louvered fins have a higher airside heat-transfer coefficient and higher air pressure drop. Fins are spaced from 3 to 20 per inch depending on the application with given load and air pressure drop. Figure 35.1 shows some fin-coil arrangements. Spiral fins are wound on the tubes under pressure and then are solder-coated for good fin bond. In plate-fin coils, tubes are mechanically expanded after fins are assembled. Fins have collars to grip the tube for good fin bond. The collar height can be adjusted to space the fins automatically. The inside surface of the tube is usually smooth. To enhance heat transfer, internal fins or turbulators are used; these are either fabricated or extruded.

35.3 COIL TYPES

35.3.1 Water Coils

Water coils are used for air cooling or heating applications. The performance of water coils depends on the air and water velocity, entering air and water temperatures, and mass flow rates. The water velocity inside the tubes usually ranges from 1 to 8 ft/s (0.3 to 2.44 m/s) depending on the design water pressure drop; the optimum water velocity is 3 to 5 ft/s (0.9 to 1.52 m/s). Water coils are provided with various water circuit arrangements. For example, a typical eight-tube-high coil can be arranged for 2, 4, 8, and 16 circuits. Figure 35.2 shows typical water

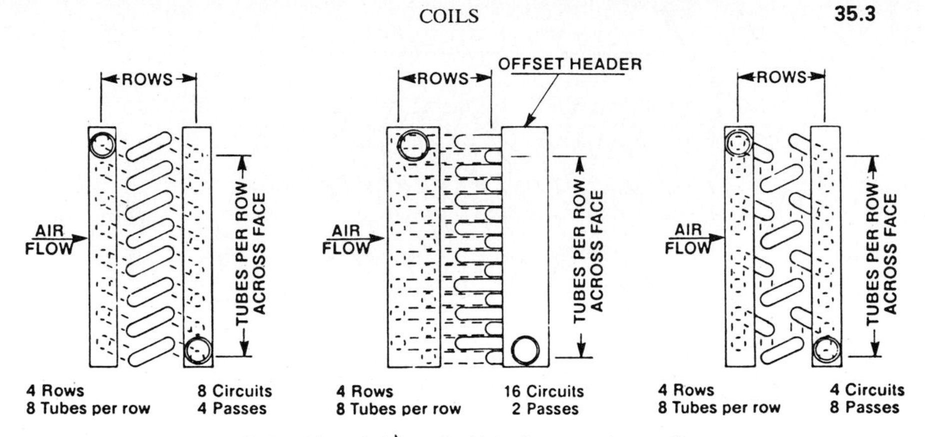

FIGURE 35.2 Water coil circuiting of eight-tube-high, four-row-deep coil.

coil circuiting of an eight-tube-high and four-row-deep coil. Coils are provided with a vent and a drain for proper functioning of the coil. Unless vented, air may be trapped in the coil, possibly causing a reduction in capacity and noise in the system. For job conditions where the waterside may be fouled, a fouling factor is used in designing the coils.

Several types of water coils are used in the field, depending on the application. Following is a description of these coils.

Standard Water Coils. Figure 35.3 shows standard water coils with brazed copper headers and return bends. Full-tube-size return bends are used to minimize fluid friction. Headers are correctly sized for each application to provide uniform distribution to all coil circuits. Each header has vent or drain plugs. These coils are constructed of various materials to suit the application. The coils are circuited for draining. However, to ensure complete drainability, coils have tubes pitched in the frame and auxiliary headers are provided for rapid drainage. Figure 35.4 shows coils with auxiliary headers for complete and rapid drainage.

Internally Cleanable Coils. Figure 35.5 shows cleanable plug coils. Coils have plugs on the supply or return connection end of each tube to permit cleaning of the tube interiors. Full-tube-size return bends are used on one end only. Each tube has a special full-size fitting with an easily removable plug with neoprene O-ring to permit insertion of cleaning tools into the tubes. Plugs may be provided on both sides of the coils for cleaning. Headers are sized for each application to provide uniform distribution to all coil circuits. Each header has vent or drain plugs. In general, coils are made of copper tubes; headers and return bends use brass plugs. Special-construction coils are also available for higher pressures and temperatures. Figure 35.6 shows a coil with a removable box-type steel header on the supply connection end. This construction offers access to the tube interiors when periodic mechanical cleaning is necessary to remove sediment deposits or scale. Tubes are rolled into a heavy steel tube plate. The box header is baffled for correct circuiting and is gasketed and bolted to the tube sheet. Each connection contains a drain or vent. The opposite end of the coil has full-size return bends so that access to the tube interior exists at only one end. Coils may have removable water box-type steel

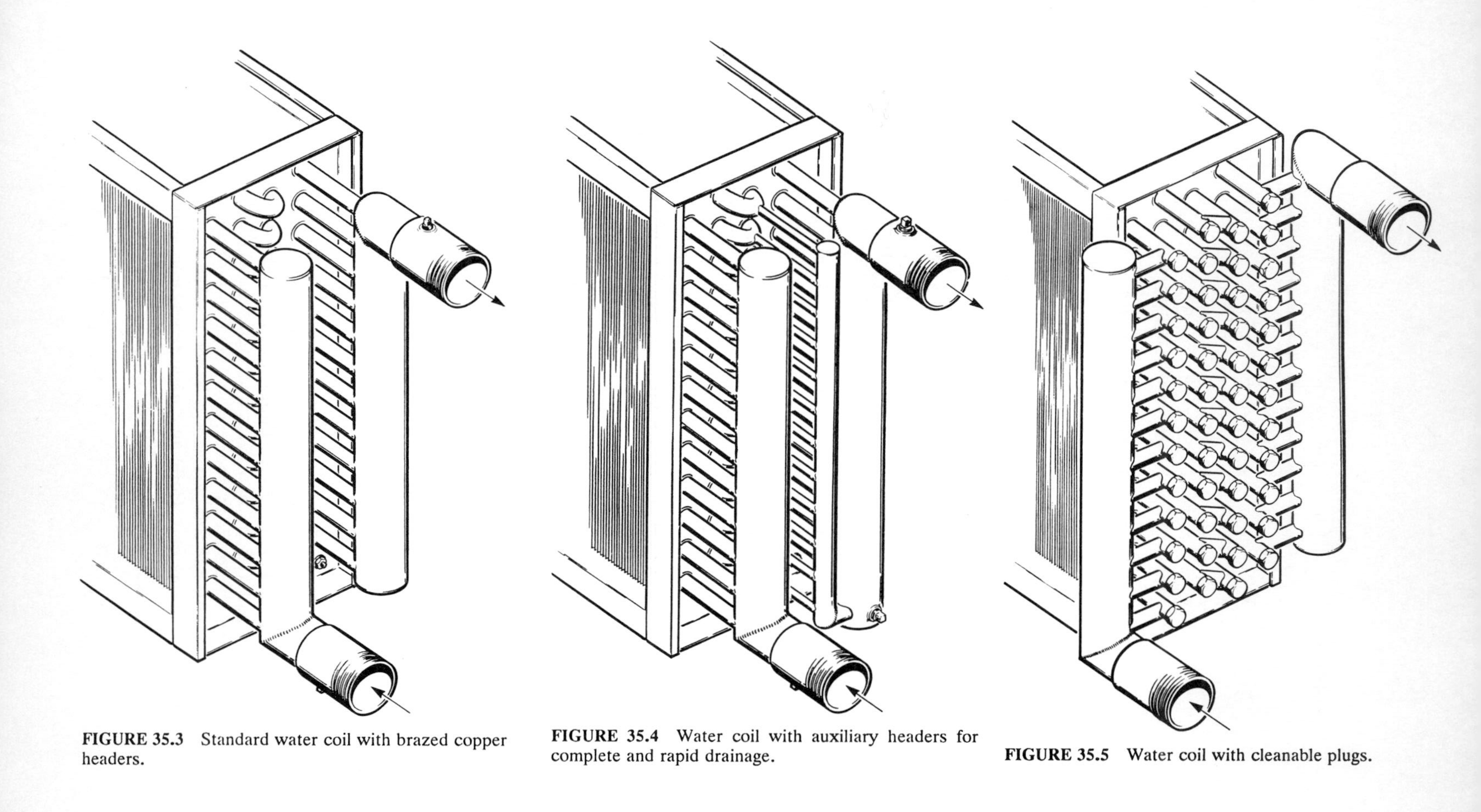

FIGURE 35.3 Standard water coil with brazed copper headers.

FIGURE 35.4 Water coil with auxiliary headers for complete and rapid drainage.

FIGURE 35.5 Water coil with cleanable plugs.

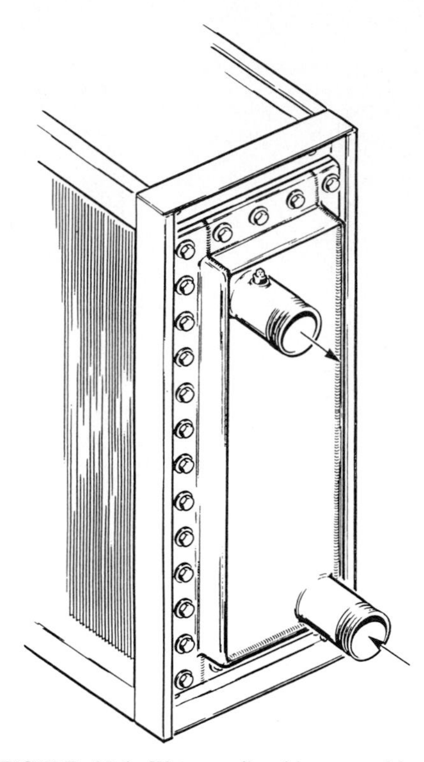

FIGURE 35.6 Water coil with removable box-type steel header.

headers on both ends of the coil. This type of coil provides straight-through mechanical cleaning of the tube interiors. Depending upon the application, these coils may be constructed of other materials.

35.3.2 Direct Expansion Coils

Refrigerant coils are more complex than water coils. Refrigerant distribution and loading are critical to coil performance. These coils are used for recirculation, flooded, and adiabatic expansion systems. Flooded and circulation coils are used mainly for low-temperature applications. Expansion coils are used in air-conditioning applications. Direct expansion coils have either a capillary tube or a thermostatic expansion valve to regulate the flow. The capillary tube system is used in small packaged units, such as room air conditioners. The bore and length of capillary tube are sized for full-load design conditions to evaporate liquid refrigerant. Capillary tube systems do not operate efficiently at other than design conditions. Thermostatic expansion valves are used for larger systems. A typical coil with a thermostatic valve assembly is shown in Fig. 35.7. The thermostatic expansion valve regulates the flow of refrigerant to the coil in direct proportion to the load. It also maintains the superheat at the coil suction outlet within the predetermined limits. The superheat is generally set at the range of 5 to 10°F (2.8 to 5.6°C).

Direct expansion coils are circuited to provide good heat transfer and oil return at reasonable pressure drop across the circuit. Circuit loading will depend on the tube diameter, circuit length, type of refrigerant, and operating conditions. For multicircuit coils, a distributor is provided to evenly distribute the refrigerant in each circuit. Table 35.1 lists the recommended circuit loading for various tube-size coils.

Some coils may have multiple thermostatic expansion valves. If the valves are located on the face of the coil, it is called "face control." If the valves are located in the (rows) depth of the coil, it is called "row control." Figure 35.8 shows a coil with both face and row control arrangements. The face of the coil is controlled by two thermostatic expansion valves. Face control is used in coils to provide equal loading on all refrigerant circuits within the coil. Row control is usually designed for special coils, and loading can be divided into several ways. Table 35.2 lists the progressive loading per row.

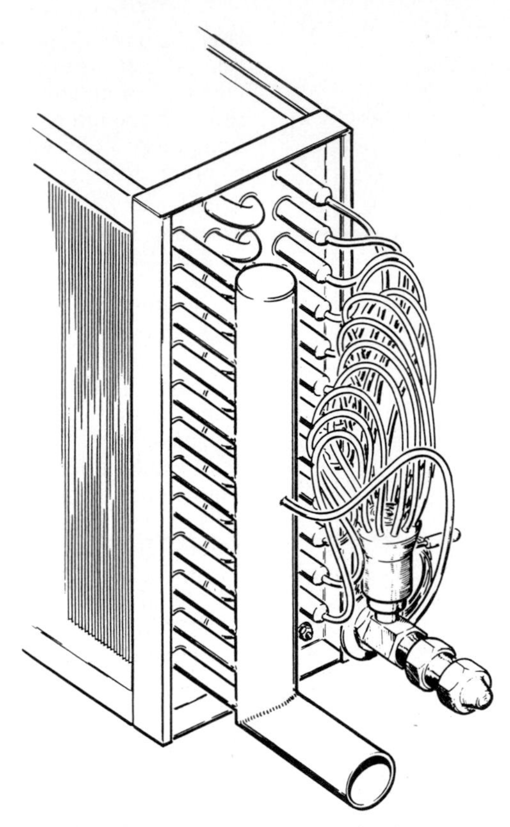

FIGURE 35.7 Direct expansion coil with thermostatic expansion valve.

TABLE 35.1 Recommended Circuit Loading, Btu/h*

Application	Refrigerant	Tube outer diameter		
		⅜ in (9.5 mm)	½ in (12.7 mm)	⅝ in (15.9 mm)
Comfort air conditioning	12	6000	12,000	18,000
	22	9000	18,000	24,000
	502	8000	16,000	22,000
Commercial refrigeration	12	2000	4,000	6,000
	22	3000	6,000	9,000
	502	2800	5,400	7,800

*For metric conversion factor, see App. B.

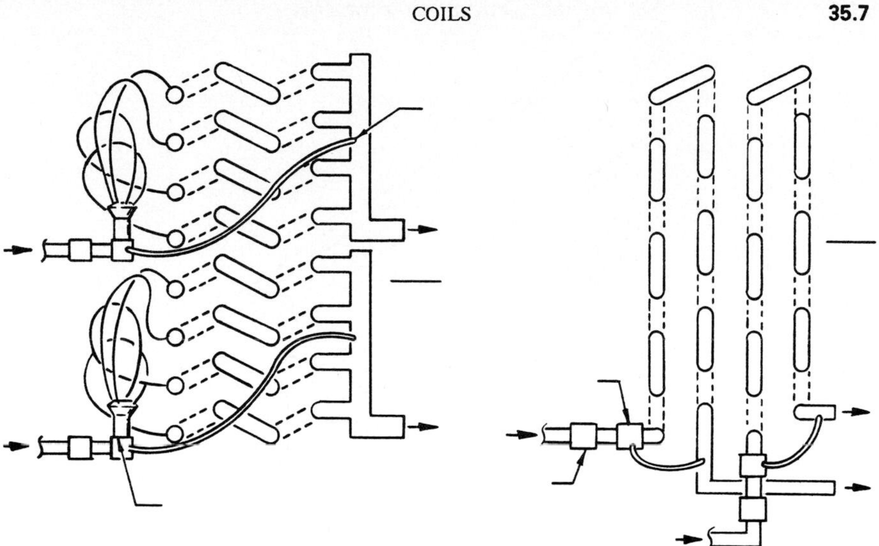

FIGURE 35.8 DX coil with face and row control arrangement.

TABLE 35.2

Rows deep	Approximate percentage of total capacity by rows							
	1	2	3	4	5	6	7	8
2	55	100						
3	40	73	100					
4	33	60	82	100				
5	29	52	71	87	100			
6	26	47	60	78	90	100		
7	24	43	58	70	81	91	100	
8	21	39	53	65	76	85	93	100

35.3.3 Steam Coils

Standard Steam Coils. Coils are made out of single tubes. Steam enters through one end of the coil, and the condensate comes out the other end. Coils are provided with tubes pitched in the casing for condensate drainage or can be mounted with the tubes vertical. Generally these coils are used with entering air temperatures above freezing. However, vertical tube coils can be used with entering air below freezing. These coils are constructed of copper, steel, and stainless-steel tubes with aluminum, copper, steel, and stainless-steel fins. Figure 35.9 shows a standard steam coil.

Distributing Steam Coils. Distributing steam coils are constructed with orificed inner steam distributing tubes, centered and supported to provide uniform steam distribution and maximum protection against freeze-ups. This design prevents freezing of the condensate, provided a sufficient amount of steam is supplied to the coil and the condensate is removed by proper trapping as fast as it is con-

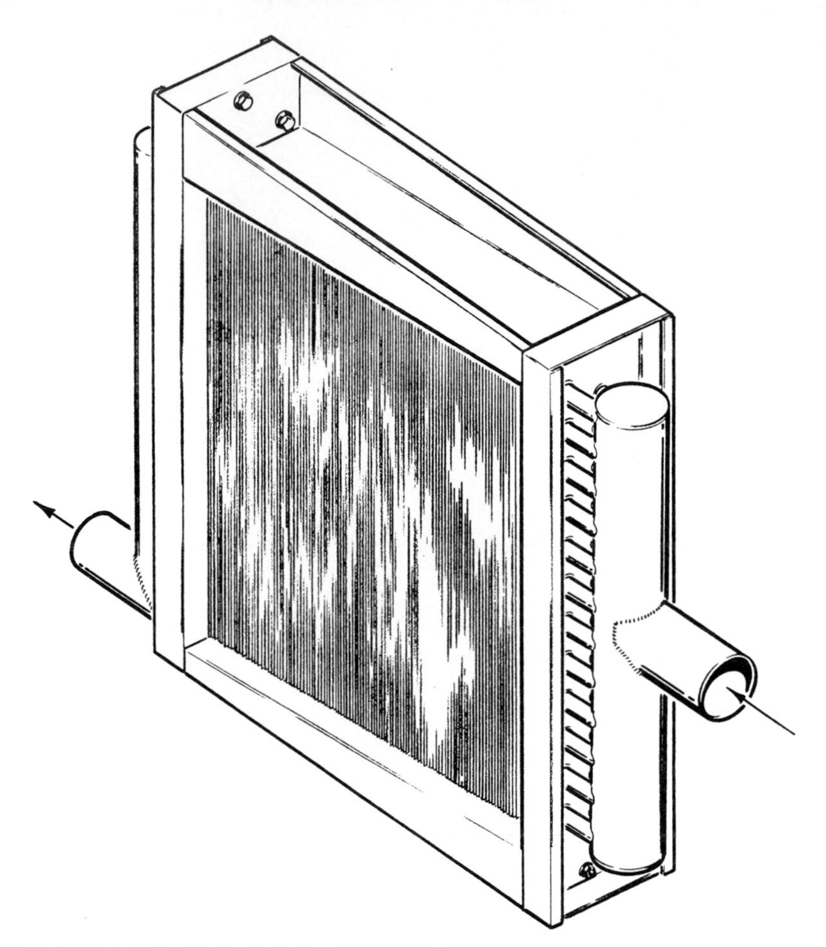

FIGURE 35.9 Standard single-tube steam coil.

densed. Coils are provided with tubes pitched in the casing. Figure 35.10 shows distributing steam coils. Coils are constructed of copper, steel, and stainless-steel tubes with aluminum, copper, steel, and stainless-steel fins.

When uniform leaving air temperatures over the entire coil face area are required, coils are designed with two steam inlet connections feeding the alternate tubes. As shown in Fig. 35.11, this design prevents stratification of heated and unheated air. By keeping both ends of the coil equally warm, maximum protection against freeze-up is provided. These coils have inner steam distributing tubes. Steam supply and return connections are on the same end of the coil, and tubes are pitched in the casing for condensate flow. Headers are sized for each application to provide uniform steam distribution to all coil circuits. Generally coils are made of copper tubes and aluminum fins. Special construction for higher-pressure applications is also available.

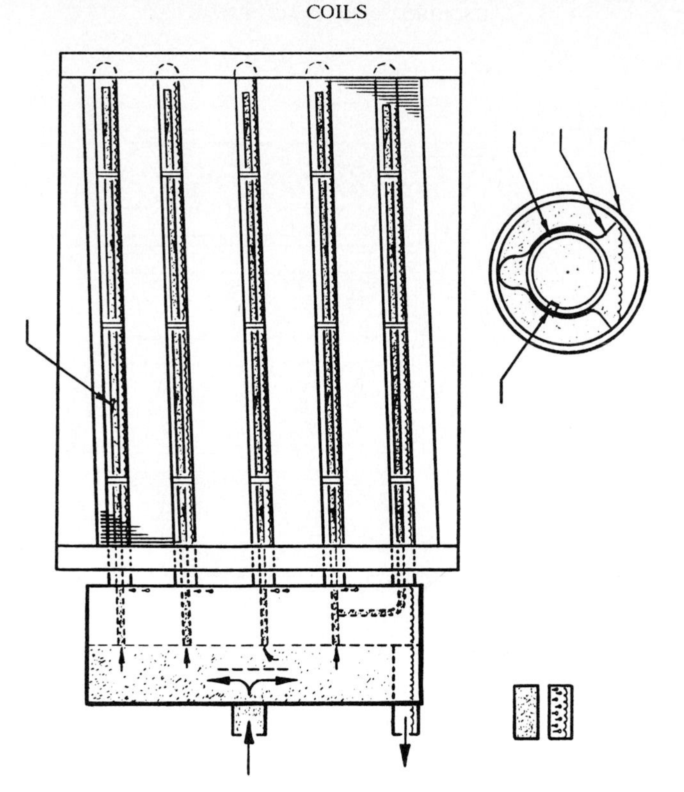

FIGURE 35.10 Distributing steam coil with orificed inner steam distributing tube.

35.4 COIL APPLICATIONS

Coils are either used by original equipment manufacturers or in-field built-up systems. Cooling coils are provided with a drain pan to catch the condensate formed during the cooling cycle. In the case of stacked coils, the condensate trough is provided in between the two coils to prevent flooding of the bottom coil. The drain pan connection should be on the downstream side of the coils and should be big enough for rapid drainage. Drain pans should be insulated to prevent sweating. Generally factory-assembled central station units incorporate these features.

The American Refrigeration Institute (ARI) standard for forced-circulation air-cooling and air-heating coils[2] covers the application range for all types of coils. These are the application and design ranges of the various types of coils.

For cooling coils:

Entry air, dry bulb	65 to 100°F (18.3 to 37.8°C)
Entry air, wet bulb	60 to 85°F (15.6 to 29.4°C)
Air face velocity	200 to 800 ft/min (61 to 244 m/min)
Fluid velocity	1 to 8 ft/s (0.3 to 2.44 m/s)

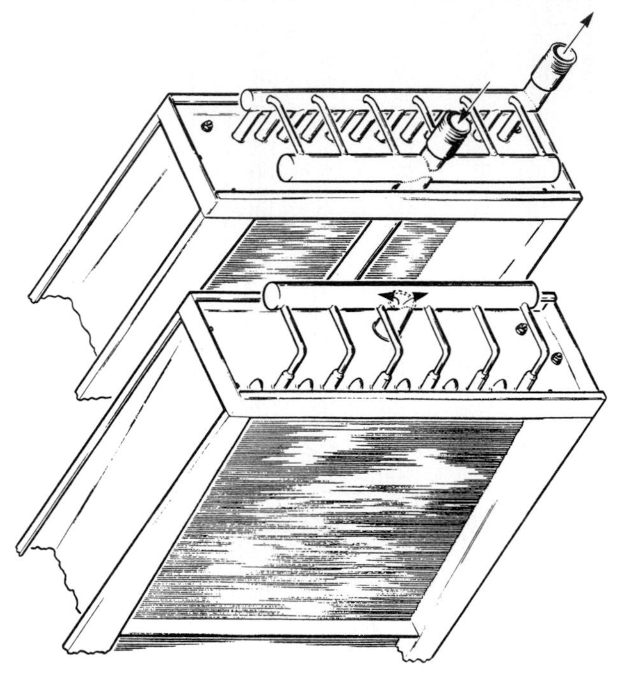

FIGURE 35.11 Distributing steam coil feeding alternative tube from both ends to prevent stratifications.

Entry fluid temperature	35 to 65°F (1.7 to 18.3°C)
Saturated suction temperature	30 to 55°F (−1.1 to 12.8°C)

For heating coils:

Entry air, dry bulb	0 to 100°F (−17.8 to 37.8°C) for hot-water coils; −20 to 100°F (−28.9 to 37.8°C) for steam
Air face velocity	200 to 1500 ft/min (61 to 457 m/min)
Fluid velocity	0.5 to 8 ft/min (0.15 to 2.44 m/s)
Entry fluid temperature	120 to 250°F (48.9 to 121.1°C)

For special applications such as low-temperature refrigeration (industrial or processing), the above-mentioned ranges may be varied.

In air dehumidifying coil applications, the air face velocity should be kept low to prevent condensed moisture from being blown off the coil. The maximum recommended face velocity is 550 ft/min (168 m/min). Over 550 ft/min (168 m/min) the condensate will be blown off the coil into the ductwork. In this case eliminators are used to prevent any water carryover. To enhance heat transfer, cooling coils are sprayed with water. In this case the leaving air temperature approaches the saturation temperature. Coils may have air bypass to control the air temperature.

35.5 COIL SELECTION

The following variables are to be considered in choosing a coil:

- Total load, Btu/h (W)
- Entering air temperatures, dry and wet bulb
- Available space for the system
- Cooling or heating media flow rate and entering temperature
- Air quantity
- Allowable air and water pressure drop
- Special coil material considerations

The total load is determined based on information available elsewhere in this handbook and on actual room load. Outdoor design temperature data are also available elsewhere in this handbook. With other known variables, a specific coil can be selected from various coil manufacturers' catalogs.

35.6 HEAT-TRANSFER CALCULATIONS

For sensible cooling coils, design problems with dry finned-tube heat exchangers require solution of the equation

$$q = U_o A \, \Delta t_m \tag{35.1}$$

where q = rate of heat transfer, Btu/h (J/s)
U_o = overall heat-transfer coefficient, Btu/(h · ft^2 · °F) [J/(s · m^2 · °C)]
A = total outside surface, ft^2 (m^2)
Δt_m = log mean temperature difference, °F (°C)

The key coil design parameter in the formula is the overall heat transfer coefficient U_o which is a function of (1) the metal thermal resistance of external fins and tube wall, (2) inside surface heat-transfer coefficient, and (3) the airside or outside surface heat-transfer coefficient. These are expressed in mathematical terms by the equation

$$U_o = \frac{1}{R_m + B/h_i + 1/h_o} \tag{35.2}$$

where U_o = overall heat-transfer coefficient of dry surface, Btu/(h · ft^2 · °F) [J/(s · m^2 · °C)]
R_m = metal thermal resistance of external fins and tube wall, ft^2/(h · °F · Btu) [m^2/(s · °C · J)]
h_i = heat-transfer coefficient of inside surface, Btu/(h · ft^2 · °F) [J/(s · m^2 · °C)]
h_o = heat-transfer coefficient of outside surfaces, Btu/(h · ft^2 · °F) [J/(s · m^2 · °C)]
B = ratio of outside and inside surfaces = A_o/A_i

A_o = total outside surface area, ft² (m²)
A_i = total inside surface area, ft² (m²)

35.7 METAL RESISTANCE OF EXTERNAL FINS AND TUBE WALL

The total metal thermal resistance R_m to heat flow through external fins and the prime tube wall may be calculated as follows:

$$R_m = R_f + R_t \tag{35.3}$$

Here the variable-fin thermal resistance R_f, based on the total surface effectiveness of dry surface, is

$$R_f = \left(\frac{1 - \eta}{\eta}\right)\frac{1}{h_o} \qquad \eta = \frac{\phi A_s + A_p}{A_o} \tag{35.4}$$

where η = total surface effectiveness
ϕ = fin efficiency
A_s = secondary fin surface, ft² (m²)
A_p = net primary surface, ft² (m²)

The tubeside resistance R_t is calculated from

$$R_t = \frac{BD_o}{24K_t} \ln \frac{D_o}{D_i} \tag{35.5}$$

where D_o = tube outside diameter, in (mm)
D_i = tube inside diameter, in (mm)
K_t = thermal conductivity of tube material, Btu/(h · ft²) [J/(s · m²)]

Metal thermal resistance is mainly dependent on the tube material, fin material, outside heat-transfer coefficient, and surface effectiveness.

35.8 HEAT-TRANSFER COEFFICIENT OF INSIDE SURFACE

The inside heat-transfer coefficient varies due to the tube diameter, mass flow rate, and physical properties of the fluid. For all water coils with smooth internal tube walls, the tubeside heat-transfer coefficient can be calculated from

$$h_i = \frac{150(1 + 0.011t_{wm})(V_w)^{0.8}}{D_i^{0.2}} \tag{35.6}$$

where t_{wm} = mean water temperature and V_w = water velocity.

For other than water, the following equation[1] should be used to calculate the inside heat-transfer coefficient:

$$h_i = 0.023\left(\frac{K}{D_i}\right)\text{Re}^{0.8}\text{PR}^{0.33} \tag{35.7}$$

where K = thermal conductivity of fluid, Btu/(h · ft^2 · °F) [J/(s · m^2 · °C)]
Re = Reynolds number
Pr = Prandtl number

Due to the two-phase flow characteristics of refrigerant coils, it is not possible to accurately calculate the inside heat-transfer coefficient as a function of flow rates. However, information is available in various handbooks for approximate data. Unit efficiency dictates that the evaporator pressure drop be kept low [perhaps 2 to 4 lb/in^2 (13.8 to 27.6 kPa)] to maintain as large a log mean temperature difference as possible for maximum capacity. Condenser refrigerant pressure drop should be limited [perhaps 10 to 15 lb/in^2 (68.9 to 103.4 kPa)], to keep the head pressure low for minimum compressor input.

35.9 HEAT-TRANSFER COEFFICIENT OF OUTSIDE SURFACE

This coefficient typically accounts for 50 to 80 percent of the overall coefficient and therefore has been the target of past design work to improve it. Initial development for improved capacity has moved from in-line to staggered-tube arrangement and from flat to corrugated fin surfaces. With every improvement an increase in coil air friction has occurred, so the type of fin surface must be evaluated before a coil is designed. The finside heat-transfer coefficient is generally determined by tests; however, some empirical data are available.

35.10 DEHUMIDIFYING COOLING COILS

When the dew point of entering air is higher than entering fluid temperature, normally the coil will remove moisture in addition to sensible cooling. In this case, the dew point temperature of air leaving a cooling coil is lower than that of the air entering the coil.

In most air-conditioning applications, air contains a mixture of water and dry air and enters the coil to lose sensible and latent heat. Latent heat occurs only in those parts of the coil where the temperature of the coil surface is below the dew point temperature of the entering air.

ARI (American Refrigeration Institute) Standard 410-81 covers in detail the procedure to rate or select dehumidifying cooling coils. The coils may be fully wet or partially wet depending on the entering air and water temperatures. Most coil manufacturers have published rating tables to select coils. These tables have been derived from test data. The method of testing coils is given in American Society of Heating, Refrigeration, and Air-Conditioning Engineers' (ASHRAE) Standard 33-78.[3] Due to the complexity of calculations, most manufacturers have computer programs to rate or select dehumidifying cooling coils. The dry portion of the coil is rated per dry-bulb temperatures, and the wet portion of the coil is rated per enthalpy of the air and fluid.

35.11 REFERENCES

1. W. H. McAdams, *Heat Transmission*, 3d ed., McGraw-Hill, New York, 1954.
2. ARI Standard 410-81, Air Conditioning and Refrigeration Institute, Arlington, VA, 1981.
3. ASHRAE Standard 33-78, American Society of Heating, Refrigeration and Air-Conditioning Engineers, Atlanta, GA, 1978.

CHAPTER 36
AIR FILTRATION AND AIR POLLUTION CONTROL EQUIPMENT

Richard D. Rivers
Environmental Quality Sciences, Inc.,
Louisville, KY

36.1 GAS PURIFICATION EQUIPMENT CATEGORIES

Gas purification devices are found in a very broad array of air- and gas-handling systems. Ultrahigh efficiency filters remove particles from the air of clean rooms where semiconductor integrated circuits are manufactured; inexpensive fiber filters protect the heat exchangers in residential air conditioners; large-scale electrostatic precipitators and scrubbers remove fly ash and sulfur dioxide from power plant stack gases. Gas-cleaning technology is usually divided into two categories, based on the purpose of the device considered:

- *Air filters:* These devices remove pollutants entering buildings and industrial processes or recirculated within them. It is common to use the terms "air cleaner," "air filter," or merely "filter" to refer to this type of device, regardless of whether air or some other gas is the working fluid.
- *Air pollution control equipment:* These devices capture pollutants emitted by industrial processes, to prevent their release into the atmosphere.

Pollutant concentrations for the two categories differ greatly. Air pollution control equipment typically must deal with pollutant concentrations 1000 to 100,000 times those found in air filtration applications. A few devices, such as systems to protect turbines and compressors from desert sandstorms, may operate at contaminant concentrations between the filtration and pollution control categories.

Gas purification devices in the above application areas are further subdivided into devices which capture *particulate* contaminants and those which capture *gaseous* contaminants. The operating principles for capture of the two contaminant types are quite different, and in only a few cases does the same device remove both gaseous and particulate contaminants.

Before we discuss the operating principles and design of gas purification equipment, it will be helpful to examine the nature of gasborne contaminants and to define some terms used in gas purification technology.

36.2 PARTICULATE CONTAMINANTS

36.2.1 Size Range of Aerosols

Both solid and liquid particles can be undesirable contaminants in gases. The range of particle sizes which can exist in a flowing gas stream is about 0.001 to 100 μm (1 μm = 1 micrometer = 10^{-6} m). Below 0.001 μm, a "particle" includes so few molecules that it behaves as a gaseous molecule. Particles in the 0.001- to 0.01-μm size range tend to agglomerate rapidly, forming larger particles. Particles larger than 100 μm are so heavy that they do not remain suspended in gases; they can be transported only by repeated bouncing from surfaces. Particles which remain in suspension for extended times are called "aerosols."

36.2.2 Aerosol Shape

Liquid aerosols in small sizes are very nearly perfect spheres. Larger liquid aerosols (droplets) may be distorted by aerodynamic and gravity forces, but are still nearly spherical. Solid aerosols may be spherical, but are often distorted or ragged in appearance. Some aerosols (e.g., soot particles) are randomly shaped chains or agglomerates of smaller particles. Aerosols which have length-to-diameter ratios greater than about 5 (asbestos, lint) are usually considered fibers and behave somewhat differently from near-spherical aerosols.

36.2.3 Aerosol Size Distribution Statistics

A few aerosols are of nearly uniform and constant size. More typically, however, an aerosol cloud includes a range of particle sizes. Statistical concepts are therefore useful in describing most aerosols and are almost necessary to calculate the performance of gas-cleaning devices. The first step in defining the statistics of an aerosol cloud is to subdivide it into size groups, each of which contains only a narrow range of sizes. This may be done with an instrument which actually separates the particles by size or one which measures an image or other size-dependent property of individual particles. The number or mass of particles in each narrow size range is determined, and a particle-size histogram (Fig. 36.1) is plotted. Aerosol histograms may have a single peak (as in Fig. 36.1) or several (Fig. 36.2). Each peak is called a "mode." The terms "unimodal," "bimodal," "trimodal," and "multimodal" are used to label aerosols with one, two, three, and many modes. Where more than one mode is present, the measured aerosol comes from more than one source or is the result of more than one production mechanism.

It is possible to divide the histogram into the components which make up each individual peak. When this is done, the histogram is usually skewed to one side of the peak, as in Fig. 36.1. If the data are replotted on a logarithmic scale for particle diameters, the histogram appears symmetric about the peak (Fig. 36.3). Such a histogram can be fitted quite closely to a common statistical function, the

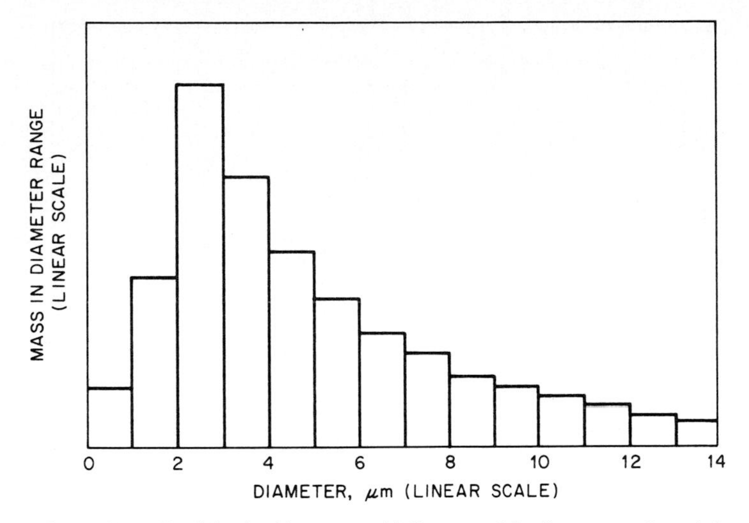

FIGURE 36.1 Particle-size histogram with linear particle-diameter scale and single peak.

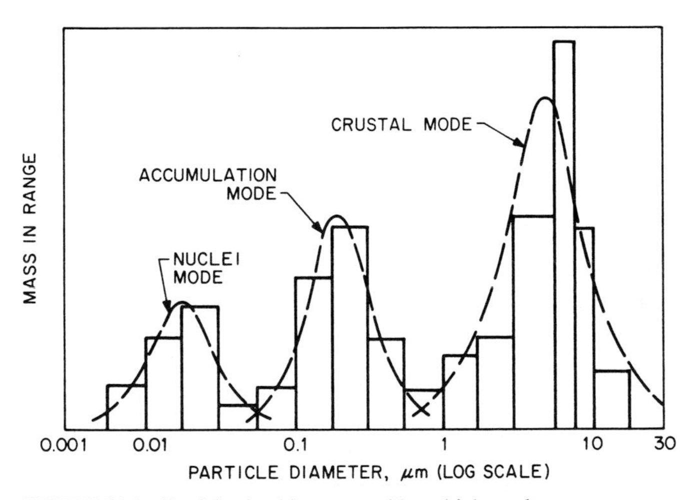

FIGURE 36.2 Particle-size histogram with multiple peaks.

"normal" or "gaussian" distribution. Such an aerosol is said to be "lognormally distributed." We can use this curve to replace the data array used to plot the histogram with just three parameters: the total count of particles, mean diameter (diameter at the peak of the smooth curve), and standard deviation of the curve (half the width between the inflection points).

The process of fitting a smooth curve to the aerosol size distribution histogram is greatly simplified by plotting the *sum* of counts for all sizes larger than a given size against the logarithm of the size, using a special log-probability graph paper.

For aerosols which are lognormally distributed, this form of plot (Fig. 36.4)

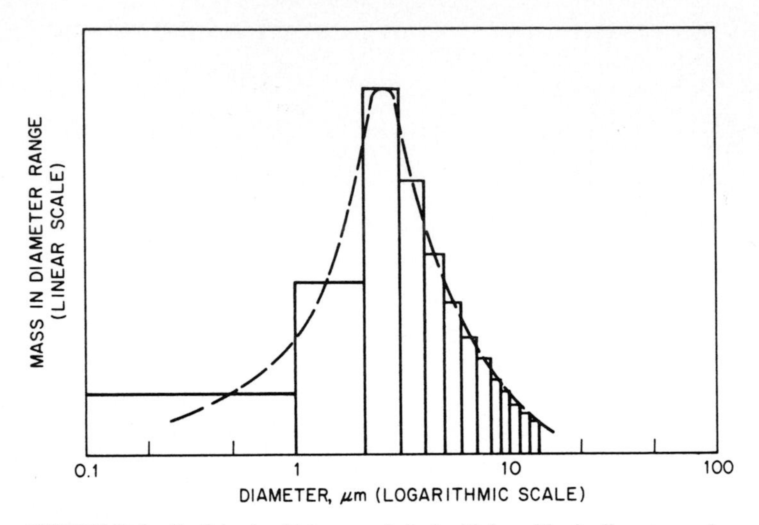

FIGURE 36.3 Particle-size histogram plotted with logarithmic diameter scale.

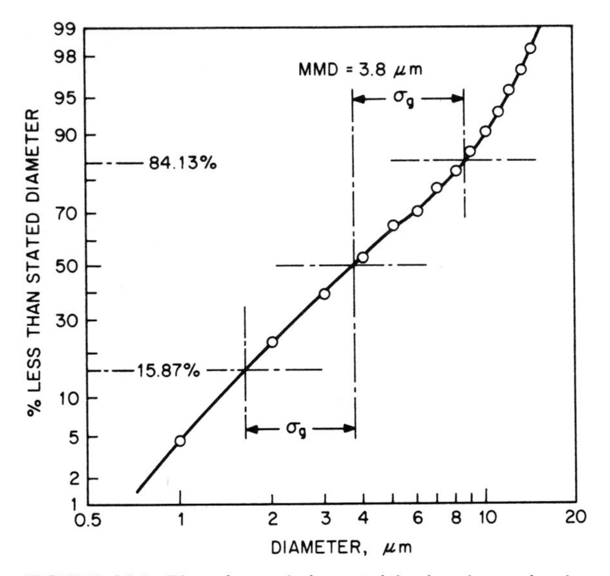

FIGURE 36.4 Plot of cumulative particle-size data using log-probability scales.

appears as a straight line. Small deviations in measured size distributions can thus be smoothed out, and the distribution parameters picked off the plot. These parameters are

$$\begin{aligned} d_g &= \text{geometric mean diameter, above which lies 50\% of distribution count} \\ d_{\sigma-} &= \text{diameter above which lies 84.13\% of distribution} \\ d_{\sigma+} &= \text{diameter above which lies 15.87\% of distribution} \\ \sigma_g &= \text{geometric standard deviation} \\ &= \left(\frac{d_{\sigma+}}{d_g}\right) = \left(\frac{d_g}{d_{\sigma-}}\right) \end{aligned} \tag{36.1}$$

The value of d_m is considerably larger if we have based the distribution histogram on the masses of particles in each diameter range, rather than the count. This distinction is very important when aerosol data are reported or used. If the distribution is indeed lognormal, the relations between mass and count distribution (geometric) mean sizes can be calculated from the Hatch-Choate equation:

$$\ln d_m = \ln d_g + 3(\ln \sigma_g)^2 \tag{36.2}$$

This conversion is useful when one needs to compute the penetration of a particulate filter or air pollution control device on a mass basis, yet only count data are available for the aerosol; or when penetration on a count basis is wanted, but only mass data are available for the aerosol.

The lognormal plot is not the only way in which aerosol distributions can be presented. A form often used is the Lundgren plot, shown in Fig. 36.5. Here the horizontal axis is the logarithm of the particle diameter, and the vertical axis is the logarithm of the term $\Delta c/\Delta \ln d$, where

$$\begin{aligned} \Delta c &= \text{particle count per m}^3 \text{ for particle size range } d_i \text{ to } d_{i+1} \\ \Delta \ln d &= \text{difference between natural logarithms of } d_i \text{ and } d_{i+1} \end{aligned}$$

Similar plots can also be made with $\Delta w/\Delta \ln d$ or $\Delta s/\Delta \ln d$ on the vertical axis. In these cases, Δw is the total mass of particles per cubic meter in the size range d_i to d_{i+1}, and Δs is the total surface of the particles in the range from d_i to d_{i+1}. This type of plot shows the modal pattern of the aerosol quite clearly. Another form, with the logarithm of particle counts greater than a given diameter plotted against the logarithm of these diameters, is used in determining the "class" of clean rooms (see Fig. 36.6).

It is possible to convert any of these plots to one of the others. Programs for hand calculators and microcomputers exist to help in this rather tedious work.[1] Further discussions of aerosol statistics are found in many texts.[2,3]

36.2.4 Aerosol Concentrations

Aerosol mass concentrations are expressed in mass of particulate per unit volume of gas. Units commonly used are μg/m³, mg/m³, g/m³, gr/ft³, and gr/1000 ft³. Table 36.1 gives some typical values for industrial-process concentrations. The outdoor mass concentration (total suspended particulates, or TSP) is measured con-

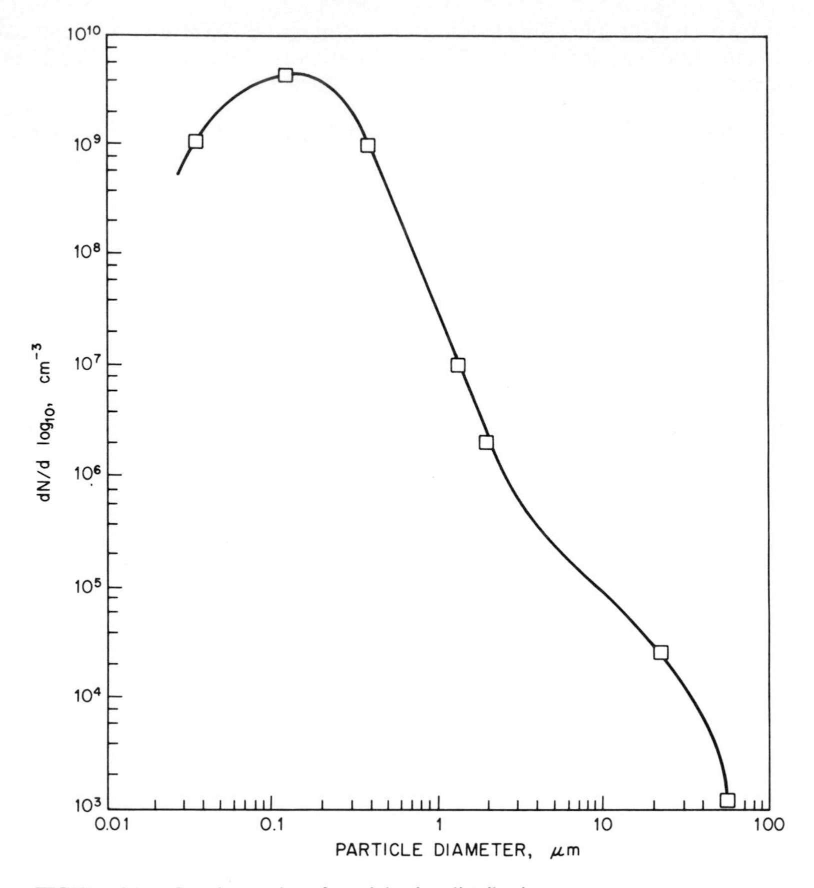

FIGURE 36.5 Lundgren plot of particle-size distribution.

tinually at hundreds of stations throughout the world. Average values obtained for 80 U.S. cities are listed in Table 36.2 and for 90 rural stations in Table 36.3. Outdoor concentrations are generally high in arid regions with agricultural activity and in cities with high automotive vehicle populations. The trend in TSP concentrations has been downward since the advent of national pollution control regulations and the shift from coal-fired residential heating. Tables 36.2 and 36.3 give average values; the percentage of each year for which TSP concentrations exceed a given level is lognormally distributed in almost all locations. Desert zones have special patterns. There seasonal winds bring sandstorms and duststorms with extremely high concentrations (see Table 36.3).

Average aerosol concentration has been found[6] to decrease with height above grade, approximately as given by Eq. (36.3):

$$C = C_o e^{-0.008h} \tag{36.3}$$

where C_o = concentration near surface, mg/m³
C = concentration at height h, mg/m³
h = height above grade, m

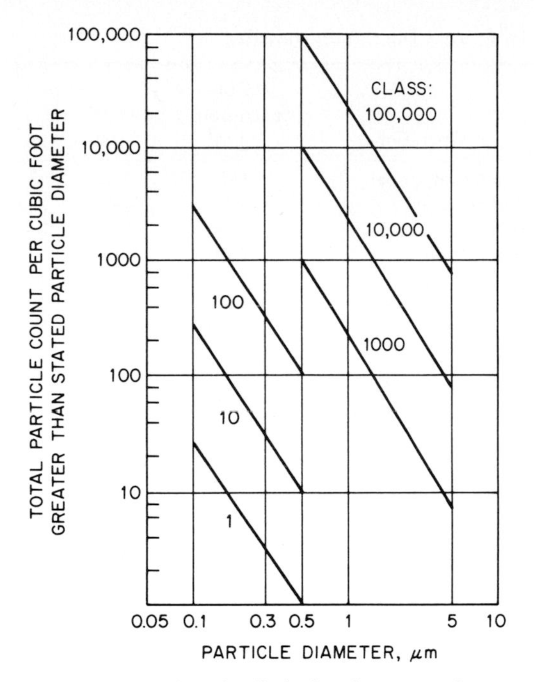

FIGURE 36.6 Particle-size limits for clean-room classes.

Indoor aerosol concentrations are the result of the combined effects of outdoor concentration, air leakage into and from the space, internal generation and deposition, air-flow patterns (especially the recirculation rate), and filtration. Calculation of indoor dust concentration is detailed in Sec. 36.4.3.

36.2.5 Aerosol Sources and Generation Modes

Outdoor aerosol particle-size distributions exhibit three modes:

- *Coarse-particle or crustal mode:* Chiefly soil particles, suspended naturally by wind erosion or by human activity in agriculture, construction, or travel on unpaved roads. Particles are produced by mechanical fracture of solids.
- *Fine-particle or accumulation mode:* Chiefly combustion products, with some production by atmospheric photochemical reactions, agglomeration, and suspension of very fine soil particles.
- *Nuclei mode:* Produced by evaporation, recondensation, and reactions of various compounds, primarily in combustion processes (including gasoline and diesel engines). Some particles in this mode originate from sea spray and radon gas. Nuclei-mode particles are short-lived, coagulating into accumulation-mode particles. Since there is essentially continuous production, a distinct nuclei mode will appear in ambient aerosols.

Indoor aerosols exist in the same three modes, in part because they arise from infiltration of outdoor aerosols into indoor spaces. However, interior generation sources often add to one or more of the modes, thus altering the overall distribu-

TABLE 36.1 Industrial Process Dust Characteristics

Service	Operation	Concentration, g/m³	MMD, μm	σ_g	d, g/cm³	Code*
Boiler fly ash	Chain grate, coal	1–5	4–20	3	0.7–3	1
	Stoker, coal	5–7	4–20	2–3	0.7–3	2
	Pulverized, coal	10–30	4–20	2–3	0.7–3	3
	Wood-fired	5–20	12	4		4
Ceramics	Frit spraying	1–8	5	2	2.7	5
Chemicals	Crushing, grinding	5–15	5–20	2	2.5	6
	Dryers, kilns	10–20	10	2	2.5	7
	Bin ventilation	1–5	5–10	2	1	8
	Materials handling	1–20	5–20	2	1	9
Coal mining and	Dedusting, cleaning	10–20	10	1.7	2	10
handling	Drying	10–20	5–11	1.7	2	11
Fertilizers	Ammoniators	1–7	5–10	2	2	12
Food	Sugar processing	1–11	5–10	2	1.7	13
	Coffee roasting	1–5	1–2	2	2	14
	Grain drying	1–5	20–30	3	2	15
	Flour handling	7–10	10	2	1.5	16
Foundries	Cupolas	2–10	1–10	2	4	17
	Electric furnaces	0.7–2	0.5–1.5	2	4.5	18
	Molding	7–10	2–10	2	2.5	19
	Shakeouts	3–12	1–10	4	2.5	20
Leatherworking	Buffing	7–11	11	2.5	0.7	21
Metalworking	Grinding	0.5–8	10–40	2.5	3–6	22
	Polishing, buffing	1–5	5–15	2.5	4	23
	Machining	1–8	1–5	2.2	3.7	24
	Welding, cutting	0.02–1.5	0.5–3	1.8	4	25
	Abrasive cleaning	7–30	5–10	2	2.5	26
	Spray painting	0.2–3	15–30	2	1.1	27
Papermaking	Paper cutting	7–10	12	4	1	28
Pharmaceuticals	Coating, packing	5–10	5–10	2	1.7	29
Plastics, rubber	Mixing	7–11	5	2	1.2	30
	Grinding, buffing	7–11	20	2.5	1.2	31
Plating	Acid mist collection	1–30	1.6–2	1.7	1	32
Printing	Press rooms	0.001–0.01	10	3	1.5	33
Rock products	Crushing, screening	10–30	5–10	2.5	2.5	34
	Dryers, kilns	7–20	10–15	2.5	2.5	35
Steelmaking	Open-hearth furnace	1–20	0.5–3	2	5.2	36
	Electric-arc furnaces	1–5	0.5–1	2	5.2	37
	Basic oxygen furnaces	5–30	0.5–1	2	5.2	38
	Scarfing	1–5	1.5	2.5	6	39
	Sintering, pelletizing	10–20	5–12	2	5.4	40
	Strip mills	1–5	5	2.5	7	41
Textiles	Carding, weaving	0.001–2	3–10	2.5	1.2	42
	Synthetic fiber drawing	0.01–2	5	2	1	43
Woodworking	Sawing, planing	7–12	10–1000	5	1	44
	Sanding	7–10	10–30	2.6	1	45
Miscellaneous	Incinerators	1–15	10–30	3	0.7–2	46
	Engine, compressor intakes	0.001–0.03	1–30	2	2.5	47

*Operation codes are used in Table 36.10.

TABLE 36.2 Total Suspended Particulate Concentrations for Major U.S. Population Centers, 1981

Population center	TSP, μg/m³	Population center	TSP, μg/m³
Birmingham, AL	111	Long Branch, Asbury Park, NJ	62
Phoenix, AZ	178	Newark, NJ	95
Tucson, AZ	53	New Brunswick, NJ	69
Anaheim, Santa Ana, CA	104	Albany, Schenectady, NY	59
Fresno, CA	109	Buffalo, NY	97
Los Angeles, Long Beach, CA	121	Nassau, Suffolk Counties, NY	56
Oxnard, Ventura, CA	90	New York, NY	68
Riverside, San Bernadino, CA	157	Rochester, NY	73
Sacramento, CA	68	Syracuse, NY	76
San Diego, CA	95	Charlotte, Gastonia, NC	67
San Francisco, Oakland, CA	56	Greensboro, Winston-Salem, NC	61
San Jose, CA	64	Raleigh, Durham, NC	53
Denver, Boulder, CO	183	Akron, OH	67
Hartford, CT	47	Cincinnati, OH	84
Wilmington, DE	65	Cleveland, OH	129
Washington, DC	67	Columbia, OH	74
Ft. Lauderdale, FL	69	Dayton, OH	77
Jacksonville, FL	79	Toledo, OH	72
Miami, FL	97	Youngstown, Warren, OH	112
Orlando, FL	67	Oklahoma City, OK	96
Tampa, St. Petersburg, FL	87	Tulsa, OK	99
West Palm Beach, FL	59	Portland, OR	114
Atlanta, GA	79	Allentown, Easton, PA	84
Honolulu, HI	51	Philadelphia, PA	75
Chicago, IL	111	Pittsburgh, PA	100
Gary, Hammond, IN	121	Scranton, Wilkes-Barre, PA	61
Indianapolis, IN	80	San Juan, PR	94
Louisville, KY	92	Providence, Pawtucket, RI	57
New Orleans, LA	82	Greenville, Spartanburg, SC	63
Baltimore, MD	90	Memphis, TN	74
Boston, MA	62	Nashville, TN	74
Springfield, Holyoke, MA	73	Austin, TX	96
Detroit, MI	116	Dallas, Ft. Worth, TX	77
Flint, MI	60	Houston, TX	159
Grand Rapids, MI	58	San Antonio, TX	73
Minneapolis, St. Paul, MN	100	Salt Lake City, UT	67
Kansas City, MO	96	Norfolk, Portsmouth, VA	64
St. Louis, MO	190	Richmond, VA	50
Omaha, NE	63	Seattle, Everett, WA	87
Jersey City, NJ	78	Milwaukee, WI	73

Note: Population center = metropolitan statistical area.
Source: Reference 4.

TABLE 36.3 Rural Aerosol Concentrations in the United States

Site	Concentration, μg/m³	Site	Concentration, μg/m³
Clanton, AL	54	Kalispell, MT	75
Florence, AL	75	North Platte, NE	66
Kenai, AK	47	Winnemuca, NV	80
Kingman, AZ	56	Northumberland, NH	46
Miami, AZ	51	Glassboro, NJ	48
Yuma Co., AZ	123	Roswell, NM	98
Dumas, AR	98	Zuni Pueblo, NM	41
Russellville, AR	78	Lake Placid, NY	33
Alturas, CA	47	Ulster Co., NY	35
Barstow, CA	152	Warsaw, NY	42
El Centro, CA	112	Brevard, NC	41
Napa, CA	59	Chapel Hill, NC	48
Visalia, CA	167	New Bern, SC	58
Clear Creek, CO	80	Sheridan Co., ND	40
Cortez, CO	49	Findlay, OH	75
La Junta, CO	97	Chillicothe, OH	67
Danbury, CT	64	Coshocton, OH	60
Seaford, DE	57	McAlester, OK	91
Merritt Is., FL	40	Corvallis, OR	35
Naples, FL	39	Bend, OR	64
Okeechobee, FL	27	Erie Co., PA	49
Panama City, FL	47	Johnstown, PA	112
Putnam Co., FL	36	York Co., PA	60
Dalton, GA	51	Sumter, SC	59
Cordele, GA	50	Huron, SD	48
Maui Co., HI	60	Dyersburg, TN	76
Blaine Co., ID	45	McMinville, TN	63
Mt. Vernon, IL	70	Bryan, TX	63
Petersburg, IL	64	Eagle Pass, TX	117
Bloomington, IN	75	Jeff Davis Co., TX	24
Estherville, IA	48	McAllen Co., TX	85
Dodge City, KS	53	Plainview, TX	150
Glasgow, KY	70	San Juan Co., UT	37
Iberville Par., LA	51	Tooele, UT	58
Millinocket, ME	38	Rutland, VT	56
Acadia National Park, ME	21	Accomack, VA	52
Garrett Co., MD	48	Martinsville, VA	56
Athol, MA	49	Aberdeen, WA	41
Plymouth, MA	41	Moses Lake, WA	75
Albion, MI	44	Mt. Vernon, WA	47
Petosky, MI	57	Weston, WV	77
Hibbing, MN	46	La Crosse, WI	51
Stearns Co., MN	52	Rhinelander, WI	40
Columbia, MO	80	Converse Co., WY	25
Big Horn Co., MT	37	Yellowstone Park, WY	13

Source: EPA-450/2-78/040 (Ref. 5).

TABLE 36.4 Recommended Models for Ambient Aerosols

	Coarse (crustal)	Fine (accumulation)	Nuclei
Aerosol mode			
MMD, μm	18	0.4	0.02
σ_g†	4.8	3.1	2.0
Density ρ, g/cm^3	2.5	1.0	1.0
Composition models			
Urban, high automotive, arid, %	13.0	86.1	0.9
Urban, low automotive, arid, %	45.0	54.5	0.5
Urban, high automotive, temperate, %	30.0	69.4	0.6
Urban, low automotive, temperate, %	25.0	74.3	0.7
Rural, arid, %	55.0	44.5	0.5
Rural, temperate, %	20.0	79.4	0.6
Dust storms,‡ %	100.0	—	—
Cigarette smoke (sidestream), %	—	80	20
Radon progeny, %	—	20	80
Street dust, %	100.0	—	—

†$\sigma_g = d_{84.13\%}/d_{50\%}$ (geometric standard deviation), where $d_p\%$ = diameter above which lies p percent of the mass.
‡Concentration is 1 to 200 mg/m^3.
Sources: References 7–11.

tion. A major indoor aerosol is tobacco smoke, which contains both liquid and solid particles.

Industrial processes can produce aerosols of virtually any size and chemical composition. Concentrations are usually high enough to allow reasonably rapid aerosol concentration and particle-size measurements to be made. Table 36.4 gives typical values for MMD (geometric mass-mean diameter), σ_g (geometric standard deviation), and density for these aerosol modes. Table 36.4 also lists models for the percentage of the three modes in aerosols in different geographical situations. These models, coupled with TSP concentration data from the Environmental Protection Agency (EPA), make it possible to characterize the outdoor aerosol with sufficient accuracy for the calculation of air filtration system performance (see Sec. 36.4.3). Some important aerosols appearing indoors are also characterized in this table.

36.2.6 Liquid Aerosols

The most common liquid aerosols are water: fog and rain. These are nontoxic but may be harmful to air filtration equipment. Electrostatic air filters, e.g., cannot operate when free water covers their high-voltage insulators; some fibrous filter mats collapse when wetted, and the pores of activated carbon can be plugged with liquid water. Hence systems with danger of heavy rain or fog exposure must be protected with louvers and/or coalescer pads (coarse fiber filters) which can capture and drain off water droplets. Liquid aerosols sometimes appear in industrial situations (e.g., paint overspray, textile fiber finishes, machine-tool coolants, ink mists). Fiber-bed plugging can be serious if the liquid viscosity is high. Reentrainment of collected liquid into the air stream is a problem if viscosity is low. Special equipment types have been developed which allow collected liquid aerosols to drain from their filter media. Volatile aerosols may be collected by an air filter, evaporate, pass through the filter, condense, and thus reappear downstream of the filter.

36.2.7 Radioactive Aerosols

Some alpha-particle emitters produce recoil effects which can dislodge collected particles from filter fibers. Intense radiation from collected particles may damage filter media, seals, and structures. Reliability requirements for air filtration systems in nuclear applications are very stringent, and disposal of collected dust and used filters poses special problems. The needs of this technology have been dealt with in extensive literature,[12–15] some of which is applicable to general filtration problems.

Radon gas creates unique air filtration problems. It enters a space as a gas, but then by radioactive decay it is converted to extremely fine (<0.02 μm) "progeny" or "daughter" particles. (See Table 36.4.)

36.2.8 Biological Aerosols

Microorganisms (bacteria, viruses, pollen, spores) have the filtration properties of other aerosols of like size and density. Densities for pollen (~ 1.0 g/cm^3) are, however, less than densities for soil particles, which range from 1.0 to 20.0 μm in diameter. Viable organisms can multiply after being captured by a filter and grow through it. In general, however, filters offer few nutrients, and the effect is not often observed.

36.2.9 Fibrous Aerosols

Lint, plant parts, hairs, and fibers of polymers, glass, and asbestos all become airborne. With the exception of asbestos and debris from raw cotton and similar vegetable fibers, these are generally nontoxic. They may, however, be very damaging to mechanisms such as computer disk drives or to processes like paint spraying or photography. The lengths of fibrous aerosols make them relatively easy to filter, but they tend to form mats over the surfaces of filter media. This causes cake formation and rapid increase of air-flow resistance. Special filters with very thin, cheap filter webs which are automatically "renewed" are used to collect the high fiber concentration found in textile and printing plants.

36.2.10 Aerosol Charging

A substantial fraction of all aerosols carry electric charges of either + or − sign. Aerosols which are freshly generated, whether by grinding, spraying, or condensation from combustion gases, carry somewhat higher charges than aged aerosols. The maximum charge which can be put on an aerosol particle is a function[2,3,9] of its diameter, its dielectric constant, and the ion density, electric field, and gas temperature in the vicinity of the particle. Charged particles move in electric fields and create fields. These effects are the basis of electrostatic precipitators and electrically augmented air filters, whose operation is described in Sec. 36.5. Charged aerosol particles may be rendered nearly neutral by flooding them with a cloud of bipolar ions, containing nearly equal numbers of positive and negative ions.

36.3 CONTAMINANT EFFECTS

The control of gaseous contaminants in ventilation air and in stack gases is based on the damage they can inflict on people, processes, materials, or the general environment. The decision to control a given contaminant therefore depends on determination of its damaging properties. For some gaseous contaminants, this determination has been quite thorough, and allowable air concentration levels in workplaces or stack emission rates are mandated by federal and local laws. For others, contaminant control levels must be based on published studies and such poorly defined quantities as odor thresholds.

36.3.1 Gaseous Contaminant Toxicity

Air pollution health effects usually depend on the cumulative dose of contaminant inhaled. However, there are upper limits to permissible concentration even for short exposure periods. Table 36.5 lists a few well-recognized toxic gaseous contaminants along with their physical and toxicological properties. The abbreviations in the table are defined as follows:

MW = molecular weight, g/(gmol)
BP = boiling point, °C at 1-atm (760-mmHg) pressure
DEN = contaminant gas density, g/m^3

Maximum allowable concentrations (MACs):

IDLH = concentration immediately dangerous to life and health
AMP = acceptable maximum peak for short-term exposure
ACC = acceptable ceiling concentration, not to be exceeded during any 8-h shift, except for periods defined by AMP
TWA8 = time-weighted average concentration, not to be exceeded in any 8-h period of 40-h week
OTH = odor threshold, concentration below which 50% of observers cannot detect contaminant's presence

Alternative terms for the above items are MAC (maximum acceptable ceiling) = AMP, STEL (short-term exposure limit) = ACC, and TLV (threshold limit value) = TWA8. The AMP, ACC, and TWA8 concentration levels are requirements of OSHA (Occupational Safety and Health Administration). Table 36.6 lists the group of contaminants defined by OSHA as "high-hazard"; these must be controlled in ways specified by OSHA. Reference 16 gives details on the con-

TABLE 36.5 Gaseous Air Pollutants: Toxicity, Odor, and Physical Properties

Compound	Allowable concentration, mg/m^3						
	IDLH	AMP	ACC	TWA8	OTH	BP, °C	MW
Acetone	48,000			2,400	47	56	58
Acrolein	13			0.25	0.35	52	56 T
Acrylonitrile	10			45	50	77	53
Allyl chloride	810		9	3	1.4	44	77
Ammonia	350		35	35	33	−33	17 T
Benzene	10,000		25	5	15	80	78 T
Benzyl chloride	50			5	0.2	179	127
Carbon dioxide	90,000		54,000	9,000	∞	−78	44
Carbon disulfide	1,500	300	90	60	0.6	46	76
Carbon monoxide	1,650		220	55	∞	313	28 T
Carbon tetrachloride	1,800	1,200	150	60	130	77	154
Chlorine	75		1.5	3	0.007	−34	71
Chloroform	4,800		9.6	240	1.5	124	119
Chloroprene	1,440		3.6	90		120	89
p-Cresol	1,100			22	0.056	305	108 T
Dichlorodifluoromethane	250,000			4,950	5,400	−30	121
Dioxane	720			360	304	101	68
Ethylene Dibromide	3,110	271	233	155		131	188
Ethylene Dichloride	4,100	818	410	205	25	84	99
Ethylene Oxide	1,400		135	90	196	10	44
Formaldehyde	124	12	6	4	1.2	97	30
n-Heptane	17,000			2,000	2.4	98	100
Hydrogen fluoride	13		5	2	2.7	19	20
Hydrogen sulfide	420	70	28	30	0.007	−60	34 T
2-Butanone (MEK)	8,850			590	30	79	72 T
Mercury	28			0.1	∞	357	201
Methanol	32,500			260	130	64	32 T

Methylene chloride	7,500		3,480	1,740	750	40	85 T
Nitric acid	250			5		84	63
Nitrogen dioxide	90		1.8	9	51	21	46
Ozone	20			2	0.2	−112	48
Parathion	20			0.11		375	291
Phenol	380		60	19	0.18	182	94 T
Phosgene	8		0.8	0.4	4	8	90
Sulfur dioxide	260			13	1.2	−10	64T
Sulfuric acid	80			1	1	270	98
Tetrachloroethane	1,050			35	24	146	108
Tetrachloroethylene	3,430	2,060	1,372	686	140	121	166
o-Toluidine	440			22	24	199	107
Toluene	7,600	1,900	1,140	760	8	111	92 T
Toluene diisocyanate	70		0.14	0.14	15	251	174
1,1,1-Trichloroethane	2,250			45	1.1	113	133
Trichloroethylene	5,413	1,620	1,080	541	120	87	131
Vinyl chloride			0.014	0.003	1,400	−14	63
p-Xylene	43,500		870	435	2	137	106 T

Notes:
IDLH, AMP, ACC, TWA8, and OTH are defined in the text.
BP = boiling point at 760 mmHg; MW = molecular weight, g/gmol.
The letter T in the rightmost column indicates that this compound has been found in tobacco smoke.
To convert from mg/m^3 to ppm, use the following equation: ppm = $9.62\, sP/[(MW)(t + 273)]$, where P = carrier gas pressure, mmHg; t = carrier gas temperature, °C; and s = contaminant concentration, mg/m^3.
Source: References 18 through 20.

TABLE 36.6 OSHA-Regulated High-Hazard Toxic Airborne Contaminants

Asbestos	4-Aminodiphenyl
4-Nitrobiphenyl	Ethyleneimine
α-Naphthylamine	β-Propiolactone
Methyl chloromethyl ether	2-Acetylaminofluorene
3,3′-Dichlorobenzidine (and its salts)	4-Dimethylaminoazobenzene
	N-Nitrosodimethylamine
bis-Chloromethyl ether	Vinyl chloride (monomer)
β-Naphthylamine	Benzene
Benzidine	Coke oven emissions

Note: 4,4-Methylene-bis(2-chloroaniline) (MOCA) was originally on this list, but was removed for legal reasons rather than because of any proof of lower carcinogenicity.
Source: Reference 16, secs. 1910.1001 through 1910.1029.

trol regulations for these and several hundred other gaseous contaminants. Medical opinion is by no means unanimous that the allowed concentrations are correct or that all significant toxic substances are on the lists.

Air known to contain the substances in Table 36.6 in clearly detectable amounts must not be recirculated. Trace quantities of some compounds in this list have been detected in tobacco smoke, and other highly suspect carcinogens are present in tobacco smoke. In most cases, however, the annoyance level of contaminants—odor, lacrimation, and reduction in visibility—rather than matters of health, will determine control levels. In general, we find that

$$\text{IDLH} > \text{AMP} > \text{ACC} > \text{TWA8} > \text{OTH}$$

There are important exceptions to this sequence, however; e.g., carbon dioxide and carbon monoxide have no odor. Such substances give no warning of their presence and so are very dangerous. Air-handling systems must provide enough makeup air to eliminate the possibility of odorless gases reaching dangerous levels.

The allowable levels of gaseous contaminant emissions from stacks are determined first by EPA regulations. Very few stack gas emissions are controlled as of this writing; Table 36.7 lists those that are. Beyond this, odors, corrosive effects, and the desire to live in peace with one's neighbors often determine which contaminants get controlled.

36.3.2 Odors

Individual humans have widely different abilities to detect the odor of any one compound. Perceived odor intensities are approximately proportional to concentration to the nth power, with $0.2 < n < 0.7$. The nose sometimes quickly loses its sensitivity to a compound on exposure; in other cases, an odor can be "learned" and hence becomes easier to detect at lower levels. Odor thresholds, the intensity of odors in a space, or the effectiveness of an odor control device must be determined by panels of observers making "blind" choices in comparison with standards. Testimonials without such experimental controls are useless. The OTH values listed in Table 36.5 have been gathered from generally respected sources, but should be taken as order-of-magnitude guidelines rather than absolute values.

TABLE 36.7 National Ambient Air-Quality Standards, 1985

Pollutant	Averaging time*	Primary standard,† μg/m³	Secondary standard,‡ μg/m³
Total suspended particulate matter	Annual	75	60
	24 h	260	150
Sulfur oxides	Annual	80	. . .
	24 h	365	. . .
	3 h	. . .	1,300
Carbon monoxide	8 h	10,000	10,000
	1 h	40,000	40,000
Nitrogen oxides (NO_x)	Annual	100	100
Ozone	1 h	240	240
Nonmethane hydrocarbons	3 h	160	160
Lead	3 months	1.5	1.5

*Not to be exceeded more than once a year.
†This standard is intended to protect public health.
‡This standard is intended to protect the general environment from damage and to avoid problems with visibility, comfort, etc.
Source: Reference 26.

Stack emissions cause no odor problems if they are diluted to the point where odorous compounds are below threshold concentrations at ground levels. Determining what emission level is permissible before control equipment is installed requires a calculation of the diffusion of the stack plume over the surrounding countryside. Standard methods have been established for doing these calculations.[22]

36.3.3 Special Contaminants: Ozone and Radon

Ozone (O_3) is a highly reactive, unstable form of oxygen (O_2). Ozone quickly disappears in a confined space unless it is continuously resupplied to the space. Office copiers, welding, electric-motor commutators, and incorrectly designed electronic air cleaners (including desktop units) all produce ozone. Ozone has a strong odor, and persons exposed to it will ordinarily complain of its presence before concentrations reach hazardous levels—but not necessarily before they develop headaches or eye irritation. Ozone is very corrosive to rubber and plastics.

Radon is present in soils, groundwater, stone, concrete, indeed virtually anything that comes out of the ground. It is a harmful gas when inhaled, but by radioactive decay it generates far more toxic "progeny." These are condensation aerosols of polonium, bismuth, and lead, with diameters from 0.01 to 0.1 μm; they are trapped deep within the human lung, where they emit beta and alpha particles. There is strong evidence that radon progeny form a significant source of lung cancer. Reference 23 reviews the current knowledge of the radon problem.

36.3.4 Material and Product Damage

Both gaseous and particulate contaminants damage structures and process materials. A common example is wall discoloration and streaking. The dark patterns on walls are the result of particle deposition under combined turbulence, thermal, and electrostatic effects. Overall discoloration of walls can be due to deposition of tar aerosols (from tobacco smoke) or reactions with gaseous contaminants, such as sulfur dioxide. Fabrics, silver, artworks, paper, photographic film, plastics, and elastomers are vulnerable to degradation by gaseous contaminants. Ozone, oxides of nitrogen, and sulfides and sulfates are especially prevalent and troublesome.

The only sure way to decrease the rate of wall discoloration is to remove the contaminants causing it from ventilation air. The ASHRAE dust spot efficiency test for particulate air filters[24] is in essence a measure of wall blackening. Measurement of other damaging effects is more difficult, for one must usually measure the long-term cumulative exposure from very low concentrations.

Some industrial operations require careful filtration of process gases and the air in the work environment. Paint spraying and high-quality printing are examples of operations where dust causes product defects. The most demanding area for this type of control is semiconductor microcircuit manufacture. Here particles with as small as a 0.01-μm diameter can produce defects and reduce yields significantly. To cope with this cleanliness requirement, "clean rooms" are constructed. A typical configuration is shown in Fig. 36.7. The activities in the room are continuously washed with a downflow of air that is essentially particle-free. The air leaving through the floor grills carries off internally generated particles which might otherwise diffuse onto the product. Similar arrangements are used in pharmaceutical production to maintain sterility and chemical purity. Figure 36.6 defines the clean-room classes established under Federal Standard 209.[25] For a room to qualify for a given class, the per-cubic-foot count of particles larger than each size must fall below the line labeled for that class. Standard 209 details procedures for the counting process.

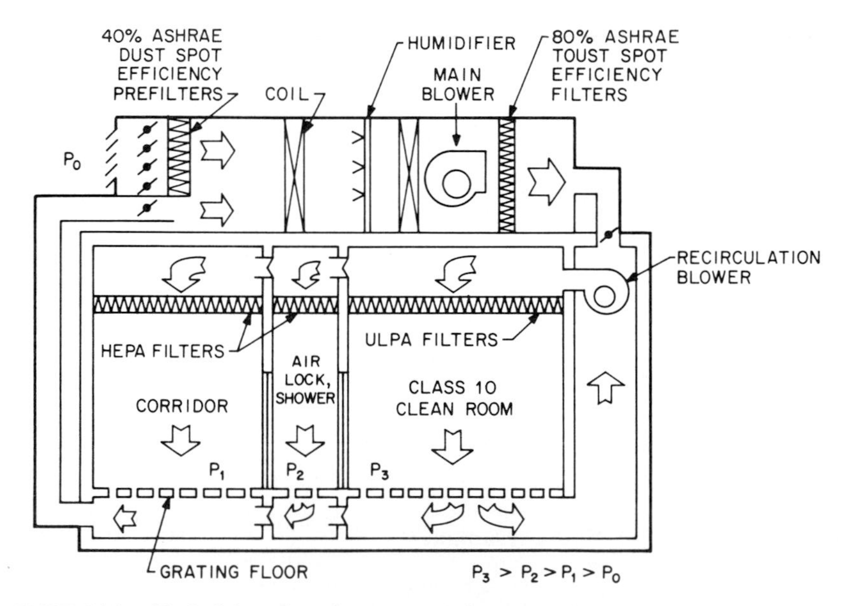

FIGURE 36.7 Typical downflow clean-room configuration. classes.

36.3.5 Particulate Contaminant Toxicity

Reference 16 lists OSHA-mandated TWA8 and ACC values for both particulate and gaseous contaminants. Among those listed are heavy metals; organotin compounds; beryllium and cadmium; silica, talc, and some other minerals; several pesticides and herbicides; and raw cotton dust. Asbestos is singled out along with some other carcinogens for prescriptive control measures, rather than allowable concentrations. This list by no means includes all materials which have been implicated by toxicity studies.

References 16 to 19 provide additional guidance with regard to allowable concentrations for particulate contaminants, and the journals listed in the bibliography (Sec. 36.11) provide periodic reviews of toxicity knowledge as well as reports of studies on specific materials. Allowable airborne levels of radioactive substances are controlled by a different set of regulations, promulgated in most of the U.S. state agencies acting in behalf of the U.S. Nuclear Regulatory Commission. Reference 26 defines these limits, which are unique in that the age of the exposed subjects is an important factor.

36.3.6 Environmental Damage

Pollutants emitted into the atmosphere are windborne away from the emission point. Both aerosols and gaseous contaminants can travel immense distances; dusts from the Sahara have been detected in North America. The major impact of industrial pollutants is relatively close to the emission point, however, for concentrations diminish as we move away from the source. Concentration at a distance from a source is a function of wind speed, stack height, stack velocity, meteorological parameters, and observer location. (See Sec. 36.4.2 for details.)

The damage caused to the environment may be due to the contaminant actually emitted, e.g., sulfur dioxide destroying plant growth near a copper smelter, or the damage may be due to some sort of interaction between the pollutant and other compounds. The creation of "smog" or "brown clouds," e.g., is a complex chemical reaction involving organic vapors from vehicles, combustion products, and energy supplied by sunlight. Damage may occur when the pollutant finally deposits on the earth's surface, as in the case of acid rain. The phenomena involved are complex, and data interpretation is frequently controversial.

Engineers concerned with air pollution control must meet at least the standards set by the EPA for emission levels. These standards appear in the Federal Register and are gathered in Reference 21. Local standards may be more stringent. In general, emission standards limit both concentration and total emission, with stricter concentration limits on larger installations. Sometimes (especially in the case of radioactive emissions) the concentration at the boundary of the owner's site is controlled. Where this is the case, calculation of dispersion from the emission point is required.

36.3.7 Visibility Problems

Aerosols reduce visibility by simultaneously absorbing and scattering light. The percentage of light transmitted by a column of polluted air is

$$T = 100e^{-NC_s x} \tag{36.4}$$

where T = transmittance, %
N = no. particles per volume of air, m^{-3}
C_s = scattering cross section for particle, m^2
x = length of air column, m

Here C_s is a complicated function of light wavelength and the diameter and index of refraction of the particles. In general, it has a maximum when the particle radius is near the wavelength of the light (about 0.4 to 0.7 μm for the visible spectrum). Since N is usually larger for small particles, aerosols with diameters near 0.4 μm are the most troublesome from the standpoint of visibility. The "blue haze" observed in sports arenas is the result of absorption and scattering of light by tobacco smoke aerosols, which are typically in this size range.

The visibility of smoke plumes emerging from a stack was at one time a measure of the acceptability of the discharge. Present-day air pollution control equipment reduces discharges to near invisibility, except for harmless water droplets, which evaporate and become invisible almost instantly. The classic Ringlemann chart method of smoke shade measurement has been shown to be very subjective and is rarely specified. Instruments which measure the light transmission across a stack by lamp or photocell systems are sometimes used to monitor the performance of an air pollution control system, but cannot measure the actual mass of aerosols emitted.

36.3.8 Nuisances

Dust kittens, cobwebs, minor eye irritation, nontoxic fallout, and temporary stimulation of coughing and sneezing may all be considered nuisances caused by air contaminants. These matters alone may dictate the quality of gas cleaning in an installation, but they are virtually impossible to quantify.

36.3.9 Flame and Explosion Hazards

Combustible gaseous contaminants can, under certain conditions, burn or explode. The minimum vapor concentration in air necessary to support combustion, called the "lower flammable limit" or "lower explosive limit" (LEL), ranges from about 1 to 20 percent by volume. Safety engineers recommend holding calculated concentrations to below one-fourth the LEL, to allow for imperfect mixing with ventilation air. Even those concentrations will usually be many times toxic or odorous levels, hence not likely to exist in spaces ventilated for human occupancy. Nevertheless, the design process should examine the possibility of reaching one-fourth the LEL and should provide sufficient ventilation or dilution air in a reliable manner.

Combustible dusts are also explosive. This is a serious problem in pollution control systems, where even if average concentrations are below the LEL, there are many opportunities for locally higher concentrations. It is important to keep runs to pollution control equipment short and clean, to prevent dust buildup. In many cases, vents are desirable to allow pressure relief when an explosion does occur. The National Fire Protection Association (NFPA) provides codes for appropriate vent designs and tables of LEL values for both gases and dusts.[27,28]

36.4 AIR QUALITY

36.4.1 Outdoor Air Quality

In the United States, outdoor air quality is regulated on a national scale by several acts of Congress from 1955 to 1977. During this period, substantial research was undertaken by the Environmental Protection Agency to determine existing levels of pollution, detrimental effects of pollution, and the technical feasibility and cost of air pollution. By 1980, it became apparent that the standards mandated by Congress were not being met. Thereafter, regulatory effort was directed in the following manner:

1. Strict control of airborne carcinogens and other pollutants of verified, high toxic risk (see Table 36.6)
2. Use of best-available control technology (BACT) for air pollution from new plants in specific industries [new source performance standards (NSPS)]
3. Maintenance of areas of the nation with high-quality air as close as possible to their present condition by prevention of significant deterioration (PSD) rules
4. Use of "offsets" and "bubbles," allowing the owner of plants in a given region to emit excess pollution from a new source, provided a greater compensating reduction is made on emissions from an existing source

This regulatory procedure makes for considerable complexity, but also allows considerable exercise of engineering skill to minimize pollution. For large facilities at least, extensive environmental impact analyses are required before an EPA construction license can be obtained. References 29 and 30 give details on these regulations and their impact on pollution control equipment design. The fact that air-quality levels and emission standards have been promulgated for very few substances does not exempt the designer of pollution control systems from paying close attention to the hazards possible from the release of other substances. References 3, 7, 11, 16, 17, 18, 29, 30, and 48 all provide guidance on the current understanding of the toxicity of air pollutants, as do the sources listed in the bibliography (Sec. 36.11).

The goals for outdoor air quality remain those established by Congress. However, only a few such goals—national ambient air-quality standards (NAAQS)—have been promulgated. (See Table 36.7) Reports[4] are made periodically showing the level of attainment of these standards throughout the United States. It is these reports which provide the estimates of existing concentrations, used in Sec. 36.2.4.

36.4.2 Dispersion of Air Contaminants

Pollutants emitted outdoors from a point source are carried away by winds. As a plume moves with the wind, air turbulence causes it to spread, both horizontally and vertically. The pollutant concentration in the plume thus (on average) decreases with the distance from the source. Plume concentrations are predictable with reasonable accuracy, provided that the meteorological parameters which define the wind and turbulence structure are known and, of course, that the physical properties of the contaminant and such items as height of release point, stack exit velocity, gas temperatures, etc., are known. Reference 22 provides a review of these relationships. With a reasonable amount of calculation one can estimate the average and peak concentrations to be expected near pollutant-emitting stacks and thus the potential for health hazard, odors, or nuisance.

Not all air pollutants are emitted from stacks. So-called fugitive emissions, which escape directly from processes without being captured by a hood, may contribute greatly to the surrounding pollution. Reference 31 shows that total material escaping from a process equipped with a hood and an air pollution control system is

$$G_t = (1 - 10^{-4}E_hE_e)G \tag{36.5}$$

where G_t = total emission to surroundings, g/s
E_h = hood capture efficiency, %
E_e = air pollution control equipment efficiency, %
G = process pollutant emission rate, g/s

Clearly, it makes no sense to invest in high-efficiency pollution control equipment and then couple it with a hood system of low efficiency. Reference 32 is a guide to proven hood design principles.

Dispersion of contaminants indoors is more difficult to predict than outdoors, for there are ordinarily many sources, many cross currents of air flow, and even the motion of people within the space affects dispersion. Most investigation of indoor contaminant dispersion has been done in connection with clean rooms. These usually have a relatively low uniform air velocity directed downward from a ceiling completely covered with air filters. Pollutants in these "laminar-flow rooms" are washed downward and away from the workspaces in the room, leaving either through low sidewall vents or a perforated floor. In more conventional rooms, the most common assumption is that pollutants are completely mixed with ventilation air.

36.4.3 Indoor Air Quality

Ventilation. Most city dwellers spend the majority of their lives indoors; indoor air quality is therefore of equal or greater importance to them than outdoor air quality. At present, there are no regulations on indoor air quality except in industrial workplaces (see Sec. 36.3.1 and Ref. 16) and implicitly in the ventilation requirements of building codes. Indoor air quality can be maintained by ventilation alone provided outdoor air for ventilation is itself acceptable. A minimum "fresh" air supply of about 5 ft^3/min (8.5 m^3/h) is needed for each occupant of a space, just to carry away carbon dioxide and supply oxygen. This cannot be made up by recirculation without the very complicated systems used in spacecraft and submersibles. Such minimal ventilation is not, however, adequate to eliminate odors, internally generated contaminants such as cigarette smoke, or in some cases radon infiltration. ASHRAE Standard 62-89 establishes higher ventilation rates, based on space use and number of occupants in the space.[33] Ventilation rates as high as 7 times the above minimum are recommended for some situations. The standard allows fresh air above the 5 ft^3/min (8.5 m^3/h) minimum to be obtained by adequate filtration processes and sets standards for the allowable contaminant content after filtration. The specified levels for particulates are fairly easy to obtain; specified gaseous contaminant levels are very difficult to obtain with the present state of the art.

Calculation of Indoor Air Quality. Both particulate and gaseous contaminant concentrations can be calculated with reasonable accuracy by use of system models, such as shown in Fig. 36.8. The steady-state concentration in the space, assuming complete mixing, is

$$C_i = \frac{G_i + 0.01(P_i Q_i + 0.01 P_2 P_1 Q_m) C_o}{Q_{v2} + Q_x + K_d A_d + Q_r(1 - 0.01 P_2)} \tag{36.6}$$

where C_i = steady-state indoor concentration, mg/m^3
C_o = outdoor concentration, mg/m^3
G_i = rate of indoor generation, mg/min
Q_i = air leakage through cracks into space, m^3/min
Q_m = makeup air flow through filtration system, m^3/min
Q_x = air leakage from space, m^3/min
Q_{v2} = exhaust air, m^3/min
Q_r = recirculation air flow, m^3/min
P_1 = penetration of contaminant through prefilter, %
P_2 = penetration of contaminant through main filter, %
P_i = penetration of outdoor contaminant through cracks, %
K_d = indoor dust deposition velocity, m/min
A_d = indoor dust deposition area, m^2

Equation (36.6) applies to each contaminant, i.e., to each gaseous contaminant compound, and to each particle size, in the case of particulate contaminants. This means that, in the particulate case, it is necessary to divide the aerosol spectrum into narrow bands and make a calculation for that narrow size range using the filter penetrations appropriate to that size. Hand calculation is extremely tedious, but computer programs are available to do the work. Reference 34 details these calculations and includes procedures for more complicated recirculation patterns.

Infiltration and Exfiltration. Equation (36.6) assumes that the amount of air entering the space through cracks and leaving, either through cracks or intentional

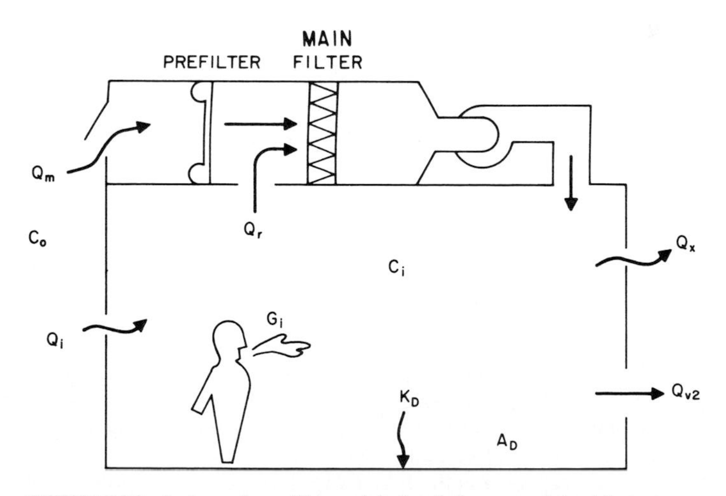

FIGURE 36.8 Indoor air-quality model. Symbols are explained in text.

exhausts, is known. The *ASHRAE Handbook, Fundamentals*,[35] provides considerable guidance on this subject. For spaces where infiltration is not allowable (such as clean rooms), the protected space must be pressurized to prevent air entry through any path other than the filter system. Pressurization of about 25 Pa (0.1 inWG) is generally adequate for this purpose.

36.5 PARTICULATE AIR FILTERS

36.5.1 Operating Principles

Fibrous Filters. Fibrous air filters have four general configurations:

1. *Panel filters*, where the filter medium is very open, with low air-flow resistance. Medium is held in relatively thin panels oriented perpendicular to the main air-flow direction, so that the medium velocity is equal to the duct velocity.
2. *Pleated-medium filters*, using denser, high-resistance medium pleated in zigzag fashion, so that considerable medium area is available and air velocity through the medium is low. The filter medium is held more or less rigidly in parallel planes by spacers, which can be corrugated paper or aluminum inserts, or ribbons or threads attached to the surface of the filter media, or corrugations of the media itself.
3. *Tubular pocket or bag filters*, where the medium is formed into tubes or pockets which are closed at one end and sealed at the opposite end into a header plate or frame. The aim of this construction is the same as in pleating—to expand the medium area and reduce medium velocity and hence pressure drop. These filters are usually collapsible and are held in extended form by their own air-flow resistance or by wire-frame supports.
4. *Roll filters*, where a flexible, compressible panel-type medium of glass or polymer fiber is supplied in the same manner as camera roll film. Clean filter medium is usually advanced into the air-filtering zone by an automatic transport mechanism, activated by a clock or sensor of resistance or medium dust load.

Fibrous filter media are nonwoven webs or blankets of glass, polymeric, metal, ceramic, or natural fibers. Fiber diameters range from about 0.2 to 100 μm. Media fibers are normally bonded together to give dimensional stability and resist compression. Practical media are made with as much open space as possible, to allow air passage at low pressure drop. They often contain no more than 1 percent solids.

Fibrous air filters capture particles by a combination of the following effects:

Aerodynamic Capture. This occurs because air flow must change direction to pass around filter fibers. Particles have greater inertia than gas molecules and thus tend to cross flow streamlines as these streamlines bend. Particles which pass close enough to a filter fiber to strike it will usually adhere to the fiber (Fig. 36.9*a*). This type of capture, the result of aerodynamic effects, is also called *impaction, impingement*, or *interception*. The effect increases capture probability as particle mass, density, and velocity increase.

Diffusional Capture. This is the result of random thermal motion of the molecules of the gas carrying the particles. Individual gas molecules impact the par-

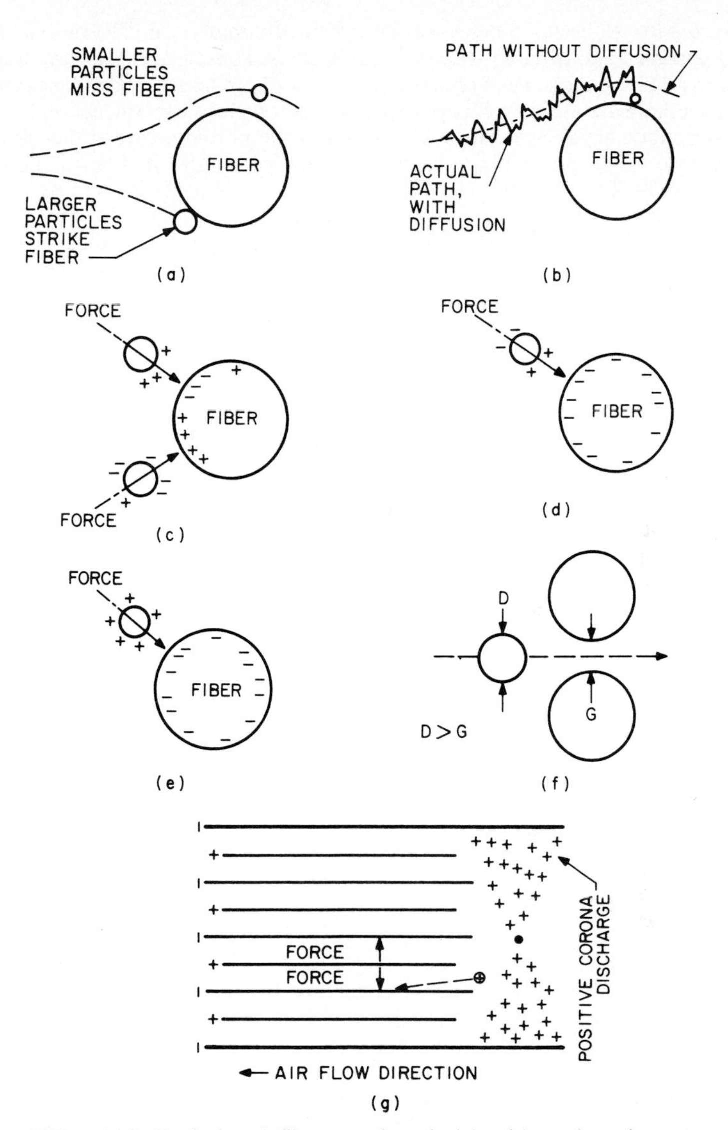

FIGURE 36.9 Particulate air filter operating principles: (*a*) aerodynamic capture; (*b*) diffusional capture; (*c*) charged-particle-induced electric force; (*d*) dielectric force with charged filter, uncharged particle; (*e*) coulomb force with charged filter, charged particle; (*f*) straining; (*g*) two-stage electrostatic air filter.

ticle from all directions. If the particle is small enough, the number of impacts in a small time interval may be predominantly from one direction. This drives the particle in erratic fashion out of its normal path (Fig. 36.9*b*). In effect, this erratic motion greatly increases the width of region in which a particle-fiber collision can occur. The effect increases as the particle size decreases and as the gas velocity decreases.

Electrostatic Effects. These are the result of unbalanced electric charges on aerosol particles and filter medium fibers or both. Some fraction of the particles in an aerosol cloud are always charged. When such a charged particle approaches a fiber, the particle induces an opposite charge on the fiber surface. These two charges produce an electric force which drives the particle toward the fiber, and they increase as the gas velocity decreases and particle size increases. The effect can be intensified by intentionally charging the aerosol particles and by impressing electric fields on the filter media (Fig. 36.9*c* through 36.9*e*).

Straining. This occurs when particles are larger than the spaces between fibers. The effect is of little importance for ventilation-type filters, except in the collection of lint and other fibrous particles. Most aerosols are much smaller than the passages through ventilation-type filter media. Membranes, however, may have openings which are smaller than many aerosols, and straining becomes a significant capture mechanism (Fig. 36.9*f*).

Membrane Filters. Membrane filters are made from very thin polymer sheets pierced by extremely small air-flow passages. They are available in several forms:

1. Collimated pore membranes, whose air passages are micrometer-diameter holes, all of nearly uniform diameter, running straight through the membrane. These membranes have a high solids content and high air-flow resistance, but can offer absolute removal of particles larger than their tube diameters.
2. Granular symmetric membranes. These are formed like sand beds or sponges on a microscopic scale.
3. Granular asymmetric membranes. These have a relatively open, spongelike structure which acts as a support for a very thin face layer containing ultrafine pores. This thin layer is the actual filter.
4. Fiber membranes, which are similar in structure to the fibrous media described above, but have considerably smaller-scale geometries.

Membranes are usually pleated and formed into panels or cylindrical cartridge filters for gas filtration use. The rather fragile nature of membranes, their cost, and the relatively high pressure drop for a given flow have, to date, limited their application to the production of ultrapure gases, to air sampling, and to fabric air pollution control equipment. Future developments might make them practical for ventilation applications. Membranes differ from usual gas filtration media in that they often have more media volume filled with solids than open space and hence provide very little space for dust storage.

Electrostatic Filters. Electrostatic air filters for ventilation service have the form shown in Fig. 36.9*g*. The filter has two stages: an ionizer to charge the aerosol and a plate package to capture it. In the ionizer stage, fine tungsten wires are suspended parallel to the grounded struts, midway between them. Each wire is maintained at a high voltage (+6 to +15 kV, depending on the strut spacing). The intense electric field near the wire ionizes oxygen molecules in that area. Positive ions of each ion pair formed are driven across the space between the ionizer wire and struts. During the transit, some ions attach themselves to passing aerosol particles. These charged particles then enter the plate package. The adjacent plates of the plate package are alternately grounded and charged, so that fields perpendicular to the air flow exist between the plates. These fields drive the charged particles to the plates. Here the particles are retained, usually with the help of a low-flammability viscous liquid adhesive on the plates. Electrostatic air cleaners can readily remove about 90 percent of atmospheric aerosols from the air flow, with an overall pressure drop of no more than 3 mmWG. This pressure drop does not increase appreciably as dust builds up on the plates. Eventually, however, the plates must be washed clean and recoated with adhesive.

In an alternative system, the plates are left dry, and collected aerosols are allowed to build up on the plates until they blow off. Released particles are agglomerates much larger than the entering aerosols and can be collected in a roll or fine-fiber-medium filter. The rate of resistance rise is much slower for electrostatically agglomerated aerosols; hence these systems have favorable maintenance characteristics.

Combined Forms. Electric fields can be maintained in fiber beds, provided the fibers do not become too conductive. Such fields can aid the collection of both charged and uncharged aerosols; the effects are much stronger for charged aerosols than for uncharged. However, when the medium becomes conductive—from humidity usually but also because ambient aerosols are conductive—the collection effectiveness drops to the level of the fiber bed alone, without electrostatic augmentation. Electrets, or "permanent" electric fields, have been impressed on fibers in air filters. This gives the same effect as an externally supplied electric field. However, once the electrets are drained down by conduction, they do not necessarily recover; hence this type of medium is very humidity-sensitive. Examples of air filter configurations are shown in Fig. 36.10.

36.5.2 Performance Characteristics

Penetration and Efficiency. Figure 36.11 shows the penetration and efficiency of a range of air filters as a function of particle diameter. The scales used in the figure (logarithmic for diameter, probability for penetration and efficiency) allow presentation of a wide range of data in a single chart. The curves were developed from test data, rather than theoretical calculations; manufacturers can supply similar curves for specific filters operating at chosen velocities. Such data are essential to the air-quality calculation described in Sec. 36.4.1. Note the penetration maxima in the region below 0.3 μm. This is the result of the competing effects of diffusion and aerodynamic forces. For low-efficiency (high-penetration) filters, another penetration peak occurs at large particle diameters, because of the poor adhesion of large particles to the filter fibers.

Pressure Drop and Service Life. The pressure drop across an air filter (usually called "filter resistance") is dependent on the quantity of gas flow through the filter, on the physical properties of the gas (density and viscosity), on the geometry of the filter, and, in the case of fibrous filters, on the amount, size, and shape of aerosol stored in the filter media. It is customary to express pressure drop as a function of the air filter average face velocity, which is air flow divided by the face area of the filter. This allows the same data to be applied to like filters of any flow capacity. Gas in passing through a filter can experience any of three flow regimes: laminar, turbulent, or an unstable, transitional regime between these two. For fine-fiber-medium filters, the flow in the fiber bed is most likely to be laminar, while the flow through entrance and exit passages is probably turbulent. Even filters which operate at duct velocity (roll filters and panel filters) will ex-

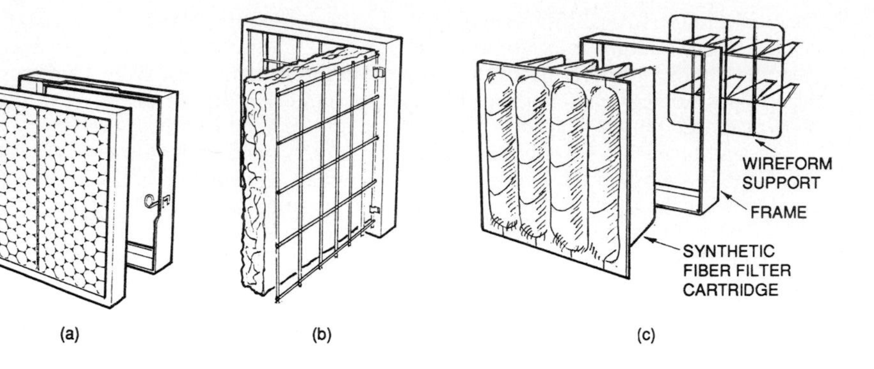

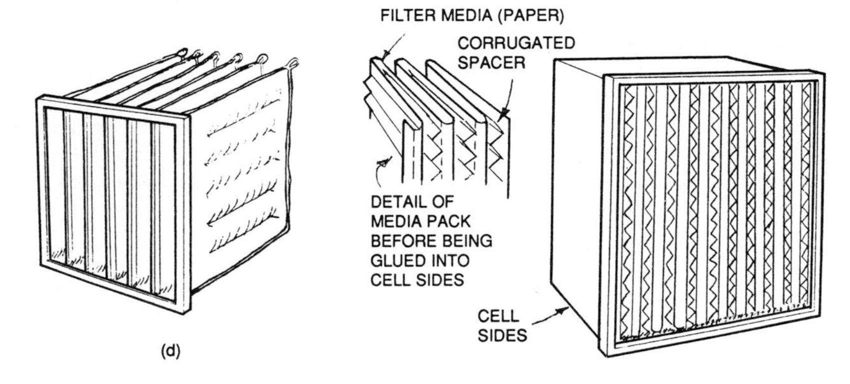

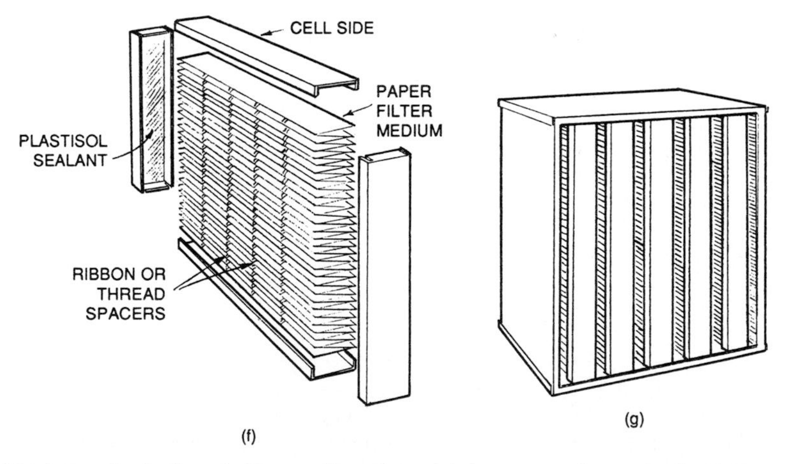

FIGURE 36.10 Particulate air filter configurations: (*a*) throwaway viscous impingement panel; (*b*) replaceable medium panel; (*c*) replaceable medium extended surface, supported; (*d*) self-supporting "soft" cartridge, extended surface; (*e*) rigid cartridge pleated medium, with separators; (*f*) rigid cartridge pleated medium, no separators; (*g*) zigzag cartridge, no separators; (*h*) roll filter; (*i*) self-cleaning viscous impingement curtain; (*j*) two-stage electrostatic air filter.

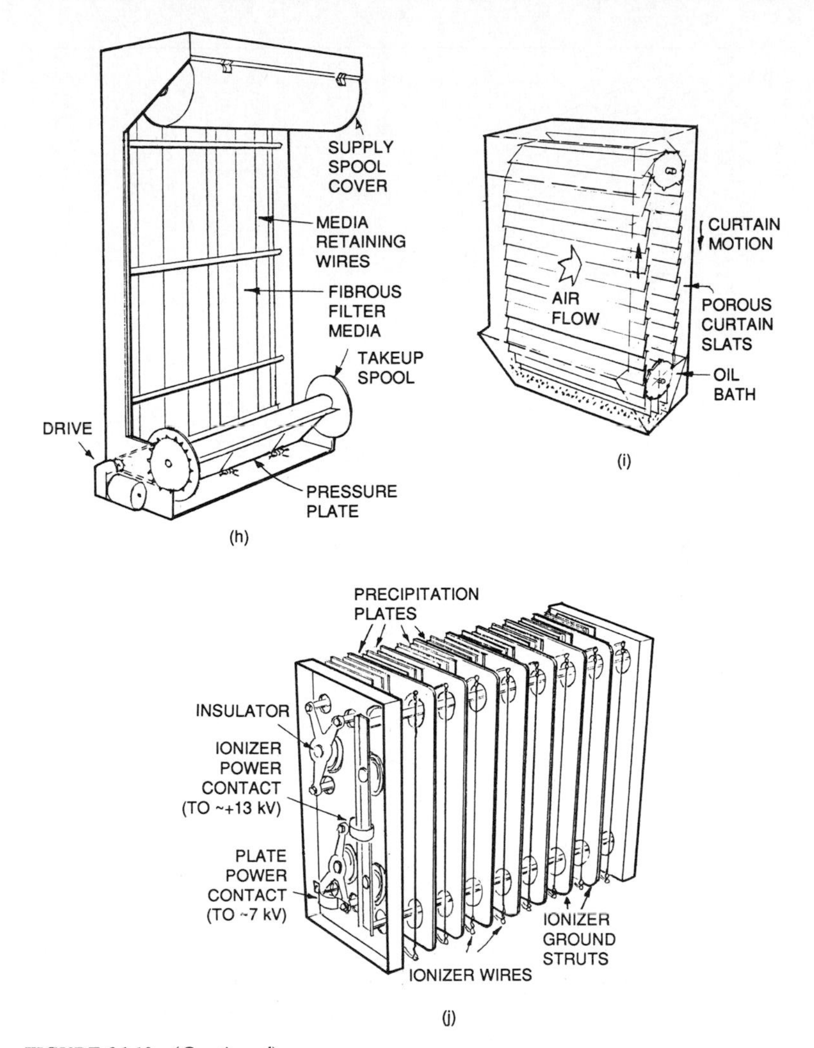

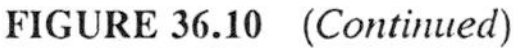

FIGURE 36.10 (*Continued*)

hibit both flow regimes, but in that case the turbulent-flow term is dominant. Filter manufacturers supply curves of filter resistance versus air flow, made in accordance with the test procedures of ASHRAE Standard 52-76.[24] Unfortunately, the standard requires this "resistance traverse" for a clean filter only; designers actually need velocity-resistance curves for aerosol-loaded filters as well. Where traverse data are available, we can solve for K_T and K_L in the general resistance expression

$$R = K_T \rho V^2 + K_L \mu V \frac{A_f}{A_m} \tag{36.7}$$

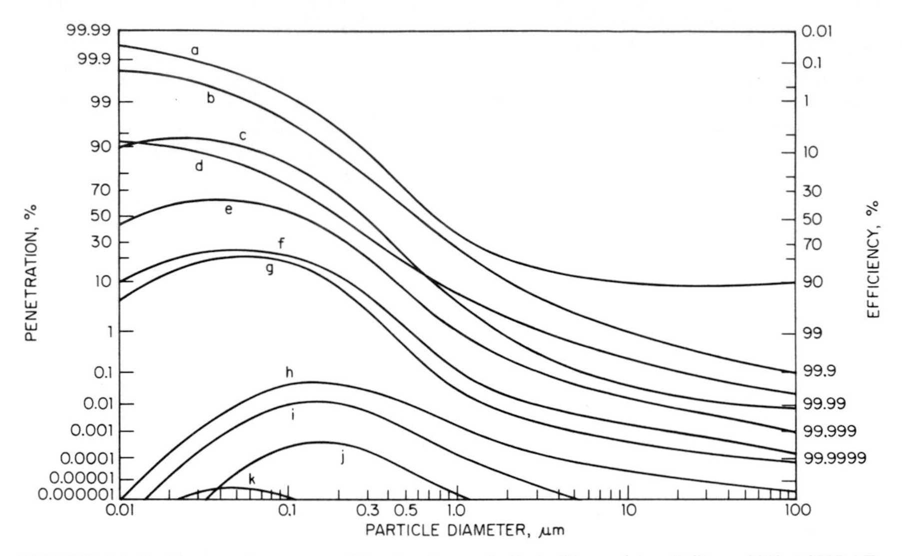

FIGURE 36.11 Penetration vs. particle size for typical air filters: (*a*) roll filter, 20% ASHRAE 52-76 dust spot efficiency (ADSE); (*b*) pleated-medium pocket filter, 45% ADSE; (*c*) pleated-medium pocket filter, 85% ADSE; (*d*) 85% ADSE electrostatic air cleaner; (*e*) pleated-medium pocket filter, 95% ADSE; (*f*) 95% ADSE pocket filter with electrostatic agglomerator; (*g*) pleated glass-fiber paper medium, 5% DOP penetration (DOPP); (*h*) standard HEPA filter, 0.03% DOPP; (*i*) HEPA filter, 0.01% DOPP; (*j*) ULPA filter, 0.001% DOPP; (*k*) membrane filter, 0.2-μm pore diameter.

Here

R = device resistance, Pa
V = device face velocity, m/s
ρ = gas density, kg/m^3
μ = gas viscosity, poise (P)
K_T = constant for device or its turbulent-flow elements (dimensionless)
K_L = constant for device or its laminar-flow elements, 1/m
A_f = device face area, m^2 (ft^2)
A_m = device medium area, m^2 (ft^2)

(If units of inches per second are used, the numerical values obtained for K_T and K_L will be different from those obtained using SI units.)

Some filtration devices (membrane filters, HEPA, and ULPA filters) operate at such low velocities that they have essentially zero values of K_T. Others (cyclones, louvers, electrostatic precipitators, and, of course, ductwork and orifices) have only turbulent elements, hence for them $K_L = 0$. Another group (fabric filters, extended-medium filters, adsorbers, panel filters, roll filters) have both laminar and turbulent elements ($K_L > 0, K_T > 0$). The value of K_T is constant for purely turbulent-flow devices, there being no dust buildup. The turbulent-flow term in the general resistance equation comes from the turning, converging, and diverging of gas flow in the filter structure, and this does not change greatly as the filter medium loads with dirt. The value of K_L is, however, strongly dependent on the amount and type of dust loaded into the device. A given mass of fine dust captured by a piece of filter medium produces a greater increase in K_L than does the same mass of coarser dust. No very satisfactory theory relating dust size and filter medium geometry has been developed.

Most designers will be content to take manufacturers' estimates of filter service life for their operating conditions. However, those who perform the indoor air-quality calculations of Sec. 36.4.1 and who want to know the service life of filters predicted by that model will need life data on the type of dust actually reaching the device. Dust-holding capacity by the ASHRAE Standard 52-76 test cannot be translated directly to dust-holding capacity in service; the ASHRAE dust-holding capacity can be taken only as a rough comparative value between filters.

Maintenance and Energy Consumption. Most air filters are disposable items; they are thrown away when the filter resistance exceeds the design value. Some, however, are cleaned and reused. Renewable filters (including electrostatic air cleaners) usually have a coating of a liquid adhesive to bind captured dust to the filter; this must be replaced after the filter is cleaned.

Systems are usually supplied with draft gauges measuring filter resistance (the static pressure drop between upstream and downstream filter faces) to indicate when replacement or renewal should occur.

The energy use by a filter is given by

$$P = \frac{0.001QR}{E_m E_d E_b} \tag{36.8}$$

where P = power consumed by system, kW
Q = air flow, m^3/s
E_m, E_d, E_b = fractional efficiencies of motor, drive, and blower at flow Q
R = filter resistance, Pa

(If Q is in ft^3/min and R is in inWG, the factor 0.001 should be replaced by 0.0001175.)

If air flow Q changes, the energy consumption drops—but only if this change is produced by a change in blower speed. If flow is controlled by dampers, these eat up the energy not used by the filter bank. Thus a low-resistance filter is useful only if the original design of the system contemplated this low resistance and a blower appropriate to the overall system resistance was chosen. The blower must be capable of supplying the terminal resistance of the filter at design air flow. Controls to regulate flow when the filter is below terminal resistance can save energy, for a low *initial* filter resistance will increase air flow and fan power, rather than reducing energy consumption.

For electrostatic filters, there will be some energy input for high-voltage power supplies. The energy used by any auxiliary blowers or motors in a system must also be included in total consumption.

Operating Limitations and Hazards. The threats to reliable air filter operation are high temperatures, excessive air velocity and filter resistance, wetting, vibration, and corrosion. Local or temporary excursions of any of these factors can cause catastrophic or gradual failures. It is therefore important to estimate the ranges of temperature and air velocity expected and to check the design of the filter for material and structural compatibility with these conditions. Air-flow distribution over filter banks should be as nearly uniform as possible. Filters that may be exposed to rain and snow should be protected by inlet louvers; air washers and cooling coils ahead of filters must never operate at velocities which allow carryover of sprays and condensate droplets. The air flow through filters provides immense potential for oxidation; construction materials must therefore be polymers, coated metals, aluminum, or stainless steel, depending on service requirements.

It is desirable for air filters to be nonflammable or at least to contribute minimally to fire and smoke hazards. For this reason, filters must usually meet one of the flammability categories specified in UL 900.[36] Class 1 filters do not contribute fuel when attacked by flame and emit only negligible amounts of smoke; class 2 filters may burn moderately or emit moderate amounts of smoke or both. UL 900 specifies tests for qualification in these classes. Tests are performed on clean filters, hence may underestimate the hazard present if dust of high flammability or smoke is trapped in a filter. Electrostatic filters are powered by high-voltage supplies (from 6 to 20 kV), some of which are capable of supplying up to 30 mA of direct current. These are quite lethal levels. High-voltage cables must be insulated with ample safety factors and be run through metal conduit. All doors to plenums accessing electrostatic filters must be provided with closures which turn off line power to power supplies, then force a delay of 15 s or so to allow the power supply capacitors to bleed to about 10 V before anyone can enter the plenum. The power supplies themselves must be able to withstand essentially steady arc-overs within the filter elements.

36.6 GASEOUS CONTAMINANT AIR FILTERS

36.6.1 Operating Principles

Removal of an unwanted gaseous contaminant from air or other carrier gas is a mass-transfer process. The contaminant molecules must be driven across an interface of such a nature that they cannot easily return. Some porous membranes have this asymmetric permeability for specified gases, and development continues to widen the number of contaminant types that can be separated by membranes.

More common, however, for air filtration applications are adsorption materials. In these the interface is a solid surface covered with vast numbers of microscopic pores which trap contaminant molecules. If the trapping process at the surface involves a chemical reaction, the process is called "chemisorption." For air pollution control, the interface is often the surface of a liquid into which the contaminant molecules dissolve. (This is called "*ab*sorption.") Ventilation air which passes through a liquid scrubber emerges very nearly saturated with the liquid vapor; even if the vapor is water, most of it must be removed, at considerable expense and complexity. Catalysts find limited use in air filtration systems, because only a few are available which operate at ambient temperatures; the cost of heating and then cooling air passing through a catalytic bed ordinarily prohibits their use. Dry adsorbers, chemisorbers, and low-temperature catalysts currently available for air filtration applications are listed in Table 36.8, with the contaminant categories for which they are suited.

Adsorption. In general, high-boiling-point contaminants condense most easily and are therefore easiest to adsorb. Molecular size and structure influence the

TABLE 36.8 Low-Temperature Adsorbers, Chemisorbers, and Catalysts

Material	Impregnant	Vapors or gases captured
	Physical adsorbers	
Activated carbon	None	Organic vapors; SO_2, H_2S, acid gases, NO_2
Activated alumina	None	Polar organic compounds*
Activated bauxite	None	Polar organic compounds
Silica gel	None	Water, polar organic compounds
Molecular sieves (zeolites)	None	Carbon dioxide, iodine
Porous polymers	None	Various organic vapors
	Chemisorbers	
Activated alumina	KNO_3	Hydrogen sulfide, sulfur dioxide
Activated carbon	I_2, Ag, S	Mercury vapor
Activated carbon†	I_2, KI_3, amines	Radioiodine, organic iodides
Activated carbon	$NaHCO_3$	Nitrogen dioxide
LiO_3, NaO_3, KO_3	. . .	Carbon dioxide
LiO_2, NaO_2, KO_2, $Ca(O_2)_2$	. . .	Carbon dioxide
Li_2O_2, Na_2O_2	. . .	Carbon dioxide
LiOH	. . .	Carbon dioxide
$NaOH + Ca(OH)_2$	. . .	Acid gases
Activated alumina, activated carbon	(Various, some proprietary)	Formaldehyde, mercury vapor
	Catalysts	
Activated carbon	None	Ozone
Activated carbon‡	Cu, Cr, Ag, NH_4	Acid gases, chemical warfare agents
Activated alumina	Pt, Rh oxides	Carbon monoxide

*Polar organics are alcohols, phenols, aliphatic and aromatic amines, etc.
†Mechanism may be isotopic exchange as well as chemisorption.
‡"ASC Whetlerite"

process, however; adsorption is generally easier for large molecules than for small, at least at the low contaminant concentrations of concern in air filtration. Low temperatures aid absorption; some substances, such as the noble gases, cannot be adsorbed effectively above cryogenic temperatures. Contaminants which are readily adsorbed will tend to block the adsorption of those which are more difficult to adsorb, and may even replace them. Water vapor (humidity) is the most common source of this interference. If the temperature of an adsorber is increased, adsorbed contaminant vapor may be driven out of pores and returned to the carrier gas stream. This process is called "desorption" or "elution."

The most important requirement for an adsorber is that it have a very large surface area covered with large numbers of pores of dimensions appropriate to the contaminant to be adsorbed. Surface areas of 50 to 2000 m^2/g are available. Beyond this, the adsorber is more effective if it exhibits the special pore size and binding properties mentioned above.

Adsorbers are made into granules, pellets, and fibers, all with the intent of allowing the formation of porous beds through which air can pass while it comes into contact with the adsorber surface. These granules must be rigid, so that the air passages are kept open, and must not break up into powder when the bed is vibrated.

Chemisorption. Most chemisorbers are adsorbers whose surfaces have been coated or "impregnated" with a chemical compound which reacts with the contaminant to be captured. In some cases, the body of the chemisorber is itself a substance which reacts with the contaminant. The reaction must convert the contaminant to a substance which is nonvolatile or of low volatility. In addition, the reaction must not yield any gaseous product which is itself unwanted. Because many chemical reactions are ionic, the presence of some condensed water is essential to the performance of some chemisorbers. Chemisorbers generally are more effective as temperatures rise, because chemical reactions are more rapid at higher temperatures. The reactant material must, of course, not react appreciably with the carrier gas or decompose or evaporate. Mechanical requirements are the same as those for adsorbers.

Catalysis. For ventilation applications, oxidation is about the only catalytic process likely to be useful. One must be careful that the reaction product is innocuous and nonodorous, or else the effluent may be more troublesome than the contaminant being controlled. Catalysts are sensitive in difficult-to-predict ways to surface adsorption of extraneous materials, "poisons." Considerable testing is necessary to determine their level of penetration and life as a function of temperature and contaminant concentration.

36.6.2 Adsorption Filter Configurations

It is most important that the adsorber prohibit air passage without contact with the adsorption medium. There must be no thin spaces or pockets in the medium body. Air channels either through the medium body or around its edges will increase penetration substantially. To prevent these failures, the edges of each medium-holding cell are covered with "baffles" to force air flow through the main medium body. Medium granules are packed tightly into the cells, by allowing the granules to fall into place rather slowly, as individual grains, usually with additional gentle vibration of the cell. For high-reliability cells, the perforated faces of each cell are drawn together after filling, to compress the medium fill. Elastomeric "springs" are sometimes inserted in cell ends to maintain pressure on the medium granules. Regardless of how deep the bed is, the granules cannot be unrestrained, for air flow will "mine" wide passages through the granules. Units for toxic contaminant capture have all metal-to-metal joints sealed by seam welding, gaskets, or sealants.

Three types of adsorption units are in general use for air filtration systems. The first, for commercial odor control applications (Fig. 36.12*a*), is made of perforated panels 13 to 25 mm in depth arranged in zigzag fashion to give bed velocities about 15 percent of duct velocities. The panels are removable so that the carbon they hold can be removed, regenerated, and put back into service. The second type, illustrated in Fig. 36.12*b*, is of sinuous form, with edge baffles to reduce bypassing. Beds are usually 50 mm deep. The third configuration, shown in Fig. 36.12*c*, has horizontal trays with baffles and controlled spacing of perforated sheets. Bed depths are from 50 to 300 mm. The perforated sheets can be lined with cloth to allow finer granule adsorption medium (down to 30 mesh) to be held. Both sinuous and horizontal panel units can be reloaded with adsorption medium by removing the entire unit from the system.

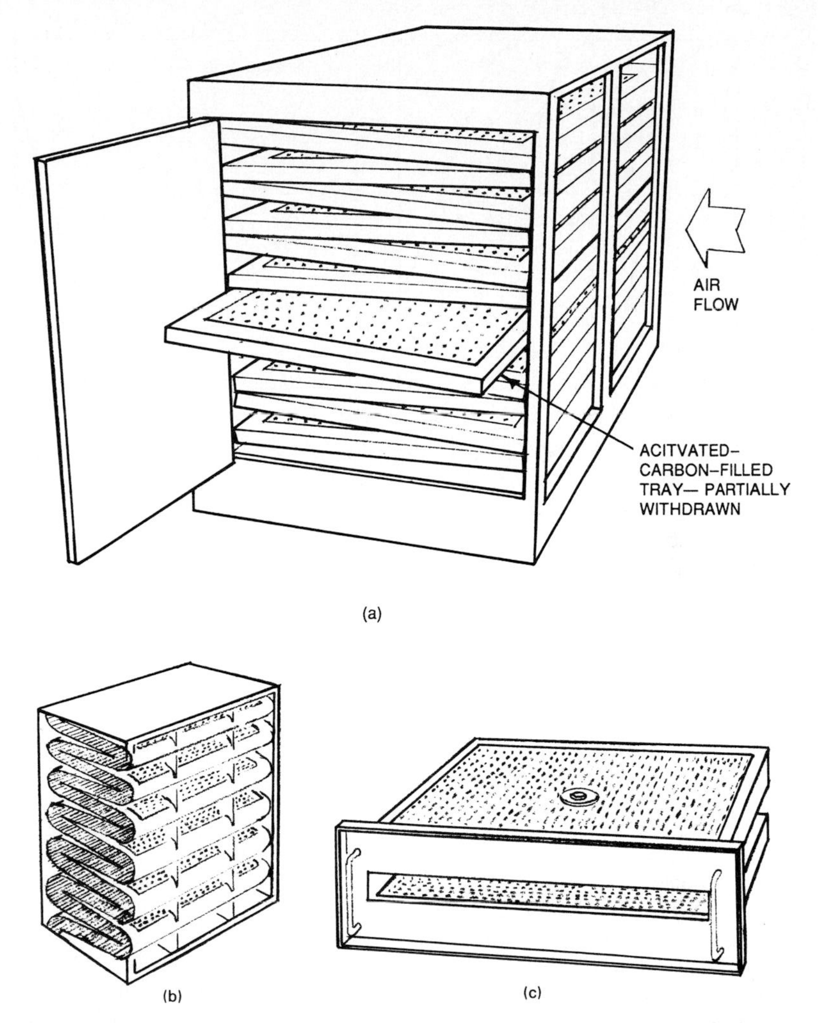

FIGURE 36.12 Adsorber configurations for air filtration: (*a*) removable panel odor filter; (*b*) sinuous-bed filter; (*c*) flat panel nuclear or toxic contaminant filter.

36.6.3 Performance Characteristics

Penetration and Breakthrough. The penetration of adsorbates through any adsorber or chemisorber bed increases as contaminants are trapped by the bed. If the bed is deep enough and if adsorption is vigorous, the unpolluted bed adsorbs essentially all the adsorbate. Penetration is very low until nearly all active adsorption pores are used up. Thereafter very little adsorbate is captured, and penetration rises to 100 percent. Because the change in penetration is often sudden, it is usually called "breakthrough." For shallower beds with low inlet concentrations and weaker adsorption, this breakthrough is less sharp. Figure 36.13 shows the typical form of the passage of an adsorption wave through an adsorber bed. For the fresh bed ($T = 0$) the adsorbate concentration in the flowing gas de-

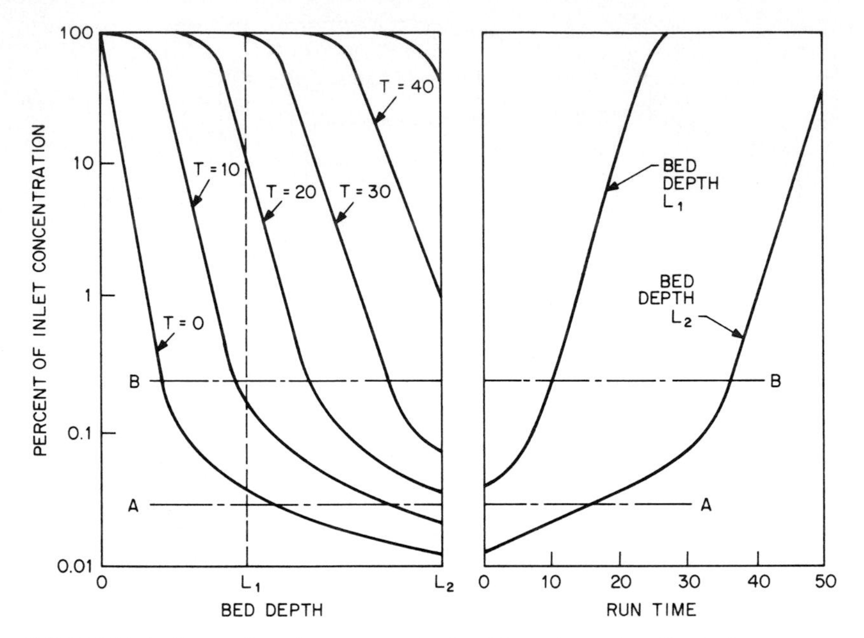

FIGURE 36.13 Penetration of gaseous contaminant through an adsorption bed.

creases nearly exponentially with bed depth until the concentration in the bulk gas stream begins to equal the concentration at the adsorber surface. The force driving adsorbate into the pores is now very small, and the adsorbate concentration in the stream decreases more slowly for a given amount of bed depth. Finally the initial portions of the bed become filled with adsorbate and useless; the concentration curve essentially starts anew at the boundary of this spent bed. (Note curves for $T = 10$ to $T = 40$ h in Fig. 36.13.)

It is not customary for suppliers of adsorbers and chemisorbers to supply this type of data for their products. They appear in published papers (e.g., Nelson and Correia[37]) and can be developed with reasonable accuracy from theory. The following must all be known to evaluate penetration test data for an adsorber:

- Carrier gas temperature, relative humidity, pressure, and velocity
- Adsorbate physical properties and concentration at bed inlet
- Adsorber bulk density, particle-size distribution, activity, and retentivity for adsorbates of interest
- Bed depth and pressure drop

Service Life. The operating parameters just listed plus the allowable bed penetration determine the service life of an adsorber bed. Two breakthrough levels are indicated in Fig. 36.13. For the bed with depth L_1, breakthrough level A is reached instantly; level B requires about 10-h exposure. For a bed of depth L_2, breakthrough at level A requires 16 h and at level B, 36 h. (Times given are merely for illustration and do not apply to any specific adsorber.) Detailed calculation of such breakthrough curves is time-consuming. A preliminary best-case analysis will frequently determine whether detailed calculation or, better, testing of the adsorber-adsorbate system is worthwhile. This calculation proceeds as follows:

If the adsorber is assumed 100 percent efficient on the adsorbate of interest, a simple mass-balance calculation shows that the maximum life of the adsorber bed is

$$t = \frac{360 f \rho_a L}{cv} \tag{36.9}$$

where t = life, h
ρ_a = adsorber bulk density, g/cm^3
L = bed depth, cm
c = adsorbate concentration upstream of bed, mg/m^3
v = carrier gas velocity through actual bed, m/s
f = ratio of adsorbate mass in bed to adsorber mass, at saturation

The term f is called the "fractional retentivity," and it is usually tabulated as a percentage. The manufacturers' literature for activated carbons give values from 2 to 40 percent for retentivity with values of 15 percent being typical. Carbon bulk densities are about 0.5 g/cm^3, bed depths are 2.5 cm, and velocities are about 0.2 m/s. Given these values, we get

$$t = \frac{(3170)(0.15)(0.5)(2.5)}{(c)(0.2)} = \frac{2972}{c} \quad \text{h}$$

A change-out period of 6 months (4380 h) might be acceptable from an economic standpoint. In that case,

$$c_{\text{max}} = \frac{2972}{4380} = 0.68 \text{ mg/m}^3$$

This is the maximum inlet concentration tolerable for these assumptions. Table 36.5 shows that this is in the order of magnitude of some odor thresholds, but considerably below most TWA8 values. Thus fixed activated carbon beds are likely to require replacement far more often than every 6 months in industrial situations where concentrations are near TWA8 levels. This is especially true where the assumptions made for this best-case analysis are unrealistically favorable.

A different limitation will exist for highly odorous or toxic adsorbates, where allowable concentrations may be quite low. In these cases, the adsorbate concentration downstream of the adsorber must be substantially lower than that upstream, or else there will be little justification for the air purification system. Reduction of the mass concentration of an odorous adsorbate by 75 percent, e.g., will be barely noticeable by most observers. Toxic adsorbates must be removed with substantial safety margins. Thus penetrations in the order of 1 to 10 percent must be achieved for toxic or odorous adsorbates throughout the useful life of the bed. Such penetrations are difficult to achieve at the low concentrations listed in the OTH column of Table 36.5 unless deep adsorber beds and long stay times are employed. This does not tend to make economic systems.

Detailed Life Calculations. Economic comparisons and, indeed, decisions on whether to use adsorption at all (instead of ventilation) depend on knowing the service life of the adsorber more accurately than through best-case calculations. The theory of adsorption is poorly developed for low concentrations and multiple adsorbates, which is usually the situation for air filtration. Some procedures require that the user know contaminant physical properties which are not generally available. A fairly simple semiempirical procedure has been proposed by Nelson and Correia.[37] They obtain this expression for the time required for penetration of a single contaminant to reach 10 percent:

$$t_{10} = \frac{280Wf}{C^{2/3}M^{1/3}Q} \qquad (36.10)$$

where W = carbon mass, g
C = inlet concentration, mg/m^3
M = molecular weight, g/(gmol)
Q = air flow, m^3/min
f = ratio of mass of contaminant adsorbed to mass of carbon at 10% penetration breakpoint, %

Nelson and Correia (Ref. 37) give a means of calculating f based on the boiling point of the adsorbate. Adsorber suppliers often provide f, based on test data. In addition, Ref. 38 indicates that interferences between contaminants are not too severe at low loadings. That paper gives a procedure for handling multiple adsorbates.

Life of Chemisorbers. Similar procedures are applicable to chemisorptive beds as well as those which operate on purely physical adsorption. For chemisorbers, the best-case analysis is even simpler, for once all the impregnant has reacted with the adsorbate, the bed is spent, or at least reduced to the performance of an unimpregnated bed. If the carrier for the impregnant is a poor physical adsorber (which is the case for activated alumina with most contaminants, e.g.), the bed is a useless energy consumer. Truly catalytic beds, of course, suffer less from these disabilities, but even they eventually become "poisoned" and must be replaced.

Inlet Concentrations. For any single gaseous contaminant, the concentration reaching the adsorber inlet can be calculated by the procedures described in Sec. 36.4.3. Indoor contaminant source terms may be derived from the data in Table 36.9. EPA annual air-quality reports, such as Refs. 4 and 5, give data for outdoor levels of the priority pollutants (Table 36.7). Others can be obtained from the periodicals listed in the bibliography (Sec. 36.11). Intakes are often located well above ground, where gaseous contaminant concentrations may actually be *higher* than at grade, where EPA samples are often located. The following expression fits typical vertical distribution patterns for gaseous contaminants:[39]

$$C = C_2 \exp\left[2.9 \times 10^{-5}(h^2 - 274h + 545)\right] \qquad (36.11)$$

where C = concentration at height h above grade
C_2 = concentration at local sampling station (assumed 2 m above grade)
h = height above grade, m

Regeneration. Activated carbon can be restored to nearly new performance by passing air, nitrogen, or steam at elevated temperatures through it. Steam is generally preferred, because it is cheaper than nitrogen, yet does not support combustion. Collected contaminants are desorbed or eluted from the carbon, leaving cleaned pores available for another adsorption cycle. The desorbed contaminants, of course, appear in the steam; unless some means of detoxification of the contaminant is provided, regeneration merely moves a contaminant from one place to another. The usual detoxification process is incineration, which may be

TABLE 36.9 Indoor Gaseous Contaminant Generation

	Activities									
	Theaters, concert halls	Meeting rooms	Trade shows	Quality restaurants	Cafeterias	Bars	Enclosed sports arenas	Corridors, foyers	Office space	Units
Occupancy factors										
Population*	1.5	0.8	0.3	0.5	0.8	1.5	1.5†	0.5	0.2	occupant/m^2
Smoking	0.0	0.5	0.2	1.2	1.5	2.5	2.0	0.2	0.7	cigarette/(h · occupant)
Human emissions										
Carbon dioxide	24	33	48	40	62	55	55	21	35	g/(h · occupant)
Methane	2	2	2	2	2	2	2.7	1	2	mg/(h · occupant)
Hydrogen	1	1	1	1	1	1	1.3	0.5	1	mg/(h · occupant)
Hydrogen sulfide	0.015	0.015	0.015	0.015	0.015	0.015	0.02	0.007	0.015	mg/(h · occupant)
Ammonia	9	18	30	18	40	35	35	7	20	mg/(h · occupant)
Smoking										
Carbon dioxide	—	0.7	0.7	0.7	0.7	0.7	0.7	0.7	0.7	g/cigarette
Carbon monoxide	—	80	80	80	80	80	80	80	80	mg/cigarette
Organics	—	15	15	15	15	15	15	15	15	mg/cigarette
Building emissions										
Formaldehyde*	0.003	0.003	0.003	0.003	0.003	0.003	0.003	0.003	0.003	mg/(min · m^2)
Ozone	—	—	0.1	—	—	—	—	—	0.1	mg/photocopy

*Per m^2 of floor area.
†Seating areas only.

economically feasible if the desorption flow is low enough. The cost of handling and transport to and from the regenerator must be added to the regeneration costs to determine whether regeneration is practical. The alternative to regeneration is to bury or burn the spent carbon and replace it. Adsorbants for critical service should always be new material, not regenerated.

Pressure Drop and Energy Consumption. Bed pressure drop can be calculated approximately if the bed depth granule diameter, percentage of solids, gas velocity, and gas physical properties are known. Suppliers of granular adsorbers provide pressure drop test data for air on their various grades. Overall pressure drop is the drop through the granular bed plus the drop through any structure used to hold the granules. Structure resistance may not be negligible; it is best to obtain actual test data on the complete adsorber unit. Pressure drop through an adsorber does not change if the adsorber is protected from dust buildup, which can be done by placing particulate filters upstream of the adsorber.

Environmental Considerations. Humidity is always a factor in gaseous contaminant air filtration. Physical adsorbers generally work best at low humidities and may fail entirely if they adsorb too much water. Chemisorbers, however, usually cannot operate at extremely low humidities. Chemical reactions tend to work better at high temperatures, hence these favor chemisorbers and catalysts. Temperature excursions above specified operating conditions, however, may drive adsorbates out of physical adsorbers and may evaporate or decompose impregnants, chemisorptive compounds, or catalysts. Activated carbon is flammable and a poor heat conductor; adsorption is a process which generates heat. Spontaneous ignition is possible when adsorbate concentrations are high. Where unusual conditions are expected, the adsorber/adsorbate system should be tested at those conditions.

Vibration of the adsorber bed can cause attrition of adsorber granules. Standard tests for attrition resistance, ignition point, and other adsorber properties have been established by the American Society of Testing and Materials (ASTM).[40]

36.7 PARTICULATE AIR POLLUTION CONTROL EQUIPMENT

Table 36.10 gives characteristics and typical application areas for various types of particulate air pollution control equipment; Figs. 36.14 and 36.15 give schematic illustrations of their operating principles. References 3, 40, 41, and 42 provide further design advice. In general, the higher the energy input to the device, the lower the penetration through it. Devices which circumvent this rule (e.g., electrostatic precipitators) will probably require high initial investment and more extensive control systems and maintenance. Equipment choices are usually based, first, on which system will provide the needed penetration level and, second, which will function reliably in the operating environment.

36.7.1 Operating Principles

Inertial (Aerodynamic) Collectors

Large-Scale Cyclones (Fig. 36.14a). Ga enters tangentially at high velocity at the top of the cyclone, changes direction at the bottom, and leaves through the

TABLE 36.10 Particulate Air Pollution Control Equipment: Characteristics and Application Areas

Control equipment type	Pressure drop,* Pa	Temperature range,† °C		Application areas (see Table 36.1 for codes)
		Low	High	
Large-scale cyclones	125–800	−40	550	4, 5, 9, 10, 15, 16, 22, 28, 31, 35, 42, 44
Small, multiple cyclones	500–1500	−40	550	1–5, 7, 9, 10, 15, 16, 22, 26, 31, 34, 35, 42
Louvers	125–500	−40	350	47
Dry dynamic collectors‡		−30	370	5–7, 10, 11, 16, 22, 23, 26, 29, 31, 34, 44, 45
Wet dynamic collectors‡		5	70	5, 7, 9, 13, 19–21, 29, 31, 45
Centrifugal mist collectors‡		5	110	24
Shaker baghouse	500–1500	−30	300	6–10, 17, 26, 35, 42, 44, 45
Collapse-cycle baghouse	500–1500	−30	300	
Reverse-pulse baghouse	500–1800	−30	300	1–10, 13, 15–17, 20–22, 26, 29, 30, 40, 44, 45
Pulsed cartridge collectors	300–1000	−20	110	5, 8, 9, 13, 22, 24, 25, 29–31, 45, 47
Granular beds		−40	800	46
Lint arrestors	180–400	−20	65	13, 28, 29, 33, 42
Low-voltage, two-stage electrostatic precipitators	125–375	−30	60	14, 22, 24, 25, 31, 43
High-voltage, one-stage electrostatic precipitators	125–375	−20	400	1–3, 17, 18, 35–41, 46
Venturi scrubbers	2000–20,000	5	90	7, 11, 12, 17, 18, 35–39, 46
Packed-bed scrubbers	250–2500	5	90	32
Impingement scrubbers	500–2000	5	90	6, 7, 9, 19–26, 29, 31, 32, 34, 35, 39, 41
Submerged-nozzle scrubbers	500–1500	5	90	6, 7, 9, 19–27, 29, 31, 32, 34, 35, 39, 41

*249 Pa = 1 inWG.

†Temperature range is determined by materials of construction, except for wet scrubbers. Values given are typical for commercially available devices. To convert to USCS units: °F = 1.8 × °C + 32.

‡These devices combine blower and collector; they usually provide from 500 to 2500 Pa external static pressure. Motors may require protection to meet listed temperatures.

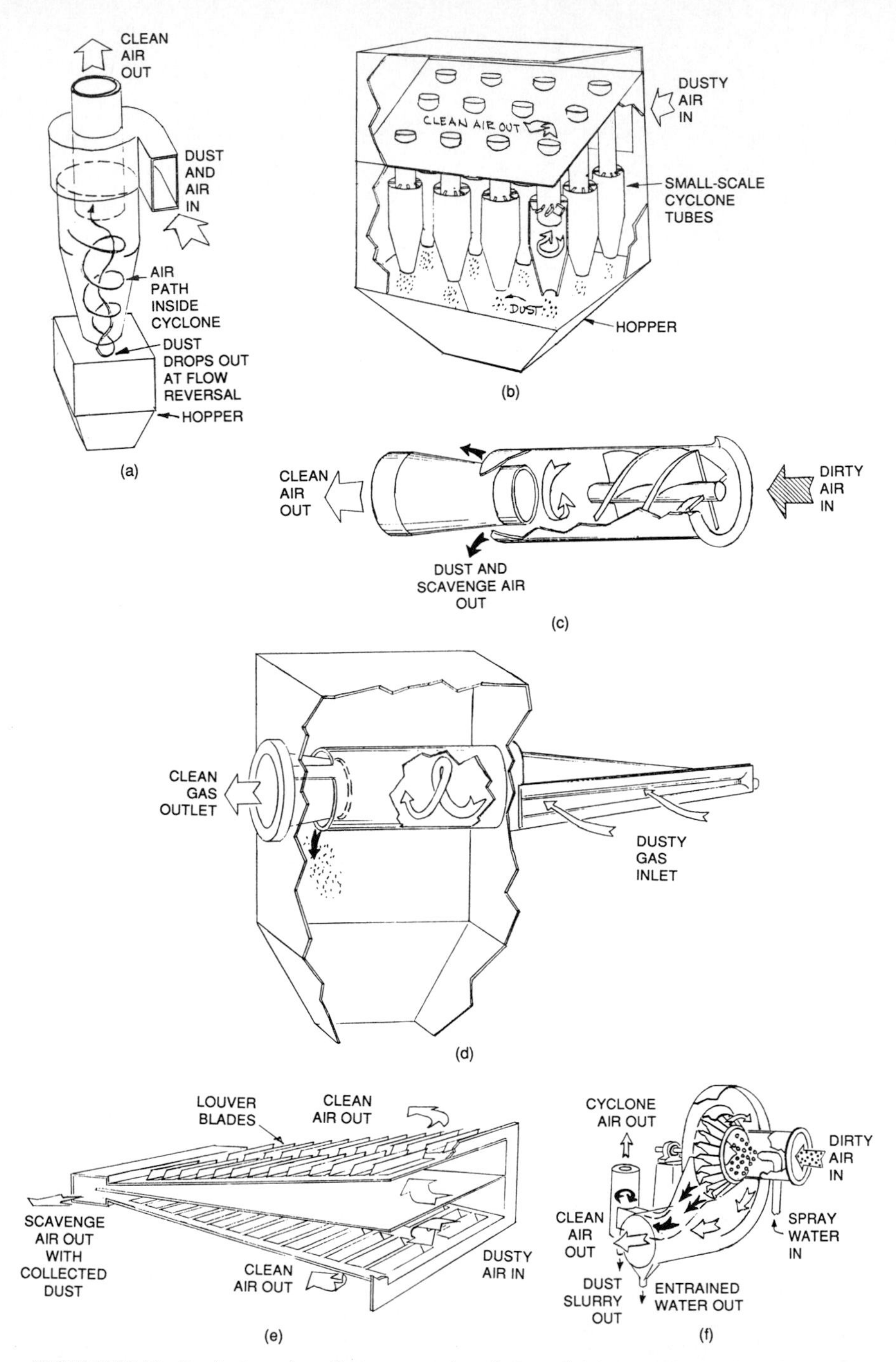

FIGURE 36.14 Particulate air pollution control equipment: (*a*) large-scale cyclone; (*b*) multiple bank of small-scale cyclones; (*c*) helical-vane cyclone; (*d*) straight-through cyclone; (*e*) louver; (*f*) dynamic blower-collector; (*g*) centrifugal mist collector; (*h*) shaker baghouse; (*i*) collapse-cycle baghouse; (*j*) reverse-pulse baghouse; (*k*) pulsed panel cartridge collector; (*l*) pulsed cylindrical element cartridge collector; (*m*) vacuum-type lint collector; (*n*) paper web lint/ink mist collector.

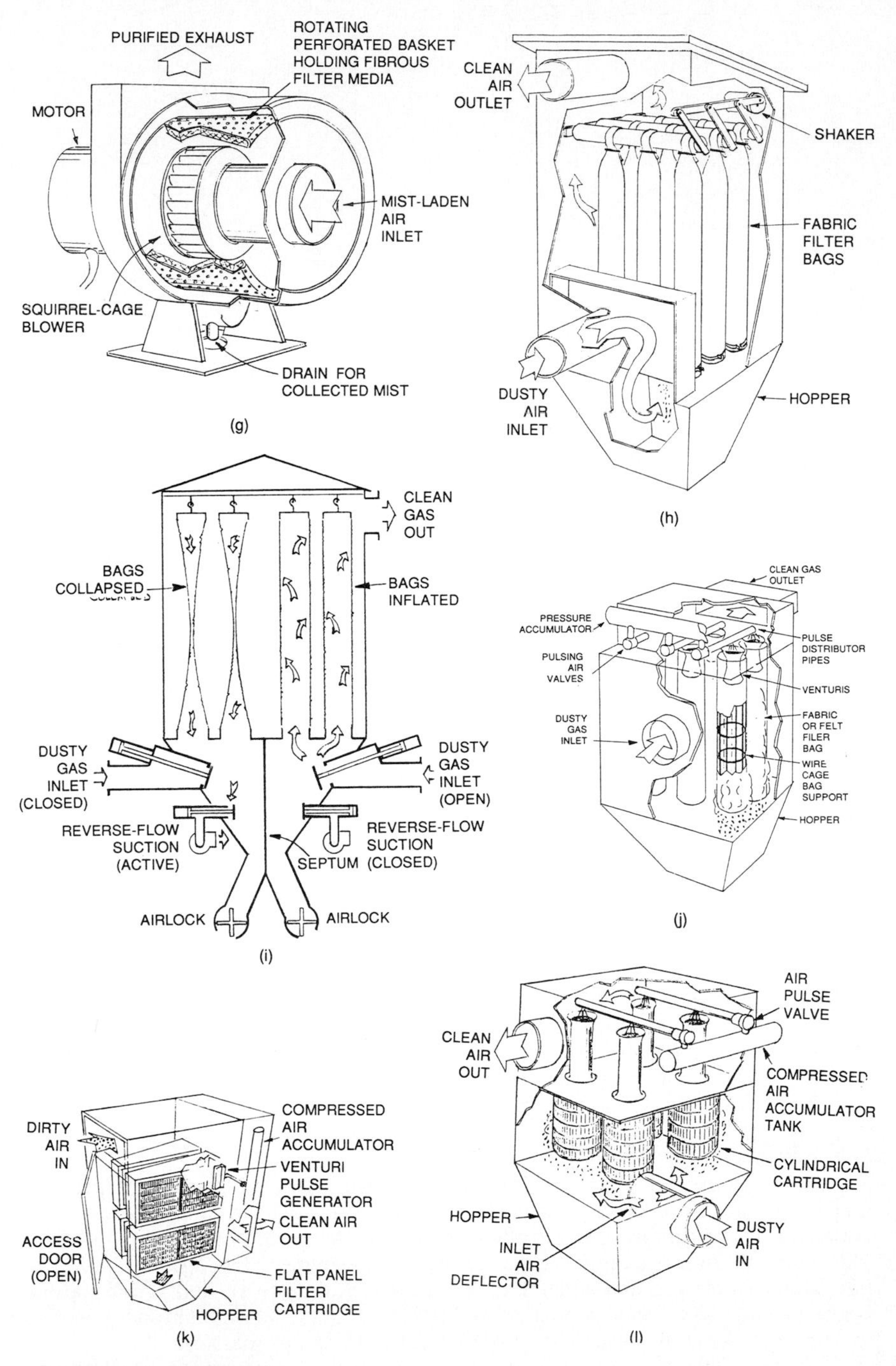

FIGURE 36.14 (*Continued*)

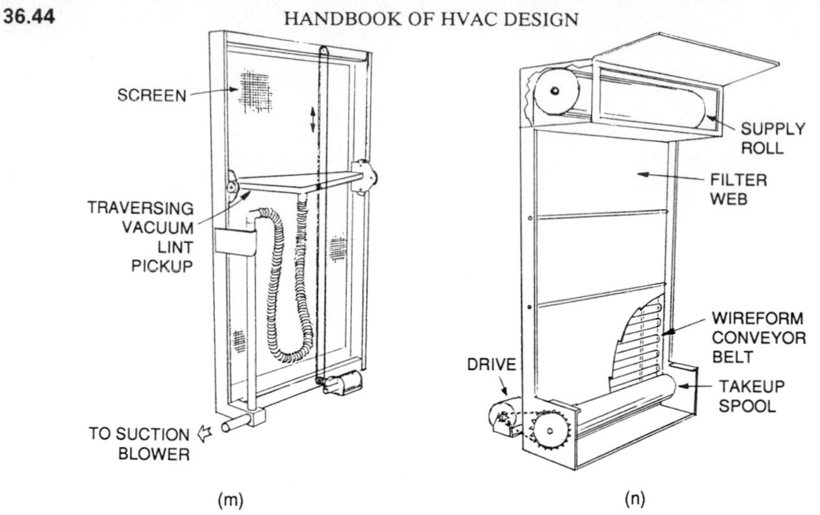

FIGURE 36.14 (*Continued*)

central cylinder. Dust is centrifuged to the cyclone walls and falls into the hopper at the gas turnaround zone. Higher velocities—up to about 30 m/s—produce lower penetrations, but these must be considered large-particle control devices, or precleaners for more effective devices.

Small-Scale Cyclones (Fig. 36.14b). Sometimes these are simply small-scale versions of Fig. 36.14*a*, with many units mounted in parallel. Spinning can be imparted to the gas by "twirlers," helical vanes at the gas inlet to each cyclone (Fig. 36.14*c*); in this case, gas entry is axial instead of tangential. In another version (Fig. 36.14*d*) the clean air passes out through a central tube, with no flow reversal. A secondary air flow (for the dust-heavy gas) greatly improves performance.

Louvers (Fig. 36.14e). Gas is forced to reverse direction by louver blades. Dust-heavy gas is drawn out a slot at the downstream end of the device.

Dynamic Collectors (Fig. 36.14f). These combine a fan and a particulate separator. The specially shaped fan blades centrifuge particles to the outside of the fan spiral, where they are skimmed off into a hopper. The device can be assisted with water sprays in the inlet.

Centrifugal Mist Collector (Fig. 36.14g). In this device, a fibrous cylindrical air filter is rotated rapidly. Air entrained by the filter exhausts radially. Entrained droplets, however, impinge on the filter fibers and drain off through a perimeter slot. The device is used to capture coolant-oil mists from machine tools.

Fabric Collectors

Shaker Baghouses (Fig. 36.14h). These operate on the same principle as a household vacuum cleaner. Gas passes through the fabric, where a cake of dust builds up, greatly improving filtration efficiency. At some point, groups of bags are shaken to dislodge the collected cake, and the cycle starts again. Units with flat fabric panels are available. Fabrics used include woven and felted natural synthetic and glass fibers as well as laminations of these and membrane materials to improve base collection efficiency and cake-release properties.

Collapse-Cycle Baghouses (Fig. 36.14i). Here the particulate cake is removed by a complete reversal of gas flow, which collapses each tube.

Reverse-Pulse Baghouses (Fig. 36.14j). In these devices, particles are collected on the outside of the fabric tubes, which are kept from collapsing by an internal wireform. When the dust cake builds to the maximum desired level, a pulse of compressed air is injected through a venturi at the top of each of a group of bags. This pulse induces gas flow into the bag, causing a momentary flow reversal to ripple down each pulsed bag and break the cake loose. The system has the advantage of having a minimum of moving parts in the gas stream.

Pulsed Cartridge Collectors (Fig. 36.14k and l). These are similar in principle to reverse-pulse baghouses, except that the collection medium is a pleated paper web instead of fabric. Both cylindrical and flat-panel configurations for the cartridges are available.

Lint Arrestors. Collectors for lint in textile and printing plants have two forms. One form makes use of an endless wire-screen band (Fig. 36.14*m*). Linty air passes through the band twice, with almost all the lint depositing on the screen during the first pass. The band moves slowly, so that the collected lint is steadily fed into the nozzle of a vacuum cleaner for transfer to a fabric collector of much smaller capacity. The second type of lint collector (Fig. 36.14*n*) makes use of an endless wire-mesh band, but this is merely a support for a very thin, porous paper web. Lint builds up on this to make an effective filter. The web and its collected lint are rolled up to form a disposable package. This form of lint collector is also effective in capturing the ink mist which often accompanies lint in printing operations.

Electrostatic Precipitators

Low-Voltage Two-Stage Precipitators. A few industrial applications (e.g., welding-fume collection) use the same type of two-stage collector described for air filtration use in Sec. 36.5. These will be made somewhat more rugged for this service and may include a deionizing stage downstream to eliminate the space-charge effects which occur when dust concentrations are high.

High-Voltage, Single-Stage Precipitators (Fig. 36.15a). These are widely used for stack gas cleaning, especially in fossil-fuel power plants. They consist of parallel grounded plates, with ionizer electrodes suspended midway between these, separated from each other by a distance about equal to the plate-to-plate spacing. The ionizer electrodes may be wires, but are often tubes which carry barbs of various forms; the ions are created at these barbs. Plate-to-plate spacings are large—typically 30 cm—and voltages high—typically 70 kV. Dust collected on the plates is removed by rapping the plates. Voltage is adjusted so that there is a controlled arcing rate between electrodes.

Wet Scrubbers

Venturi Scrubbers. A venturi scrubber (Fig. 36.15*b*) is a duct with a short converging entrance and a longer diverging exit. This shape allows the gas to accelerate to a high velocity at the smallest cross section (the throat) but to drop back to the normal duct velocity without excessive pressure losses. Water is injected into the venturi throat, through very crude, difficult-to-plug nozzles or slots. The very large difference between the water and gas velocities tears the water into a cloud of fine droplets, which, however, are not traveling as fast as the aerosol particles in the gas. The particles therefore collide and combine with the water droplets, which are subsequently removed by a cyclone or other low-

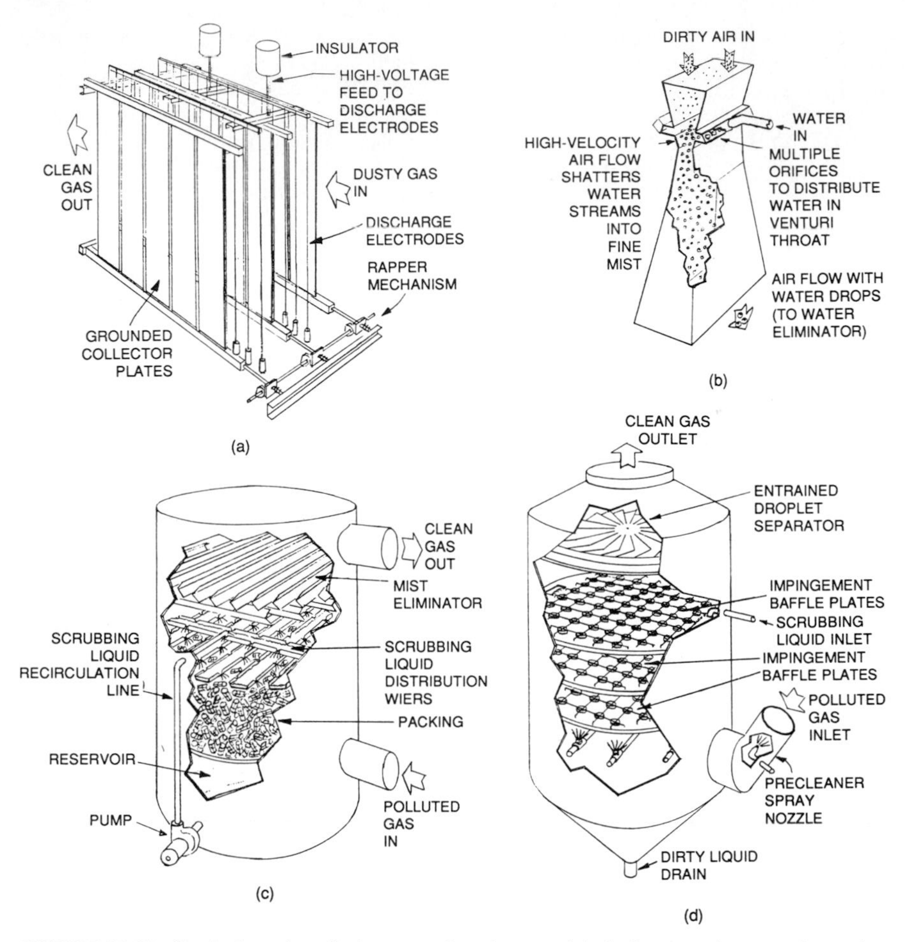

FIGURE 36.15 Particulate air pollution control equipment: (*a*) single-stage electrostatic precipitator; (*b*) venturi scrubber; (*c*) packed tower scrubber; (*d*) impingement baffle scrubber; (*e*) fiber-bed scrubber; (*f*) sinuous submerged-nozzle scrubber; (*g*) submerged jet scrubber; (*h*) granular bed-collector.

energy scrubber. Penetrations can be very low through venturi scrubbers and reliability high—but at the cost of high energy consumption.

Packed-Bed Scrubbers (Fig. 36.15c to e). In these scrubbers the gas is passed through beds of packings (marbles, rings, saddles, impingement plates, or fibers, which are continuously irrigated with a liquid spray). Particles are captured by impaction on both the packing materials and droplets of the liquid and are continuously removed in the flowing liquid. The liquid is filtered externally and recirculated.

Submerged-Nozzle Scrubbers (Fig. 36.15f). In this device, liquid-particle contact is made in a sort of upside-down waterfall formed in a half-submerged, sinuous passage. The device has neither spray nozzles nor liquid filters to maintain; the collected dust settles out in the liquid tank, from which it is periodically removed by scraping. Another form uses a submerged jet (Fig. 36.15*g*).

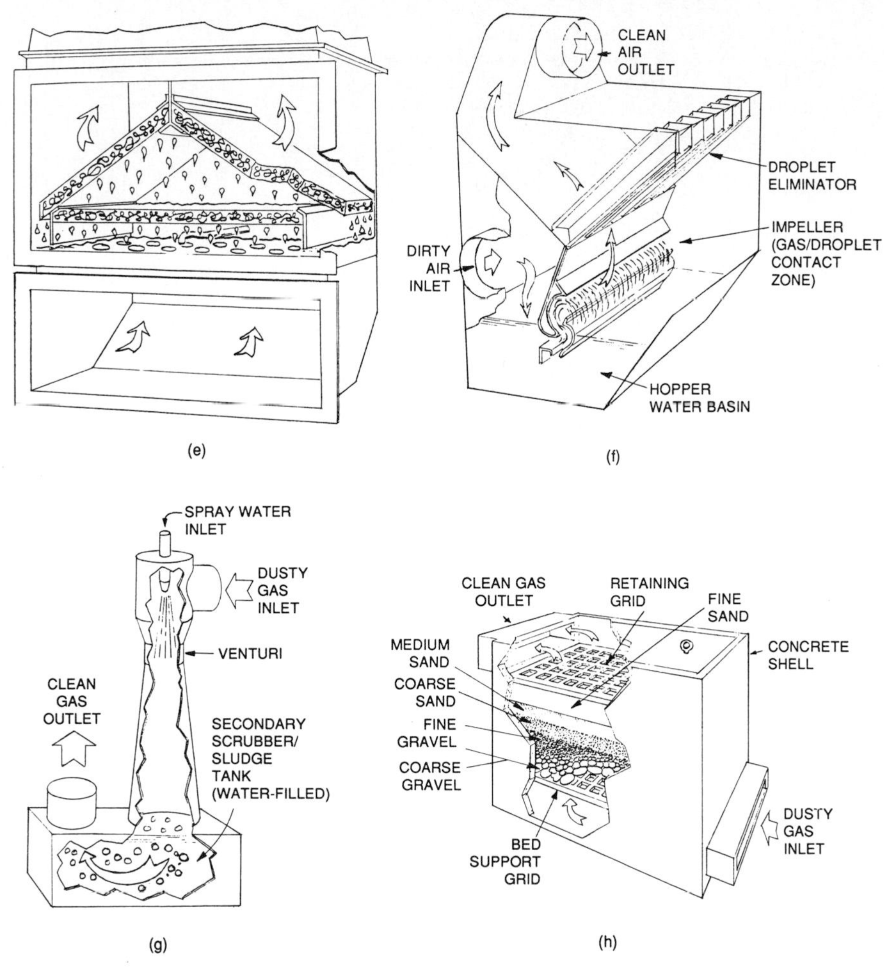

FIGURE 36.15 (*Continued*)

Granular Bed Collectors (Fig. 36.15h). In these, the gas passes through a rather deep bed made of granules with progressively smaller diameters as it moves from the inlet to outlet face of the collector. (Some reverse gradation near the outlet face may be necessary to keep the finer granules from entraining into the gas stream.) Granules are usually sand and pebbles. Collection is by impaction, sedimentation, and diffusion of dust onto the granules. This concept has been used in high-temperature systems and in highly radioactive streams.

36.7.2 Performance Characteristics

Penetration and Efficiency. Typical penetration–particle diameter curves for selected air pollution control devices are shown in Fig. 36.16. Detailed data for specific devices are usually available from device manufacturers. Such data must be

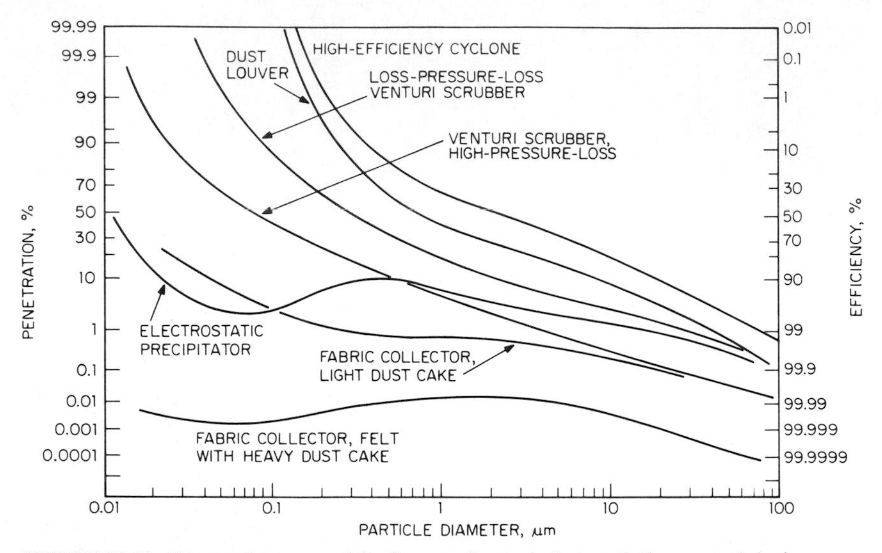

FIGURE 36.16 Penetration vs. particle diameter for typical air pollution control devices.

quite specific with regard to operating conditions during the test, giving gas constituents, temperature and pressure, and particle density. A common problem in pollution control equipment selection is to estimate the device penetration at some gas condition entirely different from the test, or for particles with density different from that used in the test.

A reasonable approximation to this can be obtained by converting the diameter scale of the diameter-penetration curve to an "interception parameter" scale. This interception parameter is, for a given device,

$$N_i = \frac{\rho_p d^2 V}{\mu} \tag{36.12}$$

where μ = gas viscosity, g/(cm · s)
ρ_p = particle density, g · s/cm
d = particle diameter, cm
V = gas velocity through device, cm/s
N_i = interception parameter

This same parameter can then be calculated for the operating conditions for which penetration data are desired, using diameters over the range of interest. One can then pick off penetrations for the values of N_i which correspond to the chosen diameters and plot a new diameter-penetration curve. The units used are immaterial but must, of course, be kept the same for all conditions examined.

Pressure Drop and Service Life. Pressure drops through pollution control devices are obtainable from manufacturers of the devices, with proper consideration of gas temperature and density and liquid flows, if any. The pressure drop through fabric collectors and pulsed cartridge collectors rises to a nearly steady-state level, which is a function of aerosol concentration, medium velocity, and frequency and vigor of the cleaning cycle. The other particulate pollution control devices listed above have essentially steady pressure drop, for they do not store dust on medium which remains continuously in the air stream. The media in fabric collectors and pulsed cartridge units do have finite service lives and must be replaced when they begin to lose strength or become clogged with dirt. For lint collectors which use replaceable filter webs, media must be replaced when the roll has run out. All pollution control equipment is subject to corrosion and abrasive wear.

Disposal of Captured Dust. Dust captured by dry collectors is usually deposited in a pyramidal hopper. Rotary "air locks" allow dust to be removed without inrush or outrush of gas. The dust removed may be bagged or dropped into containers, but care must be taken to avoid loss during this transfer. Wet collectors may transfer innocuous wastes directly to sewers, but it is more common to allow the collected material to settle in a pond and recirculate the scrubbing liquid; even water is sometimes precious. Whether the collected dust is dry, in a slurry, sludge, or pelletized form, final disposal may be simple or difficult. Innocuous dusts may be piled or buried, but there are often regulations with regard to disposal of hazardous materials. Many dusts have recovery value; this should be investigated.

Operating Limitations and Maintenance. Air pollution control equipment must be designed to withstand the temperature and pressure excursions normal to the process involved. If the gases involved are known to be corrosive, the materials chosen must withstand this corrosion for an economic equipment lifespan. Wet collectors especially require operation within fairly narrow flow limits. Fabric collectors need to be inspected internally at regular intervals to check on failed bags; regular preventive maintenance schedules should be followed on all equipment. Support structures and foundations must meet the usual American Institute of Steel Construction (AISC) requirements, and maintenance ladders and catwalks must be well thought out and designed to OSHA standards. Safety interlocks are essential on all electrostatic precipitators.

36.8 GASEOUS CONTAMINANT AIR POLLUTION CONTROL EQUIPMENT

36.8.1 Operating Principles

Equipment for control of gaseous air pollution emissions uses many of the same principles described in Sec. 36.6 for air filtration systems. The differences are matters of scale, maintenance, and operating environment. In addition, the emission from a pollution control system can sometimes be saturated with water vapor without causing problems; stack emissions need not be at air-conditioning temperature; reaction products which might not be tolerable in air for breathing may be at a low enough level to be emitted into the atmosphere and diluted. Incineration is often practical, and condensation, even at cryogenic temperatures, is sometimes used.

Air pollution control systems for control of gaseous emissions are basically chemical engineering problems. The literature in the subject is immense. References 3 and 41 through 43 provide design guidance beyond the brief outline given below, and periodicals listed in the bibliography (Sec. 36.11) give current developments.

Adsorption. The fundamentals of adsorption for stack gas cleaning are the same as for air filtration service. (See Sec. 36.6.) Dry adsorbers are less used for pollution control than for air cleaning, because temperatures are often high. Where they are used, the normal maintenance arrangement is in situ regeneration. Highly practical solvent recovery systems are built in this way. Dry adsorbers (e.g., activated carbon for sulfur dioxide capture) are sometimes used as a device to concentrate a contaminant, which is subsequently immobilized by a second system when the adsorber is regenerated. Contaminant concentrations in air pollution control systems are usually much greater than those in air filtration, and regenerative adsorption systems are therefore more likely to be practical.

Absorption and Chemisorption. Dry adsorbers impregnated with various reactants are used to capture mercury, radioiodine and its organic compounds, hydrogen sulfide, and acid gases (especially those in the highly toxic chemical-warfare-agent class).

Liquid scrubbers, both with and without added reagents, are commonplace in gaseous contaminant pollution control. Scrubbing liquids are sprayed into chambers which may be packed with rings, saddles, marbles, fibers, perforated plates, etc., to provide large liquid-gas interface areas (Fig. 36.15*c* to *e*). Bed length and flow velocities must be adjusted to give adequate contact time, yet not allow entrainment of the scrubbing liquid into the gas stream. The reaction used must not create solid precipitates which will fill up the passages through the packing. Final disposal of the reacted contaminant usually involves a secondary cycle outside the scrubber, where precipitates can be filtered, centrifuged, or settled out of the scrubber liquid. The mass-transfer area and contact time for particulate-type wet scrubbers are usually inadequate for gaseous contaminant collection.

Condensation. If the contaminant has a higher boiling point than the carrier gas, it is possible to condense it out of the carrier gas on a cool surface. This technique is often used in equipment for stack gas sampling, where flows are small and one wants to capture all condensables. The condensation of radioactive xenon, krypton, and radon in activated carbon at cryogenic temperatures is one of the few practical ways to capture these gaseous contaminants.

Catalysis. Catalysis is generally more useful in air pollution control than in air filtration because the control targets are usually a few known contaminants and higher temperatures are available. Reaction heats add to the energy available to sustain catalytic actions.

Catalytic Incineration. Catalytic incineration is sometimes a practical alternative to thermal incineration or regenerative adsorption systems. With a catalyst present, oxidation of the contaminant can be achieved at lower temperatures (300 to 600°C) than with thermal incineration (>650°C). The entire carrier gas stream must be held at these temperatures, in either case. Recuperative heat exchangers downstream of the catalyst bed or incinerator return energy from the exhaust stream for use in preheating the carrier gas or some other process. A problem with catalytic incineration is that combustion of the contaminant is sometimes incomplete and so the resulting products may be odorous or toxic.

Thermal Incineration. In thermal incineration, the gas stream is heated, usually with direct-fired natural-gas burners, to the temperature at which the contaminant oxidizes. This is a form of combustion, and once the process starts, it may be possible to reduce the fuel input—or, with very high concentrations, the reaction may be self-sustaining. The method is simple, nonspecific, generally reliable, but expensive, for the gas temperature must be raised substantially.

36.8.2 Performance Characteristics

System Penetration. Higher device penetrations are usually allowable for gaseous contaminants in air pollution control systems than in air filtration, because the chief aim is to bring the effluent down to some safe, legislated or aesthetically acceptable control level. Atmospheric dispersion from tall stacks tends to increase safety and aesthetic acceptability. Concentrations of contaminant and carrier gas constituents at the control equipment inlet must be known, along with gas temperature and pressure and many gas properties. Mass transfer, adsorption, catalysis, and combustion theory are comparatively well developed at the concentration levels met in air pollution control, and traditional chemical engineering techniques can be applied. Much engineering effort is needed for accurate prediction of performance. In general, performance for well-designed systems remains stable for a given inlet condition; the exception to this is the catalytic incinerator. These decline in performance as the catalyst gradually becomes poisoned. Systems with catalysts must operate at some minimum temperature, while adsorbers fail miserably at high temperatures.

Environmental Limitations. Temperatures, gas and liquid flows, and reagent concentrations must be accurately controlled if systems are to operate as designed. Buildup of precipitates in liquid scrubbers is not necessarily reversible; corrosion is a major problem in all systems, especially where sulfuric acid can condense out of the gas stream. The combination of heat, free water, and catalysts is inherently destructive to metals; stainless steel, monel, titanium, glass, and plastics (such as polyester/glass-fiber composites) are often needed for materials of construction. The disposal of liquid and solid wastes from control processes is a serious problem, subject to controls as stringent as those for the gasborne pollutants themselves.

Pressure Drop and Energy Consumption. Suppliers provide data on system pressure drops at operating conditions (which include liquid-flow rates for scrubber systems). Overall energy consumption calculations must include the cost of air horsepower as well as liquid pumping costs and fossil-fuel consumption. Reagent costs, catalyst or adsorber replacement, regeneration steam, heat-exchanger fan power, waste disposal, and amortization of capital are among overall life-cycle system costs.

36.9 GAS PURIFICATION EQUIPMENT: PERFORMANCE TESTING

Purchasers of pollution control equipment often require a performance guarantee, to be proved by tests on the installed equipment. The only logical test is the percentage penetration for the equipment at specified operating conditions. It is not logical to specify concentrations downstream of the equipment, for this is always a function of upstream concentration, which is beyond the control of the pollution control equipment supplier. Thus performance evaluation of air pollution control equipment requires measurement of pollutant concentrations both upstream and downstream of the tested device. This validation process is usually a small part of the cost of the installed equipment; it can be done relatively quickly and may be required by regulatory agencies. If the device performance is sensitive to changes in operating velocity, temperature, etc., tests should be run over the expected range of these parameters. A useful survey of test methods is given in Ref. 44.

Testing of installed air filtration systems has been rare, except in critical cases. Customers usually rely on tests of filter samples. These tests should be made according to industry standard procedures to allow comparison between suppliers. (See "Laboratory Tests" below.) Field testing is possible but expensive in relation to equipment cost, and there are almost no standard procedures available, except for clean rooms and nuclear safety systems.

36.9.1 Aerosol Contaminant Testing

Traditional aerosol testing methods make use of integrated samples. Known sampling flows are drawn simultaneously upstream and downstream of the gas cleaner. Filters having extremely low penetrations for the aerosol present are placed in the sampling lines, to gather as much contaminant as possible. Enough sample must be collected to allow reliable measurement of the masses of aerosol captured.

Sampling filters are never perfectly stable; some gain or loss in filter mass is to be expected when no aerosol at all is collected. If we want the aerosol mass to be at least 20 times the expected gain or loss in the sampling filter, then the sampling time t (in minutes) must be

$$t = \frac{333fW}{PCQ} \tag{36.13}$$

where f = expected sampling filter gain or loss, % (due to moisture adsorption, etc., *not* aerosol accumulation)

W = sampling filter mass, g

P = air pollution control system penetration, %

C = upstream aerosol concentration, mg/m^3

Q = sampler flow, m^3/s

Substitution of reasonable values in Eq. (36.13) shows that even for high upstream concentrations, the sampling filter must be very stable (low f) and the mass of the sampling filter kept low.

The EPA has mandated a method for particulate measurement in stacks (method 5, Ref. 21, part 60, app. A). This procedure is usually specified for validation of stack gas particulate emissions.

Equation (36.13) applies to air filtration situations as well. In this, concentrations are so low that sampling times must be on the order of days, not minutes. Sensitive detectors have been developed to overcome some of these problems. Piezoelectric and beta-ray mass sensors are available to determine mass concentrations within a few minutes, with reasonable reliability. Optical particle counters take an entirely different approach. These measure the amount of light scattered by single particles passing through the instrument. This scattered light intensity is chiefly a function of the diameter of the particles. By counting the number of light pulses in a range of size bands, the instrument constructs a histogram of the particle sizes in the aerosol, as in Fig. 36.1. Calculations from the histogram data enable determination of particulate mass concentrations, provided the aerosol density is known. Users of optical particle counters should be cautious in interpreting data from these instruments. The particle sizes they give are not necessarily the true, "geometric" diameters of the particles observed, nor do these counters directly measure the filtration properties or mass of individual particles. Particles of several different diameters may appear of equal size in an optical particle counter. They are, however, widely applied in clean rooms, where their high sensitivity is necessary and the operating environment is gentle enough for their use. References 43 and 44 examine the capabilities and limitations of aerosol detection instruments in detail.

Laboratory Tests for Air Filters. The test standards most used in the United States for air filters are ASHRAE 52-76[24] and MIL-STD-282.[46] ASHRAE 52-76 prescribes four tests for air filters for general ventilation service:

Resistance, or the pressure drop across the filter as a function of air flow through it

Dust-holding capacity, or the increase in resistance as a function of the amount of ASHRAE standard air filter test dust loaded on it

Arrestance, or the percentage (by mass) of this dust captured by the filter at sequential dust loadings

Dust spot efficiency, or a measure of the ability of the filter to remove dust which will blacken walls

The dust spot test is made by sampling air both upstream and downstream of the tested filter. These samples are passed through glass-fiber filter "targets," which become darker as aerosol builds up on them. The rate of change of light transmission of these dust spots is used to compute the dust spot efficiency.

MIL-STD-282 is used to evaluate HEPA filters. An aerosol of DOP (dioctyl phthalate, or diethylhexyl phthalate) having a mass-mean diameter near 0.3 m is fed to the filter. The total light scattered by the aerosol cloud is measured upstream and downstream of the filter. Penetrations as low as 0.002 percent are routinely measured by this method. For lower penetrations, particle count techniques[25,45] must be used.

36.9.2 Gaseous Contaminant Testing

Many detection systems are available for both integrated and instantaneous measurement of gaseous contaminant concentrations and thus the performance of gaseous contaminant air pollution control and air filtration devices. Table 36.11 lists the major types of detectors available. Their operating principles are, briefly:

Gas washers (also called "bubblers" or "impingers"): These are miniature liquid scrubbers, frequently used in series to provide adequate removal efficiency. Reagents are often used to make them efficient and specific to a chosen contaminant.

Gas adsorber tubes: These are small tubes filled with adsorber granules (usu-

TABLE 36.11 Instruments for Measuring Gaseous Contaminant Concentration

Instrument type	Sample*	Contaminants detected
Gas washers (bubblers, impingers)	IM	Acid gases, ozone, organics
Gas adsorption tubes	IM	Many, with specific tube fill
Gas detector tubes	IM	Many, with specific tube fill
Chemiluminescence	RT	Ozone, ethylene oxide, organics
Infrared spectrometry	RT	Ozone, organics, acid gases
Laser spectrometry (LIDAR)	RT	Many gaseous; also aerosols
Gas chromatography (GC)	RT	Many, with proper column and detector
Mass spectroscopy (MS) (also GC/MS)	RT	Almost all compounds
Flame ionization	RT	Many, especially with GC
Atomic adsorption	RT	Many, with proper light wavelength
Passive detector badges	IM	Growing number
Piezoelectric	RT	Limited; based on dry chemisorbers
Freeze-trapping	IM	All condensable gases
Thermal conductivity	RT	Gases with thermal properties different from those of air

*Sample codes: RT = real time, instantaneous or nearly so; IM = sample is integrated over extended period.

ally impregnated to give low penetration of a specific gas). After the sample is captured by drawing polluted air through the tube, the contaminant is desorbed by flushing with an appropriate liquid. Various liquid analysis techniques can be used to detect the amount of contaminant desorbed.

Gas detector tubes: These are small tubes filled with a chemisorber that shows a color change with a chosen contaminant. The progress of the adsorption wave down the tube can be seen and measured.

Chemiluminescent analyzers: These have a lightproof chamber through which the polluted gas passes. Another gas, selected to react with the chosen pollutant, is mixed into this flow. The reaction is of a type that emits photons, which are observed by a sensitive photocell.

Infrared spectrometry: The transmission of infrared light is measured through a very long path containing the polluted gas. The path is folded with mirrors to make a compact instrument.

Laser spectrometry: A pulse of narrow-beam laser light is directed through a region containing the contaminant. Contaminant molecules scatter the laser light (in proportion to their numbers) in various directions, including back to the source. The intensity of the returned pulse is a measure of contaminant concentration. The method is unique in its ability to measure concentrations at a distance from the instrument.

Gas chromatography: A small grab sample (of microliter size) is mixed with an inert carrier gas and drawn through a small-diameter adsorber column. Individual components of the polluted gas sample migrate down the column at different, but predictable, rates. The mass of each component present is measured as it comes out the end of the column, using one or more detection devices.

Mass spectroscopy: Individual molecules of the contaminant are injected through electrostatic and/or magnetic fields in a vacuum. The flight path of the molecule is bent, with the path radius being a function of the molecular mass. The numbers of molecules of various masses are measured electronically. The method is sometimes coupled with gas chromatography.

Flame ionization: The contaminant is passed through a colorless flame, where it is ionized and changes the conductivity of the flame region; this change can be detected electronically.

Atomic adsorption: The contaminant is passed through a colorless flame, where it is ionized. A light beam from a narrow-wavelength-range source also passes through the flame. The ionized contaminant reduces the transmission of light through the flame; this change measures contaminant concentration. There are also flameless techniques for exciting the contaminant atoms.

Passive detector badges: These diffusively chemisorb specific contaminants, which are later desorbed for analysis. No pumps are needed, hence the badges are easily worn by individuals to measure personal exposure.

Piezoelectric: A quartz crystal is coated with a chemisorber specific to the chosen contaminant. As contaminant mass is added to the chemisorber, the oscillation frequency of the crystal changes.

Freeze-trapping: A tube is immersed in a liquid-nitrogen bath. Any gas which can condense at the tube (trap) temperature will condense out when passed through it. The condensed liquids can be distilled off later for individual contaminant measurement.

Thermal conductivity: The pressure of contaminant molecules can alter the thermal conductivity of a gas, roughly in proportion to the number of contaminant molecules present. This method is not highly specific, hence it can be used only where a single contaminant is present or in conjunction with gas chromatography.

Gaseous contaminant detectors are quite sensitive to environmental conditions, and much care is needed to avoid spurious readings. The low concentrations present in air filtration situations (often parts per billion) frequently demand special procedures, such as accumulation of the contaminant on an adsorber, with subsequent desorption at higher concentration and higher temperature. (See Ref. 47.)

NIOSH[48] has validated test procedures for many gaseous contaminants. Presumably these will be updated as new techniques are shown to be reliable.

36.10 REFERENCES*

1. J. W. Johnson et al., *A Computer-Based Cascade Impactor Data Reduction System*, EPA-600/7-78/042, NTIS, 1978.
2. H. L. Green and W. R. Lane, *Particulate Clouds*, 2d ed., D. Van Nostrand, Princeton, NJ, 1964.
3. A. C. Stern (ed.), *Air Pollution*, 3d ed., Academic Press, New York, 1977.

*Note: NTIS = National Technical Information Service, Springfield, VA 22161; USGSD = Superintendent of Documents, U.S. Government Printing Office, Washington, DC 20402.

4. *National Air Quality and Emissions Trends Report, 1982*, EPA-450/4-84/002, NTIS, 1984.
5. *Air Quality Data—1977 Annual Statistics*, EPA-450/2-78/040, NTIS, 1978.
6. P. J. Lioy, R. P. Mallon, and T. J. Kneip, "Long-Term Trends in Total Suspended Particulates, Vanadium, Manganese and Lead at near Street Level and Elevated Sites in New York City," *Journal of Air Pollution Control Association*, vol. 30, 1980, p. 153.
7. National Research Council Subcommittee on Airborne Particles, *Airborne Particles*, University Park Press, Baltimore, MD, 1981.
8. International Symposium on Indoor Air Pollution, *Health and Energy Conservation*, CONF-811048-SUMS (DE84004447), NTIS, 1981.
9. B. Y. H. Liu, D. Y. H. Pui, and H. J. Fissan (eds.), *Aerosols: Proceedings of the 1st International Conference*, Elsevier, New York, 1984.
10. C. S. Dudney and P. J. Walsh, *Report of Ad Hoc Task Force on Indoor Air Pollution*, ORNL/TM-7679, NTIS, 1981.
11. National Research Council, *Indoor Pollutants*, EPA-600/6-82/001, PB82180563, NTIS, 1982.
12. M. J. First (ed.), *Proceedings 16th DOE Nuclear Air Cleaning Conference*, CONF-801038, NTIS, 1980 (vol. 3 is index to all previous conference proceedings).
13. M. J. First (ed.), *Proceedings 17th DOE Nuclear Air Cleaning Conference*, DE83009767 and DE83009768, NTIS, 1983.
14. M. J. First (ed.), *Proceedings 18th DOE Nuclear Air Cleaning Conference*, CONF-840806, NTIS, 1984.
15. C. A. Burchsted, A. B. Fuller, and J. E. Kahn, *Nuclear Air Cleaning Handbook*, 2d ed., ERDA 76-21, NTIS, Springfield, VA, 1976.
16. *Code of Federal Regulations, Title 29* (*Labor*), Part 1900.1000ff, USGSD, Washington, DC, annual.
17. *Threshold Limit Values for Chemical Substances in the Work Environment*, American Conference of Governmental Industrial Hygienists, Cincinnati, OH, 1984.
18. F. W. Makison, R. S. Stricoff, and L. J. Partridge, *NIOSH/OSHA Pocket Guide to Chemical Hazards*, DHEW-NIOSH Pub. 78-210, USGSD, Washington, D.C., 1978. See also comment, *American Industrial Hygiene Association Journal*, vol. 44, Sept. 1983, p. A-6.
19. A. Dravnieks and B. K. Krotoszynski, "Systematization of Analytical and Odor Data on Odorous Air," in *ASHRAE Symposium, Odors and Odorants: The Engineering View*, ASHRAE, Atlanta, GA, 1969.
20. C. E. Billings and L. C. Jonas, "Odor Thresholds as Compared to Threshold Limit Values," *American Industrial Hygiene Association Journal*, vol. 42, 1981, p. 479.
21. *Code of Federal Regulations, Title 40* (*Environment*), USGSD, Washington, DC, annual.
22. *UNAMAP* (*Version 5*) *User's Network for Applied Modeling of Air Pollution*, EPA-DF-83/007, PB83-244368, NTIS, Springfield, VA, 1983.
23. R. G. Sextro, "Control of Indoor Radon and Radon Progeny Concentrations," symposium paper HI85-39-3, ASHRAE, Atlanta, GA, 1985.
24. American Society of Heating, Refrigeration, and Air-Conditioning Engineers, *Methods of Testing Air-Cleaning Devices Used in General Ventilation for Removing Particulate Matter*, Standard 52-76, ASHRAE, Atlanta, GA, 1976.
25. *Clean Room and Work Station Requirements, Controlled Environments*, Federal Standard 209d. General Services Administration, Washington, DC, 1988.
26. *Code of Federal Regulations, Title 10* (*Energy*), Part 20, USGSD, Washington, DC, annual.

27. *Guide for Explosion Venting* (NFPA 68-1978), National Fire Protection Association, Quincy, MA, 1978.
28. *Fire Hazard Properties of Flammable Liquids, Gases and Volatile Solids* (NFPA 325M-1984), National Fire Protection Association, Quincy, MA, 1984.
29. *Control Techniques for Particulate Emission from Stationary Sources*, vol. 1: EPA 450/3-81-005A, PB83-127498; vol. 2: EPA 450/3-81-005B, PB83-127480; NTIS, Springfield, VA, 1981, 1983.
30. *Control Technology for Volatile Organic Emissions from Stationary Sources*, EPA 450/2-78-022, PB284804, NTIS, Springfield, VA, 1978.
31. M. J. Ellenbecker, "Evaluation of Total Airborne Emissions from Industrial Processes," *Journal of Air Pollution Control Association*, vol. 33, 1983, p. 884.
32. Committee on Industrial Ventilation, *Industrial Ventilation*, 18th ed., American Conference of Governmental Industrial Hygienists, Cincinnati, OH, 1984.
33. ASHRAE, *Ventilation for Acceptable Indoor Air Quality*, Standard 62-81, ASHRAE, Atlanta, GA, 1989.
34. R. D. Rivers, "Predicting Particulate Air Quality in Recirculatory Ventilation Systems," *ASHRAE Transactions*, vol. 88, part 1, 1982.
35. ASHRAE, *1985 ASHRAE Handbook, Fundamentals*, ASHRAE, Atlanta, GA, 1985, chap. 22.
36. *Standard for Test Performance of Filter Units* (UL 900), 4th ed., Underwriters' Laboratories, Northbrook, IL, 1983.
37. G. O. Nelson and N. Correia, "Respirator Cartridge Efficiency Studies," *American Industrial Hygiene Association Journal*, vol. 37, 1976, p. 514.
38. L. A. Jonas, E. B. Sansone, and T. S. Farris, "Prediction of Activated Carbon Performance in Binary Vapor Mixtures," *American Industrial Hygiene Association Journal*, vol. 44, 1983, p. 716.
39. C. I. Harding and T. R. Kelley, "Horizontal and Vertical Distribution of Corrosion Rates in an Industrialized Seacoast City," *Journal Air Pollution Control Association*, vol. 17, 1966, p. 545.
40. *1985 Annual Book of ASTM Standards*, sec. 15.01 (Standards D2652, D2854, D2862, D2866, D2867, D3466, D3467, D3802, D3803, D3838), American Society of Testing and Materials, Philadelphia, PA, 1985.
41. H. W. Parker, *Air Pollution*, Prentice-Hall, Englewood Cliffs, NJ, 1977.
42. P. N. Cheremisinoff and R. A. Young, *Air Pollution Control Design Handbook*, Marcel Dekker, New York, 1977.
43. J. A. Danielson (ed.), *Air Pollution Engineering Manual*, 2d ed., EPA Publication AP-40, NTIS, 1973.
44. R. R. Wilson et al., *Guidelines for Particulate Sampling in Gaseous Effluents from Industrial Processes*, EPA/600/7-79/028, PB290899/4, NTIS, Springfield, VA, 1979.
45. D. A. Lundgren et al. (eds.), *Aerosol Measurement*, University Presses of Florida, Gainesville, FL, 1979.
46. *Military Standard: Filter Units, Protective Clothing, Gas Mask Components and Related Products, Performance Test Standards*, MIL-STD-282, Naval Publications and Forms Center, Philadelphia, PA, 1956.
47. J. F. Walling et al., *Sampling Air for Gaseous Organic Chemicals Using Solid Adsorbents: Applications to Tenax*, EPA/600/4-82/059, PB82-262189, NTIS, Springfield, VA, 1982.
48. P. M. Eller (ed.), *NIOSH Manual of Analytical Methods*, 3d ed., vols. 1, 2, PB8J-179018/GAR, NTIS, Springfield, VA.

36.11 BIBLIOGRAPHY

36.11.1 Books and Monographs

Hidy, G. M.: *Aerosols: An Industrial and Environmental Science*, Academic Press, Orlando, FL, 1983.

Patty's Industrial Hygiene and Toxicology, 3d ed., Wiley-Interscience, New York, 1985; vol. 1: G. D. Clayton and F. E. Clayton (eds.), *General Principles*; vol. 2*a, b, c*: G. D. Clayton and F. E. Clayton (eds.), *Toxicology*; vol. 3*a, b*: L. J. Cralley and L. V. Cralley (eds.), *The Work Environment*.

Registry of Toxic Effects of Chemical Substances, vols. 1, 2, U.S. Dept. of Health and Human Services, HE 20.7112:980, USGSD, Washington, DC, 1980.

Sax, N. I.: *Dangerous Properties of Industrial Materials*, 7th ed., Van Nostrand-Reinhold, Florence, KY, 1988.

36.11.2 Periodicals

ASHRAE Journal, American Society Heating, Refrigeration, and Air-Conditioning Engineers, Atlanta.

Atmospheric Environment, Pergamon Press, Elmsford, NY.

American Industrial Hygiene Association Journal, AIHA, Akron, OH.

Aerosol Science and Technology, Elsevier Science Publishing Co., New York, NY.

Journal of Aerosol Science, Pergamon Press, Elmsford, NY.

Journal of the Air Pollution & Waste Management Association, A&WMA, Pittsburgh, PA.

Environmental Science and Technology, ACS, Washington, D.C.

Filtration and Separation, Uplands Press Ltd., London, England.

Chemical Engineering, McGraw-Hill, New York, NY.

NTIS Published Searches (annually NTIS), Springfield, VA.

U.S. Government Research and Development Reports (biweekly, NTIS).

CHAPTER 37
AIR-HANDLING UNITS*

James A. Reese
Tempmaster Corporation, North Kansas City, Missouri

37.1 SYSTEM DESIGN

37.1.1 Focus: Variable-Air-Volume All-Air Systems

The experience of most skilled designers indicates that for most buildings a variation on the basic variable-air-volume (VAV) all-air system will yield the best combination of comfort, first cost, and life cycle cost. Therefore, this discussion focuses on the design and selection of equipment for all-air VAV systems and deals lightly with alternative systems.

37.1.2 Comfort

With all the written and verbal discussions concerning energy, controls, building automation, environmental impact, etc., we must remember that the original objective of the building's design, and of the heating, ventilating, and air-conditioning (HVAC) system in particular, is *the comfort of the building's occupants*. It is the task of the designer to provide environmental comfort and acceptable indoor air quality (temperature, humidity, ventilation, and noise level) consistent with meeting the specific project requirements; intent of local, state, and federal codes; and first cost, life cycle cost, and energy budgets.

Temperature Control. The design goal for VAV systems is to provide inexpensive temperature control in large numbers of zones by modulating the flow of constant-temperature air in response to a local thermostat.

Comparable temperature control can be achieved at higher first cost and maintenance cost with multiple-fan coil and hydronic heat pump systems and at higher energy requirements due to mixing and/or reheat in multizone, dual-duct, and reheat systems.

Humidity Control. Precise humidity control requires reheat with constant-volume air flow or a supply air temperature reset with variable volume. How-

*All art reproduced in this chapter is courtesy of Tempmaster Corp.

ever, one of the major advantages of a VAV system is that comfortable humidity levels are easily maintained at part-load conditions—especially important with higher summer interior design temperatures. The proper design of air-handling and refrigeration equipment for humidity control is covered in Chaps. 32 through 40 and 52 of this book.

Air Movement. In perimeter zones with nonair heating or with VAV heating, loss of air circulation may occur at the changeover point. Reheating or mixing to maintain air flow should be avoided for energy cost reasons. Maintenance of minimum circulation in perimeter zones can be best accomplished by the use of central fan perimeter heating/cooling or heating-only systems or the use of appropriately selected variable-volume fan terminals, which recirculate unheated air from the core of the building at light-load conditions. Suggested systems are described in Sec. 37.2.

In interior zones, modern buildings with low lighting loads sometimes require very few cubic feet per minute of primary air per square foot. A properly designed variable-volume system using high induction slot diffusers can handle interior zone loads as low as 0.3 $ft^3/(min \cdot ft^2)$ [0.5 $m^3/(h \cdot m^2)$] and still provide adequate air circulation in the space. For lower primary air requirements or for systems using non-coanda (surface-effect) slot diffusers (e.g., perforated metal-covered diffusers and light troughers), drawthrough fan-powered terminals can be used to mix ceiling plenum air with primary air. Additional information on surface effect may be found in the American Society of Heating, Refrigeration, and Air-Conditioning Engineers' (ASHRAE) chapter on air-diffusing equipment.* Surface effect refers to the ability of air to hug the ceiling and not drop. This will provide constant air circulation and predictable diffuser performance in the interior spaces. For diffuser application rules and selection procedures, see diffuser manufacturers' catalogs and/or the ASHRAE handbooks on fundamentals* and equipment.†

Ventilation. Ventilation requirements were diminished considerably with the press for energy conservation. However, concern over indoor air quality has resulted in a reversal of this practice. With a VAV system and other systems that are designed for free cooling (up to 100 percent outside air), any percentage of outside air can be easily provided. Ventilation in the interior-cooling-only VAV system can be accomplished without the energy penalty of heating the ventilation air. It is generally accepted that if the correct amount of ventilation air is provided, regardless of the total amount of air being circulated, it will be distributed adequately since nearly all buildings require cooling year-round. When this is not the case, special provisions may be necessary to supply ventilation air directly to the required areas. And ventilation is improved in perimeter spaces when heating is provided by an air system which draws from the general building return plenum rather than the individual space.

Sound. The proper design of air distribution ductwork and the proper selection of fans, terminals, and diffusers are essential in eliminating noise problems in VAV systems. Excessive noise can make a space unusable for its original purpose. This is a common problem in poorly designed VAV systems. A solution is to use computer duct design programs combining static regain duct sizing with

---

**1985 ASHRAE Handbook, Fundamentals*, ASHRAE, Atlanta, GA, 1985.
†*1983 ASHRAE Handbook, Equipment*, ASHRAE, Atlanta, GA, 1983.

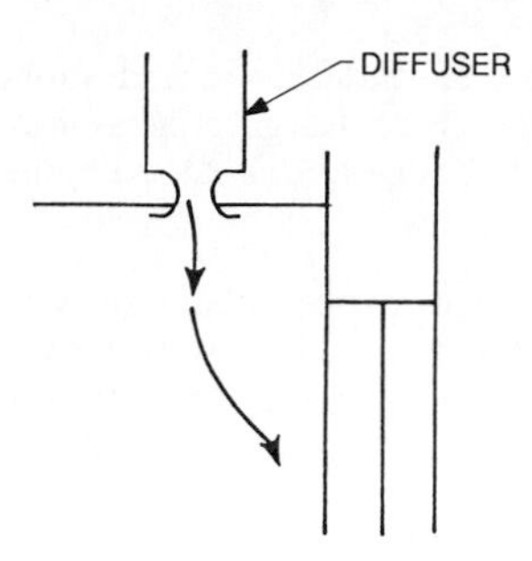

FIGURE 37.1 Skin loss greater than 400 Btu/(h · lin ft) (385 W/lin m). Vertical high-induction slot diffusers located over windows with constant volume of skin heating air.

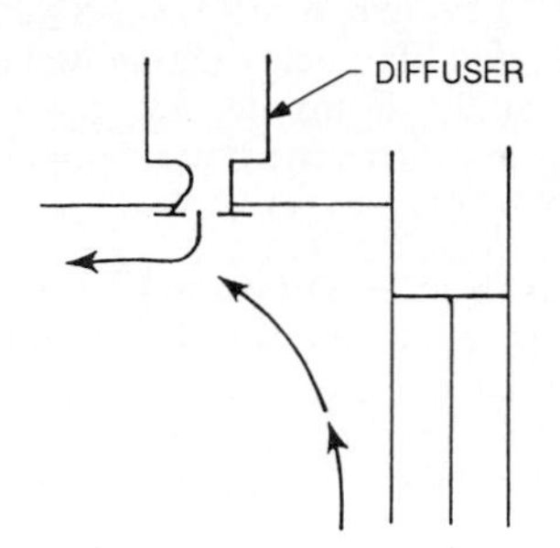

FIGURE 37.2 Skin loss 250 to 400 Btu/(h · lin ft) (240 to 385 W/lin m). Slot diffusers next to outside wall directed inward with constant volume of skin heating air.

fitting selection, VAV unit sizing, and fan and attenuation selection to achieve supply-side noise criteria requirements. See Chaps. 49 and 50 of this book for noise and vibration control.

Control of Downdraft. Control of downdraft is a major consideration in perimeter zones. This draft control can be achieved with all-air systems by the introduction of heating air as follows:

Skin loss greater than 400 Btu/(h · lin ft) (385 W/lin m) (Fig. 37.1). High induction downblow slot diffusers, located over windows, discharging a constant volume of skin heating air.

Skin loss of 250 to 400 Btu/(h · lin ft) (240 to 385 W/lin ft) (Fig. 37.2). Slot diffusers next to the outside wall, discharging inward, with a constant or variable volume of skin heating air. Air temperature is reset to approximate the skin load of the building. Variable volume can be used to trim individual zones to minimize energy waste.

Skin loss less than 250 Btu/(h · lin ft) (240 W/lin m). Downdrafts are not experienced in this skin loss range. Therefore, slot diffusers at any location can provide constant-volume or variable-volume heating air.

The current trend toward building skins with lower winter skin heating loss [<250 Btu/(h · lin ft) (240 W/lin m)] is increasing the use of systems which do not separate the heating from the cooling.

37.1.3 Budget

Attention is gradually shifting from concern with first cost only to concern with life cycle cost, or the total cost over a period of owning and operating the building. The design professional now has computer analysis tools available to make intelligent recommendations to the architect and owner:

- To influence the building envelope design (*U* factor, glass, etc.) to minimize total building owning cost.
- To compare various system concepts and equipment selections to choose the best from a comfort and owning-cost standpoint. Important cost factors, trade-

offs, and life cycle costing techniques must be well understood and applied intelligently. Computer analysis can do this in the early design stages on a cost-effective basis without demanding an overwhelming and costly engineering effort.

It is most important to establish the overall budget properly and early so that inexperienced clients will not expect more than their budgets can provide. First-cost estimates must be developed from experience and will vary from area to area and project to project, depending on the mix and cost of piping, sheet metal, and electrical labor and material.

37.1.4 Energy and Safety Codes

Energy. Reference 1 relating to air consumption in new buildings offers two alternatives:

1. Follow carefully prescribed dos and don'ts which allow very little originality or trade-offs, particularly where substantial first-cost advantage might be obtained. This is the most straightforward.
2. Demonstrate through computer simulation that the proposed design is as efficient as the design prescribed by the ASHRAE standard. This approach will generally lead to the best system.

Many energy codes follow this pattern. An alternative approach gaining popularity is the use of an energy budget. It permits any design which leads to an annual per-square-foot energy consumption of less than a prescribed amount.

Safety. There are fire, electrical, and ventilation codes to satisfy. The type of controls, e.g., might be influenced by whether low-voltage control wiring must be enclosed in conduit; whether pneumatic lines must be copper rather than plastic; etc. Fire codes can play an important role in determining the best system, type of controls, and location of air supply units.

37.1.5 Building Specifics

Building Configuration. Building configuration can have a substantial effect on load, energy consumption, and proper system design. The size, shape, and orientation; equipment room location and space; trunk duct space; and amount of glass may change the best system choice from one type to another.

Space Utilization. Use of space is another big factor. If the space is owner-occupied and can be prezoned, a system might be used that has limited flexibility but may be very energy-efficient and/or low in first cost. A speculative building, however, needs the combination of flexibility and low tenant completion costs. Therefore, system-powered boxes might be a good choice, to avoid the need for multiple trades every time a space is finished. Special-purpose buildings, such as medical offices, influence choice, as does average zone size. For example, many medical buildings require a system which can provide heating and cooling year-round to all spaces. This rules out cooling-only interior systems or locking out heating in summer.

Zone Size. Zone size can have quite an effect on system cost. A rectangular building should normally have a minimum of nine zones per floor. The number of additional zones depends on building application and tenant requirements. The finished cost per ft^3/min (m^3/h) of 250 ft^3/min (425 m^3/h) zones is often 2½ times the finished cost per ft^3/min (m^3/h) of 2000 ft^3/min zones (3400 m^3/h).

Energy Development and Cost. The availability and relative cost of different forms of energy (and local expectations of changes in cost and availability) will often determine the type of heating and cooling equipment to be used.

37.1.6 Design Checklist for Energy Efficiency

Given the design suggestions above as well as first-cost and building constraints, the crux of the HVAC system and equipment selection is the energy comparison. A well-written, sophisticated, and flexible computerized energy analysis program is a valuable tool. System energy requirements can be minimized by following this checklist:

1. Take advantage of heat produced by the sun, lights, and people in control of the perimeter heating system.
2. Allow conditioned primary air volume to vary down to shut-off in both interior and perimeter spaces at part-load conditions. Accomplish minimum air circulation in the perimeter with plenum air provided the minimum required ventilation air requirement is satisfied.
3. Mixing or reheat should be avoided, if possible. Control systems should be selected so that control miscalibration will not lead to mixing or reheat.
4. If reheat must be used, primary air temperature should be reset to minimize the reheating penalty at part-load conditions. With primary air temperatures reset, interior cooling-only terminals and diffusers will have to be oversized.
5. Take return air for perimeter heating from as close to the core of the building as possible, to maximize pickup of heat from interior loads.
6. Do not condition outside air if it can be avoided. Ventilation air should be distributed to the interior cooling-only zones of the building when codes and building design permit. Keep ventilation air off when the building is unoccupied and during warmup cycles.
7. Optimize the supply air temperature design to minimize system energy consumption as the supply air temperature is reduced, fan energy decreases, compressor energy increases, and outside air load increases.
8. Minimize face velocities on coils and filters as well as terminal and diffuser pressure drop. Use static regain computerized duct designs to reduce fan static pressure and resultant horsepower.
9. Utilize refrigeration, heating, and air-handling controls which provide maximum free cooling without heating penalty and unload and reset refrigeration and heating equipment for maximum part-load efficiency.
10. Use localized heating with deep night setbacks when the building will experience substantial periods of nonoccupancy. Turn off interior fans during unoccupied periods unless filtering or ventilation is required to prevent the buildup of air pollutants due to continuous off-gassing.
11. Maximize the efficiency of refrigeration equipment by utilizing low head and

high suction pressures. Evaporative-type air-cooled or water-cooled centrifugal or reciprocating equipment is normally the most efficient. Do not forget the effects of pumps and other refrigeration auxiliaries, however. For air-cooled machines, consider the effect of dirt on the condenser coils and the resultant degradation of efficiency.

12. Utilize energy transfer and heat recovery whenever possible. Make sure that the net heat saved is more than the energy expended to save it. Heat must be available when it is needed unless storage is provided. If you are considering heat pumps, be sure to analyze the coincidence of the amount and availability of heat when required (include the effect and cost of storage, if necessary). Pay particular attention to the effects of diversity (requirement for instantaneous peak capacity) in determining installed tonnage, cubic feet per minute (cubic meters per hour), and connected load.
13. If thermal storage is used, take advantage of the opportunity to lower the supply air cooling temperature and reduce the size and operating cost of the air distribution system.
14. Use central building automation to schedule equipment operation, provide optimized summer and winter early startup, control equipment performance at maximum efficiency, and control electrical demand.

37.1.7 Use an Energy Analysis Program

Utilizing an energy analysis program enables the professional designer to make quick and accurate estimates of the annual energy consumption of each of the HVAC systems being considered. Trade-offs can then be made and systems compared. For example, a very energy-efficient system may cost more than a less efficient system but have a desirably short payback on its cost premium. Maintenance difficulty should also be reviewed. First-cost or energy cost savings can be negated by a system that is difficult and costly to maintain.

37.2 TYPICAL SYSTEM DESIGNS

To assist the designer in choosing the best system type, control method, and equipment, various possibilities are described and illustrated in the following pages. Each of these systems can be simulated on computerized energy analysis programs and designed with the assistance of a computerized duct design program. Financial analysis is also available through computerized programs.

37.2.1 Series Fan Terminal, Skin VAV Cooling Only—Interior (Fig. 37.3)

Advantages

1. Low first cost with electric heat
2. Continuous air circulation at all times with excellent downdraft control
3. Minimal simultaneous heating and cooling, if properly controlled

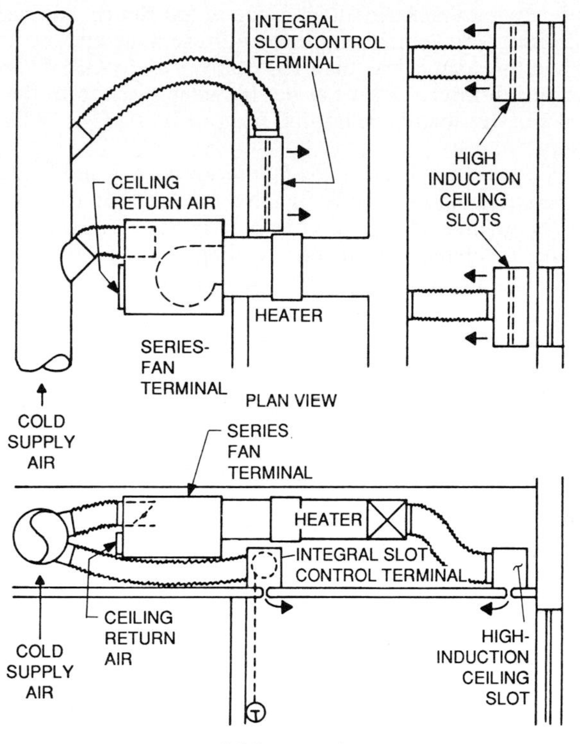

FIGURE 37.3 Series fan terminal skin VAV cooling only—interior. Skin load less than 400 Btu/(h · lin ft) (385 W/lin m). (Above these limits use a downblow diffuser above windows.) *Note*: Make runouts straight for 4 diameters upstream from box inlet.

4. Possibility of free heat transfer from interior to perimeter spaces, if fan terminal units are located near central core
5. Excellent way to use VAV primary air and maintain any design air-flow circulation rate desired in interior spaces

Application Guidelines. Install fan terminal units as close as possible to the center core, to minimize wiring or hot-water piping and reduce noise in occupied spaces.

The System. Fan terminal units provide constant air circulation year-round. Cold primary air is varied, and the unit fan pulls in return air to make up the difference. Heating is not energized until the fan terminal's cooling damper is closed.

Fan terminal units handle the skin heating and cooling load and provide downdraft protection where needed. A minimum of one unit per building face per floor is recommended to minimize simultaneous heating and cooling in perimeter spaces in winter. If the lineal wall exceeds 100 ft (30.5 m), additional units should be considered.

Where the heating transmission load exceeds 400 Btu/(h · lin ft) (385 W/lin m) including infiltration, downblow induction diffusers should be used to avoid downdraft. When this load is less than 400 Btu/(h · lin ft) (385 W/lin m), a ceiling high-induction slot diffuser, located at but blowing away from the outside wall, can be used. When this load is below 250 Btu/(h · lin ft) (240 W/lin m), any kind of diffuser is satisfactory.

Lights, people, and equipment loads in interior spaces and lights, people, and solar loads in perimeter spaces can be handled for zones up to approximately 400 ft^3/min (680 m^3/h) by integral diffuser control terminals. Conventional single-duct control terminals supplying one or more ceiling slot diffusers can be used for larger zones. Fan terminal units can be employed to increase air circulation in very low-load interior spaces.

Variable-volume constant-temperature cold air is used as the primary air for this system.

An efficient heat pump system can be made by using a central double-bundle condenser and hot-water heating coils in the fan terminal units. Thermal storage can also be used with this system but thermal insulation and isolation must be provided if the resulting supply air temperature is below 50°F (10°C).

Controls. Pneumatic, electric, or electronic controls can be used by selecting a particular zone or using the outdoor or exterior wall temperature to control the series-type fan terminal unit. A wide variety of optional velocity limits and deadband adjustments are available.

VAV terminals in perimeter spaces should be interlocked with the fan terminal unit serving the same spaces, and the control sequencing should be arranged to minimize simultaneous heating and cooling.

Central readout and reset of zone temperatures and central night setback with an electronic building automation system should be considered. Night temperature control is obtained by intermittent operation of the fan terminal and its heater.

37.2.2 Series Fan Terminal Perimeter, VAV Interior (Figs. 37.4 and 37.5)

Advantages

1. Minimum installed ductwork and low first cost in buildings with large zone sizes
2. Continuous air circulation at all times with excellent downdraft control
3. No danger of simultaneous heating and cooling
4. Low first cost with electric heat

The System. Series fan terminal units provide constant total air circulation year-round for perimeter spaces and handle the entire cooling/heating load. Primary cold air is varied, and the unit fan pulls in return air to make up the difference. Heating is not energized until the unit cooling damper is closed.

Use one fan terminal unit per zone and select the unit for any ratio of primary to total air less than unity. Where downdraft may be a problem, high-induction downblow diffusers can be used to blanket the wall with ceiling slot diffusers for the remainder of the air.

Where the heating transmission load exceeds 400 Btu/(h · lin ft) (385 W/lin m) including infiltration, high-induction downblow diffusers should be used to avoid

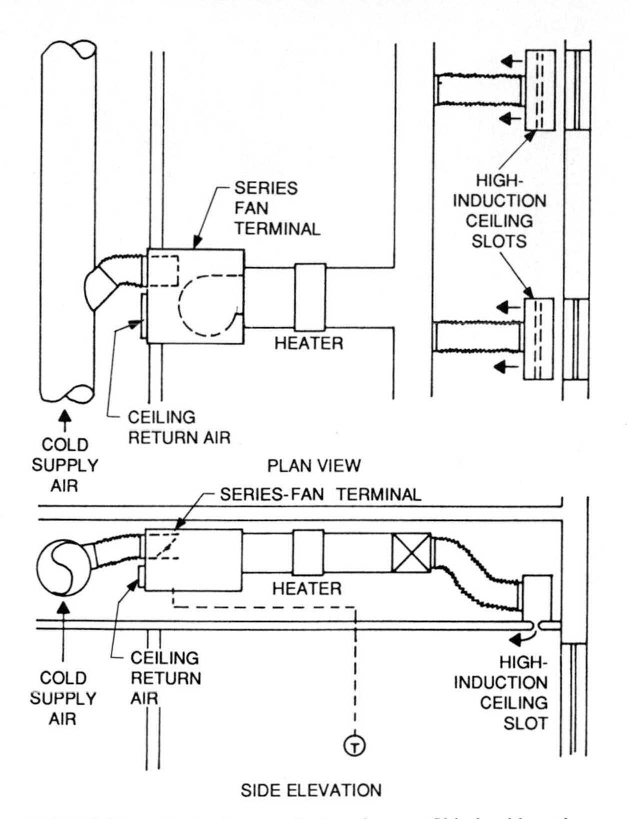

FIGURE 37.4 Series fan terminal perimeter. Skin load less than 400 Btu/(h · lin ft) (385 W/lin m). (Above these limits use a downblow diffuser above windows.) *Note*: Make runouts straight for 4 diameters upstream from box inlet.

downdraft. When this load is less than 400 Btu/(h · lin ft) (385 W/lin m), a ceiling slot diffuser located at, but blowing away from, the outside wall can be used. When this load is below 250 Btu/(h · lin ft) (240 W/lin m), downdraft is not a problem.

Lights and people loads in interior spaces may be handled by integral diffuser control terminals for zones up to approximately 400 ft^3/min (680 m^3/h) and control terminals with one or more ceiling slot diffusers for larger zones. Fan terminal units can be employed to increase air circulation in very low-load interior spaces.

Variable-volume constant-temperature cold air is used as the primary air for this system.

This arrangement can also be used for interior spaces such as conference rooms to increase air circulation rates or to provide local air filtering in conjunction with VAV primary air control (Fig. 37.5). It can also be used with low-temperature VAV primary air to achieve any desired supply air temperature to the room. Caution should be exercised to ensure that modifications are made in the terminal design to accommodate very low-temperature supply air.

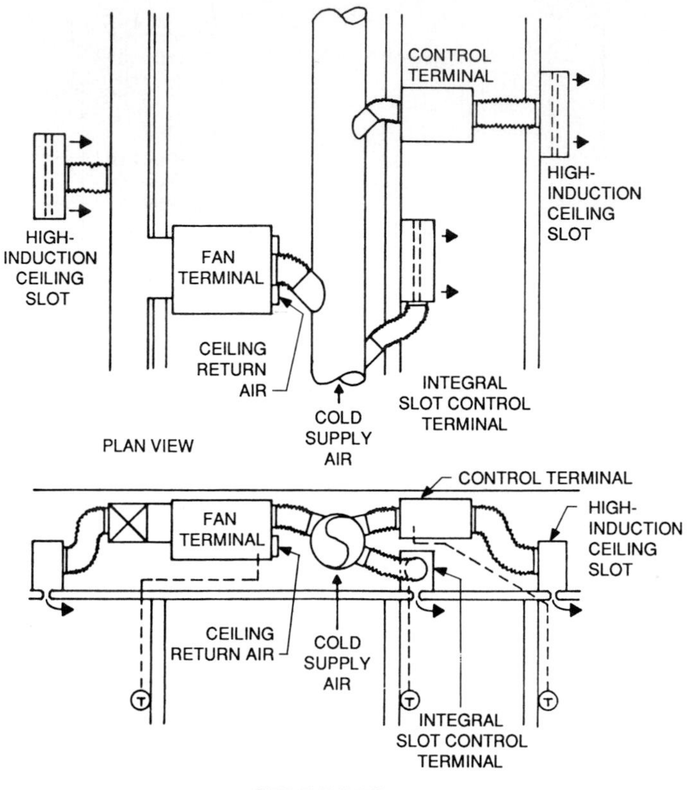

FIGURE 37.5 Fan terminal for interior space. *Note*: Make runouts straight for 4 diameters upstream from box inlet.

Controls. Controls can be pneumatic, electric, or electronic with a wide variety of optional velocity limits and dead-band adjustments.

Central readout and reset of zone temperatures and central night setback are available. Night temperature control is obtained by intermittent operation of fan terminals and their heaters.

37.2.3 Parallel Fan Terminal Perimeter, VAV Interior (Fig. 37.6)

Advantages

1. Minimum installed ductwork and low first cost in buildings with large zone sizes
2. No danger of simultaneous heating and cooling
3. Minimum fan energy because smaller fan terminal fans operate only upon a call for heat

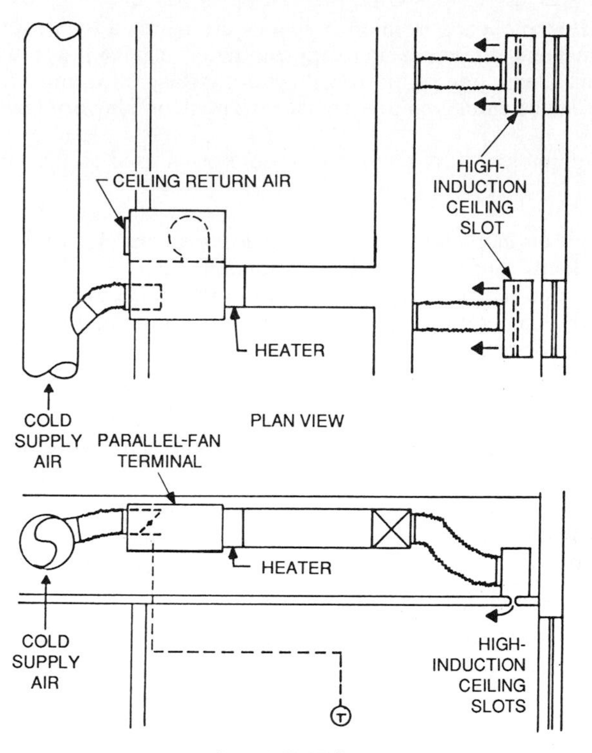

FIGURE 37.6 Parallel-fan terminal perimeter. Skin load less than 400 Btu/(h · lin ft) (385 W/lin m). *Note*: Make runouts straight for 4 diameters upstream from box inlet.

Application Guidelines. This is not recommended for applications with a high winter heating load because of low diffuser velocity in the heating season and marginal downdraft control.

The System. The parallel fan terminal unit has a unit-mounted fan installed in parallel with a primary air damper. In the cooling season, the unit fan is off and cold primary air bypasses it. Heating is energized only after the primary air damper is closed and the unit fan is up to full speed.

The parallel fan terminal perimeter system is best suited for electric heat applications for medium-size and large zones. One unit can meet the entire heating and cooling requirement if the winter heating load is relatively light. This system is not recommended where winter transmission load plus infiltration exceeds 400 Btu/(h · lin ft) (385 W/lin m). When this load is between 250 and 400 Btu/(h · lin ft) (240 and 385 W/lin m), it can be employed with a ceiling high-induction slot diffuser located at, but blowing away from, the outside wall, provided that the total air quality is held nearly constant throughout the year (select design heating cfm/design cooling cfm in perimeter spaces). When this load is below 250 Btu/(h · lin ft) (240 W/lin m), any diffuser and cooling-heating air-flow ratio are satisfactory.

Lights and people loads in interior spaces are handled by single integral slot control terminal units for zones up to approximately 400 ft^3/min (680 m^3/h) and by control terminals with one or more ceiling slot diffusers for larger zones. Series fan terminals can be employed to increase air circulation in very low-load interior spaces.

Variable-volume constant-temperature cold air is used as the primary air for this system.

Controls. Controls can be pneumatic, electric, or electronic, with a wide variety of optional velocity limits and dead-band adjustments.

Central readout and reset of zone temperatures and central night setback are available. Night temperature control is obtained by intermittent operation of fan terminal fans and heaters.

37.2.4 Variable-Volume, Double-Duct Perimeter, VAV Interior (Fig. 37.7)

Advantages

1. Low first cost when startup occupancy is low because both heating and cooling means can be installed as the tenant finish is completed
2. Minimum piping with hot-water heating
3. Possibility of no simultaneous cooling and heating
4. Reduced fan energy at light-load conditions

Application Guidelines. First, the system must be properly designed and appropriately applied to enjoy its advantages. Second, this system is not recommended for applications with a high winter heating load because of low diffuser velocity in the heating season and poor downdraft control.

The System. A skin heating-only or heating/cooling variable-volume supply unit and a perimeter and interior space cooling-only variable-volume supply unit both feed common perimeter double-duct terminal units. Diffusers, boxes, and ductwork do not have to be installed in initially unoccupied spaces.

To minimize reheating of cold "economizer" air in the winter, two separate supply fans are used and economizer dampers are put only on the interior cooling fan. Both fans must have inlet vane controls or variable-speed controls and properly located duct static pressure sensors. The supply duct pressure sensor should be located as far away from the fan as possible, to minimize overpressurization of ducts but still provide the required pressure to all ducts. Care must be taken in parallel duct systems when the solution involves using multiple sensors controlled to satisfy the one with the greater need. Maximum resetting of the hot duct temperature minimizes hot duct pressure variation and increases air circulation in perimeter spaces at light-load conditions.

The variable-volume two-fan perimeter system is not recommended where the winter transmission load plus infiltration exceeds 400 Btu/(h · lin ft) (385 W/lin m). When this load is between 250 and 400 Btu/(h · lin ft) (240 and 385 W/lin m), it can be employed with split diffusers (heating separate from cooling to maintain high slot velocity at full heating) located at, but blowing away from, the outside wall, provided that the heating diffuser section is properly matched to the heating transmission plus infiltration air flow and that the hot air temperature is reset.

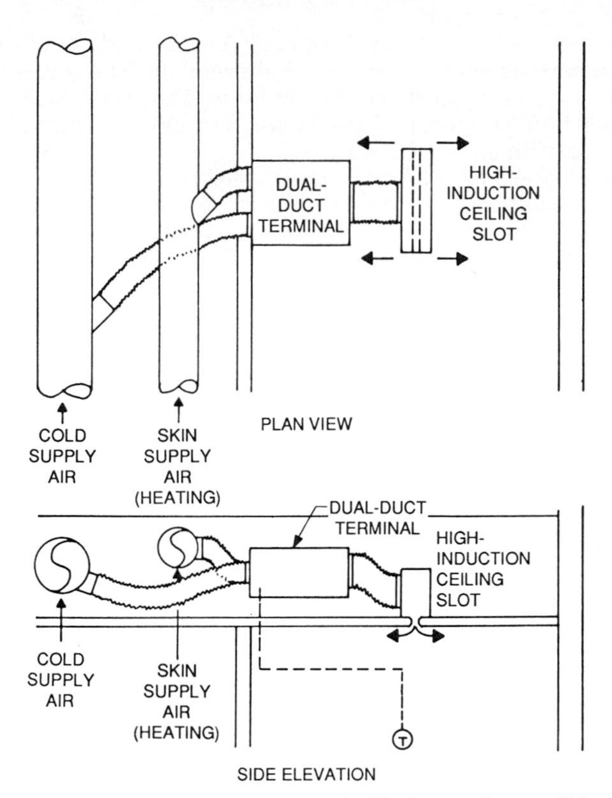

FIGURE 37.7 Variable-volume double-duct perimeter. Skin load less than 250 Btu/(h · lin ft) (240 W/lin m). Single box and diffuser for both systems. *Note*: Make runouts straight for 4 diameters upstream from box inlet.

When this load is below 250 Btu/(h · lin ft) (240 W/lin m), any high-induction slot diffuser and cooling/heating air-flow ratio are satisfactory.

Lights and people loads in interior spaces are handled by single integral diffuser control terminal units for zones up to approximately 400 ft^3/min (680 m^3/h) and by control terminal units with one or more high-induction ceiling slot diffusers for larger zones. Series fan terminal units can be employed to increase air circulation in very low-load interior spaces.

Variable-volume constant-temperature cold air is used for the interior system handling lights and people loads in both interior and perimeter spaces. Variable-volume reset temperature air is used for the skin system handling heating transmission and infiltration loads. Cooling transmission and infiltration and solar loads may be assigned to either air distribution system.

An efficient central heat pump system can be made by using a double-bundle condenser and hot-water heating coils for the skin air distribution system. Thermal storage can be included.

Controls. Controls can be pneumatic, electric, or electronic, with a wide variety of optional velocity limits and dead-band adjustments.

The simplest controls are obtained with zero minimum air flow, a fixed small dead band, and heating-only skin system air distribution. This forces the interior air distribution system to handle all cooling loads. This works well in buildings where minimum heating is required and the space requires cooling year-round except for morning warmup.

If the skin air system also has a cooling capability, interior system duct sizes are reduced, but additional controls are required to accomplish changeover.

If a relatively high minimum air quantity is required, the inlet velocity on one or both heating and cooling inlets must be measured and controlled, to avoid the possibility of prohibitive energy consumption. In such cases, it is important to reset at least the hot duct temperature to the greatest possible extent.

Central readout and reset of zone temperatures and central night setback are available. Winter temperature control is obtained by operation of the skin air system only with appropriate temperature resetting.

37.2.5 Central Constant-Volume System, Skin VAV Interior Cooling Only (Fig. 37.8)

Advantages

1. Maximum zone relocation flexibility and relatively low first cost in buildings with many small zones and high occupancy at startup
2. Minimum piping with hot-water heating
3. Continuous air circulation at all times with excellent downdraft control
4. Quiet, reliable, and easy to maintain

Application Guidelines. Heating reset schedule should be set carefully to avoid unnecessary winter heating.

The System. A skin heating-only or heating/cooling constant-volume supply unit handles the heating or heating and cooling skin transmission and infiltration load. A cooling-only variable-volume supply unit handles all lights, people, and solar loads and may also handle the skin cooling transmission and infiltration load.

To minimize reheating of cold economizer air in the winter, economizer dampers are put only on the interior variable-volume supply fan. The constant-volume perimeter system maintains air circulation through perimeter spaces even when variable-volume boxes serving these spaces are closed.

Where the heating transmission load exceeds 400 Btu/(h · lin ft) (385 W/lin m) including infiltration, high-induction downblow slot diffusers should be used to avoid downdraft. When this load is less than 400 Btu/(h · lin ft) (385 W/lin m), a high-induction ceiling slot diffuser located at, but blowing away from, the outside wall can be used. When this load is below 250 Btu/(h · lin ft) (240 W/lin m), any kind of diffuser is satisfactory.

Lights and people loads in interior spaces and lights, people, and solar loads in perimeter spaces are handled by single integral slot control terminal units for zones up to approximately 400 ft^3/min (680 m^3/h) and control terminal units with one or more high-induction ceiling slot diffusers for larger zones. Series fan terminals and parallel fan terminal boxes can be employed to increase air circulation in very low-load interior spaces.

When the skin system is heating-only, a very efficient central heat pump sys-

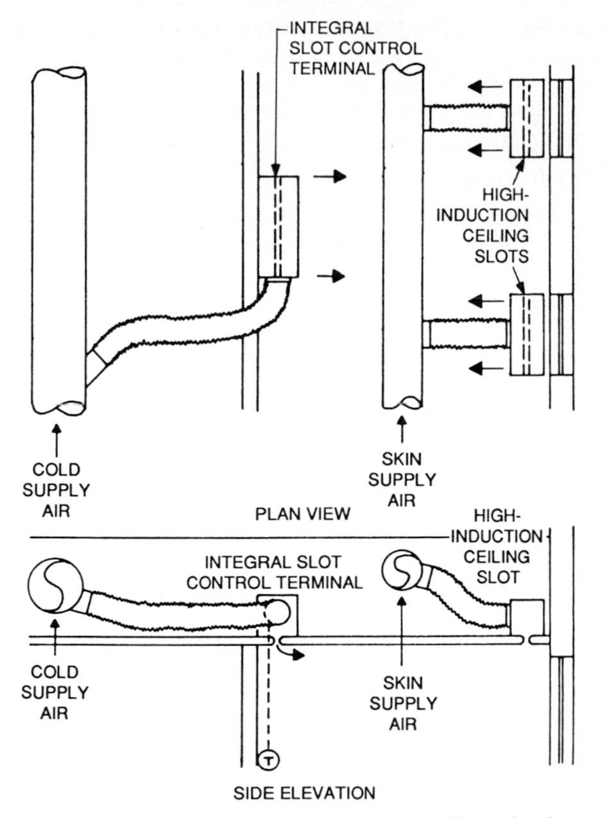

FIGURE 37.8 Constant-volume skin VAV cooling only—interior. Skin load less than 400 Btu/(h · lin ft) (385 W/lin m). (Above these limits use a downblow diffuser above windows.) *Note*: Make runouts straight for 4 diameters upstream from box inlet.

tem can be made by using an auxiliary air-cooled condenser in the main refrigeration circuit as the heating coil for the skin system. When the skin system is a heating/cooling system, a double-bundle condenser and hot-water heating coil are employed. Thermal storage can be included in either case.

Controls. VAV box controls can be pneumatic, self-powered, electric, or electronic with velocity-resetting options. The skin system is reset off the outside temperature to just handle the heating transmission plus infiltration load in winter and cooling transmission plus infiltration in summer, if cooling capability is included.

Heating energy is conserved by setting the heating reset schedule to maintain a relatively low inside temperature when lights are off. Heat from lights then brings the winter temperatures up to a comfortable level.

Central readout and reset of zone temperatures and central night setback are available. Night temperature control is obtained by intermittent or slow-speed operation of the skin fan only with appropriate temperature resetting.

37.2.6 Furnace Fan Coil Heating-Only Skin, VAV Cooling Only, Interior (Fig. 37.9)

Advantages

1. Low first cost with electric heat
2. Good downdraft control
3. Minimal simultaneous heating and cooling, if properly controlled
4. Reduced fan energy because smaller fans operate only on a call for heat
5. Possibility of free heat transfer to perimeter spaces if heating fan units are located near central core

Application Guidelines. Install furnace fan heating units as close as possible to the central core to minimize wiring or hot-water piping and reduce noise in occupied spaces.

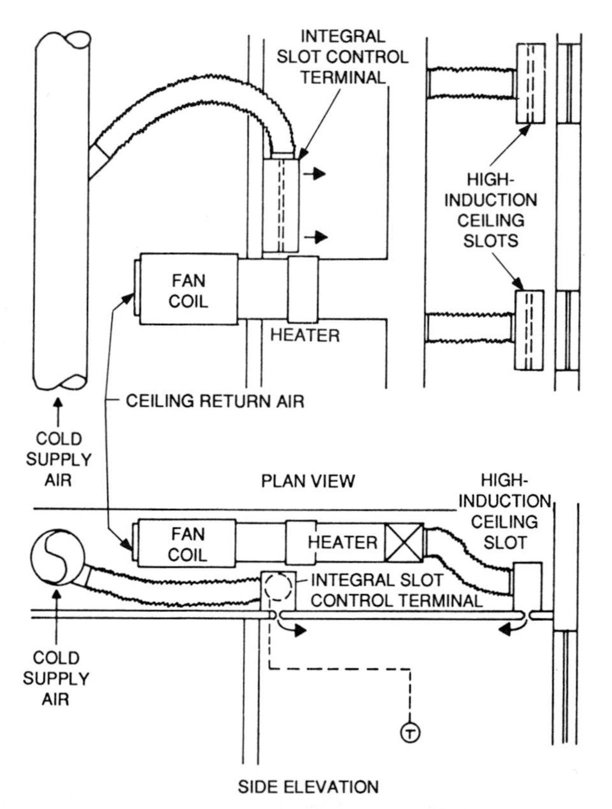

FIGURE 37.9 Fan-coil heating only, skin VAV cooling only—interior. Skin load less than 400 Btu/(h · lin ft) (385 W/lin m). (Above these limits use a downblow diffuser above windows.) *Note*: Make runouts straight for 4 diameters upstream from box inlet.

The System. Separate heating-only furnace heating fan coils per building face or per face per floor supply for constant-volume skin systems. With only one central cooling-only VAV system, this arrangement permits zoning of the skin to take advantage of solar and internally generated heat.

Where heating transmission load exceeds 400 Btu/(h · lin ft) (385 W/lin m) including infiltration, high-induction downblow slot diffusers should be used to avoid downdraft. When this load is less than 400 Btu/(h · lin ft) (385 W/lin m), a high-induction ceiling slot diffuser located at but blowing away from the outside wall can be used. When this load is less than 250 Btu/(h · lin ft) (240 W/lin m), any kind of diffuser is satisfactory.

All cooling loads in perimeter and interior spaces are handled by single integral slot control terminal units for zones up to 400 ft^3/min (680 m^3/h) and by control terminal units with one or more high-induction ceiling slot diffusers for larger zones. Series fan terminal boxes can be employed to increase air circulation in very low-load interior spaces.

Variable-volume constant-temperature air is used as the primary cooling air for this system. Heating air is at constant volume during the heating season with reset temperatures.

An efficient heat pump system can be made by using a central double-bundle condenser and hot-water heating coils in the fan coil units. Thermal storage can be included.

Controls. Pneumatic, self-powered, electric, or electronic controls can be used by selecting a particular control zone or using the outdoor or exterior wall temperature to control the heating unit. A wide variety of optional velocity limits and dead-band adjustments are available.

With electronic controls there is gradual fan turn-on, and each fan coil can be interlocked with one or more of the perimeter cooling-only VAV boxes. The heating fans are off during the cooling season. In midseason and in winter they operate to provide minimum air circulation and heating.

Central readout and reset of zone temperatures and central night setback are available. Night temperature control is accomplished by intermittent operation of furnace heating fans and heaters.

37.2.7 VAV Reheat Perimeter (Fig. 37.10)

Advantages. Minimum installed ductwork and low first cost are the benefits of this system.

Application Guidelines. The system must be properly designed, controlled, and restricted to light-load heating applications to avoid substantial energy waste. It is not recommended for applications with high winter heating load because of low diffuser velocity in the heating season and marginal downdraft control.

The System. Perimeter spaces are served by a variable-volume terminal unit which is controlled to maintain a certain minimum air flow when a downstream heating coil is on.

A VAV reheat system is not recommended where winter transmission load plus infiltration exceeds 250 Btu/(h · lin ft) (240 W/lin m). When the load is below this value, any diffuser is satisfactory.

Lights and people loads in interior spaces are handled by single induction slot

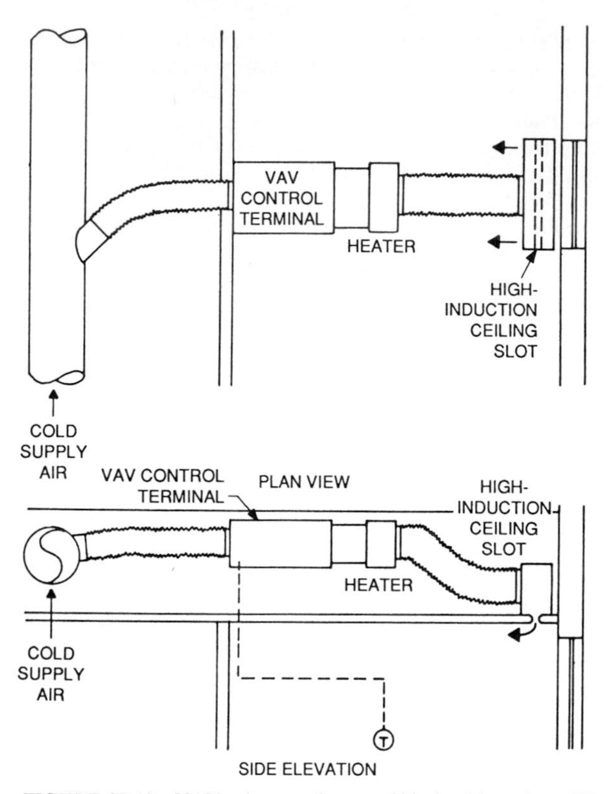

FIGURE 37.10 VAV reheat perimeter. Skin load less than 250 Btu/(h · lin ft) (240 W/lin m). *Note*: Make runouts straight for 4 diameters upstream from box inlet.

control terminal units for zones up to approximately 400 ft^3/min (680 m^3/h) and single-duct terminal units with one or more high-induction ceiling slot diffusers for larger zones. Single fan terminal boxes can be employed to increase air circulation in very low-load interior spaces.

Variable-volume temperature-reset cold air is recommended for this system. To be able to achieve substantial upward resetting of the cold supply air temperature without losing control of interior zones in winter, all boxes and diffusers serving pure interior spaces should be oversized. The amount of oversize depends on the ratio of interior to perimeter space, with the greatest amount of oversizing used where the interior is small compared with the perimeter.

Controls. Controls can be pneumatic, electric, or electronic with a wide variety of optional velocity limits and dead-band adjustments. Unless the duct static pressure is well controlled and uniform throughout the building, special controls are needed to ensure desired air flow while reheat is maintained at the design value for reheating. Duct static pressure control is discussed in Sec. 37.5. If this value is relatively low and the cold air temperature resetting is substantial, the energy wasted in reheating can be kept within reasonable bounds.

Electronic controls enjoy a special advantage in that they permit the air flow to decrease to a cooling minimum air flow which is below the heating minimum air flow. This avoids overcooling and subsequent reheating at times when the net space load is very small. In many mild-climate applications, this additional feature can almost completely eliminate the energy waste associated with reheating.

Central readout and reset of zone temperatures and central night setback are available. Night temperature control is obtained by intermittent operation of the main fan.

37.2.8 Baseboard Heating, Skin VAV Cooling-Only Interior (Fig. 37.11)

Advantages

1. Maximum zone relocation flexibility and relatively low first cost in buildings with many small zones and high occupancy at startup
2. Good downdraft control when excess heat is provided
3. Quiet operation

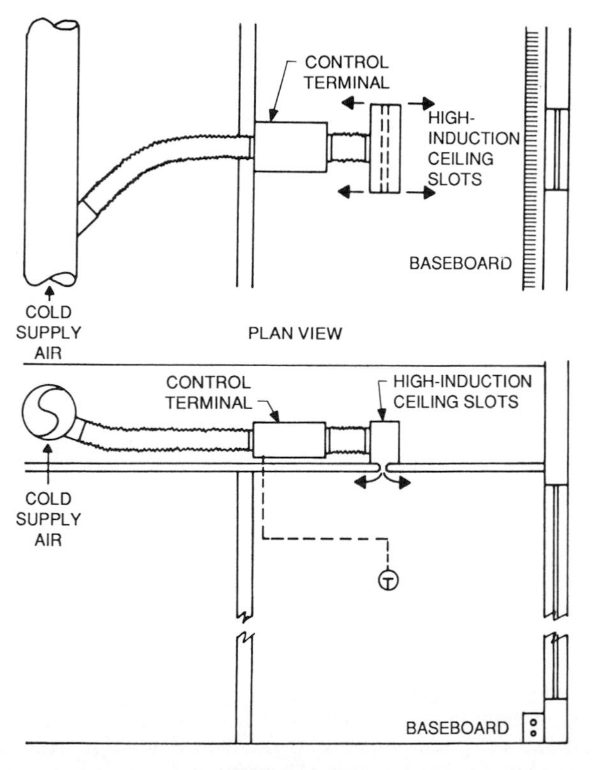

FIGURE 37.11 Baseboard heating, skin VAV cooling only—interior. *Note*: Make runouts straight for 4 diameters upstream from box inlet.

Application Guidelines. To maintain perimeter space ventilation in winter, it is usually necessary to overheat with the baseboard system and thereby force the perimeter space cooling boxes to recool with outside air. Alternatively, VAV boxes can be provided with a minimum air setting and the baseboard controlled to maintain space temperature, but this generally has higher first cost.

The System. An electric or hot-water baseboard skin radiation heating system handles the heating skin transmission and infiltration load. A cooling-only variable-volume supply unit handles all cooling and ventilation air loads. The cooling system often includes an outside air economizer.

All cooling loads are handled by single-type integral slot control terminal units for zones up to approximately 400 ft^3/min (680 m^3/h) and by control terminal units with one or more high-induction ceiling slot diffusers for larger zones. Series fan terminal boxes can be employed to increase air circulation in very low-load interior spaces.

Variable-volume constant-temperature cold air is used for the air distribution system.

Controls. VAV box controls can be pneumatic, self-powered, electric, or electronic, with velocity resetting options. A hot-water baseboard skin heating system is usually reset from the outside air temperature to provide some overheating for the sake of ventilation. Electric baseboard systems are usually broken into relatively small zones with each zone sequenced to come on as the nearest cooling box approaches some minimum position.

Central readout and reset of zone temperatures and central night setback are available. Night temperature control is accomplished by resetting the hot-water temperature or by intermittent operation of electric baseboard heaters.

37.2.9 System Fans

See Chap. 34 for additional information on fans.

Three factors can cause unsatisfactory operation in fans:

1. Surge, which can occur in all fan types
2. Paralleling, which may result from multiple fans with a characteristic dip in the curve, common with forward-curved (FC) and vane axial types
3. Resonance, which is a beat frequency produced by multiple fans operating at slightly different speeds

Fan surge (Fig. 37.12), as it is normally referred to in the industry, is a result of stall as air passes over the fan blades. This produces a static pressure and noise level fluctuation. The magnitude is on the order of 10 percent of block-tight static pressure for an airfoil centrifugal fan. While this pulsation is typically less for FC and vane axial fans, it still results in unsatisfactory operation in any type of fan. Do not choose a fan or allow a fan to operate in this area of its performance curve. Most manufacturers allow some margin of safety between their catalog cutoff limit and the point where tests indicate that surge begins.

With multiple fans (of the same zone) having a dip in their curve, paralleling can occur (Fig. 37.13). This is caused by one fan operating on one side of its peak and the other fan operating on the opposite side of the peak. This results in unstable operation and static pressure pulsation, increased horsepower, and in-

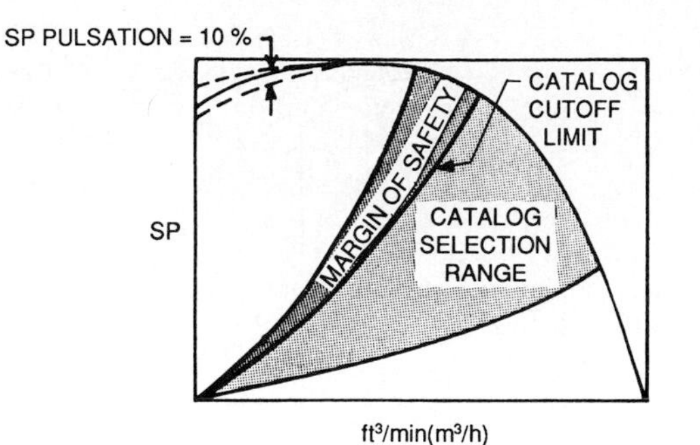

FIGURE 37.12 Fan surge.

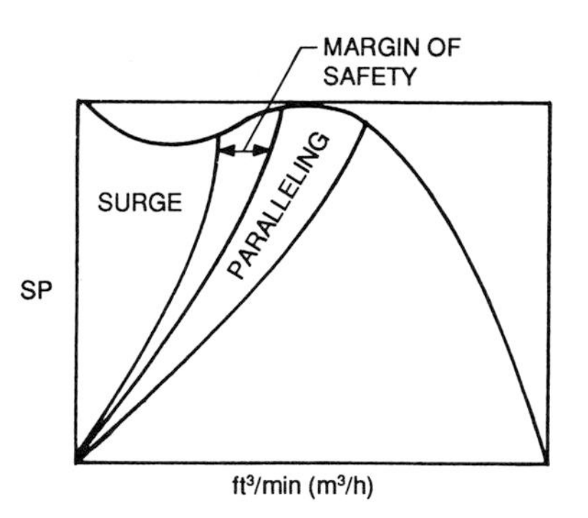

FIGURE 37.13 Multiple fans, paralleling and resonance.

creased noise level. The symptoms of paralleling are the same as those for surge, and with multiple fans of this type, it is often difficult to tell which is the cause. It is obvious that a fan cannot be allowed, and should not be selected, to operate under either of these conditions. Resonance may also result from multiple fans. The solution is to operate them at the exact same speed or speeds far enough to avoid a beat frequency.

37.3 FAN MODULATION METHODS

The following types of fan capacity modulation can be used:

- Static pressure modulation with inlet dampers and discharge dampers
- Scroll volume
- Runaround
- Speed variation
- Controllable pitch
- Inlet vanes

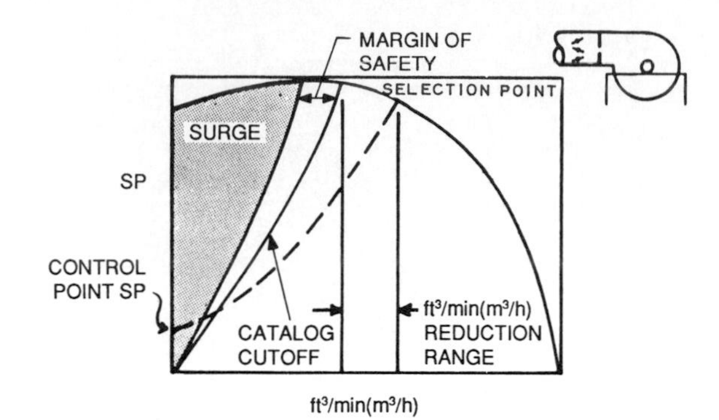

FIGURE 37.14 Static pressure control.

Inlet dampers and discharge dampers provide load modulation by adding resistance to the system, either upstream or downstream of the fan. Figure 37.14 shows the characteristic curves of an FC fan with the catalog cutoff limit and surge range indicated. A variable-volume system will require load and static pressure as indicated by the dashed line between the selection point and control point. As resistance is added to the system through dampers, the operating point moves up on the fan curve and the load is reduced. Note that the load reduction is limited to where the fan reaches the catalog cutoff limit because of fan surge. This modulation method is used widely on single FC fans in systems requiring relatively low static pressure.

Scroll volume control (Fig. 37.15) is achieved in FC fans by attaching a flat sheet inside the top of the fan housing, which in effect changes the scroll's shape. The result is a variety of performance curves, which will result in a capacity reduction. However, there is little or no horsepower savings and it does not work on BI or AF types. Its primary application is on low-pressure class 1 FC fans or to correct paralleling on multiple fans.

The next method used is the runaround (Fig. 37.16). This is a bypass duct which diverts some of the air back from the discharge of the fan to the inlet and prevents it from operating in the unstable range. Keep in mind that while the runaround prevents unstable fan operation, the duct static pressure is not being reduced and an additional set of dampers downstream from the runaround should

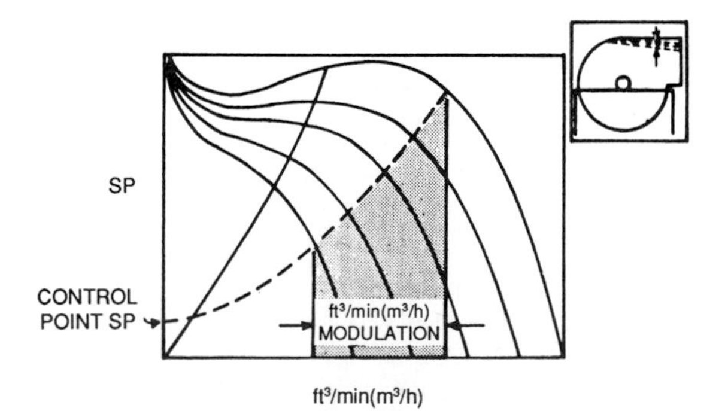

FIGURE 37.15 Scroll volume control.

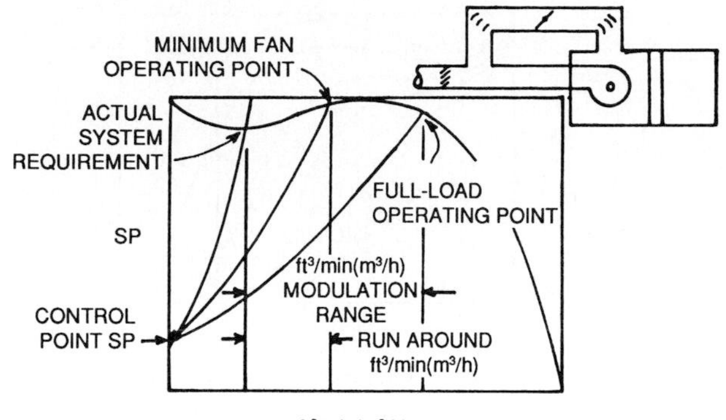

FIGURE 37.16 Runaround bypass.

be used to prevent excessive duct static pressure. Because it works well and is relatively inexpensive, the runaround method is frequently used to correct an existing problem, although there is little horsepower savings and it is rather space-consuming.

Speed variation (Fig. 37.17) is another means of controlling fan capacity. It is available through fluid or mechanical speed reducers or solid-state voltage rectifiers. This affords probably the greatest opportunity for horsepower reduction, providing the power consumed by the speed reducer does not offset this reduction. Sound level is also reduced with this method of control. However, at present it is normally higher in first cost than other options and must be justified on its operating-cost savings. It is also possible to use multispeed motors which will provide two or more steps of reduction. However, the added first cost, plus control complexity, is the main reason for its not being more popular.

Controllable-pitch axial fans (Fig. 37.18) achieve capacity reduction by changing the blade pitch. This produces a family of fan curves which provide capacity reduction. While this method of control may achieve horsepower reduction approaching that of variable speeds, the sound level is not necessarily reduced and may, in fact, be increased. This, coupled with the high sound level generated by this type of fan, requires considerable attention to acoustical treatment down-

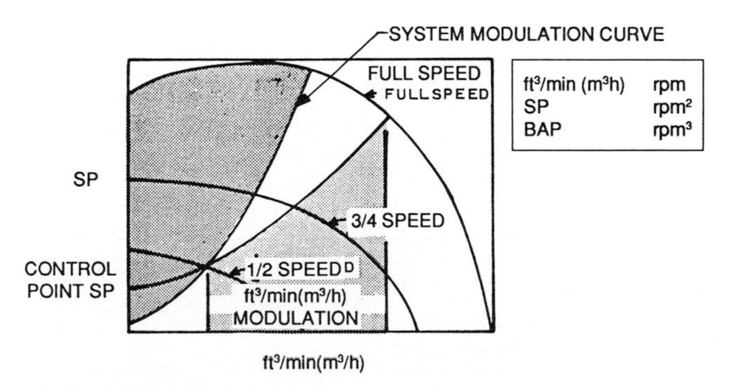

FIGURE 37.17 Speed control.

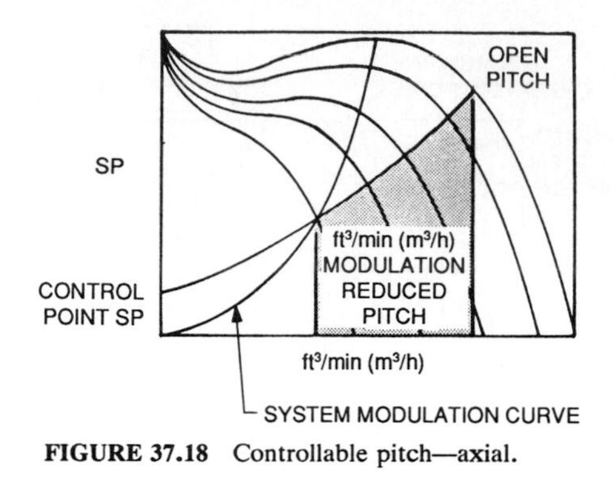

FIGURE 37.18 Controllable pitch—axial.

stream of the fan. It is also high in first cost when attenuation requirements are considered.

Inlet vanes (Fig. 37.19) are readily available from almost all manufacturers of centrifugal fans except in smaller sizes. They achieve speed reduction by creating spin in the direction of rotation of the fan, which reduces the load, static pressure, and boiler horsepower (bhp). This spin changes the surge characteristics of the fan slightly, as indicated by the bump in the surge curve. However, inlet vanes do not eliminate surge, as is sometimes assumed. Horsepower reductions are achieved, although not to the extent of variable-speed or controllable-pitch fans. Since the centrifugal fan is typically more efficient than vane axial fans, the centrifugal fan with inlet vanes frequently requires lower horsepower than the controllable-pitch vane axial. Inlet vanes may increase the sound level of centrifugal fans on the order of 5 dB, although centrifugal fan sound levels are quite low for the duty performed. Inlet vanes are relatively low in first cost, typically ranging from 20 to 50 percent of the cost of the fan, depending on fan size.

Variable-volume systems without fan control are not recommended. Even

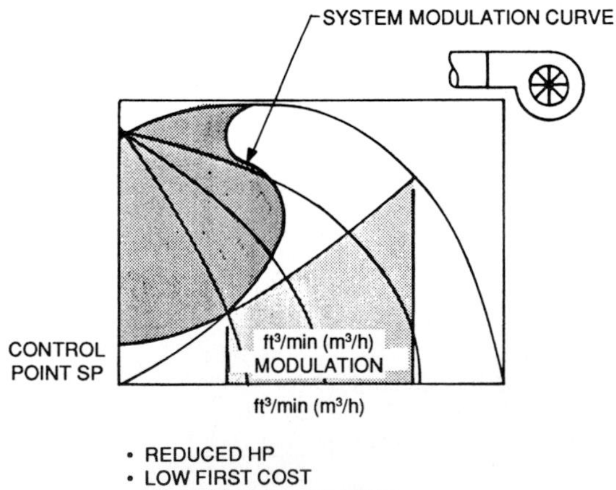

FIGURE 37.19 Inlet vanes.

though the fan can be undersized to prevent surge at the minimum load requirement, the excess pressure buildup increases noise and leakage problems.

37.4 DEVIATION FROM CATALOG RATINGS

Fans are tested, rated, and cataloged under more or less ideal conditions. The standard Air Moving and Conditioning Association (AMCA) test procedure used is shown in Fig. 37.20. Note that 10 diameters of duct are used and performance is measured 8½ diameters from the discharge. A flow straightener is also used.

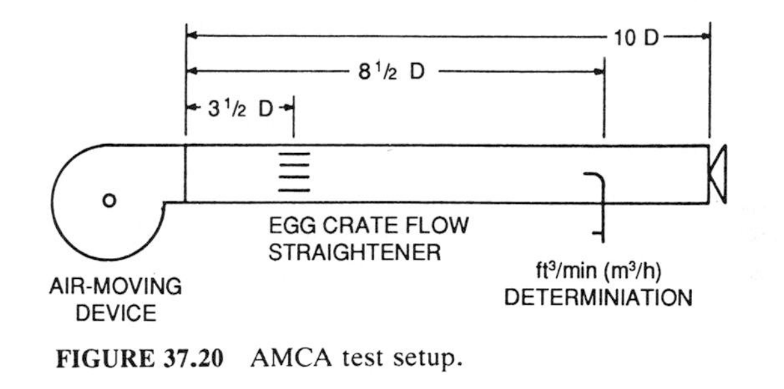

FIGURE 37.20 AMCA test setup.

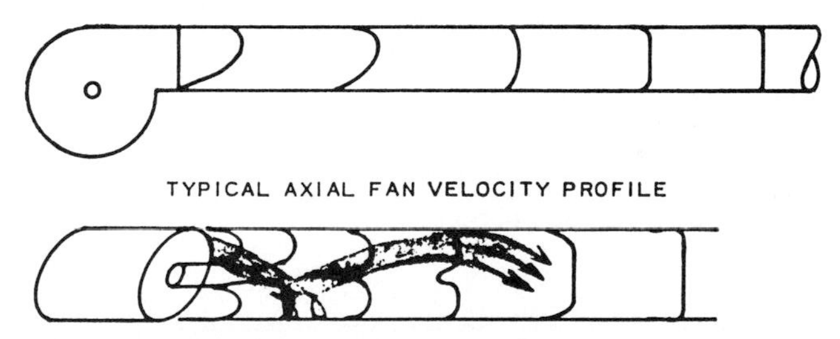

FIGURE 37.21 Fan velocity profiles.

Most fans do not have a uniform discharge velocity profile (see Fig. 37.21). Part of the high-velocity energy is recaptured as static regain and is included in catalog ratings. However, when a minimum of 5 fan diameters of duct is not provided in a given application, such as a blowthrough installation, this energy is lost and cataloged ratings must be adjusted accordingly. The typical static regain included in manufacturers' ratings which must be compensated for when not installed with same manner as rated is shown in Fig. 37.22. For example, on blowthrough applications, multiply the factor by the outlet velocity pressure and add it to the static pressure required to make the fan selection.

Be sure to consider these factors in any fan selection:

1. Whether full static regain is achieved
2. Plenum wall within ¾ wheel diameter of inlet

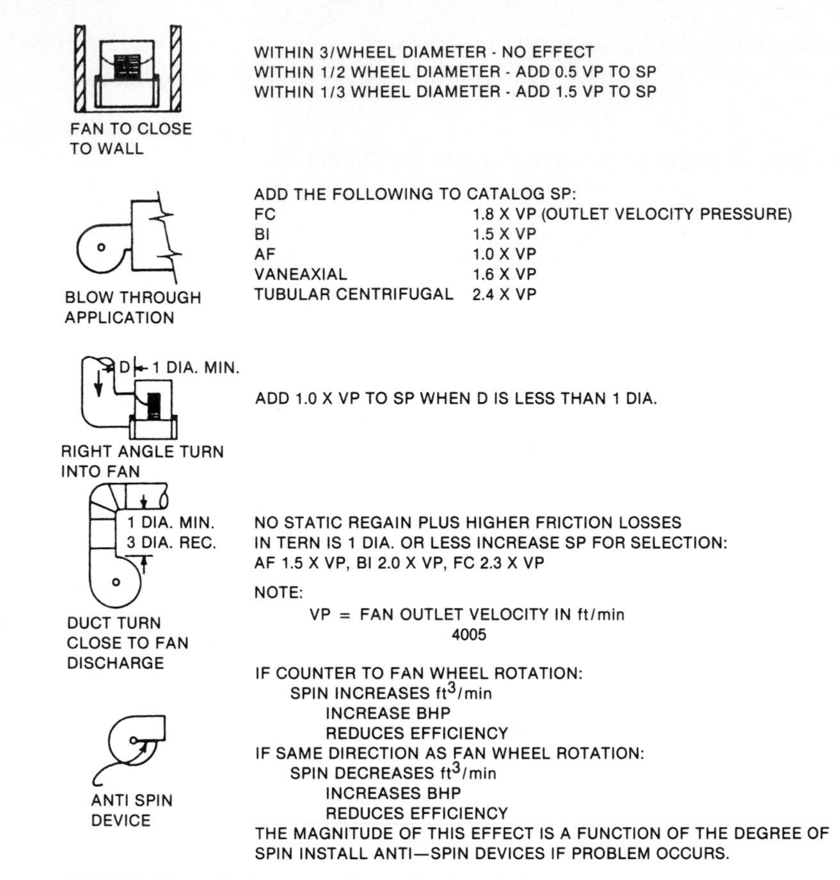

FIGURE 37.22 Performance effect of fan installation.

3. Effect of drive sheaves and belt guards close to fan inlet
4. Inlet spin
5. Nonuniform air flow to inlet
6. Inlet vane resistance

37.4.1 Example of Static Regain Effect

To illustrate the effect of static regain and how to compensate for its absence in fan selection, consider the following example (Fig. 37.23). (*Note*: For metric equivalents, BHP × 0.746 = kW, FPM (or ft/min) × 0.0058 = m/s, and inWG × 249 = Pa.)

This is required at entrance to supply duct system: 2830 ft/min velocity, 3 in WG static pressure (all at discharge). Assume a 3 percent speed increase and 13 percent horsepower increase for inlet vanes and a 2 percent speed increase and 6 percent horsepower increase for drive losses.

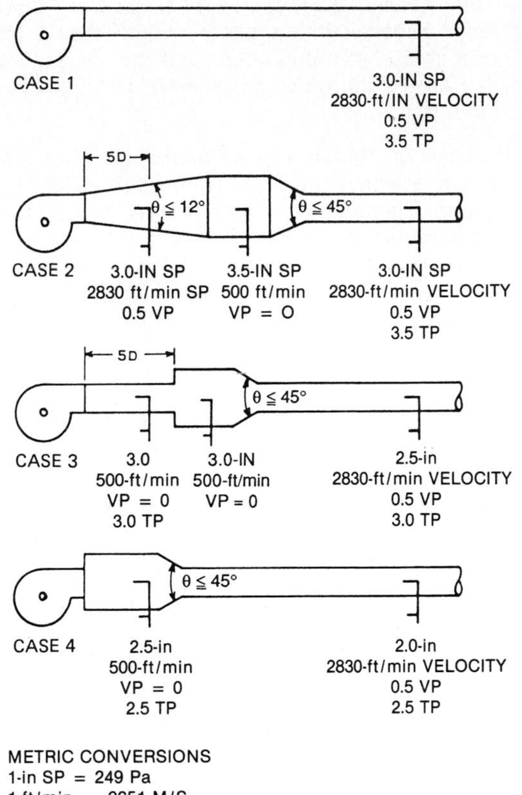

FIGURE 37.23 Effect of static regain at constant fan speed.

Case 1. Ducted discharge is same size as fan outlet (duct exceeds 8 fan diameters). The fan is applied in same manner tested, so catalog ratings may be used to determine the required speed. From the catalog, RPM (or r/min) = 900.

$$\text{RPM required} = \text{catalog RPM} \times \begin{pmatrix}\text{assumed increase}\\ \text{due to inlet vanes}\end{pmatrix} \times \begin{pmatrix}\text{assumed increase due}\\ \text{to drive losses}\end{pmatrix}$$

$$= 900 \times 1.03 \times 1.02 = 946$$

From catalog, BHP = 23.0.

$$\text{BHP (kW) required} = \text{catalog BHP} \times \text{kW} \times \begin{pmatrix}\text{assumed increase}\\ \text{due to inlet vanes}\end{pmatrix} \times \begin{pmatrix}\text{assumed increase due}\\ \text{to drive losses}\end{pmatrix}$$

$$\text{BHP required} = 23.0 \times 1.13 \times 1.06 = 27.6$$

Case 2: Discharge with Gradual Expansion and Gradual Contraction Transitions. There is no difference between this application and case 1 since no dynamic losses (ignore friction in this example) occur and the energy is transferred from velocity (leaving the fan) to static (in the plenum) and back to velocity (at the entrance to the main duct).

Case 3: Ducted Discharge to Plenum with Minimum of 3 Fan Wheel Diameters of Duct. In this case, one assumes the velocity pressure is lost by the sudden expansion into the plenum. Therefore, to obtain 3 inWG static pressure + ½ inWG velocity pressure (2830 ft/min velocity) (3½ inWG total pressure), we must select the fan to provide 3½ inWG total pressure in the plenum (3½ inWG static pressure since velocity pressure is nearly zero in the plenum).

Select fan for 3½ inWG static pressure with 2830 ft/min outlet velocity (assume airfoil is centrifugal).

From catalog, RPM = 935. So

$$\text{RPM required} = 935 \times 1.03 \times 1.02 = 982$$

From catalog, BHP = 26.0, so

$$\text{BHP required} = 26.0 \times 1.13 \times 1.06 = 31.1$$

If the fan is operated at the same speed as the fan in case 1, the static pressure at entrance to duct system would be only 2½ inWG.

Case 4: Blowthrough with Fan Discharging Directly into Plenum. This is the same as case 3 except no static regain occurs. Therefore, to achieve 3 inWG static pressure at 2830 ft/min in the downstream duct, assuming an airfoil centrifugal fan (regain effect of 1.0 × velocity pressure, from Fig. 37.22) and loss of fan discharge velocity pressure, we would select the fan for:

Actual static required in discharge plenum	3.0 inWG static pressure (747 Pa)
Absence of duct to achieve regain	0.5 inWG static pressure (125 Pa)
Discharge loss (sudden expansion)	0.5 inWG static pressure (125 Pa)
Fan selection	4.0 inWG (997 Pa)

From catalog, RPM = 970, so

$$\text{RPM required} = 970 \times 1.03 \times 1.02 = 1019$$

From catalog, BHP = 29.4, so

$$\text{BHP required} = 29.4 \times 1.13 \times 1.06 = 35.2$$

If this fan is operated at the same speed as in case 1 (900 r/min), the static pressure at entrance to the duct system would be only 2 inWG, as shown in Fig. 37.23.

This example illustrates the importance of making proper allowances for the difference between catalog ratings and the way that a fan is applied. In this case the horsepower actually required is 53 percent greater than that indicated in the catalog.

37.5 FAN CONTROL SENSOR LOCATION

A very important consideration in the design of the duct and fan system is the location of the sensor which controls fan modulation. Many possibilities exist in any given system with widely ranging results. The following considerations are important:

- Duct system configuration
- Parallel risers and trunks
- Possible use of the space
- Fan selection and surge characteristics
- Duct construction and tightness
- Sound level characteristics of the terminal unit

Now let

TP = total pressure
SP = static pressure
VP = velocity pressure
HL = head losses due to friction and air turbulence at duct fittings
VEL = velocity
u = upstream
d = downstream
ΣHL = sum of head losses from fan to a given point

Then

$$\mathrm{TP} = \mathrm{SP} + \mathrm{VP}$$
$$\mathrm{TP}_u = \mathrm{TP}_d + \mathrm{HL}$$
$$\mathrm{SP}_u + \mathrm{VP}_d = \mathrm{SP}_d + \mathrm{VP}_d + \mathrm{HL}$$
$$\mathrm{VEL} = \frac{\text{air flow}}{\text{duct area}}$$
$$\mathrm{VP, in\ WG} = \left(\frac{\mathrm{VEL}}{4005}\right)^2 \qquad \text{VEL in ft/min}$$

To illustrate, consider the high-velocity static regain design in Fig. 37.24. This is a typical result of computer-optimized design.

The fan is sized to handle 99,000 ft^3/min (168,201 m^3/h), serving seven floors of a commercial office building. The upper six floors consist of office space which is expected to have very limited use at night and on weekends. However, the bottom floor is a commercial store which may be in full operation at times when the rest of the building is not. With this design and space utilization, the three operating conditions which cover the capacity modulation range and encompass the worst situation that would probably exist are

1. Full load
2. 50 percent capacity evenly distributed
3. Only the bottom floor, with 14 percent of load in operation

First, the building is analyzed at full-load conditions at the fan in the main riser just prior to the first branch takeoff and at the bottom of the main riser just prior to the last branch takeoff.

At full load the total pressure at the fan consists of a velocity pressure of 0.4

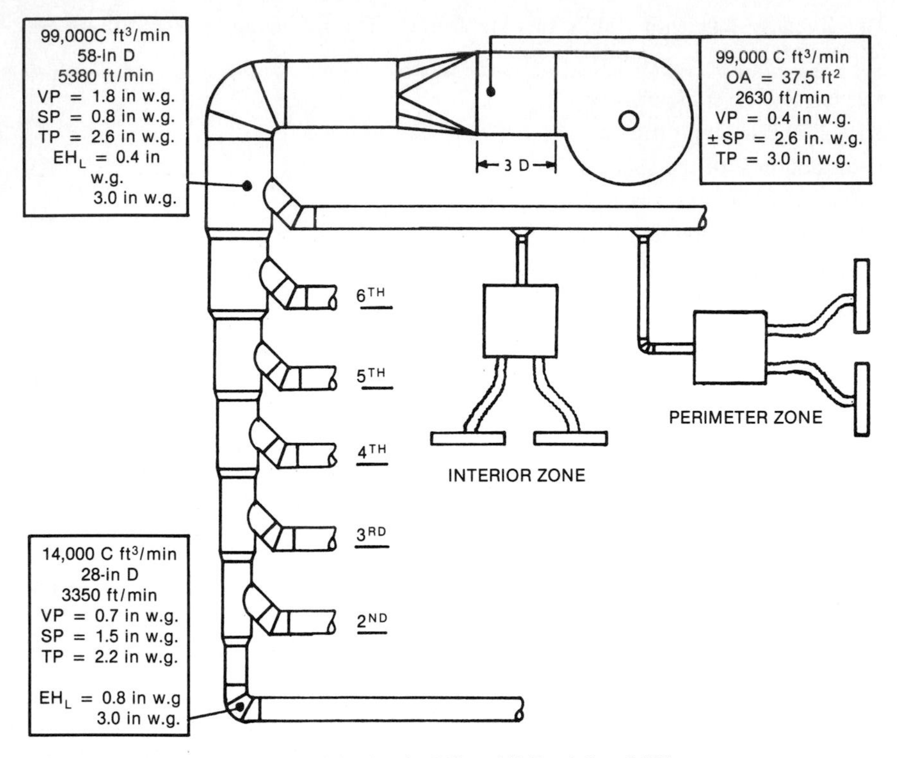

FIGURE 37.24 Fan control—full load. 1 in WG = 249 Pa; 1 ft = 0.305 m.

inWG (100 Pa) and a static pressure of 2.6 inWG (647 Pa). Just downstream in the main riser, the duct size is reduced, the velocity pressure equals 1.8 inWG (448 Pa), and the static pressure equals 0.8 inWG (200 Pa), for a total pressure of 2.6 inWG (647 Pa). The difference between the total pressure at this point and that at the fan is the head losses of 0.4 inWG (100 Pa) due to dynamic losses in the transition and elbow and the friction loss in that length of duct.

At the bottom of the riser, only 14,000 ft^3/min (23,786 m^3/h) is left in the duct with a velocity pressure of 0.7 inWG (174 Pa), a static pressure of 1.5 inWG (374 Pa), and a total pressure of 2.2 inWG (548 Pa). The difference in total pressure between this and 3.0 inWG (747 Pa) at the fan is the sum of the head losses, equal to 0.8 inWG (200 Pa). The bottom of the main riser in this example would be the proper place to locate the fan capacity control signal set at 1.5 inWG (374 Pa) static pressure. This location provides for significant capacity reduction at the fan without short-circuiting any trunk ducts at the bottom of the main riser.

At each branch takeoff, static regain occurs which increases the static pressure. However, dynamic loss occurs at the takeoff which reduces static pressure. And velocity pressure is increased through the reducer which also reduces static pressure. The objective in high-velocity static regain design is to attempt to maintain, as closely as possible, a constant static pressure throughout the main riser and trunk duct system. This means that the net reduction in static pressure due to friction and dynamic losses, plus a reduction due to a change in velocity pressure through reducers, must be offset by the static regain achieved at each takeoff.

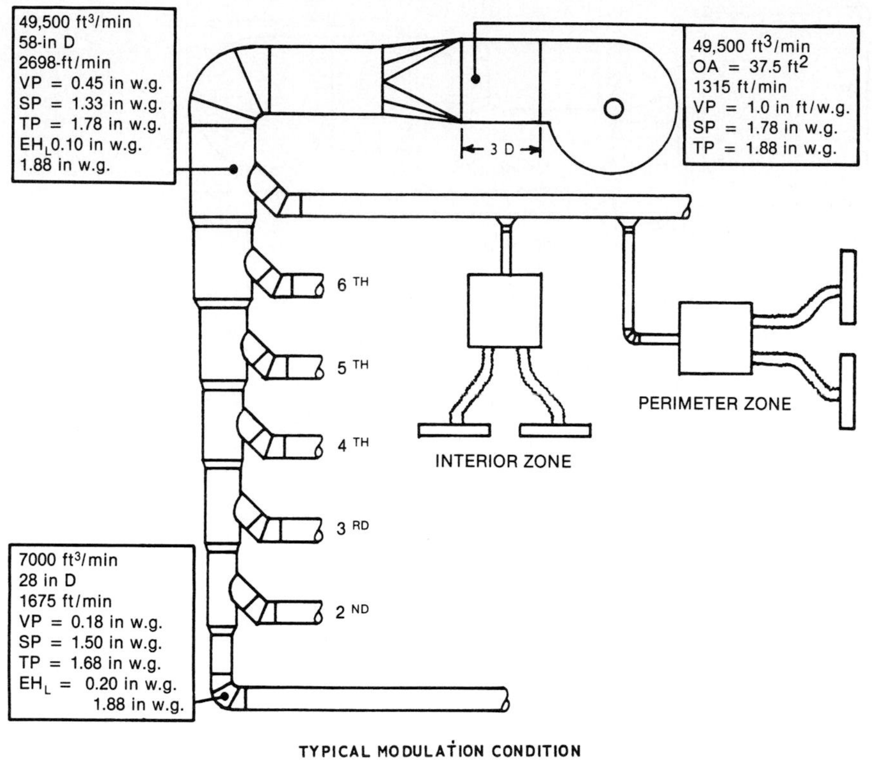

FIGURE 37.25 Fan control—50 percent of capacity. 1 in WG = 249 Pa; 1 ft = 0.305 m.

Next consider what happens when the system is at 50 percent capacity, evenly distributed to reach each floor (Fig. 37.25). This situation could exist, e.g., in a building with a lot of glass if the sun went under a cloud. It would be a relatively common occurrence. Under these conditions, the head losses and velocity pressure regain are reduced by the square of the air-flow change. The fan capacity control is set at 1.5 inWG (374 Pa) static pressure (the value required at maximum load), even though the actual requirement on the bottom floor may be less. The net effect in this case is a total pressure reduction at the fan of 1.88 inWG (468 Pa). Note that at the top and bottom of the riser the total pressure, plus the sum of the head losses, is equal to that at the fan.

Finally, consider the situation when only the first floor is in operation (Fig. 37.26). This is not the same "system" as in the previous two situations, because full flow still exists in the last section of the main riser between the first and second floors. In the next section up between the second and third floors, only half the design air flow exists. This is reduced in succeeding sections to ⅓, ¼, ⅕, ⅙, and ⅐ of the design air flow. The resulting head losses in each section are reduced by the square of the load change, and since none of the other floors is taking any air, no static regain occurs.

Enough total pressure must be produced at the fan to provide the required

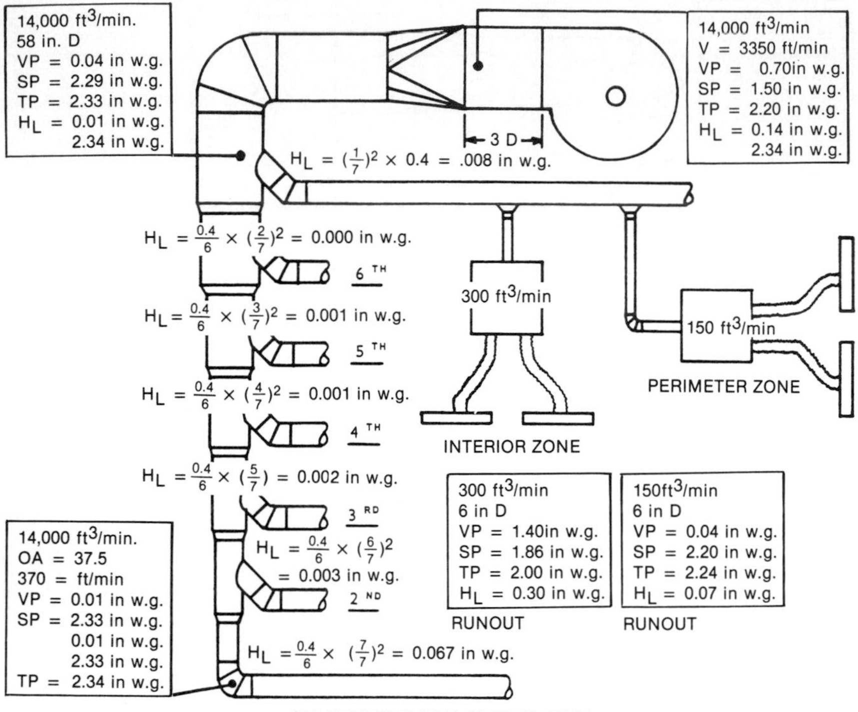

FIGURE 37.26 Fan control—14 percent of full load. 1 in WG = 249 Pa; 1 ft = 0.305 m.

static pressure and velocity pressure at the bottom of the main riser (in this example) as well as offset all the head losses to that point.

The resulting velocity pressure, static pressure, and total pressure are shown at the top of the main riser and at the fan. Note that the velocity pressure now is practically zero at these points, and thus the static pressure component is almost all the total pressure.

Under this condition, if a person came into an office on the top floor, there would be about 2.29 inWG (570 Pa) of static pressure in the entire top-floor trunk duct system. Assuming an internal office so that the terminal unit requires full load as soon as the lights are on and the people load exists, 1.86 inWG (463 Pa) of static pressure would exist at the terminal after the losses in the takeoff fitting and runout ductwork are deducted. The sound level, duct construction, and other considerations would have to be evaluated on the basis of this maximum static pressure condition. Also shown is the same situation involving a perimeter office with only half of the design terminal air flow in the absence of any solar load. Since lower losses exist in the takeoff fitting and runout, the static pressure would be higher at this terminal and the sound level would have to be evaluated on the basis of the reduced load and higher static pressure.

Figure 37.27 indicates the static pressure requirements for each of these conditions. Remember, the static pressure which the sensor is set to control must satisfy the highest static pressure requirement at that point. Thus, if the sensor is

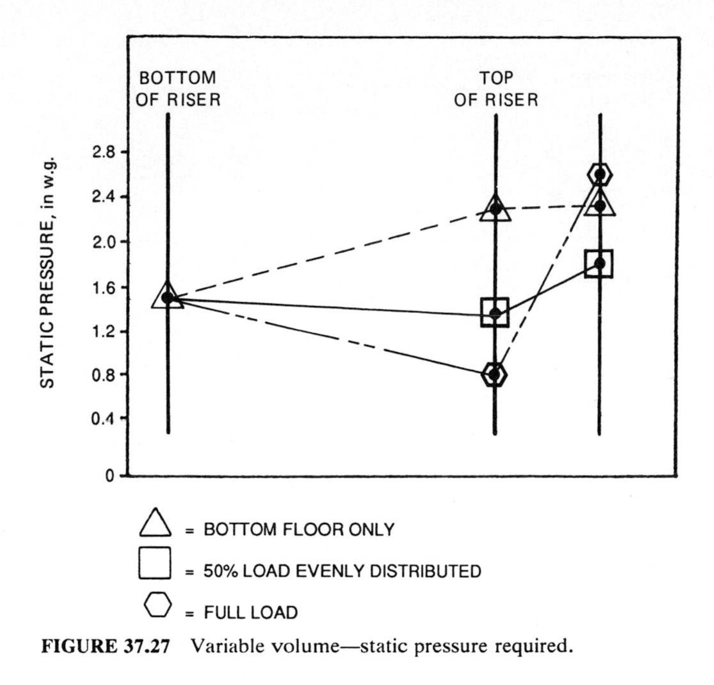

FIGURE 37.27 Variable volume—static pressure required.

located at the fan, it must be set to maintain full-load discharge static pressure of 2.6 inWG (647 Pa) at all times (in this example). This obviously would result in considerable excess pressure at all other conditions and therefore is not the most desirable control point. Locating the sensor at the top of the main riser in the high-velocity ductwork would still require a rather high static pressure at the worst condition (only the bottom floor in operation). Thus, this is obviously not a good location. Locating the sensor at the bottom of the main shaft and setting it for 1.5 inWG (374 Pa) is probably the best location for this example.

37.6 FAN SELECTION

Now consider the selection of fans for various applications. Consideration should be given to the varying system resistance in multizone and particularly double-duct systems and its effect on load, fan horsepower, and duct static pressure (see Fig. 37.28).

For example, when the average of all zone requirements is 50 percent hot deck and 50 percent cold deck, the friction loss through the parallel components (i.e., coils, dampers, and supply ducts for double duct) is reduced by the square of the air-flow change. This means that the operating point on the fan curve is changed, resulting in increased air flow.

On double-duct systems, since the same static pressure exists at the entrance to both ducts, considerable excess pressure may exist in the duct requiring less air. Dampers are sometimes installed and controlled from a remote point in the duct to prevent excessive static pressure buildup (see Fig. 37.29). The same rules for locating the sensor apply as described in Sec. 37.5.

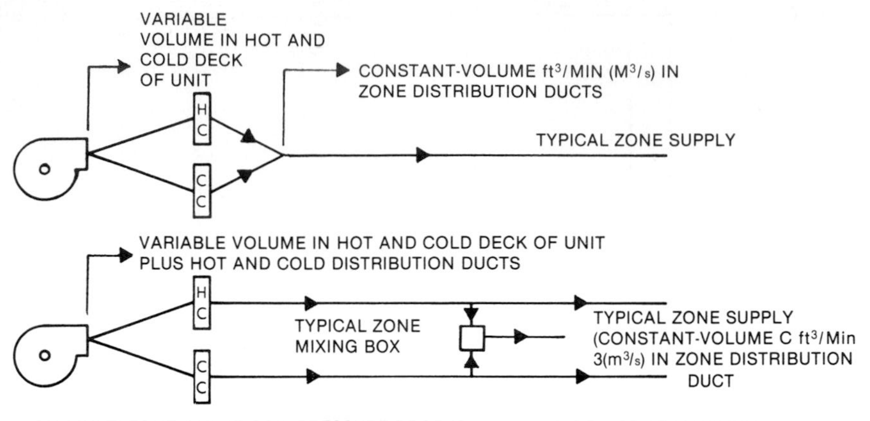

ON MULTI ZONE (M2): IF COIL PRESSURE DROP IS 1-in w.g. (249-Pa) SP AT MAXIMUM-LOAD ft³/min(m³/s), THEN COIL SP AT 50% LOAD = ¼ in w.g. (62 Pa) [SP AT MAXIMUM LOAD × (½)²], OR ¾-in w.g. (187-Pa) SP VARIATION.

ON DUAL DUCT (DD): IF COIL & TRUNK DUCT PRESSURE DROP IS 4-in w.g. (996-Pa) SP AT MAXIMUM-LOAD ft³/min (m³/s), THEN SP AT 50% LOAD = 1 in w.g. (249 Pa) [SP AT MAXIMUM LOAD × (½)²], or 3-in w.g. (747-Pa) VARIATION.

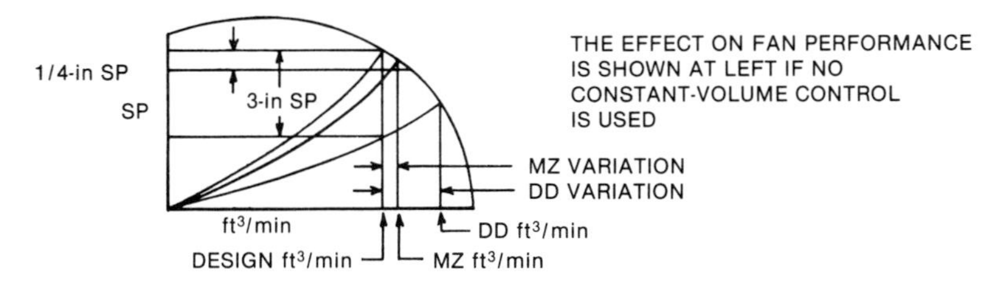

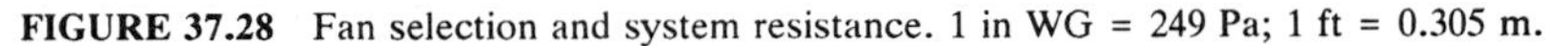

FIGURE 37.28 Fan selection and system resistance. 1 in WG = 249 Pa; 1 ft = 0.305 m.

With constant-volume systems, the rule of thumb has typically been to select the largest fan possible, which would also be the most efficient and quietest. But while the largest fan may be the most efficient, it is not the correct fan for a variable-volume system. The reason is that little load variation is possible without putting the fan into surge. This is true regardless of the sensor's location in the system. For that reason, a fan one size, or possibly even two sizes, smaller (Fig. 37.30) is required to provide the necessary load variation. While the smaller fan may be less efficient at its full-load operating point, there may be very little horsepower variation, compared with a large fan, at the typical modulated condition where the fan will operate a majority of the time.

The following illustrates fan selection for a variable-volume system. Figure 37.31 shows a 60-in (1.52-m) double-width fan which is the second largest size that could be used for the selection. It results in 71 percent static efficiency at full load and 77 percent total efficiency. The minimum system resistance curve is shown as well as the situation that could exist at the fan when only the bottom floor is in operation. These two curves, then, determine the maximum amount of excess pressure that can exist ahead of any terminal. So this gives an easy means of evaluating acceptability of the resulting sound level and duct construction criteria. Note, however, that this fan is entering the surge region at about 50,000

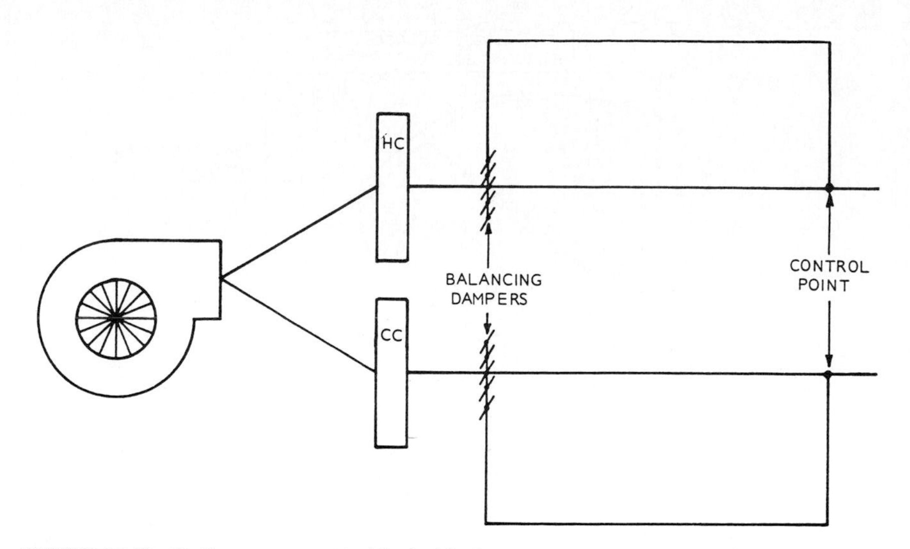

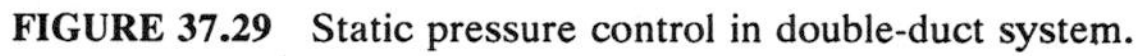

FIGURE 37.29 Static pressure control in double-duct system.

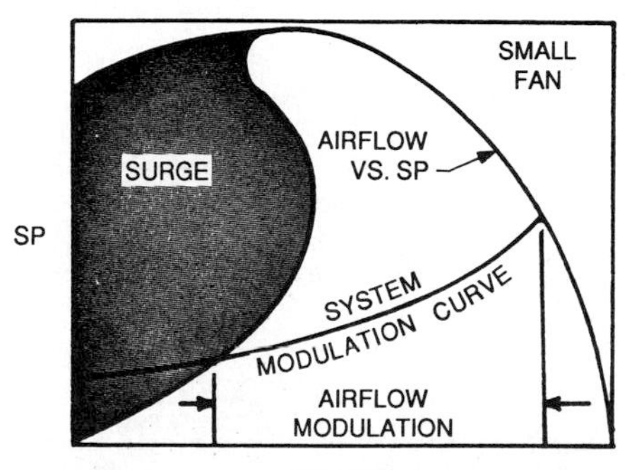

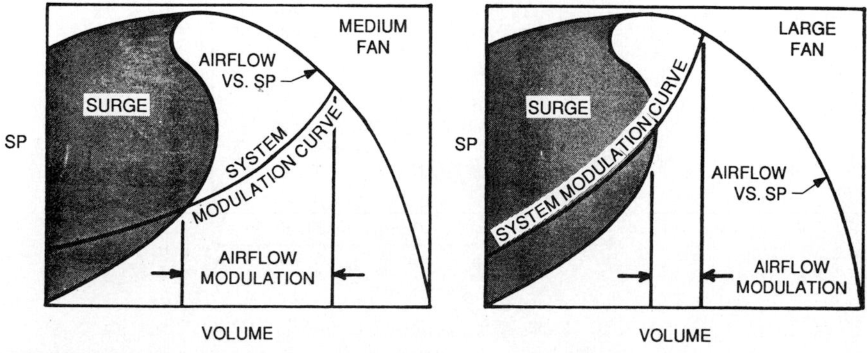

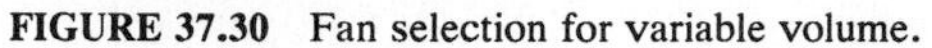

FIGURE 37.30 Fan selection for variable volume.

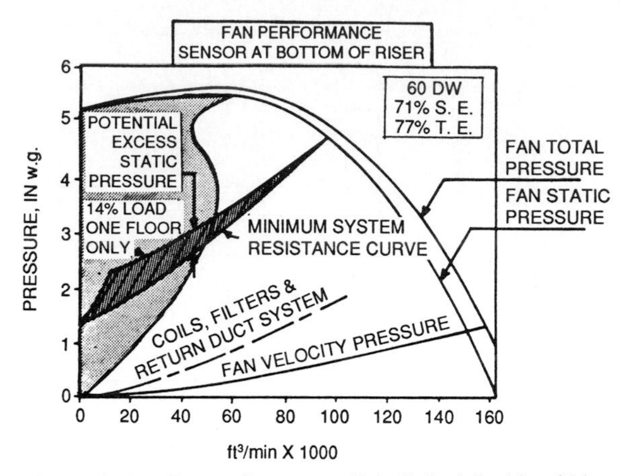

FIGURE 37.31 Fan performance—60-in (1.5-m) double width.

ft^3/min (84,950 m^3/h) with a static pressure of approximately 3.0 inWG (747 Pa) at the fan at the worst condition. The surge condition in this static pressure range is about the limit of what we could reasonably expect to get by with. If this situation could exist for extended periods with only the bottom floor in operation, a run-around duct, in conjunction with the inlet vanes, might be considered.

Figure 37.32 shows the situation with a fan two sizes smaller. This 49-in (1.245-m) double-width fan has a considerably reduced static efficiency of 54 percent, but the total efficiency of 67 percent has only dropped off by 10 points. Since total efficiency is the only true criterion in a properly designed drawthrough fan system, this selection looks better than if static efficiency had mistakenly been used for the comparison. But static efficiency is the proper criterion for comparing blowthrough applications where the velocity head is dissipated. This smaller fan allows reduction in load down to approximately 20,000

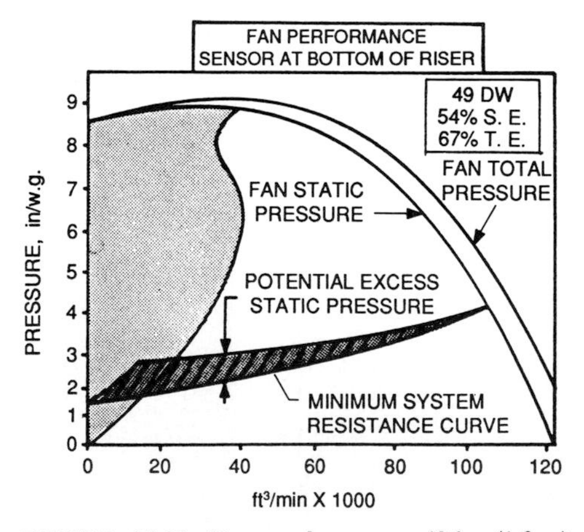

FIGURE 37.32 Fan performance—49-in (1.2-m) double width.

ft³/min (33,980 m³/h) before entering the surge range. At this operating point, the fan static pressure with inlet vane or variable-speed control is only 1½ to 2½ inWG (8 to 18 Pa). The resulting increase in noise or pulsation would probably be acceptable at these static pressures. The system designer must evaluate the potential minimum load for the system and the way in which it would be distributed in the building. She or he must then compare this value with the fan operating characteristics to determine the required method of control.

37.7 RETURN AIR FANS

Use of a return air fan should be considered whenever high return air duct losses are anticipated. In these cases, a return air fan may prove useful by providing a positive means for exhausting air. It also has the advantage of reducing the negative pressure on the suction side of the supply fan, permitting smaller return duct sizing.

Power relief fans should be used in lieu of return fans if the consideration is solely to provide sufficient building exhaust. Generally, when the return air path pressure drop is between 0.15 and 0.50 inWG (37 and 125 Pa), power relief fans give good results.

Return fan volume is calculated by subtracting the amount of air necessary to pressurize the building from the supply fan volume. Typically, this is about 10 percent of supply fan volume. Any forced exhaust should be deducted from the return air volume.

If the return air fan does not follow the supply fan (to maintain the design load differential), it could overpower the system. This would create a negative static pressure in the building and a positive static pressure at the outside air entrance (Fig. 37.33). One means of preventing this is to control the return fan from a differential-pressure controller sensing outdoor static and building pressure on

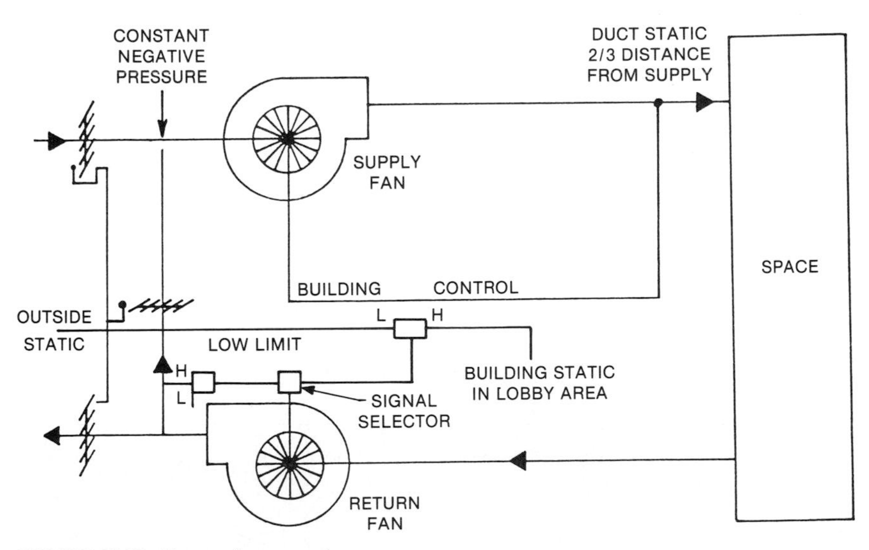

FIGURE 37.33 Return fan control.

the ground floor. A low limit control must be installed in the return/exhaust plenum to maintain a minimum positive pressure to prevent damage to ductwork on system startup. Control of the return fan from the same sensor as the supply fan has been used on small systems, but the fan and system characteristics are different so building pressure control cannot be ensured. Fan tracking systems are available to ensure that the air flow from the return system matches the volume of the supply system. However, these systems do not sense or control building pressure directly.

To summarize, remember these points:

1. Use static regain duct design for best results in large systems.
2. Understand the fan characteristics to prevent unsatisfactory operation.
3. Carefully evaluate the use of the building and the expected capacity under various anticipated operating conditions to determine fan control.
4. Consider a smaller fan for variable-volume than for constant-volume systems.
5. Locate the sensor for minimum static pressure variation in the system and to ensure proper static pressure under all operating conditions.
6. Make sure that the combination of the fan, system, and control points is compatible so the sound level and duct static pressure stay within acceptable limits.
7. Control return fans from a building pressure sensor.
8. Consider use of relief fans in lieu of return fans.

37.8 REFERENCE

1. *Energy Consumption in New Buildings*, Standard 90-75, American Society for Heating, Ventilating, and Air-Conditioning Engineers (ASHRAE), Atlanta, GA, 1981.

CHAPTER 38

AIR MAKEUP (REPLACEMENT AIR OR MAKEUP AIR)

Walter B. Schumacher
Vice President, Engineering, Aerovent, Inc., Piqua, Ohio

38.1 INTRODUCTION

"Makeup-air units," most recently called "replacement-air units," are air-handling units designed primarily to supply air to a building or space, replacing air handled through mechanical exhaust systems, process-equipment ventilation, and even gravity ventilation systems. People have long recognized the need for ventilation systems and have provided means to exhaust air from buildings in various ways, but as new technology in buildings increased and buildings became tighter and better-insulated, the exhaust systems began to work harder and became more important to plant operations. It was finally realized what all good ventilation engineers knew—that we must supply air to the buildings in order for the exhaust systems to work, rather than rely on openings throughout the buildings for this makeup air supply.

Recent experiences have only reaffirmed the fact that in order to ensure that the ventilation systems work at their designed rates, *it is necessary to provide a volume of air at least equal to that being exhausted.* As a result, a make-up (or replacement-air) system is required. Even today, however, many people do not fully recognize the need for mechanical replacement of the air being exhausted and mistakenly rely on the negative-pressure effect of the building to supply the air.

The basic rule for ventilation and replacement (or makeup) air is that for every cubic foot (meter) of air exhausted, a cubic foot (meter) of air must enter the building. Without a well-designed supply air system for this replacement air, all the air must enter whatever openings are available. This means that the air must enter through open doorways, windows, cracks, etc. During cold seasons, doors and windows are closed and, as a result, the exhaust systems become starved without the proper supply systems. During the heating season this creates cold drafts and low space temperatures along the building's interior perimeter. The beauty of a well-designed supply air system for this replacement air is that the needs of the ventilation exhaust systems are satisfied, the negative pressure in the building is overcome, and unwanted infiltration of air through the cracks is eliminated. Thus the perimeter of the building is no longer a cold and drafty perimeter, but more uniform in temperature with the central part of the space.

Usually a well-designed system is designed for 10 percent excess air, thus creating an exfiltration factor. This creates a buffer, preventing cold drafts from leaking into the building, or actually forces some small amount of air outward through the openings, and it thus helps maintain a uniform temperature throughout the complete space.

38.2 TYPES OF MAKEUP AIR (REPLACEMENT AIR)

There are several types of units that can be designed to provide the necessary replacement air. Each can be classified by:

- Type of heat source, if any
- Type of air-moving device used
- The complexity of the system

The basic types of units are:

- Ambient-air supply units
- Heated-air supply units
- Cooled-air supply units
- Combination heated- and cooled-air supply units

The basic heated-air-type units are divided into the following categories:

- Direct-fired natural gas
- Indirect-fired systems
- Steam with modulating coils
- Steam with face and bypass coils
- Modulated hot-water glycol systems
- Face and bypass hot water (hot-water or glycol systems)
- Electric-heat systems
- Heat-recovery systems

The cooling systems employed are (1) evaporative cooling with spray washers, (2) evaporative cooling with pad systems, (3) chilled-water coils, and (4) direct-expansion coils.

38.3 HEAT SOURCES

38.3.1 Direct-Fired Natural Gas

The most popular, lowest-cost, and lowest-operating-cost heat system is the direct-fired natural-gas unit (Fig. 38.1). This system usually relies on the con-

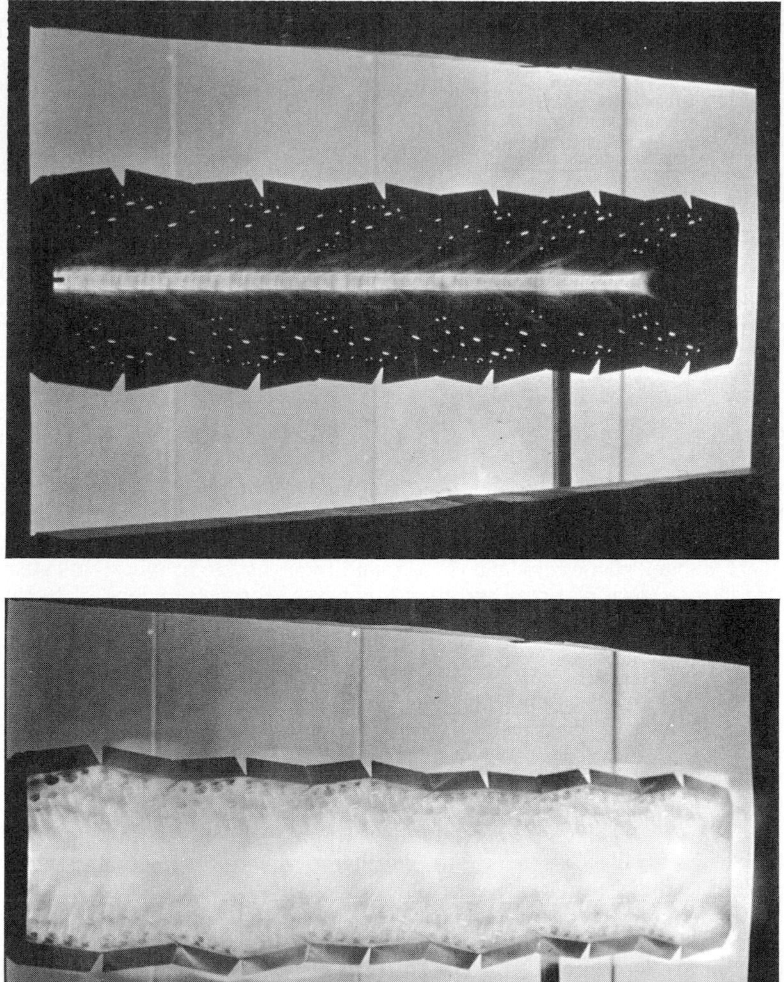

(b)

FIGURE 38.1 Direct-fired gas burner—Airflo type. (*a*) Low fire; (*b*) high fire. (*Courtesy of Maxon Corporation, Muncie, Indiana.*)

trolled flow of air across the burner at specific velocities for proper combustion. A turndown ratio of 25:1 is the common design rate. The direct-fired units also have the highest combustion efficiency. Outlet temperature control or room temperature control may be used.

38.3.2 Indirect-Fired Gas and Oil Systems

Although indirect-fired gas and oil systems are available, they are less efficient and more limited in capacity than the direct-fired gas units; they are also bulkier and require more maintenance. Some applications and codes dictate that indirect-fired units must be used, but these units are less popular than other types of units and are not as readily available. Their initial cost is also higher.

38.3.3 Steam Units

Steam units have long been popular for heating where large installations are supported by steam-generating plants and utilize surplus steam from process applications or from boilers utilizing surplus or waste fuel. The two types of systems are (1) modulated steam flow and (2) face-and-bypass systems where full steam is applied to the coil at all times and the air is passed through the coil or bypassed with a series of dampers.

The modulated-steam-coil units should be designed with the "nonfreeze"-style coil for the most freeze protection (Fig. 38.2). The temperature control system used with modulated coils should always be of the "outlet temperature," or discharge temperature, design [referred to as outlet temperature control (OTC)]. This will ensure that a warm room or area around the thermostat will not cause the steam valve to restrict the steam flow to the coil in the presence of low-temperature air, which would increase the possibility of freeze problems. Modulated coils should always be used with gravity-return condensate systems with adequate gravity head on properly sized steam traps and with vacuum breakers to protect the coil system. Vented condensate receivers and pumping systems are quite often required in order to return the condensate to the boiler.

Face-and-bypass steam units can be designed with alternative coil styles, but the nonfreeze coil is preferred for a standard system with side-by-side or over-under coil-bypass arrangements. The coil/damper must be sized for equal air-flow pressure drop across each section. (See Figs. 38.3 and 38.4.) One disadvantage of face-and-bypass systems is the heat pickup from the coil area under mild conditions due to leakage through the face damper and from wiping by air coming through the bypass area. To help overcome this problem, some manufacturers supply an integral face-and-bypass coil that uses alternate passages of coil tubes and bypass areas, with the damper blades on the inlet and discharge sides of the coil.

38.3.4 Hot-Water Systems

Hot-water systems are manufactured similarly to steam systems; however, plain water coils for modulated flow are not recommended because of the high potential for freeze problems. Ethylene glycol solutions should be used for the best freeze protection even with face-and-bypass systems. Some people argue that if the water is always kept flowing, it will not freeze, but what about times when the power fails or the units are turned off?

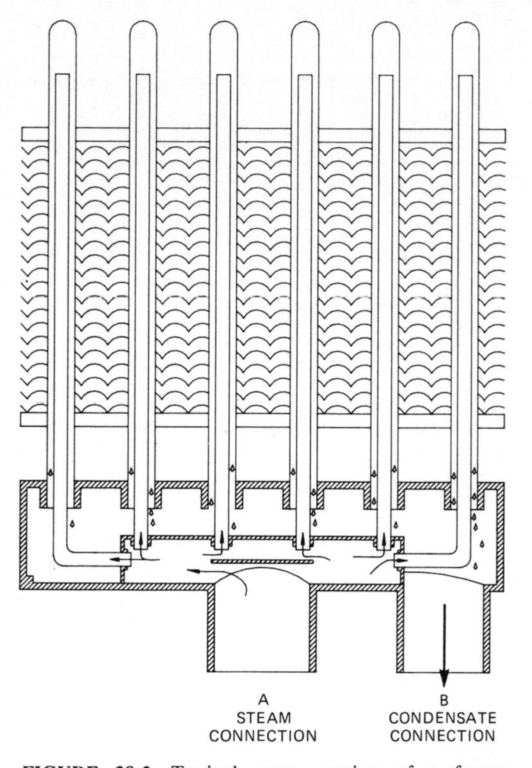

FIGURE 38.2 Typical cross section of a freeze-resistant coil. The steam supply helps keep the condensate from freezing. Vertical installation enhances drainage. *(Courtesy of Aerovent, Inc.)*

Water systems require large flow rates and properly sized piping and pumping systems. An example of this is a calculation of the amount of water that must be handled due to steam condensate versus hot water. Steam has 960 Btu/lb (2232 kJ/kg) of water when under 5 lb/in^2 (35,000 N/m^2) pressure, while a water flow of 1 gal/min (3.78L/min) releases 500 Btu/h (527 kJ/h) per degree of temperature difference. For each 1 million Btu/h (1,055,100 kJ/h), there would be 1,000,000/960 lb/h = 1041 lb/h (473 kg/h) of condensate in 5 lb/in^2 (35,000 N/m^2) of steam, or 2.08 gal/min (7.9 L/min) of water, while a hot-water system would require 1,000,000/(500 × 20) gal/min = 100 gal/min (378 L/min) of water for a 20°F (11°C) ΔT. These are standard formulas used for heat-transfer calculations. This translates to water systems handling about 50 times the amount of water in a given period of time than steam systems do.

38.3.5 Electric-Heat Units

During the fuel crisis of the 1970s, electricity became the "alternative" fuel for natural and propane gas where hydroelectric or coal-fired generation was available. Equipment is available for heating makeup air with multistaged electric

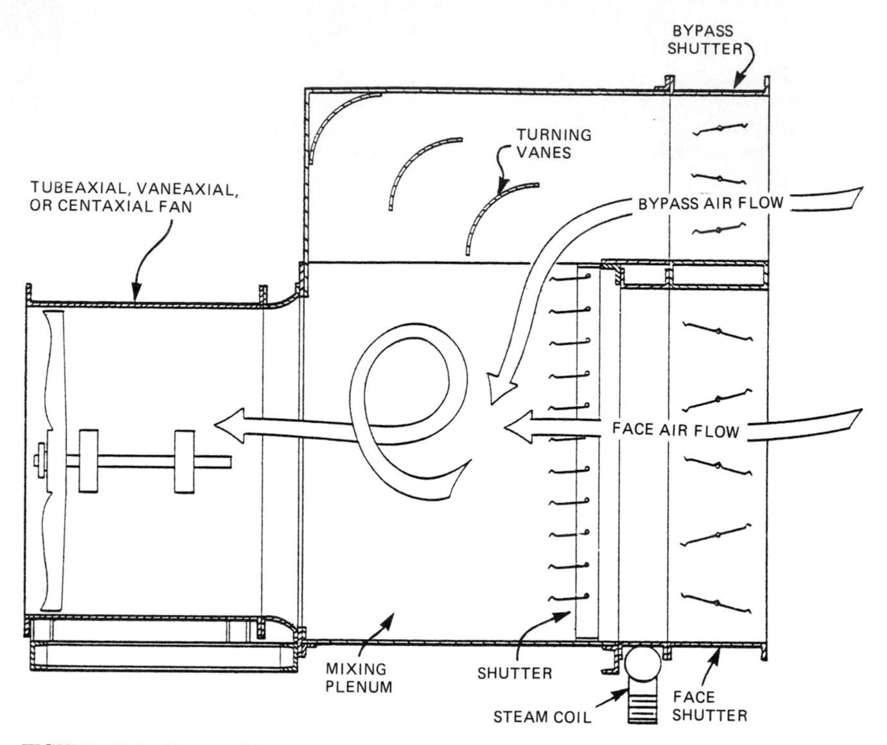

FIGURE 38.3 Face-and-bypass steam coil system with bypass damper located above face damper. Automatic damper located on coil discharge to minimize wiping of the coil. *(Courtesy of Aerovent, Inc.)*

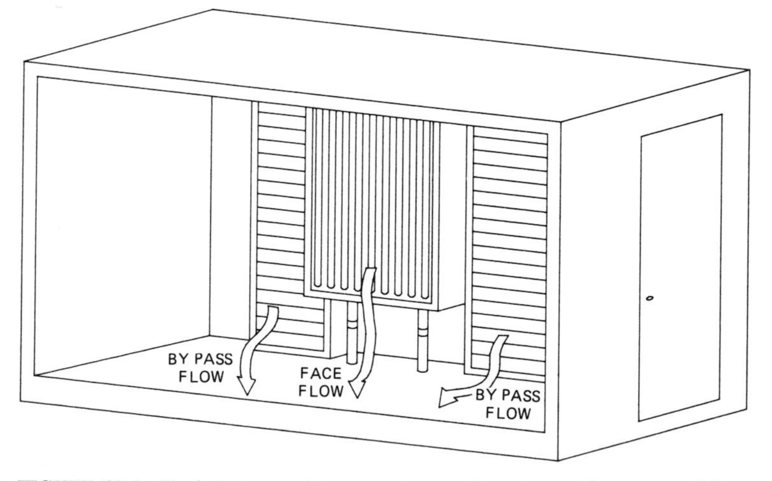

FIGURE 38.4 Typical face-and-bypass steam coil system with center coil/face damper and side-by-side bypass dampers. Promotes better blending for single- or double-inlet fans. *(Courtesy of Aerovent, Inc.)*

heaters. Reasonable temperature control can be obtained by selecting sufficient stages of heating elements. The heating sections are compact and require minimum maintenance. However, the cost of the electric service and the operating costs presently limit the applications to smaller units or to localities where no alternative fuels are available.

38.3.6 Heat Recovery

Air-to-air heat exchangers, heat-exchange pipes, rotary heat sinks, etc., are among the many types of heat-recovery devices used to provide the heat to the incoming air. These units utilize waste heat from exhaust processes, thus saving on fuel costs, but they increase the initial investment and require higher power on both the supply and exhaust systems.

Mixed air of good quality has also been used successfully in many applications (Fig. 38.5). A constant-volume supply fan receives the air from outside or the outside air mixed with the warm air from a clean or filtered exhaust system that contains no noxious fumes and is below accepted threshold limits. The exhaust system is designed with a set of dampers to direct a portion or all of the exhaust air to the supply fan to satisfy the temperature requirement, or to exhaust all or part of the air. This type of system requires no additional fuel if the exhaust tem-

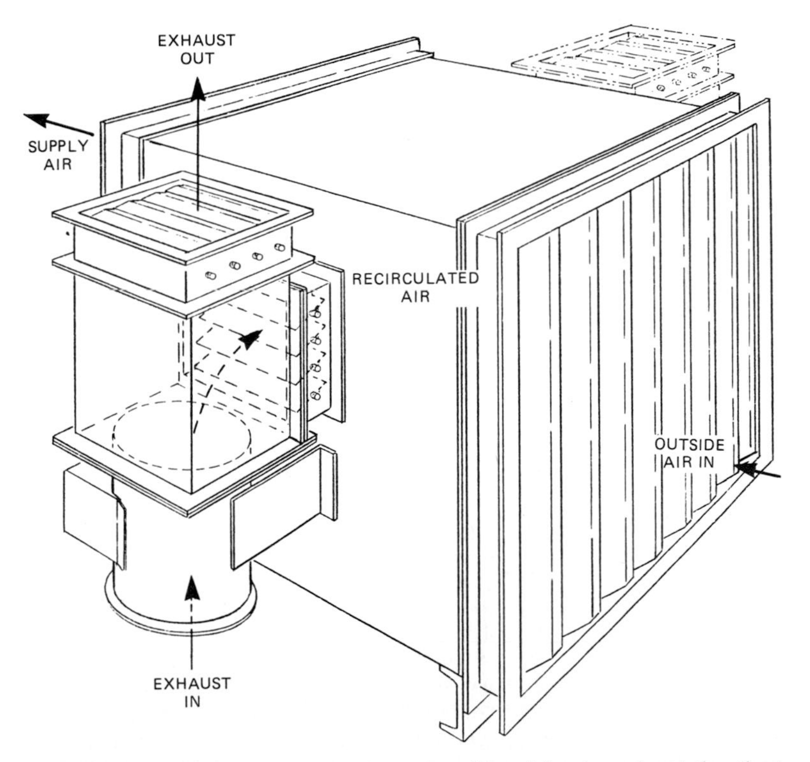

FIGURE 38.5 "Mixing box" for supply system with reclaimed or exhaust air and outside air. (*Courtesy of Aerovent, Inc.*)

perature is adequate, and the only additional cost (beyond that of an unheated air supply system) is the cost of the dampers and temperature control system.

38.3.7 Heat Requirements

The basic formula to determine the heat required for air-handling units in Btu's per hour (joules per hour) is:

$$\text{Btu/h} = \text{ft}^3/\text{min} \times 1.08 \times \text{temperature rise in °F}$$

$$(\text{J/h} = \text{m}^3/\text{min} \times 1.6 \times 10^{-5} \times \Delta°\text{C})$$

where 1.08 and 1.6×10^{-5} are constants.

EXAMPLE Find the heat required for 50,000 ft^3/min of air with a temperature rise from 0 to 92°F.

Solution

$$\text{Btu/h} = 50{,}000 \times 1.08 \times 92°\text{F} = 5{,}000{,}000$$

38.4 UNHEATED AIR AND FILTERED AIR

Plain, unheated-air supply units, filtered or unfiltered, are used to advantage in mild climates, in summertime, or when warm air can be entrained with the supply air stream (Fig. 38.6). Warm air can be entrained by controlling the supply air flow through the warm air layer near the ceiling in high bay areas or over heat-producing devices. Many installations vary the exhaust rate from summer to winter. The extra air for the increased summer ventilation rate also can be supplied with separate units to supplement the wintertime, heated-air units. In some geographic areas with mild climates or for summer-only requirements, this could be the only type of supply unit needed.

38.5 COOLING SYSTEMS

Makeup-air systems with cooling capacity are popular in many areas of the United States, particularly in the warmer climates and where temperature control for processes is important year-round. They are also used for humidity control in conjunction with reheating capabilities.

Evaporative cooling is most commonly used in areas of low ambient relative humidities; however, almost any geographic area can use evaporative cooling during the peak-temperature period of the day. Temperatures can be lowered by up to 90 percent of the wet-bulb depression. The evaporative cooling process causes an increase in the relative humidity of evaporatively cooled air. As this cool moist air mixes with the warm dry room air, it is reheated and thus the relative humidity is lowered, since no additional moisture is added in the process. The net effect is that the temperature of the conditioned air in the space is lowered with some increase in relative humidity, which results in an improvement of room comfort level over that of outside air. The two types of evaporative coolers

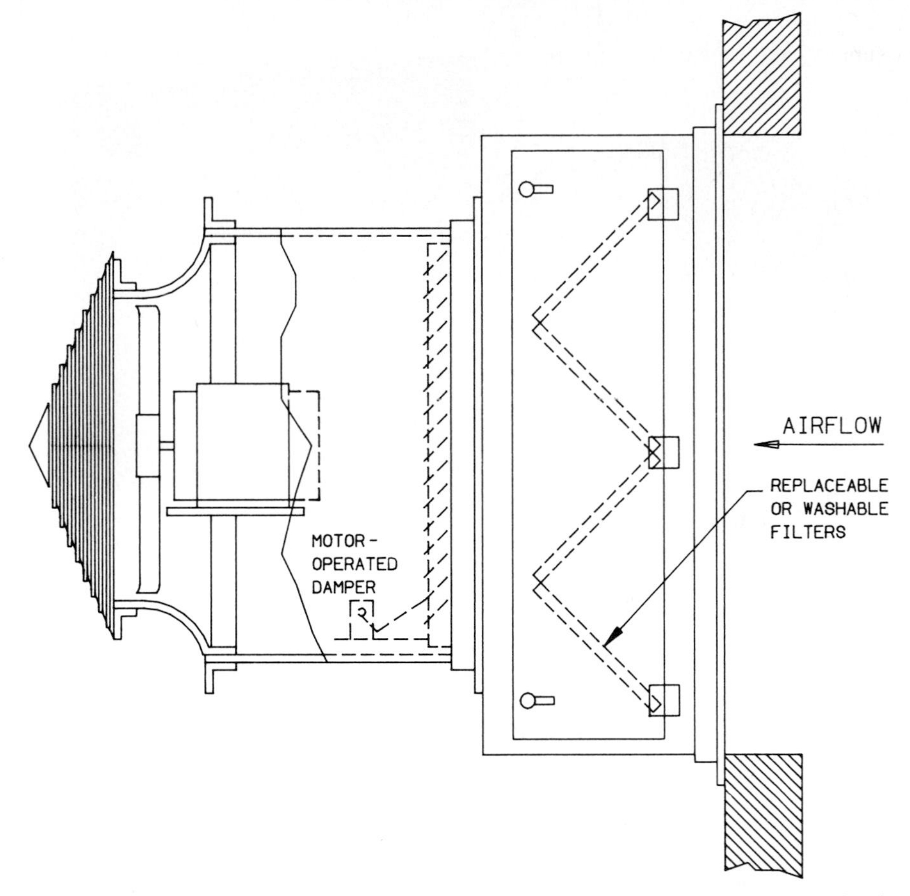

FIGURE 38.6 Simple filtered air supply unit with propeller fan, damper, and filter cabinet. Good for single-point air entry. (*Courtesy of Aerovent, Inc.*)

used are high-velocity [up to 1500 ft/min (460 m/min) air velocity] spray-washer types (Fig. 38.7) and the wetted-media or wetted-pad types (Fig. 38.8).

Basic recommendations for cooling systems:

1. A slight positive pressure, 5 to 10 percent in excess of exhaust air, is recommended for most applications. An exception is where the migration of fumes from a given area to other areas must be avoided; negative pressure must be maintained in that area. The differential pressure should be at least 0.05 inWG (1.27 mm WG) below the pressure of other areas to induce proper flow.
2. For seated workers, ducted systems with distributing directional grilles or registers are necessary.
3. Areas with low ceiling height, 8 to 10 ft (2.4 to 3.0 m), require better distribution of air than high-ceiling areas do. High ceilings require distribution to bring the cool air to individual workers or to the 8- to 10-ft station, minimizing the total volume of cool air required. (See Fig. 38.9.)

FIGURE 38.7 High-velocity evaporative cooler using high-density water spray; 800- to 1500-ft/min (240- to 460-m/min) air velocities. Inset illustrates the spray pattern. (*Courtesy of Aerovent, Inc.*)

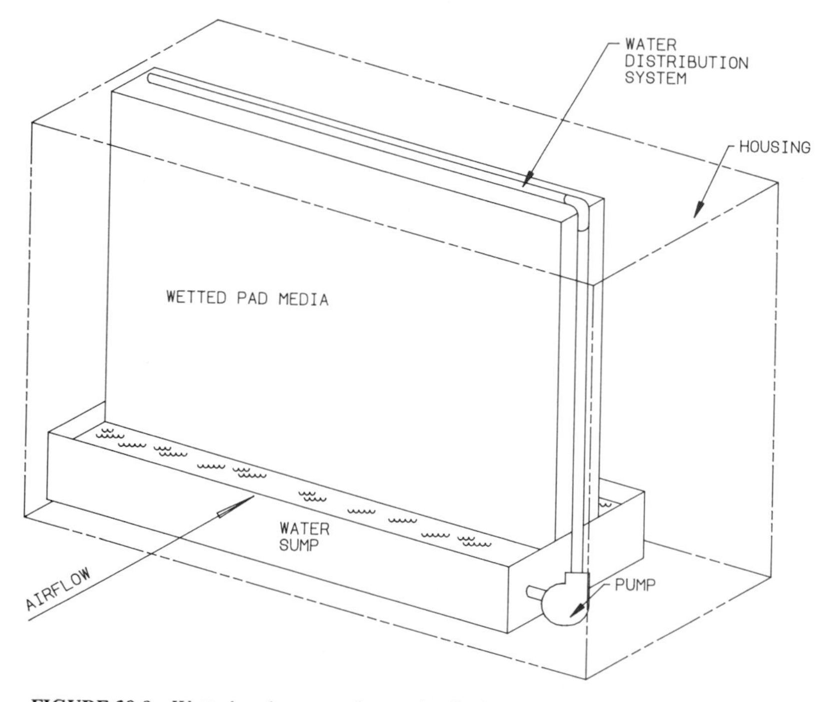

FIGURE 38.8 Wetted-pad evaporative cooler for low velocities; 200- to 600-ft/min (60- to 180-m/min) air velocities. (*Courtesy of Aerovent, Inc.*)

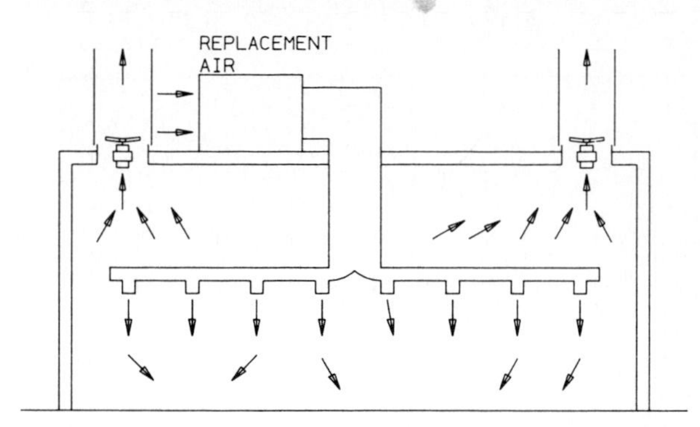

FIGURE 38.9 Typical replacement-air supply system with distribution ducts to bring air to the 8- to 10-ft-high (2.4- to 3.0-m-high) work area. Excellent distribution of air. *(Courtesy of Aerovent, Inc.)*

38.6 TYPES OF UNITS BY AIR-MOVING DEVICES

The various units previously described all contain an air-moving device—or fan. They can quite simply be divided into these categories:

1. Axial-flow air-moving device (e.g., Fig. 38.10)
   - Propeller type
   - Tubeaxial type
   - Vaneaxial type
   - In-line centrifugal type
2. Centrifugal air-moving device
   - Single-width, single-inlet configuration (Fig. 38.11)
   - Double-width, double-inlet configuration (Fig. 38.12)

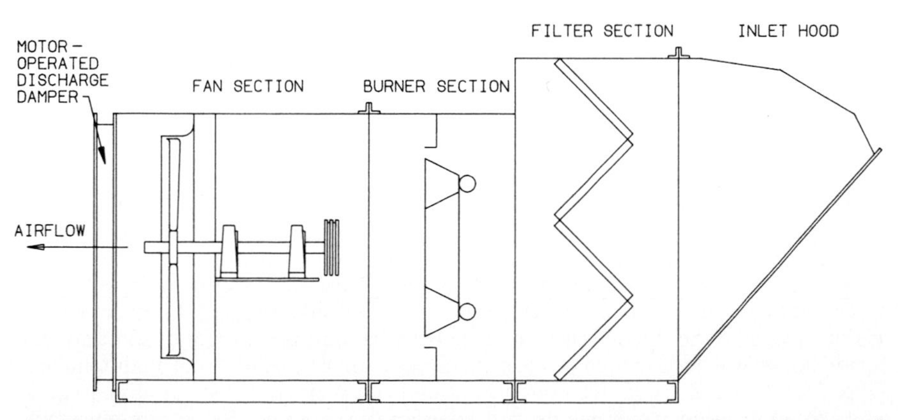

FIGURE 38.10 Propeller-fan type of gas makeup-air unit—typical cross section. *(Courtesy of Aerovent, Inc.)*

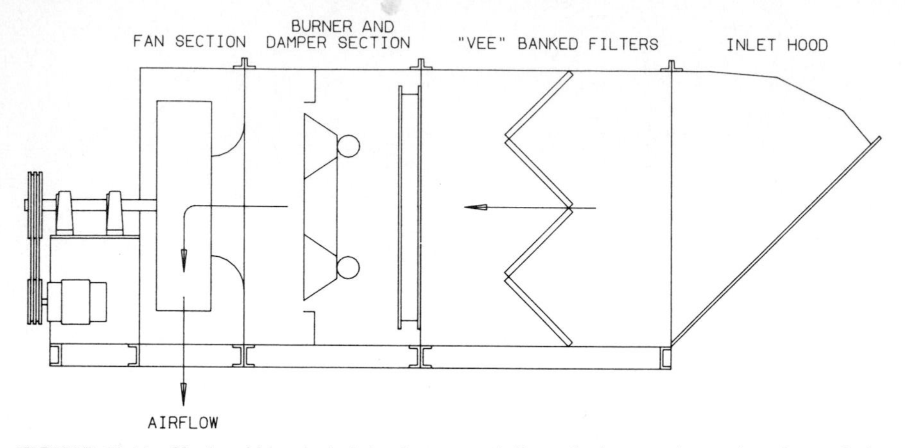

FIGURE 38.11 Single-width, single-inlet fan type of direct-fired gas makeup-air unit—typical cross section. (*Courtesy of Aerovent, Inc.*)

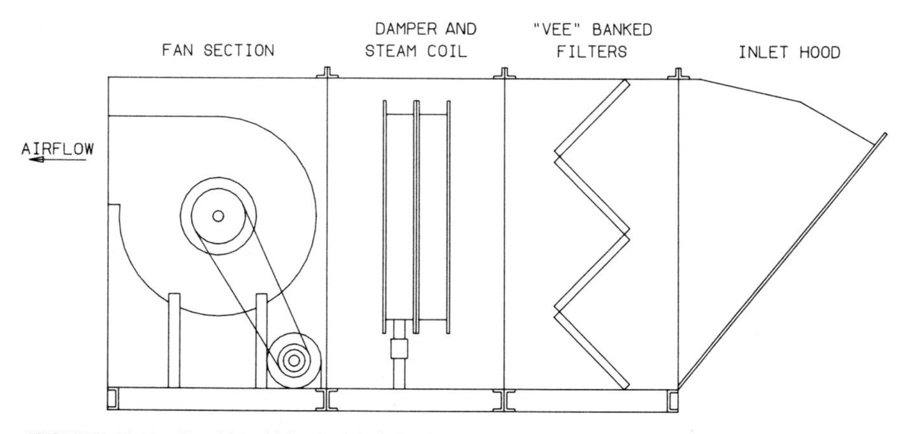

FIGURE 38.12 Double-width, double-inlet fan type of makeup-air unit with a modulated steam coil—typical cross section. (*Courtesy of Aerovent, Inc.*)

In addition to the subdivision of makeup-air (or replacement-air) units by air-moving types, they can also be grouped into various physical requirements. The basic standard packaged unit would be built for inside installation with the option of weatherproofing for outside installation. (A special construction unit with special modifications and features constitutes another subdivision.) Another subdivision would be the "penthouse" type, where the unit is self-contained in an exterior housing with a complete access corridor for convenience of maintenance (Fig. 38.13). All of the above configurations have their own distinct advantages, and the application requirements and economics would be the determining factors as to which type could or should be used. The basic propeller packaged unit

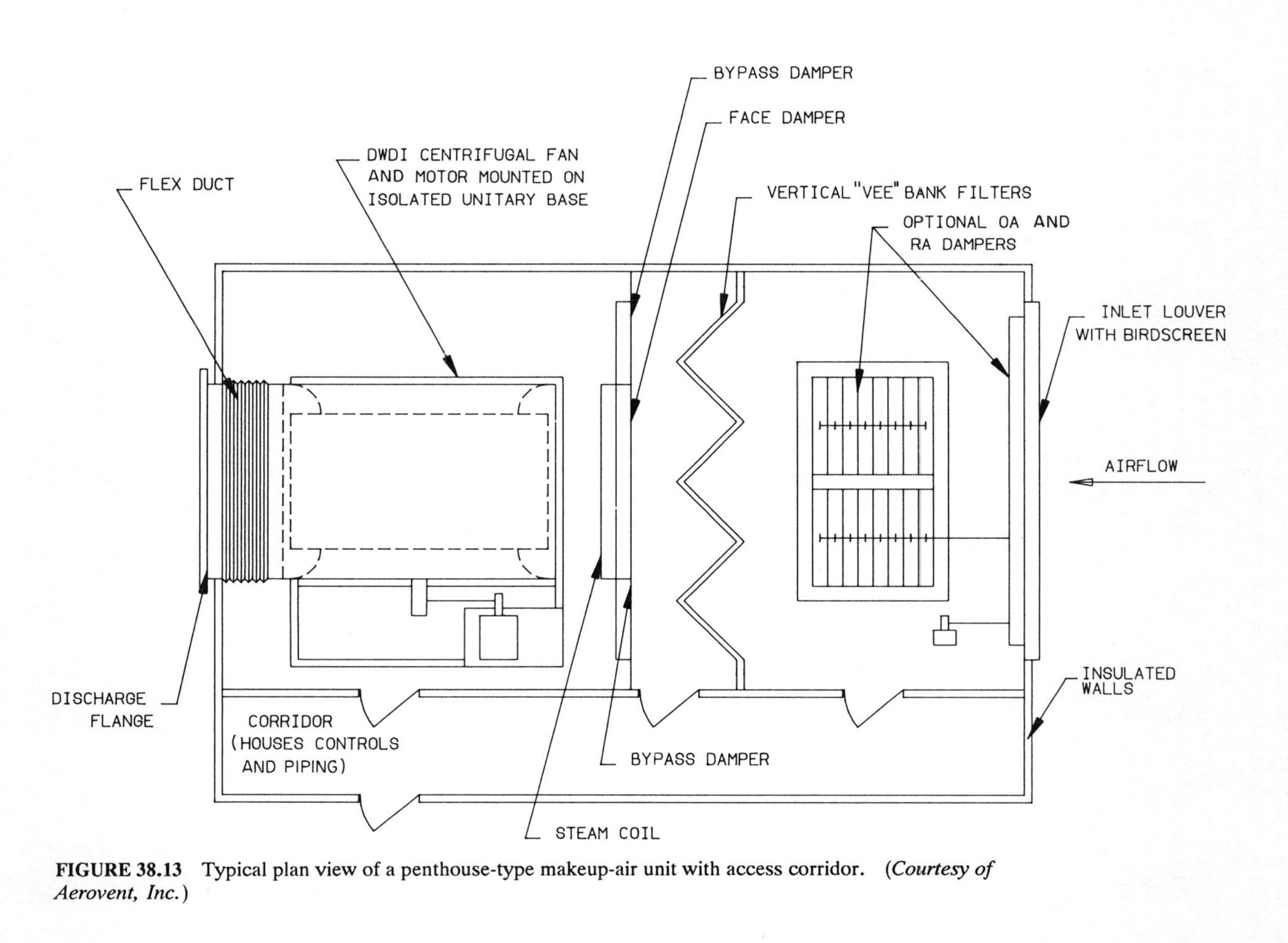

FIGURE 38.13 Typical plan view of a penthouse-type makeup-air unit with access corridor. (*Courtesy of Aerovent, Inc.*)

is the most economical unit, regardless of the basic functional requirement, while the custom-penthouse-type unit is at the most expensive end of the scale.

Where adequate space is available within the plant under the ceiling, on a mezzanine, over office cubicles, etc., the standard nonweathertight unit with an outside air inlet can be mounted inside. Where no space within the plant is available, a unit with weatherproof door, cabinets over controls, or other protection can be mounted on steelwork on the roof or adjacent to the plant with discharge ductwork to deliver the air within the plant. Sometimes extensive ducts are run across the roof with drops into the plant where required. In addition, weathertight units can be mounted on roof curbs. Penthouse units with their work corridor are usually mounted on curbs; however, they are sometimes mounted on separate platforms with the air ducted to the plant. On very large industrial buildings it is sometimes advantageous to mount units in equipment rooms or large penthouses constructed on the roof of the main plant. The most common installation on small industrial plants is weatherproof units, roof-mounted where the construction will take the added roof loading. In addition to the descriptions mentioned above, various other appurtenances can be included in the design of the units, such as a variety of filtration systems, humidification systems, return-air combinations, and various other customer options.

38.7 APPLICATION—GENERAL

The purpose of installing makeup-air (replacement-air) units would be to meet the requirements briefly reviewed in Sec. 38.1. The basic application for this type of unit is to supply air to the space to be ventilated (exhausted) so that the ventilation system will work properly. "Working properly" means that the full design air volume is supplied so that the space will not have drafts and hot or cold spots due to infiltration from the outside, which results from the excess negative pressure created by exhaust systems when they lack sufficient makeup air.

In order to determine the volume of makeup air (or supply air) required, the total ventilation (or exhaust) requirement must be known. In a mechanical ventilation system, the volume of air to be exhausted can be determined by (1) the total exhaust ventilation design volume, (2) a plant survey of the actual equipment presently installed or planned to be installed and the total published ratings or run-volume measurements of all the units, or (3) a combination of the total run volumes of installed exhaust units (fans) plus any design volumes of new units to be added and/or any volume changes to be made to installed devices. In addition to this, any process ventilation must be considered along with any combustion air requirements. All exhaust and process ventilation should be designed to comply with the requirements or recommendations of the Environmental Protection Agency (EPA), the Occupational Safety and Health Administration (OSHA), and the American Conference of Governmental Industrial Hygienists.

The total ventilation requirement is the first thing to consider for a proposed application. As discussed in Sec. 38.1, the amount of air to be supplied should be at least equal to the total ventilation requirement, and an amount up to 10 percent greater than the total ventilation requirement may be advisable.

The air must then be provided to the space in the best manner for distribution. This can vary depending on the building's structure, the location of the ventilation systems, and the location of the various work centers. One of the basic requirements that we always consider is that the supply air should be brought into the area where it will serve the best purpose. For a makeup-air unit that supplies air to contaminated and noncontaminated areas of a building, it is a good design procedure *not* to draw the supply air across or from a contaminated area, but to divide the makeup-air system so that the air is independently supplied directly to the contaminated and to the noncontaminated areas; this will minimize the possibility of supplying contaminated makeup air to other zones. The ideal situation is to have the contaminated or undesirable areas operate under a negative-pressure condition so that the air will flow to this area from the other areas in the plant. Where complicated systems such as this can exist, the use of multiple units or well-designed distribution systems should be considered. Of course, in areas where the ventilation requirement is simple, the supply system can be simple too, consisting of a single discharge point or a very short duct terminating in a distribution box with discharge grille assemblies to spread the air flow across the desired area (Figs. 38.14 and 38.15). Quite often it is cheaper to go with one larger unit and a discharge or distribution duct system than with multiple units. On the other hand, the total volume of makeup air required may be such that multiple units, even with extensive distribution ducts, are justified. Furthermore, the reliability factor should be considered; particularly in critical areas, multiple units can provide more reliability than can a single unit.

A final consideration is the makeup air's quality. Is the plain unfiltered or untempered ambient outside air satisfactory, needing only to be introduced in the proper location? Or is it necessary to "condition" this air with heating, filters, cooling, and/or humidity control?

1. *Heating.* Although perhaps the most common application is the addition of heat and filtration to the air, many systems are being designed for winter-and-summer application, and this may well require a combination of heating and ambient-air units. During mild weather conditions, no heat is necessary and the air-moving device used by the heating system might be sufficient to supply unheated ambient air. However, if the air volume required for the mild condition is greater than that needed for the heated condition, a separate ambient-air unit could be considered; the disadvantage is that this type of selection requires two distribution systems. An alternative to this is a two-speed unit, with low speed

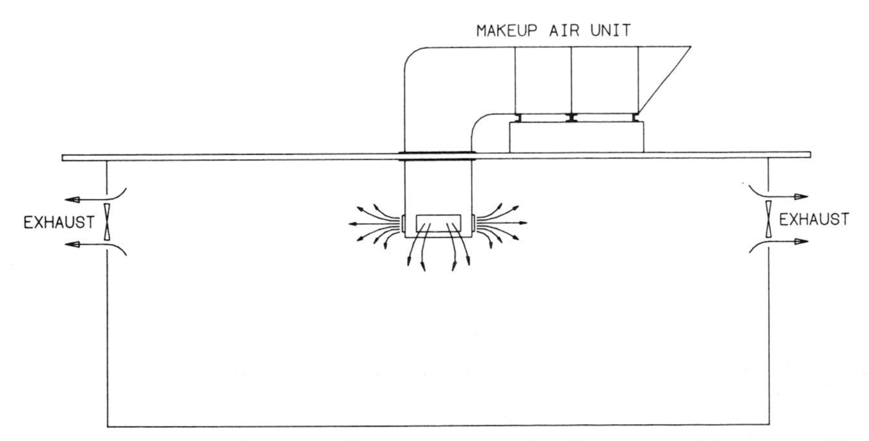

FIGURE 38.14 Makeup-air unit with short duct and a distribution box. Minimum cost with good results. (*Courtesy of Aerovent, Inc.*)

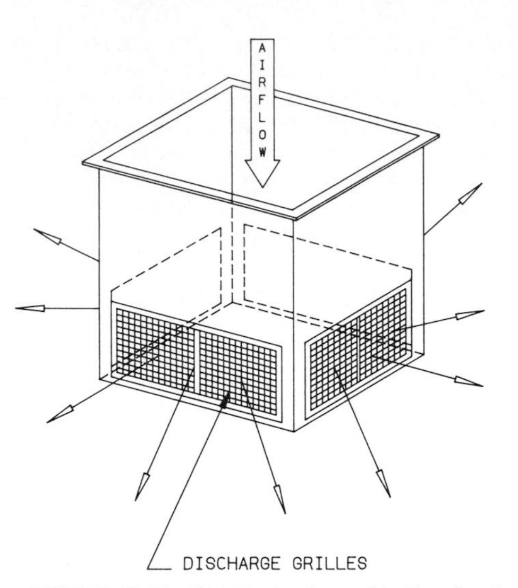

FIGURE 38.15 Distribution box with directional grilles on four sides. Allows for single-point entry of air into the building with good distribution. (*Courtesy of Aerovent, Inc.*)

for the winter (or heating) season and with high speed for summer ventilation. A bypass damper could also be installed that would open up the area around the heating device and reduce the internal resistance to air flow, thus allowing more air to be handled at a lesser resistance with a reduced power requirement.

2. *Filtration.* Does the air need to be filtered? If so, to what qualifications? To remove most of the larger airborne particles, a normal industrial 2-in-thick (5-cm-thick) washable filter or a 2-in-thick disposable-type filter can be used. For higher-quality applications, a higher-efficiency filter (such as a "bag" type) can be used by itself, or it can be used in conjunction with a prefilter of the above type for prolonged life. In addition to this, for very high quality air, filters of the high-efficiency particulate air (HEPA) type can also be used; however, they should be used with quality prefilters. Each additional requirement for higher efficiency adds additional static-pressure requirements to the air-moving device and thus increases the power requirement.

3. *Cooling and humidification.* In some areas, cooling may be required in addition to heating, for the heat load during the summertime may be such that additional quality must be induced into the ambient air to ensure the proper environment in a given work space. This can be done through evaporative or mechanical cooling. The type would depend both on the availability of various resources and on the operating cost. Evaporative cooling is the lowest-cost approach, whereas mechanical cooling is naturally the higher-cost approach. Most all areas in the United States can utilize evaporative cooling during the most severe temperature conditions (as explained in Sec. 38.5); however, when the humidity must be closely controlled, mechanical cooling becomes a requirement. In addition to cooling the air, it may be necessary to use mechanical cooling with reheating in order to reduce the relative humidity. Sometimes it is necessary to add humidification, particularly during cooler weather and during the heating season.

In summary, the choice of which makeup-air unit to install should be determined by:

- The air volume required
- The physical location where the unit will be placed
- Any cost restraints
- The physical features and degree of reliability desired
- The necessary treatment required (heating, filtration, cooling, humidification)
- The flow restriction in terms of the static-pressure requirement

The combinations that can be added to makeup-air systems seem almost endless and are governed only by the customer's needs and economic justification for these combinations. The addition of each individual option increases the cost of the equipment, the static-pressure requirement of the air-moving device, the power required by the air-moving device, and the operating cost; it also increases the need for (and cost of) maintenance. Custom design in makeup-air units is a standard practice, but when costs are of the utmost consideration, standard packaged-type units offer the most capability at the lowest cost.

38.8 APPLICATION—POSITIVE-PRESSURE HEATING

Since the late 1960s, air-makeup-style equipment has been used as the primary or total heating source for many industrial plants, warehouses, and similar-type buildings. This is referred to as "positive-pressure heating," or "fresh-air heating." Direct-fired gas units are the most common and most economical units for this purpose. The basic principle of this method is that (1) the fresh-air supply unit, by heating the air being introduced, provides the amount of heat required to overcome the building's heat loss and that (2) this required heat is obtained by proper sizing of the burner's capacity. The burner is modulated with a modulating room thermostat: a room temperature control (RTC) unit. The fresh-air units provide excellent air motion in the space being heated and minimize stratification, thus reducing the amount of heat lost through the ceiling or roof.

Presently there are two concepts used for fresh-air heating units. One is the standard makeup-air-unit approach, in which the entire volume of supply air is outside air. The other is the recirculation-type unit, which brings a given percentage of air into the building from outside and passes it across the burner while the remaining air is either recirculated or introduced from outside and bypasses the burner, the percentages being either modulated or selected (as with a day-night switch).

The sizing of the units has been simplified, dispensing with long-drawn-out calculations. The following guidelines have proved to be reliable over the last 20 years:

Building type	Air changes per hour
Extremely tight, well-insulated	¾
Tight construction, well-insulated	1
Steel construction, insulated	1½
Brick or block—tight, insulated roof	2
Brick or block—many windows, little insulation in roof	3

Where design temperatures for the heating season are above 0°F (−18°C), a 120°F (66.6°C) temperature rise for the entire air volume is standard. For design temperatures down to −20°F (−29C), a 135°F (74.9°C) temperature rise is recommended.

The above table's air-flow values are for heating only. When exhaust is present, the exhaust volume should be added to the unit's volume if it exceeds 5 percent of the calculated volume. When the makeup-air requirement equals or exceeds four air changes per hour, no additional volume should be required, but the units should be designed for a 120°F (66.6°C) temperature rise and should have room temperature control.

When a constant-volume makeup-air unit is used for heating, as is done quite often, it usually incorporates a day-night setback thermostat for the off-duty-cycle times. During the occupied cycles the unit runs continuously and monitors the temperature. During the off-duty cycle a lower temperature is selected, and the unit runs only long enough to satisfy the room thermostat at this lower setting; the unit then cycles on and off. In order to invoke greater economies with this system when there is a multiple number of units on the building, one unit is usually set as the prime unit for the heating during the off-duty cycle, and the others are set with their thermostats as backup should the one not be adequate. However, one unit is often more than adequate, and a two-speed motor on the fan device is recommended so that the unit can be operated at lower speed during the off-duty cycle for economies on power and heating.

When using recirculation-type heating units, it should be remembered that the only makeup air is that brought in from outside. When used on this cycle for heating during unoccupied times, there is no problem; however, when used during operating hours the unit should be properly controlled for the required volume. Maximum heat can also be supplied best with full outside-air volume.

Recirculation-type heating units were developed to provide constant air motion. This is the same principle used in steam units with face and bypass dampers, which can recirculate the return air all of the time. With direct-fired gas units, however, recirculating air across the burner presents a problem, because the contaminants in the return air could change properties when passing through the burner; consequently, many of the return-air design units bypass the burner with a variable volume of return air versus outside air, but they still allow the air-moving device to keep the larger volume of air in motion.

Because of its two channels of air—one coming across the burner, and the other bypassing around the burner—the mixing of the warm and cooler air becomes more of a problem with the recirculation-type unit. For this reason, centrifugal fans are most commonly used because they provide a better mixing of the two air streams as they pass through the fan. In addition to this, short runs of duct can help provide maximum air mixing before the air is discharged into the space to be conditioned. As the demand for heat becomes more severe, this mixing problem becomes more critical because there is a greater temperature differential between the heated-air channel and the return-air channel; however, recognizing this problem in advance and thus allowing adequate time for the different channels of air to mix in a distribution system does make for a very acceptable installation.

38.9 SUMMARY

Makeup air (or replacement air) has grown in popularity over the last 35 years. Almost all new industrial plants have provisions to provide this feature in their initial design. However, special functions are often deemed necessary, and it frequently happens that obsolete designs are copied from previous specifications. In either case, before excessive expenditures are made or a lot of time is spent in designing a system from scratch, I strongly suggest that one contact several manufacturers of makeup-air equipment and thoroughly explore their latest products. These units, while not on-the-shelf types, are commonly built by various manufacturers and contain many of the features that a customer needs. The manufacturers have the experience and expertise to select the proper air-moving device and the proper heating or cooling components and to recommend the necessary type of filtration. For most of the alternatives, they have standard features that would meet most of an individual customer's specifications. Also, because they have worked with many specially designed units, their input could create great savings both in the initial purchase cost of the various components and in the continuing costs of operation and maintenance.

The basic consideration in selecting equipment is that the customer's present and foreseeable future needs be properly served. For example, the units purchased should not be undersized, because when the customer's need for makeup-air increases, it will be difficult and entail excessive costs to install new equipment at that time. Rather, many of the units initially purchased can be those which have the increased capacity needed for future conversions. For instance, instead of increasing the fan speed by installing a higher-power motor, one could initially purchase the larger motor and run the fan at a reduced speed; a two-speed motor has been an excellent option in many cases. There are many reputable firms which continually manufacture makeup-air equipment and which have valuable information that they are willing to share with potential customers.

Makeup air pays for itself in improved working conditions and in the improved functioning of exhaust systems, air-handling process systems, and combustion systems and is one of the most efficient ways to distribute conditioned air throughout an industrial plant—and direct-fired gas systems are the most efficient present-day systems for converting fuel to heat. While the capacity of air-makeup-type units is usually stated in maximum Btu (J) capability, the unit seldom operates at maximum capacity unless the minimum outside air temperature is present. Normally the unit is modulated to partial heat flow, and then only for the heating season, which varies in length for different geographic locations. The actual operating costs vary with the degree-days (the unit of measure used in all heating calculations). The modulated capabilities offer more uniform conditions than do the standard-type heating devices that operate on an on-off cycle.

Makeup air is a necessity for the industrial facility and for many commercial-type applications. Quite often it is also the ideal solution for high-moisture and -condensation problems, such as in gymnasiums and arenas. The applications are endless and the results are amazing.

CHAPTER 39
DESICCANT DRYERS

Douglas Kosar
Senior Project Manager, Desiccant Technology, Gas Research Institute, Chicago, Illinois

39.1 INTRODUCTION

Desiccants exhibit an affinity so strong for moisture that they can draw water vapor directly from the surrounding air. This affinity can be regenerated repeatedly by applying heat to the desiccant material to drive off the collected moisture. Desiccants are placed in dehumidifiers, which have traditionally been used in tandem with mechanical refrigeration in specialty air-conditioning systems. The systems have been most commonly applied in atypical air-conditioning situations that involve large dehumidification load fractions. This situation often arises with the depressed humidity levels required for operations in many industries. The depressed humidity levels, below the level necessary for comfort, are generally unattainable cost-effectively with mechanical refrigeration and reheat.

In the past, desiccant dehumidification was also integrated with mechanical refrigeration in early air-conditioning approaches for comfort in businesses and homes. However, the advent of abundant and inexpensive electricity and of mass production in factory packaging made the use of electric-motor-driven vapor-compression refrigeration, by itself, the maturing market choice for cost-effective commercial and residential air-conditioning systems.

To the present day, though, in specialty applications desiccant dehumidification still holds its economic advantage over dehumidification by mechanical refrigeration and reheat. In industrial air-conditioning, numerous moisture-sensitive manufacturing and storage applications utilize desiccant dehumidification. The dramatic decrease in product rejection, and thus the direct increase in profitability of the product, yields a quick payback on the initial investment in depressed humidity control equipment.

Interest is now being revived in thermal-driven desiccant dehumidification in nonindustrial air-conditioning applications to offset rising electricity prices. Lower-cost thermal energy, including natural gas, waste heat, solar energy, and other sources, is substituted for electric energy to meet the dehumidification load on the air-conditioning system. Available desiccant dehumidification equipment has been considered too expensive, compared to assembly-line mechanical refrigeration equipment, for application outside the industrial field of use. But today, desiccant dehumidification technology, supported by ongoing research and development, is providing a cost-effective means to reduce electric air-conditioning capacity and thus to lower electric-energy costs and power-demand charges in certain nonindustrial air-conditioning situations too.

39.2 PSYCHROMETRICS

In air-conditioning applications, air must be supplied to a conditioned space in sufficient quantity and at the proper temperature and humidity to meet the thermal (or sensible heat) and dehumidification (or latent heat) load on the space. To understand air-conditioning processes well, especially those utilizing desiccant dehumidification, it is necessary to discuss the psychrometric chart. The psychrometric chart (Fig. 39.1) is a graphic representation of the condition of the air at any given point in an air-conditioning process. It relates the dry-bulb temperature (the air temperature read by a common thermostat) on the horizontal scale to the absolute moisture content of the air on the vertical scale. If the dry-bulb temperature of a given amount of air is decreased to the point that moisture will condense, it is saturated. This saturation point is the dew-point temperature; it is represented by the outer curved boundary, or 100 percent relative humidity (rh), on the chart.

The sources of a space's air-conditioning load are internal (including sensible and latent heat gains from people, lights, and equipment) and external (including sensible and latent heat gains from transmission, conduction, diffusion, and infiltration). In industrial, institutional, and commercial structures, the required ventilation air is an additional external sensible and latent load; the amount of fresh air, as determined by state and local building codes, must be treated by the air-conditioning system and introduced to the space to maintain air quality. Depending on the climatic conditions, applicable codes, and operations in the space, a range of sensible and latent heat loads can be encountered when maintaining temperature and humidity set points in the space. The fraction of the air-conditioning

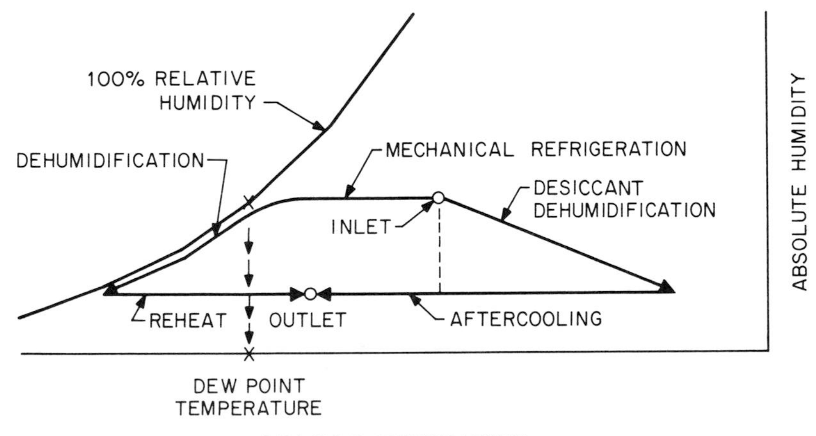

FIGURE 39.1 Psychrometric chart with cooling process paths.

load that is sensible is the sensible heat fraction (SHF); likewise, the fraction of the air-conditioning load that is latent is the latent heat fraction (LHF).

In a conventional vapor-compression air-conditioning system, air is cooled by a coil that has a surface temperature below the dew point of the air. As the air cools below its dew point, moisture condenses on the coil's surface and is drained away. The dry-bulb temperature, the absolute humidity of the passing air, and the coil's surface temperature are the primary determinants of the amounts of sensible cooling and latent cooling that result. Once again, the ratio of sensible heat removal to total cooling capacity is the SHF, and, likewise, the ratio of latent heat removal to total cooling capacity is the LHF.

To maintain the temperature and humidity set points in the space, the SHF and LHF of the system's cooling capacity must equal the SHF and LHF of the air-conditioning load. In certain air-conditioning situations, the LHF of the system's cooling capacity is not high enough (or the SHF of the system's cooling capacity is not low enough) to meet the latent load and maintain the space's humidity set point. In this case, in order to obtain the necessary latent cooling capacity, additional sensible cooling capacity must first be designed into the cooling unit. Either the original amount of air must be sensibly cooled further below the dew-point temperature or additional amounts of air must be sensibly cooled down past the dew-point temperature to increase the latent cooling capacity. The cooler air temperature or the larger amount of cooled air is now greater than is necessary to meet the sensible load, and reheating equipment must be added in order to raise the supply air's dry-bulb temperature to maintain the space's temperature.

Air-conditioning design practice limits increases in air-handling requirements, typically restricting flow rates to under 500 cubic feet per minute of standard air (SCFM) per ton [67 L/(s · kW)] of cooling. Decreasing the dew-point temperature to achieve latent cooling has its practical limitations too. While the sensible heat removal remains essentially constant per degree of dew-point temperature reduction at saturation, the amount of latent heat removal per degree of dew-point drop lessens considerably. Table 39.1 illustrates this phenomenon by comparing moisture removal for a series of dew-point depressions. If sensible cooling is already met at the dew points shown in Table 39.1, then for each additional Btu/(h · SCFM) [0.621 W/(L/s)] of latent cooling, that dew point must be lowered by the excess sensible cooling and reheating listed in Table 39.2. When cooling-coil surfaces must be held at 32°F (0°C) or lower to produce resulting air dew points of 40°F (4.4°C) or lower, condensed moisture will freeze on the coils and cannot be drained away while the system is operating. Defrosting equipment must be added to the system to eliminate ice buildup on the cooling coils; the cooling will be interrupted periodically to defrost the coils before the heat transfer and consequently the capacity are reduced substantially. Dual cooling coils (at a minimum)

TABLE 39.1 Sensible and Latent Cooling by Dew-Point Depression at Saturation Conditions

Dew-point temperature, °F(°C)	Sensible heat removal, Btu/(h · SCFM) [W/(L/s)]	Latent heat removal, Btu/(h · SCFM) [W/(L/s)]
60–59 (15.6–15)	1.08 (0.671)	1.86 (1.16)
50–49 (10–9.4)	1.08 (0.671)	1.36 (0.845)
40–39 (4.4–3.9)	1.08 (0.671)	0.97 (0.602)

TABLE 39.2 Sensible Overcooling and Reheating for Latent Cooling

Dew-point temperature, °F(°C)	Lowered dew point °F(°C)	Sensible overcooling, Btu/(h · SCFM) [W/(L/s)]	Sensible reheating, Btu/(h · SCFM) [W/(L/s)]	Total excess sensible energy, Btu/(h · SCFM) [W/(L/s)]
60 (15.6)	59.46 (15.3)	0.58 (0.36)	0.58 (0.36)	1.16 (0.72)
50 (10)	49.27 (9.6)	0.79 (0.49)	0.79 (0.49)	1.58 (0.98)
40 (4.4)	38.97 (3.9)	1.11 (0.69)	1.11 (0.69)	2.22 (1.38)

must be added to the equipment to produce continuous cooling. Finally, all heat (the amount depends on the defrosting method) added to defrost the coils (except that drained away with the melted condensate) will increase the air-conditioning load in the space.

This "overdesigned" process, shown in Fig. 39.1, yields inefficient operation, because the unnecessary sensible cooling, defrosting, and reheating increase the energy costs when meeting high design LHF loads. The initial equipment costs increase too as additional unnecessary sensible cooling, defrosting [if the coil surface's temperature is 32°F (0°C) or lower], and reheating capacity is purchased to acquire the design's latent cooling capacity. At off-design conditions when the sensible cooling requirement is lower, an even higher LHF load can exist, especially in the more humid climates. In such climates, where the outdoor dew point is generally high and close to the dry-bulb temperature, the part load condition has a larger fraction of moisture and a correspondingly smaller fraction of sensible heat. At these part load conditions, the operation to control moisture becomes increasingly inefficient and may be unable to satisfy the humidity set point in the conditioned space.

An alternative to this mismatch in air-conditioning load and system capacity is the use of a desiccant dehumidifier to meet the latent cooling need. A desiccant dehumidifier utilizes a material that sorbs water vapor directly from the air to its surface or into a chemical solution. The moisture releases heat as it is sorbed and raises the temperature of the desiccant and, in turn, the dry-bulb temperature of the dehumidified air. A desiccant dehumidifier is typically coupled with an aftercooling heat exchanger to reduce the dry-bulb temperature of the drier, but warmer, exiting air. These components can then be integrated with reduced-capacity electric vapor-compression equipment, or other air-conditioning device, for further aftercooling to meet the remaining sensible load. With this "hybrid" operation shown in Fig. 39.1, temperature and humidity can be controlled independently and without excess sensible cooling capacity and reheating and defrosting requirements.

39.3 DESICCANTS

Desiccants are solid or liquid materials. Familiar solid desiccants are silica gel or activated alumina. Common liquid desiccants are water solutions of lithium chloride or glycol. The term "desiccant" is applied to these and other materials having a large capacity for moisture relative to their own weight. The ability of desiccants to collect gases or vapors is not restricted to moisture. They can remove and exhaust various airborne pollutants, and are generally known as "sorbents" in this role. In addition, certain desiccants have biocidal effects, killing bacteria and viruses in a contacting air stream (see Sec. 39.8, Air Quality).

The weight of water-vapor uptake in a desiccant is determined by an equilibrium of the partial pressures exerted by the moisture in the sorbent and in the surrounding air. The partial pressure of moisture is termed "vapor pressure." Vapor pressure is directly related to dew-point temperature. Lower vapor pressure in a desiccant results in a depression of the dew-point temperature for surrounding air that has a higher vapor pressure.

The equilibrium water capacity of a desiccant at a given vapor pressure decreases as the desiccant's temperature increases. Combining the variables of dry-bulb temperature with vapor pressure (or dew-point temperature) into relative humidity produces a single equilibrium water-capacity curve for 0 to 100 percent relative humidity (Ref. 1). The water-vapor equilibrium curves for several solid and liquid desiccants are shown in Figs. 39.2 and 39.3, respectively.

The sorption of water vapor in a desiccant raises the temperature of the desiccant itself and then raises the dry-bulb temperature of the surrounding air. This increase results mostly from the conversion of latent to sensible energy produced by the heat of condensation from the sorbed moisture. However, a further dry-bulb temperature increase results from the additional sensible energy release due to physical, electrical, or chemical interactions of the desiccant with the sorbed moisture. This additional heat effect can vary from a fraction of the heat of con-

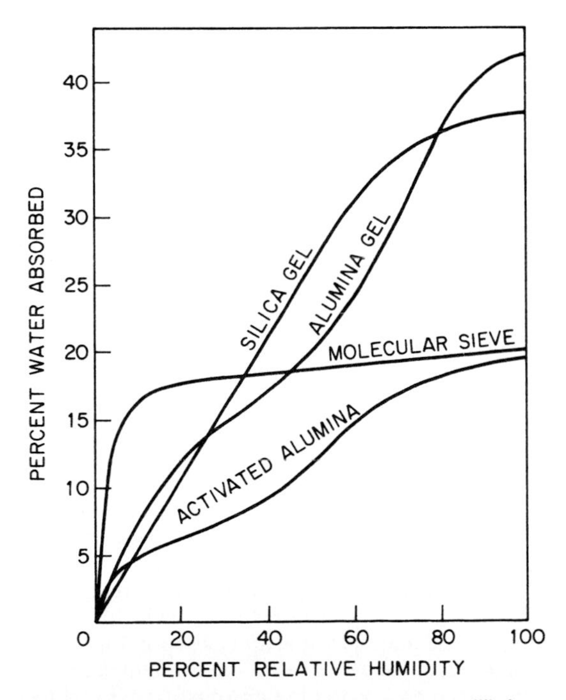

FIGURE 39.2 Solid-desiccant water-vapor equilibrium curves. (*Reprinted by permission from the ASHRAE Handbook—1985 Fundamentals.*)

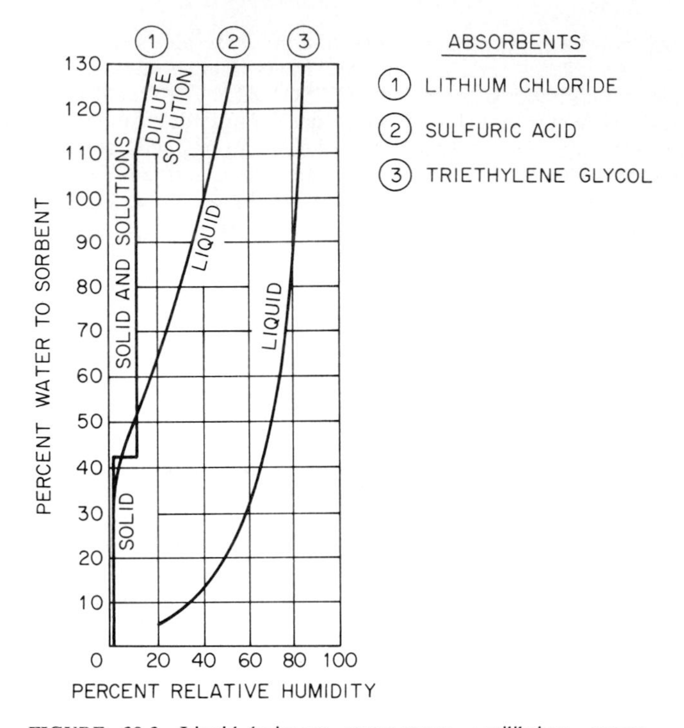

FIGURE 39.3 Liquid-desiccant water-vapor equilibrium curves. *(Reprinted by permission from the ASHRAE Handbook—1985 Fundamentals.)*

densation to a few times this heat, depending on the desiccant. In practice, however, dehumidifiers exhibit process-air dry-bulb temperature increases that typically range from about 1.1 to 2 times that associated with heat of condensation alone. The entire heat energy released, including the heat of condensation, is termed the "heat of sorption."

The most important feature of the sorption process is its reversibility. In most cases, the desorption process is accomplished by the application of heat to raise the desiccant's temperature above its temperature in the sorption process. The relative humidity of the surrounding air decreases as its dry-bulb temperature and the desiccant's temperature are increased. This lowers the sorbed water equilibrium content in the desiccant. The desiccant desorbs moisture, and water vapor is released as heat energy is transferred to the desiccant. The regenerated desiccant is now ready once again to sorb water vapor. Regeneration temperatures in dehumidifiers are generally within the 130 to 300°F (54.4 to 148.9°C) range.

A desiccant may eventually lose significant moisture-cycling capacity. Over time, thermal and mechanical degradation can account for some loss in capacity for solid desiccants, but substantial capacity reduction for solid and liquid desiccants and dehumidifiers is generally caused by contamination with airborne materials and reactive substances. Incoming air must be filtered to remove particulates. Depending on the desiccant, the application may be assessed initially for other contaminants as well. Most equipment manufacturers do offer desiccant-analysis services for their installed dehumidifiers.

Although the percentage of water sorbed in some desiccants is significantly greater than in others, the minimum dew-point temperature produced by a sorption-desorption process is not determined by the equilibrium capacity (Ref. 2). Equilibrium capacity does determine the amount of desiccant that air must contact for a given reduction in absolute humidity. But most desiccants can ideally provide minimum dew-point temperatures below freezing (32°F, or 0°C) at room-air dry-bulb temperatures (75°F, or 23.9°C).

The ability to reach this minimum dew-point temperature is limited, though, by practical dehumidifier design and operation considerations. For instance, in solid systems the transfer of moisture from the air creates a propagating concentration wave front through the material. Less than the ideal shock wave-front behavior leads to average outlet air conditions from the desiccant material above the minimum dew-point achieved. Economical sizing for dehumidification capacity typically does not allow equipment to provide air at the minimum dew-point possible for the desiccant. Also, higher regeneration temperatures lower the dew-point temperature possible for the desiccant, but material tolerances and poorer operating efficiencies often restrict their use in dehumidifiers. As an example, in certain liquid systems the regeneration temperatures are restricted to below the levels at which the concentrations created would result in the crystallization of solids in solution. Available desiccant dehumidifiers attain dew-point temperatures that generally range from 10 to −55°F (−12.2 to −48.3°C) (Ref. 3).

39.4 DEHUMIDIFIERS

Solid desiccants are placed in some form of bed to dehumidify air. The older and simpler dehumidifier designs utilize desiccant-filled shallow disks. The perforated disk rotates slowly to expose one portion of the desiccant bed to the process air stream while the other portion simultaneously passes through the regeneration air stream. A partition and flexible seals separate the process and regeneration air streams in the dehumidifier. Moist air enters the process side and passes through the desiccant-filled bed and is dehumidified. Regeneration occurs on the other side of the partition, where heated air enters from the same or opposite direction, passes through the bed, and, laden with moisture, is exhausted from the dehumidifier.

Alternatively, the solid desiccant in some dehumidifiers is contained between inner and outer closely spaced cylinders. A further variation of this design places the desiccant in a circular arrangement of compartmentalized vertical beds. A more recent dehumidifier design employs a parallel-passageway wheel to minimize pressure drop. The desiccant is impregnated on a substrate, which is typically a corrugated structure forming channels for air flow.

In all these designs, air flows through the solid desiccant contained in the perforated containers or corrugated structures, which rotate continuously (or in step fashion) from process to regeneration and back, as shown in Fig. 39.4 (and see the accompanying Fig. 39.5). Often a small amount of process air is used to cool, or purge the heat from, the desiccant immediately following regeneration. The purge air stream is then ducted to the entering regeneration air stream. A heat exchanger may be provided that transfers energy from the exiting regeneration

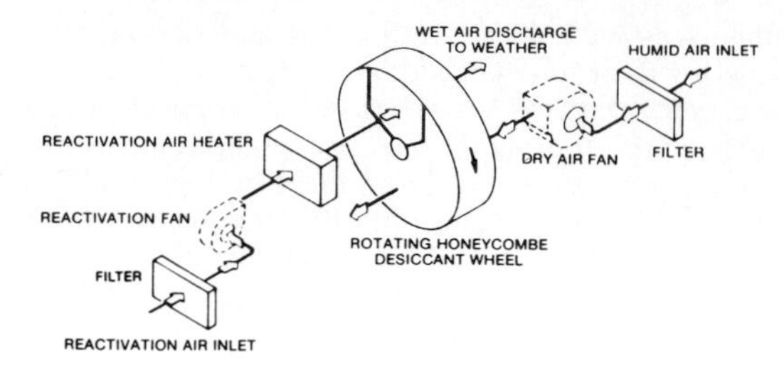

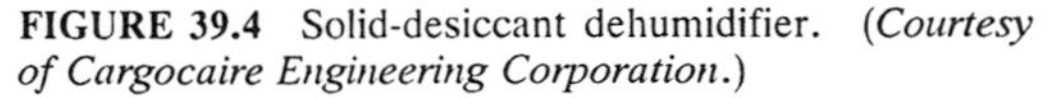
FIGURE 39.4 Solid-desiccant dehumidifier. (*Courtesy of Cargocaire Engineering Corporation.*)

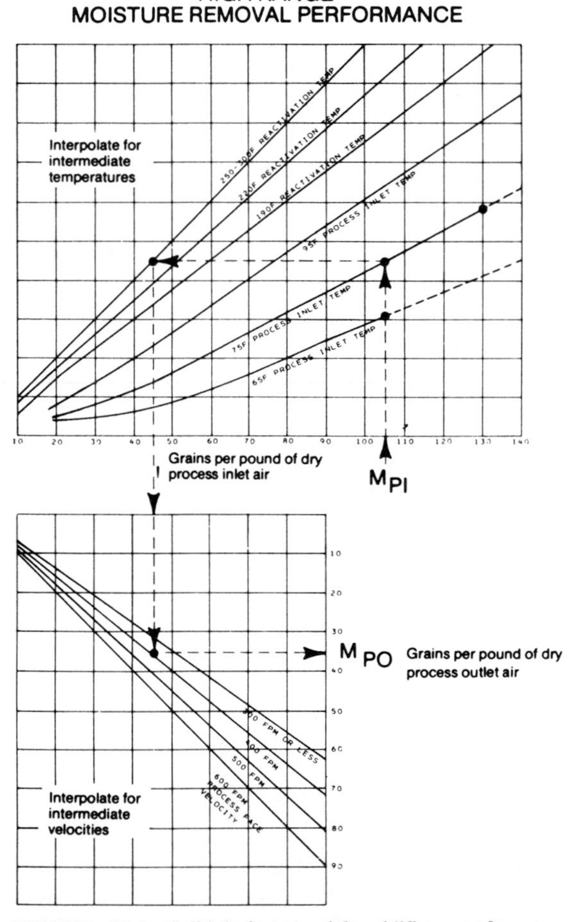

FIGURE 39.5 Solid-desiccant dehumidifier performance map. (*Courtesy of Cargocaire Engineering Corporation.*)

air stream to the entering regeneration air stream. Alternatively, a heat exchanger may cool the process air stream exiting the dehumidifier and preheat the incoming regeneration air stream in turn. Additional sensible cooling of the process air stream will be required to control the dry-bulb temperature for comfort in the building.

Liquid desiccants are enclosed in chambers to dehumidify air. The desiccant is distributed in one chamber, where it contacts the passing process air stream. As the moisture is sorbed by the desiccant, heat is released. Cooling coils in the chamber remove the heat from the desiccant and from the exiting process air stream in the chamber. (The simultaneous desiccant dehumidification and aftercooling generally follows the dashed-line path shown in Fig. 39.1.) The moisture-laden desiccant is then pumped to the other chamber, where a heating coil drives off the water in an exhaust air stream. The desiccant can now be pumped from the regeneration chamber, called the "regenerator," to be redistributed in the dehumidification chamber, called the "conditioner." An interchanger is often used to cool the warmer desiccant leaving the regenerator by exchanging heat with the cooler desiccant from the conditioner. The liquid-desiccant dehumidifier configuration is shown in Fig. 39.6 (and see the accompanying Fig. 39.7). As with other dehumidifiers, additional process-air sensible cooling may be required to provide comfortable space dry-bulb temperatures.

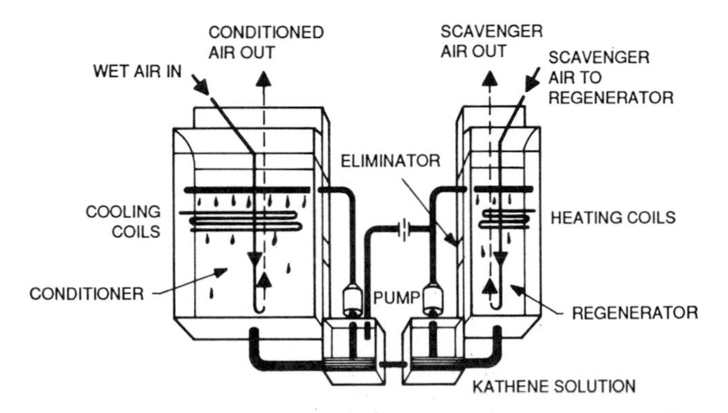

FIGURE 39.6 Liquid-desiccant dehumidifier. (*Courtesy of Kathabar Systems*.)

Dehumidifiers range in size from 25 to 100,000 SCFM (11.8 to 47,200 L/s) with moisture-removal capacities from about 0.5 to 10,000 lb/h (0.23 to 4540 kg/h), or 500 to 10,500,000 Btu/h (0.15 to 3078 kW), of latent cooling (Refs. 4 to 9). Individual solid systems are designed from 25 to 62,500 SCFM (11.8 to 29,500 L/s), and liquid systems are designed from 1000 to 100,000 SCFM (472 to 47,200 L/s). Figures 39.5 and 39.7 show representative dehumidifier performance characteristics for solid and liquid systems, respectively. The coefficients of performance (COPs) range from as low as 0.35 to 0.7. The COP is based on the latent energy of the moisture removed divided by the heat energy necessary for regeneration. Waste-heat utilization in a system design can reduce, and possibly eliminate, the purchase of thermal energy for regeneration. Manufacturers should be contacted for complete dehumidifier performance data and design specifications.

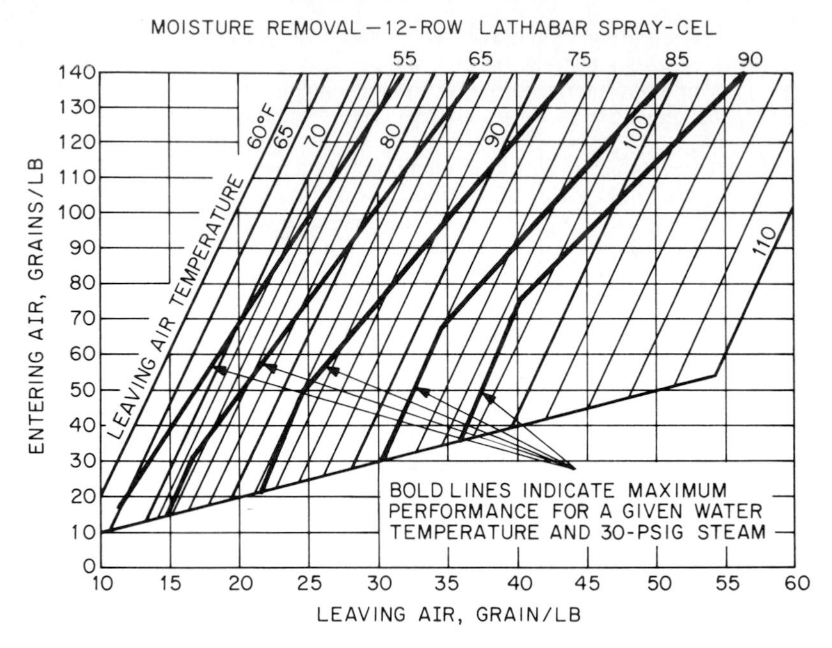

FIGURE 39.7 Liquid-desiccant dehumidifier performance map. Water temperature refers to cooling-coil water temperature and 30-psig steam refers to heating-coil steam pressure in Fig. 39.6. *(Courtesy of Kathabar Systems.)*

39.5 APPLICATIONS

In numerous industrial environments, desiccant dehumidification removes unwanted humidity from the air to promote drying and prevent condensation or moisture regain. This allows production processes to proceed efficiently with minimum quality problems. It also allows raw materials and finished goods to be stored without degradation problems. To manufacture pharmaceuticals, for example, relative humidities down to 10 percent at room temperature are required for granule and powder handling, proper compounding, and long shelf life.

Other industries—including food and beverage production, electronics manufacturing, chemicals preparation, rubbers and plastics fabrication, metals treatments, photographic production, wood processing, cosmetics manufacturing, printing and painting operations, and glass preparations—also use desiccant dehumidification to control depressed humidity levels at room-air dry-bulb temperature. Many of these same industries also use desiccant dehumidification to lower humidity levels for transporting and warehousing in order to prevent corrosion, mold and mildew formation, and other deterioration. Table 39.3 lists several industries and their temperature- and humidity-controlled processes.

The often dramatic decrease in product rejection and direct increase in profitability of the product is the primary incentive for the investment in depressed humidity control by desiccant dehumidification in the industrial sector. These depressed humidity levels are typically unattainable cost-effectively with mechanical refrigeration and reheat. The small volume of sales and specialty application

TABLE 39.3 Temperature and Humidity Set Point for Industrial Air Conditioning

Process	Dry Bulb, °F	(°C)	rh, %
Ceramics			
Refractory	110–150	(43–66)	50–90
Molding room	80	(27)	60–70
Clay storage	60–80	(16–27)	35–65
Decalcomania production	75–80	(24–27)	48
Decorating room	75–80	(24–27)	48
Cereal			
Packaging	75–80	(24–27)	45–50
Distilling			
Storage:			
Grain	6	(16)	35–40
Liquid yeast	32–33	(0–1)	
General manufacturing	60–75	(16–24)	45–60
Aging	65–72	(18–22)	50–60
Electrical products			
Electronics and X ray:			
Coil and transformer winding	72	(22)	15
Semiconductor assembly	68	(20)	40–50
Electrical instruments:			
Manufacture and laboratory	70	(21)	50–55
Thermostat assembly and calibration	75	(24)	50–55
Humidistat assembly and calibration	75	(24)	50–55
Small mechanisms:			
Close tolerance assembly	72	(22*)	40–45
Meter assembly and test	75	(24)	60–63

Process	Dry Bulb, °F	(°C)	rh, %
Pharmaceuticals			
Powder storage (prior to mfg)	†	†	
Manufactured powder storage and packing areas	75	(24)	35
Milling room	75	(24)	35
Tablet compressing	75	(24)	35
Tablet coating room	75	(24)	35
Effervescent tablets and powders	75	(24)	20
Hypodermic tablets	75	(24)	30
Colloids	75	(24)	30–50
Cough drops	75	(24)	40
Glandular products	75	(24)	5–10
Ampoule manufacturing	75	(24)	35–50
Gelatin capsules	75	(24)	35
Capsule storage	75	(24)	35
Microanalysis	75	(24)	50
Biological manufacturing	75	(24)	35
Liver extracts	75	(24)	35
Serums	75	(24)	50
Animal rooms	75–80	(24–27)	50
Small animal rooms	75–78	(24–26)	50
Photo studio			
Dressing room	72–74	(22–23)	40–50
Studio (camera room)	72–74	(22–23)	40–50
Film darkroom	70–72	(21–22)	45–55
Print darkroom	70–72	(21–22)	45–55
Drying room	90–100	(32–38)	35–45

TABLE 39.3 Temperature and Humidity Set Point for Industrial Air Conditioning (*Continued*)

Process	Dry Bulb, °F	(°C)	rh, %
Electrical Products (*Continued*)			
Switchgear:			
Fuse and cutout assembly	73	(23)	50
Capacitor winding	73	(23)	50
Paper storage	73	(23)	50
Conductor wrapping with yarn	75	(24)	65–70
Lightning arrester assembly	68	(20)	20–40
Thermal circuit breakers assembly and test	75	(24)	30–60
High-voltage transformer repair	79	(26)	5
Water wheel generators:			
Thrust runner lapping	70	(21)	30–50
Rectifiers:			
Processing selenium and copper oxide plates	73	(23)	30–40
Gum			
Manufacturing	77	(25)	33
Rolling	68	(20)	63
Stripping	72	(22)	53
Breaking	73	(23)	47
Wrapping	73	(23)	58
Lenses (optical)			
Fusing	75	(24)	45
Grinding	80	(27)	80

Process	Dry Bulb, °F	(°C)	rh, %
Photo Studio (*Continued*)			
Finishing room	72–75	(22–24)	40–55
Storage room (b/w film and paper)	72–75	(22–24)	40–60
Storage room (color film and paper)	40–50	(4–10)	40–50
Motion picture studio	72	(22)	40–55
Plastics			
Manufacturing areas:			
Thermosetting molding compounds	80	(27)	25–30
Cellophane wrapping	75–80	(24–27)	45–65
Plywood			
Hot pressing (resin)	90	(32)	60
Cold pressing	90	(32)	15–25
Rubber-dipped goods			
Manufacture	90	(32)	
Cementing	80	(27)	25–30‡
Dipping surgical articles	75–80	(24–32)	25–30‡
Storage prior to manufacture	60–75	(16–24)	40–50‡
Laboratory (ASTM Standard)	73.4	(23)	50*

*Temperature to be held constant.
†Store in sealed plastic containers in sealed drums.
‡Dew point of air must be below evaporation temperature of solvent.
Source: Reprinted by permission from the *ASHRAE Handbook—1987 HVAC Systems and Applications*.

pricing generally place a relatively high initial cost on desiccant dehumidification equipment.

The same costly, heavy-duty grade of equipment used in industry is currently applied to very few applications in the institutional and commercial sector. The economic benefits to the building's operations, not only air-conditioning, again justify the initial investment. Many hospitals and health-care centers use desiccant dehumidification to control humidity precisely and independently of temperature, and to remove microorganisms when ventilating critical areas. Municipal waterworks use it to prevent damage to cool-water piping and control system surfaces by condensation and corrosion. Enclosed ice rinks and swimming pools also apply desiccant dehumidification to control similar degrading effects of moisture on the building's structure and equipment.

Significant development is being sponsored to lower the initial cost and improve the performance of desiccant dehumidification equipment (Refs. 10 and 11). Objectives are the broader application of desiccant dehumidification as a least costly cooling energy service with value-added features such as enhanced comfort and improved air quality.

With recently developed equipment, supermarkets using desiccant dehumidification can hold relative humidity between 30 and 40 percent to save refrigerated-case operating energy costs (Refs. 12 and 13). Other developing applications in more humid climates include hotels and restaurants. Hotels have high building-maintenance costs and often premature room renovations because of mold, mildew, and corrosion damage caused by excessive humidity levels (Ref. 14). Restaurants use up to 40 percent makeup air in air-conditioning operations to compensate for kitchen exhaust and to control cooking odors (Ref. 14).

In all commercial and institutional buildings, the ventilation air requirements necessary to meet current and proposed air-quality standards (see Sec. 39.8, Air Quality) create substantial dehumidification loads, especially in the more humid climates. Desiccant dehumidification equipment to reduce air-conditioning operating energy costs while enhancing comfort and improving air quality is available for office and other building applications (Ref. 15).

To date, desiccant dehumidification has seen negligible use in the residential sector. Although desiccant dehumidification can enhance comfort and improve air quality over mechanical refrigeration alone, the current initial-cost premiums and performance levels preclude market penetration (Ref. 16).

39.6 SYSTEMS

Typically an application will require that a space be maintained at a dry-bulb temperature and relative or absolute humidity. Internal and external sensible and latent loads are quantified for interior and exterior conditions and ventilation air requirements. The summation of all the sensible and latent loads equals the total cooling load. Supply air requirements are specified, typically between 250 and 500 SCFM/ton [33.5 and 67.0 L/(s · kW)] of total cooling. To help determine which air-conditioning process provides the least costly energy service for the total cooling load, the series of graphs in Fig. 39.8 provides an approximate comparison of operating energy costs.

Parts *a* to *d* of Fig. 39.8 determine the required process outlet (supply) dry-bulb and dew-point temperatures. These are based on process inlet dry-bulb temperature and absolute humidity and on the SHF and LHF of the total design cool-

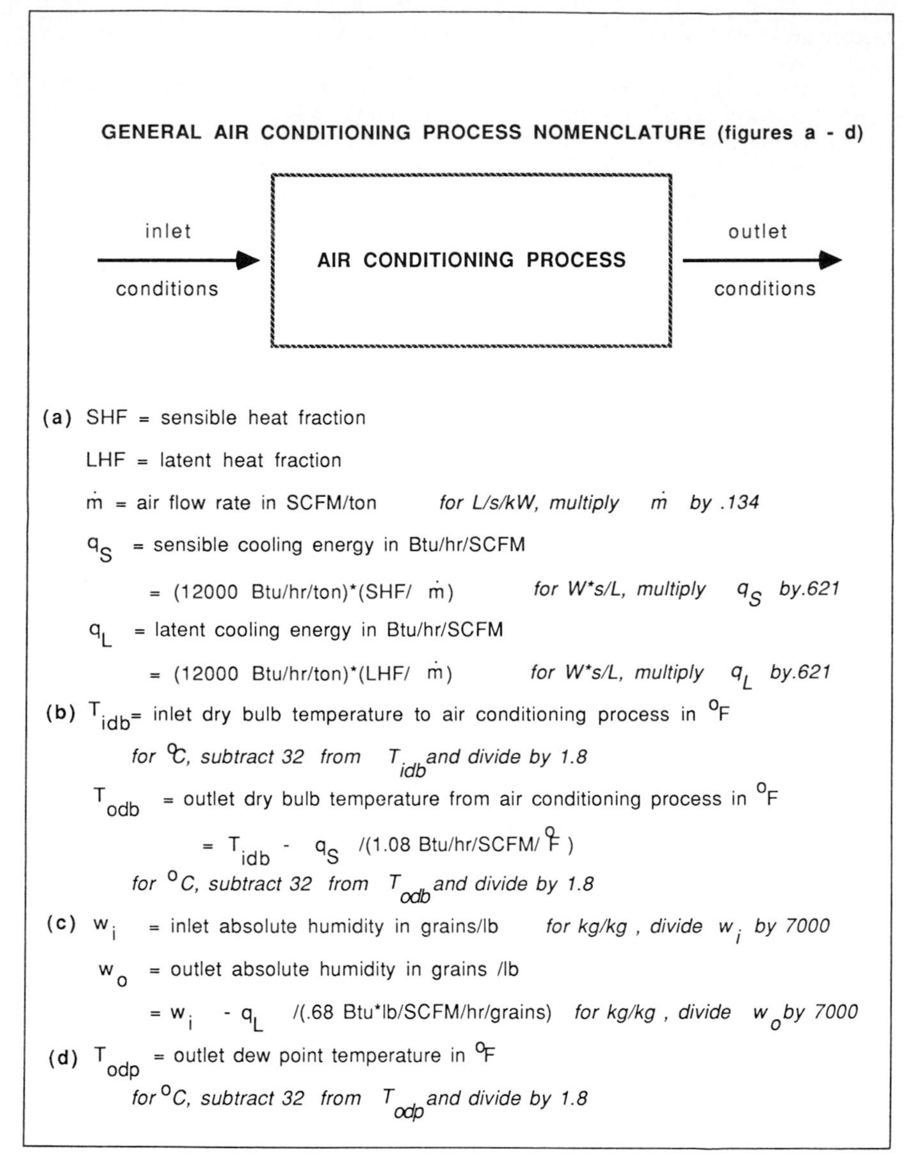

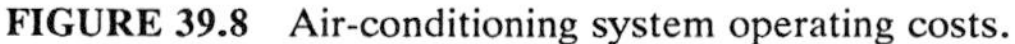

FIGURE 39.8 Air-conditioning system operating costs.

ing load. It is assumed that the process inlet air condition is a mixed ventilation (ambient) and return (space) air condition. As a guideline, for process outlet dew points less than 40°F (4.4°C), desiccant dehumidification is advisable in order to avoid excessive operating energy costs for overcooling, defrosting, and reheating. If the process outlet dew point is above 40°F (4.4°C) and greater than the process outlet dry-bulb temperature, then mechanical refrigeration alone is gen-

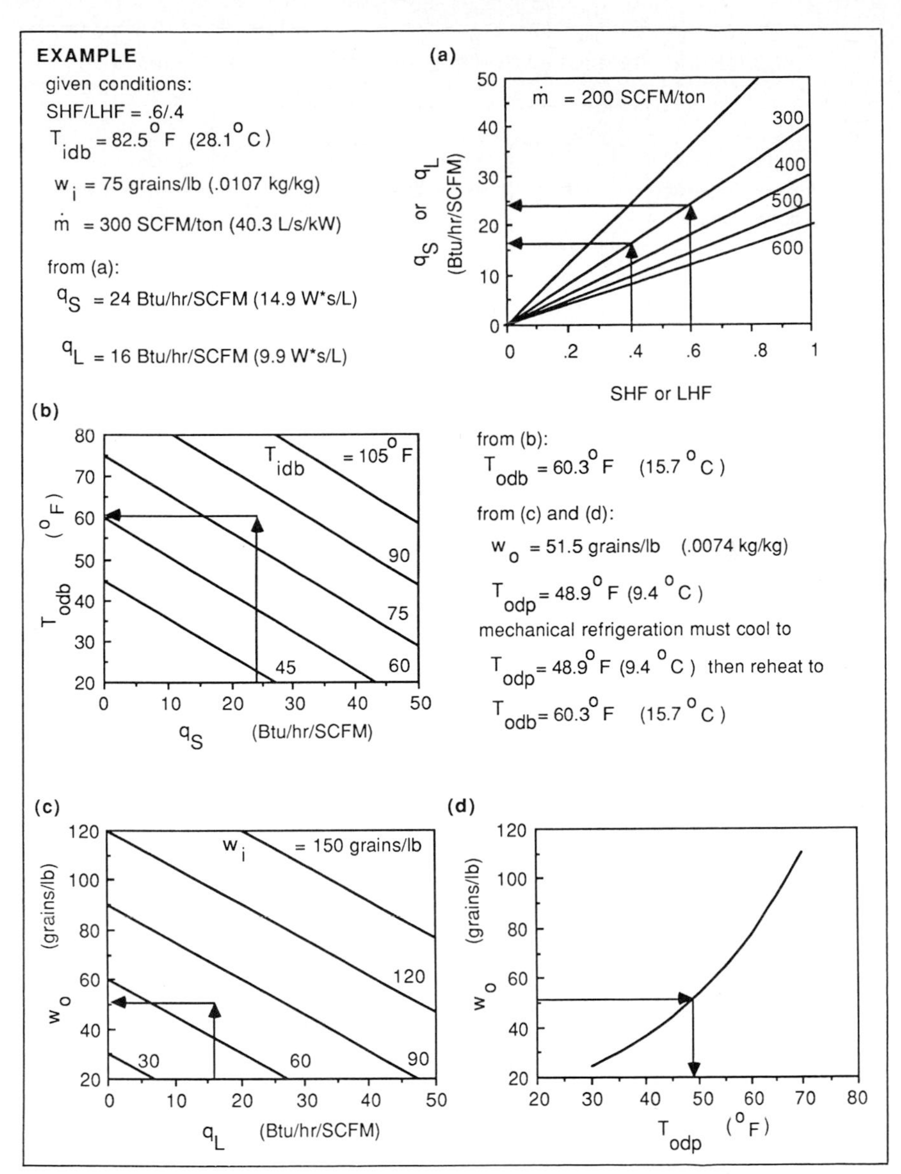

FIGURE 39.8 (*Continued*)

erally recommended. However, if the electricity price is substantially greater (4 times or more) than the price of natural gas or other thermal energy, then desiccant dehumidification may save operating costs. And finally, if the process outlet dew point is above 40°F (4.4°C) and less than the process outlet dry-bulb temperature, then desiccant dehumidification may again save operating costs.

Parts *e* to *n* of Fig. 39.8 compare operating costs. Parts *e* to *i* determine the

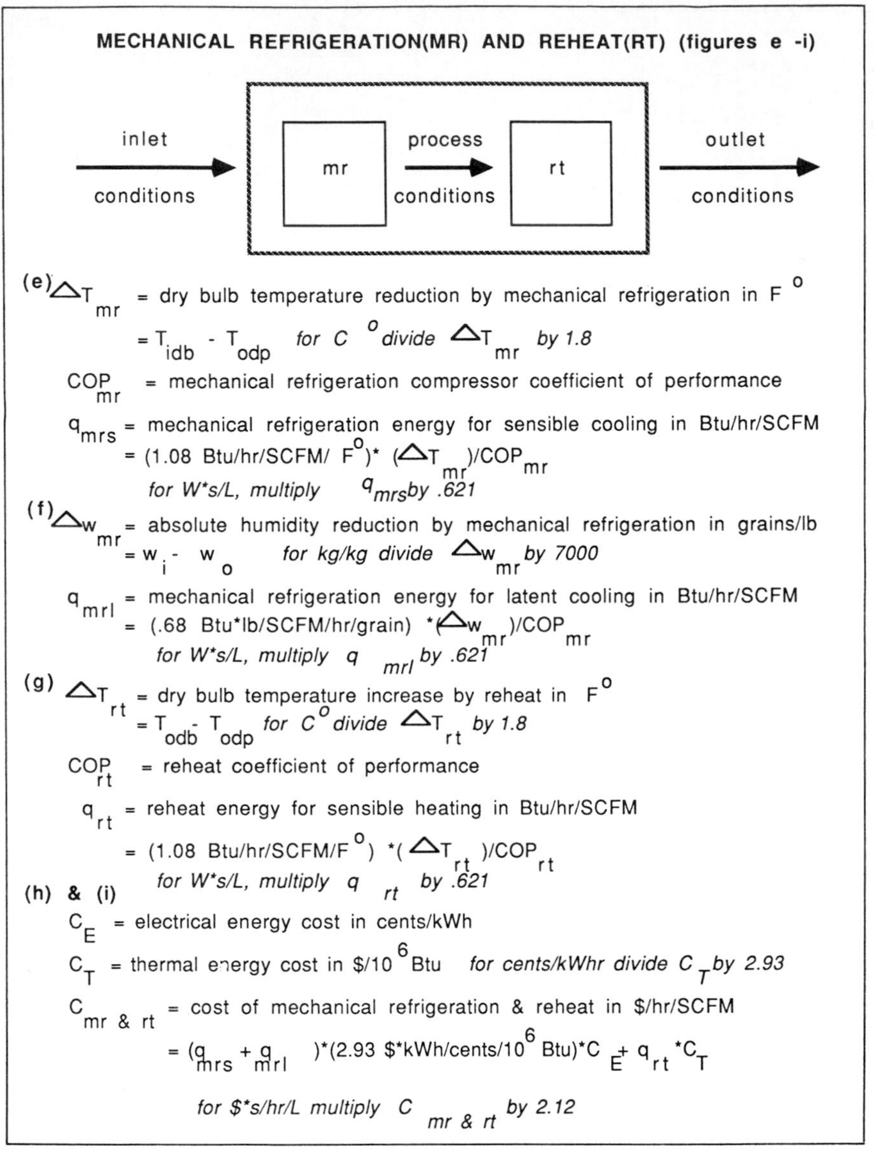

FIGURE 39.8 (*Continued*)

operating energy costs for mechanical refrigeration with reheat, and parts *j* to *n* determine the operating energy costs for desiccant dehumidification with aftercooling. Aftercooling can be accomplished with mechanical refrigeration, indirect evaporative cooling (cooling tower), air-to-air heat exchange, or a combination of any or all of these options. Part *l* in Fig. 39.8 incorporates high-COP aftercooling operations. The use of indirect evaporative cooling will depend on

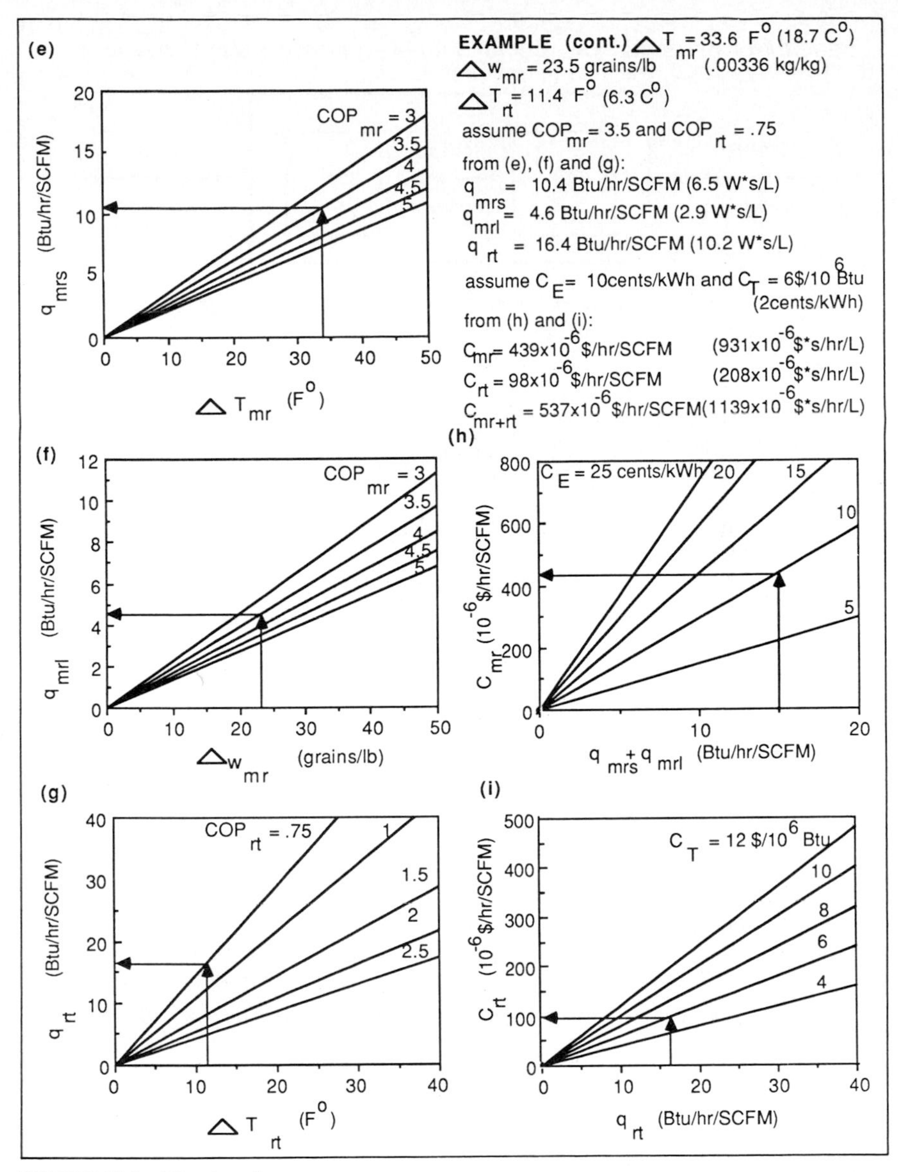

FIGURE 39.8 (*Continued*)

local weather design conditions and water hardness. The use of stationary or rotary air-to-air heat exchangers will depend on the degree of exhaust- or ambient-air heat sinks available. Desiccant dehumidification can cause sizable temperature increases in the process air stream (see Sec. 39.3, Desiccants). Part k accounts for heat-of-sorption effects. Typically, if desiccant dehumidification is to compete with mechanical refrigeration and reheat on an operating-cost basis in

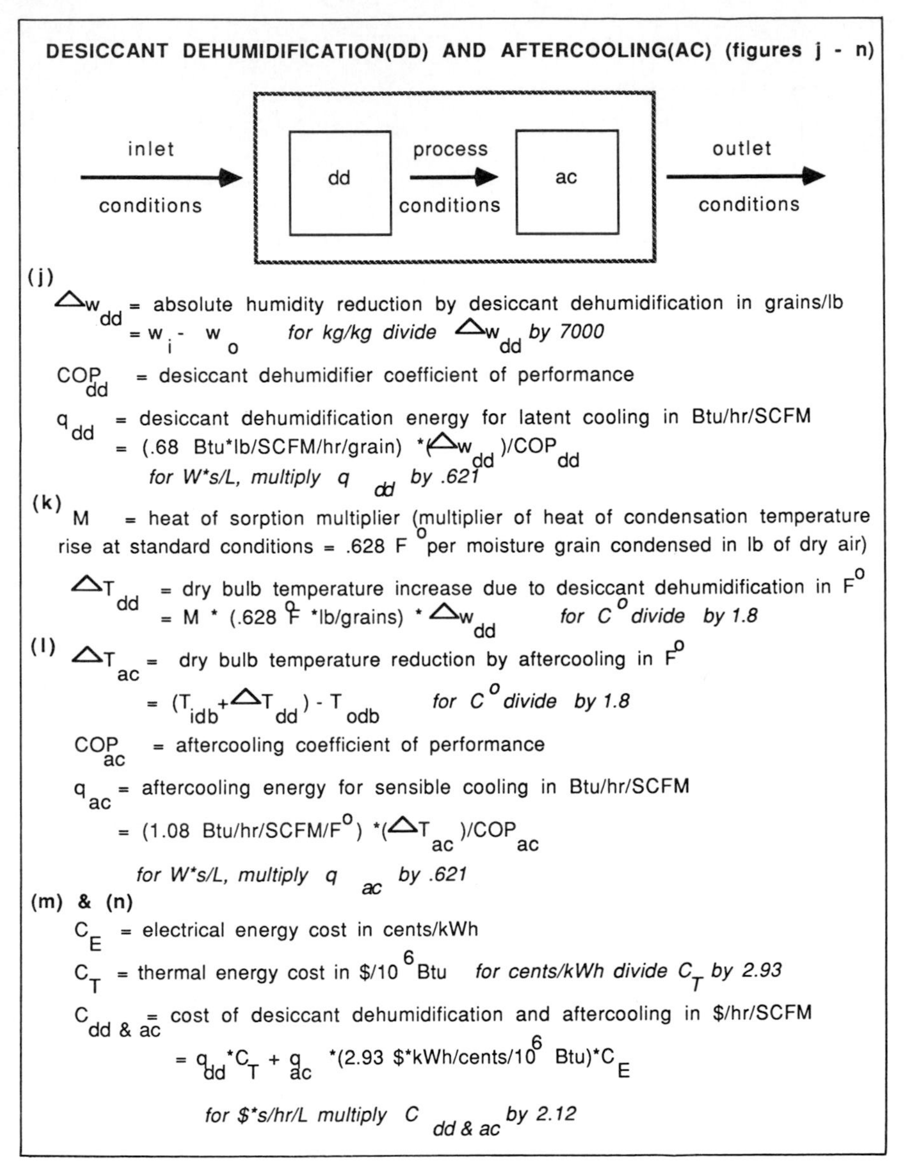

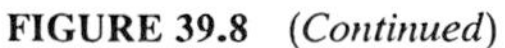

FIGURE 39.8 *(Continued)*

a higher process-outlet dew-point application, it is necessary to utilize aftercooling options that consume low electric energy. However, the operating-cost savings of the desiccant dehumidification and aftercooling designs must pay back the premiums between their initial-cost and the initial cost of the mechanical-refrigeration-and-reheat design in an acceptable time period (see Sec. 39.9, Economics).

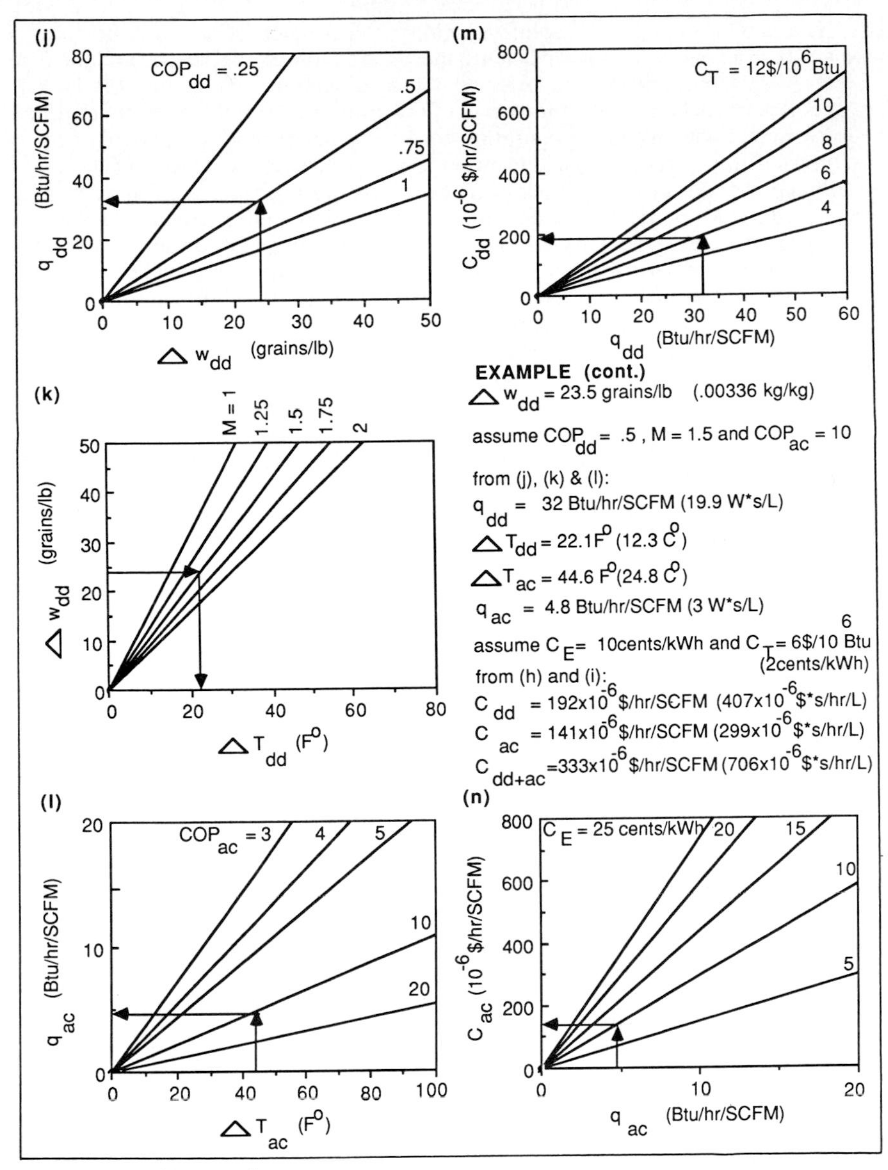

FIGURE 39.8 (*Continued*)

When desiccant dehumidification shows a potential operating-cost advantage, it must be determined whether or not deep drying of the ventilation air alone is sufficient to meet the total latent load. Parts *a, c*, and *d* in Fig. 39.8 can help one determine a new required process outlet humidity based on a process inlet absolute humidity equal to ventilation (ambient) air conditions. Performance charts for desiccant dehumidifiers, such as shown in Figs. 39.5 and 39.7, are then used to ascertain if that outlet absolute humidity can be met. If dehumidification of ventilation air alone is not sufficient, then a fraction of the return air must be mixed with the ventilation air prior to dehumidification. Minimizing the size of the desiccant dehumidifier and the air flow through it to achieve the required moisture removal lowers the operating costs. The initial cost of that equipment is minimized too. (Precooling of air entering a solid-desiccant dehumidifier, especially ambient air, can make operating- and initial-cost sense but is not addressed here; see Ref. 3 or consult manufacturers for additional information.) As with all the operating costs calculated in Fig. 39.8, the operating cost for desiccant dehumidification from part *m* is a per-unit-of-air-flow value that can then be multiplied by the actual volume processed. (New aftercooling process air conditions and operating costs must also be determined.)

The low dew-point capabilities of desiccant dehumidifiers can allow them to be designed to handle less air flow than is necessary for dehumidification by mechanical refrigeration and reheat. This is a significant operating-cost savings when the latent load dictates air-flow requirements. Similar operating-cost savings opportunities arise if the latent load dictates the cooling coil temperature of mechanical refrigeration. With desiccant dehumidifiers meeting the latent load, aftercooling with mechanical refrigeration is designed for the sensible load only, and cooling-coil surface temperatures can typically be raised for a higher compressor COP.

A general schematic of an air-conditioning system integrating desiccant dehumidification is shown in Fig. 39.9. Aftercooling is located downstream of the dehumidifier. In the case of liquid-desiccant dehumidification, some or all after

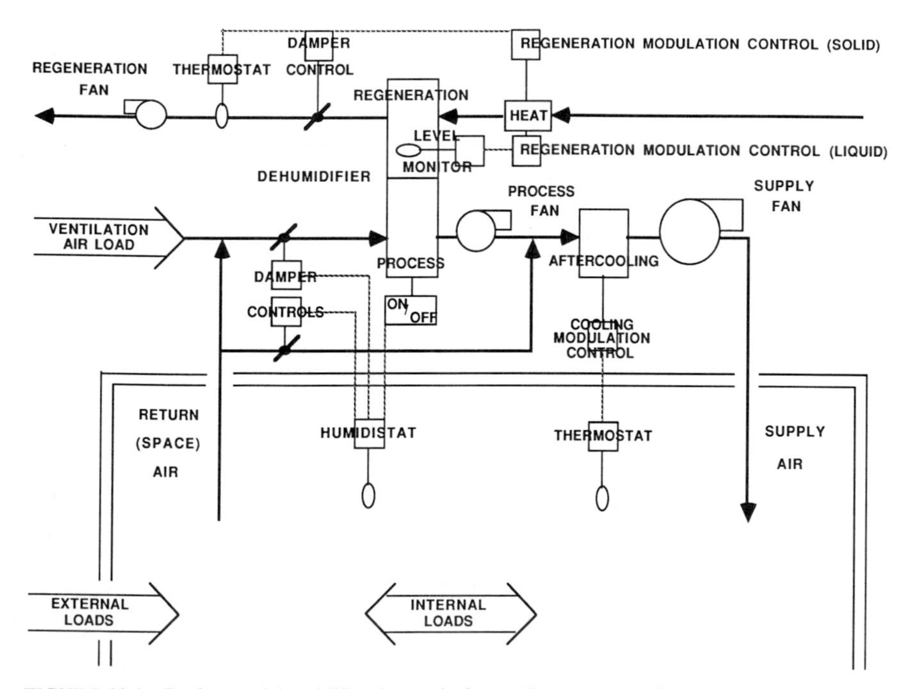

FIGURE 39.9 Desiccant dehumidification and aftercooling system and controls.

cooling is incorporated in the dehumidifier itself (see Sec. 39.4, Dehumidifiers), most often with cooling coils circulating tower water or chilled water. No reheat capabilities, only space heating, need be incorporated farther downstream.

Partial or total bypass of return air around the dehumidifier is directed by face and bypass dampers. If design conditions allow total bypass, partial bypass must be accommodated for humidity control of unventilated or unoccupied space.

Regeneration can be accomplished by direct firing of natural gas in a regeneration air stream that then contacts the desiccant. Indirect firing is required with certain desiccants, and recommended for all, to prevent physical or chemical degradation by products of combustion. Alternatively, hot-water or steam coils piped to a boiler can heat the regeneration air stream. In the case of liquid-desiccant dehumidification, regeneration is typically accomplished with hot water or steam circulating in heating coils contacting the desiccant in the regenerator. A single liquid-desiccant regenerator can serve multiple conditioners. Regenerated liquid desiccant can also be stored for later dehumidification.

Hot water or steam available from cogeneration systems, engine-driven chillers, or other processes producing waste heat can be utilized for regeneration. Solar energy collected by even low-temperature flat-plate systems can provide either preheat or all regeneration requirements. The use of expensive electricity is rarely justified for regeneration except for ease of installation and operation in small or portable applications.

39.7 CONTROLS

Air-conditioning systems operate at off-design conditions most of the time and rely on controls to regulate capacity efficiently to meet varying internal and external loads and maintain humidity and dry-bulb temperature set points in the space. Combined control is difficult with the mechanical-refrigeration approach to dehumidification. Accurate dry-bulb temperature control can be achieved with reheat. An acceptable range of absolute humidity control may be obtained with face and bypass dampers around the cooling coil. However, complex coordination of the controls may be required because of interdependent operation if, for instance, hot-refrigerant-gas bypass is utilized for reheat.

An air-conditioning system employing a desiccant dehumidifier allows independent control of the dry-bulb temperature and absolute humidity. Generally, a humidistat starts the desiccant dehumidifier when space humidity (absolute or relative) exceeds the upper limit of the set-point dead band. Operation is initiated at either maximum or minimum dehumidification capacity.

Solid-desiccant dehumidifiers can modulate the regeneration energy input to regulate the dehumidification capacity and to control space humidity. A reduction in the moisture content of the incoming process air stream and the resulting drop in moisture load increases the exiting regeneration air stream's dry-bulb temperature. Sensing a dry-bulb temperature increase, a controller throttles back the energy input to match the intended operating regeneration dry-bulb temperature. This is accomplished either by throttling the temperature by modulating heat input, or by throttling air flow by adjusting a regeneration air damper. Beds rotate slowly, on the order of 10 revolutions per hour, and humidity levels may range significantly in the space without additional controls. Some dehumidifiers also operate with variable-air-volume controls on the process side to allow more rapid response and tight space humidity control. Face and bypass dampers regulated by a space humidistat are used to control dehumidification process airflow rates and hence capacity.

Liquid-desiccant dehumidifiers maintain space humidity by controlling the solution's concentration and temperature to determine moisture removal. Upon sufficient moisture collection, a liquid-level controller, usually in the conditioner sump, activates the regenerator. As the dehumidification load lessens, the conditioner solution's level lowers. In response, the regenerator temperature controller modulates the heat input to return an intended operating-solution concentration to the conditioner. Modulating conditioner-coil coolant flow, usually based on the exiting air's dry-bulb temperature, regulates the solution's temperature.

Aftercooling must take into consideration the sensible heat load generated in the desiccant dehumidifier. Changes in the moisture load met by the dehumidifier will vary the process air dry-bulb temperature. The dry-bulb temperature is monitored in downstream ductwork or by the space thermostat and typically inputs to a control valve for modulating refrigerant or chilled-water flow in the cooling coil. Most or all of the aftercooling may be incorporated directly in the conditioner of a liquid-desiccant dehumidifier. System controls are displayed on the schematic in Fig. 39.9.

39.8 AIR QUALITY

In the more humid climates, the fresh-air requirements to meet indoor air quality create substantial dehumidification loads. Provision and treatment of adequate ventilation air to control air quality is the subject of recent revisions in air-conditioning standards. The ASHRAE standard entitled *Ventilation for Acceptable Indoor Air Quality* is proposed for modification to triple the minimum outdoor air requirements per person (Refs. 17 and 18). Although not binding, ASHRAE standards are usually adopted in state and local building codes.

The proposed increase in ventilation rates has resulted from the growing incidence of the "sick"-building syndrome as reported by the National Institute for Occupational Safety and Health (NIOSH) and other public and private organizations (Ref. 19). Nowhere is this more prevalent than in office buildings, where it is estimated that over 33 million Americans work. A recent survey of office workers showed that over half believe that better air quality would result in a more productive work environment (Ref. 20). In order to retain occupancy, office-building owners-operators may be motivated by the value to their tenants of air-conditioning service, not solely by the initial cost of the equipment (Ref. 21). With employees valued yearly at $100 or more per square foot of office-building space, even a small-percentage loss in work time due to illness or discomfort costs employers several dollars per square foot every year. In an entire office building, literally tens of thousands of dollars per year in lost employee time could be associated with inadequacies of the air-conditioning system.

The cost-effective dehumidification of ventilation air is not the only means by which desiccants can control air quality. Several industries, including chemicals preparation, rubbers and plastics fabrication, and printing and painting operations, use sorbents to perform selective separation and recovery or disposal of contaminant gases and vapors for pollution control of both the work and ambient environments (Refs. 22 to 24). Activated carbon, for example, has a high affinity and capacity for organic compounds. Certain liquid desiccants also exhibit biocidal effects, killing very high percentages of bacteria, viruses, and mold present in the process air. With the recent emphasis on pollution and air-quality control in general, this use is expanding. Desiccant-material manufacturers should be contacted for specific applications.

39.9 ECONOMICS

The analysis shown in Fig. 39.8 accounts for many of the key factors in the competitive operating-cost design of desiccant dehumidification in an air-conditioning system. Those key design parameters are:

1. Outlet dry-bulb temperature
2. Outlet dew-point temperature
3. Mechanical-refrigeration efficiency and electric-energy operating cost
4. Reheat efficiency and thermal-energy operating cost
5. Desiccant dehumidification efficiency and thermal-energy operating cost
6. Aftercooling efficiency and electric-energy operating cost

A more comprehensive and accurate seasonal analysis of the competing air-conditioning designs should be completed that will account for all electric power, off-design performance, and controls in operation. In addition, for refrigeration condensing units and aftercooling devices that utilize water, usage and sewage charges must be taken into account.

A very important and growing trend confronting electricity users in the commercial, institutional, and industrial sectors is the practice of utility companies charging for power demand, which generally peaks with air-conditioning use. A summer month that has a power peak can also set the demand charge for the entire year when this cost is "ratcheted." When analyzing annual operating costs, the appropriate local electric- and thermal-energy rate structure must be employed. Anticipated energy-cost escalation must be included for calculations in future years.

Projected equipment maintenance costs, including anticipated component replacement purchases, and life expectancy must be incorporated for life-cycle operating costs. Manufacturers can provide estimates for these values, which vary depending on the application. Harsh industrial environments, as expected, generally increase the maintenance and replacement costs and shorten the life expectancy.

Although rising electric-energy rates and demand charges are creating opportunities to save operating costs with air-conditioning designs that utilize thermal-driven desiccant dehumidification, this alternative must satisfy an application's economic criteria. Whether it be simple payback, return on investment, or life-cycle cost, all methods require that the operating costs offset the initial-cost premium of the more expensive desiccant dehumidification design. Only in low-dew-point applications below 40°F (4.4°C) does the initial cost of mechanical refrigeration and reheat generally become greater. Above that dew-point temperature, the selected economic criteria must be evaluated.

The economic evaluation can entail more than the operating, maintenance, and initial costs of the air-conditioning equipment itself. In new construction, desiccant dehumidification applications that reduce air-handling requirements use smaller ductwork and save initial building costs. Proper humidity control in existing construction lowers building maintenance costs by stopping mold and mildew formation, corrosion, and their associated deterioration of materials, equipment, and furnishings. Depressed humidity control in manufacturing and storage decreases product losses. And finally, the improvements in human comfort and well-being through humidity control and air-quality control can increase revenues as a result of greater productivity and can reduce the costs of absenteeism.

39.10 REFERENCES

1. *1985 ASHRAE Handbook, Fundamentals*, ASHRAE, Atlanta, GA, 1985, chaps. 11 and 19.
2. R. K. Collier et al., *Advanced Desiccant Materials Assessment*, Gas Research Institute (National Technical Information Service Document PB 87-172805), Chicago, IL, 1986.
3. *The Dehumidification Handbook*, Cargocaire Engineering Corporation, Amesbury, MA, 1985.
4. Assorted product literature, available from Cargocaire Engineering Corporation, P.O. Box 640, Amesbury, MA 01913. Telephone: (508) 388-0600.
5. Assorted product literature, available from Kathabar Systems of Somerset Technologies, Inc., P.O. Box 791, New Brunswick, NJ 08903. Telephone: (201) 356-6000.
6. Assorted product literature, available from Airflow Company, 295 Bailes Lane, Frederick, MD 21701. Telephone: (301) 695-6500.
7. Assorted product literature, available from Bry-Air, Inc., P.O. Box 269, Sunbury, OH 43074. Telephone: (614) 965-2974.
8. Assorted product literature, available from Miller-Picking Corporation, P.O. Box 130, Johnstown, PA 15907. Telephone: (814) 479-4023.
9. Assorted product literature, available from Niagra Blower Company, 673 Ontario Street, Buffalo, NY 14207. Telephone: (716) 875-2000.
10. *Gas Research Institute 1988–1992 Research and Development Plan and 1988 Research and Development Program*, Gas Research Institute, Chicago, IL, 1987.
11. *Solar Buildings Program Multiyear Plan—Draft*, Department of Energy, Washington, DC, 1987.
12. K. L. Bowlen et al., *Development and Field Test of a Desiccant-Based Space-Conditioning System for Supermarkets*, Gas Research Institute (NTIS Document PB89-167464), Chicago, IL, 1988.
13. "The Cold Facts About the Cost of Supermarket Humidity," Cargocaire Engineering Corporation, Amesbury, MA 1985.
14. D. R. Kosar, "The Application of Gas Fired Desiccant Dehumidification to Cooling Systems," *Proceedings of Association of Energy Engineers World Energy Engineering Congress*, Atlanta, GA, 1987.
15. "Energy Integrated Systems (a better way to air condition buildings)," Kathabar Systems of Somerset Technologies, Inc., New Brunswick, NJ, 1979.
16. D. Novosel, "Development of a Desiccant Based Environmental Control Unit," *Proceedings of ASHRAE IAQ 87: Practical Control of Indoor Air Problems*, Washington, DC, 1987.
17. *Ventilation for Acceptable Indoor Air Quality*, ASHRAE Standard 62-1981, ASHRAE, Atlanta, GA, 1981.

18. *Ventilation for Acceptable Indoor Air Quality*, ASHRAE Standard 62-1989, ASHRAE, Atlanta, GA, 1989.

19. J. Wax, "Tracking the Cause of Sick Buildings," *Heating, Air Conditioning and Refrigeration News*, Jan. 26, 1987, p. 6.

20. "Indoor Air Quality: A National Survey of Office Worker Attitudes," Honeywell, Inc., Minneapolis, MN, 1985.

21. P. G. Clark, "HVAC/R Industry: Present to the Year 2000," keynote luncheon address at HVAC & Buildings Systems Congress, Cleveland, OH, 1987.

22. Assorted product literature, available from Davison Chemicals of W. R. Grace & Company, P.O. Box 2117, Baltimore, MD 21203. Telephone: (301) 659-9000.

23. Assorted product literature, available from Linde Division of Union Carbide Corporation, P.O. Box 44, Tonowanda, NY 14151. Telephone: (716) 879-2000.

24. Assorted product literature, available from Aluminum Company of America, 1501 Alcoa Building, Pittsburgh, PA 15219. Telephone: (412) 553-4545.

39.11 BIBLIOGRAPHY

ASHRAE: *Brochure on Psychrometry*, American Society of Heating, Refrigerating and Air-Conditioning Engineers, Atlanta, GA, 1977.

———: *1987 ASHRAE Handbook, HVAC Systems and Applications*, American Society of Heating, Refrigerating and Air-Conditioning Engineers, Atlanta, GA, 1987, chaps. 28 and 51.

———: *1986 ASHRAE Handbook, Refrigeration*, American Society of Heating, Refrigerating and Air-Conditioning Engineers, Atlanta, GA, 1986, chap. 1.

———: *1983 ASHRAE Handbook, Equipment*, American Society of Heating, Refrigerating and Air-Conditioning Engineers, Atlanta, GA, 1983, chaps. 6 and 7.

Carrier Air Conditioning Company: *Handbook of Air Conditioning System Design*, McGraw-Hill, New York, 1965.

Holszauer, R.: "Industrial Dehumidification," *Plant Engineering*, July 12, 1979, pp. 86–93.

McNall, P. E.: "Indoor Air Quality," *ASHRAE Journal*, June 1986, pp. 39–42, 44, 46, 48.

Morris, R. H.: "Indoor Air Pollution: Airborne Viruses and Bacteria," *Heating/Piping/Air Conditioning*, February 1986, pp. 59–68.

"New IAQ Standard: A Preview," *Engineered Systems*, July–August 1986, p. 52.

Raffaelle, P.: "The Air Quality Issue: Will Standards Raise Costs?" *Energy User News*, March 1987, pp. 6–7.

Riggs, J. L.: *Engineering Economics*, McGraw-Hill, New York, 1982.

Schoening, N. C., and M. D. Clayton: *Electric and Gas Rates and Fuel Oil Prices in 18 Representative U.S. Cities, 1985*, Gas Research Institute, Chicago, IL, 1986.

Schoening, N. C., and J. M. Fay: *A Study of Commercial Electric Demand Charges for 28 U.S. Locations*, Gas Research Institute, Chicago, IL, 1986.

Stoecker, W. F., and J. W. Jones: *Refrigeration and Air Conditioning*, McGraw-Hill, New York, 1982.

CHAPTER 40
RECIPROCATING REFRIGERATION UNITS

Chan Madan
Continental Products Inc.,
Indianapolis, Indiana

40.1 RECIPROCATING COMPRESSORS

A "reciprocating compressor" is a single-acting piston machine driven directly by a pin and connecting rod from its crankshaft. It is a positive-displacement compressor in which an increase in the pressure of the refrigerant gas is achieved by reducing the volume of the compression chamber through work applied to the mechanism.

Various combinations of piston size (bore), length of piston travel (stroke), number of cylinders, and shaft speed make it possible to have various compressor sizes, ranging from 1/12 to 200 hp (0.06 to 149 kW).

The most commonly used reciprocating compressor is for refrigerants R-12 and R-22. For heating, ventilating, and air-conditioning (HVAC) and process cooling, the most practical refrigerant is R-22; in some instances, R-12 is utilized. Other halocarbon refrigerants used are R-502, R-114, and R-503; R-502 is used for low-temperature applications, usually below a 0°F (−17°C) suction temperature, and R-114 and R-503 are used for temperatures below −40°F (−40°C). Reciprocating compressors for ammonia (R-717) are used for refrigeration applications and are not practical for HVAC work.

40.1.1 Open-Type Compressor Units

Open-type compressors are designed to have the drive shaft extend outside the crankcase through a mechanical seal. This seal prevents outward leakage of refrigerant and oil and inward leakage of air and moisture. Figure 40.1 shows a cutaway of a typical open-drive compressor.

The drive shaft is adaptable to an electric-motor or gas-engine drive. Electric-motor drives are either belt-driven or directly coupled to the compressor by means of a flexible coupling. Gas-engine drives are usually directly coupled.

FIGURE 40.1 Open-drive compressor. (*Courtesy of Carrier Corporation.*)

40.1.2 Belt-Driven Compressor Units

Figure 40.2 shows a typical belt-driven unit. The drive consists of (1) a flywheel mounted on the compressor shaft and (2) a small pulley mounted on the motor shaft. These are interconnected by one or more V-belts. The size of the flywheel is usually fixed by the manufacturer, and the size of the motor pulley can be varied to achieve the desired speed. Speed variations are also obtained by ranging the size of the compressor flywheel and motor pulley.

The compressor is rigidly mounted on a steel base. The motor is mounted on an adjustable rail; this allows alignment of the motor pulley and tightening of the belts. Proper belt alignment and belt tension are most important for efficient compressor operation.

Belt Alignment. Belt alignment (Fig. 40.3) can be checked as follows:

1. Line up the compressor flywheel and motor pulley with a straightedge. Slide the motor pulley on the shaft to correct any parallel misalignment. For correct angular alignment, loosen the motor hold-down bolts and turn the motor frame as required.
2. When the alignment is completed, move the motor away from the compressor with the adjusting screws to tighten the belts. Tighten the belts just enough to prevent slippage.

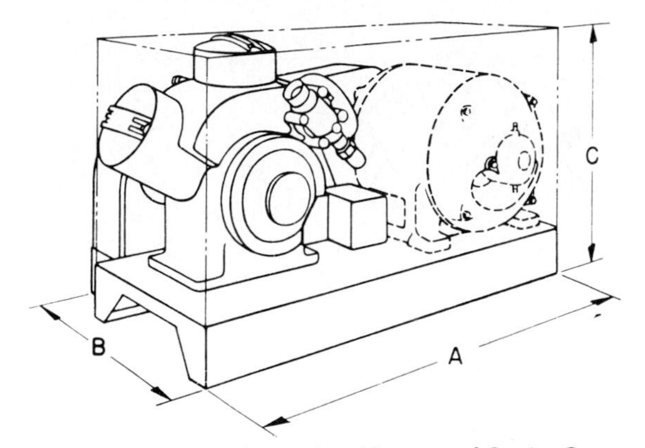

FIGURE 40.2 Belt-driven unit. (*Courtesy of Carrier Corporation.*)

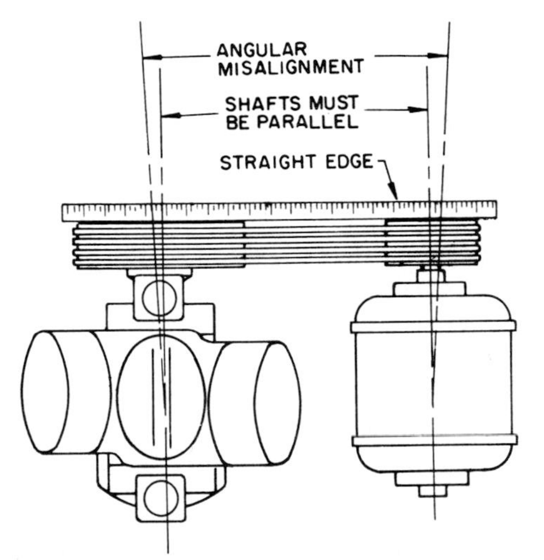

FIGURE 40.3 Correct belt alignment. (*Courtesy of Carrier Corporation.*)

Belt Tension. Belt tension can be checked by checking the amount of deflection as the belt is depressed at the center of the span. The rule of thumb is that belts deflect approximately 1 in for every 24-in span (1 cm for every 24-cm span). A longer span will deflect proportionately more.

40.1.3 Direct-Drive Compressor Units

Figure 40.4 shows a typical direct-drive unit. The compressor shaft is connected directly to an electric motor through a flexible coupling and is designed to run at motor speed; this speed is 1750 rpm for a 60-Hz power supply and 1450 rpm for a 50-Hz power supply. Two-speed or variable-speed motors are sometimes used

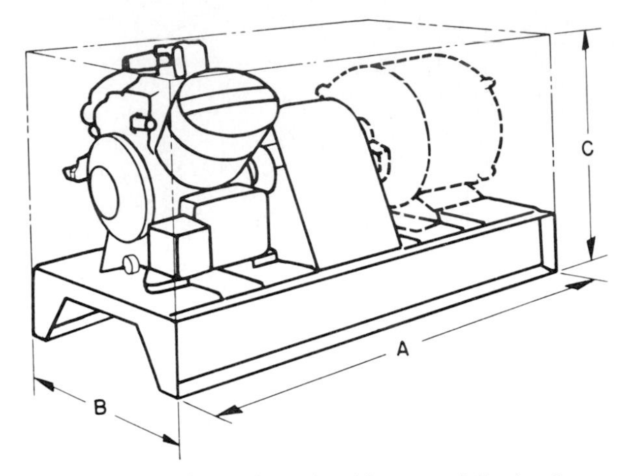

FIGURE 40.4 Direct-drive unit. *(Courtesy of Carrier Corporation.)*

for closer capacity control, but for HVAC applications this is cost-prohibitive and not commonly used.

The compressor and motor are rigidly mounted on a steel base. Proper coupling alignment is essential for trouble-free operation. The maximum permissible angular or parallel misalignment for all couplings is 0.010 in (0.25 mm). The manufacturer's recommendations are necessary for alignment. Basically, there are two alignment methods employed:

1. The dial-indicator method (Fig. 40.5)
2. The straightedge-and-caliper method (Fig. 40.6)

40.1.4 Hermetic Compressors

Hermetic compressors are also known as "sealed" or "welded" compressors or as "cans." They have the motor and compressor mounted within a common pressure vessel and sealed by welding. Figure 40.7 shows a cutaway of a typical hermetic compressor.

The compressor consists of pistons connecting rods, bearings, and a lubrication system and is driven by a crankshaft that is common to both the compressor and the motor. The motor-starter windings are of refrigerant-gas-cooled design. Most hermetic compressors are internally spring-mounted. Some have built-in suction accumulators for protection against liquid floodback. Lubrication of the compressor is usually achieved by a careful design that allows lubrication of the bearing and motor surfaces.

Hermetic compressors are commercially available in sizes from 1/12 to 24 hp (0.06 to 18 kW). Larger sizes are proprietary and are made by some manufacturers for use on their own packages.

Hermetic Motors and Motor Protection. Hermetic motors are specially designed by various manufacturers to keep the compressor losses to a minimum. This allows the compressors to operate effectively at maximum compression ratios.

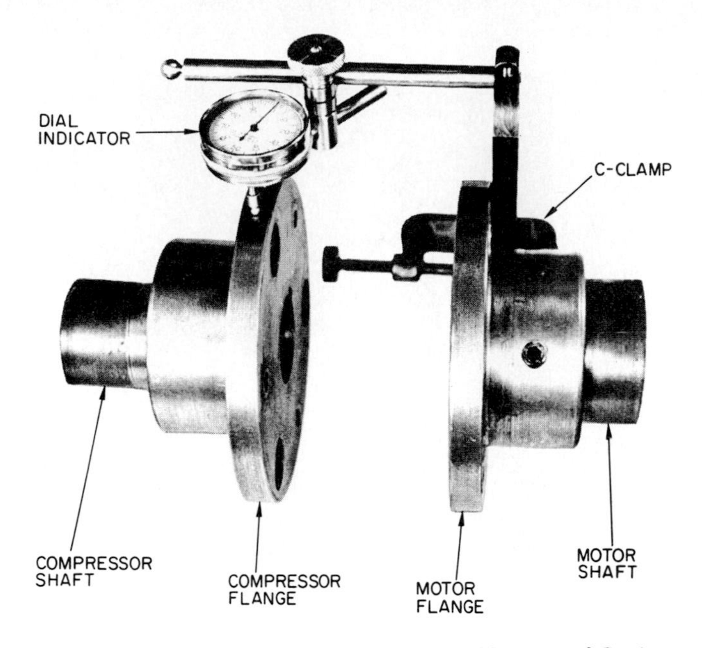

FIGURE 40.5 Alignment with a dial indicator. (*Courtesy of Carrier Corporation.*)

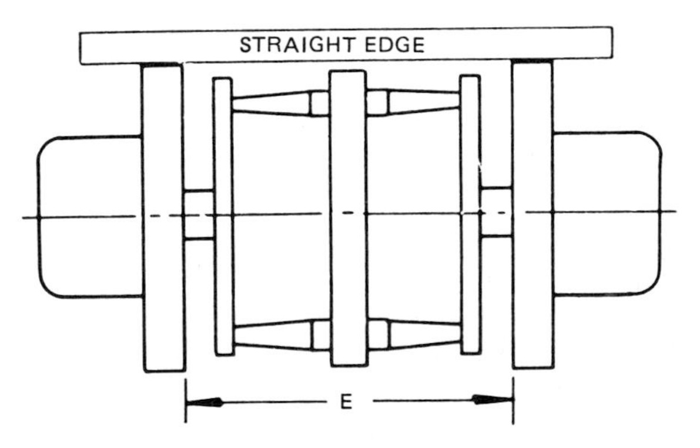

FIGURE 40.6 Alignment with a straightedge and calipers. (*Courtesy of Carrier Corporation.*)

Furthermore, these motors must have high dielectric strength, be resistant to abrasives, and be compatible with the refrigeration lubrication oil and refrigerant. Such other factors as insulation, efficiency, performance starting current, starting and breakdown torques, cost, and availability are also important.

Although the suction-gas-cooled design feature allows hermetic motors to be of considerably small sizes, it also poses a problem in protecting the motors from quick overheating or from drawing excessive current (amps). The most common method of motor protection is to have thermal overload devices embedded in the windings. These mechanisms trip when overloading, overheating, or any other abnormal condition occurs.

FIGURE 40.7 Hermetic compressor. (*Courtesy of Copeland Corporation.*)

40.1.5 Semihermetic Compressors

Semihermetic compressors are also known as "accessible" hermetic compressors. They have the motor and compressor mounted within a common pressure vessel and sealed by bolted plates so that the motor and compressor parts are accessible. In case of a compressor failure, these parts can easily be repaired or replaced, and the rebuilt compressor can be bolted back together. Except for this accessibility, all other features in the semihermetic design remain similar to the hermetic design: The compressor consists of pistons, connecting rods, and lubrication-system bearings and is driven by a crankshaft that is common to both the compressor and the motor.

Semihermetic compressors are available in the following designs for compressor-motor cooling:

1. *Air-cooled design.* This design uses air circulation for proper cooling of the compressor motor. A constant air flow is required across the compressor housing, and this is accomplished by direct impingement of air from fan discharge. Typically, air-cooled compressors are limited in size to ¼ to 3 hp (0.19 to 2.24 kW).
2. *Water-cooled design.* In this design, water coils are wrapped around the compressor housing; compressor-motor cooling is provided by circulating water. Typically, this design is limited to water-cooled condensing units and is not practical in today's HVAC market.
3. *Refrigerant-cooled design.* This design is the one most commonly used in HVAC applications. As in hermetic compressors, the motor is cooled by return suction gas. Refrigerant-cooled compressors are available in sizes from 2 to 100 hp (1.5 to 75 kW).

Semihermetic Motors and Motor Protection. Although the features of semihermetic motors are similar to those of hermetic motors, the motor protectors can be different for different manufacturers.

One manufacturer uses solid-state sensors, which are calibrated for proper motor protection. These sensors are mounted internally in the motor windings and are designed such that a change in temperature causes a change in electrical resistance through the sensor. This relationship between temperature and resistance remains stable, so the calibration of the protection system is made on the basis of a resistance reading.

The second component in this solid-state motor protector is the control module. It consists of a sealed enclosure that contains a triac, or a relay, a transformer, and several electronic components. Electrical leads from the motor's internally mounted sensors are connected to the module as shown in Fig. 40.8. As the motor temperature rises or falls, the resistance also rises or falls, and the change in resistance is transmitted from the sensors to the module, which triggers the control circuit at a predetermined setting of openings and closings. The module's voltage is determined by the voltage of the power source; 120- or 240-V modules are most commonly used.

This solid-state motor protector provides protection against high temperature due to motor overload, a locked rotor, or a loss of refrigerant charge. Another feature of the solid-state protector is the built-in time delay for a minimum of 2 min each time the power circuit is "opened." This time-delay feature also acts as a protection against short cycling.

Figure 40.9 shows a typical line-voltage circuit, which allows basic motor protection against high and low pressure and abnormal oil-pressure conditions.

Another method of compressor-motor protection is to use a compressor head sensor mounted in the discharge head of the compressor (Fig. 40.10).

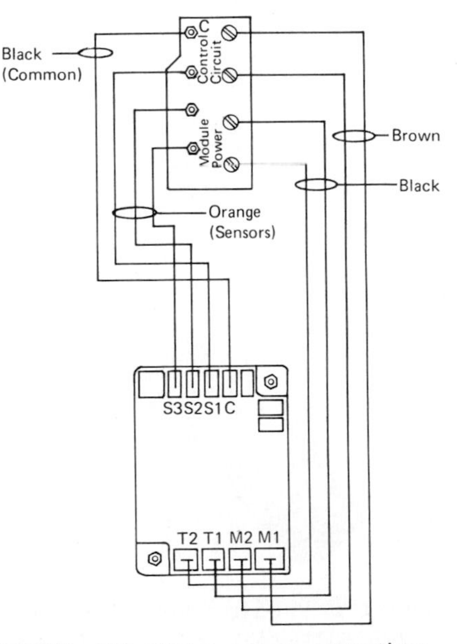

FIGURE 40.8 Motor sensor connections. (*Courtesy of Copeland Corporation.*)

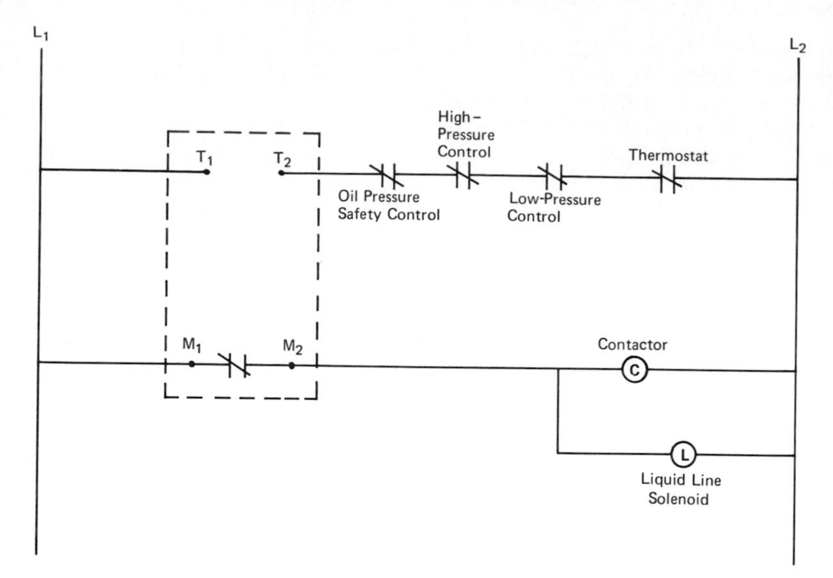

FIGURE 40.9 Line-voltage circuit. (*Courtesy of Copeland Corporation.*)

FIGURE 40.10 Compressor head sensor. (*Courtesy of Carrier Corporation.*)

Trouble-Shooting the Motor Protector and Motor Sensors. If the compressor motor is not operating, check the solid-state circuit as follows:

1. Allow at least an hour for the compressor motor to cool down and the motor protector to reset.
2. Connect a jumper wire across the control-circuit terminals on the terminal board in order to bypass the control circuit.
3. Try to start the compressor. If the compressor does not operate with the jumper wire installed, the problem is external to the solid-state motor protector. If the compressor operates with the jumper wire installed, the control module has an open circuit and needs to be replaced.

If the motor protector is still not operating, check the motor sensors. The following method is based on checking the resistance through the sensors. The sensors are very sensitive and can be easily damaged; therefore, no attempt should be made to check continuity through them.

1. Remove the electrical leads from the sensors and from the common terminals on the terminal board.
2. Apply a maximum of 6 V. Use an ohmmeter (maximum of 6 V) to read the resistance through each sensor terminal. If the resistance reading is between 500 Ω (cold) and 20,000 Ω (hot), this indicates that the sensors are good. A resistance approaching infinity would indicate an open circuit, and a reading of 0 would indicate a short. All three circuits can be checked as above. If all three resistance readings are between 500 Ω (cold) and 20,000 Ω (hot), but no one reading is higher than 10,000 Ω, then the sensors are probably not damaged.
3. If the sensors have proper resistance and the compressor will run with the control circuit in bypassed mode, but will not run if connected through the solid-state motor protector, then the protector is defective.

40.1.6 Methods of Capacity Control

Air-conditioning systems require some means of capacity control in order to meet the varying load conditions. A varying load can be caused by a simple weather change, by occupancy, or by a lighting-load change. To maintain human-comfort conditions under the above circumstances, it is important that some means of capacity control be employed. The simplest form of capacity control is on-off operation of the compressor; however, if this method is repeated too often (short cycling), it can lead to premature compressor failure.

Depending on the application, capacity control can be applied by these methods: (1) multiple compressors, (2) two-speed compressors, (3) cylinder unloading, or (4) hot-gas bypass.

Multiple-Compressors. This method falls under the category of simple on-off operation. However, systems using two to eight compressors have been applied, allowing two to eight steps of unloading. Two-compressor systems are available in a 4- to 200-hp (3- to 150-kW) range, which covers most reciprocating-compressor applications for air-conditioning. Multiple-compressor systems, especially two-compressor systems, allow redundancy. Most engineers thus prefer them over single-compressor systems.

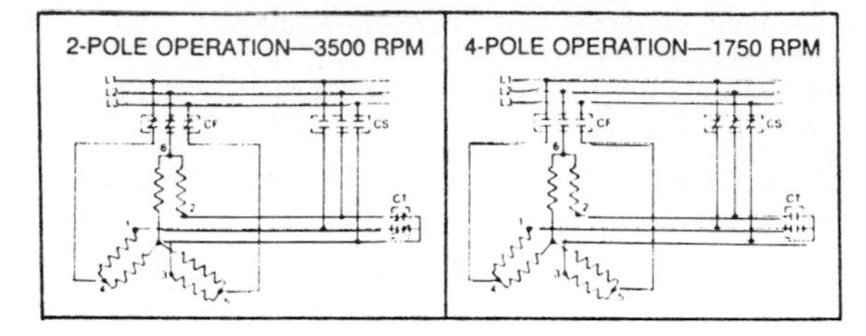

FIGURE 40.11 Two-speed compressor wiring. (*Courtesy of Bristol Compressors, Inc.*)

Two-Speed Compressors. Two-speed compressors are available in a 7.5- to 20-hp (3- to 15-kW) range. These compressors operate at either 3600- or 1800-rpm synchronous speeds and thus provide both 100 percent and 50 percent capacity control by virtue of the speed change. Figure 40.11 shows the wiring for a typical two-speed compressor.

Cylinder Unloading. Capacity control by cylinder unloading is commonly used for compressors ranging in size from 7.5 to 100 hp (5.6 to 75 kW). The unloading is achieved by preventing the suction gas from entering the unloaded cylinder or by bypassing the suction gas to the discharge chamber without compression. Various manufacturers achieve these results by employing any one of three means: (1) external capacity control valves, (2) internal capacity control valves, or (3) a suction-cutoff unloader.

1. *External capacity control valves.* External capacity control valves are usually used in high-temperature applications [maximum 25°F (−4°C)] and are not suitable for medium- or low-temperature work. A typical external capacity control operation is shown in Figs. 40.12 (normal operation) and 40.13 (unloaded position).

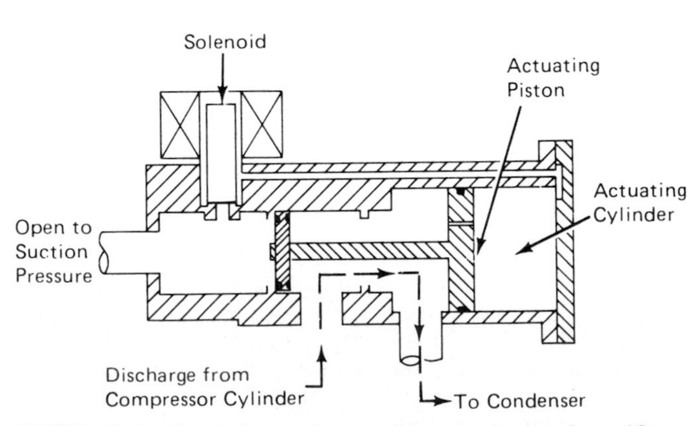

FIGURE 40.12 External capacity control—normal operation. (*Courtesy of Copeland Corporation.*)

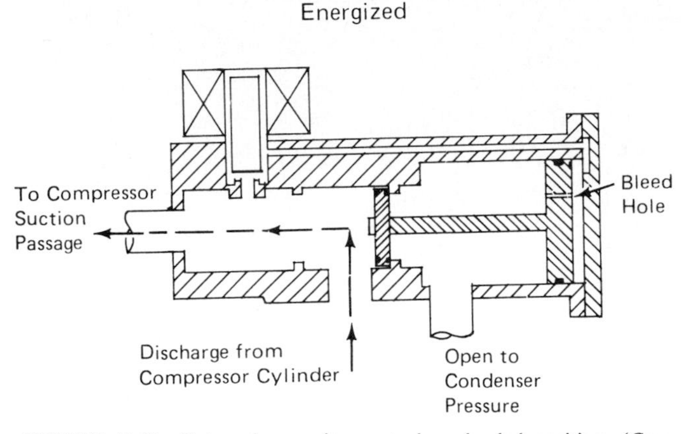

FIGURE 40.13 External capacity control—unloaded position. (*Courtesy of Copeland Corporation.*)

In this arrangement, the bypass valve connects the discharge port of the unloaded cylinder to the suction side of the compressor chamber. This allows the piston to pump suction vapor through the bypass circuit and not compress the gas. Because of this decreased volume of gas capacity, reduction is achieved.

The unloader assembly is controlled by a 120- or 240-V solenoid valve. The compressor head is divided by a web into suction and discharge chambers. One cylinder in the discharge chamber is isolated by an unloading head and gasket so that the unloading cylinder may discharge, either to the normal discharge side during the loaded cycle or to the suction side for the unloaded position.

In the normal discharge position, the compressor is fully loaded and the solenoid valve is de-energized. For reduced capacity, the solenoid valve is energized to move the actuating piston so that it allows the discharge gas to return to the suction chamber. The solenoid valve may be energized by a two-stage temperature control thermostat or by a reverse-acting low-pressure switch. To avoid rapid cycling (loading and unloading), it is necessary to have a wide differential, whether it is a temperature- or a pressure-operated control.

When the compressor operates in the unloaded position, a new refrigeration-system balance starts taking place. The suction pressure begins to rise, and the condensing pressure begins to drop. This procedure continues until a new system balance is reached.

External capacity control allows reduced power consumption because of the fact that less work is being done by the compressor during its operation in the unloaded position.

2. *Internal capacity control valves.* The internal capacity control consists of a solenoid-valve unloader plunger and a capacity-control-valve plate. A typical internal capacity control is shown in Fig. 40.14.

In the normal *loaded* position, the solenoid valve is de-energized, the needle valve is on the lower port, and the unloading-plunger chamber is exposed to the suction pressure through the suction port. Since the face of the plunger is open to the suction chamber, the gas pressure across the plunger is equalized and the plunger is held in the open position by the spring.

During the *unloaded* position, the solenoid valve is on the upper port and the unloading-plunger chamber is exposed to the discharge pressure through the dis-

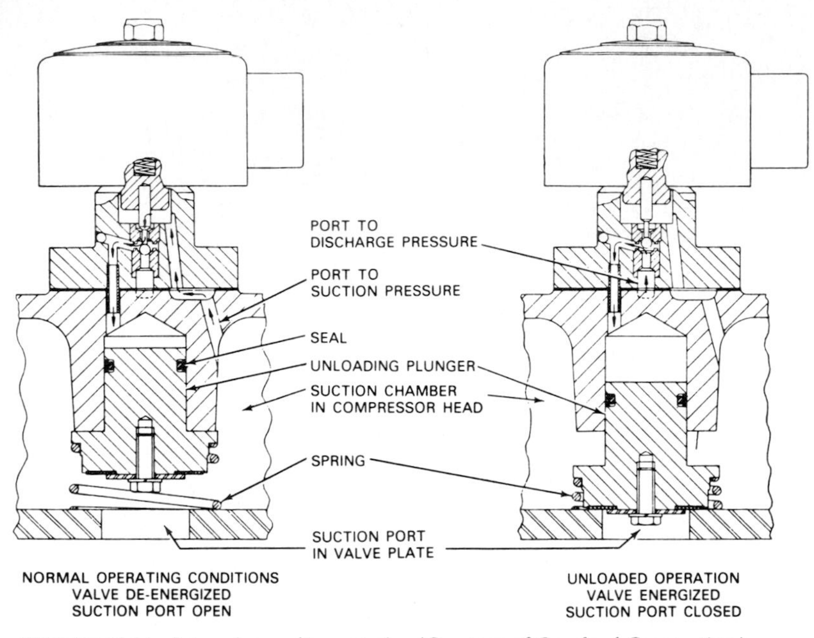

FIGURE 40.14 Internal capacity control. (*Courtesy of Copeland Corporation.*)

charge port. The differential between the discharge and suction pressures forces the plunger down, sealing the suction port in the valve plate and thus preventing the suction vapor from entering the unloading cylinder. Figure 40.15 shows the pressure-port connections to the unloading solenoid valve; the action of the valve then determines which pressure is applied to the unloading-valve plunger.

A newer discus-type compressor is shown in Fig. 40.16, and Fig. 40.17 shows a discus valve. The same basic unloading mechanism is used. The solenoid valve can be energized to unload either by temperature or pressure control. Multiple-step controllers are required for four-, six-, and eight-cylinder compressors. Again, to avoid rapid cycling, it is important to use wide differential controls.

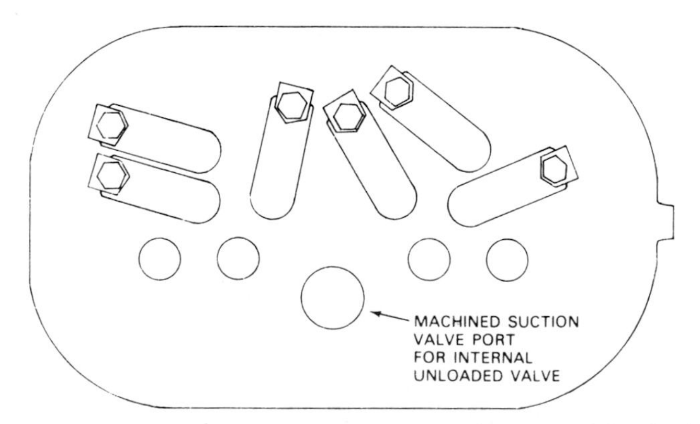

FIGURE 40.15 Pressure-port connections. (*Courtesy of Copeland Corporation.*)

FIGURE 40.16 Discus-type compressor. *(Courtesy of Copeland Corporation.)*

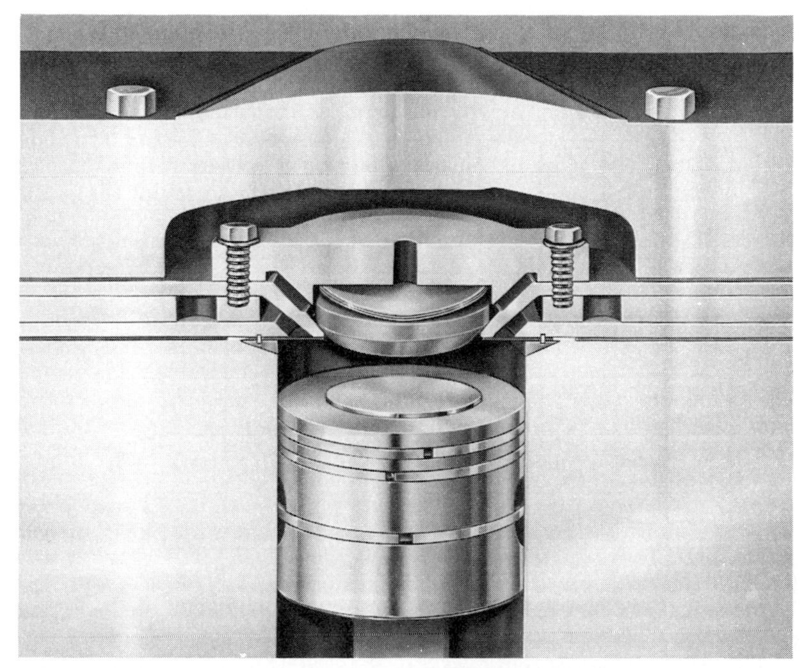

FIGURE 40.17 Discus valve. *(Courtesy of Copeland Corporation.)*

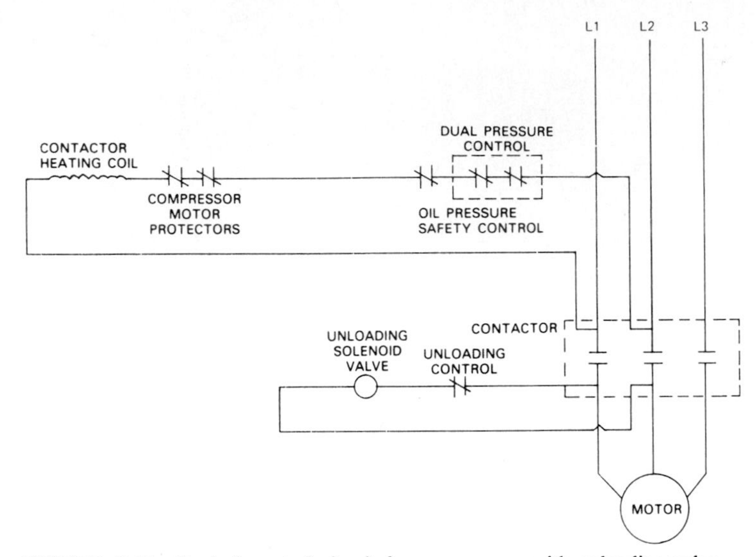

FIGURE 40.18 Typical control circuit for compressor with unloading valve—line-voltage source. (*Courtesy of Copeland Corporation.*)

A typical control circuit operating with line voltage is shown in Fig. 40.18, and Fig. 40.19 shows a control circuit operating with a separate power source or through a control transformer.

3. *Suction-cutoff unloader.* One manufacturer uses a suction-cutoff capacity control system, operated either electrically by a solenoid valve (Fig. 40.20) or

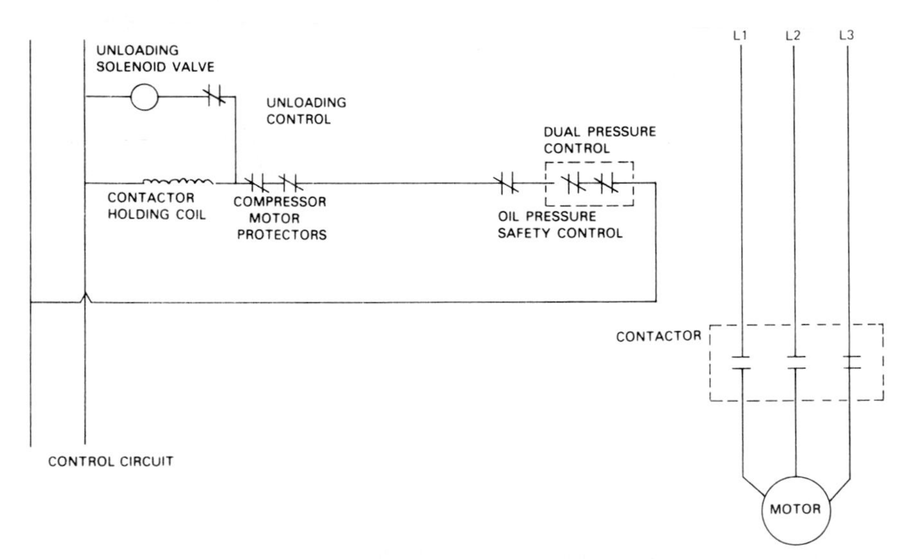

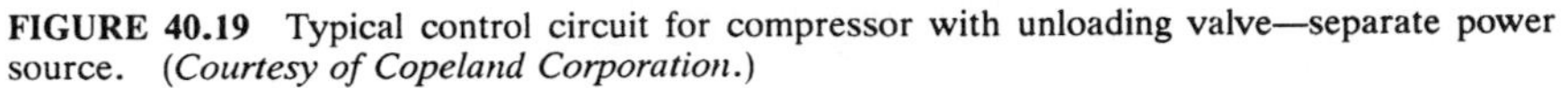

FIGURE 40.19 Typical control circuit for compressor with unloading valve—separate power source. (*Courtesy of Copeland Corporation.*)

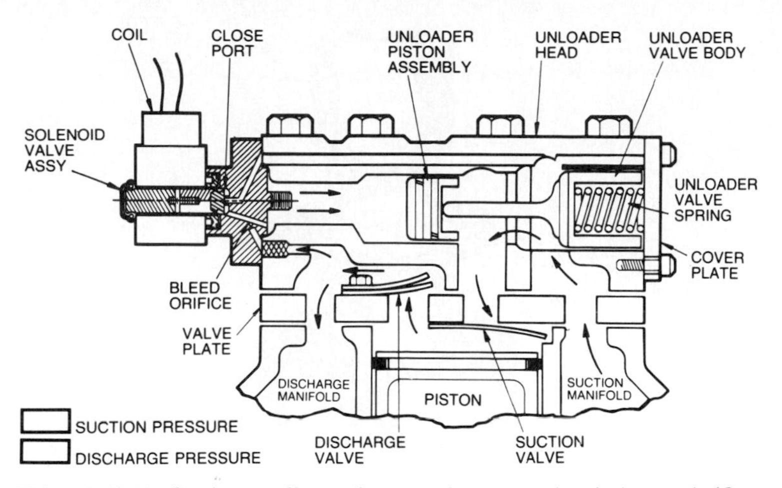

FIGURE 40.20 Suction-cutoff capacity control system—electrical control. (*Courtesy of Carrier Corporation.*)

mechanically by a pressure control valve (Fig. 40.21). Figure 40.22 shows the pressure-actuated capacity control valve in detail.

Hot-Gas Bypass. Compressor-capacity modulation by the use of hot-gas bypass is utilized whenever on-off controls are unacceptable or whenever capacity unloaders are not able to go down in capacity to match the load. Basically, hot-gas bypass operates on the principle of bypassing a part of the compressed gas in order to prevent compressor suction pressure from falling below a predetermined setting. A typical hot-gas bypass-valve hookup is shown in Fig. 40.23.

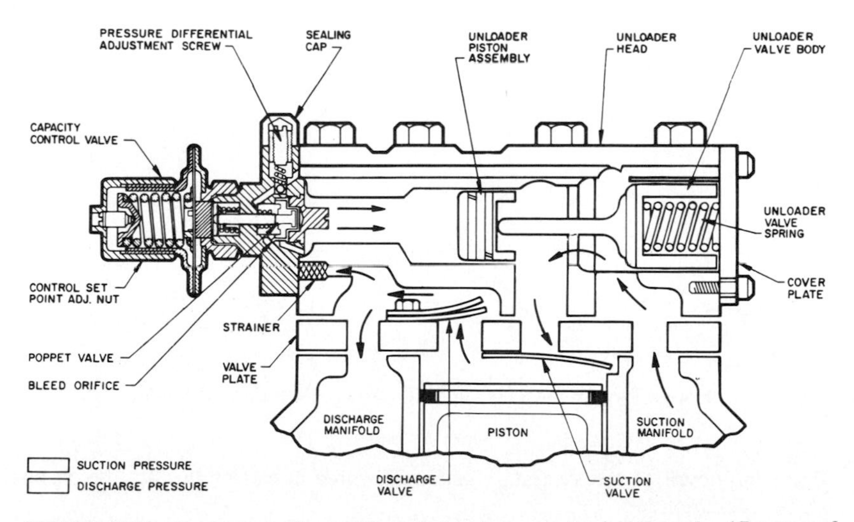

FIGURE 40.21 Suction-cutoff capacity control system—mechanical control. (*Courtesy of Carrier Corporation.*)

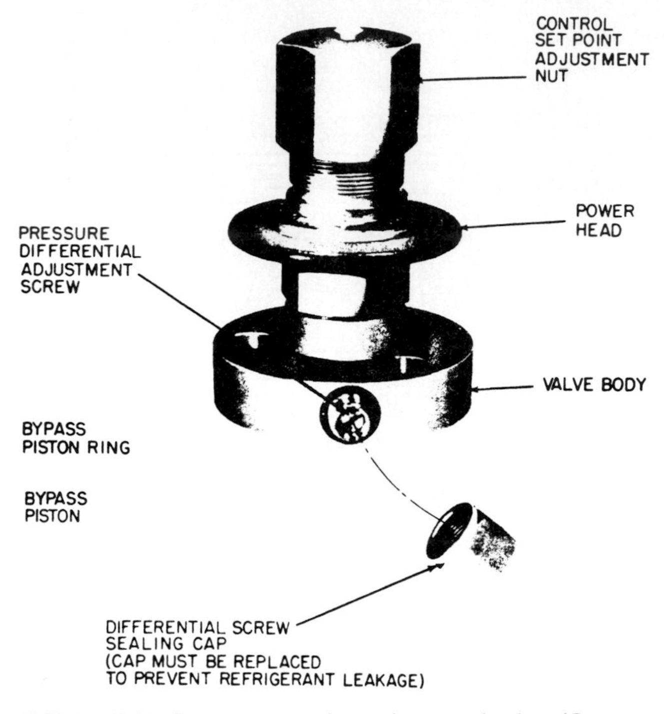

FIGURE 40.22 Pressure-actuated capacity control valve. (*Courtesy of Carrier Corporation.*)

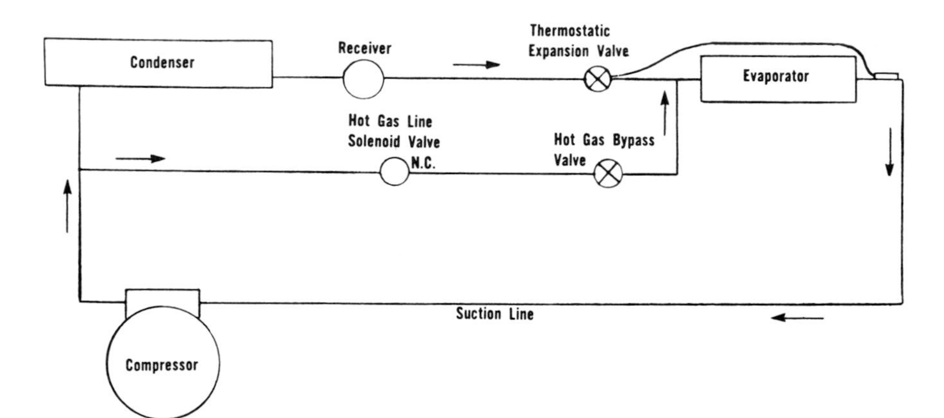

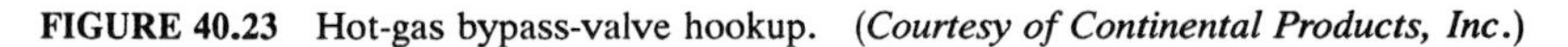

FIGURE 40.23 Hot-gas bypass-valve hookup. (*Courtesy of Continental Products, Inc.*)

For pumpdown of the system, a solenoid valve ahead of the hot-gas bypass valve is recommended.

Although bypassed hot gas can be injected directly into the suction of the compressor, this is not recommended. Instead, it should be injected after the expansion valve or at entry to the evaporator. This allows the hot gas to mix with the liquid-and-gas mixture from the expansion valve, thereby returning cool gas to the compressor. This method also allows the superheat setting to be maintained.

40.2 RECIPROCATING LIQUID CHILLER SYSTEMS

A liquid chiller system cools water, glycol, brine, alcohol, acids, chemicals, or other fluids. The most common use of a chiller system is as a water chiller for human-comfort cooling application. The chilled water generated by the chiller system is circulated through the cooling coil of a fan coil (or air-handling unit), as shown in Fig. 40.24.

The fan coil circulates air within the conditioned space. Air from the room moves over the chilled-water cooling coil of the fan coil, gets cooled and dehumidified, and returns back to the room. In this process the chilled water in the cooling coil picks up the heat and is returned back to the chiller system for cooling. As the cycle is repeated, the chiller system maintains the conditioned space at the comfort level.

Field-Assembled Liquid Chillers. Originally, liquid chillers were "field-assembled," with the components "field-matched" to develop a field-erected system. These systems were custom-built to perform a specific application. As a result, the design depended on the application, the availability of parts, the labor, and the field engineer. Some systems were well thought out, were carried out with detail and care, and performed very well for the particular application. Others were ill-conceived, had a poor choice of components, and resulted in a bad experience for the owners.

Field-assembled systems could not be pretested to check if they would per-

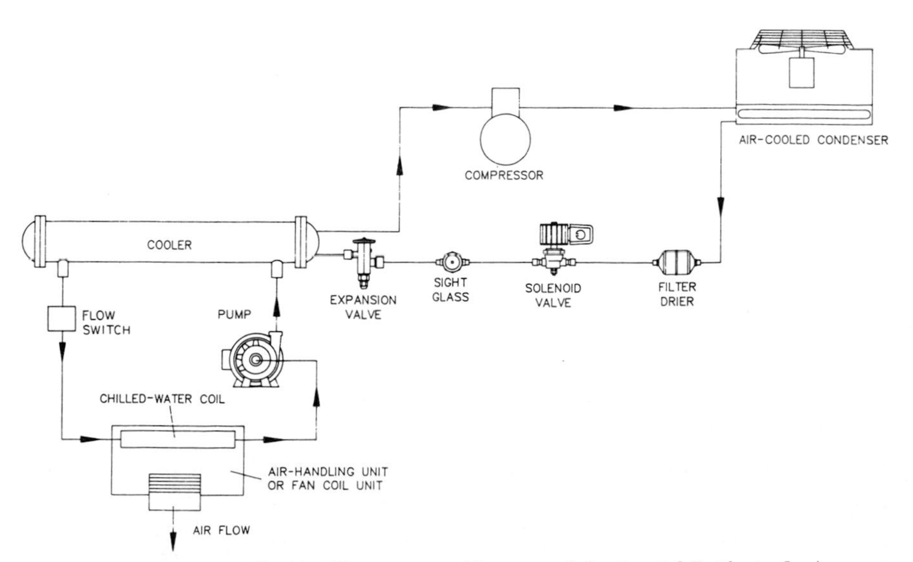

FIGURE 40.24 Typical liquid chiller system. (*Courtesy of Continental Products, Inc.*)

form properly. They depended entirely on the design concepts, the availability of matched parts, and the field experience of the labor force.

As the cost of field labor became prohibitive and as owners had poor experiences from field-erected systems, the concept of "factory packaging" became popular.

Factory-Packaged Liquid Chillers. The idea of a completely pre-engineered system is to assemble all the components on a common steel skid and to pipe, wire, pressure-test, evacuate, and charge the system with refrigerant (usually R-22). In this manner, all the system's components are preselected, heat-transfer balanced with each other, prepiped, prewired, and factory run-tested before actually being installed on the job. Factory-packaged systems, if manufactured to good engineering standards and correctly capacity-rated, are very cost-effective. They result in years of good service to the owners.

There have been many improvements from early factory packages to today's systems. Today it is reasonable to expect a reliable, fully factory-tested packaged liquid chiller system that has the following: compressor(s); condenser(s); a liquid chiller; refrigeration specialties, such as expansion valve(s), filter dryer(s), sight glass(es), and solenoid valve(s); and electrical components with power and safety controls. Chiller and condenser pumps (for water-cooled systems) and a factory-mounted and -wired flow switch are also available. Most manufacturers run-test the system before giving it its final, preshipment paint finish or cleanup. Fieldpiping of water and additional electrical components are all that are needed before the system is ready for startup.

40.2.1 Packaged Reciprocating Liquid Chiller Systems

Packaged reciprocating chillers are available with the following three types of reciprocating compressor:

1. Welded hermetic chiller available up to 50 hp (37.3 kW), with a 2- to 25-hp (1.5- to 18.6-kW) compressor
2. Semihermetic chiller available up to 400 hp (298 kW), with a 3- to 100-hp (2.24- to 75-kW) compressor
3. Open-drive chiller available up to 400 hp (298 kW), with a 2- to 150-hp (1.5- to 112-kW) compressor

Compressors can be enclosed within an acoustical enclosure to provide a dampening of sound levels.

For packaged chillers, the condensers chosen can be air-cooled (Fig. 40.25), water-cooled (Fig. 40.26), or evaporative-cooled (Fig. 40.27). Chillers with air-cooled condensers are the most commonly used because they require the least maintenance; however, the operating cost is higher than for water-cooled or evaporative-cooled systems. Evaporative-cooled packaged chillers are the most efficient and require some maintenance.

Major Components. A typical reciprocating liquid chiller system essentially consists of four components:

1. Compressor(s)—hermetic, semihermetic, or open-drive
2. Condenser(s)—air-cooled, water-cooled, or evaporative-cooled
3. Expansion valve(s)
4. Evaporator cooler(s)—direct-expansion or flooded type

FIGURE 40.25 Air-cooled chiller. *(Courtesy of Continental Products, Inc.)*

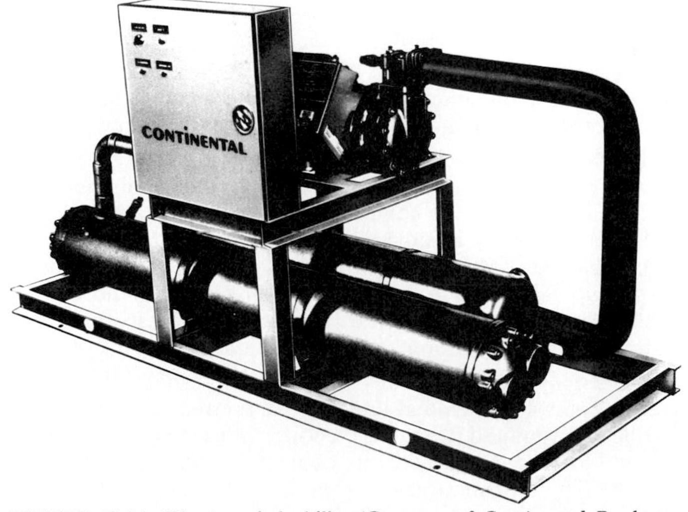

FIGURE 40.26 Water-cooled chiller. *(Courtesy of Continental Products, Inc.)*

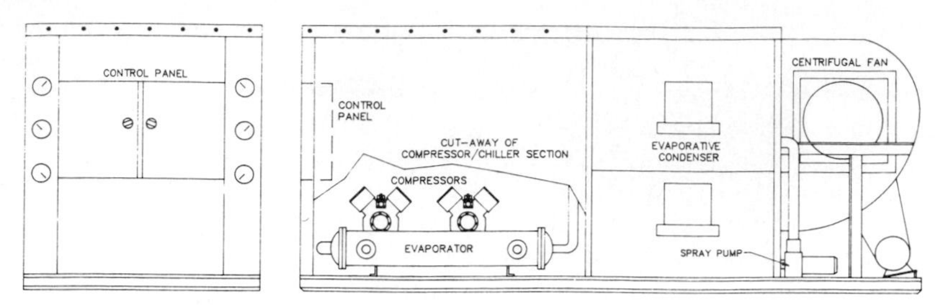

FIGURE 40.27 Evaporative-cooled chiller. (*Courtesy of Continental Products, Inc.*)

The other essential parts of the total system are the refrigerant charging valve, filter dryer, liquid solenoid valve, sight glass and moisture indicator, expansion valve, and electrical control center.

Electrical Control Center. In order to operate and manage the system, the electrical control center is an essential part of the total system. The control center includes power, operating, and safety controls, usually mounted in a common control panel.

The power controls are separated from the operating and safety controls by a divider plate or by other means. The power controls include a starting contactor (in the case of hermetic and semihermetic compressors, which have internally protected compressor motors) or NEMA-rated starters with overload protection (in the case of open-drive compressors, which use NEMA-rated electric motors).

The operating control consists of a chiller thermostat, which senses the incoming water to the cooler.

The safety controls consist of the following:

1. *High-condenser-pressure switch.* It opens if the compressor discharge pressure reaches a preset value. This control is usually of a manual-reset type; that is, the operator will have to reset the control manually to restart the system. Manual reset control is used to give the operator a chance to find out the cause of high condenser pressure.
2. *Low-refrigerant-pressure switch.* It opens when the evaporator cooler's pressure reaches a minimum safe limit preset by the manufacturer. This switch may be of a manual- or automatic-reset type; since it is also used for the pumpdown cycle, it is usually of an automatic-reset type.
3. *Oil-pressure control.* This switch, which is usually of a manual-reset type, is provided to shut down the compressor when the compressor's lubrication oil pressure drops below a minimum safe value determined by the compressor manufacturer or when sufficient oil pressure is not developed after compressor startup. This switch is used only on semihermetic and open-drive compressors, not on hermetic types. (Hermetic compressors usually have no means to sense the oil pressure, but rely on internal means of lubrication.)
4. *Freeze-protection switch.* This switch (the so-called "low chilled-liquid temperature control") operates similarly to a thermostat, sensing the temperature of the chilled liquid leaving the cooler. In case of a freeze-up condition and of liquid leaving the cooler, this controller opens the circuit to stop the total system. The minimum value is preset at the factory to prevent cooler freeze-up in case other safety controls malfunction.
5. *Low-pressure freezestat.* This control is usually of a manual-reset type and

has a 60-s built-in time delay. It senses the evaporator cooler's pressure, and if this pressure continues to drop below the preset value for a period of 60 s, the switch opens to shut down the total system. This acts as a protection against freeze-up as well as against a loss of refrigerant.

6. *Flow switch.* This control, which can be either factory-mounted or field-furnished, is needed in the chilled-water piping to protect against a cooler freeze-up in the event of no liquid flow through the system. This device is of an interlock type and is an essential protection against pump failure or other malfunctioning of the total system.
7. *Motor overload protection.* This is provided with hermetic and semi-hermetic compressors as a built-in feature, whereas for open-drive compressors the overload heaters are sized to protect the motor.
8. *Power-factor corrector capacitors.* These are for improved motor performance, reduced line losses, and lower utility costs.
9. *Indicator lights.* Various indicator lights show the system's operation and, in case of a system failure, provide diagnostics for ease of service.
10. *Pressure gauges.* High- and low-pressure and oil-pressure gauges are factory-installed and piped to the compressor.
11. *Compressor cycle meter, ammeter, and unit disconnect switches.*

Several other safety controls are available, as follows:

1. *Lock-out timer.* This control prevents the compressor from short-cycling on power interruptions to safety controls.
2. *Phase failure.* This control relay monitors the sequential loss and reverse of a three-phase power supply.
3. *Alarm-bell contacts.* These allow alarm-bell connections to the high-pressure control or to other safety controls for signaling if the unit fails on manual-reset safety controls.
4. *Low ambient controls.* These controls are specially used for air-cooled condenser chillers or evaporative-cooled chillers. The low ambient controls may consist of fan-cycling pressure controls or fan speed controls. The fan speed control usually has a solid-state controller and a single-phase condenser fan motor (which modulates the fan speed in response to condenser pressure). The low ambient controls allow operation of the chiller system on days of low ambient temperature.
5. *Relief valve.* The pressure-relief valve is set at a pressure above the high-pressure cutout to relieve the system before the system reaches its maximum design working pressure. These valves should be piped and vented outside. Fusible plugs or rupture disks can be used in some instances.
6. *High-motor-temperature protection.* This control consists of a high-temperature thermostat or thermistor. It is located in the discharge gas stream of the compressor.
7. *High oil temperature.* This controller protects the compressor when there is a loss of oil cooling or when a bearing failure results in excessive heat generation.

Other Components. Factory-mounted, -piped, and -wired pumps for chilled water and condenser water are becoming available as a part of the packaged chiller system. This has eliminated the need for field labor for plumbing, wiring, and interlocking the pumps with the chiller control panels. Factory-mounted pumps are checked for pump rotation, which is phased in with an air-cooled condenser motor, to ensure that the condenser fans operate vertically up and that the pumps operate in the correct direction.

Other accessories, such as filters, air eliminators, and storage tanks, can also be factory-mounted and -piped, eliminating the need for separate mechanical areas and the chance of incompatible field components.

40.2.2 Typical Refrigeration Cycle

In a typical chiller system (Fig. 40.24), as the water or other liquid flows through the system, the flow-switch contact is made, and if the thermostat calls for cooling and all safety devices are closed, the compressor will start. The hot gas from the compressor is discharged into the condenser (air-cooled, water-cooled, or evaporative-cooled). As it travels through the condenser, this high-pressure refrigerant cools and changes its phase to high-pressure liquid. In the case of an air-cooled condenser, the condenser rejects the heat to the air; in the case of a water-cooled or evaporative-cooled condenser, the condenser rejects the heat to the water. The high-pressure liquid refrigerant now goes through a filter dryer. Then it goes through the liquid solenoid valve (which should be open now), sight glass, and moisture indicator and into the expansion valve. The expansion valve meters the liquid refrigerant through the evaporator cooler. The cooler allows the water (or other liquid) to be cooled by the action of the evaporating liquid refrigerant. The refrigerant picks up the heat from the flowing liquid, is returned back to the suction side of the compressor as a low-pressure gas, and is then ready to be recycled again through the reciprocating compressor.

40.2.3 Evaporator Coolers

The evaporator cooler is the part of the packaged chiller system in which the refrigerant is vaporized from a liquid-and-gas mixture to an all-gaseous form, thereby producing the cooling effect on the water or other liquid that is being cooled.

Direct-Expansion Coolers. In a typical packaged chiller system, the most common evaporator cooler is of the direct-expansion type, in which the refrigerant is fed through a thermostatically controlled expansion valve. This valve meters the refrigerant in accordance with the amount of superheat in the refrigerant vapor leaving the evaporator cooler.

A direct-expansion cooler is generally made of shell-and-tube construction; the refrigerant flows through the inside of the tubes, and the liquid to be cooled flows from the outside of the tube on the shell side. A baffle arrangement is provided from the shell side to keep optimum liquid velocity across the tubes, thereby increasing the heat transfer. A proper design both of the refrigerant circuit and of the manner of feeding the refrigerant will establish proper refrigerant velocities through the tubes; these velocities must be kept to a reasonable level to ensure refrigerant oil return to the compressor with excessive refrigerant pressure drop through the tubes. Usually the liquid refrigerant is fed through the bottom of the shell (in a typical two-pass arrangement) and returned from the top. The water or other liquid to be cooled is circulated outside the tubes with the help of the baffles, which direct the flow over the tubes several times, allowing maximum heat transfer between the refrigerant and the fluid and thereby cooling the liquid as it leaves the chiller evaporator. In a typical direct-expansion cooler, mild steel shell and copper tubes are utilized with steel baffles. The refrigerant-side connections are made of copper, and the fluid-side connections are made of steel. For optimum heat transfer, the copper tube utilized has some means of creating turbulence inside the tubes; this can take the form of aluminum or of another insert inside the tubes, or the inside of the copper tube, itself, can have an enhanced surface that provides the turbulence to the refrigerant as it flows through the tubes.

The other essential parts of this cooler are: a drain plug in the shell side to allow the fluid to be drained out of the heat exchanger, an air vent on top of the shell that allows air elimination on the shell side, a means of locating the thermostat valve on the water-inlet side, and a means of locating the freezestat bulb on the water-outlet side. Couplings may be provided on both the water-inlet and water-outlet sides to measure the pressure drop through the cooler and/or to sense the temperature on the inlet and outlet of the cooler. On the refrigeration side, the expansion valve is mounted on the refrigerant inlet side, and a connection [usually a ¼-in (8-mm) fitting] is provided on the refrigerant-outlet connection. The thermostatic expansion-valve sensing bulb is mounted on the suction side, and an equalizer connection [connected to this ¼-in (8-mm) fitting] is provided to sense the vapor pressure of the refrigerant as it leaves the cooler. This also provides a means of measuring the superheat of the system. Most manufacturers recommend a superheat of 7 to 15°F (4 to 9°C) to make sure that the refrigerant, as it returns to the compressor, is in a superheated form and is not carrying any liquid refrigerant back to the compressor.

Another type of direct-expansion evaporator cooler is made of tube-and-tube construction. In this type of cooler, the refrigerant flows through the outer shell of the tube, and the liquid to be cooled flows through the inside of the tube. This type of chiller is used only for small (1- to 3-ton capacity) systems. Larger systems using tube-and-tube heat exchangers have been constructed by manifolding two or more tube-and-tube heat exchangers, but this is not recommended.

Flooded Coolers. In a typical flooded cooler, the refrigerant is vaporized on the outside of the tubes, and the water or other liquid to be cooled is circulated through the tubes, or enclosed in a shell. The operation is opposite to that of the direct-expansion-type cooler. In the flooded operation, the shell can still be made of steel, and the tubes are typically made of copper. In a typical R-22 or R-12 system, part of the shell contains the tubes; the top part of the shell, however, is left open so that the leaving suction gas does not carry liquid refrigerant to the compressor. The tubes are completely submerged in the liquid refrigerant; thus, as the water or other liquid flows through the tubes, it gets cooled, and the liquid refrigerant is boiled on the outside of the tubes. The cooler allows the water (or other liquid) to be cooled by the action of the evaporating liquid refrigerant, and now that this refrigerant has picked up the heat from the flowing liquid, it is returned back to the suction side of the compressor as a low-pressure gas, ready to be recycled again through the reciprocating compressor. Some manufacturers provide eliminators above the copper tubes to make sure that no liquid droplets leave the cooler. The heat transfer depends on how well the refrigerant is distributed across the tube bundle. To maintain the proper liquid-refrigerant level inside the shell, a level control is utilized that allows the refrigerant to be metered inside the shell in accordance with the superheat of the system. Other means to separate the liquid refrigerant from the vaporized gas can be utilized; one is to use an outside suction trap to prevent the return of excessive liquid refrigerant to the compressor.

Flooded coolers are not commonly used with reciprocating compressor systems, but they may be used with centrifugal or screw compressor packages.

40.2.4 Condensers

Condensers are heat exchangers designed to condense the high-pressure, high-temperature refrigerant discharged by the compressor. In this process, the condensers reject the heat that was picked up by the evaporator cooler or chiller. At the same time, the condensers convert the high-pressure, high-temperature gas into high-pressure, high-temperature liquid refrigerant, ready to be recycled through the expansion valve, evaporator, and back to the compressor.

There are three types of condensers: air-cooled, water-cooled, and evaporative-cooled. They are discussed in Secs. 40.2.5, 40.2.6, and 40.2.7, respectively.

Condenser-Coil Circuiting. For multiple compressors, multiple-circuit condensers are used. For example, a two-, three-, or four-compressor unit has a two-, three-, or four-circuit condenser, respectively. In this manner, each circuit is independent of the other, thus providing redundancy in case of failure. Independent refrigerant circuits also ensure that refrigerant and lubrication oil for each section of the compressor system are not mixed. Typical two- and four-circuit condenser headers are shown in Fig. 40.28.

Parallel piping of two, three, or four compressors into a common hot-gas inlet connection, with a common liquid outlet connection, is also being utilized.

Compressors are cycled on and off for capacity reduction, but the condenser-coil surface remains the same. For example, if the condenser coil is sized for three compressors and one out of three compressors is shut down, the refrigerant gas from the other two compressors will continue to pump into the same coil. Now this coil will be oversized for the amount of hot gas being pumped into it. Although the condenser fans will also be cycled to reduce the air flow, the net effect is still a larger condenser surface, resulting in subcooled liquid. A more efficient system is to feed the subcooled liquid through the expansion valve.

For parallel piping, the following need and drawback should be considered:

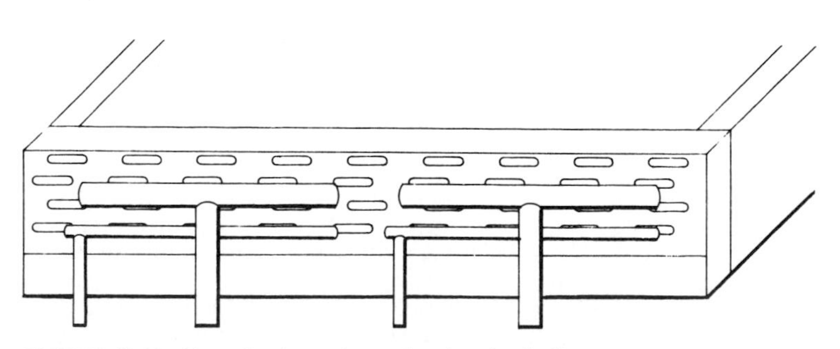

FIGURE 40.28 Two-circuit condenser headers (typical).

1. It is essential that there be some means to interconnect multiple compressors in order to maintain lubrication oil in each compressor; otherwise, one compressor may be starved of lubrication and thus become damaged.
2. Since multiple compressors have a common refrigerant circuit, even if one (hermetic or semihermetic) compressor fails or "burns out," it contaminates the complete system. To replace one compressor, the complete system has to be "cleaned," evacuated before refrigerant is charged, and put back into service. Condensers with independent refrigerant circuits do not have this problem.

Condenser Components

1. *Fans.* Condenser fans are of a propeller type and are statically and dynamically balanced for low-vibration operation. Propeller fans are made of aluminum, galvanized steel, stainless steel, or plastic materials and range in diameter from 18 to 30 in (46 to 76 cm). Direct-driven fans are mounted on the fan motor shaft, and belt-driven fans have belt-and-pulley combinations.

For belt-driven condensers, larger-diameter fans are utilized and have the advantage of fewer fans, compared to several direct-driven smaller fans. Lower speeds (400 to 700 rpm) are achieved by belt drives. The motors are typically standard open drip-proof or totally enclosed NEMA-rated, four-pole, 1800 rpm. These are readily available.

Belt-driven chiller packages are more suitable for high to medium ambient conditions where an on-off cycling of fans is not needed. Other means of low ambient control are utilized for medium to low ambient control conditions.

If belt drives are used and the fans cycle often, the belt tension needs to be checked more often than usual. Typically, an access door is provided to access the motor bearings and belts. For additional convenience, extended lubrication lines are installed with external grease fittings.

2. *Motors.* Typical direct-driven fan motors are six-pole and operate at 1100-rpm speeds. These motors range in size from ⅓ to 2 hp (0.25 to 1.5 kW) and have been specially designed for air-cooled condenser applications by various motor manufacturers. A typical condenser fan motor is of a 56-frame "totally enclosed air over" design with a built-in overload protector.

3. *Motor speed control for low ambient operation.* For multiple fans on a medium- to large-size chiller package, the fan motor can be cycled on and off by sensing condenser pressure or ambient temperature. This is adequate for medium ambient temperature operation. For lower ambients or a single-fan chiller package [3 to 9 hp (2.24 to 6.7 kW)], another choice is to modulate the fan motor speed. Typical fan-speed controllers sense condenser pressure or liquid temperature and modulate the motor in response to a rise or fall of pressure or temperature. At higher pressure or temperature, fans operate at higher speed; at lower pressure or temperature, fans modulate at lower speeds. All condenser fan motors are not suitable for fan speed control. Typically, a ball-bearing-type motor is needed to allow operation at lower speeds. Some single-phase motors, specially designed for speed control, have proved successful. Three-phase motors and controllers are being developed.

For a three-phase packaged chiller with multiple fans requiring fan speed control, a combination of three-phase and single-phase motors is used. For 208–230/3/60 power, a single-phase motor presents no problem, since the motor

FIGURE 40.29 Typical fan venturi. (*Courtesy of Continental Products, Inc.*)

can be wired to two of the three power legs. For 460/3/60 power, single-phase motors require a step-down transformer or a separate single-phase power source.

4. *Fan venturi.* Its design is critical for optimum air flow with minimum air losses as well as for low outlet noise. See Fig. 40.29.

5. *Fan guards.* Fan guards are mounted around a fan venturi and are designed to meet the standards of the Occupational Safety and Health Administration (OSHA) so as to protect against accidents as well as to allow free air circulation.

40.2.5 Air-Cooled Condensers

Air is used as the medium to cool and condense the hot refrigerant vapor. Generally, an air-cooled condenser consists of copper tubes and of copper or aluminum fins, which are expanded on the tubes for maximum heat transfer. The tubes are arranged in parallel or staggered rows and circuited for low refrigerant and air-pressure drop. The complete tube-and-fin condenser coil has a hot-gas inlet and refrigerant-liquid outlet connections. Condenser fan(s) are either directly driven or belt-driven to allow ambient air to circulate over the condenser coil.

Air is drawn from the bottom, goes over the condenser coil and condenser fan motor, and discharges upward. High-pressure, high-temperature refrigerant is being pumped through the hot-gas inlet of the condenser coil and is distributed according to the coil circuit, moving in the opposite direction of the air movement. In this process, heat transfer takes place. The cooler ambient air is circulated over the hot refrigerant. The ambient air picks up the heat, gets warm, and is discharged to the atmosphere. The hot refrigerant gas gets cooled and condenses into liquid.

Medium to Low Ambient Controls. The capacity of an air-cooled condenser is based on the temperature difference between the summer ambient-air temperature and the condensing temperature. When the packaged chiller is operated at temperature conditions lower than the design ambient-air temperature, the temperature difference between the condensing temperature and the ambient-air temperature is reduced, resulting in increased condenser capacity and lower condensing pressure.

If the ambient-air temperature falls below 60°F (15°C), the condensing pressure falls below a point where the expansion valve can no longer feed the cooler (evaporator) properly. Therefore, for 60°F (15°C) and below, it is necessary to use one or more of the following means to control the condensing pressure:

1. *Fan cycling.* By cycling one fan of a two-fan system, two fans of a three-fan system, and so on, a reasonable condensing pressure can be maintained. The cycling of fans can be in response to ambient-actuated thermostats, sensing the ambient-air temperature entering the condenser coil, or in response to the actual

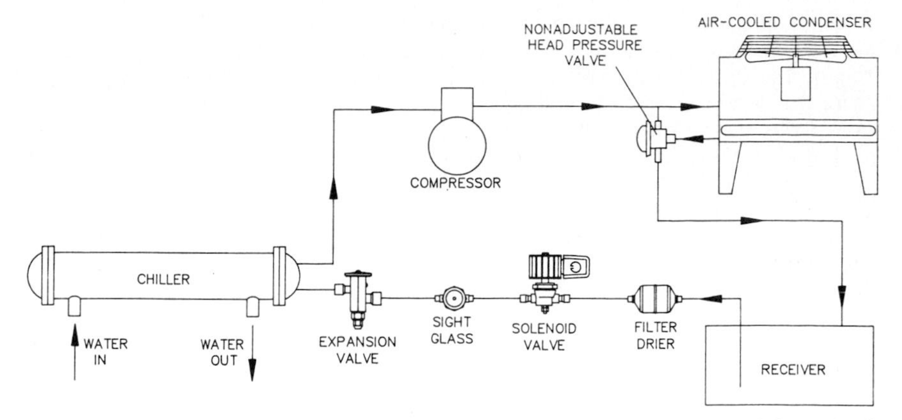

FIGURE 40.30 Typical head pressure-valve piping—nonadjustable. (*Courtesy of Continental Products, Inc.*)

condensing pressure. Fan cycling is reasonable and simple and is recommended for medium-temperature applications.

2. *Fan speed control.* By fan-speed modulation, single-fan chiller packages or multiple-fan systems can operate at medium to low ambient conditions.

3. *Flooded-head pressure control.* Flooded-head pressure control holds back enough refrigerant in the condenser coils to render some of the coil surface inactive. This reduction of the effective condensing surface results in a higher condensing pressure, thus allowing enough liquid-line pressure for normal expansion-valve operation.

Typical head pressure-valve piping is shown in Figs. 40.30 (a nonadjustable-valve combination) and 40.31 (an adjustable-valve combination). Both valves require the use of a liquid-refrigerant receiver, as shown. The capacity of the receiver is critical in that it must be large enough to hold all the refrigerant during high ambient conditions. If the receiver is too small, liquid refrigerant will be held

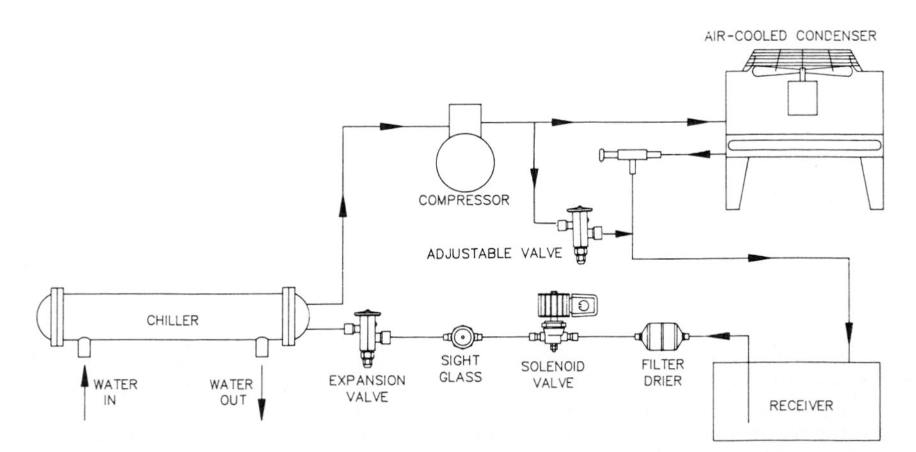

FIGURE 40.31 Typical head pressure-valve piping—adjustable. (*Courtesy of Continental Products, Inc.*)

back in the condenser during high ambients, resulting in high discharge pressures. During low ambients the receiver pressure falls until it approaches the setting of the control valve orifice. This orifice throttles, restricting the flow of liquid from the condenser. Thus the liquid refrigerant is backed up in the condensing coil, reducing the surface area.

Flooded controls can maintain operation down to −40°F (−40°C) ambient or below. Under normal summer conditions the liquid side of the valve remains open, and the hot-gas side is fully closed. Under low ambient conditions, the liquid side remains closed on startup, causing the condenser to "flood." This flooding continues until the condensing pressure reaches the valve setting [typically 180 psig (1241 kPag) for R-22, or 100 psig (689 kPag) for R-12]. Meanwhile the gas-side valve is open, allowing a portion of the hot gas to flow directly to the receiver, maintaining high pressure of the liquid for proper expansion-valve operation. Once the preset pressure is achieved, the valve modulates to maintain high head pressure, regardless of the ambients.

4. *Inlet-fan damper control.* This control modulates the air flow through the coil by the movement of dampers, in response to the condensing pressure. Usually a combination of fan cycling and damper control is utilized. Inlet dampers are mounted on the inlet of the fan and are actuated by a damper motor. Outlet dampers have also been used, but not very successfully. Experience has shown that damper control is generally not as effective as other means to achieve the same results, so its usage is limited.

40.2.6 Water-Cooled Condensers

Water-cooled condensers are of the following three types, all of which use water as the cooling medium:

1. *Shell-and-tube condensers.* These condensers, for chillers, are generally built with a steel shell and finned copper tubes. The cooling water circulates through the tubes, and the hot-gas refrigerant is on the outside of the tubes on the shell side. The condensation of refrigerant vapor takes place as the high-temperature hot gas comes in contact with the cool tube surfaces. The condenser water thus picks up the heat rejected by the compressor. Water circuiting is baffled so as to have two, three, four, or six passes.
2. *Shell-and-coil condensers.* In this arrangement, a copper or cupronickel coil is contained within a shell. This type of condenser is limited to smaller sizes and is not generally used for chiller packages.
3. *Tube-and-tube condensers.* This type (which cannot be cleaned easily) consists of two tubes, one contained within the other. The annular space is used for water flow, and the inner tube is used for refrigerant condensing. Because the refrigerant undergoes a considerable pressure drop in the single tube, the use of tube-and-tube condensers is limited to smaller chiller packages up to 10 tons.

Condenser tubes can be cleaned mechanically or chemically. In any case, it is important to have cooling-water treatment for an efficient overall chiller system. Cooling water for condensers can be obtained from a cooling tower or from city water, but because it is uneconomical to use city water, a cooling tower is commonly used. A cooling tower's water temperature can vary according to the wet-bulb temperature of the ambient air. To keep the chiller system operating at a low water temperature, a water-regulating valve is installed on the condenser's water inlet. This valve modulates the flow of water in response to the condensing temperature or pressure.

40.2.7 Evaporative-Cooled Condensers

Evaporative-cooled condensers employ a copper, stainless-steel, or steel tube condensing coil that is kept continuously wet on the outside by a water-recirculating system. Simultaneously, centrifugal or propeller fan(s) move atmospheric air over the coil. A portion of the recirculated water evaporates, reusing heat from the condensing coil and thus cooling the refrigerant to its liquid state.

A typical evaporative-cooled chiller is shown in Fig. 40.27. The complete evaporative-cooled condenser consists of the following:

- Condensing coil (usually prime surface without fins)
- Centrifugal or propeller fan(s)
- Water distribution system
- Drift eliminator
- Water makeup and drains

Since it combines principles of both air-cooled and water-cooled systems, an evaporative-cooled condenser can be considered a combination of these. The driving force is the ambient wet-bulb temperature, which is usually 15 to 25°F (8 to 14°C) lower than the ambient dry-bulb temperature. The overall effect is that an evaporative-cooled condenser operates at a much lower condensing temperature than an air- or water-cooled system. This results in the lowest compressor energy input and hence in the most efficient packaged chiller system.

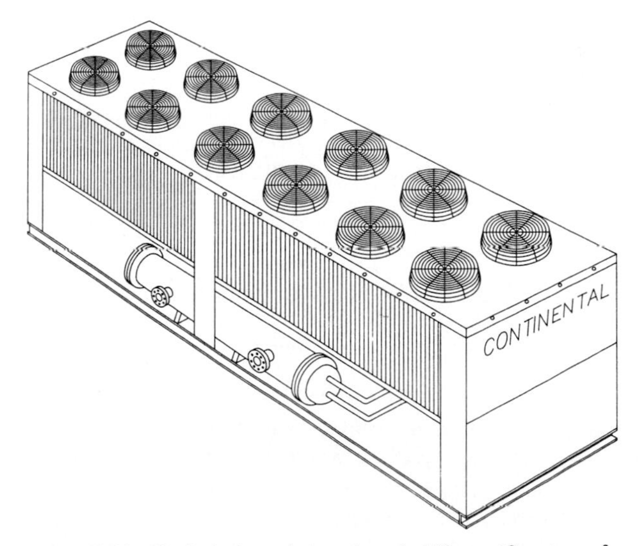

FIGURE 40.32 Typical air-cooled packaged chiller. *(Courtesy of Continental Products, Inc.)*

Water Treatment. As the water evaporates, the dissolved solids present in the water continue to concentrate and deposit on the condensing tubes, causing scale and corrosion. Combined with airborne impurities and biological contaminants, this means that the complete system can lose its high efficiency quickly and require a major overhaul. It becomes absolutely necessary to have a water-treatment program. A bleed-off valve is generally provided with the package, but additional treatment from a water-treatment specialist is needed.

40.2.8 Chiller-System Freeze Protection

If there is any danger of freezing in a closed-loop chilled-water system, it is recommended that the system be charged with a premixed industrial-grade heat-transfer fluid. Automotive antifreeze or other commercial glycols are not recommended; these may include silicates that can coat the cooler tubes, fouling the system prematurely and shortening the life of the pump seals.

For more details about heat-transfer fluids, consult your local industrial chemical supplier.

Fouling Factor. Fouling results from scaling, corrosion, sediment, and biological growth (slime, algae, etc.), and most water supplies contain dissolved or suspended materials that cause these problems. Such fouling causes thermal heat transfer to the water side of chiller systems, and the allowance for resistance to this heat transfer is called the "fouling factor."

A general practice is to allow a fouling factor of 0.00025 $(h \cdot ft^2 \cdot °F)/Btu$ $[(m^2 \cdot °C)/W]$ for coolers and water-cooled condensers. For seawater, or where the cooling water is untreated, a fouling factor of 0.001 $(h \cdot ft^2 \cdot °F)/Btu$ $[(m^2 \cdot °C)/W]$ is recommended. In this case, the use of 304 or 316 stainless steel, 90/10 cupronickel, 70/30 cupronickel, or admiralty brass tubes may be considered; the condenser heads can be made of brass or can be treated with epoxy or other protective coating.

40.2.9 Types of Refrigerant

R-22 is the most popular refrigerant for reciprocating liquid chillers.

Although R-12 and R-500 are suitable for high-condensing-temperature chillers, heat-pump chillers, or heat-recovery chillers, their use is being curtailed because of the chlorofluorocarbon (CFC) controversy.

40.2.10 Chiller Ratings

Capacity Rating Standards. Most manufacturers rate their packaged chillers according to Air Conditioning and Refrigeration Institute (ARI) Standard 590 (see Table 40.1). ARI Standard 590 is based on the following:

- Air-cooled package at an ambient-air temperature of 95°F (34°C)
- Water-cooled package at a condenser-entering water temperature of 85°F (30°C) and at a condenser-leaving water temperature of 95°F (35°C)
- Evaporative-cooled package at a dry-bulb temperature of 95°F (35°C) or at a wet-bulb temperature of 75°F (24°C)

TABLE 40.1 Typical Air-Cooled Chiller Ratings

Models MBA use a single compressor, models DBA are dual-circuit with a dual compressor, and models FBA have four compressors with four independent circuits.

Model	Tons	kW	EER	Model	Tons	kW	EER
MBA 3	2.6	3.6	8.8	DBA 52	42.4	43.8	11.6
4	3.9	5.5	9.3	62	52.4	55.0	11.4
5	4.6	6.0	9.3	70	62.3	66.2	11.3
7	6.6	8.0	9.8	75	67.7	73.0	11.1
9	8.6	9.2	11.2	80	73.2	79.8	11.0
10	10.1	10.8	11.3	90	79.7	85.4	11.2
15	13.6	14.9	11.0	100	86.3	91.0	11.4
20	16.5	16.6	11.9	110	92.8	102.5	10.9
25	21.2	21.9	11.6	120	101.0	115.8	10.5
30	24.2	26.1	11.1	FBA 130	114.1	121.9	11.2
35	31.2	33.1	11.3	140	128.3	136.4	11.3
40	36.6	39.9	11.0	160	146.3	159.6	11.0
50	43.1	45.5	11.4	180	159.4	170.8	11.2
60	49.7	57.0	10.5	200	177.8	187.5	11.4
				240	198.8	228.0	10.5

- Cooler water for all types at an entering temperature of 54°F (12°C) and at a leaving temperature of 44°F (7°C)
- Fouling factor for both the cooler and the condenser = 0.00025 $(h \cdot ft^2 \cdot °F)/Btu$ $[(m^2 \cdot °C)/W]$

ASHRAE Standard 30–77 is used for testing reciprocating liquid chillers for rating verification.

Energy Efficiency Ratio (EER)

$$\text{EER} = \frac{\text{Btu/h output}}{\text{watts input}}$$

Typically:

- The EER for air-cooled packages ranges from 8 to 12.
- The EER for water-cooled packages ranges from 9 to 13.
- The EER for evaporative-cooled packages ranges from 10 to 16.

40.2.11 Chiller Selection Guidelines

To select a packaged chiller from the manufacturer's rating table, it is necessary to know at least four of the following five items:

1. Capacity in tons or Btu/h (kW)
2. Fluid flow rate in gal/min (L/min)
3. Entering fluid temperature in °F (°C)

4. Leaving fluid temperature in °F (°C)
5. Type of fluid (water or other)

Use the following formula to calculate the fifth variable if only four are known:

$$\text{Tons} = \frac{\text{gal/min} \times \Delta T \times \text{cp} \times \text{SG}}{24}$$

where ΔT = (entering fluid temperature, °F) − (leaving fluid temperature, °F)
cp = specific heat of fluid, Btu/lb
SG = specific gravity of fluid

40.2.12 Types of Chillers

Heat-Recovery Chillers. Any HVAC or process application that has a simultaneous use for chilled and hot water is a potential heat-recovery installation. Typical applications are: buildings that need cooling on one side and heating on another; computer room cooling and reheating; restaurants; hotels; and hospitals.

Heat-recovery chillers extract heat from superheated gas vapor before it condenses in the condenser. Thus heat recovery offers "free heat" and eliminates, in certain instances, the need for separate heating equipment.

It is important that a heat-recovery heat exchanger not be oversized; otherwise, the advantage of high-temperature heat recovery is lost. A pressure-enthalpy diagram, as shown in Fig. 40.33, demonstrates the potential heat recovery for a typical chiller system.

Figure 40.34 shows a schematic of a typical heat-recovery chiller. For air-cooled chillers, a heat-recovery heat exchanger can be piped in series, as shown. For water-cooled systems, heat recovery can be in series or in parallel.

Factory-packaged heat-recovery chillers are available in sizes from 3 to 200 hp (2.24 to 150 kW).

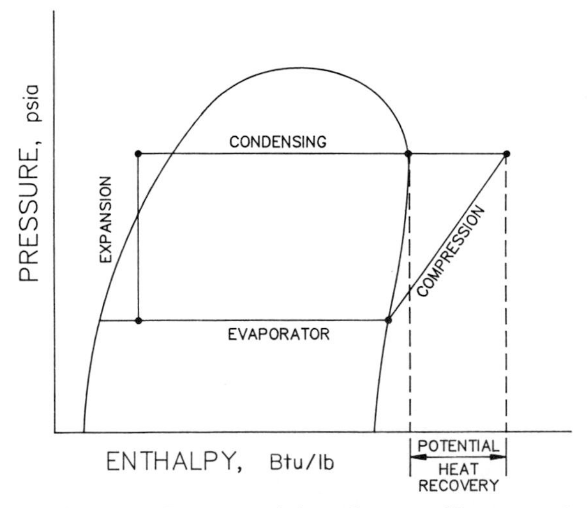

FIGURE 40.33 Pressure-enthalpy diagram. (*Courtesy of Continental Products, Inc.*)

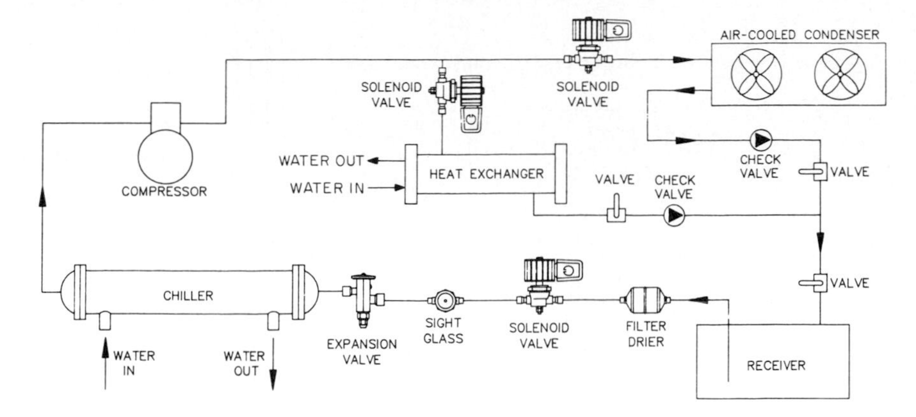

FIGURE 40.34 Typical heat-recovery chiller.

Heat-Pump Chillers. Heat-pump chillers are becoming more and more popular because of the following advantages:

1. They eliminate the use of a separate boiler or heating system.
2. They eliminate redundant piping for heating and cooling.
3. They use a two-pipe system but provide the comfort of a three-pipe system.
4. They use the same air handlers (or fan coils) for cooling and heating.
5. They use the same chilled-water pump for cooling and heating.

Heat-pump chillers utilize the same heat exchanger for cooling water as they do for heating water. The principle of operation is that, during heating, a reversing valve directs the flow of the hot-gas refrigerant from the compressor to the water heat exchanger instead of to the condenser. The heat exchanger is now being used as a condenser, and the condenser is being used as an evaporator. The gas is returned back to the compressor, through the reversing valve, with a common suction connection.

During heating, the same pump is used that circulates the water during cooling. This eliminates the need for separate hot-water and chilled-water pumps and piping. For the summer season, the same valve is reversed back to normal cooling.

Figure 40.35 shows a heat-pump chiller schematic. Air-cooled heat pumps use outside ambient air as the medium; therefore, they need wider fin spacing as well as hot-gas defrosting. Water-cooled heat pumps can use groundwater, river water, or wastewater as the medium.

Available in sizes of 5 to 200 hp (3.7 to 150 kW), heat-pump chillers are suitable for most locations with a winter-design dry-bulb temperature of 20°F (−7°C). They are also available with auxiliary electric heaters, which are useful for unexpected cold spells or as a backup.

Low-Temperature Glycol, Brine, Alcohol, and Gas Chillers. For low-temperature cooling with glycol, brine, alcohol, gases, or other fluids, several special features are necessary. Factory-packaged chillers for these applications are available. Field modifications of an HVAC chiller do not always produce the desired results.

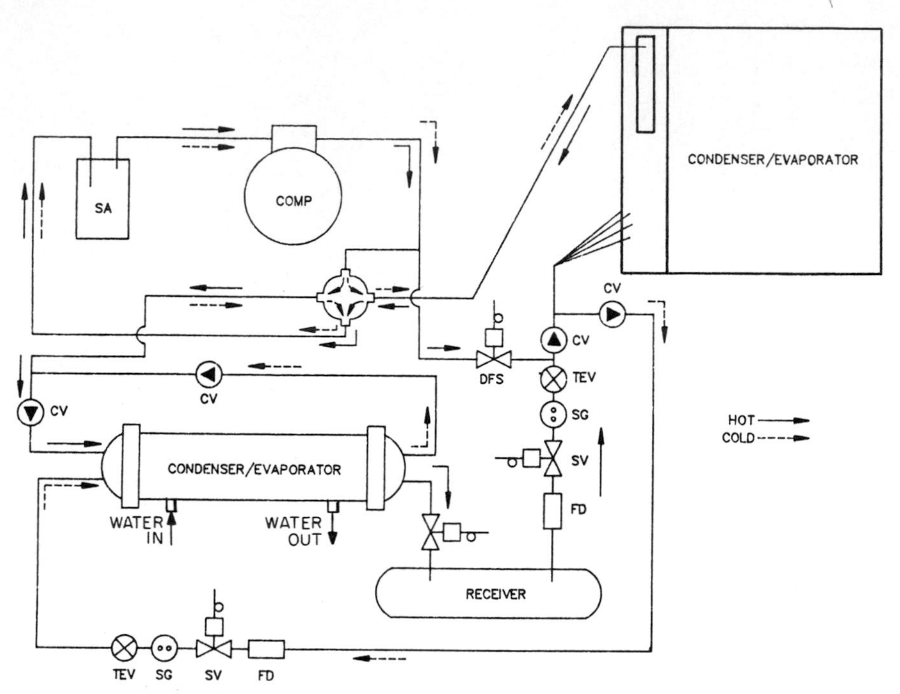

FIGURE 40.35 Heat-pump chiller—schematic.

Some of the considerations for process chillers are:

1. Type of refrigerant. R-22 is recommended.
2. Correct sizing of expansion valve.
3. Temperature controller for low temperature.
4. Low-pressure switch.
5. Low-temperature cutout.
6. Oil separator(s).
7. Suction accumulators.
8. Dual-compressor system with common dual-circuit chiller for 50 percent redundancy.
9. Dual compressor, dual condenser, dual cooler, and dual electrical components for 100 percent redundancy.
10. Primary chiller, with secondary heat exchanger, for corrosives, chemicals, food products (e.g., wines and fruit juices), gas cooling and condensation, incineration, and environmental protection.

Packaged Process Chillers. Packaged chillers used for HVAC are designed primarily for human-comfort conditions. A typical system is designed for chilled-water flow to produce a temperature difference of 10°F (5°C), cooling from 54 to 44°F (12 to 7°C) or from 52 to 42°F (11 to 6°C).

For process cooling, it is not always possible:

1. To maintain a 10°F (5°C) temperature difference.
2. To work between a 40 and 50°F (4 and 10°C) temperature range.
3. To keep a steady load.
4. To use ambient-related controls. The load may be constant year-round.
5. To use single-compressor systems.
6. To have a high return-water temperature.
7. To use standard electrical components, such as in explosion-proof atmospheres.
8. To use standard construction (steel or copper coolers or condenser) or copper or aluminum air-cooled condensers.

Typical applications for process chillers include the following:

Acid cooling
Bakeries
Breweries
Candy and fruit glazing
Chemicals and petrochemicals
Chicken and fish hatcheries
Computer and clean-room cooling
Dough mixers
Electronic-cabinet cooling
Environmental test chambers
Explosion-proof chillers
Flight simulators
Foundries
Fruit-juice cooling
Ice rinks
Laser cooling
Lobster tanks
Machine-oil cooling
Marine systems
Milk cooling
Mushroom cooling
Pharmaceuticals
Photo labs
Plastics, injection and blow molding
Plating and metal finishing
Printing plants
Pulp and paper
Shrimp freezing
Soil freezing
Solvent recovery
Steel mills
Textile plants
Welding
Wineries

CHAPTER 41
ABSORPTION CHILLERS

Nick J. Cassimatis
Gas Energy, Inc., Brooklyn, New York

41.1 INTRODUCTION

The purpose of this chapter is to provide the engineer or designer with practical information that can be used in the application of absorption chillers to air-conditioning systems.

Absorption chillers are machines that utilize heat energy directly to chill the circulating medium, usually water. The absorption cycle utilizes an absorbent (usually a salt solution) and a refrigerant (water).

Absorption chillers are usually classified according to the type of heat energy used as the input and whether it has a single-stage or two-stage generator. Absorption chillers using steam, hot water, or a hot gas as the heat energy source are referred to as "indirect-fired" machines, while those which have their own flame source are called "direct-fired" machines. Machines having one generator are called "single-stage absorption chillers," and those with two generators are called "two-stage absorption chillers."

Although small-tonnage absorption machines are produced for residential and small commercial applications in the 5- to 10-ton (17- to 144-kW) sizes, this chapter deals with large-tonnage equipment—machines ranging in capacity from 100 to 1500 tons (352 to 5280 kW).

The reader should keep in mind that the operating cycle of the single-stage absorption chiller is identical to that of the two-stage chiller except that there is no second-stage generator. The following discussion applies equally to both single-stage and two-stage chillers.

41.2 DESCRIPTION OF THE CYCLE

The absorption cycle is not much different than the more familiar mechanical refrigeration cycle.

Mechanical Refrigeration Cycle. In the mechanical refrigeration cycle (Fig. 41.1), refrigerant vapor is drawn in by the compressor (1), compressed to high temperature and high pressure, and discharged into the condenser (2). In the condenser,

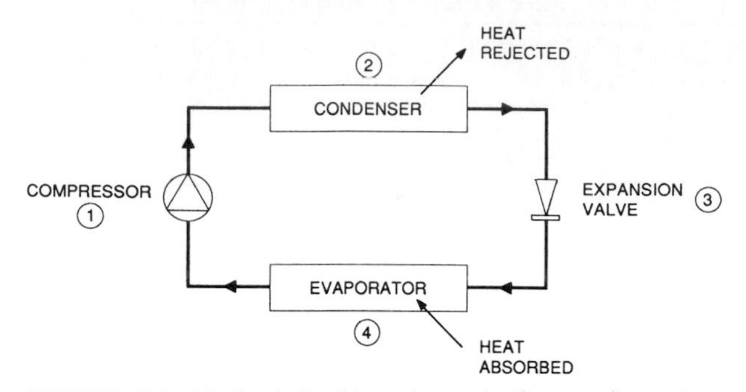

FIGURE 41.1 Mechanical refrigeration cycle. See text for explanation of circled numbers. *(Courtesy of Gas Energy, Inc.)*

the vapor is cooled and condensed to a high-pressure, high-temperature liquid by the relatively cooler water flowing through the condenser tubes.

The heat removed from the refrigerant is absorbed by the condenser water and is rejected to the atmosphere by the cooling tower.

The hot refrigerant liquid is metered through an expansion valve (3) into the low-pressure evaporator (4). The lower pressure causes some of the refrigerant to evaporate (flash), chilling the remaining liquid to a still lower temperature.

Heat is transferred from the warm system water (which is flowing through the evaporator tubes) to the cooler refrigerant. This exchange of heat causes the refrigerant to evaporate and the system water to cool.

Absorption Cycle. A parallel can be drawn between the mechanical refrigeration cycle and the absorption cycle. In the absorption cycle (Fig. 41.2), energy in the form of heat is added to the first-stage generator, and the solution pump (1) (which can be compared to the compressor of the mechanical system) pumps the lithium bromide solution from the low-pressure absorber to the relatively higher-

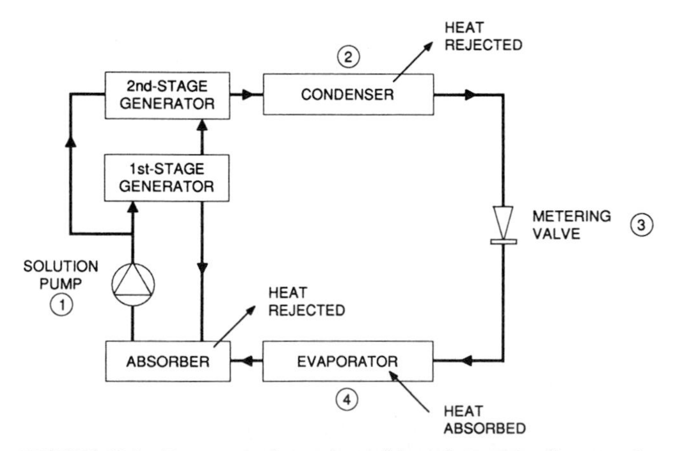

FIGURE 41.2 Two-stage absorption refrigeration cycle. See text for explanation of circled numbers. *(Courtesy of Gas Energy, Inc.)*

pressure first- and second-stage generators. In both these sections, heat is used to produce refrigerant vapor.

The refrigerant vapor is cooled and condensed into liquid by the cooling water flowing through the condenser tubes (2).

Liquid refrigerant from the condenser is metered through a metering orifice (3) (similar to the expansion valve of the mechanical system) and is pumped by the refrigerant pump to the evaporator (4), where it is sprayed over the evaporator tubes. The extremely low vacuum in the evaporator causes some of the refrigerant to evaporate.

During this process, heat is transferred from the relatively warm system water (which is flowing through the evaporator tubes) to the cooler refrigerant.

In the absorber, the spray solution mixes with the liquid-and-vapor refrigerant, and the resultant heat of absorption is removed by the condenser water (from the cooling tower) flowing through the tubes. In this system, note that heat is removed in both the condenser and the absorber sections by the cooling water and that it is finally rejected to the atmosphere by the cooling tower in the same manner as in the mechanical system.

Figure 41.3 shows schematically the arrangement of the different components of a typical two-stage absorption chiller. To make the absorption cycle more understandable, Fig. 41.4 breaks it down into six steps.

Reference 1 contains a description of the operating principles and thermodynamics of the absorption cycle.

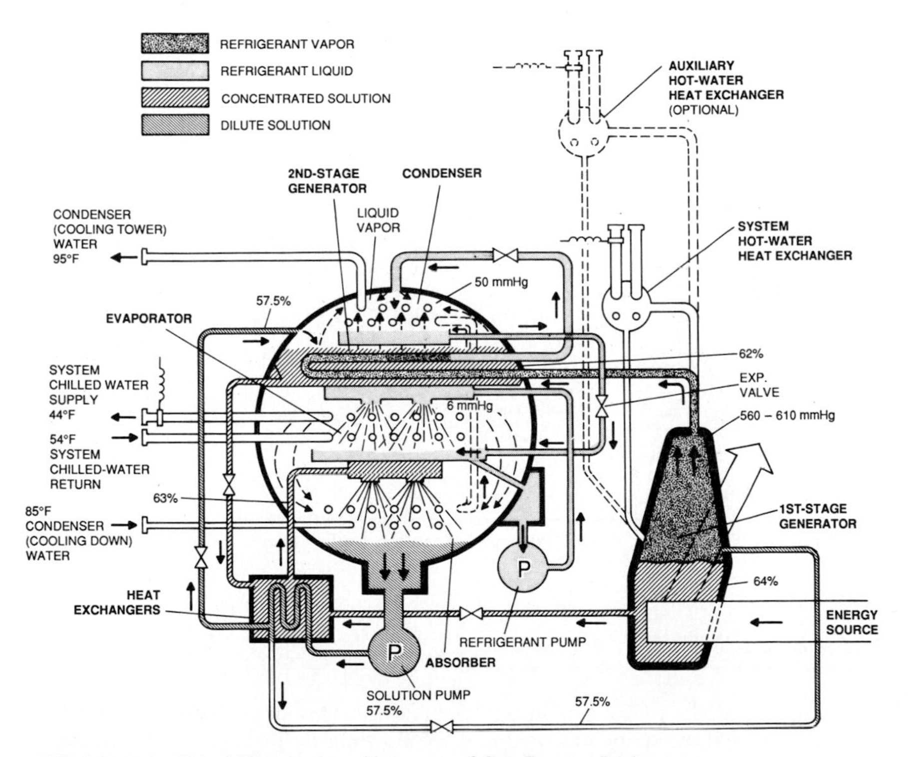

FIGURE 41.3 The chilling cycle. (*Courtesy of Gas Energy, Inc.*)

The two-stage absorption chilling cycle is continuous; however, for the sake of clarity and simplicity, it is divided into six steps.

1. Solution Pump/ Heat Exchangers

A dilute solution (57.5%) of lithium bromide and water descends from the Absorber to the Solution Pump. This flow of dilute solution is split into two streams and pumped through heat exchangers to the First Stage Generator and to the Second Stage Generator.

Parallel flow type chillers with split of solution flow virtually eliminate the possibility of crystallization (solidification) by allowing the unit to operate at much lower solution concentration and temperatures than in series flow systems.

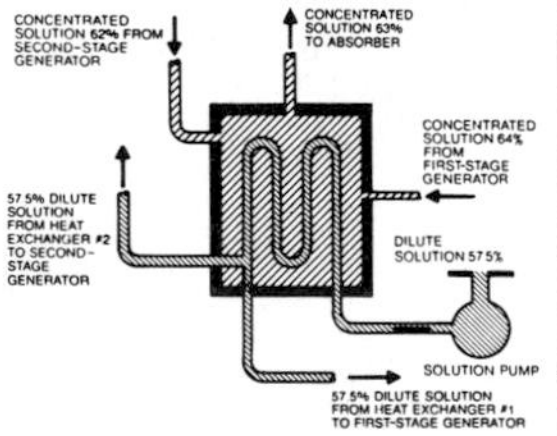

Note: There are two heat exchangers, but only one is shown for illustrative purposes.

2. First-Stage Generator

An energy source heats dilute lithium bromide solution (57.5%) coming from the Solution Pump/Heat Exchangers. This produces hot refrigerant vapor which is sent to the Second-Stage Generator, leaving a concentrated solution (64%) that is returned to the Heat Exchangers.

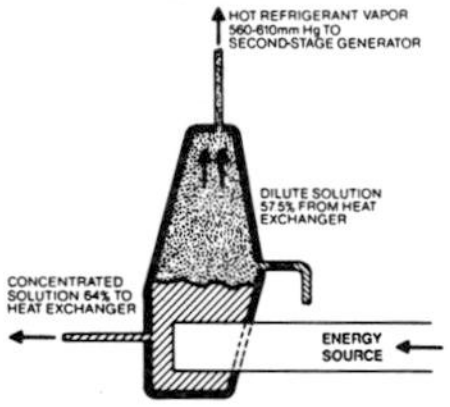

3. Second-Stage Generator

The energy source for the production of refrigerant vapor in the Second-Stage Generator is the hot refrigerant vapor produced by the First Stage Generator.

This is the heart of the efficient two-stage absorption effect. The refrigerant vapor produced in the First-Generator is increased by 40% – at no additional expense of fuel. The result is much higher efficiency than in conventional systems

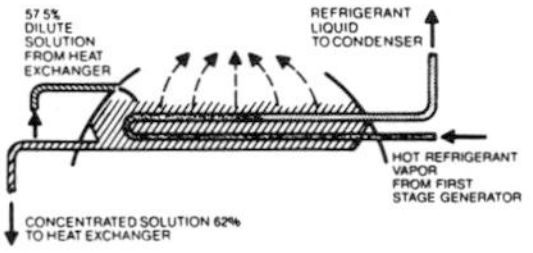

This additional refrigerant vapor (dotted arrow) is produced when dilute solution from the Heat Exchanger is heated by refrigerant vapor from the First-Generator. The additional concentrated solution that results is returned to the Heat Exchanger. The refrigerant vapor from the First-Stage Generator condenses into liquid giving up its heat, and continues to the Condenser.

4. Condenser

Refrigerant from two sources – (1) liquid resulting from the condensing of vapor produced in the First-Stage Generator, and (2) vapor (dotted arrows) produced by the Second-Stage Generator – enters the Condenser. The refrigerant vapor is condensed into liquid and the refrigerant liquid is cooled. The refrigerant liquids are combined and cooled by condenser water. The liquid then flows down to the Evaporator.

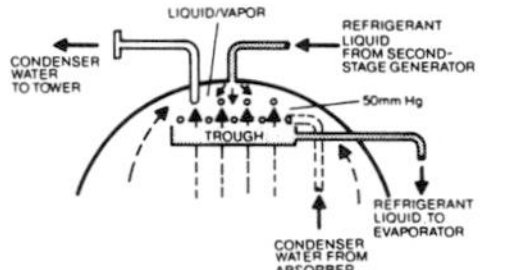

5. Evaporator

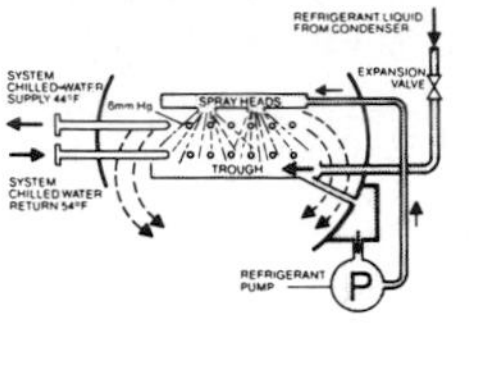

Refrigerant liquid from the Condenser passes through an Expansion Valve and flows down to the Refrigerant Pump, where it is pumped up to the top of the Evaporator. Here the liquid is sprayed out as a fine mist over the Evaporator tubes. Due to the extreme vacuum (6 mmHg) in the Evaporator, some of the refrigerant liquid vaporizes, creating the refrigerant effect. (This vacuum is created by hygroscopic action – the strong affinity lithium bromide has for water – in the Absorber directly below.)

The refrigerant effect cools returning system chilled water in the Evaporator tubes. The refrigerant liquid/vapor picks up the heat of the returning chilled water, cooling if from 54 to 44°F (12 to 7°C). The chilled water is then supplied back to the system.

6. Absorber

As refrigerant liquid/vapor descends to the Absorber from the Evaporator, concentrated solution (63%) coming from the Heat Exchanger is sprayed out into the flow of descending refrigerant. The hygroscopic action between lithium bromide and water – and the related changes in concentration and temperature – result in the creation of an extreme vacuum in the Evaporator directly above. The dissolving of the lithium bromide in water gives off heat, which is removed by condenser water entering from the Cooling Tower at 85°F (30°C) and leaving for the Condenser at 92°F (33°C) (dotted lines). The resultant dilute lithium bromide solution collects in the bottom of the Absorber, where it flows down to the Solution Pump.

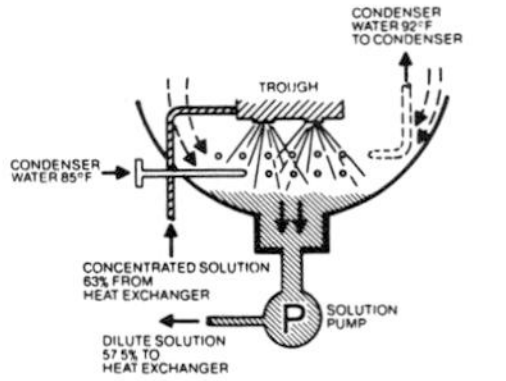

The chilling cycle is now completed and begins again at Step 1.

FIGURE 41.4 The six steps in the two-stage absorption chilling cycle. (*Courtesy of Gas Energy, Inc.*)

41.3 EQUIPMENT

A recent newcomer to the absorption field is the two-stage absorption chiller-heater. This type of equipment, manufactured primarily in Japan, is a two-stage absorption chiller that can also double as a heater.

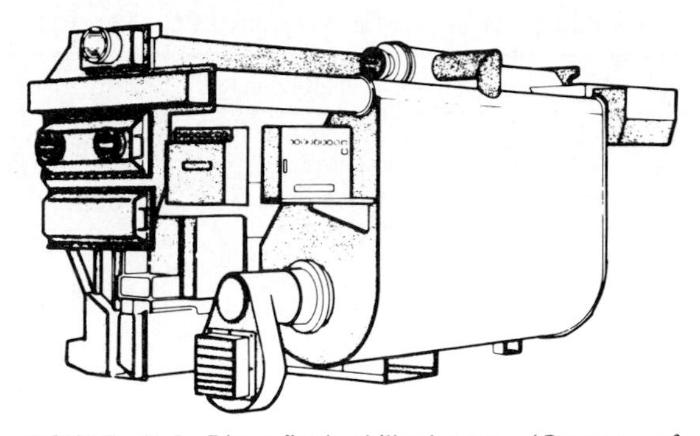

FIGURE 41.5 Direct-fired chiller-heater. (*Courtesy of Gas Energy, Inc.*)

The chiller-heater is either of the direct-fired type (Fig. 41.5) or of the heat-recovery type (Fig. 41.6), and in the heating mode it operates in much the same manner as a boiler except that it operates under vacuum. Heat applied to the first-stage generator creates a hot vapor, which is piped to the shell side of a shell-and-tube heat exchanger attached to the unit. The system water circulating through the hot-water heat-exchanger tubes is heated by the hot vapor condensing on the outside of the tubes (shell side).

The greatest advantage of chiller-heater types of machines is that one piece of equipment can meet both the cooling and heating needs of the facility and can offer substantial space savings, especially in buildings where space is at a premium. Another advantage is that under proper conditions an absorption chiller-heater can provide both chilling and heating water simultaneously.

Table 41.1 gives general performance information for single- and two-stage absorption units.

Most absorption equipment currently manufactured for use in air-conditioning is water-cooled single- or two-stage equipment using the lithium bromide–water cycle.

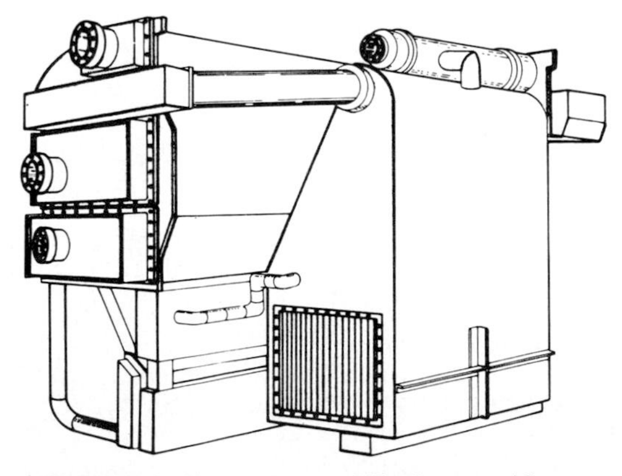

FIGURE 41.6 Heat-recovery chiller-heater. (*Courtesy of Gas Energy, Inc.*)

TABLE 41.1 Nominal Rating Conditions for Absorption Chillers

	Single-stage	Two-stage
Chilled water:		
Leaving temperature, °F (°C)	44 (6.7)	44 (6.7)
Flow rate, gal · min/ton (L/kJ)	2.4 (0.043)	2.4 (0.043)
Temperature drop, °F (°C)	10 (5.5)	10 (5.5)
Condenser water:		
Entering temperature, °F (°C)	85 (29.4)	85 (29.4)
Flow rate, gal · min/ton (L/kJ)	3.6 (0.065)	3.6 (0.065)
Temperature, °F (°C)	10 (5.5)	15 (5.5)
Steam pressure range (dry at saturated), lb/in^2 (kPa)	9–12 (62–83)	43–130 (296–896)
Fouling factor (absorber and evaporator), $h \cdot ft^2 \cdot °F/Btu$ ($m^2 \cdot K/W$)	0.0005 (0.0009)	0.0005 (0.0009)

Reference 2 provides more information on the different types of absorption equipment available.

41.4 APPLICATIONS

Since the energy input of these machines is in the form of heat, they are best-suited to applications where heat is available in a usable form and at a relatively low cost. The available heat source also dictates which type of machine is best-suited to a particular application. Table 41.2 shows different machines and their applications.

With the increased emphasis on operating costs, the many types of chillers available allow the engineer or owner to select the one that offers the lowest operating costs.

TABLE 41.2 Applications of Absorption Chillers

	Machine type		
Application	Direct-fired	Steam	Heat-recovery
Areas where electricity costs are high	✓	✓	✓
Areas where the primary fuel cost (gas, oil) is relatively low	✓	✓	
Areas where the cost of utility steam is low or where steam is available as a byproduct		✓	
Areas where heat in the form of hot gases is available as a byproduct		✓	✓

41.5 ENERGY ANALYSIS

There are many detailed computer programs that can help the engineer chose the best type of chiller for his or her particular application. However, a quick way to estimate the operating costs of the different types of chillers is as follows:

$$C = H \times T \times E \times R$$

where C = annual energy costs, dollars
H = equivalent full ton hours of operation
T = chiller rated capacity, tons
E = chiller energy usage at full load
R = local utility rate

EXAMPLE Which type of chiller will be best for a building which requires 500 tons of chilling and which operates an average of 850 full ton hours per year? The various energy costs and chiller energy usages are:

Electricity (including demand charges) = \$0.14/kWh
Natural gas = \$4.15/1000ft^3
Utility steam = \$8.75/1000lb
Electric centrifugal chiller: energy usage = 0.65 kW/ton
Two-stage steam chiller: steam rate = 9.7 lb/(ton · h)
Direct-fired chiller: gas usage = 11.73 ft^3/(ton · h) (LHV)

Solution

1. Electric centrifugal chiller:

$$C = H \times T \times E \times R = 850 \times 500 \times 0.65 \times 0.14 = \$38{,}675$$

2. Two-stage steam chiller:

$$C = H \times T \times E \times R = 850 \times 500 \times 9.7 \times 8.75/1000 = \$36{,}072$$

3. Direct-fired chiller:

$$C = H \times T \times E \times R = 850 \times 500 \times 11.73/0.90 \times 4.15/1000 = \$22{,}988$$

where the 0.90 converts the 11.73 ft^3/(ton · h) to HHV.

This simple calculation indicates quickly that a direct-fired chiller will be the best choice if the only consideration is the operating cost. Of course, the engineer must evaluate many other factors before deciding which type of chiller will be best for the building.

41.6 UNIT SELECTION

For a given application, the absorption unit selected must provide the required chilling capacity with the smallest possible machine size. Machine size is usually based on specified chilled-water flow rates and temperatures, but it can also be influenced by the flow rate and temperature rise of the condenser water.

To arrive at the best system, both the initial system cost and the annual operating cost must be carefully examined. When selecting an absorption unit, the manufacturer's procedures should be followed.

Since chiller performance improves with increased leaving chilled-water tem-

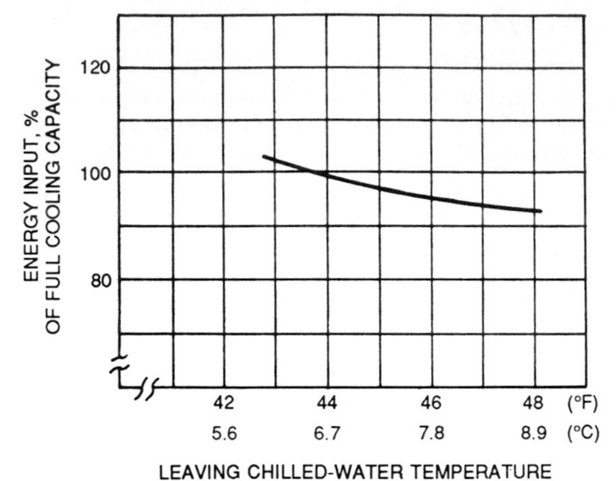

FIGURE 41.7 Energy input versus leaving chilled-water temperature. (*Courtesy of Gas Energy, Inc.*)

perature, the unit should be chosen carefully. Consideration should also be given to resetting this temperature during milder weather.

Figures 41.7 and 41.8 show the effect on capacity as the chilled-water and condenser-water temperatures vary.

The cooling tower used with absorption chillers is usually larger than that used with mechanical systems. The cooling tower cost, however, can be reduced if the chiller can operate with a greater condenser water temperature rise. Also, since chiller performance improves with lower entering condenser water temperature, the small additional cost of a slightly larger cooling tower can be offset by the

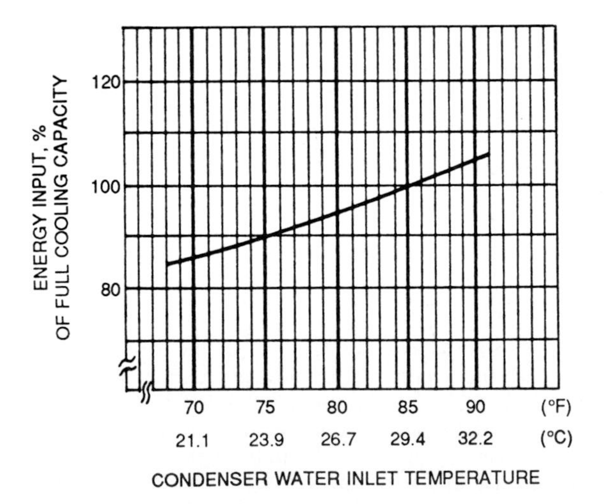

FIGURE 41.8 Energy input versus condenser water inlet temperature. (*Courtesy of Gas Energy, Inc.*)

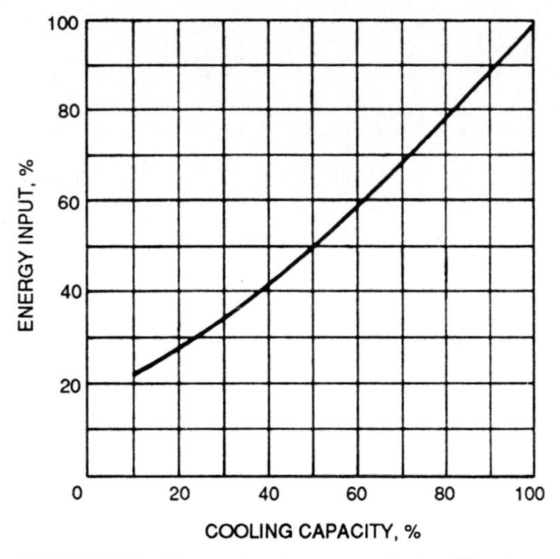

FIGURE 41.9 Energy input versus output. *(Courtesy of Gas Energy, Inc.)*

energy-cost savings. If a source of good-quality water is available, such as a river, a lake, or a well, it can be used in lieu of a cooling tower.

Another important consideration in selecting an absorption chiller is the fact that there is an almost direct relationship between load and energy input (see Fig. 41.9).

Steam absorption chillers have the additional advantage of being able to operate over a wide range of pressures (see Fig. 41.10).

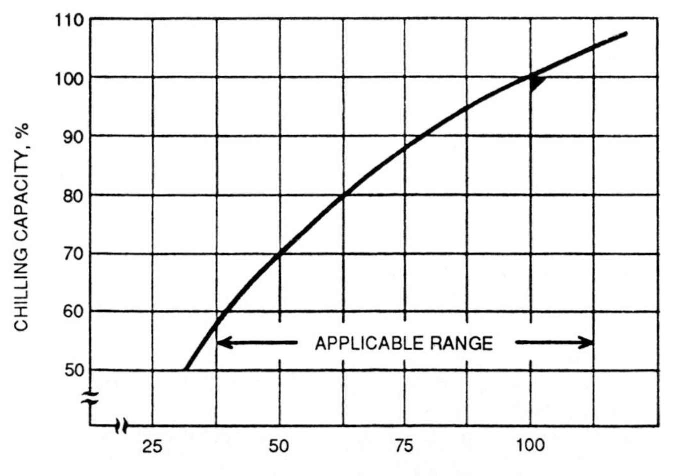

FIGURE 41.10 Effect of steam pressure on capacity. *(Courtesy of Gas Energy, Inc.)*

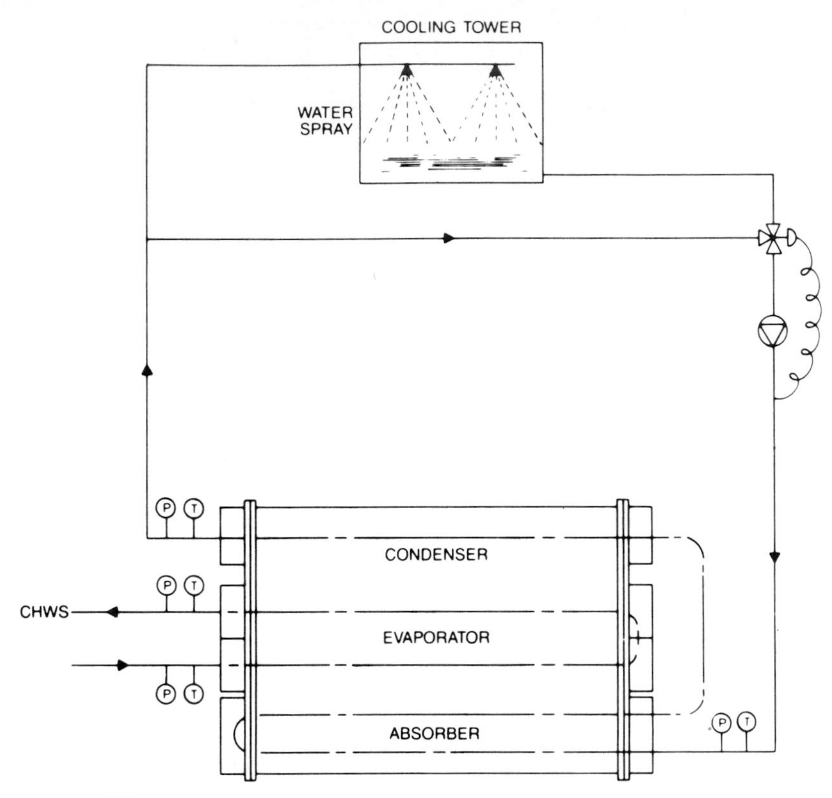

Note:
To ensure uniform and correct flow from the chiller, balancing valves should be installed downstream of the chilled water, and condenser water.

Locate pressure and temperature gauges between the chiller water outlets and the balancing valves.

FIGURE 41.11 Typical absorption chiller piping. (*Courtesy of Gas Energy, Inc.*)

The reader should consult manufacturers' engineering manuals and catalogs, which usually contain more detailed data on these machines. The final selection should be based on the best combination of system components and lowest operating costs. The designer should therefore consider carefully the chilled-water temperature, condenser water temperature, water quality, fuel supply, etc., before making the final selection.

Figure 41.11 shows a typical cooling tower piping schematic with a three-way mixing valve and bypass.

Figure 41.12 shows a typical steam piping schematic for a steam absorption chiller.

Figure 41.13 shows the basic elements of an absorption chiller used to recover the exhaust gas from a heat source such as an engine, a gas turbine, or some process.

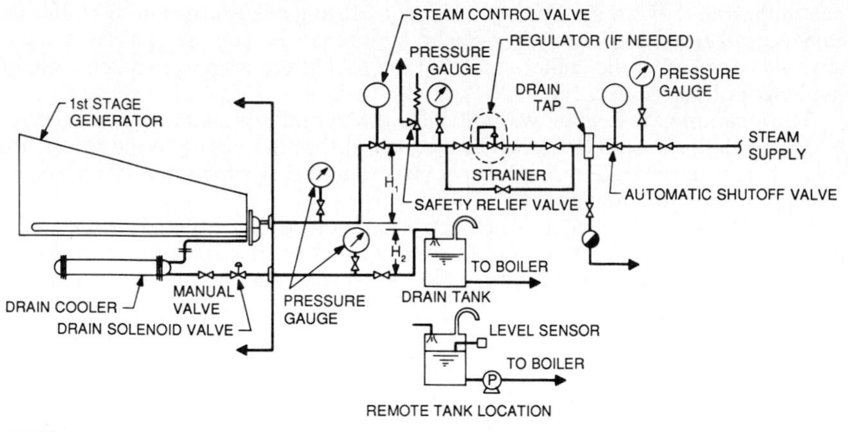

NOTES:

H_1 = 40-in (102 cm) MINIMUM TO PREVENT CONDENSATE BACKFLOW.

H_2 = 32.8-ft (10-m) MAXIMUM TO PREVENT EXCESSIVE BACK PRESSURE.

CONDENSATE LEAVES DRAIN COOLER AT APPROXIMATELY 14.3 psig (98.6 kPa),195°F (90.6°C).

MAXIMUM INLET STEAM PRESSURE 128 psig (882 kPa).

AUTOMATIC SHUTOFF VALVE TO BE FAIL SAFE TYPE (CLOSES UPON POWER FAILURE).

BOTH THE STEAM SUPPLY AND THE CONDENSATE DRAIN LINES MUST BE PROPERLY PITCHED TO PREVENT HAMMERING.

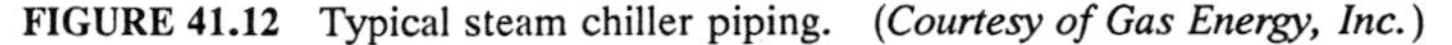

FIGURE 41.12 Typical steam chiller piping. (*Courtesy of Gas Energy, Inc.*)

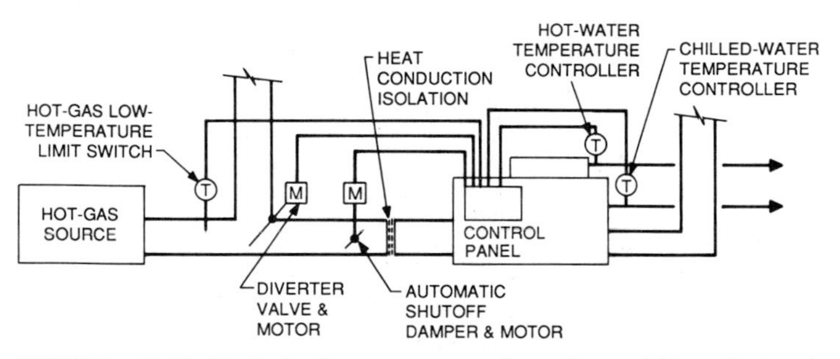

FIGURE 41.13 Typical heat-recovery absorption chiller ductwork. (*Courtesy of Gas Energy, Inc.*)

41.7 LOCATION

Because of their rather compact size, light weight, and absence of vibration, absorption chillers can be installed almost anywhere in the building: basement, mid-floor, or roof. Generally speaking, absorption chillers are not recommended for outdoor installation.

When selecting a site, consider structural support, access for service, and tube pull area. The machine-room floor should be able to support the full operating weight of the chiller. The chiller should be located so that all piping presents a neat appearance, with a minimum number of fittings and turns and with full and easy access to the chiller for service and maintenance. The tube pull area is usually on either end of the chiller, and the tube pull space is approximately equal to the length of the chiller.

The location of the chiller will influence the overall installation cost of the system. The chiller should therefore be located so that the piping to the tower, flue stack (direct-fired chillers), building services, and fuel supply and the electrical wiring are kept to a minimum.

Consider the type of fuel to be used when selecting the installation site. For instance, a gas chiller allows more flexibility in terms of its placement than oil or steam chillers do because it requires no storage tank and very little (if any) maintenance.

41.8 INSTALLATION

Since the absorption chiller is almost vibrationless, vibration eliminators are generally not required. However, vibration-eliminating mounts or pads should be considered when the machine is installed in an area where even mild noise or vibration would be a problem, such as on a floor near a conference room or sleeping area or on the roof of such an area.

Absorption chillers will operate properly and produce maximum capacity only if they are installed level. It is therefore important that the machine be leveled when it is installed in place and be checked again (and adjusted if necessary) after the piping, refrigerant, solution, and system water have been installed. Follow the manufacturer's instructions and observe the allowable tolerances.

Leave sufficient space, usually 40 to 60 in (1.0 to 1.5 m), around the chiller for service and maintenance work.

The machine room must be well lighted and properly ventilated to keep its ambient temperature under 100°F (38°C). Humidity in the machine room should also be held to a minimum to prevent corrosion and damage to electrical controls and components. Since the chiller contains large amounts of water (refrigerant), it should be protected from freezing; machine-room temperatures should be kept above 35°F (2°C).

Follow standard engineering practices in designing the piping system and other services. Adequate support must be provided for system piping so that no weight is placed on the chiller's water boxes and connecting nozzles.

41.9 INSULATION

Absorption chillers are normally shipped to the jobsite uninsulated. After the machine has been installed, piped, operated, and tested, it must be insulated.

Absorption chillers must be insulated on the cold surfaces of the evaporator, on the refrigerant lines and pump, on the chilled-water boxes, and on the refrigerant tank. This prevents them from sweating, which causes corrosion.

The first-stage generator, heat exchangers, and piping carrying hot solution must also be insulated to prevent heat loss, minimize machine-room temperature, and protect personnel against injury.

Follow the manufacturer's recommendations and local codes on the insulation material to use and on methods of application.

41.10 OPERATION AND CONTROLS

Absorption chillers meet their load variations and maintain their leaving chilled-water temperature by varying the heat input to the chiller, which in turn varies the reconcentration rate of the absorbent solution. Chiller capacity is directly proportional to solution concentration in the absorber.

The nominal operating range of most absorption chillers can be adjusted from a range of 40 to 42°F (4.4 to 5.6°C) to a range of 48 to 50°F (8.9 to 10°C). However, machine efficiency will decrease as the leaving chilled-water temperature is reset below the design temperature. The manufacturer's recommendation should be followed.

The chiller will maintain the selected leaving chilled-water temperature even though the building's load changes during the course of operation. This is usually accomplished through a temperature controller, whose sensor is located in the leaving chilled-water outlet of the chiller. The controller, in turn, regulates the heat input to the chiller by modulating the energy input of the heat source. Energy-management-type control systems can accomplish chiller control by monitoring many other system temperatures in addition to that of the leaving chilled water and by then resetting the leaving chilled-water temperature to maximize chiller efficiency.

Most of the chiller controls and components are factory-mounted, -wired, and -set—except for the temperature- and pressure-sensing devices, which are usually placed in their proper location at the jobsite.

41.10.1 Safety Controls

In addition to the operating controls, absorption chillers are equipped with factory-installed and -wired automatic safety controls. Typical safety devices are:

1. *Low-temperature cutout switch:* This is a thermostat that will stop the chiller if the evaporator temperature falls too low.
2. *Chilled-water flow switch:* This can be either of the pressure- or flow-sensitive type. It will stop the chiller if the chilled-water flow is reduced below the design level or is stopped.
3. *Cooling-water flow switch:* Similar to the chilled-water flow switch, this will stop the chiller if the flow of cooling water is interrupted or reduced below the design limits.

The above three switches will usually reset themselves when proper water-flow levels have been reestablished.

4. *Concentration limiters:* Found on some chillers, these are designed to prevent the solution from reaching a high absorbent concentration, which can lead to crystallization. This high concentration occurs when the cooling-water tem-

perature becomes too low and/or when the solution flow within the machine is reduced.

5. *First-stage generator high-pressure and high-temperature switches:* These switches are designed to interrupt operation if either the temperature or pressure within the machine exceeds the preset limits. These switches will not reset themselves; they must be reset manually by the operator after the cause has been determined and corrected.
6. *Alarms:* Some or all of the above devices are usually wired so that, when activated, they will sound an alarm to alert the operator.
7. *Rupture disk:* Some absorption chillers are fitted with a rupture disk or some other type of pressure-relief device on the chiller's generator section to protect against overpressure.
8. *Other controls:* Special machines, such as of the direct-fired or heat-recovery types, may employ additional controls to provide safe operation.

Direct-fired machines will normally be provided with all necessary controls, safeties, indicators, and monitoring devices needed for the proper and safe operation of the burner.

Heat-recovery machines will normally be provided with controls to interlock the chiller to the heat source and with an exhaust-gas low-temperature-limit switch, which will prevent the chiller from operating until the gas temperature reaches the design temperature.

41.10.2 System Interlock Controls

For proper operation, the absorption chiller must be electrically interlocked with the chilled-water and cooling-water (condenser) pumps and the cooling tower; special machines, such as the heat-recovery type, must also be interlocked to the heat source. Most chiller manufacturers provide a signal source at the chiller control panel to start the pumps when the chiller starts.

In addition to the control interlocks, most manufacturers include terminals in their control panels to permit the monitoring of machine operation from a remote location. Provision can also be made to sound an alarm or activate other warning devices when the chiller operation is abnormal or interrupted.

The installing contractor will usually be required to provide flow switches on the chilled-water and condenser water piping to prove water flow to the chiller control circuitry. If water flows are not established within a short time, the chiller will cease operation and an alarm will sound to alert the operator.

The chiller manufacturer will normally provide sufficient information about where the interlocks should be located and how they should be wired. As a rule, external controls are not provided by the chiller manufacturer.

41.10.3 Lead-Lag Controls

In some installations, even though a single chiller can meet the building's needs, two or more units may be installed to provide flexibility, economy of operation, and at least partial redundancy. In cases where two machines are installed to provide either half or part of the building's total load, lead-lag controls can be supplied to allow the operator to chose the "lead" or "lag" machine to meet the load requirements.

Lead-lag control can be achieved whether the chillers are installed in parallel or in series. In either case, a return chilled-water temperature sensor may be used to cycle either the lead or the lag unit when one unit can carry the load. Figures 41.14 and 41.15 show typical lead-lag connections.

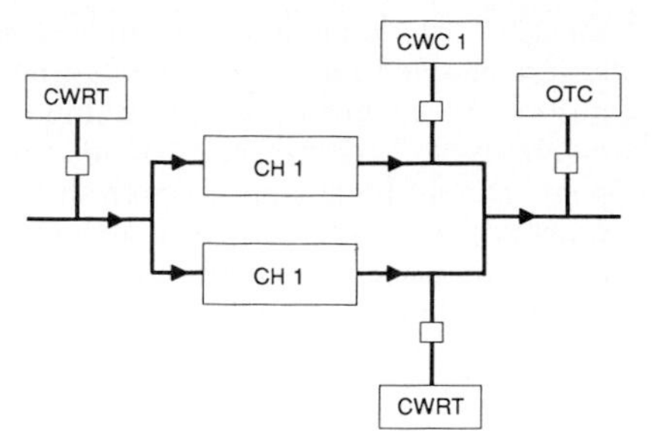

FIGURE 41.14 Chillers installed in parallel. CH = chiller; CWRT = chilled-water return temperature sensor; CWC = chilled-water temperature controller; OTC = override temperature controller. (*Courtesy of Gas Energy, Inc.*)

When the chillers are installed in parallel and one chiller is shut down, the chilled water flowing through it will bypass the operating chiller and the mixture temperature will be higher than the set point. To prevent this from happening, an override temperature controller can be used to reset the operating unit to a lower setting so that the mixture temperature is at or close to the original setting.

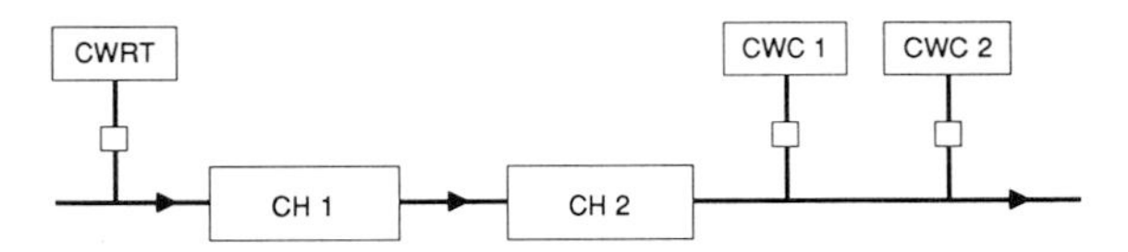

FIGURE 41.15 Chillers installed in series. See Fig. 41.14 for definitions of the abbreviations. (*Courtesy of Gas Energy, Inc.*)

41.11 OPERATION AND MAINTENANCE

Absorption chillers are simple to operate and comparatively trouble-free. Chillers presently manufactured domestically and in Japan are built to rigid standards of internal cleanliness and vacuum integrity. They are usually operated and leak-tested prior to shipment from the factory.

41.11.1 Corrosion Control and Leak Tightness

The two most important maintenance concerns for the operator of the absorption chiller are the need to maintain corrosion control and leak tightness. The useful life of the chiller is directly proportional to how tight the chiller is maintained and how good the corrosion control is.

The lithium bromide solution used as the absorbent in absorption chillers is extremely corrosive and will attack steel, copper, and copper alloys in the presence of air and at temperatures above 300°F (150°C).

All absorption chillers use some type of corrosion inhibitor to protect the internal parts of the chiller against corrosive attacks. It is very important that any inhibitor use be maintained within the limits specified by the manufacturer. A solution sample must be obtained at least twice per season and analyzed to determine the presence of adequate amounts of inhibitors. It should also be tested for alkalinity and the presence of ammonia, copper, iron, etc. Ideally these components should be absent, or present only in very small amounts. The chiller manufacturer normally provides guidelines as to the acceptable ranges within which these components can be present; usually these are given as parts per million (ppm) or milligrams per liter (mg/L).

More significant, however, than the numerical value of these components is the change that occurs from one sampling to the next. This change is a good indicator of the inhibitors' effectiveness. The laboratory test results will determine how much of which types of inhibitor should be added to maintain the solution within the manufacturer's specifications.

Failure to maintain proper inhibitor control can have a detrimental effect on the life of the chiller.

Whenever an absorption chiller is to be opened to the atmosphere for repair and maintenance, it is standard practice to break the vacuum with nitrogen and to bleed a small amount of nitrogen into the unit while the work is performed. Nitrogen is an inert gas, and its use will prevent corrosive attack while the chiller is exposed to the atmosphere.

As explained earlier, absorption chillers operate at a relatively high vacuum; thus, unless the chiller is properly maintained, air will leak into the machine.

41.11.2 Purging

In addition to air leakage, there is the constant production of noncondensable gases that are created during the operation of the chiller. To prevent the accumulation of air and noncondensable gases, absorption chillers are equipped with either automatic or manual purging systems, which must be operated regularly. If purging is not carried out regularly, such accumulation can reduce the chilling capacity, permit deterioration of the chiller's internal parts through corrosion, and even cause crystallization within the chiller.

Crystallization is the precipitation of the lithium bromide salt crystals from the solution. The precipitate forms a slushlike mixture that can plug pipelines and other fluid passages within the chiller, rendering the machine inoperable. The most common causes of crystallization are:

- Insufficient purging of the chiller
- Low condenser water temperature
- High solution concentration
- Power failure while the chiller is in operation

Present-day absorption chillers are not likely to experience crystallization if properly operated and maintained. Most manufacturers provide adequate controls to safeguard against crystallization.

On steam-type absorption chillers it is important that the steam's quality and pressure are as per the manufacturer's specifications. A condensate sample should be obtained and analyzed at least once per season.

The typical component content allowance is as follows:

pH	7 to 9 [at 75°F (24°C)]
Ni	Less than 0.17 ppm (mg/L)
Cu	Less than 0.40 ppm (mg/L)
Carbonic acid	Less than 3 ppm (mg/L)

41.11.3 The Burner and Pumps

The burner of direct-fired absorption chillers should be checked and serviced at least once per season. This service should include a complete analysis of combustion products.

The solution and refrigerant pumps of the absorption chillers should periodically be disassembled, inspected, and rebuilt if necessary. Follow the manufacturer's schedule for the recommended intervals.

41.11.4 Water Treatment and Tube Cleaning

Absorption chillers are designed for maximum efficiency and long life. However, as with any piece of equipment of comparable size and complexity, it can only deliver its design output and efficiency if it is properly operated and maintained. One of the most important elements of proper maintenance is the cleanliness of the tubes.

To prevent fouling and scaling, it is very important that the chiller's owner-operator engage the services of a reputable water-treatment specialist for both the initial charging of the system and its continuous monitoring and treatment. Improperly treated or maintained water can result in decreased efficiency, high operating costs, and premature failure of the chiller.

It is equally important to clean and inspect the tubes of the absorber, condenser, and evaporator at the frequencies recommended by the manufacturer. That is, in addition to periodic cleaning, the tubes must be inspected for wear and condition. Tube failures usually occur as a result of corrosion, erosion, and the stress corrosion and fatigue caused by thermal stress.

Eddy-current inspection of all heat-exchanger tubes is an invaluable preventive-maintenance method. It provides a quick method of determining tube condition at a reasonable cost.

As with all maintenance programs, follow the manufacturer's recommendations on inspection intervals and employ a reputable organization to perform the eddy-current testing.

41.12 REFERENCES

1. *1985 ASHRAE Handbook, Fundamentals*, ASHRAE, Atlanta, 1985, chap. 1.
2. *1983 ASHRAE Handbook, Equipment*, ASHRAE, Atlanta, chap. 14.

CHAPTER 42
CENTRIFUGAL CHILLERS

John M. Schultz
Retired Chief Engineer, Centrifugal Systems, York International Corporation, York, Pennsylvania

42.1 INTRODUCTION

A refrigeration system that uses a centrifugal compressor to cool water for air-conditioning purposes is called a "centrifugal chiller." The capacities of these machines range from 100 to 10,000 tons (350 kW to 35 MW) of refrigeration. Their reliability is high, and their maintenance requirements low, because centrifugal compression involves the purely rotary motion of only a few mechanical parts.

Most centrifugal chillers have water-cooled condensers. The source of the condenser water is usually a cooling tower, but lake or river water can also be used. Air-cooled condensers are employed in locations where cooling water is not available.

In sizes above 1300 tons (4600 kW), the major components of a water-cooled chiller (heat exchangers, compressor, etc.) must be shipped individually for mounting and connecting at the jobsite; these chillers are called "field-erected" units. Smaller sizes can be completely assembled, piped, and wired before they leave the manufacturer's plant; these units are called "factory packages."

Figure 42.1 is a photograph of a factory package which even includes a reduced-voltage motor starter. Only the water piping and power supply need to be connected to this unit in the field before startup. Figure 42.2 is a rear view of the same unit. Figure 42.3 is a photograph of a large field-erected unit.

42.2 REFRIGERATION CYCLES

Many water-cooled chillers employ a single stage of compression in the basic vapor-compression refrigeration cycle shown in Fig. 42.4. Sometimes a liquid subcooler is added between the condenser and the expansion device to improve the coefficient of performance (reduce the power requirement) of the cycle. When a subcooler is used, it is usually built into the bottom of the condenser. Two or three stages of compression are employed in some water-cooled designs and in all air-cooled units. In these cases, some improvement in the coefficient of performance can be obtained by using the intercooled refrigeration cycle of Fig.

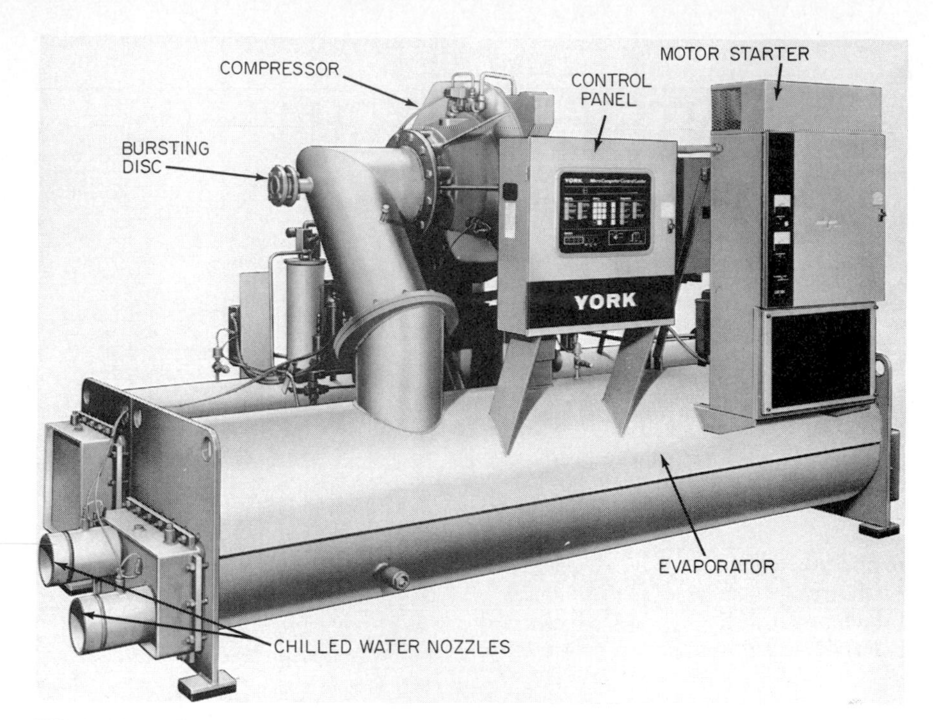

FIGURE 42.1 Factory-packaged centrifugal chiller—front view. Capacity: 350 tons (1200 kW).

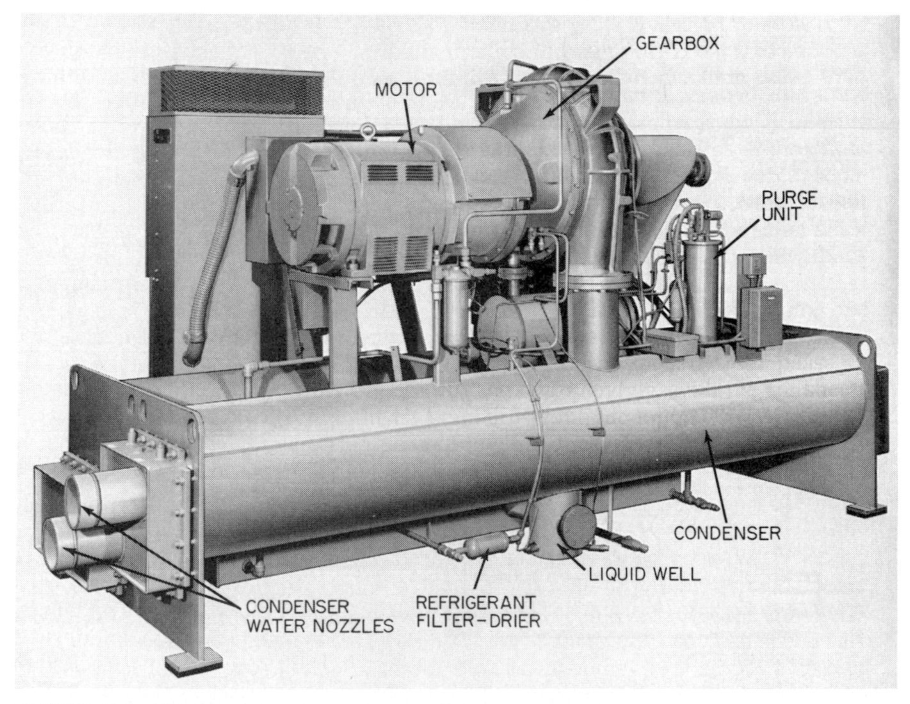

FIGURE 42.2 Factory-packaged centrifugal chiller—rear view.

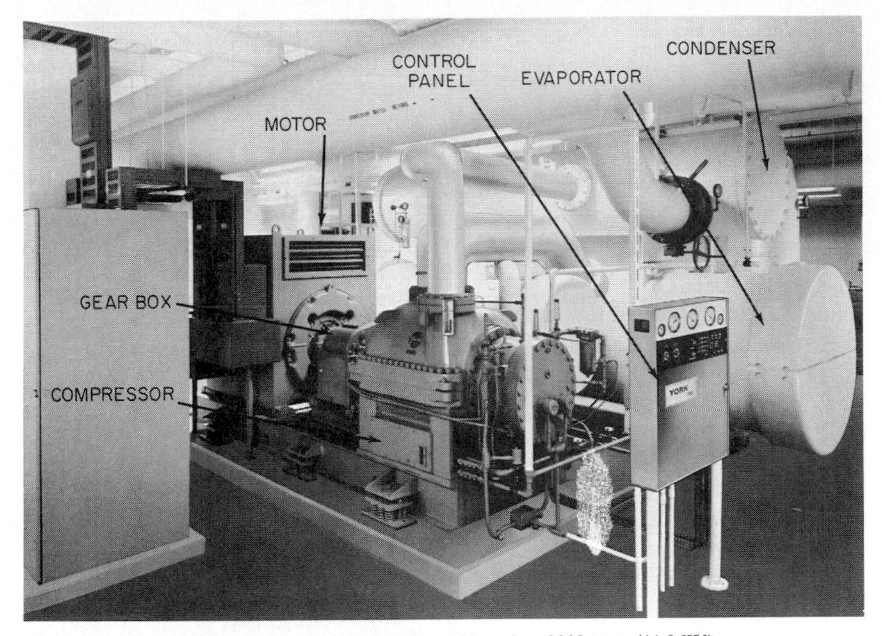

FIGURE 42.3 Field-erected centrifugal chiller. Capacity: 3000 tons (11 MW).

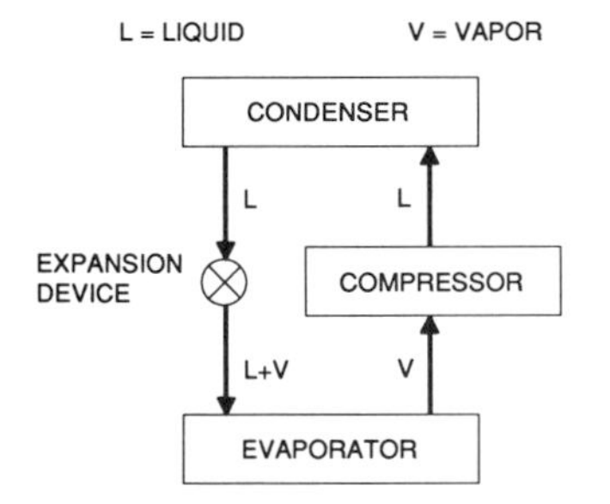

FIGURE 42.4 Single-stage vapor-compression refrigeration cycle.

42.5. This figure shows two stages of compression and one intercooler. When three compression stages are used, two intercoolers are possible.

The degree to which subcooling and intercooling reduce the power requirement of the basic refrigeration cycle depends on which refrigerant is being used and how great the temperature difference (temperature lift) is between the evaporator and the condenser. The choice of refrigerant and cycle for a particular application involves many considerations. Different manufacturers choose different combinations to achieve the same overall results. Both high-cost, low-power units and low-cost, high-power units are available from all manufacturers.

42.3 COMPONENTS

In most centrifugal chillers, the compressor is driven by a squirrel-cage induction motor, either directly or through speed-increasing gears. In factory packages, the gears are built into the compressor; in field-erected units, a separate gearbox is usually employed. Because the optimum performance of a centrifugal compres-

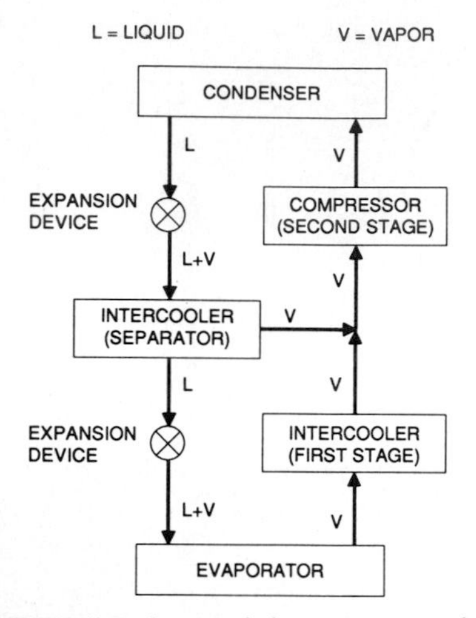

FIGURE 42.5 Intercooled vapor-compression refrigeration cycle. (*Courtesy of Gas Energy, Inc.*)

sor is strongly influenced by its rotating speed, different gear ratios are used with the same compressor for different operating conditions.

Steam turbines, gas turbines, and internal-combustion engines are also used to drive centrifugal compressors, with or without gears, generally in a field-erected configuration.

42.3.1 Motors

A motor that is built into the compressor and operates in a refrigerant atmosphere is called a "hermetic" motor. Refrigerant liquid and vapor cool the motor. No mechanical shaft seal is needed because the shaft that connects the motor to the compressor never penetrates the refrigerant envelope. Additional compressor power is required to provide the refrigeration effect that cools the hermetic motor and to overcome the windage loss that results from running the motor in a dense refrigerant atmosphere.

An "open" motor is one that operates in atmospheric air and is air-cooled. It is often attached directly to the compressor, in which case it is called a "close-coupled" motor. Open motors tend to be more efficient than hermetic motors and are easier to maintain. If the winding of a hermetic motor fails, the entire refrigerant charge may be contaminated by the burning insulation.

42.3.2 Evaporators

Horizontal shell-and-tube evaporators are used in centrifugal chillers. These are of the flooded type with liquid refrigerant surrounding the tubes. The chilled water flows inside the tubes. The tube bundle occupies only the lower part of the shell; the upper part is a space in which liquid refrigerant drops out of the vapor that is flowing to the compressor. Mist eliminators are sometimes installed above the tube bundle to help remove the entrained liquid from the vapor.

The outside surface (refrigerant side) of the evaporator tubes is finned to increase the heat-transfer area or is coated to increase the boiling heat-transfer coefficient. Sometimes the fins are modified for boiling enhancement. The inside (water side) of the tubes may be smooth, or it may have ribs or corrugations to increase the heat-transfer coefficient of the water film. Figure 42.6 shows three kinds of evaporator tube.

42.3.3 Condensers

Horizontal shell-and-tube condensers are used in water-cooled units, with the cooling water inside the tubes. The tube bundle fills most of the shell. The outside

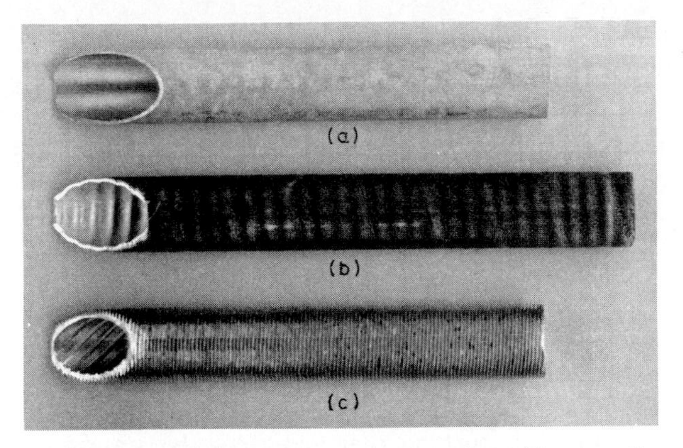

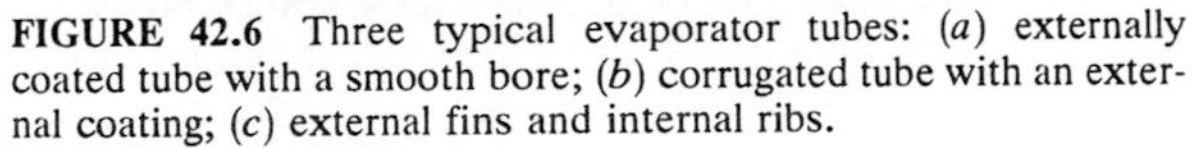

FIGURE 42.6 Three typical evaporator tubes: (*a*) externally coated tube with a smooth bore; (*b*) corrugated tube with an external coating; (*c*) external fins and internal ribs.

surface of the tubes is finned to increase the refrigerant-side heat transfer. Sometimes the fins are modified to increase the condensing coefficient. The inside of the tubes may be smooth, or it may have ribs to enhance water-side performance.

Air-cooled condensers contain coils of externally finned tubes. The refrigerant condenses inside the tubes and drains into a receiver under the coils. Fans are used to force the flow of cooling air over the outside of the tubes. Some liquid subcooling almost always occurs in air-cooled condensers.

Liquid refrigerant flows from the condenser to the evaporator through an expansion (throttling) device. The device may be a liquid-level control or float valve in the receiver of an air-cooled condenser or in a liquid well at the bottom of a water-cooled condenser. Alternatively, it may be a thermostatic expansion valve that controls the superheat of the vapor leaving the evaporator. Often, the expansion device is merely one or more orifices that have been sized and arranged so as to pass a negligible amount of vapor with the flowing liquid.

42.3.4 Purge Units

Most packaged centrifugal chillers use refrigerant 11. The evaporators of these chillers operate in a vacuum. Some air inevitably leaks into the vacuum, bringing with it a small amount of water vapor. The same thing occurs, to a lesser degree, with refrigerant 114. In the case of refrigerant 113, both the evaporator and the condenser operate in a vacuum. The leakage air and water vapor tend to collect in the condenser, from which they must be purged in order to maintain condenser heat-transfer performance and prevent acid formation in the refrigerant.

To remove these noncondensables, a purge unit is provided. This device draws refrigerant from the top of the condenser along with any air or water vapor that may be present. The purge unit compresses and condenses most of the refrigerant and returns the condensate to the system. The air does not condense and is blown off, along with a small amount of uncondensed refrigerant, to the atmosphere. Some water vapor is expelled with the air, and some condenses with the condensed refrigerant. The condensed water is removed from the condensed refrigerant either by a filter-drier cartridge or by a gravity (density) separator. Water from the latter must be drained manually.

Purge units vary in their ability to separate refrigerant from the air that they expel to the equipment room. Efficiency in this respect is desirable because of the cost of replacing the lost refrigerant and because the needless release of refrigerant to the atmosphere can be harmful to the environment.

Some purge units operate automatically, and some of these include an alarm system that indicates when an air leak is causing an excessive amount of purging.

42.3.5 Pumpout Units

Refrigerants 12, 22, 114, and 500 have standby (idle) pressures that exceed atmospheric pressure. When a chiller that uses one of these refrigerants needs to be serviced, the refrigerant must be removed from the system and stored. A pumpout unit that contains a compressor, a condenser, and a storage receiver is a requirement for servicing one of these high-pressure chillers. The pumpout unit is a part of some of the factory packages that use high-pressure refrigerants.

Pumpout provisions vary from one manufacturer to another in their ability to isolate and save refrigerant during the servicing of a unit. Efficiency in conserving refrigerant is even more important for a pumpout unit than it is for a purge unit because the potential for refrigerant loss is greater with high-pressure refrigerants than it is with low-pressure refrigerants.

42.3.6 Additional Information

Much detailed information about compressors, motors, heat exchangers, etc., can be found in Ref. 1.

42.4 CAPACITY CONTROL

The volumetric flow capacity of a centrifugal compressor must be adjusted for variations in evaporator loading and for changes in the temperature lift (compressor head) between the evaporator and the condenser. A modulating capacity control system is a part of every centrifugal chiller. Several different control methods are used. The simplest and least efficient method of capacity control is to throttle the compressor suction flow; this is done only on some small factory packages.

The remaining centrifugal compressors have a set of movable guide vanes at the inlet to the first impeller. These are called "prerotation vanes" (PRVs) because their function is to create a swirling motion (rotation) in the refrigerant's flow stream just ahead of the impeller. Figure 42.7 is a photograph of a typical PRV design. A compressor that has more than one impeller may have more than one set of PRVs. The aerothermodynamic losses associated with PRV control are less than those which are incurred with throttling control.

Figure 42.8 represents the performance of a centrifugal compressor operating at constant speed and using PRV control. Each vane position generates a line that terminates at a "surge" point above which the refrigerant alternately flows backward and forward through the compressor. This unstable operating condition,

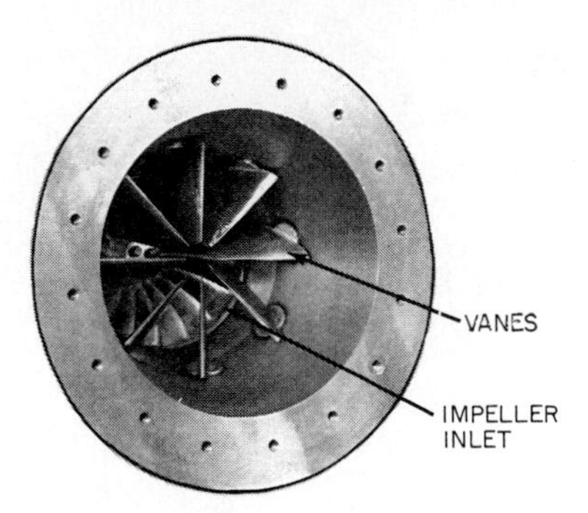

FIGURE 42.7 Prerotation vanes (PRVs).

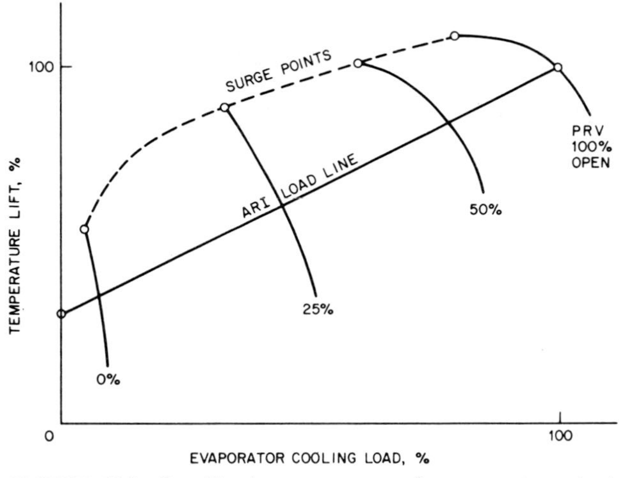

FIGURE 42.8 Centrifugal compressor performance at constant speed. (*Courtesy of Gas Energy, Inc.*)

called "surging," can be avoided by recirculating some of the compressor discharge vapor back to the compressor suction. The recirculated vapor, called "hot-gas bypass" (HGBP), adds to the evaporator flow so that the total flow is in the region of stable operation to the right of the surge points.

HGBP is inefficient because it imposes additional load on the compressor. The need for HGBP is reduced in some compressors by using diffuser controls that move the surge points to the left. Methods that can be used to reduce the size of the surge area are (1) varying the width of the radial diffusion space that surrounds the impeller and (2) throttling the flow between the impeller and the diffuser.

The most efficient way to control the capacity of a centrifugal compressor is to vary its speed. Turbine and engine drives lend themselves to this type of control rather easily. Wound-rotor motors can also be used. The most common method of varying speed, however, is to vary the frequency of the alternating current that is supplied to an induction motor that drives the compressor. Variable-frequency inverters are used to modulate the motor speeds of many centrifugal chillers with

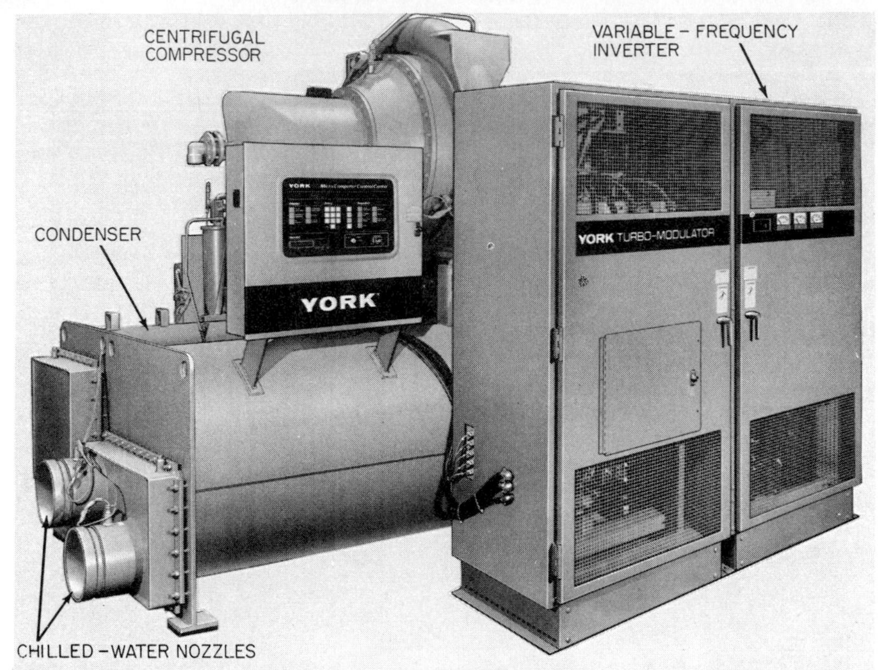

FIGURE 42.9 Variable-speed inverter drive for centrifugal package.

capacities up to 700 tons (2500 kW). The capacity control system of most inverter drives combines both speed control and PRV control to optimize the compressor efficiency at every possible operating condition. PRV control is not used with variable-speed turbine and engine drives. Figure 42.9 shows a variable-frequency inverter connected to a packaged centrifugal chiller.

42.5 POWER CONSUMPTION

Of primary importance in the design of an air-conditioning system is the optimization of initial cost and operating cost. The principal operating cost of a centrifugal chiller is the cost of the power that it consumes. At full load, the power requirement of most water-cooled, motor-driven chillers ranges from 0.5 to 0.9 kW/ton (0.14 to 0.26 kW/kW). The variation in electric power consumption is due to a number of factors that can be summarized by the equation

$$\text{kW/ton} \sim \frac{(\text{lbm}^*/\text{min/ton})(\text{temperature lift, °F})}{\begin{pmatrix}\text{compression}\\ \text{efficiency}\end{pmatrix}\begin{pmatrix}\text{mechanical}\\ \text{efficiency}\end{pmatrix}\begin{pmatrix}\text{electrical}\\ \text{efficiency}\end{pmatrix}}$$

*lbm = pounds mass

or
$$\text{kW/kW} \sim \frac{[\text{kg/(s} \cdot \text{kW)}](\text{temperature lift, °C})}{\begin{pmatrix}\text{compression}\\ \text{efficiency}\end{pmatrix}\begin{pmatrix}\text{mechanical}\\ \text{efficiency}\end{pmatrix}\begin{pmatrix}\text{electrical}\\ \text{efficiency}\end{pmatrix}}$$

The refrigerant mass flow rate {lb/min/ton [kg/(s · kW)]} depends mainly on which refrigerant and which refrigeration cycle are being used. The evaporating and condensing temperatures are of secondary importance. To the extent that these temperatures matter, reducing the difference between them (temperature lift) reduces the refrigerant flow rate.

Temperature lift is the chief factor in determining the compression work input (head) per unit of flow. It also influences the compressor's aerodynamic efficiency (compression efficiency). Lowering the temperature lift increases the compression efficiency at full load; it also increases the efficiency at part load when speed control is used, but not when PRV control is used. Temperature lift equals the difference between the water temperatures leaving the heat exchangers added to the temperature differences within the heat exchangers between the refrigerant and the leaving water:

$$\text{Temperature lift} = (\text{LCWT} - \text{LEWT}) + (\text{LEWT} - \text{ET}_p) + (\text{CT}_p - \text{LCWT})$$

where LCWT = leaving condenser water temperature, °F (°C)
LEWT = leaving evaporator water temperature, °F (°C)
ET_p = refrigerant evaporator temperature, °F (°C)
CT_p = refrigerant condenser temperature, °F (°C)

Much can be done to lower the power consumption of a centrifugal chiller by increasing the chilled-water temperature, decreasing the condenser water temperature, and using effective heat exchangers that have small leaving-temperature differences.

Compression efficiency depends primarily on the aerodynamic design of the centrifugal impeller(s) and diffuser(s). Additional considerations, when more than one stage of compression is used, are the design of the flow passages between the stages and the mixing of intercooler vapor flow(s) with interstage flow(s).

The mechanical efficiency of a centrifugal chiller is reflected by its oil-cooling requirement. The friction losses that are carried off by the lubricating oil are generated in the compressor and hermetic-motor bearings, the gearbox, and the shaft seal.

The electrical efficiency is the product of the motor's efficiency and (when used) the variable-frequency inverter's efficiency.

42.5.1 Part Load

When the evaporator's cooling load is reduced, the temperature lift is also reduced. This is because the heat-transfer rates are reduced in both heat exchangers, which lowers the leaving-temperature differences, and because (usually) the entering condenser water temperature (ECWT) or air temperature is lower. With constant-speed PRV control, the compression efficiency is reduced at part load, as are the other efficiencies as well. The net effect on electric power consumption (kW) when the leaving evaporator water temperature (LEWT) is held constant is illustrated in Fig. 42.10 for a water-cooled unit.

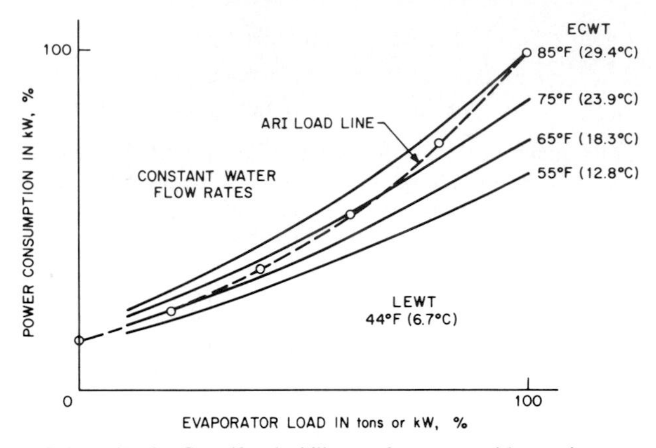

FIGURE 42.10 Centrifugal chiller performance with varying condenser water temperature. *(Courtesy of Gas Energy, Inc.)*

The influence of temperature lift on power consumption can easily be seen in Fig. 42.10. Except in the immediate vicinity of the full-load design point (100 percent load and power), the ECWT lines will be everywhere lower with variable-speed capacity control than they will be with constant-speed PRV control. The difference becomes more pronounced as the ECWT values become less than the full-load value. The reason is that compression efficiency is adversely affected by low temperature lift when PRV control is used.

Because centrifugal chillers rarely operate at full load, part-load power is considerably more significant than full-load power in determining the total annual operating cost.

42.5.2 Multiple Units

Two or more centrifugal chillers are often operated together to cool a large building. This not only provides some cooling capacity in the event of a chiller failure, it also saves power. When two machines are used, one machine alone can handle a cooling load that is less than half of the building's maximum load. The load will be high for the one operating machine, and so will be its efficiency. The power consumption will be less than it would be if the load were being handled by one machine of twice the capacity.

42.5.3 Heat Recovery

In cool weather, part of a building may need heating at the same time as another part needs cooling. In that event, the condenser heat can be applied to the heating load instead of being dissipated to the atmosphere in a cooling tower. This mode of operation is called "heat recovery." The heating water is circulated in a closed loop that is separated from the cooling tower water in order to avoid contamination of the former by the latter. The two water circuits may pass through the same "double bundle" condenser, or the heating water may have a separate "auxiliary" condenser. Factory packages with heat-recovery condensers are available with capacities up to 1300 tons (4600 kW).

For the heating to be effective, the temperature of the water leaving the condenser must be over 100°F (38°C). This makes the ECWT 95°F (35°C) or more. At the same time, the evaporator load may be only half the full-load value. Figure 42.10 shows that these conditions require considerably more power than would be the case with a normal ECWT of 70°F (21°C) at that load. The cost of this additional power is included in the economic evaluation of a heat-recovery application.

The temperature lift during heat-recovery operation is greater than it is during full-load cooling operation. Figure 42.8 indicates that at loads less than 50 percent of full load, the compressor will generally be operating in the surge area when heat-recovery temperature lift is needed. An automatic HGBP valve is required in heat-recovery units to prevent surging.

Heat-recovery operation occurs during cool weather when the cooling load can be met with an LEWT that is higher than the full-load value. By increasing the LEWT during heat-recovery operation, the temperature lift will be reduced. This will reduce the need for HGBP and its power penalty. Even without heat recovery, it always saves power to operate with the highest possible LEWT.

Because the compressor must be designed for the heat-recovery lift, its efficiency will be less during the summer cooling season than it would be if it did not have this extra capability. The cost of the extra summertime power consumption that results from the lower compression efficiency is also part of the economic analysis. With speed control, the summer power penalty is significantly less than it is with PRV control.

42.5.4 Free Cooling

In cold weather, the condenser water temperature may be less than the chilled-water temperature that is needed for the cooling load that exists at that time. Under these conditions, refrigerant will migrate from the evaporator to the condenser without the compressor running. An appreciable amount of cooling can be accomplished in this manner, particularly if the migration is made easier by opening a bypass pipe around the compressor. This mode of operation is called "free cooling" because it involves no chiller power.

In some units, more refrigerant is needed during free-cooling operation than can be tolerated during normal compressor operation. A storage tank is provided in these cases to handle the extra refrigerant.

42.6 RATINGS

A centrifugal chiller's rating is a statement of its refrigeration capacity, power consumption, and heat-exchanger water-pressure drop at a specific set of operating conditions. "Refrigeration capacity" is the net usable cooling capacity that remains after the chiller's own internal needs (hermetic-motor cooling, oil cooling, purge-unit cooling, etc.) have been satisfied. In a motor-driven unit, the "power consumption" is the total electric power that is required to operate the unit (motor, inverter, control panel, oil heater, oil pump, purge unit, condenser fans, etc.). Turbine and engine drives involve additional forms of energy consumption (steam rate, fuel flow, etc.). In these cases, only the input power to the compressor shaft may be known by the chiller manufacturer.

The Air-Conditioning and Refrigeration Institute (ARI) has established a standard set of operating conditions at which centrifugal chillers are to be rated (Ref. 2).

42.6.1 Full Load

The standard ARI conditions for full-load ratings are:

Evaporator: LEWT = 44°F (6.7°C). Water flow rate = 2.4 gal/min/ton [0.043 L/(s · kW)]. It follows from this that the temperature of the chilled water returning to the evaporator is 54°F (12.2°C).

Water-cooled condenser: ECWT = 85°F (29.4°C). Water flow rate = 3.0 gal/min/ton [0.054 L/(s · kW)]. The temperature of the cooling water leaving the condenser is about 95°F (35.0°C). The exact value depends on how much power is put into the refrigerant by the driver of the compressor and by the electrical auxiliaries of the unit.

Air-cooled condenser: Entering air temperature = 95°F (35.0°C). Barometric pressure = 29.92 inHg (101 kPa).

Heat-exchanger fouling: Water-side fouling allowance = 0.00025 h · ft^2 · °F/Btu (0.044 m^2 · °C/kW). Air and refrigerant-side fouling = 0.

42.6.2 Part Load

The standard ARI conditions for part-load ratings are the same as for full-load ratings except that the ECWT of water-cooled condensers varies linearly with load from 85°F (29.4°C) at full load to 60°F (15.6°C) at zero load. For air-cooled condensers, the entering air temperature varies linearly with load from 95°F (35.0°C) at full load to 55°F (12.8°C) at zero load. The standard part-load conditions for water-cooled condensers are indicated by the ARI load lines in Figs. 42.8 and 42.10.

The ARI standard includes a method for calculating an *integrated part-load value* (IPLV) for power consumption which reflects the typical loads and conditions that are encountered over a full operating year. The IPLV is an index of total annual power consumption.

Temperatures, flow rates, and fouling allowances other than those of the ARI standard may be needed for some applications, but the standard values are typical of most applications. Standardized rating conditions are essential if valid comparisons are to be made among the offerings of different chiller manufacturers.

42.6.3 Testing

The ARI standard includes a method for verifying the rating of a centrifugal chiller by test. The standard specifies the allowable tolerances for tested values of full-load capacity, full-load power consumption, part-load power consumption at selected capacities, and the overall IPLV. The tolerance values are a function of the chilled water temperature range at full load and a percent of full-load capacity at which the test is conducted. Typical values are 5 percent for full-load capacity and power consumption, and 10 percent for half-load power consumption and IPLV.

The ARI standard specifies the test procedures, instruments, measurements, and calculations that are needed to verify a chiller's rating within the specified tolerances. To comply at the jobsite with all of the test requirements is sometimes difficult or impractical. Factory tests of packaged units can be performed by chiller manufacturers in order to avoid the problems that may arise with field testing.

Because factory testing of individual units for individual users is time-consuming and expensive, ARI has instituted a program that certifies the performance of all a manufacturer's units of a particular type and size. Specific models from a manufacturer's product line are periodically chosen at random for testing in a factory test facility that has been certified by ARI. The tests are conducted and the results are verified by an independent testing agency under contract to ARI. The model numbers of the chillers whose ratings have been confirmed by these tests are published by ARI in a directory of certified products (Ref. 3). The directory is updated and reissued periodically to reflect the results of the continuing random test program.

When a new unit is tested at the factory, its heat-exchanger tubes are clean. This makes the water-side fouling resistance zero. To simulate the effects of rated fouling during the test, the temperature lift must be increased above its clean-tube value. To accomplish this, the chilled-water temperature is lowered, and the condenser water temperature raised, by amounts that are calculated by an equation in the ARI standard. In a field test, it is usually assumed that the water-side fouling resistances coincide with the rated fouling allowances, in which case no adjustments of the water temperatures are made. If the tubes are cleaned just before the test, however, the fouling resistances are zero and the fouling allowances must be simulated by adjusting the water temperatures.

42.7 CONTROLS

Most centrifugal chillers have electronic microprocessor-based control systems, although electromechanical and pneumatic controls are used in some applications. Microprocessor controls can be programmed to start and stop a chiller automatically at different times of the day on different days of the week. The chiller controls can also operate the cooling tower fans and the building's water pumps. At startup, the chilled-water temperature's pulldown rate and the pulldown motor current can be programmed to limit electric power demand costs.

Microprocessor control systems can be interfaced with building energy management (BEM) computer systems. The BEM system can start and stop the chiller and can vary the LEWT and the motor-current limit.

The status of any of the system-monitoring parameters (pressures, temperatures, etc.) can be obtained from an alphanumeric display on the microprocessor control panel. The set points of the controls can also be read and compared with the measured operating values. If one of the controls stops the compressor, the display indicates the cause of the shutdown and provides diagnostic information.

42.7.1 Chilled-Water Temperature

The primary function of the control system is to maintain the LEWT at a given set point by modulating the compressor's speed and/or PRV position. At loads that are less than the compressor's minimum capacity (about 10 percent of full load), the modulating controls become ineffective. Under these conditions, the LEWT falls below the control set point and the compressor is automatically stopped. After the building's load has restored the chilled-water temperature to its normal value, the compressor is automatically started again. To prevent this off-on cycling from occurring so often that it harms the motor, a time delay of about 30 minutes is interposed between starts.

If the evaporator pressure becomes too low, a low-evaporator-pressure control (or low-refrigerant-temperature control) will stop the compressor before the evaporating refrigerant can freeze the chilled water in the evaporator tubes. Because this is an abnormal condition, the compressor will not restart automatically after a low-evaporator-pressure or -temperature shutdown.

One of the causes of low evaporator pressure is low chilled-water flow. A flow switch should be interlocked with the chiller controls in order to prevent the compressor from starting or operating without sufficient chilled-water flow. Sometimes an alternate interlock is used that prevents the compressor from running unless the chilled-water pump is running. This alternate indication of chilled-water flow is less certain than a flow switch.

42.7.2 Condenser

A high-discharge-pressure cutout will stop the compressor if the condenser pressure becomes too high. A condenser water flow switch or a condenser pump interlock are used to prevent the compressor from starting or operating without sufficient condenser water flow.

If the condenser pressure becomes too low, the flow of liquid refrigerant to the evaporator may become inadequate. This can usually be prevented by cycling the cooling-tower fans, but in some cases a cooling tower bypass valve in the condenser water circuit must be used to control the minimum condenser pressure.

A cooling tower bypass is required by a heat-recovery unit. When both condenser water circuits are used simultaneously, the temperature of the tower water must be forced up to the temperature of the heating water by a controlled bypass of the cooling tower.

42.7.3 Motor

The maximum motor current can be set at a value that is less than the full-load motor current in order to reduce power costs and/or demand charges. In this control mode, the compressor capacity control is overridden by the motor-current control.

Other controls will stop the unit in the event of motor overload (excessive current), high voltage, low voltage, a momentary power fault, phase imbalance, or excessive winding temperature in a hermetic motor.

42.7.4 Compressor

Low-oil-pressure and high-oil-temperature cutouts are provided to stop the compressor if its proper lubrication is threatened. The oil-pump motor is protected by cutouts that respond to overload and/or a high winding temperature. When water-cooled oil coolers are employed, the flow of water may be controlled in order to prevent the oil temperature from becoming too low. A thermostatically controlled heater keeps the oil hot during standby (idle) periods in order to minimize the amount of refrigerant vapor that is absorbed by the oil.

A high-temperature cutout in the compressor discharge pipe stops the compressor if overheating occurs. Some chillers also have a device that stops the compressor if surging occurs.

Excessive compressor speed must be prevented if a turbine or engine drive is used. The overspeed trip device that accomplishes this is mounted on the driver, not on the compressor.

42.8 INSTALLATION

When a centrifugal chiller is installed in a building, sufficient space must be left around it and above it to allow for future maintenance and service needs. At one end of the heat exchangers, the space must be long enough to allow for cleaning and possible replacement of the heat-transfer tubes. The water piping should be arranged with access to the tubes in mind. For maximum accessibility, "marine" water boxes can be employed. These provide an access port through which the heat-transfer tubes can be serviced without disconnecting any of the water piping.

42.8.1 Noise and Vibration

The chiller is mounted on vibration isolator pads or springs to prevent the transmission of structure-borne noise and vibration to the building. Spring isolators are generally used only in upper-floor installations and other vibration-sensitive locations. When spring mountings are needed, the water piping is isolated from the chiller and the pipe hangers are isolated from the building.

Some applications are sensitive to airborne noise. In these cases, the acoustic treatment of the equipment room is based on the noise characteristics of the particular unit to be installed. The chiller manufacturer will supply the necessary information in the manner prescribed by an ARI standard for measuring and reporting machinery noise (Ref. 4). (See Chaps. 49 and 50 for further discussions of noise and vibration.)

42.8.2 Ventilation

Building codes and safety standards include ventilation requirements that are based, in part, on the amount of refrigerant charge that is contained in the largest chiller in the equipment room. Also included is a requirement that the outlet of each chiller's pressure-relief device (bursting disc or relief valve) be piped to the outdoors to prevent a large amount of refrigerant vapor from filling the equipment room during a fire or other emergency.

When an open motor is used to drive a compressor, pump, or fan in the equipment room, the effect of the motor's heat on the temperature of the air in the room is also a consideration.

42.8.3 Thermal Insulation

The temperature of the evaporator, the low-side piping on the chiller, and the chilled-water piping will generally be lower than the dew point of the air in the equipment room. These components will "sweat" unless they are thermally insulated. Factory packages can be obtained with the necessary thermal insulation already in place.

42.8.4 Instrumentation

Pressure gauges and thermometers are usually installed in the water piping to aid in operating and servicing the unit. To obtain the water pressure drop of each heat exchanger, one gauge that has valved connections to both the inlet and outlet water pipes of the exchanger is more accurate than two separate gauges.

Heat-exchanger pressure drop provides only a rough indication of the water flow rate. If the chiller's performance is to be measured in the equipment room with reasonable accuracy, flow meters must be installed in the water piping.

42.8.5 Power Supply

The motor starter may be a part of a factory-packaged chiller, or it may be mounted separately from the unit. A fused disconnect switch is needed in either case. Capacitors for power-factor correction can be used with any starter except the star-delta, open-transition type.

42.8.6 Safety Codes

The design, manufacture, and installation of centrifugal chillers are governed by an ANSI/ASHRAE safety code (Ref. 5). This code invokes other codes and standards, such as the National Electrical Code® and the ASME Boiler and Pressure Vessel Code. State and local laws and codes must also be considered.

42.9 OPERATION

For the most part, centrifugal chillers operate automatically. The only manual functions are starting the unit, setting the LEWT and the motor-current limit, and stopping the unit. In many cases, some or all of these functions are also automatic.

A human operator performs other functions, such as maintaining a log of operating information that will aid in the servicing of the unit and will even indicate the need for service in the first place. Such indications as excessive air leakage, loss of refrigerant, oil loss, insufficient oil temperature, plugging of the oil filter, fouling of the heat-exchanger tubes, internal water leakage, unusual vibration, and excessive surging can all be deduced from operator logs and observations. Of course, these indications can be automated, too, by computerized data logging and programmed analysis.

42.10 MAINTENANCE

Specific maintenance requirements vary from one chiller to another. Routine procedures for most chillers include an annual change of the compressor's lubricating oil, the oil-filter element, and the refrigerant filter-drier cartridge(s). The filter-drier cartridge of a purge unit is replaced several times per year.

The oil-filter element should be checked for bearing wear particles and for an excessive amount of iron rust particles. The compressor bearings are not usually inspected unless there is an indication of need, such as wear particles in the oil. Most rust particles originate in the heat exchangers, not in the compressor. They are an indication of air and moisture in the system. Chemical analyses of the oil and the refrigerant are also performed in order to evaluate the internal health of the unit. A check for leaks is usually made twice per year unless excessive purging or a lowering of the refrigerant liquid's level indicates an earlier need.

Control settings and instrument calibrations should be checked annually, as should the electrical resistances of the compressor's motor windings.

Heat-Exchanger Tubes. Proper maintenance of the heat-transfer tubes begins with maintaining the quality of the water that flows through the tubes. The chilled water is almost always contained in a closed circuit, so one initial chemical treatment may last for several years. The condenser water circuit is almost always open to the atmosphere and is therefore in continuous need of chemical treatment in order to prevent fouling (sedimentation, corrosion, scaling, etc.) of the inside surface of the tubes. Samples from both water circuits should be analyzed each year in order to check the chemical quality of the water. (See Chap. 53 for a discussion of water treatment.)

The water strainers are cleaned annually and the inside of the heat-transfer tubes inspected. The tubes are cleaned if the inspection indicates a need. Automatic tube-cleaning systems can be installed that will operate continuously, whenever the unit is running. Keeping the tubes clean reduces power consumption because water-side fouling increases the temperature lift.

The tubes can also be inspected for wear on the outside surface (at the tube supports) and for fatigue cracks in the tube wall. This is accomplished by passing an electromagnetic probe through each tube. The moving probe generates eddy currents in the tube wall and senses the impedance of the tube wall. The output of the sensing coil is displayed on a cathode-ray tube (CRT) or recorded on a strip chart. Fluctuations of the output signal indicate variations in the thickness or composition of the tube wall.

42.11 REFERENCES

1. *1988 ASHRAE Handbook, Equipment*, ASHRAE, Atlanta, 1988.
2. ARI Standard 550-88, *Standard for Centrifugal or Rotary Water-Chilling Packages*, ARI, Arlington, VA, 1988.
3. *Directory of Certified Applied Air-Conditioning Products*, Part X: "Certified Centrifugal or Rotary Screw Water-Chilling Packages" ARI, Arlington, VA, 1989.
4. ARI Standard 575-87, *Standard for Measuring Machinery Sound within Equipment Rooms*, ARI, Arlington, VA, 1987.
5. ANSI/ASHRAE Standard 15-89, *Safety Code for Mechanical Refrigeration*, ASHRAE, Atlanta, 1989.

CHAPTER 43
SCREW COMPRESSORS

Kenneth Puetzer
Chief Engineer, Sullair Refrigeration, Subsidiary of Sundstrand Corp., Michigan City, Indiana

43.1 INTRODUCTION

43.1.1 History

The screw compressor was originally developed in Germany in the late 1800s but was not commercially applied until the mid-1900s. Modern screw compressors for the refrigeration and air-conditioning industry use an oil-flooded arrangement.

43.1.2 Design

The screw compressor is a positive-displacement compressor that uses a rotor driving another rotor (twin) or gaterotors (single) to provide the compression cycle. Current designs consist of both the twin screw (Fig. 43.1) and the single screw (Fig. 43.2). Both designs use injected fluids to cool the compressed gas, seal the rotor(s), and lubricate the bearings. Compressor designs may incorporate an internal valve for capacity control, economizer ports for improved performance, and an internal valve for variation of internal volume ratio.

A typical screw compressor package consists of the compressor, driver, oil separator and reservoir, oil cooling, control panel, and valving (Fig. 43.3).

43.2 TWIN-SCREW COMPRESSORS

43.2.1 Design

The most common type of screw compressor used today is the twin screw. As the name implies, it uses a double set of rotors (male and female) to accomplish the compression cycle. The male rotor usually has 4 lobes while the female rotor consists of 6 lobes; this is normally referred to as a 4 + 6 arrangement. However, some compressors, especially for air-conditioning applications, are using other variations, such as 5 + 7.

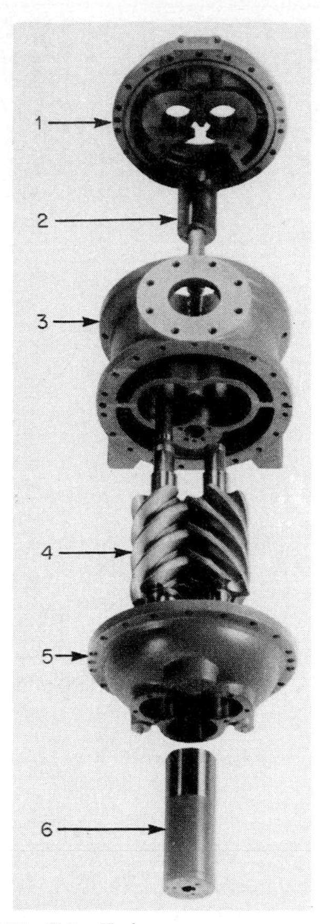

FIGURE 43.1 Twin-screw compressor. (1) Discharge housing; (2) slide valve; (3) stator; (4) male and female rotors; (5) inlet housing; (6) hydraulic capacity control cylinder. (*Sullair Refrigeration.*)

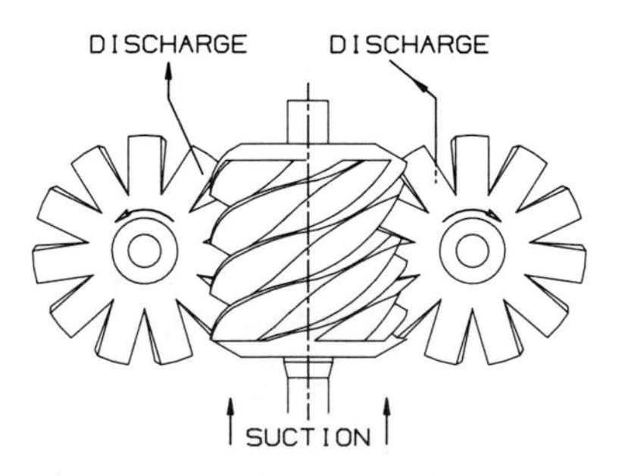

FIGURE 43.2 Single-screw compressor. (*Single Screw Compressor, Inc.*)

FIGURE 43.3 Screw compressor package. (1) Inlet check valve and strainer; (2) compressor; (3) filter; (4) capacity control actuator; (5) oil cooler; (6) oil separator; (7) discharge connection. (*Sullair Refrigeration.*)

Until the mid-1960s, the rotors were cut using a symmetrical or circular profile. This was replaced by the asymmetrical profile, which is a line-generated profile and which improved the adiabatic efficiency of the screw compressor.

43.2.2 Compression Process

The compression process starts with the rotors meshed at the inlet port of the compressor (Fig. 43.4). As the rotors turn, the lobes are separated, causing a reduction in pressure and drawing in the refrigerant. The intake cycle is completed when the lobe has turned far enough to be sealed off from the inlet port. As the lobe continues to turn, the volume trapped in the lobe between the meshing point of the rotors, the discharge housing, and the stator and rotors is continuously decreased. When the rotor turns far enough, the lobe opens to the discharge port, allowing the gas to leave the compressor.

The gas is continuously compressed until the lobes are totally meshed. This eliminates the undesirable condition in reciprocating compressors where the gas in the clearance volume between the piston and the top of the cylinder reexpands within the cylinder, resulting in reduced volumetric efficiency and an increase in horsepower (wattage).

The ratio of the volume of gas trapped after the intake cycle to the volume of gas trapped just before the lobe opens to the discharge port is known as the *built-*

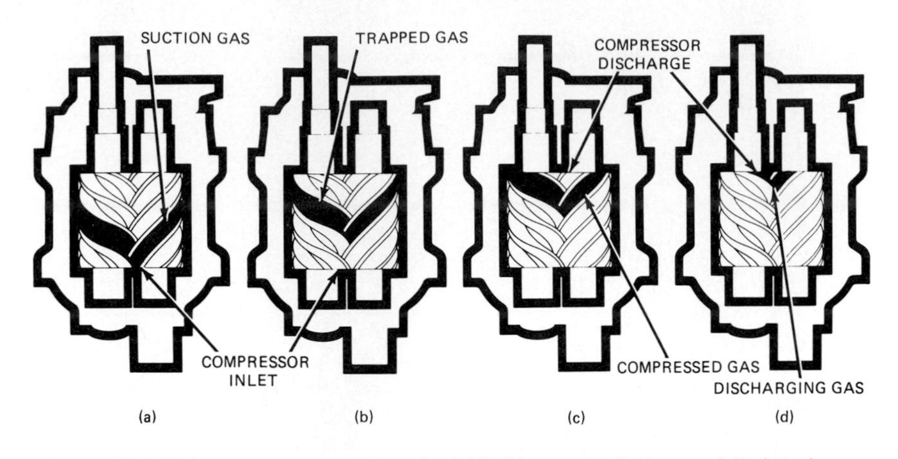

FIGURE 43.4 Twin-screw compression cycle. (*a*) Refrigerant gas is drawn axially into the compressor rotors as they turn past the intake port in the housing. (*b*) The rotors have turned past the intake port; gas is trapped in the compressor housing and rotor cavities. (*c*) As the rotors continue to turn, the lobes reduce the volume in the cavities to compress the trapped gas. (*d*) The process is completed as the compressed gas is discharged through the discharge port. (*Sullair Refrigeration.*)

in volume ratio VI. With the injection oil performing a majority of the cooling and with very low pressure differential across each lobe, the screw compressor can reach compression ratios as high as 20:1.

43.2.3 Lubrication System

The current refrigeration twin-screw compressor is designed with a relatively large oil flow of up to 100 gal/min (6.3 L/s) for large compressors. This oil flow serves three functions:

1. Lubrication of the bearings, gears, and shaft seal
2. Cooling of the gas stream
3. Formation of a film between the rotors

The twin-screw compressor requires bearings to handle both the thrust loads and the radial loads. These bearings are usually a combination either of (1) angular contact ball bearings for the thrust loads and sleeve bearings for the radial loads or of (2) tapered roller bearings for the thrust loads and straight roller bearings for the radial loads. The thrust bearings are designed to carry only the thrust loads in both directions, while the radial loads are carried by the radial bearings only. Some of the smaller compressors use a set of gears to increase the speed of the drive rotor. These gears are also lubricated by this oil.

During the compression cycle, the refrigerant gas picks up heat. The injected oil cools this gas and allows the compressor to operate under the high compression ratios. Even under the high compression ratios, the discharge temperature of a screw compressor seldom exceeds 200°F (93°C), with normal temperatures running around 160 to 180°F (71 to 82°C).

The oil also forms a film between the two rotors to allow the drive rotor to

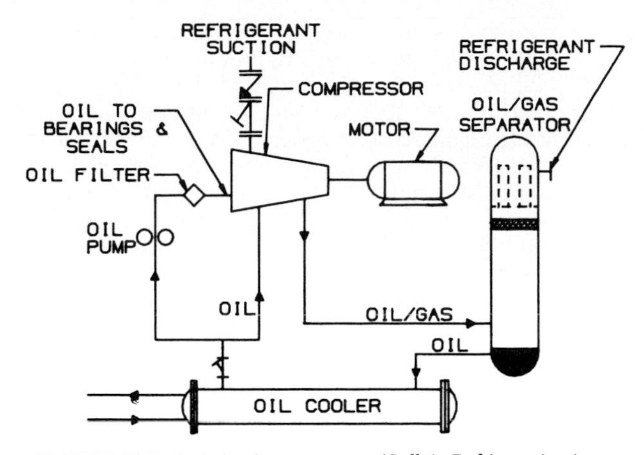

FIGURE 43.5 Lubrication system. (*Sullair Refrigeration.*)

turn the driven rotor without metal-to-metal contact. The majority of modern compressors use the male rotor to drive the female rotor. However, some of the newer compressors are using the female rotor to drive the male rotor.

With this type of lubrication system, the compressor package requires a complete lubrication system (Fig. 43.5). The system consists of a gas/oil separator and reservoir, an oil-cooling system, an optional oil pump, and an oil distribution system. Since all this oil is injected directly into the gas stream, an oil separation system must be installed.

The oil separation systems are normally designed for oil carryover to the system of less than 20 ppm (20 mg/L) under steady-state operating conditions. However, where the compressor package is built as a complete chiller package, the compressor may be designed for larger carryovers and allow the oil to circulate throughout the system and return to the compressor.

The separator (Fig. 43.6) is normally designed to accomplish the separation by several of the following methods:

1. Changes in direction of flow
2. Reduction in velocity
3. Centrifugal or mesh pad devices
4. Coalescing materials for removal of fine mist

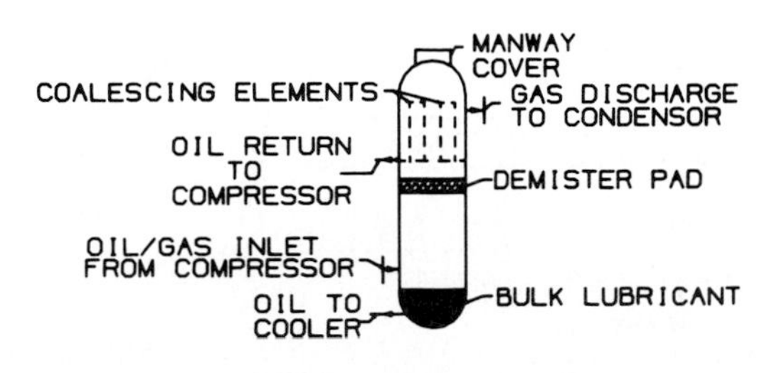

FIGURE 43.6 Gas/oil separator—breakdown and part description. (*Sullair Refrigeration.*)

The oil pump is used to overcome line and cooler pressure losses and/or to boost the oil pressure higher than the discharge pressure of the compressor. With sleeve bearings, this oil pump is normally required. However, with roller bearings, this pump may be eliminated if there is enough pressure to push the oil into the compressor and if the compressor is designed for the lower oil pressure.

The oil-cooling system is used to remove a major portion of the heat of compression. The amount of heat removed by the oil-cooling system depends on the refrigerant and the compression ratio but can be as high as 40 percent of the total heat rejected from the refrigeration system. Depending on the type of cooling system, this heat may be subtracted from the condenser's heat load.

43.2.4 Types of Cooling

The most common forms of cooling the screw compressor are:

1. Water or glycol cooling in a heat exchanger
2. Refrigerant thermosiphon cooling in a heat exchanger
3. Direct injection of refrigerant into the compressor
4. Direct-expansion cooling in a heat exchanger

The most common form of oil cooling uses a shell-and-tube heat exchanger with water or glycol in the tubes to remove the heat. The water can be supplied from either an open system (Fig. 43.7) or a closed system (Fig. 43.8). With the open system, the water can be taken from the normal plant water supply, from the sump of an evaporative condenser, or from a cooling tower. With the closed system, the water or glycol can be taken from an air-cooled heat exchanger, a closed-circuit evaporative cooler, or another heat exchanger that may be used to

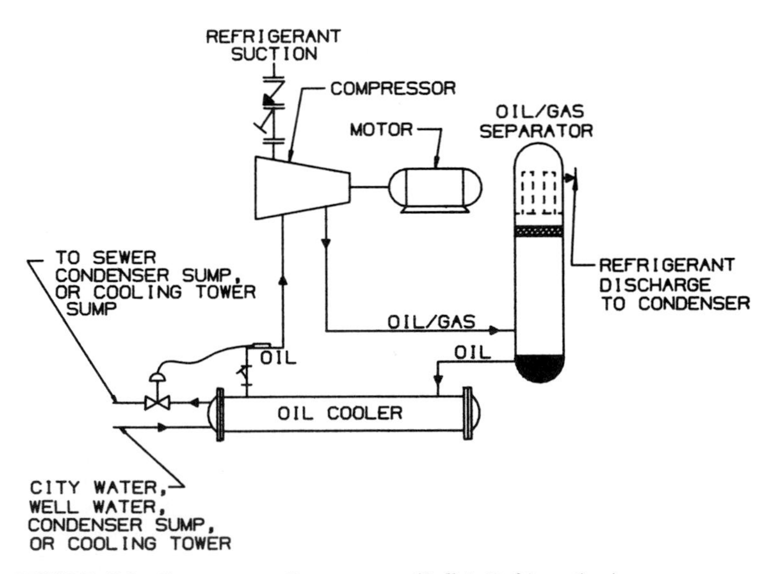

FIGURE 43.7 Open-type cooling system. (*Sullair Refrigeration.*)

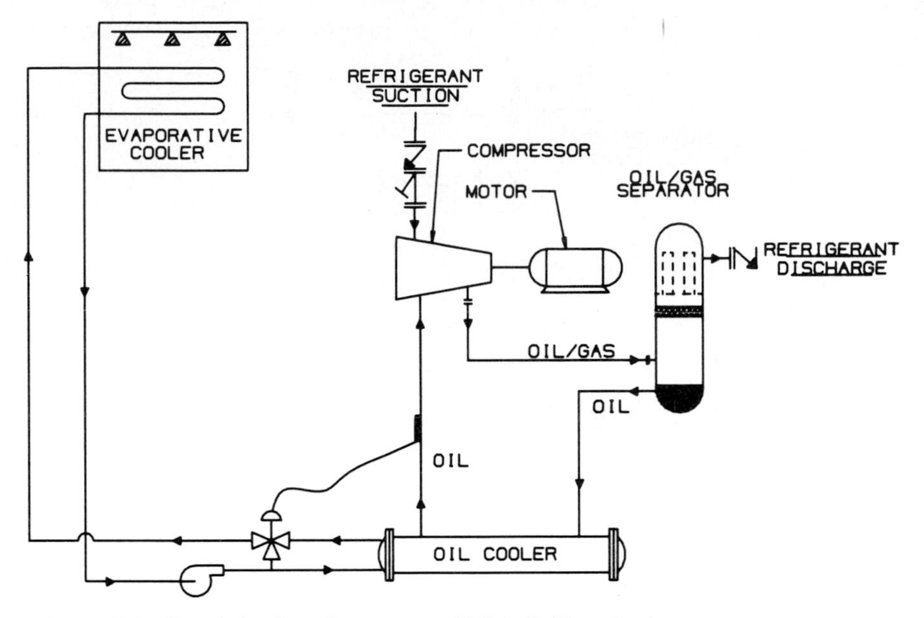

FIGURE 43.8 Closed-circuit cooling system. (*Sullair Refrigeration*.)

recover this heat for heating potable water, for boiler makeup, or for building heat. Due to the poor heat transfer, low oil flow, and large oil-temperature drop, the oil cooler is usually designed with outerfin tubes in order to enhance the heat transfer. The design is usually based on a tube-side (water-side) fouling of 0.0005 $ft^2 \cdot °F \cdot h/Btu$ ($3.69\ cm^2 \cdot °C \cdot s/cal$). Therefore, care must be taken to avoid fouling of the oil cooler. The heat removed in the oil cooler can be subtracted from the total refrigeration heat rejection when sizing the condenser.

If a source of clean water is not available or is too expensive, the cooling fluid can be replaced with refrigerant. Some applications have used a direct-expansion cooler and injected the evaporated gas into the compressor at either the suction port or an economizer port (Fig. 43.9). This procedure will add to the horsepower (wattage) of the compressor and will decrease the capacity if injected into the suction of the compressor. The condenser must be sized for the entire refrigeration heat of rejection plus the added horsepower (wattage).

The thermosiphon system (Fig. 43.10) uses the liquid from the condenser to flood the cooler and return the evaporated gas directly to the condenser. This procedure has no effect on the compressor's performance, but the entire refrigeration heat of rejection must be removed in the condenser.

The direct liquid-refrigerant injection system (Fig. 43.11) is used quite often on industrial refrigeration systems using ammonia but has seen very limited use on halocarbon systems. This system injects liquid refrigerant directly into the compressor at a point where the internal pressure is below the discharge pressure. The liquid is boiled off, cooling the gas-and-oil stream in the compressor. The liquid flow is controlled by modulating a liquid valve in response to the discharge temperature. Discharge temperatures are usually either kept around 120°F (49°C) or kept 20°F (11°C) above the condensing temperature, whichever is higher. This method has a slight effect on compressor capacity and results in a slight increase

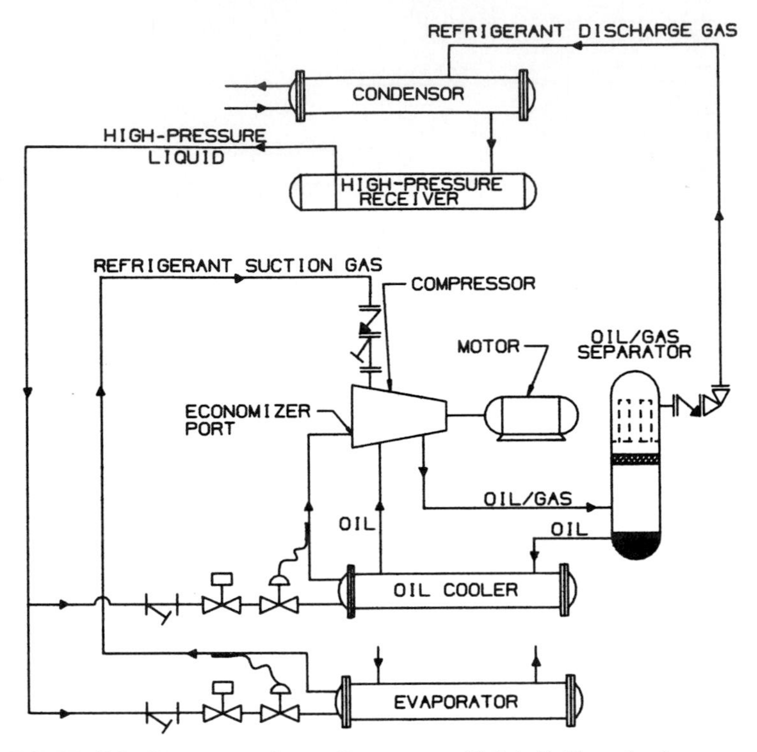

FIGURE 43.9 Direct-expansion cooling system. (*Sullair Refrigeration.*)

in compressor horsepower (wattage). The entire refrigeration heat rejection must be removed in the condenser. The main problem with this type of cooling in halocarbon systems is that the halocarbon gas is very soluble in the refrigerant oils; thus, with the low temperatures of the oil and gas and with the high gas solubility, the oil viscosity is usually reduced below the minimum viscosity level required by the compressor.

43.2.5 Capacity

The screw compressor has been built with R-22 capacities ranging from 50 to 1600 tons (176 to 5626 kW) when operating at 35°F (1.7°C) evaporating temperature and 105°F (41°C) condensing temperature. Typical performance curves are shown in Fig. 43.12. When comparing data from different manufacturers, one must make sure that the ratings are based on or corrected to the same rating conditions. Most manufacturers include correction procedures in their manuals. These correction factors consist of one or more of the following:

1. *Degrees of subcooling (C1):* Subcooling of the liquid below the condensing temperature must be accomplished by using a cooling medium other than the refrigerant listed in item 2 below.
2. *Degrees of superheating (C2):* The superheat can be either useful (if it is ac-

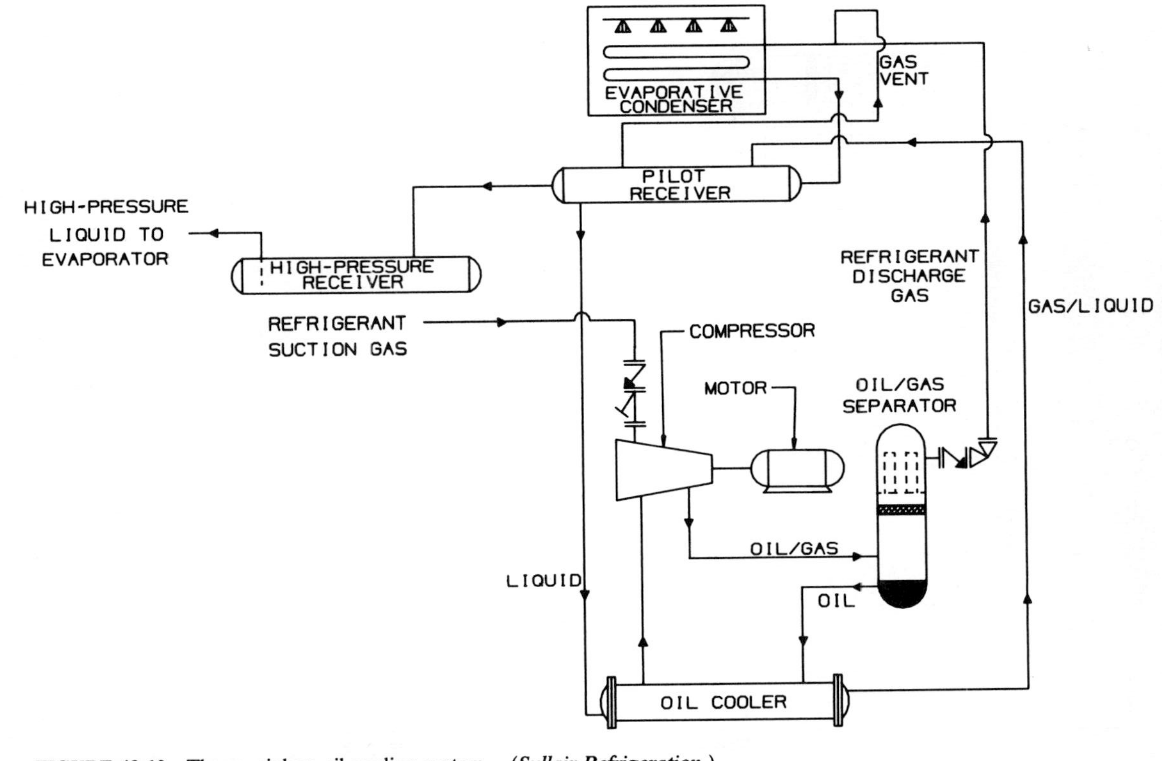

FIGURE 43.10 Thermosiphon oil-cooling system. (*Sullair Refrigeration.*)

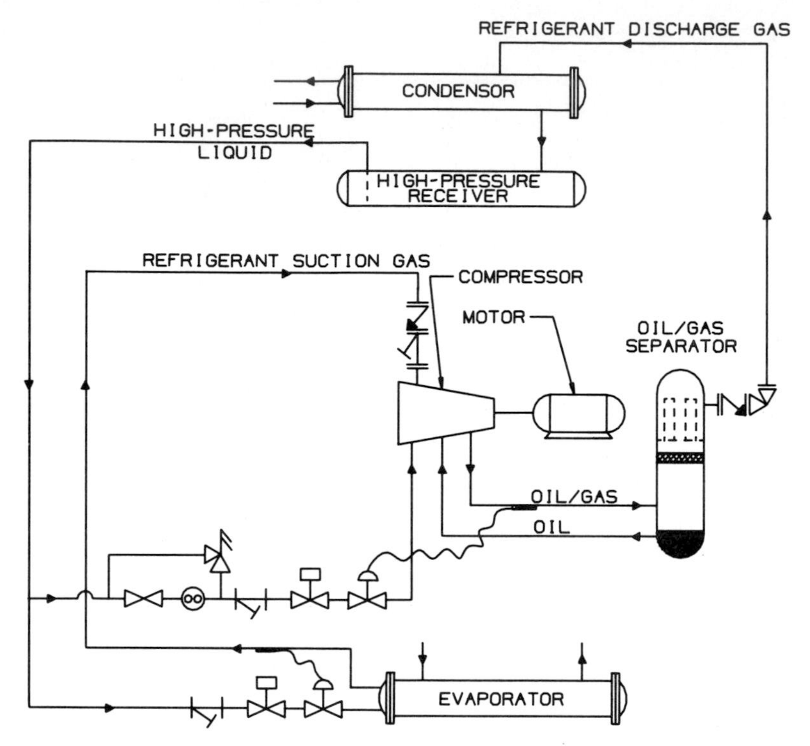

FIGURE 43.11 Liquid-injection cooling system. (*Sullair Refrigeration.*)

complished in the evaporator or by subcooling the liquid) or nonuseful (if the superheating occurs as a result of heat loss to the atmosphere or if the superheat is used to cool some other fluid).

3. *Suction pressure drop (C3) and discharge pressure drop (C4):* Most screw compressors include an isolation valve, a check valve and a strainer in the suction line, and a discharge check-valve, an oil separator and a discharge stop valve. These could amount to a pressure drop of several psi (kPa). Many compressor manufacturers rate their compressors based on the conditions at the compressor flanges rather than from the suction-valve flange to the discharge-valve flange after the separator.
4. *Speed (C5):* For speed variations less than 20 percent of the rated speed, both the capacity and the horsepower (wattage) can be ratioed directly to that speed on which the ratings are based.

Typical correction factors for the above are shown in Table 43.1. These corrections cover the majority of the required corrections, but the manufacturer's manual should be reviewed for any additional corrections.

As an example of a typical selection, we require a compressor operating with R-22 and producing 475 tons (1670 kW) when running at 35°F (1.7°C) suction and 115°F (46°C) condensing with 20°F (11°C) subcooling and 10°F (5.6°C) superheating, which is accomplished in the evaporator. Because of the motor require-

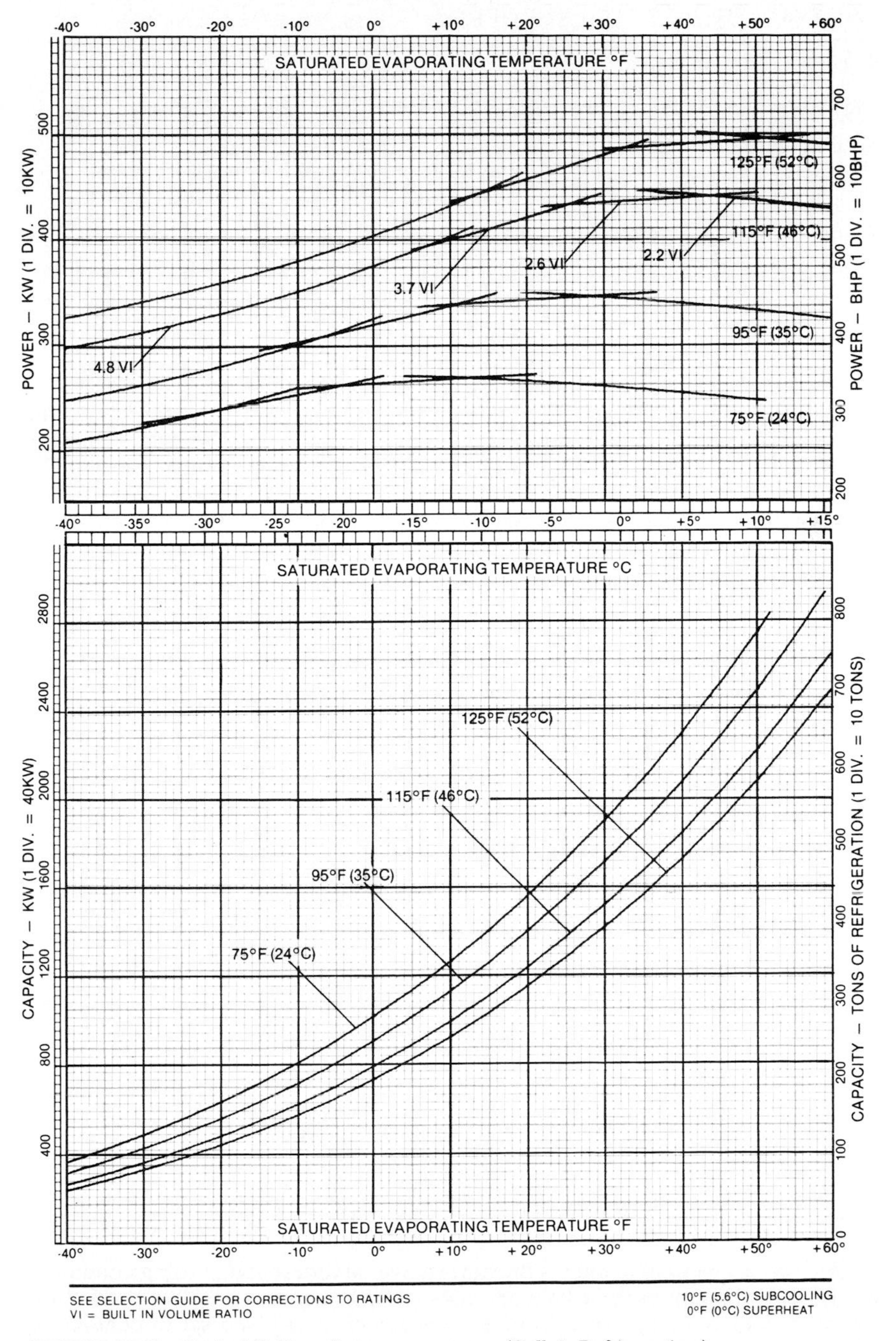

FIGURE 43.12 Typical R-22 performance curves. (*Sullair Refrigeration.*)

TABLE 43.1 Typical R-22 Correction Factors

Description	Value	Applies to
Liquid subcooling (C1)	0.0050/°F (0.009/°C)	Capacity
Suction superheat (C2)		
Useful	−0.0005/°F (−0.0009/°C)	Capacity
Nonuseful	−0.0028/°F (−0.0050/°C)	Capacity
Suction pressure drop (C3)	0.975	Capacity
Discharge pressure drop (C4)	1.02	Capacity and BHP

ments, the compressor will run at only 3500 rpm. Read up from the 35°F (1.7°C) saturated evaporating temperature line on the bottom curve of Fig. 43.12 to the 115°F (46°C) curve; at the intersection with the curve, read to the right a capacity of 480 tons (or to the left a capacity of 1688 kW). The vertical line can then be continued up to the upper curve where it intersects the 115°F (46°C) horsepower curve; at the intersection with the curve, read to the right a power requirement of 582 BHP (or to the left a power of 434 kW).

As the design parameters stated, the system will be designed for 20°F (11.1°C) subcooling. From Fig. 43.12, the compressor is rated with 10°F (5.6°C) subcooling. From Table 43.1, the correction factor is 0.005 × °F (0.009 × °C); therefore, the capacity will be corrected by a factor of 1 + 0.005 × (20°F − 10°F) = 1.05 [or 1 + 0.09 (11.1°C − 5.6°C) = 1.05]. The design also calls for a useful superheat of 10°F (5.6°C). Again from Table 43.1, the correction factor is −0.0005/°F (−0.0009/°C); therefore, the capacity will be corrected by a factor of 1 + (−0.0005 × 10) = 0.995 [1 + (0.0009 × 5.6) = 0.995].

The curve in Fig. 43.12 is based on pressures at the compressor flanges. Therefore, the suction pressure-drop correction of 0.975 and the discharge pressure-drop correction of 1.02 in Table 43.1 must be included. The compressor performance will then be as follows:

$$\text{Capacity} = 480 \times 1.05 \times 0.995 \times 0.975 \times 1.02 \times \frac{3500\ \text{rpm}}{3550\ \text{rpm}} = 491.7\ \text{tons}$$

$$\left(\text{Capacity} = 1688 \times 1.05 \times 0.995 \times 0.975 \times 1.02 \times \frac{3500\ \text{rpm}}{3550\ \text{rpm}} = 1729.1\ \text{kW}\right)$$

$$\text{Power} = 582 \times \frac{3500\ \text{rpm}}{3550\ \text{rpm}} = 573.8\ \text{BHP}$$

$$\left(\text{Power} = 434 \times \frac{3500\ \text{rpm}}{3550\ \text{rpm}} = 427.9\ \text{kW}\right)$$

The horsepower curve shows the horsepower at different operating conditions when using different built-in volume ratios (VI). The curves are drawn using the VI that will give the lowest operating horsepower for a given set of conditions. Since the screw compressor is a fixed-volume-ratio compressor, the adiabatic efficiency will peak when the compression ratio matches the internal built-in vol-

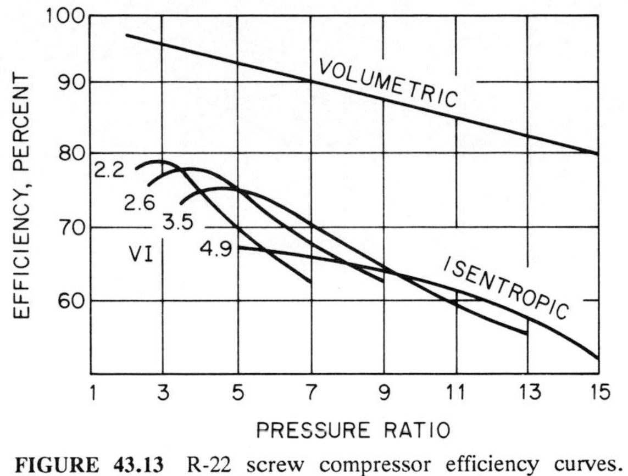

FIGURE 43.13 R-22 screw compressor efficiency curves. (*ASHRAE.*)

ume ratio. Figure 43.13 shows a typical R-22 performance curve with adiabatic efficiencies plotted for different VI. The peak adiabatic efficiency occurs at a pressure ratio as calculated in Eq. (43.1):

$$PR = (VI)^k \tag{43.1}$$

where PR = ideal pressure ratio
VI = built-in volume ratio
k = isentropic exponent

At this point the gas in the compressor will be compressed to a pressure internally (before the lobe opens to the discharge port) that exactly matches the pressure in the discharge line (Fig. 43.14, curve 1). As the lobe opens to the discharge port, the compressor will continue compression enough to push the gas into the discharge line. When operating off of this peak-efficiency point, the compressor will either be overcompressing before the lobe opens to discharge or undercompressing (Fig. 43.14). When overcompressing (curve 2), the gas will be compressed to a pressure internally (before the lobe opens to the discharge line) that is actually higher than the condensing pressure. When this occurs, the gas will create a pulse going into the oil separator when the lobe opens to the discharge line and very little additional compression will occur. Efficiency is lost because the compressor must exert work to compress the gas to a pressure higher than that required by the system. When undercompressing (curve 3), the gas will be compressed to a pressure internally (before the lobe opens to the discharge line) that is actually less than the condensing pressure. When this occurs, the gas will create a pulse going back into the compressor when the lobe opens to the discharge line. This gas, which flows back into the compressor, must be recompressed to the discharge line pressure.

When selecting a compressor with, say, design operating conditions of 35°F (1.7°C) evaporating and 115°F (46°C) condensing, the designer may select a compressor with a built-in volume ratio of 2.6 per the curve in Fig. 43.12 in order to give the lowest horsepower at design conditions. However, if the majority of the time the system would be operating at 95°F (35°C) condensing, then a better se-

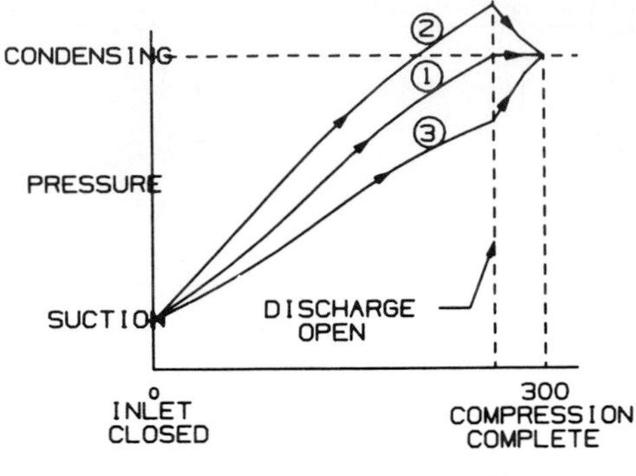

FIGURE 43.14 Internal compression curves. (1) Ideal built-in volume ratio (VI); (2) high VI—overcompression; (3) low VI—undercompression. (*Sullair Refrigeration.*)

lection may be a 2.2 VI. This decision must be made by the designer working with the compressor manufacturer.

The last items that must be calculated are the condenser heat load (CHL) and the oil-cooler heat load (OCHL). The total heat load (THL) for a screw compressor is calculated by using Eq. (43.2):

$$\begin{aligned} \text{THL} &= \text{capacity} \times 12{,}000 + \text{power} \times 2545 \text{ (Btu/h)} \\ &= \text{capacity} + \text{power (kW)} \end{aligned} \tag{43.2}$$

where THL = total heat rejection, tons (kW)
capacity = compressor capacity, tons (kW)
power = compressor power, BHP (kW)

If the compressor is using liquid injection or thermosiphon oil cooling, then the CHL is equal to the THL calculated by Eq. (43.2).

However, if the compressor is using water or glycol cooling, then OCHL can be deducted from the THL to calculate the CHL. The OCHL is calculated by Eq. (43.3):

$$\text{OCHL} = \text{THL} \times \text{OCHLM} \tag{43.3}$$

where OCHL = oil-cooler heat load, Btu/h (kW)
OCHLM = oil-cooler heat-load multiplier

The OCHLM is dependent on oil flows and on the compressor manufacturer's design parameters; therefore, these multipliers must be obtained from the manufacturer. A typical curve for these multipliers for an R-22 compressor is shown in Fig. 43.15. For our example, we would enter the curve at the 35°F (1.7°C) saturated evaporating temperature and draw a straight line up to the 115°F (46°C) saturated condensing temperature curve. At the intersection with this curve, we draw a horizontal line across and read 4.5 percent for the OCHLM. The heat loads are then calculated as follows:

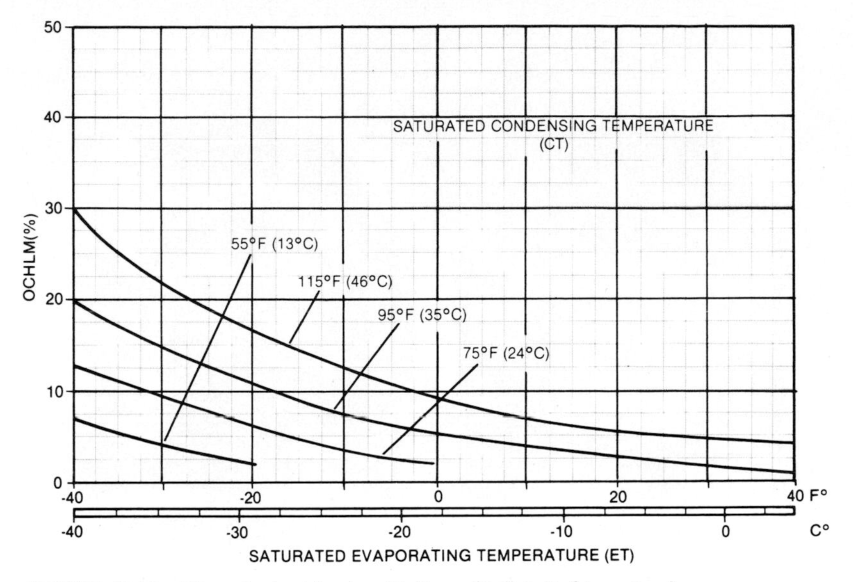

FIGURE 43.15 Oil-cooler heat-load multiplier. (*Sullair Refrigeration*.)

$$\text{THL} = 491.7 \times 12{,}000 + 573.8 \times 2545 = 7{,}360{,}721 \text{ Btu/h}$$
$$\text{OCHL} = 7{,}360{,}721 \times 0.045 = 331{,}233 \text{ Btu/h}$$
$$\text{CHL} = 7{,}360{,}721 - 331{,}233 = 7{,}029{,}288 \text{ Btu/h}$$

or in kilowatts:

$$\text{THL} = 1729.2 + 427.9 = 2157.1 \text{ kW}$$
$$\text{OCHL} = 2157.1 \times 0.045 = 97.1 \text{ kW}$$
$$\text{CHL} = 2157.1 - 97.1 = 2060.0 \text{ kW}$$

43.2.6 Capacity Control

An important advantage of screw compressors is the ability to vary compressor capacity as the load varies. This capacity reduction is accomplished in several ways:

1. Internal slide valve
2. Internal turn valve
3. Plug valves
4. Suction throttling
5. Variable speed

The internal slide valve (Fig. 43.16) is the most common form of capacity control for the larger compressors. The slide valve is usually constructed by replacing a portion of the stator with a valve shaped identical to the stator (Fig. 43.16). When the compressor is first started, the slide valve is located at minimum posi-

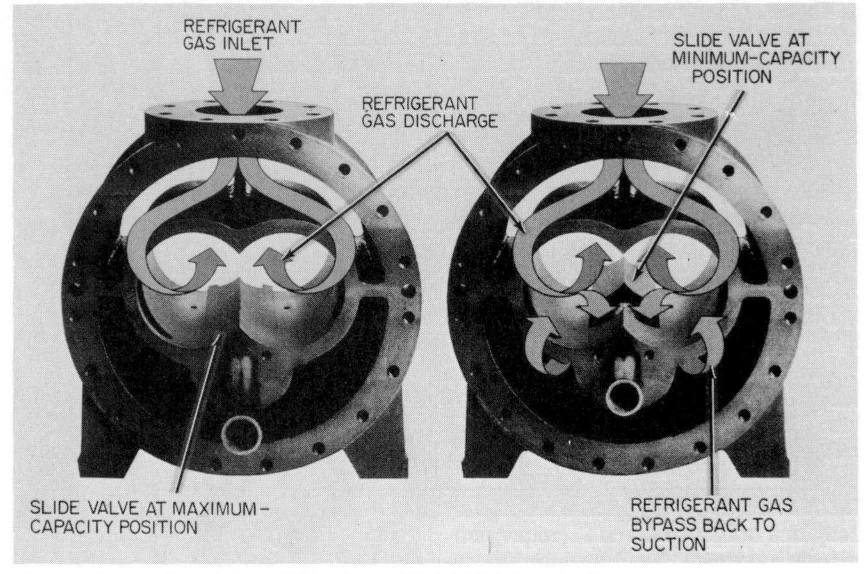

FIGURE 43.16 Compressor slide valve. (*Sullair Refrigeration.*)

tion. As the compressor loads, the slide valve is moved toward the inlet until it is against the stator when the compressor is at full load.

In order to control the position of this slide valve, a hydraulic piston (Fig. 43.17) is used. When high-pressure oil is drained through port B back to suction, the discharge gas will exert a higher pressure on the end of the slide valve than

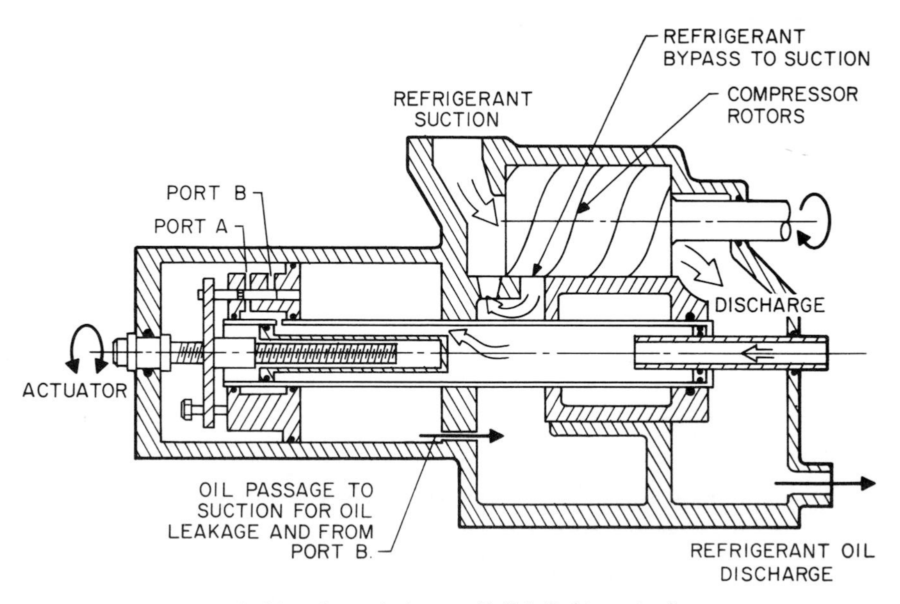

FIGURE 43.17 Internal slide valve and piston. (*Sullair Refrigeration.*)

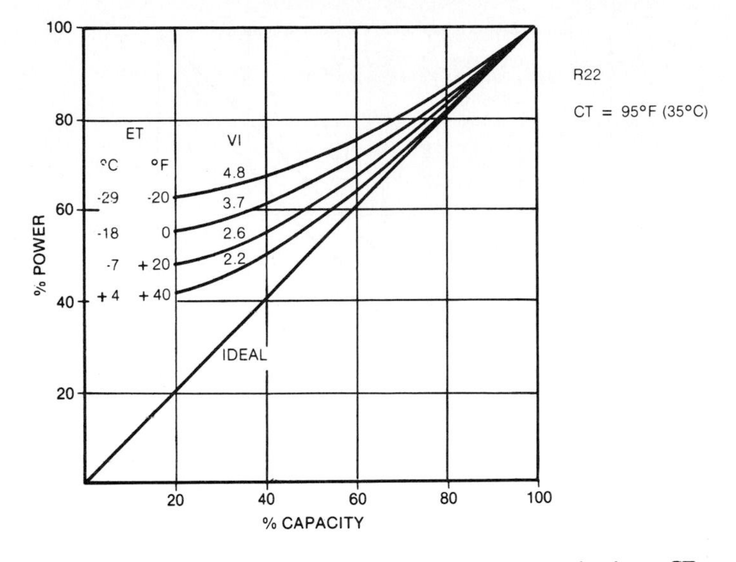

FIGURE 43.18 Part-load curves with constant condensing. CT = condensing temperature; ET = evaporating temperature; VI = volume ratio; IT = intermediate temperature. (*Sullair Refrigeration.*)

the suction gas exerts on the other end. This pressure differential causes the slide valve to move to the left, closing the gap between the stator and the slide valve and increasing the compressor capacity. To unload the compressor, high-pressure oil is injected behind the piston through port A. The resulting pressure differential across the piston overcomes the differential across the slide valve and drives the slide valve to the right. This allows more gas to flow from the rotors back to suction and less to the discharge port. A typical part-load curve for a screw compressor is shown in Fig. 43.18. This curve is based on a constant condensing temperature. Figure 43.19 shows a part-load curve based on the condensing temperature being reduced in accordance with ARI Standard 550-77.

In order to provide ease of starting, the compressor should be started at minimum position. On a normal shutdown, the compressor will always be at minimum position. However, after an abnormal or manual shutdown, the compressor may not be at minimum position and must be moved to minimum position before restarting the compressor. This is accomplished by using a motor, a spring, or an external source of high oil pressure to drive the slide valve back to minimum position.

As screw compressors became smaller, cheaper means of performing capacity control were required. This introduced the internal turn-valve design. In this design, the stator is machined with slots to allow the gas to bypass back to suction. Also included is an internal turn valve, which is located in the stator just below these slots. The turn valve consists of a hollow cylinder (Fig. 43.20), which is machined out. As the valve is rotated, the machined-out areas expose more of the slots in the stator and allow the gas to flow back to the suction side of the compressor. The turn valve is rotated by a motor, which can be driven electrically, hydraulically, or pneumatically.

Some of the newer small screw compressors are using a plug valve for capacity control. The plug valve consists of a piston which forms a portion of the stator when loaded and is removed to create a hole in the unloaded position. The piston

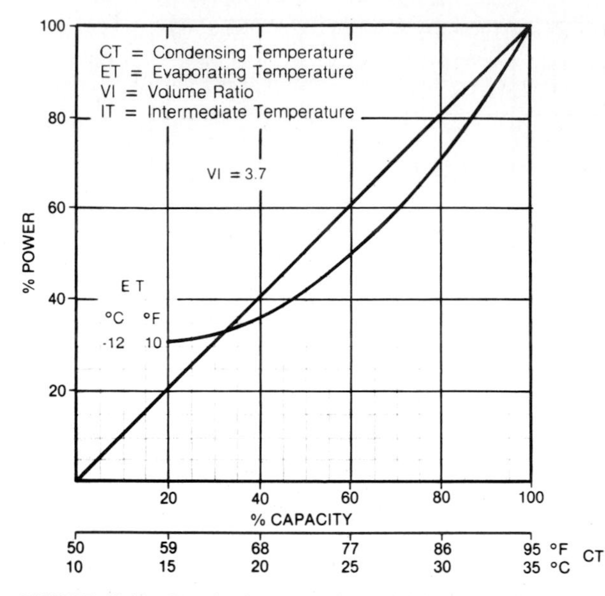

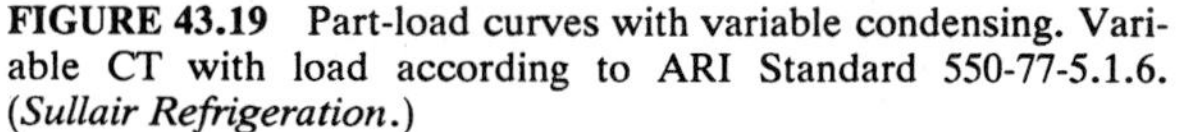

FIGURE 43.19 Part-load curves with variable condensing. Variable CT with load according to ARI Standard 550-77-5.1.6. (*Sullair Refrigeration.*)

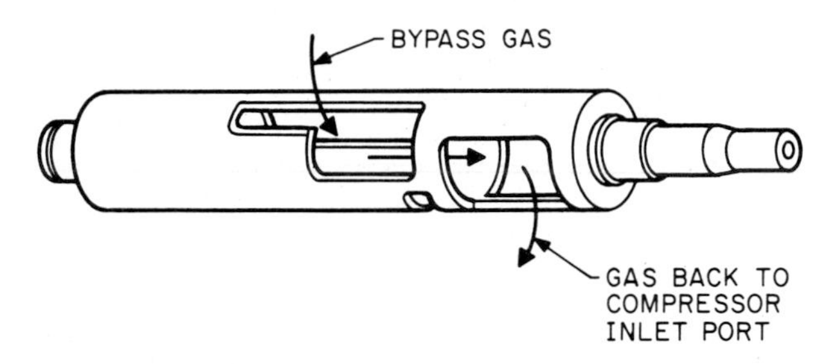

FIGURE 43.20 Internal turn-valve unloader. (*Sullair Refrigeration.*)

must be machined to the same shape and tolerance as the stator. Most applications use multiple plugs to give step-type unloading. The plugs can be moved either by hydraulics or by electric actuators.

Other screw compressors are simply installing a butterfly valve in the suction line. Under full load, the valve is wide open and, as the system calls for capacity reduction, the valve is rotated toward the closed position. As the valve closes, it creates a pressure drop, which increases the specific volume of the gas and decreases the mass flow to the compressor.

With the advances in solid-state electrical components, variable-speed drives are becoming less expensive and more efficient. As these advances (along with larger sizes) are made commercially available, they are being applied to the screw compressor as a means of capacity control.

The type of capacity control used is dependent on first cost, operating cost, size of compressor, and system operation. Therefore, the designer should work with the compressor manufacturers to determine the best choice for this application.

43.2.7 Variable Volume Ratio

Until the mid-1980s, the screw compressor was built with a fixed built-in volume ratio. This required that the compressor be picked with the correct built-in volume ratio for the design and normal operating conditions. For the industrial market, this created only a couple of percentage points of horsepower losses at full load when operating the compressor at conditions that differed from the ideal pressure ratio, and this could be almost eliminated by proper selection of the built-in volume ratio. The ideal pressure ratio is related to the built-in volume ratio by Eq. (43.1).

With increased flexibility required in industrial refrigeration systems, air-cooled screw compressor applications, and ice storage systems, a means to change the built-in volume ratio was required.

The usual method to supply the variable-volume-ratio compressor is to use a split slide valve (Fig. 43.21). However, designs using dual slide valves and plug valves have been designed. In order to decrease the built-in volume ratio, both valves are moved toward the inlet end. When reducing the capacity, the slide valve is moved toward the discharge end while the slide stop remains fixed. Under part-load conditions the slide stop and/or the slide valve may be moved in order to change the built-in volume ratio and the compressor capacity. This will

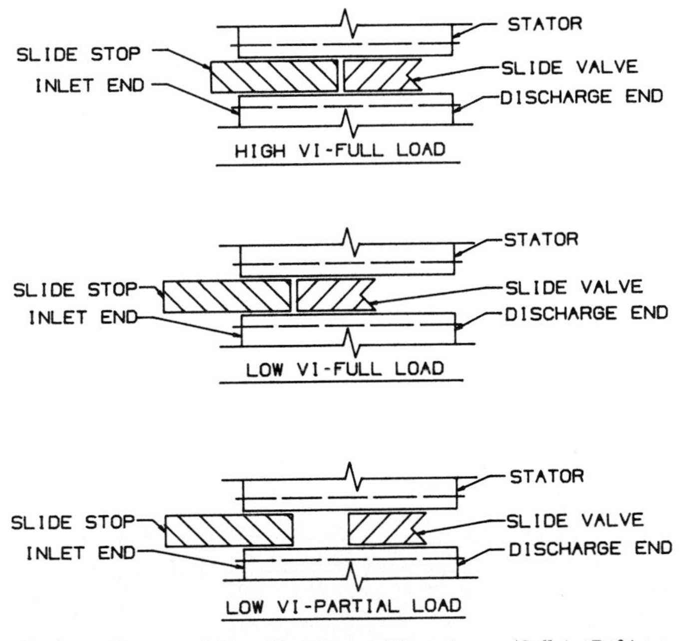

FIGURE 43.21 Variable-volume-ratio slide valves. (*Sullair Refrigeration.*)

be determined by the manufacturer, depending on the valve lengths and type of control the manufacturer is looking for. In order to know where the built-in volume ratio should be set, the compressor must constantly monitor the suction and discharge pressures and the current built-in volume ratio and capacity along with the refrigerant *k* value. This type of data monitoring requires the use of a microprocessor for control. If automatic control is not required, the slide stop could be manually set if seasonal changes or mode-of-operation changes are the only changes required.

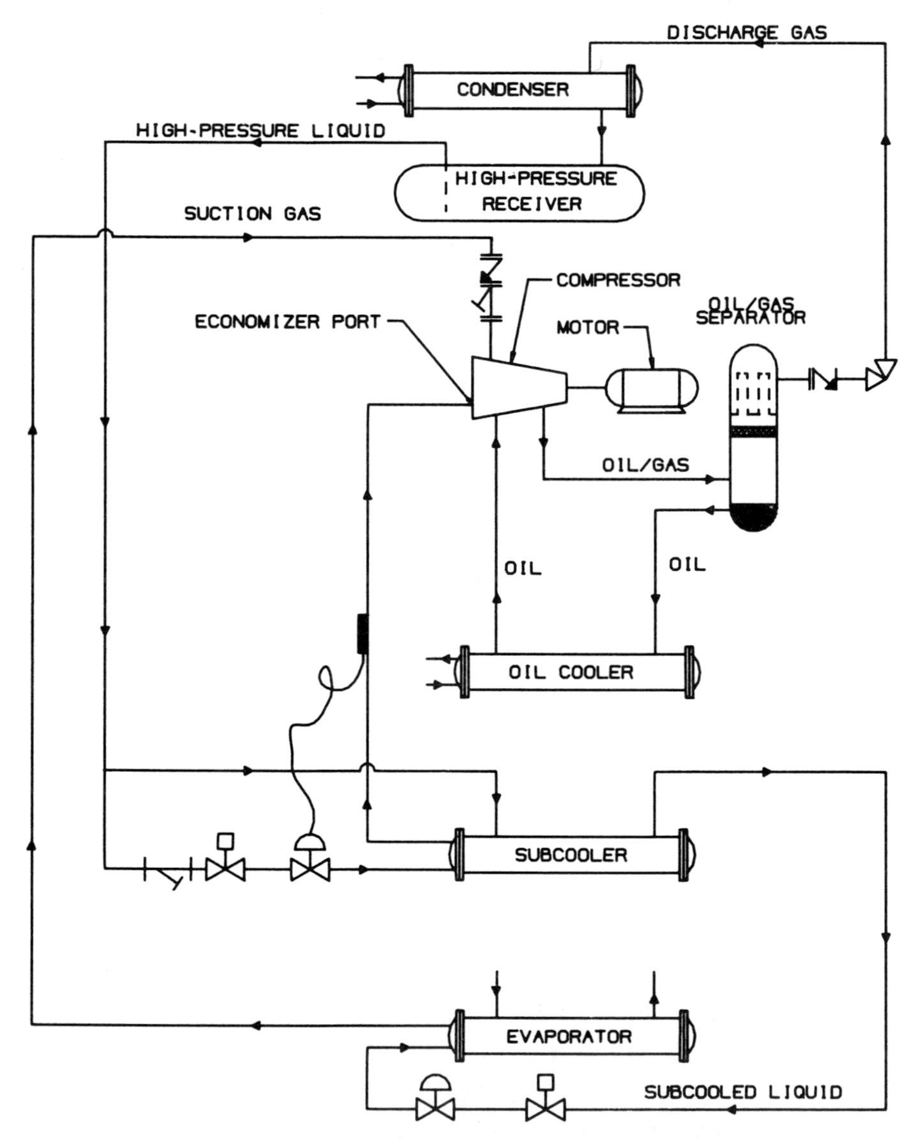

FIGURE 43.22 Direct-expansion economizer cycle. (*Sullair Refrigeration.*)

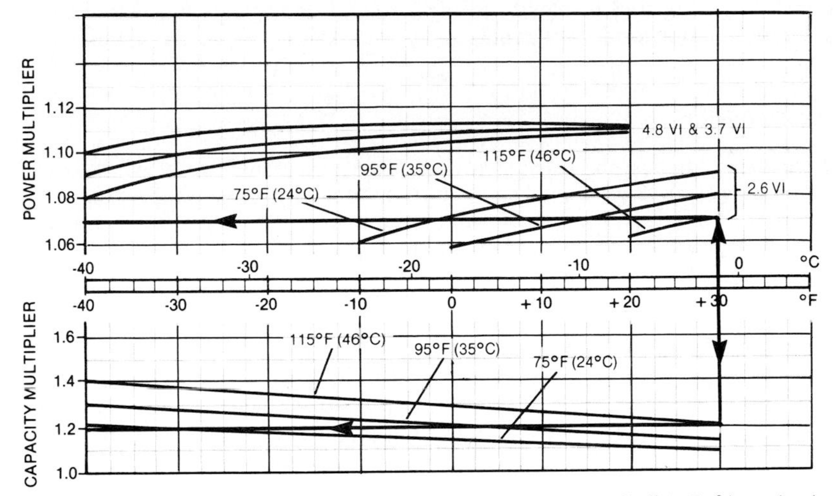

FIGURE 43.23 R-22 economizer capacity and power multipliers. *(Sullair Refrigeration.)*

43.2.8 Economizer

With the capability of cooling the gas during compression, the screw compressor can operate at compression ratios as high as 20:1 without having excess discharge temperatures. However, these pressure ratios do reduce the performance considerably. In order to overcome this, the screw compressor has been built with an economizer port. This port allows refrigerant gas to be injected into the compressor after compression has been partially performed. This gas could come from a heat exchanger used to subcool the liquid going to the main evaporator (Fig. 43.22), from a flash-type subcooler, or from an external load. A typical curve indicating capacity and power multipliers for an economizer system on an R-22 compressor is shown in Fig. 43.23. For an R-22 compressor with a 2.6 VI operating at 30°F (−1°C) evaporating temperature and 115°F (46°C) condensing temperature, the compressor capacity would be multiplied by 1.20 and the power would be multiplied by 1.07 to obtain the performance with an economizer.

As the compressor unloads to approximately 75 percent, this port pressure will reduce and will equal the suction pressure. Therefore, the system designer must make sure that, if this is a problem, proper precautions are taken, such as using a back-pressure regulator, using a shell-and-tube heat exchanger, or shutting off the economizer port. Deciding what type of system to use will depend on the refrigeration system design, and the designer should work with the compressor manufacturer to decide which is best for the system.

43.3 SINGLE-SCREW COMPRESSORS

43.3.1 Development

The single screw, or monoscrew, was originally developed in the early 1960s but was not applied to the refrigeration and air-conditioning industries until 10 years later. Today it is applied mostly to the air-conditioning market, with very little application in industrial refrigeration.

43.3.2 Design

As the name implies, the single screw consists of one main rotor working with a pair of gaterotors (Fig. 43.2). The main rotor typically has six helical flutes and a globoid (or hourglass) root profile. The gaterotors each have 11 teeth and are located on opposite sides of the main rotor.

43.3.3 Compression Process

The compression process (Fig. 43.24) is very similar to the twin screw. As the main rotor rotates, one of the teeth of the gaterotor engages a flute and draws a reduced pressure behind the tooth and within the flute. This reduced pressure draws gas through the suction connection of the compressor.

As the main rotor continues to rotate, it engages a tooth from the gaterotor on the opposite side. The gas is now trapped within the flute, the casing wall, and

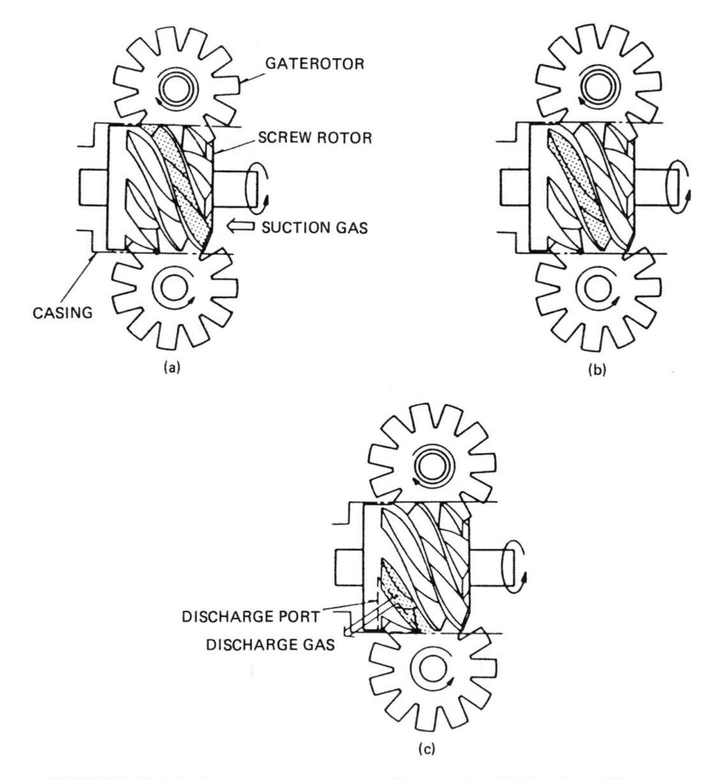

FIGURE 43.24 Monoscrew compression cycle. (*a*) Suction; (*b*) compression; (*c*) discharge. (*Single Screw Compressor, Inc.*)

the tooth of the gaterotor. Compression starts and continues until the flute opens to the discharge port.

Gas is continuously pushed out of the flute until the volume within the flute and tooth of the gaterotor is zero. As with the twin screw, the volume of gas left in the flute for reexpansion to suction is negligible.

43.3.4 Lubrication and Cooling

The lubrication system and means of cooling are identical to the systems outlined under the twin screw.

43.3.5 Capacity Control

Capacity control methods for the single screw are basically the same as for the twin screw. However, the geometry of the single screw allows two compression

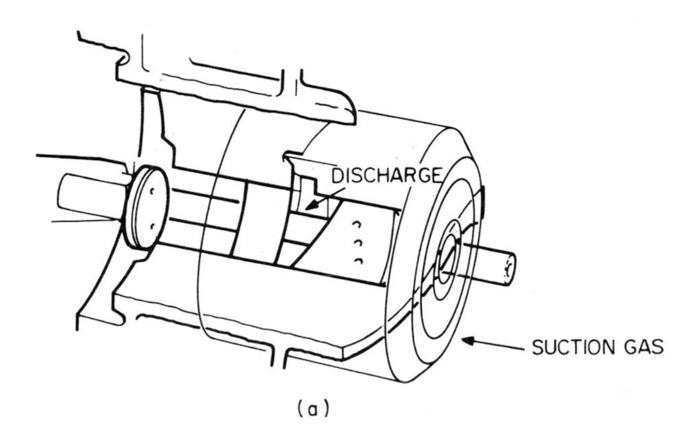

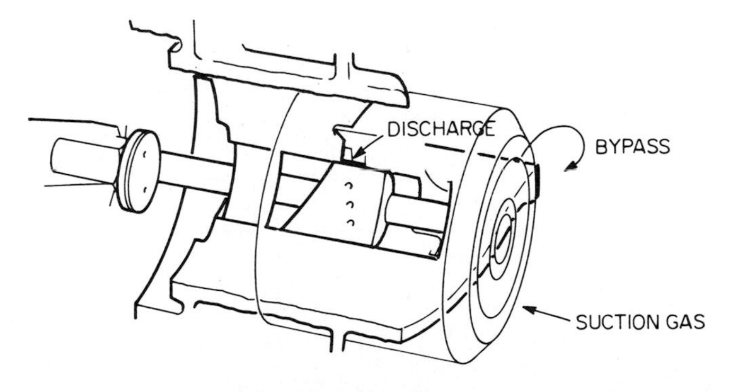

FIGURE 43.25 Single-screw axial-type capacity control mechanism. (*a*) Slide valve fully in loaded position; (*b*) slide valve in part-load position. (*Single Screw Compressor, Inc.*)

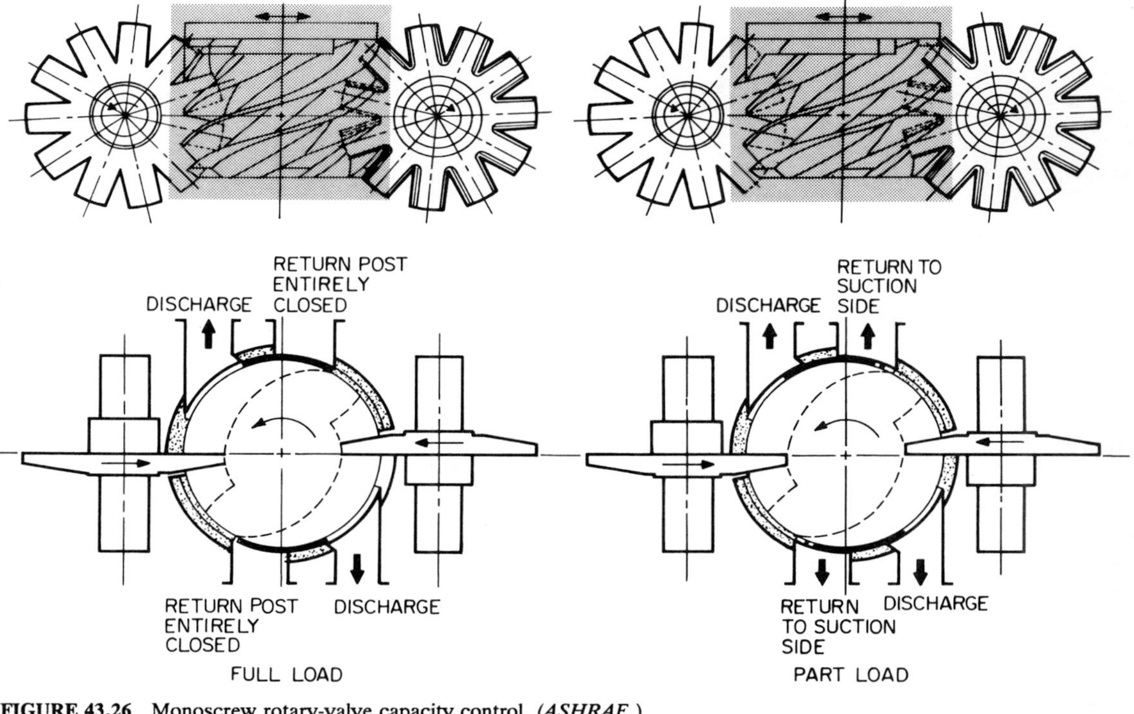

FIGURE 43.26 Monoscrew rotary-valve capacity control. (*ASHRAE.*)

processes to occur simultaneously. The unloading mechanism consists of dual slide valves which are hydraulically activated and which retard the point at which the compression process begins (Fig. 43.25).

In addition to the above methods of capacity control, the single screw can also use a rotating ring (Fig. 43.26) that is built into the stator housing and allows gas to flow from the compression chamber back to suction through the stator housing.

43.3.6 Economizer

Like the twin screw, the single screw can incorporate an economizer cycle. However, the single screw, with dual slide valves, can have one valve operating at minimum position while the second valve is operating at near maximum. This allows the economizer to operate at a lower percent capacity before the port opens up to suction pressure.

43.4 SEMIHERMETIC SCREW COMPRESSORS

Both the twin screw and the single screw are available in semihermetic designs (Fig. 43.27). With the low discharge temperatures that are inherent with screw

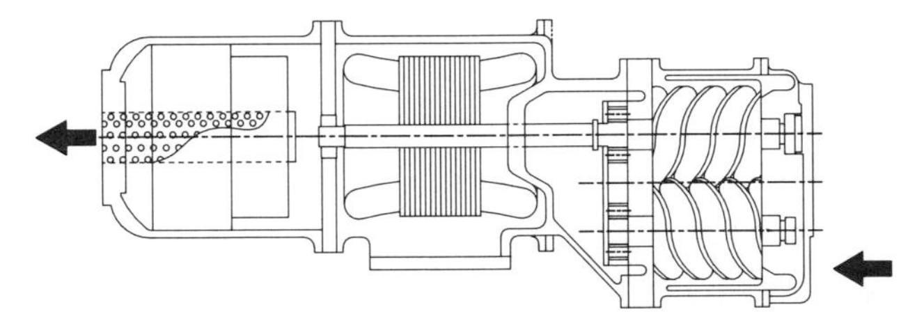

FIGURE 43.27 Semihermetic screw compressor.

compressors, the motor can be designed to be cooled by suction gas, discharge gas, intermediate gas to the economizer port, or liquid injection. Most semihermetic screw compressors are of the smaller compressor design, and the oil separator is incorporated with the compressor and motor.

CHAPTER 44
COOLING TOWERS*

John C. Hensley
Marketing Services Manager, The Marley Cooling Tower Company, Mission, Kansas

44.1 INTRODUCTION

As a general rule, air-conditioning and refrigeration systems in excess of 150- to 200-ton (528- to 704-kW) capacities make use of water as the medium for heat rejection, and the majority of such installations utilize cooling towers for the ultimate rejection of this heat to the atmosphere. In smaller systems, air-cooled heat exchangers and evaporative condensers are increasing in use, but cooling towers continue to be the method of choice where limiting the energy usage is a primary consideration.

This chapter defines the various types and configurations of cooling towers utilized for air-conditioning and refrigeration service and explains the principles by which they operate. It also discusses the environmental and energy-consuming aspects of towers and the external factors that can adversely affect their thermal performance capability.

Within reasonable limits, a cooling tower will dissipate whatever heat load is imposed on it, regardless of its size and efficiency. The tower *does not* establish the load—it merely reacts to it. However, the size and capability of the tower *does* establish the temperature level at which the heat load is dissipated, and this level, in turn, determines the operating efficiency of the system as a whole. Therefore, the student of air-conditioning is well advised to become familiar with the application of cooling towers—and with their operating characteristics.

44.2 TOWER TYPES AND CONFIGURATIONS

Cooling towers are designed and manufactured in several types and materials of construction, with numerous sizes (models) available in each type. They range in size from the very small [as little as 10 gal/min (63×10^{-5} m^3/s)] to the very large

*The contents of this chapter were adapted from *Cooling Tower Fundamentals* and *Cooling Tower Information Index*, published by The Marley Cooling Tower Company, P.O. Box 2912, Mission, KS 66201. All art used courtesy of The Marley Cooling Tower Company.

[250,000 gal/min (15 m^3/s) and larger]. Those which are normally utilized in air-conditioning service are described in this section.

Understanding the primary types and configurations, along with their advantages and limitations, can be of vital importance to the specifier and user and is essential to the full understanding of this chapter.

44.2.1 Atmospheric Type

Atmospheric towers utilize no mechanical device (fan) to create air flow through the tower. There are two main types of atmospheric towers: large and small. The large hyperbolic towers are equipped with "fill" (see Sec. 44.4); since their primary application is with electric power generation, which is beyond the scope of this book, the reader is referred to T. C. Elliott, ed., *The Standard Handbook for Powerplant Engineering* (McGraw-Hill, New York, 1989) for additional information on hyperbolic towers. The smaller atmospheric towers utilized in the air-conditioning industry are not normally equipped with fill to enhance the transfer of heat.

The small atmospheric tower shown in Fig. 44.1 derives its air flow from the natural induction (aspiration) provided by a pressure-spray type of water distribution system, and the surface contact between air and water is limited by the pressure at the sprays and by the characteristics of the nozzles.

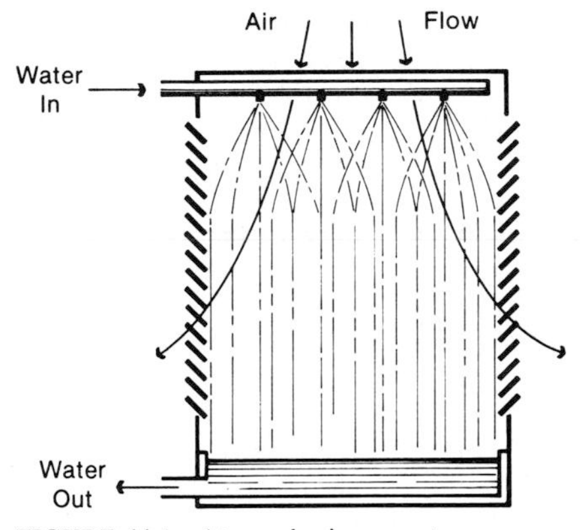

FIGURE 44.1 Atmospheric spray tower.

Although relatively inexpensive, atmospheric towers are usually applied only in very small sizes, and they tend to be energy-intensive because of the high spray pressures required. Because they are far more affected by adverse wind conditions than are other types, their use on systems requiring accurate, dependable cold-water temperatures is not recommended.

44.2.2 Mechanical-Draft Types

Mechanical-draft towers use either single or multiple fans to provide the flow of a known volume of air through the tower. Thus their thermal performance tends toward greater stability—and is affected by fewer psychrometric variables—than

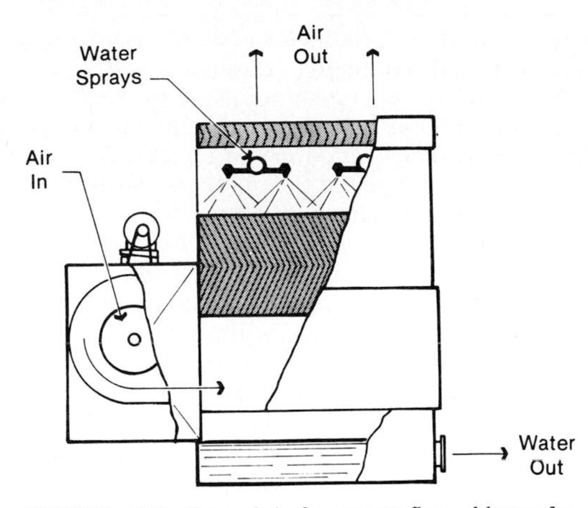

FIGURE 44.2 Forced-draft, counterflow, blower-fan tower.

that of the atmospheric towers. The presence of fans also provides a means of regulating the air flow to compensate for changing atmospheric and load conditions; this is accomplished by fan capacity manipulation and/or cycling, as described in Sec. 44.9.

Mechanical-draft towers are categorized as either forced-draft towers (Fig. 44.2), wherein the fan is located in the ambient air stream entering the tower and the air is blown through, or induced-draft towers (Fig. 44.3), wherein a fan located in the exiting air stream draws air through the tower.

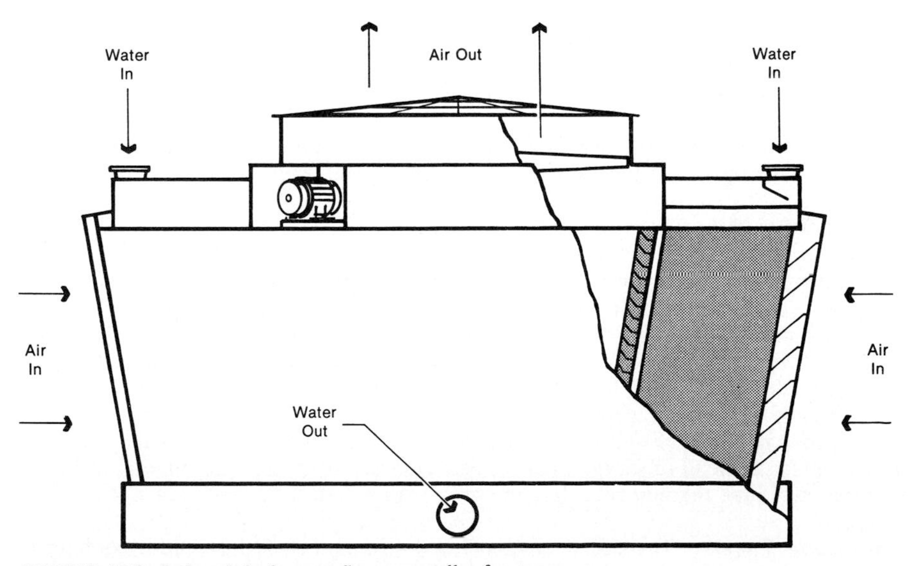

FIGURE 44.3 Induced-draft, crossflow, propeller-fan tower.

Forced-Draft Towers. These are characterized by high air-entrance velocities and low exit velocities. Accordingly, they are extremely susceptible to recirculation (see Sec. 44.5.1) and are therefore considered to have less performance stability than induced-draft towers. Of equal concern in northern climates is the fact that forced-draft fans located in the cold entering ambient air stream can become subject to severe icing, and the resultant imbalance, when moving air laden with either natural or recirculated moisture.

Usually, forced-draft towers are equipped with centrifugal blower-type fans, which, although requiring approximately twice the operating horsepower (wattage) of propeller-type fans, have the advantage of being able to operate against the high static pressures associated with ductwork. So equipped, they can be installed either indoors (space permitting) or within a specially designed enclosure that provides sufficient separation between the air-intake and -discharge locations to minimize recirculation.

Given the increasing economic pressure dictating reduced energy consumption in cooling towers and the increasing wintertime usage of cooling towers for "free cooling" (see Sec. 44.9.2), it is to be anticipated that the use of forced-draft towers, particularly those equipped with blower fans, will soon be relegated to the exceptional situation and that induced-draft configurations will prevail.

Induced-Draft Towers. These have an air-discharge velocity 3 to 4 times higher than their air-intake velocity, with the intake velocity approximating that of a 5-mph (0.22-m/s) wind. Therefore, there is little tendency for a reduced-pressure zone to be created at the air inlets by the action of the fan alone (see Sec. 44.5.1). The potential for recirculation on an induced-draft tower is not self-initiating, and it can be more easily quantified and compensated for (see Sec. 44.6) purely on the basis of ambient wind conditions.

Induced-draft towers are also typically forgiving of wintertime operation. The location of the fan within the warm air stream (the movement of which continues even with the fan off) provides excellent protection against the formation of ice on the mechanical components. As will be seen in Sec. 44.10, fans so located actually provide a means by which to facilitate deicing in extreme conditions.

44.2.3 Characterization by Air Flow

Cooling towers are also classified in terms of the relative flow relationship of the air to the water within the tower as being either counterflow or crossflow towers.

Counterflow Towers. In counterflow towers (Fig. 44.4), the air moves vertically upward through the fill, counter to the downward fall of the water. Historically, because of the need for expansive intake and discharge plenums, the use of high-pressure spray systems, and the typically higher air-pressure losses, some of the smaller counterflow towers were physically higher, required greater pump head, and utilized more fan power than their crossflow counterparts. With the advent of more sophisticated fills, fans, and spray systems, however, this operational economic gap is rapidly closing.

Although the enclosed nature of a counterflow tower tends to make it somewhat more difficult to service, it also restricts exposure of the circulating water to direct sunlight, thereby retarding the growth of algae.

Crossflow Towers. Crossflow towers (Figs. 44.3 and 44.5) have a fill configuration through which the air flows horizontally across the downward fall of the water. The water to be cooled is delivered to hot-water inlet basins located above

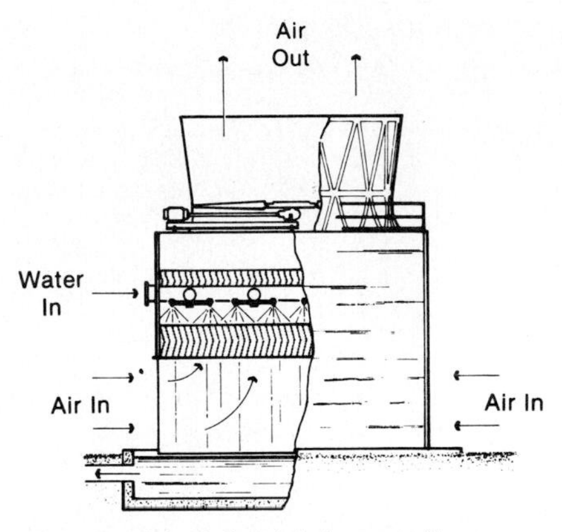

FIGURE 44.4 Induced-draft counterflow tower.

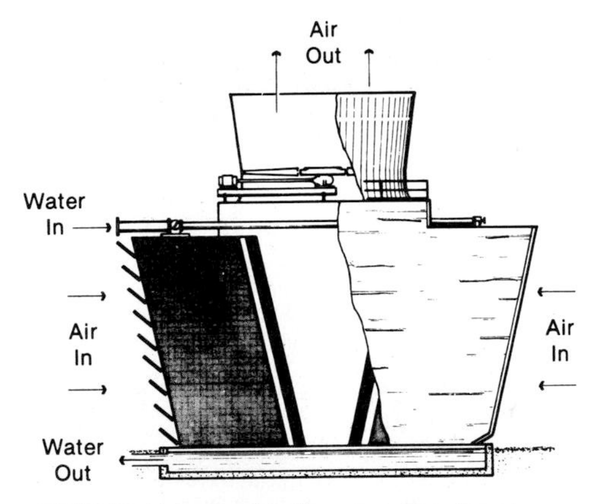

FIGURE 44.5 Double-flow crossflow tower.

the fill areas and is distributed to the fill by gravity through metering orifices in the floor of these basins. This obviates the need for a pressure-spray distribution system and places the resultant gravity system in full view for ease of maintenance.

44.2.4 Characterization by Construction

Field-Erected Towers. These are assembled primarily at the site of ultimate use. All large towers, as well as many of the smaller towers, are prefabricated, piece-marked, and shipped to the site for assembly. The labor and/or supervision for site assembly is usually provided by the cooling tower's manufacturer.

Factory-Assembled Towers. These are almost completely assembled at their point of manufacture and are shipped to the site in as few sections as the mode of transportation will permit. The relatively small tower shown in Fig. 44.6 would ship essentially intact. The larger, multicell towers (Fig. 44.7) are assembled as

FIGURE 44.6 Small factory-assembled tower.

FIGURE 44.7 Multicell factory-assembled tower.

"cells" or "modules" at the factory and are shipped with appropriate hardware for ultimate joining by the user. Factory-assembled towers are sometimes referred to as "packaged" or "unitary" towers.

44.3 TOTAL HEAT EXCHANGE

A cooling tower is a specialized heat exchanger in which two fluids (air and water) are brought into direct contact with each other to effect the transfer of heat. In the spray-filled tower shown in Fig. 44.8, this is accomplished by spraying a

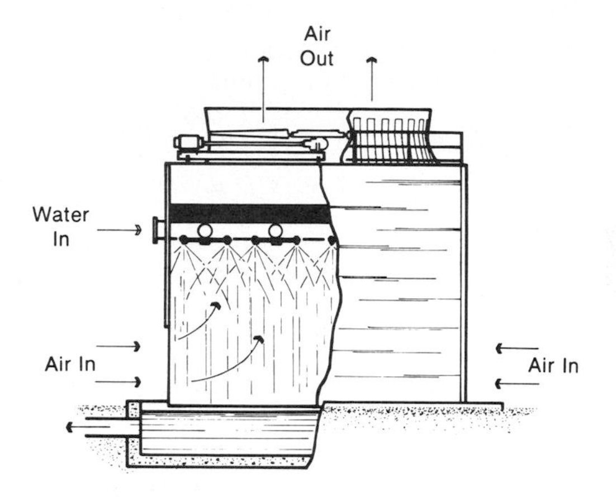

FIGURE 44.8 Spray-filled counterflow tower.

flowing mass of water into a rainlike pattern through which an upward-moving mass flow of cool air is induced by the action of the fan.

Ignoring any negligible amount of sensible heat exchange that may occur through the walls (casing) of the tower, the heat gained by the air must equal the heat lost by the water. Within the air stream, the rate of heat gain is identified by the expression $G(h_2 - h_1)$, where:

G = mass flow of dry air through the tower, lb/min (kg/min)

h_1 = enthalpy (total heat content) of entering air, Btu/lb (J/kg) of dry air

h_2 = enthalpy of leaving air, Btu/lb (J/kg) of dry air

Within the water stream, the rate of heat loss would *appear* to be $L(t_1 - t_2)$, where:

L = mass flow of water entering the tower, lb/min (kg/min)

t_1 = temperature of hot water entering the tower, °F (°C)

t_2 = temperature of cold water leaving the tower, °F (°C)

This derives from the fact that a Btu (calorie) is the amount of heat gain or loss necessary to change the temperature of 1 lb (1 g) of water by 1°F (1°C). However, because of the evaporation that takes place within the tower, the mass flow of water leaving the tower is actually less than that entering it, and a proper heat balance must account for this slight difference. Since the rate of evaporation must equal the rate of change in the humidity ratio (absolute humidity) of the air stream, the rate of heat loss represented by this change in humidity ratio can be expressed as $G(H_2 - H_1)(t_2 - 32)$, where:

H_1 = humidity ratio of entering air, lb (kg) of vapor per lb (kg) of dry air
H_2 = humidity ratio of leaving air, lb (kg) of vapor per lb (kg) of dry air
$(t_2 - 32)$ = an expression of water enthalpy at the cold-water temperature, Btu/lb (J/kg) [the enthalpy of water is considered to be zero at 32°F (0°C)]

Including this loss of heat through evaporation, the total heat balance between the air and the water, expressed as a differential equation, is

$$G\,dh = L\,dt + G\,dH(t_2 - 32)$$
$$[G\,dh = L\,dt + G\,dH(t)] \tag{44.1}$$

44.3.1 Heat Load, Range, and Water Flow Rate

The expression $L\,dt$ in Eq. (44.1) represents the heat load imposed on the tower by whatever process it is serving. However, because the mass of water per unit time is not easily measured, the heat load is usually expressed as

$$\text{Heat load} = \text{gal/min} \times R \times 8.33 = \text{Btu/min}$$
$$[\text{L/min} \times R(°\text{C}) \times 0.998 = \text{kcal/min}] \tag{44.2}$$

where gal/min (L/min) is the water flow rate through the system, R (the range) is the difference between the hot- and cold-water temperatures [in °F (°C) (see Fig. 44.9), and 8.33 = pounds per gallon of water (0.998 = kilogram per liter of water).

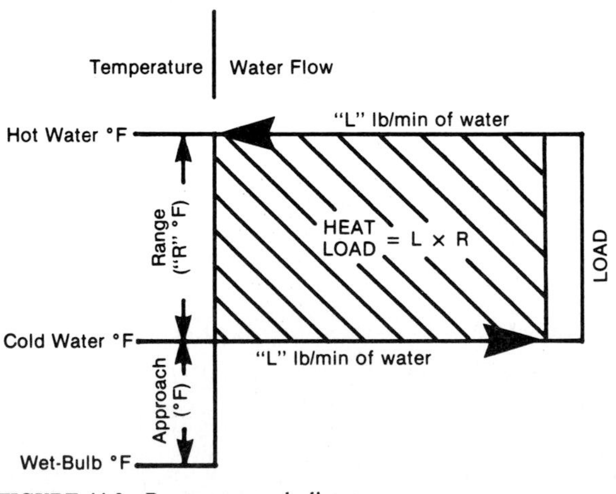

FIGURE 44.9 Range-approach diagram.

Note from Eq. (44.2) that the heat load establishes *only* an imposed temperature differential in the condenser water circuit and is unconcerned with the actual hot- and cold-water temperatures. Therefore, the mere indication of a heat load is meaningless to the application engineer attempting to size a cooling tower properly. More information of a specific nature is required.

Optimum operation of an air-conditioning system usually occurs within a relatively narrow band of condenser water flow rates and cold-water temperatures. This establishes two of the parameters required to size a cooling tower accurately—namely, the water flow rate and the cold-water temperature.

The total heat to be rejected from the system establishes a third parameter—the hot-water temperature coming to the tower. For example, let us assume that a refrigerant-compression type of air-conditioning unit of 500-ton (1760-kW) nominal capacity rejects heat to the condenser water circuit at a rate of 120,000 Btu/min (30,240 kcal/min) (including heat of compression) and performs most efficiently if supplied with 1500 gal/min (5677.5 L/min) of water at 85°F (29.4°C). With a slight transposition of Eq. (44.2), the condenser water temperature rise can be determined as

$$R = \frac{120{,}000}{1500 \times 8.33} = 9.6°\text{F} \qquad R = \frac{30{,}240}{5677.5 \times 0.998} = 5.3°\text{C}$$

Therefore, the hot-water temperature coming to the tower would be 85 + 9.6 = 94.6°F (29.4 + 5.3 = 34.7°C).

44.3.2 Wet-Bulb Temperature

Having determined that the cooling tower must be able to cool 1500 gal/min (5677.5 L/min) of water from 94.6°F (34.7°C) to 85°F (29.4°C), what conditions of the entering air must be known?

Equation (44.1) would identify enthalpy to be of prime concern, but air enthalpy is not something that is routinely measured and recorded at any geographic location. However, wet-bulb and dry-bulb temperatures are values easily measured, and a glance at Fig. 44.10 shows that lines of constant wet-bulb temperature are essentially parallel to lines of constant enthalpy. Therefore, the wet-bulb temperature is the only air parameter needed to size a cooling tower properly, and its relationship to other parameters is as shown in Fig. 44.9.

44.3.3 Enthalpy Exchange Visualized

To understand the exchange of total heat (including a slight amount of mass exchange) that takes place in a cooling tower, let's assume a tower designed to cool 120 gal/min (1000 lb/min) [454.2 L/min (543.6 kg/min)] of water from 85°F (29.4°C) to 70°F (21.1°C) at a design wet-bulb temperature of 65°F (18.3°C) and (for purposes of illustration only) a coincident dry-bulb temperature of 78°F (25.6°C). (These conditions of the air are defined as point 1 in Fig. 44.10.) Let's also assume that air is caused to move through the tower at the rate of 1000 lb/min (approximately 13,500 ft^3/min) [543.6 kg/min (382.3 m^3/min)].

Since the mass flows of air and water are equal, 1 lb (0.4536 kg) of air can be said to contact 1 lb (0.4536 kg) of water, and the psychrometric path of one such unit of air has been traced in Fig. 44.10 as it moves through the tower.

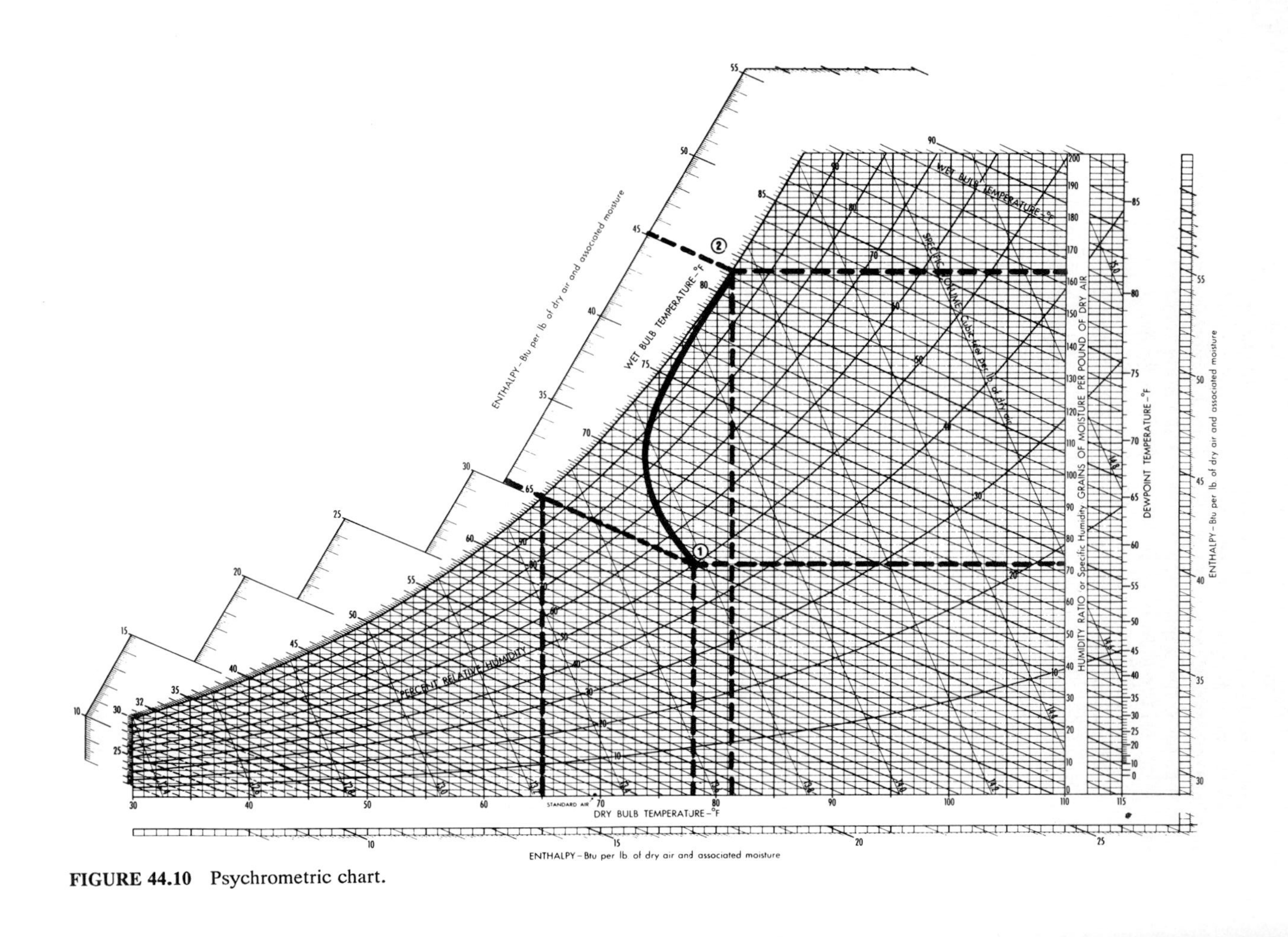

FIGURE 44.10 Psychrometric chart.

Air enters the tower at condition 1 [65°F (18.3°C) wet bulb and 78°F (25.6°C) dry bulb] and begins to gain enthalpy (total heat) and moisture content in an effort to achieve equilibrium with the water. This pursuit of equilibrium (solid line) continues until the air exits the tower at condition 2. The dashed lines identify the following changes in the psychrometric properties of this pound of air due to its contact with the water:

- The total heat content (enthalpy) increased from 30.1 Btu (7.59 kcal) to 45.1 Btu (11.37 kcal). This enthalpy increase of 15 Btu (3.78 kcal) was gained from the water. Therefore, 1 lb (0.4536 kg) of water was reduced in temperature by the required amount of 15°F (8.3°C).
- The air's moisture content increased from 72 grains (gr) [4.67 grams (g)] to 163 gr (10.56 g) (7000 gr = 1 lb). These 91 gr (5.89 g) of moisture (0.013 lb of water) were evaporated from the mass flow of water at a latent heat of vaporization of about 1000 Btu/lb (556 kcal/kg). This means that 13 of the 15 Btu (3.28 of the 3.78 kcal) removed from the water (86 percent of the total) occurred by virtue of *evaporation*.

Note: Water's latent heat of vaporization varies with temperature from about 1075 Btu/lb at 32°F (597.7 kcal/kg at 0°C) to 970 Btu/lb at 212°F (539.3 kcal/kg at 100°C). Actual values at specific temperatures are tabulated in various thermodynamics manuals.

44.3.4 Effects of Variables

Although several parameters are defined in Fig. 44.9, each of which will affect the size of a tower, understanding their effect is simplified if one thinks only in terms of (1) heat load, (2) range, (3) approach, and (4) wet-bulb temperature. If three of these parameters are held constant, changing the fourth will affect the tower size as follows:

1. Tower size varies directly and linearly with heat load (Fig. 44.11).

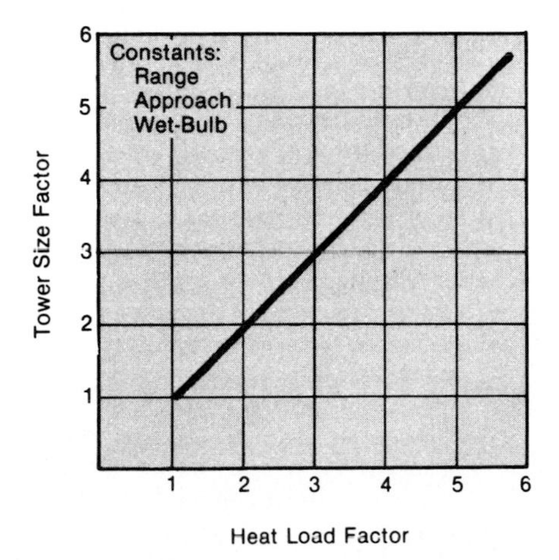

FIGURE 44.11 Effect of heat load on tower size.

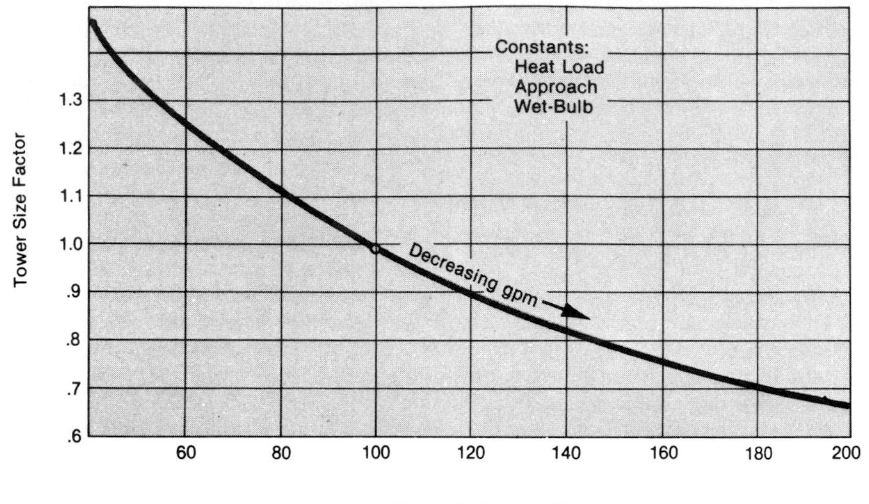

FIGURE 44.12 Effect of range on tower size.

2. Tower size varies inversely with range (Fig. 44.12). Two primary factors account for this. First, increasing the range (Fig. 44.9) also increases the temperature differential between the incoming hot-water temperature and the entering wet-bulb temperature. Second, as evidenced by Eq. (44.2), increasing the range at a constant heat load requires that the water flow rate be decreased. The resultant decrease in internal static pressure increases the flow of air through the tower.
3. Tower size varies inversely with approach (Fig. 44.13). A longer approach requires a smaller tower. Conversely, a closer approach requires an increasingly larger tower, and at a 5°F (3°C) approach the effect on tower size begins to become asymptotic. For this reason, *it is not customary in the cooling tower industry to guarantee any approach of less than 5°F (3°C).*

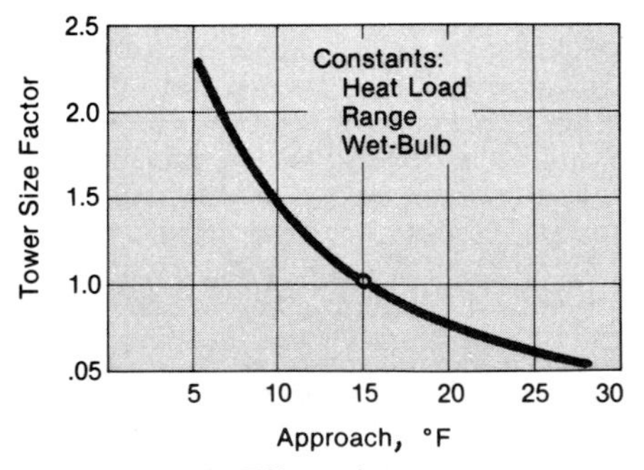

FIGURE 44.13 Effect of approach on tower size.

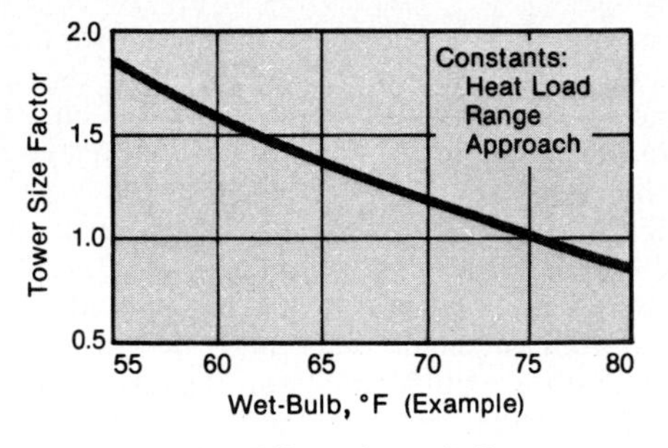

FIGURE 44.14 Effect of wet bulb on tower size.

4. Tower size varies inversely with wet-bulb temperature. When the heat-load, range, and approach values are fixed, reducing the design wet-bulb temperature increases the size of the tower (Fig. 44.14). This is because *most* of the heat transfer in a cooling tower results from evaporation, and air's ability to absorb moisture is reduced as the temperature is reduced.

44.4 COOLING TOWER FILL

Although cooling tower fill is often acceptably referred to as a "heat-transfer surface," such terminology is not true in its strictest sense. The heat-transfer surface in the classic cooling tower is actually the exposed surface of the water itself. The fill is merely a medium which causes more water surface to be exposed to the air (increasing the *rate* of heat transfer) and which increases the time of air-water contact by retarding the progress of the water (increasing the *amount* of heat transfer).

44.4.1 The Purpose of Fill

At a given rate of air moving through a cooling tower, the extent of heat transfer that can occur depends on the amount of water surface exposed to that air. In the tower shown in Fig. 44.8, the total exposure consists of the cumulative surface areas of a multitude of random-size droplets, the size of which depends on the type of nozzle utilized and on the pressure at which the water is sprayed. Higher pressures normally produce a finer spray and greater total surface-area exposure.

However, droplets contact each other readily in the overlapping spray patterns and coalesce into larger droplets, which reduces the net surface-area exposure. Consequently, predicting the thermal performance of a spray-filled tower is difficult at best and is highly dependent on good nozzle design as well as on a constant water pressure.

This relationship between water-surface exposure and heat-transfer rate is intrinsic to all types of towers, regardless of the type of water distribution system utilized. The more water surface exposed to a given flow of air, the greater will be the rate of heat transfer.

For a specific heat-dissipation problem, however, the *rate* at which enthalpy will be exchanged is important in the sense that it allows the designer to predict a finite *total* exchange of heat—with the total being a function of the period of time that the air and water are in intimate contact. Within psychrometric limits, the longer the contact period, the greater will be the total exchange of heat—and the colder will be the cold-water temperature.

In the spray-filled tower, the total time of air-water contact can only be increased by increasing the height of the tower, causing the water to fall through a greater distance. Given a tower of infinite height, the cold-water temperature produced by that tower would equal the incoming air's wet-bulb temperature, and the leaving air temperature would equal the incoming hot-water temperature. (These are the psychrometric limits previously mentioned.)

Obviously, a tower of infinite height would cost an infinite amount of money. More practically, structural limitations would begin to manifest themselves very early in the attempt toward infinite height. Early cooling tower designers quickly discovered these limitations and devised the use of fill as a far better means of increasing not only the rate of heat transfer, but its amount as well.

44.4.2 Types of Fill

The two basic types of fill utilized in present-day cooling towers are the splash type (Fig. 44.15) and the film type (Fig. 44.16). Either type may be used in towers of both counterflow and crossflow configuration, situated within the towers as shown in Figs. 44.4 and 44.5, respectively.

Both types of fill exhibit advantages in varied operating situations, assuring that neither type is likely to endanger the continued utilization of the other. Offsetting cost comparisons tend to keep the two types competitive, and the operational advantages peculiar to a specific situation are usually what tip the scales of preference. Therefore, specifiers are cautioned against excluding either type unless there are overriding reasons for doing so.

FIGURE 44.15 Splash fill installed.

FIGURE 44.16 Film fill installed.

Splash Type. Splash-type fill (Fig. 44.17) causes the flowing water to cascade through successive elevations of parallel "splashbars." Equally important is the increased time of air-water contact brought about by repeated interruption of the water's flow progress.

Since the movement of water within a cooling tower is essentially vertical, splash-type fill must obviously be arranged with the wide dimension of the

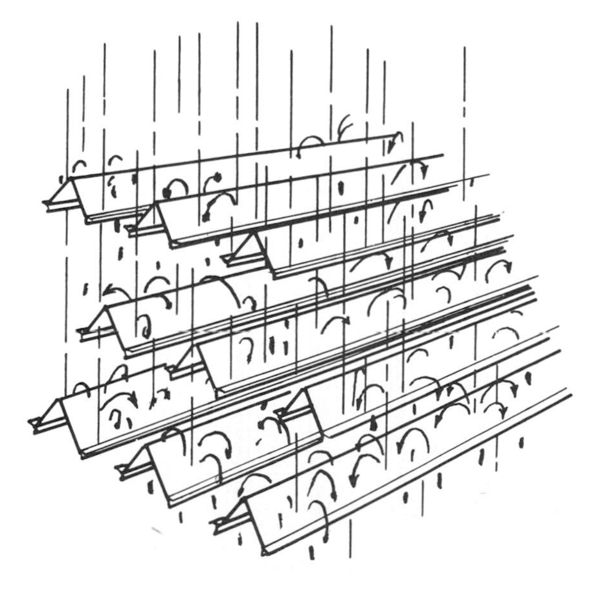

FIGURE 44.17 Splash-fill concept.

splashbars situated in a horizontal plane; otherwise, maximum retardation and breakup of the water could not be realized. When wood splashbars, for example, are utilized, they are typically ⅜ in (1 cm) thick by 1½ in (3.8 cm) wide by about 4 ft (1.2 m) long, and, as situated in the tower, only the ⅜-in (1-cm) dimension is vertical. Consequently, splash-type fill provides the least opposition to air flow in a horizontal direction, which accounts for the fact that splash-type fill is not routinely applied in counterflow towers.

Because of the water dispersal that takes place within splash-type fill, splash-filled towers are quite forgiving of the poor initial water distribution that can result from clogged or missing nozzles. The splashing action effectively redistributes the water at each level of splashbars. The relatively wide spacing of splash-type fill renders it ideal for use in a contaminated-water situation, where fills of closer spacing would be subject to clogging.

Splashbars are typically manufactured of polyvinyl chloride (PVC), polypropylene, polyethylene, or treated wood. Occasionally they are made of rolled aluminum or stainless steel to withstand high temperatures or particularly aggressive water conditions. Splashbars are normally supported by fiber-reinforced polyester (FRP) grids.

Film Type. Film-type fill has gained prominence in the cooling tower industry because of its ability to expose greater water surface within a given packed volume. Hence, film-filled towers tend to be somewhat smaller than splash-filled towers of equal performance. Film fill is equally effective in both crossflow and counterflow towers.

As can be seen in Fig. 44.18, water flows in a thin "film" over vertically oriented sheets of PVC fill, which are usually spaced approximately ¾ in (1.9 cm)

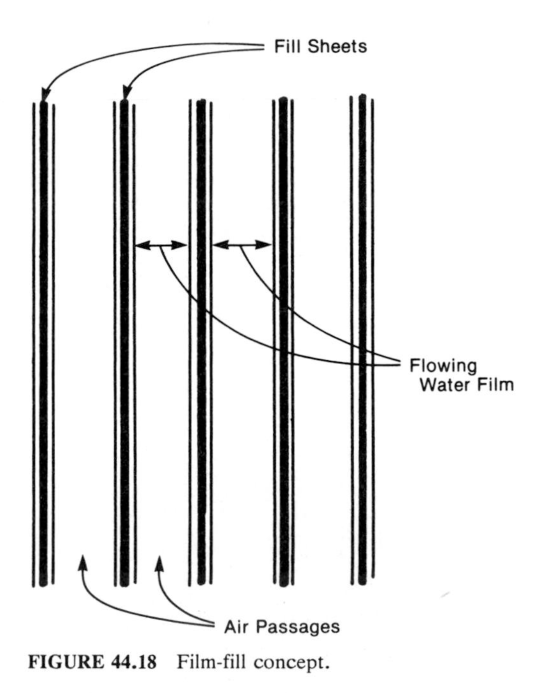

FIGURE 44.18 Film-fill concept.

apart. For purposes of clarity, the fill sheets in Fig. 44.18 are shown as if they were flat. In actuality, these sheets are usually molded into corrugated, or "chevron," patterns (Fig. 44.16) to create a certain amount of turbulence within the air stream and to extend the exposed water surface.

Unlike splash fill, film-type fill affords no opportunity for the water to redistribute itself during its vertical progress. Consequently, uniformity of the initial water distribution at the top of the fill is of prime importance, as is vigilant maintenance of the water distribution system. Areas deprived of water will become unrestricted paths for useless air flow, and thermal performance will degrade.

Where normally "clean" circulating water is anticipated, film-type fill will usually be the proper choice. However, the narrow passages afforded by the close spacing of fill sheets makes film fill very susceptible to poor water quality. High turbidity, leaves, debris, or the presence of algae, slime, or a fatty acid condition can diminish the passage size and affect the heat-transfer efficiency.

44.5 EXTERNAL INFLUENCES ON PERFORMANCE

Unlike other components comprising the total air-conditioning system, the cooling tower is normally exposed to the vagaries of wind and weather, which can affect its level of performance. Being a "breathing" device, it can also be affected by abnormalities relating to both the quality and quantity of its air supply.

Although cooling tower manufacturers strive for designs that will minimize the impact of such external forces, it is ultimately incumbent upon the site designer (1) to allocate an adequate, properly configured space for the cooling tower or (2) to adjust the thermal design conditions accordingly. This section discusses the first of these options, and Sec. 44.6 discusses the second.

44.5.1 Recirculation

The air moving through a cooling tower undergoes considerable increase in wet-bulb temperature, compared to its entering condition. Occasionally a portion of this warm, saturated, leaving air stream will be reintroduced into the tower, causing an artificial elevation of the entering wet-bulb temperature and a resultant increase in the tower's cold-water temperature. This condition is called "recirculation," and the potential for it to occur is directly related to the type of tower utilized and to its orientation with respect to the prevailing wind.

In the induced-draft tower (Fig. 44.3), discharge air leaves the tower at a velocity of approximately 2000 ft/min (10 m/s), whereas the intake velocity is usually less than 700 ft/min (3 m/s). This velocity differential essentially precludes self-imposed recirculation. However, the velocity relationships in the forced-draft tower (Fig. 44.2) are the reverse of those encountered in the induced-draft tower. Air enters the fan at a velocity exceeding 2000 ft/min (10 m/s) and leaves the tower at velocities normally less than 700 ft/min (3 m/s). The high entrance velocity evacuates a low-pressure zone at the intake, in which a part of the billowing exhaust is often recaptured.

When flowing wind encounters an obstruction, such as a cooling tower, the normal path of the wind is disrupted and a reduced-pressure zone forms on the lee side (downwind) of the obstruction. In trying to fill this "void," the wind will attempt to carry with it the tower's saturated exhaust air—which has been suitably deflected in the direction of the trouble zone. The wind's success or failure in this attempt depends, of course, on the velocity of the air leaving the tower. Higher-velocity discharges obviously suffer less downwind deflection than do lower-velocity discharges.

Whether or not the plume deflection results in recirculation also depends on the nature of the downwind tower face. If that face is an air intake, some degree of recirculation will definitely occur. If it is a cased (solid) face, there will be small likelihood of recirculation. Therefore, proper orientation of the tower with respect to the prevailing wind (coincident with the design wet-bulb temperature) is very important. Normally, seasonal shifts in wind direction are of little concern because they are usually accompanied by reductions in the wet-bulb temperature.

44.5.2 Interference

Intervening heat sources of higher enthalpy content located upwind of a cooling tower can also cause an elevation in the wet-bulb temperature of the entering air. These sources, most often, are other cooling towers, whose discharge air (depressed by the wind) "interferes" with the thermal quality of the air supplied to the tower in question. Where location of an intended cooling tower downwind of such a heat source is unavoidable, the design wet-bulb temperature should be adjusted, or higher system operating temperatures should be anticipated.

44.5.3 Air Restrictions

In residential, commercial, and small industrial installations, cooling towers are frequently shielded from view with barriers or enclosures. Usually this is done for aesthetic reasons. Quite often these barriers restrict air flow, resulting in low-pressure areas (inviting recirculation) and poor air distribution to the tower's air inlets. Sensible construction and placement of screening barriers will help minimize any negative effect on the thermal performance of the cooling tower.

Screening in the form of shrubbery, fences, or louvered walls should be placed several feet from the air inlet to allow normal air entry into the tower. When a tower is enclosed, it is desirable for the enclosure to have openings opposite each air-intake face and for the net free area of the openings to be at least equal to the gross intake area of that tower face.

Although recommended clearances vary with the type of enclosure under consideration, the basic rule of thumb for an induced-draft tower is to place the enclosure wall no closer to the tower than the gross height of the tower's air-intake opening. For forced-draft towers, this distance should be at least 3 times the air-intake height. However, since these recommended distances tend to increase with tower length and number of operating cells, screening barriers or enclosures should not be installed without prior consultation with the cooling tower manufacturer.

44.6 CHOOSING THE DESIGN WET-BULB TEMPERATURE

Selection of the design wet-bulb temperature must be made on the basis of conditions existing at the site proposed for the cooling tower. It should be that which will result in the maximum acceptable cold-water temperature at, or near, the time of peak load demand.

Performance analyses have shown that most air-conditioning installations based on wet-bulb temperatures that are exceeded in no more than 1 percent of the total hours during a normal summer have given satisfactory results. The hours in which peak wet-bulb temperatures exceed the upper 1 percent level are seldom consecutive hours and usually occur in periods of relatively short duration. The "flywheel" effect of the total water system inventory is usually sufficient to carry through the above-normal periods without detrimental results.

Air temperatures, wet-bulb as well as coincident dry-bulb, are routinely measured and recorded by the U.S. Weather Bureau, worldwide U.S. military installations, airports, and various other organizations to whom anticipated weather patterns and specific air conditions are of vital concern. Compilations of these data exist that are invaluable to both the users and designers of cooling towers.

However, it must be realized that the wet-bulb temperature determined from these publications represents the ambient for a geographic area and does not take into account localized heat sources and the potential for recirculation, which may artificially elevate that temperature at a specific site. Where local effects and anticipated recirculation are known, the design wet-bulb temperature must be increased by an appropriate amount. Where doubt exists, the design wet-bulb temperature should be increased by 1°F (0.5°C) for an induced-draft tower, or 2°F (1°C) for a forced-draft tower. (Readers interested in the derivation of this "rule" are referred to "External Influences on Cooling Tower Performance," *ASHRAE Journal*, January 1983.)

44.7 TYPICAL COMPONENTS

Although techniques and technology vary among cooling tower manufacturers, in the realm of mechanical-draft types (which predominate in air-conditioning applications) certain primary and essential components are common to all. Each, for example, will assume a modified box shape designed to contain and control a transient flow of water while, at the same time, permitting the relatively free passage of a continuous flow of air. Each will also be equipped with a fan to provide that amount of air, and virtually all will be equipped with fill to promote intimate contact between air and water.

Beyond these essential concerns, means must be provided to recover the flowing mass of water and replenish that which is lost, to assure mechanical operation within design limits, to promote thermal performance stability, to simplify maintenance, and to reduce adverse environmental impact. Components devoted to these considerations are described in this section. Fill was discussed in Sec. 44.4, and fans will be discussed in Sec. 44.9.

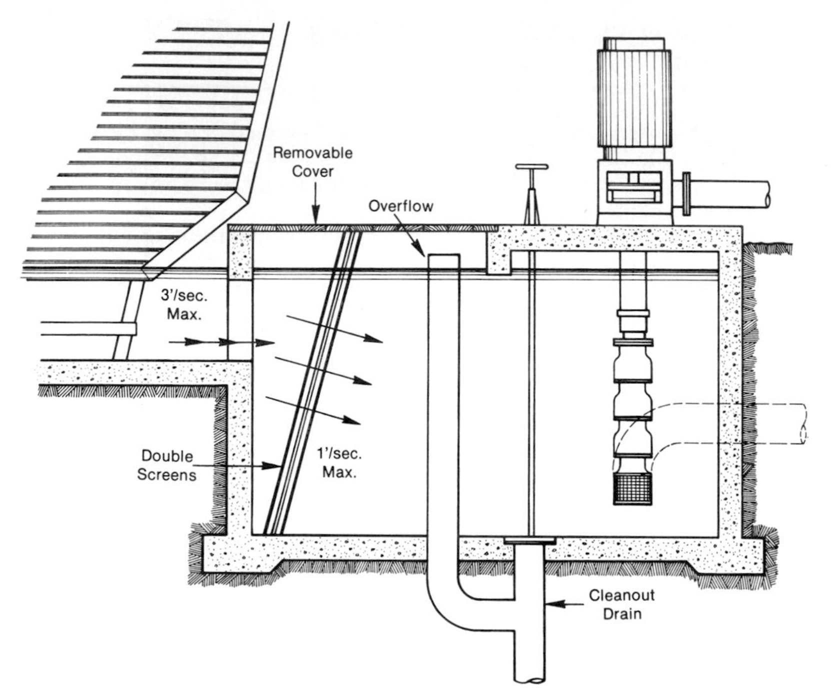

FIGURE 44.19 Typical cross section of concrete sump pit.

44.7.1 Cold-Water Basin

The cooling tower basin serves the two fundamentally important functions of (1) collecting the cold water following its transit of the tower and (2) acting as the tower's primary foundation. Because it also functions as a collecting point for foreign material washed out of the air by the circulating water, it must be accessible, cleanable, have adequate draining facilities, and be equipped with suitable screening to prevent entry of debris into the suction-side piping.

Ground-level installations, typical of many larger projects, may utilize concrete basins. Occasionally, concrete basins are also used on rooftop installations, with the basin slab either poured integrally with the roof or separated by a waterproof membrane. In all cases, concrete basins are designed by the structural engineer for construction by the general contractor, based on dimensional drawings and load schedules provided by the cooling tower manufacturer. Contiguous sumps (Fig. 44.19) provide outflow submergence and house the facilities for overflow and cleanout.

FIGURE 44.20 Steel grillage supporting tower equipped with wood cold-water collection basin.

Factory-assembled towers (Figs. 44.6 and 44.7), as well as elevated field-erected towers (Fig. 44.20), are normally equipped with basins pro-

FIGURE 44.21 Plywood cold-water basin floor. (Note depressed sump.)

vided by the cooling tower manufacturer. The materials utilized are compatible with the tower's overall construction; they include wood, steel, stainless steel, and plastic. For towers that come equipped with basins, the cooling tower manufacturer will include drain and overflow fittings, makeup valve(s), sumps and screens (Fig. 44.21), and provision for anchorage. The basin of the factory-assembled tower shown in Fig. 44.22 merely dumps by gravity into a concrete collection basin, the side curbs of which serve to support the tower.

A grillage of steel or concrete is normally utilized for support of a tower installed over a wood, steel, or plastic basin (Fig. 44.20). Grillages must be designed to withstand the total wet operating weight of the tower and connecting

FIGURE 44.22 Two-cell factory-assembled tower supported by concrete collection pit.

piping, plus the dead loads contributed by stairways, catwalks, etc. They must also accept transient loads attributable to winds, earthquakes, and maintenance traffic. The grillage members must be level and of sufficient strength to preclude excess deflection under load.

44.7.2 Louvers and Drift Eliminators

Water management and control are of prime concern to the cooling tower designer. Where air enters or leaves the tower, water also has the opportunity to escape. On the air-intake sides of induced-draft towers, louvers are devised to prevent the escape of random water droplets. Also, because a cooling tower promotes maximum contact between water and relatively high-velocity air [normally 500 to 700 ft/min (2.5 to 3.5 m/s)], water droplets become entrained in the leaving air stream. Collectively, these entrained water droplets are called "drift" (and are not to be confused with the pure water vapor with which the effluent air stream is saturated or with any droplets formed by the condensation of that vapor). The composition and quality of drift is the same as that of the circulating water flowing through the tower. Its potential for nuisance—in the spotting of cars, windows, and buildings—is considerable. Located upwind of critical areas, a cooling tower producing significant drift can pose an operating hazard.

FIGURE 44.23 Film fill with integral louvers and drift eliminators.

Drift eliminators remove entrained water from the discharge air by causing it to make sudden changes in direction. The resultant centrifugal force separates the drops of water from the air and deposits them on the eliminator surface, from which they flow back into the tower.

As with fill, PVC has become the dominant material from which to manufacture louvers and drift eliminators. They are normally formed into a honeycomb configuration with labyrinth passages and, for use with crossflow towers, can be molded integrally with the fill sheets (Fig. 44.23). Although current industry standards continue to limit allowable drift to 0.2 percent of the circulating water rate, actual drift rates seldom exceed 0.02 percent with present technology.

44.7.3 Fan Drive Mechanisms

The optimum speed of a cooling tower fan seldom coincides with the most efficient speed of the driving motor. Even in the smaller fan sizes, design speeds are usually less than 500 rpm, whereas motors are usually applied at a nominal speed of 1800 rpm. Speed reduction is therefore necessary, and it is usually accom-

FIGURE 44.24 Fan and drive mechanism mounted on support assembly.

plished either by differential gears of positive engagement or by differential pulleys connected through V-belts.

Typically, gear reduction units are of the right-angle type, coupled to the motor through an extended drive shaft (Fig. 44.24). This permits the motor to be located outside the fan cylinder where it is more accessible (Fig. 44.25) and operates in a less aggressive environment. Gear reduction units are applied throughout a wide range of operating powers, from 250 hp (175 kW) and larger down to as little as 5 hp (3 kW). V-belt drives, on the other hand, are seldom applied at powers greater than 50 hp (35 kW) in cooling towers. Prudent limiting of belt lengths requires that the motor be located in the hot, humid atmosphere within

FIGURE 44.25 Motor location outside fan cylinder.

the tower. In addition to complicating the routine maintenance of the motor, this tends to increase the frequency of necessary belt adjustment to reduce the energy loss associated with belt slippage.

44.8 MATERIALS OF CONSTRUCTION

The structures of cooling towers are typically made of wood or steel, depending on one's preference and on local building or fire codes. Casings, basins, and decks are usually of like material, with casings also being manufactured of fire-retardant glass-reinforced polyester plastic. Woods used are Douglas fir and redwood, either of which should be pretreated to prevent fungal attack. Being the single most available and most renewable structural material, wood is expected to continue as one of the predominant cooling tower materials for the foreseeable future.

Because it lends itself well to plant manufacturing techniques, steel is the principal material used in factory-assembled cooling towers. Appropriate grades of carbon steel are utilized for framing, casing, decking, hot- and cold-water basins, etc., and are usually galvanized for protection against corrosion. Minimum protection offered in the industry consists of G-90 galvanizing, ranging to G-210 galvanizing at the premium end of the scale. In addition, coatings are often applied over the galvanizing in an effort to extend its useful service life.

Although some towers in smaller sizes have been produced primarily in plastic, their susceptibility to fire and to ultraviolet deterioration has discouraged their widespread use. With technological advances in the plastics industry, such towers may well become more prevalent in the future.

Currently, and for the foreseeable future, those who seek to obtain a cooling tower offering the longest possible service life typically specify stainless steel. Such towers combine the fire resistance, corrosion immunity, and service longevity sought by the nonspeculative buyer. Furthermore, the increased use of stainless steel—plus economic pressures in the steel market as a whole—have tended to decrease the premium price differential. Therefore, the movement toward cooling towers of higher-quality construction can be expected to increase.

44.9 ENERGY MANAGEMENT AND TEMPERATURE CONTROL

Energy is consumed by a cooling tower in the operation of fans for the movement of air. It is also consumed within the system served by the cooling tower in the operation of condenser water pumps. Of these two energy-consuming aspects, fan manipulation offers the most productive means of controlling temperature—and thereby of controlling energy use.

Unless specifically permitted by the cooling tower manufacturer, *pump manipulation to vary water flow over the tower should not be used*. This is because the distribution system in a given tower is designed to produce optimum fill coverage at a specific water flow, and variations in that flow tend to reduce the tower's thermal efficiency.

Before attempting to manage energy use by the manipulation of fans, how-

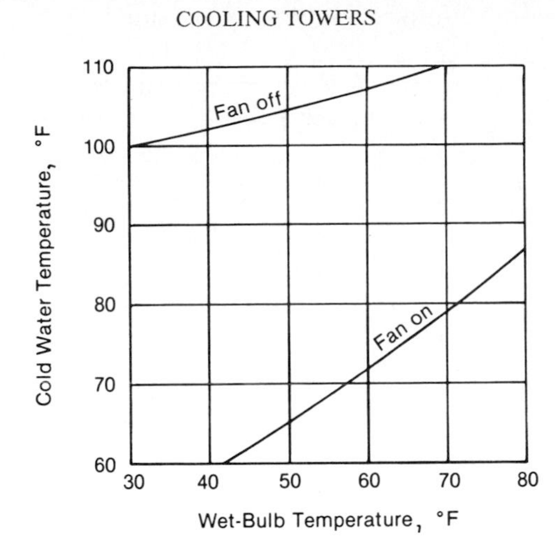

FIGURE 44.26 Typical performance curve—single-cell tower with single-speed motor.

ever, the operator should make sure that such manipulation is not premature—and therefore not self-defeating. Observing that the cold-water temperature produced by a cooling tower reduces as the outside wet-bulb temperature reduces (Fig. 44.26), many operators will begin cycling fans without regard to the effect that this may have on the overall system's energy consumption.

Bear in mind that the system compressor horsepower (wattage) is perhaps 20 times greater than that consumed by the cooling tower fans, and some air-conditioning systems reward a reduction in condenser water temperature by a commensurate reduction in compressor horsepower (wattage). Although those systems which do benefit from reduced condenser water temperature usually have some limiting temperature [perhaps 75°F (24°C)] below which no further reduction in compressor horsepower (wattage) is realized, the operator should allow that cold-water temperature from the tower to be reached before undertaking any fan manipulation.

44.9.1 Fan Cycling

The ability to achieve effective air-side control depends primarily on the number of cells comprising the tower and on the speed-change characteristics of the motors driving the fans. Figure 44.26 defines the operating modes available with a single-cell tower equipped with a single-speed motor. In this most rudimentary of cases, control of the cold-water temperature can be attempted only by cycling the fan motor on and off, and great care must be exercised to prevent an excessive number of starts from burning out the motor. (As a basic rule of thumb, 30 s/h of starting time should not be exceeded.)

Conversely, the operating characteristics of a three-cell tower equipped with single-speed and two-speed motors are shown in Figs. 44.27 and 44.28, respectively. The numbers in parentheses represent the approximate percentage of total fan power consumed in each operating mode. Note that the opportunity for both temperature control and energy management is tremendously enhanced by the use of two-speed motors. At any selected cold-water temperature, it can be seen

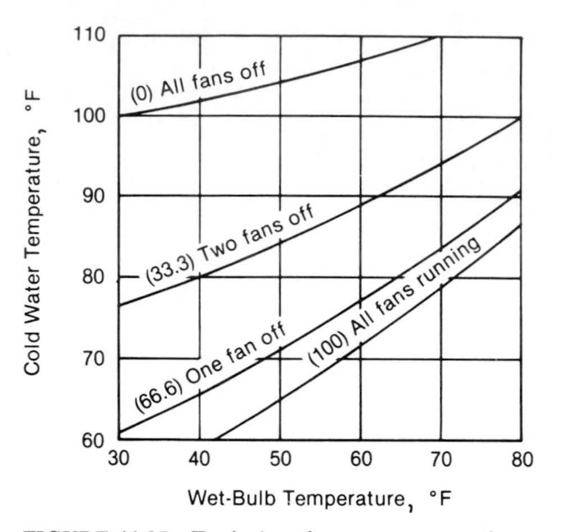

FIGURE 44.27 Typical performance curve—three-cell tower with single-speed motors.

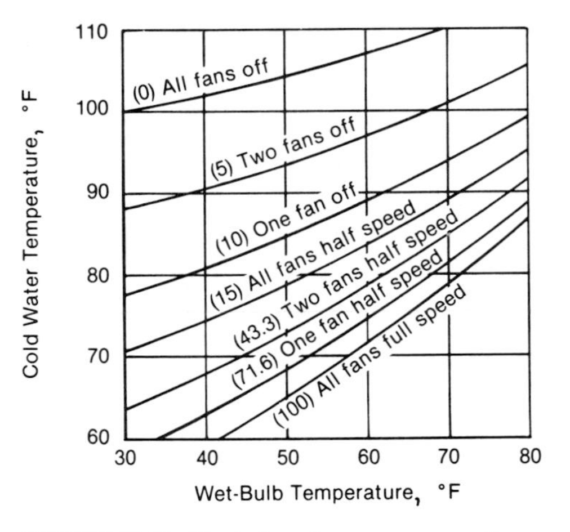

FIGURE 44.28 Typical performance curve—three-cell tower with two-speed motors.

that an increase in the number of fan-and-speed combinations causes the operational mode lines to come closer together. It follows, therefore, that the capability to modulate a fan's speed or capacity (within the range from 0 to 100 percent) would represent the ultimate in temperature control and energy management.

The technology by which to approach this ideal situation currently exists in the form of the automatic variable-pitch fans (Fig. 44.29) and electrical frequency modulating devices provided by several manufacturers. Figure 44.30 indicates the cold-water temperature control—and energy reductions—that can be achieved with automatic variable-pitch (AVP) fans. For purposes of the curve, it is assumed that the cold-water temperature (dashed line) has been set to reduce no lower than 75°F (24°C). In a reducing wet-bulb temperature, fan power (solid

FIGURE 44.29 Automatic variable-pitch fan used to manipulate air flow and fan horsepower (wattage).

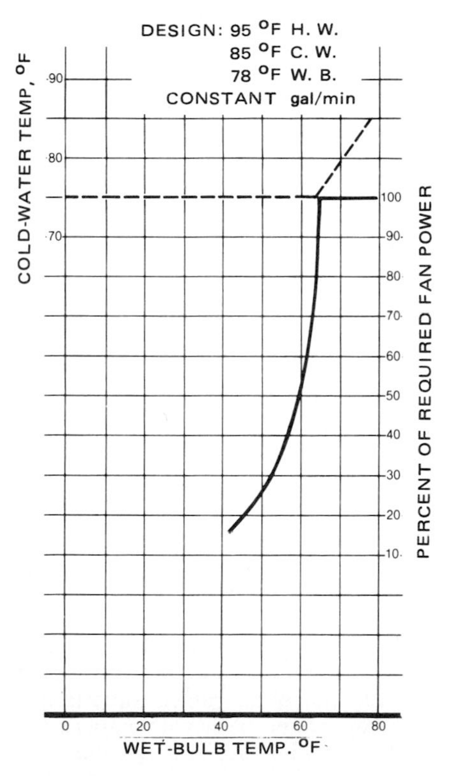

FIGURE 44.30 Typical power reduction with AVP fans.

line) will remain at 100 percent until such time as that cold-water temperature is reached, after which the fan automatically depitches to maintain that cold-water temperature. In this typical case, note that a further reduction of only 20°F (11°C) in wet-bulb temperature results in an 80 percent reduction in fan power.

Were it not for the physical characteristics of standard fans—as well as the limitations of their drive mechanisms—results similar to those shown in Fig. 44.30 could be achieved with frequency modulation control devices. Most fans have at least one critical speed that occurs between 0 and 100 percent of design rpm, and some have several. Typically, fans are designed such that their critical speeds do not coincide with the rpm's that will be produced by normal motor speed changes. They are also, of course, designed to minimize the effect of any critical speed. Nevertheless, it would be foolhardy to run the risk of protracted operation at a critical speed; therefore, it becomes necessary to predetermine the critical speeds and to prevent corresponding frequencies from being utilized. Avoiding critical speeds, however, reintroduces some miniature "notches" in an otherwise constant cold-water temperature line and, to avoid "hunting," requires that some appreciable tolerance be designed into the control mechanism.

44.9.2 Free Cooling

Because of increased computer loads, process loads, and the need for "core cooling" of larger buildings, air-conditioning systems are being used for longer periods of the year and, in many cases, year-round. The ability of cooling towers to produce water temperatures typical of chilled-water temperatures in colder months has led to an increase in the utilization of "free cooling" as the best means to reduce overall system energy demand.

Direct Free Cooling. The simplest and most thermally effective—*yet least recommended*—arrangement for free cooling is shown in Fig. 44.31, where a simple bypass system physically interconnects the chilled-water and condenser water loops into one common water path between the load and the cooling tower. The dashed lines indicate the water flow path during the free-cooling mode of operation. The absence of a heat exchanger separating the two water loops precludes the need for a temperature differential, so the load benefits from the cooling tower's full capability.

The reason why this direct system is least recommended is because the intermixing of the two water streams contaminates the "clean" chilled water with "dirty" condenser water, which is a situation most operators are reluctant to allow. Those who have been most successful in the utilization of this direct-connected system have used a "side stream" filtration arrangement to filter a portion of the total water flow continuously. In the figure, the flow from the small pump would be directed through the filter during tower operation and through an instantaneous heater during periods of shutdown.

Indirect Free Cooling. By the simple expedient of installing a heat exchanger, piped in a parallel bypass circuit with the chiller, free cooling is accomplished without concern for contamination of the chilled-water loop (Fig. 44.32). Because plate heat exchangers can function properly with only a small temperature difference [as little as 2°F (1°C), depending on size], they permit separation of the water loops with minimal sacrifice of free-cooling opportunity.

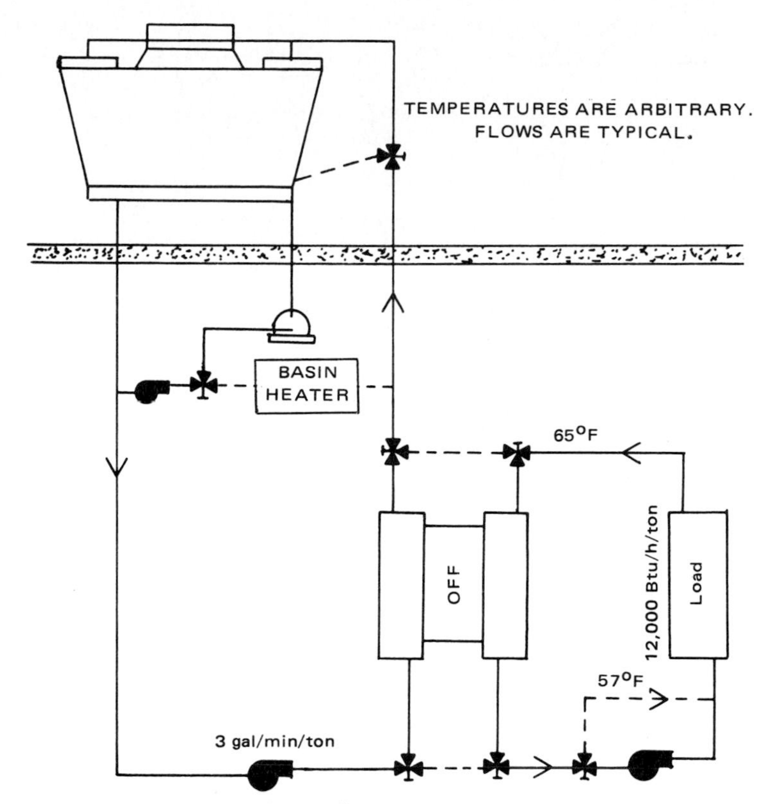

FIGURE 44.31 Direct free-cooling system.

Tower Selection for Free Cooling. One of the prime advantages of free cooling is that it can be applied to many existing systems. Most towers are capable of producing the temperatures required at some point in the year. Designers of new systems, however, have the opportunity to take free cooling into account early in the design and maximize free cooling to its fullest extent. Those who do usually choose a tower size that will produce a necessary "chilled"-water temperature at the highest possible wet-bulb temperature. This ensures that compressor energy will be avoided for the maximum number of annual operating hours.

44.10 WINTERTIME OPERATION

Although cooling towers have always been used 12 months of the year in process industries, such usage has only recently become prevalent in the air-conditioning industry. Accordingly, operators who have been accustomed to shutting down their tower in October and restarting it in May must now concern themselves with the fact that the tower may become a potential ice-maker in freezing weather.

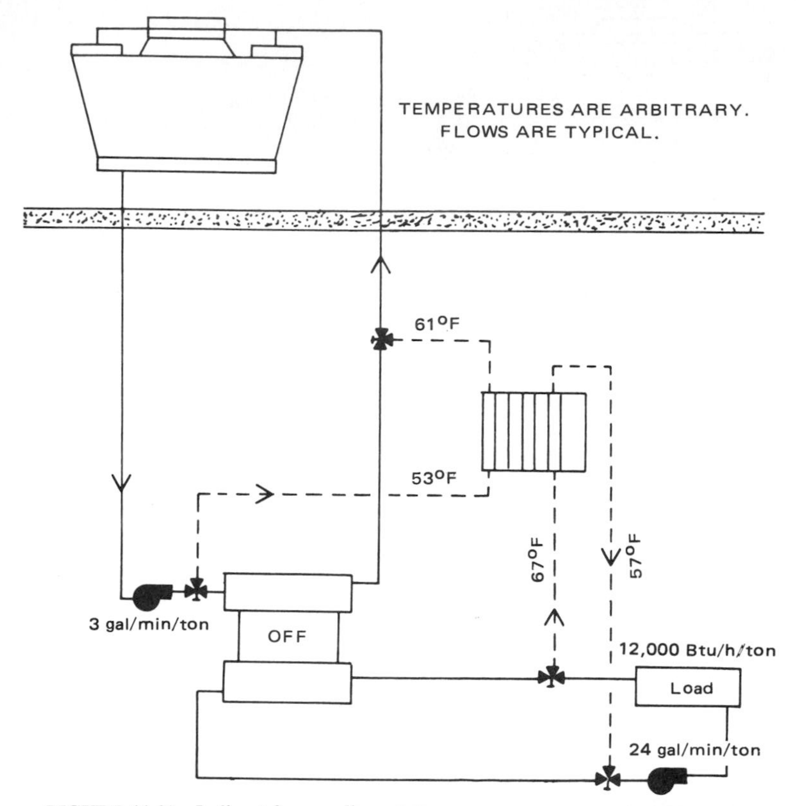

FIGURE 44.32 Indirect free-cooling system.

44.10.1 Periods of Shutdown

The most obvious time of concern, of course, is when the system is shut down for an interim period of time (i.e., nights, weekends, etc.). During these shutdown periods, no system heat is being added to the condenser water—the water is stagnant—and freezing is free to occur in the cooling tower basin as well as in all exposed lines. At these times, the operator basically has two possible courses of action: (1) Drain the system to the point where all outside components are empty of water or (2) find a way to add heat to the exposed portion of the system.

With proper forethought, system draining can become an automatic process. The "indoor tank" method shown in Fig. 44.33 allows water in the tower's cold-water basin to drain continuously into an indoor storage tank, from which the condenser water pumps take suction. The small bypass drain line connecting the main supply and return lines, just below the "roof" level, is there to ensure drainage of the tower supply line at the time of pump shutdown. This bypass line may be equipped with an automatic valve designed to open on pump shutdown, or it may merely be a relatively small open line that would allow a limited amount of warm water to bleed continuously into the tank.

Where space does not allow the use of an indoor tank, heat must be added to

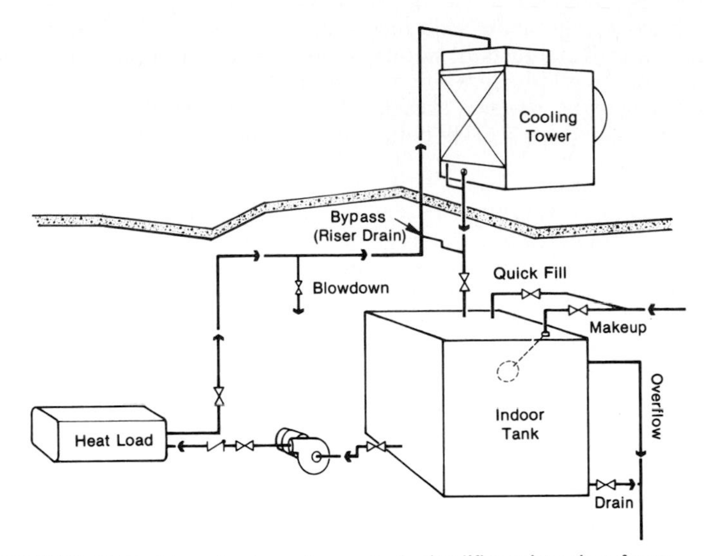

FIGURE 44.33 Use of indoor storage tank simplifies wintertime freeze protection.

the exposed system by means of some type of temperature-actuated device. Foremost among the systems used are electric immersion heaters submerged in the water of the cooling tower basin, although steam or hot-water loops are also routinely utilized. Where electric immersion heaters are used, care must be taken to ensure that the heater element is always adequately covered with water. Such heater elements operate at about 1500°F (815°C) in the open air and constitute an extremely high fire risk.

Where such devices are used in the cooling tower basin, the designer must realize that they give no protection to exposed piping. Therefore, all exposed piping must be heat-traced and insulated.

A very effective way to protect both the cooling tower basin and the primary exposed piping is shown in Fig. 44.31. Note the presence of a small bypass pump and an instantaneous heater. During shutdown periods, this pump would be temperature-actuated to pump a small amount of water through the heater and back to the cooling tower basin. Note also that a bypass valve has been added to the tower riser to divert this small flow directly into the cooling tower basin; this is to prevent the water from going over the top of the tower, where it could induce freezing on the fill. In the case of this system, the designer must remember that the exposed makeup line to the tower gets no protection from freezing and must make provisions accordingly.

44.10.2 Periods of Operation

Manipulation of the air flow (see Sec. 42.9) is an invaluable tool not only in the retardation of ice formation but also in the reduction or elimination of ice already formed. In addition to bringing less cold air into contact with the circulating water, reducing the entering air-flow velocity alters the path of the falling water, allowing it to impinge upon—and melt—ice previously formed by those random droplets which wind gusts or normal splashing may have caused to escape the protection of the relatively warm main stream of water.

The falling-water patterns associated with each type of tower (crossflow and counterflow) have much to do both with the type of ice formed and with its location.

Crossflow Towers. Crossflow towers equipped with splash-type fill tend to form louver ice, wherein water droplets generated by the splashing action may escape the fill area, impinge upon the louvers, and be frozen almost instantaneously. Usually this ice can be controlled merely by reducing the fan's speed or by turning the fan off for a short period of time. Severe cases may require that the fan be reversed for an interim period. This causes warmed air to flow outward through the louvers to effect deicing. However, since this also causes frigid air to flow downward through the fan, the amount of time in this operating mode must depend on the tendency for ice to form on the fan cylinder and mechanical equipment. Therefore, fan reversal should be limited to the minimum time required to accomplish acceptable deicing and should be monitored by the operator.

Also, the direction of a fan's rotation should not be changed without allowing the fan to come essentially to a stop; otherwise, tremendous stresses will be imposed on the entire drive train. Prudence dictates that the starter be equipped with a 2-min time delay between directional changes.

Used in crossflow configuration, PVC film-type fill also tends to limit the amount of ice formed. This is due to the absence of any splashing action within the fill. Typically, thin ice may tend to form on the lower leading edges of the fill, turn to slush as the air flow becomes blocked, and shortly disappear. This is particularly true of current fill configurations having louvers molded integrally with the fill. Given normal control measures, these towers have proved to be very civilized in their wintertime operation. This can be very important in the low-load, low-temperature situations encountered in free cooling.

Counterflow Towers. Since the fill of a counterflow tower is elevated appreciably above the cold-water basin's level, the generation of random water droplets produced by this free-fall tends to be irrespective of the type of fill utilized. Droplets that splash in an outboard direction may freeze on the louvers and lower structure.

Deicing measures for counterflow towers are similar to those utilized for crossflow towers but tend to be somewhat less effective. With the fans off, the normally vertical sides of a counterflow tower place the air inlets beyond the reach of the falling-water pattern. In this fans-off mode, only the convective heat from the water promotes deicing, and the process may take considerably longer than with a crossflow tower. Although fan reversal can also be effectively utilized for deicing of a counterflow tower, many operators are reluctant to reverse fans because of the small amount of water caused to escape the air inlets by the outward flow of air, for this water may produce sufficient ice in the immediate region of the tower to be considered hazardous, requiring separate measures for its control.

Fan Options. Single-speed motors afford the least opportunity for air-flow variation, and towers so equipped require maximum vigilance on the part of the operator to determine the proper cyclic operation of the fans to accomplish the best ice control.

Two-speed fan motors offer significantly greater operating flexibility and should be given maximum consideration in the purchase of towers for use in cold climates. To achieve a balance between cooling effect and ice control, the fans may individually be cycled back and forth between full speed and half speed, as required; this practice is limited only by the maximum allowable motor insulation temperature (which temperature could be exceeded if there were an abnormal number of speed changes per hour). In many cases, it will be found that operating all the tower fans at half speed produces the best combination of cooling effect and ice control.

Operators who wish to be relieved of the burden of selecting the proper air flow for any operating situation may choose such devices as the AVP fan (Fig. 44.29) described in Sec. 44.9.1. In most operating cases, by the time the ambient air temperature has dropped low enough to produce ice, the fan will have depitched to a much reduced air-flow capability, making the formation of ice unlikely. By the simple expedient of adjusting the cold-water temperature control point, operators of sensitive systems can easily strike a balance between maximum cooling and minimum icing.

For water treatment and blowdown rates, see Chap. 53.

CHAPTER 45
APPLICATIONS OF HVAC SYSTEMS

Ernest H. Graf, P.E.
Assistant Director, Mechanical Engineering, Giffels Associates, Inc., Southfield, Michigan

William S. Lytle, P.E.
Project Engineer, Mechanical Engineering, Giffels Associates, Inc., Southfield, Michigan

45.1 GENERAL CONSIDERATIONS

As a system design develops from concept to final contract documents, the following subjects (in Secs. 45.1.1 to 45.1.11) should be considered throughout the HVAC design period.[1] These subjects are of a general nature inasmuch as they are applicable to all HVAC designs, and they may become specific requirements inasmuch as codes are continually updated.

45.1.1 Cooling Towers and Legionnaire's Disease

Since the 1976 outbreak of pneumonia in Philadelphia, cooling towers have frequently been linked with the *Legionella pneumophila* bacteria, or Legionnaires' disease. Much is yet to be learned about this bacteria, but until it is known to be eliminated, several precautions should be taken:

1. Keep basins and sumps free of mud, silt, and organic debris.
2. Use inhibitors as recommended by water-treatment specialists. Do not overfeed, because high concentrations of some inhibitors are nutrients for microbes.
3. Do not permit the water to stagnate. The water should be circulated throughout the system for at least 1 h each day regardless of the water temperature at the tower. The water temperature in indoor piping will probably be 60°F

[1]The preliminary design, calculations, equipment, and control of heating, ventilating, and air-conditioning (HVAC) systems are discussed in other chapters.

(15.6°C) or warmer, and one purpose of circulating the water is to disperse active inhibitors throughout the system.

4. Minimize leaks from processes to cooling water, especially at food plants. Again, the processes may contain nutrients for microbes.

45.1.2 Elevator Machine Rooms

These spaces are of primary importance to the safe and reliable operation of elevators. In the United States, all ductwork or piping in these rooms must be for the sole purpose of serving equipment in these rooms unless the designer obtains permission from the authorities in charge of administering ANSI Standard 17.1, *Safety Code for Elevators and Escalators*. If architectural or structural features tend to cause an infringement of this rule, the duct or pipe must be furred in and enclosed in an approved manner.

45.1.3 Energy Conservation

A consequence of the 1973 increase in world oil prices is legislation governing the design of buildings and their HVAC systems. Numerous U.S. states and municipalities include an energy code or invoke a particular issue of ASHRAE Standard 90 as a part of their building code. Standard 90 establishes indoor and outdoor design conditions, limits the overall U-factor for walls and roofs, limits reheat systems, requires the economizer cycle on certain fan systems, limits fan motor power, requires minimum duct and pipe insulation, requires minimum efficiencies for heating and cooling equipment, etc. Certain occupancies, including hospitals, laboratories, and computer rooms, are exempt from portions of the standard.

In the interest of freedom of design, the energy codes permit trade-offs between specified criteria as long as the annual consumption of depletable energy does not exceed that of a system built in strict conformance with the standard. Certain municipalities require that the drawings submitted for building-permit purposes include a statement to the effect that the design complies with the municipality's energy code. Some states issue their own preprinted forms that must be completed to show compliance with the state's energy code.

45.1.4 Equipment Maintenance

The adage "out of sight out of mind" applies to maintenance. Equipment that a designer knows should be periodically checked and maintained may get neither when access is difficult. Maintenance instructions are available from equipment manufacturers; the system designer should be acquainted with these instructions, and the design should include reasonable access, including walk space and headroom, for ease of maintenance. Some features for ease of maintenance will increase project costs, and the client should be included in the decision to accept or reject these features.

Penthouse and rooftop equipment should be serviceable via stairs or elevators and via roof walkways (to protect the roofing). Ship's ladders are inadequate when tools, parts, chemicals, etc., are to be carried. Rooftop air handlers, especially those used in cold climates, should have enclosed service corridors. If heavy rooftop replacement parts, filters, or equipment are expected to be skidded or rolled across a roof, the architect must be advised of the loading to permit proper roof system design.

Truss-mounted air handlers, unit heaters, valves, exhaust fans, etc., should be over aisles (for servicing from mechanized lifts and rolling platforms) when catwalks are impractical. Locate isolated valves and traps within reach of building columns and trusses to provide a degree of stability for service personnel on ladders.

It is important that access to ceiling spaces be coordinated with the architect. Lay-in ceilings provide unlimited access to the space above, except possibly at lights, speakers, sprinklers, etc. When possible, locate valves, dampers, air boxes, coils, etc., above corridors and janitor closets so as to disturb the client's operations the least.

Piping-system diagrams and valve charts are important and should be provided by the construction documents. Piping should be labeled with service and flow arrows, and valves should be numbered, especially when not within easy view of the source (such as steam piping not being within easy view of the boiler).

For piping of approximately 3 in (7.5 cm) and larger, use only flanged or lugged valves when it is intended that the item immediately adjacent to the valve will be removed for servicing. Remember that wafer valves are unsuitable inasmuch as both pipe flanges are required to hold the valve in place (see Chap. 48).

Pump performance and strainer clogging can be monitored by the pressure-gauge arrangement shown in Fig. 45.1 or by installing pressure gauges upstream

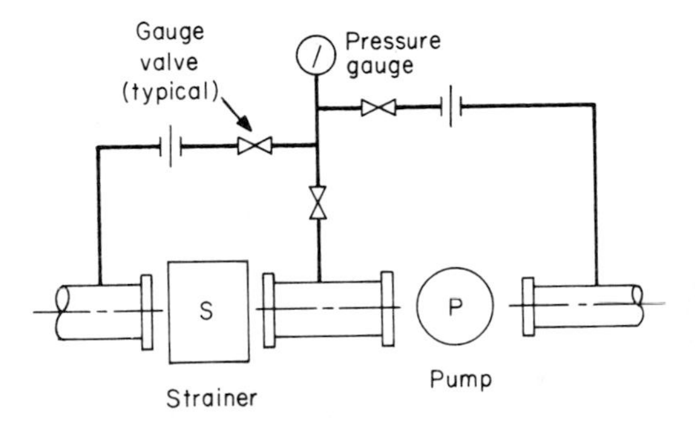

FIGURE 45.1 Multiple-point pressure gauge.

and downstream of strainers, pumps, etc. Using the readings from one gauge eliminates the suspicions caused by the inherent inaccuracies among multiple gauges. Frequently remaining serviceable for a long time, ⅜-in (10-mm) globe-pattern gauge valves are preferred to gauge cocks.

The observation of steam-trap operation can be facilitated by having a ⅜-in (10-mm) test valve at the trap discharge pipe (Fig. 45.2). With valve V-1 closed, trap leakage and cycling may be observed at an open test valve. The test valve can be used to monitor reverse-flow leaks at check valves.

45.1.5 Equipment Noise and Vibration

Noise and vibration can reach unacceptable levels in manufacturing plants as well as in offices, auditoriums, etc. Once an unacceptable level is "built in," it is

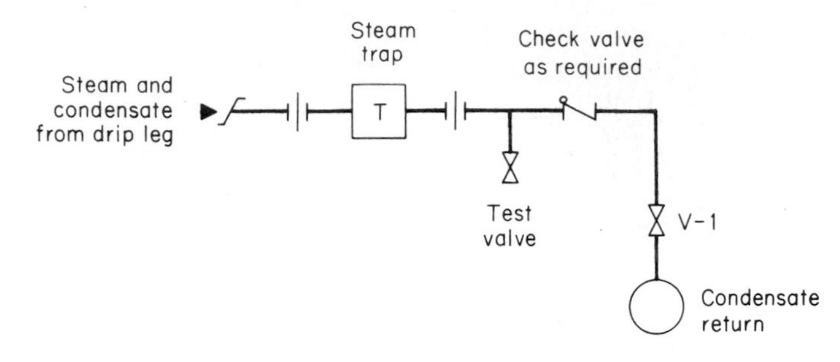

FIGURE 45.2 Test valve at steam trap.

very costly to correct. The noise and vibration control recommendations in Chaps. 49 and 50 of this book and in the *1987 ASHRAE Handbook, HVAC Systems and Applications*, should be followed. Sound and vibration specialists should be consulted for HVAC systems serving auditoriums and other sensitive areas. Fans, dampers, diffusers, pumps, valves, ducts, and pipes which have sudden size changes or interior protrusions or which are undersized can be sources of unwelcome noise.

Fans are the quietest when operating near maximum efficiency, yet even then they may require sound attenuation at the inlet and outlet. Silencers and/or a sufficient length of acoustically lined ductwork are commonly used to "protect" room air grilles nearest the fan. Noise through duct and fan sides must also be considered. In the United States, do not use acoustic duct lining in hospitals except as permitted by the U.S. Department of Health and Human Services (DHHS) Publication HRS-M-HF 84-1.

Dampers with abrupt edges and those used for balancing or throttling air flows cause turbulence in the air stream, which in turn is a potential noise source. Like dampers, diffusers (as well as registers, grilles, and slots) are potential noise sources because of their abrupt edges and integral balancing dampers. Diffuser selection, however, is more advanced in that sound criteria are readily available in the manufacturers' catalogs. Note, however, that a background noise (or "white" noise) is preferable in office spaces because it imparts a degree of privacy to conversation. Diffusers can provide this.

Pumps are also the quietest when operating near maximum efficiency. Flexible connectors will dampen vibration transmission to the pipe wall but will not stop water- or liquid-borne noise.

Valves for water, steam, and compressed-air service can be a noise source or even a source of damaging vibration (cavitation), depending on the valve pattern and on the degree of throttling or pressure reduction. Here again, the findings of manufacturers' research are available for the designer's use. (See Chap. 48 for a discussion of cavitation in valves.)

Equipment rooms with large fans, pumps, boilers, chillers, compressors, and cooling towers should not be located adjacent to sound- or vibration-sensitive spaces. General office, commercial, and institutional occupancies usually require that this equipment be mounted on springs or vibration isolation pads (with or without inertia bases) to mitigate the transfer of vibration to the building's structure. Spring-mounted equipment requires spring pipe hangers and flexible duct and conduit connections. Air-mixing boxes and variable-volume boxes are best located above corridors, toilet rooms, public spaces, etc. Roof fans, exhaust pipes from diesel-driven generators, louvers, etc., should be designed and located to minimize noise levels, especially when near residential areas.

45.1.6 Evaporative Cooling

An air stream will approach at it's wet bulb temperature a 100 percent saturated condition after intimate contact with recirculated water. Evaporative cooling can provide considerable relief without the cost of refrigeration equipment for people working in otherwise unbearably hot commercial and industrial surroundings, such as laundries, boiler rooms, and foundries. Motors and transformers have been cooled (and their efficiency increased) by an evaporatively cooled air stream.

Figure 45.3 shows the equipment and psychrometric elements of a "direct" evaporative cooler. Its greatest application is in hot, arid climates. For example, the 100°F (38°C), 15 percent relative humidity (RH) outdoor air in Arizona could be cooled to 70°F (21°C), 82 percent RH with an 88 percent efficient unit. Efficiency is the quotient of the dry-bulb conditions shown at (2), (3), and (4) in Fig. 45.3. Note that the discharge air from a direct evaporative cooler is near 100 percent humidity and that condensation will result if the air is in contact with surfaces below its dew point. The discharge dew point in the above example is 64°F (18°C).

Figure 45.4 schematically shows an "indirect" evaporative cooler. Whereas a direct evaporative cooler increases the air stream's moisture, an indirect evaporative cooler does not; that is, there is sensible cooling only at (1) to (2) in Fig. 45.4. Air is expelled externally at (5). When an indirect cooler's discharge (2) is ducted to a direct cooler's inlet, the final discharge (3) will be somewhat cooler and include less moisture than that of a direct cooler only. Various combinations of direct and indirect equipment have been used as stand-alone equipment or to augment refrigeration equipment for reduced overall operating costs. Refer to the *1988 ASHRAE Handbook, Equipment*, and the *1987 ASHRAE Handbook, HVAC Systems and Applications*.

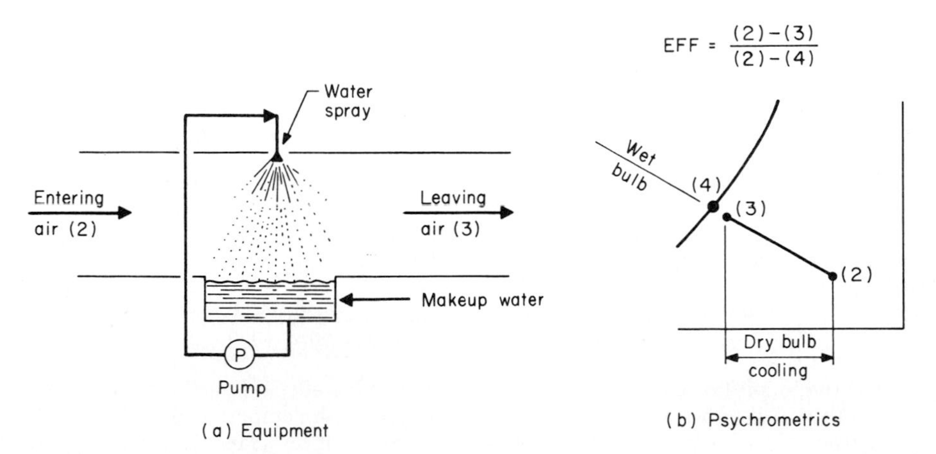

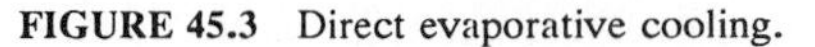

FIGURE 45.3 Direct evaporative cooling.

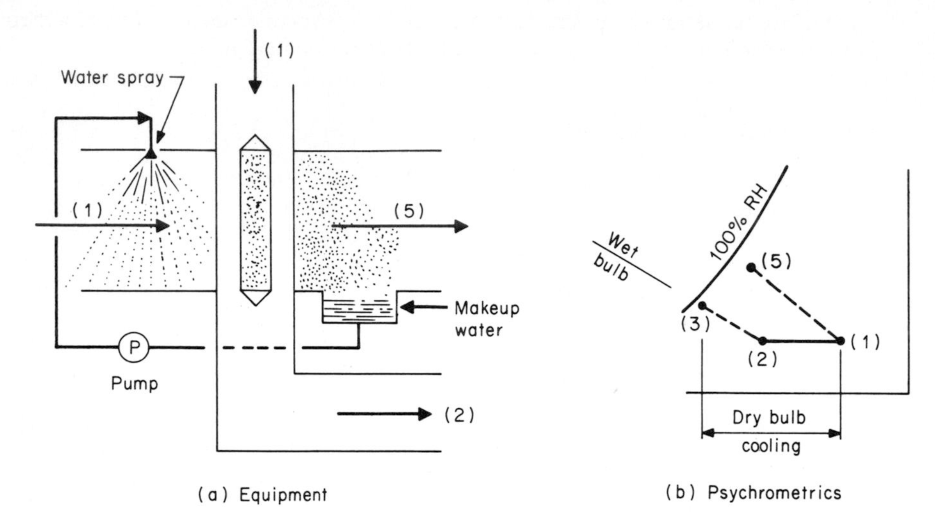

FIGURE 45.4 Indirect evaporative cooling.

Some evaporative cooling equipment operates with an atomizing water spray only, with any overspray going to the drain. Some additional air cooling is available when the water temperature is less than the air wet-bulb temperature. Evaporative cooling involves large quantities of outdoor air, and there must be provisions to exhaust the air. Evaporative cooling has also been applied to roof cooling; a roof is wetted by fine sprays, and the water evaporation causes cooler temperatures at the roof's upper and lower surfaces. The water supply for all applications must be analyzed for suitability and, as needed, treated to control scale, algae, bacteria, etc.

45.1.7 Fire and Smoke Control Dampers

Wherever practical and/or necessary, building walls and floors are made of fire-resistant material to hinder the spread of fire. Frequently, HVAC ducts must penetrate walls and floors. In order to restore the fire resistance of a penetrated wall, fire dampers or equal protection must be provided whenever a fire-resistance-rated wall, floor, or ceiling is penetrated by ducts or grilles. Fire dampers are approved devices (approved by administrators of the building code, fire marshall and/or insurance underwriter) that automatically close in the presence of higher-than-normal temperatures to restrict the passage of air and flame. Smoke dampers are approved devices that automatically close to restrict the passage of smoke.

The following are general applications for fire or smoke dampers per the National Fire Protection Association Standard NFPA-90A, 1989 edition:

- Provide 3-h fire dampers in ducts that penetrate walls and partitions which require a 3-h or higher resistance rating, provide 1½-h dampers in ducts that penetrate those requiring a rating of 2 h or higher but less than 3 h, and provide 1½-h dampers in ducts that penetrate shaft walls requiring a rating of 1 to 2 h.

- Provide fire dampers in all nonducted air-transfer openings that penetrate partitions if they require a fire-resistance rating.
- Provide smoke dampers at air-handling equipment whose capacity exceeds 15,000 ft^3/min (7080 L/s). The dampers shall isolate the equipment (including filters) from the remainder of the system except that the smoke dampers may be omitted (subject to approval by the authority having jurisdiction) when the entire air-handling system is within the space served or when rooftop air handlers serve ducts in large open spaces directly below the air handler.

Exceptions to the above are allowed when the facility design includes an engineered smoke control system. Note that schools, hospitals, nursing homes, jails, etc., may have more stringent requirements.

Dampers that "snap" closed have often incurred sufficient vacuum on the downstream side to collapse the duct (see Ref. 1). Smoke and other control dampers that close "normally" and restrict the total air flow of a rotating fan can cause pressure (or vacuum) within the duct equal to fan shutoff pressure. A fan might require a full minute after the motor is de-energized before coasting to a safe speed (pressure). Provide adequate duct construction, relief doors, or delayed damper closure (as approved by the authority having jurisdiction).

Refer to the building codes, local fire marshall rules, insurance underwriter's rules, and NFPA-90A for criteria regarding fire and smoke dampers.

45.1.8 Outdoor Air

This is needed to make up for air removed by exhaust fans; to "pressurize" buildings so as to reduce the infiltration of unwanted hot, cold, moist, or dirty outdoor air; to dilute exhaled carbon dioxide, off-gassing of plastic materials, tobacco smoke, body odors, etc.; and to replenish oxygen.

A frequently used rule of thumb to provide building pressurization is to size the return fan's air flow for 85 percent of the supply fan's, thereby leaving 15 percent for pressurization and small toilet-exhaust makeup. This is acceptable for simple, constant-volume systems and buildings. The required outdoor air can also be established by estimating the air flow through window and door cracks, open windows and doors, curtain walls, exhaust fans, etc. Building pressurization should be less than 0.15 in water gauge (WG) (4 mm WG) on ground floors that have doors to the outside so that doors do not "hang" open from outflow of air. The building's roof and walls must be basically airtight to attain pressurization. If there are numerous cracks, poor construction joints, and other air leaks throughout the walls, it is impractical to pressurize the building—and worse, the wind will merely blow in through the leaks on one side of the building and out through the leaks on the other side. Variable-air-volume (VAV) systems require special attention regarding outdoor air because as the supply fan's air flow is reduced, the outdoor and return air entering this fan tend to reduce proportionately.

The National Fire Protection Association (NFPA) standards recommend minimum outdoor air quantities for hazardous occupancies. NFPA standards are a requirement insofar as building codes have adopted them by reference. Building codes frequently specify minimum outdoor air requirements for numerous hazardous and nonhazardous occupancies. ASHRAE Standard 62 recommends minimum quantities of outdoor air for numerous activities. In the interest of energy conservation, 5 ft^3/min (2.4 L/s) per person had been considered acceptable for sedentary nonsmoking activities, but this was later determined to be inadequate. ASHRAE 62-1989 requires at least 15 ft^3/min (7.1 L/s) per person.

45.1.9 Perimeter Heating

The heat loss through outside walls, whether solid or with windows, must be analyzed for occupant comfort. The floor temperature should be no less than 65°F (18°C), especially for sedentary activities. In order to have comfortable floor temperatures, it is important that perimeter insulation be continuous from the wall through the floor slab and continue below per Refs. 2 and 3.

Walls with less than 250 Btu/h · lin ft (240 W/lin m) loss may generally be heated by ceiling diffusers that provide air flow down the window—unless the occupants would be especially sensitive to cold, such as in hospitals, nursing homes, day-care centers, and swimming pools. Walls with 250 to 450 Btu/h · lin ft (240 to 433 W/lin m) can be heated by warm air flowing down from air slots in the ceiling; the air supply should be approximately 85 to 110°F (29 to 43°C). Walls with more than 450 Btu/h · lin ft (433 W/lin m) should be heated by underwindow air supply or radiation. See Ref. 4 for additional discussion. The radiant effect of cold surfaces may be determined from the procedures in ANSI/ASHRAE Standard 55.

Curtain-wall construction, custom-designed wall-to-roof closures, and architectural details at transitions between differing materials have, at times, been poorly constructed and sealed, with the result that cold winter air is admitted to the ceiling plenum and/or occupied spaces. Considering that the infiltration rates published by curtain-wall manufacturers are frequently exceeded because of poor construction practices, it is prudent to provide overcapacity in lieu of undercapacity in heating equipment. The design of finned radiation systems should provide for a *continuous* finned element along the wall requiring heat. Do not design short lengths of finned element connected by bare pipe all within a continuous enclosure. Cold downdrafts can occur in the area of bare pipe. Reduce the heating-water supply temperature and then the finned-element size as required to provide the needed heat output and water velocity.

The surface temperatures of glass, window frames, ceiling plenums, structural steel, vapor barriers, etc., should be analyzed for potential condensation, especially when humidifiers or wet processes are installed.

45.1.10 Process Loads

Heat release from manufacturing processes is frequently a major portion of an industrial air-conditioning load. Motors, transformers, hot tanks, ovens, etc., form the process load. If all motors, etc. in large plants are assumed to be fully loaded and to be operating continuously, then invariably the air-conditioning system will be greatly oversized. The designer and client should mutually establish diversity factors that consider actual motor loads and operating periods, large equipment with motors near the roof (here the motor heat may be directly exhausted and not affect the air-conditioned zone), amount of motor input energy carried off by coolants, etc. Diversity factors could be as much as 0.5 or even 0.3 for research and development shops containing numerous machines that are used only occasionally by the few operators assigned to the shop.

45.1.11 Room Air Motion

Ideally, occupied portions [or the lower 6 ft (2 m)] of air-conditioned spaces for sedentary activities would have 20- to 40-ft/min (0.1- to 0.2-m/s) velocity of air movement, with the air being within 2°F (1°C) of a set point. It is impractical to expect this velocity throughout an entire area at all times inasmuch as air would have to be supplied at approximately a 2-ft^3/min · ft^2 (10.2-L/s · m^2) rate or higher. This rate is easily incurred by the design load of perimeter offices, laboratories, computer rooms, etc., but would only occur in an interior office when there is considerable heat-release equipment. The supply air temperature should be selected such that, at design conditions, a flow rate of at least 0.8 ft^3/min · ft^2 (4.1 L/s · m^2), but never less than 0.5 ft^3/min · ft^2 (2.5 L/s · m^2), is provided.

People doing moderate levels of work in non-air-conditioned industrial plants might require as much as a 250-ft/min (1.3-m/s) velocity of air movement in order to be able to continue working as the air temperature approaches 90°F (32°C). This would not necessarily provide a "full comfort" condition, but it would provide acceptable relief. Loose paper, hair, and other light objects may start to be blown about at air movements of 160 ft/min (0.8 m/s); see Ref. 5. Workers influenced by high ambient temperatures and radiant heat may need as much as a 4000-ft/min (20-m/s) velocity of a 90°F (32°C) air stream to increase their convective and evaporative heat loss. These high velocities would be in the form of spot cooling or of a relief station that the worker could enter and exit at will. Air movement can only compensate for, but not stop, low levels of radiant heat. Only effective shielding will stop radiant energy. Continuous air movement of approximately 300 ft/min (1.5 m/s) and higher can be disturbing to workers.

Situations involving these higher air movements and temperatures should be analyzed by the methods in Refs. 6 to 9.

45.2 OCCUPANCIES

45.2.1 Clean Rooms

For some manufacturing facilities, an interior room that is conditioned by a unitary air conditioner with 2-in- (5-cm-) thick throwaway filters might be called a "clean room"; that is, it is "clean" relative to the atmosphere of the surrounding plant. Generally, however, clean rooms are spaces associated with the microchip, laser optics, medical, etc., industries where airborne particles as small as 0.5 micrometer (μm) and less are removed. One micrometer equals one-millionth of a meter, or 0.000039 in (0.000001 m).

Clean rooms are identified by the maximum permissible number of 0.5-μm particles per cubic foot. For example, a class 100 clean room will have no more than 100 of these particles per cubic foot, a class 10 clean room no more than 10, etc. This degree of cleanliness can be attained by passing the air through a high-efficiency particulate air (HEPA) filter installed in the plane of the clean-room ceiling, after which the air continues in a downward vertical laminar flow (VLF) to return grilles located in the floor or in the walls at the floor. Horizontal laminar flow (HLF) rooms are also built wherein the HEPA filters are in one wall and the return grilles are in the opposite wall. A disadvantage with an HLF room is that downstream activities may receive contaminants from upstream activities.

An alternative to an entire space being ultraclean is to provide ultraclean chambers within a clean room (e.g., class 100 chambers in a class 10,000 room). This is feasible when a product requires the class 100 conditions for only a few operations along the entire assembly line.

The air-conditioning system frequently includes a three-fan configuration (primary, secondary, and makeup) similar to that shown in Fig. 45.5. The primary fan maintains the high air change through the room and through the final HEPA filters. The secondary fan maintains a side-stream (to the primary circuit) air flow through chilled-water or brine cooling coils, humidifiers, and heating coils. The makeup fan injects conditioned outdoor air into the secondary circuit, thereby providing clean-room pressurization and makeup for exhaust fans. Clean-room air changes are high, such that the total room air might be replaced every 7 s, and this generally results in the fan energy being the major portion of the internal heat gain. Whenever space permits, locate filters downstream of fans so as to intercept containments from the lubrication and wear of drive belts, couplings, bearings, etc.

For additional discussions, refer to the *1982 ASHRAE Handbook, Applications*, and to the latest issue of federal Standard 209, entitled *Clean Room and Work Station Requirements, Controlled Environment*.

45.2.2 Computer Rooms

These rooms are required to house computer equipment that is sensitive to swings in temperature and humidity. Equipment of this type normally requires controlled conditions 24 hours per day, 7 days per week. Computer equipment

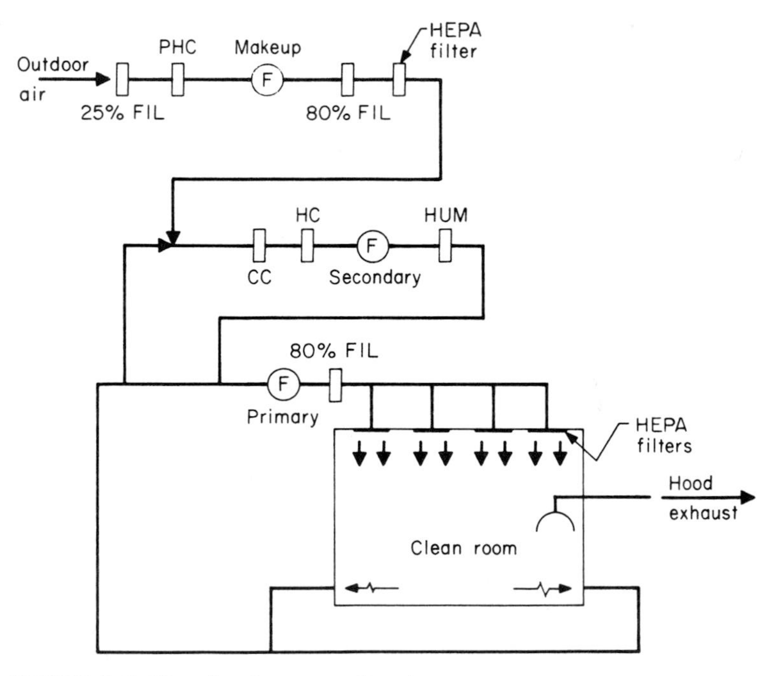

FIGURE 45.5 Three-fan clean-room air system.

can be classified as (1) data processing, (2) computer-aided design and drafting (CADD), and (3) microcomputer. Microcomputers are generally similar to standard office equipment and require no special treatment. Some CADD equipment is also microcomputer-based and falls into the same category. Data processing and larger CADD systems fall into the realm of specialized computer rooms, and these are discussed below.

Data processing and large CADD systems operate on a multiple-shift basis, requiring air-conditioning during other than normal working hours. Humidity stability is of prime importance with data processing equipment and CADD plotters. The equipment is inherently sensitive to rapid changes in moisture content and temperature.

To provide for the air-conditioning requirements of computer equipment, two components are necessary: a space to house the equipment and a system to provide cooling and humidity control. Fundamental to space construction is a high-quality vapor barrier and complete sealing of all space penetrations, such as piping, ductwork, and cables. To control moisture penetration into the space effectively, it is necessary to extend the vapor barrier up over the ceiling in the form of a plenum enclosure. Vapor-sealing the ceiling itself is not generally adequate due to the nature of its construction and to penetration from lighting and other devices.

A straightforward approach to providing conditioning to computer spaces is to use packaged, self-contained computer-room units specifically designed for the service. Controls for these units have the necessary accuracy and response to provide the required room conditions. An added advantage to packaged computer-room units is flexibility. As the needs of the computer room change and as the equipment and heat loads move around, the air-conditioning units can be relocated to suit the new configuration. The units can be purchased either with chilled-water or direct-expansion coils, as desired. Remote condensers or liquid coolers can also be provided. Large installations lend themselves quite well to heat recovery; therefore, the designer should be aware of possible potential uses for the energy.

Centrally located air-handling units external to the computer space offer benefits on large installations. More options are available with regard to introduction of ventilation air, energy recovery, and control systems. Maintenance is also more convenient where systems are centrally located. There are obvious additional benefits with noise and vibration control. Use of a centrally located system must be carefully evaluated with regard to first cost and to potential savings, as the former will carry a heavy impact.

The load in the room will be primarily sensible. This will require a fairly high air-flow rate as compared to comfort applications. High air-flow rates require a high degree of care with air distribution devices in order to avoid drafts. One way to alleviate this problem is to utilize underfloor distribution where a raised floor is provided for computer cable access. A typical computer-room arrangement is shown in Fig. 45.6. Major obstructions to air flow below the floor must be minimized so as to avoid dead spots.

In summary, important points to remember are:

1. Completely surround the room with an effective vapor barrier.
2. Provide well-sealed wall penetrations where ductwork and piping pass into computer space.
3. Provide high-quality humidity and temperature controls capable of holding close tolerances: ±1°F (0.6°C) for temperature, and ±5 percent for relative humidity.

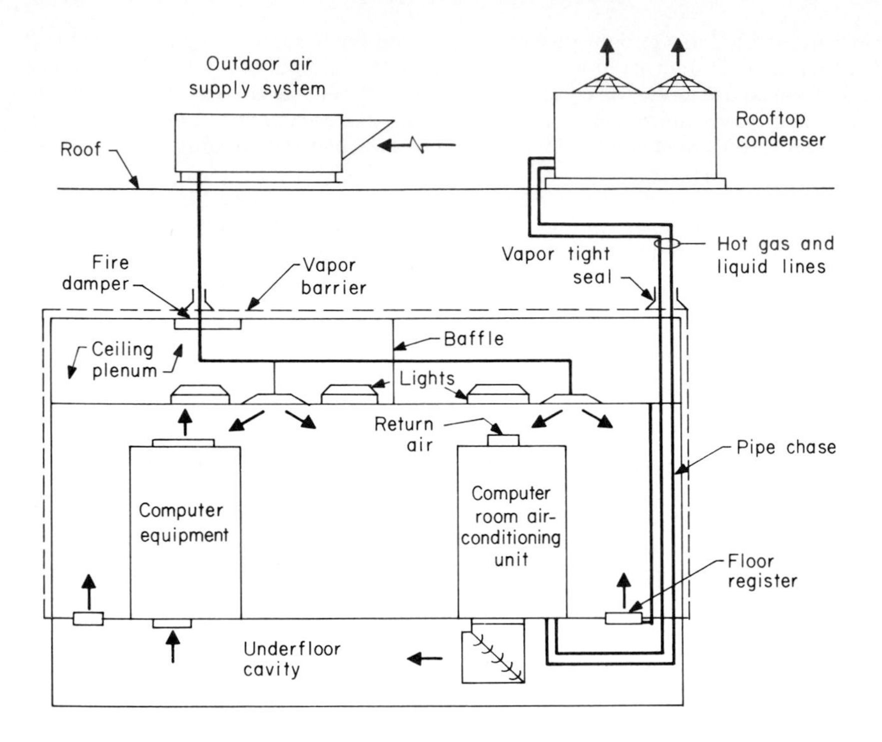

1. Locate floor registers so as to be in nontraffic areas and free from obstruction
2. Ceiling plenum baffles located where and as directed by local codes and insurance underwriters

FIGURE 45.6 Typical computer-room layout.

4. Pay close attention to air distribution, avoiding major obstructions under floors where underfloor distribution is used.
5. Be alert to opportunities for energy recovery.
6. Make sure that the chosen control parameters and design temperatures and conditions satisfy the equipment manufacturer's specifications.
7. Be attentive to operating-noise levels within the computer space.
8. If chilled water or cooling water is piped to computer-room units within the computer-room space, provide a looped- or grid-type distribution system with extra valved outlets for flexibility.

45.2.3 Offices

Cooling and heating systems for office buildings and spaces are usually designed with an emphasis on the occupants' comfort and well-being. The designer should remain aware that not only the mechanical systems but also the architectural features of the space affect the comfort of the occupants. And the designer will do well to remember that the mechanical system should in all respects be invisible to the casual observer.

The application of system design is divided into three parts: the method of energy transfer, the method of energy distribution, and the method of control. Controls are discussed in Chap. 52 and will therefore not be discussed here.

To properly apply a mechanical system to control the office environment, it is necessary to completely understand the nature of the load involved. This load will have a different character depending on the part of the office that is being served. Perimeter zones will have relatively large load swings due to solar loading and heat loss because of thermal conduction. The loading from the occupants will be relatively minor. Core zones, on the other hand, will impart more loading from building occupants and installed equipment.

For the office environment, the more common system used today is the variable-air-volume (VAV) system. This approach was originally developed as a cooling system, but with proper application of control it will serve equally well on heating. In climates where there is need for extensive heating, perimeter treatment is required to replace the skin loss of the building structure. An old but reliable method is fin-tube radiation supplied with hot water to replace the skin loss. A system that is being seen with more regularity is in the form of perimeter air supply. Care should be taken with the application of perimeter air systems to ensure that wall U-values are at least to the level of ASHRAE Standard 90. If this is not done, interior surface temperatures will be too low and the occupants in the vicinity will feel cold.

Avoid striking the surface of exterior windows with conditioned air, as this will probably cool even double-pane glass to below the dew point of the outdoor air in the summer. The result will be fogged windows and a less-than-happy client.

In the interest of economy from a final cost and operating basis, it is best to return the bulk of the air circulated to the supply fan unit. Only enough outdoor air should be made up to the building space to provide ventilation air, replace toilet exhaust, and pressurize the building. For large office systems, it is generally more practical to return spent air to the central unit or units through a ceiling plenum. If the plenum volume is excessively large, a better approach would be to duct the return air directly back to the unit. The ceiling plenum will be warmer during the cooling season when the return air is ducted, and this will require a somewhat greater room air supply because more heat will be transmitted to the room space from the ceiling rather than directly back to the coil through the return air.

Terminal devices require special attention when applied to VAV systems. At low flow rates, the diffuser will tend to dump unless care is taken in the selection to maintain adequate throw. Slot-type diffusers tend to perform well in this application, but there are other diffuser designs, such as the perforated type, that are more economical and will have adequate performance.

The air-handling, refrigeration, and heating equipment could be located either within an enclosed mechanical-equipment room or on the building roof in the form of unitary self-contained equipment. For larger systems, of 200 tons (703 kW) of refrigeration or more, the mechanical-equipment room offers distinct advantages from the standpoint of maintenance; however, the impact on building cost must be evaluated carefully. An alternate approach to the enclosed equipment room is a custom-designed factory-fabricated equipment room. These are shipped to the jobsite in preassembled, bolted-together, ready-to-run modules. For small offices and retail stores, the most appropriate approach would be roof-mounted, packaged, self-contained, unitary equipment. It will probably be found that this is the lowest in first cost, but it will not fare well in a life-cycle analysis because of increased maintenance costs after 5 to 10 years of service.

45.2.4 Test Cells

The cooling and heating of test cells poses many problems.

Within the automotive industry, test cells are used for:

- Endurance testing of transmissions and engines
- Hot and cold testing of engines
- Barometric testing and production testing

The treatment of production test cells would be very similar to the treatment of noisy areas in other parts of an industrial environment. These areas are generally a little more open in design, with localized protection to contain the scattering of loose pieces in the event of a mechanical failure of the equipment being tested. Hot and cold rooms and barometric cells are usually better left to a package purchase from a manufacturer engaged in that work as a specialty.

Endurance cells, on the other hand, are generally done as a part of the building package (Fig. 45.7). It will be found that these spaces are air-conditioned for personnel comfort during setup only. The cell would be ventilated while a test is under way. Heat gains for the nontest air-conditioned mode would be from the normal sources: ambient surroundings, lights, people, etc. Air distribution for air-conditioning would be similar to any space with a nominal loading of 200 to 400 ft^2/ton (5.3 to 10.6 m^2/kW) of refrigeration. It should be remembered, however, that sufficient outdoor air will be needed to make up for trench and floor exhaust while maintaining the cell at a negative-pressure condition relative to other areas. Consult local building codes to ensure compliance with regulations concerning exhaust requirements in areas of this nature.

During testing, as stated above, the cell would only be ventilated. Outdoor air would be provided at a rate of 100 percent in sufficient quantity to maintain reasonable conditions within the cell. Temperatures within the cell could often be in excess of 120°F (49°C) during a test. Internal-combustion engines are generally liquid-cooled, but even so, the frame losses are substantial and large amounts of outdoor air will be required in order to maintain space conditions to even these high temperature limits. In cold climates, it is necessary to temper ventilation air to something above freezing; 50°F (10°C) is usually appropriate, but each situation needs to be evaluated on its own merit. The engine losses are best obtained from the manufacturer, but in the absence of this data there is information in the *1982 ASHRAE Handbook, Applications*, that will aid in completing an adequate heat balance. The dynamometer is most often air-cooled and can be thought of as similar to an electric motor. The engine horsepower (wattage) output will be converted to electricity, which is usually fed into the building's electrical system; therefore, the dynamometer losses to the cell will be on the order of 15 to 20 percent of the engine shaft output.

The engine test cell will require a two-stage exhaust system for cooling. The first stage would be to provide low-level floor and trench exhaust to remove heavy fuel vapors and to maintain negative conditions in the cell at all times. The second stage would be interlocked with the ventilation system and would come on during testing and would exhaust at a rate about 5 to 10 percent greater than

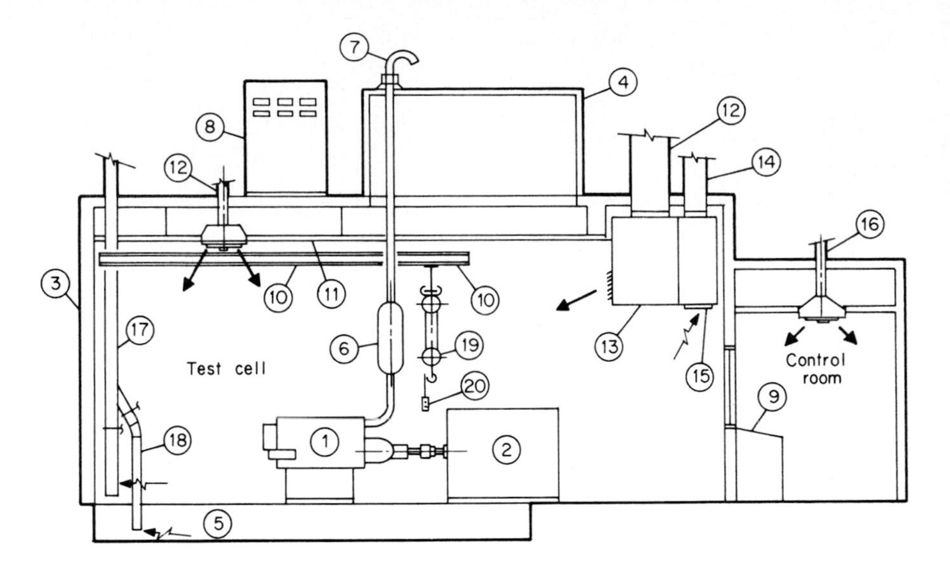

1. Engine
2. Dynamometer
3. Blast wall
4. Blast cupola
5. Fuel and service trench
6. Muffler
7. Engine exhaust
8. Dynamometer
9. Control panel
10. Crane
11. Suspended ceiling
12. Supply air (conditioned, unconditioned)
13. Supply air plenum
14. Cell exhaust
15. Exhaust plenum
16. Control room supply (conditioned)
17. Exhaust duct
18. Trench exhaust duct
19. Electric hoist
20. Hoist electric control

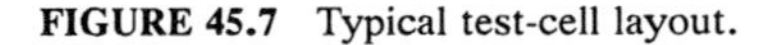

FIGURE 45.7 Typical test-cell layout.

the supply rate to maintain negative-pressure conditions. The second stage would also be activated in the event of a fuel spill to purge the cell as quickly as possible. Activation of the purge should be by automatic control in the event that excessive fumes are detected. An emergency manual override for the automatic purge should be provided. Shutdown of the purge should be manual. Consult local codes for explicit requirements.

Depending on the extent of the engine exhaust system, a helper fan may be required to preclude excessive back pressure on the engine. Where more than one cell is involved, one fan would probably serve multiple cells. Controls would need to be provided to hold the back pressure constant at the engine (Fig. 45.8).

Air-conditioning for the test cell could be via either direct-expansion or chilled-water coils. During a test, the cell conditioning would be shut down in all areas except the control room. Depending on equipment size, it usually is an advantage to have a separate system cooling the control room. One approach to heating and cooling an endurance-type test cell is shown schematically in Fig. 45.9. Local building codes and the latest volumes of NFPA should be reviewed to

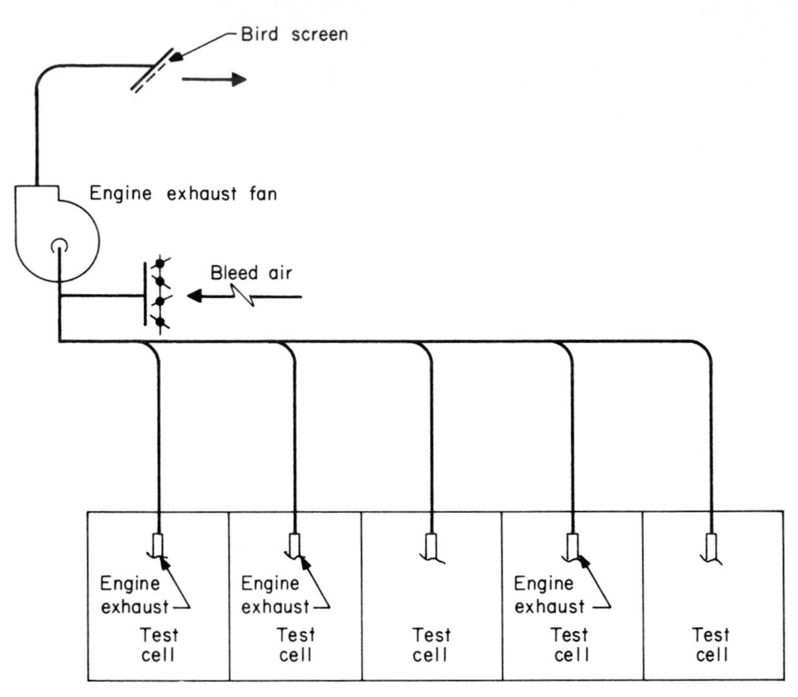

FIGURE 45.8 Engine exhaust helper fan.

ensure that local requirements are being met. Fuel vapors within the cell should be continually monitored. The cell should purge automatically in the event that dangerous concentrations are approached.

The following is suggested as the sequence of events for the control cycle of the test cell depicted in Fig. 45.9:

Setup Mode

1. AC-1 and RF-1 are running. Outdoor-air and relief-air dampers are modulated in an economizer arrangement.
2. EF-2 is controlled manually and runs at all times, maintaining negative conditions in the cell and the control room.
3. EF-1 is off and D-1 is shut.

Emergency Ventilation Mode

1. If vapors are detected, D-2 shuts and D-1 opens.
2. EF-1 starts and AC-1 changes to high-volume delivery with cooling coil shut down and outdoor-air damper open.
3. HV-1 starts and its outdoor-air damper opens.
4. System should be returned to normal manually.

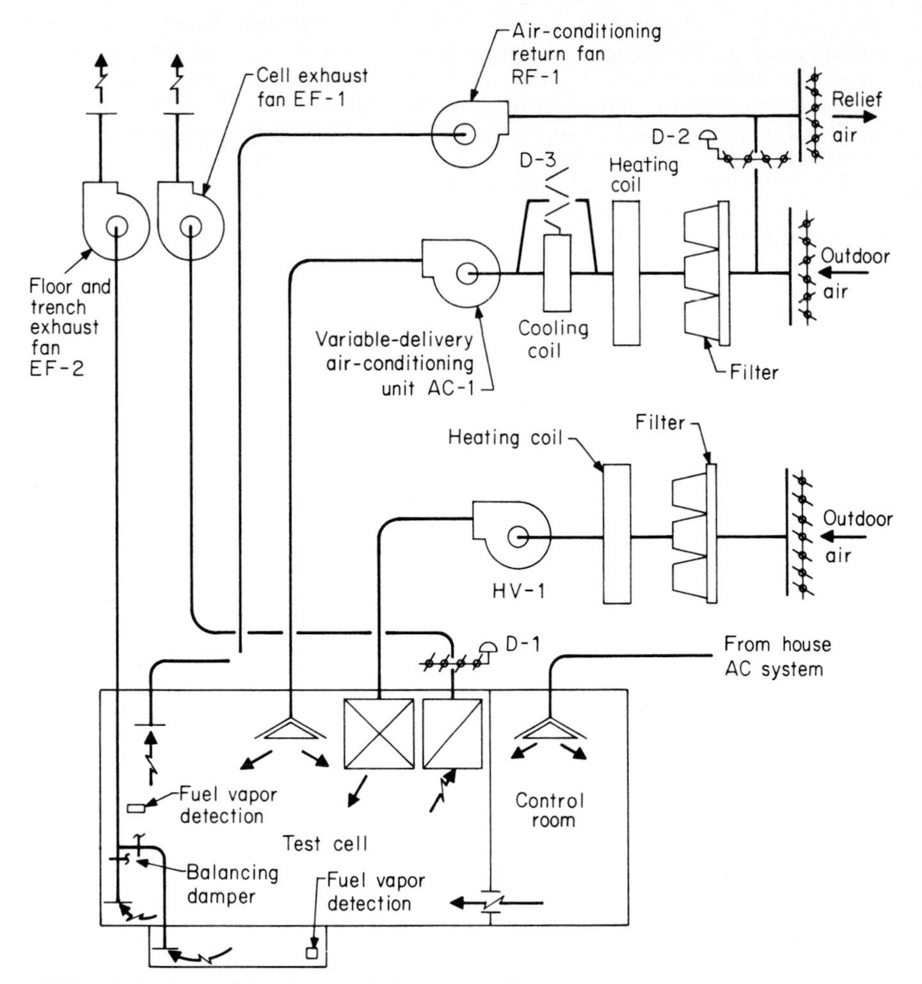

FIGURE 45.9 Test-cell heating, ventilating, and cooling.

Test Mode

1. AC-1 cooling coil shuts down.
2. AC-1 changes to high-volume delivery with outdoor-air damper fully open. D-2 closes and D-3 opens.
3. HV-1 starts and EF-1 starts.

45.3 EXHAUST SYSTEMS

One of the early considerations in the design of an exhaust (or ventilation) system should be the ultimate discharge point into the atmosphere. Most of the emissions from ventilation systems are nontoxic or inert and thus will not require a permit for installation or building operation. But should the exhaust air stream contain any of the criteria pollutants—those pollutants for which emissions and ambient concentration criteria have been established, such as CO, NO_x, SO_2, lead, particulate matter (PM), and hydrocarbons (HC)—it is likely that a permit to install the system will be required.

Once it is determined that a permit will be necessary, an emissions estimate must be made to determine estimates of both uncontrolled (before a pollution control device) and controlled emissions. The emissions estimate may be obtained from either the supplier of the equipment being contemplated for installation or from the Environmental Protection Agency (EPA) Publication AP-42, *Compilation of Air Pollutant Emission Factors*. AP-42 contains emission factors for many common industrial processes, which, when applied to process weight figures, yield emission rates in pounds (kilograms) per hour or tons per year, depending on process operating time. The permit to install an application may be obtained from the state agency responsible for enforcing the federal Clean Air Act. In most states, the Department of Environmental Protection or Department of Natural Resources will have jurisdiction. In general, the permit-to-install application requires the information and data listed in Fig. 45.10.

When designing an area or process exhaust system and a control system for the exhaust, it would be well to keep in mind that federal and local air-quality regulations may govern the type of emission control equipment installed and the maximum allowable emissions. The factors dictating what regulations apply include the type of process or equipment being exhausted, the type and quantity of emissions, the maximum emission rate, and the geographic location of the exhausted process. In order to determine what specific rules and regulations apply, the requirements of the U.S. Code of Federal Regulations Title 40 (40 CFR)

1. Applicant name and address.
2. Person to contact and telephone number.
3. Proposed facility location.
4. SIC (Standard Industrial Classification Code).
5. Amount of each air contaminant from each source in pph (pounds per hour) and tpy (tons per year) at maximum and average.
6. What federal requirement will apply to the source?
   - NESHAPS (national emission standard on hazardous air pollutants).
   - NSPS (new source performance standards).
   - PSD (prevention of significant deterioration).
   - EOP (emission offset policy).
7. Will BACT (best available control technology) be used?
8. Will the new source cause significant degradation of air quality?
9. How will the new source affect the ambient air quality standard?
10. What monitoring will be installed to monitor the process, exhaust, or control device?
11. What is the construction schedule and the estimated cost of the pollution abatement devices?

FIGURE 45.10 Commonly requested information for air-quality permit applications.

should be understood early in the project stages so that all applicable rules may be accommodated. Should the design office lack the necessary expertise in this area, a qualified consultant should be engaged. The federal government has issued a list entitled "Major Stationary Sources." The exhaust system's designer should be acquainted with this list, for it identifies the pollutant sources governed by special requirements. Several of the more common sources are listed in Fig. 45.11, and 40 CFR should be consulted for the complete listing. One of the major

1. Fossil fuel-fired generating plants greater than 250 million Btu input
2. Kraft pulp mills
3. Portland cement plants
4. Iron and steel mill plants
5. Municipal incinerators greater than 250 tons/day charging
6. Petroleum refineries
7. Fuel conversion plants
8. Chemical process plants
9. Fossil fuel boilers, or combination thereof totaling more than 250 MBtu
10. Petroleum storage and transfer units exceeding 300,000 barrel storage
11. Glass fiber processing plants

FIGURE 45.11 Major stationary sources—partial list.

sets of rules included in 40 CFR are the Prevention of Significant Deterioration (PSD) rules, which establish the extent of pollution control necessary for the major stationary sources.

If a source is determined to be "major" for any pollutant, the PSD rules may require that the installation include the best available control technology (BACT). The BACT is dependent on the energy impact, environmental impact, economic impact, and other incidental costs associated with the equipment. In addition, the following items are prerequisites to the issue of a permit for pollutants from a major source:

1. Review and compliance of control technology with the:
   a. State Air Quality Implementation Plan (SIP).
   b. New Source Performance Standards (NSPS) (see Fig. 45.12).
   c. National Emissions Standards for Hazardous Air Pollutants (NESHAPs).
   d. BACT.
2. Evidence that the source's allowable emissions will not cause or contribute to the deterioration of the National Ambient Air Quality Standard (NAAQS) or the increment over baseline, which is the amount the source is allowed to increase the background concentration of the particular pollutant.
3. The results of an approved computerized air-quality model that demonstrates the acceptability of emissions in terms of health-related criteria.
4. The monitoring of any existing NAAQS pollutant for up to 1 year or for such time as is approved.
5. Documentation of the existing (if any) source's impact and growth since August 7, 1977, in the affected area.
6. A report of the projected impact on visibility, soils, and vegetation.

1. Fossil-fuel-fired steam generators with construction commencing after 8-17-78
2. Electric utility steam generators with construction commencing after 9-18-78
3. Incinerators
4. Portland cement plants
5. Sulfuric acid plants
6. Asphalt acid plants
7. Petroleum refineries
8. Petroleum liquid storage vessels constructed after 6-1-73 and prior to 5-19-78
9. Petroleum liquid storage vessels constructed after 5-18-78
10. Sewage treatment plants
11. Phosphate fertilizer industry — wet process phosphoric acid plants
12. Steel plants — electric arc furnaces
13. Steel plants — electric arc furnaces and argon decarburization vessels constructed after 8-17-83
14. Kraft pulp mils
15. Grain elevators
16. Surface coating of metal furniture
17. Stationary gas turbines
18. Automobile and light-duty truck painting
19. Graphic arts industry — rotogravure printing
20. Pressure-sensitive tape and label surface coating operations
21. Industrial surface coating: large appliance
22. Asphalt processing and asphalt roofing manufacture
23. Bulk gasoline terminals
24. Petroleum dry cleaners

FIGURE 45.12 New Source Performance Standards—partial list.

7. A report of the projected impact on residential, industrial, commercial, and other growth associated with the area.
8. Promulgation of the proposed major source to allow for public comment. Normally, the agency processing the permit application will provide for public notice.

One of the first steps regarding potential pollutant sources is to determine the applicable regulations. For this, an emissions estimate must be made, and the "in-attainment" or "non-attainment" classification of the area in which the source is to be located must be determined. The EPA has classified all areas throughout the United States, including all U.S. possessions and territories. The area is classified as either "in-attainment" (air quality is better than federal standards) or "non-attainment" (air quality is worse than federal standards).

If the source is to be located in a non-attainment area, the PSD rules and regulations do not apply, but all sources that contribute to the violation of the NAAQS are subject to the Emissions Offset Policy (EOP). The following items must be considered when reviewing a source that is to be located in a non-attainment area:

1. The lowest achievable emission rate (LAER), which is defined as the most stringent emission limit that can be achieved in practice
2. The emission limitation compliance with the SIP, NSPS, and NESHAPS
3. The contribution of the source to the violation of the NAAQS
4. The impact on the non-attainment area of the fugitive dust sources accompanying the major source

In general, the EOP requires that for a source locating in a non-attainment area, more than equivalent offsetting emission reductions must be obtained from existing emissions prior to approval of the new major source or major modification. The "bubble" concept, wherein the total emissions from the entire facility with the new source does not exceed the emissions prior to addition of the new source, may be used to determine the emission rate. If there were emission reductions at "existing" sources, they would offset the contributions from the new source, or "offset" the new emissions. This same bubble concept may be used for sources that qualify for in-attainment or PSD review.

In the design of a polluting or pollution control facility, stack design should be considered. A stackhead rain-protection device (Figs. 45.13 and 45.17) should be

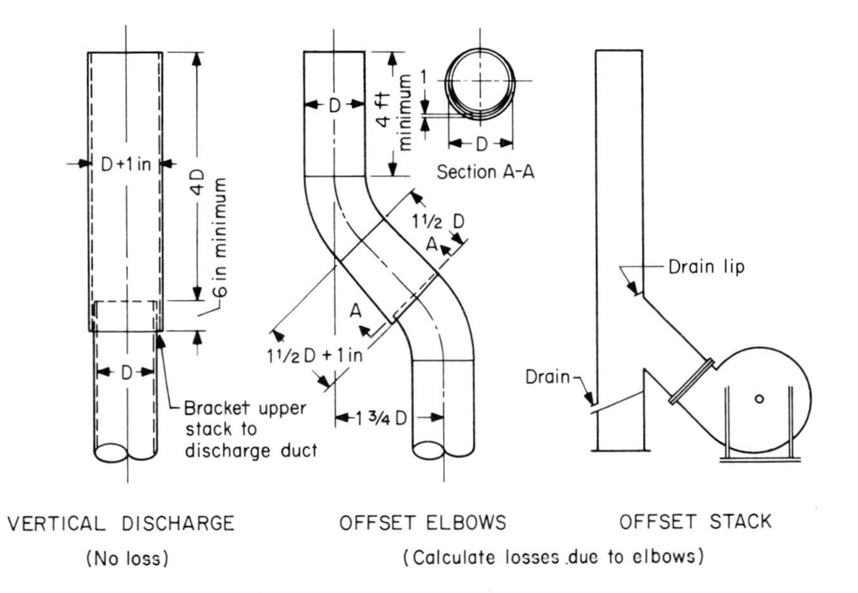

1. Rain protection characteristics of these caps are superior to a deflecting cap located 0.75D from top of stack.
2. The length of upper stack is related to rain protection. Excessive additional distance may cause "blowout" of effluent at the top gap between upper and lower sections.

FIGURE 45.13 Typical rain-protection devices. *(From Industrial Ventilation—A Manual of Recommended Practice, 18th ed., Committee on Industrial Ventilation, American Conference of Governmental Industrial Hygienists, copyright 1984, p. 6–41.)*

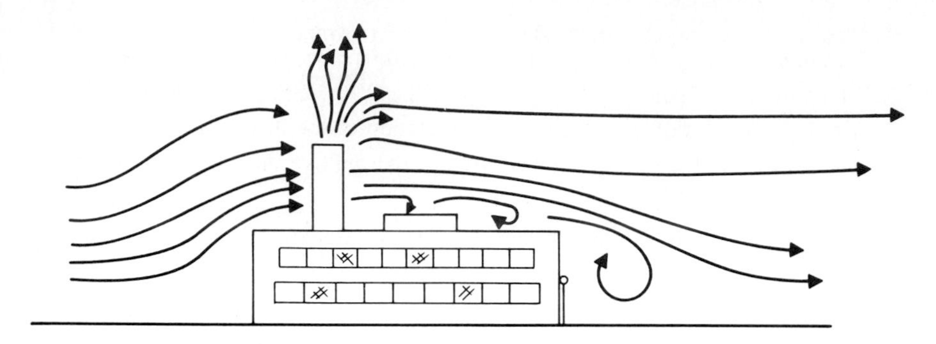

GEP stack height minimizes re-entrainment of exhaust gasses into air which might enter building ventilation system.

FIGURE 45.14 GEP stack.

used in lieu of the weather cap found on many older installations, since this cap does not allow for adequate dispersion of the exhaust gas. When specifying or designing stack heights, it should be noted that the EPA has promulgated rules governing the minimum stack height; these rules are known as "good engineering practice" (GEP). A GEP stack has sufficient height to ensure that emissions from the stack do not result in excessive concentrations of any air pollutant in the vicinity of the source as a result of atmospheric downwash, eddy currents, or wakes caused by the building itself or by nearby structures (Figs. 45.14 and 45.15). For uninfluenced stacks, the GEP height is 98 ft (30 m). For stacks on or near structures, the GEP height is (1) 1.5 times the lesser of the height or width of the structure, plus the height of the structure, or (2) such height that the owner of the building can show is necessary for proper dispersion. In addition to GEP stack height, stack exit velocity must be maintained for proper dispersion characteristics.

Figures 45.16 and 45.17 illustrate the relationship between velocity at discharge and the velocity at various distances for the weather-cap- and stackhead-type rain hoods, respectively. Maintaining an adequate exit velocity ensures that the exhaust gases will not reenter the building through open windows, doors, or mechanical ventilation equipment. Depending on normal ambient atmospheric conditions, the exit velocities may range from 2700 to 5400 ft/min (14 to 28 m/s).

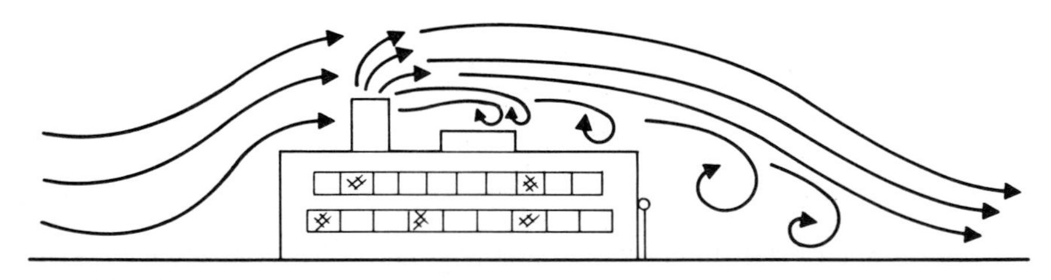

Non-GEP stack allows exhaust gasses to be entrained in building wakes and eddy currents.

FIGURE 45.15 Non-GEP stack.

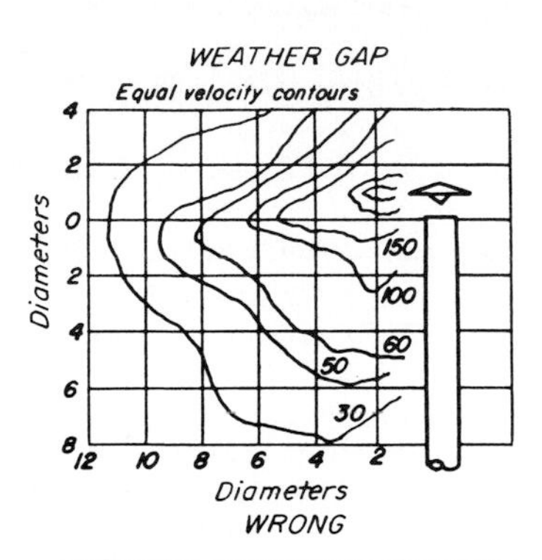

FIGURE 45.16 Weather-cap dispersion characteristics. *(From Industrial Ventilation—A Manual of Recommended Practice, 18th ed., Committee on Industrial Ventilation, American Conference of Governmental Industrial Hygienists, copyright 1984, p. 6–39.)*

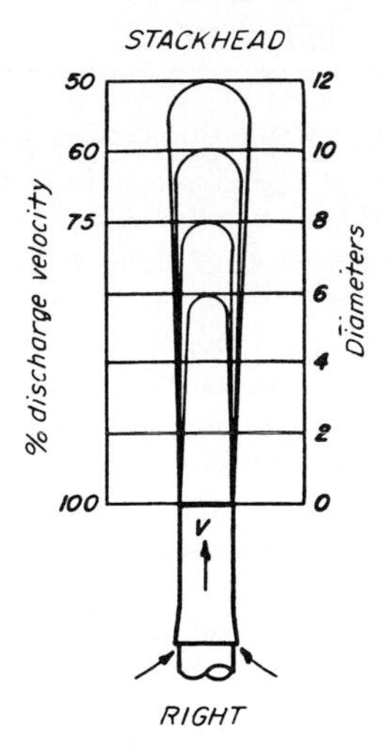

FIGURE 45.17 Stackhead dispersion characteristics. *(From Industrial Ventilation—A Manual of Recommended Practice, 18th ed., Committee on Industrial Ventilation, American Conference of Governmental Industrial Hygienists, copyright 1984, p. 6–39.)*

In practice, it has been found that 3500 ft/min (18 m/s) is a good average figure for stack exit velocity, giving adequate plume rise yet maintaining an acceptable noise level within the vicinity of the stack.

Care must be taken when designing exhaust systems handling pollutants for which no specific federal emission limit exists (noncriteria pollutants). All pollutants not included in the criteria pollutant category or the NESHAPS category are considered noncriteria pollutants. When establishing or attempting to determine acceptable concentration levels for noncriteria pollutants, the local authority responsible for regulating air pollution should be consulted since policy varies from district to district. In general, however, noncriteria pollutants' allowable emission rates are based on the American Conference of Governmental Industrial Hygienists (ACGIH) time-weighted average acceptable exposure levels.

A hazardous air pollutant is one for which no ambient air-quality standard is applicable, but which may cause or contribute to increased mortality or illness in the general population. Emission standards for such pollutants are required to be set at levels that protect the public health. These allowable pollutants' emission levels are known as NESHAPS and include levels for radon-222, beryllium, mercury, vinyl chloride, radionuclides, benzene, asbestos, arsenic, and fugitive organic leaks from equipment.

An exhaust stream that includes numerous pollutants, with some being noncriteria pollutants, can be quickly reviewed by assuming that all the exhaust consists of the most toxic pollutant compound. If the emission levels are acceptable for that review, they will be acceptable for all other compounds.

45.4 REFERENCES

1. United McGill Corporation, *Engineering Bulletin*, vol. 2, no. 9, July 1986.
2. *Energy Conservation in New Building Design*, ASHRAE Standard 90A-1980, ASHRAE, Atlanta, GA, p. 18, para. 4.4.2.4.
3. *1989 ASHRAE Handbook, Fundamentals*, ASHRAE, Atlanta, GA, 1989, p. 25–8, fig. 6.
4. Tom Zych, "Overhead Heating of Perimeter Zones in VAV Systems," *Contracting Business*, August 1985, pp. 75–78.
5. *Thermal Environmental Conditions for Human Occupancy*, ANSI/ASHRAE Standard 55-1981, ASHRAE, Atlanta, GA, p. 4, para. 5.1.3.
6. Knowlton J. Caplan, "Heat Stress Measurements," *Heating/Piping/Air Conditioning*, February 1980, pp. 55–62.
7. *Industrial Ventilation—A Manual of Recommended Practice*, 18th ed., Committee on Industrial Ventilation, American Conference of Governmental Industrial Hygienists, Lansing, MI, 1984, sec. 3, pp. 3-1 through 3-12.
8. *1987 ASHRAE Handbook, HVAC Systems and Applications*, ASHRAE, Atlanta, GA, chap. 41, pp. 41.1–41.8.
9. W. C. L. Hemeon, *Plant and Process Ventilation*, 2d ed., Industrial Press, New York, 1963, chap. 13, pp. 325–334.

CHAPTER 46
HVAC APPLICATIONS FOR COGENERATING SYSTEMS

Alan J. Smith
Senior Project Engineer, Brown & Root, Inc., Houston, Texas

46.1 INTRODUCTION

A cogeneration facility consists of equipment that uses energy to produce both electric energy and forms of useful thermal energy (such as heat or steam) for industrial, commercial, heating, or cooling purposes. Cogeneration facilities are designed as either topping-cycle or bottoming-cycle facilities. Topping-cycle facilities first transform fuel into useful electric power output; the reject heat from power production is then used to provide useful thermal energy. In contrast, bottoming-cycle facilities first apply input energy to a useful thermal process, and the reject heat emerging from the process is then used for power production. Either of these cycles can apply thermal energy to meet process heating, ventilating, and air-conditioning (HVAC) or comfort requirements for steam or for hot or chilled water.

This chapter describes the various methods of applying thermal energy from a cogeneration system to HVAC systems.

46.2 HVAC APPLICATIONS FOR THERMAL ENERGY

Feasible methods for applying thermal energy to meet process HVAC or comfort (hereafter referred to as "utility") requirements are:

- Steam generation and mechanical-drive and/or absorption chillers (Fig. 46.1)
- Hot-water generation and absorption chillers (Fig. 46.2)
- Exhaust-gas-driven chiller-heater units (Fig. 46.3)

46.2.1 Steam and Hot-Water Generation

Typical Facilities. Existing facility equipment can simplify the installation of a cogeneration system. Steam produced from cogenerated thermal energy will be

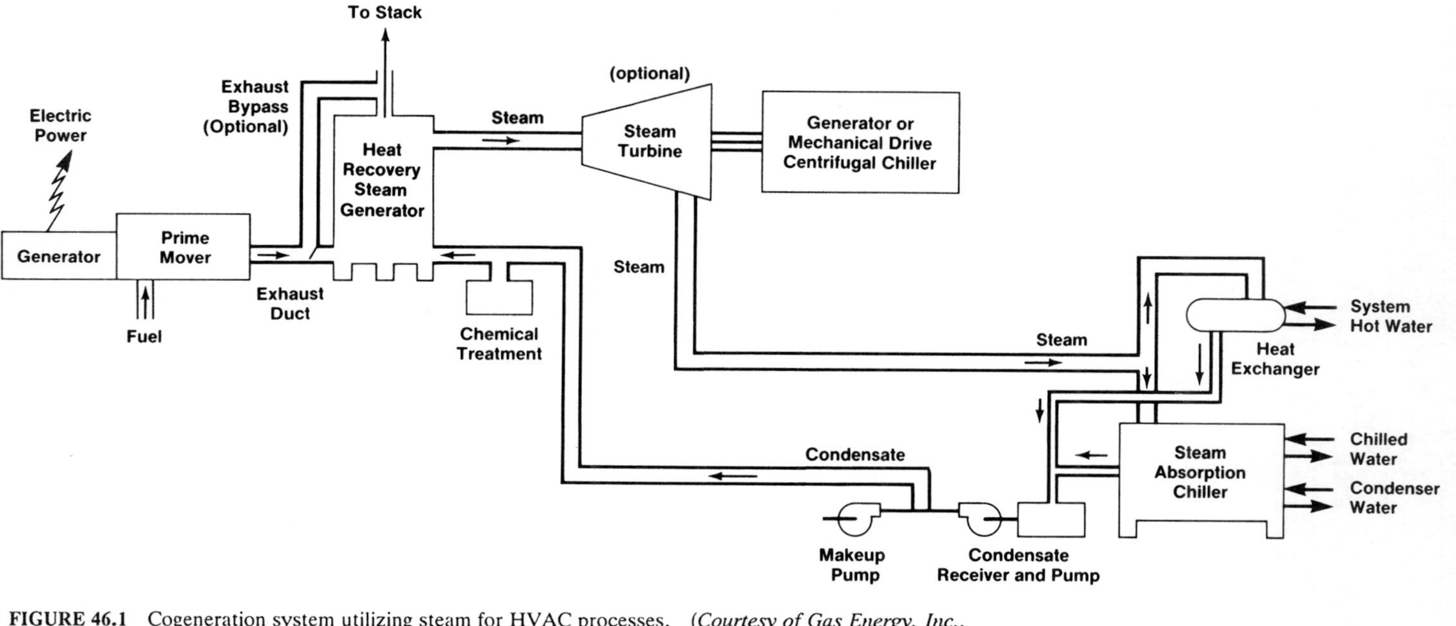

FIGURE 46.1 Cogeneration system utilizing steam for HVAC processes. (*Courtesy of Gas Energy, Inc., a subsidiary of Brooklyn Union Gas.*)

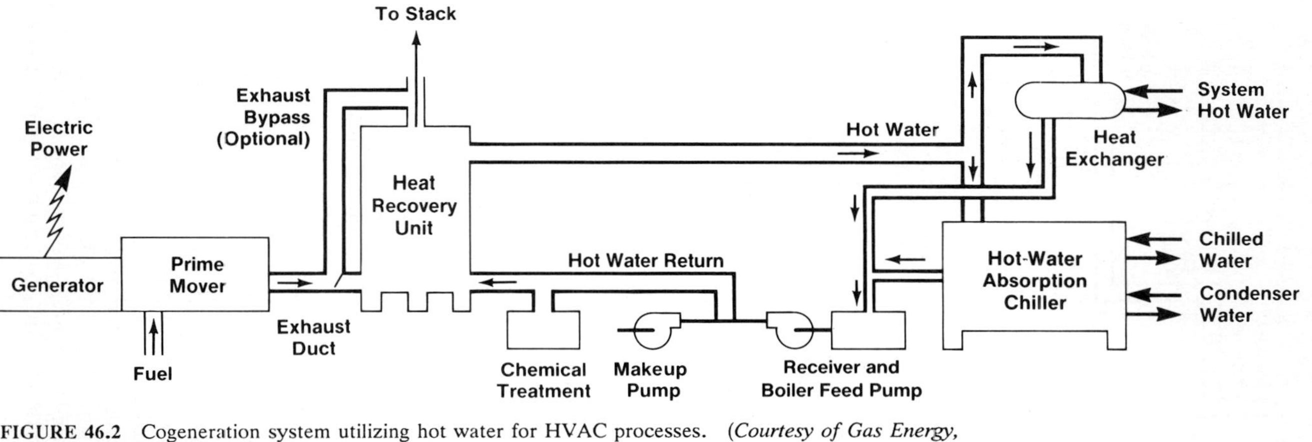

FIGURE 46.2 Cogeneration system utilizing hot water for HVAC processes. (*Courtesy of Gas Energy, Inc., a subsidiary of Brooklyn Union Gas.*)

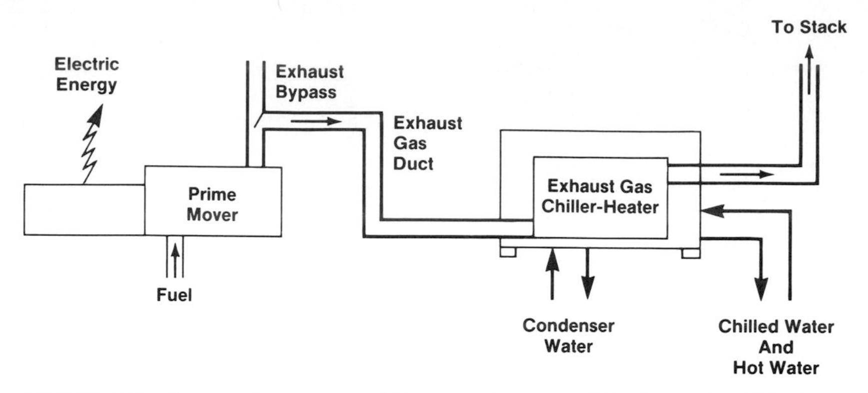

FIGURE 46.3 Cogeneration system utilizing an exhaust-gas chiller-heater for HVAC processes. (*Courtesy of Gas Energy, Inc., a subsidiary of Brooklyn Union Gas.*)

cost-effective in cases where it supplements a facility's existing steam requirements. A hot-water system rather than a steam system should be considered for facilities with requirements for hot water. The steam or hot-water generation system design should include a standby energy source to ensure that utility requirements are met if the cogeneration system operates at a reduced electrical generation level or suffers an unplanned outage. For maintenance purposes, the exhaust system design for the heat-recovery steam generator (HRSG) unit should include guillotine seal plates and a seal air blower to isolate the HRSG.

The HRSG has a minimum recommended exhaust temperature of 250°F (121°C). Maintaining this temperature protects the HRSG from water-vapor condensation and acid formation that occurs when the exhaust temperature drops below the dew point. Although an HRSG can be designed for lower exhaust temperatures, the corrosion-resistant design has minimum economic feasibility.

The HRSG normally imposes a back pressure of 6 to 8 in water gauge (inWG) (1493 to 1990 Pa) on the prime mover exhaust. This back pressure results in a horsepower penalty of approximately 17.5 hp/inWG (1.89 W/Pa) for a combustion gas turbine rated at 4900 BHP (3653.93 kW). Within the limitations specified by internal-combustion-engine vendors, exhaust-gas back pressure does not appreciably reduce the mechanical power output of the engine.

Steam or Hot-Water Generation Control. In facilities that can use a limited amount of thermal energy from a cogeneration system, the steam or hot-water production rate can be controlled by regulating the throttle of the prime mover or by bypassing the exhaust-gas heat around the heat-recovery unit and sending it up the stack. Excess steam or hot water can also be diverted to the condensers.

Using intermediate steam turbines, combined-cycle cogeneration systems (Fig. 46.1) use thermal energy not required for utility service to generate additional electricity.[1] One patented cycle varies its steam production rate by reinjecting high-pressure steam into the power turbine section of a combustion gas turbine. This procedure reduces the amount of steam that must be used by a facility and increases the electric power output of the unit.

[1]Combined-cycle systems simultaneously produce power using a fossil-fueled prime mover and a steam turbine generator unit.

Steam or Hot-Water Absorption Chiller Units. Factors that affect the selection of an absorption chiller in a cogeneration system are:

- Available steam or hot-water pressure and temperature
- Steam consumption rate
- Physical size
- Machine performance under partial-load conditions

Steam and hot-water requirements for typical units are summarized in Table 46.1. The steam consumption rate of the two-stage machine is approximately 40 percent less than that of the single-stage machines. Condenser water requirements are also reduced more than 20 percent compared to the requirements of similar amounts of single-stage absorption chillers.

Two-stage absorption machines are designed with absorbent streams using parallel or series flow (Fig. 46.4). The configuration of the parallel-flow machine results in reduced height in all machine sizes and reduced width in larger machine sizes. Either type of machine can be installed assembled in capacities up to 750 tons (2635.7 kW). Above 750 tons (2635.7 kW), the series-flow machine must be partially assembled at the installation site, while the parallel-flow machine can be transported and installed as a single unit.

The steam utilization characteristics of absorption chillers affect their sizing in cogeneration systems. The single-stage absorption machine's electricity and steam consumption rate per ton (kW) of chilled-water production decreases with reduced load down to approximately 30 percent of design capacity. At this point, consumption rises unless other cycle enhancement is added. Steam consumption curves decrease slightly at reduced-load conditions for series-flow two-stage machines. Two-stage machines using parallel flow maintain flat steam consumption curves over the entire load range.

Noncondensing Steam Turbines. Noncondensing steam turbines driving mechanical chillers can be used in series with conventional absorption chillers by matching steam flow rates and exhaust pressure from the steam turbine (Fig. 46.1). The traditional distribution of chiller capacity is one-third of the tonnage for the mechanical-drive chiller and two-thirds of the tonnage for the absorption chiller.

Typical steam inlet pressure for noncondensing steam turbines is at least 400 lb/in^2 (2757.9 × 10^3 Pa), with exhaust steam pressure approximately 150 lb/in^2 (1034.2 × 10^3 Pa). Figure 46.5 illustrates the range of inlet steam pressures and flows commonly used with noncondensing steam turbines. The typical steam consumption rate for steam turbines which power mechanical-drive centrifugal chillers is between the consumption rates of the single- and the two-stage absorption machines. Noncondensing steam turbines enhance the energy efficiency of a cogeneration cycle, because exhaust steam can be used for other heating or absorbing processes. For example, the exhaust steam can be used for a steam absorption chiller rather than being exhausted to the facility's condenser (Fig. 46.1).

46.2.2 Exhaust-Gas-Driven Chiller-Heater Units

A modification of the two-stage parallel-flow absorption chiller permits driving the chiller with high-temperature exhaust gas from a combustion gas turbine or an internal-combustion engine (Fig. 46.3). Moreover, the chillers can be purchased

TABLE 46.1 Thermal Energy Requirements for Absorption Chillers

Absorption machine type	Steam supply conditions		Hot-water supply conditions		Nominal steam consumption rate	
	psig	kPa	°F	°C	lbm/(h · ton)	kg/(s · W)
Single-stage, small	—	—	160	71.1	—	—
Single-stage	8–15	55.16–103.42	270	132.2	17.5–20	8.3–9.29
Two-stage	43–150	296.7–1034.21	300–400	148.9–204.4	9.9–12	4.6–5.5

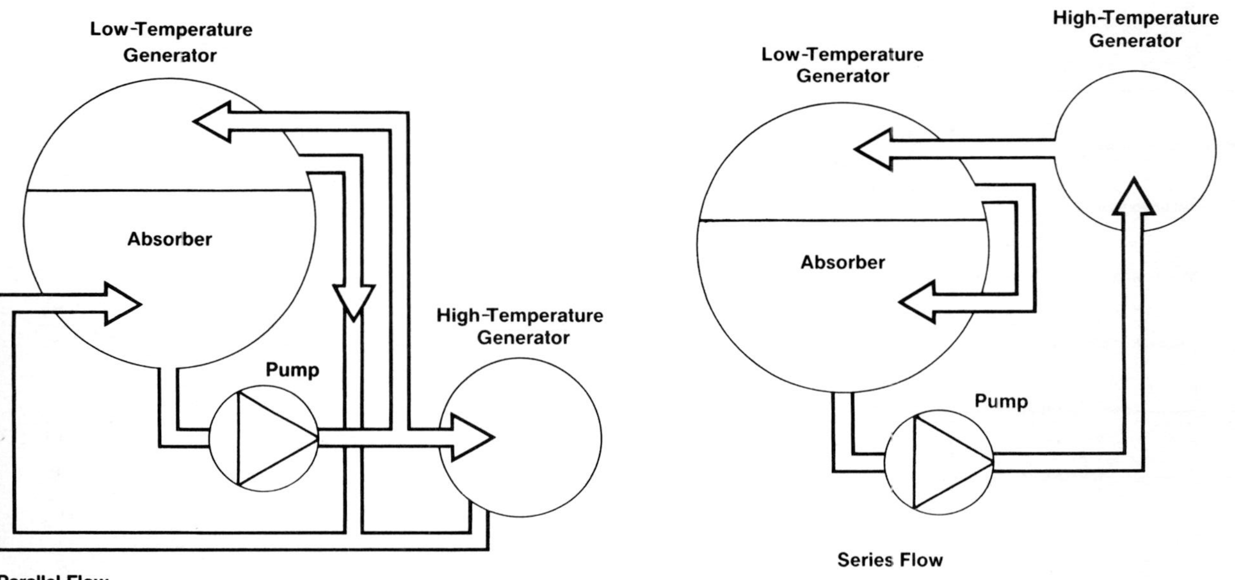

FIGURE 46.4 Comparison of parallel- and series-flow absorption chillers. (*From "Direct Fired Absorption Chiller-Heater," Hitachi Review, vol. 30, no. 1, 1981, Fig. 5. Used courtesy of Gas Energy, Inc., a subsidiary of Brooklyn Union Gas.*)

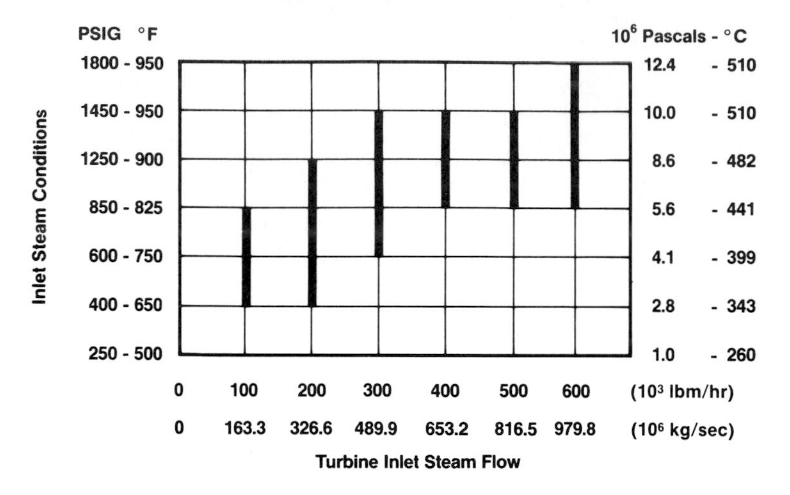

FIGURE 46.5 Range of initial steam conditions normally selected for industrial steam turbines. (*From J. M. Kovacik, "Industrial Plant Objectives and Cogeneration System Development," General Electric Corporation.*)

with an additional second-stage heat exchanger that converts the thermal energy contained in an internal-combustion engine's jacket cooling water into additional chilled-water capacity.

Exhaust-gas chillers simultaneously produce chilled water and hot water. The units can be equipped with supplemental firing (90 percent efficiency) to add energy to the exhaust gas as well as maintain utility service if the prime mover fails.

The use of exhaust-gas chillers eliminates the need for a steam or hot-water generation system and its associated condensate feedwater system. The layout space and maintenance requirements are substantially reduced, compared to the conventional steam or hot-water systems.

The exhaust-gas system design between the prime mover and the exhaust-gas chiller should include an effective bypass damper or guillotine seal plates combined with a seal air blower. The guillotine seal plates isolate the chiller from prime mover exhaust gas during chiller maintenance operations. If a bypass damper is used without seal plates, the user should verify that the damper has performed successfully in similar service. The position of the bypass damper should also be indicated directly, to aid operation by confirming the exhaust-gas flow path.

Typical heat-recovery parameters for the exhaust-gas chillers are summarized in Table 46.2. The thermal energy used by the exhaust-gas chiller and its resulting cooling capacity are then:

$$Q = MC_p(T_1 - T_2) \tag{46.1}$$

$$Q_{hx} = M_j C_{pj}(T_1 - T_2) \tag{46.2}$$

where Q = heat removed from exhaust gas
Q_{hx} = heat removed from jacket water[2]
M = exhaust-gas flow rate

[2]Jacket-water heat recovery is associated with internal-combustion engines.

TABLE 46.2 Operating Parameters for Exhaust-Gas Chillers

Parameter	Exhaust gas	Jacket water
Coefficient of performance	1.14	0.6–0.7
Interconnection efficiency	0.95	0.95
Minimum temperature	550°F (287.8°C)	180°F (82.2°C)
Stack temperature	375°F (max.) (190°C)	. . .
Jacket-water temperature difference	. . .	10–20°F (5.6–11.1°C)

Source: Courtesy of Gas Energy, Inc., a subsidiary of Brooklyn Union Gas.

M_j = jacket-water mass flow rate[2]
C_p = exhaust-gas specific heat
C_{pj} = jacket-water specific heat[2]
T_1 = entering temperature
T_2 = exiting temperature

$$\text{Cooling capacity} = \binom{\text{exhaust-gas}}{\text{cooling capacity}} + \binom{\text{jacket-water}}{\text{cooling capacity}} = (1.14 \times 0.95 \times Q) + (0.6 \times 0.95 \times Q_{hx}) \quad (46.3)$$

46.3 OPERATIONAL CRITERIA

Electricity demand and process energy demand (chilled water, hot water, and steam) establish sizing and operating criteria for a cogeneration system. These data must be examined over specific periods of time (seasonally, weekly, daily, and even hourly in some cases) to establish a specific cyclic pattern for the energy.

The specific components and sources of the demand must be known. Careful consideration should be given to the decrease in a facility's electricity requirements if electric-driven centrifugal chillers are to be replaced by steam absorption units as part of the cogeneration system.

Typical operational criteria that could result from process data are:

- The facility will be able to efficiently use thermal energy produced by the prime movers.
- The cogeneration facility will supply the base electric load.
- The cogeneration facility will engage in interchange sales.

The decision to engage in interchange sales of electricity to the interconnected utility should be studied. The capital cost associated with compliance with utility interconnection standards may exceed the revenue obtained from selling a small amount of power to the interconnecting utility.

Typical ranges for the electric power generation capacity of industrial, institutional, residential, and commercial cogeneration systems are summarized in Table 46.3. Industrial and institutional facilities can achieve significant economic benefit from cogeneration systems due to their balanced requirements for electric and thermal energy.

TABLE 46.3 Typical Cogeneration System Electric Power Generation Capacities

Application	Electrical output, kW
One- and two-family homes	5–15
Multifamily dwellings	20–5000
Office buildings	200–10,000
Local shopping centers	100–250
Distribution centers	250–2500
Regional shopping centers	5000–15,000
Industrial institutional facilities	Site-dependent

Source: Richard Stone, "Stand Alone Cogeneration By Large Building Complexes," *Energy Economics, Policy and Management* (Fairmont Press, Atlanta), vol. 62, Summer 1982.

46.4 FUEL

The selection of a cogeneration system's fuel supply and an assessment of the system's economic viability are affected by fuel supply reliability and by projections of future fuel prices. Fuel choice also affects the heat-recovery equipment design downstream of the prime mover. The HVAC unit or the HRSG heat-transfer surface design must be compatible with constituents contained in the prime mover exhaust gases.

46.4.1 Fuel Supply Reliability

Factors useful in assessing fuel supply reliability include:

- Assurances from the supplier that fuel supplies are adequate
- Identification of alternative fuel sources, including provisions to use them in the system design (No. 2 fuel oil or natural gas)
- Identification of alternative means of providing utility services (a standby electric-motor-driven chiller or steam generation from another source)

46.4.2 Fuel Price Forecasts

The economic benefit of a cogeneration system may be determined through cost comparison of the total cost associated with a cogeneration system and the cost of providing similar services using electricity purchased from the existing utility. Electricity cost projections are required in order to make this comparison. Rate structure information required for this project can be obtained from both the electric service contract between the facility and the utility and form 10K that the utility files with the Securities and Exchange Commission.

Form 10K can supply data useful in establishing a demand component and a fuel component in the rate structure, such as:

- Present and future fuel mixture
- Historical fuel cost

- Projected capital requirements

Industry trade groups and governmental organizations are also valuable sources for obtaining fossil-fuel cost, availability, and demand data. Publications prepared by the U.S. Department of Energy provide sample methodology for making these projections.

Additionally, federal regulations regarding fuel pricing can materially affect the fuel selection process. For example, the natural-gas pricing structure has changed as a result of the 1981 Federal Energy Regulatory Commission (FERC) Order 319, which authorized transportation services for up to 5 years of natural gas purchased from sources other than pipeline companies. Using this program, "high-priority users"—schools and hospitals—have achieved energy cost savings ranging from 20 to 45 percent, depending on wellhead prices and transportation costs.

46.4.3 Heat-Recovery Equipment

Fuels having large amounts of particulates or corrosive substances may require special features, such as a washing system. This would ensure proper heat transfer across surfaces inside the recovery equipment.

46.5 PRIME MOVERS

Combustion gas turbines and internal-combustion engines are the prime movers used in topping cycles. Typical thermal energy temperatures are summarized in Table 46.4.

46.5.1 Combustion Gas Turbine Generators

Combustion gas turbine generator (CGTG) units exhibit the following characteristics in a cogeneration system:

- High temperature of exhaust gas
- High quantity of exhaust gas

With thermal energy recovery, the overall cycle energy efficiency of a CGTG unit exceeds 60 percent. Typical types of heat-recovery equipment used in CGTG cogeneration systems are:

TABLE 46.4 Typical Waste-Heat Temperatures

	Gas turbine		Internal-combustion engine	
Thermal energy source	°F	°C	°F	°C
Exhaust gas	900–1000	482–537.8	1000–1200	537.8–648.9
Lube oil	165 (max.)	73.8	160–200	71.1–93.3
Jacket water	. . .	. . .	180–250	82.2–121.1

- Heat-recovery steam generator or hot-water heater
- Exhaust-gas chillers

Combustion turbines typically generate up to 10 lb/h (16,330 kg/s) of 15- to 150-psig (103.42- to 1034.2-kPa) steam per horsepower (0.7457 kW) of output. Because of the volume of excess air contained in the CGTG exhaust, it is possible to supplement the heat contained in the turbine exhaust to gain additional steam-generating capacity or cooling capacity by burning additional fuel. This supplemental firing gas has an efficiency of 90 percent.

Heat Balance. Mechanical energy makes up approximately 30 percent of a CGTG unit's heat balance under full-load conditions. Exhaust gas contains essentially the remainder of the energy, with small portions allocated to lube oil and radiation. This exhaust-gas thermal energy can be directly applied to driving an HRSG or an exhaust-gas chiller-heater.

The lube oil temperature is low and the quantity of heat is small, and thus in most cases it is not economical to recover heat from this source.

Load Control. Single- and two-shaft combustion turbines are available. The two-shaft units are designed with separate shafts for the compressor section and the power turbine section. Separate shafts permit the rotating speed of the compressor section to be controlled by the requirements of the power turbine rather than by the rotating speed of the generator.

Partial-load operating efficiencies between the single- and two-shaft types of combustion gas turbines are illustrated in Fig. 46.6. The two-shaft units are able to maintain higher exhaust temperatures, and therefore greater operating efficiency, under partial-load conditions. The two-shaft units, however, will have higher heat rates at full-load conditions. If partial-load operation of a combustion turbine is required because of cogeneration system operating criteria, consideration should be given to a two-shaft combustion turbine.

46.5.2 Internal-Combustion Engines

Internal-combustion engines exhibit the following characteristics in cogeneration systems:

- High mechanical efficiency
- More efficient operation at partial loads (Fig. 46.7)
- High-temperature exhaust gases
- Readily available maintenance services

Heat-recovery equipment used in cogeneration systems using internal-combustion engines includes:

- Water tube boilers with steam separators
- Coil-type hot-water heaters
- Steam separators for use with high-temperature cooling of engine jackets
- Exhaust-gas-driven chillers

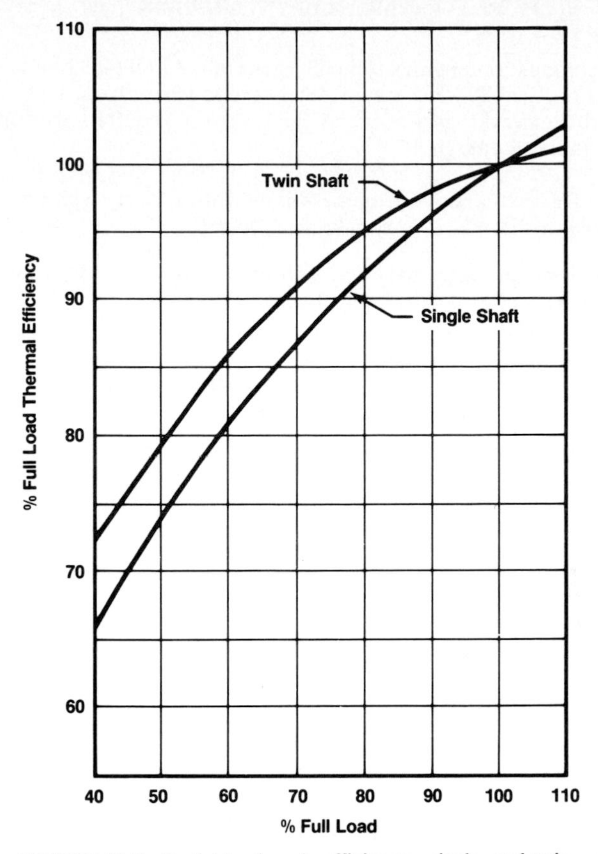

FIGURE 46.6 Partial-load cycle efficiency—single- and twin-shaft turbines. *(Courtesy of Ruston Gas Turbines Limited.)*

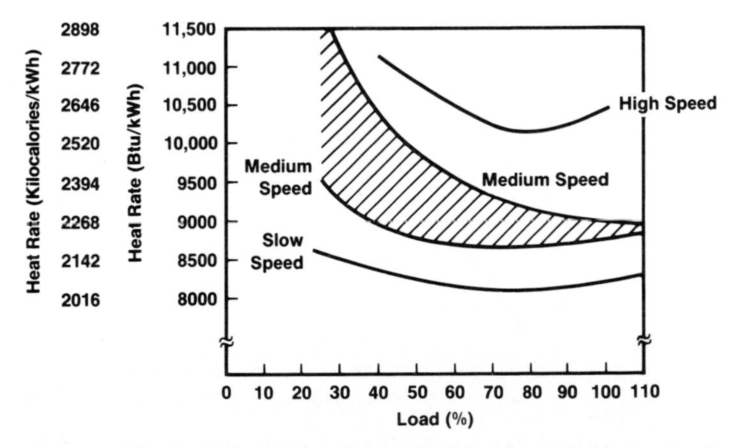

FIGURE 46.7 Typical variation of internal-combustion-engine heat rate with load. *(From Handbook of Industrial Cogeneration, Publication DE 82 009604, Department of Energy, Washington, DC, October 1981, Fig. 16-1.)*

Internal-combustion engines typically generate 3 lb/h (4899 kg/s) of 15- to 150-psig (103.42- to 1034.32-kPa) steam per horsepower output. Due to the lack of oxygen in the exhaust gas, electric heaters are required to supplement the exhaust-gas thermal energy.

Jacket-Water Heat Recovery. Cogeneration heat-recovery systems that use engine jacket-water thermal energy take four forms:

1. The heated jacket water may be routed to process needs. Engine cooling is dependent on the leak-tight integrity of this system.
2. The jacket cooling-water circuit for each engine transfers heat to an overall utilization circuit serving facility process needs. The overall utilization circuit may also be heated by the engine levels. This configuration minimizes connections to the jacket cooling-water system.
3. The recovered heat in the jacket cooling-water system is flashed to steam in an attached steam flash chamber. Water enters the engine at 235°F (112.7°C) and exits at 250°F (121.1°C). Steam is produced at 235°F (112.7°C), 8 psig (55.168 kPa). Flow must be restricted at the entrance to the steam flash chamber to maintain sufficient back pressure on the liquid coolant in the engine chambers.
4. Some engines use natural-convection ebullient cooling. A steam-and-water mixture rises through the engine to a separating tank, where the steam is released and the water is recirculated. A rapid coolant flow is required through the engine because a very small temperature rise occurs in the fluid. Back pressure must be controlled, for the steam bubbles in the engine could rapidly

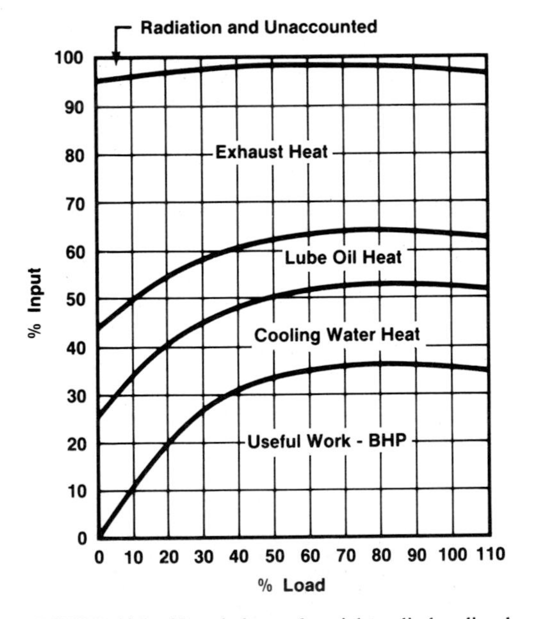

FIGURE 46.8 Heat balance for eight-cylinder diesel engine. *(From "Cogeneration," Fairbanks-Morse Engine Division, Colt Industries.)*

expand, causing the engine to overheat. This system produces 15-psig (103.42-kPa) steam at 250°F (121.1°C).

The temperature and pressure of these jacket-water heat-recovery systems make them suitable for single-stage absorption chiller application (Table 46.1).

Heat Balance. A typical heat balance for an internal-combustion engine is illustrated in Fig. 46.8. The exhaust heat makes up the largest portion of the energy.

The jacket cooling-water component of thermal energy from an internal-combustion engine contains 30 percent of the heat input (Fig. 46.8). Jacket cooling-water temperatures are summarized in Table 46.4. Some internal-combustion-engine manufacturers discourage operating with high jacket-water temperatures because special gasket and seal designs are required.

The lubricating oil system also contains usable heat (Fig. 46.8). The normal operating temperature for the system is 165°F (73.9°C). The lube oil cooling fluid may also be routed through the exhaust heat-recovery unit if process requirements specify heat at a higher temperature. By elevating the lube oil coolant temperature above 180°F (82.2°C) toward 200°F (93.3°C), special lubricants may be required to ensure an adequate useful life of the oil.

Load Control. The heat rate of an internal-combustion engine remains constant above approximately 50 percent load, as illustrated in Fig. 46.8. From the engine heat balance, energy normally being converted to mechanical energy is transferred to thermal energy below 50 percent power. Cogeneration systems are suited to using a large portion of this thermal energy.

CHAPTER 47
PUMPS FOR HEATING AND COOLING

Warren Fraser
Will Smith
Dresser Pump Division, Dresser Industries, Mountainside, New Jersey

47.1 INTRODUCTION

Pumps can be classified into two basic types—centrifugal and positive-displacement. Each operates on a different principle, and their basic design differences influence their areas of application.

The centrifugal pump produces a head differential, and thus a flow, by increasing the velocity of the liquid through the machine with a rotating vaned impeller. The conversion of a portion of the input energy to head and velocity energy by the impeller within the pump casing produces the flow and head differential from suction to discharge. It follows that the design of the impeller and casing controls the relationship between head, flow, and input horsepower (wattage) at any given speed. Since the design of the impeller and casing can be widely varied, the centrifugal pump is a very versatile machine for pumping the liquids required in most applications today.

The positive-displacement pump operates on an entirely different principle: the alternation of filling a cavity and then displacing a given volume of liquid. Pumps designated as power, steam, gear, and rotary pumps fall into this classification. Power pumps are used primarily for applications requiring small flows at high pressures. Steam-operated reciprocating pumps are economical when low rates at moderate pressures are required, such as for feed pumps in smaller plants or for utilizing waste process steam. The principal advantage of rotary pumps is their ability to deliver a constant volume of liquid against a varying discharge head.

Whether centrifugal or positive-displacement pumps are involved, it is obvious, even though often overlooked, that all the wetted parts of the pump must be of materials that are compatible with the fluid being pumped in terms of corrosion, erosion, and temperature.

47.2 CENTRIFUGAL PUMPS

47.2.1 General Theory

The application of centrifugal pumps requires a basic knowledge of some of the design parameters and the general performance characteristics of the wide range of pumps now available. Since the centrifugal pump produces head by accelerating the liquid through the impeller and then converting the velocity into head in the casing, the basic relation is $H = V^2/2g$, where H is the head in ft (m) of liquid, V is the velocity in ft/s (m/s), and g is the constant for the acceleration of gravity: 32.2 ft/s^2 (0.981 m/s^2). ["Feet (meters) of head" is a common pump term used to express foot-pound (Newton-meter) force per pound (kilogram) mass of specific energy added to the liquid.] By means of this simple power law, the performance of a given pump at any one speed can be predicted at any other speed merely by varying (1) the head produced by the square of the rotational speed, (2) the output flow or capacity directly as the change in the rotational speed, or (3) the input horsepower (wattage) required by the cube of the speed change. These relations are designated the "similarity relations" and are expressed mathematically as

$$\frac{H_1}{H_2} = \frac{N_1^2}{N_2^2} \qquad \frac{Q_1}{Q_2} = \frac{N_1}{N_2} \qquad \frac{P_1}{P_2} = \frac{N_1^3}{N_2^3} \tag{47.1}$$

where H = head of fluid of liquid being pumped, ft (m)
N = rotational speed of pump, r/min
Q = capacity of pump, gal/min (m^3/h)
P = power required, hp (kW)

There are, however, two limitations to the range throughout which the similarity relations can be applied. The first is a mechanical limitation of the maximum speed at which the pump is designed to operate. This limitation is set by the manufacturer and should not be exceeded.

The second limitation is the head required on the suction side of the pump to prevent cavitation at the impeller inlet. Adequate suction head, or "net positive suction head" (NPSH), is mandatory for satisfactory pump operation and should be carefully checked for any application, regardless of pump size or type. The suction-head limitation is usually stated by the pump manufacturer as the "required net positive suction head" (NPSHR). In practice, the "available net positive suction head" (NPSHA) at the pump suction must equal or exceed the NPSHR. If it does not, the pump's output will be reduced and the pump will be damaged mechanically.

The NPSHA at the pump suction can be determined as soon as the piping design has been completed and the absolute pressure of the source of the liquid to be pumped has been established. Since the NPSH is an atmospheric-pressure head, the friction loss through the piping must be expressed in terms of feet (meters) of head as well as in terms of the absolute pressure of the liquid source. To determine the NPSHA, use the relation

$$\begin{aligned}\text{NPSHA} &= \text{static head} + \text{atmospheric pressure} - \text{friction head loss}\\ &\quad - \text{vapor pressure}\end{aligned} \tag{47.2}$$

In an open suction system, the static head is the height in feet (meters) of the liquid's surface above (negative if below) the center of the pump-inlet connection; in a closed system, it is the positive suction system pressure expressed in feet (meters) of liquid pumped. The atmospheric pressure is the local barometric pressure expressed in feet (meters); at sea level it is 34 ft (10.4 m) of water. The friction head loss is the loss due to fluid friction of the liquid in the suction line.

EXAMPLE 47.1 Say that the NPSHA at the pump suction drawing from a cooling tower basin is as follows:

1. The static difference in elevation between the level in the basin and the pump centerline (which is positive for a submerged impeller) = +6 ft.
2. Since the water surface in the basin is open to the atmosphere, the absolute head on the surface is the head corresponding to the absolute atmospheric pressure = 33 ft.
3. The friction loss in the suction piping at, say, 2000 gal/min = 3 ft.
4. The vapor pressure of the water at the impeller at, say, 68°F = 0.8 ft.

Solution From Eq. (47.2) we have

$$\text{NPSHA} = +6 \text{ ft} + 33 \text{ ft} - 3 \text{ ft} - 0.8 \text{ ft} = 35.2 \text{ ft}$$

Thus the pump selected must be capable of pumping 2000 gal/min with an NPSHR of 35.2 ft or less.

A more generalized means to determine the NPSHR for centrifugal pumps has been developed in the form of specific-speed limitations, as published in the *Hydraulic Institute Standards*. The term "specific speed" is used here to designate the relation between capacity, head, and speed for any given design at the point of maximum efficiency.[1] It is a dimensionless index number and has no relation to the size of the pump. It does have significance, however, in determining the maximum rotational speed permissible for a given capacity and head and the NPSHA in the design stages of a pumping system. Specific speed is defined as

$$N_s = \frac{N\sqrt{Q}}{H^{0.75}} \tag{47.3}$$

where N_s = specific speed[2], gal/(min · ft) [m^3/(s · m)]
N = rotational speed of pump, r/min
Q = capacity of pump at maximum efficiency, gal/min (m^3/s)
H = total head of pump at maximum efficiency, ft (m) of liquid

With a given required flow and head and an NPSHA, it is thus possible with the Hydraulic Institute's specific-speed limitations to determine the maximum pump rotational speed. This kind of analysis is useful in determining the most economic relation between the initial pump and system costs.

The concept of specific speed is also useful in designating general performance characteristics for the entire range of good centrifugal pump designs. Figures 47.1 to 47.4 show the variation in head, capacity, input horsepower (wattage), and efficiency as well as the general impeller configuration for selected specific speeds. It is important to consider the relation between head, capacity, and input horsepower (wattage) throughout the expected range of operation so that an ade-

[1]The definition of "specific speed" is the speed of a pump for unit head and unit capacity at the point of best efficiency.

[2]While N_s is dimensionless in concept, the use of inconsistent units for flow and head results in numerical differences between the conventional and metric systems.

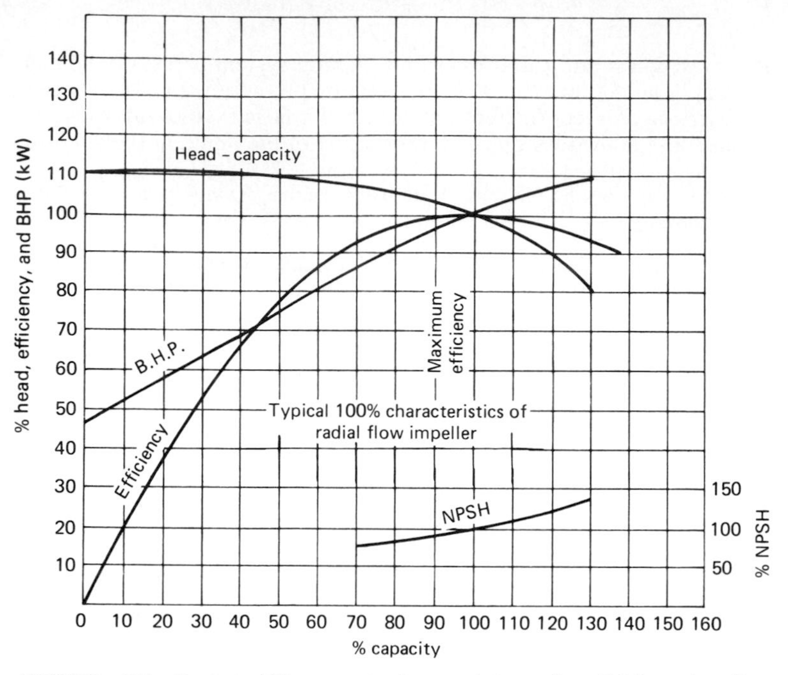

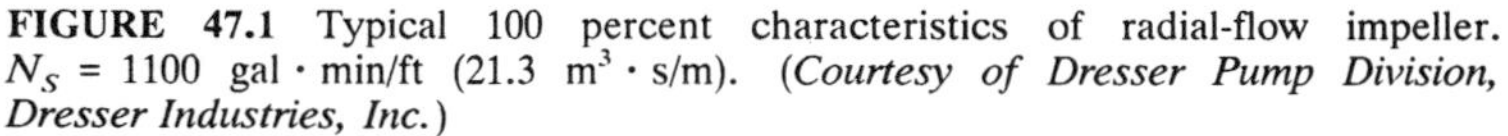

FIGURE 47.1 Typical 100 percent characteristics of radial-flow impeller. N_S = 1100 gal · min/ft (21.3 $m^3 \cdot s/m$). (*Courtesy of Dresser Pump Division, Dresser Industries, Inc.*)

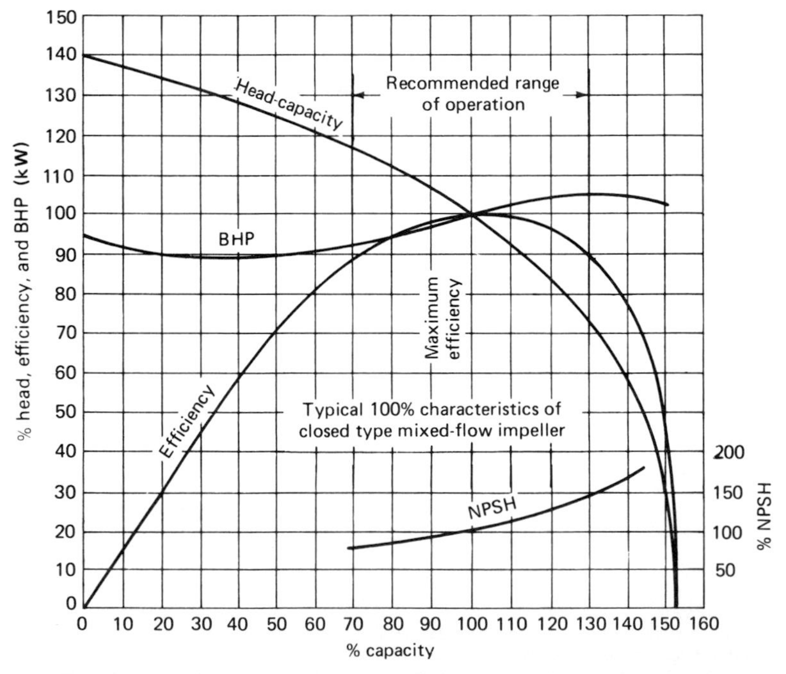

FIGURE 47.2 Typical 100 percent characteristics of closed-type mixed-flow impeller. N_S = 4000 gal · min/ft (77.4 $m^3 \cdot s/m$). (*Courtesy of Dresser Pump Division, Dresser Industries, Inc.*)

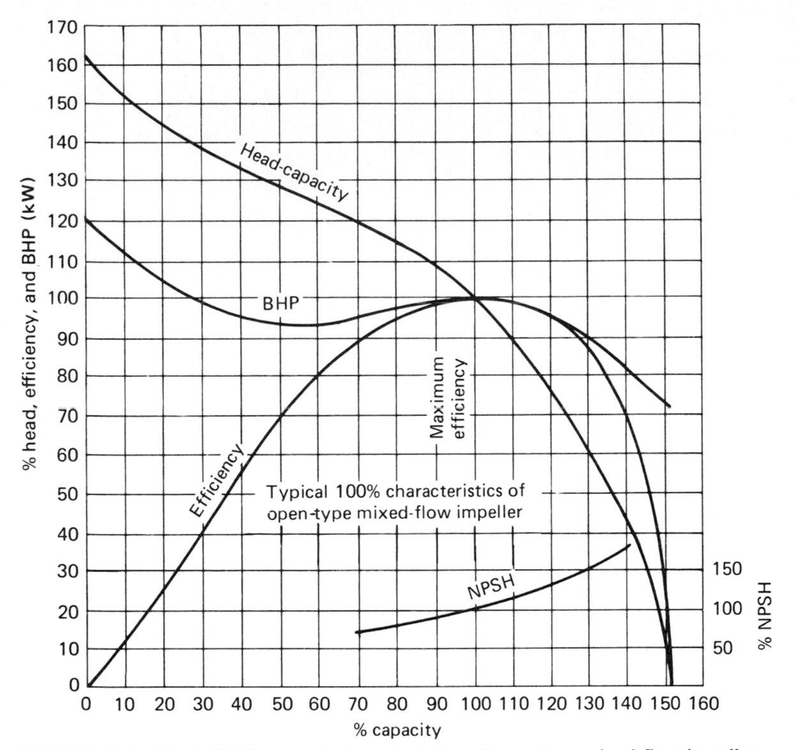

FIGURE 47.3 Typical 100 percent characteristics of open-type mixed-flow impeller. N_S = 6250 gal · min/ft (120.9 m^3 · s/m). *(Courtesy of Dresser Pump Division, Dresser Industries, Inc.)*

quately powered driver can be provided and so that the NPSHR does not exceed the NPSHA at any point. With an increase in flow, the NPSHR increases while, in most installations, the NPSHA decreases because of the increased piping losses. It follows, therefore, that the NPSHA and NPSHR should be compared not at the design point, but at the maximum expected capacity. The divergence of the NPSHR and NPSHA with increasing flows also points out the danger of overestimating the system discharge head against which the pump must operate. Referring to the pump characteristics shown in Figs. 47.1 to 47.4, it can be seen that if the head is lower than estimated, the pump will operate at an increased capacity and require an increased NPSH that was not allowed for in the pump selection.

Optimum efficiencies for centrifugal pumps are obtained in the 2000 to 3000 specific-speed range. Figure 47.5 shows the variation in pump efficiency with specific speed and rated capacity as well as typical impeller designs for the range of specific speeds.

When feasible, it is advisable to operate a centrifugal pump at or near the capacity corresponding to the point of maximum efficiency, as shown on the characteristic performance curve. This is not only to save on power costs but also to prevent reduction in the life of pump components, which is caused when operation occurs too far to the right or left of the maximum-efficiency capacity. The

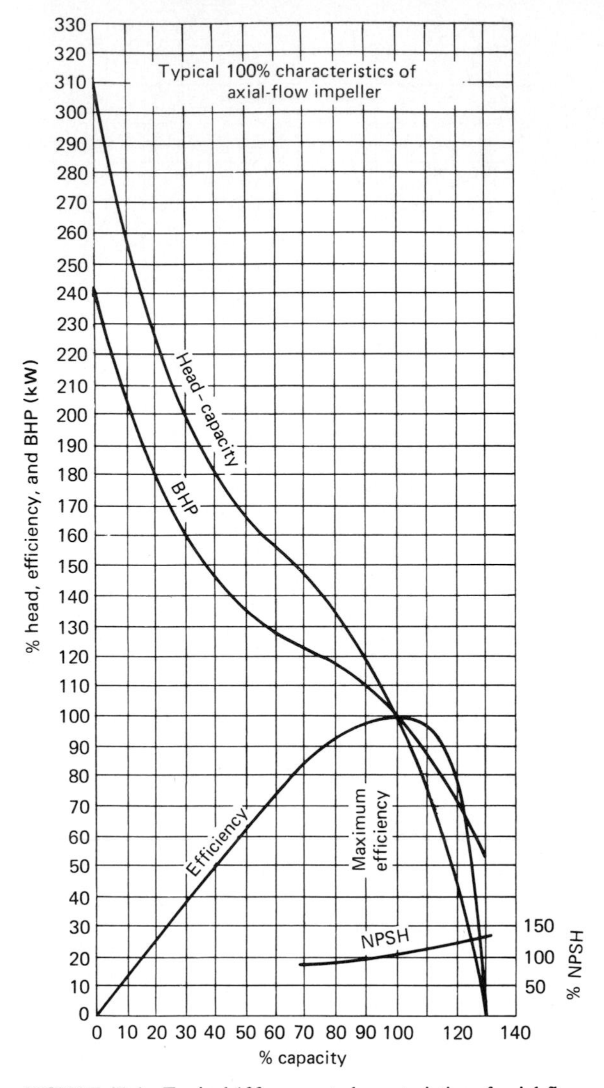

FIGURE 47.4 Typical 100 percent characteristics of axial-flow impeller. N_S = 11,000 gal · min/ft (213 m^3 · s/m). *(Courtesy of Dresser Pump Division, Dresser Industries, Inc.)*

impeller vanes are designed to operate at the flows corresponding to those at peak efficiency, and operation at higher or lesser flows means that the vane angles are mismatched to the flow angles; accelerated erosion on the impeller and casing can result, with accompanying noise, vibration, and unstable operation. This is particularly relevant where quiet or continuous operation is required. The recommended range of operation for a typical centrifugal pump is shown in Fig. 47.2.

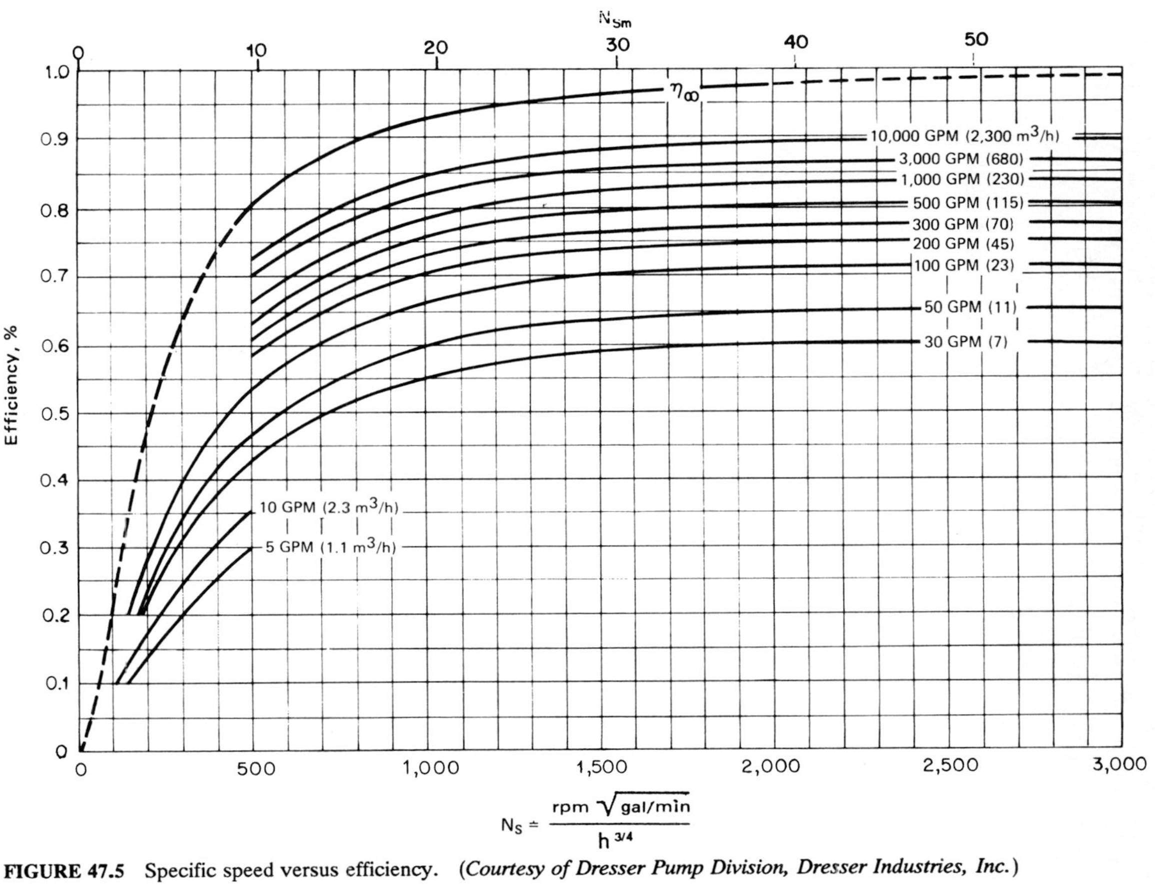

$$N_S = \frac{\text{rpm}\sqrt{\text{gal/min}}}{h^{3/4}}$$

FIGURE 47.5 Specific speed versus efficiency. (*Courtesy of Dresser Pump Division, Dresser Industries, Inc.*)

The power requirement for a centrifugal pump can be determined from the relation

$$P = \frac{Q \times H \times S}{C \times E} \tag{47.4}$$

where P = power, BHP (kW)
Q = flow rate, gal/min (m^3/h)
H = head of fluid being pumped, ft (m)
S = specific gravity of the fluid
C = a constant depending on the system of units used: 3960 for English units, and 3678 for metric units
E = pump efficiency, expressed as a decimal

47.2.2 Parallel and Series Operation

Two or more pumps can be installed in parallel to increase the capacity of any given system or to match the flow requirements of a variable-capacity system. The increase in flow or the incremental steps in pumping rates will depend on the shape of the head capacity curve, the characteristic of the system curve, and the number of pumps in operation. The system curve is the head against which the pumps must operate and may consist of static head, friction head, or a combination of static and friction head. The intersection of the pump head capacity curve and the system head determines the capacity that will be delivered by the pump(s).

An example of two centrifugal pumps operating in parallel is shown in Fig. 47.6. With pump 1 in operation, it will deliver 100 percent capacity and intersect

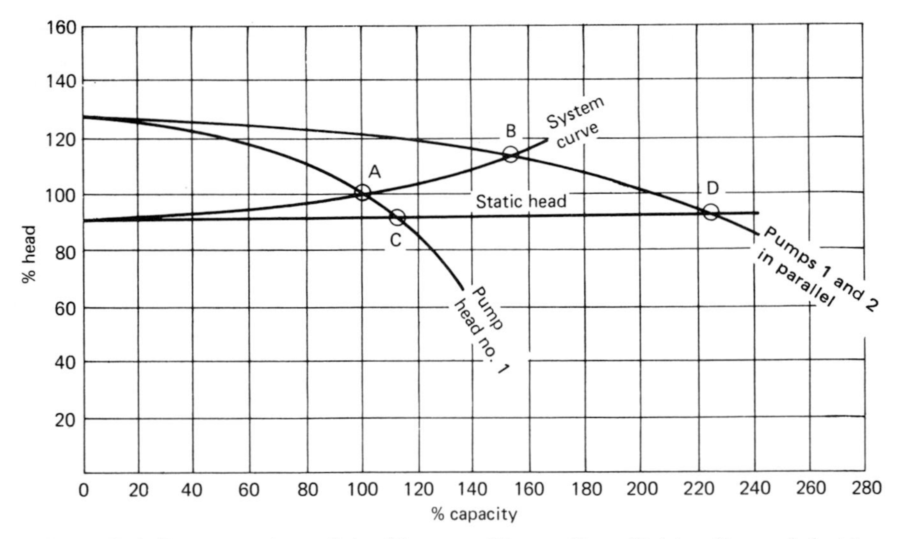

FIGURE 47.6 Two pumps in parallel. *(Courtesy of Dresser Pump Division, Dresser Industries, Inc.)*

the system curve shown at point A. The same pump operating against a pure static head will intersect at point C and deliver 114 percent capacity. Additional capacity can be produced by operating pump 2 in parallel with pump 1. The two pumps will now intersect the system curve at point B and deliver 153 percent capacity. The two pumps operating against a pure static head will deliver 228 percent capacity and intersect the static head at point D.

The same two operating in series and pumping against a pure friction curve system is shown in Fig. 47.7. In this case, pump 1 operating alone will pump 100 percent capacity and intersect the system curve at point A. When pump 2 operates in series with pump 1, the two pumps will deliver 125 percent capacity and intersect the system curve at point B.

Of course, if the system head becomes as high the shutoff head for either of the pumps operating in parallel (Fig. 47.6), that pump can contribute no capacity to the system. Figure 47.7 illustrates what happens when a second pump is

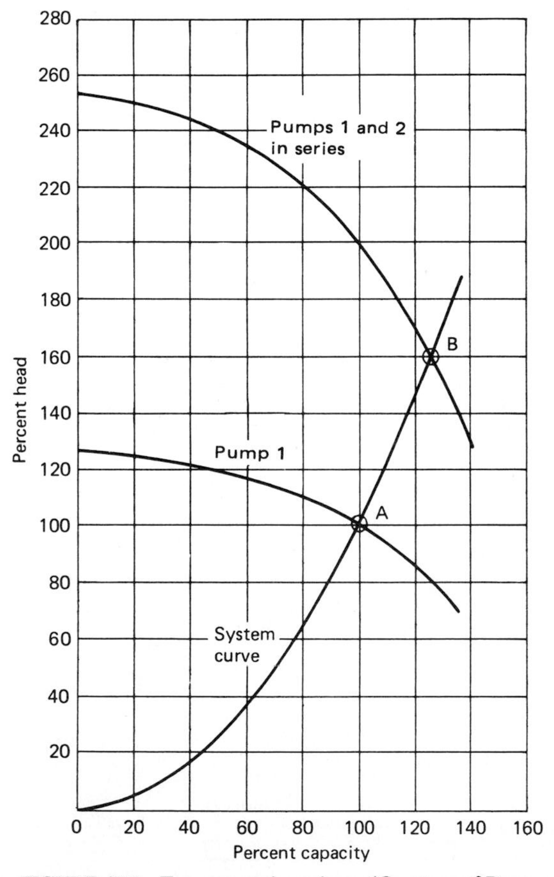

FIGURE 47.7 Two pumps in series. (*Courtesy of Dresser Pump Division, Dresser Industries, Inc.*)

started up in series, as opposed to a second pump being run in parallel as shown in the previous figure.

47.2.3 Frame-Mounted and Close-Coupled End-Suction Pumps

Both frame-mounted and close-coupled end-suction centrifugal pumps have wide application for general and specialized services. Since many of these units are used throughout the industry, a high degree of standardization has been achieved along with an interchangeability of parts to reduce the inventory of spare parts and to meet changing pumping requirements.

Figure 47.8 shows a typical frame-mounted pump, and Fig. 47.9 shows a typical close-coupled pump. As seen in these two figures, the liquid inlets of both types are identical. The frame-mounted pump is used with a separate driver coupled to the pump and mounted on a common baseplate. The close-coupled unit has the liquid inlet bolted directly onto the motor frame, with the impeller

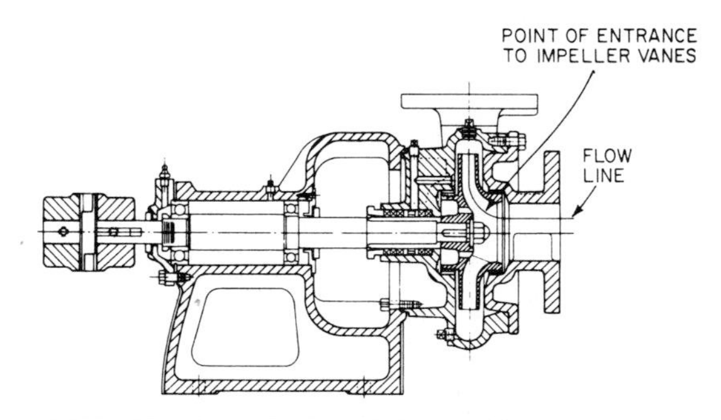

FIGURE 47.8 Frame-mounted end-suction pump. [*From Igor J. Karassik et al. (eds.),* Pump Handbook, *1st ed., McGraw-Hill, New York, © 1976. Reproduced with permission.*]

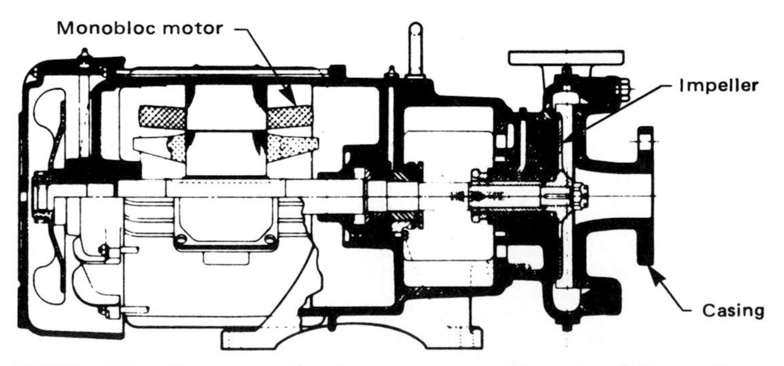

FIGURE 47.9 Close-mounted end-suction pump. (*Courtesy of Dresser Pump Division, Dresser Industries, Inc.*)

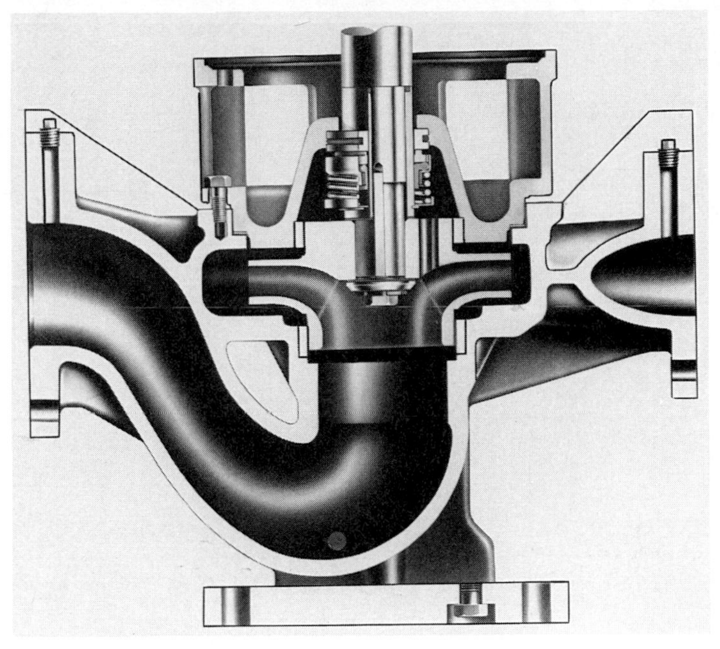

FIGURE 47.10 Vertical in-line pump.

mounted on an extension of the motor shaft. The close-coupled unit has the advantage of compactness, and it eliminates the problem of maintaining alignment between pump and driver.

Many pumps for hot-water circulation are for flow rates and heads in the range of in-line centrifugal pumps that are supported by the pipeline in which they are installed. Figure 47.10 illustrates an in-line pump.

47.2.4 Double-Suction Centrifugal Pumps

The double-suction impeller design is basically two single-suction impellers placed back to back in an integral casing. As shown in Fig. 47.11, the incoming flow is divided before entering the impeller, with one-half of the total flow entering each side of the impeller. The inlet velocities and the NPSHR are thereby reduced as compared to a single-suction impeller delivering the same total flow. In many applications, the low NPSHR will often dictate the use of a double-suction pump. The following inherent advantages in double-suction design will require careful consideration in the selection of the type of pump best-suited to a particular set of conditions:

1. Since the impeller is symmetrical about a plane normal to the axis, the hydraulic end thrust is virtually eliminated. This means that smaller bearings are needed to produce the same bearing life as does a single-suction pump of the same rating.

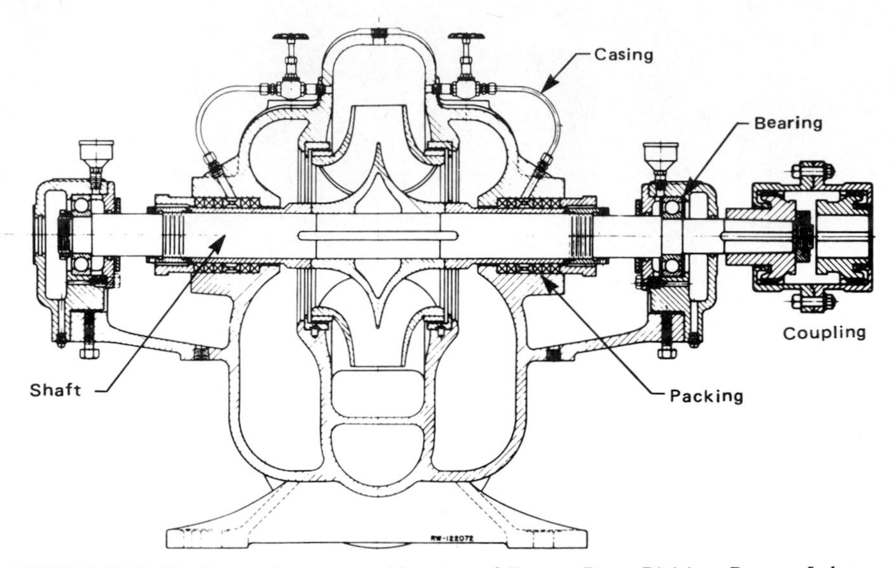

FIGURE 47.11 Double-suction pump. (*Courtesy of Dresser Pump Division, Dresser Industries, Inc.*)

2. The suction and discharge connections are in the same line, thus simplifying the piping arrangement just as with in-line pumps.
3. The rotor is readily accessible by removing the top half of the horizontally split volute casing.
4. Due to a high degree of standardization in the smaller sizes, delivery time and spare-parts inventory are reduced.

47.2.5 Vertical Multistage Pumps

The vertical multistage pump has found wide application throughout the industry primarily because of its compactness, low cost, and vertical arrangement. This type of pump is usually referred to as a "deepwell turbine pump," although its application has gone far beyond its original use as a well pump. A typical design for use in well service is shown in Fig. 47.12. An adaptation for use as a wet-pit general-water-service pump is shown in Fig. 47.13.

Figure 47.14 shows a further modification for use in condensate and process applications.

The vertical multistage pump can be provided with either an open- or closed-shaft arrangement. The open-shaft design is shown in Fig. 47.12 with the intermediate bearings supported in spiders attached at intervals to the column pipe. The bearings are lubricated by the liquid pumped and can be of any appropriate bearing material, although bronze or rubber or a combination of the two is most frequently specified. It is quite evident, however, that the open-shaft design is limited to applications pumping liquids that not only adequately lubricate the

FIGURE 47.12 Deepwell turbine pump. (*Courtesy of Dresser Pump Division, Dresser Industries, Inc.*)

bearings but also are noncorrosive and nonabrasive to the bearing material specified.

With the closed-shaft design, the shaft and bearings are enclosed in a cover tube, as shown in Fig. 47.13. The bearings are usually of bronze and are threaded externally. The cover tube is threaded internally at both ends to receive the bearings, and at the top of the column a special bearing is provided that can be adjusted to impose an axial load on the cover tube to maintain bearing alignment. The bearings are usually oil-lubricated, with the lubricant supplied from a drop-feed oiler.

There are no thrust bearings in the vertical multistage pump. The thrust of the pump is carried by the special thrust bearing provided in the motor. It is almost universal practice to furnish hollow-shaft motors with deep-well pumps so that the position of the pump rotor can be adjusted to compensate for the accumulation of dimensional tolerances and the stretch of the shaft in deep settings. Another valuable feature of the hollow-shaft motor is the adaption of a nonreverse ratchet at the top of the motor that prevents the line shaft of the pump from backing out of the threaded couplings in the event that the motor accidentally operates in a direction of rotation opposite to that for which the pump was designed.

In the case of the wet-pit pump with a short setting (Fig. 47.13) or the condensate or process pump (Fig. 47.14), a solid-shaft motor with an intermediate flanged coupling between the pump and drive is preferred. The reason for this is that the adjustable feature of the hollow-shaft motor is not required for the shorter-setting arrangements, and a more accurate alignment of pump and motor shaft can be achieved with the solid-shaft motor than with the motor that has a hollow shaft.

47.2.6 Horizontal Multistage Pumps

For pressures up to 1600 lb/in^2 (110 bar), the horizontal multistage pump (or a positive-displacement type) is the preferred selection. A typical pump (Fig. 47.15) may have 10 to 12 stages with external oil- or grease-lubricated bearings. The interstage passages from one impeller to the next can be either of the multivane diffuser design or the volute design. The multivane diffuser consists of a series of vanes that are cast integrally with an interstage diaphragm and assembled on the

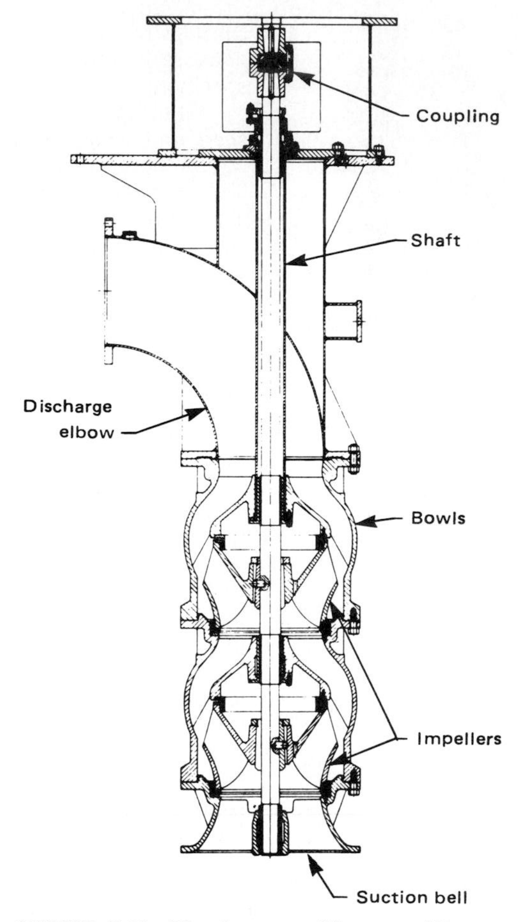

FIGURE 47.13 Wet-pit pump. *(Courtesy of Dresser Pump Division, Dresser Industries, Inc.)*

shaft between each impeller. In this design the entire rotor assembly (consisting of the shaft, impellers, and diffusers) is removed from the casing for disassembly and maintenance.

In the volute design, one-half of the volute is cast in the upper portion of the split casing, and the other half is cast in the lower portion of the casing. When the casing is bolted together, the two halves of the volute match and provide a smooth transition in the flow passage from one impeller to the next. Radial reaction on the impellers can be minimized either by staggering the volute tongues from one stage to the next or by providing symmetrical double volutes.

Both the diffuser and volute designs can be furnished with either packed boxes or mechanical seals. If the pump operates with a negative suction head, an external source of sealing water must be provided to prevent the leakage of air into the pump.

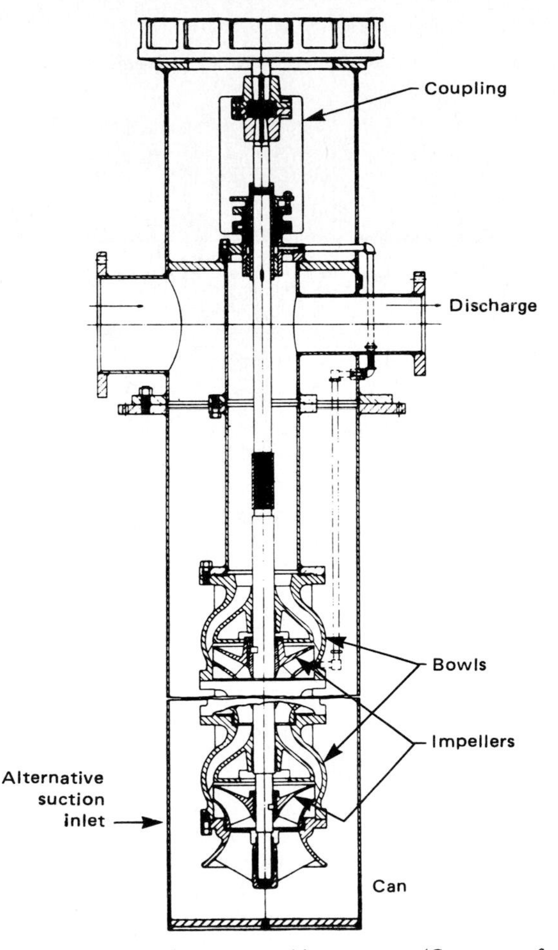

FIGURE 47.14 Can-type turbine pump. (*Courtesy of Dresser Pump Division, Dresser Industries, Inc.*)

Unlike the turbine pump for condensate or process service (Fig. 47.14), the NPSHA may limit the speed of the horizontal pump to less than the rated speed. When the NPSHA is 10 ft (3 m) or less, the vertical turbine pump (Fig. 47.14) is usually a more economical selection. The primary advantage of the horizontal multistage pump is that the shaft is supported by bearings located externally to the liquid and can be expected to operate for longer periods between bearing maintenance or replacement.

47.2.7 Submersible Pumps

The submersible pump was developed to eliminate the necessity for the long line shaft and the associated problems of alignment and adequate lubrication of the bearings in deep-well pumps. The term "submersible pump" means that while

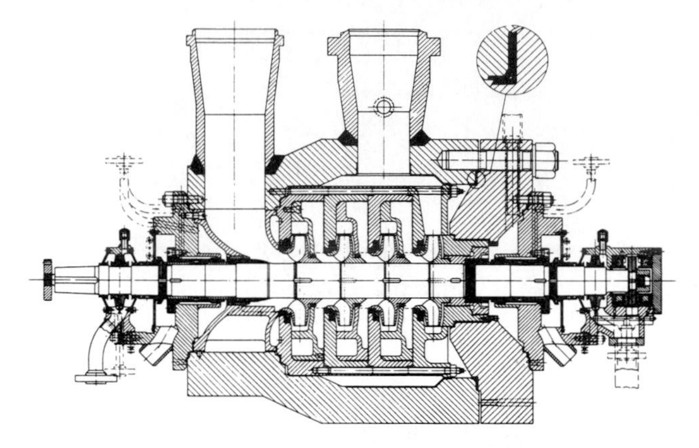

FIGURE 47.15 Horizontal multistage pump. (*Courtesy of Dresser Pump Division, Dresser Industries, Inc.*)

the pumping assembly is the same as for the standard deep-well pump, the motor is submersed in the water *below* the pumping assembly rather than at ground level.

While the line shaft has been eliminated with the reduction in first cost and installation cost, the design of the motor must incorporate special features to permit submerged operation. The first requirement is that the motor must be designed with a length-to-diameter ratio much higher than the standard motor to permit installation in a standard well diameter for which the pump was designed. The second requirement is that the bearings must operate under water for extended periods. The real problem here is the motor thrust bearing, which must be capable of carrying the full thrust developed by the pump. The line bearings are only lightly loaded. The tilted pad thrust bearing operating against a graphite thrust collar is widely used in conjunction with graphite bearings. The bearings are usually lubricated with water or a water-oil emulsion, and leakage of the lubricant from the motor is prevented by a mechanical seal.

47.2.8 Self-Priming Pumps

Unlike a positive-displacement pump, a centrifugal pump *will not start pumping unless the liquid can flow into the pump suction under a positive head*. This is an obvious disadvantage for pumps that must operate under a suction lift, and various methods have been developed to circumvent this limitation. There is no problem with the vertical deep-well or wet-well pump, as the impeller is always located beneath the surface of the water level and is self-priming. With the centrifugal pump operating on a lift, a priming device can be used to evacuate the air from the casing and thus raise the water to the impeller eye. An auxiliary vacuum pump or an ejector operating on air pressure is commonly used for this purpose.

For special applications where starting and stopping is frequent and automatic starting is necessary, self-priming pumps have been developed. A typical self-priming pump is shown in Fig. 47.16. It is well known that a small centrifugal pump can be primed if sufficient liquid is forced back from the discharge to the suction, and the self-priming design utilizes this device. Priming is accomplished by trapping a small portion of the liquid, each time the pump is shut down, in a reservoir mounted on the pump casing. On the next starting cycle, this trapped

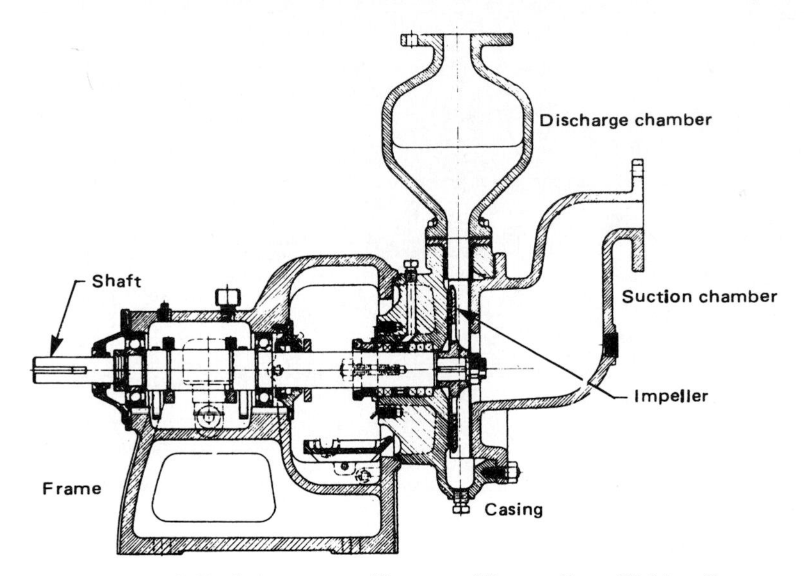

FIGURE 47.16 Self-priming pump. (*Courtesy of Dresser Pump Division, Dresser Industries, Inc.*)

liquid recirculates through the impeller, and through this pumping action and the entrainment of air bubbles the pressure in the suction pipe is reduced to a point sufficient to raise the suction liquid level to the inlet of the impeller. The pump then operates in its normal manner.

47.2.9 Axial-Flow Pumps

The axial-flow, or propeller, pump has been developed where large flows at low head are required. For process work, such as circulation in an evaporator, the axial-flow pump is usually of the type of design shown in Fig. 47.17. Where large quantities of water at low heads are required, such as in drainage service or to supply cooling water for condensers, the axial-flow wet-pit pump is widely used in place of the larger and slower-speed centrifugal pump. The mechanical construction of a typical wet-pit pump is very similar to that of the vertical deep-well pump shown in Fig. 47.13. In fact, it is an extension of the deep-well pump for use in applications requiring higher specific speeds.

By the nature of its design, the axial-flow pump requires some special precautions on its application. The first is its rather unusual head capacity and horsepower (wattage) capacity characteristics as compared to the lower-specific-speed centrifugal pump. The typical characteristics of an 11,000-specific-speed pump are shown in Fig. 47.4. Unlike the centrifugal pump, the axial-flow pump requires an increasing input horsepower (wattage) with a reduction in flow. This means that the axial-flow pump cannot operate against a closed discharge valve unless the driver is sized to deliver this increased power. Fortunately, most systems involving axial-flow pumps do not require discharge valves in their operation. For drainage work, a free discharge is usually used. For larger installations, a siphon can be used on the discharge to recover a portion of the static head, and the

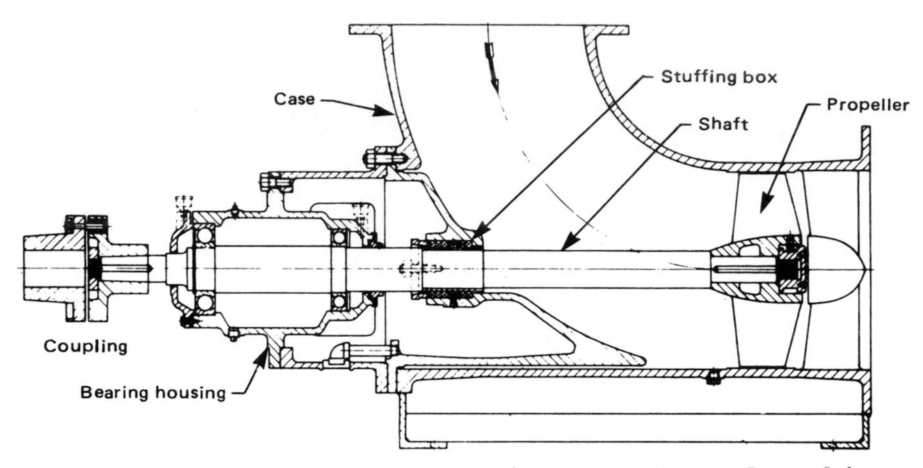

FIGURE 47.17 Elbow propeller pump. (*Courtesy of Dresser Pump Division, Dresser Industries, Inc.*)

driver has to be powered only to establish the siphon and not to operate against a closed discharge valve during startup.

The head capacity curve of the axial-flow pump exhibits a rising horsepower (wattage) toward zero capacity and an unstable portion that begins at approximately 70 percent of the rated capacity and extends to shutoff (or zero flow). Operation in this portion of the pump characteristic should be avoided to prevent unstable operation and a material reduction in bearing life.

47.2.10 Regenerative Pumps

The regenerative pump produces head by adding energy to the liquid through a series of straight vanes located only at the periphery of the impeller. As shown in Fig. 47.18, the liquid enters the impeller along the axis of the shaft and flows to the periphery of the impeller through clearances on both sides of the impeller, and the re-

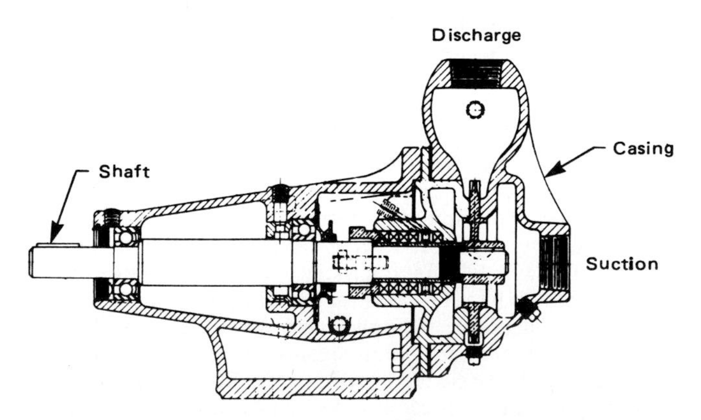

FIGURE 47.18 Regenerative pump. (*Courtesy of Dresser Pump Division, Dresser Industries, Inc.*)

peated entry and reentry of liquid between the vanes produces very small capacities at high heads. Unlike the centrifugal pump, the regenerative pump exhibits a very steep head capacity curve that is more suitable for many applications.

Operation at less than 40 percent of the rated capacity should be avoided, for the impeller clearances are subjected to accelerated water, which may increase the noise level.

47.3 POSITIVE-DISPLACEMENT PUMPS

47.3.1 Reciprocating Pumps

Reciprocating pumps operate by the alternate filling and displacement of the liquid in a pressure chamber by means of a plunger or piston. Power pumps are those reciprocating pumps driven by an external source of power, such as a motor or an engine, and direct-acting pumps are those driven by a piston connected directly to the piston or plunger in the liquid inlet of the pump. These are often referred to as "steam pumps."

The reciprocating pump is used in those applications which require low capacity at high discharge pressure or where a constant capacity is desired over a wide range of pressures. Unlike the small-capacity centrifugal pump, the reciprocating pump has a high efficiency, as the losses consist only of leakage past the valves and piston, losses through the valves, and frictional losses of the moving parts. Reciprocating pumps are self-priming, and this may be a factor in evaluating the choice between reciprocating and centrifugal pumps.

The discharge from a reciprocating pump will vary in a cyclic pattern as a result of the alternate suction and discharge strokes. For the same reason, the flow into the suction of the pump will also vary with the same cyclic pattern. These pulsations can be the source of noise and vibration in both the suction and discharge pipes. Provision should be made for the installation of cushioning chambers on both sides of the pump. Cushioning chambers usually consist of a small chamber partially filled with a gas that will effectively absorb the pressure variations and reduce the pressure pulsations in the pipe. The bladder-type cushioning chamber shown in Fig. 47.19 prevents the liquid being pumped from absorbing the gas.

As with centrifugal pumps, the suction head available and the suction head required must be carefully considered in the application of reciprocating pumps. To reduce losses, the suction pipe should be a size larger than the pump suction flange, and fittings should be kept to a minimum. Tees or short-radius elbows at the pump suction flange should be avoided in any installation. The flow entering the pump must have a uniform velocity profile across the pipe if the stated NPSHR for the pump is to be achieved. Cavitation and noise may result if the NPSHA is less than the published NPSHR performance.

An additional factor must be considered in the determination of the NPSHA for a reciprocating pump. This factor is an acceleration head loss that results from the cyclic variation in velocity of the liquid in the suction pipe. An additional head is required to maintain sufficient head above the vapor pressure of the liquid at the low-pressure point of the cycle to prevent cavitation and to provide sufficient acceleration to the liquid so that it can follow the velocity changes of the plungers during one full cycle. The magnitude of the pressure reduction resulting from this velocity change can be minimized by locating the pump as close to the source of the liquid being pumped as possible and by installing suction

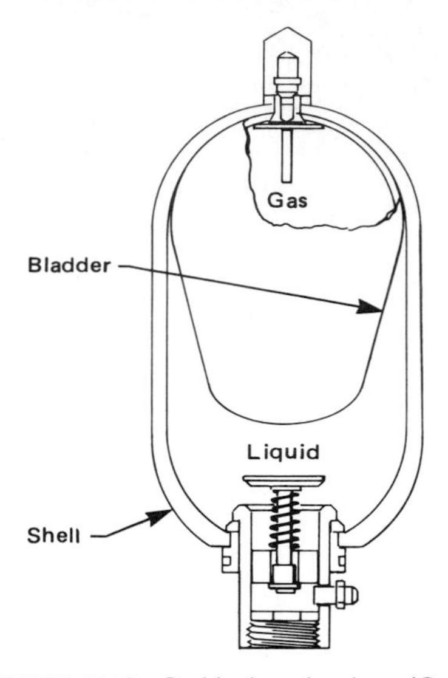

FIGURE 47.19 Cushioning chamber. (*Courtesy of Dresser Pump Division, Dresser Industries, Inc.*)

cushion chambers that will dampen pulsations. The chamber contains a volume of gas under a pressure equal to the average pressure in the suction pipe. The pressure of the gas responds to the rise and fall of the liquid pressure and will absorb energy at the high-pressure peaks and then return energy to the liquid at the low-pressure peaks. The return of energy to the liquid at the low-pressure peaks will provide additional acceleration to the liquid and minimize the possibility of intermittent cavitation from pressure fluctuations.

47.3.2 Power Pumps

Power pumps are built with the stroke of the plungers in either the horizontal or vertical position. A typical vertical design is shown in Fig. 47.20, while a horizontal design is shown in Fig. 47.21. The liquid cylinder may be either cast or forged, depending on the pressure developed. The plungers can be furnished in a variety of materials, depending on the corrosive and abrasive characteristics of the liquid pumped.

From the manufacturer's ratings of power pumps, the operating speed for any capacity up to the rated speed can be determined as follows:

$$N = \frac{QN_r}{D\eta_v}$$

where N = required pump speed, r/min
Q = required flow, gal/min (m^3/h)
D = displacement of pump at rated speed, gal/min (m^3/h)
N_r = rated speed, r/min
η_v = volumetric efficiency, expressed as a decimal

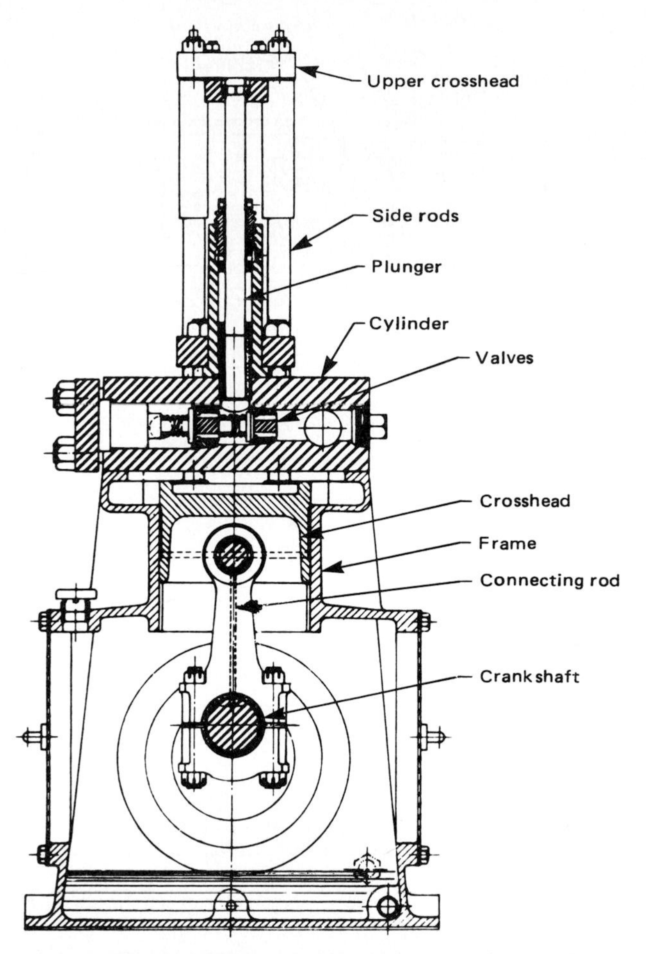

FIGURE 47.20 Vertical power pump. (*Courtesy of Dresser Pump Division, Dresser Industries, Inc.*)

The required input horsepower (wattage) can be determined from the relation

$$P = \frac{Q\,\Delta P}{C \times \eta_m}$$

where P = power required to drive the pump,[2] hp (kW)
Q = capacity of pump, gal/min (m^3/h)
ΔP = differential pressure developed by pump, lb/in^2 (bar)
C = a constant depending on the system of units used: 1714 for English units of gal/min and lb/in^2, and 36 for metric units of m^3/h and bar
η_m = approximate efficiency, as follows:

[2]The power required of a positive-displacement pump is independent of the specific gravity of the liquid being pumped.

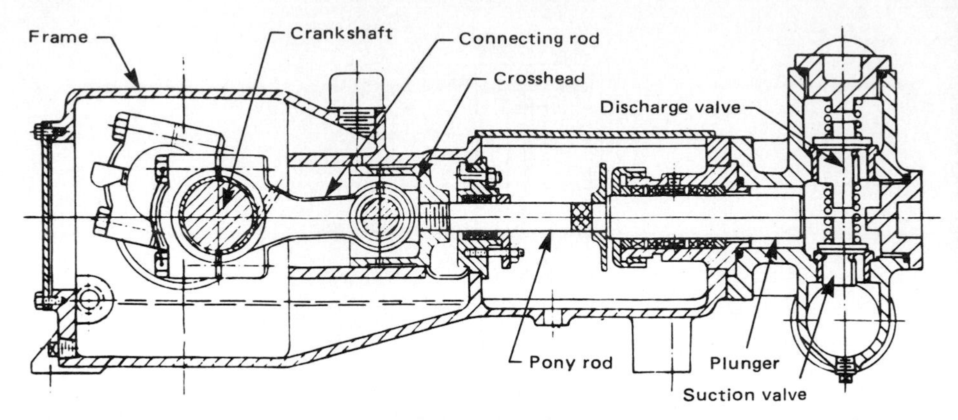

FIGURE 47.21 Horizontal power pump. *(Courtesy of Dresser Pump Division, Dresser Industries, Inc.)*

Stroke		
in	mm	η_m, %
2	50.8	82
4	101.6	85
5	127.0	87
6	152.4	87

47.3.3 Steam Pumps

Steam pumps are usually either of simplex or duplex construction. The simplex pump has a single steam cylinder, while the duplex has a pair of steam cylinders mounted on a common frame. The liquid inlet can be supplied either with a packed piston or a plunger. The packed-piston design is shown in Fig. 47.22. Of the two, the packed piston is the lighter and less expensive construction and is used for moderate pressures on clear and noncorrosive liquids. The plunger construction is used for high-pressure applications and for the more severe pumping conditions that score or erode the close clearances of the piston design.

Steam pump dimensions are specified as steam piston diameter × liquid piston diameter × stroke. Thus a 6 × 4 × 6 steam pump designates the steam piston as 6 in (152.4 mm) in diameter, the liquid piston as 4 in (101.6 mm) in diameter, and the stroke as 6 in (152.4 mm). To determine the liquid-inlet size for a given capacity, the following relations can be used:

Duplex pump:

$$D_L = 3.5\frac{\text{gal/min}}{\text{ft/min}} \qquad D_L = 50.52\,\frac{\text{m}^3/\text{h}}{\text{m/min}}$$

Simplex pump:

$$D_L = 5.0\,\frac{\text{gal/min}}{\text{ft/min}} \qquad D_L = 71.5\,\frac{\text{m}^3/\text{h}}{\text{m/min}}$$

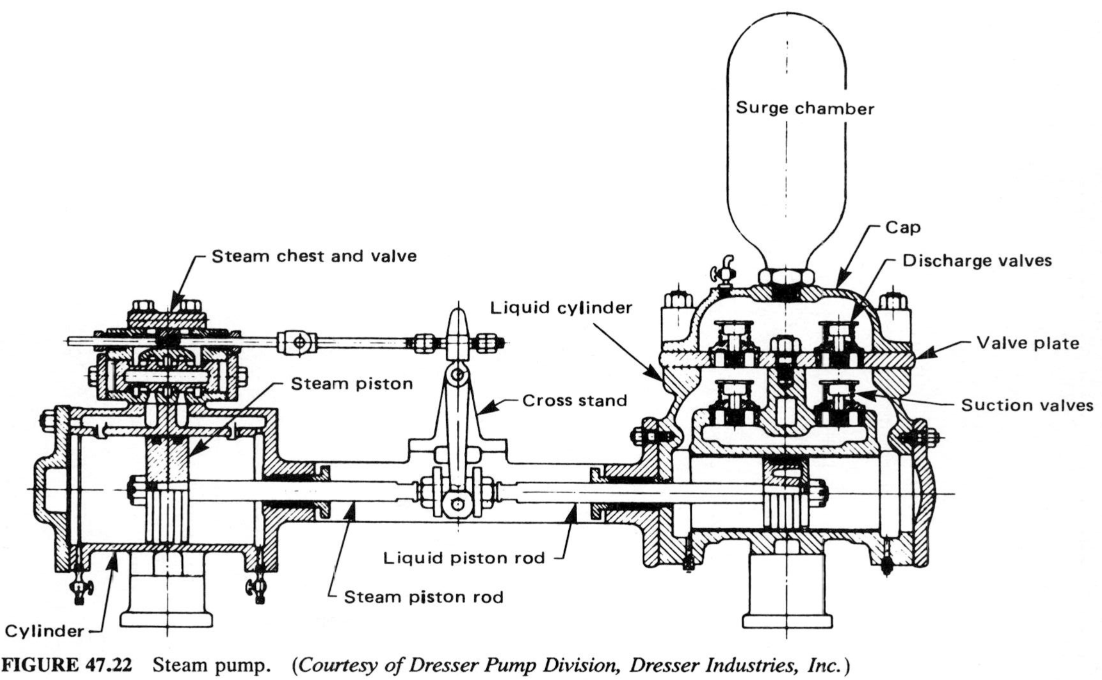

FIGURE 47.22 Steam pump. *(Courtesy of Dresser Pump Division, Dresser Industries, Inc.)*

TABLE 47.1 Direct-Acting Pump—Recommended Piston Speeds

Stroke		Strokes per minute	Speed	
in	mm		ft/min	m/min
4	101.6	71	47	14.3
6	152.4	60	60	18.3
8	203.2	51	68	20.3
10	254.0	45	75	22.9
12	304.8	40	80	24.4
18	457.2	32	95	29.0
24	609.6	26	105	32.0

where D_L is the diameter of the liquid piston in in (mm), gal/min (m^3/h) is the capacity, and ft/min (m/min) is the piston speed. The recommended piston speeds for liquids with viscosities of less than 800 SSU and various lengths of stroke are shown in Table 47.1.

To determine the size of the steam piston, the following relation can be used:

$$D_S = \frac{D_L P_L}{P_S E_m}$$

where D_S = diameter of steam piston, in (mm)
D_L = diameter of liquid piston, in (mm)
P_L = discharge liquid pressure minus the suction pressure (or plus a negative suction pressure), lb/in^2 (bar)
P_S = inlet minus the exhaust steam pressure, lb/in^2 (bar)
E_m = mechanical efficiency in decimals selected from Table 47.2 Plunger pump mechanical efficiencies are approximately 4 percent less than those listed for piston pumps.

TABLE 47.2 Efficiency versus Stroke

Stroke		Mechanical efficiency
in	mm	
3	76	0.50
4	102	0.55
5	127	0.60
6	152	0.65
7	178	0.65
8	203	0.65
10	254	0.70
12	305	0.70
18	457	0.73
24	610	0.75

47.3.4 Rotary Pumps

One common type of rotary pump is the gear pump, which is used for handling viscous liquids such as oils, tars, and nonabrasive sludges. A typical gear pump is

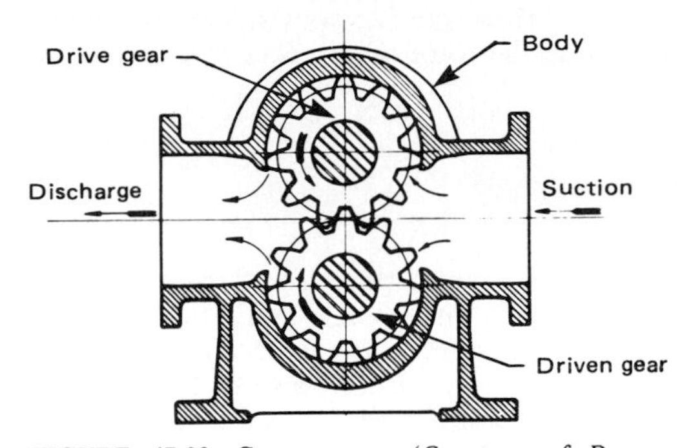

FIGURE 47.23 Gear pump. (*Courtesy of Dresser Pump Division, Dresser Industries, Inc.*)

shown in Fig. 47.23. The liquid flows into the suction and is trapped between the gear teeth and the housing. As the gears turn, the liquid is carried around to the discharge and is squeezed out from between the meshing gear teeth. Spur gears may be used, although double helical gears are usually preferred, as they run more quietly at higher speeds. The output of a gear pump is essentially constant with increasing pressure over a wide range of high efficiency.

Other types of rotary pumps are built with different devices to trap an increment of liquid on the inlet side and then discharge the liquid at a higher pressure at the discharge side. Cams, screws, sliding vanes, and rotating lobes are widely used. A typical sliding-vane pump is shown in Fig. 47.24. Sliding vanes, fitted to a slotted rotor, press out against a housing mounted eccentrically to the rotor. As the rotor turns, liquid is trapped between the rotor and the sliding vanes on the

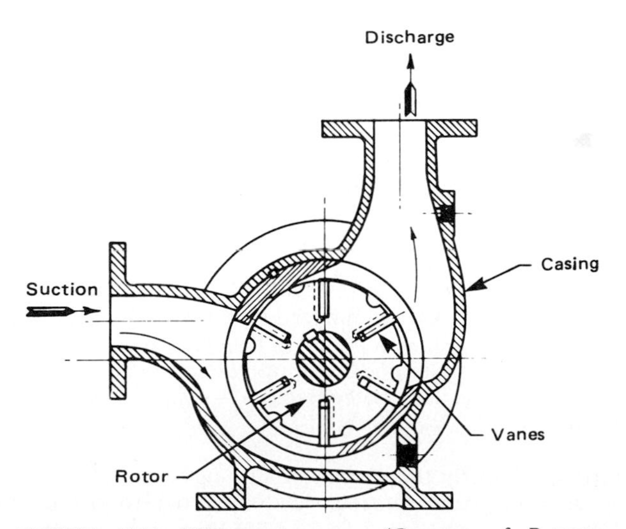

FIGURE 47.24 Sliding-vane pump. (*Courtesy of Dresser Pump Division, Dresser Industries, Inc.*)

inlet side of the pump, and then squeezed out on the discharge side as the sliding vanes are forced back into the rotor by the eccentric housing.

47.4 HEATING SYSTEMS

Heat is usually generated at a central point and transferred to one or more points of use. The transfer may be by means of a liquid (usually water), which has its temperature increased at the source and which gives up its heat at the point of use by reduction of its temperature. It may also be transferred by means of a vapor (usually steam), which changes from a liquid to a vapor at the source, giving up its heat at the point of use by condensation. Pumps may be required in both of these methods.

47.4.1 Hot-Water Systems

A centrifugal pump best meets the requirements of this service. Water is usually used in a closed circuit so that there is no static head. The only resistance to flow is that from friction in the piping and fittings, the heater, the heating coils or radiators, and the control valves. In selecting the pump, the total flow resistance at the required flow rate should be calculated as accurately as possible, with some thought as to how much variation there might be as a result of inaccuracy of calculations or changes in the circuit because of installation conditions. It is not good practice to select a pump for a head or capacity considerably higher than that required, as this is likely to result in a higher noise level as well as in increased power.

When hot water is used for radiation in a single circuit through several radiators, the water-temperature variation is usually only about 20°F (11°C) at the time of maximum requirements, so that there is not too great a difference in heat output between the first and last radiator in the circuit. With the flow rate based on water at 180 to 200°F (82 to 93°C) to heat the air to about 75°F (24°C), a 10 percent reduction in the flow would have little effect, as the actual water-temperature difference would increase to only 22°F (12°C). The reduction in the heat output of the radiator with 178°F (81°C) rather than 180°F (82°C) water would be only about 2 percent. Reference to Fig. 47.6 on the selection of pumps and the prediction of performance from the head capacity pump curves and system head flow-rate curves will show that a rather large percentage undercalculation of head loss of the circuit would be necessary to produce a flow rate 10 percent less than desired.

Greater temperature differences than 20°F (11°C) are frequently used for other radiation circuits. In such cases, a reduced flow rate would have a greater effect on temperature differential than it would in a single circuit. Whatever the condition, the pump selection should be based on full consideration of all the factors, and not on the use of so-called "safety factors"—which, however well intended, often lead to trouble.

Air in the Circuit. Initially the entire circuit will be full of air, which must be displaced by the water. Arrangements should be provided to vent most of the air before the pump is operated. Even if all the air is eliminated at the start, more will be separated from the water when it is heated, and any water added later to replace the loss will result in additional trapped air when the water is heated. Means must therefore be provided for continuous air separation. This cannot be adequately accomplished by vents at high points in the piping because the flow is usually turbulent and the air is not separated at the top of the pipe.

A separator installed before the pump intake will remove the air circulating in the system. In a heating system an air separation device is often provided at the point where the water leaves the boiler or other heating source. This is the point of the lowest pressure and highest temperature in the system and is therefore the most effective point for separation of air from the water.

If there are places in the system where the flow is not turbulent, air may accumulate and interfere with heat transfer. Automatic air vents should seldom if ever be used. If they are used, it is important that they be located only where the pressure of the water is always above that of the surrounding air, whether the pump is operating or idle; otherwise, the air vent may become an air intake.

Pump Selection. Several important factors influence the choice of a pump for a hot-water system with a number of separate heating coils, each having a separate control. Many systems in the past used three-way valves to change the flow from the coil to the bypass.

When two-way valves are used, a low-flow operation may occur for a large portion of the operating time. For this type of operation, therefore, the pump selected should have a flat performance curve so that the head rise is limited at reduced flows. A very high head rise can cause problems when many of the valves are closed. Excessive flow rates through the coils and greater pressure difference across the control valves are some of the problems that can be avoided with a flat pump curve. A centrifugal pump should not operate very long with zero flow, for it would overheat. This condition is controlled by using one or two three-way valves, a relief bypass, or a continuous small bleed between the supply and the return line. Whichever means is used to control minimum flow, the circuit must be such that it can dissipate the heat corresponding to the pump power at that operating condition, without overheating the pump.

Types of Water Circuit. There are several types of water circuit. Those shown in Figs. 47.25 and 47.26 are suitable for the smaller systems and can be used for larger systems by having several of these circuits in parallel. The one shown in Fig. 47.27 is suitable for small or very large circuits, but the reverse return would add considerably to the cost if the circuit extended in one direction instead of in a practically closed loop as shown. For the extended circuit, a simple two-pipe circuit, with proper design for balancing, could be used.

There are a number of reasons for using other circuits, particularly primary-secondary pumping where the system is more extended or complicated, such as continuous circulation branches with controlled temperature. When a coil heats air, part or all of which may be below freezing, the velocity of the water in the tubes and its temperature at any point in the coil must be such that the temperature of the inside surface of the tube is not below freezing. The circuit shown in Fig. 47.28 makes this possible.

Primary-secondary pumping permits flow rates and temperatures in branch circuits different from those in the main circuit without the flow and pressure dif-

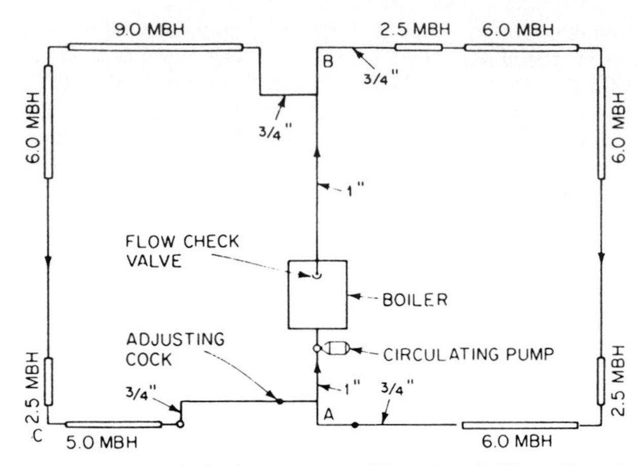

FIGURE 47.25 Series loop system. [*From Igor J. Karassik et al. (eds.),* Pump Handbook, *1st ed., McGraw-Hill, New York, © 1976. Reproduced with permission.*]

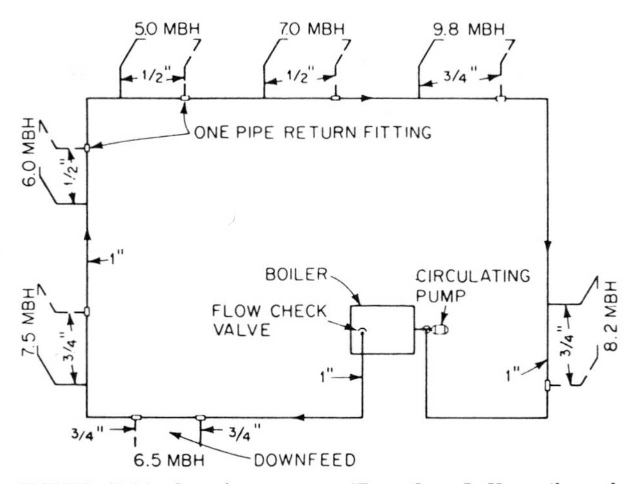

FIGURE 47.26 One-pipe system. [*From Igor J. Karassik et al. (eds.),* Pump Handbook, *1st ed., McGraw-Hill, New York, © 1976. Reproduced with permission.*]

ferences in the mains or branches having a significant effect on each other. There are many possible primary-secondary circuits to meet different requirements.

47.4.2 Steam Systems

No pumping is required with the smallest and simplest steam systems if there is sufficient level difference between the boiler and condensers (radiators, heating coils, etc.) to provide the required flow. When insufficient head exists between

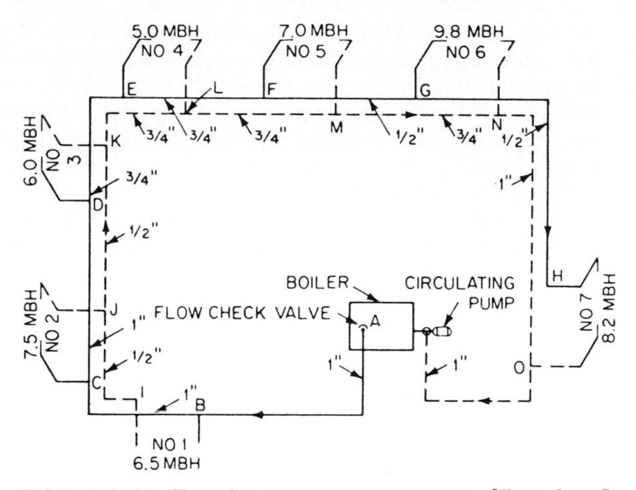

FIGURE 47.27 Two-pipe reverse-return system. [*From Igor J. Karassik et al. (eds.),* Pump Handbook, *1st ed., McGraw-Hill, New York, © 1976. Reproduced with permission.*]

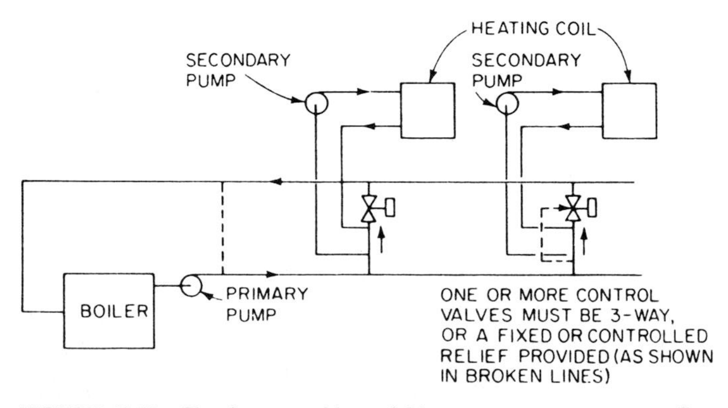

FIGURE 47.28 Circuit to provide variable temperature at constant flow rate for two or more coils. [*From Igor J. Karassik et al. (eds.),* Pump Handbook, *1st ed., McGraw-Hill, New York, © 1976. Reproduced with permission.*]

the level of the condensate in the condenser and the boiler to produce the required flow to the boiler, then a pump must be introduced to provide the required head. Since the condensate in the hot well will be at or near its saturation temperature and pressure, the only NPSH available to the pump will be the submergence less the losses in the piping between the hot well and the pump. A pump must be selected that will operate on these low values of NPSH without destructive cavitation.

In many cases, particularly for very large systems, vacuum pumps are used to remove both the condensate and the air from the condensers. This permits smaller piping for the return of condensate and air, more positive removal of condensate from condensers, and, when rather high vacuums [above 20 in (51 cm)] are possible, some control of the temperature at which the steam condenses. The use of vacuum return, particularly with higher vacuums, helps reduce the possibility of freezing heating coils exposed to outside air or exposed to stratified outside and recirculated air. Vacuum return pumps are available for handling air and water. Vacuum, condensate, and boiler-feed pumps with condensate tanks are all available in package form.

Most condensate pumps are centrifugal. Vacuum pumps may be rotary, including a rotary type with a water seal and displacement arrangement.

Where steam for heating is available at sufficient pressure to drive a direct-acting steam pump, pumping requirements can be provided at practically no energy cost. This is because direct-acting pumps utilize only the displacement energy of the steam, preserving its internal and latent thermal energy at the exhaust pressure, which energy can then be used for heating.

47.4.3 Fuel Oil

When oil burners are fairly far from the oil storage tank or when there are a number of burners at different locations in a building, a fuel-oil circulating system is required. The flow rate is relatively low [1 gal/min would provide over 8 million Btu/h (140,560 kW/h)], and a small gear pump is usually used.

47.5 COOLING SYSTEMS

47.5.1 Air-Conditioning Systems

Many air-conditioning systems produce chilled water at a central location and distribute it to air-cooling coils in various locations throughout the building or group of buildings. Centrifugal pumps are particularly well-suited for this service.

The type of circuit and the number of pumps used require an evaluation of several factors. First, the cooling requirements usually vary over a wide range.

Second, the flow rate through a chiller must be kept above the low point where freezing would be possible and below the point where tube damage would result. Some methods of chiller capacity control require a constant flow rate through the chiller.

Third, the temperature of the surface cooling the air must be low enough to control the relative humidity. This limits the use of parallel circuits through chillers when one circuit may not be in operation and permits unchilled water to mix with water of the operating chiller. Under these conditions it is difficult or impossible to attain a sufficiently low mixture temperature. It also requires greater variation of the water flow rate through the air-cooling coils, which limits the ability to control the temperature effectively.

Fourth, below-freezing air may sometimes pass over all or part of a coil. This condition would require a flow rate and water temperature adequate to keep the temperature of all water-side surfaces of tubes above freezing. Many water circuits are available to achieve the desired results. For control of the relative humidity, the air-flow circuit must also be considered. Figure 47.29 shows a circuit for cooling coils with a variable air-flow rate at constant air-leaving temperature and with two chillers in series; in addition, the two chillers are shown in parallel with a third chiller. The arrangement permits continuous flow through the coils to

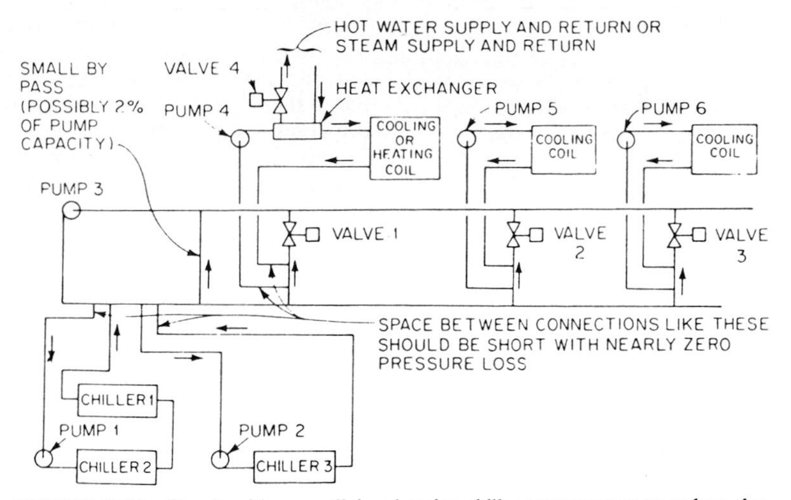

FIGURE 47.29 Circuit with a parallel-and-series chiller arrangement to reduce the possibility of freezing coils. [*From Igor J. Karassik et al. (eds.),* Pump Handbook, *1st ed., McGraw-Hill, New York, © 1976. Reproduced with permission.*]

reduce the possibility of freezing when the average temperature of the air entering the coil is above freezing but when the usual stratification results in a below-freezing temperature for some of the air entering the coil. (The word "reduce" is used because full prevention requires appropriate air-flow patterns, water velocities, and temperatures to ensure that the water side of the surface will not be below freezing at any point in the coil.) One of the coils is also arranged to add heat when the temperature of the air leaving the coil must sometimes be above that of the average air-entering temperature.

Some circuits attempt to obtain the desired results from the circuit in Fig. 47.29 with fewer pumps. However, the use of fewer pumps, although it would reduce the cost for pumps slightly, would also require three-way instead of two-way valves, would make control somewhat more complicated, and would almost certainly result in greater power consumption. The circuit shown in Fig. 47.29 permits pump heads to match the requirements exactly. It also permits stopping an individual pump when flow is not required in one of the circuits; the two-way valves 1, 2, and 3 will reduce pump circulation and the power of pump 3 at partial cooling load.

Air Separation and Removal. The methods for handling air with chilled water are about the same as those for hot water except that there is not usually a rise in temperature above that of the makeup water to produce additional separation of air. An expansion tank is required, but the reduced temperature difference, as compared to that of hot water, requires a much smaller tank size.

Condenser Water Circulation. Condenser water may be recirculated and cooled by passing through a cooling tower, or it may be pumped from a source such as a well, a lake, or the ocean.

Cooling Tower Water. Centrifugal pumps are used for the circulation of cooling tower water. The circuit is open at the tower, where the water falls or is

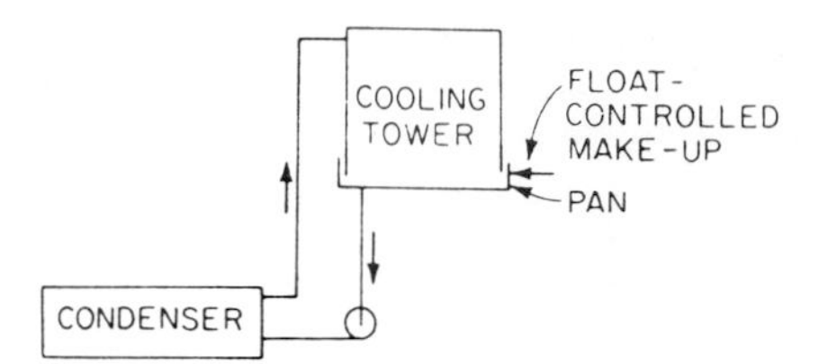

FIGURE 47.30 Cooling tower with condenser below pan water level. [*From Igor J. Karassik et al. (eds.),* Pump Handbook, *1st ed., McGraw-Hill, New York, © 1976. Reproduced with permission.*]

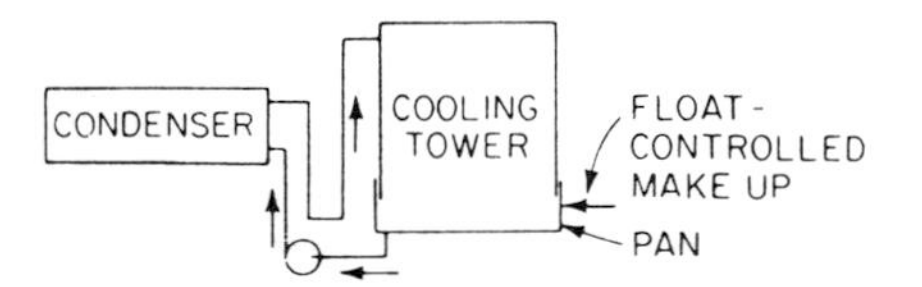

FIGURE 47.31 Cooling tower with condenser above pan water level. [*From Igor J. Karassik et al. (eds.),* Pump Handbook, *1st ed., McGraw-Hill, New York, © 1976. Reproduced with permission.*]

sprayed through the air and transfers heat to the air before the water falls to the pan at the base of the tower. A pump then circulates the water through the condenser as shown in Fig. 47.30. In this case, the pump must operate against a head equal to the resistance of the condenser and piping plus the static head required to the tower from the water level in the pan.

Figure 47.31 shows a somewhat similar circuit except that the condenser level is above the pan water level. The size of the pan of a standard cooling tower is sufficient to hold the water in the tower distribution system so that the pan will not overflow and waste water each time the pump is shut down. This capacity also ensures that the pan will have enough water to provide the required amount above the pan level immediately after startup, without waiting for the makeup that would have been needed if there was any overflow when the pump stopped. When the condenser or much of the piping is above the cooling tower pan's overflow, the amount draining when the pump is stopped would exceed the pan capacity unless means were provided to keep the condenser and lines from draining. In Fig. 47.31 it will be noted that the line from the condenser drops below the pan level before rising at the tower. This keeps the condenser from draining by making it impossible for air to enter the system. This is effective for levels of a few feet, but it is useless if the level difference approaches the barometric value. Such large level differences should be avoided, if possible, since they require special arrangements and controls.

When a cooling tower is to be used at low outside temperatures, it is necessary to avoid the circulation of any water outside unless the water temperature is well above freezing. The arrangement shown in Fig. 47.32 provides this protection. The inside tank must now provide the volume previously supplied by the pan in addition to the volume of the piping from the tower to the level of the inside tank. Condensers or piping above the new overflow level must be treated as already described and illustrated in Fig. 47.31, or additional tank volume must be provided.

The only portion of the inside tank that will be available for the water that drains down after the pump stops is that above the operating level. This operating level, or available head, is fixed by the height of liquid required to avoid cavitation or air entrainment at the inlet to the pump. The size of the pipe at the tank outlet should be determined by the velocity that can be attained from the available head, not by the pressure loss. [A good approximation is $d_p = 0.3\sqrt{\text{gal/min}/h}$, where d_p is the pipe diameter in inches (centimeters) and h is the head in feet (meters).] See Fig. 47.33.

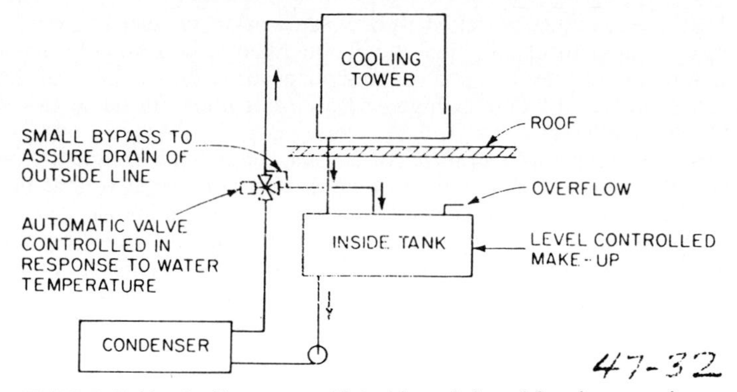

FIGURE 47.32 Cooling tower with inside tank for subfreezing operation.

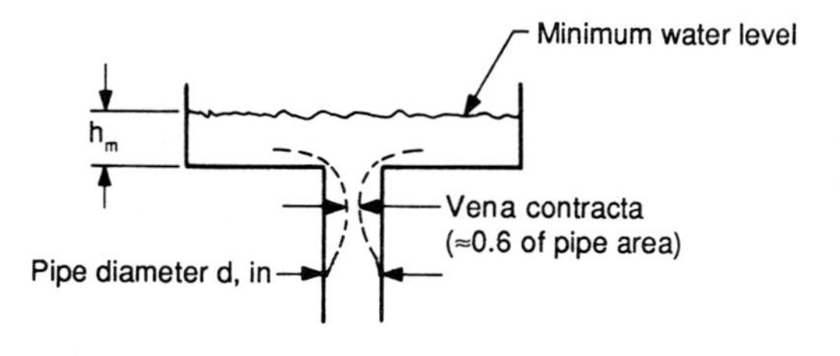

h_m = minimum height of water to prevent air entrainment (vortex) at pump section in open cooling tower pans

A = area of pipe = $1/4\ d^2/144$, ft²

A_c = area at vena contracta = $0.6\ A$, ft²

Q = required flow in pipe, gal/min

V_c = velocity at vena contracta
= flow in ft³/min/A_c

$$= \frac{Q(\text{gal/min } 0.1337 \text{ ft}^3/\text{gal})}{0.6(n/4)d^2/144}, \text{ ft/min}$$

$h_m = V_c/2g$, ft

g = 32.2, ft/s

FIGURE 47.33 Calculation of velocity at vena contracta. [*From Igor J. Karassik et al. (eds.),* Pump Handbook, *1st ed., McGraw-Hill, New York, © 1976. Reproduced with permission.*]

Well, Lake, or Ocean Water. Centrifugal pumps are used for all these services. The level from which the water is pumped is a critical factor. The level of the water in a well will be considerably lower during pumping than when the pump does not operate. In the case of pumping from a lake or from the ocean, the drawdown is usually not significant. When pumping from a pit where the water flows by gravity, there will be a drawdown that will depend on the rate of pumping. With seawater supply there will be tidal variations. A lake supply may havc seasonal level differences.

All these factors must be taken into consideration in selecting the level for mounting the pump to ensure that it will be filled with water during startup. Check or foot valves may be used for this purpose. Also, the head of the water entering the pump at the time of highest flow rates must not be so low that the NPSHR is not available.

To ensure proper pump-operating conditions, the pump is frequently mounted below the lowest level expected during zero flow conditions as well as below the lowest level expected at the greatest flow rate. These conditions may require a vertical turbine-type pump (Fig. 47.12). The motor should either be above the highest water level with a vertical shaft between the motor and the pump bowls or be of the submerged type connected directly to the pump bowls.

47.5.2 Refrigeration Systems

For refrigeration systems with temperatures near or below freezing, pumps are often required for brine or refrigerant circulation. The transfer of lubrication oil also frequently requires pumps.

Brine Circulation. The word "brine," as used in refrigeration, applies to any liquid (1) which does not freeze at the temperatures at which it will be used and (2) which transfers heat by a change solely in its temperature without a change in its physical state. As far as pumping is concerned, brine systems are very similar to systems for circulating chilled water or any liquid in a closed circuit. A centrifugal pump is the preferred choice for this service. It must be constructed of materials suitable for the temperatures and fluids encountered. For salt brines, the pump materials must be compatible with other metals in the system to avoid damage from galvanic corrosion.

Tightness is usually more important in a brine-circulating system than in a chilled-water system. This is true not only because of the higher cost of the brine but also because of problems that are caused by the entrance of minute amounts of moisture into the brine at very low temperatures.

Refrigerant Circulation. For a number of reasons, including pressure and level as well as improvement of heat transfer, refrigerant liquid may be circulated with a pump. The centrifugal pumps with mechanical shaft seals are usually preferred for this purpose.

The same liquid being pumped as a refrigerant may also serve as a brine. Whereas the fluid is all in liquid form throughout the brine circuit, it is in vapor form during part of its circulation as a refrigerant. In a refrigerant-circulating system, most of the heat transfer is by evaporation or condensation or both.

As there are changes from liquid to vapor, the liquid to be pumped must be in a saturated condition in some portion(s) of the circuit. Sufficient NPSH for the pump must be provided by the static level of liquid in the tank where the liquid is collected. The level difference required for the NPSH must provide adequate margin to compensate for any temperature rise between the tank and the pump. This is an important consideration because the liquid temperature will usually be considerably lower than that of the air surrounding the pump intake pipe.

When the pump is not operating, it may be warm and may contain much refrigerant in vapor form. It is usually necessary to provide a valved bypass from the pump discharge back to the tank so that gravity circulation can cool the pump and prevent vapor binding.

Pumps for this service may require a double seal, with the space between the seals containing circulated refrigerant oil at an appropriate pressure. This will reduce the possibility of the loss of relatively expensive refrigerant and eliminate the entrance of any air or water vapor at pressures below atmospheric. A hermetic motor may also be used for this service and thus avoid the use of seals.

Lubricating-Oil Transfer. Because the flow rates for lubricating-oil transfer are rather low, the gear pump is usually preferred. The NPSH requirement is also critical here because, although the oil itself is well below the saturation temperature at the existing pressure, it may contain liquid refrigerant in solution. Any temperature rise or pressure reduction will result in the separation of refrigerant vapor. It is important, therefore, to design the path for oil flow from the level in the tank where it is saturated with the same safeguards as are necessary for refrigerant pumping.

To reduce the oil-pumping problem, the oil can be heated to a temperature above that of the ambient air and vented to a low pressure in the refrigerant circuit. This eliminates temperature rise in the pump as well as in the suction with the corresponding reduction of NPSHA.

Usually the oil flow is intermittent, and the best results are obtained by continuous pump operation discharging to a three-way solenoid valve. The discharge would be bypassed back to the tank whenever transfer from the tank is not required. This ensures even temperature conditions and a pump free of vapor.

47.6 SELECTION

Four categories of data must be made available to the pump supplier in order to ensure optimum pump selection: (1) intended service, (2) liquid properties, (3) performance requirements, and (4) energy source (along with driver options).

Intended Service. As discussed in Secs. 47.4 and 47.5, the type of pump will generally be dictated by the service to which it is applied. In the vast majority of cases, one form or another of a centrifugal pump will be appropriate.

Liquid Properties. The thermodynamic, physical, and chemical properties of the liquid to be pumped are significant factors in determining the proper materials of construction and the design of the pump.

Performance Requirements. The design flow rate, suction, and discharge pressure are obvious requirements. Just as important, but often overlooked, are the off-design operating conditions. In most cases, system control valves alter the flow rate, the system head pressure, or both, necessitating pump operation over a significant range. Startup conditions are seldom identical to running conditions. The size, type, and hydraulic characteristic of the best pump for a particular application will be affected by this information.

Energy Source. The energy source determines the driver options and, to some extent, the pump type. Certainly for the vast majority of HVAC applications, electric power is used; thus the supply voltage, phases, and frequency, along with driver environmental exposure, are required. In those special cases where electric power is not to be used, the alternatives (such as diesel or gasoline, steam, or even wind power) will have an impact on the pump selection, including the need for speed increasers or reducers.

The horsepower (wattage) rating of the motor should be selected to equal or exceed the required horsepower of the pump throughout the expected operating range. When the operating range of the pump is not known, it is best to select a motor for the maximum horsepower shown on the pump rating curve.

The type of motor enclosure is an important consideration in the selection of motor drives for pumps. Since most pumps operate in a damp environment, it is best to select an open drip-proof or splash-proof motor. For outdoor applications, a weather-protected motor should be selected. Explosion-proof motors are required in environments where explosive gases or vapors may be present.

Internal-combustion engines are generally used for centrifugal and reciprocating pumps where electric power is not available or where standby service is required in the event of a power failure. Air-cooled gasoline or diesel engines are available for pumps requiring up to 75 hp (56 kW). For horsepowers above 75, liquid-cooled gasoline or diesel engines are required. Engines are usually rated for pump service on the severity of the duty required by the pump. The net maximum horsepower (wattage) rating of the engine should not be used for pump applications. For intermittent operation, select an engine to operate at 90 percent of the net maximum rating. For continuous pump operation, select an engine to operate at 75 percent of the net maximum rating.

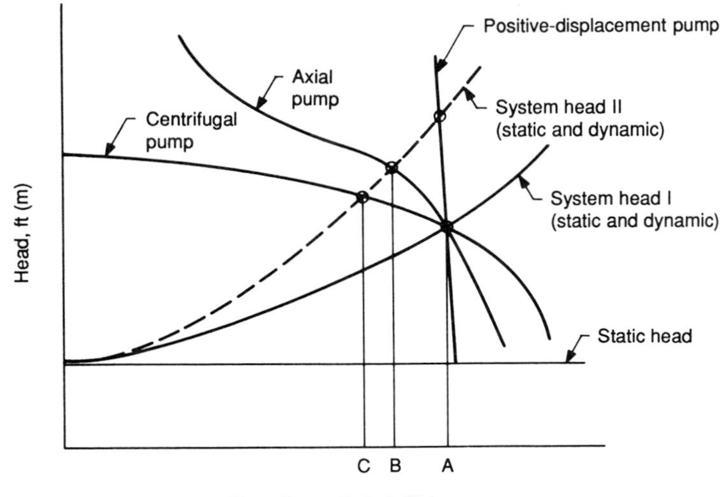

If fluid friction (dynamic contribution to system head) increases from I to II (such as from additional length of circuit, throttling of valves or reduction of pipe size(, the capacity change from initial flow A would hardly change with a positive-displacement pump, would reduce to B with an axial flow pump (high specific speed), and would reduce even further to C with a lower specific-speed centrifugal pump.

FIGURE 47.34 Interaction of system and pump head capacity characteristics.

Economic Factors. In addition to the above four factors, economic factors may bear on which pump to select. These include the first cost, the installation cost, and an evaluation of the energy and maintenance costs over the economic life of the installation. The cost of compromise should be carefully considered.

The type of pump appropriate for an application is influenced by the interaction of the system and pump head capacity characteristics, as illustrated in Fig. 47.34. The selection of a specific pump thus requires a calculation of the system head against which the pump must deliver the desired flow. Then the particular pump that meets the hydraulic performance criteria, the desired off-design point performance, and the economic evaluation standards can be selected.

EXAMPLE 47.2 Determine the total head, power requirement, and operating cost of a cooling tower pump that must meet the following criteria:

- Capacity = 100 gal/min of 60°F water
- Static rise in head = 150 ft
- Piping = 150 ft of 2-in Schedule 40 pipe with three 90° elbows and one check valve
- Static lift (the difference in level between suction and discharge) = 125 ft

Solution First, determine the pump's total head from the friction loss and the static lift:

1. *Friction loss in piping:* From Fig. 47.35 the friction loss in 100 ft of 2-in pipe at 100 gal/min is 17 ft, so the loss in 150 ft = 150/100 × 17 = 25.5 ft.

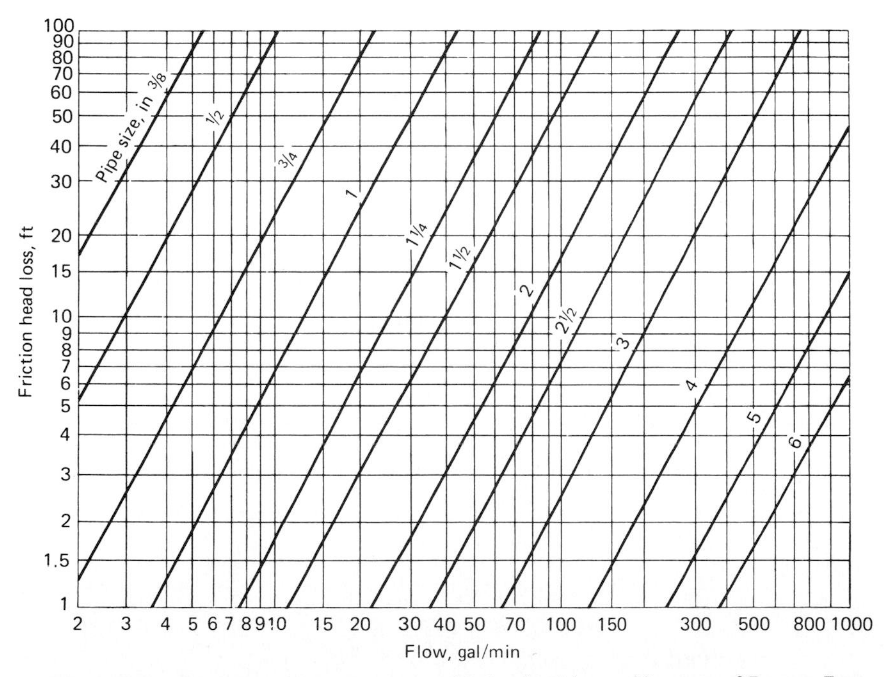

FIGURE 47.35 Friction loss for water versus friction head loss. *(Courtesy of Dresser Pump Division, Dresser Industries, Inc.)*

2. *Friction loss in elbows and check valve:*
   - From Fig. 47.36 the equivalent length of 2-in pipe for one 90° elbow is 8.5 ft, so the equivalent pipe for three elbows = 3 × 8.5 = 25.5 ft.
   - From Fig. 47.36 the equivalent length of 2-in pipe for one check valve is 19 ft.

   Thus the total equivalent length of pipe for the three elbows and one check valve = 25.5 + 19 = 44.5 ft, and the friction loss in 44.5 ft of 2-in pipe = 44.5/100 × 17 = 7.6 ft.
3. *Total friction loss:* Adding the above two, the total loss = 25.5 + 7.6 = 33.1 ft.

Therefore, the pump's total head = 33.1 ft of friction loss + 125 ft of static lift = 158.1 ft.

Second, determine the pump's power requirement. Calculate the pump's specific speed N_s from Eq. (47.3), using a maximum motor speed of 3450 rpm:

$$N_s = \frac{N\sqrt{Q}}{H^{0.75}} = \frac{\text{rpm} \times \sqrt{\text{gal/min}}}{\text{head}^{0.75}} = \frac{3450 \times \sqrt{100}}{158.1^{0.75}} = 774$$

Then note from Fig. 47.37 that the pump's ideal efficiency can be expected to be 57 percent at the best-efficiency point. Now calculate the pump's power requirement from Eq. (47.4):

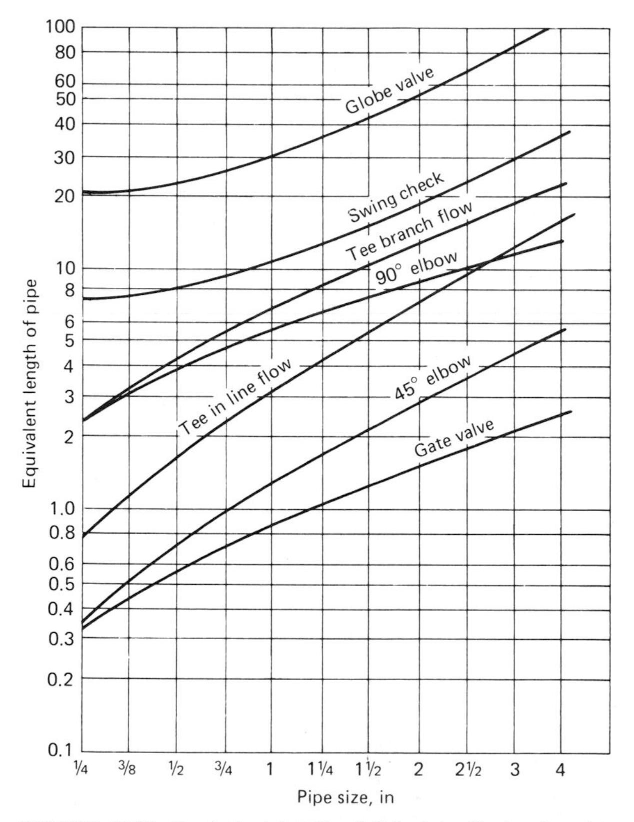

FIGURE 47.36 Equivalent length of Schedule 40 pipe for pipe fittings. (*Courtesy of Dresser Pump Division, Dresser Industries, Inc.*)

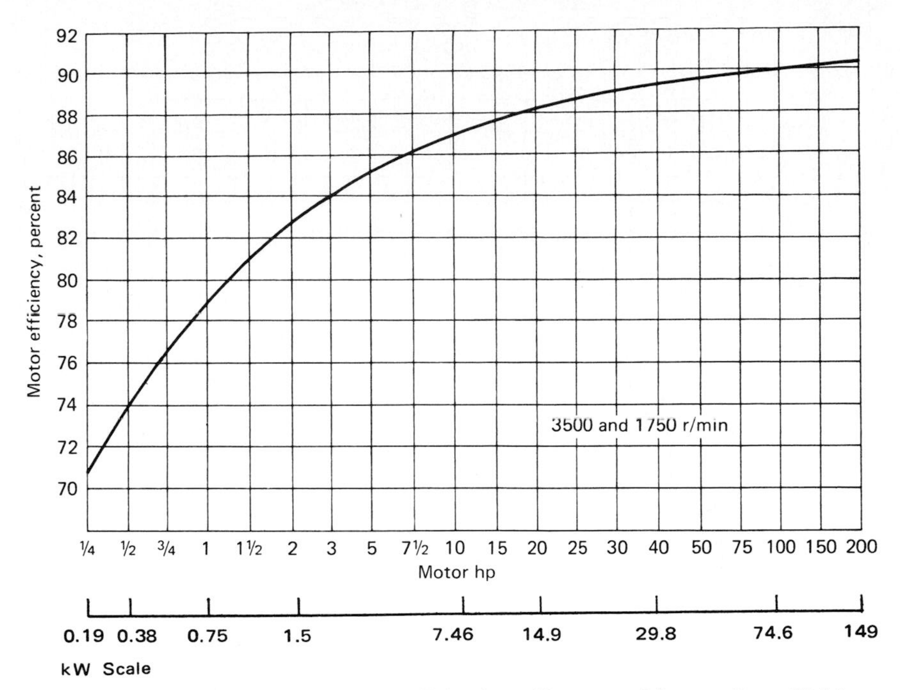

FIGURE 47.37 Typical induction motor efficiencies. (*Courtesy of Dresser Pump Division, Dresser Industries, Inc.*)

$$P = \frac{Q \times H \times S}{C \times E} = \frac{\text{gal/min} \times \text{head} \times 1.0}{3960 \times \text{efficiency}} = \frac{100 \times 158.1 \times 1.0}{3960 \times 0.57} = 7 \text{ hp}$$

Note: Figures 47.1 to 47.4 illustrate deviations of performance that should be expected for various specific-speed designs when selection is not at the ideal best-efficiency point, and Fig. 47.37 provides typical efficiencies for nominal sizes of induction motors (e.g., if a 7.5-hp motor were chosen for Example 47.2, the motor efficiency could be expected to be 86 percent).

Third, determine the cost of pump operation from Fig. 47.38:

- Gal/min × head = 100 × 158.1 = 15,810
- Combined efficiency of pump and motor = $E_p \times E_m$ = 0.58 × 0.86 = 0.499
- Pumping power = 6.5 kWh/h

Thus, if the power costs 5¢/kWh, the pumping cost = 5 × 6.5 = 32.5¢/h.

As a shortcut, Tables 47.3 to 47.6 may be used for estimating the motor horsepower (wattage) required for known flows and heads.

47.7 INSTALLATION AND OPERATION

47.7.1 Centrifugal Pumps

The best guide for the installation, operation, and maintenance of a particular pump is the manufacturer's manual. Personnel involved with these functions

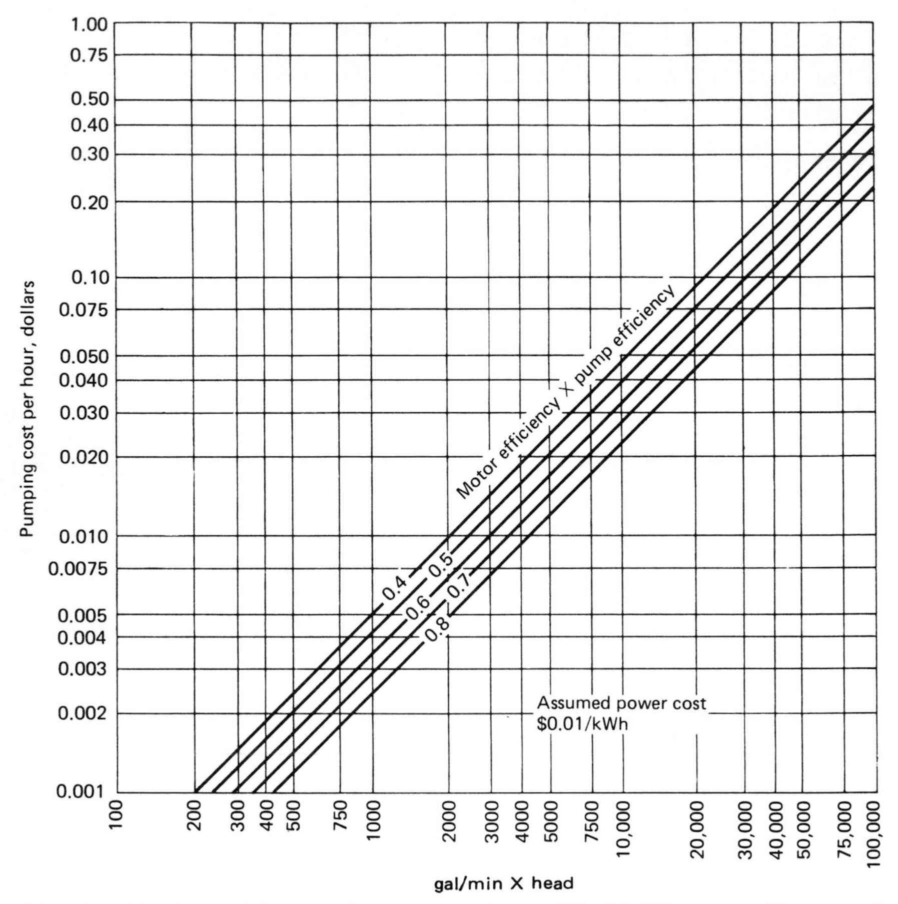

FIGURE 47.38 Determining pumping costs per hour, 60°F (15.6°C) water. (*Courtesy of Dresser Pump Division, Dresser Industries, Inc.*)

should study the information in these manuals and keep them accessible for future reference. Satisfactory performance and longevity of the pump, as well as initial warranty responsibility, often depend on following these instructions.

47.7.2 Piping

The suction piping to any pump must be carefully designed to ensure that the flow enters the pump with a minimum of rotation and swirl and with as uniform a velocity profile as practical. Short-radius elbows, tees, check valves, double elbows, and globe valves should not be used at the suction of a pump. The turbulence created by these fittings is carried directly into the pump and may result in noise and a loss in capacity and efficiency.

TABLE 47.3 Approximate Horsepower Selection Chart, End-Suction Pumps

Gallons per minute	Total head, ft																	
	20	30	40	50	60	70	80	90	100	125	150	175	200	250	300	350	400	500
10	. . .	¼	¼	½	¾	¾	¾	1	1½	2	3	5	5	5	5	15	15	20
20	¼	¼	½	¾	¾	1	1	1½	1½	2	3	5	5	5	7½	15	15	20
30	⅓	½	¾	1	1	1½	1½	1½	2	2	3	5	5	5	7½	15	15	20
40	½	¾	1	1	1	1½	1½	1½	2	3	5	5	5	7½	7½	15	15	20
50	¾	¾	1	1½	1½	1½	2	2	3	5	5	5	7½	7½	15	15	15	20
60	¾	1	1½	1½	1½	2	2	3	3	5	5	7½	7½	10	15	15	15	20
70	¾	1	1½	1½	2	2	3	3	3	5	7½	7½	7½	10	15	20	20	25
80	¾	1	1½	2	2	3	3	3	3	5	7½	7½	7½	10	15	20	20	25
90	¾	1	1½	2	2	3	3	5	5	7½	7½	7½	10	15	15	20	20	25
100	1	1	1½	2	3	3	3	5	5	7½	7½	10	10	15	20	20	20	25
120	1	1½	2	3	3	3	5	5	5	7½	7½	10	10	15	20	25	25	30
140	1	1½	2	3	3	5	5	5	7½	7½	10	10	15	20	25	25	25	30
160	1½	2	3	3	5	5	7½	7½	7½	10	10	15	15	20	25	30	30	40
180	1½	3	3	5	5	7½	7½	7½	7½	10	10	15	15	20	25	30	30	40
200	2	3	3	5	5	7½	7½	7½	7½	10	15	15	20	25	30	40	40	40
250	. . .	3	5	5	7½	7½	7½	7½	10	15	15	20	20	25	40	40	40	
300	. . .	5	5	7½	7½	7½	10	10	10	15	20	25	25	30	40	50		
350	. . .	5	5	7½	7½	7½	10	15	15	20	20	25	25	40	50	50		
400	. . .	7½	7½	7½	7½	10	15	15	15	20	20	25	30	40	50			
450	. . .	. . .	7½	7½	10	15	15	15	15	20	25	30	40	40	50			
500	. . .	. . .	. . .	10	15	15	15	15	20	20	25	30	40	50				
600	. . .	. . .	. . .	. . .	. . .	. . .	20	20	20	25	30	40	40					
700	. . .	. . .	. . .	. . .	. . .	. . .	. . .	25	. . .	40	40	50	50					
800	. . .	. . .	. . .	. . .	. . .	. . .	. . .	. . .	. . .	. . .	50	50						

Source: Dresser Pump Division, Dresser Industries, Inc.

TABLE 47.4 Approximate Horsepower Selection Chart, Double-Suction Pumps

Gallons per minute	Total head, ft																
	30	40	50	60	70	80	90	100	120	140	160	180	200	225	250	275	300
20	...	...	...	1	1	1½	1½	2	3	5	5	5	5				
40	...	1	1	1½	1½	2	2	3	5	5	5	5	7½				
60	1	1½	1½	2	2	3	3	5	5	5	7½	7½	10				
80	1½	2	2	3	3	3	5	5	7½	7½	7½	10	15				
100	2	2	2	3	3	5	5	5	7½	7½	10	10	15				
150	2	3	3	3	5	5	5	7½	7½	10	10	15	15				
200	3	3	5	5	5	7½	7½	7½	10	15	15	15	20				
250	3	3	5	5	7½	7½	10	10	15	15	15	20	20				
300	3	5	5	7½	7½	7½	10	15	15	20	20	25	25				
400	5	7½	7½	7½	10	10	15	20	20	20	25	25	30	30	40		
500	5	7½	10	10	15	15	15	20	20	25	30	30	40	50	50		
600	7½	7½	10	15	15	20	20	20	25	30	30	40	40	50	50		
800	10	10	15	15	20	20	25	25	30	40	40	50	50	60	60	75	100
1000	10	15	15	20	25	30	30	30	40	40	50	60	60	75	75	100	125
1200	15	15	20	25	25	30	40	40	40	50	60	75	75	100	100	100	125
1400	15	20	20	25	30	40	40	50	50	60	75	100	100	100	100	125	150
1600	15	25	25	30	40	40	50	50	60	75	75	100	100	100	125	150	150
1800	20	25	30	40	40	50	50	60	75	75	100	100	125	125	125	150	200
2000	30	30	40	40	40	60	60	75	75	100	100	125	125	125	150	200	200
2500	30	40	40	50	50	60	75	75	100	100	125	150	150	200	200	250	
3000	...	40	50	60	75	75	100	100	125	125	150	200	200	250	250		
3500	...	40	60	75	75	100	100	125	125	150	150	200	200				
4000	...	60	75	75	100	100	125	125	150	150	200	200					
5000	...		100	100	125	125	150	150	200	200	250						

Source: Dresser Pump Division, Dresser Industries, Inc.

TABLE 47.5 Approximate Horsepower Selection Chart, Vertical Multistage Pumps

Gallons per minute	Total Head, ft																	
	30	40	50	60	70	80	90	100	110	120	130	140	150	160	170	180	190	200
50	1	1	1	1	1½	1½	1½	1½	1½	2	2	3	3	3	3	3	5	5
75	1	1	1½	1½	2	2	3	3	3	3	3	5	5	5	5	5	5	5
100	1½	1½	2	3	3	3	3	5	5	5	5	7½	7½	7½	7½	7½	7½	7½
150	2	2	3	3	5	5	5	5	5	7½	7½	7½	7½	10	10	10	10	10
200	3	3	5	5	5	5	7½	7½	7½	10	10	10	10	15	15	15	15	15
250	3	5	5	5	7½	7½	7½	7½	7½	10	10	15	15	15	15	15	15	15
300	3	5	5	7½	7½	7½	10	10	10	15	15	15	15	20	20	20	20	20
350	5	5	7½	7½	10	10	10	15	15	15	15	20	20	20	20	20	25	25
400	5	7½	7½	10	10	15	15	15	15	20	20	20	20	20	25	25	25	25
450	5	7½	7½	10	10	15	15	15	15	20	20	20	20	25	25	25	30	30
500	5	7½	10	10	10	15	15	15	15	20	20	25	25	25	30	30	30	30
600	5	7½	10	10	15	15	20	20	20	25	25	30	30	30	40	40	40	40
700	7½	10	15	15	15	20	20	25	25	25	30	30	40	40	40	40	40	40
800	7½	10	15	15	20	20	25	25	25	30	30	40	40	40	40	40	50	50
900	10	15	15	20	20	25	25	30	30	40	40	40	40	40	50	50	50	50
1000	10	15	20	20	20	25	30	30	30	40	40	40	50	50	50	50	60	60
1100	10	15	20	20	25	30	30	40	40	40	40	50	50	50	60	60	75	75
1200	15	15	20	25	25	30	30	40	40	50	50	60	60	60	75	75	75	75
1300	15	15	20	25	30	40	40	50	50	50	50	60	60	75	75	75	100	100
1400	15	20	25	25	30	40	40	50	50	60	60	60	60	75	75	100	100	100
1500	15	20	25	30	40	40	40	50	60	60	60	75	75	75	75	100	100	100
1600	15	20	25	30	40	40	50	50	60	60	60	75	75	75	100	100	100	100
1700	20	20	25	40	40	50	50	50	60	75	75	75	75	100	100	100	100	100
1800	20	25	30	40	40	50	50	60	60	75	75	75	100	100	100	100	125	125
1900	20	25	30	40	50	50	60	60	75	75	75	100	100	100	100	125	125	125
2000	. . .	25	30	40	50	50	60	60	75	75	75	100	100	100	125	125	125	125

Source: Dresser Pump Division, Dresser Industries, Inc.

TABLE 47.6 Approximate Horsepower Selection Chart, Regenerative Pumps

Gallons per minute	Total head, ft														
	25	50	75	100	150	200	250	300	350	400	450	500	550	600	650
5	¼	½	1	1	½	3	3	3	3	3	5	10	10	15	15
10	¼	½	½	1	2	3	3	3	5	5	10	10	10	15	15
15	¼	¾	1	1½	3	3	3	5	5	10	10	10	15	15	
20	½	¾	1½	2	3	5	5	5	10	10	10	15	15	20	
25	...	1	1½	3	3	5	5	7½	10	10	15	15	15	20	
30	...	1½	1½	3	5	5	5	7½	10	10	15	15	20	20	
35	...	1½	2	3	5	5	5	7½	10	15	15	15	20		
40	...	1½	2	3	5	7½	7½	10	15	15	15	20			
50	...	...	...	...	...	7½	10	10	15	15	15	25			

Source: Dresser Pump Division, Dresser Industries, Inc.

The discharge piping should be firmly supported to minimize the load that is transferred to the pump. Excessive piping loads can distort the pump and its alignment to the driver, with resulting damage to the pump and driver bearings. In severe cases, coupling and shaft failures have resulted.

In installations where flexible pipe connections are used on the discharge side of the pump, careful consideration must be given to the fact that both the pump and the piping must be anchored rigidly to prevent movement from the piston effect of the flexible connection. Both the piping and the pump are subjected to a load equal to the cross-sectional area of the pipe times the discharge pressure of the pump in pounds per square inch (bars). In some installations where this precaution has not been considered, the resulting piston effect of a flexible pipe connection was sufficient to force the pump off its foundation.

47.7.3 Sumps

For wet- or dry-pit pumps taking suction from an open sump, the severity of the turbulence or swirl in the pump can have a marked effect on pump performance. The flow into the sump should be as straight and uniform as feasible, with as few turns or abrupt changes in cross section as possible upstream from the sump. Many examples have been reported where noise, vibration, loss of capacity, driver overload, and excessive bearing and shaft failures have been corrected by modifying the sump design. The pump manufacturer should be consulted prior to the actual construction of the sump as to the adequacy of its design for the particular pump selected for the intended service.

47.7.4 Foundations

A concrete foundation with a securely grouted pump base is the preferred arrangement for either horizontal or vertical pumps. In the case of vertical dry-pit pumps with intermediate line shafting, the cost of installation can be reduced by supporting the line shaft bearings and the driver on structural steel supports. It is essential in this type of construction not only that the structural steel supports be designed for the loads and maximum deflection furnished by the pump manufacturer, but that the natural frequency of the structure be determined prior to installation. The calculated natural frequency should be at least 100 percent higher than the operating frequency of the pump.

Some small (often fractional-horsepower) pumps are designed to be mounted and supported by the piping (they are referred to as "in-line" pumps). In these cases, special attention must be given to the pipe supports, as they must be capable of sustaining the weight of the pump and its drive motor.

47.7.5 Startup

Prior to startup on any new system, all pipe lines should be thoroughly cleaned, and it is advisable to install a temporary strainer at the pump suction to prevent any debris that may have been left in the piping from damaging the pump.

It is also important to know whether a multipump system has been designed for single-pump operation. As most centrifugal pumps have a rising-horsepower characteristic at lower discharge heads, a single pump operating on a multipump system will require more input power when operating alone than when operating with additional pumps on the line. In addition, a single pump operating on a multipump system will require more NPSH when operating alone than when operating in conjunction with additional pumps. It is always advisable, therefore, to start up a single pump with the discharge throttled to prevent operation farther out on the pump characteristic curve than that for which the system was designed.

CHAPTER 48
VALVES

Edward Di Donato
Flow Control Division, Rockwell International,
Livingston, New Jersey

48.1 INTRODUCTION

48.1.1 Valves for Heating and Cooling Systems

In the majority of piping systems, valves are seldom given the detailed attention they deserve. Valves are usually considered just another piece of plumbing. More importance is given to pumps, compressors, motors, and other more complex equipment. Yet system function depends a good deal on proper valve selection and operation since valves are one of the major controlling elements in any fluid-handling system. The control requirement may be on-off (to isolate system components); throttling (modulation of flow); reduction of fluid pressure; etc. Whatever its function, selection of the proper valve for a given service must take into consideration each operation the valve must perform and the operating conditions under which it must function.

This chapter on valving for HVAC systems is confined to the use of valves for both isolation and balancing applications and offers some considerations in valve selection.

48.1.2 Isolation Valves

The location and selection of isolation valves (or stop valves) is an important consideration in plant design. It is usually the practice to locate isolation valves in the inlet and outlet lines to all pumps, condensers, vessels, and long lengths of pipe so that they can be isolated in the event of leaks or to repair equipment.

Although gate valves, globe valves, ball valves, plug valves, and butterfly valves can be used as isolation valves, some knowledge of their operating and sealing characteristics is essential to suitable valve selection.

48.2 VALVE SEALING

In a discussion of the phenomenon of sealing in valves, it is interesting to give consideration to the mechanism of closure and how it may affect the sealing function. There are several different valve types, and one can observe significant differences between the sealing actions of the various types.

The fact that there are, indeed, widely differing kinds of valves can be attributed to the relative simplicity of the basic valve functions, which can be produced in a variety of ways. One of these functions is to serve as an intermediate piece of pipe when the valve is in the open position, permitting the contained fluid to pass through with a minimum of resistance. Conversely, in the closed position, the function is to serve as a stopper, or the equivalent of a blind flange, isolating one side of the valve from the other to prevent the flow of the contained fluid.

To judge the merits of any valve, it is usually desirable to evaluate three characteristics:

1. How well does it imitate the piece of pipe in the open position (with respect to flow and pressure loss characteristics)?
2. How well does it imitate the blind flange in the closed position?
3. How easily can it be changed from one position to the other?

48.2.1 Flow Characteristics

There are many different types of valves, and not enough space here to cover them all. The ways valves provide the shutoff and sealing function, however, are obviously limited in principle. A closure element must be moved, somehow, from a position in which its effect on flow is minimum to a position in which it completely obstructs the flow. Let's look at a few valve types and see how these movements are accomplished.

In the conduit-type gate valve (Fig. 48.1), the gate simply slides into and out of the closed position.

There is a degree of similarity between this and the action in a tapered plug valve, with the blind side of the plug sliding into and out of position (Fig. 48.2). The difference, of course, is that the sliding in this case is achieved by virtue of rotation of the plug in place, whereas in the gate valve there is straight linear motion into and out of position.

The ball valve (Fig. 48.3) is again similar, rotating as the plug does. In this case, however, the ball shape of the plug simplifies the geometry of the seal between the body and the ball and thus facilitates the use of specially designed sealing elements, as we will see a little later.

Another widely used type of gate valve has a smaller, simpler closure element (Fig. 48.4). This uses a wedging action between the gate and the body seats to provide some of the sealing force.

An obvious alternative to sliding the closure to and from its closed position is to move it directly toward and away from the body seat. This is done in globe-type valves (Fig. 48.5), and we can think of the action as replacing the sliding flange of the other types with a plug or stopper, which must be inserted into the body-seat opening. The conventional angle valves (Fig. 48.6) and the Y valves (Fig. 48.7) all use this principle.

To evaluate different types of valves, it is interesting to examine the questions asked earlier. The first was, How well does the valve imitate the piece of pipe in the open position? The sliding-closure-element types are generally superior in this respect, as they permit the flow to be straight through when the valve is in the open position. The wedge gate valve is not quite as good as the others, be-

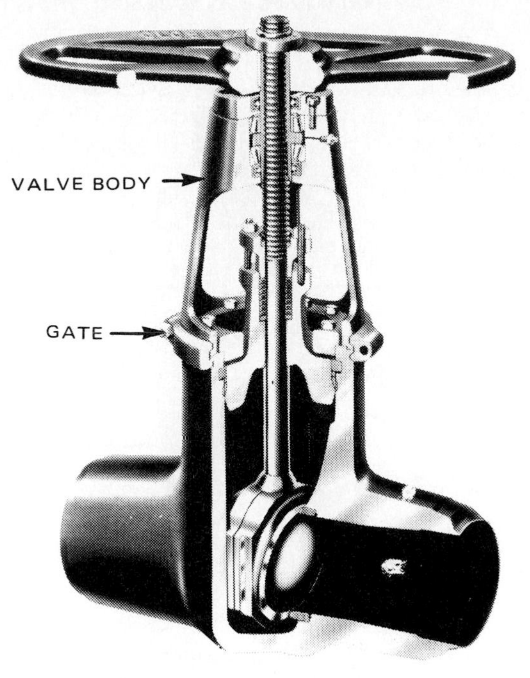

(a)

FIGURE 48.1 (*a*) Conduit-type gate valve; (*b*) standard gate valve. (*Courtesy of Nordstrom Valves, Inc., Sulphur Springs, Texas.*)

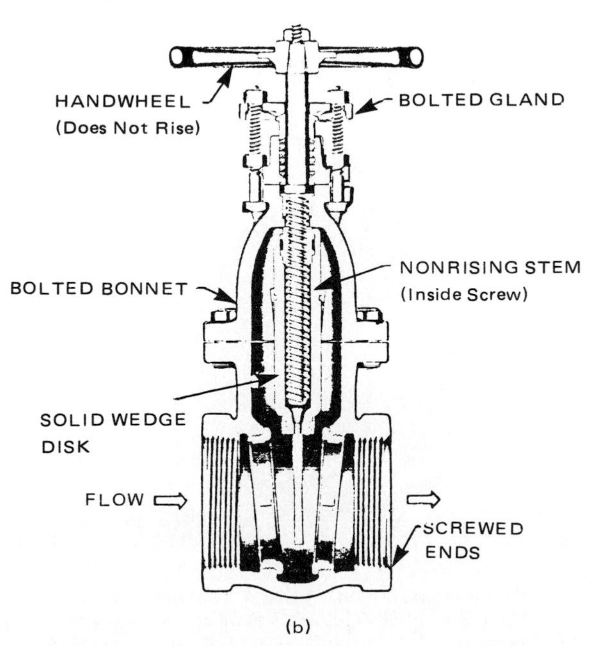

(b)

FIGURE 48.1 (*Continued*)

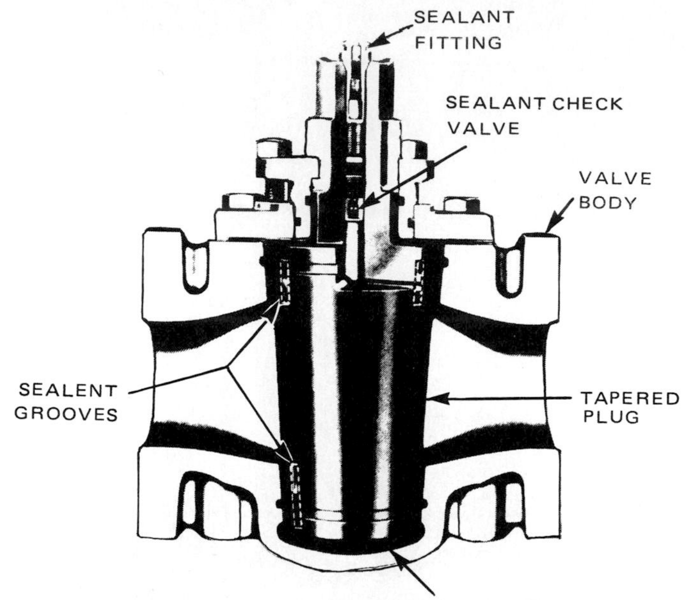

FIGURE 48.2 Tapered plug valve. (*Courtesy of Nordstrom Valves, Inc., Sulphur Springs, Texas.*)

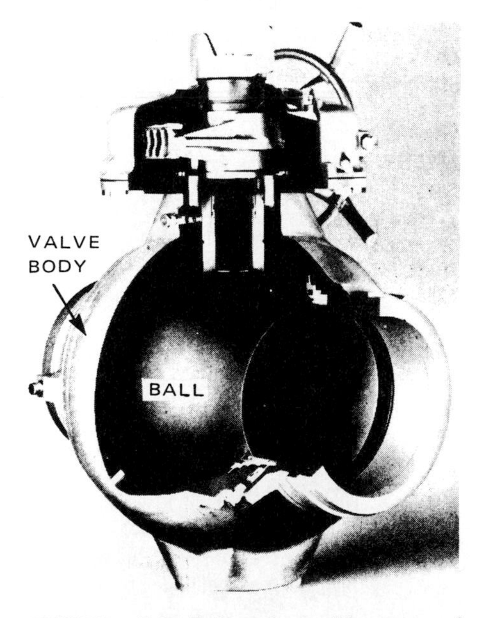

FIGURE 48.3 Ball valve. (*Courtesy of Nordstrom Valves, Inc., Sulphur Springs, Texas.*)

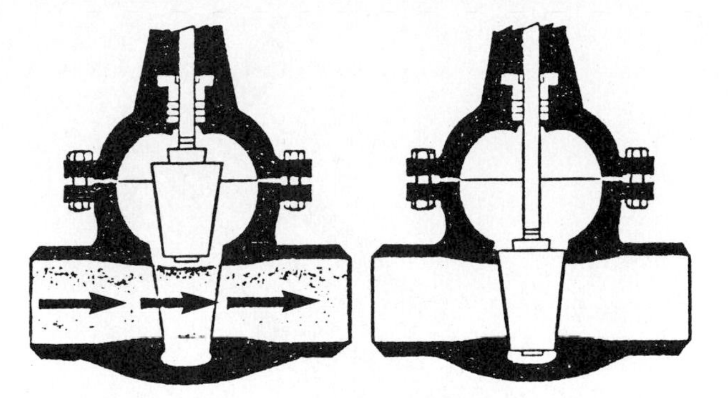

FIGURE 48.4 Gate valve, wedge-seal type. (*Courtesy of Nordstrom Valves, Inc., Sulphur Springs, Texas.*)

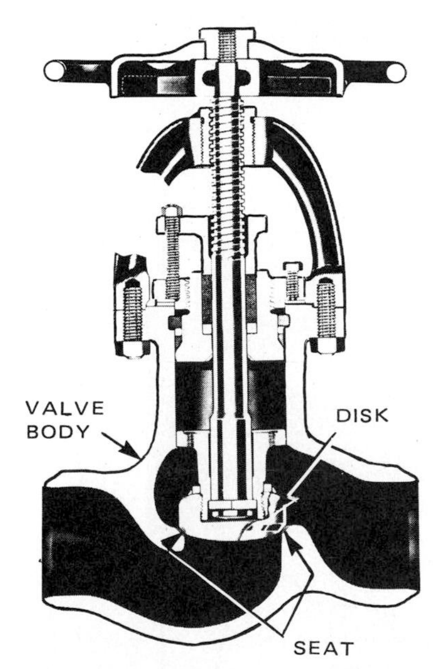

FIGURE 48.5 Globe-type valve. (*Courtesy of Nordstrom Valves, Inc., Sulphur Springs, Texas.*)

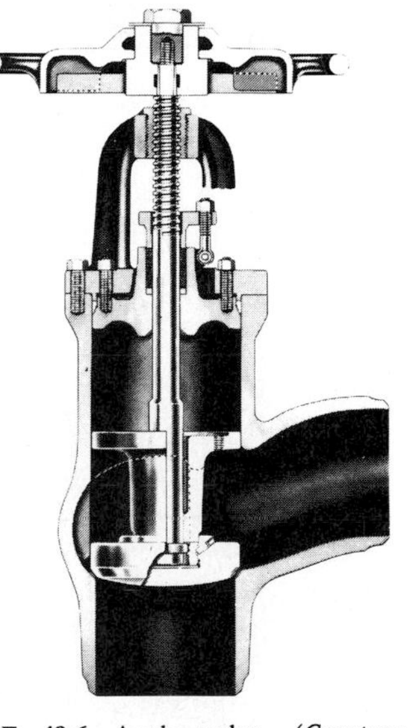

FIGURE 48.6 Angle valve. (*Courtesy of Nordstrom Valves, Inc., Sulphur Springs, Texas.*)

cause the gap between the seats, when the gate is in the open position, causes some flow turbulence that is not found in the other types.

Actually, the flow capacity of the through-conduit type of valve is somewhat better than is needed in many applications, and in such cases it can be economically attractive to use a reduced-port valve (Fig. 48.8). In gas or product transmission pipelines, however, valves are usually required to be full-port so that they will permit the passage of pipe scrapers or fluid-separating "pigs."

Conventional globe-type valves (Fig. 48.9) cannot provide a straight-through flow path for the fluid and are therefore characterized by somewhat less favor-

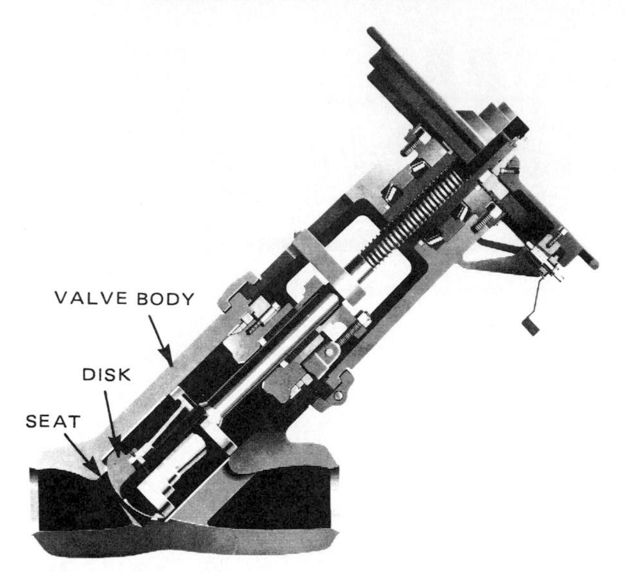

FIGURE 48.7 Y valve. (*Courtesy of Nordstrom Valves, Inc., Sulphur Springs, Texas.*)

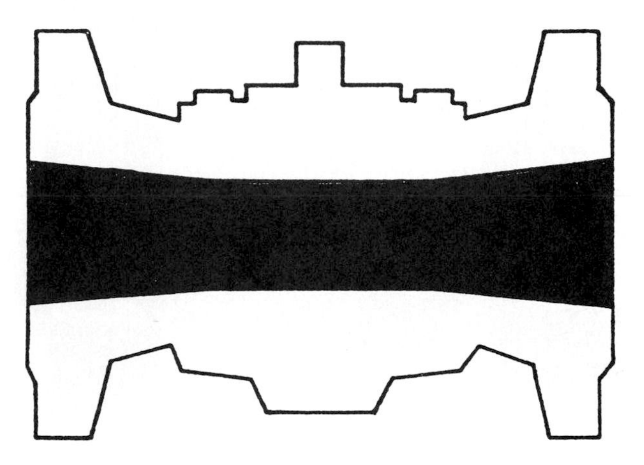

FIGURE 48.8 Reduced-port valve. (*Courtesy of Nordstrom Valves, Inc., Sulphur Springs, Texas.*)

able flow characteristics. Angle-type valves (Fig. 48.10) and Y valves (Fig. 48.11) are much better. For straight flow lines, Y valves provide the best flow capacity available in a globe-type valve; in addition, their operating characteristics are considered to be superior for some services. Where a piping arrangement requires a 90° elbow, an angle valve provides good flow capacity (Fig. 48.12) and can also save on installation costs.

Although not a through-conduit type, butterfly valves in the smaller sizes also have favorable flow characteristics (Fig. 48.13). Their pressure drops are roughly comparable to those of reduced-port plug valves.

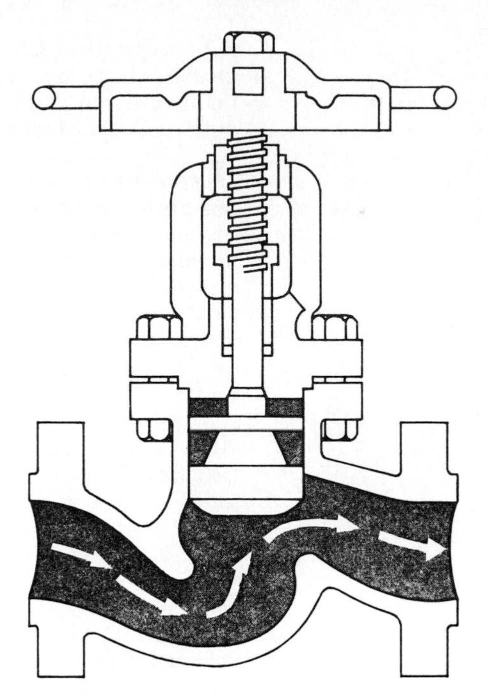

FIGURE 48.9 Conventional globe-type valve. (*Courtesy of Nordstrom Valves, Inc., Sulphur Springs, Texas.*)

Valves lacking a straight-through flow path result in a loss of energy and add to pumping power costs. However, in practical terms, the cost of pumping power must be balanced against the high capital cost of valves with very large ports. It has been found, for example, that well-streamlined plug valves with port areas 40 to 60 percent of the pipe area often provide an excellent compromise between capital cost and pumping cost.

Low-pressure butterfly valves use a flat, circular disk mounted on a shaft in the plane of the disk, and the disk rotates within the body by means of a 90° rotation of the shaft (Fig. 48.14*a*).

High-pressure butterfly valves (Fig. 48.14*b* and *c*), one of the newest types of valve available, have the circular disk plane offset from shaft so that an uninterrupted surface is available on the periphery for engagement with the body seat. Opening and closing is still accomplished by a 90° rotation of the shaft.

48.2.2 Closure Tightness

The second basic consideration is, How well does the valve imitate the blind flange? Our presumption, of course, is that with the blind flange we have assurance of a reliable closure with no leakage.

In valves, of course, we cannot always assume that there will be no leakage. Even in cases where the valve has no apparent defects, leakage can result from the presence of foreign material, such as sand or metallic corrosion products, be-

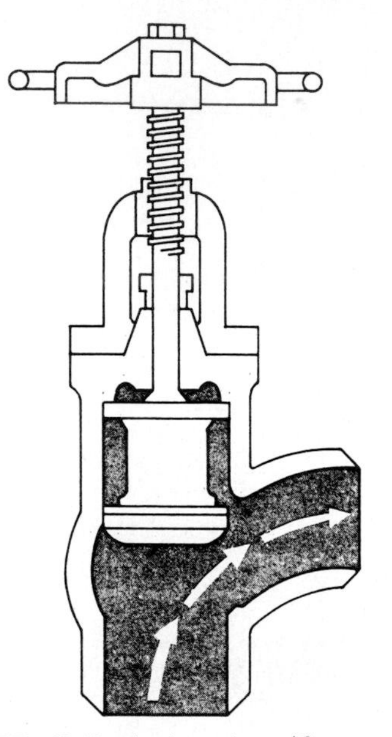

FIGURE 48.10 Angle valve. (*Courtesy of Nordstrom Valves, Inc., Sulphur Springs, Texas.*)

ing caught between the sealing elements. There are, in fact, three distinctly different kinds of sealing action utilized in industrial valves.

Metal to Metal. One is the most obvious arrangement of a metallic surface on one closure element contacting a metallic surface on the other closure element. For tight sealing, it is necessary for the surfaces to be held in intimate contact over the entire periphery of the seal. Any discontinuities in this contact, very small or even microscopic, will constitute leakage paths. Also, any imperfections in geometry that would prevent the required continuity of contact will result in leakage. Good metal-to-metal sealing, therefore, is best achieved by providing very smooth surfaces very accurately finished and by providing a relatively large mechanical force pushing the sealing elements against each other.

Sealant Material. Recognizing that all surfaces have some degree of imperfection and that even near-perfect surfaces are susceptible to damage, provision is made in some valve types for the introduction of a special sealant material into

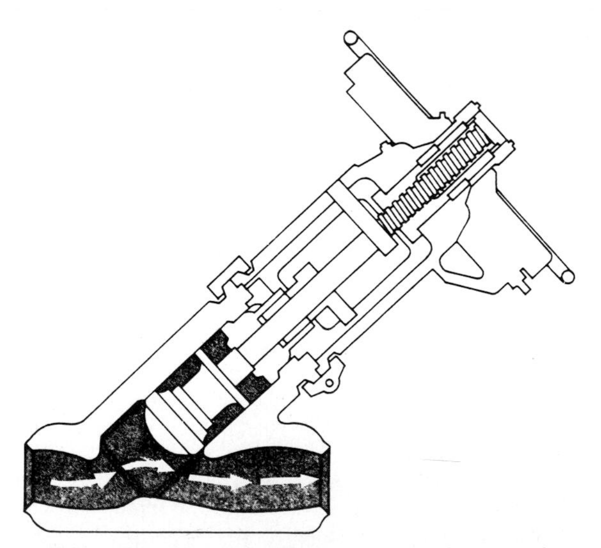

FIGURE 48.11 Y valve. (*Courtesy of Nordstrom Valves, Inc., Sulphur Springs, Texas.*)

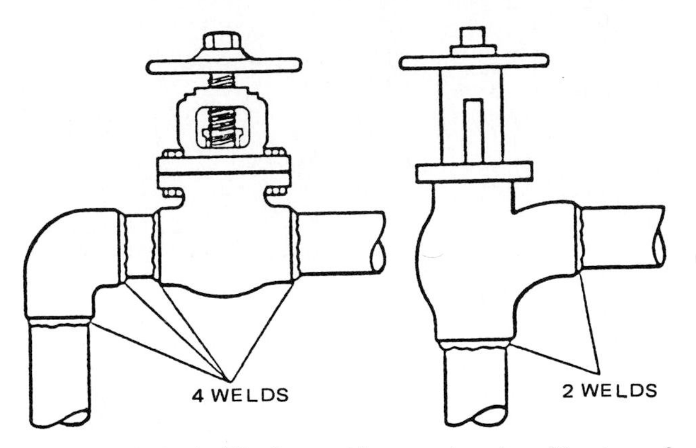

FIGURE 48.12 A 90° elbow with an angle valve. (*Courtesy of Nordstrom Valves, Inc., Sulphur Springs, Texas.*)

the sealing contact area (Fig. 48.15). This sealant reduces, and in most cases eliminates, leakage by filling the scratches and microscopic crevices in the sealing interface. In valve constructions using a sliding motion of the closure element relative to the seat, this sealant material provides helpful additional effects of lubrication of the sliding surfaces to prevent wear or damage, and it protects those important surfaces against corrosion.

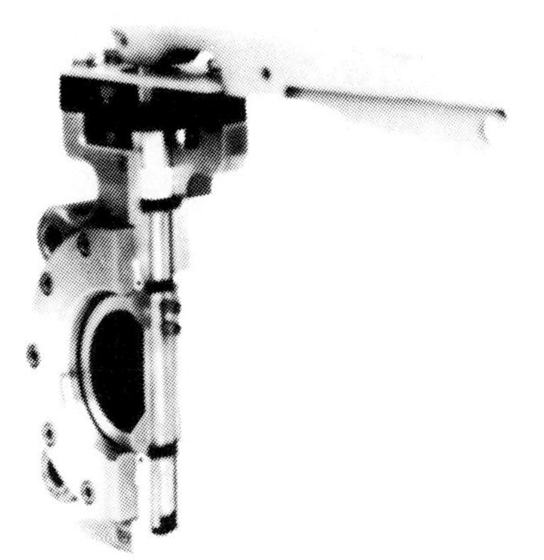

FIGURE 48.13 Butterfly valve. (*Courtesy of Nordstrom Valves, Inc., Sulphur Springs, Texas.*)

Nonmetallic Seal. This method of improving upon the reliability of tight valve sealing is the provision of a tough, resilient, nonmetallic sealant material in the seat contact area. In all such designs the basic objective is simply to provide a material that is much more deformable than the metal seat sealing face materials, with the intention that the deformable material will be squeezed into the scratches and microscopic crevices in the contacting materials and will thus prevent leakage.

Application. How well, then, do the various types of valves shut off flow in the closed position? Actually, because of the many variables that affect the sealing capability, it is not possible to evaluate any one valve design as "best" in all applications. Certain observations can be made, however, and these will help explain (in at least some cases) why some types of valves are predominantly used in some particular services and not in others.

Metal-to-metal seals without sealant or other nonmetallic backup are widely used in globe-type valves (Fig. 48.16). The lifting and reseating movement of the disks in these valves avoids a sliding contact between the sealing surfaces that

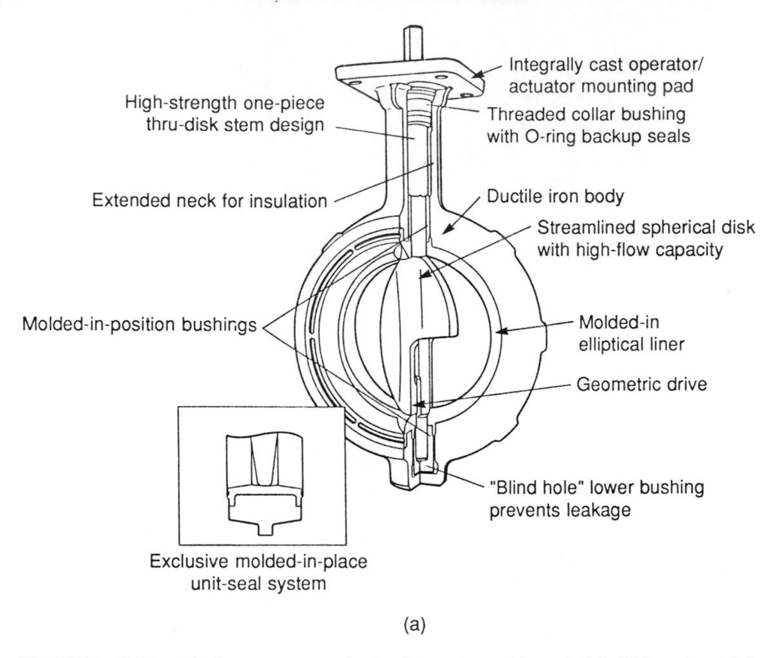

FIGURE 48.14 (*a*) Low-pressure butterfly valve. (*b*) and (*c*) Off-center high-pressure butterfly valves. (*Courtesy of Nordstrom Valves, Inc., Sulphur Springs, Texas.*)

can cause scratching and wear. The mechanism for providing this motion is also capable of applying a strong seating force to the closure element, enhancing the intimacy of mutual contact between the sealing surfaces that minimizes the leakage.

It will also be apparent that with this construction, using appropriate metals in the body, disk, and sealing surfaces, satisfactory valve performance can be anticipated over a wide range of temperatures, such as up to 1112°F (600°C) and higher.

The tapered plug valve is the most universally used example of a valve having metal-to-metal sealing enhanced by the application of a specially compounded sealant between the metal-to-metal sealing surfaces. The plug valve, in contrast to the globe-type valve, does not avoid sliding metal-to-metal contact. In fact, it maximizes this contact by having virtually the entire outside surface of the plug in contact with the corresponding internal surfaces of the body. This extensive surface contact has the desirable effect of spreading the contact load over a large bearing area, thus minimizing the contact stress. With the contact stress kept relatively low, and with the presence of sealant between the surfaces at essentially all times, the likelihood of damage to those surfaces is minimized and there is good assurance of completely tight sealing throughout the life of the valve.

The operating principle of the tapered plug valve with an integral sealant system has been applied to conduit-type gate valves (see Fig. 48.1) in small to medium sizes and medium to very high pressures. The sliding contact of the closure element (gate) against the body seal, characteristic of most conduit-type gate valves, is mitigated very usefully by the large contact area provided by the ex-

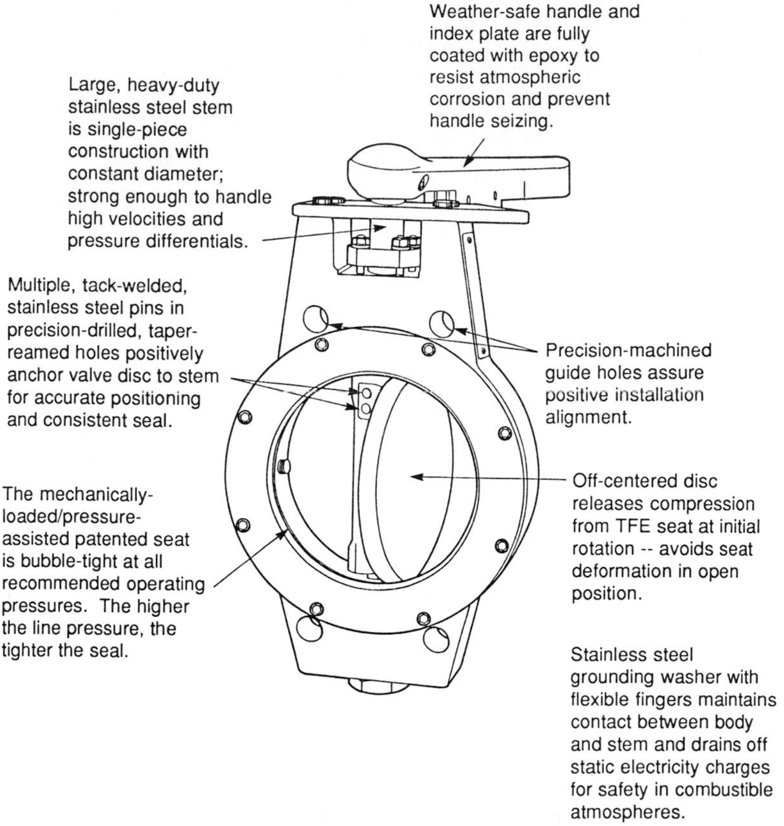

FIGURE 48.14 (*Continued*)

tended seat surfaces and, as in the case of the plug valve, by the constant presence of sealant between these parts. These mutually enhancing features—surfaces protected against wear and damage by the presence of sealant, and sealant to provide maximum probability of tight sealing—give this type of valve superior performance in many applications.

The use of a solid, nonmetallic sealant material is also helpful in improving the tight-shutoff reliability of valves. The problem of holding the sealant material in place must be faced; this is complicated, of course, by the fact that nonmetallic materials are generally quite inferior to metals in terms of mechanical strength. An example of what can happen as a result of improper application is shown in Fig. 48.17. Here we see sequentially how a simple seal can fail if improperly applied. With the elastomer sealing against a gap that opens or closes on its downstream side, the pressure can cause failure of the seal during operation of the valve. Now simply by changing the location of the elastomer so that the opening or closing gap is on the upstream side, as in Fig. 48.18, we can eliminate this problem.

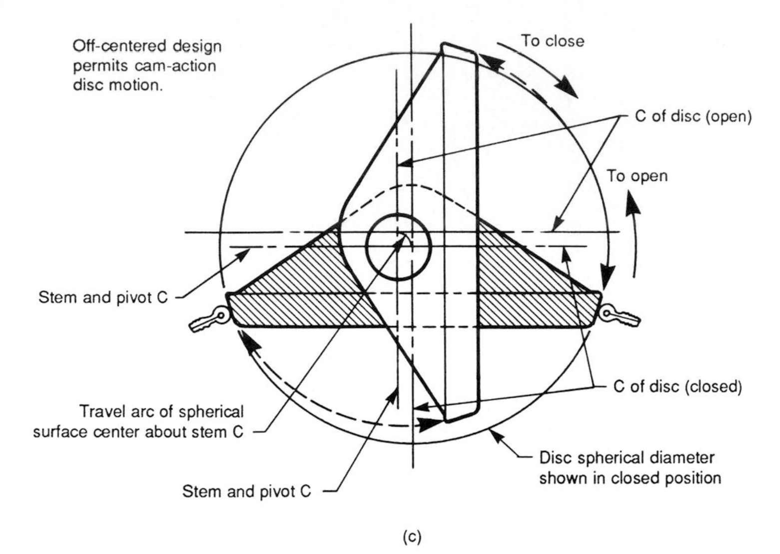

FIGURE 48.14 (*Continued*)

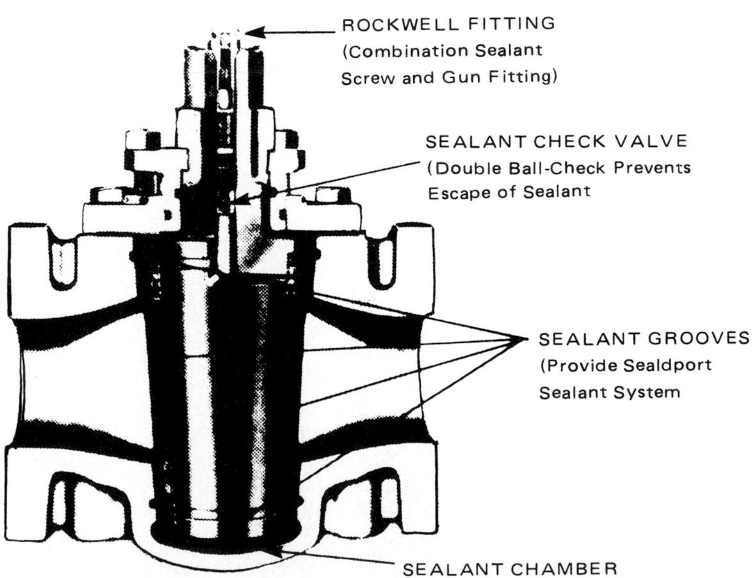

FIGURE 48.15 Sealant material used in contact area. *(Courtesy of Nordstrom Valves, Inc., Sulphur Springs, Texas.)*

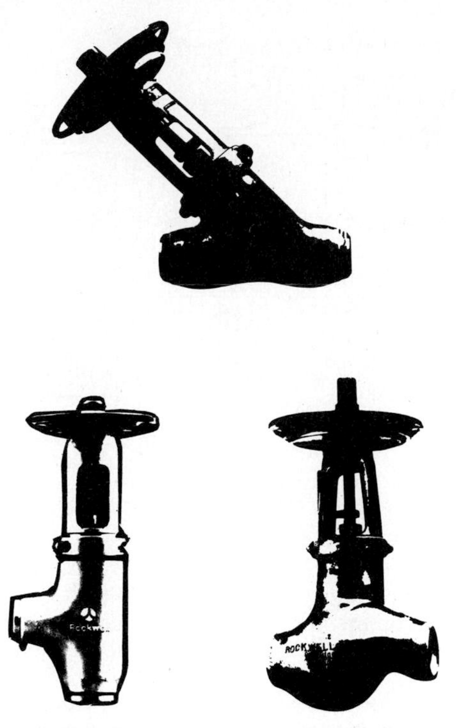

FIGURE 48.16 Metal-to-metal seals. (*Courtesy of Nordstrom Valves, Inc., Sulphur Springs, Texas.*)

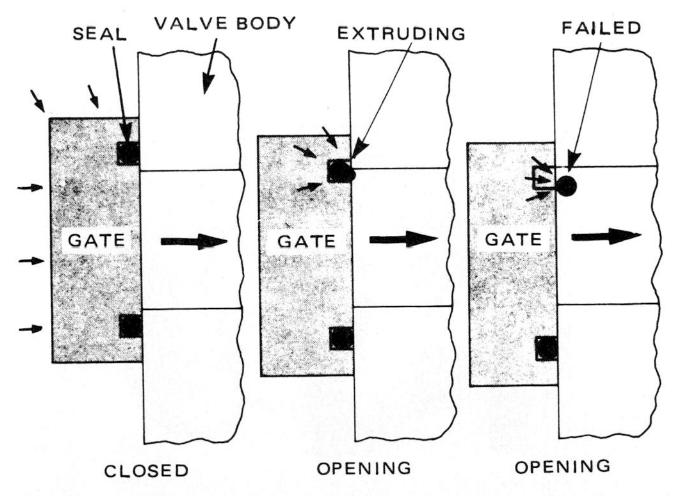

FIGURE 48.17 Improper application of sealant. (*Courtesy of Nordstrom Valves, Inc., Sulphur Springs, Texas.*)

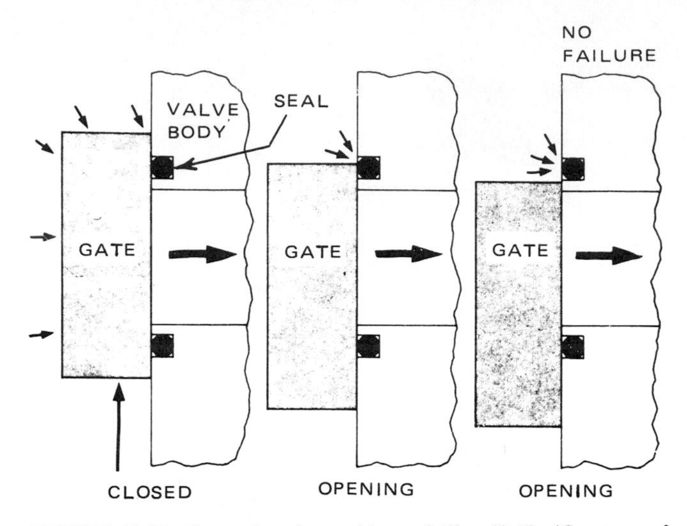

FIGURE 48.18 Correcting the problem of Fig. 48.17. (*Courtesy of Nordstrom Valves, Inc., Sulphur Springs, Texas.*)

The application of this principle is shown in Fig. 48.19. Since the valve design provides the shutoff seal on the upstream seat, the sealing action is correct, and the elastomer can function satisfactorily against openings and closings that are under very high pressure.

The need for large full-port conduit-type valves for transmission pipelines has stimulated the development of trunnion-mounted ball valves (see Fig. 48.3) with combination sealing methods. In the design illustrated in Fig. 48.20, we see the use of all the sealing methods discussed above. The basic relationship of the seat ring to the sphere is intimate metal-to-metal contact between accurately machined and highly finished surfaces. A tough, resilient nylon ring is securely retained in a groove in the seat ring, providing a positive local seal point at the line

FIGURE 48.19 Cylindrical plug valve with elastomeric seal. (*Courtesy of Nordstrom Valves, Inc., Sulphur Springs, Texas.*)

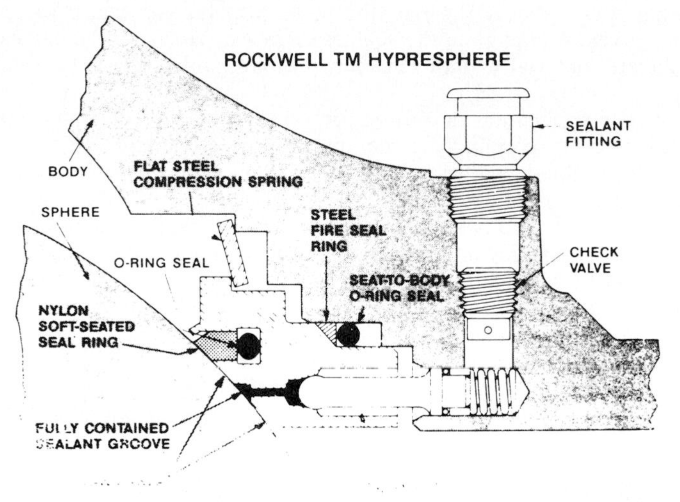

FIGURE 48.20 Sealing methods. *(Courtesy of Nordstrom Valves, Inc., Sulphur Springs, Texas.)*

of contact of the nylon ring with the surface of the sphere. And finally, means are provided for distributing a sealant material to the surface of the sphere through a groove in the seat ring. Conduit gate valves are also offered for large-diameter pipeline service and are usually provided with elastomeric seals (Fig. 48.21). Wedge-type gate valves are not generally offered for large pipeline service, but rather for general service applications, including use at elevated temperatures. Consequently, wedge-type gate valves usually have a simple metal-to-metal seating, with no elastomers or provisions for injecting sealant.

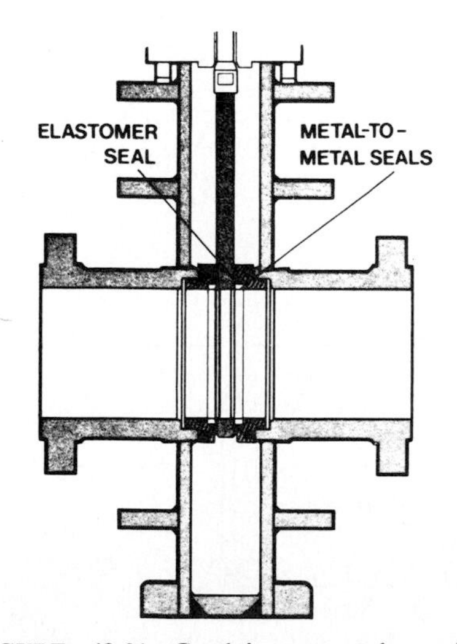

FIGURE 48.21 Conduit gate valve with elastomeric seal. (*Courtesy of Nordstrom Valves, Inc., Sulphur Springs, Texas.*)

48.2.3 Ease of Operation

The third basic consideration is: How easily can the valve be changed from one position to the other? As will be apparent, there can be a relationship between the relative difficulty of operation and the reliability of tight valve sealing.

In all metal-seated valves, for example, our primary concerns are the condition of the sealing surface and the amount of force with which the sealing surfaces are pressed together. In the globe-type valves (including angle, and Y designs), the absence of sliding action in their operation results in good preservation of the highly finished surfaces, and the large force provided gives good sealing performance. Perhaps the most troublesome cause of leakage in globe-type valves is damage to the sealing surfaces, which can be caused by foreign particulate material caught between the disk and body when the valve is closed.

The need for application of a large force in the closure of globe-type valves necessitates a correspondingly large mechanical advantage in the operating mechanism for manual operation. Globe-type valves are usually comparatively slow and difficult to operate.

Gate valves are generally more susceptible to mechanical damage of the sealing surfaces in the opening and closing operations. They are somewhat easier to operate than globe-type valves because the stem is not required to sustain the entire fluid pressure load on the disk, as is the case with globe-type valves. Instead, the fluid pressure load is sustained on gate guides and, ultimately, on the seat. The stem, then, is required only to overcome the sliding friction that results from the movement of the gate sliding on the guides or on the seat. It is apparent, of course, that such sliding under heavy load involves a considerable risk of damage to the sealing surfaces. There is therefore a direct trade-off of valve characteristics. The gate valve is easier to operate than the globe valve, but it is more susceptible to damage during operation.

In conduit-type gate valves, it can be further noted that the fluid pressure load is carried entirely on the seating surfaces, since there is no separation of these surfaces as provided in a wedge gate valve (see Fig. 48.21). Now the valve size and the amount of pressure become particularly significant, because with increasing size and increasing pressure the magnitude of the fluid pressure load increases very rapidly (Fig. 48.22). It is reasonable to expect, therefore, that as the sizes and pressures increase, gate-type valves will have increasingly serious problems related to the sliding of the gate across the valve seat.

Tapered plug valves (see Fig. 48.2) also operate by a sliding movement of the closure element (plug) across the seating surface. As mentioned earlier, the contact area in this case is quite large relative to the effective flow-port area that determines the fluid pressure load. And the presence of the plug-valve sealant between the surfaces serves as protection against wear and damage.

In the case of plug valves, the closure sliding motion is comparable to that in gate valves, but it involves rather considerably less distance because the motion is across the short dimension of a narrow opening.

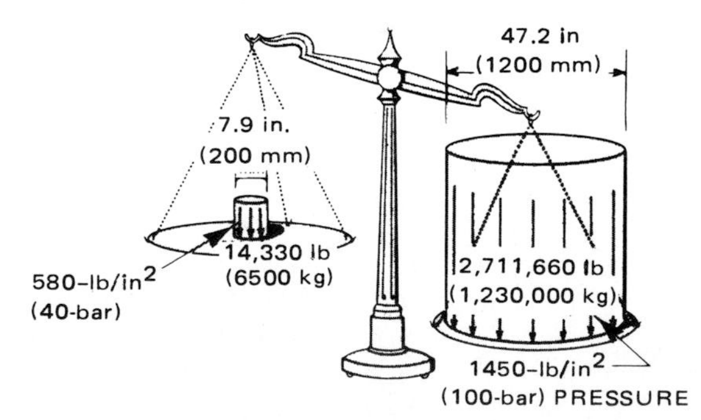

FIGURE 48.22 Fluid pressure load. (*Courtesy of Nordstrom Valves, Inc., Sulphur Springs, Texas.*)

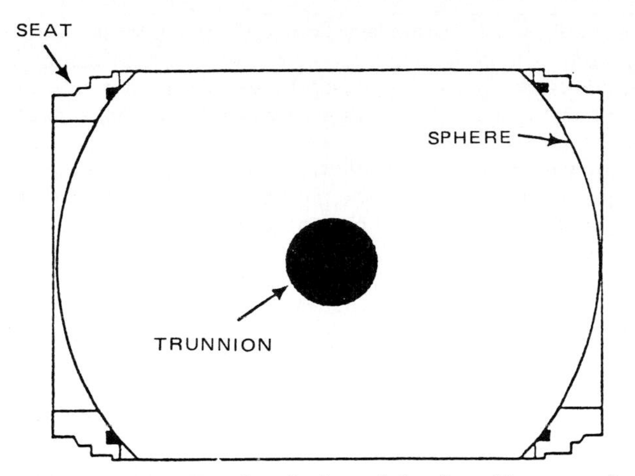

FIGURE 48.23 Trunnion in journal bearing. (*Courtesy of Nordstrom Valves, Inc., Sulphur Springs, Texas.*)

The operating effort required for ball valves is strongly influenced by a design feature that greatly reduces the force acting between the seat and the sphere. In large ball valves, the ball is pivotally supported on trunnions, which carry the fluid pressure load imposed on the ball. The trunnions are carried in low-friction journal bearings, and since the radius of these bearings (Fig. 48.23) is much less than the distance between the center of ball rotation and the seating surface, the sliding distance, and therefore the total energy required to move the valve between the open and closed positions, is considerably less than it is for any of the other valve types discussed.

There must still be contact, and therefore friction, at the seat-sphere contact area. The contact force, however, is controllable by the designer, who may thus strive for the optimum: the minimum contact force that produces good sealing between the seat and the sphere.

48.2.4 Valve Integrity in Case of Fire

One additional consideration is of basic interest in the process of evaluating the performance of the various types of valve seals. This is the matter of valve integrity when exposed to elevated temperatures resulting from a fire. Since the valve may itself be installed in a line carrying combustible fluid, a gross failure of the valve could literally add fuel to the fire. On the other hand, a valve capable of surviving such exposure could afford effective protection against such a development.

It is reasonable to assume that exposure to fire will destroy any nonmetallic sealing materials other than asbestos-containing packing, as is used in stem seals. An important question, then, is, How will the function of the valve be affected by the loss of the elastomeric seal rings or sealants used in the valve? The answers will vary with valve types and designs.

Globe-type valves will remain functional, for they do not rely on heat-sensitive materials.

Wedge gate valves will remain functional. If fire safeness is important, however, consideration should be given to the possibility of pressure buildup in the body of a closed valve (Fig. 48.24), since the double seating of this type of valve (as well as of some types of conduit gate valves) can trap fluids in the center section.

Tapered plug valves are highly resistant to fire damage, since the loss of sealant due to high temperature leaves the plug and body intact. Injection of sealant subsequent to cooling has restored fire-damaged valves to useful serviceability on many occasions.

Conduit gate valves may remain substantially functional, depending on individual design details. In general, a closed valve under differential pressure will provide good downstream protection because a loss of nonmetallic sealing materials will simply permit the gate to move slightly downstream and seat against the metal portion of the seat ring. The sealant-sealed conduit-type gate valve mentioned earlier provides excellent fire safety, in that sealant lost by reason of excessive heat can be replaced by injection after cooling.

Ball valves vary considerably in terms of their ability to function during or following exposure to fire. The construction described earlier has good fire safety because of its primary reliance on metal-to-metal ball-to-seat sealing. Some designs place primary reliance on the nonmetallic seat insert and would be expected to leak excessively if the insert were destroyed.

A specific point of concern in trunnion-mounted ball valves is the seal between the seat ring and the body. Since the ball does not press downstream against the seat as the gate in a gate valve does, loss of the elastomeric seal be-

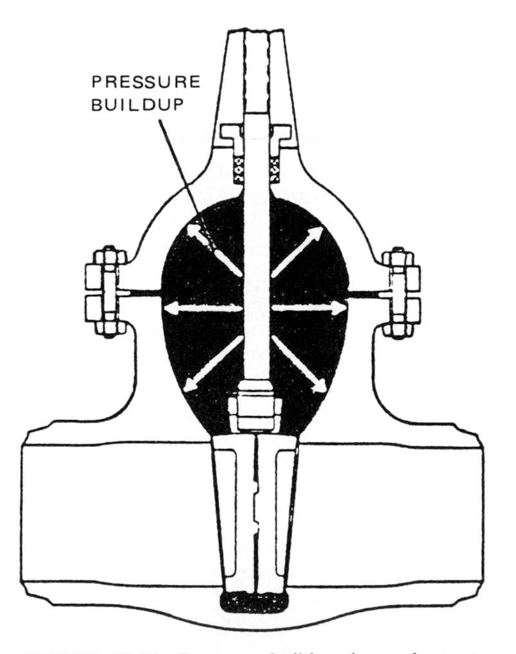

FIGURE 48.24 Pressure buildup in wedge gate valve. (*Courtesy of Nordstrom Valves, Inc., Sulphur Springs, Texas.*)

tween the seat and body can also constitute a serious failure in such a valve. In the design shown in Fig. 48.20, however, the use of a tapered metallic backup ring behind the O-ring provides effective shutoff even when the O-ring is destroyed.

The cylindrical plug valve described earlier (see Fig. 48.19) depends on the elastomer for its sealing function and is not considered to be a firesafe construction.

48.3 Isolation Valves and Balancing Valves

Isolation valves spend almost all of their service lives either open or closed, and they experience part-open operation only in those brief periods when they are moved from one position to the other. However, balancing valves are destined to spend most of their service lives partially open, throttled so as to limit the flow in a branch of a system to that which is needed for heating and cooling functions. When performing these duties, valves must imitate variable orifices instead of the straight pipe and blind flanges that isolation valves are called upon to mimic.

Ideally, each branch of a system might be sized with the proper flow capacity to suit its functional needs. Practically, most branches are oversized to various degrees to accommodate increments in standard pipe sizes and other available equipment. Sometimes systems must be sized to accommodate future increases in demand. Balancing valves are required in such systems to restore order by introducing the flow resistance that is required to approximate the ideal system. The balancing valves have a function partly comparable to that of modulating control valves; they must dissipate energy. Unlike modulating control valves, which must change position frequently to cope with the changes in system demand, balancing valves are often set to the required position and rarely adjusted after proper balance between system branches is achieved. Sometimes seasonal adjustments or changes to accommodate local variations in heating or cooling loads are necessary, but the objective in many cases is to "set it and forget it."

While the throttling duties of some balancing valves are mild, it must be recognized that energy dissipation is achieved by closing valves partially so as to produce higher-than-usual local flow velocities through orificelike restrictions. In mild cases, there may not be problems for years; in severe cases, there is a potential for dislodging or tearing away soft-seat inserts, for erosion of metal, and for noise damage problems associated with cavitation. Thus balancing-valve applications must be viewed as at least potentially more problematic than isolation-valve applications.

It may be noted that some of the potential problem areas for balancing valves may not be too serious as long as the valve is used only for throttling. Loss of a soft-seat insert or minor metal damage from erosion or cavitation might have little effect on system flow. However, if the same valve should be required to be closed and serve as an isolation valve at some later time, the damage produced by throttling might have a profound effect on tightness. Many prudent engineers include two valves in series in such cases—one for isolation and one for throttling. Where economic considerations make this a problem, valves must be selected very carefully to assure satisfaction of both required functions: isolation and balancing.

While any valve suitable for isolation can also be used for throttling or balancing just by adjusting it to a position intermediate between open and closed, proper selection of valves involves the consideration of several factors. One of these is the flow characteristic of the valve, which describes the relationship of the valve flow coefficient with the amount or percentage of valve opening. An improper flow characteristic may cause an oversensitivity of the system or branch flow to small variations in valve opening. A linear relationhip between system flow and valve position is often desirable for easy adjustment; in control-valve terms, this implies that the goal is a "linear installed characteristic." Since this relationship is highly dependent on the amount of fixed resistance that exists in system piping and other equipment in series with a balancing valve, the inherent characteristic of the valve should be such that a linear installed characteristic may be approximated.

Figure 48.25 illustrates how a linear installed characteristic may be best approximated by using a valve with an inherent characteristic that is often referred to in control-valve terminology as "equal percentage." Most gate and globe valves have inherent characteristics that are between linear and quick opening. The use of such valves in systems with much fixed resistance produces an installed characteristic that would result in little system-flow change between about 30 to 100 percent opening, so a very sensitive adjustment between closed and 30 percent open would be necessary to balance the system or branch flow. Fortunately, simple quarter-turn ball, plug, and butterfly valves produce inherent characteristics comparable to complex equal-percentage control valves. Properly

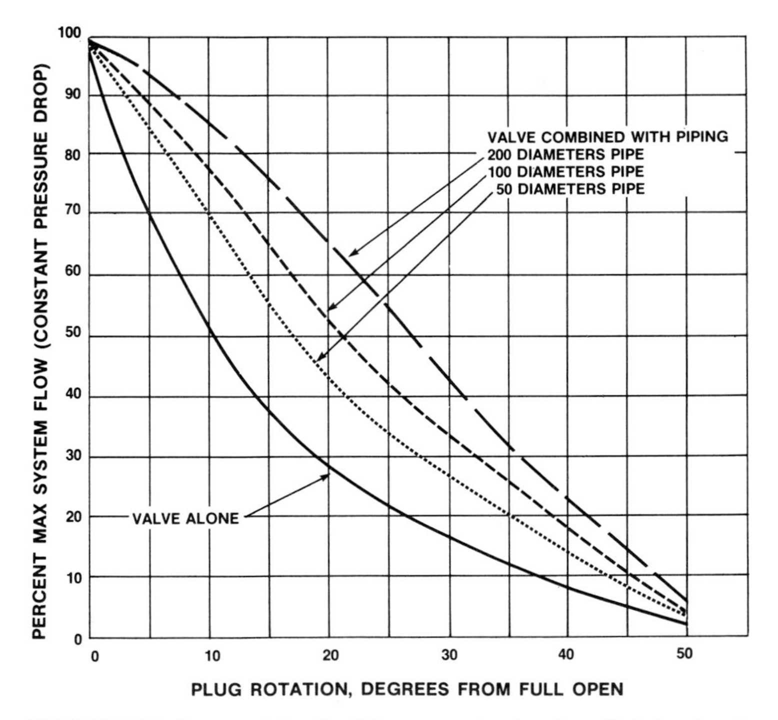

FIGURE 48.25 Representative throttling curves showing pipe effects for sizes 1 to 6 Rockwell Nordstrom plug valves. (*Courtesy of Nordstrom Valves, Inc., Sulphur Springs, Texas.*)

sized, such quarter-turn valves can provide excellent adjustability in balancing applications.

In present practice, the decision on balancing applications often rests on deciding between a plug valve and a butterfly valve. Usually, both provide inherent flow characteristics that will provide satisfactory adjustment in system or branch flow comparable to that shown in Fig. 48.25. There are no simple rules of thumb to guide this decision, but the following guidelines should be considered. (Valve manufacturers should be consulted regarding the specific performance characteristics of specific valve sizes, pressure classes, and actuator types; some characteristics that apply to one size or class may not apply to another.)

48.3.1 Pressure and Differential Pressure

Plug valves are offered in a broad range of pressure classes, with different materials and end connections. In most cases, valves are capable of operating at differential pressures equal to the body pressure rating. Some low-pressure butterfly valves are restricted to differential pressures less than their body ratings because of seat, stem, or disk design limitations. Most modern high-performance butterfly valves do not have such limitations, but soft seat inserts may not be suitable for continuous high-differential throttling.

48.3.2 Precision of Control

Manual operation is used for most balancing-valve applications. Worm-geared manual actuators are available for most plug and butterfly valves, and they are usually necessary for valves in sizes 6 to 8 in (150 to 200 mm) or larger, depending on the valve type and operating pressure. Such actuators offer excellent capability for precise adjustment. They lock in position under normal conditions. However, severe vibration (which can be produced by cavitating flow) may cause drift even with gears that are normally self-locking. This is another reason to watch out for cavitation.

In smaller sizes, the plug ball and butterfly valves may be easily operated with a lever or wrench. It should be recognized that flow through all quarter-turn valves produces a dynamic torque, usually in a direction tending to close a throttled valve. Butterfly valves have relatively low internal friction to resist flow-induced torque; to lever operating valves usually requires "index plates," which require the lever to be locked into a discrete position between open and closed. This limits the precision of the balancing adjustment. Many plug valves have sufficient internal friction to permit them to be adjusted to any position with a removable wrench without requiring a locking device to hold the plug in place. For most moderate throttling and balancing duties, this is an advantage, allowing small plug valves to be adjusted very precisely. Nevertheless, high flows and pressure drops may cause valve drift, particularly if there is vibration (again a reason to watch out for cavitation).

48.3.3 Isolation Capability of Balancing Valves

Section 48.3.2 described in detail the requirements for a successful isolation valve. This section describes the requirements for throttling and trimming. Fortunately, the plug and butterfly valves that are often used for balancing are also usually excellent as isolation valves. However, where valves used for balancing are also called upon to serve an isolation function, some secondary features must be considered.

Most butterfly valves (like most ball valves) achieve their isolation capabilities through the use of soft seats made of elastomeric or polymeric materials (such as TFE). When new and intact, most such valves easily provide drop-tight shutoff of closed valves at both low and high differential pressures. In contrast, most plug valves are of metal-to-metal seating construction but provide for sealant injection to achieve drop-tight isolation. Providing that there is no valve damage, these factors indicate an advantage to the butterfly valve, which requires no sealant injection. Still, considering the severity of some of the services encountered in balancing-valve applications, the risks of damage to soft seats must be recognized. Unless a butterfly valve is provided with a backup metal-to-metal seat, damage or destruction of the primary soft seat may produce leakage rates that are unacceptable in sensitive HVAC applications. If excessive leakage due to seat damage should require shutdown of an entire system or interruption of service to a large area, another type of valve (or two types in series) should be considered.

In such cases, the plug valve offers the advantage that sealant injection will usually provide the required shutoff requirements without system shutdown or valve removal for maintenance. Soft-sealed valves can usually be repaired, but this requires system downtime. Even where the damage to a metal plug valve is so severe as to require maintenance, sealant injection will usually permit the maintenance to be deferred until a convenient schedule period.

48.3.4 Cavitation Resistance

Cavitation is a two-stage process consisting of (1) the formation of vapor bubbles in a liquid where local static pressure falls below the vapor pressure (boiling point) and (2) the collapse of such bubbles where the pressure rises above the vapor pressure. Even where both the inlet and outlet pressures of a throttled valve are well above the vapor pressure, cavitation may occur as a result of the low pressures existing at high-velocity regions within the valve. Vapor bubbles or cavities produced within the valve will collapse either downstream of the throttling restriction in the valve or in the downstream piping.

Cavitation effects may range from hardly noticeable to violent, depending on many considerations. "Mild" cavitation may cause intense "roar" and "rocks and gravel" noises; in this region, vibration and damage to metal parts can be severe.

A detailed discussion of the cavitation resistance of specific valve types and applications is outside the scope of this chapter. Qualified consultants or valve manufacturers with good experience and test data should be involved in evaluating questionable applications. Readers should recognize that some data published in usually reputable sources may show only where "severe" cavitation occurs. Less severe conditions may produce noise and damage that would be unacceptable in some HVAC applications. The following suggestions are intended only to provide a preliminary evaluation of potentially serious or questionable applications:

Calculate the system cavitation parameter K_{sc} by the equation

$$K_{sc} = \frac{\Delta P}{P_{in} - P_{vap}}$$

where P_{in} is the inlet pressure and P_{vap} is the vapor pressure.

- If $K_{sc} = 0.1$, there should be no problem with any standard quarter-turn valve.
- If $K_{sc} = 0.2$, there should be no problem with any throttled butterfly valve except in the worst cases.
- If $K_{sc} = 0.3$, there should be no problem with any throttled plug valve except in the worst cases.

For application outside the limits given above, the valve manufacturer should be consulted.

For a more detailed description of cavitation, flow characteristics, etc., see Ref. 1.

48.4 REFERENCE

1. *Flow Manual for Quarter-Turn Valves*, Rockwell International,

CHAPTER 49
NOISE CONTROL

Martin Hirschorn
President, Industrial Acoustics Company,
Bronx, New York

49.1 INTRODUCTION

Is noise control engineering a science or an art? It is a bit of both.

Acoustic theory helps explain the acoustic world we live in and enables us to establish general design parameters for engineered noise control solutions and products, but it does not always do so very accurately. For instance, it is impossible to calculate the noise reduction of barriers, walls, enclosures, rooms, and silencers or the propagation of sound waves over open surfaces with the degree of accuracy needed for reliability. There are just too many variables. Consequently, we cannot rely on theory for more than directional indicators.

Optimum noise control solutions must therefore be based on engineered products with performance characteristics obtained from repeatable laboratory tests and/or extensive field data—because if we overdesign, it costs too much money, and if we do not adequately provide for noise control, we may have an unacceptable job. For critical jobs, where there are significant uncertainty factors, model testing is essential. This may include power plants, aviation terminals and test facilities, industrial factories, and air-handling units in high-rise buildings. Furthermore, apart from economics considerations, the structural, mechanical, aerodynamic, and thermodynamic engineering aspects of the solution to a noise control problem are often more complex than its acoustic components, so in many instances a multidisciplinary approach is essential.

This chapter is concerned primarily with basic acoustic engineering principles and how they can be applied to solve noise problems inherent in HVAC systems. The chapter first discusses the theory of sound, with emphasis on the acoustic engineering aspects, and then examines the nature of noise in HVAC systems and the means available for controlling noise.

49.2 THE NATURE OF SOUND

Sound is essentially the sensation produced through the ear by fluctuations of pressure in the adjacent air, and "noise" can be defined as sound that annoys, usually because the sound pressure level is too high. High noise levels not only interfere with direct voice communications and electronically transmitted speech; they are also considered a health hazard in both the working and living environments.

Sound waves are propagated in air as compressional waves. Although compressional waves are generally caused by vibrations of solid bodies, they can also be caused by pressure waves generated by the gas discharge of a jet engine or the subsonic velocities in an air-conditioning duct. When these waves strike solid bodies, they cause the bodies to vibrate, or oscillate.

To illustrate what happens, we can think of sound being generated by a piston oscillating back and forth in an air-filled tube (Fig. 49.1). This action of the compressor causes the air molecules adjacent to the piston to be alternately crowded together (or compressed) and then moved apart (or rarefied). The oscillation generated by the piston in this manner is referred to as "simple harmonic motion." And as shown in Fig. 49.2, a plot of the piston displacement can be presented as a sinusoidal function; that is, the sound wave generated in its purest form for a discrete sound is sinusoidal and has a frequency equal to the number of times per second that the piston moves back and forth.

49.2.1 Displacement Amplitude and Particle Velocity

Specifically, sound is transmitted through individual vibrating air particles. The vibration causes the particles to move, but they do not change their average positions if the transmitting medium itself is not in motion. The average maximum distance moved by individual particles is called the "displacement amplitude," and the speed at which they move is called the "particle velocity."

In air, the displacement amplitude may range from 4×10^{-9} in (10^{-7} mm) to a few millimeters per second. The smallest amplitude would be the lowest discernible sound, and the largest amplitude would be the loudest sound the human ear can perceive as a proper sound.

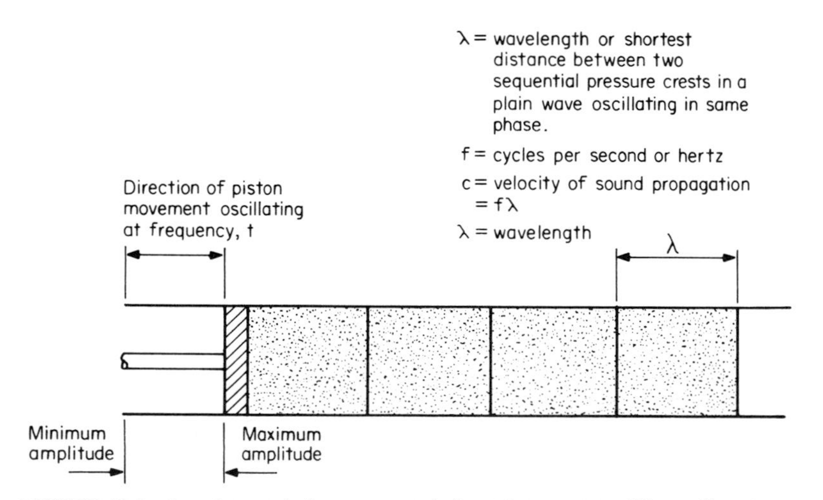

FIGURE 49.1 Sound wave being propagated through a compressible medium in a tube.

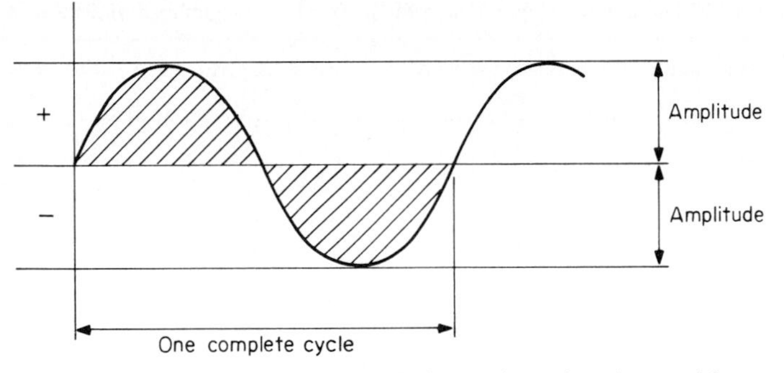

FIGURE 49.2 Sine wave of the simple harmonic motion characterizing a pure tone.

49.2.2 Frequency

The frequency of a sound wave is expressed in hertz (Hz). The range of human hearing extends from 20 to 20,000 Hz, but 12,000 to 13,000 Hz is the limit for many adults (and the exposure of teenagers to noisy rock music is likely to result in "old-age deafness" before they reach the age of 30). Figure 49.3 plots the threshold of hearing for young adults with normal hearing.

49.2.3 Wavelength

The wavelength of sound is the distance between analogous points of two successive waves. It is denoted by the Greek letter λ and can be calculated from the relationship

$$\lambda = \frac{c}{f} \tag{49.1}$$

where c is the speed of sound in ft/s (m/s), and f is the frequency in Hz.

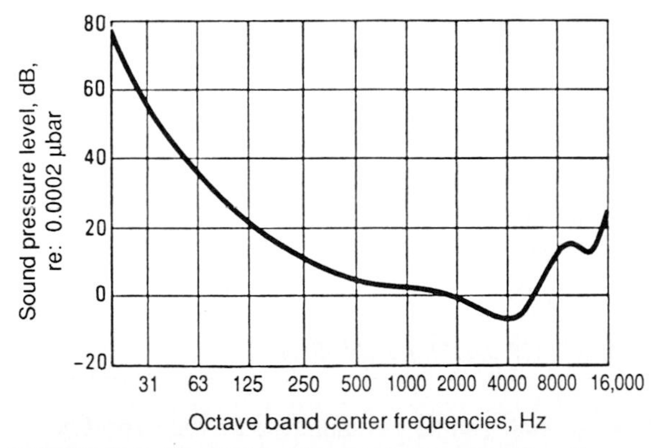

FIGURE 49.3 Threshold of hearing for young adults with normal hearing. [*C. M. Harris (ed.), Handbook of Noise Control, 2d ed., McGraw-Hill, New York, 1979, part 1, p. 8–4.*]

TABLE 49.1 Nominal One-Third Octave Band Center Frequencies and Ranges

Third octave band no.	Center frequency, Hz	Frequency range, Hz	Corresponding full octave band
14	25	22–28	Suboctave 22–45
15	—31.5—	28–36	
16	40	35–45	
17	50	45–56	1 45–89
18	—63—	56–71	
19	80	71–89	
20	100	89–112	2 89–178
21	—125—	112–141	
22	160	141–178	
23	200	178–224	3 178–355
24	—250—	224–282	
25	315	282–355	
26	400	355–447	4 354–709
27	—500—	447–563	
28	630	562–708	
29	800	708–892	5 707–1414
30	—1000—	891–1123	
31	1250	1122–1413	
32	1600	1412–1779	6 1411–2822
33	—2000—	1778–2240	
34	2500	2238–2819	
35	3150	2817–3549	7 2815–5630
36	—4000—	3547–4469	
37	5000	4465–5625	
38	6300	5621–7082	8 5617–11234
39	—8000—	7077–8916	
40	10000	8909–11225	

Note: Band numbers and center frequencies (nominal for ordinary use) are per ANSI/IEC standards. Frequency band limits are rounded to the nearest hertz.

49.2.4 Sound Level

For convenience in measuring sound without having to take data at a large number of discrete frequencies, sound levels are often measured in one-third octave bands or in full octaves, as per Table 49.1. This table shows, for instance, that the one-third octave band with a center frequency of 63 Hz has a range from 56 to 71 Hz, while the corresponding full octave has a range from 45 to 89 Hz.

49.3 THE SPEED OF SOUND IN AIR

The speed of sound in air can be calculated from the expression

$$c = \begin{cases} 49.03\sqrt{460 + °\text{F}} & \text{in English units} \\ 20.05\sqrt{273 + °\text{C}} & \text{in metric units} \end{cases} \tag{49.2}$$

TABLE 49.2 Wavelengths of Sound in Air at 70°F (21°C)
c = 1129 ft/s (344 m/s)

f, Hz	63	125	250	500	1000	2000	4000	8000
λ, ft	17.92	9.03	4.52	2.26	1.129	0.56	0.28	0.14
λ, m	5.46	2.75	1.38	0.69	0.34	0.17	0.085	0.043

where c is the speed of sound in ft/s (m/s). Note that for all practical purposes, the speed of sound in air is independent of pressure.

For example, if the temperature is 70°F (21°C), the speed of sound is

$$c = 49.03\sqrt{530} = 1129 \text{ ft/s (344 m/s)}$$

We can then use this value of c to compute the wavelength λ at various frequencies f at 70°F (21°C). At a frequency of 1000 Hz, for instance, we find from Eq. (49.1) that $\lambda = c/f = 1129/1000 = 1.129$ ft (0.344 m); likewise, at 20 Hz the wavelength would measure about 56.5 ft (17.2 m), and at 20,000 Hz it would measure about ⅔ in (17.2 mm). For 70°F (21°C), Table 49.2 gives the wavelengths of sound in air at several frequencies.

As a practical matter, because the thickness of walls and the absorptive sections of silencers are small in relation to the wavelengths of low frequencies, such structures generally attenuate sound much better in the middle and high frequencies than in the low ones. Larger and more complex structures are required for reducing low-frequency noise.

49.4 THE SPEED OF SOUND IN SOLIDS

The speed of sound in longitudinal waves in a solid bar can be shown to be

$$c_S = \sqrt{\frac{E}{\rho}} \tag{49.3}$$

where c_s is the speed of sound in solids (m/s), E is the bar's modulus of elasticity (N/m^2), and ρ is its density (kg/m^3). This obviously means that sound travels faster through media of high modulus of elasticity and of low density. Accordingly, because rubber has a much higher elasticity and lower density than steel does (as one example), a rubber insert in a steel pipe will tend to slow down sound transmission along the pipe. Table 49.3 shows the speed of sound in various media.

One can speculate that since the elasticity and density in an absolute vacuum are zero, theoretically no sound waves should be able to travel through it. An absolute vacuum may thus be the ultimate noise barrier. However, no one is yet known to have been able to come up with a practical earthborn design.

49.5 THE DECIBEL

In using the term "decibel" it is important to understand the difference between sound power levels and sound pressure levels, since both are expressed in decibels.

TABLE 49.3 Speed of Sound in Various Media (Shown in Descending Order of Magnitude)

Medium	Speed		Medium	Speed	
	ft/s	m/s		ft/s	m/s
Steel	16,500	5029	Concrete	10,600	3231
Aluminum	16,000	4877	Water	4,700	1433
Brick	13,700	4176	Lead	3,800	1158
Wood (hard)	13,000	3962	Cork	1,200–1,700	366–518
Glass	13,000	3962	Air	1,129	344
Copper	12,800	3901.	Rubber	130–490	40–149
Brass	11,400	3475			

Sound pressure levels, which can readily be measured, quantify in decibels the intensity of given sound sources. Sound pressure levels vary substantially with distance from the source, and they also diminish as a result of intervening obstacles and barriers, air absorption, wind, and other factors.

Sound power levels, on the other hand, are constants independent of distance. It is very difficult to establish the sound power level of any given source because this level cannot be measured directly, but must be calculated by means of elaborate procedures; thus, as a practical matter, sound power levels are converted to sound pressure levels, which form the basis of practically all noise control criteria.

(As one example, Sec. 49.24 illustrates how the sound power level of a fan in an HVAC system is a critical element in the silencer selection procedure to meet specified sound pressure level criteria in an office or space.)

49.5.1 Sound Power Level

The lowest sound level that people of excellent hearing can discern has an acoustic power, or sound power, of about 10^{-12} W. On the other hand, the loudest sound generally encountered is that of a jet aircraft, with a sound power of about 10^5 W. Thus the ratio of loudest to softest sounds generally encountered is 10^{17}:1.

A tenfold increase is called a "bel," so the intensity of the jet aircraft's noise can also be referred to as "17 bels." This cuts the expression of immense ranges of intensity down to manageable size. However, since the bel is still a rather large unit, it is divided into 10 subunits called "decibels" (dB). Thus the jet noise is 170 dB, and to avoid confusion with any other reference intensity, we can say that it is 170 dB with reference to 10^{-12} W.

Sound power level L_W in decibels is therefore defined as

$$L_W = 10 \log \frac{W}{10^{-12}} \qquad \text{dB re } 10^{-12} \text{ W} \tag{49.4}$$

where W is the sound power in watts. The sound power level in decibels can also be computed from

$$L_W = 10 \log W + 120 \tag{49.5}$$

TABLE 49.4 Sound Power Level L_W of Typical Sources

Source	Sound power W, W	L_W, dB re 10^{-12} W
Saturn rocket	100,000,000	200
Afterburning jet engine	100,000	170
Large centrifugal fan at 500,000 ft³/min (849,500 m³/h)	100	140
Seventy-five-piece orchestra/vaneaxial fan at 100,000 ft³/min (169,900 m³/h)	10	130
Large chipping hammer	1	120
Blaring radio	0.1	110
Centrifugal fan at 13,000 ft³/min (22,087 m³/h)	0.1	110
Automobile on highway	0.01	100
Food blenders—upper range	0.001	90
Dishwashers—upper range	0.0001	80
Voice—conversational level	0.00001	70
Quite-Duct silencer, self-noise at + 1000 ft/min (5.1 m/s)	0.00000001	40
Voice—very soft whisper	0.000000001	30
Quietest audible sound for persons with excellent hearing	0.000000000001	0

Since 10^{-12} as a power ratio corresponds to -120 dB, we can see that by definition 1 W is equivalent to a 120-dB power level. Table 49.4 shows the sound power levels of typical sources.

Note that certain older literature may contain sound power level data referenced to 10^{-13} W, an obsolete standard. Where this is the case, deduct 10 dB to convert to the current standard of 10^{-12} W.

The question now is, How does one measure sound power W? This is where another way of looking at sound power helps. As shown in Fig. 49.4, consider a

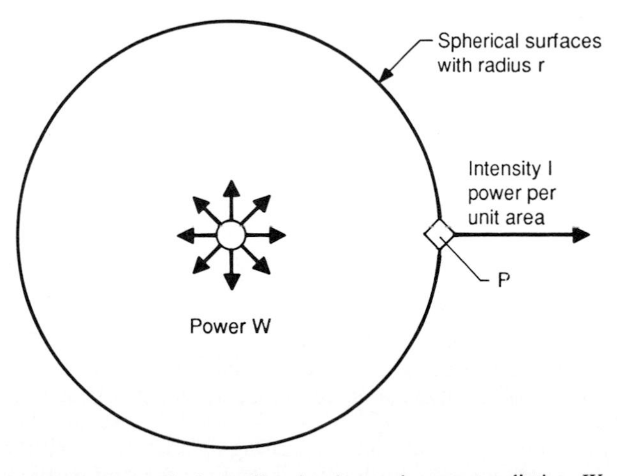

FIGURE 49.4 Ideal nondirectional sound source radiating W watts and producing a sound intensity I in watts per unit.

simple nondirectional source located at the center of a spherical surface (or at the center of a number of expanding spherical surfaces). Here the total sound power in watts is equal to sound intensity I (W/m^2) times the surface area S (m^2):

$$W = IS \tag{49.6}$$

where S for a spherical surface is $4\pi r^2$. Of course, as the sound waves move farther from their source, the surrounding spherical surface will become larger, and less power will pass through any unit element of the surface.

If the sound source is directional, the intensity will vary over the surface and the radiated power must be found by integration:

$$W = \int_s IS^{dS} \tag{49.7}$$

Since sound intensity I is rather difficult to measure, we measure sound pressure p instead. The relationship between sound pressure and intensity is

$$I = \frac{p^2_{\text{rms}}}{\rho c} \quad \text{W/m}^2 \tag{49.8}$$

where p = root-mean-square (rms) sound pressure, N/m^2
ρ = density of air, kg/m^3
c = speed of sound in air, m/s

The form of this equation will be familiar to many since it is analogous to the formula relating to electric power, voltage, and resistance:

$$P = \frac{E^2}{R}$$

where P = power, W
E = voltage, V
R = resistance, Ω

Sound intensity level L_i is defined as

$$L_i = 10 \log \frac{I}{I_{\text{ref}}} \quad \text{dB re } I_{\text{ref}} \tag{49.9}$$

where I_{ref} is 10^{-12} W/m^2.

49.5.2 Sound Pressure Level

Since sound-measuring instruments respond to sound pressure, the word "decibel" is generally associated with sound pressure level, but it is also a unit of sound power level. The square of sound pressure is proportional to, though not equal to, sound power.

Assuming a point source of sound radiating spherically in all directions, Eq. (49.6) tells us that $W = IS$. Accordingly, $10 \log W = 10 \log I + 10 \log S$, where S is the surface of the radiating sphere in ft^2 (m^2). Equation (49.5), however, tells us that 10 log $W = L_W - 120$. It can also be shown that 10 log $I = L_i - 120$. As a result, we get

$$L_W - 120 = L_i - 120 + 10 \log S$$

or

$$L_W = L_i + 10 \log S$$

Since $L_i = 10 \log (I/I_{ref})$ and $I = p^2/\rho c$, and since L_i can also be expressed as 10 log (p^2/p^2_{ref}), which is also referred to as sound pressure level L_p, then

$$L_p = 10 \log \left(\frac{p}{p_{ref}}\right)^2 = 20 \log \frac{p}{p_{ref}} \tag{49.10}$$

Accordingly,

$$L_W = L_p + 10 \log S \tag{49.11}$$

49.6 DETERMINATION OF SOUND POWER LEVELS

The concept of the imaginary radiating sphere emanating from the sound source will be referred to again in Sec. 49.8, Propagation of Sound Outdoors. Here, on the other hand, without considering imaginary spheres, we are concerned with measuring the sound power of a source that is confined within a structurally rigid space; for very large pieces of equipment and operating machinery in a plant, this approach may be the most practical way to estimate sound power.

The best method for determining the sound power level of a source is to measure it inside a good reverberant room with a truly diffuse sound field. With the sound power thus contained within the room, and with its intensity evenly distributed throughout the room, often only one sound pressure level measurement has to be taken. Then the sound power level can be calculated from $L_W = L_p + K$, where K is a constant dependent on the room volume, on the reverberation time at a given frequency or frequency band, and on the humidity.

Another method consists of containing the sound within the rigid walls of a pipe or duct equipped with an anechoic termination to minimize end reflections. Here all the sound energy must travel through the duct, and its sound field can be measured at a suitable measuring plane by averaging the sound pressure level across it. Equation (49.11) can then be used for calculating the power level of the noisemaker. Figure 49.5 illustrates such an arrangement for a ducted fan, which is also the basis of U.S. and British standards. (The 1986 U.S. standard was published jointly by ASHRAE, ANSI, and AMCA: ANSI/ASHRAE 68-1986 and ANSI/AMCA 330-86.) Although the anechoic duct method must overcome some practical difficulties, such as allowing for aerodynamically induced noise at the microphone, it clearly illustrates the relationship between measurements of sound pressure level and sound power level.

A test code of the U.S. Air Movement and Control Association (AMCA) requires the use of a reverberant or semireverberant room for determination of fan sound power levels. In such an arrangement the microphone would not be affected by aerodynamic flow. These two methods can yield comparable results,

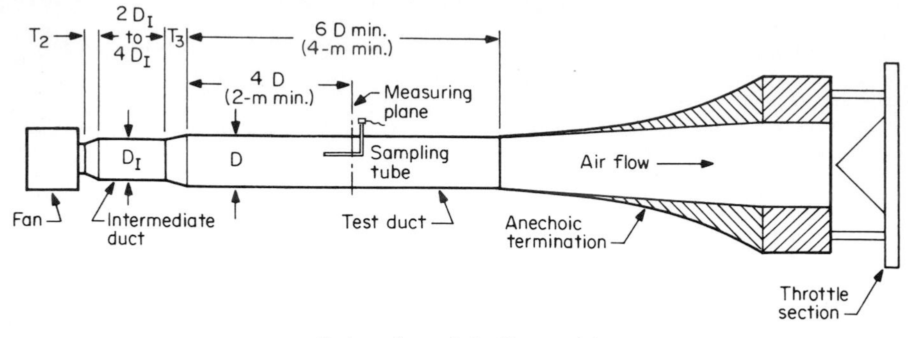

FIGURE 49.5 Anechoic duct method for fan sound power level determination. (*British Standard 848, Methods of Testing Fans, 1966, Part 2; ASHRAE/AMCA, Laboratory Method of Testing In-Duct Sound Power Measurement Procedure for Fans, 1986.*)

but relative fan sound power levels are likely to be most comparable if they have been determined under identical conditions.

In the British Standard 848, *Methods of Testing Fans*, 1966, Part 2, the sound power level L_W in each frequency band would be calculated after averaging the sound pressure level L_p across the duct area according to $L_W = L_p + 10 \log A$, where A is the cross-sectional area in ft^2 (m^2) at the plane of measurement. The U.S. standard (as in Fig. 49.5) uses $L_W = L_p + 20 \log D - 1.1$, where D is the diameter in ft (m) of the test duct.

49.7 CALCULATING CHANGES IN SOUND POWER AND SOUND PRESSURE LEVELS

49.7.1 Sound Power Level

Let L_{W1} be the sound power level corresponding to sound power W, and let L_{W2} be the sound power level twice as great, or $2W$. Then from Eq. (49.4)

$$L_{W1} = 10 \log \frac{W}{W_{\text{ref}}}$$

and
$$L_{W2} = 10 \log \frac{2W}{W_{\text{ref}}} = 10 \log \frac{W}{W_{\text{ref}}} + 10 \log 2 = L_{W1} + 3 \text{ dB}$$

Note: In eq. (49.4), $W_{\text{ref}} = 10^{-12}$ W = 1 pW (picowatt).

49.7.2 Sound Pressure Level

Assume L_{p1} to correspond to sound pressure p, and L_{p2} to sound pressure $2p$. Then from Eq. (49.10),

$$L_{p1} = 20 \log \frac{p}{p_{\text{ref}}}$$

and

$$L_{p2} = 20 \log \frac{2p}{p_{\text{ref}}} = 20 \log \frac{p}{p_{\text{ref}}} + 20 \log 2 = L_{p1} + 6 \text{ dB}$$

The addition of two equal sound pressures results in an increase of 6 dB, and the addition of two equal sound powers results in an increase of 3 dB. However, when two equal sound pressure levels are added, we are adding in effect two equal sound power levels, therefore:

$$\begin{aligned} L_{p1} + L_{p1} &= 10 \log \left(\frac{p}{p_{\text{ref}}}\right)^2 + 10 \log \left(\frac{p}{p_{\text{ref}}}\right)^2 \\ &= 10 \log \left(\frac{p}{p_{\text{ref}}}\right)^2 \times 2 \\ &= 10 \log \left(\frac{p}{p_{\text{ref}}}\right)^2 + 3 \text{ dB} \end{aligned}$$

Similarly, it can be said that when N identical sound sources are added,

$$L_p(\text{total}) = L_p(\text{single source}) + 10 \log N \tag{49.12}$$

where N is the number of sources; 10 log N is plotted as a function of N in Fig. 49.6.

Table 49.5 shows how to add two unequal decibel levels, and Fig. 49.7 presents Table 49.5 graphically. Examples:

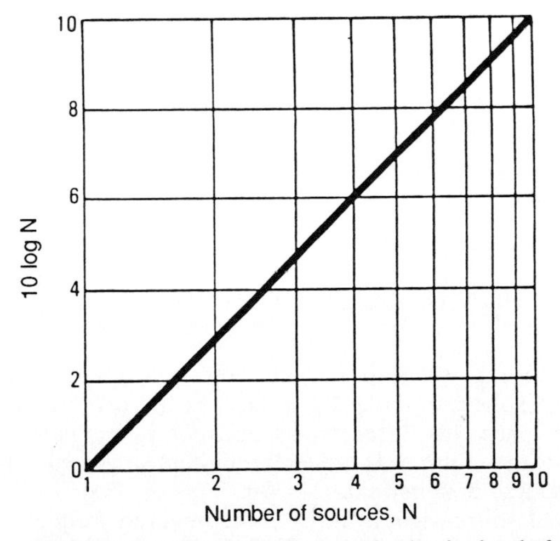

FIGURE 49.6 Predicting the combined noise level of identical sources.

TABLE 49.5 Addition of Sound Levels

Difference between the two levels, dB	Add to the higher level, dB
0	3
1	2.5
2	2
3	2
4	1.5
5	1
6	1
7	1
8	0.5
9	0.5
10 or more	0

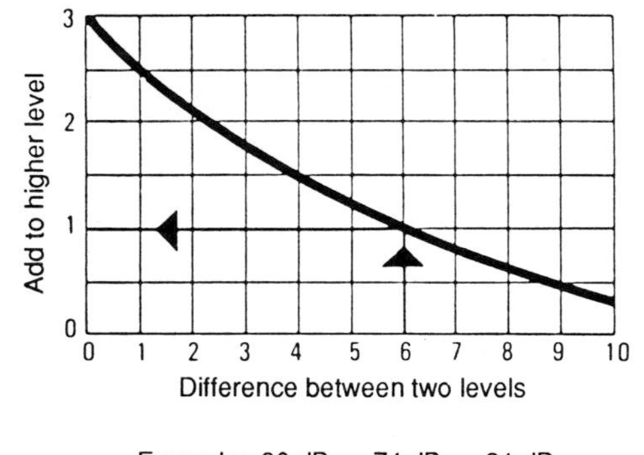

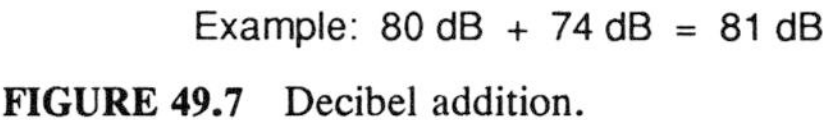

FIGURE 49.7 Decibel addition.

1. Two fans produce an L_p of 95 dB each in the fourth octave band at a given location. The combined L_p in that band would then be 98 dB.
2. If one of these fans is slowed down to produce an L_p of 90 dB, the combined L_p would then be 96 dB.

49.8 PROPAGATION OF SOUND OUTDOORS

Section 49.5.1 (and Fig. 49.4) introduced the concept of sound propagating through a series of spheres increasing in size as the distance r from its source increases. We now need to differentiate between hemispherical and spherical sound sources. If the source is considered hemispherical, the surface area $S = 2\pi r^2$; if the source is spherical, $S = 4\pi r^2$.

A fully spherical source would not be encountered frequently in a practical situation (examples would be an aircraft flying overhead, a rocket in flight, or noise emanating from the top of a tall building or a vertical stack or from a bird flying through air). If the source radiates hemispherically, as most sources do when close to the ground or to other reflective surfaces, then for a uniformly directional source (such as a siren), the relationship between sound pressure and sound power would be

$$\begin{aligned} L_p &= L_W - 10 \log 2\pi r^2 \\ &= L_W - 20 \log r - 10 \log 2\pi \\ &= \begin{cases} L_W - 20 \log r + 2.3 & \text{if } r \text{ is in feet} \\ L_W - 20 \log r - 8 & \text{if } r \text{ is in meters} \end{cases} \end{aligned} \tag{49.13}$$

For a spherical source, the relationship would be

$$\begin{aligned} L_p &= L_W - 10 \log 4\pi r^2 \\ &= L_W - 20 \log r - 20 \log 4\pi \\ &= \begin{cases} L_W - 20 \log r + 0.7 & \text{if } r \text{ is in feet} \\ L_W - 20 \log r - 11 & \text{if } r \text{ is in meters} \end{cases} \end{aligned} \tag{49.14}$$

It will be noted that the sound pressure level for a hemispherical source is 3 dB higher than for a spherical source because the same sound intensity is considered to pass through an area half the size of a full sphere.

Not all sound sources radiate uniformly. If a sound source has a marked directional characteristic, this characteristic has to be taken into account; it is called the "directivity index" (DI). Figure 49.8 illustrates how noise emanating from an opening, stack, or pipe will vary with the directivity angle.

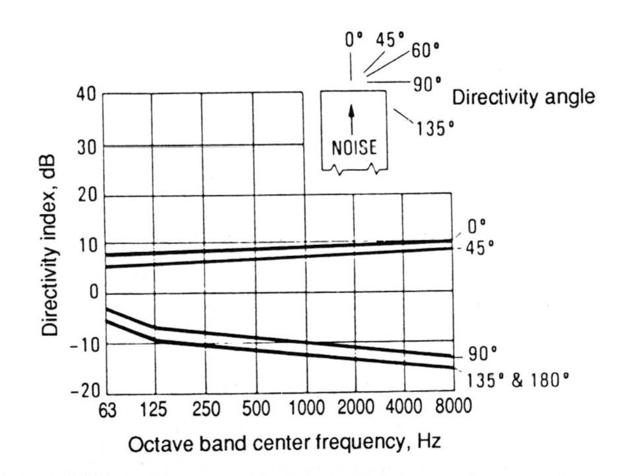

The noise emanating from an opening, stack or pipe, will vary with directivity angle between the point of measurement and the conduit centerline. Data shown for stack or pipe diameter of approximately 10-ft (3.05-m) equivalent diameter.

FIGURE 49.8 Directivity index from openings, stacks, or pipes. (*NEMA Standards Publication SM 33, Directivity in Openings, Stacks or Pipes, 1964.*)

Other factors affecting the radiation of sound might be barriers and the attenuation of sound due to atmospheric conditions (such as molecular absorption in the air, wind, and rain) and ground conditions (including grass, trees, shrubbery, snow, paving, and water). Attenuation due to such factors is generally significant in the high frequencies and over long distances and makes reliable and repeatable outdoor measurements very difficult to obtain.

For a directional noise source, we can therefore estimate sound pressure levels by modifying Eq. (49.13) as follows:

$$L_p = \begin{cases} L_W - 20 \log r + \text{DI} - A_a - A_b + 2.3 & \text{if } r \text{ is in feet} \\ L_W - 20 \log r + \text{DI} - A_a - A_b - 8 & \text{if } r \text{ is in meters} \end{cases} \tag{49.15}$$

where DI = directivity index
A_a = attenuation due to atmospheric conditions
A_b = attenuation due to barriers
r = distance from source, ft (m)

For instance, if we know or estimate the L_W of a fan (now provided by many manufacturers), we can also estimate the L_p at a distance r by taking into account directivity and the other factors indicated in Eq. (49.15).

49.9 THE INVERSE-SQUARE LAW

From Eq. (49.15) we can see that if the sound pressure level of a source is measured at two different distances from the source, the difference in sound pressure levels at those locations is

$$L_{p2} - L_{p1} = 10 \log \left(\frac{r_2}{r_1}\right)^2 = 20 \log \frac{r_2}{r_1} \tag{49.16}$$

where L_{p1} = sound pressure level at location 1, dB
L_{p2} = sound pressure level at location 2, dB
r_1 = distance from source to location 1, ft (m)
r_2 = distance from source to location 2, ft (m)

The relationship between $(L_{p2} - L_{p1})$ and r_2/r_1 is shown in Fig. 49.9.

Equation (49.16) shows that the sound pressure level varies inversely with the square of the distance from the source, with L_p decreasing by 6 dB for each doubling of distance from the source. This relationship is known as the "inverse-square law."

At locations very close to a sound source, a measurement point will be in what is known as the "near field" of the source. In the near field, neither Eq. (49.15) nor Eq. (49.16) applies, and L_p will vary substantially with small changes in position. As the distance increases, however, L_p will decrease according to the inverse-square law; Eqs. (49.15) and (49.16) will apply, and a measurement point can be said to be in the "far field" of the source.

For all practical purposes, the inverse-square law functions only in a "free field," which is defined as a space with no reflective boundaries or surfaces. Outdoors, such conditions are likely to exist only in an open field. In a reverberant field, such as might exist in the courtyard of a building or in a narrow street, the sound pressure level may decrease by a factor of less than 6 dB for each doubling

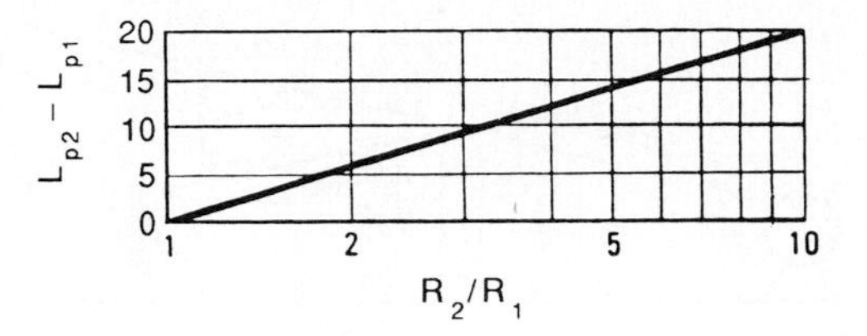

R_1 = distance from source to location 1

R_2 = distance from source to location 2

L_{p1} = sound pressure level, location 1

L_{p1} = sound pressure level, location 2

FIGURE 49.9 Inverse-square law.

of the distance. On the other hand, in a field of freshly fallen snow the decrease may be more than that predicted by the inverse-square law.

49.10 PARTIAL BARRIERS

Unobstructed sound propagates directly along a straight-line path from the source. If a barrier is interposed between that source and a receiver, some of the sound will be reflected back toward the source. These reflections can, of course, be attenuated by placing sound-absorptive surfaces on the barrier side facing the source.

Another portion of the sound emanating from the source is transmitted through the barrier (Fig. 49.10). To meet structural and wind loading criteria,

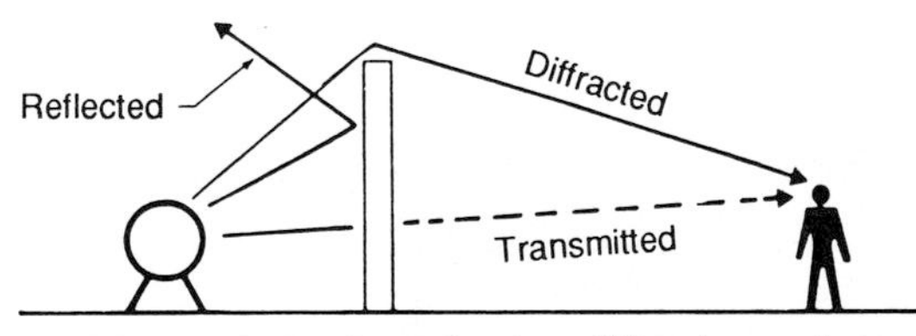

FIGURE 49.10 Barrier reflection, diffraction, and transmission.

however, most barrier designs significantly inhibit noise transmission to the extent that sound reaches the receiver primarily by diffracting over and around the barrier. As shown in Fig. 49.11, the presence of the barrier creates a "shadow zone" in which diffraction attenuates the noise reaching the receiver; the extent of this attenuation is the angle Θ between the straight and diffracted sound paths. Angle Θ (and thereby barrier attenuation) increases if the receiver or source is placed closer to the barrier or (assuming that the barrier is long enough to prevent sound from diffracting around the ends) if the barrier height is increased.

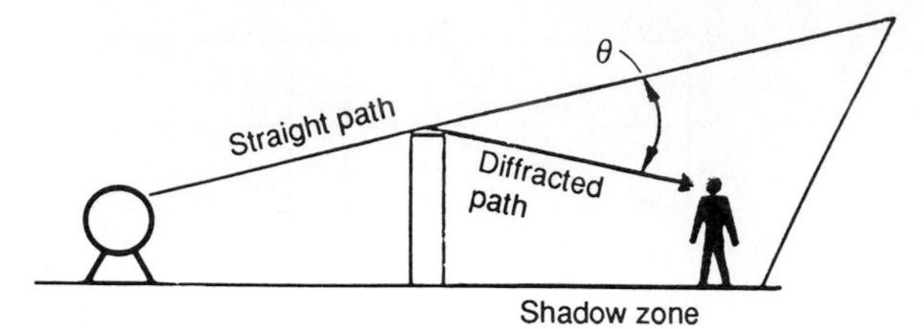

FIGURE 49.11 The shadow zone behind a barrier.

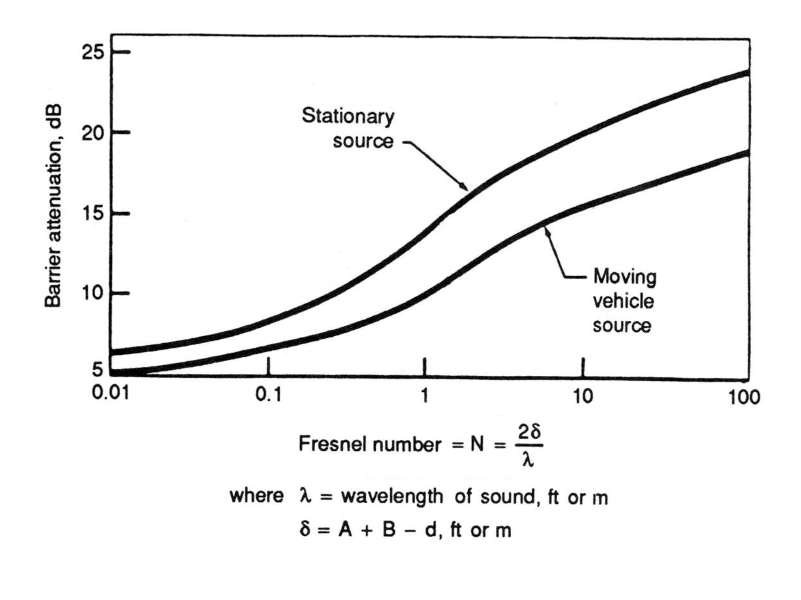

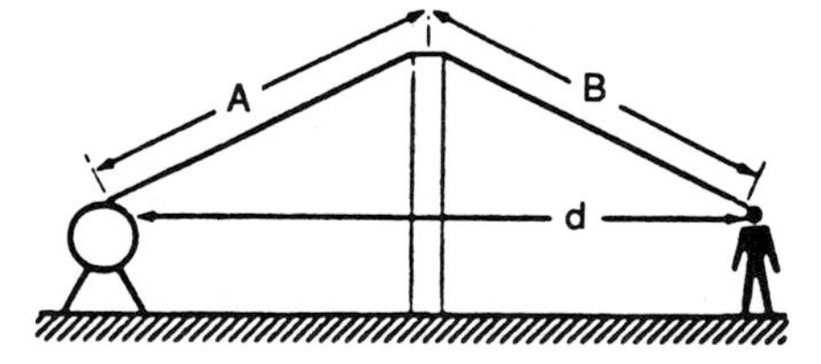

Note: N is a dimensionless number and can be used in English or metric units on a consistent basis.

FIGURE 49.12 Barrier attenuation as a function of Fresnel number. (*Z. Maekawa, "Shielding Highway Noise," Noise Control Engineering, vol. 9, no. 1, July–Aug, 1977.*)

The theoretical relationship between barrier height, source and receiver position, and barrier attenuation from diffraction can be mathematically expressed as a function of Fresnel number N, as shown in Fig. 49.12.

49.11 PROPAGATION OF SOUND INDOORS

Assume that a sound source is on the floor of an enclosed space and that there are no partitions or barriers between the source and the receiver, and assume further that none of the sound leaves the space and reaches the receiver by a flanking path. Under these conditions, the sound in the space will reach the receiver by two paths: a direct sound path and a reverberant sound path.

49.11.1 Direct Sound Path

In the far field of the source, sound from a source on or near the center of a wall or floor in a room will propagate to the receiver according to the inverse-square law:

$$L_{pd} = \begin{cases} L_W - 20 \log r + 2.3 & \text{if } r \text{ is in feet} \\ L_W - 20 \log r - 8 & \text{if } r \text{ is in meters} \end{cases} \tag{49.17}$$

where L_{pd} is the sound pressure level from direct sound.

49.11.2 Reverberant Sound Path

Reverberant sound will reach the receiver after reflecting off surfaces in the space. If the sound in the space is diffuse (essentially equal at all locations), Eq. (49.18) applies:

$$L_{pr} = \begin{cases} L_W - 10 \log A + 16.3 & \text{if } A \text{ is in sabins} \\ L_W - 10 \log A + 6.0 & \text{if } A \text{ is in metric sabins} \end{cases} \tag{49.18}$$

where L_{pr} is the sound pressure level from reverberant sound, and A is the total absorption. The "total absorption" of a surface is the product of the surface area S and the absorption coefficient α of that surface:

$$A = S\alpha \tag{49.19}$$

where the units of A are sabins if S is in ft^2, and metric sabins if S is in m^2; α is the sound absorption coefficient, the dimensionless ratio of sound energy absorbed by a given surface to that incident upon the surface (see Sec. 49.22 and Table 49.43).

Total room absorption can be calculated as follows:

$$A = \Sigma S\bar{\alpha} = S_1\alpha_1 + S_2\alpha_2 + S_3\alpha_3 + \cdots + S_n\alpha_n \tag{49.20}$$

where
A = total absorption in room, sabins (metric sabins)
S = total surface area in room, ft^2 (m^2)
$\bar{\alpha}$ = average room absorption coefficient

$S_1, S_2, S_3, \ldots, S_n$ = surface area of different segments of wall, ceiling, and floor surfaces in room

$\alpha_1, \alpha_2, \alpha_3, \ldots, \alpha_n$ = corresponding sound absorption coefficients

Reverberant sound may be reduced by adding sound-absorptive materials to reflective room surfaces. The theoretical reduction in reverberant sound due to adding sound-absorptive treatment to the surfaces of a room containing a diffuse sound field is equal to

$$\text{Reduction in reverberant sound} = 10 \log \frac{A_2}{A_1} \tag{49.21}$$

where A_1 is the total room absorption after adding sound-absorptive treatment, and A_2 is the total room absorption before adding treatment. This is illustrated in Fig. 49.13.

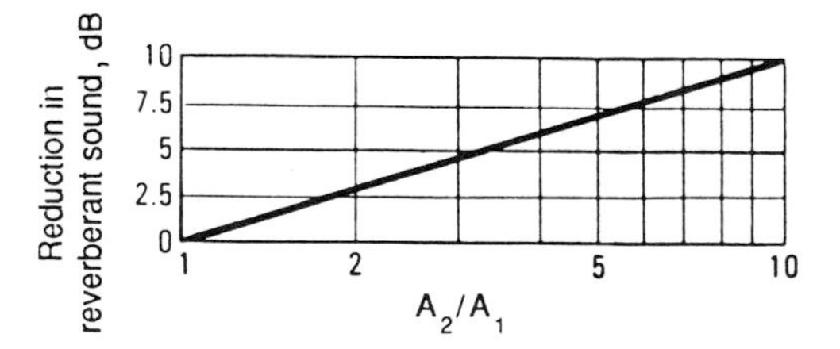

A_2 = Total room absorption after adding sound absorptive treatment

A_2 = Total room absorption before adding treatment

FIGURE 49.13 Effect of increasing room absorption.

49.11.3 Effects of Direct and Reverberant Sound

The effects of direct and reverberant sound are shown in Fig. 49.14. Direct sound predominates close to the source, but direct sound diminishes with distance. Thus, farther from the source, reverberant sound predominates; under ideal conditions this occurs when the L_p in the room levels off with increasing distance from the source.

The quantitative relationship between L_W and L_p from both direct and reflected sound paths is shown in Fig. 49.15 as a function of distance from the source and total room absorption. Add 3 dB to L_p if the source is on the wall or floor of the room, add 6 dB if the source is at the intersection of two walls (or a wall and ceiling), and add 9 dB if the source is in a corner (Ref. 1).

Note in Fig. 49.15 [and also Eq. (49.17)] that increasing the total room absorption has no effect on direct sound; accordingly, adding sound-absorptive materials to room surfaces will show maximum reduction in L_p in areas where reverberant sound predominates. Also, note that small increases in total room absorption will not produce significant decreases in sound pressure level; even in

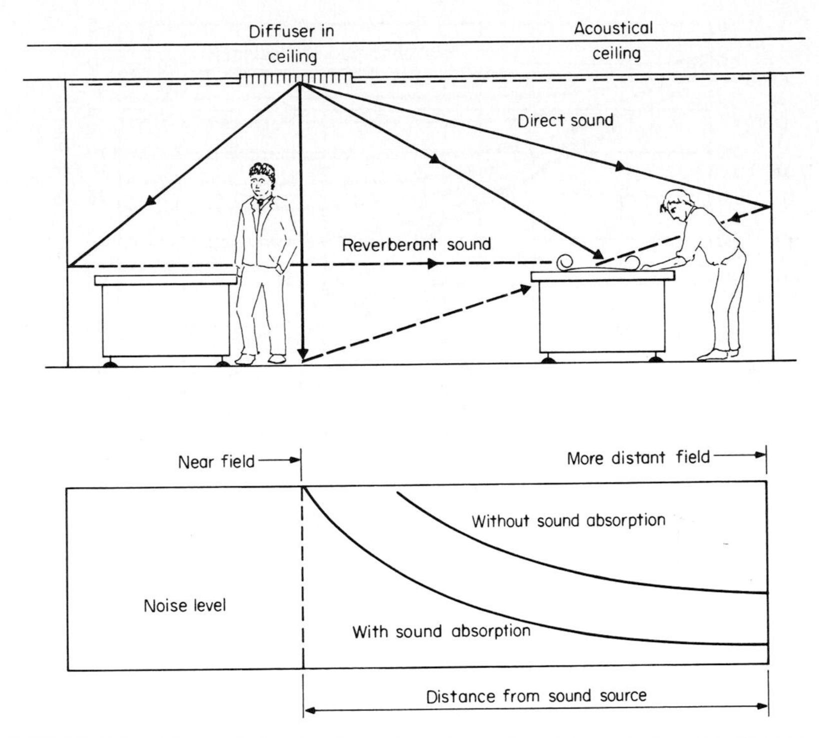

FIGURE 49.14 Effects of direct and reverberant sound on listeners in the source's near and far fields. Close to the source, direct sound predominates; at a distance, reverberant sound predominates.

a location dominated by reverberant sound, doubling thc room absorption will decrease the sound pressure level by only 3 dB.

49.12 SOUND TRANSMISSION LOSS

Figure 49.16 shows that when a sound path is broken by a partition, part of the sound is reflected, part is absorbed, and part is transmitted through the partition.

Ten times the logarithmic ratio of incident sound power to transmitted sound power is defined as "sound transmission loss" (TL). As shown in Fig. 49.17,

$$\mathrm{TL} = 10 \log \frac{W_i}{W_t} = 10 \log \frac{1}{\tau} \qquad (49.22)$$

where τ = sound transmission coefficient
W_i = incident sound power, W
W_t = transmitted sound power, W

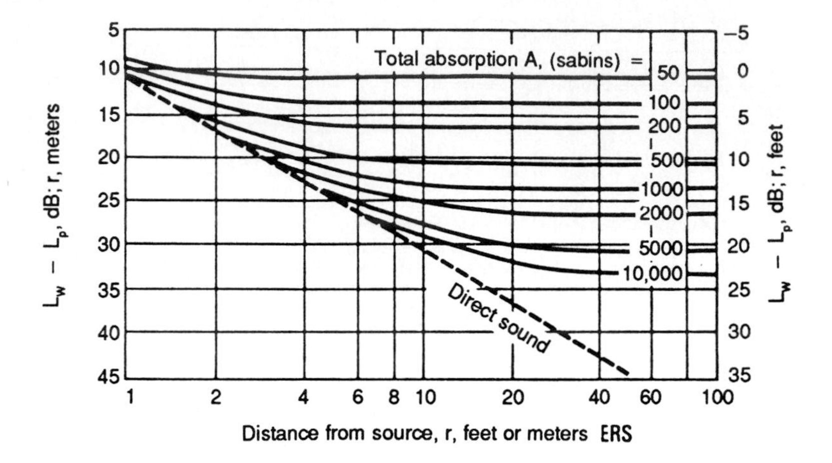

FIGURE 49.15 Effects of direct and reverberant sound in rooms. [*C. M. Harris (ed.), Handbook of Noise Control, 2d ed., McGraw-Hill, New York, 1979, part 1, p. 8–4.*]

49.12.1 The Mass Law

The mass law provides a theoretical relationship between the sound transmission loss of a single-wall (solid) partition, its weight, and the frequency of sound being transmitted through it. For normal incidence (NI), the relationship is

$$\mathrm{TL} = \begin{cases} 10 \log w + 20 \log f - 33.5 & \text{if } w \text{ is in lb/ft}^2 \\ 10 \log w + 20 \log f - 47.5 & \text{if } w \text{ is in kg/m}^2 \end{cases} \tag{49.23}$$

where w is the weight (or mass density), and f is the frequency in hertz.

Equation (49.23) tells us that for each doubling of the barrier's weight, the transmission loss increases by 6 dB. Equally, by doubling or halving the frequency, a 6-dB shift in TL occurs.

Equation (49.23) is commonly known as "the mass law," but more accurately it is an approximation. Actual data can deviate from mass law predictions by 10 dB or more, and the law generally does not apply to nonhomogeneous structures. As will be shown in Sec. 49.21.1, for example, multilayer walls or double walls separated by an air space generally provide greater TL than that predicted by the mass law.

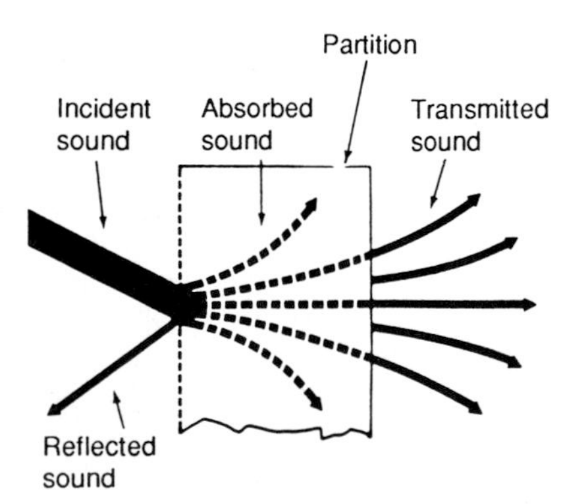

FIGURE 49.16 Effect of partitions on incident sound. (*Noise Control: A Guide for Workers and Employees, U.S. Department of Labor, Occupational Safety and Health Administration, 1980.*)

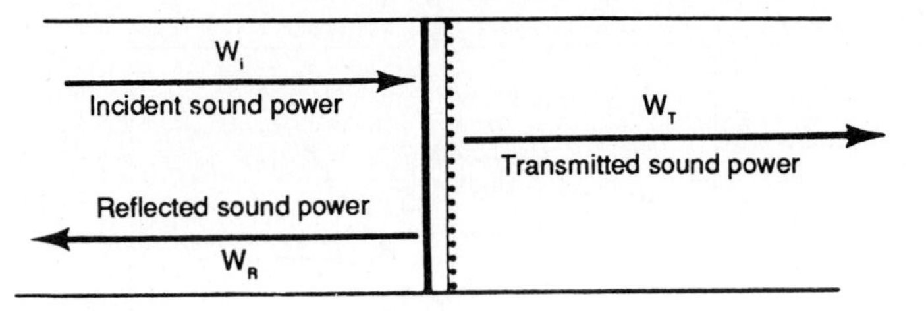

FIGURE 49.17 Incident sound power versus reflected and transmitted sound power.

49.12.2 The Effect of Openings on Partition TL

Windows, access ports, door seals, wall-to-ceiling joints, cutouts for wiring or plumbing, and other openings can significantly diminish the TL capabilities of a structure. As an example, if a 100-ft^2 (9.3-m^2) partition has a TL rating of 40 dB at a given frequency, a 1 percent [or 1-ft^2 (0.093-m^2)] opening in that partition will reduce the overall TL to 20 dB unless noise control measures are applied to the opening. The theoretical effect of openings in partitions or complete enclosures is shown in Fig. 49.18.

49.12.3 Single-Number TL Ratings: STC Ratings

For engineering rating purposes, the TL of partitions is frequently defined in terms of a single-number decibel rating known as "sound transmission class" (STC). STC ratings are determined by plotting contours of TL versus frequency in one-third octave bands from 125 to 4000 Hz and comparing the results with standard contours defined in ASTM E413 (Fig. 49.19). The TL data and STC rat-

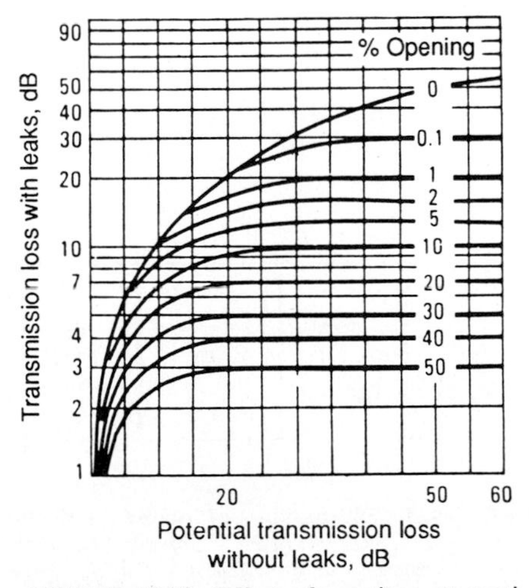

FIGURE 49.18 Effect of openings on partition TL.

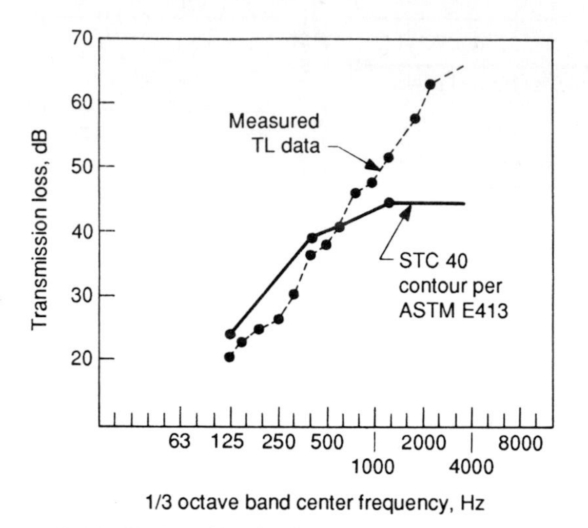

FIGURE 49.19 ASTM E413 contours for sound transmission class (STC) and noise isolation class (NIC). (*ASTM E413, Standard Classification for Determination of Sound Transmission Class, 1973.*)

ings of typical structures are listed in Table 49.39. The total deficiencies must not be greater than 32 dB, but any single band's deficiency cannot be greater than 8 dB.

49.13 NOISE REDUCTION AND INSERTION LOSS

As shown in Fig. 49.20, "noise reduction" (NR) is simply the difference in sound pressure level between any two points along the sound path from a noise source:

$$NR = L_{p1} - L_{p2} \tag{49.24}$$

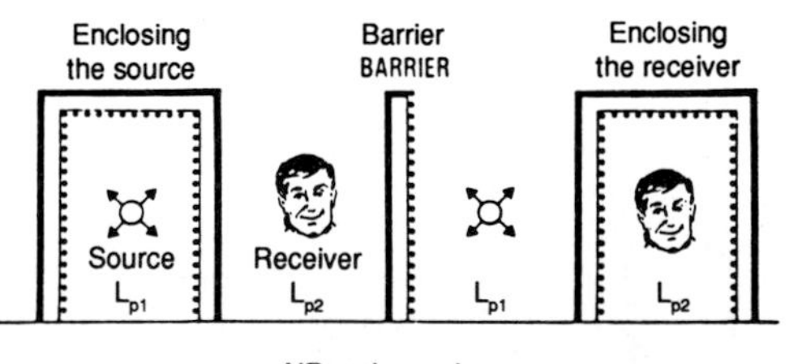

FIGURE 49.20 Illustration of noise reduction: NR = $L_{p1} - L_{p2}$. (*Lawrence G. Copley, "Control of Noise by Partitions and Enclosures," Tutorial Papers on Noise Control for Inter-Noise, Institute of Noise Control Engineers, 1972.*)

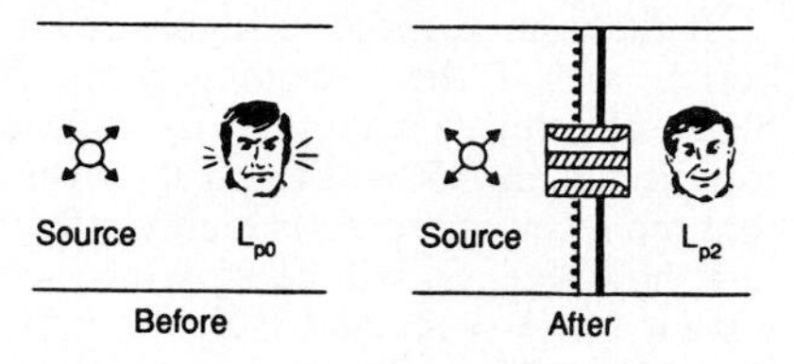

FIGURE 49.21 Illustration of insertion loss: IL = L_{p0} − L_{p2}. *(Lawrence G. Copley, "Control of Noise by Partitions and Enclosures," Tutorial Papers on Noise Control for Inter-Noise, 1972.)*

"Insertion loss" (IL), on the other hand, is the before-versus-after difference at the same measurement point, brought about by interposing a means of noise control between the source and the receiver (Fig. 49.21):

$$\text{IL} = L_{p0} - L_{p2} \qquad (49.25)$$

Like TL, NR and IL are typically rated as a function of full octave bands or one-third octave bands. The NR ratings of several types of soundproof room are listed in Table 49.42. A single-number NR rating system called "noise isolation class" (NIC) is often used for such rooms. Similar to the STC ratings described in Sec. 49.12.3, NIC ratings are established by plotting NR as a function of frequency and comparing the results against standard contours defined in ASTM E413.

49.14 THE EFFECTS OF SOUND ABSORPTION ON RECEIVING-ROOM NR CHARACTERISTICS

Figure 49.22 shows a receiver located within a room outside of which is a noise source. The relationship between the NR and TL characteristics of such a room can be shown to be represented by

$$\text{NR} = \text{TL} + 10 \log \frac{\bar{\alpha}_2 A_2}{S} \qquad (49.26)$$

where NR = $L_{p1} - L_{p2}$
L_{p1} = sound pressure level in source room, dB
L_{p2} = sound pressure level in receiving room, dB
TL = transmission loss of receiving-room walls, dB
$\bar{\alpha}_2$ = average sound absorption coefficient in receiving room
A_2 = total wall area in receiving room, ft^2 (m^2)
S = surface area separating the two rooms, ft^2 (m^2)

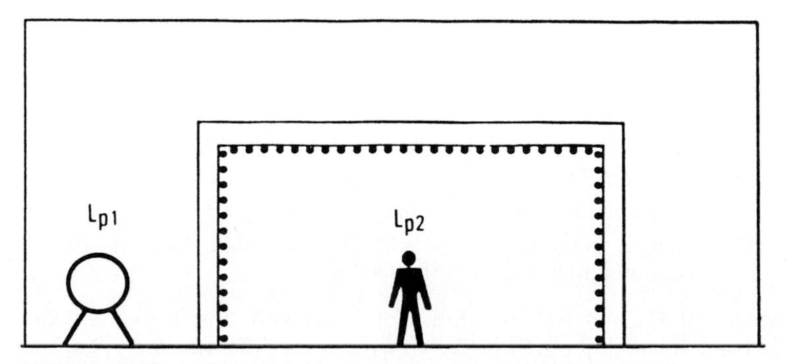

FIGURE 49.22 Noise source in outer room, and receiver in inner room.

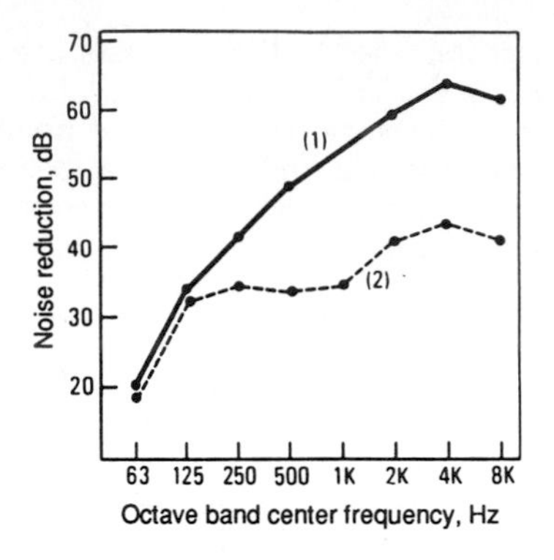

FIGURE 49.23 Additional NR demonstrated by adding sound absorption to the inside surface of a reflective receiving room. *(Martin Hirschorn, Noise Control Reference Handbook, Industrial Acoustics Company, 1982.)*

If the source room is highly reverberant and if the receiving room is highly absorptive such that $\bar{\alpha}_2$ is close to unity, then NR = TL. In the event that the receiving room is highly reflective, however, $\bar{\alpha}_2$ will be very low; for instance, if $S = A_2$ and if $\bar{\alpha}_2 = 0.01$, then NR = TL − 20 dB.

Accordingly, a highly absorptive receiving room can be seen to have a potential of 20-dB more noise reduction than a reflective receiving room with the same TL. This effect is illustrated in Fig. 49.23, which shows the NR of a 6-ft 4-in by 6-ft 0-in by 6-ft 6-in (1930- by 1829- by 1981-mm) room, which could be a fan plenum, tested with and without 2 in (51 mm) of sound-absorptive materials on the otherwise highly reflective steel inside walls. The sound absorption coefficient of a 2-in (51-mm) liner is relatively low at low frequencies, so the liner has little effect on NR. At the higher frequencies, however, NR is approximately 20 dB higher with the absorptive liner in place.

49.15 FAN NOISE

Fans are the primary source of noise generation in HVAC systems. It is always best to use fan L_W data provided by the fan manufacturer. However, if these data are not available, Eq. (49.27) can be used to predict the estimated fan L_W (dB re $1p_W$); see note on p. 49.10 (Ref. 2):

$$L_W = K_W + 10 \log \frac{Q}{Q_1} + 20 \log \frac{P}{P_1} + C + \text{BFI} \tag{49.27}$$

where K_W = specific sound power level, dB re $1p_W$ (from Table 49.6)

Q = flow rate, ft^3/min (L/s)

Q_1 = 1 when Q is in ft^3/min, 0.472 when Q is in L/s

P = fan pressure head, in WG (Pa) [in WG is "inch water gauge"]

P_1 = 1 when P is in inches WG, 249 when P is in Pa

C = correction factor for point of operation, dB

BFI = blade frequency increment to be added only to octave band containing blade pass frequency

The values of K_W and BFI are shown in Table 49.6, and Table 49.7 shows the octave band in which the BFI is likely to occur for different fan types. Values for C are given in Table 49.8.

TABLE 49.6 Specific Sound Power Levels K_W (dB re $1p_W$) and Blade Frequency Increment (BFI) for Various Types of Fans

Fan type	Wheel size	Octave band center frequency, Hz							
		63	125	250	500	1000	2000	4000	BFI
Centrifugal									
Airfoil, backward-curved,	Over 36 in (900 mm)	32	32	31	29	28	23	15	3
backward-inclined	Under 36 in (900 mm)	36	38	36	34	33	28	20	
Forward-curved	All	47	43	39	33	28	25	23	2
Radial blade and pressure	Over 40 in (1000 mm)	45	39	42	39	37	32	30	
blower	From 40 in (1000 mm) to 20 in (500 mm)	55	48	48	45	45	40	38	8
	Under 20 in (500 mm)	63	57	58	50	44	39	38	
Vaneaxial	Over 40 in (1000 mm)	39	36	38	39	37	34	32	
	Under 40 in (1000 mm)	37	39	43	43	43	41	28	6
Tubeaxial	Over 40 in (1000 mm)	41	39	43	41	39	37	34	
	Under 40 in (1000 mm)	40	41	47	46	44	43	37	5
Propeller									
Cooling tower	All	48	51	58	56	55	52	46	5

Note: These values are the specific sound power levels radiated from either the inlet or the outlet of the fan. If the total sound power level being radiated is desired, add 3 dB to the above values.

Source: *1987 ASHRAE Handbook, Systems*, ASHRAE, Atlanta, 1987, chap. 52, "Sound and Vibration Control."

TABLE 49.7 Octave Band in Which Blade Frequency Increment (BFI) Is Likely to Occur

Fan type	Octave band in which BFI occurs*
Centrifugal	
Airfoil, backward-curved, backward-inclined	250 Hz
Forward-curved	500 Hz
Radial blade and pressure blower	125 Hz
Vaneaxial	125 Hz
Tubeaxial	63 Hz
Propeller	
Cooling Tower	63 Hz

*Use for estimating purposes. For speeds of 1750 r/min (29 r/s) or more, move the BFI to the next higher octave band. Where the actual fan is known, use the manufacturer's data.

Source: *1987 ASHRAE Handbook, Systems*, ASHRAE, Atlanta, 1987, chap. 52, "Sound and Vibration Control."

Fans can generate high-intensity noise levels of a discrete tone at the blade pass frequency (BPF). The noise level's intensity will vary with the type of fan. The BPF can be established if the rpm and number of blades of the fan are known; the following equation can then be used:

$$\text{BPF} = \frac{\text{rpm} \times \text{number of blades}}{60} \quad \text{Hz} \tag{49.28}$$

For example, if the rpm is 1200 and the number of blades is 8, BPF = 160 Hz.

These discrete tones at the BPF are usually the most predominant noises emanating from large fans, but such discrete frequencies may not show up in an octave band analysis. To find them, narrower frequency ranges may have to be measured, such as one-third octave bands or even one-tenth octave bands.

BFIs vary from 2 dB for centrifugal radial-blade fans to 8 dB for centrifugal pressure-blower fans (see Table 49.6). BFIs and second harmonic frequencies generally occur in the 63- to 500-Hz region (see Table 49.7).

Once a decision has been made as to the type of fan to be used, it is best to select one that operates close to the peak of its efficiency curve. Such a fan will typically generate the lowest noise level. The correction factor *C* for off-peak operation is shown in Table 49.8.

TABLE 49.8 Correction Factor *C* for Off-Peak Operation

Static efficiency, % of peak	Correction factor, dB
90 to 100	0
85 to 89	3
75 to 84	6
65 to 74	9
55 to 64	12
50 to 54	15

Source: *1984 ASHRAE Handbook, Systems*, ASHRAE, Atlanta, GA, 1984, chap. 32, "Sound and Vibration Control."

TABLE 49.9 Calculation of Total Fan L_W in Example 49.1

	Octave band center frequency, Hz						
Calculation	63	125	250	500	1000	2000	4000
Specific fan K_W	37	39	43	43	43	41	28
10 log (Q/Q_1)+20 log (P/P_1)	55	55	55	55	55	55	55
C	0	0	0	0	0	0	0
BFI	. . .	. . .	6				
+3 dB to get total L_W (see note below Table 49.6)	3	3	3	3	3	3	3
Total L_W	95	97	107	101	101	99	86

EXAMPLE 49.1 A 35.5-in-diameter vaneaxial fan with eight blades has a 20,000-ft^3/min flow rate, develops a 4-in WG head at a speed of 1765 r/min, and operates at 95 percent of peak efficiency. Determine the fan's L_W and BPF.

Solution Calculate the fan's total L_W from Eq. (49.27):

- For K_W, Table 49.6 gives a range of octave band center frequencies for a vaneaxial fan with a diameter (or wheel size) under 40 in.
- Flow rate Q = 20,000 ft^3/min, and Q_1 = 1. Thus 10 log (Q/Q_1) = 43.
- Fan pressure head P = 4 in WG, and P_1 = 1. Thus 20 log (P/P_1) = 12.
- Correction factor C comes from Table 49.8; at 95 percent peak efficiency, C = 0.
- From Table 49.6, for vaneaxial fans, BFI = 6. Furthermore, Table 49.7 and its note show that this BFI occurs in the 250-Hz octave band.

These data are tabulated in Table 49.9, which shows the total L_W.

To calculate the fan's BPF, use Eq. (49.28). Given 1765 r/min and eight blades, BPF = (1765 × 8)/60 = 235 Hz, and Table 49.9 shows that the nearest octave band to 235 Hz in this example is 250 Hz.

49.16 COOLING TOWER NOISE

In the typical mechanically induced-draft cooling tower (Fig. 49.24), noise is generated by fan noise and water impact; at most locations of interest, however, fan noise predominates. For evaluation and control of cooling tower noise, see Refs. 3 to 6. A typical cooling tower noise control installation, consisting of air-intake and -discharge silencers, is shown in Fig. 49.25.

Cooling tower fan noise, if not available from the manufacturer, can be estimated from Eq. (49.27) and Tables 49.6 to 49.8. It should be noted, however, that the intake noise must propagate upstream against the air flow, make a 90° turn, divide as it disperses through the side of the tower, and pass through the louvers. This tortuous path results in the cooling tower fan's intake noise being less than its discharge noise. Typical fan attenuation at the air intake can amount to as much as 3, 7, 11, and 9 dB in the first four octave bands, respectively; however, in the last four bands water noise predominates. Clearly, wherever possible, data based on actual measurements and provided by the cooling tower manufacturer should be used.

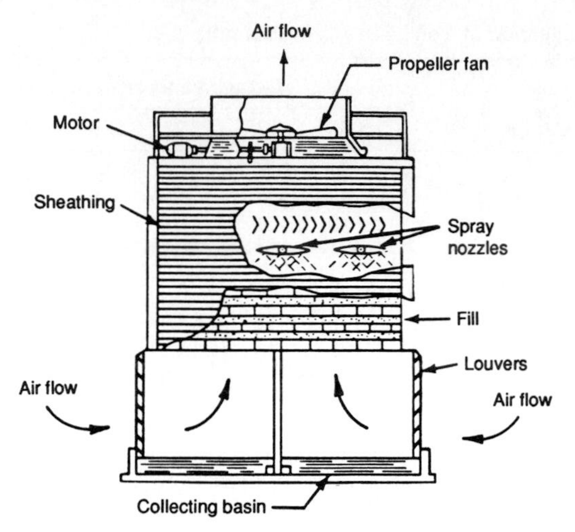

FIGURE 49.24 Mechanically induced-draft cooling tower.

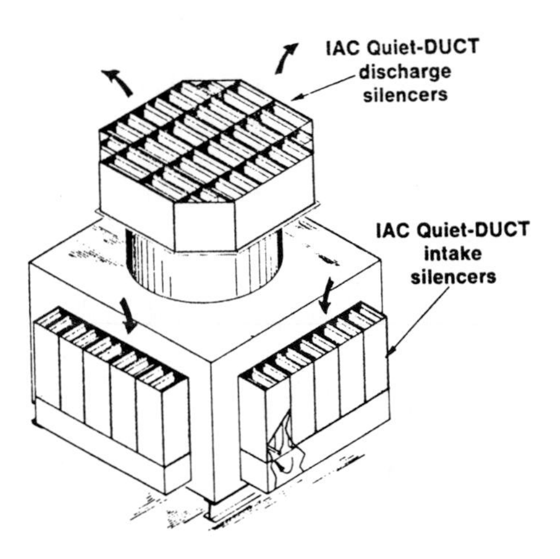

FIGURE 49.25 Silencers for cooling towers. (*Application Manual for Duct Silencers, Bulletin 1.0301.4, Industrial Acoustics Company, 1989.*)

49.17 DUCT SILENCERS—TERMINOLOGY AND TYPES

Duct silencers reduce the air-flow noise inside air-handling systems that is caused by the following:

- The fan—the air's prime mover
- The passage of air through straight ducts
- The impact of air flowing through duct components, such as elbows, branches, mixing boxes, rods, and orifices

We can generalize that any form of air movement will generate noise. If V is the velocity of air flow in a straight duct, the sound power level may be a function of V^5 to V^7, depending on the frequency and the duct component. This means that the noise generated by air flow inside a duct may increase or decrease by 15 to 21 dB every time the velocity is doubled or halved.

Six principal parameters are generally used to describe the aeroacoustic characteristics of silencers:

1. *Dynamic insertion loss (DIL):* The DIL is the difference between two sound power levels or intensity levels when measured at the same point in space before and after a silencer has been inserted between the measuring point and the noise source.
2. *Self-noise (SN):* The SN is the sound power level in decibels generated by a given volume of air flowing through a silencer of stated cross-sectional area.
3. *Air flow:* Accurate aerodynamic measurements are essential in describing any component of an air-handling system. DIL and SN data are always reported as a function of silencer face air-flow velocity.
4. *Static pressure drop:* This is generally related to silencer face velocity and volumetric air-flow capacity for a given silencer face area. For energy conservation considerations, it can also be related to the horsepower (kilowatts) required to overcome the pressure drop.
5. *Forward flow:* This applies to DIL and SN ratings with the air flow moving in the same direction as the noise propagation, such as in a fan discharge system.
6. *Reverse flow:* This applies to DIL and SN ratings with the air flow and noise propagation moving in opposite directions, such as in a fan inlet system.

There are many types of silencers, including the following:

Reactive Silencers. These have tuned cavities and/or membranes and are designed mainly to attenuate low-frequency noise in diesel, gasoline, and similar engines. Such silencers, however, are rarely used in HVAC systems.

Diffuser-Type Silencers. These are used primarily for jet engine test facilities and pneumatic cleaning nozzles in manufacturing operations. They often employ perforated "pepper pots" that slow down the flow velocities and/or prevent the generation of low-frequency noise.

Active Attenuators. Much work has been done during the last 10 years on "active" silencers. These attenuate noise by means of electronic cancellation techniques involving microphones, speakers, synchronizing sensors, and microprocessors.

Such silencers are effective at low frequencies under 300 Hz but are not suitable for broadband noise reduction without the addition of a dissipative silencer.

Moreover, their cost and maintenance requirements do not make such silencers a practical proposition. However, they might constitute an answer in unusual situations where there is no room for conventional silencers and where very low frequency noise must be controlled.

Packless Silencers. These can be used where the acoustic infill of conventional silencers could become a breeding ground for disease-carrying bacteria or where

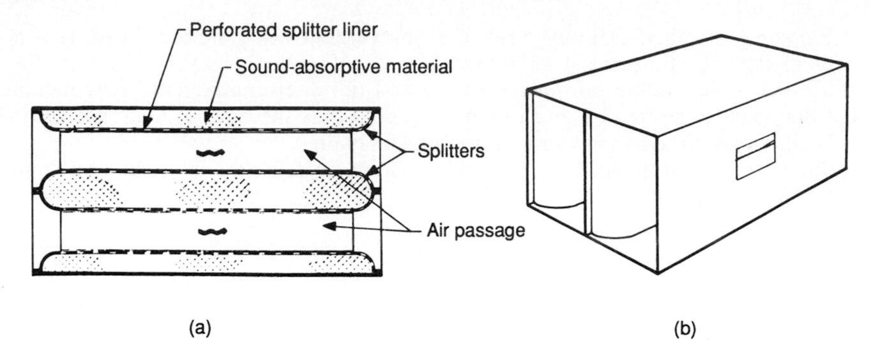

FIGURE 49.26 "Round-nosed" rectangular silencer. (*a*) Cross section; (*b*) external view. (*Application Manual for Duct Silencers, Bulletin 1.0301.4, Industrial Acoustics Company, 1989.*)

particulate matter from fiber erosion can contaminate streams of air or gas. This makes packless silencers particularly suitable for microchip manufacturers, food processing plants, hospitals, and pharmaceutical and other manufacturing plants requiring clean-room environments.

The absence of acoustic materials also reduces fire hazards where flammable materials could saturate the infill. Other applications therefore include engine test cells, kitchen exhausts, and facilities in general where fuels, grease, acids, and solvents might be carried in streams of air or gas.

Packless silencers could well become more important for general use if it becomes established that fiberglass causes lung illnesses.

Dissipative Silencers. These are widely used in HVAC duct systems. Figures 49.26 and 49.27 show the general configuration of rectangular splitter silencers. The splitter, consisting of a strong, perforated-steel envelope containing sound-

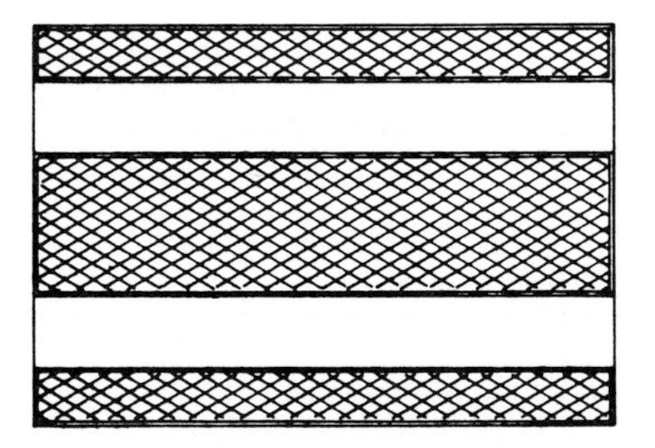

(Frequently used in Europe. Constitutes poor aerodynamic and self-noise design. See Sect. 49.23.3 and Fig. 49.59.)

FIGURE 49.27 "Flat-nosed" rectangular silencer. (*Martin Hirschorn, "The Aero-Acoustic Rating of Silencers for 'Forward' and 'Reverse' Flow of Air and Sound," Noise Control Engineering, vol. 2, no. 1, Winter 1974.*)

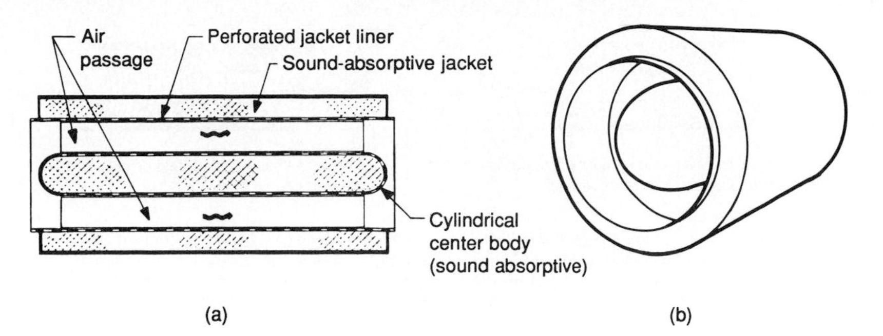

FIGURE 49.28 Cylindrical silencer. (*a*) Cross section; (*b*) external view. (*Application Manual for Duct Silencers, Bulletin 1.0301.3, Industrial Acoustics Company.*)

absorptive materials, divides the air or gas flow into smaller sound-attenuating passages. Rectangular silencers are used in rectangular ducts and are sometimes set up in very large tiers, or banks, on the intakes and exhausts of fans.

Figure 49.28 shows a tubular, or cylindrical, silencer. At first sight it looks similar in cross section to the rectangular silencer, but it consists of an outer cylindrical shell and an inner concentric bullet. Cylindrical silencers are often used in circular duct systems in conjunction with vaneaxial fans.

Dissipative silencers are available in a variety of executions, lengths, and cross sections to meet almost any noise-reduction and pressure-drop requirement of an HVAC system. The use of dissipative silencers is further discussed and illustrated in Secs. 49.23 to 49.29 in terms of applications. For discussions of the principles of silencer performance and duct break-out noise, respectively, see Secs. 49.18 and 49.19.

49.18 EFFECTS OF FORWARD AND REVERSE FLOW ON SILENCER SN AND DIL

The self-noise (SN) of a silencer varies by 7 to 26 dB for each doubling and halving of flow velocity, depending on the frequency, on the silencer's configuration, and on whether the noise and air flow are traveling in the same direction (i.e., forward or reverse flow).

As explained in Sec. 49.17, forward flow occurs if the air flow is traveling in the same direction as the sound propagation, as on the supply side of an HVAC system, and reverse flow occurs when air is traveling in a direction opposite to the direction of sound propagation, such as in a duct's return-air system. Both are illustrated in Fig. 49.29.

Figure 49.30 illustrates the effects of forward and reverse flow on silencer SN. Low-frequency SN is the greatest in the forward-flow mode, while high-frequency SN is the greatest in the reverse-flow mode.

Because of the forward- and reverse-flow phenomena, silencer performance is best rated with air flow in terms of dynamic insertion loss (DIL) determined in accordance with ASTM E477 (Ref. 7) in a reverberant room in the reverse and forward modes. The test arrangement is shown in Figs. 49.31 and 49.32. See Ref. 8.

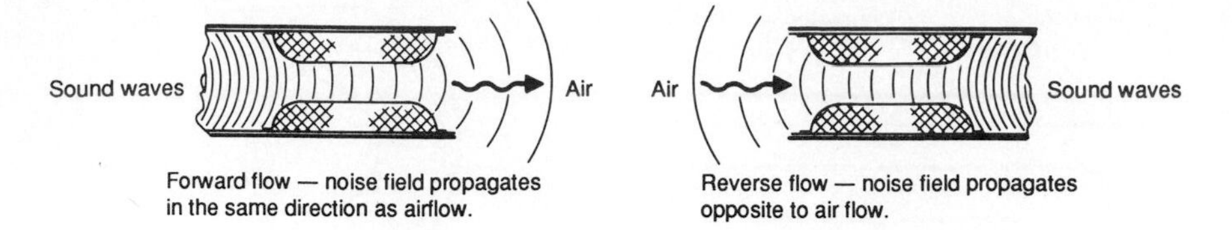

Note: If velocity of air through silencer is 70 ft/s (21.3 m/s), the speed of sound in the forward-flow direction would be 1100 + 70 = 1170 ft/s (335.3 + 21.3 = 356.6 m/s). Similarly in the reverse-flow direction, the speed of sound through the silencer would be 1100 − 70 = 1030 ft/s (335.3 − 21/3 = 314 m/s). Approximate velocity of sound at sea level = 1100 ft/s (335.3 m/s).

FIGURE 49.29 Schematic of reverse flow versus forward flow. (*Application Manual for Duct Silencers, Bulletin 1.0301.4, Industrial Acoustics Company, 1989.*)

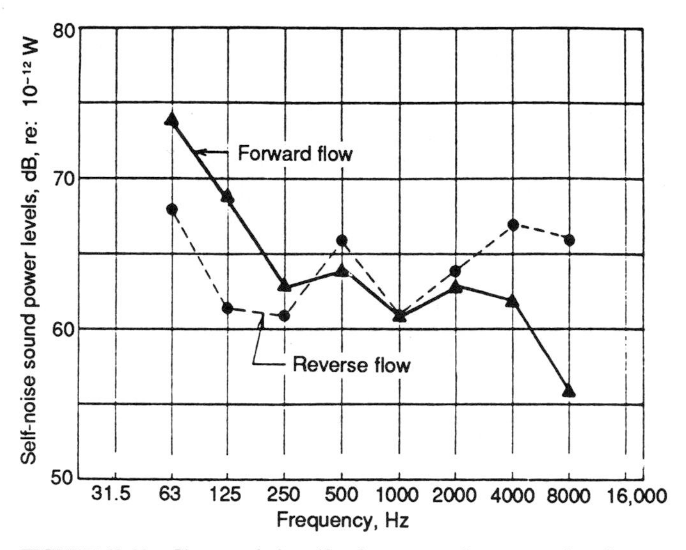

FIGURE 49.30 Characteristic self-noise spectra for rectangular silencers with 30 percent free area. (*M. Hirschorn, "Acoustic and Aerodynamic Characteristics of Duct Silencers for Airhandling Systems," ASHRAE Paper CH-81-6, 1981.*)

49.18.1 Brief Theory of the Effects of Air-Flow Direction on Silencer Performance

In examining the influence of air flow on the acoustic DIL, observers have found that air flow affects sound transmission in three major ways: (1) convection, (2) refraction, and (3) flow modification of the acoustic impedance of the duct walls. Since the third effect is rather insignificant for silencers using absorptive materials, it will not be discussed here.

Convection. The term "convection" signifies that the speed of sound in the forward direction is greater than in the reverse direction. As a result, the sound waves (previously referred to as the "noise field") maintain longer contact with

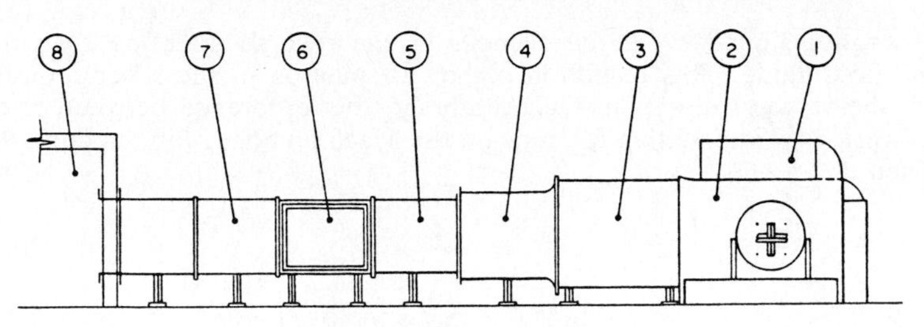

1. Air flow measurements station
2. System fan
3. System silencer
4. Signal source chamber
5. Upstream pressure test station
6. Silencer under test
7. Downstream pressure test station
8. Reverberation room

FIGURE 49.31 Typical facility for rating duct silencers with or without air flow. (*ASTM E477, Standard Method of Testing Duct Liner Materials and Prefabricated Silencers for Acoustical and Airflow Performance, American Society for Testing and Materials, 1973.*)

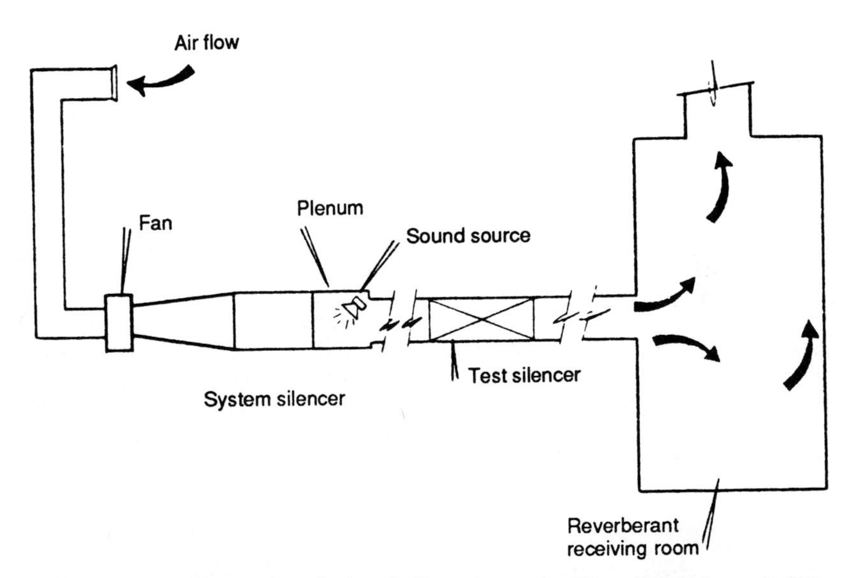

FIGURE 49.32 Schematic of the facility shown in Fig. 49.31; forward flow illustrated. (*M. Hirschorn, "Acoustic and Aerodynamic Characteristics of Duct Silencers for Airhandling Systems," ASHRAE Paper CH-81-6, 1981.*)

the absorptive boundary in the silencer in the reverse direction than in the forward-flow mode. This results in higher attenuation in the reverse direction than in the forward direction. Quantitatively, this difference between reverse- and forward-flow attenuation depends on the Mach number M in the duct, which is defined as

$$M = \frac{V}{c} \tag{49.29}$$

where V is the velocity of air, and c is the velocity of sound. At sea level, c is approximately 1100 ft/s (335.3 m/s), and V in an air-conditioning silencer might typically be on the order of 70-ft/s (21.3-m/s) throat velocity, or a Mach number of about 0.064.

This dependence on the Mach number is modified by whether the air-flow pattern in the flow sublayer close to the boundary is streamlined or turbulent. If the pattern is streamlined, the ratio between reverse- and forward-flow attenuation can be shown to be $(1 + M)/(1 - M)^1$; if the pattern is turbulent, the ratio is expected to be $(1 + M^2)/(1 - M^2)^2$. If the Mach number is about 0.064 and if the turbulent sublayer is streamlined, this would correspond to a theoretical ratio between reverse- and forward-flow attenuation of about 14 percent; however, much wider fluctuations have been measured under actual test conditions.

Where turbulent flow conditions control, the ratio between reverse- and forward-flow attenuation might then be on the order of 30 percent or more; consequently, it follows that shape and construction can have a major effect on silencer attenuation values and that it cannot be concluded that all silencers will necessarily behave alike. There is only one way to be sure that silencers will provide the performance specified, and that is on the basis of actual test data.

(It is interesting to note that if the velocity of air through a duct equals Mach 1, then theoretically no noise at all should be transmitted in the reverse-flow direction. In fact, experimental jet engine intake silencers have been constructed on this principle.)

Refraction. At higher frequencies, refraction begins to be significant, and it works in opposition to the effect of convection. That is, refraction tends to increase high-frequency attenuation in the forward-flow direction and decrease it in the reverse-flow direction. This situation is illustrated schematically in Fig. 49.33. As a sound wave travels in the forward-flow direction, there is a tendency for it to be refracted toward the boundary, which leads to smaller attenuation in the reverse-flow direction. This effect is significant only at higher frequencies when the wavelength is smaller than the cross-sectional dimensions of the duct.

It will be noted from the data in Fig. 49.55 (p. 49.79) that in the reverse-flow

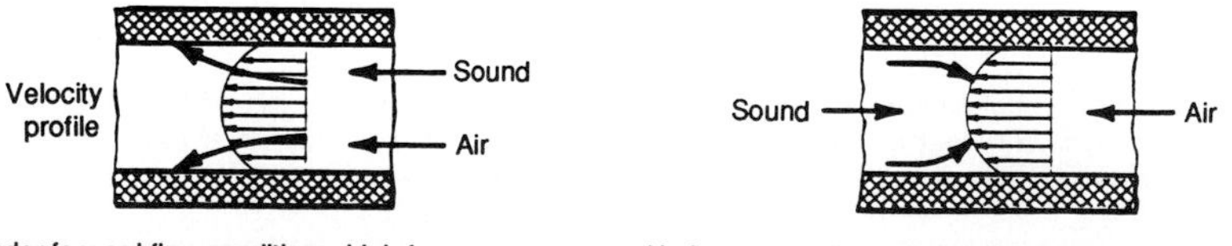

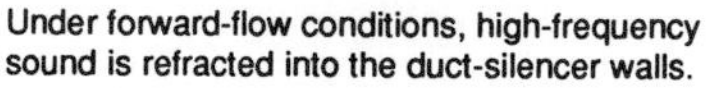

Under reverse-flow conditions, sound is refracted away from the walls and toward the center of the duct silencer.

FIGURE 49.33 The refraction of sound under forward- and reverse-flow conditions. (*Application Manual for Duct Silencers, Bulletin 1.0301.3, Industrial Acoustics Company.*)

mode, silencer attenuation falls off markedly from the sixth octave band upward and increases for the forward-flow mode (Refs. 9 to 11).

49.19 SOUND TRANSMISSION THROUGH DUCT WALLS—DUCT BREAK-OUT AND BREAK-IN NOISE

The break-out phenomenon in particular illustrates the importance of reducing fan noise by means of silencers directly after the fan. Otherwise, duct runs that lack an adequate acoustic design may radiate unacceptably high noise levels into occupied spaces.

Air ducts are commonly manufactured from light-gauge sheet materials, which provide only partial containment of the sound field within the duct. Internal noise can be transmitted into the surrounding space (break-out), and in some cases external noise can pass into the duct (break-in), which then becomes a path for noise to travel into other occupied areas.

The phenomena of break-out and break-in sound transmission are illustrated in Figs. 49.34 and 49.35.

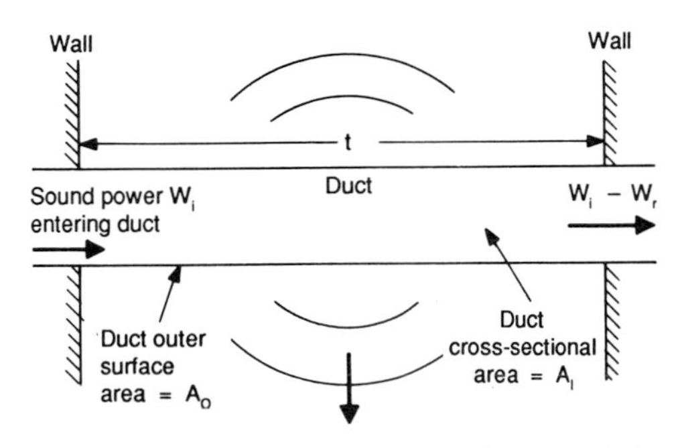

FIGURE 49.34 Break-out sound transmission through duct walls. (*ASHRAE Handbook 1987 Systems, chap. 52, "Sound and Vibration Control," American Society of Heating, Refrigerating and Air-Conditioning Engineers, Inc., Atlanta, 1987.*)

The magnitude of the sound transmission loss (TL) of a duct wall differs from that of a plenum wall panel due to the frequency-dependent nature of the sound propagation within the duct. If the cross-sectional dimensions of the duct are smaller than one-half of the wavelength, only plane waves can propagate within the duct. The vibration response of the duct walls and the pattern of radiation of sound from the duct are governed by the directional characteristics of the internal sound field. The forced response of the duct wall is proportional to the local sound pressure, which propagates in an axial direction at a speed that is equal to or greater than the speed of sound.

Practical TL curves have been developed that are divided into two regions (Ref. 2): one where plane-mode transmission within the duct predominates, and another where cross-modes prevail.

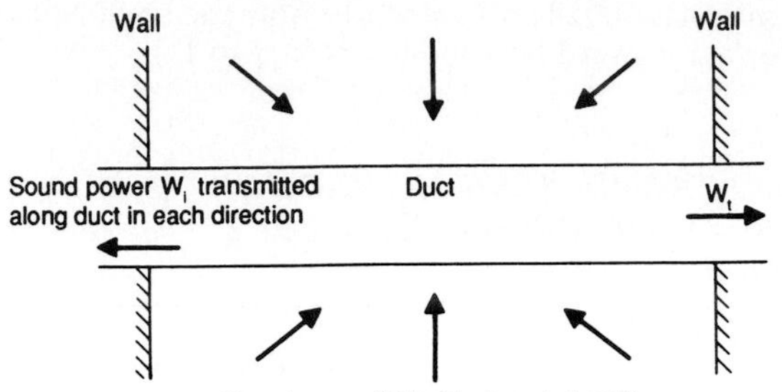

FIGURE 49.35 Break-in sound transmission through duct walls.

For break-out, the limiting frequency f_1 between these curves is given by

$$f_1 = \frac{24{,}134}{(ab)^{0.5}} \tag{49.30}$$

where a and b are the duct cross-sectional dimensions in inches; when working with metric units, convert millimeters to inches before using Eq. (49.30) (mm divided by 25.4).

For break-in, the lowest acoustic cross-mode frequency is used as the limiting frequency:

$$f_1 = \frac{6764}{a} \tag{49.31}$$

where a is the larger duct dimension in inches; when working with metric units, convert millimeters to inches before using Eq. (49.31) (mm divided by 25.4).

Below the limiting frequencies, break-out TL is given by

$$TL_{\text{out}} = 10 \log \left(\frac{fq^2}{a + b} \right) + 17 \tag{49.32}$$

and break-in TL is given by the larger of

$$\text{TL}_{\text{in}} = \text{TL}_{\text{out}} - 4 - 10 \log \frac{a}{b} + 20 \log \frac{f}{f_1} \tag{49.33}$$

or

$$\text{TL}_{\text{in}} = 10 \log \left[12l \left(\frac{1}{a} + \frac{1}{b} \right) \right] \tag{49.34}$$

where f = frequency, Hz
q = mass per unit area of duct wall, lb_m/ft^2 ($\text{kg/m}^2 \times 0.2048$)
l = duct length, ft (m)

TABLE 49.10 Examples of Duct Break-out and Break-in TL versus Frequency

Duct and TL type	Octave band center frequency, Hz							
	63	125	250	500	1000	2000	4000	8000
Rectangular,* TL_{out}	19	22	25	28	31	37	43	45
Rectangular,* TL_{in}	14	14	22	25	28	34	40	42
Circular,† TL_{out}	45	50	26	26	25	22	36	43

*Duct size: 44 by 12 in (1118 by 305 mm), 22 ga [0.034 in (0.85 mm)].
†Duct size: 26-in (660-mm) diameter, 24 ga [0.028 in (0.7 mm)].

Above the limiting frequencies, break-out TL is given by

$$TL_{out} = 20 \log qf - 31 \tag{49.35}$$

and break-in TL is given by

$$TL_{in} = TL_{out} - 3 \tag{49.36}$$

Air ducts are frequently installed above suspended ceilings or under access flooring. These confined spaces have the effect of modifying the radiating pattern of sound around the duct. Close proximity of the duct to a concrete slab will modify the response of the duct wall; the slab will also act as a reflecting plane to the overall sound radiation. Rigid partitions perpendicular to the axis of the duct may cause standing waves at frequencies where the wavelength is equal to the distance between the partitions. These standing waves can raise the local sound pressure levels in the occupied space by as much as 10 dB.

Circular ducts provide much higher TL than do rectangular ducts in the low frequencies, where most duct break-out problems occur. At higher frequencies, however, circular ducts can exhibit a resonance phenomenon at the duct's so-called "ring frequency," where the TL is sharply reduced.

Examples of break-out and break-in TL values are shown in Table 49.10 for an unlined rectangular duct made from 22-gauge [0.034-in (0.85-mm)] sheet steel and measuring 44 in (1118 mm) wide by 12 in (305 mm) deep; the break-out values for an equal-area circular-section duct 26 in (660 mm) in diameter made from spiral-wound 24-gauge [0.028-in (0.7-mm)] steel are shown for comparison. More comprehensive listings of break-out and break-in TL are shown in Tables 49.11 to 49.16.

Lagging on the outside of ductwork is often used to increase the TL values. The increase in performance due to the lagging will depend on the type and rigidity of the lagging material. A hard outer layer of sheet metal or gypsum board may not be a very effective means of reducing low-frequency noise caused by resonance effects in rectangular ducts. Limp covering materials that effectively add mass to the duct wall may improve the TL values by reducing wall response without adding stiffness. In critical situations, it may be necessary to apply panels with air spaces to the duct surfaces for maximum noise reduction.

Undoubtedly, more correlation between field and empirical data is required on break-out and break-in noise. Some acoustic consultants and engineers consider that the TL data presented here (from Ref. 2) may be overstated when translated to field installations; for instance, the break-out noise sound levels are likely to be higher than would be arrived at by using the TL figures in Tables 49.10 to

TABLE 49.11 TL_{out} versus Frequency for Various Rectangular Ducts

Duct size* in (mm)	Gauge in (mm)	63	125	250	500	1000	2000	4000	8000
		Octave band center frequency, Hz							
12 × 12 (300 × 300)	24 ga 0.028 (0.7)	21	24	27	30	33	36	41	45
12 × 24 (300 × 600)	24 ga 0.028 (0.7)	19	22	25	28	31	35	41	45
12 × 48 (300 × 1200)	22 ga 0.034 (0.85)	19	22	25	28	31	37	43	45
24 × 24 (600 × 600)	22 ga 0.034 (0.85)	20	23	26	29	32	37	43	45
24 × 48 (600 × 1200)	20 ga 0.04 (1.0)	20	23	26	29	31	39	45	45
48 × 48 (1200 × 1200)	18 ga 0.052 (1.3)	21	24	27	30	35	41	45	45
48 × 96 (1200 × 2400)	18 ga 0.052 (1.3)	19	22	25	29	35	41	45	45

*All duct lengths are 20 ft (6 m).

TABLE 49.12 TL_{in} versus Frequency for Various Rectangular Ducts

Duct size* in (mm)	Gauge in (mm)	63	125	250	500	1000	2000	4000	8000
		Octave band center frequency, Hz							
12 × 12 (300 × 300)	24 ga 0.028 (0.7)	16	16	16	25	30	33	38	42
12 × 24 (300 × 600)	24 ga 0.028 (0.7)	15	15	17	25	28	32	38	42
12 × 48 (300 × 1200)	22 ga 0.034 (0.85)	14	14	22	25	28	34	40	42
24 × 24 (600 × 600)	22 ga 0.034 (0.85)	13	13	21	26	29	34	40	42
24 × 48 (600 × 1200)	20 ga 0.04 (1.0)	12	15	23	26	28	36	42	42
48 × 48 (1200 × 1200)	18 ga 0.052 (1.3)	10	19	24	27	32	38	42	42
48 × 96 (1200 × 2400)	18 ga 0.052 (1.3)	11	19	22	26	32	38	42	42

*All duct lengths are 20 ft (6 m).

TABLE 49.13 TL_{out} versus Frequency for Various Circular Ducts*

Duct size and type	Octave band center frequency, Hz							
	63	125	250	500	1000	2000	4000	8000
8-in (200-mm) diam, 26 ga [0.022 in (0.55 mm)], long seam, length = 15 ft (4.5 m)	>45	(53)	55	52	44	35	34	26
14-in (350-mm) diam, 24 ga [0.028 in (0.7 mm)], long seam, length = 15 ft (4.5 m)	>50	60	54	36	34	31	25	38
22-in (550-mm) diam, 22 ga [0.034 in (0.85 mm)], long seam, length = 15 ft (4.5 m)	>47	53	37	33	33	27	25	43
32-in (800-mm) diam, 22 ga [0.034 in (0.85 mm)], long seam, length = 15 ft (4.5 m)	(51)	46	26	26	24	22	38	43
8-in (200-mm) diam, 26 ga [0.022 in (0.55 mm)], spiral-wound, length = 10 ft (3 m)	>48	>64	>75	>72	56	56	46	29
14-in (350-mm) diam, 26 ga [0.022 in (0.55 mm)], spiral-wound, length = 10 ft (3 m)	>43	>53	55	33	34	35	25	40
26-in (650-mm) diam, 24 ga [0.028 in (0.7 mm)], spiral-wound, length = 10 ft (3 m)	>45	50	26	26	25	22	36	43
26-in (650-mm) diam, 16 ga [0.064 in (1.6 mm)], spiral-wound, length = 10 ft (3 m)	>48	>53	36	32	32	28	41	36
32-in (800-mm) diam, 22 ga [0.034 in (0.85 mm)], spiral-wound, length = 10 ft (3 m)	>43	42	28	25	26	24	40	45
14-in (350-mm) diam, 24 ga [0.028 in (0.7 mm)], long seam with two 90° elbows, length = 15 ft (4.5 m) plus elbows	> 50	54	52	34	33	28	22	34

*In cases where background noise swamped the noise radiated from the duct walls, a lower limit in the TL is indicated by a > sign. Parentheses indicate measurements in which background noise has produced a greater uncertainty than usual in the data.

TABLE 49.14 TL_{in} versus Frequency for Various Circular Ducts*

Duct size and type	Octave band center frequency, Hz							
	63	125	250	500	1000	2000	4000	8000
8-in (200-mm) diam, 26 ga [0.022 in (0.55 mm)], long seam, length = 15 ft (4.5 m)	>17	(31)	39	42	41	32	31	23
14-in (350-mm) diam, 24 ga [0.028 in (0.7 mm)], long seam, length = 15 ft (4.5 m)	>27	43	43	31	31	28	22	35
22-in (550-mm) diam, 22 ga [0.034 in (0.85 mm)], long seam, length = 15 ft (4.5 m)	>28	40	30	30	30	22	22	40
32-in (800-mm) diam, 22 ga [0.034 in (0.85 mm)], long seam, length = 15 ft (4.5 m)	(35)	36	23	23	21	19	35	40
8-in (200-mm) diam, 26 ga [0.022 in (0.55 mm)], spiral-wound, length = 10 ft (3 m)	>20	>42	>59	>62	53	53	43	26
14-in (350-mm) diam, 26 ga [0.022 in (0.55 mm)], spiral-wound, length = 10 ft (3 m)	>20	>36	44	28	31	32	22	37
26-in (650-mm) diam, 24 ga [0.028 in (0.7 mm)], spiral-wound, length = 10 ft (3 m)	>27	38	20	23	22	19	33	40
26-in (650-mm) diam, 16 ga [0.064 in (1.6 mm)], spiral-wound, length = 10 ft (3 m)	>30	>41	30	29	29	25	38	33
32-in (800-mm) diam, 22 ga [0.034 in (0.85 mm)], spiral-wound, length = 10 ft (3 m)	>27	32	25	22	23	21	37	42
14-in (350-mm) diam, 24 ga [0.028 in (0.7 mm)], long seam with two 90° elbows, length = 15 ft (4.5 m) plus elbows	>27	37	41	29	30	25	19	31

*In cases where background noise swamped the noise radiated from the duct walls, a lower limit in the TL is indicated by a > sign. Parentheses indicate measurements in which background noise has produced a greater uncertainty than usual in the data.

TABLE 49.15 TL_{out} versus Frequency for Various Flat-Oval Ducts

Duct size*										
$a \times b$	Gauge	Octave band center frequency, Hz								
in (mm)	in (mm)	63	125	250	500	1000	2000	4000	8000
12 × 6 (300 × 150)	24 ga 0.028 (0.7)	31	34	37	40	43			
24 × 6 (600 × 150)	24 ga 0.028 (0.7)	24	27	30	33	36			
24 × 12 (600 × 300)	24 ga 0.028 (0.7)	28	31	34	37				
48 × 12 (1200 × 300)	22 ga 0.034 (0.85)	23	26	29	32				
48 × 24 (1200 × 600)	22 ga 0.034 (1.85)	27	30	33					
96 × 24 (2400 × 600)	20 ga 0.04 (1.0)	22	25	28					
96 × 48 (2400 × 1200)	18 ga 0.052 (1.3)	28	31						

*All duct lengths are 20 ft (6 m).

TABLE 49.16 TL_{in} versus Frequency for Various Flat-Oval Ducts

Duct size*									
$a \times b$	Gauge	Octave band center frequency, Hz							
in (mm)	in (mm)	63	125	250	500	1000	2000	4000	8000
12 × 6 (300 × 150)	24 ga 0.028 (0.7)	18	18	22	31	40			
24 × 6 (600 × 150)	24 ga 0.028 (0.7)	17	17	18	30	33			
24 × 12 (600 × 300)	24 ga 0.028 (0.7)	15	16	25	34				
48 × 12 (1200 × 300)	22 ga 0.034 (0.85)	14	14	26	29				
48 × 24 (1200 × 600)	22 ga 0.034 (1.85)	12	21	30					
96 × 24 (2400 × 600)	20 ga 0.04 (1.0)	11	22	25					
96 × 48 (2400 × 1200)	18 ga 0.052 (1.3)	19	28						

*All duct lengths are 20 ft (6 m).

49.16. However, in the meantime, the above procedures (including Tables 49.10 to 49.16) can be used, bearing in mind that the introduction of safety factors might be in order.

49.20 NOISE CRITERIA

Noise is unwanted or objectionable sound, and numerous standards define its limits for specific types of noisemakers, specify how the sound is to be measured, and in certain instances specify when. These standards are published by local and national government agencies, national and international standards organizations, the military, professional societies, and others. A few of these standards (or criteria) are given below.

49.20.1 dBA Criteria

One way of rating sounds is by means of the A scale, a sound-level meter weighting network that approximates the response of the human ear to sound. Both the human ear and the A-weighting network are more sensitive to high-frequency than low-frequency sound. In decibels, A-scale levels are expressed as dBA. Typical noise source dBA levels are shown in Table 49.17.

A dBA sound level can be determined in two ways: (1) by using an instrument that reads directly in dBA or (2) by applying weighting factors to measured octave band sound pressure levels (SPLs). Table 49.18 shows the A-scale weighting factors as well as an example of their use.

The A-scale weighting factors are shown in Fig. 49.36, along with responses for two other scales often found on sound-level meters. The B network is generally not used in noise control engineering. The C network measures overall sound pressure levels based on an essentially flat spectrum over the audible frequency range.

The upper part of Table 49.19 (from James Botsford, Ref. 12) permits estimates of sound pressure levels when the dBC and dBA, or C- and A-scale sound-level meter, values are known (see also Ref. 13). By taking corresponding (C − A) values from −1 to + 20, Botsford developed curves that show their average octave band relationships based on about 1000 noises. He stated that these curves can be used to predict the levels of five out of eight octave bands within 3 dB for two-thirds of all noises. The lower part of Table 49.19 evaluates his octave band relationships in terms of actual measurements.

With the A-scale weighting factors shown in Table 49.18, dBA design-guide sound pressure levels can be established on the basis of equivalent octave band levels, as shown in Table 49.20, where the speech interference levels (SILs) are the arithmetic average of the sound pressure levels at the 500-, 1000-, and 2000-Hz center frequencies (see Sec. 49.20.4).

TABLE 49.17 Typical Noise Source dBA Levels

Noise source	dBA
Noise at ear level from rustling leaves	20
Room in a quiet dwelling at midnight	32
Soft whisper at 5 ft (1.52 m)	34
Large department store	50–65
Room with window air conditioner	55
Conversational speech	60–75
Self-service grocery store	60
Busy restaurant or canteen	65
Within typing pool (nine typewriters in use)	65
Passenger car at 50 ft (15.2 m)	69
Vacuum cleaner in private home at 10 ft (3.05 m)	69
Ringing alarm clock at 2 ft (0.61 m)	80
Loudly reproduced orchestral music in large room	82
Buses, trucks, motorcycles at 50 ft (15.2 m)	82–85
Pneumatic tools at 50 ft (15.2 m)	85
Eight-hour OSHA criteria—hearing conservation programs	85
Medium-size automatic printing-press plant	86
Bulldozer at 50 ft (15.2 m)	87
Jackhammer at 50 ft (15.2 m)	88
Eight-hour OSHA criteria—engineering or administrative noise controls	90
Heavy city traffic	90
Heavy diesel-propelled vehicle at 25 ft (7.6 m)	92
Grinders	93–95
Small air compressor	94
Hammermill	96
Plastic chipper	96
Cutoff saw	97
Home lawn mower	98
Multiple spot welder	98
Turbine condenser	98
Drive gear	103
Banging of steel plate	104
High-pressure gas leak	106
Magnetic drill press	106
Air chisel	106
Positive-displacement blower	107
Large air compressor	108
Jet aircraft at 500 ft (152 m) overhead	115
Human pain threshold	120
Inside jet engine test cell	150

TABLE 49.18 A-Scale Weighting Factors

Octave band center frequency, Hz	31.5	63	125	250	500	1000	2000	4000	8000
Weighting factor	−39	−26	−16	−9	−3	0	+1	+1	−1

Example of dBA calculation from octave band levels

Octave band center frequency, Hz	63	125	250	500	1000	2000	4000	8000
SPL spectrum, dB	83	85	82	81	76	60	50	44
A-scale weighting factor	−26	−16	−9	−3	0	+1	+1	−1
Spectrum adjusted to A-scale	57	69	73	78	76	61	51	43

Logarithmic Decibel Addition

57 + 69 → 60; 73 + 78 → 79; 76 + 61 → 76; 51 + 43 → 51.5

60 + 79 → 79; 76 + 51.5 → 76

79 + 76 → 81 dBA

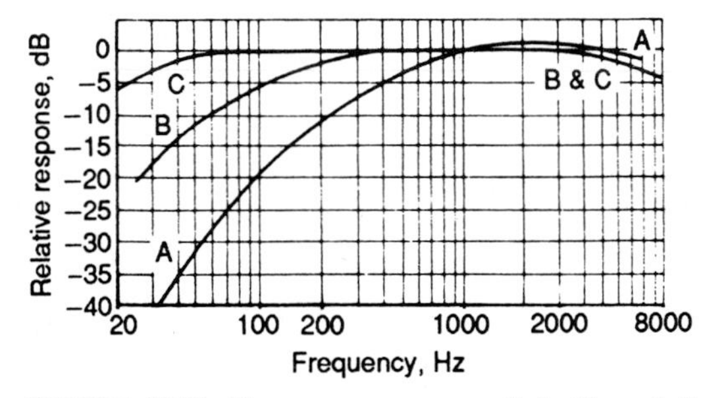

FIGURE 49.36 Frequency response of A, B, and C scales.

Figure 49.37 shows the statistical expectations of community response to noise. The parameter L_{dn} is a day-night equivalent A-weighted sound level with an additional 10-dB penalty imposed on noise exposure between 10 p.m. and 7 a.m. In addition to L_{dn}, several other terms involve the use of dBA ratings:

L_{eq} Equivalent sound level, the dBA of a steady-state sound that has the same dBA-weighted sound energy as that contained in the actual time-varying sound being measured over a specific period.

TABLE 49.19 Estimating Octave Band Values if the C- and A-Scale Sound-Level Meter Readings Are Known

Determine the C-scale minus A-scale weighting factors, and deduct from the C scale the corresponding values shown below to obtain the approximate octave band sound levels:

	Octave band center frequency, Hz								
C − A	31.5	63	125	250	500	1000	2000	4000	8000
−1	−26	−24	−23	−20	−17	−10	−6	4	−8
0	−20	−19	−17	−15	−13	−7	−6	−7	−9
2	−13	−12	−11	−10	−8	−6	−8	−11	−14
5	−9	−8	−7	−7	−8	−10	−13	−17	−22
10	−6	−5	−6	−8	−13	−17	−20	−26	−32
15	−5	−4	−6	−14	−19	−23	−28	−33	−41
20	−5	−4	−6	−19	−26	−31	−38	−44	−50

The above relationships were checked against actual measurements, and here are the results:

	Octave band center frequency, Hz							
Comparison	63	125	250	500	1000	2000	4000	8000
Generator, 40 hp (30 kW)								
Actual measurement	77	83	73	62	60	57	49	43
Botsford prediction	80	78	70	65	61	56	51	43
Difference	−3	+5	+3	−3	−1	+1	−2	0
Generator, 20 hp (15 kW)								
Actual measurement	74	77	74	69	64	59	51	41
Botsford prediction	77	75	67	62	58	53	48	40
Difference	−3	+2	+7	+7	+6	+6	+3	+1
Vaneaxial fan, 25,000 ft^3/min (11.8 m^3/s), no load*								
Actual measurement	73	83	84	89	86	82	69	63
Botsford prediction	81	82	83	85	87	85	82	79
Difference	−8	+1	+1	+4	−1	−3	−13	−16

*If this comparison is typical, the fan prediction seems close in five octave bands but is on the high side, particularly in the 4000- and 8000-Hz bands.

$L_{eq(x)}$ L_{eq} over a period of x hours. That is, if $x = 24$ h, we have $L_{eq(24)}$.

L_d The equivalent A-weighted sound level between 7 a.m. and 10 p.m. Also known as "daytime equivalent sound level."

L_n The equivalent A-weighted sound level between 10 p.m. and 7 a.m. Also known as "nighttime equivalent sound level."

L_x The time-varying dBA level that will be or is exceeded x percent of the time.

The relationships between L_d, L_n, L_{dn}, and $L_{eq(24)}$ are defined by the following equations and are summarized in Table 49.21. Typical L_{dn} levels at various locations are shown in Table 49.22.

TABLE 49.20 dBA Octave Band Design-Guide Table

		Octave band center frequency, Hz									
dBA	SIL	31.5	63	125	250	500	1000	2000	4000	8000	16,000
115	. . .	142	131	122	115	109	106	105	104	104	112
110	. . .	137	126	117	110	104	101	100	99	99	107
105	. . .	132	121	112	105	99	96	95	94	94	102
100	92	127	116	107	100	94	91	90	89	89	97
95	87	122	111	102	95	89	86	85	84	84	92
90	82	117	106	97	90	84	81	80	79	79	87
85	77	112	101	92	85	79	76	75	74	74	82
80	72	107	96	87	80	74	71	70	69	69	77
75	67	102	91	82	75	69	66	65	64	64	72
70	62	97	86	77	70	64	61	60	59	59	67

Source: M. Hirschorn, "Noise Level Criteria and Methods of Engineering Noise Control," *National Safety News*, September 1972.

$$L_{\mathrm{dn}} = 10 \log \frac{1}{24} (15 \cdot 10^{L_d/10} + 9 \cdot 10^{(L_n + 10)/10}) \tag{49.37}$$

$$L_{eq(24)} = 10 \log \frac{1}{24} (15 \cdot 10^{L_d/10} + 9 \cdot 10^{L_n/10}) \tag{49.38}$$

49.20.2 Community and Workplace Noise Regulations

Municipalities, states, and agencies of the U.S. government have established noise criteria for a broad range of conditions.

Many HVAC noise considerations relate to indoor space, but the compressors, chillers, fans, and cooling towers associated with air-conditioning systems can have a significant impact on the acoustic environment of a building. Accordingly, some of the criteria are for outdoor areas or are referenced to property boundary lines. Several are shown in the tables noted below.

Local, State, Highway, and Navy Regulations. The New York City Noise Control Code establishes three ambient noise quality zones, as shown in Table 49.23. The Chicago noise control ordinance is shown in Table 49.24, Minnesota noise control regulations are shown in Table 49.25, the Federal Highway Administration design noise levels are shown in Table 49.26, and general specifications for ships of the U.S. Navy in regard to noise levels are shown in Table 49.27.

OSHA Regulations. People working in noisy environments, such as mechanical-equipment rooms, may also be concerned about the standards set by the U.S. Occupational Safety and Health Administration (OSHA). OSHA criteria for noise levels exceeding 85 dBA for an 8-h day are listed in Table 49.28; in essence, these criteria mandate hearing-conservation measures (including annual audiometric testing and provision of hearing protectors). OSHA criteria for workplace exposures exceeding 90 dBA for an 8-h day (Table 49.29) require engineered noise control measures or administrative procedures that would limit the time employees are exposed to excessive noise levels.

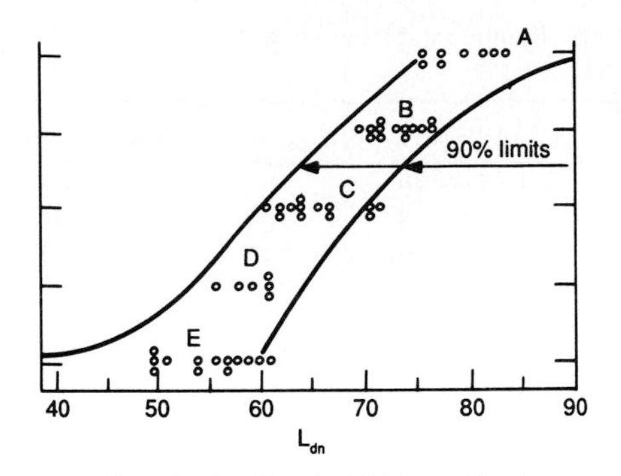

Normalized outdoor day/night sound level
of intruding noise, dB

Community reaction

A. Vigorous action
B. Several threats of legal action or strong appeals to local officials to stop noise.
C. Widespread complaints or single threat of legal action.
D. Sporadic complaints.
E. No reaction although noise is generally noticeable.

Corrections for background sound — apply for the season of operation and the ambient sound characteristics of the nearby neighborhood (see refs. 1 and 2).

Type of correction	Description	Amount correction
Seasonal	Summer (or year-round operation)	0
	Winter only (or windows always closed)	+5
Background sound	Quiet suburban or rural community (remote from large cities and from industrial activity and trucking)	-10
	Normal suburban community (not located near industrial activity)	-5
	Urban residential community (not immediately adjacent to heavily traveled roads and industrial areas)	0
	Noisy urban residential community (near relatively busy roads or industrial areas)	+5
	Very noisy urban residential community	+10

These corrections are based on reported typical residual noise levels as shown above. If measured data at the site under investigation differs significantly from the table, different corrections may be warranted. The residual sound level is that sound level exceeded 90% of the time.

Source: "Community Noise", U.S. Environmental Protection Agency Report NTID 300.3, December 1971, Washington, DC.

FIGURE 49.37 Statistical expectations of community response to noise. (*Gas Turbine Installation Sound Emissions, ANSI B-133.8, American Society of Mechanical Engineers, 1977.*)

TABLE 49.21 Relationships between L_d, L_n, L_{dn}, and $L_{eq(24)}$ Sound Levels

$L_d - L_n$	Add to L_d for L_{dn}	Add to L_d for $L_{eq(24)}$
−4	10	+2
−2	8	+1
0	6.5	0
2	5	−0.7
4	3.5	−1
6	2	−1.5
8	1	−1.7
10	0	−1.8

TABLE 49.22 Typical L_{dn} Sound Levels at Various Locations

Location	Typical L_{dn}, dBA
Wilderness ambient	35
Rural residential	40
Agricultural cropland	44
Wooded residential	51
Old urban residential	60
Urban row housing on major avenue	68
Urban high-density apartment	78
Downtown with some construction activity	79
Three-quarter mile from touchdown at major airport	86
Apartment next to freeway	88

TABLE 49.23 New York City Noise Control Code
Measured for any one hour

	Standards in $L_{eq(1)}$	
Ambient noise quality zone	7 a.m.–10 p.m.	10 p.m.–7 a.m.
Low-density residential	60 dBA	50 dBA
High-density residential	65 dBA	55 dBA
Commercial and manufacturing	70 dBA	70 dBA

Note:: All noise measurements shall be made at the property line of the impacted site. When instrumentation cannot be placed at the property line, the measurement shall be made as close thereto as is reasonable. However, noise measurements shall not be made at a distance less than 25 ft (7.6 m) from a noise source.

TABLE 49.24 Chicago Environmental Control Ordinance—Maximum Sound Pressure Levels at Residential and Business-Commercial Boundaries

Manufacturing zoning districts	Octave band frequency, Hz									
	31.5	63	125	250	500	1000	2000	4000	8000	dBA
	Residential boundaries									
Restricted	72	71	65	57	51	45	34	34	32	55
General	72	71	66	60	54	49	44	40	37	58
Heavy	75	74	69	64	58	52	47	43	40	61
	Business-commercial boundaries									
Restricted	79	78	72	64	58	52	46	41	39	62
General	79	78	73	67	61	55	50	46	43	64
Heavy	80	79	74	69	63	57	52	48	45	66

A. Districts as defined in City of Chicago Zoning Ordinance.
B. Maximum levels in "Restricted" Manufacturing Zoning Districts apply at the lot boundaries.
C. Maximum levels in "General" or "Heavy" Manufacturing Zoning Districts apply at the boundary of the Residence, Business or Commercial District, or at 125 ft (38.1 m) from the nearest property line of a plant or operation, whichever distance is greater from that plant or operation.
Note: The New York City Noise Control Code and the Chicago Environmental Control Ordinance also contain allowable noise limits for various noise sources. Reference each document for complete details.

TABLE 49.25 State of Minnesota Noise Control Regulations

NAC*	Day (7 a.m.–10 p.m.)		Night (10 p.m.–7 a.m.)	
	L_{50}	L_{10}	L_{50}	L_{10}
1	60	65	50	55
2	65	70	65	70
3	75	80	75	80

*NAC stands for Noise Area Classification system according to land activity at receiver.
Acceptable sound levels for the receiver are a function of the intended activity in that land area. The Noise Area Classifications are grouped and defined as follows:
- NAC-1: Residential areas, hotels, hospitals, schools, resorts, etc.
- NAC-2: Urban shopping areas, rapid transit terminals, finance, insurance, legal trade areas.
- NAC-3: Manufacturing areas.

HUD Site Acceptability Standards. The standards of the U.S. Department of Housing and Urban Development (HUD) are shown in Table 49.30.

49.20.3 Noise Criteria (NC) Curves

One of the most commonly used ways to rate the noisiness of an indoor space is of established octave band spectra known as "NC curves" (Ref. 2). These curves are plotted in Fig. 49.38 and tabulated in Table 49.31.

The projected or measured NC level within an occupied space is determined by the highest NC level corresponding to the sound pressure level in any octave

TABLE 49.26 Federal Highway Administration (FHWA) Design Noise Levels, dBA

$L_{eq(1)}$	L_{10}	Area
57 (exterior)	60 (exterior)	Within parks, open spaces, and other tracts of land requiring special qualities of serenity and quiet
67 (exterior)	70 (exterior)	Playgrounds, recreation and picnic areas, and outside of residences, motels, public meeting rooms, schools, churches, libraries, and hospitals
72 (exterior)	75 (exterior)	Other developed lands
52 (interior)	55 (interior)	Within residences, motels, hotels, public meeting rooms, schools, houses of worship, libraries, hospitals, and auditoriums

TABLE 49.27 General Specifications for Ships of the United States Navy, Section 073
Airborne noise levels in decibels

Ship Space Category	Octave band center frequencies, Hz									SIL value
	32	63	125	250	500	1000	2000	4000	8000	
A	115	110	105	100		SIL value requirement		85	85	64
B	90	84	79	76	73	71	70	69	68	
C	85	78	72	68	65	62	60	58	57	
D	115	110	105	100	90	85	85	85	95	
E	115	110	105	100		SIL value requirement		85	85	72
F	115	110	105	100		SIL value requirement		85	85	65

Category A: Spaces, other than category E spaces, where intelligible speech communication is necessary.

Category B: Spaces where comfort of personnel in their quarters is normally considered to be an important factor.

Category C: Spaces where it is essential to maintain especially quiet conditions.

Category D: Spaces or areas where a higher noise level is expected and where deafness avoidance is a greater consideration than intelligible speech communication.

Category E: High noise level areas where intelligible speech communication is necessary.

Category F: Topside operating stations on weather decks where intelligible speech communication is necessary.

band. A sound pressure level of 57 dB in the 63-Hz band, for example, corresponds to NC 30, whereas 57 dB in the 125-Hz band corresponds to NC 40.

If a sound pressure level spectrum in the eight octave bands were 57, 60, 62, 54, 51, 44, 38, and 32 dB, the corresponding NC levels for each octave band would therefore be 30, 45, 55, 50, 50, 45, 40, and 35. However, since the highest NC level is 55, it is also the controlling one for this single-number rating. Not

TABLE 49.28 OSHA Criteria for Hearing Conservation Programs

Employers shall administer continuing, effective hearing conservation programs wherever employee noise exposures equal or exceed an 8-hour time-weighted average of 85 dBA or, equivalently, a dose of 50% measured according to the following equation:

$$D = 100 \left(\frac{C_1}{T_1} + \frac{C_2}{T_2} + \cdots + \frac{C_N}{T_N} \right)$$

where D = workday dose, %
1, 2, 3 = periods of exposure to different levels
C = actual exposure timc at different levels
T = permissible exposure time at a given level in accordance with the following table

A-weighted sound level L, dB	Reference duration T, h	A-weighted sound level L, dB	Reference duration T, h
80	32.0	92	6.2
81	27.9	93	5.3
82	24.3	94	4.6
83	21.1	95	4.0
84	18.4	96	3.5
85	16.0	97	3.0
86	13.9	98	2.6
87	12.1	99	2.3
88	10.6	100	2.0
89	9.2	101	1.7
90	8.0	102	1.5
91	7.0	103	1.4

Examples:

1. Assume exposure of:

85 dBA for 5 hours   87 dBA for 2 hours   80 dBA for ½ hour

$$D = 100 \left(\frac{5}{16} + \frac{2}{12.1} + \frac{0.5}{32} \right) = 49.34\%$$

(acceptable, since D is less than 50%)

2. Assume exposure of:

100 dBA for 1 hour   90 dBA for 4 hours   85 dBA for 3 hours

$$D = 100 \left(\frac{1}{2} + \frac{4}{8} + \frac{3}{16} \right) = 118.75\%$$

(unacceptable, since D exceeds 50%)

Note: The exposure in example 2, when evaluated in reference to OSHA criteria for engineering or administrative controls, can be shown to be acceptable, since levels below 90 dBA do not enter into those criteria. However, exposures exceeding a 50% dose still require implementation of hearing conservation programs.

necessarily an intentional justification is the fact that spectra with strong peaks can be more objectionable than NC comparisons would indicate.

Typical NC design levels for a variety of indoor space usages are shown in Table 49.32.

NC curves have been in widespread use since the late 1950s. Since then, sim-

TABLE 49.29 OSHA Criteria for Engineering or Administrative Controls

Feasible administrative or engineering controls shall be utilized if noise dose D is greater than 1.0 in accordance with the following equation:

$$D = \frac{C_1}{T_1} + \frac{C_2}{T_2} + \frac{C_3}{T_3} + \cdots + \frac{C_N}{T_N}$$

where D = daily noise dose (must not exceed unity)
C = actual exposure time at a given noise level
T = permissible exposure time at that level in accordance with the following table

Duration per day, h	Permissible exposure "slow" response, dBA
8	90
6	92
4	95
3	97
2	100
1.5	102
1	105
0.5	110
0.25 or Less	115

Exposure to impulsive or impact noise should not exceed 140 dB peak sound pressure level.

Examples:

1. For an 8-hour day at constant noise levels, 90 dBA is the maximum allowable level.
2. Assume exposure of:

   100 dBA for 2 hours   90 dBA for 6 hours

   $$D = \frac{2}{2} + \frac{6}{8} = 1.75$$

   Engineering or administrative controls are necessary to reduce noise dose to unity.
3. Assume exposure of:

   100 dBA for 1 hour   90 dBA for 4 hours   85 dBA for 3 hours

   Exposure below 90 dBA does not contribute to OSHA noise "dose" for administrative or engineering controls to be employed. Therefore:

   $$D = \frac{1}{2} + \frac{4}{8} = 1.00$$

   (acceptable)

TABLE 49.30 U.S. Department of Housing and Urban Development (HUD) Site Acceptability Standards

L_{dn} at 6.5 ft (2 m) from building setback line nearest noise source	Acceptability
Not exceeding 65 dBA	Normally acceptable
Above 65 dBA but not exceeding 75 dBA	Normally unacceptable
Above 75 dBA	Unacceptable

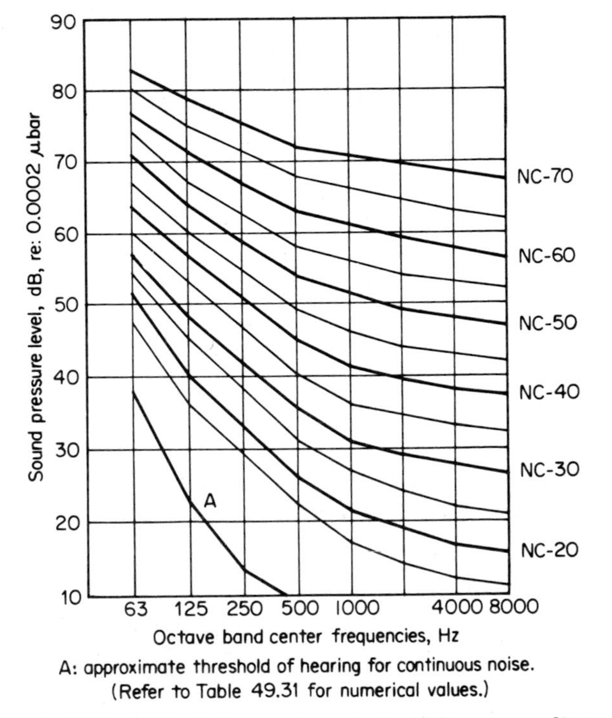

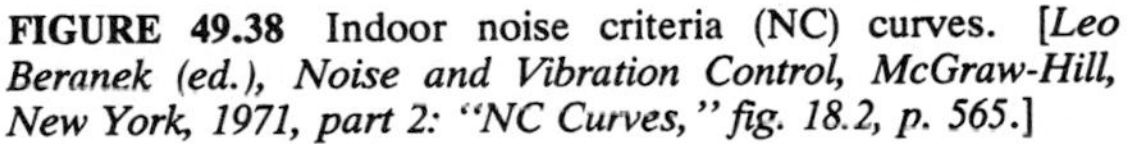
A: approximate threshold of hearing for continuous noise. (Refer to Table 49.31 for numerical values.)

FIGURE 49.38 Indoor noise criteria (NC) curves. [*Leo Beranek (ed.), Noise and Vibration Control, McGraw-Hill, New York, 1971, part 2: "NC Curves," fig. 18.2, p. 565.*]

ilar systems have been proposed: preferred noise criteria (PNC) curves, room criteria (RC) curves, and noise rating (NR) curves. These have not enjoyed the popularity of NC curves, but they have been used occasionally and should be understood.

Preferred Noise Criteria (PNC) Curves. When researchers created a spectrum corresponding to an NC curve in all octave bands, they found that the resultant sound was objectionable in terms of low-frequency rumble as well as high-frequency hissing.

TABLE 49.31 NC Curve Tabulations

Noise criteria	Octave band number 1	2	3	4	5	6	7	8
NC 20	51	41	33	26	22	19	17	16
NC 25	54	45	38	31	27	24	22	21
NC 30	57	48	41	35	31	29	28	27
NC 35	60	53	46	40	36	34	33	32
NC 40	64	57	51	45	41	39	38	37
NC 45	67	60	54	49	46	44	43	42
NC 50	71	64	59	54	51	49	48	47
NC 55	74	67	62	58	56	54	53	52
NC 60	77	71	67	63	61	59	58	57
NC 65	80	75	71	68	66	64	63	62
NC 70	83	79	75	72	71	70	69	68

TABLE 49.32 Typical Room Design NC Criteria, dB

Type of area	Low	Average	High
Residences			
Private homes (rural and suburban)	20	25	30
Private homes (urban)	25	30	35
Apartment houses, two- and three-family units	30	35	40
Hotels			
Room, suites, banquet rooms, ballrooms	30	35	40
Halls, corridors, lobbies	35	40	45
Kitchens, laundries, garages	40	45	50
Hospitals and clinics			
Private rooms	25	30	35
Operating rooms, wards	30	35	40
Laboratories, halls, lobbies, waiting rooms	35	40	45
Washrooms and toilets	40	45	50
Offices			
Boardroom	20	25	30
Conference rooms	25	30	35
Private offices, reception rooms	30	35	40
General open offices, drafting rooms	35	40	50
Halls and corridors	35	45	55
Tabulation and computation	40	50	60
Auditoriums and music halls			
Concert and opera halls, sound studios	20	22	25
Legitimate theaters, multipurpose halls	25	27	30
Movie theaters, TV audience studios, semi-outdoor amphitheaters, lecture halls	30	32	35
Lobbies	35	40	45

TABLE 49.32 Typical Room Design NC Criteria, dB (*Continued*)

Type of area	Low	Average	High
Houses of worship and schools			
Sanctuaries	20	25	30
Libraries, classrooms	30	35	40
Laboratories, recreation halls	35	40	45
Corridors, halls, kitchens	35	45	50
Public buildings			
Public libraries, museums, courtrooms	30	35	40
Post offices, banking areas, lobbies	35	40	45
Washrooms and toilets	40	45	50
Restaurants, cafeterias, lounges			
Restaurants, nightclubs	35	40	45
Cocktail lounges	35	45	50
Cafeterias	40	45	50
Stores, retail			
Clothing and department stores	35	40	45
Department stores (main floor), small stores, supermarkets	40	45	50
Sports activities indoors			
Coliseums	30	35	40
Bowling alleys, gymnasiums	35	40	45
Swimming pools	40	50	55
Transportation (rail, bus, plane)			
Ticket sales offices	30	35	40
Lounges and waiting rooms	35	45	50
Manufacturing areas			
Supervisor's office	40	45	50
Assembly lines, light machinery	45	60	70
Foundries, heavy machinery	55	65	75

The preferred noise criteria (PNC) curves developed in 1971 are generally 4 to 5 dB lower than the NC curves in the 63-Hz band; 1 dB lower in the 125-, 250-, 500-, and 1000-Hz bands; and 4 to 5 dB lower in the 2000-, 4000-, and 8000-Hz bands. PNC curves are shown in Fig. 49.39 and tabulated in Table 49.33. They are used the same way as NC curves.

Room Criteria (RC) Curves. The use of NC curves can result in undesirable rumble if the NC level is determined primarily by sound pressure levels at the lower frequencies or in undesirable hissing if the NC level is determined at higher frequencies.

To establish a more balanced sound, RC curves have been established for which the objective is to design sound spectra to within ±2 dB of an RC curve at all frequencies. In this regard, a spectrum exceeding an RC curve by more than 5 dB below 250 Hz may result in too much rumble, whereas a spectrum more than 5 dB higher at 2000 Hz would probably have an unacceptable hissing quality. RC curves are shown in Fig. 49.40. Recommended RC curve design goals are the same as those shown in Table 49.32.

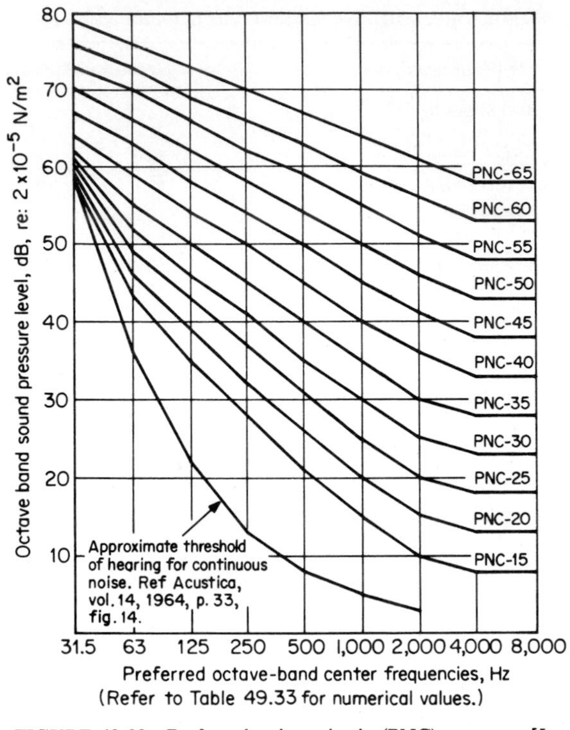

FIGURE 49.39 Preferred noise criteria (PNC) curves. [*Leo Beranek (ed.), Noise and Vibration Control, McGraw-Hill, New York, 1971, part 2: "NC Curves," fig. 18.3, p. 567.*]

TABLE 49.33 Preferred Noise Criteria (PNC) Curve Tabulations

PNC	Octave band center frequency, Hz								
curve	31.5	63	125	250	500	1000	2000	4000	8000
PNC 15	58	43	35	28	21	15	10	8	8
PNC 20	59	46	39	32	26	20	15	13	13
PNC 25	60	49	43	37	31	25	20	18	18
PNC 30	61	52	46	41	35	30	25	23	23
PNC 35	62	55	50	45	40	35	30	28	28
PNC 40	64	59	54	50	45	40	36	33	33
PNC 45	67	63	58	54	50	45	41	38	38
PNC 50	70	66	62	58	54	50	46	43	43
PNC 55	73	70	66	62	59	55	51	48	48
PNC 60	76	73	69	66	63	59	56	53	53
PNC 65	79	76	73	70	67	64	61	58	58

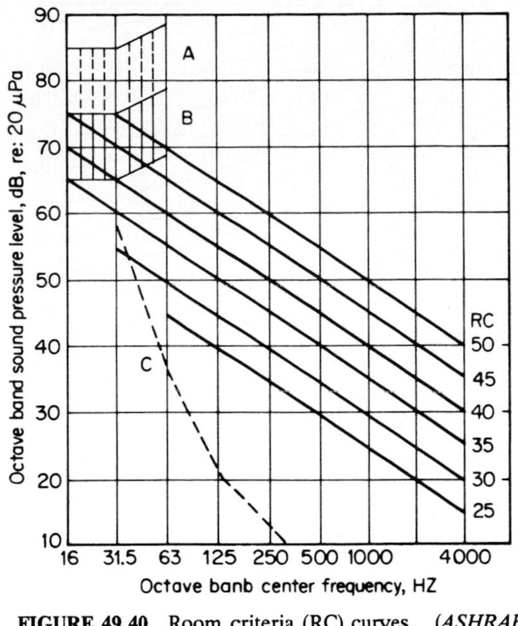

FIGURE 49.40 Room criteria (RC) curves. (*ASHRAE Handbook 1987 Systems, Chap. 52, "Sound and Vibration Control," American Society of Heating, Refrigerating and Air-Conditioning Engineers, Inc., Atlanta, 1987.*)

Noise Rating (NR) Curves. NR curves were developed by the International Organization for Standardization (ISO) in 1971 to rate noisiness with the 1000-Hz octave band as the reference point. At NR 70, for instance, the curve has a level of 70 dB at 1000 Hz.

Compared to NC curves, the NR curve generally permit higher levels at lower frequencies and lower levels at higher frequencies, although there are considerable variations in the NC/NR 20 to NC/NR 70 range. An NR level is determined by the highest level corresponding to any of the octave bands. NR curves are shown in Fig. 49.41 and tabulated in Table 49.34.

49.20.4 Speech Interference Levels

Speech interference levels (SILs) are the arithmetic average of the sound pressure levels at the 500-, 1000-, and 2000-Hz center frequencies. Table 49.35 shows the SILs and distances at which the average person would have to talk in a normal, raised, and very loud (or shouting) voice to be understood. Table 49.36 shows how the relative difficulty of telephone usage in noisy areas varies with SIL and dBA levels.

Assuming a spectral content of the NC curves in Fig. 49.38, Table 49.37 shows the approximate relationships between NC level, SIL, and dBA. These relationships will vary with other spectral shapes, however, and should be checked accordingly.

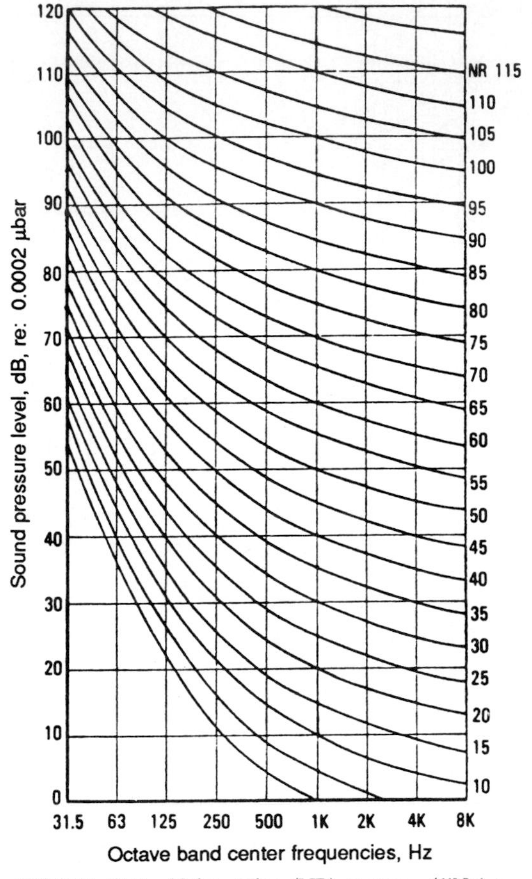

FIGURE 49.41 Noise rating (NR) curves. (*"Noise Assessment with Respect to Community Response," ISO Draft Standard 1996, November 1969.*)

49.20.5 Ambient Noise Levels as Criteria

In the absence of standards, the simplest approach to establishing criteria is to stipulate that the noise from equipment such as fans or other HVAC machinery shall not cause the ambient, or background, sound levels to increase in any octave band. This assumes, of course, that these levels are satisfactory and are not affected by other existing noisemakers.

The combined effect of ambient sound and equipment sound can be additive. Suppose the existing sound pressure level is 60 dB in a given octave band; if other equipment contributes a sound pressure level of 60 dB in the same octave band, the resultant theoretical sound pressure level would be approximately 63 dB (see Fig. 49.6). Ideally, equipment sound levels should be 10 dB below the ambient sound level so as not to raise the combined sound level.

Table 49.38 shows the octave band ambient sound levels typically found in a

TABLE 49.34 Noise Rating (NR) Curve Tabulations

NR curve	Octave band center frequency, Hz								
	31.5	63	125	250	500	1000	2000	4000	8000
0	55.4	35.5	22.0	12.0	4.8	0	−3.5	−6.1	−8.0
5	58.8	39.4	26.1	16.6	9.7	5	+1.6	−1.0	−2.8
10	62.2	43.4	30.7	21.3	14.5	10	6.6	+4.2	+ 2.3
15	65.6	47.3	35.0	25.9	19.4	15	11.7	9.3	7.4
20	69.0	51.3	39.4	30.6	24.3	20	16.8	14.4	12.6
25	72.4	55.2	43.7	35.2	29.2	25	21.9	19.5	17.7
30	75.8	59.2	48.1	39.9	34.0	30	26.9	24.7	22.9
35	79.2	63.1	52.4	44.5	38.9	35	32.0	29.8	28.0
40	82.6	67.1	56.8	49.2	43.8	40	37.1	34.9	33.2
45	86.0	71.0	61.1	53.6	48.0	45	42.2	40.0	38.3
50	89.4	75.0	65.5	58.5	53.5	50	47.2	45.2	43.5
55	92.9	78.9	69.8	63.1	58.4	55	52.3	50.3	48.6
60	96.3	82.9	74.2	67.8	63.2	60	57.4	55.4	53.8
65	99.7	86.8	78.5	72.4	68.1	65	62.5	60.5	58.9
70	103.1	90.8	82.9	77.1	73.0	70	67.5	65.7	64.1
75	106.5	94.7	87.2	81.7	77.9	75	72.6	70.8	69.2
80	109.9	98.7	91.6	86.4	82.7	80	77.7	75.9	74.4
85	113.3	102.6	95.9	91.0	87.6	85	82.8	81.0	79.5
90	116.7	106.6	100.3	95.7	92.5	90	87.8	86.2	84.7
95	120.1	110.5	104.6	100.3	97.3	95	92.9	91.3	89.8
100	123.5	114.5	109.0	105.0	102.2	100	98.0	96.4	95.0
105	126.9	118.4	113.3	109.6	107.1	105	103.1	101.5	100.1
110	130.3	122.4	117.7	114.3	111.9	110	108.1	106.7	105.3
115	133.7	126.3	122.0	118.9	116.8	115	113.2	111.8	110.4
120	137.1	130.3	126.4	123.6	121.7	120	118.3	116.9	115.6
125	140.5	134.2	130.7	128.2	126.6	125	123.4	122.0	120.7
130	143.9	138.2	135.1	132.9	131.4	130	128.4	127.2	125.9

Note: This table is not part of ISO Standard 1996, but it does correspond to the NR curves shown in Fig. 49.41.

TABLE 49.35 Approximate Speech Interference Levels (SILs)

Distance		SIL, dBA			
ft	mm	Normal	Raised	Very loud	Shouting
1	305	77	83	89	95
3	914	67	73	79	85
6	1829	61	67	73	79
12	3658	55	61	67	73

TABLE 49.36 Telephone Usage in Noisy Areas

SIL, dB	dBA	Telephone use
Less than 65	72	Satisfactory
65–80	72–87	Difficult
Above 80	87	Impossible

TABLE 49.37 Approximate Relationships between NC Level, SIL, and dBA

NC	20	30	40	50	60	70
SIL	22	32	42	51	61	71
dBA	31	40	49	58	68	77

Source: M. Hirschorn, *Noise Control Reference Handbook*, Industrial Acoustics Company, Bronx, NY, 1989.

TABLE 49.38 Estimated Outdoor Ambient Sound Levels, dB

	Octave band center frequency, Hz							
	63	125	250	500	1000	2000	4000	8000
	Octave band number							
Condition	1	2	3	4	5	6	7	8
Nighttime								
Rural*	42	37	32	27	22	18	14	12
Suburban*	47	42	37	32	27	23	19	17
Urban*	52	47	42	37	32	28	24	22
Business or commercial area	57	52	47	42	37	33	29	27
Daytime								
Business or commercial area	62	57	52	47	42	38	34	32
Industrial or manufacturing area	67	62	57	52	47	43	39	37
Within 300 ft (91 m) of continuous heavy traffic	72	67	62	57	52	48	44	42

*No nearby traffic of concern.

Note: The sound levels listed here are generally applicable for various outdoor locations and thus can be used as design criteria.

variety of outdoor environments. Indoor ambient levels can be estimated from the data in Table 49.32.

49.21 ENCLOSURE AND NOISE PARTITION DESIGN CONSIDERATIONS

49.21.1 Actual versus Predicted Sound Transmission Losses

One rule of noise control engineering is that theoretical prediction schemes for acoustic structures or silencers are useful only as guidelines. Substantial and costly errors can develop if not checked out in the laboratory and/or field. For instance, air-handling units (Sec. 49.28) for acoustically critical situations are usually checked out in specially constructed full-scale test facilities, which try to encompass all variables before assembling 91 identical units, as in one application.

The sound transmission loss (TL) of partitions is generalized by the mass law: Eq. (49.23). This equation is valid only when incident sound is normal to the par-

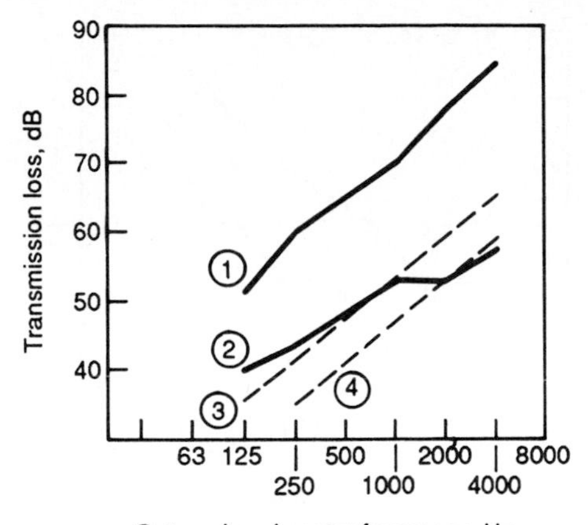

1. Measured data for two 4-in (102-mm) acoustic panels separated by 4-in (102-mm) air space. Total weight: 20 lb/ft² (98 kg/m²).
2. Measured data for single panel, 4-in (102-mm) thick. Weight: 10 lb/ft² (49 kg/m²).
3. Mass law prediction for 20 lb/ft² (98 kg/m²) panel.
4. Mass law prediction for dense concrete: 10-lb/ft² (49-kg/m²) panel.

FIGURE 49.42 Sound transmission loss of single- and double-wall acoustic panels compared to mass law predictions. (*Martin Hirschorn, Noise Control Reference Handbook, Industrial Acoustics Company, 1982.*)

tition's surface and is within a frequency range where the TL is not affected by the partition's stiffness or internal damping. For multilayer and double-wall structures separated by air spaces, the mass law is not applicable.

This disparity between theory and practice is illustrated in Fig. 49.42, which compares the measured and predicted performance levels of commercially available double- and single-wall nonabsorptive noise control partitions 4 in (102 mm) thick.

The measured TL of the 10-lb/ft² (48.8-kg/m²) single-wall partition is in some instances more than 10 dB greater than predicted by the mass law. To achieve a TL of 40 dB in the 125-Hz octave band, for example, the mass law indicates a surface density of 38 lb/ft² (185.5 kg/m²).

Figure 49.43 compares the TL of 20-lb/ft² (97.6-kg/m²) modular steel partitions 4 in (102 mm) thick separated by a 4-in (102-mm) air space with that of a 150-lb/ft² (732.3-kg/m²) concrete wall. The concrete-wall data are presented both on a calculated and measured basis. However, modular partitions, regardless of their materials, will be only as good as the construction of their joints.

Table 49.39 shows the measured TL of commercially available modular steel noise control partitions and other building materials (see Sec. 49.12.3).

49.21.2 Joints

To facilitate handling, acoustic panel components for the construction of modular soundproof rooms, machinery enclosures, and fan plenums, for instance, usually measure no more than 48 in (1219 mm) wide by 144 in (3658 mm) high. Their

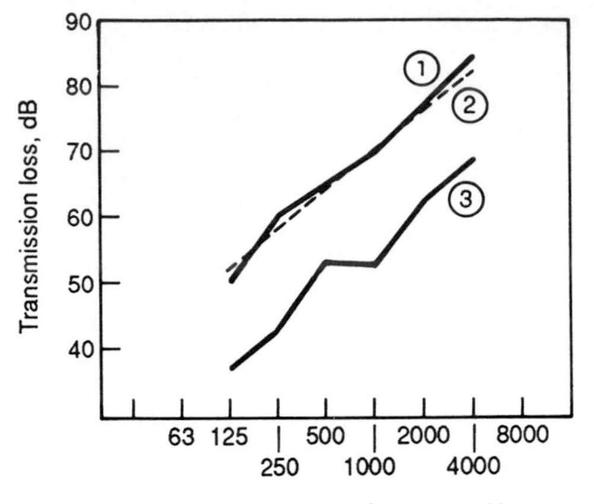

1. Measured data for two 4-in (102-mm) acoustic panels separated by 4-in (102-mm) air space. Total weight: 20 lb/ft² (98 kg/m²).
2. Mass law prediction: 150-lb/ft² (723-kg/m²) panel.
3. Measured data for dense concrete: 150-lb/ft² (723-kg/m²) panel.

FIGURE 49.43 Sound transmission loss of double-wall acoustic panel compared to single concrete wall and mass law prediction. (*Martin Hirschorn, Noise Control Reference Handbook, Industrial Acoustics Company, 1982.*)

weight ranges from 380 to 480 lb (172 to 218 kg), depending on their acoustic and structural characteristics. When employed as wall sections, they must be joined together to protect the acoustic integrity of the panels. Figure 49.18 shows that an opening as small as 0.1 percent can reduce the TL of a partition from 40 dB to 30 dB, as one example.

Also, HVAC plenums must be airtight because they can be subjected to air-pressure differentials on the order of 10 inWG (2491 Pa). In addition, they must be designed to withstand seismic upsets, and outdoor installations must be able to withstand snow covers and wind-velocity forces in excess of 100 mi/h (161 km/h).

Figure 49.44 shows several acoustic panel joint designs. Note the following about these designs:

- Batten strips can work acoustically but have limited structural strength.
- Back-to-back channel joiners have greater strength but are difficult to seal against air-flow or noise leakage.
- Tongue-in-groove joints can provide acceptable acoustic seals without using separate joiners and can provide reasonable structural strength for relatively lower pressure applications.
- H joiners, roll-formed from single lengths of steel, form center-supported box beams and constitute excellent panel joints for high-pressure applications. They also provide excellent acoustic joints.

Figure 49.45 shows the load-bearing strength of panels 4 in (102 mm) thick and 48 in (1219 mm) wide connected with H sections. A thinner tongue-in-groove

TABLE 49.39 Transmission Loss (TL) Data and STC Ratings for Commercially Available Noise Control Partitions and Commonly Used Building Materials

Product	Octave band center frequency, Hz								STC	Weight	
	63	125	250	500	1000	2000	4000	8000		lb/ft^2	kg/m^2
Acoustic partition											
Noishield regular panel	20	21	27	38	48	58	67	66	40	8.0	39.1
Noise-Lock I panel	25	27	31	41	51	60	65	66	44	9.5	46.4
Noise-Lock II, Fire-Noise-Lock panel	27	30	32	41	50	59	67	71	45	10.5	51.3
Super-Noise-Lock panel	31	34	35	44	54	63	62	68	48	15.0	73.2
Noishield hard panel	22	33	45	52	58	68	75	65	56	9.5	46.4
Noishield septum	21	19	23	35	50	60	68	72	37	9.0	43.9
Trackwall (industrial regular)	18	25	35	45	52	51	56	58	46	10.0	48.8
APR single-leaf personnel door	22	22	28	39	33	31	35	37	33	7.0	34.2
Standard Noise-Lock door	. . .	26	42	43	47	52	56	. . .	47	7.0	34.2
Single Trackwall (architectural)	. . .	40	44	48	53	53	58	. . .	51	10.0	48.8
Double Trackwall (architectural)	. . .	51	60	65	70	78	85	. . .	70	20.0	98.0
Single Trackwall (architectural, absorptive on one side)	. . .	28	40	50	53	53	58	. . .	50	14.0	68.3
Quadraseal Noise-Lock door	. . .	47	52	58	66	70	67	. . .	63	20.0	98.0
Building material											
Concrete, 12 in (305 mm) thick	30	37	43	53	53	63	69	. . .	53	150.0	732.0
Plasterboard, ⅜ in (9.5 mm) thick	. . .	12	18	22	28	32	25	. . .	26	1.6	7.8
Plaster, ½ in (12.7 mm) thick, over ⅜-in (9.5-mm) gypsum lath, both sides 2- by 4-in (51- by 102-mm) studs on 16-in (406-mm) centers	. . .	30	37	42	44	39	51	. . .	39	13.4	65.4
Galvanized steel, 22 ga (0.7 mm)	. . .	13	17	22	28	34	38	. . .	27	1.4	6.8
Solid-core wood door, normally closed	. . .	23	27	29	27	26	29	. . .	27	3.9	19.1

Back-to-back channels

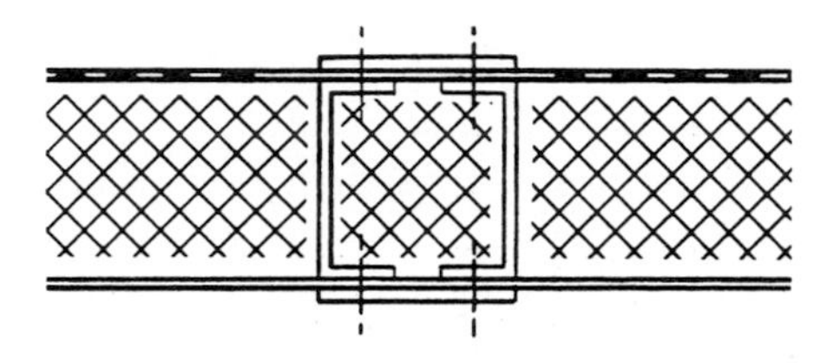

Batten strips

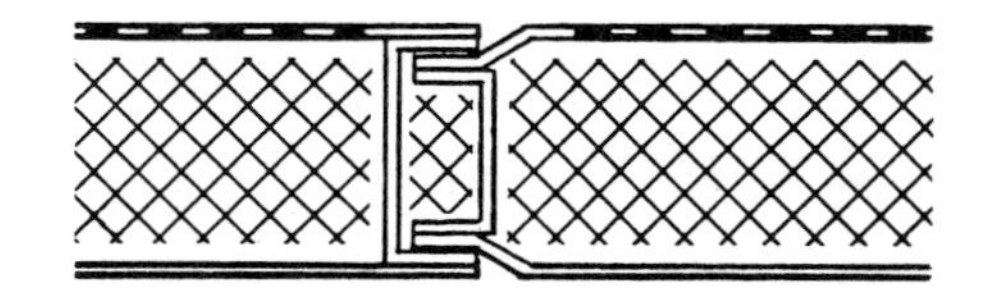

Tongue and groove joint

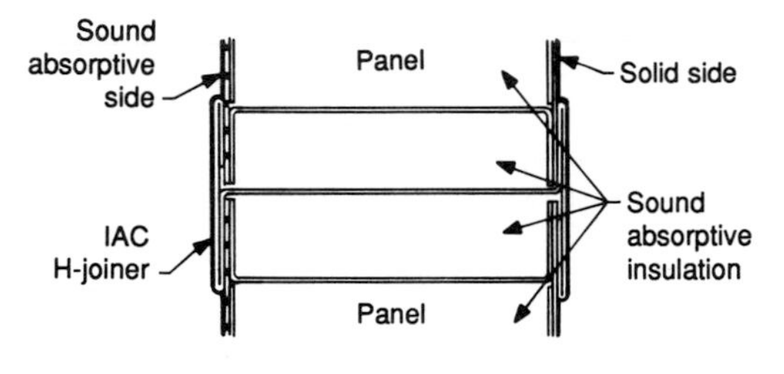

One-piece H-joiner

FIGURE 49.44 Acoustic panel joints. (*Courtesy of Industrial Acoustics Company.*)

panel 3 in (76 mm) thick and 40 in (1016 mm) wide is also shown, though 2-in (51-mm) tongue-in-groove panels are also used in low-pressure fan plenum installations. All of these are of a welded-steel acoustic plenum panel design, and all are commercially available.

The data in Fig. 49.45 indicate that the heavier and wider panels connected by H joiners can be installed in longer unsupported spans than can the narrower and thinner tongue-in-groove structures.

Adjacent wall and roof sections are not the only locations where close atten-

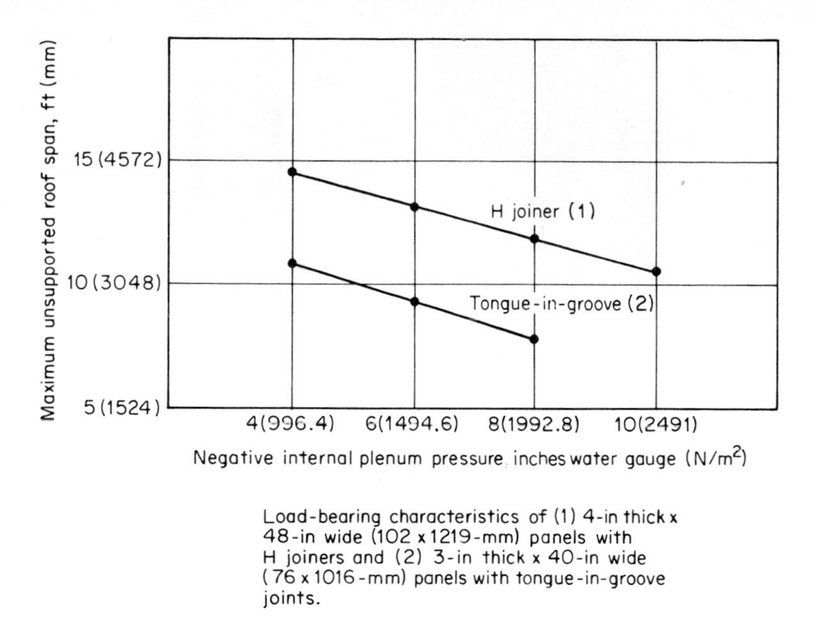

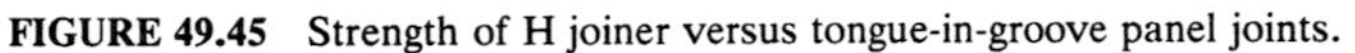

FIGURE 49.45 Strength of H joiner versus tongue-in-groove panel joints.

tion must be paid to airtight, acoustically tight joiners. Wall-to-ceiling, wall-to-floor, and corner joints are equally important, and examples of such good joint designs are shown in Fig. 49.46.

49.21.3 Windows and Seals

Here again, joints and seals play an important role. Window panes must be securely sealed within their frame perimeters, and the frame itself must be sealed in place within the wall. Desiccant should be provided within double-glazed windows to dry up moisture between the panes.

Figure 49.47 shows the sound transmission loss (TL) characteristics of 0.236- and 0.472-in- (6- and 12-mm-) thick single-glazed windows and of one double-glazed window with a 7.87-in (200-mm) air space and sound-absorbent material inside the air space.

For single-glazed windows, the data in Ref. 14 show that attenuation up to about 500 Hz increases by 3 to 5 dB when doubling the glass thickness from 6 mm to 12 mm; however, resonance and coincidence effects cause significant TL dips at specific frequencies. Coincidence effects occur because of the excitation of a panel by an airborne sound wave. A coincidence effect's frequency is reduced by increasing the thickness of the glass. Laminated windows of the same thickness will significantly reduce the reduction in TL caused by coincidence effects.

For double-glazed windows, air spaces between two glass panes will not produce significant TL improvements at spacings less than approximately 4 in (101.6 mm). The use of sound-absorptive materials inside the air space is definitely helpful, as is the use of glass panes of dissimilar thicknesses.

Figure 49.48 shows a modern acoustic window design that uses high-strength

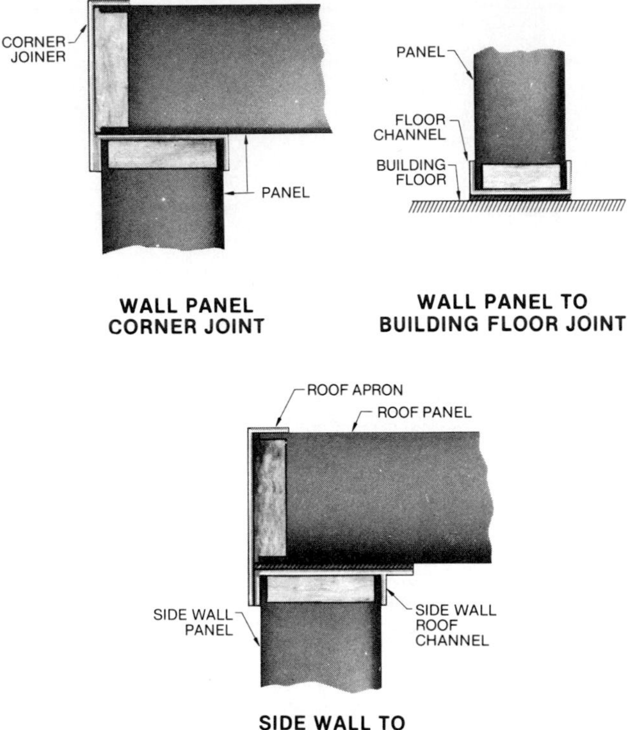

FIGURE 49.46 Wall-to-ceiling, wall-to-floor, and corner joints. (Moduline for New Construction and Renovation, *Bulletin 6.0513.0, Industrial Acoustics Company.*)

aluminum-alloy framing, tempered or laminated glass, extruded aluminum mullions, and weather-and-acoustic seals. Multiple modules as large as 48 by 48 in (1219 by 1219 mm) can be installed in lengths practically without limit to form a "window wall" (Fig. 49.49), such as in a sewage treatment plant.

Acoustically, the design shown in Fig. 49.48 exhibits a sound transmission class (STC) performance of STC 39 when single-glazed with ½-in (12.7-mm) glass, of STC 49 when double-glazed with ¼-in (6.35-mm) glass and a 7.9-in (200-mm) air space, and of STC 50 when the double glazing uses glass thicknesses of ½ in (12.7 mm) and ¼ in (6.35 mm). Capable of withstanding hurricane-force winds of up to 182 mi/h (293 km/h), the structure has a thermal U-value of 0.40 Btu · h/(ft^2 · °F) [2.27 W/(m^2 · °C)]. And it allows no water penetration when tested in accordance with ASTM Specification 331-70 under conditions equivalent to a 68-mi/h (109-km/h) wind load and a heavy rain.

49.21.4 Doors and Seals

Relying on silenced duct systems for ventilation, many acoustically designed structures use fixed (nonopening) windows. The acoustic doors are operable, of

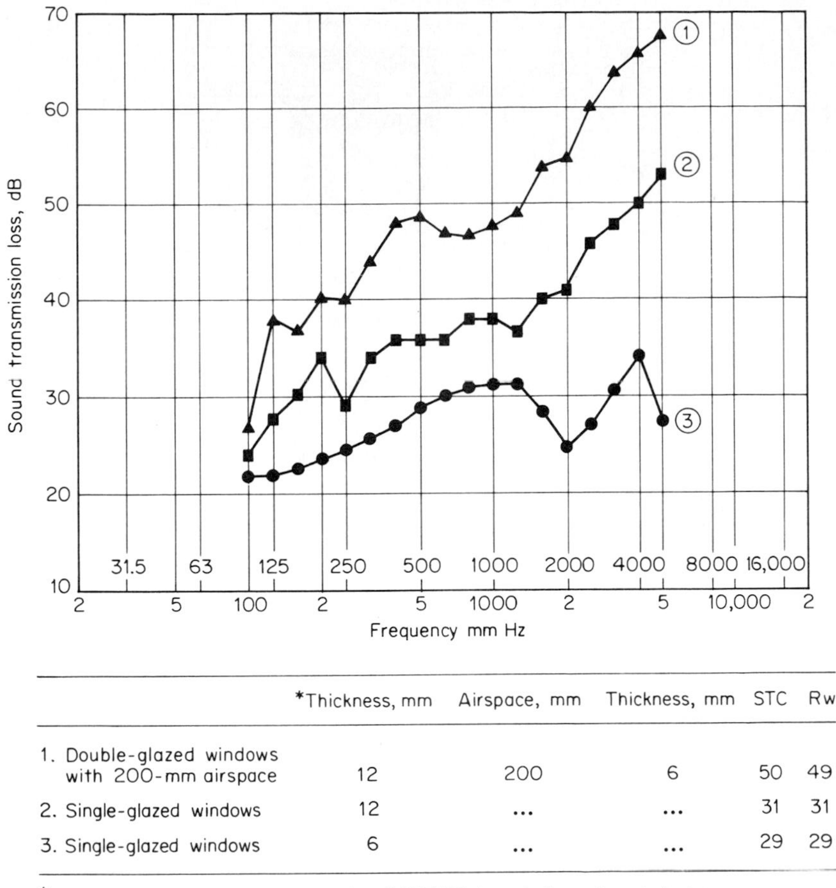

	*Thickness, mm	Airspace, mm	Thickness, mm	STC	Rw
1. Double-glazed windows with 200-mm airspace	12	200	6	50	49
2. Single-glazed windows	12	...	...	31	31
3. Single-glazed windows	6	...	...	29	29

*Multiply dimension in millimeters by 0.03937 to get dimensions in inches.

FIGURE 49.47 Sound transmission loss for double-glazed windows with air space and for single-glazed windows. (*Curve 1: Industrial Acoustics Company. Curves 2 and 3: Technical Advisory Service, Pilkington Flat Glass, Ltd., England.*)

course, but their seals must be designed to provide reliable perimeter tightness while allowing easy access.

Still another requirement in many applications is a flush sill. Sills raised to create a seal can obstruct access and may not be suitable when "barrier-free" doorways are required. "Sweep" seals, which have continuous contact with the floor, can acoustically eliminate the need for sills but may be subject to high wear and may make a door more difficult to operate. Crank-down seals can also be used, but they are inconvenient for doors requiring quick and easy access.

The state-of-the-art swinging door (Fig. 49.50 on p. 49.70) uses cam-action hinges that gradually raise and lower the leaf approximately ½ in (12.7 mm) from fully closed to fully open and back again. A simple seal at the bottom of the leaf compresses against the floor as the door closes, and it lifts up as the door opens. At the jambs and head of the door, magnetic seals are used. Self-adjusting against

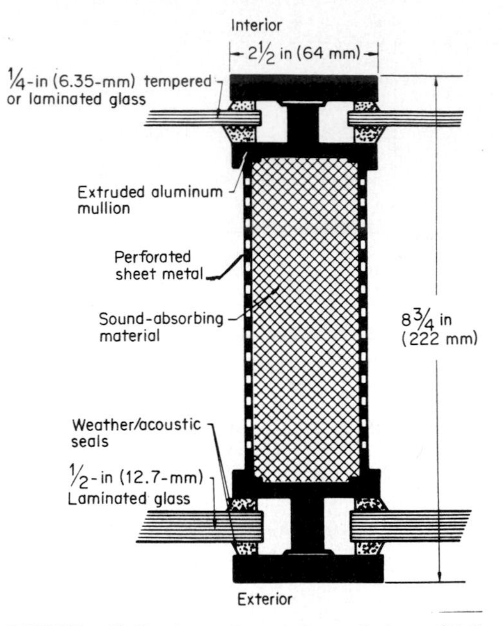

FIGURE 49.48 Acoustic window design. (IAC Noise-Lock Vision Wall Window Wall System, *Bulletin 3.0301.1, Industrial Acoustics Company.*)

the steel frame, these seals compensate for slight misalignments due to plumb tolerances or lack of absolute flatness. Note that these are double seals, providing closure at both the inside and outside surfaces of the opening.

In sliding doors, bottom closure can be accomplished by using a crank-down or pneumatically operated bottom seal or by aligning the door on an incline so that a compression seal lifts on opening and seals on closing. The best head and jamb seals for sliding doors are pneumatic devices that are actuated only when the door is in the closed position (Fig. 49.51).

The performance of commercially available acoustic doors is shown in Table 49.40. Sound transmission loss versus frequency for a range of STC contours is shown in Table 49.41.

49.21.5 Transmission Loss of Composite Structures

The transmission loss of an acoustic structure (partition or enclosure) can be no greater than the TL of its elements. If wall panels have a TL of 50 dB in a given octave band, for example, an overall TL of 50 dB could not be achieved unless all doors, windows, and joiners exhibit corresponding TL characteristics. The TL of a composite partition consisting of acoustically dissimilar elements (such as walls and doors) can be approximated by the following theoretical relationships:

$$\mathrm{TL}_c = 10 \log \frac{1}{\tau_c} \qquad (49.39a)$$

FIGURE 49.49 Acoustic "window wall." *(IAC Vision Wall, Bulletin 3.1232.1, Industrial Acoustics Company.)*

$$\tau_c = \frac{\tau_1 S_1}{S_T} + \frac{\tau_2 S_2}{S_T} + \frac{\tau_3 S_3}{S_T} + \cdots + \frac{\tau_N S_N}{S_T} \tag{49.39b}$$

where TL_c = sound transmission loss of composite partition, dB
τ_c = sound transmission coefficient of composite partition
$S_1, S_2, \ldots, S_N$ = area of individual elements in partition, ft^2 (m^2)
$\tau_1, \tau_2, \tau_3, \ldots, \tau_N$ = sound transmission coefficients of individual elements in partition
S_T = total partition area = sum of $S_1, S_2, S_3, \ldots, S_N$

EXAMPLE 49.2 A 22.3-m^2 partition consists of IAC Noishield paneling and a Noise-Lock door. For the paneling, $\tau = 1.58 \times 10^{-6}$ and $S = 20.3$ m^2; for the door, $\tau = 6.31 \times 10^{-6}$ and $S = 2$ m^2. Calculate TL_c at 2000 Hz.

Solution

$$\tau_c = \frac{1.58 \times 10^{-6} \times 20.3}{22.3} + \frac{6.31 \times 10^{-6} \times 2}{22.3} = 2.0 \times 10^{-6}$$

$$TL_c = 10 \log \frac{1}{2 \times 10^{-6}} = 57 \text{ dB}$$

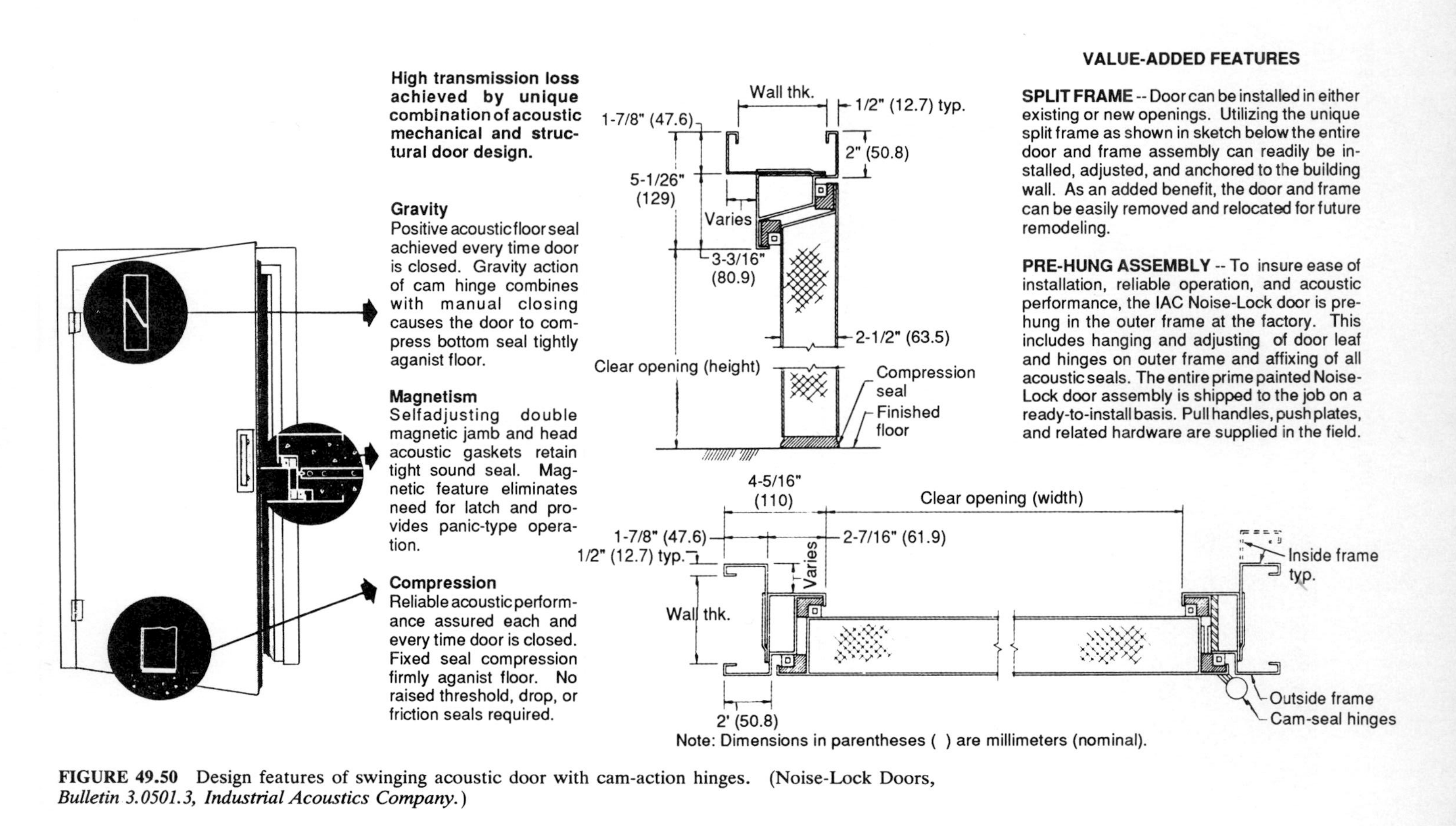

FIGURE 49.50 Design features of swinging acoustic door with cam-action hinges. (Noise-Lock Doors, *Bulletin 3.0501.3, Industrial Acoustics Company.*)

49.21.6 Flanking Paths

An acoustic "flanking path" circumvents noise control silencers or partitions. Typical examples are floors and ceilings that have a lower TL than the partitions between rooms (Fig. 49.52). Another example is break-out noise transmitted through a duct wall into the surrounding space (see Sec. 49.19). Flanking paths can dramatically diminish the effectiveness of the best silencers, acoustic enclosures, and partitions.

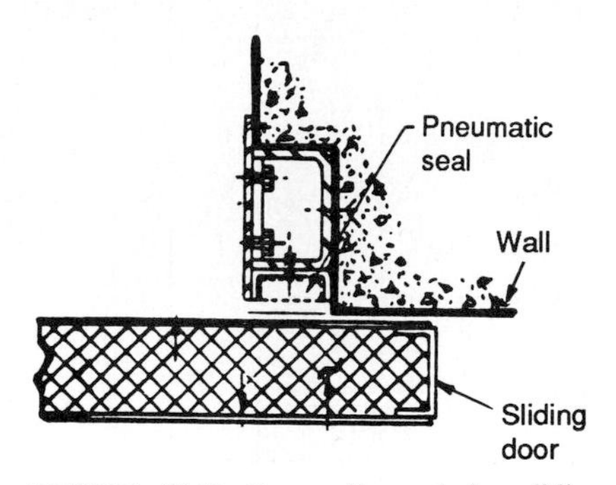

FIGURE 49.51 Pneumatic seal for sliding acoustic door. *(Courtesy of Industrial Acoustics Company.)*

49.21.7 Room Performance

Given the variables of panel and door design, joints and door seals, flanking paths, and ventilation systems, the noise reduction provided by an enclosure or room is most reliably determined by building and testing it as an entity. This has proven to be feasible for modular audiometric rooms, music practice rooms, quiet havens for noisy areas, and many others. Some are often shipped completely preassembled.

Table 49.42 shows the performance of several commercially available modular soundproof rooms. Comparing the acoustic performance of different walls, windows, and doors is meaningful only if all data are taken on the same basis.

The U.S. standard for such building components is ASTM E90, which defines specific test and calculation procedures to be followed in the determination of TL and NR ratings (Ref. 15). For complete rooms, ASTM E596 applies (Ref. 16).

49.22 SOUND ABSORPTION IN ROOMS

The theory of acoustically absorptive surfaces on indoor sound has been summarized in Secs. 49.11 and 49.14. Table 49.43 gives sound absorption coefficients of commercially available acoustic panel products and commonly used building materials.

In many instances, however, actual sound absorption coefficient data may not be readily available for the specific wall, floor, and ceiling surfaces being used. When the sound sources are HVAC terminals, Tables 49.44, 49.45, and 49.46 can be used in a general way to predict how absorption can affect sound pressure levels at a given location.

For rooms of various sizes and absorption categories ("hard," "average," and "soft"), Table 49.44 estimates the expected attenuation when a listener within the room is within 5 ft (1.5 m) of a terminal within the room. At such a close distance, the direct sound field from the terminal predominates over the reverberant sound field, which, however, decreases with increasing room size; therefore, sound-absorptive treatment provides less attenuation in smaller than in larger rooms. Table 49.45 gives examples of rooms that are typically "hard," "average," or "soft" from a sound absorption standpoint. Table 49.46 (p. 49.77)

TABLE 49.40 Sound Transmission Loss of Commercially Available Acoustic Doors

Swinging-door type	Door leaf thickness, in (mm)	Transmission loss, dB — Center frequency, Hz							
		125	250	500	1000	2000	4000	STC	UL label[a]
Magnetic D seal[b]									1 h and 1.5 h
Laboratory test	2.5 (63.5)	26	42	43	47	52	56	47	
Field performance		26	40	38	39	48	54	43	
Quadraseal, laboratory test[c]	2.5, 2.5 (63.5, 63.5)	47	52	58	66	70	67	63	
		Sound absorption coefficient						NRC	
Absorptive one side, magnetic seal, one surface perforated, 23% open area	2.5 (63.5)	0.50	0.68	1.03	1.05	1.00	0.99	0.95	

Sliding-door type	Door panel thickness, in (mm)	Transmission loss, dB — Center frequency, Hz							
		125	250	500	1000	2000	4000	STC	UL label
Standard design I, 14 ga, solid both sides[d]	4 (102)	36	41	47	53	51	58	51	1 h
Standard design II, 14 ga, solid both sides	4 (102)	36	41	47	53	51	58	51	1.5 h
Double-wall standard design, 4-in air space, solid all sides	12 (305)	53	61	65	74	78	88	70	
Absorptive one side, 14 ga solid one side, 18 ga perforated one side	4 (102)	25	36	45	53	50	56	46	

Absorptive one side with septum, 14 ga solid one side, 18 ga perforated one side[3]	4 (102)	26	43	49	54	52	58	50	
Sound absorption coefficient								NRC	
Absorptive septum design, perforated one side, 23% open area[f]	4 (102)	0.45	0.95	0.85	0.85	0.85	0.85	0.90	

Application information[g]

1. All transmission-loss tests conducted in accordance with ASTM E90-61, -66, or -70 and E413-73 by independent acoustic consultant or laboratory. Copies available on request.
2. Sound absorption test conducted in accordance with ASTM C423. Copies available on request.
3. Sound absorption is an *optional* feature with the system. It is obtained by perforating one side of the door leaf with 3⁄32-in (2.38-mm) holes on 3⁄16-in (4.76-mm) staggered centers. This feature may lower transmission loss by approximately 6 dB. Sound absorption will help reduce reverberation time and noise-level buildup on the perforated side of the door.
4. Underwriters' Laboratory fire resistance and water hose tests conducted in accordance with UL 10B procedure. Certification available on request.
5. Maximum glass area for fire-rated doors is 100 in^2 (645 cm^2), with no dimension exceeding 12 in (305 mm); ¼-in (6-mm) wire-reinforced glass must be specified.
6. UL labels available for single-leaf doors up to 3- by 7-ft (914- by 2134-mm) clear opening and for 7-ft by 7-ft 6-in (2134- by 2286-mm) double-leaf designs for flush 2½-in (63.5-mm) doors.
7. 1½-h UL label for single-leaf doors up to 3 ft 6 in by 7 ft 6 in (1067 by 2286 mm) and for double-leaf openings up to 7 ft 0 in by 7 ft 6 in (2134 by 2286 mm). Certifications are available for sizes greater than listed above.
8. UL label available for opening up to 13 ft 6 in by 12 ft 7 in (4115 by 3835 mm). Certifications also available for larger openings.
9. IAC heavy-duty industrial and commercial doors are available with sound transmission class (STC) ratings of 50 and noise isolation class (NIC) ratings of up to 75.

Note: Arithmetic average of sound absorption coefficients in ⅓ octave bands centered at 250, 500, 1000 and 2000 Hz. By convention, maximum NRC used is 0.95.

[a]Fire-rated doors use D-type compression seals and can be supplied with blast pressure ratings. Extensive laboratory testing of doors with magnetic seals and compression D-type seals, respectively, has established that the noise-reduction and sound-transmission-loss characteristics of these seals are closely comparable.

[b]See Fig. 49.50 for a description of the magnetic seal.

[c]A Quadraseal door is two parallel doors acting in tandem.

[d]14 ga = 1.9 mm; 18 ga = 1.2 mm.

[e]A septum door leaf has a solid metal sheet across the entire face as part of the acoustic infill.

[f]Higher sound-absorption data are available without septum.

[g]See ASTM E90, *Standard Method for Laboratory Measurement of Airborne Sound Transmission Loss of Building Partitions*, 1975; and ASTM E413, *Standard Classification for Determination of Sound Transmission Class*, 1973.

Source: Industrial Acoustics Company.

TABLE 49.41 Sound Transmission Loss (TL) versus Frequency for a Range of STC Contours*

Frequency, Hz															
125	160	200	250	315	400	**500**	630	800	1000	1250	1600	2000	2500	3150	4000
44	47	50	53	56	59	**60**	61	62	63	64	64	64	64	64	64
43	46	49	52	55	58	**59**	60	61	62	63	63	63	63	63	63
42	45	48	51	54	57	**58**	59	60	61	62	62	62	62	62	62
41	44	47	50	53	56	**57**	58	59	60	61	61	61	61	61	61
40	43	46	49	52	55	**56**	57	58	59	60	60	60	60	60	60
39	42	45	48	51	54	**55**	56	57	58	59	59	59	59	59	59
38	41	44	47	50	53	**54**	55	56	57	58	58	58	58	58	58
37	40	43	46	49	52	**53**	54	55	56	57	57	57	57	57	57
36	39	42	45	48	51	**52**	53	54	55	56	56	56	56	56	56
35	38	41	44	47	50	**51**	52	53	54	55	55	55	55	55	55
34	37	40	43	46	49	**50**	51	52	53	54	54	54	54	54	54
33	36	39	42	45	48	**49**	50	51	52	53	53	53	53	53	53
32	35	38	41	44	47	**48**	49	50	51	52	52	52	52	52	52
31	34	37	40	43	46	**47**	48	49	50	51	51	51	51	51	51
30	33	36	39	42	45	**46**	47	48	49	50	50	50	50	50	50
29	32	35	38	41	44	**45**	46	47	48	49	49	49	49	49	49
28	31	34	37	40	43	**44**	45	46	47	48	48	48	48	48	48
27	30	33	36	39	42	**43**	44	45	46	47	47	47	47	47	47
26	29	32	35	38	41	**42**	43	44	45	46	46	46	46	46	46
25	28	31	34	37	40	**41**	42	43	44	45	45	45	45	45	45
24	27	30	33	36	39	**40**	41	42	43	44	44	44	44	44	44
23	26	29	32	35	38	**39**	40	41	42	43	43	43	43	43	43
22	25	28	31	34	37	**38**	39	40	41	42	42	42	42	42	42
21	24	27	30	33	36	**37**	38	39	40	41	41	41	41	41	41
20	23	26	29	32	35	**36**	37	38	39	40	40	40	40	40	40
19	22	25	28	31	34	**35**	36	37	38	39	39	39	39	39	39
18	21	24	27	30	33	**34**	35	36	37	38	38	38	38	38	38
17	20	23	26	29	32	**33**	34	35	36	37	37	37	37	37	37
16	19	22	25	28	31	**32**	33	34	35	36	36	36	36	36	36
15	18	21	24	27	30	**31**	32	33	34	35	35	35	35	35	35
14	17	20	23	26	29	**30**	31	32	33	34	34	34	34	34	34

*A particular STC contour is identified by its TL value at 500 Hz.

is the same as Table 49.44 (p. 49.76), except that it also estimates attenuation characteristics in the far field up to 20 ft (6.1 m) from an HVAC terminal.

Figure 49.53 also illustrates how sound is attenuated in accordance with room size, distance from sound source, and degree of sound-absorptive treatment.

49.23 SILENCERS

Sections 49.17 and 49.18 discussed silencer terminology, types of silencers, and the effects of forward and reverse flow on silencer self-noise (SN) and dynamic

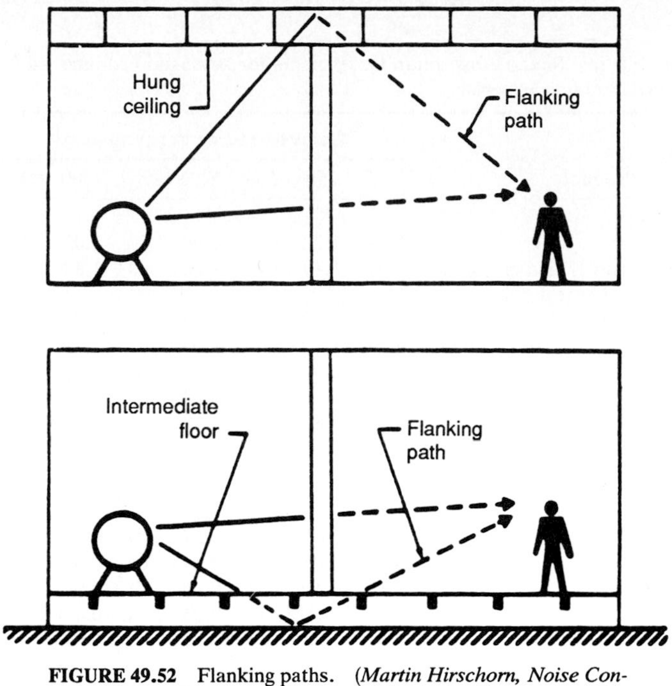

FIGURE 49.52 Flanking paths. (*Martin Hirschorn, Noise Control Reference Handbook, Industrial Acoustics Company, 1982.*)

TABLE 49.42 Noise Reduction (NR)* and Noise Isolation Class (NIC) Characteristics of Commercially Available "Quiet Rooms"

	Octave band center frequency, Hz								
Product	63	125	250	500	1000	2000	4000	8000	NIC
All-purpose room									
Single-glazed	17	20	28	31	36	35	37	45	34
Double-glazed	17	23	30	38	44	45	47	49	41
Music practice room or educational study room									
Outside to inside	···	28	37	38	48	54	58	···	45
Inside to outside	···	23	44	42	50	55	62	···	47
Room to room, 4-in (102-mm) space	···	37	72	74	90	>92	>91	···	61
Medical room									
Single-walled	33	31	39	50	57	61	68	62	53
Double-walled	37	52	64	80	93	>93	>93	>93	75
"Mini" Series 250 audiometric booth	··	18	32	38	44	51	52	50	39

*NR ratings are outside to inside unless otherwise stated. Where background noise swamped the noise exceeded from the walls, a lower limit on the NR is indicated by a > sign.

Source: ASTM E90, *Standard Method for Laboratory Measurements of the Noise Reduction of Complete Sound Isolating Enclosures*, 1978.

TABLE 49.43 Typical Sound Absorption Coefficients for Acoustic Products and Commonly Used Building Materials

Product	Octave band center frequency							
	125	250	500	1000	2000	4000	8000	NRC
Acoustic product*								
Noishield	0.89	1.20	1.16	1.09	1.01	1.03	0.93	0.95
Noise-Lock II and II, Super-Noise-Lock, and Fire-Noise Lock	0.94	1.19	1.11	1.06	1.03	1.03	1.04	0.95
Noishield Septum	0.50	0.68	1.03	1.05	1.00	0.99	···	0.95
Noise-Foil/Varitone (flush-mounted)	0.94	1.19	1.11	1.06	1.03	1.03	1.04	0.95
Trackwall	0.45	0.95	0.85	0.85	0.85	0.85	0.80	0.90
Noishield Hard	0.045	0.03	0.02	0.02	0.015	0.02	···	0.02
Noise-Foil/Varitone, 2-in (5-mm) thick (flush-mounted)	0.57	0.86	1.15	1.07	0.94	0.92	···	0.95
Building material								
Concrete block, coarse	0.36	0.44	0.31	0.29	0.39	0.25	···	0.35
Concrete block, painted	0.10	0.05	0.06	0.07	0.09	0.08	···	0.05
Brick	0.03	0.03	0.03	0.04	0.05	0.07	···	0.05
Plaster	0.14	0.10	0.06	0.05	0.04	0.03	···	0.05
Glass, heavy plate	0.18	0.06	0.04	0.03	0.02	0.02	···	0.05
Glass, window grade	0.35	0.25	0.18	0.12	0.07	0.04	···	0.15
Wood	0.15	0.11	0.10	0.07	0.06	0.07	···	0.10

*All products are 4 in (102 mm) thick unless otherwise noted. Numbers greater than 1.0 are caused by edge diffraction effects dependent on panel size and geometry.

TABLE 49.44 Room Attenuation in Direct Sound Field as a Function of Room Size and Absorption Characteristics

For listener within 5 ft (1.5 m) of terminal

Room surface		Room absorption characteristic, dB		
ft^2	m^2	Hard	Average	Soft
500	46.5	0	3	5
1,000	92.9	3	5	6
2,000	185.8	5	6	7
5,000	464.5	7	7	8
8,000	743.2	7	8	8
10,000+	929.0+	8	8	8

insertion loss (DIL). This section considers more specifically the effects of flow velocity on silencer attenuation; it also discusses the practical significance of interaction between SN and DIL, pressure drop and energy consumption, and the importance of silencer location for maximum noise reduction and minimum pressure drop. Section 49.24 will then present a procedure for selecting silencers for HVAC duct systems.

TABLE 49.45 Room Absorption Characteristics as a Function of Room Type

Type of room	Absorption rating
Radio and TV studios, theaters, lecture halls	Soft
Concert halls, stores, restaurants, offices, conference rooms, hotel rooms, school rooms, hospitals, private homes, libraries, businesses, machine rooms, churches, reception rooms	Average
Large churches, gymnasiums, factories	Hard

TABLE 49.46 Attenuation Estimates at Various Distances from Terminal Noise in Accordance with Room Sound Absorption Characteristics

Room surface		Distance*		Room absorption characteristic, dB		
ft^2	m^2	ft	m	Hard	Average	Soft
500	46.5	5	1.5	0	3	5
1,000	92.9	7	2.1	3	6	8
2,000	185.8	9	2.7	6	9	11
5,000	464.5	15	4.6	10	13	15
8,000	743.2	18	5.5	12	15	17
10,000	929.0	20	6.1	13	16	18

*Approximate distance r of listener.

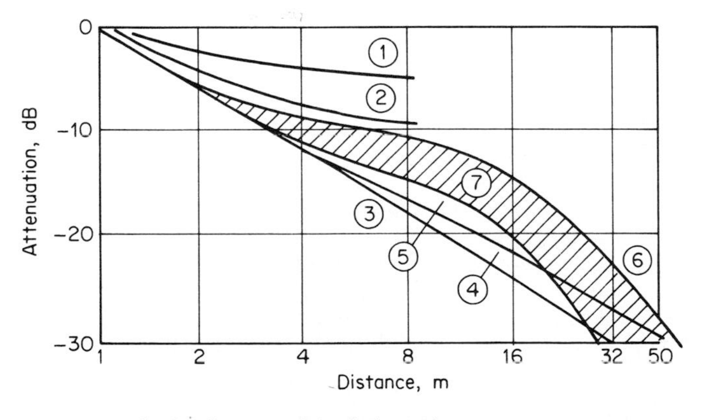

1. Small room, minimal absorption
2. Small room, strong absorption
3. Sound propagation in free field - inverse square law
4. Large office with heavily absorptive ceiling
5. Same as 4, but with reflective diffusers
6. Same as 5, but with weak absorptive ceiling
7. Range of most applications

FIGURE 49.53 The sound pressure level decreases with a doubling of the distance inside different rooms. *(Courtesy of Dr. Ing. E. Schaffert, BeSC, GmbH, Berlin.)*

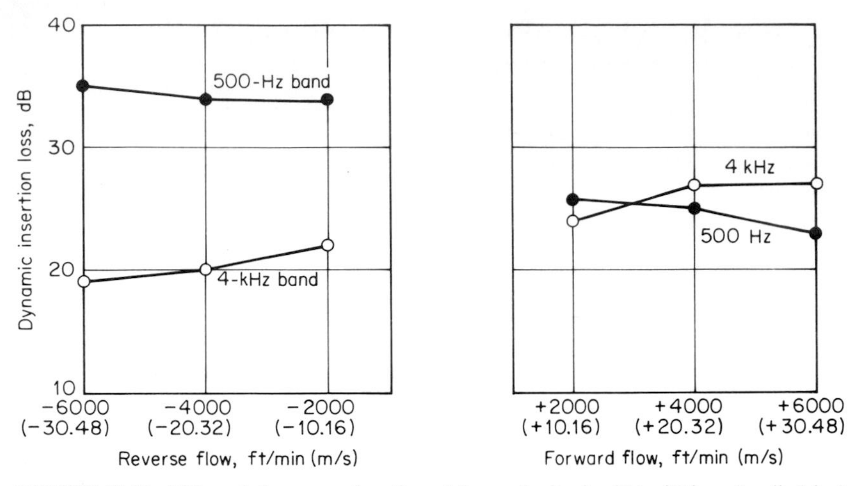

FIGURE 49.54 DIL variations as a function of face velocity for 24-in (610-mm) cylindrical silencers.

49.23.1 Specific Effects of Flow Velocity on Silencer Attenuation

Figure 49.54 shows dynamic insertion loss (DIL) variation in the 500- and 4000-Hz octave bands as a function of air flow for a 24-in (610-mm) cylindrical silencer. As the face velocity (volumetric air flow divided by silencer cross section) varies from 2000-ft/min (10.2-m/s) reverse flow to 2000-ft/min (10.2-m/s) forward flow in the 500-Hz octave band, the insertion loss varies from 34 to 26 dB, or by 8 dB. In the 4000-Hz band the corresponding change is from 22 to 24 dB.

Figure 49.55 shows the insertion loss for a rectangular silencer at 2000-ft/min (10.2-m/s) face velocity in the forward-flow and reverse-flow modes. Up to 1000 Hz the DIL is higher in the reverse-flow mode, with a difference of 6 dB in the 250-Hz band.

Figure 49.56 shows the same effects in cylindrical silencers as a function of diameter at 6000-ft/min (30.5-m/s) face velocity. (Cylindrical silencers can be designed for higher flow velocities because their pressure is significantly lower than that of rectangular silencers.)

For a cylindrical silencer with a 24-in (610-mm) diameter, Fig. 49.57 shows the change in DIL that results from using different acoustic designs while essentially maintaining the same pressure-drop characteristics (Ref. 17). (The model without a sound-absorptive jacket has a slightly lower pressure drop.)

49.23.2 Interaction of DIL with Self-Noise

The self-noise (SN) of a silencer can be thought of as a noise "floor" that may limit the ability of the silencer to achieve desired sound levels.

Figure 49.58 shows how silencer SN can preclude meeting a noise control criterion (in this case the Chicago Environmental Control Ordinance) unless the face velocity and self-noise are decreased. The condition here is particularly sensitive to flow velocity noise because of the low high-frequency sound pressure level specification.

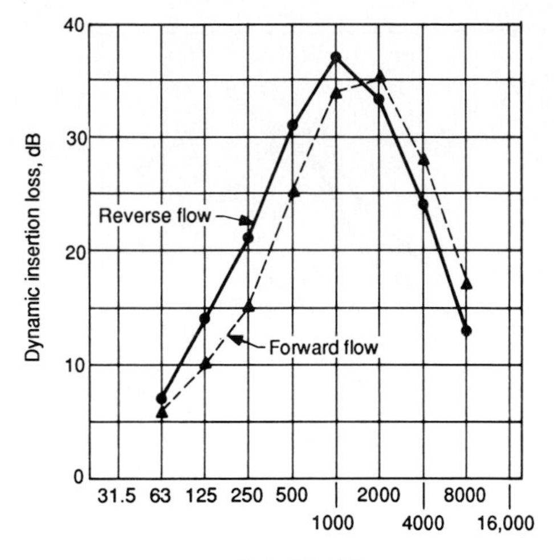

FIGURE 49.55 Characteristic DIL patterns for a dissipative rectangular silencer at 2000-ft/min (10.2-m/s) face velocity. (*Martin Hirschorn, "The Aero-Acoustic Rating of Silencers for 'Forward' and 'Reverse' Flow of Air and Sound," Noise Control Engineering, vol. 2, no. 1, Winter 1974.*)

Figure 49.59 shows the SN for two types of rectangular silencers, and Table 49.47 (p. 49.82) shows the DIL and SN data at three octave bands for several commercially available rectangular silencers.

49.23.3 Pressure Drop

To push or pull air past any ductwork component requires fan energy, which is usually described in terms of pressure head in inches water gauge (inWG) or pascals. Any duct component offers a pressure drop ΔP that must be overcome by the fan.

Obviously, the objective in any system is to keep the fan pressure head to a bare minimum. As the pressure head increases, more power is required, which increases both the initial cost and the operating costs. Most HVAC fans have a pressure head ranging from 4 to 6 inWG (996.4 to 1494.6 Pa).

Silencers are generally allowed pressure drops up to a maximum of about 0.5 inWG (124.6 Pa), though often they will be less. A silencer has three primary areas of pressure drop:

1. At the silencer inlet, the flow passage contracts to the space between the splitters (Fig. 49.26, p. 49.30). Good practice here is to design the splitters with rounded noses to minimize the abruptness of the contraction.
2. Friction losses in the splitter passages are directly proportional to silencer length.
3. An exit loss occurs as the flow expands. To minimize these losses, the

1. 12 in (305 mm) ϕ
2. 60 in (1524 mm) ϕ
3. 12 in (305 mm) ϕ
4. 60 in (1524 mm) ϕ

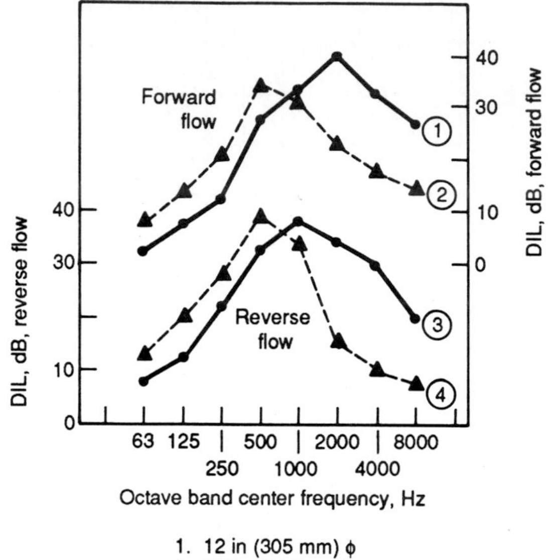

Change in DIL with diameter for cylindrical silencers at face velocity. Note that high frequency DIL decreases with diameter whereas low-frequency DIL increases. Reverse flow DIL generally is higher than forward flow in first five octaves.

FIGURE 49.56 Change in DIL with diameter for cylindrical silencers at 600-ft/min (3.05-m/s) face velocity. (*Martin Hirschorn, "The Aero-Acoustic Rating of Silencers for 'Forward' and 'Reverse' Flow of Air and Sound," Noise Control Engineering, vol. 2, no. 1, Winter 1974.*)

splitters' trailing edges can be rounded or fitted with aerodynamic tail sections.

Accordingly, the pressure drop of a silencer at a given air flow is a function of length, the free-area ratio, and the aerodynamic design of splitter leading and trailing edges.

Figure 49.60 (p. 49.83) shows a cylindrical silencer with an attached tail cone, and Fig. 49.61 illustrates the effect of the cone, which can reduce silencer pressure drop by as much as 33 percent.

At a given temperature, the pressure drop of a silencer varies as the square of the velocity through the silencer. Mathematically, this means that

$$\Delta P_2 = \Delta P_1 \left(\frac{V_2}{V_1} \right)^2 \qquad (49.40)$$

where subscript 1 represents conditions at which the pressure drop is known, and

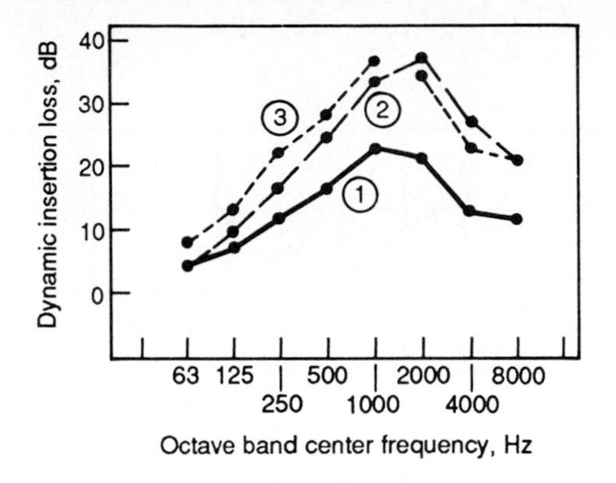

1. Type Ns — No absorptive jacket
2. Type Cs — 4-m (102-mm) absorptive jacket
3. Type FCs — 8-in (203-mm) absorptive jacket

DIL of three 24-in (610-mm) diameter cylindrical silencers at 4000-ft/min (20.3-m/s) face velocity.

FIGURE 49.57 Effect of absorptive jacket on cylindrical silencer performance. (*M. Hirschorn, "Acoustic and Aerodynamic Characteristics of Duct Silencers for Airhandling Systems," ASHRAE Paper CH-81-6, 1981.*)

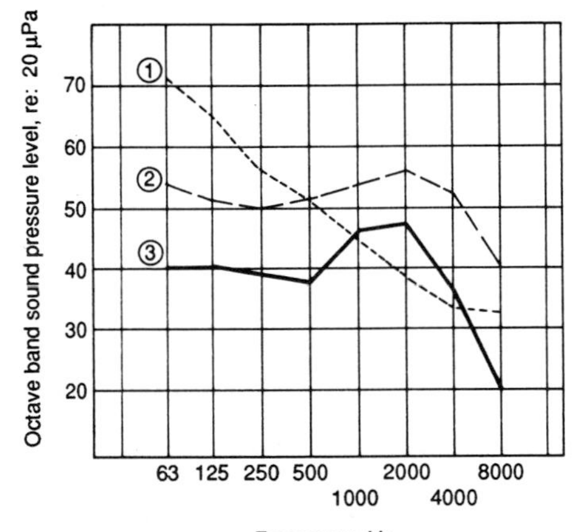

1. Chicago 55-dBA criterion
2. L_p to self-noise of 24 ft^2 (2.23 m^2) of 30 percent free area intake silencer at 1000-ft/min (5.1-m/s) flow velocity
3. L_p due to self-noise of 24 ft^2 (2.23 m^2) of 60 percent free area intake silencer at 1000-ft/min (5.1-m/s) flow velocity

FIGURE 49.58 Effect of self-noise on silencer performance with silencer in zoning boundary. (*M. Hirschorn, "Acoustic and Aerodynamic Characteristics of Duct Silencers for Airhandling Systems," ASHRAE Paper CH-81-6, 1981.*)

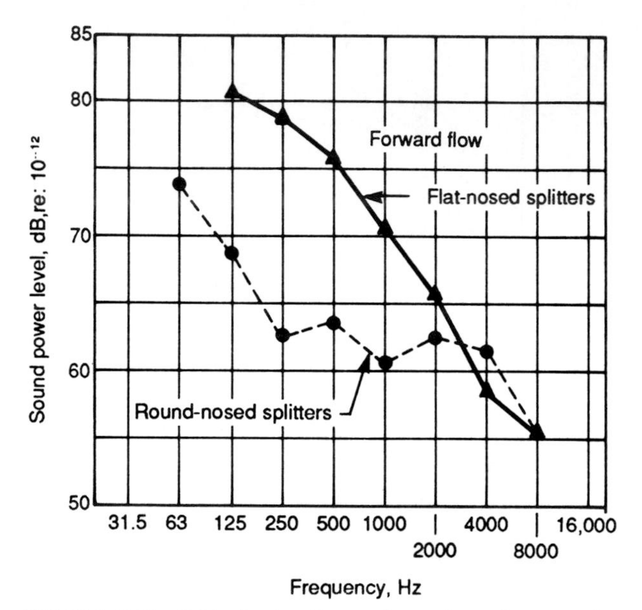

FIGURE 49.59 Self-noise for flat- and round-nosed rectangular silencers. (*Martin Hirschorn, "The Aero-Acoustic Rating of Silencers for 'Forward' and 'Reverse' Flow of Air and Sound," Noise Control Engineering, vol. 2, no. 1, Winter 1974.*)

TABLE 49.47 Dynamic Insertion Loss (DIL) and Self-Noise (SN) Data for Rectangular Silencers 3 ft (914 mm) Long

1000 ft/min (5.1 m/s) face velocity forward flow at 60°F (15.6°C)

	DIL, dB			SN, dB		
Type	250 Hz	500 Hz	2000 Hz	250 Hz	500 Hz	2000 Hz
Steel and acoustic fill standard splitter silencers						
S	16	28	35	49	47	49
Es	15	25	34	33	32	33
Ms	13	20	22	36	34	32
ML	10	17	13	30	27	28
L	9	14	23	37	32	36
Clean flow*						
Hs	14	19	28	49	47	49
Packless splitters†						
XL	17	21	11	44	46	57
XM	10	17	12	44	46	57
KL	13	12	7	36	43	46
KM	6	13	7	38	43	46

*Absorptive material protected to minimize adsorption of foreign matter.
†Reactive design using no absorptive material.

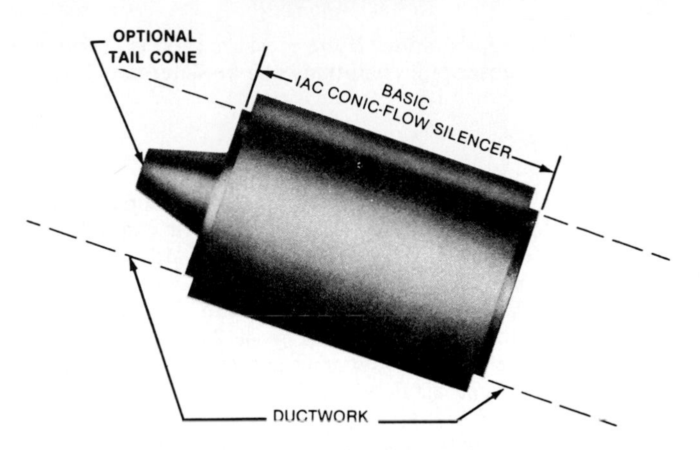

FIGURE 49.60 Silencer tail cone. (*Application Manual for Duct Silencers, Bulletin 1.0301.3, Industrial Acoustics Company.*)

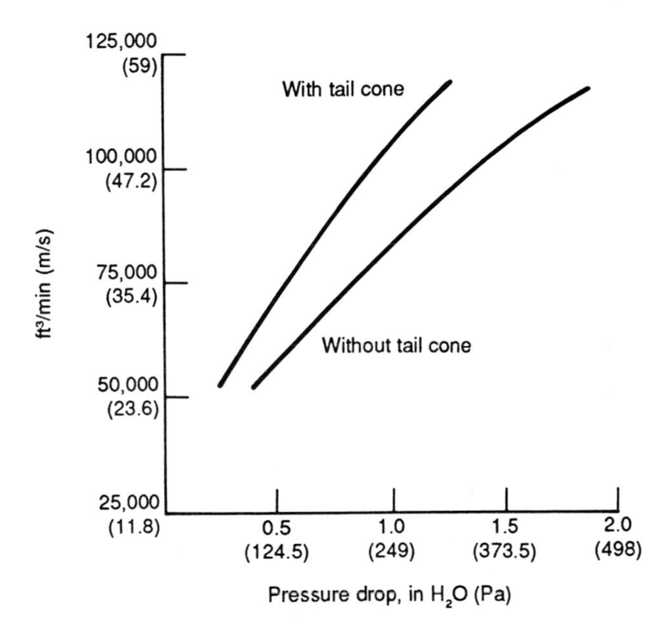

FIGURE 49.61 Pressure-drop reduction as a result of tail-cone installation. (*Application Manual for Duct Silencers, Bulletin 1.0301.3, Industrial Acoustics Company.*)

subscript 2 represents conditions at which it is to be determined. For example, if a silencer has a ΔP of 0.1 inWG (24.9 Pa) at a face velocity of 1000 ft/min (5.1 m/s), the ΔP at 2000 ft/min (10.2 m/s) will be $0.1(2000/1000)^2 = 0.1 \times 4$, or 0.4 inWG (99.64 Pa).

The pressure drop at a given temperature can be calculated if the pressure drop is known at another temperature. If the pressure drop is known at subscript 1 conditions, the ΔP at subscript 2 conditions can be determined by

$$\Delta P_2 = \Delta P_1 \frac{T_1}{T_2} \tag{49.41}$$

where T_1 and T_2 are in consistent units of absolute temperature in degrees Rankine (°R = °F + 460) or kelvin (K = °C + 273). For example, if a silencer has a ΔP_1 of 0.3 inWG (74.7 Pa) at 70°F (530°R), the ΔP_2 at 300°F (760°R) is 0.3 × (530/760) = 0.21 inWG (52.3 Pa).

Most companies in Europe design silencers with flat-nosed splitters (Fig. 49.27), as distinct from round-nosed ones (Fig. 49.26 p. 49.30). The flat-nosed configuration constitutes a poor aerodynamic and self-noise design (Fig. 49.59); its pressure drop for the same air flow can be 3 times higher than for a round-nosed silencer with an equal free area.

The significantly higher self-noise generation of flat-nosed splitters results from a higher exit velocity. Noise due to air flow is a function of its velocity V and varies from V^5 to V^7, depending on frequency and duct components.

49.23.4 Energy Consumption

The cost element in operating air-flow system components such as filters, coils, terminal devices, grilles, registers, and silencers can be determined from

$$\Delta\$ = \begin{cases} \dfrac{1.03 \times \Delta P \times Q \times U \times C}{E} & \text{in English units} \\ \dfrac{0.002434 \times \Delta P \times Q \times U \times C}{E} & \text{in metric units} \end{cases} \tag{49.42}$$

where $\Delta\$$ = annual energy cost, dollars
ΔP = pressure drop, inWG (Pa)
Q = air flow, ft³/min (m³/h)
U = utilization, percentage of total hours per year (i.e., 50 for 50 percent)
C = cost, \$/kWh
E = combined system fan and motor operating efficiency, percent (i.e., 75 for 75 percent)

For example, say that silencer A handles an air flow of 24,000 ft³/min at a pressure drop of 0.55 inWG and that silencer B handles the same air flow at a pressure drop of 0.09 inWG. If U = 50, C = \$0.05/kWh, and E = 75, the difference between the annual energy costs of silencers A and B is calculated as follows:

$$\Delta\$_A = \frac{1.03 \times 0.55 \times 24{,}000 \times 50 \times 0.05}{75} = \$453.20$$

$$\Delta\$_B = \frac{1.03 \times 0.09 \times 24{,}000 \times 50 \times 0.05}{75} = \$74.16$$

By using silencer B instead of silencer A, the saving in annual energy cost is therefore 453.20 − 74.16 = \$379.04. Installing silencer B on each floor of a 50-

story building would thus save 50 × 379.04 = $18,952 per year, and over 30 years the energy cost savings would be $568,560 without taking inflation into account. If C were $0.15/kWh instead of $0.05/kWh, the energy cost savings in 30 years would be $1,705,680, and this amount would be significantly greater if an annual inflation factor were applied.

49.23.5 Effects of Silencer Length and Cross Section

Silencer length and cross section are important because space is usually at a premium in HVAC systems. However, splitter silencers (for instance) of a given design and cross section are usually available with different free areas, with the lower pressure drop and lower attenuation usually going with the larger free area. Therefore, if cross section can be traded off for more length, the required attenuation can be obtained by selecting a longer silencer. Figure 49.62 shows the DIL as a function of length for a rectangular silencer with 30 percent free area.

The reverse trade-off is not usually practical when the length is not available, because a short silencer with high attenuation characteristics and large cross section would require long transition sections to keep the pressure drop down.

Figure 49.63 shows that when two silencers with 60 percent free area are directly installed in series, the measured DIL approximates the linearly added ratings of two silencers tested individually.

49.23.6 Impact on Silencer ΔP of Proximity to Other Elements in an HVAC Duct System

The pressure-drop and DIL characteristics of silencers in the United States are generally determined by tests conducted in accordance with ASTM E477, which

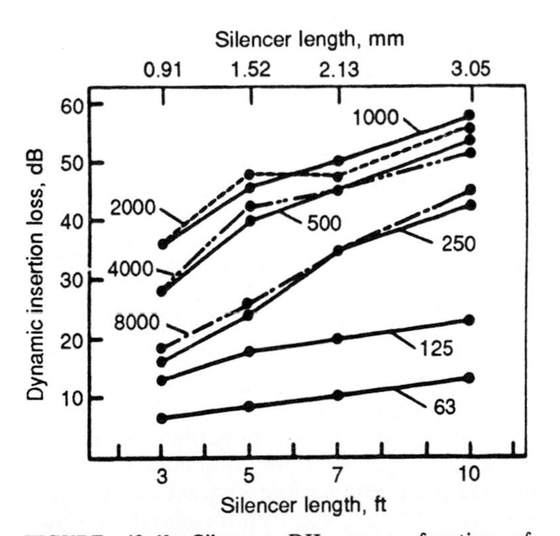

FIGURE 49.62 Silencer DIL as a function of length. (*M. Hirschorn, "Acoustic and Aerodynamic Characteristics of Duct Silencers for Airhandling Systems," ASHRAE Paper CH-81-6, 1981.*)

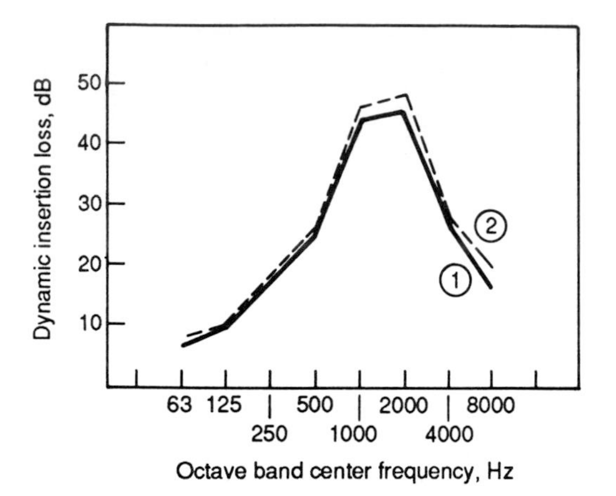

1. Measured DIL (not sensitive to spacing ranging from 24 in (610 mm) to 192 in (4880 mm) between silencers).
2. Linearly added DIL rating of two silencers tested individually.

FIGURE 49.63 Comparison of measured results and linear addition of DIL ratings for two silencers with 60 percent free area at 2000-ft/min (10.2-m/s) forward flow. (*M. Hirschorn, "Acoustic and Aerodynamic Characteristics of Duct Silencers for Airhandling Systems," ASHRAE Paper CH-81-6, 1981.*)

mandates the use of straight duct lengths "no less than five equivalent diameters from the entrance," with "the downstream duct no less than ten equivalent diameters from the exit" of the silencer under test (Ref. 7).

Table 49.48 shows the ΔP factors (multipliers relative to ASTM E477 test data) that can be applied to silencer pressure drop when a silencer is located less than 2.5D ($\sqrt{4ab/\pi}$ for a rectangular duct of dimensions a and b) upstream or downstream of other duct elements.

49.23.7 Duct Rumble and Silencer Location

In duct systems, duct rumble is generally caused by poor aerodynamic design and/or by excessive low-frequency fan noise within the duct that breaks out of the duct construction when there are no silencers close to the fan. That is, duct rumble can occur when duct sections get excited by vibrational impulses developed as a result of aerodynamic turbulence and/or by pulses generated by the prime mover (usually a fan). Even with good air-flow design, duct rumble can arise when the fan blade pass frequency or one of its harmonics coincides with the natural resonance of a given section of a duct system; therefore, rumble is most likely to occur in duct systems without silencers.

If a silencer with adequate attenuation characteristics in the critical frequencies is installed right after the fan, duct rumble should not develop or will be reduced. Preferably, the silencer installation should be made with aerodynamically well-designed transition sections; there should be no duct runs before the silencer in which resonances could develop. Because the flow velocities after a silencer

TABLE 49.48 Pressure-Drop Factors When Silencer Is Located Less than 2.5*D* Upstream or Downstream from Other Duct Elements

The air-flow and pressure-drop data for all IAC duct silencers is based on tests run in accordance with ASTM E477 and other codes that specify minimum lengths of straight duct connections upstream and downstream.

However, in practice, these corrections may vary significantly from the test procedure. The modifiers in the following are based on arrangements frequently encountered in air-handling systems and are based on conservative practice.

The factors shown apply only to silencer pressure drop and do not include the pressure drop for the other system components (elbows, transitions, etc.) adjacent to which the silencer is located.

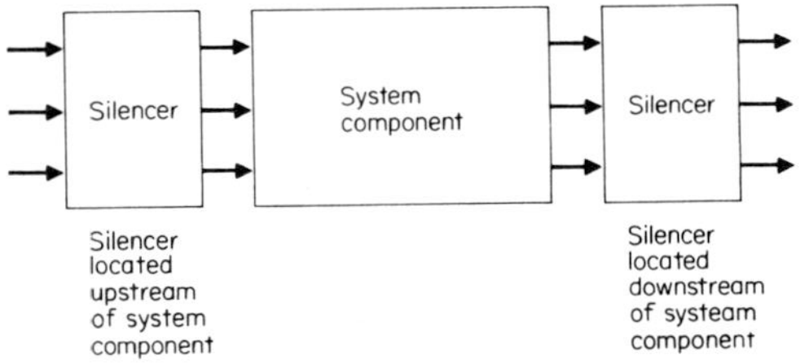

Location of silencers relative to fans	ΔP factor		
	Silencer		
	Upstream	Downstream	
Ducted centrifugal fans Discharge—Quiet-Duct Rectangular Silencers a. L_1 = one duct diameter for every 1000-ft/min (5.08-m/s) average duct velocity, including suitably designed transition section for maximum regain.	. . .	1.0	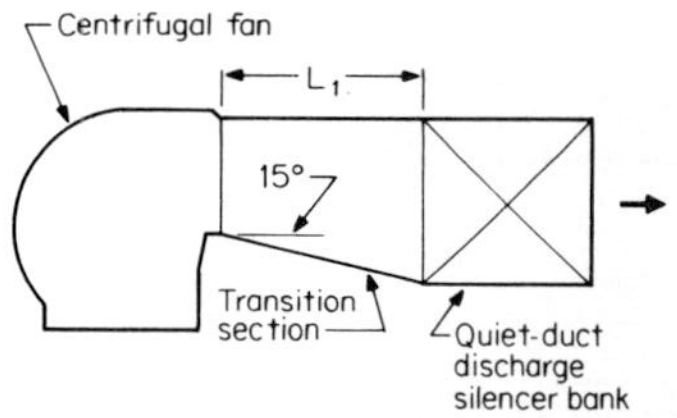 Recommended transition section arrangements between centrifugal fan and silencer bank (ducting not shown)

TABLE 49.48 Pressure-Drop Factors When Silencer Is Located Less than 2.5*D* Upstream or Downstream from Other Duct Elements (*Continued*)

Location of silencers relative to fans	ΔP factor: Silencer Upstream	ΔP factor: Silencer Downstream	
b. If space is limited, velocity distribution vanes, diffusers, or other flow equalizers will have to be provided by system designer. Allow minimum $L_1 = 0.75D$.	. . .	1.0	Intake and discharge silencers centrifugal fans (ducting not shown)
Intake—Quiet-Duct Rectangular Silencers Use minimum $L_2 = 0.75D$, including suitably designed transition sections if required.	1.0	. . .	
Ducted 50% hub-vaneaxial fans			
Discharge—Quiet-Duct Rectangular Silencers a. L_1 = one duct diameter for every 1000-ft/min (5.08-m/s) average duct velocity, including transition sections of not more than 30° included angle for maximum regain.	. . .	1.0	Recommended transition section arrangements between vane-axial fan and quiet-duct silence banks (ducting not shown)
b. When space is limited, velocity distribution vanes, diffusers, or other flow equalizers will have to be provided by system designer. Allow minimum $L_1 = 0.75D$.	. . .	1.0	
Discharge—Conic-Flow Tubular Silencers $L_1 = 0$ when fan hub is matched to silencer center body.	. . .	1.0	Conic=flow tubular silencer center-body-matched to axial fan hub (ducting not shown)
Intake—Quiet-Duct Rectangular Silencers Use minimum $L_2 = 0.75D$, including intake cones of not more than 60° included angle.	1.0	. . .	
Intake—Conic-Flow Tubular Silencers $L_2 = 0$ when fan hub is matched to silencer center body.	1.0	. . .	

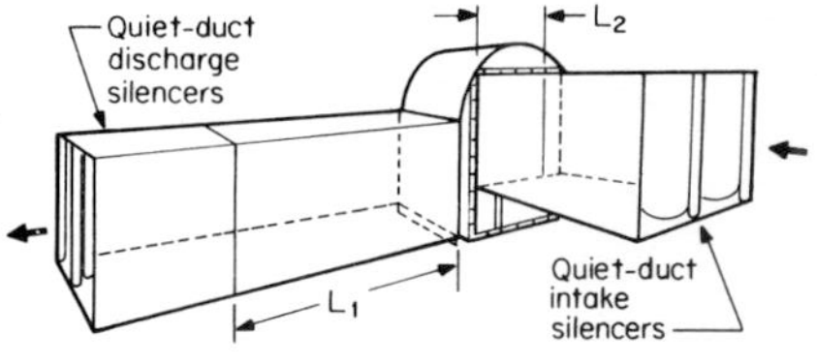

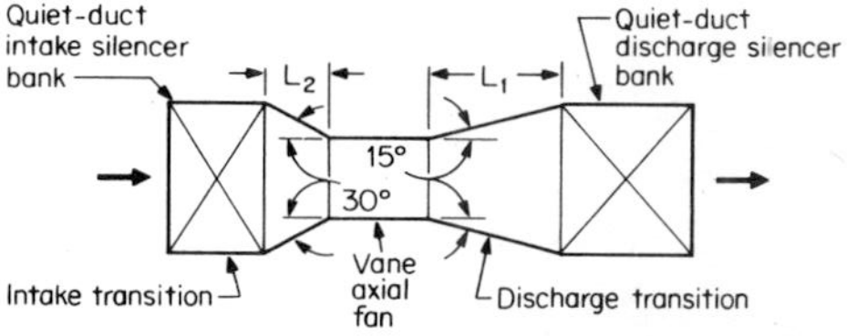

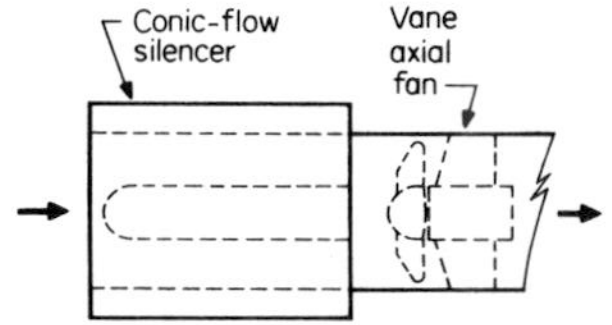

Elbows (without turning vanes)			Silencers before and after elbows
Distance of silencer from elbow:			Note: Silencer baffles should be parallel to the plane of elbow turn.
$D \times 3$	1.0	1.0	
$D \times 2$	1.5	1.5	
$D \times 1$	2.0	2.0	
Elbows (with turning vanes)			
Distance of silencer from elbow:			
$D \times 3$	1.0	1.0	
$D \times 2$	1.2	1.2	
$D \times 1$	1.75	1.75	
$D \times 0.5$	3.0	3.0	
Directly connected	4.0	Not advised	
Terminal Devices			Discharge silencer downstream of mixing box and upstream of grille
Mixing boxes, pressure reducing valves, terminal reheats, etc.	. . .	1.0	
Grilles, Registers, and Diffusers			
Deflecting Type			
Allow 24 in (610 mm) upstream and 2½D downstream of silencer.	1.0	1.0	
Nondeflecting Type			
Allow at least 12 in (305 mm).	1.0	1.0	
Transitions			Silencer between upstream and downstream transitions
With 15° included angle (7.5° slope)	1.0	1.0	
With 30° included angle (15° slope)	1.25	1.0	
With 60° included angle (30° slope)	1.5	1.0	
Coils and Filters			
Silencer downstream—12 in (305) mm from face	. . .	1.0	
Silencer upstream—24 in (610 mm) from face	1.0		

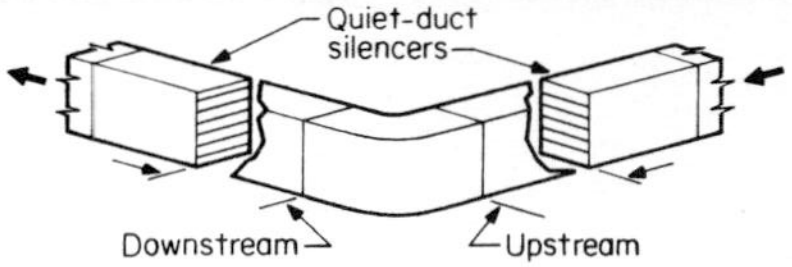

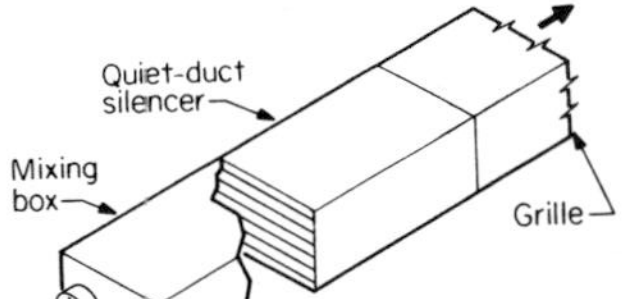

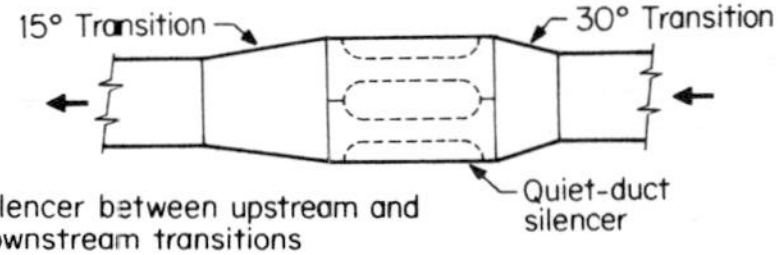

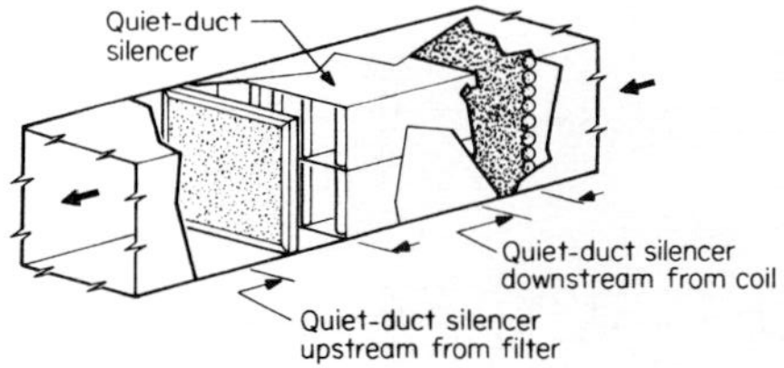

TABLE 49.48 Pressure-Drop Factors When Silencer Is Located Less than 2.5D Upstream or Downstream from Other Duct Elements (*Continued*)

Location of silencers relative to fans	ΔP factor — Silencer — Upstream	ΔP factor — Silencer — Downstream	
Cooling Towers and Condensers Type L or Type ML Silencers	2.0	2.0	Quiet-duct discharge silencers; Quiet-duct intake silencers
	This multiplier includes typical allowance for intake and discharge dump losses.		
The pressure-drop increase due to the addition of silencers to a cooling tower is partially offset by the resulting decrease in the entrance and discharge losses of the system.			
Immediately at System Entrance or Exit	Silencer at intake	Silencer at discharge	Quiet-duct intake silencers; Quiet-duct discharge silencer; Silencers immediately at intake and discharge of equipment room
Silencer type or model:			
CL, FCL	2.0	5.0	
NL	2.0	4.0	
ML	1.5	3.5	
Cs, FCs, Ns, L	1.5	3.0	
Ms	1.5	2.0	
S, Es	1.5	1.5	
The relatively higher multipliers for the lower-pressure-drop silencers, such as the CL and L types, for instance, are due to the dump losses to the atmosphere being significantly higher relative to their rated values.			
Pressure-drop factors for silencers at the entrance to the system can be materially reduced by the use of a smoothly converging bellmouth with sides having a radius equal to at least 20% of its outlet dimension.			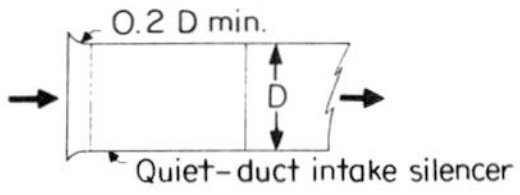 0.2 D min.; D; Quiet-duct intake silencer

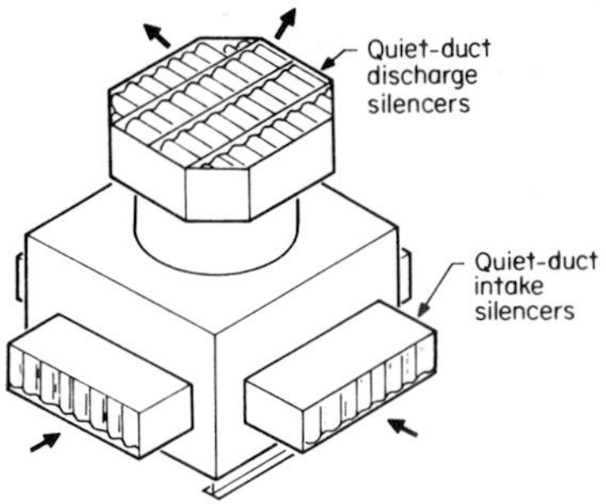

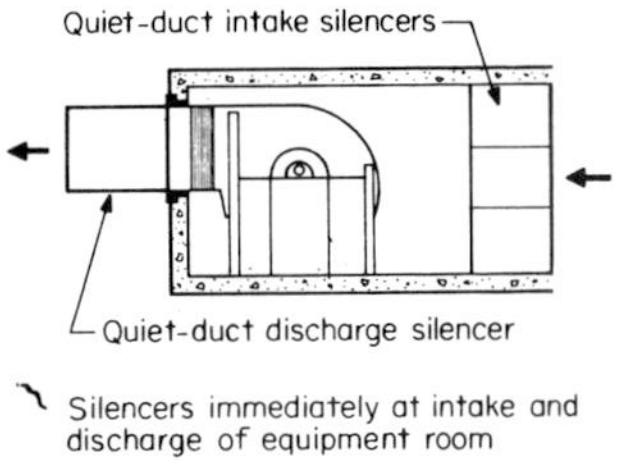

Notes applying to Table 49.48:

1. For maximum structural integrity, Quiet-Duct Silencer splitters should be installed vertically. Where vertical installation is not feasible, structural reinforcement is required for silencers wider than 24 in (610 mm).
2. Unless otherwise indicated, connecting ductwork is assumed to have the same dimensions as fan intake or discharge openings.
3. When elbows precede silencers, splitters should be parallel to the plane of elbow turn.
4. L_1 = distance from fan exhaust to entrance of discharge silencer; L_2 = distance from fan inlet to exit of intake silencer.
5. ΔP factor = multiplier relative to silencer laboratory-rated pressure-drop data.
6. D = diameter of round duct or equivalent diameter of rectangular duct.
7. Unless otherwise noted, the multipliers shown do not include the pressure drop of other components (elbows, transitions, dump

are generally low, duct rumble is not likely to develop farther downstream. Nevertheless, a smooth aerodynamic design—with no abrupt changes in cross sections and flow directions—should be maintained throughout the duct system.

Where an adequate duct silencer cannot be installed immediately after the fan inside the equipment room, some of the low-frequency noise will radiate out of the duct wall into the equipment room. In this case, the duct and the room may effectively act as a kind of low-frequency silencer with unknown acoustic performance characteristics. The equipment room itself must therefore have adequate sound transmission loss characteristics to prevent fan noise from entering critical office space. Also, a silencer must be installed in the duct immediately outside the equipment room to prevent fan noise from exiting the mechanical space into critical office-space ceiling plenums.

49.23.8 Effect of Silencer Location on Residual Noise Levels

Figure 49.64 shows a laboratory test arrangement in which a fan discharges air into a room through a duct containing a silencer together with an abrupt

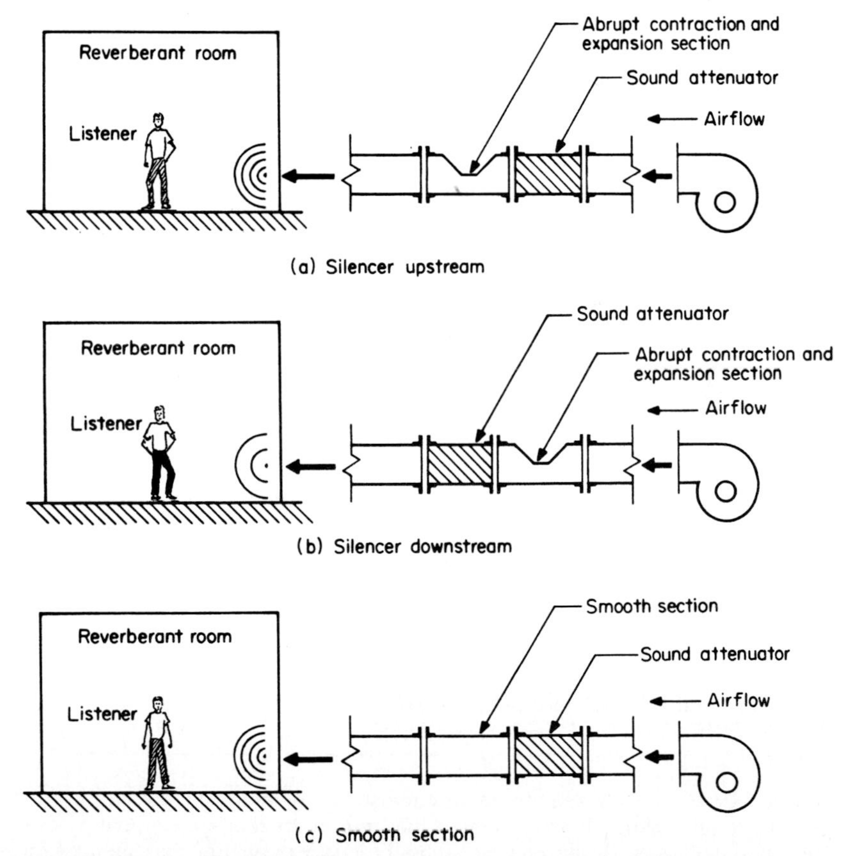

FIGURE 49.64 Installation of silencer upstream and downstream of a duct element generating high noise levels. (*Courtesy of Industrial Acoustics Company.*)

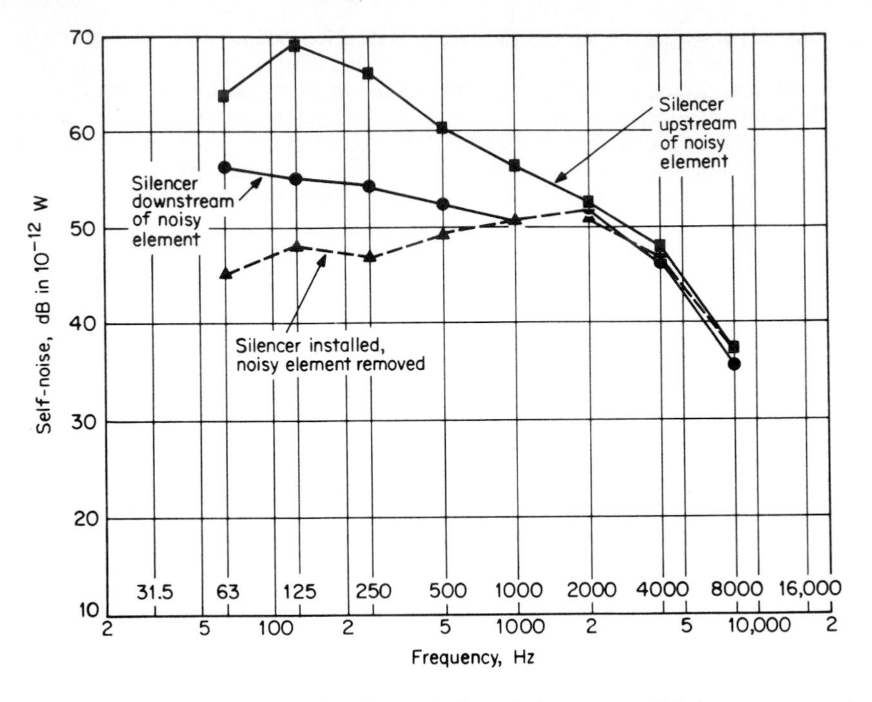

Note: Noisy element in these tests was a restricted duct section having abrupt contraction followed by sudden expansion (see Fig. 49.64).

FIGURE 49.65 Effect of silencer location on duct element noise. (*Courtesy of Industrial Acoustics Company.*)

contraction-and-expansion section. To determine the effects on residual room noise, separate tests were conducted with the silencer on each side of the contraction-expansion section. A third test was conducted with the contraction-expansion section replaced by a straight run of duct.

The results from these tests (Fig. 49.65) show how noise can be regenerated after a silencer in an aerodynamically poorly designed duct system. Clearly the silencer gives the best results in an aerodynamically smoothly designed duct system when care is taken not to include sudden contractions, expansions, or turns.

The duct system must also be designed for minimum self-noise, and silencers must be positioned between the listener and the noise source. Otherwise, the benefits of a duct silencer may be canceled out.

49.24 SYSTEMIC NOISE ANALYSIS PROCEDURE FOR DUCTED SYSTEMS

To avoid HVAC noise problems in an enclosed or open space, the noise reduction (NR) requirements (if any) to meet the criteria for that space must be determined. If silencers are required, care must be taken to ensure that silencer pressure drop is not excessive and that silencer self-noise is not so high as to prevent meeting the criteria.

Some noise reduction is inherently provided by the acoustic characteristics of ductwork. And in a room where criteria must be met, absorption and distance from terminals also influence the sound pressure level at any given location. If the ductwork (lined and unlined) and room effects are not sufficient to meet the criteria, any remaining attenuation must be provided by silencers.

49.24.1 Procedure

Silencer NR requirements can be determined by the following procedure, which consists of calculating the items in lines 1 to 11 of Table 49.58 (p. 49.101). The data used to illustrate this procedure come from Fig. 49.66, these data are summarized in Table 49.57, and the results of the calculations based on these data are given in Table 49.58.

Line 1: Noise Criteria. Using Sec. 49.20.3 (p. 49.49), establish the sound pressure level criteria in each octave band by selecting the noise criteria (NC) values from Table 49.31. From Table 49.32 an "average" criterion for a reception room

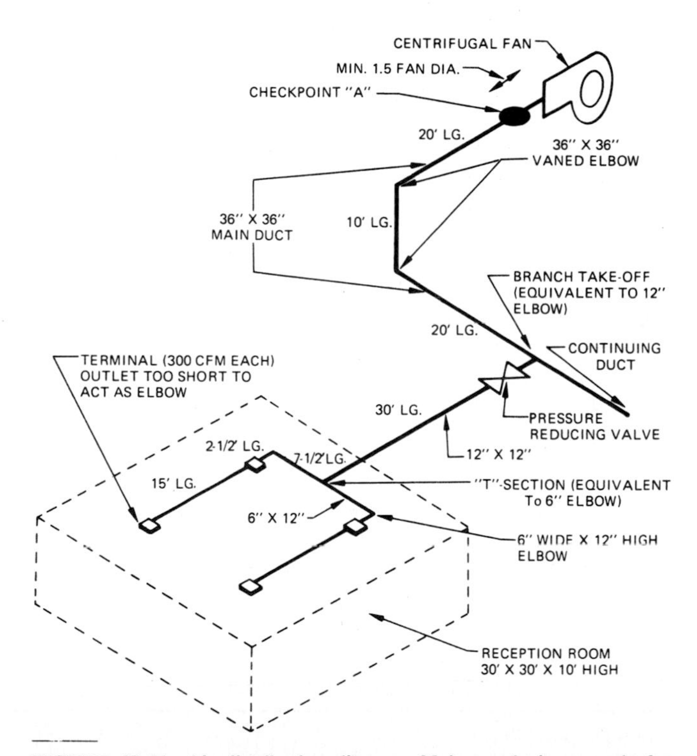

FIGURE 49.66 Air distribution diagram: Noise analysis example for ducted systems. (*Application Manual for Duct Silencers, Bulletin 1.01103, Industrial Acoustics Company.*)

is NC 35; accordingly, use the NC 35 values from Table 49.31. Enter these values on line 1 of Table 49.58.

Line 2: Room and Terminal Effect. This is a factor for converting the L_p (sound pressure level) criteria of line 1 to the L_W (sound power level) at each terminal. This factor accounts for the size and acoustic characteristics of the space as well as for the number and location of the terminals. Consider only those terminals connected directly to fans by ductwork; include the supply and return fans.

To obtain this factor, follow these steps:

1. Refer to Tables 49.32 (p. 49.55) and 49.45 (p. 49.77) to determine the characteristics of the room. For the reception room in the example, select "average" sound absorption ratings (in accordance with the criterion used in line 1).
2. Determine whether the listener is likely to be in the direct or reverberant field. This determination is made by assessing the individual's probable location in the space. For example, in a theater the key is the seat, not the lobby; in an office the key is the seated position, not the standing position.
3. Find the room absorption characteristic in either Table 49.44 or 49.46. For this example, use Table 49.46 (p. 49.77), and interpolate to find the absorption characteristic for a room surface area of 3000 ft^2; assume that the absorption characteristic is 10 dB.
4. From the value obtained in step 3, subtract the terminal effect given in Table 49.49. In the example the listener is likely to be within 11 ft of two terminals but probably not within 11 ft of more than two terminals. Therefore, subtract 3 dB from the 10 dB obtained in step 3, and enter 7 dB on line 2 of Table 49.58.

Line 3: Allowance for End Reflection. End reflection accounts for the fact that some low-frequency noise is reflected back into the duct. As shown in Table 49.50, end reflection can add significantly to system noise attenuation. Use the dimensions of the duct leading to the terminal, and interpolate as necessary. For the 6- by 12-in rectangular duct in the example, the square root of the duct area is $\sqrt{6 \times 12} = 8.5$ in; using this in Table 49.50, interpolate attenuation values of 13, 9, 5, 2, and 0 dB. Enter these values on line 3 of Table 49.58.

Line 4: Ductwork Attenuation. Determine for each octave band the attenuation provided by unlined ducts (no inner sound-absorptive lining) and, if applicable, by ducts internally lined with sound-absorptive materials. Then enter the attenuation values on line 4 of Table 49.58.

In the example there are 40 ft of 6- by 12-in duct, 40 ft of 12- by 12-in duct, and 50 ft of 36- by 36-in duct between the nearest terminal and checkpoint A. These are all unlined sheet-metal ducts, so interpolate the attenuation between the sizes listed in Table 49.51; note that this table lists the attenuation values in dB/ft. Add these calculated values for the ducts to find each band's total value, and enter the bands' totals on line 4 of Table 49.58.

TABLE 49.49 Correction Factor for Number of Terminals

Number of terminals providing primary influence	1	2	3	4	6	8
dB to be subtracted from room absorption effect	0	3	5	6	8	9

TABLE 49.50 Attenuation Due to End Reflection at Ductwork Openings
Values in decibels

Duct dimension*		Octave band number					
in	mm	1	2	3	4	5	6 and higher
5	127	17	12	8	4	1	0
10	254	12	8	4	1	0	0
20	508	8	4	1	0	0	0
40	1016	4	1	0	0	0	0
80	2032	1	0	0	0	0	0

*Diameter of round duct or square root of area of rectangular duct.

TABLE 49.51 Attenuation of Unlined Sheet-Metal Ducts
Values in dB/ft (dB/0.3 m)

	Duct size		Octave band number			
Type	in	cm	1	2	3	4 and higher
Small	6 × 6	15 × 15	0.2	0.2	0.15	0.1
Medium	24 × 24	61 × 61	0.2	0.2	0.1	0.05
Large	72 × 72	1.8 × 1.8	0.1	0.1	0.05	0.01
Round	4 to 12	10 to 30.5	0.03	0.03	0.03	0
Round	Over 12	Over 30.5	0	0	0	0

For acoustically lined duct runs, the attenuations per foot for common duct sizes are shown in Tables 49.52 and 49.53 for linings 1 and 2 in thick, respectively. Multiply the unit attenuation for each octave band by the length of the lining to get the total attenuation; however, as per the *1987 ASHRAE Handbook* (HVAC Systems and Applications, chapter 52), usable high frequency attenuation values (above 500 Hz) are limited to 10-ft (3048-mm) lengths.

To illustrate the use of Table 49.52, 10 ft of 18- by 54-in duct lined with 1.5-lb/ft^3-density fiberglass 1 in thick is estimated to provide the following attenuation:

Hertz	63	125	250	500	1000	2000	4000
Attenuation	0.4	1.1	3.1	8.4	16.5	13.2	11.6

Except for small cross-sections, the use of duct lining alone cannot sufficiently attenuate the low-frequency noise from air-handling equipment. Moreover, effective noise reduction close to the principal noise maker, the fan, is essential to prevent unacceptable duct breakout noise into offices and other rooms. The insertion loss of a 5-ft-long commercially available duct silencer is as follows:

Hertz	63	125	250	500	1000	2000	4000	8000
DIL	8	18	24	40	45	46	41	26

TABLE 49.52 Sound Attenuation in Straight, Lined, Sheet-Metal Ducts of Rectangular Cross Section in dB/ft (dB/0.3 m); 1-in (25-mm) Lining Thickness*; No Air Flow

Internal cross-sectional dimensions		Octave band center frequency, Hz						
in	mm	63	125	250	500	1000	2000	4000
4 × 4	100 × 100	0.16	0.44	1.21	3.32	9.10	10.08	3.50
4 × 6	100 × 150	0.13	0.37	1.01	2.77	7.58	8.26	3.13
4 × 8	100 × 200	0.12	0.33	0.91	2.49	6.82	6.45	2.57
4 × 10	100 × 250	0.11	0.31	0.85	2.32	6.37	5.02	2.07
6 × 6	150 × 150	0.12	0.34	0.93	2.56	7.01	7.50	3.17
6 × 10	150 × 250	0.10	0.27	0.75	2.04	5.61	5.67	2.67
6 × 12	150 × 300	0.09	0.26	0.70	1.92	5.26	4.80	2.33
6 × 18	150 × 460	0.08	0.23	0.62	1.70	4.67	2.95	1.51
8 × 8	200 × 200	0.10	0.28	0.77	2.12	5.82	6.08	2.95
8 × 12	200 × 300	0.09	0.24	0.65	1.77	4.85	4.98	2.64
8 × 16	200 × 410	0.08	0.21	0.58	1.59	4.37	3.89	2.17
8 × 24	200 × 610	0.07	0.19	0.52	1.42	3.88	2.39	1.41
10 × 10	250 × 250	0.09	0.24	0.67	1.84	5.04	5.17	2.79
10 × 16	250 × 410	0.07	0.20	0.55	1.49	4.10	4.04	2.41
10 × 20	250 × 510	0.07	0.18	0.50	1.38	3.78	3.30	2.05
10 × 30	250 × 760	0.06	0.16	0.45	1.23	3.36	2.03	1.34
12 × 12	300 × 300	0.08	0.22	0.60	1.64	4.48	4.52	2.67
12 × 18	300 × 460	0.07	0.18	0.50	1.36	3.74	3.71	2.39
12 × 24	300 × 610	0.06	0.16	0.45	1.23	3.36	2.89	1.97
12 × 36	300 × 910	0.05	0.15	0.40	1.09	2.99	1.78	1.28
15 × 15	380 × 380	0.07	0.19	0.52	1.42	3.88	3.84	2.53
15 × 22	380 × 560	0.06	0.16	0.43	1.19	3.27	3.20	2.29
15 × 30	380 × 760	0.05	0.14	0.39	1.06	2.91	2.46	1.86
15 × 45	380 × 1140	0.05	0.13	0.34	0.94	2.17	1.51	1.21
18 × 18	460 × 460	0.06	0.17	0.46	1.26	3.45	3.37	2.42
18 × 28	460 × 710	0.05	0.14	0.38	1.03	2.84	2.69	2.13
18 × 36	460 × 910	0.05	0.13	0.34	0.94	2.59	2.15	1.78
18 × 54	460 × 1370	0.04	0.11	0.31	0.84	1.65	1.32	1.16
24 × 24	610 × 610	0.05	0.14	0.38	1.05	2.87	2.73	2.26
24 × 36	610 × 910	0.04	0.12	0.32	0.87	2.39	2.24	2.02
24 × 48	610 × 1220	0.04	0.10	0.29	0.78	1.90	1.75	1.66
24 × 72	610 × 1830	0.03	0.09	0.25	0.70	1.06	1.07	1.08
30 × 30	760 × 760	0.04	0.12	0.33	0.91	2.49	2.32	2.14
30 × 45	760 × 1140	0.04	0.10	0.28	0.76	1.88	1.90	1.91
30 × 60	760 × 1520	0.03	0.09	0.25	0.68	1.35	1.48	1.57
30 × 90	760 × 2290	0.03	0.08	0.22	0.60	0.76	0.91	1.02
36 × 36	910 × 910	0.04	0.11	0.29	0.81	2.01	2.03	2.04
36 × 54	910 × 1370	0.03	0.09	0.25	0.67	1.42	1.66	1.83
36 × 72	910 × 1830	0.03	0.08	0.22	0.60	1.02	1.30	1.50
36 × 108	910 × 2740	0.03	0.07	0.20	0.54	0.57	0.80	0.98
42 × 42	1070 × 1070	0.04	0.10	0.27	0.73	1.59	1.81	1.97
42 × 64	1070 × 1630	0.03	0.08	0.22	0.60	1.11	1.47	1.75
42 × 84	1070 × 2130	0.03	0.07	0.20	0.55	0.81	1.16	1.45
42 × 126	1070 × 3200	0.02	0.06	0.18	0.49	0.45	0.71	0.94
48 × 48	1220 × 1220	0.03	0.09	0.24	0.67	1.30	1.65	1.90
48 × 72	1220 × 1830	0.03	0.07	0.20	0.56	0.92	1.35	1.70
48 × 96	1220 × 2440	0.02	0.07	0.18	0.50	0.66	1.05	1.40
48 × 144	1220 × 3660	0.02	0.06	0.16	0.45	0.37	0.65	0.91

*Based on measurements of surface-coated duct liners of 1.5-lb/ft^3 (24-kg/m^3) density. For the specific materials tested, liner density had a minor effect over the nominal range of 1.5 to lb/ft^3 (24 to 48 kg/m^3).

Source: *1987 ASHRAE Handbook, HVAC Systems and Applications*, Chap. 52.

TABLE 49.53 Sound Attenuation in Straight, Lined, Sheet-Metal Ducts of Rectangular Cross Section in dB/ft (dB/0.3 m); 2-in (50-mm) Lining Thickness*; No Air Flow

Internal cross-sectional dimensions		Octave band center frequency, Hz						
in	mm	63	125	250	500	1000	2000	4000
4 × 4	100 × 100	0.34	0.93	2.56	7.02	19.23	10.08	3.50
4 × 6	100 × 150	0.28	0.78	2.13	5.85	16.03	8.26	3.13
4 × 8	100 × 200	0.26	0.70	1.92	5.26	14.42	6.45	2.57
4 × 10	100 × 250	0.24	0.65	1.79	4.91	13.46	5.02	2.07
6 × 6	150 × 150	0.26	0.72	1.97	5.40	14.81	7.50	3.17
6 × 10	150 × 250	0.21	0.58	1.58	4.32	11.85	5.67	2.67
6 × 12	150 × 300	0.20	0.54	1.48	4.05	11.11	4.80	2.33
6 × 18	150 × 460	0.17	0.48	1.31	3.60	8.80	2.95	1.51
8 × 8	200 × 200	0.22	0.60	1.64	4.49	12.31	6.08	2.95
8 × 12	200 × 300	0.18	0.50	1.36	3.74	10.26	4.98	2.64
8 × 16	200 × 410	0.16	0.45	1.23	3.37	9.23	3.89	2.17
8 × 24	200 × 610	0.15	0.40	1.09	2.99	5.68	2.39	1.41
10 × 10	250 × 250	0.19	0.52	1.42	3.89	10.66	5.17	2.79
10 × 16	250 × 410	0.15	0.42	1.15	3.16	8.66	4.04	2.41
10 × 20	250 × 510	0.14	0.39	1.06	2.92	7.22	3.30	2.05
10 × 30	250 × 760	0.13	0.34	0.95	2.59	4.04	2.03	1.34
12 × 12	300 × 300	0.17	0.46	1.26	3.46	9.48	4.52	2.67
12 × 18	300 × 460	0.14	0.38	1.05	2.88	7.62	3.71	2.39
12 × 24	300 × 610	0.13	0.35	0.95	2.59	5.47	2.89	1.97
12 × 36	300 × 910	0.11	0.31	0.84	2.31	3.06	1.78	1.28
15 × 15	380 × 380	0.15	0.40	1.09	2.99	7.64	3.84	2.53
15 × 22	380 × 560	0.12	0.34	0.92	2.52	5.55	3.20	2.29
15 × 30	380 × 760	0.11	0.30	0.82	2.25	3.89	2.46	1.86
15 × 45	380 × 1140	0.10	0.27	0.73	2.00	2.17	1.51	1.21
18 × 18	460 × 460	0.13	0.35	0.97	2.66	5.79	3.37	2.42
18 × 28	460 × 710	0.11	0.29	0.80	2.19	3.95	2.69	2.13
18 × 36	460 × 910	0.10	0.27	0.73	2.00	2.94	2.15	1.78
18 × 54	460 × 1370	0.09	0.24	0.65	1.78	1.65	1.32	1.16
24 × 24	610 × 610	0.11	0.29	0.81	2.21	3.73	2.73	2.26
24 × 36	610 × 910	0.09	0.25	0.67	1.84	2.65	2.24	2.02
24 × 48	610 × 1220	0.08	0.22	0.61	1.66	1.90	1.75	1.66
24 × 72	610 × 1830	0.07	0.20	0.54	1.48	1.06	1.07	1.08
30 × 30	760 × 760	0.09	0.25	0.70	1.92	2.65	2.32	2.14
30 × 45	760 × 1140	0.08	0.21	0.58	1.60	1.88	1.90	1.91
30 × 60	760 × 1520	0.07	0.19	0.52	1.44	1.35	1.48	1.57
30 × 90	760 × 2290	0.06	0.17	0.47	1.28	0.76	0.91	1.02
36 × 36	910 × 910	0.08	0.23	0.62	1.70	2.01	2.03	2.04
36 × 54	910 × 1370	0.07	0.19	0.52	1.42	1.42	1.66	1.83
36 × 72	910 × 1830	0.06	0.17	0.47	1.28	1.02	1.30	1.50
36 × 108	910 × 2740	0.06	0.15	0.41	1.14	0.57	0.80	0.98
42 × 42	1070 × 1070	0.07	0.21	0.56	1.54	1.59	1.81	1.97
42 × 64	1070 × 1630	0.06	0.17	0.47	1.28	1.11	1.47	1.75
42 × 84	1070 × 2130	0.06	0.15	0.42	1.16	0.81	1.16	1.45
42 × 126	1070 × 3200	0.05	0.14	0.38	1.03	0.45	0.71	0.94
48 × 48	1220 × 1220	0.07	0.19	0.52	1.42	1.30	1.65	1.90
48 × 72	1220 × 1830	0.06	0.16	0.43	1.18	0.92	1.35	1.70
48 × 96	1220 × 2440	0.05	0.14	0.39	1.06	0.66	1.05	1.40
48 × 144	1220 × 3660	0.05	0.13	0.34	0.94	0.37	0.65	0.91

*Based on measurements of surface-coated duct liners of 1.5-lb/ft^3 (24-kg/m^3) density. For the specific materials tested, liner density had a minor effect.

Source: *1987 ASHRAE Handbook, HVAC Systems and Applications*, Chap. 52.

TABLE 49.54 Attenuation of Unlined Sheet-Metal Elbows
*Values in decibels per elbow**

Duct diameter or width†		Octave band number							
in	cm	1	2	3	4	5	6	7	8
5–10	12.7–25.4	0	0	0	0.0, 1	1, 3, 5	2, 4, 7	3, 4, 5	3
11–20	27.9–50.8	0	0	0.0, 1	1, 3, 5	2, 4, 7	3, 4, 5	3	3
21–40	53.3–101.6	0	0.0, 1	1, 3, 5	2, 4, 7	3, 4, 5	3	3	3
41–80	104.1–203.2	0, 0, 1	1, 3, 5	2, 4, 7	3, 4, 5	3	3	3	3

*Where three values are shown, use: *First*, round elbows; *second*, square elbows with vanes; *third*, square elbows without vanes, or branch takeoffs with flow diverter. Where one value is given, apply it to any of these three conditions.

†Width is dimension in plane of turn. If area transition occurs between inlet and outlet of elbow, use width of smaller cross section.

To provide equivalent attenuation in the 250-Hz octave band, a 1-in-thick lined duct 80 ft long would be required.

Line 5: Elbow Attenuation. Determine from Table 49.54 the allowances for elbows and branch takeoffs between the terminal and noise source. In the example we have one 12- by 12-in (305- × 305-mm) branch takeoff, two elbows without vanes 6 in (152 mm) wide by 12 in (305 mm) high, and two 36- by 36-in (0.91- × 0.91-m) vaned elbows between the nearest terminal and checkpoint A. Branch takeoffs may be evaluated similarly to elbows, using the width of the branch. Total the attenuation values and enter them on line 5 of Table 49.58.

Line 6: L_W Split—Branch to Terminals. Determine the allowance from Table 49.55. In the example there are four terminals; therefore, enter a value of 6 dB on line 6 of Table 49.58.

TABLE 49.55 Allowance for L_W Split—Branch to Terminals

Number of terminals	1	2	3	4	8	10	20	40	100
Allowance factor, dB	0	3	5	6	9	10	13	16	20

Line 7: L_W Split—Main Duct to Branch Ducts. Determine from Table 49.56 the allowance for the division of sound power to the duct branches located between the terminals and checkpoint A, excluding the terminal supply branch (which was accounted for in line 6). In the example the allowance is 10 dB since the area ratio

TABLE 49.56 Division of Sound Power—Main Duct to Branches

Area of branch in % of area of main duct*	⅕%	½%	1%	2%	5%	10%	20%	50%
dB to be added to branch L_W to obtain main duct L_W	27	23	20	17	13	10	7	3

*For constant-velocity systems, branch air flow can be used as percentage of main air flow; however, substantial errors can result if this practice is followed in high-velocity systems.

between the 12- by 12-in branch and the 36- by 36-in main duct is approximately 11 percent. Enter 10 dB on line 7 of Table 49.58.

Line 8: Safety Factors. Enter a safety factor of − 3 dB.

Line 9: Permissible L_W. Arithmetically add the line 1 criteria to the sum of the phenomena considered in lines 2 through 8. Make sure to subtract the 3-dB safety factor. The result is the permissible sound power level spectrum that cannot be exceeded at checkpoint A (Fig. 49.66) without exceeding the line 1 criteria.

Line 10: L_W of Fan. Enter the manufacturer's or the calculated fan sound power level in each octave band.

Line 11: DIL Required. Compare the maximum permissible sound power levels (line 9) with the predicted fan sound power levels (line 10). If the latter is greater than the former in any octave band, the difference is the dynamic insertion loss (DIL) required to meet the criteria in that band.

The data in Table 49.58 are graphically illustrated in Fig. 49.67, where the shaded area shows the required silencer DIL.

49.24.2 Silencer Selection

A silencer must be selected to provide both the necessary attenuation and DIL, and the self-noise level should be no greater than the permissible levels in line 9 of Table 49.58.

A large range of silencer types and sizes is available so that selections can be made to meet specified pressure drops.

Of course, the lower the pressure drop, the lower the operating costs, although the initial cost may be higher. (See Secs. 49.23.3, Pressure Drop, and 49.23.4, Energy Consumption.)

49.24.3 Calculating the Attenuation Effects of Lined Ducts

The sound attenuation of lined ducts may be estimated by the use of empirically derived equations that express the insertion loss of a given length of duct in terms of the duct cross section and the sound absorption properties of the lining material (Ref. 18). Absorption depends on the type of lining material, but in general it is related to density and thickness at each frequency of interest.

The equations take the form of two intersecting curves defining the low-frequency attenuation A_1 and the high-frequency attenuation A_2:

$$A_1 = \frac{t^{1.08} h^{0.356} (P/A)L \cdot f^{1.17+K_2 d}}{K_3 d^{2.3}} \quad \text{dB} \tag{49.43}$$

$$A_2 = \frac{K_4 (P/A)L \cdot f^{K_5 - 1.61 \log (P/A)}}{w^{2.5} h^{2.7}} \quad \text{dB} \tag{49.44}$$

where $K_2 = 0.19\ (0.119)$ [metric shown in parentheses]
$K_3 = 1190\ (5.46 \times 10^{-3})$

K_4 = 2.11 × 10^9 (3.32 × 10^{18})
K_5 = −1.53 (−3.79)
d = liner density, lb/ft³ (kg/m³)
t = liner thickness, in (mm)
h = the smaller inside duct dimension, in (mm)
w = the larger inside duct dimension, in (mm)
P = inside duct perimeter, in (mm)
A = inside area of duct, in² (mm²)
L = duct length, ft (m)
f = one-third octave band center frequency, Hz

The equations may be used directly to calculate attenuation in any one-third octave band, using the lower of A_1 or A_2. Due to the effect of break-out, the max-

TABLE 49.57 Summary of Acoustic Input Data for "SNAP" Calculations in Table 49.58

Line	Data
1	Sound pressure level criterion: NC 35
2	Room and terminal effect: General case = individual terminals Room size = 30 ft long, 30 ft wide, 10 ft high Number of terminals = 4 Distance from reference point to terminal 1 = 11 ft Distance from reference point to terminal 2 = 11 ft Distance from reference point to terminal 3 = 15 ft Distance from reference point to terminal 4 = 15 ft
3	Allowance for end reflection: Duct cross-sectional dimensions = 6 by 12 in
4	Ductwork attenuation—terminal to checkpoint: Duct size = 12 by 12 in; duct length = 30 ft Duct size = 36 by 36 in; duct length = 50 ft
5	Elbow attenuation—terminal to checkpoint: Branch take-off with flow diverter; width = 12 in; quantity = 1 Elbow type = square, without turning vanes; width = 6 in; quantity = 1 Elbow type = square, without turning vanes; width = 36 in; quantity = 2
6	Power-level split—branch to terminals: An allowance of 6 dB is being made to account for the sound power level division to each outlet based on four terminals (from line 2).
7	Power-level split—main duct to branch duct: Main duct = 36 by 36 in Branch duct = 12 by 12 in
8	Safety factors: A 3-dB safety factor was selected.
10	Sound power level of fan: The fan power level was input by the user.

Source: Systemic Noise Analysis Procedure, Bulletin 1.0110.3, Industrial Acoustics Company, Bronx, NY, 1976.

TABLE 49.58 Summary of Acoustic "SNAP" Calculations

		Octave band center frequency, Hz							
		63	125	250	500	1000	2000	4000	8000
		Octave band number							
Line	Calculation	1	2	3	4	5	6	7	8
1	Noise criteria at NC 35	60	53	46	40	36	34	33	32
2	Room and terminal effect	7	7	7	7	7	7	7	7
3	Allowance for end reflection	13	9	5	2	0	0	0	0
4	Ductwork attenuation	17	17	9	5	5	5	5	5
5	Elbow attenuation	0	0	7	15	25	25	19	15
6	L_W split—branch to terminals	6	6	6	6	6	6	6	6
7	L_W split—main duct to branch	10	10	10	10	10	10	10	10
	Totals	113	102	90	85	89	87	80	75
8	Safety factors	−3	−3	−3	−3	−3	−3	−3	−3
9	Permissible L_W	110	99	87	82	86	84	77	72
10	L_W of fan	102	99	98	97	96	91	87	82
11	DIL required	0	0	11	15	10	7	10	10

Source: *Systemic Noise Analysis Procedure for Air Handling Systems*, Industrial Acoustics Company, 1979.

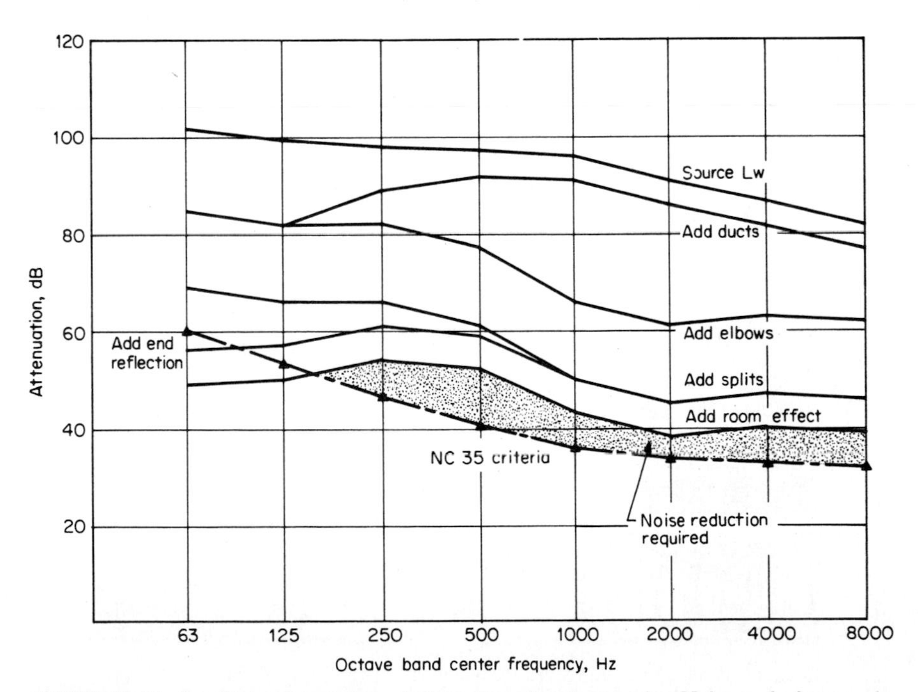

FIGURE 49.67 Graphic representation of attenuation of systemic noise: Noise analysis example for ducted system. (*Courtesy of Industrial Acoustics Company.*)

imum attenuation should be limited to 40 dB. Also, for frequencies above 1000 Hz, the lined length L should be empirically limited to 10 ft (3.05 m).

To express octave band attenuation, use the lowest of the one-third octave frequencies in the octave band of interest to calculate A_1. Use the highest of the one-third octave frequencies to calculate A_2.

49.25 ACOUSTIC LOUVERS

In an HVAC fan room, acoustic louvers can be used to allow fresh air in while reducing the fan noise outside the building. Acoustic louvers are designed to keep face velocities and pressure drops low and can occupy relatively large building wall sections. Louvers are therefore treated as wall components with openings and are therefore also acoustically rated on the basis of sound transmission loss (ASTM E90) rather than rated as silencers (ASTM E477).

A typical louver installation is shown in Fig. 49.68, and the transmission loss characteristics of commercially available acoustic louvers are listed in Table 49.59.

Systemic analysis of acoustic louvers is accomplished in seven steps (as illustrated in Table 49.62):

1. Establish the octave band sound pressure level criterion from Sec. 49.20 and Table 49.38 on p. 49.60.
2. Calculate the divergence, taking into account the face area of the louver as well as the distance from the louver to the location (external to the fan room) where the criterion must be met. Table 49.60 shows the divergence for a range of face areas and distances. Enter the table at the left in the "Distance" column and read across to the right to establish the divergence for the appropriate area; interpolate as necessary. For example, at a distance of 50 ft (15.24 m) the divergence for a 36-ft^2 (3.34-m^2) louver would be interpolated as 26 dB. (For more on divergence, see Secs. 49.8, Propagation of Sound Outdoors, and 49.9, The Inverse-Square Law.)
3. Determine the fan noise attenuation due to the sound-absorptive characteris-

FIGURE 49.68 Acoustic louver installation, SNAP Form II, Bulletin 1.0503.2. (*Courtesy of Industrial Acoustics Company.*)

TABLE 49.59 Sound Transmission Loss of Acoustic Louvers

Louver	Octave band center frequency, Hz							
	63	125	250	500	1000	2000	4000	8000
Standard model	5	7	11	12	13	14	12	9
Low-pressure-drop model	4	5	8	9	12	9	7	6

TABLE 49.60 Divergence Factor for Acoustic Louvers, dB

Distance from louver to point of interest, ft	Face area of louver, ft^2							
	12½	25	50	100	200	400	800	1600
12½	19	16	13	0	0	0	0	0
25	25	22	19	16	13	0	0	0
50	31	28	25	22	19	15	13	0
100	37	34	31	28	25	22	19	16
250	45	42	39	36	33	30	27	24
500	51	48	45	42	39	36	33	30
1000	57	54	51	48	45	42	39	36
2000	63	60	57	54	51	48	45	42

Note: When working in metric units, convert to English units before entering this table: (1) Multiply the distance in meters by 3.281 to get the distance in feet; (2) multiply the area in square meters by 10.764 to get the area in square feet.

tics of the fan room. Table 49.61 summarizes fan-room effects. Enter this table under the "Distance" column and read across to the right, interpolating as necessary to establish the room absorption for the size and type of room. For example, at a distance of 10 ft (3.05 m) a "soft" room of 2000 ft^2 (185.8 m^2) shows a 15-dB factor.

4. Add the values obtained in steps 1 to 3, and subtract from their total a 3-dB safety factor. The result is the permissible noise-source sound power level that will meet the criterion.
5. Enter the manufacturer's fan sound power level. If it is not available, calculate it by using the procedure described in Sec. 49.15.
6. Compare the values obtained in step 5 with the values obtained in step 4. If the fan sound power level is greater than the permissible sound power level in any octave band, the difference is the sound transmission loss (TL) that the louver must provide.
7. Compare the required TL to the TL of the louver to be used. Select the lowest-pressure-drop louver that will meet the requirements.

Table 49.62 shows these steps for an installation where a 24,000-ft^3/min (11.3-m^3/h) fan is located 10 ft (3.05 m) from a 36-ft^2 (3.34-m^2) air-intake louver in a 2000-ft^2 (185.8-m^2) fan plenum, which is lined with absorptive materials and can be classified as "soft." The louver is 50 ft (15.24 m) from an apartment house. The installation is close to continuous heavy traffic and the criterion is taken

(continued p. 49.106)

TABLE 49.61 Fan-Room Absorption Factors

Distance from louver to fan, ft	Room surface, ft^2																	
	500			1000			2000			4000			8000			16,000		
	Acoustic characteristic of room, dB*																	
	S	A	H	S	A	H	S	A	H	S	A	H	S	A	H	S	A	H
5	9	5	2	10	7	4	11	8	5	11	9	7	12	10	8	12	10	9
10	12	6	2	14	8	4	15	9	6	15	10	8	16	12	10	17	13	10
15	12	6	2	15	8	4	17	10	6	18	11	8	19	13	10	20	15	12
20	12	6	2	15	8	4	18	10	6	19	12	8	20	13	10	22	15	12

*S = soft (walls and ceiling covered with 4-in-thick absorptive panels); A = average (30 percent of walls and ceiling covered with 4-in-thick panels); H = hard (walls and ceiling made of concrete, brick, metal, or similar material).

Notes:

1. Bottom factors in each column are maximum for any greater distance.
2. Where direct line of sight from the noise generator to the opening exists, use one-half of the values shown in this table.
3. When working in metric units, convert to English units before entering this table: (a) Multiply the distance in meters by 3.281 to get the distance in feet; (b) multiply the area in square meters by 10.764 to get the area in square feet.

Source: *Systemic Noise Analysis Procedure*, SNAP II, Bulletin 1,0503.2, Industrial Acoustics Company, 1975.

TABLE 49.62 Systemic Noise Analysis Procedure for Louver Application

			Octave band center frequency, Hz							
			63	125	250	500	1000	2000	4000	8000
			Octave band number							
Step	Calculation	Data source	1	2	3	4	5	6	7	8
1	Criterion, dB	Table 49.38	72	67	62	57	52	48	44	42
2	Divergence, dB	Table 49.60	26	26	26	26	26	26	26	26
3	Room absorption, dB	Table 49.61	15	15	15	15	15	15	15	15
4	Total, steps 1 to 3	Steps 1 to 3	113	108	103	98	93	89	85	83
	Minus 3-dB safety factor		−3	−3	−3	−3	−3	−3	−3	−3
	Result—noise source permissible L_W, dB		110	105	100	95	90	86	82	80
5	Noise source L_W re 10^{-12} W, dB	Manufacturer's data	101	101	100	98	97	92	84	76
6	Transmission loss required, dB	Step 5 minus step 4	. . .	. . .	. . .	3	7	6	2	
7	Transmission loss of louver, dB	Table 49.59	4	5	8	9	12	9	7	6

from Table 49.38. The fan sound power level was supplied by the manufacturer. The intake louver's face area was selected on the basis of a maximum permissible pressure drop of 0.5 inWG (124.6 Pa).

49.26 HVAC SILENCING APPLICATIONS

An HVAC system may have several locations where noise control is required. Figure 49.69 shows some of the different types of noise control devices used in HVAC systems:

1. *Cylindrical silencers:* Shown also in Fig. 49.28 on p. 49.31, these control the noise of axial-flow fans.

1. Cylindrical silencers
2. Rectangular silencers
3. Fan plenums and air-handling systems
4. "Through-the-wall" vent silencers
5. "Over-the-wall" vent silencers
6. Exhaust vent silencers
7. Cooling-tower silencers
8. Noise control enclosures
9. Acoustic louvers
10. "Clean-flow" and packless silencers
11. Slim louvers

FIGURE 49.69 Typical HVAC silencing applications. (*Application Manual for Duct Silencers, Bulletin 1.0301.3, Industrial Acoustics Company.*)

2. *Rectangular silencers:* Shown also in Fig. 49.26 on p. 49.30, these are modular units often built up in banks immediately before fan intake and discharge sections with aerodynamic transition sections. Table 49.48 illustrates some ways in which rectangular silencers are used in HVAC systems. Additional applications are shown in Fig. 49.70.
3. *Acoustic plenums and quiet air-handling systems:* Acoustic plenums control fan noise in the air stream and in adjacent areas and generally provide excellent thermal insulation. Fan plenums may also include silencers; for instance, Fig. 49.71 shows a diffuser-silencer that attenuates vaneaxial fan exhaust noise while facilitating aerodynamic pressure regain. Other acoustically designed plenum housings to control fan noise in the air stream and in adjacent areas are illustrated in Figs. 49.81 and 49.82 on pp. 49.117 and 49.118.
4. (and 5 and 6) *Vent silencers:* These provide conversational privacy between rooms while allowing air circulation. They are commercially available in exhaust-vent configurations as well as in "through the wall" and "over the wall" designs behind acoustic ceilings (Figs. 49.72 and 49.73 on p. 49.109).
7. *Silencers for cooling towers and roof exhaust fans:* These silencers prevent acoustic annoyance to neighboring buildings. Figure 49.25 on p. 49.28 illustrates the use of silencers with cooling towers, and Fig. 49.70 shows other fan intake and discharge silencer arrangements.
8. *Modular air-handling units and built-up fan plenums:* Discussed in Secs. 49.28 and 49.29, these shield office occupants against machinery noise and help create quite office spaces. Figures 49.79 (on p. 49.116), 49.81, and 49.82 show several types of air-handling units used in buildings.
9. *Acoustic louvers:* These control noise, permit air flow, and provide decorative protection against weather and forced entry. They are also used as noise barriers for cooling towers and other equipment. Acoustic louvers are illustrated in Figs. 49.68 (on p. 49.102) and 49.69.
10. *Clean-flow silencers:* These offer protected acoustic infill for silencers in hospitals, clean rooms, operating suites, and pollution control research facilities.
11. *Packless silencers:* Having no acoustic fill, they can be cleaned with steam, chemicals, hot water, or vacuuming. Packless silencers are ideal for silencing microchip plants, corrosive environments, food and dairy operations, clean rooms, hospital operating rooms, and research facilities.
12. *Slim louvers:* Only 4 in (102 mm) deep, these allow for ventilation and provide noise control.

Figure 49.74*a* to *f* (beginning on p. 110) shows somc typcs of rooms in which soundproofing and the silencing of HVAC equipment are important.

49.27 SELF-NOISE OF ROOM TERMINAL UNITS

The best noise control design for an HVAC system can be undone if the self-noise from the terminal units selected exceeds the sound pressure criteria selected. (The silencer selection procedure described in Sec. 49.24.1 takes into account correction factors for terminal units, room absorption characteristics, duct end reflection, and other variables.) In general, terminal devices do not generate

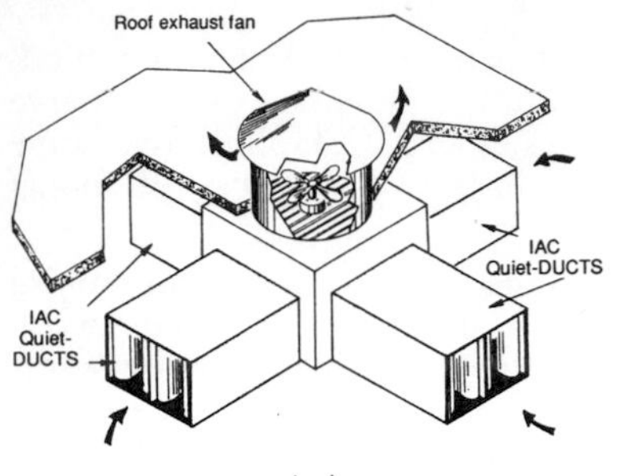

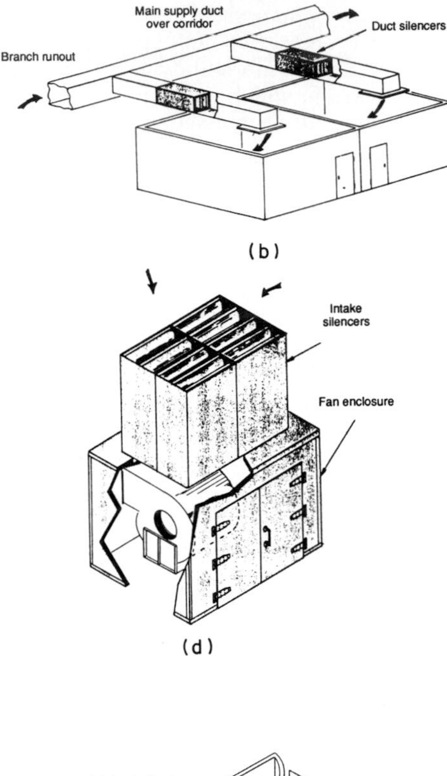

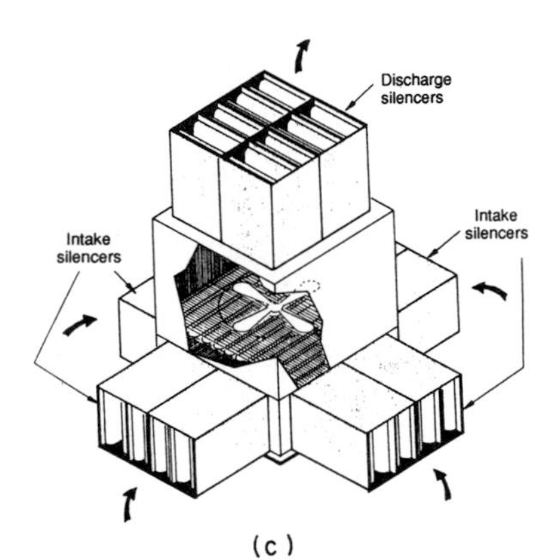

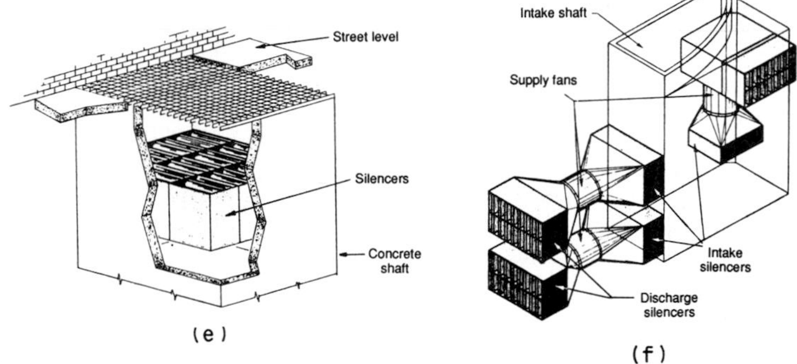

FIGURE 49.70 Rectangular silencers in ventilation HVAC systems. (*a*) Roof exhaust silencer arrangement for propeller fans; (*b*) duct silencers prevent sound transmission from room to room through connecting duct system; (*c*) silencers for air-cooled condenser; (*d*) centrifugal-fan enclosure with intake silencers; (*e*) duct silencer bank with intake or discharge shaft; (*f*) underground garage intake and discharge silencers. (Application Manual for Duct Silencers, *Bulletin 1.0301.4, Industrial Acoustics Company, 1989.*)

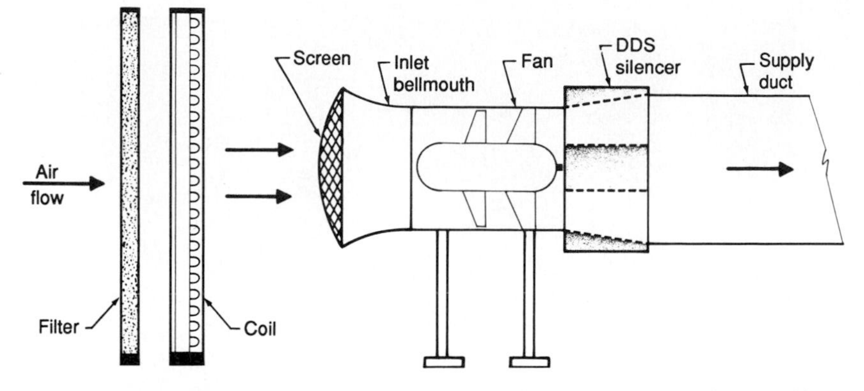

FIGURE 49.71 Diffuser-silencer for axial-flow fans. (Application Manual for Duct Silencers, *Bulletin 1.0301.4, Industrial Acoustics Company.*)

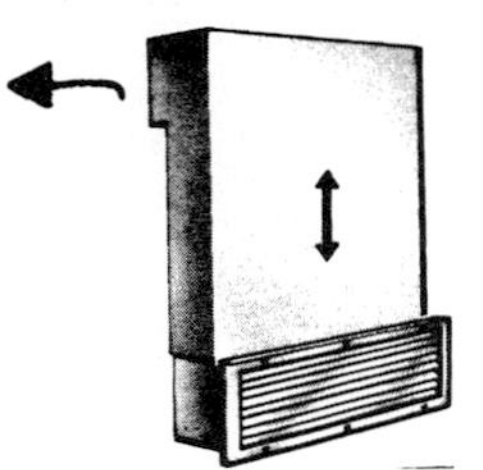

FIGURE 49.72 "Through the wall" vent silencer. (Quiet Vent Silencers for Air Transfer Systems, *Bulletin 1.0601.1, Industrial Acoustics Company.*)

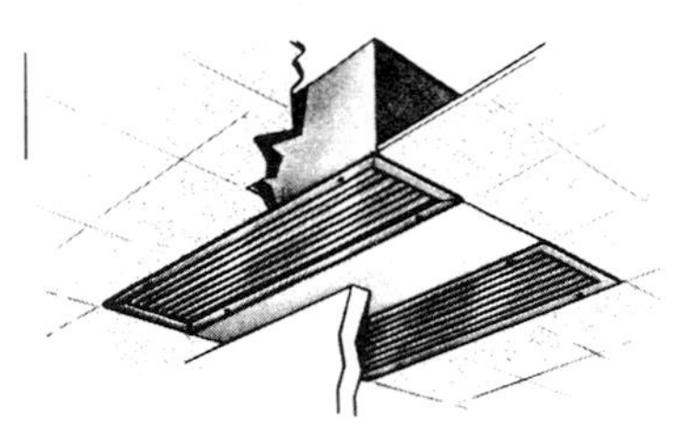

FIGURE 49.73 "Over the wall" vent silencer. (Quiet Vent Silencers for Air Transfer Systems, *Bulletin 1.0601.1, Industrial Acoustics Company.*)

significant low-frequency noise; rather, they are more likely to produce a discrete high-frequency sound.

We know that sound levels can increase with the doubling of velocities from 15 to 21 dB, depending on the frequency. Therefore, to ensure desired sound levels, velocities must be appropriately selected to ensure that the desired criterion is not exceeded by the flow noise from the terminal unit.

Manufacturers of ceiling diffusers, grills, registers, and mixing boxes now publish sound level data, usually presented in the form of an NC rating (such as NC 30 at given conditions of air-flow volume, pressure drop, and size). Such ratings usually assume a room sound absorption of 8 or 10 dB and a certain minimum distance from the sound source.

For more detailed sound pressure level L_p calculations, the sound power level L_W of a terminal device must be known; the sound pressure level at a given distance from the source can then be estimated by referring to Fig. 49.15 on p. 49.20. However, we must also know the room constant, or total room absorption A (see Sec. 49.11.2 on p. 49.17), and the directivity factor, which in this case assumes a diffuser located essentially in the middle of the ceiling. If S is the area of all the surfaces in a room and if $\bar{\alpha}$ is the average absorption coefficient of the room (taking into account also objects and people), then $A = S\,\bar{\alpha}$ (where the units of A are sabins if S is in ft^2, and metric sabins if S is in m^2) [Eq. (49.19) on

FIGURE 49.74a Anechoic chamber for testing and calibrating electronic equipment: a modular soundproof room with silenced HVAC system. (*Courtesy Industrial Acoustics Company*.)

p. 49.17]. Using Fig. 49.15 (p. 49.20), if the room constant is 1000 sabins (92.91 metric sabins) and if the distance from the source is 20 ft (6.1 m), then $L_W - L_p = 14$ dB. So if a source's L_W at a given frequency is 50 dB, the L_p would be 36 dB.

Chapter 52 of the *1987 ASHRAE Handbook, Systems*, provides the following empirical equation for "normal rooms" when there is a single source in the room:

$$L_p = \begin{cases} L_W - 5 \log V - 3 \log f - 10 \log r + 25 \text{ dB} & \text{in English units} \\ L_W - 5 \log V - 3 \log f - 10 \log r - 12 \text{ dB} & \text{in metric units} \end{cases} \quad (49.45)$$

room sound pressure level of chosen reference point, dB re 0.08 inWG (20 Pa)
L_W = sound power level of source, dB re 10^{-12} W
V = room volume, ft^3 (m^3)
f = octave band center frequency, Hz
r = distance from source to reference point, ft (m)

A general equation for an overall sound power level prediction from air-conditioning diffusers is (see Ref. 19)

$$L_W = 10 + 10S + 30 \log \xi + 60 \log u \quad (49.46)$$

where S = area of duct cross section prior to diffuser, m^2
ξ = normalized pressure drop, derived from $\Delta P / 0.5 \rho u^2$
ρ = density of air, kg/m^3
u = mean-flow velocity in duct prior to the grid, m/s
ΔP = pressure drop across diffuser, Pa

FIGURE 49.74b One of nine audiometric testing rooms in London, England.

FIGURE 49.74c One of 121 music practice rooms at the University of Western Michigan.

FIGURE 49.74d One of three studios at KFOG Radio, San Francisco.

FIGURE 49.74e Reverbrant room used for the development of air conditioning systems.

Use Eq. (49.46 on p. 49.110) with metric units only; convert English to metric units as follows:

ft × 0.3048 = m	ft^2 × 0.0929 = m^2
in^2 × 0.0006452 = m^2	inWG × 249.1 = Pa

FIGURE 49.74f Shelters for torch cutting controls in a continuous casting mill.

Equation (49.46) tells us in effect that when the velocity is doubled, the sound power level L_W increases by 18 dB. A doubling of the diffuser area will increase the L_W by 3 dB, while a doubling of the pressure drop across the diffuser (assuming that the duct velocity prior to the grid can be kept constant) will increase the diffuser's self-noise by 9 dB.

The above theory and Eq. (49.46) are intended only as guidelines to explain how the variables of pressure drop, flow velocity, and diffuser area affect the acoustic performance characteristics of diffusers; they are not intended for calculation purposes. For instance, a duct's end reflection and number of terminals would be part of an HVAC system calculation for silencer requirements, as in Sec. 49.24 beginning p. 49.92; they are not directly related to the diffuser area as given in Eq. (49.46).

Essential guidelines to terminal unit selection therefore involve attention to flow velocity and pressure drop. An increase in area is more than offset by reductions in sound power levels due to lower flow velocities and pressure drops.

Flow-noise sound levels from terminal units also depend on correct air-approach configurations (Ref. 2). Manufacturers' ratings apply only to outlets with a uniform air-velocity distribution throughout the neck of the unit. If these conditions cannot be duplicated, then increases in sound levels up to 12 dB can occur, as shown in Fig. 49.75. Also, the misalignment of flexible duct connections can increase sound levels significantly, as in Fig. 49.76.

Pressure drop is also a factor in flow-noise generation from volume control dampers. Figure 49.77 shows the decibels to be added to outlet sound for throttled damper pressure ratios close to the outlet. In acoustically critical spaces, balancing dampers, equalizers, and similar devices should never be installed directly

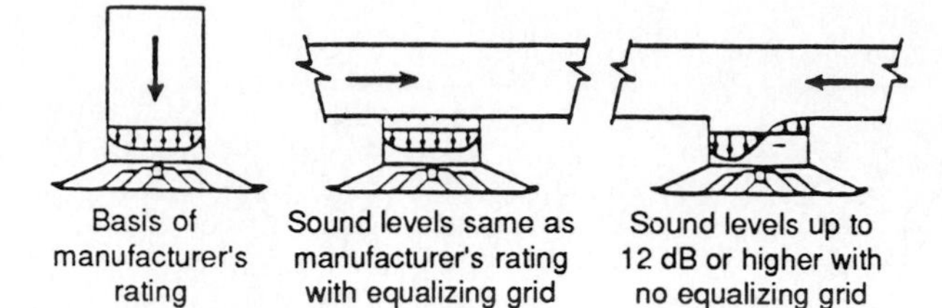

FIGURE 49.75 Proper and improper air-flow conditions to an outlet. (ASHRAE Handbook 1987 Systems, *chap. 52, "Sound and Vibration Control," American Society of Heating, Refrigerating and Air-Conditioning Engineers, Inc., Atlanta.*)

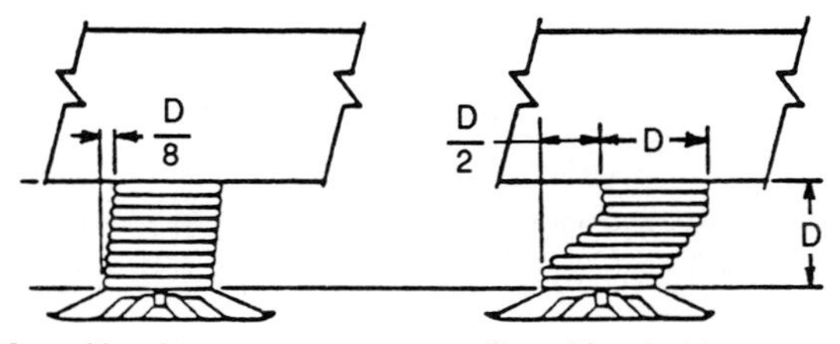

FIGURE 49.76 Effect of proper and improper alignment of flexible duct connector. (ASHRAE Handbook 1987 Systems, *chap. 52, "Sound and Vibration Control," American Society of Heating, Refrigerating and Air-Conditioning Engineers, Inc., Atlanta, 1987.*)

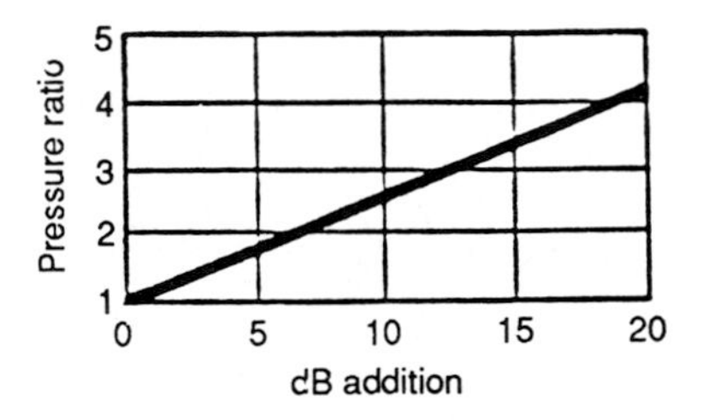

$$\text{Pressure rating (PR)} = \frac{\text{Throttled pressure (drop)}}{\text{Catalog pressure (drop)}}$$

FIGURE 49.77 Decibels to be added to outlet sound for throttled damper close to the outlet. (ASHRAE Handbook 1987 Systems, *chap. 52, "Sound and Vibration Control," American Society of Heating, Refrigerating and Air-Conditioning Engineers, Inc., Atlanta, 1987.*)

$$\text{Pressure ratio (PR)} = \frac{\text{Throttled pressure}}{\text{Minimum pressure}}$$

Location of volume damper	1.5	2	2.5	3	4	6
(A) OB damper in neck of lenear diffuser	5	9	12	15	18	24
(B) OB damper in plenum side inlet	2	3	4	5	6	9
(C) Damper in supply duct at least 5 ft (1.5 m) from plenum	0	0	0	2	3	5

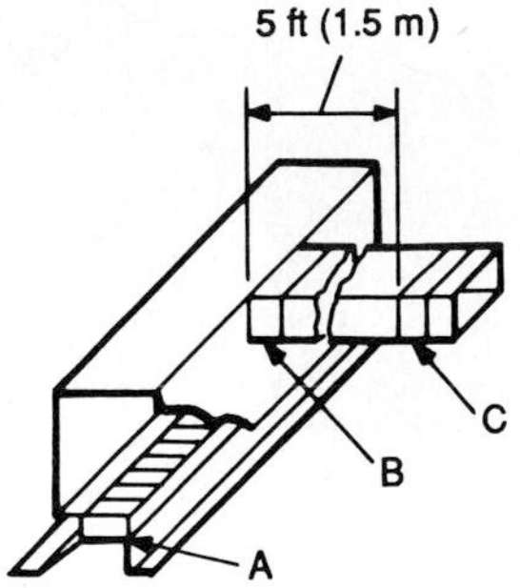

5 ft (1.5 m)
A
B
C

FIGURE 49.78 Decibels to be added to diffuser sound rating to allow for throttling of volume damper. (*ASHRAE Handbook 1987 Systems, chap. 52, "Sound and Vibration Control," American Society of Heating, Refrigerating and Air-Conditioning Engineers, Inc., Atlanta, 1987.*)

behind terminal devices or open end outlets. They should be located 5 to 10 diameters from the opening, followed by silencers or lined ducts to the terminal or open duct end. Figure 49.78 shows the decibels to be added to sound ratings for volume dampers located at different distances from diffusers and throttled at various pressure ratios.

The noise generated by multiple air-distribution devices of equal sound intensity follows the 10 log N rule (where N is the number of diffusers), which means that 3 dB must be added for two terminal units, and 10 dB for 10 units. See Fig. 49.6 on p. 49.11 for other additions.

49.28 THE USE OF INDIVIDUAL AIR-HANDLING UNITS IN HIGH-RISE BUILDINGS

During the last decade, factory-assembled and -tested individual floor air-handling units (AHUs) have been used in many high-rise office buildings. An AHU with a 20,000-ft^3/min (9.4-m^3/s) air-flow capacity, for example, can be placed on each building floor, which may have an area of 20,000 ft^2 (1858 m^2).

AHUs are often located in individual mechanical-equipment rooms in tightly configured floor layouts and may be adjacent to office spaces that must be kept quiet. NC 30 or NC 35 criteria are not uncommon.

Such individual AHUs are taking the place of large, central air-handling systems, where one system might serve a 2,000,000-ft^2 (185,800-m^2) high-rise office building from one mechanical-equipment-room floor located in the middle of the building. The core of the central air-handling system would consist of four 450,000-ft^3/min (212.25-m^3/s) supply fans, each with an 800-hp (597-kW) drive. It is clear that the energy demand and operating costs would be very high if such a

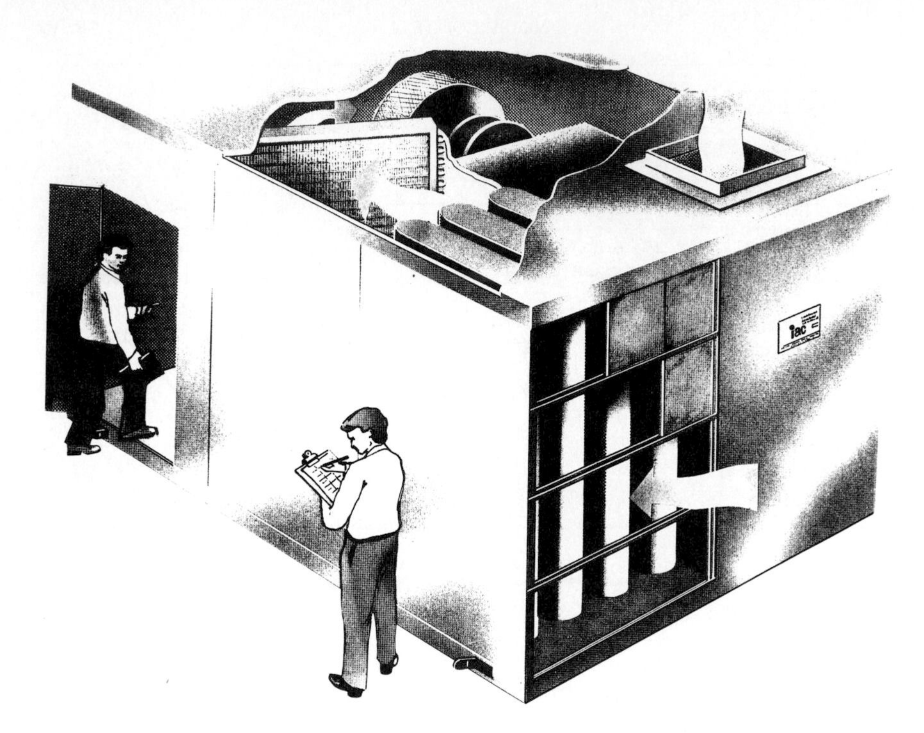

FIGURE 49.79 Quiet air-handling unit. (*Courtesy of Industrial Acoustics Company.*)

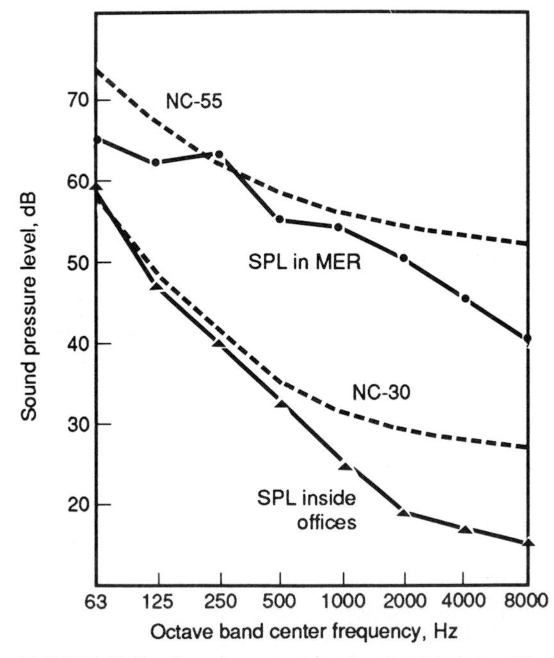

FIGURE 49.80 Sound pressure levels of quiet air-handling unit in mechanical-equipment room and outside of adjacent offices. (*Courtesy of Industrial Acoustics Company.*)

system had to be turned on to accommodate overtime work on Saturday and Sunday work in just a few offices.

Individual floor AHUs can therefore provide a high degree of flexibility and lower operating costs. A quiet AHU is illustrated in Fig. 49.79. The noise levels for such units inside a mechanical-equipment room and just outside adjacent offices are shown in Fig. 49.80.

49.29 BUILT-UP ACOUSTIC PLENUMS

The following applies to field-assembled air-handling units, as distinct from the factory-assembled and -tested AHUs discussed in Sec. 49.28. In an acoustic sound-absorbent plenum, noise from the fan inlet, fan housing, and drive motor is radiated within the plenum, resulting in a reduction of the system sound pressure levels, which at present cannot be predicted accurately.

In a typical "blow through" plenum arrangement there will be no return-air

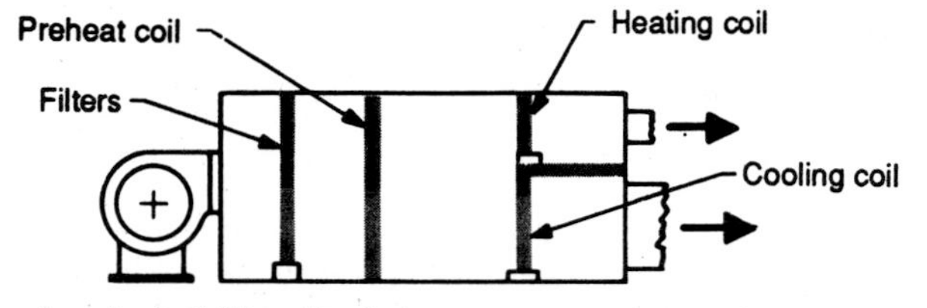

In a typical 'blow-thru' plenum arrangement, noise from the fan outlet may propagate through the plenum and then through the supply air duct system.

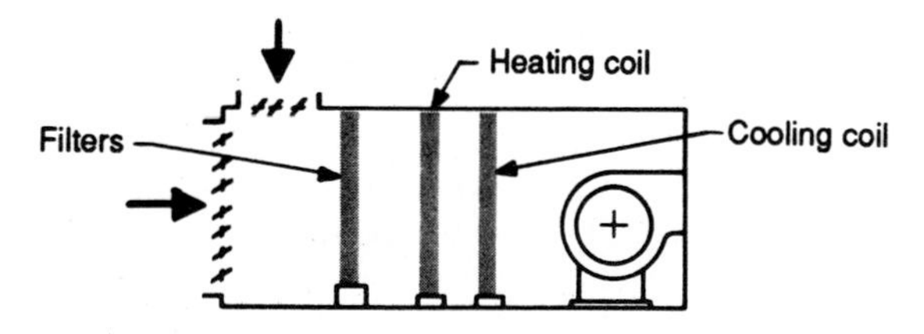

In a 'draw-thru' system, noise in the supply duct may be due to fan inlet and fan housing noise propagating through the plenum and into the return air system.

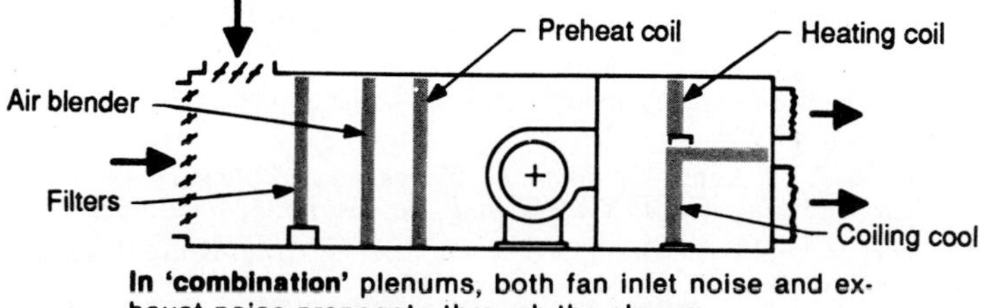

In 'combination' plenums, both fan inlet noise and exhaust noise propagate through the plenum.

FIGURE 49.81 Plenum types.

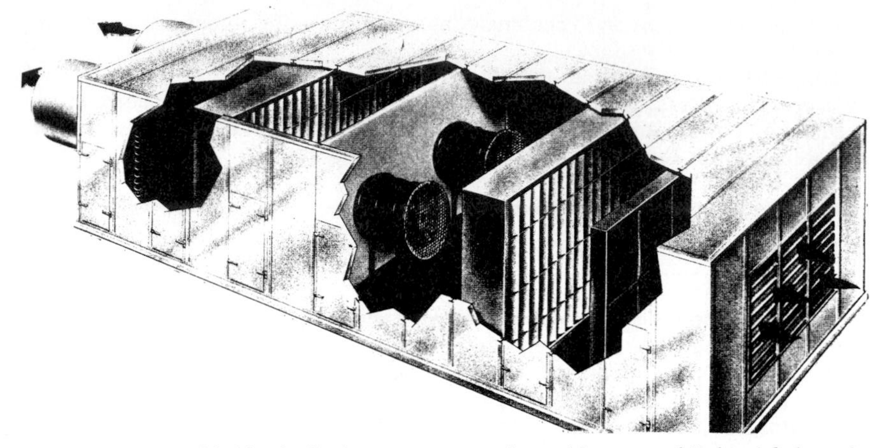

FIGURE 49.82 "Combination" plenum, cutaway view. (*Courtesy of Industrial Acoustics Company.*)

system fan attenuation. Similarly, in a "draw through" plenum there will be no supply duct system attenuation. On the other hand, in a typical "combination" plenum there will be attenuation in both systems. All three plenum types are shown in Fig. 49.81, and a cutaway view of a combination plenum is shown in Fig. 49.82.

The main parameters affecting fan plenum attenuation are the connecting duct area, the surface area of the plenum, the average absorption coefficients of the plenum walls and floors, the plenum dimensions, the angle between the fan inlet and the plenum opening, and the distance between the fan inlet and the plenum discharge. These variables are largely reflected in a formula in Ref. 2. This formula is based on dimensionally small-scale experimental models not directly related to air-conditioning fan plenums; obviously, the configuration and internal elements will vary in a full-size plenum. No full-scale data are presently available to verify this formula, which usually results in relatively high numbers.

Plenum silencer characteristics should therefore be regarded merely as *supplemental*, thus providing a safety factor for the supply-air and/or return-air duct silencers.

49.30 REFERENCES

1. C. M. Harris (ed.), *Handbook of Noise Control*, 2d ed., McGraw-Hill, New York, NY, 1979.
2. *1987 ASHRAE Handbook, Systems*, ASHRAE, Atlanta, GA, 1987, chap. 52, "Sound and Vibration Control."
3. L. Miller and F. M. Long, "*A Practical Approach to Cooling Tower Noise Evaluation,*" *Heating, Piping and Air Conditioning*, vol. 34, no. 6, June 1962.
4. H. Seelbach, Jr., and F. M. Oran, "Control of Cooling Tower Noise," *Heating, Piping and Air Conditioning*, vol. 35, issue 16, June 1963.
5. M. Hirschorn, "Silencing Cooling Towers," *Heating, Piping and Airconditioning*, vol. 24, issue 8, August 1952.

6. M. Hirschorn, "Some Considerations Involved in the Silencing of Air Conditioning and Ventilating Air Intakes, Roof Exhausters and Cooling Towers," *Noise Control*, January 1957.
7. ASTM E477, *Standard Method of Testing Duct Liner Materials and Prefabricated Silencers for Acoustical and Airflow Performance*, American Society for Testing and Materials, Philadelphia, 1973.
8. M. Hirschorn, "Acoustic and Aerodynamic Characteristics of Duct Silencers for Airhandling Systems," ASHRAE Paper CH-81-6, 1981.
9. *Application Manual for Duct Silencers*, Bulletin 1.0301.4, Industrial Acoustics Company, Bronx, NY, 1989.
10. U. Ingard, "Influence of Fluid Motion Past a Plan Boundary on Sound Reflection, Absorption and Transmission," *Journal of the Acoustical Society of America*, vol. 31, no. 7, July 1959.
11. F. Mechel, "Schalldämpfung in absorbierend ausgekleideten Kanalen mit überlagerter Luftstromung," *Proceedings of Third International Congress on Acoustics*, Stuttgart, 1959.
12. James H. Botsford, "dBA Reduction of Barriers," *Proceedings of Noise Expo*, Chicago, Illinois, 1973.
13. Laymon N. Miller, "Noise Control for Buildings and Manufacturing Plants," lecture notes, Bolt Beranek & Newman, Inc., Cambridge, Mass., 1981.
14. *Technical Advisory Service* (publication), Pilkington Flat Glass, Ltd., St. Helens, Merseyside, England, 1974.
15. ASTM E90, *Standard Method for Laboratory Measurement of Airborne Sound Transmission Loss of Building Partitions*, 1975.
16. ASTM E596, *Standard Method for Laboratory Measurements of the Noise Reduction of Complete Sound Isolating Enclosures*, 1978.
17. Martin Hirschorn, "The Aero-Acoustic Rating of Silencers for 'Forward' and 'Reverse' Flow of Air and Sound," *Noise Control Engineering*, vol. 2, no. 1, Winter 1974.
18. H. L. Kuntz and Robert M. Hoover, "The Interrelationships Between the Physical Properties of Fibrous Duct Lining Materials and Lined Duct Sound Attenuation," preliminary ASHRAE paper, 1987. See *1987 ASHRAE Handbook, Systems*, ASHRAE, Atlanta, GA, Chapter 52.
19. Leo L. Beranek (ed.), *Noise and Vibration Control*, McGraw-Hill, New York, NY, 1971.

CHAPTER 50
VIBRATION

Norman J. Mason
President, Mason Industries, Inc., Hauppauge, New York

50.1 INTRODUCTION

Chillers, pumps, blowers, cooling towers, ducts, piping, rigid electrical conduits, etc., that are rigidly bolted to a structure, transmit 100 percent of their vibratory energy. The introduction of properly selected vibration isolators will reduce this transmitted energy to the point where it is completely imperceptible or so minor as to no longer be annoying to the occupants, structurally destructive, or detrimental to critical manufacturing processes. "Vibration isolation efficiency" is defined as the percentage of vibration that is no longer transmitted to the structure because of the introduction of vibration isolators.

50.2 THEORY

The vibration problem can be approached on a theoretical basis by using Eq. (50.1), the theoretical vibration efficiency equation, which represents an isolated machine as shown in Fig. 50.1.

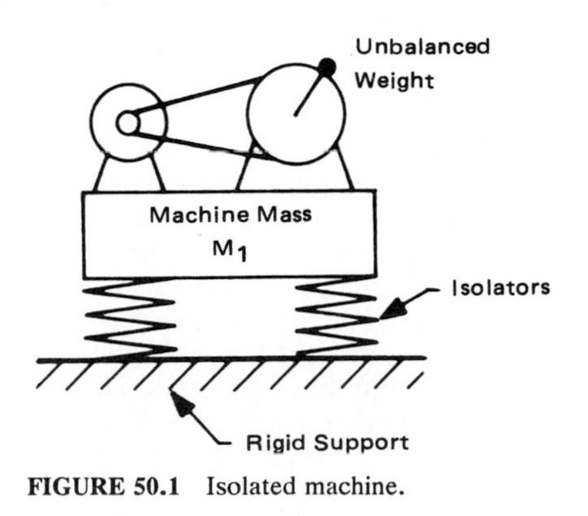

FIGURE 50.1 Isolated machine.

$$E = 100\left[1 - \frac{1}{(f_d/f_n)^2 - 1}\right] \tag{50.1}$$

where E = percentage of vibration isolated (efficiency)
f_d = disturbing frequency of the isolated machine
f_n = natural frequency of the isolated machine

The disturbing frequency should be taken as the r/min of either the equipment or the driver, whichever is lower. All equipment has some unbalance at the primary speed, and this approach is conservative, since any higher-frequency vibration used in the formula would result in overly optimistic values for the primary disturbance.

This equation states that a given percentage of the vibration can be kept out of the structure by installing the machine which is creating the vibration at a known frequency (usually the machine r/min) on a system of vibration mountings. These should resonate at a frequency that is very much lower than the disturbing frequency described above. When this ratio of disturbing frequency to the natural frequency is 3:1, about 90 percent of the transmitted vibration is theoretically eliminated. Reference to Fig. 50.1 will help to visualize the mechanical system represented by Eq. (50.1). A motor is shown driving a rotating component. Assume that there is one point of unbalance in the rotating machine and none in the motor. These two components would be kept in their relative positions by either a steel base or a steel base filled with concrete. The total weight of the system would be the weight of the motor plus the weight of the machine plus the weight of the base.

Assuming that the system has only one degree of freedom and that there are no external connections, when we push the system down it will bounce back up and continue to oscillate vertically at a specific frequency. This frequency f_n is known as the "natural" frequency because it continues to move at this rate with no further introduction of energy.

When we turn the machine on, the unbalanced force that is centrifugally generated will force the system to vibrate at the operating speed of the driven machine. This frequency is known as the "forcing" or "disturbing" frequency f_d. Examination of the efficiency equation (50.1) shows us that the larger the ratio between this forcing frequency and the natural frequency, the greater will be the degree of vibration isolation.

The only unknown in the equation is f_n. The equation used in determining this frequency is a simple relationship to the static deflection, provided the isolation material that is used has a straight-line deflection curve and virtually no damping. (Mass in no way enters the equation.) The frequency is dependent upon acceleration provided by the return spring force acting against the mass. When the deflection is directly proportionate to increased mass, as a function of the spring stiffness, mass and stiffness conveniently drop out and deflection is all that is needed to determine frequency. Static deflection refers to the actual difference in height between the unloaded spring and the spring in the deflected condition. For example, if the spring were 6 in (150 mm) tall to begin with and 4 in (100 mm) tall under load, the static deflection would be 2 in (50 mm).

If the material does not have a straight-line deflection curve, as in the case of some rubber materials in compression, or the material has appreciable dynamic stiffness (which means that in motion it acts like a stiffer material than the deflection would indicate), frequency tests can be run to determine the frequency at various loads and deflections and the curves plotted. This is normally the case for natural rubber and other elastomers and for materials such as fiberglass, felt, cork, and sisal. When using these materials, deflection is a very poor indicator of the natural frequency, and test curves must be referred to for any degree of accuracy when inserting f_n in the efficiency equation. The same applies to more exotic devices like air springs where there is no deflection under load but only a change in air pressure. Frequencies can be calculated, but test data are far more reliable.

Equation (50.2) refers to helical steel springs which are free of damping and have uniform deflection rates. The natural frequency f_n is then expressed as

$$f_n = \sqrt{\frac{188(25.4)}{d_1}} \tag{50.2}$$

where f_n = natural frequency, cycles/min (Hz)
d_1 = static deflection in the spring supports, in (mm)

In looking at this as a mechanical system, it is simpler to think in terms of the pendulum. The frequency of a pendulum is dependent on the length of the pendulum from the rotation point to the center of the bob or mass and is in no way a function of the mass. The frequency drops as the pendulum becomes longer and increases as it is shortened. Pendulum length is analogous to spring deflection. The longer the pendulum, the lower the frequency. The more static deflection, the lower the natural frequency of the system.

Returning to the efficiency equation, let us try a solution referencing a machine with a disturbing frequency of 600 cycles/min. A natural frequency of about 200 cycles/min would be required to attain a ratio of 3:1 and to arrive at 87.5 percent efficiency. Since exact numbers are not critical, it is easier to use an efficiency chart (Fig. 50.2).

This is still a theoretical discussion and not to be used for practical purposes. This is only introductory to aid in understanding.

The operating speeds or disturbing frequencies are listed across the bottom of Fig. 50.2. The vertical scale is a continuous solution of the natural frequency equation, with the deflection on the left and the corresponding natural frequencies on the right. The efficiency lines run diagonally.

To use this chart, start at the bottom with the lowest operating speed of the equipment. Move up vertically and intersect the desired efficiency line and then move over to the left to pick up the natural frequency of the mountings required for that efficiency and the corresponding spring deflection. We have outlined this procedure for 90 percent at 600 r/min in the dark lines. For all practical purposes, we would need mountings with about 1 in (25.4 mm) of static deflection in order to obtain the 90 percent efficiency.

It is important to emphasize that *commercial vibration control is not an exact science*. The weight of the equipment provided by manufacturers is approximate at best, and the location of the center of gravity is even more nebulous. Using commercially available mountings, it is impossible to select mountings for exact deflection. Vibration isolation commodities are not precision devices, and there cannot be exact cataloged springs for every load. For these reasons, deflections are generally specified as minimums. Deflections beyond those that are minimal will merely lower the natural frequency and improve performance.

Transmitted vibration is that portion of the vibration that is still sensed by the structure. At 90 percent efficiency, there is 10 percent transmission. Efficiency is basically a salesperson's term, probably developed because of the grading system

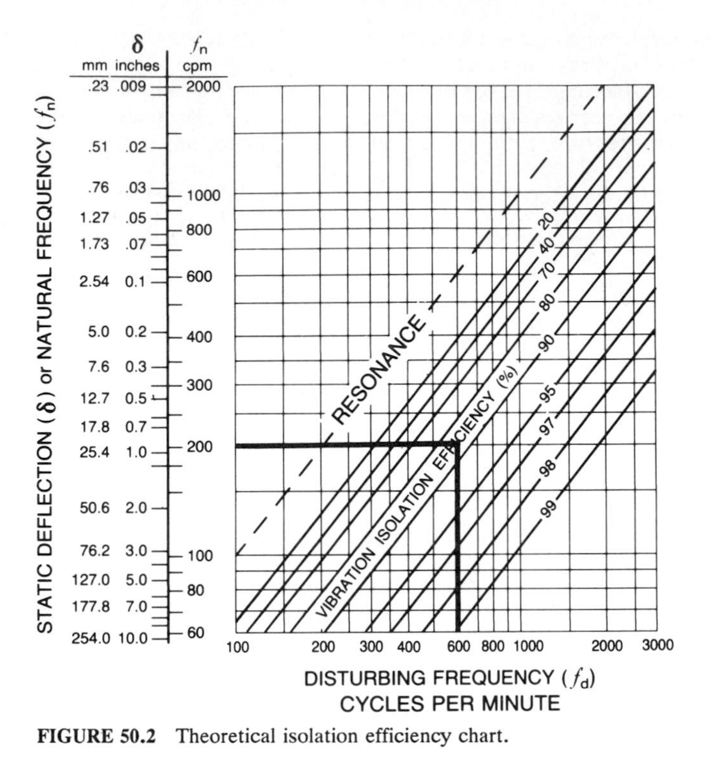

FIGURE 50.2 Theoretical isolation efficiency chart.

that we all grew up with: 85 percent fair, 90 percent good, and 95 percent excellent. In our work, these numbers can be extremely deceptive. In comparing 95 to 90 percent, we assume an improvement of only 5 percent. However, the comparison of 5 to 10 percent transmission shows that only half the remaining force is transmitted. Neither number can be considered alone as the source of the vibration, and what we have to eliminate is the deciding factor.

In broad terms a 125-hp (93-kW) pump will generate five times the vibratory energy of a 25-hp (19-kW) pump. Therefore, the isolation provided for the 125-hp (93-kW) pump must be five times more efficient to reduce the force transmission to a similar level. An 80 percent efficiency with 20 percent transmission for the 25-hp (19-kW) pump is equivalent to 96 percent efficiency with 4 percent transmission for the 125-hp (93-kW) unit.

50.3 APPLICATION

50.3.1 Basic Considerations

A completely arithmetical, rather than conventional, approach provides a better mechanical visualization of what is really going on. The results can be confirmed on the efficiency chart.

There are six basic considerations once the installation is well away from resonance ($f_d/f_n \geq 3{:}1$):

1. Efficiency is controlled by static deflection, and transmission is reduced in direct proportion to the increase in deflection.
2. If the frequency of the isolator approaches the operating speed of the equipment, resonance is approached.
3. When approaching resonance, the dynamic motion generated by the unbalanced force is amplified.
4. Dynamic motion is controlled by the unbalanced force and its relationship to the total mass. For all practical purposes, at higher frequency ratios the frequency ratio itself no longer influences motion.
5. Once the frequency ratio is 3:1 or greater, motion should only be reduced by the introduction of mass.
6. Assuming the unbalanced forces act through the center of gravity, motion is reduced in direct proportion to the increase in mass.

In reality, when a rotating machine vibrates on isolation mountings, the foundation has a small rotary motion. Since floors are much more sensitive in the vertical direction, the other modes are usually ignored and installations are always visualized moving vertically.

A spring constant is referred to as k and is defined as the number of pounds (kilograms) required to deflect the spring 1 in (25 mm). Thus a spring with a constant k of 1000 lb/in (18 kg/mm) would deflect 0.5 in (13 mm) at 500 lb (227 kg) and 1.0 in (25 mm) at 1000 lb (454 kg).

A system constant is normally defined as the number of pounds (kilograms) required to deflect all the supports of a system 1 in (25 mm) simultaneously. Using four springs with a constant k of 1000 lb/in (18 kg/mm) each, one in each corner of an installation, the system constant would be 4000 lb/in (72 kg/mm). Thus 2000 lb (907 kg) of equipment would deflect the system 0.5 in (13 mm) and 4000 lb (1814 kg), 1.0 in (25 mm).

Since the spring rate is uniform, this also means that if the upward vibratory force pulled the equipment 0.10 in (2.5 mm) up from the neutral position, it would be reducing the spring load by 0.10 of 4000 lb (1814 kg), or 400 lb (181 kg). As the vibratory motion pushed the installation 0.10 in (2.5 mm) below the neutral position, there would be an increase in the spring load of the same 400 lb (181 kg). Since the bottom of the spring is attached to the structure, this change in force of ±400 lb (181 kg) is what the structure sees as a change in the static loading. This occurs at 600 r/min. A change in static loading at a particular frequency is another definition of vibration.

This approach to the problem is best illustrated by Fig. 50.3.

If the machine is running at 600 r/min, the vibratory force transmitted would be ±400 lb (181 kg) at 600 cycles/min.

Assume that the ±400 lb (181 kg) is unacceptable in the structure. The instinctive solution would be to use a larger mass. A traditional mass ratio is three times the equipment weight, bringing the system to 16,000 lb (7258 kg) (Fig. 50.4). Continuing with a 1-in (25-mm) deflection spring grouping, the system constant would have to be raised to 16,000 lb/in (286 kg/mm) by using stiffer individual springs or clusters of four times the original spring groupings.

In basic rule 6, it was stated that the vibratory motion would be reduced in direct proportion to the increase in mass. Therefore, the amplitude of 0.10 in (2.5

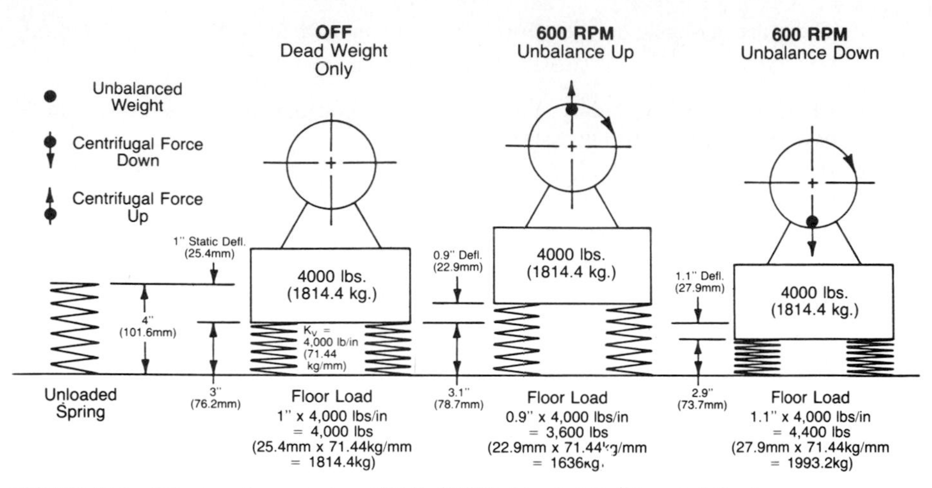

FIGURE 50.3 Vibratory transmission, 4000-lb (1800-kg) load. 1-in (25-mm) deflection.

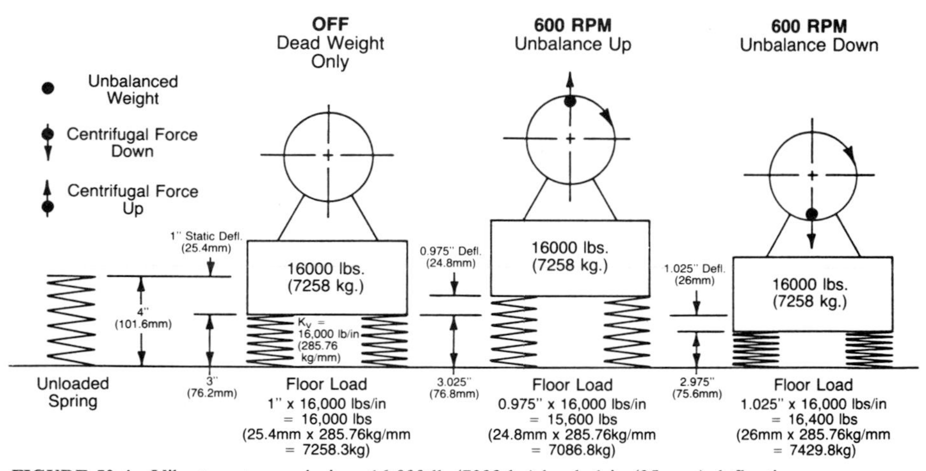

FIGURE 50.4 Vibratory transmission, 16,000-lb (7300-kg) load. 1-in (25-mm) deflection.

mm), as in Fig. 50.3, would now become 0.025 in (0.64 mm). Following the sketches across, it is found that the reduced motion is merely acting against a proportionately stiffer spring constant so that there is no reduction in vibration transmission but only in amplitude.

If the machine is running at 600 r/min, the vibratory force transmitted remains ±400 lb (181 kg) at 600 cycles/min.

The question then comes up as to how to actually reduce the transmitted vibration. Assume we wish to reduce this transmission by 75 percent so that the end result is ±100 lb (45 kg). Basic rule 1 states that efficiency is controlled by static deflection and transmission reduced in direct proportion to the increase in deflection. Rule 4 also states that, for all practical purposes, the frequency ratio no longer influences motion.

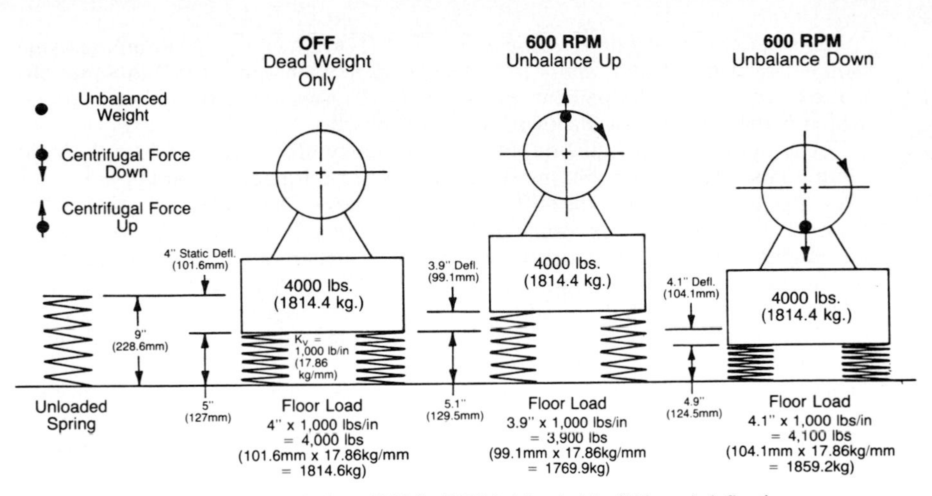

FIGURE 50.5 Vibratory transmission, 4000-lb (1800-kg) load. 4-in (100-mm) deflection.

Therefore, increase the deflection to 4 in (100 mm), as in Fig. 50.5. Since this larger deflection will give us a lower natural frequency, there will be no noticeable difference in the amplitude. The example shows that the spring constant has dropped to 1000 lb/in (18 kg/mm). Since the amplitude remains at the original ±0.10 in (2.54 mm), this amplitude multiplied by the new spring constant results in a force transmission of only ±100 lb (45 kg) at 600 cycles/min.

The problem can now be approached on the basis of reducing both amplitude and transmission by reusing the total weight of 16,000 lb (7258 kg) and providing 4-in (100-mm) deflection, as shown in Fig. 50.6. A reduction is now made in both the amplitude to 0.025 in (0.64 mm) and in the transmitted force to ±100 lb (45 kg).

The vibratory force transmitted would be ±100 lb (45.5 kg) at 600 cycles/min.

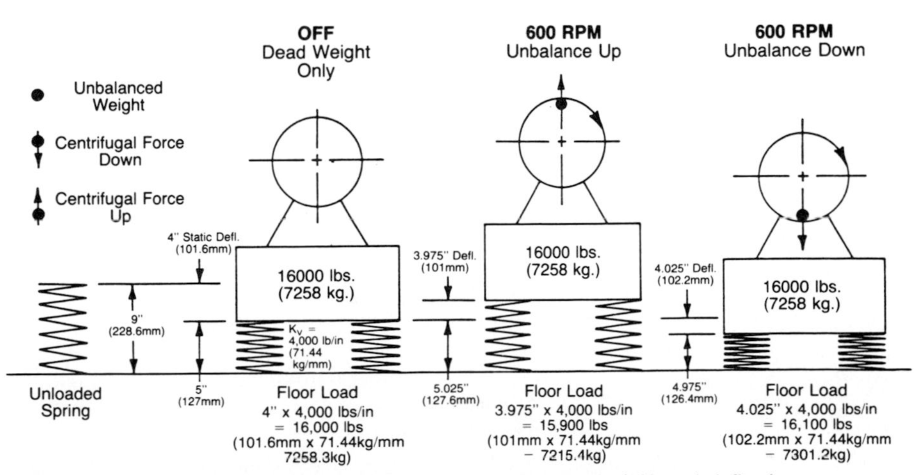

FIGURE 50.6 Vibratory transmittion, 16,000-lb (7300-kg) load. 4-in (100-mm) deflection.

This really agrees with the efficiency chart (Fig. 50.2), as a 600-r/min machine isolated by 1-in (25-mm) and 4-in (100-mm) deflection springs would show efficiencies of 90 and 97.5 percent, respectively. Transmission reduction is 10:2.5, which is the same factor 4 shown by the arithmetic.

Both the efficiency equation and the efficiency chart are based on the completely false assumption that the floor stiffness or frequency in an upper story is very high as compared to the stiffness or frequency of the isolator. In reality, the floor has a deflection of its own and a natural frequency which can be low enough to mandate the use of isolators with very much higher deflections than indicated by the chart. Figure 50.7 shows the actual conditions in a structure. Rather than a simple system with the machine or machine foundation resting on springs on a relatively unyielding support, the springs are supported by a spring board with a finite mass of its own. Schematically, this is sketched in Fig. 50.8. The machine mass rests on springs on the floor mass, and the floor stiffness is shown by a second set of springs. Although floors are supported by beams connected to vertical columns, ground-level vertical stiffness really exists only at the columns and not in between.

The worldwide structural limit on floor deflection is 1/360th of the span. In many commercial buildings design spans are at least 360 in (9144 mm), or 30 ft (9.14 m). This means that the structural engineer is allowed a floor deflection of 1 in (25 mm) at the center of the span when the floor is fully laden with both live and dead loads.

Let us make the assumption that in a particular area where a 125-hp (93 kW) pump is installed that the floor is loaded to half dead load plus live load. The floor deflection would then be 0.5 in (13 mm). If the pump is running at 1750 r/min, a quick reference to the efficiency chart shows that a mounting deflection of 0.1 in (2.54 mm) should provide 90 percent efficiency or 10 percent transmission. However, the 0.5-in (13-mm) floor deflection would be five times the deflection of the isolator. The actual efficiency would be influenced by the floor's 0.5-in (13-mm) deflection, the mass of the floor, and the floor's damping characteristics. The 90 percent theoretical efficiency could never be attained, and depending on the combination of conditions, the actual efficiency might slide down to 50 or 40 percent and not meet the requirement.

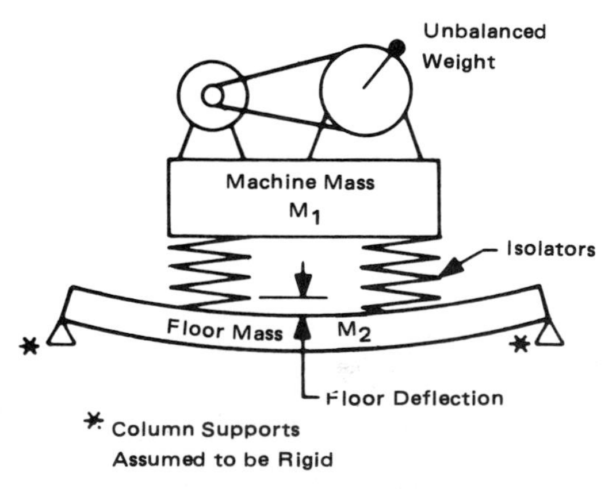

FIGURE 50.7 Actual structure conditions (floor deflection exaggerated).

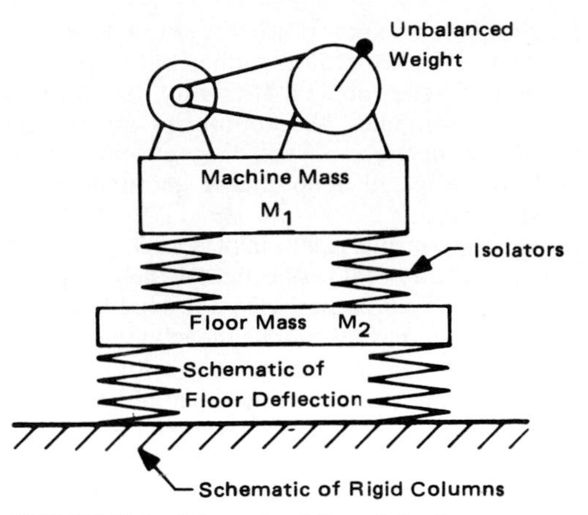

FIGURE 50.8 Schematic of floor deflection.

Therefore, rather than relying on the theoretical method, commercial selection of isolators has evolved into using isolators with deflections that equal or exceed the floor deflection to attain acceptable transmission levels. The efficiency chart should only be used as a tool to learn the subject and gain direction.

Studies show that the floor stiffness is greatly in excess of the isolator stiffness, because the mass of the floor is much greater than the mass of the machine that is to be isolated. While this may be of importance when isolating small equipment, it is certainly not significant with large pumps and chillers.

On a day-to-day basis, the cost of an involved engineering investigation of a commercial structure's stiffness and resonance, along with the possibility of error in these conclusions, dictates the continued use of the more conservative floor-deflection rather than stiffness approach. The cost of isolation is small as compared to the cost of an installed air-conditioning system. Possible savings in using lower-deflection materials are in no way proportionate to complete loss of occupancy or lower rental rates in a noisy structure or recourse to very expensive retrofits.

The recommended deflections shown in the selection guide (Table 50.6) are based on empirical data gathered through 40 years of installation experience, as well as discussions with mechanical and acoustical engineers, architects, and manufacturers. The deflections were influenced by operating speeds, size of equipment, the equipment as a vibratory source, and the sensitivity of the floor structure in terms of construction and floor span.

50.3.2 Isolation Materials

An "isolation material" can be defined broadly as any resilient material that will accept a load on a permanent basis and produce a resonant or natural frequency that is reasonably consistent and predictable. It is also important that any increase in this frequency is small as the material ages.

Vibration Pads. "Elastomeric" describes any rubberlike material. While natural rubber has the best performance characteristics, it is generally not used commercially because of aging when exposed to oxygen, ozone, or oil. A synthetic elastomer similar in properties to natural rubber but lower in cost is the oil derivative SBR (styrene-butadiene rubber). It is very commonly used where there are no specific aging requirements. The neoprenes are not quite as resilient as natural rubber or SBR, but because of their very excellent aging characteristics, better grades of pad are either all neoprene or neoprene blended with SBR or natural rubber to reduce cost.

While the selection of the proper elastomer is an important choice in terms of performance and aging, the physical properties of the compound are specifically controlled by formulation. The polymers are not used by themselves but mixed with other materials such as carbon black and clay to provide reinforcement. Larger ratios of these fillers will reduce cost but with penalty to physical properties. As a general statement, a ratio of one-third fillers to the selected polymer produces the best physical and dynamic properties. When aging is the primary concern, the formulation should contain only neoprene and not a blend of the neoprene with natural rubber or SBR. Unless a specification states exactly what the physical properties and ingredients must be, the material that is furnished will only follow the manufacturer's conscience. Unfortunately, there are no industry standards as to when a pad can be called neoprene or natural rubber. Products are available that are made with as little as 5 percent neoprene but still referred to as commercial-grade neoprene. A good guide to quality are the AASHO standards shown in Table 50.1.

Other than foams, which do not have enough capacity or stability to be used as isolators, air-free rubber materials are incompressible. When a load is applied to a pad, it changes shape but does not lose volume. The ability to change shape is controlled by the shape factor and the material's hardness. Since a pad used in compression can only change its volume by bulging, unconfined edges are referred to as "escape area," whether internal in the form of holes or external. The term "shape factor" is the ratio of the loaded area to the escape area. The lower the shape factor, the more deflection at a particular load.

Thus a 4-in (100-mm) square pad that is 1 in (25 mm) thick, covered completely either by the equipment or by a steel plate, would have a loaded area of 16 in^2 (100 cm^2). Since the perimeter is 16 in (400 mm) and the pad 1 in (25 mm) thick, the escape area is also 16 in^2 (100 cm^2). The shape factor (load area divided by escape area) would be 16/16 (100/100), or 1. Assuming that the hardness remains the same, we could increase the load-carrying capacity of this pad by using two ½-in (12.5-mm) thick pads with a steel plate separating the two layers. Since the loaded area would remain 16 in^2 but the escape would now be 16 in (400 cm) multiplied by 0.5 in (12.5 mm), or 8 in^2 (50 cm^2), the shape factor would be increased to 2 with a lower deflection for the same imposed load. These relationships are shown in shape-factor curves (Fig. 50.9). These curves are empirical based on test data, and small variations will be found from one publication to another.

Hardness is measured by a durometer (in units called "duros"), a clocklike gauge with a penetration probe on the bottom. Pads are normally used in 30 to 70 duro in 10-point increments. Since rubberlike materials cannot be exactly controlled, the normal acceptable variation is ±5 of a nominal duro. To give you some idea of the feel of these durometers, common references are rubber band stock at about 30, red erasers at about 40, white erasers at 50 to 60, and the old-fashioned hard gray erasers at 70.

Automobile tires are 50 to 70 duro. The influence of hardness on load capacity is shown in Fig. 50.10. A 70-duro material will handle about four times the load that would be carried by the same shape in 30 duro.

TABLE 50.1 Physical Properties of Structural Bearings Made from Du Pont Neoprene

Physical Property	Test Method	Performance Requirements			
		HARDNESS GRADE			
		50	60	70	
Hardness, durometer A	ASTM D 2240	50±5	60±5	70±5	
Tensile strength	ASTM D 412	2500 (1725)	2500 (1725)	2500 (1725)	psi (kPa) minimum
Elongation at break	ASTM D 412	400	300	300	% maximum
Adhesion Bond made during vulcanization	ASTM D 429	40	40	40	lbs. per inch, minimum
Low-temperature performance	(Sample first prepared 96 hr. at −20 ± 2°F (−29 ± 1°C) axial load 500 psi and strain of 20% "T" [effective thickness].) Shear resistance after 1 hr. at 25% shear strain not to exceed values shown	50 (35)	75 (52)	110 (76)	psi (kPa)
Resistance to heat	ASTM D 573				
Change in original properties after 70 hrs. at 212°F (100°C)					
Hardness		+15	+15	+15	points, maximum
Elongation		−40	−40	−40	%, maximum
Tensile Strength		−15	−15	−15	%, maximum
Resistance to oil aging*	ASTM D 471*	+80	+80	+80	%, maximum*
Change in volume after 70 hrs. immersion in ASTM Oil No. 3 at 212°F (100°C)					
Resistance to ozone	ASTM D 1149				
Condition after exposure to 100 pphm ozone in air for 100 hrs. at 100±2°F (29±1°C) (sample under 20% strain)		No Cracks	No Cracks	No Cracks	
Resistance to permanent set	ASTM D 395	35	35	35	% maximum
Compression set after 22 hrs. at 212°F (100°C)					

*This oil aging requirement is not a part of the AASHO Specification referenced. However, its inclusion is strongly recommended to assure use of a high-quality neoprene compound.

Source: American Association of State Highway Officials Standard Specification for Highway bridges, Table B.

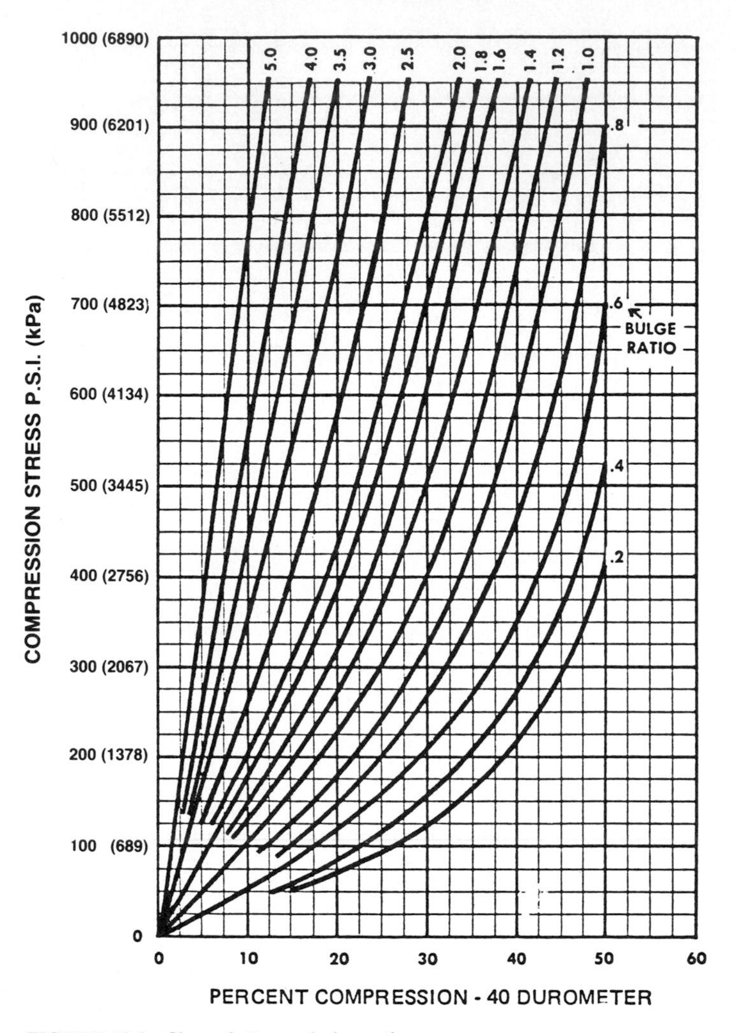

FIGURE 50.9 Shape-factor or bulge ratio.

Unfortunately, the harder the rubber material, the less it acts like a steel spring; the introduction of viscosity is similar to the introduction of a dashpot working in parallel with a spring. When the dashpot becomes large and the fluid stiff, vibratory forces are transmitted through the dashpot. It is for this reason that 70 duro is considered the extreme hardness for vibration isolation. Since 30 duro becomes uneconomical for large loadings and hard to manufacture, most pad materials fall into the 40- to 60-duro range.

Figure 50.11 is a dynamic stiffness chart based on experimental work with neoprene compounds containing no other elastomer and minimum fillers and the actual frequency at various deflections and hardness. Increased use of fillers lowers cost and quality at the expense of performance. To see the influence of the dynamic stiffness, you need only compare these frequencies and deflections with

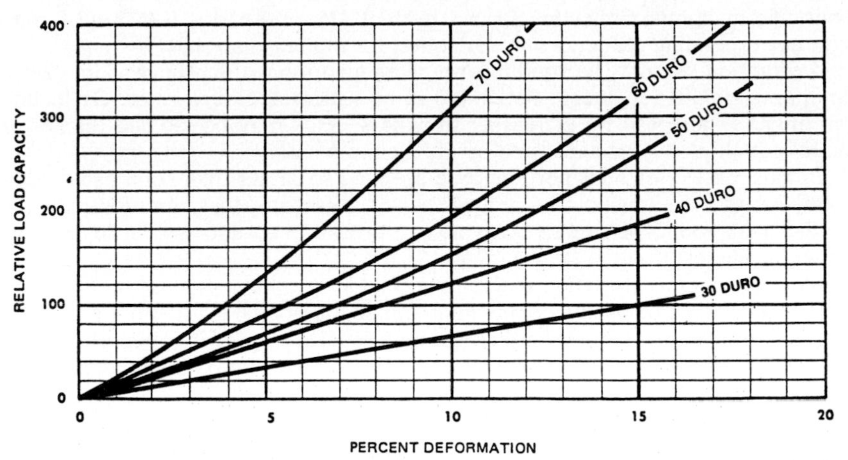

FIGURE 50.10 Influence of hardness on load capacity.

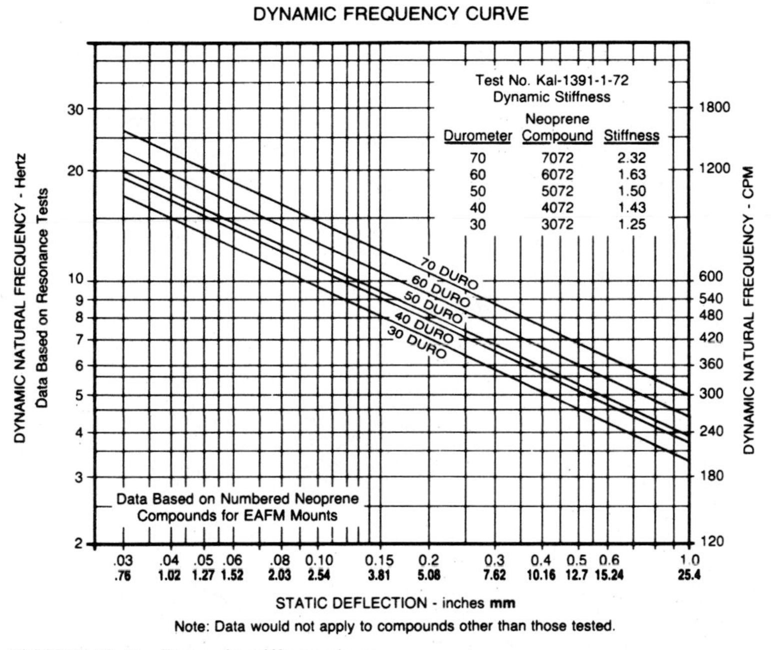

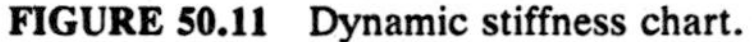

FIGURE 50.11 Dynamic stiffness chart.

the same deflections but lower frequencies shown in the efficiency chart, Fig. 50.2.

Pad deflection is limited by thickness. For the material to remain resilient and to control permanent set and creep, pad deflection should be limited to 15 percent of the thickness regardless of the rubber configuration or the rubber material.

Thus, the maximum deflection for a 1-in-thick (25-mm) pad is 0.15 in (3.8 mm) with corresponding reductions in deflection for thinner materials.

A vibration pad may be solid if for a given load it has the proper shape factor for the maximum 15 percent deflection in an acceptable durometer. Deflection can only be increased by increasing thickness. Pads may be molded thicker or made of multiple layers separated by steel shim plates.

In most cases, loadings per unit area are low, so rather than solid pads, additional escape area is needed to reduce the shape factor. Most vibration pads are molded with round or square holes and in cross-ribbed and waffle designs, as shown in Fig. 50.12.

Most commercial isolation pads are available in thicknesses up to ⅜ in (9.5 mm). They should be used at a maximum deflection of 0.06 in (1.5 mm) per layer. Dynamic frequency is 16 Hz in the best materials. A new ¾-in (19-mm) pad has been introduced recently with deflections of 0.11 in (2.8 mm) and a dynamic frequency of 12 Hz.

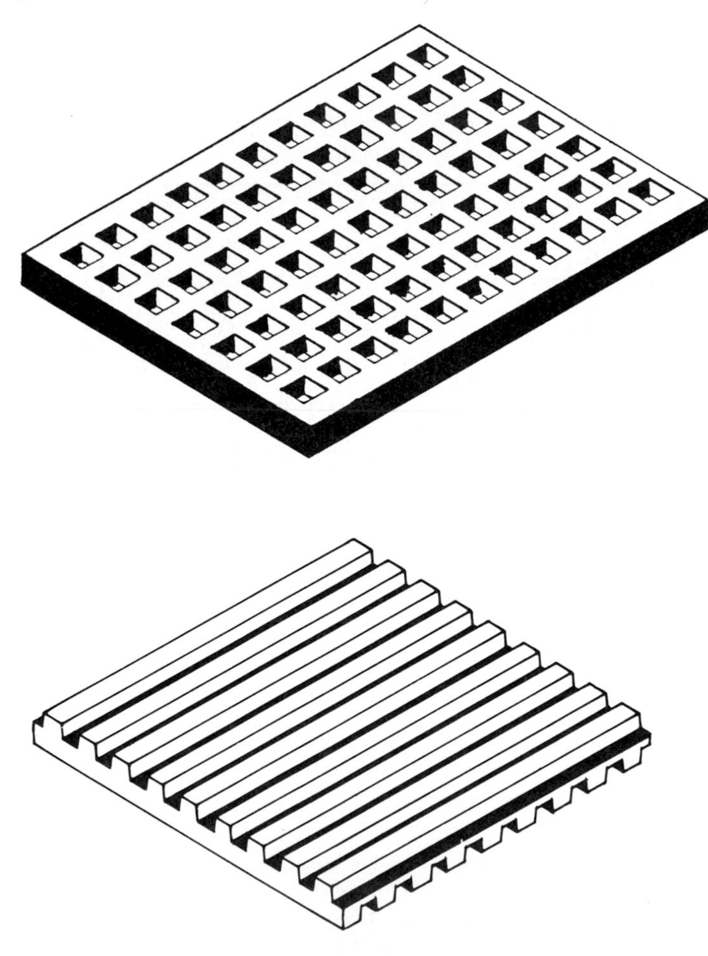

FIGURE 50.12 Typical vibration pads.

Other pad materials are cork, felt, sisal, and heavy-density precompressed fiberglass. Since the fiberglass is a fragile, spongelike material, it is normally covered with a neoprene or other coating to protect it against fraying and moisture. Fiberglass-pad frequency is not as sensitive to deflection and loading, so fiberglass pads are sometimes described as flat-frequency materials. However, this frequency is higher than neoprene or natural rubber at the same deflection. Table 50.2 shows published frequencies for 0.5-in (12.5-mm) and 1.0-in (25-mm) pads. Fiberglass pads should be avoided if there are large shear forces, such as those under a horizontal compressor.

Rubber mountings are sophisticated rubber pads. While the rubber may be loaded in shear, most commercial rubber mountings for air-conditioning applications are loaded in compression. Quality levels may be as described for pads, but the static deflections are higher only because the mountings are thicker. Rubber mountings have the advantage of provision for bolting to the equipment and to the floor when needed. Many of the newer designs have rubber under and over the base and top plates so that they can be slipped into place without bolting in stationary applications.

Springs. Spring mountings are generally required to provide the minimum deflection needed to compensate for structural flexibility. The heart of any steel spring mounting is the spring itself. It should be designed with a minimum diameter-to-deflected-height ratio of 0.8 so that the horizontal spring constant K_h is a minimum of 80 percent of the vertical K_v. Most designs end up with this 0.8 ratio, but this is not an exact rule. While this chapter is not meant to be a spring design handbook, Fig. 50.13 will give you the criteria for checking the horizontal stiffness as compared to the vertical.

An allowance of 50 percent additional capacity beyond rated load is also good practice. This means a spring rated for 2-in (50-mm) deflection would not have the coils touching (solid) before 3 in (75 mm). If it is rated for 1-in (25-mm) deflection, it should not go solid before 1½ in (37.5 mm). Overtravel allowance is needed as it is impossible to calculate exact weight distribution. Published equipment weight is inaccurate, and center-of-gravity locations are often unavailable. A 50 percent overtravel will allow for an acceptable 20 percent overload.

Isolation springs should be designed such that when the coils are touching, the elastic limit has not been exceeded so that springs will return to full height. If the spring is designed this way, it will be stressed two-thirds the elastic limit under normal load. Springs last indefinitely as vibration amplitudes are very small, and the spring movement per coil is the total amplitude divided by the number of active coils. With so little movement the stress cycle is close to zero, and these applications approach static loadings.

A simple spring can be considered a vibration isolator and often built into mechanical equipment. This low-cost method is satisfactory when thousands of springs are used in a repetitive application. When springs are used in the field, however, minimum additions to the design are normally a neoprene friction pad on the bottom to eliminate the need for bolting and to act in series with the steel coils to help eliminate high-frequency noise transmission. Since loadings are not easily or exactly determined, there must be a means of leveling, and this is usually done by means of an adjustment bolt. The bolt is often used to attach the mountings to the equipment as well.

Air Springs. The last remaining commodity of major importance is the air spring. Air springs are made of neoprene with nylon tire cord reinforcement and

TABLE 50.2 Comparison of Natural Frequencies at Given Deflections: Heavy-Density Fiberglass, AASHO Neoprene, Steel Springs

Heavy-density fiberglass					AASHO neoprene					Steel springs		
Pad thickness		Deflection		Dynamic freq., Hz	Pad thickness		15% Deflection		Dynamic freq., Hz	Deflection		Freq., Hz
in	mm	in	mm		in	mm	in	mm		in	mm	
0.5	13	0.045 to 0.090	1.1 to 2.3	22	0.3	7.6	0.045	1.1	18.0	0.045	1.1	14.8
					0.5	12.7	0.075	1.9	14.5	0.075	1.9	11.4
					0.6	15.2	0.090	2.3	13.1	0.090	2.3	10.4
1.0	25	0.09 to 0.18	2.3 to 4.6	16	1.0	25.4	0.150	3.8	10.5	0.150	3.8	8.1
					1.2	30.5	0.180	4.6	9.5	0.180	4.6	7.4

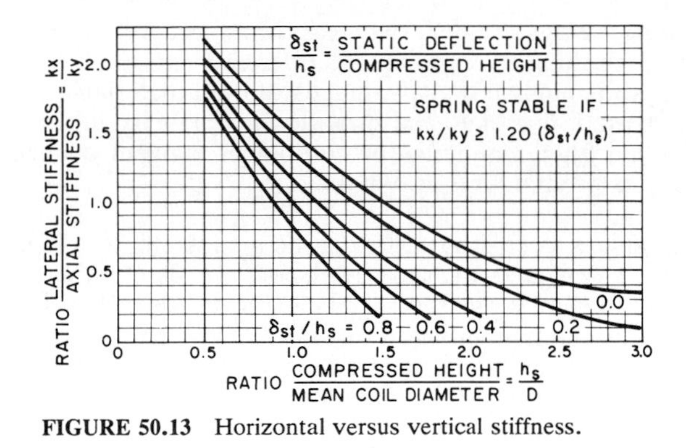

FIGURE 50.13 Horizontal versus vertical stiffness.

shaped like vertical bellows. Ethylene-propylene (EPDM) is also used for this purpose and sometimes butyl as the inner liner. Butyl is the least permeable of the rubber materials and reduces air loss.

The rolling lobe is a design variation that provides for movement by the rubber walls literally rolling down a steel stanchion rather than flexing. Both designs are equally suitable, and it is merely a matter of selecting one over the other depending on what frequency is needed. In general, rolling lobes have lower natural frequencies as compared to single-convolution and double-convolution bellows of the same height.

All air springs leak. The leakage rate is very low, but it is generally impractical to set up supervisory replenishing procedures. All air spring systems should be installed with replenishing air lines connected to height-sensitive leveling valves. If an air spring or cluster of air springs loses air and the equipment settles, air will automatically be added. Where air springs might be installed on a hot roof, the leveling system would respond and bleed small quantities of air should the air expand.

Leveling valves also level equipment that goes out of level because of external forces when the equipment is running. For example, a top horizontal fan tends to rotate away from the point of discharge. It rears up on the discharge end and settles in the back. Leveling valves automatically compensate for this and return the installation to proper elevation.

Air springs have the advantages of low frequencies and low profiles. Since there is no steel continuity, there is no noise transmission. Air spring frequency varies very little with pressure, but since the capacity is directly proportional to increased pressure, air springs need not be selected as carefully as steel springs. Most devices will handle loadings at a minimum of 25 lb/in^2 (172 kPa) and as much as 100 lb/in^2 (689 Pa), which allows for a 4:1 ratio from minimum to maximum loading on a particular mounting.

Hangers. Hangers accommodate all the above devices in modified form so that they fit within steel frames which are usually open-sided. Very simple high-frequency neoprene hangers could be pad hangers, but more often rubber elements are designed for hangers by eliminating the base plates and the tapped holes on top. The elements are molded with a projecting bushing that passes through the hanger hole to prevent the rod from rubbing. Occasionally fiberglass pads are used in lieu of neoprene elements, but they generally have even higher frequencies. Steel springs are generally fitted into neoprene cups, and the cup itself has a no-rub bushing arrangement. Rubber elements are often used in series with steel springs to combine the advantages of both materials.

It is very important to isolate and provide for flexibility in the piping so that the function of the floor mountings or equipment hangers connected to piped equipment is not interfered with or bypassed.

Connectors. Stainless steel connectors can be manufactured to specific lengths; they consist of a corrugated stainless steel body welded to the appropriate fittings. Even under moderate pressures, the closely spaced bellows become unstable and would spew out as the bellows expand. Therefore, virtually all metallic connectors have a stainless steel braid attached to the two ends to form a tube over the stainless steel corrugations. Stainless steel braid prevents elongation and adds to the radial strength.

Metallic connections are designed to allow flexing and reduce fatigue. The connector end away from the equipment should be rigidly secured so that the connector is forced to flex. This minimizes piping vibration after the anchor. Unfortunately, when any such connector is pressurized, there is a tremendous pull on the braid which makes the assembly extremely stiff. As a practical matter, it is very difficult to secure the afterend rigidly, so that in the average installation the flexible connector compensates for misalignment but does very little to reduce noise and vibration.

The next class of flexible connectors are Teflon bellows which are an improvement on the metallic connectors because the Teflon introduces a discontinuity in the metallic pipe wall. While Teflon is an excellent material, temperature and pressure ratings are often too low for high-rise structures. Teflon bellows are manufactured with built-in control rods to prevent excessive elongation. Control rods severly limit vibration and noise reduction as they bypass the flexible bellows.

Hand-built, single-arch rubber expansion joints are still being manufactured and are similar in function to Teflon. The arch in the center is all that provides the flexibility, and because of the bellowslike shape these connectors also elongate unless control rods are used or the piping is anchored. Here again the control rods tend to bypass the action of the expansion joint. The stiff walls leading up to the arch have little function other than to provide room for the steel retaining rings and the bolt heads that go between the steel rings and the arch. Applications are generally reserved to industry as they can be built up to 144 in (3.7 m) in diameter and manufactured in exotic rubber materials, particularly for high temperatures and highly corrosive chemicals.

The most recent entrant is the spherical neoprene or ethylene-propylene (EPDM) expansion joint. Unlike the three commodities described above, spherical connectors are designed on the principle of the automobile tire. The reinforcement fabric is nylon tire cord. The tire cord forms a suspension bridge from one flange to the other. When these connectors are pressurized, the nylon stretches until the stretching force equals the pressure and then the connector remains stable at that length and diameter. In most cases they can be installed without control rods.

The volumetric expansion and contraction of the connector dissipates sound energy. They do an excellent job in reducing sound transmission at blade passage frequency (number of pump buckets times shaft r/min). Unfortunately, they are too stiff to handle the primary vibration at r/min, so it is still necessary to protect the structure by isolating the pipelines with isolation hangers. The connectors take care of misalignment and virtually eliminate the high-pitched whine that normally travels through a structure. Connectors are recommended in a double-sphere configuration, for equal ends or concentric reduction. Long-radius tapered elbows save space at pump connections.

Bases. When equipment is manufactured or supplied in multiple components not connected by a common base, a base must be used to connect the two elements before the vibration isolators can be applied under the entirety. A belt pulling on a flexibly mounted motor connected to a flexibly mounted blower would pull one or the other off the isolators. Torque causes similar problems when machinery is directly connected. Bases can be constructed of structural steel, or, where additional mass is required or advisable, of steel frames filled with concrete. Most air-conditioning equipment is so well-balanced that there is no base weight criteria and rigidity is the only concern. Descriptions of these bases are covered in the selection section.

50.4 SELECTION

Following is a complete guide specification written in engineering terms. The specification selection guide provides the proper prescription for the complete isolation of a unit in terms of the type of mounting or hanger, the recommended deflection, the need for a base (if there is one), and the recommendation for a specific flexible connector.

We suggest you include all these specification paragraphs in your standard engineering specifications. In addition, prepare a drawing of standard details that becomes part of the mechanical drawing set. This eliminates constant editing on each job since the materials are not actually used unless an extra column is added to your equipment schedule, as in Table 50.3. It is this callout that defines the isolation that is to be used under each piece of equipment. Table 50.3 refers to the appropriate specification paragraph by letter with the notation as to the proper static deflection. The recommendations come right from the selection guide (Table 50.6), so the table containing them can be prepared very quickly.

The selections were based on a 30-ft (9.1-mm) floor span in the penthouse (9.1 m) and a 20-foot (6.1-m) span in other locations. Note that for pump no. 5 no isolation is called out as it is located in the basement under the garage where no one could be annoyed by the vibration. Fire pumps (no. 6) are seldom isolated.

In preparing a schedule similar to Table 50.3 there is an opportunity to consider every piece of equipment and very little possibility of skipping over some item in the rush of completing the project.

Specification A

Double-deflection neoprene mountings (Fig. 50.14) shall have a minimum static deflection of 0.35 in (8.9 mm). All metal surfaces shall be neoprene-covered to avoid corrosion and have friction pads both top and bottom so they need not be bolted to the floor. Bolt holes shall be provided for these areas where bolting is required. Steel rails (Fig. 50.15) shall be used above the mountings to compensate for the overhang on small vent sets, close-coupled pumps, etc.

Specification B

Spring-type isolators (Fig. 50.16) shall be freestanding and laterally stable without any housing and complete with ¼-in (6.4-mm) neoprene acoustical friction pads between the baseplate and the support. All mountings shall have leveling bolts that

TABLE 50.3

Fan schedule								Vibration isolation		
		Wheel diameter				Motor			Static deflection	
Fan no.	Location	in	mm	Arr.	Fan rpm	hp	kW	Spec.	in	mm
1	Penthouse	60	1524	1 SISW	503	30	22.4	B-J	1.50	38.1
2	3rd floor	49	1245	3 SISW	720	25	18.6	B-J	0.75	19.0
3	Penthouse	73	1854	3 DIDW	405	75	56.0	B-J	3.50	88.9
4	Basement	36	914	2 SISW	930	15	11.2	A-G	0.35	8.9
5	3rd floor	108	2743	3 SISW	400	125	93.2	B-J	2.50	63.5
6	3rd floor	2-27	2-686	AC unit	533	10	7.5	D	1.00	25.4
7	Penthouse	3-12	2-305	AC unit	630	5	3.7	B	0.75	19.0

Pumb Schedule					Vibration Isolation		
			Motor			Static Deflection	
Pump no.	Location	Type	hp	kW	Spec.	in	mm
1	Penthouse	Split casing	75	56.0	B-J-K	2.5	63.5
2	2nd Floor	Close coupled	1/2	0.37	B-J-K	0.75	19.0
3	3rd Floor	End suction	10	7.5	B-J-K	0.75	19.0
4	Basement	Close coupled	3	2.2	A-K	0.35	8.9
5	Basement	Split casing	50	37.2	—	—	
6	2nd Floor	Fire pump			—	—	

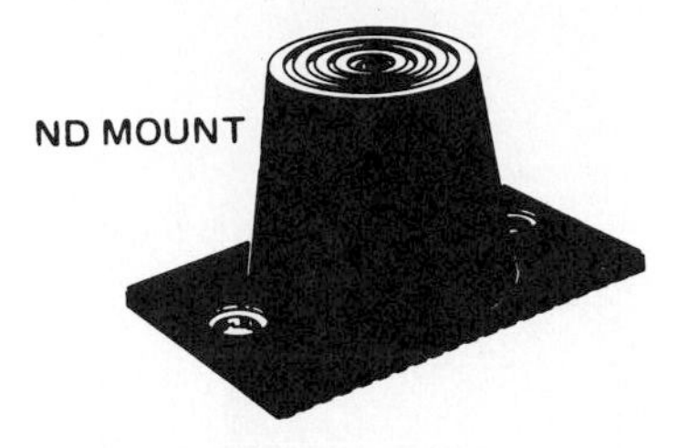

FIGURE 50.14 Double-deflection neoprene mounting.

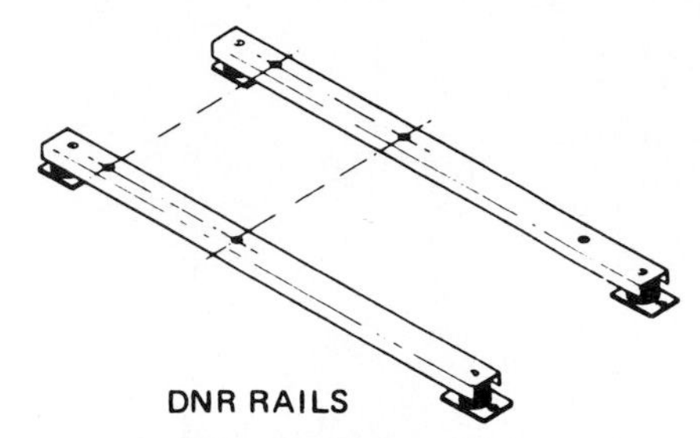

FIGURE 50.15 Steel rails.

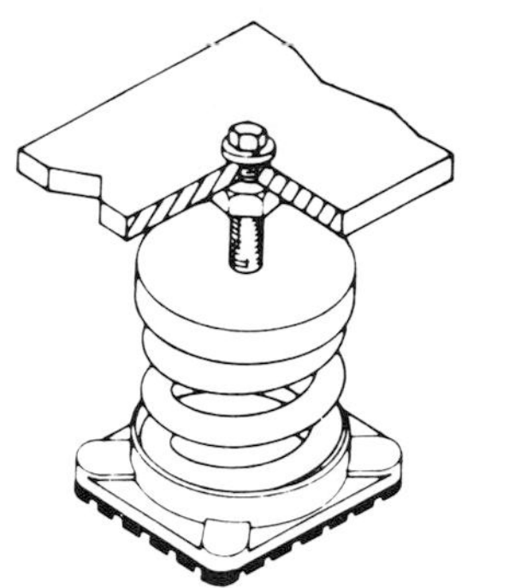

FIGURE 50.16 Spring-type isolator.

must be rigidly bolted to the equipment. Spring diameters shall be no less than 0.8 of the compressed height of the spring at rated load. Springs shall have a minimum additional travel to solid equal to 50 percent of the rated deflection. Submittals shall include spring diameters, deflections, compressed spring height, and solid spring height.

Specification BB

Air springs (Fig. 50.17) shall be manufactured with upper and lower steel sections connected by a flexible nylon-reinforced neoprene element in a multiple-bellows or rolling-lobe configuration. Air springs shall have a burst pressure that is a minimum of three times the maximum operating pressure. Air spring system elevations shall be controlled to ±⅛ in (±3.2 mm) by leveling valves connected to a 100-lb/in^2 (689-Pa) air supply. Frequency shall not exceed □ 3, □ 2, □ 1½ Hz (select frequency).

Specification C

Equipment with operating weight different from the installed weight such as chillers, boilers, etc., and equipment exposed to the wind such as cooling towers shall be mounted on spring mountings as described in engineering Specification B or BB, but a housing shall be used that includes vertical limit stops to prevent spring extension when weight is removed (Fig. 50.18). The housing shall serve as blocking during erection, and cooling tower mounts shall be located between the supporting steel and roof or the grillage and dunnage as shown on the drawings. The installed and operating heights shall be the same. A minimum clearance of ½ in (12.7 mm) shall be maintained around restraining bolts and between the housing and the spring so as not to interfere with the spring action. Limit stops shall be out of contact during normal operation. Mountings used out-of-doors shall be hot-dipped galvanized and the spring electroplated and hydrogen-relieved.

Specification D

Vibration hangers (Fig. 50.19) shall contain a steel spring seated in a neoprene cup in series with a 0.3-in (7.6-mm) deflection neoprene element. The neoprene spring cups and elements shall be molded with rod isolation bushings that pass through the hanger box. Spring diameters and hanger box lower hole sizes shall be large enough to permit the hanger rod to swing through a 30° arc before contacting the hole and short-circuiting the spring. Springs shall have a minimum additional travel to solid

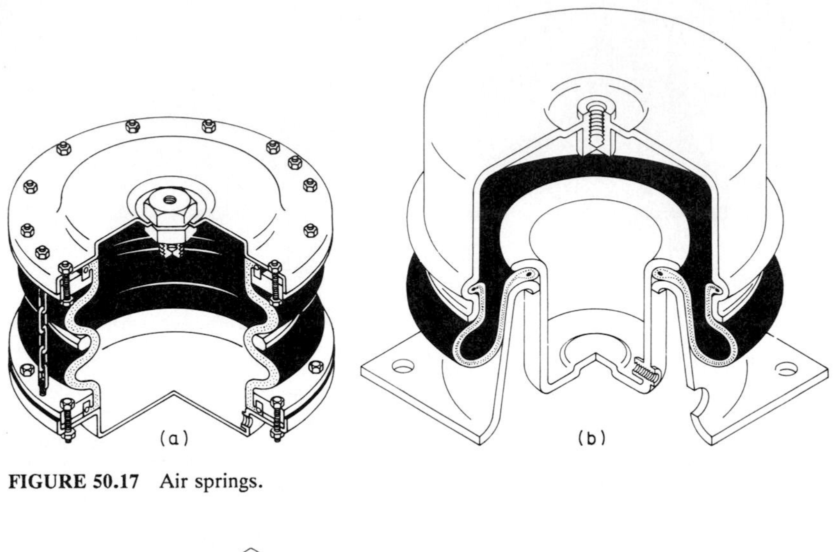

FIGURE 50.17 Air springs.

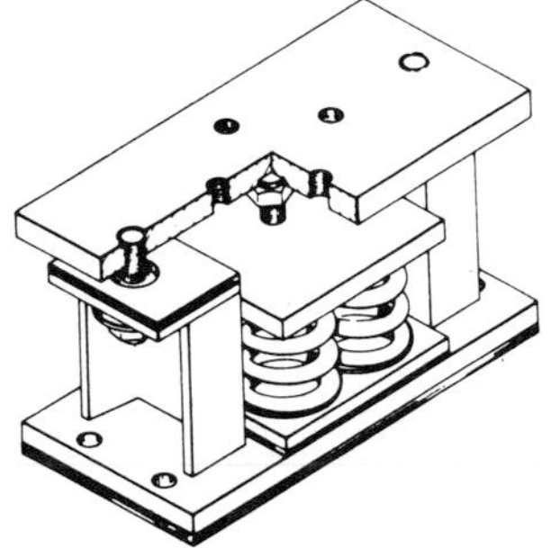

FIGURE 50.18 Housing with vertical limit stops.

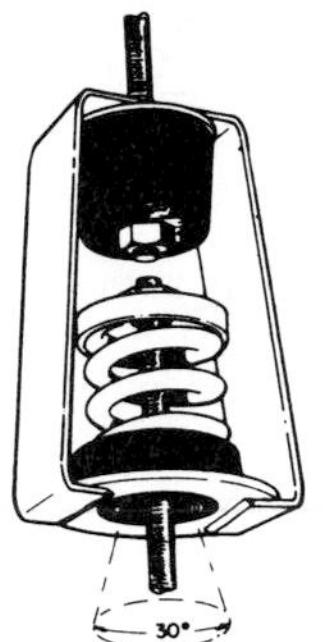

FIGURE 50.19 Vibration hanger, 30N.

equal to 50 percent of the rated deflection. Submittals shall include a scale drawing of the hanger showing the 30° capability.

Specification E

Vibration hangers (Fig. 50.20) shall be as described in engineering Specification D, but they shall be precompressed to the rated deflection so as to keep the piping or equipment at fixed elevation during installation. The hangers shall be designed with a release mechanism to free the spring after the installation is complete and the hanger is subjected to its full load. Deflection shall be clearly indicated by means of a scale. Submittals shall include a scale drawing of the hanger showing the 30° capability.

Specification F

Vibration hangers (Fig. 50.21) shall contain steel springs located in a neoprene cup manufactured with a neoprene bushing to prevent short-circuiting of the hanger rod. The cup shall contain a steel washer designed to properly distribute the load on the neoprene and prevent its extrusion. Spring diameters and hanger box lower hole sizes shall be large enough to permit the hanger rod to swing through a 30° arc before

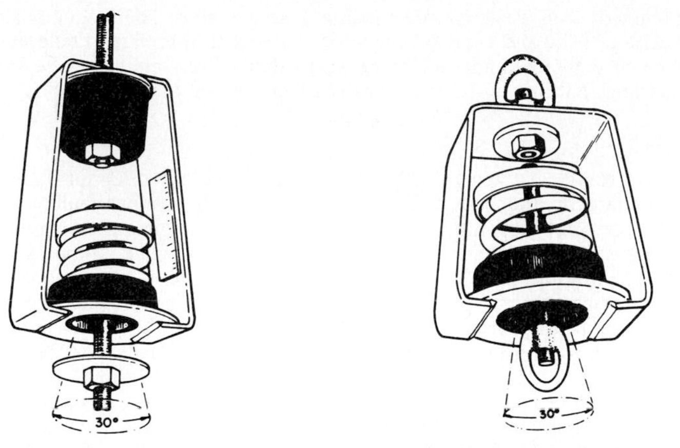

FIGURE 50.20 Vibration hanger, PC30N. **FIGURE 50.21** Vibration hanger, W30.

contacting the hole and short-circuiting the spring. Springs shall have a minimum additional travel to solid equal to 50 percent of the rated deflection. Hangers shall be provided with an eyebolt on the spring end and provision to attach the housing to the flat iron duct straps. Submittals shall include a scale drawing of the hanger showing the 30° capability.

Specification X

Air-handling equipment shall be protected against excessive displacement which might result from high air thrust in relation to the equipment weight. The horizontal thrust restraint (Fig. 50.22) shall consist of a spring element as described in Specification B, seated in a neoprene cup with the same deflection as specified for the mountings or hangers. The spring element shall be contained within a steel frame and designed so it can be preset for thrust at the factory and adjusted in the field to allow

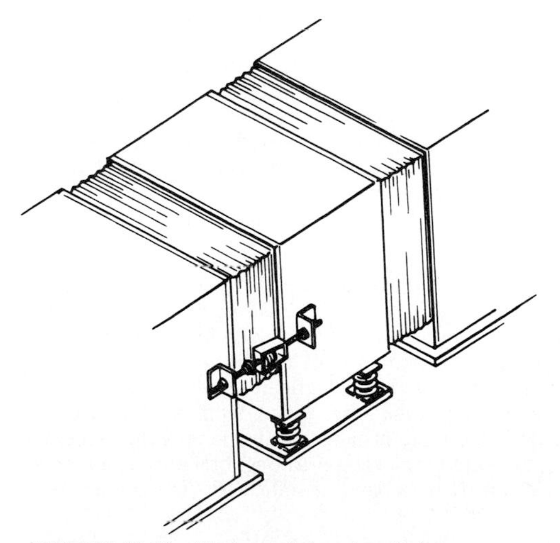

FIGURE 50.22 Horizontal thrust restraint.

for a maximum of ¼-in (6.4-mm) movement at start and stop. The assembly shall be furnished with one rod and angle brackets for attachment to both the equipment and ductwork or the equipment and the structure. Horizontal restraints shall be attached at the centerline of thrust and symmetrically on either side of the unit.

Specification Y

Curb-mounted rooftop equipment (Fig. 50.23) shall be mounted on spring isolation curbs that are directly attached to the structure. The lower member shall be a structural rectangular steel tube supporting adjustable and removable steel springs that isolate the upper floating section. The upper frame must provide continuous support for the equipment both before and after spring adjustment to allow sectional installation, and must be captive so as to resiliently resist wind and seismic forces. All directional neoprene snubber bushings shall be a minimum of ¼ in (6.4 mm) thick. Steel springs shall rest on ¼-in (6.4-mm) neoprene acoustical pads, and have a minimum deflection of □ ¾ in (19 mm), □ 1½ in (37.5 mm), □ 2½ in (62.5 mm) (check one). The ¾-in (19-mm) deflection springs must have a minimum 0.8 ratio of OD/OH (outside diameter/operating height). Other spring diameters must be a minimum of 2⅞ in (69 mm) up to 2 in (50 mm) deflection and 4 in (100 mm) thereafter. Hardware must be cadmium-plated or galvanized and the springs plated or provided with an approved rust-resistant finish.

Airtight weatherproofing shall be provided by a continuous flexible aluminum seal joined at the corners by EPDM bellows. The aluminum seal must be screwed to and provide counterflashing to the curb's waterproofing. The design shall make provision for waterproofing after the equipment is placed or for reroofing without removing the equipment. All spring locations shall have removable waterproof covers to allow for spring adjustment or replacement. Curbs shall accommodate 2 in (50 mm) of insulation. All ductwork shall be flexibly connected.

(*Note*: If additional acoustical protection is needed, please include the following paragraph.)

The floating member of the roof curb shall have perimeter angle and cross members to support two layers of ⅝-in (15.8-mm) waterproof gypsum board laid on with staggered joints. Gypsum board must surround ducts to provide a continuous sound break. This acoustical barrier shall be caulked to minimize sound transmission. Where the mechanical arrangement makes attachment to the floating member unfeasible, the barrier shall be attached at the highest practical elevation of the fixed curb with provision for 1-in- (25-mm-) thick closed-cell neoprene flexible seals around the ductwork. A 4-in (100-mm) layer of 1.5 density fiberglass shall cover the entire solid roof surface under the unit. Ductwork shall be lined with sound absorbent material and discharge ductwork painted with damping compound, provided with turning vanes, covered with two layers of ⅝-in (15.8-mm) gyp board or 2 in (50 mm) of fiberglass and vinyl wrap and supported with ¾-in (19-mm) deflection Specification D hangers for 40 ft (12.18 m). Complete instructions shall be provided by the spring isolation curb manufacturer.

Specification G

Vibration isolator manufacturer shall furnish integral structural steel bases (Fig. 50.24). Bases are preferably rectangular for all equipment. Centrifugal refrigeration machine and pump bases may be T- or L-shaped in cramped spaces.

Pump bases for split case pumps shall include supports for suction and discharge base ells. All perimeter members shall be beams with a minimum depth equal to one-tenth the longest dimension of the base. Beam depth need not exceed 14 in (356 mm) provided that the deflection and misalignment is kept within acceptable limits as determined by the manufacturer. Height-saving brackets shall be employed in all mounting locations to provide a base clearance of 1 in (25.4 mm).

Specification H

Vibration isolator manufacturer shall provide steel members welded to height-saving

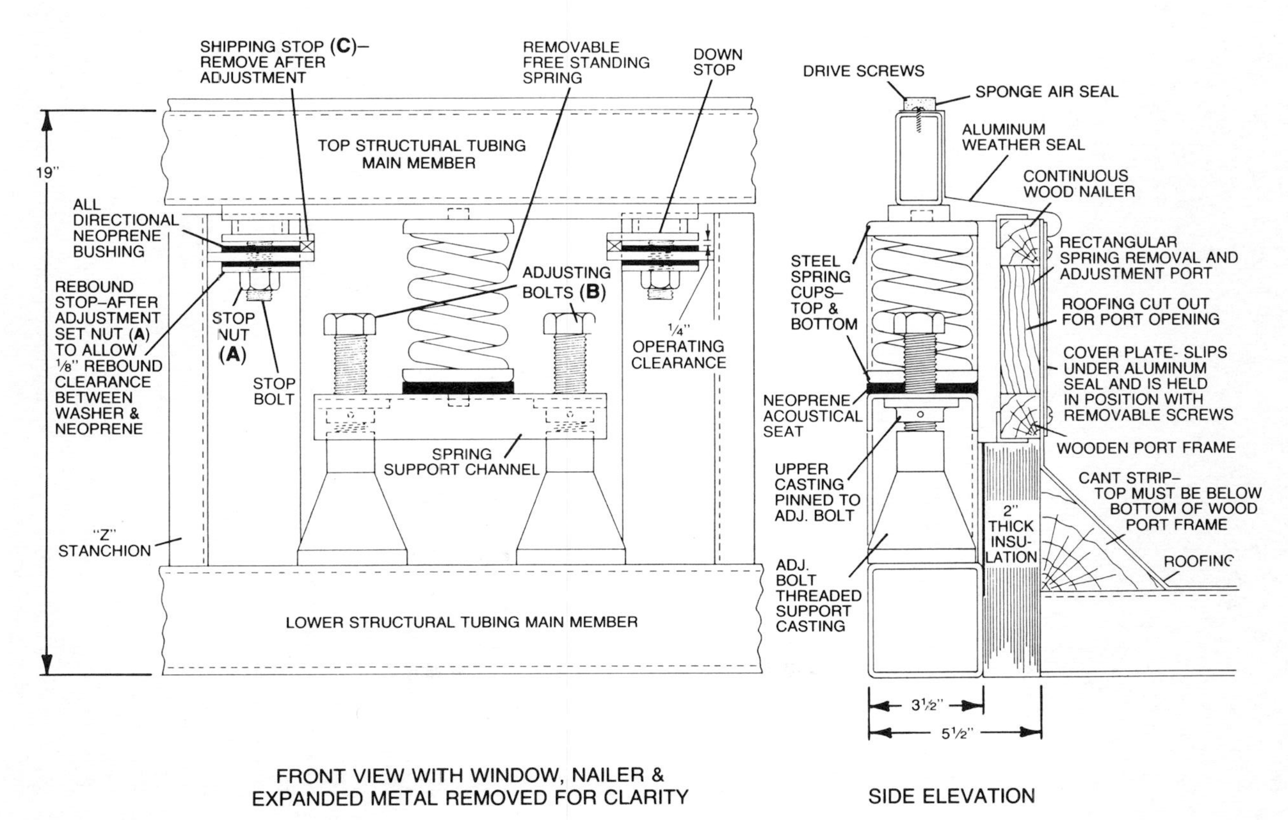

FIGURE 50.23 Curb-mounted rooftop equipment. (1 in = 25.4 mm).

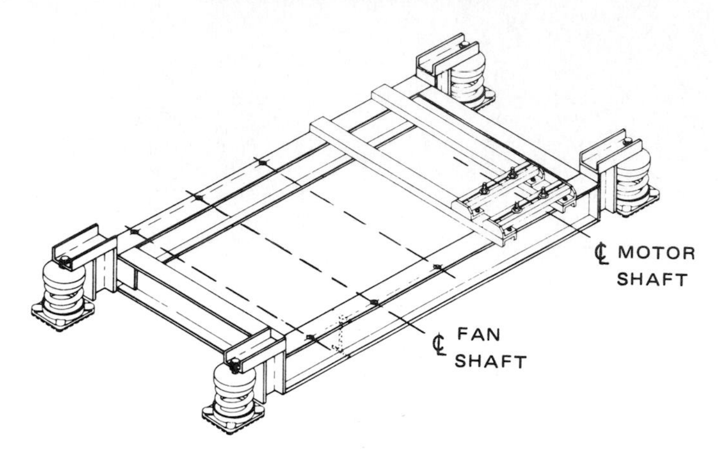

FIGURE 50.24 WF structural steel base.

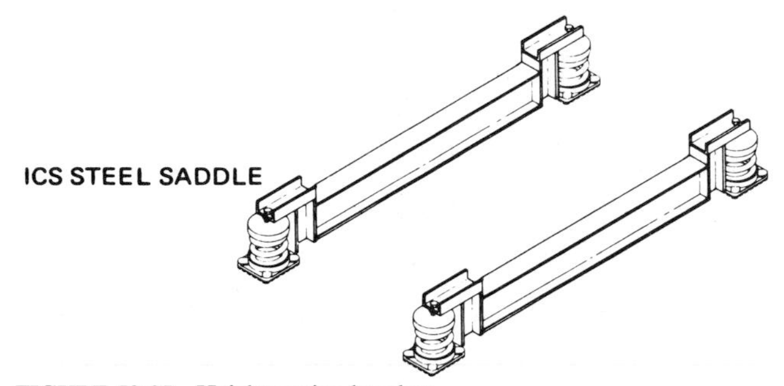

FIGURE 50.25 Height-saving bracket.

brackets (Fig. 50.25) to cradle machines having legs or bases that do not require a complete supplementary base. Members shall be sufficiently rigid to prevent strains in the equipment.

Specification J

Vibration isolator manufacturer shall furnish rectangular rolled beam or channel or formed channel concrete forms for floating foundations (Fig. 50.26). Bases for split case pumps shall be large enough to provide support for suction and discharge base ells. The base depth need not exceed 12 in (300 mm) unless specifically recommended by the base manufacturer for mass or rigidity. In general, bases shall be a minimum of one-twelfth the longest dimension of the base, but not less than 6 in (150 mm). Forms shall include minimum concrete reinforcement consisting of ½-in (12-mm) round bars or ⅜-in (10-mm) squares welded on 6-in (150-mm) centers running both ways in a layer 1½ in (38 mm) above the bottom, or additional steel as is required by the structural conditions. Forms shall be furnished with steel members to hold anchor bolt sleeves when the anchor bolts fall in concrete locations. Height-saving brackets shall be employed in all mounting locations to maintain a 1-in (25-mm) clearance below the base.

Specification K

Flexible EPDM connectors shall be used on all equipment as indicated on the draw-

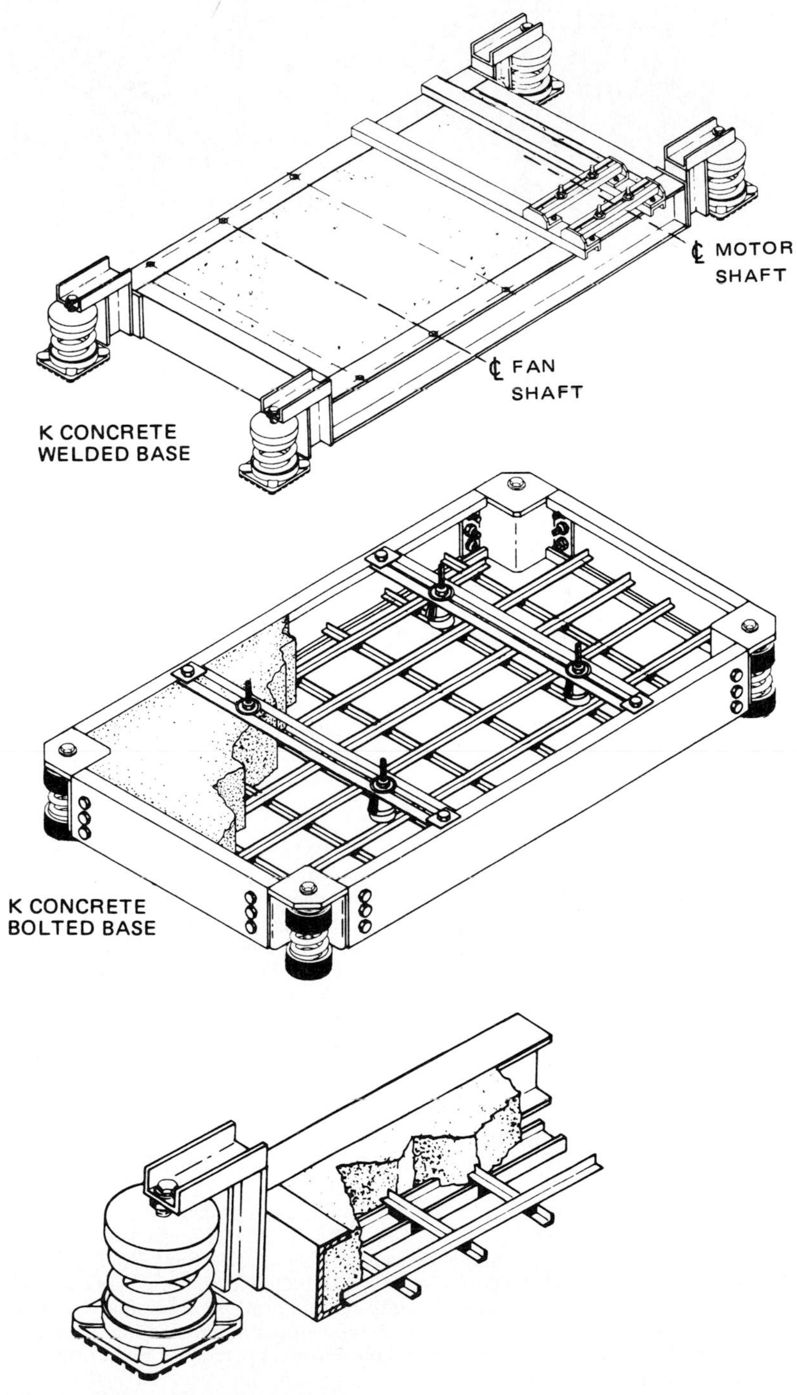

FIGURE 50.26 Floating foundations.

FIGURE 50.27 Flexible connector, double spherical.

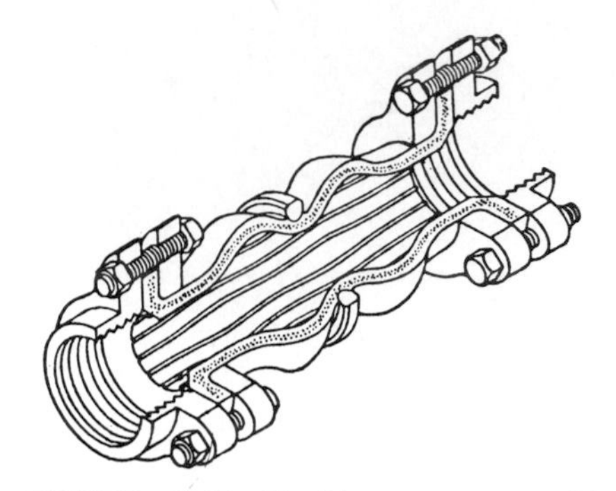

FIGURE 50.28 Flexible connector with threaded ends.

ings or on the equipment schedule. They shall be manufactured of multiple plies of frictioned nylon tire cord with an EPDM cover and liner. No steel wire or rings shall be used as internal pressure reinforcement. Straight connectors shall have two spheres (Fig. 50.27) with a centered molded-in external ductile iron ring to maintain the two spherical shapes. Two-inch and smaller sizes may have threaded ends (Fig. 50.28). Floating flanges shall have a recess to lock the bead wire in the raised-face EPDM flanges. Tapered twin-sphere connectors (Fig. 50.29) as described above shall be used where line size changes are required in straight piping runs.

FIGURE 50.29 Tapered twin-sphere flexible connector.

Flanged equipment shall be directly connected to neoprene elbows in the size range 2½ through 12 in (65 through 300 mm), if the piping makes a 90° turn and flanges are equal-sized (Fig. 50.30). Long-radius reducing EPDM elbows shall be used in place of steel or cast iron elbows at pump connections (Fig. 50.31).

All twin-sphere connectors shall be properly extended as recommended by the manufacturer to prevent additional elongation under pressure. Joints shall be designed for maximum elongation under pressure as follows (Table 50.4).

When the pressure would cause the connector to extend beyond its rated elongation, control rods shall be employed using ½-in- (13-mm-) thick bridge-bearing neoprene washer bushings designed for maximum loading of 1000 psi (Fig. 50.32).

Twin-sphere connectors shall have a minimum rating of 250 lb/in^2 (1720 kPa) at 170°F (77°C) and 165 lb/in^2 (1130 kPa) at 250°F (120°C). Elbows and reducing twin spheres shall have a minimum pressure rating of 220 lb/in^2 (1510 kPa) at 170°F (75°C) and 145 lb/in^2 (1000 kPa) at 250°F (120°C). Neoprene materials shall be limited to 220°F (104°C). Certified safety factors shall be a nominal 4:1 with minimum acceptable test results of 3.6:1. Test shall cover burst, flange leakage, extension without control rods, and tests retention at 50 percent of burst pressure without control rods.

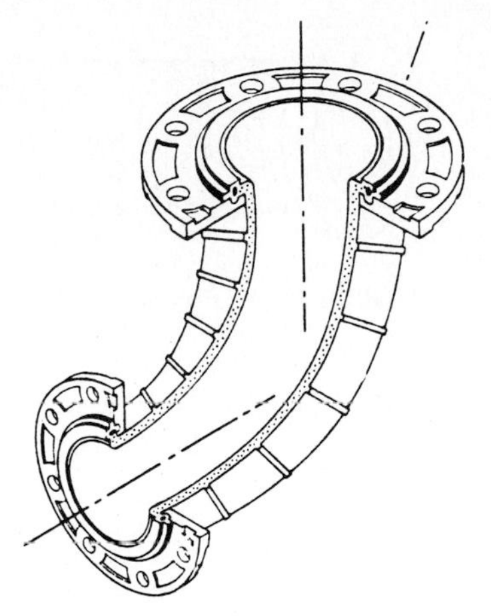

FIGURE 50.30 Neoprene elbow. **FIGURE 50.31** Reducing elbow.

TABLE 50.4 Twin-Sphere Pressure Elongation

Pipe size		Pressure		Elongation	
in	mm	lb/in^2	kPa	in	mm
1½–2½	40–65	250	1723	½	13
3–8	75–200	250	1723	¾	19
10–12	250–300	175	1206	⅞	22

Note: Based on Mason Industries Super-Flex. (Not standard for all manufacturers).

Submittals shall include two test reports by independent consultants showing minimum reduction of 20 dB in vibration accelerations and 10 dB in sound pressure levels at typical blade passage frequencies.

Connectors shall be installed on the equipment side of the shutoff valves at all times.

SPECIFICATION L

Flexible stainless steel hose (Fig. 50.33) shall have stainless steel braid and carbon steel fittings. Sizes 3 in (76.2 mm) and larger shall be flanged. Smaller sizes may have male nipples. Lengths shall be as tabulated (see Table 50.5).

Hoses shall be installed on the equipment side of the shutoff valves horizontally and parallel to the equipment shafts wherever possible.

SPECIFICATION M

Where piping passes through equipment walls, floors, or ceilings, the vibration isolator manufacturer shall provide a split seal (Fig. 50.34), consisting of two bolted pipe halves with ¾ in (19 mm) or thicker neoprene sponge bonded to the inner faces. The seal shall be tightened

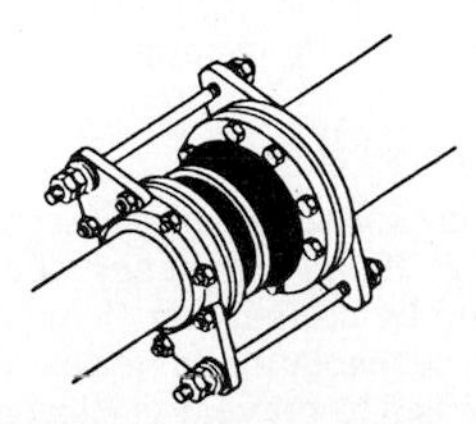

FIGURE 50.32 Connector with control rods.

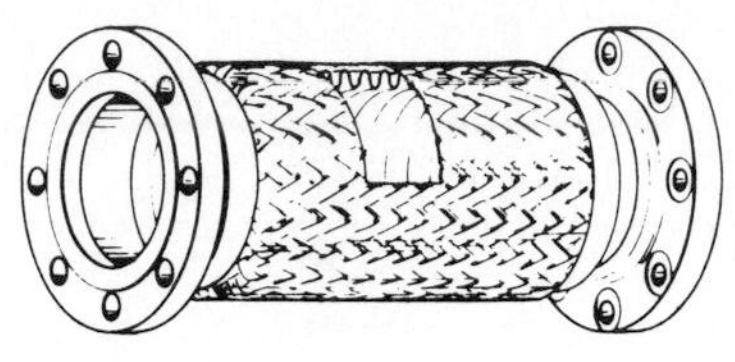

FIGURE 50.33 Flexible stainless steel hose.

TABLE 50.5 Stainless Steel General Usage Lengths

Flanged			Male nipples		
Pipe diameter, in (mm)	×	Length, in (mm)	Pipe diameter, in (mm)	×	Length, in (mm)
3 (75)	×	14 (350)	½ 13)	×	9 (230)
4 (100)	×	15 (380)	¾ (20)	×	10 (255)
5 (125)	×	19 (480)	1 (25)	×	11 (280)
6 (150)	×	20 (500)	1¼ (30)	×	12 (305)
8 (200)	×	22 (560)	1½ (40)	×	13 (330)
10 (250)	×	26 (660)	2 (50)	×	14 (355)
12 (300)	×	28 (715)	2½ (65)	×	18 (460)
14 (350)	×	30 (760)			
16 (400)	×	32 (815)			

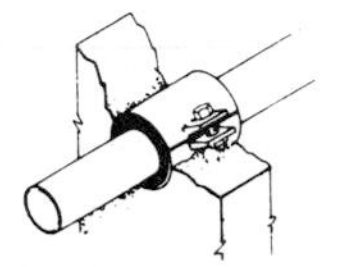

FIGURE 50.34 Split seal.

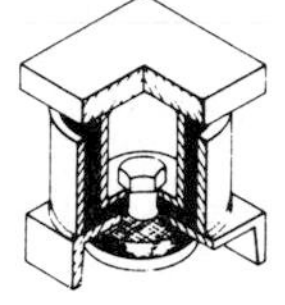

FIGURE 50.35 Acoustical pipe anchor.

around the pipe to eliminate clearance between the inner sponge face and the piping. Concrete may be packed around the seal to make it integral with the floor, wall, or ceiling if the seal is not already in place around the pipe prior to the construction of the building member. Seals shall project a minimum of 1 in (25 mm) past either face of the wall. Where temperature exceeds 240°F (116°C), 10 = lb/ft^3 (16-kg/m^3) density fiberglass may be used in lieu of the sponge.

Specification N

Vibration isolator manufacturer shall provide an all-directional acoustical pipe anchor (Fig. 50.35), consisting of a telescopic arrangement of two sizes of steel tubing separated by a minimum ½-in (12-mm) thickness of heavy-duty neoprene and duck or neoprene isolation material. Vertical restraints shall be provided by similar material arranged to prevent vertical travel in either direction. Allowable loads on the iso-

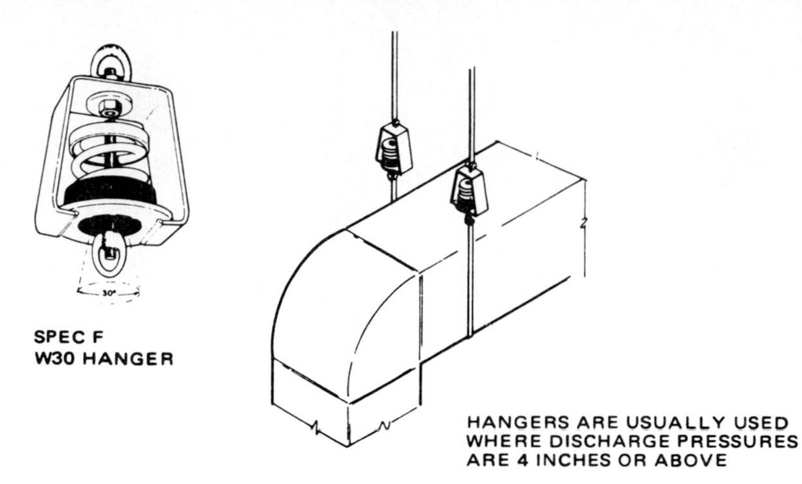

FIGURE 50.36 Hanger and floor support.

lation material shall not exceed 500 psi (3445 kPa), and the design shall be balanced for equal resistance in any direction.

Specification: Duct Isolation

All discharge runs for a distance of 50 ft (15 m) from the connected equipment shall be isolated from the building structure by means of Specification F hangers or Specification C floor supports (Fig. 50.36). Spring deflections shall be a minimum of 0.75 in (19 mm).

Specification: Horizontal Pipe Isolation (Fig. 50.37)

The first three pipe hangers in the main lines near the mechanical equipment shall be as described in Specification E. Horizontal runs in all other locations throughout the building shall be isolated by hangers as described in Specification D. Floor-supported piping shall rest on isolators as described in Specification C. Heat exchangers shall be considered part of the piping run. All Specification E hangers or the first three Specification C mounts as noted above will have the same static deflection as specified for the mountings under the connected equipment. [*Note*: If piping is connected to equipment located in basements and hangs from ceiling under occupied spaces, the first three hangers shall have 0.75-in (19-mm) deflection for pipe sizes up to and including 3 in (75 mm), 1.5-in (38-mm) deflection for pipe sizes up to and including 6 in (150 mm), and 2.5-in (64-mm) deflection thereafter.] All other hangers and mounts will have a minimum steel spring deflection of 0.75 in (19 mm). Hangers shall be located as close to the overhead supports as practical.

Equipment room seals. All piping passing through the equipment room walls, floors, or ceilings shall be protected against sound leakage by means of an acoustical wall seal as described in Specification M.

Specification: Riser Isolation

Risers shall be suspended from or supported from Specification D hangers or Specification B mountings and the piping anchored or guided with Specification N anchors (Fig. 50.38), all as indicated on the riser drawings. Wherever possible, anchors shall be located in the center of the pipe run.

Steel spring deflection shall be a minimum of 0.75 in (19 mm) except in those ex-

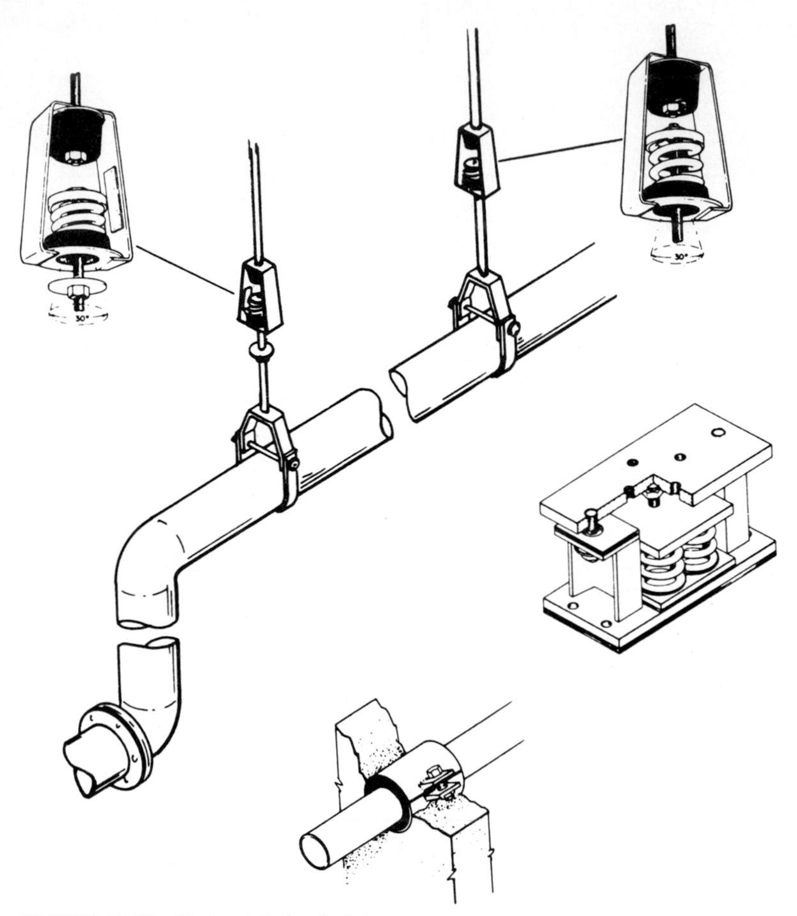

FIGURE 50.37 Horizontal pipe isolators.

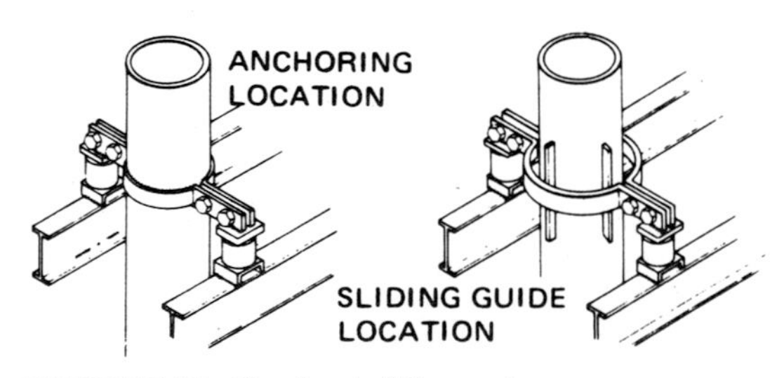

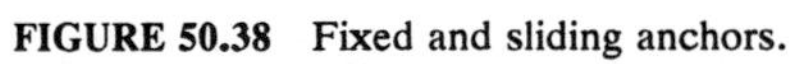

FIGURE 50.38 Fixed and sliding anchors.

pansion locations where additional deflection is required to limit deflection or load changes to ±25 percent of the initial deflection. (See Table 50.6, next page.)

50.5 SEISMIC PROTECTION OF RESILIENTLY MOUNTED EQUIPMENT

50.5.1 Theory

There is an increased awareness throughout the world of the danger of installing flexibly mounted equipment in seismic zones without proper restraint. In addition air-conditioning systems installed in military installations suffer the effects of nearby bomb blasts. These forces are similar to those introduced by earthquakes.

When an earthquake or bomb blast occurs, the ground is accelerated through a range of frequencies. The damage is not so much the force of the quake as single or multiple impulses but the introduction of a particular ground frequency that has sufficient energy to vibrate and ultimately resonate either a total building or the machinery therein. The large resonant displacements cause overstressing and failure.

A response spectrum is a curve drawn on log-log paper describing an earthquake or bomb blast. Curve shapes vary depending on intensity, soil frequencies, elevations in a structure, etc. Figure 50.39 is a typical curve and not to be used for a specific application.

The horizontal axis references the resonance or natural frequency of the object affected by the quake. The vertical axis shows the velocity attained by this object during the quake. The diagonal axis running up toward the right-hand corner reads the maximum accelerations to which the object is subjected. The axis at right angles to this will read the displacement of the object in relation to the ground or rigid support. Superimposed on these scales are the response curves. The three curves indicate 0, 1, and 2 percent of critical damping. Higher levels are most unusual. This particular curve is referred to as a 0.5*g* response spectrum as the accelerations of these three curves become asymptotic at a 0.5*g* level above a resonance of 40 Hz. Now let us see how equipment at different natural frequencies would act during an earthquake described by these curves. Reference the 0 damping curve since this would show the most extreme condition.

If the object to be studied has a natural frequency of 1 Hz (Fig. 50.40), start at the bottom of the chart at 1 Hz and move vertically to intersect the response curve at point *A*. Moving to the extreme right from point *A*, the velocity is 55 in (1375 mm)/s. By following a displacement line diagonally down to the left, we find displacement is 9 in (225 mm). By following an acceleration line down to the right, we find acceleration at 0.9*g*. If we then consider a 2-Hz natural frequency (Fig. 50.41) and work from point *B*, the maximum velocity would be 62 in (1550 mm)/s, displacement of 5 in (125 mm), and a maximum acceleration of 2*g*. At 5-Hz natural frequency (Fig. 50.42), point *C*, velocity is 38 in (950 mm)/s, displacement is 1.2 in (30 mm), and acceleration is 3*g*. At 25 Hz (Fig. 50.43), point *D*, the area is asymptotic with a velocity of 1.5 in (37.5 mm)/s, displacement 0.01 in (0.25 mm), and an acceleration of 0.6*g*. Notice that the values vary widely, as stated earlier, depending on the natural frequency of the object exposed to this particular quake.

Going back to reference point *A*, the displacement is 9 in (225 mm). While the *g* force is low, if something is vibrating at an amplitude of 9 in (225 mm) it might

TABLE 50.6 Specification Selection Guide

This Table is to be used with Vibration Control Engineering Specifications for HVAC Equipment in Office Buildings, Colleges, Theatres and Similar Structures	DEFLECTION AND MOUNTING CRITERIA FOR 4"-6"(100-150mm) THICK SOLID CONCRETE FLOORS (Note 9)									
	Ground Supported Slab or Basement		20'(6.1m) Floor Span Possible Floor Defl 0.67"(17mm)		30'(9.1m) Floor Span Possible Floor Defl 1.0"(25mm)		40'(12.2m) Floor Span Possible Floor Defl 1.33"(34mm)		50'(15.2m) Floor Span Possible Floor Defl 1.67"(42mm)	
	Engr Spec	Min. Static Deflection in (mm) (Note 1)	Engr Spec	Min. Static Deflection in (mm) (Note 1)	Engr Spec	Min. Static Deflection in (mm) (Note 1)	Engr Spec	Min. Static Deflection in (mm) (Note 1)	Engr Spec	Min. Static Deflection in (mm) (Note 1)
REFRIGERATION MACHINES (Note 2)										
Absorption Machines	A-K	0.35 (9)	C-K	0.75 (19)	C-K	0.75 (19)	C-K	1.50 (38)	C-K	1.50 (38)
Centrifugal Chillers or Heat Pumps										
Cooler Condenser Mounted Hermetic-Compressors	A-K	0.35 (9)	C-K	0.75 (19)	C-K	1.50 (38)	C-K	1.50 (38)	C-H-K	2.50 (64)
Cooler Condenser Mounted Hermetic-Compressors	A-K	0.35 (9)	C-K	0.75 (19)	C-K	1.50 (38)	C-K	1.50 (38)	C-K	2.50 (64)
Open Type Compressors (Note 4)	A-G-K	0.35 (9)	C-G-K	0.75 (19)	C-G-K	1.50 (38)	C-G-K	1.50 (38)	C-G-K	2.50 (64)
Refrigeration Reciprocating Compressors										
500 RPM to 750 RPM	B	0.75 (19)	B	1.50 (38)	B	1.50 (38)	B-H	2.50 (64)	B-H	3.50 (90)
751 RPM and over	B	0.75 (19)	B	0.75 (19)	B	1.50 (38)	B-H	2.50 (64)	B-H	3.50 (90)
Reciprocating Chillers or Heat Pumps										
500 RPM to 750 RPM	C-K	0.75 (19)	C-K	1.50 (38)	C-K	1.50 (38)	C-H-K	2.50 (64)	C-H-K	3.50 (90)
751 RPM and over	C-K	0.75 (19)	C-K	0.75 (19)	C-H-K	1.50 (38)	C-H-K	2.50 (64)	C-H-K	3.50 (90)
PACKAGED STEAM GENERATORS (Boilers) (Note 2)	A-L	0.35 (9)	C-L	0.75 (19)	C-L	0.75 (19)	C-L	1.50 (38)	C-L	2.50 (64)
PUMPS (Note 2)										
Close Coupled										
Thru 5 HP (4kw)	A-J-K	0.35 (9)	B-J-K	0.75 (19)	B-J-K	0.75 (19)	B-J-K	1.50 (38)	B-J-K	1.50 (38)
7½ HP (4kw) and larger	B-J-K	0.75 (19)	B-J-K	0.75 (19)	B-J-K	1.50 (38)	B-J-K	1.50 (38)	B-J-K	2.50 (64)
Base mounted (Note 3)										
Up to 60 HP (45kw)	B-J-K	0.75 (19)	B-J-K	0.75 (19)	B-J-K	1.50 (38)	B-J-K	1.50 (38)	B-J-K	2.50 (64)
75 HP (56kw) and larger	B-J-K	0.75 (19)	B-J-K	1.50 (38)	B-J-K	2.50 (64)	B-J-K	2.50 (64)	B-J-K	3.50 (90)
FACTORY ASSEMBLED H & V UNITS										
Curb Mounted Rooftop Units (Note 10)	–	–	Y	1.50 (38)	Y	1.50 (38)	Y	2.50 (64)	Y	2.50 (64)
Suspended Units (Note 11)										
Thru 5 HP (4kw)	D	1.00 (25)	D	1.00 (25)	D	1.00 (25)	D	1.00 (25)	D	1.00 (25)
7½ HP (6kw) and larger - 275 RPM to 400 RPM	D	1.50 (38)	D	1.50 (38)	D	1.50 (38)	D	1.50 (38)	D	1.50 (38)
7½ HP (6kw) and larger - 401 RPM and over	D	1.00 (25)	D	1.00 (25)	D	1.00 (25)	D	1.50 (38)	D	2.50 (64)
Floor Mounted Units (Note 11)										
Thru 5 HP (4kw)	A	0.35 (9)	B	0.75 (19)	B	0.75 (19)	B	0.75 (19)	B	0.75 (19)
7½ HP (6kw) and larger - 275 RPM to 400 RPM	A	0.35 (9)	B-H	1.50 (38)	B-H	1.50 (38)	B-H	1.50 (38)	B-H	1.50 (38)
7½ HP (6kw) and larger - 401 RPM and over	A	0.35 (9)	B	0.75 (19)	B	0.75 (19)	B-H	1.50 (38)	B-H	2.50 (64)
50 HP (37kw) and larger - 401 RPM and over	A	0.35 (9)	B	0.75 (19)	B-H	1.50 (38)	B-H	2.50 (64)	B-H	3.50 (90)
AIR COMPRESSORS (Note 2)										
Tank Type	B-J-L	0.75 (19)	B-J-L	0.75 (19)	B-J-L	1.50 (38)	B-J-L	2.50 (64)	B-J-L	3.50 (90)
V-W Type	B-J-L	0.75 (19)	B-J-L	0.75 (19)	B-J-L	1.50 (38)	B-J-L	2.50 (64)	B-J-L	3.50 (90)
Horiz, Vert, 1 or 2 Cylinder (Note 5)										
275 RPM to 499 RPM	B-J-L	2.50 (64)	B-J-L	2.50 (64)	B-J-L	2.50 (64)	B-J-L	3.50 (90)	B-J-L	3.50 (90)
500 RPM to 800 RPM	B-J-L	1.50 (38)	B-J-L	1.50 (38)	B-J-L	2.50 (64)	B-J-L	3.50 (90)	B-J-L	3.50 (90)
	Specification should read "J" type inertia bases with sufficient mass to limit motion to a theoretical double amplitude of 0.03"(0.76mm).									
BLOWERS										
Utilities Sets										
Floor Mounted (Note 7)	A	0.35 (9)	Spec B for 0.75"(19mm) deflection and Spec B-H for over 0.75"(19mm) deflection with deflection from Blower Minimum Deflection Guide, but not to exceed 2.5"(64mm).							
Roof Mounted			Spec B-J with defl. from Blower Minimum Deflection Guide. If roof will not handle concrete base load use Spec C for 0.75"(19mm) defl. and Spec C-H for over 0.75"(19mm) defl.							
Suspended Unit (Note 7)			Spec D with deflection from Blower Minimum Deflection Guide, but not to exceed 2.5"(64mm).							
Centrifugal Blowers (Note 8)	A-J	0.35 (9)	Spec B-J with deflection from Blower Minimum Deflection Guide.							
Fan Heads (Note 6)										
Floor Mounted	A-X	0.35 (9)	Spec B-X if 0.75"(19mm) deflection or Spec B-H-X for deflection of 1.5"(38mm) to 4.5"(115mm) from Blower Minimum Deflection Guide.							
Suspended Units			Spec D-X with deflection from Blower Minimum Deflection Guide.							
Tubular Centrifugal and Axial Fans (Note 6) (Note 1b)										
Suspended Units	Spec D with deflection from Blower Minimum Deflection Guide, Spec D-X for over 4"(100mm) statics.									
Floor Mounted with Motor on/in Fan Casing	A	0.35 (9)	Spec B for 0.75"(19mm) deflection and Spec B-H for over 0.75"(19mm) deflection with defl. from Blower Minimum Defleciton Guide. Spec B-J or B-X for over 4"(120mm) statics.							
Floor Mounted Arrangement 1 or any Separately Mounted Motor	A-J	0.35 (9)	Spec B-J with deflection from Blower Minimum Deflection Guide.							
COOLING TOWERS AND CONDENSING UNITS	A-K	0.35 (9)	Spec C with deflection from Blower Minimum Deflection Guide.							

TABLE 50.6 Specification Selection Guide (*Continued*)

Blower Minimum Deflection Guide

When blowers are 60 HP (45kw) or larger, select deflection requirements for next larger span. A minimum of 2.5"(64mm) should be used unless larger deflections are called for on the chart or these large fans are located in the lowest sub-basement or on a slab on grade.

FAN SPEED RPM	Required Deflection for Ground Supported Slab or Basement	Required Deflection for 20'(6.1m) Floor Span	Required Deflection for 30'(9.1m) Floor Span	Required Deflection for 40'(12.2m) Floor Span	Required Deflection for 50'(15.2m) Floor Span
500 and up	0.35" (9mm)	0.75" (19mm)	1.50" (38mm)	2.50" (64mm)	3.50" (90mm)
375 - 499	0.35" (9mm)	1.50" (38mm)	2.50" (64mm)	3.50" (90mm)	3.50" (90mm)
300 - 374	0.35" (9mm)	2.50" (64mm)	2.50" (64mm)	2.50" (64mm)	3.50" (90mm)
225 - 299	0.35" (9mm)	3.50" (90mm)	3.50" (90mm)	3.50" (90mm)	3.50" (90mm)
175 - 224	0.35" (9mm)	3.50" (90mm)	4.50" (115mm)	4.50" (115mm)	4.50" (115mm)

NOTES

1. a) Minimum deflections called for in this specification are not 'nominal' but certifiable minimums. The 0.75"(19mm), 1.5"(38mm), 2.5"(64mm), 3.5"(89mm) and 4.5"(115mm) minimums should be selected from manufacturers nominal 1"(25mm), 2"(50mm), 3"(75mm), 4"(100mm) and 5"(125mm) series respectively. Full deflection specifications are seldom, if ever, met because of infinite load and limited spring selection. The 0.75"(19mm), 2.5"(64mm), 3.5"(89mm) and 4.5"(115mm) minimums are attainable and should be enforced.

   b) Air Springs should be substituted for steel springs in critical areas, particularly under centrifugal compressors, screw compressors and vane axial fans 50 HP (37kw) and larger. Substitute Spec. BB for B when doing so.

2. If flexible spherical connectors are never used, omit the letters K or L from the Engineering Spec. If stainless steel hose is required, substitute L for K. If stainless steel hose is not to be used with air compressors, omit L.

3. Vacuum, Condensate or Boiler Feed Pumps shall be mounted with their tanks on a common J base with deflections as specified for base mounted pumps.

4. The base described in Spec. G is used under the drive side. Individual mountings as described in Spec. C are used under the Cooler and Condenser.

5. This type of compressor is highly unbalanced and sometimes requires inertia bases weighing 5 to 7 times the equipment weight to reduce motion to acceptable limits.

6. Spec. X is located after Spec. F.

7. Limit deflection for utility sets 18"(457mm) wheel diameter and smaller to 1½"(38mm).

8. FLOATING CONCRETE INERTIA BASES. Floating concrete inertia bases do not reduce vibration transmitted to the structure through the mountings. These bases will reduce vibratory motion, provide a very rigid machine base and minimize spring reactions to fan thrust.

9. LIGHT FLOOR CONSTRUCTION. When floors or roofs are lighter than 4"(100mm) solid concrete, it is desirable to introduce a localized mass under the vibration mountings in the form of a sub-base. This sub-base should be 12"(300mm) thick and 12"(300mm) longer and wider than the mechanical equipment above it. When this mass is provided, the 30'(9m) minimum static deflection requirements will suffice even in longer bays. The mass is also useful for unusually large bays over 50'(15.2m).

   When floors are lighter than the 4"(100mm) concrete or the location is in a particularly sensitive area and the mass described above cannot be introduced, select deflection requirements for the next larger span.

10. Spec. Y is located after F and X.

11. If blower section only, use Spec. for Fan Heads under Blowers.

very well snap or fall off the spring. This point can be emphasized by referring to a system with a resonant frequency of 0.4 Hz. Following the same procedure as for points A, B, and C, but only referring to displacement, we can see that with 0 damping the displacement would be in the order of 20 in (508 mm). It is this resonant displacement that causes failure and certainly makes any type of flexible connection impractical.

To protect resiliently mounted equipment against these events, it is most important to introduce a resilient snubbing device (Fig. 50.44) to limit the deceleration forces and control amplitudes. These snubbing devices are built with air

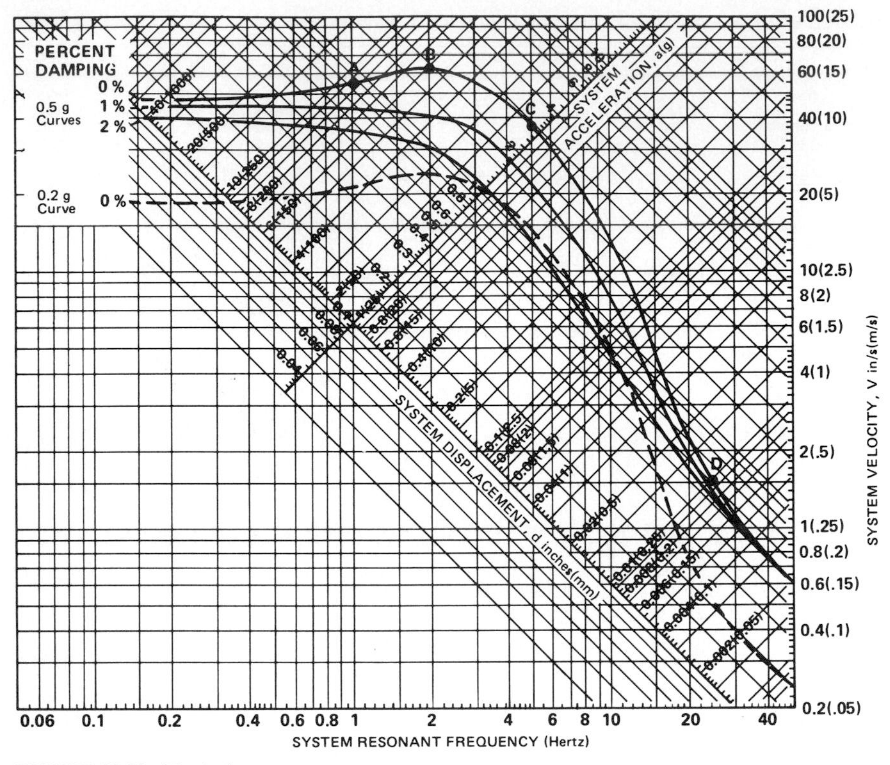

FIGURE 50.39 Typical response spectrum.

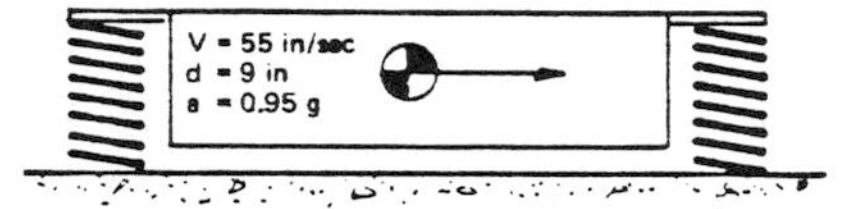

FIGURE 50.40 Response diagram, 1-Hz system.

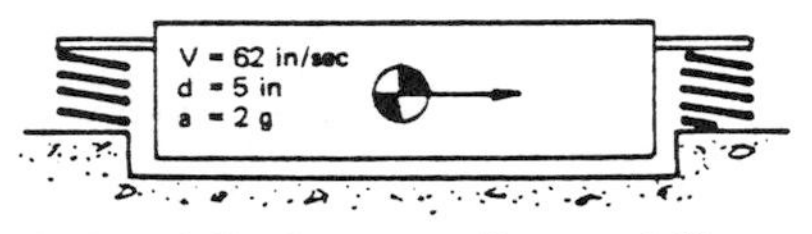

FIGURE 50.41 Response diagram, 2-Hz system.

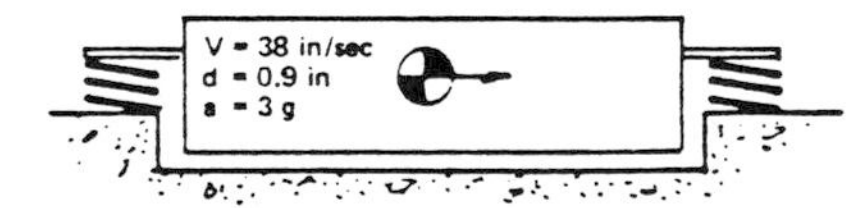

FIGURE 50.42 Response diagram, 5-Hz system.

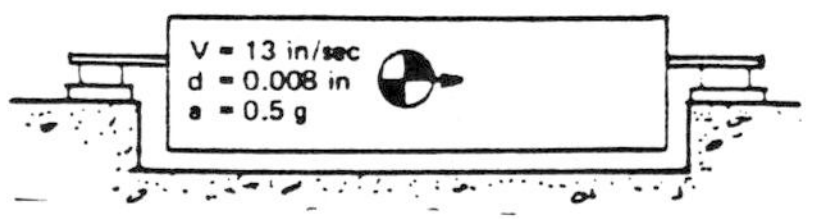

FIGURE 50.43 Response diagram, 25-Hz system.

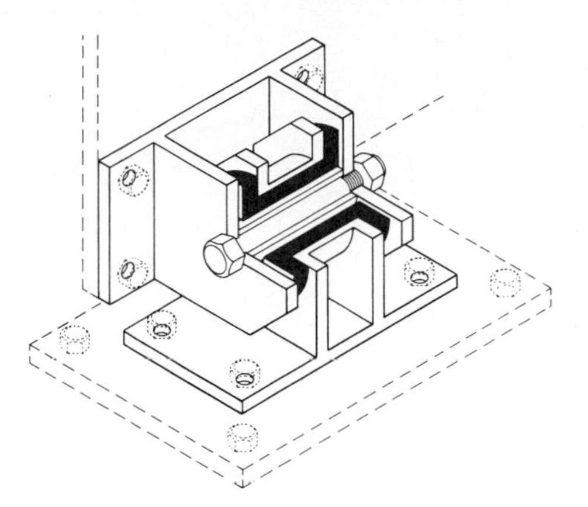

FIGURE 50.44 Snubbing device.

clearances so that during normal operation (no earthquake) they do not interfere with the vibration isolation.

Most older codes call for snubbing in terms of static capability only. The supplier need only submit calculations showing that if forces act through the center of gravity in the various modes, the anchored or snubbing device will withstand some defined static force such as 0.5*g*, 1*g*, etc. The better method (that will ultimately become the standard) is a dynamic study since an earthquake is not a static event.

All equipment has a fragility level. By definition a fragility level means it will stand a force of so many *g*'s in any direction and still remain operative. Normal transportation subjects equipment to as much as 5*g* and on rare occasions even 10*g*. Therefore an arbitrary protective standard of a maximum exposure of 4*g* is acceptable or some lower *g* force if a manufacturer advises that its equipment is even more fragile. Seismic design is then limited to the fragility level for that particular piece of equipment. A computer analysis is made in all modes, and it is demonstrated that the resiliently restrained device will not only stay in place but forces to the equipment will be 4*g* or less. The equipment will then remain operative during and after the quake.

The following typical specifications describe this type of snubbing as well as the computer program that can be specified along with it. In this specification the seismic zones are as shown in the U.S. seismic zone map from the uniform building code (Fig. 50.45). Unfortunately, zones are not described in terms of *g* levels but only in terms of damage.

The zones on this map are described numerically from zero through three. Zone zero is an area where no damage results from earthquake; zone one, minor damage; zone two, moderate damage; and zone three, major damage. Thus, in all the zones other than zero, it becomes apparent that there should be some seismic provisions, and while the response spectra would have much lower values in zone one than in zone three, the basic approach to the problem remains the same. It is interesting to note how small an area of the United States is actually zone zero.

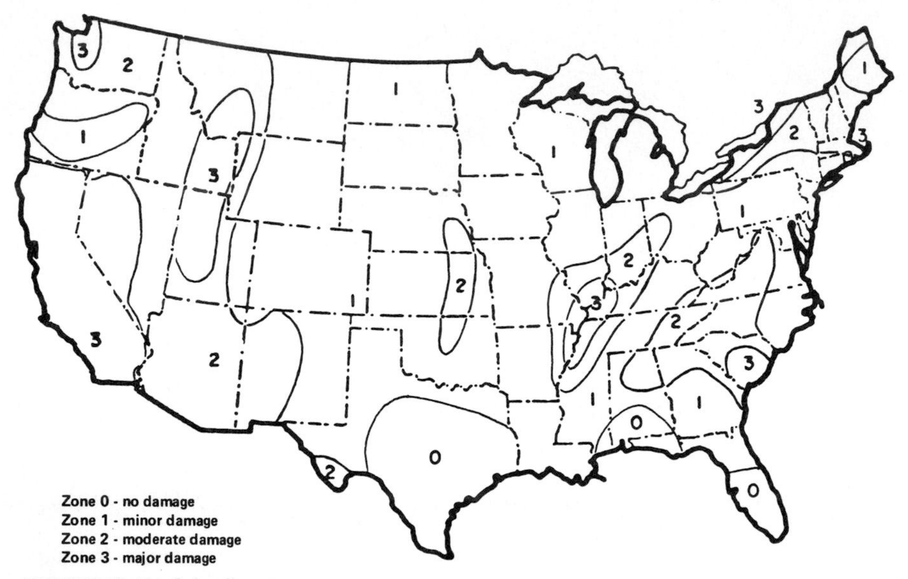

FIGURE 50.45 Seismic zone map.

50.5.2 Seismic Specifications

Specification I is suggested for those buildings where seismic response curves are available for all levels. This is usually the case for major buildings in zone three and equipment at military installations.

Specification II is included for more modest structures where this design data is not available and for possible application to large structures in zones one and two. Specification II is a definite improvement on the present static approach to the problem and would certainly represent a best-efforts basis on the part of the consultant.

A mechanical consultant has the additional option of using Specification I and providing typical response curves as suggested by a seismic authority rather than specific response curves calculated for that particular building. The seismic specifications are supplementary to the vibration control requirements.

SPECIFICATION I: SEISMIC SNUBBER SPECIFICATION USING RESPONSE SPECTRA DATA

All vibration-isolated equipment shall be mounted on rigid steel frames or concrete bases as described in the vibration control specifications unless the equipment manufacturer certifies direct attachment capability. Each spring-mounted base shall have a minimum of four all-directional seismic snubbers that are double-acting and located close to the vibration isolators if possible to facilitate attachment both to the base and to the structure (Fig. 50.46). The snubbers shall consist of interlocking steel members restrained by shock-absorbent rubber materials compounded to bridge-bearing specifications as tabulated in Table 50.7.

Elastomeric materials shall be replaceable and a minimum of ¾ in (19 mm) thick. A minimum air gap of ⅛ in (3 mm) shall be incorporated in the snubber design in all directions before contact is made between the rigid and resilient surfaces. Snubbers shall be installed with factory-set clearances. Submittals shall include load deflection

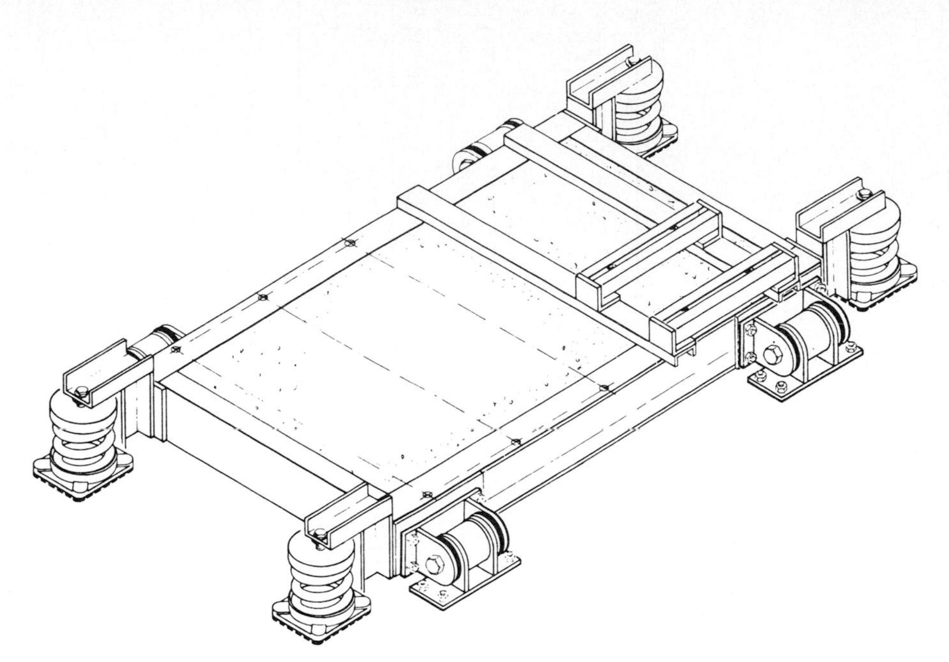

FIGURE 50.46 Spring-mounted base with snubber.

curves up to ½-in (12.7-mm) deflection in the *X, Y,* and *Z* planes. Destruction tests shall be conducted in an independent laboratory or under the signed supervision of an independent registered engineer. The snubber assemblies shall be bolted to the test machine as the snubber is normally installed. Test reports shall certify that neither the neoprene elements nor the snubber body sustained any obvious deformation after release of load at ½-in (12.7-mm) deflection.

The selection of the particular seismic snubber shall be based upon a complete dynamic analysis furnished by the vendor and based on the seismic response data.

The computer report shall include the following information along the *X, Y,* and *Z* axes.

1. The six natural frequencies of the system both with and without snubbing
2. The most probable movements (root mean square or rms values) at all mounting or combination mounting and snubber locations as well as remote source points such as duct, pipe, and electrical connections and the machine extremities
3. Maximum accelerations at the center of each significant system element such as the motor, fan, compressor, heat exchanger, or pumps as well as the mounting or combination mountings and snubber locations, all expressed in *g* units
4. The most probable force (rms value) at each mounting or mounting and snubber location expressed in pounds (kilograms)

It is the intent of the specification that the acceleration shall not exceed 4*g* or the excursion from center, ⅝ in (16 mm) at the snubbers.

When the analysis shows that levels higher than 4*g* occur at the snubbers because of the maximum allowable amplitude, the vendor shall suggest an alternate mount system based upon an additional dynamic analysis that will maintain the 4*g* level. Either the higher *g* level or the alternate method shall be accepted at the option of the consultant.

TABLE 50.7

Original physical properties				Tests for aging				
(a)	(b)		(b)	(c)			(d)	(e)
	Tensile strength (min)			Oven aging 70 h/212°F (100°C)				
Durometer	lb/in^2	(kg/cm^2)	Elongation at break (min)	Hardness (max)	Tensile strength (max)	Elongation at break (max)	Ozone 1 ppm in air by vol. 20% strain 100°F (38°C)	Compression set 22 h 158°F (70°C) Method B
40 ± 5	2000	(140)	450%	+15%	±15%	−40%	No cracks	30% (max)
50 ± 5	2500	(175)	400%	+15%	±15%	−40%	No cracks	25% (max)
60 ± 5	2500	(175)	350%	+15%	±15%	−40%	No cracks	25% (max)

Sources: (a)ASTM D-676; (b)ASTM D-412; (c)ASTM D-573; (d)ASTM D-1149; (e)ASTM D-395.

The analysis shall be the readout of a computer program with equipment modeled as a single three-dimensional rigid body composed of several rigidly attached lumped masses. For the purpose of the analysis at each support, the nonlinear snubber-isolator spring combination may be replaced by a single equivalent linear spring which is dependent on the displacement amplitude. The system must be conservative and have six natural modes and frequencies.

The final values shall be obtained as a combination of model responses in the form of a "most probable" value (rms of six modes) and an "upperbound" value sum of absolute values of six modes.

SPECIFICATION II: SEISMIC SNUBBER SPECIFICATION WITHOUT SPECTRUM DATA

All vibration-isolated equipment shall be mounted on rigid steel frames or concrete bases as described in the vibration control specifications unless the equipment manufacturer certifies direct attachment capability. Each spring-mounted base shall have a minimum of four all-directional seismic snubbers that are double-acting and located close to the vibration isolators if possible to facilitate attachment both to the base and to the structure. The snubbers shall consist of interlocking steel members restrained by shock-absorbent rubber materials compounded to bridge-bearing specifications as tabulated in Table 50.7.

Elastomeric materials shall be replaceable and a minimum of ¾ in (19 mm) thick. Snubbers shall be manufactured with an air gap between hard and resilient material of not less than ⅛ in (3 mm) nor more than ¼ in (6.4 mm). Snubbers shall be installed with factory-set clearances.

The capacity of the seismic snubber at ⅜-in (9.5-mm) deflection shall be three to four times the load assigned to the mount grouping in its immediate area. Submittals shall include load deflection curves up to ½-in (13-mm) deflection in the *X*, *Y*, and *Z* planes. Tests shall be conducted in an independent laboratory or under the signed supervision of an independent registered engineer. The snubber assemblies shall be bolted to the test machine as the snubber is normally installed. Test reports shall certify that neither the neoprene elements nor the snubber body sustained any obvious deformation after release of load.

50.6 ACOUSTICAL ISOLATION BY MEANS OF VIBRATION-ISOLATED FLOATING FLOORS

50.6.1 Theory and Methods

After the vibration problem is completely addressed, we may still be troubled by sound transmission through walls, floors, and ceilings. This occurs when a building does not have enough sound transmission resistance to handle airborne noise as opposed to mechanically transmitted noise. Airborne noise is sound radiating from equipment by air pulsation. This changing air pressure moves the floors and walls much like the cone in a loudspeaker. When building components are not sufficiently massive, the amplitude remains large enough to broadcast the sound on the other side of the architectural component with enough intensity to be annoying. The problem could be solved by increasing the mass of the components, but this is normally impractical as even 12-in (300-mm) concrete construction will transmit appreciable noise if the sound levels are high enough in the equipment room.

A second method has been made more practical. This is the floating-floor system, used above or below equipment rooms depending on the need for quiet space. In most cases, the equipment rooms are in the penthouse and the major

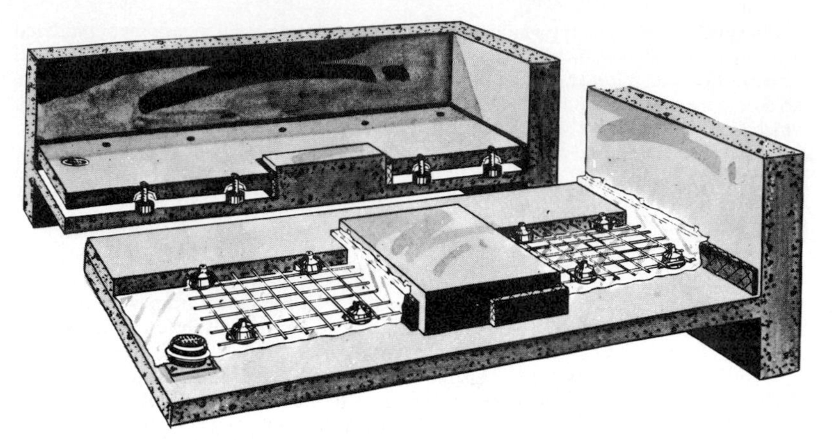

FIGURE 50.47 Floating-floor system.

problem is below. A floating floor consists of a secondary concrete floor, normally 4 in (100 mm) thick, separated from the main building structure by an air gap and supported by resilient elements (Fig. 50.47). The introduction of this construction can raise the theoretical value of a normal 6-in (150-mm) concrete floor from an STC of 54 to an STC of 79 if all flanking paths are removed.

50.6.2 Specification

A typical specification using a lift slab method is as follows.

A. Scope:

1. Isolate floating floors from building structure by means of jack-up isolators and perimeter isolation board.

B. Materials:

1. Plastic sheeting: thickness 6 mil. (1.52 mm).
2. Isolators (Fig. 50.48): Bell-shaped cast iron housing complete with ¾-in (19-mm) minimum-diameter steel jackscrews and 2-in- (50-mm-) thick bridge-bearing grade Du Pont neoprene isolators having AASHO Table B properties as in Table 50.7. Deflections shall not exceed 0.3 in or the dynamic frequency 10 Hz throughout the load range. Light-walled metal housings or isolation materials requiring protective coatings are not acceptable.
3. Perimeter isolation board: ¾-in- (19-mm-) thick 10-lb/ft^3 (16-kg/m^3) fiberglass or ½-in (12.5-mm) neoprene sponge.
4. Caulking compound: Pouring grade, nonhardening, drying or bleeding.
5. Floating-floor drains (Fig. 50.49): Cast iron pipe buckets with cast iron grills and large flanges to cover structural openings complete with waterproofing clamping ring. Upper member shall float with floating floor and sound leakage shall be prevented by an interlocking water trap. Drains shall have weep holes where indicated on drawings.
6. Riser seals (Fig. 50.50): Steel cylinders containing neoprene sponge for both structural and floating floors.

C. Floor system construction procedure:

1. Provide full-sized 10-in- (250-mm-) high concrete pedestals supported by the main structure for all mechanical equipment 75 hp (60 kW) and over. Individual pedestals may be used for chiller legs.

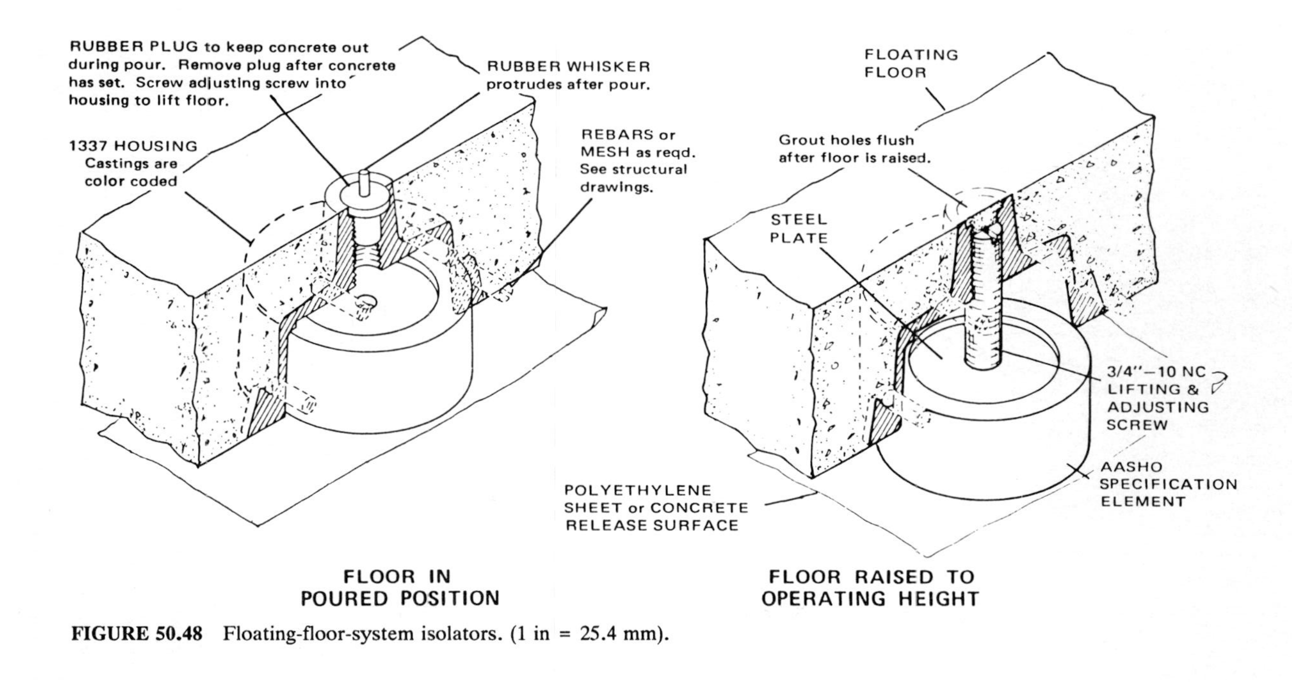

FIGURE 50.48 Floating-floor-system isolators. (1 in = 25.4 mm).

FIGURE 50.49 Floating-floor drain.

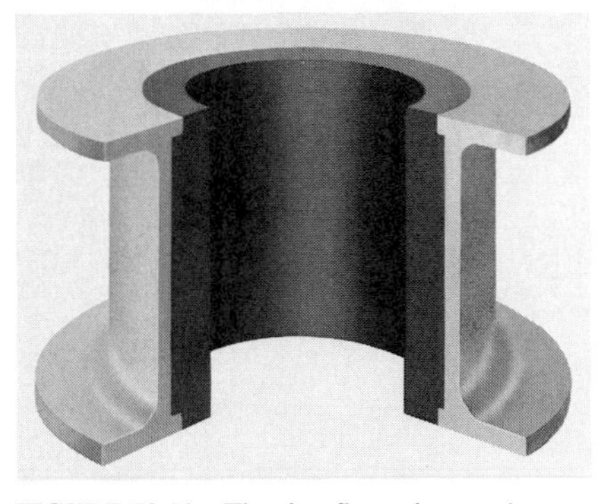

FIGURE 50.50 Floating-floor riser seal.

2. Set and waterproof drains and lower pipe seals in keeping with waterproofing specifications.
3. Cement perimeter isolation board around all pedestals, walls, columns, curbs, etc.
4. Cover entire floor area with 6-mil (1.52-mm) plastic sheeting and carry sheeting up perimeter isolation board.
5. Tape all joints with waterproof tape.
6. Place bell-shaped castings on a maximum of 48-in (1200-mm) centers in strict accordance with approved drawings prepared by the isolation manufacturer.
7. Place reinforcing as shown on the drawings and pour floor monolithically.
8. After concrete is completely cured, roll all heavy equipment across the floor and place on concrete pedestals or housekeeping pads called for in step 1.
9. Raise floor 2 in (50 mm) by means of the jackscrews. If construction sequence dictates raising the floor before placing machinery, heavy planking must be used to protect floor while machinery is being rolled into position.
10. Caulk perimeter isolation board in all locations.
11. Grout jackscrews holes.

D. Submittals:
   1. Load and deflection curves of all isolators
   2. Certification of the neoprene compound to the listed AASHO specifications
   3. Test data verifying 10-Hz maximum dynamic frequency

4. Acoustical test data from an independent laboratory showing minimum STC and INR improvements of 25 and 44, respectively, using a 4-in (100-mm) concrete floating floor and a 2-in (50-mm) air gap without fiberglass infill
5. Floating-floor structural test data with specified or lesser reinforcement

E. Manufacturer:
   1. The setting of all isolation materials and raising of the floors shall be performed by, or under, the supervision of the isolation manufacturer.

Acoustical consultants and vibration control manufacturers can offer similar assistance for increasing sound transmission losses through walls and ceilings.

CHAPTER 51
ENERGY CONSERVATION PRACTICE

Nils R. Grimm, P.E.
Section Manager—Mechanical, Sverdrup Corporation, New York, New York

51.1 INTRODUCTION

Energy conservation means many things to the design engineer. For instance:

- At one end of the scale it is the design of a system for new or retrofit projects that will have the *lowest* energy consumption over the operating life of the facility while meeting the owner's or user's needs. This is energy conservation in its pure sense, where costs are secondary to energy savings.
- At the other end of the scale it is the design of a system for new or retrofit* projects that will *minimize* energy consumption at *lowest* first cost of the project while meeting the owner's or user's needs. This is not pure energy conservation, since energy savings are secondary to costs. The prime consideration here is minimum *initial* cost; energy and maintenance cost are not included in the cost evaluation.
- Between these two extremes lies the area of design which offers the greatest challenge to the design engineer with respect to energy conservation. That is, to design the most efficient (minimized-energy-consumption) system for new or retrofit projects having the lowest life-cycle costs over the operating life of the facility and while meeting the owner's or user's needs.

The last concept of energy conservation, evaluated on life-cycle costs (LCC), will be discussed in this chapter.

*Retrofitting an existing building or facility for energy conservation means adding insulation, weatherstripping, storm windows, or replacement windows with insulated glass, or undertaking any other kind of remodeling that contributes to the prevention of unwanted heat loss or gain.

51.2 GENERAL

In new or retrofit energy conservation building design, innovation should be encouraged. However, any innovation will fail, no matter how beneficial from an energy conservation point of view, if it cannot be easily integrated into conventional construction practices and conform to established owner-user preferences, financing methods, building codes, and standards.

Though the design engineer uses the same procedures and information whether designing for energy conservation or not, there is significantly greater care and effort necessary in energy-saving design. Special attention must be given to the following factors:

- Overall values of the coefficient of heat transfer U for walls, floors, roofs, and glass
- Maximum percent fenestration (glass) area
- Building orientation with respect to fenestration per exposure
- Hours of operation of each space and area on weekdays, Saturdays, Sundays, and holidays
- Zoning of heating, ventilating, and air-conditioning (HVAC) systems
- System efficiencies at full load and at partial loads
- Ability to control, reset, start, stop, and reduce loads
- Heat recovery and heat storage
- Use of nondepletable energy sources
- Lighting illumination and fixture efficiencies
- Electrical motor efficiencies

Whether it is a new or retrofit project, reduction in one or more of the following general categories is required to reduce the energy consumed:

- Hours of system operation
- Air-conditioning loads
- Heating loads
- Ventilation and/or exhaust loads
- Domestic hot-water loads
- Lighting loads
- Off-peak loads

In addition, demand limiting and improvements in system efficiency and heat recovery are required.

Demand limiting and shifting electric loads to off-peak periods generally do not reduce the total energy required for the facility. They do reduce peak electric load, and therefore the utility or cogeneration plant energy requirement.

Of all the above energy-reduction items, it is the hours of operation that will usually have the most significant impact on energy conservation. Put another way, the energy consumption of an inefficient mechanical, plumbing, electrical, or process system that is turned off when not needed will generally be less than that of the most efficient system that is unnecessarily left on.

51.3 DESIGN PARAMETERS

Of all energy conservation factors, the major one determining the annual energy consumption of a facility is how that facility is used. This is more important than the type or capacity of the HVAC systems, boilers, chillers, and processes and the amount of glass or insulation or lighting.

It is therefore essential, if not mandatory, for the design engineer to have a definitive work schedule for each activity to be performed in the facility before energy conservation options can be considered. This schedule is part of the project design program, a topic discussed in various books (see list given in Preface), and should include the following items for each space and area:

- A detailed description of the work being performed.
- The type of process equipment and heating and cooling.
- The number of working staff or personnel by shifts for weekdays, Saturdays, Sundays, and holidays.
- The percent of equipment operating in a given hour and the average percent of full capacity for all the equipment by shifts for weekdays, Saturdays, Sundays, and holidays. If this information is not available, then the percent of maximum capacity of each operating piece of equipment for each hour of each shift for weekdays, Saturdays, Sundays, and holidays will be required.

The project annual energy budget must be determined. This establishes the maximum annual energy in Btu/ft^2 (MJ/m^2) expected to be consumed by the project.

The energy budget depends on the type of facility (such as office, hospital, institution, or warehouse). The owner or user usually establishes the energy budget. If it is not available, the engineer should establish a budget for submission to the owner or user for approval before starting the design.

It is the designer's responsibility to select and design a totally integrated system whose annual energy consumption will not exceed the project's energy budget. If the project is a new facility, the design engineer can initiate the energy conservation design. However, if the project is a retrofit, an energy audit of the existing facilities must be performed before the design engineer can start the energy conservation design.

51.3.1 Energy Audit

The purpose of the typical energy audit is threefold:

- To learn how much energy is being used annually and for what purpose.
- To identify areas of potential energy saving (heat or cooling reclamation) and areas of energy waste.
- To obtain data required to prepare plans and specifications to reduce, reclaim, or eliminate the waste identified in the audit.

It is general practice to set priorities for the recommendations of the energy audit, starting with the most cost-effective and progressing down to the least cost-effective options. Before proposing or making any modifications to a particular system, the designer should carefully study all possible effects on the total facility. For instance, a reduction in energy usage for one or more subsystems may result in an increase in the total facility energy consumption.

A typical energy-audit scope of work can be prepared for residential, commercial, institutional, or industrial facilities by selecting applicable items from the following procedures:

Utility Consumption. Obtain annual and daily records of the quantities and cost of each type of energy:

- Oil (by grade)
- Gas (natural and propane)
- Coal (by type and grade)
- Electricity

If this information is available by function, system, and process (such as office, cafeteria, or manufacturing), it should be recorded as such.

Identify all equipment observed to be idling for extended periods of time. Determine which could be turned off when not needed.

Insulation. Identify areas of damaged or missing insulation on piping systems, ductwork, and equipment.

Is the insulation type and thickness in walls and roof and on piping, ductwork, and equipment in compliance with current energy conservation standards? If not, will it be cost-effective to replace with insulation of the appropriate thickness and type or to add new insulation over the existing insulation?

If there is indication that the building, piping, ductwork, and equipment insulation may be inadequate, an infrared energy survey of the facilities should be performed to identify the hot spots (areas of greatest energy loss).

Fenestration. Is the percent of glass area high (25 percent or more of the total wall area)? Is there large glass exposure to the west and north? Is the glazing single-pane?

If the facility is fully air-conditioned, especially with large western glass exposures, the cost-effectiveness of replacing single-pane glazing with tinted Thermopane®, retrofitting shading devices in the summer, and reducing the glass area should be evaluated.

Infiltration. Is caulking around windows and exterior door frames in good condition? All defective or questionable caulking should be removed and replaced.

Is there weatherstripping around windows and exterior doors? Is it in good condition? If it is defective, it should be removed and replaced. If missing, it should be installed.

Broken windows should be replaced.

Do all building personnel entrances that are used daily have vestibules with double doors? If not, is it cost-effective to provide them? Especially in areas that have long winters, it is good practice to provide vestibules on all frequently used doors.

Do loading docks have shrouds or air-curtain fans? If not, is it cost-effective to retrofit the doors with them?

Ventilation. Is outside air set at minimum volume?

Is it cost-effective from an energy standpoint to recirculate all but the minimum ft^3/min (m^3/s) of outside air (that required to replenish oxygen and dilute unfilterable gases, e.g., carbon monoxide and carbon dioxide) through a filtering system using high-efficiency particle filters (to remove the particulate matter) in series with gas sorbers (to remove the pollutant gases).

Typical sorbers contain gas absorption materials and oxidizers such as activated charcoal and aluminum impregnated with potassium permanganate, depending on the particular gases present or anticipated in the air stream.

Exhaust and Makeup. Identify all systems that exhaust moderate to large volumes of air and fumes to the atmosphere. Can these quantities be reduced? Will it be cost-effective to recover the thermal energy being exhausted?

Identify areas and systems where the actual makeup air is excessive or deficient when compared to the required makeup air requirements. Determine the most cost-effective way to correct the makeup air volumes to the design specifications for all excessive and deficient areas or systems.

Air Systems. Is the time interval between morning startup of air-handling units and the start of the workday as short as possible but long enough to develop an acceptable temperature for arriving employees?

Is the time interval between shutdown of the refrigeration or heating system (depending on the mode of operation) and the end of the workday as long as possible but short enough to maintain an acceptable temperature at the close of the workday? For a discussion on determining the optimum startup and shutdown time periods, see Sec. 51.3.7.

Night Setback. Do the heating coil controls of the air-handling and heating and ventilating units have night setback controls that close outside air dampers and reset the thermostat downward when the facility is unoccupied?

Is the night setback temperature in the unoccupied area at least 10°F (5.5°C) lower than the nominal (occupied) space temperature? Maximum setback should maintain at least 40°F (4.5°C), however.

If any air-handling and heating and ventilating units do not have night setback controls, the cost-effectiveness of adding them should be evaluated.

If the present night setbacks are not set to maintain temperatures in the 40 to 55°F (4.5 to 12.8°C) range, the reason should be determined. If there is no valid reason, they should be adjusted to do so.

Cooldown Cycle (Cooling Mode). Do the air-handling units have a cooldown control cycle? Does that cycle close the outside air damper (assuming the building is unoccupied), de-energize the heating cycle, reset the cooling thermostat (to the occupied settings), and energize the cooling cycle?

If there are air-handling units that are normally operated 12 h or less a day without a cooldown cycle, the cost-effectiveness of adding a cooldown cycle should be evaluated.

Warmup Cycle (Heating Mode). Do the air-handling and heating and ventilating units have a warmup control cycle? Does the cycle close the outside air damper (assuming the building is unoccupied), reset the heating thermostat (to occupied setting), and de-energize the cooling and ventilating cycles?

If there are air-handling or heating and ventilating units that do not have a warmup cycle, the cost-effectiveness of adding them should be evaluated.

Low-Leakage Dampers. Do air-handling, heating and ventilating, and makeup air units have low-leakage outside air dampers? For energy conservation, a low-leakage damper is one having a maximum leakage rate less than 1 percent of the full flow in ft^3/min (m^3/s).

If there are air-handling, heating and ventilating, and makeup air units that do not have low-leakage outside air dampers and whose outside air dampers are normally closed when the building or facility they serve is unoccupied, the cost-effectiveness of retrofitting them, with respect to the energy saved, should be evaluated.

Coils (Heating and Cooling). Are coil surfaces clean? Are there any blockages or restrictions to uniform air flow across the coil face area?

Is the water side of the coil clean?

Are there any plugged tubes or indication that a coil has been frozen and repaired? It is not uncommon for repairs to frozen coils to seriously reduce heat transferability (efficiency). If this is the case, the cost-effectiveness of replacing all repaired coils should be evaluated.

All coils with dirty air side and fouled water side heat-transfer surfaces should be cleaned.

All coils found with blockages or restrictions to uniform air flow should be evaluated to determine if it will be cost-effective to correct this situation at this time.

Preheat Coils. Does the air-handling system have preheat coils? Can any of them be shut off?

If reheat is required for few zones, can variable-air-volume boxes that bypass air to the return be retrofitted to replace the reheat coils?

Is there reclaimable waste heat that could be used as an energy source for the zones that must have reheat?

Is it possible to reduce the heating-medium temperature and still maintain leaving air temperatures?

All reheat coils that are not needed should be valved off. Those reheat coils where valve turnoff is questionable—and where there is no possibility of freezing—should be shut off and their zones monitored to determine whether they can be permanently valved off.

For those instances where there are a few zones requiring reheat, the cost-effectiveness of replacing the reheat coils with variable-air-volume boxes that bypass air to the return should be evaluated.

For those instances where there are a significant number of reheat points and there is a source of waste heat that can be recovered, the cost-effectiveness of retrofitting a waste-heat recovery system for the reheater should be evaluated.

Ductwork. For comments on duct and equipment insulation, see Insulation, above.

Is there indication that the ductwork is not tight? For low-velocity systems, the leakage rate should not be greater than 7½ percent of the supply fan ft^3/min (m^3/s). For high-velocity (or medium-velocity) systems, the leakage rate should not be greater than 5 percent of the supply fan cfm (m^3/s).

Are there indications of restrictions or poorly installed ductwork? Can the supply and return fans' static pressure (total pressure for axial flow fans) be significantly reduced by modifying the ductwork?

If any of these conditions is found in the existing duct systems, the cost-effectiveness of modifying the duct systems to correct it should be studied.

Types of Systems. If the air-conditioning system is constant-volume with terminal reheat or dual ducts, the cost-effectiveness of retrofitting it to a variable-air-volume (VAV) system should be evaluated. If the air-conditioning system is multizone, it may be cost-effective to retrofit it to a VAV system.

When it is not cost-effective to retrofit a dual-duct system to a VAV system, the hot-deck automatic control should remain closed during the cooling mode. Under these conditions the hot-deck temperature will be adequate for the commercial reheat requirements, even though it will be equal only to the mixed-air temperature plus temperature increase caused by the heat added by the supply fan (which is minor).

Where reheating cannot be eliminated, are the leaving air temperatures of the coils as low as possible, yet high enough to maintain space conditions?

Can the speeds of the air-handling system supply and return fans speed be reduced by replacing the drive pulleys and belts and rebalancing the system? If the answer is "yes" or "maybe," then a study should be made to determine if the changes will be cost-effective.

Do the air-conditioning systems that serve areas that must maintain design temperatures and relative humidity 365 days a year (computer facilities, constant-temperature rooms, calibration laboratories, etc.) have some means to utilize the cooler ambient temperatures during the spring, fall, and winter months to reduce the annual energy costs?

If not, will it be cost-effective to retrofit the existing systems to have a water-cooled condenser with dry coolers as described under Condenser Water/Precooling Recovery in Sec. 51.3.6?

Liquid Refrigeration Chillers. Is the chilled-water supply temperature set at the highest temperature possible but low enough to maintain space temperatures under maximum load conditions? If not, it should be reset.

Are the automatic controls capable of resetting the chilled-water supply temperature higher as the cooling load decreases?

Is the refrigerant compressor operating at the highest suction pressure and the lowest head pressure possible, yet able to maintain the required chilled-water supply temperature under maximum load conditions? If not, this should be corrected.

Are the automatic controls for the cooling tower capable of resetting (lowering) the condenser water supply temperature as the ambient wet-bulb temperature drops?

Are the evaporator and condenser tube surfaces clean, maximizing heat-transfer efficiencies? If not, they should be cleaned.

If the present automatic control cannot reset the chilled-water supply temperature higher as the cooling load decreases, or reset (lower) the condenser water supply temperature from the cooling tower as the wet-bulb temperature drops, determine the cost-effectiveness of modifying the controls to provide these capabilities.

For facilities that have a year-round cooling requirement that cannot be met by using 100 percent outside air (economizer cycle) and have a chilled-water system with water-cooled condenser (with cooling tower or spray pound), evaluate the cost-effectiveness of the following:

- Reclaiming and reusing the heat of the condenser by providing a double bundle condenser
- Installing a plate exchanger piped into the condenser and chilled-water system. Depending on the ambient wet-bulb temperature, the plate exchanger will provide the chilled-water supply temperature and still maintain separate chilled and condenser water piping systems.

Refrigerant Compressors for DX Air-Handling Units. Is the suction pressure set at the highest temperature possible, yet able to maintain space temperatures under maximum load conditions? If not, it should be.

Is the refrigerant compressor operating at the lowest head pressure possible, yet able to maintain the required suction pressure? If not, it should be.

If the condenser is air-cooled, is the automatic control for the condenser fans capable of maintaining the lowest head pressure recommended by the compressor manufacturer while maintaining the required (compressor) suction pressure?

If the condenser is water-cooled, can the automatic control for the cooling tower reset (lower) the condenser water supply temperature as the ambient wet-bulb temperature drops, corresponding to the lowest head pressure recommended by the compressor manufacturer, and still maintain the required (compressor) suction pressure?

If the condenser is an evaporative condenser, can the automatic controls for the spray pump and condenser fan lower the condensing temperature (as the ambient wet-bulb and dry-bulb temperatures decrease) to the lowest head pressure recommended by the compressor manufacturer and yet maintain the required (compressor) suction pressure?

If the automatic controls do not maintain the lowest recommended condensing temperature, the cost-effectiveness of modifying the condenser controls should be determined.

Cooling Towers. Will it be cost-effective to reduce the blowdown (makeup water) requirements by changing or modifying water treatment?

Can the makeup water required because of drift loss be reduced by modifying or adding drift eliminators on the existing towers? Will the change be cost-effective?

Do the towers have two-speed fan motors? If not, is the energy saved enough to justify the cost of modifying the fan motors and tower controls?

Can the tower fan volume be reduced and still supply condenser water at the required temperature under design load conditions? If not, will it be cost-effective to provide this feature?

Is the automatic temperature control for the cooling tower capable of resetting (lowering) the condenser water supply temperature as the ambient wet-bulb temperature drops? If not, will it be cost-effective to retrofit the control?

Boilers. Are tubes and breeching clean?

Is the flue gas continually analyzed and the air/fuel ratio adjusted for maximum combustion efficiency, corresponding to swings in heating load?

Is heat recovered from the flue gas to preheat the combustion air or for some other preheat service?

Is the stack gas temperature as low as possible, i.e., approximately 50°F (10°C) above the lowest combustion gas dew point?

Is the breeching installed properly? Is breeching the correct size for the maximum firing rate? Is breeching pitched up toward the stack or chimney connection without restrictions?

Are the stack diameter and height adequate for the maximum firing rate of the connected boilers?

Are the burner flame shape and capacity correct for the dimensions of the combustion chamber? Is the burner type the most efficient for the boiler? Is the burner the correct size (neither undersized nor grossly oversized) for the boiler and load?

Will it be cost-effective from an energy standpoint to modify any or all of the above items?

Waste Heat and Heat Recovery. Identify areas and systems where heat can be reclaimed or recovered.

Is there a requirement for chilled water or process cooling water during the heating season? If there is, will it be cost-effective to preheat the ventilation or makeup air (outside air) and precool the chilled water or process cooling water as it is returned to the chiller by retrofitting a water-to-water and water-to-air heat-recovery system? See Fig. 51.2 and the related discussion in Sec. 51.3.6.

If the electric transformers are located indoors, will it be cost-effective to reclaim the heat generated by them?

If there are large computer rooms that operate 24 h a day or throughout the night, do they have the ability to utilize the lower-temperature ambient air to reduce the refrigeration energy demand? If not, the cost-effectiveness of retrofitting them to provide this capability should be evaluated.

In areas where ceiling height is greater than 12 ft (3.5 m), is there temperature stratification near the ceiling with a temperature difference greater than 10°F (5.5°C) during the heating season? If so, the cost-effectiveness of reclaiming this wasted heat should be evaluated. Two types of heat recovery systems are discussed under Heat Recovery by Recirculating Warm Stratified Air in Sec. 51.3.6.

Hydronic Systems. Identify leaks in condenser water, chilled water, hot water, process water, etc.

Are three-way valves used to automatically control the heating and cooling coil capacities? If two-way valves are used to automatically control the heating and cooling coil capacities, are variable-speed pumps used? If three-way or two-way automatic coil control valves with constant-speed pumps are used, will it be cost-effective to retrofit the system to one using variable-speed pumps with two-way automatic control valves?

Is the water treatment optimum to provide maximum heat-transfer efficiency within the boilers, coils, and heat exchangers and minimize corrosion and fouling of the water distribution system? Refer to Chap. 53, "Water Conditioning," for a discussion on water treatment. If the water treatment is not optimum, the cost-effectiveness of providing one that is should be evaluated.

Is the hot-water supply temperature to the fin pipe radiators automatically reset on the basis of ambient temperature?

For comments on piping and equipment insulation see Insulation, above in this section.

All leaks in the valves, equipment, and piping system should be repaired.

If the facility has a three-pipe (independent hot-water and chilled-water supply pipes and a common return pipe) distribution system, is it feasible to retrofit a two-pipe or four-pipe distribution system? If so, which is more cost-effective?

Steam Systems. Identify leaks in steam and condensation piping systems. This is especially critical for vacuum steam heating systems. Identify malfunctioning and leaking steam traps. All leaks in the piping system, valves, equipment, and malfunctioning steam traps should be repaired.

Is any condensate wasted that is suitable to be returned to the boiler?—i.e., uncontaminated? Would it be cost-effective to return it?

If high-pressure steam [at least 125 lb/in^2 (8.5 bar)] is available, will it be cost-effective to use steam-driven turbine pumps and fans, since turbines can operate as a pressure-reducing valve to supply the low-pressure [under 15 lb/in^2 (1 bar)] needs?

Is the boiler feedwater treatment optimum to provide maximum heat-transfer efficiency within the boilers, coils, and heat exchangers and minimize corrosion and fouling of the steam and condensation piping distribution systems? If the water treatment is not optimum, the cost-effectiveness of optimizing it should be evaluated.

For comments on piping and equipment insulation, see Insulation, above in this section.

Self-contained automatic radiator control valves should be retrofitted on all steam radiators and fin pipe convectors that do not already have them.

Process Equipment. Is there cost justification for:

- Replacing old equipment with new equipment requiring less energy?
- Replacing an obsolete, inefficient process and equipment with a modern process using less energy?

For batch-type processes, is it cost-effective to shut off equipment between batches?

Is the equipment startup period (the time it takes for the process to reach operating conditions) as short as possible? If the startup period is long, can the equipment be modified to shorten it? Will the modification be cost-effective?

Automatic Space Controls. Were the controls calibrated recently? If they have not been calibrated within the past 5 years, they should be recalibrated.

Are the space air-conditioning thermostats set for 78°F (25.5°C) dry-bulb temperature for comfort cooling and at the highest temperature at which the process and/or equipment can operate?

Are the space-heating thermostats set for 68°F (20°C) dry-bulb temperature for comfort heating and at the lowest temperature at which the process and/or equipment can operate?

Do thermostats reset at night or when the space is unoccupied? Can thermostats be reset by unauthorized personnel?

Are the air-handling units that have the economy cycle (provision to use 100 percent outside air for cooling) provided with enthalpy control?

Are the radiators controlled via hand valves?

Will it be cost-effective from an energy standpoint to modify any or all of the above items?

Does the facility have an energy management system? If it does, is it functionally satisfactory? If it does not, will it be cost-effective to install one?

Are the perimeter radiation hot-water supply temperature set points as low as possible for ambient air temperatures but high enough to maintain space conditions? Is the hot-water supply temperature to the heating coils as low as possible and yet able to maintain space and/or leaving air conditions? Are the controls set to prevent, or at least minimize, the effect of the perimeter system bucking the interior system in the cooling or heating mode? If the controls are not so set, they should be adjusted or modified so they will not waste energy.

Solar. Is the site's geographical location favorable for the application of solar collectors? If it is, will it be cost-effective to heat the domestic hot water or to preheat the process water?

Domestic Hot Water. Are flow restrictors installed at lavatory, bathtub, and shower fixtures?

What temperature is the hot-water supply set at? If the system supplies predominantly toilets and showers and the hot-water supply temperature is above 110°F (43°C), determine if it is cost-effective to install booster heaters locally at the equipment or fixtures that require higher temperatures and reduce the supply hot water to the 105 to 110°F (40.5 to 43°C) range.

Determine if the domestic hot-water heater is oversized. If so, is it cost-effective to reduce its capacity to match the connected load?

Does the domestic hot-water system have recirculating pumps? Do they run continuously? If so, evaluate the cost-effectiveness of shutting off the pumps after normal working hours and, if needed, installing supplementary domestic hot-water heaters for the toilets that are used during those hours.

Identify and fix all leaking fixtures, valves, and fittings.

Identify areas of damaged insulation, or those lacking insulation. Evaluate the cost-effectiveness of replacing and providing insulation where appropriate.

Is the geographical location favorable for the application of solar collectors? If it is, will it be cost-effective to install solar systems to preheat or heat the domestic hot water?

Compressed-Air Systems. Identify all leaks in compressed-air piping, valves, and fittings.

Determine if compressed-air supply pressure can be lowered. If so, the pressure control should be reset.

For central systems, determine if the compressed-air supply pressure was set for equipment in one or two areas where the required volume of compressed air is a small percentage of the plant's (volume) capacity. If so, determine if it is cost-effective to lower the supply pressure of the central system and install local air compressors in areas having equipment requiring higher pressures.

Is rejected heat from intercoolers, aftercoolers, and ventilation air reclaimed? If not, is it cost-effective to reclaim and use it?

Is the intake air to the compressor intake filter unrestricted at the pressure and quantity specified by the compressor manufacturer? If not, evaluate the cost-effectiveness of modifying the intake system to comply with the manufacturer's requirements.

Is the intake air to the compressor clean and at the lowest temperature possible? If not, will the increase in efficiency (reduction in energy consumption) produced by modifying the intake system justify the cost?

Lighting and Power. Identify areas with excessive illumination levels and areas where illumination levels can be reduced if task lighting is provided. Identify areas where lights are left on when not needed. Are there enough switches to permit leaving the lights on only in areas where persons are working (after the normal working day, etc.) and shutting off all other fixtures except for security requirements? Are the light fixtures wired to permit reducing the general illumination level by switching off alternate fixtures and reducing the number of active tubes in fluorescent fixtures? Are all fluorescent fixtures of the energy-efficient type with energy-saving ballasts?

Is exterior lighting (building, parking lot, advertising, etc.) controlled by timers or photocells?

Is the present lighting fixture maintenance program adequate to maintain maximum illuminating output? If not, determine if it will be cost-effective to increase or revise the maintenance program.

Are high-efficiency electric motors used? Are the electric motors oversized? Oversized motors operate at a lower power factor.

Determine the overall power factor for the installation. If the power factor is low (according to the electric utility standard), will it be cost-effective to provide power factor correction equipment?

From the electric utility billing criteria and the facility's hourly electric load profile, determine if it will be cost-effective to install demand-limiting equipment.

Determine the cost-effectiveness of the following:

- Reducing the illumination levels by adding task lighting where necessary.
- Retrofitting additional switches to permit shutting off lights in areas and rooms not used.
- Retrofitting the fixture circuits to permit switching off alternate fixtures and tubes.
- Relamping the facility with the most energy-efficient fixtures and bulbs for the type of work being performed in each area.
- Replacing the electric motors with those having the highest efficiency and power factor available.

51.3.2 Design

General. Though the energy required for a process normally does not vary with the seasons of the year, the energy consumed by HVAC systems does. On an annual basis, most of the energy use for building HVAC systems occurs when ambient temperatures are moderate and the systems are operating at part load. Only a small fraction of the annual hours of operation of HVAC equipment occurs when ambient summer and winter temperatures are at or near their respective design values. The designer should (from an energy consumption standpoint) be more concerned about minimizing energy consumption at various part-load conditions throughout the year than at the design heating and cooling loads.

The designer must consider carefully energy consumption of equipment that operates most of the time at or close to full load. Typically lights, fans, and pumps, before the energy crisis in the 1970s, were operated constantly and at full load. In many air-conditioned offices and institutional buildings, under such conditions, the HVAC fans and pumps on an annual basis use more energy than the central air-conditioning chillers.

Energy can be saved if the designer carefully considers the following:

1. Operating HVAC systems at part load, especially fans and pumps.
2. Selecting variable-capacity fans and pumps, capable of varying their capacities to meet their respective part-load requirements (this is usually required of systems employing variable-speed drives and/or multiple units).
3. Using high-efficiency motors.
4. Designing HVAC systems that can isolate areas having relatively constant occupancy during the normal working day from those having only part-time occupancy (such as conference rooms, auditoriums, etc.), spaces that are used 24 h a day (such as computer rooms, constant-temperature rooms, and calibration labs), and areas that are used after the normal working day and on Saturdays, Sundays, and holidays. The systems should be designed and zoned so that only the areas occupied or requiring constant exhaust, temperature, or humidity will be operating. All other cooling and exhaust systems will be off, and heating system temperatures should be reset as low as possible.
5. The lighting system should be designed to provide the minimum acceptable level of general illumination and task lighting for the working area. High-efficiency lamps and low-energy ballasts should be installed, and available daylight should be used whenever possible. Lighting circuitry should be designed to permit turning off lights in unoccupied areas and reducing lighting level for off-hours housecleaning.
6. Domestic hot-water temperature should be set as low as possible. Local generation of domestic hot water to eliminate long runs of recirculating piping should be evaluated. Water conservation fixtures should be used.
7. The design pressure in plant compressed-air systems should be as low as possible.

Components. To assist the designer in selecting the proper components for an integrated energy-efficient design that will minimize energy usage and meet the project's energy budget, the following guidance is offered. This list of components must not be considered all-inclusive. Innovations and additions should be encouraged.

Utilities. When determining the most appropriate fuel or fuels to be used, the following should be considered:

- Present and long-term availability and costs of oil, coal, gas, and electricity available at the project site.
- The various grades of oil and coal available.
- For coal, the costs of unloading, storing, handling coal; controlling air pollution (particulate matter); and ash handling and disposal must be considered.
- In locations where natural gas is available, the cost-effectiveness of using dual fuels—especially oil and gas—should not be overlooked.

Alternative Energy Sources. To reduce dependence on oil and electricity (generated by burning oil), alternative energy sources such as coal; methane gas from wells, landfill, and sewage treatment plants; wood; hydropower; sun; wind; and tidal motion (to name the most common) should be considered. Although installations using one or more of these alternative energy sources have been successful, specific environmental, meteorologic, and site-related conditions must be favorable. When site conditions are favorable, alternative energy sources should be compared with oil, gas, coal, and electricity to determine those most cost-effective.

Transmission Values. When the maximum transmission values U are not determined by the user or owner, they should be selected by the engineer to minimize energy consumption. The author has used the values in Table 51.1 as the basis for his designs and as the base U values for calculating the cost-effectiveness of using lower U values in combination with additional insulation and triple-pane glazing.

The U values in Table 51.1 are selected on the basis of heating degree-days. However, the author suggests that the design U values, for projects where the air-conditioning load is predominant, should be based on the lower of two values, one based on the actual heating degree-days and the other based on one of the following conditions:

- When the summer air-conditioning design ambient temperature is above 95°F (35°C) 2½ percent of the time and the cooling season is at least 4 months long, the U values corresponding to 3001 to 4000 (1671 to 2220) heating degree-days should be used. (Values in parentheses are Celsius degree-days; others are Fahrenheit degree-days.)
- When the summer air-conditioning design ambient temperature is between 90 and 95°F (32 and 35°C) 2½ percent of the time and the cooling season is at least 3 months, the U values corresponding to 2001 to 3000 (1111 to 1670) heating degree-days should be used.
- When the summer air-conditioning design ambient temperature is below 90°F (32°C) 2½ percent of the time and the cooling season is at least 4 months, the U values corresponding to less than 1000 (560) heating degree-days should be used.
- For all other conditions, the U values should be selected on the basis of the actual heating degree-days.

If there is any question on the selection of a particular value, the decision should be based on a life-cycle cost analysis.

Fenestration. Traditionally the architect is the one who determines the glass area of a building. However, in order to design a facility that will meet the established energy budget, it is the engineer who must determine the maximum percentage of glass area that can be permitted in conjunction with the wall construction that will not exceed the overall design U_o value (see preceding discussion under Transmission Values).

The overall U_o value is determined by the following equation:

$$U_o = \frac{U_w A_w + U_g A_g + U_d A_d + \cdots}{A_o} \tag{51.1}$$

where U_o = average or combined transmission of the gross exterior wall, floor, or roof-ceiling assembly area, Btu/(h · ft² · °F) [W/(m² · K)]

A_o = gross exterior wall, floor, or roof-ceiling assembly area, ft² (m²)

U_w = thermal transmission of the components of the opaque wall, floor, or roof-ceiling assembly area, Btu/(h · ft² · °F) [W/(m² · K)]

A_w = opaque wall, floor, or roof-ceiling assembly area, ft² (m²)

U_g = thermal transmission of the glazing (window or skylight area), Btu/(h · ft² · °F) [W/(m² · K)]

A_g = glazing area (finished opening), ft² (m²)

U_d = thermal transmission of the door or similar opening, Btu/(h · ft² · °F) [W/(m² · K)]

A_d = door area (finished opening), ft² (m²)

TABLE 51.1 Maximum Heat Transmission Values*

Heating degree-days, °F days (°C days)	Gross wall† U_o (note 1)	Gross wall† U_o (note 2)	Walls U_w (note 3)	Ceiling/roof U_r (note 4)	Floor U_f (note 5)	Floor U_f (note 6)
Less than 1000 (less than 560)	0.31 (1.760)	0.38 (2.15)	0.15 (0.853)	0.05 (0.284)	0.10 (0.568)	0.29 (1.647)
1000–2000 (561–1110)	0.23 (1.306)	0.38 (2.15)	0.15 (0.853)	0.05 (0.284)	0.08 (0.454)	0.24 (1.363)
2001–3000 (1111–1670)	0.18 (1.022)	0.36 (2.048)	0.10 (0.568)	0.04 (0.227)	0.07 (0.397)	0.21 (1.192)
3001–4000 (1671–2220)	0.16 (0.909)	0.36 (2.048)	0.10 (0.568)	0.03 (0.170)	0.07 (9.397)	0.18 (1.022)
4001–6000 (2221–3330)	0.13 (0.738)	0.31 (1.760)	0.08 (0.454)	0.03 (0.170)	0.05 (0.284)	0.14 (0.794)
6001–8000 (3331–4440)	0.12 (0.683)	0.28 (1.590)	0.07 (0.397)	0.03 (0.170)	0.05 (0.284)	0.12 (0.683)
Over 8001 (over 4441)	0.10 (0.568)	0.28 (1.590)	0.07 (0.397)	0.03 (0.170)	0.05 (0.284)	0.10 (0.568)

*Heat transmission values are expressed in English units, Btu/(ft^2 · h · °F), and, in parentheses, in SI units, W/(m^2 · K).

†Gross wall values include all doors and windows, window frames, metal ties through walls, structural steel members that protrude through all insulation to the exterior or adjacent to the exterior and continuous concrete or masonry walls or floors that extend from inside heated spaces through the building envelope to the exterior, e.g., fire walls that extend above the roof and concrete floor slabs that extend beyond the exterior wall to form a balcony or terrace.

Note 1: These gross wall U_o values are used for all new construction and major alteration of facilities other than hospitals and medical and dental clinics.

Note 2: These gross wall U_o values are to be used for hospitals and medical and dental clinics. The maximum U_o value will put a limitation on the allowable percentage of glass area to gross wall area in a building. Insulating glass will allow higher percentage of glass area than single glass.

Note 3: Wall U_w value is the thermal transmittance of all elements of the opaque wall area. U_w values are to be used for upgrading existing facilities where the alteration of walls and resizing of window glazing to meet gross wall values is not cost-effective.

Note 4: Ceiling/roof U_r values are for ceiling and roof areas where adequate space exists for insulation to be applied above the ceiling or below the roof structure. Built-up roof assemblies and ceiling assemblies in which the finished interior surface is essentially the underside of the roof deck shall have a maximum U_r value of 0.05 (0.284) for any heating degree-day area.

On existing buildings, use the maximum U_r value practical to accommodate the existing roof conditions where the life-cycle cost analysis indicates a higher life-cycle cost to implement U_r values required by Table 51.1. Examples of costs encountered on existing buildings related to implementing U_r values required by Table 51.1 are as follows: (*a*) cost of providing structural support to accommodate additional dead loads of new insulation and roofing system, and additional live loads from greater accumulations of snow (snow will melt more slowly because of increased insulation); (*b*) cost of raising roof curbs; (*c*) cost of raising cap flashings; (*d*) cost of raising roof drains.

Note 5: Floor U_f values are for floors of heated space over unheated areas such as garages, crawl spaces, and basements without a positive heat supply to maintain a minimum temperature of 50°F (10°C).

Note 6: Floor U_f values are for slab-on-grade insulation around the perimeter of the floor.

Source: *Department of Defense Construction Criteria*, document DOD 4270.1-M, Office of the Deputy Assistant Secretary of Defense (Installations), Washington, DC, 15 Dec. 1983, chap. 8, table 8-1, p. 8-8.

From Eq. (51.1) it can be seen that the percentage of glass can be maximized without increasing the design U_o by selecting the lowest economical U_w (by changing the wall construction or adding insulation) and using triple Thermopane® glazing.

Although maximizing the percentage of glazing can have aesthetic, daylighting, and passive solar heating benefits, it generally increases wall construction costs.

Insulation. In residential facilities, most of the energy is used for environmental control. In such facilities the thermal (insulation) quality of the buildings and the severity of the weather become a predominant influence on energy consumption. Other major factors are how the systems perform with respect to space temperatures and hours of operation. In facilities such as these, the insulation thickness has a direct effect on reducing the amount of energy consumed. The more insulation, the less energy required to maintain space conditions.

In nonresidential facilities, energy usage is more complex. It is influenced by the function of the particular building, type and sophistication of control systems, type of fan and pump operation (constant speed or variable speed), hours of operation, ventilating rate, and thermal (insulation) quality of the building. Buildings such as these are relatively insensitive to energy savings resulting from insulation thickness alone. The primary reason for this is that, during the cooling season, most of the air-conditioning energy is used to offset heat gains from people, lights, and equipment, which are the same for facilities in Fairbanks, Alaska, or Miami, Florida. Another reason is that energy loss through exterior areas (building skin and roof) is a small percentage of the heating and cooling load; this is especially true in high-rise office buildings and institutions.

Selecting the optimum insulation thickness and type is important, since it can improve system efficiencies and reduce the amount of energy needed to maintain the same environmental condition or process load—or increase the energy available to maintain environmental conditions or process load.

The optimum insulation thickness is the thickness which will result in the lowest total of the cost of energy lost and the cost of insulation and installation. The method and procedure to calculate the optimum insulation thickness can be found in standard design handbook sources such as Ref. 1.

If the analytical method is not used to determine the optimum insulation thickness, the author recommends the following thickness guidelines. At the very least, they can be used as a basis for comparison of insulation thicknesses and types.

1. Duct insulation—outside air, supply, and return ductwork; plenums and casing of HVAC units
   a. Indoors
      (1) Blanket-type flexible fibrous-glass insulation, minimum density 1 lb/ft^3 (16 kg/m^3), minimum thickness 2 in (50.8 mm)
      (2) Rigid-type fibrous-glass insulation, minimum density 3 lb/ft^3 (48 kg/m^3), minimum thickness 2 in (50.8 mm)
   b. Outdoors—polyurethane or polyisocyanate board, minimum density 1.7 lb/ft^3 (27.2 kg/m^3), minimum thickness 3 in (76.2 mm)
2. Equipment
   a. Pumps, chilled, dual-temperature, and hot water
      (1) Cellular glass insulation, minimum thickness 2 in (50.8 mm)
      (2) Fibrous-glass insulation, minimum density 6 lb/ft^3 (96.1 kg/m^3), minimum thickness 2 in (50.8 mm)

(3) Polyurethane or polyisocyanate, minimum density 1.7 lb/ft^3 (27.2 kg/m^3), minimum thickness 2 in (50.8 mm)

b. Expansion tanks, condensate receivers, hot-water storage tanks, and converters
   (1) Cellular glass, minimum thickness 4 in (101.6 mm)
   (2) Fibrous-glass insulation, minimum density 6 lb/ft^3 (96.1 kg/m^3), minimum thickness 4 in (101.6 mm)
   (3) Calcium silicate, minimum thickness 4 in (101.6 mm)
   (4) Polyurethane or polyisocyanate, minimum thickness 2 in (50.8 mm)

c. Chillers
   (1) Polyurethane or polyisocyanate, minimum thickness 2 in (50.8 mm)
   (2) Plastic foam, minimum thickness 2 in (50.8 mm)

d. Piping systems—chilled water, dual temperature, hot-water heating, and domestic hot water
   (1) Fibrous glass, minimum density 3 lb/ft^3 (48 kg/m^3)
   (2) Pipes less than 3 in (76.2 mm) in diameter, minimum thickness 1 in (25.4 mm)
   (3) Pipes 3 in (76.2 mm) and 4 in (101.6 mm) in diameter, minimum thickness 1½ in (38.1 mm)
   (4) Pipes 5 in (127 mm) and larger, minimum thickness 2 in (50.8 mm)
   (5) Pipes less than 3 in (76.2 mm) in diameter, minimum thickness ¾ in (19 mm)
   (6) Pipes 3 in (76.2 mm) and 4 in (101.6 mm) in diameter, minimum thickness 1 in (25.4 mm)
   (7) Pipes 5 in (127 mm) and larger, minimum thickness 1½ in (38.1 mm)

e. Steam, condensate, and boiler-feed piping
   (1) Fibrous glass, minimum density 3 lb/ft^3 (48 kg/m^3); minimum thickness of pipe insulation is listed in Table 51.2
   (2) Calcium silicate, minimum thickness ½ inch (12.7 mm) greater than those listed in Table 51.2

For outdoor insulation it is a good rule of thumb to increase thickness by 1 in (25.4 mm) over that of indoor insulation. Similarly, when chilled-water and cooling-water piping as well as air-conditioning ducts must be routed through hot areas such as boiler rooms and laundries, additional insulation thickness should be considered.

When selecting insulation thicknesses for energy conservation purposes, the

TABLE 51.2 Minimum Thickness of Fibrous Glass Pipe Insulation
(Not exposed to weather)

Maximum temperature, °F (°C)	Nominal pipe sizes, in (mm)				
	Up to 1.25 (31.75)	1.5–2.5 (38.1–68.5)	3–4 (76.2–101.6)	5–6 (127–152.4)	8 (203.2) and larger
Up to 299 (148.3)	1 (25.4)	1.5 (38.1)	2 (50.8)	2.5 (63.5)	3 (76.2)
300–499 (148.9–259.4)	1.5 (38.1)	2.5 (63.5)	3 (76.2)	3.5 (88.9)	4 (101.6)
Condensate and boiler feedwater	1 (25.4)	1 (25.4)	1.5 (38.1)	2 (50.8)	2 (50.8)

engineer must not overlook the fact that it may not be cost-effective to insulate piping and ductwork for liquids or gases in the temperature range of 55 to 120°F (12.8 to 48.9°C). As the temperature difference between the liquid or gas stream and the surrounding space decreases, so does the possibility of saving energy. A point is reached where this temperature difference is so small that heat loss or gain without insulation will not increase the annual energy requirements.

Infiltration. Infiltration is air flowing into a building or space through cracks around windows, doors, and skylights and through minute passageways and cracks within wall, floor, and roof structures. Infiltration always results in an additional heating load and an additional sensible and latent cooling load when portions of a building are under negative pressure because of stack effect in high-rise buildings or insufficient tempered makeup air. The heating and/or cooling infiltration load can be calculated from the formulas in standard handbook sources such as Refs. 2 and 3.

With present technology, it is not economically feasible to design a commercial, institutional, or industrial facility for zero infiltration. However, the engineer should select exterior wall components that will minimize the infiltration load. This will be cost-effective at the point when energy saved (by reducing the infiltration load) over the life of the facility is greater than the total cost of reducing the infiltration load.

The following ways of controlling the infiltration load are suggested to the designer:

1. Reduce the pressure differential across exterior doors and windows.
   a. For exterior personnel entrances, provide vestibules with exterior and interior doors or revolving doors. The vestibules should have cabinet heaters and ducted tempered supply (pressurization) air. With revolving doors (four-section), tempered supply air can be ducted through the top of the two completely closed-in sections. The supply air volume should be automatically controlled to maintain the inside pressure equal to or slightly greater than the outside pressure.
   b. Generally, space restraints preclude vestibules at loading-dock doors. However, air-curtain-type door heaters, mounted over the door and discharging downward, with shrouds (flexible closure pieces) to seal the space between the door opening and the truck or trailer body, have proven to be effective.
   c. Infiltration at window areas can be controlled to acceptable levels by pressurizing the space (when the space does not have to be maintained under a negative pressure) by returning slightly less air ft^3/min (m^3/s) than one supplies to the space and selecting tightly closing, well-made window assemblies and hardware with good-quality seals around the perimeter and especially at all points where the sashes slide against the frame or past another sash.
2. Provide good-quality weather stripping seals around the perimeter of all doors.
3. Provide good-quality heavy building paper between the sheathing and exterior siding on all wood-constructed exterior walls.
4. Seal all exterior brick walls.

Ventilation. From an energy standpoint, when HVAC systems are operating in either the cooling or heating mode, ventilation air should be kept at the minimum quantity required to replenish the oxygen and dilute pollutants and contaminants in the indoor air to an acceptable level.

The American Society of Heating, Refrigerating, and Air Conditioning Engineers (ASHRAE) has recently revised its recommended outside ventilation air quantities upward in order to achieve acceptable indoor air quality by dilution with outdoor air only. (See Ref. 4 for details.) For a typical office, ASHRAE is recommending a four-fold increase from 5 $ft^3/(min \cdot person)$ [2.51 $L/(s \cdot person)$] to 20 $ft^3/(min \cdot person)$ [10 $L/(s \cdot person)$].

Before designing an HVAC system with these higher outdoor air ventilating quantities (dilution only), the engineer should evaluate the cost-effectiveness of (1) adding only the minimum quantity of outside air required to replenish oxygen and dilute unfilterable gases and (2) removing or reducing contaminants and pollutants in the return air by filtration. This procedure will create, at least, the same indoor-air quality as if higher outdoor-air quantities were used.

The major design parameters for a typical dilution-removal filtration system are:

- The outdoor air quality will be set at the minimum required to replenish the oxygen (O_2) and dilute unfilterable gases, namely carbon monoxide (CO) and carbon dioxide (CO_2).
- The mixed air stream (outdoor and return air) first passes through a roughing filter, 2 in (50 mm) thick, in series with a 90 percent (minimum) ASHRAE 52-1976 efficiency and a filter at least 6 in (152 mm) deep.
- The air stream then passes through gas sorbers capable of removing a broad range of gases and vapors commonly found in a particular indoor and outdoor environment.
- The sorbers usually contain gas adsorbers and oxidizers, such as activated charcoal and alumina impregnated with potassium permanganate, depending on the gases present at the site or anticipated in the air stream.
- Odoroxidant media should be suitable for removing odorous, irritating, acidic gases from air by reacting chemically with the sorbed gases to prevent later desorption.
- The sorbers should be selected with sufficient capacity to remain active (effective) for a minimum service life of 4370 h (24 h per day for 6 months).
- The velocity through the sorber collection bed should provide a minimum residence time of approximately 0.06 s.
- High circulation rates (6 to 10 changes of the volume of air in each space per hour) are required to obtain effective mixing of the air within each space to capture and remove sufficient quantities of indoor contamination to provide the required indoor air quality.
- The filtered (supply) air should be discharged from diffusers that direct the air in a plug (flow predominately in one direction) or horizontal laminar flow pattern so the contaminants will be swept along with the flow across an occupied space to return-air intakes on the opposite side of the space.
- Since the static pressure drop across the combined high-efficiency particle filter and gas sorber is normally about 2 in water (497 Pa) and high circulation rates are required, it is not uncommon for this type of filtering system to have its own fan system and operate either in conjunction with the building HVAC system or independent of it.
- When this type of filtering system is integrated with air-handling equipment containing cooling coils, the sorber section must be located downstream of the cooling coils and coil condensate drain pan to ensure that microbiological con-

taminants living on wet surfaces are removed before the air is distributed to the occupied spaces.

Exhaust and Makeup. For energy conservation, the engineer should determine the minimum exhaust air quantity for each system consistent with applicable codes and good engineering practices. To achieve this goal, the designer should evaluate each exhaust system with respect to the following items:

- When codes and good engineering practice permit shutting off the exhaust system when a facility is not occupied or a process is not operating, the designer should design a dedicated exhaust system that can be independently taken out of service. Though starting and stopping the exhaust system can be done manually, more energy will be saved if it is automatically done.
- Where the applicable codes do not mandate the exhaust air quantity for a particular type of space activity, it should be equal to the ventilation air quantity recommended for the activity in Ref. 4. Though not as current, Ref. 3 is also used.
- The industrial exhaust hoods should be as close to the source (oxygen or exhaust air) as possible to minimize the exhaust air volume.
- Push-pull exhaust systems should be considered for large tanks and vats.
- When possible, all tanks and vats should be provided with covers to reduce emission of vapors and odors.
- Generally, recirculating systems with adequate filtration should be used instead of exhausting air to the outside, whenever the particular industrial process or equipment and good engineering permit it.
- Low-volume, high-velocity exhaust systems should be used whenever possible to control dust from portable hand tools and machining operations.
- The industrial exhaust system should conform to the recommended practices set forth in the latest edition of Ref. 5.
- Will it be cost-effective to reduce the industrial exhaust air quantities by selecting less toxic or less hazardous materials or modernizing the process or equipment?

Once the exhaust air quantities have been established, the makeup air should be equal to the total exhaust air quantity unless there are specific areas that must be maintained at a negative or positive pressure. When there are equipment, processes, or areas that must be maintained at a negative or positive pressure, they should be enclosed in the smallest envelope possible and their makeup air should be supplied from a separate zone or unit.

The designer should evaluate the cost-effectiveness of recovering the heating or cooling energy in the exhaust air to heat or cool the makeup air. Thermal wheels, parallel-plate heat exchangers, coil runaround cycles, and heat-pipe recovery systems are discussed in Sec. 51.3.6.

Low-Leakage Dampers. High-performance, low-leakage dampers should be used for outside air, relief air, and return air and for mixing hot and cold air streams. The energy that can be saved by using high-performance, low-leakage dampers instead of standard dampers is apparent when one compares the leakage rates at the same difference pressure drop across fully closed dampers.

Typical leakage rates are

Low-leakage dampers	Less than 1 percent of full flow
Standard dampers	5 percent minimum to 25 percent of full flow (depending on the quality of the manufacturing)

The static pressure drop across a fully open high-performance, low-leakage damper or standard damper is so small compared to the total system static pressure that, in general, there is no noticeable effect on the system energy usage with either type (in the fully open position).

The engineer may want to evaluate the cost-effectiveness of low-leakage dampers for other damper duties.

Coils (Heating and Cooling). From an energy conservation point of view, the engineer can reduce the energy used by the fans by selecting coils having minimum resistance to air flow. The heat-transfer surfaces on both air and fluid sides must be kept clean at all times with adequate water treatment on the fluid side and periodic cleaning on the air side. The following parameters can be used to select coils with low resistance to air flow:

For cooling coils:

Minimum velocity	400 ft/min (2 m/s)
Maximum velocity	500 ft/min (2.5 m/s)
Fan spacing	6 to 10 fins/in (0.24 to 0.39 fins/mm)

For heating coils:

Velocity range	500 to 800 ft/min (2.5 to 4 m/s)
Maximum fin spacing	14 fins/in (0.55 fins/mm)
Heat-recovery coils	Maximum velocity and fin spacing should be determined to maximize the energy recovered and minimize the cost of recovering it

The author acknowledges the manufacturers' position that for the same cooling load, coils with fin spacings of 6 to 10 fins/in (0.24 to 0.39 fins/mm) will probably require one and possibly two additional rows compared to a coil having 14 fins/in (0.55 fins/mm) and there will be no apparent difference in static pressure drop across the coil. That position is valid only when the coil is clean, however. With 14 fins/in (0.55 fins/mm) and wet fin surfaces, the particulate matter that passes through the filters will adhere to the wet fins and, in a relatively short time, will reduce the already narrow spacing between these fins. It has been the author's experience that in a short time, the static pressure drop across wet coils with 14 fins/in (0.55 fins/mm) becomes greater than for coils with 6 to 10 fins/in (0.24 to 0.39 fins/mm) even with the additional row or two. Considering the length of time between coil cleaning, the greater fin spacing can save energy.

Heating coils, on the other hand, always have dry fin surfaces and in general have no more than two rows. Under these conditions, the static pressure drop with 14 fins/in (0.55 fins/mm) will not result in a noticeable increase in fan energy.

Ductwork. Energy savings result when fans operate at the lower system pressures allowed by larger ductwork. The system design must provide operation cost savings (over the service life of the equipment) that more than offset the increased construction costs of the ductwork.

For a discussion of commercial and industrial system duct sizing procedures, the reader is referred to Chap. 4 of this book. For industrial exhaust system ducting, design in accordance with the procedures set forth in Ref. 5.

All duct seams of commercial and institutional duct systems should be taped and the maximum system leakage should not exceed the following:

- For a low-velocity system, 7½ percent of the fan ft^3/min (m^3/s).
- For a medium- or high-velocity system, 5 percent of the fan ft^3/min (m^3/s).

All duct seams of industrial exhaust systems should be welded, brazed, or soldered, depending on the system temperature and duct material.

All hot and cold ducts should be insulated. See previous discussion of duct insulation in this section.

51.3.3 Types of Systems

General. When selecting air-handling units (especially for HVAC application) the engineer must always remember that probably less than 5 percent of the actual hours of heating or cooling system operation will be at the respective design load. The remainder of the time, the system will be operating at part load.

Though the actual part-load capacity and corresponding percent of operation time should be determined for each system, Table 51.3 can be used to estimate

TABLE 51.3 Heating or Cooling Operating Time at Various Loads for Typical HVAC Systems

Percent of full load	Percent of operating time
75–100	10
50–75	50
25–50	30
0–25	10

the order of magnitude of a typical heating or cooling HVAC system. From Table 51.3, it is apparent that the energy used (especially by fan and pump motors) in the 25 to 75 percent full-load range is extremely important. It is in this load range that one should concentrate on maximizing the system efficiency and minimizing the horsepower, and not at the design load. As a general rule, more energy can be saved by reducing the fan ft^3/min (m^3/s) and pump gal/min (m^3/s) than by reducing the supply air or water temperature to meet part-load conditions.

In order to conserve energy, areas and processes that are used after normal business hours should have their own HVAC and exhaust systems. Typically, these are the areas that must maintain design temperatures and relative humidity conditions 365 days a year (computer facilities, constant-temperature rooms, calibration laboratories, etc.). Also, auditoriums, cafeterias, conference rooms, and meeting rooms that are frequently used after normal business hours should be included in this category. The facilities that serve or support these areas, such as lobbies, corridors, toilets, lounges, and lunch rooms, should also be designed to operate independently of the main building system if one is to minimize energy costs and usage.

Though the engineer has six basic types of HVAC air-distribution systems from which to select that most appropriate for a design, those that can vary the air and liquid volume in accordance with variations in load generally have the lowest energy consumption.

Basic Systems. The six basic systems and their variations are:

1. *Single duct:* This is usually a low-velocity distribution system. The unit consists of filters, cooling and heating coils, supply fan, and sometimes a return fan.

The fans are generally centrifugal type, constant or variable volume. If the fans are variable volume, centrifugal or axial flow, they can be controlled by inlet vanes or variable-speed motors. Axial-flow fans, depending on size, can also be controlled by varying the pitch of the blades.

This system is suitable for single-zone application. When more than one zone is required, terminal reheats have been used to provide zone control. However, even when waste or reclaimed heat is then used for the reheat energy, it still may not have the lowest life-cycle costs.

2. *Dual duct:* This is usually a high-velocity supply and low-velocity return duct distribution system. The unit consists of filters, cooling and heating coils, and supply and return fans. The supply distribution mains consist of hot and cold ducts with mixing boxes at each zone. The ductwork from the mixing boxes to the diffusers is low-velocity.

The system is extremely flexible with respect to future modifications and has good temperature controls.

The size of the cold duct main should be based on the maximum building peak cooling load. The cold branch mains on a floor should be sized on the maximum simultaneous internal- and external-exposure peak loads of areas they serve. The hot duct is usually sized between 50 and 75 percent of the air capacity of the cold duct.

For energy conservation, the fans are generally airfoil variable-volume, centrifugal- or axial-flow types. Variable-frequency speed control is used on both types. Axial-flow fans are also available with adjustable-pitch blades.

The hot deck coil control valve should be closed during the cooling mode to conserve energy.

Even with these energy conservation measures, this system's energy consumption is relatively high.

3. *Multizone:* This is a low-velocity duct distribution system. The unit consists of filters, cooling and heating coils, hot and cold automatic modulating coil discharge air dampers, supply fan, and sometimes a return fan.

Depending on the size of the unit, six to ten zones with controls are common. The zone controls available with this type of unit are satisfactory for comfort air conditioning (such as in an office environment) but usually not for critical areas (such as laboratories).

The fans are centrifugal, constant-speed type.

This system varies each zone supply temperature by modulating its respective hot and cold deck dampers, as required, to satisfy the particular zone space temperature set point. It is not adaptable to varying the supply air volume. In some comfort air-conditioning installations, energy can be saved during the cooling cycle by automatically closing the heating coil control valve during this mode of operation.

This system is generally relatively expensive to install and modify.

Even with the energy conservation measures noted above, the energy usage of this system will be higher than that of a variable air volume system.

A recent variation to the standard multizone uses individual zone heating and cooling coils instead of a common hot and cold deck with individual zone mixing dampers. The elimination of simultaneous heating and cooling and air-stream mixing losses can result in significant energy saving. Energy consumption of this unit can be as much as 40 percent less than a multizone unit with common hot and cold deck with individual zone mixing dampers. Only package rooftop units in the 15- to 37-ton refrigeration (52.8- to 130.1-kW) range are currently manufactured in this type.

These units are available with gas-fired heat, electric heat, or hot-water/glycol heat and direct-expansion cooling coils with multiple reciprocating compressors and air-cooled condensers. When high indoor relative humidity (in humid weather and during part load) is a concern, a direct expansion cooling coil in the outside air stream can be provided with this type of unit.

4. *Variable air volume:* This is usually a high-velocity supply, low-velocity return duct distribution system. The unit consists of filters, cooling and heating coils, and supply and return fans. Return fans have been omitted on smaller systems.

Fans are variable-volume, centrifugal- or axial-flow type. Depending on fan size, the air volume can be varied by variable-frequency control or variable inlet vanes on smaller systems, or by variable blade pitch only on larger axial-flow fans.

The supply distribution main consists of a single duct with VAV boxes at the beginning of each zone duct. The ductwork leaving the VAV boxes to the diffusers is low-velocity.

The system is extremely flexible with respect to future modifications and has good temperature controls.

Care must be exercised in selecting the type of diffusers and controls. See discussion on VAV systems in Sec. 51.3.7.

The size of the main supply duct should be based on the maximum building peak cooling load. The branch mains on a floor should be sized on the maximum simultaneous interior and exterior exposure peak loads of the areas they serve.

For the commercial office, this system generally has the lowest energy usage and construction costs. However, there have been problems when VAV systems were used to air-condition laboratories and good-quality automatic temperature controls were not employed.

5. *Fan coil unit:* Each unit usually consists of a filter, combination heating and cooling coil, centrifugal fan, and supply and return grilles. Though not common, units are available with separate heating and cooling coils.

Although ceiling-mounted units are available, fan coil units are generally located at the floor against the exterior walls, preferably under the windows.

Since these units generally have no provision for ventilation air (that is, they recirculate 100 percent of the supply air), they are used in conjunction with single-duct, dual-duct, multizone, or variable-air-volume systems. The fan coil units are sized to handle the exterior (solar, transmission, and infiltration) cooling and heating load and the interior cooling load for the first 10 to 15 ft (3 to 4.6 m) from the exterior wall. The interior system will provide the ventilation air for the exterior zones. This combined system significantly reduces the size of the distribution ductwork and the associated construction cost, since the ducted system serves only the interior loads and ventilation air requirements. The system combined with a VAV interior system is used most often in modern offices and is among the lowest energy users.

Units are available that have provision for ventilation air. They are generally self-contained, packaged heat pumps with their own air-cooled direct-expansion compressor, cooling coil, and supplementary electric heat. They are predominantly used in schools, motels, and hotels. If there are extended periods during the heating and cooling seasons when the spaces served are not occupied, energy usage is reasonable. However, in areas where the ambient heating design temperature is 12°F (−11°C) or lower and there are 5000 (2780) degree-days or more, energy usage is generally high, since under these conditions the heating is mostly electric.

The self-contained heat-pump units are thermostatically controlled. The other unit capacities can be regulated by varying the water flowing through the coil with an automatic temperature-controlled water-regulating valve or by varying the fan speed. Though varying the fan speed requires constant flow through the coil, and the choice of pump size is therefore restricted and the possibility of saving pump energy by reducing the flow is eliminated, it is economical and is the method most often provided for these units.

6. *Induction unit:* This is a constant-volume, low- or high-velocity system. It consists of a centrally located unit that filters, cools, and dehumidifies the primary air and induction units located generally at the floor along the walls. Each induction unit consists of a primary air plenum (which is sound-attenuated), primary air nozzle, mixing chamber, heating coil, and return and discharge grilles.

The primary air is ducted to each induction unit. At each induction unit the primary air flow enters the primary air plenum and leaves through the primary air nozzle at high velocity, inducing return air from the space to flow into the mixing chamber and mix with the primary air. The mixed air leaves the unit and enters the conditioned space.

The primary air provides the ventilation air and cooling requirements of the conditioned spaces. The heating coil in the return air stream provides the heating requirements.

Though this system was popular before the energy crisis and provides good temperature control, it is seldom selected any more for new facilities because of its high energy use.

51.3.4 Chillers

Centrifugal. To minimize energy use, the following guidelines should be considered:

- For commercial and institutional applications, the number and size of the refrigeration units should be determined so that the number of units on line (operating) will have the lowest kilowatts per ton (kW/W) ratio—in the range of 75 to 25 percent of design load—since approximately 80 percent of the hours of operation will be in this load range. If units have a significantly lower kilowatts per ton ratio in the 75 to 50 percent of design load range, they should be selected since approximately 50 percent of the hours of operation will occur in this load range. See the general discussion of this in Sec. 51.3.3 for typical part-load operation.
- For industrial or other applications where the cooling load does not vary appreciably with the ambient weather conditions, the number and size of the refrigeration units should be chosen to produce the lowest kilowatts per ton (kW/W) over the duration of the cooling load.
- Select chilled-water supply temperatures at the highest possible temperature

that will maintain space design temperature and humidity under maximum load conditions.

- Select refrigerant compressors to operate at the highest suction pressure and the lowest head pressure possible and still maintain the required supply chilled-water temperature under maximum load conditions.
- Select refrigerant compressors that can maximize the energy reduction possible with lower condenser water-supply temperatures under part-load conditions.
- Provide automatic controls that can reset the supply chilled-water temperature to the highest level under part-load operation and still maintain space design temperature and humidity conditions.

For a discussion on heat recovery with double bundle condensers see Sec. 51.3.6.

Heat-transfer surfaces must be kept clean at all times with adequate water treatment and periodic cleaning.

Absorption. When waste heat [preferably steam around 12 lb/in^2 (0.8 bar)] is available and chilled water is required, absorption refrigeration units should seriously be considered to save energy and improve the overall plant efficiency.

However, when steam or hot water must be generated expressly for absorption units, the engineer must evaluate the following before selecting the type of refrigeration units:

- The water rate for a single-stage absorption unit for 12-lb/in^2 (0.8-bar) steam of about 18 to 20 lb/h of steam per ton of refrigeration (2.3 to 2.6 kg/kW), or its equivalent hot-water value, is not energy-efficient. Furthermore, the heat rejection to the cooling tower is about 200 percent greater than that of an electric-driven compressor unit for the same refrigeration capacity.
- The water rate for a two-stage absorption unit with 125- to 150-lb/in^2 (8.6- to 10.3-bar) steam entering the first stage is about 12 to 14 lb/h of steam per ton of refrigeration (1.5 to 1.8 kg/kW), which indicates a significant reduction in steam energy, or its equivalent high-temperature water at 355°F (179°C), for the same refrigeration capacity. However, the lithium bromide refrigerant solution used in absorption units is extremely corrosive at the elevated temperatures at which the first stage operates. Although manufacturers of two-stage units profess that corrosion will not be a problem if their water treatment requirements are strictly adhered to, it is the author's experience and position that corrosion and/or the potential corrosion-related problems are a major concern and repair expense for users of two-stage units.
- Guidelines for selecting the number and size of absorption units are similar to those noted under the heading Centrifugal, above.

Direct-Expansion Evaporators—Screw Compressors and Reciprocating Compressors

- Generally screw compressors are more economical above 100 tons (350 kW) of refrigeration, whereas reciprocating compressors are more economical below that capacity.
- The same criteria described under the heading Centrifugal should be used to select units with water-cooled condensers.
- If the condenser is *air-cooled*, the same criteria described under the heading

Centrifugal should be used, except for the automatic controls for the condenser fans. These should be able to lower the condensing temperature (head pressure) as the ambient dry-bulb temperature drops to the lowest recommended by the compressor manufacturer, yet maintain the required (compressor) suction pressure.

- If the condenser is *evaporative*, the same criteria described under Centrifugal should be used except for the automatic controls for the evaporative condenser fans. These should automatically control the spray pump and condenser fan, so as to lower the condensing temperature (head pressure) as the ambient dry- and/or wet-bulb temperature decreases to the lowest recommended by the compressor manufacturer, yet maintain the required (compressor) suction pressure.

Cooling Tower. For energy conservation, towers should be selected in conjunction with the refrigeration unit to produce the lowest kilowatt per ton of refrigeration (kW/W) ratio. To achieve this goal, the following guidelines should be considered:

- Induced-draft towers should be selected over forced-draft towers since they require significantly less fan horsepower (kW) for the same cooling requirement.
- Hyperbolic natural-draft cooling towers are without question the most energy-efficient. However, their minimum effective size is approximately 250,000 gal/min (15.8 m^3/s), which is far greater than the central refrigeration plant requirements we are concerned with in this book.
- Though it is possible to design a natural-draft tower (without mechanical fans) in the capacity range we would need, it would be inefficient and would need a large amount of space. However, if space is available, natural-draft cooling towers, as well as spray pounds, should be considered.
- If the project is located near a river, lake, or other large body of water, it should be considered as a source of condenser water before a mechanical-draft cooling tower is selected.
- Groundwater has been used for precooling and condenser water. However, requirements for recharging wells and restrictions on groundwater contamination generally make this source of condenser water uneconomical.
- The three major cooling tower parameters are:

  Ambient wet-bulb temperature: This temperature should be selected with care, since the wet-bulb temperature of the air entering the tower is the basis for the thermal design of any evaporative-type cooling tower.

  Range: This is the difference in temperature between hot water entering the tower [condenser water return (CWR)] and the cold water leaving the tower [condenser water supply (CWS)]. Of these two temperatures, the tower size is primarily affected by the CWS temperature.

  Approach: This is the difference between the cold-water temperature leaving the tower and the entering air wet-bulb temperature. The approach is important for two primary reasons: first, it sets the CWS temperature; the lower this temperature is, the lower the refrigeration unit kilowatts per ton of refrigeration (kW/W) ratio will be. Second, it fixes the size and efficiency of the cooling tower. Although increasing the tower efficiency will measurably decrease the approach, there are limits. In practice it is the tower size that is significantly increased to achieve the lower approach requirements. The closest approach that can be achieved is 5°F (2.8°C).

- It is generally more cost-effective to increase the tower size to obtain lower CWS than to increase the refrigeration unit kilowatt per ton (kW/W) ratio.
- Towers should be selected to minimize the drift and evaporation losses.
- Automatic temperature controls capable of resetting (lowering) the condenser water supply temperature as the ambient wet-bulb temperature drops should be provided.
- Tower fan motors should be two-speed to improve part-load efficiency.
- The heat-transfer surfaces must be kept clean at all times with adequate water treatment and periodic cleaning.

Several heat-recovery systems using cooling towers are discussed in Sec. 51.3.6.

51.3.5 Boilers

To minimize energy usage the following guidelines should be considered:

- For comfort heating, the number and size of the boilers should be determined so that the number of units on line (operating) will be close to their maximum efficiency point at part loads ranging from 75 to 25 percent of design load, since approximately 80 percent of the hours of operation will be in this load range. If significantly higher efficiencies can be obtained by selecting boilers operating in the 75 to 50 percent of design load range, then the boiler size and number should be determined at this load range, since they will be operating in this range approximately 50 percent of the time. See the general discussion in Sec. 51.3.3 for typical part-load application. For process boilers, the number and size of the boilers should be determined to maximize the plant efficiency.
- Boilers should be selected for lead-lag control with low fire rates to minimize the on and off cycling of the lead boiler.
- Boiler insulation type and thickness should be selected to minimize the cooldown and radiation heat loss. The burner flame shape and heat output must be selected to match the dimensions of the combustion chamber.
- The burner controls for multiple boilers should be capable of the following:

  Automatically cycling the boiler on and off (lead-lag control) and modulating its firing rate in accordance with load swings.

  Continually monitoring the flue gas for excess O_2 and CO content and excess temperature.

  Automatically adjusting the firing rate according to the operating parameters (flue gas O_2, CO, and temperature) and actual plant load for the highest obtainable combustion efficiency.
- The combustion air volume should be set at the lowest safe maintainable value that the boiler controls can operate at.
- The stack gas temperature should be as low as possible, approximately 50°F (10°C) above the lowest combustion-gas dew point.

- The breeching and stack should be designed to minimize the resistance to combustion gases at maximum load and still maintain sufficient draft to discharge the stack gases into the atmosphere safely under minimum load.
- The breeching should be insulated to reduce flue gas temperature loss. For guidance on insulation thickness see Insulation in Sec. 51.3.2.

For a discussion of waste heat boilers and heat recovery see Sec. 51.3.6.

51.3.6 Waste Heat and Heat Recovery

Energy that is discharged into the atmosphere is wasted. However, if this energy (or part of it) is recovered and reused, it is identified as "waste heat" or "recovered heat." For the purposes of this chapter, the terms are synonymous.

The following schemes have been used successfully to recover and reuse energy that would have been wasted:

Waste-Heat Boilers. These are economically attractive for most operations with a steam or hot-water requirement and flue gases from process equipment, incinerator, gas turbine, or engine exhausts at temperatures between 500 and 2500°F (260 and 1370°C). Large waste-heat steam boilers typically include evaporator, superheater, and economizer sections. Fire-tube boilers are available up to 60,000 lb/h of steam (nominally 17,500 kW). Water-tube boilers are available from 60,000 to 200,000 lb/h of steam (nominally 17,500 to 58,500 kW). In evaluating air-to-air or air-to-water heat-recovery units with dirty flue or exhaust gases, the potential fouling of the heat-transfer surfaces and the additional cleaning costs must be considered.

Economizers. These should be considered for all steam heating and process boilers having a capacity greater than 2000 lb/h of steam (nominally 590 kW) that are used more than 30 percent of the time.

Economizers are generally used to preheat the boiler feed water. Typically, economizers will:

- Reduce the flue gas temperature from the 400 to 800°F (200 to 425°C) range to the 100 to 300°F (40 to 150°C) range.
- Increase feed-water temperature by 70 to 100°F (21 to 38°C).
- Improve boiler efficiency 5 to 10 percent.

When selecting economizers, the engineer must make sure that the temperature of the flue gases leaving the economizer is above the lowest flue gas dewpoint temperature, to prevent excessive corrosion of the downstream breeching and stack.

The engineer should not overlook the possibility of preheating domestic hot water, process water, etc., by using economizers in the breeching of hot-water boilers.

Thermal Wheel. The thermal wheel normally handles clean and filtered air with temperatures ranging from ambient to 500°F (260°C) for commercial and institutional use. The wheels that recover sensible and latent energy operate at lower temperatures. High-temperature wheels can operate up to 1500°F (815°C).

A motor slowly rotates the wheel at 1 to 3 r/min between the supply and exhaust air streams, which are moving in opposite directions. Regeneration is continuous as energy is captured while the hotter air stream moves through narrow spaces (between the corrugated metal plates of the rotor) in the hot section, is stored briefly, and is transferred to the cooler air stream in the cold section as the wheel rotates through it.

Heat-recovery wheels typically recover 60 to 80 percent of the sensible heat difference between the upstream exhaust air and the outside air. Special heat-recovery wheels are manufactured with properties like those of a desiccant dryer. They will recover 60 to 80 percent of the total (sensible and latent) energy difference between the upstream exhaust air and the outside air.

Some codes do not permit the use of heat wheels where there is a possibility of short-circuiting of hazardous or toxic exhaust air into the outside air (ventilation or makeup air). With a purge section, the cross-contamination of exhaust air should not exceed 0.1 percent of the exhaust volume. Most manufacturers recommend that both air streams be filtered with 2-in-thick (50-mm) roughing filters.

During the cooling season, the thermal wheel can be used to precool the outside air whenever the exhaust air is cooler than the ambient air. Heat wheels can be used to transfer total energy from the outside air to the exhaust air whenever the total energy of the exhaust air is less than that of the outside air. This lowers the sensible and latent enthalpies of the incoming outside air.

Parallel-Plate Heat Exchangers. These are available in counterflow and crossflow air-to-air duct exchanger designs. Since this type of heat exchanger must be located where both supply and exhaust ducts can be connected to it, its primary service has been to preheat combustion air or oven makeup air with the boiler flue gas or hot oven exhaust.

The plates are generally dimpled to keep them separate, to provide air passages, and to create air turbulence to improve heat transfer. There are no moving parts; hot exhaust gases on one side transfer energy through the thin plate walls separating the air streams to the cooler supply on the other side.

Plate-type heat exchangers typically recover a maximum of 40 to 50 percent of the sensible heat difference between the upstream exhaust air and the outside air. Unlike thermal wheels, plate-type heat exchangers cannot recover sensible heat.

There is no contamination of outside air by exhaust air with plate heat exchangers since the air streams are physically separated.

Coil Runaround Cycle (Air-to-Water–Water-to-Air). This system is typically used to preheat outside ventilation air or makeup air during the heating season. It can also be used during the cooling season to precool the outside ventilation air or makeup air whenever the exhaust air is colder than the outside air.

The system consists of a heat-recovery coil located in the exhaust duct and a preheat coil located in the outside or makeup air duct, with a runaround piping system connecting the coils, complete with circulating pump, expansion tank, generally a glycol solution (for freeze protection), and automatic controls. With the proper selection of materials, this system can recover heat from high-temperature industrial exhaust systems. The distance between the heat-recovery coil and the preheat coil is limited only by the temperature loss and energy consumed by the hydraulic system.

The coil runaround cycle typically recovers 30 to 45 percent of the sensible heat difference between the upstream exhaust air and the outside air. The coil runaround cycle cannot recover latent heat.

Heat-Pipe Recovery System. In its simplest form, this system makes use of a self-contained refrigeration cycle. Thermal energy (warm air) is applied to either end of the pipe and vaporizes the refrigerant at that end. The refrigerant vapor then travels to the other end of the pipe where cooler air is applied, condensing the refrigerant gas and absorbing the latent heat of condensation. The condensed liquid refrigerant then flows back to the evaporator section (hot side). As long as there is a temperature difference between the evaporation (hot) end and the condensing (cold) end, the transfer of energy will be continuous.

Although a heat-pipe recovery unit from the outside looks like an ordinary plate-fin water or steam coil, its major difference is the absence of return bends or headers. Each heat pipe is independent; there are no connections between them.

The heat pipe typically recovers about 50 to 80 percent of the sensible heat difference between the upstream exhaust air and the outside air.

Since the heat-pipe recovery unit will transfer recovered heat in any direction, it can also be used to precool the outside air during the cooling season, when the exhaust air is cooler than the ambient air, by reversing the slope of the heat pipes. There is no contamination of outside air with exhaust air, since the air streams are physically separate.

Generally, this type of heat-recovery unit consists of multiple rows of heat pipes with typically 14 fins/in (0.55 fins/mm), adding about 1 in of water (249 Pa) to both the exhaust air and outside air duct system static pressure requirements. The additional friction must be accounted for in the fan selection and in the energy analysis.

Each heat-pipe recovery unit must be capable of changing the slope of the heat pipe if either of the following is required:

- Preheating the outside air (ventilation or makeup air) during the heating season and precooling the outside air during the cooling season.
- Preventing frost from forming on the leaving side of the exhaust-air heat pipes when the ambient air is significantly below freezing.

The heat-recovery capacity can be regulated by changing the slope (orientation) of the pipes. Lowering the evaporator (hot) section below the horizontal will increase the gravity return flow of the condensed refrigerant to the evaporator section and thereby increase the amount of heat recovered. The ability to change the slope is required to prevent frost formation on the downstream side of the exhaust-duct heat-pipe external fins in cold climates and to change over from preheating the outside air (heating mode) to precooling the outside air (cooling mode).

If the exhaust air is grease-laden, it may be desirable to provide an automatic spray-type cleaning system for the exhaust heat pipe's external, closely spaced fins. The cleaning solution and spray pattern should be selected to effectively remove the grease buildup.

Heat Recovery by Recirculating Warm Stratified Air. Where the ceiling height is greater than 12 ft (3.5 m) in areas that are not heated with infrared heaters or radiant floor panels, the cost-effectiveness of recirculating the warm stratified air at the ceiling or under the roof to the occupied zone should be evaluated.

There are two principal recirculating methods, both requiring the installation of propeller fans uniformly spaced throughout the area. One method uses ducts to deliver the warm air to the occupied zone, the other does not.

The ducted system consists of one or more ducts about 10 in (254 mm) in diameter with a low-volume fan mounted at one end. Each duct is mounted vertically, extending from about 6 to 12 in (152 to 395 mm) above the floor to the zone of warm air. The fans discharge into their respective ducts, circulating the warm air to the floor. These fans are quiet, relatively low volume units [250 to 400 ft^3/min (0.12 to 0.19 m^3/s)]. The fans can be controlled manually or automatically by thermostats at the ceiling or under the roof. The thermostats are set to stop the fans when the temperature drops to 75°F (24°C) and start the fans when the temperature rises above 85°F (29°C).

The nonducted system consists of one or more vertically discharging high-volume, variable-speed propeller fans, mounted at the ceiling or under the roof. The fans recirculate the warm air vertically down into the occupied zone. The fans are relatively quiet, low-speed (150 to 350 r/min), high-volume [10,000 to 18,000 ft^3/min (4.7 to 8.5 m^3/s)] units and have three to six fan blades with diameters usually ranging from about 36 to 54 in (0.915 to 1.37 m). The fans are usually manually controlled from wall-mounted variable-speed controllers in the occupied zone. The fans are started, speed-adjusted, and stopped from their respective controllers as required to maintain the design space temperature. Automatic starting and stopping ability, although not usually furnished with the fans, can be provided by installing thermostats as described previously for the ducted system; however, speed selection is still manual.

Heat Pump. On small projects, when seasonal heating and cooling are required, the heat pump should be evaluated for energy conservation.

Water source units are preferred for heating the perimeter spaces of buildings which have interior spaces that must be cooled concurrently. Under these conditions, the heat rejected from the interior space can be piped to heat the exterior space.

Air-to-air heat pumps are generally not cost-effective from an annual energy standpoint for locations with heating design ambient temperatures (97½ percent basis) equal to or less than 12°F (−11°C) and equal to or greater than 5000 (2780) heating degree-days, 65°F (18°C) basis. However, this may not be true for those locations in which 30 percent or more of the total annual heating hours below 65°F (18.3°C) occur from April through October in the northern hemisphere and November through April in the southern hemisphere.

Condenser-Water Heat Recovery. Though there is the potential to recover the heat in the condenser water leaving a refrigeration unit, it is generally not cost-effective for the following reasons:

- The possibility of reclaiming heat from the air-conditioning refrigeration unit condenser normally is limited to the cooling season.
- The relatively low temperature of the hottest condenser water is 95 to 103°F (35 to 39.5°C), and the corresponding differential between the condenser water supply and return temperature is 10 to 18°F (5.5 to 10°C).
- The temperature of the potable water to the domestic hot-water heater during the summer will normally range from 50 to 70°F (10 to 21°C), depending on the source. If the source is surface water, it will be around 70°F (21°C). If it is from a well, it will be closer to 50°F (10°C), depending on the aquifer where the well water is taken from. Though not the norm, there are geographic areas where the temperature of water obtained from wells exceeds 100°F (38°C). When this occurs, the portion of hot well water that is to be used as cold water can be a

source of energy to preheat the domestic cold-water supply to the domestic hot-water heater, since the domestic cold-water temperature should be around 60°F (15.5°C) or lower.

- The heat exchanger will not be economical because of the low mean effective temperature available and the close approach.
- The demand for hot water is not constant throughout the day. For instance, in a typical office, it peaks at noon and just before the close of business. Typically, the heat from the condenser will not be available for the peak occurring at the close of business, since the refrigeration unit would normally be off at that time. The shutdown period is discussed later, in Sec. 51.5.

Using condenser water for the reheat coils is generally not cost-effective for the following reasons:

- The possibility of reclaiming heat from the air-conditioning refrigeration unit condensers normally is limited to the cooling season. This will require another source of hot water for the reheat coils when the air-conditioning system is operating on the economy cycle.
- To minimize the fouling of the reheat coils by condenser water coming directly in, since the condenser water flows in an open circuit designed to reject heat in a cooling tower, a plate heat exchanger should be provided. This will isolate the circuit flowing through the cooling tower and the circuit flowing through the boilers and will minimize the water treatment requirements. It will, however, increase the cost since it requires a secondary pumping circuit in addition to the heat exchanger.
- The relatively low temperature available from the condenser water without increasing the chiller kW/ton (kW/W) ratio is 95 to 103°F (35 to 39.5°C), and the corresponding 10 to 18°F (5.5 to 10°C) differential between the condenser-water supply and return temperatures will require reheat coils with several rows to transfer the heat. This will not only increase the system static pressure and fan air power requirements, but will also increase the required pipe sizes and flow requirements, which will increase the pumping power.

Condenser-Water Off-Season Cooling. Facilities that have a year-round cooling requirement that cannot be met with 100 percent outside air (economizer cycle) and have a chilled-water system with a water-cooled condenser (with cooling towers or a spray pound) are prime candidates for this system. In it, a plate heat exchanger is installed in parallel with the chiller. The condenser water supply is connected to one side of the plate heat exchanger and the chilled-water supply is connected to the other.

When this cooling mode is activated, the refrigeration chillers will be out of service (valved off). The separate chilled-water and condenser-water circuits in the plate exchanger will prevent fouling of the chilled-water cooling coils with dirt, mud, organic slime, and algae in the condenser water.

The tower must be winterized.

There will be a loss of 2 to 3°F (1.1 to 1.7°C); i.e., the chilled water leaving the heat exchanger will be 2 to 3°F (1.1 to 1.7°C) higher than the condenser water entering it.

The climatic weather data for the geographic area of the site must be studied to determine the period of the year when this system will satisfy the chilled-water or process-cooling-water requirements.

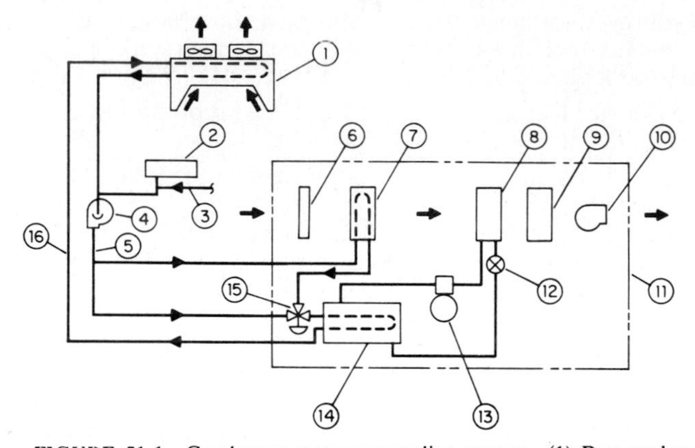

FIGURE 51.1 Condenser water precooling system. (1) Dry cooler; (2) expansion tank; (3) glycol solution makeup; (4) glycol solution circulation pump; (5) condenser water supply (glycol solution); (6) filter; (7) precooling coil; (8) direct-expansion cooling coil; (9) heating coil; (10) supply fan; (11) air-conditioning unit; (12) refrigeration expansion valve; (13) refrigeration compressor; (14) refrigeration condenser; (15) three-way mixing valve; (16) condenser water return (glycol solution).

Condenser Water/Precooling Recovery. Direct expansion (DX) air-conditioning units with water-cooled condensers that operate 24 h a day, 365 days a year are prime candidates for precooling the return air with condenser water.

In addition to the standard DX air-conditioning unit supply fan, filter, heating coil, DX cooling coil, compressor, expansion valve, and water-cooled condenser, this system contains a precooling coil between the filter and the DX cooling coil and an automatic three-way mixing valve to modulate the flow of condenser water through the precooling coil. A dry cooler is used instead of a cooling tower. This closed-circuit condenser water system is complete with variable-capacity dry cooler fans, circulating pump, expansion tank, and provision for makeup water. The condenser water is a glycol solution to prevent freeze-up in cold weather (see Fig. 51.1).

The typical annual energy savings of this system over a standard DX system operating 24 h a day, 365 days a year, accounting for the reduction in heat-transfer efficiency because of the glycol solution, are listed below for representative cities in the United States.

City	Approximate energy savings, %
Atlanta, GA	12
St. Louis, MO, and Columbus, OH	20
Portland, OR, and New York, NY	22
Chicago, IL, Detroit, MI, and Syracuse, NY	25
Minneapolis, MN, and Milwaukee, WI	30

The savings are also applicable for condenser-water precooling recovery systems in other geographic locations having operating hours, ambient weather conditions, and elevations similar to the cities listed.

The following is a typical sequence of operation for this system (see Fig. 51.1):

- When the ambient air is 65°F (18.3°C) and above, the three-way valve will modulate fully closed to the precooling coil (no flow through the precooling coil). In this mode, the unit will operate like a standard DX system. However, its efficiency will be less because of the glycol solution's adverse effect on heat transfer and pumping head.
- When the ambient air is below 65°F (18.3°C) but above 35°F (1.7°C), the three-way valve will modulate fully open to the precooling coil. (The condenser water will flow through the equipment in series, first through the precooling coil and then through the refrigeration condenser.) The automatic temperature controls will cycle the refrigeration compressor through its steps of unloading as required to maintain the design room space conditions. As the ambient air drops from 65°F (18.3°C) to 35°F (1.7°C), the cooling load on the DX system will be less and less. As a rule of thumb, the reduction in compressor operating time vs. ambient air temperature is as follows:

Ambient air temperature, °F (°C)	Reduction in compressor operation, %
65 (18.3)	0
60 (15.5)	15
55 (12.8)	30
50 (10)	45
45 (7.2)	60
40 (4.4)	80
35 (1.7)	100 (Compressor off)

- When the ambient air is 35°F (1.7°C) and below, the refrigerant compressor will be off and the three-way valve will modulate the flow through the precooling coil as required to maintain the design room space conditions. When the condenser water supply temperature drops to a preset low temperature, an aquastat through a controller will automatically cycle the dry cooler fans as required to maintain the minimum set condenser water temperature.

Water-to-Water–Water-to-Air Heat Recovery. When there is a requirement for chilled water and/or process cooling water and a simultaneous requirement to heat makeup and/or ventilation air, this system should be considered. With this system, the chilled water or process cooling water return is pumped through a closed circuit to a plate heat exchanger (precooled) before it is returned to the chiller. The heat extracted from the chilled water or process cooling water is pumped through a preheat coil in the makeup air or ventilation air duct, then back to the plate heat exchanger. See Fig. 51.2 for the schematic flow diagram of this system. The control sequence (also shown in Fig. 51.2) is as follows:

- When ambient air drops below 40°F (4.5°C), thermostat T1, through controller C-1, will automatically start pumps P-1 and P-2.
- Thermostat T2 through controller C-1 will modulate three-way mixing valve V-1 as required to maintain its set point.
- When the ambient temperature rises above 40°F (4.5°C), thermostat T1,

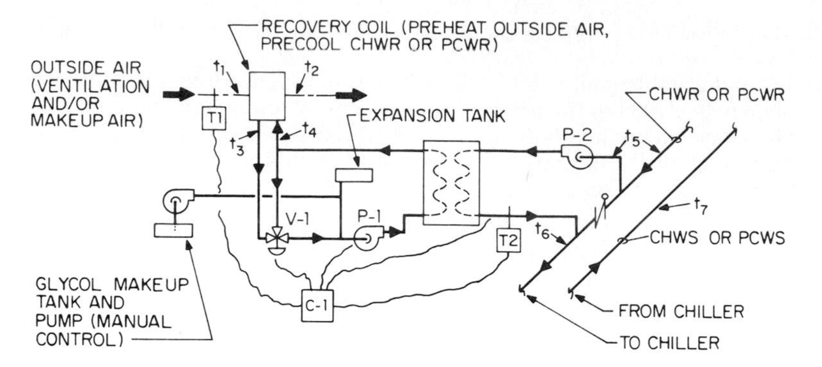

FIGURE 51.2 Water-to-water–water-to-air heat-recovery system—schematic flow diagram. Outside air temperature, t_1; preheated outside air temperature, t_2; glycol cooling supply temperature, t_3; glycol cooling return temperature, t_4; chilled-water supply, CHWS; chilled-water return, CHWR; process cooling water supply, PCWS; process cooling water return, PCWR; CHWR or PCWR temperature, t_5; precooled CHWR or PCWR temperature, t_6; CHWS or PCWS, t_7; controller, C-1; three-way mixing valve, C-1. P-1 = heat-recovery circulating pump; P-2 = CHWR or PCWR energy-recovery circulating pump; T-1 = duct thermostat; T-2 = leaving heat exchanger CHWR or PCWR acquastat. Temperatures: $t_1 < t_2$, at least 20°F (11°C) less; $t_3 < t_4$, approximately 10°F (6°C) less; $t_6 < t_7$, t_6 will equal t_7 when the ambient air temperature is about 35°F (19.5°C) below the CHWS or PCWS temperature.

through controller C-1, will stop pumps P-1 and P-2 and deactivate this control system.

- *Note:* The installation of the recovery coil will increase the system static pressure at which the supply fan must operate. Therefore, all costs to modify the fan drive and motor or even to replace the fan must not be overlooked.

Transformer Heat Recovery. Indoor electric transformers serving industrial plants or office buildings with at least 500,000 ft^2 (46,450 m^2) of floor space are a potential source of heat recovery. This is especially true if the transformers are centrally located.

The magnitude of this source of potential heat is approximately 2 percent of the kilowatt load on the transformers. The actual heat loss must be obtained from the transformer manufacturer.

The following items should be checked before this potential energy source is seriously considered:

- Approval from the utility company may be required. (Not all utility companies permit transformer heat recovery.)
- The load on the transformers must be high enough and occur simultaneously with the heating load.
- The heat-recovery system must not introduce pipes and coils with fluid within the transformer vaults.

Thermal-Storage Heat Recovery. When recoverable energy will not be available at the time it is needed, temporary-heat-storage equipment must be provided. The additional cost to provide, operate, and maintain the required thermal storage facilities (including the cost, if any, of the loss of rentable space that the storage tanks occupy) must be included in the life-cycle costs of this system.

Thermal storage can yield substantial ownership and operating savings, especially in geographic locations which have electric utility time-of-day rates, with high on-peak demand charges. These conditions usually provide adequate economic incentive for installing thermal-storage systems.

The cost-effectiveness of thermal storage should also be considered for industrial batch-type processes where thermal energy would otherwise be wasted to the atmosphere. Heat is recovered and stored during the processing of one batch and, at a later time, used to process another batch.

Though most commercial thermal-storage facilities are for chilled water, hot-water and steam storage are also feasible.

There are two main reasons why storing heat energy for commercial facilities is usually not cost-effective: (1) the costs associated with minimizing the temperature loss through the storage tanks (of the stored heating energy to the surrounding storage facilities) are high and (2) the energy used to produce heat is generally from fossil fuels (oil and coal), which are not affected by time-of-day rates and high on-peak demand charges.

The following rules of thumb are offered for preliminary guidance:

- Long storage times (greater than 24 h) between storing energy and using energy are generally not cost-effective.
- The higher the peak energy demand and the shorter the duration of the peak demand, the more cost-effective thermal storage will be.
- Chilled-water storage tanks generally range in capacity from 350 to 400 gal/ton of refrigeration (0.385 to 0.44 m^3/kW).
- Ice-bank stored cooling systems generally range from 35 to 40 gal/ton of refrigeration (0.0385 to 0.044 m^3/kW).
- It is important to design chilled-water storage tanks for maximum stratification. That is, the water returned to a tank should not mix directly with the chilled water already in the tank. This maintains design stored-chilled-water temperature as long as possible and ensures that water is drawn from the coldest portion.

Hydronic Systems. From an energy point of view, the engineer should design the piping distribution system for system dynamic heads so that the pump uses less energy. Of course, the savings in energy costs must be greater than the construction costs of the larger pipe sizes needed to ensure the lower heads.

For a discussion of commercial and industrial system pipe sizing procedures, the reader is referred to Chap. 3.

To maximize the heat-transfer efficiencies of boilers, chillers, coils, and heat exchangers and to minimize corrosion and fouling of the distribution system and components, the engineer should provide the most cost-effective water treatment that the project budget will permit. The reader is referred to Chap. 53 for a discussion of water treatment.

Insulation for energy conservation was previously discussed under Insulation in Sec. 51.3.2.

Steam Systems. From an energy point of view, the engineer should select the lowest steam pressure that will provide the required energy. On projects that require high- and low-pressure steam, the engineer should evaluate the cost-effectiveness of using steam-turbine-driven pumps and/or fans instead of pressure-reducing valves (PRVs) to reduce the high pressure to low. The energy for the steam-driven pumps and/or fans can be considered free, since it would not be available if PRVs were used to reduce the steam pressure. When the high-pressure steam is 125 lb/in^2 (8.5 bar) or greater and the low pressure is 15 lb/in^2 (1 bar) or less, this can be an extremely effective way of saving energy.

In the design of steam distribution systems, the location of steam traps is important not only from a functional point of view but also from that of operations and maintenance. Adequate access and valving is required at all trap locations for proper maintenance of stream traps to minimize the waste of energy. In addition, all critical traps should have valved bypasses so that the trap can be isolated and serviced at any time without interrupting steam flow.

For a discussion of commercial and industrial system pipe sizing procedures, the reader is referred to Chap. 3.

Insulation for energy conservation is discussed under Insulation in Sec. 5.3.2.

To maximize the heat-transfer efficiencies of boilers, coils, and heat exchangers and to minimize corrosion and fouling of the distribution system and equipment, the engineer should provide the most cost-effective water treatment that the project budget will permit. The reader is referred to Chap. 53 for a discussion of water treatment.

All steam radiators and finned pipe convectors should be controlled with self-contained automatic radiator control valves to save energy.

51.3.7 Automatic Temperature Controls

In this section, the design and functional requirements of an energy-efficient automatic temperature control system are described. In Sec. 51.5, the function of an automatic temperature control system as part of an energy management system in a building or complex is discussed.

Facilities with centrally located automatic temperature controls should be designed as an integral part of an energy management system, or, at the very least, they should have the following minimum controls and station functions that could be transmitted to an energy management system in the future:

1. Start and stop, turn on and off, open and close equipment
2. Alarms
   - *a*. Abnormal equipment status
   - *b*. Safety
   - *c*. High or low level, temperature, pressure, etc.
3. Analog monitoring
   - *a*. Temperature
   - *b*. Energy consumption (all types)
   - *c*. Pressure
4. Reset set points of local control systems
5. Change mode from warmup to cooldown, day to night, enthalpy control on and off, smoke purge on and off, etc., plus alternative modes
6. Computer programming for calculations and table lookup functions for maintenance, operation monitoring, energy management, and system use optimization
   - *a*. Fan system start and stop (clock control)
   - *b*. Enthalpy economizer control
   - *c*. Heating and cooling plant optimization and nonautomatic plant operation with centralized manual control

d. Load shedding based on demand monitoring
e. Logging and profiling

For spaces that do not require constant temperature and humidity control year-round, the optimum energy-efficient automatic temperature control system will automatically maintain design environmental conditions except if the space is unoccupied or the process is not operating, when it will maintain minimum heating with no cooling.

Though allowance must be made for the geographic location of the facility and the type of system, the following rules of thumb generally hold:

- Each degree Fahrenheit (1.8°C) that the space temperature set point can be lowered during the heating season represents a saving in energy of 3 to 5 percent.
- Each degree Fahrenheit (1.8°C) that the set point can be raised during the cooling season represents a saving in energy of 5 to 10 percent.

To design an energy-saving control system, the following controls and/or operating procedures should be used, as applicable to a specific control system:

1. Startup and Shutdown time periods should be optimized. Since substantial energy can be saved annually by minimizing the startup time period and maximizing the shutdown time period, optimization of these time periods is important.

The optimum startup or shutdown point is when the space temperature at the end of these time periods is a few degrees above the design space temperature during the cooling mode or a few degrees below the design space temperature during the heating mode. Under these conditions, one can be sure that the startup or shutdown time cannot be decreased without seriously undercooling or underheating (depending on the mode of operation), and the maximum energy will be saved.

If the space temperature at the end of the startup or shutdown periods is not a few degrees higher than the design space temperature during the cooling mode or a few degrees lower than the design space temperature during the heating mode, then the startup and/or shutdown time should be adjusted as follows:

Cooling mode

- *Startup time:* If the space temperature at the end of this period is lower than the design cooling temperature, the cooldown period is too long and should be shortened. On the other hand, if the space temperature at the end of this period is 5°F (2.8°C) or more higher than the design cooling temperature, the cooldown period is too short and should be lengthened.
- *Shutdown time:* If the space temperature at the end of this period is at or lower than the design cooling temperature, the warmup period is too short and should be lengthened. On the other hand, if the space temperature at the end of this period is 5°F (2.8°C) or more higher than the design cooling temperature, the warmup period is too long and should be shortened.

Heating mode

- *Startup time:* If the space temperature at the end of this period is at or higher than the design heating temperature, the warmup period is too long and should be shortened. On the other hand, if the space temperature at the end of this period is 5°F (2.8°C) or more lower than the design heating temperature, the warmup period is too short and should be lengthened.

- *Shutdown time:* If the space temperature at the end of this period is at or higher than the design heating temperature, the cooldown period is too short and should be lengthened. On the other hand, if the space temperature is 5°F (2.8°C) or more lower than the design heating temperature, the cooldown period is too long and should be shortened.

There are computer programs available to determine the optimum time, taking into account variables such as ambient temperature and enthalpy (cooling mode only); wind speed and direction (heating mode primarily); design space temperature and humidity (cooling mode only); U values and surface areas of exterior walls, partitions, glass, skylights, floors, ceiling, and roof; lights (normally off during cooling startup mode); power or process loads (normally off during cooling startup mode); and people (usually negligible during cooling startup). However, it is prudent to verify their accuracy by using a manual trial-and-error solution.

The computer program optimum startup and shutdown time periods should be checked manually for at least the following seasons and load conditions.

In the cooling season:

Design cooling day

Average cooling day

Minimum cooling day (with mechanical cooling operations)

In the heating season:

Design heating day

Average heating day

Minimum heating day

In intermediate seasons (based on local weather statistics):

Warmest day

Average day

Coolest day

The optimum startup and shutdown time periods for the above-noted conditions can be approximated by:

- Selecting a unit of time based on computer optimum data, past experience, or current operating practices in an existing, similar building.
- Recording the latest local weather forecast, as close to the starting time as possible, with respect to temperature, humidity, wind direction, wind speed, barometric pressure, barometer rise or fall, and precipitation for the next 12 to 24 h.
- Recording the time the startup and shutdown time periods begin and end.
- Recording the space temperature conditions at the end of the startup and shutdown time periods.

The space temperature conditions should be compared with those in the above discussion on startup and shutdown time periods to determine if the optimum time period has been reached.

This trial-and-error procedure should be repeated until the optimum startup and shutdown time periods are achieved for the three load conditions, i.e., cooling, heating, and intermediate seasons.

Though this procedure takes at least a year to determine the seasonal effect on the optimum startup and shutdown time periods, the potential savings in energy over the life of the equipment will more than justify the effort.

Once the optimum times are established, not only can the computer program be verified but facilities that do not have a computer optimization program can be manually controlled. With the optimum startup and shutdown times established, operators of facilities without computer optimization programs can, depending on the seasonal mode of operation, select the appropriate startup and shutdown time periods by listening to the local weather report. If the facility has an energy management system, the various times can be programmed by mode of operation into the system, including local weather data recording.

2. Control should be calibrated annually in order to maintain optimum performance and reliability.

3. The space air-conditioning thermostat during the occupied period should be set at 78°F (25.5°C) dry-bulb for comfort cooling and at the highest temperature at which the process and/or equipment can operate.

4. The space heating thermostat during the occupied period should be set at 68°F (20°C) dry-bulb for comfort heating and at the lowest temperature that the process and/or equipment can operate.

5. Thermostats should have night (unoccupied) reset function.

6. Thermostats should be the type that unauthorized personnel cannot reset.

7. The perimeter radiation hot-water supply temperature set points should be set as low as possible, in accordance with ambient air temperatures, to maintain the space design temperature during the occupied period and reset to maintain the night setback temperature during the unoccupied period.

8. The thermostat control in the hot-water supply to the heating coils should be set as low as possible to maintain the design space conditions during the occupied period and reset to maintain the night setback temperature during the unoccupied period.

9. The thermostat setting and the control system should be selected to prevent the perimeter system from bucking the interior system in the heating or cooling mode.

10. All air-handling and heating and ventilating unit heating-coil controls should include night setback controls that close outside air dampers and reset the thermostat downward when the facility is unoccupied. The night setback temperature in the unoccupied area should be at least 10°F (5.5°C) lower than the normal (occupied) space temperature. The maximum setback should maintain at least 40°F (4.5°C) in the space.

11. During the cooling season and at other times when there is no freeze-up potential, all heating, ventilating, and air-conditioning units should be provided with controls that will automatically stop the fans, close the heating and/or cooling control valves, and close the outside air dampers for spaces that are not occupied.

12. All exhaust fans that do not operate continuously should be provided with an automatic motor-operated damper in the ductwork at the exterior wall that will automatically close when the fan is not running.

13. Automatic modulating two-way control valves should be used to control heating and cooling coils in conjunction with variable-speed circulating pumps when the cost of the energy saved for the estimated service life of the system is equal to or greater than the cost to install and maintain the variable-speed control.

14. An economy cycle should be provided on all air-conditioning systems, except those using room fan coil units, in geographic locations where the winter design temperature is 35°F (2°C) or less (on a 97.5 percent basis). The system should be designed so that 100 percent outside air may be used, when the outside air temperature is sufficiently low [approximately 65°F (18°C)], to provide all the cooling needed or to reduce the load on the air-conditioning refrigeration equipment when enthalpy control is provided. An economy cycle in areas where the winter design temperature is above 35°F (2°C) may not be cost-effective.

15. Enthalpy control should be provided on all air-conditioning systems that are equipped with the economy cycle. This control should be capable of sensing the enthalpy of the return air and outside air streams and modulate in sequence the outside air, relief air, and return air dampers from minimum to 100 percent outside air as needed to minimize the load on the refrigeration compressor.

16. From an energy standpoint, when return fans are used with VAV systems, the controls selected to vary the supply and return fans' capacity must be accurate and reliable to assure that the design ft^3/min (m^3/s) differential between the supply and return fans is maintained over the total fan operating range. If the return fan exhausts more air than required, the difference between the air supplied to a space and the air exhausted from it will be made up by infiltration. This undesirable condition will increase the heating and cooling energy requirements.

17. When dual-duct or multizone distribution (including multizone units with individual zone heating and cooling coils) is selected for commercial office air conditioning, energy can be saved by providing controls that will automatically close the hot deck and/or zone heating coil valves during the cooling mode. The temperature of the hot duct will be adequate for the reheat requirements for comfort air conditioning, even though it will be equal to the mixed air temperature plus the temperature rise from heat added by the supply fan (which is minor).

18. In dual-duct systems, separate air-handling units for each exposure can virtually eliminate the energy waste due to the reheat effect and should be included when they are cost-effective. Controls must be provided that will automatically select the highest cold-duct temperature and the lowest hot-duct temperature required by any zone, rather than the extreme temperatures that might be required when one unit serves all five zones (north, south, east, west, and interior).

19. Chillers should be provided with controls capable of automatically resetting the supply chilled-water temperature higher as the cooling load decreases, while maintaining the space design temperature.

20. Cooling tower controls should have the following energy-saving features:

- Automatically supplying the lowest condenser water temperature (from the tower) that the actual ambient wet-bulb temperature will permit, down to the limit set by the refrigeration equipment manufacturer.
- Automatically varying the tower fan capacity on a signal from the control thermostat in the condenser water supply line.

21. During the cooling season, in some geographic locations, refrigeration energy can be saved by flushing the building with 100 percent cool, dry outside air at night when the building is unoccupied.

22. On small systems, a time clock that can activate the control circuits to start and stop equipment, reset temperatures, etc. will generally be cost-effective.

23. In domestic hot-water systems, flow restrictors should be provided on all lavatories, bathtubs, and shower fixtures to conserve energy.

The design domestic hot-water supply temperature for lavatories, bathtubs, showers, slop sinks, and service sinks for offices, dormitories, hotels, etc. should not be greater than 110°F (43°C), preferably 105°F (40.5°C). Where higher temperatures are required for kitchens and laundries, local booster heaters should be provided to increase the 105° to 110°F (40.5° to 43°C) domestic hot-water temperature range to 140° to 180°F (60° to 82°C) range.

When it is cost-effective to design a centrally generated domestic hot-water and recirculating system for facilities that are not occupied 24 h a day, the recirculating pump should automatically shut off when the facility is unoccupied.

24. In compressed-air systems, energy can be saved by selecting the lowest compressed-air volume and discharge pressure. Therefore the engineer should review the manufacturer's compressed-air requirements (volume and pressure), minutes of continuous use, and frequency of use for each piece of equipment using compressed air before establishing the compressed-air system design plant capacity, discharge pressure, and system diversity.

Sometimes the central compressed-air system supply pressure is determined by equipment in one or two areas, whose compressed-air volume requirement is a small percentage of the total plant capacity. In such cases, the cost-effectiveness of lowering the central plant discharge pressure and providing local air compressors sized for the specific higher-pressure requirement of equipment in these areas should be evaluated.

When a central compressed-air system is proposed, the cost-effectiveness of reclaiming the heat rejected from the intercoolers, aftercoolers, and ventilation air should be evaluated. This is especially true when large centrifugal and screw air compressors are proposed. To obtain the maximum compressor efficiency (lowest energy consumption), the intake air to the compressor must be unrestricted, with the intake at the location where the coldest and cleanest air can be obtained. It is a good policy to confirm the maximum intake air temperature, ft^3/min (L/s), and pressure in inches of water (Pa) at the intake connection of the compressor with the compressor manufacturer before making the final selection.

25. To minimize the energy use for lighting and power, the following should be considered:

- The general illumination level should be at the minimum recommended by good engineering standards, with local task lighting where required.
- Enough switches should be provided to permit leaving the lights on only in areas where personnel are working (after the normal working day etc.) and shutting off all other fixtures except for security requirements.
- When fluorescent fixtures are used, they should be the energy-efficient type with energy-saving ballasts.
- When color identification is not critical, high-pressure sodium-vapor fixtures may be a good choice.

- When color identification is critical, metal halide fixtures may be a good choice.
- Exterior lighting (building, parking lot, advertising, etc.) should be automatically controlled by timers and photocells.
- A maintenance program for the lighting fixtures should be mandatory.
- Whenever possible, only high-efficiency and high-power-factor electric motors should be used.
- Electric motors should not be oversized. Oversized motors generally operate at a lower power factor and efficiency.
- Evaluate the cost-effectiveness of providing power-factor-correction equipment.
- Evaluate the cost-effectiveness of providing demand-limiting equipment.
- When the architectural features of the building permit, the benefits of daylighting should be integrated with artificial lighting.

26. Though solar designs do not reduce the building and system energy requirements, they do replace energy from nonrenewable sources (oil and coal) with renewable energy, thereby saving nonrenewable energy.

The cost-effectiveness of solar heating and cooling should be evaluated for projects in geographic areas where the hours of sunshine coincidental with clear sky (little or no cloud cover) are high.

Active solar heating with flat-plate collectors has been used successfully for heating and preheating domestic hot water, space heating with hot water, heating thermal-storage tanks, and preheating process water.

Active solar cooling with concentrating or evacuated-tube collectors to obtain hot-water or steam temperatures high enough for absorption chillers has also been used. In conjunction with low-pressure steam or hot water, this type of solar-collector system can be used year-round.

In geographic areas that have favorable conditions for passive solar energy systems or features, the following have been most successful in residences:

- Orienting the building to maximize the heat gain (solar radiation) in winter with south-facing windows in the northern hemisphere and north-facing windows in the southern hemisphere.
- In the hotter climates, using solar control devices such as overhangs, louvers, and vertical fins to control the amount of direct solar radiation that windows and other openings receive to reduce the air-conditioning load.

51.4 LIFE-CYCLE COSTING*

51.4.1 General

The purpose of this section is to provide a guide to life-cycle cost (LCC) techniques for evaluating building designs for energy conservation. It is important

*This section was edited, with permission, from H. E. Marshall and R. T. Ruegg, "Life-Cycle Costing Guide for Energy Conservation in Buildings," chap. 9 in D. Watson (ed.), *Energy Conservation in Buildings*, McGraw-Hill, New York, NY, 1979.

that the cumulative effects of all direct and indirect energy conservation options are accounted for in the LCC analysis.

Life-cycle cost analysis is a variation of benefit-cost analysis, a technique for evaluating programs or investments by comparing all present and future expected benefits with all present and future costs. To be worthwhile economically, the long-run benefits or cost savings produced by an investment must exceed the long-run costs. (More extensive treatments of cost-benefit and LCC analysis are in Refs. 6 to 8.)

LCC analysis, as applied to energy conservation in buildings, is the evaluation of the net effect over time of reducing fuel costs by purchasing, installing, maintaining, operating, repairing, and replacing energy-conserving features. The results of LCC analysis may be expressed as (1) the total of conservation investment and energy consumption costs, (2) the net savings from the investment in energy conservation, or (3) the ratio of savings to costs. The choice will depend in part on the preference of the analyst and in part on the nature of the investment problem. The net savings from energy conservation are computed from

$$S = E - (A + M + R) \tag{51.2}$$

where S = net savings (or losses) from energy conservation
E = energy cost savings (benefits)
A = acquisition and installation costs
M = maintenance and operating costs
R = repair and replacement costs

A positive value for S indicates that the energy-conserving feature results in net savings and is, therefore, economically efficient; a negative value indicates that it results in net losses and is, therefore, uneconomical.

LCC analysis may be used to address two types of economic efficiency choices: (1) how much of a single energy conservation feature to use (if at all) and (2) how much of each of several energy conservation features to use in combination. By comparing the net life-cycle effects of successively increasing amounts of a given energy conservation feature, it is possible to determine which level of investment in this feature is most economical. The optimal combination of energy conservation features can be determined by substituting among alternatives until each is being used to the level at which its additional contribution to energy cost reduction per extra dollar spent is just equal to that for all the other alternatives. (For a discussion of the determination of the optimal input combinations to minimize the cost of producing a given output or maximize the output for a given cost, see Ref. 9.)

51.4.2 Discounting, Taxes, and Inflation

The results obtained by LCC analysis and cost-benefit analysis are usually expressed in either "present value" terms or in "uniform annual value" terms. "Present value" means that all past, present, and future dollars of expenditures, receipts, and savings—that is, all cash flows—are converted to an equivalent value in today's dollars. "Uniform annual value" means that all past, present, and future cash flows are converted to an equivalent level amount recurring yearly.

It is important to note that the present value of net costs and cost savings from an investment is not found by merely summing the cash flows over the expected life. Nor is the uniform annual value found by dividing cumulative net cash flows by the number of years of expected life (that is, the uniform annual value is not the same as the average yearly value). This is because the value of money is *time-dependent*. The time dependency of value reflects not only inflation, which may erode the buying power of the dollar, but also the yield of money invested over time, regardless of inflation or deflation. Hence, to evaluate the profitability of investing in energy conservation—either to determine the desirability of a single investment or to compare alternative investments—it is necessary to adjust for the differences in the timing of expenditures and cost savings.

The conversion of differently timed cash flows to a common time equivalent may be done by a technique called "discounting." This technique relies on the application of interest (discount) formulas or, to simplify the calculation, discount factors already calculated from the formulas, to adjust the cash flows.*

To apply the discount formulas or factors, it is necessary to select a discount rate. The discount rate should indicate one's time preference for money. For example, if a person had an annual discount rate of 10 percent (for example, if 10 percent interest could be earned in a risk-free savings account at the bank), a given amount of money this year would be worth 10 percent more next year. This person should therefore be indifferent to a choice between a given amount of money now and 10 percent more than that amount a year from now.

A discount rate may be either "nominal" or "real." A nominal discount rate reflects both the effects of inflation and the real earning power of money invested over time. A lower real rate, reflecting only the real earning power of money, is appropriate for evaluating investments if inflation is removed from the cash flows prior to discounting; that is, if they are stated in constant dollars.

The discount rate may be based on any of several different measures, such as the rate of return which could be realized from the next best available investment, the interest rate on savings accounts, or the cost of borrowing. *There may be a strong subjective element in the specification of the discount rate*. The choice of a rate will likely vary depending on the investor's financial position and concern for the timing of expenditures and receipts (time preference) (Ref. 12). The approach generally taken is to base the rate on a consideration of the factors at hand, and to test the outcome for sensitivity to the use of alternative discount rates where there is great uncertainty as to the correct choice.†

Table 51.4 provides several simple illustrations of the discounting of costs and savings typically associated with investments in energy conservation. The illustrations are based on a discount rate of 10 percent and a period of 10 years. The

*A familiar application of an interest or discounting formula is the use of the uniform capital recovery (UCR) formula to amortize the principal of a mortgage loan over a specified number of years at a given interest rate. This formula and the following five additional interest formulas are those most frequently used in investment analysis: (1) single compound amount (SCA) formula, used to find the future value of a present amount; (2) the single present worth (SPW) formula, used to find the present value of a future amount; (3) the uniform compound amount (UCA) formula, used to find the future value of a series of uniform annual amounts; (4) the uniform sinking fund (USF) formula, used to find the annual amount which will result in a given total value at a future time; and (5) the uniform present worth (UPW) formula, used to find the present value of a series of uniform annual amounts. A detailed explanation of discounting formulas and tables of discount factors calculated from the discount formulas for a range of years and discount rates is available in most engineering economics textbooks. See, for example, Refs. 10 and 11.

†Where there is uncertainty as to the correct value of one or more important input parameters in an evaluation, such as the discount rate, it is useful to determine whether the outcome would change significantly if alternative values were used for the input parameters. Sensitivity analysis can be used to provide additional information for making economic choices. For a description and illustration of sensitivity analysis and the mathematics of probability in economic studies, see Ref. 11, pp. 251–330.

TABLE 51.4 Discounting Cash Flows from an Energy Conservation Investment*

Type of cost or saving (1)	Cash-flow diagram; ↓ expense, ↑ savings (2)	Present value equivalent P (3)	Annual value equivalent A (4)
Purchase and installation of an energy-conserving Feature	S_1 ↓ 1 2 ... 10 years; $10,000	P_f = \$10,000	A_f = \$10,000 (0.1628) = \$1,628 (from i = 10%, N = 10 in UCR)
Repair and replacement of parts	S_1 ↓ 1 ... 5 ... 10 years; $500	P_r = \$500 (0.6209) = \$310 (from i = 10%, N = 5 in SPW)	A_r = \$500 (0.6209) × (0.1628) = \$51 (from i = 10%, N = 5 in SPW; i = 10%, N = 10 in UCR)
Annual fuel savings*	S_1 \$1200 \$1200 \$1200; ↓ 1 ↑ 2 ↑ 3 ↑ ... 10 ↑ years; \$1200	P_s = \$1200 (6.144) = \$7373 (from i = 10%, N = 10 in UPW)	A_s = \$1200
Net total savings		$P_n = P_s - (P_f + P_r)$	$A_n = A_s - (A_r)$
(fuel savings less costs)		= −\$2937	= −\$479

Note: S_1 = starting time (the present)
P = present value equivalent
A = annual value equivalent
F = future value equivalent
f = first costs
r = repair and replacement costs
s = fuel savings
N = net of total costs and savings
i = discound rate

$$\text{UCR (uniform capital recovery formula)} = A = P\frac{i(1+i)^N}{(1+i)^N - 1}$$

$$\text{SPW (single present worth formula)} = P = F\frac{1}{(1+i)^N}$$

$$\text{UPW (uniform present worth formula)} = P = A\frac{(1+i)^N - 1}{i(1+i)^N}$$

In the last three formulas, i = discount rate per period and N = number of interest periods.
*Assumes no change in fuel prices. To include fuel price escalation, the formula becomes:

$$P = C\sum_{i=1}^{N}\left(\frac{1+e}{1+i}\right)^j$$

where C = fuel cost savings at outset and e = fuel price escalation rate.

first column describes the type of cost or saving. The second column uses a cash-flow diagram to describe the timing of the cash outflows and inflows. The horizontal line with arrows represents a time scale progressing from left to right, on which S_1 (for start) indicates the present, the number on the scale indicates the number of years, each downward arrow represents an expenditure, and each upward arrow represents a saving. The third column shows the present value equivalent, and the fourth column shows the annual value equivalent of each cost or saving.

Once the various cash flows have been discounted to a present value or an annual value, they may be combined to provide a net measure of the economic impact of an investment. In column 3 of Table 51.4, for example, the present value cost of $10,000 for purchasing and installing the energy conservation feature plus the present value cost of $310 for repair and replacement equals $10,310. The present value fuel savings total $7373. Net savings of −$2937 result. This is equivalent to a net loss of $479 per year in terms of annual value (A_n from column 4). Hence, this investment is not worthwhile even though net savings in undiscounted terms amount to $1500 (i.e., a total of $10,500 for purchase, installation, repair, and replacement subtracted from $12,000 in aggregate fuel savings for 10 years equals $1500).

Depending upon the degree of accuracy desired in an evaluation, it may be important to consider the impact of taxes. By affecting revenues and costs, taxes can dramatically alter the profitability of an investment. (For a discussion of estimating the impact of taxes on investment decisions, see Ref. 11, pp. 337–382.) Potentially important tax effects include deductions from taxable income of depreciation allowances on capital expenditures, investment tax credits which directly offset tax liabilities, property taxes on capital investments, and the loss of deductions from taxable income when current operating expenses are reduced by fuel savings.

It is usually not necessary to increase the estimates of cash flows to include inflation in each item of cost or savings. Inflation can often be handled in an LCC analysis by making the simplifying assumption that all costs and revenues, except fuel costs, inflate at the same general rate, and that they therefore remain constant in real terms. (For a discussion of the circumstances under which the assumption of evenly inflating costs is not appropriate, see Ref. 10, App. G, pp. 542–552.) Because fuel prices are a dominant factor in the analysis of energy conservation investments, and because they are widely expected to change at a rate faster than overall prices in the economy, it is important to adjust estimates of future fuel cost savings to reflect their expected differential rate of price change, i.e., the rate of increase or decrease over and above the general rate of inflation. (Guidelines for using differential rates of fuel price increases are discussed in Ref. 7.) With these assumptions, all future cash flows can be evaluated with a "real" discount rate. (It should be noted that the treatment of inflation in economic analysis is different from the treatment of inflation for budgeting. To develop reliable budgets, it is essential to take into account the inflation that can occur in planned costs during the lag between the time of the preparation of the economic analysis and the time of actually spending or obligating funds.)

It is not always necessary to go through an elaborate LCC analysis before investing in energy conservation. In some cases, where first costs are low and the potential for energy conservation is high, it will not be necessary to make an explicit evaluation. Weatherstripping around poorly fitting (or leaking) windows and doors is an example of an inexpensive approach to energy conservation which can generally be undertaken with little doubt as to its favorable impact on life-cycle costs.

In cases where first costs are high or significant costs and savings are unevenly distributed over time, it is often advisable to do an LCC analysis. *Not all energy-conserving features will be economical to apply*. Their cost-effectiveness will depend particularly upon climatic conditions, their purchase and installation costs, their durability and maintainability, their ability to save energy, and the present and future prices of fuel. LCC analysis, appropriately used, can result in substantial savings both in energy conservation investments and in building costs in general.

51.4.3 Related Methods of Evaluation

There are several other methods of evaluating the economic efficiency of investment in energy conservation which are closely related to LCC analysis. Popular among these are the "payback method" and the "internal rate-of-return method."

The *payback method* measures the elapsed time between the point of the initial investment and the point at which accumulated savings, net of other accumulated costs, are sufficient to offset the initial investment. (Although costs and savings should be discounted in calculating the payback period, in practice they are frequently left undiscounted.) Shorter payback periods are generally preferred to longer payback periods.

The popularity of the payback method probably reflects the fact that it is an easily understood concept and that it emphasizes the rapid recovery of the initial investment at a time when many organizations appear to place great emphasis on flexibility in investment strategy. However, the payback method has the weakness of failing to measure cash flows that occur beyond the point at which the initial investment costs are recovered. It is possible for a project with a longer payback period to yield higher net savings than a project with a shorter payback period. Hence, use of the payback method may lead to uneconomic conservation investments.

The algebraic formula for determining discounted payback is

$$C - \sum_{j=1}^{Y} \frac{B_j - K_j}{(1 + i)^j} = 0$$

where C = initial investment cost
Y = number of years elapsed until the present value of cumulative net yearly savings just offsets the initial investment
B_j = cost savings or benefits in year j
K_j = costs in year j
i = discount rate

The objective is to find the number of years Y that solves the equation, given values of the other variables. This may be done by trial and error. Alternatively, for the special case in which the net yearly savings, $B_j - K_j$, is equal to a constant A, the following expression of the payback equation can be used:

$$Y = \frac{-\ln (1 - iC/A)}{\ln (1 + i)} \qquad (51.3)$$

where $A = B_j - K_j$.

The *internal rate-of-return method* finds the rate of return that an investment is expected to yield. The rate of return is expressed as that compound interest rate for which the total discounted benefits become just equal to total discounted costs. The rate of return is generally calculated by a process of trial and error in which various interest rates are used to discount costs and benefits to present values. These discounted costs and benefits are compared with each other until that interest rate is found for which costs and benefits are equal and net benefits are, therefore, zero.

As an illustration, let us find the internal rate of return on an investment which requires an initial, one-time cost of $10,000 and yields a yearly recurring savings of $3000 for 10 years. The initial investment of $10,000 is already in present value terms. We need now to calculate the net present value *P* of the $3,000 in yearly savings for various interest rates. First, let us calculate the value of *P* for, say, a compound interest rate of 25 percent. At this interest rate, the present value equivalent of the $3,000 for 10 years is equal to $10,713. Subtracting the present value cost of $10,000 from the present value savings yields a net present value savings of $713. The fact that $713 exceeds zero means that 25 percent is less than the internal rate of return on this investment. Trying now a higher compound interest rate of 30 percent gives a present value savings over the 10-year period of $9276. Net present value savings are now equal to −$724, an amount $724 less than the $10,000 cost. Since an interest rate of 30 percent results in net losses, this rate must be greater than the internal rate of return on this investment. Thus, we can conclude that the rate of return is bracketed by 25 and 30 percent. By interpolation, we can now estimate that the investment yields an internal rate of return of a little over 27 percent. The investment would be considered worthwhile if the 27 percent rate exceeds the rate of return which the investor could get from alternative investments.

This method of evaluation usually results in a measure consistent with an LCC approach and is somewhat more reliable than the payback method. However, the internal rate-of-return method does have the disadvantage of giving either no solution or multiple solutions under certain conditions.

The payback method, internal rate-of-return method, and LCC method have particular advantages and disadvantages. Each will serve as a useful tool for investment decisions in certain cases.* For most problems of making economically efficient decisions in energy conservation, the LCC method will provide an adequate measure.

51.5 ENERGY MANAGEMENT SYSTEMS

An energy management system (EMS) will enable management to control the operation and maintenance of the equipment while producing energy and work savings for heating, ventilating, and air conditioning; process equipment; lighting; chillers; boilers; etc.

One only has to consider the fact that the major energy consumers in a typical institutional, retail, or commercial office building are the HVAC and lighting systems (see summary in Table 51.5) to see the potential cost-effectiveness of an EMS system is in monitoring and controlling the energy usage of a facility.

The U.S. government has classified EMS into five categories associated with the total number of points connected to a system. They are:

1. Large EMS in excess of 2000 points

*For a discussion of the suitability of various evaluation methods for treating different kinds of investment decisions, see Ref. 13.

TABLE 51.5 Comparative Energy Use by System and Building Type

Zone*	Heating and ventilation†	Cooling and ventilation†	Lighting†	Power and process†	Domestic hot water†
Schools:					
A	4	3	1	5	2
B	1	4	2	5	3
C	1	4	2	5	3
Colleges:					
A	5	2	1	4	3
B	1	3	2	5	4
C	1	5	2	4	3
Office buildings:					
A	3	1	2	4	5
B	1	3	2	4	5
C	1	3	2	4	5
Commercial stores:					
A	3	1	2	4	5
B	2	3	1	4	5
C	1	3	2	4	5
Auditoriums:					
A	3	2	1	4	5
B	1	3	2	4	5
C	1	3	2	4	5
Hospitals:					
A	4	1	2	5	3
B	1	3	4	5	2
C	1	5	3	4	2

*Climatic zone A: Less than 2500 (1400) degree-days [°F (°C)]; climatic zone B: 2500–5500 (1400–3000) degree-days; climatic zone C: 5500–9500 (3000–5300) degree-days.

†The numbers 1 through 5 are in the order of descending energy demand; that is, 1 represents the greatest and 5 the least.

2. Medium EMS with 500 to 2500 points
3. Small distributed EMS with 200 to 1000 points
4. Small centralized EMS with 50 to 600 points
5. Micro EMS with less than 125 points

A typical EMS system is generally configured into a network with control functions at multiple locations and a central point of operation control and supervision. Typically an EMS consists of a central control unit (CCU), memory, storage devices, input-output devices, a central communications controller (CCC), data transmission medium (DTM), field interface devices (FIDs), multiplexers, instruments, and controls. The primary task of the CCU is to automatically perform monitoring and control functions. The control functions require the execution of algorithms used to predict environmental conditions and rate of power consumption, calculate equipment operating set points, and produce supervisory control signals to operate equipment in real time. The CCU also accommodates the operator interface required for EMS operation. Functions of the FID are to collect data and issue control commands to the local equipment, the FID's data environment. FIDs contain a microcomputer that performs local control, monitoring, and communications functions. Data from the FID are transferred via DTM to the CCU and CCC, where they are utilized for supervisory control and monitoring functions, calculations, and alarm reporting.

51.5.1 Components

Major components of a typical EMS are:

- The central control unit is a minicomputer or microcomputer with memory for the operating system software, command software, and implementation of application programs. Computations and logical decision functions for central supervisory monitoring and control are performed in this unit. Data and programs are stored in and retrieved from the memory or mass-storage devices. The CCU has input-output (I/O) ports for specific equipment such as printers and cathode-ray tube (CRT) consoles. It also has direct memory access (DMA) controllers for high-speed data transfer between the CCU and mass-storage devices, such as magnetic tape systems and disk systems. During normal operation, the CCU coordinates operation of all other EMS components, except safety interlocks.
- The central communications controller is a minicomputer with sufficient memory to (1) execute the software required to reformat, transfer, and perform error checks on data between the CCU and FIDs and (2) provide limited backup in the event of CCU failure. This unit is usually provided with a medium-to-large EMS.
- The communication link termination (CLT) is the communications equipment interface between field equipment and EMS equipment in the central control room.
- The operator's console includes a color CRT terminal and keyboard. It is the operator interface that accepts operator commands, displays alarms and data, and shows diagrammatically the systems connected to the EMS.
- The system terminal is an alphanumeric CRT terminal or a printer used to load the operating system, develop programs, run diagnostics, and support background processing. It is not intended for use by the system operator.
- Alarm and logging printers provide a permanent copy of alarm system operations and historical data.
- The disk storage system provides high-density random-access mass storage in one or more disk drives.
- A modem allows communication over the data transmission medium.
- The CCC bulk loader is a mass-storage device used to load EMS software into the CCC when the CCU is not in service. This unit is usually provided in a medium-to-large EMS.
- The flexible disk storage system is a medium-density random-access storage device, with removable storage media.
- The system real-time clock (RTC), with battery backup, synchronizes system clocks at regular intervals.

- The failover controller switches the CCU, CCC, and peripheral equipment, in the event of CCU or CCC failure, into a backup mode of operation. The failover controller should be provided on a medium-to-large EMS only.
- The magnetic tape system is a high-density serial mass-storage device.
- Field interface devices are microcomputer-based devices with memory, I/O, communications, and power supply. An FID provides an interface to the monitored and controlled data environment, performs calculations and logical operations, accepts and processes system commands, and is capable of stand-alone operation in the event of CCU or DTM failure.
- A multiplexer (MUX) is a device which combines data from a number of points in the data environment and communicates with its associated FID. It also demultiplexes commands received from the FID. The MUX is functionally part of the FID and can be in the same enclosure or remotely located.
- The memory-protection power supply is an independent backup power supply for the CCU and CCC that maintains volatile memory contents for at least 20 min.
- The data terminal cabinet (DTC) is the interface point between FID and MUX and the data environment.
- The dial-up telephone modem automatically dials and answers calls for the EMS for remote diagnostics and data acquisition.
- The data transmission medium is a communications link such as voice-grade telephone lines (wire lines), optical fibers, coaxial cable, or microwave radio.
- The FID test set is a unit consisting of a FID, a MUX, and data environment simulator panel. It is required for the central control room of large and medium systems.
- The FID-MUX portable tester is a unit for locally programming and bulk-loading diagnostic tests for FID and MUX functions.
- The power line conditioner (PLC) is a constant-voltage transformer used in the central control room, FID, and MUX to regulate voltage and provide noise isolation.

51.5.2 Software Programs

Software for energy management includes the following:

- Automatic demand control and load-shedding program
- Duty cycle program
- Time clock start and stop program
- Optimal start and stop program
- Event-printing program
- Event-initiated program
- Control interpreter language program
- Holiday schedule program
- Lighting control (on-off or stepped, timed, or light-level-sensed) program
- Routine program (aids scheduling of preventive maintenance of equipment)

51.5.3 Functions

The energy management system should include the following functions:

- Remote start-stop, on-off, and open-close control of equipment.
- Alarms:

  Abnormal equipment status
  Safety
  High or low level, temperature, pressure, etc.
- Analog monitoring:

  Temperatures
  Energy consumption (all types)
  Pressures
- Set-point reset of local control systems.
- Mode change for warmup-cooldown, day-night, enthalpy control, smoke purge, etc.
- Computer programming for calculations and table lookup functions for maintenance, operations monitoring, energy management, and system use optimization, including:

  Fan system start and stop (clock control and optimal control)
  Enthalpy economizer control
  Chiller plant optimization and nonautomatic plant operation with centralized manual control
  Load shedding based on demand monitoring
  Logging and profiling
- Chiller plant: nonautomatic, centralized manual control. Provides a combination of start-stop, status, alarm, water temperature, flow, and Btu indications for each machine including auxiliaries (pumps, valves).
- Condenser water and cooling towers: nonautomatic, centralized manual control.
- Plumbing system:

  Alarms and status points
  Start and stop control for domestic water pumps
- Emergency generator and electrical alarms.
- Lighting control by zones and floors.
- Air-handling units: monitoring and control points.
- Miscellaneous fans: start and stop control and status indication.
- Monitor space temperatures (interior and perimeter) plus outside air wet- and dry-bulb temperatures.
- Isolating damper control.

- Perimeter hot-water system: reset modulating valves on heat exchangers from outdoor air temperature.

51.5.4 Optional Security and Fire Alarm System

The security and fire alarm system should offer the following features:

- Limit intrusion and protect life and property in emergency situations with the following equipment:

  Intrusion alarm
  Smoke alarm
  Sprinkler alarm
  Fire alarm
  Equipment tamper alarms
  Special area monitoring units
- Centralized control of door and gate locking and unlocking.
- Emergency communications:

  Emergency reporting telephone system
  Firefighters' communications
  Emergency sound system
  Security intercom
  Complete radio communications
  Visual communications via closed-circuit television
  Stairway emergency communications

51.5.5 Selecting an EMS

Though it may not be cost-effective to install an EMS as described in this section for buildings under 50,000 ft^2 (4600 m^2), a system using a simple time clock to start and stop equipment can provide an economical alternative to minimize energy usage.

When selecting an EMS the designer should be concerned not only that the system offers all the functions the owner and user require but also that the vendor provides timely, effective customer services by trained, knowledgeable personnel; detailed manuals to train the owners and operating personnel; system programs that prompt the operator at each operational step, suggesting the next logical course of action; and manuals that guide the owner and user personnel step by step through the programming system. The buyer should also make sure that the EMS protects against loss of memory during power outages and can be upgraded economically in the future.

In addition, each component should be evaluated for ease of on-line programming (when applicable), modular maintenance, rugged construction, flexibility, and reliability of performance in the environment in which it will be operating.

Finally, the buyer should make sure that failure of the central computer data transmission system should not cause total system shutdown. The HVAC, refrigeration, boilers, pumps, process, etc. should be operational (stand alone) when the EMS is down.

The following are examples of medium-to-large EMS systems:

- The EMS at the Arnold Engineering Development Center (AEDC) in Tullahoma, TN, was installed by the U.S. Air Force to control energy costs. The EMS computers continuously monitor and control AEDC energy supply and usage. Because AEDC's major energy cost is incurred when the facility exceeds peak power levels set by the Tennessee Valley Authority (TVA), a prime objective of the EMS is load shedding. If AEDC power requirements during testing reach the peak, the system will transmit an alarm so that a dispatcher can temporarily shut down other equipment, such as HVAC systems, to stabilize the load. This EMS is designed to be a fully automated system. It controls and monitors most of ADEC's HVAC system—50 million Btu of heating and 24,000 tons of cooling in 48 buildings. The system also controls the main pumping station, which has six 2000-hp motors and supplies 105,000 gpm of water.
- In a commercial installation in a 40-building office complex in Michigan, an EMS starts and stops all major equipment daily. Programmed for holiday service, it turns on and off parking lot and driveway lights, allowing for variations in sunrise and sunset and for holidays, and controls HVAC and other operations.
- The EMS in the Silverdome Sports Center in Pontiac, MI, with more than 80,000 seats, schedules the operation of blowers, heaters, air conditioners, and other mechanical devices, including those necessary to keep the domed air-supported nylon roof inflated and in position. It also monitors the fire and security call stations.

REFERENCES

1. *1985 ASHRAE Handbook, Fundamentals*, ASHRAE, Atlanta, GA, 1985, chap. 20, "Thermal Insulation and Vapor Retarders."
2. Ibid., chap. 22, "Ventilation and Infiltration."
3. Carrier Corp., *Handbook of Air Conditioning System Design*, pt. 1, chap. 6, McGraw-Hill, New York, NY, 1965.
4. "Ventilation for Acceptable Indoor Air Quality," ANSI/ASHRAE Standard 62-1981R, ASHRAE, Atlanta, GA.
5. *Industrial Ventilation, A Manual of Recommended Practice*, Committee on Industrial Ventilation of American Conference of Governmental Hygienists, Lansing, MI.
6. A. K. Dasgupta and D. W. Pearce, *Cost-Benefit Analysis: Theory and Practice*, Barnes and Noble, New York, NY, 1972.
7. Reynolds, Smith and Hills, Architects-Engineers-Planners, Inc., *Life-Cycle Costing Emphasizing Energy Conservation: Guidelines for Investment Analysis*, rev. ed., Manual 76/130, Energy Research and Development Administration, Washington, DC, May 1977.
8. R. T. Ruegg et al., *Life Cycle Costing, A Guide for Selecting Energy Conservation Projects for Public Buildings*, Building Science Series 113, National Bureau of Standards, Gaithersburg, MD, May 1978.
9. E. Mansfield, *Microeconomics: Theory and Applications*, W. W. Norton, New York, NY, 1970, pp. 148–156.

10. G. W. Smith, *Engineering Economy: Analysis of Capital Expenditures*, 2d ed., Iowa State University Press, Ames, IA, 1977.
11. E. L. Grant and W. Grant Ireson, *Principles of Engineering Economy*, Ronald Press, New York, NY, 1970.
12. J. J. Mutch, *Residential Water Heating, Fuel Conservation, Economics, and Public Policy*, Th 751222, M18, Rand Corp. for National Science Foundation, 1977, Washington, DC, app. B, pp. 69–71.
13. R. T. Ruegg, "Economics of Waste Heat Recovery," in *Waste Heat Guidebook*, K. G. Kreider and M. B. McNeil (eds.), Handbook 121, National Bureau of Standards, Gaithersburg, MD, February 1974, pp. 99–105.

CHAPTER 52

AUTOMATIC TEMPERATURE, PRESSURE, FLOW CONTROL SYSTEMS

Edward B. Gut
Donald H. Spethman
Honeywell, Inc., Arlington Heights, Illinois

52.1 CONTROL BASICS

52.1.1 Control Systems

Elements of Control Systems. Control loops consist of several elements and are used to match equipment capacity to load by changing system variables. Figure 52.1 is a block diagram of a control loop and shows the relation of the elements.

The controlled variable is the condition being controlled; for HVAC systems this is typically temperature, humidity, or pressure. A sensor is the device that measures a variable and transmits its value to the controller. The controller com-

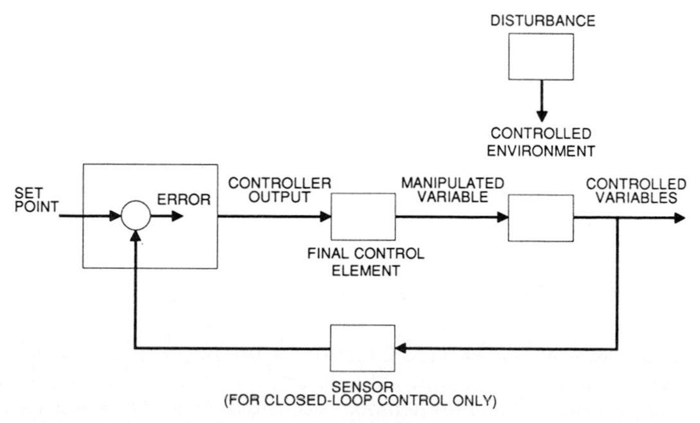

FIGURE 52.1 Basic elements of a control loop.

pares the value of the variable with the set point or desired value, and outputs a signal based on the difference between the variable and the set point.

The final control element responds to the controller signal and varies the control agent. Control elements may be valves, dampers, electric relays, or electronic motor speed controllers, and agents may be air, water, steam, or electricity. The process plant is the equipment being controlled and whose output is the controlled variable. It may be a coil, fan, steam generator, or heat exchanger.

Types of Control Loops. There are two basic types of control loops, open loop and closed loop. With open-loop control, the system sensor measures a variable external to the system yet has some relation to the controlled variable. An example is sensing outdoor temperature to control heat flow into a building to maintain indoor temperature. Thus, a fixed relationship between outdoor temperature and required heat input is assumed and the control system programmed accordingly.

Closed-loop control pertains when the system sensor measures the controlled variable, resulting in variations in the control agent to maintain the desired value of the controlled variable. Closed-loop control is also called "feedback control," and results of a corrective action are fed back within the controlled system, therefore providing true control of the controlled variable.

52.1.2 Modes of Feedback Control

Feedback-controlled systems are categorized by the type of corrective action a controller is designed to output. For all types, the set point is the desired value of the controlled variable to which the controller is set. The control point is the actual value of the controlled variable as maintained by the controller's action.

Two-Position Control. The final control element may be in one or the other position, i.e., maximum or minimum, except for the brief time when it changes positions. There are two values of the controlled variable which establish the position of the controlled element: set point and differential. Differential is the smallest range through which the controlled variable must pass to move the control element from one position to the other. Figure 52.2 shows a temperature con-

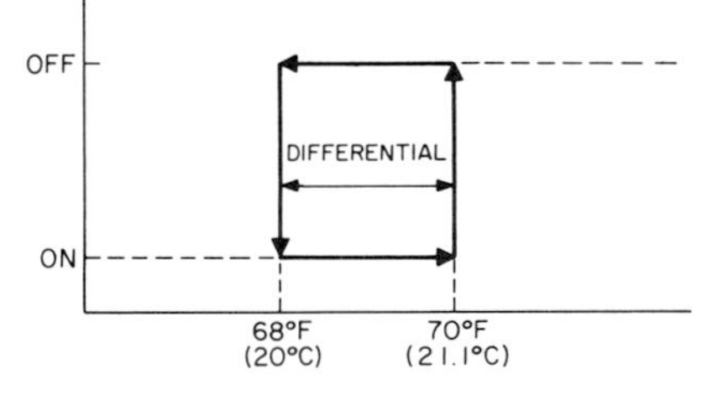

FIGURE 52.2 Two-position control.

troller or thermostat with a 70°F (21°C) set point. At 70°F (21°C) this electric thermostat would open its contacts and stop a burner. For the thermostat contacts to close, turning on the burner, the temperature must drop below the set point by the amount of the differential, 2°F (1.1°C) in this example. Differential may be subtracted from or added to set point depending on controller design.

Two-position control is a low-cost device and provides acceptable control of slow-reacting systems that minimum-lag between controller outputs and control-element response. Fast-reacting systems may overshoot excessively and be unstable. Examples of two-position control are domestic hot-water heaters, residential space-temperature controls, and HVAC system electric preheat elements.

Timed Two-Position Control. The final control element may be in one of two positions, as for a two-position control, but a timer is incorporated in the controller so that it responds to the average value of the controlled variable rather than the peak fluctuations. Timed two-position control greatly reduces the variations or swings in the control variable by anticipating controlled-variable changes due to control-system action.

A typical example of timed two-position control is residential space-temperature control. The thermostat has an electric heating element that is energized during the on period. The heat from the element warms the temperature sensor more quickly than the rising space temperature, shortening the on time and reducing temperature overshoot. During the off period the sensor heater is also off, allowing the sensor to respond directly to space temperature. This results in a relatively constant cycle time with a variable on-off ratio dependent on space load.

Timed two-position control is low-cost and may be applied to slow-reacting systems that have some lag between controller output and control-element response. The timer will anticipate the response and minimize variations in the controlled variable.

Proportional Control. *A proportional controller* has a linear relationship between the value of the incoming sensor signal and the controller's output. The relationship is generally adjustable in the controller but once adjusted remains fixed during operation. There is therefore only one value of the final control element for each value of the controlled variable within the operating range of a proportional control system.

The variation in control variable required to move the final control element through its operating range is the throttling range of the control system and is expressed in the measuring units of the controlled variable. The variable in the sensor signal required to operate a proportional controller through its range is called the "proportional band" and is expressed as a percentage of sensor span.

The "set point" of a proportional controller is defined as the sensor input which results in the controller output at the midpoint of its range.

"Offset" is the difference between the set point and the controlled variable at any instant. Sometimes offset is also referred to as "deviation," "droop," or "drift." Offset results from the fixed linear relationship between control input, sensor signal, and output. Therefore, under full-load conditions, control input must be offset by one-half the proportional band for the controller to output a signal at one extreme of its range. Similarly, at minimum load the offset will be one-half the proportional band (see Figure 52.3).

Proportional control is used with slow stable systems, allowing narrow throttling ranges and therefore small offset. Fast-reacting systems require large throttling ranges to avoid instability and cycling out the controlled variable. This, of course, increases offset.

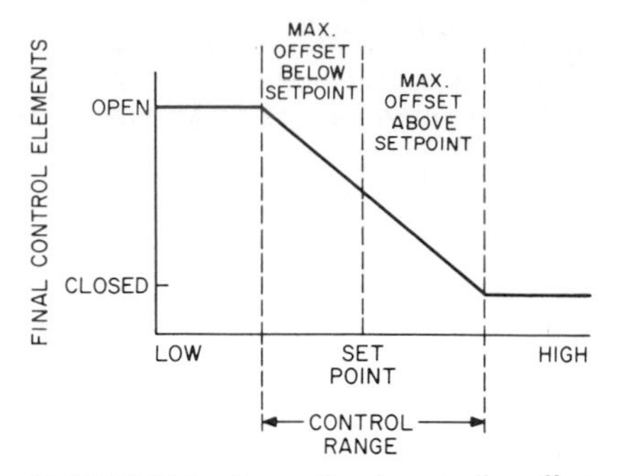

FIGURE 52.3 Proportional-controller offset.

Proportional-plus-Reset Control. A proportional-plus-reset controller has proportional action plus an automatic means of resetting the set point to eliminate offset. This controller action is also called "proportional-plus-integral," or PI, control.

A PI controller's initial output signal has a fixed relationship to a changed sensor input signal, the same as a proportional controller, but then continues to change until the control variable equals set point. The rate at which this additional change occurs is called the "reset rate" or "repeats per minute" and is the number of times the original proportional change in controller output is repeated per minute. The reset rate may also be expressed as reset or integral time, which is the amount of time for the controller to change its output as much as the first proportional change.

PI control may be applied in fast-acting systems that require large proportional bands for stability but where the resultant offset between set point and control point is undesirable due to comfort and/or energy-conservation considerations. Typical applications are mixed-air, duct-static, chiller-discharge, and coil-discharge control.

Proportional-plus-Rate Control. Also called "proportional-plus-derivative," or PD, control, this control mode adds to proportional control an automatic means of varying controller output based on changes in deviation or difference between the set point and the control variable.

When deviation increases, rate action adds to the controller's output, causing the final control element to respond an additional amount to stabilize the controlled variable more quickly than proportional control alone can. Conversely, when deviation decreases, rate action subtracts from the controller's output. When there is no change in deviation, rate action stops and the deviation is determined by the proportional band of the controller.

Proportional-plus-Integral-plus-Derivative (PID) Control. This combination of control modes is useful for controlling fast-acting systems that tend to be unstable such as duct static-pressure control. For these applications the controller may be set with a large proportional band for system stability, a slow reset to eliminate deviation from set point yet retain stability, and derivative action to speed control response when the system is upset due to changes.

Adaptive Control. A PID controller must be properly tuned to the system it is controlling to achieve stable and accurate control. The proportional, integral, and derivative parameters of the controllers are dependent on the characteristics of the system being controlled and can be time-consuming to establish. Even then they will be optimum for one operating condition and compromised for the remaining operating range.

Adaptive control is the ability of a controller to adapt to the system it is controlling by determining the ideal PID parameters and adjusting itself according. Two types of adaptive control have been developed, self-tuning and model reference.

A self-tuning controller begins with initial PID parameters. With input from the controller's output and the control-variable value, it establishes new parameters. After a few cycles of control-system operations, the controller determines the optimum parameters. The process continues as the system operates, so every time the system changes, the controller reestablishes the parameters so that they are optimum for every condition of system operation.

The model-reference controller compares its output with that of a fixed model and develops the PID parameters to achieve control-system operation for the model. While the model may not be exactly the same as the actual system, it is very close and allows the controller to develop the parameter values quickly.

Floating Control. A floating control outputs a corrective signal when the difference between the set point and sensor signal is greater than a set amount or differential. The output signal will increase or decrease a final control element depending on if the controlled variable is below or above set point. If the difference is less than the differential, the controller output is zero and the final control element remains in the position it was last driven to. Floating controls may be applied to systems that react quickly with little lag and have slow load changes.

Time-Proportioning Control. Time proportioning is a method of controlling electric heating elements. The final control element is either on or off, but the ratio of on-to-off time is varied depending on system load, therefore varying the energy inputs.

The sum of on and off time, or the total time per cycle, is constant. Time-proportioning control is also called "average-position control" and is a relatively low-cost way to simulate proportional control when large electrical loads are controlled.

52.1.3 Flow-Control Characteristics

Flow Control. Proper volume or flow control of one form or another is essential to the successful operation of most HVAC systems. Usually the flow of water, steam, and/or air is controlled to modulate system outputs or capacity as required by changing loads. As in other control loops, a sensor measures the control variable and a controller compares the sensor signal to a set point and outputs a corrective signal as required to a final control element. For water and steam flow the final control element is a valve, and for air flow a damper.

The flow-control characteristics of valves and dampers are designated in terms of the flow versus opening based on a constant pressure drop across the element. The three common characteristics are quick opening, linear, and equal percentage.

As shown in Fig. 52.4, quick opening provides for more percentage of full flow than when the valve or damper is opened. Linear characterization has the same

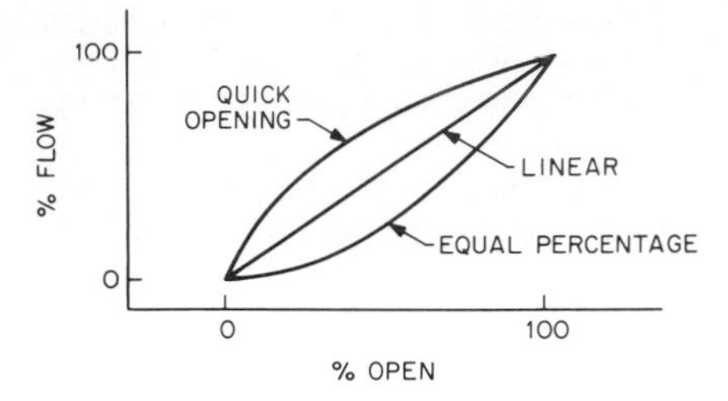

FIGURE 52.4 Flow-control characteristics.

percentage of full flow as when the valve is open, while equal percentage increases flow by an equal percentage over the previous value for each equal increment of opening. In other words, a 10 percent change in opening from 20 to 30 percent increases the flow by the same percentage of flow at 20 percent opening as an increase in opening from 70 to 80 percent would increase flow from the 70 percent opening position. These different characteristics are required to match the control needs of water, steam, and air flow.

Pressure drop across a valve or damper in a system rarely stays constant. Therefore actual opening-flow characteristics vary from manufacturer's ratings, which are based on constant pressure drop. The amount of this variation depends on how much the pressure drop changes and is determined by overall system design. The pressure drop is minimum when the valve or damper is full open and increases as the valve or damper closes. When fully closed, the entire pressure drop is across the valve or damper.

For the valve or damper to provide approximately its design characteristic, the design or full-open pressure drop should be a fairly large percentage of the total system drop. As a high pressure drop consumes energy, consideration should be given to design or control a system to provide a more constant pressure drop, allowing the valve or damper to be sized for a lower pressure drop at full flow.

Control of Water Flow. One of the primary uses of water-flow control is to modulate the capacity of a heating or cooling coil. However, the capacity of a coil is not linear with water flow; instead, as the flow is reduced, more energy is transferred from the water, partly offsetting the reduction in flow. Figure 52.5 shows the relationship of capacity versus flow for a heating or cooling coil. This nonlinearity is primarily a consideration with hot-water coils due to the large temperature difference between the water and air flow through a coil. For hot-water coils, this nonlinear variation may be reduced by designing the coil for a higher water-temperature drop or by reducing water temperature as system load decreases.

Since hot-water coils have a significantly nonlinear relationship between heat transfer and water flow, equal-percentage valves are used for coil water-flow control, resulting in a more linear relationship between valve position and coil heat output.

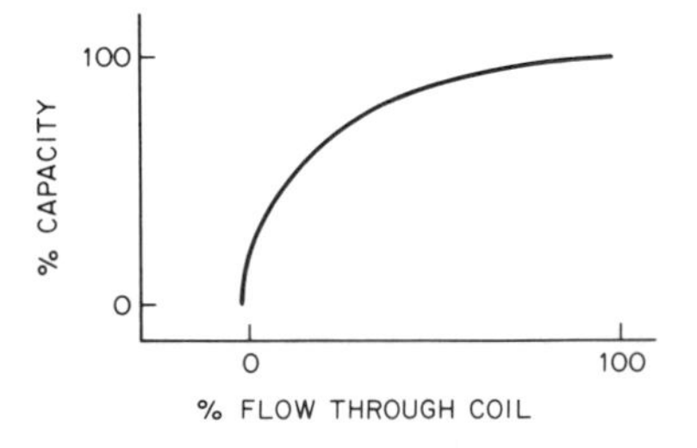

FIGURE 52.5 Heating- or cooling-coil capacity versus flow.

The capacity of a water valve is determined by the pressure drop across the valve and is independent of the supply pressure. Valve capacity is rated by a flow coefficient, or C_v, which is defined as the amount of water in gal/min (m^3/h) that will flow through an open valve at 1 lb/in^2 (1 atm, 101.325 kPa) pressure drop. For valves in systems, the pressure drop increases as the valve closes, offsetting part of the desired flow reduction. To minimize this, valves should be sized so that they constitute about 25 to 50 percent of the system resistance that a valve controls.

Valve pressure-drop changes can be minimized by providing a system bypass valve to maintain total system flow even when control valves close. Or system flow may be modulated by an automatic flow-control valve in series with the pump, or the pump may be operated at varying speeds based on system pressures near the far end of the piping circuit.

Control of Steam Flow. Control of steam flow is usually applied to modulate the heat output of a steam-to-water and steam-to-air heat exchanger.

For one-pipe steam systems, line-size, two-position valves are used to ensure proper flow of steam and simultaneous drainage of condensate. Two-pipe steam systems may be controlled by two-position or modulating valves which must be sized properly for good control. Since output of a steam heat exchanger is linear with steam flow, valves with linear flow-opening characteristics should be used for modulating control.

The capacity of a steam valve is determined by the pressure drop across it and the inlet pressure. Valves for two-position applications are sized to provide the required full flow with minimum pressure drop and to be able to close against system pressure.

Modulating steam valves must be sized to only full-load flows, which may be less than full heat-exchanger flow, to avoid system instability due to excessive capacity. Since steam valve capacity depends on pressure drop and inlet pressure, it is important that valve inlet and outlet pressures are kept fairly constant to maintain a linear relationship between valve opening and heat-exchanger output.

Supply pressures can be controlled by automatic pressure-reducing valves in the supply lines or by a narrow differential controller. The effect of variations in return pressures can be minimized by sizing the valve so that the outlet pressure is near its minimum value or at a pressure resulting in critical velocity in the fully open valve, whichever is higher.

Critical velocity in a valve is the velocity at which an increase in pressure drop will not result in an increase in velocity or flow through the valve. This occurs when outlet pressure is about 58 percent of inlet pressure. For some applications with large-capacity modulating, two steam valves in parallel may be used for better full-range control. The valves should be sized so that one valve has about one-third full-load capacity and the other valve about two-thirds full-load capacity. The valves are operated in sequence so that the smaller valve controls during low loads and the larger valves operates when the smaller valve is fully open.

Control of Air Flow. Air flow in HVAC systems is controlled in an on-off mode or modulating mode. The on-off mode is generally used to allow outside air into a building when desired such as during occupied times and to prevent outside air from entering at other times. Modulating air flow is used to blend air from more than one source to achieve a desired temperature or to vary the volume of air delivered to match load requirements.

Dampers are used to control air flow and are produced in two basic designs, parallel-blade and opposing-blade configurations (see Fig. 52.6). The opening-flow characteristics of these configurations for constant pressure drop and for various ratios of system pressure drop without the damper to damper pressure drop at full open flow is shown in Figs. 52.7 and 52.8.

However, as with valves, dampers installed in systems have varying pressure drops as they modulate, being minimum when full open and maximum when

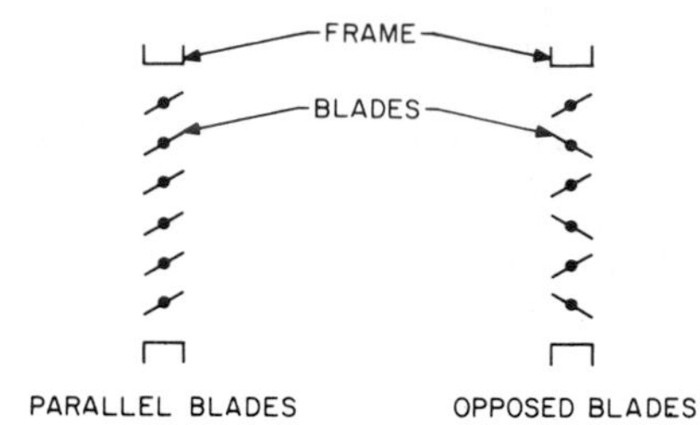

FIGURE 52.6 Damper-blade configurations.

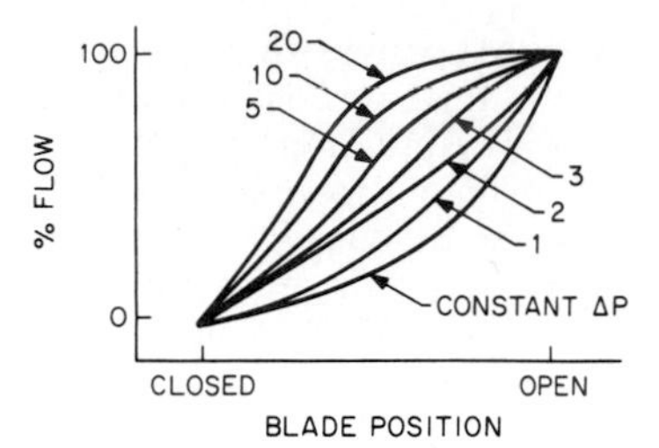

FIGURE 52.7 Characteristics of parallel-blade dampers. Curves other than the constant-ΔP curve represent ratios of system pressure drop to open-damper pressure drop at full flow.

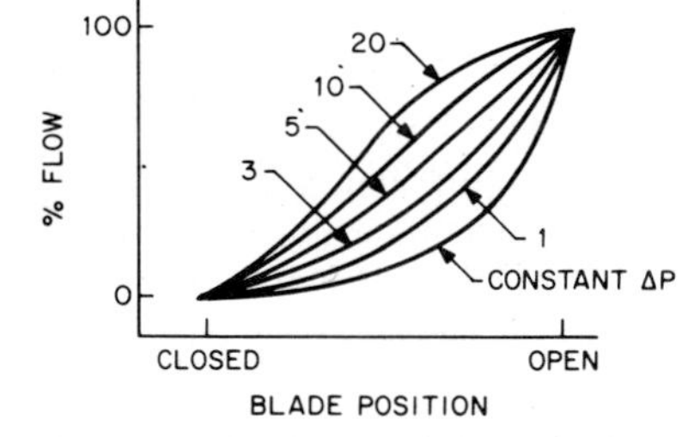

FIGURE 52.8 Characteristics of opposed-blade dampers. Curves other than the constant-ΔP curve represent ratios of system pressure drop to open-damper pressure drop at full flow.

closed. For two-position applications, dampers should be selected on the basis of full-flow pressure drop, leakage, and closed-pressure differential ability. Modulating characteristics are not important.

Typical modulating applications are mixed air, face and bypass, and volume control. Mixed air, or control of outside, return, and exhaust air, requires the coordination of three dampers for modulating outside and return air to maintain a constant supply volume and for modulating exhaust-air volume as outside air varies. Face and bypass control is used to vary the amount of air through and around a coil to vary the temperature of the total air flow after the coil. The face damper controls air flow through the coil, and the bypass damper the air flow around the coil. The dampers are arranged so that when one opens the other closes, and the sum of the air flow through both dampers is constant. To achieve this relationship it is important that both dampers are selected for linear control.

Volume control of air flow may be used to maintain static pressure in a duct or space or to match space- or zone-conditioning needs. Variable air flow is achieved by changing duct system resistance to air flow or by diverting air flow through an alternative or bypass route. Dampers should be selected to provide equal changes in air flow for equal changes in control variables, which may be temperature, pressure, or flow volume in these specified systems, for stable control over the full operating range.

52.2 CONTROL EQUIPMENT TYPES

The elements of a control loop are divided into four categories: sensors, controllers, final control elements, and auxiliary equipment, and may be pneumatic, electric, or electronic.

52.2.1 Sensors

The controlled variable of a system is measured by a sensor. A sensor output signal, whether pneumatic or electric, may change electrical resistance depending on the value of the sensed variable. The usual pneumatic sensor-signal range is 3 to 15 lb/in^2 (20.7 to 103.4 kPa), while electric sensors output 2 to 20 V dc or 4 to 20 mA. Resistance sensors have a nominal resistance of 500 Ω. Temperature-sensing elements are usually bimetal, rod and tube, sealed bellows, and resistance.

Bimetal is the oldest and most common type of temperature-sensing element. Its operation is based on the principal that the change in size with the change in temperature is different for different metals. Combining two metals, one with a large expansion coefficient and one with a small coefficient, into a strip, the strip will deflect with temperature changes due to the different amounts of expansion (see Fig. 52.9). The amount of deflection is proportional to temperature and can

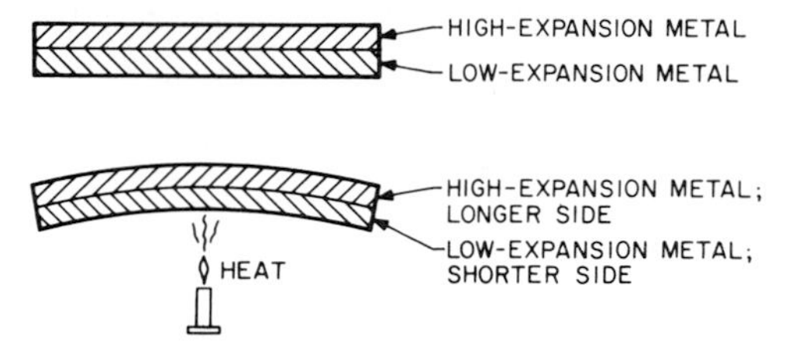

FIGURE 52.9 Bimetal strip.

therefore be used to measure or sense temperature and generate a proportional pneumatic or electric signal. Bimetal strips may be used as straight elements or may be U-shaped or spiral-wound depending on the space available and the temperature-deflection characteristics desired.

Rod-and-tube elements also use the different expansion rates of metals to generate movement with temperature changes. However, they are constructed with a low-expansion rod, high-expansion tube (see Fig. 52.10) and are usually used for insertion directly into the medium, such as water, steam, or air.

Sealed bellows (see Fig. 52.11) consist of a capsule and bellow evacuated of air and filled with a vapor or liquid. As a vapor or liquid changes pressure or volume with temperature changes, the bellows moves, providing an indication of sensed temperature. The variation of sealed bellows is the remote-bulb element (see Fig. 52.12). A bulb is attached to the bellows assembly by a capillary tube so that temperature changes at the bulb result in pressure changes which are transmitted to the bellows, resulting in movement corresponding to temperature at the bulb.

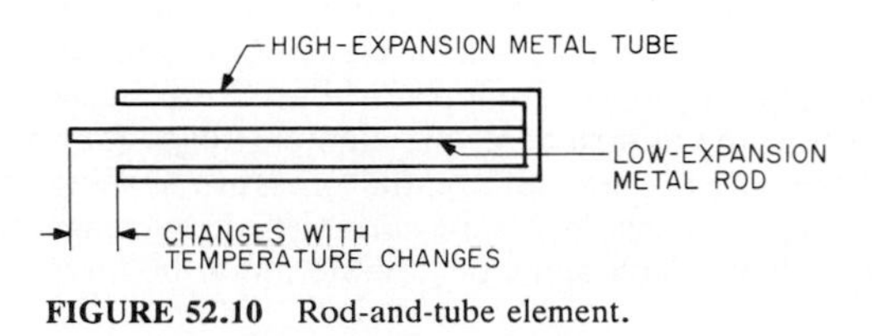

FIGURE 52.10 Rod-and-tube element.

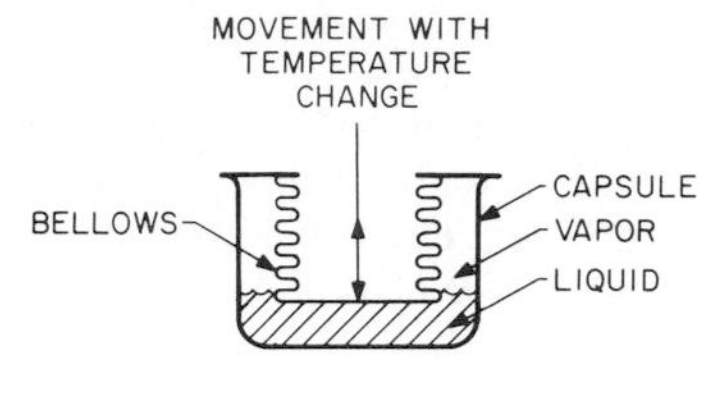

FIGURE 52.11 Sealed bellows.

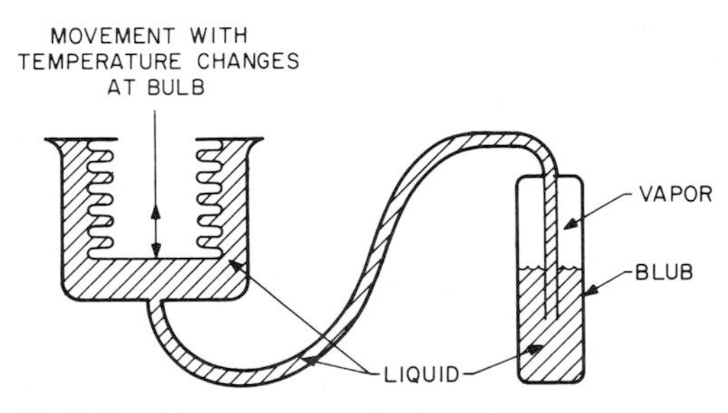

FIGURE 52.12 Remote-bulb element.

Resistance elements consist of an element with a known temperature life and resistance characteristic. The element may be wired, wound on the bobbin, or a thermister which is a semiconductor, whose resistance depends on temperature.

Humidity-sensing elements are hydroscopic or electric. The hydroscopic elements are based on the fact that certain materials change size as they absorb or release moisture. Typical materials are hair, wood, leather, or nylon whose size changes due to moisture absorption or release, based on the moisture content of surrounding air, will indicate humidity of the air. This size change is used to develop a pneumatic signal proportional to humidity or to turn an electric switch on and off.

Electric humidity-sensing elements are constructed to provide either a resistance change with ambient humidity changes or a capacitor change and are generally used with electronic controllers.

Dew-point sensors are constructed by winding two wires around a hollow tube impregnated with lithium chloride. The conductivity of the lithium chloride varies as it absorbs or releases moisture to the surrounding air. Electric power supplied to the two wires around the sleeve will flow through the lithium chloride at a rate depending on its conductivity, which varies with dew points. As the electricity flows through the wires, the temperature of the cavity of the tube is elevated and is a measure of dew point (see Fig. 52.13). The cavity temperature may be sensed with any temperature sensor that will fit inside the tube.

Pressure sensors may be high-range (pounds per square inch, or psi) or low-range (inches of water). High-range sensor elements usually are Bourdon tubes, bellows, or diaphragms to provide movement based on pressure. If one side of the element is open to the atmosphere, the element responds to sensed pressure above or below atmospheric. For differential pressure sensing, both sides of an element are connected to sense pressure variables. Low-range pressure sensors generally use large slack diaphragms or flexible metal bellows to transduce low

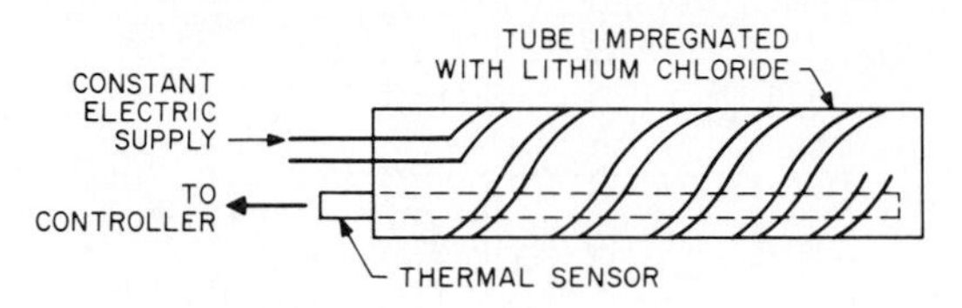

FIGURE 52.13 Dew-point sensor.

pressures into usable forces for indicating pressure. Outputs of pressure sensors may be pneumatic, electric analog, or electric on-off.

Pneumatic air-velocity sensors are of the differential-pressure or of the deflected-jet type. The differential-pressure types use a restriction in the air stream, such as an orifice plate, or sense static and total pressure to generate differential pressures that represent air velocity (see Fig. 52.14). The deflected-jet type has a small air jet flowing across the measured air stream from an emitter tube. The air is captured in a collector tube and generates a recovery pressure (see Fig. 52.15). When the velocity of the measured air stream is low, most of the air jet depinges on the collector tube and the recovery pressure is high. As the air-stream velocity increases, the air jet is deflected and recovery pressure diminishes. The recovery pressure is, therefore, a direct indication of air-stream velocity.

Electric air-velocity sensors use a heated wire or thermistor placed in the air stream. The amount of current required to maintain the wire or thermistor temperature varies with the cooling effect of differing air velocities and, therefore, is a measure of air velocity. A reference wire or thermistor shielded from the air stream compensates for varying air temperatures.

Water-flow sensors may be differential-pressure types, such as orifice plates, pitot tubes, or flow nozzles, that have limited range or vortex-shedding, turbine, or magnetic types that have greater range but are more expensive.

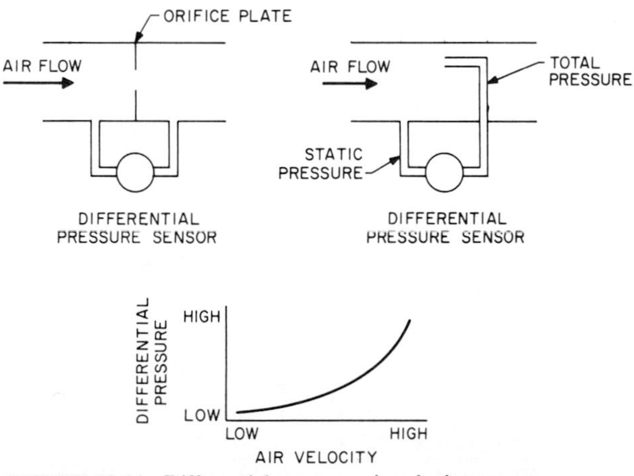

FIGURE 52.14 Differential-pressure air-velocity sensors.

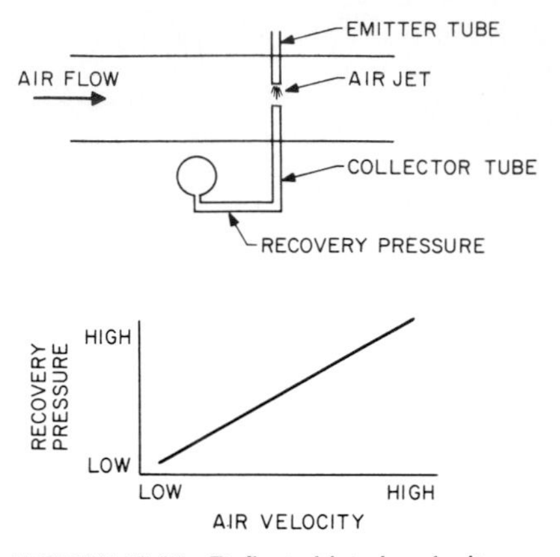

FIGURE 52.15 Deflected-jet air-velocity sensor.

Other sensing elements including smoke and high-temperature detectors, specific-gravity, current, CO, and CO_2 sensors are often used for complete control of HVAC systems.

52.2.2 Controllers

Controllers provide these set-point and for some the proportional-band integral and derivative parameters of a control loop. They compare the sensor signal with the set point and output a corrective signal as determined by the controlled settings. This signal may be *direct-acting*, increasing with sensor-signal increases, or *reverse-acting*, decreasing with sensor-signal increasing. Controllers may incorporate a sensing element for sensing and controlling in one device. Proportional controllers may also be designed to use remote sensors and are called sensor-controller systems.

Controllers may be pneumatic or electric. Pneumatic controllers receive a sensor signal and output a proportional signal between 3 and 13 lb/in^2 (20.6 and 270 kPa). The controller may be a nonrelay or relay type. Nonrelay types use a restricted supply air, bleeding varying amounts to the atmosphere to generate a corrective output signal (see Fig. 52.16). Since the capacity of the output signal is restricted, amplification should be limited to small volume-control elements or where long response times are acceptable. Relay-type controllers incorporate a capacity amplifier for the corrective signal for greater output volume.

Electric controllers also may have integral or remote sensors. Outputs are two-position to cycle equipment, floating to open, hold, or close a final-control

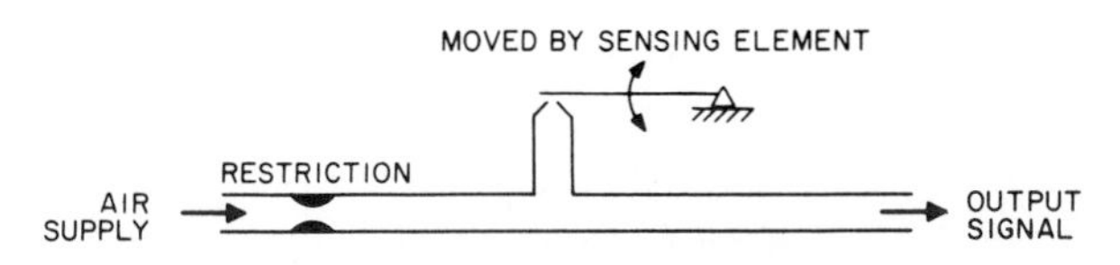

FIGURE 52.16 Nonrelay pneumatic controller.

element or proportioning to position a final-control element. Proportioning controllers may be electronic and analog or digital.

Electronic and analog controllers are similar to pneumatic controllers. That is, their response to a sensor signal is fixed by their design and only their parameters, such as set-point, direct- or reverse-action, proportional-band, and if included integral- and derivative-timing, are adjustable. Digital controllers are microprocessor-based, and their response to a sensor signal is programmable. This provides great flexibility for the application of a digital controller and allows control strategy changes after installation. Digital controllers generally are able to handle a number of control loops independently or with some interaction, as desired, and use the signal from a single sensor in any control loop.

Pneumatic and electric controllers may also provide indication and/or recording of the value of the sensed variable for visual checks or for a history of system operation. Transducers may be used with controllers to convert sensor signals and controller outputs from pneumatic to electric, or vice versa, as required by the controller or final controlled elements.

52.2.3 Final-Control Elements

Final-control elements are valves, dampers, electric heaters, relays, and motors for fans, pumps, burners, refrigeration, and other HVAC equipment. All these elements may be operated on-off or to position, while valves and dampers may also be used with floating-control and proportional-control modes. Final-control elements may be normally open, that is, open with no controller signal, or normally closed.

Pneumatic valves and damper operators have a flexible diaphragm or bellows attached to a valve stem or damper linkage (see Fig. 52.17). Movement is opposed by a compression spring, while a pneumatic controller signal is connected to the operator and generates a force depending on the pressure of the signal in the area of the diaphragm. When the signal pressure multiplied by the area exceeds the force of the spring, the operator moves, also moving the valve or damper until the spring force and controller signal's generated force are in balance. When the controller signal reduces, the spring causes the operator to retract. By selection of springs, various operator position-controller signal characteristics can be attained. Since operator position depends on the balance between the diaphragm and spring force, any external force from a valve or damper will offset the operator. For control systems requiring accurate synchronization of final-control elements, this may be a problem. For precise positioning, a positive positioner is used. It senses controller input signal and operator position and feeds or bleeds air to or from the operator to position it regardless of external load.

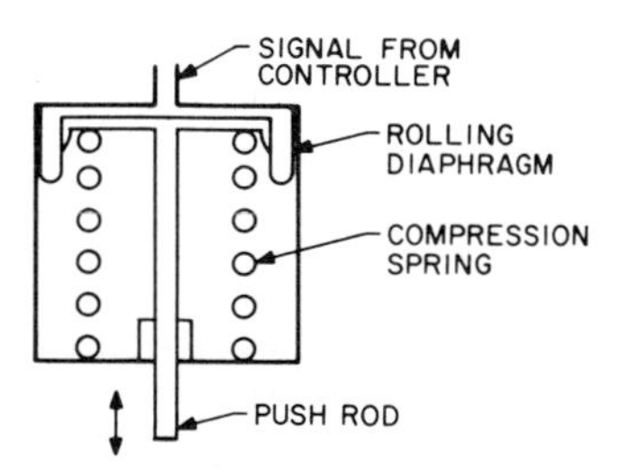

FIGURE 52.17 Pneumatic valve or damper operation.

Electric motors are unidirectional, spring-return, or reversible. Unidirectional motors are for two-position operation: opening a valve or damper in half a revolution and closing it in the second half. Once initiated, the motor continues through half a revolution. When it receives a second signal, the controller continues through the next half-revolution cycle.

Spring-return motors are also used for two-position operation. A control signal drives the motor to one end of its movement and holds it there. When the controller is satisfied and ends its output, the motor is driven back by an internal spring which was wound during its initial movement.

Reversible motors are used with floating- or proportional-control modes. The motor can be operated in either direction, depending on the controller signal; it stops when the signal stops. For proportional control, a potentiometer on the motor shaft is used to signal the motor position to the controller.

52.2.4 Auxiliary Equipment

Many control systems require auxiliary equipment for complete system operation. For pneumatic control systems, these include:

Air-compressor systems with dryers and filters to provide clean dry air at the proper pressures to power the system

Pneumatic-electric relays for switching electric loads with pneumatic signals and electric-pneumatic relays for switching pneumatic lines with electric signals

Two-position relays for converting proportional pneumatic signals to two-position and proportional relays for reversing signals, selecting the higher or lower of two or more signals, averaging two signals, adding or subtracting a constant from a signal, and amplifying signal pressure or air-flow capacity

Switching relays to divert signals automatically or manually

Gradual switches to manually vary air pressure in a circuit

Electric systems utilize transformers to provide required voltage, relays to switch electric loads larger than a controller's capacity, potentiometers for manual positioning of proportional control devices or for remote set-point adjustments, manual on-off switches, and auxiliary switches on dampers and valves for control of sequence operation.

Other auxiliary devices are common to pneumatic and electric systems. These include step controllers for operating a number of electric switches by a proportional operator to control stages of electric heating or refrigeration. Power controllers may be solid-state, saturable-core, or variable autotransformers and are used to control electric resistance heaters with a proportional pneumatic or electric control signal. Clocks and timers are used to control apparatus or control-system sequences based on time of day or elapsed time.

52.2.5 Pneumatic, Electric, Electronic Comparisons

A pneumatic control system may be as shown in Fig. 52.18. Advantages of pneumatic controls are the inherent modulation sensors and controller signals and the low cost of modulating operators. The system is explosionproof, and the control elements require little maintenance and are easy to troubleshoot. Disadvantages are the need for an air-compressor system which is too expensive for small sys-

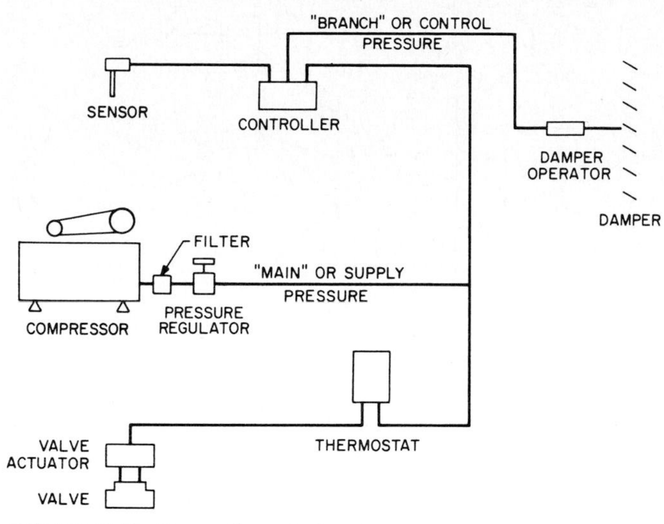

FIGURE 52.18 Pneumatic control system.

tems. Compressed air must be piped to all controls, increasing installation costs, and transducers are required for interfacing to automation systems.

Electric control systems, as shown in Fig. 52.19, can be installed wherever

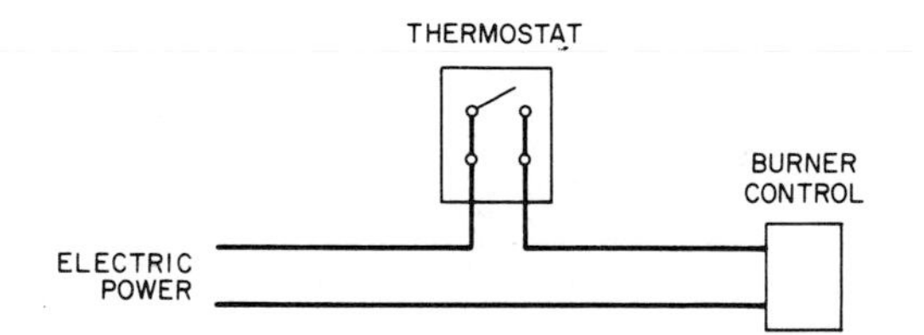

FIGURE 52.19 Electric two-position control system.

electric power is available and are low-cost for small simple systems. However, modulating operators are expensive, and explosionproof housings are required in hazardous areas. Electronic modulating controls allow remote sensors and set-point adjustment, provide high accuracy, and readily interface with automation systems. They are higher in cost and require more skilled personnel for trouble-shooting.

Direct digital controllers (see Fig. 52.20), offer many advantages. They have very high accuracy, so control-loop accuracy is limited only by the sensor and final-control element. They are capable of complex control algorithms which may easily be changed by reprogramming. This allows flexible building operation during construction, startup, occupancy, full-occupancy, and expansion phases. An entire building can be controlled from one location, and building by energy-management strategies can be accomplished. Direct digital controllers are higher

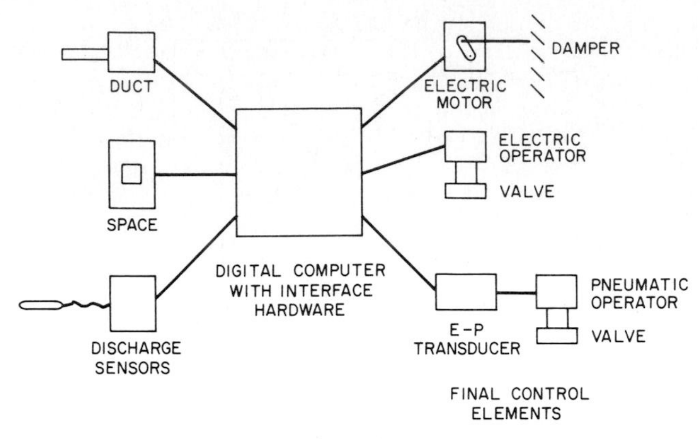

FIGURE 52.20 Direct digital control system.

in cost but can handle multiple control loops and require ordinary maintenance personnel skills.

52.3 CONTROL APPLICATIONS

These are defined by general functions in the total HVAC system.

52.3.1 Boiler Control

The control of steam or water boilers involves three types of functionality: flame safeguard, load control, and control of excess air. Steam boilers also include water-level control. The means of accomplishing each of these functions is influenced by the size of the boiler involved. In general, small boilers have a single control package which accomplishes flame safeguard and load control with no need for excess air control. Large boilers can have different control packages for all three functions. The application of flame-safeguard control is very dependent upon boiler and burner design and therefore is normally supplied as a complete package by the boiler-burner manufacturer. The type of fuel(s) selected, the size of design load, and the type of approval required are the primary decisions of the HVAC designer that establish the type of boiler controls that are appropriate. On larger-size installations, the method of load control, the use of multiple boilers, and the cost effectiveness of appropriate types of excess air control are additional considerations for the HVAC designer. This section explains means of accomplishing the three basic types of boiler-control functionality. It also explains an auxiliary function of monitoring smoke control.

Flame-Safeguard Control. The objective of flame-safeguard control is to ensure that safe conditions exist for initiating and sustaining combustion. On small- to medium-size [up to 400,000 Btu/h (422,000 kJ/h)] burners, the flame-safeguard-control function is provided by a package called a primary control. The primary control starts the burner in the proper sequence, proves that combustion air is available, purges the combustion chambers and proves the burner flame is established, and supervises the flame during burner operation. It causes safety shutdown on failure to ignite the pilot or main burner or on loss of flame. In addition, the primary control checks itself against unsafe failure. Typically, a check for flame-simulating failure and a continuity check of safety-switch circuitry is made. The ultimate in self-checking capability is a dynamic self-check system that checks its internal circuitry about one to four times each second during operation.

On larger-size systems over 400,000 Btu/h (422,000 kJ/h), additional timing functions are required, and the package is called a flame-safeguard-programming control. Added functions include purging with a minimum flow rate before light-off, sequencing of ignition and pilot light-off, proof of pilot before main valve opening, and sometimes it includes interrupting the pilot to prove the main flame before the normal "run" period. The function of proving the flame is common to all sizes of burner in both primary control and programming types of flame-safeguard controllers. However, there are two general approaches to sensing flame. These approaches are flame-rod sensing and optical detection. (This ignores thermal sensing which is only appropriate for domestic-size burners.) Flame-rod sensing depends upon the ability of a flame to conduct current due to the ionized combustion gases in a flame. Flame-rod rectification that passes current in only one direction is used because it provides protection against the high-resistance leakage to ground giving a false indication that flame is present. This means of sensing is suitable for gas. It is not usually suitable for oil because of the tendency to form carbon and ash deposits on the rod. Optical sensing is implemented in three methods that sense visible, infrared, or ultraviolet (uv) light. (See Fig. 52.21 for characteristics of all four methods of sensing.) The visible-light optical method uses a rectifying photocell, typically cadmium sulfide, and is suitable only on oil because gas flames may generally not give sufficient visible light. The rectification method is used because it again gives shorted lead protection. The infrared optical detection is accomplished by use of a lead sulfide cell and is suitable for gas or oil flames. A sensing design is used that reacts to flickering flame and not to a steady signal from a hot refractory surface. The ultraviolet detection can sight either a gas or oil flame, but it must be positioned so that it does not see ignition spark or hot refractory over 2500°F (1371°C) or is masked by oil vapor that absorbs uv.

Firing-Rate Control. Control of a boiler to meet variations in load is sensed by temperature or pressure and accomplished in three general ways: on-off control of fixed firing rate, high-low firing rate, and modulated firing rate. The on-off control is simply accomplished by a pressure or temperature controller connected to the primary controller to start and stop combustion as required to meet the load. This is the normal method of control on smaller boilers (under 400,000 Btu/h (422,000 kJ/h) where the start-stop sequence is simpler and a primary control is used rather than a programmer. High-low firing provides two rates according to heat demand. The high and low fire rates are obtained by simultaneously positioning the air damper and fuel valve to the proper flow rate of air and fuel. On two-stage burners, a second fuel valve is opened to provide high fire. High-low firing is used on large commercial and industrial boilers to provide a safe light-off. The low rate is used for safe light-off and low loads, while the high rate is used for heavy loads. A modulated firing rate to meet load variations is accomplished by proportional control of fuel and air between low and high fire rates. The methods of accomplishing this range from simple linkages between air and fuel valves to

	Flame rod	Rectifying photocell	Infrared (lead sulfide) detector	Ultraviolet detector
Will supervise oil	No	Yes	Yes	Yes
Will supervise gas	Yes	No	Yes	Yes
Will detect ignition spark	No	No	No	Yes
Type of signal	dc	dc	ac	dc
Light responsive	None	Visible	Infrared	Ultraviolet
Max. ambient temperature at the sensor or cell	500°F (260°C)[a]	165°F (74°C)	125°F (52°C)	125–250°F (52–121°C)[b]
False flame signals by:				
Inductive pickup	No	No	No[c]	No[d]
Capacitive pickup	No	No	No[e]	No[d]
Refractory glow	No	Yes	No[f]	No[g]
Checkout tests required:				
Hot refractory saturation	No	No	Yes	No
Hot refractory hold-in	No	Yes	Yes	No
Ignition interference test	Yes	No	No	Yes[h]
Pilot turndown test	Yes	Yes	Yes	Yes

[a]Maximum insulator temperature is 500°F (260°C). Maximum rod temperatures are 2000°F (1093°C) for Jellif Alloy "K," 2462°F (1350°C) for Kanthal A-1, and 2600°F (1427°C) for Globar.

[b]C7012E and F—135°F (57°C); C7012A and C—135°F (57°C); C7044A—212°F (100°C); C7027A—215°F (102°C); C7035A—250°F (121°C).

[c]Leads must be run alone in grounded conduit all the way to the wiring subbase.

[d]C7027 and C7035 should use No. 14 TW leadwire when running long distances of up to 1100 ft (3609 m). C7012E should use coaxial cable; RG62U will provide satisfactory performance with lengths of the order of 500 ft (1641 m).

[e]Leads must be BX cable, shielded cable, or twisted pair. Conduit must be rigid or fastened securely to minimize vibration.

[f]Will not sense the hot refractory alone, but turbulent hot air, steam, smoke, or oil spray may cause the radiation to fluctuate, stimulating flame.

[g]An ultraviolet detector will respond to hot refractory at different temperatures: C7076 maximum hot refractory at 2200°F (1204°C); C7012 maximum hot refractory at 2300°F (1260°C); and C7027, C7035, and C7044 maximum hot refractory at 2800°F (1538°C).

[h]An ultraviolet detector will sense ignition spark; ignition spark response test is required.

FIGURE 52.21 Flame-detector comparison chart. (*Courtesy of Honeywell, Inc.*)

complex metering and flow-control systems. The accuracy of the fuel-air-ratio control influences efficiency. Modulated control of low firing rates reduces stack temperatures which improves efficiency for those low loads. Modulated control is appropriate for applications requiring close control of temperature or pressure. It may also be appropriate for larger boilers with significant load variations where improved seasonal efficiency can be accomplished and justifies the added cost.

52.3.2 Control of Excess Air

The efficiency of a boiler is dependent upon the efficiency of combustion and the thermal efficiency of the boiler. Both of these efficiencies are dependent upon the air-fuel ratio. Too little air causes incomplete combustion and a sharp drop in efficiency, while too much air causes excessive thermal losses. The theoretical optimum air quantity is described as 0 percent excess air. See Fig. 52.22 for a characteristic of efficiency as a function of excess air. Both O_2 and CO_2 in stack gas are indicators of the quantity of excess air, as shown in Fig. 52.23. As can be seen, O_2 is a more sensitive indicator of excess air below 60 percent. The quantity of combustibles such as H_2 and CO in stack gas varies with excess air also;

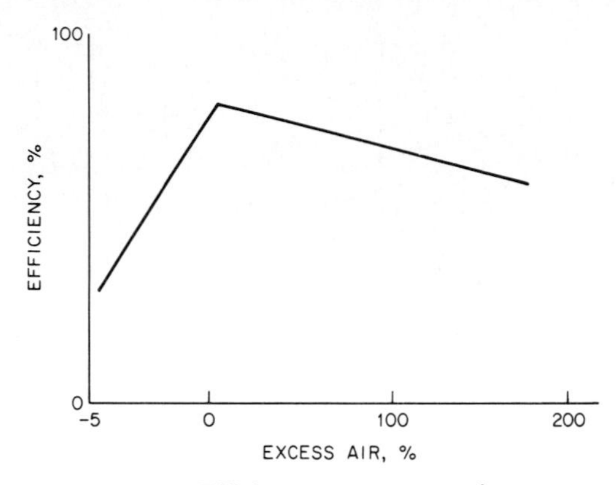

FIGURE 52.22 Efficiency versus excess air.

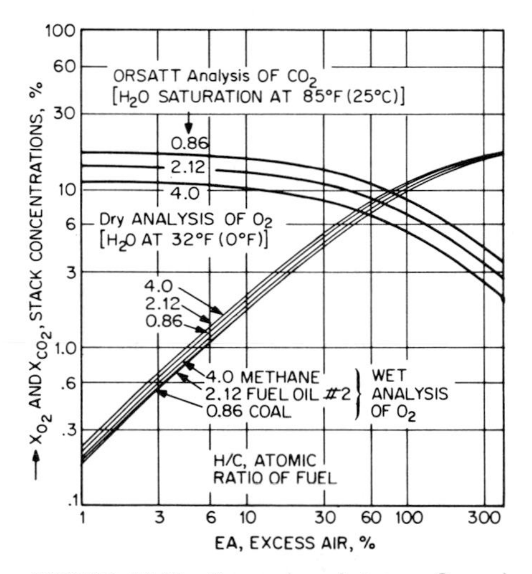

FIGURE 52.23 Comparison between O_2 and CO_2: stack-gas analysis versus excess air. (*Courtesy of Honeywell, Inc.*)

however, their characteristics vary with fuel type and boiler design and can also indicate other combustion abnormalities. Measurement of oxygen in stack gas is therefore the preferred means today of controlling excess air to improve efficiency. The zirconium oxide oxygen sensor used with microprocessor-based controllers is the most common type of excess-air controller. The control of air quantity is to maintain a minimum practical percentage of oxygen (typically 1 to 5 percent) in the stack gas, which accomplishes maximum efficiency. The other factors which can guide efficiency improvement are carbon monoxide content and temperature of the flue gas. Monitoring of these conditions may be justified. The stack temperature in a given installation increases with firing rate and with fouling of the heat exchanger. Increases in stack temperature both decrease efficiency and increase the effect on excess air in decreasing efficiency. See Fig.

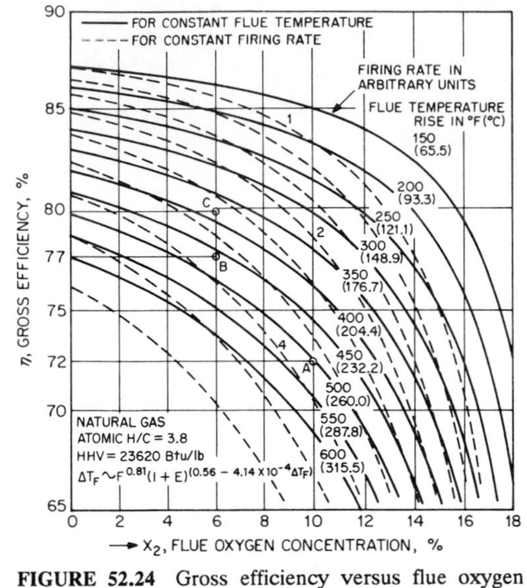

FIGURE 52.24 Gross efficiency versus flue oxygen concentration. (*Courtesy of Honeywell, Inc.*)

52.24 for these characteristics for natural gas. Example point A shows 10 percent oxygen and 500°F (260°C) stack yields with 72 percent efficiency. Example points B and C at 6 percent oxygen show how a decrease in the firing rate and flue temperature increases efficiency from 77 to 80 percent. Increases of stack-gas temperature for a given load can be an indication of the need to clean a heat exchanger. Monitoring of CO in conjunction with O_2 control of excess air can be used as an indicator of combustion abnormalities. The improvements in efficiency from control of excess air can range from 3 to 10 percent. Initial cost and maintenance requirements must be compared to the savings for any specific system considered.

Smoke Monitoring and Control. A smoke-detector system is used to ensure compliance with antipollution regulations. This is typically accomplished by continuous sensing of smoke capacity in a smoke stack or boiler breeching. The sensor and light source are installed on opposite sides of the chamber. Opacity is measured in Ringelmanns, with 0 meaning clear and 5 meaning 100 percent opacity. Both alarm action and shutdown action with a persisting alarm are used. Induction and recoding of readings are also used.

52.3.3 HVAC Fan Systems

As described in the chapters that cover fans and air-handling systems, HVAC fan system basic categories are single- and dual-duct, while zone temperatures are maintained by constant-volume–variable-temperature or variable-volume–constant-temperature discharge air. There are many terminal unit and fan configurations within these categories, and this section will discuss control techniques that may be used to achieve desired system operation.

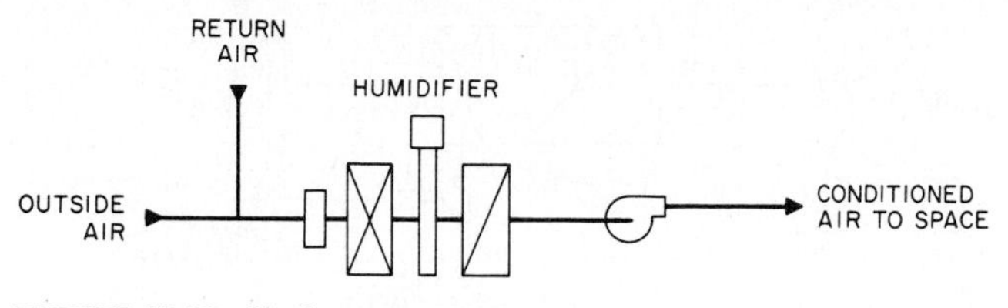

FIGURE 52.25 Single-zone system.

Single Duct. Single-duct systems contain the primary heating and cooling coils in a series air-flow path and use a single-duct distribution system to supply all terminal units. Single-zone systems (see Fig. 52.25) are applicable where load variations occur uniformly in the zone or where the load is constant, as the load is sensed at only one point in the zone. Control of a single-zone system may be by varying cooling or heating or reheating to change zone-discharge air temperature to maintain space temperature while discharge air volume is constant. Humidity may also be controlled based on zone requirements. Since system control is directly based on space temperature and humidity, close system control may be achieved. If reheat is not used, summer dehumidification is not possible.

Multiple-zone systems (see Fig. 52.26) distribute conditioned air to many

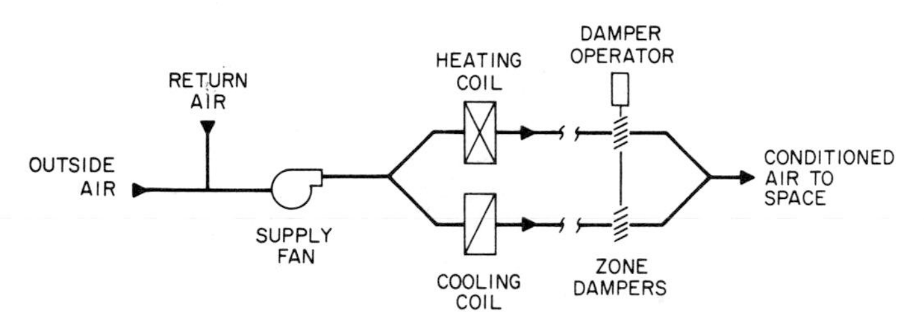

FIGURE 52.26 Multiple-zone system.

zones from one air handler. Zone-temperature control is achieved by controlling each zone's discharge-air temperature. The following are basic single-duct terminal units controlled circuits.

Constant-volume reheat (see Fig. 52.27): The balance damper may be set manually or controlled by a fixed set point, mechanical volume regulator, or air-flow controller. The space thermostat responds to space temperature to control the amount of reheat, which may be hot water or electric heat. Close-temperature and dew-point control may be achieved, but energy is wasted due to reheating cooled air.

Variable-volume, cooling only, pressure-dependent (see Fig. 52.28): Maximum and minimum air volumes are set with mechanical stops, while the air volume between these limits is controlled by the space thermostat signal to the damper operator. This system is low-cost and avoids reheat energy waste. Discharge volume is affected by duct static pressure, so terminal units affect one another as they change air volume. This interaction can result in loss of temperature control but is minimal with duct static pressures less than 1.0 in (2.5 cm) of water.

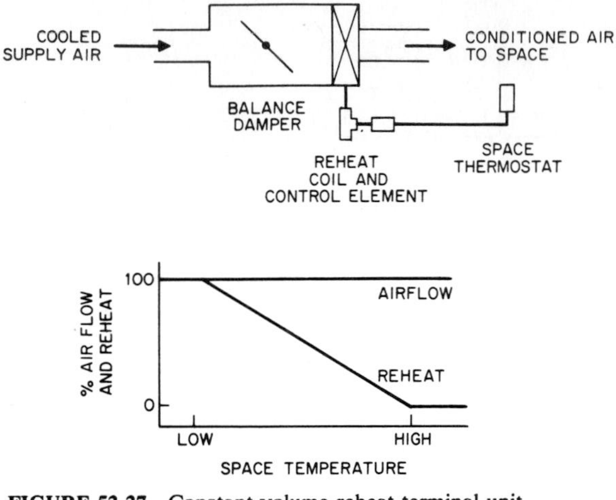

FIGURE 52.27 Constant-volume reheat terminal unit.

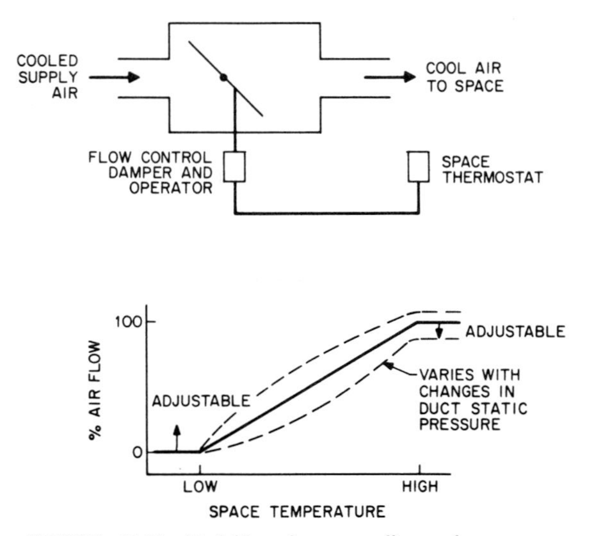

FIGURE 52.28 Variable-volume, cooling only, pressure-dependent terminal unit.

Variable-volume, cooling only, pressure-independent (see Fig. 52.29): Maximum and minimum air volumes are adjustable set-point limits of the air-velocity controller. The thermostat signal resets the air-velocity controller set point to maintain space temperature. The velocity controller will maintain the required air velocity, and therefore the discharge volume, regardless of duct static-pressure variations, avoiding interaction between zones.

Variable-volume, heating and cooling, pressure-independent (see Fig. 52.30): The thermostat signal resets the air-velocity-controller set point, between maximum and minimum limits, as required by the space cooling load. For heating, air ve-

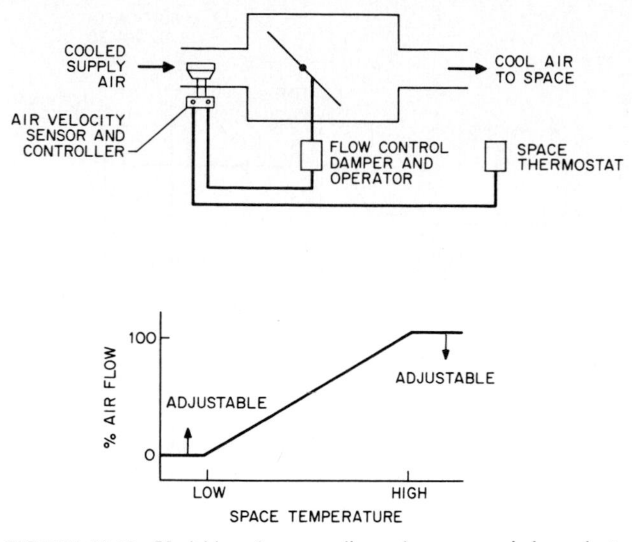

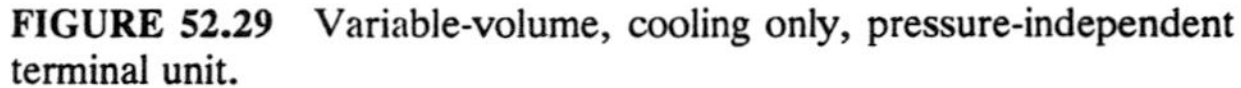

FIGURE 52.29 Variable-volume, cooling only, pressure-independent terminal unit.

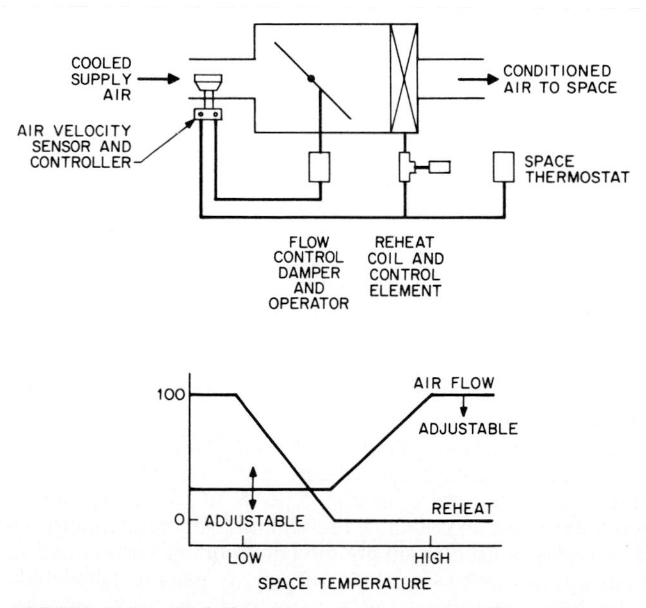

FIGURE 52.30 Variable-volume, heating and cooling, pressure-independent terminal unit.

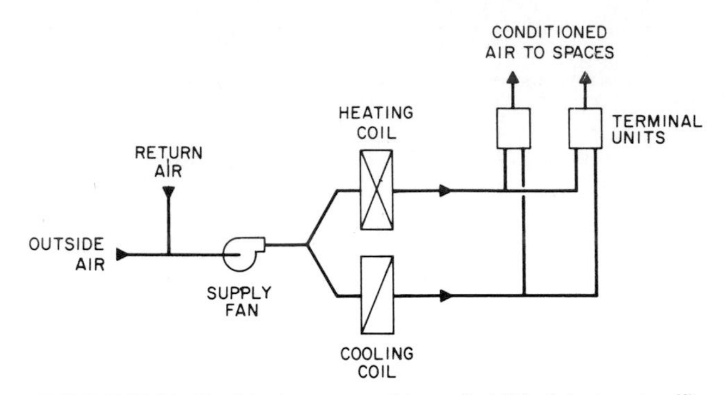

FIGURE 52.31 Double-duct system (also called "dual-duct system").

locity is maintained at minimum and reheat is modulated as required. The reheating of cooled air is minimized, reducing energy requirements.

Double Duct (Dual Duct). Double-duct (dual-duct) systems (see Fig. 52.31) condition all the air in a central system and distribute it to condition spaces through two-duct systems, one for cooled air and one for heated air. In each space, a terminal unit blends the two air supplies to maintain space temperature.

Constant-volume, pressure-dependent (see Figure 52.32): The damper and damper operators are connected to respond to thermostat signals in opposite directions; i.e., when one opens the other closes. As room temperature decreases, the thermostat signal changes, causing the hot supply damper to open and the cold supply damper to close. This varies the discharge temperature while maintaining the discharge volume. Supply volume will change with duct-pressure changes, upsetting discharge temperature and space control.

Variable-volume, pressure-independent (see Fig. 52.33): Air-velocity controllers provide independently adjustable velocity set-point limits and proper velocity controls, regardless of supply-duct-pressure variations. The hot and cold air can be controlled to avoid or minimize mixing, dependent on minimum ventilation needs, and therefore minimize simultaneous heating and cooling.

Air Handlers. Air handlers deliver conditioned air to a supply-duct system. This air may be filtered, heated, cooled, humidified, dehumidified, and a source of outside ventilation air. The conditioned air may be delivered at a constant volume or a variable volume, depending on the terminal units used. A return or relief fan may also be used to ensure positive ventilation. Air-handling systems consist of four basic sections: mixed-air, conditioning, fan, and terminal (see Fig. 52.34).

The mixed-air section is a plenum where recirculated or return air and fresh or outside air are mixed. A filter removes dust, dirt, and other particles from the air before it enters the conditioning section. Special requirements of kitchens, laboratories, and other areas do not allow recirculating air from conditioned spaces so that outside air only is used to supply the conditioning section and all air for the spaces exhausted. Systems using all outside air are usually called makeup-air systems.

Outside and return air are mixed in the mixed-air section, and the quantity of

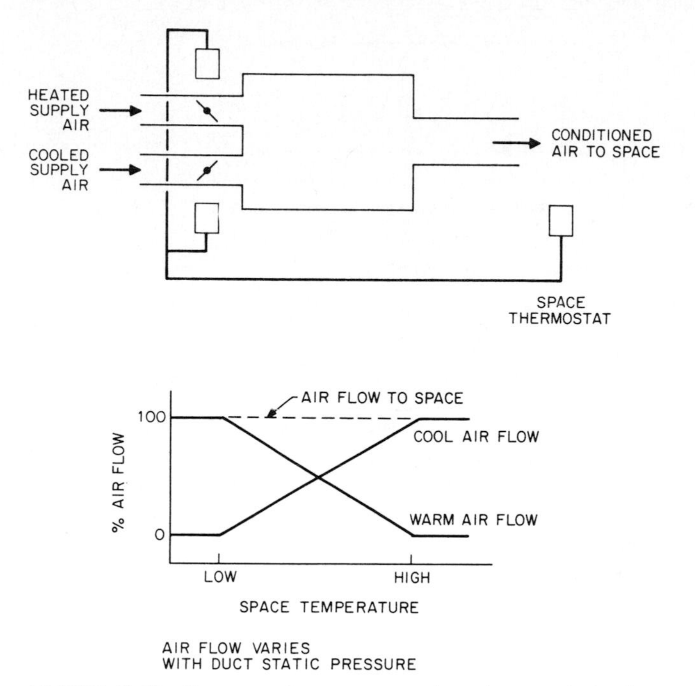

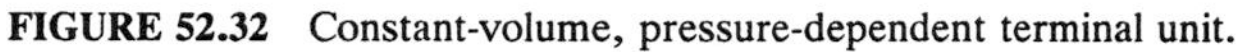

FIGURE 52.32 Constant-volume, pressure-dependent terminal unit.

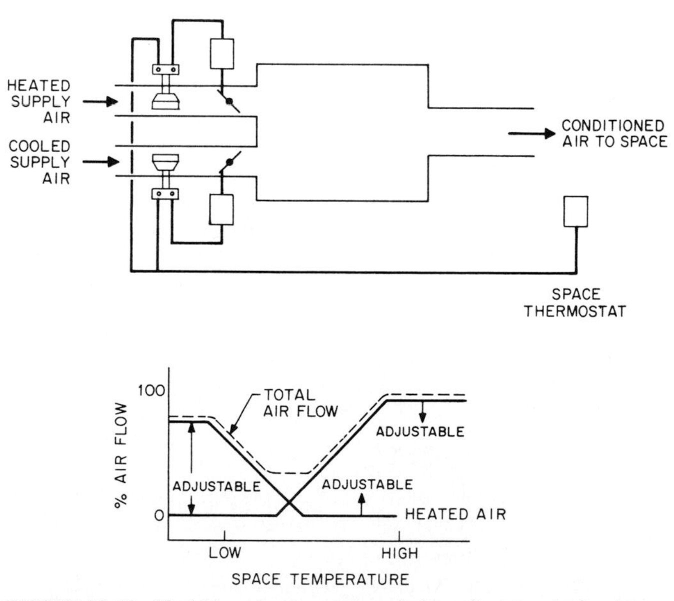

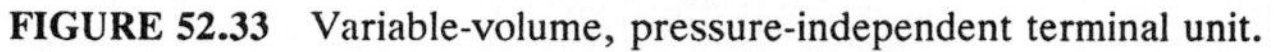

FIGURE 52.33 Variable-volume, pressure-independent terminal unit.

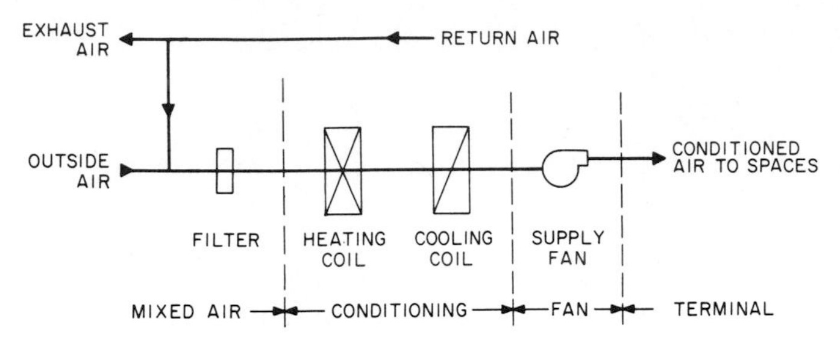

FIGURE 52.34 Air-handler sections.

outside air is usually controlled in one of three ways: fixed minimum, manually adjustable, and automatically controlled.

Fixed minimum is used only on constant-volume systems (see Fig. 52.35) to meet ventilation requirements of local codes. The outside damper may be fixed to allow the required quantity of outside air into the system or the outside damper may be motorized to open to the desired position during building occupancy only, avoiding unnecessary ventilation during morning warmup or cooldown operation.

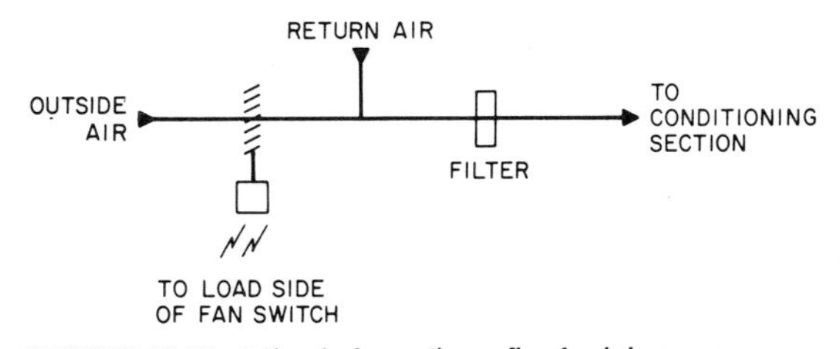

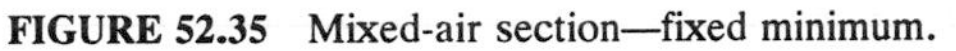

FIGURE 52.35 Mixed-air section—fixed minimum.

Manually adjustable outside-air systems, also used only on constant-volume systems (see Fig. 52.36), provide a positioning switch to vary the amount of outside air introduced into a building. An additional damper is required in the return-air duct. The outside- and return-air dampers are connected so that as one opens the other closes. Therefore, as the quantity of outside air is increased, the amount of return air is decreased, and vice versa, maintaining a constant total

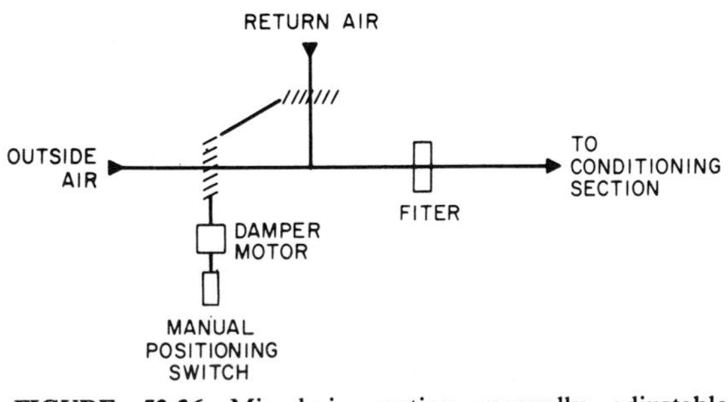

FIGURE 52.36 Mixed-air section—manually adjustable minimum.

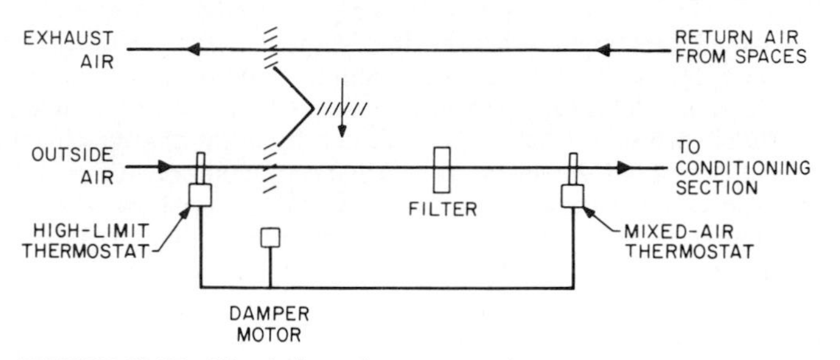

FIGURE 52.37 Mixed-air section—economizer system.

quantity of mixed air. This system provides the capability of using outside air to provide space cooling when outside temperatures are below inside temperatures. Note that a low-temperature limit control is required to prevent the low freezing air from entering water or steam coils.

Automatically controlled mixed-air sections are called "economizer" systems (see Fig. 52.37). These may be used with constant- and variable-volume systems and automatically control the proportion of outside and return air to provide space cooling when outside temperatures permit. The return and outside dampers are controlled by a duct thermostat, typically the conditioning-section discharge thermostat, to provide a heating, ventilation, refrigeration, and conditioning sequence.

When outside temperatures are cold, dampers are positioned to allow only the minimum amount of outside air for ventilation. As outside temperatures increase and the need for heating stops, more outside air is introduced and less return air is recirculated. As outside temperature continues to increase, more outside air is used until there is full outside air flow and no return air is circulated. When outside temperature rises further so that ventilation no longer satisfies space cooling needs, the outside air damper closes to minimum and the return air damper opens. This limit on ventilation may be accomplished by an outside air temperature controller, sensing dry-bulb temperature, or by an enthalpy controller which compares outside- and return-air enthalpy and selects the source with minimum enthalpy. An enthalpy controller is useful in areas where outside air moisture content is high when dry-bulb temperature is low.

The conditioning section consists of heating and cooling coils and for some systems a humidifier as required to condition the supply air to meet space needs. Heating may be by steam, hot water, or electric coils and control by modulating or two-position space or duct thermostat to vary steam, hot water, or electrical flows through the coil. See Fig. 52.38.

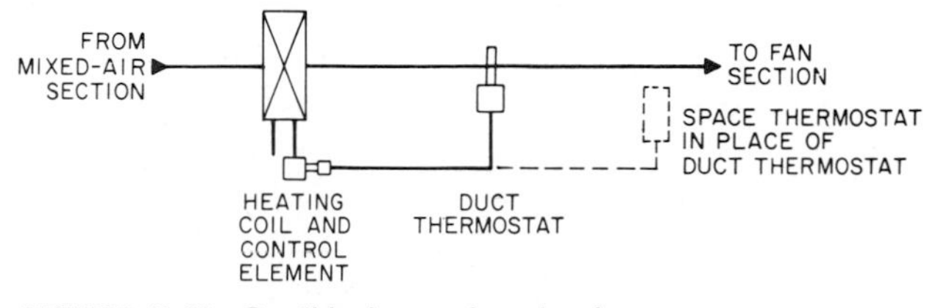

FIGURE 52.38 Conditioning section—heating.

Cooling is done by a chilled-water or direct-expansion coil. Chilled-water coils may be controlled by modulating or two-position space or duct thermostat, similar to hot-water coils. Direct-expansion coils have refrigerant flowing through them to directly absorb heat from the air as the refrigerant changes from liquid to gas in the coil. Control is by cycling the refrigeration compressor or opening and closing a solenoid valve to vary refrigerant flows. A two-position space or return-air-duct thermostat is used. Discharge temperature control results in short-cycling the compressor or a wide temperature swing in discharge air temperature and is therefore not recommended.

Face and bypass dampers may also be used with either type of cooling coil to control duct air temperature (see Fig. 52.39). For direct-expansion coils, the compressor must be turned off when the face damper closes to prevent frost buildup on the coil. Humidifiers are used when moisture must be added to condition spaces (see Fig. 52.40). A humidifier water or steam valve is controlled by a space or return-air duct humidity controller. The humidifier must be turned off when the supply fan is off to avoid introducing liquid water in the duct.

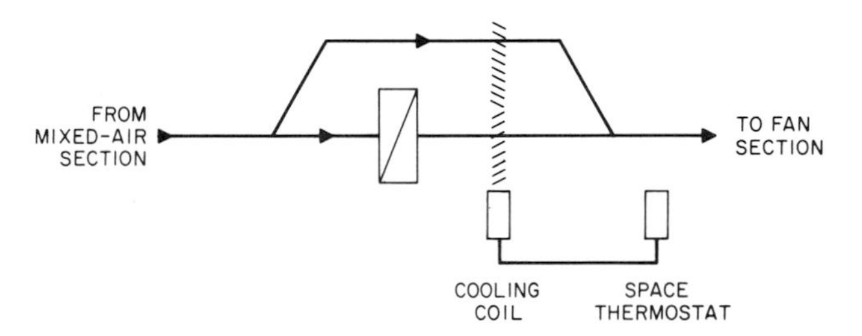

FIGURE 52.39 Conditioning section—face and bypass dampers.

The fan section consists of one or more supply fans and may be in a draw-through or blow-through configuration (see Fig. 52.41). For constant-volume systems, the supply fan is selected and adjusted by motor-fan pulley selection to deliver the design volume of supply air.

For variable-volume systems (see Fig. 52.42), duct static pressure at the pressure pickup is controlled at just enough pressure to ensure full air delivery at the last terminal unit. Fan discharge pressure and volume will vary to maintain con-

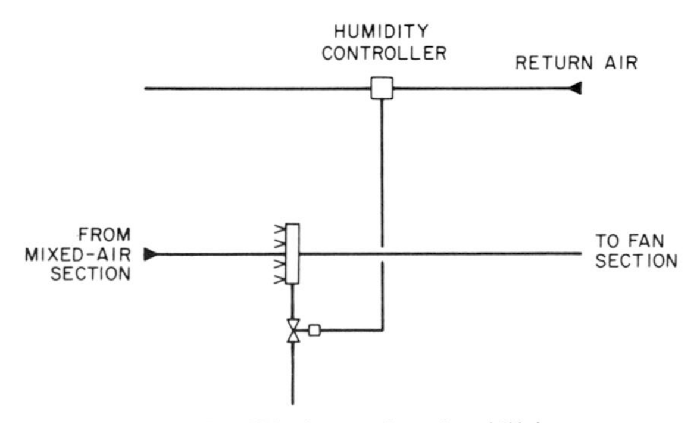

FIGURE 52.40 Conditioning section—humidifying.

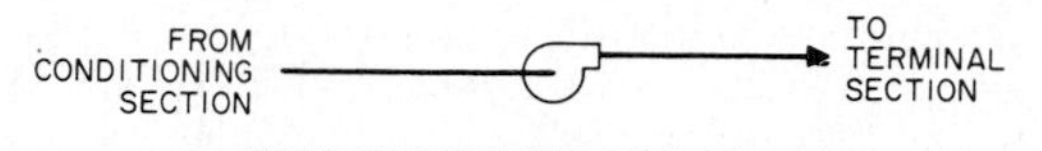

DRAW-THROUGH FAN CONFIGURATION

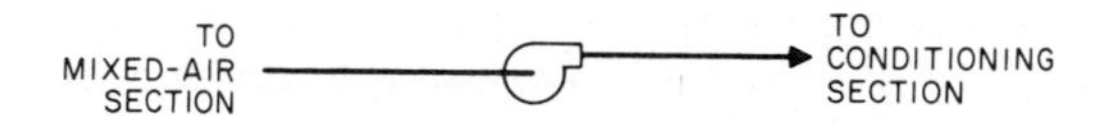

BLOW-THROUGH FAN CONFIGURATION

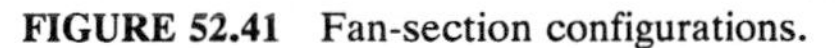

FIGURE 52.41 Fan-section configurations.

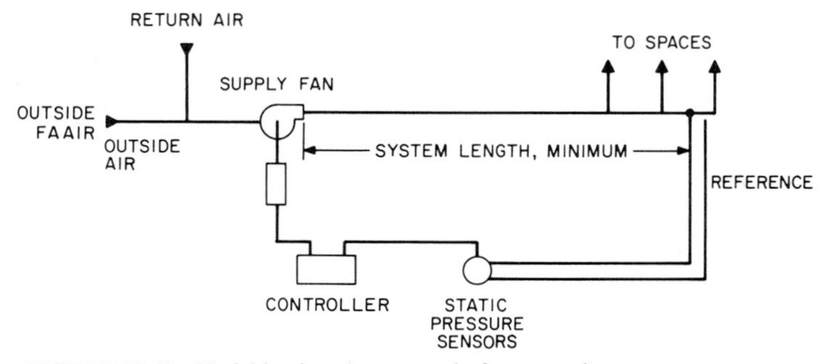

FIGURE 52.42 Variable-air-volume supply-fan control.

stant pressure at this point. Duct static pressure responds very quickly to changes in fan output; that is, the process has a short time constant. Therefore, for stable operations, a wide throttling range or proportional band is required. This results in a wide variation in duct static pressure, causing high static pressure during minimum-volume operation and excessive energy consumption. A proportional-integral-derivative controller should therefore be used for duct static pressure control of variable-air-volume (VAV) systems. The controller may be pneumatic, electronic, or direct digital, but the PID mode of control will provide stable, responsive control with no offset between set point and control point.

Some supply-duct systems split or fork into two or more trunks (see Fig. 52.43).

The static-pressure pickups should be near the end of each branch and a selector relay used to determine which branch has the lowest pressure. The fan is controlled based on the needs of that branch.

Duct systems may contain fire or smoke dampers between the fan and static-pressure pickups (see Fig. 52.44). If the damper should close, the downstream static pressure will decrease, resulting in full fan volume. Since the damper is closed, excessive duct pressure will develop between the fan and damper, possibly damaging the duct or damper. A high limit control at the fan discharge will prevent this. The control may be used to turn the fan off when a certain limit is reached or to modulate the fan at the limit.

Duct static pressure may be sensed by a single pitot tube if there is adequate straight duct upstream of the pitot tube to ensure an accurate, stable signal. Many times the duct configuration does not provide this, so a multiple pitot tube station with air-flow straighteners must be used.

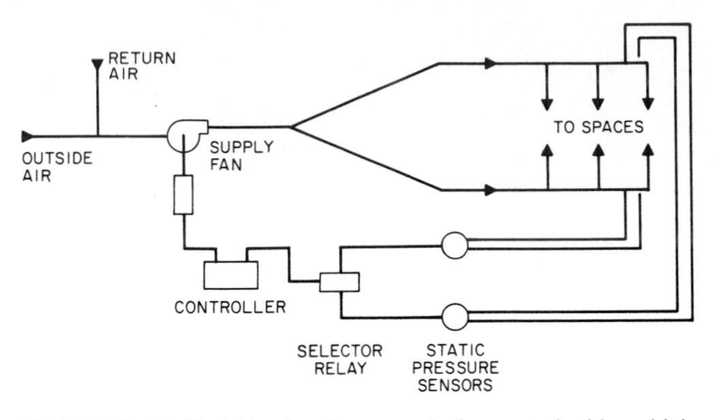

FIGURE 52.43 Variable-air-volume supply-fan control with multiple ducts.

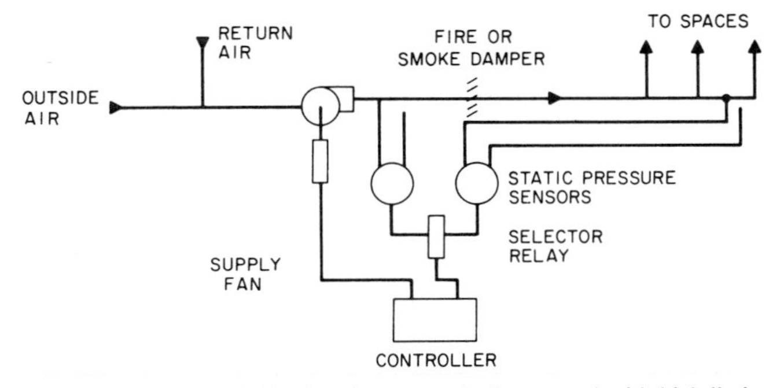

FIGURE 52.44 Variable-air-volume supply-fan control with high limit.

Variable-volume return fans are similar in design to VAV supply fans. They are controlled to return air from conditioned spaces according to the volume of air delivered by these supply fans. There are three basic methods of VAV return-fan control: slaving, direct building pressure, and tracking. Each has cost and operational advantages and disadvantages.

Slaving is the most basic control method. The supply-fan control signal is simply applied to the return fan and the fans are operated together, or slaved (see Fig. 52.45). Since return fans are usually smaller than supply fans or may be a different type and operate at different discharge pressures, the air volume of supply and return fans is not the same at a given control signal. This difference in volume results in changing amounts of ventilation air and building pressurization, wasting energy and causing drafts through doors and windows.

This problem may be minimized by adjusting the return-fan modulation mechanism so that the correct volume is returned at minimum control signal and at maximum control signal. Additionally, fans should be selected to have similar modulation characteristics.

Return-fan slaving is applicable to systems that modulate over a minimal range, typically less than 50 percent, have matched supply and return fans, and are less than about 20,000 ft^3/min (5600 m^3/min). If the conditioned spaces have

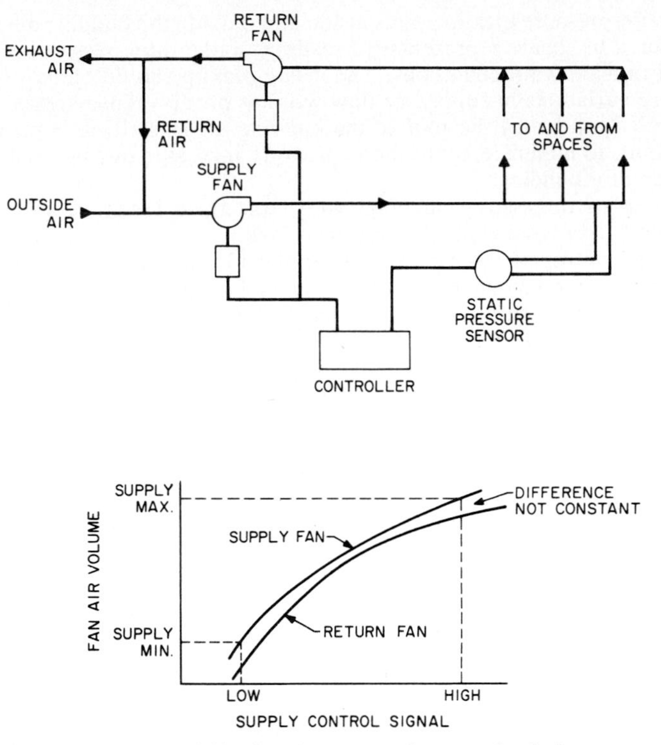

FIGURE 52.45 Variable-air-volume return-fan control—slaving.

exhaust fans such as for kitchens or toilets, this exhaust-air volume must be taken into account when setting up the return fans and must not change during operation or building pressurization will be upset.

Direct-building static pressure controls the return fan directly from the static pressure in the occupied space (see Fig. 52.46). This technique assures constant building pressurization but varies the amount of minimum ventilation air depending on changes in building exfiltration. These changes may be due to outside doors or windows being open or closed or exhaust fans being on or off.

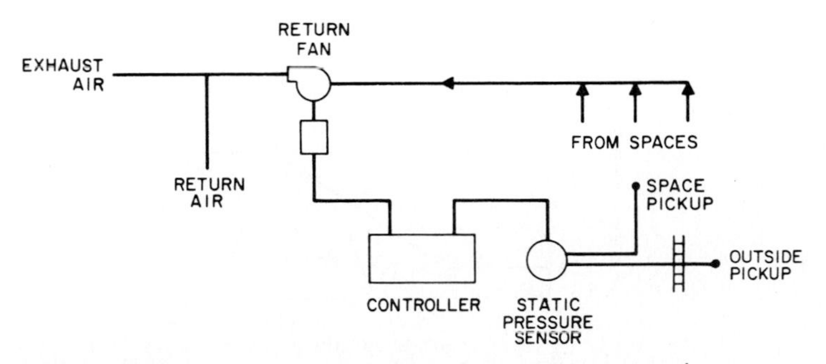

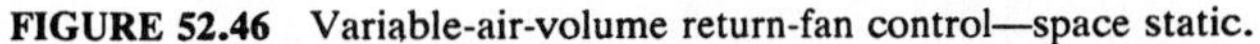

FIGURE 52.46 Variable-air-volume return-fan control—space static.

The static-pressure pickup points inside and outside the building must be carefully selected to ensure representative readings under all modes of system operation and outside wind conditions. The inside pickup should be in a large open area where variations in supply air flow will not produce false signals. The outside pickup must be on the roof of the building, at least 10 ft (3 m) above the highest point, to minimize atmospheric pressure increases due to wind impinging on the side of a building.

Low two- or three-story buildings with large open areas are candidates for direct-building pressure control of returned fans.

Tracking control of returned fans is applicable to almost all systems. With tracking, return-fan volume is controlled to equal supply-fan volume minus an allowance for building exfiltration and exhaust fans. This difference is maintained constant for all modes of operation (see Fig. 52.47).

Air-flow stations are used to sense supply and return air-flow volumes. These stations consist of multiple pitot tubes plus air-flow straighteners and provide an accurate and stable velocity-pressure signal. The signal from the return-fan air-flow station is amplified by a transmitter and input to a controller. The controller outputs a signal to the return fan to maintain volume at a set point.

Similarly, a supply air-flow station and transmitter provide a signal representing supply air-flow volume. This signal is input to the return-fan controller to change its set point based on supply air-flow volume. Therefore, as supply air

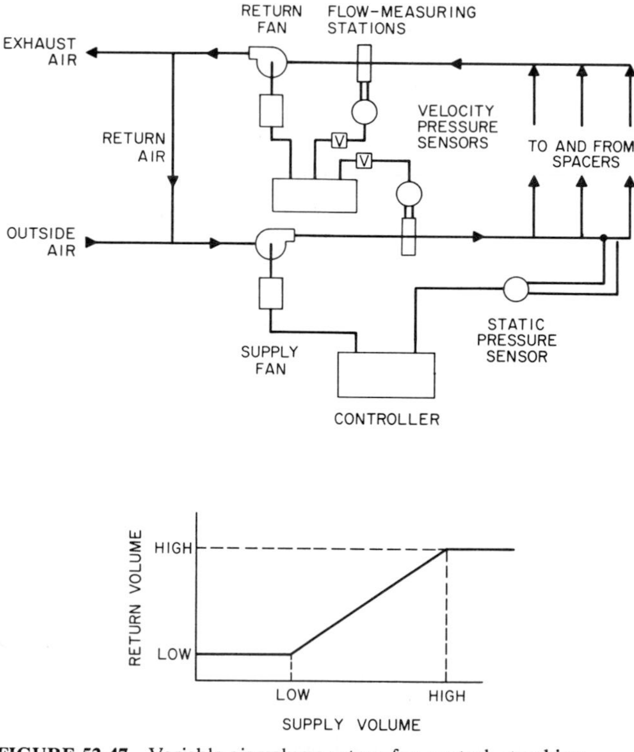

FIGURE 52.47 Variable-air-volume return-fan control—tracking.

volume is modulated, return air volume is modulated, maintaining a constant differential volume.

Tracking control may be used for all sizes of system and fan types as it is based on supply and return air volumes. However, air volume is measured by velocity pressure which has a squared relationship with air velocity. The tracking error introduced by this nonlinear relationship is minimal for systems with less than 50 percent turndown. However, systems with more than 50 percent turndown must have square root extractors between the transmitter and controller to linearize the signal.

Tracking-controls systems may be pneumatic, electronic, or direct digital. The control principles are the same. As for supply- and static-pressure control, return air flow is a fast process, so PID control modes must be used for stable, responsive, and accurate operation.

Some large systems with a larger range between full and minimum air volume use two or more fans to save energy and/or building space. These fans are staged; that is, one fan is used when air-volume requirements are low. When one fan can no longer deliver the required volume, the next fan is turned on and two fans operate in parallel; similarly, a third or more fans may be used.

Total air volume should be used as the criterion for turning fans on or off (see Fig. 52.48). When the first fan reaches full volume, the pneumatic-electric (PE)

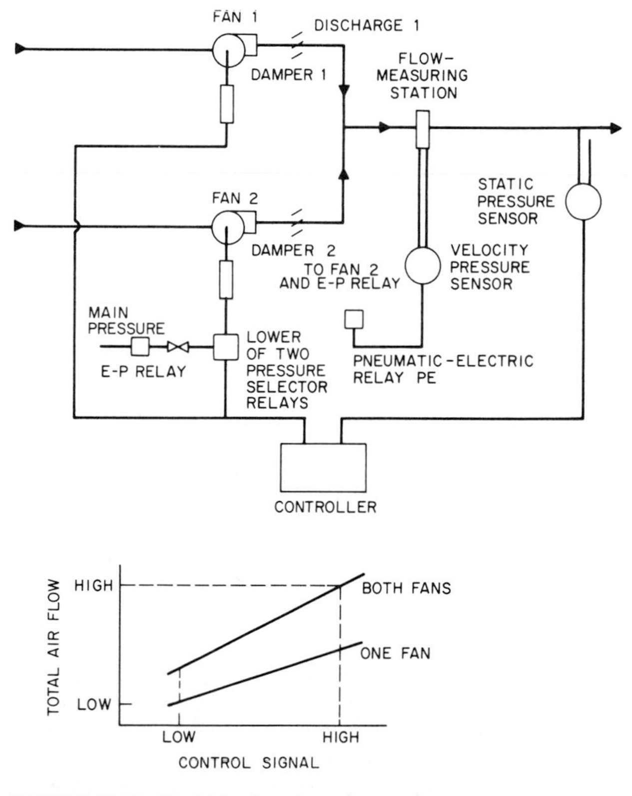

FIGURE 52.48 Variable-air-volume fan staging.

relay turns on fan 2. Fan 2 starts in a minimum-volume position and discharge damper 2 opens. An electric-pneumatic (EP) relay is also turned on and applies main pressure through a restriction to a lower of a two pressures relay. The relay transmits the slowly rising pressure to fan 2, increasing its output volume. When the desired total output is achieved, the fan-volume signal decreases the fan 1 output until fans 1 and 2 operate at partial volume, controlled by the volume signal, and their total output meets system needs. When required, fans 1 and 2 will increase output to full volume and additional fans 3 turned on. As volume needs are reduced, fan 3, then fan 2 are turned off. This fan-staging technique is applicable to multiple supply and return fans and may be accomplished with pneumatic, electronic, or direct digital control.

Some buildings have exhaust fans that are turned on and off during occupancy. Examples are kitchen, laboratory, and paint-spray-booth exhaust fans. To maintain constant building pressurization even though fans are turned on and off requires a signal to the return-fan controller to reset its set point equal to the volume of exhaust-fan discharge.

52.3.4 Refrigeration Control

The use of refrigeration for air conditioning can generally be categorized as either central-plant or packaged direct-expansion (DX) systems. Central plant normally means central water chillers and water-cooled condensers with distribution of chilled water to fan-system coils. A central plant also normally means evaporative cooling towers to reject condenser water heat. Packaged DX systems normally mean a fan system with direct-expansion cooling coils to cool air and compressors with air-cooled condensers. The control considerations for each of these situations is covered below. An optimizing control consideration is that the energy required to do a ton of cooling is directly related to the pressure change that must be accomplished by the compressor. Therefore, raising the evaporator (suction) pressure and lowering the condenser (head) pressure decrease the energy consumption per unit of cooling. Control strategies to minimize energy consumption are based upon raising the chilled-water supply temperature and lowering the condenser-water temperature to minimize energy and still meet load needs.

Liquid-Chiller Control. Control of capacity to meet the cooling load is the primary operating control of the liquid chiller. Normally this is done by a discharge chilled-water sensor controller whose set point can be automatically changed. The use of discharge control gives fast response to load changes and maintains the temperature that is of primary interest. The set point can be changed as appropriate to meet load with minimum energy consumption. Reset of supply-water temperature should be from an indicator of load. Either using return-water temperature or raising water temperature to just satisfy the greatest load are two ways to do this. The preferred method of load-reset is by monitoring the chilled-water valve position on major fan systems and resetting to keep the greatest demand valve almost full open. This approach ensures that all loads are met, whereas use of return-water temperature does not recognize individual load diversity. Where chilled-water constant-speed pumping horsepower is more than 25 to 33 percent of chiller horsepower, the increases in chilled-water temperature could force the use of more pumping energy than was saved in reducing compressor energy. In these cases, chilled-water reset should not be used or should be limited to reset only when flow is below a break-even point of chiller versus pump energy.

Close control to set point may be obtained with the proportional-integral-derivative control algorithm as the primary operating controller. The capacity of centrifugal chillers is normally varied in a proportional manner by vane position or speed control. The capacity of reciprocating chillers is normally varied in stages by unloading cylinders or cycling multiple compressors. Safety limits appropriate to the machine are normally included with the chiller and many include some or all of the following:

1. Low evaporator temperature or pressure limit
2. High condenser-pressure limit
3. High motor-temperature limit
4. Proper oil temperature and pressure limits

Other control or monitoring functions to be considered as part of building automation include the following:

1. Discharged chilled-water temperature and set point
2. Return chilled-water temperature
3. Chiller flow
4. Chiller energy use

Evaporative Cooling-Tower Control. Control of cooling-tower capacity is by keeping on enough fan capacity to meet the heat-rejection load, thereby limiting how high the condenser water temperature can go. Also, under lower ambient conditions, bypassing some hot condenser return water is necessary to limit how low the condenser water temperature can go. See Fig. 52.49 for the schematic of this control scheme. For minimum energy usage, the set point for condenser water temperature should be as low as can be safely used by the refrigeration system and as can be provided by outside-air conditions. The design approach value of the evaporative cooling tower represents how close the cooled-water temperature may come to outside-air temperature at design conditions. When outside air is at lower wet bulb then design, the tower can cool water to wet-bulb temperature

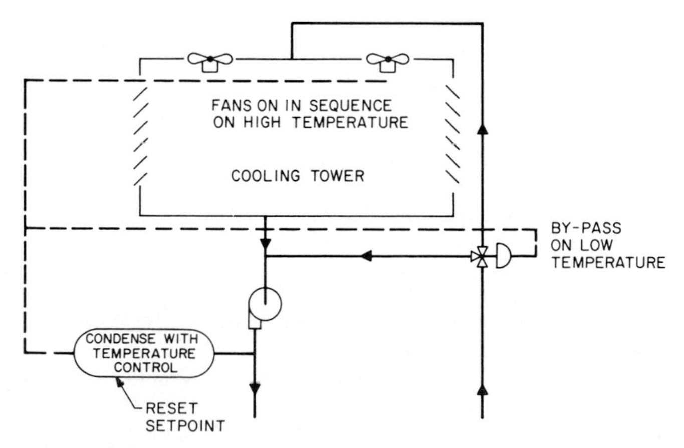

FIGURE 52.49 Cooling-tower control.

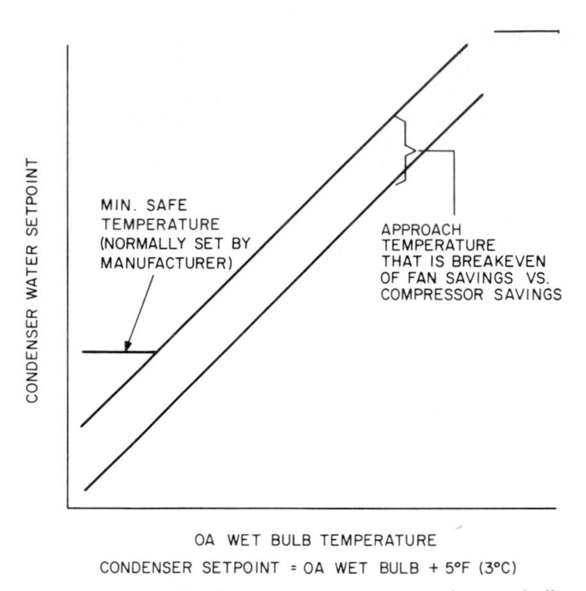

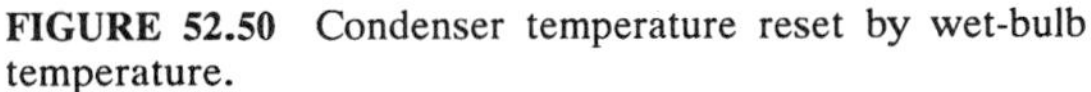

FIGURE 52.50 Condenser temperature reset by wet-bulb temperature.

plus approach value, but it cannot cool it down to wet-bulb temperature. Therefore, the set point of the controller should be at the lowest attainable temperature to both save chiller energy and not waste fan energy trying to reach an unobtainable value. See Fig. 52.50 for the reset schedule of condenser-water set point as a function of outside-air wet-bulb temperature.

Packaged Direct-Expansion (DX) Systems. A packaged DX system is typically a rooftop-mounted fan system that includes a DX air-cooling system, use of outside air for ventilation, and, in some cases, a means of heating air. Normally control of such a system is applied by the unit manufacturer to vary heat, ventilation, and cooling capacity as needed to meet a load. In the case of single-space zone control, this can be done by a space thermostat. In the case of a multiple-zone application, such as VAV zones or reheat zones, a discharge-air controller is used to control capacity of the system. Control of DX cooling capacity is typically accomplished by multiple compressor-coil sets that are turned on and off as needed. A coil is fed liquid refrigerant through a thermal expansion valve that limits the volume of refrigerant to the quantity that can be evaporated by the existing load. Other means of changing capacity are by unloading compressor cylinders and shutting off coils with solenoid valves. These approaches require application care and control that ensure proper operating conditions for all stages of capacity control.

The control of system capacity is from a space or discharge controller that responds to thermal load and calls for more or fewer stages of capacity. Minimum off-time control is used to prevent short cycling that could damage equipment. Other safety controls associated with refrigeration control are normally included, such as:

1. Low suction-pressure limit
2. High head-pressure limit

The principles of optimizing energy usage by resetting the discharge temperature can be applied to package DX systems also. The normal means is to reset the discharge-air temperature control up under a light-load condition. This can be accomplished by a set-point adjustment in the original equipment manufacturer (OEM) control circuit, which is reset by the field-applied building automation control or energy management systems. The air-cooled condenser associated with the normal DX unit is already minimizing energy by giving the lowest condenser pressures that are available and safe. Other controller monitoring functions for potential field tie-in to building automation include the following:

1. Fan start-stop
2. Discharge-air temperature and set point
3. Mixed-air temperature and set point
4. Minimum outside-air-position set point
5. Maximum outside-air enthalpy override
6. Zone temperature and set point
7. Filter maximum pressure-drop status

52.3.5 Central Heating and Cooling Plants

The central plant is one that generates chilled water and/or hot water or steam for distribution to the point of usage. The energy source may be electricity or a fuel-combustion source. Central plants offer the opportunity to minimize energy use by matching plant to load and using heat-pump cycle and thermal storage. The general objective of total plant optimization is accomplished by control strategies that select the most efficient plant lineups to meet a measured or projected load. On a project without thermal-storage equipment, the loads to be considered are the instantaneous loads for heating and cooling. In projects with thermal-storage equipment, the loads to be considered are the 24-h load profiles of heating and cooling. The optimizing control tactics to build a total plant optimizing strategy include the following:

1. Supply chilled water at a temperature level that minimizes chiller and pumping energy while meeting the greatest demand-load needs. (This was covered in Sec. 52.3.4 under Liquid-Chiller Control.)
2. Select the chiller or chiller combination that meets a required load at minimum cost. The information used to make this decision should reflect the influence of actual refrigerant head pressures.
3. When a heat-pump cycle is available to use rejected heat, it should be capacity-controlled to satisfy heating load when it is larger than cooling load, otherwise it should be controlled to satisfy the cooling load.
4. When thermal storage is available, daytime rejected heat is stored and/or nighttime low-cost cooling is stored to minimize energy costs. Storage also reduces the size of cooling-generating equipment.
5. Select boilers as required to meet load at minimum cost.

The specifics of control in each situation are now discussed.

Multiple Boilers and Heat Exchangers. The basis of capacity control of a single boiler was covered in Sec. 52.3.1. With multiple-boiler control, the primary decision is when to change the number of boilers on-line. This can be done when the temperature or pressure controller indicates a need for more capacity because one boiler is not carrying the load. Alternatively, an optimizing-control approach can also be used to make the decision of one or two boilers based on minimum operating cost. On any boiler with constant O_2 control, an increased firing rate causes decreased efficiency because a higher flue-temperature rise results in higher thermal loss. However, adding a boiler on-line adds standing losses due to the auxiliaries and the thermal losses through the added casing, piping, and breeching of the second boiler. In any specific case, the second boiler should be used when sharing the load increases efficiency to a greater degree than standby and startup loss of the second boiler detract from efficiency. This calculated changeover loading can be established by initial analysis using the type of information seen in Fig. 52.24 of Sec. 52.3.1. Assuming overhead standing losses of 2 to 3 percent of full load, the calculated load characteristics of one versus two identical boilers can be seen in Fig. 52.51. For very large boilers, the added so-

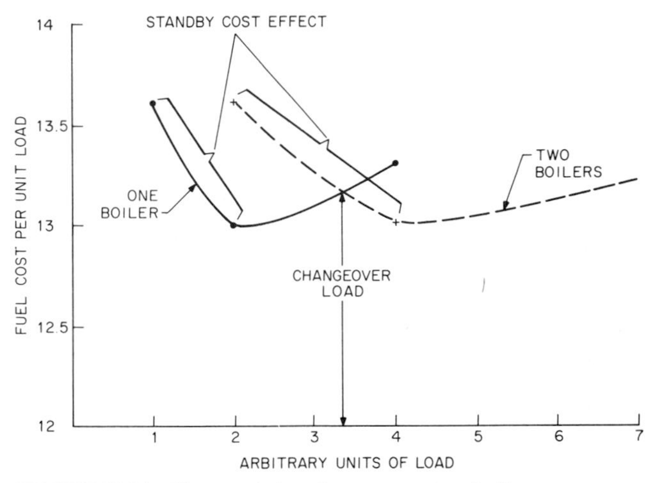

FIGURE 52.51 Characteristics of one versus two boilers.

phistication of dynamically monitoring efficiency and using measured performance to make these analyses and decisions might be justified. However the decision is reached, the control is then implemented by measuring heating load or projected load values and giving a turn-on or turn-off decision based on the changeover values established by analysis.

Multiple heat exchangers would normally be associated with separate load circuits and would be on, if that load was present. If there were parallel heat exchangers on the same load pump, they would normally both have secondary flow but would have sequenced control of the primary (source) flow to improve controllability.

Multiple Chillers. The goal of selecting from alternatives of chiller lineups is to minimize the operating cost to meet an existing load. The energy requirements of a centrifugal chiller are a function of load and the evaporator and condenser pressures which establish refrigerant lift (also called "refrigerant head"). The condenser-water supply temperature from the cooling tower and the chilled-water supply temperatures establish these conditions and can be used as an indicator of the refrigerant head. See Fig. 52.52 for the pressure and temperature relationships as a function of load. It is to be noted that the indicated head consisting of condenser supply temperature and discharge chilled-water temperature can be used and the other variations are a function of load. The energy required versus load characteristics for different indicated heads are the variables that can be used to guide selection decisions. See Fig. 52.53 for an example of the characteristics of one versus two identical chillers for energy per unit load as a function of load and indicated refrigerant head. The basis of changeover from one chiller to two is a changeover-load value reset by indicated refrigerant head and subject to any equipment limitations. See Fig. 52.54 for an example of the changeover schedule. For larger systems and dissimilar chillers, a more sophisticated approach that monitors actual performance and dynamically updates curves may be justified. This would include a more sophisticated analysis of the minimum energy combination to meet different loads. Actual implementation of the optimized selection decision is by measuring load and comparing it to the

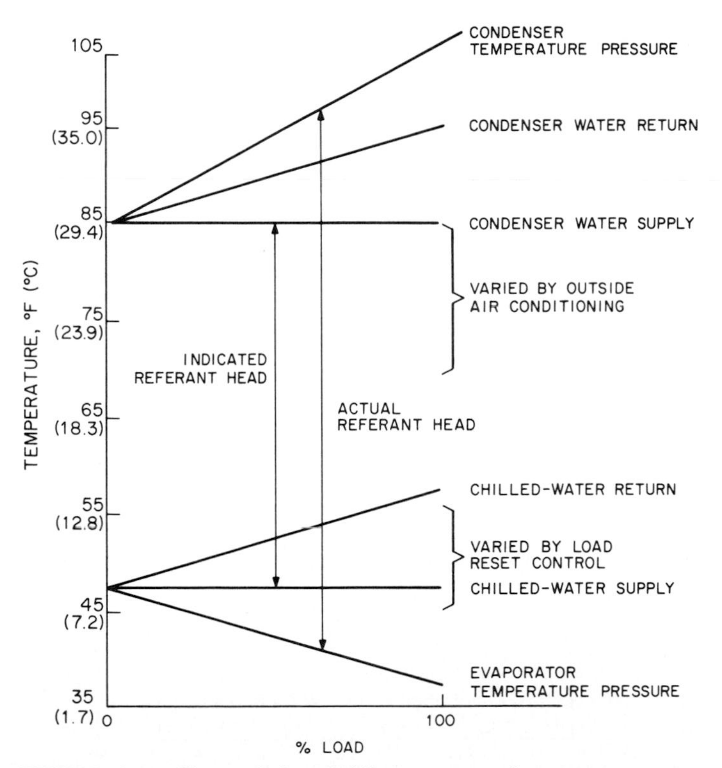

FIGURE 52.52 Characteristics of chiller pressure and temperatures versus load.

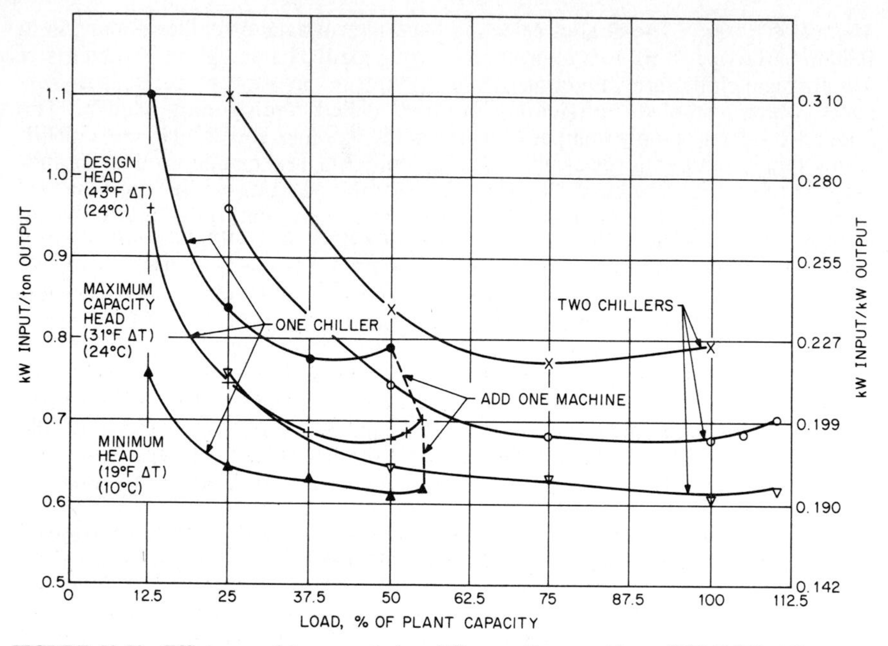

FIGURE 52.53 Efficiency of two equal-size chillers. • One machine—43°F (24°C) ΔT; x two machines—43°F (24°C) ΔT; + one machine—31°F (17°C) ΔT; ○ two machines—31°F (17°C) ΔT; ▲ one machine—19°F (10°C) ΔT; ▼ two machines—19°F (10°C) ΔT.

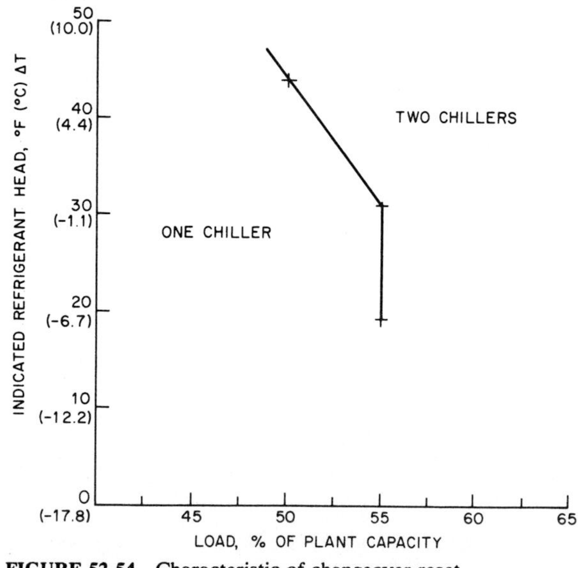

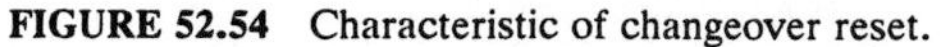

FIGURE 52.54 Characteristic of changeover reset.

changeover value and either giving operator messages or equipment start-stop commands.

Heat-Pump-Cycle Chillers. The function of the heat-pump cycle of chillers is to transfer heat from one system or section of the building to another. The equipment involved is a double-bundle condenser and the controls to operate it. A normal application is to have interior-zone cooling accomplished by heat-pump-cycle chillers that reject heat to perimeter heating loads and only reject excess heat tc a cooling tower. Then fuel heat is used only if rejected-heat supply is inadequate. See Fig. 52.55 for the system and control sequence. A further refinement in an optimized system with chilled-water reset is that after outside-air cooling is shut off to gain heat, the chilled-water reset action is removed to gain more rejected heat before the use of fuel heat. The interaction of heating needs limiting free cooling is a significant optimizing action that will maintain the balance of free heating versus free cooling over a broad range of outside-air temperatures. Figure 52.56 gives an example of instantaneous load and source relationships showing that between 25 and 50°F (−3.9 and 10°C) outside air, free cooling is limited by this action.

Thermal Storage. The function of storage is to save heating or cooling for future use. A normal application is to have enough storage to meet the load changes that occur in a 24-h diurnal cycle. In cooling-load seasons, the storage of low-cost nighttime cooling can save energy and demand charges. Another benefit may be in decreased design size of chilled-water generating equipment. In heating-load seasons, the storage of daytime excess rejected refrigeration heat or solar heat can be used for nighttime heating loads. The storage mechanisms may be water tanks or ice bins for cooling and may be water tanks or thermal conducting fluids for heat. The efficiency of storage depends upon insulation, and in the case of water storage, it depends upon minimizing the mixing of return and storage water which decreases the stored-energy temperature differential. The primary control functions are to charge and discharge storage at the proper times and rates. An

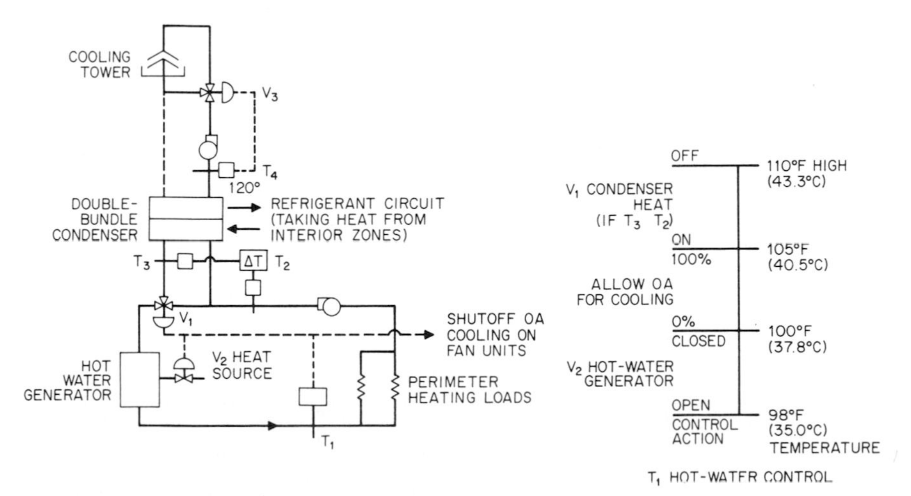

FIGURE 52.55 Heat-pump-cycle controls and sequence.

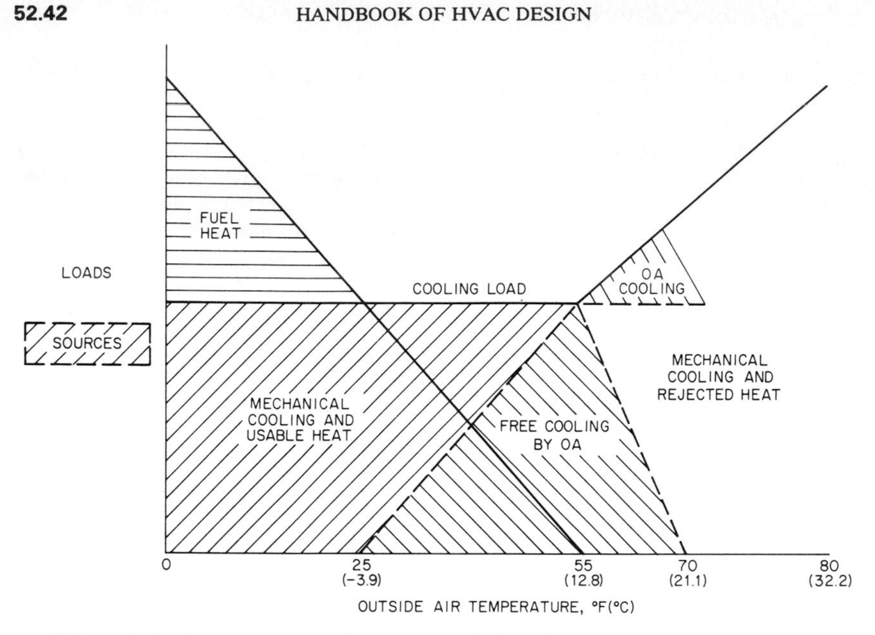

FIGURE 52.56 Instantaneous load and source relationships.

example of the physical arrangements of segmented water storage is shown in Fig. 52.57. The three basic modes of control are charge or discharge and instantaneous cooling. Besides the three basic control modes, it would also be possible to simultaneously do instantaneous cooling and either charge or discharge. In this type of operation with simultaneous charge and load, low differential pressure on loads would limit the amount of charging flow. In the case of simultaneous stor-

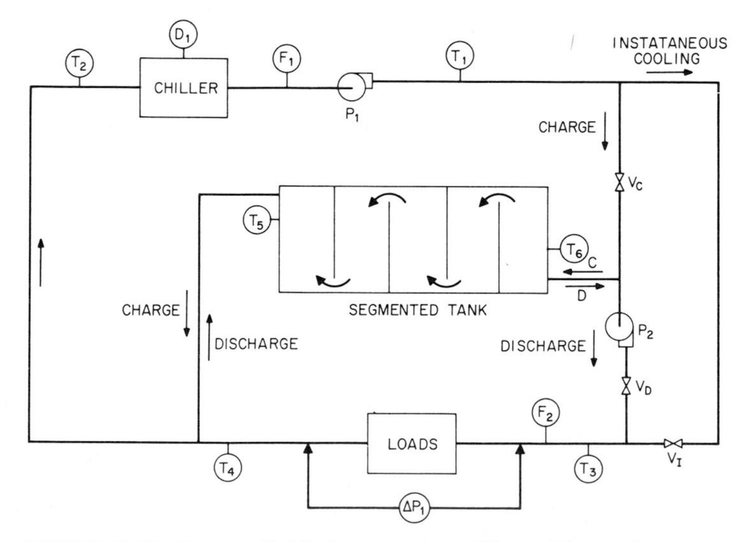

FIGURE 52.57 Segmented chilled-water storage. (*Notes*: All controls are not used in all sequences.)

age discharge and instantaneous cooling, the supplemental stored cooling would be mixed with limited instantaneous cooling to meet load requirements. Proper combinations of control modes depend upon relative storage size, relative costs of stored versus instantaneous cooling, and electrical rate structures. The timing of the storage cycle is normally when cooling generation is lowest in cost, such as during nighttime at lowest condenser-water conditions and possibly with lower time-of-day electricity rates. The cooling rate should be at the most efficient rate that satisfies charging-quantity requirements within the limits of the time span available.

The simpliest control needs occur when storage size is large enough to satisfy the loads through the peak demand period of a design day and stored cooling is less costly than instantaneous cooling. In this case, charging control can be a fixed time schedule to start charging which is established by an analysis of maximum charge requirements and the time span of low-cost cooling availability. The stopping of charging would then be done when storage was filled. Discharge is controlled to meet load conditions with the chiller operated, only if necessary, to supplement the use of stored cooling. The sequence of control for this full-cooling storage scheme is as follows (refer to Fig. 52.57, control symbol numbers):

Charge cycle: Schedule start. Valves V_D and V_I are closed, and V_C is throttled to maintain charging flow rate at F_1. T_1 controls chiller capacity to maintain 40°F (5°C) chilled water supply (CHWS) temperature. When T_5 is 40°F (5°C), charge cycle is stopped.

Discharge cycle: Scheduled load time and T_6 is under 42°F (6°C). Valves V_C and V_I are closed, chiller and pump P_1 are off. Pump P_2 is on.

Instantaneous cooling cycle: Schedule load time and T_6 is over 42°F (6°C). Pump P_2 is off; V_C and V_D are closed. Pump P_1 is on and V_I is open, and T_1 controls chiller capacity to maintain reset CHWS temperature.

If stored energy costs are larger than instantaneous energy costs because of losses and/or no time-of-day rates, then the primary savings from storage are the savings of demand charges. In this case, the only cooling that should be stored would be what was required to meet the demand limit the next day. The size of storage and the demand-control objectives establish the control strategies for this situation. The control objectives are first to limit demand for the billing period and second to minimize energy costs. If the size of storage is sufficient for a demand day's use, and demand savings with no daytime generation offsets the cost of storage loss, then charging should be only at night and only for the quantity that is needed for the next day. Discharge is then controlled to meet load. In this case, the sequence of the control is the same as above for full storage and lower stored-cooling costs, except that rather than filling the storage tank, a measured quantity is stored in the tank. This quantity is measured by chiller delivery calculation using T_1, T_2, and F_1 readings and integrated with time. The quantity used by the load the previous day can be the basis of establishing what is needed for storage for the next day. This usage can be calculated with T_3, T_4, and F_2 readings and integrated with time.

If the size of storage is not sufficient for a demand day, then its discharge should be controlled as a supplement to limit the instantaneous cooling to the demand limit. This can be done by limiting chiller capacity and controlling the mixing of stored and instantaneous cooling to meet chilled-water needs in sequence with further reductions in chiller cooling capacity. The charging could be done at night on a scheduled basis to fill the storage. The sequence of control is as follows:

Discharge and instantaneous cooling: Pump P_1 on, V_C closed, and V_I open. T_1 controls chiller capacity subject to the demand limit control D_1. T_3 controls V_I and V_D to maintain CHWS temperature as necessary.

Night charge: Scheduled start continuing until T_5 equals 40°F (5°C). Pump P_1 is on. F_1 controls V_C to maintain charging flow rate. Pump P_2 off, and V_D and V_I are closed. T_1 controls chiller capacity to maintain 40°F (5°C) CHWS temperature.

Simultaneous-day charge: Used only if low loads increase daytime cooling costs significantly above stored cooling costs. Pump P_1 is on, and V_I open. Pump P_2 is off, and V_D is closed. T_1 controls chiller capacity to maintain 40°F (5°C) CHWS temperature. Differential T_1 controls V_D to limit charge rate and maintain adequate supply to loads.

The decisions described above may be established by design analysis to provide the simplest appropriate control sequence. On large jobs, it may be justified to have the sophistication of on-line calculation for dynamic analysis and improved decisions under varying conditions.

Stored heating can be treated simply by storing excess rejected heat to full storage capacity and using it to supply all heating needs before fueled heat is used. This is the same as the simplest cooling-storage control sequence described above. The more accurate design decision making and choice of control sequence by cost comparisons can be done by calculating the cost of stored rejected heat as equal to the increased energy cost to run the higher condenser temperatures for heating. These decisions can also be design-analyzed to select the control sequence or they can be on-line dynamic analyses and control decisions.

52.3.6 Water-Distribution Control

Differential pressures in a water-distribution system are determined by the pump characteristic curve and the system resistance curve (see Fig. 52.58). These curves intersect at the operating point which denotes pump pressure differential and system flow. Water-distribution systems usually consist of a pump, heating or cooling source, piping, and a number of coils, each controlled to satisfy zone loads. As these loads change, the volume of water through an individual coil is

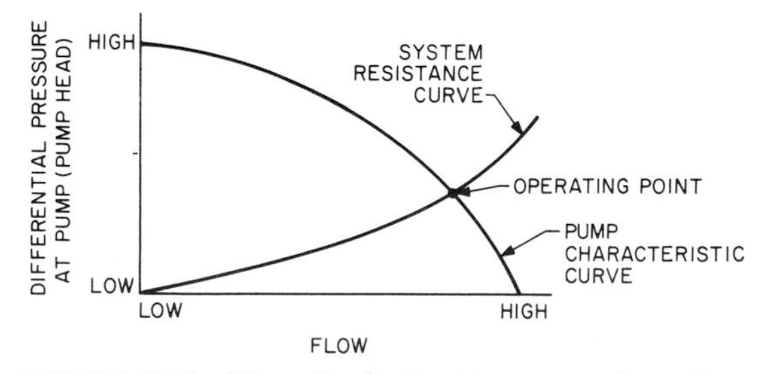

FIGURE 52.58 Water-distribution-system operating point.

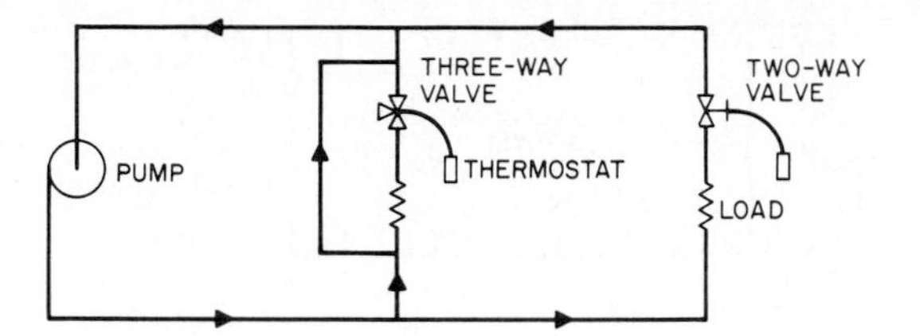

FIGURE 52.59 Water-distribution-system coil-control flow control.

changed either by bypassing some water around the coil or by restricting the flow of water through it (see Fig. 52.59).

Coils modulated by bypass or three-way valves theoretically have no effect on the system resistance. This is true if the resistance of the bypass, coil, and valve together is equal for all valve positions. Generally, this is essentially so, so the result is little or no change in system resistance and the pump operates at the full-load operating point, even when the HVAC system is lightly loaded. This presents no control problem but consumes excessive pumping energy.

Coils modulated by restricting flow or by two-way valves greatly affect system resistance. With all valves open, resistance is minimum. As valves close, water flow decreases and system resistance increases. Finally, when all valves are closed, system resistance is infinite and the system is dead-ended. Figure 52.60 illustrates how various system resistances establish various system operating points. Note that as system resistance increases, system flow decreases and pump pressure differential increases. However, pump horsepower decreases.

The changes in pump pressure differential may be a control problem for two reasons. First, modulating control valves provide their design-stem position-flow characteristic when operated at a fixed pressure differential. While some variation in pressure differential, and therefore valve flow characteristic, is acceptable, excessive variations result in poor control and system instability. Second, valves are capable of closing against certain maximum pressure-differential pressures. Higher differential pressures will force the valve open, resulting in undesired flow through the coil.

Pressure-differential changes across coil valves can be minimized by adding a pump bypass valve to the distribution system. The bypass valves may be close to the pump and control to maintain constant system differential pressure near the pump (see Fig. 52.61). This configuration limits coil-control-valve pressure dif-

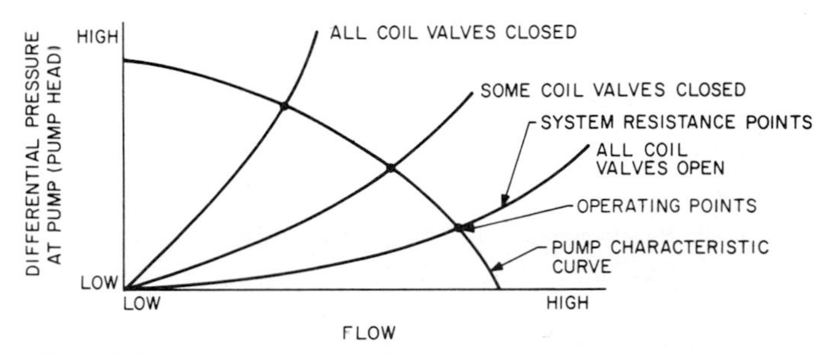

FIGURE 52.60 Effect of coil-valve operation on system curve.

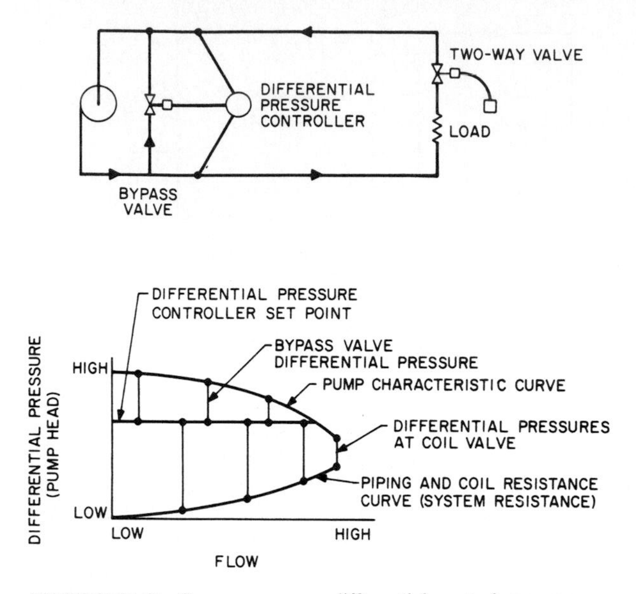

FIGURE 52.61 Bypass pressure-differential control at pump.

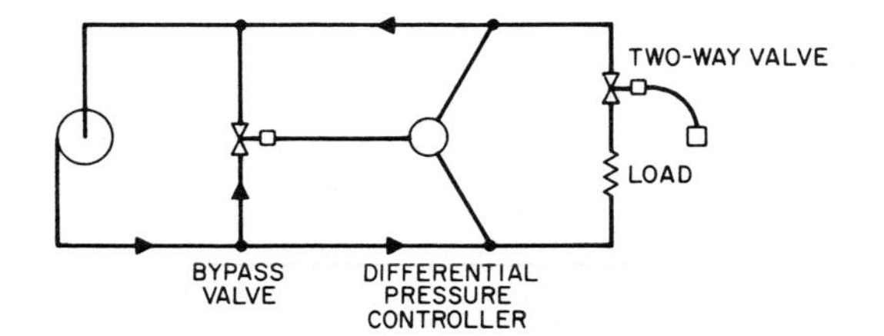

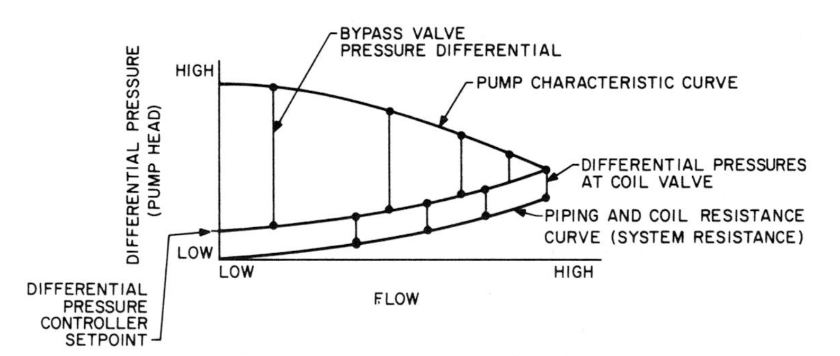

FIGURE 52.62 Bypass pressure-differential control at far end of system.

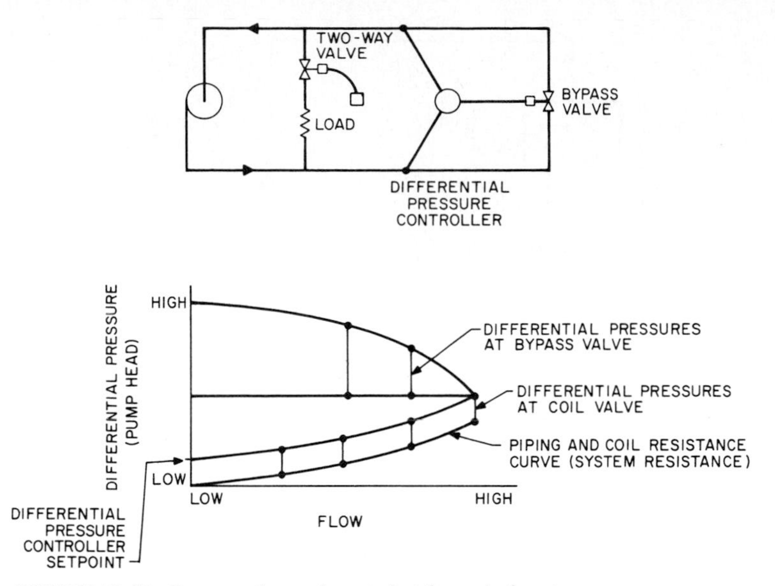

FIGURE 52.63 Bypass valve and control at far end of system.

ferential somewhat. A refinement is to sense in control-system differential pressure near the far end of the distribution system (see Fig. 52.62), keeping valve differential pressure constant.

With both methods, the bypass valve must be capable of closing against full pump differential pressure. Locating the valve at the far end of the distribution system maintains full flow and pressure drop through the distribution piping, reducing the close-off capability requirement of the bypass valve (see Fig. 52.63).

Bypass control of differential pressure is effective in limiting variations in pressure drop across control valves. However, the pump always operates at full load, wasting energy and all except design-load conditions.

System pressure differential may be controlled and pump energy reduced, at hard-load conditions, in several ways. A two-way valve, controlling the pressure differential at the end of the system, may be placed at the pump discharge (see Fig. 52.64). As the coil valves throttle, reducing system flow and, therefore, system pressure drop, the pump-discharge valve is throttled to keep the sensed differential pressure constant. The pump operates along its operating curve, consuming less energy, as its volume is reduced. Another method is to incorporate two pumps in parallel and stage their operation to meet system needs (see Fig. 52.65). When the coil control valves are throttled, one pump can supply the re-

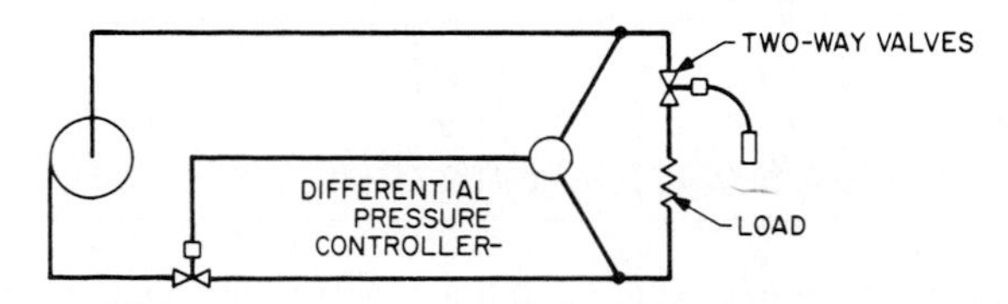

FIGURE 52.64 Pump-discharge control valve.

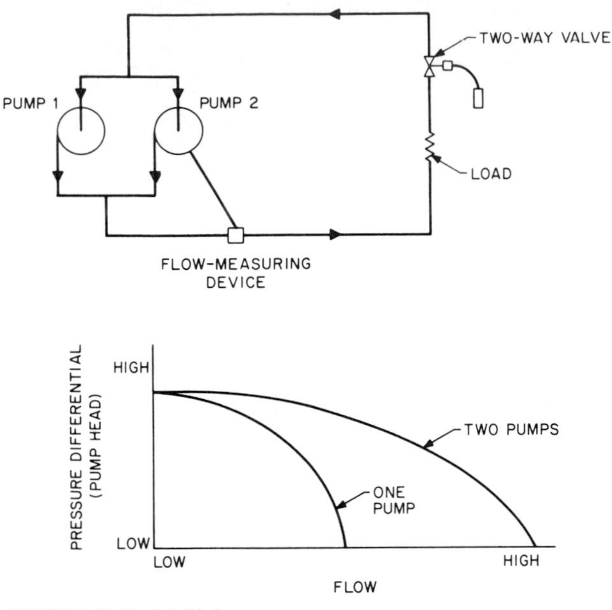

FIGURE 52.65 Multiple pumps.

quired volume. As the coil control valves open, more volume is required until the full capacity of the first pump is reached, at which time the second pump is turned on. Both pumps operate to satisfy the load. When the coil control valves throttle and system flow is reduced to below the capacity of the first pump, the second pump is turned off. Note that the pumps are turned on and off by system flow, not by differential pressure. Differential-pressure variations with two pumps in parallel is less than with one large pump. Variation may be further reduced by increasing the number of pumps. Staging pumps also reduce power under part-load conditions when fewer than all pumps are required. Also, if one pump should fail, the remaining pumps can handle most of the load.

Variable-speed operation of the pumps may also be used to patrol system pressure differential and save energy (see Fig. 52.66). The speed of the pump is controlled by system pressure differential as sensed at the far end of the system. This allows the pump to operate at the speed which will just supply the pressure required by the coil and control valve.

Considerable energy savings can be achieved by variable-speed pumping. Since a much larger reduction in coil flow is required to achieve an incremental reduction in coil capacity and since system pumping losses decrease as the square of the flow reduction, large pumping-energy reduction results when coil capacity is modulated, which is at all except design conditions.

52.4 BUILDING AUTOMATION CONTROL

"Building automation control" can be defined as the combination of central and remote equipment that aids in the operation and management of building ser-

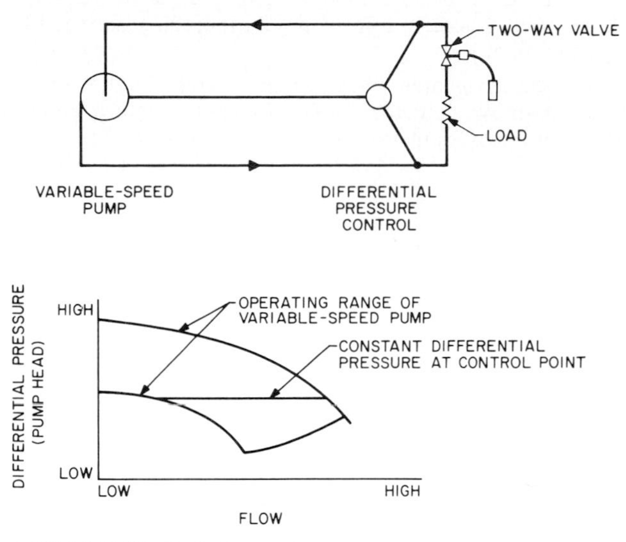

FIGURE 52.66 Variable-speed pump.

vices. These services can include HVAC, utilities, lighting, and fire and security systems. Operating aids include automatic control (based on schedules and measured conditions), log and display information on building systems, and operator command capability. Management aids include reports that summarize building services furnished, energy used, and cost or performance indices. Central equipment includes the CRT or display, printers, and keyboards that provide operator interface. Remote equipment includes data control panels (DCPs) [otherwise known as field interface devices (FIDs) of multiplexers (MUXs)] and the sensors and actuators that mount in the controlled system. A building automation system is also known as a "facility management system" (FMS) or as an "energy monitoring and control system" (EMCS).

State-of-the-art building automation system products are designed by manufacturers with unique communications protocols and tasks dependent on the system architecture, size range, and product features. The design of a specific building project and its building automation system by the mechanical system designer or other consulting engineer is primarily a matter of defining the system variables to be controlled and the functionality of the building automation system. Examples of system variables and functionality include the following:

- Points to be monitored for analog values or digital status.
- Points to be operator-commanded by analog control value or digital start-stop.
- HVAC systems to be controlled by standard energy management programs including definition of the sensor and control or actuator points to be used with the programs.
- Systems to be controlled by direct digital control (DDC) with sequence of operation and the sensors and actuators identified.
- Operator interface type of equipment and features (CRT in color or black and white, graphic creation, logs required, remote dial-up, etc.).

- Security alarm points, access control points including card readers, and TV monitoring requirements.
- Fire alarm and fire management with smoke control. Establish the type and location of fire and smoke detectors, type of alarm signaling, control means for smoke control, and provisions for elevator interlock and fire department manual override control.

52.4.1 Building Automation System Types

The specific designs of building automation systems have been rapidly changing to use the latest in electronic solid-state technology. The use of microprocessors has become standard practice which makes building automation system product designs both hardware- and software-dependent. The location of where functionality is implemented within a system is now a product-design decision that establishes the type of "architecture" and some of the features of the system. Two general types of architecture are centralized and distributed, with intermediate variations dependent upon multiple levels of intelligence and location of functionality.

Specific functionality requires specific data at the same location in the system to support implementation. In many cases, this requires the communication of data from one location in a system to another. Data types can be viewed as status, control, qualification, and information. Status data describes the state of the building at a given point in time. Examples are whether the fan is on or off and the temperatures of a room and the air supplied to the room. Generally, status is a measured condition, though it can also be a value calculated from measured conditions. An example of this is measured energy in Btu's calculated from process water flow and temperature-difference measurements. It is to be noted that status data originates at remote sensor locations and is first put into the system memory at DCPs.

Control data indicates what conditions should be, and this data is normally commands that are put into memory to be executed by the system. Examples are operator commands and set points entered by keyboard, and energy management program calculated set points for startup times. Control data from operators is first put into the system memory at the operator interface processor. Energy management strategy (EMS) or DDC program calculated control data is first put into the system memory at the physical location where the program is executed. This could be at remote DCPs or at the central operator interface processor.

Qualification data is normally the constants applied against status data by calculations that provide control data. Examples are room-temperature limits and occupancy times that are used with room- and outside-air temperature status to calculate an optimum start-control time. Qualification data is normally put into the system from the central operator interface processor.

Compiling status and control data for use in analyzing and recording the operation of the building leads to a fourth type of data, information. This includes historical data which is stored for later display or processing. A significant aspect of this type of data is that it requires a large amount of system storage resources as compared to the single piece of status data that it may be related to. Information-type data may be the result of any or all other types of data and could be generated either remotely or centrally. Normally, information is used at the operator interface processor and is stored there or in mass storage which is accessed through the operator interface processor area.

Centralized System Equipment. In a centralized system, most or all of the data processing and control algorithm execution is done at the central processor. On a very small system, all sensors and actuators could be hard-wired to the central processor. On a normal- to larger-size system, sensors and actuators are connected to multiple DCPs that communicate with the central processor. See Fig. 52.67 for an example of a centralized system with DCPs. The data used in this system is essentially all resident in the central processor. This enables fast response to operator inquiries and rapid calculation of global strategies. It also lends itself to mass storage of information. This system does require that all status and control data be communicated from and to remote DCPs. This could cause delays in updating information in the central processor dependent upon communication load and speed. The centralized system does not lend itself to direct digital control which should be done in remote DCPs for fast speed of response and improved reliability.

Distributed Systems. In distributed systems, control functionality plus status, control, and sometimes qualification types of data are placed in the DCP. The qualification-type data is particularly necessary in providing DDC as part of a building automation system. The qualification data is the DDC program including control logic and parameters. A completely distributed system is one in which each DCP has the ability to operate independently. The operator interface function and all DCPs can communicate in a peer or a master-slave fashion. See Fig. 52.68 for an example of the completely distributed system with DDC. The data in this system is essentially all resident in the remote DCPs, except for any information data in mass storage at the operator interface. This enables fast response

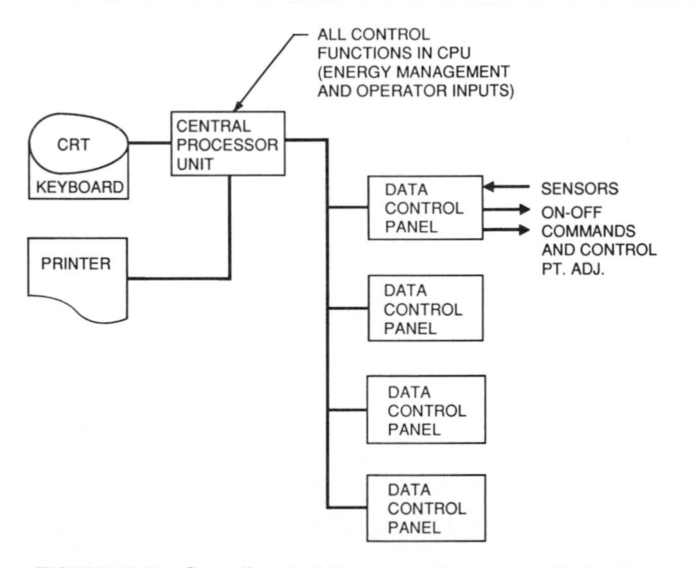

FIGURE 52.67 Centralized building automation system with data control panels.

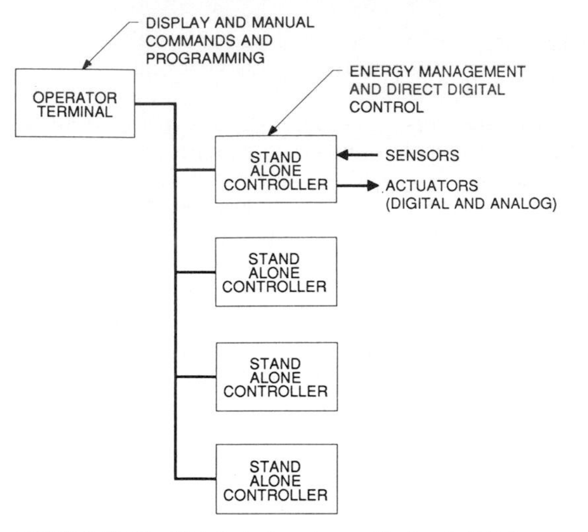

FIGURE 52.68 Distributed system with direct digital control.

of DDC control and any local EMS programs. The response to operator inquiries and commands is dependent upon the loading and speed of the communications.

Operator Interface Equipment. The broad definition of operator interface function includes all information storage, processing, and display or printing plus all operator input such as alarm acknowledgement, commands, parameter setting, or programming. The provision for several types of operators can be handled by multiple interface terminals with appropriate types of information segregated to different terminals. An example is an HVAC operator on one terminal and a security operator on another terminal. An operator-level distinction can also be made on a given terminal by distinction in the level of access given to the operator who has signed onto a terminal. For instance, the lowest level of operator could read status, acknowledge alarms, and issue normal commands. A higher level of operator would be able to reprogram parameters and issue override commands.

Communications. The communication of information between parts of the system is accomplished by one of two general methods (types of protocol): a master-slave or a peer type of protocol. The master-slave protocol may be used with either centralized or distributed systems but is normally associated with a centralized system. The peer protocol is appropriate to a distributed type of architecture.

Various types of information to be communicated require different predefined message structures so that sender and receiver can properly coordinate different types of communications. Examples of message types include the following:

- Group poll demand: response to get information on all points in a group.
- Point command
- Point global information
- Point status-demand response

- Program file header
- Program file item

Different types of communications are used in different situations. The most common is a dedicated bus controlled by a protocol that time-shares the use of the bus for different communication tasks. In this situation, the bus is in continual use to give the fastest possible total system response. The physical communication media could be two-wire electrical pulses of voltage or it could be fiber-optic cable with electro-optic coupling of pulsed light signals. In some cases, communication is over a phone line by the use of modems which buffer (store) and retransmit at a speed and protocol suitable for the type of phone line involved. With dedicated phone lines this can be done on a continual basis under control of the systemwide communication protocol. It can also be done on a demand basis by a dial-up over nondedicated phone lines. In this case, the demand for a communication could be one-way or two-way if remote alarms are to be automatically reported.

Another form of shared communications is to use one channel on a multiple-channel local-area network (LAN). A LAN is normally a coaxial cable like closed-circuit TV (CCTV) which carries multiple channels of high-frequency signals. These are primarily used for high-speed communication of masses of data between computers. A modem is required for each connection to a channel on this shared cable. This type of system is most appropriate where multiple types of systems use computers in multiple buildings or locations.

52.4.2 Automation-System Applications

The application of a building automation system (BAS) [also called a facility management system (FMS) or an energy monitoring and control system (EMCS)] considers what building services are to be monitored or automated, the objectives to be attained, the functions to do, and the interface to the building systems to be controlled. The way the building is used and operated should help establish the way the building systems are zoned or subdivided. The basic use and nature of a facility and/or the operating philosophy of the management may tend to emphasize the importance of a type of building service. Examples are:

Prestige office building emphasizes quality of HVAC results. (This would stress monitoring and management reports.)

High-rise building with heavy occupancy emphasizes life safety with fire and smoke control.

High energy cost and minimal operating staff emphasizes automatic control and energy management programs.

Widely dispersed multiple buildings with central operating management responsibility emphasizes automatic control with exception reporting and maintenance scheduling.

HVAC Scheduling and Control. The BAS control of HVAC has in the past been by supervision and optimization of the local loop hardware control of the individual HVAC subsystems. These control strategies are generally called energy management strategies (EMS). Examples are the optimized start-stop of fan systems and the optimizing reset of supply temperature levels. See Fig. 52.69 for the

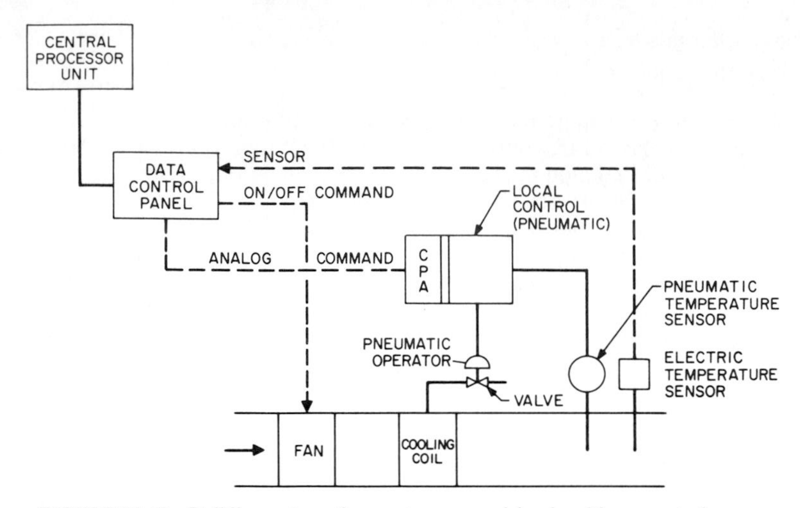

FIGURE 52.69 Building automation system supervising local loop control.

hardware configuration of a building automation system which supervises hardware local loop control. Other energy management strategies such as enthalpy, duty cycle, electric demand control, and chiller selection also provide optimizing control of the HVAC subsystems. More recently, building automation systems have also included direct digital control that provides the local loop control function on individual subsystems such as HVAC fan systems. The same BAS functions of supervising and optimizing local control are done in these building automation systems with DDC. See Fig. 52.70, which shows a BAS with DDC configuration. Note the elimination of the local controllers and interface hardware on the system with DDC. The application of different standard energy man-

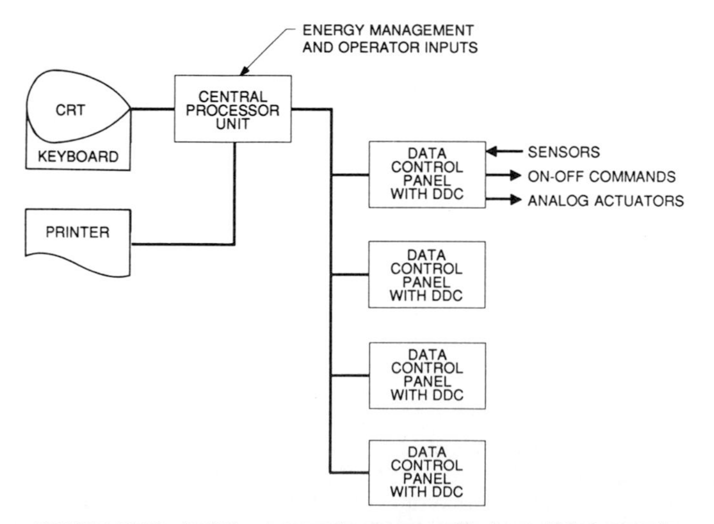

FIGURE 52.70 Building automation system with direct digital control.

agement strategies to appropriate HVAC systems include the following considerations:

1. Optimum start-stop is appropriate to systems that have unoccupied periods when the spaces go beyond comfort limits of temperature. The strategy uses space and outside-air temperature inputs to calculate the proper lead time that provides comfort conditions at the time of change and occupancy. It also prevents wasteful use of ventilation air when preconditioning is being done before occupancy.

2. Enthalpy control or dry-bulb economizer cutout is appropriate on fan systems that can use up to 100 percent outside air for cooling. An enthalpy control makes better decisions and saves cooling costs compared to a dry-bulb economizer in all climates but the very dry ones. The enthalpy benefit is related to the amount of time the outside air is lower in dry-bulb temperature but higher in enthalpy (total heat) than the return air.

3. Load reset of supply-air temperature level is appropriate for the type of fan systems that reheat cool supply air or those that mix heated and cooled air. Load reset benefit is in minimizing the energy wasted by bucking heating and cooling. Load reset of chilled-water supply temperature is appropriate for chiller systems with constant water flow or with variable-speed pumping or where the chiller horsepower is more than three or four times the chilled-water pumping horsepower. In these cases, the raising of the chiller evaporator temperature decreases power required by the chiller more than the pumping energy is increased. Load reset of VAV air supply is appropriate if chilled water is also reset to obtain chiller savings from the elevated temperature levels in the VAV cooling coil, and if limits are placed on the supply-air reset action to prevent forcing the use of more fan energy than is saved in the chiller.

4. Zero energy band (ZEB) is the strategy of having a band of acceptable room temperature in which neither heating nor cooling is added. Heating is added below the band and cooling is added above the band. This strategy is appropriate for HVAC systems that add both heating and cooling to a space. Implementation on a single-zone system is by sequencing of heating, ventilation, and cooling actuators. On multiple-zone systems, the heating and cooling supply temperatures are reset with ventilation controlled in the ZEB range. On VAV fan systems, the reset action should be limited to the break-even air volume at which increases in temperature cause more fan costs than there are chiller savings.

5. Duty cycling is appropriate to constant-volume fan systems that do not have load reset. Duty cycling reduces the fan energy and outside air-conditioning energy needed. With modulating-temperature feedback, maximum savings are made consistent with maintaining comfort.

6. Electric demand control is appropriate where there are demand charges and some ability to limit the use of electricity during peak demand periods. The benefit is a reduction in the demand charges. In the case of "ratchet" type demand charges, savings can be effective for extended periods of time up to a year. The use of thermal cooling storage can be very effective in limiting electric demand on heavy cooling days. In addition, it can take advantage of time-of-day rate structures that give reduced night rates.

7. Chiller plant optimization can consist of three separate strategies: chilled-water reset (previously described under load reset, item 3), selection of multiple chillers, and reset of condenser water control of tower fans. The optimized selection of multiple chillers is appropriate whenever multiple chillers exist in a single chilled-water system (plant). The benefit is saving money by meeting required load at the lowest cost. The basis of optimized selection of identical-type chillers is relatively straightforward. The basis of selection of dissimilar types of chillers is much more complex. Both are related to part-load characteristics as a function of refrigerant lift (condenser temperature minus evaporator temperature). Reset condenser water control of tower fans is appropriate on large cooling towers with multiple fans. The benefit is saving fan energy when outside-air wet-bulb temperatures are high enough that minimum condenser water temperatures cannot be obtained. At this time fans would be run with no benefit. In these cases, the attainable condenser temperature is made the set point for controlling fan operation, thereby saving fan energy.

HVAC Monitoring and Logging. The use of monitoring and logging is an operating aid to both operators and managers. The operator can use alarms and alarm summary logs to flag conditions that require attention. Operator alarm examples include:

- Space-temperature and humidity limits
- Discharge-temperature limits including those related to adjustable set points
- Chilled-water supply temperature limits
- Air filter differential alarm

Complete status information is available to help the operator analyze abnormal conditions. Operator status examples include:

- All-point status summary
- Selected HVAC system or group display log
- Selected point trend log.

Management aids are primarily in the form of energy audit reports that periodically give a summary of energy used and some measure of performance. In some cases, the quality of the HVAC services provided may also be monitored by summarizing occurrences of conditions being out of normal range. These reports may be custom-configured to meet specific project circumstances. Remote dial-up capability can give flexibility to monitor from an operator's terminal at remote locations either permanently or on a scheduled off-hour basis.

Direct Digital Control. A definition of direct digital control (DDC) is "the use of a computer (or microprocessor) to periodically use measured inputs to calculate a control signal and update the position of a proportioning actuator." The use of intelligent DCPs has enabled the practical implementation of DDC in BAS. The advantages of DDC include:

1. The elimination of controller, relay, and interface hardware
2. The flexibility of software which allows easy interface between building automation systems, energy management strategies, and local control
3. The use of more sophisticated control algorithms such as proportional-integral-derivative and adaptive control

The application considerations for DDC are basically the same as with control by discrete components. It is a matter of establishing the desired sequence of control for a fan system or other HVAC subsystem. In addition to the ease of interface between the BAS and the local control, DDC allows more generous use of operator selection of manual automatic modes of control. The considerations unique to DDC are that DDC cycle time must be fast enough to meet system control response needs and that recovery from power failure must be provided. A 2- to 4-s cycle time is fast enough to control the fastest HVAC processes. The use of battery backup for memory-containing operating parameters is an acceptable approach. The continued operation of local control by DDC when there is no communication with the central system is another important requirement. This control should carry on with the last communicated parameters or go to prespecified default values that are in memory.

Lighting Control. The control of lighting can be both an energy-saving and a labor-saving strategy. The control of lighting can be a scheduled function or controlled by ambient light level to complement natural lighting. Schedules can be for hours of occupancy with additional off-hour housekeeping times that are staggered space by space.

Maintenance Scheduling. Equipment run time is the basis of scheduling preventive maintenance on most motor-driven equipment. Pressure drop is the basis of scheduling some filter and heat-exchanger maintenance. A BAS can give maintenance management aid by preprogrammed instructions that associate with the task that is scheduled and report what skill level and materials are needed.

Fire Alarm and Smoke Control. The integration of a fire alarm system and the BAS allows automatic activation of smoke-control systems. This is particularly appropriate to high-rise buildings and is known as "high-rise fire management." Application considerations include the type of fire alarm system, the type and location of operators' terminal(s), compatibility of smoke control and HVAC zoning, and suitability of communications for speed of response and reliability. In many cases, the building codes may require fire department approval and a U.L.-approved-type fire alarm system.

The elements of a fire alarm system can be classified as detection devices and a processor-control panel that provides notification and control action. The fire and smoke detectors are connected to special data-gathering panels that not only monitor alarm status but also electrically supervise the integrity of the wiring to the detectors. There are different notification means appropriate to different sizes of systems. On larger sizes of systems that incorporate smoke control, it is appropriate to have prerecorded evacuation directions for each fire-smoke zone. In many buildings, it is also appropriate to automatically send notification to the fire department. The control actions in a high-rise building include initiating smoke control and override of automatic elevators to return them to the ground floor for use by fire fighters. All fire or smoke alarms should be recorded and require an operator acknowledgment.

Operator's Terminal. The operating people responsible for reacting to a fire condition are the ones who should have a fire management operator's terminal. This terminal reports and can acknowledge all alarms and can provide manual override commands to smoke-control equipment. Either this terminal or a fire fighter's control panel should provide to the fire fighters the means to manually control fans and dampers used in fire management and smoke control. If notification to occupants is by prerecorded messages via an interior communications system, the fire fighter's panel should have this manually controlled communications capability. If the HVAC operator is different than the fire management operator, then system information should be segregated to two different terminals. The command priorities appropriate to give emergency override would be given to the fire fighter's emergency panel.

Fire-Control Zoning. This is related to the fire compartmentation of the building. In addition, the smoke-control zoning must be the same as the fire zoning but is also related to the HVAC air-distribution zoning. If the HVAC zones are not the same as fire zones, then smoke-control dampers or other arrangements must be made to provide smoke-control zones that are the same as fire zones. The strategy of smoke control is to stop the supply of air and to exhaust air from the fire zone. It is also to stop the exhaust and use supply air to pressurize zones that are adjacent to the fire zone. The benefit of this strategy is to contain smoke in the fire zone. See Fig. 52.71 for an example of smoke control.

Communications Speed of Response and Reliability. The requirements in a fire management system for speed of response and reliability are greater than in a normal building automation system. The normal expectation in a fire management system is for alarm response time of 5 s or less, while the normal updating of analog information for HVAC can be as slow as 60 s. The reliability of com-

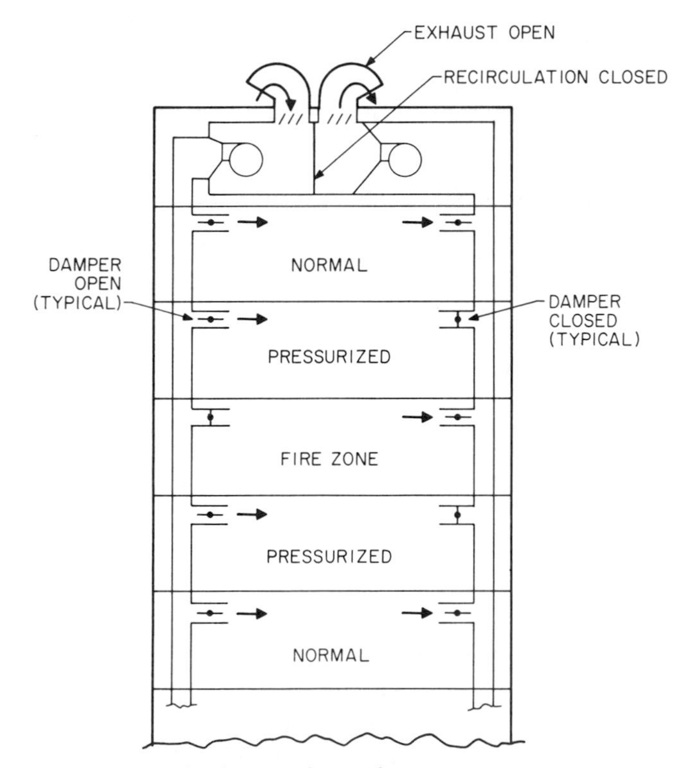

FIGURE 52.71 Smoke-control example.

munications for fire management requires dual communications cables to give automatic standby communication if one cable is nonfunctional.

Security Alarm and Access Control. The application of security systems may involve perimeter, area, or object protection. It may also involve remote camera and audio surveillance, personnel identification, and access control. The use of perimeter, area, or object protection is accomplished by some form of intrusion sensing and some type of alarm system. The types of alarm systems include:

- Local alarm in the vicinity of the protected area.
- Central-station alarm relayed to a monitoring facility owned and manned by a protection agency. (If to a police department, the scheme is called a "direct alarm.")
- Proprietary central alarm relayed to a monitoring facility owned, manned, and operated by the proprietor.

The types of perimeter sensing include:

- Outdoor taut-wire detection used on fences
- Indoor window and door switches
- Indoor photoelectric beam
- Indoor invisible light beam

The types of area-protection sensing include:

- Audio detection of abnormal noises
- Electronic vibration detection of intrusion attempts through walls
- Motion detection by distortion of an inaudible wave pattern.

Types of object protection include capacitance detection of anyone approaching or touching a protective object which is acting as an antenna.

Access may be manually controlled from a central security terminal, with personnel identification accomplished by TV camera and/or audio communication. Access may also be automatically controlled by the use of personnel identification, such as a card reader input, compared to computer files of authorized entrants. In an automatic-access control system, different zones of access may be implemented. Also, logs of who entered when and who made unauthorized entry attempts may be maintained. The hardware implementation of high-security sensors includes not only sensor-line supervision but also sensor-line monitoring that detects attempts to tamper with sensor circuitry.

52.5 SELECTION

Selection is normally based upon considerations of functions required or desired and upon the economic considerations. These considerations in total must be specific to a project. However, defining an approach to considering functionality and economics can be generalized and is given below.

52.6 TOTAL BUILDING FUNCTION

The objectives of the owner for the building are the most basic criteria that will influence the total selection process. Examples of owners and their objectives are as follows:

1. Owner-occupant with objective of having a prestigious showplace.
2. Owner-occupant with objective of a high-quality environment with a reasonable life-cycle cost.
3. Owner-occupant with objective of a utilitarian environment and minimum first cost with operating-cost improvements that pay for themselves in 3 years.
4. Owner-lessor with objective of high-quality environment and cost-effective management of building services.
5. Owner-developer with objective of lowest first cost and short-term ownership.

The first step, then, is to establish relative priorities of first cost, operating cost, and quality of environment.

52.6.1 Type of Building and System Zoning

The building structure and zoning help establish the type of HVAC system and the zoning. High-rise buildings tend to use centralized systems with large zones of water distribution and have large built-up fan systems with field-applied controls. Low-rise buildings tend to use packaged rooftop fan systems with local control by OEM controls.

52.6.2 Types of Occupancy and Use

Twenty-four-hour usage of buildings such as hospitals and hotels does not have some optimizing opportunities that exist for buildings which have periods when they are unoccupied. Buildings with different occupancy schedules in different areas require consideration of system zoning based on occupancy. The zoning may require both fan-system and central-plant segmentation and control for effective operating-cost control.

52.6.3 Accuracy Requirements

The accuracy of control can have an effect on operating costs and upon the first cost of system and controls. The accuracy needs for control of space temperature and humidity may relate to process or equipment being used in the space. Examples of high-accuracy uses may be hospital operating rooms or manufacturing constant-temperature rooms. The accuracy needs for intermediate control such as mixed-air and supply-air control relate to their effect on operating costs. Inaccuracies in fixed set-point control of supply temperatures will usually detract some from efficient operation. Load reset control of supply set points usually compensate for inaccuracies in the basic discharge control. Measurements for energy calculation and optimizing decision making need relatively high accuracy to be effective. Once total system accuracy needs are established, they should be specified and individual high-accuracy sensors should be identified.

52.6.4 Economic Justification

When the functional definition is completed of what is required and what is desired, then the cost-benefit aspects of each function can be considered.

System Size and Number of Control Points. Total system size will not have a great effect on the per unit cost of local control. However, the size can have a significant effect on the type of a building automation system and upon the per-unit cost of it. In considering size, use a minimum required number of points and the maximum number including expansion to bracket the types of system that should be considered.

Degree of Automatic Control Function. The use of energy management programs will automatically do much of the optimizing control functions that are sometimes done by operators. If added points of measurement and control are required for a specific EMS function, this should be considered on a cost-benefit basis. An example of this is the added flow and temperature measurements and cost to accomplish chiller selection versus the estimated energy-savings benefit. Another aspect is the addition of automatic valves to automatically change the chiller line up to save operator labor.

Alternative Approach Assessment. When a minimum and a maximum approach are considered, two complete scenarios can be defined. These would include size and functionality considerations. It should also include the benefits of programs and features that are part of a candidate system, even if they were not required. The final selection should also consider long-term reliability of both product and the supplier service to keep a control system running properly.

CHAPTER 53
WATER CONDITIONING

Richard T. Blake
Chief Chemist, Metropolitan Refining Co., Inc., Long Island City, New York

53.1 INTRODUCTION

It is the object of this chapter to discuss state-of-the-art technology of water treatment for commercial and industrial heat-transfer equipment, with specific emphasis on heating, ventilating, and air-conditioning (HVAC) systems. Since water treatment for industrial processes requires a specific design for each process, it is beyond the scope of this chapter to cover all aspects of industrial water treatment. For fuller coverage of industrial water treatment, see the bibliography (Sec. 53.10) at the end of the chapter.

53.2 WHY WATER TREATMENT?

Water treatment for corrosion and deposit control is a specialized technology. Essentially, it can be understood when one first recognizes why treatment is necessary to prevent serious failures and malfunction of equipment which uses water as a heat-transfer medium. This is seen more easily when one observes the problems water can cause, the mechanism by which water causes these problems, which leads to solutions, and the actual solutions or cures available.

Water is a universal solvent. Whenever it comes into contact with a foreign substance, there is some dissolution of that substance. Some substances dissolve at faster rates than others, but in all cases a definite interaction occurs between water and whatever it contacts. It is because of this interaction that problems occur in equipment such as boilers or cooling-water systems in which water is used as a heat-transfer medium. In systems open to the atmosphere, corrosion problems are made worse by additional impurities picked up by the water from the atmosphere.

Most people have seen the most obvious examples of corrosion of metals in contact with water and its devastating effect. Corrosion alone is the cause of failure and costly replacement of equipment and is itself a good reason why water treatment is necessary.

53.2.1 Cost of Corrosion

The direct losses due to corrosion of metals for replacement and protection are reported to be $10 to $15 billion annually; over $5 billion is spent for corrosion-resistant metallic and plastic equipment, almost $3 billion for protective coatings, and over $340 million for corrosion inhibitors.[1] All this is just to minimize the losses due to corrosion. Typical examples of these losses resulting from failures of piping, boiler equipment, and heat-exchanger materials because of corrosion and deposits are depicted in this chapter. Only with correct application of corrosion inhibitors and water treatment will HVAC equipment, such as heating boilers and air-conditioning chillers and condensers, provide maximum economical service life. However, even more costly than failures and replacement costs, and less obvious, is the more insidious loss in energy and operating efficiency due to corrosion and deposits.

In heat-transfer equipment, corrosion and deposits will interfere with the normal efficient transfer of heat energy from one side to the other. The degree of interference with this transfer of heat in a heat exchanger is called the *fouling factor*. In the condenser of an air-conditioning machine, a high fouling factor causes an increase in condensing temperature of the refrigerant gas and thus an increase in energy requirements to compress the refrigerant at that higher temperature. The manufacturer's recommended design fouling factor for air-conditioning chillers and condensers is 0.0005. This means that the equipment cannot tolerate deposits with a fouling factor greater than 0.0005 without the efficiency of the machine being seriously reduced.

Figure 53.1 graphically illustrates the effect of scale on the condensing temperature of a typical water-cooled condenser. From this graph, we see that the condensing temperature increases in proportion to the fouling factor. An increase in condensing temperature requires a proportionate increase in energy or compressor horsepower to compress the refrigerant gas. Thus the fouling factor affects the compressor horsepower and energy consumption, as shown in Fig. 53.2. Condenser tubes are quickly fouled by a hard water supply which deposits calcium carbonate on the heat-transfer surface. The explanation of the mechanism of this type of fouling is given in a later section.

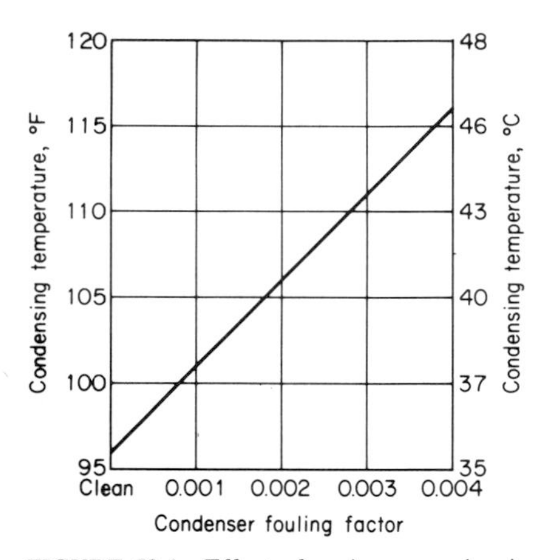

FIGURE 53.1 Effect of scale on condensing temperature. (*From Carrier System Design Manual, part 5, "Water Conditioning," Carrier Corporation, Syracuse, NY, 1972, p. 5-2. Used with permission.*)

Table 53.1 lists the fouling factors of various thicknesses of a calcium carbonate type of scale deposit most frequently found on condenser watertube surfaces where no water treatment or incorrect treatment is applied.

The additional energy consumption required to compensate for a calcium carbonate type of scale on condenser tube surfaces of a refrigeration machine is illustrated in Fig. 53.3. The graph shows that a scale thickness of only 0.025 in (0.635 mm) [fouling factor of 0.002] will result in a 22 percent increase in energy consumption, which is indeed wasteful.

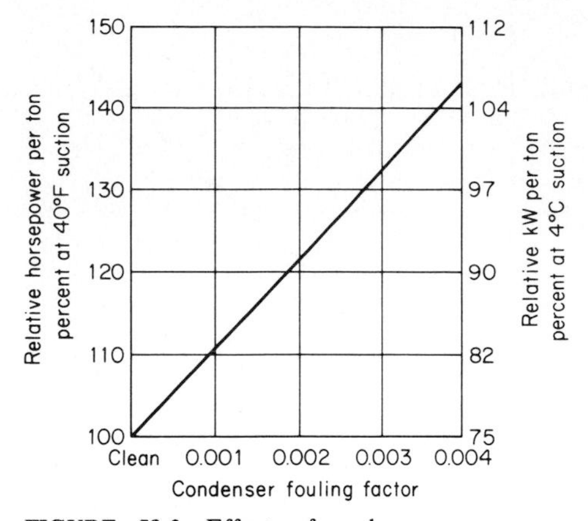

FIGURE 53.2 Effect of scale on compressor horsepower. (*From Carrier System Design Manual, part 5, "Water Conditioning," Carrier Corporation, Syracuse, NY, 1972, p. 5-2. Used with permission.*)

TABLE 53.1 Fouling Factor of Calcium Carbonate Type of Scale

Approximate thickness of calcium carbonate type of scale, in (mm)	Fouling factor
0.000	Clean
0.006 (0.1524)	0.0005
0.012 (0.3048)	0.0010
0.024 (0.6096)	0.0020
0.036 (0.9144)	0.0030

Source: *Carrier System Design Manual*, part 5, "Water Conditioning," Carrier Corp., Syracuse, NY, 1972, p. 5–3. Used with permission.

53.2.2 Cost of Scale and Deposits

The actual cost of scale is even more surprising. For example, a 500-ton air-conditioning plant operating with a scale deposit of 0.025 in (0.635 mm) of a calcium carbonate type will increase energy requirements by 22 percent if the same refrigeration load is maintained and cost $2870 in additional energy consumption required for only 1 month (720 h) of operation. This is based on an efficient electric-drive air-conditioning machine's requiring 0.75 kW/(h · ton) of refrigeration for compressor operation. The average cost for this energy in early 1984 was 5.0 cents/kWh.

With proper care and attention to water treatment, wasteful use of energy can be avoided. Likewise, in a boiler operation for heating or other purposes, an insulating scale deposit on the heat-transfer surfaces can substantially increase energy requirements.

Boiler scale or deposits can consist of various substances including iron, silica, calcium, magnesium, carbonates, sulfate, and phosphates. Each of these, when deposited on a boiler tube, contributes in some degree to the insulation of

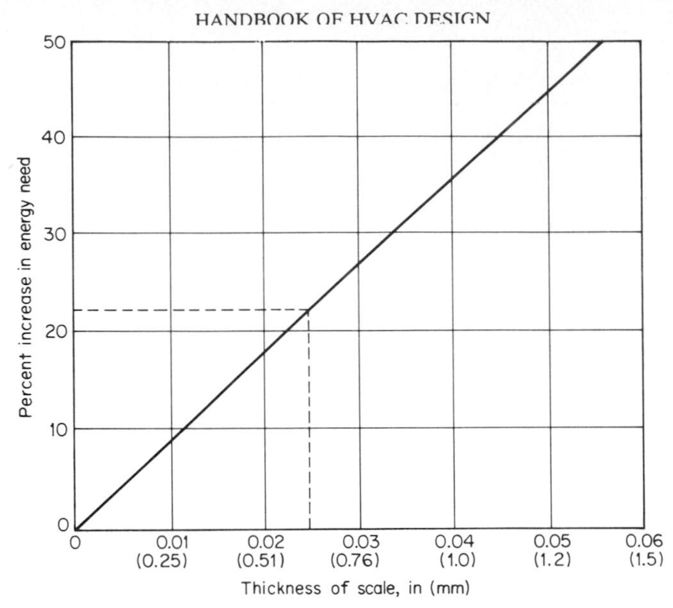

FIGURE 53.3 Effect of condenser tube scale on energy consumption, $K = 1.0$ Btu/(h · ft^3 · °F). *Example*: Scale that is 0.025 in (0.6 mm) thick requires 22 percent increase in energy.

the tube. That is, the deposits reduce the rate of heat transfer from the hot gases or fire through the boiler metal to the boiling water.

When this occurs, the temperature of the boiler tube metal increases. The scale coating offers a resistance to the rate of heat transfer from the furnace gases to the boiler water. This heat resistance results in a rapid rise in metal temperature to the point at which the metal bulges and eventual failure results. This is the most serious effect of boiler deposits, since failure of such tubes causes boiler explosions. Figure 53.4 shows a boiler tube blister caused by a scale deposit.

Table 53.2 shows the average loss of energy as a result of boiler scale. A normal scale of only 1/16-in (1.588-mm) thickness can cause an energy loss of 4 percent. For example, a loss of 4 percent in energy as a result of a scale deposit can

FIGURE 53.4 Boiler tube blister. *(Courtesy of Metropolitan Refining Co., Inc.)*

TABLE 53.2 Boiler Scale Thickness vs. Energy Loss

Normal scale, calcium carbonate type, in (mm)	Dense scale (iron silica type)	Energy loss, %
1/32 (0.794)	1/64 (0.397)	2
1/16 (1.588)	1/32 (0.794)	4
3/32 (2.381)	3/64 (1.191)	6
1/8 (3.175)	1/16 (1.588)	8
3/16 (4.763)	3/32 (2.381)	12
1/4 (6.350)	1/8 (3.175)	16

mean that 864 gal (3270.6 L) more of No. 6 fuel oil than is normally used would be required for the operation of a steam boiler at 100 boiler hp (bhp) (1564.9 kg/h) for 1 month (720 h).

53.3 WATER CHEMISTRY

Water and its impurities are responsible for the corrosion of metals and formation of deposits on heat-transfer surfaces, which in turn reduce efficiency and waste energy. Having seen the effects of corrosion and deposits, let us see how they can be prevented. The path to their prevention can best be approached through understanding their basic causes, why and how they occur.

Water, the common ingredient present in heat-transfer equipment such as boilers, cooling towers, and heat exchangers, contains many impurities. These impurities render the water supply more or less corrosive and/or scale-forming.

53.3.1 Hydrologic Cycle

The hydrologic cycle (Fig. 53.5) consists of three stages: evaporation, condensation, and precipitation. This cycle begins when surface waters on the earth are heated by the rays of the sun, vaporized, and raised into the troposphere, a thin layer of air and moisture approximately 7 mi (11 Km) thick which surrounds the earth. Clouds of condensed moisture form in the troposphere, and when carried over land by the wind, they contact cold-air currents. This causes precipitation of

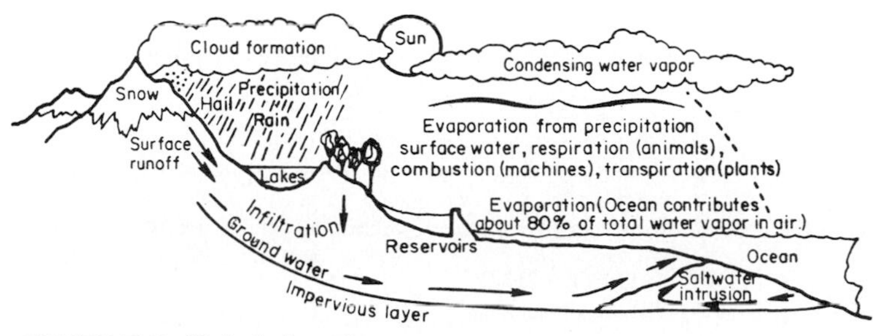

FIGURE 53.5 Hydrologic cycle.

rain or snow. In this manner, water returns to the earth's surface, only to repeat the cycle.

Throughout the hydrologic cycle, the water absorbs impurities. While falling through the atmosphere, water dissolves the gases, oxygen, nitrogen, carbon dioxide, nitrogen oxides, sulfur oxides, and many other oxides present in the atmosphere in trace amounts.

The quantity of these gases in the atmosphere depends on the location. For example, in large urban areas rainwater contains high concentrations of carbon dioxide, sulfur oxides, and nitrogen oxides. In rural areas, water contains lesser amounts of these gases. A study made by Gene E. Likens of Cornell University noted that for the past 20 years the acidity of our rainfall has steadily increased.[2] This is caused by the increased amounts of sulfur and nitrogen oxide gases that pollute the atmosphere.

53.3.2 Water Impurities

In contact with the earth surface, rainwater will tend to dissolve and absorb many of the minerals of the earth. The more acidic the rainfall, the greater the reaction with the earth's minerals. This reaction includes hydrolysis and hydration. As water passes over and through gypsum, calcite, dolomite, and quartz rock, it will dissolve calcium, silica, and magnesium minerals from these rocks (Table 53.3). In similar manner, other minerals present in the earth's crust can be dissolved and taken up by the water. Table 53.4 shows some of the minerals present in the earth's surface which by reaction with water become impurities in water. Water accumulates on the earth's surface in lakes, rivers, streams, and ponds and can be collected in reservoirs. These surface water supplies usually contain fewer minerals but are more likely to contain dissolved gases.

Underground water supplies are a result of surface waters' percolating through the soil and rock. The water supplies usually contain large quantities of minerals and not much dissolved gases, although there are numerous exceptions to this general rule. Table 53.5 lists the various sources of water. Figures 53.6 through 53.10 show typical analyses of surface waters and underground well waters.

A brief observation of the analyses of these different water supplies shows that the natural impurities and mineral content do indeed vary with location. In fact, many well water supplies in a very proximate location exhibit vast differences in mineral content. Let us examine each of the basic impurities of water to see how they contribute to corrosion and deposits.

TABLE 53.3 Reactions of Water with Minerals

Hydrolysis is the chemical reaction between water and minerals in which the mineral dissolves in the water:

$NaCl$	+	$H_2O \rightarrow Na^+$	+	Cl^-	+	H_2O
Sodium chloride	+	Water = Sodium ion in solution	+	Chloride ion in solution	+	Water

Hydration is the absorption of water by minerals, changing the nature of the mineral:

$CaSO_4$	+	$2H_2O \rightarrow CaSO_4 \cdot 2H_2O$
Calcium sulfate	+	Water = Calcium sulfate hydrate

TABLE 53.4 Mineral Groups

Silicates	Quartz, augite, mica, chert, feldspar, hornblend
Carbonates	Calcite, dolomite, limestone
Halides	Halite, fluorite
Oxides	Hematite, ice, magnetite, bauxite
Sulfates	Anhydrite, gypsum
Sulfides	Galena, pyrite
Natural elements	Copper, sulfur, gold, silver
Phosphates	Apatite

TABLE 53.5 Sources of Water

Surface water	Lakes and reservoirs of fresh water
Groundwater	Water below the land surface caused by surface run-off drainage and seepage
Water table	Water found in rock saturated with water just above the impervious layer of the earth
Wells	Water-bearing strata of the earth—water seeps and drains through the soil surface, dissolving and absorbing minerals of which the earth is composed (thus the higher mineral content of well water)

53.3.3 Dissolved Gases

Oxygen. One of the gases in the atmosphere is oxygen which makes up approximately 20 percent of air. Oxygen in water is essential for aquatic life; however, it is the basic factor in the corrosion process and is, in fact, one of the essential elements in the corrosion process of metals. Therefore, dissolved oxygen in water is important to us in the study of corrosion and deposits.

Carbon Dioxide. Carbon dioxide is present in both surface and underground water supplies. These water supplies absorb small quantities of carbon dioxide from the atmosphere. Larger amounts of carbon dioxide are absorbed from the decay of organic matter in the water and its environs. Carbon dioxide contributes significantly to corrosion by making water acidic. This increases its capability to dissolve metals. Carbon dioxide forms the mild carbonic acid when dissolved in water, as follows:

$$\underset{\text{Carbon dioxide}}{CO_2} \; \underset{+}{+} \; \underset{\text{Water}}{H_2O} \; \underset{=}{\rightarrow} \; \underset{\text{Carbonic acid}}{H_2CO_3}$$

Sulfur Oxides. Sulfur oxide gases are present in the atmosphere as a result of sulfur oxides absorbed from the atmosphere, in which they are present as pollutants from the combustion of fuels containing sulfur, such as coal and fuel oil. In large urban areas, the quantity of sulfur oxides that are absorbed by surface water supplies and aerated waters used in cooling towers can be significant. Also

METROPOLITAN REFINING COMPANY INC.
50-23 23rd STREET • LONG ISLAND CITY • N.Y. 11101

EFFICIENCY THROUGH CHEMISTRY
METROPOLITAN EFFICIENCY PRODUCTS ®

CERTIFICATE OF ANALYSIS

NEW YORK CITY

DATE
SAMPLING DATE
REPRESENTATIVE

ANALYSIS NUMBER		339568		
SOURCE		CITY		
pH		6.9		
P ALKALINITY	$CaCO_3$, mg/L			
FREE CARBON DIOXIDE	CO_2, mg/L			
BICARBONATES	$CaCO_3$, mg/L	12.		
CARBONATES	$CaCO_3$, mg/L			
HYDROXIDES	$CaCO_3$, mg/L			
M (Total) ALKALINITY	$CaCO_3$, mg/L	12.		
TOTAL HARDNESS	$CaCO_3$, mg/L	16.		
SULFATE	SO_4, mg/L			
SILICA,	SiO_2, mg/L	1.5		
IRON	Fe, mg/L	Trace		
CHLORIDE	NaCl, mg/L	13.		
ORGANIC INHIBITOR	mg/L			
PHOSPHATE	PO_4, mg/L			
CHROMATE	Na_2CrO_4, mg/L			
NITRITE	$NaNO_2$, mg/L			
ZINC	Zn, mg/L			
SPECIFIC CONDUCTANCE	mmhos/cm			
TOTAL DISSOLVED SOLIDS	mg/L	33.5		
SUSPENDED MATTER				
BIOLOGICAL GROWTHS				
SPECIFIC GRAVITY @ 15.5°/15.5°C				
FREEZING POINT °C/°F				
	% BY WEIGHT			

NOTES:
1. ANALYTICAL RESULTS EXPRESSED IN MILLIGRAMS PER LITRE (mg/L) ARE EQUIVALENT TO PARTS PER MILLION (ppm). DIVIDE BY 17.1 TO OBTAIN GRAINS PER GALLON (gpg).
2. CYCLES OF CONCENTRATION = CHLORIDES IN SAMPLE / CHLORIDES IN MAKEUP

TREATMENT	TREATMENT CONTROL	FOUND	RECOMMENDED	FOUND	RECOMMENDED

REMARKS:

RTB:MW

SPEEDIPLY® PAT'D MCP® PAT'D MBF 28 FORM 1078-1

FIGURE 53.6 New York City (Croton Reservoir) water analysis. *(Courtesy of Metropolitan Refining Co., Inc.)*

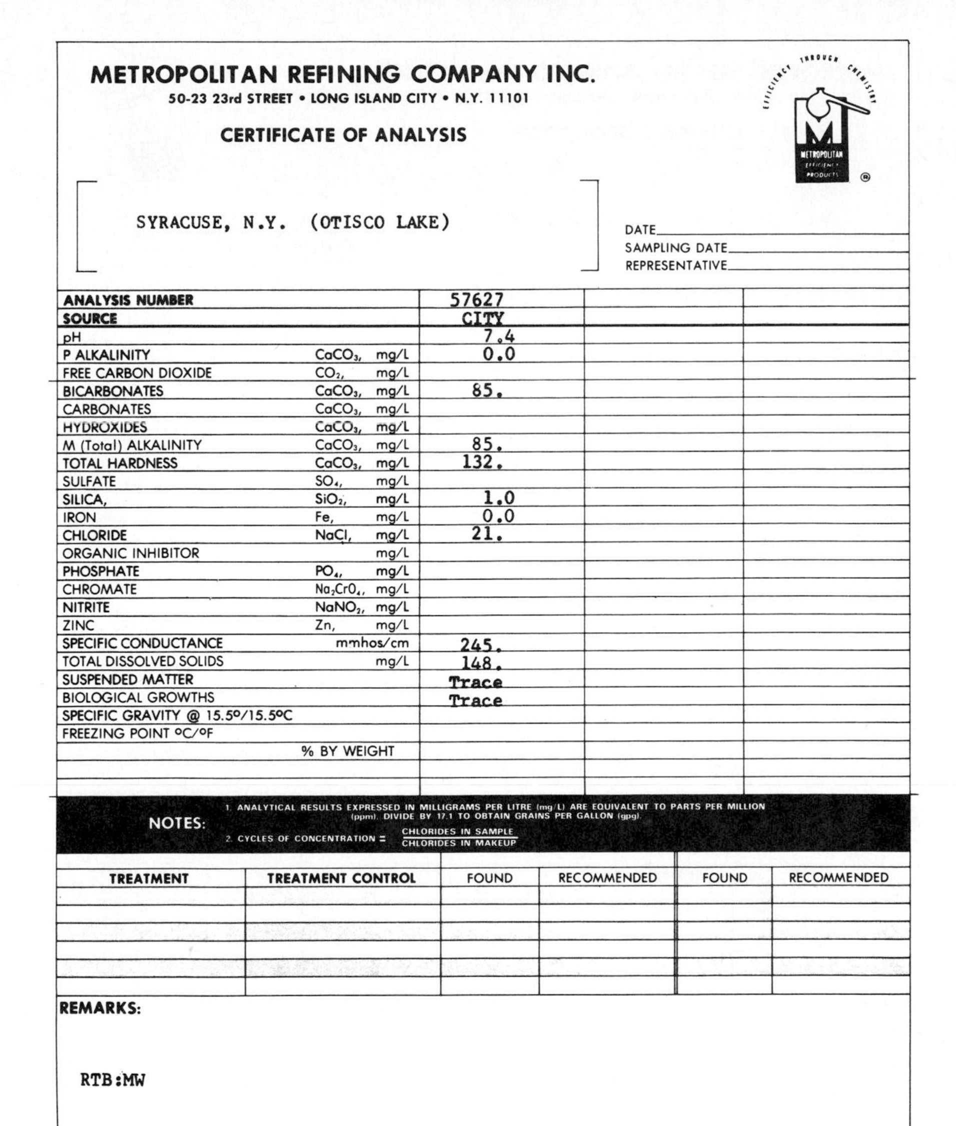

METROPOLITAN REFINING COMPANY INC.
50-23 23rd STREET • LONG ISLAND CITY • N.Y. 11101

CERTIFICATE OF ANALYSIS

SYRACUSE, N.Y. (OTISCO LAKE)

DATE
SAMPLING DATE
REPRESENTATIVE

ANALYSIS NUMBER			57627		
SOURCE			CITY		
pH			7.4		
P ALKALINITY	$CaCO_3$,	mg/L	0.0		
FREE CARBON DIOXIDE	CO_2,	mg/L			
BICARBONATES	$CaCO_3$,	mg/L	85.		
CARBONATES	$CaCO_3$,	mg/L			
HYDROXIDES	$CaCO_3$,	mg/L			
M (Total) ALKALINITY	$CaCO_3$,	mg/L	85.		
TOTAL HARDNESS	$CaCO_3$,	mg/L	132.		
SULFATE	SO_4,	mg/L			
SILICA,	SiO_2,	mg/L	1.0		
IRON	Fe,	mg/L	0.0		
CHLORIDE	NaCl,	mg/L	21.		
ORGANIC INHIBITOR		mg/L			
PHOSPHATE	PO_4,	mg/L			
CHROMATE	Na_2CrO_4,	mg/L			
NITRITE	$NaNO_2$,	mg/L			
ZINC	Zn,	mg/L			
SPECIFIC CONDUCTANCE		mmhos/cm	245.		
TOTAL DISSOLVED SOLIDS		mg/L	148.		
SUSPENDED MATTER			Trace		
BIOLOGICAL GROWTHS			Trace		
SPECIFIC GRAVITY @ 15.5°/15.5°C					
FREEZING POINT °C/°F					
		% BY WEIGHT			

NOTES: 1. ANALYTICAL RESULTS EXPRESSED IN MILLIGRAMS PER LITRE (mg/L) ARE EQUIVALENT TO PARTS PER MILLION (ppm). DIVIDE BY 17.1 TO OBTAIN GRAINS PER GALLON (gpg).
2. CYCLES OF CONCENTRATION = CHLORIDES IN SAMPLE / CHLORIDES IN MAKEUP

TREATMENT	TREATMENT CONTROL	FOUND	RECOMMENDED	FOUND	RECOMMENDED

REMARKS:

RTB:MW

SPEEDIPLY® PAT'D MCP® PAT'D MBF 28 FORM 1078-I

FIGURE 53.7 Water analysis of Syracuse, NY (Otisco Lake). *(Courtesy of Metropolitan Refining Co., Inc.)*

METROPOLITAN REFINING COMPANY INC.

50-23 23rd STREET • LONG ISLAND CITY • N.Y. 1110J

CERTIFICATE OF ANALYSIS

WASHINGTON, D.C. (POTOMAC RIVER)

DATE

SAMPLING DATE

REPRESENTATIVE

ANALYSIS NUMBER		20197		
SOURCE		CITY		
pH		7.7		
P ALKALINITY	$CaCO_3$, mg/L			
FREE CARBON DIOXIDE	CO_2, mg/L			
BICARBONATES	$CaCO_3$, mg/L	90.		
CARBONATES	$CaCO_3$, mg/L			
HYDROXIDES	$CaCO_3$, mg/L			
M (Total) ALKALINITY	$CaCO_3$, mg/L	90.		
TOTAL HARDNESS	$CaCO_3$, mg/L	140.		
SULFATE	SO_4, mg/L			
SILICA,	SiO_2, mg/L	7.0		
IRON	Fe, mg/L	0.0		
CHLORIDE	NaCl, mg/L	41.		
ORGANIC INHIBITOR	mg/L			
PHOSPHATE	PO_4, mg/L			
CHROMATE	Na_2CrO_4, mg/L			
NITRITE	$NaNO_2$, mg/L			
ZINC	Zn, mg/L			
SPECIFIC CONDUCTANCE	mmhos/cm			
TOTAL DISSOLVED SOLIDS	mg/L	195.		
SUSPENDED MATTER				
BIOLOGICAL GROWTHS				
SPECIFIC GRAVITY @ 15.5°/15.5°C				
FREEZING POINT °C/°F				
	% BY WEIGHT			

NOTES:

1. ANALYTICAL RESULTS EXPRESSED IN MILLIGRAMS PER LITRE (mg/L) ARE EQUIVALENT TO PARTS PER MILLION (ppm). DIVIDE BY 17.1 TO OBTAIN GRAINS PER GALLON (gpg)
2. CYCLES OF CONCENTRATION = $\frac{\text{CHLORIDES IN SAMPLE}}{\text{CHLORIDES IN MAKEUP}}$

TREATMENT	TREATMENT CONTROL	FOUND	RECOMMENDED	FOUND	RECOMMENDED

REMARKS:

RTB:MW

SPEEDIPLY• PAT'D MCP• PAT'D MBF 28 FORM 1078-I

FIGURE 53.8 Potomac River (Washington, DC) water analysis. *(Courtesy of Metropolitan Refining Co., Inc.)*

METROPOLITAN REFINING COMPANY INC.

50-23 23rd STREET • LONG ISLAND CITY • N.Y. 11101

CERTIFICATE OF ANALYSIS

EFFICIENCY THROUGH CHEMISTRY — METROPOLITAN EFFICIENCY PRODUCTS ®

JAMAICA, N. Y. (WELLS)

DATE

SAMPLING DATE

REPRESENTATIVE

ANALYSIS NUMBER		38140		
SOURCE		CITY WATER		
pH		7.0		
P ALKALINITY	$CaCO_3$, mg/L	0.0		
FREE CARBON DIOXIDE	CO_2, mg/L			
BICARBONATES	$CaCO_3$, mg/L	30.		
CARBONATES	$CaCO_3$, mg/L			
HYDROXIDES	$CaCO_3$, mg/L			
M (Total) ALKALINITY	$CaCO_3$, mg/L	30.		
TOTAL HARDNESS	$CaCO_3$, mg/L	60.		
SULFATE	SO_4, mg/L			
SILICA,	SiO_2, mg/L	14.3		
IRON	Fe, mg/L	0.07		
CHLORIDE	NaCl, mg/L	29.		
ORGANIC INHIBITOR	mg/L			
PHOSPHATE	PO_4, mg/L			
CHROMATE	Na_2CrO_4, mg/L			
NITRITE	$NaNO_2$, mg/L			
ZINC	Zn, mg/L			
SPECIFIC CONDUCTANCE	mmhos/cm	154.		
TOTAL DISSOLVED SOLIDS	mg/L	106.		
SUSPENDED MATTER				
BIOLOGICAL GROWTHS				
SPECIFIC GRAVITY @ 15.5°/15.5°C				
FREEZING POINT °C/°F				
	% BY WEIGHT			

NOTES:
1. ANALYTICAL RESULTS EXPRESSED IN MILLIGRAMS PER LITRE (mg/L) ARE EQUIVALENT TO PARTS PER MILLION (ppm). DIVIDE BY 17.1 TO OBTAIN GRAINS PER GALLON (gpg).
2. CYCLES OF CONCENTRATION = $\frac{\text{CHLORIDES IN SAMPLE}}{\text{CHLORIDES IN MAKEUP}}$

TREATMENT	TREATMENT CONTROL	FOUND	RECOMMENDED	FOUND	RECOMMENDED

REMARKS:

RTB:MW

SPEEDIPLY® PAT'D MCP® PAT'D MBF 28 FORM 1078-I

FIGURE 53.9 Water analysis of Jamaica, NY (wells). (*Courtesy of Metropolitan Refining Co., Inc.*)

METROPOLITAN REFINING COMPANY INC.
50-23 23rd STREET • LONG ISLAND CITY • N.Y. 11101

CERTIFICATE OF ANALYSIS

EFFICIENCY THROUGH CHEMISTRY
METROPOLITAN EFFICIENCY PRODUCTS ®

YELLOW SPRINGS, OHIO (WELLS)

DATE
SAMPLING DATE
REPRESENTATIVE

ANALYSIS NUMBER		47588		
SOURCE		CITY WATER		
pH		7.6		
P ALKALINITY	$CaCO_3$, mg/L	0.0		
FREE CARBON DIOXIDE	CO_2, mg/L			
BICARBONATES	$CaCO_3$, mg/L	300.		
CARBONATES	$CaCO_3$, mg/L			
HYDROXIDES	$CaCO_3$, mg/L			
M (Total) ALKALINITY	$CaCO_3$, mg/L	300.		
TOTAL HARDNESS	$CaCO_3$, mg/L	454.		
SULFATE	SO_4, mg/L			
SILICA,	SiO_2, mg/L	9.5		
IRON	Fe, mg/L	0.0		
CHLORIDE	NaCl, mg/L	58.		
ORGANIC INHIBITOR	mg/L			
PHOSPHATE	PO_4, mg/L			
CHROMATE	Na_2CrO_4, mg/L			
NITRITE	$NaNO_2$, mg/L			
ZINC	Zn, mg/L			
SPECIFIC CONDUCTANCE	mmhos/cm	840.		
TOTAL DISSOLVED SOLIDS	mg/L	514.		
SUSPENDED MATTER		Abs.		
BIOLOGICAL GROWTHS		Abs.		
SPECIFIC GRAVITY @ 15.5°/15.5°C				
FREEZING POINT °C/°F				
	% BY WEIGHT			

NOTES:
1. ANALYTICAL RESULTS EXPRESSED IN MILLIGRAMS PER LITRE (mg/L) ARE EQUIVALENT TO PARTS PER MILLION (ppm). DIVIDE BY 17.1 TO OBTAIN GRAINS PER GALLON (gpg).
2. CYCLES OF CONCENTRATION = $\frac{\text{CHLORIDES IN SAMPLE}}{\text{CHLORIDES IN MAKEUP}}$

TREATMENT	TREATMENT CONTROL	FOUND	RECOMMENDED	FOUND	RECOMMENDED

REMARKS:

RTB:MW

SPEEDIPLY® PAT'D MCP® PAT'D MBF 28 FORM 1078-I

FIGURE 53.10 Water analysis of Yellow Springs, OH (wells). *(Courtesy of Metropolitan Refining Co., Inc.)*

when dissolved in water, sulfur oxides form acids which create a corrosive atmosphere.

$$\underset{\text{Sulfur trioxide}}{SO_3} + \underset{\text{Water}}{H_2O} \rightarrow \underset{\text{Sulfuric acid}}{H_2SO_4}$$

Nitrogen Oxides. Nitrogen oxides are also present in the atmosphere both naturally and from pollutants created by the combustion process. These, too, form acids when absorbed by water and contribute to the corrosion process.

$$\underset{\text{Nitrogen}}{3NO_2} + \underset{\text{Water}}{H_2O} \rightarrow \underset{\text{Nitric acid}}{2HNO_3} + \underset{\text{Nitric oxide}}{NO}$$

Hydrogen Sulfide. The odor typical of rotten eggs which is found in some waters is due to the presence of hydrogen sulfide. This gas comes from decaying organic matter and from sulfur deposits. Hydrogen sulfide forms when acidic water reacts with sulfide minerals such as pyrite, an iron sulfide commonly called "fool's gold":

$$\underset{\text{Ferric sulfide}}{FeS} + \underset{\text{Acid in solution}}{2H^+} \rightarrow \underset{\text{Iron in solution}}{Fe^{2+}} + \underset{\text{Hydrogen sulfide}}{H_2S}$$

Hydrogen sulfide reacts with water to form hydrosulfuric acid, a slightly acidic solution. Its presence in water is also due to the decomposition of organic matter and protein which contain sulfur. Hydrogen sulfide is also a constituent of sewer gas, marsh gas, and coal gas. It can be present in water and also comes from these sources. Because of its acidic reaction in water, hydrogen sulfide is very corrosive and must be removed or neutralized.

53.3.4 Dissolved Minerals

Alkalinity. Alkalinity is the quantity of dissolved alkaline earth minerals expressed as calcium carbonate. It is the measured carbonate and bicarbonate minerals calculated as calcium carbonate since that is the primary alkaline earth mineral contributing to alkalinity. Alkalinity is also measured and calculated as the hydroxide when that is present. All natural waters contain some quantity of alkalinity. It contributes to scale formation because its presence encourages deposition of calcium carbonate, or lime scale.

pH Value. The *quality* of alkalinity, or the measure of the relative strength of acidity or alkalinity of a water, is the pH value, a value calculated from the hydrogen-ion concentration in water. The pH scale ranges from 0 to 14. A pH of 7.0 is neutral. It indicates a balance between the acidity and alkalinity. As the pH decreases to zero, the alkalinity decreases and the acidity increases. As the pH increases to 14, the alkalinity increases and the acidity decreases.

The pH scale (Fig. 53.11) is used to express the strength or intensity of the acidity or alkalinity of a water solution. This scale is logarithmic so that a pH change of 1 unit represents a tenfold increase or decrease in the strength of acidity or alkalinity. Hence water with a pH value of 4.0 is 100 times more acid in

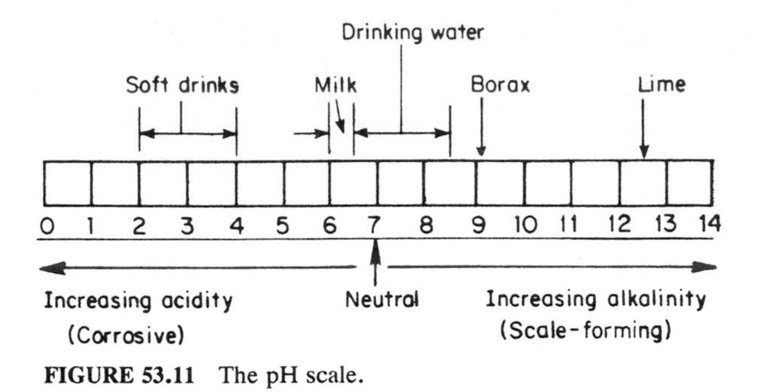

FIGURE 53.11 The pH scale.

strength than water with a pH value of 6.0. Water is corrosive if the pH value is on the acidic side. It will tend to be scale-forming if the pH value is alkaline.

Hardness. Hardness is the total calcium, magnesium, iron, and trace amounts of other metallic elements in water which contribute to the hard feel of water. Hardness is also calculated as calcium carbonate, because it is the primary component contributing to hardness. Hardness causes lime deposits or scale in equipment.

Silica. Silica is dissolved sand or silica-bearing rock such as quartz through which the water flows. Silica is the cause of very hard and tenacious scales that can form in heat-transfer equipment. It is present dissolved in water as silicate or suspended in very fine, invisible form as colloidal silica.

Iron, Manganese, and Alumina. Iron, manganese, and alumina are dissolved or suspended metallic elements present in water supplies in varying quantities. They are objectionable because they contribute to a flat metallic taste and form deposits. These soluble metals, when they react with oxygen in water exposed to the atmosphere, form oxides which precipitate and cause cloudiness, or "red water." This red color, particularly from iron, causes staining of plumbing fixtures, sinks, and porcelain china and is a cause of common laundry discoloration.

Chlorides. Chlorides are the sum total of the dissolved chloride salts of sodium, potassium, calcium, and magnesium present in water. Sodium chloride, which is common salt, and calcium chloride are the most common of the chloride minerals found in water. Chlorides do not ordinarily contribute to scale since they are very soluble. Chlorides are corrosive, however, and cause excessive corrosion when present in large volume, as in seawater.

Sulfates. Sulfates are the dissolved sulfate salts of sodium, potassium, calcium, and magnesium in the water. They are present due to dissolution of sulfate-bearing rock such as gypsum. Calcium and magnesium sulfate scale is very hard and difficult to remove and greatly interferes with heat transfer.

Total Dissolved Solids. The total dissolved solids (TDS) reported in water analyses are the sum of dissolved minerals including the carbonates, chlorides, sulfates, and all others that are present. The dissolved solids contribute to both scale formation and corrosion in heat-transfer equipment.

Suspended Matter. Suspended matter is finely divided organic and inorganic substances found in water. It is caused by clay silt and microscopic organisms which are dispersed throughout the water, giving it a cloudy appearance. The measure of suspended matter is turbidity. Turbidity is determined by the intensity of light scattered by the suspended matter in the water.

53.4 CORROSION

Corrosion is the process whereby a metal through reaction with its environment undergoes a change from the pure metal to its corresponding oxide or other stable combination. Usually, through corrosion, the metal reverts to its naturally occurring state, the ore. For example, iron is gradually dissolved by water and oxidized by oxygen in the water, forming the oxidation product iron oxide, commonly called rust.

This process occurs very rapidly in heat-transfer equipment because of the presence of heat, corrosive gases, and dissolved minerals in the water, which stimulate the corrosion process.

The most common forms of corrosion found in heat-transfer equipment are

1. General corrosion
2. Oxygen pitting
3. Galvanic corrosion
4. Concentration cell corrosion
5. Stress corrosion
6. Erosion-corrosion
7. Condensate grooving

53.4.1 General Corrosion

General corrosion is found in various forms in heat-transfer equipment. In a condenser water or cooling tower circuit, it can be seen as an overall deterioration of the metal surface with an accumulation of rust and corrosion products in the piping and water boxes. On copper condenser tubes, it is observed most frequently as a surface gouging or a uniform thinning of the tube metal.

In boilers, general corrosion is observed in the total overall disintegration of the tube metal surface in contact with the boiler water. (See Figs. 53.12 and 53.13.)

FIGURE 53.12 General corrosion on condenser tube. *(Courtesy of Metropolitan Refining Co., Inc.)*

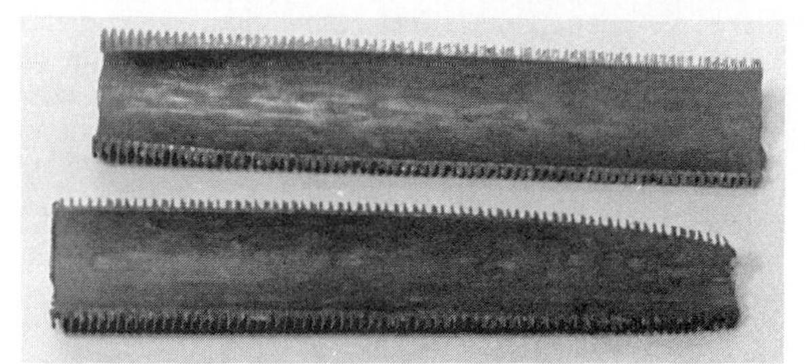

FIGURE 53.13 Pitting corrosion on condenser tubes. (*Courtesy of Metropolitan Refining Co., Inc.*)

General corrosion occurs when the process takes place over the entire surface of the metal, resulting in a uniform loss of metal rather than a localized type of attack. It is often, but not always, accompanied by an accumulation of corrosion products over the surface of the metal (Fig. 53.14).

Iron and other metals are corroded by the metal going into solution in the water. It is necessary, therefore, to limit corrosion of these metals by reducing the activity of both hydroxyl ions and hydrogen ions, i.e., by maintaining a neutral environment.

Another important factor in the corrosion process is dissolved oxygen. The evolution of hydrogen gas in these reactions tends to slow the rate of the corrosion reaction and indeed, in many instances, to stop it altogether by forming an inhibiting film on the surface of the metal which physically protects the metal from the water.

Accumulation of rust and corrosion products is further promoted by the presence of dissolved oxygen. Oxygen reacts with the dissolved metal, eventually forming the oxide which is insoluble and in the case of iron builds up a voluminous deposit of rust. Since the role of dissolved oxygen in the corrosion process

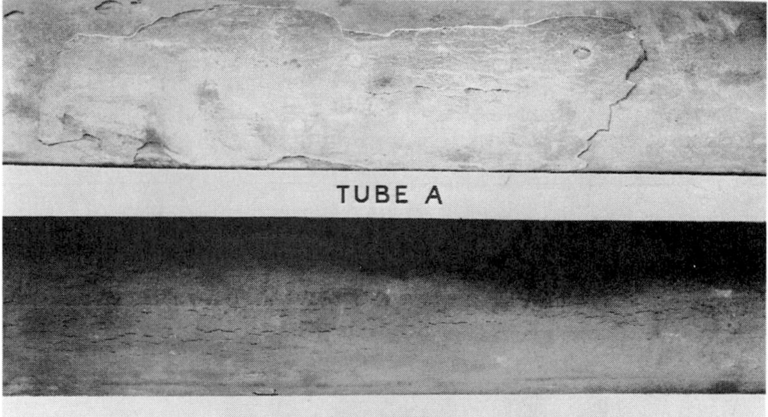

FIGURE 53.14 Boiler tube corrosion. (*Courtesy of Babcock & Wilcox Co.*)

is important, removal of dissolved oxygen is an effective procedure in preventing corrosion.

53.4.2 Oxygen Pitting

The second type of corrosion frequently encountered in heat-transfer equipment is pitting. Pitting is characterized by deep penetration of the metal at a small area on the surface with no apparent attack over the entire surface as in general corrosion.

The corrosion takes place at a particular location on the surface, and corrosion products frequently accumulate over the pit. These appear as a blister, tubercle, or carbuncle, as in Fig. 53.15.

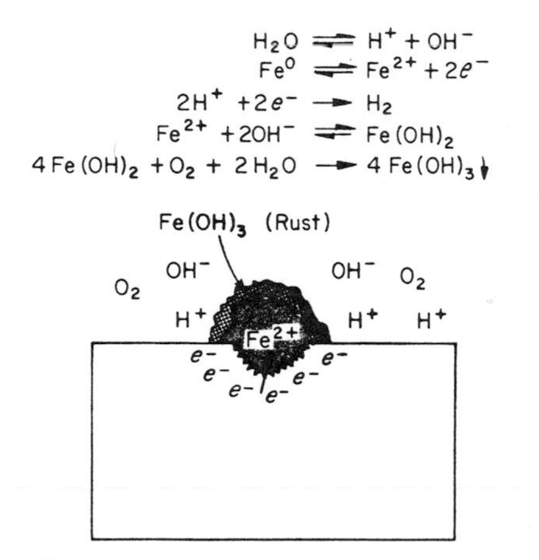

FIGURE 53.15 Reactions forming blisters over pit.

Oxygen pitting is caused by dissolved oxygen. It differs from localized pitting due to other causes, such as deposits of foreign matter, which is discussed in Sec. 53.4.4. Following are examples of pitting caused by dissolved oxygen (Figs. 53.16 and 53.17).

Oxygen pitting occurs in steam boiler systems where the feedwater contains dissolved oxygen. The pitting is found on boiler tubes adjacent to the feedwater entrance, throughout the boiler, or in the boiler feedwater line itself.

One of the most unexpected forms of oxygen pitting is commonly found in boiler feedwater lines following a deaerator. It is mistakenly believed that mechanically deaerated boiler feedwater will completely prevent oxygen pitting. However, quite to the contrary, water with a low concentration of dissolved oxygen frequently is more corrosive than that with a higher dissolved oxygen content. This is demonstrated by the occurrence of oxygen pitting in boiler feedwater lines carrying deaerated water.

Mechanical deaerators are not perfect, and none can produce a feedwater with zero oxygen. The lowest guaranteed dissolved oxygen content that deaerators produce is 0.0005 cm^3/L. This trace quantity of dissolved oxygen is sufficient to cause severe pitting in feedwater lines or in boiler tubes adjacent to the feedwater entrance. This form of pitting is characterized by deep holes scattered over the surface of the pipe interior with little or no accumulation of corrosion products or rust, since there is insufficient oxygen in the environment to form the ferric oxide rust. (See Fig. 53.18.)

53.4.3 Galvanic Corrosion

Corrosion can occur when different metals come in contact with one another in water. When this happens, an electric current is generated similar to that of a

FIGURE 53.16 Pitting on boiler tube. (*Courtesy of Metropolitan Refining Co., Inc.*)

FIGURE 53.17 Blister over pits on boiler tubes. (*Courtesy of Babcock & Wilcox Co.*)

FIGURE 53.18 Pitting in boiler feedwater line. (*Courtesy of Metropolitan Refining Co., Inc.*)

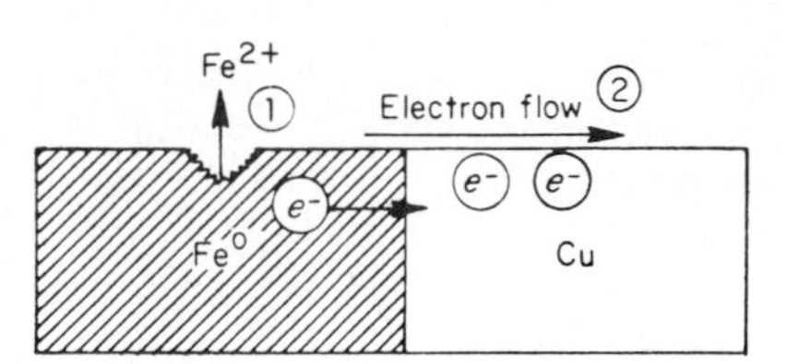

FIGURE 53.19 Galvanic corrosion caused by dissimilar-metal couple. (1) Iron going into solution loses two electrons: $Fe^0 \rightarrow Fe^{2+} + 2e^-$; (2) electrons flow to copper, the less reactive metal.

storage battery. The more active metal will tend to dissolve in the water, thereby generating an electric current (an electron flow) from the less active metal. This current is developed by a coupling of iron and copper, as in Fig. 53.19.

This tendency of a metal to give up electrons and go into solution is called the "electrode potential." This potential varies greatly among metals since the tendency of different metals to dissolve and react with the environment varies.

In galvanic corrosion, commonly called "dissimilar-metal corrosion," there are four essential elements:

1. A more reactive metal called the "anode"
2. A less reactive metal called the "cathode"
3. A water solution environment called the "electrolyte"
4. Contact between the two metals to facilitate electron flow

The rate of galvanic corrosion is strongly influenced by the electrode potential difference between the dissimilar metals. The galvanic series is a list of metals in order of their activity, the most active being at the top of the list and the least active at the bottom. The farther apart two metals are on this list, the greater will be the reactivity between them and, therefore, the faster the anodic end will corrode. The galvanic series is shown in Fig. 53.20.

If one or more of these four essential elements are eliminated, the corrosion reactions will be disrupted and the rate of corrosion slowed or halted altogether.

One method of preventing this type of corrosion is to eliminate contact of dissimilar metals in HVAC equipment by using insulating couplings or joints, such as a dielectric coupling which interferes with the electron flow from one metal to the other. Other forms of protection involve the removal of dissolved oxygen and use of protective coatings and inhibitors which provide a barrier between the corroding metal and its environment.

53.4.4 Concentration Cell Corrosion

Concentration cell corrosion is a form of pitting corrosion that is a localized type of corrosion rather than a uniform attack. It is frequently called "deposit corro-

Corroded end (anodic, or least noble)

Magnesium alloys (1)
Zinc (1)
Beryllium
Aluminum alloys (1)
Cadmium
Mild steel, wrought iron
Cast iron, flake or ductile
Low-alloy high-strength steel
Nickel-resist, types 1 & 2
Naval bronze (CA464), yellow bronze (CA268), aluminum bronze (CA687), Red bronze (CA230), Admiralty bronze (CA443) manganese bronze
Tin
Copper (CA102, 110), silicon bronze (CA655)
Lead-tin solder
Tin bronze (G & M)
Stainless steel, 12–14% chromium (AISI Types 410, 416)
Nickel silver (CA 732, 735, 745, 752, 764, 770, 794)
90/10 Copper-nickel (CA 706)
80/20 Copper-nickel (CA 710)
Stainless steel, 16–18% chromium (AISI Type 430)
Lead
70/30 Copper-nickel (CA 715)
Nickel-aluminum bronze
Inconel* alloy 600
Silver braze alloys
Nickel 200
Silver
Stainless steel, 18 chromium, 8 nickel (AISI Types 302, 304, 321, 347)
Monel* Alloys 400, K-500
Stainless steel, 18 chromium, 12 nickel-molybdenum (AISI Types 316, 317)
Carpenter 20† stainless steel, Incoloy* Alloy 825
Titanium, Hastelloy‡ alloys C & C 276, Inconel* alloy 625
Graphite, graphitized cast iron

Protected end (cathodic, or most noble)

* International Nickel Trademark.
† Union Carbide Corp. Trademark.
‡ The Carpenter Steel Co. Trademark.

FIGURE 53.20 Galvanic series of metals and alloys in flowing aerated seawater at 40 to 80°F (4 to 27°C). (*From* 1976 ASHRAE Handbook, Systems, *chap. 36, "Corrosion and Water Treatment," p. 36.2. Used with permission.*)

sion" or "crevice corrosion" since it occurs under deposits or at crevices of a metal joint.

Deposits of foreign matter, dirt, organic matter, corrosion products, scale, or any substance on a metal surface can initiate a corrosion reaction as a result of differences in the environment over the metal surface. Such differences may either be differences of solution ion concentration or dissolved oxygen concentration.

With concentration cell corrosion, the corrosion reaction proceeds as in galvanic corrosion since this differential also forms an electrode potential difference. This can best be prevented by maintaining clean surfaces.

53.4.5 Stress Corrosion

Stress corrosion is a combination of exposure of a metal to a corrosive environment and application of stress on the metal. It is frequently seen on condenser tubes and boiler tubes in the area where the tubes are rolled into the tube sheets. In steam boilers, stress corrosion has been referred to as "necking and grooving." It is seen as a circumferential groove around the outside of a firetube where it enters the tube sheet. Figure 53.21 shows this type of corrosion.

The corrosion failure is a result of a corrosive environment and stresses and strains at the point of failure. Usually it occurs at the hottest end of the tube at the beginning of the first pass against the firewall. It concentrates at the tube end because of strains from two sources. First, when tubes are rolled in, stresses are placed on the metal, expanding the metal to fit the tube sheet. Second, when a boiler is fired, the heat causes rapid expansion of the tube, and consequently strains are greatest at the tube ends, which are fixed in the tube sheets. This actually causes a flexing and bowing of the tube, and sometimes the expansion is so severe that the tubes loosen in the sheets. During this bending of the tube, the natural protective iron oxide film forming at the tube ends tends to tear or flake off, exposing fresh steel to further attack. Eventually, the tube fails due to both corrosion and stress.

FIGURE 53.21 Necking and grooving on boiler firetube. *(Courtesy of Metropolitan Refining Co., Inc.)*

Stress corrosion can also occur on condenser tubes and heat-exchanger tubes from heat expansion that causes stresses in the metal at tube supports or tube sheets. This problem is reduced by more gradual firing practices in boilers, which allow more gradual temperature changes, and by using proper inhibitors to correct the corrosive environment.

53.4.6 Erosion-Corrosion

"Erosion-corrosion" is the gradual wearing away of a metal surface by both corrosion and abrasion. It is also commonly called "impingement corrosion."

Water moving rapidly through piping can contain entrained air bubbles and suspended matter, sand, or other hard particulates. This is not uncommon in cooling tower waters where such particles are washed from the atmosphere. These abrasive particles remove natural protective oxide films present on the surface of the metal and cause general corrosion of the exposed metal. The higher the velocity of the impinging stream, the greater the rate of erosion-corrosion.

53.4.7 Condensate Grooving

Condensate grooving is a particular phenomenon of steam condensate line corrosion in HVAC equipment. It is found in steam condensate piping on all types of equipment, heat exchangers, steam-turbine condensers, unit heaters, steam absorption condensers, radiators, or any type of unit utilizing steam as a heat-transfer medium.

Condensate grooving is a direct chemical attack by the steam condensate on the metal over which it flows and is identified by the typical grooves found at the bottom of the pipe carrying the condensate. This is shown in Fig. 53.22.

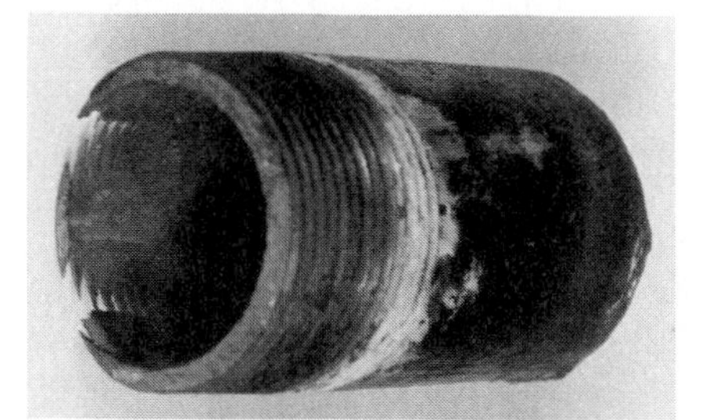

FIGURE 53.22 Steam condensate line corrosion. (*Courtesy of Metropolitan Refining Co., Inc.*)

The primary cause of condensate grooving is carbon dioxide. The dissolved carbon dioxide forms a mild carbonic acid. The methods available to prevent this type of corrosion include removal of bicarbonate and carbonate alkalinity from the boiler makeup water (dealkalinization) and use of carbonic acid neutralizers and filming inhibitors.

53.5 SCALE AND SLUDGE DEPOSITS

The most common and costly water-caused problem encountered in HVAC equipment is scale formation. The high cost of scale formation stems from the significant interference with heat transfer caused by water mineral scale deposits.

53.5.1 Mineral Scale and Pipe Scale

At this point, we should differentiate between mineral scale and pipe scale. *Mineral scale* is formed by deposits of the more insoluble minerals present in water, the heat-transfer medium (Fig. 53.23). *Pipe scale* (Fig. 53.24) is the natural iron oxide coating or corrosion products that form on the interior of piping which flake off and appear as a scale.

FIGURE 53.23 Pipe scale and iron corrosion products. *(Courtesy of Metropolitan Refining Co., Inc.)*

FIGURE 53.24 Mineral scale deposits of water minerals. *(Courtesy of Metropolitan Refining Co., Inc.)*

Mineral scale in steam boilers, heat exchangers, and condensers consists primarily of calcium carbonate, the least soluble of the minerals in water. Other scale components, in decreasing order of occurrence, are calcium sulfate, magnesium carbonate, iron, silica, and manganese. Present also in some scales are the hydroxides of calcium, magnesium, and iron as well as the phosphates of these minerals, where phosphates and alkalinity are used as a corrosion or scale inhibitor. Sludge is a softer form of scale and results when hard-water minerals reacting with phosphate and alkaline treatments forming a soft, pastelike substance rather than a hard, dense material. In most cases, scales contain a complex mixture of mineral salts because scale forms gradually and deposits the different minerals in a variety of forms.

The major cause of mineral scale is the inverse solubility of calcium and magnesium salts. Most salts or soluble substances, such as table salt or sugar, are more soluble in hot water than in cold.

Calcium and magnesium salts, however, dissolve more readily and in greater quantity in cold water than in hot, hence inverse solubility. This unique property is responsible for the entire problem of mineral scale on heat-transfer surfaces in HVAC equipment. From this property alone, we can readily understand why mineral scale forms on hot-water generator tubes, condenser tubes, boiler tubes, etc. It is simply the fact that the hottest surface in contact with the water is the tube surface of this type of equipment.

In condenser water systems using recirculating cooling tower water or once-through cooling water, the water temperature is much lower than that in steam boiler or hot-water systems. At these lower temperatures most of the scale-forming minerals will remain in solution, but the tendency will be to deposit calcium carbonate on the heat-transfer surfaces where there is a slight rise in temperature.

The primary factors which affect this tendency are:

1. Alkalinity
2. Hardness
3. pH
4. Total dissolved solids

The higher the alkalinity of a water, the higher the bicarbonate and/or carbonate content. As these minerals approach saturation, they tend to come out of solution.

Likewise, a higher concentration of hardness will increase the tendency of calcium and magnesium salts to come out of solution. The pH value reflects the ratio of carbonate to bicarbonate alkalinity. The higher the pH value, the greater the carbonate content of the water. Since calcium carbonate and magnesium carbonate are less soluble than the bicarbonate, they will tend to precipitate as the pH value and carbonate content increase.

Also affecting this tendency are the total dissolved solids and temperature. The higher the solids content, the greater the tendency to precipitate the least soluble of these solids. The higher the temperature, the greater the tendency to precipitate the calcium and magnesium salts because of their property of inverse solubility.

53.5.2 Langelier Index

The Langelier index is a calcium carbonate saturation index that is very useful in determining the scaling or corrosive tendencies of a water. It is based on the assumption that a water with a scaling tendency will tend to deposit a corrosion-inhibiting film of calcium carbonate and hence will be less corrosive, whereas a water with a nonscaling tendency will tend to dissolve protective films and be more corrosive. This is not entirely accurate since other factors are involved in corrosion, as we have seen in Sec. 53.4 on corrosion, but it is an extremely valuable index in determining a tendency of a water.

In the 1950s, Eskell Nordell arranged five basic variables into an easy-to-use chart to quickly determine the pH of saturation of calcium carbonate and the Langelier index.[3] This index is based on the pH of saturation of calcium carbonate.

The pH of saturation of calcium carbonate is the theoretical pH value of a particular water if that water is saturated with calcium carbonate. As the actual pH of a recirculating water approaches or even exceeds the pH of saturation of calcium carbonate, the tendency is to form a scale of calcium carbonate. If the actual pH is well below the pH of saturation of calcium carbonate, the tendency is to dissolve minerals and therefore to be corrosive. The Langelier index, therefore, is determined by comparing the actual pH of a recirculating water with the pH of saturation of calcium carbonate.

To determine the Langelier index, the actual pH of the water must be measured, and the pH of saturation of calcium carbonate, called the pHs, is calculated from a measure of the total alkalinity, hardness, total dissolved solids, and temperature.

A useful shortcut calculation of pHs can be made for cold well or municipal water supplies that are used for once-through cooling or service water. The reason why this rapid calculation is valid is that these supplies are usually consistent in temperature [49 to 57°F (10 to 14°C)] and total dissolved solids (50 to 300 mg/L). If a water supply has these characteristics, the following formula can be used (see Fig. 53.25).

$$\text{pHs @ 50°F (10°C)} = 11.7 - (C + D)$$

Likewise for hot-water supplies at 140°F (60°C), a short-form calculation of the pH of saturation of calcium carbonate can be done with the following formula:

$$\text{pHs @ 140°F (60°C)} = 10.8 - (C + D)$$

Once the pH of saturation of calcium carbonate has been calculated, the Langelier saturation index (SI) can be determined from the formula

$$\text{SI} = \text{pH} - \text{pHs}$$

where pH = actual measured pH of the water and pHs = pH of saturation of calcium carbonate as calculated from Fig. 53.25. Figure 53.26 can also be used to determine the pH of saturation.

A positive index indicates scaling tendencies; a negative one, corrosion tendencies. A very handy guide in predicting the tendencies of a water by using the Langelier saturation index is shown in Table 53.6.

53.5.3 Ryznar Index

Another useful tool for determining the tendencies of a water is the Ryznar index. This index is also based on the pH of saturation of calcium carbonate and was intended to serve as a more accurate index of the extent of scaling or corrosion in addition to the tendency. This index is calculated as follows:

$$\text{Ryznar index} = 2(\text{pHs}) - \text{pH}$$

where pHs = pH of saturation of calcium carbonate, as calculated from Fig. 53.25, and pH = actual measured pH of the water. Table 53.7 can be used to determine the tendency and extent of corrosion or scaling with the Ryznar index.

A

Total solids (mg/L)	*A*
50–300	0.1
400–1000	0.2

B

Temperature °F	(°C)	*B*
32– 34	(0– 1.1)	2.6
36– 42	(2.2– 5.5)	2.5
44– 48	(6.7– 8.9)	2.4
50– 56	(10.0–13.3)	2.3
58– 62	(14.4–16.7)	2.2
64– 70	(17.8–21.1)	2.1
72– 80	(22.2–26.7)	2.0
82– 88	(27.8–31.1)	1.9
90– 98	(27.8–31.1)	1.8
100–110	(37.8–43.3)	1.7
112–122	(44.4–50.0)	1.6
124–132	(51.1–55.6)	1.5
134–142	(56.7–63.3)	1.4
148–160	(64.4–71.1)	1.3
162–178	(72.2–81.1)	1.2

C

Calcium hardness (mg/L of $CaCO_3$)	*C*
10– 11	0.6
12– 13	0.7
14– 17	0.8
18– 22	0.9
23– 27	1.0
28– 34	1.1
35– 43	1.2
44– 55	1.3
56– 69	1.4
70– 87	1.5
88– 110	1.6
111– 138	1.7
139– 174	1.8
175– 220	1.9
230– 270	2.0
280– 340	2.1
350– 430	2.2
440– 550	2.3
560– 690	2.4
700– 870	2.5
800–1000	2.6

D

M Alkalinity (mg/L of $CaCo_3$)	*D*
10– 11	1.0
12– 13	1.1
14– 17	1.2
18– 22	1.3
23– 27	1.4
28– 35	1.5
36– 44	1.6
45– 55	1.7
56– 69	1.8
70– 88	1.9
89– 110	2.0
111– 139	2.1
140– 176	2.2
177– 220	2.3
230– 270	2.4
280– 350	2.5
360– 440	2.6
450– 550	2.7
560– 690	2.8
700– 880	2.9
890–1000	3.0

$pH_s = (9.3 + A + B) - (C + D)$
$SI = pH - pH_s$
If index is 0, water is in chemical balance.
If index is positive, scale-forming tendencies are indicated.
If index is negative, corrosive tendencies are indicated.

FIGURE 53.25 Data for calculations of the pH of saturation of calcium carbonate. (*From Eskell Nordell, Water Treatment for Industrial and Other Uses, 2d ed., © 1961 by Litton Educational Publishing Inc., reprinted with permission of Van Nostrand Reinhold Co.*)

Let us see how these indices can help us in analyzing a particular water supply. Figure 53.8 depicts an analysis report on the Washington, DC, water supply. The Langelier saturation index at 50°F (10°C) is determined by using this analysis and the data shown on Fig. 53.25 as follows:

$$\begin{aligned} \text{pHs} &= 9.3 + A + B - (C + D) \\ &= 9.3 + 0.1 + 2.3 - (1.8 + 2.0) \\ &= 8.2 \end{aligned}$$

and

$$\text{SI} = \text{pH} - \text{pHs} = 7.3 - 8.2 = -0.9$$

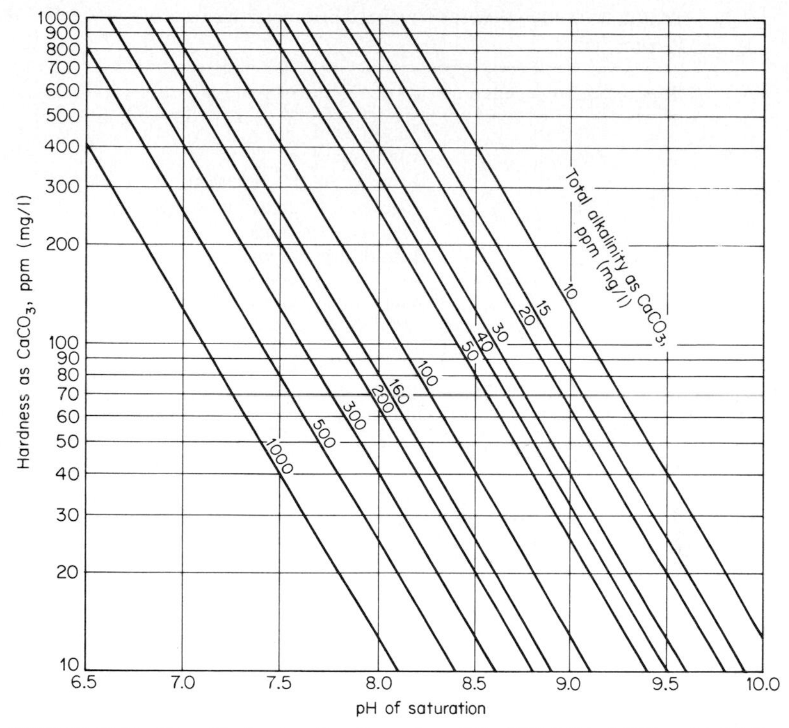

FIGURE 53.26 The pH of saturation for waters 49 to 57°F (10 to 14°C) and total dissolved solids of 50 to 300 mg/L.

TABLE 53.6 Prediction of Water Tendencies by the Langelier Index

Langelier saturation index	Tendency of water
2.0	Scale-forming and for practical purposes noncorrosive
0.5	Slightly corrosive and scale-forming
0.0	Balanced, but pitting corrosion possible
−0.5	Slightly corrosive and non-scale-forming
−2.0	Serious corrosion

Source: *Carrier System Design Manual*, part 5, "Water Conditioning," Carrier Corporation, Syracuse, NY, 1972, pp. 5–12.

From Table 53.6, according to the Langelier saturation index this water supply is somewhat more than "slightly corrosive and non-scale-forming."

To learn more about this water, the Ryznar index (RI) can be calculated in the same manner:

$$RI = 2(pHs) - pH = 16.4 - 7.3 = 9.1$$

According to Table 53.6, this water supply tendency indicates "intolerable cor-

TABLE 53.7 Prediction of Water Tendencies by the Ryznar Index

Ryznar stability index	Tendency of water
4.0–5.0	Heavy scale
5.0–6.0	Light scale
6.0–7.0	Little scale or corrosion
7.0–7.5	Significant corrosion
7.5–9.0	Heavy corrosion
9.0+	Intolerable corrosion

Source: *Carrier System Design Manual*, part 5, "Water Conditioning," Carrier Corporation, Syracuse, NY, 1972, pp. 5–14.

rosion." The Ryznar index, being more quantitative, indicates that the degree of corrosion would be greater than we would anticipate from the tendency shown by the qualitative Langelier saturation index.

In an examination of a water supply, both the Langelier and the Ryznar indices are used to determine the scale-forming or corrosion tendencies.

In open cooling tower condenser water systems and steam boilers, however, there is a constant accumulation of minerals as a result of evaporation of pure water, such as distilled water, and makeup water containing the various mineral impurities. Therefore, in these systems the pH, concentration of hardness, total dissolved solids, and alkalinity are constantly changing, making a study of the Langelier and Ryznar indices relatively complex and subject to gross inaccuracies.

53.5.4 Boiler Scale

Scale in boilers is a direct result of precipitation of the calcium, magnesium, iron, and silica minerals present in the boiler feedwater. Scale can be prevented by removing a portion of the scale-forming ingredients prior to the boiler with external water-softening equipment or within the boiler itself with internal boiler water treatment.

One of the most troublesome deposits frequently encountered in steam boilers is iron and combinations of iron with calcium and phosphate used in boiler water treatment. These sticky, adherent sludge deposits are caused by excessive amounts of iron entering the boiler with the feedwater. The iron is in the form of iron oxide or iron carbonate corrosion products. It is a result of corrosion products from the sections prior to the boiler, such as steam and condensate lines, condensate receivers, deaerators, and boiler feedwater lines. A program for preventing scale deposits must include treatment to prevent this troublesome type of sludge deposit.

53.5.5 Condensate Scale

In recirculating cooling tower condenser water systems for air conditioning and refrigeration chillers, scale deposits are a direct result of precipitation of the calcium carbonate, calcium sulfite, or silica minerals due to such an overconcentration of these minerals that their solubility or pH of saturation is exceeded and the minerals come out of solution. Scale in this equipment can include foreign substances such as corrosion products, organic matter, and mud or dirt. These are usually called "foulants" rather than "scale." Treatment to prevent mineral scale should, therefore, include sufficient dilution of the recirculating water to prevent the concentration of minerals from approaching the saturation point, pH control to prevent the pH from reaching the pH of saturation of calcium carbonate, and chemical treatments to inhibit and control scale crystal formation.

53.6 FOULANTS

In addition to water mineral scale, other deposits of mud, dirt, debris, foreign matter, and organic growth are a recurrent problem in recirculating water systems. Deposits of foreign matter plug narrow passages, interfere with heat transfer, and foul heat-transfer surfaces, causing inefficient performance of the equipment and high energy consumption.

53.6.1 Mud, Dirt, and Clay

Open recirculating cooling tower systems are most subject to deposits of mud, dirt, and debris. A cooling tower is a natural air washer with water spraying over slats and tower fill washing the air blown through either naturally or assisted by fans.

Depending on the location, all sorts of airborne dust and debris end up in cooling tower recirculating water systems. These vary from fine dust particles to pollen, weeds, plant life, leaves, tree branches, grass, soil, and stones.

The fine particles of dust and dirt tend to collect and compact in the condenser water system, especially in areas of low circulation. At heat-transfer surfaces, the dust and dirt can deposit and compact into a sticky mud and seriously interfere with operating efficiency.

Muddy foulants are a common occurrence and form with the combination of airborne particles, corrosion products, scale, and organic matter. Very rarely can one identify a foulant as a single compound because it is usually a complex combination of all these things.

In closed recirculating water systems, foulants are not nearly as varied and complex as in open systems, but they are just as serious when they occur. Deposits in closed systems are usually caused by dirt or clay entering with the makeup water or residual construction debris. A break in an underground water line can result in dirt, sand, and organic matter being drawn into a system and is a common source of fouling.

Makeup water containing unusual turbidity or suspended matter is usually treated at the source by coagulation, clarification, and filtration so as to maintain its potability. Suspended matter and turbidity, therefore, are not common in makeup water in HVAC systems since the makeup water usually comes from a municipal or local source, over which there is a water authority responsible for delivery of clear, potable water.

Where a private well water, pond, or other nonpublic source of water is available for use as makeup water to recirculating water systems and boilers, it should be carefully examined for turbidity and suspended matter. The suspended matter measured as turbidity should be no more than the maximum of 1 turbidity unit for drinking water recommended by the Environmental Protection Agency. When the supply is excessively turbid, some form of clarification such as coagulation, settling, filtration, and/or fine strainers should be used to remove the suspended matter and reduce the turbidity to below 1 unit.

The more common problem with suspended matter and turbidity results from makeup water that is temporarily or occasionally dirty. This may occur when the local water authority is cleaning sections of a distribution main or installing new mains or when water mains are cut into during some nearby construction project. This kind of work creates a disturbance of the water mains, causing settled and lightly adherent pipeline deposits to break off and be flushed into the water supply. These deposits consist mostly of iron oxide corrosion products and dirt, clay, or silt.

53.6.2 Black Mud and Mill Scale

One of the most common and difficult foulants found in closed systems is a black mud made up of compacted, fine, black magnetic iron oxide particles. This black mud not only deposits at heat-transfer surfaces, but also clogs or blocks narrow passages in unit heaters, fan-coil units, and cooling, reheat, and heating coils in air-handling units. This black mud is a result of wet very fine particles of black magnetic iron oxide being compacted into a dense adherent mud.

The interior of black iron piping, commonly used for recirculating water, has a natural black iron oxide protective coating ordinarily held intact by oil-based inhibitors used to coat the pipe to prevent corrosion during storage and layup. This natural iron oxide protective coating is called *mill scale*, a very general term which can be applied to any form of pipe scale or filings washed off the interior of the pipe. This mill scale film becomes disturbed and disrupted during construction due to the constant rough handling, cutting, threading, and necessary battering of the pipe. After construction, the recirculating water system is filled and flushed with water, which removes most of the loosened mill scale along with any other construction debris. However, very fine particles of magnetic iron oxide will continue to be washed off the metal surface during operation, and in many instances this washing persists for several years before it subsides. Mill scale plugging can be a serious problem. It is best alleviated in a new system by thorough cleaning and flushing with a strong, low-foaming detergent-dispersant cleaner. This, however, does not always solve the problem. Even after a good cleanout, gradual removal of mill scale during ensuing operation can continue.

53.6.3 Boiler Foulants

In steam boilers, foulants other than mineral scale usually consist of foreign contaminants present in the feedwater. These include oil, clay, contaminants from a process, iron corrosion products from the steam system, and construction debris in new boiler systems. Mud or sludge in a boiler is usually a result of scale-forming minerals combined with iron oxide corrosion products and treatment chemicals. Such foulants are controlled by using proper dispersants which prevent adherence on heat-transfer surfaces.

In heating boilers, the most frequent foulants other than sludge are oil and clay. Oil can enter a boiler system through leakage at oil lubricators, fuel oil preheaters, or steam heating coils in fuel oil storage tanks. When oil enters a boiler, it causes priming and foaming by emulsifying with the alkaline boiler water. Priming is the bouncing of the water level that eventually cuts the boiler off at low water due to the very wide fluctuation of this level. Oil can also carbonize at hot boiler tubes, causing not only serious corrosion from concentration corrosion cells but also tube ruptures as a result of overheating due to insulating carbon deposits. Whenever oil enters a boiler system, it must be removed immediately to prevent these problems. This is easily done by boiling out with an alkaline detergent cleaner for boilers.

Clay is a less frequent foulant in boilers, but it, too, can form insulating deposits on tube surfaces. Clay enters a boiler with the boiler makeup water that is either turbid or contaminated with excessive alum, used as a coagulant in the clarification process. Clay can be dispersed with the use of dispersants in the internal treatment of the boiler, but makeup water should be clear and free of any turbidity before it is used as boiler feedwater. Where turbidity and clay are a constant problem, filtration of the boiler feedwater is in order.

53.6.4 Construction Debris

All new systems become fouled and contaminated with various forms of foreign matter during construction. It is not uncommon to find these in the interior of HVAC piping and heat exchangers: welding rods, beads, paper bags, plastic wrappings, soft drink can rings, pieces of tape, insulation wrappings, glass, and any other construction debris imaginable.

It is necessary not only to clean out construction debris from the interior of HVAC systems prior to initial operation, but also to clean the metal surfaces of oil and mill scale naturally present on the pipe interior. This oil and mill scale, as has been shown, can seriously foul and plug closed systems and cause boiler tube failures, if the oil is carbonized during firing. Every new recirculating water system and boiler must be cleaned thoroughly with a detergent-dispersant type of cleaner or, as in steam boilers, with an alkaline boilout compound. This initial cleanout will remove most of the foulants and prevent serious operational difficulties.

53.6.5 Organic Growths

Organic growths in HVAC equipment are usually found in open recirculating water systems such as cooling towers, air washers, and spray coil units. Occasionally closed systems become fouled with organic slimes due to foreign contamination. Open systems are constantly exposed to the atmosphere and environs which contain not only dust and dirt but also innumerable quantities of microscopic organisms and bacteria. Cooling tower waters, because they are exposed to sunlight, operate at ideal temperatures, contain mud as a medium and food in the form of inorganic and organic substances, and are a most favorable environment for the abundant growth of biological organisms. Likewise, air washers and spray coil units, as they wash dust and dirt from the atmosphere, collect microscopic organisms which tend to grow in the recirculating water due to the favorable environment. The organisms that grow in such systems consist primarily of algae, fungi, and bacterial slimes.

53.6.6 Algae

Algae are the most primitive form of plant life and together with fungus form the family of thallus plants. Algae are widely distributed throughout the world and consist of many different forms. The forms found in open recirculating water systems are the blue-green algae, green algae, and brown algae. The blue-green algae, the simplest form of green plants, consist of a single cell and hence are called unicellular. Green algae are the largest group of algae and are either unicellular or multicellular. Brown algae are also large, plantlike organisms that are multicellular.

Large masses of algae can cause serious problems by blocking the air in cooling towers, plugging water distribution piping and screens, and accelerating corrosion by concentration cell corrosion and pitting. Algae must be removed physically before a system can be cleaned since the mass will provide a continuous source of material for reproduction and biocides will be consumed only at the surface of the mass, leaving the interior alive for further growth.

53.6.7 Fungi

Fungi are also a thallus plant similar to the unicellular and multicellular algae. They require air, water, and carbohydrates for growth. The source of carbohydrates can be any form of carbon. Fungi and algae can grow together; the algae living within the fungus mass are furnished with a moist, protected environment, while the fungus obtains carbohydrates from the algae.

53.6.8 Bacteria

Bacteria are microscopic unicellular living organisms that exhibit both plant and animal characteristics. They exist in rod-shaped, spiral and spherical forms. There are many thousands of strains of bacteria, and all recirculating waters contain some bacteria. The troublesome ones, however, are bacterial slimes, iron bacteria, sulfate-reducing bacteria, and pathogenic bacteria.

Pathogenic bacteria are disease-bearing bacteria. Cooling tower waters, having ideal conditions for the growth of bacteria and other organisms, can promote the growth of pathogenic bacteria. In isolated instances, pathogenic bacteria have been found growing in cooling tower waters. Therefore, it is as important to keep these systems free of bacterial contamination, to inhibit growth of pathogenic bacteria, as it is to prevent growth of slime-forming and corrosion-promoting bacteria.

53.7 PRETREATMENT EQUIPMENT

Prior to internal treatment of HVAC equipment, it is frequently necessary to use mechanical equipment to remove from the feedwater supply damaging impurities such as dissolved oxygen, excess hardness, or suspended solids.

The choice of proper equipment and its need can be determined by studying the quality and quantity of makeup water used in a boiler, condenser water system, and an open or a closed recirculating water system.

53.7.1 Water Softeners

Hardness in the makeup water is the cause of scale formation. In equipment using large volumes of a hard water, a substantial amount of scale can form on heat-transfer surfaces in a short time. In these circumstances, it may be economical to remove the hardness from the water supply before it is used in the equipment.

Determining the Need for Water Softeners. In open cooling tower condenser water, evaporative condensers, and surface spray units, removal of the hardness with pretreatment equipment is usually not economical. These systems operate at pH values close to the neutral point, where the hardness can be kept in solution with the aid of antiscalants and pH control chemical treatments. This treatment will be required even if most of the hardness is removed. Therefore, it is not economically advantageous to install water softeners in this equipment unless the makeup water is so hard that it cannot be used at all in the particular system without some form of pretreatment to at least partially soften the water.

Waters with a hardness of more than 300 or 400 mg/L require some form of pretreatment to partially soften the water so that it can be used for cooling tower makeup water. Likewise, it is not economical to install water-softening equipment on closed chilled-water and low-temperature hot-water heating systems since such systems use very little makeup water and internal treatment with antiscalants can prevent scale on heat-transfer surfaces.

With steam and high-temperature hot-water boilers, however, removal of hardness from the makeup water is frequently required. The determining factor usually is the quality of available makeup water. With high-temperature hot-water boilers, if the hardness of the makeup water exceeds 100 mg/L, the initial fill and makeup water should be softened to remove the hardness and prevent excessive scale or sludge deposits on heat-transfer surfaces.

With steam boilers, the determining factors are both the hardness and the amount of makeup water used. In low-pressure heating applications where steam is used for heating only and possibly small amounts of humidification requiring less than 10 percent of the steam generated, the boiler feedwater will consist of 90 percent or more of return steam condensate. In instances such as this, the makeup water will not require external hardness removal since the small amount of hardness entering the system can be controlled with internal treatment.

In steam boiler systems that use more than 10 percent raw makeup water, it may prove economical and practical to install a water softener. Usually if the makeup water in these systems exceeds 100 mg/L hardness and the amount of makeup water is more than 1000 gal/day (3785 L/day), a water softener will be required. This can be justified by comparing the cost of internally treating the boiler water to control scale and sludge deposits with the operating cost of an external water softener.

Another useful guideline is that if the hardness of water entering a boiler exceeds 1000 gr/h, a water softener usually is required.

Grains per hour are determined by multiplying the hardness in grains per gallon (17.1 mg/L per 1.0 gr/gal) by the makeup rate in gallons per hour:

$$\text{Grains per hour} = \text{hardness, gr/gal} \times \text{makeup, gal/h}$$

Ion-Exchange Water Softeners. The water softener used for boiler makeup water is a synthetic zeolite softener containing an ion-exchange water softener resin. This ion-exchange resin adsorbs calcium and magnesium ions from the water passing over the resin bed. The resin at the same time releases sodium, hence the term "ion exchange." Figure 53.27 shows a typical ion-exchange water softener used for boilers.

The size of the softener required depends on the rate of makeup water and the amount of hardness to be removed. The softener should have a minimum delivery rate of 6⅔ gal/min (25.2 L/min) per 100 bhp [3450 lb/h (1564.9 kg/h) steam rate].

53.7.2 Dealkalizer

Another ion-exchange water conditioner that may be required is the dealkalizer. A steam boiler that operates with a makeup water containing excessive quantities of carbonate and bicarbonate alkalinity not only will develop excess alkalinity in the boiler (causing priming, foaming, and carryover), but also will generate large quantities of carbon dioxide as a result of decomposition of the carbonates and bicarbonates. This results in an acid steam condensate, as noted previously, causing severe corrosion of steam and condensate return lines.

The alkalinity in these cases can be reduced by 90 percent by passing the makeup water through a dealkalizer following the water softener. Usually a dealkalizer cannot be economically justified unless the total alkalinity as calcium carbonate exceeds 100 mg/L and the makeup rate exceeds 100 gal/h (378.5 L/h). With lesser quantities, the effects of the carbon dioxide generated can be con-

FIGURE 53.27 Ion-exchange water softener. *(Courtesy of Ermco Inc.)*

trolled by the use of steam and return line treatments. The economics of using a dealkalizer can be determined by comparing the costs of the steam and return line treatments with the costs of installing and operating a dealkalizer.

In installations where the steam comes in direct contact with food products, the U.S. Food and Drug Administration permits use of certain treatments for control of carbon dioxide corrosion under limited conditions. When it is not possible to control the treatment within these specified limitations, installation of a dealkalizer would be justified to remove 90 percent of the alkalinity to reduce the excessive corrosion tendencies of the steam and return condensate.

The dealkalizer contains an ion-exchange resin similar to the water softener with the capability of exchanging carbonate, bicarbonate, sulfate, and other anions for chloride, hence the name "chloride anion dealkalizer." Figure 53.28 shows a typical installation of a chloride anion exchange unit.

The dealkalizer installed must be sized to remove the alkalinity from the makeup water and to deliver dealkalized water at a rate of 6⅔ gal/min (25.2 L/min) per 100 bhp [3450 lb/h (1565 kg/h) steam rate] so that it would be able to deliver the maximum amount of makeup water required at any given moment of operation. It usually accompanies the ion-exchange water softener and is installed as a complete packaged softener-dealkalizer. With this equipment, alkalinity is reduced and corrosion of steam and return condensate lines can be controlled.

FIGURE 53.28 Dealkalizer. (*Courtesy of Cochrane Division, Crane Co.*)

53.7.3 Deaerators

To prevent serious corrosion and pitting of steam boiler feedwater lines and boiler tubes, it is necessary to remove the dissolved oxygen from the boiler feedwater, which may be done by the use of chemical treatment. In many installations, however, it is neither practical nor economical to use chemical treatment alone. In these circumstances, it is necessary to remove most of the dissolved oxygen mechanically by using feedwater heaters or deaerators followed by small quantities of treatment to remove the last traces of corrosive gases.

Whenever a steam boiler system is open to the atmosphere through vented condensate receivers, feedwater tanks, etc., the air absorbed will result in high quantities of dissolved oxygen. This increases in direct proportion to the amount of makeup water used because the cold raw makeup water, high in dissolved oxygen, not only will increase the dissolved oxygen content, but also will lower the temperature of the feedwater in the return condensate tank, enabling more oxygen from the atmosphere to be dissolved in the feedwater. This happens because the solubility of oxygen in water is inversely proportional to temperature. Figure 53.29 shows the solubility of oxygen with respect to temperature.

The feedwater temperature of low-pressure heating boilers operating with theoretically all return condensate will remain close to 180 to 200°F (82 to 93°C) since little or no cold-water makeup is used. In systems such as this, a deaerator or feedwater heater is not usually required. Hence the current design custom is to avoid the use of preheaters or deaerators with low-pressure heating boilers. However, if the low-pressure steam heating boiler were to operate with any steam loss such as at steam humidifiers in air-handling units or steam tables or pressure cookers in a cafeteria or restaurant, the cold-water makeup replacing that loss increases the dissolved oxygen content. In these cases, it is essential that a mechanical preheater or deaerator be used to remove the dissolved oxygen from the feedwater. Without it, the cost for chemical oxygen scavengers alone is excessive.

The simplest mechanical deaerator is the open feed-water heater or preheater. This consists of open or closed steam coils placed in the vented condensate tank.

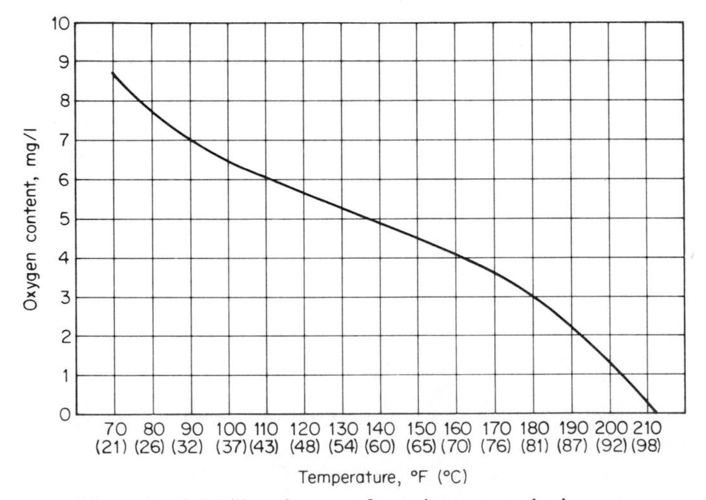

FIGURE 53.29 Solubility of oxygen from air at atmospheric pressure.

The coils are thermostatically controlled to maintain the temperature of the feed water at 180 to 200°F (80 to 95°C) or the highest possible temperature without causing boiler feed-water pump difficulties such as cavitation or steam shock.

The deaerator is a more complex device that utilizes steam injection to scrub the incoming makeup water and condensate of the dissolved corrosive gases, oxygen and carbon dioxide. In the deaerator, water is sprayed over inert packing such as glass beads or plastic fill or is trickled through baffles or trays to break up the water and provide for intimate mixing of the feed water with the incoming steam. The steam is injected counter to the flow of the incoming water which drives the gases, oxygen and carbon dioxide, upward out the vent. The vent releases only noncondensible gases and steam losses are at a minimum.

Some deaerators can produce a feed water with dissolved oxygen content as low as 0.005 cm^3L. To completely protect the equipment, the last trace of oxygen must be removed by superimposing a chemical oxygen scavenger, also packaged deaerating heaters can provide a feed-water quality of 0.03 cm^3/L dissolved oxygen that is more than adequate for HVAC boiler systems. These packages include a factory-assembled unit complete with deaerataing heater, controls, boiler feed-water pumps, and level controls. The deaerating heater consists of a storage or collection tank for condensate and raw makeup water and the deaerator section. The feed water is pumped from the storage section through a spray manifold into the top of the deaerating section. Steam is injected through a sparger into the bottom of the deaerating section and is bubbled through the feed water. The steam drives out the oxygen and carbon dioxide gases through the vent. See Chap. 7 for a full discussion of aerators.

It is important to use the available equipment to remove dissolved corrosive gases from boiler feed water for both low- and high-pressure boilers when any amount of the steam produced is used for any purpose where some cold-water makeup is required.

53.7.4 Abrasives Separators

In closed hot-water, dual-temperature chilled and so-called secondary recirculating water systems, one of the major problems is the suspended matter. Suspended matter not only causes deposit corrosion and fouling of heat-transfer surfaces, but also can seriously damage mechanical seals and shafts on pumps. Hard and abrasive particles will score shafts and mechanical seals, causing leaks and premature failures.

Mechanical seal failures are frequently attributed to the water treatment used in the recirculating water when they are actually caused by abrasives, finely divised iron particles, grit, sand, and other foreign particles. Water treatment and other minerals dissolved in the water will not cause seal failures unless excessive operating temperatures are encountered. This causes flashing of the water lubricant at the seal interface, resulting in precipitation of dissolved minerals which end up scoring the seal surface in the absence of the water lubricant. This occurs at temperatures in excess of 160°F (71°C). See Ref. 4.

To minimize these failures, very inexpensive abrasives separators are used. These not only remove the suspended abrasive particles from the water lubricant at the seal interface, but also force clean, clear water at the pump discharge pressure back into the seal cavity, preventing flashing due to excessive temperatures.

A typical abrasives separator is pictured in Fig. 53.30. The device shown removes suspended abrasive particles by centrifugal force as the water piped from

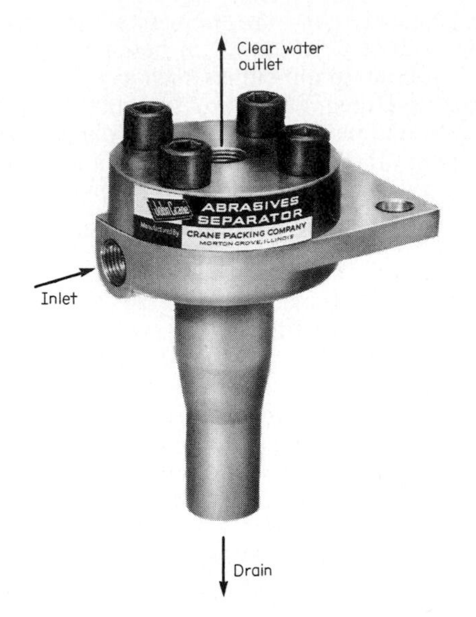

FIGURE 53.30 Abrasives separator. (*Courtesy of Crane Packing Co.*)

the pump discharge enters the separator and is rotated in the cone-shaped bore at a high velocity developed by the pressure differential. Clear water is taken from the middle of the top outlet and piped to the pump housing for flushing the seal faces. This clear flush water under pump discharge pressure prevents flashing at the seal surfaces.

53.7.5 Strainers and Filters

It is common practice to install strainers prior to pumps on open recirculating condenser water systems to protect the pump internals, vanes, shaft, and impellers from large damaging flakes of rust, suspended dirt, or other foreign particles that can enter an open system. Similarly, strainers should be installed on closed systems since they are frequently plagued with suspended black magnetic iron oxide mill scale as well as foreign particles.

The bucket strainer can be used as a coarse filter to remove larger particles before the pumps and heat exchangers. Iron and steel particles, however, can be so small that they pass through the finest mesh screen. Magnetic inserts attached to the strainer's bucket catch the fine iron particles which can then be cleaned out by flushing with a high-pressure hose. Figure 53.31 pictures a typical magnetic insert strainer bucket used for this purpose.

53.7.6 Free Cooling

One of the most interesting applications of strainers and/or filters on open cooling tower water systems is described in U.S. Patent 3,995,443 dated December 5,

FIGURE 53.31 Strainer with magnetic inserts. (*Courtesy of Hayward Manufacturing Co.*)

1976. This patent describes a process which is commonly called "free cooling" or "crossover cooling." During seasons of the year when the cooling tower water temperature can be held at 60°F (15.5°C) or below, the tower water is diverted to the chilled-water circuit bypassing the refrigeration machine. In this way, natural cooling is provided by naturally cool cooling tower water rather than by artifically chilled water. This procedure can provide substantial savings in energy since the refrigeration machine does not have to operate during the off-peak spring and fall months in certain areas of the country.

The dangers of diverting cooling tower water to the chilled-water circuit are obvious. Suspended matter, airborne particles, dirt foulants, mud, slime, etc., commonly present in open cooling tower waters can seriously foul chilled-water circuits. These systems have much narrower passageways in fan-coil units, terminal units, air-handling units, etc., with delicate controls, often needle valves, which are readily fouled by even the slightest amount of suspended matter. Therefore, it is necessary to maintain the cooling tower water free of suspended matter and foulants if such a procedure is to be successfully applied.

Methods of keeping the system clean include use of modern polymeric dispersants and deposit inhibitors along with very fine separators, strainers, or sand filters. The latter system uses highly efficient heat exchangers in place of the strainers and frequently solves the contamination problem by isolating the closed circuit from the open circuit in "free-cooling" systems.

53.7.7 Gadgets

From time to time over the past 100 years, various forms of gadgets have appeared for which exaggerated claims have been made. Some are said to eliminate and prevent all forms of corrosion and deposits in recirculating water systems and boilers without the use of chemical treatment.

Today several gadgets are being marketed which allegedly work by electromagnetism or mysterious electronic field forces to eliminate and prevent scale, corrosion, and even biological growths in recirculating water systems. The sources cited in Ref. 5 provide the scientific basis for a true refutation of the gadget makers' claims. In short, there is no simple, effective substitute for true chemical and biological control of the chemical and biological substances present in water that are responsible for corrosion and deposits. If you are still in doubt after reading the claims for any device, review Refs. 12 through 18.

53.8 TREATMENT OF SYSTEMS

53.8.1 General

The chemicals, equipment, and method of treatment required to optimize a water treatment system to reduce the corrosion, scaling, and fouling to the accepted minimum must be tailored not only to each system but also to each geographic location.

To obtain the optimum treatment of each system, it is strongly recommended that a reputable water treatment company and/or water treatment consultant be retained to select the chemicals, equipment, and method of treatment for each system required.

53.8.2 Boiler Water Systems

General. Internal treatment of HVAC boilers is required to prevent the problems of corrosion, pitting, scale deposits, and erratic boiler operation due to priming, foaming, and carryover. To prevent these problems, correct blowdown and treatment must be applied.

Blowdown. Blowdown of a boiler is the spontaneous removal of some concentrated boiler water from the boiler under pressure. The recommended maximum concentrations of these impurities which must be controlled are outlined in Table 53.8. These limits are for boilers with an operating pressure up to 250 psig (1724 kPa) and are used as a guide only. Actual operating experience will determine the true limits with any specific boiler operation.

TABLE 53.8 Maximum Concentration of Boiler Water Solids for Boilers up to 250 lb/in^2 (1124 kPa)*

Silica	150 mg/L
Suspended solids	600 mg/L
Total dissolved solids	3000 mg/L

*Adapted from American Boiler and Affiliated Industries Manufacturers Association and American Society of Heating, Refrigeration, and Air-Conditioning Engineers' guidelines.

The number of times that the solids in the makeup water have been accumulated in the boiler water is called the "cycles of concentration."

To determine the required amount of blowdown, it is necessary to examine the makeup water analysis and compare it to the maximum allowable concentration of solids as outlined in Table 53.8.

The maximum cycles of concentration permitted for each of the items listed in Table 53.8 is determined by dividing the maximum values in this table by the amount of each given in the makeup water analysis.

Let us examine the water analysis of Omaha, NE (Fig. 53.32), and compare it with the maximum concentration of solids allowed in a boiler (Table 53.8), to determine the maximum cycles of concentration allowed:

Analytical results	Maximum cycles of concentration
Silica, mg/L:	
$\dfrac{\text{Maximum}}{\text{Makeup water}} = \dfrac{150 \text{ mg/L}}{7.7 \text{ mg/L}} =$	19.5
Suspended solids (hardness), mg/L:	
$\dfrac{\text{Maximum}}{\text{Makeup water}} = \dfrac{600 \text{ mg/L}}{159 \text{ mg/L}} =$	3.8
Total dissolved solids, mg/L:	
$\dfrac{\text{Maximum}}{\text{Makeup water}} = \dfrac{3000 \text{ mg/L}}{414 \text{ mg/L}} =$	7.2

From the above we see that the maximum allowable cycles of concentration are 3.8. The lowest value obtained is used, for if this value were exceeded, difficulty with that particular impurity would result. If the makeup water were softened, hardness would be removed and no longer considered a limiting factor. In this case, the maximum allowable cycles of concentration would then be increased to 7.2, the next lowest value.

After the maximum allowable cycles of concentration in a boiler is ascertained, the blowdown rate required to maintain the solids accumulation below this maximum level can be calculated. Blowdown is used to remove accumulated boiler water solids.

The amount of solids present in the concentrated boiler blowdown water is equal to the amount of solids in the makeup water multiplied by the cycles of concentration. This can be expressed mathematically as BCX, where

B = blowdown, gal (L)

C = cycles of concentration

X = total solids concentration of makeup water, ppm (mg/L or g/gal)

The amount of solids entering the boiler with the makeup is expressed mathematically as MX, where

M = makeup water, gal (L)

METROPOLITAN REFINING COMPANY INC.

50-23 23rd STREET • LONG ISLAND CITY • N.Y. 11101

CERTIFICATE OF ANALYSIS

EFFICIENCY THROUGH CHEMISTRY

METROPOLITAN EFFICIENCY PRODUCTS ®

OMAHA, NE.

DATE

SAMPLING DATE

REPRESENTATIVE

ANALYSIS NUMBER				
SOURCE	AVERAGE ANALYSIS FLORENCE PLANT			
pH		9.2		
P ALKALINITY	$CaCO_3$, mg/L	10.		
FREE CARBON DIOXIDE	CO_2, mg/L			
BICARBONATES	$CaCO_3$, mg/L	52.		
CARBONATES	$CaCO_3$, mg/L	20.		
HYDROXIDES	$CaCO_3$, mg/L			
M (Total) ALKALINITY	$CaCO_3$, mg/L	72.		
TOTAL HARDNESS	$CaCO_3$, mg/L	159.		
SULFATE	SO_4, mg/L	181.		
SILICA,	SiO_2, mg/L	7.7		
IRON	Fe, mg/L			
CHLORIDE	NaCl, mg/L	23.		
ORGANIC INHIBITOR	mg/L			
PHOSPHATE	PO_4, mg/L			
CHROMATE	Na_2CrO_4, mg/L			
NITRITE	$NaNO_2$, mg/L			
ZINC	Zn, mg/L			
SPECIFIC CONDUCTANCE	mmhos/cm	678.		
TOTAL DISSOLVED SOLIDS	mg/L	414.7		
SUSPENDED MATTER				
BIOLOGICAL GROWTHS				
SPECIFIC GRAVITY @ 15.5°/15.5°C				
FREEZING POINT °C/°F				
	% BY WEIGHT			

NOTES:
1. ANALYTICAL RESULTS EXPRESSED IN MILLIGRAMS PER LITRE (mg/L) ARE EQUIVALENT TO PARTS PER MILLION (ppm). DIVIDE BY 17.1 TO OBTAIN GRAINS PER GALLON (gpg).
2. CYCLES OF CONCENTRATION = $\frac{\text{CHLORIDES IN SAMPLE}}{\text{CHLORIDES IN MAKEUP}}$

TREATMENT	TREATMENT CONTROL	FOUND	RECOMMENDED	FOUND	RECOMMENDED

REMARKS:

R. T. Blake

SPEEDIPLY• PAT'D MCP• PAT'D MBF 28 FORM 1078-I

FIGURE 53.32 Water analysis of Omaha, NE. (*Courtesy of Metropolitan Refining Co., Inc.*)

Since blowdown is designed to maintain a specific level of cycles of concentration, that level can be kept consistent only if the amount of solids leaving the boiler is precisely equal to the amount of solids entering the boiler. This is expressed mathematically as

$$\underset{\text{(solids leaving)}}{BCX} = \underset{\text{(solids entering)}}{MX}$$

Solving this mathematical equation for blowdown B, we obtain

$$B = \frac{M}{C}$$

This formula is used to determine a blowdown rate with respect to the makeup rate. In percent, it can be expressed as

$$\%\ \text{Blowdown} = \frac{100}{C}$$

In actual practice, however, it is not usually possible to measure the blowdown rate even though it is possible to calculate the amount required. Therefore, to determine if the blowdown rate is sufficient, the cycles of concentration are measured through the use of a simple chloride test.

Chlorides are the most soluble minerals and are always present in the makeup water in some degree. In addition, chlorides are only added to a boiler with the makeup water and not with treatment or from any other source. The cycles of concentration are found by comparing the chlorides of the makeup water with the chlorides of the boiler water:

$$C = \frac{\text{chlorides in boiler}}{\text{chlorides in makeup}}$$

This very simple and practical test is used by operating engineers to control the blowdown rate.

Scale Control. After the maximum allowable cycles of concentration are determined and a blowdown rate is established to prevent accumulation of minerals beyond the maximum allowable limit, treatment to prevent deposits and maintain precipitated solids in suspension must be considered.

As outlined previously, the hardness minerals, calcium and magnesium, are precipitated in the boiler water and tend to build a scale on the heat-transfer surfaces unless some treatment is used. Without treatment, these minerals will eventually precipitate as the insoluble carbonate and sulfate salts.

Similarly, silica and complex silicates will form hard, dense scales if silica is present in excess of its solubility. Some treatment for preventing these hard, dense scales includes phosphate to preferentially precipitate the calcium as phosphate, and in the presence of excess alkalinity phosphate is precipitated as calcium hydroxy phosphate, also know as apatite $[Ca_3(PO_4)_2 \cdot Ca(OH)_2]$. Magnesium with hydroxide alkalinity present in the boiler water will form the hydroxide precipitating as brucite $(MgOH_2)$. In these forms, the particles are more easily dispersed and held in suspension.

Nonphosphate treatments may consist of carbonate and silicate salts to preferentially precipitate calcium carbonate and magnesium silicate, which are more readily dispersed and held in suspension by polymers. The formulated boiler water treatment may include soluble organic polymers such as lignins, tannins, and polyelectrolytes that promote formation of the insoluble precipitate within the boiler water rather than on the heat-transfer surface. These materials act as nucleating sites, or places for the soluble ions to meet and form the insoluble particle dispersed throughout the boiler water. Some polymers are called "polyelectrolytes" because of their many positive and negative electrolytic sites on the polymer chain.

The polyelectrolytes such as polyacrylamides, polyacrylates, polymethacrylates, polymerized phosphonates, polymaleates, and polystyrene sulfonates also distort the crystal growth of the scale particle, rendering it less adhesive to heat-transfer surfaces and more readily dispersed with reduced tendency to compact into a dense scale. Some polymers, such as the polyacrylates, also sequester hardness similar to the chelates and act to even redissolve existing deposits. These polymers therefore act to prevent and remove scale by a threefold mechanism of dispersion, crystal distortion, and sequestration.

Another type of scale inhibitor is the chelating agent. Chelants are organic materials capable of solubilizing calcium, magnesium, and iron, preventing formation of the insoluble salts.

Both EDTA and NTA (nitrilotriacetic acid) are used in boiler water treatment as chelates to prevent scale formation and in some cases to remove existing scale deposits.

The use of these materials in HVAC equipment is limited because they require close control and are particularly corrosive to iron when they are not controlled. Some proprietary formulations contain small amounts of chelants to provide very efficient scale control. Combinations of chelating agents and polymers have been widely applied with excellent results.

Corrosion Control—Environmental Stablizers. Corrosion in boilers may be controlled with environmental stablizers and corrosion inhibitors. Environmental stablizers are materials used to correct the corrosive environment by neutralizing the acid or removing the corrosive gases. Corrosion of iron is greatly influenced by the pH value of the water in contact with the iron. The lower the pH value, the higher the corrosion rate; and the higher the pH value, the lower the corrosion rate. At a pH of 11.0, the corrosion rate of iron will be virtually nil.

Treatment of boiler water to control general corrosion therefore includes an alkaline substance to raise the pH value to 10.5 to 11.5.

Carbon dioxide gas is very corrosive to steel and copper and can cause failure of unit heaters and condensate handling lines and equipment.

The carbon dioxide generated from the carbonates and bicarbonates naturally present in the boiler makeup water can be neutralized by using neutralizing amines, mild alkaline materials related to ammonia but much milder than it. Amines used for this purpose are morpholine, cyclohexylamine, and diethylaminoethanol. These materials, liquid at room temperature, boil and vaporize at approximately the same temperature as water, thereby going off with the steam, a mixture known as an "azeotrope." As the steam condenses into water, the amines also condense, rendering that initial condensate alkaline and much less corrosive to the metal surface. The flowing condensate containing the neutralizing amine will then retain higher pH values, even with the adsorption of carbon dioxide gases. For the flowing condensate, a pH range of 7.0 to 9.0 is the most favorable for good corrosion control of all metals in heating systems including iron, steel, brass, and copper.

As we have seen, dissolved oxygen in boiler water causes localized corrosion called pitting, and to prevent pitting, the oxygen must be removed. This is another way to alter or stabilize the environment. Deaerators are used for this purpose, but no deaerator is perfect, and the best ones produce a boiler feed water with a dissolved oxygen content of 0.0005 cm^3/L. This very low oxygen content can cause serious pitting failures especially in boiler feed-water lines and at boiler tubes adjacent to the feed-water entrance to the boiler. To prevent this, chemical oxygen scavengers are used to absorb the oxygen from the water and to ensure a completely oxygen-free environment. The most common oxygen scavengers are sodium sulfite and hydrazine. These materials have a strong affinity for oxygen and will absorb it from the water. Hydrazine is no longer widely used as an oxygen scavenger due to its toxicity. Other oxygen scavengers currently being used as a replacement for hydrazine are diethylhyroxyl amine and sodium erythorbate.

The maintenance of a slight excess of the oxygen scavenger in the water provides assurance that there is no dissolved oxygen present. Complete water treatment formulations containing sodium sulfite for this purpose will also include a catalyst, ensuring that the reaction between the dissolved oxygen and the oxygen scavenger is instantaneous, even in cold water.

Corrosion Inhibitors. Corrosion inhibitors are substances which do not necessarily alter the environment, or conditions involving the corrosion process, but act as a barrier between the corrosive medium and corroding metal surface. Physical barriers such as protective coatings and galvanizing immediately come to mind as a common application of a corrosion inhibitor. These physical barriers actually separate the corrosive atmosphere containing water, oxygen, and acid gases from the base metal.

Corrosion inhibitors can be added to water which form a protective film on the metal surface, acting as a barrier to the corrosion process, i.e., inhibiting the corrosion reaction. Such barriers form by a chemical reaction between the metal surface and the inhibitor or by a physical attraction and adsorption on the metal surface. With this type of inhibitor, the film is not visible and nonaccumulative. The thickness is only the thickness of one molecule of the inhibiting film, hence it is called a "monomolecular film."

With a film of this thickness, there is no interference with heat transfer, and therefore the inhibitors are found to be very effective in heat-transfer equipment.

Inorganic inhibitors used in boilers for this purpose are molybdates, nitrites, borates, silicates, and phosphates. Some organic inhibitors used are phosphonates, polyacrylates, organic polyphosphate esters, and nitrogen-containing organics such as mercaptobenzothiozole, triazoles, fatty acid amides, and amines.

The films formed can be either adsorbed films, as in some organic inhibitors, or a chemically formed reaction product of the inhibitor and metal surface.

The inhibiting film may also be a combination of both adsorption and reaction. The result is a reduction of the corrosion rate and passivity. Passivity is described by Uhlig as a state of an active metal in which reactivity is substantially reduced, resisting corrosion, or when its electrochemical behavior becomes that of a less active metal.[7] That is, the metal becomes passive or is passivated. Inhibitors in this sense are also called "passivators."

In steam and return condensate lines, film-forming inhibitors used are organic amines or amides such as octadecylamine and mixtures of octadecanol and stearamide. These inhibitors form insoluble films on the steam and condensate piping. They steam distill from the boiler; i.e., they carry over with the steam and deposit in the steam and condensate system. The disadvantage of such inhibitors is that this film is not self-limiting, and heavy deposits plugging steam traps, strainers, etc., can result unless particular care is taken to avoid this problem. The filming-type steam and condensate line corrosion inhibitors are employed most successfully in systems where there is little or no return condensate. In heating boilers where most condensate is returned, the neutralizing amines are more suitable. Combinations of neutralizing and filming amines are also used.

Priming, Foaming, and Carryover. Priming of a boiler water is the bumping and bouncing of the water level of the boiler during operation. Foaming, however, is a less violent activity. It consists of the formation of small bubbles in the surface of the boiling water like the soap foam in a washer.

Carryover essentially is the contamination of the steam with boiler water. It is a result of priming and foaming and can be a more subtle entrapment of boiler water with the steam, causing steam contamination without the evidence of priming and/or foaming.

The causes of priming, foaming, and carryover can be many and varied. Most frequently, they are caused by contamination of the boiler water with oil or other foreign substances. Other causes are excessive solids accumulation due to the lack of blowdown, high alkalinity, overtreatment, or mechanical malfunction.

Adequate blowdown and certain antifoam treatments can reduce the problems of priming, foaming, and carryover.

Treatment of Low-Pressure Steam Heating Boilers. Most low-pressure steam heating boilers have a unique type of operation in that *all* the steam produced is for heating purposes *only*. Therefore, little makeup water is required as well as little accumulation of solids. The only makeup water is for slight loss at vents, leaks, or overflow at condensate receivers.

Without makeup water and accumulation of minerals, even with a very hard water supply, problems of scale formation are significantly reduced. However, corrosion under these conditions can become aggravated. The treatment program will usually consist of a corrosion inhibitor for the boiler and neutralizing or filming amine for the steam and condensate system.

This program, however, is limited to "closed" systems where all steam is returned to the boilers as condensate and where *no* steam is used for humidification or cooking, as in cafeterias or restaurants. Whenever steam is consumed, makeup water will tend to accumulate minerals in the boiler, and other treatment programs will have to be used, as described below.

The corrosion inhibitors most widely used in low-pressure steam boilers are sodium molybdate and sodium nitrite.[6,7]

To be certain that these inhibitors provide a continuous protective film and that no area of the metallic surfaces remains exposed, certain minimum levels must be maintained. Inhibitors are required to provide buffering, so that the pH can be maintained at 7.0 to 10.0 for maximum effectiveness.

For treating the steam and condensate of low-pressure heating systems, the neutralizing amines can be added gradually at regular intervals in sufficient quantity to maintain the pH value of the condensate at 7.5 to 8.5. The neutralizing amines should not be used in systems containing nitrites or where steam is used for humidification.

Treatment of High- and Low-Pressure Process-Steam Boilers. Because high-pressure steam systems operate at higher boiler water temperatures than low-pressure steam heating boilers, they cannot be treated with the corrosion inhibitors mentioned above. The method of treatment of high-pressure steam boilers involves heat-stable substances such as environmental stabilizers that alter the condition responsible for corrosion and deposits.

With low-pressure steam boilers operating with some makeup water due to steam losses at steam humidifiers, kitchen steam tables, dishwashers, or for any other purpose, the conditions favoring use of corrosion inhibitors are altered.

With makeup water required because of steam losses, minerals present in the makeup will tend to accumulate. This requires increased blowdown to reduce the mineral buildup and additional treatment to replace that lost in blowdown. Likewise, a different type of treatment including scale inhibitors will be required to prevent hardness in the makeup water from forming scale. Therefore, such systems must be treated as high-pressure steam boilers with environmental stabilizers, rather than with corrosion inhibitors.

Treatment for corrosion control will consist of adjustment of the alkalinity and pH value with caustic soda to maintain the pH value at 10.5 to 11.5 consistent with efficient operating conditions.

The boiler feed water should be deaerated to remove dissolved oxygen and to prevent oxygen pitting, followed by superimposing of an oxygen scavenger to remove the last traces of dissolved oxygen. Materials used for this purpose are sodium sulfite at levels of 20 to 59 mg/L or the organic materials diethylhydroxylamine and sodium erythorbate.

In most HVAC applications, sodium sulfite is preferred because of cost and ease of storage, handling, and control. Organic materials are preferred when dissolved solids are objectionable, such as in electrode boilers. These do not add to the mineral content of the boiler water. Diethylhydroxylamine, however, has the disadvantage of forming ammonia or amines which are corrosive to copper, brass, and other copper alloys commonly used in heating systems. The choice of oxygen scavengers, therefore, must be carefully considered.

For control of scale and deposits, boiler water treatment includes a zeolite softener on makeup water when required, followed by internal treatment with phosphates, chelating agents, polymers, and/or dispersants.

The basic program includes a phosphate polymer combination used to maintain a sodium phosphate concentration in the boiler water at 20 to 60 mg/L, ensuring that all the calcium has precipitated as phosphate.

The nonphosphate program includes polymers, phosphonates, or combinations of polymers, phosphonates, and chelants with the polymer in the range of 2 to 100 mg/L, depending on the type of polymer and formulation used. Chelates should not be used in excess of 2 to 3 mg/L free chelate because corrosion of iron can be accelerated with free chelates in boiler water.

Antifoam agents are used in the boiler water treatment formulations to prevent foaming and carryover. They enhance more efficient nucleate boiling and prevent foaming and carryover. Antifoam agents for boilers include polyalkalyene glycols and polyamides used at 10 to 100 mg/L.

Finally, the treatment program for high-pressure steam boilers and low-pressure process boilers should include a steam and return condensate line corrosion inhibitor. This should be either a neutralizing amine or a mixture of neutralizing amines to maintain the pH of the condensate at 7.0 to 9.0. Where the steam loss is high and the use of neutralizing amines is too costly, a film-forming inhibitor may be used to protect the condensate piping.

Test Controls. The test control of water treatment is one of the most important aspects of the water treatment program. Without proper testing, more harm than good can result.

A typical analysis of a low-pressure steam boiler using a sodium chromate inhibitor is shown in Fig. 53.33. With high-pressure steam boilers and low-pressure process boilers, the tests must be made more frequently. Testing 3 times a day

METROPOLITAN REFINING COMPANY INC.
50-23 23rd STREET • LONG ISLAND CITY • N.Y. 11101

CERTIFICATE OF ANALYSIS

EFFICIENCY THROUGH CHEMISTRY — METROPOLITAN EFFICIENCY PRODUCTS ®

DATE
SAMPLING DATE
REPRESENTATIVE

ANALYSIS NUMBER		341247		
SOURCE		LP STEAM BOILER		
pH		7.9		
P ALKALINITY	$CaCO_3$, mg/L	0.0		
FREE CARBON DIOXIDE	CO_2, mg/L			
BICARBONATES	$CaCO_3$, mg/L	250.		
CARBONATES	$CaCO_3$, mg/L			
HYDROXIDES	$CaCO_3$, mg/L			
M (Total) ALKALINITY	$CaCO_3$, mg/L	250.		
TOTAL HARDNESS	$CaCO_3$, mg/L	86.		
SULFATE	SO_4, mg/L			
SILICA,	SiO_2, mg/L			
IRON	Fe, mg/L			
CHLORIDE	NaCl, mg/L	35.		
ORGANIC INHIBITOR	mg/L			
PHOSPHATE	PO_4, mg/L			
CHROMATE	Na_2CrO_4, mg/L	2800.		
NITRITE	$NaNO_2$, mg/L			
ZINC	Zn, mg/L			
SPECIFIC CONDUCTANCE	mmhos/cm	3831.		
TOTAL DISSOLVED SOLIDS	mg/L	3448.		
SUSPENDED MATTER				
BIOLOGICAL GROWTHS				
SPECIFIC GRAVITY @ 15.5°/15.5°C				
FREEZING POINT °C/°F				
	% BY WEIGHT			

NOTES:
1. ANALYTICAL RESULTS EXPRESSED IN MILLIGRAMS PER LITRE (mg/L) ARE EQUIVALENT TO PARTS PER MILLION (ppm). DIVIDE BY 17.1 TO OBTAIN GRAINS PER GALLON (gpg).
2. CYCLES OF CONCENTRATION = CHLORIDES IN SAMPLE / CHLORIDES IN MAKEUP

TREATMENT	TREATMENT CONTROL	FOUND	RECOMMENDED	FOUND	RECOMMENDED
--	CHROMATE	2800	2000 - 2500		
--	pH	7.9	7.0 - 10.0		

REMARKS:

R. V. Blake

SPEEDIPLY• PAT'D MCP• PAT'D MBF 28 FORM 1078-1

FIGURE 53.33 Analysis of low-pressure steam boiler water. *(Courtesy of Metropolitan Refining Co., Inc.)*

may be required if load variations are such that conditions can change that rapidly. Daily testing with field test kits is more common and the preferred frequency for best results.

A typical test analysis of a high-pressure steam boiler water, with recommended controls, is shown in Fig. 53.34.

METROPOLITAN REFINING COMPANY INC.

50-23 23rd STREET • LONG ISLAND CITY • N.Y. 11101

CERTIFICATE OF ANALYSIS

EFFICIENCY THROUGH CHEMISTRY
METROPOLITAN EFFICIENCY PRODUCTS ®

DATE
SAMPLING DATE
REPRESENTATIVE

ANALYSIS NUMBER		374129		
SOURCE		HIGH PRESSURE BOILER		
pH		11.3		
P ALKALINITY	$CaCO_3$, mg/L	153.		
FREE CARBON DIOXIDE	CO_2, mg/L			
BICARBONATES	$CaCO_3$, mg/L			
CARBONATES	$CaCO_3$, mg/L	122.		
HYDROXIDES	$CaCO_3$, mg/L	92.		
M (Total) ALKALINITY	$CaCO_3$, mg/L	214.		
TOTAL HARDNESS	$CaCO_3$, mg/L	0.0		
SULFATE	SO_4, mg/L			
SILICA,	SiO_2, mg/L	23.		
IRON	Fe, mg/L			
CHLORIDE	NaCl, mg/L	841.		
ORGANIC INHIBITOR	mg/L			
PHOSPHATE	PO_4, mg/L	50.		
CHROMATE	Na_2CrO_4, mg/L			
NITRITE	$NaNO_2$, mg/L			
ZINC	Zn, mg/L			
SPECIFIC CONDUCTANCE	mmhos/cm	2244.		
TOTAL DISSOLVED SOLIDS	mg/L	1818.		
SUSPENDED MATTER				
BIOLOGICAL GROWTHS				
SPECIFIC GRAVITY @ 15.5°/15.5°C				
FREEZING POINT °C/°F				
	% BY WEIGHT			

NOTES:
1. ANALYTICAL RESULTS EXPRESSED IN MILLIGRAMS PER LITRE (mg/L) ARE EQUIVALENT TO PARTS PER MILLION (ppm). DIVIDE BY 17.1 TO OBTAIN GRAINS PER GALLON (gpg).
2. CYCLES OF CONCENTRATION = CHLORIDES IN SAMPLE / CHLORIDES IN MAKEUP

TREATMENT	TREATMENT CONTROL	FOUND	RECOMMENDED	FOUND	RECOMMENDED
	pH	11.3	10.5 - 11.5		
	P ALKALINITY	153.	200 - 400		
	PHOSPHATE	50.	30 - 50		
	SULFITE	35.	30 - 50		
	TOTAL DIS. SOLIDS	1818.	3000 MAX.		

REMARKS:

SPEEDIPLY PAT'D MCP PAT'D MBF 28 FORM 1078-I

FIGURE 53.34 Analysis of high-pressure steam boiler water. *(Courtesy of Metropolitan Refining Co., Inc.)*

Field test kits are available for easy testing of boiler water samples. Figure 53.35 depicts a typical drop test for chloride and sulfite used for boiler water.

More elaborate test cabinets are used in the large boiler room where a watch engineer is on duty to make tests at frequent intervals. Figure 53.36 illustrates this type of equipment.

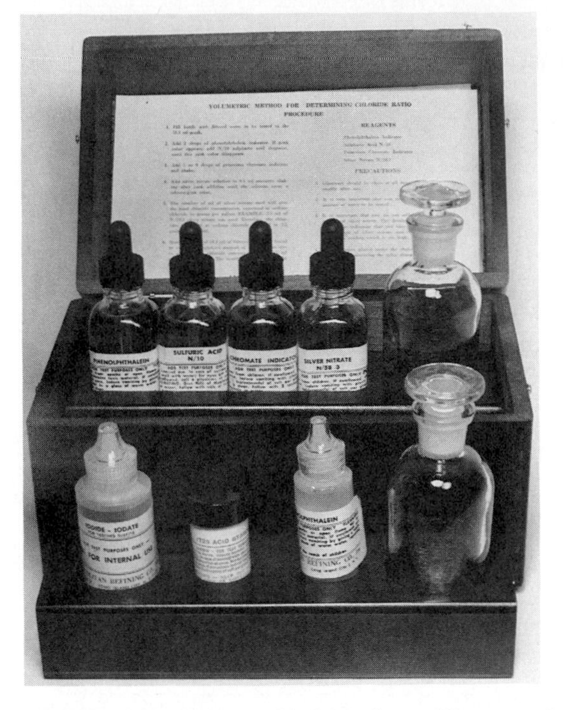

FIGURE 53.35 Field test kit for boilers. (*Courtesy of Metropolitan Refining Co., Inc.*)

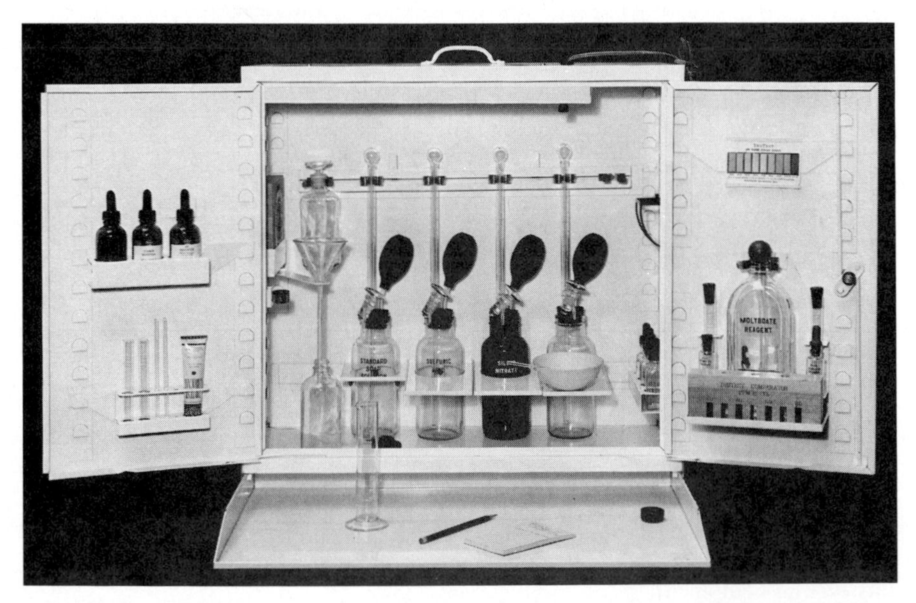

FIGURE 53.36 Test cabinet for water analyses. (*Courtesy of Tru Test Laboratories, Inc.*)

Feed Methods. Water treatment is applied to boilers by various means, ranging from shock dosage to highly sophisticated chemical proportioning pumps and controllers. These methods are briefly described:

Shock Feed Methods. Shock feeders are most convenient and effective with low-pressure heating boilers which require only periodic addition of a corrosion inhibitor and perhaps weekly addition of a steam condensate neutralizing amine. One of the handier devices for this purpose is the force hand pump which adds treatment directly to the boiler through a ¾-in (20-mm) drain cock. Figure 53.37 illustrates this application.

Another shock device is the pot type or bypass feeder. This kind of feeder should be installed on the makeup water line to the boiler to inject treatment directly into the boiler with the pressure of the boiler feedwater pump. If it is installed on a raw makeup water line, an approved backflow preventer must be used to avoid the possibility of raw water contamination with boiler water treatment. A typical installation is shown in Fig. 53.38.

Proportioning Feed Methods. With high-pressure steam boilers and low-pressure process boilers, it is desirable to feed treatment to the boiler continuously in proportion to actual need. Proportioning feed systems will feed treatment based on feedwater flow.

Basic Proportioning Feed System. A simple proportioning feed system consists of a chemical feed pump interlocked with the boiler feedwater pump to transfer treatment directly from the drum into the feedwater tank prior to the boiler. This method is ideal where a single liquid boiler treatment formulation is used that contains all the required corrosion and deposit inhibitors *without* phosphates.

The feed system illustrated in Fig. 53.39 is applicable to any size boiler system provided the above conditions are met.

Proportioning Water Treatment System for Steam Boilers without Deaerator. This more complex program requires a more sophisticated feed system. The treatments with alkalinity and phosphates must be added directly to the boiler to prevent deposits in the feedwater line and damage to feedwater pumps. Figure 53.40 outlines this method of feeding treatment directly into each boiler in proportion to feedwater flow.

Proportional Water Treatment System for Steam Boilers with Deaerators. Where deaerators are installed, it is best to feed treatments for corrosion and pitting directly to the storage section of the deaerator. Some deaerator manufactur-

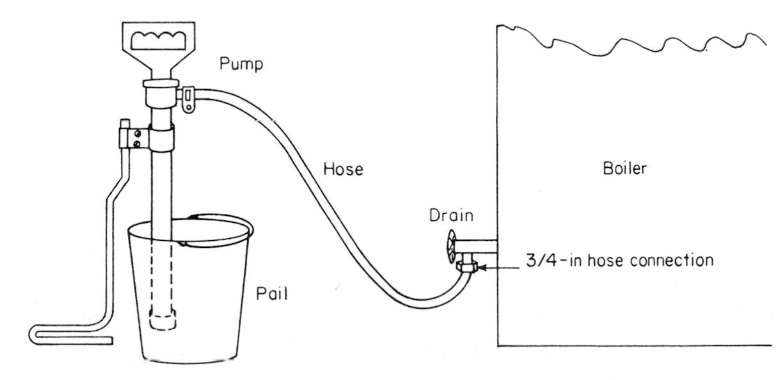

FIGURE 53.37 Hand pump for feeding inhibitors into boilers.

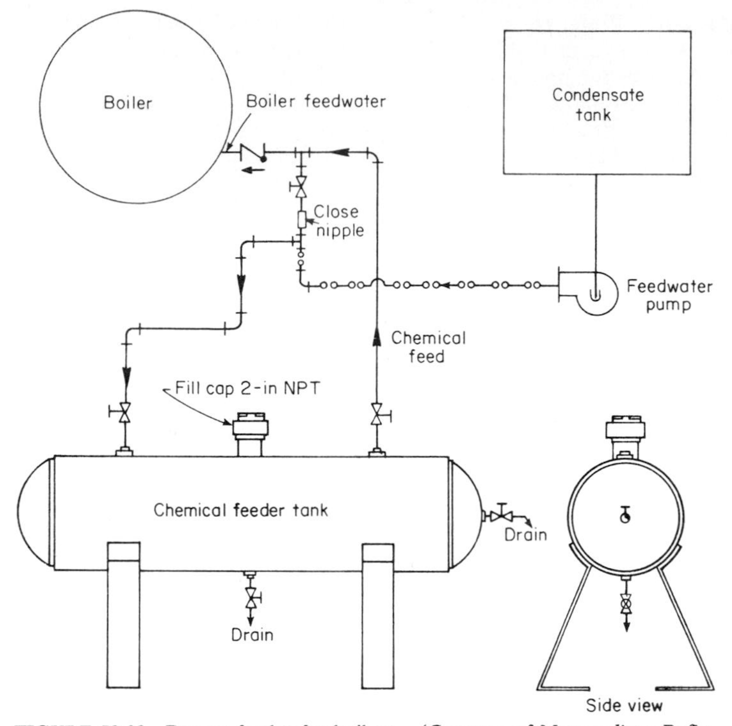

FIGURE 53.38 Bypass feeder for boilers. *(Courtesy of Metropolitan Refining Co., Inc.)*

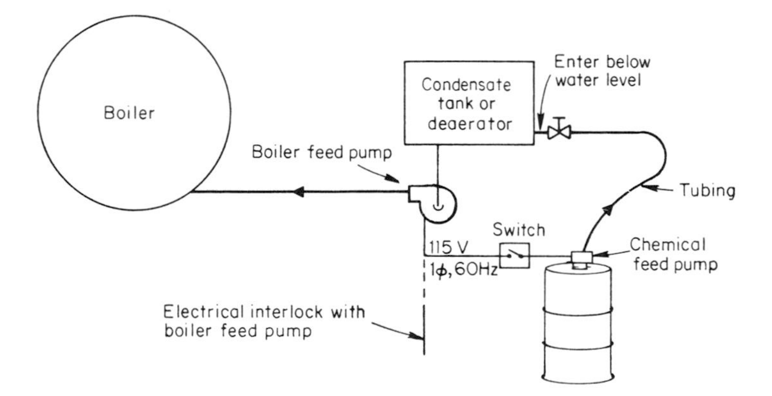

FIGURE 53.39 Simple proportioning water treatment feed. *(Courtesy of Metropolitan Refining Co., Inc.)*

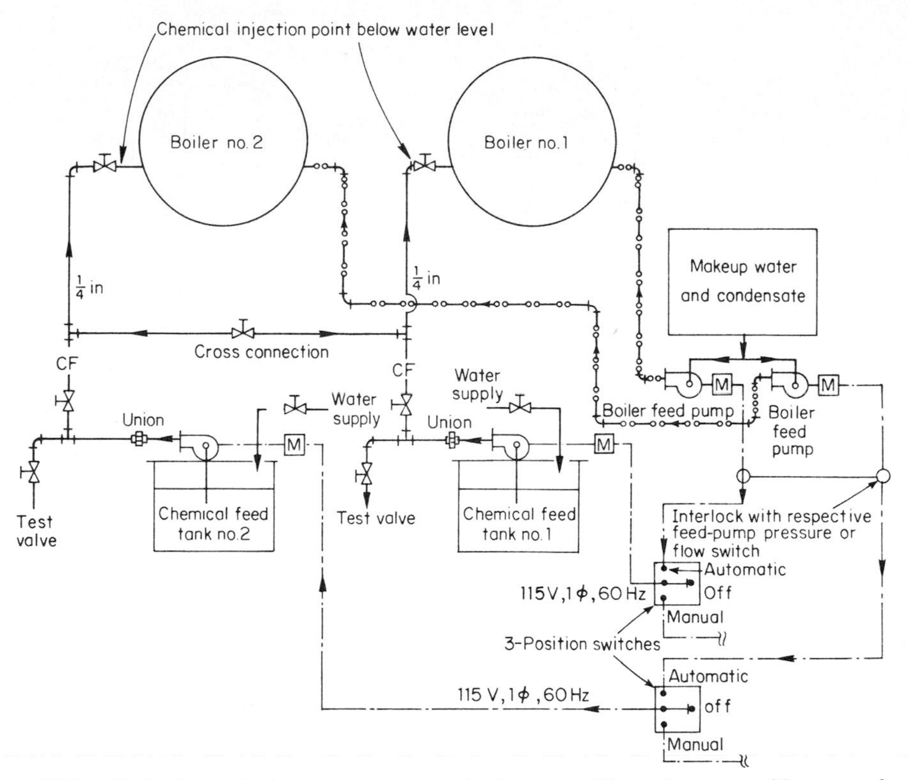

FIGURE 53.40 Proportioning water treatment feed system without deaerator. (*Courtesy of Metropolitan Refining Co., Inc.*)

ers warn against feed of treatment to the storage section of the deaerator if phosphates and other alkaline compounds are used, because they can cause precipitation of deposits prior to the boiler and can damage the deaerator tank, boiler feedwater pump, and lines. In this case, the phosphate and alkaline treatments should be fed directly to the boiler, and only the oxygen scavenger and amine, or other non-deposit-forming materials, should be fed into the preboiler section. When this is explained to the deaerator manufacturer, the objection is usually removed. Figure 53.41 describes this method.

Automatic Blowdown and Feed Controllers. Since ASME (American Society of Mechanical Engineers) code restrictions prohibit installing automatic valves on the bottom blowdown line, use of controllers and automatic valves is restricted to the side or top continuous blowdown line. The side or top blowdown is strictly used for controlling dissolved solids, because it will not remove suspended solids settled in the bottom of the boiler. Even when top blowdown controllers are installed, the bottom blowdown still must be administered to remove the settled and suspended solids. The top blowdown must never be used as a substitute for the bottom blowdown.

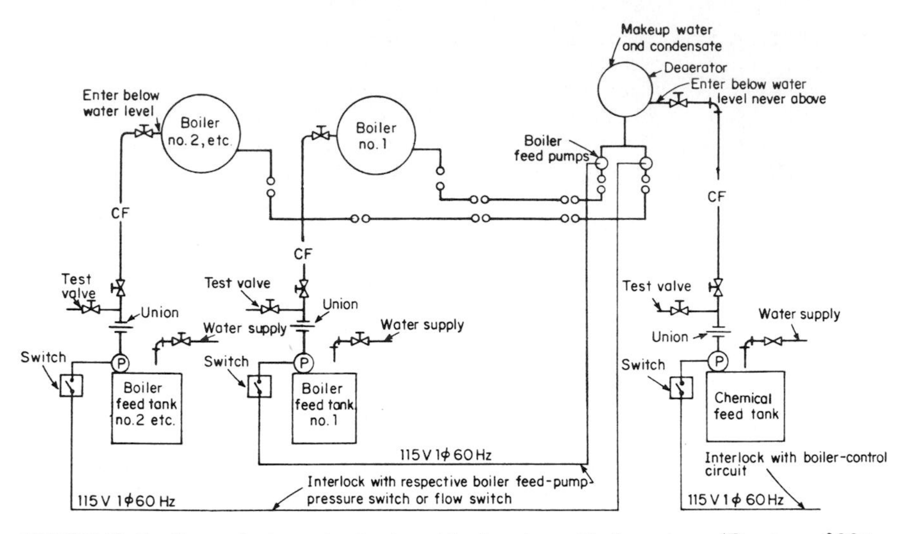

FIGURE 53.41 Proportioning water treatment feed system with deaerator. (*Courtesy of Metropolitan Refining Co., Inc.*)

53.8.3 Treatment for Open Recirculating Water Systems

The major problems encountered in open cooling water systems are corrosion and deposits of scale, dirt, mud, and organic slime or algae growths.

In open systems, there is a loss of water caused by both evaporation and windage drift. The water lost by drift will be of the same quality as the recirculating water; i.e., it will contain the same amount of dissolved minerals and impurities.

The water lost by evaporation, however, will be of a different quality. Evaporation from the recirculated water will be pure water vapor similar to water produced by distillation containing, theoretically, no minerals or dissolved solids unless droplets of water spray are carried over with the water vapor. This pure water vapor leaves the dissolved minerals behind in the recirculating water since they do not vaporize with the water.

The water lost by evaporation is replaced by makeup water which contains dissolved minerals and impurities. As a result, there is an accumulation of minerals that constantly increases as the water in the system makeup water is introduced. Eventually the recirculating water will become so saturated with minerals that the most insoluble salts will come out of solution and form a scale on heat-transfer surfaces or within other parts of the system. To avoid this, the mineral content of the recirculating water must be controlled and limited. This is done by bleed-off.

Bleed-off. Bleed-off is the continuous removal of a small quantity of water concentrated with minerals from an open recirculating cooling water system. The water lost by windage drift also contains concentrated minerals and can be considered as bleed-off water. The term "zero bleed-off" refers to the most ideal situation where the windage drift is such that the minerals and suspended matter lost with the drift water droplets are sufficient to eliminate any need for an additional continuous bleed-off. In most cases, however, an additional bleed-off is required to limit the solids accumulations.

The most troublesome mineral scale is calcium carbonate since it is the least soluble of the salts present in the recirculating water. Inhibitors are used to increase the solubility of calcium carbonate and some other minerals, but even when they are used, it is necessary to limit excessive concentration of minerals by bleed-off.

Bleed-off also limits concentrations of alkalinity, total hardness, and silica.

Alkalinity should be limited to prevent precipitation of calcium carbonate scale. Alkalinity is present in the form of bicarbonates and carbonates which combine with calcium and magnesium to form calcium carbonate and magnesium carbonate. (Calcium carbonate, being less soluble than magnesium carbonate, will form first.)

Total hardness should be limited to prevent calcium sulfate scale. Sulfates are naturally present in the recirculating water or are added in the form of sulfuric acid for pH control.

When the concentration of calcium hardness exceeds the solubility of calcium sulfate, it will come out of solution. Inhibitors can be used to extend this solubility by forming a supersaturated solution. But bleed-off must still be used to maintain the maximum limit. Silica must be limited to prevent silica scale, and the limit is the solubility of silica at the temperatures encountered.

The maximum levels of alkalinity hardness and silica that are suggested with scale inhibitors present are shown in Table 53.9.

TABLE 53.9 Maximum Concentration of Mineral Solids for Cooling Towers, Evaporative Condensers, Air Washers, and Spray Coil Units

Total alkalinity, calcium carbonate	500 mg/L
Total hardness as calcium carbonate	1200 mg/L
Silica as silicon dioxide	150 mg/L

Alkalinity should be limited not only to prevent precipitation of calcium carbonates, but also to prevent high pH conditions which may be damaging to some components of the system such as galvanized steel, brass, or cooling tower lumber.

With a total alkalinity of 500 mg/L, the pH value of a cooling tower water is expected to be approximately 8.7. This is based on the average pH of cooling tower water, as shown in Fig. 53.42. This chart shows the expected average pH value which will vary according to atmospheric conditions. For example, if the atmosphere has a considerable quantity of carbon dioxide and/or sulfur dioxide, these gases will tend to neutralize the alkalinity of the cooling tower water. For this reason, the alkalinity of the cooling tower water is not always a mathematical total of alkalinity from the makeup water; therefore, the condition may vary from location to location, and the maximum limitation of 500 mg/L of alkalinity as calcium carbonate is used only as a guide.

The limitation of total hardness as calcium carbonate to 1200 mg/L is based on the solubility of calcium sulfate which is the next least soluble of the calcium salts.

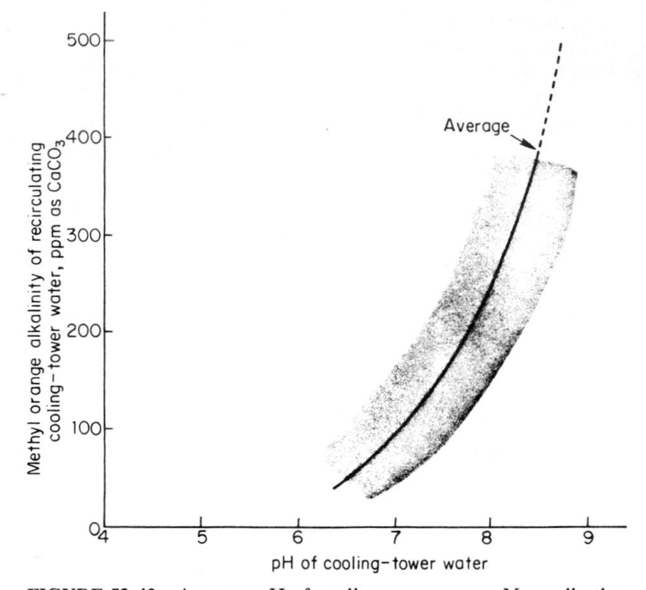

FIGURE 53.42 Average pH of cooling tower water. Normally the pH level for 90 percent of towers is within shaded area. (*From Carrier System Design Manual, part 5, "Water Conditioning," Carrier Corporation, Syracuse, NY, 1972, p. 5–15. Data from Betz Laboratories.*)

If calcium sulfate exceeds 1200 mg/L (calculated as calcium carbonate), calcium sulfate will tend to precipitate or a supersaturated solution will form. Scale inhibitors can extend this solubility beyond this natural maximum, but the guideline of 1200 mg/L maximum is used to limit the cycles of concentration when required.

The silica limitation is based on the solubility of silicon dioxide (SiO_2) at the temperatures encountered in HVAC cooling-water systems. Silica is soluble up to 150 mg/L. Beyond this, it will tend to come out of solution. Other factors, such as presence of scale inhibitors and other mineral salts, will have an influence on this, but this limitation is used as a guide in determining the maximum cycles of concentration where silica is a factor.

Once the maximum limitations for certain dissolved minerals permitted in a recirculating cooling water system have been established, it is necessary to limit these minerals by bleed-off. See Table 53.9 for maximum concentrations.

The maximum cycles of concentration recommended for each of the items listed in Table 53.9 are determined by dividing the maximum values in this table by the amount of each found in the makeup water analysis. To demonstrate how this is done, let us examine a city water analysis from Harrisburg, PA (Fig. 53.43), and determine the maximum cycles of concentration to be recommended in a cooling tower system water.

METROPOLITAN REFINING COMPANY INC.

50-23 23rd STREET • LONG ISLAND CITY • N.Y. 11101

CERTIFICATE OF ANALYSIS

HARRISBURG, PA.

DATE

SAMPLING DATE

REPRESENTATIVE

ANALYSIS NUMBER		359065		
SOURCE		CITY		
pH		7.5		
P ALKALINITY	$CaCO_3$, mg/L			
FREE CARBON DIOXIDE	CO_2, mg/L			
BICARBONATES	$CaCO_3$, mg/L	167.		
CARBONATES	$CaCO_3$, mg/L			
HYDROXIDES	$CaCO_3$, mg/L			
M (Total) ALKALINITY	$CaCO_3$, mg/L	167.		
TOTAL HARDNESS	$CaCO_3$, mg/L	149.		
SULFATE	SO_4, mg/L			
SILICA,	SiO_2, mg/L	10.		
IRON	Fe, mg/L			
CHLORIDE	NaCl, mg/L	187.		
ORGANIC INHIBITOR	mg/L			
PHOSPHATE	PO_4, mg/L			
CHROMATE	Na_2CrO_4, mg/L			
NITRITE	$NaNO_2$, mg/L			
ZINC	Zn, mg/L			
SPECIFIC CONDUCTANCE	mmhos/cm	818.		
TOTAL DISSOLVED SOLIDS	mg/L	501.		
SUSPENDED MATTER				
BIOLOGICAL GROWTHS				
SPECIFIC GRAVITY @ 15.5°/15.5°C				
FREEZING POINT °C/°F				
	% BY WEIGHT			

NOTES:
1. ANALYTICAL RESULTS EXPRESSED IN MILLIGRAMS PER LITRE (mg/L) ARE EQUIVALENT TO PARTS PER MILLION (ppm). DIVIDE BY 17.1 TO OBTAIN GRAINS PER GALLON (gpg).
2. CYCLES OF CONCENTRATION = $\frac{\text{CHLORIDES IN SAMPLE}}{\text{CHLORIDES IN MAKEUP}}$

TREATMENT	**TREATMENT CONTROL**	FOUND	RECOMMENDED	FOUND	RECOMMENDED

REMARKS:

R. T. Blake

SPEEDIPLY® PAT'D MCP® PAT'D MBF 28 FORM 1078-

FIGURE 53.43 City water analysis of Harrisburg, PA. *(Courtesy of Metropolitan Refining Co., Inc.)*

Shown below is a comparison of each of the limiting impurities with that impurity in the makeup water, to determine the maximum cycles of concentration.

Analytical results	Maximum cycles of concentration
Alkalinity	
$\frac{\text{Maximum}}{\text{Makeup water}} = \frac{500 \text{ mg/L}}{167 \text{ mg/L}} =$	3.0
Hardness	
$\frac{\text{Maximum}}{\text{Makeup water}} = \frac{1200 \text{ mg/L}}{149 \text{ mg/L}} =$	8.0
Silica	
$\frac{\text{Maximum}}{\text{Makeup water}} = \frac{150 \text{ mg/L}}{10 \text{ mg/L}} =$	15.0

From the above we see that the maximum cycles of concentration recommended are 3.0.

The lowest value obtained is used, for if it were exceeded, difficulty with that particular impurity would result. In this case, it would be the alkalinity, an excess causing precipitation of carbonate salts or excessive pH conditions. If a neutralizing acid were used in a system with this water, the alkalinity would be reduced and no longer considered a limiting factor. In that case, the maximum cycles of concentration would be increased to 8.0, the next lowest value.

After the maximum cycles of concentration recommended in an open recirculating cooling water system have been determined, the bleed-off rate required to maintain the solids accumulation below this maximum level can be calculated. The mathematical formula for the bleed-off rate with respect to the evaporation rate of an open cooling water system is

$$B = \frac{E}{C - 1}$$

where B = bleed-off rate, gal/min (L/min)

E = evaporation rate, gal/min (L/min)

C = cycles of concentration

The purpose of bleed-off is to remove dissolved solids in order to maintain a maximum level, determined by the maximum recommended cycles of concentration. To maintain this constant maximum level of solids, the amount of solids entering the system must be equal to the amount of solids leaving the system. This can be expressed mathematically as

$$\underset{\text{(solids leaving)}}{BCX} = \underset{\text{(solids entering)}}{MX} \tag{53.1}$$

where C = cycles of concentration
B = bleed-off rate, gal/min (L/min)
M = makeup water, gal/min (L/min)
X = solids concentration of makeup water, ppm (mg/L or g/gal)

The makeup rate to an open cooling water system is proportional to the load on the system and is equivalent to the evaporation rate plus any losses of drift, overflow or bleed-off. This is expressed as

$$M = E + B \tag{53.2}$$

where M = total makeup rate and E = evaporation rate. Substituting in Eq. (53.1), we obtain

$$BCX = (E + B)X$$

Solving for B gives

$$B = \frac{E}{C - 1}$$

The bleed-off is a total of all water losses from the recirculating systems such as leaks at pumps, overflow, windage drift, and actual bleed. All these together should be sufficient to keep the cycles of concentration below the recommended maximum.

To determine if the total bleed-off from a system is adequate, the cycles of concentration are measured by a simple chloride test. Therefore, a measure of the chlorides in the recirculating water will tell very accurately how much the makeup water has concentrated in that system.

The cycles of concentration are found by comparing the chlorides of the makeup water with the chlorides of the recirculating water:

$$\text{Cycles of concentration} = \frac{\text{chlorides in recirculating water}}{\text{chlorides in makeup water}}$$

Scale Control. Bleed-off is necessary to prevent deposits of mineral salts as a scale. Scale inhibitors must be used, however, to minimize bleed-off water losses and to permit operation at as high a solids concentration as possible, as outlined in Table 53.9. Without scale inhibitors, it would not be possible to accumulate bicarbonate alkalinity and hardness beyond limits which are much lower than those outlined in Table 53.9.

Acid Feed. One method of inhibiting scale is by using acid to control the Langelier saturation index. Acids neutralize the bicarbonate alkalinity. Acids can be used to maintain the pH value as close to 7.0 as possible with the alkalinity below the saturation level. This method, however, requires very careful pH and alkalinity control and, most important, careful and controlled feed of acid to avoid excessive corrosive conditions.

Water Softeners. Another way to control scale in open recirculating water systems is by using ion-exchange water softeners. By reducing the hardness, higher cycles of concentration can be established without exceeding the pH of saturation.

However, to be effective, this method must include simultaneous feed of acid to reduce the alkalinity. Therefore, it offers no further advantage over alkalinity reduction without softening.

Ion-exchange softeners operate at less than 100 percent efficiency throughout the softening cycle. Therefore, trace quantities of hardness bypass the softeners before regeneration.

Scale Inhibitors. The most economical and effective method to control scale is with scale inhibitors with or without alkalinity reduction. Scale inhibitors, when added to the recirculating water at very low levels, 1 to 20 mg/L, will reduce the scaling tendency. This is accomplished by holding the scale-forming minerals in solution beyond their saturation level, forming supersaturated solutions, by interfering with the growth of scale crystals and formation of scale, or by dispersing the particles of scale and preventing them from adhering to heat-transfer surfaces. In practice, scale inhibitors most likely act in all three ways, thereby preventing scale formation of carbonate, silicate, and sulfate salts of calcium and magnesium.

Sodium hexametaphosphate, when added to water at only 2 to 3 mg/L, was found to inhibit precipitation of calcium carbonate from supersaturated solutions. This phenomenon was called "threshold treatment" since the polyphosphate holds calcium carbonate in solution at the "threshold" of precipitation. This was later applied to cooling water treatment when polyphosphate became widely used as an inhibitor for scale as well as corrosion.

Polyphosphates are used in the range of 0.5 to 5 mg/L to effectively inhibit scale in condenser water systems. There are, however, some disadvantages to polyphosphate scale inhibitors. These products tend to hydrolyze and revert to the orthophosphate form, which is ineffective as a threshold scale inhibitor. In fact, precipitates of calcium and magnesium orthophosphates can form, causing soft scales on heat-transfer surfaces called "sludge." In addition, overfeed of polyphosphates in excess of 20 mg/L can result in the formation of insoluble calcium polyphosphate sludge. Finally, the ability of polyphosphates to hold calcium carbonate in solution diminishes significantly with high pH conditions (above 8.0) in recirculating cooling water systems.

Other scale inhibitors developed in the 1960s and 1970s have proved more effective than polyphosphates without the above disadvantages.

Other materials which have been used to prevent precipitation of scale-forming minerals include starch, tanning (a natural derivation of wood and bark), lignin (a wood derivative), and other synthetic organic polymers. They control scale formation by interfering with the growth of the scale crystal and by dispersion of scale particles.

The low-molecular-weight polyacrylates, polymethacrylates, polymaleates, polyhydroxy alkyl acrylates, and polyacrylamides are the most effective of the synthetic polymers.

Other water-soluble polymers that exhibit similar scale-inhibiting properties are carboxymethyl cellulose, hydroxyethyl cellulose, polystyrene sulphonates, polyvinyl alcohol,and polyamino esters.

Some of the most widely used organic substances for both scale and corrosion control are the organic phosphonates. In particular, amino methylene phosphonate is used at very low levels, 1 to 2 mg/L, as a threshold treatment to prevent precipitation of calcium carbonate and calcium sulfate.

Other equally effective phosphonates, are phosphous Butane tricarboxylic acid, hydroxyethylidene diphosphonate, and poly amino substituted phosphonates. Organic phosphate esters have also been reported to inhibit scale and corrosion.

All these polymers, phosphonates, and organic phosphates have the distinct advantage of excellent scale-inhibiting properties over a wide range of pH and temperature conditions without the disadvantages of inorganic polyphosphates mentioned above.

In actual applications, combinations of these with corrosion inhibitors and dispersants are used in the proprietary formulations available for both corrosion and deposit control.

Corrosion Inhibitors. Just as in boilers, corrosion in open cooling water systems is controlled with both environmental stabilizers and corrosion inhibitors. In open cooling water systems, however, oxygen cannot be economically removed with oxygen scavengers because these systems are constantly aerated. Therefore, corrosion inhibitors that are effective in oxygen-containing environments must be utilized.

Also unlike boilers, recirculating cooling water systems contain various types of metals and alloys, and inhibitors and environmental conditions maintained must be compatible with these multimetallic systems. A particular problem associated with these systems is galvanic corrosion caused by bimetallic couples.

Treatment of recirculating cooling water must first include control of the pH value for both scale and corrosion control. A pH value within the range of 6.5 to 8.5 is usually maintained depending on the type of corrosion and scale inhibitor used. Lower pH values tend to render the recirculating water excessively corrosive, while higher pH values will result in both amphoteric metal corrosion (as with zinc, brass, and aluminum, and scale-forming conditions.

In areas where the atmospheric environment is such that acid conditions develop in open cooling water systems, neutralizing substances, such as caustic soda and sodium carbonate (soda ash), can be used to maintain the desired pH value in the neutral range. These conditions can be found in large urban areas or at locations subject to acid fumes from an adjacent boiler stack or incinerator.

Conversely, in areas where the makeup water is excessively alkaline, it may be necessary to add acid to maintain the desired pH range for control of scale, as previously outlined. Within the pH range of 6.5 to 8.5, corrosion is controlled by corrosion inhibitors added to the recirculating water. Inhibitors are substances which do not necessarily alter the environment, but do act as a barrier between the corrosive medium and the metal surface. These materials, when added to the recirculating water, form a protective barrier on the metal surface either by chemical reaction with the metal surface or by physical or chemical adsorption on the metal surface. An actively corroding metal can be rendered passive through the use of inhibitors that react in this manner.

Corrosion Testing. The relative performance of the types of programs available is based on actual field experience and corrosion monitoring. This can be determined by inserting corrosion test coupons in the open recirculating cooling water system. This corrosion test method has been described by the National Association of Corrosion Engineers (NACE) and is consistent with the ASTM standard "Corrosivity Testing of Industrial Cooling Water (Coupon Test Method)."[8]

The test coupon is placed in the recirculating water system in a test coupon "rack" described in Fig. 53.44. The corrosion test report will include the calculated corrosion rate in mils per year or micrometers per year and other pertinent data such as depth of pits, variations, and types of corrosion deposits.

The corrosion rate of mild steel for a 30-day exposure time in an open recirculating cooling water system is generally found to be rated in accordance with Table 53.10. The corrosion rate on mild steel can also be evaluated based on the lifespan of a standard 6-in (152.4-mm) Schedule 40 steel pipe assuming failure would occur first at a threaded joint (see Table 53.11).

Likewise the corrosion rate on standard 16-gauge copper condenser tubing,

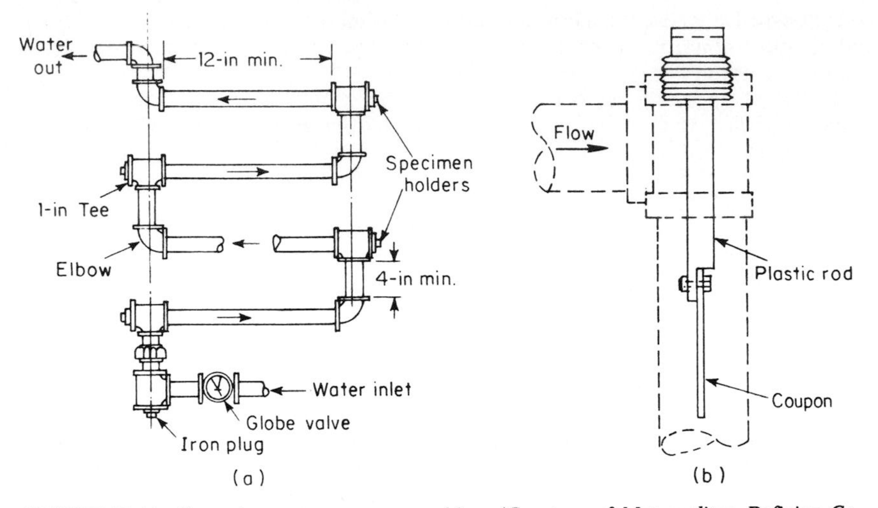

FIGURE 53.44 Corrosion test coupon assembly. (*Courtesy of Metropolitan Refining Co., Inc.*)

TABLE 53.10 Mild-Steel Corrosion Rates, 30-Day Exposure

Corrosion rate, mils/(MDD)*	Corrosion control
>5 (>27.3)	Poor
2–5 (10.9–27.3)	Good
0–2 (0–10.9)	Excellent

*MDD = milligrams per square deciliter per day.
Source: See Ref. 9.

TABLE 53.11 Evaluation of Corrosion Rates for Mild Steel

Corrosion rate, mils/yr (MDD)*	Half-life of standard 6-in (152.4-mm) Schedule 40 pipe, years	Corrosion control
0–2 (0–10.9)	>70	Excellent
2–5 (10.9–27.3)	70–28	Good
5–8 (27.3–43.7)	28–17½	Fair
8–10 (43.7–54.6)	17½–14	Poor
>10 (>54.6)	<14	Intolerable

*MDD = milligrams per square deciliter per day.

based on the expected lifespan of the tubing (assuming the corrosion is uniform), can be evaluated by using Table 53.12.

Control of Organic Growths. Organisms are present in the atmosphere attached to duct and dirt, and particles are swept into the recirculation water system with the air motion.

TABLE 53.12 Evaluation of Corrosion Rates for Standard Copper Condenser Tubing

Corrosion rate, mils/yr (MDD)*	Expected life of standard 16-gauge copper condenser tubing, years	Corrosion control
0–1 (0–6.2)	>65	Excellent
1–2 (6.2–12.4)	65–32½	Good
2–3 (12.4–18.6)	32½–21⅔	Fair
3–4 (18.6–24.8)	21⅔–16¼	Poor
>4 (>24.8)	<16¼	Intolerable

*MDD = milligrams per square decimeter per day.

These organisms, causing slime, fungi, and algae, may be controlled by substances capable of killing bacteria without harming the materials of construction or damaging the environment. These substances are called "microbicides," "biocides," "slimicides," "algicides," and "fungicides." They fall into the general classification called "pesticides" which are controlled and registered by the Environmental Protection Agency.

Pesticides used in cooling water treatment are more commonly referred to as "microbicides." The most widely used microbicide for recirculating water systems is chlorine. Chlorine used in excess will damage wood and organic fill in cooling towers. Usually it is only used in systems large enough to justify equipment for its controlled feed.

Where chlorine is used on a continuous basis, the concentration of free chlorine should be maintained at 0.3 to 0.5 mg/L to minimize the attack on materials of construction.

For cleaning purposes, shock feed of up to 50 mg/L can be used, provided this high chlorine content is held for no more than 8 h and the system is thoroughly flushed and drained, to remove dead organic matter and excess chlorine.

The use of chlorine on an intermittent basis has been found to be effective in inhibiting the growth of a particular biological organism while at the same time minimizing the disadvantages of continuous chlorine feed.

Chlorine is a very strong oxidizing agent, and so it is excessively corrosive to metals and damaging to cooling tower lumber and organic fill. Furthermore, it is very difficult to control the feed of chlorine at levels which will be effective in limiting organic growth while not attacking the materials of construction.

For these reasons, nonoxidizing, less corrosive, and less harmful microbicides are used in cooling towers, air washers, and other open recirculating water systems for HVAC. The most widely used microbicides in this type of equipment are the quaternary ammonium compounds and organosulfur compounds. Chlorophenols had been used in the past; however, the federal EPA has restricted the use of these types of pesticides. Other effective microbicides are organotin copper compounds used with the quaternary ammonium compounds, 2-dibromo-3-nitrilo-proprionamide, mixed isothiazouins, and 1-bromo-3-chloro dimethyl hydantoin.

The nonoxidizing microbicides are best applied to clean systems to inhibit, i.e., prevent, an organic growth rather than to kill or remove an existing growth. Sterilizing dosages of these biocides can be used to kill an organism, but usually penetrating agents are needed to assist the biocides in penetrating the mat of the organism, especially when the growth is of the algae type which forms layers of thick mats on the surface of the equipment.

Cleaning a contaminated system requires a significant amount of physical flushing and removal of organic matter to prevent reinfestation. Care must be taken to circulate the solution through the many nooks and crannies in these systems where growths can adhere and eventually grow to reinfest a system. Table 53.13 lists several of the more common microbicides and their effective sterilizing dosage and inhibiting dosage for most of the organisms found in HVAC open recirculating water systems.

A *sterilizing dosage* is one from which a complete kill can be expected within a few hours, provided the sterilizing solution comes into contact with the organism. An *inhibiting dosage* is the level at which the growth of the organism can be prevented once the dosage is removed from the system.

The manner of feeding biocides to a system is very important. Often the continuous feed at low inhibiting dosages is not effective and is very costly. Organisms tend to build up an immunity to a single biocide, and low dosages may even encourage the growth of some organisms. Therefore, a shock dosage at a level high enough for a bacterial kill is more effective than a continuous dosage at a low level. In addition, occasional change of the type of biocide used will prevent growth of an immunity.

TABLE 53.13 Common Microbicides Used in Treating Open Recirculating Water Systems

	Sterilizing dosage, ppm (mg/L)	Inhibiting dosage, ppm (mg/L)	Organisms controlled
Quaternary ammonium compounds			
n-Alkyl (60% C_{14}, 30% C_{16}, 5% C_{12}, 5% C_{18}) Dimethyl benzyl ammonium chloride	100–200	5–50	Algae, mold, and bacteria
n-Alkyl (98% C_{12}, 2% C_{14}) Dimethyl 1-naphthylmethyl ammonium chloride	100–200	2–20	Algae and bacterial slimes
Poly (oxyethylene) (dimethylimino) ethylene (dimethylimino) ethylene dichloride)	50–100	1–10	Algae, fungus, and bacterial slimes
Mixture of alkyl dimethyl (benzyl ammonium chlorides and ethylbenzyl ammonium chlorides)	50–100	1–10	Algae, fungus, and bacterial slimes
Organosulfur compounds			
Dimethyl dithiocarbonate salts of sodium or potassium	50–100	10–20	Bacterial and fungal slimes and molds
Methylene bis thiocyanate	50–100	2–10	Bacterial slime, fungi, and algae
N-methyl dithiocarbonate salts of sodium or potassium	50–100	10–20	Bacterial and fungal slimes, molds, and algae

Treatment of HVAC systems should include periodic shock dosages of a biocide only when experience indicates a need.

Bacteria Test. Test methods are available for monitoring the recirculating water to predetermine the need for biocides. These methods include test strips or dipsticks used to obtain a total bacteria count. Tests made periodically during operation of the system will provide a history that will indicate if there is an increase in the bacteria count during any particular season, showing the possible need of a biocide.

These tests are also used to determine the effectiveness of a biocide program and indicate when treatment should be changed or altered.

Mud and Dirt Control. Deposits of mud, dirt, and foreign suspended matter washed out of the atmosphere into open recirculating water systems can be as troublesome as scale and must be controlled for efficient operation of heat-transfer equipment. In the past, treatment for mud, dirt, and silt consisted only of physical removal with filters and separators or manual cleaning.

Now soluble polymers used in water clarification are applied to cooling waters with significant success. These polymers are high-molecular-weight polyelectrolytes that attract small dirt particles, forming a larger mass. This process is called "coagulation." The larger mass, called an "agglomerate," has a tendency to "float" or remain nonadherent to piping surfaces, and it does not compact into a mud. Coagulant-type polyelectrolytes can be used for mud and dirt control at dosages as low as 0.5 to 1 mg/L.

Another type of mud and dirt control agent is the dispersant. Dispersants have the opposite effect and prevent formation of mud deposits because of their ability to "repel" particles. Dispersants prevent the formation of larger agglomerates and thus their compacting into a mud. This causes the particles to be held in suspension, so they are more readily removed with normal bleed-off.

Coagulants usually are best applied in a shock manner on a periodic basis rather than continuously, while dispersants are applied on a continuous basis for maximum results. These may be included in the proprietary corrosion and deposit inhibitor formulation.

Cooling Tower—Evaporative Condenser Treatment. A complete treatment program for condenser water systems for HVAC equipment should include all the materials required to prevent corrosion and deposits. In addition, parameters should be given for bleed-off control to avoid excessive accumulation of minerals in the cooling tower and evaporative condenser circuit.

Air Washer and Surface Spray Treatment. Treatment of air washers and surface spray units will be similar to that of open cooling tower and evaporative condenser systems. The essential difference is that the former systems are subject to a wide variation of conditions with respect to temperature and humidity.

Test Controls. Simple, easy-to-use test kits (similar to that shown in Fig. 53.47) are available for regular testing of inhibitor levels and chlorides for cycles of concentration to control bleed-off. The tests should be made daily to ensure adequate control. Additional tests are made by the water service company to verify the accuracy of field test results.

Feed Methods. Treatments may be fed into systems simply by shock feed through any convenient opening, usually in the tower or condenser pan. This requires frequent additions at high levels to ensure maintenance of minimum levels of inhibitor in the recirculating water. For example, to maintain a minimum of 300 mg/L of a corrosion inhibitor in a cooling tower will require shock dosages at levels up to 664 mg/L every day. Figure 53.45 shows the treatment level after each initial shock dosage every 24 h in a 1000-gal (3800-L) system operating at 100-ton

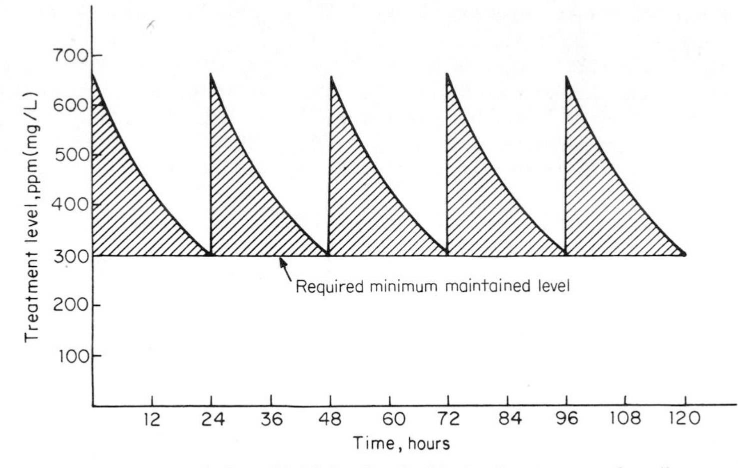

FIGURE 53.45 Variation of inhibitor level with shock treatment of cooling towers.

(352-kW) capacity, 24 h/day, 7 cycles of concentration, 0.5 gal/min (1.9 L/min) bleed-off rate.

As shown in Fig. 53.45 this method can be very costly and wasteful. The shaded area indicates overtreatment required to maintain the minimum level of 300 mg/L. To conserve treatment, therefore, proportional feed methods should be used.

Shock methods of feed are required, however, for certain types of treatment such as cleaners and biocides. These are fed at high levels on an intermittent basis to effect a cleanup action or biological kill, rather than to continuously maintain minimum levels.

1. *Cannister-type feeders:* To obtain a more continuous or gradual feed, the cannister-type feeder is used. This feeder comes in various designs. All consist of a can, or reservoir, containing treatment that is gradually dissolved or diluted so that it is fed slowly over a long time.

2. *Proportioning feed methods:* A more accurate method of feeding treatment is by means of a proportioning chemical feed pump. Proportioning feed pumps add treatment directly in proportion to the loss of treatment from the system, thereby maintaining a precise and consistent minimum treatment level. This avoids the excessive losses of shock feed or less accurate canister feeders.

A simple proportioning feed method utilizes a chemical feed pump interlocked with the recirculating water pump on the cooling tower, evaporative condenser, or spray cooler. The pump adds treatment directly from the drum or mixing tank. Figure 53.46 outlines this method of feed.

This system automatically injects the necessary treatment into the recirculating condenser water line, based on an assumed continuous-load factor. Sufficient water is bled from the system on a continuous basis when the condenser water pump is on. This maintains the cycles of concentration at a safe

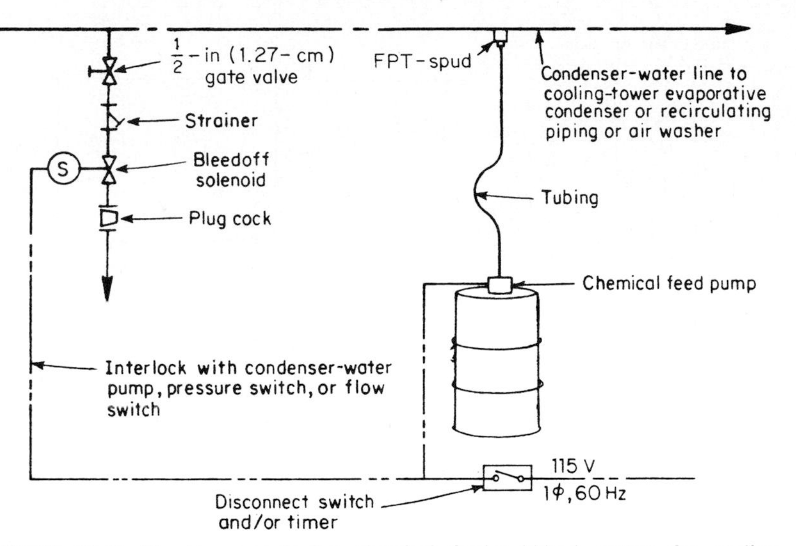

FIGURE 53.46 Simple proportioning chemical feed-and-bleed system for cooling towers. (*Courtesy of Metropolitan Refining Co., Inc.*)

operating level. Treatment is then fed continuously in proportion to the bleed rate so as to make up for loss of treatment in the bleed-off.

One of the most basic, fundamental, and universally adaptable fully automatic systems uses the water flow meter as a control. Since the rate of makeup water is proportional to evaporation and the actual load on a condenser water system, this is a fully automatic control system, independent of the quality of makeup water, that requires no additional instrumentation for accuracy other than pH control, if needed.

With this system, sufficient water can be bled automatically from the recirculating water to maintain the maximum cycles of concentration permitted. Treatment is then fed in direct proportion to bleed to replace any that is lost in the bleed-off. Figure 53.47 outlines this type of system.

Another widely used automatic proportioning feed-and-bleed system uses a total dissolved solids (TDS) analytical instrument as a control. This is commonly called the "conductivity controller" or "TDS controller." Since the dissolved mineral solids in a water are directly proportional to the electrical conductivity, the conductivity measurement can be used to determine any variation in dissolved solids. Because the dissolved solids content increases in proportion to the evaporation and load on the cooling system, this system is fully automatic.

In this system, bleed-off is activated when the TDS controller detects high TDS content so that the maximum permissible cycles of concentration are maintained. Treatment is then fed in direct proportion to bleed, to replace any that is lost in the bleed-off. Figure 53.48 outlines a system of this type.

Another controller used for HVAC systems, the pH controller is needed where pH control is required because of atmospheric conditions and/or the quality of makeup water or the type of program used. The controller automatically maintains the pH value of the recirculating water at the desired level and, when necessary, activates a chemical feed pump to add acid or alkali (Fig. 53.49). Fig-

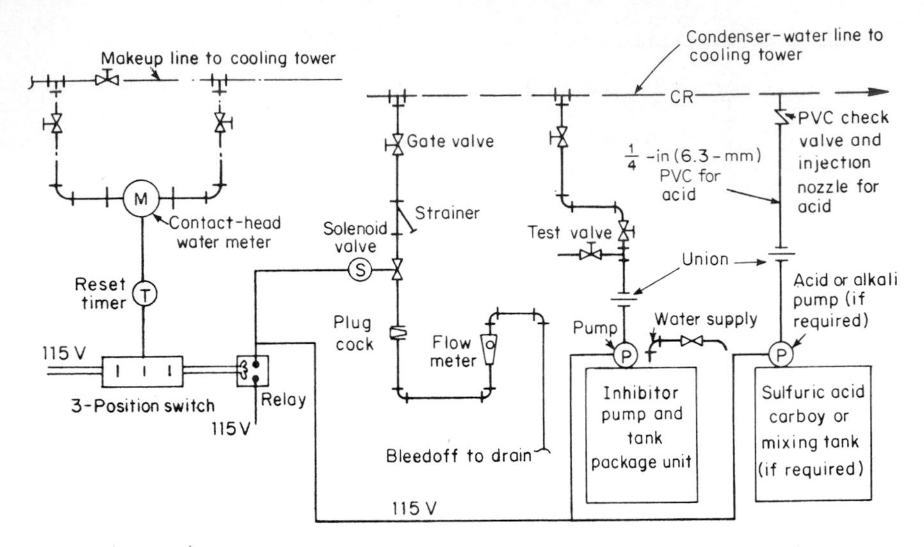

FIGURE 53.47 Water-meter-activated water treatment feed-and-bleed system for condenser water systems. *(Courtesy of Metropolitan Refining Co., Inc.)*

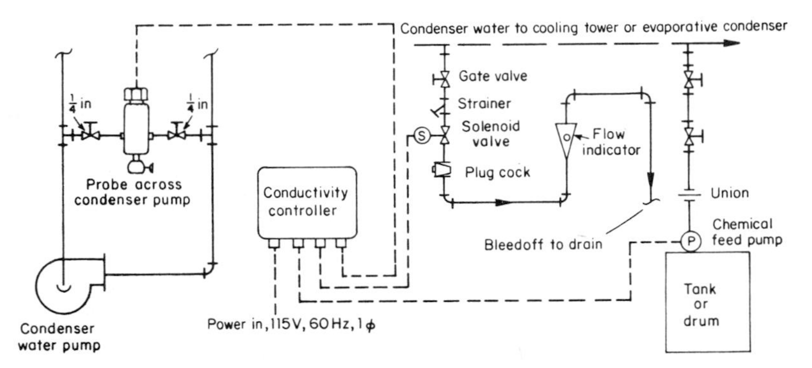

FIGURE 53.48 Conductivity-controlled automatic water treatment feed-and-bleed system. *(Courtesy of Metropolitan Refining Co., Inc.)*

ure 53.50 shows the type of TDS and pH controllers, chemical feed pumps, and package mixing tanks available.

53.8.4 Treatment of Closed Recirculating Water Systems

General. The need for treatment in closed recirculating water systems has frequently been questioned. It is often observed that because such systems are closed, they do not take on the impurities, excess minerals, and dissolved corrosive gases which are the prime causes of scale and corrosion.

Operating experience, however, indicates that much more makeup is used in closed systems than one would anticipate. For example, a very slow drip past

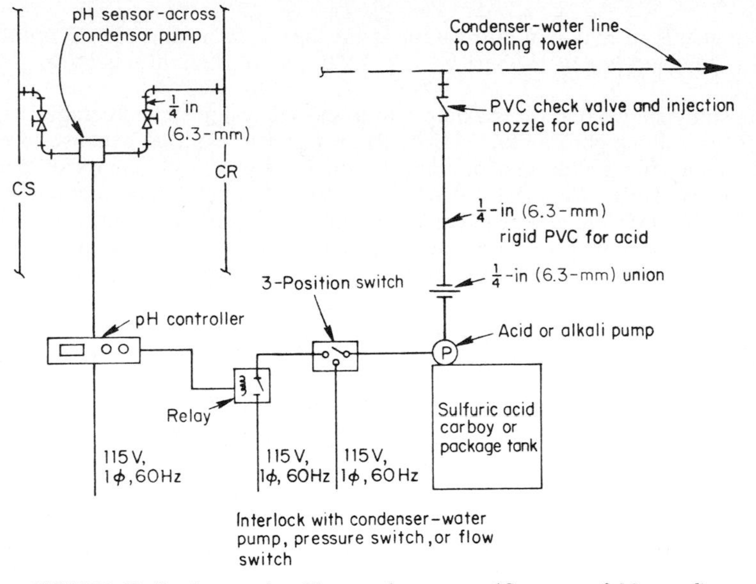

FIGURE 53.49 Automatic pH control system. (*Courtesy of Metropolitan Refining Co., Inc.*)

FIGURE 53.50 Automatic water treatment system. (*Courtesy of Electrosystems, Inc.*)

packing glands to maintain lubrication at the rate of 1 oz/min (30 mL/min) will require a makeup rate of 337.5 gal/month (1260 L/month), or 4050 gal/year (15,000 L/year).

One study shows that the makeup rate to closed systems can average 100 percent of the volume per month.[10] The makeup water is required because of water losses at packing glands and draining for maintenance and repairs of pumps, valves, or controls. With makeup water, dissolved minerals and oxygen are introduced. The calcium and magnesium hardness will deposit at heat-transfer surfaces and will accumulate as more makeup is introduced. All closed systems must "breathe," and expansion tanks are a source of atmospheric air and corrosive gases.

Galvanic couples of copper, iron, brass, and aluminum are commonly used in closed systems without dielectric fittings which result in a higher rate of corrosion of the less noble metal at the couple.

Therefore, it is generally accepted that treatment of these systems is required to prevent problems of corrosion and deposits as well as to maintain clean, efficient heat-transfer surfaces.[11]

Before closed recirculating water systems are treated to prevent scale and corrosion, it is essential to clean them thoroughly. All types of debris may enter a system during construction, and these must be removed. In addition, protective oil-based coatings and mill scale must be removed from the interior of the piping to avoid corrosion and damage from these sources. Black magnetic iron mill scale is a particular problem in closed systems and must be removed not only to prevent corrosion and heat-insulating deposits but also to avoid damage to pumps, packing, and mechanical seals.

Materials most effectively used for such cleanouts are low-foaming detergents built with strong dispersants to prevent resettling.

Dispersants are an important part of the closed system cleaner because they help to remove fine, black magnetic iron oxide particles by suspending them throughout the body of the recirculating cleaning solution and preventing them from settling and plugging.

The detergent-dispersant cleaner should be added to the recirculating water at the dosage recommended by the manufacturer for 3 to 4 h. Then drain, refill, and recirculate as fast as possible, all at the same time, to obtain maximum flow through all portions of the system and to avoid pockets and dead-end locations where sediment can be trapped.

The system should be flushed out until all traces of residual detergent and suspended matter are gone.

After cleanout, closed systems must be treated for control of corrosion and deposits. First, the makeup water should be analyzed to determine if pretreatment, such as a water softener, is required. Then treatment must be added to prevent further corrosion and deposits.

Closed systems are filled with fresh water and corrosion and deposit inhibitors or, under freezing conditions, with a solution of water and an inhibited glycol antifreeze compound. Antifreeze solutions are being used with increasing frequency to protect exposed piping of closed hydronic hot-water or chilled-water systems from freezing and rupturing. Where antifreeze solutions are required, an inhibited ethylene glycol could be used at a maximum concentration of 30 percent. Specific HVAC-grade glycol antifreeze should be used, not automotive-grade. Samples of solutions should be analyzed on a periodic basis to ensure that the inhibitor has not been depleted or decomposed.

Closed systems filled with raw water must be treated further for corrosion and deposit control.

High-Temperature Hot-Water Heating Systems. High-temperature hot-water (HTHW) heating systems are defined as those which operate above 350°F (178°C) and 450 lb/in^2 (3.00 kPa). These systems are frequently used to provide central heating for large complexes or campuslike situations.

Unless zeolite-softened water is used in these systems, the high temperature encountered, combined with the large volume of water in the system, will cause scale deposits to accumulate at the heat-transfer surfaces of the HTHW boiler. The ratio of water volume to heat-transfer surface area is such that even the initial scale deposits would be sufficient to cause significant interference with heat transfer.

Additional treatment should include an oxygen scavenger such as diethylhydroxyl amine or sodium sulfite. The latter is preferable in systems containing nonferrous metals. Caustic soda should be used to maintain the pH value of the water at 8.0 to 10.0.

Dispersants and scale inhibitors are used to maintain clean heat-transfer surfaces. These consist of thermally stable inhibitors such as polyacrylates, polymethacrylates, and phosphonates.

Medium-Temperature Hot-Water Systems. Medium-temperature hot-water (MTHW) heating systems are defined as those which operate at 250°F (120°C) up to 350°F (177°C) with pressures above 30 lb/in^2 (206 kPa). This kind of system is treated in the same way as HTHW systems. Corrosion and pitting are controlled by oxygen scavengers, sodium sulfite, or diethylhydroxyl amine, and the pH value is maintained at 8.0 to 10.0 with caustic soda or a similar alkali.

The system should be filled with zeolite-softened water to avoid scale deposits on heat-transfer surfaces. Scale inhibitor and dispersant should be added in the form of thermally stable inhibitors, as described above.

Low-Temperature Hot-Water Systems. Low-temperature hot-water (LTHW) systems are defined as those which operate below 250°F (120°C) with maximum water pressure of 30 lb/in^2 (206 kPa). In low-temperature systems, bimetallic couples of steel, copper, brass, and aluminum are frequently used.

The recirculating water in an LTHW system is best treated with corrosion inhibitors such as a sodium nitrite-borax combination. This inhibitor and similar inhibitors are not used at temperatures above 250°F (120°C) because of their questionable stability at higher temperatures. Apparently the film-forming mechanism is severely restricted at higher temperatures so that there is spotty pitting and even decomposition of the inhibitor in the case of nitrite inhibitors.

Sodium nitrite-borax inhibitor systems and other nitrite-based compositions are used at 3000 to 4000 mb/L at temperatures above 180°F, while 1000 to 2000 mg/L is sufficient for systems operating below 180°F. With nitrite-based treatments, it is important to use specific copper and brass inhibitors such as mercaptobenzothiozole or phenyl triazoles to minimize copper attack and copper-induced pitting of steel.

Because of concern about the toxic nature of chromates and the possibility that leaks from closed systems could cause staining, nitrite-based inhibitors are used more and more frequently.

Other corrosion inhibitors that are available in proprietary blends include phosphate organics, amides, silicates, benzoates, phosphonates, and molybdates, as well as various combinations of these with other organics. The concentrations recommended by the suppliers of these inhibitors should be followed to obtain good corrosion control.

Inhibitors for control of scale and deposit are used along with the corrosion inhibitors and are usually included in the proprietary blend.

Scale inhibitors for LTHW heating systems include sodium polyacrylates, polymethacrylates, polymaleates, sulfonated polystyrene, carboxy methyl cellulose lignins, and phosphonates. Some of these materials also act as dispersants which prevent sedimentation of fine particulate matter and provide excellent scale and deposit control. When these inhibitors are used, it is not necessary to soften the water used to fill LTHW heating systems which require makeup water only to replace normal losses resulting from drips, leaks, repairs, etc.

Chilled-Water Systems. Chilled-water systems operate below 140°F (60°C), usually in the range of 44 to 54°F (10 to 12°C) for comfort cooling. Sodium nitrite-borax inhibitors and other nitrite-based blends are used in the range of 500 to 1000 ppm with the pH maintained within the range of 7.0 to 10.0.

It is important that nitrite-based inhibitors include nonferrous-metal inhibitors to reduce nitrite attack on solder and other nonferrous metals. Among these inhibitors are borax, sodium benzoate, phosphates, and copper inhibitors, mercaptobenzothiazole, or phenyl triazoles. Chilled-water systems invariably contain components of multimetallic construction which include iron, steel, copper, brass, and aluminum and which require balanced formulated inhibitors, to ensure that all metals are protected as well as the galvanic couples of these metals.

Other inhibitors for chilled-water systems are organics blended with phosphates, silicates, or molybdate, sodium benzoate, organic phosphonates, and organic amines such as methyl glycine and imidazoline.

The corrosion inhibitor or inhibitor blend may be combined with dispersants and deposit inhibitors for a complete treatment. The dispersants are the same as those used in hot-water systems.

Closed-Circuit Coolers. Closed-circuit coolers are closed condenser water systems which are combined with an evaporative cooler. The closed condenser water circuit is treated the same as the closed chilled-water circuit with formulated inhibitors for control of corrosion and deposits.

Solar Heat-Exchange Systems. Another type of closed system being used with growing frequency is the energy-saving solar heat-exchange circuit. These systems also require treatment for control of corrosion and deposits. The most effective treatments are those used for chilled- and hot-water heating circuits.

Test Controls. The tests should be made for inhibitor level and pH. The testing service provided by the inhibitor supplier should include the recommended inhibitor level as the sample analysis.

Method of Feed. Treatments are applied to closed systems by several convenient means. They may be added directly through any opening into the system, such as at the expansion tank or open reservoir. Force pumps (similar to Fig. 53.37) are used to inject treatment directly through a drain valve.

FIGURE 53.51 Chemical feed pumps and mixing tanks with control panel skid-mounted. (*Courtesy of Neptune Chemical Pump Co.*)

In most instances, however, it is more convenient to install a bypass feeder across the recirculating pump as shown in Fig. 53.51.

With a differential in pressure between the discharge side and the suction side of the recirculating pump, the water in the system is forced through the bypass feeder, so the treatment is injected directly into the system.

The maintenance of a treatment program for closed systems is relatively simple, and the benefits obtained in corrosion protection and efficient operation are substantial.

53.9 REFERENCES

1. R. H. Hausler, "Economics of Corrosion Control," Materials Performance, National Association of Corrosion Engineers, Houston, June 1978, p. 9.
2. Gene E. Likens, "Acid Precipitation," *Chemical and Engineering News*, Nov. 22, 1976, p. 29.
3. Eskell Nordell, *Water Treatment for Industrial and Other Uses*, Reinhold Publishing Co., New York, 1951, p. 222.
4. Sidney Sussman, "Causes and Cures of Mechanical Shaft Seal Failures in Water Pumps," *Heating/Piping/Air Conditioning*, September 1963.

5. Hugh P. Goddard (ed.), "Materials Performance," vol. 13, no. 4, p. 9, April 1974. Reprinted by permission.
6. W. A. Kilbaugh and F. J. Pocock, "Pointers on the Care of Low Pressure Steam Steel Boilers," *Heating/Piping/Air Conditioning*, February 1962.
7. H. F. Hinst, "12 Ways to Avoid Boiler Tube Corrosion," Babcock and Wilcox Co. Reprint.
8. *Cooling Water Treatment Manual*, National Association of Corrosion Engineers, Houston, 1971, pp. 12–16.
9. *Carrier System Design Manual*, part 5, "Water Conditioning," Carrier Corp., Syracuse, NY, 1972, pp. 5–24.
10. Sidney Sussman, "Hot Water Heating Needs Water Treatment," *Air Conditioning, Heating and Ventilating*, August 1965.
11. *1980 ASHRAE Handbook and Product Directory, Systems*, ASHRAE, Atlanta, 1980, chap. 36, "Corrosion and Water Treatment," p. 36.16.
12. R. Eliassen and H. H. Uhlig, "So-called Electrical and Catalytic Treatment of Water for Boilers," *JAWWA*, vol. 44, pp. 576–582, 1952.
13. B. Q. Welder and E. P. Partridge, "Practical Performance of Water Conditioning Gadgets," *Ind. Eng. Chem.*, vol. 46, pp. 954–960, 1954.
14. R. Eliassen and R. T. Skrinde, "Experimental Evaluation of Water Conditioning Performance," *JAWWA*, vol. 49, pp. 1179–1190, 1957.
15. A. M. Henricks, "Water Conditioning Gadgets: Fact or Fancy," Paper presented at South Central Regional Meeting of NACE, Oklahoma City, October 1–4, 1957.
16. R. Eliassen, R. T. Skrinde, and W. B. Davis, "Experimental Performance of Miracle Water Conditioners," *JAWWA*, vol. 50, pp. 1371–1384, 1958.
17. "Federal Trade Commission Decision on Evis Water Conditioner Claims," *JAWWA*, vol. 51, pp. 708–709, 1959.
18. "Why Be a Gadget Sucker?" *Corrosion*, vol. 16, no. 7, p. 7, 1960.

53.10 BIBLIOGRAPHY

American Public Health Association (APHA): *Standard Methods for the Examination of Water and Waste Water*, APHA, Washington, DC, 1980.

American Water Works Association: *Water Quality and Treatment*, McGraw-Hill, New York, 1971.

Atkinson, J. J. N., and H. Van Droffelaar: *Corrosion and Its Control*, National Association of Corrosion Engineers, Houston, 1982.

Betz Handbook of Industrial Water Conditioning, Betz Laboratories, Trevose, PA, 1980.

Blake, Richard T.: *Water Treatment for HVAC and Potable Water Systems*, McGraw-Hill, New York, 1980.

Butler, G., and H. C. K. Ison: *Corrosion and Its Prevention in Waters*, Litton Educational Publishing Co. (Reinhold), Chicago, 1966.

Drew—Principles of Industrial Water Treatment, Drew Chemical Corporation, Boonton, NJ, 1983.

Kemmer, Frank N. (ed.): *The Nalco Water Handbook*, McGraw-Hill, New York, 1979.

McCoy, James W.: *Chemical Analysis of Industrial Water*, Chemical Publishing Co., New York, 1969.

McCoy, James W.: *The Chemical Treatment of Cooling Water*, Chemical Publishing Co., New York, 1974.

NACE Cooling Water Treatment Manual, National Association of Corrosion Engineers, Houston, 1971.

Nathan, C. C. (ed.): *Corrosion Inhibitors*, National Association of Corrosion Engineers, Houston, 1973.

Nordell, Eskel: *Water Treatment for Industrial and Other Uses*, Litton Educational Publishing Co. (Reinhold), Chicago, 1961.

Pincus, L. I.: *Practical Boiler Water Treatment*, McGraw-Hill, New York, 1962.

Powell, Sheppard T.: *Water Conditioning for Industry*, McGraw-Hill, New York, 1954.

Speller, Frank N.: *Corrosion Causes and Prevention*, McGraw-Hill, New York, 1951.

Uhlig, H. H. (ed.): *The Corrosion Handbook*, John Wiley & Sons, New York, 1948.

Uhlig, H. H.: *Corrosion and Corrosion Control*, John Wiley & Sons, New York, 1971.

APPENDIX A
ENGINEERING GUIDE FOR ALTITUDE CORRECTIONS

A.1 INTRODUCTION

Altitude affects the operation of air-conditioning equipment in the following areas:

1. Psychrometric air properties
2. Air density
3. Temperature level of steam-heating equipment

A.1.1 Effect on Psychrometric Air Properties

Altitude affects the psychrometric properties of air such as enthalpy, dew-point temperature, and specific humidity. This is best shown visually by the construction of an altitude psychrometric chart using the following relationship:

$$W = \frac{0.622P''}{P - P''}$$

where W = specific humidity, lb water vapor/lb dry air (kg water vapor/kg of dry air)
P'' = partial pressure of water vapor at the dew-point temperature, psia (kPa absolute)
P = barometric pressure, psia (kPa absolute)

By assuming various values of the dew-point temperature and calculating values of W corresponding to the barometric pressure, the saturation curve may be drawn for air at elevations above sea level. This curve shows that air at elevation has a higher specific humidity than air at sea level for a given dry-bulb temperature and percent relative humidity. The higher specific humidity causes the

Note: The material in this appendix, including tables and illustrations, is taken from Carrier Corporation's *Engineering Guide for Altitude Effects*, 1967, courtesy of Carrier Corporation. Edited for this handbook by Nils R. Grimm.

enthalpy of air to be higher for a given dry-bulb temperature and percent relative humidity or for a given value of dry-bulb and wet-bulb temperature.

A.1.2 Effect on Air Flow

The density of air at altitude varies inversely as the absolute pressure from the perfect gas law:

$$p = \frac{P}{RT}$$

where p = density, lb/ft^3 (kg/m^3)
P = pressure, lb/ft^2 (Pa)
R = gas constant for air = 53.3 (287.05)
T = absolute temperature, °F (°R or K)

Therefore, to maintain the same mass flow rate at altitude as at sea level, the air-volume flow rate must be increased at a rate inversely proportional to the air-density ratio. If the air-volume flow rate is held constant, the mass flow rate decreases at a rate directly proportional to the air-density ratio.

A.1.3 Effect on Temperature Level of Equipment Using Steam

Although altitude causes no change in the saturation temperature-pressure relationship, there is a change in the temperature corresponding to a certain gauge pressure. For example, at sea level the saturation temperature of 5 psig (0.34 bar) steam is 227°F (108.5°C), while at 7500 ft (2285.9 m) the saturation temperature is 216.3°F (102.4°C). The effect of altitude on temperature level should be considered in steam unit heaters, heating and ventilating units, and absorption refrigeration machines.

In the range of 0 to 10,000 ft (0 to 3050 m), the effect of altitude on the thermal properties of air (such as viscosity, thermal conductivity, and specific heat) is very small and can be disregarded.

A.2 ADJUSTMENT DATA FOR VARIOUS KINDS OF AIR-CONDITIONING EQUIPMENT

A ready reference index (Table A.1) is provided to allow quick reference to correction factors for various air-conditioning products. *For some products, it is impractical to give the correction factor. In such cases, refer to the applicable section of this appendix referenced in the table for the procedure or recommendations for correcting for altitude effects.*

A.2.1 Open and Hermetic Compressors

Compressor capacity is expressed as a function of saturated discharge temperature and saturated suction temperature. Regardless of compressor type, the ca-

TABLE A.1 Ready Reference Index

Effect of altitude on air-conditioning equipment capacity

Product name	Altitude correction factors — Elevation, ft (m) 2500 (760)	5000 (1500)	7500 (2300)	10,000 (1005)
Reciprocating compressors	None	None	None	None
Condensing units (water-cooled)	None	None	None	None
Air-cooled condensers	0.95	0.90	0.85	0.80
Evaporative condensers	1.00	1.01	1.02	1.03
Liquid coolers	None	None	None	None
Absorption refrigeration machines	See Sec. A.2.6, p. A.7			
Centrifugal refrigeration machines (open and hermetic type)	None	None	None	None
Reciprocating liquid chilling packages with air-cooled condensers	0.98	0.97	0.95	0.93
Induction-room terminals:				
Chilled-water	0.93	0.86	0.80	0.74
Hot water—gravity and forced air	0.95	0.90	0.85	0.81
Steam heating	See Sec. A.2.11, p. A.16			
Electric heating	See Sec. A.2.13, p. A.17			
Air terminal units and outlets	See Sec. A.2.14, p. A.17			
Condensing units (air-cooled)	0.98	0.97	0.95	0.93
Central station air-handling units:				
Chilled-water and direct-expansion	See Sec. A.2.15, p. A.19			
Steam and hot water	See Secs. A.2.11, p. A.16, and A.2.15, p. A.19			
Chilled-water fan-coil units:				
Chilled-water	See Sec. A.2.9, p. A.8			
Steam and hot water	See Secs. A.1.3, p. A.2, and A.2.15, p. A.19			
Direct-expansion fan-coil units:				
Direct-expansion, total capacity SHF 0.40–0.95	0.97	0.95	0.93	0.91
Direct-expansion, total capacity SHF 0.95–1.00	0.93	0.86	0.79	0.73
Direct-expansion, sensible capacity SHF 0.40–0.95	0.92	0.85	0.78	0.71
Hot water	0.95	0.90	0.85	0.81
Steam	See Sec. A.1.3, p. A.2			
Room fan-coil units:				
Chilled-water, total capacity SHF 0.40–0.95	0.97	0.95	0.93	0.91
Chilled-water, total capacity SHF 0.95–1.00	0.93	0.86	0.80	0.74
Chilled-water, sensible capacity SHF 0.40–0.95	0.92	0.85	0.78	0.71
Hot water	0.95	0.90	0.85	0.81
Steam	See Sec. A.2.11, p. A.16			
Central station air-handling units with sprayed coil:				
Chilled-water and direct-expansion	See Sec. A.2.17, p. A.20			
Steam and hot water	See Secs. A.2.11, p. A.16, and A.2.15, p. A.19			

TABLE A.1 Ready Reference Index (*Continued*)

Effect of altitude on air-conditioning equipment capacity

Product name		Altitude correction factors — Elevation, ft (m) 2500 (760)	5000 (1500)	7500 (2300)	10,000 (1005)
Unit Heaters:					
Steam		See Sec. A.2.11, p. A.16			
Gas-fired		See Sec. A.2.15, p. A.19			
Hot water		0.95	0.90	0.85	0.81
Heating and ventilating units:					
Steam		See Sec. A.1.3, p. A.2			
Hot water		0.95	0.90	0.85	0.81
Roof-mounted air-conditioning units:					
Direct-expansion, total capacity	SHF 0.40–0.95	0.98	0.96	0.94	0.92
Direct-expansion, total capacity	SHF 0.95–1.00	0.96	0.92	0.88	0.84
Direct-expansion, sensible capacity	SHF 0.40–0.95	0.92	0.85	0.78	0.71
Gas-fired heating		See Sec. A.2.16, p. A.19			
Package air-conditioning unit with air-cooled condenser:					
Direct-expansion, total capacity	SHF 0.40–0.95	0.98	0.96	0.94	0.92
Direct-expansion, total capacity	SHF 0.95–1.00	0.96	0.92	0.88	0.84
Direct-expansion, sensible capacity	SHF 0.40–0.95	0.92	0.85	0.78	0.71
Package air-conditioning unit with water-cooled condenser:					
Direct-expansion, total capacity	SHF 0.40–0.95	0.99	0.985	0.98	0.97
Direct-expansion, total capacity	SHF 0.95–1.00	0.96	0.93	0.90	0.87
Direct-expansion, sensible capacity	SHF 0.40–0.95	0.93	0.87	0.80	0.74
Package air-conditioning unit with air- or water-cooled condenser:					
Hot water		0.95	0.90	0.85	0.80
Steam		See Secs. A.1.3, p. A.2, and A.2.11, p. A.16			
Gas-fired furnace		See Sec. A.2.16, p. A.19			
Centrifugal fans		See Sec. A.2.18, p. A.23			
Motors		See Sec. A.2.20, p. A.24			
Pumps		See Sec. A.2.19, p. A.24			

Altitude thermal capacity = sea-level thermal capacity × altitude correction factor.

pacity is the same at altitude as at sea level for the same saturated suction and discharge temperatures. At altitude, however, there is a change in the gauge pressure corresponding to the saturated discharge and saturated suction temperatures. For example, the gauge pressure at 5000 ft (1500 m) of elevation is higher by 2.6 psi (0.179 bar) than the gauge pressure at sea level (see Table A.2).

A.2.2 Water-Cooled Condensing Units

Since air is not used in these condensers, operation at altitude causes no change in capacity.

A.2.3 Air-Cooled Condensers

When an air-cooled condenser is operated at altitude, the actual air volume is maintained constant (for the same air temperature and fan speed) but the air-weight flow decreases. This causes the capacity of the condenser to decrease.

To obtain the capacity of altitude for these condensers, multiply the published values of capacity by the following factors:

Elevation, ft (m)	Altitude capacity factors
2,500 (760)	0.95
5,000 (1500)	0.90
7,500 (2300)	0.85
10,000 (3050)	0.80

A.2.4 Evaporative Condensers

There is a slight increase in the capacity of an evaporative condenser at altitude when operating at the same fan speed as at sea level.

To obtain the capacity at altitude for the above units, multiply the published values of capacity by the following factors:

Elevation, ft (m)	Altitude capacity factors
2,500 (760)	1.00
5,000 (1500)	1.01
7,500 (2300)	1.02
10,000 (3050)	1.03

A.2.5 Liquid Coolers

Since the heat exchange is between water and refrigerant, altitude operation causes no change in capacity.

TABLE A.2 Altitude Pressure for Air

Altitude	Barometric pressure		Gauge-pressure correction
ft	in Hg	(psia)	psia
0	29.92	(14.70)	
500	29.38	(14.40)	-0.30
1,000	28.86	(14.19)	-0.51
1,500	28.33	(13.91)	-0.79
2,000	27.82	(13.58)	-1.12
2,500	27.32	(13.41)	-1.29
3,000	26.82	(13.20)	-1.50
3,500	26.33	(12.92)	-1.78
4,000	25.84	(12.70)	-2.00
4,500	25.37	(12.44)	-2.26
5,000	24.90	(12.23)	-2.57
5,500	24.43	(12.01)	-2.69
6,000	23.98	(11.78)	-2.92
6,500	23.53	(11.55)	-3.15
7,000	23.09	(11.33)	-3.37
7,500	22.65	(11.10)	-3.60
8,000	22.22	(10.92)	-3.78
8,500	21.80	(10.70)	-4.00
9,000	21.39	(10.50)	-4.20
9,500	20.98	(10.30)	-4.40
10,000	20.58	(10.10)	-4.60
m	mmHg	(bar)	bars
0	759.98	(1.014)	0
152.4	746.25	(0.993)	-0.021
304.8	733.04	(0.979)	-0.035
457.2	719.58	(0.959)	-0.055
609.6	706.63	(0.937)	-0.077
762	693.93	(0.925)	-0.089
914.4	681.23	(0.910)	-0.104
1066.7	668.78	(0.891)	-0.123
1219.1	656.34	(0.876)	-0.138
1371.5	644.4	(0.859)	-0.155
1523.9	632.46	(0.843)	-0.171
1676.3	620.52	(0.828)	-0.186
1828.7	609.09	(0.812)	-0.202
1981.1	597.66	(0.797)	-0.217
2133.5	586.49	(0.781)	-0.233
2285.9	575.31	(0.766)	-0.248
2438.3	564.39	(0.753)	-0.261
2590.7	553.72	(0.738)	-0.276
2743.1	543.31	(0.724)	-0.290
2895.4	532.89	(0.710)	-0.304
3047.9	522.73	(0.697)	-0.317

A.2.6 Absorption Refrigeration Machine

The capacity of the absorption refrigeration machine is a function of the absolute pressure of the steam supplied to the machine. Since the capacity of the above unit is expressed as a function of gauge pressure, it is necessary to correct for the reduction in absolute pressure at altitude. The correction of capacity for operation at altitude may be obtained by subtracting from the actual gauge pressure the pressure correction tabulated below to obtain a pseudo gauge pressure. Use this pseudo gauge pressure to obtain absorption-machine capacity:

Elevation, ft (m)	Gauge-pressure correction, psig (bar)
2,500 (760)	−1.3 (−0.0897)
5,000 (1500)	−2.6 (−0.179)
7,500 (2300)	−3.6 (−0.248)
10,000 (3050)	−4.6 (−0.317)

A.2.7 Air-Cooled Condensing Units

Graphical plots of air-cooled condenser capacity at sea level and at various altitudes in combination with compressor capacity show that the decrease in condensing-unit capacity at altitude is small compared to the reduction in capacity of the air-cooled condenser alone. To obtain the capacity at altitude for air-cooled condensing units, multiply the published values of capacity by the following factors:

Elevation, ft (m)	Altitude capacity factors
2,500 (760)	0.98
5,000 (1500)	0.97
7,500 (2300)	0.95
10,000 (3050)	0.93

A.2.8 Liquid Chilling Units with Air-Cooled Condensers

Graphical plots made of air-cooled condensers' capacity at sea level and at various altitudes in combination with condenserless liquid chilling units show that the decrease in capacity at altitude is small compared to the reduction in capacity of the air-cooled condenser alone.

To obtain the capacity at altitude for the above units, multiply the published values of capacity by the following factors:

Elevation, ft (m)	Altitude capacity factors
2,500 (760)	0.98
5,000 (1500)	0.97
7,500 (2300)	0.95
10,000 (3050)	0.93

A.2.9 Chilled-Water Central Station Air-Handling Units; Chilled-Water Fan-Coil Units

Altitude capacity may be compared with sea-level capacity on the basis of constant air-volume flow (varying-weight flow) or constant air-weight flow (varying-volume flow).

For the above units the comparison is most conveniently made on the basis of constant air-weight flow. This shows that the capacity of the units increases if the cooling coil is dehumidifying. The comparison is based on the same entering-air dry-bulb and wet-bulb temperature, the same water flow, and the same entering-water temperature. Therefore, to obtain the same capacity from the unit operating at altitude with the same air-weight flow, the same entering dry-bulb and wet-bulb temperature, and same entering-water temperature, a smaller amount of water is required.

The following procedure is recommended to determine the gallons per minute (gpm) required at altitude when the sensible heat factor (SHF) is 0.95 or less.

Given: entering-air dry-bulb and wet-bulb temperatures, leaving-air dry-bulb and wet-bulb temperatures, entering-water temperature, and standard cubic feet per minute (cfm).

1. Calculate the load using the formula

$$\text{GTH} = 4.45 \times \text{cfm}(h_1 - h_2) \qquad (A.1)$$

where GTH = grand total heat, Btu/h

h_1 = entering-air enthalpy, Btu/lb at altitude entering-air wet-bulb temperature. Use altitude psychrometric charts, Figs. A.1 to A.4.

h_2 = leaving-air enthalpy, Btu/lb at altitude leaving-air wet-bulb temperature. Use altitude psychrometric charts, Figs. A.1 to A.4.

cfm = rate of air flow, standard cfm. If cfm is given at altitude, convert to standard cfm in order to use Eq. (A.1). (Standard cfm = altitude cfm × air-density ratio at altitude.)

2. Using the load calculated in step 1, the same standard air cfm, and the entering-air enthalpy at sea level (with the specified entering wet-bulb temperature), calculate the value of the leaving-air enthalpy at sea level as follows:

$$h_{2s} = h_{1s} - \frac{\text{GTH}}{4.45 \times \text{cfm}} \qquad (A.2)$$

where h_{1s} = entering-air enthalpy at sea level and specified altitude entering wet-bulb temperature, Btu/lb

h_{2s} = leaving-air enthalpy at sea level, Btu/lb

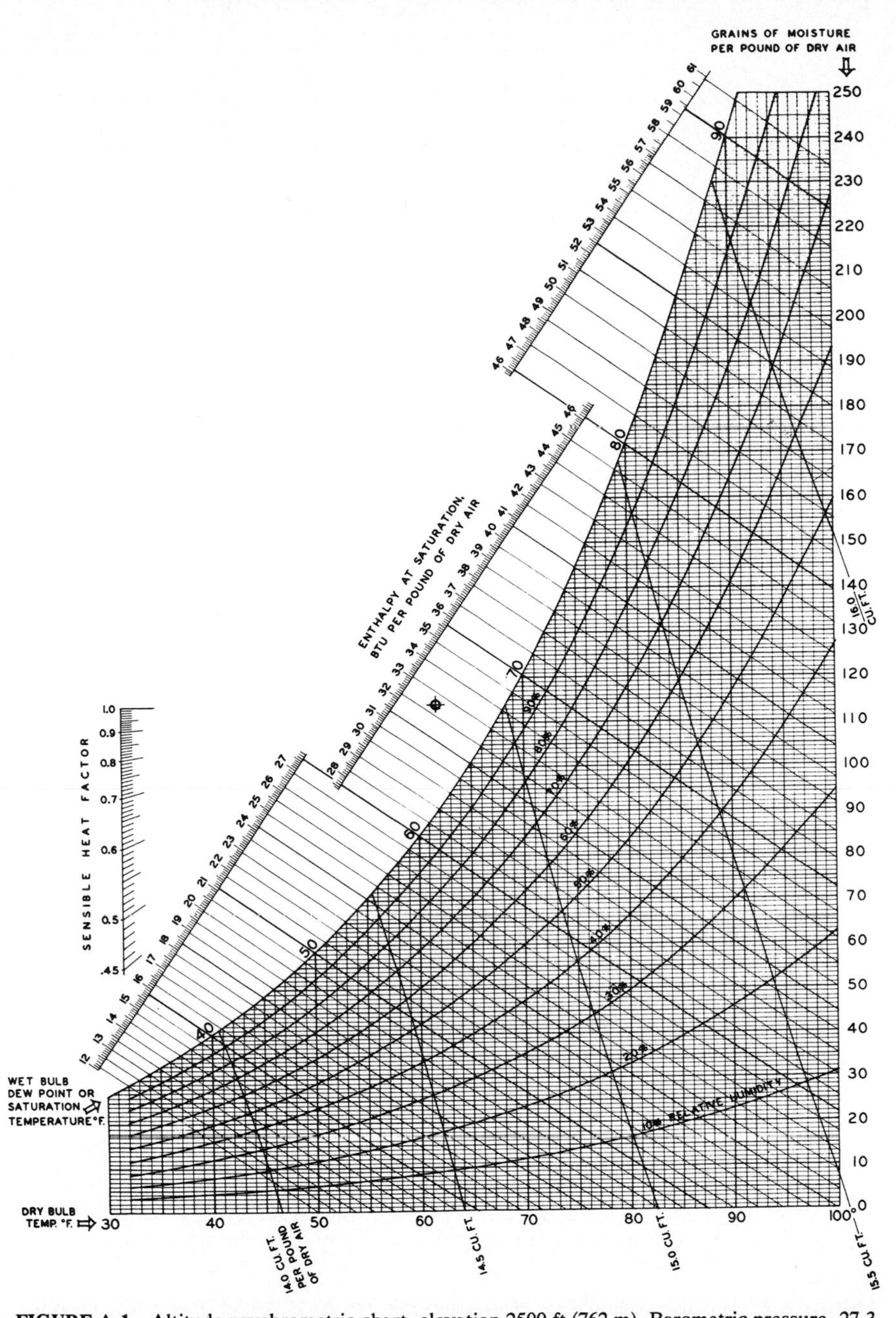

FIGURE A.1 Altitude psychrometric chart, elevation 2500 ft (762 m). Barometric pressure, 27.3 in Hg = 13.4 psia (93 kPa); normal temperatures. Metric conversions: °C = (°F − 32)/1.8; kJ (0.0004299) = Btu/lb; kg/kg (6997.9) = grains of moisture/lb of dry air; m /kg (16.01846) = ft /lb of dry air.

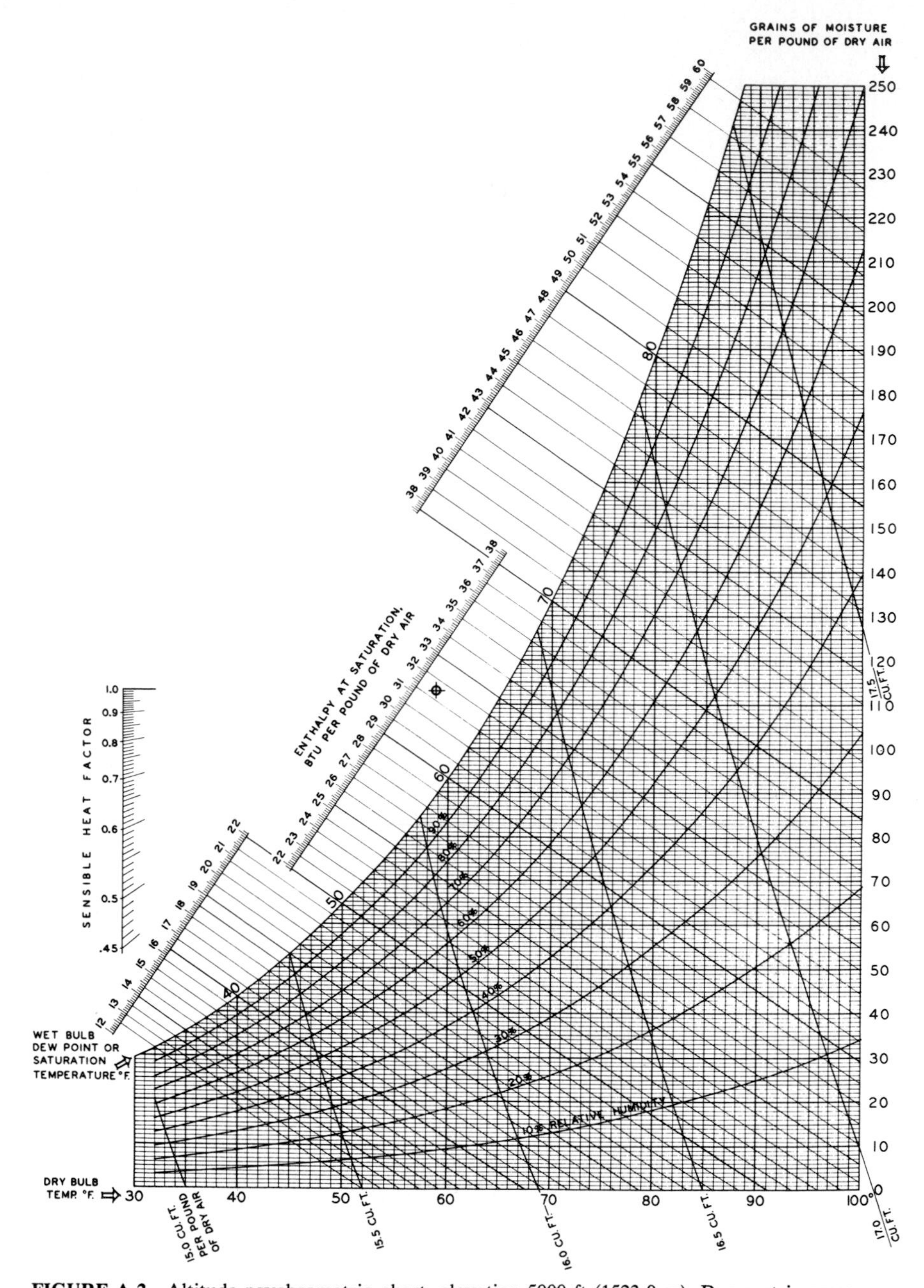

FIGURE A.2 Altitude psychrometric chart, elevation 5000 ft (1523.9 m). Barometric pressure, 24.9 in Hg = 12.23 psia (84 kPa); normal temperatures. For metric conversion factors, see Fig. A.1.

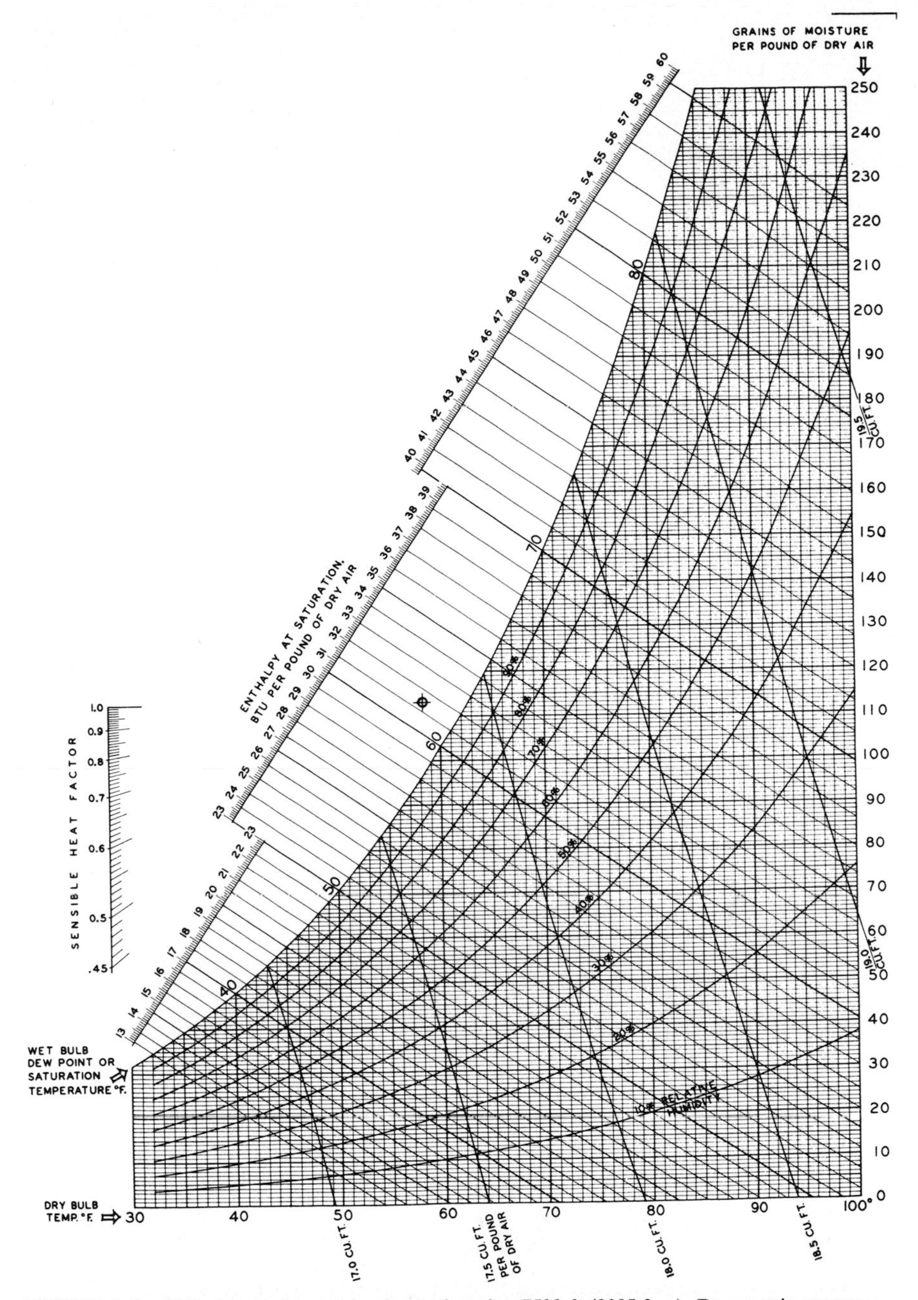

FIGURE A.3 Altitude psychrometric chart, elevation 7500 ft (2285.9 m). Barometric pressure, 22.7 in Hg = 11.1 psia (76.5 kPa); normal temperatures. For metric conversion factors, see Fig. A.1.

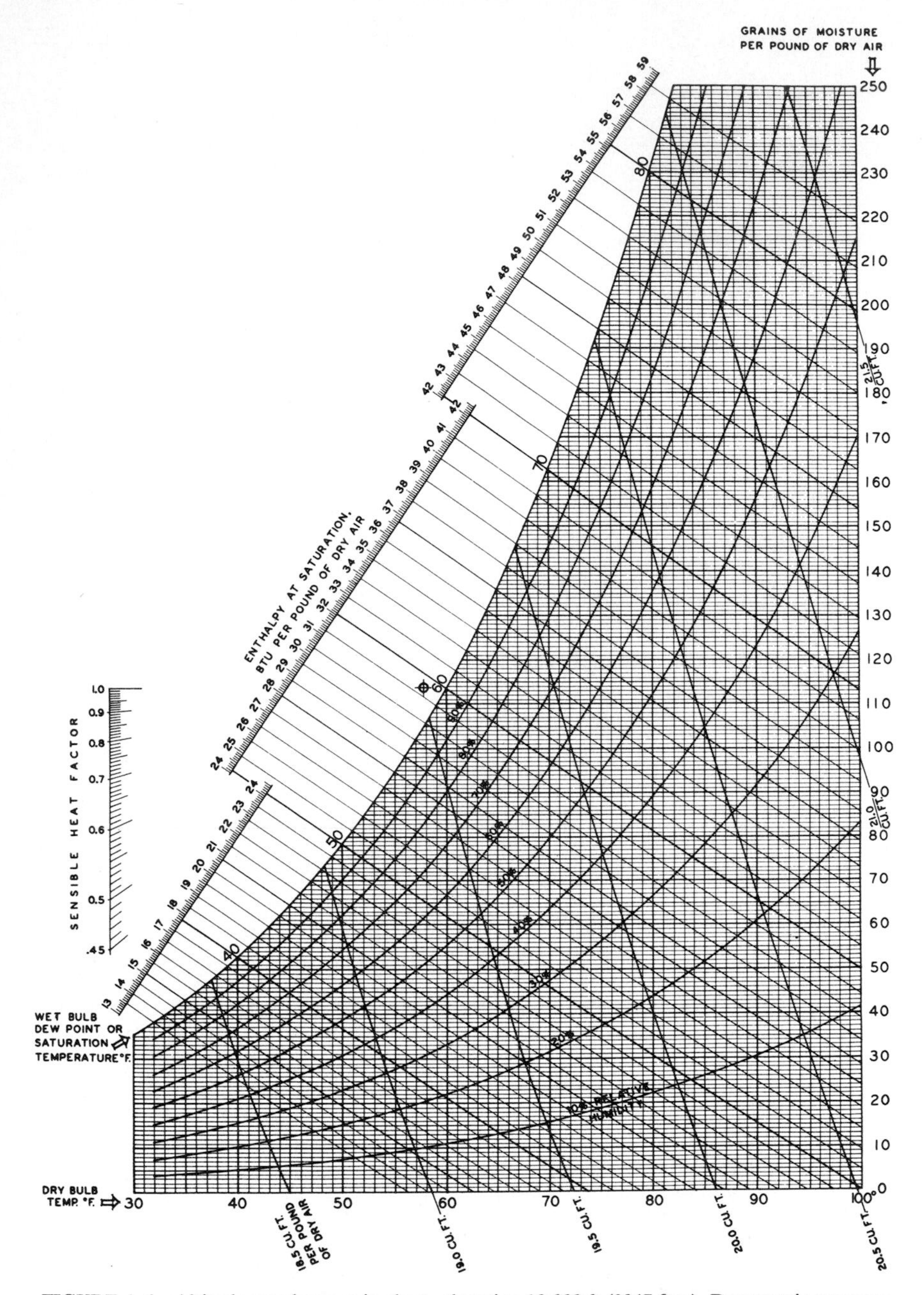

FIGURE A.4 Altitude psychrometric chart, elevation 10,000 ft (3047.9 m). Barometric pressure, 20.6 in Hg = 10.1 psia (69.63 kPa); normal temperatures. For metric conversion factors, see Fig. A.1.

The saturation temperature which corresponds to the above value of enthalpy is the leaving-air wet-bulb temperature for sea-level operation.

3. From the manufacturer's performance chart, determine the sea-level adp using the entering wet-bulb temperature specified, the assumed coil surface and standard air velocity, and the sea-level leaving wet-bulb temperature determined from step 2.

4. Determine the gallons/minute required at sea level from published aparatus dewpoint (adp) ratings using the adp from step 3, the load determined from step 1, and the design entering-water temperature.

5. Multiply the sea-level gpm determined from step 4 by the following factors:

Elevation, ft (m)	Altitude sensible heat factor, 0.40–0.95
2,500 (760)	0.97
5,000 (1500)	0.94
7,500 (2300)	0.91
10,000 (3050)	0.88

6. Determine actual leaving dry-bulb temperature at altitude by the formula

$$t_2 = t_s + \mathrm{BF}(t_1 - t_s) \tag{A.3}$$

where t_s = adp at altitude – saturation temperature corresponding to h_s in the formula

$$h_s = h_1 - \frac{h_1 - h_2}{1 - \mathrm{BF}} \tag{A.4}$$

t_1 = entering-air dry-bulb temperature, °F (°C)
t_2 = leaving-air dry-bulb temperature, °F (°C)
BF = bypass factor. Determine from the manufacturer's performance chart using standard air cfm.
h_s = enthalpy at t_s, Btu/lb (W/kg). Use altitude psychrometric charts, Figs. A.1 to A.4.
h_1 = entering-air enthalpy at altitude entering-air wet-bulb temperature, Btu/lb (W/kg). Use altitude psychrometric charts, Figs. A.1 to A.4.
h_2 = leaving-air enthalpy at altitude leaving-air wet-bulb temperature, Btu/lb (W/kg). Use altitude psychrometric charts, Figs. A-1 to A-4.

7. For chilled-water units operating with an SHF of 0.95 or greater, consider the unit to be doing all sensible cooling and to have the same capacity at altitude as at sea level. Follow the procedure for determining the gpm for a dry coil at sea level as given in the manufacturer's published data.

A.2.10 Direct-Expansion Central Station Air-Handling Units

Altitude capacity may be compared with sea-level capacity on the basis of constant air-volume flow (varying-weight flow) or constant air-weight flow (varying-volume flow).

For the above units the comparison is most conveniently made on the basis of constant air-weight flow. This shows that there is an increase in the capacity of these units if the cooling coil is dehumidifying. The comparison is based on the same air-weight flow at sea level and altitude, the same entering dry-bulb and wet-bulb temperature, the same refrigerant, and the same suction temperature. Therefore, to obtain the same capacity from the unit operating at altitude with the same air-weight flow, the same entering dry-bulb and wet-bulb temperature, and the same refrigerant, a higher saturated suction temperature is required at altitude than at sea level.

The following procedure is recommended to determine the required suction temperature at altitude when the sensible heat factor is 0.95 or less.

Given: entering-air dry-bulb and wet-bulb temperatures, leaving-air dry-bulb and wet-bulb temperatures, and standard cfm.

1. Calculate the load using the following formula:

$$\text{GTH} = 4.45 \times \text{cfm}(h_1 - h_2) \qquad (A.1)$$

where GTH = grand total heat, Btu/h

h_1 = entering-air enthalpy, Btu/lb at altitude entering-air wet-bulb temperature. Use altitude psychrometric charts, Figs. A.1 to A.4.

h_2 = leaving-air enthalpy, Btu/lb at altitude leaving-air wet-bulb temperature. Use altitude psychrometric charts, Figs. A.1 to A.4.

cfm = rate of air flow, standard cfm. If cfm is given at altitude, convert to standard cfm in order to use Eq. (A.1). (Standard cfm = altitude cfm × air-density ratio at altitude.)

2. Using the load calculated in step 1, the same standard cfm, and the entering-air enthalpy at sea level (with the specified entering wet-bulb temperature), calculate the value of the leaving-air enthalpy at sea level as follows:

$$h_{2s} = h_{1s} - \frac{\text{GTH}}{4.45 \times \text{cfm}} \qquad (A.2)$$

where h_{1_s} = entering-air enthalpy at sea level and specified altitude entering wet-bulb temperature, Btu/lb

h_{2s} = leaving-air enthalpy at sea level, Btu/lb as calculated

The saturation temperature which corresponds to the above value of enthalpy is the leaving-air wet-bulb temperature for sea-level operation.

English units should be used with Eq. (A.2). To obtain metric equivalents:

(1)*(2)	
W(3.413)	= Btu/h

$$\frac{\text{kJ}}{\text{kg}}(0.0004299) = \text{Btu/lb}$$

$$\frac{\text{m}^3}{\text{s}}(2118.9) = \text{cfm}$$

$$\frac{\text{L}}{\text{s}}(2.1189) = \text{cfm}$$

3. From the manufacturer's performance chart, determine the sea-level adp using the entering wet-bulb temperature specified, the assumed coil surface and standard air velocity, and the sea-level leaving wet-bulb temperature determined from step 2.

4. From the published adp ratings, using the load and adp determined above, obtain the value of the suction temperature required for operation at sea level.

5. To obtain the value of the suction temperature required for operation at altitude, increase the sea-level suction temperature obtained from step 4 by the following number of degrees:

Elevation, ft (m)	Altitude sensible heat factor, 0.40–0.95
2,500 (760)	0.6
5,000 (1500)	1.3
7,500 (2300)	2.0
10,000 (3050)	2.6

6. Determine actual leaving dry-bulb temperature at altitude by the formula

$$t_2 = t_s + \text{BF}(t_1 - t_s) \tag{A.3}$$

where t_s = adp at altitude-saturation temperature corresponding to h_s in the formula

$$h_s = h_1 - \frac{h_1 - h_2}{1 - \text{BF}} \tag{A.4}$$

t_1 = entering-air dry-bulb temperature, °F (°C)
t_2 = leaving-air dry-bulb temperature, °F (°C)
BF = bypass factor. Determine from the manufacturer's performance chart using standard air cfm.
h_s = enthalpy at t_s, Btu/lb (W/kg). Use altitude psychrometric charts, Figs. A.1 to A.4.
h_1 = entering-air enthalpy at altitude entering-air wet-bulb temperature, Btu/lb (W/kg). Use altitude psychrometric charts, Figs. A.1 to A.4.
h_2 = leaving-air enthalpy at altitude leaving air wb, Btu/lb (W/kg). Use altitude psychrometric charts, Figs. A.1 to A.4.

Maximum Standard Air Coil Face Velocities. To be consistent with established limits of standard air face velocity (to avoid problems with water carryover), multiply the maximum allowable air face velocity at sea level by the following factors to establish the maximum standard air velocity at altitude:

Elevation, ft (m)	Maximum standard air face velocity factor	Maximum actual air face velocity factor
2,500 (760)	0.95	1.05
5,000 (1500)	0.91	1.10
7,500 (2300)	0.87	1.15
10,000 (3050)	0.83	1.20

For example, the limiting standard air face velocity is 700 ft/min (3.05 m/s). The maximum allowable standard air face velocity at 7500 ft (300 m) would be 0.87 × 700 = 609 ft/min (0.87 × 3 = 2.61 m/s). The maximum allowable actual air face velocity at 7500 ft (300 m) would be 1.15 × 700 = 805 ft/min (1.15 × 3.05 = 3.507 m/s).

A.2.11 Induction Units; Central Station Air-Handling Units; Fan-Coil Units; Room Fan-Coil Units; Heating and Ventilating Units; Unit Heaters; Packaged Air-Conditioning Units

Steam Heating. Operation of the above units at altitude with a steam-heating coil requires a correction for the reduction in absolute steam pressure and a correction for the reduced air-weight flow at the same fan speed. The effect of these two corrections is combined in the values shown in the second table below.

To obtain the heating capacity at altitude for the above units, multiply the published values of capacity at the altitude cfm by the following factors:

Elevation,	Altitude capacity factors							
	Steam pressure, psig (bars)							
ft (m)	2	(0.138)	10	(0.69)	50	(3.45)	100	(6.9)
2,500 (760)	0.92	(0.0634)	0.93	(0.0611)	0.94	(0.0648)	0.94	(0.0648)
5,000 (1500)	0.85	(0.0586)	0.87	(0.0600)	0.88	(0.0607)	0.88	(0.0607)
7,500 (2300)	0.78	(0.0538)	0.80	(0.0552)	0.82	(0.0566)	0.83	(0.0572)
10,000 (3050)	0.71	(0.0490)	0.74	(0.0510)	0.77	(0.0551)	0.78	(0.0538)

A.2.12 Induction Units; Fan-Coil Units; Room Fan-Coil Units; Unit Heaters; Packaged Air-Conditioning Units.

Hot Water. Operation at altitude affects the capacity of the above units only in the reduction of air-weight flow through the unit. To obtain the heating capacity at altitude for these units, multiply the published value of capacity at the altitude cfm by the following factors:

Elevation, ft (m)	Altitude capacity factors
2,500 (760)	0.95
5,000 (1500)	0.90
7,500 (2300)	0.85
10,000 (3050)	0.81

A.2.13 Induction Units

Electric Heat. Operation at altitude does not affect the capacity of the above units since the capacity is determined only by the power consumption of the electric heating elements. It is necessary, however, to increase the minimum required actual primary air volume over that published in order to maintain the same minimum air-weight flow. Failure to compensate for the reduction in air-weight flow may trip the heater element's thermal overload.

Minimum Required Primary Air; Altitude cfm (m^3/s)

	Elevation, ft (m)			
Heater wattage	2500 (760)	5000 (1500)	7500 (2300)	10,000 (3050)
500	39 (18.4)	43 (20.3)	47 (22.2)	51 (24.1)
1000	77 (36.3)	85 (40.1)	93 (43.9)	102 (48.1)
1500	110 (51.9)	121 (57.1)	132 (62.3)	146 (68.9)
2000	148 (69.8)	162 (76.5)	178 (84.0)	197 (93.0)
2500	181 (85.4)	198 (93.4)	218 (102.9)	240 (113.3)
3000	219 (108.3)	241 (113.7)	264 (124.6)	291 (137.3)
3500	252 (118.9)	277 (143.5)	304 (143.5)	335 (158.1)
4000	290 (136.9)	319 (150.6)	350 (165.2)	385 (181.7)
4500	323 (156.4)	354 (167.1)	382 (180.3)	429 (202.5)
5000	362 (170.8)	396 (186.9)	437 (206.2)	489 (226.5)

Note: 1000 L/s = 1 m^3/s.

A.2.14 Air Terminal Units and Outlets

When the sea-level cfm has been determined for a terminal unit or outlet, the required unit discharge air volume at altitude is equal to the sea-level (standard) cfm (m^3/s) divided by the air-density ratio (see Table A.3). Where volume regulators are part of the terminal, they should be set accordingly.

For example, if a terminal unit or outlet is operating at 5000 ft (1500 m) of elevation and the design air flow required is 100 standard cfm (0.0472 m^3/s), the unit or outlet should be set to discharge an actual air volume of

$$\frac{100}{0.832} = 120 \text{ cfm} \quad \frac{0.0472}{0.832} = 0.0567 \text{ m}^3/\text{s}$$

TABLE A.3 Altitude Density for Air

Altitude	Air-density ratio	Density	Specific volume
ft	At 70°F	lb/ft^3 at 70°F	ft^3/lb at 70°F
0	1.000	0.0750	13.33
500	0.982	0.0737	13.58
1,000	0.964	0.0722	13.86
1,500	0.947	0.0710	14.10
2,000	0.930	0.0697	14.36
2,500	0.913	0.0684	14.61
3,000	0.896	0.0672	14.88
3,500	0.880	0.0660	15.17
4,000	0.864	0.0648	15.45
4,500	0.848	0.0635	15.76
5,000	0.832	0.0625	16.00
5,500	0.817	0.0613	16.32
6,000	0.801	0.0601	16.66
6,500	0.786	0.0590	16.98
7,000	0.772	0.0579	17.28
7,500	0.757	0.0567	17.66
8,000	0.743	0.0557	17.95
8,500	0.729	0.0545	18.38
9,000	0.715	0.0536	18.69
9,500	0.701	0.0526	19.05
10,000	0.687	0.0515	19.45
m	At 21.1°C	kg/m^3 at 21.1°C	m^3/kg at 21.1°C
0	1.000	1.201	0.8326
152.4	0.982	1.181	0.8467
304.8	0.964	1.157	0.8643
457.2	0.947	1.137	0.8795
609.6	0.930	1.116	0.8961
762	0.913	1.096	0.9124
914.4	0.896	1.076	0.9294
1066.7	0.880	1.067	0.9372
1219.1	0.864	1.038	0.9634
1371.5	0.848	1.017	0.9833
1523.9	0.832	1.001	0.999
1676.9	0.817	0.982	1.0183
1828.7	0.801	0.963	1.0384
1981.1	0.786	0.945	1.0582
2133.5	0.772	0.927	1.0787
2285.9	0.757	0.908	1.1013
2438.3	0.743	0.892	1.1211
2590.7	0.729	0.873	1.1455
2743.1	0.715	0.859	1.1641
2895.4	0.701	0.843	1.1862
3047.9	0.687	0.825	1.2121

This allows the correct air-weight flow to be supplied to the conditioned space.
Where sound-level guides are available for outlet selection, altitude cfm should be used.

A.2.15 Central Station Air-Handling Units; Chilled-Water Fan-Coil Units; Heating and Ventilating Units

Hot Water. When operated at constant fan speed, the above units have a reduced capacity at altitude due to the reduced air-weight flow through the units. To obtain the heating capacity at altitude, the following procedure is recommended:

1. Multiply the altitude cfm (m^3/s) by the air-density ratio to determine the standard cfm.
2. Divide the standard cfm (m^3/s) by the coil face area to determine standard air face velocity.
3. Use this face velocity with the specified gpm to determine the heat transfer index from the published curves and gpm for the coil surface required.
4. Calculate the unit capacity to verify that it equals or exceeds the specified capacity from

$$Q = \text{HTI} \times 1000(T_w - T_a) \tag{A.5}$$

where Q = unit capacity, Btu/h
HTI = heat transfer index
T_w = entering-water temperature, °F
T_a = entering-air temperature, °F

English units should be used with Eq. (A.5). To obtain metric equivalents:

$$\text{W}\ (3.413) = \text{Btu/h}$$

$$°\text{C}(1.8) + 32 = °\text{F}$$

A.2.16 Unit Heaters and Duct Furnaces; Roof-Mounted Heating Units; Furnaces

Gas-Fired. Operation at altitude affects the unit capacity because the flue passages allow a constant volume of combustion air to pass through the unit regardless of altitude. At altitude, therefore, the combustion air-weight flow through the unit is reduced, causing a reduction in unit capacity.

The recommendation of the American Gas Institute for derating gas-fired heating units for altitude operation is as follows: "Ratings need not be corrected for elevations up to 2000 ft (610 m). For elevations above 2000 ft (610 m), ratings should be reduced 4 percent for each 1000 ft (305 m) above sea level."

The gas flow rate should be adjusted to the reduced capacity by one of the following methods of control:

1. Reduce manifold pressure by changing the setting of the pressure-regulating valve. This provides only a partial degree of gas-flow control but may be suf-

ficient to provide the proper gas flow when operating at low altitudes without any change of the orifice size.

2. Change to a different-size orifice to provide proper gas flow. This change is usually necessary for operation at higher altitudes. Consult the manufacturer's engineering department for correct orifice size for the application. After changing to the correct orifice size, it may also be necessary to vary the setting of the pressure regulator to obtain the proper gas-flow rate.

For gas-fired duct furnaces, an accessory modulating gas valve is available for installation downstream from the pressure-regulating valve. This varies the gas-flow rate to the furnace at the required rate for heating.

A.2.17 Direct-Expansion Fan-Coil Units; Chilled-Water Room Fan-Coil Units; Roof-Mounted Air-Conditioning Unit; Package Air-Conditioning Unit

General. In using the altitude factors to determine unit capacity for the above units, refer first to the published ratings. If the sensible heat factor (SHF) shown for the design conditions is less than 0.95, the unit is dehumidifying at altitude. In these cases the published values of total capacity and sensible capacity should be multiplied by the factors under the 0.40 to 0.95 sensible heat factor columns shown in various of the unnumbered tables in this appendix. If the sensible heat factor shown for the design conditions is greater than 0.95, the unit still may be dehumidifying at altitude due to the higher specific humidity of altitude air for a given design dry-bulb and wet-bulb temperature. To verify whether or not the unit is dehumidifying at altitude, the adp should be calculated from

$$h_s = h_1 - \frac{h_1 - h_2}{1 - \text{BF}} \tag{A.4}$$

where h_s = enthalpy at t_s, Btu/lb. Use altitude psychrometric charts, Figs. A.1 to A.4.

h_1 = entering-air enthalpy at altitude entering-air wet-bulb temperature, Btu/lb. Use altitude psychrometric charts, Figs. A.1 to A.4.

h_2 = leaving-air enthalpy at altitude leaving-air wet-bulb temperature, Btu/lb. Use altitude psychrometric charts, Figs. A.1 to A.4.

BF = bypass factor. Determine from published data using standard air cfm.

t_s = adp at altitude − saturation temperature corresponding to h_s

The actual air-condition line should then be drawn on the altitude psychrometric chart from the entering-air conditions to the adp.

If the slope of this line is less than 0.95 m, multiply the published values of total capacity and sensible capacity by the factors under the 0.40 to 0.95 sensible heat factor columns shown in various of the unnumbered tables of this appendix. If the slope of the conditions line is in the range of 0.95 to 1.00, multiply the published values of capacity by the factors under the 0.95 to 1.00 sensible heat factor columns in tables already noted.

Direct-Expansion Fan-Coil Units. For these units, the comparison of altitude capacity with sea-level capacity is made on the basis of constant air-volume flow (varying-weight flow).

To obtain the total heat capacity at altitude for these units, multiply the published values of total heat capacity at the altitude cfm (m^3/s), entering-air wet-bulb temperature, and coil-refrigerant temperature by the following factors:

Elevation, ft (m)	Altitude total capacity factors	
	Altitude SHF 0.40–0.95	Altitude SHF 0.95–1.00
2,500 (760)	0.97	0.93
5,000 (1500)	0.95	0.86
7,500 (2300)	0.93	0.79
10,000 (3050)	0.91	0.73

To obtain the sensible heat capacity at altitude for these units, multiply the published values of sensible heat capacity at the altitude cfm (m^3/s), entering-air wet-bulb temperature, and coil-refrigerant temperature by the following factors:

Elevation, ft (m)	Altitude sensible capacity factors for altitude SHF 0.40–0.95
2,500 (760)	0.92
5,000 (1500)	0.85
7,500 (2300)	0.78
10,000 (3050)	0.71

Chilled-Water Room Fan-Coil Units. For these units, the comparison of altitude capacity with sea-level capacity is made on the basis of constant air-volume flow (varying-weight flow).

To obtain the total heat capacity at altitude for these units, multiply the published values of total heat capacity at the entering-air wet-bulb temperature, gpm (m^3/s), and entering-water temperature by the following factors:

Elevation, ft (m)	Altitude total capacity factors	
	Altitude SHF 0.40–0.95	Altitude SHF 0.95–1.00
2,500 (760)	0.97	0.93
5,000 (1500)	0.95	0.86
7,500 (2300)	0.93	0.80
10,000 (3050)	0.91	0.74

To obtain the sensible heat capacity at altitude for these units, multiply the published values of sensible heat capacity at the altitude cfm (m^3/s), entering-air wet-bulb temperature, and coil-refrigerant temperature by the following factors:

Elevation, ft (m)	Altitude sensible capacity factors for altitude SHF 0.40–0.95
2,500 (760)	0.92
5,000 (1500)	0.85
7,500 (2300)	0.78
10,000 (3050)	0.71

Direct-Expansion Roof-Mounted Air-Conditioning Unit; Direct-Expansion Package Air-Conditioning Unit. Since the above units are rated in terms of cfm, the comparison of altitude capacity with sea-level capacity is made on the basis of constant air-volume flow (varying-weight flow).

To obtain the total heat capacity at altitude for these units, multiply the published values of total heat capacity at the altitude cfm (m^3/s), entering-air wet-bulb temperature, and condensing temperature (for water-cooled units) or temperature of air entering condenser (for air-cooled units) by the following factors:

Unit with Air-Cooled Condenser

Elevation, ft (m)	Altitude total capacity factors	
	Altitude SHF 0.40–0.95	Altitude SHF 0.95–1.00
2,500 (760)	0.98	0.96
5,000 (1500)	0.96	0.92
7,500 (2300)	0.94	0.88
10,000 (3050)	0.92	0.84

Unit with Water-Cooled Condenser

Elevation, ft (m)	Altitude total capacity factors	
	Altitude SHF 0.40–0.95	Altitude SHF 0.95–1.00
2,500 (760)	0.99	0.96
5,000 (1500)	0.985	0.93
7,500 (2300)	0.98	0.90
10,000 (3050)	0.97	0.87

To obtain the sensible heat capacity at altitude for these units, multiply the published values of sensible heat capacity at the altitude cfm (m^3/s), entering-air wet-bulb temperature, and condensing temperature by the following factors:

Unit with Air-Cooled Condenser

Elevation, ft (m)	Altitude sensible capacity factors	
	Altitude SHF 0.40–0.95	Altitude SHF 0.95–1.00
2,500 (760)	0.92	0.96
5,000 (1500)	0.85	0.92
7,500 (2300)	0.78	0.88
10,000 (3050)	0.71	0.84

Unit with Water-Cooled Condenser

Elevation, ft (m)	Altitude sensible capacity factors	
	Altitude SHF 0.40–0.95	Altitude SHF 0.95–1.00
2,500 (760)	0.93	0.96
5,000 (1500)	0.87	0.93
7,500 (2300)	0.80	0.90
10,000 (3050)	0.74	0.87

For all fan-coil units having a belt-driven evaporator fan, the fan speed may be increased at altitude to circulate a larger air volume through the evaporator. This increases the unit capacity so that the altitude capacity factors shown are increased. For cases where the evaporator fan speed is increased, capacity values must be obtained from the unit manufacturer for each specific application.

A.2.18 Centrifugal Fans

Altitude affects both the fan static pressure and brake horsepower as stated by the fan law: "For constant capacity and speed, the horsepower and pressure vary directly as the air density (directly as the barometric pressure and inversely as the absolute temperature)."

Fan tables and curves are based on air at standard conditions. For operation at altitude a correction should be applied. With a given capacity [cfm (m^3/s)] and total static pressure at altitude operating conditions, the adjustments are made as follows:

1. Determine the air-density ratio (Table A.2).
2. Calculate the equivalent static pressure by dividing the given static pressure by the air-density ratio.
3. Enter the fan tables at the given capacity [cfm (m^3/s)] and the equivalent static pressure to obtain fan speed and a pseudo brake horsepower. The speed is correct as determined.

4. Multiply the pseudo brake horsepower by the air-density ratio to find the true brake horsepower at the operating conditions.
5. Determine which class of fan is required by referring to the manufacturer's fan catalog, using the equivalent static pressure and the fan outlet velocity.

In determining the required fan horsepower for equipment where values of horsepower (sea level) are published as a function of external static pressure, the adjusted horsepower must be determined from adjusted values of total static pressure (not external pressure) following the procedure given above.

Refer to Sec. A.4, "System Pressure Loss" for static pressure adjustment. Where unit static pressure drop is not available, estimate or obtain it from the manufacturer.

A.2.19 Pumps

Altitude affects the operation of pumps installed in an open system because it reduces the available net positive suction heat (NPSH). The available NPSH must always be equal to or greater than the required NPSH in order to produce flow through the pump. A lack of sufficient NPSH allows the liquid to flash into vapor inside the pump, often resulting in unstable flow and cavitation.

Refer to Carrier Corp.'s *System Design Manual*, part 8, chap. 1, or Chap. 47 of this book for the procedure to calculate the available NPSH.

Pumps installed in a closed system are selected on a standard gpm (m^3/s) versus head losses, and the effect of NPSH on flow to the pump suction need not be considered.

A.2.20 Motors

Since the effectiveness of cooling air depends on its density, motor cooling is decreased with altitude. To compensate for this decrease, it is necessary to provide additional margin for the increase in motor temperature. The decrease in cooling effectiveness may be offset by one or more of the following:

1. Elimination of the service factor
2. Provision of a higher-temperature insulation
3. Special motor design
4. Use of a larger-size motor.

Up to 3300 ft (1005 m) of elevation, all motors meet their rated temperature guarantees. Between 3300 and 9900 ft (1005 and 3020 m) of elevation, motors operate satisfactorily if not operated above nameplate rating when applied voltage is constant. If a variation in voltage of the magnitude of ±10 percent exists and the motor is to be operated above 3300 ft (1005 m), consult the motor manufacturer for recommendations. It is recommended that the motor manufacturer be consulted for all applications above 9900 ft (3020 m).

A.3 LOAD CALCULATION

The following adjustments are required to calculate loads at high altitude:

1. The design outside- and room-air moisture content must be adjusted to the new elevation by one of the following methods:
   a. If dry-bulb temperature and percent relative humidity are given, divide the specific humidity at sea level by the air-density ratio.
   b. If dry-bulb and wet-bulb temperatures are given, obtain the specific humidity at altitude from the formula

$$W_1 = W_0 + \frac{P_0 - P_1}{P_1} W_s \qquad (A.6)$$

where W_0 = specific humidity at altitude for specified dry-bulb and wet-bulb temperatures, lb/lb of dry air (kg/kg of dry air)
W_s = specific humidity at sea level and saturated wet-bulb temperature, lb/lb of dry air (kg/kg of dry air)
P_0 = barometric pressure, psia (kPa absolute)
P_1 = altitude pressure, psia (kPa absolute)

Note: The values of specific humidity given in table 1, part 1, chap. 2 of Carrier Corp.'s *System Design Manual* are already corrected for the altitude condition.

2. Solar heat-gain corrections should be made for altitude. Refer to part 1, chap. 3, table 6 of Carrier Corp.'s *System Design Manual.*
3. Because of the increased moisture content of the air, the effective sensible heat factor must be corrected. Refer to part 1, chap. 8, table 66 of Carrier Corp.'s *System Design Manual.*

A.4 SYSTEM PRESSURE LOSS

A.4.1 Air Friction through Air-Conditioning Equipment (Filters, Heating Coils, Cooling Coils, etc.)

To obtain correct values of air friction at altitude and also at temperatures other than 70°F (21.1°C), the following procedure is recommended:

1. Correct the altitude cfm (m^3/s) to standard cfm (m^3/s) by multiplying the altitude cfm by the air-density ratio (Table A.3).
2. From published data determine the value of air friction based on standard cfm flow.
3. Divide the standard air friction obtained from step 2 above by the air-density ratio to obtain the air friction at altitude and temperature.

A.4.2 Air Friction through Ducts

1. Use the altitude cfm (m^3/s) to obtain a value of air friction for standard air from duct design manuals such as *ASHRAE Handbook of Fundamentals*, chap. 12, "Duct Design."
2. To correct for nonstandard air density at altitude, multiply the value of air friction obtained from step 1 by the following factors:

Elevation, ft (m)	Temp. corr. factor F_1	Temperature, °F (°C)	Temp. corr. factor F_2
2,000 (610)	0.944	0 (−17.8)	1.120
4,000 (1220)	0.890	50 (10)	1.031
6,000 (1830)	0.838	100 (37.8)	0.957
8,000 (2440)	0.788	150 (65.6)	0.894
10,000 (3050)	0.742	200 (93.5)	0.840
		250 (121.1)	0.792
		300 (143.9)	0.749

Duct air friction (at altitude and temperature) = duct air friction (standard air) × F_1 × F_2

BIBLIOGRAPHY

Carrier Corporation: *System Design Manual*, McGraw-Hill, New York, 1960.

APPENDIX B
METRIC CONVERSION TABLES

NOTE: Most of the text in this handbook provides SI metric conversions. However, editorial factors necessitated omissions in some tables. Listed below are conversions most likely encountered in HVAC.

In the table below the first two digits of each numerical entry represents a power of 10. For example, the entry "−2 2.54" expresses 2.54×10^{-2}. Standard abbreviations are used as appropriate.

To convert from	To	Multiply by
	Area	
acre	m^2	+03 4.046
circular mil	m^2	−10 5.067
in^2	m^2	−04 6.4512
ft^2	m^2	−02 9.290
mi^2	m^2	+06 2.589
$yard^2$	m^2	−01 8.361
	Density	
lb/in^3	kg/m^2	+04 2.768
lb/ft^3	kg/m^2	+01 1.602
	Energy	
Btu (mean)	joule	+03 1.056
Btu/lb	J/kg	+03 2.324
ft · lb	joule	+00 1.356
	Energy/area time	
$Btu/(ft^2 \cdot s)$	W/m^2	+04 1.135
	Force	
lb	N	+00 4.448
oz	N	−01 2.780
	Length	
ft	m	−01 3.048
in	m	−02 2.54

To convert from	To	Multiply by
	Length (*cont.*)	
mil	m	−05 2.540
mi (U.S. statute)	m	+03 1.609
yd	m	−01 9.144
	Mass	
oz	kg	−02 2.835
lb	kg	−01 4.536
ton (short)	kg	+02 9.072
	Power	
Btu/h	W	−01 2.931
(ft · lb)/hr	W	−04 3.766
hp [550 (ft · lb)/s]	W	+02 7.457
	Pressure	
bar	Pa	+05 1.00
ft H_2O	Pa	+03 2.989
inHg	Pa	+03 3.386
lb/in^2 (psi)	Pa	+03 6.895
	Speed	
ft/s	m/s	−01 3.048
mi/h	m/s	−01 4.470
	Temperature	
°F	°C	(°F−32)/1.8
	Viscosity	
cSt	m^2/s	−06 1.00
cP	(Pa · s)	−03 1.00
	Volume	
barrel (42 gal)	m^3	−01 1.590
fluid oz (U.S.)	m^3	−05 2.597
ft^3	m^3	−02 2.832
gal (U.S. liquid)	m^3	−03 3.785
gal (U.S. liquid)	L	00 3.785
in^3	m^3	−05 1.639
ppm	mg/L (of H_2O)	00 1.000

INDEX

- A-scale, weighting factors, **49**.42 to **49**.46
- A-style watertube steam boilers, **18**.6, **18**.8
- ABMA (*see* American Boiler Manufacturers Association)
- Abrasives separators, **53**.37 to **53**.38
- Absorption:
  - in air pollution control, **36**.50
  - definition of, **36**.32
  - effects of on energy conservation, **51**.26
- Absorption chillers, **41**.1 to **41**.18
  - controls for, **41**.13, **41**.14
  - definition of, **41**.1
  - installation of, **41**.11 to **41**.12
  - location of, **41**.9 to **41**.11
  - maintenance of, **41**.14 to **14**.18
  - operation of, **41**.12, **41**.14
  - unit selection of, **41**.7, **41**.9
- Absorption cycle, description of, **41**.2 to **41**.3
- Absorption refrigeration machines, effects of altitude on, **A**.7
- Access controls, 52.59
- Accessories and components for chimneys, **30**.23, **30**.25
- Accumulation mode aerosols, **36**.7
- Accuracy requirements of building controls, **52**.60 to **52**.61
- ACGIH (*see* American Conference of Governmental Industrial Hygienists)
- ACI (*see* American Concrete Institute)
- Acid brick, use of in chimneys, **30**.58 to **30**.59
- Acid dew points:
  - effects of on chimneys and flues, **30**.55 to **30**.59
  - in incinerators, **30**.51 to **30**.52
- Acid feed in open recirculating water systems, **53**.59
- Acoustic louvers, **49**.102 to **49**.107
- Acoustic plenums:
  - built-up, **49**.117 to **49**.118
  - and quiet air-handling systems, **49**.107
- Acoustic program, **2**.6
- Acoustical isolation by floating floors, **50**.41 to **50**.45
- Active attenuators, **49**.29
- Active solar energy systems, definition of, **9**.1
- Adaptive controls, **52**.4 to **52**.5
- Adsorption:
  - in air filters, **36**.32 to **36**.34
  - in air pollution control, **36**.50
- Adsorption filters:
  - configurations of, **36**.34 to **36**.35
  - performance characteristics of, **36**.35 to **36**.40
- Aerodynamic capture by fibrous filters, **36**.24
- Aerosol concentrations in rural U.S. (table), **36**.10
- Aerosol contaminant testing, **36**.52 to **36**.53
- Aerosols:
  - concentrations of, **36**.5 to **36**.7
  - generation modes of, **36**.7
  - shape of, **36**.2
  - size distribution statistics for, **36**.2 to **36**.5
  - size range of, **36**.2
  - sources of, **36**.7
- AHU (*see* Air-handling units)
- Air:
  - density of, **34**.3, **34**.4
  - in hot-water heating systems, **47**.26 to **47**.27
  - motion of in rooms, **45**.9
  - movement of around buildings, **30**.76 to **30**.82
  - quantities of outdoors, **45**.7 to **45**.8
- Air baffles, in valance units, **17**.2
- Air changes per hour in replacement-air systems (table), **38**.18
- Air conditioning:
  - effects of on boiler loads, **18**.34 to **18**.35
  - form for recording data, **1**.6, **1**.8
- Air conditioning equipment, effects of altitude on, **A**.2 to **A**.24
- Air conditioning load program, **2**.7
- Air conditioning systems, **47**.30 to **47**.34
  - controls for, **39**.21 to **39**.22
  - use of dessicant dryers in, **39**.13 to **39**.21
- Air conditioning units, effects of altitude on, **A**.16, **A**.17, **A**.19, **A**.20 to **A**.23, **A**.25 to **A**.26
- Air contaminants, dispersion of, **36**.21 to **36**.22
- Air curtains, **16**.2, **16**.5
- Air ducts, **1**.13
- Air filters:
  - definition of, **36**.1
  - gaseous contaminants, **36**.32 to **36**.40
  - laboratory testing for, **36**.53
  - particulate, **36**.24 to **36**.32
- Air filtration, **36**.1 to **36**.58
- Air flow:
  - control of, **52**.7 to **52**.8
  - definition of, **49**.29
  - effects of altitude on, **A**.2
- Air friction, effects of altitude on, **A**.25 to **A**.26
- Air handlers, **52**.24, **52**.26 to **52**.34
- Air heaters for boilers, maintenance of, **28**.10
- Air interlocks, **22**.2
- Air makeup, **38**.1 to **38**.19
- Air movement, indoor, **37**.2
- Air Movement and Control Association (AMCA): test code for fans, **34**.2
- Air pollution control devices, **36**.1 to **36**.58
  - definition of, **36**.1
  - efficiency of, **36**.47 to **36**.48
  - performance characteristics of, **36**.47 to **36**.49, **36**.51
- Air preheaters, **26**.1 to **26**.2
- Air quality, **36**.21 to **36**.24
  - effect of dessicant dehumidifiers on, **39**.22 to **39**.23
- Air restrictions on cooling towers, **44**.18 to **44**.19
- Air separation and removal in air-conditioning systems, **47**.31
- Air springs, **50**.15, **50**.17
- Air supply in boiler rooms, **23**.1 to **23**.3
- Air systems, energy audits of, **51**.5
- Air terminal units and outlets, effects of altitude on, **A**.17 to **A**.19
- Air vents for steam heating systems, **5**.13, **5**.17
- Air washers, treatment of, **53**.65
- Air-cooled chillers (dwg.), **40**.19

Air-cooled condensers:
for chillers, **40**.26 to **40**.28
effects of altitude on, **A**.5
Air-cooled condensing units, effects of altitude on, **A**.7
Air-flow stations, **52**.32
Air-handling units (AHU), **37**.1 to **37**.38
effects of altitude on, **A**.14 to **A**.19, **A**.20
in high-rise buildings, **49**.115 to **49**.117
typical system designs for, **37**.6 to **37**.21
Air-insulated type of low-heat chimneys, **30**.12 to **30**.13
Air-moving devices, types of, **38**.11 to **38**.14
Air-quality permit applications, forms for, **45**.18
Air-to-air heat pumps, **8**.10
Air-to-water heat pumps, **8**.10
Algae, **53**.32
Algicides, **53**.62 to **53**.63
Alkali materials for incinerator chimneys, **30**.45
Alkalies in refuse, **30**.49 to **30**.50
Alkalinity, **53**.13
All-air systems, **32**.1 to **32**.12, **37**.1 to **37**.38
Alternative energy sources, effects of on energy conservation, **51**.13
Altitude, effect on steam systems, **3**.3
Altitude corrections, engineering guide for, **A**.1 to **A**.26
Altitude density for air (table), **A**.18
Altitude effect on hydronic systems, **3**.1
Altitude pressure for air (table), **A**.6
Altitude psychrometric charts, **A**.9 to **A**.12
Alumina in water, **53**.14
Ambient aerosols, recommended models for (table), **36**.11
Ambient noise levels as criteria, **49**.58, **49**.60
Ambient temperatures, in door heating, **16**.8 to **16**.9
Ambient wet-bulb temperatures, effects of on energy conservation, **51**.27
AMCA (*see* Air Movement and Control Association)
American Boiler Manufacturers Association (ABMA): *Industry Standards and Engineering Manual*, **18**.15
American Concrete Institute (ACI) standards, **30**.7, **30**.33, **30**.36
American Conference of Governmental Industrial Hygienists (ACGIH), **45**.23
American Society of Civil Engineers (ASCE): *Design and Construction of Steel Chimney Liners*, **30**.36
American Society of Heating, Refrigerating, and Air Conditioning Engineers (ASHRAE), **6**.1
American Society of Mechanical Engineers (ASME):
Boiler and Pressure Vessel Code, **18**.1
Power Boiler Code, **18**.11
Power Test Code, **18**.15
Analog controllers, **52**.13
Anemostats (dwg.), **12**.17
Angle valves, **48**.2, **48**.5, **48**.6, **48**.8
Anodes, **53**.19
Antifreeze in brine systems, **31**.8 to **31**.10
Appliance outlet temperatures, effects of on chimneys, **30**.59, **30**.60
Approach, effects of on energy conservation, **51**.27 to **51**.28
Approach assessment, alternative to, **52**.61
ASCE (*see* American Society of Civil Engineers)
ASHRAE (*see* American Society of Heating, Refrigerating, and Air Conditioning Engineers)
Ash test, **19**.3
Atmospheric burners for watertube boilers, **18**.26 to **18**.27
Atmospheric equipment for burners, **21**.13 to **21**.14
Atmospheric type of cooling towers, **44**.2
Atomic adsorption, testing with, **36**.55
Automatic blowdown and feed controllers, **53**.53
Automatic control function, degree of, **52**.61
Automatic controls for fin-tube heaters, **15**.14, **15**.17
Automatic temperature controls, effects of on energy conservation, **51**.38 to **51**.44
Auxiliary control equipment, **52**.14
Available net positive suction head (NPSHA), **47**.2
Axial fan noise, **34**.23 to **34**.24
Axial-flow air-moving devices, **38**.11
Axial-flow fans, **34**.5 to **34**.9
definition of, **34**.3
Axial-flow pumps, **47**.17 to **47**.18

Backward-curved blades for fans, **34**.10
Backward-curved impellers, **34**.10
BACT (*see* Best available control technology)
Bacteria, **53**.32
Bacteria test for open recirculating water systems, **53**.65
Bag filters, **36**.24
Balanced temperatures, definition of, **8**.9
Balanced-pressure steam traps, **5**.16 to **5**.17
Balancing valves, **48**.18 to **48**.20
isolation capability of, **48**.21
Ball valves, **48**.2, **48**.4
operation of in fires, **48**.17
Barometric pressure versus altitude, effects of on chimneys, **30**.59, **30**.60
BAS (*see* Building automation systems)
Baseboard heaters, **8**.6 to **8**.7, **15**.3, **15**.6, **15**.7
ratings for, **15**.9 to **15**.10, **15**.11, **15**.12
skin VAV cooling-only interior systems, **37**.19 to **37**.20
Bases for vibration control, **50**.19
Belt-driven compressor units, **40**.2 to **40**.3
Best available control technology (BACT), **45**.19
BFI (*see* Blade frequency increment)
Bimetal sensors, **52**.9
Bimetallic thermostatic steam traps, **5**.17 to **5**.18
Biocides, **53**.63
Biological aerosols, **36**.12
Bird screens for fans, **34**.5
Black mud, **53**.30
Blade frequency increment (BFI), **49**.24 to **49**.26
Blade pass frequency (BPF), **49**.24, **49**.26
Bleed-off of open recirculating cooling water systems, **53**.54 to **53**.59
Blow-through plenums, **49**.117
Blowdown in watertube steam boilers, **18**.14
Blowdown of boilers, **28**.6 to **28**.9, **53**.40 to **53**.43
Blower fan unit heaters (dwg.), **12**.6
Blowers, **34**.1 to **34**.40
Boiler blowdown, **28**.6 to **28**.9, **53**.40 to **53**.43
Boiler capacities, for snow melting, **10**.5
Boiler casing and insulation for watertube steam boiler furnaces, **18**.12

Boiler construction for cast-iron boilers, **18**.30
Boiler controls, **22**.5
Boiler design for watertube steam boilers, **18**.11 to **18**.15
Boiler feed pumps, inspection of, **28**.3
Boiler feedwater, use of deaerators with, **24**.1
Boiler foulants, **53**.30 to **53**.31
Boiler room logs, **28**.5
Boiler rooms, ventilation of, **23**.1 to **23**.3
Boiler scale, **53**.28
Boiler water systems, water treatment for, **53**.40 to **53**.43
Boilers, **18**.1 to **18**.47
 controls for, **52**.16 to **52**.18
 definition of, **18**.1
 design for sootblowers in, **27**.1
 dry storage of, **28**.8 to **28**.9
 economizers for, **25**.1
 effects of on energy conservation, **51**.28 to **51**.29
 electric, **29**.1 to **29**.18
  classification of, **29**.1 to **29**.2
 electrode, **29**.5 to **29**.10
 energy audits of, **51**.8 to **51**.9
 feedwater treatment for, **28**.6
 inspection of, **28**.1 to **28**.3
 maintenance of, **28**.7 to **28**.11
 number of required, **18**.33
 operation of, **28**.3 to **28**.7
 relief valves for, **28**.5 to **28**.6
 safety valves for, **28**.5 to **28**.6
 shutdowns of, **28**.7 to **28**.9
 startup of, **28**.3 to **28**.4
 system pressures for, **18**.36, **18**.37
 types of, **18**.1 to **18**.2
 water level controls for, **22**.4 to **22**.5
 wet storage of, **28**.8
Boilout, **28**.4 to **28**.5
Boosters, **20**.9
Borosilicate block, use of in chimneys, **30**.58
Bottoming-cycle facilities, definition of, **46**.1
BPF (*see* Blade pass frequency)
Breeching ductwork, **23**.2
Breechings:
 definition of, **30**.24
 for factory-built listed chimneys, **30**.24 to **30**.30
 field-installed factory-built insulated carbon-steel types, **30**.28 to **30**.29
 insulated carbon-steel factory-built types, **30**.27
 joints for, **30**.29
 for medium-heat appliance chimneys, **30**.27
 requirements for, **30**.25 to **30**.26
 steel, **30**.29
Brine, definition of, **31**.1, **47**.34
Brine circulation, **47**.34
Brine systems, **31**.8 to **31**.10
 design of, **31**.8 to **31**.10
Bubblers, testing with, **36**.53
Bucket steam traps, **5**.18 to **5**.19
Budgets, for air-handling systems, **37**.3 to **37**.4
Building appliance chimneys, **30**.56
Building automation controls, **52**.48 to **52**.59
Building automation systems (BAS), **52**.50 to **52**.53
Building configurations, effect of on air-handling units, **37**.4
Building function for control systems, **52**.60
Building heating, for variable-air-volume systems, **32**.10
Building loads, and heat pumps, **8**.8 to **8**.10
Building static pressure, for variable-air-volume systems, **32**.10
Building types and uses, **52**.60
Buildings, heat loss calculations in, **12**.10
Burner controls, inspection of, **28**.2 to **28**.3
Burners, **21**.1 to **21**.16
 for absorption chillers, **41**.17
 for watertube boilers, **18**.26 to **18**.27
 (*See also* Gas burners; Oil burners; Solid-fuel burners)
Butterfly valves, **20**.9, **48**.6, **48**.9
Bypass boilers for incinerators, **30**.51
Bypass chimneys for incinerators, **30**.51
Bypass feeders, for boilers, **53**.51, **53**.52

C-scale, weighting factors, **49**.42, **49**.45
Cabinet convectors, **15**.1 to **15**.3
 heating elements for, **15**.7 to **15**.9
Cabinet unit heaters, **13**.1 to **13**.21
 applications of, **13**.17, **13**.19 to **13**.21
 definition of, **13**.1
 for heating and cooling, **13**.3 to **13**.5
 for heating only, **13**.2 to **13**.3
 selection of, **13**.5 to **13**.17
CAD Interface with Ultra Edition Load Design and Duct Design Program, **2**.6 to **2**.7
CADD (*see* Computer-aided design and drafting)
Calcium aluminate, use of in chimneys, **30**.56 to **30**.57
Calcium carbonate:
 effect of in recirculating water systems, **53**.55
 saturation index for (table), **53**.25
Calcium salts, inverse solubility of, **53**.24
Calculation books, **1**.10
Calculations:
 for designs, **2**.1 to **2**.9
 preparation of, **1**.10 to **1**.11
Cannister-type feeders for open recirculating water systems, **53**.66
Capacities of cast-iron boilers, **18**.32
Capacity, of fans, **34**.2
Capacity controls:
 for centrifugal chillers, **42**.6 to **42**.8
 for compressors: methods of, **40**.9 to **40**.16
 for screw compressors, **43**.23 to **43**.25
Carbon dioxide, **53**.7
 removal of from feedwater, **24**.1 to **24**.7
Carbon monoxide, for burners, **21**.16
Carrier programs, **2**.7 to **2**.8, **4**.4
Carrying and guide bars for plate-and-frame heat exchangers, **14**.12
Carryover of boiler water, **53**.46
Cash flows, **51**.45 to **51**.48
Casing corrosion in HTW generators, **18**.22
Cast-iron boilers, **18**.29 to **18**.32
 design of, **18**.30 to **18**.31
 types of, **18**.29 to **18**.30
 uses of, **18**.30
Catalysis, in air pollution control, **36**.50
Catalysts, in air filters, **36**.32 to **36**.34
Catalytic incineration, in air pollution control, **36**.50
Cathode-ray tube (CRT) consoles, **51**.52
Cathodes, **53**.19
Cavitation resistance in valves, **48**.22

CCC (*see* Central communications controller)
CCU (*see* Central control unit)
Centifugal fans, definition of, **34**.3
Central communications controller (CCC), **51**.51 to **51**.53
Central constant-volume system, skin VAV interior cooling only systems, **37**.14 to **37**.15
Central control unit (CCU), **51**.51 to **51**.53
Central cooling plants, **52**.39 to **52**.41
Central equipment:
for dual-duct air systems, **32**.11
for induction unit air systems, **32**.7 to **32**.8
for multizone air systems, **32**.4 to **32**.5
for single-zone air systems, **32**.1 to **32**.3
for variable-air-volume systems, **32**.9 to **32**.10
Central heating plants, **52**.37 to **52**.38
Central hot-water systems, **8**.4
Centralized electric heating systems, **8**.4
Centralized system equipment, **52**.51
Centrifugal air-moving devices, **38**.11
Centrifugal blower fan unit heaters, **12**.4
Centrifugal chillers, **42**.1 to **42**.17
controls for, **42**.13 to **42**.14
definition of, **42**.1
effects of on energy conservation, **51**.25 to **51**.26
installation of, **42**.14 to **42**.16
maintenance of, **42**.16 to **42**.17
ratings of, **42**.11 to **42**.13
testing of, **42**.12 to **42**.13
Centrifugal fan noise, **34**.22 to **34**.23
Centrifugal fans, **34**.9 to **34**.13
effects of altitude on, **A**.23 to **A**.24
Centrifugal mist collectors of pollutants, **36**.44
Centrifugal pumps, **47**.1 to **47**.19
installation and operation of, **47**.39 to **47**.40
theory of, **47**.1 to **47**.8
Centrifugal type impellers, **34**.9 to **34**.13
CFR (*see* U.S. Code of Federal Regulations)
CGTG (*see* Combustion gas turbine generators)
Charge cycles, **52**.43
Charged aerosols, **36**.12 to **36**.13
Charging valves, for direct expansion systems, **33**.8
Check valves, for direct expansion systems, **33**.8
Chemical composition of gaseous fuels (table), **19**.4
Chemicals, effects of on unlined steel stacks, **30**.73 to **30**.74
Chemiluminescent analyzers, testing with, **36**.54
Chemisorbers, in air filters, **36**.32 to **36**.34
Chemisorption:
in air pollution control, **36**.50
definition of, **36**.32
Chilled water, definition of, **31**.1
Chilled water systems, **31**.1 to **31**.7, **31**.10 to **31**.14
applications of, **31**.1 to **31**.2
design of, **31**.2 to **31**.6
installation of, **31**.6 to **31**.7
temperatures for, **31**.5, **31**.12
water treatment for, **53**.72
Chilled water temperature controls, **42**.13 to **42**.14
Chilled water units, effects of altitude on, **A**.8 to **A**.13, **A**.21
Chiller Economics Program, **2**.6
Chiller-heater equipment, **41**.3 to **41**.5
Chiller-system, freeze protection, **40**.30
Chillers:
effects of on energy conservation, **51**.25
liquid reciprocating, **40**.17 to **40**.35
ratings for, **40**.30 to **40**.31
selection guidelines for, **40**.31 to **40**.32
types of, **40**.32 to **40**.35
Chimney systems:
balancing of, **30**.64, **30**.66
characteristics of, **30**.55 to **30**.59
multiple systems, **30**.66 to **30**.69
Chimneys:
as boiler gas outlets, **23**.2
for building heating appliances: tests for, **30**.20 to **30**.21
corrosion resistance of, **30**.3
design of, **30**.59 to **30**.82
effects of winds on, **30**.74 to **30**.82
factory-built prefabricated, **30**.1 to **30**.84
for incinerators, **30**.43 to **30**.59
NFPA standards for, **30**.2 to **30**.4
problems in incinerators, **30**.49 to **30**.51
selection chart for, **30**.8, **30**.9
sizing of, **30**.59 to **30**.71
system losses in, **30**.63 to **30**.64
UL standards for, **30**.2 to **30**.4
for wood-burning appliances, **30**.14 to **30**.15
Chlorides in water, **53**.14
Chlorine, use of as in microbicides, **53**.63
CICIND (*see* International Committee on Industrial Chimneys)
Circulation ratio of water and steam, **18**.10
Classical type precast-concrete chimneys, **30**.38 to **30**.39
Classroom air conditioners, **13**.5
Clay, **53**.29 to **53**.30
Clean-flow silencers, **49**.107
Clean rooms, **45**.9 to **45**.10
contaminant dispersions in, **36**.22
Clearances for electric boilers, **29**.16 to **29**.17
Close-coupled motors, **42**.4
Closed-circuit coolers, water treatment for, **53**.72
Closed-loop controls, definition of, **52**.2
Closed recirculating water systems, treatment of, **53**.68 to **53**.72
Closure tightness of valves, **48**.7 to **48**.15
CLT (*see* Communication link termination)
Cluster-supported chimneys, **30**.33, **30**.37
Coal:
composition of (table), **19**.5
corrosion problems of in low-heat chimneys, **30**.18
Coal-fired units, preheaters for, **26**.1
Coarse-particle aerosols, **36**.7
Codes, for HVAC systems, **1**.2, **1**.4
Coefficient of performance, **8**.8
Cogenerating facilities, definition of, **46**.1
Cogenerating systems, **46**.1 to **46**.15
fuel for, **46**.10 to **46**.11
for thermal energy, **46**.1 to **46**.9
Cogeneration systems:
operational criteria for, **46**.9 to **46**.10
prime movers for, **46**.11 to **46**.15
Coil runaround cycle (air-to-water-water-to-air), effects of on energy conservation, **52**.30
Coil Selection Programs, **2**.5
Coils (heating and cooling):
effects of on energy conservation, **51**.21
energy audits of, **51**.6

Coils, **35**.1 to **35**.14
  applications of, **35**.9 to **35**.10
  arrangement of, **35**.1 to **35**.2
  construction of, **35**.1 to **35**.2
  heat-transfer calculations for, **35**.11 to **35**.12
  for plate-and-frame heat exchangers, **14**.13
  selection of, **35**.11
  types of, **13**.2 to **13**.3, **35**.2 to **35**.9
Cold-water basins for cooling towers, **44**.20 to **44**.22
Collapse-cycle baghouses, **36**.44
Combination plenums, **49**.117, **49**.118
Combined collector-storage systems, in solar systems, **9**.3
Combined thermosiphon/refractory-lined type of low-heat chimneys, **30**.13, **30**.14, **30**.16
Combustible dusts, **36**.20
Combustible gaseous contaminants, **36**.20
Combustion, principles of, **21**.3, **21**.6
Combustion fuel-air-ratio controls for burners, **21**.13 to **21**.14
Combustion gas turbine generators (CGTG), for cogeneration systems, **46**.11 to **46**.12
Combustion problems in incinerators, **30**.48
Combustion products, **21**.6
Comfort, indoor, **37**.1 to **37**.3
Commercial and industrial hot-water boilers, **29**.4
Commercial watertube boilers, **18**.26 to **18**.27
Commercial/industrial chimneys, **30**.3 to **30**.4
Communication link termination (CLT), **51**.52
Communications equipment, speed of in fire control systems, **52**.58
Communications in building automation, **52**.52 to **52**.53
Community noise regulations, **49**.46 to **49**.49
Components of centrifugal chillers, **42**.3 to **42**.6
Components of systems, effects of on energy conservation, **51**.13
Composite structures, transmission losses of, **49**.68 to **49**.69
Compressed-air systems, energy audits of, **51**.11
Compression process in single-screw compressors, **43**.22 to **43**.23
Compression process in twin-screw compressors, **43**.3 to **43**.4
Compressor controls for centrifugal chillers, **42**.14
Compressor motors, troubleshooting of, **40**.9
Compressors:
  for direct expansion systems, **33**.5
  methods of capacity control of, **40**.9 to **40**.16
  open and hermetic: effects of altitude on, **A**.2, **A**.5
  reciprocating, **40**.1 to **40**.16
  screw, **43**.1 to **43**.25
Computer calculations:
  cooling loads, **2**.3 to **2**.8
  heating loads, **2**.3 to **2**.8
Computer programs, **2**.3 to **2**.8
  role of, **2**.1
Computer rooms, HVAC systems for, **45**.10 to **45**.12
Computer-aided design and drafting (CADD), equipment for, **45**.11
Concentration cell corrosion, **53**.19 to **53**.21
Concept design procedures, **1**.4 to **1**.10
Concept phase, **1**.2 to **1**.4
  form for recording data, **1**.5
  form for summarizing design data, **1**.3
Condensate, determining load of, **5**.23
Condensate grooving, **53**.22
Condensate return in steam heating systems, **5**.5
Condensate scale, **53**.28 to **53**.29
Condensation:
  in air pollution control, **36**.50
  cycle of, **53**.5
Condenser-coil circuiting for chillers, **40**.24 to **40**.25
Condenser components for chillers, **40**.25
Condenser controls for centrifugal chillers, **42**.14
Condenser pressure controls, for direct expansion systems, **33**.7
Condenser water circulation in air-conditioning systems, **47**.31 to **47**.34
Condenser-water heat recovery systems, effects of on energy conservation, **51**.32 to **51**.33
Condenser-water off-season cooling, effects of on energy conservation, **51**.33
Condenser-water/precooling recovery, effects of on energy conservation, **51**.34 to **51**.35
Condensers:
  for centrifugal chillers, **42**.4 to **42**.5
  for chillers, **40**.24
  for direct expansion systems, **33**.5
Conduit gate valves, operation of in fires, **48**.17
Conduit-type gate valves, **48**.2, **48**.3, **48**.16
Cone-jet deflectors for heaters (dwg.), **12**.15
Connectors:
  for listed gas appliances, **30**.5
  for listed Type L appliances, **30**.6 to **30**.7
  for vibration control, **50**.18 to **50**.19
Construction debris, **53**.31
Construction of fans, **34**.25 to **34**.31
Consumer Products Safety Commission (CPSC): wood-burning appliance chimney tests, **30**.17
Contaminants:
  damage of materials by, **36**.17 to **36**.18
  effects of, **36**.13 to **36**.20
Contractor Estimating Program, **2**.6
Control equipment:
  for hot-water systems, **6**.12
  types of, **52**.8 to **52**.16
Control loops, types of, **52**.2
Control over unit heater operation, **12**.18 to **12**.19
Control points, number of, **52**.61
Control systems:
  basics of, **52**.1 to **52**.8
  for door heating, **16**.6 to **16**.7
  economic justification of, **52**.61
  for electrode boilers, **29**.10 to **29**.14
  elements of, **52**.1 to **52**.2
  selection of, **52**.59
  for solid-fuel-fired boilers, **21**.12 to **21**.15
Controllers, **52**.12 to **52**.13
Controls:
  for absorption chillers, **41**.13, **41**.14
  for air conditioning systems, **39**.21 to **39**.22
  applications of, **52**.16 to **52**.48
  automatic: energy audits of, **51**.10 to **51**.11
  for centrifugal chillers, **42**.13 to **42**.14
  for electrode boilers, **29**.9
  for fan systems, **34**.17 to **34**.20
  for jet electrode steam boilers, **29**.14
  operating (*see* Operating controls)
  for resistance steam boilers, **29**.2
  safety (*see* Safety controls)

Controversial test standards for wood-burning appliance chimneys, **30**.17
Convection, relation to silencer performance, **49**.32
Convection heating surfaces of watertube steam boiler furnaces, **18**.12
Convection heating units, metallic heating elements, **8**.6
Convection zone of watertube steam boilers, **18**.9 to **18**.10
Convectors:
architectural enclosures for, **15**.4, **15**.12
production enclosures for, **15**.1 to **15**.3, **15**.4, **15**.5, **15**.6
ratings for, **15**.9 to **15**.10, **15**.11, **15**.12
Conventional globe-type valves, **48**.5, **48**.9
Conversion factors for chimney design, **30**.61, **30**.62
Cooldown cycle (cooling mode), energy audits of, **51**.5
Coolers, evaporator, **40**.22 to **40**.24
Cooling:
types of for screw compressors, **43**.6 to **43**.8
use of replacement air in, **38**.16 to **38**.17
Cooling coils:
dehumidifying, **35**.13 to **35**.14
selection for cabinet unit heaters, **13**.15, **13**.16, **13**.17, **13**.18
Cooling elements, selection of for valance units, **17**.6, **17**.8 to **17**.9
Cooling loads:
computer calculations of, **2**.3 to **2**.8
factors in, **2**.1 to **2**.3
manual calculations of, **2**.3
Cooling operations, with valance units, **17**.2 to **17**.4
Cooling systems, **47**.30 to **47**.35
for replacement air systems, **38**.8 to **38**.11
Cooling tower water, **47**.31 to **47**.33
average pH of (graph), **53**.56
Cooling towers, **44**.1 to **44**.33
components of, **44**.19 to **44**.24
effects of on energy conservation, **51**.27 to **51**.28
effects of variables on, **44**.11 to **44**.13
energy audits of, **51**.8
evaporative condenser treatment for, **53**.65
heat exchange in, **44**.7 to **44**.13
and Legionnaire's disease, **45**.1 to **45**.2
nature of fill, **44**.13 to **44**.17
noise of, **49**.27 to **49**.28
performance of, **44**.17 to **44**.18
size of, **44**.11 to **44**.13
types and configurations of, **44**.1 to **44**.7
water cleanliness in, **53**.38 to **53**.39
Cornice heaters, **8**.7
Corrosion, **53**.15 to **53**.22
in chimneys servicing solid-fuel appliances, **30**.18
control of environmental stabilizers for, **53**.44 to **53**.45
costs of, **53**.2 to **53**.3
effects of, **53**.15 to **53**.16
effects on low-heat chimneys, **30**.17 to **30**.18
elimination of in steam plants, **24**.1 to **24**.7
erosion, **53**.22
impingement, **53**.21 to **53**.22
stress, **53**.21
Corrosion control in absorption chillers, **41**.14, **41**.16
Corrosion inhibitors, **53**.45
in open recirculating water systems, **53**.61 to **53**.62
Corrosion resistance of chimneys, **30**.3
Corrosion-resistant materials for fans, **34**.29 to **34**.31
Corrosion testing in open recirculating water systems, **53**.61 to **53**.62
Corrosion tests for chimneys, **30**.23
Cost estimates, **1**.4 to **1**.5, **1**.10
Costs:
of centrifugal chillers, **42**.8 to **42**.11
comparison of different unit heating systems, **12**.2 to **12**.4
of deposits, **53**.3 to **53**.5
effect of on air-handling units, **37**.4
electric heating versus oil heating (graph), **8**.2, **8**.3
of fuels for gas-fired systems, **7**.7 to **7**.8
life-cycle costing, **51**.44 to **51**.50
of scale, **53**.3 to **53**.5
Counterflow cooling towers, **44**.4, **44**.32
Cove heaters, **8**.7
CPSC (*see* Consumer Products Safety Commission)
Crevice corrosion, **53**.19 to **53**.21
Critical velocity, definition of, **52**.7
Cross-flow fans, definition of, **34**.5
Crossflow cooling towers, **44**.4 to **44**.5, **44**.32
CRT (*see* Cathode-ray tube consoles)
Crustal mode aerosols, **36**.7
Cylinder unloading for compressors, **40**.10

D-style watertube steam boilers, **18**.6
Dampers:
for air flow control, **52**.7 to **52**.8
for fire and smoke, **45**.6 to **45**.7
Data control panels (DCPs), **52**.49, **52**.50
Data terminal cabinet (DTC), **51**.53
Data transmission medium (DTM), **51**.51 to **51**.53
dBA criteria, **49**.42, **49**.46
DCP (*see* Data control panels)
DDC (*see* Direct digital controls)
Deaerators, **24**.1 to **24**.7, **53**.36 to **53**.37
maintenance of, **28**.9 to **28**.10
packed column, **24**.3
pressurized, **24**.2 to **24**.3
Dealkalinizers, **53**.34 to **53**.35
Decentralized electric heating systems, **8**.2
Decibels, measurements using, **49**.5 to **49**.9
Deflection (*see* Vibration deflection)
Deflection curve, **50**.2 to **50**.3
Deflectors for louver blades in unit heaters (dwg.), **12**.18
Dehumidification in valance units, **17**.4
Dehumidifiers, **39**.7 to **39**.9
applications of, **39**.10, **39**.13
Dehumidifying cooling coils, **35**.13 to **35**.14
Delivery systems for door heating, **16**.2
Density of air, **34**.3, **34**.4
Density versus temperature, effects of on chimneys, **30**.64
Deposit corrosion, **53**.19 to **53**.21
Deposits, costs of, **53**.3 to **53**.5
Design, effects of on energy conservation, **51**.12 to **51**.22
Design calculations, **2**.1 to **2**.9
Dessicant, definition of, **39**.4
Dessicant dehumidification, **39**.1 to **39**.25
economics of, **39**.23 to **39**.24

Dessicant dehumidifiers:
- controls for, **39**.21 to **39**.22
- effect of on air quality, **39**.22 to **39**.23

Dessicant dryers, **39**.1 to **39**.25
- use of in systems for, **39**.13 to **39**.21

Dessicants, **39**.4 to **39**.7
Dew-point sensors, **52**.10
Dew points of corrosives in flue gases, **30**.47
Diffuser-type silencers, **49**.29
Diffusers for chilled water storage systems, **31**.13
Diffusional capture by fibrous filters, **36**.24 to **36**.25
Digital controllers, **52**.13
DIL (*see* Dynamic insertion losses)
Direct acting pumps, **47**.24
Direct and reverberant sound, effects of, **49**.18 to **49**.19
Direct digital controllers, comparisons with, **52**.15 to **52**.16
Direct digital controls (DDC), **52**.49, **52**.50, **52**.56 to **52**.57
Direct drive compressor units, **40**.3 to **40**.4
Direct expansion coils, **35**.5 to **35**.7
Direct expansion coolers, **40**.22 to **40**.23
Direct expansion evaporators, effects of on energy conservation, **51**.26 to **51**.27
Direct expansion systems (DX), **33**.1 to **33**.11
- design of, **33**.10 to **33**.11
- equipment for, **33**.4 to **33**.9
- operation of, **33**.1 to **33**.4
- uses of, **33**.9 to **33**.10

Direct-fired chiller-heaters, **41**.4
Direct-fired gas for door heating, **16**.6
Direct-fired natural gas heating, **38**.2 to **38**.4
Direct-forced circulation systems, in solar systems, **9**.3
Direct free cooling with cooling towers, **44**.28
Direct memory access (DMA), **51**.52
Direct sound path indoors, **49**.17
Dirt, effects of, **53**.29 to **53**.30
Dirt controls in open recirculating water systems, **53**.65
Discharge and instantaneous cooling, **52**.44
Discharge cycles, **52**.43
Discharge lines, **20**.2 to **20**.9
Discounting, **51**.46 to **51**.47
Discus-type compressors, **40**.12 to **40**.14
Displacement amplitude, **49**.2
Disposal of captured dust, **36**.49
Dissipative silencers, **49**.30 to **49**.31
Dissolved gases, **53**.7, **53**.13
Dissolved minerals in water, **53**.13 to **53**.15
Dissolved solids in water, **53**.14
Distributed systems, **52**.51 to **52**.52
Distributing steam coils, **35**.7 to **35**.9
Distribution systems, **1**.13 to **1**.14
- for chilled water, **31**.3 to **31**.4
- for solar systems, **9**.2 to **9**.5

Disturbing frequencies, **50**.2 to **50**.3
DMA (*see* Direct memory access)
Dock seals, **16**.11
Dock shelters, **16**.11
Door heaters, **16**.1 to **16**.7
- design of, **16**.8 to **16**.9
- installation of, **16**.11
- selection of, **16**.7 to **16**.11

Door heating, **16**.1 to **16**.7
- available equipment, **16**.2 to **16**.6
- characteristics of loads, **16**.1 to **16**.2

Door heating (*Cont.*):
- comparison of alternatives for, **16**.10 to **16**.11
- control systems for, **16**.6 to **16**.7
- delivery systems for, **16**.2
- load calculations for, **16**.9 to **16**.10
- worksheet for, **16**.12 to **16**.17

Door switches for door heating, **16**.6, **16**.7
Doors and seals, transmission losses in, **49**.66 to **49**.68, **49**.70 to **49**.74
Double duct (dual duct) fan systems, **52**.24, **52**.26
Double-suction centrifugal pumps, **47**.11 to **47**.12
Double-wall construction of watertube steam boiler furnaces, **18**.13 to **18**.14
Double-wall single-flue stacks, **30**.33, **30**.36
Double-wall steel stack with single flue, **30**.32 to **30**.33
Double-width, double-inlet configuration air moving devices, **38**.11 to **38**.13
Downcomer tubes in watertube steam boilers, **18**.11
Downdraft, control of, **37**.3
Drafts of chimneys, theoretical, **30**.59
Draw-through plenums, **49**.117
Drift eliminators for cooling towers, **44**.22
Drip legs, definition of, **5**.21
Drum internals (steam quality) in watertube steam boilers, **18**.14
Drums for watertube steam boilers, **18**.11
Dry storage of boilers, **28**.8 to **28**.9
DTC (*see* Data terminal cabinet)
DTM (*see* Data transmission medium)
Dual duct air systems, **32**.11
Dual duct systems, effects of on energy conservation, **51**.23
Duct break-in noise, **49**.35 to **49**.42
Duct break-out noise, **49**.35 to **49**.42
Duct Design Program, **4**.4
Duct rumble, relation to silencers, **49**.86, **49**.91
Duct silencers:
- terminology for, **49**.28 to **49**.29
- types of, **49**.29 to **49**.31

Duct sizing, **4**.1 to **4**.4
- computer programs for, **4**.3 to **4**.4

Duct velocities (table), **4**.2
Duct walls, sound transmission through, **49**.35 to **49**.42
Ducts, air, **1**.13
Ductwork:
- for dual-duct air systems, **32**.11 to **32**.12
- effects of on energy conservation, **51**.21 to **51**.22
- energy audits of, **51**.6

Ductwork attenuation, **49**.94 to **49**.95, **49**.96, **49**.97, **49**.101
Ductwork systems:
- for air systems, **32**.10 to **32**.11
- for induction unit air systems, **32**.8
- for multizone systems, **32**.5 to **32**.6
- for single-zone air systems, **32**.2, **32**.3

Duplex pumps, **47**.22
DX (*see* Packaged direct-expansion systems)
DX coils (*see* Evaporators [DX coils])
DX Piping Design Program, **3**.5
DX systems (*see* Direct expansion systems)
Dynamic collectors of pollutants, **36**.44
Dynamic insertion losses (DIL), **49**.29, **49**.31, **49**.32, **49**.78
- effects of cross section, **49**.85
- effects of silencer length, **49**.85

Dynamic insertion losses (DIL) (*Cont.*):
interaction of with self-noise, **49**.78 to **49**.79, **49**.81, **49**.82

Earthquakes, effects of in vibration isolation, **50**.33 to **50**.41
Economic justification of control systems, **52**.61
Economics of dessicant dehumidification, **39**.23 to **39**.24
Economizer systems, **52**.27
Economizers, **25**.1
effects of on energy conservation, **51**.29
maintenance of for boilers, **28**.11
for screw compressors, **43**.21
Eddy zones around buildings, **30**.76 to **30**.82
Effective projected radiant surface (EPRS), **18**.15
Efficiency:
of air pollution control devices, **36**.47 to **36**.48
of cast-iron boilers, **18**.31 to **18**.32
of filters, **36**.27
Elbow attenuation, **49**.98, **49**.101
Elbow propeller pumps, **47**.18
Electrets in air filters, **36**.27
Electric air-velocity sensors, **52**.11
Electric boilers, **29**.1 to **29**.18
classification of, **29**.1 to **29**.2
electric wiring for, **29**.17 to **29**.18
installation of, **29**.15 to **29**.18
Electric circuit design, **8**.11
Electric coils for cabinet unit heaters, **13**.3
Electric control systems, comparisons with, **52**.15
Electric controllers, **52**.12
Electric heat output for snow-melting, **10**.6 to **10**.7
Electric heat units, for replacement air systems, **38**.5, **38**.7
Electric heat tracing (*see* Heat tracing)
Electric heaters, **7**.2
additional length or width of (table), **7**.5 to **7**.6
number and size of, **7**.3 to **7**.8
overhead, **7**.2
portable, **7**.2
spacing and arrangement of, **7**.3 to **7**.6
watt density increase of, **7**.4 to **7**.5
Electric heating coils for cabinet unit heaters, **13**.15
Electric heating sections for door heating, **16**.6
Electric heating systems, **8**.1 to **8**.11
centralized or decentralized, **8**.2 to **8**.4
energy needed versus air flow and temperature rise (graph), **8**.5
specifying of, **8**.10 to **8**.11
Electric heating versus gas heating, energy cost comparison (graph), **8**.3
Electric heating versus oil heating, energy cost comparison (graph), **8**.2
Electric humidity-sensing elements, **52**.10
Electric infrared heaters, **8**.7
Electric unit heaters, **12**.7, **12**.11
types of, **8**.6
Electric wiring for electric boilers, **29**.17 to **29**.18
Electrical codes, **8**.11
for electric boilers, **29**.17
Electrical control center for liquid chillers, **40**.20
Electrical snow melting, **10**.6 to **10**.7
Electrode boilers, **29**.2, **29**.5
types of, **29**.5 to **29**.15
Electrode steam boilers (dwgs.), **29**.10 to **29**.13
Electrolytes, **53**.19
Electronic controllers, **52**.13
Electrostatic effects in filters, **36**.26
Electrostatic filters, **36**.26 to **36**.27
Electrostatic precipators, **36**.45
Element sizes in valance units, **17**.5 to **17**.6
Elevator machine rooms, **45**.2
EMCS (*see* Energy monitoring and control systems)
Emergency ventilation mode of test cells, **45**.16
Emissions Offset Policy (EOP), **45**.20
EMS (*see* Energy management strategy)
EMS (*see* Energy management systems)
Enclosure and noise partitions, design considerations for, **49**.60 to **49**.71
Enclosures for valance units, **17**.2
End reflections, effect of on noise, **49**.94, **49**.95, **49**.101
End-suction pumps:
close-coupled, **47**.10 to **47**.11
frame-mounted, **47**.10 to **47**.11
Energy analysis of absorption chillers, **41**.6 to **41**.7, **41**.8
Energy Analysis Programs, **2**.4
Energy audit, **51**.3 to **51**.12
Energy codes for air-handling units, **37**.4
Energy conservation:
definition of, **51**.1
design parameters for, **51**.3 to **51**.44
effects of on energy conservation, **51**.12 to **51**.22
in HVAC systems, **45**.2
Energy conservation practice, **51**.1 to **51**.57
Energy consumption:
in adsorption filters, **36**.40
in air-flow systems, **49**.84 to **49**.85
in air pollution control, **36**.51
of filters, **36**.31
Energy efficiency, design checklist for air-handling units, **37**.5 to **37**.6
Energy management:
of cooling towers, **44**.24 to **44**.25
software programs for, **51**.53
Energy management strategy (EMS), **52**.50
Energy management systems (EMS), **51**.50 to **51**.56
components of, **51**.52 to **51**.53
examples of, **51**.56
functions of, **51**.54 to **51**.55
optional security and fire alarm systems for, **51**.55
selection of, **51**.55 to **51**.56
Energy monitoring and control systems (EMCS), **52**.49, **52**.53
Engineering guide for altitude corrections, **A**.1 to **A**.26
Entering air temperature, **12**.13
Enthalpy, definition of, **5**.1
Enthalpy exchange visualized for cooling towers, **44**.9 to **44**.11
Entropy, definition of, **5**.1
Environmental considerations of adsorption filters, **36**.40
Environmental damage by pollutants, **36**.19
Environmental limitations, in air pollution control, **36**.51
Environmental Protection Agency (EPA), **45**.18
Environmental stabilizers, corrosion control of, **53**.44 to **53**.45

EOP (*see* Emissions Offset Policy)
EPA (*see* Environmental Protection Agency)
EPRS (*see* Effective projected radiant surface)
Equal/Friction Duct Design Program, **4**.3
Equal-friction method, **4**.1
Equal-velocity method, **4**.1
Equipment Estimating Program, **2**.8
Equipment location, **1**.11
Equipment maintenance, **45**.2 to **45**.3
Equipment noise and vibration, **45**.3 to **45**.5
Equipment selection, **1**.11
Equipment Selection Program, **2**.8
Equivalent length of pipe, determination of, **5**.6
Equivalent static pressure of fans, **34**.20, **34**.22
Erosion-corrosion, **53**.21 to **53**.22
Ethylene glycol, **9**.6
Evaporation, cycle of, **53**.5
Evaporative condensers, effects of altitude on, **A**.5
Evaporative-cooled condensers, **40**.29 to **40**.30
Evaporative cooling, **45**.5 to **45**.6
Evaporative cooling-tower controls, **52**.35 to **52**.36
Evaporator coolers, **40**.22 to **40**.24
Evaporators for centrifugal chillers, **42**.4, **42**.5
Evaporators (DX coils) for direct expansion systems, **33**.5
Excess air controls for boilers, **52**.18 to **52**.20
Exfiltration of outside air, **36**.23 to **36**.24
Exhaust and makeup:
  effects of on energy conservation, **51**.20
  energy audits of, **51**.5
Exhaust systems, **45**.17 to **45**.23
Exhaust-gas-driven chiller-heater units in cogeneration systems, **46**.5, **46**.8 to **46**.9
Expansion tanks in hot-water systems, **6**.10 to **6**.12
Expansion valves for direct expansion systems, **33**.5
Expansion vessels for HTW generators, **18**.24 to **18**.25
Explosion hazards, effects of contaminants, **36**.20
Explosive atmospheres, nit heaters for, **12**.22
External capacity control valves for compressors, **40**.10 to **40**.11
External fins, metal thermal resistance of, **35**.12
External headers for cast-iron boilers, **18**.31

Fabric collectors of pollutants, **36**.44
Facility management systems (FMS), **52**.49
Factory areas, location of unit heaters in (dwg.), **12**.23
Factory-assembled cooling towers, **44**.5 to **44**.6
Factory-built chimneys and vents, standards for, **30**.2 to **30**.4
Factory-built prefabricated vents and chimneys, **30**.1 to **30**.84
Factory-packaged liquid chillers, **40**.18
Fan coil units, effects of on energy conservation, **51**.24 to **51**.25
Fan cycling:
  for chillers, **40**.26 to **40**.27
  in cooling towers, **44**.25 to **44**.28
Fan drive mechanisms for cooling towers, **44**.22 to **44**.23
Fan guards, for chillers, **40**.26
Fan noise:
  axial, **34**.23 to **34**.24
  centrifugal, **34**.22 to **34**.23
Fan options for cooling towers, **44**.32 to **44**.33
Fan speed control for chillers, **40**.27
Fan static pressure, definition of, **34**.2
Fan System Economics, **2**.4
Fan systems, **34**.13 to **34**.20, **52**.20 to **52**.34
  capacity control for, **34**.17 to **34**.20
  with heat exchange, **34**.16
  with mass exchange, **34**.16
  matching of fans for, **34**.15
  pressure of gases in, **34**.1 to **34**.2
  relation to fans, **34**.17
Fan velocity pressure, definition of, **34**.2
Fan venturi, for chiller condensers, **40**.26
Fan-coil units for heating and cooling, **13**.3 to **13**.4
Fans, **34**.1 to **34**.40
  for air-handling systems, **37**.20 to **37**.21
  for air-handling units and modulation methods for, **37**.21 to **37**.25
  capacity of, **34**.2
  for chiller condensers, **40**.25
  construction of, **34**.25 to **34**.31
  deviation from catalog ratings, **37**.25 to **37**.28
  laws for, **34**.20, **34**.21
  laws of and sound power levels, **34**.24 to **34**.25
  location of, **34**.14 to **34**.15
  location of control sensors for, **37**.28 to **37**.33
  manufacturers' figures for sound power levels, **49**.99
  noise of, **34**.22 to **34**.24, **49**.24 to **49**.27
  power formulas for, **34**.3
  ratings of from published data, **34**.36 to **34**.40
  return air, **37**.37 to **37**.38
  selection of, **34**.31 to **34**.40, **37**.33 to **37**.37
  specifications for, **34**.31 to **34**.35
  system matching of, **34**.15
  types of, **34**.3 to **34**.13
Feed methods, for open recirculating water systems, **53**.65, **53**.68
  for water treatment, **53**.51 to **53**.54
Feedback controls, **52**.2 to **52**.5
Feedback systems for burners, **21**.15
Feedwater:
  removal of carbon dioxide from, **24**.1 to **24**.7
  removal of oxygen from, **24**.1 to **24**.7
Feedwater controls, **22**.4 to **22**.5
Feedwater treatment for boilers, **28**.6
Fenestration:
  effects of on energy conservation, **51**.14, **51**.16 to **51**.18
  energy audits of, **51**.4
Fibrous aerosols, **36**.12
Fibrous filters, **36**.24
FID (*see* Field interface devices)
Field-assembled liquid chillers, **40**.17
Field-erected cooling towers, **44**.5
Field interface devices (FID), **51**.51 to **51**.53, **52**.49
Fill in cooling towers, **44**.13 to **44**.17
Film-type fill, **44**.16 to **44**.17
Filter-driers for direct expansion systems, **33**.7
Filtered air for replacement air systems, **38**.8, **38**.9
Filters:
  combined forms of, **36**.27, **36**.28, **36**.29
  energy consumption of, **36**.31
  hazards of, **36**.31 to **36**.32
  maintenance of, **36**.31
  operating limitations of, **36**.31 to **36**.33

Filters (*Cont.*):
performance characteristics of, **36**.27, **36**.29 to **36**.32
service life of, **36**.27, **36**.29 to **36**.31
for water systems, **53**.38
Filtration, use of replacement air in, **38**.16
Fin-tube heaters:
applications of, **15**.13 to **15**.14
automatic controls for, **15**.14, **15**.17
Fin-tube radiation, heating elements for, **15**.7 to **15**.9
Fin-tube radiator units, **15**.1 to **15**.3
Fin-tube walls for watertube steam boiler furnaces, **18**.11
Final-control elements, **52**.13 to **52**.14
Fine-particle aerosols, **36**.7
Finned surfaces for plate-and-frame heat exchangers, **14**.13 to **14**.15
Fins, types of, **15**.7 to **15**.8
Fire alarms, **52**.57
Fire clay castable refractory, use of in chimneys, **30**.57
Fire-control zoning, **52**.58
Fire dampers, **45**.6 to **45**.7, **52**.29
Firebox boilers, **18**.27 to **18**.29
Fireplace chimneys, **30**.69 to **30**.71
Fires, effect of on valves, **48**.16 to **48**.18
Firetube boilers, **18**.2
Firetube heating boilers, operating pressures of, **18**.5 to **18**.6
Firing rate controls, **22**.4
controls for boilers, **52**.17 to **52**.18
Fixed-flow equipment for burners, **21**.14
Flame hazards, effects of contaminants, **36**.20
Flame ionization, testing with, **36**.55
Flame safeguards, **22**.1
controls for boilers, **52**.16 to **52**.17
Flame stability, **21**.6
Flammable atmospheres, unit heaters for, **12**.22
Flanking paths, transmission losses of, **49**.71, **49**.75
Flash point, definition of, **19**.1
Flash tanks in steam heating systems, **5**.26
Float-and-thermostatic steam traps, **5**.19 to **5**.20
Floating controls, **52**.5
Floating floors, acoustical isolation by, **50**.41 to **50**.45
Flooded coolers, **40**.23 to **40**.24
Flooded-head pressure controls for chillers, **40**.27
Floor deflection, limits to, **50**.9
Flow control equipment:
for burners, **21**.13 to **21**.14
characteristics of, **52**.5 to **52**.8
Flow control valves for fuel oil piping, **20**.9
Flow of fuel oil, **20**.2 to **20**.3
Flow rate of steam (tables), **5**.9 to **5**.10
Flow velocity, effects of on silencer attenuation, **49**.78, **49**.79, **49**.80, **49**.81
Flue gas, **18**.9
Flue gas fluid dynamic behavior in heat-recovery boilers, **18**.38, **18**.41 to **18**.42, **18**.43, **18**.44
Flue liners, deterioration of, **30**.55
FMS (*see* Facility management systems)
Foaming of boiler water, **53**.46
Forced-circulation boilers with release drums, **18**.21
Forced-circulation systems, definition of, **6**.1
Forced-convection burners for watertube boilers, **18**.27
Forced-draft design of firetube boilers, **18**.4, **18**.5
Forced-draft type of cooling towers, **44**.3, **44**.4
Forward and reverse flow, effects of on silencers, **49**.31 to **49**.35
Forward flow, **49**.29
Forward-curved blades, for fans, **34**.10
Foulants, **53**.29 to **53**.32
Foundations for pumps, **47**.45
Four-pass design of firetube boilers, **18**.3 to **18**.4
Four-pipe multiple-coil systems for fan-coil units, **13**.4 to **13**.5
Four-pipe single-coil systems for fan-coil units, **13**.4 to **13**.5
Free cooling:
of centrifugal chillers, **42**.11
in cooling towers, **44**.28 to **44**.29
Free-standing chimneys, **30**.23
Free-standing roof-mounted refractory-lined chimneys, **30**.33, **30**.36
Freeze and thaw tests for chimneys, **30**.23
Freeze protection in steam heating systems, **5**.24
Freeze-trapping, testing with, **36**.55
Frequency of sound waves, **49**.3
Fresh-air heating, **38**.17 to **38**.19
Friction factor for piping, effects of on chimneys, **30**.64, **30**.65
Friction losses, in pipe fittings (graph), **20**.4
Front and rear walls of watertube steam boiler furnaces, **18**.12
Fuel-air-ratio control systems, for burners, **21**.15
Fuel for cogeneration systems, **46**.10 to **46**.11
Fuel costs for gas-fired systems, **7**.7 to **7**.8
Fuel interlocks, **22**.1 to **22**.2
Fuel oil:
flow of, **20**.2 to **20**.3
grades of (table), **19**.2
heaters of, **20**.7 to **20**.8
for heating systems, **47**.30
pressure of, **20**.2 to **20**.3
pump suction (graphs), **20**.5, **20**.6
pumping of, **20**.3
temperature of, **20**.2 to **20**.3
viscosity of, **20**.2 to **20**.3
Fuel oil burners, preheaters for, **26**.1
Fuel oil piping, **20**.1 to **20**.9
flow control valves for, **20**.9
valves for, **20**.8 to **20**.9
Fuel oil pumps, **20**.2 to **20**.3
Fuel oil unit heaters, **12**.6 to **12**.7
Fuel oils, characteristics of, **21**.3, **21**.4 to **21**.5
Fuel system controls, **22**.3
Fuel systems for boilers, maintenance of, **28**.11
Fueled heaters, **7**.2
Fuels, **19**.1 to **19**.5
for cast-iron boilers, **18**.31
chemical composition of (table), **19**.4
control of, **21**.6 to **21**.7
effects of on energy conservation, **51**.13
heating value of, **19**.1, **19**.3
specific gravity and heating value of (table), **19**.3
Fugitive emissions, **36**.22
Functions of energy management systems, **51**.54 to **51**.55
Fungi, **53**.32
Fungicides, **53**.63
Furnace design (six-wall cooling) for watertube steam boiler furnaces, **18**.11

Furnace fan coil heating-only skin,
VAV cooling only, **37**.16 to **37**.17
Furnace floors for watertube steam boiler furnaces, **18**.11 to **18**.12
Furnace heat release for watertube steam boiler furnaces, **18**.15 to **18**.16
Furnaces:
effects of altitude on, **A**.19 to **A**.20
of watertube steam boilers, **18**.8 to **18**.9
Fusible plugs for direct expansion systems, **33**.8

Gadgets for water systems, **53**.39 to **53**.40
Galvanic corrosion, **53**.17, **53**.19
Galvanic series:
anodic (table), **53**.20
definition of, **53**.19
Gas (high-efficiency) unit heaters, **12**.6
Gas (natural or propane) unit heaters, **12**.6
Gas adsorber tubes, testing with, **36**.53 to **36**.54
Gas burners, **21**.1 to **21**.2
Gas chromatography, testing with, **36**.54
Gas compressors, **20**.9
Gas detector tubes, testing with, **36**.54
Gas-fired boilers, safety shutoff valves for, **22**.3
Gas-fired heaters, **7**.2
placement of, **7**.8 to **7**.9
Gas-fired infrared units, **7**.6 to **7**.10
Gas-fired propeller unit heaters (dwg.), **12**.6
Gas-fired systems:
design of total-heating, **7**.7
placement of thermostats for, **7**.8
Gas heating systems:
direct-fired natural gas, **38**.2 to **38**.4
indirect-fired, **38**.4
Gas heating versus electric heating, energy cost comparison (graph), **8**.3
Gas piping, **20**.9 to **20**.12
components for, **20**.11 to **20**.12
pressure regulators for, **20**.11 to **20**.12
sizing of, **20**.10 to **20**.11
typical installation (dwg.), **20**.10
Gas pressure regulators, criteria for selection, **20**.12
Gas purification equipment, **36**.1 to **36**.2
performance testing of, **36**.51 to **36**.55
Gas strainers, **20**.9
Gas train components, **20**.11 to **20**.12
Gas unit heaters, connections for (dwg.), **12**.9
Gas vents:
NFPA standards for, **30**.2 to **30**.4
UL standards for, **30**.2 to **30**.4
Gas washers, testing with, **36**.53
Gaseous contaminant air filters, **36**.32 to **36**.40
Gaseous contaminant air pollution control, equipment for, **36**.49 to **36**.51
Gaseous contaminant testing, **36**.53 to **36**.55
Gaseous contaminants:
indoor generation of (table), **36**.39
toxicity of, **36**.13 to **36**.16
Gases:
characteristics of (table), **21**.2
dissolved, **53**.7, **53**.13
Gaskets for plate-and-frame heat exchangers, **14**.10 to **14**.11
Gear pumps, **47**.24 to **47**.25
GEP (*see* Good Engineering Practice)
Globe-type valves, **48**.2, **48**.5
operation of in fires, **48**.17
Glycol cooling for screw compressors, **43**.6 to **43**.7
Glycol sludge in solar systems, **9**.9
Good Engineering Practice (GEP), **45**.22
Government Economics Program, **2**.5
Granular bed collectors of pollutants, **36**.47
Gravity hot-water systems, **6**.1

Hangers for vibration control, **50**.17 to **50**.18
Hardness of water, **53**.14
Hardware of vibration pads, effect of on load capacity (graph), **50**.10, **50**.13
Hazards of filters, **36**.31 to **36**.32
Headers:
for coil design for plate-and-frame heat exchangers, **14**.15
for shell-and-tube heat exchangers, **14**.8
Heads for plate-and-frame heat exchangers, **14**.11 to **14**.12
Heat balance for internal-combustion engines, **46**.15
Heat demand in heat tracing, calculations of, **11**.4
Heat exchange:
in cooling towers, **44**.7 to **44**.13
fan systems with, **34**.16
Heat-exchanger tubes, maintenance of, **42**.17
Heat exchangers, **14**.1 to **14**.17, **52**.38
maintenance of, **14**.16 to **14**.17
for snow melting: sizing of, **10**.6 to **10**.7
in solar systems, **9**.7
Heat input areas of watertube steam boilers, **18**.8 to **18**.9
Heat load in cooling towers, **44**.8, **44**.9
Heat losses:
in buildings (calculations of), **12**.10
from insulated metal pipes (table), **11**.5
Heat of combustion, definition of, **19**.1
Heat of sorption, definition of, **39**.6
Heat-pipe recovery systems, effects of on energy conservation, **51**.31
Heat-pump chillers, **40**.33
Heat-pump-cycle chillers, **52**.41
Heat pumps, **8**.8
and building loads, **8**.8 to **8**.10
effects of on energy conservation, **51**.32
types of, **8**.10
Heat recovery, energy audits of, **51**.9
Heat-recovery boilers, **18**.36 to **18**.42
design considerations for, **18**.38, **18**.39 to **18**.40
Heat-recovery chiller-heaters, **41**.4
Heat-recovery chillers, **40**.32
Heat-recovery equipment for cogeneration systems, **46**.11
Heat-recovery steam generator (HRSG) units, **46**.4
Heat-recovery units, for incineration of hospital waste, **30**.52 to **30**.54
Heat recovery by recirculating warm stratified air, effects of on energy conservation, **51**.31 to **51**.34
Heat recovery equipment, inspection of, **28**.3
Heat recovery of centrifugal chillers, **42**.10 to **42**.11
Heat recovery units, for replacement air, **38**.7 to **38**.8
Heat requirements, for air-handling units, **38**.8
Heat-resistant materials for fans, **34**.25 to **34**.29
Heat sources, **38**.2 to **38**.8

Heat tracing, **11**.1 to **11**.7
accessory and control equipment for, **11**.7
applications for, **11**.2
calculations of in heat demand, **11**.4
definition of, **11**.1
design for, **11**.4 to **11**.7
examples of (dwgs.), **11**.3
specifications for, **11**.4 to **11**.7
wrapping factor for (table), **11**.6
Heat-transfer calculations for coils, **35**.11 to **35**.12
Heat-transfer coefficient of inside surfaces, **35**.12 to **35**.13
Heat-transfer coefficient of outside surfaces, **35**.13
Heat-transfer coefficients (table), **12**.12
Heat transfer elements, in valance units, **17**.2
Heat-transfer media, for solar systems, **9**.6
Heat-transfer plates for heat exchangers, **14**.9 to **14**.10
Heater motors, **12**.2
Heaters:
effects of altitude on, **A**.19 to **A**.20
of fuel oil, **20**.7 to **20**.8
gas-fired: placement of, **7**.8 to **7**.9
number and size of, **7**.3 to **7**.4
thermostats for, **12**.19
Heating:
form for recording data, **1**.7, **1**.9
perimeter, **45**.8
use of replacement air in, **38**.15 to **38**.16
Heating, cooling, and ventilation selections, for cabinet unit heaters, **13**.15, **13**.17
Heating, ventilating, and air-conditioning systems (*see* HVAC systems)
Heating boilers, loads for, **18**.33 to **18**.34
Heating costs, comparison of unit heating systems, **12**.2 to **12**.4
Heating elements:
in cabinet convectors, **15**.7 to **15**.9
for fin-tube radiation, **15**.7 to **15**.9
for valance units: selection of, **17**.9 to **17**.10
Heating load program, **2**.7
Heating loads:
computer calculations of, **2**.3 to **2**.8
factors in, **2**.1 to **2**.3
manual calculations of, **2**.3
Heating loads of plastics in incinerators, **30**.48
Heating operations in valance units, **17**.5
Heating sections for door heating, **16**.4 to **16**.6
Heating surfaces for watertube steam boiler furnaces, **18**.15
Heating systems, **47**.26 to **47**.30
selection of, **8**.1 to **8**.4
Heating units, selection of, **15**.10 to **15**.13
Heating value of fuels, **19**.1 (table), **19**.3
HEPA (*see* High-efficiency particulate air)
Hermetic compressors, **40**.4 to **40**.5
Hermetic motors for centrifugal chillers, **42**.4
HGBP (*see* Hot gas bypass)
High-efficiency particulate air (HEPA), **45**.9
High-limit controls, **22**.2
High-pressure boilers, **18**.6
High-pressure butterfly valves, **48**.7, **48**.10
High-pressure pipe sizing (tables), **5**.13 to **5**.16
High-pressure process-steam boilers, treatment of, **53**.46 to **53**.47
High-rise building pressures, effects of wind on, **30**.74 to **30**.76
High-temperature detectors, **52**.12
High-temperature hot-water heating systems, water treatment for, **53**.71
High-temperature water (HTW) systems, **6**.2, **18**.16
High-velocity horizontal air stream heaters, **12**.14
High-voltage single-stage precipitators, **36**.45
Highway noise regulations, **49**.46, **49**.50
HLF (*see* Horizontal laminar flow)
Horizontal delivery unit heaters, **12**.5, **12**.13
Horizontal laminar flow (HLF), **45**.9
Horizontal multistage pumps, **47**.13 to **47**.15, **47**.16
Horizontal power pumps, **47**.22
Horsepower, estimates for pumps (table), **47**.41 to **47**.44
Hospital waste:
incineration of, **30**.52 to **30**.55
with bypass and heat-recovery systems, **30**.53
with heat-recovery but no bypass, **30**.54
with heat-recovery units, **30**.52 to **30**.53
with scrubbers and a bypass system, **30**.54
with scrubbers but no bypass chimneys, **30**.54 to **30**.55
without waste heat recovery or scrubbers, **30**.52
Hot gas bypass (HGBP), **42**.7
Hot gas bypass controls for direct expansion systems, **33**.6 to **33**.7
Hot gas bypasses for compressors, **40**.15 to **40**.17
Hot gas double risers for direct expansion systems, **33**.7
Hot gas mufflers for direct expansion systems, **33**.8
Hot water, energy audits of domestic units, **51**.11
Hot water absorption chiller units in cogeneration systems, **46**.5, **46**.6 to **46**.7
Hot water boiler systems, selection of, **18**.32 to **18**.36
Hot water boilers:
pressure drop in (graph), **29**.16
resistance, **29**.3 to **29**.5
Hot water cogeneration, systems for, **46**.1, **46**.4
Hot water coils:
for cabinet unit heaters, **13**.3
for door heating, **16**.4 to **16**.6
Hot water demands (table), **18**.35
Hot water generation controls in cogeneration systems, **46**.4
Hot water generators, **18**.16, **18**.17
Hot water heating, radiation in, **15**.1 to **15**.17
Hot water heating systems, **38**.4 to **38**.6, **47**.26 to **47**.28
Hot water selection for cabinet unit heaters, **13**.7, **13**.9 to **13**.15
Hot water systems, **6**.1 to **6**.12
central, **8**.4
classes of, **6**.1 to **6**.2
design of, **6**.2 to **6**.3
Hot water unit heaters, **12**.5 to **12**.6
connections for (dwg.), **12**.8
HRSG (*see* Heat-recovery steam generator)
HTW (*see* High-temperature water systems)
Hub ratio of fans, **34**.5
HUD site noise standards, **49**.49, **49**.53
Human hearing, range of, **49**.3
Humidification, use of replacement air in, **38**.16 to **38**.17

Humidity control, indoor, **37**.1 to **37**.2
Humidity-sensing elements, **52**.10
HVAC controls, **52**.53 to **52**.56
HVAC logging, **52**.56
HVAC monitoring, **52**.56
HVAC systems:
applications for cogenerating systems, **46**.1 to **46**.15
applications of, **45**.1 to **45**.24
characteristics of, **1**.1
codes for, **1**.2, **1**.4
design of, **1**.1 to **1**.2
energy conservation in, **45**.2
for occupancies, **45**.9 to **45**.17
regulations for, **1**.2, **1**.4
silencing applications in, **49**.106 to **49**.113
Hydraulic Institute's standards for pumps, **47**.3
Hydrochloric acid in incinerators, **30**.47
Hydrologic cycle, **53**.5 to **53**.6
Hydrolysis, definition of, **53**.6
Hydronic cabinet unit heaters, **13**.1 to **13**.21
Hydronic systems, **3**.1 to **3**.3
effects of on energy conservation, **51**.37 to **51**.38
energy audits of, **51**.9 to **51**.10
Hydronics, **15**.1
Hydrostatic pressure tests and startup for plate-and-frame heat exchangers, **14**.12 to **14**.13
Hydrostatic tests of boilers, **28**.1

I/O (*see* Input-output ports)
Impaction in filters, **36**.24
Impingement corrosion, **53**.21 to **53**.22
Impingement in filters, **36**.24
Impingers, testing with, **36**.53
In-line centrifugal type air-moving devices, **38**.11
In-line pumps, **47**.11
Incineration of general waste, chemical changes in flue gases from, **30**.46 to **30**.47
Incineration systems, temperatures for, **30**.45 to **30**.46
Incinerator operators, precautions for, **30**.50
Incinerators:
chimney problems with, **30**.49 to **30**.51
chimneys for, **30**.43 to **30**.59
design features to handle downtime of waste heat boilers, **30**.51
disruptive effects on from carbon monoxide, **30**.50 to **30**.51
for hospital waste, **30**.52 to **30**.55
increased combustion temperatures for, **30**.50
maintaining flue surface temperatures above acid dew points, **30**.51 to **30**.52
Incomplete combustion in incinerators, **30**.48
Indirect-fired gas for door heating, **16**.6
Indirect-fired gas heating systems, **38**.4
Indirect-fired oil heating systems, **38**.4
Indirect forced-circulation systems in solar systems, **9**.3, **9**.4
Indirect free cooling with cooling towers, **44**.29
Indoor air, **37**.2
Indoor air quality, **36**.22 to **36**.24
Indoor gaseous contaminants, generation of (table), **36**.39
Induced-draft type of cooling towers, **44**.3, **44**.4
Induction unit air systems, **32**.6 to **32**.8
Induction units:
effects of altitude on, **A**.17
effects of on energy conservation, **51**.25
Industrial air conditioning, settings for (tables), **39**.11, **39**.12
Industrial process dust, characteristics of (table), **36**.8
Industrial watertube boilers, **18**.6 to **18**.26
Industrial watertube HTW generators, **18**.16, **18**.18
Inertial (aerodynamic) collectors of pollutants, **36**.40, **36**.44
Infiltration:
effects of on energy conservation, **51**.18
energy audits of, **51**.4
Infiltration load for door heating, **16**.7 to **16**.8
Infiltration of outside air, **36**.23 to **36**.24
Inflation, **51**.46 to **51**.48
Infrared (radiant) snow melting, **10**.7 to **10**.8
Infrared heaters, electric, **8**.7
Infrared heating, **7**.1 to **7**.10
applications of, **7**.1
physiology of, **7**.2 to **7**.3
Infrared spectrometry, testing with, **36**.54
Inhibiting dosage for open circulating water systems, **53**.64
Input-output (I/O) ports, **51**.52
Insertion losses of sound, **49**.23
Inspection of boilers, **28**.1 to **28**.3
Installation:
of absorption chillers, **41**.11 to **41**.12
of centrifugal chillers, **42**.14 to **42**.16
of chilled water storage systems, **31**.13
of electric boilers, **29**.15 to **29**.18
of pumps, **47**.39 to **47**.46
Instantaneous cooling cycles, **52**.43
Instantaneous cycles, **52**.43
Instrumentation of centrifugal chillers, **42**.15 to **42**.16
Insulated double-wall steel stacks, **30**.32 to **30**.33
Insulated metal pipes, heat losses from (table), **11**.5
Insulated refractory-lined type of low-heat chimneys, **30**.13, **30**.14
Insulation:
of absorption chillers, **41**.12
effects of on chimney life spans, **30**.55 to **30**.56
energy audits of, **51**.4
Integrated heat recovery, **8**.8
Interception in filters, **36**.24
Interlocks for boilers, inspection of, **28**.2 to **28**.3
Intermittent pilot ignition systems, **12**.20
Internal capacity control valves, **40**.11 to **40**.12
Internal-combustion engines for cogeneration systems, **46**.12 to **46**.15
Internal floating head removable tube bundles, **14**.5 to **14**.6
Internal push-nipple boilers, **18**.31
Internal rate-of-return method, **51**.49 to **51**.50
Internally cleanable coils, **35**.3 to **35**.5
International Committee on Industrial Chimneys (CICIND): code for concrete chimneys, **30**.36, **30**.37
Inverse-square law, relation to sound, **49**.14 to **49**.15
Ion-exchange water softeners, **53**.34
Iron in water, **53**.14
Isolation efficiency chart, **50**.3
Isolation materials, **50**.9 to **50**.19
Isolation valves, **48**.1, **48**.18 to **48**.20
Isolators (*see* Vibration isolators)

Jacket-water heat recovery in cogeneration systems, **46**.14 to **46**.15
Jackson type of precast-concrete chimneys, **30**.41 to **30**.42
Jet electrode steam boilers, **29**.14 to **29**.15
Jet-induced circulation HTW generators, **18**.16 to **18**.17, **18**.19

Laboratory tests for air filters, **36**.53
LAER (*see* Lowest achievable emission rate)
Lake water for cooling systems, **47**.33 to **47**.34
Langeller index, **53**.24 to **53**.25
Large jet-induced circulation multidrum HTW generator units, **18**.19 to **18**.21
Large-scale cyclones, **36**.40, **36**.44
Laser spectrometry, testing with, **36**.54
Latent cooling capacity, **39**.3 to **39**.4
Latent heat fraction (LHF), definition of, **39**.3
Layouts on architectural drawings, **1**.13
LCC (*see* Life-cycle costing)
Lead-lag controls for absorption chillers, **41**.14, **14**.15
Leak protection, in solar systems, **9**.9
Leak tightness in absorption chillers, **41**.14, **41**.16
Legionnaire's disease and cooling towers, **45**.1 to **45**.2
LHF (*see* Latent heat fraction)
Life calculations of adsorption filters, **36**.37 to **36**.38
Life-cycle costing (LCC), **51**.44 to **51**.50
Life Cycle Costs Program, **2**.7 to **2**.8
Lighting, energy audits of, **51**.11 to **51**.12
Lighting controls, **52**.57
Limit devices, **22**.3
Lined ducts, attenuation effects of, **49**.99, **49**.100, **49**.102
Liners (*see* Flue liners)
Lint arrestors, **36**.45
Liquid aerosol, **36**.11 to **36**.12
Liquid-chiller controls, **52**.34 to **52**.35
Liquid chiller systems, reciprocating, **40**.17 to **40**.35
Liquid chillers, safety controls for, **40**.20 to **40**.22
Liquid chilling units with air-cooled condensers, effects of altitude on, **A**.7 to **A**.8
Liquid coolers, effects of altitude on, **A**.5
Liquid dessicants, use of in dehumidifiers, **39**.9, **39**.10
Liquid-expansion steam traps, **5**.18
Liquid receivers for direct expansion systems, **33**.9
Liquid refrigeration chillers, energy audits of, **51**.7 to **51**.8
Liquid scrubbers, testing with, **36**.53
Load analysis of boiler systems, **18**.32 to **18**.38
Load calculations, effects of altitude on, **A**.24 to **A**.25
Load controls for internal-combustion engines, **46**.15
Load Design Programs, **2**.4
Load Estimating for Residential Program, **2**.7
Load on the system, definition of, **8**.1
Local noise regulations, **49**.46, **49**.48, **49**.49
Locating unit heaters, **12**.20 to **12**.24
Location of equipment, **1**.11
Logs for boiler rooms, **28**.5
Louver blades on deflectors for unit heaters (dwg.), **12**.18
Louvers:
  in air pollution control equipment, **36**.44
  for cooling towers, **44**.22
  for gas-fired systems, **7**.7
Low-heat appliance chimneys:
  connectors for, **30**.17
  in current use, **30**.10 to **30**.18
Low-leakage dampers:
  effects of on energy conservation, **51**.20 to **51**.21
  energy audits of, **51**.6
Low-pressure heating boilers, **18**.5 to **18**.6
Low-pressure process-steam boilers, treatment of, **53**.46 to **53**.47
Low-pressure steam heating boilers, treatment of, **53**.46
Low-pressure steam pipe sizing (tables), **5**.8 to **5**.12
Low-temperature alcohol chillers, **40**.33 to **40**.34
Low-temperature brine chillers, **40**.33 to **40**.34
Low-temperature gas chillers, **40**.33 to **40**.34
Low-temperature glycol chillers, **40**.33 to **40**.34
Low-temperature hot-water systems, water treatment for, **53**.71 to **53**.72
Low-temperature water (LTW) systems, **6**.1 to **6**.2, **18**.16
Low-voltage two-stage precipitators, **36**.45
Low-water cutoffs, **22**.2
Lower explosive limit, definition of, **36**.20
Lower flammable limit, definition of, **36**.20
Lowest achievable emission rate (LAER), **45**.21
LTW (*see* Low-temperature water systems)
Lubricating-oil transfer, **47**.35
Lubrication systems for twin-screw compressors, **43**.4 to **43**.6

Mach numbers, relation to sound attenuation, **49**.34
Magnesium salts, inverse solubility of, **53**.24
Maintenance:
  of absorption chillers, **41**.14 to **41**.18
  of air pollution control devices, **36**.49
  of boilers, **28**.7 to **28**.11
  of centrifugal chillers, **42**.16 to **42**.17
  of equipment, **45**.2 to **45**.3
  of filters, **36**.31
  of heat exchangers, **14**.16 to **14**.17
  of oil burners, **28**.5
  of unit heating systems, **12**.3 to **12**.4
Maintenance scheduling, **52**.57
Makeup air (*see* Replacement air)
Makeup water in HTW generators, **18**.22
Manganese in water, **53**.14
Manual method for duct sizing, **4**.2 to **4**.3
Masonry chimney liner dimensions with circular equivalent, effects of on chimneys, **30**.64, **30**.65
Mass exchange, fan systems for, **34**.16
Mass flow for incinerator chimneys, effects of on chimneys, **30**.62
Mass flow input ratio, effects of on chimneys, **30**.61, **30**.62
Mass flow of combustion products, effects of on chimneys, **30**.61 to **30**.62
Mass law of sound transmission losses (TL), **49**.20
Mass spectroscopy, testing with, **36**.55

Materials and equipment for solar systems, **9**.9 to **9**.10
Materials for cooling towers, **44**.24
Maximum instantaneous demand, effects of on boiler loads, **18**.34, **18**.35
Maximum temperatures, design for, **2**.2
Mean radiant temperature (MRT), **7**.3
Mechanical components for hot-water systems, **6**.12
Mechanical-draft types of cooling towers, **44**.2 to **44**.4
Mechanical refrigeration cycle, **41**.1 to **41**.2
Mechanical requirements for electric boilers, **29**.16
Mechanical strength tests for chimneys, **30**.23
Medium-heat appliance chimneys, **30**.21 to **30**.24
Medium-high temperature (MTW) systems, **18**.16
Medium-pressure pipe sizing (tables), **5**.13 to **5**.16
Medium-temperature hot-water systems, water treatment for, **53**.71
Medium-temperature water (MTW) systems, **6**.2
Medium to low ambient controls for chillers, **40**.26
Membrane construction of watertube steam boiler furnaces, **18**.14
Membrane filters, **36**.26
Membrane tube walls for watertube steam boiler furnaces, **18**.11
Metal thermal resistance of external fins and tube walls, **35**.12
Metal-to-metal valve seals, **48**.8, **48**.13
Method of feed for closed recirculating water systems, **53**.72 to **53**.73
Metric conversion tables, **B**.1 to **B**.2
Microbicides, **53**.63 to **53**.64
Mill scale, **53**.30
Mill-type buildings, location of unit heaters in (dwg.), **12**.21
Mineral groups (table), **53**.7
Mineral scale, **53**.22 to **53**.24
Mineral solids, maximum concentration of (table), **53**.55
Minimum clearance in valance units, **17**.6
Minimum outdoor air for variable-air-volume systems, **32**.10
Minimum temperatures, design for, **2**.2
Mixed air systems, **38**.7 to **38**.8
Mixed-flow fans, definition of, **34**.3
Modified Scotch marine type boilers, **18**.2
Modified thermosiphon type of low-heat chimneys, **30**.11 to **30**.12
Modular air-handling units, **49**.107
Modulating gas shutoff valves, **20**.9
Modulating steam valves, **52**.7
Mollier diagram, definition of, **5**.1 (graph), **5**.2
Monitor-type buildings, location of unit heaters in (dwg.), **12**.23, **12**.24
Monitoring of chilled water systems, **31**.7
Motor controls for centrifugal chillers, **42**.14
Motors:
  for centrifugal chillers, **42**.4
  for chiller condensers, **40**.25 to **40**.26
  effects of altitude on, **A**.24
  heater, **12**.2
Mounting height of unit heaters, **12**.13 to **12**.15, **12**.20
MRT (*see* Mean radiant temperature)
MTW (*see* Medium-temperature water systems)
Mud, **53**.29 to **53**.30
Mud control in open recirculating water systems, **53**.61
Multiple boilers, **52**.38
Multiple chillers, **52**.39 to **52**.41
Multiple-compressors, **40**.9
Multiple flues:
  for incinerators, **30**.52
  in a structural-steel windshield, **30**.33
Multiple heat-transfer fluids for solar systems, **9**.7
Multiple units of centrifugal chillers, **42**.10
Multiplexers (MUX), **51**.53, **52**.49
Multizone air systems, **32**.4 to **32**.6
Multizone systems, effects of on energy conservation, **51**.23 to **51**.24
MUX (*see* Multiplexers)

NAAQS (*see* National Ambient Air Quality Standard)
Narrow buildings, location of unit heaters in (dwg.), **12**.22
National Ambient Air Quality Standard (NAAQS), **45**.19
National Electrical Code (NEC), requirements for electric boilers, **29**.17
National Emission Standards for Hazardous Air Pollutants (NESHAPS), **45**.19, **45**.23
National Fire Protection Association (NFPA), gas vent and chimney standards, **30**.2 to **30**.4
National Fire Protection Association (NFPA) standards, **45**.6, **45**.7 to **45**.8
National Institute for Occupational Safety and Health (NIOSH), **39**.22
Natural frequencies, **50**.2 to **50**.3
Natural gas burners, preheaters for, **26**.1
Natural gas unit heaters, **12**.6
Navy noise regulations, **49**.46, **49**.50
NC (*see* Noise criteria)
NEC (*see* National Electrical Code)
Negative pressures, effect of on unit heating systems, **12**.3
NESHAPS (*see* National Emission Standards for Hazardous Air Pollutants)
Net costs, **51**.45 to **51**.48
Net positive suction head (NPSH), **47**.2
New Source Performance Standards (NSPS), **45**.19
NFPA (*see* National Fire Protection Association)
Night charges, **52**.44
Night setback, energy audits of, **51**.5
NIOSH (*see* National Institute for Occupational Safety and Health)
Noise:
  analysis procedure for ducted systems, **49**.92 to **49**.102
  of cooling towers, **49**.27 to **49**.28
  criteria for, **49**.42 to **49**.50
  definition of, **49**.1 to **49**.2
  in equipment, **45**.3 to **45**.5
  of fans, **49**.24 to **49**.27
  in hydronic systems, **3**.2
Noise and vibration of centrifugal chillers, **42**.15
Noise control, **49**.1 to **49**.119
Noise criteria (NC):
  curves, **49**.49, **49**.50, **49**.51, **49**.53 to **49**.55
  for silencers, **49**.93 to **49**.94
Noise reduction (NR), **49**.22 to **49**.23
  effects of sound absorption on, **49**.23 to **49**.24

Noncondensing steam turbines in cogeneration systems, **46**.5, **46**.6
Nondirectional sound sources, **49**.7
Nonmetallic sealant materials, **48**.11, **48**.14
Nonmetallic seals for valves, **48**.9
Nonremovable (fixed-tubesheet) tube bundles, **14**.2 to **14**.3
NPSH (*see* Net positive suction head)
NPSHA (*see* Available net positive suction head)
NPSHR (*see* Required net positive suction head)
NR (*see* Noise reduction)
NSPS (*see* New Source Performance Standards)
Nuclei mode aerosols, **36**.7
Nuisances due to pollutants, **36**.20

Occupancies, HVAC systems for, **45**.9 to **45**.17
Occupational Safety and Health Administration (OSHA), permissible contaminant concentrations set by, **36**.13 to **36**.17
Ocean water for cooling systems, **47**.33 to **47**.34
Odors, intensities of, **36**.14 to **36**.17
Offices, HVAC systems for, **45**.12 to **45**.14
Oil burners, **21**.2 to **21**.7
  maintenance of, **28**.5
Oil heating systems, indirect-fired, **38**.4
Oil heating versus electric heating, energy cost comparison (graph), **8**.2
Oil piping, **20**.1 to **20**.9
Oil-fired unit heaters, **12**.6 to **12**.7
  connections for (dwg.), **12**.10
Once-through orifice-controlled forced-circulation boilers—drumless, **18**.21
One-pump design for HTW generators, **18**.22
Open and hermetic compressors, effects of altitude on, **A**.2, **A**.5
Open recirculating water systems, treatment of, **53**.54 to **53**.68
Open-loop controls, definition of, **52**.2
Open-type compressor units, **40**.1, **40**.2
Openings, effects of on sound transmission losses of partitions, **49**.21 to **49**.22
Operating controls, **22**.1 to **22**.5
Operating Cost Analysis Program, **2**.8
Operating limitations:
  of air pollution control devices, **36**.49
  of filters, **36**.31 to **36**.32
Operating pressure for boilers, **18**.34 to **18**.36
Operation:
  of absorption chillers, **41**.12, **41**.14
  of boilers, **28**.3 to **28**.7
  of centrifugal chillers, **42**.16
  of pumps, **47**.39 to **47**.46
Operator interface equipment, **52**.52
Operator's terminals in fire control systems, **52**.57 to **52**.58
Optional security and fire alarm systems for energy management systems, **51**.55
Organic growths, **53**.31 to **53**.32
  control of in open recirculating water systems, **53**. 62, **53**.64
Orificed HTW generators, **18**.22
OSHA (*see* Occupational Safety and Health Administration)
OSHA noise regulations, **49**.46, **49**.51, **49**.52
Outdoor air, quantities of, **45**.7 to **45**.8
Outdoor air quality, **36**.21 to **36**.22
Outdoor temperatures in door heating, **16**.9
Outside design data, form for recording, **1**.12
Overhead electric heaters, **7**.2
Oxygen, **53**.7
  for burners, **21**.16
  removal of from feedwater, **24**.1 to **24**.7
Oxygen pitting, **53**.17, **53**.18, **53**.19
Ozone, effects of, **36**.17

Packaged boilers, **18**.1 to **18**.2
  selection of, **18**.2 to **18**.3
Packaged direct-expansion (DX) systems, **52**.36 to **52**.37
Packaged firetube boilers, design of, **18**.3 to **18**.5
Packaged process chillers, **40**.34 to **40**.35
Packaged reciprocating liquid chillers systems, **40**.18 to **40**.22
Packaged steam generators, **18**.2
Packed column deaerators, **24**.3
Packed floating tubesheet removable tube bundles, **14**.4 to **14**.5
Packed-bed scrubbers, **36**.46
Packless silencers, **49**.29 to **49**.30, **49**.107
Panel joints, transmission losses in, **49**.61 to **49**.62, **49**.64 to **49**.65, **49**.66
Pant-leg air curtains, **16**.2 to **16**.3
Parallel and series operation of centrifugal pumps, **47**.8 to **47**.10
Parallel fan terminal perimeters, VAV interior systems, **37**.10 to **37**.12
Parallel-plate heat exchangers, effects of on energy conservation, **51**.30
Part load of centrifugal chillers, **42**.9 to **42**.10
Particle velocity, **49**.2
Particulate air filters, **36**.24 to **36**.32
Particulate air pollution control, equipment for, **36**.40 to **36**.44
Particulate contaminants, **36**.2 to **36**.13
  toxicity of, **36**.19
Partitions, effect of openings in on sound transmission losses, **49**.21 to **49**.22
Passive detector badges, testing with, **36**.55
Passive solar energy systems, definition of, **9**.1
Payback estimates, **1**.5, **1**.10
Payback method, **51**.49
Pendulums, frequency of, **50**.3
Penetration:
  for air pollution control devices, **36**.47 to **36**.48
  of filters, **36**.27
Perimeter heating, **45**.8
Periods of operation of cooling towers, **44**.31 to **44**.33
Permissible sound levels, **49**.99
Pesticides, **53**.62 to **53**.63
pH of saturation of calcium carbonate, **53**.25 to **53**.26
pH value, **53**.13 to **53**.14
Physical and chemical behavior of gases in heat-recovery boilers, **18**.38, **18**.41
Pickup allowances, effects of on boiler loads, **18**.34, **18**.35
Pickup load, **10**.2 to **10**.4
PID (*see* Proportional-plus-integral-plus-derivative controls)
Piezoelectric devices, testing with, **36**.55
Pipe, determination of equivalent length of, **5**.6
Pipe scale, **53**.22 to **53**.24
Pipe sizing, **3**.1 to **3**.6
  criteria for in steam heating systems, **5**.5 to **5**.7
  in hot-water systems, **6**.9 to **6**.10
  for solar systems, **9**.5

Piping, **1**.13, **20**.1 to **20**.12
for chilled water systems, **31**.5 to **31**.6
Piping arrangements:
for fin-tube heaters, **15**.14, **15**.15 to **15**.17
in steam heating systems, **5**.5
Piping layouts:
for hot-water systems, **6**.3 to **6**.6
for snow melting, **10**.4
Piping losses, effects of on boiler loads, **18**.34, **18**.35
Piping supports in steam heating systems, **5**.24 to **5**.25
Pitting, **53**.17, **53**.18, **53**.19
Plastic coatings for fan parts, **34**.30
Plastics, problems with incineration of, **30**.48
Plate-and-frame heat exchangers, **14**.8 to **14**.16
PLC (*see* Power line conditioner)
Pleated-medium filters, **36**.24
Plug valves, **48**.16
Plume concentrations, nature of, **36**.21
PNC (*see* Preferred noise criteria)
Pneumatic air-velocity sensors, **52**.11
Pneumatic control systems, comparisons with, **52**.14 to **52**.15
Pneumatic control valves, **20**.9
Pneumatic controllers, **52**.12, **52**.13
Pollutants:
control of, **21**.6
environmental damage by, **36**.19
relation to winds on chimneys, **30**.74 to **30**.82
Portable electric heaters, **7**.2
Positive-displacement pumps, **47**.19 to **47**.26
Positive-pressure heating, **38**.17 to **38**.19
Potassium silicate, use of in chimneys, **30**.57 to **30**.58
Power, energy audits of, **51**.11 to **51**.12
Power consumption of centrifugal chillers, **42**.8 to **42**.11
Power formulas for fans, **34**.3
Power line conditioner (PLC), **51**.53
Power pumps, **47**.20 to **47**.22
Power supplies for centrifugal chillers, **42**.16
Precast chimneys, **30**.37 to **30**.43
limitations and advantages of, **30**.38
Precast-concrete chimneys, types of, **30**.38 to **30**.43
Precast reinforced-concrete chimneys, **30**.33
Precast systems, review of, **30**.42 to **30**.43
Precipitation, cycle of, **53**.5 to **53**.6
Preferred noise criteria (PNC), curves of, **49**.53, **49**.56
Preheat coils, energy audits of, **51**.6
Preheaters:
air, **26**.1 to **26**.2
for coal-fired units, **26**.1
for fuel oil burners, **26**.1
for natural gas burners, **26**.1
recuperative, **26**.1 to **26**.2
regenerative, **26**.1 to **26**.2
types of, **26**.1 to **26**.2
Preliminary design phase, **1**.3 to **1**.4
Prerotation vanes (PRVs), **42**.6 to **42**.8
Present value, definition of, **51**.45
Pressure and differential pressure in valves, **48**.20 to **48**.21
Pressure conditions in steam heating systems, **5**.4 to **5**.5
Pressure drop:
in adsorption filters, **36**.40
Pressure drop (*Cont.*):
of air pollution control devices, **36**.48 to **36**.49, **36**.51
in ductwork in relation to silencers, **49**.79, **49**.80, **49**.83 to **49**.84
in filters, **36**.27, **36**.29 to **36**.31
in hot-water boilers (graph), **29**.16
in hot-water systems, **6**.6 to **6**.9
Pressure gauges for direct expansion systems, **33**.9
Pressure of fuel oil, **20**.2 to **20**.3
Pressure of gases in fan systems, **34**.1 to **34**.2
Pressure regulators for gas piping, **20**.11 to **20**.12
Pressure sensors, **52**.10 to **52**.11
Pressure vessels:
inspection of, **28**.1
for watertube steam boilers, **18**.11
Pressure-enthalpy diagram (graph), **40**.32
Pressure-reducing valves (PRV), **22**.3
in steam heating systems, **5**.26
Pressures and ratings for jet electrode steam boilers, **29**.15
Pressurized deaerators, operation of, **24**.2 to **24**.3
Pressurized or ventilated annular areas in steel stacks, **30**.33
Prime movers for cogeneration systems, **46**.11 to **46**.15
Priming of boiler water, **53**.46
Process chillers, applications for, **40**.35
Process equipment, energy audits of, **51**.10
Process loads, **45**.8
Process-steam boilers, **18**.6
Propeller fan unit heaters, **12**.4
Propeller fans, **34**.5, **34**.6
Propeller type air-moving devices, **38**.11
Propellor pumps (*see* Axial-flow pumps)
Proportional controls, **52**.3
Proportional water treatment system for steam boilers with deaerators, **53**.51, **53**.53
Proportional-plus-integral-plus derivative (PID) controls, **52**.4
Proportional-plus-rate controls, **52**.4
Proportional-plus-reset controls, **52**.4
Proportioning feed methods:
for boilers, **53**.51 to **53**.53
for open recirculating water systems, **53**.65 to **53**.68
Proportioning water treatment system for steam boilers without deaerators, **53**.51, **53**.52
PRV (*see* Pressure-reducing valves)
PRVs (*see* Prerotation vanes)
Psychrometric air properties, effects of altitude on, **A**.1 to **A**.2
Psychrometric charts, **39**.2
Psychrometric density charts, **34**.4
Psychrometrics, **39**.2 to **39**.4
Pulsed cartridge collectors of pollutants, **36**.45
Pump selection, hot-water systems, **47**.27
Pump suction, fuel oil (graphs), **20**.5 to **20**.6
Pumping considerations for solar systems, **9**.7 to **9**.8
Pumping of fuel oil, **20**.3
Pumping requirements in hot-water systems, **6**.6 to **6**.9
Pumpout units for centrifugal chillers, **42**.6
Pumps:
for absorption chillers, **41**.17
centrifugal (*see* Centrifugal pumps)
economic factors for, **47**.37

Pumps (*Cont.*):
effects of altitude on, **A**.24
energy sources for, **47**.35 to **47**.36
for fuel oil, **20**.2 to **20**.3
for heating and cooling, **47**.1 to **47**.46
horsepower estimates for (table), **47**.41 to **47**.44
installation and operation of, **47**.39 to **47**.46
intended service for, **47**.35
liquid properties of, **47**.35
performance requirements of, **47**.35
piping for, **47**.40
positive-displacement (*see* Positive-displacement pumps)
selection of, **47**.35 to **47**.39
Purge units for centrifugal chillers, **42**.5 to **42**.6
Purging of absorption chillers, **41**.16 to **41**.17
Purifier cartons, **18**.14

Quietness of unit heaters, **12**.15 to **12**.17, **12**.19

Radiant convector wall panels, **8**.7
Radiation:
for hot water heating, **15**.1 to **15**.17
for steam heating, **15**.1 to **15**.17
Radiators, **15**.1 to **15**.3
Radioactive aerosols, **36**.12
Radon, effects of, **36**.17
Rain caps on chimneys, **30**.81 to **30**.82
Rain-protection devices (dwgs.), **45**.21
Range, effects of on energy conservation, **51**.27
Ratings:
for baseboard heaters, **15**.9 to **15**.10, **15**.11, **15**.12
of centrifugal chillers, **42**.11 to **42**.13
for convectors, **15**.9 to **15**.10, **15**.11, **15**.12
for electrode boilers, **29**.9
of fans from published data, **34**.36 to **34**.40
RC (*see* Room criteria)
Reactive silencers, **49**.29
Real-time clock (RTC), **51**.52
Reciprocating compressors, **40**.1 to **40**.16
effects of on energy conservation, **51**.26
Reciprocating liquid chiller systems, **40**.17 to **40**.35
Reciprocating pumps, **47**.19 to **47**.20
Reciprocating refrigeration units, **40**.1 to **40**.35
Recirculating air curtains, **16**.3, **16**.5
Recirculation in cooling towers, **44**.17 to **44**.18
Recovery/time in door heating, **16**.9
Rectangular insulated refractory-lined type of low-heat chimneys, **30**.13 to **30**.14, **30**.15
Rectangular silencers, **49**.107, **49**.108
Recuperative preheaters, **26**.1 to **26**.2
Refraction, relation to sound attenuation, **49**.34
Refractories, effects of halogens and alkalies on, **30**.51
Refractories in chimneys, deterioration of, **30**.56 to **30**.59
Refractory in watertube steam boiler furnaces, **18**.15
Refrigerant circulation, **47**.34 to **47**.35
Refrigerant compressors for DX air-handling units, energy audits of, **51**.8
Refrigerant piping for direct expansion systems, **33**.6
Refrigerant systems, **3**.4 to **3**.5
Refrigerants:
characteristics of, **33**.2 to **33**.4

Refrigerants (*Cont.*):
for screw compressors, **43**.7 to **43**.8, **43**.10
types of for chillers, **40**.30
Refrigeration, cycles of, **42**.1, **42**.3
Refrigeration controls, **52**.34 to **52**.37
Refrigeration cycles (graph), **33**.3
typical cycles, **40**.22
Refrigeration systems, **47**.34 to **47**.35
Refrigeration units, reciprocating, **40**.1 to **40**.35
Regeneration of adsorption filters, **36**.38, **36**.40
Regenerative preheaters, **26**.1 to **26**.2
Regenerative pumps, **47**.18 to **47**.19
Regulations for HVAC systems, **1**.2, **1**.4
Relief valves, **22**.2
for boilers, **28**.5 to **28**.6
for direct expansion systems, **33**.8
for hot-water boilers, **29**.4
Replacement air, **38**.1 to **38**.19
types of, **38**.2
Replacement air systems:
applications of, **38**.14 to **38**.17
cooling systems for, **38**.8 to **38**.11
Required net positive suction head (NPSHR), **47**.2
Required sound levels, **49**.99
Residential hot-water boilers, **29**.3 to **29**.4
Residential low-heat appliance chimney, **30**.7 to **30**.10
Resistance boilers, **29**.1, **29**.3 to **29**.5
Resistance elements, **52**.10
Resistance hot-water boilers, **29**.3 to **29**.5
Resistance loss coefficients, effects of on chimneys, **30**.63, **30**.64
Resistance steam boilers, **29**.2 to **29**.3
ratings available, **29**.3
selection of, **29**.3
Return air fans, **37**.37 to **37**.38
Reverberant sound path indoors, **49**.17 to **49**.18
Reverse flow, **49**.29
Reverse-pulse baghouses, **36**.45
Reversible motors, **52**.14
Richter type of precast-concrete chimneys, **30**.39 to **30**.41, **30**.49
Riser tubes in watertube steam boilers, **18**.10 to **18**.11
Rod-and-tube elements, **52**.9
Roll filters, **36**.24
Roof-mounted tower-supported chimneys, **30**.33
Room air motion, **45**.9
for variable-air-volume systems, **32**.10
Room and terminal effects on silencers, **49**.94, **49**.101
Room criteria (RC), curves for, **49**.55, **49**.57
Room performance, transmission losses of, **49**.71, **49**.75
Room terminal units, self-noise of, **49**.107, **49**.109 to **49**.110, **49**.112 to **49**.115
Rooms, sound absorption in, **49**.71, **49**.74, **49**.76 to **49**.77
Rotary pumps, **47**.24 to **47**.26
Rotary sootblowers, **27**.2
RTC (*see* Real-time clock)
Ryznar index, **53**.25, **53**.27 to **53**.28

Safety checks for boilers, **28**.5 to **28**.6
Safety codes:
for air-handling units, **37**.4
for centrifugal chillers, **42**.16

Safety controls, **22**.1 to **22**.5
for absorption chillers, **41**.13
for liquid chillers, **40**.20 to **40**.22
Safety devices, **22**.1 to **22**.2
Safety limits of boilers, **28**.2 to **28**.3
Safety shutoff valves, **22**.3
Safety valves, **22**.2
for boilers, **28**.5 to **28**.6
for resistance steam boilers, **29**.2
Saturated steam (table), **5**.3
SCA (*see* Single compound amount)
Scale:
control of, **53**.43 to **53**.44
costs of, **53**.3 to **53**.5
Scale control in open recirculating water systems, **53**.59
Scale deposits, **53**.22 to **53**.29
Scale inhibitors in open recirculating water systems, **53**.60 to **53**.61
Scheduling of HVAC systems, **52**.53 to **52**.56
Screw compressors, **43**.1 to **43**.25
capacity control of, **43**.15 to **43**.19
effects of on energy conservation, **51**.26
performance of, **43**.8, **43**.10, **43**.11 to **43**.15
types of cooling for, **43**.6 to **43**.8
Scrubbers, **36**.45 to **36**.46
role of in hospital waste incineration, **30**.53 to **30**.55
Sealant materials for valves, **48**.8 to **48**.9, **48**.12
Sealed bellows, **52**.9
Security alarms, **52**.59
Seismic protection of resiliently mounted equipment, **50**.33 to **50**.41
Selection:
of control systems, **52**.59
of energy management systems, **51**.55 to **51**.56
of equipment, **1**.11
of fans, **34**.31 to **34**.40
of heating units, **15**.10 to **15**.13
of hot-water system pumps, **47**.27
of packaged boilers, **18**.2 to **18**.3
of pumps, **47**.35 to **47**.39
of silencers, **49**.99
of vibration isolators, **50**.19 to **50**.33
Self-noise (SN), **49**.29
of room terminal units, **49**.107, **49**.109 to **49**.110, **49**.112 to **49**.115
of silencers, **49**.31
Self-priming pumps, **47**.16 to **47**.17
Semihermetic compressors, **40**.6 to **40**.9
Semihermetic motors, protection of, **40**.7 to **40**.9
Semihermetic screw compressors, **43**.25
Sensible heat fraction (SHF), definition of, **39**.2 to **39**.3
Sensors, **52**.9 to **52**.12
Series and parallel operation of centrifugal pumps, **47**.8 to **47**.10
Series fan terminal perimeters, VAV interior systems, **37**.8 to **37**.10
Series fan terminals, skin VAV cooling only—interior, **37**.6 to **37**.8
Service life:
of adsorption filters, **36**.36 to **36**.38
of air pollution control devices, **36**.48 to **36**.49
of filters, **36**.27, **36**.29 to **36**.31
Shaker baghouses, **36**.44
Shape-factor of vibration pads (graph), **50**.12
Shell-and-coil condensers, **40**.28
Shell-and-tube condensers, **40**.28
Shell-and-tube heat exchangers, **14**.1 to **14**.2
tubes for, **14**.6 to **14**.7
SHF (*see* Sensible heat fraction)
Shock feed methods for boiler water treatment, **53**.51, **53**.52
Shutdown periods for cooling towers, **44**.30 to **44**.31
Shutoff valves for direct expansion systems, **33**.8
Shutting down of boilers, **28**.7 to **28**.9
Sight glasses for direct expansion systems, **33**.8
Signal temperatures for corrosion in incinerators, **30**.47 to **30**.48
SIL (*see* Speech interference levels)
Silencers, **49**.74, **49**.78 to **49**.92
for cooling towers and roof exhaust fans, **49**.107
effects of forward and reverse flow on self-noise type, **49**.31 to **49**.35
effects of location on residual noise levels, **49**.91 to **49**.98
NR requirements for, **49**.93 to **49**.99
pressure-drop and DIL characteristics of, **49**.85 to **49**.86, **49**.87 to **49**.90
relation to duct rumble, **49**.86, **49**.91
relation to pressure drops in ductwork, **49**.79, **49**.80, **49**.83 to **49**.84
selection of, **49**.99
(*See also* Duct silencers)
Silencing applications in HVAC systems, **49**.106 to **49**.113
Silica in water, **53**.14
Simple harmonic motion, **49**.2, **49**.3
Simplex pumps, **47**.22
Simultaneous-day charges, **52**.44
Single compound amount (SCA), **51**.46
Single duct fan systems, **52**.21
Single duct systems, effects of on energy conservation, **51**.23
Single present worth (SPW), **51**.46
Single-screw compressors, **43**.21 to **43**.25
Single-wall chimneys lined with acid-resistant borosilicate block, **30**.33, **30**.35
Single-wall insulated but unlined stacks, **30**.32
Single-wall refractory-lined free-standing chimneys, **30**.33 to **30**.34
Single-width, single-inlet configuration air-moving devices, **38**.11 to **38**.13
Single-zone constant-volume air systems, **32**.1 to **32**.2
Single-zone constant-volume air systems with reheat, **32**.2 to **32**.4
SIP (*see* State Air Quality Implementation Plan)
Sizing:
of chimneys, **30**.59 to **30**.71
of electric heating systems, **8**.10 to **8**.11
of gas piping, **20**.10 to **20**.11
for low-pressure steam pipe (tables), **5**.8 to **5**.12
for medium- and high-pressure steam pipe (tables), **5**.13 to **5**.16
of pipe in hot-water systems, **6**.9 to **6**.10
of pipe for steam heating systems, **5**.5 to **5**.7
of steam traps, **5**.22
Slaving of fans, **52**.30 to **52**.31
Sliding-vane pumps, **47**.25
Slim louvers, **49**.107
Slimicides, **53**.63 to **53**.64
Sludge deposits, **53**.22 to **53**.29
Small-scale cyclones, **36**.44

Smoke control dampers, **45**.6 to **45**.7, **52**.29
Smoke detectors, **52**.12, **52**.20, **52**.57
SN (*see* Self-noise)
Snow, characteristics of, **10**.2, **10**.4
Snow melting:
  boiler capacities for, **10**.5
  determination of loads for, **10**.2 to **10**.4
  electric heat output for, **10**.6 to **10**.7
  electrical, **10**.6 to **10**.7
  infrared (radiant), **10**.7 to **10**.8
  piping layouts for, **10**.4
  sizing of heat exchangers for, **10**.4 to **10**.5
  systems for, **10**.1 to **10**.8
Software programs for energy management, **51**.53
Solar collectors, energy audits of, **51**.11
Solar distribution systems, general design, **9**.5
Solar heat-exchange systems, water treatment for, **53**.72
Solar heating systems:
  glycol sludge in, **9**.9
  heat exchangers for, **9**.7
  heat-transfer media for, **9**.6
  leak protection in, **9**.9
  materials and equipment for, **9**.9 to **9**.10
  multiple heat-transfer fluids for, **9**.7
  pumping considerations for, **9**.7 to **9**.8
  storage subsystems for, **9**.9
Solar hot-water operating temperatures, **9**.5 to **9**.6
Solar space heating, **9**.1 to **9**.10
Solenoid valves for direct expansion systems, **33**.8
Solid and gaseous fuels for low-heat chimneys, tests for, **30**.18
Solid dessicants, use of in dehumidifiers, **39**.7 to **39**.9
Solid-fuel boilers, **18**.42 to **18**.44, **18**.47
Solid-fuel burners, **21**.7 to **21**.15
Solid-Pac insulated type of low-heat chimneys, **30**.10
Sootblowers, **27**.1 to **27**.2
  rotary, **27**.2
  types of, **27**.2
  ultrasonic, **27**.2
Sorbents, **39**.5
Sound:
  absorption of, **49**.23 to **49**.24
  definition of, **49**.1
  indoor, **37**.2 to **37**.3
  levels of, **49**.4
  mass law for transmission losses, **49**.20
  nature of, **49**.1 to **49**.4
  partial barriers to, **49**.15 to **49**.17
  propagation of indoors, **49**.17 to **49**.19
  propagation of outdoors, **49**.12 to **49**.14
  speed of in air, **49**.4 to **49**.5
  speed of in solids, **49**.5
  transmission losses of, **49**.19 to **49**.22, **49**.60 to **49**.65
Sound absorption in rooms, **49**.71, **49**.74, **49**.76 to **49**.77
Sound attenuation, **49**.94 to **49**.95, **49**.96, **49**.97, **49**.98, **49**.101
  safety factors in, **49**.99
Sound emissions of unit heaters, **12**.16
Sound measurements, standards for, **12**.17
Sound power, calculation of changes in levels of, **49**.10
Sound power levels, **49**.6 to **49**.8
  definition of, **49**.6
  determination of, **49**.9 to **49**.10
Sound power levels and fan laws, **34**.24 to **34**.25
Sound pressure, calculation of changes in levels of, **49**.10 to **49**.12
Sound pressure levels, **49**.6, **49**.8 to **49**.9
Sound transmission losses, actual versus predicted, **49**.60 to **49**.65
Sound transmission through duct walls, **49**.35 to **49**.42
Sound waves:
  frequency of, **49**.3
  wavelengths of, **49**.3
Space utilization, effect of on air-handling units, **37**.4
Spark-resistant construction for fans, **34**.31
Spec Writers Program, **2**.6
Specific gravity, of fuels (table), **19**.3
Specific speed, definition of, **47**.3
Specification Writing and Word Processing Program, **2**.8
Specifications for fans, **34**.31 to **34**.35
Speech interference levels (SIL), **49**.57, **49**.59, **49**.60
Splash type fill, **44**.15 to **44**.16
Split-branch to terminals, attenuation in, **49**.98
Split-main duct to branch ducts, attenuation of, **49**.98
Spray electrode steam boilers (*see* Jet electrode steam boilers)
Spring constant, definition of, **50**.5
Spring-return motors, **52**.14
Springs, **50**.15, **50**.17
  deflection of, **50**.3, **50**.6 to **50**.9
SPW (*see* Single present worth)
Stacks:
  as boiler gas outlets, **23**.2
  definition of, **30**.30
  freestanding individual steel stacks, **30**.31
  standards for, **30**.31
  steel, **30**.30 to **30**.33
  typical installations of, **30**.33
Standard steam coils, **35**.7
Standard water coils, **35**.3
Standards:
  American/British for sound power levels, **49**.9
  for chimneys and gas vents, **30**.2 to **30**.4
  for sound measurements, **12**.17
Startup of boilers, **28**.3 to **28**.4
Startup of pumps, **47**.45 to **47**.46
State Air Quality Implementation Plan (SIP), **45**.19
State noise regulations, **49**.46, **49**.49
Static pressure drop, **49**.29
Static regain effect in fans, **37**.26 to **37**.28
Static regain method, **4**.1
Steam, **5**.1 to **5**.26
  flow rate of (tables), **5**.9 to **5**.10
  nature of, **5**.1 to **5**.5
Steam absorption chiller units in cogeneration systems, **46**.5, **46**.6 to **46**.7
Steam boiler systems, selection of, **18**.32 to **18**.36
Steam boilers, resistance, **29**.2 to **29**.3
Steam boilers (*See also* Watertube steam boilers)
Steam coils, **35**.7 to **35**.9
  for cabinet unit heaters, **13**.2
  for door heating, **16**.4 to **16**.6
Steam flow, control of, **52**.7

Steam generation, cogeneration systems for, **46**.1, **46**.4
Steam heating, radiation in, **15**.1 to **15**.17
Steam heating systems, **5**.4 to **5**.26, **47**.28 to **47**.30
Steam/hot-water horizontal delivery unit heater (dwg.), **12**.5
Steam pipe, capacities for low-pressure systems (tables), **5**.11 to **5**.12
Steam pipe sizing (tables), **5**.6 to **5**.8
Steam plants, elimination of corrosion in, **24**.1 to **24**.7
Steam production in watertube boilers, **18**.10 to **18**.11
Steam pumps, **47**.22 to **47**.26
Steam selection for cabinet unit heaters, **13**.5 to **13**.7, **13**.8, **13**.9
Steam separation in watertube steam boilers, **18**.14
Steam separators in steam heating systems, **5**.26
Steam systems, **3**.3 to **3**.4
  effects of on energy conservation, **51**.37 to **51**.38
  energy audits of, **51**.10
Steam traps, **5**.14 to **5**.21
  location of, **5**.21 to **5**.22
  selection of, **5**.22 to **5**.23
  sizing of, **5**.22
  test valves at (dwg.), **45**.4
Steam unit heaters, **12**.5 to **12**.6
  connections for (dwg.), **12**.9
Steam unit heating systems, **38**.4 to **38**.6
Steam valves, **52**.7
Steel chimneys, deterioration of, **30**.55 to **30**.56
Steel stacks, **30**.30 to **30**.33
  chemical loading for, **30**.71 to **30**.74
Sterilizing dosage for open recirculating water systems, **53**.64
Stokers, **21**.7, **21**.9 to **21**.11
Stop valves, **48**.1
Storage subsystems for solar systems, **9**.9
Strainers:
  for direct expansion systems, **33**.9
  maintenance of, **28**.10
  for steam heating systems, **5**.25 to **5**.26
  for water systems, **53**.38
Straining in filters, **36**.26
Stratified chilled-water storage systems, **31**.10 to **31**.14
Stress corrosion, **53**.21
Strip doors, **16**.11
Structural bearings made from Du Pont Neoprene (table), **50**.11
Submerged-nozzle scrubbers, **36**.46
Submersible pumps, **47**.15 to **47**.16
Suction lines, **20**.1 to **20**.2
  sizing of, **3**.4
Suction piping for direct expansion systems, **33**.7
Suction throttling of screw compressors, **43**.15, **43**.20
Suction-cutoff unloaders for compressors, **40**.14 to **40**.15
Sulfates in water, **53**.14
Sulfur in fuel, **19**.1, **19**.3
Sulfur oxides, **53**.7
  effects of on unlined steel stacks, **30**.71 to **30**.73
Sulfuric acid in incinerators, **30**.46 to **30**.47
Summer temperatures, design for, **2**.2 to **2**.3
Sumps, **47**.45
Surface spray treatments, **53**.65
Surging, in centrifugal chillers, **42**.7
Suspended matter in water, **53**.15
Suspended particulate concentrations for major U.S. population centers (table), **36**.9
System constant, definition of, **50**.5
System controls for snow melting, **10**.8
System expansion in HTW generators, **18**.24
System interlock controls for absorption chillers, **41**.13 to **41**.14
System penetration in air pollution control, **36**.51
System pressure losses, effects of altitude on, **A**.25 to **A**.26
System pressures for boilers, **18**.36, **18**.37
System size, economics of, **52**.61
Systems, effects of on energy conservation, **51**.22 to **51**.25

Tangent tube walls for watertube steam boiler furnaces, **18**.11
Tap dryness, **18**.11 to **18**.12
Tapered plug valves, **48**.2, **48**.4, **48**.16
  operation of, **48**.10 to **48**.11
    in fires, **48**.17
Taxes, **51**.48
TDS (*see* Total dissolved solids)
Temperature control:
  of cooling towers, **44**.24 to **44**.28
  indoor, **37**.1
Temperature controls, effects of on energy conservation, **51**.38 to **51**.44
Temperature gauges for direct expansion systems, **33**.9
Temperature level of equipment using steam, effects of altitude on, **A**.2
Temperatures:
  for chilled water systems, **31**.5, **31**.12
  of fuel oil, **20**.2 to **20**.3
  for incineration systems, **30**.45 to **30**.46
  maximum, **2**.2
  minimum, **2**.2
  summer, **2**.2 to **2**.3
  of valance units and height above ground, **17**.6
  winter, **2**.2 to **2**.3
Test cells, **45**.14 to **45**.17
Test controls:
  for open recirculating water systems, **53**.65
  for water treatment, **53**.68
Testing of centrifugal chillers, **42**.12 to **42**.13
Theoretical draft per unit of chimney height, **39**.59, **30**.60
Thermal conductivity, testing with, **36**.55
Thermal incineration in air pollution control, **36**.50
Thermal insulation of centrifugal chillers, **42**.15
Thermal resistance of external fins and tube walls, **35**.12
Thermal storage, **52**.41 to **52**.44
Thermal-storage heat recovery, effects of on energy conservation, **51**.36 to **51**.37
Thermal tests for chimneys, **30**.22 to **30**.23
Thermal wheels, effects of on energy conservation, **51**.29 to **51**.30
Thermodynamic steam traps, **5**.20 to **5**.21
Thermosiphon systems, design of, **9**.2
Thermosiphon type of low-heat chimneys, **30**.11
Thermostats:
  for door heating, **16**.6, **16**.7

Thermostats (*Cont.*):
for gas-fired systems, **7**.8
for heaters, **12**.19
Thermosyphon cooling for screw compressors, **43**.7, **43**.9
Throat tile, **18**.15
Time-proportioning controls, **52**.5
Timed two-position controls, **52**.3
TL (*see* Sound, transmission losses of)
Topping-cycle facilities, definition of, **46**.1
Total dissolved solids (TDS) in analytical instruments, **53**.68 to **53**.71
Total pressure, definition of, **34**.2
Totally immersed electrode boilers, **29**.5, **29**.8, **29**.9
Tower selection for free cooling with cooling towers, **44**.29
Toxic atmospheres, location of unit heaters in, **12**.23
Toxicity:
of gaseous contaminants, **36**.14 to **36**.16
of particulate contaminants, **36**.19
TRACE II Input/Edit Program, **2**.5
Tracking-control systems, **52**.33
Trane programs, **2**.4 to **2**.7
for duct sizing, **4**.3
Transducers, **52**.13
Transformer heat recovery, effects of on energy conservation, **51**.36
Transmission values, effects of on energy conservation, **51**.14, **51**.15
Transmitted vibrations, definition of, **50**.3
Treatment of water systems, **53**.40 to **53**.73
Truncone deflectors for unit heaters (dwg.), **12**.16
Trunnion-mounted ball valves, **48**.14
Tube axial fans, **34**.7
Tube bundles:
internal floating head removable, **14**.5 to **14**.6
nonremovable (fixed-tubesheet), **14**.2 to **14**.3
packed floating tubesheet: removable, **14**.4 to **14**.5
Tube cleaning for absorption chillers, **41**.17 to **41**.18
Tube joints in heat exchangers, **14**.7 to **14**.8
Tube pitch of watertube steam boiler furnaces, **18**.12
Tube sizes and orifices in HTW generators, **18**.21 to **18**.22
Tube walls, metal thermal resistance of, **35**.12
Tube-and-tube condensers, **40**.28 to **40**.29
Tubeaxial type air-moving devices, **38**.11
Tubes:
for coil design for plate-and-frame heat exchangers, **14**.15 to **14**.16
for shell-and-tube heat exchangers, **14**.6 to **14**.7
and tube attachments for watertube steam boilers, **18**.11
Tubular pocket filters, **36**.24
Tubular units, **7**.2
Turndown ratio for electrode boilers, **29**.8
Twin-screw compressors, **43**.1 to **43**.21
design of, **43**.1 to **43**.3
Two-fan systems, **34**.15 to **34**.16
Two-pipe double-coil systems for fan-coil units, **13**.4
Two-pipe single-coil systems for fan-coil units, **13**.4
Two-position controls, **52**.2 to **52**.3
Two-pump design for HTW generators, **18**.22 to **18**.24
Two-speed compressors, **40**.10
Two-stage absorption chiller-heaters, **41**.3 to **41**.5
Type B gas vents, **30**.4 to **30**.5
Type BW gas vents, **30**.5, **30**.6
Type L venting systems, **30**.5 to **30**.6
Types of systems, energy audits of, **51**.7

U-tube removable tube bundles, **14**.3 to **14**.4
UCA (*see* Uniform compound amount)
UCR (*see* Uniform capital recovery)
UL (*see* Underwriters' Laboratories)
UL 103 equilibrium temperature test for flue gases, **30**.19 to **30**.20
UL 103 thermal shock test for flue gases, **30**.18 to **30**.19
ULC (*see* Underwriters' Laboratories of Canada)
Ultrasonic sootblowers, **27**.2
Underground breechings, **30**.29, **30**.30
Underwriters' Laboratories (UL), gas vent and chimney standards, **30**.1 to **30**.4
Underwriters' Laboratories (UL), Type HT test, **30**.16
Underwriters' Laboratories of Canada (ULC), type S629 test, **30**.16 to **30**.17
Unheated air for replacement air systems, **38**.8
Unidirectional motors, **52**.13
Uniform annual value, definition of, **51**.45
Uniform capital recovery (UCR), **51**.46
Uniform compound amount (UCA), **51**.46
Uniform present worth (UPW), **51**.46
Uniform sinking fund (USF), **51**.46
Uninsulated double-wall steel stacks, **30**.32
Unit air heaters, mounting height for, **12**.13 to **12**.15
Unit heater connections, **12**.7 to **12**.10
Unit heater operation, control over, **12**.18 to **12**.19
Unit heaters:
classification of, **12**.4 to **12**.7
definition of, **12**.1
heating systems and equipment for, **12**.1 to **12**.25
location of, **12**.20 to **12**.24
mounting height of, **12**.20
quietness of, **12**.15 to **12**.17, **12**.19
replacement versus repairs of, **12**.24 to **12**.25
selection of, **12**.11, **12**.13
sound emissions of, **12**.16
Unit heating systems:
effect of negative pressures on, **12**.3
maintenance of, **12**.3 to **12**.4
Unit ventilators, for heating, cooling, and ventilating, **13**.5
U.S. Code of Federal Regulations (CRF), **45**.18 to **45**.19
Unlined chimneys, deterioration of, **30**.55 to **30**.56
Unlined steel stacks, chemical loading for, **30**.71 to **30**.74
Updraft construction of firetube boilers, **18**.4 to **18**.5
UPW (*see* Uniform present worth)

Urethane asphalt membranes, use of in chimneys, **30**.58
USF (*see* Uniform sinking fund)
Utility consumption, audits of, **51**.4

Valance cooling, **17**.1 to **17**.4
Valance heaters, **8**.7
Valance heating, **17**.1 to **17**.2, **17**.5
Valance terminal unit, definition of, **17**.1
Valance units:
  construction of, **17**.1 to **17**.2
  cooling operations with, **17**.2, **17**.4
  design of, **17**.5 to **17**.10
  heating operations in, **17**.5
  minimum clearance in, **17**.6
  operation of, **17**.2, **17**.4
  selection of cooling elements for, **17**.6, **17**.8 to **17**.9
  selection of heating elements for, **17**.9 to **17**.10
  temperatures of in relation to height above ground, **17**.6
Valve sealing, **48**.1 to **48**.18
Valves, **48**.1 to **48**.22
  closure tightness of, **48**.7 to **48**.15
  flow characteristics of, **48**.2 to **48**.7
  for fuel oil piping, **20**.8 to **20**.9
  integrity of in case of fire, **48**.16 to **48**.18
  operation of, **48**.9 to **48**.15
  for screw compressor controls, **43**.15 to **43**.19
  sealant materials for, **48**.8 to **48**.9
Vane axial fans, **34**.7
Vaneaxial type air-moving devices, **38**.11
Varatrain (Static Regain) Duct Design Program, **4**.3
Variable-air-volume (VAV) systems, **32**.8 to **32**.11, **37**.1 to **37**.38, **45**.7, **45**.13, **52**.29
Variable-flow equipment for burners, **21**.14
Variable-immersion electrode boilers, **29**.9 to **29**.14
Variable-inlet-vane (VIV) controls for fans, **34**.18
Variable-pitch controls for fans, **34**.18 to **34**.19
Variable speed drives for screw compressors, **43**.15, **43**.18
Variable-volume double-duct perimeters, **37**.12 to **37**.14
Variable volume ratio for screw compressors, **43**.19 to **43**.20
VAV reheat perimeter systems, **37**.17 to **37**.19
VAV systems (*see* Variable-air-volume systems)
Vent silencers, **49**.107, **49**.109
Ventilation:
  of boiler rooms, **23**.1 to **23**.3
  of centrifugal chillers, **42**.15
  effects of on energy conservation, **51**.18 to **51**.20
  energy audits of, **51**.4 to **51**.5
  indoors, **36**.22 to **36**.23, **37**.2
Ventilators for electric heating systems, **8**.6
Venting in hot-water systems, **6**.10 to **6**.12
Venting systems, **30**.4 to **30**.7
  NFPA standards for, **30**.4 to **30**.7
  UL standards for, **30**.4 to **30**.7
Vents, factory-built prefabricated, **30**.1 to **30**.84
Venturi scrubbers, **36**.45 to **36**.46
Vertical delivery unit heaters, **12**.14, **12**.15
Vertical laminar flow (VLF), **45**.9, **45**.10
Vertical multistage pumps, **47**.12 to **47**.13
Vertical turbine pumps, **47**.15
Vessel design codes for resistance steam boilers, **29**.2 to **29**.3
Vestibules, **16**.11
Vibration, **50**.1 to **50**.45
  in equipment, **45**.3 to **45**.5
  theory of, **50**.1 to **50**.4
Vibration control:
  applications of, **50**.4 to **50**.19
  theory versus practice, **50**.3
Vibration deflection, **50**.6
Vibration isolation efficiency, definition of, **50**.1
Vibration isolators, **50**.1 to **50**.45
  seismic protection of, **50**.33 to **50**.41
  seismic specifications for, **50**.38 to **50**.41
  selection of, **50**.19 to **50**.33
  specifications for, **50**.19 to **50**.33
Vibration pads, **50**.9, **50**.15 to **50**.16
Viscosity of fuel oil, **20**.2 to **20**.3
Viscosity of oil, **19**.3
Visibility problems due to aerosols, **36**.19 to **36**.20
VIV (*see* Variable-inlet-vane controls)
VLF (*see* Vertical laminar flow)
Voltage classifications for electrode boilers, **29**.2

Wall panels, radiant convectors, **8**.7
Wall-fin elements, ratings of (table), **6**.3
Warehouses, location of unit heaters in (dwg.), **12**.22
Warm-air systems, **8**.4 to **8**.6
Warmup cycle (heating mode), energy audits of, **51**.5
Waste heat:
  effects of on energy conservation, **51**.29 to **51**.38
  energy audits of, **51**.9
Waste-heat boilers, effects of on energy conservation, **51**.29
Water:
  analyses of (tables), **53**.8 to **53**.12
  dissolved minerals in, **53**.13 to **53**.15
  for electrode boilers, **29**.5, **29**.6, **29**.7
  reactions with minerals, **53**.6
Water analysis for boilers, **53**.40 to **53**.43, **53**.48
Water chemistry, **53**.5 to **53**.15
Water circuits, types of, **47**.27 to **47**.28, **47**.29
Water coils, **35**.2 to **35**.5
Water conditioning, **53**.1 to **53**.73
  pretreatment equipment for, **53**.32 to **53**.40
Water cooling for screw compressors, **43**.6 to **43**.7
Water damage in steam heating systems, **5**.23 to **5**.24
Water drainback systems, for solar systems, **9**.6 to **9**.7
Water flow, control of, **52**.6 to **52**.7
Water flow rate in cooling towers, **44**.8, **44**.9
Water hammering, **5**.23 to **5**.24
Water impurities, **53**.6 to **53**.7
Water level controls for boilers, **22**.4 to **22**.5
Water Piping Design Program, **3**.2 to **3**.3
Water quality for electric boilers, **29**.18
Water requirements for jet electrode steam boilers, **29**.14 to **29**.15
Water softeners, **53**.33 to **53**.34
  in open recirculating water systems, **53**.59

Water sources (table), **53**.7
Water systems:
  gadgets for, **53**.39 to **53**.40
  treatment of, **53**.40 to **53**.72
Water treatment:
  for absorption chillers, **41**.17
  for boiler water systems, **53**.40 to **53**.53
  for condensers, **40**.30
  need for, **53**.1 to **53**.5
  in steam heating systems, **5**.24
  test controls of, **53**.47 to **53**.50
Water valves, **52**.6
Water-cooled chillers (dwg.), **40**.19
Water-cooled condensers, **40**.28 to **40**.29
Water-cooled condensing units, effects of altitude on, **A**.5
Water-distribution controls, **52**.44 to **52**.48
Water-flow quantity in HTW generators, **18**.25 to **18**.26
Water-to-water heat pumps, **8**.10
Water-to-water/water-to-air heat recovery, effects of on energy conservation, **51**.35 to **51**.36
Watertube boilers:
  commercial, **18**.26 to **18**.27
  for high-temperature water systems, **18**.16
  industrial, **18**.6 to **18**.26
Watertube steam boilers:
  principles of operation of, **18**.8 to **18**.11
  types of, **18**.6, **18**.8
  uses of, **18**.6, **18**.7
Watt density:
  for electric heaters (table), **7**.4
  increase of, **7**.4 to **7**.5
Wavelengths of sound waves, **49**.3
Wedge gate valves, operation of in fires, **48**.17
Wedge-seal gate valves, **48**.2, **48**.5
Wedge-type gate valves, **48**.15
Well water for cooling systems, **47**.33 to **47**.34
Wet scrubbers, **36**.45
Wet storage of boilers, **28**.8
Wet-base boilers, **18**.31
Wet-bulb temperatures:
  in cooling towers, **44**.9
  selection of for cooling towers, **44**.19
Wet-pit pumps, **47**.13, **47**.14
White noise, definition of, **45**.4
Windows and seals, transmission losses in, **49**.65 to **49**.66, **49**.67, **49**.68, **49**.69
Winds:
  effects of on cooling towers, **44**.18
  effects of on high-rise building pressures, **30**.74 to **30**.76
Winter temperatures, design for, **2**.2 to **2**.3
Wintertime operation of cooling towers, **44**.29 to **44**.33
Wood:
  composition of (table), **19**.5
  corrosion problems of in low-heat chimneys, **30**.18
Woodburning appliance chimneys, **30**.2 to **30**.3
  tests for, **30**.16 to **30**.17
Workplace noise regulations, **49**.46 to **49**.49
Wrapping factor for heat tracing (table), **11**.6

Y valves, **48**.6, **48**.8

ZEB (*see* Zero energy band)
Zero energy band (ZEB), **52**.55
Zinc chloride:
  for incinerator chimneys, **30**.45
  in incinerators, **30**.47
Zone size, effect of on air-handling units, **37**.5
Zoning of building systems, **52**.60